Because LEARNING EVOLVES, So Does Your Classroom

Learning
REVOLUTIONIZED

Up-to-Date Presentation

Reorganized and reimagined, the Fourteenth Edition of *Brock Biology of Microorganisms* provides the most up-to-date, accurate, and approachable introduction to the study of microbiology for today's students. Recognizing that mastering the principles of the dynamic field of microbiology today requires understanding the underlying molecular biology, the Fourteenth Edition provides both the foundation for the science and the science itself.

Brock Biology of Microorganisms guides students through the six major themes of microbiology in the 21st century as outlined by the American Society of Microbiology Conference on Undergraduate Education (ASMCUE): Evolution, Cell Structure and Function, Metabolic Pathways, Information Flow and Genetics, Microbial Systems, and The Impact of Microorganisms. Reflecting how the "omics" revolution has transformed all of biology, an undercurrent of genomics supports content in every chapter of the Fourteenth Edition.

New Dynamic Art

An enhanced and revised art program in the Fourteenth Edition provides the consistency and context students need to visualize microbiology. With new illustrations, significant art style updates, and nearly 200 new color photos, the Fourteenth Edition beautifully presents microbiology as the visual science it is today.

Consistent and Concise This new figure covers billions of years of evolution in a concise and effective manner. The appealing and engaging art program depicts biological elements consistently.

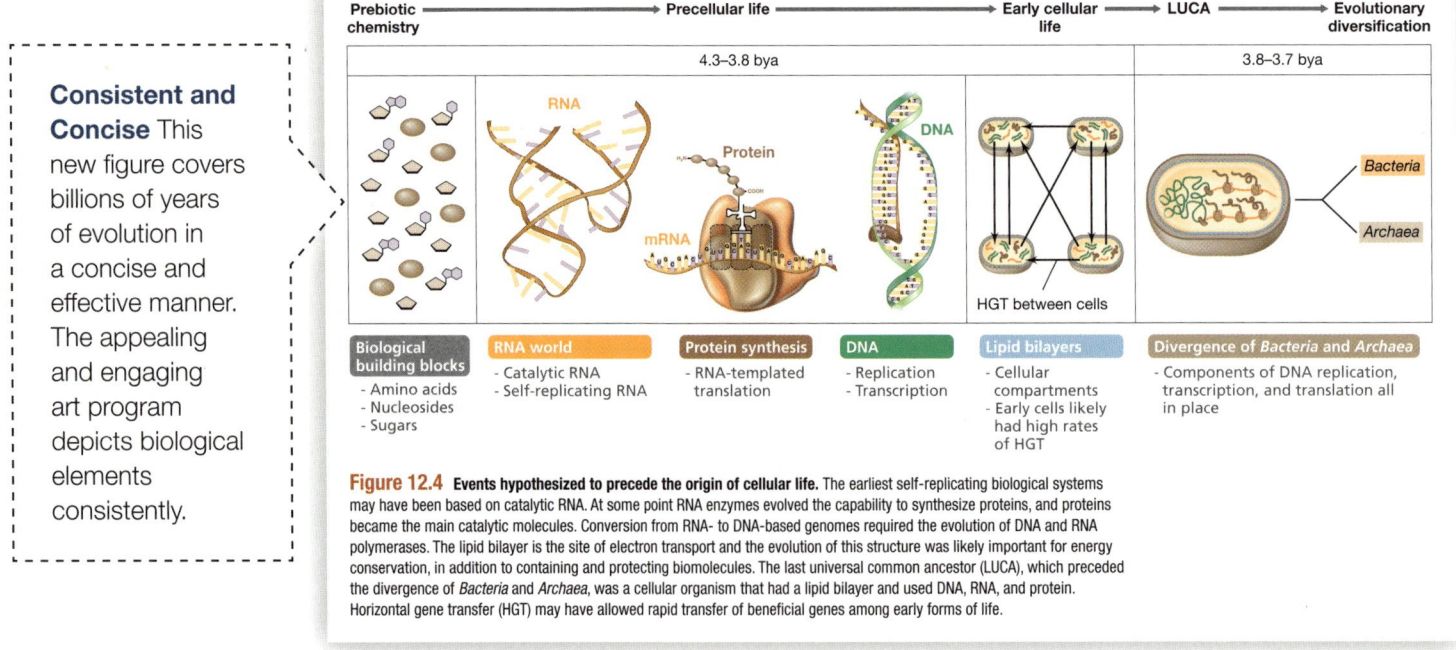

Figure 12.4 Events hypothesized to precede the origin of cellular life. The earliest self-replicating biological systems may have been based on catalytic RNA. At some point RNA enzymes evolved the capability to synthesize proteins, and proteins became the main catalytic molecules. Conversion from RNA- to DNA-based genomes required the evolution of DNA and RNA polymerases. The lipid bilayer is the site of electron transport and the evolution of this structure was likely important for energy conservation, in addition to containing and protecting biomolecules. The last universal common ancestor (LUCA), which preceded the divergence of *Bacteria* and *Archaea*, was a cellular organism that had a lipid bilayer and used DNA, RNA, and protein. Horizontal gene transfer (HGT) may have allowed rapid transfer of beneficial genes among early forms of life.

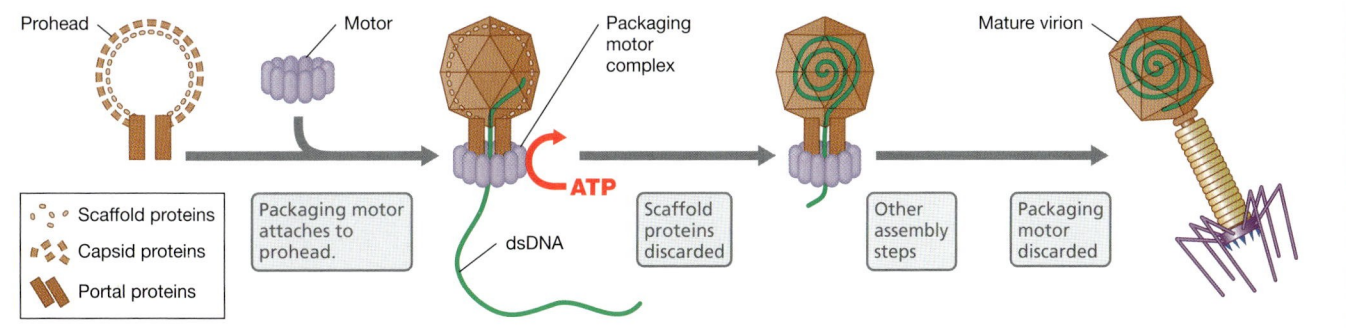

Figure 8.14 **Packaging of DNA into a T4 phage head.** Proheads are assembled from capsid and portal proteins, both of which remain in the mature virion. As the head fills with DNA, it expands and becomes more angular. Once the head is filled, the packaging motor detaches and the tail components are added.

Clear Step Text and Arrows

The virion is a key player in viral replication, an important aspect of microbiology courses. This figure clearly depicts the three relevant protein types that are integral to virion formation. Using didactic artwork along with clear step text and arrows, students can easily follow the steps of virion formation.

New Figures Depict Key Topics

The processes of transcription and translation are key to understanding protein synthesis. DNA, RNA, and protein are depicted in this new figure with visual prominence and are uniquely distinguished from each other. Step text and arrows guide student understanding.

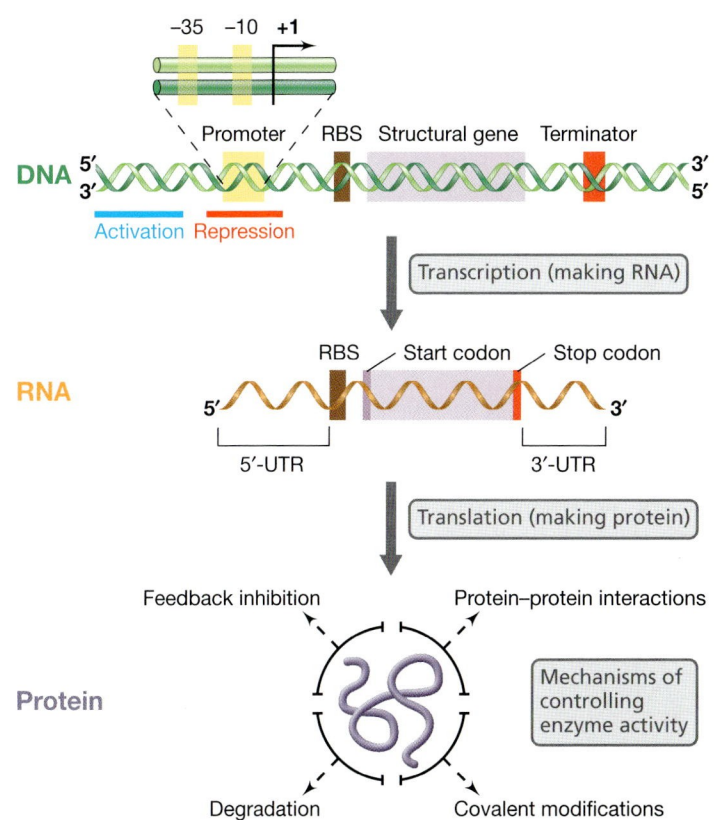

Figure 7.1 **Gene expression and regulation of protein activity.** The promoter and terminator as well as regions involved in transcriptional activation and repression are indicated. The 5′ untranslated region (5′-UTR) is a short region between the start of transcription and the start of translation, while the 3′ untranslated region (3′-UTR) is a short region between the stop codon and the transcription terminator. These are the regions where translational regulation often occurs. Mechanisms for regulating protein activity after translation are shown at the bottom.

Learning
REVOLUTIONIZED

microbiology**now**

New MicrobiologyNow chapter openers showcase current microbiology research, contextualizing chapter content and adding interest and relevancy for today's microbiology students. MicrobiologyNow research-based coaching activities are also assignable within MasteringMicrobiology®, encouraging and developing students' critical thinking skills.

32 · Eukaryotic Pathogens: Fungal and Parasitic Diseases

microbiology**now**

Deadly Fungi

People usually associate fungi with decaying organic matter, or in a medical context, with superficial fungal infections such as athlete's foot. But fungi can cause serious, even deadly infections, as we were reminded when pathogenic fungi triggered a widespread outbreak of fungal meningitis in the United States in 2012.[1]

Glucocorticoids such as methylprednisolone are often prescribed for pain relief, especially in adults suffering chronic lower back pain. The drug is typically injected directly into the outermost part of the spinal canal (epidural injection). Several lots of methylprednisolone formulated by a small pharmaceutical company in Massachusetts were found to be contaminated with fungi, including the mold *Exserohilum rostratum* (photo). *E. rostratum* is common in soil and can naturally infect several human tissues, in particular the cornea, lungs, and lining of the heart (pericardium). By early December of 2012 there were 590 fungal infections—mostly cases of meningitis—linked to injections of the contaminated drug, with 37 of these being fatal.

Using standard epidemiological techniques, the Centers for Disease Control and Prevention (Atlanta, Georgia, USA) and a specially formed multistate response team quickly linked the contaminated medication to the Massachusetts company and more specifically, to particular batches of the drug produced by this company. *E. rostratum* was found in sealed vials of methylprednisolone, and the nonpathogenic yeast *Rhodotorula* and mold *Rhizopus* were found in some others.

This outbreak of fungal disease underscores the critical importance of ensuring that drugs are both sterile and uncontaminated with foreign substances, especially drugs used for injections. In addition, the quick action by public health authorities to identify the disease source and prevent further infections is a testimony to the effectiveness of this branch of the healthcare system in the United States.

[1]Smith R.M., et al. 2012. Fungal infections associated with contaminated methylprednisolone injections—Preliminary report. *N. Engl. J. Med.* DOI: 10.1056/NEJMoa1213978.

EXPLORE THE MICROBIAL WORLD

Culturing the Uncultured

General ecologists refer to two distinct categories of ecological niche, the **fundamental niche** and the **realized niche**. The fundamental niche refers to the range of environments in which a species will be sustained when it is not resource-limited, such as may result from competition with other species. By contrast, the realized niche refers to the range of natural environments supporting a species when it is confronted with factors such as resource limitation, predation, and competition from other species. This distinction between fundamental and realized niches yields a clue as to why some microorganisms are so difficult to isolate from the environment.

Establishing laboratory conditions that fall within the fundamental niche will be sufficient to maintain an organism in pure culture, but those conditions will likely not be adequate to selectively enrich for that species when it must compete with other organisms present in the sample inoculum. Since the realized niche of most microorganisms is unknown, there has been an increasing emphasis on developing high-throughput methods for cultivation, using robotics to set up many cultivation attempts in parallel. High-throughput methods allow for simultaneous investigation of many alternative growth conditions in an attempt to replicate the realized niche or, alternatively, allow the organism to occupy its fundamental niche by relieving it from competition. The latter approach is more commonly employed, since fewer media need to be prepared to sustain growth.

Methods for collecting and depositing a single cell in nutrient media are well developed, including the established dilution methods, as well as the more recent applications of flow cytometric cell sorting and laser tweezers. When high-throughput cultivation is coupled with molecular probes to screen for growth of novel organisms identified previously by culture-independent methods (Section 18.5), there has been increasing success in isolating some interesting bacteria from the uncultured majority in nature. In fact, this general method was used for the successful isolation of one of the most abundant organisms on Earth, *Pelagibacter ubique* (**Figure 1**).[1] As discussed in Section 19.10, this bacterium is a species of highly successful marine bacterioplankton (suspended bacterial cells) that thrive on the very dilute pool of dissolved organic matter present in the open oceans.

Dedicated patience is needed in any cultivation effort, as the discovery of slow-growing or dormant organisms may require months of incubation. Also, many or perhaps even most microorganisms in nature are adapted to extremely low nutrient concentrations, being inhibited by nutrient concentrations used to grow organisms commonly studied in the laboratory. In addition, others may depend on complex interspecies relationships that cannot be replaced by a defined growth medium. Thus, high-throughput methods are ideally suited to surveying various combinations of resources to find the set that best supports the laboratory culture of the most interesting and ecologically relevant organisms from nature.

Culture-independent methods have identified over 50 major divisions (phyla) within the domain *Bacteria* alone. Remarkably, only 12 phyla were known in 1987! Among the major bacterial divisions so far identified, only about half now have cultured representatives. Additionally, even when cultures are available for a division, they are generally few in number and therefore do not fully encompass the phylogenetic diversity within the group. Thus, the challenge of obtaining representative cultures is one of both *breadth* of coverage (obtaining at least one member of each division) and *depth* of coverage (developing a culture collection that spans the phylogenetic diversity of each division). The relatively poor representation of the natural diversity of *Bacteria* in culture collections today (☞ Figure 15.1) also applies to microbial eukaryotes and the *Archaea*. However, this situation is not so much a problem as it is a remarkable opportunity for a new generation of microbiologists interested in microbial diversity. We now have the understanding and technology necessary to more fully explore the remarkable diversity of microbial life through culture-based analyses.[1]

[1]Rappé, M.S., S.A.Connon, K.L. Vergin, & S.J. Giovannoni. 2002. Cultivation of the ubiquitous SAR11 marine bacterioplankton clade. *Nature 418:* 630-633.

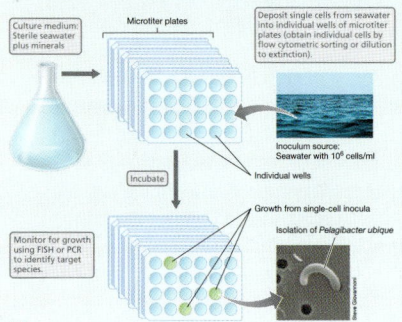

Figure 1 Methodological pipeline for high-throughput cultivation of previously uncultured microorganisms. The method shown here was used to isolate *Pelagibacter ubique*, one of the most abundant bacteria on Earth. Following the addition of filter-sterilized seawater and low nutrient concentrations to the individual wells, pure cultures of *Pelagibacter* and other novel marine *Bacteria* were obtained.

EXPLORE THE MICROBIAL WORLD

Explore the Microbial World features examine specific topics, helping students focus on the big picture of microbiology and exciting their scientific curiosity. Examples include: "The Black Death Decoded," "Genomics: One Cell at a Time," and "Culturing the Uncultured."

CULTIVATE UNDERSTANDING
with MasteringMicrobiology®

MicroLab Tutors
MicroLab Tutors prepare students for lab by introducing and assessing their understanding of lab concepts and techniques. Certain topics also include molecular animation, encouraging students to visualize what's happening at the cellular level.

Lab Technique Videos
Lab Technique Videos demonstrate specific lab techniques in 3–5 minutes. These videos cover important procedures and labs, such as aseptic technique, Gram staining, ELISA, and identification of an unknown, helping students prepare for wet lab and review techniques on their own.

Pre-Lab Quiz with Lab Technique Video: Gram Stain

MasteringMicrobiology®

Microbiology Lab | MicroLab Tutor: Gram Stain

Item Type: Coaching Activities | Difficulty: 2 | Time: 18m | Contact the Publisher | Manage this Item: Standard View

MicroLab Tutor: Gram Stain

Introduction

The Gram stain was developed in the 1880s by Dr. Hans Christian Gram, a Danish bacteriologist whose goal was to find a better staining technique for bacteria in sputum from pneumonia patients.

We still use this technique today to help us to distinguish bacteria from human cellular material in samples such as sputum and cervical smears. We also use this technique to classify bacteria into one of two large groups: Gram-positive bacteria, which have a thick peptidoglycan wall, and Gram-negative bacteria, which have only one or two layers of peptidoglycan covered by an outer membrane that is mostly lipid.

Let's watch as the Gram stain is performed on two different bacteria and then answer the questions that follow.

Part A

Arrange the steps of the Gram staining procedure in their correct order.

Gram-positive

Complexes of crystal violet and Gram's iodine

MicroFlix®
MicroFlix are 3D movie-quality animations with self-paced coaching activities and gradable quizzes that help students master the three toughest topics in microbiology: metabolism, DNA replication, and immunology. Additional 3D BioFlix® animations help students review relevant concepts from general biology.

MicroLab Practicals
MicroLab Practical assessments provide extra practice analyzing and interpreting lab results—putting knowledge into practice.

Additional resources include **Clinical Case Study Coaching Activities** and **MicroCareers Coaching Activities**.

MasteringMicrobiology Study Area
The MasteringMicrobiology Study Area is rich with media assets to give students extra practice on their own time. It includes chapter quizzes, quantitative questions, MCAT prep questions, animations, and more.

Dynamic, Interactive Classroom Tools
Learning Catalytics, now available within MasteringMicrobiology
This student engagement, assessment, and classroom intelligence system allows students to use their laptops, smartphones, or tablets to respond to questions in class. Learning Catalytics includes questions specific to *Brock Biology of Microorganisms*.

BEST SUPPORT for Instructors and Students

FOR INSTRUCTORS

Instructor Resource DVD

978-0-321-92834-4 / 0-321-92834-2

This cross-platform set of DVDs offers a wealth of media resources including all the art from the book in both JPEG and PowerPoint formats, PowerPoint lecture outlines, a computerized Test Bank, and answer keys all in one convenient location. The animations help bring lectures to life, while the selected step-edit figures help break down complicated processes.

Instructor Manual / Test Bank

By W. Matthew Sattley, Christopher A. Gulvik, and Sherry L. Seston

978-0-321-92832-0 / 0-321-92832-6

The Instructor Manual provides chapter summaries that help with class preparation as well as the answers to the end-of-chapter review and application questions. The Test Bank contains 3000 questions for use in quizzes, tests, and exams.

MasteringMicrobiology®

www.masteringmicrobiology.com

Here you'll find a host of assets including the following assignable assessments: chapter-specific Reading Questions, Test Bank questions, Quantitative Questions, and MCAT prep questions that emulate the appropriate content sections, cognitive skills, and question types included on the MCAT.

Pearson Custom Library

The Pearson Custom Library gives you the freedom to create your own customized textbook that meets your course needs. Select the book chapters you need, in the sequence you want. Delete chapters you don't use: Your students pay only for the material you choose. You're in control. Find out more at www.pearsoncustom.com.

Additional supplements:

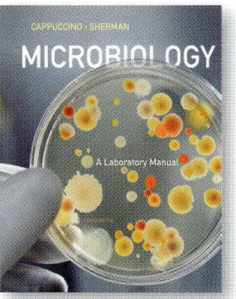

Microbiology: A Laboratory Manual

Tenth Edition

By James Cappuccino and Natalie Sherman

978-0-321-84022-6 / 0-321-84022-4

Versatile, comprehensive, and clearly written, this competitively priced laboratory manual can be used with any undergraduate microbiology text—and now features brief clinical applications for each experiment, MasteringMicrobiology quizzes that correspond to each experiment, and a new experiment on handwashing.

Laboratory Experiments in Microbiology

Tenth Edition

By Ted R. Johnson and Christine L. Case

978-0-321-79438-3 / 0-321-79438-9

Techniques in Microbiology: A Student Handbook

By John M. Lammert

978-0-132-24011-6 / 0-132-24011-4

FOR STUDENTS

Books a la Carte

978-0-321-92835-1 / 0-321-92835-0

This edition features the same content as *Brock Biology of Microorganisms*, Fourteenth Edition, in a convenient, three-hole-punched, loose-leaf version. Books a la Carte also offers a great value for your students—this format costs 35% less than a new textbook.

MasteringMicrobiology Study Area

The MasteringMicrobiology Study Area is rich with media assets to give students extra practice on their own time. It includes chapter quizzes, quantitative questions, MCAT prep questions, animations, and more.

BROCK BIOLOGY OF MICRO ORGANISMS

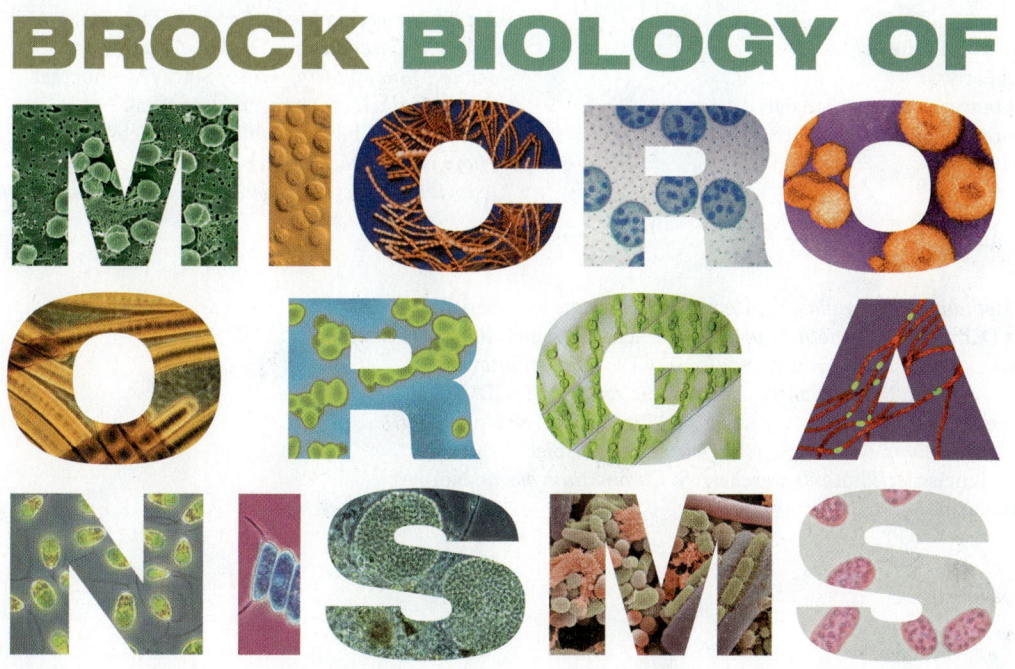

FOURTEENTH EDITION

MICHAEL T. MADIGAN
Southern Illinois University Carbondale

JOHN M. MARTINKO
Southern Illinois University Carbondale

KELLY S. BENDER
Southern Illinois University Carbondale

DANIEL H. BUCKLEY
Cornell University

DAVID A. STAHL
University of Washington Seattle

PEARSON

Boston   Columbus   Indianapolis   New York   San Francisco   Upper Saddle River
Amsterdam   Cape Town   Dubai   London   Madrid   Milan   Munich   Paris   Montréal   Toronto
Delhi   Mexico City   São Paulo   Sydney   Hong Kong   Seoul   Singapore   Taipei   Tokyo

Senior Acquisitions Editor: Kelsey Churchman
Associate Editor: Nicole McFadden
Assistant Editor: Ashley Williams
Director of Development: Barbara Yien
Senior Art Development Editor: Elisheva Marcus
Managing Editor: Michael Early
Production and Design Manager: Michele Mangelli
Production Supervisor: Karen Gulliver
Copyeditor: Anita Wagner
Proofreader: Martha Ghent
Indexer: Sallie Steele

Art Coordinator: Jean Lake
Director, Media Development: Lauren Fogel
Media Producer: Joe Mochnick
Art: Imagineering Media Services, Inc.
Text and Cover Design: Tandem Creative, Inc.
Senior Manufacturing Buyer: Stacey Weinberger
Senior Marketing Manager: Neena Bali
Compositor: Cenveo® Publisher Services
Photo Image Lead: Donna Kalal
Associate Project Manager, Text Permissions: Michael Farmer

Cover photo credits: **M:** *Methanococcus jannaschii,* Electron Microscope Lab, Berkeley; **I:** Cyanobacteria, Thomas D. Brock; **C:** *Crenothrix polyspora,* Michael Wagner; **R:** *Volvox carteri,* Aurora M. Nedelcu; **O:** Arenavirus, Charles Humphrey/PHIL/CDC; **O:** *Oscillatoria,* Richard W. Castenholz; **R:** *Candida,* Maxine Jalbert, Dr. Leo Kaufman/PHIL/CDC; **G:** *Micrasterias,* blickwinkel/Alamy; **A:** *Anabaena,* Alicia M. Muro-Pastor; **N:** *Dunaliella,* Arthur M. Nonomura; **I:** *Scenedesmus,* Ralf Wagner; **S:** *Pleurocapsa,* Daniel H. Buckley; **M:** Oral bacteria, Steve Gschmeissner/Photo Researchers; **S:** *Chromatium okenii,* Norbert Pfennig and Michael T. Madigan.

Credits can be found on page 961.

Library of Congress Cataloging-in-Publication Data
Madigan, Michael T., 1949-
Brock biology of microorganisms / Michael T. Madigan. . . [et al.]. — Fourteenth edition.
    pages cm
Includes index.
ISBN 978-0-321-89739-8
1. Microbiology. I. Title.
QR41.2.B77 2015
579—dc23

2010044962

ISBN 10: 0-321-89739-0 (student edition)
ISBN 13: 978-0-321-89739-8 (student edition)
3 4 5 6 7 8 9 10—RRD—17 16 15 14

www.pearsonhighered.com

About the Authors

Michael T. Madigan received his B.S. in Biology and Education from Wisconsin State University–Stevens Point (1971) and his M.S. (1974) and Ph.D. (1976) in Bacteriology from the University of Wisconsin–Madison. His graduate research was on the hot spring bacterium *Chloroflexus* in the laboratory of Thomas Brock. Following three years as a postdoctoral scientist at Indiana University, Mike moved to Southern Illinois University Carbondale, where he taught courses in introductory microbiology and bacterial diversity as a professor of microbiology for 33 years. In 1988 Mike was selected as the Outstanding Teacher in the College of Science and in 1993, the Outstanding Researcher. In 2001 he received the SIUC Outstanding Scholar Award. In 2003 he received the Carski Award for Distinguished Undergraduate Teaching from the American Society for Microbiology, and he is an elected Fellow of the American Academy of Microbiology. Mike's research is focused on bacteria that inhabit extreme environments, and for the past 15 years he has studied Antarctic microbiology. In addition to research papers, he has edited a major treatise on phototrophic bacteria and served for 10 years as chief editor of the journal *Archives of Microbiology*. He currently serves on the editorial board of the journals *Environmental Microbiology* and *Antonie van Leeuwenhoek*. Mike's other interests include forestry, swimming, reading, and caring for his dogs and horses. He lives on a quiet lake with his wife, Nancy, four shelter dogs (Gaino, front in photo, deceased 30 September 2013; Pepto, back in photo; Peanut; and Merry), and three horses (Eddie, Gwen, and Festus).

John M. Martinko received his B.S. in Biology from Cleveland State University. He then worked at Case Western Reserve University, conducting research on the serology and epidemiology of *Streptococcus pyogenes*. His doctoral work at the State University of New York at Buffalo investigated antibody specificity and antibody idiotypes. As a postdoctoral fellow, he worked at Albert Einstein College of Medicine in New York on the structure of major histocompatibility complex proteins. Since 1981, he has been in the Department of Microbiology at Southern Illinois University Carbondale where he was Associate Professor and Chair, and Director of the Molecular Biology, Microbiology, and Biochemistry Graduate Program. His research interests centered on the structure–function relationships of immune system proteins, including immunoglobulins, T cell receptors, and major histocompatibility proteins. His teaching interests include an advanced course in immunology as well as immunology and inflammation instruction to medical students. For his educational efforts, he won the 2007 Southern Illinois University Outstanding Teaching Award. He has been active in a number of educational outreach programs for pre-university students and teachers. He has also been a faculty member at Bard College in its innovative Citizen Science program, an interactive laboratory, computer, and problem-based-learning science curriculum that introduces freshmen students to critical thinking through the discovery and application of scientific principles. He was the Chair of the Institutional Animal Care and Use Committee at SIUC and continues to act as a consultant in the area of animal care. He is also an avid golfer and cyclist. John lives in Carbondale with his wife Judy, a high school science teacher.

Kelly S. Bender received her B.S. in Biology from Southeast Missouri State University (1999) and her Ph.D. (2003) in Molecular Biology, Microbiology, and Biochemistry from Southern Illinois University Carbondale. Her dissertation work focused on the genetics of perchlorate-reducing bacteria. During her postdoctoral fellowship, Kelly worked on the genetic regulation of sulfate-reducing bacteria in the laboratory of Judy Wall at the University of Missouri–Columbia. She also completed a transatlantic biotechnology fellowship at Uppsala University in Sweden researching regulatory small RNAs in bacteria. In 2006, Kelly returned to her alma mater, Southern Illinois University Carbondale, as an Assistant Professor in the Department of Microbiology and was tenured and promoted to Associate Professor in 2012. Her lab studies a range of topics including the regulation of stress responses by small RNAs, microbial community dynamics of sites impacted by acid mine drainage, and the bioremediation of uranium by metal- and sulfate-reducing bacteria. Kelly teaches courses in microbial genetics and molecular biology, has served on numerous federal grant review panels, and is an active member of the American Society for Microbiology. Her other interests include biking, cooking, and spending time with family, friends, and her miniature schnauzer, Pepper.

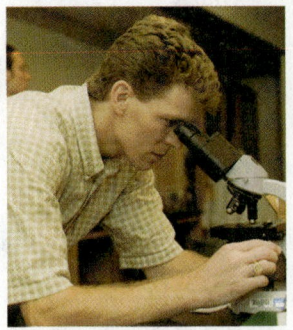

Daniel H. Buckley is an Associate Professor at Cornell University in the Department of Crop and Soil Sciences. He earned his B.S. in Microbiology (1994) at the University of Rochester and his Ph.D. in Microbiology (2000) at Michigan State University. His graduate research focused on the ecology of soil microbial communities and was conducted in the laboratory of Thomas M. Schmidt in affiliation with the MSU Center for Microbial Ecology. Dan's postdoctoral research examined linkages between microbial diversity and biogeochemistry in marine microbial mats and stromatolites and was conducted in the laboratory of Pieter T. Visscher at the University of Connecticut. Dan joined the Cornell faculty in 2003. His research program investigates the ecology and evolution of microbial communities in soils with a focus on the causes and consequences of microbial diversity. He has taught both introductory and advanced courses in microbiology, microbial diversity, and microbial genomics. He received a National Science Foundation Faculty Early Career Development (CAREER) award in 2005 for excellence in integrating research and education. He has served as Director of the Graduate Field of Soil and Crop Sciences at Cornell and Codirector of the Marine Biological Laboratory Microbial Diversity Summer Course in Woods Hole, Massachusetts. He currently serves on the editorial boards of *Applied and Environmental Microbiology* and *Environmental Microbiology*. Dan lives in Ithaca, New York, with his wife Merry and sons Finn and Colin. Dan enjoys running and a variety of outdoor sports but, most of all, catching critters down at the creek with his boys.

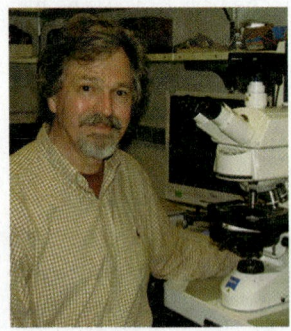

David A. Stahl received his B.S. degree in Microbiology from the University of Washington, Seattle, and completed graduate studies in microbial phylogeny and evolution with Carl Woese in the Department of Microbiology at the University of Illinois at Urbana-Champaign. Subsequent work as a postdoctoral fellow and research associate with Norman Pace, then at the National Jewish Hospital in Colorado, involved early applications of 16S rRNA-based sequence analysis to the study of natural microbial communities. In 1984 Dave joined the faculty at the University of Illinois with appointments in Veterinary Medicine, Microbiology, and Civil Engineering. In 1994 he moved to the Department of Civil Engineering at Northwestern University, and in 2000 returned to the University of Washington as professor in the Departments of Civil and Environmental Engineering and Microbiology. Dave is known for his work in microbial evolution, ecology, and systematics, and received the 1999 Bergey Award and the 2006 ASM Procter & Gamble Award in Applied and Environmental Microbiology. He is a fellow in the American Academy of Microbiology and a member of the National Academy of Engineering. His main research interests surround the biogeochemistry of nitrogen and sulfur and the microbial communities that sustain the associated nutrient cycles. His laboratory was first to culture ammonia-oxidizing *Archaea*, a group believed to be the key mediators of this process in the nitrogen cycle. Dave has taught several courses in environmental microbiology, was one of the founding editors of the journal *Environmental Microbiology,* and has served on many advisory committees. Outside the lab, Dave enjoys hiking, bicycling, spending time with family, reading a good science fiction book, and—with his wife Lin—renovating an old farmhouse on Bainbridge Island.

Dedications

Michael T. Madigan
dedicates this book to the memory of his old friend, Snuffy. I sure miss those long walks, just you and me.

John M. Martinko
dedicates this book to his mother Lottie, who inspired all her children to achieve and excel.

Kelly S. Bender
dedicates this book to the memory of her grandmother, Alberta, whose biggest regret in life was not being able to attend school past the fifth grade.

Daniel H. Buckley
dedicates this book to Merry. Thanks for sharing this adventure and all the others.

David A. Stahl
dedicates this book to his wife, Lin. My love, and one that helps me keep the important things in perspective.

Preface

Because learning evolves, so do we. Streamlined, completely up to date, and written with deference to the history of microbiology and excitement for the future, the Fourteenth Edition of *Brock Biology of Microorganisms* is the strongest yet. For three generations, students and instructors have relied on the accuracy, authority, consistency, and up-to-date science of *Brock Biology of Microorganisms* to learn the core principles of microbiology and pique their interest in the future of the discipline. With the Fourteenth Edition, students will benefit from the book's emphasis on cutting-edge research, its seamless integration of and introduction to modern molecular microbiology, and a visually stunning art program revision. Additionally, for the first time, *Brock Biology of Microorganisms* is supported by MasteringMicrobiology, Pearson's online homework, tutorial, and assessment system.

Experienced authors Madigan, Martinko, and Stahl welcome two new coauthors on the Fourteenth Edition: Kelly S. Bender and Daniel H. Buckley. Significantly revising the coverage of molecular biology and microbial genetics, Kelly has been highly praised for both her undergraduate teaching and graduate student mentoring at Southern Illinois University. At Cornell, Dan participates in the Cornell Institute for Biology Teachers' summer workshop for high school science teachers and codirects the world-famous Summer Course in Microbial Diversity at Woods Hole. Both new coauthors have greatly strengthened the core mission of *Brock Biology of Microorganisms*: to remain the best learning resource for today's microbiology students and instructors.

What's New in the 14th Edition?

Reorganized and reimagined, the Fourteenth Edition guides students through the six major themes of microbiology in the 21st century as outlined by the American Society of Microbiology Conference on Undergraduate Education (ASMCUE): Evolution, Cell Structure and Function, Metabolic Pathways, Information Flow and Genetics, Microbial Systems, and the Impact of Microorganisms. With enhanced and revised artwork and nearly 200 new color photos, *Brock Biology of Microorganisms* beautifully presents microbiology as the visual science it is. The new chapter-opening features, titled "MicrobiologyNow," engage students in cutting-edge research relevant to each chapter's content and connect to assignable and assessable MasteringMicrobiology coaching activities. "Explore the Microbial World" features focus on specific topics that help students get a feel for the "big picture" in microbiology while simultaneously fueling their scientific curiosity.

Genomics, and all of the various "omics" it has spawned, support content in every chapter of *Brock Biology of Microorganisms*, reflecting how the omics revolution has transformed all of biology. Gone is the day of microbiology as a descriptive science. Mastering the principles of the dynamic field of microbiology today requires understanding the underlying molecular biology. As authors, we are well aware of this, and have written *Brock Biology of Microorganisms* in a way that provides both the foundation for the science and the science itself. The result is a truly robust and eminently modern treatment of microbiology.

To strengthen the learning experience, additional Mastering-Microbiology content includes chapter-specific reading quizzes, MicroLab Tutorials, MicrobiologyNow coaching activities, Clinical Case and MicroCareer coaching activities, animation quizzes, MCAT Prep questions, and many additional study and assessment tools. Collectively, the content and presentation of *Brock Biology of Microorganisms*, coupled with the powerful learning tools of MasteringMicrobiology, create an unparalleled educational experience in microbiology.

Revision Highlights

Chapter 1

- Chapter 1 has been revised to provide an up-to-date and succinct introduction to and overview of microbiology, including basic elements of cell structure and the phylogenetic tree of life.
- The power of genomics for solving mysteries in microbiology is revealed in a new Explore the Microbial World feature, "The Black Death Decoded" on forensic studies of victims of Europe's "Black Death" outbreak over 650 years ago.

Chapter 2

- Coverage of microbial cell structure and function now combines material on *Bacteria* and *Archaea* with that of microbial eukaryotes, giving students a thorough introduction to comparative cell structure and providing the instructor with all of the tools necessary for effective classroom presentations.

Chapter 3

- The essential features of microbial metabolism necessary for understanding how microorganisms transform energy are laid out in a logical sequence with an appropriate level of detail on metabolic diversity for introductory students. Newly rendered art makes mastering key metabolisms a more visual and engaging experience.

Chapter 4

- The basic principles of molecular microbiology are heavily revised and beautifully presented early in the text to provide a useful background for students as they proceed through the book.
- Beautiful new art stresses consistency and simplicity to make complex molecular concepts easy to learn, retain, and apply.

Chapter 5

- To close Unit 1, this chapter draws on the previous four chapters in describing the end result of molecular biology and physiology: cell division and population growth.
- The chapter now incorporates the essentials of microbial growth control to allow instructors to better tie important practical content to the basic science of the growth process itself.

Chapter 6

- Complete coverage of microbial genomics and the omics revolution that is driving the science of microbiology today appears much earlier in the book than in the previous edition. Coverage of the technology, biology, and evolution of genomes is laid out in a new and exciting fashion.
- Marvel at the power of genomics in a new Explore the Microbial World feature on single-cell genomics: "Genomics, One Cell at a Time."

Chapter 7

- Chapter 7 contains major updates on the regulation of gene expression—one of the hottest areas in microbiology today—including expanded coverage of cell sensing capacities and signal transduction.
- Explore new aspects of gene regulation, including the importance of small RNAs and the regulation of special events in model bacteria such as sporulation in *Bacillus*, cell differentiation in *Caulobacter*, and heterocyst formation in the nitrogen-fixing cyanobacterium *Anabaena*.

Chapter 8

- The basic principles of virology are presented without extraneous detail and use bacteriophage T4 as a model for depicting key virological concepts.
- New coverage of the virosphere and viral ecology reveals the overwhelming genetic diversity of viruses.

Chapter 9

- Coverage of viral genomes and diversity now directly follows the basic virology chapter to better link the two closely related topics.
- New coverage of the evolution of viral genomes and a new organization that more directly contrasts the biology of DNA and RNA viruses support a more consistent and conceptual understanding of viral diversity.

Chapter 10

- Coverage of the fundamental principles of the genetics of *Bacteria* and *Archaea* is now strategically located in the book to better incorporate supporting concepts from molecular microbiology, growth, regulation, and virology.

Chapter 11

- Complete coverage of the molecular biology of gene cloning and other major genetic manipulations forms a prelude to coverage of the application of these methods in the fast-moving field of biotechnology.
- Enter the world of synthetic biology and learn how this hot new area promises yet another revolution in biology.

Chapter 12

- Microbial evolution and systematics benefits from a major revision that focuses on the mechanisms of microbial evolution, including the importance of genomic evolution and horizontal gene transfer.
- Consider how metabolic interdependencies in microbial communities may have evolved in a fascinating new Explore the Microbial World feature, "The Black Queen Hypothesis."

Chapter 13

- Microbial metabolic diversity is now presented in a single chapter to better compare and contrast the key metabolisms of *Bacteria* and *Archaea* and to emphasize how "the unity of biochemistry" has pervaded microbial metabolism.
- Metabolic diversity is now strategically positioned to naturally segue into the new chapter on bacterial functional diversity.

Chapters 14 and 15

- Chapter 14, "Functional Diversity of *Bacteria*," now explores bacterial diversity with respect to the ecological, physiological, and morphological characteristics of well-known bacteria. Chapter 15, "Diversity of *Bacteria*," presents the diversity of bacterial life in a truly phylogenetic context. New, colorful, and easy-to-follow phylogenetic trees summarize bacterial diversity in both chapters.

Chapter 16

- Archaeal diversity is revised with a stronger phylogenetic thread and new coverage of the more recently discovered archaeal phyla of *Thaumarchaeota*, *Nanoarchaeota*, and *Korarchaeota*.
- Learn how the previously unrecognized *Thaumarchaeota* are probably the most common *Archaea* on Earth and review the physicochemical limits to life, all of which are currently defined by species of *Archaea*.

Chapter 17

- Eukaryotic microbial diversity benefits from new phylogenetic coverage and a chapter prelude on the importance of endosymbioses in the evolution of eukaryotic cells.
- Many new color micrographs portray the beauty and diversity of eukaryotic microbial life.

Chapter 18

- The modern tools of the microbial ecologist are described with examples of how each has sculpted the science. Also, learn how the omics revolution has provided a new window for simplifying complex problems in microbial ecology.
- In the new Explore the Microbial World feature "Culturing the Uncultured," discover how novel ecological methods have yielded laboratory cultures of the marine bacterium *Pelagibacter*, the most abundant organism on Earth.

Chapter 19

- The properties and microbial diversity of the major microbial ecosystems are compared and contrasted in an exciting new way.
- New environmental census data for freshwater habitats and the microbial ecology of arid landscapes highlight new material in this chapter, along with fresh coverage of the link between marine microorganisms and climate change.

Chapter 20

- Chapter 20 includes new coverage of the remarkable abilities of microorganisms to respire solid metal oxides in the iron and manganese cycles.
- Learn how humans are profoundly affecting the nitrogen and carbon cycles, including inorganic nutrient overloads and other forms of pollution, and how all of this feeds back into climate change.

Chapter 21

- A new chapter on the "built environment" shows how humans create new microbial habitats through construction of buildings, supporting infrastructure, and habitat modification. Witness the substantial positive and negative effects microorganisms have on important human infrastructure including wastewater treatment, microbial mining and acid mine drainage, corrosion of metals, the biodeterioration of stone and concrete, and the problem of pathogens in drinking water.

Chapter 22

- Here you will find expanded coverage of how microorganisms profoundly affect the physiology of plants and animals through symbiotic associations, including the dynamic topic of the human microbiome and its relationship to health and disease.
- Discover how a common mechanism used by bacteria and fungi to form symbiotic associations with the roots of plants provides the plants with key nutrients.

Chapter 23

- Major topics in human microbiology including the normal microflora, pathogenesis, and host factors in infection and disease are presented in a style that unites these concepts and reveals how they tip the balance toward health or disease.

Chapter 24

- Chapter 24 is designed to be the straightforward and concise overview of immunology that many instructors use to teach the fundamental concepts of the science.
- This chapter is loaded with practical information on vaccines, inflammation, and allergic responses in an easily teachable format.

Chapter 25

- Built on the shoulders of the previous chapter, Chapter 25 offers a more complete picture of immune mechanisms, with an emphasis on the molecular and cellular interactions that control innate and adaptive immunity.

Chapter 26

- This is a brief chapter that considers immunology from a completely molecular perspective, including the important receptor–ligand interactions that trigger the immune response and the genetics of the key proteins that drive adaptive immunity.

Chapter 27

- Reorganized and up to date, Chapter 27 describes the role of the clinical microbiologist and introduces the tools used to identify and track infectious diseases in clinical laboratories.
- New coverage of antimicrobial agents and their clinical usage underscores the important role of both drug therapy and drug resistance in medicine today.

Chapter 28

- A revised discussion of epidemiology introduces the concept of the reproduction number (R) and its implications for disease spread and control by herd immunity.
- Find up-to-date coverage of emerging infectious diseases and current pandemics, including those of HIV/AIDS, cholera, and influenza, and the role of the epidemiologist in public health microbiology.

Chapter 29

- Coverage of diseases transmitted from person to person is reorganized and illustrated with dozens of new color photos showing symptoms and treatments. To better consolidate material that fits a common theme, infectious diseases in this and each of the next three chapters are presented by taxonomy.

Chapter 30

- Bacterial and viral diseases transmitted by insect vectors or from soil are consolidated and illustrated by dozens of new color photos.
- This chapter contains new coverage of important viral diseases such as yellow fever and dengue fever and the bacterial diseases anthrax, tetanus, and gas gangrene.

Chapter 31

- Common-source diseases linked to contaminated food and water are now consolidated to better emphasize their similar modes of transmission. Coverage within the chapter is by taxonomy—bacterial versus viral—and illustrated with nearly 30 new color photos.
- Find new coverage of the potentially fatal foodborne infection caused by the intracellular bacterium *Listeria*.

Chapter 32

- All infectious diseases caused by eukaryotic microorganisms— fungi and parasites—are consolidated into one chapter to retain the taxonomic theme of medical microbiology. The visual experience is bolstered by 35 new color photos showing the pathogens and disease symptoms. Coverage of fungal and microbial parasitic diseases is expanded; the chapter also includes first-time coverage of major helminthic infections.

Other Learning Tools

- Two appendices, including a primer on bioenergetic calculations and a list of higher order taxa described in *Bergey's Manual of Systematic Bacteriology*; a glossary; and a thorough index round out the learning package in *Brock Biology of Microorganisms*, 14e.

Acknowledgments

A textbook is a complex entity and emerges from the contributions of a large book team. Besides the authors, the team is composed of folks both inside and outside of Pearson. Acquisitions Editor Kelsey Churchman and Associate Editor Nicole McFadden, both of Pearson, were the workhorses in editorial. Kelsey paved the way for the Fourteenth Edition of *Brock Biology of Microorganisms* (*BBOM* 14e) and skillfully maneuvered the book around the challenges inherent in any major textbook project. Nicole ran the day-to-day operations of the *BBOM* team, expertly processing the manuscript and keeping all facets of the project on track. The authors thank Kelsey and Nicole for their dedication to *BBOM* and professionalism in seeing the Fourteenth Edition through to completion.

The *BBOM* 14e production and design team in San Francisco was composed of Michele Mangelli (Mangelli Productions), Yvo Riezebos (Tandem Creative, Inc.), and Elisheva Marcus (Pearson). Michele managed the production team and kept everyone on mission and on budget. The artistic magic of Yvo is clearly visible in the beautiful text and cover designs of *BBOM* 14e. Elisheva (Ellie) was our art development editor and created the new art look for *BBOM* 14e, one that readers will immediately appreciate for its clarity, consistency, and modern style. Ellie's strong background in both science and art will be obvious throughout the book and her contributions were a major improvement to this edition. Thank you, Michele, Yvo, and Ellie. The authors also thank the art team at Imagineering (Toronto) for helping the authors convey a strong educational message and for their excellent suggestions for art improvements.

Other important people in the production team included Karen Gulliver, Jean Lake, Kristin Piljay, Betsy Dietrich, and Martha Ghent. Karen was our excellent and highly efficient production editor; Karen kept pages moving smoothly and tolerated the authors' many requests. Jean was our art coordinator, tracking and routing art and handling interactions between the art studio, art reviewers, and the authors, to ensure quality control and a timely schedule. Betsy and Martha worked with Jean and Karen to ensure an art program and text free of both bloopers and subtle errors. Kristin was our photo researcher who helped the authors acquire photos that meet the exacting standards of *BBOM*. The authors are extremely grateful to Karen, Jean, Kristin, Betsy, and Martha for transforming literally thousands of pages of text and art manuscript into the superb learning tool you have in front of you.

Special thanks go out to two other members of the production team. Anita Wagner was our absolutely spectacular copyeditor; the authors could not have had a more skillful and effective person in this key position on the book team. Anita improved the accuracy, clarity, and consistency of the text and rendered her editorial services in a style that the authors found both helpful and time saving. Elizabeth McPherson (University of Tennessee) was our accuracy checker; her eagle eye, extensive knowledge of all areas of microbiology, and prompt service ensured the authority of the final product.

Thanks are also due to Joe Mochnick of the Pearson media team, as well as Ashley Williams for conducting reviews and managing the supplements to this text. And, since no book is worth a hoot if it does not do well in the marketplace, many thanks go to Neena Bali, who leads the marketing efforts for this title.

The authors also wish to acknowledge the excellent contributions of Matt Sattley (Indiana Wesleyan University), who composed the Instructor's Manual that accompanies *BBOM* 14e, and Christopher Gulvik (Georgia Institute of Technology) and Sherry L. Seston (Alverno College), who revised the test bank questions for this edition. We are also grateful to the top-notch educators who constructed the MasteringMicrobiology program that accompanies this text; these include: Ann Paterson, Narveen Jandu, Jennifer Hatchel, Susan Gibson, Ines Rauschenbach, Lee Kurtz, Vicky McKinley, Clifton Franklund, Benjamin Rohe, Ben Rowley, and Helen Walter. And finally, special thanks go out to Nicolás Pinel (Institute for Systems Biology) for producing the beautiful overview figures that summarize the microbial diversity of major microbial habitats.

No textbook in microbiology could be published without reviewing of the manuscript and the gift of new photos from experts in the field. We are therefore extremely grateful for the kind help of the many individuals who provided manuscript reviews and new photos. These include:

Jill Banfield, *University of California, Berkeley*
Dennis A. Bazylinski, *University of Nevada Las Vegas*
J. Thomas Beatty, *University of British Columbia*
Jayne Belnap, *US Geological Survey*
Karim Benzerara, *Centre National de la Recherche Scientifique, France*
Odile Berge, *INRA-PACA, France*
Robert Blankenship, *Washington University St. Louis*
F. C. Boogerd, *VU University Amsterdam, The Netherlands*
Yan Boucher, *University of Alberta, Canada*
Don Bryant, *Penn State University*
Richard W. Castenholz, *University of Oregon*
Clara Chan, *University of Delaware*
Todd Ciche, *Michigan State University*
David P. Clark, *Southern Illinois University*
J. Collier, *Université de Lausanne*
Patricia Dominguez-Cuevas, *Newcastle University, England*
Cheryl Drake, *Memorial Health System, Springfield, Illinois*
Kimberley D. Ellis, *Tufts University School of Medicine*
David Emerson, *Bigelow Laboratory*
Jeff Errington, *Newcastle University, England*

Katharina Ettwig, *Radboud University, The Netherlands*
Teresa Fischer, *Indian River State College*
Derek J. Fisher, *Southern Illinois University*
Rachel Foster, *Max Planck Institute for Marine Microbiology, Germany*
Jed Fuhrman, *University of Southern California*
Sandra Gibbons, *University of Illinois at Chicago* and *Moraine Valley Community College*
Steve Giovannoni, *Oregon State University*
Eric Grafman, *Centers for Disease Control Public Health Image Library*
Claudia Gravekamp, *Albert Einstein College of Medicine*
A.D. Grossman, *Massachusetts Institute of Technology*
Ricardo Guerrero, *University of Barcelona, Spain*
Daniel P. Haeusser, *University of Houston–Downtown*
Markus Huettel, *Florida State University*
Michael Ibba, *The Ohio State University*
Vaughn Iverson, *University of Washington*
Shawna Johnston, *University of Calgary, Canada*
Megan Kempher, *Southern Illinois University*
Phil Kirchberger, *University of Alberta, Canada*
Susan Koval, *University of Western Ontario, Canada*
F. Leng, *Florida International University*
James Little, *Emory University*
Huub Loozen, *Merck Sharp & Dohme, The Netherlands*
Nicole B. Lopanik, *Georgia State University*
Derek R. Lovely, *University of Massachusetts*
Fritz E. Lower, *Southern Illinois University School of Medicine*
Thomas C. Marlovits, *Research Institute of Molecular Pathology, Austria*
Ann G. Matthysse, *University of North Carolina at Chapel Hill*
Carmody McCalley, *University of Arizona*
Vicky McKinley, *Roosevelt University*
Mary Ann Moran, *University of Georgia*
Alicia María Muro-Pastor, *Instituto Bioquímica Vegetal y Fotosíntesis, Spain*
Alison E. Murray, *Desert Research Institute*
Gerard Muyzer, *University of Amsterdam, The Netherlands*
Jeffrey Nash, *Udon Thani Rajabhat University, Thailand*
Lars Peter Nielsen, *Aarhus University, Denmark*
Sean O'Connell, *Western Carolina University*
Norman Pace, *University of Colorado*
Ann V. Paterson, *Williams Baptist College*
C. O. Patterson, *Texas A&M University*
Jennifer Pett-Ridge, *Lawrence Livermore National Laboratory*
Niels Peter Revsbech, *University of Aarhus, Denmark*
Virginia Rich, *University of Arizona*
D. Rudner, *Harvard Medical School*

Verena Salman, *University of North Carolina*
Karin Sauer, *Binghamtom University*
Bernhard Schink, *University of Konstanz, Germany*
Gerald Schönknecht, *Oklahoma State University*
Matt Schrenk, *East Carolina University*
Kimberley Seed, *Tufts University School of Medicine*
Christine Sharp, *Wairakei Research Center, New Zealand*
Nancy L. Spear, *Murphysboro, Illinois*
S. R. Spilatro, *Marietta College*
K.O. Stetter, *Universität Regensburg, Germany*
Matthew Stott, *GNS Science, New Zealand*
Matthew Sullivan, *University of Arizona*
Andreas Teske, *University of North Carolina*
Tim Tolker-Nielsen, *University of Copenhagen, Denmark*
Tjisse van der Heide, *University of Groningen, The Netherlands*
Laura van Niftrik, *Radboud University, The Netherlands*
Claire Vieille, *Michigan State University*
Michael Wagner, *University of Vienna, Austria*
Susan C. Wang, *Washington State University*
David Ward, *Montana State University*
Peter K. Weber, *Lawrence Livermore National Laboratory*
James Weisshaar, *University of Wisconsin*
Angel White, *Oregon State University*
Kenneth H. Williams, *Lawrence Berkeley National Laboratory*
Mark Young, *Montana State University*
Davide Zannoni, *University of Bologna, Italy*
Lanying Zeng, *Texas A & M University*
Steve Zinder, Cornell University

As hard as a publishing team may try, no textbook can ever be completely error-free. Although we are confident the reader will be hard pressed to find errors in *BBOM* 14e, any errors that do exist, either of commission or omission, are the responsibility of the authors. In past editions, users have been kind enough to contact us when they spot an error so we can fix it in a subsequent printing. Users should feel free to continue to contact the authors directly about any errors, concerns, questions, or suggestions they have about the book. We are always happy to hear from users; your comments help make the book stronger.

Michael T. Madigan (madigan@micro.siu.edu)

John M. Martinko (martinko@micro.siu.edu)

Kelly S. Bender (bender@micro.siu.edu)

Daniel H. Buckley (dbuckley@cornell.edu)

David A. Stahl (dastahl@u.washington.edu)

Brief Contents

Contents

CHAPTER 14

Functional Diversity of *Bacteria* 433

microbiology**now**
One Culture Away 433

CHAPTER 15

Diversity of *Bacteria* 479

microbiology**now**
Discovering Novel Microbial Phyla 479

UNIT 4 Microbial Ecology and Environmental Microbiology

CHAPTER 19

Microbial Ecosystems 597

microbiology**now**
Living in a World of Extreme Energy Limitation 597

CHAPTER 20

Nutrient Cycles 631

microbiology**now**
Microbial Power Lines 631

CHAPTER 21

Microbiology of the Built Environment 649

microbiology**now**
Your Subway System: What's in the Air? 649

CHAPTER 22

Microbial Symbioses 669

microbiology**now**
A Symbiotic Trio Sustains Seagrass Ecosystems 669

UNIT 5 Pathogenicity and Immunology

CHAPTER 23

Microbial Interactions with Humans 705

CHAPTER 24

Immunity and Host Defense 731

CHAPTER 25

Immune Mechanisms 753

CHAPTER 26

Molecular Immunology 773

UNIT 6 Infectious Diseases and Their Transmission

CHAPTER 30

Vectorborne and Soilborne Bacterial and Viral Diseases 885

microbiology**now**
Vampire Bats and Rabies 885

CHAPTER 31

Water and Food as Vehicles of Bacterial Diseases 903

microbiology**now**
Prison Pruno Packs a Life-Threatening Punch 903

CHAPTER 32

Eukaryotic Pathogens: Fungal and Parasitic Diseases 923

microbiology**now**
Deadly Fungi 923

CHAPTER

1 · Microorganisms and Microbiology

microbiology**now**

Microbial Life Is Everywhere

As you embark on your journey to the microbial world, you will be astounded to learn where microorganisms live in nature. In brief, they live everywhere, including locations too harsh for macroorganisms. For example, a research team studying the permanently ice-covered Lake Vida in the McMurdo Dry Valleys of Antarctica (top photo) found living bacteria immersed in a subfreezing salt solution at −13°C! These hardy microorganisms were discovered by microbiologists wearing protective clothing to prevent contamination during the drilling process (bottom photos).

The Lake Vida bacteria, a metabolic group called psychrophiles (a term that means "cold loving"), were shown to carry out various metabolic reactions at the temperature of their icy home. Specific genes isolated from the various Lake Vida bacteria were used to classify the organisms, and future studies of their genetic blueprints—their genomes—should help to reveal the genetic secrets that allow these organisms to thrive in the constant cold.

Lake Vida is unusual even for Antarctic lakes, since its ice cover extends all the way to the bottom. Sunlight, only available 6 months of the year, cannot penetrate deeply into the lake. So the Lake Vida bacteria are probably metabolizing and growing, albeit extremely slowly, on organic carbon that was trapped in the ice at the time the lake became ice-sealed, millennia ago.

Microbiologists study bacteria from extreme environments to reveal the environmental limits to life and to search for unique products that might benefit humans or our planet. But in addition to contributing to basic and applied science, the Lake Vida bacteria are models for the kinds of life forms that could inhabit other icy worlds, such as Mars, or Jupiter's moon, Europa.

Murray, A.E., et al. 2012. Microbial life at −13°C in the brine of an ice-sealed Antarctic lake. *Proc. Natl. Acad. Sci. (USA)*. *109*: 20626–20631.

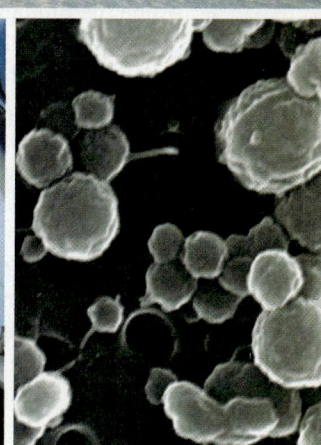

1

I • Introduction and Major Themes of Microbiology

1.1 What Is Microbiology About and Why Is It Important?

The science of microbiology is all about **microorganisms** and how they work, especially the bacteria, a very large group of very small cells (**Figure 1.1**) that have enormous basic and practical importance. Microbiology is also about diversity and evolution of microbial cells, about how different kinds of microorganisms arose and why. Microbiology embraces ecology, so it is also about where microorganisms live on Earth, how they associate and cooperate with each other, and what they do in the world at large, in soils and waters and in animals and plants.

The science of microbiology revolves around two interconnected themes: (1) understanding the nature and functioning of the microbial world, and (2) applying our understanding of the microbial world for the benefit of humankind and planet Earth. As a *basic* biological science, microbiology uses microbial cells to probe the fundamental processes of life. In so doing, microbiologists have developed a sophisticated understanding of the chemical and physical basis of life and have learned that all cells share much in common. As an *applied* biological science, microbiology is at the forefront of many important breakthroughs in human and veterinary medicine, agriculture, and industry. From infectious diseases to soil fertility to the fuel you put in your automobile, microorganisms affect the everyday lives of humans in both beneficial and detrimental ways.

Microorganisms existed on Earth for billions of years before plants and animals appeared, and we will see later that the genetic and physiological diversity of microbial life dwarfs that of the plants and animals. Although microorganisms are the smallest forms of life (Figure 1.1), collectively they constitute the bulk of biomass on Earth and carry out many necessary chemical reactions for higher organisms. In the absence of microorganisms, higher life forms would never have appeared and could not be sustained. Indeed, the very oxygen we breathe is the result of past microbial activity. Moreover, humans, plants, and animals are intimately dependent on microbial activities for the recycling of key nutrients and for degrading organic matter. It is thus safe to say that no other life forms are as important as microorganisms for the support and maintenance of life on Earth.

This chapter begins our journey into the microbial world. Here we will begin to discover what microorganisms are and what they do and explore their evolutionary history and impact on planet Earth. We will also place microbiology in historical context, as a process of scientific discovery. From the landmark contributions of both early microbiologists and scientists practicing today, the microbial world will begin to unfold.

MINIQUIZ
- If microbial life had not evolved, would you be here today? Give one good reason why or why not.
- Why are microbial cells useful tools for basic science? Why are microorganisms important to humans?

1.2 Structure and Activities of Microbial Cells

Microbial cells are living compartments that interact with their environment and with other cells in dynamic ways. In Chapter 2 we will examine the structure of cells in detail and relate specific structures to specific functions. Here we present a snapshot of

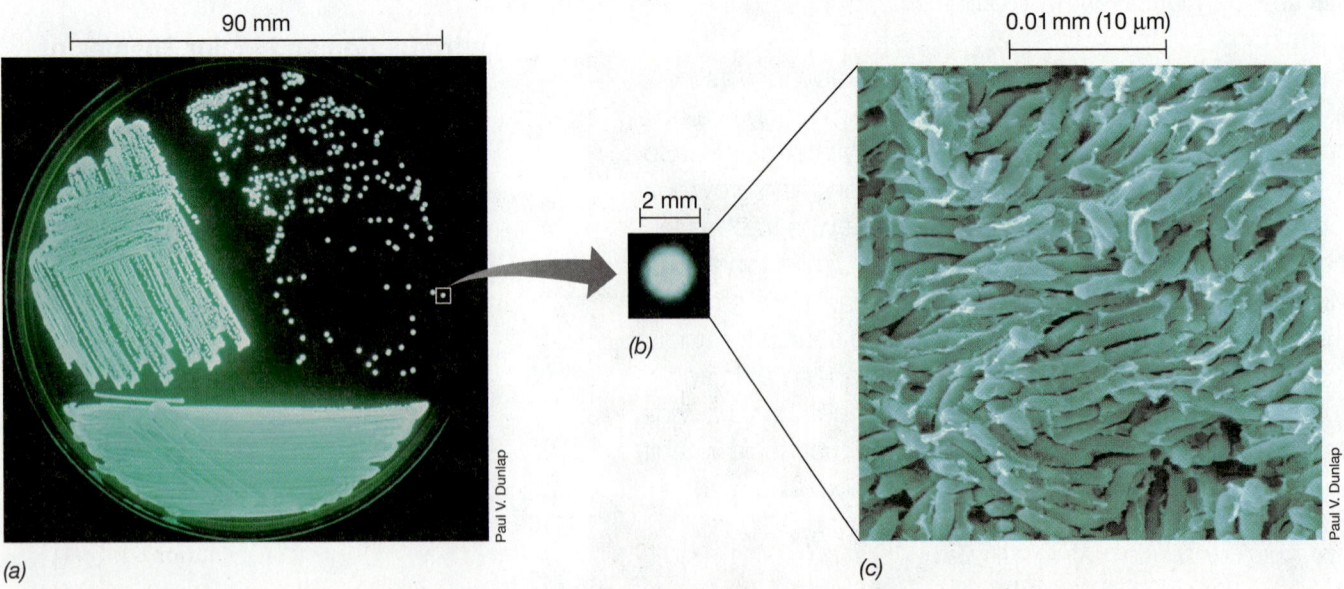

Figure 1.1 Microbial cells. *(a)* Bioluminescent (light-emitting) colonies of the bacterium *Photobacterium* grown in laboratory culture on a Petri plate. *(b)* A single colony can contain more than 10 million (10^7) individual cells. *(c)* Scanning electron micrograph of cells of *Photobacterium*.

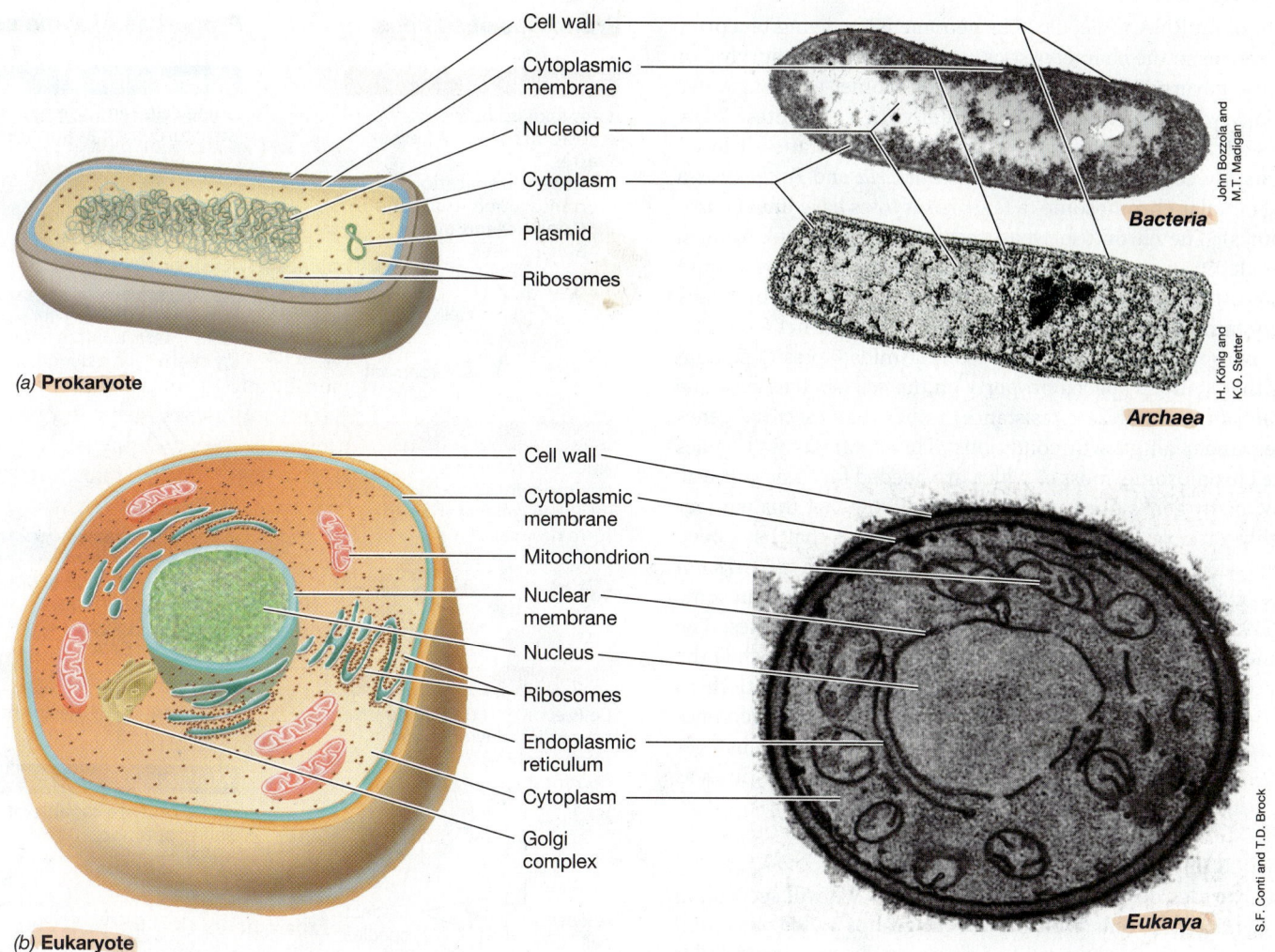

Figure 1.2 Microbial cell structure. *(a)* (Left) Diagram of a prokaryotic cell. (Right) Electron micrograph of *Heliobacterium modesticaldum* (*Bacteria*, cell is about 1 μm in diameter) and *Thermoproteus neutrophilus* (*Archaea*, cell is about 0.5 μm in diameter). *(b)* (Left) Diagram of a eukaryotic cell. (Right) Electron micrograph of a cell of *Saccharomyces cerevisiae* (*Eukarya*, cell is about 8 μm in diameter).

microbial structure and activities. We purposely exclude viruses in this discussion because although they resemble cells in many ways, viruses are not cells but instead a special category of microorganism. We consider the structure, diversity, and activities of viruses in Chapters 8 and 9.

Elements of Microbial Structure

All cells have much in common and contain many of the same components (**Figure 1.2**). All cells have a permeability barrier called the **cytoplasmic membrane** that separates the inside of the cell, the **cytoplasm**, from the outside. The cytoplasm is an aqueous mixture of **macromolecules**—proteins, lipids, nucleic acids, and polysaccharides—small organic molecules (mainly precursors of macromolecules), various inorganic ions, and **ribosomes**, the cell's protein-synthesizing structures. The **cell wall** lends structural strength to a cell; it is a relatively permeable structure located outside the membrane and is a much stronger layer than the membrane itself. Plant cells and most microorganisms have cell walls, whereas animal cells, with rare exceptions, do not.

Examination of the internal structure of cells reveals two patterns, called **prokaryote** and **eukaryote**. Prokaryotes include the *Bacteria* and the *Archaea* and consist of small and structurally rather simple cells (Figure 1.2a). Eukaryotes are typically much larger than prokaryotes and contain an assortment of membrane-enclosed cytoplasmic structures called **organelles** (Figure 1.2b). These include, most prominently, the DNA-containing nucleus but also mitochondria and chloroplasts, organelles that specialize in supplying the cell with energy, and various other organelles. Eukaryotic microorganisms include algae, protozoa and other protists, and the fungi. The cells of plants and animals are also eukaryotic. Despite the clear-cut *structural* differences between prokaryotes and eukaryotes (Figure 1.2), the word "prokaryote" does not imply *evolutionary* relatedness. As we will see in the next section, although species of *Bacteria* and *Archaea* may *look* similar, they are not closely related in an evolutionary sense.

Genes, Genomes, Nucleus, and Nucleoid

The life processes of a cell are controlled by its complement of genes, its **genome**. A gene is a segment of DNA that encodes a protein or an RNA molecule. The genome is the living blueprint of an organism; the characteristics, activities, and very survival of a cell are governed by its genome. The genomes of prokaryotes and eukaryotes are organized differently. In eukaryotes, DNA is present as linear molecules within the membrane-enclosed **nucleus**. By contrast, the genome of *Bacteria* and *Archaea* is a closed circular chromosome (a few prokaryotes have linear chromosomes). The chromosome aggregates within the cell to form the **nucleoid**, a mass visible in the electron microscope (Figure 1.2a). Most prokaryotes have only a single chromosome, but many also contain one or more small circles of DNA distinct from that of the chromosome, called *plasmids*. Plasmids typically contain genes that confer a special property on the cell (such as a unique metabolism, or antibiotic resistance) rather than essential genes needed under all growth conditions. This contrasts with genes on the chromosome, most of which are needed for basic survival.

How many genes does a cell have? We know that this number is highly variable because of the many genomes that have been sequenced. The genome of the model bacterium *Escherichia coli* is fairly typical in size; it is a single circular chromosome of 4,639,221 base pairs of DNA arranged into 4288 genes. The genomes of a few prokaryotes are three times this size while the genomes of some others contain as few as one-twentieth as many genes. Eukaryotic cells typically have much larger genomes than do prokaryotes. A human cell, for example, contains over 1000 times as much DNA as a cell of *E. coli* and about 7 times as many genes.

Activities of Microbial Cells

What activities do microbial cells carry out? We will see that in nature, microbial cells typically live in groups called *microbial communities*. **Figure 1.3** considers some of the ongoing cellular activities within the microbial community. All cells show some form of **metabolism** by taking up nutrients from the environment and transforming them into new cell materials and waste products. During these transformations, energy is conserved that can be used by the cell to support synthesis of new structures. Production of these new structures culminates in the division of the cell to form two cells. In microbiology, we use the word **growth** to refer to the increase in cell number as a result of cell division.

During metabolism and growth, both genetic and catalytic events occur in cells; biological information flow is initiated and metabolic pathways are engaged. On the genetic side, the cell's genome is replicated, and the proteins needed to support growth under a given set of conditions are biosynthesized in the sequential processes of *transcription* and *translation* (Figure 1.3). These events require that the cell's catalytic machinery—its **enzymes**—carry out reactions that supply the energy and precursors necessary for the biosynthesis of all cell components. Catalytic and genetic events in a microbial cell are coordinated and highly regulated to ensure that new cell materials are made in the proper order and concentrations and that the cell remains optimally tuned to its surroundings.

Many microbial cells are capable of **motility**, typically by self-propulsion (Figure 1.3). Motility allows cells to move away from unfavorable conditions and to exploit new resources or growth opportunities. Some microbial cells undergo **differentiation**, which

Properties of *all* cells:

Metabolism

Cells take up nutrients, transform them, and expel wastes.
1. **Genetic** (replication, transcription, translation)
2. **Catalytic** (energy, biosyntheses)

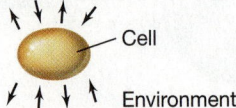

Growth

Nutrients from the environment are converted into new cell materials to form new cells.

Evolution

Cells evolve to display new properties. Phylogenetic trees capture evolutionary relationships.

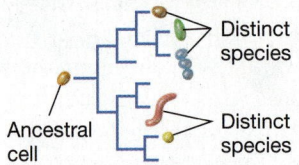

Properties of *some* cells:

Differentiation

Some cells can form new cell structures such as a spore.

Communication

Cells interact with each other by chemical messengers.

Genetic exchange

Cells can exchange genes by several mechanisms.

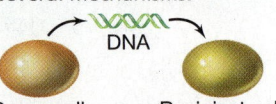

Motility

Some cells are capable of self-propulsion.

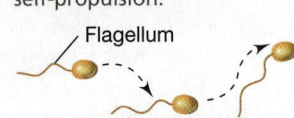

Figure 1.3 The properties of microbial cells. Major activities ongoing in cells in the microbial community are depicted.

may result in the formation of modified cells specialized for growth, dispersal, or survival. Cells respond to chemical signals in their environment, including those produced by other cells of either the same or different species, and these signals often trigger new cellular activities. Microbial cells thus exhibit intercellular **communication**; they are "aware" of their neighbors and can respond accordingly. Many prokaryotic cells can also transfer genes to or accept genes from neighboring cells, either of the same species or of a different species, in the process of **genetic exchange**.

Evolution (Figure 1.3) is the process of descent with modification in which genetic variants (mutants) are selected based on their reproductive fitness. Although we learn from elementary biology that evolution is a very slow process, evolution in microbial cells can be very rapid when selective pressure is strong. For example, witness today how genes encoding antibiotic resistance in pathogenic (disease-causing) bacteria have been selected and widely distributed by the indiscriminate use of antibiotics in human and veterinary medicine. Genetic exchange between prokaryotic cells, which is independent of evolution (Figure 1.3), can also significantly accelerate the adaptation of cells to new habitats or to rapidly changing conditions.

Not all of the processes depicted in Figure 1.3 occur in all cells. Metabolism, growth, and evolution, however, are universal. We now take a peek at the results of microbial evolution in the form of the enormous diversity of the microbial world that has been unveiled by modern microbiology.

MINIQUIZ

- What important functions do the following play in a cell: cytoplasmic membrane, ribosomes, cell wall?
- What type of cells have a nucleus? Nucleoid? What is a cell's genome and why is it important?
- What do the terms "growth" and "motility" mean in microbiology?

1.3 Evolution and Diversity of Microbial Cells

Microorganisms were the first entities on Earth that showed the properties we associate with life. How did microbial cells originate and how are extant microbial cells related to one other?

The First Cells and the Beginnings of Evolution

Because all cells are constructed in similar ways, it is thought that all cells have descended from a common ancestral cell, the *last universal common ancestor* (LUCA). After the first cells arose from nonliving materials, a process that occurred over hundreds of millions of years, their subsequent growth formed cell populations and these began to interact with other cell populations to form microbial communities. Along the way, evolution and genetic exchange served up variants that could be selected for improvements that made their success and survival more probable. Today we see the grand result of these processes, ongoing for nearly 4 billion years.

Life on Earth through the Ages

Earth is 4.6 billion years old and evidence shows that microbial cells first appeared between 3.8 and 3.9 billion years ago (**Figure 1.4**). During the first 2 billion years of Earth's existence, its atmosphere was anoxic (O_2 was absent), and only nitrogen (N_2), carbon dioxide (CO_2), and a few other gases were present. Only microorganisms capable of anaerobic metabolisms could survive under these conditions. The evolution of phototrophic microorganisms—organisms that harvest energy from sunlight—occurred within 1 billion years of the formation of Earth. The first phototrophs were relatively simple ones, such as purple or green bacteria and other anoxygenic (non-oxygen-evolving) phototrophs (**Figure 1.5a**). Cyanobacteria (oxygen-evolving phototrophs) (Figure 1.5b) evolved from anoxygenic phototrophs nearly a billion years later and began the slow process of oxygenating Earth's atmosphere. Triggered by increases in O_2 in the atmosphere, multicellular life forms eventually evolved and continued to increase in complexity, culminating in the plants and animals we know today. But plants and animals have only existed for about half a billion years. The timeline of life on Earth (Figure 1.4a) shows that 80% of life's history was exclusively microbial, and thus in many ways, Earth can be considered a microbial planet.

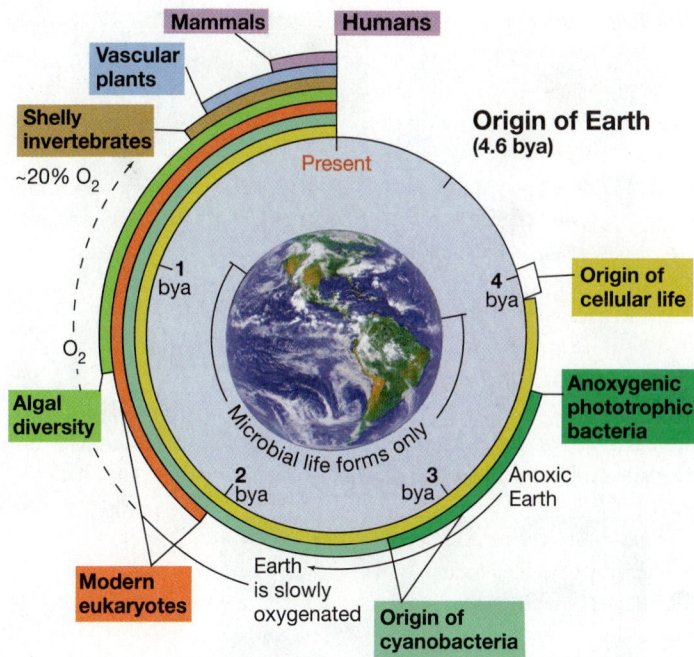

(a)

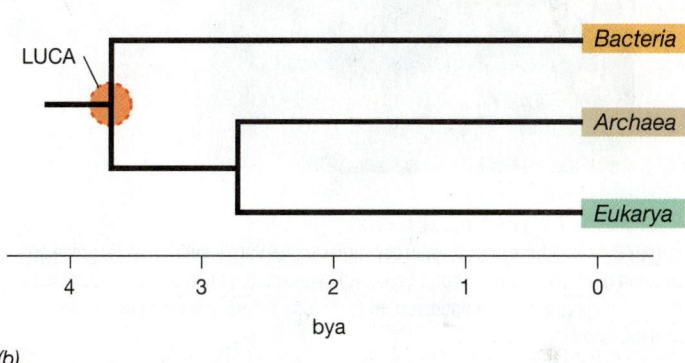

(b)

Figure 1.4 A summary of life on Earth through time and origin of the cellular domains. *(a)* Cellular life was present on Earth by 3.8 billion years ago (bya). Cyanobacteria began the slow oxygenation of Earth about 3 bya, but current levels of O_2 in the atmosphere were not achieved until 500–800 million years ago. Eukaryotes are nucleated cells (Figure 1.2*b*) and include both microbial and multicellular organisms. *(b)* The three domains of cellular organisms are *Bacteria, Archaea,* and *Eukarya. Archaea* and *Eukarya* diverged long before nucleated cells with organelles ("modern eukaryotes" in part a) appear in the fossil record. LUCA, last universal common ancestor.

As evolutionary events unfolded, three major lineages of microbial cells—the *Bacteria*, the *Archaea*, and the *Eukarya* (Figure 1.4*b*)—were distinguished; microbial *Eukarya* were the ancestors of the plants and animals. These major cell lineages are called **domains**. Over enormous periods of time, natural selection filled every suitable environment on Earth with microorganisms whose ancestry can be traced back to one of these three domains.

Microbial Diversity

Assessing the phylogenetic history of the microbial world—and thus revealing its true diversity—had to wait until tools were available that could do the job. Unlike plants and animals for which bones, fossils, leaves, and the like can be used to help reconstruct

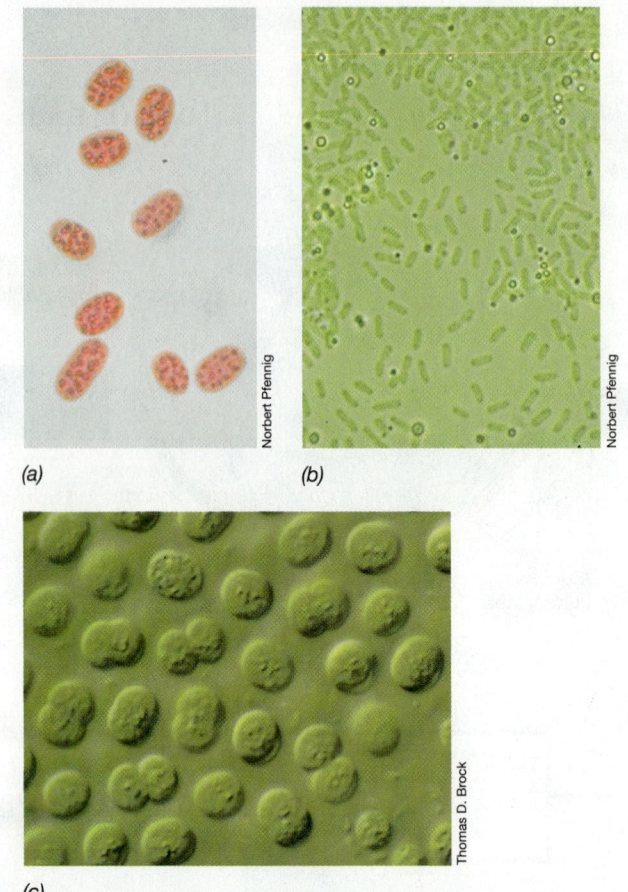

(a)

(b)

Norbert Pfennig

Norbert Pfennig

(c)

Thomas D. Brock

Figure 1.5 Phototrophic microorganisms. *(a)* Purple sulfur and *(b)* green sulfur bacteria (both anoxygenic phototrophs). *(c)* Cyanobacteria (oxygenic phototrophs). Purple and green bacteria appeared on Earth long before oxygenic phototrophs evolved (see Figure 1.4a).

phylogenies, such remains were unavailable to guide the construction of a microbial evolutionary tree. However, discoveries made in the past 40 years or so have clearly shown that each cell contains a record of its evolutionary history embedded in its genes. For reasons that will be presented in later chapters, genes that encode *ribosomal RNAs* have emerged as excellent barometers of microbial diversity. Ribosomal RNAs are components of ribosomes (Figure 1.2), the structures that synthesize new proteins as part of the process of translation. The technology for revealing the phylogeny of a microorganism from its ribosomal RNA genes is well developed, and from only a few cells, a phylogenetic tree that reveals the position of any organism relative to its neighbors can be constructed (**Figure 1.6a**).

As the ribosomal RNA phylogenetic tree of life has been fleshed out (Figure 1.6b), it has shown that thousands of species of *Bacteria* and *Archaea* exist as well as hundreds of species of microbial *Eukarya* (the tree in Figure 1.6b shows only a few landmark lineages). The tree of life has also revealed two important facts that were previously unsuspected: (1) *Bacteria* and *Archaea* are phylogenetically distinct despite sharing many structural features (Figure 1.2a), and (2) *Archaea* are more closely related to *Eukarya* than to *Bacteria*. From the last universal common

ancestor of all cells (Figure 1.4b), evolution proceeded along two paths to form the domains *Bacteria* and *Archaea*. At some later time, the domain *Archaea* diverged to distinguish the *Eukarya* from the *Archaea* (Figures 1.4b and 1.6b).

The tools for generating microbial phylogenies on pure cultures of microorganisms (Figure 1.6a) have been adapted for use in natural environments in order to probe the diversity of microbial communities. These techniques have greatly improved our picture of microbial diversity and have led to the staggering conclusion that most microorganisms that exist on Earth have yet to be brought into laboratory culture! It now appears that our understanding of microbial diversity is still in its infancy. Nevertheless, the universal tree of life provides us with a roadmap to guide future work on microbial diversity and has unveiled the previously hidden concept of three evolutionary domains of life.

MINIQUIZ

• How old is Earth and when did cells first appear on Earth?

• Why were cyanobacteria so important in the evolution of life on Earth?

• How can the phylogenetic history of microorganisms be determined?

• Name the three domains of life.

1.4 Microorganisms and Their Environments

In nature, microbial cells live in association with other cells. A *population* is a group of cells derived from a single parental cell by successive cell divisions. The immediate environment in which a microbial population lives is called its **habitat**. Populations of cells interact with other populations in **microbial communities** (**Figure 1.7**). The abundance and diversity of any microbial community is strongly controlled by the *resources* (foods) available and *conditions* (temperature, pH, presence or absence of oxygen, and so on) that prevail in that community.

Microbial Ecosystems

Microbial populations can interact with each other in beneficial, neutral, or harmful ways. For example, the metabolic waste products of one group of organisms can be nutrients or even poisons to other groups of organisms. Habitats differ markedly in their characteristics, and a habitat that is favorable for the growth of one organism may be harmful for another. Collectively, we call all the living organisms, together with the physical and chemical components of their environment, an **ecosystem**. Major microbial ecosystems are *aquatic* (oceans, ponds, lakes, streams, ice, hot springs), *terrestrial* (surface soils, deep subsurface), and *higher organisms* (in or on plants and animals).

An ecosystem is greatly influenced by microbial activities. Microorganisms carrying out metabolic processes remove nutrients from the ecosystem and use them to build new cells. At the same time, they excrete waste products back into the

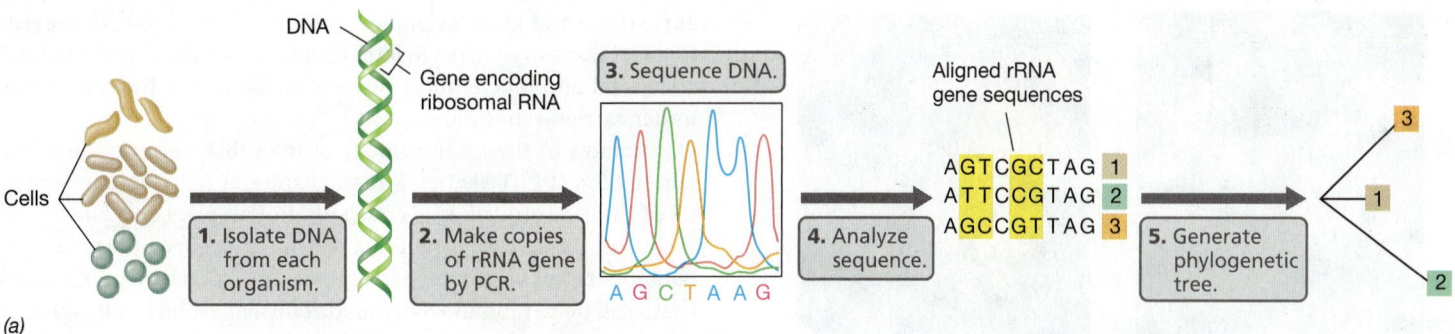

(a)

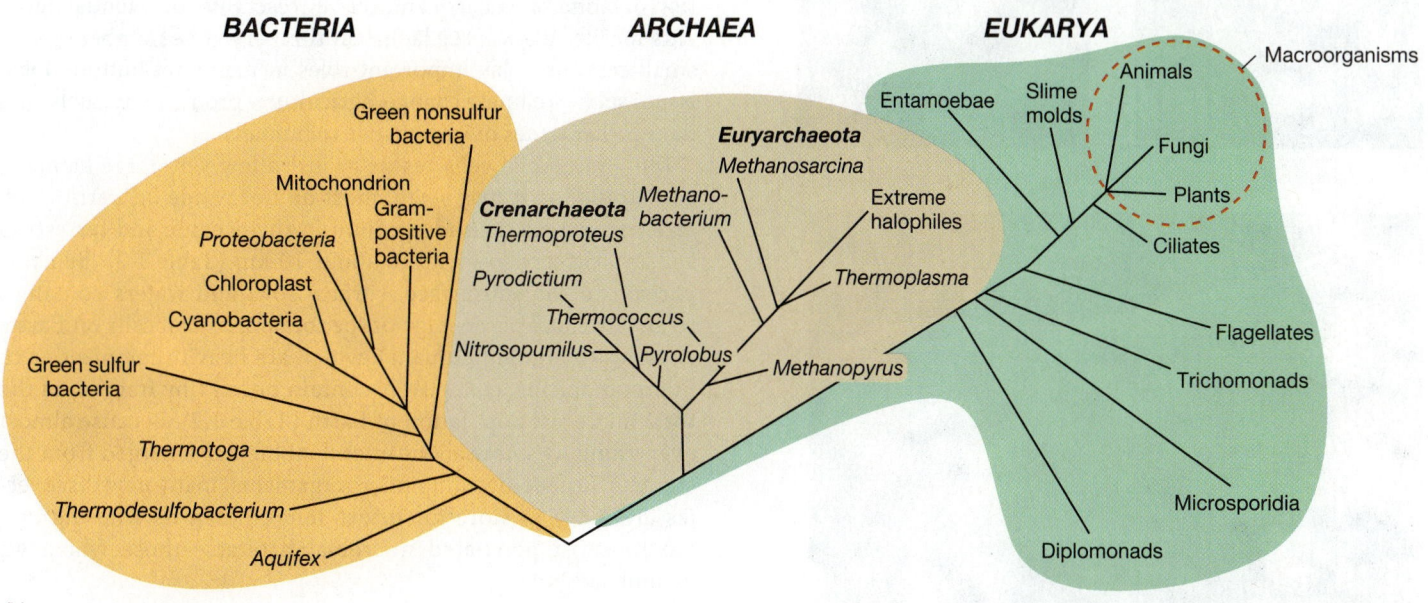

(b)

Figure 1.6 Evolutionary relationships and the phylogenetic tree of life. *(a)* The technology behind ribosomal RNA gene phylogenies. 1. DNA is extracted from cells. 2. Copies of the gene encoding rRNA are made by the polymerase chain reaction (PCR; ⇄ Section 11.3). 3, 4. The gene is sequenced and the sequence aligned with sequences from other organisms. A computer algorithm makes pairwise comparisons at each base and generates a phylogenetic tree, 5, that depicts evolutionary relationships. In the example shown, the sequence differences are highlighted in yellow and are as follows: organism 1 versus organism 2, three differences; 1 versus 3, two differences; 2 versus 3, four differences. Thus organisms 1 and 3 are closer relatives than are 2 and 3 or 1 and 2. *(b)* The phylogenetic tree of life. The tree shows the three domains of organisms and a few representative groups in each domain.

environment. Thus, microbial ecosystems expand and contract, depending on the resources and conditions available and the different populations of organisms they can support. Over time, the metabolic activities of microorganisms can gradually change their ecosystems, both chemically and physically. For example, molecular oxygen (O_2) is a vital nutrient for some microorganisms but a poison to others. If aerobic (oxygen-consuming) microorganisms remove O_2 from a habitat, rendering it anoxic (O_2-free), the changed conditions may favor the growth of anaerobic microorganisms that were formerly present in the habitat but unable to grow. In other words, as resources and conditions change in a microbial habitat, cell populations rise and fall, changing the makeup of the community and redefining the ecosystem. In later chapters we will return to a consideration of the ways in which microorganisms affect animals, plants, and the whole global ecosystem. This is the science of **microbial ecology**, perhaps the most exciting subdiscipline of microbiology today.

Microorganisms in Natural Environments

Microorganisms are present everywhere on Earth that will support life. These include habitats we are all familiar with—soil, water, animals, and plants—as well as virtually any structures made by humans. In the human body alone, microbial cells outnumber our body cells by a factor of ten. Sterility (the absence of life forms) in any natural sample is extremely rare.

In some microbial habitats higher organisms cannot survive because the habitat is too hot or too cold, too acidic or too caustic, too salty or otherwise osmotically stressing, or includes enormous pressures. Although one might predict that such "extreme environments" would pose challenges for any life forms, these punishing habitats are often teeming with microorganisms. Such

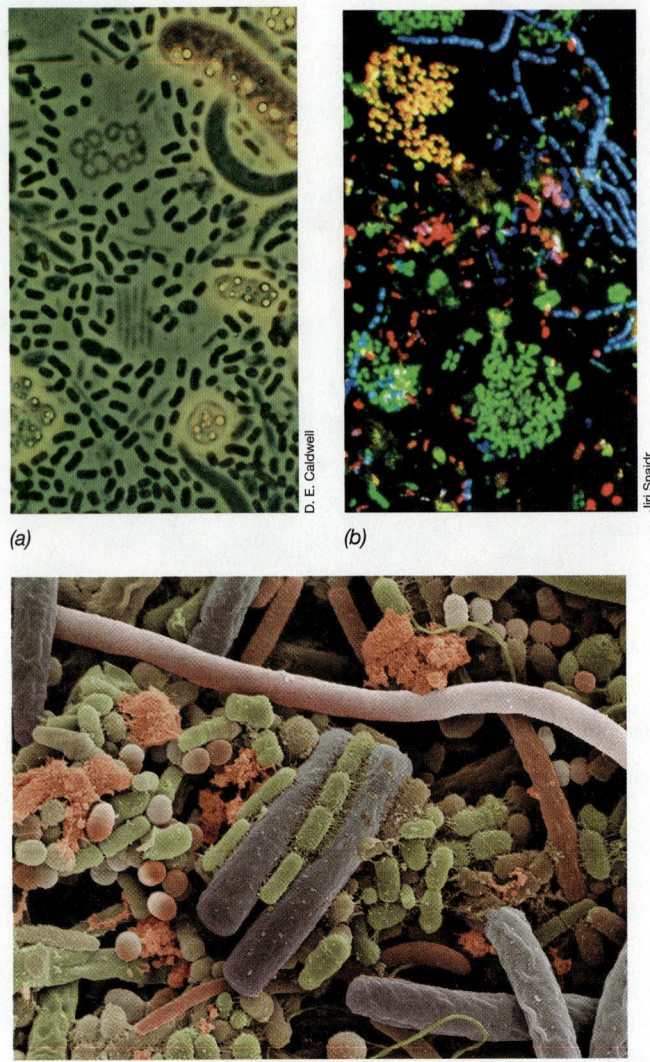

(a)

D. E. Caldwell

(b)

Jin Snaidr

(c)

Figure 1.7 Microbial communities. *(a)* A bacterial community that developed in the depths of a small Michigan lake, showing cells of various green and purple (large cells with sulfur granules) phototrophic bacteria. *(b)* A bacterial community in a sewage sludge sample. The sample was stained with a series of dyes, each of which stained a specific bacterial group. From *Journal of Bacteriology 178:* 3496–3500, Fig. 2b. © 1996 American Society for Microbiology. *(c)* Scanning electron micrograph of a microbial community scraped from a human tongue.

microorganisms are called **extremophiles** and include a large and remarkable group of mainly *Bacteria* and *Archaea*, whose collective properties define the physiochemical limits to life (Table 1.1).

Extremophiles abound in such harsh environments as volcanic hot springs; on or in the ice covering lakes (see page 1), glaciers, or the polar seas; in extremely salty bodies of water; in soils and waters having a pH as low as 0 or as high as 12; and in the deep sea or deep in the earth where pressures can exceed atmospheric by over 1000 times. Interestingly, these prokaryotes do not just *tolerate* their particular environmental extreme but actually *require* it in order to grow. That is why they are called extremophiles (the suffix *-phile* means "loving"). Table 1.1 summarizes the current "record holder" extremophiles, lists the term used to describe

each class, and gives examples of their habitats. We will revisit many of these organisms in later chapters and discover the special structural and biochemical properties that allow them to thrive under extreme conditions.

Estimates of the total number of microbial cells on Earth are about 2.5×10^{30} (Table 1.2). The total amount of carbon present in all of these microbial cells equals that of all plants on Earth, and plant carbon far exceeds animal carbon. Moreover, the collective contents of nitrogen and phosphorus in microbial cells exceed that of all plant biomass by 10-fold. Thus, microbial cells, small as they are, are not inconsequential; they constitute the major fraction of biomass on Earth and are key reservoirs of essential nutrients for life. We will see later that this very large number of very small cells also play important roles in many hot-button global issues such as climate change, agricultural productivity, fuels, and many other issues of importance to humans.

Most microbial cells reside in just a few very large habitats, and strange as it may seem, most do not reside on Earth's *surface* but instead lie underground in the oceanic and terrestrial subsurfaces at depths up to about 10 km (Table 1.2). By comparison to the subsurface, surface soils and waters contain a relatively small percentage of the total microbial cells on Earth. Animals (including humans), which are heavily colonized with microorganisms, collectively contain only a tiny fraction of the total microbial population of Earth (Table 1.2). Because almost everything we know about microbial life has emerged from the study of surface-dwelling microorganisms, many new discoveries are likely in store for future microbiologists who dig into Earth's most populated microbial habitats—those which we cannot see.

MINIQUIZ

- How does a microbial community differ from a microbial population?
- What is a habitat? How can microorganisms change the characteristics of their habitats?
- What is an extremophile?
- Where do most microorganisms live in nature?

1.5 The Impact of Microorganisms on Humans

Through the years microbiologists have made great strides in discovering how microorganisms work, and application of this knowledge has greatly advanced human health and welfare. Besides understanding microorganisms as agents of disease, microbiology has made great advances in understanding the important role that microorganisms play in food and agriculture, and microbiologists have been able to exploit microbial activities to produce valuable human products, generate energy, and clean up the environment.

Microorganisms as Agents of Disease

The statistics summarized in **Figure 1.8** show how microbiologists and clinical medicine have combined to conquer infectious diseases

Table 1.1 Classes and examples of extremophiles[a]

Extreme	Descriptive term	Genus/species	Domain	Habitat	Minimum	Optimum	Maximum
Temperature							
High	Hyperthermophile	*Methanopyrus kandleri*	Archaea	Undersea hydrothermal vents	90°C	106°C	122°C[b]
Low	Psychrophile	*Psychromonas ingrahamii*	Bacteria	Sea ice	−12°C	5°C	10°C
pH							
Low	Acidophile	*Picrophilus oshimae*	Archaea	Acidic hot springs	−0.06	0.7[c]	4
High	Alkaliphile	*Natronobacterium gregoryi*	Archaea	Soda lakes	8.5	10[d]	12
Pressure	Barophile (piezophile)	*Moritella yayanosii*	Bacteria	Deep ocean sediments	500 atm	700 atm[e]	>1000 atm
Salt (NaCl)	Halophile	*Halobacterium salinarum*	Archaea	Salterns	15%	25%	32% (saturation)

[a]The organisms listed are the current "record holders" for growth in laboratory culture at the extreme condition listed.
[b]Anaerobe showing growth at 122°C only under several atmospheres of pressure.
[c]*P. oshimae* is also a thermophile, growing optimally at 60°C.
[d]*N. gregoryi* is also an extreme halophile, growing optimally at 20% NaCl.
[e]*M. yayanosii* is also a psychrophile, growing optimally near 4°C.

in the past 100 years. At the beginning of the twentieth century, the major causes of human death were infectious diseases caused by bacterial and viral **pathogens**. In those days children and the aged in particular succumbed in large numbers to microbial diseases. Today, however, infectious diseases are much less deadly, at least in developed countries. Control of infectious disease has come from a combination of advances including our increased understanding of disease processes, improved sanitary and public health practices, active vaccine campaigns, and the widespread use of antimicrobial agents, such as antibiotics. As we will see in the second part of this chapter, the development of microbiology as a science can be traced to pioneering studies of infectious disease.

Although many infectious diseases are now controlled, many others can still be a threat, particularly in developing countries. For example, diseases such as malaria, tuberculosis, cholera, African sleeping sickness, measles, pneumonia and other respiratory diseases, and diarrheal syndromes are still common in developing countries. Moreover, humans worldwide are under threat from diseases that could quickly emerge, such as bird or swine flu, or Ebola hemorrhagic fever; these are primarily animal diseases that under certain circumstances can be transmitted to humans and spread quickly through a population. Thus, microorganisms can still be serious health threats to humans in all parts of the world.

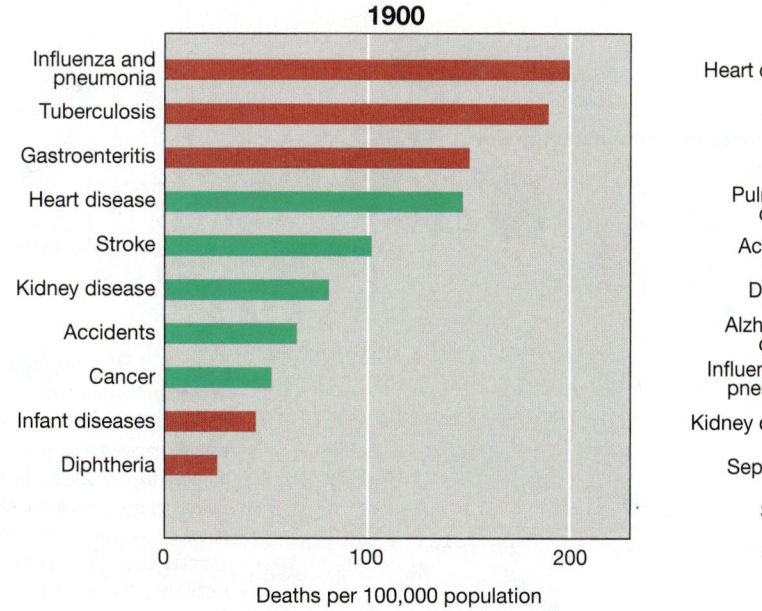
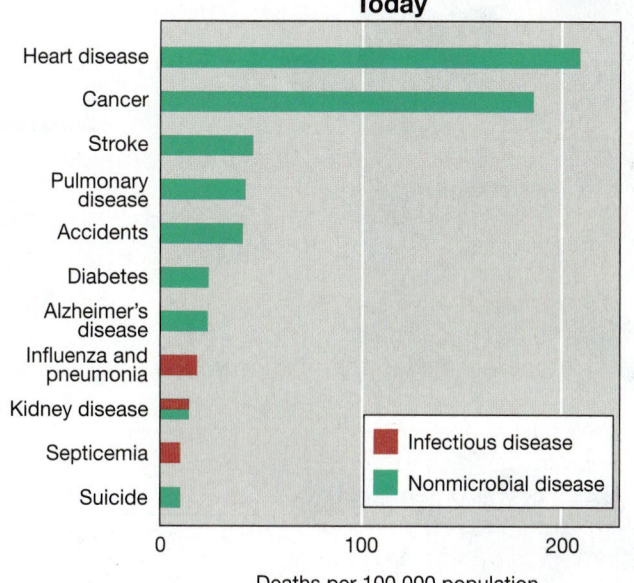

Figure 1.8 Death rates for the leading causes of death in the United States: 1900 and today. Infectious diseases were the leading causes of death in 1900, whereas today they account for relatively few deaths. Kidney diseases can be caused by microbial infections or systemic sources (diabetes, cancers, toxicities, metabolic diseases, etc.). Data are from the United States National Center for Health Statistics and the Centers for Disease Control and Prevention.

Table 1.2 Distribution of microorganisms in and on Earth[a]

Habitat	Percent of total
Marine subsurface	66
Terrestrial subsurface	26
Surface soil	4.8
Oceans	2.2
All other habitats[b]	1.0

[a]Data compiled by William Whitman, University of Georgia, USA; refer to total numbers (estimated to be about 2.5×10^{30} cells) of *Bacteria* and *Archaea*. This enormous number of cells contains, collectively, about 5×10^{17} grams of carbon.
[b]Includes, in order of decreasing numbers: freshwater and salt lakes, domesticated animals, sea ice, termites, humans, and domesticated birds.

Although we must appreciate the powerful threat posed by pathogenic microorganisms, in reality, most microorganisms are not harmful to humans. In fact, most microorganisms cause no harm but instead are beneficial, and in many cases are even essential, to human welfare and the functioning of the planet. We turn our attention to these microorganisms now.

Microorganisms, Agriculture, and Human Nutrition

Agriculture benefits from the cycling of nutrients by microorganisms. For example, a number of major crop plants that feed humans and domesticated animals are legumes. Legumes live in close association with bacteria that form structures called *nodules* on their roots. In the nodules, these bacteria convert atmospheric nitrogen (N_2) into ammonia (NH_3, the process of *nitrogen fixation*) that the plants use as a nitrogen source for growth (**Figure 1.9**).

Nitrogen fixation also eliminates the need for farmers to apply costly and polluting nitrogen fertilizers. Other bacteria cycle sulfur, oxidizing toxic sulfur compounds such as hydrogen sulfide (H_2S) into sulfate (SO_4^{2-}), which is nontoxic and an essential plant nutrient (Figure 1.9c).

Also of major agricultural importance are microorganisms that inhabit the *rumen* of ruminant animals, such as cattle and sheep. The rumen is a microbial ecosystem in which large populations of microorganisms digest and ferment the polysaccharide cellulose, the major component of plant cell walls (Figure 1.9d). Without these symbiotic microorganisms, ruminants could not thrive on cellulose-rich (but otherwise nutrient-poor) food, such as grass and hay. Many domesticated and wild herbivorous mammals—including deer, bison, camels, giraffes, and goats—are also ruminants.

The human gastrointestinal (GI) tract lacks a rumen, and microbial numbers comparable to those in the rumen (about 10^{11} microbial cells per gram of contents) occur only in the colon (large intestine). The colon (**Figure 1.10**) follows the stomach and small intestine in the digestive tract, but unlike the rumen, the colon lacks significant numbers of cellulose-degrading microorganisms. Microbial cell numbers are low in the highly acidic (pH 2) stomach (about 10^4 per gram) but increase to about 10^8 per gram near the end of the small intestine (pH 5) and then reach maximal numbers in the colon (pH 7) (Figure 1.10). Microorganisms in the colon assist in digestive processes by synthesizing certain vitamins and other essential nutrients but also compete for space and resources with pathogenic microorganisms that may enter the GI tract in contaminated food and water. Thus by their sheer numbers alone, the colonic microflora help prevent pathogens from gaining a foothold.

In addition to benefiting agriculture, microorganisms also have negative effects on the industry. Microbial diseases of plants and

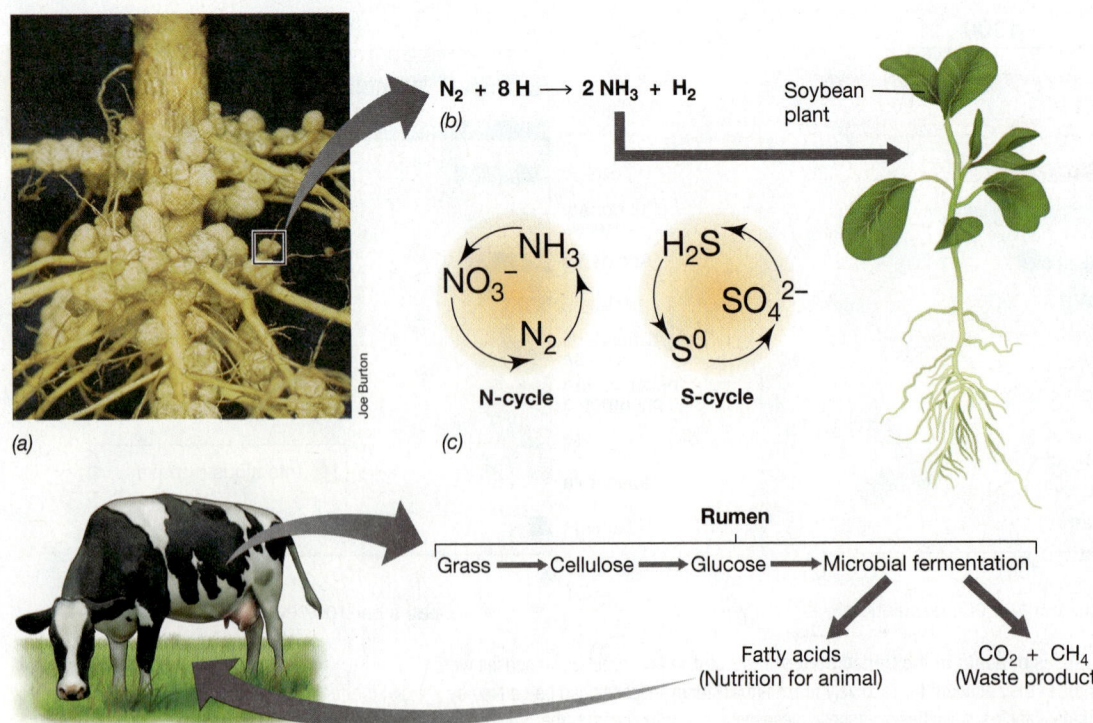

N$_2$ + 8 H $\longrightarrow$ 2 NH$_3$ + H$_2$
(b)

Soybean plant

N-cycle

S-cycle

(a)

(c)

Joe Burton

Rumen

Grass $\longrightarrow$ Cellulose $\longrightarrow$ Glucose $\longrightarrow$ Microbial fermentation

Fatty acids (Nutrition for animal)

CO_2 + CH_4 (Waste products)

(d)

Figure 1.9 **Microorganisms in modern agriculture.** *(a, b)* Root nodules on this soybean plant contain bacteria that fix molecular nitrogen (N_2) for use by the plant. *(c)* The nitrogen and sulfur cycles, key nutrient cycles in nature. *(d)* Ruminant animals. Microorganisms in the rumen of the cow convert cellulose from grass into fatty acids that can be used by the animal. The other products are not so desirable, as CO_2 and CH_4 are the major gases that cause global warming.

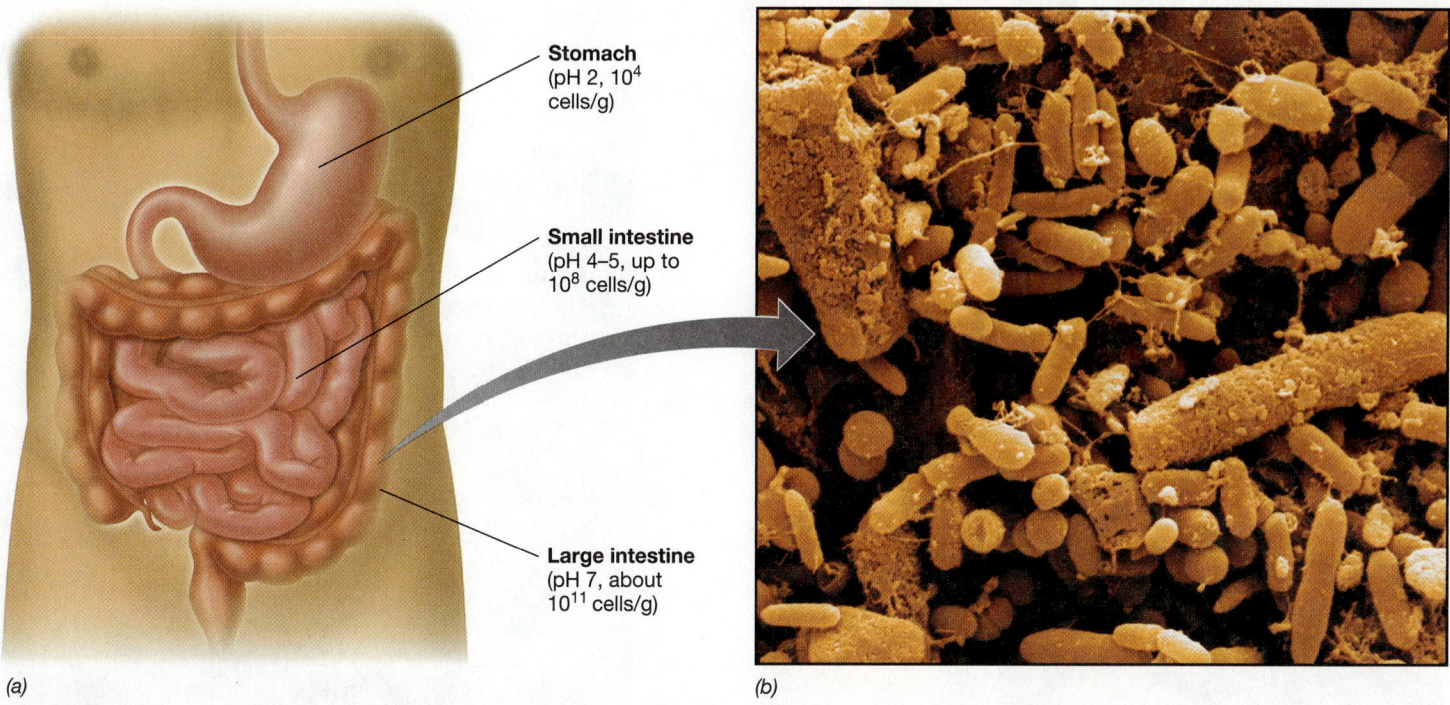

Figure 1.10   The human gastrointestinal tract. *(a)* Diagram of the human GI tract showing the major organs. *(b)* Scanning electron micrograph of microbial cells in the human colon (large intestine). Cell numbers in the colon can reach as high as 10^{11} per gram. As well as cell *numbers*, the microbial *diversity* in the colon is also quite high.

animals used for human food cause major economic losses every year. Occasionally a food product causes serious human disease, such as when pathogenic *Escherichia coli* or *Salmonella* is transmitted from infected meat, or when microbial pathogens are ingested with contaminated fresh fruits and vegetables. Microorganisms thus significantly impact the agriculture industry in both positive and negative ways.

Microorganisms and Food, Energy, and the Environment

Microorganisms play important roles in the food industry, including food spoilage, safety, and production. Food spoilage alone results in huge economic losses each year, and the canning, frozen food, and dried-food industries were developed as means to preserve foods that would otherwise undergo microbial spoilage. Food safety requires the constant monitoring of food products to ensure that they are free of pathogens and the tracking of disease outbreaks to identify the source(s) of the pathogens. Fresh foods such as meats, fruits, and vegetables are most vulnerable to microbial contamination and typically have brief "shelf lives" for the very reason that contamination is virtually impossible to prevent.

Although food safety is a major problem in the food industry, not all microorganisms

in foods cause harm to the food products or to those who eat them. Many are desirable or even essential, such as those that grow in fermented foods (**Figure 1.11**). For example, many dairy products depend on the activities of microorganisms to produce key acids characteristic of the products, such as in the fermentations that yield cheeses, yogurt, and buttermilk. Sauerkraut, pickles, and some sausages are also subject to microbial fermentations. Moreover, baked goods and alcoholic beverages rely on the fermentative activities of yeast, which generate carbon dioxide (CO_2)

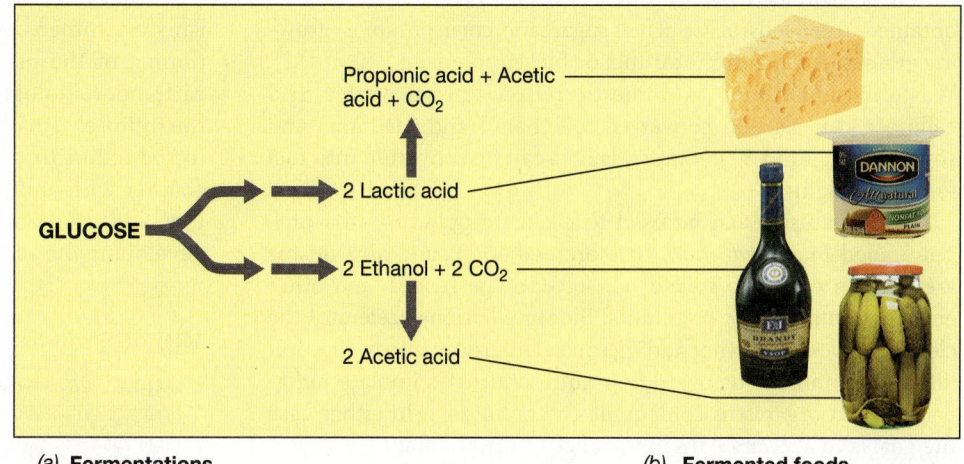

(a) Fermentations

GLUCOSE

2 Lactic acid → Propionic acid + Acetic acid + CO_2

2 Ethanol + 2 CO_2

2 Acetic acid

(b) Fermented foods

Figure 1.11   Fermented foods. *(a)* Major fermentations in various fermented foods. It is the fermentation product (ethanol, or lactic, propionic, or acetic acids) that both preserves the food and renders in it a characteristic flavor. *(b)* Photo of several fermented foods showing the characteristic fermentation product in each.

(a)

(b)

Figure 1.12 Ethanol as a biofuel. *(a)* Major crop plants used as feedstocks for biofuel ethanol production. Top: switchgrass, a source of cellulose. Bottom: corn, a source of cornstarch. Both cellulose and starch are composed of glucose, which is fermented to ethanol by yeast. *(b)* An ethanol plant in the United States. Ethanol produced by fermentation is distilled and then stored in the tanks.

to raise the dough and alcohol as a key ingredient, respectively (Figure 1.11). These products of fermentation are not only desirable chemicals but also function to preserve the food product from deleterious microbial growth.

Some microorganisms produce *biofuels.* For example, natural gas (methane, CH_4) is a product of the anaerobic metabolism of a group of *Archaea* called *methanogens.* Ethyl alcohol (ethanol), which is produced by the microbial fermentation of glucose obtained from feedstocks such as sugarcane, corn, or rapidly growing grasses, is a major motor fuel or fuel supplement (**Figure 1.12**). Waste materials such as domestic refuse, animal wastes, and cellulose can also be converted to ethanol and methane; and soybeans (Figure 1.9) contain oils that can be converted into fuel for diesel engines.

Microorganisms can be used to clean up pollution in a process called *bioremediation.* In bioremediation, microorganisms are used to consume spilled oil, solvents, pesticides, and other environmentally toxic pollutants. Bioremediation accelerates the cleanup process by either adding special microorganisms to a polluted environment or by adding nutrients that stimulate indigenous microorganisms to degrade the pollutants. In either case the goal is to accelerate disappearance of the pollutant.

Microorganisms can also be harnessed to produce commercially valuable products. In *industrial microbiology,* naturally occurring microorganisms are grown on a massive scale to make large amounts of products of relatively low value, such as antibiotics,

enzymes, and certain chemicals. By contrast, *biotechnology* employs genetically engineered microorganisms to synthesize products of high value, such as insulin or other human proteins, usually on a small scale. Genomics has greatly enhanced both industrial microbiology and biotechnology by making it possible to inspect the genome of virtually any organism for genes of potential commercial interest.

As the above discussion shows, the influence of microorganisms on humans is great and their activities essential for functioning of the planet. Or, as the eminent French chemist and early microbiologist Louis Pasteur so aptly put it: "The role of the infinitely small in nature is infinitely large." We continue our introduction to the microbial world in the second half of this chapter with an historical overview of the contributions of Pasteur and a few other key scientists who were instrumental in developing the science of microbiology.

MINIQUIZ

- List two ways in which microorganisms are important in the food and agricultural industries.
- Give some examples of biofuels. How can nitrogen fixation in root nodules help biofuel production?
- What is biotechnology and how can it improve the lives of humans?

II • Microbiology in Historical Context

The future of any science is rooted in its past accomplishments. Although microbiology claims very early roots, the science did not really develop in a systematic way until the nineteenth century because technology such as microscopes and culturing techniques had to catch up with the already strong scientific curiosity. In the past 150 years or so, microbiology has moved forward in a way unprecedented by any other biological science and has spawned several new fields in modern biology. We retrace some highlights in the history of microbiology now and describe a few of the major contributors.

1.6 The Discovery of Microorganisms

Although the existence of creatures too small to be seen with the naked eye had been suspected for centuries, their discovery had to await invention of the microscope. The English mathematician and natural historian Robert Hooke (1635–1703) was an excellent microscopist. In his famous book *Micrographia* (1665), the first book devoted to microscopic observations, Hooke illustrated, among many other things, the fruiting structures of molds (**Figure 1.13**). This was the first known description of microorganisms.

The first person to see bacteria, the smallest microbial cells, was the Dutch draper and amateur microscopist Antoni van Leeuwenhoek (1632–1723). Van Leeuwenhoek constructed extremely simple microscopes containing a single lens to examine various natural substances for microorganisms (**Figure 1.14**). These microscopes were crude by today's standards, but by careful manipulation and focusing, van Leeuwenhoek was able to see bacteria. He discovered bacteria in 1676 while studying pepper–water infusions, and reported his observations in a series of letters to the prestigious Royal Society of London, which published them in English translation in 1684. Drawings of some of van Leeuwenhoek's "wee animalcules," as he referred to them, are shown in Figure 1.14*b*, and a photo taken through a van Leeuwenhoek microscope is shown in Figure 1.14*c*.

Because experimental tools to study microorganisms were crude at this time, little progress in understanding the nature and importance of bacteria was made for the next 150 years. However, in the mid-nineteenth century, microbiology reawakened. A major contributor during this period was the German–Polish scientist, Ferdinand Cohn. Cohn (1828–1898) was trained as a botanist, and his interests in microscopy led him to the study of unicellular algae and later to bacteria, including the large sulfur bacterium *Beggiatoa* (**Figure 1.15**). Cohn was particularly interested in heat resistance in bacteria, which led to his discovery that some bacteria form *endospores*. We now know that bacterial endospores are formed by differentiation from the mother (vegetative) cell and are extremely heat-resistant structures. Cohn described the life cycle of the endospore-forming bacterium *Bacillus* (vegetative cell → endospore → vegetative cell) and showed that vegetative cells but not endospores were killed by boiling.

Cohn also laid the groundwork for a system of bacterial classification and devised many highly effective methods for preventing the contamination of culture media, such as the use of cotton for

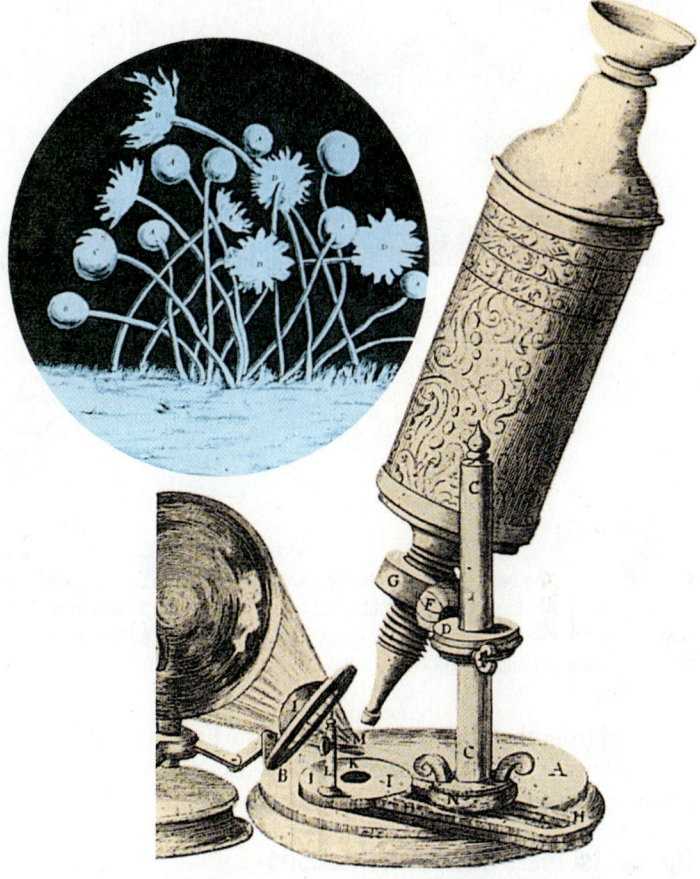

Figure 1.13 Robert Hooke and early microscopy. A drawing of the microscope used by Robert Hooke in 1664. The lens was fitted at the end of an adjustable bellows (G) and light focused on the specimen by a separate lens (1). Inset: Hooke's drawing of a bluish mold he found degrading a leather surface; the round structures contain spores of the mold.

closing flasks and tubes. These methods were later adopted by Robert Koch, the first medical microbiologist, and allowed him to make rapid progress in the isolation and characterization of several disease-causing bacteria. Cohn was also a contemporary of Louis Pasteur, and we consider the contributions of Pasteur and Koch in the next two sections.

MINIQUIZ

- What prevented the science of microbiology from developing before the era of Hooke and van Leeuwenhoek?
- What major discovery emerged from Cohn's study of heat resistance in microorganisms?

1.7 Pasteur and Spontaneous Generation

Major advances in microbiology were made in the nineteenth century because of the interest in two major questions of the day: (1) Does spontaneous generation occur? and (2) What is the nature of infectious disease? Answers to these seminal questions

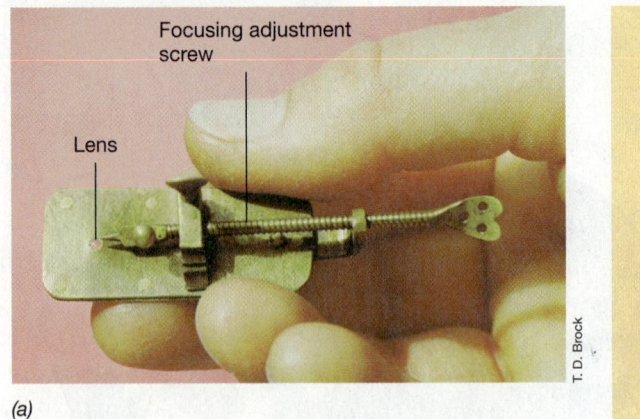

Focusing adjustment screw

Lens

(a)

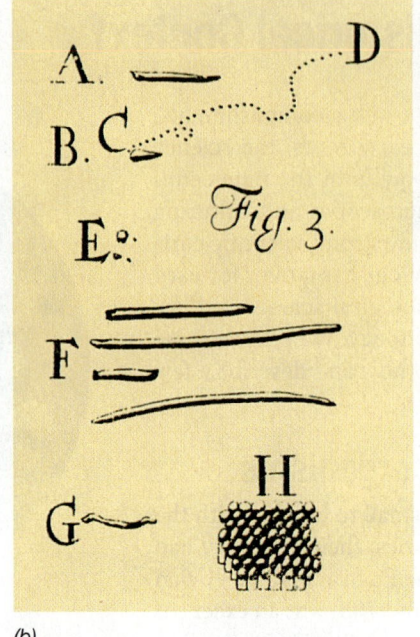

(b)

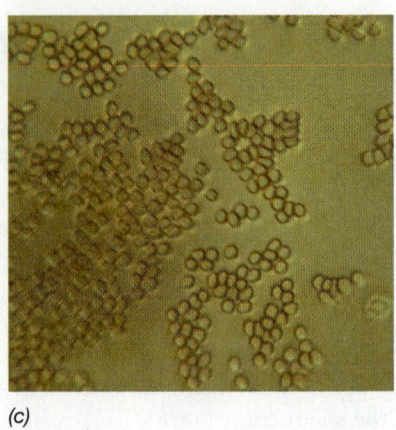

(c)

Figure 1.14 The van Leeuwenhoek microscope. *(a)* A replica of Antoni van Leeuwenhoek's microscope. *(b)* Van Leeuwenhoek's drawings of bacteria, published in 1684. Even from these simple drawings we can recognize several shapes of common bacteria: A, C, F, and G, rods; E, cocci; H, packets of cocci. *(c)* Photomicrograph of a human blood smear taken through a van Leeuwenhoek microscope. Red blood cells are clearly apparent.

emerged from the work of two giants in the fledgling field of microbiology: the French chemist Louis Pasteur and the German physician Robert Koch. We begin with the work of Pasteur.

Optical Isomers and Fermentations

Pasteur was a chemist by training and was one of the first to recognize the significance of *optical isomers*. A molecule is optically active if a pure solution or crystal diffracts light in only one

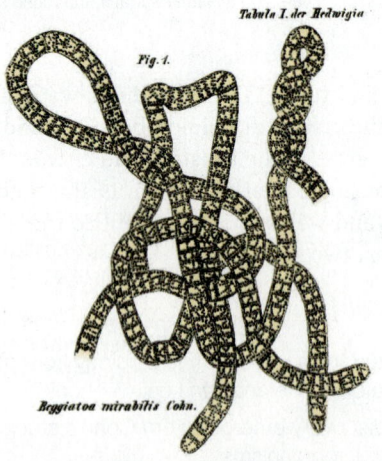

Figure 1.15 Drawing by Ferdinand Cohn of the large filamentous sulfur-oxidizing bacterium *Beggiatoa*. The small granules inside the cells consist of elemental sulfur, produced from the oxidation of hydrogen sulfide (H_2S). Cohn was the first to identify the granules as sulfur in 1866. A cell of *Beggiatoa* is about 15 µm in diameter. *Beggiatoa* moves on solid surfaces by a gliding mechanism, and in so doing, cells often twist about one another. Compare this drawing with Winogradsky's drawings of *Beggiatoa* in Figure 1.24b.

direction. Pasteur studied crystals of tartaric acid that he separated by hand into those that bent a beam of polarized light to the left and those that bent the beam to the right. Pasteur discovered that the mold *Aspergillus* metabolized D-tartrate, which bent light to the right, but did not metabolize its optical isomer, L-tartrate (**Figure 1.16**). The fact that a living organism could discriminate between optical isomers was not lost on Pasteur, and he began to suspect that some chemical activities were actually catalyzed by microorganisms and that these could be distinct from purely chemical reactions.

Pasteur initiated studies on the mechanism of the alcoholic fermentation, which in the mid-nineteenth century was assumed to be a strictly chemical process. The yeast cells in the fermenting broth were thought to be some sort of chemical substance formed by the fermentation. However, microscopic observations and other simple but rigorous experiments convinced Pasteur that the alcoholic fermentation was catalyzed by living microorganisms, the yeast cells. From these foundational studies, Pasteur began a series of classic experiments on spontaneous generation, experiments that are forever linked to his name and to the science of microbiology.

Spontaneous Generation

The concept of **spontaneous generation** had existed since biblical times and its basic tenet can be easily grasped. If food or some other perishable material is allowed to stand for some time, it putrefies. When examined microscopically, the putrefied material is teeming with microorganisms. From where do these organisms arise? Some people said they developed from seeds or germs that entered the food from air. Others said they arose spontaneously

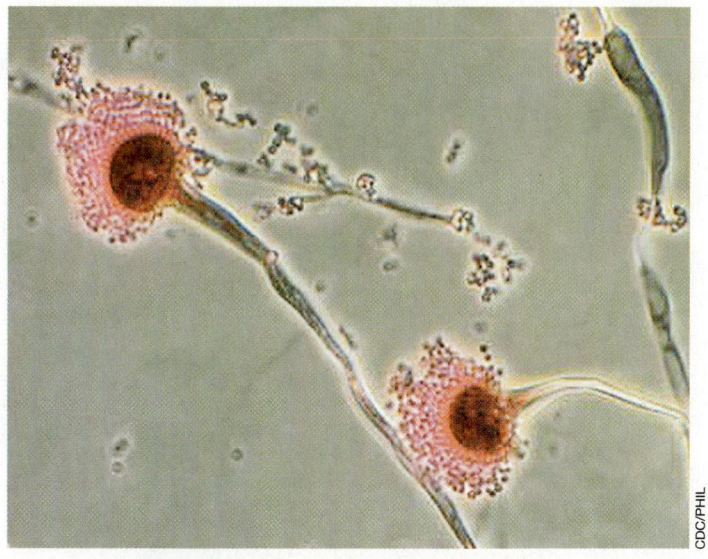

CDC/PHIL

(a)

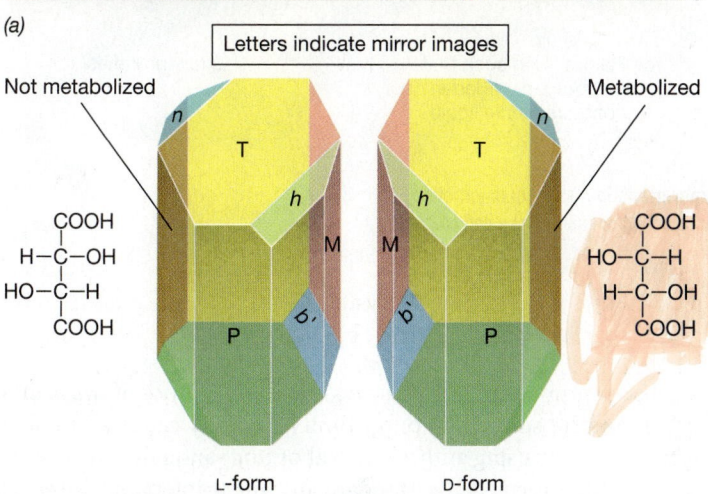

Letters indicate mirror images

Not metabolized

Metabolized

```
COOH
H–C–OH
HO–C–H
COOH
```

```
COOH
HO–C–H
H–C–OH
COOH
```

L-form

D-form

(b)

Figure 1.16 Louis Pasteur and optical isomers. *(a)* Light micrograph of cells of the mold *Aspergillus*. *(b)* Pasteur's drawings of crystals of tartaric acid. Left-handed L-form crystals bend light to the left, and right-handed crystals bend light to the right. Note that the two crystals are mirror images of one another, a hallmark of optical isomers. Pasteur found that only D-tartrate was metabolized by *Aspergillus*.

from nonliving materials, that is, by *spontaneous generation*. Who was right? Keen insight was necessary to solve this controversy, and this was exactly the kind of problem that appealed to Louis Pasteur.

Pasteur became a powerful opponent of spontaneous generation. Following his discoveries on the tartaric acid and alcoholic fermentations, Pasteur predicted that microorganisms in putrefying materials were descendants of cells that entered from the air or cells that had been on the decaying materials to begin with. Pasteur further reasoned that if food were treated in such a way as to destroy all living organisms present—that is, if it were rendered **sterile**—and then protected from further contamination, it should not putrefy.

Pasteur used heat to kill contaminating microorganisms, and he found that extensive heating of a nutrient solution followed

by sealing kept it from putrefying. Proponents of spontaneous generation criticized these experiments by declaring that "fresh air" was necessary for the phenomenon to occur. In 1864 Pasteur countered this objection simply and brilliantly by constructing a swan-necked flask, now called a *Pasteur flask* (**Figure 1.17**). In such a flask nutrient solutions could be heated to boiling and sterilized. However, after the flask was cooled, air could reenter but the bend in the neck prevented particulate matter (including microorganisms) from entering the nutrient solution and initiating putrefaction. Nutrient solutions in such flasks remained sterile indefinitely.

The microbial growth was observed only after particulate matter from the neck of the flask was allowed to enter the liquid in the flask (Figure 1.17c), which settled the spontaneous generation controversy forever. Pasteur's work on spontaneous generation naturally led to the development of effective sterilization procedures that were eventually standardized and carried over into both basic and applied microbiological research as well as clinical medicine. The food industry also benefited from the work of Pasteur, as his principles were quickly adapted for the preservation of milk and many other foods by heat treatment (pasteurization).

Other Accomplishments of Pasteur

Pasteur went on from his famous work on spontaneous generation to many other triumphs in microbiology and medicine. Some highlights include his development of vaccines for the diseases anthrax, fowl cholera, and rabies. Pasteur's work on rabies was his most famous success, culminating in July 1885 with the first administration of a rabies vaccine to a human, a young French boy named Joseph Meister who had been bitten by a rabid dog. In those days, a bite from a rabid animal was invariably fatal. News spread quickly of the success of Meister's vaccination, and of one administered shortly thereafter to a young shepherd boy, Jean-Baptiste Jupille (**Figure 1.18a**). Within a year several thousand people bitten by rabid animals had traveled to Paris to be treated with Pasteur's rabies vaccine.

Pasteur's fame from his rabies research was legendary and led the French government to establish the Pasteur Institute in Paris in 1888 (Figure 1.18b). Originally established as a clinical center for the treatment of rabies and other contagious diseases, the Pasteur Institute today is a major biomedical research center focused on antiserum and vaccine research and production. The medical and veterinary breakthroughs of Pasteur were not only highly significant in their own right but helped solidify the concept of the germ theory of disease, whose principles were being developed at about the same time by a second giant of this era, Robert Koch.

MINIQUIZ

- Define the term sterile. How did Pasteur's experiments using swan-necked flasks defeat the theory of spontaneous generation?
- Besides ending the controversy over spontaneous generation, what other accomplishments do we credit to Pasteur?

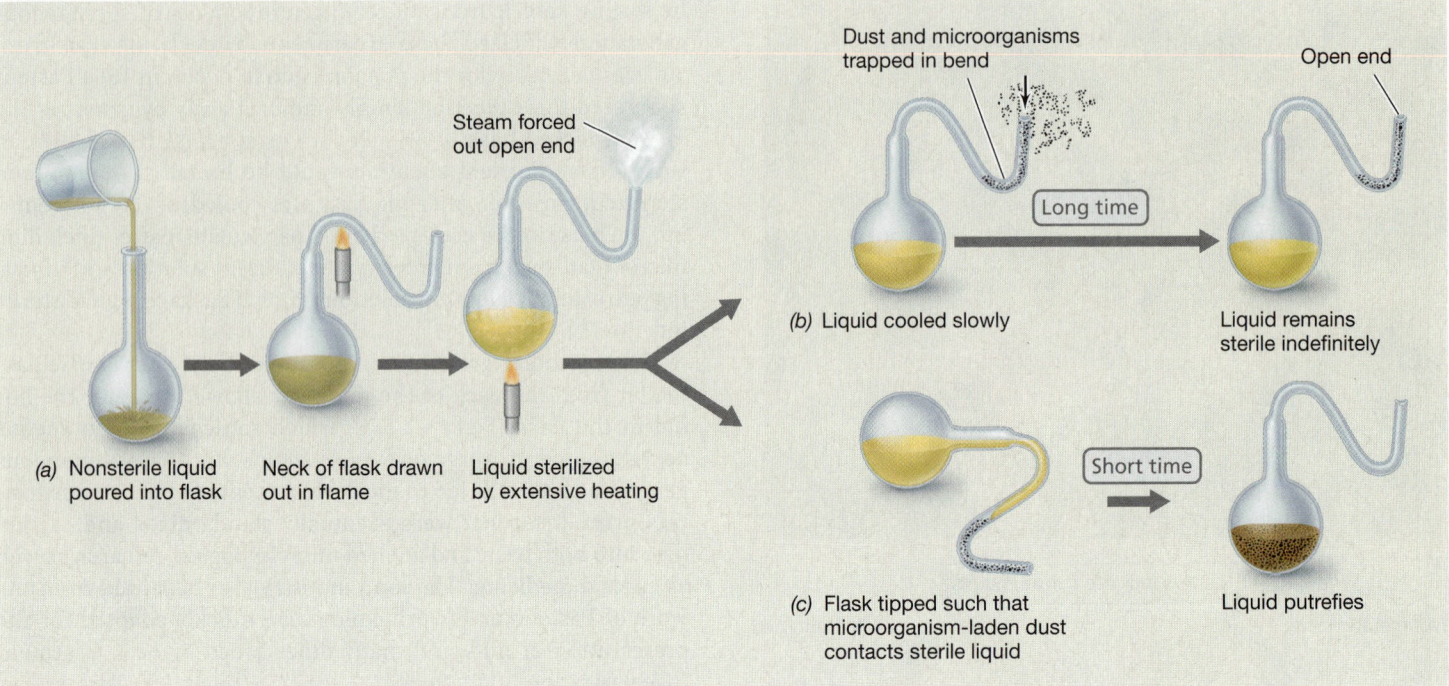

Figure 1.17 **The defeat of spontaneous generation: Pasteur's swan-necked flask experiment.** In *(c)* the liquid putrefies because microorganisms enter with the dust. The bend in the flask allowed air to enter (a key objection of Pasteur's sealed flasks) but prevented microorganisms from entering.

1.8 Koch, Infectious Disease, and Pure Cultures

Proof that some microorganisms cause disease provided the greatest impetus for the development of microbiology as an independent biological science. Even as early as the sixteenth century it was thought that something that induced disease could be transmitted from a diseased person to a healthy person. After the discovery of microorganisms, it was widely believed that they were responsible, but definitive proof was lacking. Improvements in sanitation by the Hungarian physician Ignaz Semmelweis (attempts to control hospital-associated infections, 1847) and the British physician Joseph Lister (introduction of aseptic techniques for surgeries, 1867) provided indirect evidence for the importance of microorganisms in causing human diseases. But it was not until the work of the German physician Robert Koch (1843–1910) (**Figure 1.19**) that the concept of infectious disease was developed and given direct experimental support.

The Germ Theory of Disease and Koch's Postulates

In his early work Koch studied anthrax, a disease of cattle and occasionally of humans. Anthrax is caused by an endospore-forming bacterium called *Bacillus anthracis.* By careful microscopy and staining, Koch established that the bacteria were always present in the blood of an animal that was succumbing to the disease. However, Koch reasoned that the mere *association* of

the bacterium with the disease was not actual proof of *cause and effect*, and he seized the opportunity to study cause and effect experimentally using anthrax and laboratory animals. The results of this study formed the standard by which infectious diseases have been studied ever since.

Koch used mice as experimental animals. Using appropriate controls, Koch demonstrated that when a small drop of blood from a mouse with anthrax was injected into a healthy mouse, the latter quickly developed anthrax. He took blood from this second animal, injected it into another, and again observed the characteristic disease symptoms. However, Koch carried this experiment a critically important step further. He discovered that the anthrax bacteria could be grown in nutrient fluids *outside the host* and that even after many transfers in laboratory culture, the bacteria still caused the disease when inoculated into a healthy animal.

On the basis of these experiments and others on the causative agent of tuberculosis, Koch formulated a set of rigorous criteria, now known as **Koch's postulates,** for definitively linking cause and effect in an infectious disease. Koch's postulates, summarized in **Figure 1.20**, stressed the importance of *laboratory culture* of the putative infectious agent followed by introduction of the suspected agent into virgin animals and recovery of the pathogen from diseased or dead animals. With these postulates as a guide, Koch, his students, and those that followed them discovered the causative agents of most of the important infectious diseases of humans and domestic animals. These discoveries also led to the

(a)

(b)

M.T. Madigan

Figure 1.18 **Louis Pasteur and some symbols of his contributions to microbiology.** *(a)* A French 5-franc note honoring Pasteur. The shepherd boy Jean-Baptiste Jupille is shown killing a rabid dog that had attacked children. Pasteur's rabies vaccine saved Jupille's life. In France, the franc preceded the euro as a currency. *(b)* Part of the Pasteur Institute, Paris, France. Today this structure, built for Pasteur by the French government, houses a museum that displays some of the original swan-necked flasks used in his experiments and a chapel containing Pasteur's crypt.

development of successful treatments for the prevention and cure of many of these diseases, thereby greatly improving the scientific basis of clinical medicine and human health and welfare (Figure 1.8).

The modern genomic era has also weighed in on the question of cause and effect in infectious diseases by developing molecular methods for identifying potential pathogens. Using these methods, a pathogen can be identified even if it cannot be cultured, or even if the pathogen itself is long since dead (see Explore the Microbial World, "The Black Death Decoded"). Such methods have revolutionized infectious disease diagnosis and treatment.

Koch, Pure Cultures, and Microbial Taxonomy

The second of Koch's postulates states that the suspected pathogen must be isolated and grown away from other microorganisms

Figure 1.19 **Robert Koch.** The German physician and microbiologist is credited with founding medical microbiology and formulating his famous postulates.

in laboratory culture (Figure 1.20); in microbiology we say that such a culture is *pure*. To accomplish this important goal, Koch and his associates developed several simple but ingenious methods of obtaining and growing bacteria in **pure culture**, and many of these methods are still used today.

Koch started by using natural surfaces such as a potato slice to obtain pure cultures, but quickly developed more reliable and reproducible growth media employing liquid nutrient solutions solidified with gelatin, and later with agar, an algal polysaccharide with excellent properties for this purpose. Along with his associate Walther Hesse, Koch observed that when a solid surface was incubated in air, masses of bacterial cells called *colonies* developed, each having a characteristic shape and color (Figure 1.21). He inferred that each colony had arisen from a single bacterial cell that had grown to yield the mass of cells. Koch reasoned that each colony harbored a population of identical cells, that is, a *pure culture*, and Koch quickly realized that solid media provided an easy way to obtain pure cultures. Richard Petri, another associate of Koch's, developed the transparent double-sided "Petri dish" in 1887 and this quickly became the standard tool for obtaining pure cultures.

KOCH'S POSTULATES

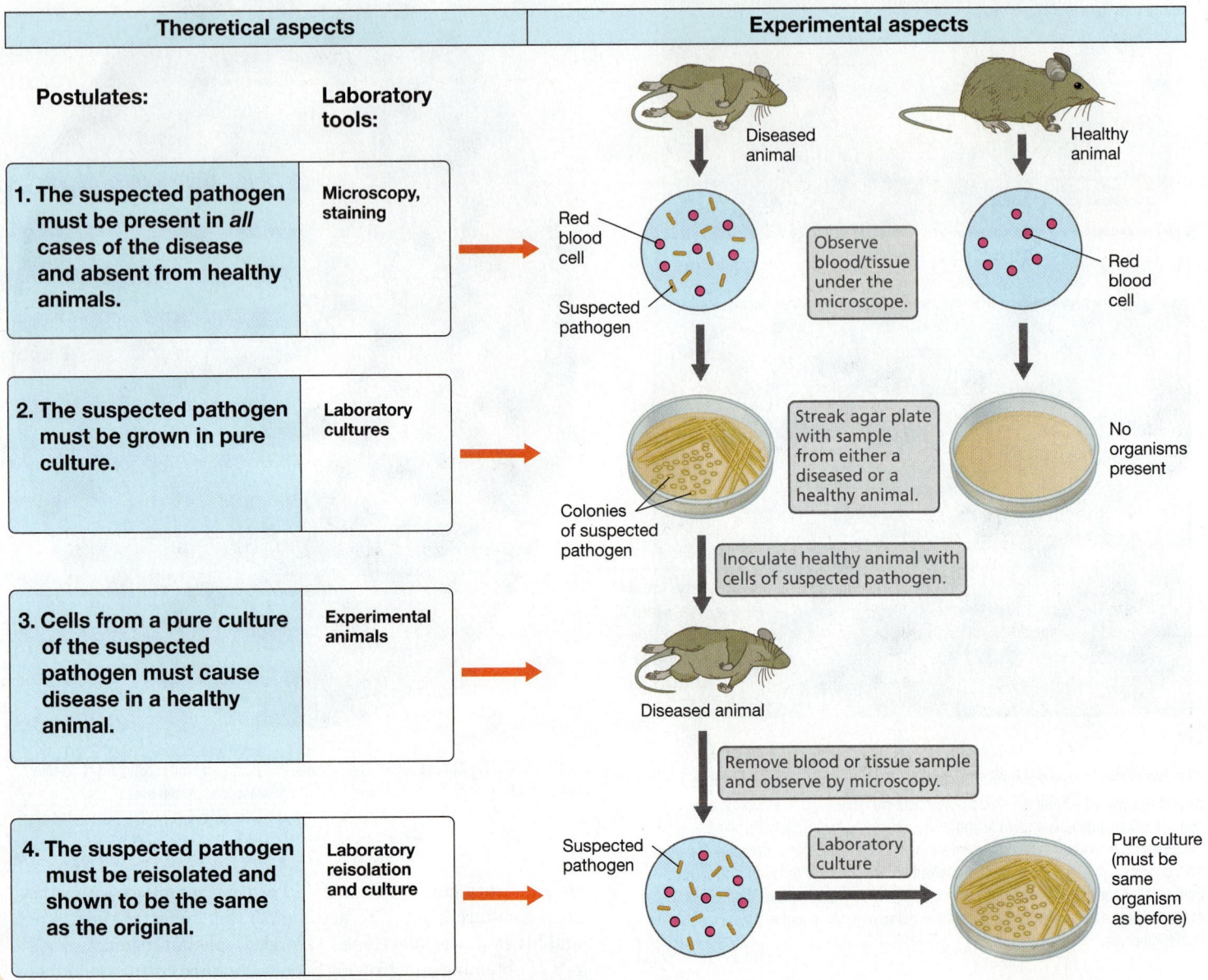

Figure 1.20 Koch's postulates for proving cause and effect in infectious diseases. Note that following isolation of a pure culture of the suspected pathogen, the cultured organism must both initiate the disease and be recovered from the diseased animal. Establishing the correct conditions for growing the pathogen is essential; otherwise it will be missed.

Koch was keenly aware of the implications his pure culture methods had for classifying microorganisms. He observed that colonies that differed in color and size (Figure 1.21) bred true and that cells from different colonies typically differed in size and shape and often in their nutrient requirements as well. Koch realized that these differences were analogous to the criteria taxonomists had established for the classification of larger organisms, such as plant and animal species, and he suggested that the different types of bacteria should be considered as "species, varieties, forms, or other suitable designation." Such insightful thinking was important for the rapid acceptance of microbiology as a new biological science, rooted as biology was in classification during Koch's era.

Koch and Tuberculosis

Koch's crowning scientific accomplishment was his discovery of the causative agent of tuberculosis. At the time Koch began this work (1881), one-seventh of all reported human deaths were caused by tuberculosis (Figure 1.8). There was a strong suspicion that tuberculosis was a contagious disease, but the suspected agent had never been seen, either in diseased tissues or in culture. Following his successful studies of anthrax, Koch set out to demonstrate the cause of tuberculosis, and to this end he brought together all of the methods he had so carefully developed in his previous studies with anthrax: microscopy, staining, pure culture isolation, and an animal model system (Figure 1.20).

EXPLORE THE
MICROBIAL WORLD

It is sometimes impossible to satisfy Koch's postulates, and in such cases, genomics may be able to link cause with effect in a different way. Thousands of microbial genomes have been sequenced and have revealed that pathogens often contain "signature" genes that can be used to positively identify them in a clinical specimen without the need for laboratory culture. This technology has greatly improved both the rapidity and accuracy of disease diagnoses. Although such genomic methods have been used primarily to diagnose disease in sick but still living patients, the technology has also been used to solve ancient medical mysteries where both sick patients and a recoverable pathogen are long gone. An excellent example is research that revealed the causative agent of the "Black Death."

The Black Death swept through Europe in the middle of the fourteenth century from origins near the Crimean Peninsula (present-day Ukraine). It has long been thought that the Black Death was a massive outbreak of bubonic plague, a typically fatal disease whose causative agent, *Yersinia pestis* (**Figure 1**), was discovered by the Swiss microbiologist Alexandre Yersin in 1894 and later linked to the disease through animal model studies. However, in the case of the Black Death, the connection to *Y. pestis* was uncertain for at least two major reasons. First, this deadly and widespread disease outbreak (the Black Death killed nearly one-third of the European population) occurred 660 years ago, and second, historical descriptions of victims' symptoms were often ambiguous, leaving open the possibility that other pathogens could have been responsible. Genomic studies confirmed that the Black Death was a severe outbreak of plague, and the published study[1] has become a model for how genomics can contribute to disease sleuthing.

How was the Black Death–plague link confirmed? At the height of the Black Death outbreak in the year 1349, a new cemetery was dug in East Smithfield, England. According to burial records, the cemetery was prepared specifically to inter Black Death victims, and in a little more than a year, the graveyard was filled with nearly 2500 bodies. No further burials occurred. A team of scientists examined corpses removed from the East Smithfield cemetery knowing in hindsight that all of the bodies were of Black Death victims. Because of this, the scientists could rule out other causes of death.[1]

Bubonic plague is an infection of the lymphatic system caused by cells of *Y. pestis* transmitted to a person from the bite of an infected flea. The bacteria multiply in the lymph nodes, forming painful swellings called *buboes,* and from there cells travel throughout the body and cause tissues to hemorrhage and then blacken (thus the term "Black Death") (**Figure 2**). Using tooth and bone samples from corpses disinterred from East Smithfield and employing a method for *Y. pestis* "DNA capture" developed from previous genomic studies of the pathogen, the international research team[1] fished out enough ancient DNA to reconstruct the genome of the bacterium that caused the Black Death. By comparing this genome with that from isolates of *Y. pestis* obtained from recent localized outbreaks, the mystery behind this devastating medieval disease was solved: The Black Death was indeed bubonic plague.

Further analyses of the Black Death *Y. pestis* genome showed that the Black Death strain was the ancestor of all modern *Y. pestis* strains and that the genomes of modern strains have evolved very little from the Black Death strain during the intervening 660 years. This points to the heightened importance of other factors—extremely poor sanitation, an influx of rats (rats harbor

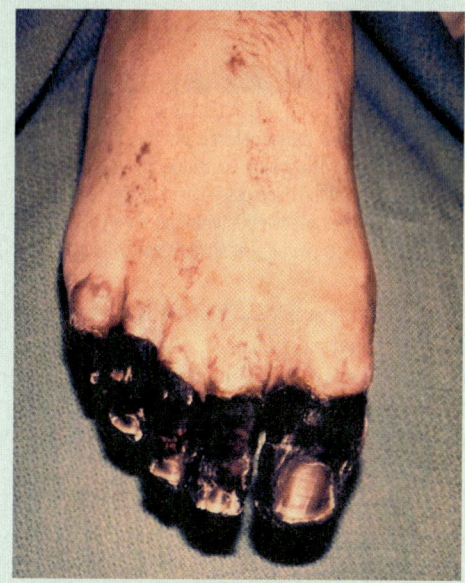

Figure 2 **Symptoms of bubonic plague.** Blackened skin on the toes of this plague victim comes from internal bleeding (hemorrhaging) due to systemic infection with *Yersinia pestis.*

the fleas that carry *Y. pestis*), and poor nutrition—in intensifying the Black Death outbreak compared with less pervasive waves of plague that visited Europe in earlier times. Indeed, the Black Death was the most devastating plague pandemic the world has ever seen. And because the Black Death touched such a wide geographic area, the Black Death strain of *Y. pestis* was able to infect a very large population of fleas and rats. From these sources, this ferocious pathogen became firmly entrenched and has periodically reemerged to trigger localized outbreaks of bubonic plague, all of which can be traced back to the Black Death bacterium that took its gruesome toll more than half a century ago.

In the United States, a handful of plague cases are observed each year. However, plague today carries two concerns. In addition to dealing with the natural disease, we must also be on guard against the use of *Y. pestis* as an agent of bioterrorism!

[1]Bos, K.I., et al. 2011. A draft genome of *Yersina pestis* from victims of the Black Death. *Nature* 478: 506–510.

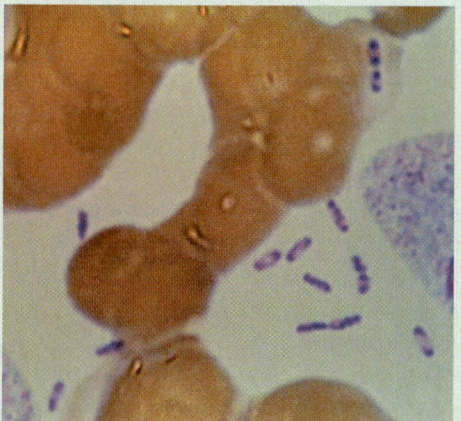

Figure 1 **Light micrograph of cells of the bacterium *Yersinia pestis* in a blood smear. This bacterium is the causative agent of bubonic plague.**

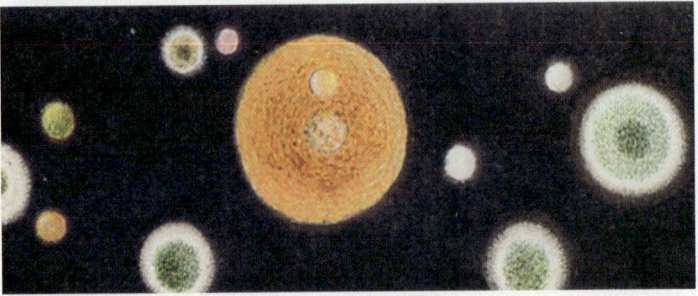

Figure 1.21 **A hand-colored photograph taken by Walther Hesse of colonies formed on agar.** The colonies include those of molds and bacteria obtained during Hesse's studies of the microbial content of air in Berlin, Germany, in 1882. From Hesse, W. 1884. "Ueber quantitative Bestimmung der in der Luft enthaltenen Mikroorganismen." *Mittheilungen aus dem Kaiserlichen Gesundheitsamte. 2*: 182–207.

The bacterium that causes tuberculosis, *Mycobacterium tuberculosis*, is very difficult to stain because *M. tuberculosis* cells contain large amounts of a waxlike lipid in their cell walls. Nevertheless, Koch devised a staining procedure for *M. tuberculosis* cells in lung tissue samples. Using this method, he observed the blue, rod-shaped cells of *M. tuberculosis* in tubercular tissues but not in healthy tissues (**Figure 1.22**). Obtaining cultures of *M. tuberculosis* was not easy, but eventually Koch succeeded in growing colonies of this organism on a solidified nutrient solution containing blood serum. Under the best of conditions, *M. tuberculosis* grows slowly in culture, but Koch's persistence and patience eventually led to pure cultures of this organism from human and animal sources.

From this point Koch used his postulates (Figure 1.20) to obtain definitive proof that the organism he had isolated was the cause

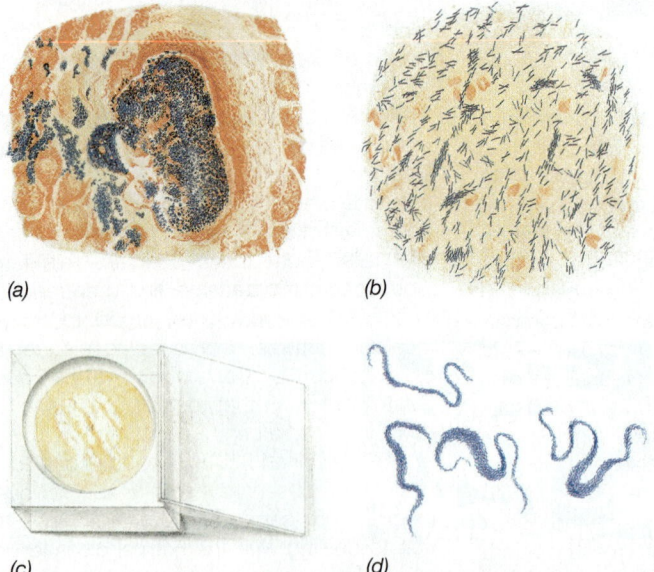

(a)   (b)

(c)   (d)

Figure 1.22 **Robert Koch's drawings of *Mycobacterium tuberculosis*.** *(a)* Section through infected lung tissue showing cells of *M. tuberculosis* (blue). *(b) M. tuberculosis* cells in a sputum sample from a tubercular patient. *(c)* Growth of *M. tuberculosis* on a glass plate of coagulated blood serum stored inside a glass box to prevent contamination. *(d) M. tuberculosis* cells taken from the plate in part c and observed microscopically; cells appear as long cordlike forms. Original drawings from Koch, R. 1884. "Die Aetiologie der Tuberkulose." *Mittheilungen aus dem Kaiserlichen Gesundheitsamte 2*: 1–88.

of the disease tuberculosis. Guinea pigs can be readily infected with *M. tuberculosis* and eventually succumb to systemic tuberculosis. Koch showed that tuberculous guinea pigs contained masses of *M. tuberculosis* cells in their lungs and that pure cultures obtained from such animals transmitted the disease to healthy animals. In this way, Koch successfully satisfied all four of his postulates, and the cause of tuberculosis was understood. Koch announced his discovery of the cause of tuberculosis in 1882, and for this accomplishment he was awarded the 1905 Nobel Prize for Physiology or Medicine. Koch had many other triumphs in the growing field of infectious diseases, including the discovery of the causative agent of cholera (the bacterium *Vibrio cholerae*) and developing methods to diagnose infection with *M. tuberculosis* (the tuberculin skin test).

MINIQUIZ -

• How do Koch's postulates ensure that cause and effect of a given disease are clearly differentiated?

• What advantages do solid media offer for the isolation of microorganisms?

• What is a pure culture?

1.9 The Rise of Microbial Diversity

As microbiology entered the twentieth century, its initial focus on basic principles, methods, and medical aspects broadened to include studies of the microbial diversity of soil and water and the metabolic processes that microorganisms carried out in these habitats. Major contributors in this era included the Dutchman Martinus Beijerinck and the Russian Sergei Winogradsky.

Martinus Beijerinck and the Enrichment Culture Technique

Martinus Beijerinck (1851–1931) was a professor at the Delft Polytechnic School in Holland and was originally trained in botany, so he began his career in microbiology studying plants. Beijerinck's greatest contribution to the field of microbiology was his clear formulation of the **enrichment culture technique**. In enrichment cultures, microorganisms are isolated from natural samples using highly selective nutrient and incubation conditions to favor a particular metabolic group of organisms. Beijerinck's skill with the enrichment method was readily apparent when, following Winogradsky's discovery of the process of nitrogen fixation, he isolated the aerobic nitrogen-fixing bacterium *Azotobacter* from soil (**Figure 1.23**). Nitrogen-fixing bacteria can use atmospheric nitrogen (N_2) to make important nitrogenous substances in the cell, such as amino acids to make proteins and nucleotides to make nucleic acids.

Using the enrichment culture technique, Beijerinck isolated the first pure cultures of many soil and aquatic microorganisms, including sulfate-reducing and sulfur-oxidizing bacteria, nitrogen-fixing root nodule bacteria (Figure 1.9), lactic acid bacteria, green algae, various anaerobic bacteria, and many others. In addition, in his classic studies of "mosaic disease" of tobacco, Beijerinck used selective filters to show that the infectious agent in this disease (a virus) was smaller than a bacterium and that it

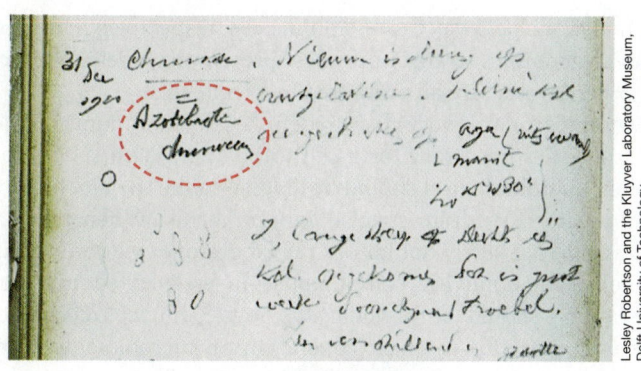

(a)

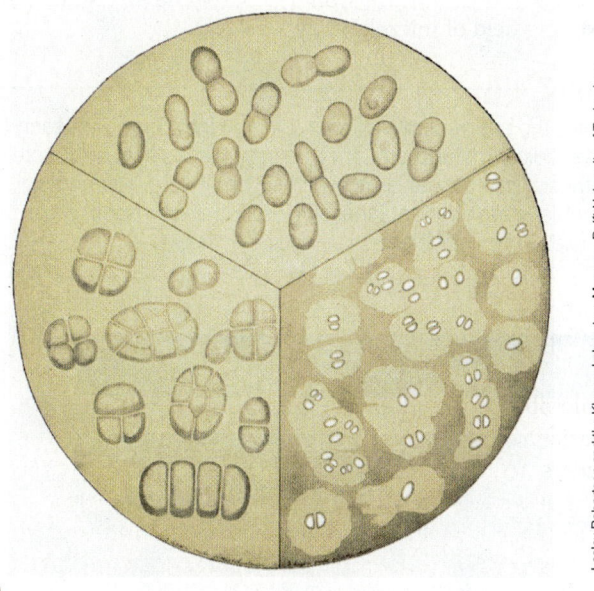

(b)

Lesley Robertson and the Kluyver Laboratory Museum, Delft University of Technology

Figure 1.23 Martinus Beijerinck and *Azotobacter*. *(a)* A page from the laboratory notebook of M. Beijerinck dated 31 December 1900 describing the aerobic nitrogen-fixing bacterium *Azotobacter chroococcum* (name circled in red). Compare Beijerinck's drawings of pairs of *A. chroococcum* cells with the photomicrograph of cells of *Azotobacter* in Figure 14.32. *(b)* A painting by M. Beijerinck's sister, Henriëtte Beijerinck, showing cells of *Azotobacter chroococcum*. Beijerinck used such paintings to illustrate his lectures.

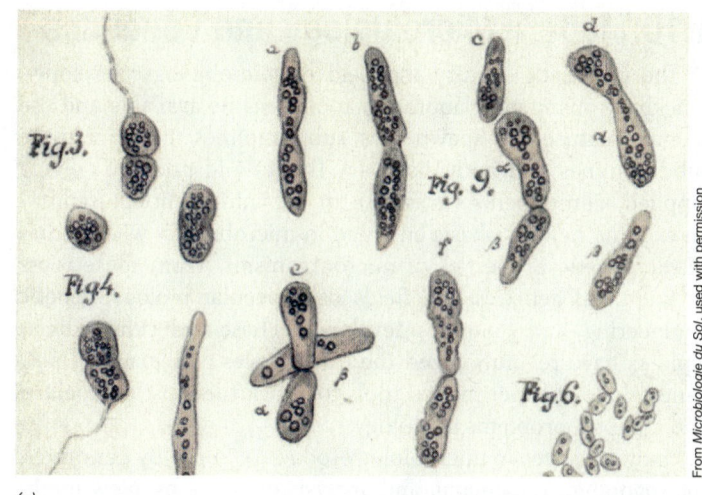

(a)

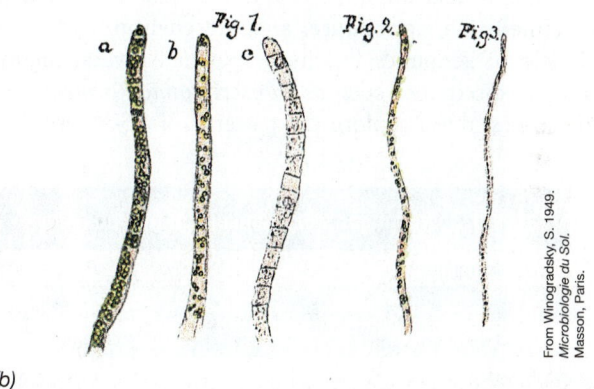

(b)

From *Microbiologie du Sol*, used with permission

From Winogradsky, S. 1949. *Microbiologie du Sol.* Masson, Paris.

Figure 1.24 Sulfur bacteria. The original drawings were made by Sergei Winogradsky in the late 1880s and then copied and hand-colored by his wife Hèléne. *(a)* Purple sulfur phototrophic bacteria. Figures 3 and 4 show cells of *Chromatium okenii* (compare with photomicrographs of *C. okenii* in Figures 1.5*a* and 1.7*a*). *(b) Beggiatoa*, a sulfur chemolithotroph (compare with Figures 1.15 and 14.27).

somehow became incorporated into cells of the living host plant. In this insightful work, Beijerinck not only described the first virus, but also the basic principles of virology, which we present in Chapter 8.

Sergei Winogradsky, Chemolithotrophy, and Nitrogen Fixation

Like Beijerinck, Sergei Winogradsky (1856–1953) was interested in the bacterial diversity of soils and waters and was highly successful in isolating several notable bacteria from natural samples. Winogradsky was particularly interested in bacteria that cycle nitrogen and sulfur compounds, such as the nitrifying bacteria and the sulfur bacteria (**Figure 1.24**). He showed that these bacteria catalyze specific chemical transformations in nature and proposed the important concept of **chemolithotrophy**, the oxidation of *inorganic* compounds to yield energy. Winogradsky further

showed that these organisms, which he called *chemolithotrophs* (meaning, literally, "earth eaters"), are widespread in nature and obtain their carbon from CO_2. Winogradsky thus revealed that, like photosynthetic organisms, chemolithotrophic bacteria are *autotrophs*.

Winogradsky performed the first isolation of a nitrogen-fixing bacterium, the anaerobe *Clostridium pasteurianum*, and as just mentioned, Beijerinck used this discovery to guide his isolation of *aerobic* nitrogen-fixing bacteria years later (Figure 1.23). Winogradsky lived to be almost 100, publishing many scientific papers and a major monograph, *Microbiologie du Sol* (*Soil Microbiology*). This work, a milestone in microbiology, contains drawings of many of the organisms Winogradsky studied during his lengthy career (Figure 1.24).

MINIQUIZ
- What is meant by the term "enrichment culture"?
- What is meant by the term "chemolithotrophy"? In what way are chemolithotrophs like plants?

1.10 Modern Microbiology and Genomics

In the twentieth century the field of microbiology developed quickly, as many new laboratory tools became available and the science matured to spawn new subdisciplines. Most of these subdisciplines had both discovery (basic) and problem-solving (applied) components (Table 1.3). In the mid-twentieth century an exciting new emphasis emerged in microbiology with studies of the genetic properties of microorganisms. From these roots in microbial genetics, the fields of molecular biology, genetic engineering, and genomics developed. These molecular subdisciplines have revolutionized the life sciences and spawned new generations of experimental tools to attack the most compelling and complex problems in biology.

Many advances in microbiology today are fueled by **genomics**, the mapping, sequencing, and analysis of genomes. New methods for DNA sequencing and improved computational capacities have unleashed huge amounts of genomic data for attacking problems in medicine, agriculture, and the environment. The fast-paced field of genomics has itself spawned several highly focused new subdisciplines such as *transcriptomics, proteomics,* and *metabolomics*; these explore the patterns of RNA, protein, and metabolic pathway expression in cells, respectively. The concepts of genomics, transcriptomics, proteomics, metabolomics, and other "omics" are all presented in Chapter 6.

Genomics today is very close to defining the minimum complement of genes necessary for a cell to be alive. With such information, microbiologists should be able to define the biochemical prerequisites for life in precise genetic terms. When that day arrives, and it is likely not far off, the laboratory creation of a living cell from nonliving components—in essence, spontaneous generation—should be possible. Obviously, much exciting science is in store for the next generation of microbiologists, and your continued journey through this book will position you to both understand it and appreciate it. Good luck, and welcome to the exciting field of microbiology!

MINIQUIZ

- Identify the subdiscipline of microbiology that deals with each of these topics: metabolism, enzymology, nucleic acid and protein synthesis, microorganisms and their natural environments, microbial classification, inheritance of characteristics, gene complements of different organisms.

Table 1.3 The major subdisciplines of microbiology

Subdiscipline	Focus
I. Basic emphases[a]	
Microbial physiology	Nutrition, metabolism
Microbial genetics	Genes, heredity, and genetic variation
Microbial biochemistry	Enzymes and chemical reactions in cells
Microbial systematics	Classification and nomenclature
Virology	Viruses and subviral particles
Molecular biology	Nucleic acids and protein
Microbial ecology	Microbial diversity and activity in natural habitats; biogeochemistry
Genomics	Genome sequencing and comparative analyses
II. Applied emphases[a]	
Medical microbiology	Infectious disease
Immunology	Immune systems
Agricultural/soil microbiology	Microbial diversity and processes in soil
Industrial microbiology	Large-scale production of antibiotics, alcohol, and other chemicals
Biotechnology	Production of human proteins by genetically engineered microorganisms
Aquatic microbiology	Microbial processes in waters and wastewaters, drinking water safety

[a]None of these subdisciplines are devoted entirely to basic science or applied science. However, the subdisciplines listed in I tend to be more focused on discovery and those in II more focused on solving problems or making commercial products.

BIG IDEAS

1.1 • Microorganisms are single-celled microscopic organisms that are essential for the well-being and functioning of other life forms and the planet. As a science, microbiology has its basic and applied components, which generate new knowledge and solve problems, respectively.

1.2 • Like houses, cells are constructed of many parts, all of which interact to yield the intact living organism. Prokaryotic and eukaryotic cells differ in cellular architecture, and an organism's characteristics are defined by its complement of genes—its genome. Many activities are carried out by all cells, including metabolism, growth, and evolution.

1.3 • Diverse microbial populations were widespread on Earth for billions of years before higher organisms appeared, and cyanobacteria in particular were important because they oxygenated the atmosphere. *Bacteria*, *Archaea*, and *Eukarya* are the major phylogenetic lineages (domains) of cells.

1.4 • Microorganisms live in populations that interact with other populations to form microbial communities. The activities of microorganisms in microbial communities can greatly affect the chemical and physical properties of their habitats. Microbial biomass on Earth exceeds that of higher organisms, although most microbial cells actually reside in the terrestrial and oceanic deep subsurface.

1.5 • Microorganisms can be both beneficial and harmful to humans, although many more microorganisms are beneficial (or even essential) than are harmful. Agriculture, food, energy, and the environment are all impacted in major ways by microorganisms.

1.6 • Robert Hooke was the first to describe microorganisms, and Antoni van Leeuwenhoek was the first to describe bacteria. Ferdinand Cohn founded the field of bacteriology and discovered bacterial endospores.

1.7 • Louis Pasteur devised ingenious experiments proving that living organisms do not arise spontaneously from nonliving matter. Pasteur developed many concepts and techniques central to the science of microbiology, including sterilization, and developed a number of key vaccines for humans and other animals.

1.8 • Robert Koch developed a set of criteria called Koch's postulates for linking cause and effect in infectious diseases. Koch also developed the first reliable and reproducible means for obtaining and maintaining microorganisms in pure culture.

1.9 • Martinus Beijerinck and Sergei Winogradsky explored soil and water for microorganisms that carry out important natural processes, such as nutrient cycling and the biodegradation of particular substances. Out of their work came the enrichment culture technique and the concepts of chemolithotrophy and nitrogen fixation.

1.10 • In the middle to latter part of the twentieth century, various basic and applied subdisciplines of microbiology emerged. These paved the way for the current era of molecular microbiology, with genomic sciences currently on center stage.

MasteringMicrobiology®

Review what you know and challenge what you have learned with MasteringMicrobiology! Access study materials, chapter quizzes, animations, and microbiology lab tutorials in the Study Area to ensure that you have mastered this chapter's content.

KEY TERMS

Cell wall a rigid layer present outside the cytoplasmic membrane; it confers structural strength on the cell and prevents osmotic lysis

Chemolithotrophy a form of metabolism in which energy is generated from the oxidation of inorganic compounds

Communication interactions between cells using chemical signals

Cytoplasm the fluid portion of a cell, bounded by the cytoplasmic membrane

Cytoplasmic membrane a semipermeable barrier that separates the cell interior (cytoplasm) from the environment

Differentiation modification of cellular components to form a new structure, such as a spore

Domain one of the three main evolutionary lineages of cells: the *Bacteria*, the *Archaea*, and the *Eukarya*

Ecosystem organisms plus their nonliving environment

Enrichment culture technique a method for isolating specific microorganisms from nature using specific culture media and incubation conditions

Enzyme a protein (or in some cases an RNA) catalyst that functions to speed up chemical reactions

Eukaryote a cell having a membrane-enclosed nucleus and various other membrane-enclosed organelles; *Eukarya*

Evolution descent with modification leading to new forms or species

Extremophiles microorganisms that inhabit environments unsuitable for higher life forms, such as environments that are extremely hot or cold, or environments that are acidic, alkaline, or extremely salty

Genetic exchange the transfer of genes or the acceptance of genes between prokaryotic cells

Genome an organism's full complement of genes

Genomics the mapping, sequencing, and analysis of genomes

Growth in microbiology, an increase in cell number with time

Habitat the environment in which a microbial population resides

Koch's postulates a set of criteria for proving that a given microorganism causes a given disease

Macromolecules a polymer of monomeric units that includes the proteins, nucleic acids, polysaccharides, and lipids

Metabolism all biochemical reactions in a cell

Microbial community two or more populations of cells that coexist and interact in a habitat

Microbial ecology the study of microorganisms in their natural environments

Microorganism a microscopic organism consisting of a single cell or cell cluster or a virus

Motility the movement of cells by some form of self-propulsion

Nucleoid the aggregated mass of DNA that makes up the chromosome of prokaryotic cells

Nucleus a membrane-enclosed structure in eukaryotic cells that contains the cell's DNA genome

Organelles a bilayer-membrane-enclosed structure such as the mitochondrion, found in eukaryotic cells

Pathogen a disease-causing microorganism

Prokaryote a cell that lacks a membrane-enclosed nucleus and other organelles; *Bacteria* or *Archaea*

Pure culture a culture containing a single kind of microorganism

Ribosomes a structure composed of RNAs and proteins upon which new proteins are made

Spontaneous generation the hypothesis that living organisms can originate from nonliving matter

Sterile free of all living organisms (cells) and viruses

REVIEW QUESTIONS

1. What are the two major themes of microbiology and how do they differ in their focus? (Section 1.1)

2. How can prokaryotic and eukaryotic cells be distinguished? List the major activities carried out by cells, and in each case, describe why the activity occurs. (Section 1.2)

3. Why did the evolution of cyanobacteria change Earth forever? How many domains of life are there and how are they related? (Section 1.3)

4. What is an ecosystem? What effects can microorganisms have on their ecosystems? (Section 1.4)

5. How would you convince a friend that microorganisms are much more than just agents of disease? (Section 1.5)

6. For what contributions are Robert Hooke and Antoni van Leeuwenhoek most remembered in microbiology? What time period were these scientists active? (Section 1.6)

7. Explain the principle behind the Pasteur flask in studies on spontaneous generation. Why were the results of this experiment inconsistent with the theory of spontaneous generation? (Section 1.7)

8. What is a pure culture and how can one be obtained? Why are pure cultures important for medical microbiology and other areas of microbiology? (Section 1.8)

9. What are Koch's postulates and how did they influence the development of microbiology? Why are Koch's postulates still relevant today? (Section 1.8)

10. What were the major microbiological interests of Martinus Beijerinck and Sergei Winogradsky? It can be said that both men discovered nitrogen fixation. Explain. (Section 1.9)

11. Select one major subdiscipline of microbiology from each of the two major categories of Table 1.3. Why do you think the subdiscipline is "basic" or "applied"? (Section 1.10)

APPLICATION QUESTIONS

1. Pasteur's experiments on spontaneous generation contributed to the methodology of microbiology, understanding of the origin of life, and techniques for the preservation of food. Explain briefly how Pasteur's experiments affected each of these topics.

2. Describe the lines of proof Robert Koch used to definitively associate the bacterium *Mycobacterium tuberculosis* with the disease tuberculosis. How would his proof have been flawed if any of the tools he developed for studying bacterial diseases had not been available for his study of tuberculosis?

3. Imagine that all microorganisms suddenly disappeared from Earth. From what you have learned in this chapter, why do you think that animals would eventually disappear from Earth? Why would plants disappear? By contrast, if all higher organisms suddenly disappeared, what in Figure 1.4*a* tells you that a similar fate would not befall microorganisms?

2 · Microbial Cell Structure and Function

microbiology**now**

Archaeal Tortoise and Hare

Motility is important for microorganisms because the ability to move allows cells to explore new habitats and exploit their resources. Motility has been studied for over 50 years in the flagellated bacterium *Escherichia coli*. It is with *E. coli* that scientists first discovered that the bacterial flagellum functions by rotating and that when speed is expressed in terms of body lengths traveled per second, swimming *E. coli* cells are actually moving faster than the fastest animals.

Studies of the archaeon *Halobacterium* showed that its flagella also rotate but that they were thinner than their bacterial counterparts and were composed of a protein distinct from flagellin, the protein that makes up bacterial flagella. Moreover, observations of swimming cells showed that *Halobacterium* was a slowpoke, moving at less than one-tenth the speed of *E. coli*. This raised the interesting question of whether this was true of all *Archaea;* are they naturally joggers instead of sprinters?

Microbiologists recently zeroed in on the movements of swimming *Archaea* and showed that *Halobacterium* was the slowest of all species examined.[1] By contrast, cells of the archaeon *Methanocaldococcus* (cells with flagellar tufts in photo) swam nearly 50 times faster than cells of *Halobacterium* and 10 times faster than cells of *E. coli*. Astonishingly, *Methanocaldococcus* moves at nearly 500 cell lengths per second, which makes it the fastest organism on Earth!

The thin diameter of the archaeal flagellum obviously does not mandate a slow swimming speed as some had predicted from the *Halobacterium* work. Instead, swimming speeds of *Archaea* can and do vary greatly.[1] Indeed, the existence of both a "tortoise" and a "hare" within the *Archaea* shows that there is still much to learn about the structure and function of microbial cells.

[1]Herzog, B., and R. Wirth. 2012. Swimming behavior of selected species of *Archaea*. *Appl. Environ. Microbiol. 78*: 1670–1674.

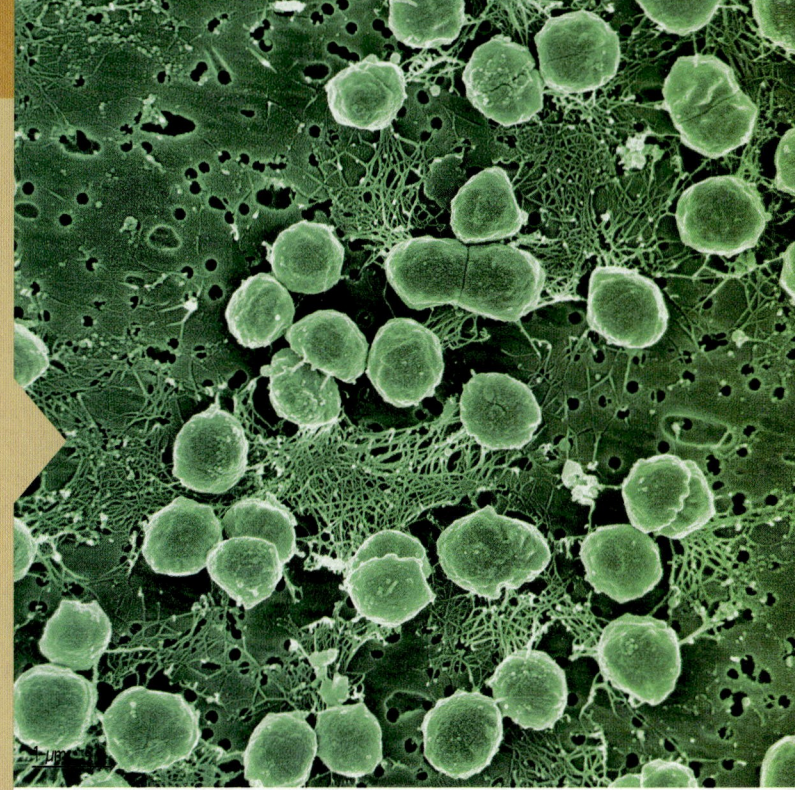

I • Microscopy

Historically, the science of microbiology has taken its greatest leaps forward as new tools for the study of microorganisms are developed and old tools improve. The microscope is the microbiologist's oldest and most basic tool for studying microbial structure. Many types of microscopy are used and some are extremely powerful. So as a prelude to our study of cell structure, let's first take a look at some common tools for visualizing cells with a goal of understanding how they work and what they can tell us.

2.1 Discovering Cell Structure: Light Microscopy

To see microorganisms, one needs a microscope of some sort, either a *light* microscope or an *electron* microscope. In general, light microscopes are used to examine cells at relatively low magnifications, and electron microscopes are used to examine cells and cell structures at very high magnification.

All microscopes employ lenses that magnify the image. Magnification, however, is not the limiting factor in our ability to see small objects. It is instead **resolution**—the ability to distinguish two adjacent objects as distinct and separate—that governs our ability to see the very small. Although magnification can be increased virtually without limit, resolution cannot, because resolution is a function of the physical properties of light.

We begin with the light microscope, for which the limits of resolution are about 0.2 μm (μm is the abbreviation for micrometer, 10^{-6} m). We then proceed to the electron microscope, for which resolution is considerably greater.

The Compound Light Microscope

The light microscope uses visible light to illuminate cell structures. Several types of light microscopes are used in microbiology: *bright-field*, *phase-contrast*, *differential interference contrast*, *dark-field*, and *fluorescence*.

With the bright-field microscope, specimens are visualized because of the slight differences in contrast that exist between them and their surroundings, differences that arise because cells absorb or scatter light to varying degrees. The modern compound light microscope contains two lenses, *objective* and *ocular*, that function in combination to form the image. The light source is focused on the specimen by the condenser (**Figure 2.1**). Bacterial cells are typically difficult to see well with the bright-field microscope because the cells themselves lack significant contrast with their surrounding medium. Cells visualized by a form of light microscopy called phase-contrast (Section 2.2; see inset Figure 2.1) overcome these limitations. Pigmented microorganisms are also an exception because the color of the organism itself adds contrast, which makes them easier to visualize by bright-field optics (**Figure 2.2**).

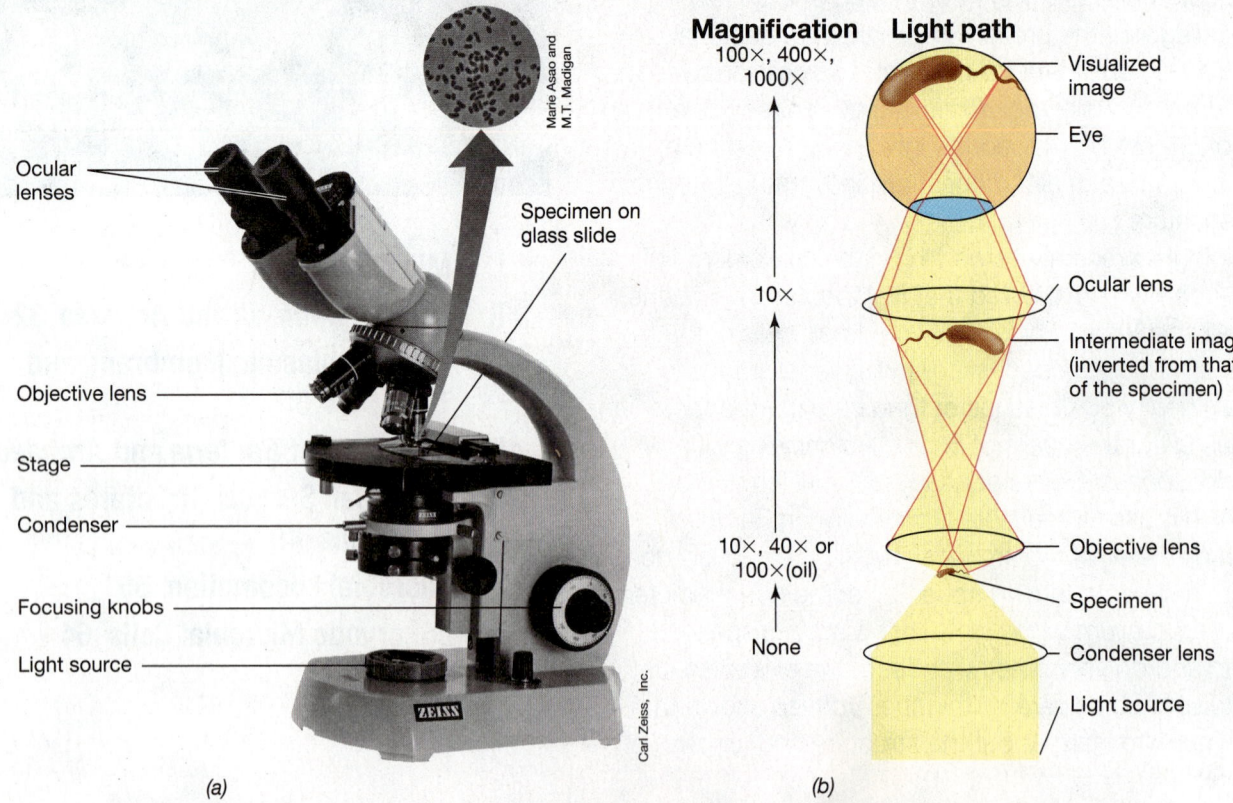

Figure 2.1 Microscopy. *(a)* A compound light microscope (inset photomicrograph of unstained cells taken through a phase-contrast light microscope). *(b)* Path of light through a compound light microscope. Besides 10×, ocular lenses are available in 15–30× magnifications. Figure 2.5 compares cells visualized by bright field with those by phase contrast.

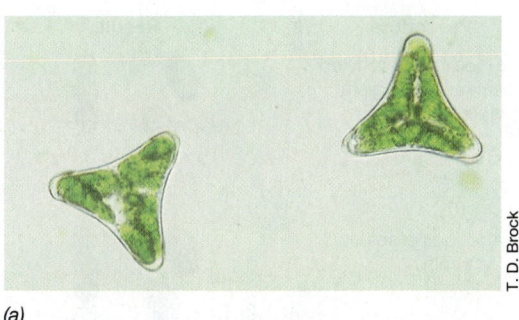

(a)

T. D. Brock

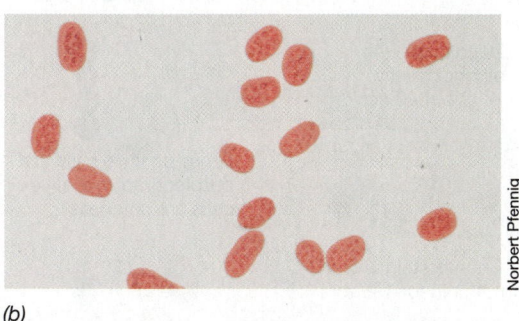

(b)

Norbert Pfennig

Figure 2.2 Bright-field photomicrographs of pigmented microorganisms.
(a) A green alga (eukaryote). The green structures are chloroplasts. *(b)* Purple
phototrophic bacteria (prokaryote). The algal cell is about 15 μm wide, and the
bacterial cells are about 5 μm wide.

For cells lacking pigments there are several ways to boost contrast, and we consider these methods in the next section.

Magnification and Resolution

The total magnification of a compound light microscope is the *product* of the magnification of its objective and ocular lenses (Figure 2.1*b*). Magnifications of about 2000× are the upper limit for light microscopes, and at magnifications above this, resolution does not improve. Resolution is a function of the wavelength of light used and a characteristic of the objective lens known as its *numerical aperture*, a measure of light-gathering ability. There is a correlation between the magnification of a lens and its numerical aperture; lenses with higher magnification typically have higher numerical apertures. The diameter of the smallest object resolvable by any lens is equal to 0.5λ/numerical aperture, where λ is the wavelength of light used. This formula reveals that resolution is highest when blue light is used to illuminate a specimen (blue light has shorter wavelengths than white or red light) and the objective has a very high numerical aperture.

As mentioned, the highest resolution possible in a compound light microscope is about 0.2 μm. What this means is that two objects that are closer together than 0.2 μm cannot be resolved as distinct and separate. Microscopes used in microbiology have ocular lenses that magnify 10–20× and objective lenses that magnify 10–100× (Figure 2.1*b*). At 1000×, objects 0.2 μm in diameter can just be resolved. With the 100× objective, and with certain other objectives of very high numerical aperture, an optical grade oil is placed between the microscope slide and the objective. Lenses on which oil is used are called *oil-immersion* lenses. Immersion oil increases the light-gathering ability of a lens

by allowing some of the light rays emerging from the specimen at angles (that would otherwise be lost to the objective lens) to be collected and viewed.

MINIQUIZ --

• Define the terms magnification and resolution.

• What is the upper limit of magnification for a bright-field microscope? Why is this so?

2.2 Improving Contrast in Light Microscopy

In light microscopy, improving contrast improves the final image. Staining is a quick and easy way to improve contrast, but there are many other ways to do this.

Staining: Increasing Contrast for Bright-Field Microscopy

Dyes can be used to stain cells and increase their contrast so that they can be more easily seen in the bright-field microscope. Dyes are organic compounds, and each class of dye has an affinity for specific cellular materials. Many dyes used in microbiology are positively charged, and for this reason they are called *basic dyes*. Examples of basic dyes include methylene blue, crystal violet, and safranin. Basic dyes bind strongly to negatively charged cell components, such as nucleic acids and acidic polysaccharides. Because cell surfaces tend to be negatively charged, these dyes also combine with high affinity to the surfaces of cells, and hence are very useful general-purpose stains.

To perform a simple stain one begins with dried preparations of cells (**Figure 2.3**). A clean glass slide containing a dried suspension of cells is flooded for a minute or two with a dilute solution of a basic dye, rinsed several times in water, and blotted dry. Because their cells are so small, it is common to observe dried, stained preparations of *Bacteria* or *Archaea* with a high-power (oil-immersion) lens.

Differential Stains: The Gram Stain

Stains that render different kinds of cells different colors are called *differential* stains. An important differential-staining procedure used in microbiology is the **Gram stain** (**Figure 2.4**). On the basis of their reaction in the Gram stain, *Bacteria* can be divided into two major groups: **gram-positive** and **gram-negative**. After Gram staining, gram-positive bacteria appear purple-violet and gram-negative bacteria appear pink (Figure 2.4*b*). The color difference in the Gram stain arises because of differences in the cell wall structure of gram-positive and gram-negative cells, a topic we will consider later. After staining with a basic dye such as crystal violet that renders cells purple in color, treatment with ethanol decolorizes gram-negative cells but not gram-positive cells. Following counterstaining with a different-colored stain, typically the red stain safranin, the two cell types can be distinguished microscopically by their different colors (Figure 2.4*b*).

The Gram stain is the most common staining procedure used in microbiology, and it is often done to begin the characterization of a

UNIT 1

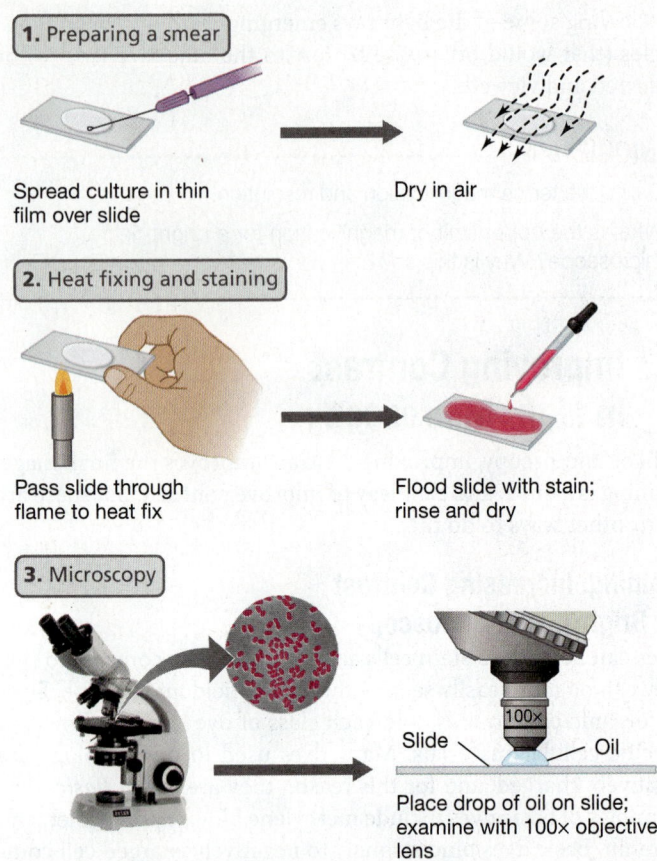

1. Preparing a smear

Spread culture in thin film over slide

Dry in air

2. Heat fixing and staining

Pass slide through flame to heat fix

Flood slide with stain; rinse and dry

3. Microscopy

Slide — 100× — Oil

Place drop of oil on slide; examine with 100× objective lens

Figure 2.3 Staining cells for microscopic observation. Stains improve the contrast between cells and their background. Center: Same cells as shown in Figure 2.1 inset but stained with a basic dye.

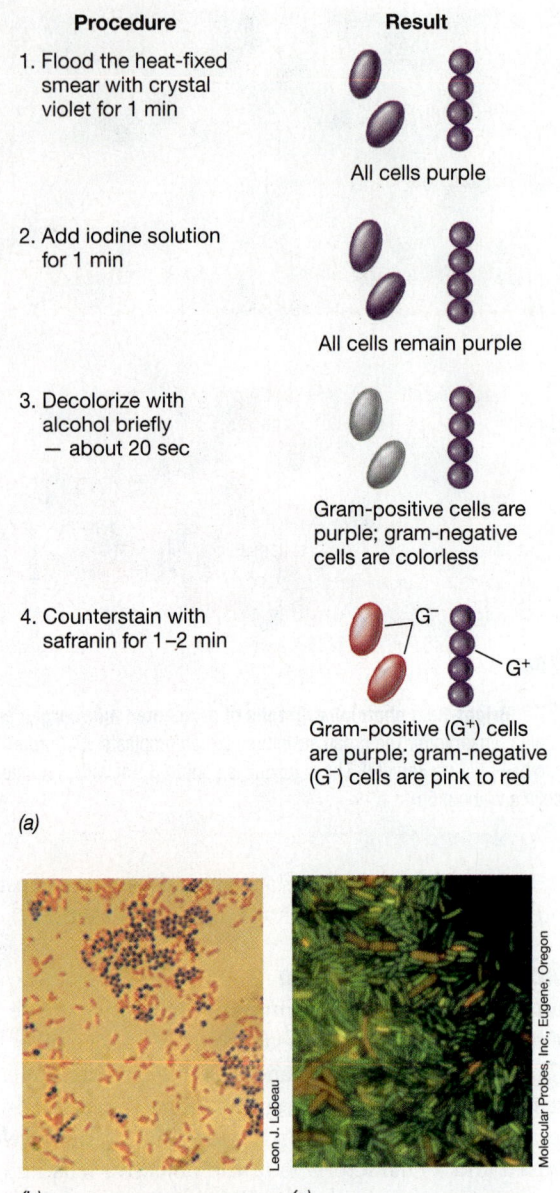

Procedure	Result
1. Flood the heat-fixed smear with crystal violet for 1 min	All cells purple
2. Add iodine solution for 1 min	All cells remain purple
3. Decolorize with alcohol briefly — about 20 sec	Gram-positive cells are purple; gram-negative cells are colorless
4. Counterstain with safranin for 1–2 min	Gram-positive (G⁺) cells are purple; gram-negative (G⁻) cells are pink to red

(a)

(b) Leon J. Lebeau

(c) Molecular Probes, Inc., Eugene, Oregon

Figure 2.4 The Gram stain. *(a)* Steps in the procedure. *(b)* Microscopic observation of gram-positive (purple) and gram-negative (pink) bacteria. The organisms are *Staphylococcus aureus* and *Escherichia coli*, respectively. *(c)* Cells of *Pseudomonas aeruginosa* (gram-negative, green) and *Bacillus cereus* (gram-positive, orange) stained with a one-step fluorescent staining method. This method allows for differentiating gram-positive from gram-negative cells in a single staining step.

newly isolated bacterium. If a fluorescent microscope is available, the Gram stain can be reduced to a one-step procedure; gram-positive and gram-negative cells fluoresce different colors when treated with a special chemical (Figure 2.4c).

Phase-Contrast and Dark-Field Microscopy

Although staining is widely used in light microscopy, staining kills cells and can distort their features. Two forms of light microscopy improve image contrast of unstained (and thus live) cells. These are phase-contrast microscopy and dark-field microscopy (**Figure 2.5**). The phase-contrast microscope in particular is widely used in teaching and research for the observation of living preparations.

Phase-contrast microscopy is based on the principle that cells differ in refractive index (a factor by which light is slowed as it passes through a material) from their surroundings. Light passing through a cell thus differs in phase from light passing through the surrounding liquid. This subtle difference is amplified by a device in the objective lens of the phase-contrast microscope called the *phase ring*, resulting in a dark image on a light background (Figure 2.5b; see also inset to Figure 2.1). The ring consists of a phase plate that amplifies the variation in phase to produce the higher-contrast image.

In the dark-field microscope, light reaches the specimen from the sides only. The only light that reaches the lens is that scattered by the specimen, and thus the specimen appears light on a dark

background (Figure 2.5c). Resolution by dark-field microscopy is often better than by light microscopy, and some objects can be resolved by dark-field that cannot be resolved by bright-field or even by phase-contrast microscopes. Dark-field microscopy is a particularly good way to observe microbial motility, as bundles of flagella (the structures responsible for swimming motility) are often resolvable with this technique (see Figure 2.50a).

Fluorescence Microscopy

The fluorescence microscope is used to visualize specimens that fluoresce, emitting light of one color after absorbing light

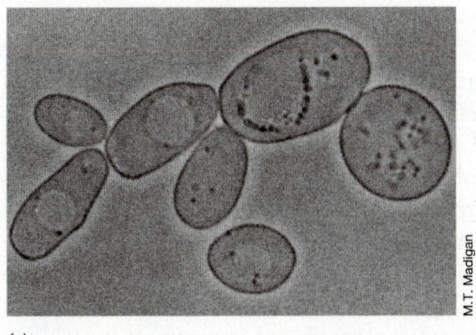
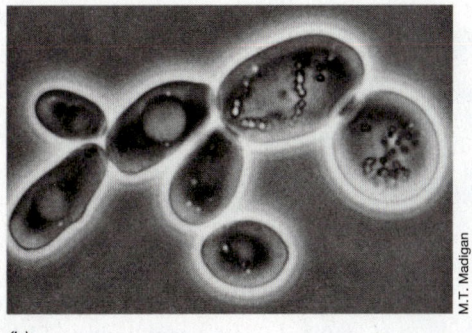
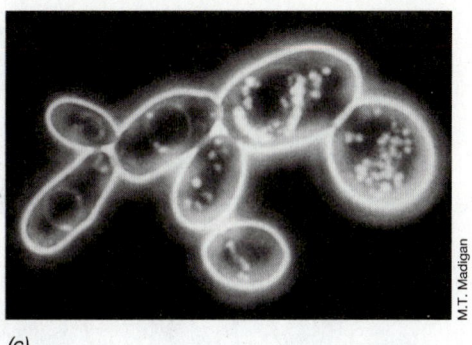

(a)   *(b)*   *(c)*

Figure 2.5 Cells visualized by different types of light microscopy. The same field of cells of the yeast *Saccharomyces cerevisiae* visualized by *(a)* bright-field microscopy, *(b)* phase-contrast microscopy, and *(c)* dark-field microscopy. Cells average 8–10 μm wide.

of another color (**Figure 2.6**). Cells fluoresce because they either contain naturally fluorescent substances such as chlorophyll or other fluorescing components (autofluorescence, Figure 2.6*a*, *b*), or because they have been stained with a fluorescent dye (Figure 2.6*c*). DAPI (4′,6-*di*amidino-2-*p*henyl*i*ndole) is a widely used fluorescent dye. DAPI stains cells bright blue because it complexes with the cell's DNA (Figure 2.6*c*). DAPI can be used to visualize cells in their natural habitats, such as soil, water, food, or a clinical specimen. Fluorescence microscopy using DAPI is therefore widely used in clinical diagnostic microbiology and also in microbial ecology for enumerating bacteria in a natural environment or in a cell suspension.

MINIQUIZ ---

- What color will a gram-negative cell be after Gram staining by the conventional method?
- What major advantage does phase-contrast microscopy have over staining?
- How can cells be made to fluoresce?

2.3 Imaging Cells in Three Dimensions

Thus far we have considered forms of microscopy in which the rendered images are two-dimensional. How can this limitation be overcome? We will see in the next section that the scanning electron microscope offers one solution to this problem, but certain forms of light microscopy can also improve the three-dimensional perspective of an image.

Differential Interference Contrast Microscopy

Differential interference contrast (DIC) microscopy is a form of light microscopy that employs a polarizer in the condenser to produce polarized light (light in a single plane). The polarized light then passes through a prism that generates two distinct beams. These beams pass through the specimen and enter the objective lens where they are recombined into one. Because the two beams pass through substances that differ in refractive index, the combined beams are not totally in phase but instead interfere with each other, and this effect enhances subtle differences in cell structure. Thus, by DIC microscopy, cellular structures such as the nucleus of eukaryotic cells (**Figure 2.7** or endospores, vacuoles, and inclusions of bacterial cells, appear more three-dimensional. DIC microscopy is typically used on unstained cells as it can reveal internal cell structures that are nearly invisible by bright-field without the need for staining (compare Figure 2.5*a* with Figure 2.7).

Confocal Scanning Laser Microscopy

A confocal scanning laser microscope (CSLM) is a computer-controlled microscope that couples a laser to a fluorescent microscope. The laser generates a bright three-dimensional

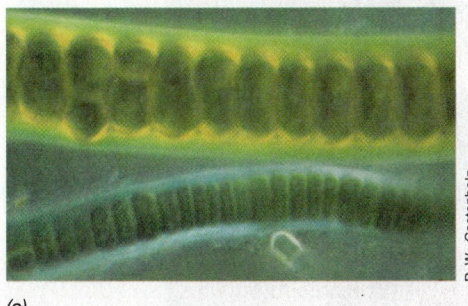
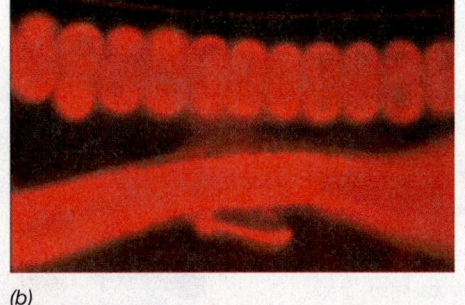
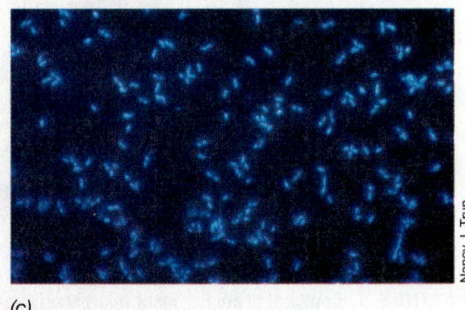

(a)   *(b)*   *(c)*

Figure 2.6 Fluorescence microscopy. *(a, b)* Cyanobacteria. The same cells are observed by bright-field microscopy in part a and by fluorescence microscopy in part b. The cells fluoresce red because they contain chlorophyll *a* and other pigments. *(c)* Fluorescence photomicrograph of cells of *Escherichia coli* made fluorescent by staining with the fluorescent dye DAPI, which binds to DNA.

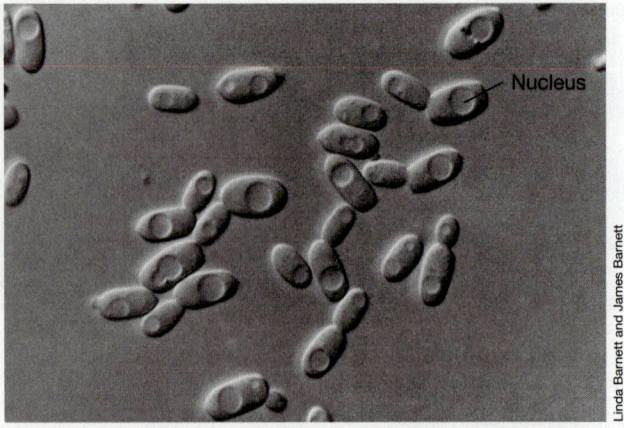

Linda Barnett and James Barnett

Figure 2.7 Differential interference contrast microscopy. Cells of the yeast *Saccharomyces cerevisiae* are given a three-dimensional effect by this form of microscopy. The yeast cells are about 8 μm wide. Note the clearly visible nucleus and compare to the bright-field image of yeast cells in Figure 2.5a.

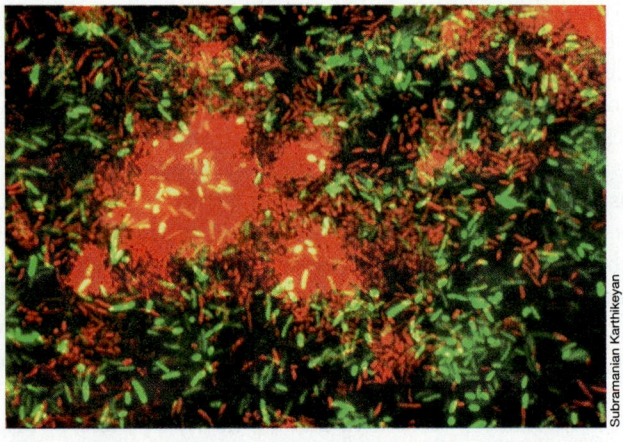

(a)

Subramanian Karthikeyan

(b)

Gernot Arp and Christian Boeker, Carl Zeiss, Jena

Figure 2.8 Confocal scanning laser microscopy. *(a)* Confocal image of a microbial biofilm community. The green, rod-shaped cells are *Pseudomonas aeruginosa* experimentally introduced into the biofilm. Cells of different colors are present at different depths in the biofilm. *(b)* Confocal image of a filamentous cyanobacterium growing in a soda lake. Cells are about 5 μm wide.

image and allows the viewer to access several planes of focus in the specimen (**Figure 2.8**). To do this, the laser beam is precisely adjusted such that only a particular layer within a specimen is in perfect focus at one time. By precisely illuminating only this single plane, the CSLM eliminates stray light from other focal planes. Thus, when observing a relatively thick specimen such as a bacterial biofilm (Figure 2.8a), not only can cells on the surface of the biofilm be observed, as would be the case with conventional light microscopy, but cells in the various layers can also be observed by adjusting the plane of focus of the laser beam. Using CSLM it has been possible to improve on the 0.2-μm resolution of the compound light microscope to a limit of about 0.1 μm.

Cells in CSLM preparations can be stained with fluorescent dyes to make them more distinct (Figure 2.8a). Alternatively, false color can be added to unstained preparations such that different layers in the specimen have different colors (Figure 2.8b). A CLSM employs a computer to assemble digital images for subsequent image processing. Images obtained from the different layers can then be digitally reconstructed to yield a three-dimensional image of the entire specimen.

CSLM is widely used in microbial ecology, especially for identifying specific populations of cells in a microbial habitat or for resolving the different components of a structured microbial community, such as a biofilm (Figure 2.8a) or a microbial mat. In general, CSLM is particularly useful anywhere thick specimens need to be examined for their microbial content with depth.

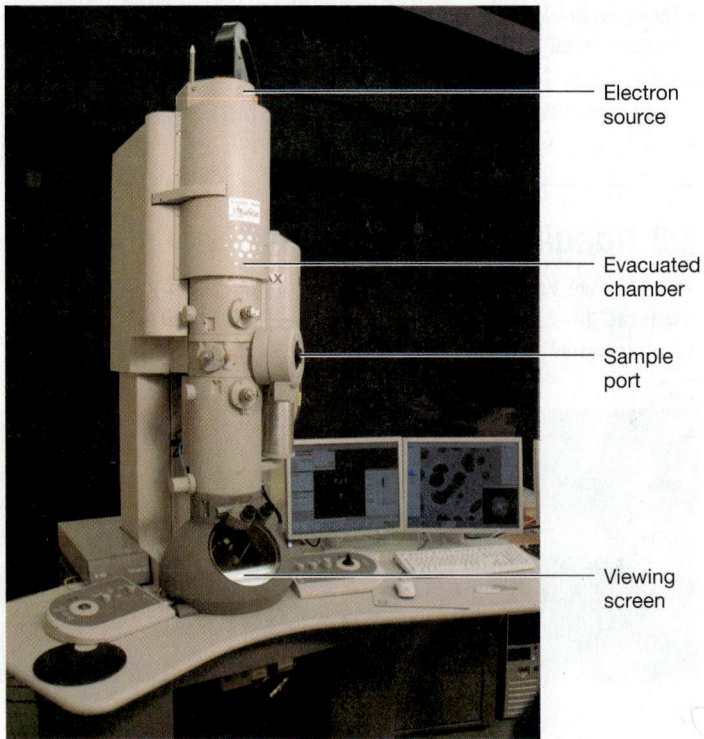

Electron source

Evacuated chamber

Sample port

Viewing screen

Figure 2.9 The electron microscope. This instrument encompasses both transmission and scanning electron microscope functions.

2.4 Probing Cell Structure: Electron Microscopy

Electron microscopes use electrons instead of visible light (photons) to image cells and cell structures. In the electron microscope, electromagnets function as lenses, and the whole system operates in a vacuum (**Figure 2.9**). Electron microscopes are fitted with cameras to allow a photograph, called an *electron micrograph*, to be taken. Two types of electron microscopy are in routine use in microbiology: transmission and scanning.

Transmission Electron Microscopy

The *transmission electron microscope* (TEM) is used to examine cells and cell structure at very high magnification and resolution. The resolving power of a TEM is much greater than that of the light microscope, even allowing one to view structures at the molecular level (**Figure 2.10**). This is because the wavelength of electrons is much shorter than the wavelength of visible light, and, as we have learned, wavelength affects resolution (Section 2.1). For example, whereas the resolving power of a light microscope is about 0.2 *micrometer*, the resolving power of a TEM is about 0.2 *nanometer*, a thousandfold improvement. With such powerful resolution, objects as small as individual protein and nucleic acid molecules can be visualized by transmission electron microscopy (Figure 2.10).

Unlike photons, electrons are very poor at penetrating; even a single cell is too thick to penetrate with an electron beam. Consequently, to view the internal structure of a cell, *thin sections* of the cell are needed, and the sections must be stabilized and stained with various chemicals to make them visible. A single bacterial cell, for instance, is cut into extremely thin (20–60 nm) slices, which are then examined individually by TEM (Figure 2.10*a*). To obtain sufficient contrast, the sections are treated with stains such as osmic acid, or permanganate, uranium, lanthanum, or lead salts. Because these substances are composed of atoms of high atomic weight, they scatter electrons well and thus improve contrast. If only the *external* features of an organism are to be

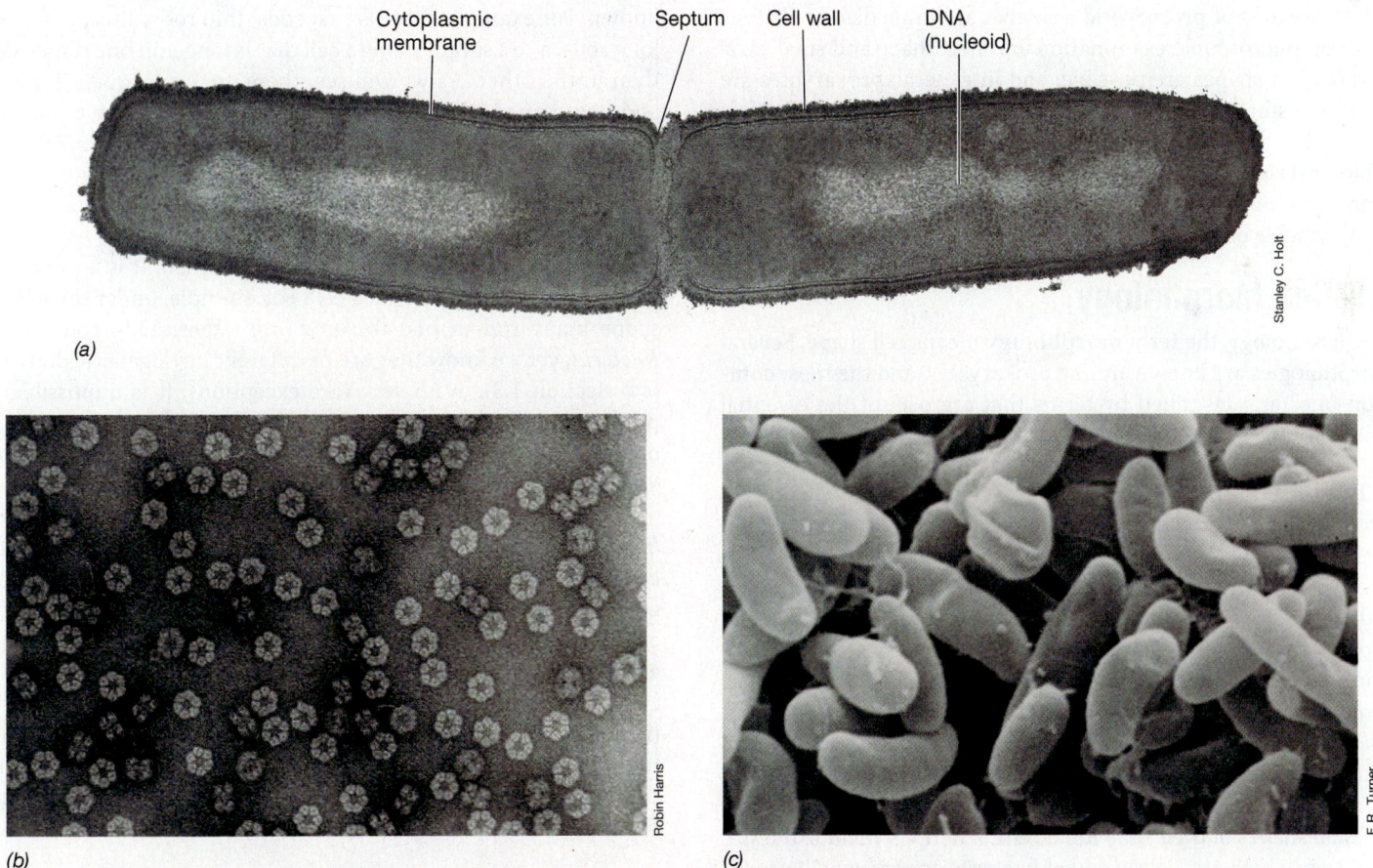

Cytoplasmic membrane   Septum   Cell wall   DNA (nucleoid)

(a)   Stanley C. Holt

(b)   Robin Harris   *(c)*   F.R. Turner

Figure 2.10 Electron micrographs. *(a)* Micrograph of a thin section of a dividing bacterial cell, taken by transmission electron microscopy (TEM). The cell is about 0.8 μm wide. *(b)* TEM of negatively stained molecules of hemoglobin. Each hexagonal-shaped molecule is about 25 nanometers (nm) in diameter and consists of two doughnut-shaped rings, a total of 15 nm wide. *(c)* Scanning electron micrograph of bacterial cells. A single cell is about 0.75 μm wide.

observed, thin sections are unnecessary. Intact cells or cell components can be observed directly in the TEM by a technique called *negative staining* (Figure 2.10*b*).

Scanning Electron Microscopy

For optimal three-dimensional imaging of cells, a *scanning electron microscope* (SEM) is used (Figure 2.9). In scanning electron microscopy, the specimen is coated with a thin film of a heavy metal, typically gold. An electron beam then scans back and forth across the specimen. Electrons scattered from the metal coating are collected and projected on a monitor to produce an image (Figure 2.10*c*). In the SEM, even fairly large specimens can be observed, and the depth of field (the portion of the image that remains in sharp focus) is extremely good. A wide range of magnifications can be obtained with the SEM, from as low as 15× up to about 100,000×, but only the *surface* of an object is typically visualized.

Electron micrographs taken by either TEM or SEM are originally taken as black-and-white images. Although the original image contains the maximum amount of scientific information that is available, color is often added to electron micrographs by manipulating them in a computer. However, such false color does not improve resolution of a micrograph. Its primary value is to increase the artistic value of the image for public consumption in the mass media. The maximum scientific content and detail in an electron micrograph is set at the moment the micrograph is taken, and thus false color will be used sparingly in electron micrographs in this book so as to present the micrographs in their original scientific context.

MINIQUIZ
- What is an electron micrograph? Why do electron micrographs have greater resolution than light micrographs?
- What type of electron microscope would be used to view a cluster of cells? What type would be used to observe internal cell structure?

II • Cells of *Bacteria* and *Archaea*

Two features of prokaryotic cells that are immediately obvious upon microscopic examination are their shape and small size. A variety of shapes are possible, and in general, prokaryotes are extremely small relative to eukaryotic cells. Cell shape can be useful for distinguishing different cells and undoubtedly has some ecological significance, but cell shape rarely has phylogenetic relevance. By contrast, the typically small size of prokaryotes affects many aspects of their biology.

2.5 Cell Morphology

In microbiology, the term **morphology** means cell shape. Several morphologies are known among prokaryotes, and the most common ones are described by terms that are part of the essential lexicon of the microbiologist.

Major Cell Morphologies

Examples of bacterial morphologies are shown in **Figure 2.11**. A cell that is spherical or ovoid in morphology is called a *coccus* (plural, cocci). A cylindrically shaped cell is called a *rod* or a *bacillus*. Some rods form spiral shapes and are called *spirilla*. The cells of some prokaryotes remain together in groups or clusters after cell division, and the arrangements are often characteristic. For instance, some cocci form long chains (for example, the bacterium *Streptococcus*), others occur in three-dimensional cubes (*Sarcina*), and still others in grapelike clusters (*Staphylococcus*).

A few bacterial groups are immediately recognizable by the unusual shapes of their individual cells. Examples include the spirochetes, which are tightly coiled bacteria; appendaged bacteria, which possess extensions of their cells as long tubes or stalks; and filamentous bacteria, which form long, thin cells or chains of cells (Figure 2.11).

The cell morphologies described here should only be considered representative; many variations of these morphologies are known. For example, there are fat rods, thin rods, short rods, and long rods, a rod simply being a cell that is longer in one dimension than in the other. As we will see, there are even square bacteria and star-shaped bacteria! Cell morphologies thus form a continuum, with some shapes, such as rods, being very common, whereas others are more unusual.

Morphology and Biology

Although cell morphology is easily determined, it is a poor predictor of other properties of a cell. For example, under the microscope many rod-shaped *Archaea* look identical to rod-shaped *Bacteria*, yet we know they are of different phylogenetic domains (⮂ Section 1.3). With very rare exceptions, it is impossible to predict the physiology, ecology, phylogeny, pathogenic potential, or virtually any other property of a prokaryotic cell by simply knowing its morphology.

Why is a cell the shape it is? Although we know something about *how* cell shape is controlled, we know little about *why* a particular cell evolved the morphology it has. Several selective forces undoubtedly help shape the morphology of a given species. Some examples of these might include optimization for nutrient uptake (small cells and others with high surface-to-volume ratios, such as appendaged cells), swimming motility in viscous environments or near surfaces (helical- or spiral-shaped cells), gliding motility (filamentous bacteria), and so on. Morphology is not a trivial feature of a microbial cell but instead a genetically encoded property that maximizes fitness of the organism for success in its particular habitat.

MINIQUIZ
- How do cocci and rods differ in morphology?
- Is cell morphology a good predictor of other properties of the cell?

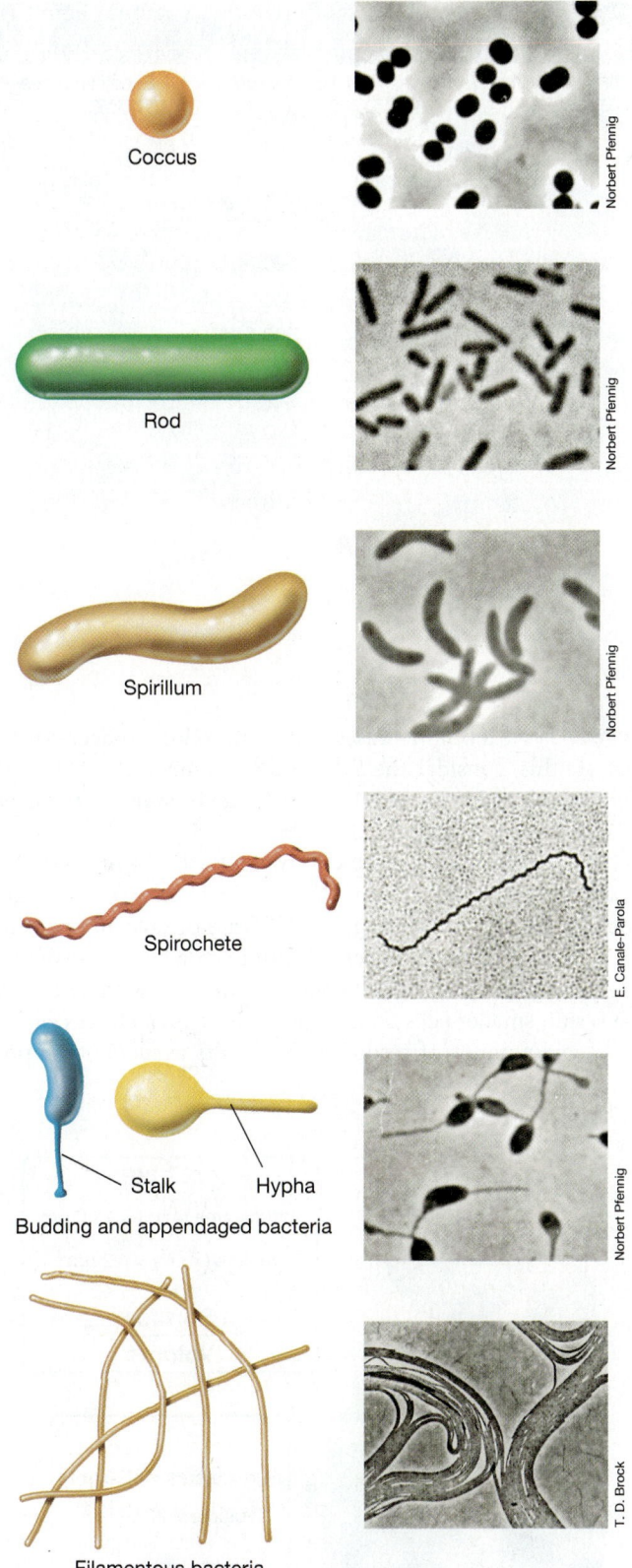

Figure 2.11   Cell morphologies. Beside each drawing is a phase-contrast photomicrograph of cells showing that morphology. Coccus (cell diameter in photomicrograph, 1.5 μm); rod (1 μm); spirillum (1 μm); spirochete, (0.25 μm); budding (1.2 μm); filamentous (0.8 μm). All photomicrographs are of species of *Bacteria*. Not all of these morphologies are known among the *Archaea*.

2.6 Cell Size and the Significance of Being Small

Prokaryotes vary in size from cells as small as about 0.2 μm in diameter to those more than 700 μm in diameter (**Table 2.1**). The vast majority of rod-shaped prokaryotes that have been cultured are between 0.5 and 4 μm wide and less than 15 μm long. But a few very large prokaryotes, such as *Epulopiscium fishelsoni*, are known, with cells longer than 600 μm (0.6 millimeter) (**Figure 2.12**). This bacterium, phylogenetically related to the endospore-forming bacterium *Clostridium* and found in the gut of tropical marine fish called surgeonfish, contains multiple copies of its genome. The many copies are apparently necessary because the volume of an *Epulopiscium* cell is so large (Table 2.1) that a single copy of its

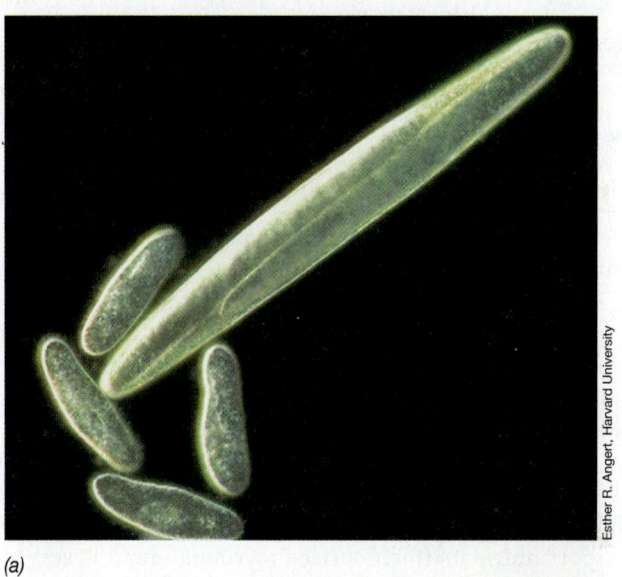

(a)

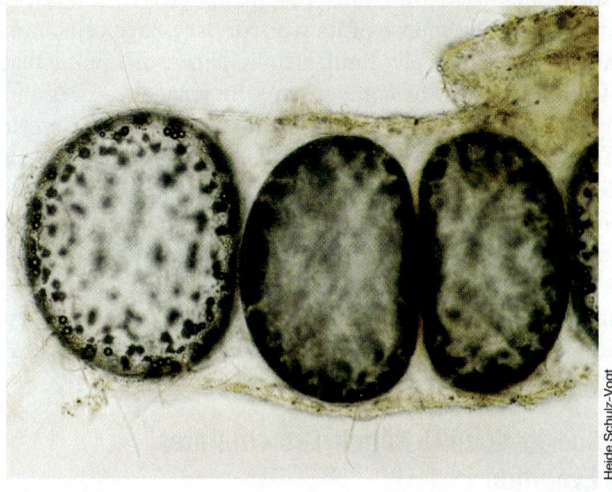

(b)

Figure 2.12   Some very large prokaryotes. Dark-field photomicrograph of two giant prokaryotes, species of *Bacteria*. *(a) Epulopiscium fishelsoni.* The rod-shaped cell is about 600 μm (0.6 mm) long and 75 μm wide and is shown with four cells of the protist *Paramecium* (a eukaryote), each of which is about 150 μm long. *(b) Thiomargarita namibiensis*, a large sulfur chemolithotroph and currently the largest known prokaryote. Cell widths vary from 400 to 750 μm.

Table 2.1 Cell size and volume of some cells of *Bacteria*, from the largest to the smallest

Organism	Characteristics	Morphology	Size[a] (μm)	Cell volume (μm^3)	E. coli volumes
Thiomargarita namibiensis	Sulfur chemolithotroph	Cocci in chains	750	200,000,000	100,000,000
Epulopiscium fishelsoni[a]	Chemoorganotroph	Rods with tapered ends	80×600	3,000,000	1,500,000
Beggiatoa species[a]	Sulfur chemolithotroph	Filaments	50×160	1,000,000	500,000
Achromatium oxaliferum	Sulfur chemolithotroph	Cocci	35×95	80,000	40,000
Lyngbya majuscula	Cyanobacterium	Filaments	8×80	40,000	20,000
Thiovulum majus	Sulfur chemolithotroph	Cocci	18	3,000	1,500
Staphylothermus marinus[a]	Hyperthermophile	Cocci in irregular clusters	15	1,800	900
Magnetobacterium bavaricum	Magnetotactic bacterium	Rods	2×10	30	15
Escherichia coli	Chemoorganotroph	Rods	1×2	2	1
Pelagibacter ubique[a]	Marine chemoorganotroph	Rods	0.2×0.5	0.014	0.007
Mycoplasma pneumoniae	Pathogenic bacterium	Pleomorphic[b]	0.2	0.005	0.0025

[a]Where only one number is given, this is the diameter of spherical cells. The values given are for the largest cell size observed in each species. For example, for *T. namibiensis*, an average cell is only about 200 μm in diameter. But on occasion, giant cells of 750 μm are observed. Likewise, an average cell of *S. marinus* is about 1 μm in diameter. The species of *Beggiatoa* here is unclear and *E. fishelsoni*, *Magnetobacterium bavaricum*, and *P. ubique* are not formally recognized names in taxonomy.
[b]*Mycoplasma* is a bacterium that lacks a cell wall and can thus take on many shapes (*pleomorphic* means "many shapes").

Source: Data obtained from Schulz, H.N., and B.B. Jørgensen. 2001. *Ann. Rev. Microbiol. 55:* 105–137.

genome would be insufficient to support its transcriptional and translational demands.

Cells of the largest known prokaryote, the sulfur chemolithotroph *Thiomargarita* (Figure 2.12*b*), are even larger than those of *Epulopiscium*, about 750 μm in diameter; such cells are just visible to the naked eye. Why these cells are so large is not well understood, although for sulfur bacteria a large cell size may be a mechanism for storing inclusions of sulfur (an energy source). It is hypothesized that the upper size limit for prokaryotic cells results from the decreasing ability of larger and larger cells to transport nutrients (their surface-to-volume ratio is very small; see the next subsection). Since the metabolic rate of a cell varies inversely with the square of its size, for very large cells, nutrient uptake would eventually limit metabolism to the point that the cell would no longer be competitive with smaller cells.

Very large cells are uncommon in the prokaryotic world. In contrast to *Thiomargarita* or *Epulopiscium* (Figure 2.12), the dimensions of an average rod-shaped prokaryote, the bacterium *E. coli*, for example, are about 1 × 2 μm; these dimensions are typical of the cells of most prokaryotes. By contrast, eukaryotic cells can be 2 to more than 600 μm in diameter, although very small eukaryotes are uncommon, most being 8 μm in diameter or greater. In general, then, it can be said that prokaryotes are very small cells compared with eukaryotes.

Surface-to-Volume Ratios, Growth Rates, and Evolution

There are significant advantages to being small. Small cells have more surface area relative to cell volume than do large cells; that is, they have a higher *surface-to-volume ratio*. Consider a coccus. The volume of a coccus is a function of the cube of its radius ($V = \frac{4}{3}\pi r^3$), while its surface area is a function of the square of the radius ($S = 4\pi r^2$). Therefore, the S/V ratio of a coccus is $3/r$

(Figure 2.13). As a cell increases in size, its S/V ratio decreases. To illustrate this, consider the S/V ratio for some of the cells of different sizes listed in Table 2.1: *Pelagibacter ubique*, 22; *E. coli*, 4.5; and *E. fishelsoni* (Figure 2.12*a*), 0.05.

The S/V ratio of a cell affects several aspects of its biology, including even its evolution. Because how fast a cell can grow depends, among other things, on the rate of nutrient exchange, the higher S/V ratio of smaller cells supports a faster rate of nutrient exchange per unit of cell volume compared with larger cells. As a result, smaller cells tend to grow faster than larger cells, and for a given amount of resources (nutrients available to support

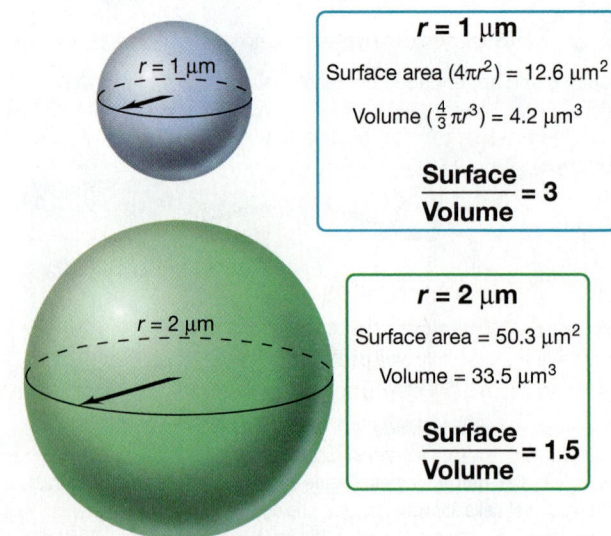

Figure 2.13 Surface area and volume relationships in cells. As a cell increases in size, its S/V ratio decreases.

growth), a larger population of small cells than of large cells is possible. This in turn can affect evolution.

Each time a cell divides, its chromosome replicates. As DNA is replicated, occasional errors, called *mutations*, occur. Because mutation rates appear to be about the same in all cells, large or small, the more chromosome replications that occur, the greater the total number of mutations in the cell population. Mutations are the "raw material" of evolution; the larger the pool of mutations, the greater the evolutionary possibilities. Thus, because prokaryotic cells are quite small and are also genetically haploid (allowing mutations to be expressed immediately), they have, in general, the capacity for more rapid growth and faster evolution than larger, genetically diploid cells. In the latter, not only is the S/V ratio smaller, but the effects of a mutation in one gene can be masked by a second, unmutated gene copy. These fundamental differences in size and genetics between prokaryotic and eukaryotic cells is a major reason why prokaryotes adapt rapidly to changing environmental conditions and more easily exploit new habitats than eukaryotic cells. We will illustrate this concept in later chapters when we consider, for example, the enormous diversity of prokaryotes (Chapters 13–16) and the rapidity of their evolution (⮌ Section 12.6).

Lower Limits of Cell Size

From the foregoing, one could imagine that smaller and smaller bacteria would have greater and greater selective advantages in nature. However, this is not true, as there are lower limits to cell size. If one considers the volume needed to house the essential components of a free-living cell—proteins, nucleic acids, ribosomes, and so on—a structure 0.1 μm in diameter or less is insufficient to do the job, and structures 0.15 μm in diameter are marginal. Thus, structures observed in natural samples that are 0.1 μm or even smaller and "look" like bacterial cells are almost certainly not so. Despite this, some very small prokaryotic cells are known and many have been grown in the laboratory. Open ocean water, for example, contains 10^5–10^6 prokaryotic cells per milliliter, and these tend to be very small cells, 0.2–0.4 μm in diameter. We will see later that many pathogenic bacteria are also very small. When the genomes of these pathogens are examined, they are found to be highly streamlined and missing many genes whose functions are supplied to them by their hosts.

MINIQUIZ

- What physical property of cells increases as cells become smaller?
- How can the small size and haploid state of prokaryotes accelerate their evolution?
- What are the approximate limits to how small a cell can be? Why is this so?

III • The Cytoplasmic Membrane and Transport

We now consider the structure and function of one of a cell's most critical structures, the cytoplasmic membrane. The cytoplasmic membrane plays many roles, chief among them as the "gatekeeper" for dissolved substances that enter and exit the cell.

2.7 Membrane Structure

The **cytoplasmic membrane** surrounds the cytoplasm and separates it from the environment. If the cytoplasmic membrane is compromised, the integrity of the cell is destroyed, the cytoplasm leaks into the environment, and the cell dies. The cytoplasmic membrane is structurally weak and confers little protection from osmotic lysis, but it is an ideal structure for its major function on the cell: selective permeability.

Composition of Membranes

The general structure of the cytoplasmic membrane is a phospholipid bilayer. Phospholipids are composed of both hydrophobic (fatty acid) and hydrophilic (glycerol–phosphate) components (**Figure 2.14**). As phospholipids aggregate in an aqueous solution, they naturally form bilayers. In a phospholipid membrane, the fatty acids point inward toward each other to form a hydrophobic environment, and the hydrophilic portions remain exposed to the external environment or the cytoplasm (Figure 2.14*b*). Common fatty acids in the cytoplasmic membrane include those with 14 to 20 carbon atoms.

The cytoplasmic membrane is only 8–10 nanometers wide but is still visible in the transmission electron microscope, where it appears as two dark lines separated by a light line (Figure 2.14*c*). This *unit membrane*, as it is called (because each phospholipid leaf forms half of the "unit"), consists of a phospholipid bilayer with proteins embedded in it (**Figure 2.15**). Although in a diagram the cytoplasmic membrane appears rigid, it is actually somewhat fluid, having a consistency approximating that of a low-viscosity oil. Thus, some freedom of movement exists for proteins embedded in the membrane. The cytoplasmic membranes of some *Bacteria* are strengthened by sterol-like molecules called *hopanoids*. Sterols are rigid and planar molecules that function to strengthen the membranes of eukaryotic cells, and hopanoids serve a similar function in *Bacteria*.

Membrane Proteins

The protein content of the cytoplasmic membrane is quite high, and membrane proteins typically have hydrophobic surfaces in regions that span the membrane and hydrophilic surfaces in regions that contact the environment and the cytoplasm (Figures 2.14 and 2.15). The *outer* surface of the cytoplasmic membrane faces the environment and in gram-negative bacteria interacts with a variety of proteins that bind substrates or process larger molecules for transport into the cell (periplasmic proteins, Section 2.11). The *inner* surface of the cytoplasmic membrane touches the cytoplasm and interacts with proteins and other molecules in this milieu.

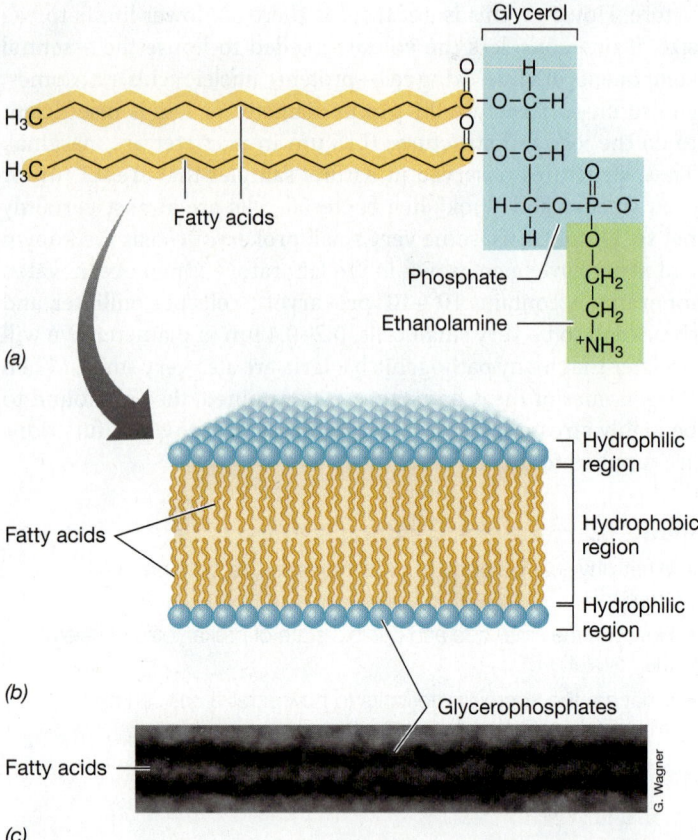

(a)

(b)

(c)

Figure 2.14 Phospholipid bilayer membrane. *(a)* Structure of the phospholipid phosphatidylethanolamine. *(b)* General architecture of a bilayer membrane; the blue spheres depict glycerol with phosphate and/or other hydrophilic groups. *(c)* Transmission electron micrograph of a membrane. The light inner area is the hydrophobic region of the model membrane shown in part b.

Many membrane proteins are firmly embedded in the membrane and are called *integral* membrane proteins. Other proteins have one portion anchored in the membrane and extramembrane regions that point into or out of the cell (Figure 2.15). Still other proteins, called *peripheral* membrane proteins, are not membrane-embedded but nevertheless remain associated with membrane surfaces. Some of these peripheral membrane proteins are lipoproteins, molecules that contain a lipid tail that anchors the protein into the membrane. Peripheral membrane proteins typically interact with integral membrane proteins in important cellular processes such as energy metabolism and transport. Membrane proteins that need to interact with each other in some process are typically grouped together into clusters to allow them to remain adjacent to one another in the semifluid environment of the membrane.

Archaeal Membranes

In contrast to the lipids of *Bacteria* and *Eukarya* in which *ester* linkages bond fatty acids to glycerol, the lipids of *Archaea* contain *ether* bonds between glycerol and their hydrophobic side chains (Figure 2.16). Archaeal lipids thus lack fatty acids, per se, although the hydrophobic side chains play the same functional role as fatty acids. Archaeal lipids are formed from multiple units of the five-carbon hydrocarbon isoprene (Figure 2.16c).

The cytoplasmic membrane of *Archaea* is formed from either glycerol diethers, which have 20-carbon side chains (the 20-C unit is called a *phytanyl* group composed of 5 isoprene units), or diglycerol tetraethers, which have 40-carbon side chains (**Figure 2.17**). In the tetraether lipid, the ends of the phytanyl side chains that point inward from each glycerol molecule are covalently linked. This forms a lipid *monolayer* instead of a lipid *bilayer* membrane (Figure 2.17d, e). In contrast to lipid bilayers, lipid monolayer membranes are extremely resistant to heat and are therefore widely distributed among hyperthermophilic *Archaea*, organisms that grow best at temperatures above 80°C. Membranes with a mixture of bilayer and monolayer character are also possible, with some of the opposing hydrophobic groups covalently bonded and others not.

Many archaeal lipids contain rings within the hydrocarbon side chains. For example, *crenarchaeol*, a lipid widespread among species of *Thaumarchaeota*, a major phylum of *Archaea*, contains four 5-carbon (cyclopentyl) rings and one 6-carbon (cyclohexyl) ring (Figure 2.17c). Rings in the hydrocarbon side chains affect the chemical properties of the lipids and thus overall membrane function. Sugars can also be present in archaeal lipids. For example, the predominant membrane lipids of many *Euryarchaeota*, a major group of *Archaea* that includes the methanogens and extreme halophiles (⟳ Figure 1.6b), are glycerol diether glycolipids.

Despite the differences in chemistry between the cytoplasmic membranes of *Archaea* and organisms in the other domains, the fundamental construction of the archaeal cytoplasmic membrane—inner and outer hydrophilic surfaces and a hydrophobic interior—is the same as that of membranes in *Bacteria* and *Eukarya*. Evolution has selected this design as the best solution to the main function of the cytoplasmic membrane—permeability—and we consider this problem now.

MINIQUIZ

- Draw the basic structure of a lipid bilayer and label the hydrophilic and hydrophobic regions.
- How are the membrane lipids of *Bacteria* and *Archaea* similar, and how do they differ?

2.8 Membrane Function

The cytoplasmic membrane has several functions. First and foremost, the membrane is a permeability barrier, preventing the passive leakage of solutes into or out of the cell (**Figure 2.18**). Second, the membrane is an anchor for many proteins. Some of these are enzymes that function in energy conservation, and others transport solutes into and out of the cell. The cytoplasmic membrane is a major site of energy conservation in the prokaryotic cell. The membrane can exist in an energetically charged form in which protons (H^+) are separated from hydroxyl ions (OH^-) across the membrane surface (Figure 2.18c). Charge separation forms an energized state, analogous to the potential energy present in a charged battery. This energy source, called the *proton motive force*, is responsible for driving many energy-requiring functions in the cell, including many transport reactions, swimming motility, and the biosynthesis of ATP.

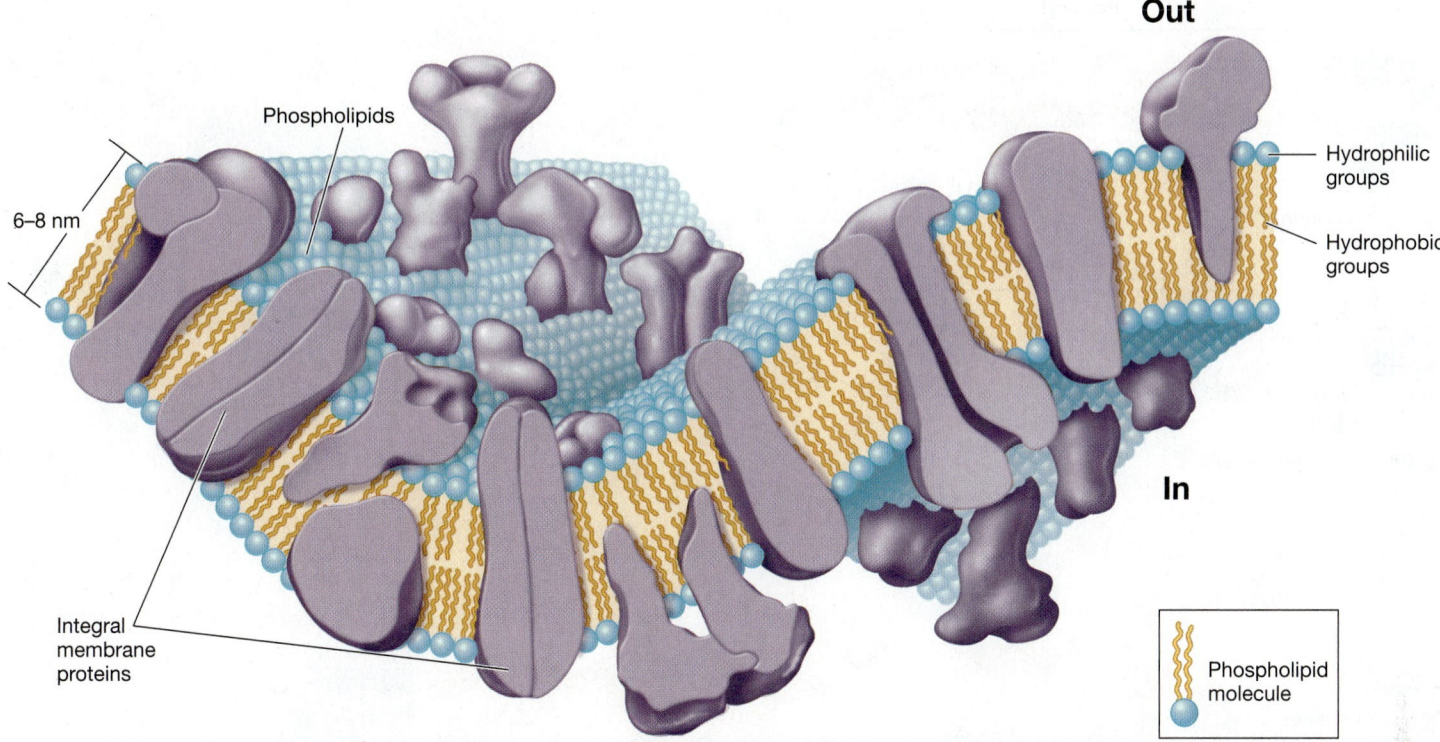

Out

Phospholipids

6–8 nm

Hydrophilic groups

Hydrophobic groups

In

Integral membrane proteins

Phospholipid molecule

Figure 2.15 Structure of the cytoplasmic membrane. The inner surface (**In**) faces the cytoplasm and the outer surface (**Out**) faces the environment. Phospholipids compose the matrix of the cytoplasmic membrane with proteins embedded or surface associated. The general architecture of the cytoplasmic membrane is similar in both prokaryotes and eukaryotes, although there are chemical differences.

Permeability

The cytoplasm is a solution of salts, sugars, amino acids, nucleotides, and many other substances. The hydrophobic portion of the cytoplasmic membrane (Figures 2.14 and 2.15) is a tight barrier to diffusion of these substances. Although some small hydrophobic molecules pass the cytoplasmic membrane by diffusion, polar and charged molecules do not diffuse but instead must be transported. Even a substance as small as a proton (H^+) cannot diffuse across the membrane. The relative permeability of the membrane to some biologically relevant substances is shown in **Table 2.2**. As

can be seen, most substances cannot diffuse into the cell and thus must be transported.

One substance that does freely pass the membrane in both directions is water, a molecule that is somewhat polar but sufficiently small to pass between phospholipid molecules in the lipid bilayer (Table 2.2). In addition to water that enters by diffusion, membrane proteins called *aquaporins* function to accelerate the movement of water across the membrane. For example, aquaporin AqpZ of *Escherichia coli* imports water to or exports water from the cytoplasm, depending on osmotic conditions.

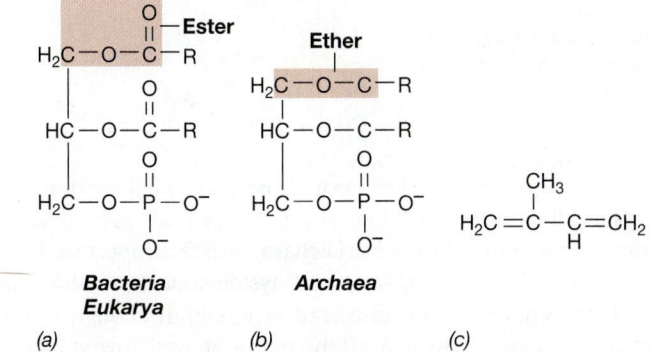

Figure 2.16 General structure of lipids. *(a)* The ester linkage and *(b)* the ether linkage. *(c)* Isoprene, the parent structure of the hydrophobic side chains of archaeal lipids. By contrast, in lipids of *Bacteria* and *Eukarya*, the side chains are composed of fatty acids (see Figure 2.14*a*).

Table 2.2 Comparative permeability of membranes to various molecules

Substance	Rate of permeability[a]	Potential for diffusion into a cell
Water	100	Excellent
Glycerol	0.1	Good
Tryptophan	0.001	Fair/Poor
Glucose	0.001	Fair/Poor
Chloride ion (Cl⁻)	0.000001	Very poor
Potassium ion (K⁺)	0.0000001	Extremely poor
Sodium ion (Na⁺)	0.00000001	Extremely poor

[a]Relative scale—permeability with respect to permeability to water given as 100. Permeability of the membrane to water may be affected by aquaporins.

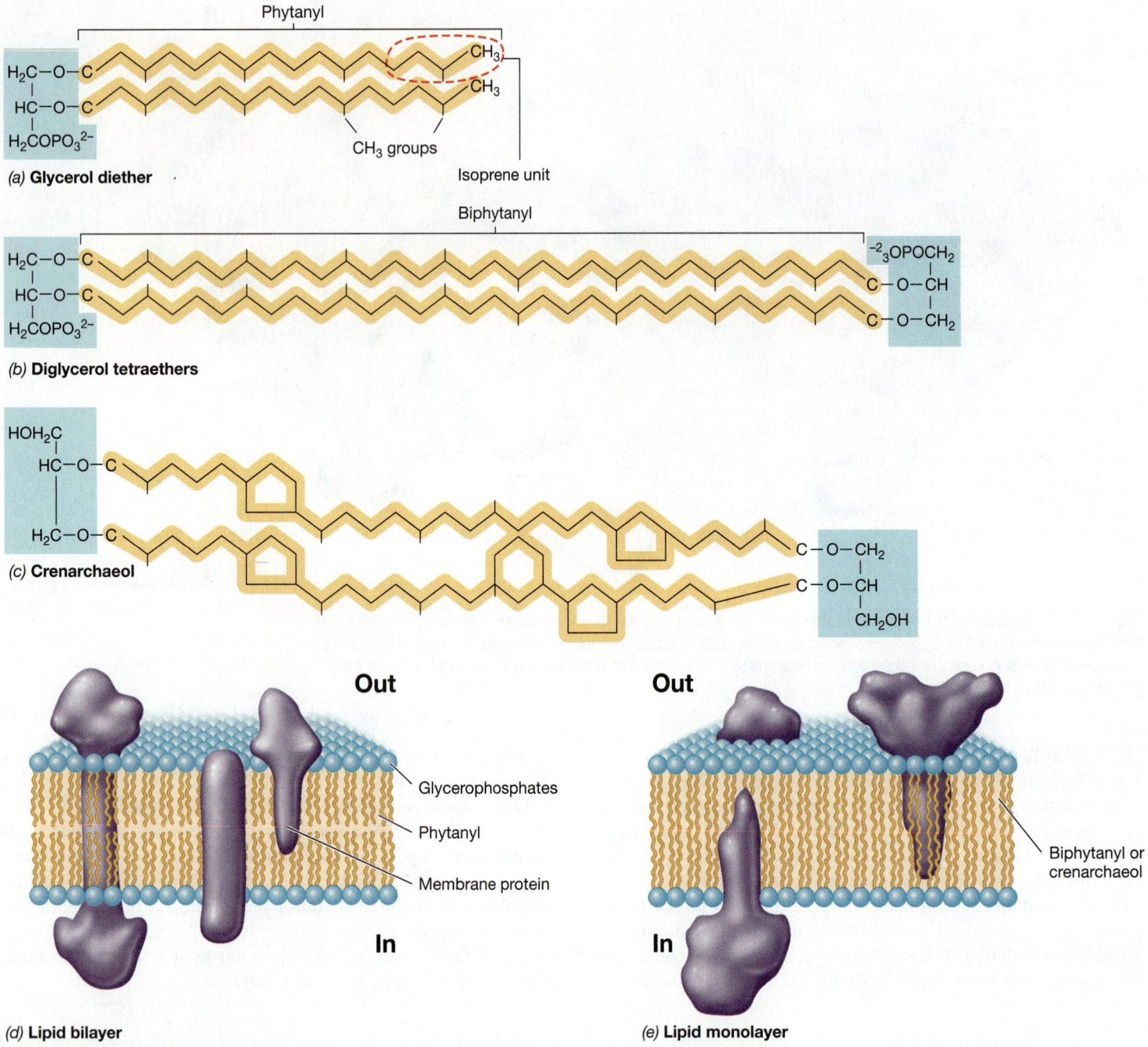

(a) **Glycerol diether**

(b) **Diglycerol tetraethers**

(c) **Crenarchaeol**

(d) **Lipid bilayer**

(e) **Lipid monolayer**

Figure 2.17 **Major lipids of *Archaea* and the architecture of archaeal membranes.** *(a, b)* Note that the hydrocarbon of the lipid is bonded to the glycerol by an ether linkage in both cases. The hydrocarbon is phytanyl (C_{20}) in part a and biphytanyl (C_{40}) in part b. *(c)* A major lipid of *Thaumarchaeota* is crenarchaeol, a lipid containing 5- and 6-carbon rings. *(d, e)* Membrane structure in *Archaea* may be bilayer or monolayer (or a mix of both).

Transport Proteins

Transport proteins do more than just carry solutes across the membrane—they function to *accumulate* solutes against the concentration gradient. The necessity for carrier-mediated transport is easy to understand. If diffusion were the only mechanism by which solutes entered a cell, the intracellular concentration of nutrients would never exceed that of the external concentration, which for most nutrients in nature is often quite low (**Figure 2.19**). This would be insufficient for cells to carry out biochemical reactions. Transport reactions move nutrients from low concentration to high concentration, and as we will see in the next section, this costs the cell energy.

Transport systems show several characteristic properties. First, in contrast with diffusion, transport systems show a *saturation effect*. If the concentration of a substrate is high enough to saturate the transporter, which often occurs at very low substrate concentrations, the rate of uptake becomes maximal and the addition of more substrate does not increase the rate (Figure 2.19). This feature of transport proteins is essential for concentrating nutrients from very dilute environments. A second characteristic

Functions of the cytoplasmic membrane

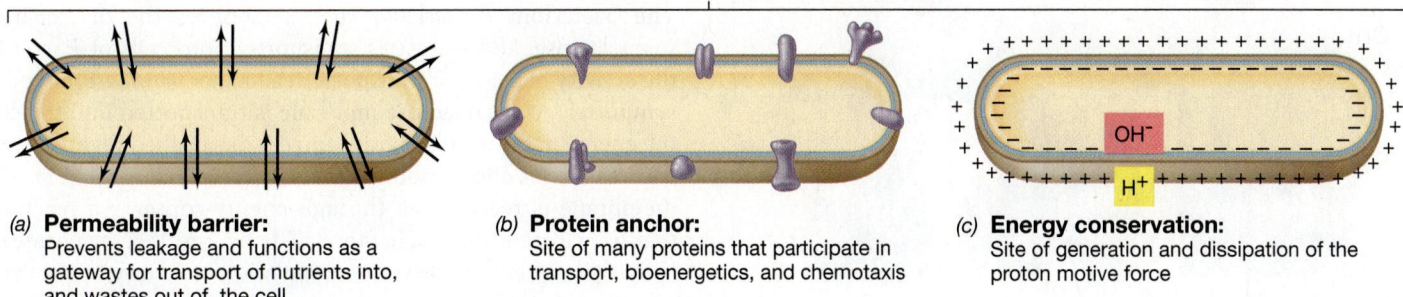

(a) **Permeability barrier:**
Prevents leakage and functions as a gateway for transport of nutrients into, and wastes out of, the cell

(b) **Protein anchor:**
Site of many proteins that participate in transport, bioenergetics, and chemotaxis

(c) **Energy conservation:**
Site of generation and dissipation of the proton motive force

Figure 2.18 **The major functions of the cytoplasmic membrane.** Although structurally weak, the cytoplasmic membrane has many important cellular functions.

of carrier-mediated transport is *high specificity*. Many transport proteins carry only a single kind of molecule, whereas a few carry a related class of molecules, such as several different sugars or several different amino acids. This economizing reduces the need for separate transport proteins for each different amino acid or sugar. A third major characteristic of transport systems is that their synthesis is often *highly regulated* by the cell. That is, the specific complement of transporters present in the cytoplasmic membrane of a cell is a function of both the nature and concentration of resources in its environment. Some nutrients are transported by one transporter when present at high concentration and by a separate, usually higher-affinity transporter, when present at very low concentration.

MINIQUIZ

- Why can a cell not depend on simple diffusion as a means of acquiring its nutrients?
- Why is physical damage to the cytoplasmic membrane potentially lethal for the cell?

2.9 Nutrient Transport

To fuel metabolism and support growth, cells need to import nutrients and export wastes on a continuous basis. To fulfill these requirements, several transport mechanisms exist in prokaryotes, each with its own unique features.

Transport Events and Transporters

At least three transport mechanisms have been well characterized in prokaryotes. **Simple transport** consists only of a membrane-spanning transport protein, **group translocation** employs a series of proteins in the transport event, and **ABC transport systems** consist of three components: a substrate-binding protein, a membrane-integrated transporter, and an ATP-hydrolyzing protein (**Figure 2.20**). All of these systems drive the actual transport event using the energy of the proton motive force or ATP or some other energy-rich organic compound.

Membrane transporters are typically composed of 12 polypeptides that weave back and forth through the membrane to form a

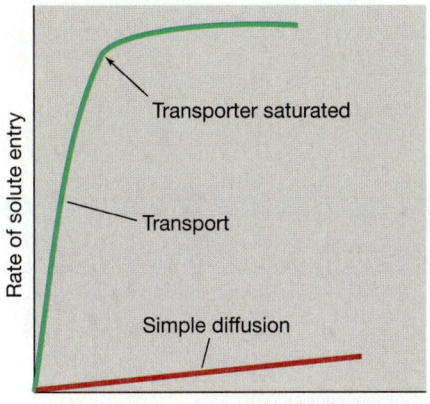

Figure 2.19 **Transport versus diffusion.** In transport, the uptake rate shows saturation at relatively low external concentrations.

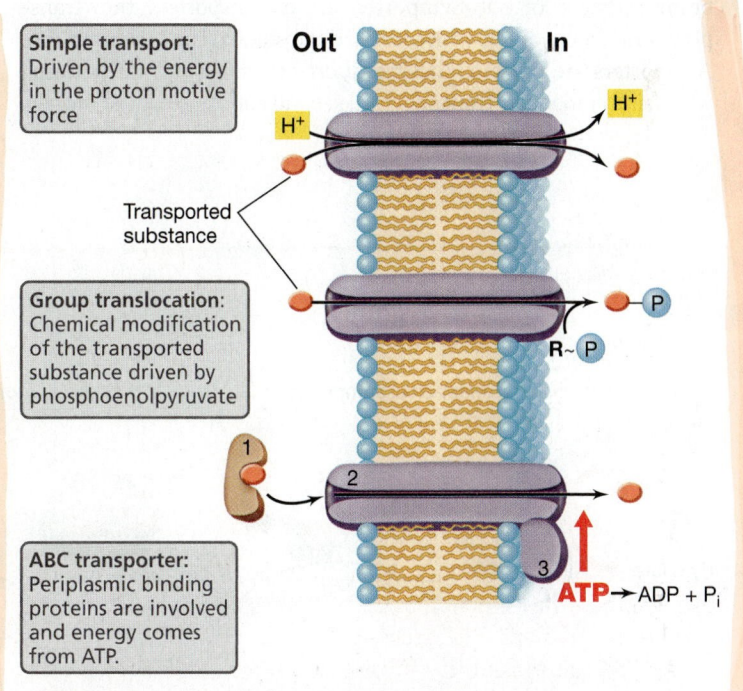

Simple transport: Driven by the energy in the proton motive force

Group translocation: Chemical modification of the transported substance driven by phosphoenolpyruvate

ABC transporter: Periplasmic binding proteins are involved and energy comes from ATP.

Figure 2.20 **The three classes of transport systems.** Note how simple transporters and the ABC system transport substances without chemically modifying them, whereas group translocation results in chemical modification (in this case phosphorylation) of the transported substance. The three proteins of the ABC system are labeled 1, 2, and 3.

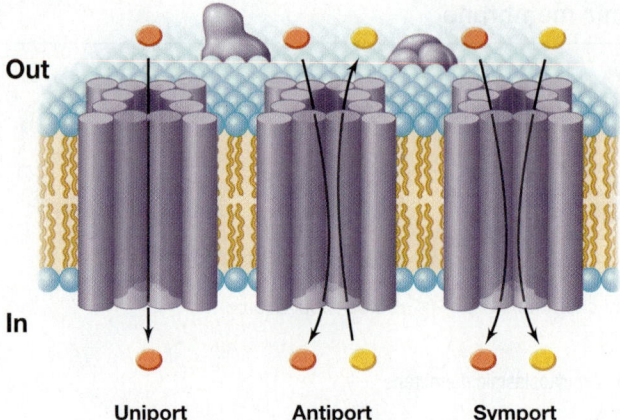

Uniport   Antiport   Symport

Figure 2.21 *Structure of membrane-spanning transporters and types of transport events.* Membrane-spanning transporters are made of 12 α-helices (each shown as a cylinder) that aggregate to form a channel through the membrane. Shown are examples of three different transport events: uniport, antiport, and symport. Red discs represent the transported molecule; yellow discs represent the cotransported molecule.

channel; it is through this channel that the solute is actually carried into the cell. Transport requires that a conformational change occur in the transport protein following binding of its solute. Like a gate swinging open, the conformational change then brings the solute into the cell.

Regardless of transport *mechanism*, actual transport *events* are of three types—*uniport, symport,* and *antiport*— and each event is catalyzed by a protein called a *porter* (**Figure 2.21**). Uniporters are proteins that carry a substance unidirectionally across the membrane, either in or out. Symporters are cotransporters; they transport a molecule along with a second substance, typically a proton. Antiporters are proteins that transport one substance into the cell while simultaneously transporting a second substance out of the cell.

Simple Transporters and Group Translocation

The bacterium *Escherichia coli* metabolizes the disaccharide sugar lactose, and lactose is transported into cells of *E. coli* by the activity of a simple transporter called *lac permease*, a type of symporter. As each lactose molecule is transported into the cell, the energy in the proton motive force is diminished slightly by the cotransport of one proton into the cytoplasm (Figure 2.21). The membrane is reenergized through energy-conserving reactions that we will describe in Chapter 3. The net result of lac permease activity is the energy-driven accumulation of lactose in the cytoplasm against the concentration gradient. Once in the cytoplasm, lactose is broken down and used for ATP synthesis and to make new carbon skeletons.

Group translocation differs from simple transport in two ways: (1) the transported substance is chemically modified during the transport process, and (2) an energy-rich organic compound rather than the proton motive force drives the transport event. One of the best-studied group translocation systems transports the sugars glucose, mannose, and fructose in *E. coli*. These compounds are phosphorylated during transport by the *phosphotransferase system*. The phosphotransferase system consists of a family of proteins that work in concert; five proteins are necessary to transport any given sugar. Before the sugar is transported, the proteins in the phosphotransferase system are themselves alternately phosphorylated and dephosphorylated in a cascading fashion until the actual transporter, Enzyme II$_c$, phosphorylates the sugar during the transport event (**Figure 2.22**). A protein called *HPr*, the enzyme that phosphorylates HPr (Enzyme I), and Enzyme II$_a$ are all cytoplasmic proteins. By contrast, Enzyme II$_b$ lies on the inner surface of the membrane and Enzyme II$_c$ is an integral membrane protein.

HPr and Enzyme I are nonspecific components of the phosphotransferase system and participate in the uptake of several different sugars. Several different versions of Enzyme II exist, one for

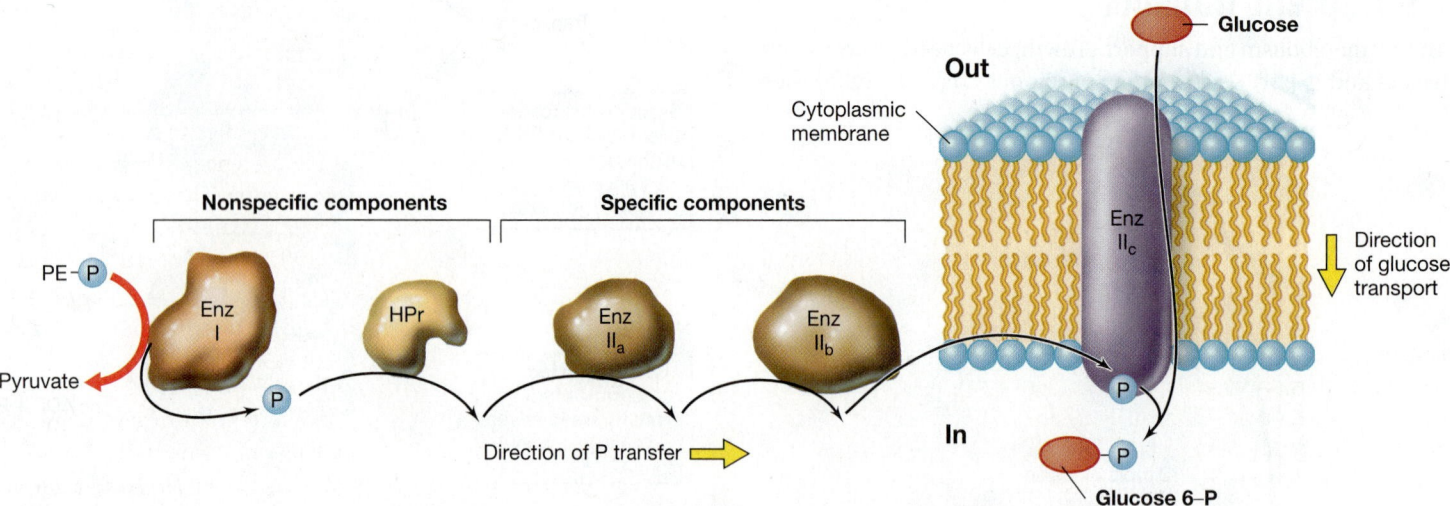

Figure 2.22 *Mechanism of the phosphotransferase system of Escherichia coli.* For glucose uptake, the system consists of five proteins: Enzyme (Enz) I, Enzymes II$_a$, II$_b$, and II$_c$, and HPr. A phosphate cascade occurs from phosphoenolpyruvate (PE-P) to Enzyme II$_c$ and the latter actually transports and phosphorylates the sugar. Proteins HPr and Enz I are nonspecific and transport any sugar. The Enz II components are specific for each particular sugar.

each different sugar transported (Figure 2.22). Energy to drive the phosphotransferase system comes from phosphoenolpyruvate, an energy-rich intermediate in glycolysis (⮂ Section 3.8).

Periplasmic Binding Proteins and the ABC System

We will learn shortly that gram-negative bacteria contain a region called the *periplasm* that lies between the cytoplasmic membrane and a second membrane layer called the *outer membrane*, part of the gram-negative cell wall (Section 2.11). The periplasm contains many different proteins, several of which function in transport and are called *periplasmic binding proteins*. Transport systems that employ periplasmic binding proteins along with a membrane transporter and ATP-hydrolyzing proteins are called ABC transport systems, the "ABC" standing for *ATP-binding cassette*, a structural feature of proteins that bind ATP (**Figure 2.23**). More than 200 different ABC transport systems have been identified in prokaryotes. ABC transporters exist for the uptake of organic compounds such as sugars and amino acids, inorganic nutrients such as sulfate and phosphate, and certain metals.

A characteristic property of periplasmic binding proteins is their high substrate affinity. These proteins can bind their substrate(s) even when the substrate is present at extremely low concentration; for example, less than 1 micromolar (10^{-6} M). Once its substrate is bound, the periplasmic binding protein interacts with its respective membrane transporter to transport the substrate into the cell driven by the energy in ATP (Figure 2.23).

Even though gram-positive bacteria lack a periplasm, they also have ABC transport systems. In gram-positive bacteria, however, substrate-binding proteins (the functional equivalent of periplasmic binding proteins) are anchored to the external surface of the cytoplasmic membrane. Once they bind their substrate, these proteins interact with a membrane transporter to catalyze the ATP-driven uptake of the substrate.

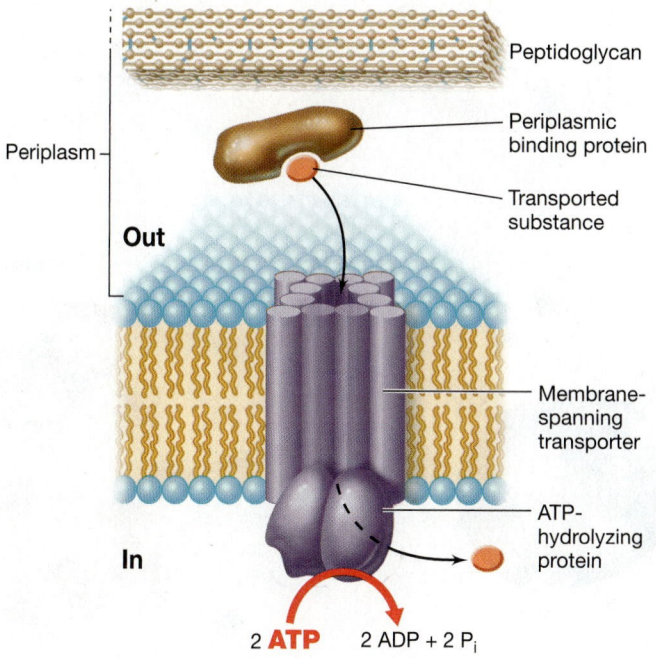

Figure 2.23   Mechanism of an ABC transporter. The periplasmic binding protein has high affinity for substrate, the membrane-spanning proteins form the transport channel, and the cytoplasmic ATP-hydrolyzing proteins supply the energy for the transport event.

MINIQUIZ -

- Compare and contrast simple transporters, the phosphotransferase system, and ABC transporters in terms of (1) energy source, (2) chemical alterations of the solute during transport, and (3) number of proteins required.

- Which major characteristic of periplasmic-binding proteins makes them ideally suited for organisms living in very nutrient-poor environments?

IV • Cell Walls of *Bacteria* and *Archaea*

The cytoplasm of prokaryotic cells maintains a high concentration of dissolved solutes, and this creates a significant osmotic pressure—about 2 atm (203 kPa)—in a typical cell. This is about the same as the pressure in an automobile tire. To withstand these pressures and prevent bursting (cell lysis), most cells of *Bacteria* and *Archaea* contain a wall. Besides protecting against osmotic lysis, cell walls also confer shape and rigidity on the cell. Knowledge of cell wall structure and function is important not only for understanding how prokaryotic cells work, but also because certain antibiotics target cell wall synthesis, leaving the cell susceptible to lysis. Since human cells lack cell walls, these antibiotics are of obvious benefit for treating bacterial infections.

2.10 Peptidoglycan

As we saw earlier, species of *Bacteria* can be divided into two major groups, called *gram-positive* and *gram-negative*. The distinction

between gram-positive and gram-negative bacteria is based on the Gram stain reaction (Section 2.2), and differences in cell wall structure are at the heart of the Gram stain reaction. The surface of gram-positive and gram-negative cells as viewed in the electron microscope differs markedly, as shown in **Figure 2.24**. The gram-negative cell wall, or *cell envelope* as it is often called, consists of at least two layers, whereas the gram-positive cell wall is typically much thicker and consists primarily of a single type of molecule.

We focus here on the polysaccharide component of the cell walls of *Bacteria*, both gram-positive and gram-negative. In the next section, we describe the special wall components present in gram-negative *Bacteria*. In Section 2.12, we describe the cell walls of *Archaea*.

Chemistry of Peptidoglycan

The walls of *Bacteria* have a rigid layer that is primarily responsible for the strength of the wall. This rigid layer, called

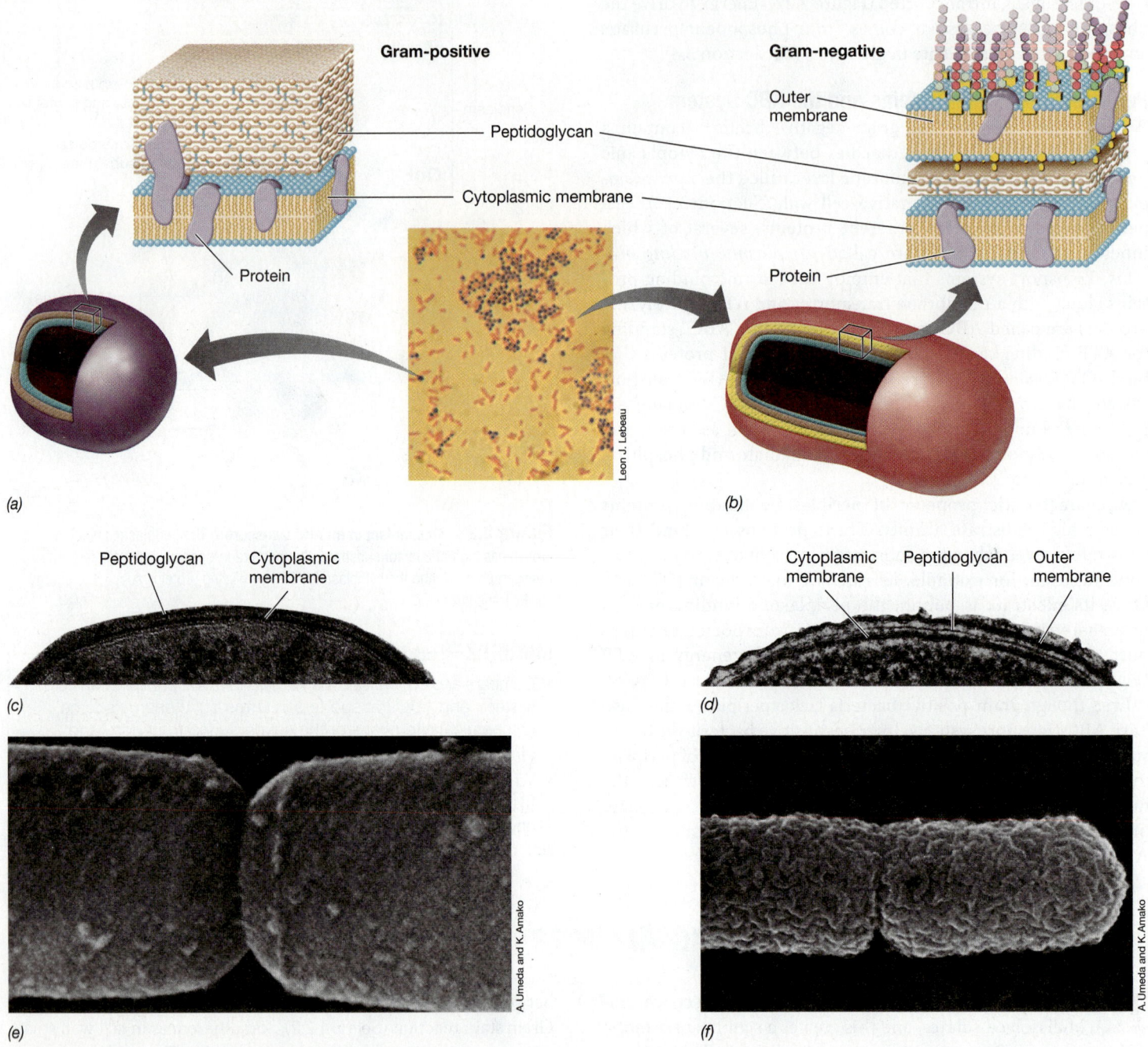

Figure 2.24 Cell walls of *Bacteria*. *(a, b)* Schematic diagrams of gram-positive and gram-negative cell walls. The Gram stain photo in the center shows cells of *Staphylococcus aureus* (purple, gram-positive) and *Escherichia coli* (pink, gram-negative). *(c, d)* Transmission electron micrographs (TEMs) showing the cell wall of a gram-positive bacterium and a gram-negative bacterium. *(e, f)* Scanning electron micrographs of gram-positive and gram-negative bacteria, respectively. Note differences in surface texture. Each cell in the TEMs is about 1 μm wide.

peptidoglycan, is a polysaccharide composed of two sugar derivatives—*N-acetylglucosamine* and *N-acetylmuramic acid*—and a few amino acids, including L-alanine, D-alanine, D-glutamic acid, and either L-lysine or a structurally similar molecule, diaminopimelic acid (DAP). These constituents are connected to form a repeating structure called the *glycan tetrapeptide* (**Figure 2.25**).

Long chains of peptidoglycan are biosynthesized adjacent to one another to form a sheet surrounding the cell. The individual chains are connected by cross-links of amino acids. The glycosidic bonds connecting the sugars in the glycan strands are covalent bonds, but these provide rigidity in only one direction. Only after cross-linking is peptidoglycan strong in both the X and Y directions (**Figure 2.26**). Cross-linking occurs to different degrees

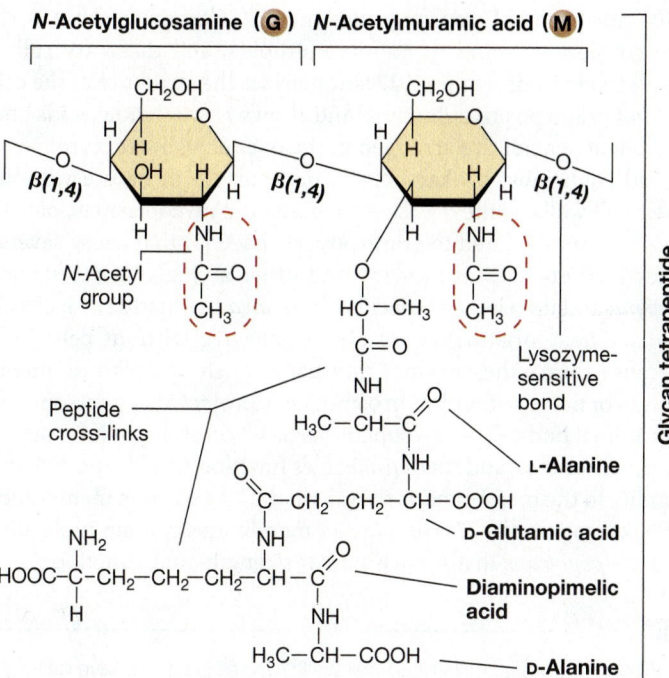

Figure 2.25 **Structure of the repeating unit in peptidoglycan, the glycan tetrapeptide.** The structure given is that found in *Escherichia coli* and most other gram-negative *Bacteria*. In some *Bacteria*, other amino acids are present as discussed in the text.

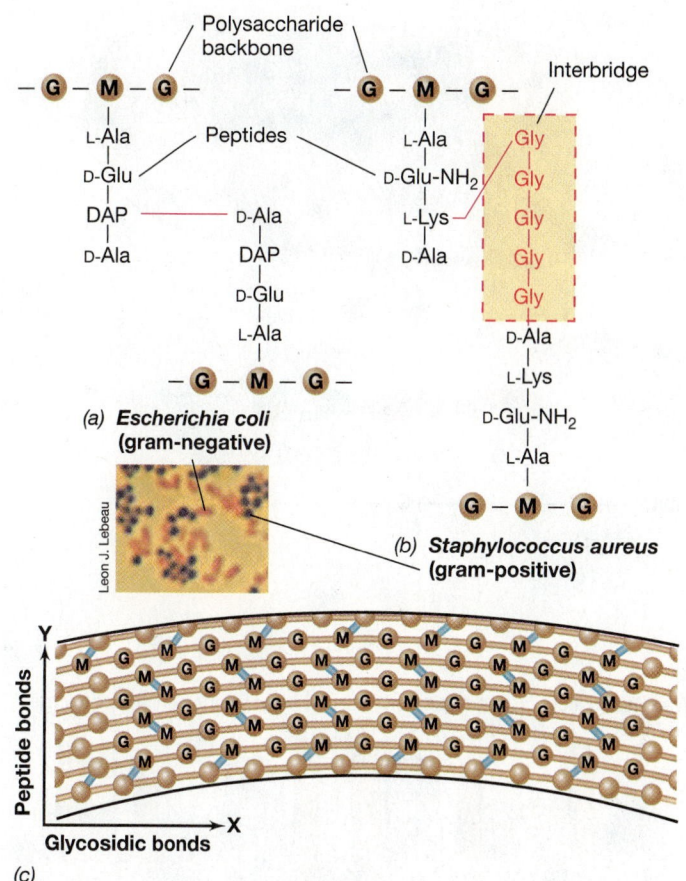

Figure 2.26 **Peptidoglycan in *Escherichia coli* and *Staphylococcus aureus*.** *(a)* No interbridge is present in *E. coli* peptidoglycan nor that of other gram-negative *Bacteria*. *(b)* The glycine interbridge in *S. aureus* (gram-positive). *(c)* Overall structure of peptidoglycan. G, *N*-acetylglucosamine; M, *N*-acetylmuramic acid. Note how glycosidic bonds confer strength on peptidoglycan in the X direction whereas peptide bonds confer strength in the Y direction.

in different species of *Bacteria*, and the more extensive the cross-linking, the greater the rigidity.

In gram-negative bacteria, the peptidoglycan cross-link is formed by a peptide bond from the amino group of DAP of one glycan chain to the carboxyl group of the terminal D-alanine on the adjacent glycan chain (Figure 2.26). In gram-positive bacteria, the cross-link often occurs through a short peptide interbridge, the kinds and numbers of amino acids in the interbridge varying from species to species. In the gram-positive bacterium *Staphylococcus aureus*, whose cell wall chemistry is well understood, the interbridge consists of five glycine residues (Figure 2.26b). The overall structure of peptidoglycan is shown in Figure 2.26c.

Peptidoglycan can be destroyed by certain agents. One such agent is the enzyme *lysozyme*, a protein that cleaves the β-1,4-glycosidic bonds between *N*-acetylglucosamine and *N*-acetylmuramic acid in peptidoglycan (Figure 2.25), thereby weakening the wall. When this occurs, water can enter the cell and cause cell lysis. Lysozyme is present in animal secretions including tears, saliva, and other body fluids, and functions as a major line of defense against bacterial infection. When we consider peptidoglycan biosynthesis in Chapter 5 we will see that the antibiotic penicillin also targets peptidoglycan, but in a different way from that of lysozyme. Whereas lysozyme destroys preexisting peptidoglycan, penicillin instead prevents its biosynthesis, leading to a weakened molecule and osmotic lysis.

Peptidoglycan is found only in *Bacteria*; *N*-acetylmuramic acid and the amino acid analog DAP have never been found in the cell walls of *Archaea* or *Eukarya*. However, not all *Bacteria* examined have DAP in their peptidoglycan; some have lysine

instead. An unusual feature of peptidoglycan is the presence of two amino acids of the D stereoisomer, D-alanine and D-glutamic acid. Proteins, by contrast, are always constructed of L-amino acids. More than 100 chemically distinct peptidoglycans have been described that vary in their peptide cross-links and/or interbridge. By contrast, the glycan portion of all peptidoglycans is constant; only *N*-acetylglucosamine and *N*-acetylmuramic acid are present and are connected in β-1,4 linkage (Figures 2.25 and 2.26).

The Gram-Positive Cell Wall

As much as 90% of the gram-positive cell wall is composed of peptidoglycan. And, although some bacteria have only a single layer of peptidoglycan, many gram-positive bacteria have several sheets of peptidoglycan stacked one upon another (Figure 2.26a). It is thought that peptidoglycan is synthesized by the cell in "cables" about 50 nm wide, with each cable containing several cross-linked glycan strands (**Figure 2.27a**). As the peptidoglycan is synthesized, the cables become cross-linked to form an even stronger cell wall structure.

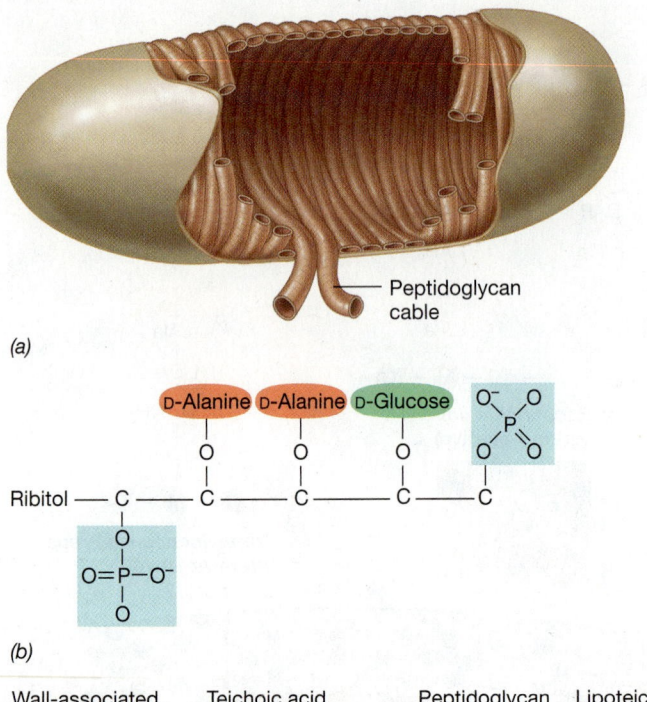

(a)

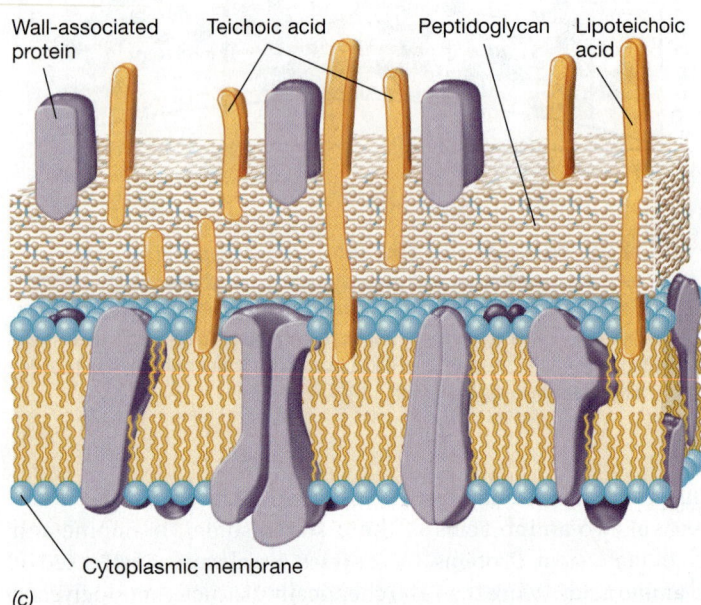

(b)

(c)

Figure 2.27 Structure of the gram-positive bacterial cell wall. *(a)* Schematic of a gram-positive rod showing the internal architecture of the peptidoglycan "cables." *(b)* Structure of a ribitol teichoic acid. The teichoic acid is a polymer of the repeating ribitol unit shown here. *(c)* Summary diagram of the gram-positive bacterial cell wall.

Many gram-positive bacteria have acidic molecules called **teichoic acids** embedded in their cell wall. The term "teichoic acids" includes all cell wall, cytoplasmic membrane, and capsular polymers composed of glycerol phosphate or ribitol phosphate. These polyalcohols are connected by phosphate esters and typically contain sugars or D-alanine (Figure 2.27b). Teichoic acids are covalently bonded to muramic acid in the wall peptidoglycan. Because the phosphates are negatively charged, teichoic acids are in part responsible for the overall negative electrical charge of the cell surface. Teichoic acids also function to bind Ca^{2+} and

Mg^{2+} for eventual transport into the cell. Certain teichoic acids are covalently bound to membrane lipids, and these are called *lipoteichoic* acids. Figure 2.27 summarizes the structure of the cell wall of gram-positive *Bacteria* and shows how teichoic acids and lipoteichoic acids are arranged in the overall wall structure.

Although most prokaryotes cannot survive in nature without their cell walls, some do. These include the mycoplasmas, pathogenic bacteria related to gram-positive bacteria that cause several infectious diseases of humans and other animals, and *Thermoplasma* and its relatives, species of *Archaea* that naturally lack cell walls. These organisms are able to survive without cell walls because they either contain unusually tough cytoplasmic membranes or because they live in osmotically protected habitats such as the animal body. Most mycoplasmas have sterols in their cytoplasmic membranes, and these molecules function to add strength and rigidity to the membrane as they do in the cytoplasmic membranes of eukaryotic cells. *Thermoplasma* membranes contain molecules called *lipoglycans* that serve a similar strengthening function.

MINIQUIZ

- Why do bacterial cells need cell walls? Do all bacteria have cell walls?
- Why is peptidoglycan such a strong molecule?
- What does the enzyme lysozyme do?

2.11 LPS: The Outer Membrane

In gram-negative bacteria, only a small amount of the total cell wall consists of peptidoglycan, as most of the wall is composed of the **outer membrane**. This layer is effectively a second lipid bilayer, but it is not constructed solely of phospholipid and protein, as is the cytoplasmic membrane (Figure 2.15). Instead, the outer membrane also contains polysaccharide, and the lipid and polysaccharide are linked to form a complex. Because of this, the outer membrane is often called the **lipopolysaccharide** layer, or simply **LPS** for short.

Chemistry and Activity of LPS

The structure of LPS from several bacteria is known. As seen in **Figure 2.28**, the polysaccharide portion of LPS consists of two components, the *core polysaccharide* and the *O-specific polysaccharide*. In *Salmonella* species, where LPS has been well studied, the core polysaccharide consists of ketodeoxyoctonate (KDO), various seven-carbon sugars (heptoses), glucose, galactose, and *N*-acetylglucosamine. Connected to the core is the O-specific polysaccharide, which typically contains galactose, glucose, rhamnose, and mannose, as well as one or more dideoxyhexoses, such as abequose, colitose, paratose, or tyvelose. These sugars are connected in four- or five-membered sequences, which often are branched. When the sequences repeat, the long O-specific polysaccharide is formed.

The relationship of the LPS layer to the overall gram-negative cell wall is shown in **Figure 2.29**. The lipid portion of the LPS, called *lipid A*, is not a typical glycerol lipid (see Figure 2.14a), but instead the fatty acids are connected through the amine groups from a disaccharide composed of glucosamine phosphate. The disaccharide is attached to the core polysaccharide through KDO

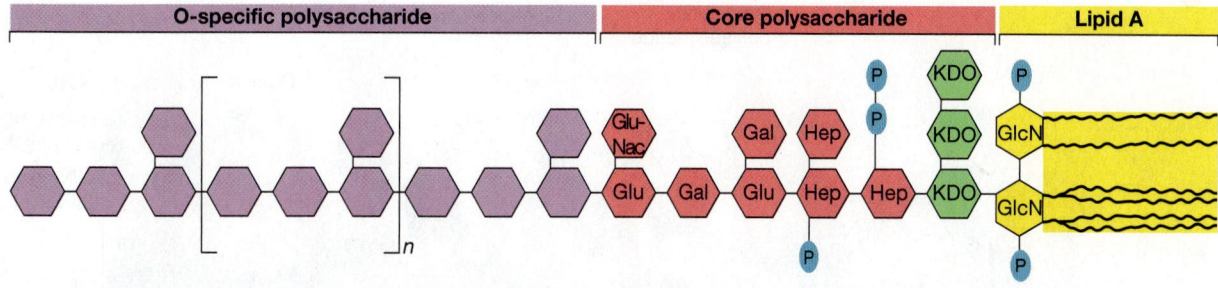

Figure 2.28 Structure of the lipopolysaccharide of gram-negative *Bacteria*. The chemistry of lipid A and the polysaccharide components varies among species of gram-negative *Bacteria*, but the major components (lipid A–KDO–core–O-specific) are typically the same. The O-specific polysaccharide is highly variable among species. KDO, ketodeoxyoctonate; Hep, heptose; Glu, glucose; Gal, galactose; GluNac, *N*-acetylglucosamine; GlcN, glucosamine; P, phosphate. Glucosamine and the lipid A fatty acids are linked through the amine groups. The lipid A portion of LPS can be toxic to animals and comprises the endotoxin complex. Compare this figure with Figure 2.29 and follow the LPS components by their color-coding.

(Figure 2.28). Fatty acids commonly found in lipid A include caproic (C_6), lauric (C_{12}), myristic (C_{14}), palmitic (C_{16}), and stearic (C_{18}) acids. LPS replaces much of the phospholipid in the outer half of the outer membrane and functions as an anchor tying the outer membrane to peptidoglycan. Thus, although the outer membrane is technically a lipid bilayer, its structure is distinct from that of the cytoplasmic membrane.

Besides its importance in rendering strength to the gram-negative cell, an important biological activity of LPS is its toxicity to animals. Common gram-negative pathogens for humans include species of *Salmonella*, *Shigella*, and *Escherichia*, among many others, and some of the gastrointestinal symptoms these pathogens elicit are due to toxic outer membrane components. Toxicity is associated with the LPS layer, in particular, lipid A. The term *endotoxin* refers to this toxic component of LPS. Some endotoxins cause violent symptoms in humans, including gas, diarrhea, and vomiting, and the endotoxins produced by *Salmonella* and enteropathogenic strains of *E. coli* transmitted in contaminated foods are classic examples of this (∞ Sections 23.10 and 31.10).

The Periplasm and Porins

Although permeable to small molecules, the outer membrane is impermeable to proteins and other very large molecules. In fact, one of the major functions of the outer membrane is to prevent proteins whose activities occur outside the cytoplasmic membrane from diffusing away from the cell. These proteins are present in a region called the **periplasm**. This space, located between the outer surface of the cytoplasmic membrane and the inner surface of the outer membrane, is about 15 nm wide (Figure 2.29). The periplasm is gel-like in consistency because of the high concentration of proteins there.

Depending on the organism, the periplasm can contain several different classes of proteins. These include hydrolytic enzymes, which function in the initial degradation of food molecules; binding proteins, which begin the process of transporting substrates (Section 2.9); and chemoreceptors, which are proteins that govern the chemotaxis response (Section 2.19). Most of these proteins reach the periplasm by way of a protein-exporting system present in the cytoplasmic membrane (∞ Section 4.14).

The outer membrane is relatively permeable to small molecules (even hydrophilic molecules) because of proteins called *porins*

that function as channels for the entrance and exit of solutes (Figure 2.29*a, c*). Several porins are known, including both specific and nonspecific classes. Nonspecific porins form water-filled channels through which any small substance can pass. By contrast, specific porins contain a binding site for only one or a small group of structurally related substances. Structurally, porins are transmembrane proteins made up of three identical subunits. Besides the channel present in each barrel of the porin, the barrels of the three porin proteins associate in such a way that a small hole about 1 nm in diameter is formed in the outer membrane through which very small molecules can travel (Figure 2.29*c*).

Relationship of Cell Wall Structure to the Gram Stain

The structural differences between the cell walls of gram-positive and gram-negative *Bacteria* are responsible for differences in the Gram stain reaction. Recall that in the Gram stain, an insoluble crystal violet–iodine complex forms inside the cell. This complex is extracted by alcohol from gram-negative but not from gram-positive bacteria (Section 2.2). As we have seen, gram-positive bacteria have very thick cell walls consisting primarily of peptidoglycan. During Gram staining, the gram-positive cell wall is dehydrated by the alcohol, causing the pores in the walls to close and preventing the insoluble crystal violet–iodine complex from escaping. By contrast, in gram-negative bacteria, alcohol readily penetrates the lipid-rich outer membrane and extracts the crystal violet–iodine complex from the cell. After alcohol treatment, gram-negative cells are nearly invisible unless they are counterstained with a second dye, a standard procedure in the Gram stain (Figure 2.4).

MINIQUIZ

- What chemical components are found in the outer membrane of gram-negative bacteria?
- What is the function of porins and where are they located in a gram-negative cell wall?
- What component of the gram-negative cell has endotoxin properties?
- Why does alcohol readily decolorize gram-negative but not gram-positive bacteria?

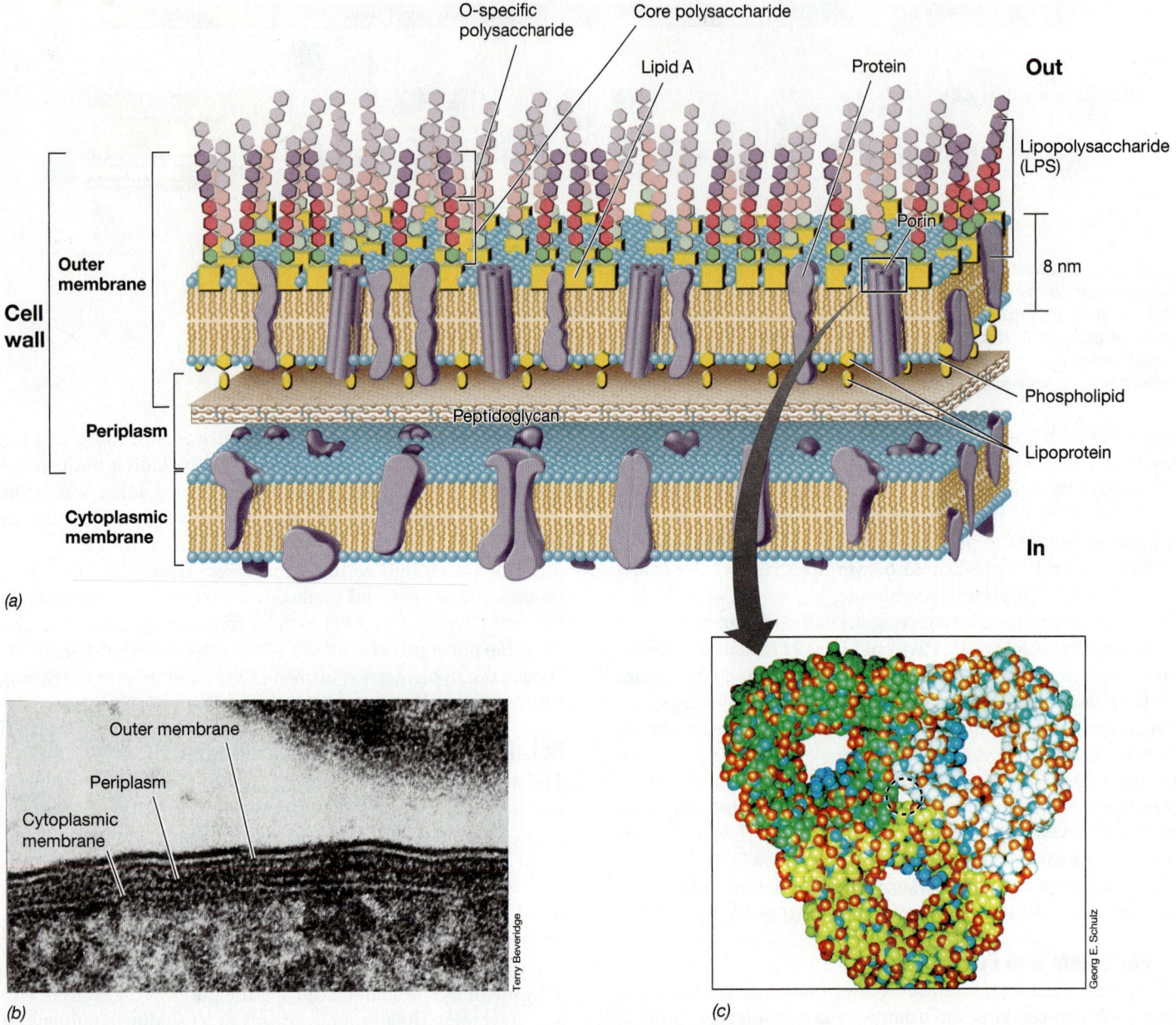

Figure 2.29 **The gram-negative cell wall.** *(a)* Arrangement of lipopolysaccharide, lipid A, phospholipid, porins, and lipoprotein in the outer membrane. See Figure 2.28 for details of the structure of LPS. *(b)* Transmission electron micrograph of a cell of *Escherichia coli* showing the cytoplasmic membrane and wall. *(c)* Molecular model of porin proteins. Note the four pores present, one within each of the proteins forming a porin molecule and a smaller central pore (circled) between the porin proteins. The view is perpendicular to the plane of the membrane.

2.12 Archaeal Cell Walls

Peptidoglycan, a key biomarker for *Bacteria*, is absent from the cell walls of *Archaea*. An outer membrane is typically lacking in *Archaea* as well. Instead, a variety of cell wall types are found in *Archaea*, including those containing polysaccharides, proteins, or glycoproteins.

Pseudomurein and Other Polysaccharide Walls

The cell walls of certain methanogenic *Archaea* contain a molecule that is remarkably similar to peptidoglycan, a polysaccharide called *pseudomurein* (the term "murein" is from the Latin word for "wall" and was an old term for peptidoglycan) (**Figure 2.30**). The backbone of pseudomurein is formed from alternating repeats of *N*-acetylglucosamine (also present in peptidoglycan) and *N*-acetyltalosaminuronic acid; the latter replaces the *N*-acetylmuramic acid of peptidoglycan. Pseudomurein also differs from peptidoglycan in that the glycosidic bonds between the sugar derivatives are β-1,3 instead of β-1,4, and the amino acids are all of the L stereoisomer (Figure 2.30). It is thought that peptidoglycan and pseudomurein either arose by convergent evolution after *Bacteria* and *Archaea* had diverged or, more likely, by evolution from a common polysaccharide present in

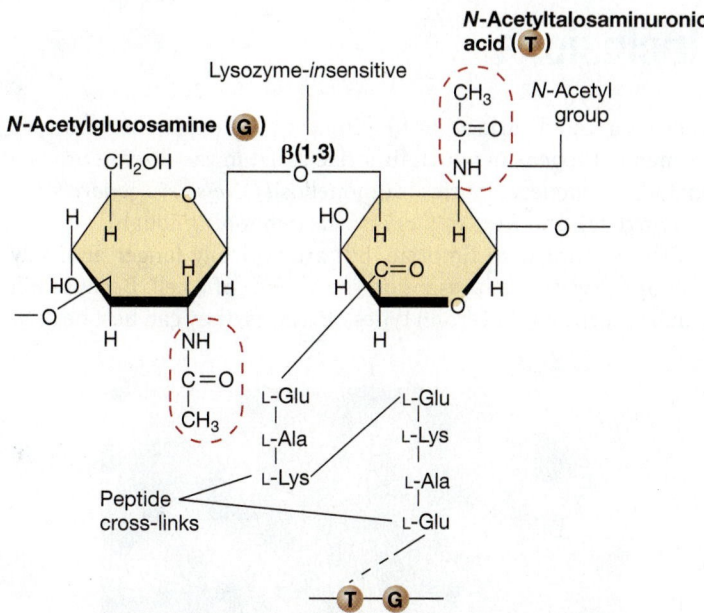

Figure 2.30 Pseudomurein. Structure of pseudomurein, the cell wall polymer of *Methanobacterium* species. Note the similarities and differences between pseudomurein and peptidoglycan (Figure 2.25).

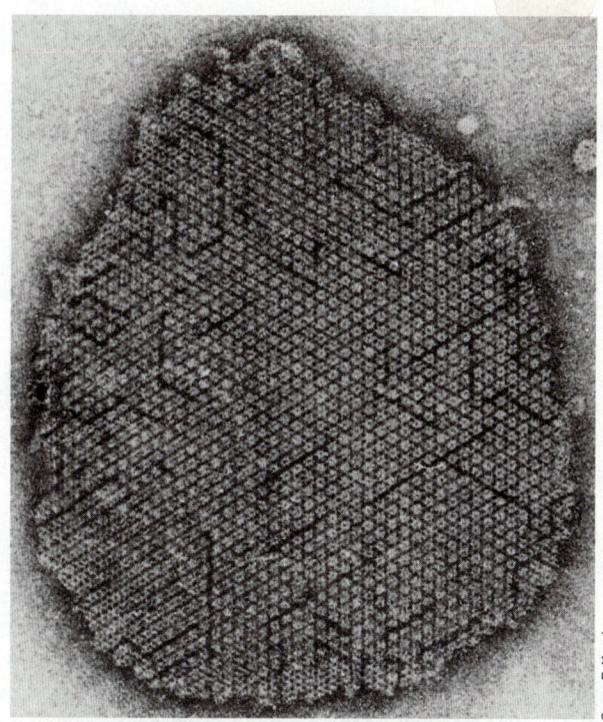

Susan F. Koval

Figure 2.31 The S-layer. Transmission electron micrograph of a portion of an S-layer showing the paracrystalline structure. Shown is the S-layer from *Aquaspirillum* (a species of *Bacteria*), and this S-layer shows hexagonal symmetry common in S-layers of *Archaea*.

the cell walls of the common ancestor of the domains *Bacteria* and *Archaea*.

Cell walls of some other *Archaea* lack pseudomurein and instead contain other polysaccharides. For example, *Methanosarcina* species have thick polysaccharide walls composed of polymers of glucose, glucuronic acid, galactosamine uronic acid, and acetate. Extremely halophilic (salt-loving) *Archaea* such as *Halococcus*, which are related to *Methanosarcina*, have similar cell walls that are also highly sulfated. The negative charges on the sulfate ion (SO_4^{2-}) bind Na^+ present in the habitats of *Halococcus*—salt evaporation ponds and saline seas and lakes—at high levels. The sulfate–sodium complex helps stabilize the *Halococcus* cell wall in such strongly ionic environments.

S-Layers

The most common type of cell wall in *Archaea* is the paracrystalline surface layer, or **S-layer** as it is called. S-layers consist of interlocking molecules of protein or glycoprotein (**Figure 2.31**). The paracrystalline structure of S-layers can form various symmetries, including hexagonal, tetragonal, or trimeric, depending upon the number and structure of the subunits of which it is composed. S-layers have been found in representatives of all major lineages of *Archaea* and also in several species of *Bacteria* (Figure 2.31).

The cell walls of some *Archaea*, for example the methanogen *Methanocaldococcus jannaschii*, consist only of an S-layer. Thus, S-layers are sufficiently strong to withstand osmotic pressures without any other wall components. However, in many organisms S-layers are present in addition to other cell wall components, usually polysaccharides. For example, in *Bacillus brevis*, a species of *Bacteria*, an S-layer is present along with peptidoglycan.

However, when an S-layer is present along with other wall components, the S-layer is always the *outermost* wall layer, the layer that is in direct contact with the environment.

Besides serving as protection from osmotic lysis, S-layers may have other functions. For example, as the interface between the cell and its environment, it is likely that the S-layer functions as a selective sieve, allowing the passage of low-molecular-weight solutes while excluding large molecules or structures (such as viruses). The S-layer may also function to retain proteins near the cell surface, much as the outer membrane (Section 2.11) does in gram-negative bacteria.

We thus see several cell wall structures in species of *Archaea*, varying from those that closely resemble peptidoglycan to those that totally lack polysaccharide. But with rare exception, all *Archaea* contain a cell wall of some sort, and as in *Bacteria*, the archaeal cell wall functions to prevent osmotic lysis and gives the cell its shape. Because they lack peptidoglycan, *Archaea* are naturally resistant to lysozyme (Figure 2.30) and the antibiotic penicillin, agents that either destroy peptidoglycan or interrupt its biosynthesis (Section 2.10).

MINIQUIZ

- How does pseudomurein resemble peptidoglycan? How do the two molecules differ?
- What is the composition of an S-layer?
- Why are *Archaea* insensitive to penicillin?

V • Other Cell Surface Structures and Inclusions

In addition to cell walls, cells of *Bacteria* and *Archaea* may have other layers or structures in contact with the environment and often contain one or more types of cellular inclusions. We examine some of these here.

2.13 Cell Surface Structures

Many prokaryotes secrete sticky or slimy materials on their cell surface that consist of either polysaccharide or protein. These are not considered part of the cell wall because they do not confer significant structural strength on the cell. The terms "capsule" and "slime layer" are used to describe these layers.

Capsules and Slime Layers

The terms capsule and slime layer are often used interchangeably, but the two terms do not refer to the same thing. Traditionally, if the layer is organized in a tight matrix that excludes small particles, such as India ink, it is called a **capsule**. Such a structure is readily visible by light microscopy if cells are treated with India ink and can also be seen in the electron microscope (**Figure 2.32**). By contrast, if the layer is more easily deformed, it will not exclude particles and is more difficult to see; this form is called a *slime layer*. Capsules typically adhere firmly to the cell wall, and some are even covalently linked to peptidoglycan. Slime layers, by contrast, are loosely attached and can be lost from the cell surface.

Outer surface layers have several functions. Surface polysaccharides assist in the attachment of microorganisms to solid surfaces. As we will see later, pathogenic microorganisms that enter the body by specific routes usually do so by first binding specifically to surface components of host tissues; this binding is often mediated by bacterial cell surface polysaccharides (⇄ Section 23.1). When the opportunity arises, bacteria of all types typically bind to solid surfaces, often forming a thick layer of cells called a *biofilm*. Extracellular polysaccharides play a key role in the development and maintenance of biofilms as well.

Besides attachment, outer surface layers can have other functions. These include acting as virulence factors in certain bacterial diseases and preventing the cell from suffering dehydration. For example, the causative agent of the diseases anthrax and bacterial pneumonia—*Bacillus anthracis* and *Streptococcus pneumoniae*, respectively—each contain a thick capsule of either protein (*B. anthracis*) or polysaccharide (*S. pneumoniae*). Encapsulated cells of these bacteria avoid destruction by the host immune system because the immune cells that would otherwise recognize these pathogens as foreign and destroy them are blocked from doing so because of the bacterial capsule. In addition to this role in disease, outer surface layers of virtually any type bind water and likely protect the cell from desiccation in periods of dryness.

Fimbriae and Pili

Fimbriae and pili are filamentous proteins that extend from the surface of a cell and can have many functions. *Fimbriae* (**Figure 2.33**) enable cells to stick to surfaces, including animal tissues in the case of pathogenic bacteria, or to form pellicles (thin sheets of cells on a liquid surface) or biofilms on solid surfaces. Notorious human pathogens in which fimbriae assist in the disease process include *Salmonella* species (salmonellosis), *Neisseria gonorrhoeae* (gonorrhea), and *Bordetella pertussis* (whooping cough).

Pili are similar to fimbriae, but are typically longer and only one or a few pili are present on the surface of a cell. Because pili can be receptors for certain types of viruses, they can best be seen

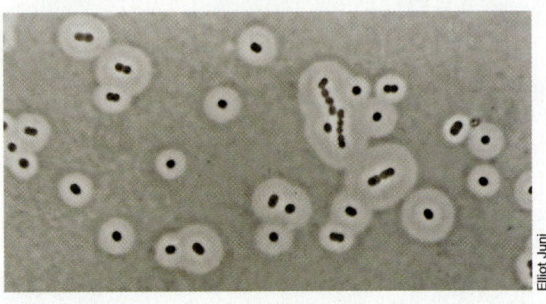

(a) Elliot Juni

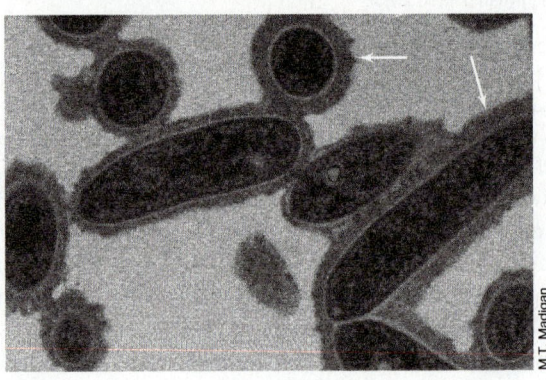

(b) M.T. Madigan

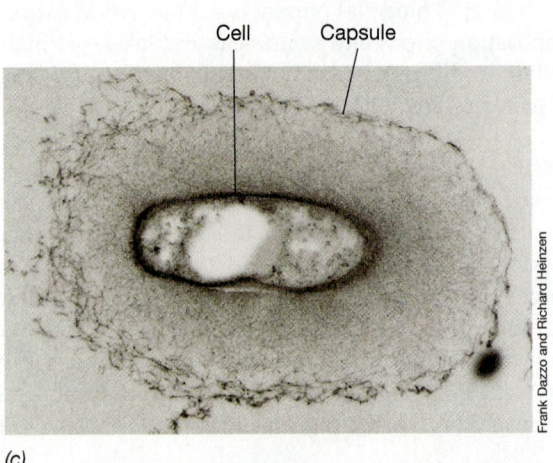

Cell   Capsule

(c) Frank Dazzo and Richard Heinzen

Figure 2.32 Bacterial capsules. *(a)* Capsules of *Acinetobacter* species observed by phase-contrast microscopy after negative staining with India ink. India ink does not penetrate the capsule and so the capsule appears as a light area surrounding the cell, which appears black. *(b)* Transmission electron micrograph of a thin section of cells of *Rhodobacter capsulatus* with capsules (arrows) clearly evident; cells are about 0.9 μm wide. *(c)* Transmission electron micrograph of *Rhizobium trifolii* stained with ruthenium red to reveal the capsule. The cell is about 0.7 μm wide.

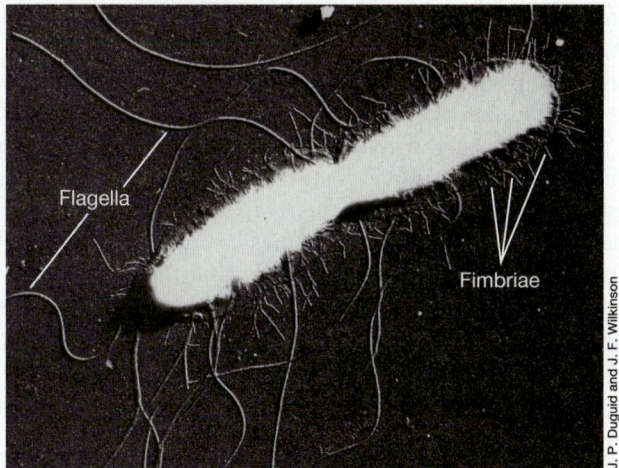

Figure 2.33 Fimbriae. Electron micrograph of a dividing cell of *Salmonella typhi*, showing flagella and fimbriae. A single cell is about 0.9 μm wide.

under the electron microscope when they become coated with virus particles (**Figure 2.34**). Many classes of pili are known, distinguished by their structure and function. Two very important functions of pili include facilitating genetic exchange between cells in a process called *conjugation* and enabling the adhesion of pathogens to specific host tissues that they subsequently invade. The latter function has been best studied in gram-negative pathogens such as *Neisseria*, species of which cause the diseases gonorrhea and meningitis, but pili are also present on certain gram-positive pathogens such as *Streptococcus pyogenes*, the bacterium that causes both strep throat and scarlet fever.

One important class of pili, called *type IV pili*, assist cells in adhesion but also allow for an unusual form of cell motility called *twitching motility*. Type IV pili are present only at the poles of those rod-shaped cells that contain them. Twitching motility is a type of gliding motility, movement along a solid surface (Section 2.18). In twitching motility, extension of pili followed by their retraction drags the cell along a solid surface, with energy supplied by ATP. Certain species of *Pseudomonas* and *Moraxella* are well known for their twitching motility.

Type IV pili have also been implicated as key colonization factors for certain human pathogens, including *Vibrio cholerae* (cholera) and *Neisseria gonorrhoeae* (gonorrhea). The twitching

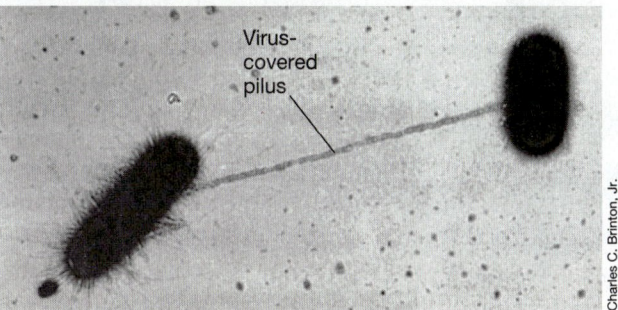

Figure 2.34 Pili. The pilus on an *Escherichia coli* cell that is undergoing conjugation (a form of genetic transfer) with a second cell is better resolved because viruses have adhered to it. The cells are about 0.8 μm wide.

motility of these pathogens presumably assists the organism to locate specific sites for attachment to initiate the disease process. Type IV pili are also thought to mediate genetic transfer by the process of transformation in some bacteria, which, along with conjugation and transduction, are the three known means of horizontal gene transfer in prokaryotes (Chapter 10).

MINIQUIZ --

- Could a bacterial cell dispense with a cell wall if it had a capsule? Why or why not?

- How do fimbriae differ from pili, both structurally and functionally?

2.14 Cell Inclusions

Inclusions are often present in prokaryotic cells. Inclusions function as energy reserves and/or carbon reservoirs or have special functions. Inclusions can often be seen directly with the light microscope and are usually enclosed by single-layer (nonunit) membranes that partition the inclusion off in the cell. Storing carbon or other substances in an insoluble form is advantageous for cells because it reduces the osmotic stress that would occur should the same amount of the substance be dissolved in the cytoplasm.

Carbon Storage Polymers

One of the most common inclusion bodies in prokaryotic organisms is **poly-β-hydroxybutyric acid (PHB)**, a lipid that is formed from β-hydroxybutyric acid units. The monomers of PHB polymerize by ester linkage and then the polymer aggregates into granules; the latter can be seen by either light or electron microscopy (**Figure 2.35**).

The monomer in the polymer is usually hydroxybutyrate (C_4) but can vary in length from as short as C_3 to as long as C_{18}. Thus, the more generic term *poly-β-hydroxyalkanoate* (PHA) is often used to describe this class of carbon- and energy-storage polymers. PHAs are synthesized by cells when there is an excess of carbon and are broken down as carbon or energy sources when conditions warrant. Many *Bacteria* and *Archaea* produce PHAs.

Another storage product is *glycogen*, which is a polymer of glucose and, like PHA, is a reservoir of both carbon and energy and is produced when carbon is in excess. Glycogen resembles starch, the major storage reserve of plants, but differs slightly from starch in the manner in which the glucose units are linked together.

Polyphosphate, Sulfur, and Carbonate Minerals

Many microorganisms accumulate inorganic phosphate (PO_4^{3-}) in the form of granules of *polyphosphate* (**Figure 2.36a**). These granules can be degraded and used as sources of phosphate for nucleic acid and phospholipid biosyntheses and in some organisms can be used directly to make the energy-rich compound ATP. Phosphate is often a limiting nutrient in natural environments. Thus, if a cell happens upon an excess of phosphate, it is advantageous to be able to store it as polyphosphate for future use.

Many gram-negative prokaryotes can oxidize reduced sulfur compounds, such as hydrogen sulfide (H_2S); these organisms are the "sulfur bacteria," discovered by the great microbiologist Sergei

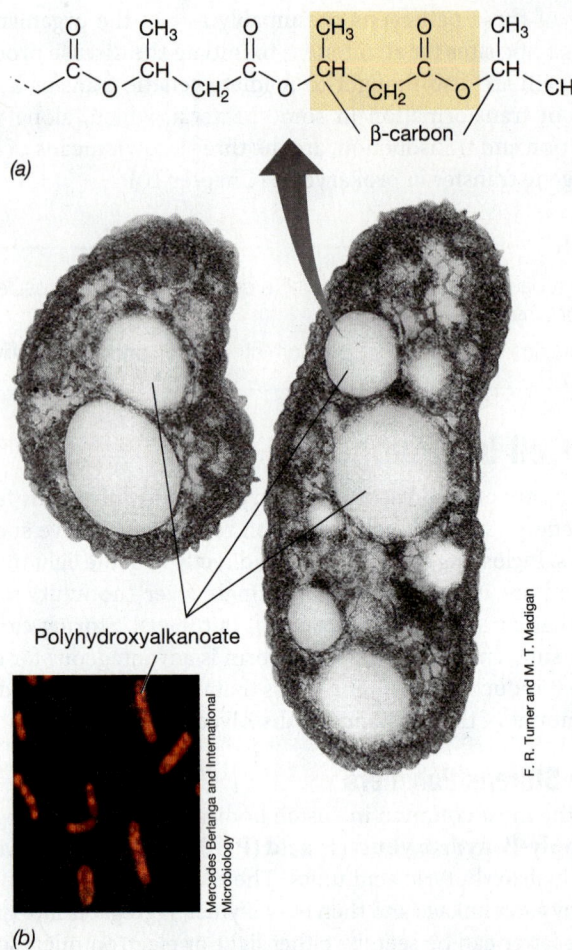

(a)

β-carbon

Polyhydroxyalkanoate

(b)

Mercedes Berlanga and International Microbiology

F. R. Turner and M. T. Madigan

Figure 2.35 Poly-β-hydroxyalkanoates. *(a)* Chemical structure of poly-β-hydroxybutyrate, a common PHA. A monomeric unit is shown in color. Other PHAs are made by substituting longer-chain hydrocarbons for the –CH₃ group on the β-carbon. *(b)* Electron micrograph of a thin section of cells of a bacterium containing granules of PHB. Color photo: Nile red–stained cells of a PHA-containing bacterium.

Winogradsky (⮑ Section 1.9). The oxidation of sulfide is linked to the need for electrons to drive either reactions of energy metabolism (chemolithotrophy) or CO_2 fixation (autotrophy). In either case, *elemental sulfur* (S^0) from the oxidation of sulfide may accumulate in the cell in microscopically visible granules (Figure 2.36b). This sulfur remains as long as the source of reduced sulfur from which it was derived is still present. However, as the reduced sulfur source becomes limiting, the sulfur in the granules is oxidized to sulfate (SO_4^{2-}), and the granules slowly disappear as this reaction proceeds. Interestingly, although sulfur globules appear to reside in the cytoplasm, they are actually present in the periplasm (Section 2.11). In these cells the periplasm expands outward to accommodate the growing globules as H_2S is oxidized to S^0 and then contracts inward as S^0 is oxidized to SO_4^{2-}.

Filamentous cyanobacteria (see Figure 2.55) have long been known to form carbonate minerals on the external surface of their cells. However, some cyanobacteria also form carbonate minerals *inside* the cell, as cell inclusions. For example, the unicellular cyanobacterium *Gleomargarita* forms intracellular granules of benstonite, a carbonate mineral that contains barium, strontium, and magnesium (Figure 2.37). The microbiological process of forming minerals is called *biomineralization*. It is unclear exactly why this particular mineral is formed by this cyanobacterium, although it may serve the cell as ballast to maintain cells in their habitat, deep in an alkaline lake in Mexico. The biomineralization of several different minerals is catalyzed by various prokaryotes (⮑ Section 13.21), but only in the case of *Gleomargarita* and magnetosomes (to be discussed next) do we see the process yield intracellular inclusions.

Magnetic Storage Inclusions: Magnetosomes

Some bacteria can orient themselves within a magnetic field because they contain **magnetosomes**. These structures are intracellular particles of the iron oxide mineral magnetite—Fe_3O_4

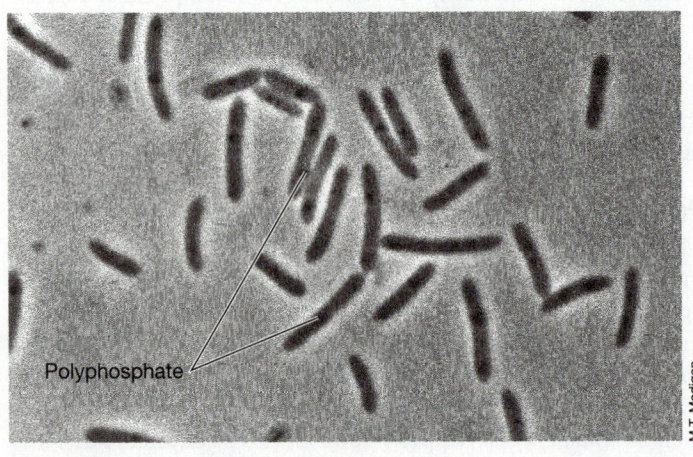

(a)

Polyphosphate

M.T. Madigan

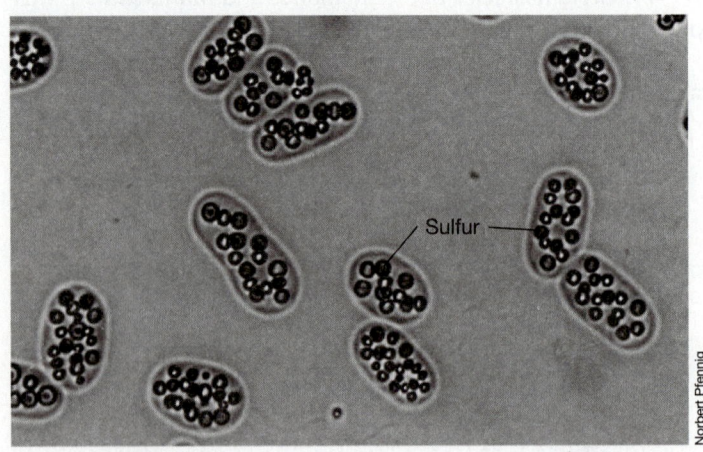

(b)

Sulfur

Norbert Pfennig

Figure 2.36 Polyphosphate and sulfur storage products. *(a)* Phase-contrast photomicrograph of cells of *Heliobacterium modesticaldum* showing polyphosphate as dark granules; a cell is about 1 μm wide. *(b)* Bright-field photomicrograph of cells of the purple sulfur bacterium *Isochromatium buderi*. The intracellular inclusions are sulfur globules formed from the oxidation of hydrogen sulfide (H_2S). A cell is about 4 μm wide.

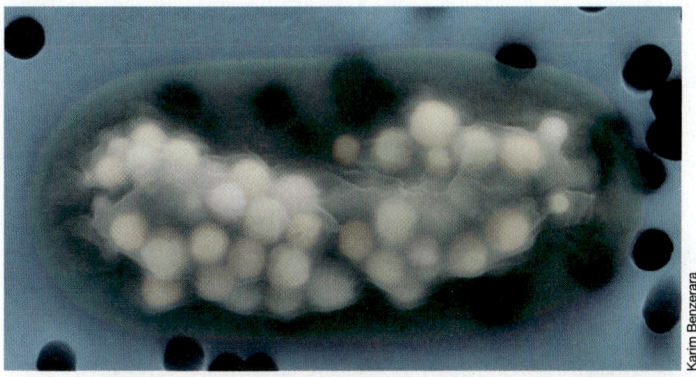

Figure 2.37 Biomineralization by a cyanobacterium. Electron micrograph of a cell of the cyanobacterium *Gleomargarita* containing granules of the mineral benstonite [(Ba,Sr,Ca)$_6$Mg(CO$_3$)$_{13}$]. A cell is about 2 μm wide.

(**Figure 2.38**). In some magnetotactic bacteria the sulfur-containing mineral greigite (Fe$_3$S$_4$) is formed. Both magnetite and greigite are magnetic minerals. Magnetosomes impart a magnetic dipole on a cell, allowing it to orient itself in a magnetic field. Bacteria that produce magnetosomes exhibit *magnetotaxis*, the process of migrating along Earth's magnetic field lines. Magnetosomes have been found in several aquatic organisms that grow best at low O$_2$ concentrations. It has thus been hypothesized that one function

of magnetosomes may be to guide these primarily aquatic cells downward (the direction of Earth's magnetic field) toward the sediments where O$_2$ levels are lower. One greigite-producer is a sulfate-reducing bacterium, and these organisms are obligate anaerobes. Remaining in anoxic zones would be especially important for such magnetotactic species.

An individual magnetosome is enclosed by a thin membrane composed of phospholipid, protein, and glycoprotein (Figure 2.38*b*, *c*). Although this membrane is not a true unit (bilayer) membrane as is the cytoplasmic membrane, magnetosome membrane proteins are functional, as they catalyze Fe^{3+} precipitation during magnetosome synthesis. A similar nonunit membrane surrounds granules of PHAs and sulfur globules, as well. The morphology of magnetosomes appears to be species-specific, varying in shape from square to rectangular to spike-shaped. No magnetosome-containing *Archaea* have yet been discovered.

MINIQUIZ --

- Under what growth conditions would you expect PHAs or glycogen to be produced?

- Why would it be impossible for gram-positive bacteria to store sulfur as gram-negative sulfur-oxidizing chemolithotrophs can?

- How are magnetosomes and the *Gleomargarita* inclusions similar and how do they differ?

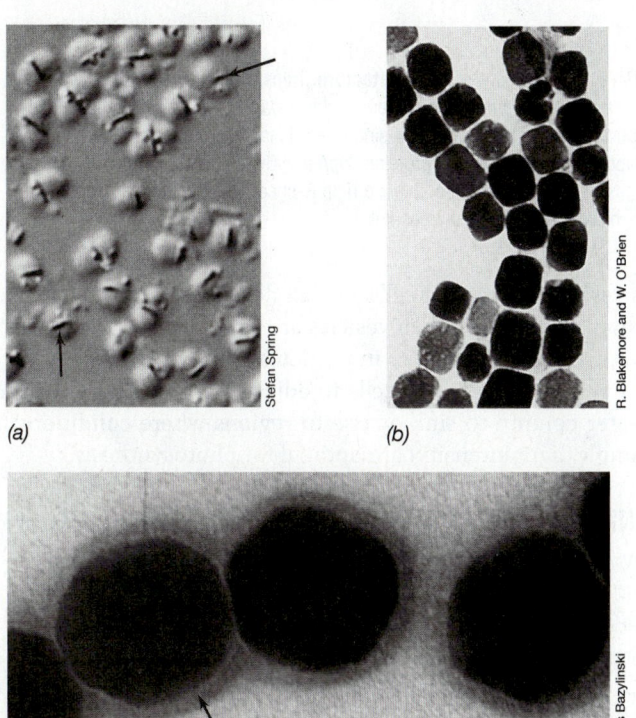

(a)   *(b)*

(c)

Figure 2.38 Magnetotactic bacteria and magnetosomes. *(a)* Differential interference contrast micrograph of coccoid magnetotactic bacteria; note chains of magnetosomes (arrows). A cell is 2.2 μm wide. *(b)* Magnetosomes isolated from the magnetotactic bacterium *Magnetospirillum magnetotacticum*; each particle is about 50 nm wide. *(c)* Transmission electron micrograph of magnetosomes from a magnetic coccus. The arrow points to the membrane that surrounds each magnetosome. A single magnetosome is about 90 nm wide.

2.15 Gas Vesicles

Some prokaryotes are *planktonic*, meaning that they live within the water column of lakes and the oceans. Many planktonic organisms can float because they contain **gas vesicles**, structures that confer buoyancy that allow the cells to position themselves in a water column in a particular location.

The most dramatic examples of gas-vesiculate bacteria are cyanobacteria that form massive accumulations called *blooms* in lakes or other bodies of water (**Figure 2.39**). Cyanobacteria are oxygenic phototrophic bacteria (↻ Sections 1.3, 13.4, and 14.3). Gas-vesiculate cells rise to the surface of the lake and are blown by winds into dense masses. Several other primarily aquatic

Figure 2.39 Buoyant cyanobacteria. Flotation of gas-vesiculate cyanobacteria that formed a bloom in a freshwater lake, Lake Mendota, Madison, Wisconsin (USA).

prokaryotes have gas vesicles, and the property is found in both *Bacteria* and *Archaea* but not microbial eukaryotes.

Gas Vesicle Structure

Gas vesicles are conical-shaped structures made of protein; they are hollow yet rigid and of variable length and diameter (**Figure 2.40**). Gas vesicles in different species vary in length from about 300 to more than 1000 nm and in width from 45 to 120 nm, but the vesicles of a given species are of constant size. Gas vesicles may number from a few to hundreds per cell and are impermeable to water and solutes but permeable to gases. The presence of gas vesicles in cells can be detected either by light microscopy, where clusters of vesicles, called *gas vacuoles*, appear as irregular bright inclusions (Figure 2.40*a*), or by transmission electron microscopy of cell thin sections (Figure 2.40*b*).

Gas vesicles are composed of two different proteins. The major protein, called *GvpA*, forms the watertight vesicle shell itself and is a small, hydrophobic, and very rigid protein, copies of which align to form the parallel "ribs" of the vesicle. The rigidity is essential for the structure to resist the pressures exerted on it from outside. The minor protein, called *GvpC*, functions to strengthen the shell of the gas vesicle by cross-linking and binding the ribs at an angle to group several GvpA molecules together (**Figure 2.41**).

The composition and pressure of the gas inside a gas vesicle is that in which the organism is suspended. Because an inflated gas vesicle has a density only one-tenth that of the cell proper, inflated

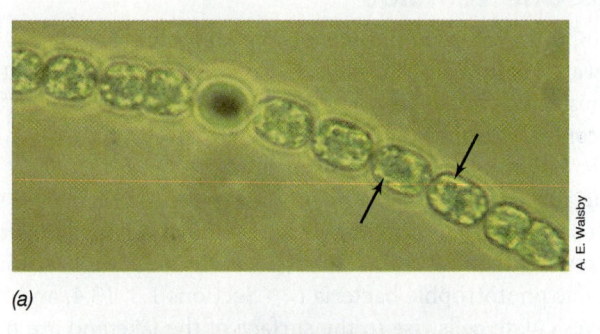

(a)

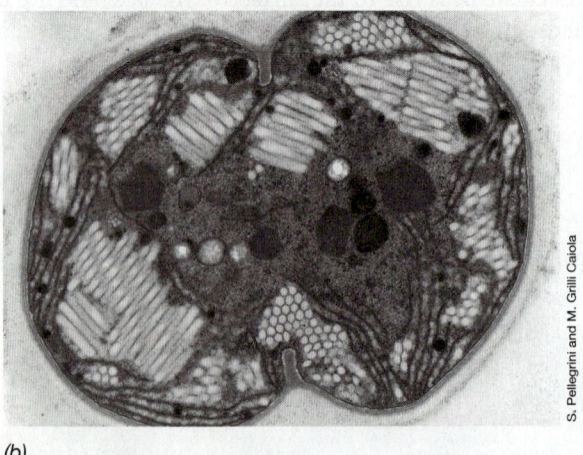

(b)

Figure 2.40 Gas vesicles of the cyanobacteria *Anabaena* and *Microcystis.* *(a)* Phase-contrast photomicrograph of *Anabaena*. Clusters of gas vesicles form phase-bright gas vacuoles (arrows). *(b)* Transmission electron micrograph of *Microcystis*. Gas vesicles are arranged in bundles, here seen in both longitudinal and cross section. Both cells are about 5 μm wide.

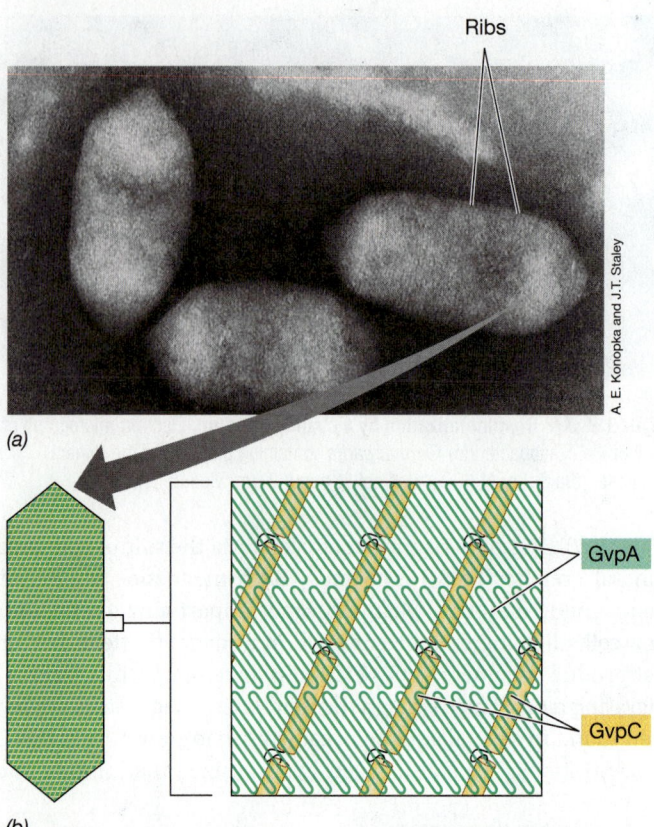

Ribs

(a)

GvpA

GvpC

(b)

Figure 2.41 Gas vesicle architecture. Transmission electron micrograph of gas vesicles purified from the bacterium *Ancylobacter aquaticus* and examined in negatively stained preparations. A single vesicle is about 100 nm in diameter. *(b)* Model of how gas vesicle proteins GvpA and GvpC interact to form a watertight but gas-permeable structure. GvpA, a rigid β-sheet, makes up the rib, and GvpC, an α-helix structure, is the cross-link.

gas vesicles decrease a cell's overall density and thereby increase its buoyancy; then, when vesicles are collapsed, buoyancy is lost. Phototrophic prokaryotes in particular can benefit from gas vesicles because they allow cells to adjust their vertical position in a water column to sink or rise to regions where conditions (for example, light intensity) are optimal for photosynthesis.

MINIQUIZ

- What gas is present in a gas vesicle? Why might a cell benefit from controlling its buoyancy?
- How are the two proteins that make up the gas vesicle, GvpA and GvpC, arranged to form such a water-impermeable structure?

2.16 Endospores

Certain species of *Bacteria* produce structures called **endospores** (**Figure 2.42**) during a process called *sporulation*. Endospores (the prefix *endo* means "within") are highly differentiated cells that are extremely resistant to heat, harsh chemicals, and radiation. Endospores function as survival structures and enable the organism to endure unfavorable growth conditions, including but not limited to extremes of temperature, drying, or nutrient depletion. Endospores can thus be thought of as the dormant stage of

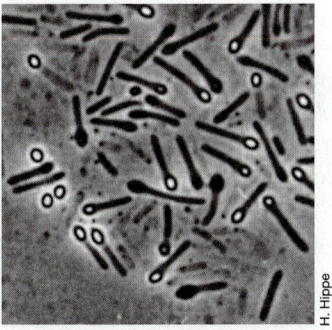
(a) **Terminal endospores**

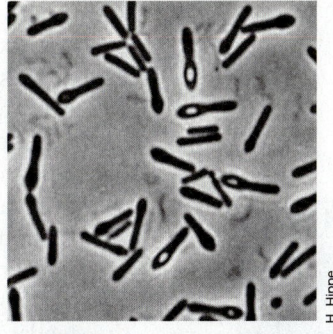
(b) **Subterminal endospores**

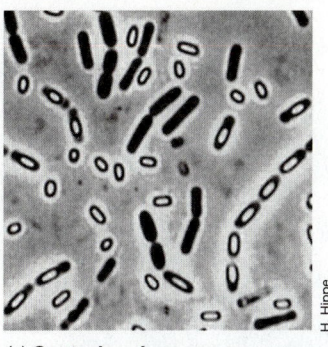
(c) **Central endospores**

Figure 2.42 **The bacterial endospore.** Phase-contrast photomicrographs illustrating endospore morphologies and intracellular locations in different species of endospore-forming bacteria. Endospores appear bright by phase-contrast microscopy.

a bacterial life cycle: vegetative cell → endospore → vegetative cell. Endospores are also easily dispersed by wind, water, or through the animal gut. Endospore-forming bacteria are commonly found in soil, and species of *Bacillus* are the best-studied representatives.

Endospore Formation and Germination

During endospore formation, a vegetative cell is converted into a nongrowing, heat-resistant, and light-refractive structure (**Figure 2.43**). Cells do not sporulate when they are actively growing but only when growth ceases owing to the exhaustion of an essential nutrient. Thus, cells of *Bacillus*, a typical endospore-forming bacterium, cease vegetative growth and begin sporulation when, for example, a key nutrient such as carbon or nitrogen becomes limiting.

An endospore can remain dormant for years but can convert back to a vegetative cell rapidly. This process involves three steps: *activation*, *germination*, and *outgrowth* (**Figure 2.44**). Activation occurs when endospores are heated for several minutes at an elevated but sublethal temperature. Activated endospores are then conditioned to germinate when supplied with certain nutrients, such as certain amino acids. Germination, typically a rapid process (on the order of minutes), involves loss of refractility of the endospore, increased ability to be stained by dyes, and loss of resistance to heat and chemicals. The final stage, outgrowth, involves visible swelling due to water uptake and synthesis of RNA, proteins, and DNA. The vegetative cell emerges from the broken endospore and begins to grow, remaining in vegetative growth until environmental signals once again trigger sporulation.

Endospore Structure

Endospores are visible by light microscopy as strongly refractile structures (Figure 2.42). Endospores are impermeable to most dyes, so occasionally they are seen as unstained regions within cells that have been stained with basic dyes such as methylene blue. To stain endospores, special stains and procedures must be used. In the classical endospore-staining protocol, the stain malachite green is used and is infused into the spore with steam.

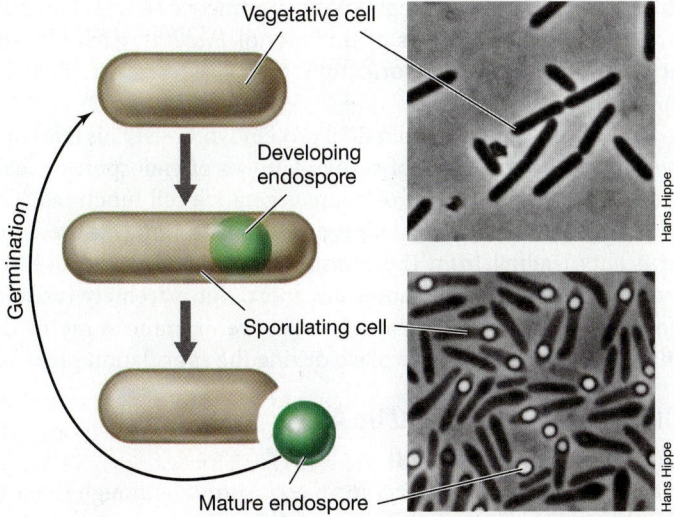

Figure 2.43 **The life cycle of an endospore-forming bacterium.** The phase-contrast photomicrographs are of cells of *Clostridium pascui*. A cell is about 0.8 μm wide.

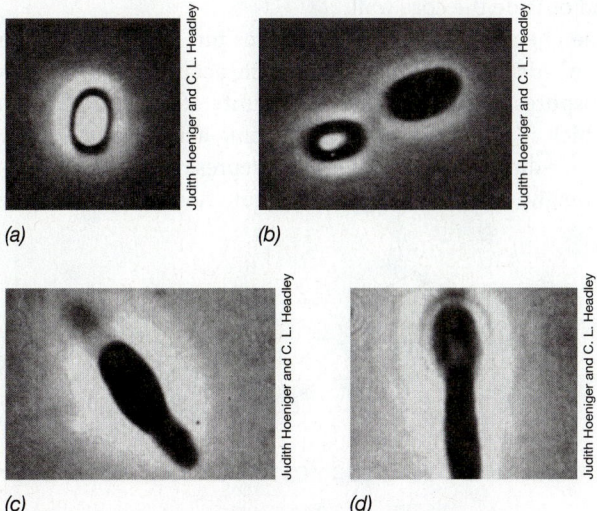

Figure 2.44 **Endospore germination in *Bacillus*.** Conversion of an endospore into a vegetative cell. The series of phase-contrast photomicrographs shows the sequence of events starting from (a) a highly refractile free endospore. (b) Activation: Refractility is being lost. (c, d) Outgrowth: The new vegetative cell is emerging.

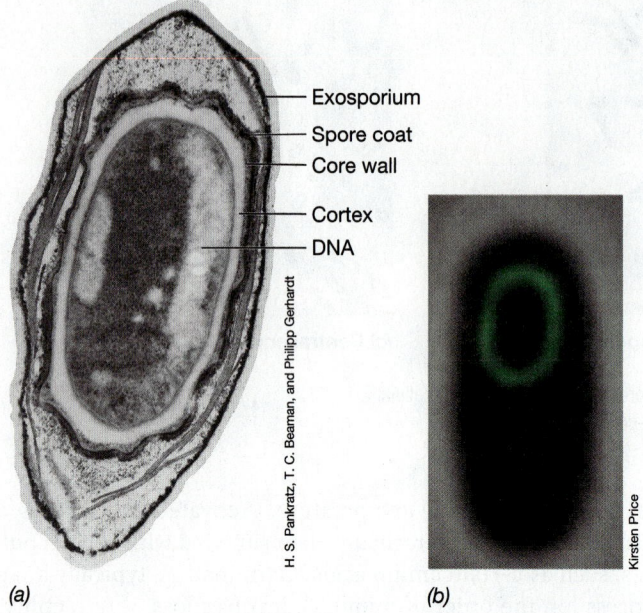

(a)

(b)

H. S. Pankratz, T. C. Beaman, and Philipp Gerhardt

Kirsten Price

Figure 2.45 Structure of the bacterial endospore. *(a)* Transmission electron micrograph of a thin section through an endospore of *Bacillus megaterium*. *(b)* Fluorescent photomicrograph of a cell of *Bacillus subtilis* undergoing sporulation. The green color is a dye that specifically stains a sporulation protein in the spore coat.

The structure of the endospore as seen with the electron microscope differs distinctly from that of the vegetative cell (**Figure 2.45**). The endospore contains many layers absent from the vegetative cell. The outermost layer is the *exosporium*, a thin protein covering. Moving inward there are several *spore coats*, composed of layers of spore-specific proteins (Figure 2.45*b*). Below the spore coat is the *cortex*, which consists of loosely cross-linked peptidoglycan, and inside the cortex is the *core*, which contains the core wall, cytoplasmic membrane, cytoplasm, nucleoid, ribosomes, and other cellular essentials. Thus, the endospore differs structurally from the vegetative cell primarily in the kinds of structures found outside the core wall.

One chemical found in endospores but not vegetative cells is **dipicolinic acid** (**Figure 2.46**), which accumulates in the core. Endospores also contain large amounts of calcium (Ca^{2+}), most of which is complexed with dipicolinic acid (Figure 2.46*b*). The calcium–dipicolinic acid complex represents about 10% of the dry weight of the endospore and functions to bind free water

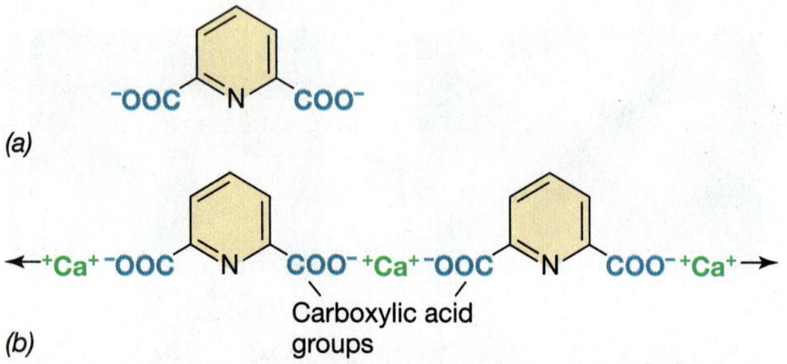

Figure 2.46 Dipicolinic acid (DPA). *(a)* Structure of DPA. *(b)* How Ca^{2+} cross-links DPA molecules to form a complex.

within the endospore, thus helping to dehydrate it. In addition, the complex inserts between bases in DNA, which helps stabilize DNA against heat denaturation.

The core of the endospore differs significantly from the cytoplasm of the vegetative cell that produced it. The core of an endospore contains less than one quarter of the water found in the vegetative cell, and thus the consistency of the core cytoplasm is that of a gel. Dehydration of the core greatly increases the heat resistance of macromolecules within the spore. Some bacterial endospores survive heating to temperatures as high as 150°C, although 121°C, the standard for microbiological sterilization (121°C is autoclave temperature, ⮌ Section 5.17), kills the endospores of most species. Dehydration has also been shown to confer resistance in the endospore to toxic chemicals, such as hydrogen peroxide (H_2O_2), and causes enzymes in the core to become inactive. In addition to the low water content of the endospore, the pH of the core is about one unit lower than that of the vegetative cell cytoplasm.

The endospore core contains high levels of *small acid-soluble spore proteins* (SASPs). These proteins are only made during the sporulation process and have at least two functions. SASPs bind tightly to DNA in the core and protect it from potential damage from ultraviolet radiation, desiccation, and dry heat. Ultraviolet resistance is conferred when SASPs change the molecular structure of DNA from the normal "B" form to the more compact "A" form. A-form DNA better resists pyrimidine dimer formation by UV radiation, which can cause mutations (⮌ Section 10.4), and resists the denaturing effects of dry heat. In addition, SASPs function as a carbon and energy source for the outgrowth of a new vegetative cell from the endospore during germination.

The Sporulation Cycle

Sporulation is an example of cellular differentiation (⮌ Figure 1.3), and many genetically directed changes in the cell occur during the conversion from vegetative growth to sporulation. The structural changes occurring in sporulating cells of *Bacillus* are shown in **Figure 2.47**. Sporulation can be divided into several stages. In *Bacillus subtilis*, which has been studied in detail, sporulation takes about 8 hours and begins with asymmetric cell division (Figure 2.47). Genetic studies of mutants of *Bacillus*, each blocked at one of the stages of sporulation, indicate that more than 200 spore-specific genes exist.

Sporulation requires differential protein synthesis. This is accomplished by the activation of several families of endospore-specific genes and the turning off of many vegetative cell functions. The proteins encoded by sporulation-specific genes catalyze the series of events leading from the moist, metabolizing, vegetative cell to the relatively dry, metabolically inert, but extremely resistant endospore (**Table 2.3**). In Section 7.11 we examine some of the molecular events that take place during the sporulation process.

Diversity and Phylogenetic Aspects of Endospore Formation

Nearly 20 genera of *Bacteria* form endospores, although the process has only been studied in detail in a few species of *Bacillus* and *Clostridium*. Nevertheless, most of the secrets behind endospore biology, such as the formation of calcium–dipicolinate complexes

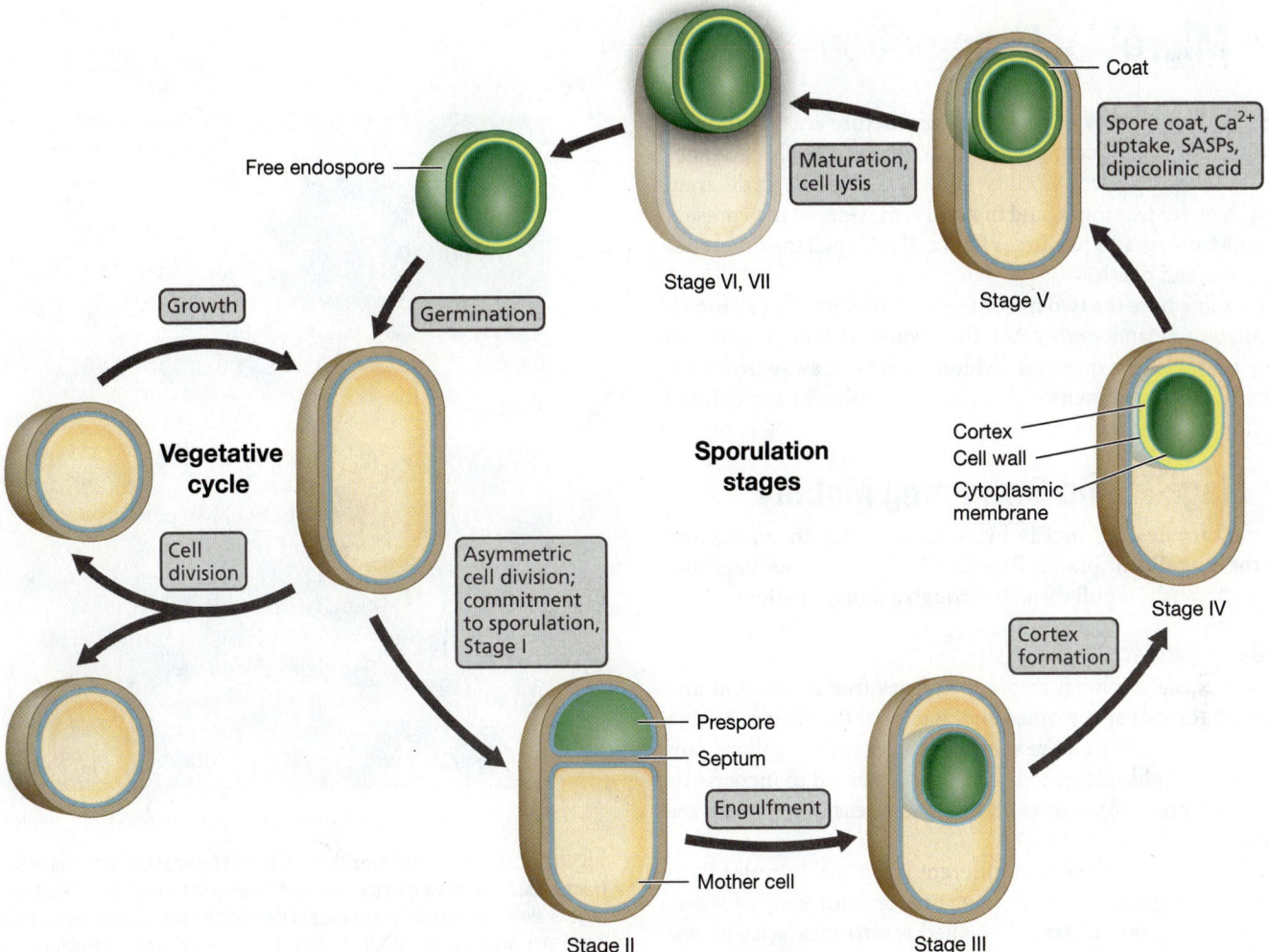

Figure 2.47 Stages in endospore formation. The stages are defined from genetic and microscopic analyses of sporulation in *Bacillus subtilis*, the model organism for studies of sporulation.

Table 2.3 Differences between endospores and vegetative cells		
Characteristic	*Vegetative cell*	*Endospore*
Microscopic appearance	Nonrefractile	Refractile
Calcium content	Low	High
Dipicolinic acid	Absent	Present
Enzymatic activity	High	Low
Respiration rate	High	Low or absent
Macromolecular synthesis	Present	Absent
Heat resistance	Low	High
Radiation resistance	Low	High
Resistance to chemicals	Low	High
Lysozyme	Sensitive	Resistant
Water content	High, 80–90%	Low, 10–25% in core
Small acid-soluble spore proteins	Absent	Present

and the production of endospore-specific SASPs, seem universal. From a phylogenetic perspective, the capacity to produce endospores is found only in a particular sublineage of the gram-positive bacteria. Despite this, the physiologies of endospore-forming bacteria are highly diverse and include anaerobes, aerobes, phototrophs, and chemolithotrophs. In light of this physiological diversity, the actual triggers for endospore formation may vary with different species and could include signals other than simple nutrient starvation, the major trigger for endospore formation in *Bacillus*. No *Archaea* have been shown to form endospores, suggesting that the capacity to produce endospores evolved sometime after the prokaryotic lineages diverged about 3.5 billion years ago (⮂ Figure 1.4b).

MINIQUIZ -

• What is dipicolinic acid and where is it found?

• What are SASPs and what is their function?

• What is formed when an endospore germinates?

VI • Microbial Locomotion

We finish our survey of prokaryotic structure and function by examining cell locomotion. Many microbial cells can move under their own power. Motility allows cells to reach different parts of their environment, and in nature, movement may present new opportunities and resources for a cell and spell the difference between life and death.

We examine here the two major types of prokaryotic cell movement, *swimming* and *gliding*. We then consider how motile cells are able to move in a directed fashion toward or away from particular stimuli (phenomena called *taxes*) and present examples of these simple behavioral responses.

2.17 Flagella and Swimming Motility

Many prokaryotes are motile by swimming due to a structure called the **flagellum** (plural, flagella) (**Figure 2.48**). The flagellum functions to push or pull the cell through a liquid medium.

Flagella of *Bacteria*

Bacterial flagella are long, thin appendages free at one end and attached to the cell at the other end. Bacterial flagella are so thin (15–20 nm, depending on the species) that a single flagellum cannot be seen by light microscopy unless it is stained to increase its diameter (Figure 2.48). However, flagella are easily seen with the electron microscope (**Figure 2.49**).

Flagella can attach to cells in different locations. In **polar flagellation**, the flagella are attached at one or both ends of a cell. Occasionally a group of flagella (called a *tuft*) may arise at one end of the cell, a type of polar flagellation called *lophotrichous* (Figure 2.48c). Tufts of flagella can often be seen in unstained cells by dark-field or phase-contrast microscopy (**Figure 2.50**). When a tuft of flagella emerges from both poles of the cell, flagellation is called *amphitrichous*. In **peritrichous flagellation** (Figures 2.48a and 2.49b), flagella are inserted at many locations around the cell surface. The type of flagellation—polar or peritrichous—is a characteristic used in the classification of bacteria.

Flagellar Structure

Flagella are not straight but helical. When flattened, flagella show a constant distance between adjacent curves, called the *wavelength*, and this wavelength is characteristic for the flagella of any

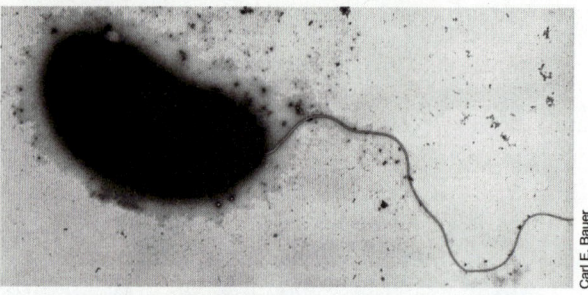

(a)

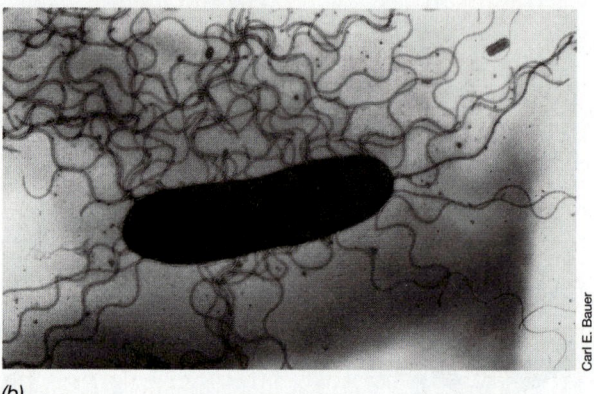

(b)

Figure 2.49 Bacterial flagella as observed by negative staining in the transmission electron microscope. *(a)* A single polar flagellum. *(b)* Peritrichous flagella. Both micrographs are of cells of the phototrophic bacterium *Rhodospirillum centenum*, which are about 1.5 μm wide. Cells of *R. centenum* are normally polarly flagellated but under certain growth conditions form peritrichous flagella. See Figure 2.59b for a photo of colonies of *R. centenum* cells that move toward an increasing gradient of light (phototaxis).

given species. The filament of a bacterial flagellum is composed of many copies of a protein called *flagellin*. The shape and wavelength of the flagellum are in part determined by the structure of the flagellin protein and also to some extent by the direction of rotation of the filament. The amino acid sequence of flagellin is highly conserved in species of *Bacteria*, suggesting that flagellar motility evolved early and has deep roots within this domain.

A flagellum consists of several components and rotates, much like a propeller on a boat motor. The base of the flagellum is structurally different from the filament. There is a wider region at the base of the filament called the *hook*. The hook consists of a single type of protein and connects the filament to the flagellum motor in the base (**Figure 2.51**).

The flagellum motor is anchored in the cytoplasmic membrane and cell wall. The motor consists of a central rod that passes through a series of rings. In gram-negative bacteria, an outer ring, called the *L ring*, is anchored in the lipopolysaccharide layer. A second ring, called the *P ring*, is anchored in the peptidoglycan layer of the cell wall. A third set of rings, called the *MS* and *C rings*, are located within the cytoplasmic membrane and the cytoplasm, respectively (Figure 2.51a). In gram-positive bacteria, which lack an outer membrane, only the inner pair of rings

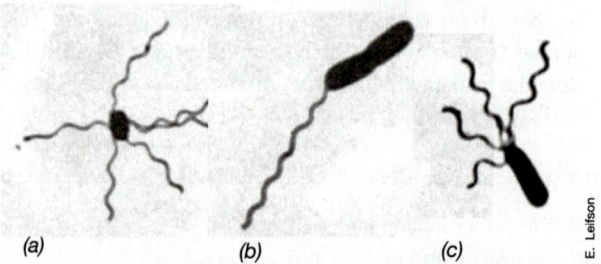

(a)   (b)   (c)

Figure 2.48 Bacterial flagella. Classic light photomicrographs taken by Einar Leifson of bacteria containing different arrangements of flagella. Cells are stained with the Leifson flagella stain. *(a)* Peritrichous. *(b)* Polar. *(c)* Lophotrichous.

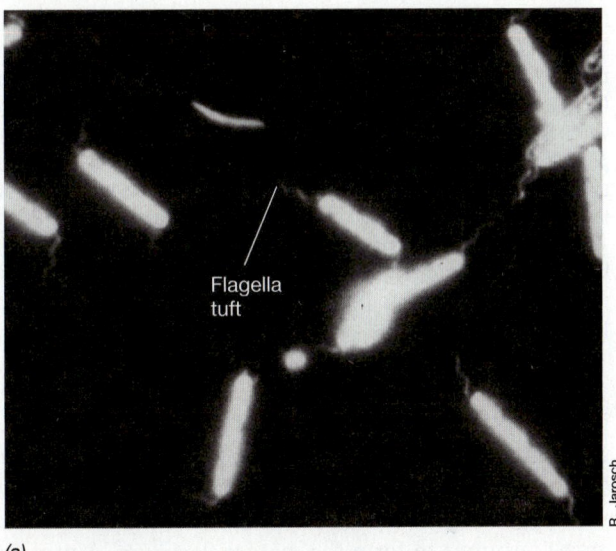

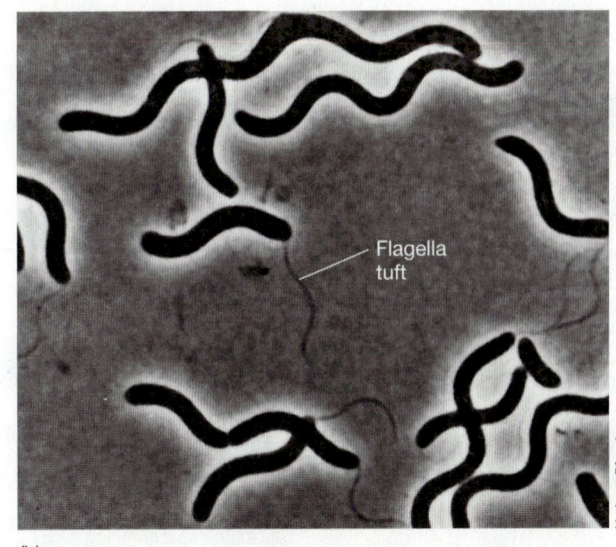

(a)

(b)

R. Jarosch

Norbert Pfennig

Figure 2.50 Bacterial flagella observed in living cells. (a) Dark-field photomicrograph of a group of large rod-shaped bacteria with flagellar tufts at each pole (amphitrichous flagellation). A single cell is about 2 μm wide. (b) Phase-contrast photomicrograph of cells of the large phototrophic purple bacterium *Rhodospirillum photometricum* with a tuft of lophotrichous flagella that emanate from one of the poles. A single cell measures about 3 × 30 μm.

is present. Surrounding the inner ring and anchored in the cytoplasmic membrane are a series of proteins called *Mot proteins*. A final set of proteins, called *Fli proteins* (Figure 2.51*a*), function as the motor switch, reversing the direction of rotation of the flagella in response to intracellular signals.

Flagellar Movement

The flagellum is a tiny rotary motor. How does this motor work? Rotary motors contain two main components: the *rotor* and the *stator*. In the flagellar motor, the rotor consists of the central rod and the L, P, C, and MS rings. Collectively, these structures make up the **basal body**. The stator consists of the Mot proteins that surround the basal body and function to generate torque.

Rotation of the flagellum is imparted by the basal body. The energy required for rotation of the flagellum comes from the proton motive force (Section 2.8). Proton movement across the cytoplasmic membrane through the Mot complex drives rotation of the flagellum, and about 1000 protons are translocated per rotation of the flagellum; a model for how this works is shown in Figure 2.51*b*. In this proton turbine model, protons flowing through channels in the Mot proteins exert electrostatic forces on helically arranged charges on the rotor proteins. Attractions between positive and negative charges would then cause the basal body to rotate as protons flow though the Mot proteins.

Archaeal Flagella

As for *Bacteria*, flagellar motility is widespread among species of *Archaea*; major genera of methanogens, extreme halophiles, thermoacidophiles, and hyperthermophiles (Figure 1.6*b*) are all capable of swimming motility. Archaeal flagella are roughly half the diameter of bacterial flagella, measuring only 10–13 nm in width (**Figure 2.52**), but impart movement to the cell by rotating, as do flagella in *Bacteria*. However, unlike *Bacteria*, in which a single

type of protein makes up the flagellar filament, several different flagellin proteins are known from *Archaea*, and their amino acid sequences and genes that encode them bear little relationship to those of bacterial flagellin.

Studies of swimming cells of the extreme halophile *Halobacterium* show that they swim at speeds only about one-tenth that of cells of *Escherichia coli*. Whether this holds for all *Archaea* is unknown, but the significantly smaller diameter of the archaeal flagellum compared with the bacterial flagellum would naturally reduce the torque and power of the flagellar motor such that slower swimming speeds are not surprising. Moreover, from biochemical experiments with *Halobacterium* it appears that archaeal flagella are powered directly by ATP rather than by the proton motive force, the source of energy for the flagella of *Bacteria* (Figure 2.51*b*). If this holds for the flagella of all motile *Archaea*, it would mean that the flagellar motors of *Archaea* and *Bacteria* employ fundamentally different energy-coupling mechanisms. Combined with the clear differences in flagellar protein structure between *Archaea* and *Bacteria*, this suggests that, as for endospores, flagellar motility evolved separately as prokaryotes diverged over 3.5 billion years ago (Figure 1.4*b*).

Flagellar Synthesis

Several genes encode the motility proteins of *Bacteria*. In *Escherichia coli* and *Salmonella enterica* serovar Typhimurium, for which motility studies have been extensive, over 50 genes are linked to motility. These genes encode the structural proteins of the flagellum and motor apparatus, of course, but also proteins that export the structural proteins through the cytoplasmic membrane to the outside of the cell, and proteins that regulate the many biochemical events surrounding the synthesis of new flagella.

A flagellar filament grows not from its base, as does an animal hair, but from its tip. The MS ring is synthesized first and inserted

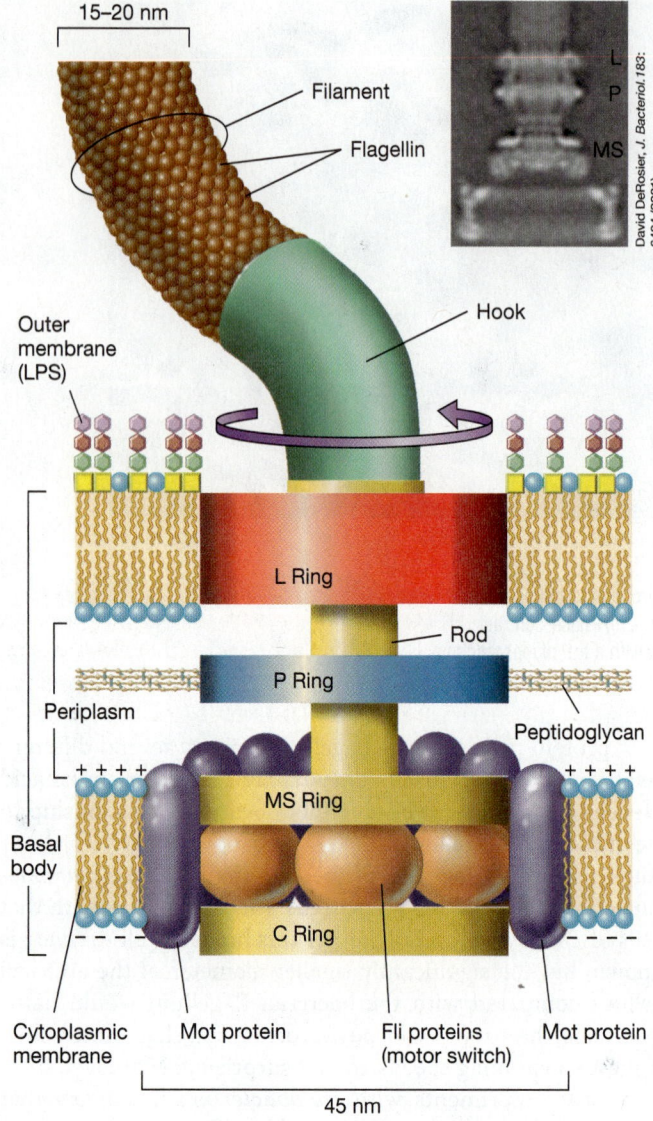

15–20 nm

Filament

Flagellin

Hook

Outer membrane (LPS)

L Ring

Rod

P Ring

Periplasm

Peptidoglycan

MS Ring

Basal body

C Ring

Cytoplasmic membrane

Mot protein

Fli proteins (motor switch)

Mot protein

45 nm

(a)

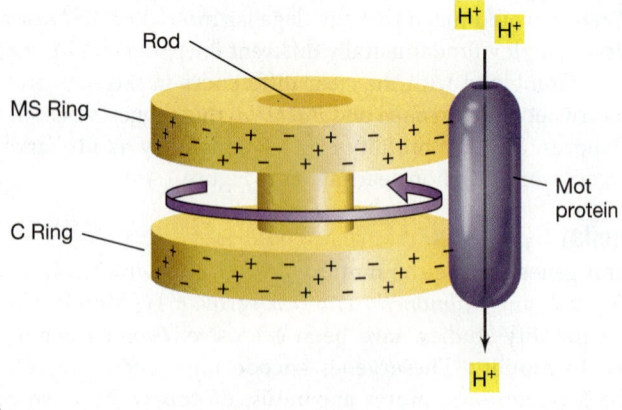

Rod

MS Ring

C Ring

Mot protein

H⁺

(b)

L
P
MS

David DeRosier, J. Bacteriol.183: 6404 (2001)

Figure 2.51 Structure and function of the flagellum in gram-negative Bacteria. (a) Structure. The L ring is embedded in the LPS and the P ring in peptidoglycan. The MS ring is embedded in the cytoplasmic membrane and the C ring in the cytoplasm. A narrow channel exists in the rod and filament through which flagellin molecules diffuse to reach the site of flagellar synthesis. The Mot proteins function as the flagellar motor, whereas the Fli proteins function as the motor switch. The flagellar motor rotates the filament to propel the cell through the medium. Inset: transmission electron micrograph of a flagellar basal body from *Salmonella enterica* with the various rings labeled. (b) Function. A "proton turbine" model has been proposed to explain rotation of the flagellum. Protons, flowing through the Mot proteins, exert forces on charges present on the C and MS rings, thereby spinning the rotor.

into the cytoplasmic membrane. Then other anchoring proteins are synthesized along with the hook before the filament forms (**Figure 2.53**). Flagellin molecules synthesized in the cytoplasm pass up through a 3-nm channel inside the filament and add on at the terminus to form the mature flagellum. A protein "cap" is present at the end of the growing flagellum. Cap proteins assist flagellin molecules that have diffused through the filament channel to assemble in the proper fashion at the flagellum terminus (Figure 2.53). Approximately 20,000 flagellin protein molecules are needed to make one filament. The flagellum grows more or less continuously until it reaches its final length. Broken flagella still rotate and can be repaired with new flagellin units passed through the filament channel to replace the lost ones.

Cell Speed and Motion

In *Bacteria*, flagella do not rotate at a constant speed but instead increase or decrease their rotational speed in relation to the strength of the proton motive force. Flagella can rotate at up to 300 revolutions per second and propel cells through a liquid at up to 60 cell lengths/sec. By contrast, the fastest known animal, the cheetah, moves at a maximum rate of about 25 body lengths/sec. Thus, when size is accounted for, a bacterial cell swimming at 60 cell lengths/sec is actually moving over twice as fast as the fastest animal!

The swimming motions of polarly and lophotrichously flagellated organisms differ from those of peritrichously flagellated organisms, and these can be distinguished microscopically (**Figure 2.54**). Peritrichously flagellated organisms typically move in a straight line in a slow, deliberate fashion. Polarly flagellated organisms, on

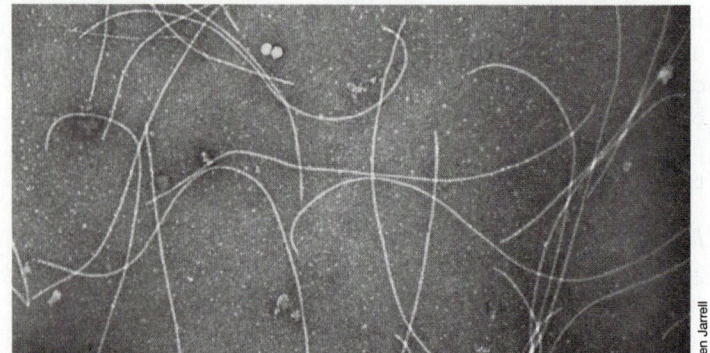

Ken Jarrell

Figure 2.52 Archaeal flagella. Transmission electron micrograph of flagella isolated from cells of the methanogen *Methanococcus maripaludis*. A single flagellum is about 12 nm wide.

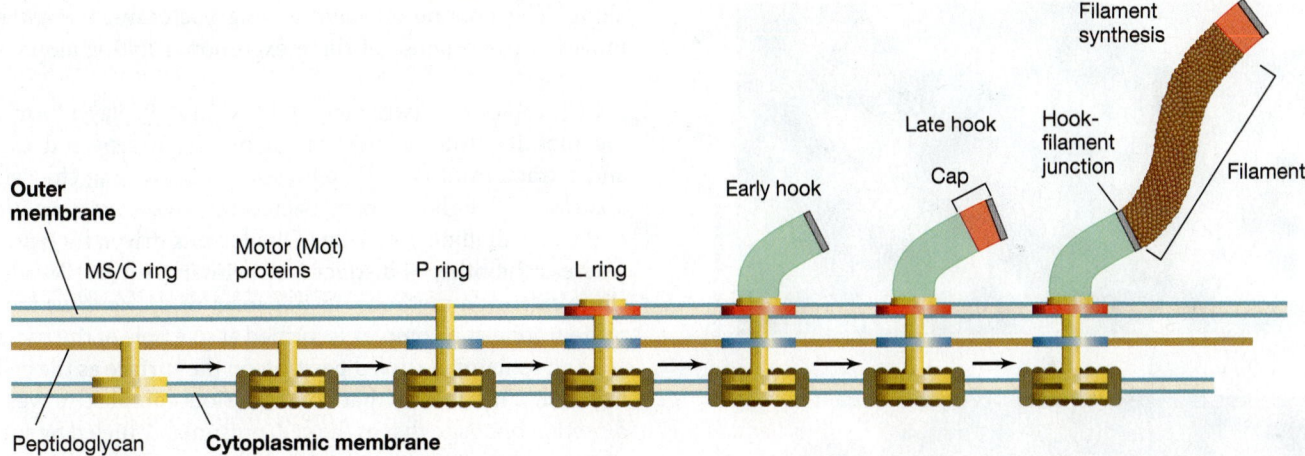

Figure 2.53 Flagella biosynthesis. Synthesis begins with assembly of MS and C rings in the cytoplasmic membrane, followed by the other rings, the hook, and the cap. Flagellin protein flows through the hook to form the filament and is guided into position by cap proteins.

the other hand, move more rapidly, spinning around and seemingly dashing from place to place. The different behavior of flagella on polar and peritrichous organisms, including differences in reversibility of the flagellum, is illustrated in Figure 2.54.

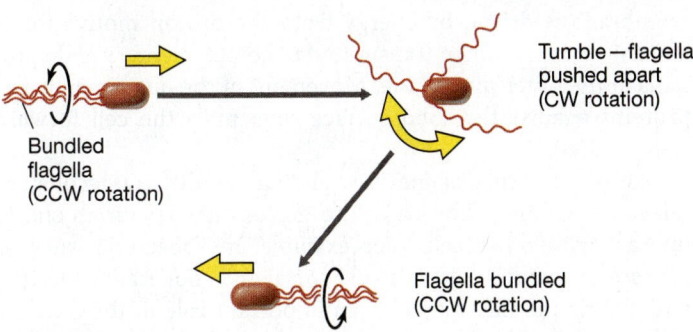

(a) **Peritrichous**

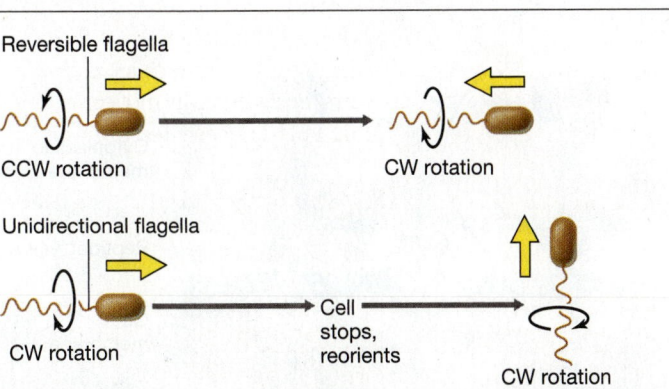

(b) **Polar**

Figure 2.54 Movement in peritrichously and polarly flagellated prokaryotes. (a) Peritrichous: Forward motion is imparted by all flagella rotating counterclockwise (CCW) in a bundle. Clockwise (CW) rotation causes the cell to tumble, and then a return to counterclockwise rotation leads the cell off in a new direction. (b) Polar: Cells change direction by reversing flagellar rotation (thus pulling instead of pushing the cell) or, with unidirectional flagella, by stopping periodically to reorient, and then moving forward by clockwise rotation of its flagella. The yellow arrows show the direction the cell is traveling.

Swimming speed is a genetically governed property because different motile species, even different species that are the same cell size, can swim at different maximum speeds. When assessing the capacity of a laboratory culture of a bacterium for swimming motility and swimming speed, observations should be made on young cultures. In old cultures, otherwise motile cells often stop swimming and the culture may appear to be nonmotile.

MINIQUIZ

- Cells of *Salmonella* are peritrichously flagellated, those of *Pseudomonas* polarly flagellated, and those of *Spirillum* lophotrichously flagellated. Using a sketch, show how each organism would appear in a flagella stain.
- Compare the flagella of *Bacteria* and *Archaea* in terms of their structure and function.

2.18 Gliding Motility

Some prokaryotes are motile but lack flagella. Most of these non-swimming yet still motile bacteria move by *gliding*. Unlike flagellar motility, in which cells stop and then start off in a different direction, gliding motility is a slower and smoother form of movement and typically occurs along the long axis of the cell.

Diversity of Gliding Motility

Gliding motility is widely distributed among *Bacteria* but has been well studied in only a few groups. The gliding movement itself—up to 10 μm/sec in some gliding bacteria—is considerably slower than propulsion by flagella but still offers the cell a means of moving about its habitat.

Gliding prokaryotes are filamentous or rod-shaped cells, and the gliding process requires that the cells be in contact with a solid surface (**Figure 2.55**). The morphology of colonies of a typical gliding bacterium are distinctive, because cells glide out and move away from the center of the colony (Figure 2.55c). Perhaps the best-known gliding bacteria are the filamentous cyanobacteria (Figure 2.55a, b), certain gram-negative *Bacteria* such as

(a)

Richard W. Castenholz

(b)

Richard W. Castenholz

(c)

Mark J. McBride

(d)

Mark J. McBride

Figure 2.55 **Gliding bacteria.** *(a, b)* The large filamentous cyanobacterium *Oscillatoria* has cells about 35 μm wide. *(b) Oscillatoria* filaments gliding on an agar surface. *(c)* Masses of the bacterium *Flavobacterium johnsoniae* gliding away from the center of the colony (the colony is about 2.7 mm wide). *(d)* Nongliding mutant strain of *F. johnsoniae* showing typical colony morphology of nongliding bacteria (the colonies are 0.7–1 mm in diameter). See also Figure 2.56.

Myxococcus and other myxobacteria, and species of *Cytophaga* and *Flavobacterium* (Figure 2.55c, d). No gliding *Archaea* are known.

Mechanisms of Gliding Motility

More than one mechanism is responsible for gliding motility. Cyanobacteria glide by secreting a polysaccharide slime from pores onto the outer surface of the cell. The slime contacts both the cell surface and the solid surface against which the cell moves. As the excreted slime adheres to the surface, the cell is pulled

along. The nonphototrophic gliding bacterium *Cytophaga* also moves at the expense of slime excretion, rotating along its long axis as it does.

Cells capable of "twitching motility" also display a form of gliding motility using a mechanism by which repeated extension and retraction of type IV pili (Section 2.13) drag the cell along a surface. The gliding myxobacterium *Myxococcus xanthus* has two forms of gliding motility. One form is driven by type IV pili, whereas the other is distinct from either the type IV pili or the slime extrusion methods. In this form of *M. xanthus* motility, a protein adhesion complex is formed at one pole of the rod-shaped cell and remains at a fixed position on the surface as the cell glides forward. This means that the adhesion complex moves in the direction opposite that of the cell, presumably fueled by some sort of cytoplasmic motility engine.

Neither slime extrusion nor twitching is the mechanism of gliding in other gliding bacteria. In the genus *Flavobacterium* (Figure 2.55c), for example, no slime is excreted and cells lack type IV pili. Instead of using one of these gliding mechanisms, the movement of proteins on the *Flavobacterium* cell surface supports gliding motility in this organism. Specific motility proteins anchored in the cytoplasmic and outer membranes are thought to propel cells of *Flavobacterium* forward by a ratcheting mechanism (**Figure 2.56**). Movement of gliding-specific proteins in the cytoplasmic membrane is driven by energy from the proton motive force, and this motion is then transmitted to complementary glide proteins in the outer membrane. Movement of the outer membrane proteins against the solid surface then pulls the cell forward (Figure 2.56).

Like other forms of motility, gliding motility has ecological relevance. Gliding allows a cell to exploit new resources and to interact with other cells. For example, myxobacteria, such as *Myxococcus xanthus*, have a very social and cooperative lifestyle, and gliding motility may play an important role in the intimate cell-to-cell interactions necessary to complete their life cycle (⮂ Section 14.19).

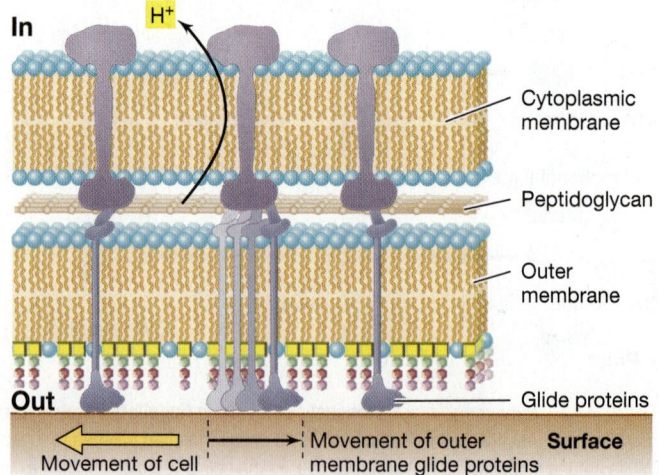

Figure 2.56 **Gliding motility in *Flavobacterium johnsoniae.*** Tracks (yellow) exist in the peptidoglycan that connect cytoplasmic proteins to outer membrane glide proteins and propel the glide proteins along the solid surface. Note that the glide proteins and the cell proper move in opposite directions.

MINIQUIZ
- How does gliding motility differ from swimming motility in both mechanism and requirements?
- Contrast the mechanism of gliding motility in a filamentous cyanobacterium and in *Flavobacterium*.

2.19 Chemotaxis and Other Taxes

Prokaryotes often encounter gradients of physical or chemical agents in nature and have evolved means to respond to these gradients by moving either toward or away from the agent. Such a directed movement is called a *taxis* (plural, taxes). **Chemotaxis**, a response to chemicals, and **phototaxis**, a response to light, are two well-studied taxes. Here we discuss these taxes in a general way. In Section 7.8 we examine the molecular mechanism of chemotaxis and its regulation in *Escherichia coli* as a model for all taxes in *Bacteria*.

Chemotaxis has been well studied in swimming bacteria, and much is known at the genetic level concerning how information on the chemical state of the environment is communicated to the flagellum. Our discussion here will thus deal solely with swimming bacteria. However, some gliding bacteria (Section 2.18) are also chemotactic, and there are phototactic movements in filamentous cyanobacteria (Figure 2.55a, b). In addition, many species of *Archaea* are also chemotactic and many of the same types of proteins that control chemotaxis in *Bacteria* are present in motile *Archaea* as well.

Chemotaxis in Peritrichously Flagellated *Bacteria*

Much research on chemotaxis has been done with the peritrichously flagellated bacterium *E. coli*. To understand how chemotaxis affects the behavior of *E. coli*, consider the situation in which a cell encounters a gradient of some chemical in its environment

(Figure 2.57). In the absence of the gradient, cells move in a random fashion that includes *runs*, in which the cell is swimming forward in a smooth fashion, and *tumbles*, when the cell stops and jiggles about. During forward movement in a run, the flagellar motor rotates counterclockwise. When flagella rotate clockwise, the bundle of flagella pushes apart, forward motion ceases, and the cells tumble (Figure 2.57).

Following a tumble, the direction of the next run is random. Thus, by means of runs and tumbles, the cell moves about its environment in a random fashion but does not really go anywhere. However, if a gradient of a chemical attractant is present, these random movements become biased. If the organism senses that it is moving toward higher concentrations of the attractant, runs become longer and tumbles are less frequent. The result of this behavioral response is that the organism moves up the concentration gradient of the attractant (Figure 2.57b). If the organism senses a repellent, the same general mechanism applies, although in this case it is the *decrease* in concentration of the repellent (rather than the *increase* in concentration of an attractant) that promotes runs.

How are chemical gradients sensed? Prokaryotic cells are too small to sense a gradient of a chemical along the length of a single cell. Instead, while moving, cells monitor their environment by sampling chemicals periodically and comparing the concentration with that sensed a few moments before. Bacterial cells thus respond to *temporal* rather than *spatial* differences in the concentration of a chemical as they swim. Sensory information is fed through an elaborate cascade of proteins that eventually affect the direction of rotation of the flagellar motor. The attractants and repellents are sensed by a series of membrane proteins called *chemoreceptors*. These proteins bind the chemicals and begin the process of sensory transduction to the flagellum (⟳ Section 7.8). Chemotaxis can thus be considered a type of sensory response system, analogous to sensory responses in the nervous system of animals.

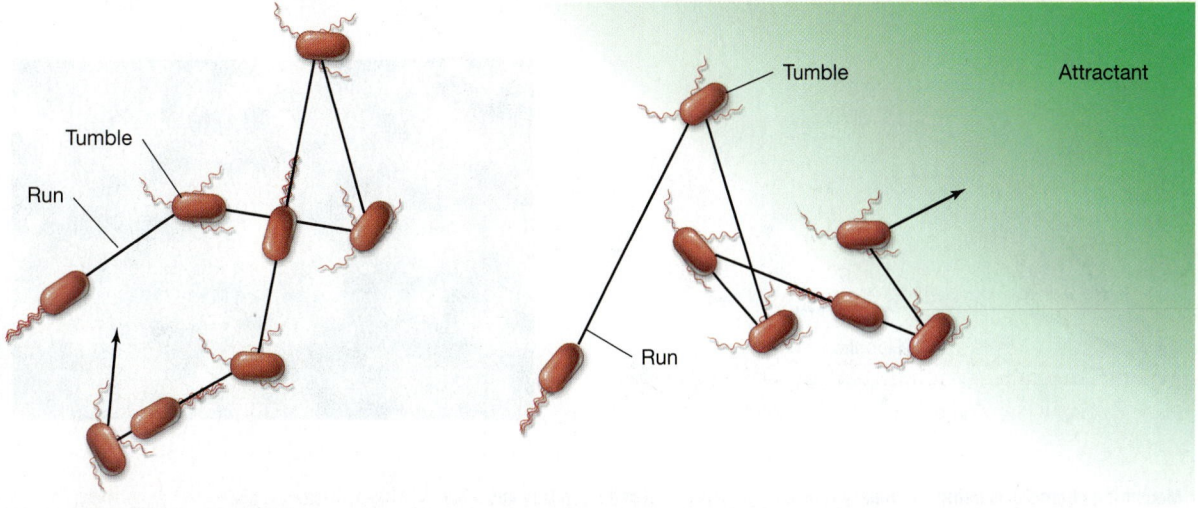

(a) **No attractant present: Random movement**   (b) **Attractant present: Directed movement**

Figure 2.57 **Chemotaxis in a peritrichously flagellated bacterium such as *Escherichia coli*.** *(a)* In the absence of a chemical attractant, the cell swims randomly in runs, changing direction during tumbles. *(b)* In the presence of an attractant, runs become biased, and the cell moves up the gradient of the attractant. The attractant gradient is depicted in green, with the highest concentration where the color is most intense.

Chemotaxis in Polarly Flagellated *Bacteria*

Chemotaxis in polarly flagellated cells is similar to that in peritrichously flagellated cells such as *E. coli*, but there are some differences. Many polarly flagellated bacteria, such as *Pseudomonas* species, can reverse the direction of rotation of their flagella and in so doing immediately reverse their direction of movement (Figure 2.54*b*). However, some polarly flagellated bacteria, such as the phototrophic purple bacterium *Rhodobacter sphaeroides*, have flagella that rotate only in a clockwise direction. How do such cells change direction, and are they chemotactic?

In cells of *R. sphaeroides*, which have only a single flagellum inserted subpolarly, rotation of the flagellum stops periodically. When it stops, the cell becomes reoriented in a random way (Figure 2.54*b*). As the flagellum begins to rotate again, the cell moves in a new direction. Nevertheless, cells of *R. sphaeroides* are strongly chemotactic to certain organic compounds and also show tactic responses to oxygen and light. *R. sphaeroides* cannot reverse its flagellar motor and tumble as *E. coli* can, but cells do maintain runs as long as they sense an increasing concentration of attractant. If the cells sense a decreasing concentration of attractant, movement ceases. By such starting and stopping, a cell eventually finds the path of increasing attractant and maintains a run until either its chemoreceptors are saturated or it begins to sense a decrease in the level of attractant.

Measuring Chemotaxis

Bacterial chemotaxis can be demonstrated by immersing a small glass capillary tube containing an attractant into a suspension of motile bacteria that does not contain the attractant. From the tip of the capillary, a gradient forms into the surrounding medium, with the concentration of chemical gradually decreasing with distance from the tip (**Figure 2.58**). When an attractant is present, chemotactic bacteria will move toward it, forming a swarm around the open tip (Figure 2.58*c*) with many of the bacteria swimming into the capillary itself. Of course, because of random movements some chemotactic bacteria will swim into the capillary even if it contains a solution of the same composition as the medium (control solution, Figure 2.58*b*). However, when an attractant is present, the number of bacteria within the capillary will be many times higher than external cell numbers. If the capillary is removed after a time period and the cells within the capillary are counted and compared with that of the control, attractants can easily be identified (Figure 2.58*e*).

If the inserted capillary contains a repellent, just the opposite occurs; the cells sense an increasing gradient of repellent and the appropriate chemoreceptors affect flagellar rotation to gradually move the cells away from the repellent. In this case, the number of bacteria within the capillary will be fewer than in the control (Figure 2.58*d*). Using this capillary method, it is possible to screen chemicals to see if they are attractants or repellents for a given bacterium.

Chemotaxis can also be observed under a microscope. Using a video camera that captures the position of bacterial cells with time and shows the motility tracks of each cell, it is possible to see the chemotactic movements of cells (Figure 2.58*f*). This method has been adapted to studies of chemotaxis of bacteria in natural environments. In nature it is thought that the major chemotactic

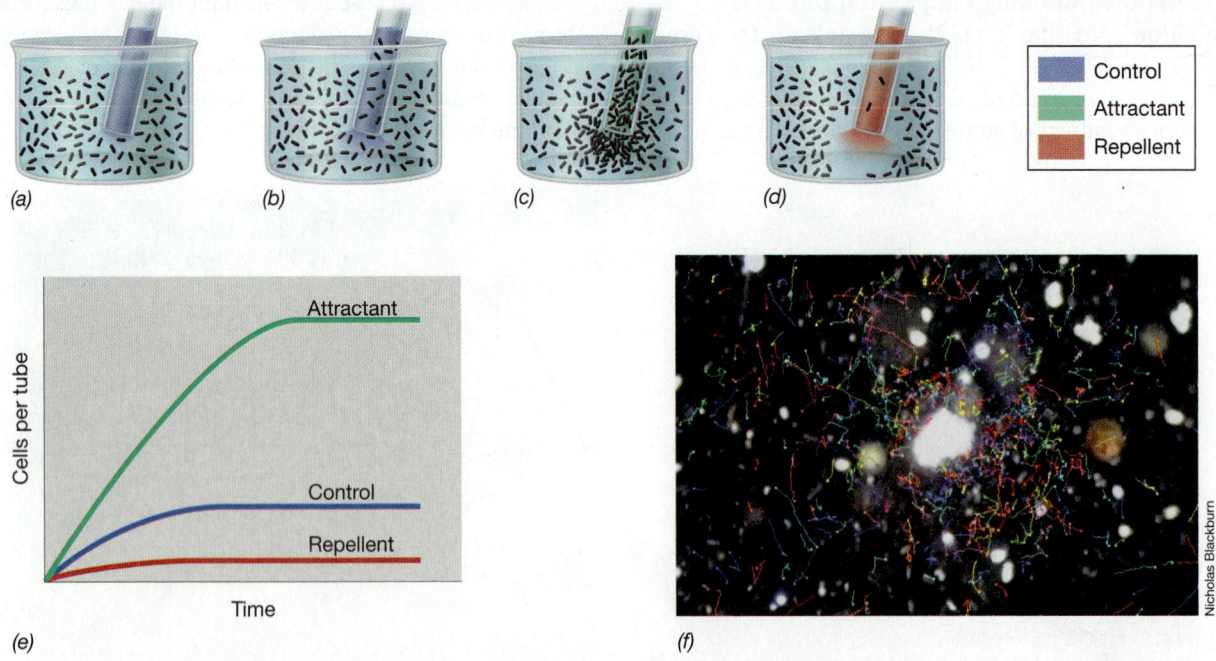

■ Control
■ Attractant
■ Repellent

(a)　(b)　(c)　(d)

Attractant

Cells per tube

Control

Repellent

Time

(e)

Nicholas Blackburn

(f)

Figure 2.58 Measuring chemotaxis using a capillary tube assay. *(a)* Insertion of the capillary into a bacterial suspension. As the capillary is inserted, a gradient of the chemical begins to form. *(b)* Control capillary contains a salt solution that is neither an attractant nor a repellent. Cell concentration inside the capillary becomes the same as that outside. *(c)* Accumulation of bacteria in a capillary containing an attractant. *(d)* Repulsion of bacteria by a repellent. *(e)* Time course showing cell numbers in capillaries containing various chemicals. *(f)* Tracks of motile bacteria in seawater swarming around an algal cell (large white spot, center) photographed with a tracking video camera system attached to a microscope. The bacterial cells are showing positive aerotaxis by moving toward the oxygen-producing algal cell. The alga is about 60 μm in diameter.

agents for bacteria are nutrients excreted from larger microbial cells or from live or dead macroorganisms. Algae, for example, produce both organic compounds and oxygen (O_2, from photosynthesis) that can trigger chemotactic movements of bacteria toward the algal cell (Figure 2.58f).

Phototaxis

Many phototrophic microorganisms can move toward light, a process called *phototaxis*. The advantage of phototaxis for a phototrophic organism is that it allows it to orient itself most efficiently to receive light for photosynthesis. This can be shown if a light spectrum is spread across a microscope slide on which there are motile phototrophic purple bacteria. On such a slide the bacteria accumulate at wavelengths at which their photosynthetic pigments absorb (**Figure 2.59**; ⮌ Sections 13.1–13.4 cover photosynthesis). These pigments include, in particular, bacteriochlorophylls and carotenoids.

Two different light-mediated taxes are observed in phototrophic bacteria. One, called *scotophobotaxis*, can be observed

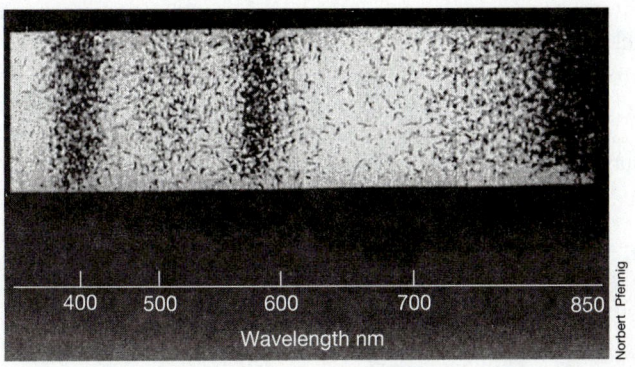

(a)

Light

(b)

0   1   2
Time (h)

only microscopically and occurs when a phototrophic bacterium happens to swim outside the illuminated field of view of the microscope into darkness. Entering darkness negatively affects photosynthesis and thus the energy state of the cell and signals the cell to tumble, reverse direction, and once again swim in a run, thus reentering the light. Scotophobotaxis is presumably a mechanism by which phototrophic purple bacteria avoid entering darkened habitats when they are moving about in illuminated ones, and this likely improves their competitive success.

True phototaxis differs from scotophobotaxis; in phototaxis, cells move up a gradient of light from lower to higher intensities. Phototaxis is analogous to chemotaxis except that the attractant in this case is light instead of a chemical. In some species, such as the highly motile phototrophic purple bacterium *Rhodospirillum centenum* (Figure 2.49), entire colonies of cells show phototaxis and move in unison toward the light (Figure 2.59b).

Several components of the regulatory system that govern chemotaxis also control phototaxis. This connection has emerged from the study of mutants of phototrophic bacteria defective in phototaxis; such mutants show defective chemotaxis systems as well. A *photoreceptor*, a protein that functions similar to a chemoreceptor but senses a gradient of light instead of chemicals, is the initial sensor in the phototaxis response. The photoreceptor then interacts with the same cytoplasmic proteins that control flagellar rotation in chemotaxis, maintaining the cell in a run if it is swimming toward an increasing intensity of light. Thus, although the stimulus in chemotaxis and phototaxis is different—chemicals versus light—the response following reception of the stimulus is controlled by a common series of proteins. Section 7.8 discusses the activities of these proteins in more detail.

Other Taxes

Other bacterial taxes, such as movement toward or away from oxygen (*aerotaxis*, see Figure 2.58f) or toward or away from conditions of high ionic strength (*osmotaxis*), are known among various swimming prokaryotes. In some gliding cyanobacteria an unusual taxis, *hydrotaxis* (movement toward water), has also been observed. Hydrotaxis allows gliding cyanobacteria that inhabit dry environments, such as desert soils, to glide toward a gradient of increasing hydration.

From the study of microbial taxes, it is obvious that motile prokaryotes are "tuned in" to the chemical and physical state of their habitats. And from a mechanistic standpoint, it is interesting that these cells process the results of their environmental samplings through a common system that ultimately controls flagellar activity. By being able to move toward or away from various stimuli, prokaryotic cells improve their chances of competing for resources and avoiding the harmful effects of substances that could damage or kill them.

Figure 2.59 Phototaxis of phototrophic bacteria. *(a)* Scotophobic accumulation of the phototrophic purple bacterium *Thiospirillum jenense* at wavelengths of light at which its pigments absorb. A light spectrum was displayed on a microscope slide containing a dense suspension of the bacteria; after a period of time, the bacteria had accumulated selectively and the photomicrograph was taken. The wavelengths at which the bacteria accumulated are those at which the photosynthetic pigment bacteriochlorophyll *a* absorbs (compare with Figure 13.3*b*). *(b)* Phototaxis of an entire colony of the purple phototrophic bacterium *Rhodospirillum centenum*. These strongly phototactic cells move in unison toward the light source at the top. See Figure 2.49 for electron micrographs of flagellated *R. centenum* cells.

MINIQUIZ

- Define the word chemotaxis. How does chemotaxis differ from aerotaxis?
- What causes a run versus a tumble?
- How can chemotaxis be measured quantitatively?
- How does scotophobotaxis differ from phototaxis?

VII • Eukaryotic Microbial Cells

Compared with prokaryotic cells, microbial eukaryotes typically have structurally more complex and much larger cells (⮌ Figure 1.2). We finish up our study of microbial cell structure and function with a consideration of structure/function issues in microbial eukaryotes, common models for the study of eukaryotic biology. Microbial eukaryotes include the fungi, the algae, and the protozoa and other protists. We cover the diversity of microbial eukaryotes in Chapter 17.

2.20 The Nucleus and Cell Division

Eukaryotic cells vary in the complement of organelles they contain, but a unit membrane–enclosed nucleus is universal and a hallmark of the eukaryotic cell. Mitochondria are nearly universal among eukaryotic cells, while pigmented chloroplasts are found only in phototrophic cells. Other structures include the Golgi complex, lysosomes, endoplasmic reticula, and microtubules and microfilaments (**Figure 2.60**). Some microbial eukaryotes have flagella or cilia—organelles of motility—while others do not. Microbial eukaryotes may also have extracellular components,

such as a cell wall in fungi and algae (most protozoans lack a cell wall).

Nucleus

The **nucleus** contains the chromosomes of the eukaryotic cell. DNA within the nucleus is wound around basic (positively charged) proteins called **histones**, which tightly pack the negatively charged DNA to form nucleosomes (**Figure 2.61b**) and from them, chromosomes. The nucleus is enclosed by a pair of membranes, each with its own function, separated by a space. The inner membrane is a simple sac while the outer membrane is in many places continuous with the endoplasmic reticulum. The inner and outer nuclear membranes specialize in interactions with the nucleoplasm and the cytoplasm, respectively. The nuclear membranes contain pores (Figures 2.60 and 2.61a), formed from holes where the inner and outer membranes are joined. The pores allow transport proteins to import and export other proteins and nucleic acids into and out of the nucleus, a process called *nuclear transport*.

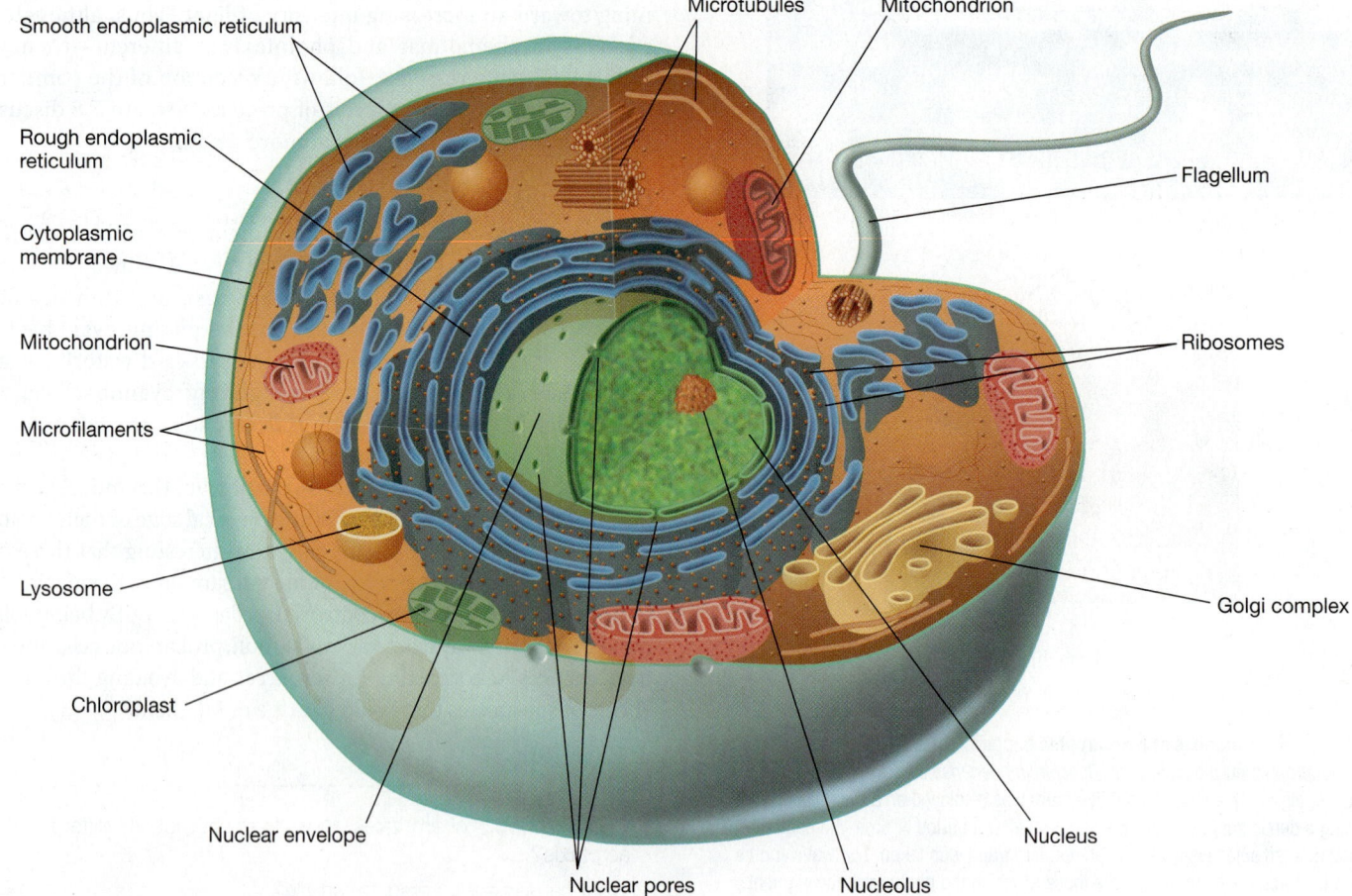

Smooth endoplasmic reticulum

Rough endoplasmic reticulum

Cytoplasmic membrane

Mitochondrion

Microfilaments

Lysosome

Chloroplast

Nuclear envelope

Nuclear pores

Nucleolus

Nucleus

Golgi complex

Ribosomes

Flagellum

Mitochondrion

Microtubules

Figure 2.60 Cutaway schematic of a microbial eukaryote. Although all eukaryotic cells contain a nucleus, not all organelles and other structures shown are present in all microbial eukaryotes. Not shown is the cell wall, found in fungi, algae, plants, and a few protists.

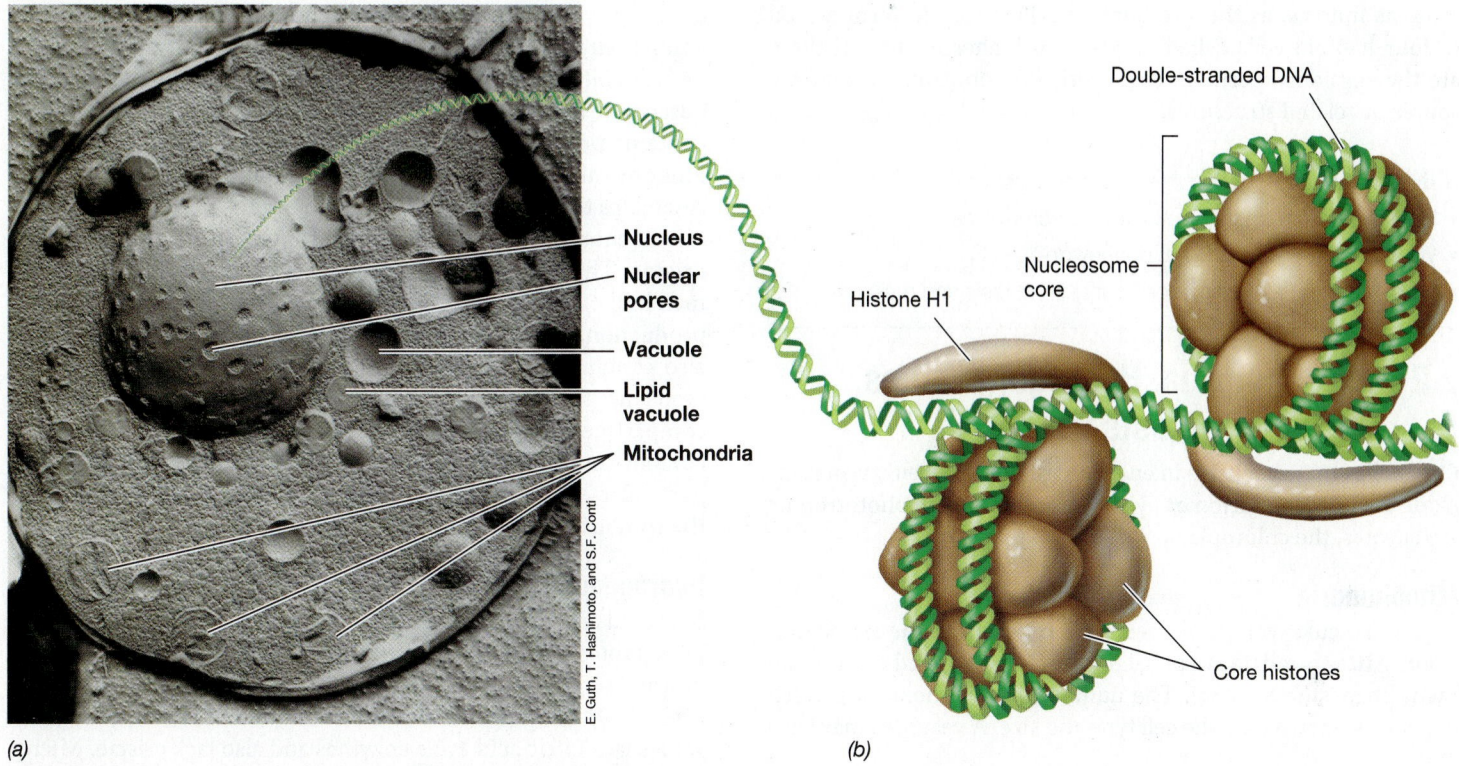

(a)   (b)

E. Guth, T. T. Hashimoto, and S.F. Conti

Figure 2.61 **The nucleus and DNA packaging in eukaryotes.** *(a)* Electron micrograph of a yeast cell prepared in such a way as to reveal a surface view of the nucleus. The cell is about 8 μm wide. *(b)* Packaging of DNA around histone proteins to form a nucleosome. Nucleosomes are arranged along the DNA strand like beads on a string and aggregate to form chromosomes during the process of mitosis (see Figure 2.62).

Within the nucleus is found the *nucleolus* (Figure 2.60), the site of ribosomal RNA (rRNA) synthesis. The nucleolus is rich in RNA, and ribosomal proteins synthesized in the cytoplasm are transported into the nucleolus and combine with rRNA to form the small and large subunits of eukaryotic ribosomes. These are then exported to the cytoplasm, where they associate to form the intact ribosome and function in protein synthesis.

Cell Division

Eukaryotic cells divide by a process in which the chromosomes are replicated, the nucleus disassembled, the chromosomes segregated into two sets, and a nucleus reassembled in each daughter cell. Many microbial eukaryotes can exist in either of two genetic states: *haploid* or *diploid*. Diploid cells have two copies of each chromosome whereas haploid cells have only one. For example, the brewer's yeast *Saccharomyces cerevisiae* can exist in the haploid state (16 chromosomes) as well as in the diploid state (32 chromosomes). However, regardless of its genetic state, during cell division the chromosome number is first doubled and later halved to give each daughter cell its correct complement of chromosomes. This is the process of **mitosis**, unique to eukaryotic cells. During mitosis, the chromosomes condense, divide, and are separated into two sets, one for each daughter cell (**Figure 2.62**).

In contrast to mitosis, **meiosis** is the process of conversion from the diploid to the haploid stage. Meiosis consists of two cell divisions. In the first meiotic division, homologous chromosomes segregate into separate cells, changing the genetic state from diploid to haploid. The second meiotic division is essentially the

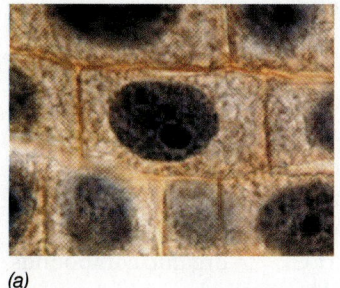   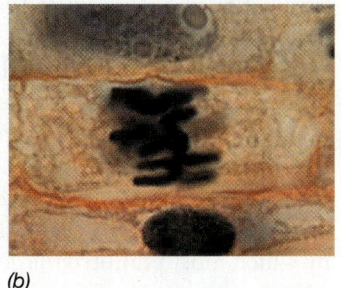   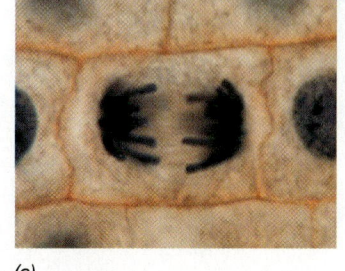   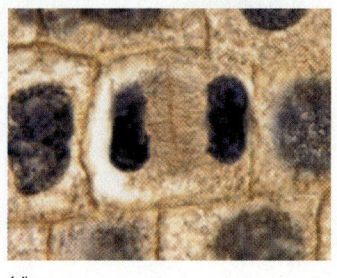

(a)   (b)   (c)   (d)

Figure 2.62 **Light micrograph of plant cells undergoing mitosis.** *(a)* Interphase, distinct chromosomes are not apparent. *(b)* Metaphase. Homologous chromosomes are lining up along the cell center. *(c)* Anaphase. Homologous chromosomes are pulling apart. *(d)* Telophase. Chromosomes have separated into the newly forming daughter cells.

same as mitosis, as the two haploid cells divide to form a total of four haploid cells called *gametes.* In higher organisms these are the eggs and sperm; in eukaryotic microorganisms, they are spores or related structures.

MINIQUIZ --

- How is DNA arranged in eukaryotic chromosomes?
- What are histones and what do they do?
- What are the major differences between mitosis and meiosis?

2.21 Mitochondria, Hydrogenosomes, and Chloroplasts

Organelles that specialize in energy metabolism in eukaryotes include the mitochondrion or hydrogenosome, and in phototrophic eukaryotes, the chloroplast.

Mitochondria

In aerobic eukaryotic cells, respiration occurs in the mitochondrion. **Mitochondria** are of bacterial dimensions and can take on many shapes (**Figure 2.63**). The number of mitochondria per cell depends somewhat on the cell type and size. A yeast cell may have

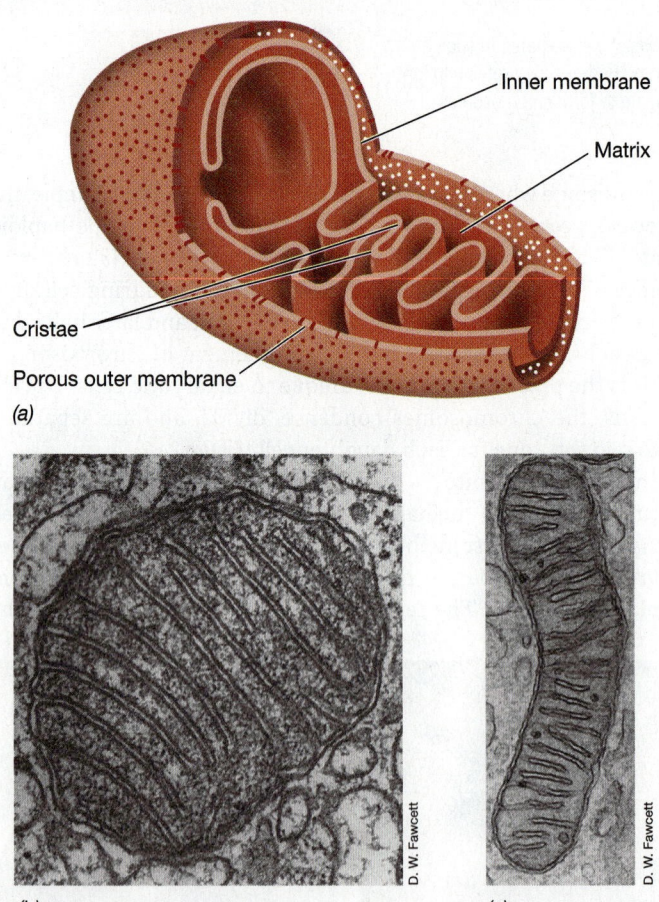

Figure 2.63 Structure of the mitochondrion. *(a)* Diagram showing the overall structure of the mitochondrion; note the inner and outer membranes. *(b, c)* Transmission electron micrographs of mitochondria from rat tissue showing the variability in morphology; note the cristae.

only a few mitochondria per cell (Figures 2.60 and 2.61*a*), while an animal cell may have a thousand. The mitochondrion is enclosed by a double membrane system. Like the nuclear membrane, the outermost mitochondrial membrane is relatively permeable and contains pores that allow the passage of small molecules. The inner membrane is less permeable and its structure more closely resembles that of the cytoplasmic membrane of *Bacteria.*

Mitochondria also contain folded internal membranes called **cristae**. These membranes, formed by invagination of the inner membrane, contain the enzymes needed for respiration and ATP production, the major function of the mitochondrion. Cristae also contain transport proteins that regulate the passage of key molecules such as ATP into and out of the *matrix*, the innermost compartment of the mitochondrion (Figure 2.63*a*). The matrix contains enzymes for the oxidation of organic compounds, in particular, enzymes of the citric acid cycle, the major pathway for the combustion of organic compounds to CO_2 (↩ Section 3.12).

Hydrogenosomes

Some eukaryotic microorganisms are killed by O_2 and, like many prokaryotes, live an anaerobic lifestyle. Such cells lack mitochondria and some of them contain structures called **hydrogenosomes** (**Figure 2.64**). Although similar in size to mitochondria, hydrogenosomes lack citric acid cycle enzymes and also lack cristae. Microbial eukaryotes that contain hydrogenosomes carry out a strictly fermentative metabolism. Examples include the human parasite *Trichomonas* (↩ Sections 17.3 and 32.4) and various protists that inhabit the rumen of ruminant animals (↩ Sections 1.5 and 22.7) or anoxic muds and lake sediments.

The major biochemical reaction in the hydrogenosome is the oxidation of the compound pyruvate to H_2, CO_2, and acetate (Figure 2.64*b*). Some anaerobic eukaryotes have H_2-consuming methanogens in their cytoplasm. These *Archaea* consume the H_2 and CO_2 produced by the hydrogenosome and use it to form methane (CH_4). Because hydrogenosomes cannot respire, they cannot oxidize the acetate produced from pyruvate oxidation as mitochondria do. Acetate is therefore excreted from the hydrogenosome into the cytoplasm of the host cell (Figure 2.64*b*).

Chloroplasts

Chloroplasts are the chlorophyll-containing organelles of phototrophic microbial eukaryotes and function to carry out photosynthesis. Chloroplasts are relatively large and readily visible with the light microscope (**Figure 2.65**), and their number per cell varies among species.

Like mitochondria, chloroplasts have a permeable outer membrane and a much less-permeable inner membrane. The inner membrane surrounds the **stroma**, analogous to the matrix of the mitochondrion (Figure 2.65*c*). The stroma contains the enzyme *ribulose bisphosphate carboxylase* (RubisCO), the key enzyme of the **Calvin cycle**, the series of biosynthetic reactions by which most phototrophs convert CO_2 to organic compounds (↩ Section 13.5). The permeability of the outer chloroplast membrane allows glucose and ATP produced during photosynthesis to diffuse into the cell cytoplasm where they can be used in biosynthesis.

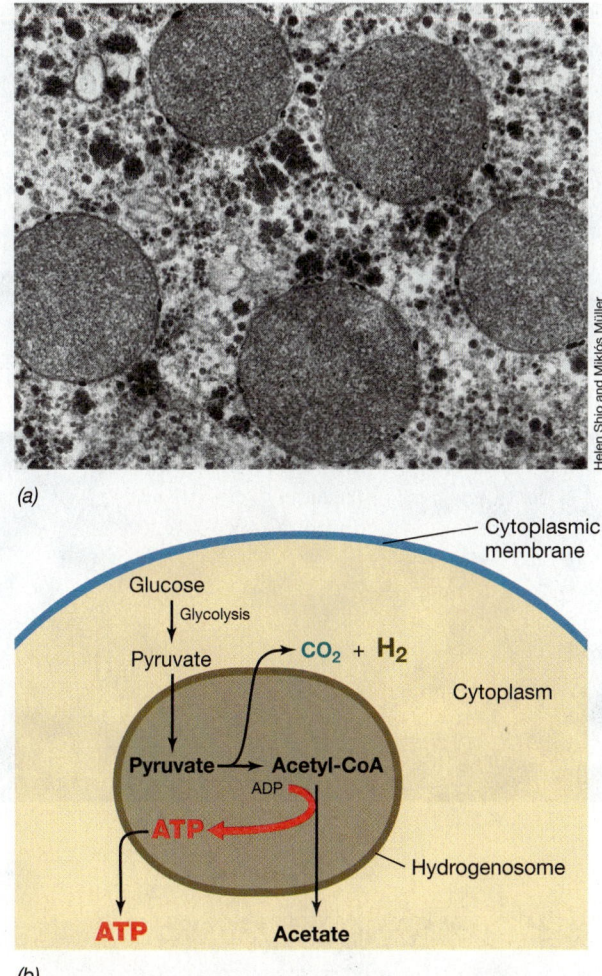

Figure 2.64 **The hydrogenosome.** (a) Electron micrograph of a thin section through a cell of the anaerobic protist *Trichomonas vaginalis* showing five hydrogenosomes in cross section. Compare their internal structure with that of mitochondria in Figure 2.63. (b) Biochemistry of the hydrogenosome. Pyruvate is taken up by the hydrogenosome, and H_2, CO_2, acetate, and ATP are produced.

Chlorophyll and all other components needed for the light reactions of photosynthesis in chloroplasts are located in a series of flattened membrane discs called **thylakoids** (Figure 2.65c). Like the cytoplasmic membrane, the thylakoid membrane is highly impermeable and its major function is to form the light-driven proton motive force (Figure 2.18c) that results in ATP synthesis.

Organelles and Endosymbiosis

On the basis of their relative autonomy, size, and morphological resemblance to bacteria, it was hypothesized over 100 years ago that mitochondria and chloroplasts were descendants of respiratory and photosynthetic bacterial cells, respectively. By associating with nonphototrophic eukaryal hosts, the latter gained a new form of energy metabolism while the symbiotic bacterial cells received a stable and supportive growth environment inside the host. Gradually, over time these originally free-living symbionts became an intimate part of the eukaryotic cell. This idea of symbiotic bacteria as the ancestors of the mitochondrion, hydrogenosome, and chloroplast is called the **endosymbiotic hypothesis**

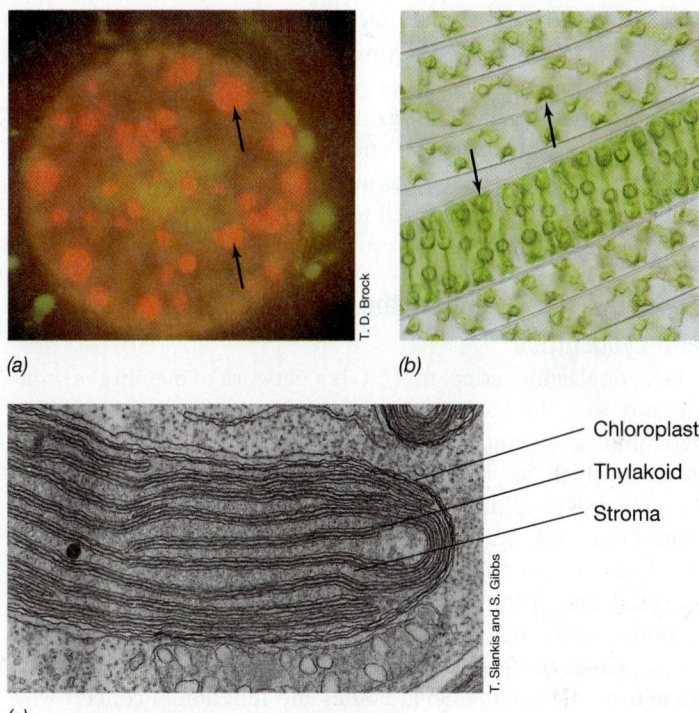

Figure 2.65 **Chloroplasts of a diatom and a green alga cell.** (a) Fluorescence photomicrograph of a diatom shows chlorophyll fluorescence (compare with Figure 2.6); arrows, chloroplasts. The cell is about 40 μm wide. (b) Phase-contrast photomicrograph of the filamentous green alga *Spirogyra* showing the characteristic spiral-shaped chloroplasts (arrows) of this phototroph. A cell is about 20 μm wide. (c) Transmission electron micrograph showing a chloroplast of a diatom; note the thylakoids.

("endo" means "inside") of eukaryotic cell origin (⟲ Sections 12.3 and 17.1) and is now well accepted in biology.

Several lines of evidence support the endosymbiotic hypothesis. These include in particular the fact that mitochondria, hydrogenosomes, and chloroplasts contain their own genomes and ribosomes. The genomes are arranged in a circular fashion as for bacterial chromosomes, and the sequence of genes that encode ribosomal RNA (⟲ Figure 1.6a) from organelles clearly points to their bacterial origin. Thus, the eukaryotic cell is a genetic chimera containing genes from two domains of life: host cell (*Eukarya*) genes and endosymbiont (*Bacteria*) genes.

MINIQUIZ -
- What key reactions occur in the mitochondrion and chloroplast, and what key product is made there?
- Compare and contrast pyruvate metabolism in the mitochondrion and the hydrogenosome.
- What is the endosymbiotic hypothesis and what evidence is there to support it?

2.22 Other Major Eukaryotic Cell Structures

Besides the nucleus and the mitochondrion (or hydrogenosome), and chloroplasts in photosynthetic cells, other cytoplasmic structures may be present in microbial eukaryotes. These include the endoplasmic reticulum, the Golgi complex, lysosomes, a variety of tubular structures, and structures that impart motility. However, unlike mitochondria and chloroplasts, these structures lack DNA and are not of endosymbiotic origin. Cell walls are also present in certain microbial eukaryotes and function as they do in prokaryotic cells to provide shape and protect the cell from osmotic lysis. The exact structure of the cell wall varies with the organism, but various polysaccharides and proteins are commonly observed.

Endoplasmic Reticulum, the Golgi Complex, and Lysosomes

The endoplasmic reticulum (ER) is a network of membranes continuous with the nuclear membrane. Two types of endoplasmic reticulum exist: *rough*, which contains attached ribosomes, and *smooth*, which does not (Figure 2.60). Smooth ER participates in the synthesis of lipids and in some aspects of carbohydrate metabolism. Rough ER, through the activity of its ribosomes, is a major producer of glycoproteins and also produces new membrane material that is transported throughout the cell to enlarge the various membrane systems before cell division.

The Golgi complex is a stack of membranes (**Figure 2.66**) that arise from preexisting Golgi bodies and function in concert with the ER. In the Golgi, products of the ER are chemically modified and sorted into those destined for secretion versus those that will function in other membranous structures in the cell. Many of the modifications are glycosylations (addition of sugar residues) that convert the proteins into various glycoproteins that can then be targeted to specific locations in the cell.

Lysosomes (Figure 2.60) are membrane-enclosed compartments that contain digestive enzymes that hydrolyze proteins, fats, and polysaccharides. The lysosome fuses with food that enters the cell in vacuoles and then releases its digestive enzymes, which break down the foods for biosynthesis and energy generation. Lysosomes also function in hydrolyzing damaged cellular components and recycling these materials for new biosyntheses.

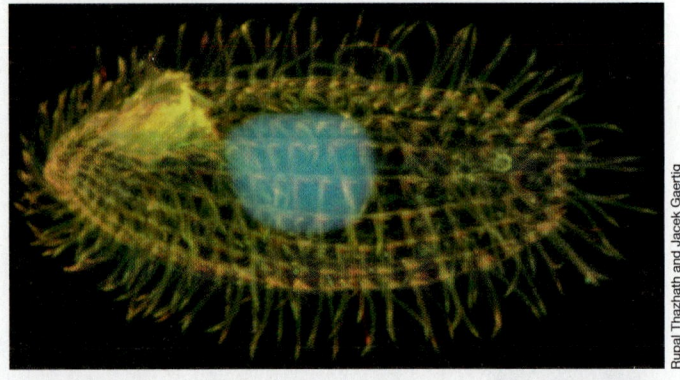

(a)

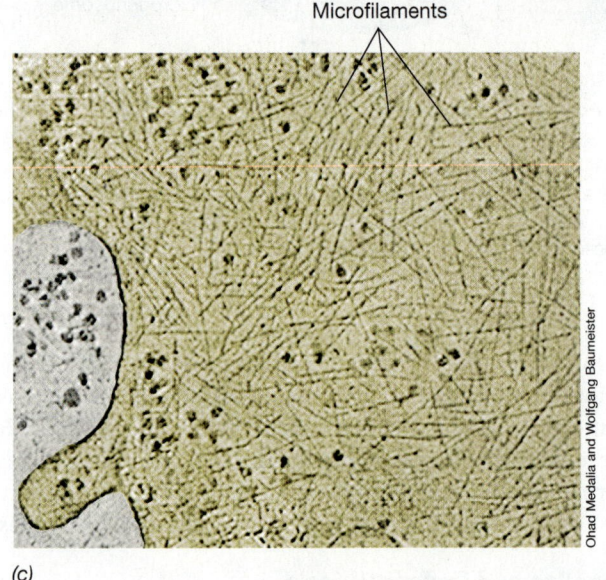

(b)

Microfilaments

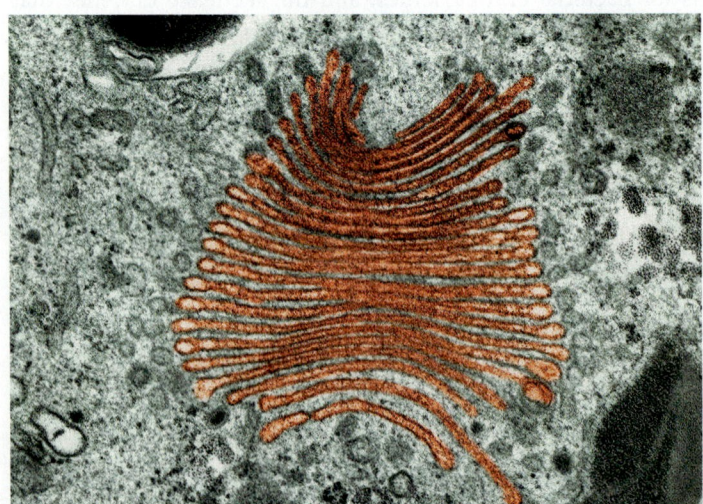

(c)

Figure 2.67 **Tubulin and microfilaments.** *(a)* Fluorescence photomicrograph of a cell of *Tetrahymena* labeled with antitubulin antibodies (red/green) and with DAPI, which stains DNA (blue, nucleus). A cell is about 10 μm wide. *(b)* An animal cell showing the role of tubulin (green) in separating chromosomes during metaphase of mitosis. *(c)* Electron microscopic image of the cellular slime mold *Dictyostelium discoideum* showing the network of actin microfilaments that along with microtubules functions as the cell cytoskeleton. Microfilaments are about 7 nm in diameter. Homologs of tubulin and microfilaments are present in *Bacteria* in the form of the proteins FtsZ and MreB, respectively (⮌ Section 5.3).

Figure 2.66 **The Golgi complex.** Transmission electron micrograph of a portion of an eukaryotic cell showing the Golgi complex (colored in gold). Note the multiple folded membranes of the Golgi complex (membrane stacks are 0.5–1.0 μm in diameter).

The lysosome thus allows the cell's lytic activities to be partitioned away from the cytoplasm proper. Following hydrolysis of macromolecules in the lysosome, the resulting nutrients pass from the lysosome into the cytoplasm for use by cytoplasmic enzymes.

Microtubules, Microfilaments, and Intermediate Filaments

Just as buildings are supported by structural reinforcement, the large size of eukaryotic cells and their ability to move requires structural reinforcement. This internal support network consists of *microtubules, microfilaments,* and *intermediate filaments*; together, these structures form the cell **cytoskeleton** (Figure 2.60).

Microtubules are tubes about 25 nm in diameter containing a hollow core and are composed of the proteins α-*tubulin* and β-*tubulin*. Microtubules have many functions including maintaining cell shape and cell motility by cilia and flagella (**Figure 2.67a**), moving chromosomes during mitosis (Figure 2.67b), and in movement of organelles within the cell. **Microfilaments** (Figure 2.67c) are smaller, about 7 nm in diameter, and are polymers of two intertwined strands of the protein actin. Microfilaments function in maintaining or changing cell shape, in cell motility by cells that move by amoeboid movement, and during cell division. **Intermediate filaments** are fibrous keratin proteins that form into fibers 8–12 nm in diameter and function in maintaining cell shape and positioning organelles in the cell.

Flagella and Cilia

Flagella and cilia are present on many eukaryotic microorganisms and function as organelles of motility, allowing cells to move by swimming. Motility has survival value, as the ability to move allows motile organisms to move about their habitat and exploit new resources. *Cilia* are essentially short flagella that beat in synchrony to propel the cell—usually quite rapidly—through the medium. *Flagella*, by contrast, are long appendages present singly or in groups that propel the cell along—typically more slowly than by cilia—through a whiplike motion (**Figure 2.68a**). The flagella of

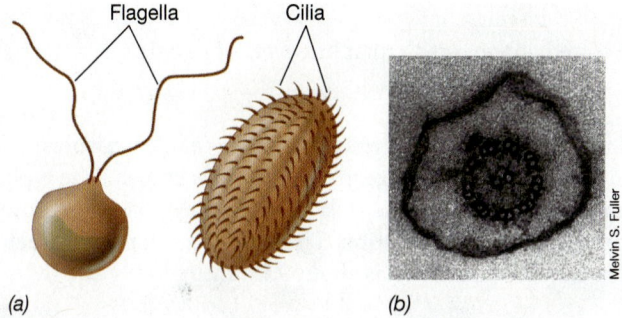

Figure 2.68 **Motility organelles in eukaryotic cells: Flagella and cilia.** *(a)* Flagella can be present as single or multiple filaments. Cilia are structurally very similar to flagella but much shorter. Eukaryotic flagella move in a whiplike motion. *(b)* Cross section through a flagellum of the fungus *Blastocladiella* showing the outer sheath, the outer nine pairs of microtubules, and the central pair of microtubules.

eukaryotic cells are structurally quite distinct from bacterial flagella and do not rotate as do bacterial flagella (Section 2.17).

In cross section, cilia and flagella appear similar. Each contains a bundle of nine pairs of microtubules surrounding a central pair of microtubules (Figure 2.68b). A protein called *dynein* is attached to the microtubules and uses ATP to drive motility. Movement of flagella and cilia is similar. In both cases, movement is the result of the coordinated sliding of microtubules against one another in a direction toward or away from the base of the cell. This movement confers the whiplike motion on the flagellum or cilium that results in cell propulsion.

MINIQUIZ

- Why are the activities in the lysosome best partitioned away from the cytoplasm proper?
- How is the cell's cytoskeleton held together?
- From a functional standpoint, how does the flagellum of eukaryotic and prokaryotic cells differ?

BIG IDEAS

2.1 • Microscopes are essential for studying microorganisms. Bright-field microscopy, the most common form of microscopy, employs a microscope with a series of lenses to magnify and resolve the image.

2.2 • An inherent limitation of bright-field microscopy is the lack of contrast between cells and their surroundings. This problem can be overcome by the use of stains or by alternative forms of light microscopy, such as phase contrast or dark field.

2.3 • Differential interference contrast (DIC) microscopy and confocal scanning laser microscopy allow enhanced

three-dimensional imaging or imaging through thick specimens.

2.4 • Electron microscopes have far greater resolving power than do light microscopes, the limits of resolution being about 0.2 nm. The two major forms of electron microscopy are transmission, used primarily to observe internal cell structure, and scanning, used to examine the surface of specimens.

2.5 • Prokaryotic cells can have many different shapes; rods, cocci, and spirilla are common cell morphologies. Morphology is a poor predictor of other cell properties and is a genetically directed characteristic that has evolved to best serve the ecology of the cell.

2.6 • Prokaryotes are typically smaller than eukaryotes, although some very large prokaryotes are known. The typical small size of prokaryotic cells affects their physiology, growth rate, ecology, and evolution. The lower limit for the diameter of a coccus-shaped cell is about 0.15 μm.

2.7 • The cytoplasmic membrane is a highly selective permeability barrier constructed of lipids and proteins that form a bilayer, hydrophobic inside and hydrophilic outside. In contrast to *Bacteria* and *Eukarya*, where fatty acids are ester-linked to glycerol, *Archaea* contain ether-linked lipids and some form monolayer instead of bilayer membranes.

2.8 • The major functions of the cytoplasmic membrane are permeability, transport, and energy conservation. To accumulate nutrients against the concentration gradient, transport systems are used that are characterized by their specificity and saturation effect.

2.9 • At least three types of nutrient transporters are known: simple, group translocation, and ABC systems. Transport requires energy from either an energy-rich compound such as ATP or from the proton motive force in order to accumulate solutes against the concentration gradient.

2.10 • Peptidoglycan is a polysaccharide found only in *Bacteria* that consists of an alternating repeat of *N*-acetylglucosamine and *N*-acetylmuramic acid, the latter cross-linked by tetrapeptides in adjacent strands. The enzyme lysozyme and the antibiotic penicillin both target peptidoglycan, leading to cell lysis.

2.11 • Gram-negative *Bacteria* have an outer membrane consisting of LPS, protein, and lipoprotein. Porins allow for permeability across the outer membrane. The gap between the outer and cytoplasmic membranes is called the periplasm and contains proteins involved in transport, sensing chemicals, and other important cell functions.

2.12 • Cell walls of *Archaea* are of several types, including pseudomurein, various polysaccharides, and S-layers, which are composed of protein or glycoprotein. As for *Bacteria*, the walls of *Archaea* protect the cell from osmotic lysis.

2.13 • Many prokaryotic cells contain capsules, slime layers, pili, or fimbriae. These structures have several functions, including attachment, genetic exchange, and twitching motility.

2.14 • Prokaryotic cells can contain inclusions of sulfur, polyphosphate, or carbon polymers, or minerals that form magnetic particles (magnetosomes). These substances function as storage materials or in magnetotaxis.

2.15 • Gas vesicles are gas-filled structures that confer buoyancy on cells. Gas vesicles are composed of two different proteins arranged to form a gas-permeable but watertight structure.

2.16 • The endospore is a highly resistant and differentiated bacterial cell produced by certain gram-positive *Bacteria*. Endospores are dehydrated and contain calcium dipicolinate and small acid-soluble proteins, absent from vegetative cells. Endospores can remain dormant indefinitely but can germinate quickly when conditions warrant.

2.17 • Swimming motility is due to flagella. The flagellum is composed of several proteins and is anchored in the cell wall and cytoplasmic membrane. In *Bacteria*, the flagellum filament is made of the protein flagellin and rotates at the expense of the proton motive force. The flagella of *Archaea* and *Bacteria* differ in structure and how energy is coupled to rotation.

2.18 • Bacteria that move by gliding do not employ rotating flagella but instead creep along a solid surface by employing one of several different mechanisms including polysaccharide excretion, twitching, or rotating glide proteins.

2.19 • Motile bacteria respond to chemical and physical gradients in their environment by controlling the lengths of runs and frequency of tumbles. Tumbles are controlled by the direction of rotation of the flagellum, which in turn is controlled by a network of sensory and response proteins.

2.20 • Microbial eukaryotes contain various organelles including the nucleus, which is universal; mitochondria or hydrogenosomes; and chloroplasts. The nucleus contains the cell's chromosomes in the form of linear DNA wrapped around histone proteins. Microbial eukaryotes divide following the process of mitosis and may undergo meiosis as well if a haploid/diploid life cycle occurs.

2.21 • The mitochondrion and hydrogenosome are energy-generating organelles of eukaryotic cells. Mitochondria carry out aerobic respiration, whereas hydrogenosomes ferment pyruvate to H_2, CO_2, and acetate. Chloroplasts are the site of photosynthetic energy production and CO_2 fixation in eukaryotic cells. These organelles were originally free-living *Bacteria* that established permanent residence inside cells of *Eukarya* (endosymbiosis).

2.22 • Endoplasmic reticula are membranous structures in eukaryotes that either contain attached ribosomes (rough ER) or do not (smooth ER). Flagella and cilia are means of motility while lysosomes specialize in breaking down large molecules. Microtubules, microfilaments, and intermediate filaments function as internal cell scaffolds.

KEY TERMS

ABC (ATP-binding cassette) transport system a membrane transport system consisting of three proteins, one of which hydrolyzes ATP; the system transports specific nutrients into the cell

Basal body the "motor" portion of the bacterial flagellum, embedded in the cytoplasmic membrane and wall

Calvin cycle the series of biosynthetic reactions by which most photosynthetic organisms convert CO_2 to organic compounds

Capsule a polysaccharide or protein outermost layer, usually rather slimy, present on some bacteria

Chemotaxis directed movement of an organism toward (positive chemotaxis) or away from (negative chemotaxis) a chemical gradient

Chloroplast the photosynthetic organelle of phototrophic eukaryotes

Cristae the internal membranes of a mitochondrion

Cytoplasmic membrane the permeability barrier of the cell, separating the cytoplasm from the environment

Cytoskeleton the cellular scaffolding typical of eukaryotic cells in which microtubules, microfilaments, and intermediate filaments define the cell's shape

Dipicolinic acid a substance unique to endospores that confers heat resistance on these structures

Endospore a highly heat-resistant, thick-walled, differentiated structure produced by certain gram-positive *Bacteria*

Endosymbiotic hypothesis the idea that mitochondria and chloroplasts originated from *Bacteria*

Flagellum a long, thin cellular appendage that rotates (in prokaryotic cells) and is responsible for swimming motility

Gas vesicles gas-filled cytoplasmic structures bounded by protein and conferring buoyancy on cells

Gram-negative a bacterial cell with a cell wall containing small amounts of peptidoglycan, and an outer membrane containing lipopolysaccharide, lipoprotein, and other complex macromolecules

Gram-positive a bacterial cell whose cell wall consists chiefly of peptidoglycan; it lacks the outer membrane of gram-negative cells

Gram stain a differential staining procedure that stains cells either purple (gram-positive cells) or pink (gram-negative cells)

Group translocation an energy-dependent transport system in which the substance transported is chemically modified during the process of being transported by a series of proteins

Histones highly basic proteins that compact and wind DNA in the nucleus of eukaryotic cells

Hydrogenosome an organelle of endosymbiotic origin present in certain microbial eukaryotes that oxidizes pyruvate to H_2, CO_2, acetate, and couples this to ATP synthesis

Intermediate filament a filamentous polymer of fibrous keratin proteins, supercoiled into thicker fibers, that functions in maintaining cell shape and the positioning of certain organelles in the eukaryotic cell

Lipopolysaccharide (LPS) a combination of lipid with polysaccharide and protein that forms the major portion of the outer membrane in gram-negative *Bacteria*

Lysosome an organelle containing digestive enzymes for hydrolysis of proteins, fats, and polysaccharides

Magnetosome a particle of magnetite (Fe_3O_4) enclosed by a nonunit membrane in the cytoplasm of magnetotactic *Bacteria*

Meiosis the nuclear division that halves the diploid number of chromosomes to the haploid

Microfilament a filamentous polymer of the protein actin that helps maintain the shape of a eukaryotic cell

Microtubule a filamentous polymer of the proteins α-tubulin and β-tubulin that functions in eukaryotic cell shape and motility

Mitochondrion the respiratory organelle of eukaryotic organisms

Mitosis nuclear division in eukaryotic cells in which chromosomes are replicated and partitioned into two daughter cells during cell division

Morphology the *shape* of a cell—rod, coccus, spirillum, and so on

Nucleus the organelle that contains the eukaryotic cell's chromosomes

Outer membrane a phospholipid- and polysaccharide-containing unit membrane that lies external to the peptidoglycan layer in cells of gram-negative *Bacteria*

Peptidoglycan a polysaccharide composed of alternating repeats of *N*-acetylglucosamine and *N*-acetylmuramic acid arranged in adjacent layers and cross-linked by short peptides

Periplasm a gel-like region between the outer surface of the cytoplasmic membrane and the inner surface of the lipopolysaccharide layer of gram-negative *Bacteria*

Peritrichous flagellation having flagella located in many places around the surface of the cell

Phototaxis movement of an organism toward light

Pili thin, filamentous structures that extend from the surface of a cell and, depending on type, facilitate cell attachment, genetic exchange, or twitching motility

Polar flagellation having flagella emanating from one or both poles of the cell

Poly-β-hydroxybutyric acid (PHB) a common storage material of prokaryotic cells consisting of a polymer of β-hydroxybutyrate or another β-alkanoic acid or mixtures of β-alkanoic acids

Resolution the ability to distinguish two objects as distinct and separate when viewed under the microscope

S-layer an outermost cell surface layer composed of protein or glycoprotein present on some *Bacteria* and *Archaea*

Simple transport system a transporter that consists of only a membrane-spanning protein and is typically driven by energy from the proton motive force

Stroma the lumen of the chloroplast, surrounded by the inner membrane

Teichoic acid a phosphorylated polyalcohol found in the cell wall of some gram-positive *Bacteria*

Thylakoid a membrane layer containing the photosynthetic pigments in chloroplasts

REVIEW QUESTIONS

1. What is the difference between magnification and resolution? Can either increase without the other? (Section 2.1)

2. What is the function of staining in light microscopy? What is the advantage of phase-contrast microscopy over bright-field microscopy? What is the advantage of DIC microscopy over bright-field microscopy? (Sections 2.2 and 2.3)

3. What is the major advantage of electron microscopes over light microscopes? What type of electron microscope would be used to view the three-dimensional features of a cell? (Section 2.4)

4. What are the major morphologies of prokaryotes? Draw cells for each morphology you list. (Section 2.5)

5. How large can a prokaryote be? How small? Why is it that we likely know the lower limit more accurately than the upper limit? What are the dimensions of the rod-shaped bacterium *Escherichia coli*? (Section 2.6)

6. Describe in a single sentence the structure of a unit membrane. (Section 2.7)

7. Describe the major structural differences between membranes of *Bacteria* and *Archaea*. (Section 2.7)

8. Explain in a single sentence why ionized molecules do not passively move through the cytoplasmic membrane. How do such molecules get through the cytoplasmic membrane? (Section 2.8)

9. Cells of *Escherichia coli* take up lactose via lac permease, glucose via the phosphotransferase system, and maltose via an ABC-type transporter. For each of these sugars describe: (1) the components of the transport system and (2) the source of energy that drives the transport event. (Section 2.9)

10. Why is the rigid layer of the bacterial cell wall called peptidoglycan? What are the structural reasons for the rigidity that is conferred on the cell wall by the peptidoglycan structure? (Section 2.10)

11. List several functions of the outer membrane in gram-negative *Bacteria*. What is the chemical composition of the outer membrane? (Section 2.11)

12. What cell wall polysaccharide common in *Bacteria* is absent from *Archaea*? What is unusual about S-layers compared to other cell walls of prokaryotes? What types of cell walls are found in *Archaea*? (Section 2.12)

13. What function(s) do polysaccharide layers outside the cell wall have in prokaryotes? (Section 2.13)

14. What types of cytoplasmic inclusions are formed by prokaryotes? How does an inclusion of poly-β-hydroxybutyric acid differ from a magnetosome in composition and metabolic role? (Section 2.14)

15. What is the function of gas vesicles? How are these structures made such that they can remain gas tight? (Section 2.15)

16. In a few sentences, indicate how the bacterial endospore differs from the vegetative cell in structure, chemical composition, and ability to resist extreme environmental conditions. (Section 2.16)

17. Define the following terms: mature endospore, vegetative cell, and germination. (Section 2.16)

18. Describe the structure and function of a bacterial flagellum. What is the energy source for the flagellum? How do the flagella of *Bacteria* differ from those of *Archaea* in both size and composition? (Section 2.17)

19. Contrast the mechanism for motility in *Flavobacterium* from that in *Escherichia coli*. (Sections 2.17 and 2.18)

20. In a few sentences, explain how a motile bacterium is able to sense the direction of an attractant and move toward it. (Section 2.19)

21. In the experiment described in Figure 2.58, what is the control and why is it essential? (Section 2.19)

22. List at least three features of eukaryotic cells that clearly differentiate them from prokaryotic cells. What are histones and what do they do? (Section 2.20)

23. How are the mitochondrion and the hydrogenosome similar structurally? How do they differ? How do they differ metabolically? (Section 2.21)

24. What major physiological processes occur in the chloroplast? (Section 2.21)

25. What evidence exists that supports the idea that the major organelles of eukaryotes were once *Bacteria*? (Section 2.21)

26. What are the functions of the following eukaryotic cell structures: endoplasmic reticulum, Golgi complex, and lysosomes? (Section 2.22)

APPLICATION QUESTIONS

1. Calculate the size of the smallest resolvable object if 600-nm (red) light is used to observe a specimen with a 100× oil-immersion lens having a numerical aperture of 1.32. How could resolution be improved using this same lens?

2. Calculate the surface-to-volume ratio of a spherical cell 15 μm in diameter and of a cell 2 μm in diameter. What are the consequences of these differences in surface-to-volume ratio for cell function?

3. Assume you are given two cultures, one of a species of gram-negative *Bacteria* and one of a species of *Archaea*. Discuss at least four different ways you could tell which culture was which.

4. Calculate the amount of time it would take a cell of *Escherichia coli* (1 × 2 μm) swimming at maximum speed (60 cell lengths per second) to travel all the way up a 3-cm-long capillary tube containing a chemical attractant.

5. Assume you are given two cultures of rod-shaped bacteria, one gram-positive and the other gram-negative. How could you differentiate them using (a) light microscopy; (b) electron microscopy; (c) chemical analyses of cell walls; and (d) phylogenetic analyses?

3 · Microbial Metabolism

microbiology**now**

A Metabolic Surprise

Students often get the impression that when it comes to metabolism, everything is known and nothing new will ever be discovered. This is especially so when they study classical metabolic pathways such as the citric acid cycle (Krebs cycle), an important series of reactions found in all cells and whose details will be covered in this chapter. It is just "another boring metabolic pathway" whose biochemistry was worked out years ago, right?

Wrong. For years, microbiologists have been baffled by the absence of two key enzymes of the citric acid cycle (CAC) in certain prokaryotes, in particular, the cyanobacteria. Cyanobacteria (photo) are oxygenic phototrophs whose photosynthetic activities oxygenated Earth billions of years ago and made the evolution of higher life forms possible. But the absence from cyanobacteria of the CAC enzymes α-ketoglutarate dehydrogenase and succinyl-CoA synthetase (enzymes that work in concert to convert α-ketoglutarate into succinate) has long pegged these organisms as "having an incomplete CAC." Is this really true?

A group of microbiologists at Penn State University (USA) reinvestigated this puzzling situation and, using a combination of genomics and biochemistry, discovered a new CAC paradigm.[1] It turns out that cyanobacteria *do* run a complete CAC, but convert α-ketoglutarate to succinate using two novel enzymes previously unknown to biology. For some reason(s), evolution has selected these enzymes rather than the canonical ones to complete the CAC in cyanobacteria and a few other prokaryotes where genes encoding these enzymes have also been discovered.

Besides solving a major metabolic mystery, this research shows the power of combining genomic analyses with biochemistry and good scientific intuition. The study also reminds us of the importance of understanding classic metabolisms as a basis for discovering new metabolisms in the microbial world.

[1]Zhang, S., and D.A. Bryant. 2011. The tricarboxylic acid cycle in cyanobacteria. *Science 334:* 1551–1553.

I • Laboratory Culture of Microorganisms

To cultivate microorganisms in the laboratory it is necessary to supply them with all of the nutrients they require. Nutrient requirements vary widely, and knowledge of the principles of microbial nutrition is necessary for the successful culture of microorganisms. We focus here on some general principles of microbial nutrition and then expand on this in Chapter 13 where the broad metabolic diversity of the microbial world will unfold.

3.1 Cell Chemistry and Nutrition

Different organisms need different complements of nutrients, and not all nutrients are required in the same amounts. Some nutrients, called *macronutrients*, are required in large amounts, while others, called *micronutrients*, are required in just trace amounts.

Chemical Makeup of a Cell

All microbial nutrients are compounds constructed from the chemical elements. However, just a handful of elements dominate living systems and are essential: hydrogen (H), oxygen (O), carbon (C), nitrogen (N), phosphorus (P), sulfur (S), and selenium (Se). In addition to these, at least 50 other elements, although not required, are metabolized in some way by microorganisms (Figure 3.1).

Besides water, which makes up 70–80% of the wet weight of a microbial cell (a single cell of *Escherichia coli* weighs just 10^{-12} g), cells consist primarily of macromolecules—proteins, nucleic acids, lipids, and polysaccharides; the building blocks (monomers) of these macromolecules are the amino acids, nucleotides, fatty acids, and sugars, respectively. Proteins dominate the macromolecular composition of a cell, making up 55% of total cell dry weight. Moreover, the diversity of proteins exceeds that of all other macromolecules combined. Interestingly, as important as DNA is to a cell, it contributes a very small percentage of a cell's dry weight; RNA is far more abundant (Figure 3.1c).

The data shown in Figure 3.1 are from actual analyses of cells of *E. coli*; comparable data vary a bit from one microorganism to the next. But in any microbial cell, carbon and nitrogen are important

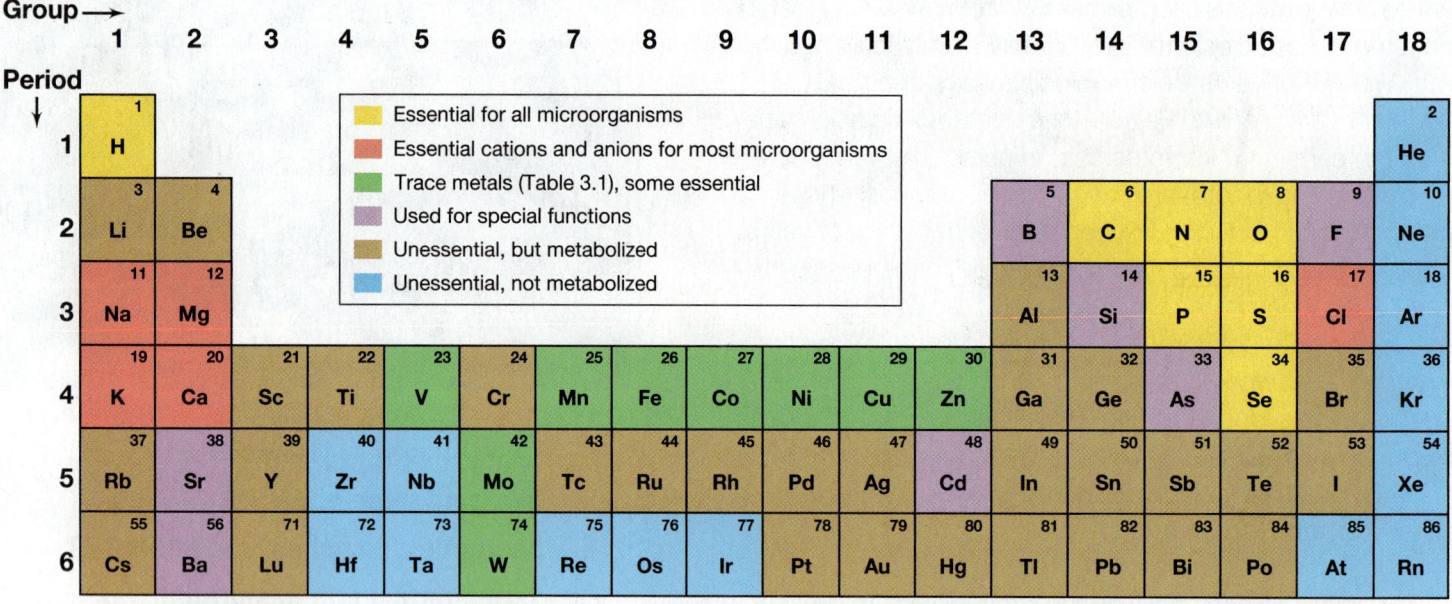

(a)

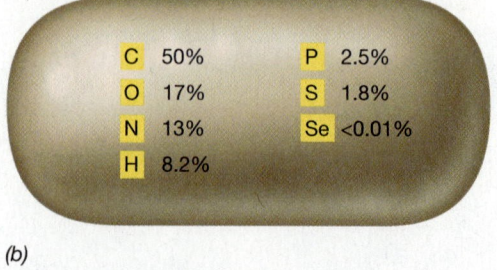

(b)

Essential elements as a percent of cell dry weight

C	50%	P	2.5%
O	17%	S	1.8%
N	13%	Se	<0.01%
H	8.2%		

Macromolecular composition of a cell

Macromolecule	Percent of dry weight
Protein	55
Lipid	9.1
Polysaccharide	5.0
Lipopolysaccharide	3.4
DNA	3.1
RNA	20.5

(c)

Figure 3.1 Elemental and macromolecular composition of a bacterial cell. *(a)* A microbial periodic table of the elements. With the exception of uranium, which can be metabolized by some prokaryotes, elements in period 7 or beyond in the complete periodic table of the elements are not known to be metabolized. *(b)* Contributions of the essential elements to cell dry weight. *(c)* Relative abundance of macromolecules in a bacterial cell. Data in *(b)* from *Aquat. Microb. Ecol. 10:* 15–27 (1996) and in *(c)* from Escherichia coli *and* Salmonella typhimurium: *Cellular and Molecular Biology.* ASM, Washington, DC (1996).

macronutrients, and thus we begin our study of microbial nutrition with these key elements.

Carbon, Nitrogen, and Other Macronutrients

All cells require carbon, and most prokaryotes require *organic* (carbon-containing) compounds as their source of carbon. About 50% of the dry weight of a bacterial cell is carbon (Figure 3.1*b*). Carbon is obtained from amino acids, fatty acids, organic acids, sugars, nitrogen bases, aromatic compounds, and countless other organic compounds. Some microorganisms are autotrophic and these organisms build their cellular structures from carbon dioxide (CO_2).

A bacterial cell is about 13% nitrogen, which is present in proteins, nucleic acids, and several other cell constituents. The bulk of nitrogen available in nature is as ammonia (NH_3), nitrate (NO_3^-), or nitrogen gas (N_2). Virtually all prokaryotes can use NH_3 as their nitrogen source, many can also use NO_3^-, and some can use organic nitrogen sources, such as amino acids. N_2 can only be used as an N source by nitrogen-fixing prokaryotes (Section 3.17).

In addition to C, N, and O and H (from H_2O), many other macronutrients are needed by cells but typically in smaller amounts (Figure 3.1*b*). Phosphorus is required for nucleic acids and phospholipids and is usually supplied to a cell as phosphate (PO_4^{2-}). Sulfur is present in the amino acids cysteine and methionine and also in several vitamins, including thiamine, biotin, and lipoic acid and is commonly supplied to cells as sulfate (SO_4^{2-}). Potassium (K) is required for the activity of several enzymes, whereas magnesium (Mg) is required to stabilize ribosomes, membranes, and nucleic acids and is also required for the activity of many enzymes. Calcium (Ca) and sodium (Na) are essential nutrients for only a few organisms, sodium in particular for marine microorganisms.

Micronutrients: Trace Metals and Growth Factors

Microorganisms require several metals for growth, typically in very small amounts, and these are part of their *micronutrient* requirements (Figure 3.1*a*). Chief among the required metals is iron (Fe), which plays a major role in cellular respiration. Iron is a key component of cytochromes and of iron–sulfur proteins that function in electron transport reactions (Section 3.10). Besides iron, many other metals are required or otherwise metabolized by microorganisms (Figure 3.1*a*). Collectively these micronutrients are called *trace elements* or *trace metals*. Trace elements typically function as cofactors for enzymes. **Table 3.1** lists the major trace elements and examples of enzymes in which each plays a role.

Growth factors are organic micronutrients (Table 3.1). Common growth factors include the vitamins, but amino acids, purines, pyrimidines, or various other organic molecules may be growth factors for one or another microorganism. Vitamins are the most commonly required growth factors and a few common ones are included in Table 3.1. Most vitamins function as coenzymes, which are nonprotein components of enzymes (Section 3.5). Vitamin requirements vary among microorganisms, ranging from none to several. Lactic acid bacteria, which include the genera *Streptococcus*, *Lactobacillus*, and *Leuconostoc* (⇆ Section 15.6), are notorious for their many vitamin requirements, which are even more extensive than those of humans (see Table 3.2).

Table 3.1 Micronutrients needed by microorganisms[a]

I. Trace Elements		II. Growth Factors	
Element	**Function**	**Growth factor**	**Function**
Boron (B)	Autoinducer for quorum sensing in bacteria; also found in some polyketide antibiotics	PABA (*p*-aminobenzoic acid)	Precursor of folic acid
Cobalt (Co)	Vitamin B_{12}; transcarboxylase (only in propionic acid bacteria)	Folic acid	One-carbon metabolism; methyltransfers
Copper (Cu)	In respiration, cytochrome *c* oxidase; in photosynthesis, plastocyanin, some superoxide dismutases	Biotin	Fatty acid biosynthesis; some CO_2 fixation reactions
Iron (Fe)[b]	Cytochromes; catalases; peroxidases; iron–sulfur proteins; oxygenases; all nitrogenases	B_{12} (Cobalamin)	One-carbon metabolism; synthesis of deoxyribose
Manganese (Mn)	Activator of many enzymes; component of certain superoxide dismutases and of the water-splitting enzyme in oxygenic phototrophs (photosystem II)	B_1 (Thiamine)	Decarboxylation reactions
Molybdenum (Mo)	Certain flavin-containing enzymes; some nitrogenases, nitrate reductases, sulfite oxidases, DMSO–TMAO reductases; some formate dehydrogenases	B_6 (Pyridoxal)	Amino acid/keto acid transformations
Nickel (Ni)	Most hydrogenases; coenzyme F_{430} of methanogens; carbon monoxide dehydrogenase; urease	Nicotinic acid (Niacin)	Precursor of NAD^+
		Riboflavin	Precursor of FMN, FAD
Selenium (Se)	Formate dehydrogenase; some hydrogenases; the amino acid selenocysteine	Pantothenic acid	Precursor of coenzyme A
Tungsten (W)	Some formate dehydrogenases; oxotransferases of hyperthermophiles	Lipoic acid	Decarboxylation of pyruvate and α-ketoglutarate
Vanadium (V)	Vanadium nitrogenase; bromoperoxidase	Vitamin K	Electron transport
Zinc (Zn)	Carbonic anhydrase; nucleic acid polymerases; many DNA-binding proteins		

[a]Not all trace elements or growth factors are needed by all organisms.
[b]Iron is typically needed in larger amounts than the other trace metals shown.

3.2 Media and Laboratory Culture

A **culture medium** is a nutrient solution used to grow microorganisms. Because laboratory culture is required for the detailed study of any microorganism, careful attention must be paid to the selection and preparation of media for laboratory culture to be successful.

Classes of Culture Media

Two broad classes of culture media are used in microbiology: defined media and complex media. **Defined media** are prepared by adding precise amounts of pure inorganic or organic chemicals to distilled water. Therefore, the *exact composition* of a defined medium (in both a qualitative and quantitative sense) is known. Of major importance in any culture medium is the carbon source because all cells need large amounts of carbon to make new cell material (Figure 3.1). The particular carbon source and its concentration depend on the organism to be cultured. **Table 3.2**

lists recipes for four culture media. Some defined media, such as the one listed for *Escherichia coli*, are considered "simple" because they contain only a single carbon source. In this medium, cells of *E. coli* make all organic molecules from glucose.

For culturing many microorganisms, knowledge of the exact composition of a medium is not essential. In these instances complex media may suffice and may even be advantageous. **Complex media** are made from digests of microbial, animal, or plant products, such as casein (milk protein), beef (beef extract), soybeans (tryptic soy broth), yeast cells (yeast extract), or any of a number of other highly nutritious substances. Such digests are commercially available in dehydrated form and need only be hydrated to form a culture medium. However, the disadvantage of a complex medium is that its nutritional composition is not known precisely. An *enriched medium*, used for the culture of nutritionally demanding (fastidious) microorganisms, many of which are pathogens, starts as a complex medium and then is embellished with additional highly nutritious substances such as serum or blood.

Culture media are sometimes prepared that are selective or differential (or both), especially media used in diagnostic microbiology. A *selective medium* contains compounds that inhibit the growth of some microorganisms but not others. For example, selective media are available for the isolation of certain pathogens, such as strains of *Salmonella* or *Escherichia coli* that cause foodborne illnesses. A *differential medium* is one in which an indicator, typically a dye, is added, which reveals by a color change whether a

Table 3.2 Examples of culture media for microorganisms with simple and demanding nutritional requirements[a]

Defined culture medium for Escherichia coli	Defined culture medium for Leuconostoc mesenteroides	Complex culture medium for either E. coli or L. mesenteroides	Defined culture medium for Thiobacillus thioparus
K_2HPO_4 7 g	K_2HPO_4 0.6 g	Glucose 15 g	KH_2PO_4 0.5 g
KH_2PO_4 2 g	KH_2PO_4 0.6 g	Yeast extract 5 g	NH_4Cl 0.5 g
$(NH_4)_2SO_4$ 1 g	NH_4Cl 3 g	Peptone 5 g	$MgSO_4$ 0.1 g
$MgSO_4$ 0.1 g	$MgSO_4$ 0.1 g	KH_2PO_4 2 g	$CaCl_2$ 0.05 g
$CaCl_2$ 0.02 g	Glucose 25 g	Distilled water 1000 ml	KCl 0.5 g
Glucose 4–10 g	Sodium acetate 25 g	pH 7	$Na_2S_2O_3$ 2 g
Trace elements (Fe, Co, Mn, Zn, Cu, Ni, Mo) 2–10 μg each	Amino acids (alanine, arginine, asparagine, aspartate, cysteine, glutamate, glutamine, glycine, histidine, isoleucine, leucine, lysine, methionine, phenylalanine, proline, serine, threonine, tryptophan, tyrosine, valine) 100–200 μg of each		Trace elements (as in first column)
Distilled water 1000 ml			Distilled water 1000 ml
pH 7	Purines and pyrimidines (adenine, guanine, uracil, xanthine) 10 mg of each		pH 7
	Vitamins (biotin, folate, nicotinic acid, pyridoxal, pyridoxamine, pyridoxine, riboflavin, thiamine, pantothenate, *p*-aminobenzoic acid) 0.01–1 mg of each		Carbon source: CO_2 from air
	Trace elements (as in first column) 2–10 μg each		
	Distilled water 1000 ml		
	pH 7		

(a)

(b)

[a]The photos are tubes of (a) the defined medium described, and (b) the complex medium described. Note how the complex medium is colored from the various organic extracts and digests that it contains. Photo credits: Cheryl L. Broadie and John Vercillo, Southern Illinois University at Carbondale.

particular metabolic reaction has occurred during growth. Differential media are useful for distinguishing bacteria and are widely used in clinical diagnostics and systematic microbiology. Differential and selective media are further discussed in Chapter 27.

Nutritional Requirements and Biosynthetic Capacity

Of the four recipes in Table 3.2, three are defined and one is complex. The complex medium is easiest to prepare and supports growth of both *Escherichia coli* and *Leuconostoc mesenteroides*, the examples used in the table. By contrast, the simple defined medium supports growth of *E. coli* but not of *L. mesenteroides*. Growth of the latter in a defined medium requires the addition of several nutrients not needed by *E. coli*. The nutritional needs of *L. mesenteroides* can be satisfied by preparing either a highly supplemented defined medium, a rather laborious undertaking because of all the individual nutrients that need to be added (Table 3.2), or by preparing a complex medium, a much less demanding operation.

The fourth medium listed in Table 3.2 supports growth of the sulfur bacterium *Thiobacillus thioparus*; this medium would not support growth of any of the other organisms. This is because *T. thioparus* is both a chemolithotroph and an autotroph and thus has no organic carbon requirements. *T. thioparus* derives all of its carbon from CO_2 and obtains its energy from the oxidation of the sulfur compound thiosulfate ($Na_2S_2O_3$). Thus, *T. thioparus* has the greatest biosynthetic capacity of all the organisms listed in the table, surpassing even *E. coli* in this regard.

The take-home lesson from Table 3.2 is that different microorganisms can have vastly different nutritional requirements. For successful cultivation, it is necessary to understand an organism's nutritional requirements and then supply it with the nutrients it needs in both the proper form and amount.

Laboratory Culture

Once a culture medium has been prepared and made sterile to render it free of all life forms, organisms can be inoculated and the culture can be incubated under conditions that will support growth (**Figure 3.2**). In a laboratory situation, inoculation will typically be with a pure culture and into either liquid or solid culture medium. Liquid culture media are solidified with agar, typically at 1–2%. Solid media immobilize cells, allowing them to grow and form visible, isolated masses called *colonies* (Figure 3.2). Microbial colonies are of various shapes and sizes depending on the organism, the culture conditions, the nutrient supply, and other physiological parameters. Some microorganisms produce pigments that cause the entire colony to be colored (Figure 3.2). Colonies permit the microbiologist to visualize the composition and presumptive purity of a culture. Plates that contain more than one kind of colony signal a contaminated culture.

Culture media must be sterilized before use, and sterilization is achieved by heating the medium in an *autoclave*. We

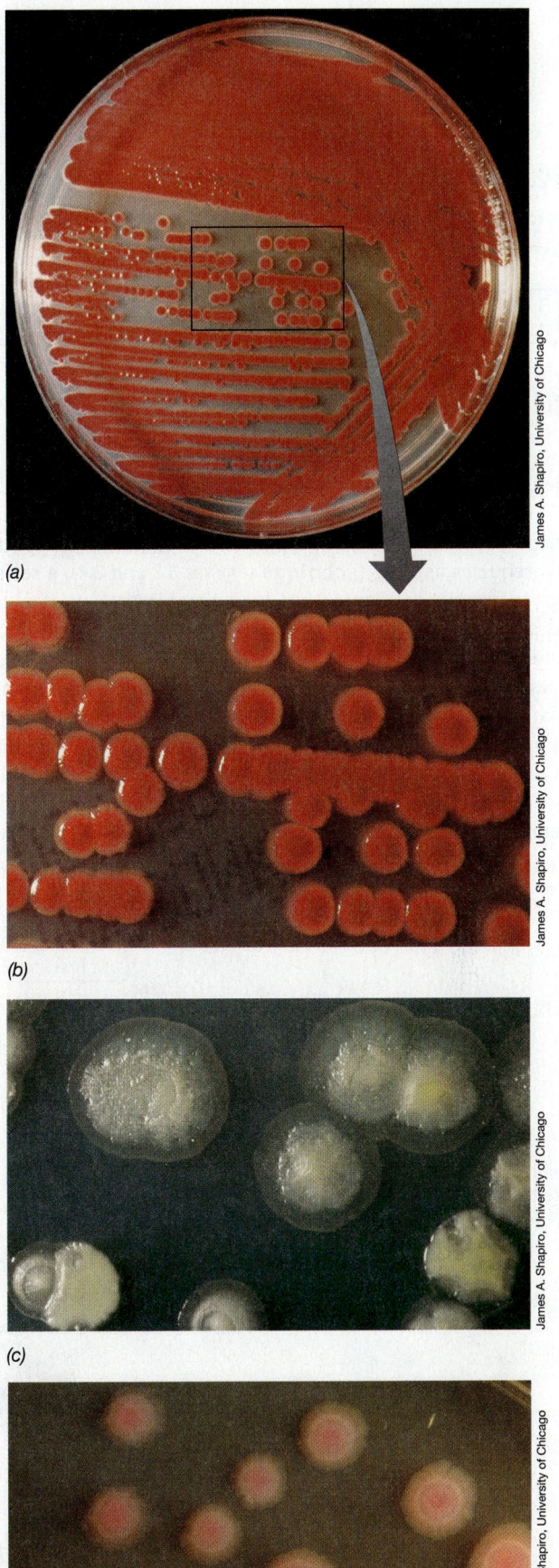

(a)

(b)

(c)

(d)

James A. Shapiro, University of Chicago

Figure 3.2 Bacterial colonies. Colonies are visible masses of cells formed from the division of one or a few cells and can contain over a billion (10^9) individual cells. *(a) Serratia marcescens*, grown on MacConkey agar. *(b)* Close-up of colonies outlined in part a. *(c) Pseudomonas aeruginosa*, grown on trypticase soy agar. *(d) Shigella flexneri*, grown on MacConkey agar.

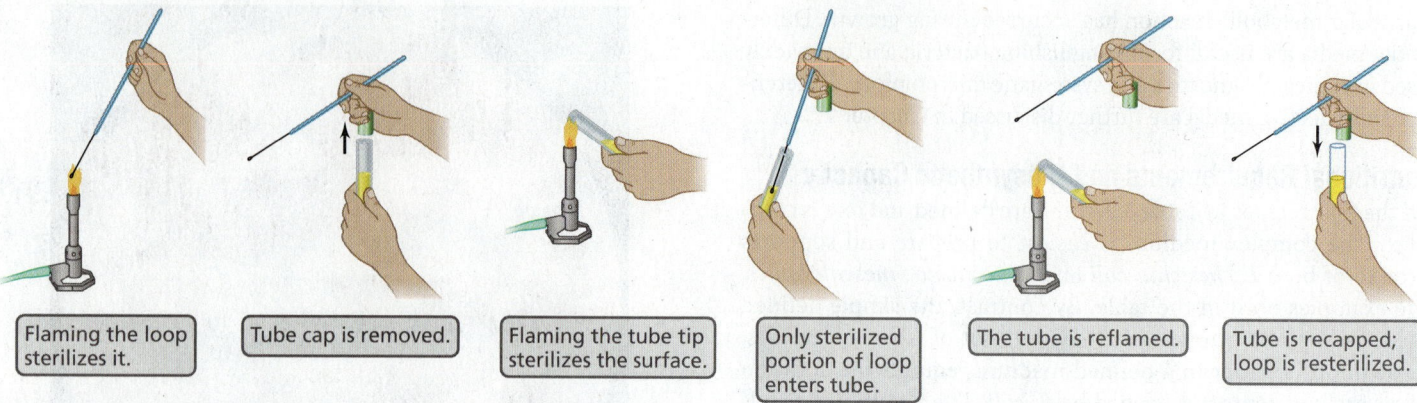

| Flaming the loop sterilizes it. | Tube cap is removed. | Flaming the tube tip sterilizes the surface. | Only sterilized portion of loop enters tube. | The tube is reflamed. | Tube is recapped; loop is resterilized. |

Figure 3.3 Aseptic transfer. After the tube is recapped at the end, the loop is resterilized before being taken out of service.

discuss the operation and principles of the autoclave in Section 5.17, along with other methods for sterilization. Once a sterile culture medium has been prepared, it is ready for inoculation. This requires **aseptic technique** (Figures 3.3 and 3.4), a series of steps to prevent contamination during manipulations of cultures and sterile culture media. A mastery of aseptic technique is required for maintaining pure cultures, as airborne contaminants are virtually everywhere (Figures 3.3 and 3.4). Picking an isolated colony and restreaking it is the main method for obtaining pure cultures from liquid samples containing several different organisms and is a common procedure in the microbiology laboratory. Other techniques for obtaining pure cultures have been developed that are especially suited for particular groups

of bacteria with unusual growth requirements and these will be discussed in Section 18.2.

MINIQUIZ

• Why would a complex culture medium for *Leuconostoc mesenteroides* be easier to prepare than a chemically defined medium?

• In which medium shown in Table 3.2, defined or complex, do you think *E. coli* would grow the fastest? Why? *E. coli* will not grow in the medium described for *Thiobacillus thioparus*. Why?

• What is meant by the word sterile? Why is aseptic technique necessary for successful cultivation of pure cultures in the laboratory?

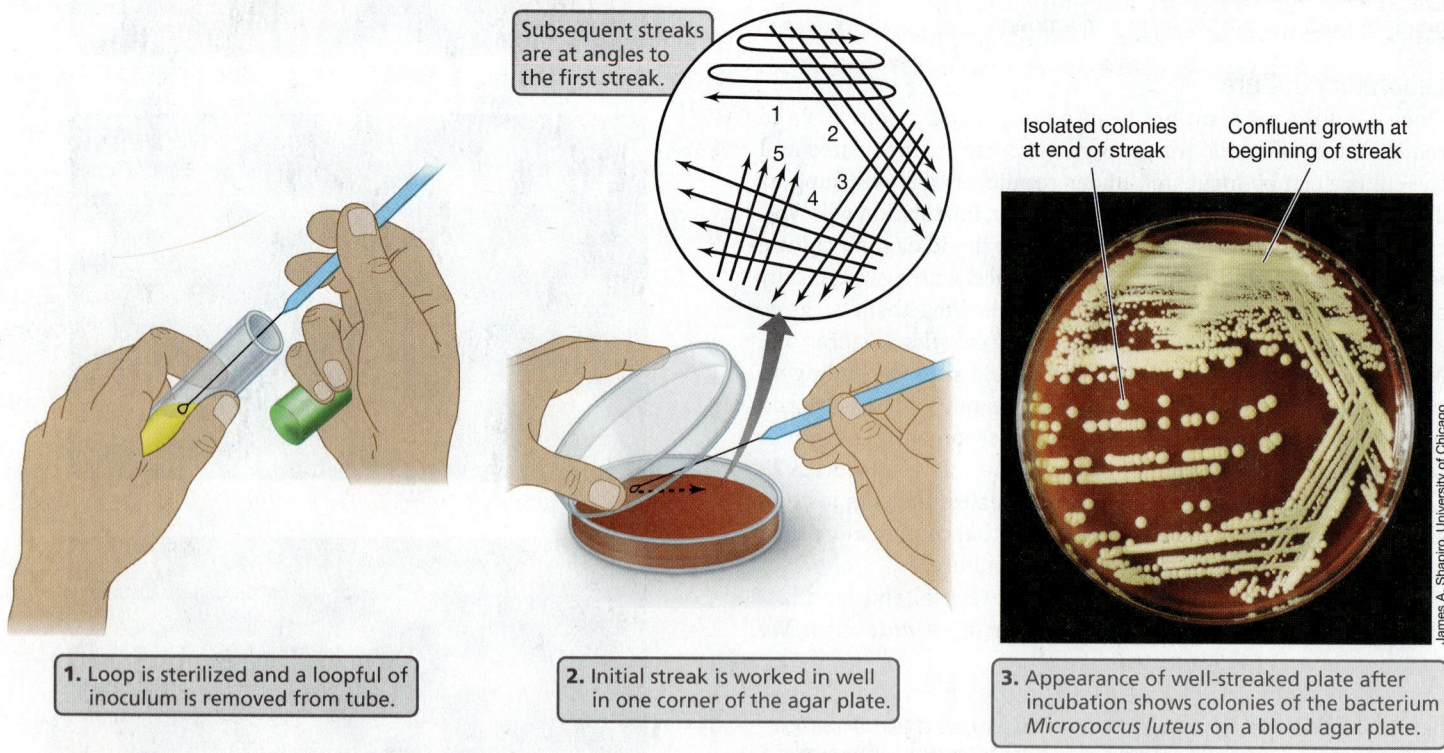

Subsequent streaks are at angles to the first streak.

Isolated colonies at end of streak

Confluent growth at beginning of streak

James A. Shapiro, University of Chicago

1. Loop is sterilized and a loopful of inoculum is removed from tube.

2. Initial streak is worked in well in one corner of the agar plate.

3. Appearance of well-streaked plate after incubation shows colonies of the bacterium *Micrococcus luteus* on a blood agar plate.

Figure 3.4 Making a streak plate to obtain pure cultures. The plate cover should be opened just enough to permit streaking manipulations.

II • Energetics, Enzymes, and Redox

All microorganisms must conserve some of the energy released in their energy-yielding reaction(s) in order to grow. Here we discuss the principles of energy conservation including the different energy classes of microorganisms, and use some simple laws of chemistry and physics to guide our understanding of bioenergetics.

3.3 Energy Classes of Microorganisms

Energy-yielding reactions are a part of metabolism called **catabolism**. Here we discuss the various energy classes of microorganisms, pointing out their similarities and differences. The terms used to describe the energy classes of microorganisms are important ones and will appear many times in this book.

Chemoorganotrophs and Chemolithotrophs

Organisms that conserve energy from chemicals are called *chemotrophs*, and those that use *organic* chemicals are called **chemoorganotrophs** (Figure 3.5). Most microorganisms that have been brought into laboratory culture are chemoorganotrophs. Many different organic chemicals can be used by one or another microorganism, and in virtually all cases, energy is conserved from the *oxidation* of the compound. The conserved energy is trapped in the cell in the form of energy-rich bonds of the compound adenosine triphosphate (ATP).

Some microorganisms can obtain energy from an organic compound only in the presence of oxygen; these organisms are called *aerobes*. Others can obtain energy only in the absence of oxygen (*anaerobes*). Still others can break down organic compounds in either the presence or absence of oxygen (*facultative aerobes*). We will deal with these options in more detail in Section 5.16.

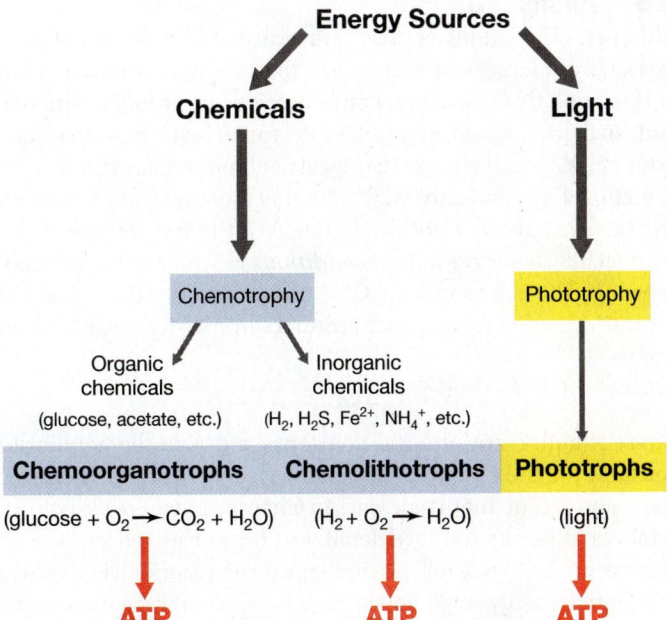

Energy Sources

Chemicals — Light

Chemotrophy — Phototrophy

Organic chemicals (glucose, acetate, etc.) — Inorganic chemicals (H_2, H_2S, Fe^{2+}, NH_4^+, etc.)

Chemoorganotrophs — **Chemolithotrophs** — **Phototrophs**

(glucose + $O_2 \rightarrow CO_2 + H_2O$) — ($H_2 + O_2 \rightarrow H_2O$) — (light)

ATP — ATP — ATP

Figure 3.5 **Metabolic options for conserving energy.** All three energy-conserving metabolisms are found within the microbial world.

Many *Bacteria* and *Archaea* can tap the energy available from the oxidation of *inorganic* compounds. This form of metabolism is called *chemolithotrophy* and was discovered by the Russian microbiologist Sergei Winogradsky (⟲ Section 1.9). Organisms that carry out chemolithotrophic reactions are called **chemolithotrophs** (Figure 3.5). Several inorganic compounds can be oxidized, for example, H_2, H_2S (hydrogen sulfide), NH_3 (ammonia), and Fe^{2+} (ferrous iron). Related groups of chemolithotrophs typically specialize in the oxidation of a related group of inorganic compounds, and thus we have the "sulfur" bacteria, the "iron" bacteria, the "nitrifying" bacteria, and so on.

The capacity to conserve energy from the oxidation of inorganic chemicals is a good metabolic strategy because competition from chemoorganotrophs is not a concern. In addition, many of the inorganic compounds oxidized by chemolithotrophs, for example H_2 and H_2S, are actually the waste products of chemoorganotrophs. Thus, many chemolithotrophs have evolved strategies for exploiting resources that chemoorganotrophs cannot, and so it is not unusual for species of these two physiological groups to live in close association with one another.

Phototrophs

Phototrophic microorganisms contain pigments that allow them to convert light energy into chemical energy and thus unlike chemotrophs, **phototrophs** do not require chemicals as a source of energy. This is a significant metabolic advantage because competition between phototrophs and chemotrophs for energy sources is not an issue and at least some sunlight is available in most microbial habitats.

Two major forms of phototrophy are known in prokaryotes. In one form, called *oxygenic* photosynthesis, oxygen (O_2) is produced. Among microorganisms, oxygenic photosynthesis is characteristic of cyanobacteria, which are prokaryotes, and algae, which are eukaryotes. The other form, *anoxygenic* photosynthesis, occurs in the purple and green bacteria and the heliobacteria (all *Bacteria*), and does not yield O_2. However, both oxygenic and anoxygenic phototrophs show great similarities in the mechanism for ATP synthesis, a result of the fact that oxygenic photosynthesis evolved from the simpler anoxygenic form some 3 billion years ago (⟲ Sections 1.3 and 12.2).

Heterotrophs and Autotrophs

Regardless of how a microorganism conserves energy, we have already seen that all cells require carbon in large amounts in order to make new cell materials (Figure 3.1). If an organism is a **heterotroph**, its cell carbon is obtained from some organic chemical. An **autotroph**, by contrast, uses carbon dioxide (CO_2) as its carbon source. Chemoorganotrophs are by definition also heterotrophs. By contrast, most chemolithotrophs and phototrophs are autotrophs. Autotrophs are also called *primary producers* because they synthesize new organic matter from CO_2. Virtually all organic matter on Earth has been synthesized by primary producers, in particular, the phototrophs.

3.4 Bioenergetics

Energy is defined as the ability to do work. In microbiology, energy transformations are measured in kilojoules (kJ), a unit of heat energy. All chemical reactions in a cell are accompanied by *changes* in energy, energy being either required for a reaction to occur or released as a reaction occurs. To identify which reactions release energy and which require energy, we need to understand some bioenergetic principles, and we focus on these here.

Basic Energetics

Although in any chemical reaction some energy is lost as heat, in microbiology we are interested in **free energy** (abbreviated **G**), which is the energy available to do work. The *change* in free energy during a reaction is expressed as $\Delta G^{0\prime}$, where the symbol Δ is read as "change in." The "0" and "prime" superscripts indicate that the free-energy value is for standard conditions: pH 7, 25°C, 1 atmosphere of pressure, and all reactants and products at molar concentrations.

Consider the reaction

$$A + B \rightarrow C + D$$

If $\Delta G^{0\prime}$ for this reaction is *negative* in arithmetic sign, then the reaction will proceed with the *release* of free energy, energy that the cell may conserve as ATP. Such energy-yielding reactions are called **exergonic**. However, if $\Delta G^{0\prime}$ is *positive*, the reaction *requires* energy in order to proceed. Such reactions are called **endergonic**. Thus, exergonic reactions *release* energy whereas endergonic reactions *require* energy.

Free Energy of Formation and Calculating Free Energy Changes ($\Delta G^{0\prime}$)

To calculate the free-energy yield of a reaction, one first needs to know the free energy of its reactants and products. This is the free energy of formation (G_f^0), the energy released or required during the formation of a given molecule from the elements. **Table 3.3** gives a few examples of G_f^0. By convention, the free energy of formation of the *elements* in their elemental and electrically neutral form (for instance, C, H_2, N_2) is zero. However, the free energies of formation of *compounds* are not zero. If the formation of a compound from its elements proceeds exergonically, then the G_f^0 of the compound is negative (energy is released). If the reaction is endergonic, then the G_f^0 of the compound is positive (energy is required).

For most compounds G_f^0 is negative. This reflects the fact that compounds tend to form spontaneously (that is, with energy being released) from their elements. However, the positive G_f^0 for nitrous oxide (N_2O) (+104.2 kJ/mol, Table 3.3) indicates that this compound does not form spontaneously. Instead, over time it decomposes spontaneously to yield N_2 and O_2. The free energies of formation of more compounds of microbiological interest are given in Appendix 1.

Table 3.3 Free energy of formation for a few compounds of biological interest

Compound	Free energy of formation (G_f^0)[a]
Water (H_2O)	−237.2
Carbon dioxide (CO_2)	−394.4
Hydrogen gas (H_2)	0
Oxygen gas (O_2)	0
Ammonium (NH_4^+)	−79.4
Nitrous oxide (N_2O)	+104.2
Acetate ($C_2H_3O_2^-$)	−369.4
Glucose ($C_6H_{12}O_6$)	−917.3
Methane (CH_4)	−50.8
Methanol (CH_3OH)	−175.4

[a]The free energy of formation values are in kJ/mol. See Table A1.1 in Appendix 1 for a more complete list of free energies of formation.

Using free energies of formation, it is possible to calculate $\Delta G^{0\prime}$ of a given reaction. For the reaction $A + B \rightarrow C + D$, $\Delta G^{0\prime}$ is calculated by subtracting the sum of the free energies of formation of the reactants (A + B) from that of the products (C + D). Thus

$$\Delta G^{0\prime} = G_f^0[C + D] - G_f^0[A + B]$$

The value obtained for $\Delta G^{0\prime}$ tells us whether the reaction is exergonic or endergonic. The phrase "products minus reactants" is a simple way to recall how to calculate changes in free energy during chemical reactions. However, before free-energy calculations can be made, it is first necessary to balance the reaction. Appendix 1 details the steps in balancing reactions and calculating free energies for any hypothetical reaction.

$\Delta G^{0\prime}$ versus ΔG

Although calculations of $\Delta G^{0\prime}$ are reasonable estimates of actual free-energy changes, in some cases they are not. We will see later in this book that the actual concentrations of products and reactants in nature, which are rarely at the molar levels used in calculations of $\Delta G^{0\prime}$, can change the results of bioenergetic calculations, sometimes in significant ways. What is most relevant to a bioenergetic calculation is not $\Delta G^{0\prime}$, but ΔG, the free-energy change that occurs *under the actual conditions* in which the organism is growing. The equation for ΔG takes into account the actual concentrations of reactants and products in the organism's habitat and is

$$\Delta G = \Delta G^{0\prime} + RT \ln K$$

where R and T are physical constants and K is the equilibrium constant for the reaction (Appendix 1). We distinguish between $\Delta G^{0\prime}$ and ΔG in important ways in Chapter 13 where we consider catabolic diversity in more detail. But for our purposes here, the expression $\Delta G^{0\prime}$ will tell us whether a given reaction yields energy or requires energy, and knowing this is sufficient for a basic understanding of energy flow in microbial systems. Only reactions that are exergonic yield energy that can be conserved by the cell, and this will be our focus in the next few sections.

3.5 Catalysis and Enzymes

Free-energy calculations reveal only whether energy is released or required in a given reaction. The value obtained says nothing about the *rate* of the reaction. If the rate of a reaction is very slow, it may be of no value to a cell. For example, consider the formation of water from O_2 and H_2. The energetics of this reaction are quite favorable: $H_2 + \frac{1}{2}O_2 \rightarrow H_2O$, $\Delta G^{0\prime} = -237\,kJ$. However, if O_2 and H_2 were mixed in a sealed bottle, no measurable amount of water would form, even after years. This is because the bonding of O_2 and H_2 to form H_2O requires that their chemical bonds first be broken. The breaking of these bonds requires some energy, and this energy is called **activation energy**.

Activation energy is the energy required to bring all molecules in a chemical reaction into the reactive state. For a reaction that proceeds with a net release of free energy (that is, an exergonic reaction), the situation is as diagrammed in **Figure 3.6**. Although the activation energy barrier is virtually insurmountable in the absence of a catalyst, in the presence of the proper catalyst, this barrier is greatly reduced. The concept of activation energy naturally leads us to consider the topic of catalysis and enzymes.

Enzymes

A **catalyst** is a substance that lowers the activation energy of a reaction (Figure 3.6), thereby increasing the reaction rate. Catalysts facilitate reactions but are not consumed or transformed by them. Moreover, catalysts do not affect the energetics or the equilibrium of a reaction; catalysts affect only the *rate* at which reactions proceed. Biological catalysts are called **enzymes**.

Most cellular reactions do not proceed at significant rates without catalysis. Enzymes are proteins (or in a few cases, RNAs) that are highly specific for the reactions they catalyze. That is, each enzyme catalyzes only a single type of chemical reaction, or in the case of a few enzymes, a single class of closely related reactions. This specificity is a function of the precise three-dimensional structure of the enzyme molecule. In an enzyme-catalyzed reaction, the enzyme (E) combines with the reactant, called a *substrate* (S), forming an enzyme–substrate complex (E—S). Then, as the reaction proceeds, the *product* (P) is released and the enzyme is returned to its original state:

$$E + S \leftrightharpoons E{-}S \leftrightharpoons E + P$$

The enzyme is generally much larger than the substrate(s), and the portion of the enzyme to which substrate binds is the enzyme's *active site*; the entire enzymatic reaction, from substrate binding to product release, may take only a few milliseconds.

Many enzymes contain small nonprotein molecules that participate in catalysis but are not themselves substrates. These small molecules can be divided into two classes based on the way they associate with the enzyme: *prosthetic groups* and *coenzymes*. Prosthetic groups bind very tightly to their enzymes, usually covalently and permanently. The heme group present in cytochromes such as cytochrome *c* (Section 3.10) is an example of a prosthetic group. **Coenzymes**, by contrast, are loosely bound to enzymes, and a single coenzyme molecule may associate with a number of different enzymes. Most coenzymes are derivatives of vitamins, and NAD^+, a derivative of the vitamin niacin (Table 3.1), is a good example.

Enzyme Catalysis

To catalyze a specific reaction, an enzyme must do two things: (1) bind its substrate and (2) position the substrate relative to specific amino acids in the enzyme's active site. The enzyme–substrate complex (**Figure 3.7**) serves both functions by aligning reactive groups and placing strain on specific bonds in the substrate(s). This reduces the activation energy required to make the reaction proceed from substrate(s) to product(s). This is shown in Figure 3.7 for the enzyme lysozyme, an enzyme whose substrate is the polysaccharide backbone of the bacterial cell wall polymer, peptidoglycan (↩ Figure 2.25).

The reaction depicted in Figure 3.7 is exergonic because the free energy of formation of the substrates is greater than that of the products. By contrast, some enzymes catalyze reactions that require energy, effectively converting energy-poor substrates into energy-rich products. In these cases, however, not only must an activation energy barrier (Figure 3.6) be overcome, but sufficient free energy must also be put into the reaction to raise the energy level of the substrates to that of the products. This is done by *coupling* the energy-requiring reaction to an energy-yielding one, such as the hydrolysis of ATP, the overall reaction proceeding with a free energy change that is either negative or near zero.

Theoretically, all enzymes are reversible in their activity. However, enzymes that catalyze highly exergonic or highly endergonic reactions typically function in only one direction. If a particularly

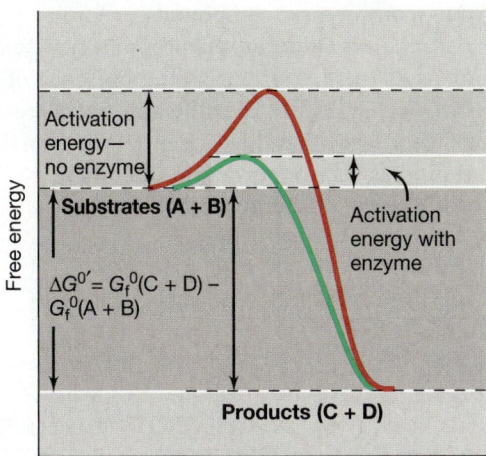

Figure 3.6 Activation energy and catalysis. Even chemical reactions that release energy may not proceed spontaneously until they are activated. Once the reactants are activated, the reaction proceeds spontaneously. Catalysts such as enzymes lower the required activation energy.

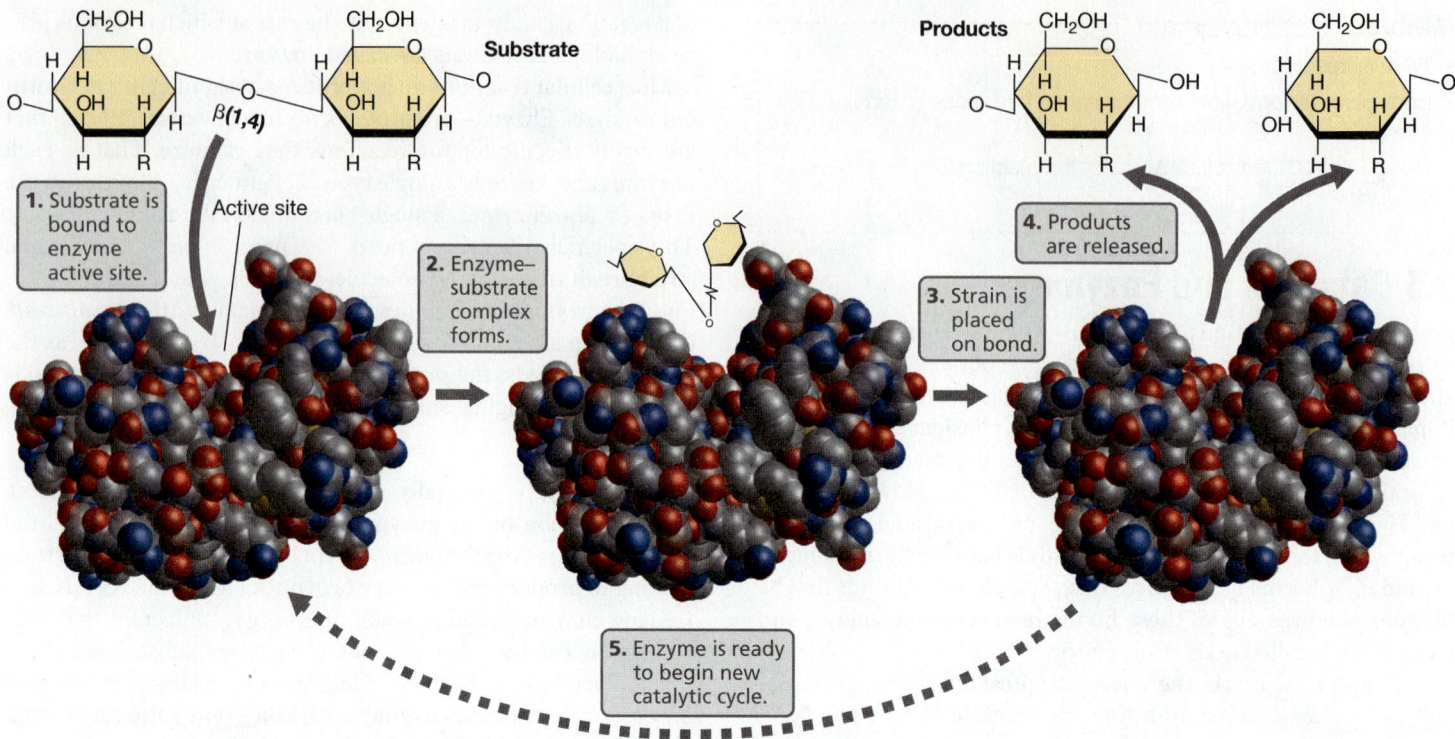

Figure 3.7 **The catalytic cycle of an enzyme.** The enzyme depicted here, lysozyme, catalyzes the cleavage of the β-1,4-glycosidic bond in the polysaccharide backbone of peptidoglycan. Following binding in the enzyme's active site, strain is placed on the bond, and this favors breakage. Space-filling model of lysozyme courtesy of Richard Feldmann.

exergonic or endergonic reaction needs to be reversed, a different enzyme usually catalyzes the reverse reaction.

MINIQUIZ

- What is the function of a catalyst? What are enzymes made of?
- Where on an enzyme does the substrate bind?
- What is activation energy?

3.6 Electron Donors and Acceptors

Cells conserve energy released from catabolic reactions by coupling it to the synthesis of energy-rich compounds, such as ATP. Reactions that release sufficient energy to form ATP are typically of an oxidation–reduction type. An *oxidation* is the removal of an electron (or electrons) from a substance, and a *reduction* is the addition of an electron (or electrons) to a substance. The term *redox* is commonly used as an abbreviation to indicate oxidation–reduction.

Redox Reactions

Redox reactions occur in pairs. For example, hydrogen gas (H_2) can release electrons and protons and become oxidized (**Figure 3.8**). However, electrons cannot exist alone in solution; they must be part of atoms or molecules. Thus the oxidation of H_2 is only a *half reaction*, a term that implies the need for a second half reaction. This is because for any substance to be oxidized, another substance must be reduced.

The oxidation of H_2 can be coupled to the reduction of many different substances, including oxygen (O_2), in a second half reaction.

This reductive half reaction, when coupled to the oxidation of H_2, yields the overall balanced net reaction (Figure 3.8). In reactions of this type, we refer to the substance *oxidized* (in this case, H_2) as the **electron donor**, and the substance *reduced* (in this case, O_2) as the **electron acceptor**. Electron donors are also commonly called *energy sources*. Many potential electron donors exist in nature, including a wide variety of organic and inorganic compounds. Many electron acceptors exist as well, including O_2, many oxidized compounds of nitrogen and sulfur, such as NO_3^- and SO_4^{2-}, and many organic compounds.

If a redox reaction is to occur, the presence of a suitable electron acceptor is just as important as the presence of a suitable electron donor. Lacking one or the other, the overall reaction cannot proceed. We will see that the concept of electron donors and electron acceptors is very important in microbiology and underlies virtually all aspects of energy metabolism.

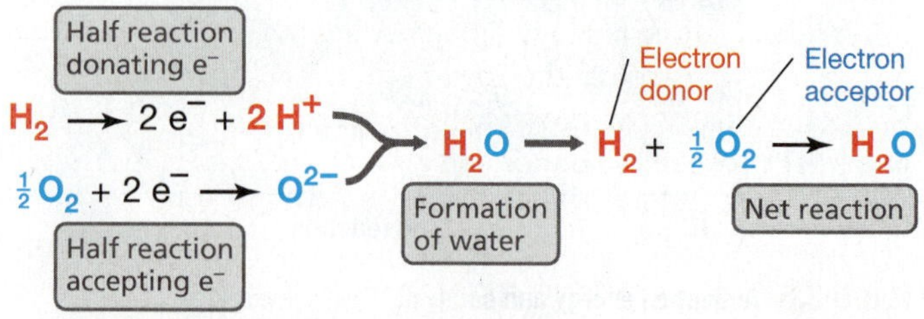

Half reaction donating e⁻

$$H_2 \longrightarrow 2\,e^- + 2\,H^+$$

$$\tfrac{1}{2}\,O_2 + 2\,e^- \longrightarrow O^{2-}$$

Half reaction accepting e⁻

Formation of water

$$H_2O$$

Electron donor / Electron acceptor

$$H_2 + \tfrac{1}{2}\,O_2 \longrightarrow H_2O$$

Net reaction

Figure 3.8 **Example of an oxidation–reduction reaction.** Each half reaction makes up half of the net reaction.

Reduction Potentials and Redox Couples

Substances differ in their inherent tendency to donate or accept electrons. This tendency is expressed as their **reduction potential** (E_0', standard conditions), measured in volts (V) in reference to that of a standard substance, H_2 (**Figure 3.9**). By convention, reduction potentials are given for half reactions written as *reductions*, with reactions at pH 7 because the cytoplasm of most cells is neutral, or nearly so.

A substance can be either an electron donor or an electron acceptor under different circumstances, depending on the substances with which it reacts. The constituents on each side of the arrow in half reactions are called a *redox couple*, such as $2\,H^+/H_2$, or $\frac{1}{2}\,O_2/H_2O$ (Figure 3.8). By convention, in writing a redox couple, the *oxidized* form of the couple is always placed on the left, before the forward slash, followed by the *reduced* form after the forward slash. In the example of Figure 3.8, the E_0' of the $2\,H^+/H_2$ couple is -0.42 V and that of the $\frac{1}{2}\,O_2/H_2O$ couple is $+0.82$ V. We will learn shortly that these values mean that O_2 is an excellent electron *acceptor* and H_2 is an excellent electron *donor*.

In reactions between two redox couples, the *reduced* substance of the couple whose E_0' is more *negative* donates electrons to the *oxidized* substance of the couple whose E_0' is more *positive*. Thus, in the couple $2\,H^+/H_2$, H_2 has a greater tendency to donate electrons than the tendency of $2\,H^+$ to accept them, and in the couple $\frac{1}{2}\,O_2/H_2O$, H_2O has a very weak tendency to donate electrons, whereas O_2 has a great tendency to accept them. It then follows that in a reaction of H_2 and O_2, H_2 will be the electron *donor* and become oxidized, and O_2 will be the electron *acceptor* and become reduced (Figure 3.8).

As previously mentioned, all half reactions are written as reductions. However, in an actual reaction between two redox couples, the half reaction with the more negative E_0' proceeds as an oxidation and is therefore written in the opposite direction. For example, in the reaction between H_2 and O_2 shown in Figure 3.8, H_2 is oxidized and is written in the reverse direction from its formal half reaction.

The Redox Tower and Its Relationship to $\Delta G^{0\prime}$

A convenient way of viewing electron transfer reactions is to imagine a vertical tower (Figure 3.9). The tower represents the range of reduction potentials possible for redox couples in nature, from those with the most negative E_0' on the top to those with the most positive E_0' at the bottom, and is thus a *redox tower*. The *reduced* substance in the redox couple at the top of the redox tower has the greatest tendency to donate electrons, whereas the *oxidized* substance in the redox couple at the bottom of the redox tower has the greatest tendency to accept electrons.

Using the tower analogy, imagine electrons from an electron donor near the top of the tower falling and being "caught" by electron acceptors at various levels. The difference in reduction potential between the donor and acceptor redox couples is quantified as $\Delta E_0'$. The further an electron drops before it is caught by an acceptor, the greater is the $\Delta E_0'$ between the two redox couples and the greater is the amount of energy released in the net reaction. That is, $\Delta E_0'$ is proportional to $\Delta G^{0\prime}$ (Figure 3.9). Oxygen (O_2), at the bottom of the redox tower, is the strongest electron acceptor of any significance in nature. In the middle of

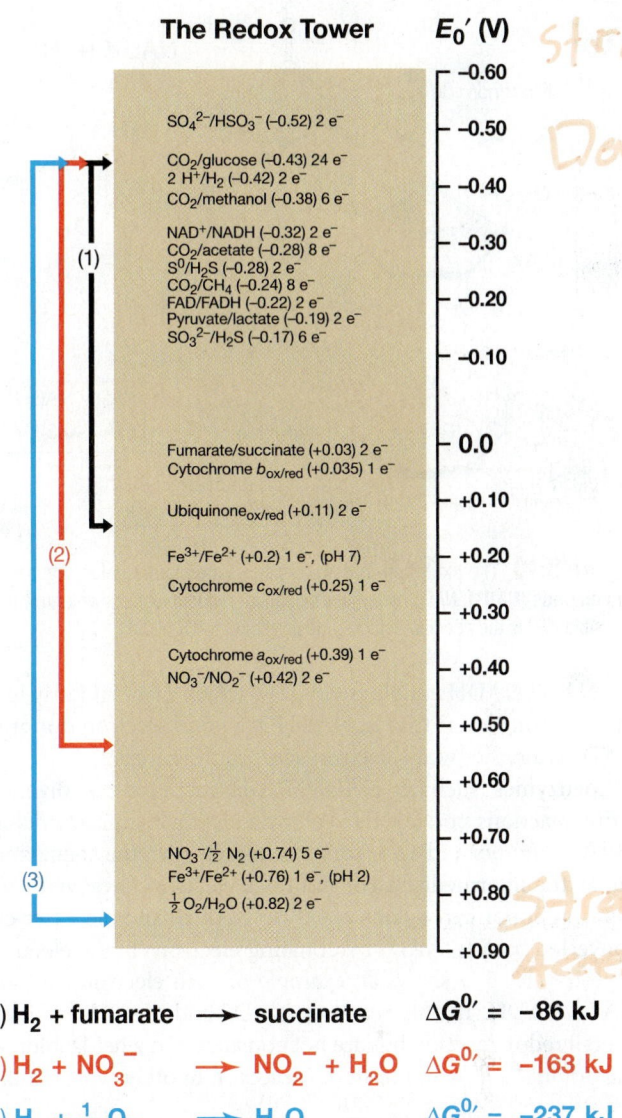

The Redox Tower E_0' (V)

strong Donors

SO_4^{2-}/HSO_3^- (-0.52) 2 e^-

CO_2/glucose (-0.43) 24 e^-
$2\,H^+/H_2$ (-0.42) 2 e^-
CO_2/methanol (-0.38) 6 e^-

$NAD^+/NADH$ (-0.32) 2 e^-
CO_2/acetate (-0.28) 8 e^-
S^0/H_2S (-0.28) 2 e^-
CO_2/CH_4 (-0.24) 8 e^-
$FAD/FADH$ (-0.22) 2 e^-
Pyruvate/lactate (-0.19) 2 e^-
SO_3^{2-}/H_2S (-0.17) 6 e^-

Fumarate/succinate ($+0.03$) 2 e^-
Cytochrome $b_{ox/red}$ ($+0.035$) 1 e^-

Ubiquinone$_{ox/red}$ ($+0.11$) 2 e^-

Fe^{3+}/Fe^{2+} ($+0.2$) 1 e^-, (pH 7)

Cytochrome $c_{ox/red}$ ($+0.25$) 1 e^-

Cytochrome $a_{ox/red}$ ($+0.39$) 1 e^-
NO_3^-/NO_2^- ($+0.42$) 2 e^-

$NO_3^-/\frac{1}{2}\,N_2$ ($+0.74$) 5 e^-
Fe^{3+}/Fe^{2+} ($+0.76$) 1 e^-, (pH 2)
$\frac{1}{2}\,O_2/H_2O$ ($+0.82$) 2 e^-

Strong Acceptors

(1) H_2 + fumarate $\longrightarrow$ succinate $\Delta G^{0\prime} = -86$ kJ

(2) H_2 + NO_3^- $\longrightarrow$ NO_2^- + H_2O $\Delta G^{0\prime} = -163$ kJ

(3) H_2 + $\frac{1}{2}\,O_2$ $\longrightarrow$ H_2O $\Delta G^{0\prime} = -237$ kJ

Figure 3.9 The redox tower. Redox couples are arranged from the strongest donors at the top to the strongest acceptors at the bottom. Electrons can be "caught" by acceptors at any intermediate level as long as the donor couple is more negative than the acceptor couple. The greater the difference in reduction potential between electron donor and electron acceptor, the more free energy is released. Note the differences in energy yield when H_2 reacts with three different electron acceptors, fumarate, nitrate, and oxygen.

the redox tower, redox couples can be either electron donors or acceptors depending on whom they react with. For instance, the $2\,H^+/H_2$ couple (-0.42 V) can react with the fumarate/succinate couple ($+0.03$ V), the NO_3^-/NO_2^- couple ($+0.42$ V), or the $\frac{1}{2}\,O_2/H_2O$ ($+0.82$ V) couple, with increasing amounts of energy being released, respectively (Figure 3.9).

Electron Carriers and NAD/NADH Cycling

Redox reactions in microbial cells are mediated by small molecules. A very common redox intermediary is the coenzyme nicotinamide adenine dinucleotide (NAD^+); the reduced form is written NADH (**Figure 3.10**). $NAD^+/NADH$ is an electron plus proton carrier, transporting 2 e^- and 2 H^+ at the same time. The reduction potential of

NAD⁺

NADH + H⁺

Nicotinamide

Ribose

Ribose Adenine

NAD⁺/ NADH
$E_0' = -0.32$ V

In NADP⁺, this OH has a phosphate attached.

Figure 3.10 The oxidation–reduction coenzyme nicotinamide adenine dinucleotide (NAD⁺). NAD⁺ undergoes oxidation–reduction as shown and is freely diffusible. "R" is the adenine dinucleotide portion of NAD⁺.

the NAD⁺/NADH couple is −0.32 V, which places it fairly high on the electron tower; that is, NADH is a good electron donor while NAD⁺ is a rather weak electron acceptor (Figure 3.9).

Coenzymes such as NAD⁺/NADH increase the diversity of redox reactions possible in a cell by allowing chemically dissimilar electron donors and acceptors to interact, with the coenzyme acting as the intermediary. For example, electrons removed from an electron donor can reduce NAD⁺ to NADH, and the latter can be converted back to NAD⁺ by donating electrons to the electron acceptor. **Figure 3.11** shows an example of such electron shuttling by NAD⁺/NADH. In this reaction, NAD⁺ and NADH facilitate the overall redox reaction but are not consumed in a net fashion as are the original donor and terminal acceptor. In other words, although

a cell requires relatively large amounts of a primary electron donor (the substance that was oxidized to yield NADH) and a terminal electron acceptor (such as O_2), it needs only a tiny amount of NAD⁺ and NADH because they are constantly being recycled (Figure 3.11).

NADP⁺ is a related redox coenzyme in which a phosphate group is added to NAD⁺. NADP⁺/NADPH typically participate in redox reactions distinct from those that use NAD⁺/NADH, most commonly in anabolic (biosynthetic) reactions in which oxidations and reductions occur (Sections 3.14–3.16).

MINIQUIZ

- In the reaction $H_2 + \frac{1}{2} O_2 \rightarrow H_2O$, what is the electron donor and what is the electron acceptor?
- Why is nitrate (NO_3^-) a better electron acceptor than fumarate?
- Is NADH a better electron donor than H_2? Is NAD⁺ a better acceptor than 2 H⁺? How do you determine this?

3.7 Energy-Rich Compounds

Energy released from redox reactions must be conserved by the cell if it is to be used to drive energy-requiring cell functions. In living organisms, chemical energy released in redox reactions is conserved primarily in phosphorylated compounds. The free energy released upon removal (hydrolysis) of the phosphate in these *energy-rich compounds* is significantly greater than that of the average covalent bond in the cell, and it is this released energy that is used by the cell.

Phosphate can be bonded to organic compounds by either *ester* or *anhydride* bonds, as illustrated in **Figure 3.12**. However, not all phosphate bonds are energy-rich. As seen in the figure, the $\Delta G^{0'}$ of hydrolysis of the phosphate ester bond in glucose 6-phosphate is only −13.8 kJ/mol. By contrast, the $\Delta G^{0'}$ of hydrolysis of the phosphate anhydride bond in phosphoenolpyruvate is −51.6 kJ/mol,

NAD⁺ reduction

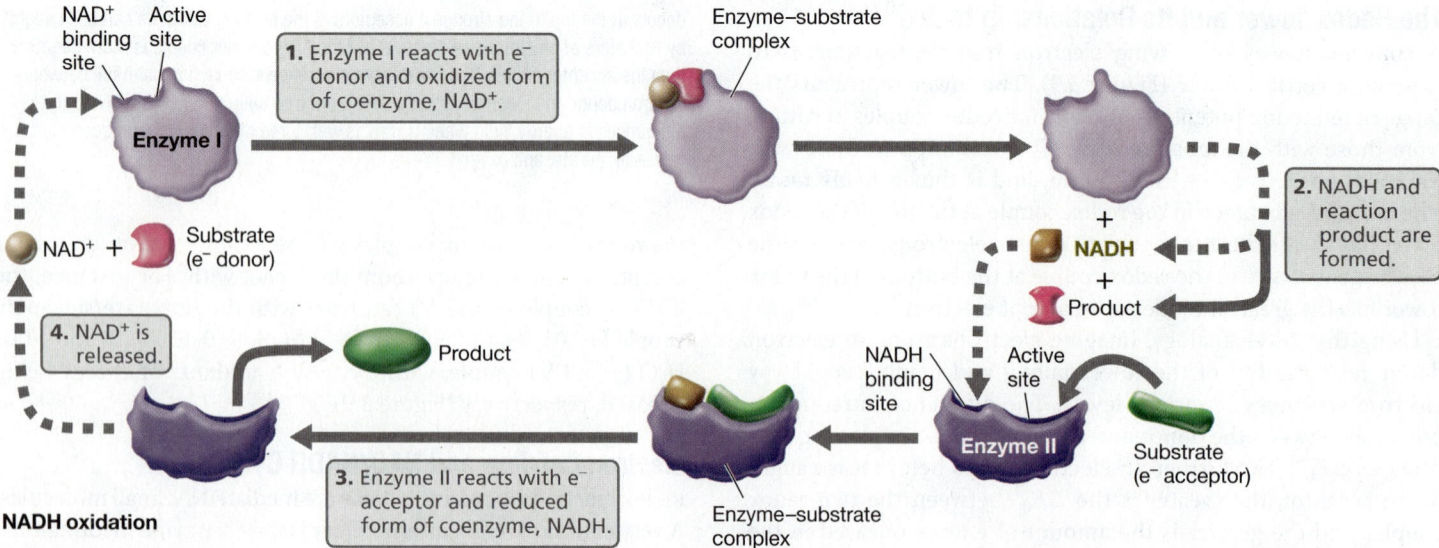

NAD⁺ binding site
Active site

Enzyme–substrate complex

1. Enzyme I reacts with e⁻ donor and oxidized form of coenzyme, NAD⁺.

Enzyme I

NAD⁺ + Substrate (e⁻ donor)

2. NADH and reaction product are formed.

+
NADH
+
Product

4. NAD⁺ is released.

Product

NADH binding site
Active site

Enzyme II

Substrate (e⁻ acceptor)

3. Enzyme II reacts with e⁻ acceptor and reduced form of coenzyme, NADH.

Enzyme–substrate complex

NADH oxidation

Figure 3.11 NAD⁺/NADH cycling. A schematic example of redox reactions in which two different enzymes are linked by their requirement for either NAD⁺ or NADH.

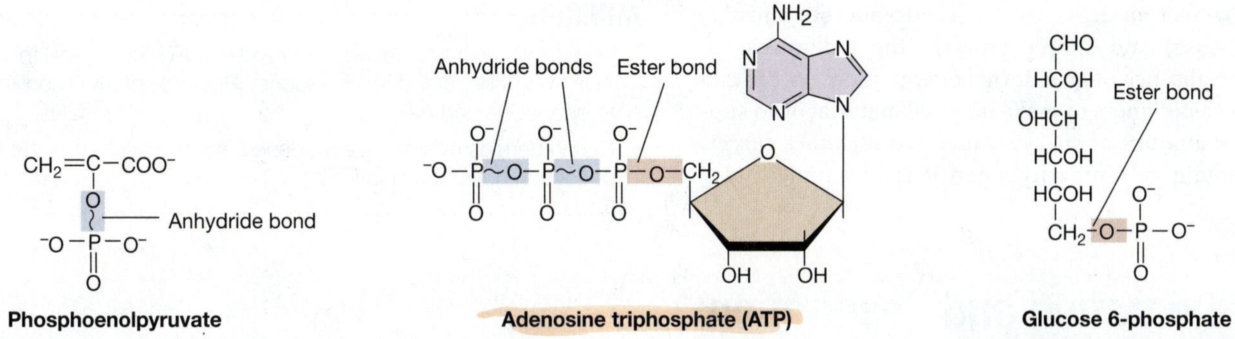

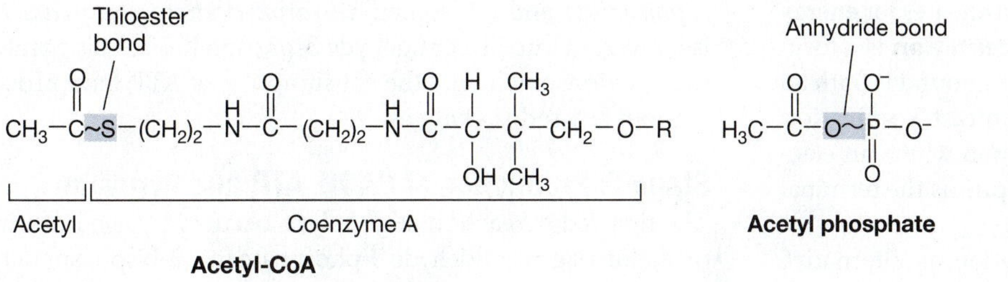

Compound	$G^{0'}$ kJ/mol
$\Delta G^{0'}$ > 30kJ	
Phosphoenolpyruvate	−51.6
1,3-Bisphosphoglycerate	−52.0
Acetyl phosphate	−44.8
ATP	−31.8
ADP	−31.8
Acetyl-CoA	−35.7
$\Delta G^{0'}$ < 30kJ	
AMP	−14.2
Glucose 6-phosphate	−13.8

Figure 3.12 Phosphate bonds in compounds that conserve energy in bacterial metabolism. Notice, by referring to the table, the range in free energy of hydrolysis of the phosphate bonds highlighted in the compounds. The "R" group of acetyl-CoA is a 3′-phospho ADP group.

almost four times that of glucose 6-phosphate. Although theoretically either compound could be hydrolyzed in energy metabolism, cells typically use a small group of compounds whose $\Delta G^{0'}$ of hydrolysis is greater than -30 kJ/mol as energy "currencies" in the cell. Thus, phosphoenolpyruvate is energy-rich whereas glucose 6-phosphate is not.

Adenosine Triphosphate

The most important energy-rich phosphate compound in cells is **adenosine triphosphate (ATP)**. ATP consists of the ribonucleoside adenosine to which three phosphate molecules are bonded in series. ATP is the prime energy currency in all cells, being generated during exergonic reactions and consumed in endergonic reactions. From the structure of ATP (Figure 3.12), it can be seen that only two of the phosphate bonds ($ATP \rightarrow ADP + P_i$ and $ADP \rightarrow AMP + P_i$) are phosphoanhydrides and thus have free energies of hydrolysis greater than -30 kJ. By contrast, AMP is not energy-rich because its free energy of hydrolysis is only about half that of ADP or ATP (Figure 3.12).

Although the energy released in ATP hydrolysis is -32 kJ, a caveat must be introduced here to define more precisely the energy requirements for the synthesis of ATP. In an actively growing *Escherichia coli* cell, the ratio of ATP to ADP is maintained at about 7:1, and this affects the energy requirements for ATP synthesis. In an actively growing cell, the actual energy expenditure (that is, the ΔG, Section 3.4) for the synthesis of 1 mole of ATP is on the order of -55 to -60 kJ. Nevertheless, for the purposes of learning and applying the basic principles of bioenergetics, we will assume that reactions conform to "standard conditions" ($\Delta G^{0'}$), and thus we will assume that the energy required for synthesis or hydrolysis of ATP is 32 kJ/mol.

Coenzyme A

Cells can use the free energy available in the hydrolysis of energy-rich compounds other than phosphorylated compounds. These include, in particular, derivatives of *coenzyme A* (for example, acetyl-CoA; see structure in Figure 3.12). Coenzyme A derivatives contain *thioester* bonds. Upon hydrolysis, these yield sufficient free energy to drive the synthesis of an energy-rich phosphate bond. For example, in the reaction

$$\text{Acetyl-S-CoA} + H_2O + ADP + P_i \rightarrow \text{acetate}^- + \text{HS-CoA} + ATP + H^+$$

the energy released in the hydrolysis of coenzyme A is conserved in the synthesis of ATP. Coenzyme A derivatives (acetyl-CoA is just one of many) are especially important to the energetics of anaerobic microorganisms, in particular those whose energy metabolism depends on fermentation (see Table 3.4). We will return to the importance of coenzyme A derivatives in bacterial bioenergetics many times in Chapter 13.

Energy Storage

ATP is a dynamic molecule in the cell; it is continuously being broken down to drive anabolic reactions and resynthesized at the expense of catabolic reactions. For longer-term energy storage, microorganisms produce insoluble polymers that can be catabolized later for the production of ATP.

Examples of energy storage polymers in prokaryotes include glycogen, poly-β-hydroxybutyrate and other polyhydroxyalkanoates, and elemental sulfur, stored from the oxidation of H_2S by sulfur chemolithotrophs. These polymers are deposited within the cell as granules that can be seen with the light or electron microscope ($\mathcal{e}\!\!\rightarrow$ Section 2.14). In eukaryotic microorganisms, starch (polyglucose) and simple fats are the major reserve materials. In the absence of an external energy source, a cell can break down these polymers to make new cell material or to supply the very low amount of energy, called *maintenance energy*, needed to maintain cell integrity when it is in a nongrowing state.

III • Fermentation and Respiration

Fermentation and respiration are two major strategies for energy conservation in chemoorganotrophs. **Fermentation** is a form of anaerobic catabolism in which an organic compound is both an electron donor and an electron acceptor. By contrast, **respiration** is the form of aerobic or anaerobic catabolism in which an electron donor is oxidized with O_2 or an O_2 substitute as the terminal electron acceptor.

One can look at fermentation and respiration as alternative metabolic options. When O_2 is available, respiration will take place because, as we will see, much more ATP is produced in respiration than in fermentation. But if conditions will not support respiration, fermentation can supply enough energy for an organism to thrive. We begin by examining a major metabolic pathway for microbial fermentations, the glycolytic pathway.

3.8 Glycolysis

A nearly universal pathway for the catabolism of glucose is **glycolysis**, which breaks down glucose into pyruvate. Glycolysis is also called the *Embden–Meyerhof–Parnas pathway* for its major discoverers. Whether glucose is fermented or respired, it travels through this pathway. In fermentation, ATP is synthesized by **substrate-level phosphorylation**. In this process, ATP is synthesized directly from energy-rich intermediates during steps in the catabolism of the fermentable substrate (**Figure 3.13***a*). This is in contrast to **oxidative phosphorylation**, which occurs in respiration; ATP is synthesized here at the expense of the proton motive force (Figure 3.13*b*).

The fermentable substrate in a fermentation is both the electron donor and electron acceptor; not all compounds can be fermented, but sugars, especially hexoses such as glucose, are excellent fermentable substrates. The fermentation of glucose through the glycolytic pathway can be divided into three stages, each requiring several independent enzymatic reactions. Stage I comprises "preparatory" reactions; these are not redox reactions and do not release energy but instead form a key intermediate of the pathway. In Stage II, redox reactions occur, energy is conserved, and two molecules of pyruvate are formed. In Stage III, redox balance is achieved and fermentation products are formed (**Figure 3.14**).

Stage I: Preparatory Reactions

In Stage I, glucose is phosphorylated by ATP, yielding glucose 6-phosphate. The latter is then isomerized to fructose 6-phosphate, and a second phosphorylation leads to the production of fructose 1,6-bisphosphate. The enzyme aldolase then splits fructose 1,6-bisphosphate into two 3-carbon molecules, *glyceraldehyde*

3-phosphate and its isomer, *dihydroxyacetone phosphate*, which is converted into glyceraldehyde 3-phosphate. To this point, all of the reactions, including the consumption of ATP, have proceeded without any redox changes.

Stage II: Production of NADH, ATP, and Pyruvate

The first redox reaction of glycolysis occurs in Stage II during the oxidation of glyceraldehyde 3-phosphate to 1,3-bisphosphoglyceric acid. In this reaction (which occurs twice, once for each of the two molecules of glyceraldehyde 3-phosphate produced from glucose), the enzyme glyceraldehyde-3-phosphate dehydrogenase reduces its coenzyme NAD^+ to NADH. Simultaneously, each glyceraldehyde 3-phosphate molecule is phosphorylated by the addition of a molecule of inorganic phosphate. This reaction, in which inorganic phosphate is converted to organic form, sets the stage for energy conservation. ATP formation is possible because 1,3-bisphosphoglyceric acid is an energy-rich compound (Figure 3.12). ATP is then synthesized when (1) each molecule

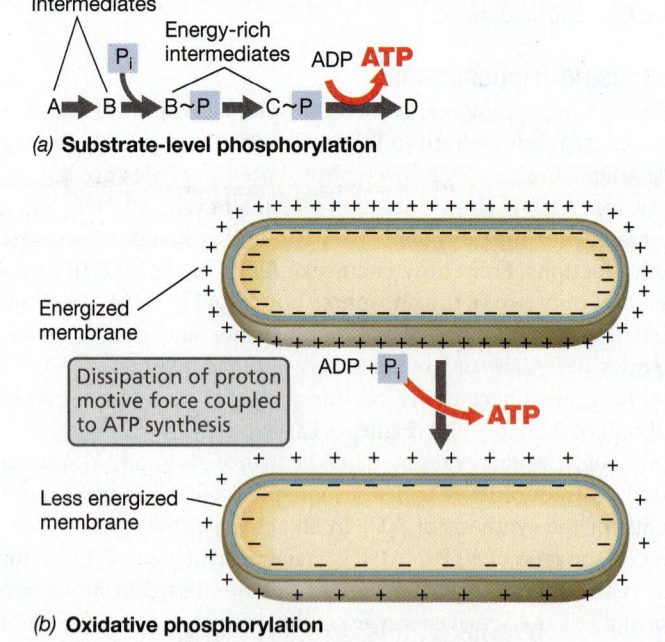

(a) **Substrate-level phosphorylation**

(b) **Oxidative phosphorylation**

Figure 3.13 Energy conservation in fermentation and respiration. *(a)* In fermentation, substrate-level phosphorylation produces ATP. *(b)* In respiration, the cytoplasmic membrane, energized by the proton motive force, dissipates energy to synthesize ATP from ADP + P_i by oxidative phosphorylation.

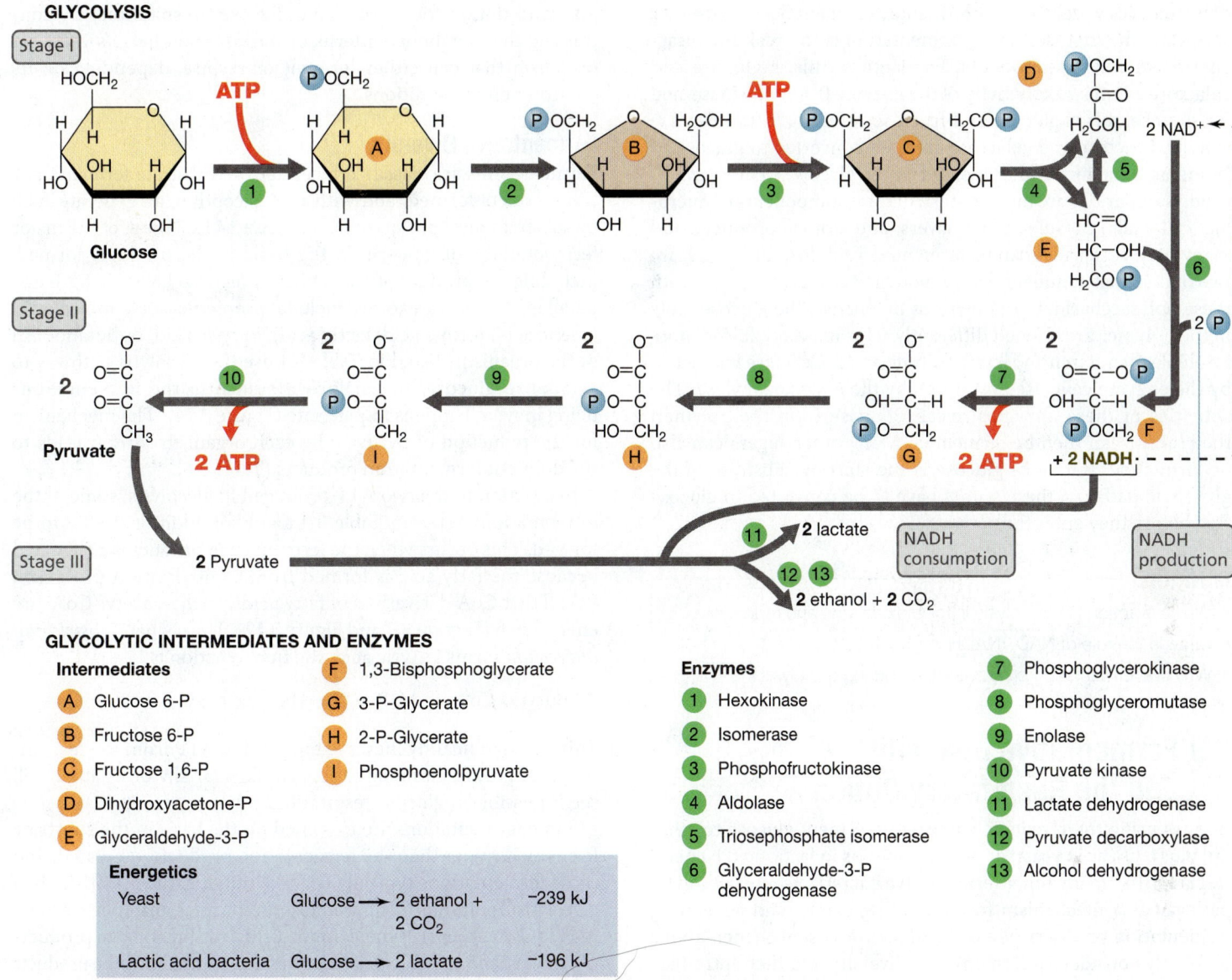

GLYCOLYTIC INTERMEDIATES AND ENZYMES

Intermediates		Enzymes	
A Glucose 6-P	F 1,3-Bisphosphoglycerate	1 Hexokinase	7 Phosphoglycerokinase
B Fructose 6-P	G 3-P-Glycerate	2 Isomerase	8 Phosphoglyceromutase
C Fructose 1,6-P	H 2-P-Glycerate	3 Phosphofructokinase	9 Enolase
D Dihydroxyacetone-P	I Phosphoenolpyruvate	4 Aldolase	10 Pyruvate kinase
E Glyceraldehyde-3-P		5 Triosephosphate isomerase	11 Lactate dehydrogenase
		6 Glyceraldehyde-3-P dehydrogenase	12 Pyruvate decarboxylase
			13 Alcohol dehydrogenase

Energetics

Yeast	Glucose ⟶ 2 ethanol + 2 CO_2	−239 kJ
Lactic acid bacteria	Glucose ⟶ 2 lactate	−196 kJ

Figure 3.14 Embden–Meyerhof–Parnas pathway (glycolysis). (Top) The sequence of reactions in the catabolism of glucose to pyruvate and then on to fermentation products. Pyruvate is the end product of glycolysis, and fermentation products are made from it. (Bottom) Intermediates, enzymes, and contrasting fermentation balances of yeast and lactic acid bacteria.

of 1,3-bisphosphoglyceric acid is converted to 3-phosphoglyceric acid, and (2) each molecule of phosphoenolpyruvate is converted to pyruvate (Figure 3.14).

During Stages I and II of glycolysis, *two* ATP molecules are consumed and *four* ATP molecules are synthesized (Figure 3.14). Thus, the net energy yield in glycolysis is *two molecules of ATP per molecule of glucose fermented*.

Stage III: Redox Balance and the Production of Fermentation Products

During the formation of two 1,3-bisphosphoglyceric acid molecules, two NAD^+ are reduced to NADH (Figure 3.14). However, recall that NAD^+ is only an electron shuttle, not a net (terminal) acceptor of electrons. Thus, the NADH produced in glycolysis

must be oxidized back to NAD^+ in order for another round of glycolysis to occur, and this is accomplished when pyruvate is reduced by NADH to fermentation products (Figure 3.14). For example, in fermentation by yeast, pyruvate is reduced to ethanol (ethyl alcohol) with the subsequent production of carbon dioxide (CO_2). By contrast, lactic acid bacteria reduce pyruvate to lactate. Many other possibilities for pyruvate reduction exist depending on the organism (see next section), but the end result is the same: NADH is reoxidized to NAD^+, and this allows earlier reactions of the pathway that require NAD^+ to continue.

Catabolism of Other Sugars and Polysaccharides

Many microorganisms can ferment disaccharides. For example, lactose (milk sugar) and sucrose (table sugar) are common disaccharides widely used by fermentative anaerobes. With either substrate, the first step in its fermentation is to break the disaccharide into its components. For lactose, this is glucose and galactose as a result of activity of the enzyme β-galactosidase, and for sucrose, this is glucose and fructose resulting from invertase activity. Fructose and galactose are then converted to glucose by isomerase enzymes and fermented by the glycolytic pathway.

Polysaccharides are important structural components of microbial cell walls, capsules, slime layers, and storage products, and many polysaccharides can be fermented. Cellulose and starch are two of the most abundant natural polysaccharides. Although both these polysaccharides are polymers of glucose, the glucose units in the polymer are bonded differently. This makes cellulose more insoluble than starch and less rapidly digested. Cellulose is attacked by the enzyme cellulase and starch by the enzyme amylase. The activities of these enzymes release glucose from the polymer; the glucose can then be fermented. Many other sugars can also be fermented. But since glucose is the starting substrate of the glycolytic pathway, these sugars have to be converted to glucose first before they enter the pathway.

MINIQUIZ -

- Which reactions in glycolysis involve oxidations and reductions?
- What is the role of $NAD^+/NADH$ in glycolysis?
- Why are fermentation products made during glycolysis?

3.9 Fermentative Diversity and the Respiratory Option

Besides using the glycolytic pathway to ferment glucose to ethanol plus CO_2, as in yeast, or to lactic acid, as in lactic acid bacteria (Figure 3.14), many other fermentative bacteria use the glycolytic pathway as a mechanism for conserving energy and generating fermentation products. We conclude our focus on fermentations by briefly considering fermentative diversity and then introduce

a second option for catabolizing glucose—respiration—by contrasting the metabolic patterns of the common baker's yeast, an organism that can either ferment or respire, depending on its environmental conditions.

Fermentative Diversity

Fermentations are classified by either the substrate fermented or the products formed, and with rare exception, all generate ATP by substrate-level phosphorylation. Table 3.4 lists some of the major fermentations of glucose on the basis of the products formed, including the production of alcohol or lactic acid, as we have just detailed. Other categories include propionic acid, mixed acid (acetic acid, formic acid, lactic acid), butyric acid, or butanol. All of the organisms listed in Table 3.4 use the glycolytic pathway to catabolize glucose, the major difference in the fermentations being in what happens to pyruvate (Figure 3.14). The mechanism for the reduction of pyruvate by each organism is what leads to the different fermentation products (Table 3.4).

In addition to the two ATP produced in glycolysis, some of the fermentations listed in Table 3.4 allow for additional ATP to be formed. This occurs when the fermentation product is a fatty acid because the fatty acid is formed from a coenzyme-A precursor. Recall that CoA derivatives of fatty acids, such as acetyl-CoA, are energy-rich (Section 3.7 and Figure 3.12). Thus, when *Clostridium butyricum* forms butyric acid, the final reaction is

$$\text{Butyryl-CoA} + \text{ADP} + P_i \rightarrow \text{butyric acid} + \text{ATP} + \text{CoA}$$

This can significantly increase the yield of ATP from the fermentation of glucose, although the yield falls far shy of what we will see is possible in glucose respiration.

Some fermentations are classified on the basis of the substrate fermented rather than the fermentation products generated, and these fermentations typically occur through pathways other than glycolysis. For instance, some endospore-forming anaerobic bacteria (genus *Clostridium*) ferment amino acids, the breakdown products of proteins, and others ferment purines and pyrimidines, the products

Table 3.4 Common bacterial fermentations and some of the organisms carrying them out

Type	Reaction	Organisms
Alcoholic	Hexose[a] → 2 ethanol + 2 CO_2	Yeast, *Zymomonas*
Homolactic	Hexose → 2 lactate⁻ + 2 H^+	*Streptococcus*, some *Lactobacillus*
Heterolactic	Hexose → lactate⁻ + ethanol + CO_2 + H^+	*Leuconostoc*, some *Lactobacillus*
Propionic acid	3 Lactate⁻ → 2 propionate⁻ + acetate⁻ + CO_2 + H_2O	*Propionibacterium*, *Clostridium propionicum*
Mixed acid[b,c]	Hexose → ethanol + 2,3-butanediol + succinate²⁻ + lactate⁻ + acetate⁻ + formate⁻ + H_2 + CO_2	Enteric bacteria including *Escherichia*, *Salmonella*, *Shigella*, *Klebsiella*, *Enterobacter*
Butyric acid[c]	Hexose → butyrate⁻ + 2 H_2 + 2 CO_2 + H^+	*Clostridium butyricum*
Butanol[c]	2 Hexose → butanol + acetone + 5 CO_2 + 4 H_2	*Clostridium acetobutylicum*
Caproate/Butyrate	6 Ethanol + 3 acetate⁻ → 3 butyrate⁻ + caproate⁻ + 2 H_2 + 4 H_2O + H^+	*Clostridium kluyveri*
Acetogenic	Fructose → 3 acetate⁻ + 3 H^+	*Clostridium aceticum*

[a]Glucose is the starting substrate for glycolysis. However, many other C_6 sugars (hexoses) can be fermented following their conversion to glucose.

[b]Not all organisms produce all products. In particular, butanediol production is limited to only certain enteric bacteria. The reaction is not balanced.

[c]Other products include some acetate and a small amount of ethanol (butanol fermentation only).

than ferment it, and the major product is CO_2 (from activities of the citric acid cycle, see Figure 3.22). Only when conditions are anoxic do yeasts switch to fermentation.

This fact has practical significance. Since the brewer and baker need the *products* of yeast fermentation rather than yeast cells themselves, care must be taken to ensure that the yeast is forced into a fermentative lifestyle. For example, when grapes are squeezed to make wine, the yeast at first respire, making the juice anoxic. Following this, the vessel is sealed against the introduction of air and fermentation begins. Yeast also serves as the leavening agent in bread, although here it is not the alcohol that is important, but CO_2, the *other* product of the alcohol fermentation (Table 3.4). The CO_2 raises the dough, and the alcohol produced along with it is volatilized during the baking process. We discuss fermented foods in more detail in Chapter 31.

MINIQUIZ

- Which fermentation products are produced by *Lactobacillus* and which by *Clostridium* species? Which would you find in fermented milk products, such as yogurt?
- Which yeast fermentation product is the desired agent in bread and what is its function in bread-making?

3.10 Respiration: Electron Carriers

Fermentation is an anaerobic process and releases only a small amount of energy. By contrast, if pyruvate is fully oxidized to CO_2 rather than reduced to some fermentation product, a far higher yield of ATP is possible. Oxidation using O_2 as the terminal electron acceptor is called *aerobic respiration*; oxidation using other acceptors under anoxic conditions is called *anaerobic respiration* (Section 3.13).

Our discussion of respiration covers both carbon transformations and redox reactions and focuses on two issues: (1) how electrons are transferred from the primary electron donor to the terminal electron acceptor and how this process is coupled to energy conservation, and (2) the pathway by which organic carbon is oxidized into CO_2. We begin with a consideration of electron transport, the series of reactions that lead to the proton motive force.

NADH Dehydrogenases and Flavoproteins

Electron transport occurs in the membrane, and several types of oxidation–reduction enzymes participate in electron transport. These include *NADH dehydrogenases, flavoproteins, iron–sulfur proteins*, and *cytochromes*. Also participating are nonprotein electron carriers called *quinones*. The carriers are arranged in the membrane in order of increasingly more positive reduction potential, with NADH dehydrogenase first and the cytochromes last (Figure 3.9).

NADH dehydrogenases are proteins bound to the inside surface of the cytoplasmic membrane and have an active site that binds NADH. The $2\ e^- + 2\ H^+$ from NADH are transferred from the dehydrogenase to a flavoprotein, the next carrier in the chain. This forms NAD^+ that is released from the dehydrogenase and can react with another enzyme (Figure 3.11).

Figure 3.15 Common food and beverage products resulting from the alcoholic fermentation of *Saccharomyces cerevisiae*.

Barton Spear

of nucleic acid breakdown. Some fermentative anaerobes even ferment aromatic compounds. In many cases, these fermentations are carried out by a single group of anaerobic bacteria; in a few cases, only a single bacterium is known to ferment a particular substance. These bacteria are metabolic specialists, having evolved the capacity to ferment a substrate not catabolized by other bacteria. Although they may seem to be metabolic oddballs, these and other fermentative bacteria are of great ecological importance in degrading the remains of dead plants, animals, and other microorganisms in anoxic environments in nature. We investigate the principles behind some of these unusual fermentations in Chapter 13.

Saccharomyces cerevisiae: Fermentation or Respiration?

During glycolysis, glucose is consumed, ATP is made, and fermentation products are generated. For the organism the crucial product is ATP; fermentation products are merely waste products. However, fermentation products are not waste products to humans. Instead, they are the foundation of the baking and fermented beverage industries (**Figure 3.15**) and are key ingredients in many fermented foods. In the baking and alcohol industries, the metabolic capacities of the key player, the baker's and brewer's yeast *Saccharomyces cerevisiae*, are on center stage. However, *S. cerevisiae* can carry out two modes of glucose catabolism, *fermentation*, as we have discussed, and *respiration*, which we will consider next.

As a rule, cells carry out that form of metabolism that most benefits them energetically. The energy available from a molecule of glucose is much greater if it is respired to CO_2 than if it is fermented. This is because unlike CO_2, organic fermentation products such as ethanol still contain a significant amount of free energy. Thus, when O_2 is available, yeast respire glucose rather

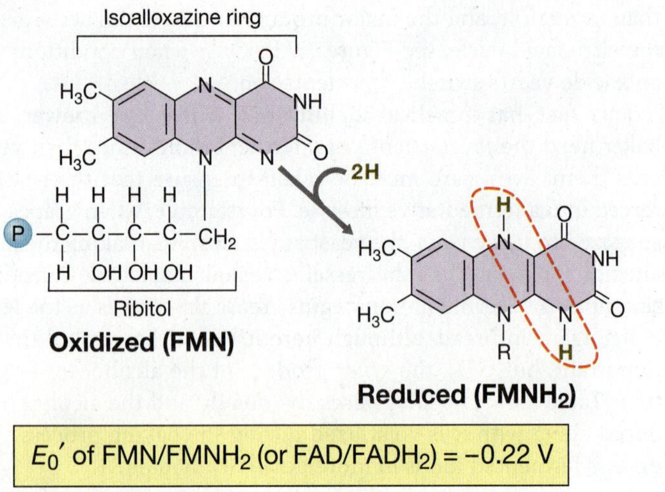

E_0' of FMN/FMNH$_2$ (or FAD/FADH$_2$) = −0.22 V

Figure 3.16   Flavin mononucleotide (FMN), a hydrogen atom carrier. The site of oxidation–reduction (dashed red circle) is the same in FMN and the related coenzyme flavin adenine dinucleotide (FAD, not shown). FAD contains an adenosine group bonded through the phosphate group on FMN.

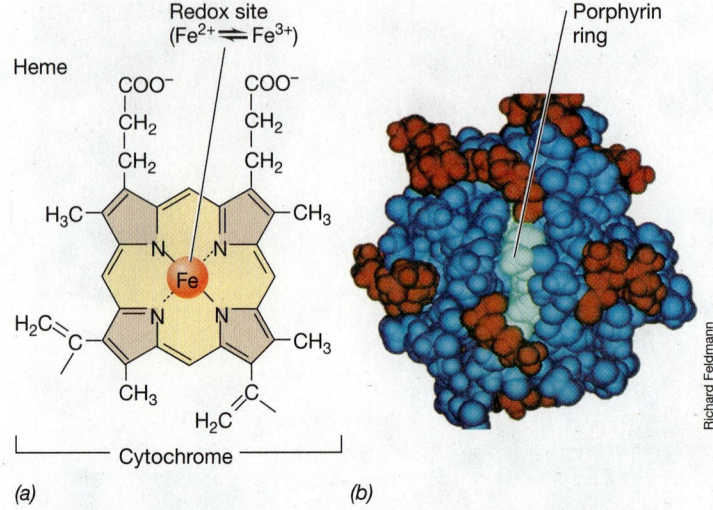

(a)   *(b)*

Figure 3.17   Cytochrome and its structure. *(a)* Structure of heme, the iron-containing portion of cytochromes. Cytochromes carry electrons only, and the redox site is the iron atom, which can alternate between the Fe^{2+} and Fe^{3+} oxidation states. *(b)* Space-filling model of cytochrome *c*; heme (light blue) is covalently linked via disulfide bridges to cysteine residues in the protein (dark blue). Cytochromes are tetrapyrroles, composed of four pyrrole rings.

Flavoproteins contain a derivative of the vitamin riboflavin (**Figure 3.16**). The flavin portion, which is bound to a protein, is a prosthetic group (Section 3.5) that is reduced as it accepts 2 e$^-$ + 2 H$^+$ and oxidized when 2 e$^-$ are passed on to the next carrier in the chain. Note that flavoproteins *accept* 2 e$^-$ + 2 H$^+$ but *donate* only electrons. We will consider what happens to the 2 H$^+$ later. Two flavins are commonly found in cells, flavin mononucleotide (FMN, Figure 3.16) and flavin adenine dinucleotide (FAD). In the latter, FMN is bonded to ribose and adenine through a second phosphate. Riboflavin, also called vitamin B$_2$, is a source of the parent flavin molecule in flavoproteins and is a required growth factor for some organisms (Table 3.1).

Cytochromes, Other Iron Proteins, and Quinones

The cytochromes are proteins that contain heme prosthetic groups (**Figure 3.17**). Cytochromes undergo oxidation and reduction through loss or gain of a single electron by the iron atom in the heme of the cytochrome:

Cytochrome—Fe^{2+} ⇆ cytochrome—Fe^{3+} − e$^-$

Several classes of cytochromes are known, differing widely in their reduction potentials (Figure 3.9). Different classes of cytochromes are designated by letters, such as cytochrome *a*, cytochrome *b*, cytochrome *c*, and so on, depending upon the type of heme they contain. The cytochromes of a given class in one organism may differ slightly from those of another, and so there are designations such as cytochromes a_1, a_2, a_3, and so on among cytochromes of the same class. Cytochromes of different classes also differ in their reduction potentials (Figure 3.9). Occasionally, cytochromes form complexes with other cytochromes or with iron–sulfur proteins. An important example is the cytochrome bc_1 complex, which contains two different *b*-type cytochromes and one *c*-type cytochrome. The cytochrome bc_1 complex plays an important role in energy metabolism, as we will see later.

In addition to the cytochromes, in which iron is bound to heme, one or more proteins with nonheme iron typically participate in electron transport chains. These proteins contain prosthetic groups made up of clusters of iron and sulfur atoms, with Fe$_2$S$_2$ and Fe$_4$S$_4$ clusters being the most common (**Figure 3.18**). *Ferredoxin*, a common nonheme iron–sulfur protein, has an Fe$_2$S$_2$ configuration. The reduction potentials of iron–sulfur proteins vary over a wide range depending on the number of iron and sulfur atoms present and how the iron centers are embedded in the protein. Thus, different iron–sulfur proteins can function at different locations in the electron transport chain. Like cytochromes, nonheme iron–sulfur proteins carry electrons only.

Quinones (**Figure 3.19**) are hydrophobic molecules that lack a protein component. Because they are small and hydrophobic, quinones are free to move about within the membrane. Like the flavins (Figure 3.16), quinones accept 2 e$^-$ + 2 H$^+$ but transfer only 2 e$^-$ to the next carrier in the chain; quinones typically participate as links between iron–sulfur proteins and the first cytochromes in the electron transport chain.

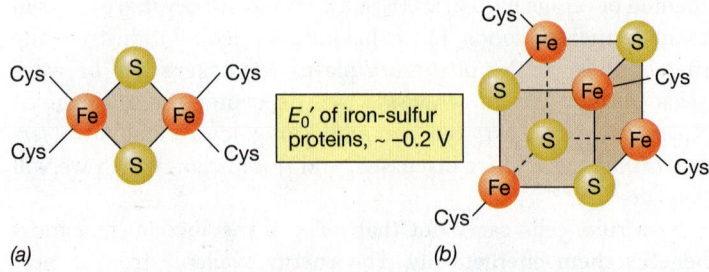

E_0' of iron-sulfur proteins, ~ −0.2 V

(a)   *(b)*

Figure 3.18   Arrangement of the iron–sulfur centers of nonheme iron–sulfur proteins. *(a)* Fe$_2$S$_2$ center. *(b)* Fe$_4$S$_4$ center. The cysteine (Cys) linkages are from the protein portion of the molecule.

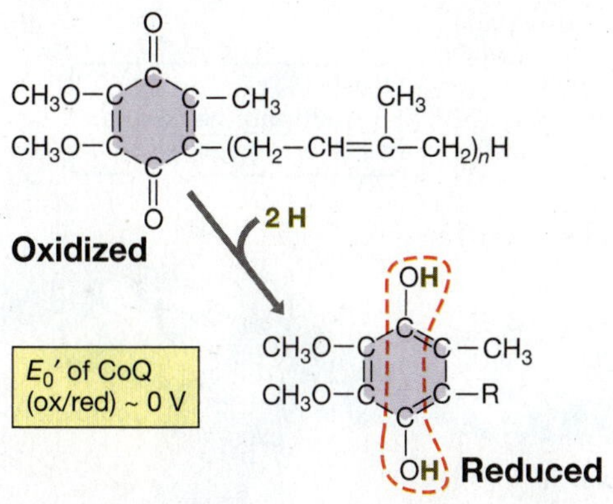

Figure 3.19 **Structure of oxidized and reduced forms of coenzyme Q, a quinone.** The five-carbon unit in the side chain (an isoprenoid) occurs in multiples, typically 6–10. Oxidized quinone requires 2 e^- and 2 H^+ to become fully reduced (dashed red circles).

E_0' of CoQ (ox/red) ~ 0 V

MINIQUIZ

- In what major way do quinones differ from other electron carriers in the membrane?
- Which electron carriers described in this section accept 2 e^- + 2 H^+? Which accept electrons only?

3.11 Respiration: The Proton Motive Force

The conservation of energy in respiration is linked to an energized state of the membrane (Figure 3.13b), and this energized state is established by electron transport. To understand how electron transport is linked to ATP synthesis, we must first understand how the electron transport system is organized in the cytoplasmic membrane. The electron transport carriers we just discussed (Figures 3.16–3.19) are oriented in the membrane in such a way that, as electrons are transported, protons are separated from electrons. Two electrons plus two protons enter the electron transport chain from NADH (through NADH dehydrogenase) to initiate the process. Carriers in the electron transport chain are arranged in the membrane in order of their *increasingly positive* reduction potential, with the final carrier in the chain donating the electrons plus protons to a terminal electron acceptor such as O_2.

During electron transport, H^+ are extruded to the outer surface of the membrane. These originate from two sources: (1) NADH and (2) the dissociation of H_2O into H^+ and OH^- in the cytoplasm. The extrusion of H^+ to the environment results in the accumulation of OH^- on the inside of the membrane. However, despite their small size, neither H^+ nor OH^- can diffuse through the membrane because they are charged and highly polar (⮌ Section 2.8). As a result of the separation of H^+ and OH^-, the two sides of the membrane differ in both charge and pH; this forms an *electrochemical potential* across the membrane. This potential, along with the difference in pH across the membrane, is called the **proton motive force (pmf)** and causes the membrane to be energized, much like a battery (Figure 3.13b). Some of the potential energy in the pmf is then conserved in the formation of ATP. However, besides driving ATP synthesis, the pmf can also be tapped to do other forms of work for the cell, such as transport reactions, flagellar rotation, and other energy-requiring reactions in the cell.

Figure 3.20 shows a bacterial electron transport chain, one of many different carrier sequences known. Nevertheless, three features are characteristic of *all* electron transport chains regardless of which specific carriers they contain: (1) the carriers are arranged in order of increasingly more positive E_0', (2) there is an alternation of electron-only and electron-plus-proton carriers in the chain, and (3) the net result is reduction of a terminal electron acceptor and generation of a proton motive force.

Generation of the Proton Motive Force: Complexes I and II

The proton motive force develops from the activities of flavins, quinones, the cytochrome bc_1 complex, and the terminal cytochrome oxidase. Following the oxidation of NADH + H^+ to form $FMNH_2$, 4 H^+ are released to the outer surface of the membrane when $FMNH_2$ donates 2 e^- to a series of nonheme iron proteins (Fe/S), forming the group of electron transport proteins called *Complex I* (Figure 3.20). These groups are called *complexes* because each consists of several proteins that function as a unit. For example, Complex I in *Escherichia coli* contains 14 separate proteins. Complex I is also called *NADH: quinone oxidoreductase* because the overall reaction is one in which NADH is oxidized and quinone is reduced. Two H^+ from the cytoplasm are taken up by coenzyme Q when it is reduced by the Fe/S protein in Complex I (Figure 3.20).

Complex II simply bypasses Complex I and feeds electrons from $FADH_2$ directly into the quinone pool. Complex II is also called the *succinate dehydrogenase complex* because of the specific substrate, succinate (a product of the citric acid cycle, Section 3.12), that it oxidizes. However, because Complex II bypasses Complex I (where electrons enter at a more negative reduction potential), fewer protons are pumped per 2 e^- that enter at Complex II than enter at Complex I (Figure 3.20); this reduces the ATP yield by one per two electrons consumed.

Complexes III and IV: bc_1 and a-Type Cytochromes

Reduced coenzyme Q (QH_2) passes electrons one at a time to the cytochrome bc_1 complex (*Complex III*, Figure 3.20). Complex III consists of several proteins that contain two different b-type hemes (b_L and b_H), one c-type heme (c_1), and one iron–sulfur center. The bc_1 complex is present in the electron transport chain of almost all organisms that can respire and also plays a role in photosynthetic electron flow in phototrophic organisms (⮌ Sections 13.3 and 13.4).

The major function of the cytochrome bc_1 complex is to move e^- from quinones to cytochrome c. Electrons travel from the bc_1 complex to cytochrome c, located in the periplasm. Cytochrome c functions as a shuttle to transfer e^- to the high-redox-potential cytochromes a and a_3 (*Complex IV*, Figure 3.20). Complex IV functions as the terminal oxidase and reduces O_2 to H_2O in the final step of the electron transport chain. Complex IV also pumps protons to the outer surface of the membrane, thereby increasing the strength of the proton motive force (Figure 3.20).

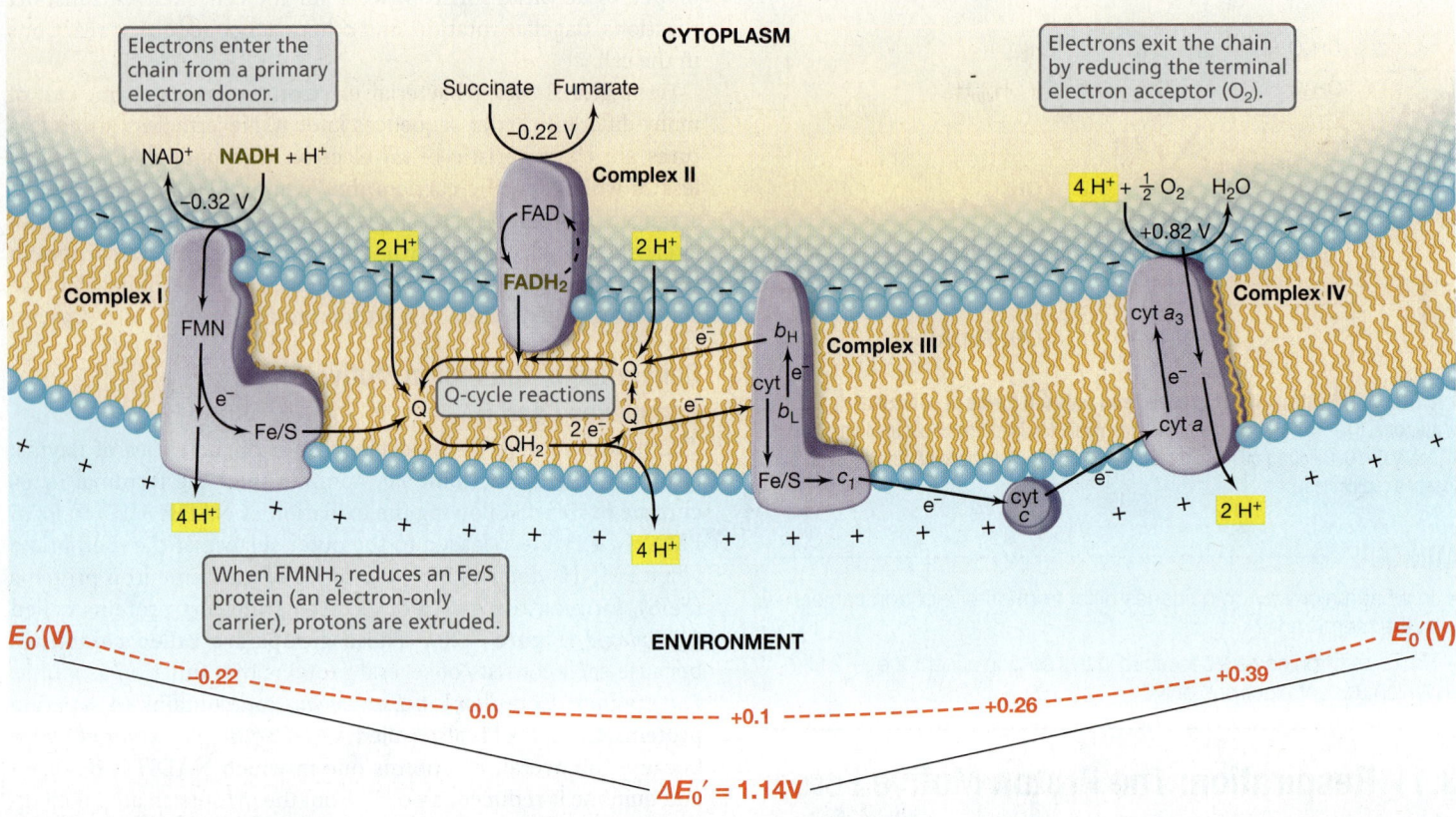

Figure 3.20 Generation of the proton motive force during aerobic respiration. The orientation of electron carriers in the membrane of *Paracoccus denitrificans*, a model organism for studies of respiration. The + and − charges at the edges of the membrane represent H$^+$ and OH$^-$, respectively. Abbreviations: FMN, flavin mononucleotide; FAD, flavin adenine dinucleotide; Q, quinone; Fe/S, iron–sulfur– protein; cyt *a, b, c*, cytochromes (b_L and b_H, low- and high-potential *b*-type cytochromes, respectively). At the quinone site, electrons are recycled from Q to bc_1 from reactions of the "Q cycle." Electrons from QH_2 can be split in the bc_1 complex between the Fe/S protein and the *b*-type cytochromes. Electrons that travel through the cytochromes reduce Q (in two, one-electron steps) back to QH_2, thus increasing the number of protons pumped at the Q-bc_1 site. Electrons that travel to the Fe/S protein proceed to reduce cytochrome c_1, and from there cytochrome *c*. Complex II, the succinate dehydrogenase complex, bypasses Complex I and feeds electrons directly into the quinone pool at a more positive E_0' than NADH (see the electron tower in Figure 3.9).

Besides transferring e$^-$ to cytochrome *c*, the cytochrome bc_1 complex can also interact with quinones in such a way that on average, two additional H$^+$ are pumped at the Q-bc_1 site. This happens in a series of electron exchanges between cytochrome bc_1 and Q, called the Q cycle. Because quinone and bc_1 have roughly the same E_0' (near 0 V, Figure 3.9), quinone molecules can alternately become oxidized and reduced using electrons fed back to quinones from the bc_1 complex. This mechanism allows on average a total of 4 H$^+$ (instead of 2 H$^+$) to be pumped to the outer surface of the membrane at the Q-bc_1 site for every 2 e$^-$ that enter the chain in Complex I (Figure 3.20). This strengthens the proton motive force, and as we will see now, it is the proton motive force that drives ATP synthesis.

ATP Synthase

How does the proton motive force generated by electron transport (Figure 3.20) actually drive ATP synthesis? Interestingly, a strong parallel exists between the mechanism of ATP synthesis and the mechanism behind the motor that drives rotation of the bacterial flagellum (⮌ Section 2.17). In analogy to how dissipation of the pmf applies torque that rotates the bacterial flagellum,

the pmf also creates torque in a large membrane protein complex that makes ATP. This complex is called **ATP synthase**, or **ATPase** for short.

ATPases consist of two components, a multiprotein complex called F$_1$ that sticks into the cytoplasm and carries out ATP synthesis, and a membrane-integrated component called F$_0$ that carries out the ion-translocating function (**Figure 3.21**). ATPase catalyzes a reversible reaction between ATP and ADP + P$_i$ as shown in the figure. The structure of ATPase proteins is highly conserved throughout all the domains of life, indicating that this mechanism of energy conservation was a very early evolutionary invention.

F$_1$ and F$_0$ are actually two rotary motors. The movement of H$^+$ through F$_0$ into the cytoplasm is coupled to the rotation of its *c* proteins. This generates a torque that is transmitted to F$_1$ via the coupled rotation of the γε subunits (Figure 3.21). The rotation causes conformational changes in the β subunits of F$_1$ and this allows them to bind ADP + P$_i$. ATP is synthesized when the β subunits return to their original conformation. As this occurs, the free energy of their rotated state is released and is coupled to ATP synthesis. Quantitative measures of the number of H$^+$ consumed by ATPase per ATP produced yield a number between 3 and 4.

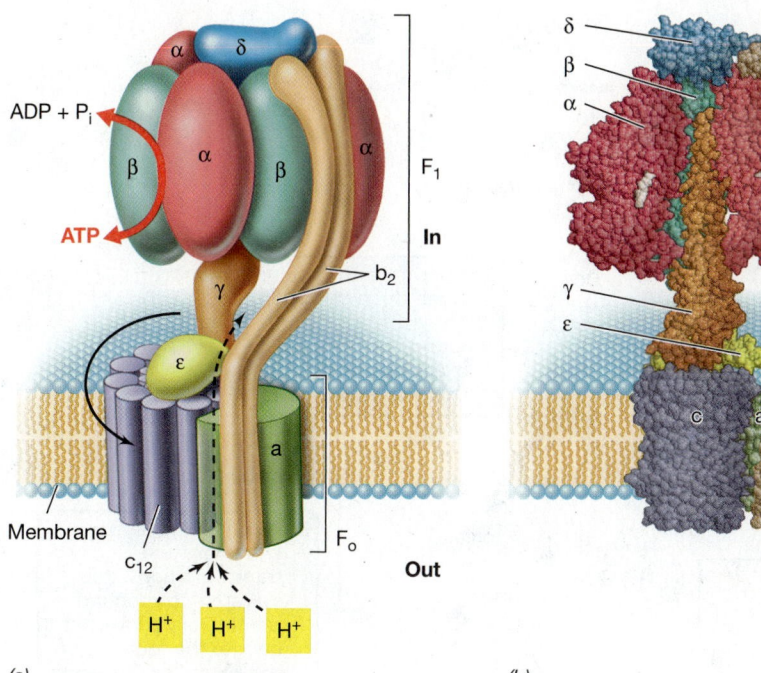

(a)

(b)

Siegfried Engelbrecht-Vandré

Figure 3.21 Structure and function of the reversible ATP synthase (ATPase) in *Escherichia coli*. *(a)* Schematic. F_1 consists of five different polypeptides forming an $\alpha_3\beta_3\gamma\epsilon\delta$ complex, the stator. F_1 is the catalytic complex responsible for the interconversion of ADP + P_i and ATP. F_o, the rotor, is integrated in the membrane and consists of three polypeptides in an ab_2c_{12} complex. As protons enter, the dissipation of the proton motive force drives ATP synthesis (3 H^+/ATP). *(b)* Space-filling model. The color-coding corresponds to the art in part *a*. Since proton translocation from outside the cell to inside the cell leads to ATP synthesis by ATPase, it follows that proton translocation from inside to outside in the electron transport chain (Figure 3.20) represents work done on the system and a source of potential energy.

Reversibility of ATPase

ATPase is reversible. The hydrolysis of ATP supplies torque for $\gamma\epsilon$ to rotate in the opposite direction from that in ATP synthesis, and this pumps H^+ from the cytoplasm to the environment through F_o (Figure 3.21). The net result in this case is *generation of* instead of *dissipation of* the proton motive force. Reversibility of the ATPase explains why strictly fermentative bacteria that lack electron transport chains and are unable to carry out oxidative phosphorylation still contain ATPases. Many important reactions in the cell, such as flagellar rotation and some forms of transport, are coupled to energy from the pmf rather than directly from ATP. Thus, the ATPase of organisms incapable of respiration, such as the strictly fermentative lactic acid bacteria, functions unidirectionally to generate this required pmf from ATP formed during substrate-level phosphorylation in fermentation.

MINIQUIZ

- How do electron transport reactions generate the proton motive force?
- What is the ratio of H^+ extruded per NADH oxidized through the electron transport chain of *Paracoccus* shown in Figure 3.20? At which sites in the chain is the proton motive force being established?
- What structure in the cell converts the proton motive force to ATP? How does it function?

3.12 Respiration: Citric Acid and Glyoxylate Cycles

Now that we have a feeling for how ATP synthesis is coupled to electron transport, we need to consider another important aspect of respiration—the production of CO_2. Our focus here will be on the citric acid cycle (Krebs cycle), a key pathway in virtually all cells, and the glyoxylate cycle, a variation on the citric acid cycle necessary when two-carbon electron donors are respired.

Respiration of Glucose

The early biochemical steps in the respiration of glucose are the same as those of glycolysis; all steps from glucose to pyruvate (Figure 3.14) are the same. However, whereas in fermentation pyruvate is reduced and converted into products that are subsequently excreted, in respiration, pyruvate is oxidized to CO_2. The pathway by which pyruvate is oxidized to CO_2 is called the **citric acid cycle** (Figure 3.22).

In the citric acid cycle, pyruvate is first decarboxylated, leading to the production of CO_2, NADH, and the energy-rich substance *acetyl-CoA*. The acetyl group of acetyl-CoA then combines with the four-carbon compound oxaloacetate, forming the six-carbon compound citric acid. A series of reactions follow, and two additional CO_2 molecules, three more NADH, and one FADH are formed. Ultimately, oxaloacetate is regenerated to return as an acetyl acceptor, thus completing the cycle (Figure 3.22).

CO_2 Release and Electron Transport: The Connection

How are reactions of the citric acid cycle and the electron transport chain connected? The oxidation of pyruvate to CO_2 requires the concerted activity of the citric acid cycle and the electron transport chain. For each pyruvate molecule oxidized through the citric acid cycle, three CO_2 molecules are produced (Figure 3.22). Electrons released during the oxidation of intermediates in the citric acid cycle are transferred to NAD^+ to form NADH, or in one reaction, to FAD to form $FADH_2$. The combined reactions of the citric acid cycle and electron transport chain allow for the complete oxidation of glucose to CO_2 along

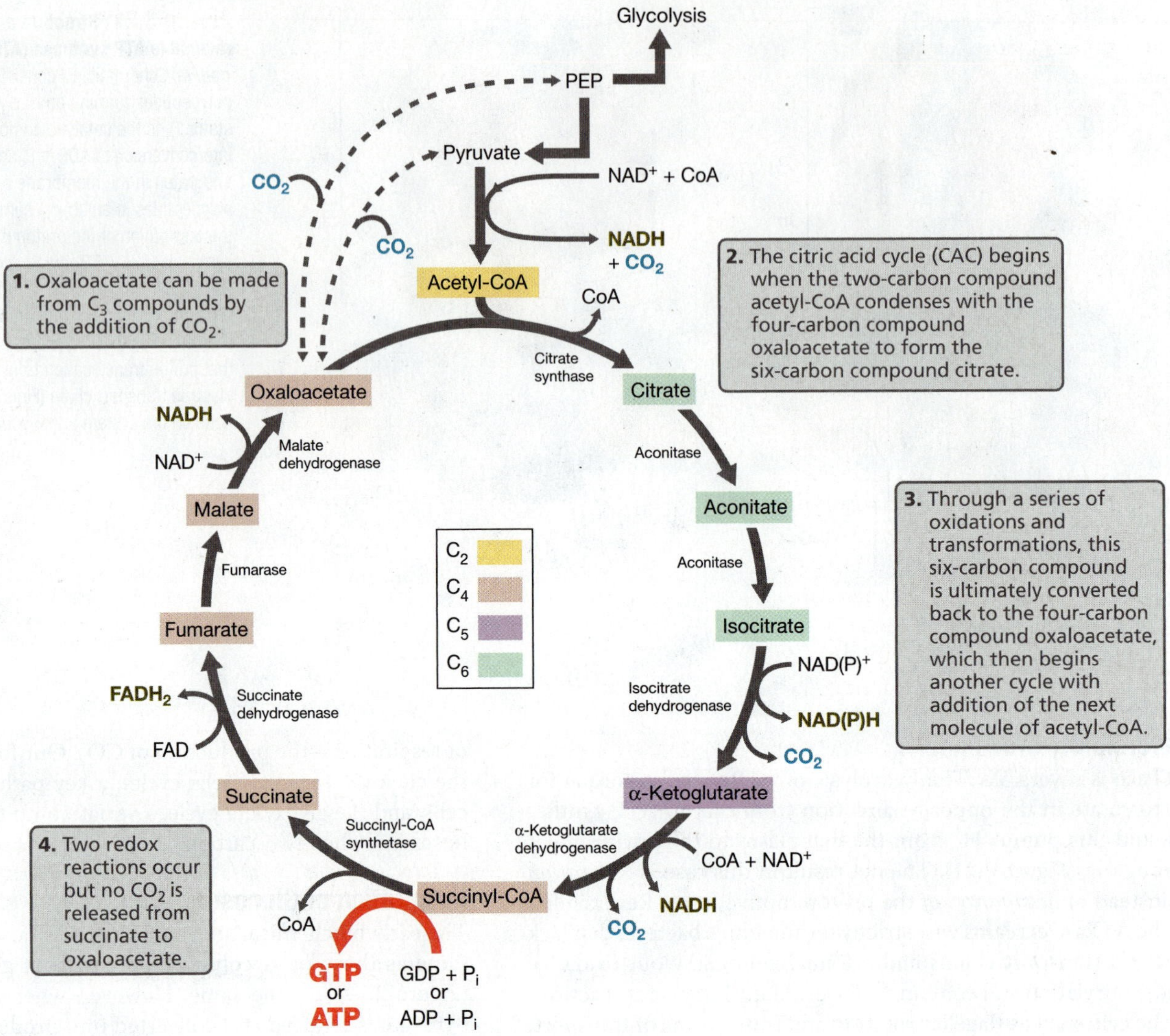

Glycolysis

PEP

Pyruvate

CO_2

CO_2

NAD$^+$ + CoA

NADH
+ CO_2

Acetyl-CoA

CoA

1. Oxaloacetate can be made from C$_3$ compounds by the addition of CO_2.

2. The citric acid cycle (CAC) begins when the two-carbon compound acetyl-CoA condenses with the four-carbon compound oxaloacetate to form the six-carbon compound citrate.

Citrate synthase

Citrate

NADH

NAD$^+$

Oxaloacetate

Malate dehydrogenase

Aconitase

Aconitate

Malate

Aconitase

Fumarase

C$_2$
C$_4$
C$_5$
C$_6$

Isocitrate

3. Through a series of oxidations and transformations, this six-carbon compound is ultimately converted back to the four-carbon compound oxaloacetate, which then begins another cycle with addition of the next molecule of acetyl-CoA.

Fumarate

NAD(P)$^+$

FADH$_2$

Isocitrate dehydrogenase

NAD(P)H

FAD

Succinate dehydrogenase

CO_2

Succinate

α-Ketoglutarate

Succinyl-CoA synthetase

α-Ketoglutarate dehydrogenase

CoA + NAD$^+$

4. Two redox reactions occur but no CO_2 is released from succinate to oxaloacetate.

CoA

Succinyl-CoA

NADH

CO_2

GTP GDP + P$_i$
or   or
ATP ADP + P$_i$

(a) **The citric acid cycle**

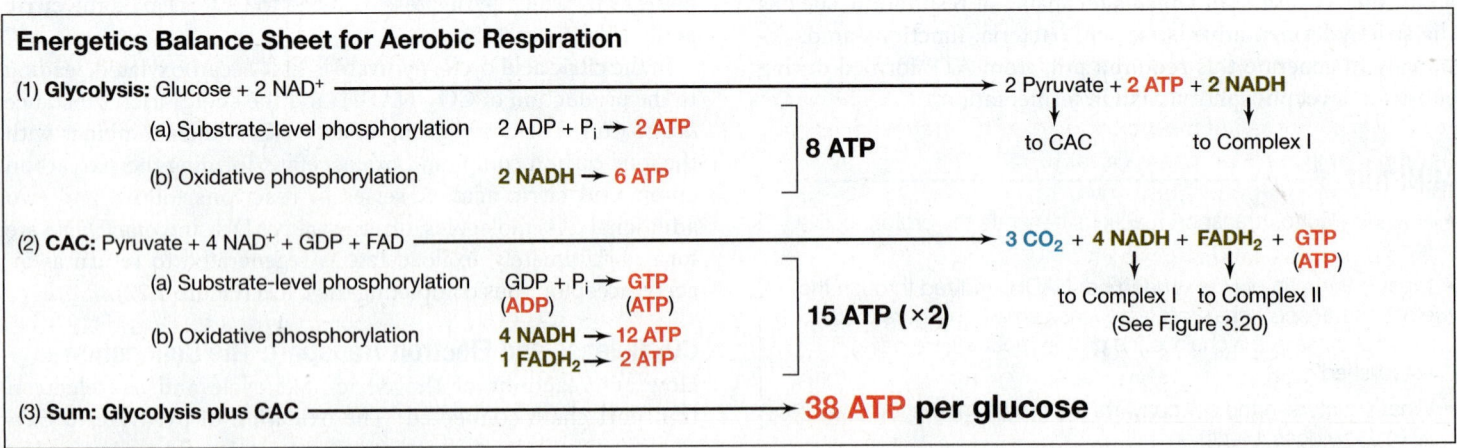

Energetics Balance Sheet for Aerobic Respiration

(1) **Glycolysis:** Glucose + 2 NAD$^+$ ⟶ 2 Pyruvate + **2 ATP** + **2 NADH**

    (a) Substrate-level phosphorylation   2 ADP + P$_i$ ⟶ **2 ATP**   ⎱ **8 ATP**   to CAC   to Complex I

    (b) Oxidative phosphorylation   **2 NADH** ⟶ **6 ATP**

(2) **CAC:** Pyruvate + 4 NAD$^+$ + GDP + FAD ⟶ **3 CO_2** + **4 NADH** + **FADH$_2$** + **GTP** **(ATP)**

    (a) Substrate-level phosphorylation   GDP + P$_i$ ⟶ **GTP**   ⎱ **15 ATP (×2)**   to Complex I   to Complex II
                                                          (ADP)   **(ATP)**   (See Figure 3.20)

    (b) Oxidative phosphorylation   **4 NADH** ⟶ **12 ATP**
                                       1 FADH$_2$ ⟶ **2 ATP**

(3) **Sum: Glycolysis plus CAC** ⟶ **38 ATP per glucose**

(b) **Energy yield from the citric acid cycle**

Figure 3.22 The citric acid cycle. *(a)* The citric acid cycle (CAC) begins when the two-carbon compound acetyl-CoA condenses with the four-carbon compound oxaloacetate to form the six-carbon compound citrate. Through a series of oxidations and transformations, citrate is converted to two CO_2 and the acetyl acceptor molecule, oxaloacetate. *(b)* The overall balance sheet of fuel (NADH/FADH$_2$) for the electron transport chain and CO_2 generated in the citric acid cycle. NADH and FADH$_2$ feed into electron transport chain Complexes I and II, respectively (Figure 3.20).

with a much greater yield of energy. Whereas only *2 ATP* are produced per glucose fermented in alcoholic or lactic acid fermentations (Figure 3.14 and Table 3.4), a total of *38 ATP* can be made by aerobically respiring the same glucose molecule to $CO_2 + H_2O$ (Figure 3.22*b*).

Biosynthesis and the Citric Acid Cycle

Besides its role in combusting pyruvate to CO_2, the citric acid cycle plays another important role in the cell. The cycle is composed of several key intermediates, small amounts of which are drawn off during growth for biosynthetic purposes. Particularly important in this regard are α-ketoglutarate and oxaloacetate, which are precursors of several amino acids (Section 3.15), and succinyl-CoA, needed to form cytochromes, chlorophyll, and several other tetrapyrroles (compounds composed of four pyrrole rings; see Figure 3.17). Oxaloacetate is also important because it can be converted to phosphoenolpyruvate, a precursor of glucose. In addition, acetate provides the starting material for fatty acid biosynthesis (Section 3.16, and see Figure 3.30). The citric acid cycle thus plays two major roles in the cell: *energy conservation* and *biosynthesis*. Much the same can be said about the glycolytic pathway, as certain intermediates from this pathway can be drawn off for biosynthetic needs as well (Sections 3.14 and 3.15).

The Glyoxylate Cycle

Citrate, malate, fumarate, and succinate are common natural products, and organisms that use these C_4 or C_6 compounds as energy sources use the citric acid cycle for their catabolism. By contrast, two-carbon compounds such as acetate cannot be used as growth substrates by the citric acid cycle alone. This is because the citric acid cycle can continue to operate only if oxaloacetate is regenerated at each turn of the cycle; any siphoning off of oxaloacetate (or any other citric acid cycle intermediates) for biosynthesis would starve the cycle of what it needs to continue functioning (Figure 3.22). Thus, when acetate is used as an electron donor, a variation on the citric acid cycle called the **glyoxylate cycle** (Figure 3.23) is employed, so named because the C_2 compound glyoxylate is a key intermediate. The glyoxylate cycle is composed of most citric acid cycle reactions plus two additional enzymes: *isocitrate lyase*, which splits isocitrate into succinate and glyoxylate, and *malate synthase*, which converts glyoxylate and acetyl-CoA to malate (Figure 3.23). The splitting of isocitrate yields succinate, which can be used for biosynthesis, and glyoxylate. The latter combines with acetyl-CoA (C_2) to yield malate (C_4). From malate, the acceptor molecule oxaloacetate is produced and can enter a new round of acetyl-CoA oxidation in the citric acid cycle (Figure 3.22).

Three-carbon compounds such as pyruvate or compounds that are converted to pyruvate (for example, lactate or carbohydrates) also cannot be catabolized through the citric acid cycle alone. But here the glyoxylate cycle is unnecessary because any shortage of citric acid cycle intermediates is corrected by synthesizing oxaloacetate from pyruvate or phosphoenolpyruvate. This occurs by the addition of CO_2 to pyruvate or phosphoenolpyruvate by the enzymes *pyruvate carboxylase* or *phosphoenolpyruvate carboxylase*, respectively (Figure 3.22).

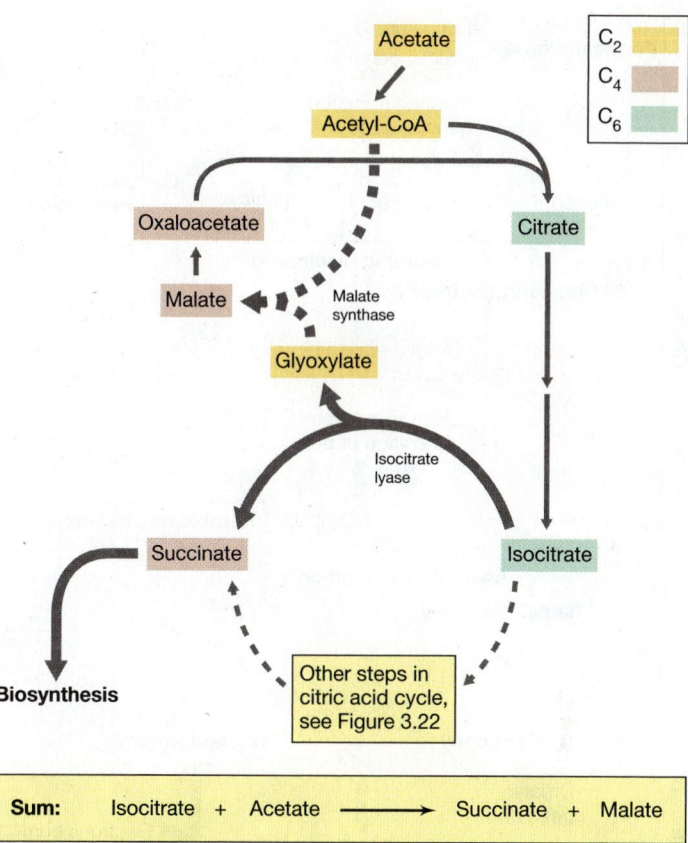

Figure 3.23 **The glyoxylate cycle.** These reactions occur in conjunction with the citric acid cycle when cells grow on two-carbon electron donors, such as acetate. The glyoxylate cycle regenerates oxaloacetate (from malate) to continue the citric acid cycle.

MINIQUIZ

- How many molecules of CO_2 and pairs of electrons are released per pyruvate that enters the citric acid cycle?
- What two major roles do the citric acid cycle and glycolysis have in common?
- Why is the glyoxylate cycle necessary for growth on acetate but not on succinate?

3.13 Catabolic Diversity

Thus far in this chapter we have dealt only with catabolism by chemoorganotrophs. We now briefly consider catabolic diversity and some of the alternatives to fermentation or respiration. These include *anaerobic respiration*, *chemolithotrophy*, and *phototrophy* (Figure 3.24).

Anaerobic Respiration

Under anoxic conditions, electron acceptors other than oxygen support respiration in certain prokaryotes. This is called **anaerobic respiration**. Some of the electron acceptors used in anaerobic respiration include nitrate (NO_3^-, reduced to nitrite, NO_2^-, by *Escherichia coli* or to N_2 by *Pseudomonas* species), ferric iron (Fe^{3+}, reduced to Fe^{2+} by *Geobacter* species), sulfate (SO_4^{2-}, reduced to hydrogen sulfide, H_2S, by *Desulfovibrio* species), carbonate (CO_3^{2-}, reduced to methane, CH_4, by methanogens or to acetate

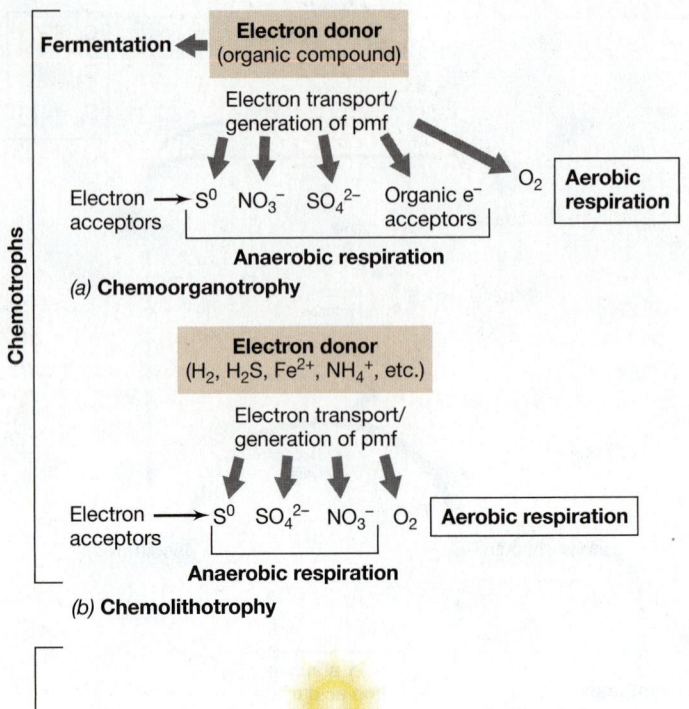

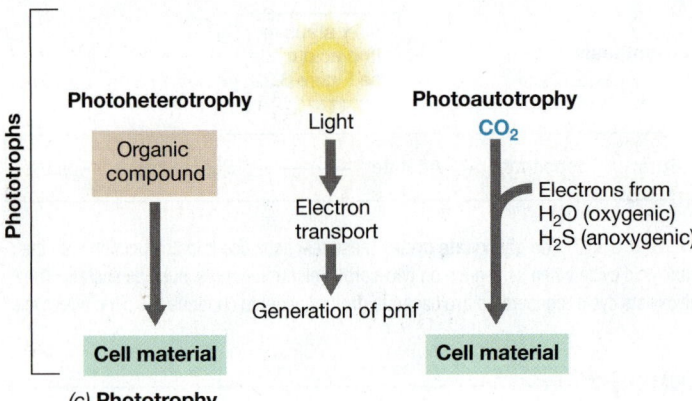

(a) **Chemoorganotrophy**

(b) **Chemolithotrophy**

(c) **Phototrophy**

Figure 3.24 Catabolic diversity. *(a)* Chemoorganotrophs. *(b)* Chemolithotrophs. *(c)* Phototrophs. Note the importance of the formation of the proton motive force driven by electron transport in both forms of respiration and in photosynthesis.

by acetogens), and even certain organic compounds, such as the citric acid cycle intermediate fumarate.

Because of the positions of these alternative electron acceptors on the redox tower (none has an E_0' as positive as the O_2/H_2O couple; Figure 3.9), less energy is conserved when they are reduced compared to when O_2 is reduced (recall that $\Delta G^{0'}$ is proportional to $\Delta E^{0'}$; Section 3.4 and Figure 3.9). Nevertheless, because O_2 is often limiting or even totally absent in many microbial habitats, anaerobic respirations can be very important means of energy generation. As in aerobic respiration, anaerobic respirations require electron transport, generate a proton motive force, and employ ATPase to make ATP (Sections 3.10–3.12).

Chemolithotrophy and Phototrophy

Organisms able to use *inorganic* chemicals as electron donors are called *chemolithotrophs* (Section 3.3). Examples of relevant inorganic electron donors include H_2S, hydrogen gas (H_2), Fe^{2+}, and NH_3.

Chemolithotrophic metabolisms are typically aerobic and begin with the oxidation of the inorganic electron donor by an electron transport chain. This results in a proton motive force, as we have already considered for the oxidation of organic electron donors by chemoorganotrophs (Figure 3.20). However, another important distinction between chemolithotrophs and chemoorganotrophs is their source of carbon for biosynthesis. Chemoorganotrophs are heterotrophs and thus use organic compounds (glucose, acetate, and the like) as carbon sources. By contrast, chemolithotrophs use carbon dioxide (CO_2) as a carbon source and are therefore autotrophs.

In the process of photosynthesis, carried out by *phototrophs*, light is used in place of a chemical to generate a proton motive force. During phototrophic metabolism, ATP is synthesized from ATPase activity during **photophosphorylation**, the light-driven analog of oxidative phosphorylation (Section 3.8). Most phototrophs assimilate CO_2 as their carbon source and are therefore *photoautotrophs*. However, some phototrophs use organic compounds as carbon sources with light as the energy source; these are the *photoheterotrophs* (Figure 3.24).

The PMF and Catabolic Diversity

With the exception of fermentation, in which substrate-level phosphorylation occurs (Section 3.8), all other mechanisms for energy conservation employ the proton motive force. Whether electrons come from the oxidation of organic or inorganic chemicals or from light-driven processes, in all forms of respiration and photosynthesis, energy conservation is linked to the establishment of a pmf and its dissipation by ATPase to form ATP (Figure 3.24). Respiration and anaerobic respiration can thus be viewed as variations on a theme of *different electron acceptors*. Likewise, chemoorganotrophy, chemolithotrophy, and photosynthesis are variations on a theme of *different electron donors*. Electron transport and the pmf link all of these processes, bringing these seemingly quite different forms of energy metabolism into a common focus. We pick up on this theme and explore it in more depth in Chapter 13.

MINIQUIZ -
- In terms of their electron donors, how do chemoorganotrophs differ from chemolithotrophs?
- What is the carbon source for autotrophic organisms?
- Why can it be said that the proton motive force is a unifying theme in most of bacterial metabolism?

IV · Biosyntheses

We close this chapter with a brief consideration of biosynthesis. Our focus here will be an overview of the biosynthesis of the building blocks of the four classes of macromolecules—sugars (polysaccharides), amino acids (proteins), nucleotides (nucleic acids), and fatty acids (lipids). Collectively, these biosyntheses are that part of metabolism called **anabolism**. We will also touch on polysaccharide and lipid biosyntheses here, and see how some prokaryotes can assimilate gaseous nitrogen (N_2) as their source of cell nitrogen.

3.14 Sugars and Polysaccharides

Polysaccharides are key components of microbial cell walls and cells often store carbon and energy reserves in the form of the polysaccharides glycogen or starch (Chapter 2). How are such large molecules made?

Polysaccharide Biosyntheses and Gluconeogenesis

Polysaccharides are synthesized from *activated* forms of glucose, either uridine diphosphoglucose (UDPG; **Figure 3.25a**) or adenosine diphosphoglucose (ADPG). UDPG is the precursor of several glucose derivatives needed for the biosynthesis of structural polysaccharides in the cell, such as *N*-acetylglucosamine and *N*-acetylmuramic acid in peptidoglycan or the lipopolysaccharide component of the gram-negative outer membrane (↺ Sections 2.10 and 2.11). Storage polysaccharides are made by adding activated glucose to preexisting polymer. For example, glycogen is synthesized as ADPG + glycogen → ADP + glycogen-glucose.

When a cell is growing on a hexose such as glucose, obtaining glucose for polysaccharide synthesis is obviously not a problem. But when the cell is growing on other carbon compounds, glucose must be synthesized. This process, called *gluconeogenesis*, uses phosphoenolpyruvate, one of the intermediates of glycolysis, as starting material and travels backwards through the glycolytic pathway to form glucose (Figure 3.14). Phosphoenolpyruvate can be synthesized from oxaloacetate, a citric acid cycle intermediate (Figure 3.22). An overview of gluconeogenesis is shown in Figure 3.25*b*.

Pentose Metabolism and the Pentose Phosphate Pathway

Pentoses are formed by the removal of one carbon atom from a hexose, typically as CO_2. The pentoses needed for nucleic acid synthesis, ribose (in RNA) and deoxyribose (in DNA), are formed

as shown in Figure 3.25*c*. The enzyme ribonucleotide reductase converts ribose into deoxyribose by reduction of the hydroxyl (—OH) group on the 2′ carbon of the 5-carbon pentose ring. This reaction occurs after, not before, synthesis of nucleotides. Thus, *ribo*nucleotides are biosynthesized, and some of them are later reduced to *deoxy*ribonucleotides for use as precursors of DNA.

Pentoses are made from hexose sugars, and the major pathway for this process is the **pentose phosphate pathway** (**Figure 3.26**). In this pathway, glucose is oxidized to CO_2, NADPH, and the key intermediate, *ribulose 5-phosphate*; from the latter, various pentose derivatives are formed. When pentoses are used as electron donors, they feed directly into the pentose phosphate pathway, typically becoming phosphorylated to form ribose phosphate or a related compound before being further catabolized (Figure 3.26).

Besides its importance in pentose metabolism, the pentose phosphate pathway is also responsible for producing many important nonpentose sugars in the cell, including C_4–C_7 sugars. These sugars can eventually be converted to hexoses for either catabolic purposes or for biosynthesis (Figure 3.26). A final important aspect of the pentose phosphate pathway is that it generates NADPH, a coenzyme used for many reductive biosyntheses, in particular, as a reductant for the production of deoxyribonucleotides (Figure 3.25*c*). Although many cells have an exchange mechanism for converting NADH into NADPH, the pentose phosphate pathway is the major means for the direct synthesis of this important coenzyme.

MINIQUIZ

• What form of activated glucose is used in the biosynthesis of glycogen by bacteria?

• What is gluconeogenesis?

• What functions does the pentose phosphate pathway play in the cell?

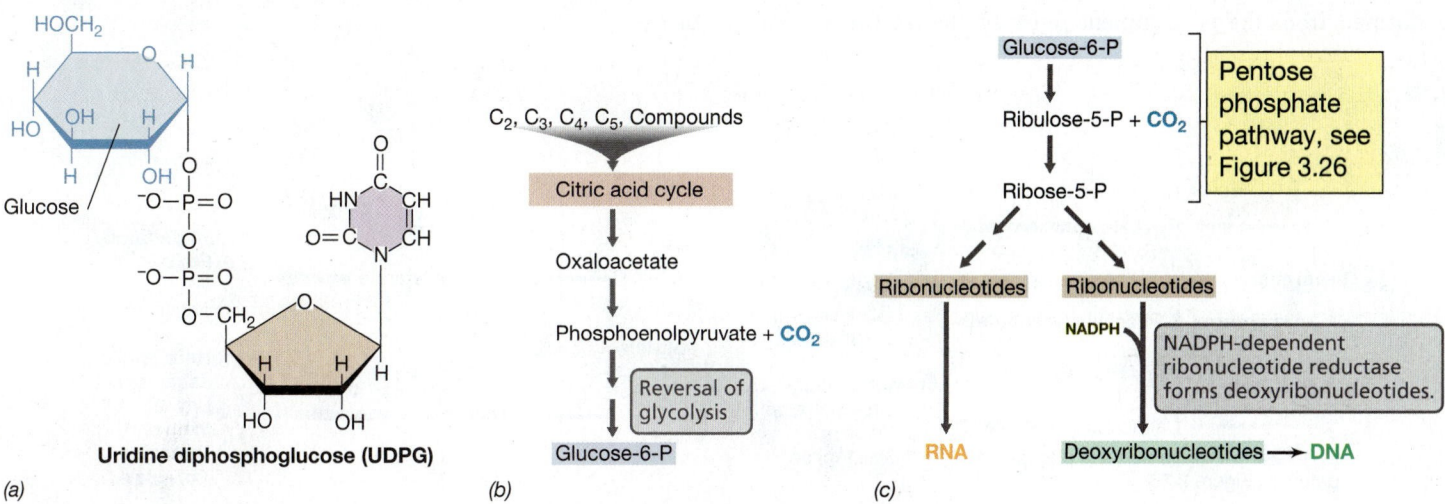

Figure 3.25 **Sugar metabolism.** *(a)* Polysaccharides are synthesized from activated forms of hexoses such as UDPG. *(b)* Gluconeogenesis. When glucose is needed, it can be biosynthesized from other carbon compounds, generally by the reversal of steps in glycolysis. *(c)* Pentoses for nucleic acid synthesis are formed by decarboxylation of hexoses such as glucose 6-phosphate. Note how the precursors of DNA are produced from the precursors of RNA by the enzyme ribonucleotide reductase.

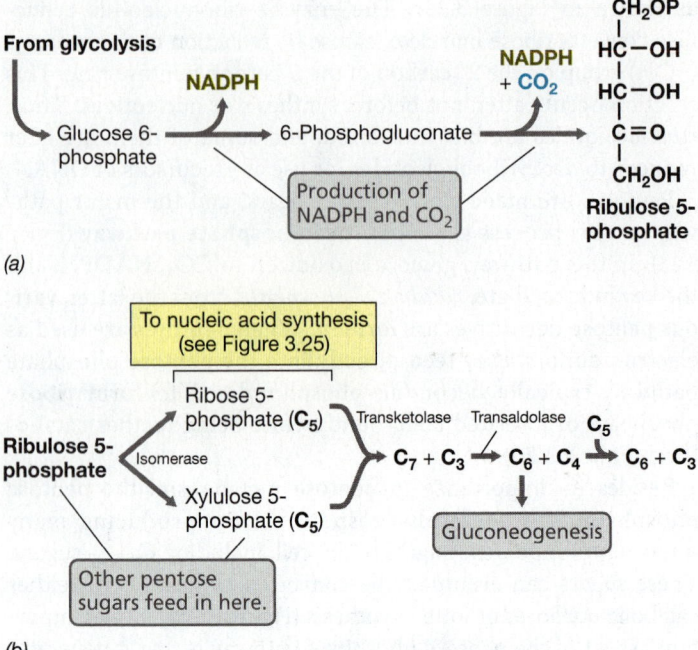

(a)

(b)

Figure 3.26 Pentose phosphate pathway. This pathway generates pentoses from other sugars for biosynthesis and also functions to catabolize pentose sugars. *(a)* Production of the key intermediate, ribulose 5-phosphate. *(b)* Other reactions in the pentose phosphate pathway.

3.15 Amino Acids and Nucleotides

The monomers in proteins and nucleic acids are the amino acids and nucleotides, respectively. Their biosyntheses are typically multistep biochemical pathways that we need not consider here. Instead, we identify the key carbon skeletons needed for the biosynthesis of amino acids and nucleotides and summarize the mechanism by which they are made.

Monomers of Proteins: Amino Acids

Organisms that cannot obtain some or all of their amino acids preformed from the environment must synthesize them from glucose or other sources. Amino acids are grouped into structurally related *families* that share several biosynthetic steps. The carbon skeletons for amino acids come almost exclusively from intermediates of glycolysis or the citric acid cycle (**Figure 3.27**).

The amino group of amino acids is typically derived from some inorganic nitrogen source in the environment, such as ammonia (NH_3). Ammonia is most often incorporated during formation of the amino acids glutamate or glutamine by the enzymes *glutamate dehydrogenase* and *glutamine synthetase*, respectively (**Figure 3.28**). When NH_3 is present at high levels, glutamate dehydrogenase or other amino acid dehydrogenases are used. However, when NH_3 is present at low levels, glutamine synthetase, with its energy-consuming reaction mechanism (Figure 3.28*b*) and correspondingly high affinity for substrate, is employed.

Once ammonia is incorporated into glutamate or glutamine, the amino group can be transferred to form other nitrogenous compounds. For example, glutamate can donate its amino group to oxaloacetate in a transaminase reaction, producing α-ketoglutarate and aspartate (Figure 3.28*c*). Alternatively, glutamine can react with α-ketoglutarate to form two molecules of glutamate in an aminotransferase reaction (Figure 3.28*d*). The end result of these types of reactions is the shuttling of ammonia into various carbon skeletons from which further biosynthetic reactions occur to form all 22 of the amino acids needed to make proteins (⮂ Figure 4.30) and other nitrogen-containing biomolecules.

Monomers of Nucleic Acids: Nucleotides

The biochemistry behind purine and pyrimidine biosynthesis is quite complex. Purines are constructed literally atom by atom from several carbon and nitrogen sources, including even CO_2 (**Figure 3.29**). The first key purine, inosinic acid (Figure 3.29*b*), is the precursor of the purine nucleotides adenine and guanine. Once these are synthesized (in their triphosphate forms) and attached to ribose, they are ready to be incorporated into DNA (following ribonucleotide reductase activity, Figure 3.25*c*) or RNA.

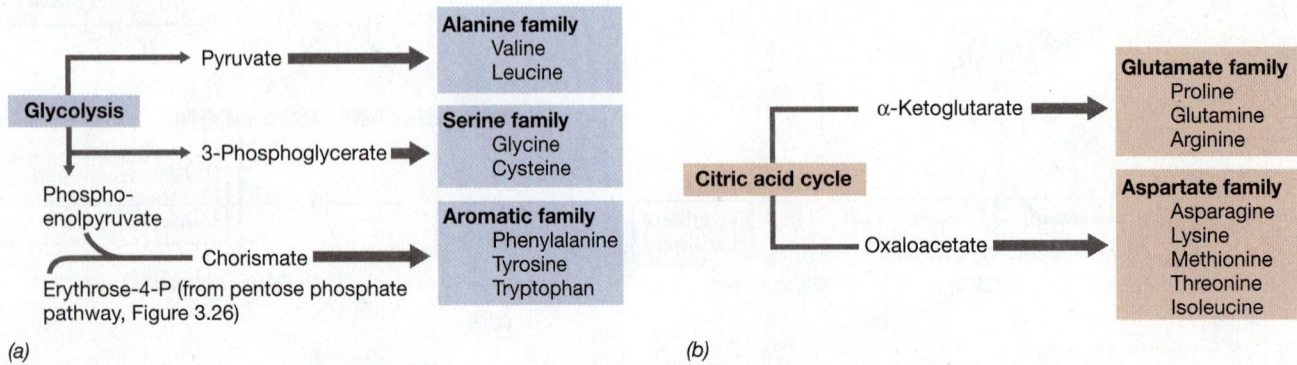

(a)   (b)

Figure 3.27 Amino acid families. Glycolysis *(a)* and the citric acid cycle *(b)* provide the carbon skeletons for most amino acids. Synthesis of the various amino acids in a family may require many steps starting with the parent amino acid (shown in bold as the name of the family).

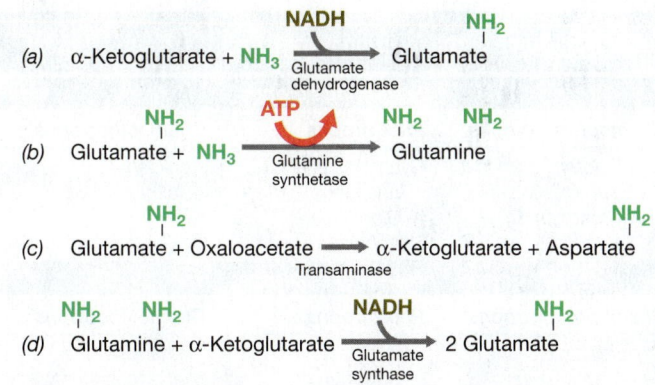

Figure 3.28 Ammonia incorporation in bacteria. Ammonia (NH_3) and the amino groups of all amino acids are shown in green. Two major pathways for NH_3 assimilation in bacteria are those catalyzed by the enzymes *(a)* glutamate dehydrogenase and *(b)* glutamine synthetase. *(c)* Transaminase reactions transfer an amino group from an amino acid to an organic acid. *(d)* The enzyme glutamate synthase forms two glutamates from one glutamine and one α-ketoglutarate.

Like the purine ring, the pyrimidine ring is also constructed from several sources (Figure 3.29*c*). The first key pyrimidine is the compound uridylate (Figure 3.29*d*), and from this the pyrimidines thymine, cytosine, and uracil are derived. Structures of all of the purines and pyrimidines are shown in the next chapter (↺ Figure 4.1).

3.16 Fatty Acids and Lipids

Lipids are major components of the cytoplasmic membrane of all cells and the outer membrane of gram-negative bacteria; lipids can also be carbon and energy reserves (↺ Figure 2.35). Fatty acids are major components of microbial lipids. However, recall that fatty acids are found only in *Bacteria* and *Eukarya*. *Archaea* do not contain fatty acids in their lipids but instead have hydrophobic isoprenoid side chains that play a similar structural role (↺ Figure 2.17). The biosynthesis of these side chains is distinct from that of fatty acids and will not be covered here.

Fatty Acid Biosynthesis

Fatty acids are biosynthesized two carbon atoms at a time by the activity of a protein called *acyl carrier protein* (ACP). ACP holds the growing fatty acid as it is being constructed and releases it once it has reached its final length (**Figure 3.30**). Although fatty acids are constructed *two* carbons at a time, each C_2 unit originates from the *three*-carbon compound malonate, which is attached

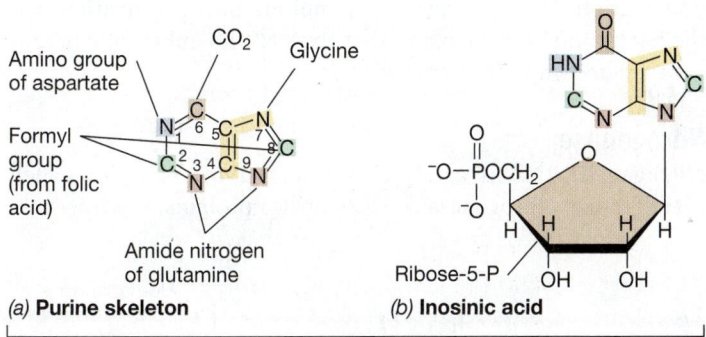

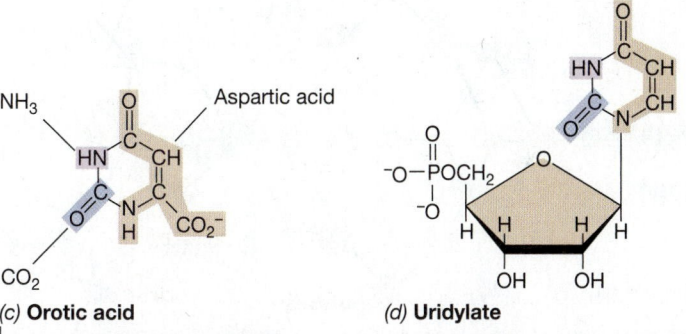

Figure 3.29 Biosynthesis of purines and pyrimidines. *(a)* Components of the purine skeleton. *(b)* Inosinic acid, the precursor of all purine nucleotides. *(c)* Components of the pyrimidine skeleton, orotic acid. *(d)* Uridylate, the precursor of all pyrimidine nucleotides. Uridylate is formed from orotate following a decarboxylation and the addition of ribose 5-phosphate.

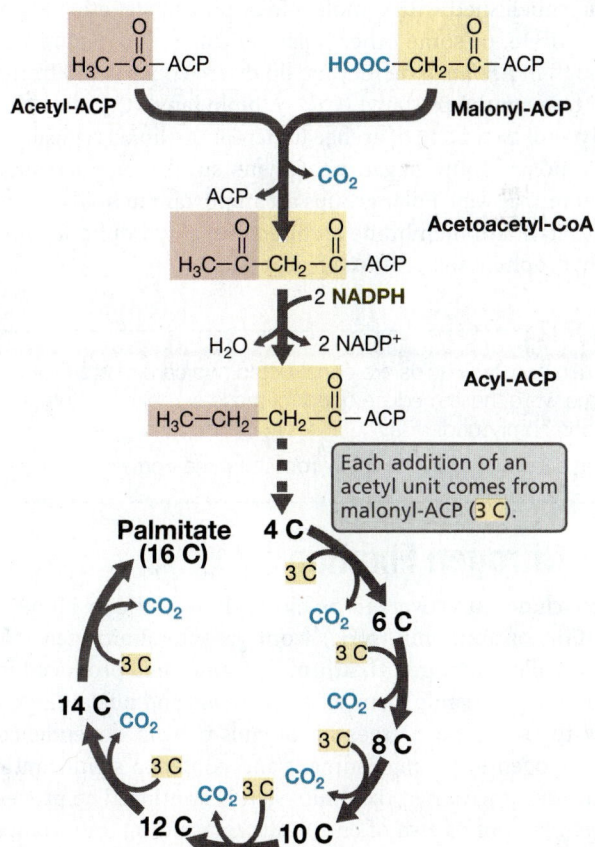

Figure 3.30 The biosynthesis of the C_{16} fatty acid palmitate. The condensation of acetyl-ACP and malonyl-ACP forms acetoacetyl-CoA. Each successive addition of an acetyl unit comes from malonyl-ACP.

to the ACP to form malonyl-ACP. As each malonyl residue is donated, one molecule of CO_2 is released (Figure 3.30).

The fatty acid composition of cells varies from species to species and can also vary in a pure culture due to differences in growth temperature. Growth at low temperatures promotes the biosynthesis of shorter-chain fatty acids whereas growth at higher temperatures promotes longer-chain fatty acids (⟳ Sections 5.12 and 5.13). The most common fatty acids in lipids of *Bacteria* are those with chain lengths of C_{12}–C_{20}.

In addition to saturated, even-carbon-number fatty acids, fatty acids can also be unsaturated, branched, or have an odd number of carbon atoms. Unsaturated fatty acids contain one or more double bonds in the long hydrophobic portion of the molecule. The number and position of these double bonds is often species-specific or group-specific, and double bonds are typically formed by desaturation of a saturated fatty acid. Branched-chain fatty acids are biosynthesized using a branched-chain initiating molecule, and odd-carbon-number fatty acids (for example, C_{13}, C_{15}, C_{17}, etc.) are biosynthesized using an initiating molecule that contains a propionyl (C_3) group.

Lipid Biosynthesis

In the assembly of lipids in cells of *Bacteria* and *Eukarya*, fatty acids are first added to a molecule of glycerol. For simple triglycerides (fats), all three glycerol carbons are esterified with fatty acids. To form complex lipids, one of the carbon atoms in glycerol is embellished with a molecule of phosphate, ethanolamine, carbohydrate, or some other polar substance (⟳ Figure 2.14a). Although in *Archaea*, membrane lipids are constructed from isoprene to form the phytanyl (C_{15}) or biphytanyl (C_{30}) side chains, the glycerol backbone of archaeal membrane lipids typically contains a polar group (sugar, phosphate, sulfate, or polar organic compound) as well. Polar groups are important in lipids for forming the standard membrane architecture: a hydrophobic interior with hydrophilic surfaces (⟳ Figure 2.17).

MINIQUIZ

- Explain how fatty acids are constructed two carbon atoms at a time while the immediate donor of these carbons is a three-carbon compound.
- What differences exist in lipids from the three domains of life?

3.17 Nitrogen Fixation

We conclude our coverage of biosyntheses by considering the formation of ammonia (NH_3) from gaseous dinitrogen (N_2), a process called **nitrogen fixation**. The ammonia produced is assimilated into organic form in amino acids and nucleotides. The ability to fix nitrogen frees an organism from dependence on fixed nitrogen in its environment and confers a significant ecological advantage when fixed nitrogen is limiting. The process of nitrogen fixation is also of enormous agricultural importance, as it supports the nitrogen needs of key crops, such as soybeans.

Only certain species of *Bacteria* and *Archaea* can fix nitrogen, and a list of some important nitrogen-fixing organisms is given in **Table 3.5**. Some nitrogen-fixing bacteria are *free-living* and carry

Table 3.5 Some nitrogen-fixing organisms[a]

Free-living aerobes		
Chemoorganotrophs	*Phototrophs*	*Chemolithotrophs*
Azotobacter, Azomonas, Azospirillum, Klebsiella,[b] Methylomonas	Cyanobacteria (e.g., Anabaena, Nostoc, Gloeothece, Aphanizomenon)	Alcaligenes Acidithiobacillus
Free-living anaerobes		
Chemoorganotrophs	*Phototrophs*	*Chemolithotrophs*[c]
Clostridium Desulfotomaculum	Purple bacteria (e.g., Chromatium, Rhodobacter) Green bacteria (e.g., Chlorobium) Heliobacteria	Methanosarcina Methanococcus Methanocaldococcus
Symbiotic		
With leguminous plants	*With nonleguminous plants*	
Soybeans, peas, clover, etc. with Rhizobium, Bradyrhizobium, Sinorhizobium	Alder, bayberry, autumn olive, many other bushy plants, with the actinomycete Frankia	

[a]Only some common genera are listed in each category; many other nitrogen-fixing genera are known.
[b]Nitrogen fixation occurs only under anoxic conditions.
[c]All are *Archaea*.

out the process completely independently. By contrast, others are *symbiotic* and fix nitrogen only in association with certain plants (⟳ Section 22.0). However, in symbiotic nitrogen fixation it is the bacterium, not the plant, that fixes N_2; no eukaryotic organisms are known to fix nitrogen.

Nitrogenase

Nitrogen fixation is catalyzed by an enzyme complex called **nitrogenase**. Nitrogenase consists of two proteins, *dinitrogenase*

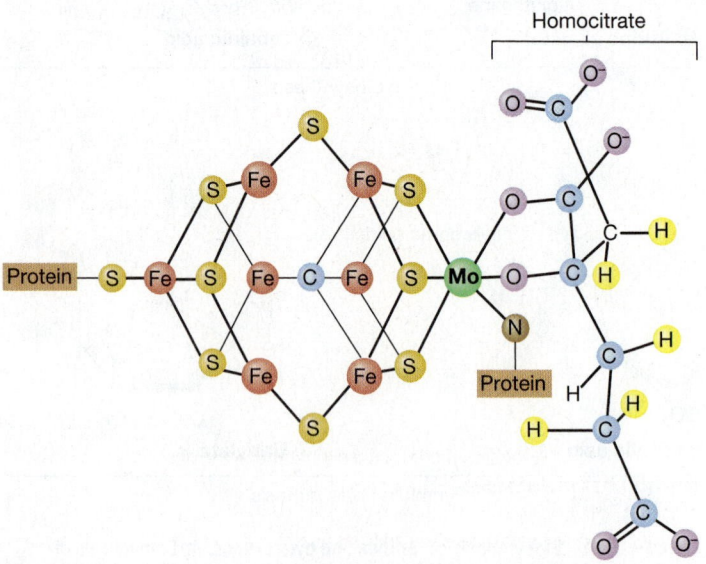

Figure 3.31 FeMo-co, the iron–molybdenum cofactor from nitrogenase. On the left side is the Fe_7S_8 cube that binds to Mo along with O atoms from homocitrate (right side, all O atoms shown in purple) and N and S atoms from dinitrogenase.

and *dinitrogenase reductase*. Both proteins contain iron, and dinitrogenase contains molybdenum as well. The iron and molybdenum in dinitrogenase are part of the enzyme cofactor called the *iron–molybdenum cofactor* (*FeMo-co*), and reduction of N_2 occurs at this site. The composition of FeMo-co is $MoFe_7S_8 \cdot$ homocitrate (**Figure 3.31**). Two "alternative" nitrogenases are known that lack molybdenum. These contain either vanadium (V) plus iron or iron-only in their cofactors and are made by certain nitrogen-fixing bacteria when molybdenum is absent from their environment (Section 14.12).

With one exception, nitrogen-fixing *Archaea* produce nitrogenases with iron as the only metal in the enzyme. Nitrogen-fixing *Archaea* appear limited to a few methane-producing species (methanogens), at least one of which can grow and fix N_2 at very high temperatures. The species *Methanosarcina barkeri*, a metabolically versatile methanogen (Section 16.2), contains genes encoding molybdenum and vanadium nitrogenases as well as an iron-only nitrogenase, and so it likely contains the full suite of nitrogenase proteins.

Nitrogen fixation is inhibited by oxygen (O_2) because dinitrogenase reductase is irreversibly inactivated by O_2. Nevertheless, many nitrogen-fixing bacteria are obligate aerobes. In these

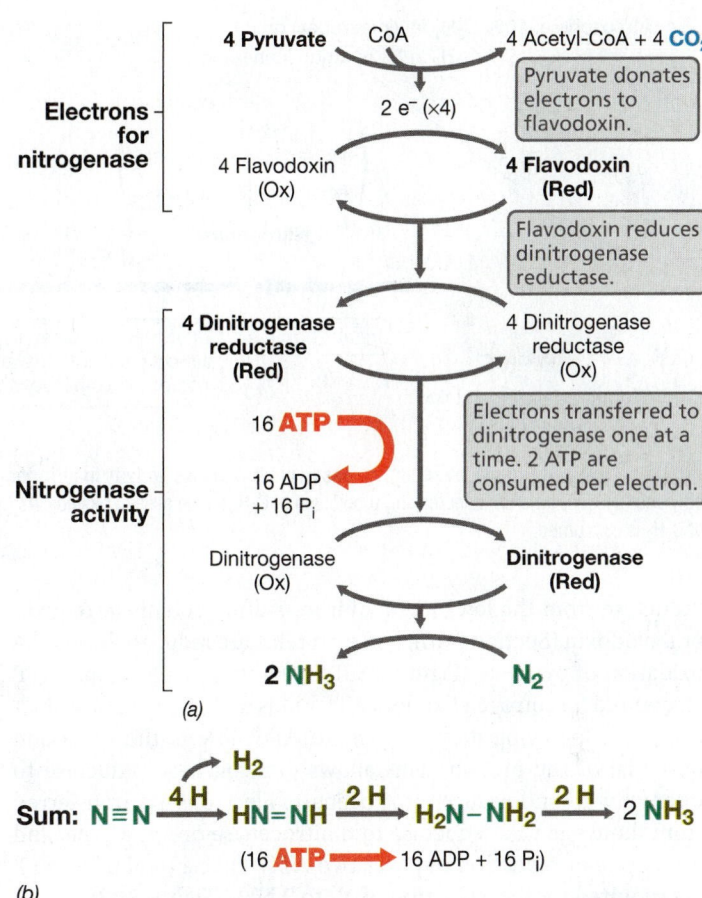

(a)

$$\text{Sum: } N \equiv N \xrightarrow[]{4\,H} \overset{H_2}{HN = NH} \xrightarrow{2\,H} H_2N - NH_2 \xrightarrow{2\,H} 2\,NH_3$$

$$(16\ \textbf{ATP} \longrightarrow 16\ ADP + 16\ P_i)$$

(b)

Figure 3.33 **Biological nitrogen fixation by nitrogenase.** The nitrogenase complex is composed of dinitrogenase and dinitrogenase reductase.

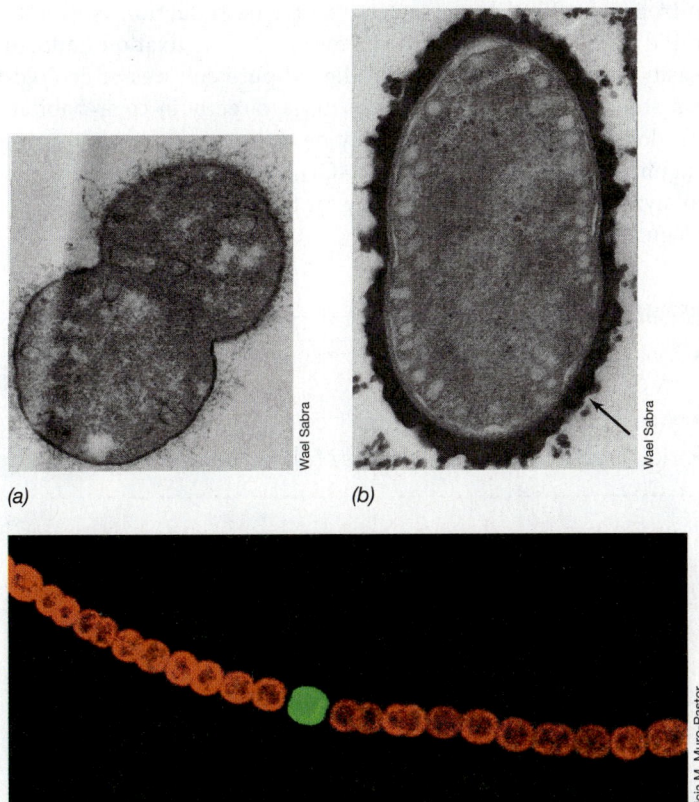

(a)   (b)

(c)

Figure 3.32 **Protection of nitrogenase in *Azotobacter vinelandii* and in the cyanobacterium *Anabaena*.** *(a)* Transmission electron micrograph of nitrogen-fixing cells of *A. vinelandii* grown with 2.5% O_2; very little slime is evident. *(b)* Cells grown in air (21% O_2). Note the extensive darkly staining slime layer (arrow). The slime retards diffusion of O_2 into the filament, thus preventing nitrogenase inactivation by O_2. *(c)* Fluorescence photomicrograph of a cell of the filamentous cyanobacterium *Anabaena* showing a single heterocyst (green). The heterocyst is a differentiated cell that specializes in nitrogen fixation and protects nitrogenase from O_2 inactivation.

organisms, nitrogenase is protected from oxygen inactivation by a combination of the rapid removal of O_2 by respiration and the production of O_2-retarding slime layers (**Figure 3.32a, b**). In heterocystous cyanobacteria, nitrogenase is protected by its localization in a differentiated cell called a *heterocyst* (Figure 3.32c; Section 14.3). Inside the heterocyst, conditions are anoxic, while in neighboring vegetative cells, conditions are just the opposite because oxygenic photosynthesis is occurring. Oxygen production is shut down in the heterocyst, thus protecting it as a dedicated site for N_2 fixation.

Electron Flow in Nitrogen Fixation

Owing to the stability of the triple bond in N_2, its activation and reduction is very energy demanding. Six electrons are needed to reduce N_2 to NH_3, and the successive reduction steps occur directly on nitrogenase with no free intermediates accumulating (**Figure 3.33**). Although only *six* electrons are necessary to reduce N_2 to two NH_3, *eight* electrons are actually consumed in the process, two electrons being lost as H_2 for each mole of N_2 reduced. For unknown reasons, H_2 evolution is an obligatory step in nitrogen fixation and occurs in the first round of the nitrogenase reduction cycle. Following this, N_2 is reduced in successive steps and ammonia is the released product (Figure 3.33).

The sequence of electron transfer in nitrogenase is as follows: electron donor → dinitrogenase reductase → dinitrogenase → N_2. The electrons for N_2 reduction are transferred to dinitrogenase

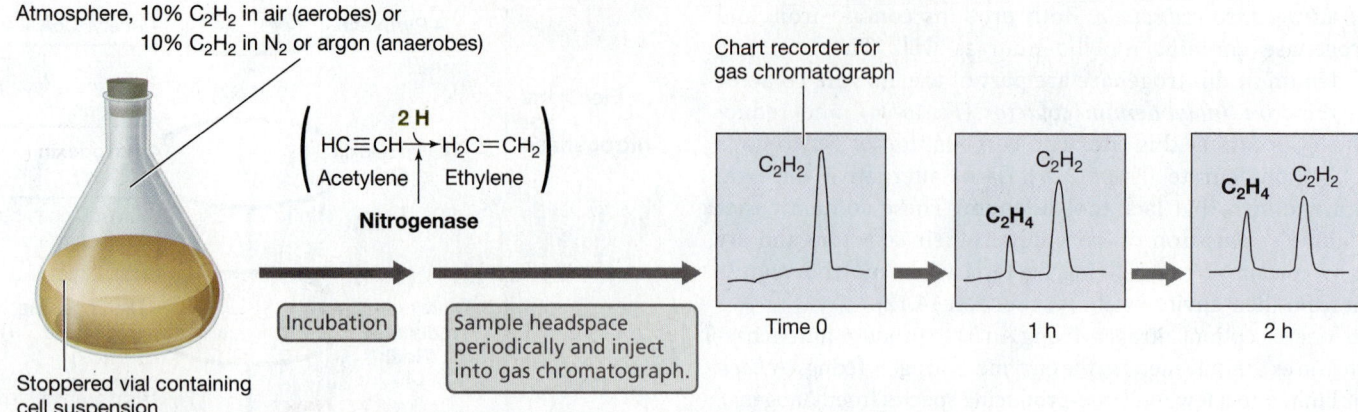

Figure 3.34   The acetylene reduction assay of nitrogenase activity in nitrogen-fixing bacteria. The results show no ethylene (C_2H_4) at time 0 but increasing production of C_2H_4 as the assay proceeds. As C_2H_4 is produced, a corresponding amount of C_2H_2 is consumed.

reductase from the low-potential iron–sulfur proteins ferredoxin or flavodoxin (Section 3.10); these proteins are reduced during the oxidation of pyruvate (Figure 3.33). In addition to electrons, ATP is required for nitrogen fixation. ATP binds to dinitrogenase reductase, and, following its hydrolysis to ADP, lowers the reduction potential of the protein. This allows dinitrogenase reductase to interact with and reduce dinitrogenase. Electrons are transferred from dinitrogenase reductase to dinitrogenase one at a time, and each cycle of reduction requires two ATP. Thus a total of 16 ATP are required for the reduction of N_2 to 2 NH_3 (Figure 3.33).

Assaying Nitrogenase: Acetylene Reduction

Nitrogenases are not entirely specific for N_2 and also reduce other triply bonded compounds, such as acetylene (HC≡CH). The reduction of acetylene by nitrogenase is only a two-electron process, and *ethylene* (H_2C=CH_2) is the final product. However, the reduction of acetylene to ethylene provides a simple and rapid method for measuring nitrogenase activity (**Figure 3.34**). This technique, known as the *acetylene reduction assay*, is widely used in microbiology to detect and quantify nitrogen fixation.

Although the reduction of acetylene is taken as strong proof of N_2 fixation, definitive proof requires an isotope of nitrogen, $^{15}N_2$, as a tracer. If a culture or natural sample is enriched with $^{15}N_2$ and incubated, the production of $^{15}NH_3$ is firm evidence of nitrogen fixation. Nevertheless, acetylene reduction is a more rapid and sensitive method for measuring N_2 fixation and can easily be used in laboratory studies of pure cultures or ecological studies of nitrogen-fixing bacteria directly in their habitat. To do this, a sample, which may be soil, water, or a culture, is incubated in a vessel with HC≡CH, and the gas phase is later analyzed by gas chromatography for the presence of H_2C=CH_2 (Figure 3.34).

MINIQUIZ --

- Write a balanced equation for the reaction catalyzed by nitrogenase.

- What is FeMo-co and what does it do?

- How is acetylene useful in studies of nitrogen fixation?

BIG IDEAS

3.1 • Cells are primarily composed of the elements H, O, C, N, P, and S. The compounds found in a cell are obtained from or formed from nutrients present in the environment. Nutrients required in large amounts are called macronutrients while those required in very small amounts, such as trace elements or growth factors, are micronutrients.

3.2 • Culture media supply the nutritional needs of microorganisms and are either defined or complex. Other media, such as selective, differential, and enriched media, are used for specific purposes. Many microorganisms can

be grown in liquid or solid culture media, and pure cultures can be maintained if aseptic technique is practiced.

3.3 • All microorganisms conserve energy from either the oxidation of chemicals or from light. Chemoorganotrophs use organic chemicals as their electron donors, while chemolithotrophs use inorganic chemicals. Phototrophic organisms convert light energy into chemical energy (ATP) and include both the oxygenic and anoxygenic phototrophs.

3.4 • Chemical reactions in the cell are accompanied by changes in energy, expressed in kilojoules. Reactions either release or consume free energy, or they are energy neutral. $\Delta G^{0\prime}$ is a measure of the energy released or consumed in a reaction under standard conditions and reveals which reactions can be used by an organism to conserve energy.

3.5 • Enzymes are protein catalysts that speed up the rate of biochemical reactions by activating the substrates that bind to their active site. Enzymes are highly specific in the reactions they catalyze, and this specificity resides in the three-dimensional structures of the polypeptides that make up the proteins.

3.6 • Oxidation–reduction reactions require electron donors and electron acceptors. The tendency of a compound to accept or release electrons is expressed by its reduction potential ($E_0{}'$). Redox reactions in a cell employ intermediaries such as $NAD^+/NADH$ as electron shuttles.

3.7 • The energy released in redox reactions is conserved in compounds that contain energy-rich phosphate or sulfur bonds. The most common of these compounds is ATP, the prime energy carrier in the cell. Longer-term storage of energy is linked to the formation of polymers, which can be consumed to yield ATP.

3.8 • The glycolytic pathway is used to break glucose down to pyruvate and is a widespread mechanism for energy conservation by fermentative anaerobes. The pathway releases only a small amount of ATP and generates fermentation products (ethanol, lactic acid, and so on) characteristic of the organism. For each glucose fermented by yeast in glycolysis, 2 ATP are produced.

3.9 • Many different fermentations of glucose use the glycolytic pathway, the differences lying in the nature of the fermentation products. Extra ATP can be obtained from glucose fermentation if fatty acids derived from coenzyme-A derivatives are fermentation products. Yeast has two options for glucose catabolism: fermentation and respiration.

3.10 • Electron transport chains are composed of membrane-associated proteins that are arranged in order of their increasing $E_0{}'$ values and function in an integrated fashion to carry electrons from the primary electron donor to the terminal electron acceptor (O_2 in aerobic respiration).

3.11 • When electrons are transported through an electron transport chain, protons are extruded to the outside of the membrane, forming the proton motive force. Key electron carriers include flavins, quinones, the cytochrome bc_1 complex, and other cytochromes. The cell uses the proton motive force to make ATP through the activity of ATPase.

3.12 • Respiration offers an energy yield much greater than that of fermentation. The citric acid cycle generates CO_2 and electrons for the electron transport chain and is also a source of biosynthetic intermediates. The glyoxylate cycle is necessary for the catabolism of two-carbon electron donors, such as acetate.

3.13 • When conditions are anoxic, several terminal electron acceptors can substitute for O_2 in anaerobic respiration. Chemolithotrophs use inorganic compounds as electron donors, whereas phototrophs use light energy. The proton motive force supports energy generation by ATPase in all forms of respiration and photosynthesis.

3.14 • Polysaccharides are important structural components of cells and are biosynthesized from activated forms of their monomers. Gluconeogenesis is the production of glucose from nonsugar precursors.

3.15 • Amino acids are formed from carbon skeletons to which ammonia is added from either glutamate, glutamine, or a few other amino acids. Nucleotides are biosynthesized using carbon from several different sources.

3.16 • Fatty acids are synthesized from the three-carbon precursor malonyl-ACP and fully formed fatty acids are attached to glycerol to form lipids. Only the lipids of *Bacteria* and *Eukarya* contain fatty acids, typically C_{12}–C_{18} in length.

3.17 • The reduction of N_2 to NH_3 is called nitrogen fixation and is catalyzed by the enzyme nitrogenase. Nitrogenase is composed of two proteins, dinitrogenase and dinitrogenase reductase. Nitrogenase can be assayed using the triply bonded compound acetylene as a surrogate of N_2, which nitrogenase reduces to ethylene.

MasteringMicrobiology®   **Review what you know and challenge what you have learned with MasteringMicrobiology!** Access study materials, chapter quizzes, animations, and microbiology lab tutorials in the Study Area to ensure that you have mastered this chapter's content.

REVIEW OF KEY TERMS

Activation energy the energy required to bring the substrate of an enzyme to the reactive state

Adenosine triphosphate (ATP) a nucleotide that is the primary form in which chemical energy is conserved and utilized in cells

Anabolic reactions (Anabolism) the sum total of all biosynthetic reactions in the cell

Anaerobic respiration a form of respiration in which oxygen is absent and alternative electron acceptors are reduced

Aseptic technique manipulations to prevent contamination of sterile objects or microbial cultures during handling

ATPase (ATP synthase) a multiprotein enzyme complex embedded in the cytoplasmic membrane that catalyzes the synthesis of ATP coupled to dissipation of the proton motive force

Autotroph an organism capable of biosynthesizing all cell material from CO_2 as the sole carbon source

Catabolic reactions (Catabolism) biochemical reactions leading to energy conservation (usually as ATP) by the cell

Catalyst a substance that accelerates a chemical reaction but is not consumed in the reaction

Chemolithotroph an organism that can grow with inorganic compounds as electron donors in energy metabolism

Chemoorganotroph an organism that obtains its energy from the oxidation of organic compounds

Citric acid cycle a cyclical series of reactions resulting in the conversion of acetate to two molecules of CO_2

Coenzyme a small and loosely bound nonprotein molecule that participates in a reaction as part of an enzyme

Complex medium a culture medium composed of chemically undefined substances such as yeast and meat extracts

Culture medium an aqueous solution of various nutrients suitable for the growth of microorganisms

Defined medium a culture medium whose precise chemical composition is known

Electron acceptor a substance that can accept electrons from an electron donor, becoming reduced in the process

Electron donor a substance that can donate electrons to an electron acceptor, becoming oxidized in the process

Endergonic requires energy

Enzyme a protein that can speed up (catalyze) a specific chemical reaction

Exergonic releases energy

Fermentation anaerobic catabolism in which an organic compound is both an electron donor and an electron acceptor and ATP is produced by substrate-level phosphorylation

Free energy (G) energy available to do work; $G^{0'}$ is free energy under standard conditions

Glycolysis a biochemical pathway in which glucose is fermented, yielding ATP and various fermentation products; also called the Embden–Meyerhof–Parnas pathway

Glyoxylate cycle a modification of the citric acid cycle in which isocitrate is cleaved to form succinate and glyoxylate during growth on two-carbon electron donors such as acetate

Heterotroph an organism that uses organic compounds as a carbon source

Nitrogenase the enzyme complex required to reduce N_2 to NH_3 in biological nitrogen fixation.

Nitrogen fixation the reduction of N_2 to NH_3 by the enzyme nitrogenase

Oxidative phosphorylation the production of ATP from a proton motive force formed by electron transport of electrons from organic or inorganic electron donors

Pentose phosphate pathway a series of reactions in which pentoses are catabolized to generate precursors for nucleotide biosynthesis or to synthesize glucose

Photophosphorylation the production of ATP from a proton motive force formed from light-driven electron transport

Phototrophs organisms that use light as their source of energy

Proton motive force (pmf) a source of energy resulting from the separation of protons from hydroxyl ions across the cytoplasmic membrane, generating a membrane potential

Reduction potential (E_0') the inherent tendency, measured in volts under standard conditions, of a compound to donate electrons

Respiration the process in which a compound is oxidized with O_2 (or an O_2 substitute) as the terminal electron acceptor, usually accompanied by ATP production by oxidative phosphorylation

Substrate-level phosphorylation the production of ATP by the direct transfer of an energy-rich phosphate molecule from a phosphorylated organic compound to ADP

REVIEW QUESTIONS

1. Why are carbon and nitrogen macronutrients but cobalt is a micronutrient? (Section 3.1)

2. Why would the following medium not be considered a chemically defined medium: glucose, 5 grams (g); NH_4Cl, 1 g; KH_2PO_4, 1 g; $MgSO_4$, 0.3 g; yeast extract, 5 g; distilled water, 1 liter? (Section 3.2)

3. What is aseptic technique and why is it necessary? (Section 3.2)

4. To which energy class does *Escherichia coli* belong? *Thiobacillus thioparus*? How do the contents of Table 3.2 tell you this? (Section 3.3)

5. Describe how you would calculate $\Delta G^{0'}$ for the reaction: glucose $+ 6 O_2 \rightarrow 6 CO_2 + 6 H_2O$. If you were told that this reaction is highly *exergonic*, what would be the arithmetic sign (negative or positive) of the $\Delta G^{0'}$ you would expect for this reaction? (Section 3.4)

6. Distinguish between $\Delta G^{0'}$, ΔG, and G_f^0. (Section 3.4)

7. Why are enzymes needed by the cell? (Section 3.5)

8. The following is a series of coupled electron donors and electron acceptors (written as donor/acceptor). Using just the data in Figure 3.9, order this series from most energy yielding to least energy yielding: H_2/Fe^{3+}, H_2S/O_2, methanol/NO_3^- (producing NO_2^-), H_2/O_2, Fe^{2+}/O_2, NO_2^-/Fe^{3+}, and H_2S/NO_3^-. (Section 3.6)

9. What is the reduction potential of the $NAD^+/NADH$ couple? (Section 3.6)

10. Why is acetyl phosphate considered an energy-rich compound but glucose 6-phosphate is not? (Section 3.7)

11. How is ATP made in fermentation and in respiration? (Section 3.8)

12. Where in glycolysis is NADH produced? Where is NADH consumed? (Section 3.8)

13. Besides lactic acid and ethanol, list some other fermentation products that can be produced when glucose is fermented through glycolysis. (Section 3.9)

14. List some of the important electron carriers found in electron transport chains. (Section 3.10)

15. What is meant by proton motive force and how is it generated? (Section 3.11)

16. How is rotational energy in the ATPase used to produce ATP? (Section 3.11)

17. How much more ATP is possible in respiration than in fermentation? Write one sentence that accounts for this difference. (Section 3.12)

18. Why can it be said that the citric acid cycle plays two major roles in the cell? (Section 3.12)

19. What is the major difference between respiration and anaerobic respiration? Which metabolic option would yield more energy, and why? (Section 3.13)

20. Which two major catabolic pathways supply carbon skeletons for sugar and amino acid biosyntheses? (Sections 3.14 and 3.15)

21. Describe the process by which a fatty acid such as palmitate (a C_{16} straight-chain saturated fatty acid) is synthesized in a cell. (Section 3.16)

22. What is the reaction carried out by the enzyme nitrogenase? How might the ability to fix nitrogen help a bacterium be more competitive in its environment? (Section 3.17)

APPLICATION QUESTIONS

1. Design a defined culture medium for an organism that can grow aerobically on acetate as a carbon and energy source. Make sure all the nutrient needs of the organism are accounted for and in the correct relative proportions.

2. *Desulfovibrio* can grow anaerobically with H_2 as electron donor and SO_4^{2-} as electron acceptor (which is reduced to H_2S). Based on this information and the data in Table A1.2 (Appendix 1), indicate which of the following components could not exist in the electron transport chain of this organism and why: cytochrome *c*, ubiquinone, cytochrome c_3, cytochrome aa_3, ferredoxin.

3. Again using the data in Table A1.2, predict the sequence of electron carriers in the membrane of an organism growing aerobically and producing the following electron carriers: ubiquinone, cytochrome aa_3, cytochrome *b*, NADH, cytochrome *c*, FAD.

4. Explain the following observation in light of the redox tower: Cells of *Escherichia coli* fermenting glucose grow faster when NO_3^- is supplied to the culture (NO_2^- is produced) and then grow even faster (and stop producing NO_2^-) when the culture is highly aerated.

4 · Molecular Microbiology

microbiology**now**

The Essence of Life: Molecular Microbiology

As you have no doubt discovered, microorganisms possess an astounding range of metabolic capabilities. The genetic blueprint of individual cells is responsible for the distinctive attributes observed for all life forms. While this repository of information must be protected and passed on from generation to generation, the information must also "come alive" to allow cells to perform an ingenious array of fascinating activities. This flow of essential biological information—from fairly inert DNA to the synthesis of proteins and enzymes critical for cellular survival—is known as the central dogma of life.

Molecular microbiology has been the cornerstone for understanding the individual steps of the central dogma: DNA replication, transcription of DNA into RNA, and translation of RNA into proteins. With the advent of cutting-edge techniques, new discoveries regarding these essential biological processes are still occurring. For example, microbiologists can now pinpoint the location of specific molecules in living cells using fluorescent labels and super-resolution fluorescence microscopy. The photo here illustrates the use of fluorescence microscopy and protein tagging in actively growing *Escherichia coli* cells to actually visualize RNA polymerases and ribosomes, two cellular machines essential to the central dogma, in action. The resulting image shows that the majority of ribosomes, the "protein factories," are located at the ends of the cell and in regions where septa form during cell division (top photo, in green), while RNA polymerases are associated with the chromosomal DNA in the nucleoid region, which is localized in the center of the cell (middle photo, in red). Overlapping the two images (bottom photo) allows us to see that spatial organization of biological information flow does indeed exist in bacterial cells, despite their lacking organelles.

Bakshi, S., A. Siryaporn, M. Goulian, and J. C. Weisshaar. 2012. Superresolution imaging of ribosomes and RNA polymerase in live *Escherichia coli* cells. *Molecular Microbiology* 85: 21–38.

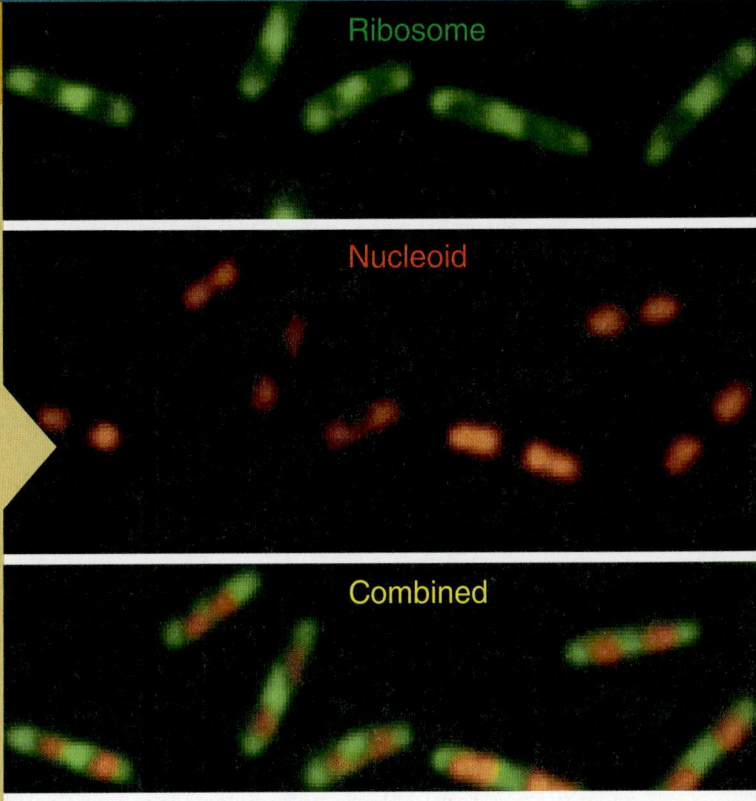

Ribosome

Nucleoid

Combined

Central to life is the flow of information. What instructs the cell to reproduce itself and survive in a given environment and what are the processes that dictate the output of cells? Cells can be considered both chemical machines and coding devices. As chemical machines, cells transform their vast array of macromolecules into new cells. As coding devices, they store, process, and use genetic information. Genes, the mechanisms by which they are transferred to new cells, and their expression are the foundation of molecular biology and the central dogma of life. This chapter highlights the genetic blueprint of cells and the steps the cell takes to transform this information into macromolecules that carry out cellular functions. Our focus here is on these processes as they occur in *Bacteria*, particularly in *Escherichia coli*, a species of *Bacteria* that is the model organism for molecular biology. This bacterium remains the best characterized of any organism, prokaryote or eukaryote.

I • The Blueprint of Life: Structure of the Bacterial Genome

4.1 Macromolecules and Genes

The functional unit of genetic information is the **gene**. All life forms, including microorganisms, contain genes. Physically, genes are located on chromosomes or other large molecules known collectively as **genetic elements**. These elements make up the total complement of genetic information, or the **genome**, in a cell or virus. In modern biology, cells can be characterized in terms of their complement of genes. Thus, if we wish to understand how microorganisms function we must understand how genes encode information.

Genetic information in cells is embedded in the **nucleic acids** DNA and RNA. **Deoxyribonucleic acid**, **DNA**, carries the genetic blueprint for the cell while **ribonucleic acid**, **RNA**, is an intermediary molecule that converts this blueprint into defined amino acid sequences in proteins. Genetic information resides in the sequence of monomers in the nucleic acids. Thus, in contrast to polysaccharides and lipids that are typically composed of long repeating units, nucleic acids are **informational macromolecules**. Because the sequence of monomers in proteins is determined by the sequence of the nucleic acids that encode them, proteins are also informational macromolecules.

The monomers of nucleic acids are called **nucleotides**, consequently, DNA and RNA are **polynucleotides**. A nucleotide has three components: a pentose sugar (either ribose in RNA or deoxyribose in DNA), a nitrogen base, and a molecule of phosphate, PO_4^{3-}. The general structures of nucleotides of both DNA and RNA are very similar (**Figure 4.1**). The nitrogen bases are either **purines** (*adenine* and *guanine*), which contain two fused heterocyclic rings, or **pyrimidines** (*thymine, cytosine,* and *uracil*), which contain a single six-membered heterocyclic ring (Figure 4.1*a*). Guanine, adenine, and cytosine are present in both DNA and RNA. With minor exceptions, thymine is present only in DNA and uracil is present only in RNA.

The nitrogen bases are attached to the pentose sugar by a glycosidic linkage between carbon atom 1 of the sugar and a nitrogen atom in the base, either nitrogen 1 (in pyrimidine bases) or 9 (in purine bases). A nitrogen base attached to its sugar, but lacking phosphate, is called a **nucleoside**. Nucleotides are nucleosides plus one or more phosphates (Figure 4.1*b*). Nucleotides play other roles in addition to their role in nucleic acids. Nucleotides, especially adenosine triphosphate (ATP) and guanosine triphosphate (GTP), are important molecules in energy conservation (⟳ Section 3.7). Other nucleotides or derivatives function in redox reactions, as carriers of sugars in polysaccharide synthesis, or as regulatory molecules.

The Nucleic Acids: DNA and RNA

The nucleic acid backbone is a polymer of alternating sugar and phosphate molecules. The nucleotides are covalently bonded by

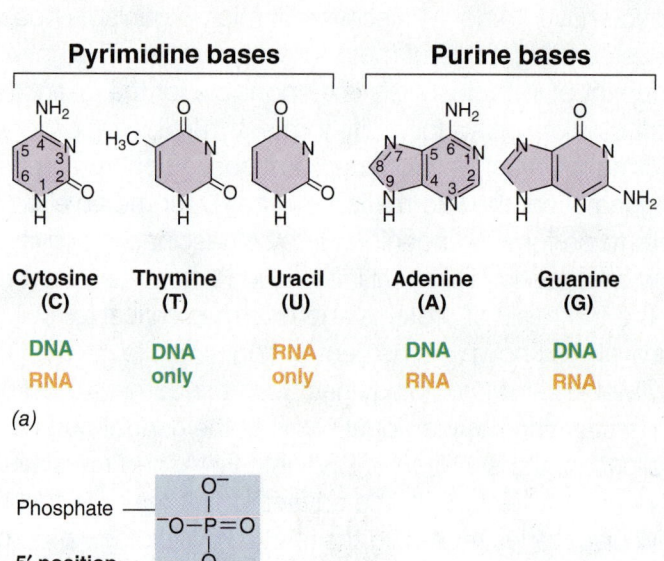

(a)

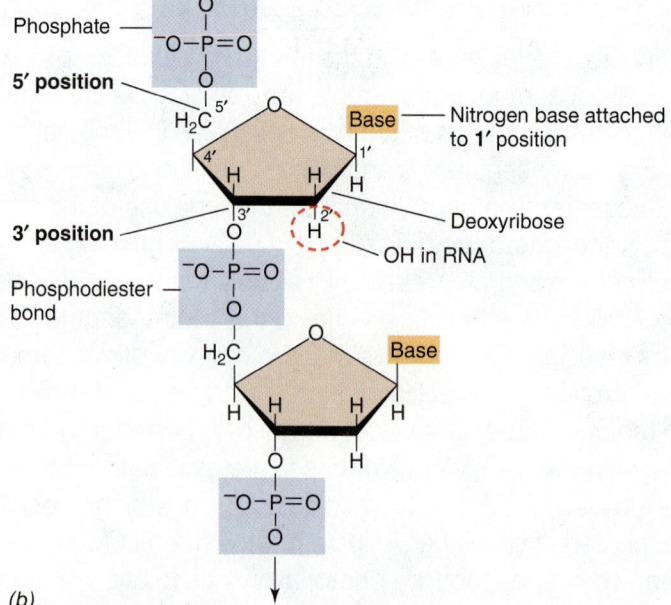

(b)

Figure 4.1 Components of the nucleic acids. (a) The nitrogen bases of DNA and RNA. Note the numbering system of the rings. In attaching itself to the 1′ carbon of the sugar phosphate, a pyrimidine base bonds through N-1 and a purine base bonds at N-9. (b) Part of a DNA chain. The numbers on the sugar of the nucleotide contain a prime (′) after them because the rings of the nitrogen bases are also numbered. In DNA, a hydrogen is present on the 2′-carbon of the pentose sugar. In RNA, an OH group occupies this position. The nucleotides are linked by a phosphodiester bond.

phosphate between the 3'- (3 prime) carbon of one sugar and the 5'-carbon of the next sugar. (Numbers with prime marks refer to positions on the sugar ring; numbers without primes to positions on the rings of the bases.) The phosphate linkage is called a **phosphodiester bond** because the phosphate connects two sugar molecules by an ester linkage (Figure 4.1b). The *sequence* of nucleotides in a DNA or RNA molecule is its **primary structure** and the sequence of bases constitutes the genetic information.

In the genome of cells, DNA is *double-stranded*. Each chromosome consists of two strands of DNA, with each strand containing hundreds of thousands to several million nucleotides linked by phosphodiester bonds. The strands are held together by hydrogen bonds that form between the bases in one strand and those of the other strand. When located adjacent to one another, purine and pyrimidine bases can form hydrogen bonds (**Figure 4.2**). Hydrogen bonding is most thermodynamically stable when guanine (G) bonds with cytosine (C) and adenine (A) bonds with thymine (T). Specific base pairing, A with T and G with C, ensures that the two strands of DNA are *complementary* in base sequence; that is, wherever a G is found in one strand, a C is found in the other, and wherever a T is present in one strand, its complementary strand contains an A.

Genes and the Steps in Information Flow

When genes are expressed, the genetic information stored in DNA is transferred to ribonucleic acid (RNA). While several classes of RNA exist in cells, three main types of RNA take part in protein synthesis. **Messenger RNA (mRNA)** is a single-stranded molecule that carries the genetic information from DNA to the ribosome, the protein-synthesizing machine. **Transfer RNAs (tRNAs)** convert the genetic information in the nucleotide sequences of RNA into a defined sequence of amino acids in proteins. **Ribosomal**

RNAs (rRNAs) are important catalytic and structural components of the ribosome. The molecular processes of genetic information flow can be divided into three stages (**Figure 4.3**):

1. **Replication**. During replication, the DNA double helix is duplicated, producing two copies. Replication is carried out by an enzyme called *DNA polymerase*.

2. **Transcription**. The transfer of genetic information from DNA to RNA is called transcription. Transcription is carried out by an enzyme called *RNA polymerase*.

3. **Translation**. Synthesis of a protein, using the genetic information in mRNA, is called translation.

The three steps shown in Figure 4.3 are characteristic of all cells and form the central dogma of molecular biology: DNA → RNA → protein. Many different RNA molecules can be transcribed from a relatively short region of the long DNA molecule. In eukaryotes, each gene is transcribed to yield a single mRNA, whereas in prokaryotes, a single mRNA molecule may carry the genetic information from several genes; that is, several protein-coding regions. A linear correspondence exists between the base sequence of a gene and the amino acid sequence of a polypeptide. Each group of three bases on an mRNA molecule encodes a single amino acid, and each such triplet of bases is called a **codon**. Codons are translated into sequences of amino acids by ribosomes (which themselves consist of proteins and RNA), tRNA, and helper proteins called *translation factors*. While the central dogma is invariant in cells, we will see later that some viruses (which are not cells, ⟳ Section 1.2) violate this process in interesting ways (Chapters 8 and 9).

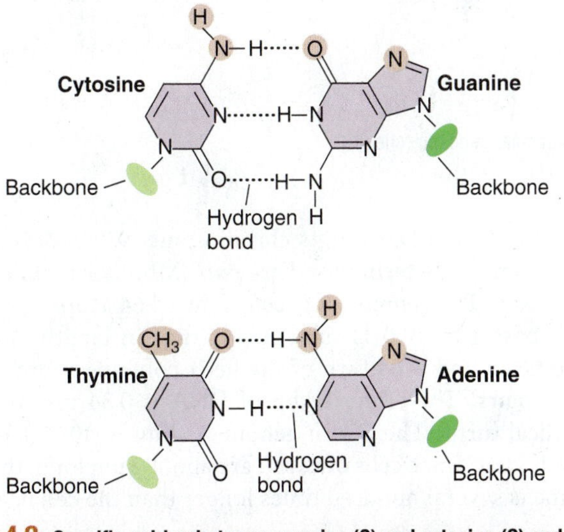

MINIQUIZ

- What is a genome?
- What components are found in a nucleotide? How does a nucleo*side* differ from a nucleo*tide*?
- What three informational macromolecules are involved in genetic information flow?
- In all cells there are three processes involved in genetic information flow. What are they?

4.2 The Double Helix

In all cells DNA exists as a double-stranded molecule with two polynucleotide strands whose base sequences are **complementary**. The complementarity of DNA arises because of specific base pairing by hydrogen bonds: adenine always pairs with thymine, and guanine always pairs with cytosine. Each adenine–thymine base pair has *two* hydrogen bonds, while each guanine–cytosine base pair has *three*. This makes GC pairs stronger than AT pairs. The two strands of the double-stranded DNA molecule are arranged in an **antiparallel** fashion (**Figure 4.4**; DNA strands appear throughout in two shades of green). Thus, the strand on the left runs 5' to 3' from top to bottom, whereas the complementary strand runs 5' to 3' from bottom to top. Although individual hydrogen bonds are very weak, the large number of such bonds between the base pairs of a long DNA molecule confers considerable stability on the molecule, sufficient to hold the two strands together.

Figure 4.2 Specific pairing between guanine (G) and cytosine (C) and between adenine (A) and thymine (T) via hydrogen bonds. These are the typical base pairs found in double-stranded DNA. Atoms that are found in the major groove of the double helix and that interact with proteins are highlighted in pink. The deoxyribose phosphate backbones of the two strands of DNA are also indicated. Note the different shades of green for the two strands of DNA, a convention used throughout this book.

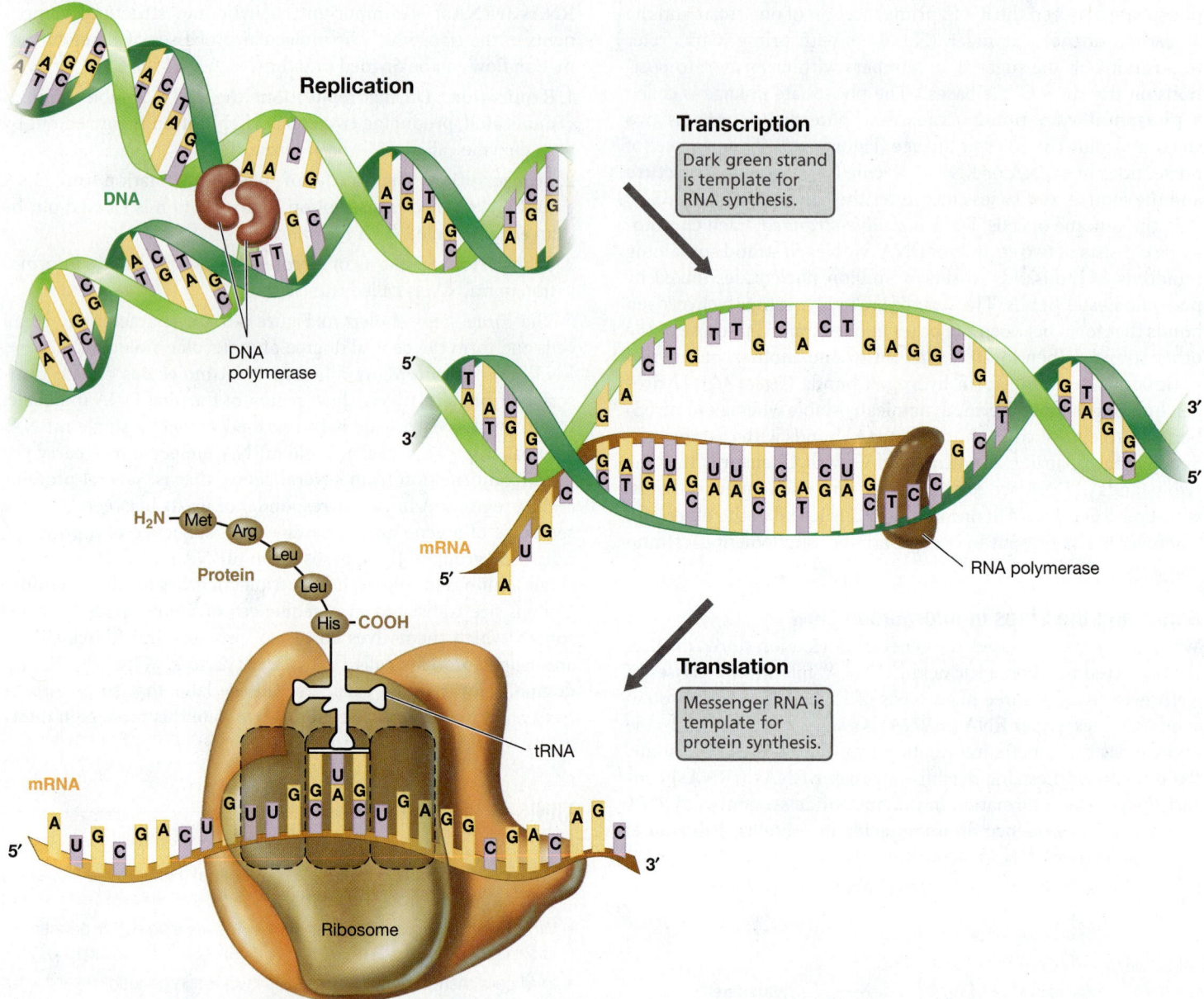

Figure 4.3 **Synthesis of the three types of informational macromolecules.** Note that for any particular gene only one of the two strands of the DNA double helix is transcribed.

The two strands of DNA are wrapped around each other to form a double helix (**Figure 4.5**). The helix forms two distinct grooves, the *major groove* and the *minor groove*. Most proteins that interact specifically with DNA bind in the major groove, where there is plenty of space. Because the double helix is a regular structure, some atoms of each base are always exposed in the major groove (and some in the minor groove). Key regions of nucleotides that are important in interactions with proteins are shown in Figure 4.2.

Size and Shape of DNA Molecules

The size of a DNA molecule is expressed as the number of nucleotide bases or base pairs per molecule. Thus, a DNA molecule with 1000 bases is 1 kilobase (kb) of DNA. If the DNA is a double helix, then *kilobase pairs* (kbp) is used. Thus, a double helix 5000 base pairs in size would be 5 kbp. The bacterium *Escherichia coli* has

about 4640 kbp of DNA in its chromosome. When dealing with large genomes, the term *megabase pair* (Mbp) for a million base pairs is used. The genome of *E. coli* is thus 4.64 Mbp.

Each base pair is 0.34 nanometer (nm) in length along the double helix, and each turn of the helix contains approximately 10 base pairs. Therefore, 1 kbp of DNA is 0.34 μm long with 100 helical turns. The *E. coli* genome is thus 4640×0.34 μm = 1.58 mm long. Since cells of *E. coli* are about 2 μm long, the chromosome is several hundred times longer than the cell itself!

Supercoiling of DNA

Considering the above calculations, how is it possible to pack so much DNA into such a tiny space? The solution is the imposition of a "higher-order" structure on the DNA, in which the double-stranded DNA is further twisted in a process called *supercoiling*.

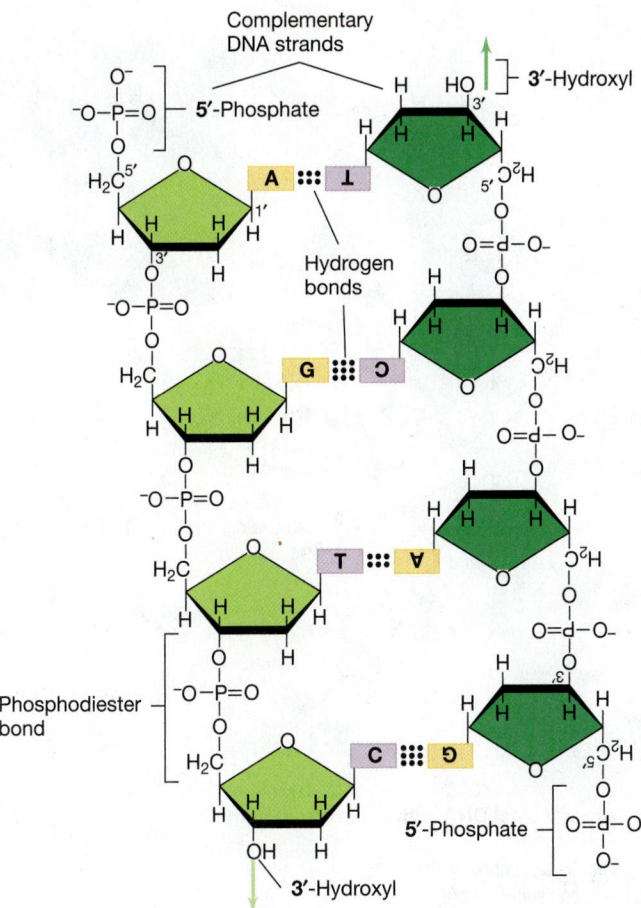

Figure 4.4 **DNA structure.** Complementary and antiparallel nature of DNA. Note that one chain ends in a 5′-phosphate group, whereas the other ends in a 3′-hydroxyl. The purple bases represent the pyrimidines cytosine (C) and thymine (T), and the yellow bases represent the purines adenine (A) and guanine (G).

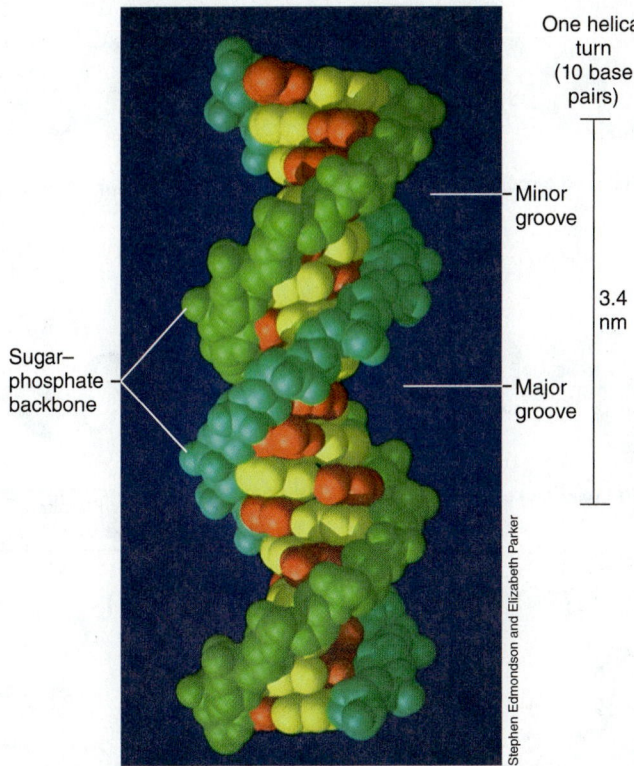

Stephen Edmondson and Elizabeth Parker

Figure 4.5 **A computer model of a short segment of DNA showing the overall arrangement of the double helix.** One of the sugar–phosphate backbones is shown in blue and the other in green. The pyrimidine bases are shown in red and the purines in yellow. Note the locations of the major and minor grooves (compare with Figure 4.2). One helical turn contains 10 base pairs.

Figure 4.6 shows how supercoiling occurs in a circular DNA molecule. If a circular DNA molecule is linearized, any supercoiling is lost and the DNA becomes "relaxed." When relaxed, a DNA molecule has exactly the number of turns of the helix predicted from the number of base pairs.

Supercoils are inserted or removed in DNA by enzymes called *topoisomerases*. The activity of supercoiling puts the DNA molecule under torsion, much like the tension added to a rubber band when it is twisted. DNA can be supercoiled in either a positive or a negative manner. In positive supercoiling the double helix is overwound (contains more than the natural number of turns), whereas in negative supercoiling the double helix is underwound (contains fewer than the natural number of turns). Negative supercoiling results when the DNA is twisted about its axis in the opposite sense from the right-handed double helix. Negatively supercoiled DNA is the form predominantly found in nature. In the *Escherichia coli* chromosome, more than 100 supercoiled domains are thought to exist, each stabilized by specific proteins bound to the DNA. Inserting supercoils into DNA requires energy from ATP, whereas releasing supercoils does not. In *Bacteria* and most *Archaea*, **DNA gyrase** is a type II topoisomerase that inserts negative supercoils into DNA by making double-strand

breaks (**Figure 4.7**). We will see later that *Archaea* that live at very high temperatures have chromosomes that are positively supercoiled and this feature helps to maintain DNA structure at such temperatures (ɘ Section 16.13). Some antibiotics inhibit the activity of DNA gyrase. These include the quinolones (such as nalidixic acid), the fluoroquinolones (such as ciprofloxacin), and novobiocin.

MINIQUIZ
- What does antiparallel mean in terms of the structure of double-stranded DNA?
- Define the term complementary when used to refer to two strands of DNA.
- What makes GC pairs stronger than AT pairs?
- Why is supercoiling important? What enzyme facilitates supercoiling?

4.3 Genetic Elements: Chromosomes and Plasmids

Structures containing genetic material (DNA in most organisms, but RNA in some viruses) are called *genetic elements*. The main genetic element in prokaryotes is the **chromosome**. Other genetic elements may be found and play important roles in gene function

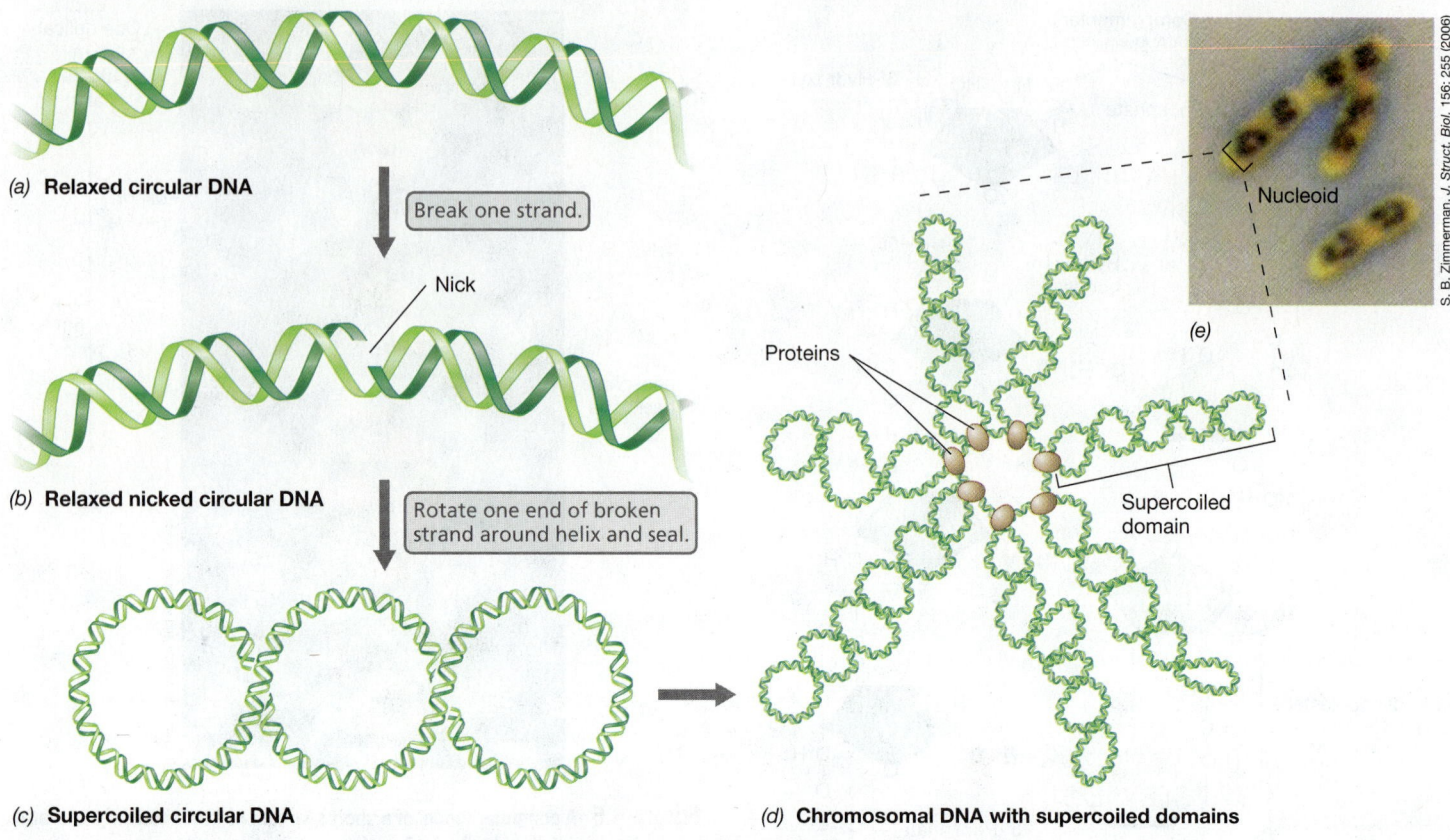

(a) **Relaxed circular DNA**

Break one strand.

Nick

(b) **Relaxed nicked circular DNA**

Rotate one end of broken strand around helix and seal.

(c) **Supercoiled circular DNA**

Proteins

Supercoiled domain

Nucleoid

(e)

(d) **Chromosomal DNA with supercoiled domains**

S. B. Zimmerman, *J. Struct. Biol.* 156: 255 (2006)

Figure 4.6 Supercoiled DNA. *(a–c)* Relaxed, nicked, and supercoiled circular DNA. A nick is a break in a phosphodiester bond of one strand. *(d)* In fact, the double-stranded DNA in the bacterial chromosome is arranged not in one supercoil but in several supercoiled domains, as shown here. *(e)* Simultaneous phase-contrast and fluorescence image of *E. coli* illustrating the location of the nucleoid within growing cells. Cells were treated with a fluorescent dye specific for DNA and the color was inverted to show the nucleoids as black.

in both prokaryotes and eukaryotes (**Table 4.1**). These include *virus genomes, plasmids, organellar genomes,* and *transposable elements.* A typical prokaryote has a single circular DNA chromosome containing all (or most) of the genes found inside the cell. Although a single chromosome is the rule among prokaryotes, there are exceptions. A few prokaryotes contain two or even three chromosomes. Eukaryotic genomes have multiple chromosomes. Also, the DNA in all known eukaryotic chromosomes is linear in contrast to most prokaryotic chromosomes, which are circular DNA molecules.

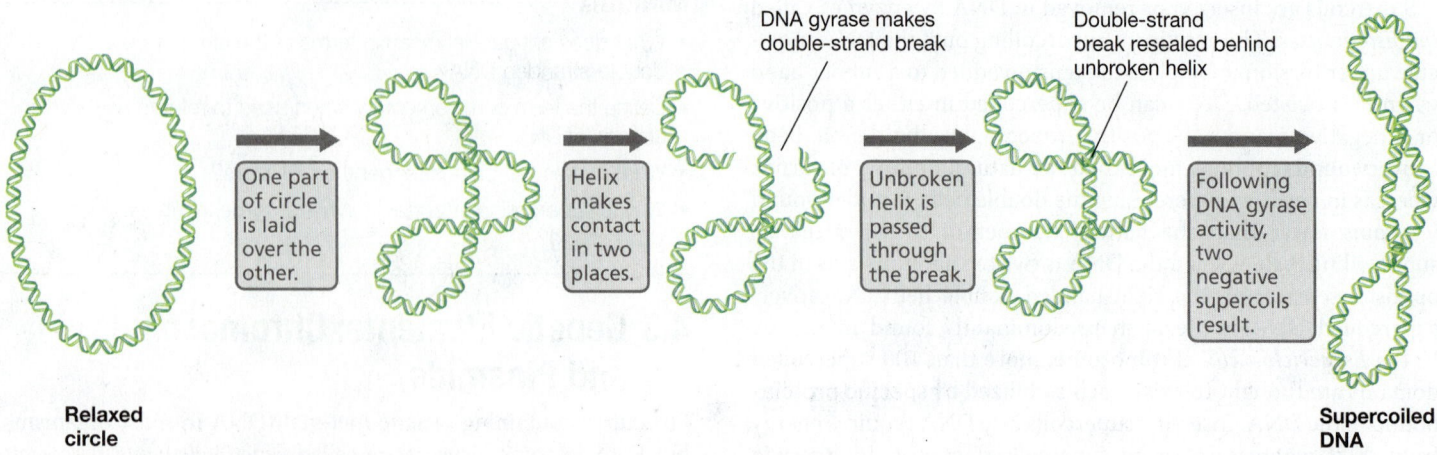

DNA gyrase makes double-strand break

Double-strand break resealed behind unbroken helix

One part of circle is laid over the other.

Helix makes contact in two places.

Unbroken helix is passed through the break.

Following DNA gyrase activity, two negative supercoils result.

Relaxed circle

Supercoiled DNA

Figure 4.7 DNA gyrase. Introduction of negative supercoiling into circular DNA by the activity of DNA gyrase (topoisomerase II), which makes double-strand breaks.

Table 4.1 Kinds of genetic elements

Organism	Element	Type of nucleic acid	Description
Prokaryote	Chromosome	Double-stranded DNA	Extremely long, usually circular
Eukaryote	Chromosome	Double-stranded DNA	Extremely long, linear
All organisms	Plasmid[a]	Double-stranded DNA	Relatively short circular or linear, extrachromosomal
All organisms	Transposable element	Double-stranded DNA	Always found inserted into another DNA molecule
Mitochondrion or chloroplast	Organellar genome	Double-stranded DNA	Medium length, usually circular
Virus	Virus genome	Single- or double-stranded DNA or RNA	Relatively short, circular or linear

[a]Plasmids are uncommon in eukaryotes.

Although considered microorganisms, viruses are not cells, but instead depend on cells for their replication. Nevertheless, viruses contain genomes, of *either* DNA or RNA, that control their replication and transfer from cell to cell. Both linear and circular viral genomes are known. In addition, the nucleic acid in viral genomes may be single-stranded or double-stranded. **Plasmids** are genetic elements that replicate separately from the chromosome. The great majority of plasmids are double-stranded DNA, and although most plasmids are circular, some are linear. Plasmids are typically much smaller than chromosomes. **Transposable elements** are segments of DNA that can move from one site on a DNA molecule to another, either on the same molecule or on a different DNA molecule. Transposable elements do not exist as separate molecules of DNA but instead are always found inserted into other DNA molecules. Chromosomes, plasmids, virus genomes, and any other type of DNA molecule may act as a host for a transposable element. Transposable elements are found in both prokaryotes and eukaryotes and play important roles in genetic variation (ᗡᖷ Section 10.11).

Arrangement of Genes on the *Escherichia coli* Chromosome

Many bacterial genomes including that of *Escherichia coli* have been completely sequenced, thus revealing the number and location of the genes they possess. The strain of *E. coli* whose chromosome was originally sequenced, strain MG1655, is a derivative of *E. coli* K-12, the traditional strain used for genetics. A corresponding genetic map of the 4,639,675 bp chromosome is presented in **Figure 4.8**, with only a few of the several thousand genes in the *E. coli* chromosome depicted. Map distances are given in 100 kilobase pairs of DNA. Genomic analysis has revealed 4288 possible protein-encoding genes that account for 88% of the *E. coli* genome. Approximately 1% of the genome are genes encoding tRNAs and rRNAs. This is in contrast to eukaryotic genomes that typically contain much more DNA than is needed to encode all the proteins required for cell function. For instance, in the human genome only about 3% of the total DNA actually encodes proteins. The "extra" DNA in eukaryotes is present as intervening DNA between coding sequences (which is removed after transcription) or as repetitive sequences, some of which are repeated hundreds or thousands of times.

Genetic mapping of the genes that encode the enzymes that function in the same biochemical pathway in *E. coli* has shown that these genes are often clustered. On the genetic map in Figure 4.8, a few such clusters are shown. Notice, for instance, the *gal*, *trp*, and *his* gene clusters. Each of these gene clusters constitutes an **operon** that is transcribed into a single mRNA encoding several individual proteins. Genes for many other biochemical pathways in *E. coli* are not clustered. For example, genes for maltose degradation (*mal* genes, Figure 4.8) are scattered throughout the chromosome. Sequence analysis of the *E. coli* chromosome has shown that over 70% of the 2584 predicted or known transcriptional units contain only a single gene, and only 6% of operons have four or more genes. Some coding sequences are on one strand of the chromosome while others are on the opposite strand, and genomic analysis has shown that there are about equal numbers of genes on both strands. In contrast to prokaryotes, eukaryotes do not contain operons.

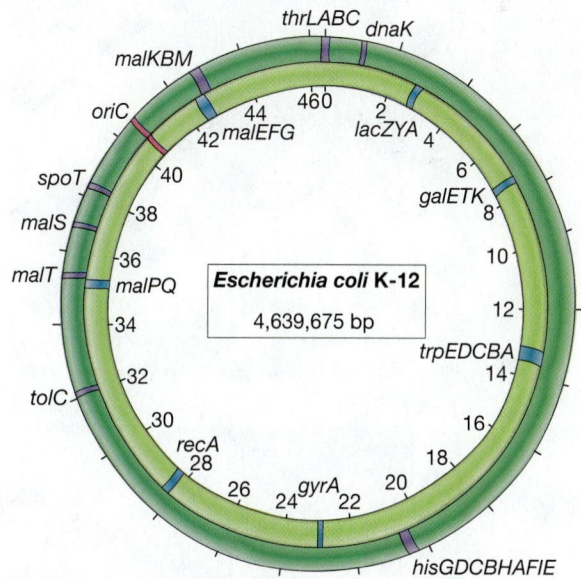

Figure 4.8 **The chromosome of *Escherichia coli* strain K-12.** Map distances are given in 100 kilobases of DNA. The chromosome contains 4,639,675 base pairs and 4288 open reading frames (genes). Depending on the DNA strand, the locations of a few genes and operons are indicated. Replication (Figure 4.3) proceeds in both directions from the origin of DNA replication, *oriC*, indicated in red.

General Principles of Plasmids

Many prokaryotic cells contain other genetic elements, in particular, plasmids, in addition to the chromosome. While plasmids possess their own origin of replication, they do rely on chromosomally encoded enzymes for their replication. Most plasmids are usually expendable since they rarely contain genes required for growth under all conditions. By contrast, essential genes reside on chromosomes. Unlike viruses, plasmids do not have an extracellular form and exist inside cells as free DNA. Thousands of different plasmids are known. Indeed, over 300 different naturally occurring plasmids have been isolated from strains of *Escherichia coli* alone.

Almost all known plasmids consist of double-stranded DNA. Most are circular, but many linear plasmids are also known. Naturally occurring plasmids vary in size from approximately 1 kbp to more than 1 Mbp. Typical plasmids are circular double-stranded DNA molecules less than 5% the size of the chromosome (**Figure 4.9**). Most plasmid DNA isolated from cells is supercoiled, this being the most compact form that DNA takes within the cell (Figure 4.6). Some bacteria may contain several different types of plasmids. For example, *Borrelia burgdorferi* (the Lyme disease pathogen, ↩ Section 30.4) contains 17 different circular and linear plasmids!

Cellular replication enzymes also replicate plasmids. The genes encoded by a plasmid function to direct the initiation of replication and partitioning of replicated plasmids between daughter cells. Different plasmids may be present in cells in different numbers called *copy number*. Some plasmids are present in only one to three copies, whereas others may be present in over 100 copies. Copy number is controlled by genes on the plasmid and by interactions between the host and the plasmid.

Types of Plasmids

Although by definition plasmids do not encode functions essential to the host, plasmids may carry genes that profoundly influence host cell physiology. Among the most widespread and well-studied groups of plasmids are the resistance plasmids, usually just called *R plasmids*, which confer resistance to antibiotics or other growth inhibitors. In general, the resistance genes encode proteins that either inactivate the antibiotic or protect the cell by some other mechanism. Several antibiotic resistance genes can be encoded on a single R plasmid; alternatively, a cell with multiple resistance may contain several different R plasmids. Plasmid R100, for example, is a 94.3-kbp plasmid (**Figure 4.10**) that encodes resistance to sulfonamides, streptomycin, spectinomycin, fusidic acid, chloramphenicol, and tetracycline. Plasmid R100 also encodes resistance to mercury. Pathogenic bacteria resistant to antibiotics are of considerable medical significance, and their increasing incidence is correlated with the increasing use of antibiotics for treating infectious diseases in humans and animals (↩ Section 27.17).

Pathogenic microorganisms possess a variety of characteristics that enable them to colonize hosts and establish infections. Two major characteristics of the virulence (disease-causing ability) of pathogens are often plasmid encoded: (1) the ability of the pathogen to attach to and colonize specific host tissue

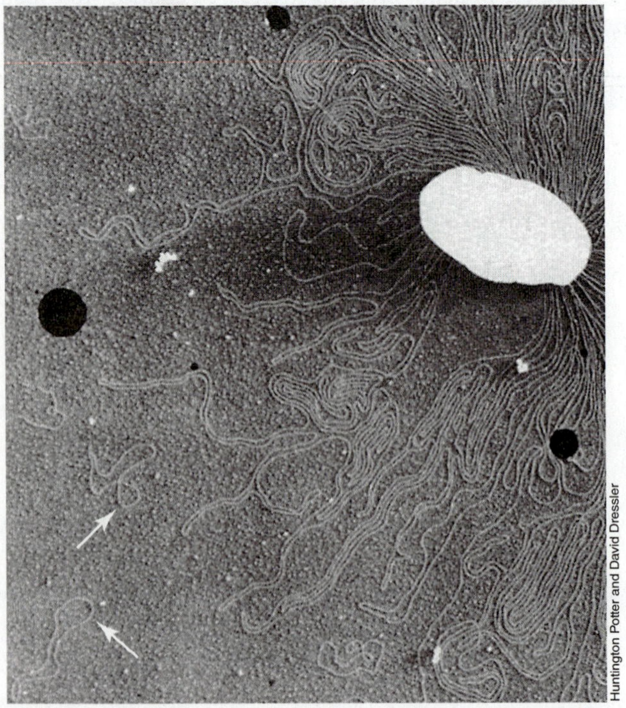

Figure 4.9 **The bacterial chromosome and bacterial plasmids, as seen in the electron microscope.** The plasmids (arrows) are the circular structures and are much smaller than the main chromosomal DNA. The cell (large, white structure) was broken gently so the DNA would remain intact.

Huntington Potter and David Dressler

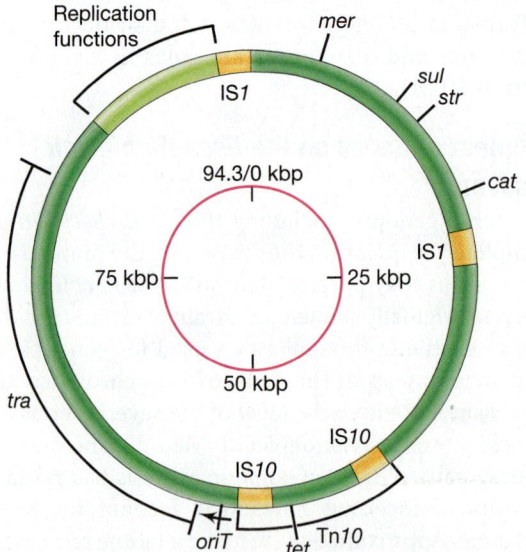

Figure 4.10 **Genetic map of the resistance plasmid R100.** The inner circle shows the size in kilobase pairs. The outer circle shows the location of major antibiotic resistance genes and other key functions: *mer*, mercuric ion resistance; *sul*, sulfonamide resistance; *str*, streptomycin resistance; *cat*, chloramphenicol resistance; *tet*, tetracycline resistance; *oriT*, origin of conjugative transfer; *tra*, transfer functions. The locations of insertion sequences (IS) and the transposon Tn*10* are also shown. Genes for plasmid replication are found in the region from 88 to 92 kbp.

and (2) the production of toxins, enzymes, and other molecules that cause damage to the host. Many bacteria also produce proteins that inhibit or kill closely related species or even different strains of the same species. These agents, called **bacteriocins**, are analogous to antibiotics, but have a narrower spectrum of activity than antibiotics. The genes encoding bacteriocins and the proteins needed for processing and transporting them and for conferring immunity on the producing organism are usually found on plasmids. For example, *E. coli* produces bacteriocins called *colicins* that bind to specific receptors on the surface of susceptible cells and kill them by disrupting membrane function. Other colicins are nucleases that degrade the DNA or RNA of susceptible strains.

In some cases plasmids encode properties fundamental to the ecology of the bacterium. For example, the ability of *Rhizobium* to interact with plants and form nitrogen-fixing root nodules requires certain plasmid functions (⮂ Section 22.3). Other plasmids confer special metabolic properties on bacterial cells, such as the ability to degrade toxic pollutants. Some special properties conferred by plasmids are summarized in Table 4.2.

MINIQUIZ

- What defines a chromosome in prokaryotes?
- What are viruses and plasmids?
- Approximately how large is the *Escherichia coli* genome in base pairs? How many genes does it contain?
- What properties does an R plasmid confer on its host cell?

Table 4.2 Examples of traits conferred by plasmids in prokaryotes

Traits	Organisms
Antibiotic production	*Streptomyces*
Conjugation	Wide range of bacteria
Metabolic functions	
Degradation of octane, camphor, naphthalene	*Pseudomonas*
Degradation of herbicides	*Alcaligenes*
Formation of acetone and butanol	*Clostridium*
Lactose, sucrose, citrate, or urea utilization	Enteric bacteria
Pigment production	*Erwinia, Staphylococcus*
Gas vesicle production	*Halobacterium*
Resistance	
Antibiotic resistance	Wide range of bacteria
Resistance to toxic metals	Wide range of bacteria
Virulence	
Tumor production in plants	*Agrobacterium*
Nodulation and symbiotic nitrogen fixation	*Rhizobium*
Bacteriocin production and resistance	Wide range of bacteria
Animal cell invasion	*Salmonella, Shigella, Yersinia*
Coagulase, hemolysin, enterotoxin	*Staphylococcus*
Toxins and capsule	*Bacillus anthracis*
Enterotoxin, K antigen	*Escherichia coli*

II • Transmission of Genetic Information: DNA Replication

DNA replication is necessary for cells to divide, whether to reproduce new organisms, as in unicellular microorganisms, or to produce new cells as part of a multicellular organism. To successfully transmit genetic information from a mother cell to an identical daughter cell, DNA replication must be highly accurate. This process requires the activities of a host of special enzymes.

4.4 Templates and Enzymes

As we have seen, DNA exists in cells as a double helix with complementary base pairing (Figures 4.3 and 4.4). If the double helix is opened up, a new strand can be synthesized as the complement of each parental strand. As shown in **Figure 4.11**, replication is a **semiconservative** process, meaning that the two resulting double helices consist of one new strand and one parental strand. The DNA strand that is used to make a complementary daughter strand is called the template strand, and in DNA replication, each parental strand is a template for one newly synthesized strand (Figure 4.11).

The precursor of each new nucleotide in the DNA strand is a deoxynucleoside 5'-triphosphate. During insertion, the two terminal phosphates are removed and the innermost phosphate is then covalently bound to a deoxyribose of the growing chain (**Figure 4.12**). This addition of the incoming nucleotide requires the

presence of a free hydroxyl group, which is available only at the 3' end of the molecule. This leads to the important principle that DNA replication always proceeds *from the 5' end to the 3' end*, the 5'-phosphate of the incoming nucleotide being attached to the 3'-hydroxyl of the previously added nucleotide.

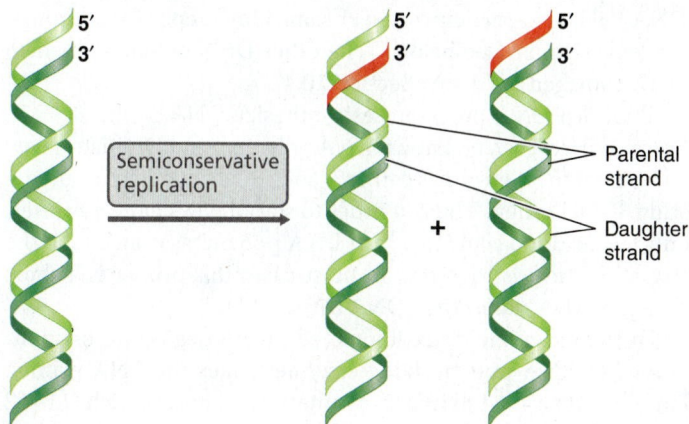

Figure 4.11 **Overview of DNA replication.** DNA replication is a semiconservative process in all cells. Note that the new double helices each contain one new daughter strand (shown topped in red) and one parental strand.

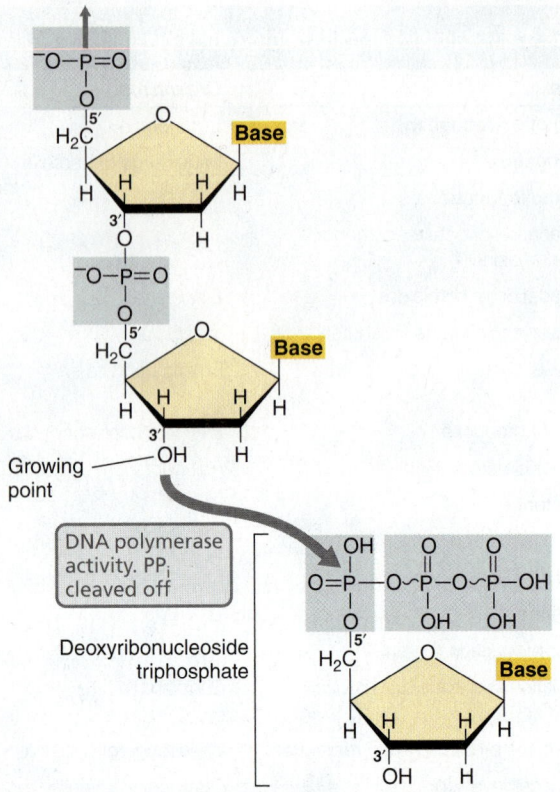

Figure 4.12 Extension of a DNA chain by adding a deoxyribonucleoside triphosphate at the 3′ end. Growth proceeds from the 5′-phosphate to the 3′-hydroxyl end. DNA polymerase catalyzes the reaction. The four precursors are deoxythymidine triphosphate (dTTP), deoxyadenosine triphosphate (dATP), deoxyguanosine triphosphate (dGTP), and deoxycytidine triphosphate (dCTP). Upon nucleotide insertion, the two terminal phosphates of the triphosphate are split off as pyrophosphate (PP$_i$). Thus, two energy-rich phosphate bonds are consumed when adding each nucleotide.

Enzymes that catalyze the addition of deoxynucleotides are called **DNA polymerases**. Several such enzymes play a role in replication, each with a specific function. There are five different DNA polymerases in *Escherichia coli*, DNA polymerases I, II, III, IV, and V. DNA polymerase III (DNA Pol III) is the primary enzyme for replicating chromosomal DNA. DNA polymerase I (DNA Pol I) also participates in chromosomal replication, though to a lesser extent (see below). The other DNA polymerases help repair damaged DNA (⟳ Section 10.4).

All known DNA polymerases synthesize DNA in the 5′ → 3′ direction. However, no known DNA polymerase can initiate a new chain; all of these enzymes can only add a nucleotide onto a preexisting 3′-OH group. Thus, in order to start a new chain, a **primer**, a nucleic acid molecule to which DNA polymerase can attach the first nucleotide, is required. In most cases this primer is a short stretch of RNA rather than DNA (**Figure 4.13**).

When the double helix is opened at the beginning of replication, an RNA-polymerizing enzyme makes the RNA primer. This enzyme, called **primase**, synthesizes a short stretch (11–12 nucleotides) of RNA that is complementary in base pairing to the template strand DNA. At the growing end of this RNA primer is a 3′-OH group to which DNA polymerase adds the first deoxyribonucleotide. Continued extension of the molecule thus

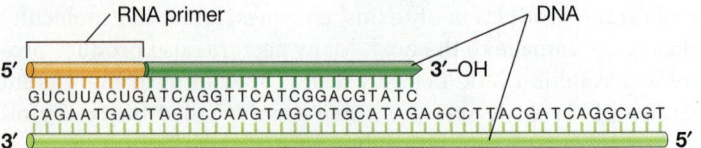

Figure 4.13 The RNA primer. Structure of the RNA–DNA hybrid formed during initiation of DNA synthesis. Orange depicts the RNA primer.

occurs as DNA rather than RNA. The newly synthesized molecule has a structure like that shown in Figure 4.13. The primer is eventually removed and replaced with DNA, as described in the next section.

MINIQUIZ -

• To which end (5′ end or 3′ end) of a newly synthesized strand of DNA does DNA polymerase add a base?

• Why is a primer required for DNA replication? What is the primer made of?

4.5 The Replication Fork

Much of our understanding of the details of DNA replication has been obtained from studying the bacterium *Escherichia coli*; however, DNA replication is probably quite similar in all *Bacteria*. By contrast, although most species of *Archaea* have circular chromosomes, many events in DNA replication resemble those in eukaryotic cells more than those in *Bacteria*, a reflection of the phylogenetic affiliation between *Archaea* and *Eukarya* (Figure 1.6*b*).

Initiation of DNA Synthesis

Before DNA polymerase can synthesize new DNA, the double helix must be unwound to expose the template strands. The zone of unwound DNA where replication occurs is called the **replication fork**. The enzyme **DNA helicase** unwinds the double helix, using energy from ATP, and exposes a short single-stranded region (**Figure 4.14**). Helicase moves along the DNA and separates the strands just in advance of the replication fork. The single-stranded region is immediately covered with copies of single-strand binding protein to stabilize the single-stranded DNA and prevent the double helix from re-forming. Unwinding of the double helix by helicase generates positive supercoils ahead of the advancing replication fork. To counteract this, DNA gyrase travels along the DNA ahead of the replication fork and inserts negative supercoils to cancel out the positive supercoiling.

Bacteria possess a single location on the chromosome where DNA synthesis is initiated, the origin of replication (*oriC*). This is a specific DNA sequence of about 250 bases that is recognized by initiation proteins, in particular a protein called DnaA (**Table 4.3**), which binds to this region and opens up the double helix. Next to assemble is the helicase (known as DnaB), which is helped onto the DNA by the helicase loader protein (DnaC). Two helicases are loaded, one onto each strand, facing in opposite directions. Next, two primase and then two DNA polymerase enzymes are loaded onto the DNA behind the helicases. Initiation of DNA

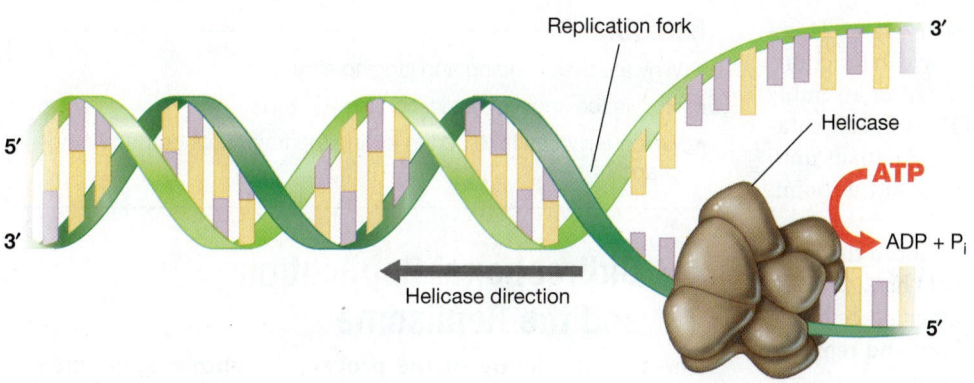

Figure 4.14 **DNA helicase unwinding a double helix.** In this figure, the helicase is denaturing or pulling the two antiparallel strands of DNA apart beginning from the right and moving to the left.

replication then begins on the two single strands. As replication proceeds, the replication fork appears to move along the DNA (Figure 4.14).

Leading and Lagging Strands

Figure 4.15 shows details of DNA replication at the replication fork. An important distinction in replication between the two DNA strands can be made due to the fact that replication always proceeds from 5′ to 3′ (5′ → 3′, always adding a new nucleotide to the 3′-OH of the growing chain). On the strand growing from the 5′-PO_4^{2-} to the 3′-OH, called the **leading strand**, DNA synthesis occurs *continuously* because there is always a free 3′-OH at the

replication fork to which a new nucleotide can be added. But on the opposite strand, called the **lagging strand**, DNA synthesis occurs *discontinuously* because there is no 3′-OH at the replication fork to which a new nucleotide can attach (Figure 4.15). The 3′-OH on this strand is located at the opposite end, away from the replication fork. Therefore, on the lagging strand, RNA primers must be synthesized by primase multiple times to provide free 3′-OH groups for DNA Pol III. By contrast, the leading strand is primed only once, at the origin. As a result, the lagging strand is made in short segments, called *Okazaki fragments*, named after their discoverer, Reiji Okazaki. These lagging strand fragments are joined together later to yield a continuous strand of DNA.

Synthesis of the New DNA Strands

After synthesizing the RNA primer, primase is replaced by DNA Pol III. This enzyme is actually a complex of several proteins (Table 4.3), including the polymerase core enzyme itself. Each molecule of polymerase is held on the DNA by a sliding clamp, which encircles and slides along the single template strands of DNA. Consequently, the replication fork contains two polymerase core

Enzyme	Encoding genes	Function
DNA gyrase	*gyrAB*	Replaces supercoils ahead of replisome
Origin-binding protein	*dnaA*	Binds origin of replication to open double helix
Helicase loader	*dnaC*	Loads helicase at origin
Helicase	*dnaB*	Unwinds double helix at replication fork
Single-strand binding protein	*ssb*	Prevents single strands from annealing
Primase	*dnaG*	Primes new strands of DNA
DNA polymerase III		Main polymerizing enzyme
Sliding clamp	*dnaN*	Holds Pol III on DNA
Clamp loader	*holA–E*	Loads Pol III onto sliding clamp
Dimerization subunit (Tau)	*dnaX*	Holds together the two core enzymes for the leading and lagging strands
Polymerase subunit	*dnaE*	Strand elongation
Proofreading subunit	*dnaQ*	Proofreading
DNA polymerase I	*polA*	Excises RNA primer and fills in gaps
DNA ligase	*ligA, ligB*	Seals nicks in DNA
Tus protein	*tus*	Binds terminus and blocks progress of the replication fork
Topoisomerase IV	*parCE*	Unlinking of interlocked circles

Table 4.3 Major enzymes involved in DNA replication in *Bacteria*

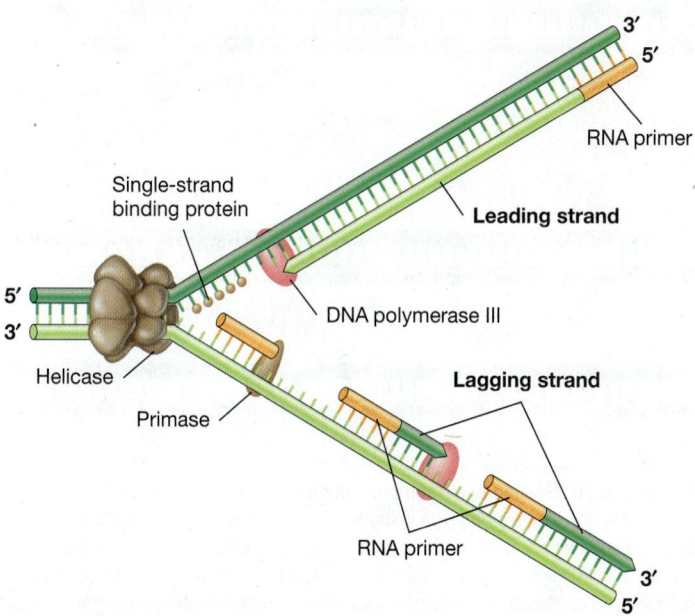

Figure 4.15 **Events at the DNA replication fork.** Note the polarity and antiparallel nature of the DNA strands.

enzymes and two sliding clamps, one set for each strand. However, there is only a single clamp-loader complex, which functions to assemble the two sliding clamps onto the DNA. After assembly on the lagging strand, the elongation activity of DNA Pol III, catalyzed by DnaE, then adds deoxyribonucleotides sequentially until it reaches previously synthesized DNA (**Figure 4.16**). At this point, DNA Pol III stops.

The next enzyme to take part, DNA Pol I, has more than one enzymatic activity. Besides synthesizing DNA, Pol I has a $5' \rightarrow 3'$ exonuclease activity that removes the RNA primer preceding it (Figure 4.16). When the primer has been removed and replaced with DNA, DNA Pol I is released. The very last phosphodiester bond is made by an enzyme called **DNA ligase**. This enzyme seals nicks in DNAs that have an adjacent $5'$-PO_4^{2-} and $3'$-OH (something that DNA Pol III is unable to do), and along with DNA Pol I, it also participates in DNA repair. DNA ligase is also important for sealing genetically manipulated DNA during molecular cloning (⇄ Section 11.4).

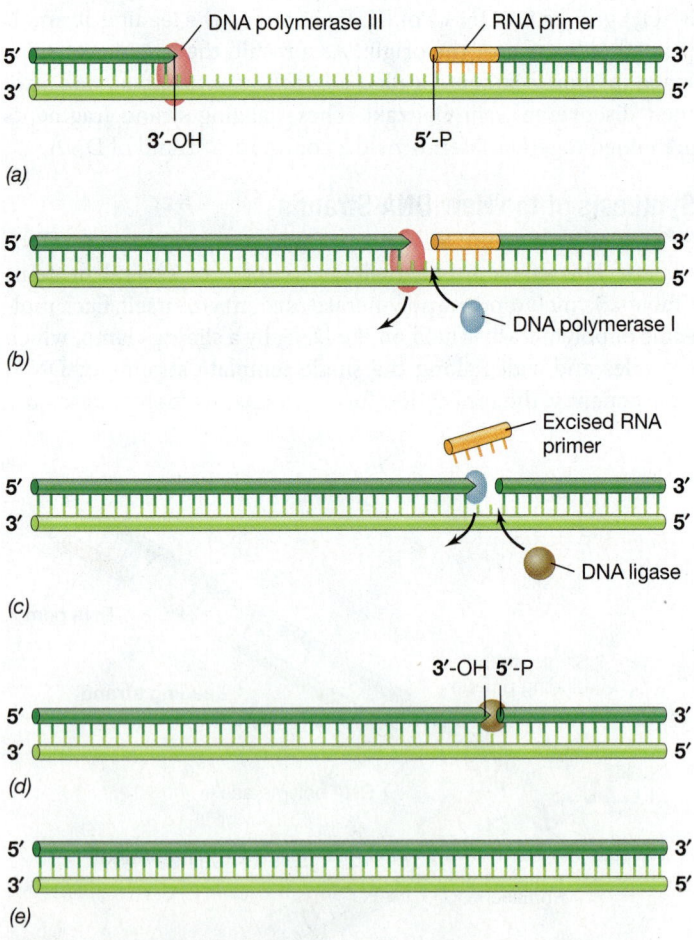

Figure 4.16 Sealing two fragments on the lagging strand. *(a)* DNA polymerase III is synthesizing DNA in the $5' \rightarrow 3'$ direction toward the RNA primer of a previously synthesized fragment on the lagging strand. *(b)* On reaching the fragment, DNA polymerase III leaves and is replaced by DNA polymerase I. *(c)* DNA polymerase I continues synthesizing DNA while removing the RNA primer from the previous fragment, and DNA ligase replaces DNA polymerase I after the primer has been removed. *(d)* DNA ligase seals the two fragments together. *(e)* The final product, complementary and antiparallel double-stranded DNA.

MINIQUIZ

- Why are there leading and lagging strands?
- How is the origin of replication recognized?
- Which enzymes take part in joining the fragments of the lagging strand?

4.6 Bidirectional Replication and the Replisome

The circular nature of the prokaryotic chromosome creates an opportunity for speeding up the replication process. In *Escherichia coli*, and probably in all prokaryotes with circular chromosomes, DNA replication is *bidirectional* from the origin of replication, as shown in **Figure 4.17**. There are thus *two* replication forks on each chromosome, each moving in opposite directions. These are held together by the two Tau protein subunits. In circular DNA, bidirectional replication leads to the formation of characteristic shapes called theta structures (Figure 4.17).

During bidirectional replication, synthesis occurs in both a leading and lagging fashion on each template strand, allowing DNA to replicate as rapidly as possible (Figure 4.17). While DNA Pol III can add nucleotides to a growing DNA strand at the rate of about 1000 per second, chromosome replication in *E. coli* still takes about 40 min. Interestingly, under the best growth conditions, *E. coli* can grow with a doubling time of about 20 min. The solution to this conundrum is that cells of *E. coli* growing at doubling times shorter than 40 min contain multiple DNA replication forks. That is, a new round of DNA replication begins before the last round has been completed. We consider this problem in more detail in Chapter 5 (⇄ Figure 5.4).

The Replisome

Figure 4.15 shows the differences in replication of the leading and the lagging strands and the enzymes that participate in the process. From such a schematic drawing it might appear that each replication fork contains several different proteins all working independently. Actually, this is not the case. Instead, replication proteins aggregate to form a large replication complex called the **replisome** (**Figure 4.18**). The lagging strand of DNA actually loops out to allow the replisome to move smoothly along both strands, and the replisome literally pulls the DNA template through it as replication occurs. Therefore, it is the DNA, rather than DNA polymerase, that moves during replication. Note also how helicase and primase form a subcomplex, called the *primosome*, which aids their working in close association during the replication process.

In summary, in addition to DNA Pol III, the replisome contains several key replication proteins: (1) DNA gyrase, which removes supercoils; (2) DNA helicase and primase (the primosome), which unwind and prime the DNA; and (3) single-strand binding protein, which prevents the separated template strands from re-forming a double helix (Figure 4.18). Table 4.3 summarizes the properties of proteins essential for DNA replication in *Bacteria*.

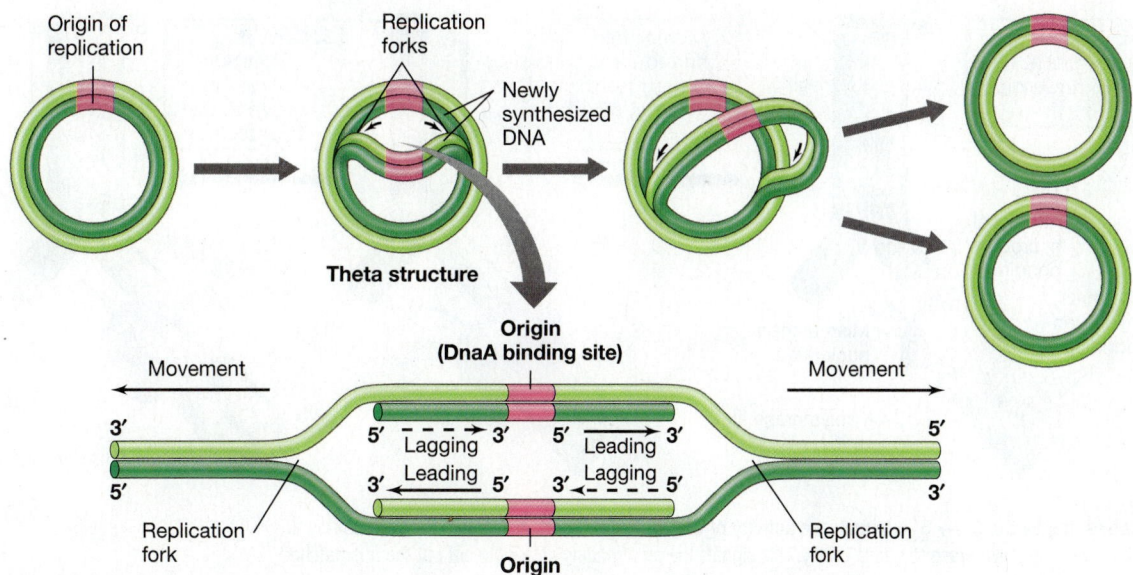

Figure 4.17 Replication of circular DNA: the theta structure. In circular DNA, bidirectional replication from an origin forms an intermediate structure resembling the Greek letter theta (θ). Inset shows dual replication forks in the circular chromosome. In *Escherichia coli*, the origin of replication is recognized by a specific protein, DnaA. Note that DNA synthesis is occurring in both a leading and a lagging manner on each of the new daughter strands. Compare this figure with the description of the replisome shown in Figure 4.18.

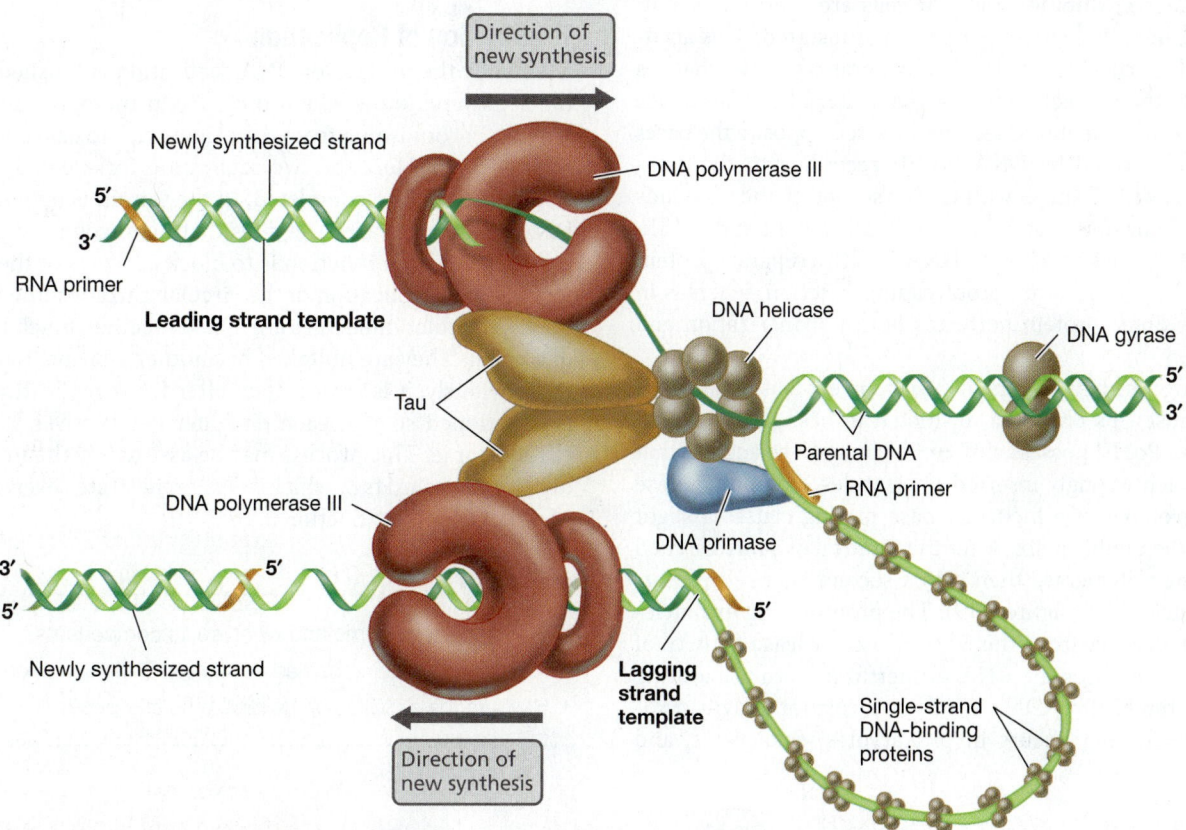

Figure 4.18 The replisome. The replisome consists of two copies of DNA polymerase III and DNA gyrase, plus helicase and primase (together forming the primosome), and many copies of single-strand DNA-binding protein. The Tau subunits hold the two DNA polymerase assemblies and helicase together. Just upstream of the rest of the replisome, DNA gyrase removes supercoils in the DNA to be replicated. Note that the two polymerases are replicating the two individual strands of DNA in opposite directions. Consequently, the lagging-strand template loops around so that the whole replisome moves in the same direction along the chromosome.

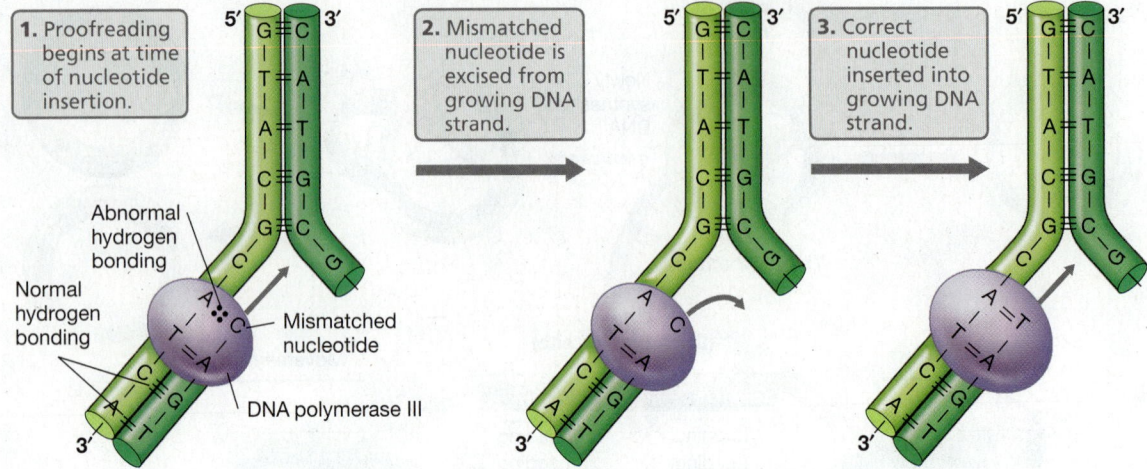

Figure 4.19 **Proofreading by the 3′ → 5′ exonuclease activity of DNA polymerase III.** A mismatch in base pairing at the terminal base pair causes the polymerase to pause briefly. This signals the proofreading activity to cut out the mismatched nucleotide, after which the correct base is inserted by the polymerase activity.

Fidelity of DNA Replication: Proofreading

DNA replicates with a remarkably low error rate. Nevertheless, when errors do occur, a mechanism exists to detect and correct them. Errors in DNA replication introduce *mutations*, changes in DNA sequence. Mutation rates in cells are remarkably low, between 10^{-8} and 10^{-11} errors per base pair inserted. This accuracy is possible partly because DNA polymerases get two chances to incorporate the correct base at a given site. The first chance comes when complementary bases are inserted opposite the bases on the template strand by DNA Pol III according to the base-pairing rules, A with T and G with C. The second chance depends upon a second enzymatic activity of both DNA Pol I and Pol III, called *proofreading* (**Figure 4.19**). In DNA Pol III, a separate protein subunit, DnaQ, performs the proofreading function, whereas in DNA Pol I, a single protein performs both polymerization and proofreading.

Proofreading activity occurs if an incorrect base has been inserted because this creates a mismatch in base pairing. Both DNA Pol I and Pol III possess a 3′ → 5′ exonuclease activity that can remove such wrongly inserted nucleotides. The polymerase senses the error because incorrect base pairing causes a slight distortion in the double helix. After the removal of a mismatched nucleotide, the polymerase then gets a second chance to insert the correct nucleotide (Figure 4.19). The proofreading exonuclease activity is distinct from the 5′ → 3′ exonuclease activity of DNA Pol I that removes the RNA primer from both the leading and lagging strands. Only DNA Pol I has this latter activity. Exonuclease proofreading occurs in prokaryotes, eukaryotes, and viral DNA replication systems. However, many organisms have additional mechanisms for reducing errors made during DNA replication that operate after the replication fork has passed by. We will discuss some of these in Chapter 10.

Termination of Replication

Eventually the process of DNA replication is finished. How does the replisome know when to stop? On the opposite side of the circular chromosome from the origin is a site called the *terminus of replication*. Here the two replication forks collide as the new circles of DNA are completed. In the terminus region are several DNA sequences called *Ter* sites that are recognized by a protein called Tus, whose function is to block progress of the replication forks. When replication of the circular chromosome is complete, the two circular molecules are linked together, much like the links of a chain. They are unlinked by another enzyme, topoisomerase IV. Obviously, it is critical that, after DNA replication, the DNA is partitioned so that each daughter cell receives a copy of the chromosome. This process may be assisted by the important cell division protein FtsZ, which helps orchestrate several key events of cell division (↩ Section 5.2).

MINIQUIZ --

- What is the replisome and what are its components?
- How is proofreading carried out during DNA replication?
- How are the activities of the replisome stopped?

III • RNA Synthesis: Transcription

Transcription is the synthesis of ribonucleic acid (RNA) using DNA as a template. There are three key differences in the chemistry of RNA and DNA: (1) RNA contains ribose instead of deoxyribose; (2) RNA contains uracil instead of thymine; and (3) except in certain viruses, RNA is not double-stranded. The change from deoxyribose to ribose affects the chemistry of a nucleic acid; enzymes that act on DNA usually have no effect on RNA, and vice versa. However, the change from thymine to uracil does not affect base pairing, as these two bases pair with adenine equally well.

While RNA predominantly exists in single strands, molecules typically fold back upon themselves in regions where complementary base pairing is possible. The term **secondary structure** refers to this folding, and the term *primary structure* refers to the nucleotide sequence, as it does for DNA. Secondary structure leads to highly folded and twisted RNA molecules whose biological function depends critically on their final three-dimensional shape.

RNA plays several important roles in the cell. As we have seen (Figure 4.3), three major types of RNA participate in protein synthesis: *messenger* RNA (mRNA), *transfer* RNA (tRNA), and *ribosomal* RNA (rRNA). Several other types of RNA are also known but mostly function in regulation (Chapter 7). All RNA molecules result from the transcription of DNA. It should be emphasized that RNA operates at two levels, genetic and functional. At the genetic level, mRNA carries genetic information from the genome to the ribosome. In contrast, rRNA has both a functional and a structural role in ribosomes, and tRNA has an active role in carrying amino acids for protein synthesis. Indeed, some RNA molecules including rRNA have enzymatic activity. Here we focus on how RNA is synthesized in *Bacteria*, once again using *Escherichia coli* as our model organism.

4.7 Transcription

Transcription is catalyzed by the enzyme **RNA polymerase**. Like DNA polymerase, RNA polymerase forms phosphodiester bonds but in this case between the ribonucleotides rATP, rGTP, rCTP, and rUTP rather than deoxyribonucleotides. Polymerization is driven by energy released from the hydrolysis of two energy-rich phosphate bonds of the incoming ribonucleoside triphosphates. The mechanism of RNA synthesis is much like that of DNA synthesis (Figure 4.12): During elongation of an RNA chain, ribonucleoside triphosphates are added to the 3′-OH of the ribose of the preceding nucleotide. Thus chain growth is 5′ to 3′ and the newly synthesized strand of RNA is antiparallel to the DNA template strand it was transcribed from. The overall process of RNA synthesis is illustrated in **Figure 4.20**.

RNA polymerase uses DNA as a template, but only one of the two strands is transcribed for any given gene. Nevertheless, genes are present on both strands of DNA and thus DNA sequences on both strands are transcribed, although at different locations. Unlike DNA polymerase, RNA polymerase can initiate new strands of RNA on its own; no primer sequence is necessary. As the newly made RNA dissociates from the DNA, the opened DNA closes back into the original double helix. Transcription stops at specific sites called *transcription terminators*. Unlike DNA replication, which copies entire genomes, transcription copies much smaller units of DNA, often as little as a single gene. This system allows the cell to transcribe different genes at different frequencies, depending on the needs of the cell for different proteins. In other words, gene expression is regulated. As we shall see in Chapter 7,

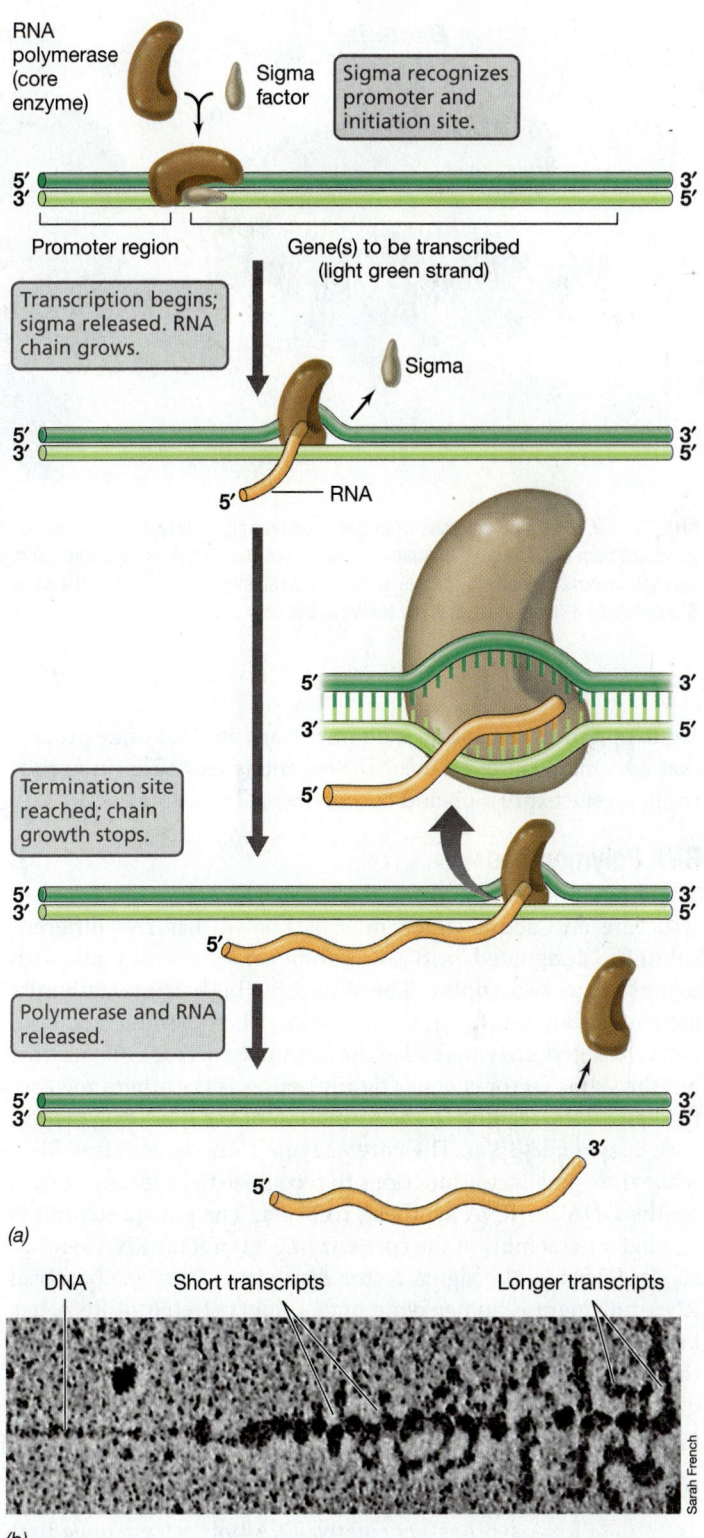

(a)

(b)

Figure 4.20 **Transcription.** *(a)* Steps in RNA synthesis. The initiation site (promoter) and termination site are specific nucleotide sequences on the DNA. RNA polymerase moves down the DNA chain, temporarily opening the double helix and transcribing one of the DNA strands. *(b)* Electron micrograph illustrates transcription along a gene on the *Escherichia coli* chromosome. Transcription is proceeding from left to right, with the shorter transcripts on the left becoming longer as transcription proceeds.

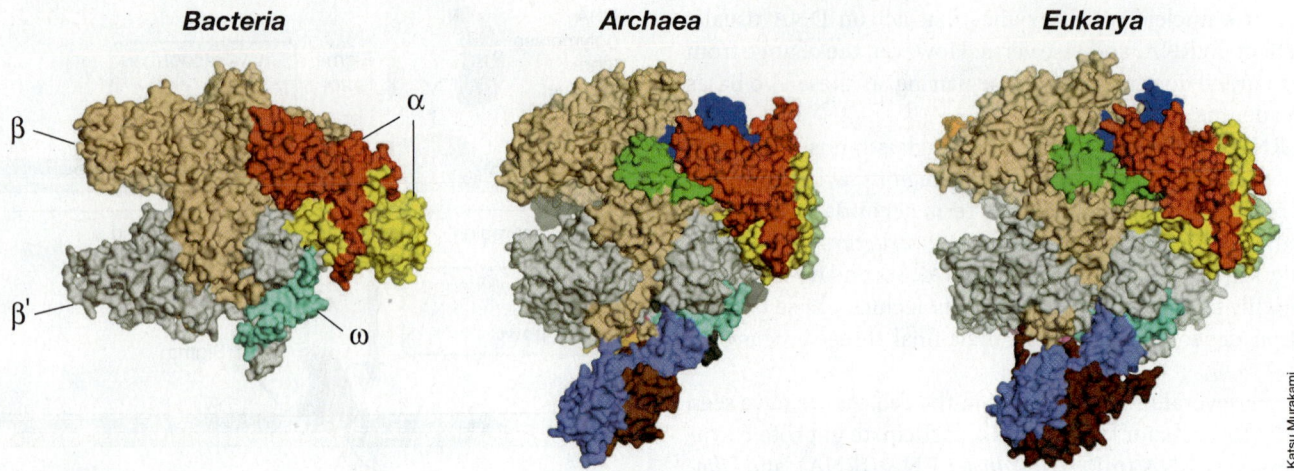

Figure 4.21 RNA polymerase from the three domains. Surface representation of multi-subunit cellular RNA polymerase structures from *Bacteria* (left, *Thermus aquaticus* core enzyme), *Archaea* (center, *Sulfolobus solfataricus*), and *Eukarya* (right, *Saccharomyces cerevisiae* RNA Pol II). Orthologous subunits are depicted by the same color. A unique subunit in the *S. solfataricus* RNA polymerase is not shown in this view.

regulation of transcription is an important and elaborate process that uses many different mechanisms and is very efficient at controlling gene expression and conserving cell resources.

RNA Polymerases

The RNA polymerase from *Bacteria*, which has the simplest structure and about which most is known, has five different subunits, designated β, β', α, ω (omega), and σ (sigma), with α present in two copies. The β and β' (beta prime) subunits are similar but not identical (**Figure 4.21**). The subunits interact to form the active enzyme, called the RNA polymerase holoenzyme, but the sigma factor is not as tightly bound as the others and easily dissociates, leading to the formation of the RNA polymerase core enzyme, $\alpha_2\beta\beta'\omega$. The core enzyme alone synthesizes RNA while the sigma factor functions to recognize the appropriate site on the DNA for RNA synthesis to begin. The omega subunit is needed for assembly of the core enzyme but not for RNA synthesis. In *Bacteria*, the sigma factor dissociates from the bacterial RNA polymerase holoenzyme once a short stretch of RNA has been formed (Figure 4.20). Elongation of the RNA molecule is then catalyzed by the core enzyme alone (Figure 4.20). Sigma is only needed to form the initial RNA polymerase–DNA complex at the promoter.

Promoters

To initiate RNA synthesis correctly, RNA polymerase must first recognize the initiation sites on the DNA called **promoters** (Figure 4.20). In *Bacteria*, promoters are recognized by the sigma subunit of RNA polymerase. Once RNA polymerase has bound to a promoter, transcription can proceed (Figure 4.20). In this process, the DNA double helix at the promoter is opened up by the RNA polymerase to form a transcription bubble. As the polymerase moves, it unwinds the DNA in short segments. This transient unwinding exposes the template strand and allows it to be copied into the RNA complement. Thus, promoters can

be thought of as structures that effectively "point" RNA polymerase in one direction or the other along the DNA. If a region of DNA has two nearby promoters pointing in opposite directions, then transcription from one will proceed in one direction (on one of the DNA strands) while transcription from the other promoter will proceed in the opposite direction (on the other strand).

Sigma Factors and Consensus Sequences

Promoters are specific DNA sequences that bind RNA polymerase, and **Figure 4.22** shows the sequence of several promoters from *Escherichia coli*. All these sequences are recognized by the same major sigma factor in *E. coli* called σ^{70} (the superscript 70 indicates the size of this protein, 70 kilodaltons); although these sequences are not identical, sigma recognizes two highly conserved shorter sequences within the promoter. These conserved sequences are upstream of the transcription start site. One is 10 bases before the transcription start, the −10 region, or *Pribnow box*. Although promoters differ slightly, comparison of many −10 regions gives the consensus sequence: TATAAT. The second conserved region is about 35 bases upstream from the start of transcription. The consensus sequence in the −35 region is TTGACA (Figure 4.22). Again, most promoters differ slightly, but are very close to consensus.

In *E. coli*, promoters that are most like the consensus sequence are usually more effective in binding RNA polymerase. Such promoters are called *strong promoters* and are very useful in genetic engineering, as discussed in Chapter 11. While most genes in *E. coli* require the standard sigma factor, σ^{70} (RpoD), for transcription, several alternative sigma factors exist that recognize different consensus sequences (**Table 4.4**). Each alternative sigma factor is specific for a group of genes required under special circumstances and thus essential for regulating gene expression. Consequently, it is possible to control the expression of different gene families by regulating the presence or absence of the

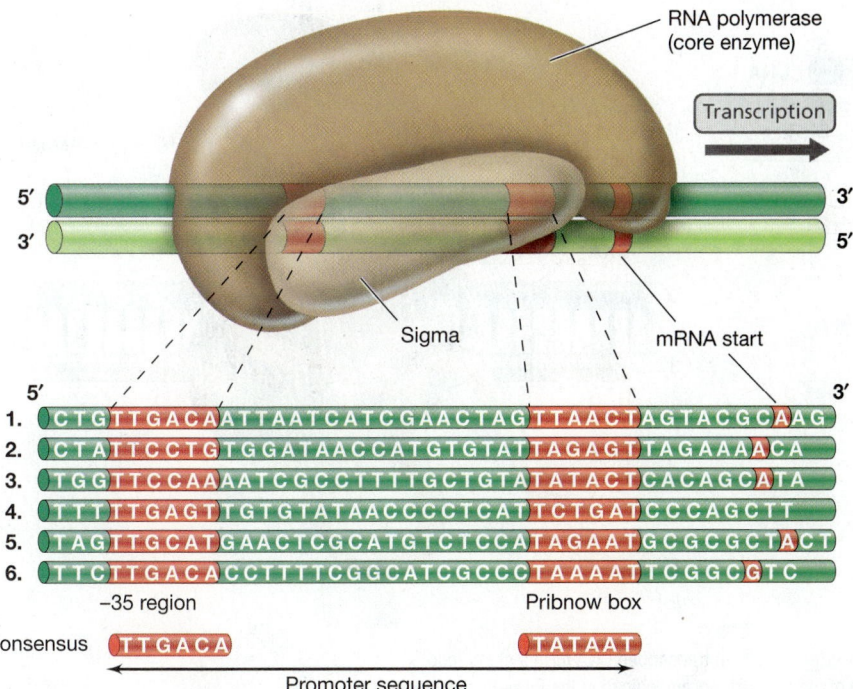

Figure 4.22 The interaction of RNA polymerase with a bacterial promoter. Shown below the RNA polymerase and DNA are six different promoter sequences identified in *Escherichia coli*. The contacts of the RNA polymerase with the −35 region and the Pribnow box (−10 sequence) are shown. Transcription begins at a unique base just downstream from the Pribnow box. Below the actual sequences at the −35 and Pribnow box regions are consensus sequences derived from comparing many promoters. Note that although sigma recognizes the promoter sequences on the 5′ → 3′ (dark green) strand of DNA, the RNA polymerase core enzyme will actually transcribe the light green strand (that runs 3′ → 5′) because core enzyme works only in a 5′ → 3′ direction.

terminate transcription at the correct position. **Termination** of RNA synthesis is governed by specific base sequences on the DNA. In *Bacteria* a common termination signal on the DNA is a GC-rich sequence containing an inverted repeat with a central nonrepeating segment. When such a DNA sequence is transcribed, the RNA forms a stem–loop structure by intra-strand base pairing (**Figure 4.23**). Stem–loops followed by a run of adenines in the DNA template (and therefore a run of uridines in the mRNA) are effective transcription terminators. This is due to the formation of a stretch of U:A base pairs that holds the RNA and DNA template together. This structure is very weak as U:A base pairs have only two hydrogen bonds each. The RNA polymerase pauses at the stem–loop, and the DNA and RNA dissociate at the run of uridines, terminating transcription.

The other mechanism for transcription termination in *Bacteria* uses a specific protein factor, known as Rho. Rho does not bind to RNA polymerase or to the DNA, but binds tightly to RNA and moves down the chain toward the RNA polymerase–DNA complex. Once RNA polymerase has paused at a Rho-dependent termination site (a specific sequence on the DNA template), Rho causes both the RNA and RNA polymerase to be released from the DNA, thus terminating transcription.

corresponding sigma factor, and this occurs by changing the rate of either synthesis or degradation of the sigma factor.

Termination of Transcription

In a growing bacterial cell, only those genes that need to be expressed are usually transcribed. Therefore, it is important to

MINIQUIZ

- In which direction along the DNA template strand does transcription proceed, and what enzyme catalyzes the reaction?
- What is a promoter? What protein recognizes the promoters in *Escherichia coli*?
- How can the expression of gene families be controlled as a group?
- What type of structures can lead to transcription termination?

4.8 The Unit of Transcription

Genetic information is organized into transcriptional units. These are segments of DNA that are transcribed into a single RNA molecule. Each transcription unit is bound by sites where transcription is initiated and terminated. Some units of transcription include only a single gene. Others contain two or more genes. The latter are said to be *cotranscribed*, yielding a single RNA molecule.

Ribosomal and Transfer RNAs and RNA Longevity

Most genes encode proteins, but others encode nontranslated RNAs, such as ribosomal RNA or transfer RNA. There are several different types of rRNA in an organism. *Bacteria* and *Archaea* produce three types: 16S rRNA, 23S rRNA, and 5S rRNA (with a ribosome having one copy of each; Section 4.14). As shown

Table 4.4 Sigma factors in *Escherichia coli*

Name[a]	Upstream recognition sequence[b]	Function
σ^{70} RpoD	TTGACA	For most genes, major sigma factor for normal growth
σ^{54} RpoN	TTGGCACA	Nitrogen assimilation
σ^{38} RpoS	CCGGCG	Stationary phase, plus oxidative and osmotic stress
σ^{32} RpoH	TNTCNCCTTGAA	Heat shock response
σ^{28} FliA	TAAA	For genes involved in flagella synthesis
σ^{24} RpoE	GAACTT	Response to misfolded proteins in periplasm
σ^{19} FecI	AAGGAAAAT	For certain genes in iron transport

[a]Superscript number indicates size of protein in kilodaltons. Many factors also have other names, for example, σ^{70} is also called σ^{D}
[b]N = any nucleotide.

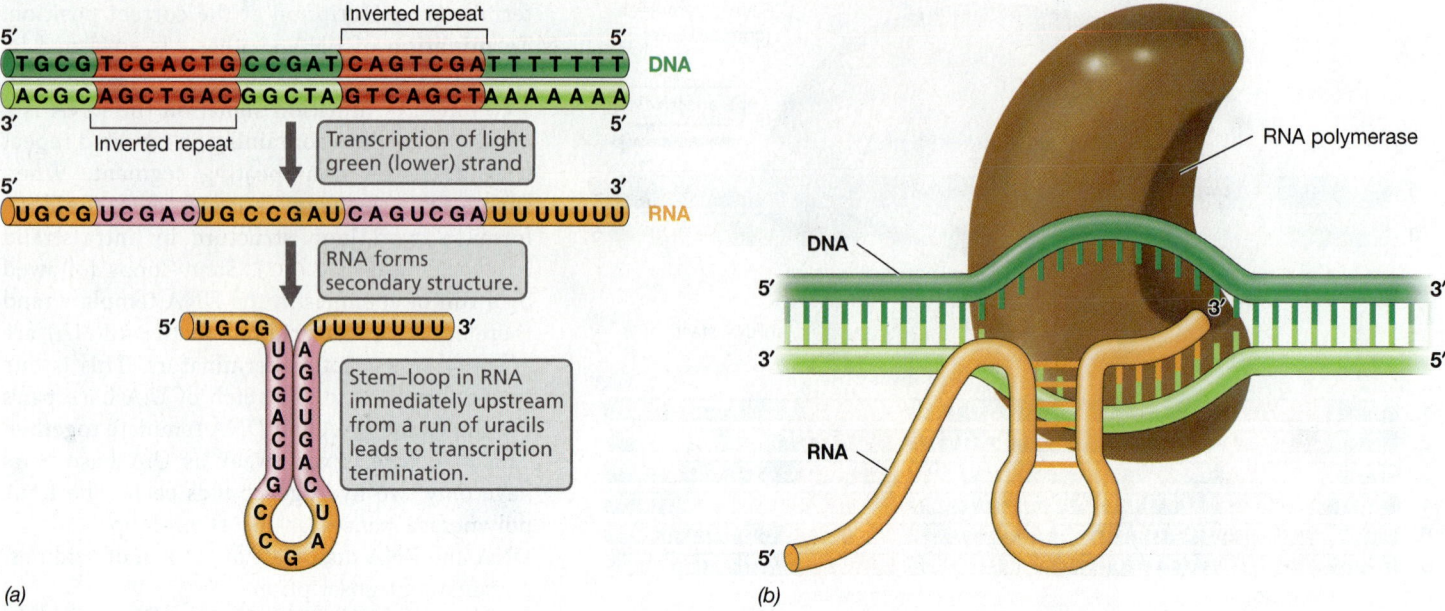

Figure 4.23 Inverted repeats and transcription termination. *(a)* Inverted repeats in transcribed DNA form a stem–loop structure in the RNA that terminates transcription when followed by a run of uracils. *(b)* Diagram indicating the formation of the terminator stem–loop in the RNA within the RNA polymerase.

in **Figure 4.24**, transcription units exist that contain one gene for each of these rRNAs, and these genes are cotranscribed. Therefore, the unit of transcription for most rRNAs is longer than a single gene. In prokaryotes tRNA genes are often cotranscribed with each other or even, as shown in Figure 4.24, with genes for rRNA. These cotranscribed transcripts are processed by specific proteins in the cell that cut them into individual units, yielding mature (functional) rRNAs or tRNAs.

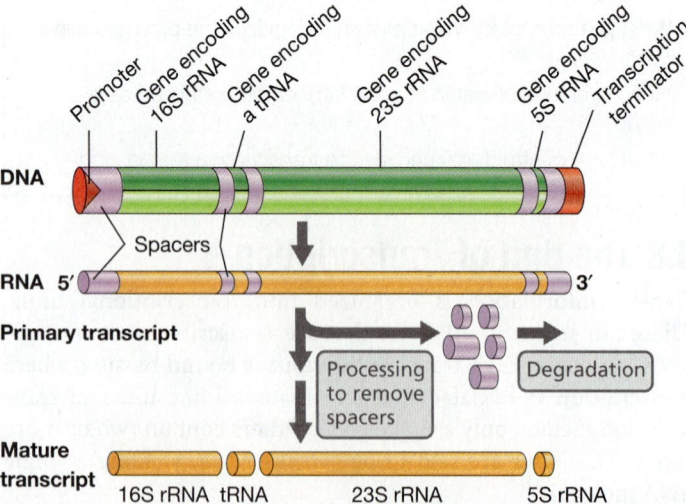

Figure 4.24 A ribosomal rRNA transcription unit from *Bacteria* and its subsequent processing. In *Bacteria*, all rRNA transcription units have the genes in the order 16S rRNA, 23S rRNA, and 5S rRNA (shown approximately to scale). Note that in this particular transcription unit the spacer between the 16S and 23S rRNA genes contains a tRNA gene. In other transcription units this region may contain more than one tRNA gene. Often one or more tRNA genes also follow the 5S rRNA gene and are cotranscribed. *Escherichia coli* contains seven rRNA transcription units.

In prokaryotes, most messenger RNAs have a short half-life (on the order of a few minutes), after which they are degraded by enzymes called *ribonucleases*. This is in contrast to rRNA and tRNA, which are stable RNAs. This stability can be ascribed to the highly folded secondary structures of tRNAs and rRNAs that prevent them from being degraded by ribonucleases. By contrast, mRNA does not form such structures and is susceptible to ribonuclease attack. The rapid turnover of prokaryotic mRNAs permits the cell to quickly adapt to new environmental conditions and halt the translation of mRNAs whose products are no longer needed.

Polycistronic mRNA and the Operon

In prokaryotes, genes that encode several enzymes of a particular metabolic pathway, for example, the biosynthesis of a particular amino acid, are often clustered together. RNA polymerase proceeds through such clusters and transcribes the entire group of genes into a single, long mRNA molecule. An mRNA encoding such a group of cotranscribed genes is called a *polycistronic mRNA* (**Figure 4.25**). Polycistronic mRNAs contain multiple *open-reading frames*, portions of the mRNA that actually encode amino acids (Section 4.11). When this mRNA is translated, several polypeptides are synthesized sequentially by the same ribosome.

A group of related genes that are transcribed together to give a single polycistronic mRNA is known as an operon. Assembling genes for the same biochemical pathway or genes needed under the same conditions into an operon allows their expression to be coordinated. Often, transcription of an operon is controlled by a specific region of the DNA just upstream of the protein-coding region of the operon. This is considered in more detail in Chapter 7.

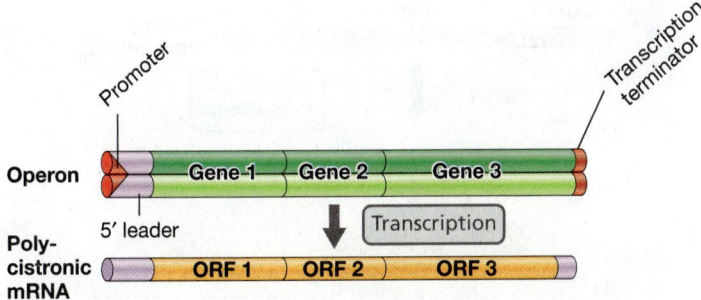

Figure 4.25 **Operon and polycistronic mRNA structure in prokaryotes.** Note that a single promoter controls the three genes within the operon and that the polycistronic mRNA molecule contains an open-reading frame (ORF) corresponding to each gene.

MINIQUIZ --
- What is the role of messenger RNA (mRNA)?
- What is a transcription unit? What is a polycistronic mRNA?
- What are operons and why are they useful to prokaryotes?

4.9 Transcription in *Archaea* and *Eukarya*

So far we have focused on transcription in *Bacteria* using *Escherichia coli* as a model system. Although in both *Archaea* and *Eukarya* the overall flow of genetic information from DNA to RNA is the same as in *Bacteria*, some details differ, and in eukaryotic cells the presence of a nucleus complicates the routing of genetic information. While *Archaea* lack a nucleus, many of their molecular properties resemble *Eukarya* more closely than *Bacteria*. These shared central dogma features confirm that these two domains are more closely related to each other than either is to *Bacteria* (⌘ Section 1.3). However, *Archaea* also share transcriptional similarities to *Bacteria*, such as operons. Units of transcription in eukaryotes include only one gene. Here we discuss key elements of transcription in *Archaea* and *Eukarya* that differ from those of *Bacteria*.

Archaeal and Eukaryotic RNA Polymerases

Archaeal and eukaryotic RNA polymerases are more similar and structurally more complex than those of *Bacteria*. *Archaea* contain only a single RNA polymerase that most closely resembles eukaryotic RNA polymerase II. The archaeal RNA polymerase typically has 11 or 12 subunits, while eukaryotic RNA polymerase II has 12 or more subunits. This is in sharp contrast to the RNA polymerase from *Bacteria* that is made up of only four different subunits plus the sigma (recognition) subunit (Figure 4.21).

We learned in Section 4.7 the importance of the promoter to transcription. The structure of archaeal promoters resembles those of eukaryotic promoters recognized by eukaryotic RNA polymerase II more than the promoters of *Bacteria*. *Eukarya* differ from both *Archaea* and *Bacteria* by having multiple RNA polymerases. Within the nucleus three separate RNA polymerases exist and transcribe different categories of genes. Mitochondria and chloroplasts also possess specific RNA polymerases, but, not surprisingly, considering the phylogenetic connections between *Bacteria* and eukaryotic cell organelles (⌘ Figure 1.6b), these are more closely related to the RNA polymerase of *Bacteria*.

Promoters and Terminators in *Archaea* and *Eukarya*

Three main recognition sequences are part of the promoters in both prokaryotic domains, and these sequences are recognized by a series of proteins called *transcription factors* that are similar in *Eukarya* and *Archaea*. The most important recognition sequence in archaeal and eukaryotic promoters is the 6- to 8-base-pair "TATA" box, located 18–27 nucleotides upstream of the transcriptional start site (**Figure 4.26**). This is recognized by the *TATA-binding protein* (TBP). Upstream of the TATA box is the *B recognition element* (BRE) sequence that is recognized by transcription factor B (TFB). In addition, an initiator element sequence is located at the start of transcription. Once TBP has bound to the TATA box and TFB has bound to the BRE, then archaeal RNA polymerase can bind and initiate transcription. This process is similar in eukaryotes except that several additional transcription factors are required.

Less is known about transcription termination in *Archaea* and *Eukarya* than in *Bacteria* (Section 4.7). Some archaeal genes have inverted repeats followed by an AT-rich sequence similar to those found in many bacterial transcription terminators. However, such termination sequences are not found in all archaeal genes. One other type of suspected transcription terminator lacks inverted repeats but contains repeated runs of thymines. In some way, this signals the archaeal termination machinery to terminate transcription. In eukaryotes, termination differs depending on the RNA polymerase and often requires a specific termination factor protein. No Rho-like proteins (Section 4.7) have been found in either *Archaea* or *Eukarya*.

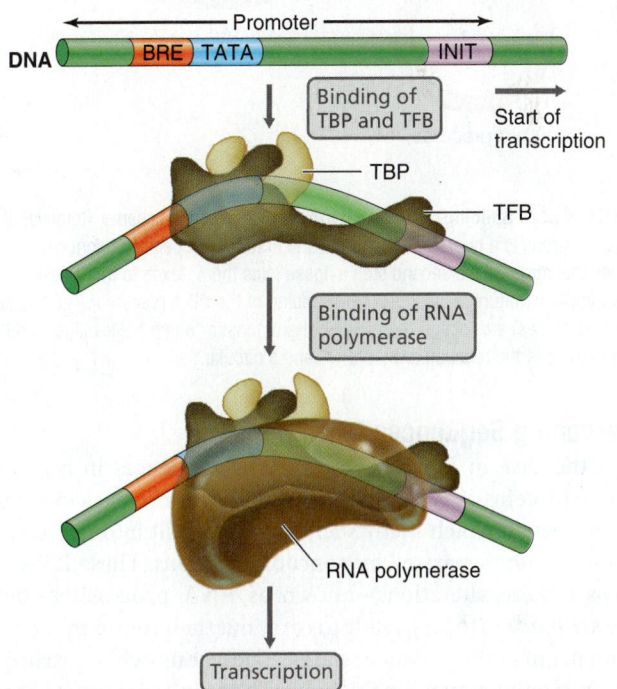

Figure 4.26 **Promoter architecture and transcription in *Archaea*.** Three promoter elements are critical for promoter recognition in *Archaea*: the initiator element (INIT), the TATA box, and the B recognition element (BRE). The TATA-binding protein (TBP) binds the TATA box; transcription factor B (TFB) binds to both BRE and INIT. Once both TBP and TFB are in place, RNA polymerase binds.

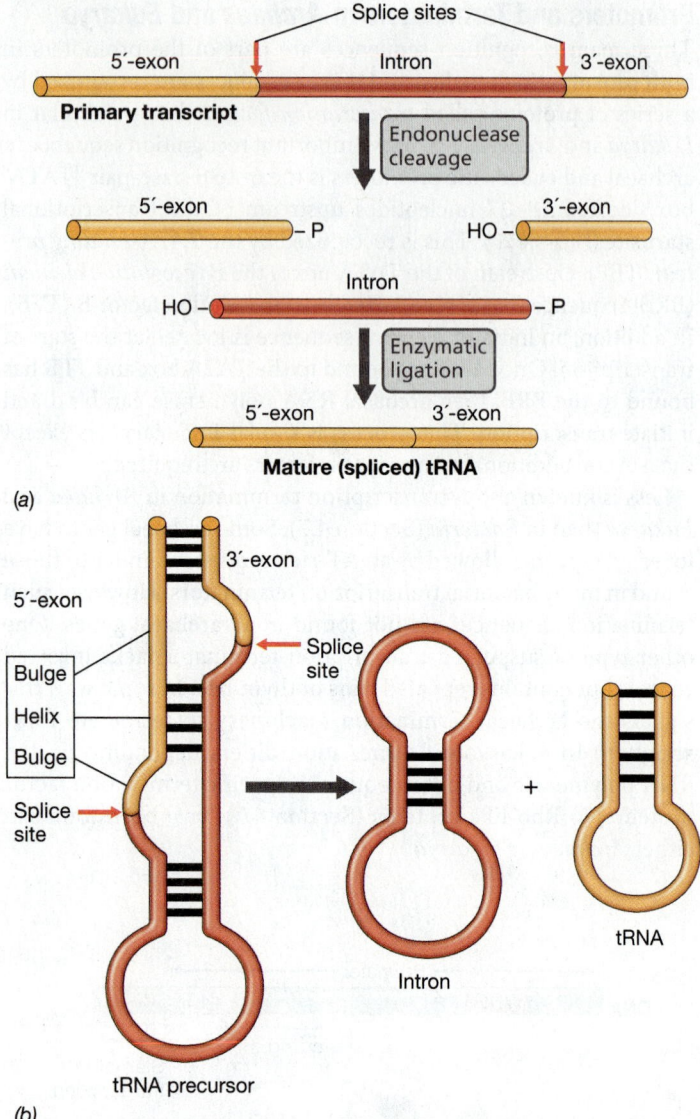

Figure 4.27 Splicing of archaeal introns. *(a)* Reaction scheme. Removal of archaeal introns is a two-step reaction. In the first step a specific endonuclease excises the intron. In the second step a ligase joins the 5′-exon to the 3′-exon, generating the mature, spliced tRNA. *(b)* Folding of the tRNA precursor. The two splice sites (red arrows) are recognized by their characteristic "bulge-helix-bulge" motifs. The products of the reaction are the tRNA and a circular intron.

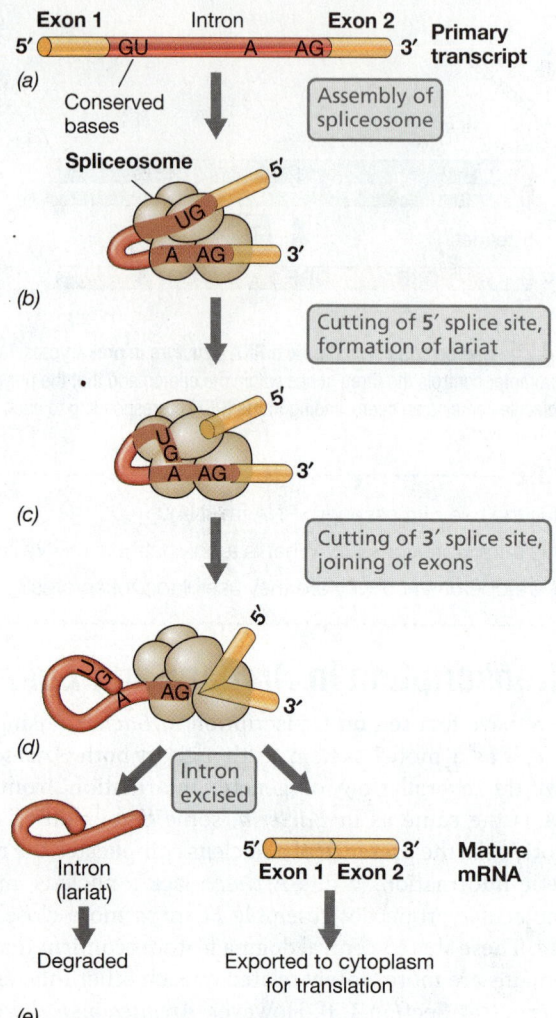

Figure 4.28 Activity of the spliceosome. Removal of an intron from the primary transcript of a protein-coding gene in a eukaryote. *(a)* A primary transcript containing a single intron. The sequence GU is conserved at the 5′ splice site, and AG is conserved at the 3′ splice site. There is also an interior A that serves as a branch point. *(b)* Several small ribonucleoprotein particles (shown in brown) assemble on the RNA to form a spliceosome. Each of these particles contains distinct small RNA molecules that take part in the splicing mechanism. *(c)* The 5′ splice site has been cut with the simultaneous formation of a branch point. *(d)* The 3′ splice site has been cut and the two exons have been joined. Note that overall, two phosphodiester bonds were broken, but two others were formed. *(e)* The final products are the joined exons (the mRNA) and the released intron.

Intervening Sequences in *Archaea*

As is the case in *Bacteria*, intervening sequences in genes that encode proteins are extremely rare in *Archaea*. This is in contrast to *Eukarya*, in which many such genes are split into two or more coding regions separated by noncoding regions. These RNA molecules require alterations—known as **RNA processing**—before they are *mature*; that is, ready to carry out their role in the cell. The segments of coding sequence are called **exons**, while **introns** are the intervening noncoding regions. The term **primary transcript** refers to the RNA molecule that is originally transcribed before the introns are removed to generate the final mRNA, consisting solely of the exons. Several tRNA- and rRNA-encoding genes of *Archaea* possess introns that must be removed after transcription to generate the mature tRNA or rRNA. These introns were

named archaeal introns because they are processed by a different mechanism than are typical eukaryotic introns. Archaeal introns are excised by a specific ribonuclease that recognizes exon–intron junctions (**Figure 4.27**). In a few cases, tRNAs in *Archaea* are assembled by splicing together segments from two or three different primary transcripts.

RNA Processing in Eukaryotes

Most primary transcripts in eukaryotes contain introns and thus require RNA processing before they can function in the cell. The process by which introns are removed and exons are joined in eukaryotes is called *splicing* (**Figure 4.28**). RNA splicing occurs in the nucleus and is done by a large macromolecular complex

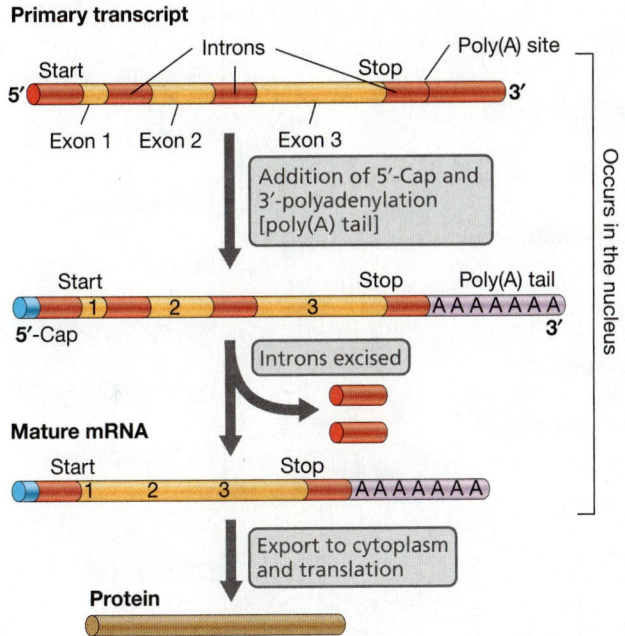

Figure 4.29 Processing of the primary transcript into mature mRNA in eukaryotes. The processing steps include adding a cap at the 5′ end, removing the introns, and clipping the 3′ end of the transcript while adding a poly(A) tail. All these steps are carried out in the nucleus. The location of the start and stop codons to be used during translation are indicated.

called the **spliceosome**. The proteins of the spliceosome cut out the intron and join the flanking exons together to form a contiguous protein-coding sequence in the mature mRNA. Many genes, especially in higher animals and plants, have multiple introns, so it is clearly important that they should all be recognized and removed by the spliceosome to generate the final mature mRNA (**Figure 4.29**).

There are two other steps in the processing of mRNA in *Eukarya* that are unique to this domain. Both steps take place in the nucleus prior to splicing. The first, called *capping*, occurs before transcription is complete. Capping is the addition of a methylated guanine nucleotide at the 5′-phosphate end of the mRNA (Figure 4.29). The cap nucleotide is added in reverse orientation relative to the rest of the mRNA molecule and is needed to initiate translation. The second consists of trimming the 3′ end of the primary transcript and adding 100–200 adenylate residues, called the *poly(A) tail* (Figure 4.29). The tail recognition sequence, AAUAAA, is located close to the 3′ end of the primary transcript. The poly(A) tail stabilizes mRNA and must be removed before the mRNA can be degraded.

MINIQUIZ --

- What three major components make up an archaeal promoter?
- What effect does the presence of a nucleus have on genetic information flow in eukaryotes?
- What steps are involved in processing of eukaryotic RNA?

IV · Protein Synthesis

Once transcription has occurred and mRNAs are made, the transcripts are translated into protein. This requires many proteins, other RNAs, and a key cellular structure, the ribosome. How these interact to produce a cell's array of proteins is what we consider now.

4.10 Polypeptides, Amino Acids, and the Peptide Bond

Proteins play major roles in cell function. Two major classes of proteins are *catalytic proteins* (enzymes) and *structural proteins*. **Enzymes** are the catalysts for chemical reactions that occur in cells. Structural proteins are integral parts of the major structures of the cell: membranes, walls, ribosomes, and so on. Regulatory proteins control most cell processes by a variety of mechanisms, including binding to DNA and affecting transcription. However, all proteins show certain basic features in common.

Proteins are polymers of **amino acids**. All amino acids contain an amino group ($-NH_2$) and a carboxylic acid group ($-COOH$) that are attached to the α-carbon (**Figure 4.30a**). Linkages between the carboxyl carbon of one amino acid and the amino nitrogen of a second (with elimination of water) are known as **peptide bonds** (**Figure 4.31**). Two amino acids bonded by peptide linkage constitute a *dipeptide*; three amino acids, a *tripeptide*; and so on. When many amino acids are linked, they form a **polypeptide**. A protein consists of one or more polypeptides. The number of amino acids

differs greatly from one protein to another, from as few as 15 to as many as 10,000.

Each amino acid has a unique side chain (abbreviated R). These vary considerably, from as simple as a hydrogen atom in the amino acid glycine to aromatic rings in phenylalanine, tyrosine, and tryptophan (Figure 4.30b). Amino acids exist as pairs of **enantiomers**. These are optical isomers that have the same molecular and structural formulas, except that they are mirror images and are designated as either D or L, depending on whether a pure solution rotates light to the right or left, respectively. Cellular proteins are composed of only L-amino acids. Nevertheless, D-amino acids are occasionally found in cells, most notably in the cell wall polymer peptidoglycan (⟲ Section 2.10) and in certain peptide antibiotics (⟲ Section 27.14). Cells can interconvert enantiomers by enzymes called *racemases*.

The chemical properties of an amino acid are governed by its side chain. Amino acids with similar chemical properties are grouped into related "families" (Figure 4.30b). For example, the side chain may contain a carboxylic acid group, as in aspartic acid or glutamic acid, rendering the amino acid acidic. Others contain additional amino groups, making them positively charged and basic. Several amino acids contain hydrophobic side chains and are called nonpolar amino acids. Cysteine contains a sulfhydryl group ($-SH$). Using their sulfhydryl groups, two cysteines can form a disulfide linkage ($R-S-S-R$) that connects two polypeptide chains.

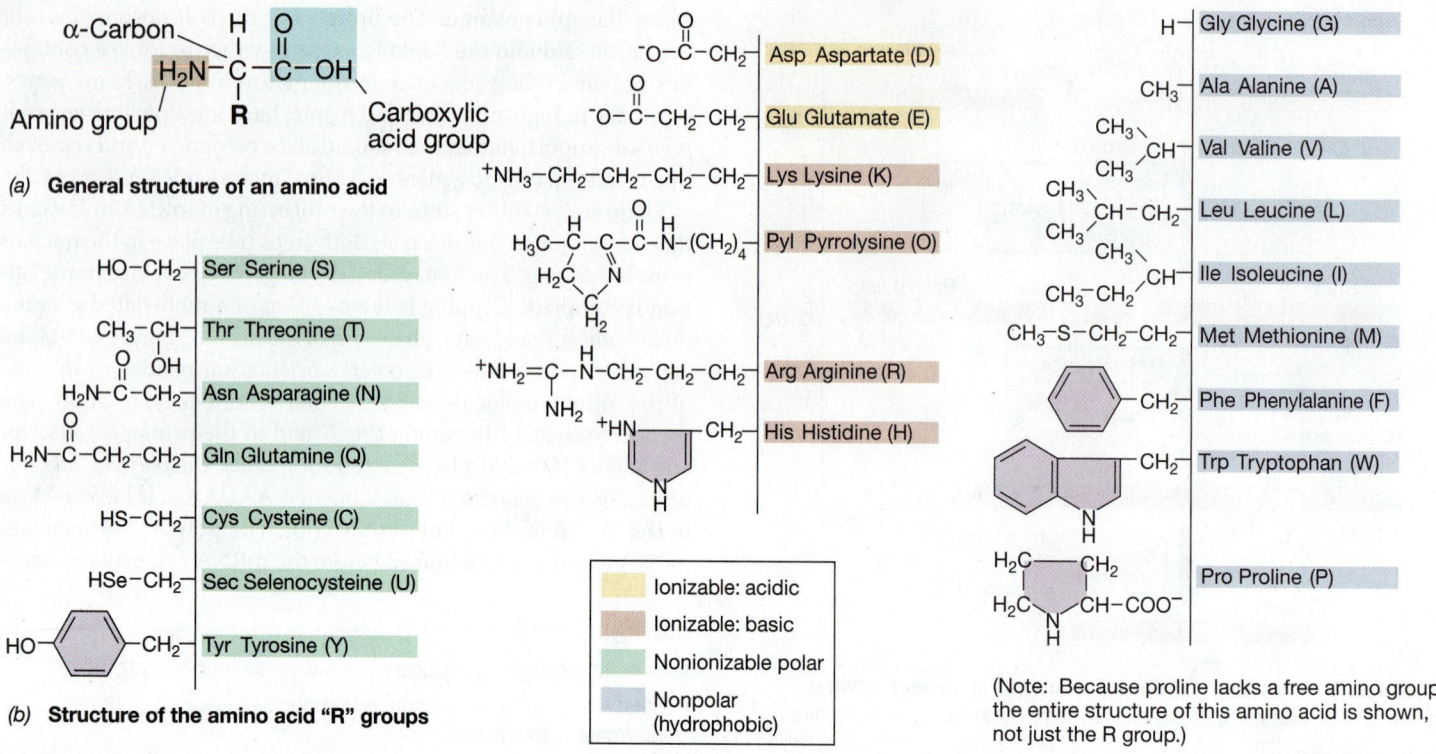

(a) **General structure of an amino acid**

(b) **Structure of the amino acid "R" groups**

- Ionizable: acidic
- Ionizable: basic
- Nonionizable polar
- Nonpolar (hydrophobic)

(Note: Because proline lacks a free amino group, the entire structure of this amino acid is shown, not just the R group.)

Figure 4.30 **Structure of the 22 genetically encoded amino acids.** *(a)* General structure. *(b)* R group structure. The three-letter codes for the amino acids are to the left of the names, and the one-letter codes are in parentheses to the right of the names. Pyrrolysine has thus far been found only in certain methanogenic *Archaea* (⟳ Section 16.2).

The diversity of chemically distinct amino acids makes possible an enormous number of unique proteins with widely different biochemical properties. If one assumes that an average polypeptide contains 300 amino acids, then 22^{300} different polypeptide sequences are theoretically possible. No cell has anywhere near this many different proteins. A cell of *Escherichia coli* contains around 2000 different kinds of proteins, with the exact number being highly dependent on the resources (nutrients) and growth conditions employed.

The linear sequence of amino acids in a polypeptide is its *primary structure*. This ultimately determines the further folding of the polypeptide, which in turn determines its biological activity (Section 4.14). The two ends of a polypeptide are designated as the "C-terminus" and "N-terminus" depending on whether a free carboxylic acid group or a free amino group is found (Figure 4.31).

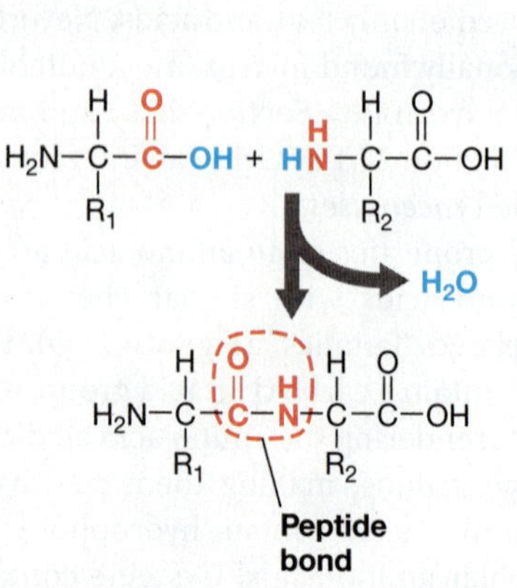

MINIQUIZ --

- Draw the structure of a dipeptide containing the amino acids alanine and tyrosine. Outline the peptide bond.
- Which enantiomeric form of amino acids is found in proteins?
- Glycine does not have two different enantiomers; why?

4.11 Translation and the Genetic Code

As we have seen, in the first two steps of biological information transfer—replication and transcription—nucleic acids are synthesized on nucleic acid templates. The last step, *translation*, also uses a nucleic acid as template, but in this case the product is a polypeptide rather than a nucleic acid. The heart of biological information transfer is the correspondence between the nucleic acid template and the amino acid sequence of the polypeptide formed. This correspondence is rooted in the **genetic code**. An RNA triplet of three bases, called a *codon*, encodes each specific amino acid. The 64 possible codons (four bases taken three at a time = 4^3) of mRNA are shown in **Table 4.5**. The genetic code is written as RNA rather than as DNA because it is mRNA that is actually translated. Note that in addition to the codons for amino acids, there are also specific codons for starting and stopping translation. Here we focus on translation in *Bacteria*, but it

Figure 4.31 **Peptide bond formation.** R_1 and R_2 refer to the side chains of the amino acids. Note that following peptide bond formation, a free OH group is present at the C-terminus for formation of the next peptide bond.

Table 4.5 The genetic code as expressed by triplet base sequences of mRNA

Codon	Amino acid	Codon	Amino acid	Codon	Amino acid	Codon	Amino acid
UUU	Phenylalanine	UCU	Serine	UAU	Tyrosine	UGU	Cysteine
UUC	Phenylalanine	UCC	Serine	UAC	Tyrosine	UGC	Cysteine
UUA	Leucine	UCA	Serine	UAA	None (stop signal)	UGA	None (stop signal)
UUG	Leucine	UCG	Serine	UAG	None (stop signal)	UGG	Tryptophan
CUU	Leucine	CCU	Proline	CAU	Histidine	CGU	Arginine
CUC	Leucine	CCC	Proline	CAC	Histidine	CGC	Arginine
CUA	Leucine	CCA	Proline	CAA	Glutamine	CGA	Arginine
CUG	Leucine	CCG	Proline	CAG	Glutamine	CGG	Arginine
AUU	Isoleucine	ACU	Threonine	AAU	Asparagine	AGU	Serine
AUC	Isoleucine	ACC	Threonine	AAC	Asparagine	AGC	Serine
AUA	Isoleucine	ACA	Threonine	AAA	Lysine	AGA	Arginine
AUG (start)[a]	Methionine	ACG	Threonine	AAG	Lysine	AGG	Arginine
GUU	Valine	GCU	Alanine	GAU	Aspartic acid	GGU	Glycine
GUC	Valine	GCC	Alanine	GAC	Aspartic acid	GGC	Glycine
GUA	Valine	GCA	Alanine	GAA	Glutamic acid	GGA	Glycine
GUG	Valine	GCG	Alanine	GAG	Glutamic acid	GGG	Glycine

[a]AUG encodes *N*-formylmethionine at the beginning of polypeptide chains of *Bacteria*.

is important to note that the translation machinery of *Archaea* and *Eukarya* are more closely related to each other than to that of *Bacteria*.

Properties of the Genetic Code

There are 22 amino acids that are encoded by the genetic information carried on mRNA (some other amino acids are formed by modification of these after translation). Consequently, because there are 64 codons, many amino acids can be encoded by more than 1 codon. Although knowing the codon at a given location unambiguously identifies the corresponding amino acid, the reverse is not true. Knowing the amino acid does not mean that the codon at that location is known. A code such as this that lacks one-to-one correspondence between "word" (that is, the amino acid) and code (codon) is called a *degenerate code*. However, knowing the DNA sequence and the correct reading frame, one can specify the amino acid sequence of a protein. This permits the determination of amino acid sequences from DNA base sequences and is at the heart of genomics (Chapter 6).

A codon is recognized by specific base pairing with a complementary sequence of three bases called the **anticodon**, which is part of tRNAs. If this base pairing were always the standard pairing of A with U and G with C, then at least one specific tRNA would be needed to recognize each codon. In some cases, this is true. For instance, there are six different tRNAs in *Escherichia coli* for the amino acid leucine, one for each codon (Table 4.5). By contrast, some tRNAs can recognize more than one codon. Thus, although there are two lysine codons in *E. coli*, there is only one lysyl tRNA, whose anticodon can base-pair with either AAA or AAG. In these special cases, tRNA molecules form standard base pairs at only the first two positions of the codon while tolerating irregular base pairing at the third position. This phenomenon

is called **wobble** and is illustrated in **Figure 4.32**, where a pairing between G and U (rather than G with C) is illustrated at the wobble position.

Several amino acids are encoded by multiple codons, and in most cases, the multiple codons are closely related in base sequence (Table 4.5). One might assume that such multiple codons would be used at equal frequencies. However, this is not so, and genomic sequence data have revealed organism-specific **codon bias**. In other words, some codons are greatly preferred over others even though they encode the same amino acid. Codon bias is correlated with a corresponding bias in the concentration of different tRNA molecules. Thus a tRNA corresponding to a rarely used codon will be in relatively short supply. Codon bias must be taken into consideration during genetic engineering. For example, a gene from one organism whose codon usage differs

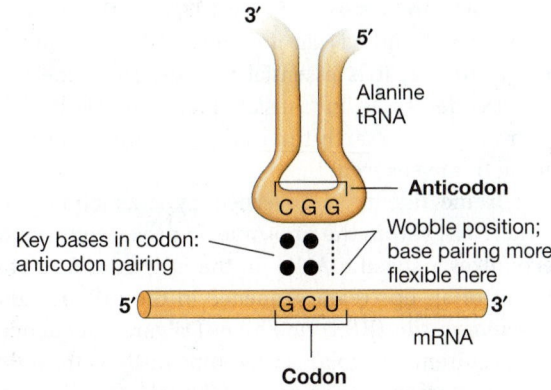

Figure 4.32 The wobble concept. Base pairing is more flexible for the third base of the codon than for the first two. Only a portion of the tRNA is shown here.

dramatically from that of another may not be translated efficiently if the gene is cloned into the latter using genetic engineering techniques (Chapter 11).

Open Reading Frames

One common method of identifying protein-encoding genes is to examine each strand of the DNA sequence for **open reading frames (ORFs)**. If an mRNA can be translated, it contains an open reading frame: a start codon (typically AUG) followed by a number of codons and then a stop codon in the same reading frame as the start codon. In practice, only ORFs long enough to encode a functional polypeptide are accepted as true coding sequences. Although most functional proteins are at least 100 amino acids in length, a few protein hormones and regulatory peptides are much shorter. Consequently, it is not always possible to tell from sequence data alone whether a relatively short ORF is merely due to chance or encodes a genuine, albeit short, polypeptide.

Using computational methods, a long DNA base sequence can be scanned to look for open reading frames. In addition to looking for start and stop codons, the search may include promoters and ribosome-binding sequences as well. The search for ORFs is very important in genomics (Chapter 6). If an unknown piece of DNA has been sequenced, the presence of an ORF implies that it can encode protein.

Start and Stop Codons and Reading Frame

Messenger RNA is translated beginning with its **start codon** (AUG, Table 4.5), which encodes a chemically modified methionine in *Bacteria* called *N-formylmethionine*. Although AUG at the *beginning* of a coding region encodes *N*-formylmethionine, AUG *within* the coding region encodes methionine. Two different tRNAs are involved in this process (Section 4.13). Conversely, *Archaea* and *Eukarya* insert a regular methionine as the first amino acid in a polypeptide.

With a triplet code it is critical for translation to begin at the correct nucleotide. If it does not, the whole reading frame of the mRNA will be shifted and thus an entirely different protein will be made. Alternatively, if the shift introduces a stop codon into the reading frame, the polypeptide will terminate prematurely. By convention, the reading frame that when translated yields the polypeptide encoded by the gene is called the *0 frame* (zero frame). As can be seen in **Figure 4.33**, the other two possible reading frames (−1 and +1) do not encode the same amino acid sequence. Therefore, it is essential that the ribosome finds the correct start codon to begin translation and, once it has, that the mRNA is translocated exactly three bases at a time. How is the correct reading frame ensured?

Reading frame fidelity is governed by interactions between mRNA and rRNA within the ribosome. In prokaryotes, ribosomal RNA recognizes a specific AUG on the mRNA as a start codon with the aid of an upstream sequence in the mRNA called the *ribosome-binding site* (RBS) or Shine–Dalgarno sequence. This alignment requirement explains why some mRNAs from *Bacteria* can use other start codons, such as GUG. However, even these unusual start codons direct the incorporation of *N*-formylmethionine as the initiator amino acid.

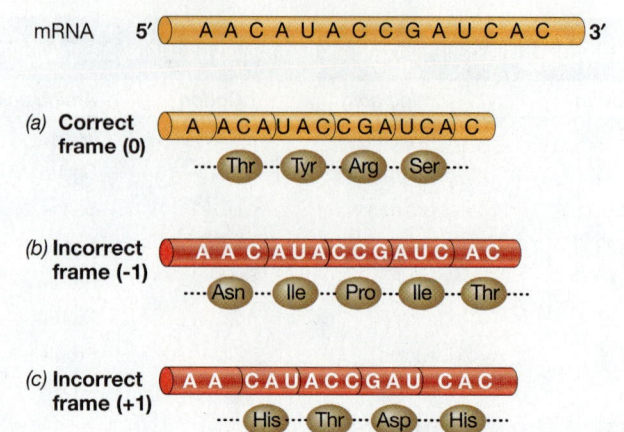

Figure 4.33 Possible reading frames in an mRNA. An interior sequence of an mRNA is shown. *(a)* The amino acids that would be encoded if the ribosome is in the correct reading frame (designated the "0" frame). *(b)* The amino acids that would be encoded by this region of the mRNA if the ribosome were in the −1 reading frame. *(c)* The amino acids that would be encoded if the ribosome were in the +1 reading frame.

A few codons do not encode any amino acid. These codons (UAA, UAG, and UGA, Table 4.5) are the **stop codons**, and they signal the termination of translation of a protein-coding sequence on the mRNA. Stop codons are also called **nonsense codons**, because they interrupt the "sense" of the growing polypeptide when they terminate translation. A few exceptions to this rule exist. For example, animal (but not plant) mitochondria use the codon UGA to encode tryptophan instead of using it as a stop codon (Table 4.5), while the genus *Mycoplasma* (*Bacteria*) and the genus *Paramecium* (*Eukarya*) use certain nonsense codons to encode amino acids. These organisms simply have fewer nonsense codons because one or two of them are used as sense codons (๛ Section 6.5).

In a few rare cases, nonsense codons encode unusual amino acids rather than one of the 20 common amino acids. These exceptions are selenocysteine and pyrrolysine, the 21st and 22nd genetically encoded amino acids (Figure 4.30). Both selenocysteine and pyrrolysine are encoded by stop codons (UGA and UAG, respectively). Both have their own tRNAs that contain anticodons that read these stop codons. Most stop codons in organisms that use selenocysteine and pyrrolysine do indeed indicate stop. However, occasional stop codons are recognized as encoding selenocysteine or pyrrolysine. This switch is controlled by a recognition sequence just downstream of the now coding stop codon. Selenocysteine and pyrrolysine are both relatively rare. Most organisms, including plants and animals, have a few proteins that contain selenocysteine. Pyrrolysine is rarer still. It has been found in certain *Archaea* and *Bacteria* but was first discovered in species of methanogenic *Archaea*.

MINIQUIZ

- What are start codons and stop codons? Why is it important for the ribosome to read "in frame"?
- What is codon bias?
- If you were given a nucleotide sequence, how would you find ORFs?

4.12 Transfer RNA

A transfer RNA carries the anticodon that base-pairs with the codon on mRNA. In addition, each tRNA is specific for the amino acid that corresponds to its own anticodon (that is, its *cognate* amino acid). The tRNA and its specific amino acid are linked by specific enzymes called **aminoacyl-tRNA synthetases**. For each amino acid, a separate aminoacyl-tRNA synthetase exists that specifically binds to both the amino acid and tRNAs possessing corresponding anticodons. These enzymes ensure that each tRNA receives its correct amino acid, so they must recognize both a specific tRNA and its cognate amino acid.

General Structure of tRNA

There are about 60 different tRNAs in bacterial cells and 100–110 in mammalian cells. Transfer RNA molecules are short, single-stranded molecules that contain extensive secondary structure and are 73–93 nucleotides long. Certain bases and secondary structures are constant for all tRNAs, whereas other parts are variable. Transfer RNA molecules also contain some purine and pyrimidine bases that are chemically modified from the standard bases found in RNA. These modifications are made to the bases after transcription. These unusual bases include pseudouridine (ψ), inosine, dihydrouridine (D), ribothymidine, methyl guanosine, dimethyl guanosine, and methyl inosine. The mature and active tRNA also contains extensive double-stranded regions within the molecule. This secondary structure forms by internal base pairing when the single-stranded molecule folds back on itself (**Figure 4.34**).

The structure of a tRNA can be drawn in a cloverleaf fashion, as shown in Figure 4.34*a*. Some regions of tRNA secondary structure are named after the modified bases found there (for example, the TψC and D loops) or after their functions (for example, the anticodon loop and acceptor stem). The three-dimensional structure of a tRNA is shown in Figure 4.34*b*. Note that bases that appear widely separated in the cloverleaf model may actually be much closer together when viewed in three dimensions. This close proximity allows some of the bases in one loop to base-pair with bases in another loop.

The Anticodon and the Amino Acid–Binding Site

One of the key variable parts of the tRNA molecule is the *anticodon*, the group of three bases that recognizes the codon on the mRNA. The anticodon is found in the anticodon loop (Figure 4.34). The three nucleotides of the anticodon recognize the codon by specifically pairing with its three bases. By contrast, other portions of the tRNA interact with both the rRNA and protein components of the ribosome, nonribosomal translation proteins, and the aminoacyl-tRNA synthetase enzyme.

At the 3′ end (acceptor stem) of all tRNAs are three unpaired nucleotides. The sequence of these three nucleotides is always cytosine-cytosine-adenine (CCA), and they are absolutely essential for function. Curiously, however, in most organisms the 3′ CCA is not encoded in the tRNA gene on the chromosome. Instead, each nucleotide is added one by one by a protein called *CCA-adding enzyme*, using CTP and ATP as substrates. The cognate amino acid is then covalently attached to the terminal adenosine of the CCA end of its corresponding tRNA by an ester linkage to the ribose sugar. As we shall see, from this location on the tRNA, the amino acid is incorporated into the growing polypeptide chain on the ribosome by a mechanism described in the next section.

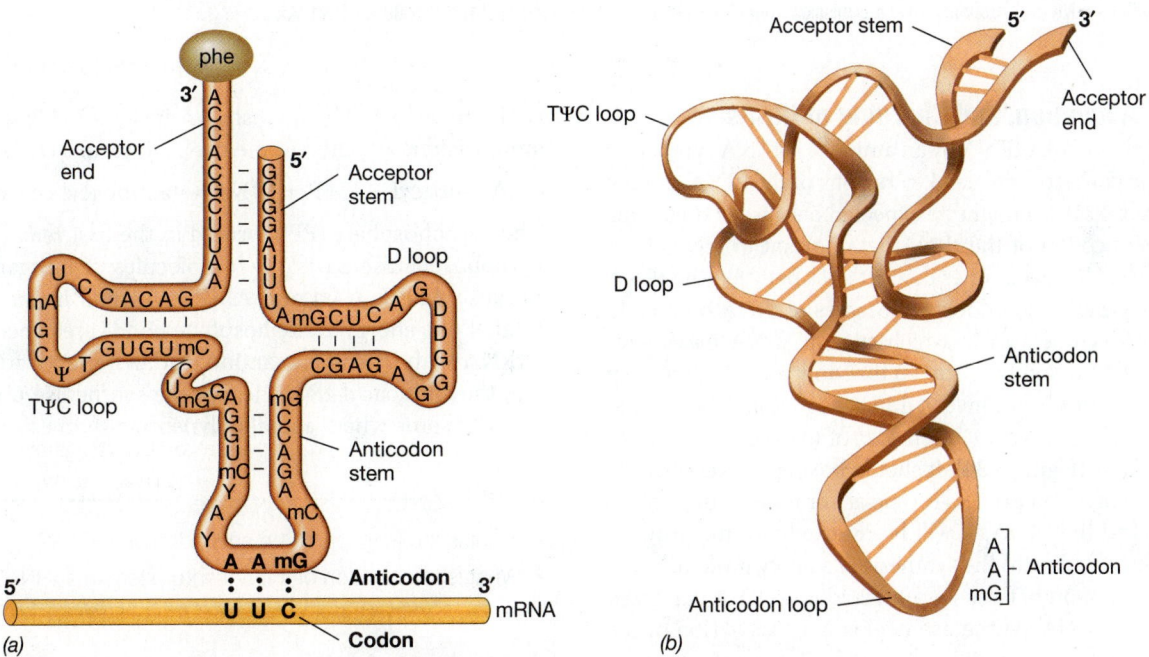

Figure 4.34 Structure of a transfer RNA. *(a)* The conventional cloverleaf structural drawing of yeast phenylalanine tRNA. The amino acid is attached to the ribose of the terminal A at the acceptor end. A, adenine; C, cytosine; U, uracil; G, guanine; T, thymine; ψ, pseudouracil; D, dihydrouracil; m, methyl; Y, a modified purine. *(b)* In fact, the tRNA molecule folds so that the D loop and TψC loops are close together and associate by hydrophobic interactions.

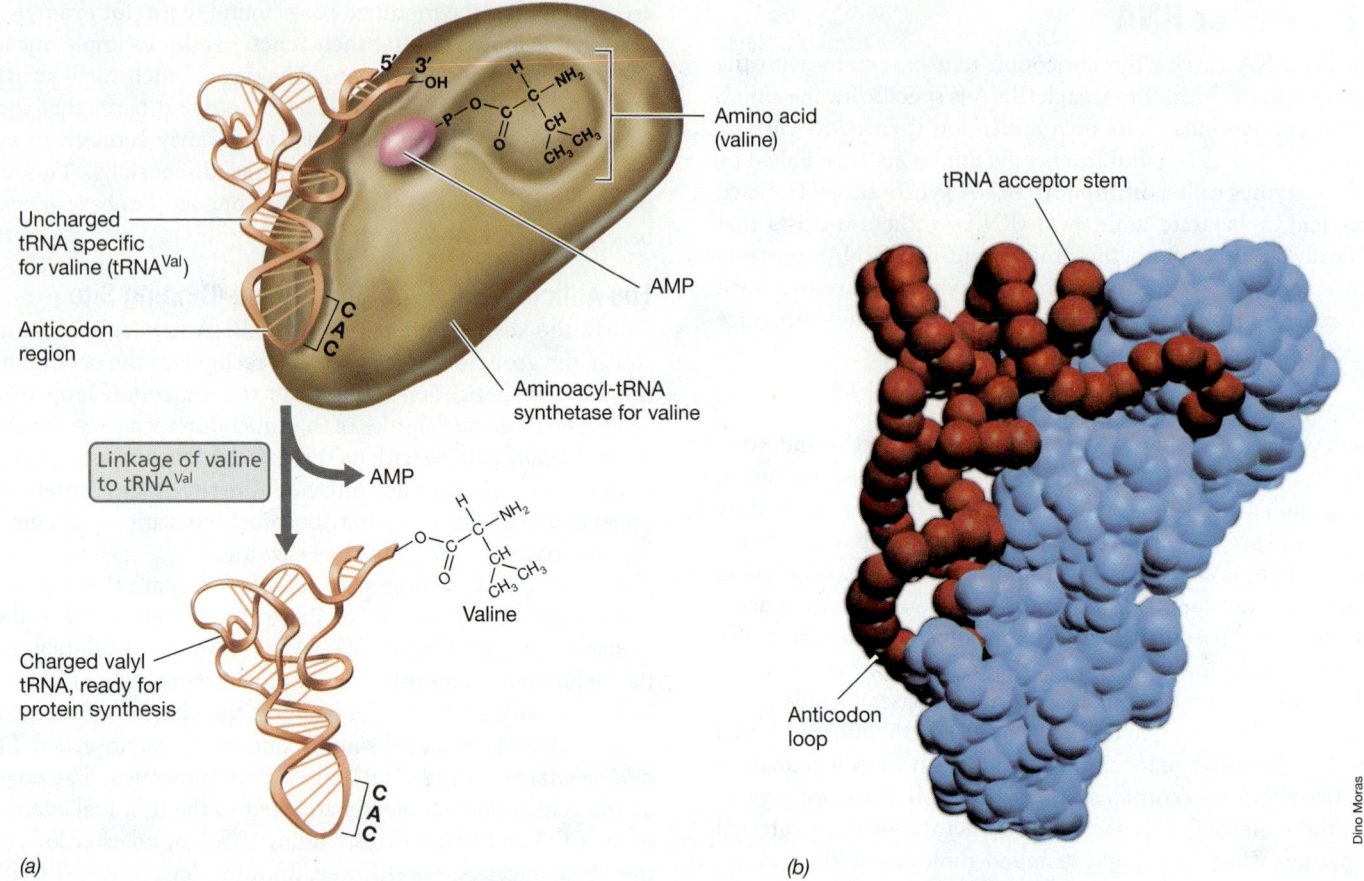

Figure 4.35 Aminoacyl-tRNA synthetase. *(a)* Mode of activity of an aminoacyl-tRNA synthetase. Recognition of the correct tRNA by a particular synthetase involves contacts between specific nucleic acid sequences in the D loop and acceptor stem of the tRNA and specific amino acids of the synthetase. In this diagram, valyl-tRNA synthetase is shown catalyzing the final step of the reaction, where the valine in valyl-AMP is transferred to tRNA. *(b)* A computer model showing the interaction of glutaminyl-tRNA synthetase (blue) with its tRNA (red). Reprinted with permission from M. Ruff et al. 1991. *Science* 252: 1682–1689. © 1991, AAAS.

Recognition, Activation, and Charging of tRNAs

Recognition of the correct tRNA by an aminoacyl-tRNA synthetase involves specific contacts between key regions of the tRNA and the synthetase (**Figure 4.35**). As might be expected because of its unique sequence, the anticodon of the tRNA is important in recognition by the synthetase. However, other contact sites between the tRNA and the synthetase are also important. Studies of tRNA binding to aminoacyl-tRNA synthetases in which specific tRNA bases have been changed by mutation have shown that only a small number of key nucleotides in tRNA are involved in recognition. These other key recognition nucleotides are often part of the acceptor stem or D loop of the tRNA (Figure 4.34). It should be emphasized that the fidelity of this recognition process is crucial, for if the wrong amino acid is attached to the tRNA, it will be inserted into the growing polypeptide, likely leading to the synthesis of a faulty protein.

The specific reaction between amino acid and tRNA catalyzed by the aminoacyl-tRNA synthetase begins with activation of the amino acid by reaction with ATP:

$$\text{Amino acid} + \text{ATP} \leftrightarrow \text{aminoacyl—AMP} + \text{P—P}$$

The aminoacyl-AMP intermediate formed normally remains bound to the tRNA synthetase until collision with the appropriate tRNA molecule. Then, as shown in Figure 4.35*a*, the activated amino acid is attached to the tRNA to form a *charged tRNA*:

$$\text{Aminoacyl—AMP} + \text{tRNA} \leftrightarrow \text{aminoacyl—tRNA} + \text{AMP}$$

The pyrophosphate (PP_i) formed in the first reaction is split by a pyrophosphatase, giving two molecules of inorganic phosphate. Because ATP is used and AMP is formed in these reactions, a total of two energy-rich phosphate bonds are expended to charge a tRNA with its cognate amino acid. After activation and charging, the aminoacyl-tRNA leaves the synthetase until it is bound by a ribosome where actual polypeptide synthesis occurs.

MINIQUIZ -
- What is the function of the anticodon of a tRNA?
- What is the function of the acceptor stem of a tRNA?

4.13 Protein Synthesis

It is vital for proper functioning of proteins that the correct amino acids be inserted at their correct locations in the polypeptide chain. This is the task of the protein-synthesizing machinery, the ribosome. Although protein synthesis is a continuous process, it can be broken down into a number of steps: *initiation, elongation,* and *termination.* In addition to mRNA, tRNA, and ribosomes, the process requires a number of proteins designated initiation, elongation, and termination factors. The energy-rich compound guanosine triphosphate (GTP) provides the necessary energy for the process.

Ribosomes

Ribosomes are the sites of protein synthesis. A cell may have many thousand ribosomes, the number increasing at higher growth rates. Each ribosome consists of two subunits. Prokaryotes have 30S and 50S ribosomal subunits that yield intact 70S ribosomes. The S-values are *Svedberg units,* which refer to the sedimentation coefficients of ribosomal subunits (30S and 50S) or intact ribosomes (70S) when subjected to centrifugal force in an ultracentrifuge. (Although larger particles do have larger S-values, the relationship is not linear and thus S-values cannot simply be summed.)

Each ribosomal subunit contains specific ribosomal RNAs and ribosomal proteins. The 30S subunit contains 16S rRNA and 21 proteins, and the 50S subunit contains 5S and 23S rRNA and 31 proteins. Thus, in *Escherichia coli*, there are 52 distinct ribosomal proteins, most present at one copy per ribosome. The ribosome is a dynamic structure whose subunits alternately associate and dissociate and also interact with many other proteins. Several proteins that are essential for ribosome function interact with the ribosome at various stages of translation. These are regarded as "translation factors" rather than "ribosomal proteins" per se.

Initiation of Translation

In *Bacteria*, such as *E. coli*, initiation of protein synthesis begins with a free 30S ribosomal subunit (**Figure 4.36**). From this, an initiation complex forms consisting of the 30S subunit, mRNA, formylmethionine tRNA, and several initiation proteins called IF1, IF2, and IF3. GTP is also required for this step. Next, a 50S ribosomal subunit is added to the initiation complex to form the active 70S ribosome. At the end of the translation process, the ribosome separates again into 30S and 50S subunits.

Just preceding the start codon on the mRNA is a sequence of three to nine nucleotides called the ribosome-binding site (RBS in Figure 4.36) that helps bind the mRNA to the ribosome. The ribosome-binding site is toward the 5′ end of the mRNA and is complementary to base sequences in the 3′ end of the 16S rRNA. Base pairing between these two molecules holds the ribosome– mRNA complex securely together in the correct reading frame. Polycistronic mRNA has multiple RBS sequences, one upstream of each coding sequence. This allows bacterial ribosomes to translate several genes on the same mRNA because the ribosome can find each initiation site within a message by binding to its RBS.

Translational initiation always begins with a special initiator aminoacyl-tRNA binding to the start codon, AUG. In *Bacteria* this is formylmethionyl-tRNA. After polypeptide completion, the formyl group is removed. Consequently, the N-terminal amino acid of the completed protein will be methionine. However, in many proteins this methionine is removed by a specific protease.

Elongation, Translocation, and Termination

The mRNA threads through the ribosome primarily bound to the 30S subunit. The ribosome contains other sites where the tRNAs interact. Two of these sites are located primarily on the 50S subunit, and they are termed the A site and the P site (Figure 4.36). The A site, the acceptor site, is the site on the ribosome where the incoming charged tRNA first attaches. Loading of a tRNA into the A site is assisted by the elongation factor EF-Tu.

The P site, or peptide site, is the site where the growing polypeptide chain is held by the previous tRNA. During peptide bond formation, the growing polypeptide chain moves to the tRNA at the A site as a new peptide bond is formed. Several nonribosomal proteins are required for elongation, especially the elongation factors, EF-Tu and EF-Ts, as well as more GTP (to simplify Figure 4.36, the elongation factors are omitted and only part of the ribosome is shown).

Following elongation, the tRNA holding the polypeptide is translocated (moved) from the A site to the P site, thus opening the A site for another charged tRNA (Figure 4.36). Translocation requires the elongation factor EF-G and one molecule of GTP for each translocation event. At each translocation step the ribosome advances three nucleotides, exposing a new codon at the A site. Translocation pushes the now empty tRNA to a third site, called the E site. It is from this exit site that the tRNA is actually released from the ribosome. As one might expect, the precision of the translocation step is critical to the accuracy of protein synthesis. The ribosome must move *exactly* one codon at each step. Although mRNA appears to be moving through the ribosome complex, in reality, the ribosome is moving along the mRNA. Thus, the three sites on the ribosome shown in Figure 4.36 are not static locations but are moving parts of a biomolecular machine.

Several ribosomes can translate a single mRNA molecule simultaneously, forming a complex called a *polysome* (**Figure 4.37**). Polysomes increase the speed and efficiency of translation simultaneously because each ribosome in a polysome complex makes a complete polypeptide. Note in Figure 4.37 how ribosomes in the polysome complex that are closest to the 5′ end (the beginning) of the mRNA molecule have short polypeptides attached to them because only a few codons have been read, while ribosomes closest to the 3′ end of the mRNA have nearly finished polypeptides.

Protein synthesis terminates when the ribosome reaches a stop codon (nonsense codon). No tRNA binds to a stop codon. Instead, specific proteins called *release factors* (RFs) recognize the stop codon and cleave the attached polypeptide from the final tRNA, releasing the finished product. Following this, the ribosomal subunits dissociate, and the 30S and 50S subunits are then free to form new initiation complexes and repeat the process.

Role of Ribosomal RNA in Protein Synthesis

Ribosomal RNA plays vital roles in all stages of protein synthesis, from initiation to termination. By contrast, the role of the many

TRANSLATION: Initiation

TRANSLATION: Elongation

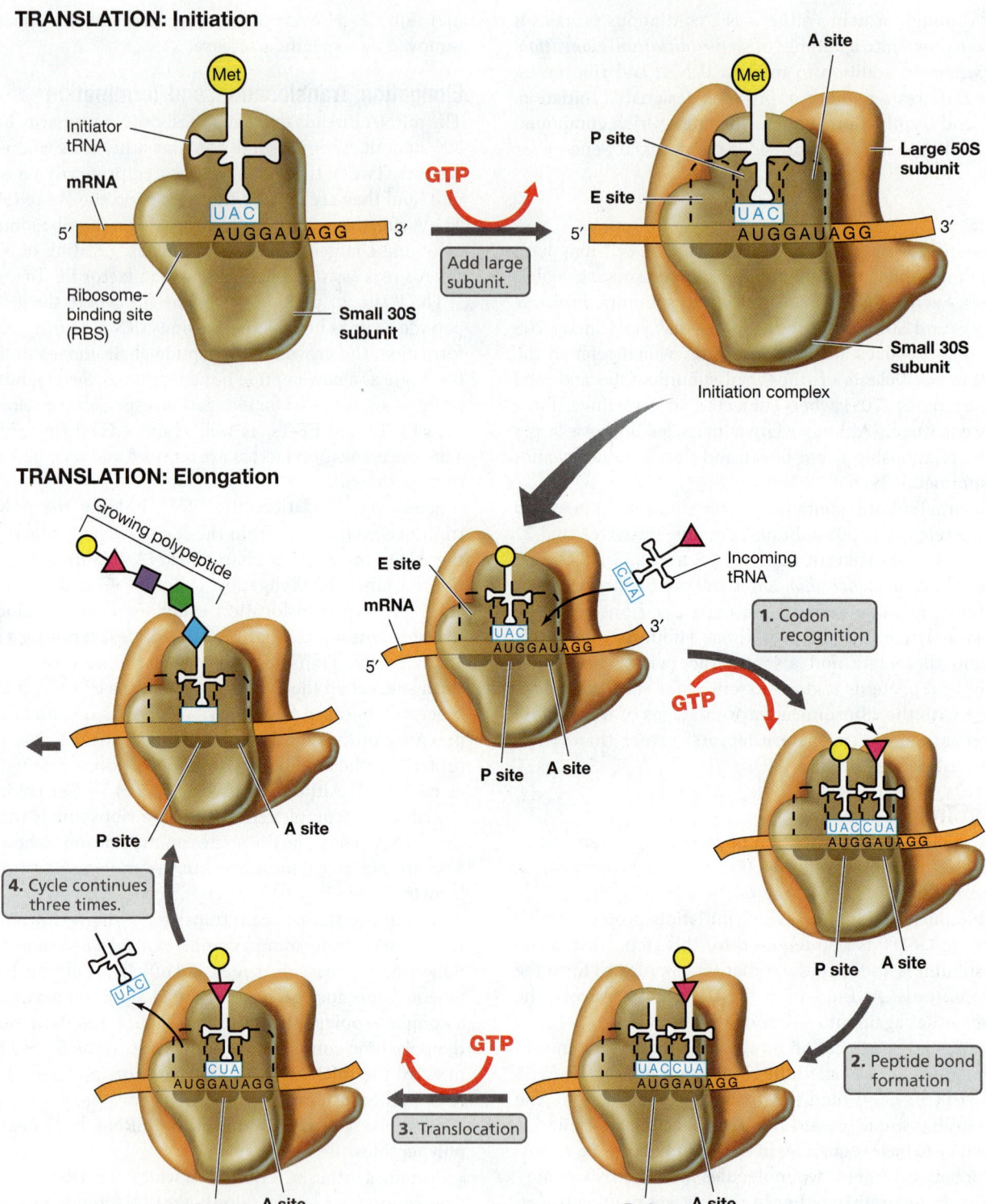

Figure 4.36 The ribosome and protein synthesis. *Initiation* of protein synthesis. The mRNA and initiator tRNA, carrying *N*-formylmethionine ("Met"), bind first to the small subunit of the ribosome. Initiation factors (not shown) use energy from GTP to promote the addition of the large ribosomal subunit. The initiator tRNA starts out in the P site. *Elongation* cycle of translation. 1. Elongation factors (not shown) use GTP to install the incoming tRNA into the A site. 2. Peptide bond formation is then catalyzed by the 23S rRNA. 3. Translocation of the ribosome along the mRNA from one codon to the next requires hydrolysis of another GTP. The outgoing tRNA is released from the E site. 4. The next charged tRNA binds to the A site and the cycle repeats. The genetic code, expressed in the language of mRNA, is shown in Table 4.5.

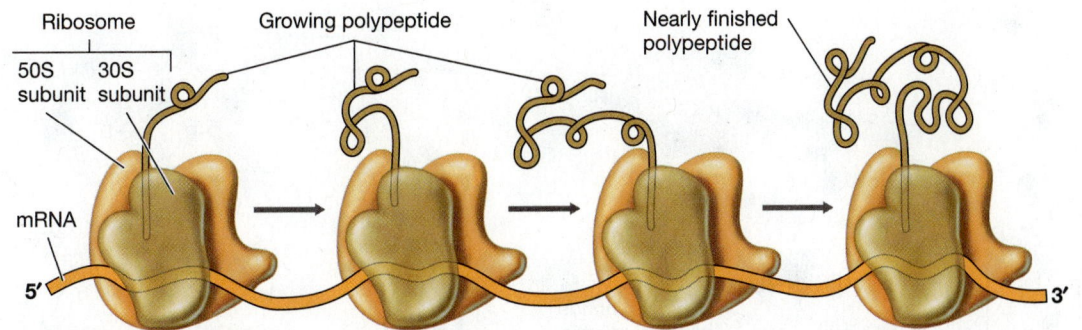

Figure 4.37 **Polysomes.** Translation by several ribosomes on a single messenger RNA forms the polysome. Note how the ribosomes nearest the 5′ end of the message are at an earlier stage in the translation process than ribosomes nearer the 3′ end, and thus only a relatively short portion of the final polypeptide has been made.

proteins in the ribosome is to form a scaffold to position key sequences in the ribosomal RNAs.

In *Bacteria* it is clear that 16S rRNA participates in initiation through base pairing with the RBS on the mRNA. Other mRNA–rRNA interactions also occur during elongation. On either side of the codons in the A and P sites, the mRNA is held in position by binding to 16S rRNA and ribosomal proteins. Ribosomal RNA also plays a role in ribosome subunit association, as well as in positioning tRNA in the A and P sites on the ribosome (Figure 4.36). Although charged tRNAs that enter the ribosome recognize the correct codon by codon–anticodon base pairing, they are also bound to the ribosome by interactions of the anticodon stem–loop of the tRNA with specific sequences within 16S rRNA. Moreover, the acceptor end of the tRNA (Figure 4.36) base-pairs with sequences in 23S rRNA.

In addition to all of this, the actual formation of peptide bonds is catalyzed by rRNA. The peptidyl transferase reaction occurs on the 50S subunit of the ribosome and is catalyzed by the 23S rRNA itself, rather than by any of the ribosomal proteins. The 23S rRNA also plays a role in translocation, and the EF proteins interact specifically with 23S rRNA. Thus, besides its role as the structural backbone of the ribosome, ribosomal RNA plays a major catalytic role in the translation process.

Freeing Trapped Ribosomes

A defective mRNA that lacks a stop codon causes a problem in translation. Such a defect may arise, for example, from a mutation that removed the stop codon, defective synthesis of the mRNA, or when partial degradation of an mRNA occurs before it is translated. If a ribosome reaches the end of an mRNA molecule and there is no stop codon, release factor cannot bind and the ribosome cannot be released from the mRNA. The ribosome is effectively "trapped."

Bacterial cells contain a small RNA molecule, called *tmRNA*, that frees stalled ribosomes (**Figure 4.38**). The "tm" in its name refers to the fact that tmRNA mimics both a tRNA, in that it carries the amino acid alanine, and mRNA, in that it contains a short stretch of RNA that can be translated. When tmRNA collides with a stalled ribosome, it binds alongside the defective mRNA. Protein synthesis can then proceed, first by adding the alanine on the

tmRNA and then by translating the short tmRNA message. The tmRNA contains a stop codon that allows release factor to bind and disassemble the ribosome. The protein made as a result of this rescue operation is defective and is subsequently degraded. This is accomplished by a short sequence of amino acids encoded by tmRNA and added to the end of the defective protein; the sequence is a signal for a specific protease to degrade the protein. Thus, through the activity of tmRNA, stalled ribosomes are not inactivated but instead are freed up to participate in protein synthesis once again.

MINIQUIZ
- What are the components of a ribosome? What functional roles does rRNA play in protein synthesis?
- How is a completed polypeptide chain released from the ribosome?
- How does tmRNA free stalled ribosomes?

4.14 Protein Folding and Secretion

For a protein to function properly it must fold correctly after it is synthesized and end up in the correct location in the cell. While many proteins exist inside the cell, some must be transported

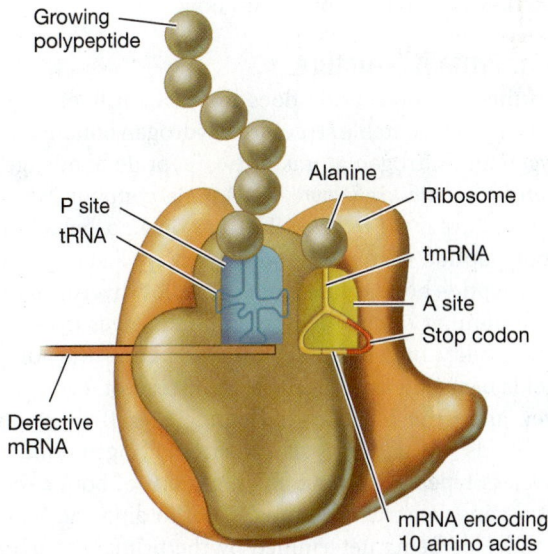

Figure 4.38 **Freeing of a stalled ribosome by tmRNA.** A defective mRNA lacking a stop codon stalls a ribosome that has a partly synthesized polypeptide attached to a tRNA (blue) in the P site. Binding of tmRNA (yellow) in the A site releases the polypeptide. Translation then continues up to the stop codon provided by the tmRNA.

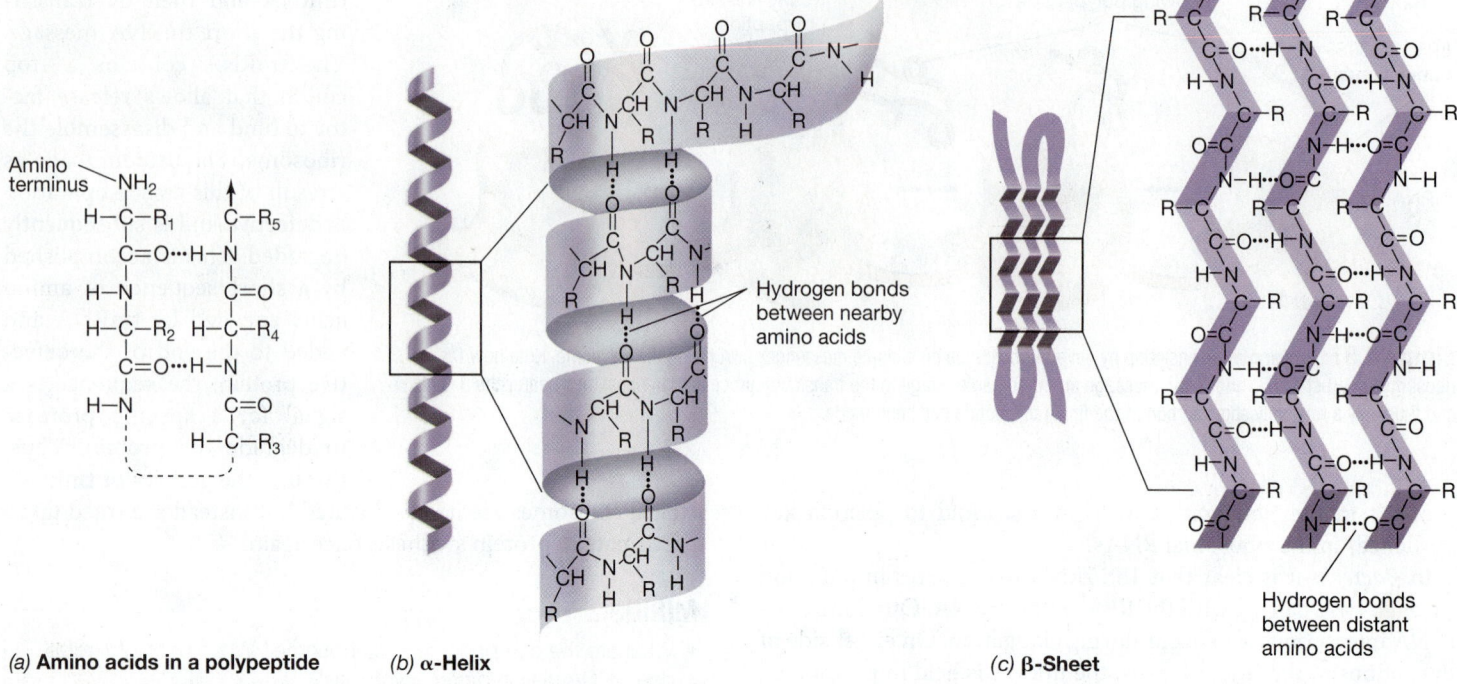

(a) **Amino acids in a polypeptide**   (b) **α-Helix**   (c) **β-Sheet**

Figure 4.39 **Secondary structure of polypeptides.** (a) Hydrogen bonding in protein secondary structure. R represents the side chain of the amino acid. (b) α-Helix secondary structure. (c) β-Sheet secondary structure. Note that the hydrogen bonding is between atoms in the peptide bonds and does not involve the R groups.

outside of the cytoplasmic membrane into the periplasm or into the inner or outer membranes to facilitate processes such as ion, sugar, and electron transport. Other proteins such as toxins and extracellular enzymes (exoenzymes) must be secreted from the cell entirely to be active in the environment. How does the cell determine the final conformation and location of proteins? We consider these two related processes now.

Levels of Protein Structure

Once formed, a polypeptide does not remain linear; instead it folds to form a more stable structure. Hydrogen bonding, between the oxygen and nitrogen atoms of two peptide bonds, generates the *secondary structure* (**Figure 4.39a**). One common type of secondary structure is the *α-helix*. To envision an α-helix, imagine a linear polypeptide wound around a cylinder (Figure 4.39b). This positions peptide bonds close enough to allow hydrogen bonding. The large number of such hydrogen bonds gives the α-helix its inherent stability. In *β-sheet* secondary structure, the polypeptide chain folds back and forth upon itself instead of forming a helix. However, as in the α-helix, the folding in a β-sheet positions peptide bonds so that they can undergo hydrogen bonding (Figure 4.39c). Polypeptides can contain regions of both α-helix and β-sheet secondary structure, the type of folding and its location in the molecule being determined by the primary structure and the available opportunities for hydrogen bonding.

Interactions between the R groups of the amino acids in a polypeptide generate two further levels of structure. **Tertiary structure** depends largely on hydrophobic interactions, with lesser contributions from hydrogen bonds, ionic bonds, and disulfide bonds. The tertiary folding generates the overall three-dimensional

shape of each polypeptide chain (**Figure 4.40**). Many proteins consist of two or more polypeptide chains. **Quaternary structure** refers to the number and type of polypeptides in the final protein. In proteins with quaternary structure, each polypeptide is called a *subunit* and has its own primary, secondary, and tertiary structure. Both tertiary and quaternary structure may be stabilized by disulfide bonds between adjacent sulfhydryl groups from cysteine residues (Figure 4.40). If the two cysteine residues are located in different polypeptides, the disulfide bond covalently links the two molecules. Alternatively, a single polypeptide chain can fold and bond to itself if a disulfide bond can form within the molecule.

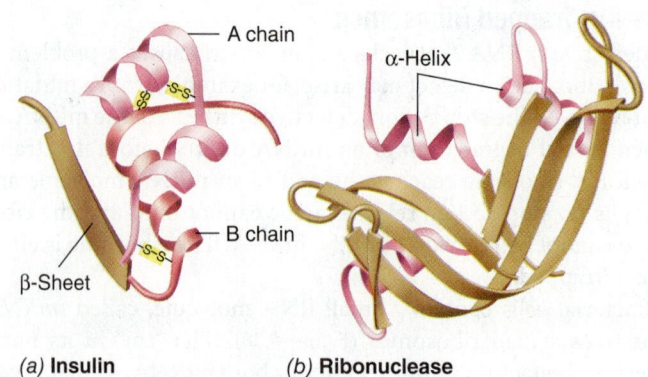

(a) **Insulin**   (b) **Ribonuclease**

Figure 4.40 **Tertiary structure of polypeptides.** (a) Insulin, a protein containing two polypeptide chains; note how the B chain contains both α-helix and β-sheet secondary structure and how disulfide linkages (S–S) help in dictating folding patterns (tertiary structure). (b) Ribonuclease, a large protein with several regions of α-helix and β-sheet secondary structure.

When proteins are exposed to extremes of heat or pH or to certain chemicals that affect their folding, they may undergo **denaturation**. This results in the polypeptide chain unfolding. When this occurs, the secondary, tertiary, and quaternary structure of the protein is destroyed along with its biological properties. However, because peptide bonds are not broken, the denatured polypeptide retains its primary structure. Depending on the severity of the denaturing conditions, the polypeptide may properly refold after the denaturant is removed. But if refolding is not correct, the protein cannot function.

Chaperonins Assist Protein Folding

Most polypeptides fold spontaneously into their active form while they are being synthesized. However, some do not and require assistance from other proteins called **chaperonins** (also known as **molecular chaperones**) for proper folding or for assembly into larger complexes. The chaperonins themselves do not become part of the assembly but only assist in folding. Indeed, one important function of chaperonins is to prevent improper aggregation of proteins. Chaperonins are widespread in all domains of life, and their sequences are highly conserved among all organisms.

Four key chaperonins in *Escherichia coli* are the proteins DnaK, DnaJ, GroEL, and GroES. DnaK and DnaJ are ATP-dependent enzymes that bind to newly formed polypeptides and prevent them from folding too quickly, which would increase the risk of improper folding (**Figure 4.41**). If the DnaKJ complex is unable to fold the protein properly, it may transfer the partially folded protein to the two multi-subunit proteins GroEL and GroES. The protein first enters GroEL, a large barrel-shaped protein that uses the energy of ATP hydrolysis to fold the protein properly. GroES assists in this (Figure 4.41). It is estimated that roughly 100 or so of the several thousand proteins of *E. coli* need help in folding from the GroEL–GroES complex, and of these, approximately a dozen are essential for survival of the bacteria.

In addition to folding newly synthesized proteins, chaperonins can also refold proteins that have partially denatured in the cell. A protein may denature for many reasons, but often it is because the organism has temporarily experienced high temperatures. Chaperonins are thus one type of *heat shock protein*, and their synthesis is greatly accelerated when a cell is stressed by excessive heat (⮂ Section 7.10). The heat shock response is an attempt by the cell to refold its partially denatured proteins for reuse before proteases recognize them as improperly folded and destroy them, freeing their amino acids to make new proteins.

Protein Secretion

Many proteins are located in the cytoplasmic membrane, in the periplasm (of gram-negative cells), or even outside the cell proper. Such proteins must get from their site of synthesis on ribosomes into or through the cytoplasmic membrane. Proteins called *translocases* transport specific proteins through and into prokaryotic membranes. For example, the Sec system both exports unfolded proteins and inserts integral membrane proteins into the cytoplasmic membrane, while the Tat system transports proteins folded in the cytoplasm through the membrane. To completely secrete proteins outside of the cell, gram-negative cells must employ additional translocases to transport proteins through the

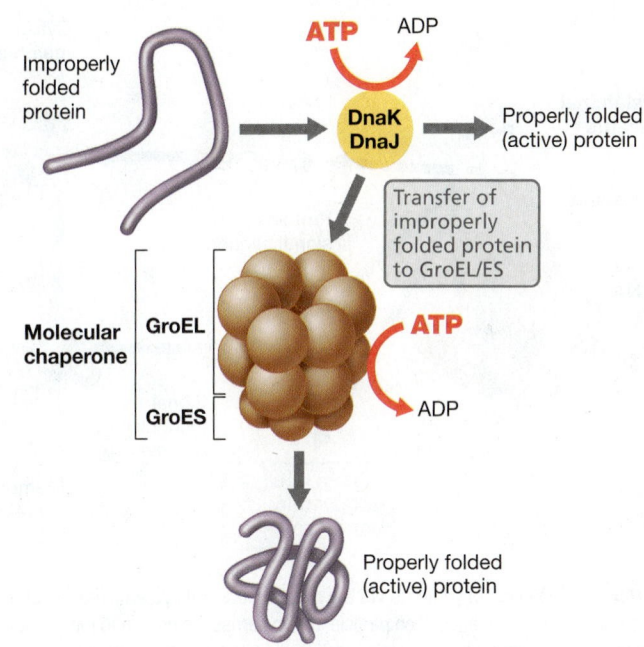

Figure 4.41 The activity of molecular chaperones. An improperly folded protein can be refolded by either the DnaKJ complex or by the GroEL–GroES complex. In both cases, energy for refolding comes from ATP.

outer membrane. At least six different types of secretion systems have been identified, some of which are employed by pathogenic bacteria to excrete toxins or harmful proteins into the host during infection.

Most proteins that must be transported into or through membranes are synthesized with an amino acid sequence of 15–20 residues, called the **signal sequence**, at the beginning (N-terminus, Figure 4.31) of the protein molecule. Signal sequences are variable, but they typically contain a few positively charged amino acids at the beginning, a central region of hydrophobic residues, and then a more polar region at their end. The signal sequence is so called because it "signals" the cell's secretory system that this particular protein is to be exported and also helps prevent the protein from completely folding, a process that could interfere with its secretion. Because the signal sequence is the first part of the protein to be synthesized, the early steps in export may actually begin before the protein is completely synthesized (**Figure 4.42**).

Proteins to be exported are recognized by either the *SecA protein* or the *signal recognition particle* (*SRP*) (Figure 4.42). Typically, SecA binds proteins that are to be exported into the periplasm whereas the SRP binds proteins that are destined to be inserted into the membrane but not released on the other side. SRPs are found in all cells. In *Bacteria*, they contain a single protein and a small noncoding RNA molecule (4.5S RNA). Both SecA and the SRP deliver proteins to the membrane secretion complex, and after crossing the membrane (Sec-mediated) or inserting into the membrane (SRP-mediated), the signal sequence is removed by a protease.

Proteins threaded through the cytoplasmic membrane in an unfolded state by the Sec system fold afterward (Figure 4.42). However, there are a few proteins, such as proteins required for energy metabolism that function in the periplasm, for example,

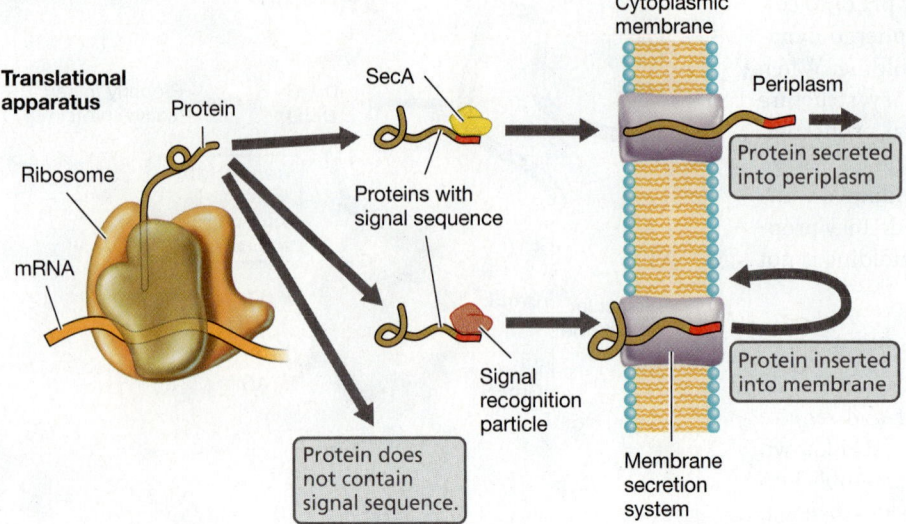

Figure 4.42 Export of proteins via the major secretory system. The signal sequence is recognized either by SecA or by the signal recognition particle, which carries the protein to the membrane secretion system. The signal recognition particle binds proteins that are inserted into the membrane, whereas SecA binds proteins that are secreted across the cytoplasmic membrane.

iron–sulfur proteins and several other redox-coupled proteins (♻ Section 3.10), that must be transported outside the cell after they have already folded. Usually this is because they contain small cofactors that must be inserted into the protein as it folds into its final form. Such proteins fold in the cytoplasm and then are exported by a transport system distinct from Sec, called the *Tat protein export system*. Tat stands for "twin arginine translocase" because the transported proteins have a short signal sequence containing a pair of arginines. This signal sequence on a folded protein is recognized by the TatBC proteins, which carry the protein to TatA, the membrane transporter. Once the protein has been transported to the periplasm using energy supplied by the proton motive force (♻ Section 3.11), the signal sequence is removed by a protease.

Types I through VI Secretion Systems

Several additional systems are used by gram-negative bacteria to deliver proteins into the outer membrane or through the outer membrane to the outside of the cell. These mechanisms are the types I–VI secretion systems. Each of these systems is composed of a large complex of proteins that form a channel through one or more membranes for the secreted molecule to travel through (Figure 4.43).

These diverse systems can be grouped into one- and two-step types. The types II and V systems are considered two-step mechanisms because they depend on the Sec or Tat system to transport either the secreted protein or a portion of the channel through the inner membrane. A second group of transporters move proteins through the outer membrane. Types I, III, IV, and VI are one-step systems because they form channels through both membranes and do not require Sec or Tat.

To inject toxin proteins into host cells, secretion systems of types III, IV, and VI also include structures on the outside of the cell that allow injection or insertion of the secreted protein into another cell. The entire type III structure has been termed the "injectisome" for its similarity to a syringe in both structure and function (Figure 4.43).

MINIQUIZ
- Define the terms primary, secondary, and tertiary structure with respect to proteins. How does a polypeptide differ from a protein?
- What does a molecular chaperone do?
- Why do some proteins have a signal sequence? What is the signal recognition particle?
- Why is it important for gram-negative bacteria to have additional secretion pathways?

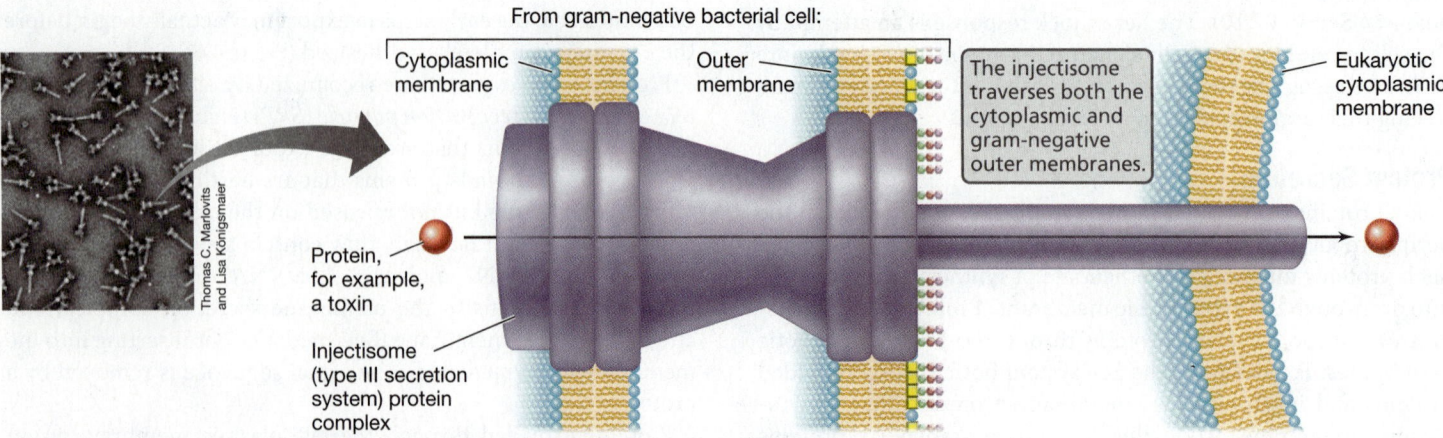

Figure 4.43 Secretion of molecules in gram-negative bacteria using the type III "injectisome" system. The protein complex that makes up the injectisome. Inset: Electron micrograph of purified injectisomes from *Salmonella enterica* serovar Typhimurium.

BIG IDEAS

4.1 • The informational content of a nucleic acid is determined by the sequence of nitrogen bases along the polynucleotide chain. Both RNA and DNA are informational macromolecules, as are the proteins they encode. The three key processes of macromolecular synthesis are: (1) DNA replication; (2) transcription (RNA synthesis); and (3) translation (protein synthesis).

4.2 • DNA is a double-stranded molecule that forms into a helix. The two strands in the double helix are complementary and antiparallel. Very long DNA molecules can be packaged into cells because they are supercoiled by enzymes called topoisomerases, such as DNA gyrase.

4.3 • In addition to the chromosome, other genetic elements can exist in cells. Plasmids are DNA molecules that exist separately from the chromosome and may confer a selective growth advantage under certain conditions. Viruses contain an RNA or DNA genome, and transposable elements exist as a part of other genetic elements. *Escherichia coli* is the prime model organism in biology.

4.4 • Both strands of the DNA helix are templates for the synthesis of new strands (semiconservative replication). The new strands are elongated by addition of deoxyribonucleotides to the 3′ end. DNA polymerases require a primer made of RNA by the enzyme primase.

4.5 • DNA synthesis begins at a site called the origin of replication. The double helix is unwound by helicase and is stabilized by single-strand binding protein. Extension of the DNA occurs continuously on the leading strand but discontinuously on the lagging strand, resulting in Okazaki fragments on the lagging strand that must be joined together.

4.6 • Starting from a single origin on a circular chromosome, two replication forks synthesize DNA simultaneously in both directions until the forks meet at the terminus region. The proteins at the replication fork form a large complex known as the replisome. Most errors in base pairing that occur during replication are corrected by the proofreading functions of DNA polymerases.

4.7 • In *Bacteria*, promoters are recognized by the sigma subunit of RNA polymerase. Alternative sigma factors allow joint regulation of large families of genes in response to growth conditions. Transcription by RNA polymerase continues until specific sites called transcription terminators are reached. These terminators function at the level of RNA.

4.8 • The unit of transcription in prokaryotes often contains more than a single gene that is transcribed into a single mRNA molecule that contains information for more than one polypeptide. A cluster of genes that are transcribed together from a single promoter constitute an operon.

4.9 • The transcription apparatus and the promoter architecture of *Archaea* and *Eukarya* have many features in common, although the components are usually relatively more simple in *Archaea*. In contrast, the processing of eukaryotic primary transcripts is unique and has three distinct steps: splicing, capping, and adding a poly(A) tail.

4.10 • Polypeptide chains contain 22 different genetically encoded amino acids that are linked via peptide bonds. The primary structure of a protein is its amino acid sequence, but the higher-order structure (folding) of the polypeptide determines its cellular function.

4.11 • The genetic code is expressed as RNA, and a single amino acid may be encoded by several different but related codons. In addition to the stop (nonsense) codons, there is also a specific start codon that signals the initiation of translation.

4.12 • Enzymes called aminoacyl-tRNA synthetases attach amino acids to their cognate tRNAs. One or more tRNAs exist for each amino acid incorporated into polypeptides by the ribosome.

4.13 • Translation occurs on the ribosome and requires mRNA and aminoacyl-tRNAs. The ribosome has three sites: acceptor, peptide, and exit. During each step of translation, the ribosome advances one codon along the mRNA, and the tRNA in the acceptor site moves to the peptide site. Protein synthesis terminates when a stop codon, which does not have a corresponding tRNA, is reached.

4.14 • Proteins must be properly folded in order to function correctly, and molecular chaperones assist with this process. Many proteins also need to be transported into or through the cytoplasmic membrane. These proteins contain a signal sequence that is recognized by the cellular translocases. Additional secretion systems are employed by gram-negative bacteria to secrete proteins into or through the outer membrane.

MasteringMicrobiology®   **Review what you know and challenge what you have learned with MasteringMicrobiology!** Access study materials, chapter quizzes, animations, and microbiology lab tutorials in the Study Area to ensure that you have mastered this chapter's content.

REVIEW OF KEY TERMS

Amino acid one of the 22 different monomers that make up proteins; chemically, a two-carbon carboxylic acid containing an amino group and a characteristic substituent on the alpha carbon

Aminoacyl-tRNA synthetase an enzyme that catalyzes attachment of an amino acid to its cognate tRNA

Anticodon a sequence of three bases in a tRNA molecule that base-pairs with a codon during protein synthesis

Antiparallel in reference to double-stranded DNA, the two strands run in opposite directions (one runs $5' \rightarrow 3'$ and the complementary strand $3' \rightarrow 5'$)

Bacteriocin a toxic protein secreted by bacteria that inhibits or kills other, related bacteria

Chaperonin or molecular chaperone a protein that helps other proteins fold or refold from a partly denatured state

Chromosome a genetic element, usually circular in prokaryotes, carrying genes essential to cellular function

Codon a sequence of three bases in mRNA that encodes an amino acid

Codon bias nonrandom usage of multiple codons encoding the same amino acid

Complementary nucleic acid sequences that can base-pair with each other

Denaturation loss of the correct folding of a protein, leading (usually) to protein aggregation and loss of biological activity

DNA (deoxyribonucleic acid) a polymer of deoxyribonucleotides linked by phosphodiester bonds that carries genetic information

DNA helicase an enzyme that uses ATP to unwind the double helix of DNA

DNA gyrase an enzyme found in most prokaryotes that introduces negative supercoils in DNA

DNA ligase an enzyme that seals nicks in the backbone of DNA

DNA polymerase an enzyme that synthesizes a new strand of DNA in the $5' \rightarrow 3'$ direction using an antiparallel DNA strand as a template

Enantiomer a form of a molecule that is the mirror image of another form of the same molecule

Enzyme a protein or an RNA that catalyzes a specific chemical reaction in a cell

Exons the coding DNA sequences in a split gene (contrast with intron)

Gene a segment of DNA specifying a protein (via mRNA), a tRNA, an rRNA, or any other noncoding RNA

Genetic code the correspondence between nucleic acid sequence and amino acid sequence of proteins

Genetic element a structure that carries genetic information, such as a chromosome, a plasmid, or a virus genome

Genome the total complement of genes contained in a cell or virus

Informational macromolecule any large polymeric molecule that carries genetic information, including DNA, RNA, and protein

Introns the intervening noncoding DNA sequences in a split gene (contrast with exons)

Lagging strand the new strand of DNA that is synthesized in short pieces and then joined together later

Leading strand the new strand of DNA that is synthesized continuously during DNA replication

Messenger RNA (mRNA) an RNA molecule that contains the genetic information to encode one or more polypeptides

Nonsense codon another name for a stop codon

Nucleic acid DNA or RNA

Nucleoside a nitrogenous base (adenine, guanine, cytosine, thymine, or uracil) plus a sugar (either ribose or deoxyribose) but lacking phosphate

Nucleotide a monomer of a nucleic acid containing a nitrogenous base (adenine, guanine, cytosine, thymine, or uracil), one or more molecules of phosphate, and a sugar, either ribose (in RNA) or deoxyribose (in DNA)

Open reading frame (ORF) a sequence of DNA or RNA that could be translated to give a polypeptide

Operon a cluster of genes that are cotranscribed as a single messenger RNA

Peptide bond a type of covalent bond linking amino acids in a polypeptide

Phosphodiester bond a type of covalent bond linking nucleotides together in a polynucleotide

Plasmid an extrachromosomal genetic element that has no extracellular form

Polynucleotide a polymer of nucleotides bonded to one another by covalent bonds called phosphodiester bonds

Polypeptide a polymer of amino acids bonded to one another by peptide bonds

Primary structure the precise sequence of monomers in a macromolecule such as a polypeptide or a nucleic acid

Primary transcript an unprocessed RNA molecule that is the direct product of transcription

Primase the enzyme that synthesizes the RNA primer used in DNA replication

Primer an oligonucleotide to which DNA polymerase attaches the first deoxyribonucleotide during DNA synthesis

Promoter a site on DNA to which RNA polymerase binds to commence transcription

Protein a polypeptide or group of polypeptides forming a molecule of specific biological function

Purine one of the nitrogenous bases of nucleic acids that contain two fused rings; adenine and guanine

Pyrimidine one of the nitrogenous bases of nucleic acids that contain a single ring; cytosine, thymine, and uracil

Quaternary structure in proteins, the number and types of individual polypeptides in the final protein molecule

Replication synthesis of DNA using DNA as a template

Replication fork the site on the chromosome where DNA replication occurs and where the enzymes replicating the DNA are bound to untwisted, single-stranded DNA

Replisome a DNA replication complex that consists of two copies of DNA polymerase III, DNA gyrase, helicase, primase, and copies of single-strand binding protein

Ribosomal RNA (rRNA) types of RNA found in the ribosome; some participate actively in protein synthesis

Ribosome a cytoplasmic particle composed of ribosomal RNA and protein, whose function is to synthesize proteins

RNA (ribonucleic acid) a polymer of ribonucleotides linked by phosphodiester bonds that plays many roles in cells, in particular, during protein synthesis

RNA polymerase an enzyme that synthesizes RNA in the $5' \rightarrow 3'$ direction using a complementary and antiparallel DNA strand as a template

RNA processing the conversion of a primary transcript RNA to its mature form

Secondary structure the initial pattern of folding of a polypeptide or a polynucleotide, usually dictated by opportunities for hydrogen bonding

Semiconservative replication DNA synthesis yielding two new double helices, each consisting of one parental and one progeny strand

Signal sequence a special N-terminal sequence of approximately 20 amino acids that signals that a protein should be exported across the cytoplasmic membrane

Spliceosome a complex of ribonucleoproteins that catalyze the removal of introns from primary RNA transcripts

Start codon a special codon, usually AUG, that signals the start of a protein

Stop codon a codon that signals the end of a protein

Termination stopping the elongation of an RNA molecule at a specific site

Tertiary structure the final folded structure of a polypeptide that has previously attained secondary structure

Transcription the synthesis of RNA using a DNA template

Transfer RNA (tRNA) a small RNA molecule used in translation that possesses an anticodon at one end and has the corresponding amino acid attached to the other end

Translation the synthesis of protein using the genetic information in RNA as a template

Transposable element a genetic element able to move (transpose) from one site to another on host DNA molecules

Wobble a less rigid form of base pairing allowed only in codon–anticodon pairing

REVIEW QUESTIONS

1. Describe the central dogma of molecular biology. (Section 4.1)

2. Genes were discovered before their chemical nature was known. First, define a gene without mentioning its chemical nature. Then name the chemicals that compose a gene. (Section 4.1)

3. DNA molecules that are AT-rich separate into two strands more easily when the temperature rises than do DNA molecules that are GC-rich. Explain this based on the properties of AT and GC base pairing. (Section 4.2)

4. Describe how DNA, which is many times the length of a cell when linearized, fits into the cell. (Section 4.2)

5. What is the size of the *Escherichia coli* chromosome and about how many proteins can it encode? What other genetic elements might be present in *E. coli*? (Section 4.3)

6. What are R plasmids and why are they of medical concern? (Section 4.3)

7. With reference to DNA, what is meant by the terms semiconservative, complementary, and antiparallel? (Section 4.4)

8. A structure commonly seen in circular DNA during replication is the theta structure. Draw a diagram of the replication process and show how a theta structure could arise. (Sections 4.5 and 4.6)

9. Why are errors in DNA replication so rare? What enzymatic activity, in addition to polymerization, is associated with DNA polymerase III and how does it reduce errors? (Section 4.6)

10. Do genes for tRNAs have promoters? Do they have start codons? Explain. (Sections 4.7 and 4.11)

11. The start and stop sites for mRNA synthesis (on the DNA) are different from the start and stop sites for protein synthesis (on the mRNA). Explain. (Sections 4.7 and 4.11)

12. What is an operon and why is it beneficial to link the expression of certain genes? (Section 4.8)

13. Why do eukaryotic mRNAs have to be "processed" whereas most prokaryotic RNAs do not? (Section 4.9)

14. Why are amino acids so named? Write a general structure for an amino acid. What is the importance of the R group to final protein structure? Why does the amino acid cysteine have special significance for protein structure? (Section 4.10)

15. What is "wobble" and what makes it necessary in protein synthesis? (Sections 4.11 and 4.12)

16. What are aminoacyl-tRNA synthetases and what types of reactions do they carry out? How does a synthetase recognize its correct substrates? (Section 4.12)

17. The enzyme activity that forms peptide bonds on the ribosome is called peptidyl transferase. Which molecule catalyzes this reaction? (Section 4.13)

18. Define the types of protein structure: primary, secondary, tertiary, and quaternary. Which of these structures are altered by denaturation? (Section 4.14)

19. Sometimes misfolded proteins can be correctly refolded, but sometimes they cannot and are destroyed. What kinds of proteins are involved in refolding misfolded proteins? What kinds of enzymes are involved in destroying misfolded proteins? (Section 4.14)

20. How does a cell know which of its proteins are designed to function outside of the cell? (Section 4.14)

APPLICATION QUESTIONS

1. The genome of the bacterium *Neisseria gonorrhoeae* consists of one double-stranded DNA molecule that contains 2220 kilobase pairs. Calculate the length of this DNA molecule in centimeters. If 85% of this DNA molecule is made up of the open reading frames of genes encoding proteins, and the average protein is 300 amino acids long, how many protein-encoding genes does *Neisseria* have? What kind of information do you think might be present in the other 15% of the DNA?

2. Compare and contrast the activity of DNA and RNA polymerases. What is the function of each? What are the substrates of each? What is the main difference in the behavior of the two polymerases?

3. What would be the result (in terms of protein synthesis) if RNA polymerase initiated transcription one base upstream of its normal starting point? Why? What would be the result (in terms of protein synthesis) if translation began one base downstream of its normal starting point? Why?

4. In Chapter 10 we will learn about mutations, inheritable changes in the sequence of nucleotides in the genome. By inspecting Table 4.5, discuss how the genetic code has evolved to help minimize the impact of mutations.

5 · Microbial Growth and Control

microbiology**now**

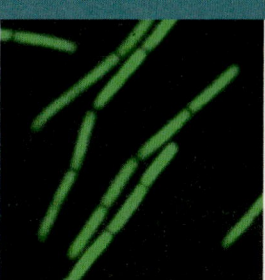
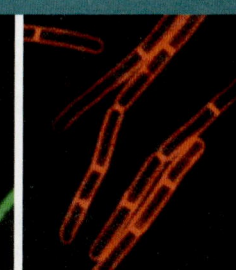

Did the First Cells on Earth Have Cell Walls?

There are many different cell shapes in the bacterial world: rods, cocci, spirilla, and more. What shape were the first cells? The peptidoglycan-containing cell wall is the hallmark of cells of *Bacteria,* as it defines a cell's morphology and prevents its osmotic lysis. But did the first cells on Earth have cell walls?

The rod-shaped bacterium *Bacillus subtilis* has been used as a model for the study of bacterial cell shape, growth, and morphogenesis. Cells of *B. subtilis* are relatively large and easy to visualize by fluorescence microscopy (top photos, L to R: DNA stain, green fluorescent protein, and membrane stain). Moreover, the genetics of this bacterium is well understood; this allows researchers to generate various mutants.

Mutant strains of *B. subtilis* that lack a cell wall, called *L-forms,* can be generated and grown in osmotically protected culture media. Remarkably, conversion of the wild type to the L-form requires only two mutations.[1] L-forms grow not by the usual binary fission process of rod-shaped bacteria, but instead by releasing small vesicles that slowly enlarge and eventually generate vesicles of their own (bottom photo). And all of this happens independently of the major cell-division and cytoskeletal proteins of bacterial cells, FtsZ and MreB.

Earth's first cells almost certainly did not look like the morphologically diverse *Bacteria* and *Archaea* we know today, but instead more like the L-forms of *B. subtilis* shown here. Lacking a cell wall would have allowed early cells to fuse and readily exchange genes. With the emergence of a peptidoglycan cell wall, a barrier to major genetic exchange would have been established, but the wall would have allowed cells to explore osmotically unprotected habitats and evolve diverse cell shapes best suited to exploiting the resources in these habitats.

[1]Errington, J. 2013. L-form bacteria, cell walls and the origins of life. *Open Biology 3:* 120143.

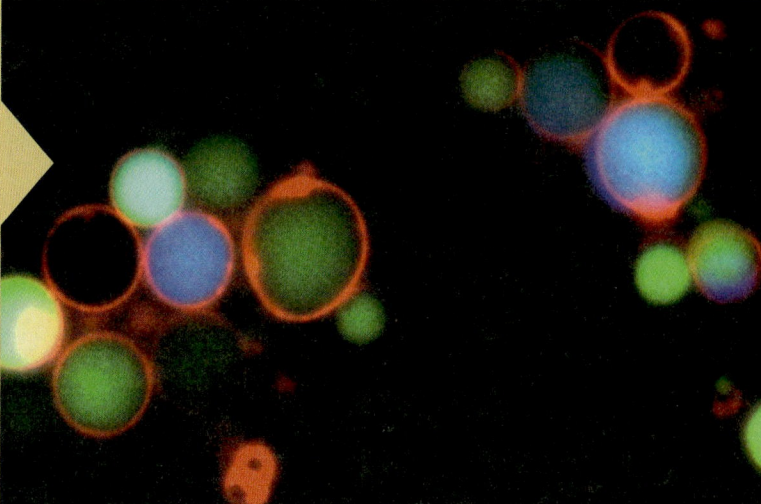

I • Bacterial Cell Division

In previous chapters we discussed cell structure and function (Chapter 2) and the principles of microbial nutrition and metabolism (Chapter 3). In Chapter 4 we learned the important molecular processes that encode the structures and metabolic processes of cells. Now we will consider how all of these come together to yield new cells during microbial growth.

Growth is the result of cell division and is the ultimate process in the life of a microbial cell. Knowledge of how bacteria grow has given us new insight into cell division in higher organisms and is useful for designing methods to control microbial growth.

5.1 Binary Fission

In microbiology, **growth** is defined as *an increase in the number of cells.* Microbial cells have a finite life span, and a species is maintained only as a result of continued growth of its population. As macromolecules accumulate in the cytoplasm of a cell, they assemble into major cell structures, such as the cell wall, cytoplasmic membrane, flagella, ribosomes, enzyme complexes, and so on, eventually leading to the process of cell division itself. In a growing culture of a rod-shaped bacterium such as *Escherichia coli*, cells elongate to approximately twice their original length and then form a partition that constricts the cell into two daughter cells (**Figure 5.1**). This process is called **binary fission** ("binary" to express the fact that two cells have arisen from one). This partition is called a *septum* and results from the inward growth of the cytoplasmic membrane and cell wall from opposing directions; septum formation continues until the two daughter cells are pinched off. There are some variations in this general pattern of binary fission. In some bacteria, such as *Bacillus subtilis*, a septum forms without cell wall constriction, while in the budding bacterium *Caulobacter*, constriction occurs but no septum is formed. But in all cases, when one cell eventually separates to form two cells, we say that one *generation* has occurred, and the time required for this process is called the **generation time** (Figure 5.1 and see Figure 5.10).

During one generation, all cellular constituents increase proportionally, and cells are said to be in *balanced growth.* Each daughter cell receives a chromosome and sufficient copies of ribosomes and all other macromolecular complexes, monomers, and inorganic ions to exist as an independent cell. Partitioning of the replicated DNA molecule between the two daughter cells depends on the DNA remaining attached to the cytoplasmic membrane during division, with constriction leading to separation of the chromosomes, one to each daughter cell (see Figure 5.3).

The generation time in a given bacterial species is highly variable and is dependent on nutritional and genetic factors, and temperature. Under the best nutritional conditions, the generation time of a laboratory culture of *E. coli* is about 20 min. A few bacteria can grow even faster than this, with the record holder having an 6-min generation time. Many bacteria grow much slower than this, with generation times of hours or days being more common. In nature, microbial cells probably grow much slower than their

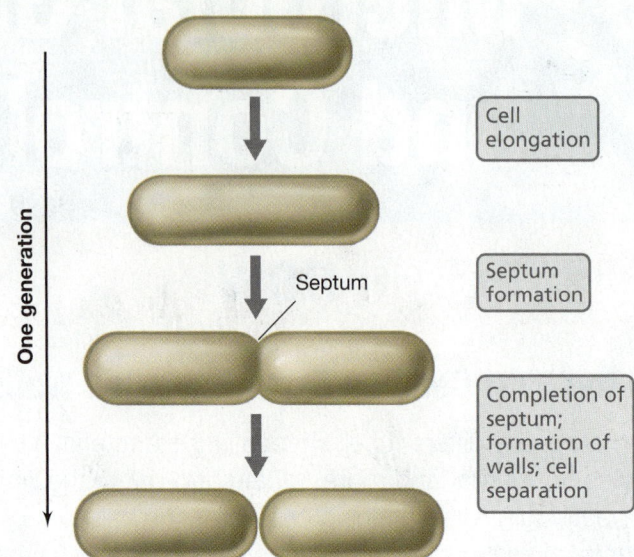

Figure 5.1 **Binary fission in a rod-shaped prokaryote.** Cell numbers double every generation.

maximum rates observed in the laboratory. This is because the conditions and resources necessary for optimal growth in the laboratory may not be present in a natural habitat, and unlike pure culture growth, microorganisms in nature live with other species in microbial communities and thus must compete with their neighbors for resources and space.

MINIQUIZ

- Summarize the steps that lead up to binary fission in a bacterium like *Escherichia coli*.

- Define the term generation. What is meant by the term generation time?

5.2 Fts Proteins and Cell Division

A series of proteins present in all *Bacteria* are essential for cell division. These proteins are called *Fts proteins* and a key one, **FtsZ**, plays a crucial role in the binary fission process. FtsZ is related to tubulin, the important cell-division protein in eukaryotes (⟳ Section 2.22), and is also found in most but not all *Archaea*. Other Fts proteins are found only in *Bacteria* and not in *Archaea*, so our discussion here will be restricted to the *Bacteria*. The gram-negative *Escherichia coli* and the gram-positive *Bacillus subtilis* have been the model bacterial species for the study of cell division events.

The Divisome

Fts proteins interact in the cell to form a division apparatus called the **divisome**. In rod-shaped cells, formation of the divisome begins with the attachment of molecules of FtsZ in a ring precisely around the center of the cell; this ring will become the

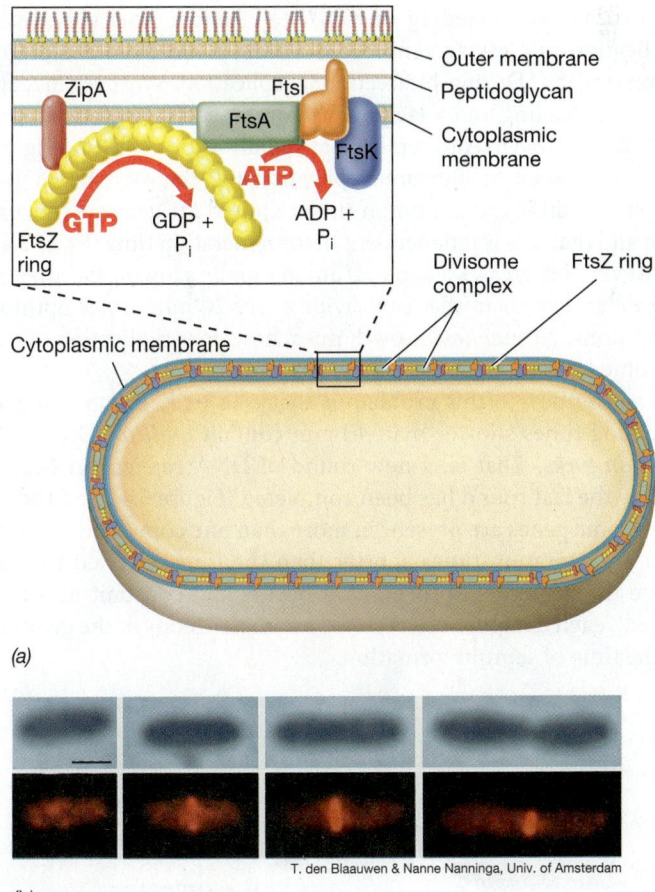

(a)

(b)

T. den Blaauwen & Nanne Nanninga, Univ. of Amsterdam

Figure 5.2 The FtsZ ring and cell division. *(a)* Cutaway view of a rod-shaped cell showing the ring of FtsZ molecules around the division plane. Blowup shows the arrangement of individual divisome proteins. ZipA is an FtsZ anchor, FtsI is a peptidoglycan biosynthesis protein, FtsK assists in chromosome separation, and FtsA is an ATPase. *(b)* Appearance and breakdown of the FtsZ ring during the cell cycle of *Escherichia coli*. Microscopy: upper row, phase contrast; bottom row, cells stained with a specific reagent against FtsZ. Cell division events: first column, FtsZ ring not yet formed; second column, FtsZ ring appears as nucleoids start to segregate; third column, full FtsZ ring forms as cell elongates; fourth column, breakdown of the FtsZ ring and cell division. Marker bar in upper left photo, 1 μm.

cell-division plane. In a cell of *E. coli* about 10,000 FtsZ molecules polymerize to form the ring, and the ring attracts other divisome proteins, including *FtsA* and *ZipA* (**Figure 5.2**). ZipA is an anchor that connects the FtsZ ring to the cytoplasmic membrane and stabilizes it. FtsA, a protein related to actin, an important cytoskeletal protein in eukaryotes (⟳ Section 2.22), also helps to connect the FtsZ ring to the cytoplasmic membrane and recruit other divisome proteins. The divisome forms about three-quarters of the way into the cell division cycle. However, before the divisome forms, the cell is already elongating and DNA replication has begun (see Figure 5.3).

The divisome also contains Fts proteins needed for peptidoglycan synthesis, such as FtsI (Figure 5.2). FtsI is one of several *penicillin-binding proteins* present in the cell. Penicillin-binding proteins are so named because their activities are inhibited by the antibiotic penicillin (Section 5.4). The divisome orchestrates synthesis of new cytoplasmic membrane and cell wall material, called

the *division septum*, at the center of a rod-shaped cell until it reaches twice its original length. The elongated cell then divides, yielding two daughter cells (Figure 5.1).

DNA Replication, Min Proteins, and Cell Division

DNA replicates before the FtsZ ring forms (**Figure 5.3**) because the ring forms in the space between the duplicated nucleoids; before the nucleoids segregate, they effectively block formation of the FtsZ ring. The proteins MinC, MinD, and MinE interact to help guide FtsZ to the cell midpoint. MinD forms a spiral structure on the inner surface of the cytoplasmic membrane and helps to localize MinC to the cytoplasmic membrane. The MinD spiral oscillates back and forth along the long axis of the growing cell and functions to *inhibit* cell division by preventing the FtsZ ring from forming (Figure 5.3). Simultaneously, however, MinE also oscillates from pole to pole, and as it does, it functions to sweep MinC and MinD aside. Hence, because MinC and MinD dwell longer at the poles than anywhere else during their oscillation cycle, the center of the cell will have, on average, the lowest concentration of these proteins. As a result, the cell center becomes the most permissive site for FtsZ binding and so the FtsZ ring forms there. In this unusual series of events, the Min proteins ensure that the divisome forms only at the *cell center* and not at the cell poles (Figure 5.3).

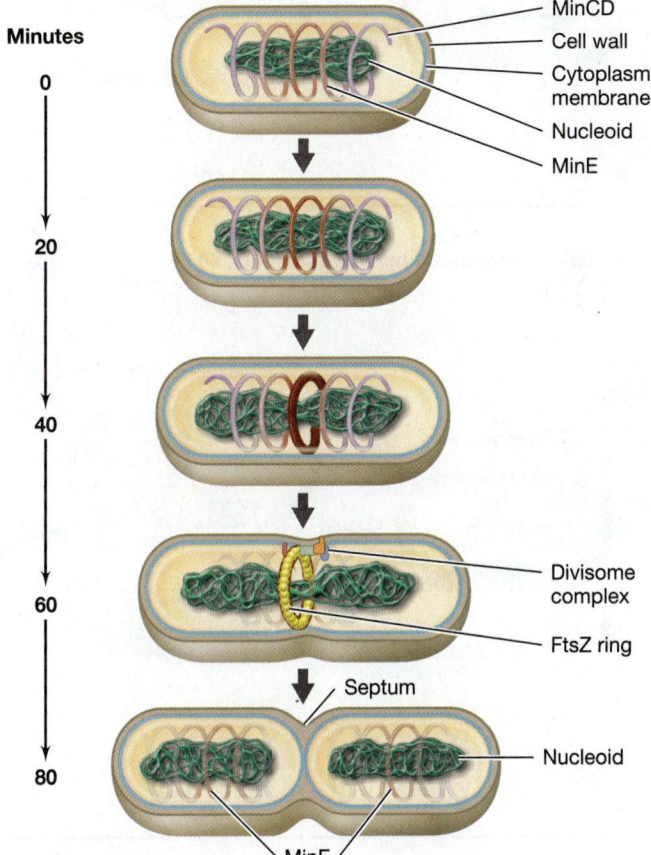

Figure 5.3 DNA replication and cell-division events. The protein MinE directs formation of the FtsZ ring and divisome complex at the cell-division plane. Shown is a schematic for cells of *Escherichia coli* growing with a doubling time of 80 min. MinC and MinD are most abundant at the cell poles.

As cell elongation continues and septum formation begins, the two copies of the chromosome are pulled apart, each to its own daughter cell (Figure 5.3). The Fts protein *FtsK* and several other proteins assist in this process. As the cell constricts, the FtsZ ring begins to depolymerize, triggering the inward growth of wall materials to form the septum and seal off one daughter cell from the other. The enzymatic activity of FtsZ also hydrolyzes guanosine triphosphate (GTP, an energy-rich compound) to yield the energy necessary to fuel the polymerization and depolymerization of the FtsZ ring (Figures 5.2 and 5.3).

There is great practical interest in understanding the details of bacterial cell division because such knowledge could lead to the development of new drugs that target specific steps in the growth of pathogenic bacteria. Like penicillin (a drug that targets bacterial cell wall synthesis), drugs that interfere with the function of specific Fts or other bacterial cell-division proteins could have broad applications in clinical medicine.

Genome Replication in Fast-Growing Cells

As we learned in Chapter 4, the circular nature of the chromosome of *Escherichia coli* and most other prokaryotes creates an opportunity for speeding up DNA replication. This is because replication of circular genomes is *bidirectional* from the origin of replication. During bidirectional replication, synthesis occurs in both a leading and a lagging fashion on each template strand, and this allows DNA to replicate as rapidly as possible (↩ Figure 4.17). Studies of chromosome replication in *E. coli* have shown that about 40 min is the minimum time required for genome replication and that this is independent of the generation time (**Figure 5.4**). However, this creates a conundrum in rapidly growing cultures of *E. coli*, an organism that can divide every 20 min under optimal conditions. At such fast growth rates, how does replication of the genome keep up with that of the cell proper?

The solution to this problem is that cells of *E. coli* growing at doubling times shorter than 40 min contain *multiple DNA replication forks*. That is, a new round of DNA replication begins before the last round has been completed (Figure 5.4), and therefore, some genes are present in more than one copy. This ensures that at generation times shorter than the time required to replicate the genome (a process that occurs at a constant maximal speed), each daughter cell receives a complete copy of the genome at the time of septum formation.

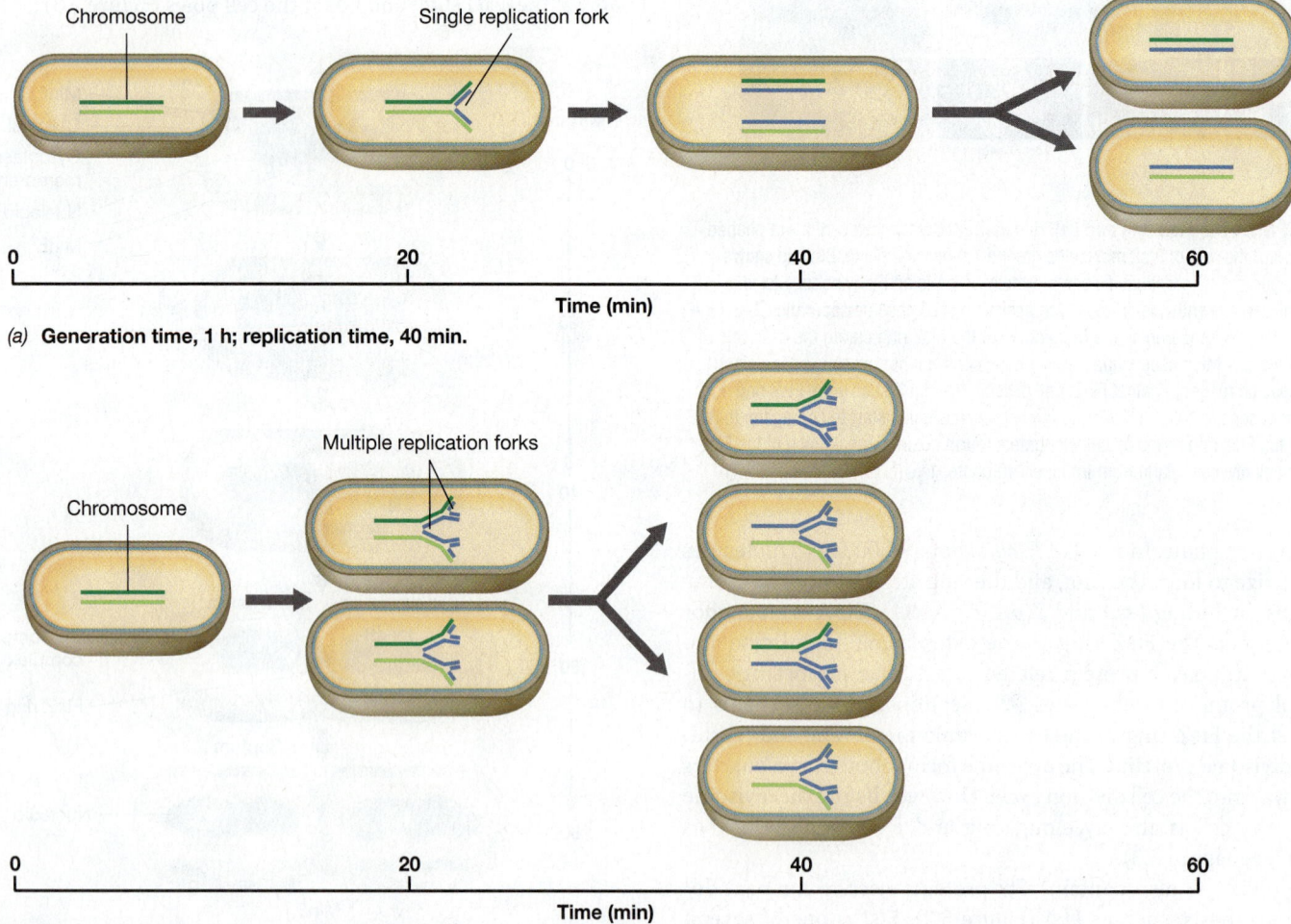

(a) **Generation time, 1 h; replication time, 40 min.**

(b) **Generation time, 20 min; replication time, 40 min.**

Figure 5.4 **Genome replication in cells of *Escherichia coli* growing at 60 min or 20 min generation times.** In cells doubling every 20 min, multiple replication forks are needed to ensure that each daughter cell gets a complete copy of the genome, which takes 40 minutes to replicate.

MINIQUIZ

- What is the divisome?
- How does FtsZ find the cell midpoint of a rod-shaped cell?
- Explain how the minimum generation time for the bacterium *Escherichia coli* can be less than the time needed to replicate its chromosome.

5.3 MreB and Cell Morphology

Just as specific proteins direct cell *division* in prokaryotes, other specific proteins direct cell *shape*. Interestingly, these shape-determining proteins show significant homology to key cytoskeletal proteins in eukaryotic cells. Like eukaryotes, prokaryotes also contain a cell cytoskeleton, and one that is both dynamic and multifaceted.

Cell Shape and MreB

The major shape-determining factor in *Bacteria* is a protein called *MreB*. MreB forms a simple cytoskeleton in *Bacteria* and in a few species of *Archaea*. MreB forms a helix of filaments around the inside of the cell, just below the cytoplasmic membrane (**Figure 5.5**). The MreB cytoskeleton presumably defines cell shape by recruiting other proteins that function in cell wall growth to group into a specific pattern. Inactivation of the gene encoding MreB in rod-shaped bacteria causes the cells to become coccus-shaped. Moreover, most naturally coccoid bacteria lack the MreB gene and thus do not make MreB. This indicates that the "default" morphology for a bacterium is most likely the sphere. Variations in the arrangement of MreB filaments in cells of nonspherical bacteria are probably responsible for the different common morphologies of prokaryotic cells (⇄ Figure 2.11).

How does MreB define a cell's shape? The helical structures formed by MreB (Figure 5.5a) are not static, but instead can rotate within the cytoplasm of a growing cell. Newly synthesized peptidoglycan (Section 5.4) is associated with the MreB helices at points where the helices contact the cytoplasmic membrane (Figure 5.5a). It is thought that MreB localizes the synthesis of new cell wall to specific locations along the long axis of a rod-shaped cell during growth. This allows new cell wall to form at several points along the cell rather than from a single location at the FtsZ site outward, as in spherical bacteria (see Figure 5.3). By rotating within the cell cylinder and initiating cell wall synthesis where it contacts the cytoplasmic membrane, MreB directs new wall synthesis in such a way that a rod-shaped cell elongates only along its long axis.

Crescentin

Caulobacter crescentus, a vibrio-shaped species of *Proteobacteria* (⇄ Section 7.12 and 14.21), produces a shape-determining protein called *crescentin* in addition to MreB. Copies of crescentin protein organize into filaments about 10 nm wide that localize onto the concave face of the curved cell. The arrangement and localization of crescentin filaments are thought to impart the characteristic curved morphology to the *C. crescentus* cell (Figure 5.5c). *Caulobacter* is an aquatic bacterium that undergoes a life cycle in which swimming cells, called *swarmers*, eventually form

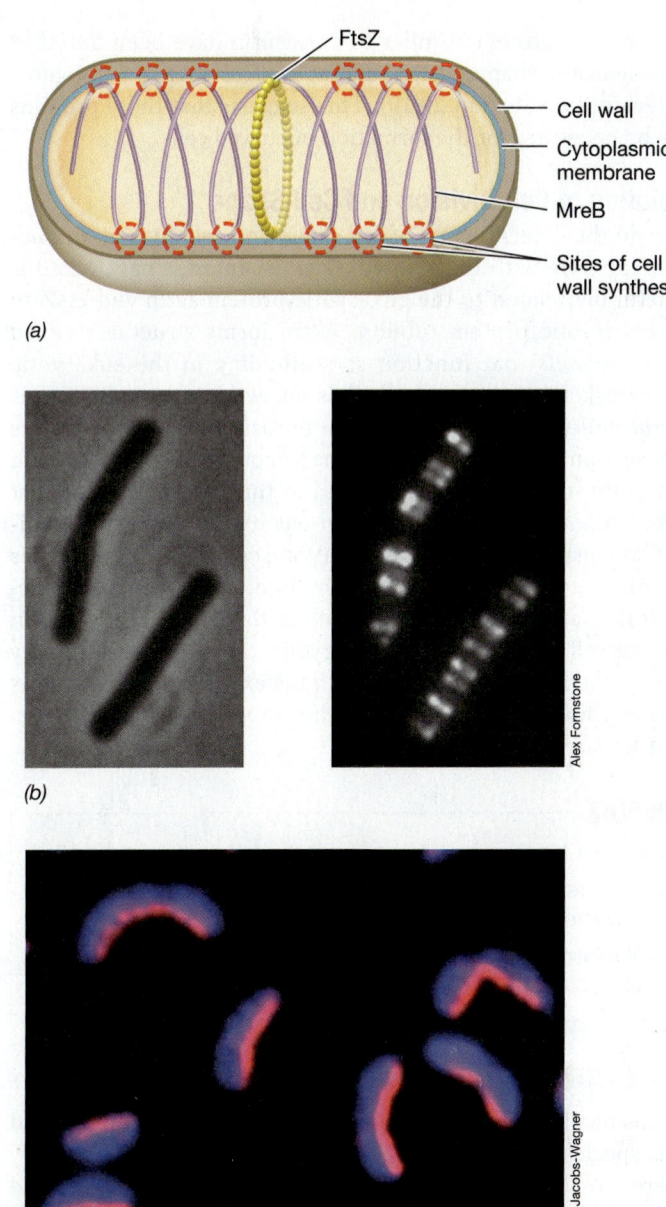

(a)

(b)

(c)

Alex Formstone

Christine Jacobs-Wagner

Figure 5.5 **MreB and crescentin as determinants of cell morphology.** *(a)* The cytoskeletal protein MreB is an actin analog that winds as a coil through the long axis of a rod-shaped cell, making contact with the cytoplasmic membrane in several locations (red dashed circles). These are sites of new cell wall synthesis. *(b)* Photomicrographs of the same cells of *Bacillus subtilis*. Left, phase contrast; right, fluorescence. The cells contain a substance that makes the MreB protein fluoresce, shown here as bright white. *(c)* Cells of *Caulobacter crescentus*, a naturally curved (vibrio-shaped) cell. Cells are stained to show the shape-determining protein crescentin (red), which lies along the concave surface of the cell, and with DAPI, which stains DNA and thus the entire cell blue.

a stalk and attach to surfaces. Attached cells then undergo cell division to form new swarmer cells that are released to colonize new habitats. The steps in this life cycle are highly orchestrated at the genetic level, and *Caulobacter* has been used as a model system for the study of gene expression in cellular differentiation (⇄ Section 7.12). Although crescentin seems to be unique to

Caulobacter, proteins similar to crescentin have been found in other helically shaped cells, such as *Helicobacter*, a pathogenic bacterium (ᴄᴅ Section 29.10). This suggests that these proteins may be necessary for the formation of curved cells.

Evolution of Cell Division and Cell Shape

How do the determinants of cell shape and cell division in *Bacteria* compare with those in eukaryotes? Interestingly, MreB is structurally related to the eukaryotic protein actin and FtsZ to the eukaryotic protein tubulin. Actin forms structures called *microfilaments* that function as scaffolding in the eukaryotic cell cytoskeleton and in cell division, whereas tubulin forms *microtubules* that are important in mitosis and other processes (ᴄᴅ Section 2.22). In addition, the shape-determining protein crescentin in *Caulobacter* is related to the keratin proteins that make up *intermediate filaments* in eukaryotic cells. Intermediate filaments form part of the eukaryotic cytoskeleton, and genes encoding similar proteins have been found in some other *Bacteria*. It thus appears that several proteins that control cell division and the cell cytoskeleton in eukaryotic cells have evolutionary roots in the *Bacteria*. However, with the exception of FtsZ, genes encoding homologs of these proteins appear to be absent from most *Archaea*.

MINIQUIZ

- How does MreB control the shape of a rod-shaped bacterium?
- What protein is thought to control the shape of cells of *Caulobacter*?
- What relationships exist between cytoskeletal proteins in *Bacteria* in eukaryotes?

5.4 Peptidoglycan Biosynthesis

In cells of all species of *Bacteria* that contain peptidoglycan, and most species do, preexisting peptidoglycan has to be temporarily severed to allow newly synthesized peptidoglycan to be inserted during the growth process. In cocci, new cell wall material grows out in opposite directions from the FtsZ ring (**Figure 5.6**), whereas as we have just seen, in rod-shaped cells, new cell wall grows at several locations along the length of the cell (Figure 5.5*a*). In either case, how does new peptidoglycan get made and how does it get outside the cytoplasmic membrane, where the peptidoglycan layer resides?

Biosynthesis of Peptidoglycan

Peptidoglycan can be thought of as a stress-bearing fabric, much like a thin sheet of rubber. Synthesis of new peptidoglycan during growth requires the controlled cutting of preexisting peptidoglycan along with the simultaneous insertion of peptidoglycan precursors. A lipid carrier molecule called *bactoprenol* plays a major role in the latter process. Bactoprenol is a hydrophobic C_{55} alcohol that bonds to a *N*-acetylglucosamine/*N*-acetylmuramic acid/pentapeptide peptidoglycan precursor (**Figure 5.7**). Bactoprenol transports peptidoglycan precursors across the cytoplasmic membrane by rendering them sufficiently hydrophobic to pass through the membrane interior.

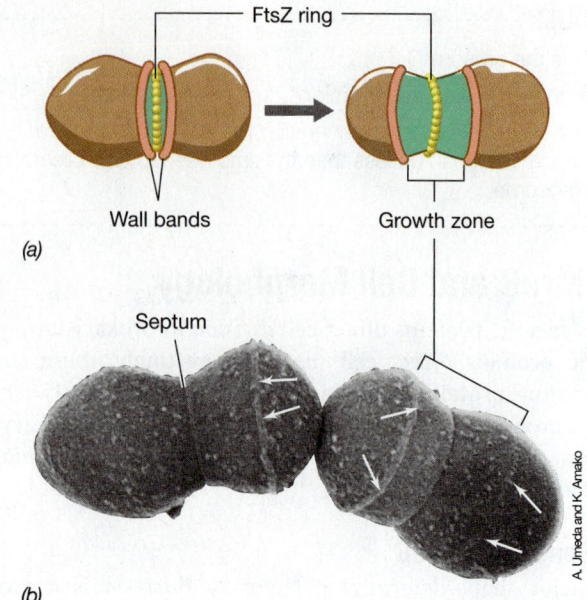

(a)

(b)

Figure 5.6 **Cell wall synthesis in gram-positive *Bacteria*.** *(a)* Localization of cell wall synthesis during cell division. In cocci, cell wall synthesis (shown in green) is localized at only one point (compare with Figure 5.5*a*). *(b)* Scanning electron micrograph of cells of *Streptococcus hemolyticus* showing wall bands (arrows). A single cell is about 1 μm in diameter.

Once in the periplasm, bactoprenol interacts with enzymes called *transglycosylases* that insert cell wall precursors into the growing point of the cell wall and catalyze glycosidic bond formation (**Figure 5.8**). Prior to this, small gaps in the existing peptidoglycan are made by enzymes called *autolysins*, enzymes that function to hydrolyze the bonds that connect *N*-acetylglucosamine and *N*-acetylmuramic acid in the peptidoglycan backbone. New cell wall material is then added across the gaps (Figure 5.8*a*). The junction between new and old peptidoglycan forms a ridge on the cell surface of gram-positive bacteria that can be observed as a *wall band* (Figure 5.6*b*). It is essential that peptidoglycan synthesis be a highly coordinated process. New tetrapeptide units must be spliced into existing peptidoglycan immediately after autolysin activity in order to prevent a breach in peptidoglycan integrity at the splice point; a breach could cause spontaneous cell lysis, called *autolysis*.

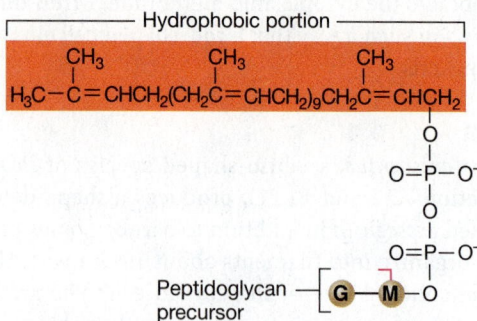

Figure 5.7 **Bactoprenol (undecaprenol diphosphate).** This highly hydrophobic molecule carries cell wall peptidoglycan precursors through the cytoplasmic membrane.

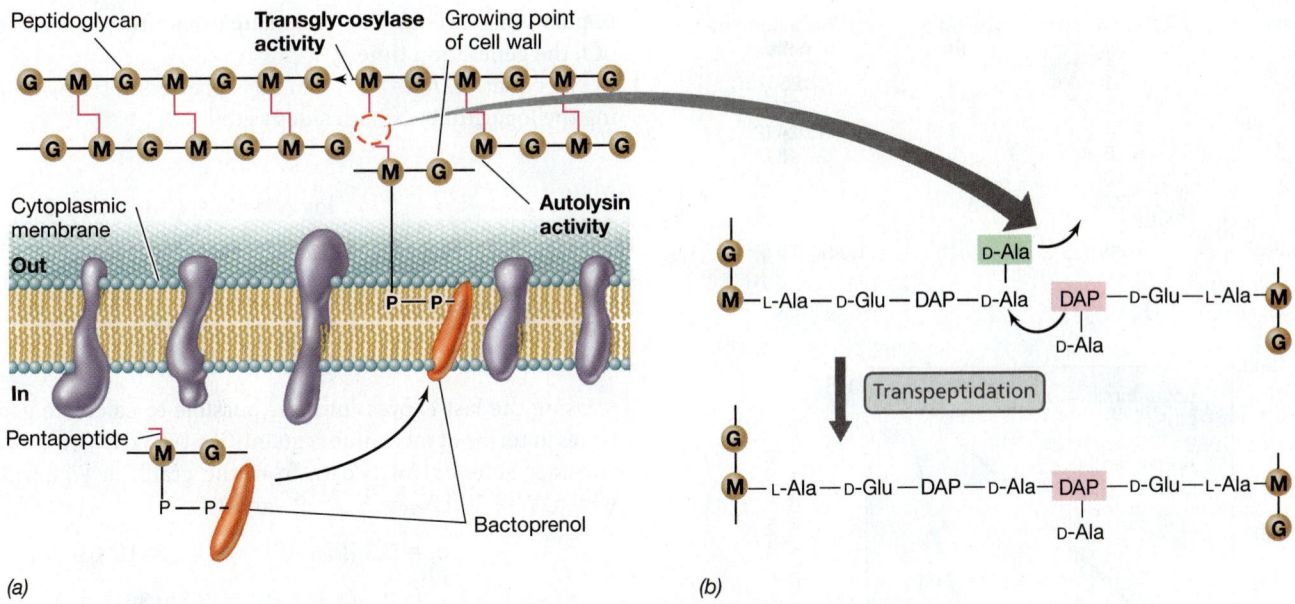

Figure 5.8 Peptidoglycan synthesis. *(a)* Transport of peptidoglycan precursors across the cytoplasmic membrane to the growing point of the cell wall. Autolysin breaks glycolytic bonds in preexisting peptidoglycan, while transglycosylase synthesizes them, linking old peptidoglycan with new. *(b)* The transpeptidation reaction that leads to the final cross-linking of two peptidoglycan chains. Penicillin inhibits this reaction.

Transpeptidation

The final step in cell wall synthesis is **transpeptidation**. Transpeptidation forms the peptide cross-links between muramic acid residues in adjacent glycan chains (c⊃ Section 2.10 and Figures 2.25 and 2.26). In gram-negative bacteria such as *Escherichia coli*, cross-links form between diaminopimelic acid (DAP) on one peptide and D-alanine on the adjacent peptide. Although there are two D-alanine residues at the end of the peptidoglycan precursor, only one remains in the final molecule as the other is removed during transpeptidation (Figure 5.8*b*). This reaction is exergonic (energy-releasing, c⊃ Section 3.4) and supplies the energy necessary to drive transpeptidation forward. In *E. coli*, the protein FtsI (Figure 5.2*a*) functions as a transpeptidase.

Transpeptidation is medically noteworthy because it is the reaction inhibited by the antibiotic penicillin. Several penicillin-binding proteins have been identified in bacteria, including FtsI (Figure 5.2*a*). When penicillin is bound to penicillin-binding proteins, the proteins are inactivated. In the absence of transpeptidation in an otherwise growing cell, the continued activity of autolysins (Figure 5.8) so weakens the peptidoglycan that the cell eventually bursts.

MINIQUIZ

- What are autolysins and why are they necessary?
- What is the function of bactoprenol?
- What is transpeptidation and why is it important?

II · Population Growth

Recall that microbial growth is defined as an increase in the *number* of cells in a population. So we now move on from considering the growth and division events in an individual cell to consider the dynamics of growth in bacterial populations.

5.5 Quantitative Aspects of Microbial Growth

During cell division, one cell becomes two. During the time that it takes for this to occur (the generation time), both total cell *number* and *mass* double (Figure 5.1). As we will see, cell numbers in a growing bacterial culture can quickly become very large, and so we turn our attention here to dealing with these large numbers in a quantitative manner.

Plotting Growth Data

A growth experiment beginning with a single cell having a generation time of 30 min is presented in **Figure 5.9**. This pattern of population increase, where the number of cells doubles in a constant time interval, is called **exponential growth**. When the cell number from such an experiment is graphed on arithmetic (linear) coordinates as a function of time, one obtains a curve with a continuously increasing slope (Figure 5.9*b*). By contrast, when the cell number is plotted on a logarithmic ($\log_{10}$) scale as a function of time (a *semilogarithmic* graph), as shown in Figure 5.9*b*, the points fall on a straight line. This straight-line function reflects the fact that the cells are growing exponentially and the population is doubling in a constant time interval.

Time (h)	Total number of cells	Time (h)	Total number of cells
0	1	4	256 (2^8)
0.5	2	4.5	512 (2^9)
1	4	5	1,024 (2^{10})
1.5	8	5.5	2,048 (2^{11})
2	16	6	4,096 (2^{12})
2.5	32	.	.
3	64	.	.
3.5	128	10	1,048,576 (2^{20})

(a)

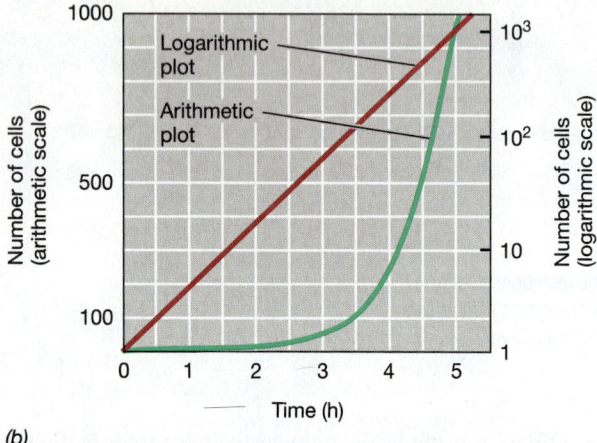

(b)

Figure 5.9 The rate of growth of a microbial culture. *(a)* Data for a population that doubles every 30 min. *(b)* Data plotted on arithmetic (left ordinate) and logarithmic (right ordinate) scales.

Semilogarithmic graphs are also convenient for estimating the generation time of a culture from growth data, since generation times may be inferred directly from the graph as shown in **Figure 5.10**. For example, when two points on the curve that represent one cell doubling on the Y axis are selected and vertical lines drawn from them to intersect the X axis, the time interval measured on the X axis is the generation time (Figure 5.10*b*).

The Mathematics of Growth and Growth Expressions

The increase in cell number in an exponentially growing bacterial culture can be expressed with simple mathematics based on a geometric progression of the number 2. As one cell divides to become two cells, we express this as $2^0 \rightarrow 2^1$. As two cells become four, we express this as $2^1 \rightarrow 2^2$, and so on (Figure 5.9*a*). A fixed relationship exists between the initial number of cells in a culture and the number present after a period of exponential growth, and this relationship can be expressed as

$$N = N_0 2^n$$

where N is the final cell number, N_0 is the initial cell number, and n is the number of generations during the period of exponential growth. The generation time (g) of the exponentially growing population is t/n, where t is the duration of exponential growth expressed in days, hours, or minutes. From a knowledge of the initial and final cell numbers in an exponentially growing cell

population, it is possible to calculate n, and from n and knowledge of t, the generation time, g.

The equation $N = N_0 2^n$ can be expressed in terms of n by taking the logarithms of both sides as follows:

$$N = N_0 2^n$$

$$\log N = \log N_0 + n \log 2$$

$$\log N - \log N_0 = n \log 2$$

$$n = \frac{\log N - \log N_0}{\log 2} = \frac{\log N - \log N_0}{0.301}$$

$$= 3.3(\log N - \log N_0)$$

Using the last expression, it is possible to calculate generation times in terms of measurable quantities, N and N_0. As an example, consider actual growth data from the graph in Figure 5.10*b*, in which $N = 10^8$, $N_0 = 5 \times 10^7$, and $t = 2$:

$$n = 3.3 [\log(10^8) - \log(5 \times 10^7)]$$

$$= 3.3 (8 - 7.69) = 3.3(0.301) = 1$$

Thus, in this example, $g = t/n = 2/1 = 2$ h. If exponential growth continued for another 2 h, the cell number would be 2×10^8. Two hours later the cell number would be 4×10^8, and so on. Besides determining the generation time of an exponentially growing culture by inspection of graphical data (Figure 5.10*b*), g can also be calculated directly from the slope of the straight-line function obtained in a semilogarithmic plot of exponential growth. The slope is equal to 0.301 n/t (or 0.301/g). In the above example, the slope would thus be 0.301/2, or 0.15. Since g is equal to 0.301/slope, we arrive at the same value of 2 for g. The term 0.301/g is called the *specific growth rate*, abbreviated k.

Other useful growth expressions can be calculated from these data. For example, the reciprocal of the generation time, called the *division rate*, is abbreviated v. The division rate is equal to 1/g and has units of reciprocal hours (h^{-1}). That is to say, whereas g is a measure of the *time* it takes for a population to double in cell number, v is a measure of the *number of generations* per unit of time in an exponentially growing culture. The slope of the line relating log cell number to time (Figure 5.10) is equal to v/3.3. Armed with knowledge of n and t, one can calculate g, k, and v for different microorganisms growing under different conditions. This is often useful for optimizing culture conditions for a newly isolated organism and also for testing the positive or negative effect of some treatment on a bacterial culture. For example, comparison with an unamended control allows factors that stimulate or inhibit growth to be identified by measuring their effect on the various growth parameters discussed here.

Consequences of Exponential Growth

During exponential growth, the increase in cell number is initially rather slow but increases at an ever faster rate. In the later stages of exponential growth, this results in an explosive increase in cell numbers. For example, in the experiment shown in Figure 5.9, the rate of cell production in the first 30 min of growth is 1 cell per 30 min. However, between 4 and 4.5 h of growth, the rate of cell production is 256 cells per 30 min, and between 5.5 and

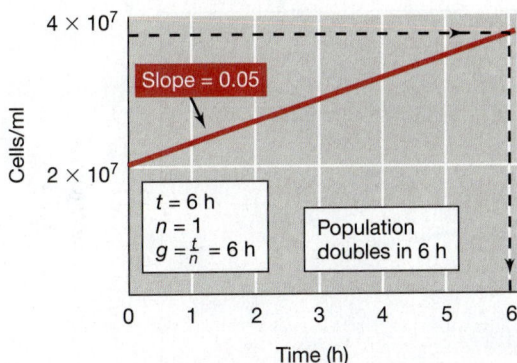

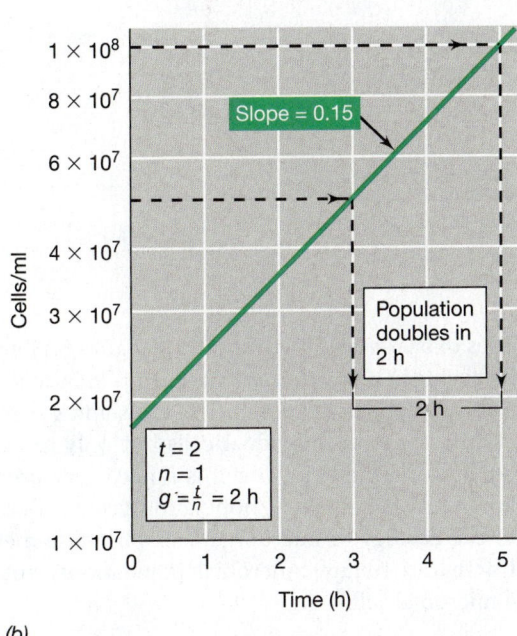

Figure 5.10 Calculating microbial growth parameters. Method of estimating the generation times (*g*) of exponentially growing populations with *g* of *(a)* 6 h and *(b)* 2 h from data plotted on semilogarithmic graphs. The slope of each line is equal to 0.301/*g*, and *n* is the number of generations in the time *t*. All numbers are expressed in scientific notation; that is, 10,000,000 is 1×10^7, 60,000,000 is 6×10^7, and so on.

6 h of growth it is 2048 cells per 30 min (Figure 5.9). Because of this, cell numbers in laboratory cultures of bacteria can quickly become very large, with final population sizes of $> 10^9$ cells/ml not uncommon.

Besides being a theoretical construct, exponential growth can have implications in everyday life. Consider something as mundane as the spoilage of milk. The lactic acid bacteria responsible for the soured flavor of spoiled milk contaminate the milk during its collection and exist in fresh, pasteurized milk in low numbers; these organisms grow slowly at refrigerator temperature (4°C) but much faster at room temperature. If a bottle of fresh milk is left to stand at room temperature overnight, some lactic acid is made, but not enough to affect milk quality. However, if week-old milk, which now contains a week's worth of bacterial growth and thus much higher cell numbers, is left standing under the same conditions, a huge amount of lactic acid is made, and spoilage results.

5.6 The Growth Cycle

The data presented in Figures 5.9 and 5.10 reflect only part of the growth cycle of a microbial population, the part called *exponential growth*. For several reasons, an organism growing in an enclosed vessel, such as a tube or a flask (a **batch culture**), cannot grow exponentially indefinitely. Instead, a typical *growth curve* for the population is obtained, as illustrated in **Figure 5.11**. The growth curve describes an entire growth cycle, and includes lag, exponential, stationary, and death phases.

Lag Phase

When a microbial culture is inoculated into fresh media, growth begins only after a period of time called the *lag phase*. This interval may be brief or extended, depending on the history of the inoculum and nature of the medium and growth conditions. If an exponentially growing culture is transferred into the same medium under the same conditions of growth (temperature, aeration, and the like), there will be essentially no lag and exponential growth begins immediately. However, if the inoculum is taken from an old culture there is usually a lag because the cells are depleted of various essential constituents and time is required for their biosynthesis. A lag is also encountered when the inoculum is of low viability (few live cells) or contains cells that have been damaged but not killed by some stressor, such as high or low temperature, radiation, or toxic chemicals.

A lag is also observed when a microbial culture is transferred from a rich culture medium to a poorer one (a medium downshift); for example, from a complex medium to a defined medium (Section 3.2). To grow in any culture medium the cells must have a complete complement of enzymes for synthesis of the essential metabolites not present in that medium. Hence, upon a medium downshift, essential metabolites must be biosynthesized, and time is needed for synthesis of the new enzymes required and for these to produce a small pool of each metabolite.

Exponential Phase

As we saw in Section 5.5, during exponential growth, the cell population doubles at regular intervals for a brief or extended period, depending on the available resources and other factors. Exponentially growing cells are said to be in the *exponential phase* of growth. Exponential phase cells are typically in their healthiest state and are thus most desirable for studies of their enzymes or other cell components.

Rates of exponential growth vary greatly. The rate of exponential growth is influenced by environmental conditions (temperature, composition of the culture medium), as well as by genetic

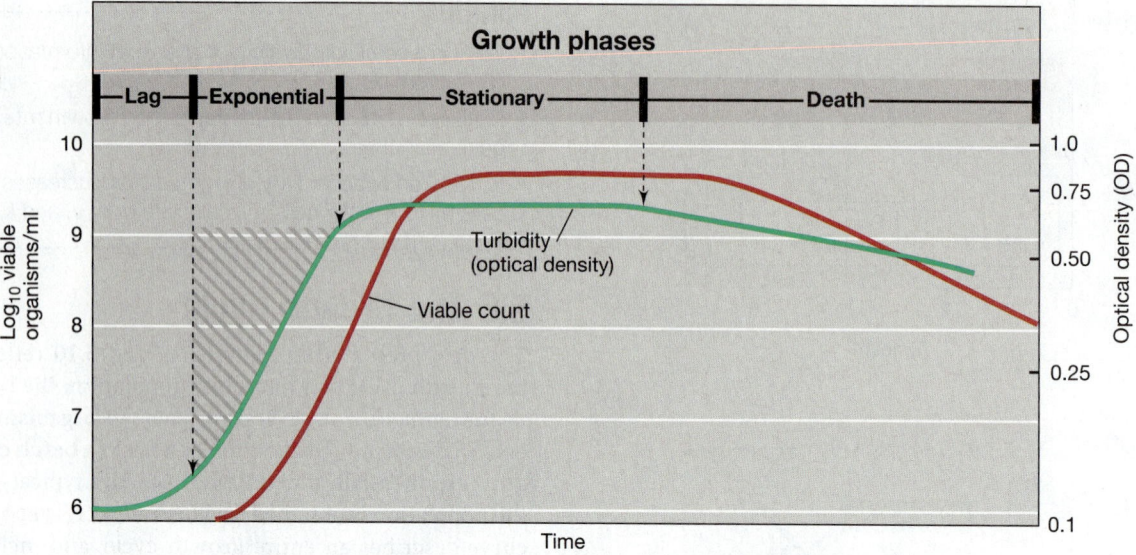

Figure 5.11 Typical growth curve for a bacterial population. A viable count measures the cells in the culture that are capable of reproducing. Optical density (turbidity), a quantitative measure of light scattering by a liquid culture, increases with the increase in cell number.

characteristics of the organism itself. In general, prokaryotes grow faster than eukaryotic microorganisms, and small eukaryotes tend to grow faster than large ones. This should remind us of the previously discussed concept of surface-to-volume ratio. Recall that small cells have an increased capacity for nutrient and waste exchange compared with larger cells, and this metabolic advantage can greatly affect their growth and other properties (Section 2.6).

Stationary and Death Phases

In a batch culture, exponential growth cannot be maintained indefinitely. Consider the fact that a single cell of a bacterium weighing one-trillionth (10^{-12}) of a gram and growing exponentially with a 20-min generation time would produce, if allowed to grow exponentially in batch culture for 48 h, a population of cells that weighed 4000 times the weight of Earth! Obviously this is impossible, and growth becomes limited in such cultures because either an essential nutrient in the culture medium is depleted or the organism's waste products accumulate. When exponential growth ceases for one (or both) of these reasons, the population enters *stationary phase* (Figure 5.11).

In the stationary phase, there is no net increase or decrease in cell number and thus the growth rate of the population is zero. Despite growth arrest, energy metabolism and biosynthetic processes in stationary phase cells may continue, but typically at a greatly reduced rate. Some cells may even divide during stationary phase but no net increase in cell number occurs. This is because some cells in the population grow while others die, the two processes balancing each other out (cryptic growth). Sooner or later, however, the population will enter the *death phase* of the growth cycle, which, like the exponential phase, occurs as an exponential function (Figure 5.11). Typically, however, the rate of cell death is much slower than the rate of exponential growth and viable cells may remain in a culture for months or even years.

The phases of bacterial growth shown in Figure 5.11 are reflections of the events in a *population* of cells, not in individual cells. Thus, the terms lag phase, exponential phase, and so on have no meaning with respect to individual cells but only to cell populations. Growth of an individual cell is a necessary prerequisite for population growth. But it is population growth that is most relevant to the ecology of microorganisms, because measurable microbial activities require microbial populations, not just an individual microbial cell.

MINIQUIZ

- In which phase of the growth curve do cells divide in a constant time period?
- Under what conditions would a lag phase not occur?
- Why do cells enter stationary phase?

5.7 Continuous Culture

Up to this point our discussion of population growth has been confined to batch cultures. The environment in a batch culture is constantly changing because of nutrient consumption and waste production. It is possible to circumvent these changes in a *continuous culture device*. Unlike a batch culture, which is a *closed* system, a continuous culture is an *open* system. In the continuous culture growth vessel a known volume of fresh medium is added at a constant rate while an equal volume of spent culture medium (which also contains cells) is removed at the same rate. Once in equilibrium, the growth vessel volume, cell number, and nutrient/waste product status remain constant, and the culture attains *steady state*.

The Chemostat

The most common type of continuous culture is the **chemostat**, a device wherein both growth rate (how *fast* the cells divide) and

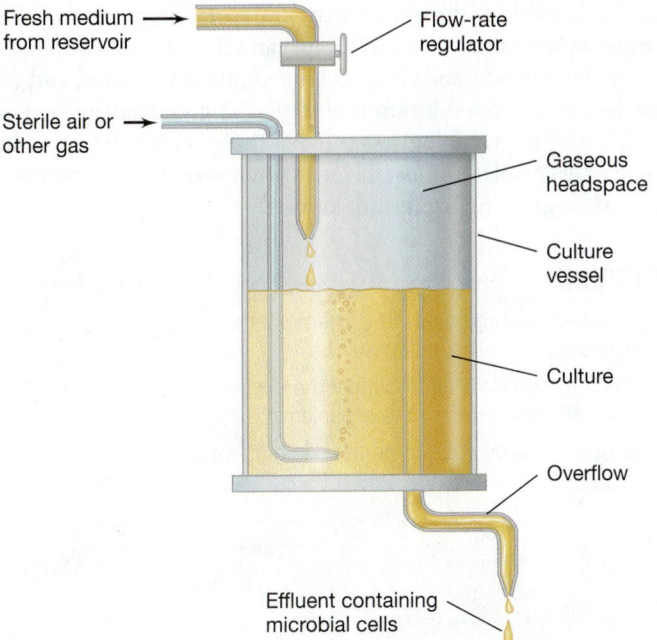

Figure 5.12 **Schematic for a continuous culture device (chemostat).** The population density is controlled by the concentration of limiting nutrient in the reservoir, and the growth rate is controlled by the flow rate. Both parameters can be set by the experimenter.

cell density (how *many* cells per ml are obtained) can be controlled independently (**Figure 5.12**). Two factors govern growth rate and cell density, respectively: (1) the *dilution rate*, which is the rate at which fresh medium is pumped in and spent medium is removed; and (2) the *concentration of a limiting nutrient*, such as a carbon or nitrogen source, present in the sterile medium entering the chemostat vessel.

In a batch culture, the nutrient concentration affects both growth rate and growth yield (**Figure 5.13**). At very low concentrations of a given nutrient, the growth rate is submaximal because

the nutrient cannot be transported into the cell fast enough to satisfy metabolic demand. At higher nutrient levels, the maximal growth rate may be obtained but the cell density can continue to increase in proportion to the concentration of nutrients in the medium (Figure 5.13). In a chemostat, by contrast, growth rate and growth yield are controlled independently: the growth rate by the dilution rate and the cell yield by the concentration of a limiting nutrient.

Varying Chemostat Parameters

The effects on bacterial growth of varying the dilution rate and concentration of growth-limiting nutrient in a chemostat are shown in **Figure 5.14**. As seen, there are rather wide limits over which the dilution rate controls growth rate, although at both very low and very high dilution rates, the steady state breaks down. At too *high* a dilution rate, the organism cannot grow fast enough to keep up with its dilution and is washed out of the chemostat. By contrast, at too *low* a dilution rate, cells may die from starvation because the limiting nutrient is not being added fast enough to support minimal cell metabolism. However, between these limits, different growth rates can be achieved by simply varying the dilution rate.

Cell density in a chemostat is controlled by a limiting nutrient, just as it is in a batch culture (Figure 5.13). If the concentration of this nutrient in the incoming medium is increased at a constant dilution rate, cell density will increase but growth rate will remain the same. Thus, by varying the chemostat dilution rate and nutrient level, one can establish dilute (for example, 10^5 cells/ml), moderate (for example, 10^7 cells/ml), or dense (for example, 10^9 cells/ml) cell populations growing at any specific growth rate.

Experimental Uses of the Chemostat

A practical advantage to the chemostat is that a cell population can be maintained in the exponential growth phase for long

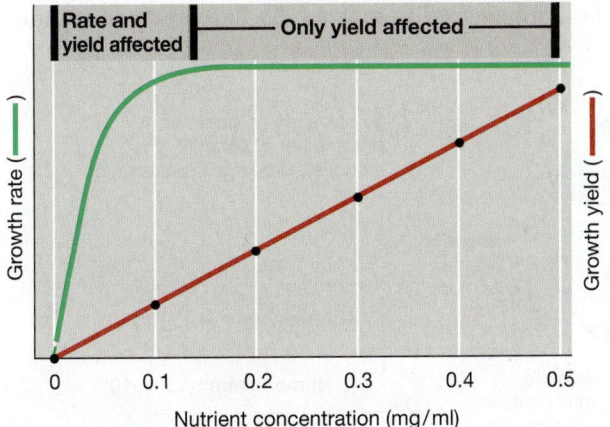

Figure 5.13 **The effect of nutrients on growth.** Relationship between nutrient concentration, growth rate (green curve), and growth yield (red curve) in a batch culture (closed system). Only at low nutrient concentrations are both growth rate and growth yield affected.

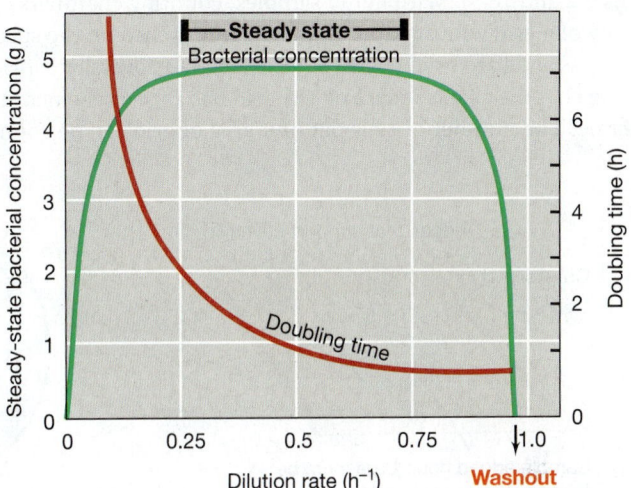

Figure 5.14 **Steady-state relationships in the chemostat.** The dilution rate is determined from the flow rate and the volume of the culture vessel. Thus, with a vessel of 1000 ml and a flow rate through the vessel of 500 ml/h, the dilution rate would be 0.5 h^{-1}. Note that at high dilution rates, growth cannot balance dilution, and the population washes out. Note also that although the population density remains constant during steady state, the growth rate (doubling time) can vary over a wide range.

periods, days or even weeks. Exponential phase cells are usually most desirable for physiological experiments, and such cells can be available at any time when grown in a chemostat. Moreover, repetition of experiments can be done with the knowledge that each time the cell population will be as close to being the same as possible. After a sample is removed from the chemostat, a period of time is required for the vessel to return to its original volume and for steady state to be reachieved. Once this has occurred, the vessel can be sampled again.

The chemostat has been used in microbial ecology as well as in microbial physiology. For example, because the chemostat can mimic the low substrate concentrations that are often found in nature, it is possible to ask which organisms in mixed cultures of known composition can best survive nutrient limitations. This can be done by monitoring changes in the microbial community as a function of varying nutrient conditions. Chemostats have also been used for the enrichment and isolation of bacteria from nature. From a natural sample, one can select a stable population under the nutrient and dilution-rate conditions chosen and then slowly increase the dilution rate until a single organism remains. In this way, microbiologists studying the growth rates of various soil bacteria isolated a bacterium with a 6-min doubling time—the fastest-growing bacterium known!

MINIQUIZ

- How do microorganisms in a chemostat differ from microorganisms in a batch culture?
- What happens in a chemostat if the dilution rate exceeds the maximal growth rate of the organism?
- Do pure cultures have to be used in a chemostat?

III • Measuring Microbial Growth

Population growth is measured by monitoring changes in cell number or changes in the level of some cellular component as a proxy for cell number. These include protein, nucleic acids, or the dry weight of the cells themselves. We consider here two common measures of cell growth: cell counts and turbidity, the latter of which is a function of cell mass.

5.8 Microscopic Counts

A total count of microbial numbers in a culture or natural sample can be done by simply observing and enumerating the cells present. The most common total count method is the *microscopic cell count*. Microscopic counts can be performed either on samples dried on slides or on liquid samples. Dried samples can be stained to increase contrast between cells and their background (⊃ Sections 2.2 and 18.3). With liquid samples, counting chambers consisting of a grid with squares of known area etched on the surface of a glass slide are used (**Figure 5.15**). When the coverslip is placed on the chamber, each square on the grid has a precisely measured volume. The number of cells per unit area of grid can be counted under the microscope, giving a measure of the number of cells per small chamber volume. The number of cells per milliliter of suspension is calculated by employing a conversion factor based on the volume of the chamber sample (Figure 5.15).

Cells in liquid samples can also be counted in a flow cytometer. This is a machine that employs a laser beam and complex electronics to count individual cells. Flow cytometry is rarely used for the routine counting of microbial cells, but has applications in the medical field for counting and differentiating blood cells and other cell types from clinical samples. It has also been used in microbial ecology to separate different types of cells for isolation purposes (⊃ Section 18.10).

Caveats of Microscopic Counting

Microscopic counting is a quick and easy way of estimating microbial cell numbers. However, it has several limitations that restrict its usefulness to rather specific applications. For example, without special staining techniques (⊃ Section 18.3), dead cells cannot be distinguished from live cells, and precision is difficult to

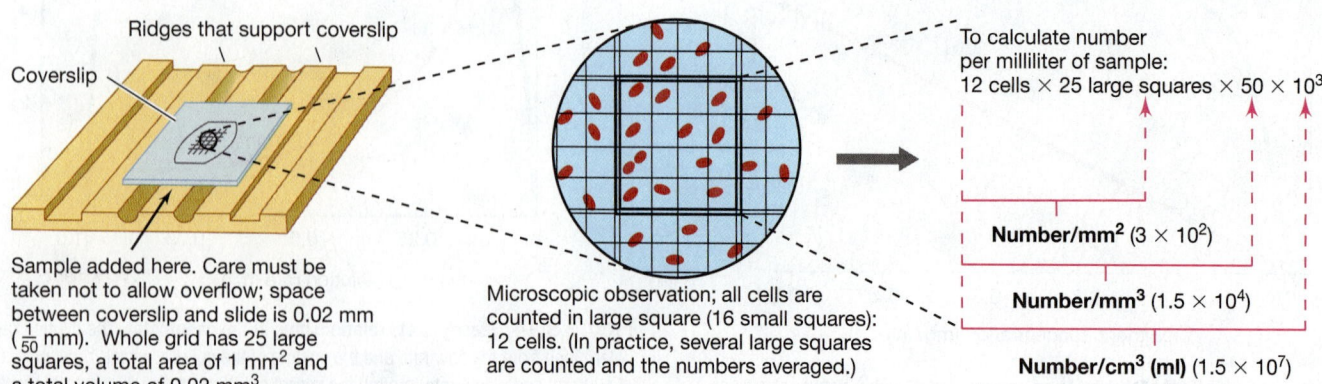

Ridges that support coverslip

Coverslip

Sample added here. Care must be taken not to allow overflow; space between coverslip and slide is 0.02 mm ($\frac{1}{50}$ mm). Whole grid has 25 large squares, a total area of 1 mm^2 and a total volume of 0.02 mm^3.

Microscopic observation; all cells are counted in large square (16 small squares): 12 cells. (In practice, several large squares are counted and the numbers averaged.)

To calculate number per milliliter of sample:
12 cells × 25 large squares × 50 × 10^3

Number/mm^2 (3 × 10^2)

Number/mm^3 (1.5 × 10^4)

Number/cm^3 (ml) (1.5 × 10^7)

Figure 5.15 **Direct microscopic counting procedure using the Petroff–Hausser counting chamber.** A phase-contrast microscope is typically used to count the cells to avoid the necessity for staining.

achieve, even if replicate counts are made. Moreover, small cells are often difficult to see under the microscope, which can lead to erroneous counts, and cell suspensions of low density (less than about 10^6 cells/milliliter) will have few if any cells in a microscope field unless the sample is first concentrated and resuspended in a small volume. Finally, motile cells must be killed or otherwise immobilized before counting, and debris in the sample may easily be mistaken for microbial cells.

Microscopic Cell Counts in Microbial Ecology

Despite its many potential caveats, microbial ecologists often use microscopic cell counts on natural samples. But they do so using stains to visualize the cells, often very powerful stains that yield phylogenetic or other key information about the cells, such as their metabolic properties.

The stain DAPI (⟳ Section 2.2 and Figure 2.6*c*) stains all cells in a sample because it reacts with DNA. By contrast, fluorescent stains that are highly specific for certain organisms or groups of related organisms can be prepared by attaching the fluorescent dyes to specific nucleic acid probes. For example, phylogenetic stains that stain only species of *Bacteria* or only species of *Archaea* can be used in combination with nonspecific stains to determine the proportion of each domain present in a given sample; the use of these stains will be discussed in Section 18.4. Other fluorescent probes target genes that encode enzymes linked to specific metabolic processes; if a cell is stained by one of these probes, a key metabolic property can be inferred that may reveal the cell's ecological role in the microbial community. In all of these cases, if cells in the sample are present in only low numbers, for example, in a sample of ocean water, this limitation can be overcome by first concentrating the cells on a filter and then counting them after staining.

Because they are easy to do and often yield very useful information, microscopic cell counts are very common in ecological studies of natural microbial environments. We pursue this theme in more detail in Chapter 18.

MINIQUIZ -
- What are some of the problems that can arise when unstained preparations are enumerated in microscopic counts?
- Using microscopic techniques, how could you tell whether *Archaea* were present in an alpine lake where total cell numbers were only 10^5/ml?

5.9 Viable Counts

A **viable** cell is one that is able to divide and form offspring, and in most cell-counting situations, these are the cells we are most interested in. For these purposes, one would use a **viable count**, also called a **plate count** because agar plates are required. The assumption made in a viable count is that each viable cell will grow and divide to yield one colony, and hence, colony numbers are a reflection of cell numbers.

Methods for Viable Counts

There are at least two ways of performing a plate count: the *spread-plate* method and the *pour-plate* method (**Figure 5.16**). In the spread-plate method, a volume (usually 0.1 ml or less) of an appropriately diluted culture is spread over the surface of an agar

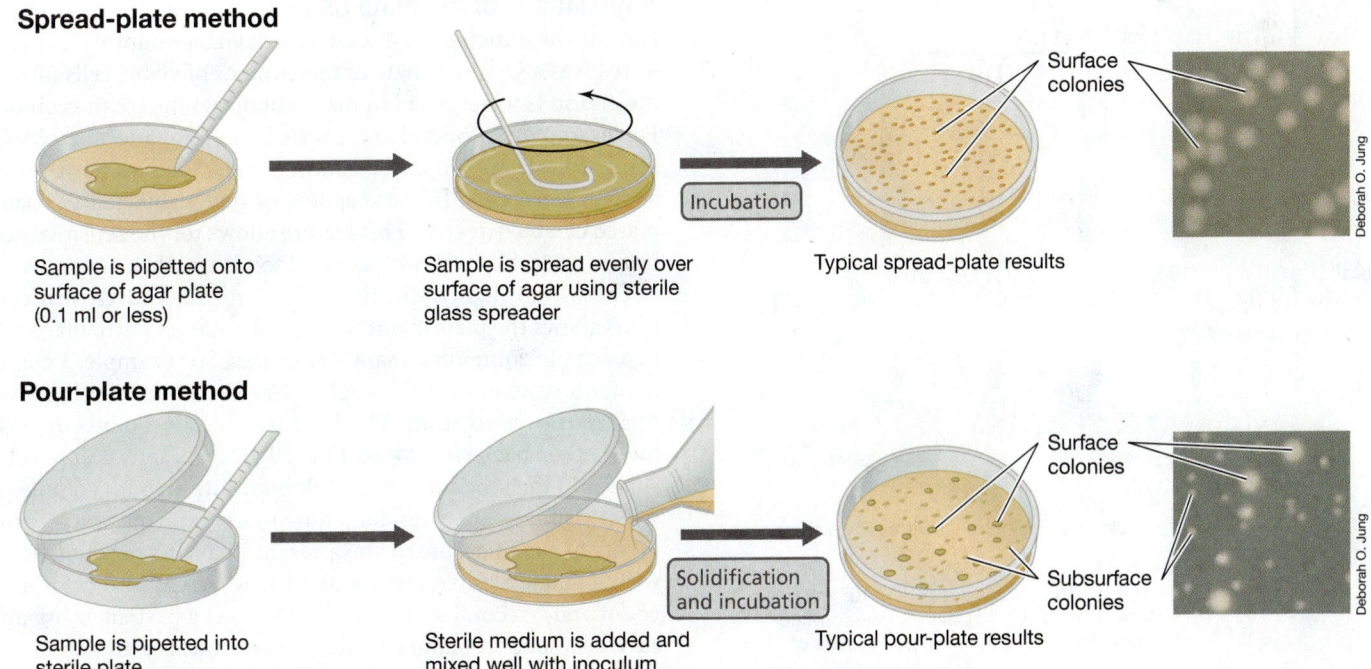

Spread-plate method

Sample is pipetted onto surface of agar plate (0.1 ml or less) → Sample is spread evenly over surface of agar using sterile glass spreader → Incubation → Typical spread-plate results → Surface colonies

Pour-plate method

Sample is pipetted into sterile plate → Sterile medium is added and mixed well with inoculum → Solidification and incubation → Typical pour-plate results → Surface colonies / Subsurface colonies

Figure 5.16 Two methods for the viable count. In the pour-plate method, colonies form within the agar as well as on the agar surface. On the far right are photos of colonies of *Escherichia coli* formed from cells plated by the spread-plate method (top) or the pour-plate method (bottom).

plate using a sterile glass spreader. In the pour-plate method, a known volume (usually 0.1–1.0 ml) of culture is pipetted into a sterile Petri plate. Molten agar medium, tempered to just above gelling temperature, is then added and mixed by gently swirling the plate on the benchtop. With both the spread-plate and pour-plate methods, it is important that the number of colonies developing on or in the medium not be too many or too few. On crowded plates some cells may not form colonies, and some colonies may fuse, leading to erroneous measurements. If the number of colonies is too small, the statistical significance of the calculated count will be low. The usual practice, which is most valid statistically, is to count colonies only on plates that have between 30 and 300 colonies.

To obtain the appropriate colony number, the sample to be counted must almost always be diluted. Because one may not know the approximate viable count ahead of time, it is usually necessary to make more than one dilution. Several 10-fold dilutions of the sample are commonly used (Figure 5.17). To make a 10-fold (10^{-1}) dilution, one can mix 0.5 ml of sample with 4.5 ml of diluent, or 1.0 ml of sample with 9.0 ml of diluent. If a 100-fold (10^{-2}) dilution is needed, 0.05 ml can be mixed with 4.95 ml of diluent, or 0.1 ml with 9.9 ml of diluent. Alternatively, a 10^{-2} dilution can be achieved by making two successive 10-fold dilutions. With dense cultures, such *serial* dilutions are needed to reach a suitable dilution for plating to yield countable colonies. Thus, if a 10^{-6} ($1/10^6$) dilution is needed, it can be achieved by making three successive 10^{-2} ($1/10^2$) dilutions or six successive 10^{-1} dilutions (Figure 5.17).

Sources of Error in Plate Counting

The number of colonies obtained in a viable count experiment depends not only on the inoculum size and the viability of the culture, but also on the culture medium and the incubation conditions. The colony number can also change with the length of incubation. For example, if a mixed culture is counted, the cells deposited on the plate will not all form colonies at the same rate; if a short incubation time is used, fewer than the maximum number of colonies will be obtained. Furthermore, the size of colonies may vary. If some tiny colonies develop, they may be missed during the counting. With pure cultures, colony development is a more synchronous process and uniform colony morphology is the norm.

Viable counts can be subject to rather large errors for several reasons. These include plating inconsistencies, such as inaccurate pipetting of a liquid sample, a nonuniform sample (for example, a sample containing cell clumps), insufficient mixing, heat intolerance (if pour plates are used), and many other factors. Hence, if accurate counts are to be obtained, great care and consistency must be taken in sample preparation and plating, and replicate plates of key dilutions must be prepared. Also, if two or more cells are in a clump, they will grow to form only a single colony. So if a sample contains many cell clumps, a viable count of that sample may be erroneously low. Data from such samples are often expressed as the number of *colony-forming units* obtained rather than the actual number of viable cells, because a colony-forming unit may contain one or more cells.

Applications of the Plate Count

Despite the difficulties associated with viable counting, the procedure gives a good estimate of the number of viable cells in a sample and so is widely used in many subdisciplines of microbiology. For example, in food, dairy, medical, and aquatic microbiology, viable counts are employed routinely. The method has the virtue of high sensitivity, because as few as one viable cell per sample plated can be detected. This feature allows for the sensitive detection of microbial contamination of foods or other materials.

The use of highly selective culture media and growth conditions allows the plate count to be used to target particular species in a sample containing many organisms. For example, a complex medium containing 10% NaCl is very useful in isolating species of *Staphylococcus* from skin, because the salt inhibits growth of most other bacteria (⮌ Section 29.9). In practical applications such as in the food industry, viable counting on both complex and selective media allows for both quantitative and qualitative assessments of the microorganisms present in a food product. That is, with a single sample one medium may be employed for a total count and a second medium used to target a particular organism, such as a specific pathogen. Targeted counting is also common in wastewater and other water analyses. For instance, enteric bacteria such as *Escherichia coli* originate from feces and are easy to target using selective media; if enteric bacteria are detected in a water sample from a swimming site, for example, their presence is a signal that the water is unsafe for human contact.

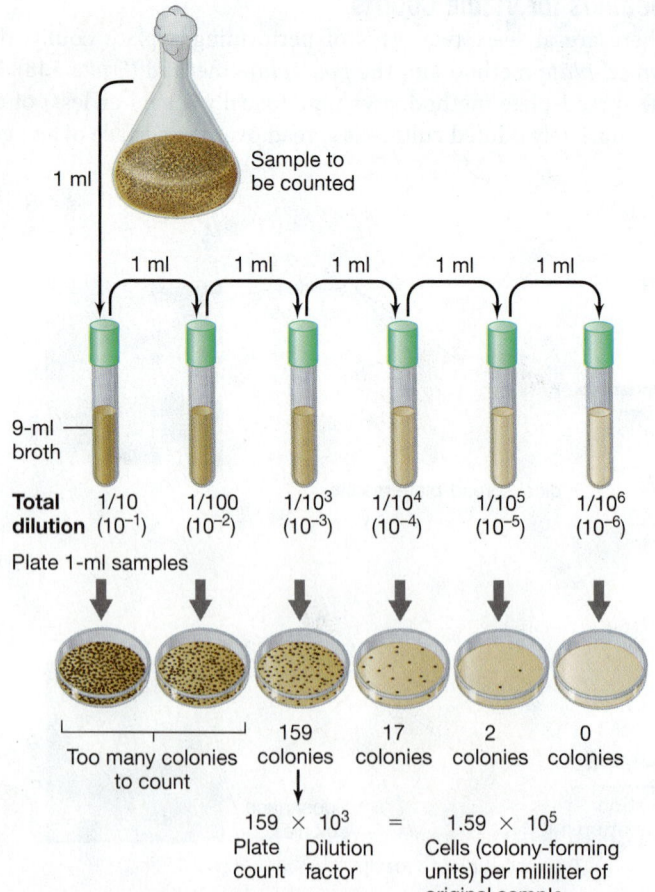

Figure 5.17 Procedure for viable counting using serial dilutions of the sample and the pour-plate method. The sterile liquid used for making dilutions can simply be water, but a solution of mineral salts or actual growth medium may yield a higher recovery. The dilution factor is the reciprocal of the dilution.

The Great Plate Count Anomaly

Direct microscopic counts of natural samples typically reveal far more organisms than are recoverable on plates of any single culture medium. Thus, although a very sensitive technique, plate counts can be highly unreliable when used to assess total cell numbers of natural samples, such as soil and water. Some microbiologists have referred to this as "the great plate count anomaly."

Why do plate counts show lower numbers of cells than direct microscopic counts? One obvious factor is that microscopic methods count dead cells, whereas by definition, viable methods do not. More important, however, is the fact that different organisms, even those present in a very small natural sample, may have vastly different requirements for nutrients and growth conditions in laboratory culture (ↄ Sections 3.1 and 3.2). Thus, one medium and set of growth conditions can only be expected to support the growth of one subset of the total microbial community. If this subset makes up, for example, 10^6 cells/g of a total viable community of 10^9 cells/g, the plate count will reveal only 0.1% of the viable cell population, a vast underestimation of the actual number and physiological types of organisms present in the sample.

Plate count results thus carry a large caveat. Targeted plate counts using highly selective media, as in, for example, the microbial analysis of sewage or food, can often yield quite reliable data, since the physiology of the targeted organisms are known. By contrast, "total" cell counts of the same samples using a single medium and set of growth conditions may be, and usually are, underestimates of actual cell numbers by one to several orders of magnitude.

MINIQUIZ

- Why is a viable count more sensitive than a microscopic count? What major assumption is made in relating plate count results to cell number?
- Describe how you would dilute a bacterial culture by 10^{-7}.
- Explain the "great plate count anomaly."

5.10 Spectrophotometry

During exponential growth, all cellular components increase in proportion to the increase in cell numbers. One such component is cell mass itself. Cells scatter light, and a rapid and useful method of estimating cell mass is *turbidity*. A cell suspension looks cloudy (turbid) to the eye because cells scatter light that passes through the suspension. The more cells that are present, the more light is scattered, and hence the more turbid the suspension. Because cell mass is proportional to cell number, turbidity can be used to estimate cell numbers and is a widely used technique in microbiology.

Optical Density

Turbidity is measured with a spectrophotometer, an instrument that passes light through a cell suspension and measures the unscattered light that emerges (**Figure 5.18**). A spectrophotometer employs a prism or diffraction grating to generate incident light of a specific wavelength (Figure 5.18*a*). Commonly used wavelengths for bacterial turbidity measurements include 480 nm

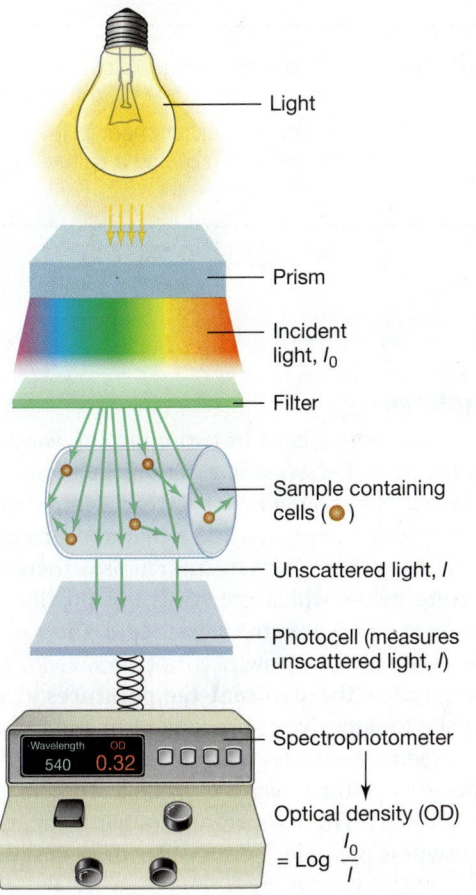

(a)

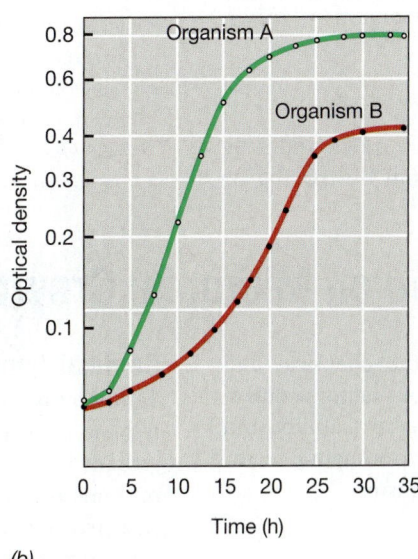

(b)

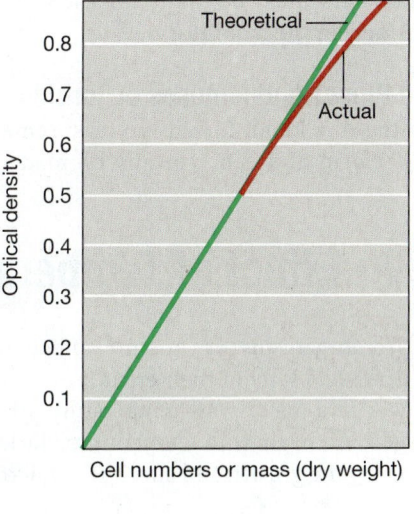

(c)

Figure 5.18 Turbidity measurements of microbial growth. *(a)* Measurements of turbidity are made in a spectrophotometer. The photocell measures incident light unscattered by cells in suspension and gives readings in optical density units. *(b)* Typical growth curve data for two organisms growing at different growth rates. For practice, calculate the generation time of the two cultures using the formula $n = 3.3(\log N - \log N_0)$ where N and N_0 are two different OD readings with a time interval t between the two (Section 5.5). Which organism is growing faster, A or B? *(c)* Relationship between cell number or dry weight and turbidity readings. Note that the one-to-one correspondence between these relationships breaks down at high turbidities.

(blue), 540 nm (green), 600 nm (orange), and 660 nm (red). Sensitivity is best at shorter wavelengths, but measurements of dense cell suspensions are more accurate at longer wavelengths. The unit of turbidity is *optical density* (*OD*) at the wavelength specified, for example, OD_{540} for measurements at 540 nm (Figure 5.18). The term *absorbance* (A), for example A_{540}, is also a commonly used unit, but it should be understood that it is the *scattering* of light, not the *absorbance* of light, that is actually being assessed in the spectrophotometer.

Relating Optical Density to Cell Numbers

For unicellular organisms, optical density is proportional, within certain limits, to cell number. Turbidity readings can therefore be used as a substitute for total or viable counting methods. However, before this can be done, a standard curve must be prepared that relates cell number (microscopic or viable count), dry weight, or protein content to turbidity. As can be seen in such a plot, proportionality only holds within limits (Figure 5.18c). At high cell densities, light scattered away from the spectrophotometer's photocell by one cell can be scattered back toward the photocell by another. To the photocell, this is as if light had never been scattered in the first place. At such high cell densities, the one-to-one correspondence between cell number and turbidity deviates from linearity, and OD measurements become less accurate. However, up to this limit, turbidity measurements can be highly accurate measures of cell number or dry weight. Also, because different organisms differ in size and shape, equal cell numbers of two different bacterial species will not necessarily yield the same OD value. Thus, to relate OD to actual cell numbers, a standard curve relating these two parameters must be made for each different organism grown routinely in the laboratory.

The Pluses and Minuses of Turbidimetric Growth

On the one hand, turbidity measurements are quick and easy to perform and can typically be made without destroying or significantly disturbing the sample. For these reasons, turbidity measurements are widely employed to monitor growth of pure cultures of *Bacteria*, *Archaea*, and many microbial eukaryotes. With turbidimetric assays, the same sample can be checked repeatedly and the measurements plotted on a semilogarithmic plot versus time (Section 5.5). From these, it is easy to calculate the generation time and other parameters of the growing culture (Figure 5.18b).

On the other hand, turbidity measurements can sometimes be problematic. Although many microorganisms grow in even suspensions in liquid medium, many do not. Some bacteria routinely form small to large clumps, and in such instances, OD measurements may be quite inaccurate as a measure of total microbial mass. In addition, many bacteria grow in films on the sides of tubes or other growth vessels, mimicking in laboratory culture a common form of growth in nature (see Explore the Microbial World, "Stick or Swim"). Hence for OD measurements to be an accurate reflection of cell mass (and thus cell numbers) in a liquid culture, clumping and biofilms have to be minimized. This can often be accomplished by stirring, shaking, or in some way keeping the cells well mixed during the growth process to prevent the formation of cell aggregates and the sticking of swimming cells to surfaces, the first step in biofilm formation. Some bacteria are just naturally planktonic—staying well suspended in liquid medium for long periods—and do not form biofilms. But if a solid surface is available, most motile bacteria will eventually develop a static biofilm, and accurately quantifying cell numbers by turbidity in these can be difficult or even impossible.

MINIQUIZ -

- List two advantages of using turbidity as a measure of cell growth.
- Describe how you could use a turbidity measurement to tell how many colonies you would expect from plating a culture of a given OD.

IV • Effect of Temperature on Microbial Growth

Microorganisms are greatly affected by the chemical and physical state of their environment, and four factors control growth in a major way: temperature, pH, water availability, and oxygen. We begin with temperature, the key environmental factor affecting the growth and survival of microorganisms.

5.11 Temperature Classes of Microorganisms

At either too cold or too hot a temperature, microorganisms will not be able to grow and may even die. The minimum and maximum temperatures supporting growth vary greatly among different organisms and usually reflect the temperature range and average temperature of the environments the organisms inhabit.

Cardinal Temperatures

Temperature affects microorganisms in two opposing ways. As temperatures rise, the rate of enzymatic reactions increases and growth becomes faster. However, above a certain temperature, proteins or other cell components may be denatured or otherwise irreversibly damaged. For every microorganism there is a *minimum* temperature below which growth is not possible, an *optimum* temperature at which growth is most rapid, and a *maximum* temperature above which growth is not possible. These three temperatures, called the **cardinal temperatures** (Figure 5.19), are characteristic for any given microorganism and can differ dramatically between species. For example, some organisms have growth temperature optima near 0°C while for others, this can be higher than 100°C. The temperature range throughout which microbial growth is possible is even wider than this, from

EXPLORE THE
MICROBIAL WORLD

In this chapter we have discussed several ways in which microbial growth can be measured, including microscopic methods, viable counts, and measurements of light scattering (turbidity) by cells suspended in a liquid culture. The turbidimetric measures of bacterial growth assume that cells remain evenly distributed in their liquid growth medium. Under these conditions, the optical density of a culture is proportional to the log of the number of cells in suspension (**Figure 1**). This floating lifestyle, called *planktonic,* is the way some bacteria, for example, organisms that inhabit the water column of a lake, actually live in nature. However, many other microorganisms are *sessile,* meaning that they grow attached to a surface. These attached cells can then develop into **biofilms**.

Humans encounter bacterial biofilms on a daily basis, for example, when cleaning out a pet's water bowl that has been sitting unattended for a few days or when you sense with your tongue the "film" that develops on your unbrushed teeth.

A biofilm is an attached polysaccharide matrix containing embedded bacterial cells. Biofilms form in stages: (1) reversible attachment of planktonic cells, (2) irreversible attachment of the same cells, (3) cell growth and production of polysaccharide, and (4) further development to form the tenacious and nearly impenetrable mature biofilm. In the early stages of biofilm formation, the attachment of bacterial cells to a surface triggers biofilm-specific gene expression. Genes that encode proteins that produce cell surface polysaccharides are transcribed, and the increased amount of slime facilitates attachment of more cells.

The motility structures of swimming bacteria—flagella—are necessary for initially establishing the biofilm state. Thin hairlike structures called type IV pili (⮂ Section 2.13), which resemble flagella but do not rotate like flagella do, are crucial for biofilm maturation. Eventually, through growth and recruitment, entire microbial communities develop within the slimy polysaccharide matrix.

Bacterial biofilms can dramatically affect humans. For example, bacterial infections are often linked to pathogens that develop in biofilms during the disease process. The genetic disease cystic fibrosis (CF) is characterized by development of a biofilm containing *Pseudomonas aeruginosa* and other bacteria in the lungs of CF patients (**Figure 2**). The biofilm matrix, which contains alginate and other polysaccharides as well as bacterial DNA, greatly reduces the ability of antimicrobial agents, such as antibiotics, to penetrate, and thus bacteria within the biofilm are little affected by the drugs. Bacterial biofilms have also been implicated in difficult-to-treat infections of implanted medical devices, such as replacement heart valves and artificial joints.

Biofilms are also a major problem in industry. Microbial biofilms can cause fouling of equipment and the contamination of products, especially if the liquid is nutrient-rich, such as milk. Biofilms can also do long-term damage to water distribution facilities and other public utilities (⮂ Sections 21.10 and 21.11). Biofilms that develop in bulk storage containers, such as fuel storage tanks, can contaminate the fuel and cause souring from chemicals, such as hydrogen sulfide (H_2S), excreted by the biofilm bacteria.

Biofilms are a common form of bacterial growth in nature. Not only does the biofilm offer protection from harmful chemicals, the thick matrix of the biofilm provides a barrier to grazing by protists and prevents bacterial cells from being washed away into a less-favorable habitat. So, while optical densities give us a laboratory picture of the perfectly suspended bacterial culture, in the "real" world bacterial growth in the biofilm state is often observed.

We examine biofilms in more detail in our focus on surfaces as microbial habitats in Sections 19.4 and 19.5.

Deborah O. Jung

OD_{540} 0   0.18   0.45   0.68

Figure 1 Liquid cultures of *Escherichia coli.* In these cultures cells are in a planktonic state and are evenly suspended in the medium. The increasing (left to right) optical density (OD_{540}) of each culture is shown below the tube. Optical density is a measure of light scattering and was measured at 540 nm here as described in Figure 5.18a. Although it is shown growing in suspension here, *E. coli* can also form biofilms. Attachment of *E. coli* cells is facilitated by their Type I fimbriae and conjugative pili (⮂ Section 2.13).

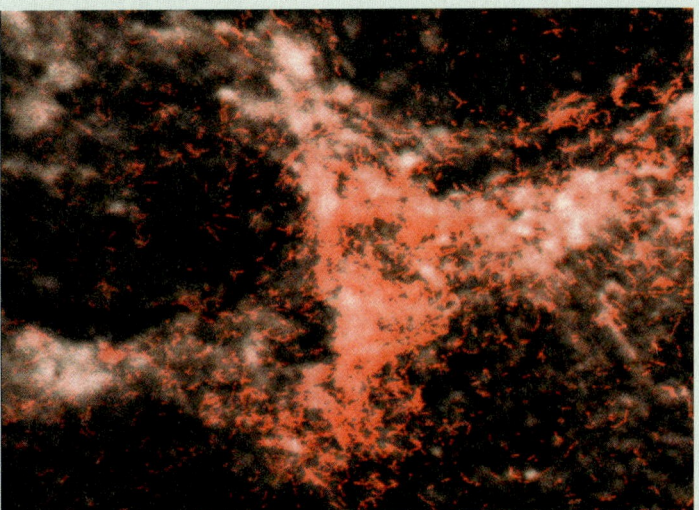

Søren Molin

Figure 2 Fluorescently stained cells of *Pseudomonas aeruginosa.* The cells were from a sputum sample of a cystic fibrosis patient. The red cells are *P. aeruginosa* and the white material is alginate, a polysaccharide-like material that is produced by cells of *P. aeruginosa.*

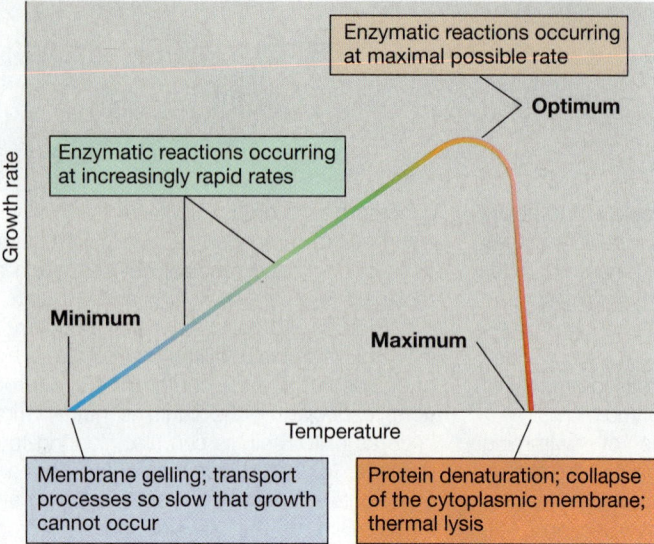

Figure 5.19 **The cardinal temperatures: minimum, optimum, and maximum.** The actual values may vary greatly for different organisms (see Figure 5.20).

as low as −15°C to at least 122°C. However, no single organism can grow over this whole temperature range, as the range for any given organism is typically less than 40°C.

The maximum growth temperature of an organism reflects the temperature above which denaturation of one or more essential cell components, such as a key enzyme, occurs. The factors controlling an organism's minimum growth temperature are not as clear. However, the cytoplasmic membrane must remain in a semifluid state for nutrient transport and bioenergetic functions to take place. That is, if an organism's cytoplasmic membrane stiffens to the point that it no longer functions properly in

transport or can no longer develop or consume a proton motive force, the organism cannot grow. In contrast to the minimum and maximum, the growth temperature *optimum* reflects a state in which all or most cellular components are functioning at their maximum rate and typically lies closer to the maximum than to the minimum (see Figure 5.20).

Temperature Classes of Organisms

Although there is a continuum of organisms, from those with very low temperature optima to those with high temperature optima, it is possible to distinguish four broad classes of microorganisms in relation to their growth temperature optima: **psychrophiles**, with low temperature optima; **mesophiles**, with midrange temperature optima; **thermophiles**, with high temperature optima; and **hyperthermophiles**, with very high temperature optima (**Figure 5.20**).

Mesophiles are widespread in nature and are the most commonly studied microorganisms. Mesophiles are found in warm-blooded animals and in terrestrial and aquatic environments in temperate and tropical latitudes. Psychrophiles and thermophiles are found in unusually cold and unusually hot environments, respectively. Hyperthermophiles are found in extremely hot habitats such as hot springs, geysers, and deep-sea hydrothermal vents.

Escherichia coli is a typical mesophile, and its cardinal temperatures have been precisely defined. The optimum temperature for most strains of *E. coli* is near 39°C, the maximum is 48°C, and the minimum is 8°C. Thus, the temperature *range* for *E. coli* is about 40 degrees, near the high end for prokaryotes (Figure 5.20).

We now consider the interesting cases of microorganisms that inhabit very low or very high temperature environments. We examine some of the physiological problems they face and some of the biochemical solutions they have evolved to survive under these extreme conditions.

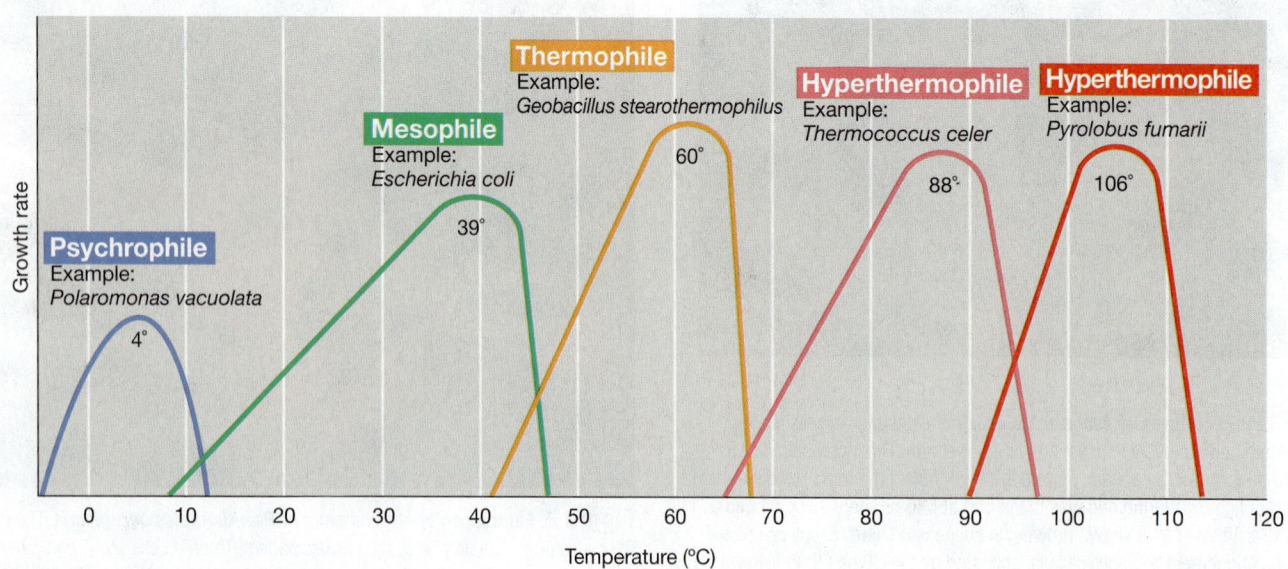

Figure 5.20 **Temperature and growth response in different temperature classes of microorganisms.** The temperature optimum of each example organism is shown on the graph.

5.12 Microbial Life in the Cold

Because humans live and work on the surface of Earth where temperatures are generally moderate, it is natural to consider very hot and very cold environments as "extreme." However, many microbial habitats are very hot or very cold, and organisms that inhabit these environments are called *extremophiles* (⮌ Section 1.4 and Table 1.1). We consider the biology of these fascinating organisms here and in the next section.

Cold Environments

Much of Earth's surface is cold. The oceans, which make up over half of Earth's surface, have an average temperature of 5°C, and the depths of the open oceans have constant temperatures of 1–3°C. Vast land areas of the Arctic and Antarctic are permanently frozen or are unfrozen for only a few weeks in summer (**Figure 5.21**). These cold environments support diverse microbial life, as do glaciers where the network of liquid water channels that run through and under the glacier are teeming with microorganisms. Even in solidly frozen materials there remain small pockets of liquid water where solutes have concentrated and microorganisms can metabolize and grow slowly (⮌ Chapter 1, page 1).

In considering cold environments, it is important to distinguish between environments that are *constantly* cold and those that are only *seasonally* cold. The latter, characteristic of temperate climates, may have summer temperatures as high as 40°C. A temperate lake, for example, may have ice cover in the winter,

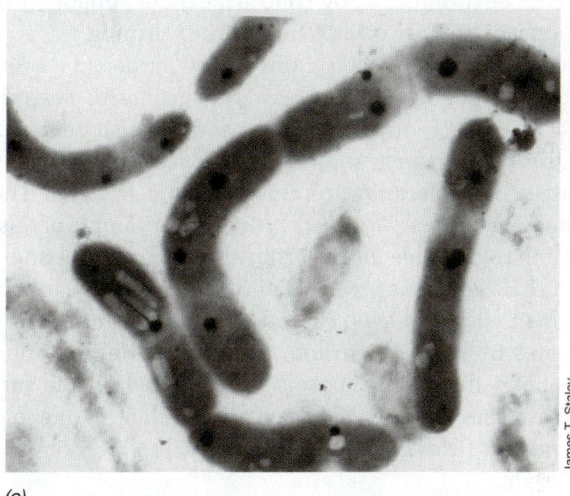

(a)

(b)

(c)

(d)

Figure 5.21 Antarctic microbial habitats and microorganisms. *(a)* A core of frozen seawater from McMurdo Sound, Antarctica. The core is about 8 cm wide. Note the dense coloration due to pigmented microorganisms. *(b)* Phase-contrast micrograph of phototrophic microorganisms from the core shown in part a. Most organisms are either diatoms or green algae (both eukaryotic phototrophs). *(c)* Transmission electron micrograph of *Polaromonas*, a gas vesiculate bacterium that lives in sea ice and grows optimally at 4°C. *(d)* Lake Bonney, McMurdo Dry Valleys, Antarctica. Although the lake is permanently ice-covered, the water column under the ice remains near 0°C and contains a diverse array of prokaryotes and microbial eukaryotes.

but the time that the water remains at 0°C is relatively brief. By contrast, Antarctic lakes contain a permanent ice cover several meters thick (Figure 5.21*d*), and the water column below the ice in these lakes remains at 0°C or colder year round. Marine sediments are also constantly cold. It is thus not surprising that the best examples of cold-active *Bacteria* and *Archaea* have emerged from these two environments.

Psychrophilic and Psychrotolerant Microorganisms

A psychrophile is an organism with an optimal growth temperature of 15°C or lower, a maximum growth temperature below 20°C, and a minimal growth temperature at 0°C or lower. Organisms that grow at 0°C but have optima of 20–40°C are called **psychrotolerant**. Psychrophiles are found in environments that are constantly cold and may be killed by warming, even to as little as 20°C. For this reason, their laboratory study requires that great care be taken to ensure that they never warm up during sampling, transport to the laboratory, isolation, or other manipulations. Seasonally cold environments, by contrast, cannot support true psychrophiles because they cannot survive the warming.

Psychrophilic algae and bacteria often grow in dense masses within and under sea ice (frozen seawater that forms seasonally) in polar regions (Figure 5.21*a, b, c*), and can also be found on the surfaces of permanent snowfields and glaciers where they impart a distinctive coloration to the surface (**Figure 5.22***a*). The common snow alga *Chlamydomonas nivalis* is an example of this, its spores being responsible for the brilliant red color of the snow surface (Figure 5.22*b*). This green alga grows within the snow as a green-pigmented vegetative cell and then sporulates. As the snow dissipates by melting, erosion, and ablation (evaporation and sublimation), the spores become concentrated on the surface. Related species of snow algae contain different carotenoid pigments, and thus fields of snow algae can also be green, orange, brown, or purple.

Several psychrophilic bacteria have been isolated, and some of these show very low growth temperature optima. A species of the sea ice bacterium *Psychromonas* grows at −12°C, the lowest temperature for any known bacterium. However, the lower temperature limit for bacterial growth is probably closer to −20°C. Even at this cold temperature, pockets of liquid water can exist, and studies have shown that enzymes from cold-active bacteria still function under such conditions. Growth rates at such cold temperatures would likely be extremely low, with doubling times of months, or even years. But if an organism can grow, even if only at a very slow rate, it can remain competitive and maintain a population in its habitat.

Psychrotolerant microorganisms are more widely distributed in nature than are psychrophiles and can be isolated from soils and water in temperate climates as well as from meat, dairy products, cider, vegetables, and fruit stored at standard refrigeration temperatures (4°C). Although psychrotolerant microorganisms grow at 0°C, most do not grow well, and one must often wait several weeks before visible growth is seen in laboratory cultures. By contrast, the same organism cultured at 30 or 35° may show growth rates similar to that of many mesophiles. Various *Bacteria*, *Archaea*, and microbial eukaryotes are psychrotolerant.

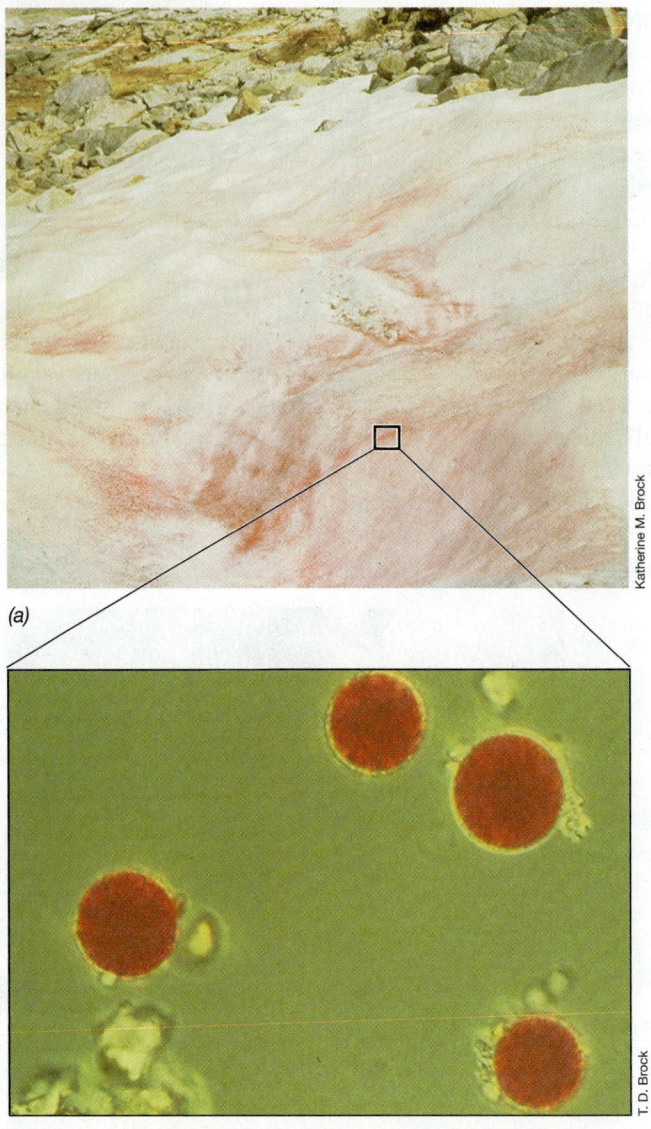

(a)

(b)

Figure 5.22 Snow algae. *(a)* Snow bank in the Sierra Nevada, California, with red coloration caused by the presence of snow algae. Pink snow such as this is common on summer snow banks at high altitudes throughout the world. *(b)* Photomicrograph of red-pigmented spores of the snow alga *Chlamydomonas nivalis*. The spores germinate to yield motile green algal cells. Some strains of snow algae are true psychrophiles but many are psychrotolerant, growing best at temperatures above 20°C. From a phylogenetic standpoint, *C. nivalis* is a green alga, and these organisms are covered in Section 17.16.

Molecular Adaptations That Support Psychrophily

Psychrophiles produce enzymes that function—often optimally—in the cold and that may be denatured or otherwise inactivated at even very moderate temperatures. The molecular basis for this is not entirely understood, but is clearly linked to protein structure. Several cold-active enzymes whose structure is known show a greater content of α-helix and lesser content of β-sheet secondary structure (⟳ Section 4.14) than do enzymes that show little or no activity in the cold. Because β-sheet secondary structures tend to be more rigid than α-helices, the greater α-helix content of cold-active enzymes allows these proteins greater flexibility for catalyzing their reactions at cold temperatures. Cold-active enzymes also tend to have greater polar and lesser hydrophobic amino acid content (Figure 4.30 for structures of amino acids) and lower numbers of weak bonds, such as hydrogen and ionic bonds, compared with the corresponding enzyme from mesophiles. Collectively, these molecular features are likely to keep cold-active enzymes flexible and functional under cold conditions.

Another characteristic feature of psychrophiles is that their cytoplasmic membranes remain functional at low temperatures. Cytoplasmic membranes from psychrophiles tend to contain a higher content of unsaturated and shorter-chain fatty acids, and this helps the membrane remain in a semifluid state at low temperatures to carry out important transport and bioenergetic functions. Some psychrophilic bacteria even contain *polyunsaturated* fatty acids, and unlike monounsaturated or saturated fatty acids that tend to stiffen at low temperatures, these remain flexible even at very cold temperatures.

Other molecular adaptations to cold temperatures include "cold-shock" proteins and cryoprotectants, and these are not limited to psychrophiles. Cold-shock proteins are even present in *Escherichia coli* and have several functions that include maintaining other proteins in an active form under cold conditions or binding specific mRNAs and facilitating their translation. The latter include, in particular, mRNAs that encode other cold-functional proteins, most of which are not produced when the cell is growing at a higher temperature. Cryoprotectants include dedicated antifreeze proteins or specific solutes, such as glycerol or certain sugars that are produced in large amounts at cold temperatures; these agents help prevent the formation of ice crystals that can puncture the cytoplasmic membrane. Highly psychrophilic bacteria often produce abundant levels of exopolysaccharides as well, and these are thought to have cryoprotectant properties too.

Although freezing temperatures may prevent microbial growth, they do not necessarily cause death. Some psychrophiles have been shown to metabolize at temperatures far lower than those supporting growth, and microbial respiration (as measured by CO_2 production) has been measured in tundra soils at nearly −40°C. Hence, enzymes continue to function at temperatures far below those that support cell growth. The medium in which cells are suspended also affects their sensitivity to freezing temperatures, and this has been exploited for the preservation of bacterial cells in microbial culture collections. For example, cells suspended in growth medium containing 10% dimethyl sulfoxide (DMSO) or glycerol and frozen at −80°C (ultracold-freezer) or −196°C (liquid nitrogen) remain viable in a frozen state for years.

MINIQUIZ --

- How do psychrotolerant organisms differ from psychrophilic organisms?
- What molecular adaptations to cold temperatures are seen in the cytoplasmic membrane of psychrophiles? Why are they necessary?

5.13 Microbial Life at High Temperatures

Microbial life flourishes in high-temperature environments, from sun-heated soils and pools of water to boiling hot springs, and the organisms that live in these environments are typically highly adapted to their environmental temperature. We examine them now (Sections 15.18 and 15.19, and Chapter 16).

Thermal Environments

Organisms whose growth temperature optimum exceeds 45°C are called *thermophiles* and those whose optimum exceeds 80°C are called *hyperthermophiles* (Figure 5.20). The surface of soils subject to full sunlight can be heated to above 50°C at midday, and some surface soils may become warmed to even 70°C. Fermenting materials such as compost piles and silage can also reach temperatures of 70°C. Thermophiles abound in such environments. The most extreme high-temperature environments in nature, however, are hot springs, and these are home to a huge diversity of thermophiles and hyperthermophiles.

Many terrestrial hot springs have temperatures at or near boiling while those at the bottom of the ocean, called *hydrothermal vents*, can have temperatures of 350°C or greater. Hot springs can be found throughout the world, but they are especially abundant in the western United States, New Zealand, Iceland, Japan, Italy, Indonesia, Central America, and central Africa. The largest concentration of hot springs in the world is in Yellowstone National Park, Wyoming (USA). Although some hot springs vary widely in temperature, many are nearly constant, varying less than 1–2°C over many years. In addition, different springs have different chemical compositions and pH values. Above 65°C, only prokaryotes are present (Table 5.1), but the diversity of *Bacteria* and *Archaea* is often extensive.

Hyperthermophiles in Hot Springs

A variety of hyperthermophiles are typically present in boiling hot springs (Figure 5.23), including both chemoorganotrophic and chemolithotrophic species. Growth rates of hyperthermophiles can be studied very simply in the field by immersing a microscope slide into a spring and then retrieving it a few days later; microscopic examination reveals microcolonies of prokaryotes that have developed from single cells that attached to and grew on the glass surface (Figure 5.23b). Simple ecological studies such as this have shown that microbial growth rates are often quite high, even in boiling springs; doubling times as short as 1 h have been recorded.

Cultures of diverse hyperthermophiles have been obtained, and a variety of morphological and physiological types of both *Bacteria* and *Archaea* are known. Some hyperthermophilic *Archaea* have growth-temperature optima above 100°C, while no species of *Bacteria* are known that grow above 95°C. Growing laboratory cultures of organisms with optima above the boiling point requires pressurized vessels that permit temperatures in the growth medium to rise above 100°C. The most heat-tolerant organisms known inhabit hydrothermal vents, with the most thermophilic example thus far being the methane-producing archaeon *Methanopyrus*, capable of growth at 122°C.

Table 5.1 Presently known upper temperature limits for growth of living organisms

Group	Upper temperature limits (°C)
Macroorganisms	
Animals	
Fish and other aquatic vertebrates	38
Insects	45–50
Ostracods (crustaceans)	49–50
Plants	
Vascular plants	45 (60 for one species)
Mosses	50
Microorganisms	
Eukaryotic microorganisms	
Protozoa	56
Algae	55–60
Fungi	60–62
Prokaryotes	
Bacteria	
Cyanobacteria	73
Anoxygenic phototrophs	70–73
Chemoorganotrophs/chemolithotrophs	95
Archaea	
Chemoorganotrophs/chemolithotrophs	122

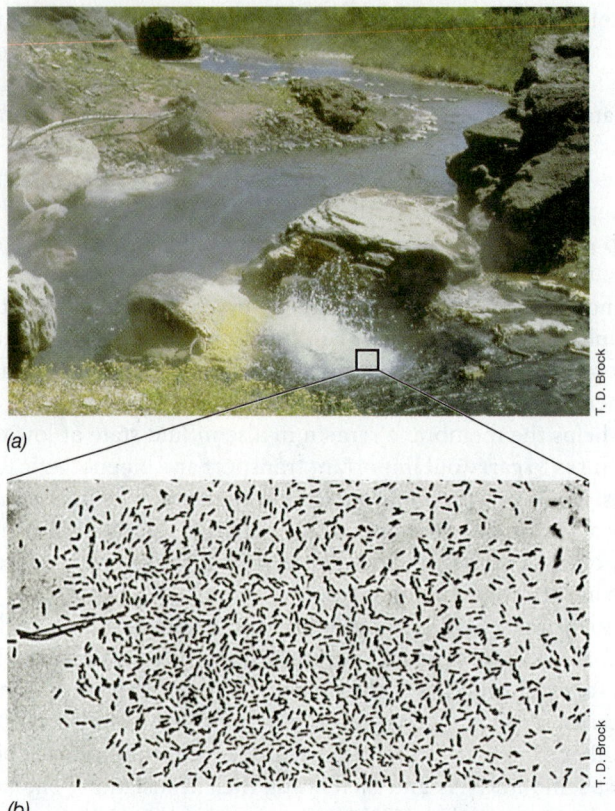

(a)

(b)

Figure 5.23 Growth of hyperthermophiles in boiling water. *(a)* Boulder Spring, a small boiling spring in Yellowstone National Park. This spring is superheated, having a temperature 1–2°C above the boiling point. The mineral deposits around the spring consist mainly of silica and sulfur. *(b)* Photomicrograph of a microcolony of prokaryotes that developed on a microscope slide immersed in such a boiling spring.

Thermophiles

Many thermophiles (optima 45–80°C) are also present in hot springs, but many are found elsewhere as well. As boiling water leaves a hot spring, it gradually cools, setting up a thermal gradient. Along this gradient, microorganisms become established, with different species growing in the different temperature ranges (**Figure 5.24**). By studying the species distribution along such natural thermal gradients, it has been possible to determine the upper temperature limits for each major group of microorganism (Table 5.1). From this information we can conclude that (1) prokaryotic organisms are able to grow at far higher temperatures than are eukaryotes, (2) the most thermophilic of all prokaryotes are certain species of *Archaea*, and (3) nonphototrophic organisms can grow at higher temperatures than can phototrophs.

Thermophilic prokaryotes have also been found in artificial thermal environments, such as hot water heaters, which typically operate at 60–80°C. Organisms resembling *Thermus aquaticus*, a common hot spring thermophile, have been isolated from both domestic and industrial hot water heaters. Electric power plants, hot water discharges, and other artificial thermal sources also provide sites where thermophiles can flourish. Many of these organisms can be easily isolated on complex media incubated at the temperature of the habitat from which the sample originated.

Protein Stability at High Temperatures

How do thermophiles and hyperthermophiles survive at high temperature? First, their enzymes and other proteins are much more heat-stable than are those of mesophiles and actually function *optimally* at high temperatures. Amazingly, however, studies of several heat-stable enzymes have shown that they often differ very little in amino acid sequence from heat-sensitive forms of the same enzyme from mesophiles. Apparently, critical amino acid substitutions at only a few locations in the enzyme allow the protein to fold in a unique, heat-stable way.

Heat stability of proteins from hyperthermophiles is also bolstered by an increased number of ionic bonds between basic and acidic amino acids and their often highly hydrophobic interiors. These combine to make a protein more resistant to unfolding. Finally, solutes such as di-inositol phosphate, diglycerol phosphate, and mannosylglycerate are produced at high levels in certain hyperthermophiles, and these are also thought to help stabilize their proteins against thermal denaturation.

Enzymes from thermophiles and hyperthermophiles have many commercial uses. Such enzymes can catalyze biochemical reactions at high temperatures and are in general more stable than enzymes from mesophiles, thus prolonging the shelf life of purified enzyme preparations. A classic example of this is the DNA polymerase isolated from *T. aquaticus. Taq polymerase*, as this enzyme is known, is used to automate the repetitive steps in the polymerase chain reaction (PCR), a technique for making multiple copies of a DNA sequence and

Nancy L. Spear

Figure 5.24 **Growth of thermophilic cyanobacteria in a hot spring in Yellowstone National Park.** Characteristic V-shaped pattern (shown by the dashed white lines) formed by cyanobacteria at the upper temperature for phototrophic life, 70–73°C, in the thermal gradient formed from a boiling hot spring. The pattern develops because the water cools more rapidly at the edges than in the center of the channel. The spring flows from the back of the picture toward the foreground. The light-green color is from a high-temperature strain of the cyanobacterium *Synechococcus*. As water flows down the gradient, the density of cells increases, less thermophilic strains enter, and the color becomes more intensely green.

a mainstay of modern biology (⊂⊃ Section 11.3). Several other uses of heat-stable enzymes and other heat-stable cell products are also known or are being developed for specific industrial applications.

Membrane Stability at High Temperatures

Besides enzymes and other macromolecules in the cell, the cytoplasmic membranes of thermophiles and hyperthermophiles must be heat-stable. Heat naturally peels apart the lipid bilayer that makes up the cytoplasmic membrane. In thermophiles, this is counteracted by constructing membranes with more long-chain and saturated fatty acid content and lower unsaturated fatty acid content than is found in the cytoplasmic membranes of mesophiles. Saturated fatty acids form a stronger hydrophobic environment than do unsaturated fatty acids, and longer-chain fatty acids have a higher melting point than shorter-chain fatty acids; collectively, these increase membrane stability.

Hyperthermophiles, most of which are *Archaea*, do not contain fatty acids in their membranes but instead have C_{40} hydrocarbons composed of repeating units of isoprene (⊂⊃ Figures 2.16c and 2.17) bonded by ether linkage to glycerol phosphate. In addition, however, the architecture of the cytoplasmic membranes of hyperthermophiles takes a unique twist: The membrane forms a lipid *monolayer* rather than a lipid *bilayer* (⊂⊃ Figure 2.17e). The monolayer structure covalently links one side of the membrane with the other and this prevents the membrane from melting at the high growth temperatures of hyperthermophiles. We consider other aspects of heat stability in hyperthermophiles, including that of DNA stability, in Chapter 16.

MINIQUIZ -

- Which domain of prokaryotes includes species with optima of >100°C? What special techniques are required to culture them?

- How does the membrane structure of hyperthermophilic *Archaea* differ from that of *Escherichia coli* and why is this structure helpful for growth at high temperature?

- What is *Taq* polymerase and why is it important?

V • Other Environmental Effects on Microbial Growth

As we have seen, temperature has a major effect on the growth of microorganisms. But many other factors do as well, chief among them being pH, osmolarity, and oxygen.

5.14 Effects of pH on Microbial Growth

Acidity or alkalinity of a solution is expressed by its **pH** on a logarithmic scale in which neutrality is pH 7 (**Figure 5.25**). pH values less than 7 are *acidic* and those greater than 7 are *alkaline*. In analogy to a temperature range, every microorganism has a pH range, typically about 2–3 pH units, within which growth is possible. Also, each organism shows a well-defined pH optimum, where growth occurs best. Most natural environments have a pH between 3 and 9, and organisms with pH growth optima in this range are most common. Terms used to describe organisms that grow best in particular pH ranges are shown in **Table 5.2**.

Acidophiles

Organisms that grow optimally at a pH value in the range termed *circumneutral* (pH 5.5 to 7.9) are called **neutrophiles** (Table 5.2).

By contrast, organisms that grow best below pH 5.5 are called **acidophiles**. There are different classes of acidophiles, some growing best at moderately acidic pH and others at very low pH. Many fungi and bacteria grow best at pH 5 or even below, while a more restricted number grow best below pH 3. An even more restricted group grow best below pH 2 and those with pH optima below 1 are extremely rare. Most acidophiles cannot grow at pH 7 and many cannot grow at pH values more than two units above their optimum.

A critical factor governing acidophily is the stability of the cytoplasmic membrane. When the pH is raised to neutrality, the cytoplasmic membranes of strongly acidophilic bacteria are destroyed and the cells lyse. This indicates that these organisms are not just acid-*tolerant* but that high concentrations of protons are actually *required* for cytoplasmic membrane stability. For example, the most acidophilic prokaryote known is *Picrophilus oshimae*, a species of *Archaea* that grows optimally at pH 0.7 and 60°C (the organism is also a thermophile). Above pH 4, cells of *P. oshimae* spontaneously lyse. As one would expect, *P. oshimae* inhabits extremely acidic thermal soils associated with volcanic activity.

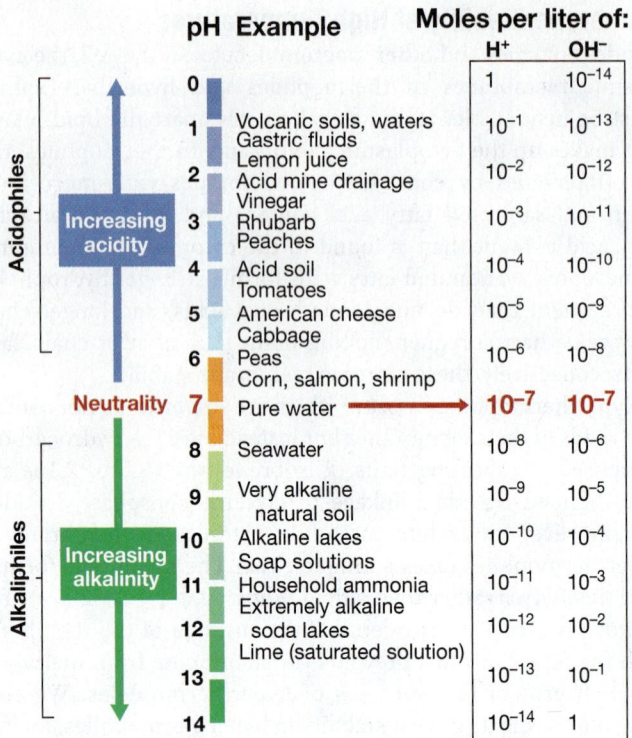

Figure 5.25 The pH scale. Although some microorganisms can live at very low or very high pH, the cell's internal pH remains near neutrality.

Table 5.2 Relationships of microorganisms to pH		
Physiological class (optima range)	**Approximate pH optimum for growth**	**Example organism[a]**
Neutrophile (pH > 5.5 and <8)	7	*Escherichia coli*
Acidophile (pH < 5.5)	5	*Rhodopila globiformis*
	3	*Acidithiobacillus ferrooxidans*
	1	*Picrophilus oshimae*
Alkaliphile (pH ≥ 8)	8	*Chloroflexus aurantiacus*
	9	*Bacillus firmus*
	10	*Natronobacterium gregoryi*

[a]*Picrophilus* and *Natronobacterium* are *Archaea*; all others are *Bacteria*.

Alkaliphiles

A few extremophiles have very high pH optima for growth, sometimes as high as pH 10, and some of these can still grow, albeit poorly, at even higher pH. Microorganisms showing growth pH optima of 8 or higher are called **alkaliphiles**. Alkaliphilic microorganisms are typically found in highly alkaline habitats, such as soda lakes and high-carbonate soils. The most well-studied alkaliphilic prokaryotes are certain *Bacillus* species, such as *Bacillus firmus*. This organism is alkaliphilic, but has an unusually broad range for growth, from pH 7.5 to 11. Some extremely alkaliphilic bacteria are also halophilic (salt-loving), and most of these are *Archaea* (⟲ Section 16.1). Some phototrophic purple bacteria (⟲ Section 14.4) are also strongly alkaliphilic. Certain alkaliphiles have industrial uses because they produce hydrolytic exoenzymes, such as proteases and lipases. Exoenzymes are those that are excreted from the cell, and in the case of alkaliphiles, their exoenzymes must function well at alkaline pH. These enzymes are produced commercially on a large scale and added as supplements to laundry detergents to remove protein and fat stains from clothing.

Alkaliphiles are of interest for several reasons, particularly how they manage bioenergetics. Imagine how a cell might generate a proton motive force (⟲ Section 3.11) when the external surface of its cytoplasmic membrane is so alkaline. One strategy for circumventing this problem in *B. firmus* is the use of sodium (Na⁺) rather than H⁺ to fuel transport reactions and motility; that is, a sodium motive instead of a proton motive force. Remarkably, however, a proton motive force is coupled to ATP synthesis in *B. firmus*, even though the external membrane surface is highly

alkaline. Exactly how this can happen is unclear, although it is thought that hydrogen ions are in some way kept very near the outer surface of the cytoplasmic membrane such that they cannot spontaneously combine with hydroxyl ions to form water.

Cytoplasmic pH and Buffers

The optimal pH for growth of an organism refers to the *extracellular* environment only; the *intracellular* pH must remain near neutrality in order to prevent the destruction of macromolecules. DNA is acid-labile and RNA is alkaline-labile, and thus a cell must maintain these key macromolecules in a stable state. Nevertheless, measurements of cytoplasmic pH in some strong acidophiles and alkaliphiles have shown a range of pH values from as low as just under pH 5 to a high of just over pH 9. If these are not the lower and upper limits of cytoplasmic pH, respectively, they are extremely close to the limits.

To prevent major shifts in pH during microbial growth in batch cultures, *buffers* are commonly added to culture media along with the nutrients required for growth. However, any given buffer works over only a relatively narrow pH range. Hence, different buffers are used for different pH classes of microorganism. Near neutral pH, potassium phosphate (KH_2PO_4) or sodium bicarbonate ($NaHCO_3$) is often employed. In addition, a suite of organic molecules called Good buffers (named for the chemist who invented them) have been designed, each of which buffers best in a specific pH range. These can be used in growth medium or for other buffering needs. In the final analysis, the best buffer for the growth of any given organism must usually be determined empirically. Various buffers are also widely used for assaying enzymes in vitro. The buffer keeps the enzyme solution at optimal pH during the assay, thus ensuring that the enzyme remains catalytically active and unaffected by any protons or hydroxyl ions generated in the enzymatic reaction.

MINIQUIZ

- How does the concentration of H⁺ change when a culture medium at pH 5 is adjusted to pH 9?
- What terms are used to describe organisms whose growth pH optimum is very high? Very low?

5.15 Osmolarity and Microbial Growth

Water is the solvent of life, and water availability is an important factor affecting the growth of microorganisms. Water availability not only depends on how moist or dry an environment is but is also a function of the concentration of solutes (salts, sugars, or other substances) dissolved in the water that is present. Solutes bind water, making it less available to organisms. Hence, for organisms to thrive in high-solute environments, physiological adjustments are necessary.

Water availability is expressed in terms of **water activity** (a_w), the ratio of the vapor pressure of air in equilibrium with a substance or solution to the vapor pressure of pure water. Values of a_w vary between 0 and 1; some a_w values are listed in **Table 5.3**. Water diffuses from regions of high water concentration (low solute concentration) to regions of lower water concentration (higher solute concentration) in the process of osmosis. The cytoplasm of a cell typically has a higher solute concentration than the environment, so the tendency for water is to diffuse into the cell. Under such conditions, the cell is said to be in *positive water balance*, which is the normal state of the cell. However, when a cell finds itself in an environment where the solute concentration exceeds that of the cytoplasm, water will flow out of the cell. If a cell has no strategy to counteract this, it will become dehydrated and unable to grow.

Halophiles and Related Organisms

In nature, osmotic effects are of interest mainly in habitats with high concentrations of salts. Seawater contains about 3% NaCl plus small amounts of many other minerals and elements. Microorganisms that inhabit marine environments almost always have a NaCl requirement and typically grow optimally at the water activity of seawater (**Figure 5.26**). Such organisms are called **halophiles**. Although halophiles require at least some NaCl for growth, the NaCl optimum observed varies with the organism and is habitat dependent. For example, marine microorganisms typically grow best with 1–4% NaCl, organisms from hypersaline environments

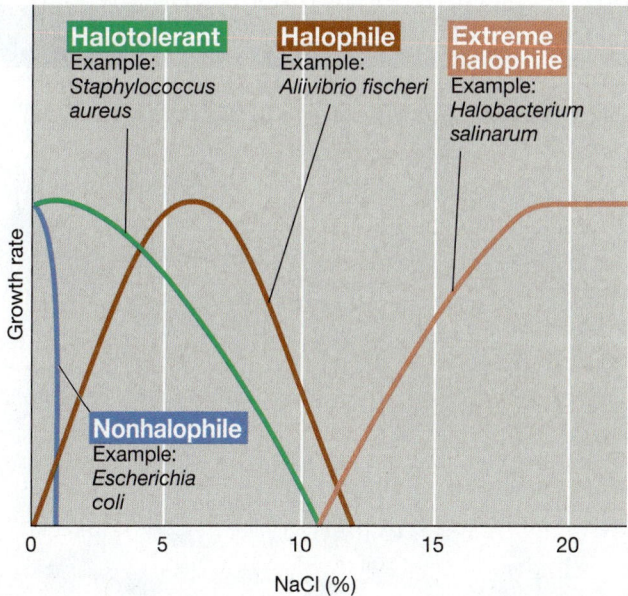

Figure 5.26 **Effect of NaCl concentration on growth of microorganisms of different salt tolerances or requirements.** The optimum NaCl concentration for marine microorganisms such as *Aliivibrio fischeri* is about 3%; for extreme halophiles, it is between 15 and 30%, depending on the organism.

(environments that are more salty than seawater), 3–12%, and organisms from extremely hypersaline environments require even higher levels of NaCl. In addition, the requirement for NaCl by halophiles is absolute and cannot be replaced by other salts, such as potassium chloride (KCl), calcium chloride ($CaCl_2$), or magnesium chloride ($MgCl_2$).

In contrast to halophiles, **halotolerant** organisms can tolerate some level of dissolved solutes but grow best in the absence of the added solute (Figure 5.26). Halophiles capable of growth in very salty environments are called **extreme halophiles** (Figure 5.26). These organisms require very high levels of NaCl, typically 15–30%, for optimum growth and are often unable to grow at all at NaCl concentrations below this. Organisms able to live in environments high in sugar are called **osmophiles** and those able to grow in very dry environments (made dry by lack of water rather than from dissolved solutes) are called **xerophiles**. Examples of these various classes of organisms are given in **Table 5.4**.

Compatible Solutes

When an organism is transferred from a medium of high a_w to one of low a_w, it maintains positive water balance by increasing its internal solute concentration. This is possible by either pumping solutes into the cell from the environment or by synthesizing a cytoplasmic solute (Table 5.4). In either case, the solute must not inhibit cellular processes in any significant way. Such compounds are called **compatible solutes**, and are typically highly water-soluble organic molecules including sugars, alcohols, or amino acid derivatives (Table 5.4). Glycine betaine, a highly soluble analog of the amino acid glycine, is widely distributed among halophilic bacteria (Table 5.4). Other common compatible solutes include sugars such as sucrose and trehalose, dimethylsulfoniopropionate, produced by marine algae, and glycerol, a common

Water activity (a_w)	Material	Example organisms[a]
1.000	Pure water	*Caulobacter, Spirillum*
0.995	Human blood	*Streptococcus, Escherichia*
0.980	Seawater	*Pseudomonas, Vibrio*
0.950	Bread	Most gram-positive rods
0.900	Maple syrup, ham	Gram-positive cocci such as *Staphylococcus*
0.850	Salami	*Saccharomyces rouxii* (yeast)
0.800	Fruit cake, jams	*Saccharomyces bailii, Penicillium* (fungus)
0.750	Salt lakes, salted fish	*Halobacterium, Halococcus*
0.700	Cereals, candy, dried fruit	*Xeromyces bisporus* and other xerophilic fungi

Table 5.3 Water activity of several substances

[a]Selected examples of prokaryotes or fungi capable of growth in culture media adjusted to the stated water activity.

Table 5.4 Compatible solutes of microorganisms

Organism	Major cytoplasmic solute(s)	Minimum a_w for growth	
Nonphototrophic *Bacteria*/freshwater cyanobacteria	Amino acids (mainly glutamate or proline[a])/sucrose, trehalose[b]	0.98–0.90	Sucrose
Marine cyanobacteria	α-Glucosylglycerol[b]	0.92	
Marine algae	Mannitol,[b] various glycosides, dimethylsulfoniopropionate	0.92	Dimethylsulfoniopropionate
Salt lake cyanobacteria	Glycine betaine	0.90–0.75	Glycine betaine
Halophilic anoxygenic phototrophic purple *Bacteria*	Glycine betaine, ectoine, trehalose[b]	0.90–0.75	Ectoine
Extremely halophilic *Archaea* and some *Bacteria*	KCl	0.75	
Dunaliella (halophilic green alga)	Glycerol	0.75	
Xerophilic and osmophilic yeasts	Glycerol	0.83–0.62	Glycerol
Xerophilic filamentous fungi	Glycerol	0.72–0.61	

[a]See Figure 4.30 for the structures of amino acids.
[b]Structures not shown. Like sucrose, trehalose is a C_{12} disaccharide; glucosylglycerol is a C_9 alcohol; mannitol is a C_6 alcohol.

solute in xerophilic fungi, organisms that grow at the lowest water activities known (Table 5.4). In contrast to these organic solutes, the compatible solute of extremely halophilic *Archaea*, such as *Halobacterium*, and a few extremely halophilic *Bacteria*, is KCl (↪ Section 16.1).

The concentration of compatible solute in a cell is a function of the levels of solute in its environment, and adjustments are made in response to the challenge from external solutes. However, in any given organism the maximal level of compatible solute is a genetically encoded characteristic. As a result, different organisms have evolved to thrive in different ranges of water activity (Tables 5.3 and 5.4). In fact, organisms designated as *nonhalotolerant, halotolerant, halophilic,* or *extremely halophilic* (Figure 5.26) are a reflection of the genetic capacity of the organisms in each group to produce or accumulate compatible solutes.

MINIQUIZ -
- What is the a_w of pure water?
- What are compatible solutes, and when and why are they needed by the cell? What is the compatible solute of *Halobacterium*?

5.16 Oxygen and Microbial Growth

For many microorganisms, oxygen (O_2) is an essential nutrient; they are unable to metabolize or grow without it. Other organisms, by contrast, cannot grow in the presence of O_2 and may even be killed by it. We therefore see, just as we did for other environmental factors considered in this chapter, classes of microorganisms based on their needs or tolerance of O_2.

Oxygen Classes of Microorganisms

Microorganisms can be grouped according to their relationship with O_2 as outlined in **Table 5.5**. **Aerobes** can grow at full oxygen tensions (air is 21% O_2) and respire O_2 in their metabolism. **Microaerophiles**, by contrast, are aerobes that can use O_2 only when it is present at levels reduced from that in air (microoxic conditions). This is because of the limited capacity of these organisms to respire or because they contain some O_2-sensitive molecule such as an O_2-labile enzyme. Many aerobes are **facultative**, meaning that under the appropriate nutrient and culture conditions they can grow in the absence of O_2.

Some organisms cannot respire oxygen and are called **anaerobes**. There are two kinds of anaerobes: **aerotolerant anaerobes**, which can tolerate O_2 and grow in its presence even though they

Table 5.5 Oxygen relationships of microorganisms

Group	Relationship to O_2	Type of metabolism	Example[a]	Habitat[b]
Aerobes				
Obligate	Required	Aerobic respiration	*Micrococcus luteus* (B)	Skin, dust
Facultative	Not required, but growth better with O_2	Aerobic respiration, anaerobic respiration, fermentation	*Escherichia coli* (B)	Mammalian large intestine
Microaerophilic	Required but at levels lower than atmospheric	Aerobic respiration	*Spirillum volutans* (B)	Lake water
Anaerobes				
Aerotolerant	Not required, and growth no better when O_2 present	Fermentation	*Streptococcus pyogenes* (B)	Upper respiratory tract
Obligate	Harmful or lethal	Fermentation or anaerobic respiration	*Methanobacterium formicicum* (A)	Sewage sludge, anoxic lake sediments

[a]Letters in parentheses indicate phylogenetic status (B, *Bacteria*; A, *Archaea*). Representatives of either domain of prokaryotes are known in each category. Most eukaryotes are obligate aerobes, but facultative aerobes (for example, yeast) and obligate anaerobes (for example, certain protozoa and fungi) are known.
[b]Listed are typical habitats of the example organism; many others could be listed.

cannot respire, and **obligate anaerobes**, which are inhibited or even killed by O_2 (Table 5.5). *Anoxic* (O_2-free) microbial habitats are common in nature and include muds and other sediments, bogs, marshes, water-logged soils, intestinal tracts of animals, sewage sludge, the deep subsurface of Earth, and many other environments. As far as is known, obligate anaerobiosis is characteristic of only three groups of microorganisms: a wide variety of *Bacteria* and *Archaea*, a few fungi, and a few protozoa. Some of the best-known obligate anaerobes are *Clostridium*, a genus of gram-positive endospore-forming *Bacteria*, and the methanogens, a group of methane-producing *Archaea*. Among obligate anaerobes, the sensitivity to O_2 varies greatly. Many clostridia, for example, although requiring anoxic conditions for growth, can tolerate traces of O_2 or even full exposure to air. Others, such as the methanogens, are killed rapidly by O_2 exposure.

Culture Techniques for Aerobes and Anaerobes

For the growth of aerobes, it is necessary to provide extensive aeration. This is because the O_2 that is consumed by the organisms during growth is not replaced fast enough by simple diffusion from the air. Therefore, forced aeration of liquid cultures is needed and can be achieved by either vigorously shaking the flask or tube on a shaker or by bubbling sterilized air into the medium through a fine glass tube or porous glass disc.

For the culture of anaerobes, the problem is not to provide O_2, but to exclude it. Bottles or tubes filled completely to the top with culture medium and fitted with leakproof closures provide suitably anoxic conditions for organisms that are not overly sensitive to small amounts of O_2. A chemical called a *reducing agent* may be added to such vessels to remove traces of O_2 by reducing it to water (H_2O). An example is thioglycolate, which is present in thioglycolate broth, a medium commonly used to test an organism's requirements for O_2 (**Figure 5.27**). Thioglycolate broth is a complex medium containing a small amount of agar, making the medium viscous but still fluid. After thioglycolate reacts with O_2 throughout the tube, O_2 can penetrate only near the top of the tube where the medium contacts air. Obligate aerobes grow only at the top of such tubes. Facultative organisms grow throughout the tube

but grow best near the top. Microaerophiles grow near the top but not right at the top. Anaerobes grow only near the bottom of the tube, where O_2 cannot penetrate. The redox indicator dye *resazurin* is present in thioglycolate broth to signal oxic regions; the dye is pink when oxidized and colorless when reduced and so

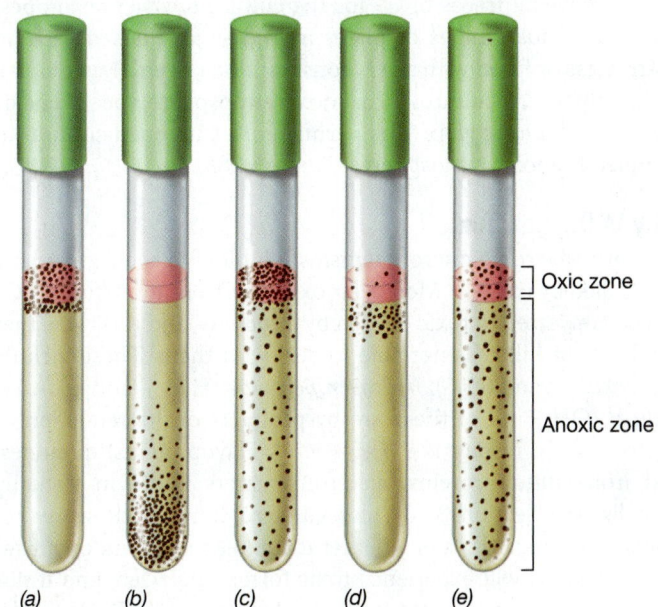

(a) (b) (c) (d) (e)

Oxic zone
Anoxic zone

Figure 5.27 Growth versus O_2 concentration. From left to right, aerobic, anaerobic, facultative, microaerophilic, and aerotolerant anaerobe growth, as revealed by the position of microbial colonies (depicted here as black dots) within tubes of thioglycolate broth culture medium. A small amount of agar has been added to keep the liquid from becoming disturbed. The redox dye resazurin, which is pink when oxidized and colorless when reduced, has been added as a redox indicator. *(a)* O_2 penetrates only a short distance into the tube, so obligate aerobes grow only close to the surface. *(b)* Anaerobes, being sensitive to O_2, grow only away from the surface. *(c)* Facultative aerobes are able to grow in either the presence or the absence of O_2 and thus grow throughout the tube. However, growth is better near the surface because these organisms can respire. *(d)* Microaerophiles grow away from the most oxic zone. *(e)* Aerotolerant anaerobes grow throughout the tube. Growth is not better near the surface because these organisms can only ferment.

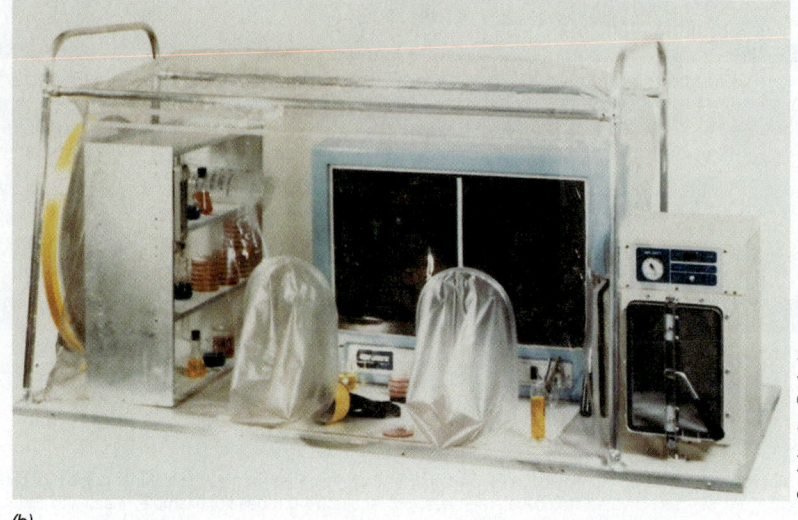

(a)   (b)

Figure 5.28 Incubation under anoxic conditions. *(a)* Anoxic jar. A chemical reaction in the envelope in the jar generates $H_2 + CO_2$. The H_2 reacts with O_2 in the jar on the surface of a palladium catalyst to yield H_2O; the final atmosphere contains N_2, H_2, and CO_2. *(b)* Anoxic glove bag for manipulating and incubating cultures under anoxic conditions. The airlock on the right, which can be evacuated and filled with O_2-free gas, serves as a port for adding and removing materials to and from the glove bag.

gives a visual assessment of the degree of penetration of O_2 into the medium (Figure 5.27).

To remove all traces of O_2 for the culture of strict anaerobes, one can incubate tubes or plates in a glass jar flushed with an O_2-free gas or fitted with an O_2 consumption system (**Figure 5.28a**). For manipulating cultures in an anoxic atmosphere, special enclosures called *anoxic glove bags* permit work with open cultures in completely anoxic atmospheres (Figure 5.28*b*).

Why Is Oxygen Toxic?

Why are anaerobic microorganisms inhibited in their growth or even killed by oxygen? Molecular oxygen (O_2) is not toxic, but O_2 can be converted to toxic oxygen by-products, and it is these that can harm or kill cells not able to deal with them. These include *superoxide anion* (O_2^-), *hydrogen peroxide* (H_2O_2), and *hydroxyl radical* (OH·). All of these are by-products of the reduction of O_2 to H_2O in respiration (**Figure 5.29**). Flavoproteins, quinones, and iron–sulfur proteins, electron carriers found in virtually all cells (↶ Section 3.10), also catalyze some of these reductions. Thus, regardless of whether it can respire O_2, an organism exposed to O_2 will experience toxic forms of oxygen, and if not destroyed, these molecules can wreak havoc in cells. For example,

superoxide anion and OH· are strong oxidizing agents that can oxidize macromolecules and any other organic compounds in the cell. Peroxides such as H_2O_2 can also damage cell components but are not as toxic as O_2^- or OH·. It should thus be clear that a major requirement for inhabiting an oxic world is to keep toxic oxygen molecules under control. We turn now to how this is done.

Superoxide Dismutase and Other Enzymes That Destroy Toxic Oxygen

Superoxide anion and H_2O_2 are the most abundant toxic oxygen species, and cells have enzymes that destroy these compounds (**Figure 5.30**). The enzymes catalase and peroxidase attack H_2O_2,

$$H_2O_2 + H_2O_2 \rightarrow 2\,H_2O + O_2$$
(a) Catalase

$$H_2O_2 + \textbf{NADH} + H^+ \rightarrow 2\,H_2O + NAD^+$$
(b) Peroxidase

$$O_2^- + O_2^- + 2\,H^+ \rightarrow H_2O_2 + O_2$$
(c) Superoxide dismutase

$$4\,O_2^- + 4\,H^+ \rightarrow 2\,H_2O + 3\,O_2$$
(d) Superoxide dismutase/catalase in combination

$$O_2^- + 2\,H^+ + \text{rubredoxin}_{\text{reduced}} \rightarrow H_2O_2 + \text{rubredoxin}_{\text{oxidized}}$$
(e) Superoxide reductase

Figure 5.30 Enzymes that destroy toxic oxygen species. *(a)* Catalases and *(b)* peroxidases are porphyrin-containing proteins, although some flavoproteins may consume toxic oxygen species as well. *(c)* Superoxide dismutases are metal-containing proteins, the metals being copper and zinc, manganese, or iron. *(d)* Combined reaction of superoxide dismutase and catalase. *(e)* Superoxide reductase catalyzes the one-electron reduction of O_2^- to H_2O_2.

Reactants		Products	
$O_2 + e^- \rightarrow$		O_2^-	(superoxide)
$O_2^- + e^- + 2\,H^+ \rightarrow$		H_2O_2	(hydrogen peroxide)
$H_2O_2 + e^- + H^+ \rightarrow$		$H_2O + $ OH·	(hydroxyl radical)
OH· $ + e^- + H^+ \rightarrow$		H_2O	(water)

Outcome:
$$O_2 + 4\,e^- + 4\,H^+ \rightarrow 2\,H_2O$$

Figure 5.29 Four-electron reduction of O_2 to H_2O by stepwise addition of electrons. All the intermediates formed are reactive and toxic to cells; water is not.

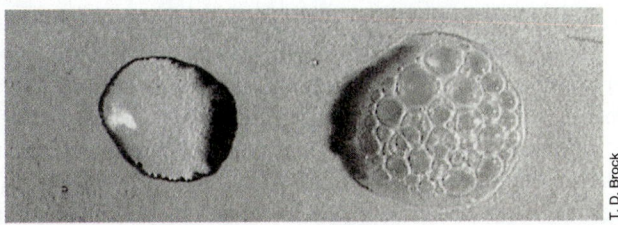

Figure 5.31 Method for testing a microbial culture for the presence of catalase. A heavy loopful of cells from an agar culture was mixed on a slide (right) with a drop of 30% hydrogen peroxide. The immediate appearance of bubbles is indicative of the presence of catalase. The bubbles are O_2 produced by the reaction $H_2O_2 + H_2O_2 \rightarrow 2 H_2O + O_2$.

forming O_2 and H_2O, respectively (Figure 5.30 and **Figure 5.31**). Superoxide anion is destroyed by the enzyme *superoxide dismutase*, an enzyme that generates H_2O_2 and O_2 from two molecules of O_2^- (Figure 5.30c). Superoxide dismutase and catalase (or peroxidase) thus work in series to convert O_2^- to harmless products (Figure 5.30d).

Aerobes and facultative aerobes typically contain both superoxide dismutase and catalase. Superoxide dismutase is an essential enzyme for aerobes. Some aerotolerant anaerobes lack superoxide dismutase and use protein-free manganese complexes instead to carry out the dismutation of O_2^- to H_2O_2 and O_2. Such a system is not as efficient as superoxide dismutase, but it is sufficient to protect the cells from O_2^- damage. In some strictly anaerobic *Archaea* and *Bacteria*, superoxide dismutase is absent and instead the enzyme *superoxide reductase* functions to remove O_2^-. Unlike superoxide dismutase, superoxide reductase reduces O_2^- to H_2O_2 without the production of O_2 (Figure 5.30e), thus avoiding exposure of the organism to O_2.

MINIQUIZ

- How does an obligate aerobe differ from a facultative aerobe?
- How does a reducing agent work? Give an example of a reducing agent.
- How does superoxide dismutase or superoxide reductase protect a cell?

VI • Control of Microbial Growth

Thus far in this chapter we have discussed microbial growth with a focus on *promoting* growth. We close this chapter by considering the opposite side of the coin, microbial *growth control*. Many aspects of microbial growth control have significant practical applications. For example, we wash fresh produce to remove attached microorganisms and we inhibit microbial growth on body surfaces by washing. However, neither of these processes kills or removes all microorganisms. Only **sterilization**—the killing or removal of all microorganisms (including viruses)—ensures that this is the case. In certain circumstances, sterility is not attainable or practical. In others, however, sterilization is absolutely essential. We review methods of growth control now.

5.17 General Principles and Growth Control by Heat

Microorganisms and their effects can be controlled in many cases by simply limiting or inhibiting growth of the cells. Methods for inhibiting microbial growth include **decontamination**, the treatment of an object or surface to make it safe to handle, and **disinfection**, a process that directly targets pathogens although it may not eliminate all microorganisms. Decontamination could be as simple as wiping off food utensils to remove food fragments (and their attached organisms) before using them, while disinfection requires agents called *disinfectants* that actually kill microorganisms or severely inhibit their growth. A solution of bleach (sodium hypochlorite), for example, is an effective disinfectant for a wide variety of applications.

Physical methods of microbial growth control are used in industry, medicine, and in the home to achieve decontamination, disinfection, and sterilization. Heat, radiation, and filtration are the most common of these methods. Perhaps the most widespread growth control method is heat. Factors that affect a microorganism's susceptibility to heat include the temperature and duration of the heat treatment and whether the heat is moist or dry.

Heat Sterilization

All microorganisms have a maximum growth temperature beyond which growth is impossible, typically because one or more key cellular structures are destroyed or a key protein is denatured (Figure 5.19). The effectiveness of heat as a sterilant is measured by the time required for a 10-fold reduction in the viability of a microbial population at a given temperature. This is called the *decimal reduction time* or *D*. The relationship between *D* and temperature is exponential, as the logarithm of *D* plotted against temperature yields a straight line (**Figure 5.32**). Moreover, heat killing is an exponential (first-order) function, proceeding more rapidly as the temperature rises. The type of heat is also important: Moist heat has better penetrating power than dry heat and, at a given temperature, produces a faster reduction in the number of living organisms.

Determination of a decimal reduction time requires a large number of viable count measurements (Section 5.9). An easier way to characterize the heat sensitivity of an organism is to measure its *thermal death time*, the time it takes to kill all cells at a given temperature. To determine the thermal death time, samples of a cell suspension are heated for different times, mixed with culture medium, and incubated. If all the cells have been killed, no growth is observed in the incubated samples. However, unlike a decimal reduction time measurement that is independent of the original cell number, the thermal death time is greatly affected by population size; a longer time is required to kill all cells in a large population than in a small one.

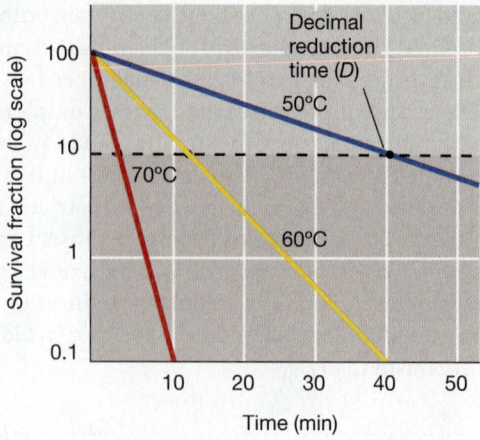

(a)

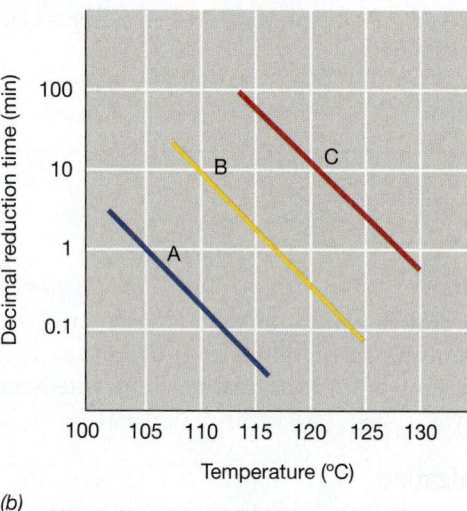

(b)

Figure 5.32 The effect of temperature on the heat killing of microorganisms. (a) The decimal reduction time (D) is the time at which only 10% of the original population of a given organism (in this case, a mesophile) remains viable at a given temperature. For 70°C, D = 3 min; for 60°C, D = 12 min; for 50°C, D = 42 min. (b) D values for model organisms of different temperature classes: A, mesophile; B, thermophile; C, hyperthermophile.

The presence of endospore-forming bacteria in a heat-treated sample can influence both the decimal reduction time and thermal death time. The heat resistance of vegetative cells and endospores from the same organism differs considerably. Recall that the mature endospore is very dehydrated and contains special chemicals, such as calcium dipicolinate, and proteins, such as small acid-soluble spore proteins (SASPs), that help confer heat stability on the structure (Section 2.16). One cannot be sure that endospores have been killed unless autoclave temperatures (at least 121°C) are reached for at least 15 min. The decimal reduction time is also a function of the inherent heat resistance of the microorganisms present; as one would expect, thermophiles and hyperthermophiles are more resistant than are mesophiles (Figure 5.32b).

The medium in which heating takes place also influences the killing of both vegetative cells and endospores. Microbial death is more rapid at acidic pH, and acid foods such as tomatoes, fruits, and pickles are much easier to sterilize than neutral pH foods such as corn and beans. High concentrations of sugars, proteins,

and fats decrease heat penetration and usually increase the resistance of organisms to heat, whereas high salt concentrations may either increase or decrease heat resistance, depending on the organism. Dry cells and endospores are more heat resistant than moist ones; consequently, heat sterilization of dry objects such as endospores always requires higher temperatures and longer heat application times than sterilization of wet objects such as liquid bacterial cultures.

The Autoclave and Pasteurization

The **autoclave** is a sealed heating device that uses steam under pressure to kill microorganisms (**Figure 5.33**). Killing of heat-resistant endospores requires heating at temperatures above the boiling point of water at 1 atm. The autoclave places steam under a pressure of 1.1 kg/cm^2 (15 lb/in^2), which yields a temperature of 121°C. At 121°C, the time to achieve sterilization of small amounts of endospore-containing material is about 15 min (Figure 5.33b). If the object to be autoclaved is bulky or large volumes of liquids are to be sterilized, heat transfer to the interior is retarded, and thus the total heating time must be extended. Note that it is not the *pressure* inside the autoclave that kills the microorganisms but the high *temperatures* that are achieved when steam is placed under pressure.

Pasteurization uses precisely controlled heat to significantly reduce the total number of microorganisms found in milk and other liquids that would be destroyed if autoclaved. The process, named for Louis Pasteur (Section 1.7), was first used for controlling the spoilage of wine. Pasteurization does not kill all organisms and is therefore not a method of sterilization. Pasteurization does, however, reduce the *microbial load*, the number of viable microorganisms in a sample. At temperatures and times used for pasteurization of food products such as milk, all known pathogenic bacteria that can be transmitted in infected milk, especially the organisms causing tuberculosis, brucellosis, Q fever, and typhoid fever, are killed. In addition, by decreasing the overall microbial load, pasteurization retards the growth of spoilage organisms, increasing the shelf life of perishable liquids (Section 31.6).

To achieve pasteurization, the liquid is passed through a tubular heat exchanger. Careful control of flow rate and the size and temperature of the heat source raises the temperature of the liquid to 71°C for 15 seconds (or even higher temperatures for shorter time periods; see Figure 5.32), after which it is rapidly cooled. This process is called *flash pasteurization*. Ultrahigh-temperature pasteurization of milk requires heat treatment at 135°C for 1 min. Milk can also be pasteurized by heating it in large vats to 63–66°C for 30 min. However, this *bulk pasteurization* method is less satisfactory because the milk heats and cools slowly, thus altering the taste of the final product, and it is a less efficient process.

MINIQUIZ

- Why is heat an effective sterilizing agent?
- What steps are necessary to ensure the sterility of material contaminated with bacterial endospores?
- Distinguish between the sterilization of microbiological media and the pasteurization of dairy products.

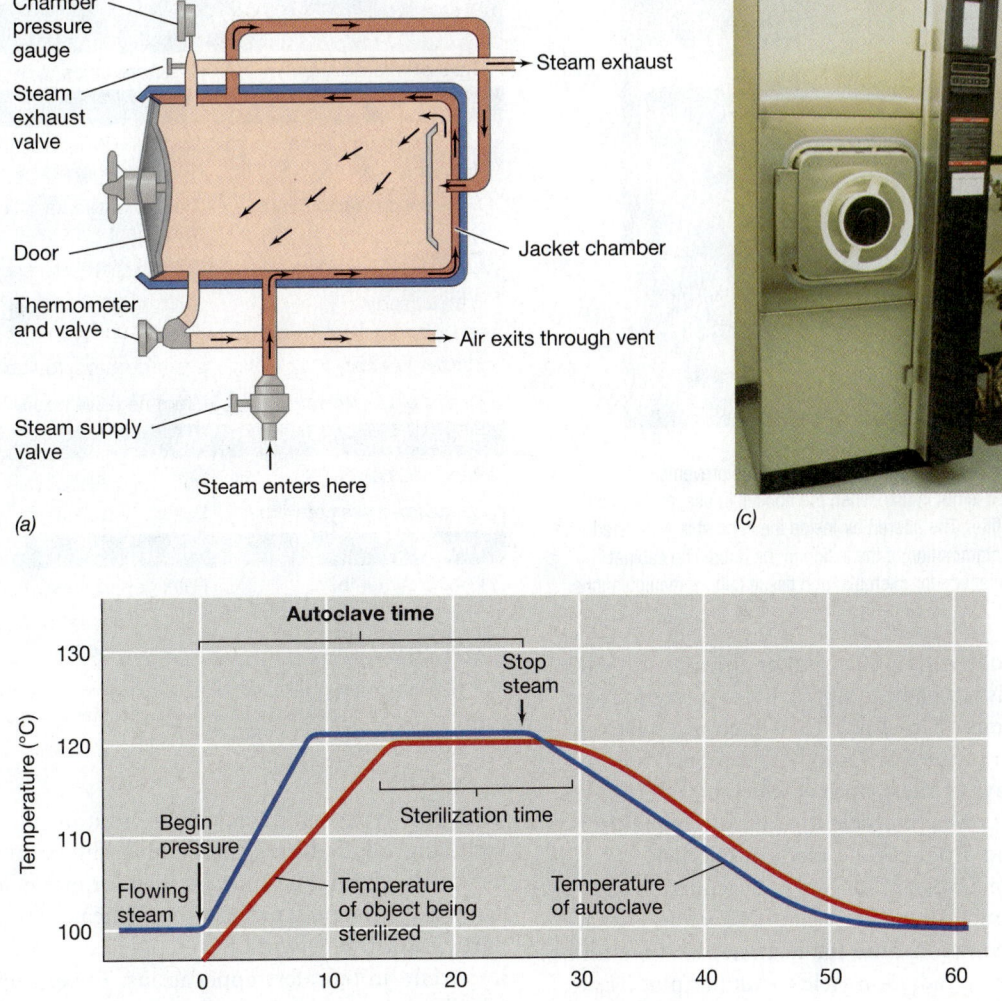

J. Martinko

(a)

(b)

(c)

Figure 5.33   The autoclave and moist heat sterilization. *(a)* The flow of steam through an autoclave. *(b)* A typical autoclave cycle. The temporal heating profile of a fairly bulky object is shown. The temperature of the object rises and falls more slowly than the temperature of the autoclave. The temperature of the object must reach the target temperature and be held for 10–15 min to ensure sterility, regardless of the temperature and time recorded in the autoclave. *(c)* A modern research autoclave. Note the pressure-lock door and the automatic cycle controls on the right panel. The steam inlet and exhaust fittings are on the right side of the autoclave.

5.18 Other Physical Control Methods: Radiation and Filtration

Heat is just one form of energy that can sterilize or reduce microbial load. Ultraviolet (UV) radiation, X-rays, and gamma rays are also effective killing agents. However, each type of energy has a different mode of action and killing efficacy and thus their applications can vary widely.

Ultraviolet and Ionizing Radiation

Ultraviolet radiation between 220 and 300 nm is absorbed by DNA and can cause mutations or other serious effects on DNA that lead to death of the exposed organism (⟳ Section 10.4). UV radiation is useful for disinfecting surfaces and air, and is widely used to decontaminate and disinfect the work surface of laboratory laminar flow hoods equipped with a "germicidal" UV light (**Figure 5.34**) and also to disinfect air circulating in hospital and food preparation rooms. However, UV radiation has very poor penetrating power, limiting its use to the disinfection of exposed surfaces or air rather than bulk objects such as canned foods or surgical clothing.

Ionizing radiation is electromagnetic radiation of sufficient energy to produce ions and other reactive molecular species from molecules with which the radiation particles collide. Ionizing radiation generates higher-energy electrons, hydroxyl radicals (OH·), and hydride radicals (H·), and each of these can damage macromolecules and kill irradiated cells (Section 5.16).

The unit of ionizing radiation is the *roentgen*, and the standard for biological applications such as sterilization is the *absorbed*

J. Martinko

Figure 5.34 **A laminar flow hood.** An ultraviolet light source prevents contamination of the hood when it is not in use. When the hood is in use, air is drawn into the cabinet through a HEPA filter. The filtered air inside the cabinet is exhausted out of the cabinet, preventing contamination of the inside of the hood. The cabinet provides a contaminant-free workspace for microbial and tissue culture manipulations.

radiation dose, measured in *rads* (100 erg/g) or *grays* (1 Gy = 100 rad). Ionizing radiation is typically generated from X-ray sources or the radioactive nuclides ^{60}Co and ^{137}Cs, which are relatively inexpensive by-products of nuclear fission. These nuclides produce X-rays or gamma rays (γ-rays), both of which have sufficient energy and penetrating power to efficiently kill microorganisms in bulk items such as food products and medical supplies.

Table 5.6 shows the dose necessary for a 10-fold reduction ($D10$) in number of selected microorganisms. The $D10$ value is analogous to the decimal reduction time for heat sterilization, and the killing curve of ionizing radiation yields a similar plot (**Figure 5.35**; compare with Figure 5.32). As is also true of heat treatments, killing endospores with ionizing radiation is more difficult than killing vegetative cells, and viruses are more difficult to kill than bacteria (Table 5.6). In addition, microorganisms in general are much more resistant to ionizing radiation than are multicellular organisms. For example, the lethal radiation dose for humans can be as low as 10 Gy if delivered over a short time period.

In the United States, the Food and Drug Administration (FDA) has approved the use of radiation for sterilization of such diverse items as surgical supplies, plastic labware, drugs, and even tissue grafts. Certain foods and food products such as fresh produce, poultry, meat products, and spices are also routinely irradiated to ensure that they are sterile or at least free of pathogens and also free of insects.

Filter Sterilization

Heat is an effective way to decontaminate most liquids, but heat-sensitive liquids and gases must be sterilized by other methods. To do this, the liquid or gas is passed through a filter with pores sufficiently small to trap any cells that may be present. For sterilization, a filter with pores of average size 0.2 µm is desirable; however, even such tiny holes will not trap most viruses. Commonly used filter pore sizes for the filter sterilization of small volumes, such as laboratory solutions, are 0.45 µm and 0.2 µm.

Table 5.6 Radiation sensitivity of some representative microorganisms

Type of microorganism	Characteristics	$D10^a$ (Gy)
Bacteria		
Clostridium botulinum	Gram-positive anaerobe; forms endospores	3300
Deinococcus radiodurans	Gram-negative, radiation-resistant coccus	2200
Lactobacillus brevis	Gram-positive, rod-shaped	1200
Bacillus subtilis	Gram-positive aerobe; forms endospores	600
Escherichia coli	Gram-negative, rod-shaped	300
Salmonella typhimurium	Gram-negative, rod-shaped	200
Fungi		
Aspergillus niger	Common mold	500
Saccharomyces cerevisiae	Baker's and brewer's yeast	500
Viruses		
Foot-and-mouth	Pathogen of cloven-hoofed animals	13,000
Coxsackie	Human pathogen	4500

^{a}$D10$ is the amount of radiation necessary to reduce the initial population or activity level 10-fold (1 logarithm). Gy, grays. 1 Gy is equivalent to 100 rads. The lethal dose for humans is 10 Gy.

Several types of filters are in routine use in microbiology, including depth filters, membrane filters, and nucleopore filters. A depth filter is a fibrous sheet or mat made from a random array of overlapping paper or borosilicate (glass) fibers that traps particles in the network of fibers (**Figure 5.36a**). Depth filters are important in biosafety applications. For example, manipulations of cell cultures, microbial cultures, and growth media require that contamination of both the operator and the experimental materials be minimized. These operations can be efficiently performed

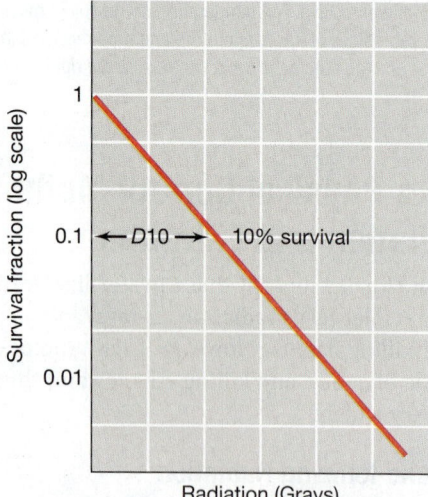

Figure 5.35 **Relationship between the survival fraction and the radiation dose of a microorganism.** The $D10$, which is the decimal reduction dose, can be interpolated from the data as shown.

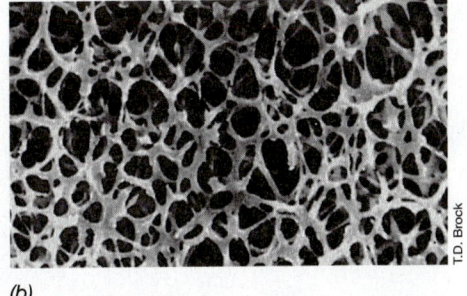

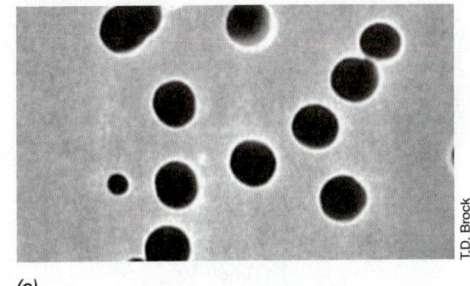

(a)   *(b)*   *(c)*

Figure 5.36 Microbiological filters. Scanning electron micrograph showing the structure of *(a)* a depth filter, *(b)* a conventional membrane filter, and *(c)* a nucleopore filter.

in a biological safety cabinet with airflow, both in and out of the cabinet, directed through a depth filter called a **HEPA filter**, or *h*igh-*e*fficiency *p*articulate *a*ir filter (Figure 5.34). HEPA filters typically remove 0.3-μm or larger particles from an airstream with an efficiency of greater than 99.9%.

Membrane filters are the most common type of filters used for liquid sterilization in the microbiology laboratory (Figure 5.36*b* and **Figure 5.37**). Membrane filters are composed of high tensile strength polymers such as cellulose acetate, cellulose nitrate, or polysulfone, manufactured in such a way as to contain a large number of tiny pores. Sterile membrane filter assemblies for the sterilization of relatively small volumes of liquids such as growth media are routinely used in research and clinical laboratories. Filtration is accomplished by using a syringe or a pump to force the liquid through the filtration apparatus into a sterile collection vessel (Figure 5.37).

Another type of membrane filter is the nucleopore filter (**Figure 5.38**). Nucleopore filters are made from a 10-μm-thick polycarbonate film that is treated with radiation and then etched with a chemical, yielding very uniform holes (Figure 5.36*c*). Nucleopore filters are commonly used to isolate specimens for scanning electron microscopy. Microorganisms are removed from a liquid or a natural sample, such as lake water, and concentrated on the filter where they can be observed directly with the microscope (Figure 5.38*a*).

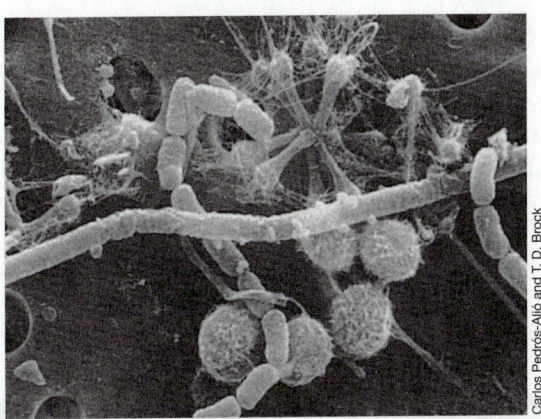

(a)

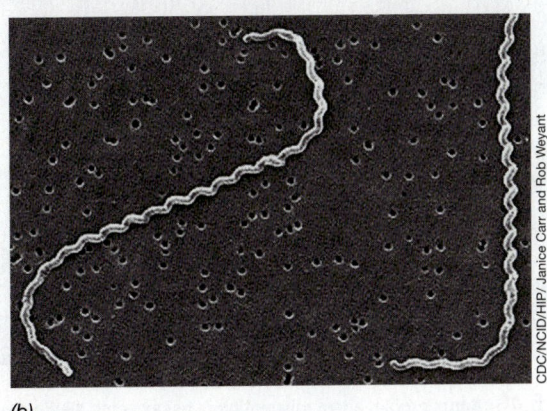

(b)

Figure 5.38 Scanning electron micrographs of bacteria trapped on nucleopore membrane filters. *(a)* Aquatic bacteria and algae. The pore size is 5 μm. *(b) Leptospira interrogans.* The bacterium is about 0.1 μm in diameter and up to 20 μm in length. The pore size of the filter is 0.2 μm.

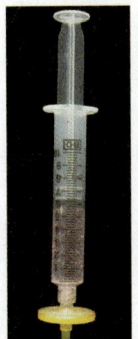

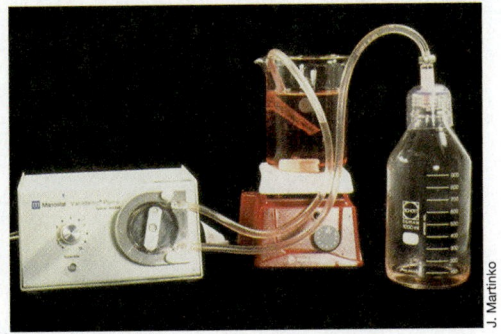

Figure 5.37 Membrane filters. Disposable, presterilized, and assembled membrane filter units. Left: a filter system designed for small volumes. Right: a filter system designed for larger volumes.

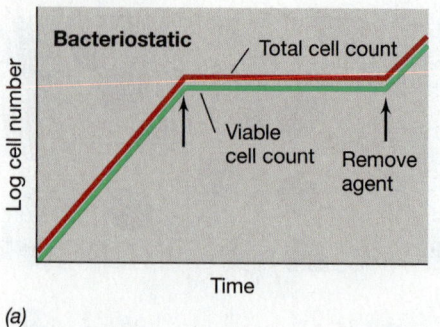
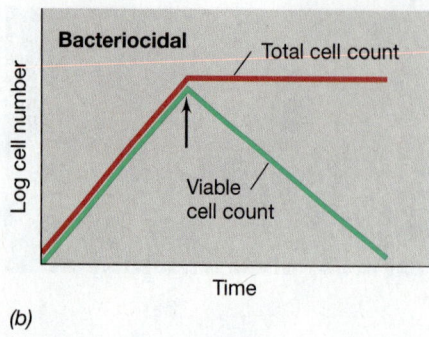
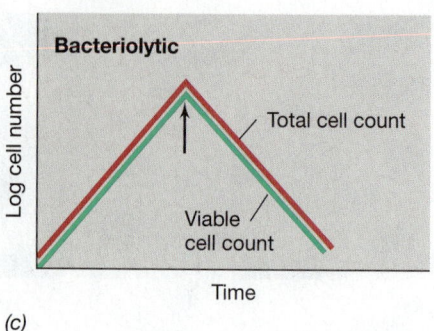

Figure 5.39 Different types of antimicrobial agents. *(a)* Bacteriostatic agents inhibit but do not kill. *(b)* Bacteriocidal agents kill. *(c)* Bacteriolytic agents lyse cells. At the time indicated by the arrow, a growth-inhibitory concentration of each antimicrobial agent was added to an exponentially growing culture. The turbidity and viable counts shown are characteristic of each type of agent.

5.19 Chemical Control of Microbial Growth

Chemicals are routinely used to control microbial growth, and an **antimicrobial agent** is a natural or synthetic chemical that kills or inhibits the growth of microorganisms. Agents that actually kill organisms are called *-cidal* agents, with a prefix indicating the type of microorganism killed. Thus, **bacteriocidal**, **fungicidal**, and **viricidal** agents kill bacteria, fungi, and viruses, respectively. Agents that do not kill but only inhibit growth are called *-static* agents, and include **bacteriostatic**, **fungistatic**, and **viristatic** compounds.

Effect of Antimicrobial Agents on Growth

Antibacterial agents are classified as bacteriostatic, bacteriocidal, or bacteriolytic (cell lysing) by observing their effects on bacterial cultures using viable and turbidimetric growth assays (**Figure 5.39**). Bacteriostatic agents are typically inhibitors of some important biochemical process, such as protein synthesis, and bind relatively weakly; if the agent is removed, the cells can resume growing. Many antibiotics fall into this category. Bacteriocidal agents, by contrast, bind tightly to their cellular targets and by definition kill the cell. However, the dead cells are not lysed, and total cell numbers, reflected in the turbidity of the culture, remain constant (Figure 5.39*b*). Formaldehyde would be an example of a bacteriocidal agent. Bacteriolytic agents kill cells by lysing them and releasing their cytoplasmic contents. Lysis decreases both the viable cell number and the total cell number (Figure 5.39*c*). An example of a bacteriolytic agent would be a detergent that ruptures the cytoplasmic membrane.

Assaying Antimicrobial Activity

Antimicrobial activity is measured by determining the smallest amount of the agent needed to inhibit the growth of a test organism, a value called the **minimum inhibitory concentration (MIC)**. To determine the MIC of a given agent against a given organism growing in a liquid medium (**Figure 5.40**), a series of tubes

Figure 5.40 Antimicrobial agent susceptibility assay using dilution methods. The assay defines the minimum inhibitory concentration (MIC). A series of increasing concentrations of antimicrobial agent is prepared in the culture medium. Each tube is inoculated with a specific concentration of a test organism, followed by a defined incubation period. Growth, measured as turbidity, occurs in those tubes with antimicrobial agent concentrations below the MIC.

Minimum inhibitory concentration

T. D. Brock

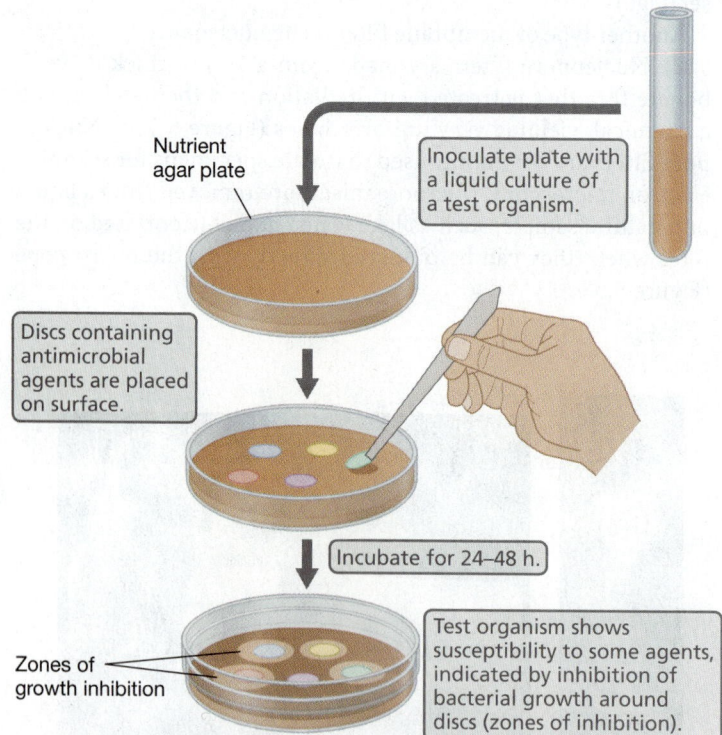

Nutrient agar plate

Inoculate plate with a liquid culture of a test organism.

Discs containing antimicrobial agents are placed on surface.

Incubate for 24–48 h.

Zones of growth inhibition

Test organism shows susceptibility to some agents, indicated by inhibition of bacterial growth around discs (zones of inhibition).

Figure 5.41 Antimicrobial agent susceptibility assay using diffusion methods. The antimicrobial agent diffuses from paper discs into the surrounding agar, inhibiting growth of susceptible microorganisms.

Table 5.7 Antiseptics, sterilants, disinfectants, and sanitizers[a]

Agent	Mode of action	Use
Antiseptics (germicides)		
Alcohol (60–85% ethanol or isopropanol in water)	Lipid solvent and protein denaturant	Topical antiseptic
Phenol-containing compounds (hexachlorophene, triclosan, chloroxylenol, chlorhexidine)	Disrupts cytoplasmic membrane	Soaps, lotions, cosmetics, deodorants, topical disinfectants; paper, leather, and textile industries
Cationic detergents, especially quaternary ammonium compounds (benzalkonium chloride)	Disrupts cytoplasmic membrane	Soaps, lotions, topical disinfectants; metal and petroleum industries
Hydrogen peroxide (3% solution)	Oxidizing agent	Topical antiseptic
Iodophors (Betadine®)	Iodinates proteins, rendering them nonfunctional; oxidizing agent	Topical antiseptic
Octenidine	Cationic surfactant, disrupts cytoplasmic membrane	Topical antiseptic
Sterilants, disinfectants, and sanitizers		
Alcohol (60–85% ethanol or isopropanol in water)	Lipid solvent and protein denaturant	General purpose disinfectant for virtually any surface
Cationic detergents (quaternary ammonium compounds, Lysol® and many related disinfectants)	Interact with phospholipids	Disinfectant/sanitizer for medical instruments, food and dairy equipment
Chlorine gas	Oxidizing agent	Disinfectant for drinking water and electrical/nuclear cooling towers
Chlorine compounds (chloramines, sodium hypochlorite, sodium chlorite, chlorine dioxide)	Oxidizing agent	Disinfectant/sanitizer for medical instruments, food/dairy equipment, and in water purification
Copper sulfate	Protein precipitant	Algicide in swimming pools
Ethylene oxide (gas)	Alkylating agent	Sterilant for temperature-sensitive materials such as plastics
Formaldehyde	Alkylating agent	Dilute (3% solution) as surface disinfectant/sterilant; concentrated (37% solution) as sterilant
Glutaraldehyde	Alkylating agent	Disinfectant or sterilant as 2% solution
Hydrogen peroxide	Oxidizing agent	Vapor used as sterilant
Iodophors (Wescodyne®)	Iodinates proteins; oxidizing agent	General disinfectant
OPA (ortho-phthalaldehyde)	Alkylating agent	Powerful disinfectant used for sterilizing medical instruments
Ozone	Strong oxidizing agent	Disinfectant for drinking water
Peroxyacetic acid	Strong oxidizing agent	Disinfectant/sterilant
Phenolic compounds	Protein denaturant	General purpose disinfectant
Pine oils (Pine-Sol®) (contains phenolics and detergents)	Protein denaturant	General purpose disinfectant for household surfaces

[a]Alcohols, hydrogen peroxide, and iodophors can be antiseptics, disinfectants, sanitizers, or sterilants depending on concentration, length of exposure, and form of delivery.

is inoculated with the test organism and a given amount of the agent added. After incubation, the tubes are scored for growth (turbidity), and the MIC is the lowest concentration of agent that completely inhibits the growth of the test organism.

Antimicrobial activity can also be assessed using solid media (**Figure 5.41**). Known amounts of an antimicrobial agent are added to filter-paper discs and the discs arranged on the surface of a uniformly inoculated agar plate. During incubation, the agent diffuses from the disc into the agar, establishing a gradient; the farther the chemical diffuses away from the filter paper, the lower is the concentration of the agent. A *zone of inhibition* is created with a diameter proportional to the amount of antimicrobial agent added to the disc, the solubility of the agent, the diffusion coefficient, and the overall effectiveness of the agent. The disc diffusion assay is routinely used to test clinically isolated pathogens for their antibiotic susceptibility (⮌ Section 27.5).

Chemical Antimicrobial Agents
Several antimicrobial agents are used to prevent growth of human pathogens on inanimate surfaces and on external body surfaces. These include sterilants, disinfectants, sanitizers, and antiseptics (**Table 5.7**).

Sterilants (also called *sterilizers* or *sporicides*) destroy all microorganisms, including endospores. Chemical sterilants are used for decontamination or sterilization in situations where it is impractical to use heat or radiation. Hospitals and laboratories, for example, must routinely decontaminate and sterilize heat-sensitive materials such as thermometers, lensed instruments, polyethylene tubing, catheters, and reusable medical equipment such as respirometers. *Cold sterilization,* a process by which gases such as ethylene oxide, formaldehyde, or peroxyacetic acid are used to treat objects within an enclosed device that resembles an autoclave, is the usual method employed. Liquid sterilants such as a sodium hypochlorite or amylphenol are used for instruments that cannot withstand high temperatures or gas.

Disinfectants are chemicals that kill microorganisms but not necessarily endospores and are used on inanimate objects. For example, phenol and cationic detergents are used to disinfect floors, tables, bench tops, walls, and so on (Table 5.7) and are important for infection control in hospitals and other medical settings. **Sanitizers**, by contrast, are less harsh chemicals than disinfectants and function to reduce microbial numbers but not necessarily sterilize the object. Sanitizers are widely used in the food industry to treat surfaces such as mixing and cooking equipment, dishes, and utensils, and are also used for dry handwashing when water is unavailable. **Antiseptics**, often called **germicides**, are chemicals that kill or inhibit the growth of microorganisms but are sufficiently nontoxic to animals to be applied to living tissues. Most germicides are used for handwashing or for treating surface wounds (Table 5.7). Certain antiseptics are also effective disinfectants. Ethanol, for example, can be both an antiseptic and a disinfectant, depending on the concentration and exposure time employed.

Several factors affect the efficacy of any chemical antimicrobial agent. For example, many antimicrobial agents are bound and inactivated by organic matter; thus, disinfecting a kitchen countertop littered with spilled foods is more difficult than disinfecting a clean countertop. Furthermore, bacteria often form biofilms, covering surfaces of tissue or medical devices with layers of microbial cells embedded in polysaccharides. Biofilms may slow or even completely prevent penetration of antimicrobial agents, reducing or negating their effectiveness.

Only sterilants are effective against endospores; these structures are extremely resistant because their spore coats prevent the penetration of most chemical agents (⟳ Section 2.16). Likewise, the bacterium *Mycobacterium tuberculosis*, the causal agent of tuberculosis, is resistant to common disinfectants because of the waxy nature of its cell wall (⟳ Sections 15.11 and 29.4). Thus, the ultimate efficacy of antiseptics, disinfectants, sterilants, and other antimicrobial compounds must be determined empirically and under the actual conditions of use. Only by actually testing the chemical and assaying for microbial growth both before and after treatment can one be confident that the agent is working as it should.

MINIQUIZ -

- Distinguish between the antimicrobial effects of -static, -cidal, and -lytic agents.
- Describe how the minimum inhibitory concentration of an antibacterial agent is determined.
- Distinguish between a sterilant, a disinfectant, and an antiseptic. What is cold sterilization?

BIG IDEAS

5.1 • Microbial growth is defined as an increase in cell numbers and is the final result of the doubling of all cell components prior to actual division that yields two daughter cells. Most microorganisms grow by binary fission.

5.2 • Cell division and chromosome replication are coordinately regulated, and the Fts proteins are keys to these processes. With the help of MinE, FtsZ defines the cell division plane and helps assemble the divisome, the protein complex that orchestrates cell division.

5.3 • MreB helps define cell shape, and in rod-shaped cells, MreB forms a cytoskeletal coil that directs cell wall synthesis along the long axis of the cell. The protein crescentin plays an analogous role in *Caulobacter*, leading to formation of a curved cell. The eukaryotic shape and cell division proteins actin and tubulin have prokaryotic counterparts.

5.4 • During bacterial growth, new cell peptidoglycan is synthesized by the insertion of new glycan tetrapeptide units into preexisting peptidoglycan. Bactoprenol facilitates transport of these units through the cytoplasmic membrane. Transpeptidation completes the process of cell wall synthesis by cross-linking adjacent ribbons of peptidoglycan at muramic acid residues.

5.5 • Microbial cells undergo exponential growth, and a semilogarithmic plot of cell numbers with time can reveal the doubling time of the population. Simple mathematics can be used to calculate various growth expressions from cell numbers data. Key expressions here are n, the number of generations; t, time; and g, generation time. The generation time is expressed as $g = t/n$.

5.6 • Microorganisms show a characteristic growth pattern when inoculated into a fresh culture medium. There is usually a lag phase and then growth commences in an exponential fashion. As essential nutrients are depleted or toxic products build up, growth ceases and the population enters the stationary phase. Further incubation can lead to cell death.

5.7 • The chemostat is an open system used to maintain cell populations in exponential growth for extended periods. In a chemostat, the rate at which a culture is diluted with fresh growth medium controls the doubling time of the population, while the cell density (cells/ml) is controlled by the concentration of a growth-limiting nutrient dissolved in the fresh medium.

5.8 • Cell counts can be done under the microscope using counting chambers. Microscopic counts measure the total number of cells in the sample and are useful for assessing the total cell numbers in a microbial habitat. Certain stains can be used to target specific cell populations in a sample.

5.9 • Viable cell counts (plate counts) measure only the living population present in the sample with the assumption that each colony originates from the growth and division of a single cell. Depending on the growth medium and conditions employed, plate counts can be fairly accurate assessments or can be highly unreliable.

5.10 • Turbidity measurements are an indirect but very rapid and useful method of measuring microbial growth. However, in order to relate a turbidity value to a direct cell number, a standard curve plotting these two parameters against one another must first be established.

5.11 • Temperature is a major environmental factor controlling microbial growth. An organism's cardinal temperatures describe the minimum, optimum, and maximum temperatures at which it grows. Microorganisms can be grouped by their cardinal temperature from cold-loving to heat-loving as psychrophiles, mesophiles, thermophiles, and hyperthermophiles.

5.12 • Organisms with temperature optima below 20°C are called psychrophiles, and the most extreme representatives inhabit constantly cold environments. Psychrophiles have evolved macromolecules that remain flexible and functional at cold temperatures, but that can be unusually sensitive to warm temperatures.

5.13 • Organisms with growth temperature optima between 45 and 80°C are called thermophiles while those with optima greater than 80°C are hyperthermophiles. These organisms inhabit hot environments that can have temperatures even above 100°C. Thermophiles and hyperthermophiles produce heat-stable macromolecules.

5.14 • The acidity or alkalinity of an environment can greatly affect microbial growth. Some organisms grow best at low or high pH (acidophiles and alkaliphiles, respectively), but most organisms grow best between pH 5.5 and 8. The internal pH of a cell must stay relatively close to neutral to prevent the destruction of DNA or RNA.

5.15 • The water activity of an aqueous environment is controlled by its dissolved solute concentration. To survive in high-solute environments, organisms produce or accumulate compatible solutes to maintain the cell in positive water balance. Some microorganisms grow best at reduced water potential and some even require high levels of salts for growth.

5.16 • Aerobes require O_2 to live, whereas anaerobes do not and may even be killed by O_2. Facultative organisms can live with or without O_2. Special techniques are needed to grow aerobic and anaerobic microorganisms. Several toxic forms of oxygen can form in the cell, but enzymes are present that neutralize most of them. Superoxide is a major toxic form of oxygen.

5.17 • Sterilization is the killing of all organisms and viruses, and heat is the most widely used method of sterilization. An autoclave employs moist heat under pressure, achieving temperatures above the boiling point of water. Pasteurization does not sterilize liquids, but it reduces microbial load, kills most pathogens, and inhibits the growth of spoilage microorganisms.

5.18 • Radiation can effectively inhibit or kill microorganisms. Ultraviolet radiation is used for decontaminating surfaces and air. Ionizing radiation is used for sterilization and decontamination where penetration is required. Filters remove microorganisms from air or liquids. Membrane filters are used for sterilization of heat-sensitive liquids, and nucleopore filters are used to isolate specimens for electron microscopy.

5.19 • Chemicals are commonly used to control microbial growth. Chemicals that kill organisms are called -cidal agents while those that arrest growth but do not kill are called -static agents. Antimicrobial agents are tested for efficacy by determining their ability to inhibit growth in vitro. Sterilants, disinfectants, and sanitizers are used to decontaminate nonliving material, while antiseptics and germicides are used to reduce the microbial load on living tissues.

MasteringMicrobiology®

Review what you know and challenge what you have learned with MasteringMicrobiology! Access study materials, chapter quizzes, animations, and microbiology lab tutorials in the Study Area to ensure that you have mastered this chapter's content.

REVIEW OF KEY TERMS

Acidophile an organism that grows best at low pH; typically below pH 5.5

Aerobe an organism that can use O_2 in respiration; some require O_2

Aerotolerant anaerobe a microorganism unable to respire O_2 but whose growth is unaffected by it

Alkaliphile an organism that has a growth pH optimum of 8 or higher

Anaerobe an organism that cannot use O_2 in respiration and whose growth is typically inhibited by O_2

Antimicrobial agent a chemical compound that kills or inhibits the growth of microorganisms

Antiseptic (germicide) a chemical agent that kills or inhibits growth of microorganisms and is sufficiently nontoxic to be applied to living tissues

Autoclave a sealed heating device that destroys microorganisms with temperature and steam under pressure

Bacteriocidal agent an agent that kills bacteria

Bacteriostatic agent an agent that inhibits bacterial growth

Batch culture a closed-system microbial culture of fixed volume

Binary fission cell division following enlargement of a cell to twice its minimum size

Biofilm an attached polysaccharide matrix containing bacterial cells

Cardinal temperatures the minimum, maximum, and optimum growth temperatures for a given organism

Chemostat a device that allows for the continuous culture of microorganisms with independent control of both growth rate and cell number

Compatible solute a molecule that is accumulated in the cytoplasm of a cell for adjustment of water activity but that does not inhibit biochemical processes

Decontamination treatment of a surface or object to make it safe to handle

Disinfectant an antimicrobial agent used only on inanimate objects

Disinfection rendering a surface or object free of all pathogen microorganisms

Divisome a complex of proteins that directs cell division processes in prokaryotes

Exponential growth growth of a microbial population in which cell numbers double within a specific time interval

Extreme halophile a microorganism that requires very large amounts of NaCl, usually greater than 10% and in some cases near to saturation, for growth

Facultative with respect to O_2, an organism that can grow in either its presence or absence

FtsZ a protein that forms a ring along the mid-cell division plane to initiate cell division

Fungicidal agent an agent that kills fungi

Fungistatic agent an agent that inhibits fungal growth

Generation time the time required for a population of microbial cells to double

Germicide (antiseptic) a chemical agent that kills or inhibits growth of microorganisms and is sufficiently nontoxic to be applied to living tissues

Growth an increase in cell number

Halophile a microorganism that requires NaCl for growth

Halotolerant a microorganism that does not require NaCl for growth but can grow in the presence of NaCl, in some cases, substantial levels of NaCl

HEPA filter a *h*igh-*e*fficiency *p*articulate *a*ir filter that removes particles, including microorganisms, from intake or exhaust air flow

Hyperthermophile a prokaryote that has a growth temperature optimum of 80°C or greater

Mesophile an organism that grows best at temperatures between 20 and 40°C

Microaerophile an aerobic organism that can grow only when O_2 tensions are reduced from that present in air

Minimum inhibitory concentration (MIC) the minimum concentration of a substance necessary to prevent microbial growth

Neutrophile an organism that grows best at neutral pH, between pH 5.5 and 8

Obligate anaerobe an organism that cannot grow in the presence of O_2

Osmophile an organism that grows best in the presence of high levels of solute, typically a sugar

Pasteurization the heat treatment of milk or other liquids to reduce its total number of microorganisms

pH the negative logarithm of the hydrogen ion (H^+) concentration of a solution

Plate count a method of counting viable cells; the number of colonies on a plate is used as a measure of cell numbers

Psychrophile an organism with a growth temperature optimum of 15°C or lower and a maximum growth temperature below 20°C

Psychrotolerant capable of growing at low temperatures but having an optimum above 20°C

Sanitizer an agent that reduces microorganisms to a safe level, but may not eliminate them

Sterilant (sterilizer, sporicide) a chemical agent that destroys all forms of microbial life

Sterilization the killing or removal of all living organisms and viruses

Thermophile an organism whose growth temperature optimum lies between 45 and 80°C

Transpeptidation formation of peptide cross-links between muramic acid residues in peptidoglycan synthesis

Viable capable of reproducing

Viable count a measurement of the concentration of live cells in a population

Viricidal agent an agent that stops viral replication and activity

Viristatic agent an agent that inhibits viral replication

Water activity the ratio of the vapor pressure of air in equilibrium with a solution to the vapor pressure of pure water

Xerophile an organism that is able to live, or that lives best, in very dry environments

REVIEW QUESTIONS

1. Describe the key molecular processes that occur when a cell grows and divides. (Section 5.1)

2. Describe the role of proteins present at the divisome. Does the FtsZ ring form before or after chromosome replication? (Section 5.2)

3. How do cells of *Escherichia coli* that carry a mutation in *mreB* (the gene encoding the protein MreB) look different microscopically from wild-type (unmutated) cells? What is the reason for this? (Section 5.3)

4. Describe how new peptidoglycan subunits are inserted into the growing cell wall. How does the antibiotic penicillin kill bacterial cells, and why does it kill only growing cells? (Section 5.4)

5. What is the difference between the specific growth rate (k) of an organism and its generation time (g)? (Section 5.5)

6. Describe the growth cycle of a population of bacterial cells from the time this population is first inoculated into fresh medium. (Section 5.6)

7. How does a chemostat regulate growth rate and cell numbers independently? (Section 5.7)

8. What is the difference between a total cell count and a viable cell count? (Sections 5.8 and 5.9)

9. How can turbidity be used as a measure of cell numbers? (Section 5.10)

10. Examine the graph describing the relationship between growth rate and temperature (Figure 5.19). Give an explanation, in biochemical terms, of why the optimum temperature for an organism is usually closer to its maximum than its minimum. (Section 5.11)

11. Describe a habitat where you would find a psychrophile, and one where you would find a hyperthermophile. How can these organisms survive under such harsh conditions? (Sections 5.12 and 5.13)

12. Concerning the pH of the environment and of the cell, in what ways are acidophiles and alkaliphiles different? In what ways are they similar? (Section 5.14)

13. Write an explanation in molecular terms for how a halophile is able to make water flow into the cell while growing in a solution high in NaCl. (Section 5.15)

14. Contrast an aerotolerant and an obligate anaerobe in terms of sensitivity to O_2 and ability to grow in the presence of O_2. How does an aerotolerant anaerobe differ from a microaerophile? (Section 5.16)

15. Compare and contrast the enzymes catalase, superoxide dismutase, and superoxide reductase as regards their substrates and products. (Section 5.16)

16. Contrast the terms thermal death time and decimal reduction time. How would the presence of bacterial endospores affect either value? (Section 5.17)

17. Describe the principle of the autoclave. How does it differ from simple boiling? Microbial culture media does not boil in the autoclave; why not? (Section 5.17)

18. Describe the effects of a lethal dose of ionizing radiation at the molecular level. (Section 5.18)

19. What type of filter would be used to filter sterilize a heat-sensitive liquid? (Section 5.18)

20. Describe the procedure for obtaining the minimum inhibitory concentration (MIC) for a chemical that is bacteriocidal for *Escherichia coli*. (Section 5.19)

21. Contrast the action of disinfectants and antiseptics. Disinfectants are not used on living tissue; why not? (Section 5.19)

APPLICATION QUESTIONS

1. Calculate g and k in a growth experiment in which a medium was inoculated with 5×10^6 cells/ml of *Escherichia coli* cells and, following a 1-h lag, grew exponentially for 5 h, after which the population was 5.4×10^9 cells/ml.

2. *Escherichia coli* but not *Pyrolobus fumarii* will grow at 40°C, while *P. fumarii* but not *E. coli* will grow at 110°C. What is happening (or not happening) to prevent growth of each organism at the nonpermissive temperature?

3. In which direction (into or out of the cell) will water flow in cells of *Escherichia coli* (an organism found in your large intestine) suddenly suspended in a solution of 20% NaCl? What if the cells were suspended in distilled water? If growth nutrients were added to each cell suspension, which (if either) would support growth, and why?

6 · Microbial Genomics

microbiology**now**

Genomics and New *Archaea*

Until recently, three phyla of *Archaea* were known, the *Euryarchaeota, Crenarchaeota*, and *Nanoarchaeota*. Curiously, every cultured species was isolated from an extreme environment, habitats that were either strictly anoxic or punishingly hot, salty, or acidic. This led many microbiologists to conclude that *Archaea* were mainly extremophiles and that they did not inhabit oceans, lakes, and soil in significant numbers. But microbial ecologists began to question this assumption when, using fluorescent microscopy, *Archaea* only marginally affiliated with *Crenarchaeota* were detected in marine and freshwater samples. Who were these organisms, and how were they making a living?

A group of microbiologists at the University of Washington in Seattle had a hunch about the metabolism of these *Archaea* and set about to isolate these organisms from marine samples (photo). With persistence, patience, and good scientific intuition, the group successfully isolated *Nitrosopumilus*, the first ammonia-oxidizing (nitrifying) archaeon known (photo inset). Although many species of *Bacteria* can nitrify, *Nitrosopumilus* can oxidize the trace amounts of ammonia found in open ocean waters, something nitrifying *Bacteria* cannot. With pure cultures of these organisms in hand, their phylogeny was more deeply explored using the powerful tools of genomics. Were nitrifying *Archaea* really just "highly divergent *Crenarchaeota*"?

Genomics is able to answer such questions, and careful analyses of the genomes of two archaeal nitrifiers[1] clearly showed that they formed their own phylum, now called the *Thaumarchaeota*. Genomic analyses allowed the entire gene set of these *Archaea* to be compared with that of all other *Archaea*. Besides revealing a fourth phylum of *Archaea*, genomics unveiled the metabolic peculiarities of *Thaumarchaeota* and this in turn provided a window into the ecological role they may play in their nutrient-deficient habitats.

[1]Spang, A., et al., 2010. Distinct gene set in two different lineages of ammonia-oxidizing *Archaea* supports the phylum *Thaumarchaeota*. *Trends in Microbiol. 18:* 331–340.

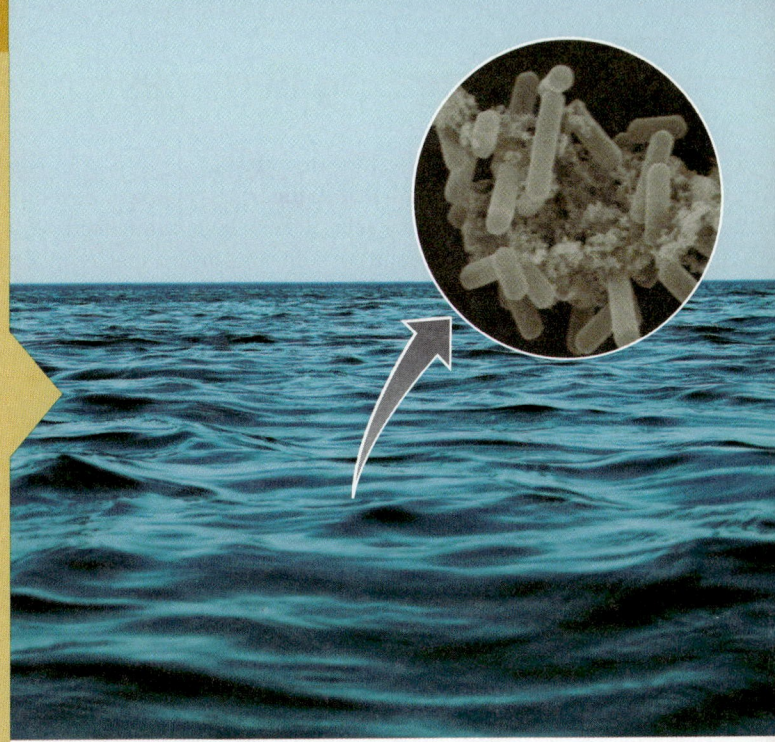

The **genome** is an organism's entire complement of genetic information, including genes that encode proteins, RNAs, and regulatory sequences, and any noncoding DNA that may be present. The genome sequence of an organism not only reveals its genes but also yields important clues to how the organism functions and its evolutionary history. Genomics encompasses not only the complete sequencing of the DNA and identification of all the genes but also the study of gene expression at a genome-wide level. The traditional approach to studying gene expression was to focus on a single gene or group of related genes. In the genomics era, the expression of all or most of an organism's genes can be examined in a single experiment.

Advances in genomics rely heavily on improvements in molecular technology and computing power. Major advances include the automation of DNA sequencing, the miniaturization of analytical procedures, and the development of powerful computational methods for analysis of DNA and protein sequences. New advances arrive every year and have driven down the cost and increased the speed at which genomes are analyzed. Here we focus on microbial genomes, some techniques used to analyze these genomes, and what microbial genomics has revealed thus far.

I • Investigating Genomes

The word **genomics** refers to the discipline of mapping, sequencing, analyzing, and comparing genomes. Several thousand genomes from prokaryotes have been sequenced, including those from multiple strains of some important species of *Bacteria* and *Archaea*. Because new advances in DNA sequencing appear frequently, the number of sequenced genomes will continue to grow rapidly. Today the major bottleneck in genomics is the analysis and visualization of vast amounts of nucleic acid sequence data. Nevertheless, genome sequences continue to provide new insights in areas as diverse as medicine and microbial evolution.

6.1 Introduction to Genomics

The first genome sequenced was the 3569-nucleotide RNA genome of the virus MS2 (cə Section 9.8) in 1976. The first DNA genome sequenced was the 5386-nucleotide sequence of the single-stranded DNA virus φX174 (cə Section 9.3) in 1977. The first bacterial genome sequenced was the 1,830,137-base-pair (bp) chromosome of *Haemophilus influenzae* published in 1995. The DNA sequences of several thousand prokaryotic genomes are now available in public databases (for an up-to-date list search http://www.genomesonline.org/), and Table 6.1 lists some representative examples. These include species of both *Bacteria* and *Archaea*, and both circular and linear genomes. Although rare, linear chromosomes are present in several *Bacteria*, including *Borrelia burgdorferi*, the agent of Lyme disease, and the important antibiotic-producing genus *Streptomyces*. Bacterial genomes range in size from approximately 0.5 to 13 megabase pairs (Mbp) and encode roughly 500 to 10,000 protein-coding genes, respectively.

The genomes of many higher organisms, including the haploid human genome, which contains about 3 billion bp but only around 25,000 protein-coding genes, have also been sequenced. The largest genomes so far sequenced, in terms of their total number of genes, are those of the black cottonwood tree (a species of poplar) with around 45,000 genes and the protozoan *Trichomonas* with an estimated 60,000 protein-encoding genes; both have many more genes than do humans.

The genomes of many pathogens have been sequenced. In some cases multiple strains of a pathogen that vary in virulence have been compared in the hope of revealing which genes are medically relevant. Moreover, hyperthermophiles (cə Section 5.12) have important uses in biotechnology because their enzymes are heat-stable. Originally the needs of the biomedical and biotechnology industries greatly affected the choice of organisms to sequence. However, genome sequencing is now so routine that projects are less governed by medical or technological needs. Indeed, a recent trend is to sequence and compare several different strains of the same organism to get a picture of genes held in common versus those that are optional (core/pan genome, Section 6.13). The genome list in Table 6.1 also includes widely studied model organisms such as *Bacillus subtilis* (sporulation), *Escherichia coli* (general biology and model gram-negative bacterium), and *Pseudomonas aeruginosa* (pathogen and model gram-negative bacterium).

MINIQUIZ
- How many genes are in the human genome?
- Name some organisms whose genomes are larger than the human genome.

6.2 Sequencing Genomes

In biology the term **sequencing** refers to determining the precise order of subunits in a macromolecule. In the case of DNA (or RNA), the sequence is the *order* in which the nucleotides are aligned. DNA sequencing technology is advancing so quickly that two or three new methods appear every year, although only a few gain widespread acceptance or stand the test of time. This is well illustrated by the drop in cost of sequencing 1 megabase of DNA. Between 2001 and 2011 this dropped 10,000-fold! Table 6.2 summarizes the sequencing methods discussed here.

First-Generation DNA Sequencing: The Sanger Dideoxy Method

The first widely used method for sequencing DNA was the dideoxy method invented by the British scientist Fred Sanger, who won a Nobel Prize for this accomplishment. Although superseded for genome sequencing by newer technologies, the dideoxy method is still used for some sequencing applications. Sanger introduced several important concepts that are still used in many newer sequencing

Table 6.1 Select prokaryotic genomes[a]

Organism	Lifestyle[b]	Size (bp)	ORFs[c]	Comments
Bacteria				
Hodgkinia cicadicola	E	143,795	169	Degenerate cicada endosymbiont
Carsonella ruddii	E	159,662	182	Degenerate psyllid endosymbiont
Buchnera aphidicola BCc	E	422,434	362	Aphid endosymbiont
Mycoplasma genitalium	P	580,070	470	Smallest nonsymbiotic bacterial genome
Borrelia burgdorferi	P	910,725	853	Spirochete, linear chromosome, causes Lyme disease
Rickettsia prowazekii	P	1,111,523	834	Obligate intracellular parasite, causes epidemic typhus
Treponema pallidum	P	1,138,006	1041	Spirochete, causes syphilis
Methylophilaceae family, strain HTCC2181	FL	1,304,428	1354	Marine methylotroph, smallest free-living genome
Aquifex aeolicus	FL	1,551,335	1544	Hyperthermophile, autotroph
Prochlorococcus marinus	FL	1,657,990	1716	Commonest marine oxygenic phototroph
Streptococcus pyogenes	FL	1,852,442	1752	Causes strep throat and scarlet fever
Thermotoga maritima	FL	1,860,725	1877	Hyperthermophile
Chlorobaculum tepidum	FL	2,154,946	2288	Model green phototrophic bacterium
Deinococcus radiodurans	FL	3,284,156	2185	Radiation resistant, multiple chromosomes
Synechocystis sp.	FL	3,573,470	3168	Model cyanobacterium
Bdellovibrio bacteriovorus	FL	3,782,950	3584	Predator of other prokaryotes
Caulobacter crescentus	FL	4,016,942	3767	Complex life cycle
Bacillus subtilis	FL	4,214,810	4100	Gram-positive genetic model
Mycobacterium tuberculosis	P	4,411,529	3924	Causes tuberculosis
Escherichia coli K-12	FL	4,639,221	4288	Gram-negative genetic model
Escherichia coli O157:H7	FL	5,594,477	5361	Enteropathogenic strain of E. coli
Bacillus anthracis	FL	5,227,293	5738	Pathogen, biowarfare agent
Pseudomonas aeruginosa	FL	6,264,403	5570	Metabolically versatile opportunistic pathogen
Streptomyces coelicolor	FL	8,667,507	7825	Linear chromosome, produces antibiotics
Bradyrhizobium japonicum	FL	9,105,828	8317	Nitrogen fixation, nodulates soybeans
Sorangium cellulosum	FL	13,033,799	9367	Myxobacterium, forms multicellular fruiting bodies
Archaea				
Nanoarchaeum equitans	P	490,885	552	Smallest nonsymbiotic cellular genome
Thermoplasma acidophilum	FL	1,564,905	1509	Thermophile, acidophile
Methanocaldococcus jannaschii	FL	1,664,976	1738	Methanogen, hyperthermophile
Pyrococcus horikoshii	FL	1,738,505	2061	Hyperthermophile
Halobacterium salinarum	FL	2,571,010	2630	Extreme halophile, bacteriorhodopsin
Sulfolobus solfataricus	FL	2,992,245	2977	Hyperthermophile, sulfur chemolithotroph
Haloarcula marismortui	FL	4,274,642	4242	Extreme halophile, bacteriorhodopsin
Methanosarcina acetivorans	FL	5,751,000	4252	Acetate using methanogen

[a]Information on prokaryotic genomes can be found at http://cmr.jcvi.org, a website maintained by The J. Craig Venter Institute, Rockville, MD, and at http://www.genomesonline.org.
[b]E, endosymbiont; P, parasite; FL, free-living.
[c]Open reading frames. Genes encoding known proteins are included, as well as ORFs that could encode a protein greater than 100 amino acid residues. Smaller ORFs are not included unless they show similarity to a gene from another organism or unless the codon bias is typical of the organism being studied.

schemes. These include sequencing by DNA synthesis rather than breakdown, using dideoxynucleotides to block chain extension, and using labeled precursors for detection.

In sequencing by DNA synthesis, short DNA oligonucleotides (typically 10–20 nucleotides) with defined sequences are used as primers. These are synthesized artificially. **Primers** are short segments of DNA or RNA that initiate the synthesis of new strands of nucleic acid. During DNA replication in vivo, RNA primers are used (⟳ Section 4.4), but in biotechnology, DNA primers are used because they are more stable than RNA primers.

Table 6.2 DNA sequencing methods

Generation	Method	Features
First generation	Sanger dideoxy method (radioactivity or fluorescence; DNA amplification)	Read length: 700–900 bases Used for the human genome project
Second generation	454 Pyrosequencing (fluorescence; DNA amplification; massively parallel)	Read length: 400–500 bases Used to sequence genome of James Watson (completed 2007)
	Illumina/Solexa method (fluorescence; DNA amplification; massively parallel)	Read length: 50–100 bases Giant panda genome (2009; Beijing Genome Institute) Denisovan genome (2010)
	SOLiD method (fluorescence; DNA amplification; massively parallel)	Read length 50–100 bases
Third generation	HeliScope Single Molecule Sequencer (fluorescence; single molecule)	Read length: up to 55 bases Fossil DNA accuracy greatly improved
	Pacific Biosciences SMRT (fluorescence; single molecule; zero mode waveguide)	Read length: 2500–3000 bases
Fourth generation	Ion torrent (electronic—pH; DNA amplification)	Read length: 100–200 bases Sequenced genome of Intel cofounder Gordon Moore (originator of Moore's law), 2011
	Oxford nanopore (electronic—current; single molecule; real time)	Read length: thousands of bases Portable MinION unit is approximately the size of a flash drive

In the Sanger procedure the sequence is determined by making a copy of the original single-stranded DNA using the enzyme DNA polymerase. As previously discussed (↩ Section 4.4), this enzyme adds deoxyribonucleoside triphosphates to a growing DNA chain. However, in Sanger sequencing, small amounts of the corresponding dideoxyribonucleotide are included in each of four incubation mixtures, one for each of the four bases—adenine,

Normal deoxyribonucleotide   **Dideoxyribonucleotide**

(a)

Direction of chain growth

No free **3′**-OH; replication will stop at this point

(b)

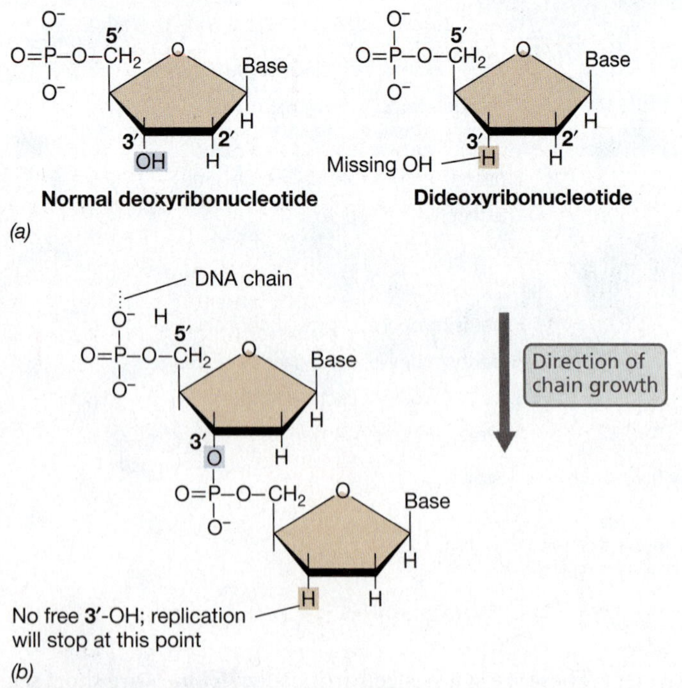

Figure 6.1 **Dideoxynucleotides and Sanger sequencing.** *(a)* A normal deoxyribonucleotide has a hydroxyl group on the 3′-carbon and a dideoxyribonucleotide does not. *(b)* Elongation of the chain terminates where a dideoxyribonucleotide is incorporated.

guanine, cytosine, and thymine (**Figure 6.1**). The dideoxy analog is a specific *chain-termination* reagent, and because it lacks a 3′-hydroxyl, it prevents further elongation of the chain after its insertion. Because dideoxyribonucleotides insert randomly, DNA chains of varying length are obtained and are separated on the basis of size by gel electrophoresis (Figure 6.1).

Originally four separate reactions (and four separate gel lanes) were used for each sequence determination, one for fragments ending with each of the four bases. The positions of the bands were located by using labeled precursors (originally radioactive but now fluorescent). By aligning the four dideoxyribonucleotide lanes and noting the vertical position of each fragment relative to its neighbor, the DNA sequence is read directly from the gel (Figure 6.2).

Automated DNA-sequencing systems use fluorescent dye-labeled primers (or nucleotides) instead of radioactive ones. The products are separated by electrophoresis in a capillary tube and the bands are scanned by a fluorescence-detecting laser. Since each of the four bases uses a fluorescent label of a different color, all four reactions are run in a single lane; the results are then analyzed by computer (**Figure 6.2**).

Shotgun Sequencing

Shotgun sequencing refers to the *preparation* of DNA for sequencing, not the sequencing itself. Most genomic sequencing projects employ shotgun sequencing. Analysis of a genome usually begins with the construction of a **genomic library**—the molecular cloning of DNA fragments that cover the entire genome (↩ Section 11.4). In the shotgun approach, the entire genome, cleaved into fragments, is cloned. The fragments are then sequenced. At this point the order and orientation of the DNA fragments are unknown. The sequences are analyzed by a computer that searches for overlapping sequences and assembles the sequenced fragments in the correct order. By its very nature, much sequencing in the shotgun method is redundant. To ensure

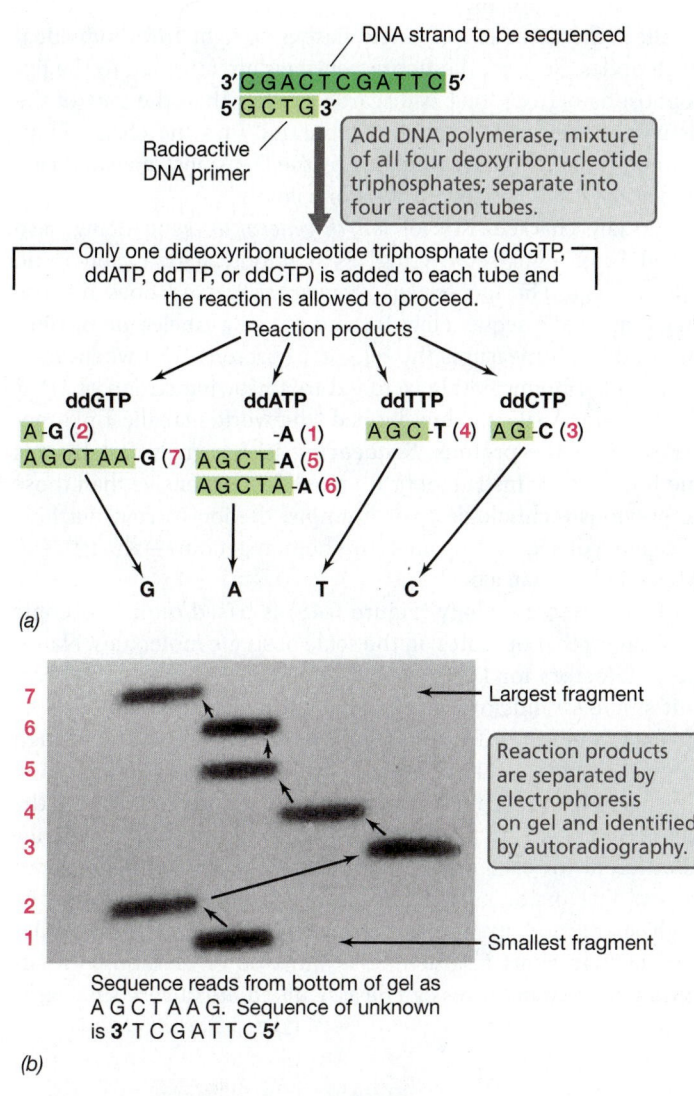

(a)

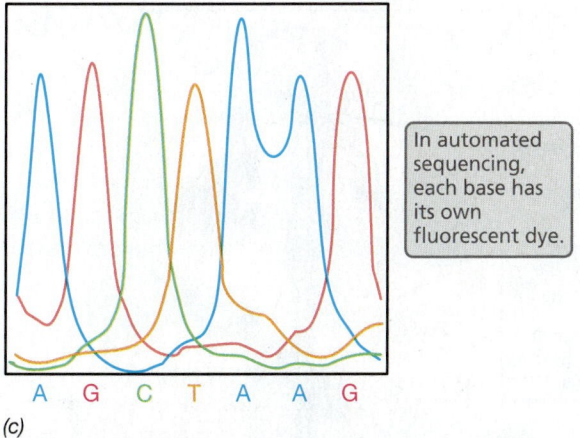

(b)

Sequence reads from bottom of gel as
A G C T A A G. Sequence of unknown
is **3′** T C G A T T C **5′**

(c)

Figure 6.2 DNA sequencing using the Sanger method. *(a)* Note that four different reactions must be run, one with each dideoxynucleotide. Because these reactions are run in vitro, the primer for DNA synthesis can be DNA. *(b)* A portion of a gel containing the reaction products from part a. *(c)* Results of sequencing the same DNA as shown in parts a and b, but using an automated sequencer and fluorescent labels. The DNA fragments are separated by size on a single capillary column and each fluorescently labeled dideoxyribonucleotide detected with a laser detector.

full coverage of a genome it is necessary to sequence a very large number of clones, many of which are identical or nearly identical. Typically, 7–10 replicate sequences (referred to as *7- to 10-fold coverage*) are obtained for any given part of the genome. This greatly reduces errors because the redundancy in sequencing allows for a consensus nucleotide to be selected at any ambiguous point.

For shotgun sequencing to be successful, the cloning must be efficient (many clones are needed) and, as far as possible, the cloned DNA fragments should be randomly generated. This may be done by enzymatic digestion of the DNA or by physical methods. The DNA fragments can be purified by size using gel electrophoresis (♻ Section 11.1) before cloning and sequencing.

Second-Generation DNA Sequencing

The term "generation" in DNA sequencing refers to successive major changes in technology that confer significant increases in speed combined with a drop in the cost of sequencing. The defining characteristic of *second-generation* sequencing is the use of *massively parallel methods*. In other words, very large numbers of samples are sequenced side-by-side in the same machine. Two major requirements for this are miniaturization and increased computing power. Second-generation methods generate sequence data 100 times faster than earlier methods. The three most widely used second-generation methods are 454 Life Sciences pyrosequencing, Illumina/Solexa sequencing, and the SOLiD/Applied Biosystems method.

In the 454 system, the DNA is broken into single-stranded segments a few hundred bases long. Each fragment is then attached to a small bead. The DNA is amplified by the polymerase chain reaction (PCR, ♻ Section 11.3), resulting in each bead carrying several identical copies of the DNA. Using robotics, the beads are then put into a fiber-optic plate containing more than a million wells, each of which holds just one bead.

Like Sanger sequencing (Figure 6.2), pyrosequencing uses synthesis of a complementary strand by DNA polymerase (**Figure 6.3**). However, instead of chain termination, each time a deoxyribonucleotide is incorporated, a molecule of pyrophosphate is released. This provides the energy necessary for activity of the light-emitting enzyme *luciferase*, which is included in each well. The four nucleotides flow sequentially over the plate in a fixed order. The one that yields a light pulse identifies which base was inserted. The Illumina/Solexa method resembles Sanger sequencing in using DNA synthesis and nucleotide chain terminators. However, in the Illumina system, the terminators are deoxy (rather than dideoxy) ribonucleotides and can be reversibly incorporated. In addition, each of the four different deoxyribonucleotides carries its own fluorescent tag that functions as a blocking group for the 3′-OH, thus causing chain termination.

Third- and Fourth-Generation DNA Sequencing

The key characteristic of *third-generation* sequencing is the sequencing of *single molecules* of DNA. There are two main approaches: one based on microscopy and the other on nanotechnology. In the *HeliScope Single Molecule Sequencer*, single-stranded fragments of DNA about 32 bases long are attached in an array on a glass slide. As the complementary strand is synthesized,

UNIT 2

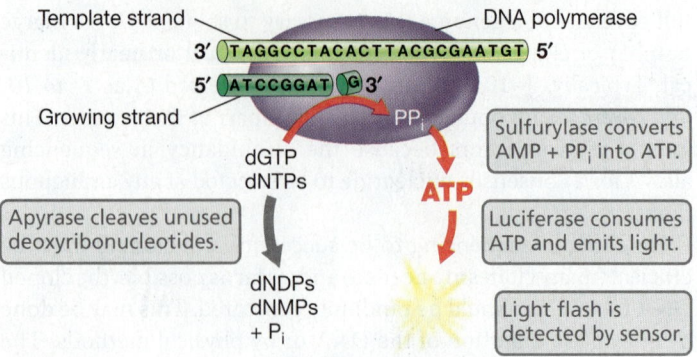

Figure 6.3 Mechanism of pyrosequencing. Whenever a new deoxyribonucleotide is inserted into the growing strand of DNA (red arrows), pyrophosphate (PP$_i$) is released and is used to make ATP from AMP by the enzyme sulfurylase. The ATP is consumed by the enzyme luciferase, which releases light. Unused deoxyribonucleotides are degraded by the enzyme apyrase (gray arrow).

fluorescent tags on the incoming nucleotides are monitored on a microscope. The machine can monitor a billion DNA fragments simultaneously. A computer then assembles the fragments into a complete sequence.

Pacific Biosciences SMRT (single-molecule real-time) sequencing uses a technique known as *zero-mode waveguides*. In this method, DNA polymerase extends a growing chain by adding deoxyribonucleotides tagged with four different fluorescent dyes. Incoming deoxyribonucleotides emit a flash of light as they are linked in place. Two novel features are critical for sequencing single molecules. First, reactions are carried out inside nanocontainers (the zero-mode waveguides). These are tiny cylindrical metal wells 20 nm wide that reduce background light sufficiently

to allow the detection of single flashes of light from individual nucleotides. Second, the fluorescent tags are attached to the pyrophosphate group that is discarded rather than the part of the deoxyribonucleotide that is incorporated into the chain. Thus, colored tags are not accumulated by the DNA; instead each reaction releases a microscopic burst of color.

The key characteristic of *fourth-generation* sequencing, also called "post light sequencing," is that optical detection is no longer used. The *ion torrent sequencing method* does not use single-molecule sequencing. Instead of using labeled deoxyribonucleotides, it measures the release of protons (H$^+$) whenever a new deoxyribonucleotide is added to a growing strand of DNA (**Figure 6.4a**). A silicon chip dubbed "the world's smallest pH meter" detects the protons. Sequencing is extremely fast by this method and the instruments are much less expensive than those of previous technologies. For example, the ion torrent machine is capable of sequencing an entire human genome—almost 3000 Mbp—in less than a day!

Nanopore technology (Figure 6.4b) is based on microscopic machinery that operates on the scale of single molecules. Nanopore detectors for DNA are extremely narrow pores that permit single strands of DNA to pass through one at a time. The Oxford Nanopore Technologies system passes DNA through nanoscale biological pores made from a protein (Figure 6.4b). As the DNA molecule transits the pore, a detector records the change in electrical current across the nanopore. This change is different for each of the four bases or combinations of those bases. The major advantages of nanopore technology are its high speed and its ability to sequence long DNA molecules (rather than short fragments like most other methods). In addition, many nanopores can be assembled into a very small area

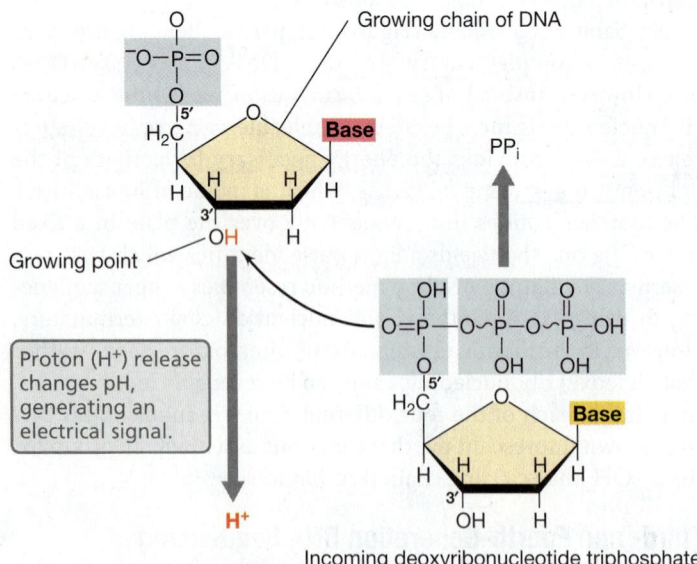

(a) **Ion torrent semiconductor sequencing**

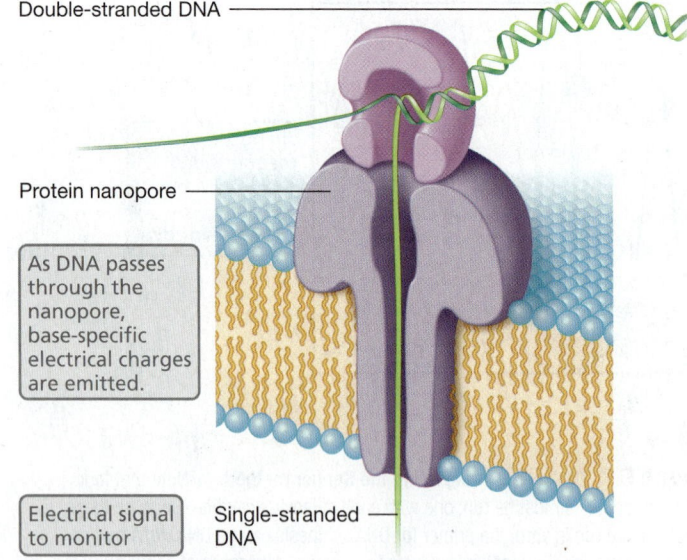

(b) **Nanopore sequencing**

Figure 6.4 Fourth-generation sequencing. (a) The ion torrent semiconductor sequencing system is based on the release of protons (H$^+$) each time a new deoxyribonucleotide is inserted into a growing strand of DNA. The resulting pH change is measured by an electrode. (b) In nanopore sequencing, a DNA double helix is converted to a single strand for passage through the pore. As the DNA transits the nanopore, it causes changes in electric charge that are base-specific.

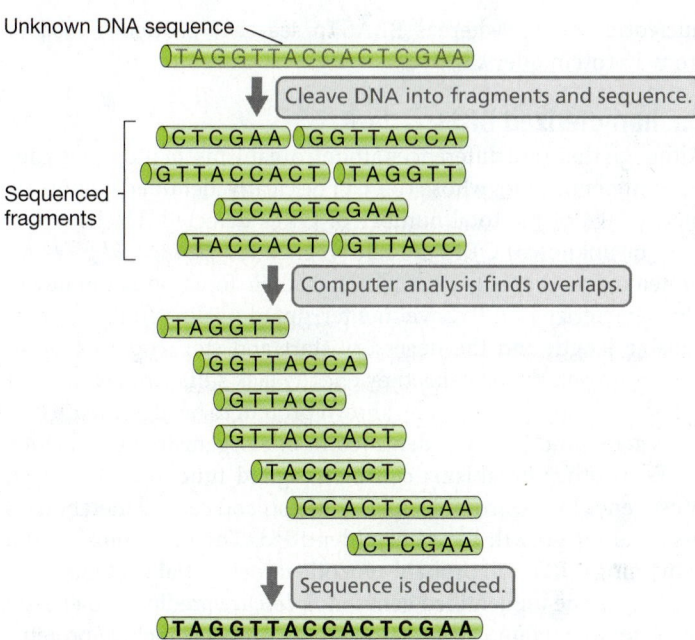

Figure 6.5 Computer assembly of DNA sequence. Most DNA sequencing methods generate vast numbers of short sequences (30 to several hundred bases) that must be assembled. The computer searches for overlaps in the short sequences and then arranges them to form a single overall sequence.

on a chip, upon which many long fragments of DNA can be sequenced simultaneously.

Genome Assembly

Regardless of how DNA is sequenced, the sequences must be assembled before they can be analyzed. Genome *assembly* consists of putting the fragments in the correct order and eliminating overlaps. In practice a computer examines many short DNA fragments that have been sequenced and deduces their order from the overlaps (**Figure 6.5**). Assembly generates a genome suitable for *annotation*, the process of identifying genes and other functional regions in the genome (discussed in the next section).

Sometimes sequencing and assembly does not yield a complete genome sequence and gaps in the genome remain. In such situations, a variety of approaches are used to obtain individual sequences that cover the gaps. Some genome projects have the goal of obtaining a *closed genome*, meaning that the entire genome sequence is determined. Other projects stop at the *draft stage*, dispensing with sequencing the small gaps. Because sequencing and assembly are heavily automated procedures, whereas gap closure is not, obtaining a closed genome is much more expensive and time consuming than a draft genome and typically needs much more human input to complete the job.

MINIQUIZ

- What is shotgun sequencing?
- What are the defining characteristics of third- and fourth-generation sequencing?
- What is accomplished during genome assembly?

6.3 Bioinformatics and Annotating Genomes

After sequencing and assembly, the next step is *genome annotation*, the conversion of raw sequence data into a list of the genes present and other functional sequences in the genome. **Bioinformatics** refers to the use of computers to store and analyze the sequences and structures of nucleic acids and proteins. Improved sequencing methods (Section 6.2) are now generating data faster than it can be properly analyzed. Thus, at present, annotation is the "bottleneck" in genomics.

Most genes encode proteins, and in most microbial genomes, especially those of prokaryotes, the great majority of the genome consists of coding sequences. Because genes in the genomes of microbial eukaryotes typically have fewer intervening sequences (introns, ↩ Section 4.9) than plant and animal genomes, and prokaryotes have almost none, microbial genomes essentially consist of a series of **open reading frames** (**ORFs**) separated by short regulatory regions and transcriptional terminators. Recall that an open reading frame is a sequence of DNA or RNA that can be translated to yield a polypeptide (↩ Section 4.11).

How Does the Computer Find an ORF?

A *functional ORF* is one that actually encodes a protein. The simplest way to find potential protein-encoding genes is to have a computer search the genome sequence for ORFs (**Figure 6.6**). Although any given gene is always transcribed from one strand, both strands are transcribed in some part of the genome (in all but the smallest plasmid or viral genomes). Thus, computer inspection of both strands of DNA is required.

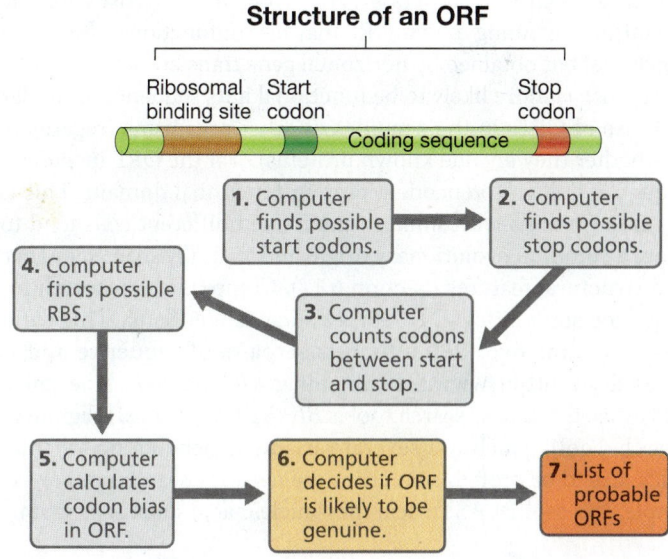

Figure 6.6 Computer identification of possible ORFs. The computer scans the DNA sequence looking first for start and stop codons. It then counts the number of codons in each uninterrupted reading frame and rejects those that are too short. The probability of a genuine ORF is made stronger if a likely ribosomal binding site (RBS) is found the correct distance in front of the reading frame. Codon bias calculations are used to test if an ORF complies with the codon usage of the organism being examined.

Table 6.3 Examples of codon bias

Arginine codon[a]	Usage of each arginine codon (%)		
	Escherichia coli	Fruit fly	Human
AGA	1	10	22
AGG	1	6	23
CGA	4	8	10
CGC	39	49	22
CGG	4	9	14
CGU	49	18	9

[a]Arginine has six codons; see Table 4.5.

The first step in finding an ORF is to locate *start* and *stop* codons in the sequence (⟲ Section 4.11 and Table 4.5). However, in-frame start and stop codons appear randomly with reasonable frequency. Thus, further clues are needed. Most proteins contain 100 or more amino acids, so most functional ORFs are longer than 300 nucleotides (100 codons). However, ignoring ORFs shorter than 100 codons will miss some genuine but short genes. In *Bacteria*, translation begins at start codons located immediately downstream of a ribosome-binding sequence (Shine–Dalgarno site) on the mRNA (⟲ Section 4.13). Thus, locating potential ribosome-binding sequences helps decide both whether an ORF is functional and which start codon is actually used.

More than one codon exists for many of the 20 common amino acids (⟲ Table 4.5), and some codons are used more frequently than others. The latter is known as **codon bias** (codon usage) and differs greatly between organisms. For example, **Table 6.3** shows the different usage of the six arginine codons in *Escherichia coli* compared to their usage in humans and fruit flies. If the codon bias in a given ORF differs greatly from the consensus for the organism containing it, that ORF may be nonfunctional or may be functional but obtained by horizontal gene transfer (Section 6.12).

An ORF is more likely to be functional if its sequence is similar to those of ORFs in the genomes of other organisms (regardless of whether they encode known proteins) or if the ORF includes a sequence known to encode a protein functional domain. This is because proteins with similar functions in different cells tend to share a common evolutionary origin and typically share sequence and structural features (Section 6.11). Computers can search for sequence similarities in databases such as GenBank. This database contains over 200 billion base pairs of sequence and is located at http://www.ncbi.nlm.nih.gov/Genbank/. The most widely used database search tool is *BLAST* (*Basic Local Alignment Search Tool*), which has several variants depending on whether nucleic acid or protein sequences are used for searching. For example, the tool BLASTn searches nucleic acid databases from a nucleotide query, whereas BLASTp searches protein databases from a protein query.

Uncharacterized ORFs

Although there are differences among organisms, in most genomes the number of genes whose role can be clearly identified is approximately 70% of the total number of ORFs detected. Uncharacterized (or unknown) ORFs are said to encode *hypothetical proteins*, proteins that probably exist although their function is unknown. Uncharacterized ORFs have uninterrupted reading frames of reasonable length and the necessary start and stop codons (Figure 6.6). However, the proteins they encode lack sufficient amino acid sequence homology with any known protein to be identified.

As gene functions are identified in one organism, homologous ORFs in other organisms can be assigned functions. However, most genes for macromolecular synthesis and central metabolism essential for growth have been identified. Therefore most of the remaining ORFs will probably encode nonessential proteins.

Many of the unidentified genes in *E. coli* are predicted to encode regulatory or redundant proteins. These probably include proteins needed only under special conditions or "backups" for key enzymes. However, even in well-studied organisms such as *E. coli*, the precise functions of many genes are often unpredictable. Some gene identifications merely assign genes to a family or a general function (such as "transporter"). By contrast, other genes are completely unknown and have only been predicted using bioinformatics. Moreover, some annotations are actually incorrect. In fact, it has been estimated that as many as 10% of genes in databases are incorrectly annotated.

Noncoding RNA

In addition to protein-encoding genes, some genes encode RNA molecules that are not translated. Such genes therefore lack start codons and may well have multiple stop codons within the gene. In addition, these genes lack codon bias; consequently, they will not be recognized by programs that search for ORFs. Some noncoding RNAs are easy to detect because they are well characterized and are highly conserved. These include tRNAs and rRNAs. However, many noncoding regulatory RNA molecules (⟲ Section 7.14) are conserved only in three-dimensional structure, with little sequence homology. Identifying these during genome annotation remains a challenge.

MINIQUIZ
- What is an open reading frame (ORF)? What is a hypothetical protein?
- How can protein homology assist in genome annotation?
- What is the major problem in identifying genes encoding nontranslated RNA?

II • Microbial Genomes

6.4 Genome Size and Content

Following sequencing, assembly, and annotation, comparative genomics can be used to compare genomes as to their size, organization, and gene content. The Microbes Online website (http://www.microbesonline.org) has nearly 4000 microbial genomes available for viewing.

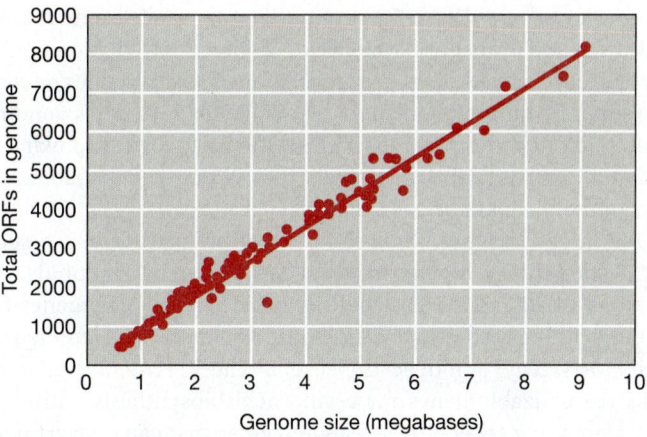

Figure 6.7 Correlation between genome size and ORF content in prokaryotes. Analyses of 115 completed prokaryotic genomes including species of both *Bacteria* and *Archaea*. Data from *Proc. Natl. Acad. Sci. (USA)101*: 3160–3165 (2004).

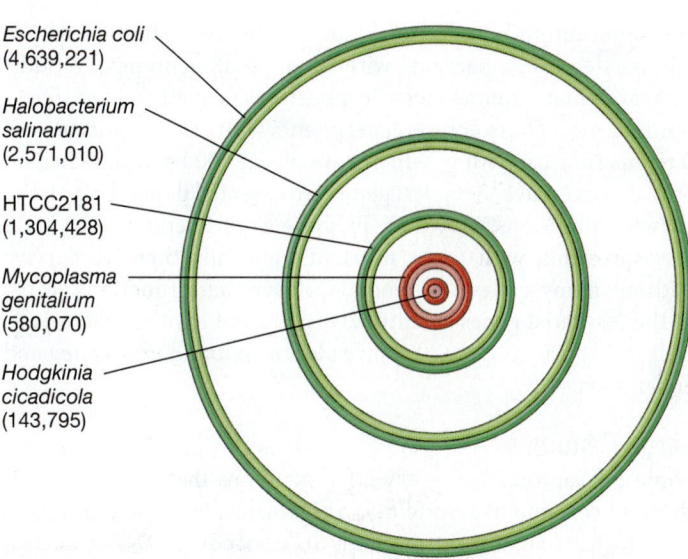

Figure 6.8 Genome size comparison. The circular genomes of several prokaryotes are shown drawn to scale. The number of nucleotides is shown next to the names. Green circles indicate free-living organisms whereas red circles indicate parasites (*Mycoplasma*) and insect symbionts (*Hodgkinia*).

Size Range of Prokaryotic Genomes

Genomes of species of both *Bacteria* and *Archaea* show a strong correlation between genome size and open reading frame (ORF) content (**Figure 6.7**). Regardless of the organism, each megabase pair of DNA in a prokaryote encodes about 1000 ORFs. As the size of prokaryotic genomes increases, the gene number also increases proportionally. This contrasts markedly with the genomes of eukaryotes, in which noncoding DNA (introns, ⮌ Section 4.9) may be a large fraction of the genome, especially in organisms with large genomes.

Analyzing genomic sequences can shed light on fundamental biological questions. For example, how many genes are necessary for a cell to exist? The record for the smallest genome for a free-living organism belongs to a species of *Bacteria* referred to as strain HTCC2181, whose genome contains 1,304,428 bp and 1354 genes. This undercuts the previous record holder, *Pelagibacter ubique*, a marine heterotroph, by a mere 4331 bp, suggesting that this is close to the practical limit for independent life. HTCC2181 is an as yet uncultured methylotrophic bacterium (methylotrophs are organisms that catabolize one-carbon compounds, such as methanol) that is common in marine coastal ecosystems.

Several other free-living *Bacteria* and *Archaea* are known that have genomes of around 1400 genes (Table 6.1). These organisms are extremely efficient in their use of DNA. They have few or no introns, inteins, or transposons and have very short intergenic spaces. The largest prokaryotic genomes contain over 10,000 genes and are primarily from soil bacteria, such as the myxobacteria, who undergo complex life cycles (⮌ Section 14.19). **Figure 6.8** shows five selected circular prokaryotic genomes drawn to scale to give a visual impression of how variable prokaryotic genomes can be.

Perhaps surprisingly, genomic analyses have shown that autotrophic organisms need only a few more genes than heterotrophs (⮌ Section 3.3). For example, the methanogen *Methanocaldococcus jannaschii* (*Archaea*) is an autotroph whose genome contains only 1738 ORFs. This enables it to be not only free-living, but also to rely on CO_2 as its sole carbon source. *Aquifex aeolicus* (*Bacteria*) is also an autotroph and contains the smallest known

genome of any autotroph at just 1.5 megabase pairs (Table 6.1). Both *Methanocaldococcus* and *Aquifex* are also hyperthermophiles, growing optimally at temperatures above 80°C. Thus, large genomes are not necessary to support autotrophic and extreme lifestyles.

Small Genomes

The smallest cellular genomes belong to prokaryotes that are parasitic or endosymbiotic (cells that live inside other cells). Genome sizes for obligately parasitic prokaryotes range from 490 kbp for *Nanoarchaeum equitans* (*Archaea*) to 4400 kbp for *Mycobacterium tuberculosis* (*Bacteria*). The genomes of several prokaryotes, including *N. equitans*, *Mycoplasma*, *Chlamydia*, and *Rickettsia*, are smaller than the largest known viral genome, that of *Mimivirus* with 1.2 Mbp (⮌ Section 9.2). The degenerate cicada endosymbiont *Hodgkinia* has a tiny genome, less than 150 kbp (Figure 6.8; see also Figure 6.14).

Genomes smaller than 1.2 Mbp are all found in bacteria that are dependent on other cells for some aspect(s) of their existence. Mycoplasmas, with genomes of just over 500 kbp and just under 500 genes, have the smallest genomes among parasitic bacteria (Figure 6.8; see also Figure 6.14). Excluding endosymbionts, the smallest prokaryotic genome is that of *N. equitans*, which is some 90 kbp smaller that that of *Mycoplasma genitalium* (Table 6.1). Despite this, the genome of *N. equitans* actually contains more genes than *M. genitalium*. This is because the *N. equitans* genome is extremely compact with almost no noncoding DNA. *N. equitans* is a hyperthermophile and a parasite of another hyperthermophile, the archaeon *Ignicoccus* (⮌ Section 16.7). *N. equitans* lacks virtually all genes that encode metabolic proteins and presumably depends on its host for most catabolic as well as anabolic functions.

Using *Mycoplasma*, which has around 500 genes, as a starting point, it has been estimated that around 250–300 genes are the minimum number possible for a viable cell. These estimates rely partly on comparisons with other small genomes. In addition, systematic mutagenesis has been performed to identify essential genes. For example, experiments with *Escherichia coli* and *Bacillus subtilis*, both of which have about 4000 genes, indicated that approximately 300–400 genes are essential depending on the growth conditions. However, in these experiments the bacteria were provided with many nutrients, allowing them to survive without many genes that encode biosynthetic functions. Most of the "essential genes" identified are present in other bacteria as well and approximately 70% have also been found in *Archaea* and eukaryotes.

Large Genomes

Some prokaryotes have very large genomes that are as large as those of certain eukaryotic microorganisms. Because eukaryotes tend to have significant amounts of noncoding DNA and prokaryotes do not, some prokaryotic genomes actually have more genes than microbial eukaryotes, despite having less DNA. For example, the genome of *Bradyrhizobium japonicum*, a bacterium that forms nitrogen-fixing root nodules on leguminous plants such as soybeans, has 9.1 Mbp of DNA and 8300 ORFs, whereas the genome of the baker's yeast *Saccharomyces cerevisiae*, a eukaryote, has 12.1 Mbp of DNA and only 5800 ORFs (see Table 6.5). The soil bacterium *Myxococcus xanthus* also has 9.1 Mbp of DNA, whereas many of its close relatives have genomes approximately half this size. It has been hypothesized that multiple duplications of substantial segments of genomic DNA might account for such very large genomes.

The largest prokaryotic genome known at present is that of *Sorangium cellulosum*, a species of the myxobacteria (🔗 Section 14.19). With just over 13 Mbp on a single circular chromosome, its genome is roughly three times larger than that of *Escherichia coli*. The *Sorangium* genome has a relatively large proportion of noncoding DNA for a bacterium—14.5%—and consequently has fewer coding sequences than might have been expected—only 9400. Nonetheless, it has more DNA than several eukaryotes including yeast and the protozoans *Cryptosporidium* and *Giardia* (see Table 6.5). The complex regulation needed for the social lifestyle of *Sorangium* is seen in its massive number of eukaryotic-type protein kinases (enzymes that phosphorylate other proteins to control their activity). It has 317 kinases, over twice that of any other genome, including eukaryotes.

In contrast to *Bacteria*, the largest genomes found in *Archaea* thus far are around 5 Mbp (Table 6.1). Overall, prokaryotic genomes range in size from those of large viruses to those of eukaryotic microorganisms.

Gene Content of Bacterial Genomes

The complement of genes in a particular organism reveals its capabilities. Conversely, genomes are molded by adaptation to an organism's lifestyle. Comparative analyses are useful when searching for genes that encode enzymes that probably exist because of the lifestyle of an organism. *Thermotoga maritima* (*Bacteria*), for example, is a hyperthermophile found in hot marine sediments, and laboratory studies have shown that it can catabolize a large

number of sugars. **Figure 6.9** summarizes some of the metabolic pathways and transport systems of *T. maritima* that were deduced from analysis of its genome. About 7% of its genes encode proteins for the metabolism of sugars. As expected, its genome is also rich in genes for transport, particularly for carbohydrates and amino acids. All this suggests that *T. maritima* exists in an environment rich in organic material.

One might imagine, for instance, that obligate parasites such as the spirochete *Treponema pallidum* (the agent of syphilis, 🔗 Sections 14.20 and 29.12) would require relatively few genes for amino acid biosynthesis because amino acids can be supplied by their hosts. This is indeed the case, as the *T. pallidum* genome lacks recognizable genes for amino acid biosynthesis, although genes encoding several proteases, enzymes that can convert peptides taken up from the host into free amino acids, are found. In contrast, the free-living bacterium *Escherichia coli* has 131 genes for amino acid biosynthesis and metabolism and the soil bacterium *Bacillus subtilis* has over 200.

A functional analysis of genes and their activities in several bacteria is given in **Table 6.4**. Thus far, a distinct pattern of gene distribution in prokaryotes has emerged. Metabolic genes are typically the most abundant class in prokaryotic genomes, although genes for protein synthesis overtake metabolic genes on a percentage basis as genome size decreases (Table 6.4 and **Figure 6.10**). Although many genes can be dispensed with, genes that encode the protein-synthesizing apparatus cannot. Thus, the smaller the genome the greater the percentage of genes that encode translational processes. The vital functions of DNA replication and transcription are encoded by genes that make up only a minor fraction of a typical prokaryotic genome.

The percentage of an organism's genes devoted to a particular function is to some degree a function of genome size. This is summarized for a large number of bacterial genomes in Figure 6.10. Core cellular processes, such as protein synthesis, DNA replication, and energy production, show only minor variations in gene number with genome size. Consequently, the relative percentage of such genes is large in organisms with small genomes. Conversely, large genomes contain more genes for regulation than small genomes. These extra regulatory systems allow the cell to adapt with more flexibility to diverse environmental situations.

Organisms with large genomes can also afford to encode many specialized metabolic genes. This likely makes these organisms more competitive in their habitats, which, for many prokaryotes with very large genomes, is soil. Here, carbon and energy sources vary greatly and are often scarce or available only intermittently (🔗 Section 19.1). A cell with a large genome that encodes multiple metabolic options would thus be strongly selected for in such a habitat. As an example of this, all of the prokaryotes listed in Table 6.1 whose genomes are larger than 6 Mbp inhabit soil.

Analyses of gene categories have also been done for several *Archaea*. On average, *Archaea* devote a higher percentage of their genomes to energy and coenzyme production than do *Bacteria* (this result is undoubtedly skewed a bit due to the large number of novel coenzymes produced by methanogenic *Archaea* [🔗 Section 13.20]). On the other hand, *Archaea* appear to contain fewer

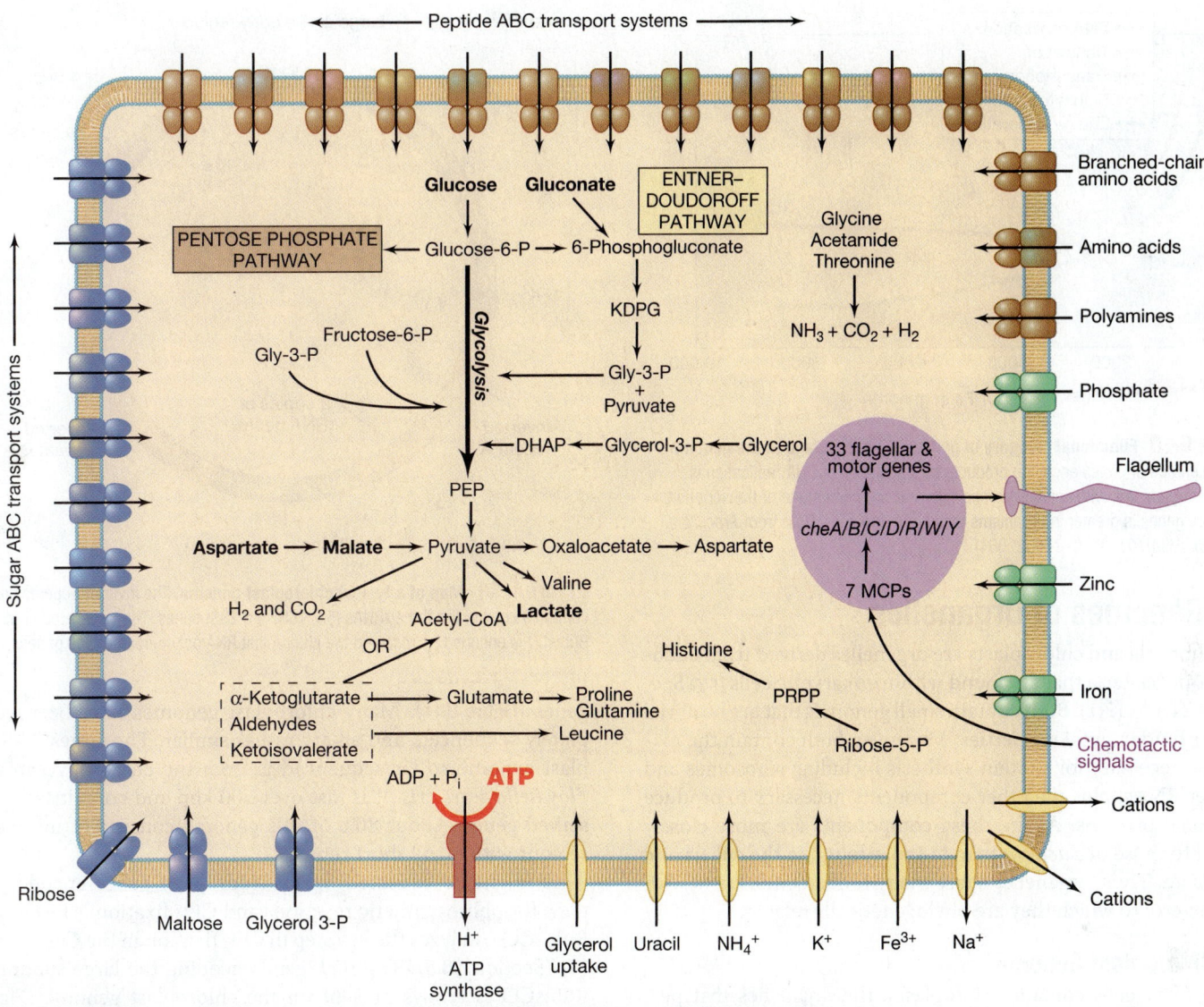

Figure 6.9 Overview of metabolism and transport in *Thermotoga maritima*. The figure summarizes the metabolic capabilities of this organism. These include some of the pathways for energy production and the metabolism of organic compounds, including transport proteins that were identified from analysis of the genomic sequence. Gene names are not shown. The genome contains several ABC-type transport systems, 12 for carbohydrates, 14 for peptides and amino acids, and still others for ions. These are shown as multi-subunit structures in the figure. Other types of transport proteins have also been identified and are shown as ovals. Chemotaxis and flagellar genes are highlighted in purple, and a few aspects of sugar metabolism are also shown. Data adapted from original figure published by The Institute for Genomic Research (TIGR, Rockville, MD).

genes for carbohydrate metabolism or membrane functions, such as transport and membrane biosynthesis, than do *Bacteria*. However, this conclusion is compromised because the corresponding pathways have been less studied in *Archaea* than in *Bacteria* and many of the relevant archaeal genes are probably still unidentified.

Both *Archaea* and *Bacteria* have relatively large numbers of genes whose functions are unknown or that encode only hypothetical proteins, although more uncertainty exists for the *Archaea* than the *Bacteria*. This may well be due to the fewer genome sequences available from species of *Archaea* than from *Bacteria*.

MINIQUIZ

- What lifestyle is typical of prokaryotes that have genomes smaller than those of certain viruses?

- Approximately how many protein-encoding genes will be found in a prokaryotic genome of 4 Mbp?

- Which is likely to have more genes, a prokaryote with 8 Mbp of DNA or a eukaryote with 10 Mbp? Explain.

- What category of genes do prokaryotes contain the most of on a percentage basis?

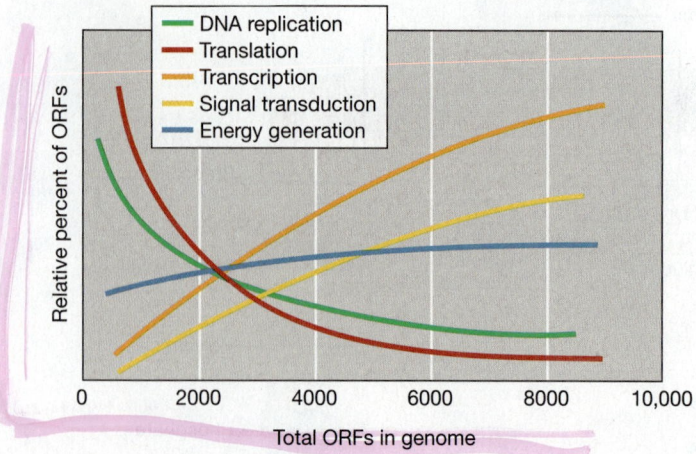

Figure 6.10 **Functional category of genes as a percentage of the genome.** The percentage of genes encoding products for translation or DNA replication is greater in organisms with small genomes, whereas the percentage of transcriptional regulatory genes is greater in organisms with large genomes. Data from *Proc. Natl. Acad. Sci. (USA)101:* 3160–3165 (2004).

6.5 Genomes of Organelles

Mitochondria and chloroplasts are organelles derived from endo-symbiotic bacteria that are found within eukaryotic cells (⮂ Sections 2.21 and 17.1). Both contain small genomes that are bacterial in their fundamental properties. Moreover, both contain the machinery necessary for protein synthesis including ribosomes and transfer RNAs, plus the other components necessary to produce functional proteins. Again, these components are more closely related to those of *Bacteria* than to those found in the eukaryotic cytoplasm. Thus, organelles share many fundamental traits with the *Bacteria* to which they are phylogenetically related.

The Chloroplast Genome

Green plant cells contain chloroplasts, the organelles that perform photosynthesis (⮂ Section 13.1). All known chloroplast genomes are circular DNA molecules, and each chloroplast contains several identical copies of the genome. The typical chloroplast genome is about 120–160 kbp and contains two inverted repeats of 6–76 kbp that each encode copies of the three rRNA

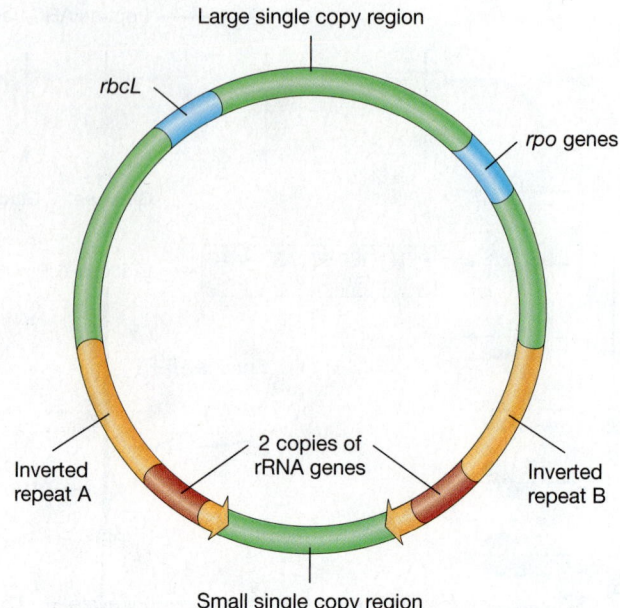

Figure 6.11 **Map of a typical chloroplast genome.** The inverted repeats each contain a copy of the three rRNAs (5S, 16S, and 23S) genes. The large subunit of RubisCO is encoded by *rbcL* and the chloroplast RNA polymerase by *rpo* genes.

genes (**Figure 6.11**). Many chloroplast genomes have been completely sequenced, and all are rather similar. The largest chloroplast genome so far sequenced is from the chlorophycean alga *Floydiella terrestris*. It is just over 500 kbp and contains 97 conserved genes. About 80% of this genome consists of intergenic regions with many short repeats.

As might be expected, many chloroplast genes encode proteins for photosynthetic reactions and CO_2 fixation. The enzyme RubisCO catalyzes the key step in CO_2 fixation in the Calvin cycle (⮂ Section 13.5). The *rbcL* gene encoding the large subunit of RubisCO is always present on the chloroplast genome (Figure 6.11), whereas the gene for the small subunit, *rbcS*, resides in the plant cell nucleus and its protein product must be imported from the cytoplasm into the chloroplast after synthesis.

The chloroplast genome also encodes rRNA used in chloroplast ribosomes, tRNA used in translation, several proteins used

---

Table 6.4 Gene function in bacterial genomes

	Percentage of genes		
Functional categories	Escherichia coli (4.64 Mbp)[a]	Haemophilus influenzae (1.83 Mbp)[a]	Mycoplasma genitalium (0.58 Mbp)[a]
Metabolism	21.0	19.0	14.6
Structure	5.5	4.7	3.6
Transport	10.0	7.0	7.3
Regulation	8.5	6.6	6.0
Translation	4.5	8.0	21.6
Transcription	1.3	1.5	2.6
Replication	2.7	4.9	6.8
Other, known	8.5	5.2	5.8
Unknown	38.1	43.0	32.0

[a]Chromosome size, in megabase pairs. Each organism listed contains only a single circular chromosome.

in transcription and translation, as well as some other proteins. Some proteins that function in the chloroplast are encoded by nuclear genes. These are thought to be genes that migrated to the nucleus as the chloroplast evolved from an endosymbiotic cell into a photosynthetic organelle. Introns (⮌ Section 4.9) are common in chloroplast genes and are primarily of the self-splicing type.

Mitochondrial Genomes and Proteomes

Mitochondria, the organelles that produce energy by respiration, are found in most eukaryotic organisms (⮌ Sections 2.21 and 17.1). Mitochondrial genomes primarily encode proteins for oxidative phosphorylation and, as for chloroplasts, also encode proteins, rRNAs, and tRNAs for protein synthesis. However, most mitochondrial genomes encode far fewer proteins than those of chloroplasts.

Several hundred mitochondrial genomes have been sequenced. The largest mitochondrial genome has 62 protein-encoding genes, but others encode as few as three proteins. The mitochondria of almost all mammals, including humans, encode only 13 proteins plus 22 tRNAs and 2 rRNAs. Figure 6.12 shows a map of the 16,569-bp human mitochondrial genome. The mitochondrial genome of the yeast *Saccharomyces cerevisiae* is larger than that of humans (85,779 bp), but has only 8 protein-encoding genes. Besides the genes encoding the RNA and proteins, the genome of yeast mitochondria contains large stretches of adenine/thymine (AT)-rich DNA that has no apparent function.

The mitochondrial genomes of plants are much larger than those of animal cells, and mostly range from around 300 kbp to 2000 kbp. Despite this, they only have about 50 highly conserved genes, most of which encode components of the respiratory chain and translation apparatus. The variation in size is due to large amounts of noncoding DNA. The mitochondrial genomes of different species of the flowering plant genus *Silene* vary amazingly in size. The largest two are approximately 7 and 11 Mbp, making them larger than most bacterial genomes!

Unlike chloroplast genomes, which are all single, circular DNA molecules, the genomes of mitochondria are quite diverse. For example, some mitochondrial genomes are linear, including those of certain algae, protozoans, and fungi. In other cases, such as the yeast *S. cerevisiae*, genetic analysis implies that the mitochondrial genome is circular, but the physical form consists of large linear molecules containing multiple genome copies. (Certain viruses, such as bacteriophage T4, also have genomes that are genetically circular but physically linear; ⮌ Section 8.6.) Finally, the mitochondria of many fungi and flowering plants contain small circular or linear plasmids in addition to the main mitochondrial genome.

Mitochondria require many more proteins than they encode. In particular, far more proteins are needed for translation than are encoded by the organelle genome. The proteins needed for many organelle functions are encoded by nuclear genes. The yeast mitochondrion contains as many as 800 different proteins (its proteome, Section 6.8). However, only eight of them are encoded by the yeast mitochondrial genome, the remaining proteins being encoded by nuclear genes (Figure 6.13). The genes for most organelle proteins are present in the nucleus, transcribed in the nucleus, and translated on the 80S ribosomes in the eukaryotic cytoplasm. The proteins are then transported into the organelles. The nuclear-encoded proteins required for translation and energy generation in mitochondria are more closely related to counterparts in the *Bacteria* than to those of the eukaryotic cytoplasm, consistent with the evolutionary history of the mitochondrion.

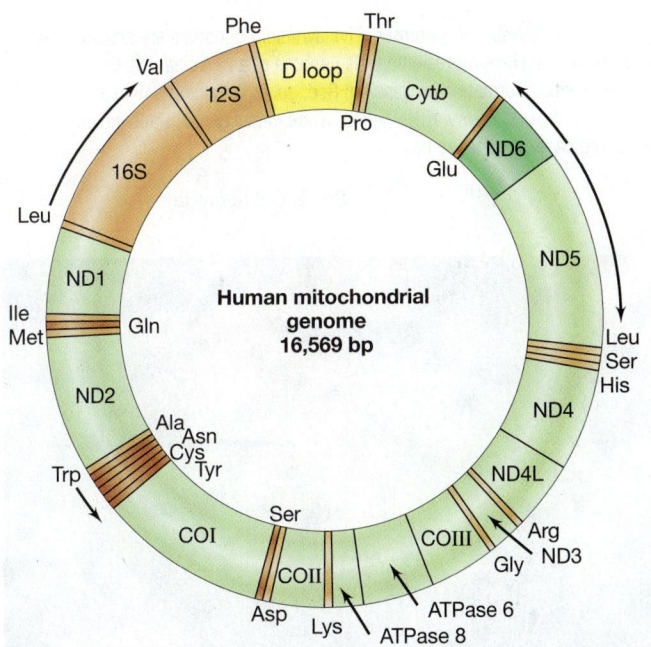

Figure 6.12 Map of the human mitochondrial genome. The genome encodes rRNAs, 22 tRNAs, and several proteins. Arrows show direction of transcription for genes of a given color, and the three-letter amino acid designations for tRNA genes are also shown. The 13 protein-encoding genes are in green. Cyt*b*, cytochrome *b*; ND1–6, components of the NADH dehydrogenase complex; COI–III, subunits of the cytochrome oxidase complex; ATPase 6 and 8, polypeptides of the mitochondrial ATPase complex. The two promoters are in the region called the D loop, which is also involved in DNA replication.

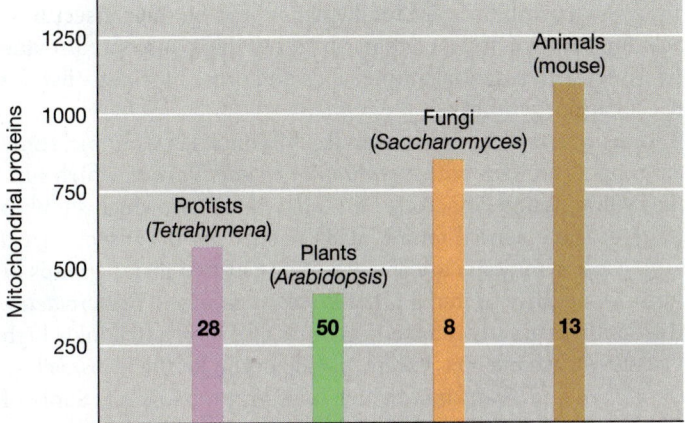

Figure 6.13 Mitochondrial proteomes. The number of proteins located in the mitochondria of different groups of model eukaryotes. The number is the best estimate because some proteins are present in very low amounts. The numbers in each colored bar are the number of proteins encoded on each organism's mitochondrial genome.

Variability in the Genetic Code

The original belief that all cells use the same genetic code led to the genetic code being regarded as universal (🔗 Table 4.5). However, later discoveries showed that some mitochondria and a few cells use slight variations of the "universal" genetic code. Alternative genetic codes were first discovered in the genomes of animal mitochondria. These modified codes typically use stop codons as sense codons. For example, animal (but not plant) mitochondria use the codon UGA to encode tryptophan instead of using it as a stop codon. Yeast mitochondria also use UGA for tryptophan but in addition use the four CUN codons (N being any nucleotide) as threonine instead of leucine. These changes may have arisen from selection pressure for smaller genomes; for example, by inhabiting an environment where many necessary nutrients were already available. Thus, the 22 tRNAs produced in mitochondria are insufficient to read the universal genetic code, even when using wobble pairing (🔗 Figure 4.32). Therefore, base pairing between the anticodon and the codon is even more flexible in mitochondria than it is in cells.

Several organisms are also known that use slightly different genetic codes. For example, in the genus *Mycoplasma* (*Bacteria*) and the genus *Paramecium* (*Eukarya*), certain stop codons encode amino acids. Consequently, these organisms have fewer stop codons. Certain fungi use the leucine codon CUG to encode serine. Curiously, however, these have become slightly ambiguous, as CUG is translated 97% of the time as serine and 3% of the time as leucine.

Symbionts and Organelles

Many insects and some other invertebrates, including certain nematodes and molluscs, contain symbiotic bacteria within their cells. Some of these bacterial symbionts are no longer capable of independent existence and show major reductions in the size of their genomes (🔗 Section 22.9). Symbiont genomes range from the same size as in free-living bacteria down to around 140 kbp for *Tremblaya* and *Hodgkinia* (Table 6.1 and Figure 6.8), the two smallest examples known (**Figure 6.14**). Thus, the genomes of some symbionts contain fewer genes than do some organelles and viruses. Such symbionts are totally dependent on their insect host cells for survival and nutrients. In turn, the symbionts provide the insect with essential amino acids and other nutrients that the insect cannot synthesize.

Some insects have two bacterial symbionts. For example, some leafhoppers contain both *Baumannia cicadellinicola*, which supplies vitamins and cofactors, plus *Sulcia muelleri*, which supplies many of the essential amino acids needed by the insect (**Figure 6.15**). Most symbionts are species of one of two major groups of the gram-negative bacterial phyla *Proteobacteria* and *Bacteroidetes*. Most of these greatly reduced genomes also have a strikingly high AT content, around 80%, except, paradoxically, for the two smallest, *Tremblaya* and *Hodgkinia*, which have around 40% AT. Some of these highly reduced genomes have apparently lost several genes regarded as essential for replication, such as the gene encoding the key cell division protein FtsZ (🔗 Section 5.2). So how these symbionts manage to replicate remains unknown.

The symbionts discussed differ from mitochondria and chloroplasts in several ways. Symbionts are restricted to a few tissues,

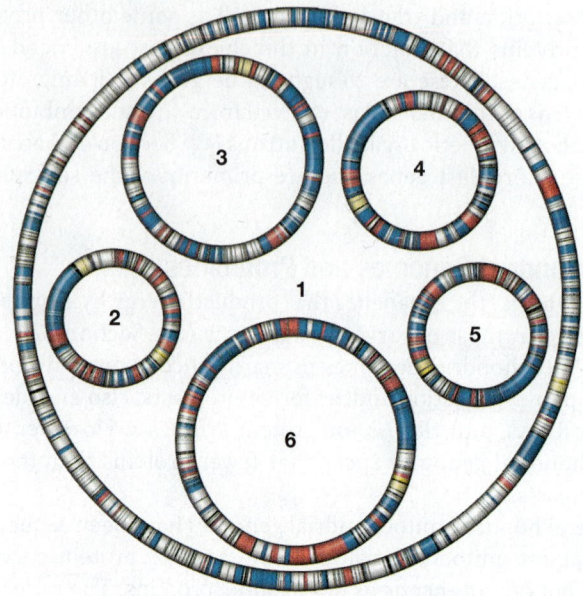

1. ***Mycoplasma genitalium***
   (*Mollicutes*)
   580.1 kbp
   GC: 31.7%
2. ***Tremblaya***
   (*Betaproteobacteria*)
   138.9 kbp
   GC: 58.8%
3. ***Zinderia***
   (*Betaproteobacteria*)
   208.5 kbp
   GC: 13.5%
4. ***Carsonella***
   (*Gammaproteobacteria*)
   159.6 kbp
   GC: 16.6%
5. ***Hodgkinia***
   (*Alphaproteobacteria*)
   143.7 kbp
   GC: 58.4%
6. ***Sulcia***
   (*Bacteroidetes*)
   245.5 kbp
   GC: 22.4%

Figure 6.14 **Symbiont genomes.** Five symbiont genomes are shown drawn to scale inside the circle representing the genome of a *Mycoplasma*. Blue: genes encoding genetic information processing; Red: genes encoding amino acid and vitamin biosyntheses; Yellow: rRNA genes; White: other genes; Gaps indicate noncoding DNA. Kbp, kilobase pairs.

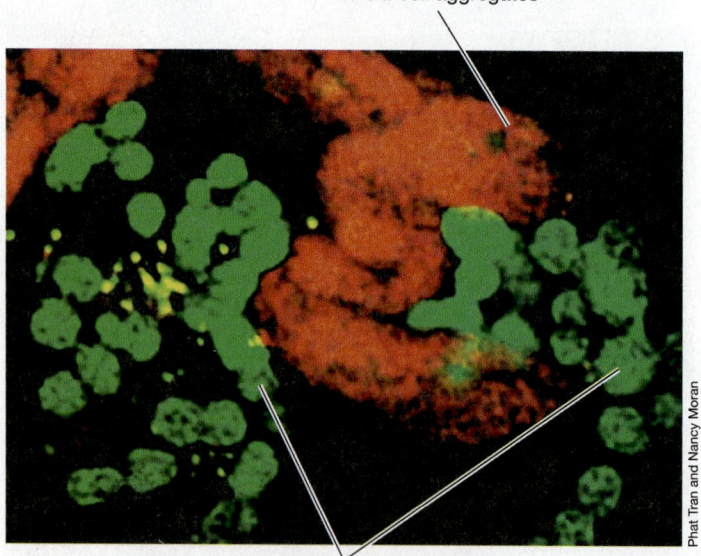

Sulcia cell aggregates

Baumannia cell aggregates

Figure 6.15 **Two endosymbionts, *Sulcia* and *Baumannia*, both inhabit the same insect cells.** Fluorescence in situ hybridizations were performed using probes that hybridize selectively to the rRNA of *Baumannia* (green) and of *Sulcia* (red).

even in a particular host organism. There is little evidence for transfer of symbiont genes to the host cell nucleus, and proteins made in the host cytoplasm do not enter the symbionts to perform vital functions. Nonetheless, some symbionts are absolutely required for host survival and cannot themselves survive outside the host. This leaves us with an important question to which there is presently no final answer: Where is the line between a symbiont and an organelle?

MINIQUIZ ---

- What is unusual about the genes that encode mitochondrial proteins?
- What do chloroplast genomes typically encode?
- What is unusual about the genomes of insect symbionts?

6.6 Eukaryotic Microbial Genomes

The genomes of many microbial eukaryotes have been sequenced (Table 6.5), and their size varies widely. Certain single-celled protozoans, including the free-living ciliate *Paramecium* (40,000 genes) and the pathogen *Trichomonas* (60,000 genes), have significantly more genes than do humans (Table 6.5). Indeed, *Trichomonas* pres-

ently holds the record for largest gene number of any organism. This is puzzling because *Trichomonas* is a human parasite, and such organisms typically have small genomes relative to comparable free-living organisms because parasites rely on their host for some or even many functions (Sections 6.3 and 6.5).

Genomes of Microbial Parasites

Apart from *Trichomonas*, parasitic eukaryotic microorganisms typically have genomes of 10–30 Mbp containing between 4000 and 11,000 genes. For example, *Trypanosoma brucei*, the agent of African sleeping sickness, has 11 chromosomes, 35 Mbp of DNA, and almost 11,000 genes. The most important eukaryotic parasite is *Plasmodium*, which causes malaria (⮂ Section 17.5). The four species of *Plasmodium* that infect humans have genomes ranging from 23 to 27 Mbp, comprising 14 chromosomes with around 5500 genes. About half of these genes have introns and about a third encode conserved hypothetical proteins of unknown function. The free-living social amoeba *Dictyostelium* has about 12,500 genes (but note that *Dictyostelium* has both single-celled and multicellular phases in its life cycle, ⮂ Section 17.8), and the pathogenic amoeba, *Entamoeba histolytica*, the causative agent of amebic dysentery, has approximately 10,000 genes.

Table 6.5 Some eukaryotic nuclear genomes[a]

Organism	Comments	Lifestyle[b]	Genome size (Mbp)	Haploid chromosomes	ORFs
Nucleomorph of *Bigelowiella natans*	Degenerate endosymbiotic nucleus	E	0.37	3	331
Encephalitozoon cuniculi	Smallest known eukaryotic genome, human pathogen	P	2.9	11	2,000
Cryptosporidium parvum	Parasitic protozoan	P	9.1	8	3,800
Plasmodium falciparum	Malignant malaria	P	23	14	5,300
Saccharomyces cerevisiae	Yeast, a model eukaryote	FL	12.1	16	5,800
Ostreococcus tauri	Marine green alga, smallest free-living eukaryote	FL	12.6	20	8,200
Aspergillus nidulans	Filamentous fungus	FL	30	8	9,500
Giardia lamblia	Flagellated protozoan, causes acute gastroenteritis	P	12	5	9,700
Dictyostelium discoideum	Social amoeba	FL	34	6	12,500
Drosophila melanogaster	Fruit fly, model organism for genetic studies	FL	180	4	13,600
Caenorhabditis elegans	Roundworm, model for animal development	FL	97	6	19,100
Mus musculus	Mouse, a model mammal	FL	2,500	23	25,000
Homo sapiens	Human	FL	2,850	23	25,000
Arabidopsis thaliana	Model plant for genetics	FL	125	5	26,000
Oryza sativa	Rice, the world's most important crop plant	FL	390	12	38,000
Paramecium tetraurelia	Ciliated protozoan	FL	72	>50	40,000
Populus trichocarpa	Black poplar, a tree	FL	500	19	45,000
Trichomonas vaginalis	Flagellated protozoan, human pathogen	P	160	6	60,000

[a]All data are for the haploid nuclear genomes of these organisms in megabase pairs. For most large genomes, both size and ORFs listed are best estimates due to large numbers of repetitive sequences and/or introns in the genomes.
[b]E, endosymbiont; P, parasite; FL, free-living.

The smallest eukaryotic cellular genome known belongs to *Encephalitozoon cuniculi,* an intracellular pathogen of humans and other animals that causes lung infections. *E. cuniculi* lacks mitochondria, and although its haploid genome contains 11 chromosomes, the genome size is only 2.9 Mbp with approximately 2000 genes (Table 6.5); this is smaller than many prokaryotic genomes (Table 6.1). As for prokaryotes, the smallest eukaryotic genome belongs to an endosymbiont. Known as a *nucleomorph,* this is the degenerate remains of a eukaryotic endosymbiont found in certain green algae that have acquired photosynthesis by secondary endosymbiosis (♻ Section 17.1). Nucleomorph genomes range from about 0.45 to 0.85 Mbp.

The Yeast Genome

Of single-celled eukaryotes, the yeast *Saccharomyces cerevisiae* is most widely used as a model organism and is also extensively used in both baking and brewing. The haploid yeast genome contains 16 chromosomes ranging in size from 220 kbp to 2352 kbp. The total yeast nuclear genome (excluding the mitochondria and some plasmid and virus-like genetic elements) is 13,400 kbp. Yeast chromosome XII contains a stretch of approximately 1260 kbp containing 100–200 repeats of yeast rRNA genes. In addition to multiple copies of the rRNA genes, the yeast nuclear genome has approximately 300 genes for tRNAs (only a few are identical) and nearly 100 genes for other types of noncoding RNA. Yeast has approximately 6000 ORFs, which is fewer than that of some genomes of *Bacteria* (Tables 6.1 and 6.5). About two-thirds of the yeast ORFs encode proteins whose functions are known.

How many yeast genes are actually essential? This question can be tackled by systematically inactivating each gene in turn with *knockout mutations* (mutations that completely inactivate genes, ♻ Section 11.5). Knockout mutations cannot normally be obtained in essential genes in a haploid organism. However, yeast can be grown in both diploid and haploid states (♻ Section 17.13). By generating knockout mutations in diploid cells and then investigating whether they can also exist in haploid cells, it is possible to determine whether a particular gene is essential for cell viability. Using knockout mutations, it has been shown that around 900 yeast ORFs (17%) are essential. Note that this number of essential genes is much greater than the approximately 300 genes (Section 6.4) estimated to be the minimal number required in prokaryotes. Because eukaryotes are more complex than prokaryotes, a larger minimal gene complement would be expected.

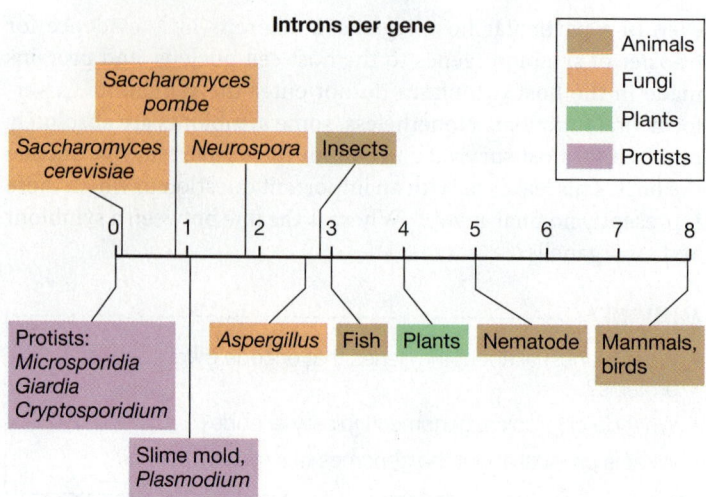

Figure 6.16 Intron frequency in different eukaryotes. The average number of introns per gene is shown for a range of eukaryotic organisms.

Being a eukaryote, the yeast genome contains introns (♻ Section 4.9). However, the total number of introns in the protein-encoding genes of yeast is a mere 225. Most yeast genes with introns have only a single small intron near the 5′ end of the gene. This situation differs greatly from that seen in more complex eukaryotes (**Figure 6.16**). For example, in the worm *Caenorhabditis elegans,* the average gene has five introns, and in the fruit fly *Drosophila,* the average gene has four. Introns are also common in the genes of plants, averaging around four per gene. The model higher plant *Arabidopsis* averages five introns per gene, and over 75% of *Arabidopsis* genes have introns. In humans almost all protein-encoding genes have introns, and it is common for a single gene to have 10 or more. Moreover, human introns are typically much longer than human exons, the DNA that actually encodes proteins. Indeed, exons make up only about 1% of the human genome, whereas introns account for 24%.

MINIQUIZ

- What is the size range of eukaryotic genomes?
- How does this compare with that of prokaryotes?
- How can you show whether a gene is essential?
- What is unusual about the genome of the eukaryote *Encephalitozoon*?

III • Functional Genomics

Despite the major effort required to generate an annotated genome sequence, the net result is simply a "list of parts." To understand how a cell functions, we need to know more than which genes are present. We must also investigate both gene expression (transcription) and the function of the final gene product. In analogy to the term "genome," the entire complement of RNA produced under a given set of conditions is known as the **transcriptome**. Similar terminology is applied to the products of translation, metabolism, and other related areas by adding the suffix "omic." **Table 6.6** summarizes the "omics" terminology used in this chapter.

6.7 Microarrays and the Transcriptome

Transcriptomics refers to the global study of transcription and is done by monitoring the total RNA generated under chosen

UNIT 2

Table 6.6 Omics terminology

DNA	**Genome** the total complement of genetic information of a cell or a virus **Metagenome** the total genetic complement of all the cells present in a particular environment **Epigenome** the total number of possible epigenetic changes **Methylome** the total number of methylated sites on the DNA (whether epigenetic or not)
RNA	**Transcriptome** the total RNA produced in an organism under a specific set of conditions
Protein	**Proteome** the total set of proteins encoded by a genome **Translatome** the total set of proteins present under specified conditions **Interactome** the total set of interactions between proteins (or other macromolecules)
Metabolites	**Metabolome** the total complement of small molecules and metabolic intermediates **Glycome** the total complement of sugars and other carbohydrates
Organisms	**Microbiome** the total complement of microorganisms in an environment (including those associated with a higher organism) **Virome** the total complement of viruses in an environment **Mycobiome** the total complement of fungi in a natural environment

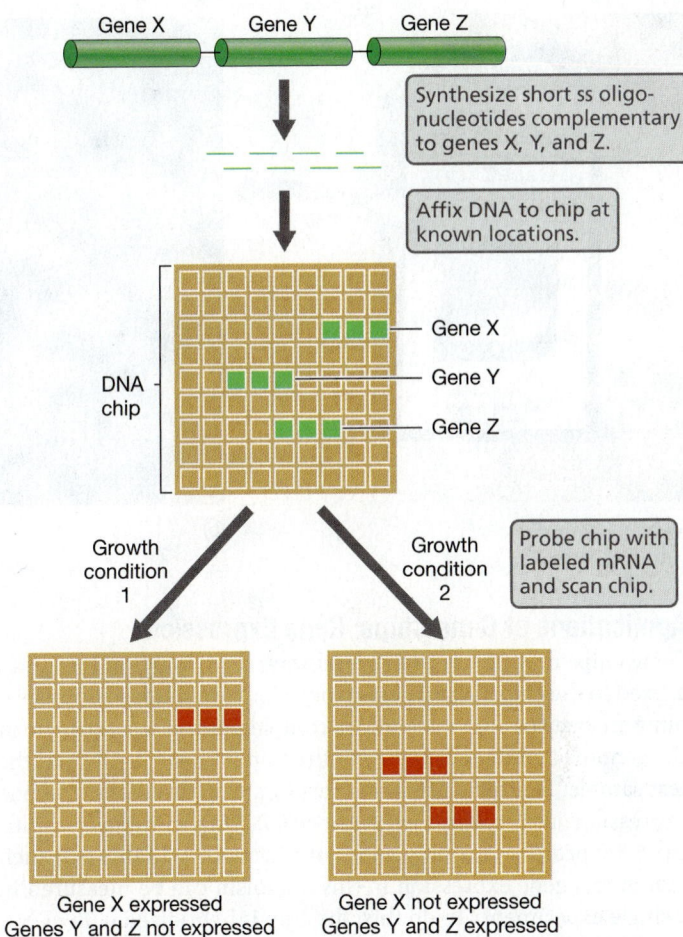

Figure 6.17 Making and using microarrays. Short single-stranded oligonucleotides corresponding to each gene in an organism are synthesized and affixed at known locations to make a microarray. The chip is assayed by hybridizing fluorescently labeled mRNA obtained from cells grown under a specific condition to the DNA probes on the chip and then scanning the chip with a laser.

growth conditions. In the case of genes whose role is still unknown, discovering the conditions under which they are transcribed may yield clues to their function. Two main approaches are used: microarrays, which depend on RNA–DNA hybridization, and RNA-Seq, which depends on second-generation (or later) sequencing.

Microarrays and the DNA Chip

Microarrays are small solid supports to which genes or, more often, segments of genes are fixed and arrayed spatially in a known pattern; they are often called **gene chips** (**Figure 6.17**). Microarray technology requires RNA–DNA hybridization. When DNA is denatured (that is, the two strands are separated), the single strands can form hybrid double-stranded molecules with other single-stranded DNA or RNA molecules by complementary or almost complementary base pairing (⮌ Section 11.2). This process is called *nucleic acid hybridization*, or **hybridization** for short, and is widely used in detecting, characterizing, and identifying segments of DNA or RNA. Segments of single-stranded nucleic acids whose identity is already known and that are used in hybridization are called **nucleic acid probes** or, simply, probes. To allow detection, probes are made radioactive or labeled with fluorescent dyes. By varying the conditions, it is possible to adjust the "stringency" of the hybridization such that complementary base pairing must be exact, or nearly so; this helps to avoid nonspecific pairing between sequences that are only partly complementary.

In a microarray, the gene segments may be synthesized by the polymerase chain reaction (PCR, ⮌ Section 11.3), or alternatively, oligonucleotides are designed and synthesized for each gene based on the genomic sequence. Once attached to the solid support, the DNA segments can be hybridized with RNA from cells grown under specific conditions and scanned and analyzed by computer. Hybridization between a specific RNA and a DNA segment on the chip indicates that the gene has been transcribed (Figure 6.17; see also Figure 6.18b). When studying protein-encoding genes, messenger RNA must be measured. In practice, mRNA is present at levels too low for direct use. Consequently, the mRNA sequences must first be amplified. This is done by using a modified version of PCR after converting the RNA to the *complementary DNA* (cDNA, ⮌ Section 11.3).

Photolithography, a process used to produce computer chips, is also used to produce microarray chips. Chips are typically about 1 to 2 cm and are inserted into a plastic holder that can easily be manipulated (**Figure 6.18a**); each chip can hold thousands of different DNA fragments. In practice, each gene is usually represented more than once in the array to increase reliability. Whole genome arrays contain DNA segments that cover the entire genome of an organism. For example, a chip that covers the entire human genome (Figure 6.18a) can analyze over 47,000 human transcripts and has room for 6500 additional oligonucleotides for use in clinical diagnostics.

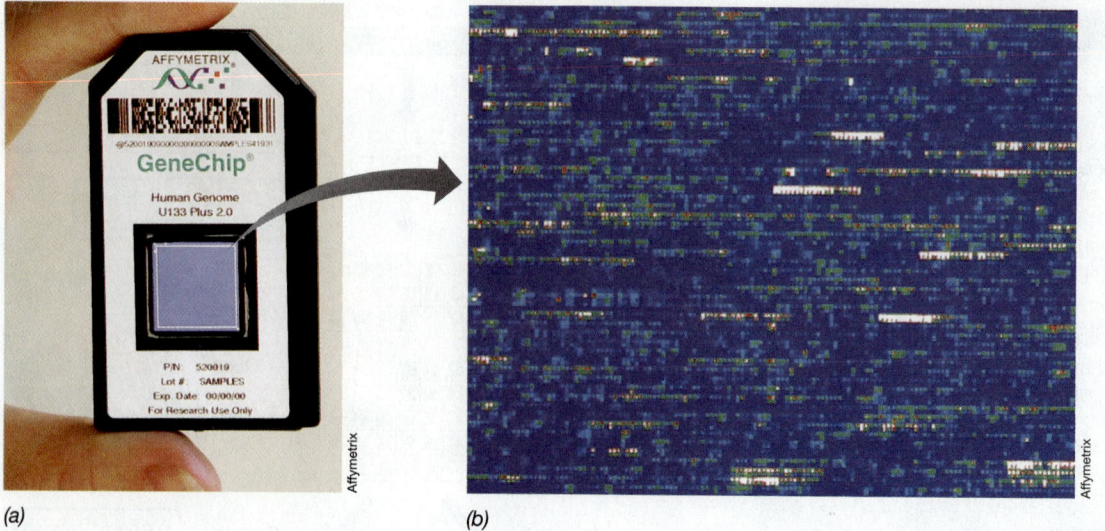

(a)

(b)

Affymetrix

Figure 6.18 **Using gene chips to assay gene expression.** (a) The human genome chip contains over 40,000 gene fragments. Blowup from part a to part b indicates location of actual microarray. (b) A hybridized yeast chip shows fragments from a quarter of the genome of the baker's yeast, *Saccharomyces cerevisiae*. Each gene is present in several copies and has been probed with fluorescently labeled cDNA (derived from mRNA) from yeast cells grown under a specific condition. The background of the chip is blue. Locations where the cDNA has hybridized are indicated by a gradation of colors up to a maximum number of hybridizations, which shows as white. Because the location of each gene on the chip is known, when the chip is scanned, it reveals which genes were expressed.

Applications of Gene Chips: Gene Expression

Gene chips may be used in several ways depending on the genes affixed to the chip. Global gene expression is monitored by assembling an array of oligonucleotides complementary to each gene in the genome and then using the entire population of mRNA as the test sample. Figure 6.18b shows part of a chip used to assay gene expression in *Saccharomyces cerevisiae*. This chip easily holds the 6000 protein-encoding genes of *S. cerevisiae* (Table 6.5) such that global gene expression in this organism can be measured in a single experiment. To do this, the chip is hybridized with cRNA or cDNA derived from mRNA obtained from yeast cells grown under specific conditions. To visualize binding, the nucleic acids are tagged with a fluorescent dye, and the chip is scanned with a laser fluorescence detector. A distinct hybridization pattern is observed, depending upon which DNA sequences correspond to which mRNAs (Figure 6.18b). The fluorescence intensity measures gene expression in a quantitative manner that allows the computer to list which genes were expressed and at what level. This reveals the *transcriptome* of the target organism grown under specified conditions (Table 6.6).

The *S. cerevisiae* gene chip has been used to study metabolic control in this important industrial organism. Yeast can grow by fermentation or by respiration. Transcriptome analysis reveals which genes are shut down and which are turned on when yeast cells are switched from fermentative (anaerobic) to respiratory (aerobic) metabolism or vice versa. Such experiments show that yeast undergoes a major metabolic "reprogramming" during the switch from anaerobic to aerobic growth. Genes that control production of ethanol (a key fermentation product) are strongly repressed, whereas citric acid cycle functions (needed for aerobic growth) are strongly activated. Overall, over 700 genes are turned on and over 1000 turned off during this metabolic transition. Moreover, by using a microarray, the expression pattern of genes of unknown function is also revealed, yielding clues to their possible role.

Applications in Identification

Microarrays can also be used to identify microorganisms. In this case the array contains a set of characteristic DNA sequences from each of a variety of organisms or viruses. Such an approach can be used to differentiate between different species or even closely related strains of the same species by differences in their hybridization patterns. This allows very rapid identification of pathogenic viruses or bacteria from clinical samples or detection of these organisms in other substances, such as food. For example, identification (ID) chips have been used in the food industry to detect specific pathogens, such as *Escherichia coli* O157:H7. A related use of DNA microarrays is to compare genes in closely related organisms. This can reveal how pathogenic bacteria may have evolved from their nonpathogenic relatives.

In environmental work, microarrays called *phylochips* have been used to assess microbial diversity. These contain oligonucleotides complementary to the 16S rRNA of different bacterial species, a molecule widely used in prokaryotic systematics (Chapter 12). After extracting bulk DNA or RNA from an environment, the presence or absence of each species can be assessed by the hybridization response on the chip (⮌ Section 18.6).

DNA chips can also be designed to identify higher organisms. A commercial chip, called the *FoodExpert-ID*, contains 88,000 gene segments from vertebrates and is used in the food industry to monitor food purity. This chip can confirm that the meat listed on a food label is as advertised, and can also detect foreign animal meats that may have been added as supplements to or substitutes for the stated ingredients. The FoodExpert-ID can also be used to detect vertebrate by-products in animal feed, a growing concern with the advent of prion-mediated infections such as mad cow disease (⮌ Section 9.13).

RNA-Seq Analysis

RNA-Seq analysis is a method in which all the RNA molecules from a cell are sequenced. Provided that the genome sequence is available for comparison, this will reveal not only which genes were transcribed but how many copies of each RNA were made. RNA-Seq is used both to measure expression of mRNA and to identify and characterize small noncoding RNAs. RNA-Seq requires high-throughput sequencing (second- or third-generation sequencing, Section 6.2) and is complicated by the fact that the

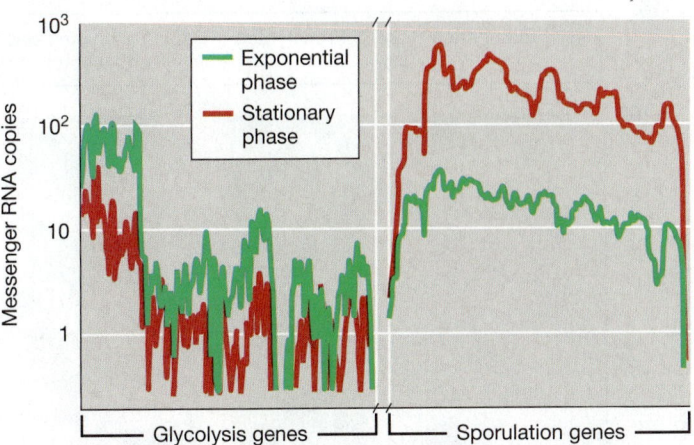

Figure 6.19 RNA-Seq analysis. The transcriptome of cultures of a *Clostridium* species grown for 4.5 h (cells in exponential phase) or 14 h (cells in stationary phase). Two genomic regions are shown: (1) ~5.4-kb segment surrounding the *gap-pgk-tpi* glycolytic operon, and (2) ~1.2-kb segment surrounding the *cotJC-cotJB* sporulation operon. Production of endospores is triggered by nutrient starvation (⮌ Section 2.16). Data from Wang, Y., X. Li, Y. Mao, and H.P. Blaschek. 2011. Single-nucleotide resolution analysis of the transcriptome structure of *Clostridium beijerinckii* NCIMB 8052 using RNA-Seq. *BMC Genomics 12:* 479–489.

most abundant RNA in a cell is ribosomal RNA (rRNA). Nevertheless, methods are available to remove rRNA or enrich mRNA from a total RNA pool. In addition, recent improvements in sequencing technology may allow sequencing without needing to remove rRNA.

RNA-Seq is beginning to overtake microarray analysis as the method of choice for global studies of gene expression. For example, **Figure 6.19** shows a comparison by RNA-Seq of cultures of a species of *Clostridium* in exponential and stationary phase. Clostridia are gram-positive rod-shaped bacteria that can produce endospores, the highly resistant and dormant stage of the cell's life cycle (⮌ Section 2.16). As one might predict, transcription of genes of the glycolytic pathway (the major means by which the organism makes ATP) is elevated during exponential growth, whereas expression of sporulation genes increases in stationary phase, when nutrients become limiting. RNA-Seq is also being used for microbial community analysis and can provide information on relative transcription levels when a genome sequence is not available for comparison. In this case the sequences detected must be identified by homology with sequences present in data banks.

As we will see in Section 6.10, *metagenomics* is the genomic analysis of pooled DNA or RNA from an environment. Metagenomic analysis using RNA-Seq has been exploited for the laboratory culture of bacteria from natural samples that previously proved intractable to laboratory culture. This was done by using RNA-Seq to reveal which genes were being transcribed at high levels by a particular microbial community. Sequence analysis then identified the proteins corresponding to the most prevalent mRNAs. This allowed the researchers to deduce which nutrients the bacteria in the sample might be using given the most likely enzyme activities of these proteins. Culture media were then devised using this information as a guide and previously uncultured bacteria were successfully cultured.

MINIQUIZ

- Why is it useful to survey expression of the entire genome under particular conditions?
- What do microarrays tell you that studying gene expression by assaying individual enzymes cannot?
- What technological advances does RNA-Seq depend on?

6.8 Proteomics and the Interactome

The genome-wide study of the structure, function, and activity of an organism's *proteins* is called **proteomics**. The number and types of proteins present in a cell change in response to an organism's environment or other factors, such as developmental cycles. As a result, the term **proteome** has unfortunately become ambiguous. In its wider sense, a proteome refers to *all* the proteins encoded by an organism's genome. In its narrower sense, however, it refers to those proteins present in a cell *at any given time*. The term *translatome* is sometimes used for the latter situation, that is, to refer to every protein made under specific conditions.

Methods in Proteomics

The first major approach to proteomics began with the advent of two-dimensional (2D) polyacrylamide gel electrophoresis. This technique can separate, identify, and measure all the proteins present in a cell sample. A 2D gel separation of proteins from *Escherichia coli* is shown in **Figure 6.20**. In the first dimension (the horizontal dimension in Figure 6.20), the proteins are separated by differences in their isoelectric points, the pH at which the net charge on each protein reaches zero. In the second dimension, the proteins are denatured in a way that gives each amino acid residue a fixed charge. The proteins are then separated by size (in much the same way as for DNA molecules; ⮌ Section 11.1).

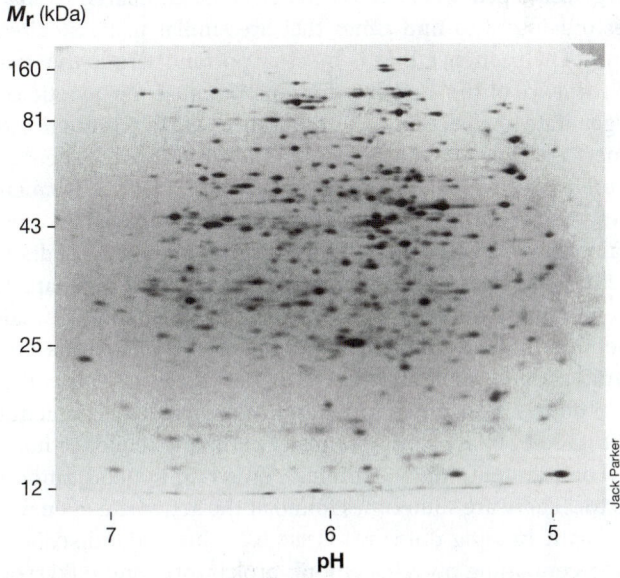

Figure 6.20 Two-dimensional polyacrylamide gel electrophoresis of proteins from *Escherichia coli*. Each spot on the gel is a different protein, radioactively labeled to allow for visualization and quantification. Denatured proteins were separated in the horizontal direction by isoelectric focusing and in the vertical direction by their mass (M_r; in kilodaltons). The largest proteins are at the top of the gel.

In studies of *E. coli* and a few other organisms, hundreds of proteins separated in 2D gels have been identified by biochemical or genetic means, and their presence or absence has been studied under various growth conditions. Using 2D gels, the presence of a particular protein under different growth conditions can be measured and related to environmental conditions. One method of connecting an unknown protein with a particular gene using the 2D gel system is to elute the protein from the gel and sequence a portion of it, usually from its amino-terminal end. Alternatively, eluted proteins can be identified by a technique called *mass spectrometry* (Section 6.9), usually after preliminary digestion to give a characteristic set of peptides. The sequence information obtained by either technique may be sufficient to completely identify the protein. Alternatively, partial sequence data may allow the design of oligonucleotide probes or primers to locate the gene encoding the protein from genomic DNA by hybridization or PCR. Then, after sequencing of the DNA, the gene may be identified.

Today, liquid chromatography is increasingly used to separate protein mixtures. In high-pressure liquid chromatography (HPLC), the sample is dissolved in a suitable liquid and forced under pressure through a column packed with a stationary phase material that separates proteins by variations in their chemical properties, such as size, ionic charge, or hydrophobicity. As the mixture travels through the column, it is separated by interaction of the proteins with the stationary phase. Fractions are collected at the column exit. The proteins in each fraction are digested by proteases and the peptides identified by mass spectrometry.

Comparative Genomics and Proteomics

Although proteomics often requires intensive experimentation, computational techniques can also help. Once the sequence of an organism's genome is obtained, it can be compared to that of other organisms to find genes that are similar to those already known. The sequence that is most important here is the *amino acid sequence* of the encoded proteins. Because the genetic code is degenerate (↩ Section 4.11), differences in DNA sequence may not necessarily lead to differences in the amino acid sequence.

Proteins with greater than 50% sequence identity frequently have similar functions. Proteins with identities above 70% almost certainly have similar functions. Many proteins consist of distinct structural modules, called *protein domains*, each with characteristic functions. Such regions include metal-binding domains, nucleotide-binding domains, or domains for certain classes of enzyme activity, such as helicase or dehydrogenase. Identification of domains of known function in a protein may reveal much about its role, even in the absence of complete sequence homology. For example, many proteins contain the metal zinc as a cofactor. These are sometimes found at the active site of enzymes or in DNA-binding domains. **Figure 6.21** shows the distribution of zinc-containing proteins among prokaryotes and eukaryotes. Whereas both groups synthesize many zinc-containing enzymes, the use of zinc-containing transcription factors is predominantly a eukaryotic trait.

Structural proteomics refers to the proteome-wide determination of the three-dimensional (3D) structures of proteins. At

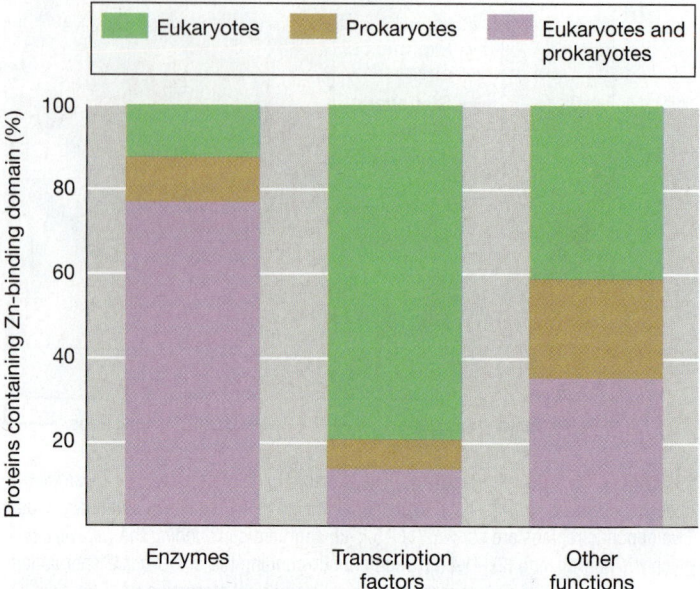

Figure 6.21 Comparative proteomics. The sequences of proteins containing zinc-binding-sequence domains from 40 *Bacteria*, 12 *Archaea*, and 5 eukaryotes were compared by functional category. Zinc-containing proteins comprise 5–6% of the total proteins in prokaryotes and 8–9% in eukaryotes and many are enzymes Eukaryotes also contain many unique zinc-containing transcription factors.

present, it is not possible to predict the 3D structure of proteins directly from their amino acid sequences. However, structures of unknown proteins can often be modeled if the 3D structure is available for a protein with 30% or greater identity in amino acid sequence.

Coupling proteomics with genomics has yielded important clues about how gene expression in different organisms correlates with environmental stimuli. Not only does such information have important basic science benefits, but it also has potential applications. These include advances in medicine, environmental protection, and agriculture. In all of these areas, understanding the link between the genome and the proteome and how it is regulated can help in fighting disease and pollution as well as yield major benefits for agricultural productivity.

The Interactome

By analogy with the terms "genome" and "proteome," the **interactome** is the complete set of *interactions* among the macromolecules within a cell (**Figure 6.22**). Originally, the term interactome applied to the interactions between proteins, many of which assemble into complexes. However, it is also possible to consider interactions between different classes of macromolecules, such as the protein–RNA interactome.

Interactome data are typically expressed in the form of network diagrams, with each node representing a protein and connecting lines representing the interactions. Diagrams of whole interactomes can be extremely complex and thus more focused interactomes, such as the motility protein network from the bacterium *Campylobacter jejuni* (Figure 6.22), are more instructive. This figure shows the core interactions between well-known components of the chemotaxis system (↩ Sections 2.19 and 7.8), including all other proteins that interact with these.

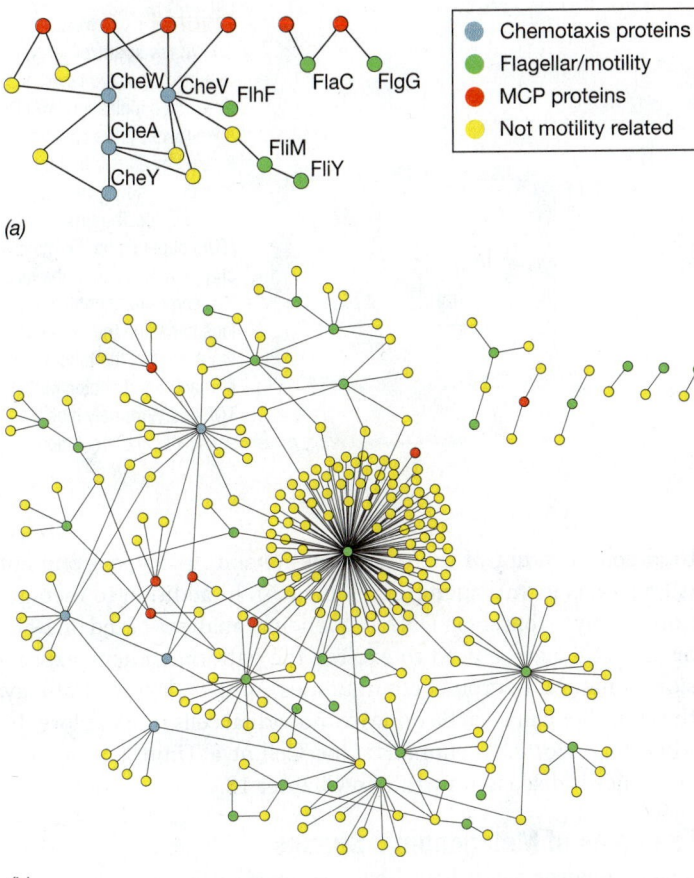

(a)

(b)

Figure 6.22 Motility protein interactome for *Campylobacter jejuni*. This network illustrates the way in which interactome data are depicted. *(a)* A subsection of the network highlighting the well-known proteins of the chemotaxis signal transduction pathway (CheW, CheA, and CheY) and their partners. MCP, methyl-accepting chemotaxis proteins (🔗 Section 7.8). *(b)* High-confidence interactions between all proteins known to have roles in motility. Note the six small networks that fall outside the single large network.

MINIQUIZ

• Why is the term "proteome" ambiguous, whereas the term "genome" is not?

• What are the most common experimental methods used to survey the proteome?

• What is the interactome?

6.9 Metabolomics and Systems Biology

The **metabolome** is the complete set of *metabolic intermediates* and other small molecules produced in an organism. Metabolomics has lagged behind other "omics" largely due to the immense chemical diversity of small metabolites that can be present in cells. This makes systematic screening technically challenging. Early attempts used nuclear magnetic resonance (NMR) analysis of extracts from cells labeled with ^{13}C-glucose (^{13}C is a heavy isotope of carbon, most of which is ^{12}C). However, this method is limited in sensitivity, and the number of compounds that can be simultaneously identified in a mixture is too low for resolution of complete cell extracts.

New Mass Spectrometry Techniques: MALDI-TOF

The most promising approach to metabolomics is the use of newly developed methods of mass spectrometry. This approach is not limited to particular classes of molecules and can be extremely sensitive. The mass of ^{12}C is defined as exactly 12 molecular mass units (daltons). However, the masses of other atoms, such as ^{14}N or ^{16}O, are not exact integers. Mass spectrometry using extremely high mass resolution allows the unambiguous determination of the molecular formula of any small molecule. Clearly, isomers will have the same molecular formula, but they can be distinguished by their different fragmentation patterns during mass spectrometry. The same approach is used to identify the peptide fragments from digested proteins during proteome analyses (Section 6.8). In this case, identifying several oligopeptides allows the identity of the parent protein to be deduced provided that its amino acid sequence is known.

In the MALDI (*m*atrix-*a*ssisted *l*aser *d*esorption *i*onization) version of mass spectrometry, the sample is ionized and vaporized by a laser (**Figure 6.23**). The ions generated are accelerated along the column toward the detector by an electric field. The time of flight (TOF) for each ion depends on its mass/charge ratio—the smaller this ratio, the faster the ion moves. The detector measures the TOF for each ion and the computer calculates the mass and hence the molecular formula. The combination of these two techniques is known as *MALDI-TOF*.

Metabolome analysis is especially useful for the study of plant biochemistry, since plants produce several thousand different metabolites—more than for most other types of organism. These include many so-called *secondary metabolites*, chemicals such as scents, flavors, alkaloids, and pigments, many of which are commercially important. Metabolomic investigations have monitored the levels of several hundred metabolites in the model plant *Arabidopsis*, and significant changes were observed in the levels of many of these metabolites in response to changes in temperature. Future directions for metabolomics include assessing the effect of disease on the metabolome of various human organs and tissues. Such results may greatly improve our understanding of how the human body combats infectious and noninfectious disease and identify key compounds important in host defenses. Such compounds could possibly be developed as drugs for clinical treatment of particular diseases.

Systems Biology

The term **systems biology** has been widely used in recent years to refer to the integration of different fields of research to give an overview of an organism or cell or even entire species or ecosystems. Systems biology integrates all of the "omics" we have studied here: genomics, transcriptomics, proteomics, metabolomics, and so on (**Figure 6.24**). The ability to store and analyze large amounts of biological information by computer is essential to systems biology, and the understanding of entire biological systems is evolving in parallel with computer power and storage capabilities.

The basic strategy of systems biology is to compile a series of "omics" data and then build a computational model of the system under study (Figure 6.24). Such models may allow for the prediction of behavior or properties of a particular organism that were not

UNIT 2

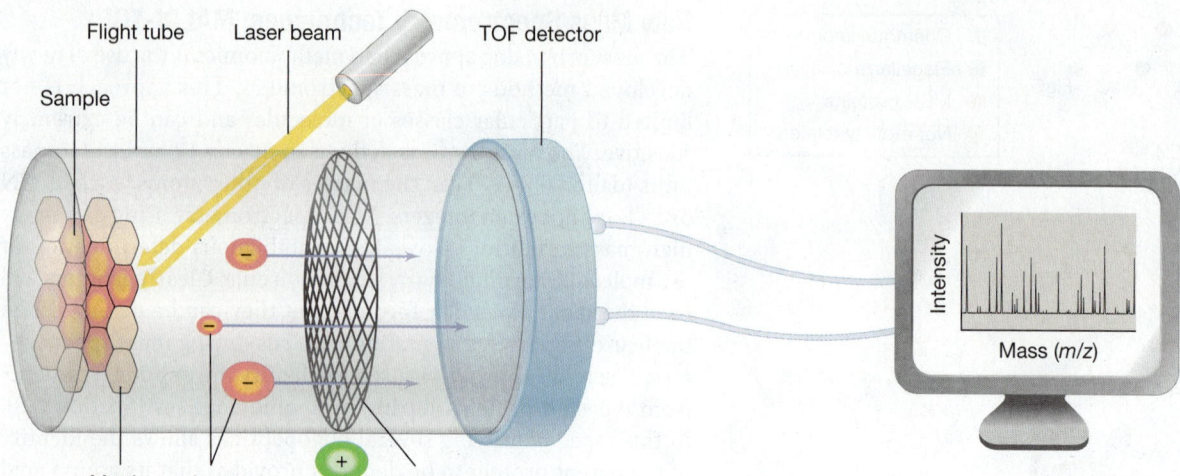

Figure 6.23   MALDI-TOF mass spectrometry. In matrix-assisted laser desorption ionization (MALDI) spectroscopy, the sample is ionized by a laser and the ions travel down the tube to the detector. The time of flight (TOF) depends on the mass/charge (m/z) ratio of the ion. The computer identifies the ions based on their time of flight; that is, the time it takes to reach the detector. MALDI-TOF has extremely high sensitivity and resolution.

obvious from the original observations. These are referred to as the *emergent properties* of an organism. It is predicted that an understanding of the emergent properties of an organism will provide much deeper insight into the overall biology of that organism than can any single "omics" study by itself.

MINIQUIZ

• What techniques are used to monitor the metabolome?

• What is a secondary metabolite?

• Why does systems biology depend on computing power? What is an "emergent property"?

6.10 Metagenomics

Microbial communities contain many species of *Bacteria* and *Archaea*, most of which have never been cultured or formally identified. **Metagenomics**, also called *environmental genomics*, analyzes pooled DNA or RNA from an environmental sample containing organisms that have not been isolated and identified. Just as the total gene content of an organism is its genome, so the total gene content of the organisms inhabiting an environment is known as its **metagenome** (Table 6.6). In addition to metagenome analyses based on DNA sequencing, analyses based on RNA or proteins may be used to explore the patterns of gene expression in natural microbial communities. With today's technology, these studies can even be done on individual cells (see Explore the Microbial World, "Genomics, One Cell at a Time"). Single-cell genomics is discussed further in Chapter 18.

Examples of Metagenomic Studies

Several environments have been surveyed by large-scale metagenome sequencing projects. Extreme environments, such as acidic runoff waters from mines, tend to have low species diversity. Consequently it has been possible to isolate community DNA and assemble much of it into nearly complete individual genomes. Conversely, complex environments such as fertile soils or aquatic environments are much more challenging, and complete genome assemblies here are much more difficult. Nonetheless, a surprising finding that has emerged from metagenomic studies thus far is that most genes in the environment belong not to cellular organisms but to viruses. This is discussed further in Chapter 9 where we consider the genomics and phylogeny of viruses.

Even if complete genomes cannot be assembled, much useful information can be derived from metagenomic surveys. For example, environments can be analyzed for the presence and distribution of different taxonomic groups of bacteria. These vary greatly in relative abundance in different environments and **Figure 6.25** illustrates this for major subgroups of *Proteobacteria* (Chapter 15) at a sampling site close to Hawaii in the Pacific Ocean. Light, oxygen, nutrients, and temperature all change with depth, and these factors can be correlated with which proteobacterial subgroups are most competitive at each depth (Figure 6.25). One curious observation that has emerged from such metagenomic studies is that much cellular DNA in natural habitats does not belong to living cells. For example, about 50–60% of the DNA in the oceans is extracellular DNA found in deep-sea sediments. Presumably this is DNA deposited when dead organisms from the upper layers of the ocean sink to the bottom and eventually lyse. Because nucleic acids are major repositories of phosphate, this DNA is a major contributor to the global phosphorus cycle.

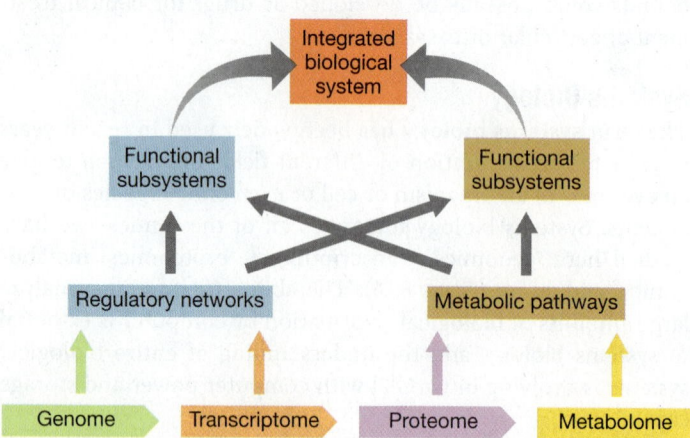

Figure 6.24   The components of systems biology. The results of various "omics" analyses are combined and successively integrated into higher-level views of the entire biology of an organism.

EXPLORE THE MICROBIAL WORLD

Genomics, One Cell at a Time

Modern genome analysis has been driven by scaling up the number of samples run simultaneously and also by scaling down the sample size. The reduction in sample size needed to do genomics has led to the capability of analyzing even single cells—a technique called *single-cell genomics*—and some amazing results have emerged.

Single cells can be isolated by using various physical techniques and then subjected to genomics procedures (**Figure 1**). Genome sequencing and transcriptome and proteome analyses have all been performed with single bacterial cells. Sequencing DNA from single cells relies on a highly modified version of PCR known as *multiple displacement amplification (MDA)* (↩ Section 18.11 and Figure 18.32). This technique amplifies the femtogram (10^{-15} g) quantities of DNA present in a single bacterial cell into the micrograms of DNA required for sequencing (a billionfold amplification). Likewise, RNA can be analyzed by RNA-Seq or following amplification by a modified version of PCR. Single-cell proteomic analyses are more tricky, but analyses that employ very sensitive fluorescence methods are available for this purpose.

Single cells have been isolated from soil and various other habitats and their DNA sequenced. Using single-cell genomics, metabolic genes present in an environment can be not only identified but actually assigned to particular species. Thus, single-cell genomics can reveal which organisms in a microbial community are degrading which nutrients. For example, single-cell genomics has been used to analyze hydrocarbon degradation by bacteria in polluted environments, leading to a better understanding of which organisms are

doing what in the overall process. Similarly, plasmids and viruses can be allocated to their correct host when a single cell is sequenced.

A surprising finding in single-cell studies has been that protein and transcript levels vary greatly from cell to cell in an actively growing pure culture of a bacterium, presumably as a result of transcription and translation occurring in bursts, rather than smoothly. This is especially true for proteins expressed at low levels. Consequently, and counterintuitively, for individual genes in a single cell there is little correlation between copy numbers of an mRNA and its corresponding protein at any given time point. This is partly due to the difference in the average lifetime of protein and mRNA molecules. Whereas most proteins survive longer than a cell generation, mRNA in bacteria is generally degraded within two or three minutes of its synthesis. Thus, mRNA levels at any given time are determined by the transcription rate in the preceding few minutes, whereas protein levels reflect synthesis over the course of an hour or so.

Single-cell genomics has a bright future for probing many important facets of an organism's biology on an individual cell rather than a cell population basis. The method has already challenged previous assumptions about the biochemical uniformity of cells in exponential-phase cultures, and it is likely that many other questions will emerge for which single-cell omics technologies will be ideally suited to yield answers not obtainable with mass cultures. Single-cell genomics is also an excellent example of how scientific methods designed with one goal in mind (that is, the genomic analysis of a *population* of cells) can be modified by creative scientists to ask scientific questions never before thought possible.

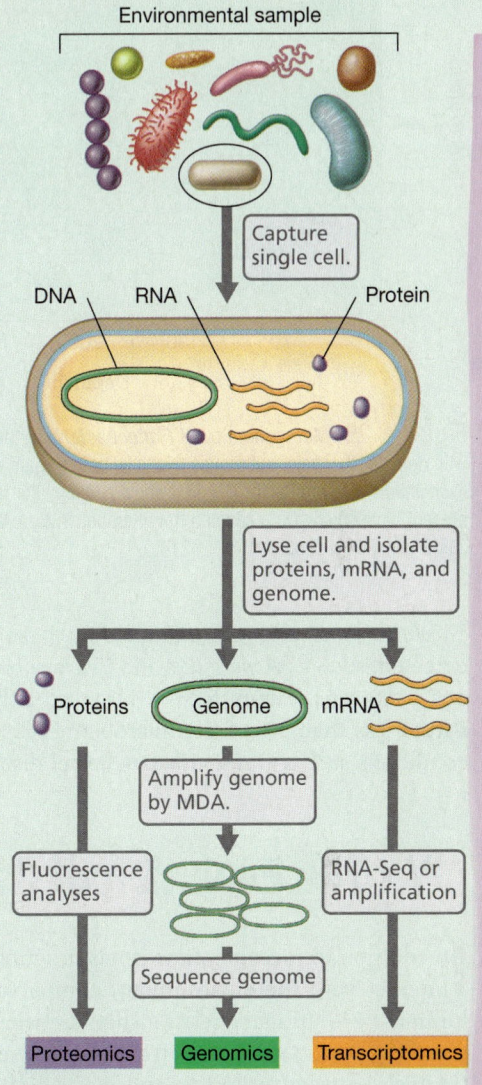

Figure 1 Single-cell genomics. A single cell isolated from an environmental sample can be the source of a diversified "omics" study.

Metagenomics and "Biome" Studies

The human body is estimated to contain about 10 trillion (10^{13}) cells, but each of us also carries around ten times more prokaryotic cells than human ones. This collection of prokaryotic cells is called the human *microbiome*. Most of these prokaryotes inhabit the gut and the majority belong to two bacterial groups, the *Bacteroidetes* and the *Firmicutes* (Chapter 15). A fascinating finding is that the composition of the gut microbiome correlates with obesity in both humans and experimental mouse models. The higher the proportion of *Firmicutes* (mostly *Clostridium* and relatives),

the fatter the human or mouse. A suggested mechanism is that species of *Firmicutes* convert more dietary fiber into short-chain fatty acids that can be absorbed by the host. Hence, the host gets more fat from the same amount of food. Moreover, although an important model organism in biology, the bacterium *Escherichia coli* comprises only about 1% of the total gut bacterial population.

Recent surveys of the human and mouse gut microbiome have also revealed several species of fungi (**Figure 6.26**), previously undetected; these make up what is called the *mycobiome* (the prefix "myco" means "fungal"). Many of these are common yeasts, such as

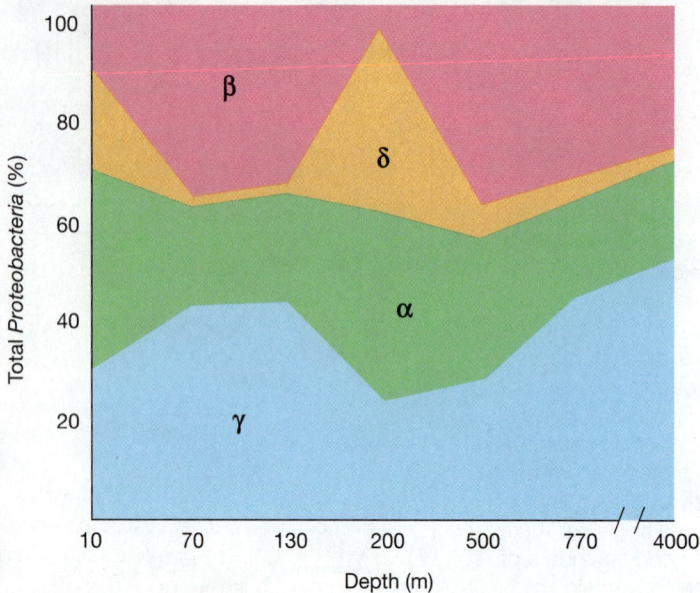

Figure 6.25 **Metagenomics of *Proteobacteria* in the ocean.** The distribution with depth of the major subgroups (alpha α, beta β, gamma γ, and delta δ) of *Proteobacteria* in the Pacific Ocean is shown. Many other types of bacteria are also present (not shown). Data adapted from Kembel, S.W., J.A. Eisen, K.S. Pollard, and J.L. Green. 2011. *PLoS One 6:* e23214.

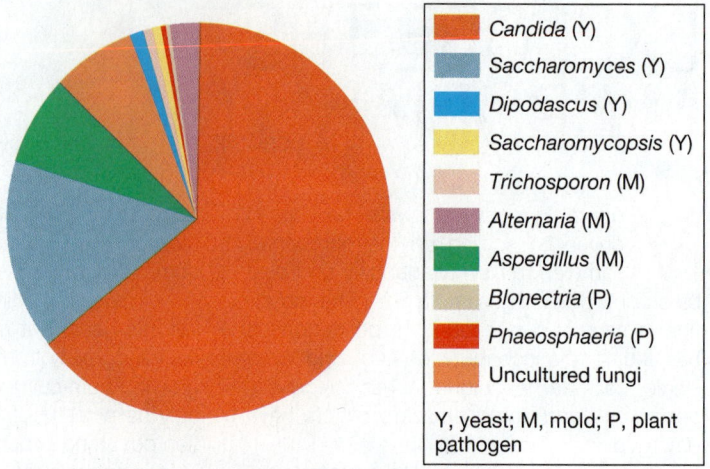

Figure 6.26 **The mouse mycobiome.** The data shown are for the fungal population of the mouse intestine. The pie chart shows the most common fungi present are yeasts. M, mold; Y, yeast; P, plant pathogen. Data adapted from Iliev, I.D., et al. *Science 336:* 1314–1317 (2012).

Saccharomyces and *Candida*, although some of the gut fungi detected, such as *Aspergillus* and *Trichosporon*, are potential serious pathogens (Figure 6.26). Moreover, although gut fungi constitute less than 1% of the microbiome, it is known that certain conditions such as inflammatory bowel disease correlate strongly with specific fungal populations. Thus metagenomics holds great promise for probing possible connections between specific microbial populations and specific diseases in humans and other animals.

MINIQUIZ

- What is a metagenome?
- How is a metagenome analyzed?
- How do the human microbiome and mycobiome differ?

IV • The Evolution of Genomes

In addition to revealing how genes function and how organisms interact with the environment, comparative genomics can illuminate evolutionary relationships between organisms. Reconstructing evolutionary trees from genome sequences helps to distinguish between primitive and derived characteristics and can resolve ambiguities in phylogenetic trees based on analyses of a single gene, such as an rRNA gene (Section 12.4). Genomics is also a link to understanding early life forms and may eventually help answer the most fundamental of all questions in biology: How did life originate?

6.11 Gene Families, Duplications, and Deletions

Genomes from both prokaryotic and eukaryotic sources often contain multiple copies of genes that are related in sequence due to shared evolutionary ancestry; such genes are called *homologous genes*, or **homologs**. Groups of gene homologs are called **gene families**. Not surprisingly, larger genomes tend to contain more individual members from a particular gene family.

Paralogs and Orthologs

Comparative genomics shows that many genes have arisen by *duplication* of other genes. Such homologs may be subdivided, depending on their origins. Genes whose similarity is the result of gene duplication at some time in the evolution of an organism are called **paralogs**. Genes found in one organism that are similar to genes in another organism because of descent from a common ancestor are called **orthologs** (**Figure 6.27**). Orthologs are often not identical because of divergent evolution in lineages following speciation. An example of paralogous genes are those encoding several variant lactate dehydrogenase (LDH) enzymes in humans. These variants, called *isoenzymes*, are structurally distinct yet all highly related and carry out the same enzymatic reaction. By contrast, the corresponding LDH from the lactic acid bacterium, *Lactobacillus*, is orthologous to all of the human LDH isoenzymes. Thus, gene families contain both paralogs and orthologs.

Gene Duplication

It is widely thought that gene duplication is the mechanism by which most new genes evolve. If a segment of duplicated DNA

UNIT 2

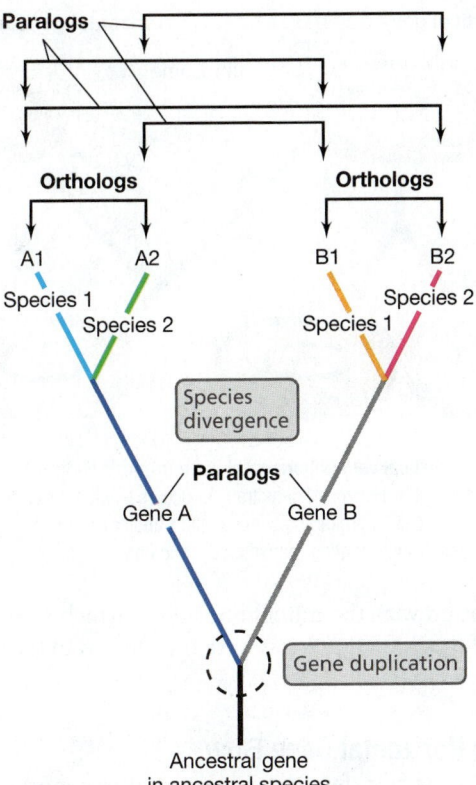

Figure 6.27 Orthologs and paralogs. This family tree depicts an ancestral gene that duplicated and diverged into two paralogous genes, A and B. Later, the ancestral species diverged into species 1 and species 2, both of which have genes for A and B (designated A1 and B1 and A2 and B2, respectively). Each such pair are paralogs. However, because species 1 and 2 are now separate species, A1 is an ortholog of A2 and B1 is an ortholog of B2.

is long enough to include an entire gene or group of genes, the organism with the duplication has multiple copies of these particular genes. After duplication, one of the duplicates is free to evolve while the other copy continues to supply the cell with the original function (Figure 6.28a). In this way, evolution can "experiment" with one copy of the gene. Such gene duplication events, followed by diversification of one copy, are thought to be the major events that fuel microbial evolution. Genomic analyses have revealed many examples of protein-encoding genes that were clearly derived from gene duplication. Figure 6.28b shows this for the enzyme RubisCO, a key enzyme of autotrophic metabolism (↺ Section 13.5). Here an ancestral gene gave rise to enzymes with different but related catalytic activities.

Duplications of genetic material may include just a handful of bases or even whole genomes. For example, comparison of the genomes of the yeast *Saccharomyces cerevisiae* and other fungi suggests that the ancestor of *Saccharomyces* duplicated its entire genome. This was followed by extensive deletions that eliminated much of the duplicated genetic material. Analysis of the genome of the model plant *Arabidopsis* suggests that there were one or more whole genome duplications in the ancestor of the flowering plants, as well.

Did bacterial genomes evolve by whole genome duplication? The distribution of duplicated genes and gene families in the genomes of bacteria suggests that many frequent but relatively small duplications have occurred. For example, the soil bacterium *Myxococcus* has a genome of 9.1 Mbp. This is approximately twice that of the genomes of its close relatives. Among a group of gram-negative bacteria called the *Alphaproteobacteria*, genome sizes range from 1.1–1.5 Mbp for parasitic species to 4 Mbp for free-living *Caulobacter*, and up to 7–9 Mbp for plant-associated bacteria

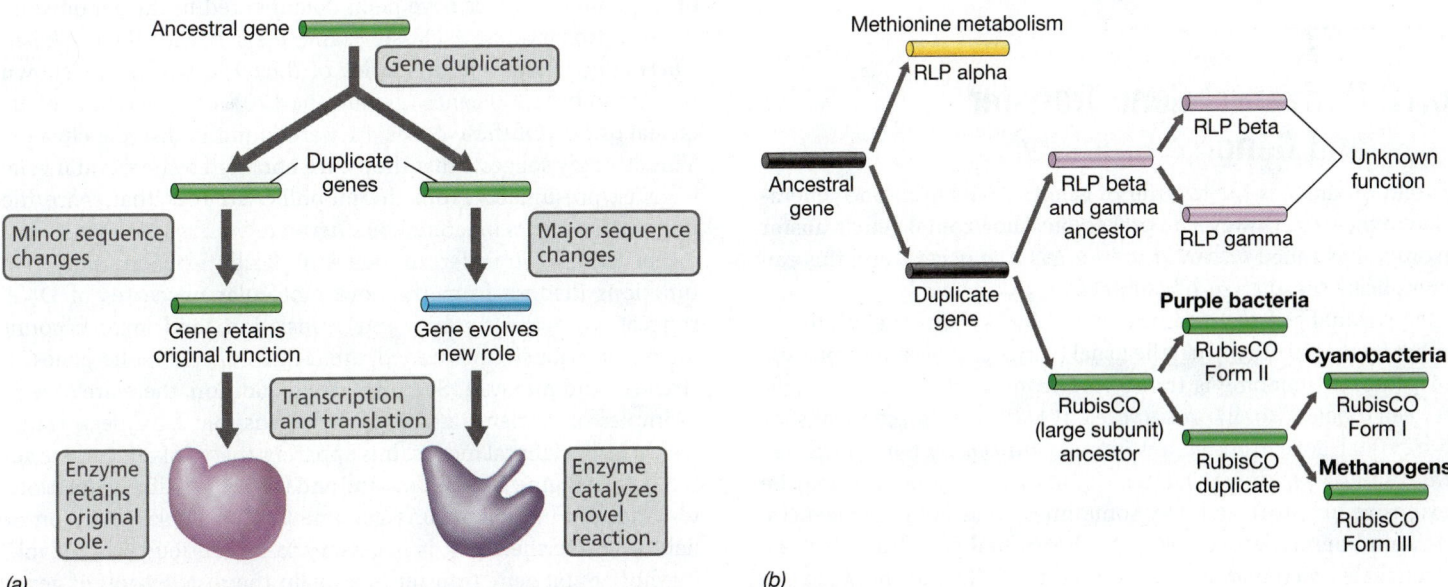

Figure 6.28 Evolution by gene duplication. *(a)* The principle of gene duplication. After duplication, the "spare" copy of a gene is free to evolve a new function. *(b)* The RubisCO (*rbcL*) family of genes. The large subunit of the enzyme RubisCO that fixes CO_2 during photosynthesis has split into three closely related forms (I, II, and III) that all retain the original function (green bars). However, RubisCO is in turn derived from an ancestral gene (black bars) of unknown function that divided to produce a gene encoding an enzyme in methionine metabolism (yellow bar) and several genes whose function is still unknown (purple bars). RLP, RubisCO-like protein.

such as *Rhizobium* (Table 6.1). However, in all of these cases gene distribution analysis points to frequent small-scale gene duplications rather than entire genome duplications. Conversely, in bacteria that are parasitic, frequent successive deletions have eliminated genes no longer needed for a parasitic lifestyle, leading to their unusually small genomes (Section 6.4, Table 6.1, and Figures 6.8 and 6.14).

Gene Analysis in Different Domains

The comparison of genes and gene families is a major task in comparative genomics. Because chromosomes from many different microorganisms have already been sequenced, such comparisons can be easily done, and the results are often surprising. For instance, genes in *Archaea* that are active in DNA replication, transcription, and translation are more similar to those in *Eukarya* than to those in *Bacteria*. Unexpectedly however, many other genes in *Archaea*, for example, those encoding metabolic functions other than information processing, are more similar to those in *Bacteria* than those in *Eukarya*. The analytical tools of bioinformatics allow genetic relationships between any organisms to be deduced quickly and at the single gene, gene group, or entire genome level. The results obtained thus far generally support the phylogenetic picture of life deduced originally by comparing rRNA sequences (⇄ Section 12.4) and suggest that many genes in all organisms have common evolutionary roots. However, these analyses have also revealed instances of horizontal gene transfer, an important issue to which we now turn.

MINIQUIZ -
- What is a homologous gene?
- What is a gene family?
- Contrast gene paralogs with gene orthologs.

6.12 Horizontal Gene Transfer and Genome Stability

Evolution dictates the transfer of genetic traits from one generation to the next. However, in prokaryotes, **horizontal gene transfer** (sometimes called *lateral gene transfer*) also occurs, and this can complicate the analysis of genomes.

Horizontal gene transfer refers to transfer from one cell to another by means other than the usual (vertical) inheritance process in which the genome is transferred from mother cell to daughter cell (**Figure 6.29**). In prokaryotes, at least three mechanisms for horizontal gene transfer are known: *transformation, transduction,* and *conjugation* (Chapter 10). Horizontal gene flow may be extensive in nature and may sometimes cross even phylogenetic domain boundaries. However, for horizontal gene transfer to be detectable by comparative genomics, the difference between the organisms must be rather large. For example, several eukaryotic genes have been found in *Chlamydia* and *Rickettsia*, both bacterial pathogens of humans. In particular, two genes encoding histone H1-like proteins have been found in the *Chlamydia trachomatis* genome, suggesting horizontal transfer from a eukaryotic source, possibly even its human host. Note that this is the reverse

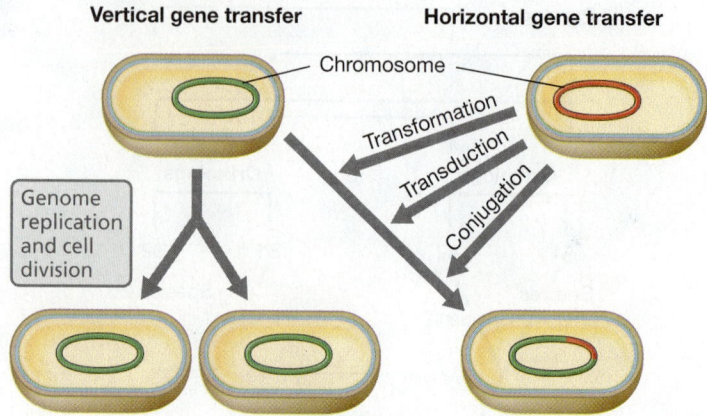

Figure 6.29 **Vertical versus horizontal gene transfer.** Vertical gene transfer occurs when cells divide. Horizontal gene transfer occurs when a donor cell contributes genes to a recipient cell. In prokaryotes, horizontal transfer occurs through one of the three mechanisms: transformation, transduction, and conjugation.

of the situation with the mitochondrion in which genes from the ancestor of the mitochondrion were transferred to the eukaryotic nucleus (Section 6.5).

Detecting Horizontal Gene Flow

Horizontal gene transfers can be detected in genomes once the genes have been annotated (Section 6.3). The presence of genes that encode proteins typically found only in distantly related species is one signal that the genes originated from horizontal transfer. However, another clue to horizontally transferred genes is the presence of a stretch of DNA whose guanosine/cytosine (GC) content or codon bias differs significantly from the rest of the genome (Figure 6.29). With these clues, many likely examples of horizontal transfer have been documented in the genomes of various prokaryotes. A classic example exists with the organism *Thermotoga maritima*, a species of *Bacteria*, which was shown to contain over 400 genes (greater than 20% of its genome) of archaeal origin. Of these genes, 81 were found in discrete clusters. This strongly suggests that they were obtained by horizontal gene transfer, presumably from thermophilic *Archaea* that share the hot environments inhabited by *Thermotoga*.

Horizontally transferred genes typically encode metabolic functions distinct from the core molecular processes of DNA replication, transcription, and translation, and may account for the previously mentioned similarities of metabolic genes in *Archaea* and *Bacteria* (Section 6.4). In addition, there are several examples of virulence genes of pathogens that have been transferred by horizontal means. It is apparent that prokaryotes are actively exchanging genes in nature, and the process likely functions to "fine-tune" an organism's genome to a particular situation or habitat. Nevertheless, it is necessary to be cautious when invoking horizontal gene transfer to explain the distribution of genes in a given organism. For example, when the human genome was first sequenced, over 200 genes were identified as being horizontal transfers from prokaryotes. However, when more eukaryotic genomes became available for comparison, homologs were found for most of these genes in many eukaryotic lineages. Consequently, it now seems that most of these genes are in fact of eukaryotic origin. Only about a dozen human genes are now accepted as strong candidates for having relatively recent prokaryotic origins. The phrase "relatively recent" here refers to genes transferred from prokaryotes after separation of the major eukaryotic lineages (♻ Section 12.4), not to genes of possible ancient prokaryotic origin that are shared by eukaryotes as a whole.

Genome Evolution and Mobile Elements

The term "mobile DNA" refers to segments of DNA that move from one location to another within host DNA molecules (♻ Section 10.11). Most mobile DNA consists of *transposable elements*, but insertion sequences and integrated virus genomes are also common. Such mobile elements can play important roles in genome evolution (**Figure 6.30**).

Transposons are common forms of mobile DNA that move between different host DNA molecules, including chromosomes, plasmids, and viruses, by the activity of an enzyme called *transposase* (♻ Section 10.11). In doing so they may pick up and horizontally transfer genes for various characteristics, including resistance to antibiotics and production of toxins. However, transposons may also mediate a variety of large-scale chromosomal changes (Figure 6.30). Bacteria that are undergoing rapid evolutionary change often contain relatively large numbers of mobile elements, especially insertion sequences, simple transposable elements whose genes encode only transposition. Recombination among identical elements generates chromosomal rearrangements such as deletions,

inversions, or translocations, and these provide a source of genomic diversity upon which natural selection can act. Thus, chromosomal rearrangements that accumulate in bacteria during stressful growth conditions are often flanked by repeats or insertion sequences.

Conversely, once a species settles into a stable evolutionary niche, most mobile elements are apparently lost. For example, genomes of species of *Sulfolobus* (*Archaea*) have unusually high numbers of insertion sequences and show a high frequency of gene translocations. By contrast, *Pyrococcus* (*Archaea*) shows an almost complete lack of insertion sequences and a correspondingly low number of gene translocations. This suggests that for whatever reason(s), perhaps because of fluctuations in conditions in their habitats, the genomes of *Sulfolobus* species are more dynamic than the more stable genome of *Pyrococcus*.

Chromosomal rearrangements due to insertion sequences have apparently contributed to the evolution of several bacterial pathogens. In *Bordetella*, *Yersinia*, and *Shigella*, the more highly pathogenic species show a much greater frequency of insertion sequences. For example, *Bordetella bronchiseptica* has a genome of 5.3 Mbp but carries no known insertion sequences. Its more pathogenic relative, *Bordetella pertussis*, the causative agent of whooping cough (♻ Section 29.3), has a smaller genome (4.1 Mbp) but has more than 260 insertion sequences. Comparison of these genomes suggests that the insertion sequences are responsible for major genome rearrangements, including the deletions that reduced the genome size in *B. pertussis*.

Insertion sequences also play a role in assembling genetic modules to generate novel plasmids. For example, 46% of the 220-kbp virulence megaplasmid of the pathogenic bacterium *Shigella flexneri* consists of insertion sequence DNA. In addition to full-length insertion sequences, there are also many fragments in this plasmid that imply multiple ancestral rearrangements.

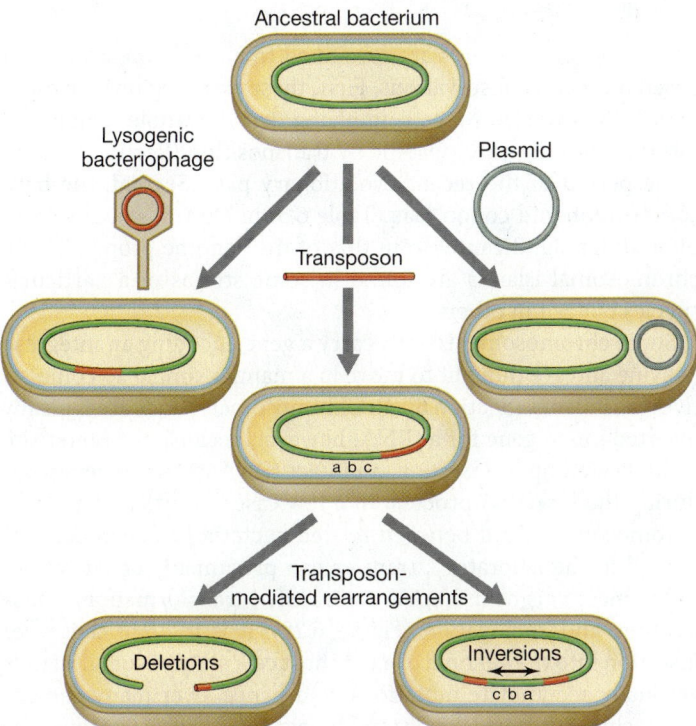

Figure 6.30 Mobile elements promote genome evolution. A variety of mobile genetic elements can move from one organism to another, thus adding genes to the genome of the recipient. The most common of these are plasmids, bacteriophages, and transposons. In the latter case, chromosomal rearrangements, such as deletions and inversions of DNA neighboring the transposon, may be mediated by the activity of the transposase.

MINIQUIZ -

- Which class of genes is rarely transferred horizontally? Why?
- List the major mechanisms by which horizontal gene transfer occurs in prokaryotes.
- How might transposons be especially important in the evolution of pathogenic bacteria?

6.13 Core Genome versus Pan Genome

One of the most important concepts to emerge from comparing the genome sequences of multiple strains of the same species is the distinction between the **pan genome** and the **core genome**. The *core* genome is that shared by all strains of a given species, whereas the *pan* genome includes the core plus all of the optional extras present in one or more strains but not all strains of that species (**Figure 6.31**). As we have seen, horizontal gene transfer of entire genetic elements such as plasmids, viruses, or transposable elements is possible. Consequently, there may be major differences in the total amount of DNA and the suite of accessory capabilities (virulence, symbiosis, or biodegradation) between strains of a single bacterial species. In other words, one could say that the core genome is typical of the species as a whole, whereas the other components of the pan genome, frequently

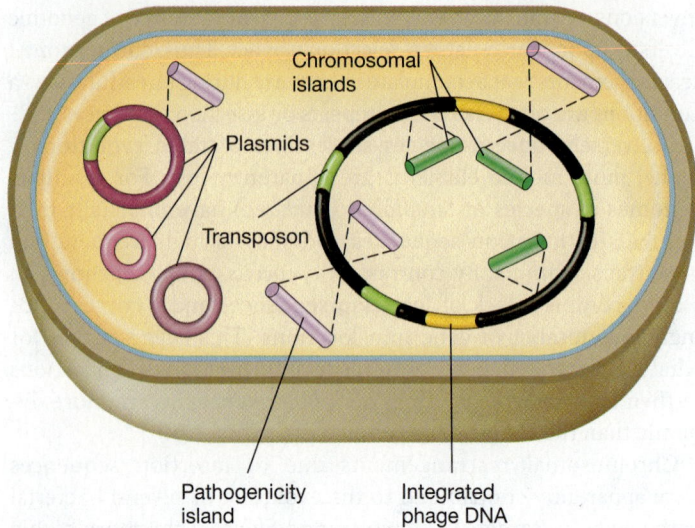

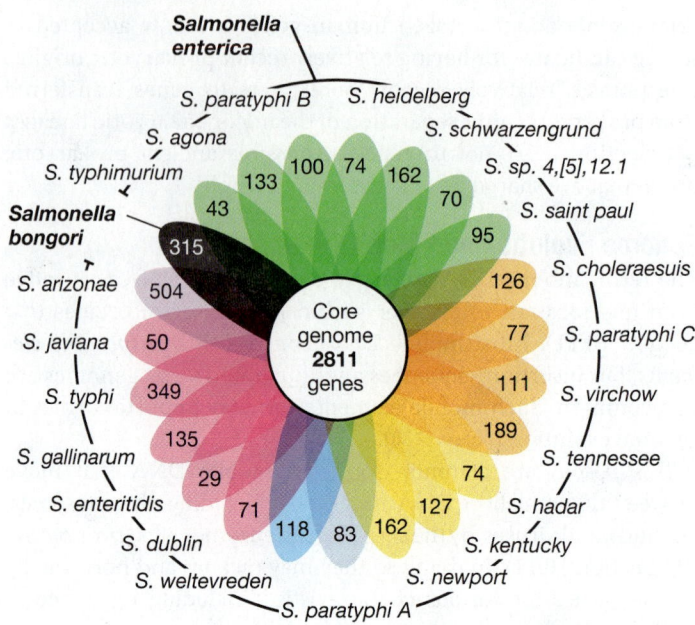

Figure 6.31 **Pan genome versus core genome.** The core genome is represented by the black regions of the chromosome and is present in all strains of a species. The pan genome includes elements that are present in one or more strains but not in all strains. Each colored wedge indicates a single insertion. Where two wedges emerge from the same location, they represent alternative islands that can insert at that site. However, only one insertion can be present at a given location. Plasmids, like the chromosome, may have insertions that are not present in all strains.

Figure 6.32 **Flowerplot of the *Salmonella enterica* pan genome.** A "flowerplot" of gene families in serovars (strains) of the gram-negative pathogenic bacterium *Salmonella enterica* (the names surrounding the flowerplot are immunologically unique serovars [S.] of *S. enterica*). The figure presents the average number of gene families found in each genome as being unique to each serovar. *Salmonella bongori* is a species distinct from *S. enterica*. Serovar 4,[5],12.i is recently identified and not yet named. Data from Jacobsen, A., R.S. Hendriksen, F.M. Aaresturp, D.W. Ussery, and C. Friis. 2011. The *Salmonella enterica* pan-genome. *Microb Ecol 62*: 487–504.

including mobile elements, are restricted to particular strains within a species.

It is difficult to define the size of the pan genome precisely because it increases as the genomes of more strains of a species are sequenced. In some cases, such as the enteric bacteria *Escherichia coli* and *Salmonella enterica*, many different isolates have been found that carry a wide range of different plasmids, transposons, and the like. Consequently the pan genome is extremely large. **Figure 6.32** illustrates the pan genome for serovars (strains) of the important human pathogen *Salmonella enterica* depicted in a "flowerplot" schematic.

Chromosomal Islands

Comparison of the core and pan genomes of particular bacteria or of genomes of particular species with their close relatives sometimes reveals extra blocks of genetic material that are part of the chromosome, rather than being plasmids or integrated viruses. These so-called **chromosomal islands** contain clusters of genes for specialized functions that are not needed for simple survival (Figure 6.31). Consequently, two strains of the same bacterial species may show significant differences in genome size.

Not surprisingly, chromosomal islands in pathogenic bacteria have drawn the most attention. However, chromosomal islands are also known that encode the biodegradation of pollutants such as aromatic hydrocarbons and herbicides. In addition, many of the genes essential for the symbiotic relationship of rhizobia with the root nodules of plants (🔗 Section 22.3) are carried in chromosomal islands. Perhaps the most unique chromosomal island is the magnetosome island of the bacterium *Magnetospirillum*; this DNA fragment carries genes that encode the formation of magnetosomes, intracellular magnetic particles used to orient the organism in a magnetic field and influence the direction of its movement (🔗 Section 2.14).

Chromosomal islands are presumed to have a "foreign" origin based on several observations. First, these extra regions are often flanked by inverted repeats, implying that the whole region was inserted into the chromosome by transposition (Section 6.12) at some period in the recent evolutionary past. Second, the base composition and codon bias (Table 6.3) in chromosomal islands often differ significantly from that of the genome proper. Third, chromosomal islands are found in some strains of a particular species but not in others.

Some chromosomal islands carry a gene encoding an integrase enzyme and are thought to move in a manner similar to conjugative transposons (Section 6.12). Chromosomal islands are typically inserted into a gene for a tRNA; however, because the target site is duplicated upon insertion, an intact tRNA gene is regenerated during the insertion process. In a few cases, transfer of a whole chromosomal island between related bacteria has been demonstrated in the laboratory; transfer can presumably occur by any of the mechanisms of horizontal transfer: transformation, transduction, and conjugation (Figure 6.29). It is thought that after insertion into the genome of a new host cell, chromosomal islands gradually accumulate mutations, and hence, over many generations, chromosomal islands tend to lose their ability to move.

Pathogenicity Islands and the Evolution of Virulence

Comparison of the genomes of pathogenic bacteria with those of their harmless relatives often reveals chromosomal islands that encode *virulence factors*, special proteins or other molecules or structures that help initiate disease (Chapter 23). Some virulence genes

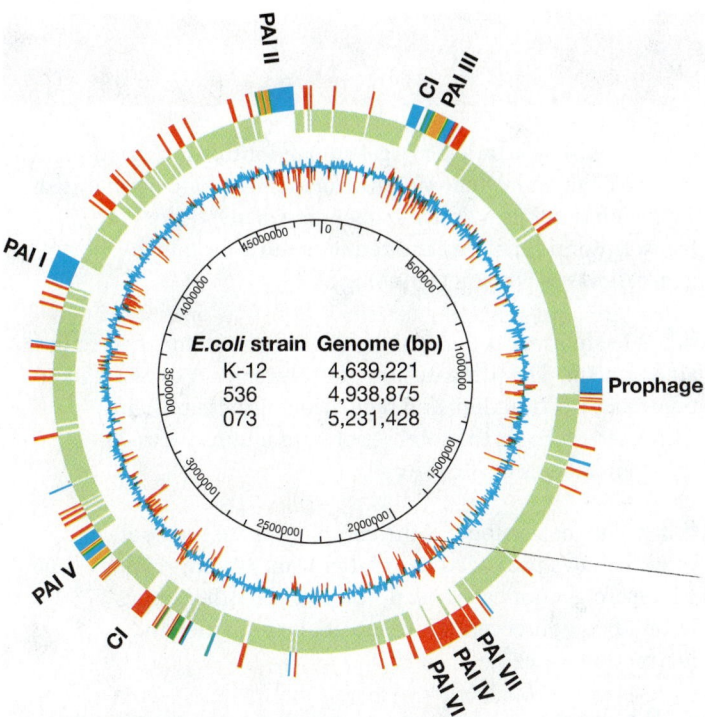

Figure 6.33 **Pathogenicity islands in *Escherichia coli*.** Genetic map of *E. coli* strain 536, a urinary tract pathogen, compared with a second pathogenic strain (073) and the nonpathogenic strain K-12. The pathogenic strains contain pathogenicity islands, and thus their chromosomes are larger than that of K-12. Inner circle, nucleotide base pairs. Jagged circle, DNA GC distribution; regions where GC content varies dramatically from the genome average are in red. Outermost circle, three-way genomic comparison: green, genes common to all strains; red, genes present in the pathogenic strains only; blue, genes found only in strain 536; orange, genes of strain 536 present in a different location in strain 073. Some very small inserts deleted for clarity. PAI, pathogenicity island; CI, chromosomal island. Prophage, DNA from a temperate bacteriophage. Note the correlation between genomic islands and skewed GC content. Data adapted from *Proc. Natl. Acad. Sci. (USA)* 103: 12879–12884 (2006).

are carried on plasmids or lysogenic bacteriophages (⟲ Sections 8.8 and 10.7); however, many others are clustered in chromosomal regions called **pathogenicity islands** (Figure 6.31 and Figure 6.33).

Pathogenicity islands are the best known of the chromosomal islands. Although pathogenicity islands are regarded as a subclass of chromosomal islands, genetically related islands that share homologous genes for integration and conjugation may carry virulence genes in some bacteria but biodegradation genes in others. For example, the identity and chromosomal location of most genes of pathogenic strains of *Escherichia coli* correspond to those of the harmless laboratory strain *E. coli* strain K-12, as would be expected. However, most pathogenic strains contain pathogenicity islands of considerable size that are not present in the *E. coli* K-12 chromosome (Figure 6.33). Consequently, two strains of the same bacterial species may show significant differences in genome size due to the presence or absence of the island. Thus, as shown in Table 6.1, the enterohemorrhagic strain *E. coli* O157:H7 contains 20% more DNA and genes than *E. coli* K-12.

Small pathogenicity islands that encode a series of virulence factors are present in certain strains of the gram-positive pathogenic bacterium *Staphylococcus aureus* and can be moved between cells by temperate bacteriophages (⟲ Section 10.7). The islands are smaller than the phage genome, and when the islands excise from the chromosome and replicate, they induce the formation of defective phage particles that carry the genes for the islands but are too small to carry the phage genome. In this way, strains of *S. aureus* that lack the islands can quickly obtain them and become more effective pathogens.

MINIQUIZ -
- What is the difference between core genome and pan genome?
- What is a chromosomal island and how can one be identified as being of foreign origin?
- What is a pathogenicity island and how does one move between bacterial species?

BIG IDEAS

6.1 • Small viruses were the first organisms whose genomes were sequenced, but now many prokaryotic and eukaryotic cellular genomes have been sequenced.

6.2 • DNA sequencing technology is advancing very quickly. The original Sanger method is rarely used and there are now four successive generations of sequencing technology. Advances in technology have greatly increased the speed of DNA sequencing. Shotgun techniques employ random cloning and sequencing of small genome fragments followed by computer-generated assembly of the genome.

6.3 • Computer analysis of sequencing data is a vital part of genomics. Computational tools are used to store and analyze the sequences and the structures of biological macromolecules.

6.4 • Sequenced prokaryotic genomes range in size from 0.15 to 13 Mbp. The smallest prokaryotic genomes are smaller than those of the largest viruses, whereas the largest have more genes than some eukaryotes. Gene content in prokaryotes is typically proportional to genome size. Many genes can be identified by their sequence similarity to genes found in other organisms. However, a significant percentage of sequenced genes are of unknown function.

6.5 • Virtually all eukaryotic cells contain mitochondria, and in addition, plant cells contain chloroplasts. Both organelles contain circular DNA genomes that encode rRNAs, tRNAs, and a few proteins needed for energy metabolism. Although the genomes of the organelles are independent of the nuclear genome, the organelles themselves are not. Many genes in the nucleus encode proteins required for organelle function.

6.6 • The complete genomic sequence of many microbial eukaryotes has been determined. The genome of the yeast *Saccharomyces cerevisiae* encodes about 6000 proteins, of which only about 900 appear essential. Relatively few of the protein-encoding genes of yeast contain introns. The number of genes in microbial eukaryotes ranges from 2000 (less than many prokaryotes) to 60,000 (more than twice as many as humans).

6.7 • Microarrays consist of genes or gene fragments attached to a solid support in a known pattern; mRNA is then hybridized with the DNA to determine patterns of gene expression. The arrays are large enough for the transcription pattern of an entire genome (the transcriptome) to be analyzed. RNA-Seq requires massive sequencing of cDNA for transcriptomic analyses and requires third- or fourth-generation sequencing technology.

6.8 • Proteomics is the analysis of all the proteins present in an organism. The ultimate aim of proteomics is to understand the structure, function, and regulation of these proteins. The interactome is the total set of interactions between macromolecules inside the cell.

6.9 • The metabolome is the complete set of metabolic intermediates produced by an organism. Systems biology uses data from genomics, transcriptomics, and other omics to build computer models of molecular activities and interactions in cells.

6.10 • Most microorganisms in the environment have never been cultured. Nonetheless, analysis of DNA samples has revealed enormous sequence diversity in most habitats. The concept of the metagenome embraces the total genetic content of all the organisms in a particular habitat.

6.11 • Genomics can be used to study the evolutionary history of an organism. Organisms contain gene families, genes with related sequences. If these arose because of gene duplication, the genes are said to be paralogs; if they arose by speciation, they are called orthologs.

6.12 • Organisms may acquire genes from other organisms in their environment by horizontal gene transfer, and such transfer may even cross phylogenetic domain boundaries. Mobile DNA elements, including transposons, integrons, and viruses, are important in genome evolution and often carry genes encoding antibiotic resistance or virulence factors.

6.13 • Comparison of the genomes of multiple strains of the same bacterial species shows a conserved component (the core genome) plus many variable genetic modules only present in certain members of the species (the pan genome). Many bacteria contain relatively large inserts of foreign origin known as chromosomal islands. These contain clusters of genes that encode specialized metabolic functions or pathogenesis and virulence factors (pathogenicity islands).

MasteringMicrobiology® **Review what you know and challenge what you have learned with MasteringMicrobiology!** Access study materials, chapter quizzes, animations, and microbiology lab tutorials in the Study Area to ensure that you have mastered this chapter's content.

REVIEW OF KEY TERMS

Bioinformatics the use of computational tools to acquire, analyze, store, and access DNA and protein sequences

Chromosomal island a bacterial chromosome region of foreign origin that contains clustered genes for some extra property such as virulence or symbiosis

Codon bias the relative proportions of different codons encoding the same amino acid; it varies in different organisms. Same as codon usage

Core genome the part of a genome shared by all strains of a species

Gene chip small solid supports to which genes or portions of genes are affixed and arrayed spatially in a known pattern (also called microarrays)

Gene family genes related in sequence to each other because of common evolutionary origin

Genome the total complement of genetic information of a cell or a virus

Genomic library a collection of cloned DNA fragments that cover an entire genome

Genomics the discipline that maps, sequences, analyzes, and compares genomes

Homologs genes related in sequence to an extent that implies common genetic ancestry; includes both orthologs and paralogs

Horizontal gene transfer the transfer of genetic information between organisms as opposed to transfer from parent to offspring

Hybridization the joining of two single-stranded nucleic acid molecules by complementary base pairing to form a double-stranded hybrid DNA or DNA–RNA molecule

Interactome the total set of interactions between proteins (or other macromolecules) in an organism

Metabolome the total complement of small molecules and metabolic intermediates of a cell or organism

Metagenome the total genetic complement of all the cells present in a particular environment

Metagenomics the genomic analysis of pooled DNA or RNA from an environmental sample containing organisms that have not been isolated; same as environmental genomics

Microarray small, solid supports to which genes or portions of genes are affixed and arrayed spatially in a known pattern (also called gene chips)

Nucleic acid probe a labeled strand of nucleic acid that can be used to hybridize with a complementary strand of nucleic acid in a mixture

Open reading frame (ORF) a sequence of DNA or RNA that could be translated to give a polypeptide

Ortholog a gene in one organism that is similar to a gene in another organism because of descent from a common ancestor (see also *paralog*)

Pan genome the totality of the genes present in the different strains of a species

Paralog a gene whose similarity to one or more other genes in the same organism is the result of gene duplication (see also *ortholog*)

Pathogenicity island a bacterial chromosome region of foreign origin that contains clustered genes for virulence

Primer an oligonucleotide to which DNA polymerase attaches the first deoxyribonucleotide during DNA synthesis

Proteome the total set of proteins encoded by a genome or the total protein complement of an organism

Proteomics the genome-wide study of the structure, function, and regulation of the proteins of an organism

Sequencing deducing the order of nucleotides in a DNA or RNA molecule by a series of chemical reactions

Shotgun sequencing sequencing of DNA from previously cloned small fragments of a genome in a random fashion; shotgun sequencing is followed by computational methods to reconstruct the entire genome sequence

Systems biology the integration of data from genomics and other "omics" areas to build an overall picture of a biological system

Transcriptome the complement of all RNA produced in an organism under a specific set of conditions

REVIEW QUESTIONS

1. Why do dideoxynucleotides function as chain terminators? (Section 6.1)

2. Give an example of a first-, a second-, and a third-generation sequencing system. (Section 6.2)

3. What characteristics are used to identify open reading frames using sequence data? (Section 6.3)

4. What is the relationship between genome size and open reading frame content of prokaryotic genomes? (Section 6.4)

5. As a proportion of the total genome, which class of genes predominates in organisms with a small genome? In organisms with a large genome? (Section 6.4)

6. Which genomes are larger, those of chloroplasts or those of mitochondria? Describe one unusual feature each for the chloroplast and mitochondrial genomes. (Section 6.5)

7. How does your genome compare with that of yeast in overall size and gene number? (Section 6.6)

8. Distinguish between the terms genome, proteome, and transcriptome. (Sections 6.7 and 6.8)

9. What does a 2D protein gel show? How can the results of such a gel be correlated with protein function? (Section 6.8)

10. Why is investigation of the metabolome lagging behind that of the proteome? (Section 6.9)

11. What are the objectives of systems biology? (Section 6.9)

12. How can gene expression be measured in uncultured bacteria? (Section 6.10)

13. Most of the genetic information on our planet does not belong to cellular organisms. Discuss. (Section 6.10)

14. What is the major difference in how duplications have contributed to the evolution of prokaryotic versus eukaryotic genomes? (Section 6.11)

15. Explain how horizontally transferred genes can be detected in a genome. (Section 6.12)

16. Explain how transposable elements promote the genome evolution of *Bacteria*. (Section 6.12)

17. Explain how chromosomal islands might move between different bacterial hosts. (Section 6.13)

18. What are pathogenicity islands and why are they important? (Section 6.13)

APPLICATION QUESTIONS

1. Apart from genome size, what factors make complete assembly of a eukaryotic genome more difficult than assembly of a prokaryotic genome?

2. Describe how one might determine which proteins in *Escherichia coli* are repressed when a culture is shifted from a minimal medium (containing only a single carbon source) to a rich medium containing many amino acids, bases, and vitamins. How might one study which genes are expressed during each growth condition?

3. The gene encoding the beta subunit of RNA polymerase from *Escherichia coli* is said to be orthologous to the *rpoB* gene of *Bacillus subtilis*. What does that mean about the relationship between the two genes? What protein do you suppose the *rpoB* gene of *B. subtilis* encodes? The genes for the different sigma factors of *E. coli* are paralogous. What does that say about the relationship among these genes?

7 · Metabolic Regulation

microbiology**now**

Luminescence or Lethality?

All human cells contain the same genetic blueprint, but what makes a brain cell different from a liver cell? This cellular differentiation is controlled by the elegant manner in which a cell deploys its genome—what is commonly known as *regulation of gene expression*. While cellular differentiation is primarily associated with multicellular organisms, some prokaryotes are also able to change their morphology or switch from benign to pathogenic forms by altering gene expression.

Photorhabdus luminescens, a bioluminescent bacterium that colonizes the gut of a worm, provides an astonishing example of this pathogenic switch. While one might assume that colonization by *Photorhabdus* would be detrimental to the worm, the relationship between these two organisms is strikingly beneficial. In the nonpathogenic stage, the bacteria reside harmlessly in the worm intestine. When the worm prepares to reproduce, it enters a host insect and regurgitates its intestinal bacteria. Through altered gene expression, some *Photorhabdus* cells then switch to the pathogenic mode and secrete insecticidal toxins that kill the host insect and enzymes that dissolve its body, releasing nutrients for both the worm and the bacterium.

Using fluorescent gene reporters, microbiologists have been able to visualize this elegant switch in a pure culture of *Photorhabdus*.[1] Cells that are in the beneficial colonization mode tend to form smaller colonies (top photo, green), whereas colonies of pathogenic cells containing the insecticidal toxins are larger (red). Researchers have also shown that both phenotypes can be exhibited within a single isolated colony! This is seen as black sectors within the large green colony in the bottom photo.

Research with *Photorhabdus* thus clearly illustrates how a simple change in gene expression can trigger a major switch in the phenotype of a bacterium, dramatically changing its lifestyle.

[1]Somvanshi, V.S., et al. 2012. A single promoter inversion switches *Photorhabdus* between pathogenic and mutualistic states. *Science 337:* 88–92.

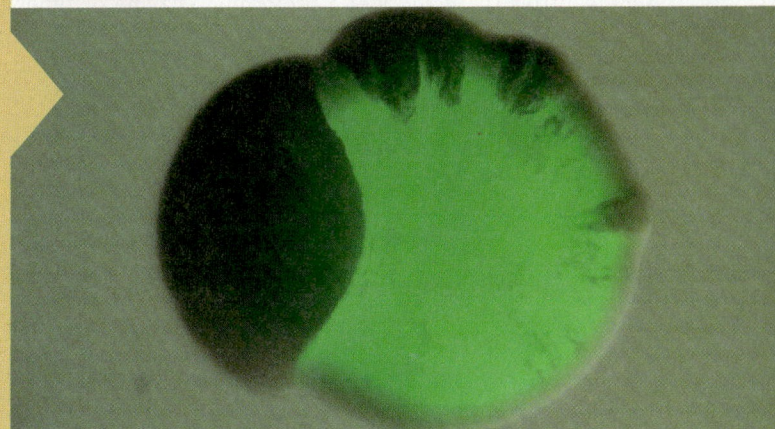

To efficiently orchestrate the numerous reactions that occur in a cell and to maximally use available resources, cells must *regulate* the types, amounts, and activities of proteins and other macromolecules they make. Regulation occurs at all molecular levels in the cell. After DNA is transcribed into RNA, the information is translated to yield specific protein. Collectively, these processes are called **gene expression**. Once proteins have been translated, additional mechanisms can be used to regulate their activity. This chapter focuses on the systems that control gene expression and the activity of proteins.

I · Overview of Regulation

Some proteins and RNA molecules are needed in the cell at about the same level under all growth conditions. The expression of these molecules is said to be *constitutive*. However, more often a particular protein or RNA is needed under some conditions but not others. For instance, enzymes required for using the sugar lactose are useful only if lactose is available to the cell. Microbial genomes encode many more proteins than are actually present in the cell under any particular growth condition (Chapter 6). Thus, regulation is a major process in all cells and helps to conserve energy and resources.

Cells use two major approaches to regulate protein function. One controls the *activity* of a preformed enzyme or other protein, and the second controls the *amount* of an enzyme or other protein. The activity of a protein can be regulated only after it has been synthesized (that is, post-translationally). Regulating the activity of an enzyme in the cell is typically very rapid (taking seconds or less), whereas synthesizing an enzyme is relatively slow (taking several minutes). After synthesis of an enzyme begins, it takes some time before it is present in amounts sufficient to affect metabolism. Conversely, after synthesis of an enzyme stops, a considerable time may elapse before the enzyme is sufficiently diluted that it no longer affects metabolism. However, working together, regulation of enzyme activity and of enzyme synthesis efficiently controls cell metabolism.

7.1 Major Modes of Regulation

Most bacterial genes are transcribed into messenger RNA (mRNA), which in turn is translated into protein, as we discussed in Chapter 4. The components of a typical gene and the points within the flow of genetic information where the amount of a specific gene product (RNA or protein) and its corresponding activity can be controlled are summarized in **Figure 7.1**. The amount of protein synthesized can be regulated at either the level of transcription, by varying the amount of mRNA made, or

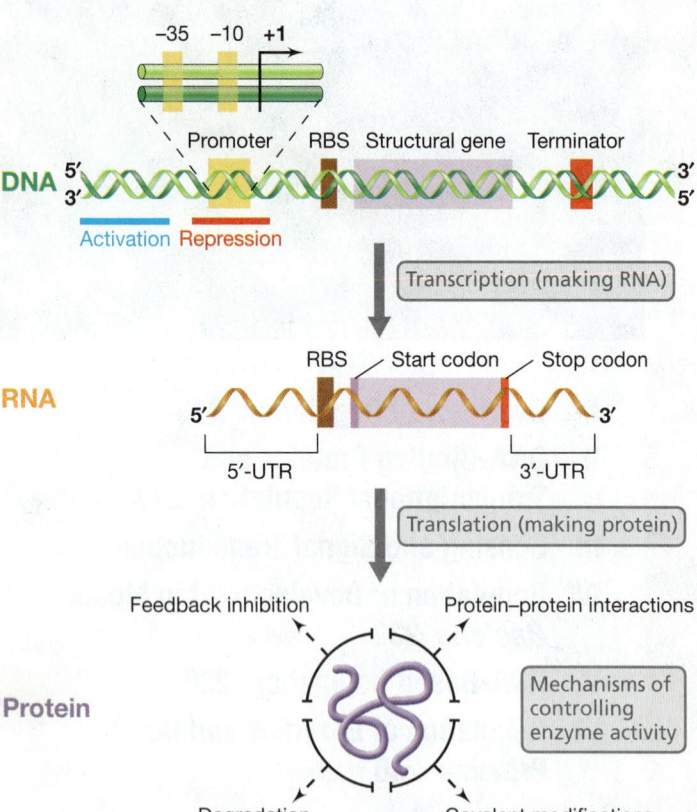

Figure 7.1 Gene expression and regulation of protein activity. The promoter and terminator as well as regions involved in transcriptional activation and repression are indicated. The 5′ untranslated region (5′-UTR) is a short region between the start of transcription and the start of translation, while the 3′ untranslated region (3′-UTR) is a short region between the stop codon and the transcription terminator. These are the regions where translational regulation often occurs. Mechanisms for regulating protein activity after translation are shown at the bottom.

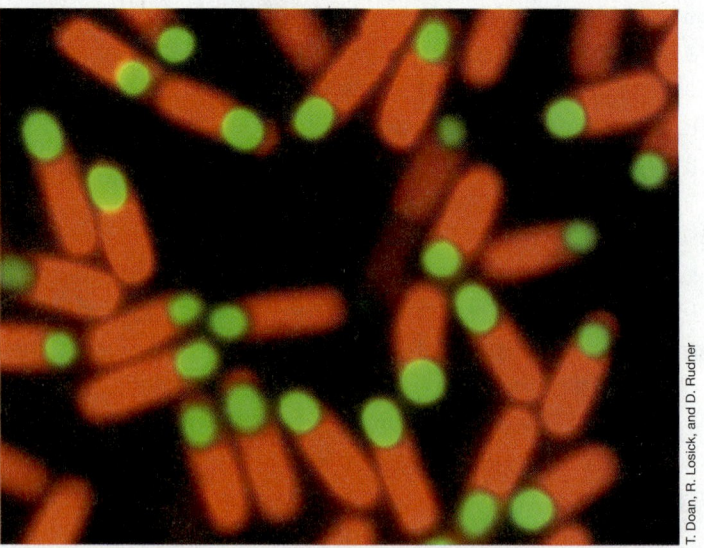

Figure 7.2 Gene expression in *Bacillus* during sporulation using the green fluorescent protein (GFP). During endospore formation, alternative sigma factors are localized to specific regions of the cell (Section 7.11). σF linked to GFP indicates expression and activity of the protein in the developing endospore (at one end of each cell). σE linked to a reporter protein that fluoresces red indicates expression and activity of the protein throughout the mother cell prior to endospore formation. Regions correspond to the model depicted in Figure 7.25*b*.

at the level of translation, by translating or not translating the mRNA. The structural gene encodes the gene product and its expression is controlled by sequences in the upstream region (⮂ Section 4.7). Note that the sequences that determine the beginning and end of transcription are distinct from those that determine the beginning and end of translation. They are separated by small spacer regions, the 5' and 3' *untranslated regions* (5'-UTR and 3'-UTR). After translation, other regulatory processes such as feedback inhibition, covalent modifications, degradation, and interactions with other proteins can further regulate the activity of some proteins.

To monitor gene expression levels corresponding to specific proteins, *reporter genes* can be employed (⮂ Section 11.6; **Figure 7.2**). Reporter genes encode a protein product that is easy to detect and assay and thus can be fused to other genes or regulatory elements to monitor gene expression. The **green fluorescent protein (GFP)**, which fluoresces bright green when exposed to a specific wavelength of light, is commonly used for monitoring gene expression. If the open-reading frame for GFP is linked directly to a regulatory region or fused to the end of a separate gene of interest, the level of fluorescence can be correlated to the level of gene expression. Figure 7.2 illustrates the use of GFP and its derivatives to monitor expression of the alternative sigma factors necessary for sporulation in *Bacillus* (Section 7.11). The use of fluorescent proteins fused to the promoter of two separate sigma factors allows the cellular location of each sigma factor to be determined. Expression of σF, as indicated by green fluorescence, is localized at the end of the cell where endospore development occurs. The expression of σE, which is necessary to activate gene transcription throughout the mother cell, can be seen as red fluorescence throughout the rest of the cell (Figure 7.2).

MINIQUIZ

- What steps in the synthesis of protein might be subject to regulation?
- Which is likely to be more rapid, the regulation of activity or the regulation of synthesis? Why?
- What mechanisms can be used to regulate the activity of some proteins?

II • DNA-Binding Proteins and Transcriptional Regulation

As we have said, the amount of a protein present in a cell may be controlled at the level of transcription, at the level of translation, or, occasionally, by protein degradation. Our discussion begins with control at the level of transcription because this is the major means of regulation in prokaryotes.

7.2 DNA-Binding Proteins

For a gene to be transcribed, RNA polymerase must recognize a specific promoter on the DNA and begin its activity (⮂ Section 4.7). Small molecules often take part in regulating this process. However, they rarely do so directly. Instead, they typically influence the binding of certain proteins, called *regulatory proteins*, to specific sites on the DNA. This event regulates gene expression by turning transcription either on or off.

Interaction of Proteins with Nucleic Acids

Interactions between proteins and nucleic acids are central to replication, transcription, and translation, and also to the regulation of these processes. Protein–nucleic acid interactions may be specific or nonspecific, depending on whether the protein attaches anywhere along the nucleic acid or binds to a specific site. Most DNA-binding proteins interact with DNA in a sequence-specific manner. Specificity is provided by interactions between specific amino acid side chains of the proteins and specific chemical groups on the nitrogenous bases and the sugar–phosphate backbone of the DNA. Because of its size, the *major groove* of DNA is the main site of protein binding, and Figure 4.2 identified atoms of the bases in the major groove that are known to interact with proteins. To achieve high specificity, the binding protein must interact simultaneously with several nucleotides.

We have already described a structure in DNA called an *inverted repeat* (⮂ Figure 4.23a). Such inverted repeats are frequently the locations at which regulatory proteins bind specifically to DNA (**Figure 7.3**). Note that this interaction does not require the formation of stem–loop structures in the DNA. DNA-binding proteins are often homodimeric, meaning they are composed of two identical polypeptide subunits, each subdivided into **domains**—regions of the protein with a specific structure and

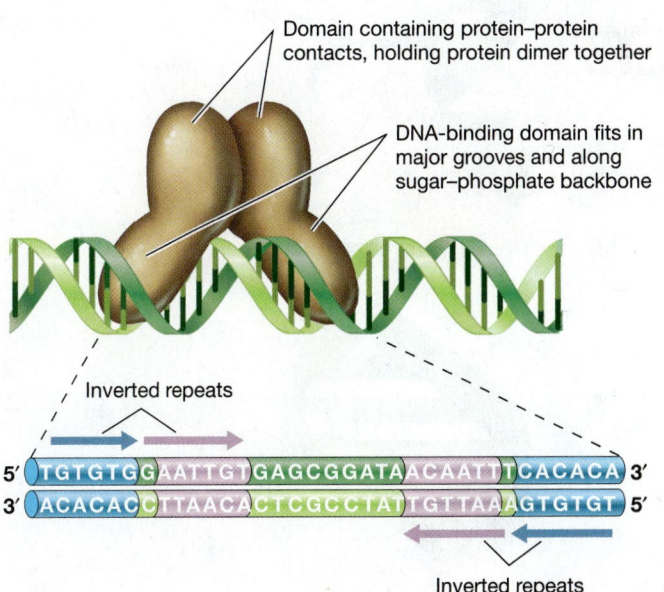

Domain containing protein–protein contacts, holding protein dimer together

DNA-binding domain fits in major grooves and along sugar–phosphate backbone

Inverted repeats

```
5′ TGTGTGGAATTGTGAGCGGATAACAATTTCACACA 3′
3′ ACACACCTTAACACTCGCCTATTGTTAAAGTGTGT 5′
```

Inverted repeats

Figure 7.3 DNA-binding proteins. Many DNA-binding proteins are dimers that combine specifically with two sites on the DNA. The specific DNA sequences that interact with the protein are inverted repeats. The nucleotide sequence of the operator gene of the lactose operon (Section 7.3) is shown, and the inverted repeats, which are sites at which the *lac* repressor makes contact with the DNA, are shown in purple and blue boxes.

function. Each subunit has a domain that interacts specifically with a region of DNA in the major groove. When protein dimers interact with inverted repeats on DNA, each subunit binds to one of the inverted repeats. The dimer as a whole thus binds to both DNA strands (Figure 7.3).

Structure of DNA-Binding Proteins

DNA-binding proteins in both prokaryotes and eukaryotes possess several classes of protein domains that are critical for proper binding to DNA. One of the most common is the *helix-turn-helix* structure (**Figure 7.4a**). This consists of two segments of polypeptide chain that have α-helix secondary structure connected by a short sequence forming the "turn." The first helix is the *recognition helix* that interacts specifically with DNA. The second helix, the *stabilizing helix*, stabilizes the first helix by interacting with it by way of hydrophobic interactions. The turn linking the two helices consists of three amino acid residues, the first of which is typically a glycine. Sequences are recognized by noncovalent interactions, including hydrogen bonds and van der Waals contacts, between the recognition helix of the protein and specific chemical groups in the sequence of base pairs on the DNA.

Many different DNA-binding proteins from *Bacteria* contain the helix-turn-helix structure. These include many repressor proteins, such as the *lac* and *trp* repressors of *Escherichia coli* (Section 7.3 and see Figure 7.4), and some proteins of bacterial viruses, such as the bacteriophage lambda repressor (Figure 7.4b). Indeed, over 250 different proteins with this motif bind to DNA to regulate transcription in *E. coli*. Two other types of protein domains are commonly found in DNA-binding proteins. One of these, the *zinc finger*, is frequently found in regulatory proteins in eukaryotes and, as its name implies, binds a zinc ion. The other protein domain commonly found in DNA-binding proteins is the *leucine zipper*, which contains regularly spaced leucine residues that function to hold two recognition helices in the correct orientation to bind DNA.

Once a protein binds at a specific site on the DNA, various outcomes are possible. Some DNA-binding proteins are enzymes that catalyze a specific reaction on the DNA, such as transcription. In other cases, however, the binding event either blocks transcription (*negative regulation*, Section 7.3) or activates it (*positive regulation*, Section 7.4).

MINIQUIZ

- What is a protein domain?
- Why are most DNA-binding proteins specific to certain chemical groups within the DNA?

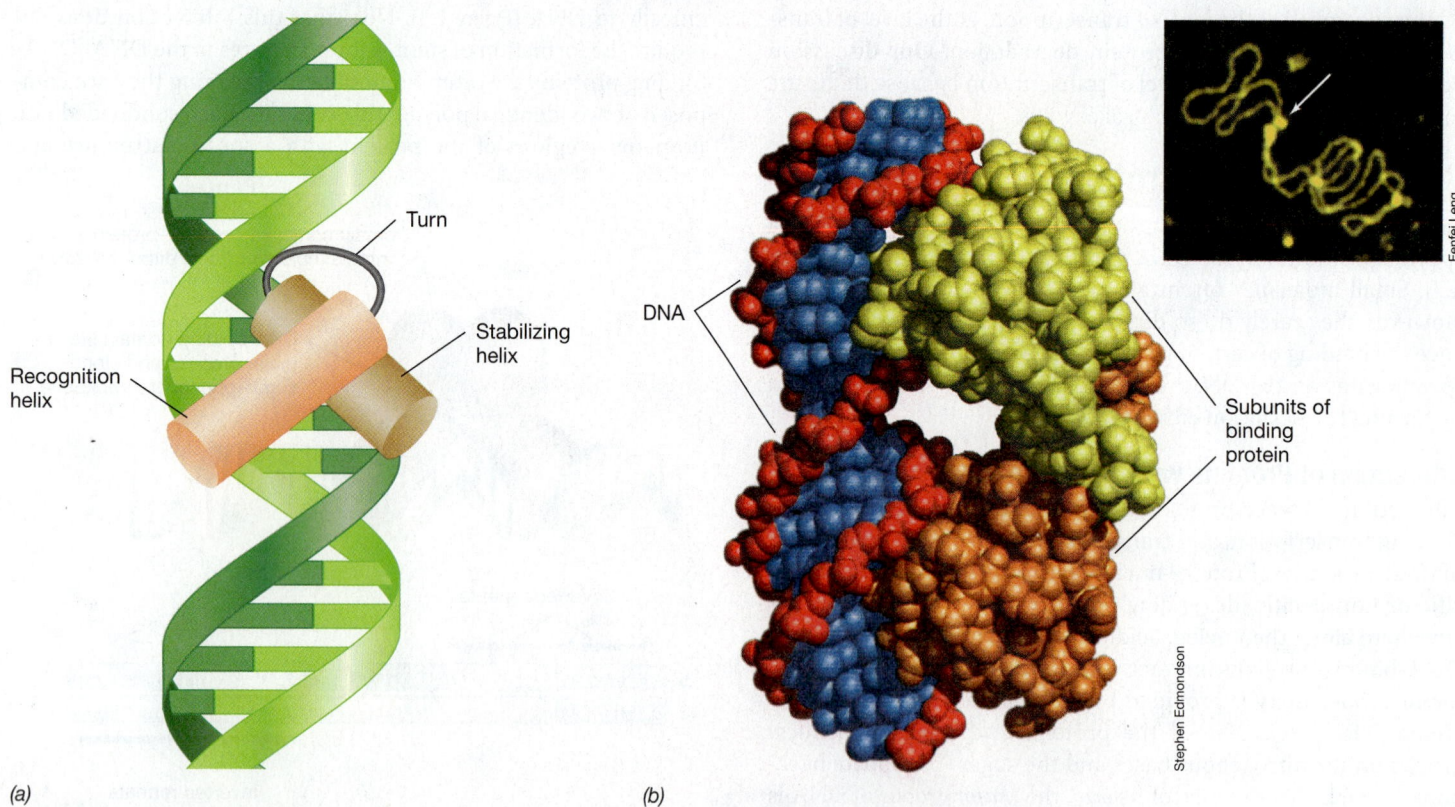

(a) *(b)*

Figure 7.4 The helix-turn-helix structure of some DNA-binding proteins. *(a)* A simple model of the helix-turn-helix structure within a single protein subunit. *(b)* A computer model of both subunits of the bacteriophage lambda repressor bound to its operator. The DNA is red and blue. One subunit of the dimeric repressor is shown in brown and the other in yellow. Each subunit contains a helix-turn-helix structure. The coordinates used to generate this image were downloaded from the Protein Data Base (http://www.pdb.org). Inset: Atomic force microscopy showing copies of the LacI repressor protein (arrow) bound to multiple operator sites on a DNA molecule.

7.3 Negative Control: Repression and Induction

Transcription is the first step in biological information flow; because of this, it is simple and efficient to control gene expression at this point. If one gene is transcribed more frequently than another, there will be more of its mRNA available for translation and therefore a greater amount of its protein product in the cell. We begin with the processes of repression and induction, simple forms of regulation that govern gene expression at the level of transcription. Here we deal with **negative control** of transcription, control that *prevents* transcription.

Enzyme Repression and Induction

Often the enzymes that catalyze the synthesis of a specific product are not made if the product is already present in the medium in sufficient amounts. For example, in *Escherichia coli* and many other *Bacteria*, the enzymes needed to synthesize the amino acid arginine are made only when arginine is absent from the culture medium; an excess of arginine decreases the synthesis of these enzymes. This is called enzyme **repression**.

As can be seen in **Figure 7.5**, if arginine is added to a culture growing exponentially in a medium devoid of arginine, growth continues at the previous rate, but production of the enzymes for arginine synthesis stops. Note that this is a *specific* effect, as the synthesis of all other enzymes in the cell continues at the previous rate. This is because the enzymes affected by a particular repression event make up only a tiny fraction of the entire complement of proteins in the cell. Enzyme repression is widespread in bacteria as a means of controlling the synthesis of enzymes required for the production of amino acids and the nucleotide precursors purines and pyrimidines. In most cases, the final product of a particular biosynthetic pathway represses the enzymes of the pathway. This ensures that the organism does not waste energy and nutrients synthesizing unneeded enzymes.

Enzyme **induction** is conceptually the opposite of enzyme repression. In enzyme induction, an enzyme is made only when its substrate is *present*. Enzyme repression typically affects biosynthetic (anabolic) enzymes. In contrast, enzyme induction usually affects degradative (catabolic) enzymes. To illustrate induction, consider the utilization of the sugar lactose as a carbon and energy source by *Escherichia coli*, the enzymes for which are encoded by the *lac* operon (⟳ Section 4.3). **Figure 7.6** shows the induction of β-galactosidase, the enzyme that cleaves lactose into glucose and galactose. This enzyme is required for *E. coli* to grow on lactose. If lactose is absent, the enzyme is not made, but synthesis begins almost immediately after lactose is added. The three genes in the *lac* operon encode three proteins, including β-galactosidase, that are induced simultaneously upon adding lactose. This type of control mechanism ensures that specific enzymes are synthesized only when needed.

Inducers and Corepressors

The substance that induces enzyme synthesis is called an *inducer* and a substance that represses enzyme synthesis is called a *corepressor*. These substances, which are normally small molecules, are collectively called *effectors*. Interestingly, not all inducers and corepressors are actual substrates or end products of the enzymes involved. For example, structural analogs may induce or repress even though they are not substrates of the enzyme. Isopropylthiogalactoside (IPTG), for instance, is an inducer of β-galactosidase even though IPTG cannot be hydrolyzed by this enzyme. In nature, however, inducers and corepressors are probably normal cell metabolites. Detailed studies of lactose utilization in *E. coli* have shown that the actual inducer of β-galactosidase is not lactose, but its isomer allolactose, which is made from lactose.

Mechanism of Repression and Induction

How can inducers and corepressors affect transcription in such a specific manner? They do this indirectly by binding to specific

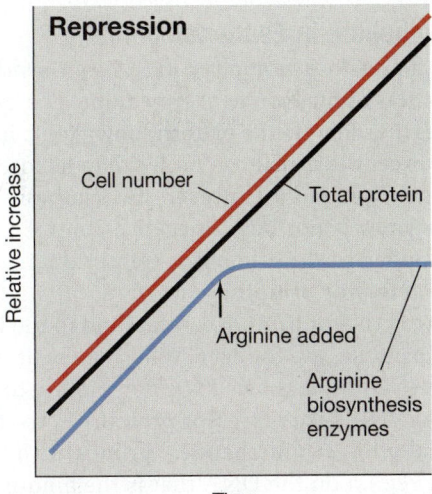

Figure 7.5 Enzyme repression. In a growing bacterial culture, the addition of arginine to the medium specifically represses production of enzymes needed to make arginine. Net protein synthesis is unaffected.

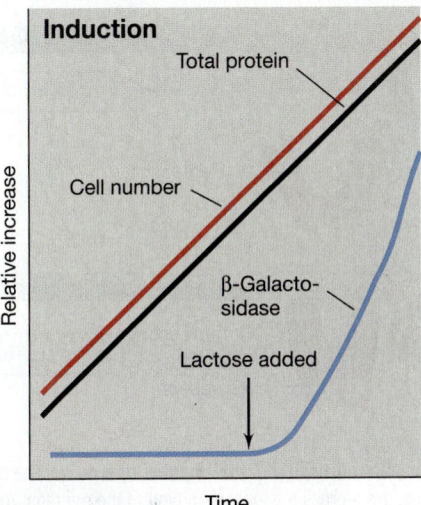

Figure 7.6 Enzyme induction. In a growing bacterial culture, the addition of lactose to the medium specifically induces synthesis of the enzyme β-galactosidase. Net protein synthesis is unaffected.

UNIT 2

DNA-binding proteins, which, in turn, affect transcription. For an example of a repressible enzyme, we consider the arginine operon (Figure 7.5). **Figure 7.7a** shows transcription of the arginine genes, which proceeds when the cell needs arginine. However, when arginine is plentiful, it acts as a corepressor. As Figure 7.7b shows, arginine binds to a specific **repressor protein**, the *arginine repressor*, present in the cell. The repressor protein is **allosteric**; that is, its conformation is altered when the effector molecule binds to it (Section 7.17).

By binding its effector, the repressor protein is *activated* and can then bind to a specific region of the DNA near the promoter of the gene called the *operator*. This region gave its name to the **operon**, a cluster of consecutive genes whose expression is under the control of a single operator (⮒ Section 4.3). All of the genes in an operon are transcribed as a single unit yielding a single mRNA (⮒ Section 4.8). The operator is located downstream of the promoter where synthesis of mRNA is initiated (Figure 7.7). If the repressor binds to the operator, transcription is physically blocked because RNA polymerase can neither bind nor proceed. Hence, the polypeptides encoded by the genes in the operon cannot be synthesized. If the mRNA is polycistronic (⮒ Section 4.8), all the polypeptides encoded by this mRNA will be repressed.

Enzyme induction may also be controlled by a repressor. In this case, the repressor protein is *active* in the absence of the inducer, completely blocking transcription. When the inducer is added, it combines with the repressor protein and inactivates it; inhibition is overcome and transcription can proceed (**Figure 7.8**).

All regulatory systems employing repressors have the same underlying mechanism: prevention of mRNA synthesis by the activity of specific repressor proteins that are themselves under the control of specific small effector molecules. And, as previously noted, because the repressor's role is to stop transcription, regulation by repressors is called *negative control*. One point to note is that genes are not turned on and off completely like light switches. DNA-binding proteins vary in concentration and affinity and thus control is quantitative. Even

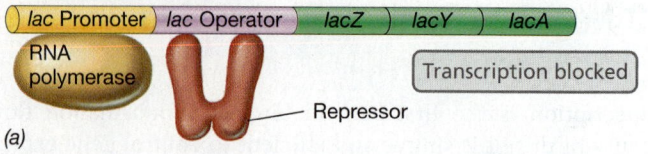

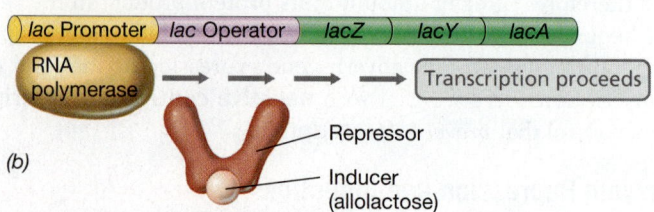

Figure 7.8 Enzyme induction in the lactose operon. *(a)* A repressor protein bound to the operator blocks the binding of RNA polymerase. *(b)* An inducer molecule binds to the repressor and inactivates it so that it no longer can bind to the operator. RNA polymerase then transcribes the DNA and makes an mRNA for that operon. For the *lac* operon, the sugar allolactose is the inducer that binds to the lactose repressor.

when a gene is "fully repressed" there is often a very low level of basal transcription.

MINIQUIZ

• Why is "negative control" so named?
• How does a repressor inhibit the synthesis of a specific mRNA?

7.4 Positive Control: Activation

Negative control relies on a protein (the repressor) to repress mRNA synthesis. By contrast, in **positive control** of transcription the regulatory protein is an *activator* that activates the binding of RNA polymerase to DNA. An example of positive regulation is the catabolism of the disaccharide sugar maltose in *Escherichia coli*.

Maltose Catabolism in *Escherichia coli*

The enzymes for maltose catabolism in *E. coli* are synthesized only after the addition of maltose to the medium. The expression of these enzymes thus follows the pattern shown for β-galactosidase in Figure 7.6 except that maltose rather than lactose is required to induce gene expression. However, the synthesis of maltose-degrading enzymes is not under negative control, as in the *lac* operon, but under positive control; transcription requires the binding of an **activator protein** to the DNA.

The maltose activator protein cannot bind to the DNA unless it first binds maltose, the inducer. When the maltose activator protein binds to DNA, it allows RNA polymerase to begin transcription (**Figure 7.9**). Like repressor proteins, activator proteins bind specifically to certain chemical groups within the DNA. However, the region on the DNA that is the binding site of the activator is not called an operator (Figures 7.7 and 7.8), but instead an *activator-binding site* (Figure 7.9). Nevertheless, the genes controlled by this activator-binding site are still called an operon.

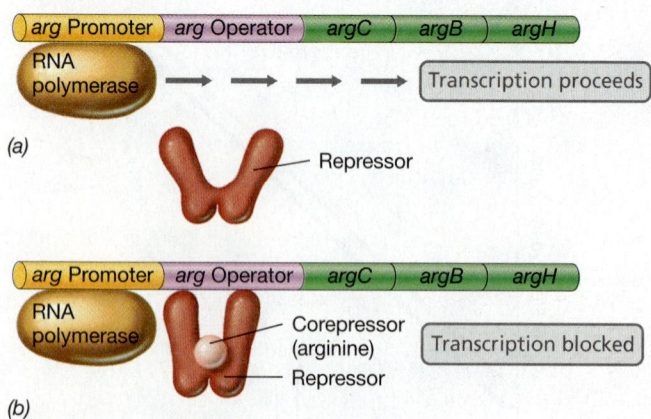

Figure 7.7 Enzyme repression in the arginine operon. *(a)* The operon is transcribed because the repressor is unable to bind to the operator. *(b)* After a corepressor (small molecule) binds to the repressor, the repressor binds to the operator and blocks transcription; mRNA and the proteins it encodes are not made. For the *argCBH* operon, the amino acid arginine is the corepressor that binds to the arginine repressor.

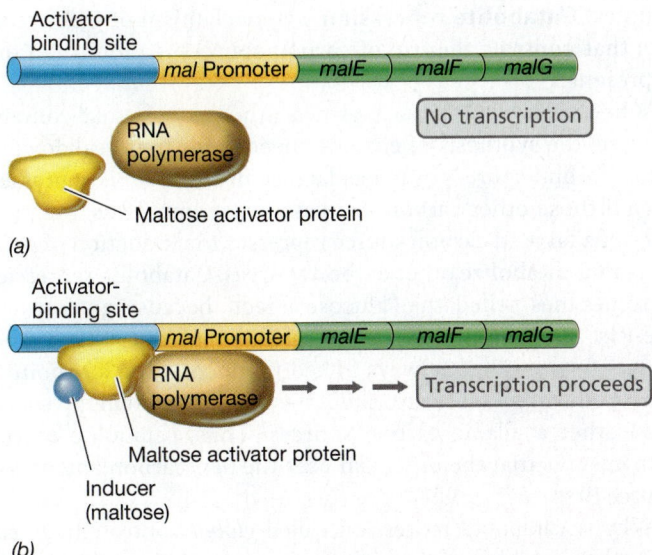

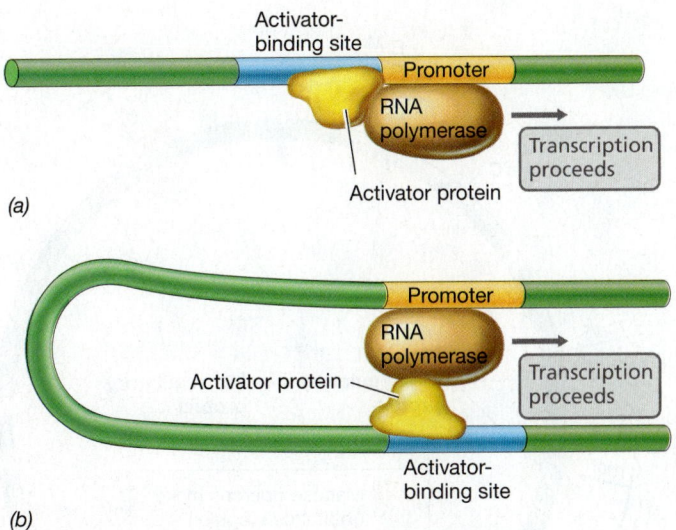

(a)

(b)

Figure 7.9 Positive control of enzyme induction in the maltose operon.
(a) In the absence of an inducer, neither the activator protein nor the RNA polymerase can bind to the DNA. *(b)* An inducer molecule (for the *malEFG* operon it is the sugar maltose) binds to the activator protein (MalT), which in turn binds to the activator-binding site. This recruits RNA polymerase to bind to the promoter and begin transcription.

(a)

(b)

Figure 7.11 Activator protein interactions with RNA polymerase. *(a)* The activator-binding site is near the promoter. *(b)* The activator-binding site is several hundred base pairs from the promoter. In this case, the DNA must be looped to allow the activator and the RNA polymerase to contact.

Binding of Activator Proteins

The promoters of positively controlled operons have nucleotide sequences that bind RNA polymerase weakly and are poor matches to the consensus sequence (⇔ Section 4.7). Thus, even with the correct sigma (σ) factor, the RNA polymerase has difficulty binding to these promoters. The role of the activator protein is to help the RNA polymerase recognize the promoter and begin transcription. For example, the activator protein may modify the structure of the DNA by bending it (**Figure 7.10**), allowing the RNA polymerase to make necessary contacts with nucleotides in the

promoter region to begin transcription. Alternatively, the activator protein may interact directly with the RNA polymerase. This can happen either when the activator-binding site is close to the promoter (**Figure 7.11a**) or when it is several hundred base pairs away from the promoter, a situation in which DNA looping is required to make the necessary contacts between protein and nucleic acid (Figure 7.11*b*).

Many genes in *E. coli* have promoters under positive control and many have promoters under negative control. In addition, many operons have promoters with multiple types of control and some have more than one promoter, each with its own control system! Thus, the simple picture outlined above does not hold for all operons. Multiple control features are common in the operons of virtually all prokaryotes, and thus their overall regulation may require a network of interactions.

Operons versus Regulons

In *E. coli*, the genes required for maltose utilization are spread out over the chromosome in several operons, each of which has an activator-binding site to which a copy of the maltose activator protein can bind (**Figure 7.12**). Therefore, the maltose activator protein actually controls the transcription of more than one operon. When more than one operon is under the control of a single regulatory protein, these operons are collectively called a **regulon**. Therefore, the enzymes for maltose utilization are encoded by the maltose regulon.

Regulons are known for operons under negative control as well. For example, the arginine biosynthetic enzymes (Section 7.3) are encoded by the arginine regulon, whose operons are all under the control of the arginine repressor protein (only one of the arginine operons was shown in Figure 7.7). In regulon control, a specific DNA-binding protein binds only at those operons it controls regardless of whether it is functioning as an activator or repressor; other operons are not affected.

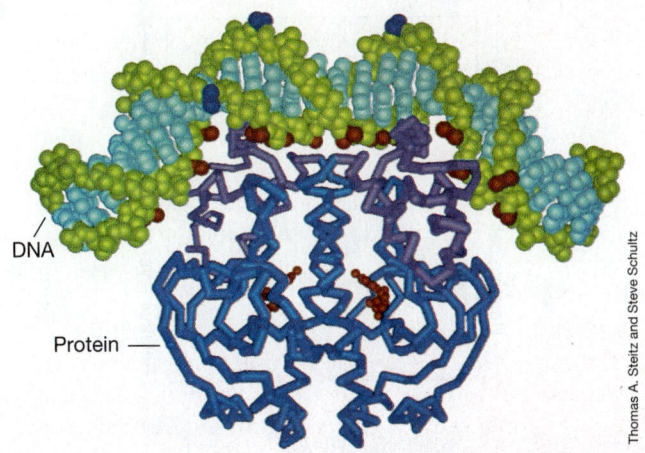

Figure 7.10 Computer model of a positive regulatory protein interacting with DNA. This model shows the cyclic AMP receptor protein (CRP), a regulatory protein that controls several operons. The α-carbon backbone of this protein is shown in blue and purple. The protein is binding to a DNA double helix (green and light blue). Note that binding of the CRP protein to DNA has bent the DNA.

Thomas A. Steitz and Steve Schultz

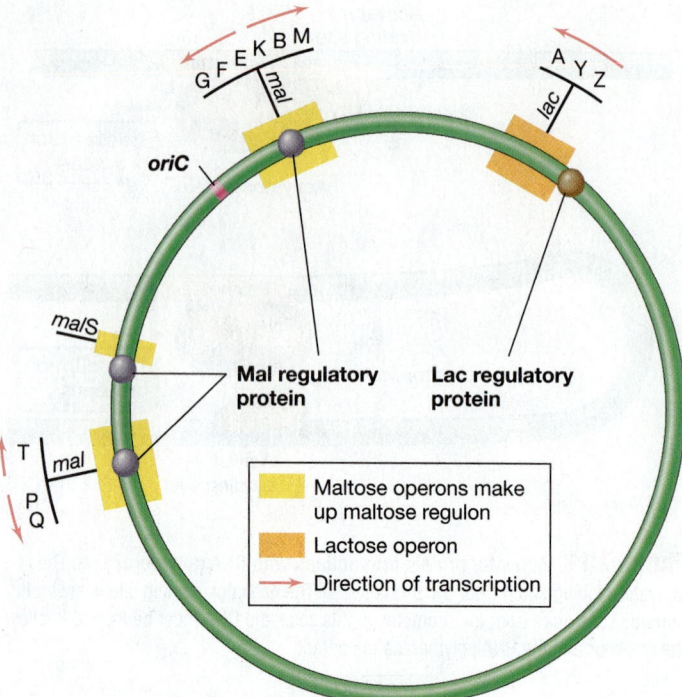

Figure 7.12 Maltose regulon of *Escherichia coli*. The genes and operons required for maltose utilization (*mal*) are dispersed throughout the *E. coli* genome and regulated by the same maltose regulatory protein. Note that the Lac repressor protein binds only to the *lac* operon, which is only located at one position on the chromosome, while the Mal repressor protein binds to multiple operons (the *mal* regulon).

MINIQUIZ

• Compare and contrast the activities of an activator protein and a repressor protein.

• Distinguish between an operon and a regulon.

7.5 Global Control and the *lac* Operon

An organism often needs to regulate many unrelated genes simultaneously in response to a change in its environment. Regulatory mechanisms that respond to environmental signals by regulating the expression of many different genes are called *global control systems*. Both the lactose operon and the maltose regulon respond to global controls in addition to their own controls discussed in Sections 7.3 and 7.4. We begin our consideration of global regulation by revisiting the *lac* operon and seeing how cells respond when given more than one sugar.

Catabolite Repression

We have not yet considered the possibility that bacteria might be confronted with several different utilizable carbon sources. For example, *Escherichia coli* can use many different sugars. When given several sugars, including glucose, do cells of *E. coli* use them simultaneously or one at a time? The answer is that *glucose is always used first*. It would be wasteful to induce enzymes for using other sugars when glucose is available, because *E. coli* grows faster on glucose than on other carbon

sources. **Catabolite repression** is a mechanism of global control that controls the use of carbon sources if more than one is present.

When cells of *E. coli* are grown in a medium that contains glucose, the synthesis of enzymes needed for the breakdown of other carbon sources (such as lactose or maltose) is repressed, even if those other carbon sources are present. Thus, the presence of a favored carbon source represses the induction of pathways that catabolize other carbon sources. Catabolite repression is sometimes called the "glucose effect" because glucose was the first substance shown to cause this response. But catabolite repression is not always linked to glucose; the key point is that the favored substrate is a better carbon and energy source than other available carbon sources. Thus, catabolite repression ensures that the organism uses the *best* carbon and energy source first.

Why is catabolite repression called *global* control? In *E. coli* and other organisms for which glucose is the best energy source, catabolite repression prevents expression of most other catabolic operons as long as glucose is present. Dozens of catabolic operons are affected, including those for lactose, maltose, a host of other sugars, and most other commonly used carbon and energy sources for *E. coli*. In addition, genes for the synthesis of flagella are controlled by catabolite repression because if bacteria have a good carbon source available, there is no need to swim around in search of nutrients.

One consequence of catabolite repression is that it may lead to two exponential growth phases, a situation called *diauxic growth*. If two usable energy sources are available, the cells grow first on the better energy source. Growth stops when the better source is depleted, but then following a lag period, it resumes on the other energy source. Diauxic growth is illustrated in **Figure 7.13** for a culture of *E. coli* grown on a mixture of glucose and lactose. The cells grow more rapidly on glucose than on lactose. Although glucose and lactose are both excellent energy sources for *E. coli*, glucose is superior, and growth is faster.

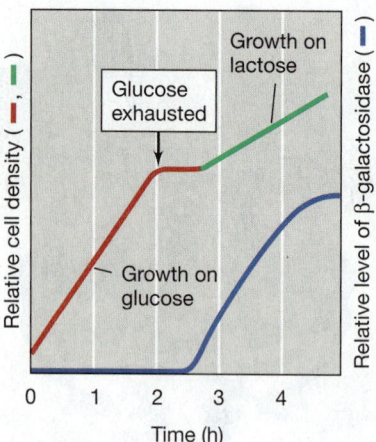

Figure 7.13 Diauxic growth of *Escherichia coli* on a mixture of glucose and lactose. The presence of glucose represses the synthesis of β-galactosidase, the enzyme that cleaves lactose into glucose and galactose. After glucose is depleted, there is a lag during which β-galactosidase is synthesized. Growth then resumes on lactose but at a slower rate, as indicated by the green line.

The proteins of the *lac* operon, including the enzyme β-galactosidase, are required for using lactose and are induced in its presence (Figures 7.6 and 7.8). But the synthesis of these proteins is also subject to catabolite repression. As long as glucose is present, the *lac* operon is not expressed and lactose is not used. However, when glucose is depleted, catabolite repression is abolished, the *lac* operon is expressed, and the cells grow on lactose.

Cyclic AMP and Cyclic AMP Receptor Protein

Despite its name, catabolite repression relies on an activator protein and is actually a form of positive control (Section 7.4). The activator protein is called the *cyclic AMP receptor protein* (*CRP*). A gene that encodes a catabolite-repressible enzyme is expressed only if CRP binds to DNA in the promoter region. This allows RNA polymerase to bind to the promoter. CRP is an allosteric protein and binds to DNA only if it has first bound a small molecule called *cyclic adenosine monophosphate* (*cyclic AMP* or *cAMP*) (**Figure 7.14**). Like many DNA-binding proteins (Section 7.2), CRP binds to DNA as a dimer.

Cyclic AMP is a key molecule in many metabolic control systems, both in prokaryotes and eukaryotes. Because it is derived from a nucleic acid precursor, it is a **regulatory nucleotide**. Other regulatory nucleotides include cyclic guanosine monophosphate (cyclic GMP; important mostly in eukaryotes), cyclic di-GMP (important in biofilm formation; Section 7.9), and guanosine tetraphosphate (ppGpp; important in the stringent response). Cyclic AMP is synthesized from ATP by an enzyme called *adenylate cyclase*. However, glucose inhibits the synthesis of cyclic AMP and also stimulates cyclic AMP transport out of the cell. When glucose enters the cell, the cyclic AMP level is lowered, CRP cannot bind DNA, and RNA polymerase fails to bind to the promoters of operons subject to catabolite repres-

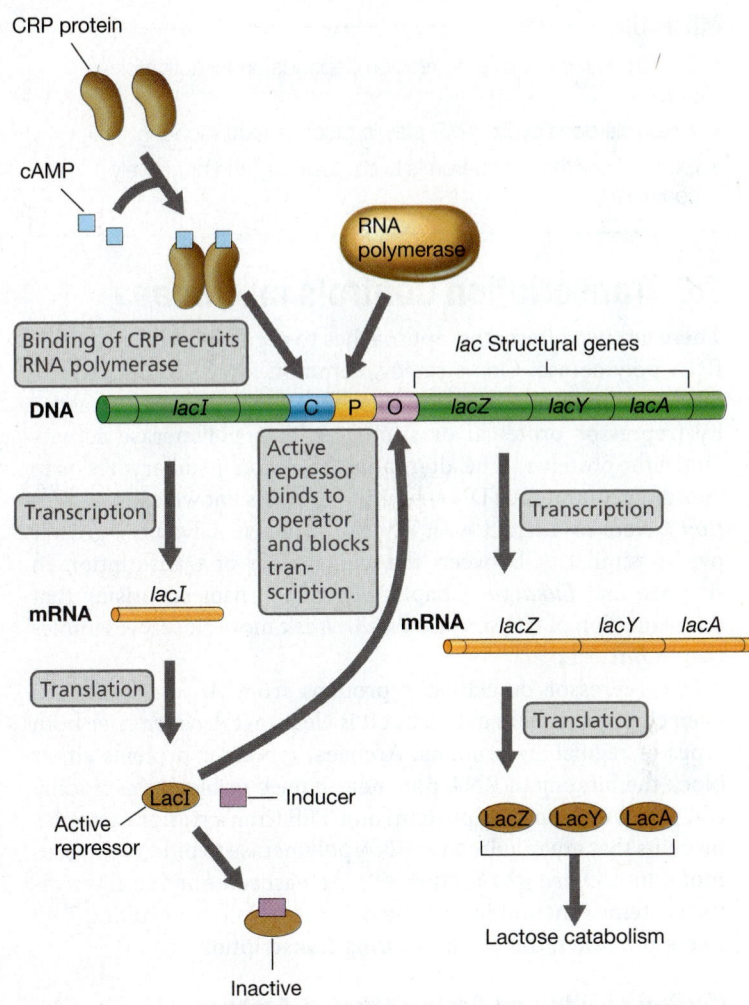

Figure 7.15 Overall regulation of the *lac* system. The *lac* operon consists of *lacZ*, encoding β-galactosidase, *lacY*, encoding lactose permease, and *lacA*, encoding lactose acetylase. The LacI repressor protein is encoded by a separate gene, *lacI*. LacI binds to the operator (O) unless the inducer is present. CRP binds to the C site when activated by cyclic AMP and recruits RNA polymerase to bind to the promoter (P) . For the *lac* operon to be transcribed by RNA polymerase, the LacI repressor must be absent (that is, inducer must be present) and cyclic AMP levels must be high (due to the absence of glucose), thus allowing CRP to bind.

sion. Thus, catabolite repression is an indirect result of the presence of a better energy source (glucose); the direct cause of catabolite repression is a low level of cyclic AMP.

Let us return to the *lac* operon and include catabolite repression. The entire regulatory region of the *lac* operon is diagrammed in **Figure 7.15**. For *lac* genes to be transcribed, two requirements must be met: (1) The level of cyclic AMP must be high enough for the CRP protein to bind to the CRP-binding site (positive control), and (2) lactose or another suitable inducer must be present so that the lactose repressor (LacI protein) does not block transcription by binding to the operator (negative control). If these two conditions are met, the cell is signaled that glucose is absent and lactose is present; then and only then does transcription of the *lac* operon begin.

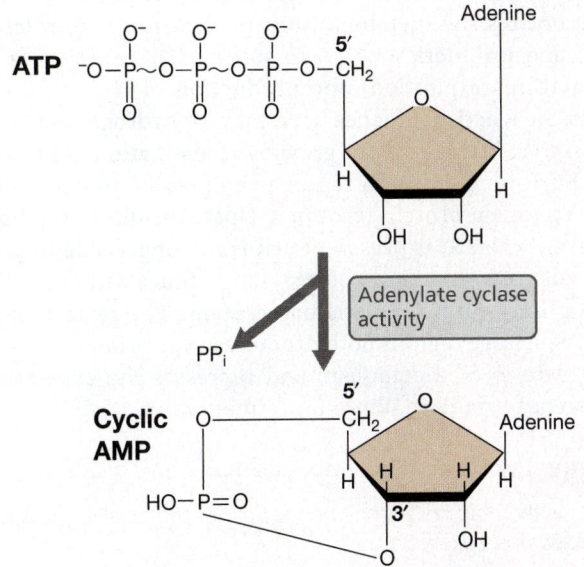

Figure 7.14 Cyclic AMP. Cyclic adenosine monophosphate (cyclic AMP) is made from ATP by the enzyme adenylate cyclase.

7.6 Transcription Controls in *Archaea*

There are two alternative approaches to regulating the activity of RNA polymerase. One strategy, common in *Bacteria*, is to use DNA-binding proteins that either block RNA polymerase activity (repressor proteins) or stimulate RNA polymerase activity (activator proteins). The alternative, common in eukaryotes, is to coordinate numerous DNA-binding proteins known as *transcription factors* to interact with RNA polymerase. Given the greater overall similarity between the mechanism of transcription in *Archaea* and *Eukarya* (Chapter 4), it is perhaps surprising that the regulation of transcription in *Archaea* more closely resembles that of *Bacteria*.

Few repressor or activator proteins from *Archaea* have yet been characterized in detail, but it is clear that *Archaea* have both types of regulatory proteins. Archaeal repressor proteins either block the binding of RNA polymerase itself or block the binding of TBP (TATA-binding protein) and TFB (transcription factor B), proteins that are required for RNA polymerase to bind to the promoter in *Archaea* (⮂ Section 4.9). At least some archaeal activator proteins function in just the opposite way, by recruiting TBP to the promoter, thereby facilitating transcription.

Control of Nitrogen Assimilation in *Archaea*

A good example of an archaeal repressor is the NrpR protein from the methanogen *Methanococcus maripaludis*. NrpR represses genes encoding nitrogen assimilation functions (**Figure 7.16**), such as those for nitrogen fixation (⮂ Section 3.17) and glutamine synthesis (⮂ Section 3.15). When organic nitrogen is plentiful in the *M. maripaludis* cell, NrpR represses nitrogen assimilation genes. However, if the level of nitrogen becomes limiting, α-ketoglutarate accumulates to high levels. This occurs because α-ketoglutarate, a citric acid cycle intermediate, is also a major acceptor of ammonia during nitrogen assimilation.

When levels of α-ketoglutarate rise, this signals the cell that ammonia is limiting and that additional pathways need to be activated for obtaining ammonia, such as nitrogen fixation or the high-affinity nitrogen assimilation enzyme glutamine synthetase. Elevated levels of α-ketoglutarate function as an inducer by binding to the NrpR protein. In this state, NrpR loses its affinity for the promoter regions of its target genes and no longer blocks transcription from promoters. In this respect, the NrpR protein resembles the LacI repressor and similar proteins of *Bacteria* (Section 7.3).

Positive Archaeal Regulatory Systems

Other archaeal proteins regulate transcription in a positive manner. Thus their binding in the promoter region increases transcription. Some of these transcription activators are related

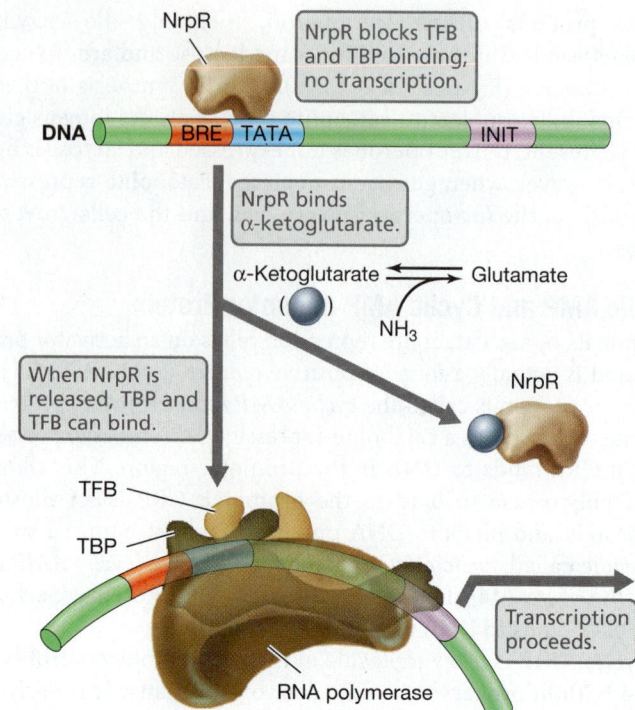

Figure 7.16 Repression of genes for nitrogen metabolism in *Archaea*. The NrpR protein of *Methanococcus maripaludis* acts as a repressor. It blocks the binding of the TFB and TBP proteins, which are required for promoter recognition, to the BRE site and TATA box, respectively. If there is a shortage of ammonia, α-ketoglutarate is not converted to glutamate. The α-ketoglutarate accumulates and binds to NrpR, releasing it from the DNA. Now TBP and TFB can bind. This in turn allows RNA polymerase to bind and transcribe the operon.

to bacterial proteins, whereas others appear to be unique to the *Archaea*. The SurR protein of the hyperthermophilic archaeon *Pyrococcus furiosus* is an example of a regulatory protein that functions either as an activator or as a repressor, depending on the location of its binding site within the promoter region. SurR controls the metabolic shift of *P. furiosus* from fermentation and production of H_2 to sulfur (S^0) reduction (a form of anaerobic respiration) and production of H_2S. When S^0 is absent, SurR activates genes necessary for hydrogenase production so that *P. furiosus* can grow by fermentation. At the same time, SurR functions as a repressor to prevent transcription of genes encoding proteins that participate in sulfur metabolism. However, when S^0 is present, SurR is no longer able to bind to DNA due to the oxidation of cysteine residues within the DNA-binding motif of this regulatory protein. The release of SurR from regulatory regions both promotes expression of genes that participate in S^0 metabolism and represses the expression of hydrogenase genes required for fermentation.

III • Sensing and Signal Transduction

Prokaryotes regulate cell metabolism in response to many different environmental fluctuations, including changes in temperature, pH, oxygen or nutrient availability, and even changes in the number of other cells present. Therefore, there must be mechanisms by which cells receive signals from the environment and transmit them to the specific target to be regulated. Some of those signals are small molecules that enter the cell and function as effectors. However, in many cases the external signal is not transmitted directly to the regulatory protein but instead is detected by a sensor that transmits it to the rest of the regulatory machinery, a process called **signal transduction**.

7.7 Two-Component Regulatory Systems

Because most signal transduction systems contain two parts, they are called **two-component regulatory systems**. Characteristically, such systems consist of a specific **sensor kinase protein** usually located in the cytoplasmic membrane, and a **response regulator protein**, present in the cytoplasm.

A kinase is an enzyme that phosphorylates compounds, typically using phosphate from ATP. Sensor kinases detect a signal from the environment and phosphorylate themselves (a process called autophosphorylation) at a specific histidine residue (**Figure 7.17**). Sensor

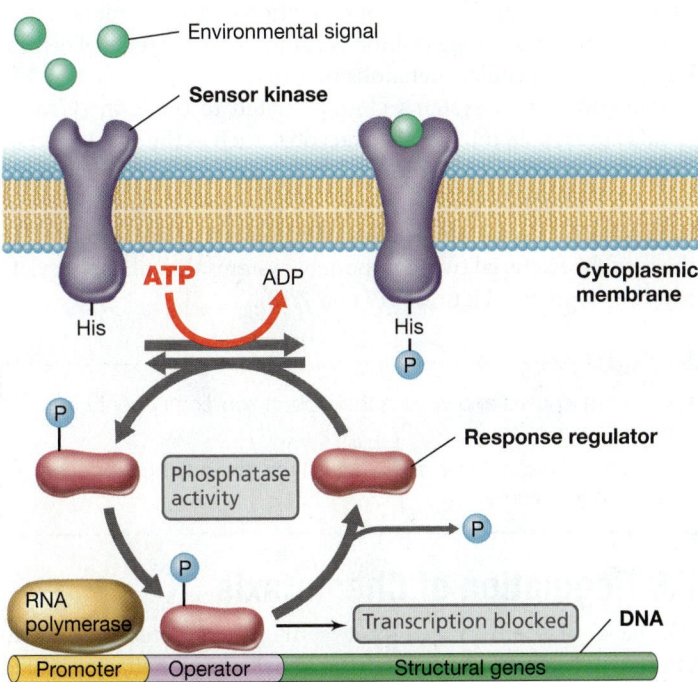

Figure 7.17 The control of gene expression by a two-component regulatory system. One component is a sensor kinase in the cytoplasmic membrane that phosphorylates itself in response to an environmental signal. The phosphoryl group is then transferred to the second component, a response regulator. The phosphorylated form of the response regulator then binds to DNA. In the system shown here, the phosphorylated response regulator is a repressor protein. The phosphatase activity of the response regulator slowly releases the phosphate from the response regulator and resets the system.

kinases thus belong to the class of enzymes called *histidine kinases*. The phosphate is then transferred from the sensor to another protein inside the cell, the response regulator. The latter is typically a DNA-binding protein that regulates transcription in either a positive or a negative fashion. In the example shown in Figure 7.17, regulation is negative; the phosphorylated response regulator functions as a repressor that binds DNA, thereby blocking transcription. Once dephosphorylated, the response regulator is released and transcription is permitted.

A balanced regulatory system must have a *feedback loop*, that is, a way to complete the regulatory circuit and terminate the response. This resets the system for another cycle. This feedback loop employs a phosphatase, an enzyme that removes the phosphate from the response regulator at a constant rate. The response regulator itself often catalyzes this reaction, although in some cases separate proteins are needed (Figure 7.17). Phosphatase activity is typically slower than phosphorylation. However, if phosphorylation ceases due to reduced sensor kinase activity, phosphatase activity eventually returns the response regulator to the fully nonphosphorylated state and the system is reset.

Examples of Two-Component Regulatory Systems

Two-component systems regulate a large number of genes in many different bacteria. Interestingly, two-component systems are either extremely rare or absent in *Archaea* and in *Bacteria* that live as parasites of higher organisms. A few key examples of two-component systems include those that respond to phosphate limitation, nitrogen limitation, and osmotic pressure.

In *Escherichia coli* almost 50 different two-component systems are present, and several are listed in **Table 7.1**. In one example, the osmolarity of the environment controls the relative levels of the proteins OmpC and OmpF in the *E. coli* outer membrane. OmpC and OmpF are *porins*, proteins that allow metabolites to cross the outer membrane of gram-negative bacteria (Section 2.11). If osmotic pressure is *low*, the synthesis of OmpF, a porin with a larger pore, increases; if osmotic pressure is *high*, OmpC, a porin with a smaller pore, is made in larger amounts. EnvZ, a cytoplasmic membrane sensor histidine kinase, detects changes in osmotic pressure. When a shift occurs, EnvZ autophosphorylates itself and transfers its phosphate group to OmpR, the response regulator of this system (**Figure 7.18**). Under conditions of *low* osmotic pressure, phosphorylated OmpR (OmpR-P) *activates* transcription of the *ompF* gene. Conversely, when osmotic pressure is *high*, OmpR-P *represses* transcription of *ompF* gene and activates transcription of *ompC* instead (Figure 7.18). The expression of *ompF* is also regulated by an additional control mechanism: regulatory RNA, and we discuss this in Section 7.14.

Two-Component Systems with Multiple Regulators

Some signal transduction systems have more than one regulatory element and their activities can quickly become quite complex. For instance, in the Ntr regulatory system, which regulates nitrogen assimilation in many *Bacteria*, the response regulator is an activator called *nitrogen regulator I* (*NRI*). NRI activates

Table 7.1 Examples of two-component systems that regulate transcription in *Escherichia coli*

System	Environmental signal	Sensor kinase	Response regulator	Activity of response regulator[a]
Arc system	Oxygen	ArcB	ArcA	Repressor/activator
Nitrate and nitrite respiration (Nar)	Nitrate and nitrite	NarX	NarL	Activator/repressor
		NarQ	NarP	Activator/repressor
Nitrogen utilization (Ntr)	Shortage of organic nitrogen	NRII (= GlnL)	NRI (= GlnG)	Activator of promoters requiring RpoN/σ^{54}
Pho regulon	Inorganic phosphate	PhoR	PhoB	Activator
Porin regulation	Osmotic pressure	EnvZ	OmpR	Activator/repressor

[a]Note that many response regulator proteins act as both activators and repressors depending on the genes being regulated. Although ArcA can function as either an activator or a repressor, it functions as a repressor on most operons that it regulates.

transcription from promoters recognized by RNA polymerase using the alternative sigma factor σ^{54} (RpoN) (⮌ Section 4.7). The sensor kinase in the Ntr system is a protein called *nitrogen regulator II* (*NRII*), which functions as both a kinase and a phosphatase. The activity of NRII is in turn regulated by another protein called *PII*, whose own activity is regulated by the addition or removal of uridine monophosphate (UMP) groups. Under nitrogen starvation conditions, UMP is added to PII, and

the resulting PII–UMP complex promotes the kinase activity of NRII and results in the phosyphorylation of NRI. Conversely, removal of UMP from PII promotes the phosphatase activity of NRII.

The *Nar regulatory system* (Table 7.1) is another example of a two-component regulatory system with multiple regulators; this system controls a set of genes that allow the use of nitrate (NO_3^-) or nitrite (NO_2^-) (or both) as alternative electron acceptors during anaerobic respiration (⮌ Section 13.17). The Nar system contains two different sensor kinases and two different response regulators. In addition, all of the genes regulated by this system are in turn controlled by the FNR protein (*f*umarate *n*itrite *r*egulator), which is a global regulator for genes of anaerobic respiration (see Table 7.2). This type of regulation in which a hierarchy of systems act in cascading fashion is common for systems of central importance to cellular metabolism.

Two-component systems closely related to those in *Bacteria* are also present in microbial eukaryotes, such as the yeast *Saccharomyces cerevisiae*, and even in plants. However, most eukaryotic signal transduction pathways rely on phosphorylation of serine, threonine, and tyrosine residues of proteins that are unrelated to those of the bacterial two-component systems that phosphorylate histidine residues (Figures 7.17 and 7.18).

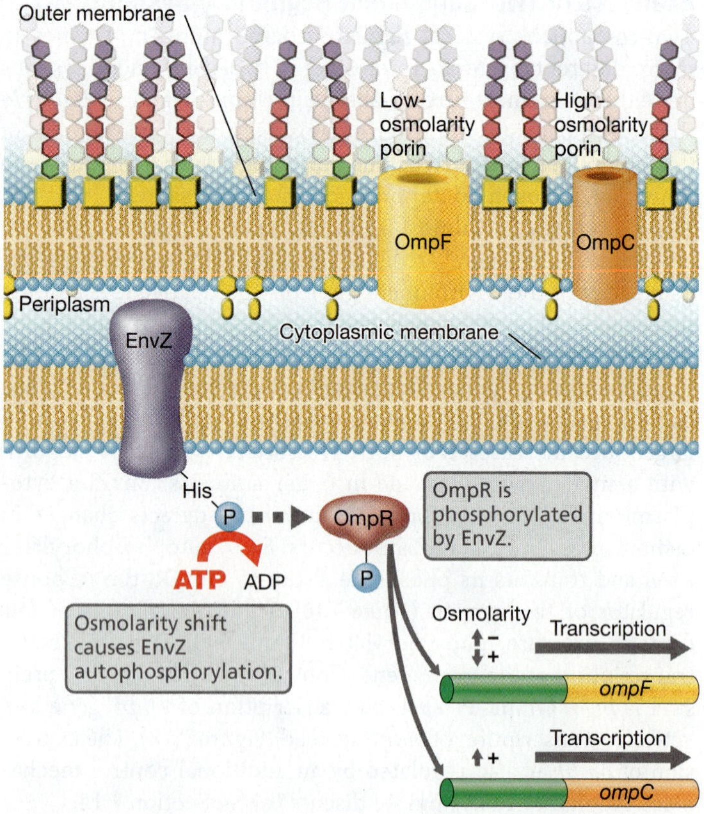

Figure 7.18 Regulation of outer membrane proteins in *Escherichia coli*. The inner membrane histidine kinase EnzZ autophosphorylates itself under osmotic pressure changes and then activates the transcriptional regulator OmpR by phosphorylation. OmpR-P binds upstream of the *ompF* gene and activates transcription under low osmotic pressure, but conversely represses transcription of *ompF* under high osmotic pressure. OmpR-P only activates transcription of the *ompC* gene under conditions of high osmolarity.

MINIQUIZ

- What are kinases and what is their role in two-component regulatory systems?
- What are phosphatases and what is their role in two-component regulatory systems?

7.8 Regulation of Chemotaxis

We have previously seen that some prokaryotes can move toward attractants and away from repellents, a behavior called *chemotaxis* (⮌ Section 2.19). We noted that prokaryotes are too small to sense spatial gradients of a chemical, but they can respond to temporal gradients. That is, they can sense the *change* in concentration of a chemical over time rather than the absolute concentration of the chemical stimulus. Prokaryotes use a modified two-component system to sense temporal changes in attractants or repellents and process this information to regulate flagellar rotation. Note that chemotaxis uses a two-component system to regulate the activity of preexisting flagella rather than to control the transcription of genes encoding flagella.

Response to Signal

The mechanism of chemotaxis depends upon a signal cascade of multiple proteins. Several sensory proteins reside in the cytoplasmic membrane and sense the presence of attractants and repellents. These sensor proteins are not themselves sensor kinases but interact with cytoplasmic sensor kinases. These sensory proteins allow the cell to monitor the concentration of various substances over time.

The sensory proteins are called *methyl-accepting chemotaxis proteins* (MCPs). *Escherichia coli* possesses five different transmembrane protein MCPs, each specific for certain compounds. For example, the Tar MCP of *E. coli* senses the attractants aspartate and maltose and the repellents cobalt and nickel. MCPs bind attractants or repellents directly or in some cases indirectly through interactions with periplasmic binding proteins. Binding of an attractant or repellent triggers interactions with cytoplasmic proteins that eventually affect flagellar rotation.

MCPs make contact with the cytoplasmic proteins CheA and CheW (**Figure 7.19**). CheA is the sensor kinase for chemotaxis. When an MCP binds a chemical, it changes conformation and, with help from CheW, leads to the autophosphorylation of CheA

to form CheA-P. Attractants *decrease* the rate of autophosphorylation, whereas repellents *increase* this rate. CheA-P then passes the phosphate to CheY (forming CheY-P); this is the response regulator that controls flagellar rotation. CheA-P can also transfer the phosphate to CheB, which plays a role in adaptation described later.

Controlling Flagellar Rotation

CheY is a central protein in the system because it governs the direction of rotation of the flagellum. Recall that if rotation of the flagellum is counterclockwise, the cell will continue to move in a run, whereas if the flagellum rotates clockwise, the cell will tumble (∞ Section 2.19). Once CheY is phosphorylated, CheY-P interacts with the flagellar motor to induce clockwise flagellar rotation, which causes tumbling (Figure 7.19). When unphosphorylated, CheY cannot bind to the flagellar motor and the flagellum rotates counterclockwise; this causes the cell to run. Another protein, CheZ, dephosphorylates CheY, returning it to the form that allows runs instead of tumbles. Because repellents increase the level of CheY-P, they lead to tumbling, whereas attractants lead to a lower level of CheY-P and smooth swimming (runs).

Adaptation

Once an organism has successfully responded to a stimulus, it must stop responding and reset the sensory system to await further signals. This is known as *adaptation*. During adaptation of the chemotaxis system, a feedback loop resets the system. This relies on the response regulator CheB, mentioned earlier.

As their name implies, MCPs can be methylated. When MCPs are fully methylated they no longer respond to attractants, but are more sensitive to repellents. Conversely, when MCPs are unmethylated they respond highly to attractants, but are insensitive to repellents. Varying the methylation level thus allows adaptation to sensory signals. This is accomplished by methylation and demethylation of the MCPs by CheR and phosphorylated CheB (CheB-P), respectively (Figure 7.19).

If the level of an attractant remains high, the rate of CheA autophosphorylation is low. This leads to unphosphorylated CheY and CheB. Consequently, the cell swims smoothly. Methylation of the MCPs increases during this period because CheB-P is not present to rapidly demethylate them. However, MCPs no longer respond to the attractant when they become fully methylated. Therefore, if the level of attractant remains high but constant, the cell begins to tumble. Eventually, CheB becomes phosphorylated and CheB-P demethylates the MCPs. This resets the receptors and they can once again respond to further increases or decreases in level of attractants. Therefore the cell stops swimming if the attractant concentration is constant. It only continues to swim if even higher levels of attractant are encountered.

The course of events is just the opposite for repellents. Fully methylated MCPs respond best to an increasing gradient of repellents and send a signal for cell tumbling to begin. The cell then moves off in a random direction while MCPs are slowly demethylated. With this mechanism for adaptation, chemotaxis successfully achieves the ability to monitor small changes in the concentrations of both attractants and repellents over time.

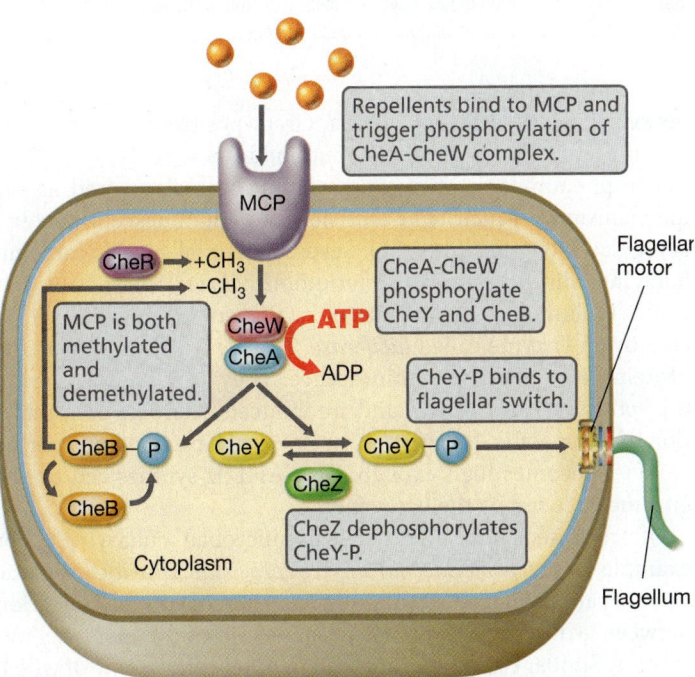

Figure 7.19 Interactions of MCPs, Che proteins, and the flagellar motor in bacterial chemotaxis. The methyl-accepting chemotaxis protein (MCP) forms a complex with the sensor kinase CheA and the coupling protein CheW. This combination triggers autophosphorylation, which can then phosphorylate the response regulators CheB and CheY. Phosphorylated CheY (CheY-P) binds to the flagellar motor switch. CheZ dephosphorylates CheY-P. CheR continually adds methyl groups to the MCP. CheB-P (but not CheB) removes them. The degree of methylation of the MCPs controls their ability to respond to attractants and repellents and leads to adaptation.

Other Taxes

In addition to chemotaxis, several other forms of taxis are known, for example, *phototaxis* (movement toward light) and *aerotaxis* (movement toward oxygen) (⮌ Section 2.19). Many of the cytoplasmic Che proteins that function to control flagellar activity in chemotaxis also play a role in these other taxes. For example, in phototaxis, a light sensor protein replaces the MCPs of chemotaxis, and in aerotaxis, a redox protein monitors levels of oxygen. These sensors then interact with cytoplasmic Che proteins to direct runs or tumbles. Thus several different kinds of environmental signals converge on the same flagellar control system, and this allows the cell to economize on its regulatory systems.

MINIQUIZ

- What are the primary response regulator and the primary sensor kinase for regulating chemotaxis?
- Why is adaptation during chemotaxis important?
- How does the response of the chemotaxis system to an attractant differ from its response to a repellent?

7.9 Quorum Sensing

Many prokaryotes respond to the presence in their surroundings of other cells of their species, and in some, regulatory pathways are controlled by the density of cells of their own kind. This is a phenomenon called **quorum sensing** (the word "quorum" in this sense means "sufficient numbers").

Mechanism of Quorum Sensing

Quorum sensing is a mechanism to assess population density. Many bacteria use this approach to ensure that sufficient cell numbers are present before initiating activities that require a certain cell density to work effectively. For example, a pathogenic (disease-causing) bacterium that secretes a toxin will have no effect as a single cell; production of toxin by one cell alone would merely waste resources. However, if a sufficiently large population of cells is present, the coordinated expression of the toxin may successfully cause disease and release resources from the host that can be used by the pathogen.

Quorum sensing is widespread among gram-negative bacteria but is also found in gram-positive bacteria. Each species that employs quorum sensing synthesizes a specific signal molecule called an **autoinducer**. This molecule usually diffuses freely across the cell envelope in either direction. Because of this, the autoinducer reaches high concentrations inside the cell only if there are many cells nearby, each making the same autoinducer. Inside the cell, the autoinducer binds to a specific transcriptional activator protein or a sensor kinase of a two-component system, ultimately triggering transcription of specific genes (**Figure 7.20b**).

While several different classes of autoinducers exist, the first to be identified were the *acyl homoserine lactones* (AHLs) (Figure 7.20a). Several different AHLs, with acyl groups of different lengths, are found in different species of gram-negative bacteria. In addition, many gram-negative bacteria make autoinducer 2 (AI-2; a cyclic furan derivative). This is apparently used as a common autoinducer

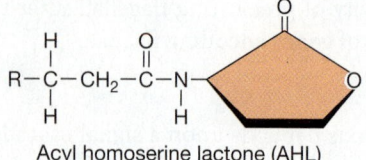

Acyl homoserine lactone (AHL)

(a)

(b)

Figure 7.20 Quorum sensing. *(a)* General structure of an acyl homoserine lactone (AHL). Different AHLs are variants of this parent structure. R = alkyl group (C_1–C_{17}); the carbon next to the R group is often modified to a keto group (C=O). *(b)* A cell capable of quorum sensing expresses AHL synthase at basal levels. This enzyme makes the cell's specific AHL. When cells of the same species reach a certain density, the concentration of AHL rises sufficiently to bind to the activator protein, which activates transcription of quorum-specific genes.

between many species of bacteria. Gram-positive bacteria generally use certain short peptides as autoinducers.

The phenomenon of quorum sensing was discovered as the mechanism by which light emission in bioluminescent bacteria is regulated (⮌ Section 15.4). Several bacterial species can emit light, including the marine bacterium *Aliivibrio fischeri*. **Figure 7.21** shows bioluminescent colonies of *A. fischeri*. The light is generated by an enzyme called *luciferase*. The *lux* operons encode the proteins needed for bioluminescence. They are under control of the activator protein LuxR and are induced when the concentration of the specific *A. fischeri* AHL, *N*-3-oxohexanoyl homoserine lactone, becomes high enough. This AHL is synthesized by the enzyme encoded by the *luxI* gene.

Quorum sensing also occurs in microbial eukaryotes. For example, in the yeast *Saccharomyces cerevisiae*, specific aromatic alcohols are produced as autoinducers and control the transition between growth of *S. cerevisiae* as single cells and as elongated filaments. Similar transitions are seen in other fungi, some of which cause disease in humans. An example is *Candida*, whose quorum sensing is mediated by the long-chain alcohol farnesol. As the concentration of farnesol increases in this dimorphic fungus, the transition from budding yeast to elongated hyphae is prevented.

Virulence Factors

Various genes are controlled by quorum sensing, including some in pathogenic bacteria. For example, Shiga toxin–producing *Escherichia coli*, such as the notorious foodborne

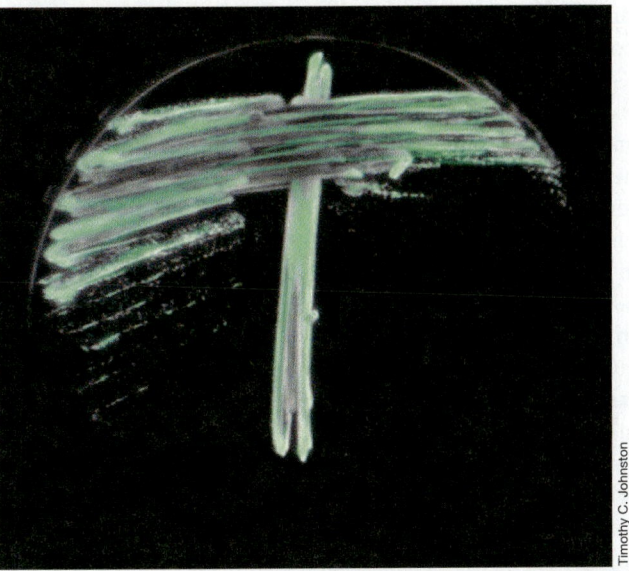

Figure 7.21 **Bioluminescent bacteria producing the enzyme luciferase.** Cells of the bacterium *Aliivibrio fischeri* were streaked on nutrient agar in a Petri dish and allowed to grow overnight. The photograph was taken in a darkened room using only the light generated by the bacteria.

pathogen *E. coli* O157: H7 (꩜ Section 31.11), produces an AHL called AI-3 that induces virulence genes. As the *E. coli* population increases in the intestine, bacterial cells produce AI-3 while host intestinal cells produce the stress hormones epinephrine

and norepinephrine. All three of these signal molecules bind to two separate sensor kinases in the *E. coli* cytoplasmic membrane, resulting in the phosphorylation and activation of two transcriptional activator proteins (**Figure 7.22***a*). These proteins activate transcription of genes encoding motility functions and secretion of the enterotoxin as well as genes encoding proteins that form lesions on the host intestinal mucosa. This is a rare example of a system that senses both bacterial and eukaryotic chemical signals to regulate gene expression.

The pathogenesis of *Staphylococcus aureus* (꩜ Section 29.9) requires, among many other things, the production and secretion of small extracellular peptides that damage host cells or that interfere with the host's immune system. The genes encoding these virulence factors are under the control of a quorum-sensing system that uses a small peptide called the *autoinducing peptide* (AIP), encoded by the *argD* gene, as the autoinducer. After synthesis of ArgD (pre-AIP), the membrane-bound ArgB protein trims the peptide into its active AIP form and secretes the small peptide outside of the cell (Figure 7.22*b*). As the cell density of *S. aureus* increases, so does the concentration of AIP. ArgC is a membrane-bound sensor kinase that binds to AIP, resulting in autophosphorylation. ArgC-P transfers its phosphate to the transcriptional activator ArgA. ArgA-P increases transcription of *argABCD* genes that encode the signal transduction system as well as an RNA molecule that controls production of a range of virulence proteins.

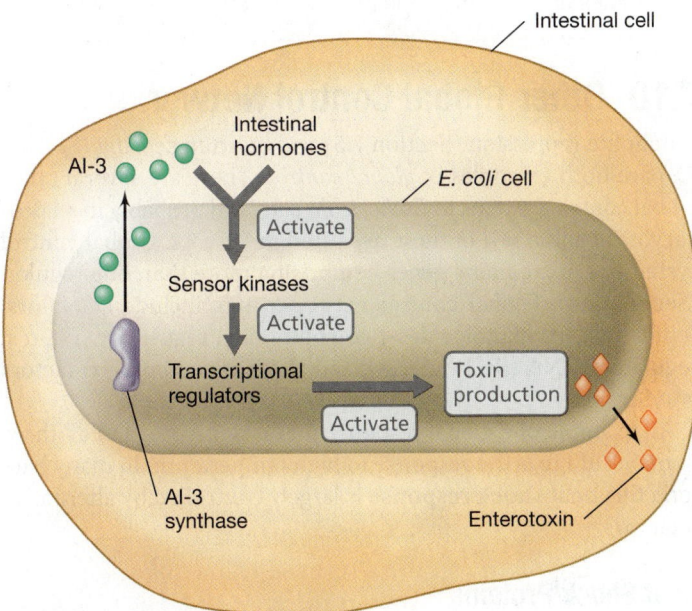

(a) **Virulence factor production in Shiga toxin-producing *Escherichia coli***

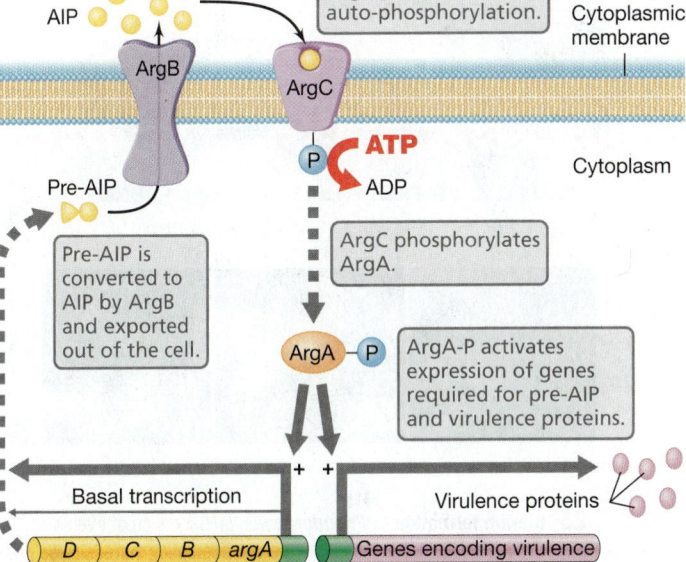

(b) **Virulence factor production in *Staphylococcus***

Figure 7.22 **Quorum sensing regulation of virulence factors.** *(a)* As the bacterial population increases, AI-3 produced by *E. coli* and epinephrine and norepinephrine produced by the intestinal cell accumulate and bind to sensor kinases, initiating a cascade of events necessary for virulence factor production. *(b)* Basal level transcription of the *argABCD* operon in *Staphylococcus* leads to production of ArgD, the pre-autoinducing peptide (AIP). ArgB trims ArgD into the functional AIP and exports it out of the cell. As the cell population increases, the AIP concentration increases and binds to ArgC, leading to autophosphorylation of ArgC. ArgC-P then activates the transcriptional activator ArgA by transfer of a phosphate group. ArgA-P increases transcription of the *argABCD* operon as well as activating the transcription of an RNA that leads to the production of virulence proteins.

Some eukaryotes produce molecules that specifically interfere with bacterial quorum sensing. Thus far, most of these have been furanone derivatives containing a halogen atom. These components mimic the AHLs or AI-2 and disrupt bacterial behavior that relies on quorum sensing. Quorum-sensing disruptors have been proposed as potential drugs to disperse bacterial biofilms and prevent the expression of virulence genes.

Biofilm Formation

Numerous signals, including cell-to-cell communication, lead to bacteria transitioning from growing freely suspended in liquid (planktonic growth) to growing in a semisolid matrix called a *biofilm* (⟳ Section 19.4 and Explore the Microbial World, "Stick or Swim," in Chapter 5). *Pseudomonas aeruginosa* forms a biofilm by producing specific polysaccharides that subsequently increase its pathogenicity and prevent the penetration of antibiotics. Quorum sensing triggers expression of a subset of the genes necessary for biofilm formation (**Figure 7.23**). Cells of *P. aeruginosa* possess two separate quorum-sensing systems, Las and Rhl, that respond to specific AHLs and activate the transcription of genes encoding exopolysaccharide synthesis as cell number increases.

Intracellular signaling also plays a role in *P. aeruginosa* biofilm formation. One of the secondary messages important in biofilm architecture is the regulatory nucleotide *cyclic di-guanosine monophosphate* (c-di-GMP). While regulatory nucleotides play an important role in all domains of life (Section 7.5), cyclic di-GMP

is only produced by prokaryotes. In fact, prokaryotic genomes encode various proteins that both synthesize and catabolize cyclic di-GMP. The synthesis or degradation of cyclic di-GMP depends on both environmental and cellular cues, and its synthesis leads to numerous physiological changes and the expression of virulence genes. Effector proteins that bind cyclic di-GMP participate in diverse activities such as exopolysaccharide production, motility, transcriptional regulation, and protein localization (both secretion and cell surface). Cyclic di-GMP also binds small regulatory RNA molecules called riboswitches (Section 7.15).

In many bacteria the formation of biofilms is triggered by the accumulation of cyclic di-GMP in the cell. Biofilm formation in *P. aeruginosa*, a notorious biofilm producer, is assisted by the synthesis of an exopolysaccharide called Pel. Pel is produced by the cyclic di-GMP receptor protein PelD and functions as both a primary scaffold for the microbial community and a mechanism for resisting antibiotics. Similarly, expression of flagellum biosynthesis genes in *P. aeruginosa* is under positive control of the cyclic di-GMP-binding protein FleQ. Flagella help mediate the attachment of *P. aeruginosa* cells during the initial stages of biofilm formation.

MINIQUIZ

- What properties are required for a molecule to function as an autoinducer?

- How do the autoinducers used in quorum sensing by gram-negative bacteria differ from those used by gram-positive bacteria?

- Besides autoinducer synthesis, what intracellular molecule promotes biofilm formation in many bacteria?

7.10 Other Global Control Networks

Catabolite repression (Section 7.5) and quorum sensing (Section 7.9) are both examples of *global control*. There are several other global control systems in *Escherichia coli* (and probably in all prokaryotes), and a few of these are listed in **Table 7.2**. Global control systems regulate many genes comprising more than one regulon (Section 7.4). Global control networks may include activators, repressors, signal molecules, two-component regulatory systems, regulatory RNA (Section 7.14), and alternative sigma (σ) factors (⟳ Section 4.7).

An example of a global response that is widespread in all three domains of life is the response to high temperature. In many bacteria this **heat shock response** is largely controlled by alternative σ factors.

Heat Shock Proteins

Most proteins are relatively stable, even to small increases in temperature. However, some proteins are less stable at elevated temperatures and tend to unfold (denature). Improperly folded proteins are recognized by protease enzymes and are degraded. Consequently, cells that experience heat stress induce the synthesis of a set of proteins—the **heat shock proteins**—that help counteract the damage. Heat shock proteins assist the cell in recovering from stress. They are induced not only by heat but

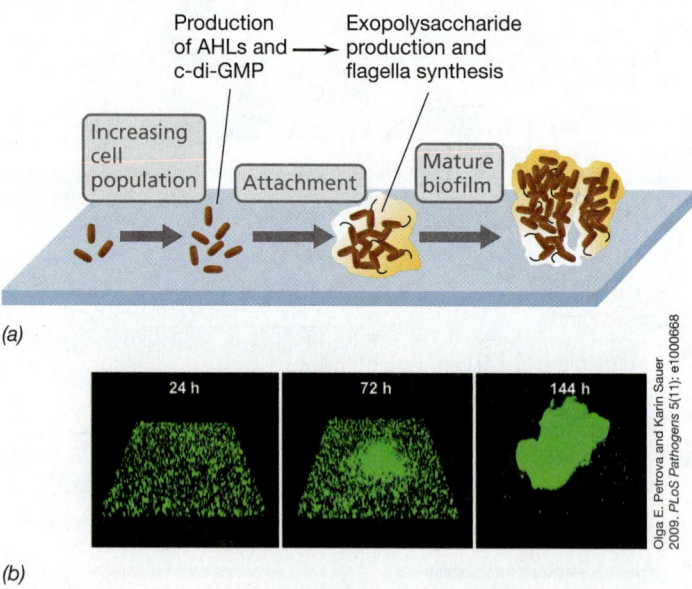

(a)

(b)

Olga E. Petrova and Karin Sauer 2009. *PLoS Pathogens* 5(11): e1000668

Figure 7.23 Biofilm formation in *Pseudomonas*. *(a)* Sequence of events leading to biofilm formation in *P. aeruginosa*. As the cell population increases, so does production of the signal molecules AHL (acyl homoserine lactones) and c-di-GMP. These signal molecules participate in activating the synthesis of exopolysaccharides and flagella necessary for full biofilm formation. *(b)* Confocal scanning laser microscopy showing the progression of *P. aeruginosa* biofilm formation over a 144-h period. Cells are stained with the LIVE/DEAD viability stain, which stains live cells green (⟳ Figure 18.7). Each rectangular pattern of cells is about 0.2 mm wide. The mature biofilm is about 0.1 mm wide and 60 μm high. Data adapted from Petrova, O.E., and K. Sauer. 2009. A novel signaling network essential for regulating *Pseudomonas aeruginosa* biofilm development. *PLoS Pathogens* 5(11): e1000668.

Table 7.2 Examples of global control systems known in *Escherichia coli*[a]

System	Signal	Primary activity of regulatory protein	Number of genes regulated
Aerobic respiration	Presence of O_2	Repressor (ArcA)	>50
Anaerobic respiration	Lack of O_2	Activator (FNR)	>70
Catabolite repression	Cyclic AMP level	Activator (CRP)	>300
Heat shock	Temperature	Alternative sigma factors (RpoH and RpoE)	36
Nitrogen utilization	NH_3 limitation	Activator (NRI)/ alternative sigma factor (RpoN)	>12
Oxidative stress	Oxidizing agents	Activator (OxyR)	>30
SOS response	Damaged DNA	Repressor (LexA)	>20

[a]For many of the global control systems, regulation is complex. A single regulatory protein can play more than one role. For instance, the regulatory protein for aerobic respiration is a repressor for many promoters but an activator for others, whereas the regulatory protein for anaerobic respiration is an activator protein for many promoters but a repressor for others. Regulation can also be indirect or require more than one regulatory protein. Many genes are regulated by more than one global system.

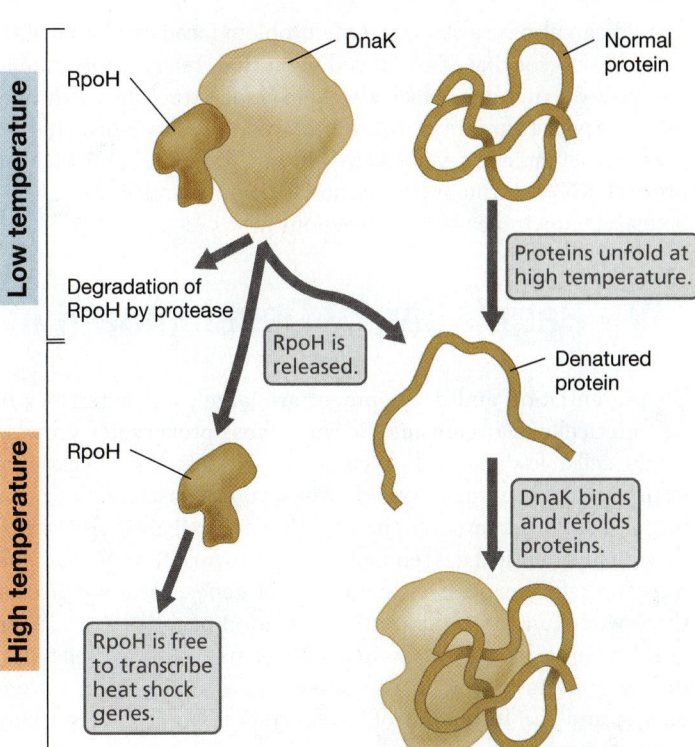

Figure 7.24 Control of heat shock in *Escherichia coli*. The RpoH alternative sigma factor is broken down rapidly by proteases at normal temperatures. This is stimulated by binding of the DnaK chaperonin to RpoH. At high temperatures, some proteins are denatured, and DnaK recognizes, binds, and refolds these unfolded polypeptide chains. This removes RpoH from DnaK, which slows the degradation rate. When the level of RpoH rises, the heat shock genes are transcribed.

also by several other stress factors that the cell may encounter. These include exposure to high levels of certain chemicals, such as ethanol, and exposure to high doses of ultraviolet (UV) radiation.

In *E. coli* and in most prokaryotes examined, there are three major classes of heat shock protein: Hsp70, Hsp60, and Hsp10. We have encountered these proteins before, although not by these names (↪ Section 4.14 and Figure 4.41). The Hsp70 protein of *E. coli* is DnaK, which prevents aggregation of newly synthesized proteins and stabilizes unfolded proteins. Major representatives of the Hsp60 and Hsp10 families in *E. coli* are the proteins GroEL and GroES, respectively. These are *molecular chaperones* that catalyze the correct refolding of misfolded proteins. Another class of heat shock proteins includes various proteases that degrade denatured or irreversibly aggregated proteins.

Heat Shock Response

In many bacteria, such as *E. coli*, the heat shock response is controlled by the alternative σ factors RpoH (σ^{32}) and RpoE (**Figure 7.24**). RpoH controls expression of heat shock proteins in the cytoplasm, and RpoE regulates the expression of a different set of heat shock proteins that function in the periplasm and cell envelope. RpoH is normally degraded within a minute or two of its synthesis. However, when cells suffer a heat shock, degradation of RpoH is inhibited and its level therefore increases. Consequently, transcription of those operons whose promoters are recognized by RpoH increases too. The rate of degradation of RpoH depends on the level of free DnaK protein, which inactivates RpoH. In unstressed cells, the level of free DnaK is relatively high and the

level of intact RpoH is correspondingly low. However, if heat begins to unfold proteins, DnaK binds preferentially to unfolded proteins and so is no longer free to promote degradation of RpoH. Thus, the more denatured proteins there are, the lower the level of free DnaK and the higher the level of RpoH; the result is heat shock gene expression. When the stress situation has passed, for example, upon a temperature downshift, RpoH is once again inactivated by DnaK, and the synthesis of heat shock proteins is greatly reduced.

Because heat shock proteins perform vital functions in the cell, there is always a low level of these proteins present, even under optimal conditions. However, the rapid synthesis of heat shock proteins in stressed cells emphasizes their importance for surviving exposure to excessive heat, chemicals, or physical agents. Such stresses can generate large amounts of inactive proteins that need to be refolded (and in the process, reactivated) or degraded to release free amino acids for the synthesis of new proteins.

There is also a heat shock response in *Archaea*, even in species that grow best at very high temperatures. An analog of the bacterial Hsp70 is found in many *Archaea* and is structurally quite similar to that found in gram-positive species of *Bacteria*. Hsp70 is also present in eukaryotes. In addition, other types of heat shock proteins are present in *Archaea* that are unrelated to stress proteins of *Bacteria*.

Cold can also be a stressor. One problem faced by all cells during cold shock is that RNA, including mRNA, tends to form stable secondary structures, especially stem–loop structures, that may interfere with translation. To counter this, cold shock proteins are present that include several RNA-binding proteins. Some of these prevent RNA secondary structure formation and others (RNA helicases) unwind base-paired regions in RNA.

MINIQUIZ

- What triggers the heat shock response?
- Why do cells have more than one type of σ factor?
- Why might the proteins induced during heat shock not be needed during cold shock?

IV • Regulation of Development in Model *Bacteria*

Differentiation and development are largely characteristics of multicellular organisms. Because most prokaryotes grow as single cells, few show differentiation. Nonetheless, occasional examples among single-celled prokaryotes illustrate the basic principle of differentiation, namely that one cell gives rise to two genetically identical descendants that perform different roles and must therefore express different sets of genes. Here we discuss three well-studied examples: the formation of endospores in the gram-positive bacterium *Bacillus*; the formation of two cell types, motile and stationary, in the gram-negative bacterium *Caulobacter*; and the formation of heterocysts in the nitrogen-fixing cyanobacterium *Anabaena*.

Although forming just two different cell types may seem superficially simple, the regulatory systems that control these processes are highly complex. There are three major phases for the regulation of differentiation: (1) triggering the response, (2) development of the differentiated cell, and (3) reciprocal communication between the two differentiating and undifferentiated cells.

7.11 Sporulation in *Bacillus*

Many microorganisms, both prokaryotic and eukaryotic, respond to adverse conditions by converting growing cells, called *vegetative cells*, into spores (⮌ Section 2.16). Once favorable conditions return, the spore germinates and the microorganism returns to its normal lifestyle. Among the *Bacteria*, the genus *Bacillus* is well known for the formation of *endospores*, that is, spores formed inside a mother cell. Prior to endospore formation, the cell divides asymmetrically. The smaller cell develops into the endospore, which is surrounded by the larger mother cell. Once development is complete, the mother cell bursts, releasing the endospore.

Endospore Formation

Endospore formation in *Bacillus subtilis* is triggered by unfavorable conditions, such as starvation, desiccation, or growth-inhibitory temperatures. Various conditions in the environment are monitored by a group of five sensor kinases. These function via a phosphotransfer relay system whose mechanism resembles that of a two-component regulatory system (Section 7.7) but is considerably more complex (**Figure 7.25**). The net result of multiple adverse conditions is the successive phosphorylation of several proteins called *sporulation factors*, culminating with sporulation factor Spo0A. When Spo0A is highly phosphorylated, sporulation proceeds. Spo0A controls the expression of several genes. The product of one of these, SpoIIE, is responsible for removing the phosphate from SpoIIAA. This allows SpoIIAA in turn

to remove an anti-sigma factor, SpoIIAB, and liberate the sigma factor, σF, as discussed below.

Once triggered, endospore development is controlled by four different σ factors, two of which, σF and σG, activate genes needed inside the developing endospore itself, and two of which, σE and σK, activate genes needed in the mother cell surrounding the endospore (Figure 7.25*b*). The sporulation signal, transmitted via Spo0A, activates σF in the smaller cell that is destined to become the endospore (σF is already present but is inactive, as it is bound by an anti-sigma factor, Figure 7.25*a*). Once free, σF binds to RNA polymerase and promotes transcription (inside the prespore) of genes whose products are needed for the next stage of sporulation. These include the gene encoding the sigma factor σG and the genes for proteins that cross into the mother cell and activate σE. Active σE is required for transcription of yet more genes inside the mother cell, including the gene for σK. The sigma factors σG (in the endospore) and σK (in the mother cell) are required for transcription of genes needed even later in the sporulation process (Figure 7.25). Eventually, the many spore coats and other unique structures typical of the endospore (⮌ Section 2.16 and Table 2.3) are formed, and the mature spore is released.

Nutrients for Endospore Formation

Nutrient limitation is a common trigger of sporulation in *Bacillus* (⮌ Section 2.16). In such a case, how do cells obtain sufficient nutrients to complete the formation of endospores? One fascinating aspect of the regulation of endospore formation is another regulatory event in which sporulating cells cannibalize their own species. Those cells in which Spo0A has already become activated secrete a protein that lyses nearby cells whose Spo0A protein has not yet become activated. This toxic protein is accompanied by a second protein that delays sporulation of neighboring cells. Cells committed to sporulation also make an antitoxin protein to protect themselves against the effects of their own toxin. When lysed, their sacrificed sister cells are used as a source of nutrients for developing endospores. Shortages of certain nutrients, in particular phosphate, increase transcription of the toxin-encoding gene.

MINIQUIZ

- How are different sets of genes expressed in the developing endospore and the mother cell?
- What is an anti-sigma factor and how can its effect be overcome?

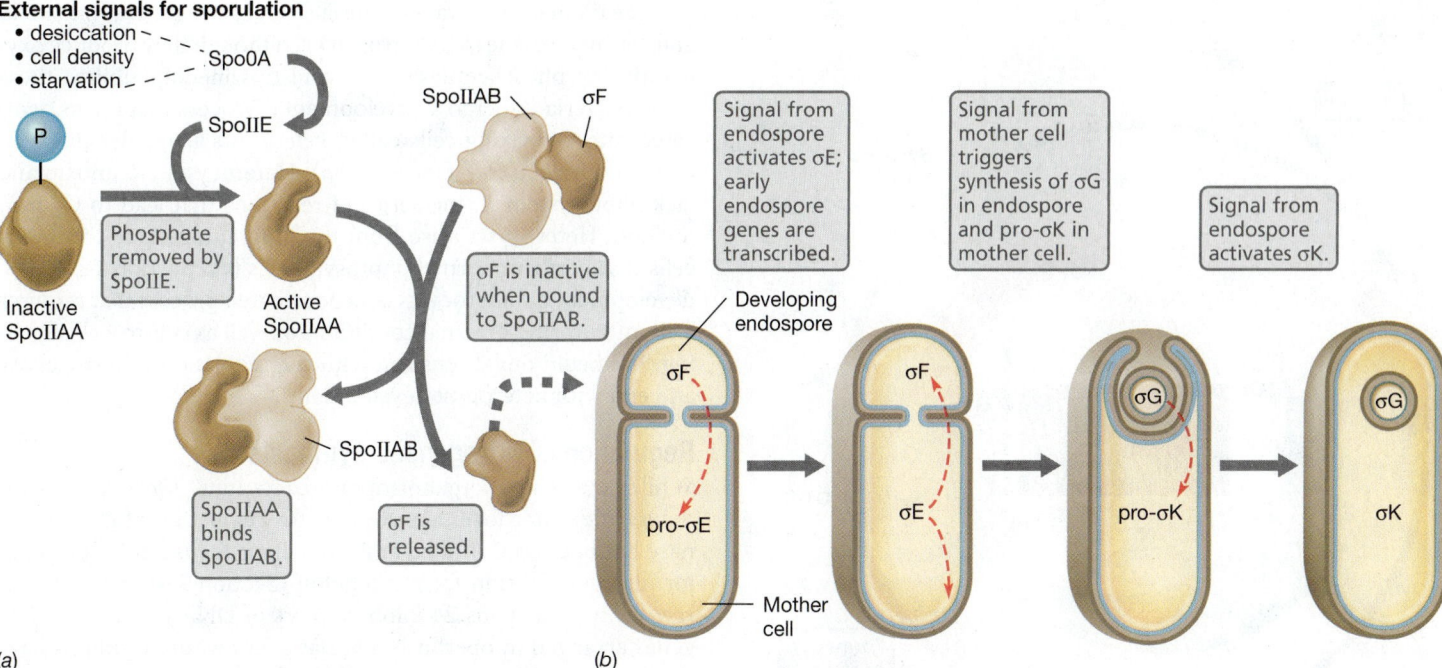

External signals for sporulation
- desiccation
- cell density
- starvation

Spo0A

SpoIIE

P

Inactive
SpoIIAA

Phosphate
removed by
SpoIIE.

Active
SpoIIAA

SpoIIAB   σF

σF is inactive
when bound
to SpoIIAB.

SpoIIAB

SpoIIAA
binds
SpoIIAB.

σF is
released.

Signal from
endospore
activates σE;
early
endospore
genes are
transcribed.

Signal from
mother cell
triggers
synthesis of σG
in endospore
and pro-σK in
mother cell.

Signal from
endospore
activates σK.

Developing
endospore

σF

pro-σE

σF

σE

σG

pro-σK

σG

σK

Mother
cell

(a)   (b)

UNIT 2

Figure 7.25 Control of endospore formation in *Bacillus*. After an external signal is received, a cascade of sigma (σ) factors controls differentiation. *(a)* Active SpoIIAA binds the anti-σ factor SpoIIAB, thus liberating the first σ factor, σF. *(b)* σF initiates a cascade of sigma factors, some of which already exist and need to be activated, others of which are not yet present and whose genes must be expressed. These σ factors then promote transcription of genes needed for endospore development.

7.12 *Caulobacter* Differentiation

The gram-negative bacterium *Caulobacter* provides another example in which a cell divides into two genetically identical daughter cells that are structurally distinct and perform different roles and express different sets of genes. *Caulobacter* is a species of *Proteobacteria* that is common in aquatic environments, typically in waters that are nutrient-poor (oligotrophic, ↩ Section 19.8). In the *Caulobacter* life cycle, free-swimming (swarmer) cells alternate with cells that lack flagella and are attached to surfaces by a stalk with a holdfast at its end. The role of the swarmer cells is dispersal, as swarmers cannot divide or replicate their DNA. Conversely, the role of the stalked cell is reproduction (**Figure 7.26**).

Regulatory Features

The *Caulobacter* cell cycle is controlled by three major regulatory proteins whose concentrations oscillate in succession. Two of these are the transcriptional regulators, GcrA and CtrA. The third is DnaA, a protein that functions both in its normal role in initiating DNA replication and also as a transcriptional regulator. Each of these regulators is active at a specific stage in the cell cycle, and each controls many other genes that are needed at that particular stage in the cycle.

CtrA is activated by phosphorylation in response to external signals. Once phosphorylated, CtrA-P activates genes needed for the synthesis of the flagella and other functions in swarmer cells. Conversely, CtrA-P represses the synthesis of GcrA and also inhibits the initiation of DNA replication by binding to and blocking the origin of replication (Figure 7.26). As the cell cycle proceeds, CtrA is degraded by a specific protease; as a consequence, levels of DnaA

rise. The absence of CtrA-P allows access to the chromosomal origin of replication, and, as in all *Bacteria*, DnaA binds to the origin and triggers the initiation of DNA replication (↩ Section 4.5). In addition, *Caulobacter* DnaA activates several other genes needed for chromosomal replication. The level of DnaA then falls due to protease degradation, and the level of GcrA rises. The GcrA regulator promotes the elongation phase of chromosome replication, cell division, and the growth of the stalk on the immobile daughter cell. Eventually, GcrA levels fall and high levels of CtrA reappear (in the daughter cell destined to swim away) (Figure 7.26).

Caulobacter as a Model for the Eukaryotic Cell Cycle

Both external stimuli and internal factors such as nutrient and metabolite levels result in the precise coordination of morphological and metabolic events within the *Caulobacter* cell cycle. Since its genome has been sequenced and good genetic systems for gene transfer and analysis are available, differentiation in *Caulobacter* has been used as a model system for studying cell developmental processes in other organisms as well. This focus is due to the strict cell cycle followed by *Caulobacter*, which resembles that of eukaryotic cells in many respects. In fact, terminology used to describe the eukaryotic cell cycle has been adapted to the *Caulobacter* system.

In eukaryotic cells, phase G1 of cell division is where growth and normal metabolic events occur while in phase G2 the cell prepares for subsequent mitotic events, which occur in the M phase. Between G1 and G2 is the S phase, where DNA replication occurs. In the *Caulobacter* life cycle there is no mitosis, of course, but analogs of the G1, G2, and S phases are apparent (Figure 7.26), and these make this bacterium an excellent model for studying cell division events in higher organisms.

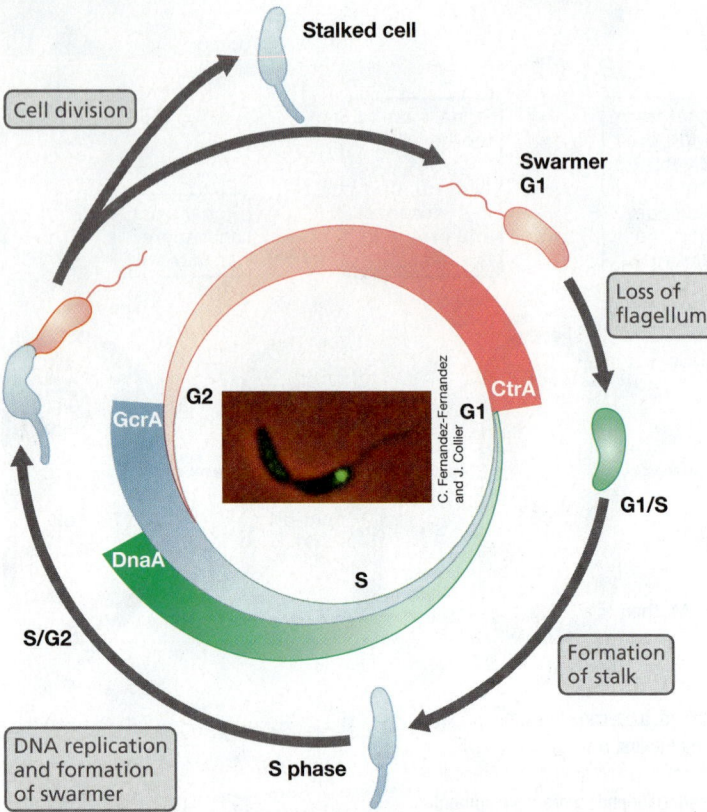

Figure 7.26 Cell cycle regulation in *Caulobacter*. Three global regulators, CtrA, DnaA, and GcrA, oscillate in levels through the cycle as shown. G1 and G2 are the two growth phases and S is the DNA synthesis phase. In G1 swarmer cells, CtrA represses initiation of DNA replication and expression of GcrA. At the G1/S transition, CtrA is degraded and DnaA levels rise. DnaA binds to the origin of replication and initiates replication (see inset photo). GcrA also rises and activates genes for cell division and DNA synthesis. At the S/G2 transition, CtrA levels begin to rise again and shut down GcrA expression. GcrA levels slowly decline in the stalked cell but are rapidly degraded in the swarmer. CtrA is degraded in the stalked cell. Inset: Using a fusion to the green fluorescent protein as a reporter (Section 7.1), a subunit of DNA polymerase is localized in the end of the stalked *Caulobacter* cell where DNA replication occurs. Each cell of the dividing *Caulobacter* pair is about 2 μm long.

MINIQUIZ --

- Why are the levels of DnaA protein controlled during the *Caulobacter* cell cycle?

- When do the regulators CtrA and GcrA carry out their main roles during the *Caulobacter* life cycle?

7.13 Nitrogen Fixation, Nitrogenase, and Heterocyst Formation

Nitrogen fixation is the process of reducing N_2 to NH_3 for incorporation into biological molecules. Nitrogen fixation is catalyzed by the enzyme *nitrogenase*, which is composed of two proteins, *dinitrogenase* and *dinitrogenase reductase* (⮑ Section 3.17). Because the process of N_2 fixation is highly energy demanding, the synthesis and activity of nitrogenase and the many other enzymes required for N_2 fixation are highly regulated. Nitrogenase is also inactivated by oxygen; therefore, it would be wasteful to synthesize the enzyme under aerobic conditions.

Some filamentous cyanobacteria, such as the genera *Anabaena* and *Nostoc*, are able to fix nitrogen even though they produce oxygen during photosynthesis. To avoid this incompatibility, these cyanobacteria undergo a developmental process that forms dedicated nitrogen-fixing cells called *heterocysts* at regular intervals along a filament (see Figure 7.28a). Heterocysts are anoxic and lack photosystem II, the series of reactions that lead to O_2 production. Heterocysts arise from the differentiation of vegetative cells that perform normal photosynthesis (⮑ Section 14.2). The development of heterocysts is a coordinated process that requires both monitoring external conditions as well as cell-to-cell signaling. We begin our discussion with the regulation of nitrogenase and end with heterocyst development.

Regulation of Nitrogenase Synthesis

While the chemoorganotrophic bacterium *Klebsiella pneumoniae* does not form heterocysts, the regulation of its nitrogenase has been well studied and will be our focus here. The genes for nitrogen fixation form a regulon (Section 7.4) called the *nif regulon*, which spans 24 kilobase pairs of DNA and contains 20 genes arranged in operons such that genes whose products have similar functions are cotranscribed (**Figure 7.27**). In addition to nitrogenase structural genes, the genes for FeMo-co synthesis (FeMo-co is a cofactor needed for nitrogenase functions, ⮑ Section 3.17), genes controlling the electron transport proteins, and a number of regulatory genes are also present in the *nif* regulon. Within the regulon, dinitrogenase is encoded by the *nifD* and *nifK* genes, while dinitrogenase reductase is encoded by the *nifH* gene. FeMo-co is encoded by several genes, including *nifN, V, Z, W, E, B,* and *Q.*

Nitrogenase is subject to strict regulatory controls. Nitrogen fixation is repressed by O_2 and by fixed forms of nitrogen, including NH_3, NO_3^-, and certain amino acids. A major part of this regulation occurs in the expression of *nif* structural genes, whose transcription is activated by the *positive* regulator NifA (Figure 7.27). By contrast, NifL is a *negative* regulator of *nif* gene expression and contains a molecule of FAD (recall that FAD is a redox coenzyme for flavoproteins, ⮑ Section 3.10) that functions as an O_2 sensor. In the presence of sufficient O_2, NifL FAD is oxidized and the protein can then repress transcription of other *nif* genes; this prevents synthesis of the oxygen-labile nitrogenase.

Ammonia prevents nitrogen fixation through a second protein, called NtrC, whose genes are not part of the *nif* operon. NtrC activity is regulated by the nitrogen status of the cell. When NH_3 is limiting, NtrC is active and promotes transcription of *nifA*. This encodes NifA, the nitrogen fixation activator protein, and *nif* gene transcription begins. The NH_3 produced by nitrogenase (⮑ Figure 3.33) does not itself prevent enzyme synthesis because it is incorporated into amino acids and used in biosynthesis soon after it is made. But when NH_3 is in excess (as in natural environments or culture media high in NH_3), nitrogenase synthesis is repressed. In this way, ATP is not wasted in making ammonia when it is already available in ample amounts.

Heterocyst Formation

Heterocyst formation in filamentous cyanobacteria requires numerous morphological and metabolic changes that are regulated

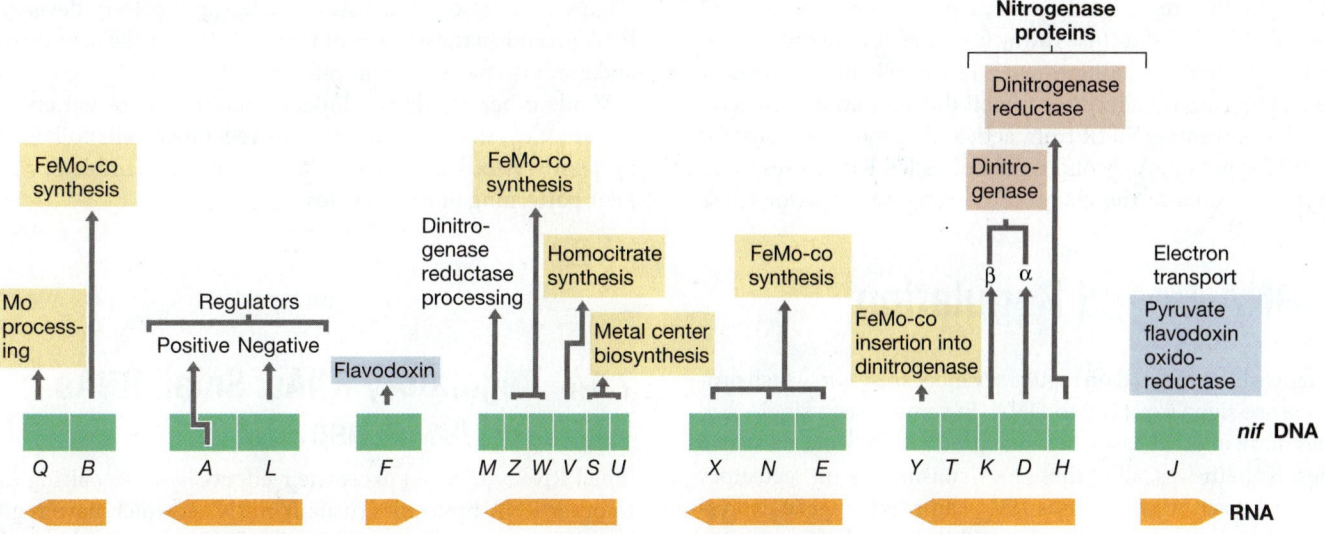

Figure 7.27 The *nif* regulon in *Klebsiella pneumoniae*, the best-studied nitrogen-fixing bacterium. The function of the *nifT* gene product is unknown. The mRNA transcripts are shown below the genes; arrows indicate the direction of transcription. Proteins that catalyze FeMo-co synthesis are shown in yellow.

by a network of systems that sense both external conditions and intracellular signaling molecules. These processes include formation of a thickened envelope to prevent O_2 diffusion into the cell, inactivation of photosystem II, expression of nitrogenase, and "patterning" of heterocyst differentiation along the filament (**Figure 7.28a**). Because nutrients can be exchanged between heterocysts and adjacent vegetative cells, other regulatory steps must be initiated to prevent nearby vegetative cells from undergoing the developmental process.

The cascade of events leading to heterocyst formation is initiated by nitrogen limitation, which is sensed as an elevation in levels of α-ketoglutarate, the acceptor molecule for formation of glutamate, in the cell. When the cell is nitrogen starved, α-ketoglutarate accumulates and activates the transcriptional global regulator NtcA. NtcA then activates transcription of the *hetR* gene, which encodes HetR, the major transcriptional regulator controlling heterocyst formation. HetR activates a cascade of genes necessary for differentiation of the heterocyst, expression of cytochrome *c* oxidases to remove O_2, as well as expression of the *nif* operon (Figure 7.27) for synthesis of nitrogenase (Figure 7.28c).

Heterocyst development in *Anabaena* is triggered by nitrogen starvation, and only specific cells within the filament form heterocysts. Interestingly, this occurs in a fairly consistent pattern (Figure 7.28a) and is under strict control. Intercellular connections between cells in an *Anabaena* filament allow vegetative cells to provide fixed carbon to the heterocyst (as an electron donor for N_2 fixation) in exchange for fixed nitrogen. However, the cell connections also allow for intercellular communication by regulatory

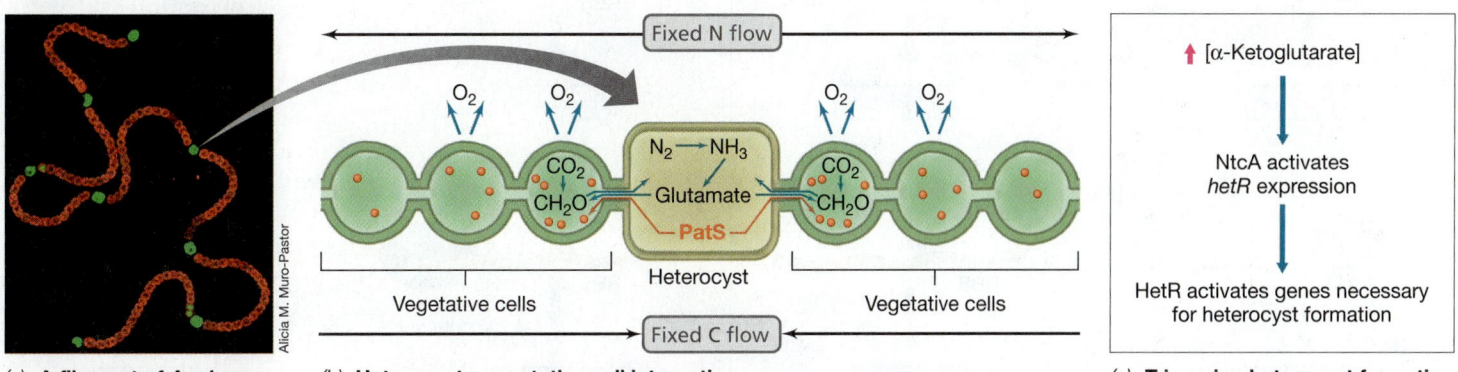

(a) **A filament of *Anabaena***

(b) **Heterocyst—vegetative cell interactions**

(c) **Triggering heterocyst formation**

Figure 7.28 Regulation of heterocyst formation. *(a)* Fluorescence microscopy showing *Anabaena* filaments expressing the green fluorescent protein linked to heterocyst-specific genes; vegetative cells are red from chlorophyll *a* fluorescence. *(b)* Molecule dispersion in heterocysts. Fixed carbon from photosynthesis in the vegetative cells is transferred to the heterocyst, while fixed nitrogen produced in the heterocyst is shared with the vegetative cells. The protein PatS, which is synthesized by heterocysts, is also dispersed to neighboring vegetative cells where it inhibits expression of genes necessary for heterocyst formation. *(c)* Cascade of events in the activation of genes necessary for heterocyst formation. The cascade is initiated by an increase in α-ketoglutarate concentration.

molecules. In this regard, differentiating cells produce a small peptide called PatS that diffuses away from the developing heterocyst to form a gradient along the vegetative cells in the filament (Figure 7.28b). PatS is believed to inhibit differentiation in vegetative cells by preventing HetR from activating genes necessary for heterocyst formation. A second regulator called PatA, a response regulator analogous to the chemotaxis response regulator CheY (Figure 7.19), also participates in heterocyst pattern development. PatA promotes the activity of HetR, decreases the activity of PatS, and may also participate in cell division.

While other regulatory links in heterocyst formation are still under study, the differentiation of vegetative cells to heterocysts in heterocystous cyanobacteria is a unique example of multicellular patterning in prokaryotes.

V • RNA-Based Regulation

Thus far we have focused on regulatory mechanisms in which proteins sense signals or bind to DNA. In some cases a single protein does both; in other cases, separate proteins carry out these two activities. Nonetheless, all of these mechanisms rely on regulatory *proteins*. However, in some cases *RNA* can regulate gene expression, both at the level of transcription and at the level of translation.

RNA molecules that are not translated to give proteins are collectively known as **noncoding RNA (ncRNA)**. This category includes the rRNA and tRNA molecules that participate in protein synthesis and the RNA present in the signal recognition particle that catalyzes some types of protein secretion (⮂ Section 4.14). Noncoding RNA also includes small RNA molecules necessary for RNA processing, especially the splicing of mRNA in eukaryotes. *Small RNAs* (sRNAs) that range from approximately 40–400 nucleotides long and regulate gene expression are widely distributed in both prokaryotes and eukaryotes. In *Escherichia coli*, for example, a number of sRNA molecules regulate various aspects of cell physiology in response to environmental or cellular signals by binding to other RNAs or in some cases to other small molecules; the end result is control of gene expression.

7.14 Regulatory RNAs: Small RNAs and Antisense RNA

Small RNAs (sRNAs) exert their effects by base-pairing directly to other RNA molecules, usually mRNAs, which have regions of complementary sequence. This binding immediately modulates the rate of target mRNA translation because a ribosome cannot translate double-stranded RNA. Thus, sRNAs provide an additional mechanism to regulate a protein's synthesis once its corresponding mRNA has already been transcribed.

Mechanisms of sRNA Activity

Small RNAs alter the translation of their mRNA target by four distinct mechanisms (**Figure 7.29**). Some sRNAs will base-pair to their target mRNA, changing its secondary structure to either block a previously accessible ribosome-binding site (RBS) (⮂ Section 4.11) or to open up a previously blocked RBS, allowing access for the ribosome. These two events decrease or increase expression of the protein encoded by the target mRNA, respectively. The other two mechanisms of sRNA interaction affect mRNA

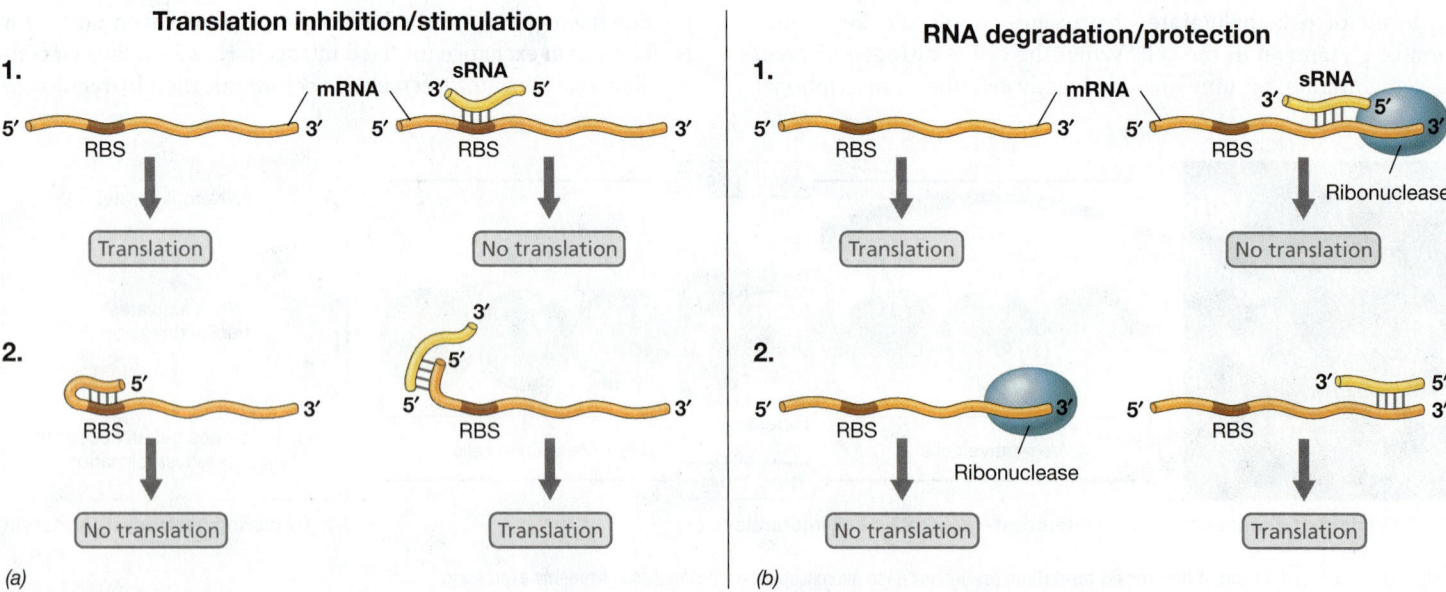

Figure 7.29 Small RNA mechanisms for modulating the translation of mRNA. *(a)* Binding of a ribosome to mRNA requires that the ribosome-binding site (RBS) of the mRNA be single-stranded. Binding of an sRNA to the RBS (shown in 1) can prevent translation, while the binding of an sRNA to an mRNA whose RBS has secondary structure (shown in 2) can stimulate translation. *(b)* Ribonuclease degrades RNA. Ribonuclease binding to partially double-stranded RNA results in RNA degradation (shown in 1), while sRNA binding at the ribonuclease binding site (shown in 2) can protect the mRNA from degradation.

stability; binding of the sRNA to its target can either increase or decrease degradation of the transcript by bacterial ribonucleases, thus modulating protein expression. Increased degradation of an mRNA prevents the synthesis of new protein molecules encoded by that mRNA. Alternatively, increasing the stability of mRNA will lead to higher corresponding protein levels in the cell (Figure 7.29).

Types of Small RNA

Small RNAs that are made by transcribing the nontemplate strand of the same gene that yielded the target mRNA are called *anti-sense* small RNAs and are thus complementary in base sequence. Transcription of antisense RNA is often enhanced under conditions in which its target genes need to be turned off. For example, the RyhB antisense RNA of *Escherichia coli* is transcribed when iron is limiting for growth. RyhB antisense RNA binds to several distinct target mRNAs that encode proteins needed for iron metabolism or that use iron as cofactors. Binding of RyhB sRNA blocks the RBS of the mRNA and thus inhibits translation (Figure 7.29). The base-paired RyhB/mRNA molecules are then degraded by ribonucleases, in particular, ribonuclease E. This forms part of the mechanism by which *E. coli* and related bacteria respond to a shortage of iron. Other responses to iron limitation in *E. coli* include transcriptional controls by repressor and activator proteins (Sections 7.3 and 7.4) that decrease and increase, respectively, the capacity of cells to take up iron or to tap into intracellular iron reserves.

Other sRNAs called *trans-sRNAs* are encoded in intergenic regions and can be spatially separated from their mRNA target. As such, these sRNAs usually have limited complementarity to their target molecule and may only base-pair with a 5- to 11-nucleotide stretch. The binding of trans-sRNAs to their targets often depends on a small protein called Hfq (**Figure 7.30**) that binds to both RNA molecules to facilitate their interaction. Hfq forms hexameric rings with RNA-binding sites on both surfaces. Hfq and functionally similar proteins are called *RNA chaperones*, as they help small RNA molecules, including many sRNAs, maintain their correct structure (Figure 7.30).

Small RNAs do not always work by affecting mRNA. For example, replication of the high copy number plasmid ColE1 in *Escherichia coli* is regulated by an sRNA that primes DNA synthesis on the plasmid and its antisense partner that blocks initiation of DNA synthesis. The level of the antisense RNA determines how often replication is initiated. Some sRNAs also bind to proteins and modulate their activity.

MINIQUIZ

- How do sRNAs alter the translation of target mRNAs?
- Why do trans-sRNAs often require a chaperone protein?

7.15 Riboswitches

RNA can carry out many roles once thought to be limited to proteins. In particular, RNA can specifically recognize and bind other molecules, including low-molecular-weight metabolites. It is important to emphasize that such binding does not require complementary base pairing (as does binding of the sRNAs described in the previous section) but results from the folding of the RNA into a specific three-dimensional structure that recognizes the target molecule, much as a protein enzyme recognizes its substrate. RNA molecules that are catalytically active are called *ribozymes*. Other RNA molecules resemble repressors and activators in binding small metabolites and regulating gene expression; these are the **riboswitches**.

Riboswitch RNAs contain regions upstream of the coding sequences that can fold into specific three-dimensional structures that bind small molecules. These recognition domains are riboswitches and exist as two alternative structures, one with the small molecule bound and the other without (**Figure 7.31**). Alternation between the two forms of the riboswitch thus depends on the presence or absence of the small molecule, which in turn controls

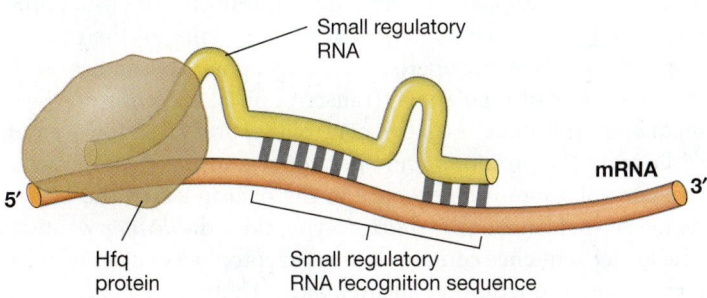

Figure 7.30 **The RNA chaperone Hfq holds RNAs together.** Binding of sRNA to mRNA often requires the Hfq protein. Small RNA molecules usually have several stem–loop structures. One consequence is that the complementary base sequence that recognizes the mRNA is noncontiguous.

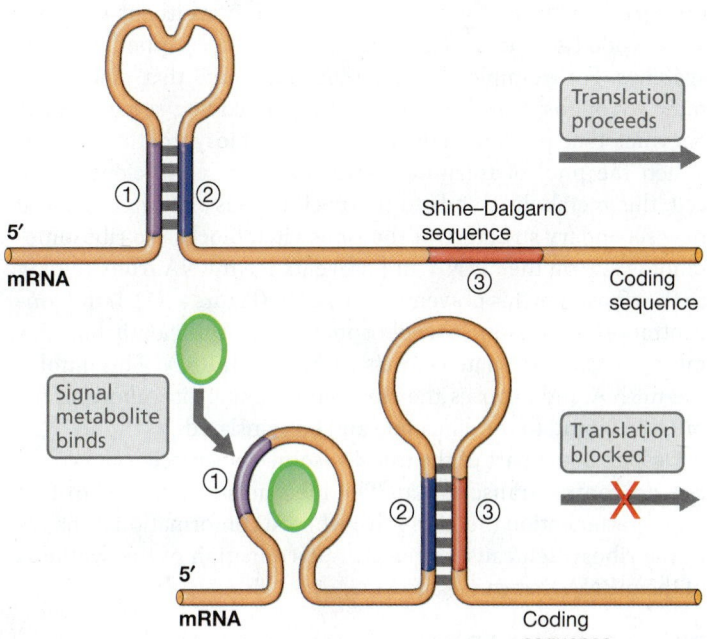

Figure 7.31 **Regulation by a riboswitch.** Binding of a specific metabolite alters the secondary structure of the riboswitch domain, which is located in the 5′ untranslated region of the mRNA, preventing translation. Numbers indicate regions within the riboswitch that can base-pair together. The Shine–Dalgarno site is where the ribosome binds the RNA.

UNIT 2

Table 7.3 Riboswitches in biosynthetic pathways of *Escherichia coli*

Type	Example
Vitamins	Cobalamin (B_{12}), tetrahydrofolate (folic acid), thiamine
Amino acids	Glutamine, glycine, lysine, methionine
Nitrogen bases of nucleic acids	Adenine, guanine (purine bases)
Others	Flavin mononucleotide (FMN), S-adenosylmethionine (SAM), glucosamine 6-phosphate (peptidoglycan precursor), cyclic di-GMP (biofilm signaling molecule)

expression of the mRNA. Riboswitches have been found that control the synthesis of enzymes in biosynthetic pathways for various vitamins, a few amino acids, some nitrogen bases, and for a precursor in peptidoglycan synthesis (**Table 7.3**).

Mechanism of Riboswitches

Earlier in this chapter we discussed the regulation of gene expression by negative control of transcription (Section 7.3). In this process, a specific metabolite interacts with a specific repressor protein to prevent transcription of genes encoding enzymes for the biosynthetic pathway of the metabolite. In contrast to this, in a riboswitch, there is no regulatory protein. Instead, the metabolite binds directly to the riboswitch at the 5′ end of the mRNA. Riboswitches usually exert their control after the mRNA has already been synthesized. Therefore, most riboswitches control *translation* of the mRNA, rather than its *transcription* (Figure 7.31).

The metabolite that is bound by the riboswitch is typically the product of a biosynthetic pathway whose constituent enzymes are encoded by the mRNAs that carry the corresponding riboswitches. For example, the thiamine riboswitch that binds thiamine pyrophosphate lies upstream of the coding sequences for enzymes that participate in the thiamine biosynthetic pathway. When the pool of thiamine pyrophosphate is sufficient in the cell, this metabolite binds to its specific riboswitch mRNA. The new secondary structure of the riboswitch blocks the ribosome-binding site on the mRNA and prevents the mRNA from binding to the ribosome; this prevents translation (Figure 7.31). If the concentration of thiamine pyrophosphate drops sufficiently low, this molecule can dissociate from its riboswitch mRNA. This unfolds the mRNA and exposes the ribosome-binding site, allowing the mRNA to bind to the ribosome and be translated.

Despite being part of the mRNA, some riboswitches nevertheless do control transcription. The mechanism is similar to that seen in attenuation (Section 7.16) where a conformational change in the riboswitch causes premature termination of the synthesis of the mRNA that carries it.

Riboswitches and Evolution

How widespread are riboswitches and how did they evolve? Thus far riboswitches have been found only in some bacteria and a few plants and fungi. Some scientists believe that riboswitches are remnants of the RNA world, a period eons ago before cells, DNA, and protein, when it is hypothesized that catalytic RNAs were the only self-replicating life forms. In such an environment, riboswitches may have been a primitive mechanism of metabolic control—a simple means by which RNA life forms could have controlled the synthesis of other RNAs. As proteins evolved, riboswitches might have been the first control mechanisms for their synthesis as well. If this is true, the riboswitches that remain today may be the last vestiges of this simple form of control because, as we have seen in this chapter, metabolic regulation is almost exclusively carried out by way of regulatory *proteins*.

MINIQUIZ
- What happens when a riboswitch binds the small metabolite that regulates it?
- What are the major differences between a repressor protein and a riboswitch in the control of gene expression?

7.16 Attenuation

Attenuation is a form of transcriptional control in *Bacteria* (and likely in *Archaea* as well) that functions by premature termination of mRNA synthesis. That is, in attenuation, control is exerted *after* the initiation of transcription but *before* its completion. Consequently, the number of completed transcripts from an operon is reduced, even though the number of initiated transcripts is not.

The basic principle of attenuation is that the first part of the mRNA to be made, called the *leader*, can fold into two alternative secondary structures. In this respect, the mechanism of attenuation resembles that of riboswitches (Figure 7.31). In attenuation, one mRNA secondary structure allows continued synthesis of the mRNA, whereas the other secondary structure causes premature termination. Folding of the mRNA depends either on events at the ribosome or on the activity of regulatory proteins, depending on the organism. The best examples of attenuation are the regulation of genes controlling the biosynthesis of certain amino acids in gram-negative *Bacteria*. The first to be described was in the tryptophan operon in *Escherichia coli*, and we focus on it here. Because the processes of transcription and translation are spatially separated in eukaryotes, attenuation control is absent from *Eukarya*.

Attenuation in the Tryptophan Operon

The tryptophan operon contains structural genes for five proteins of the tryptophan biosynthetic pathway plus the usual promoter and regulatory sequences at the beginning of the operon (**Figure 7.32**). Like many operons, the tryptophan operon has more than one type of regulation. Transcription of the entire tryptophan operon is under negative control (Section 7.3). However, in addition to the promoter and operator regions needed for negative control, there is a sequence in the operon called the *leader sequence* that encodes a short polypeptide, the *leader peptide*. The leader sequence contains tandem tryptophan codons near its terminus and functions as an attenuator (Figure 7.32).

The basis of control of the tryptophan attenuator is as follows. If tryptophan is plentiful in the cell, there will be plenty of charged tryptophan tRNAs and the leader peptide will be synthesized. Synthesis of the leader peptide results in termination of

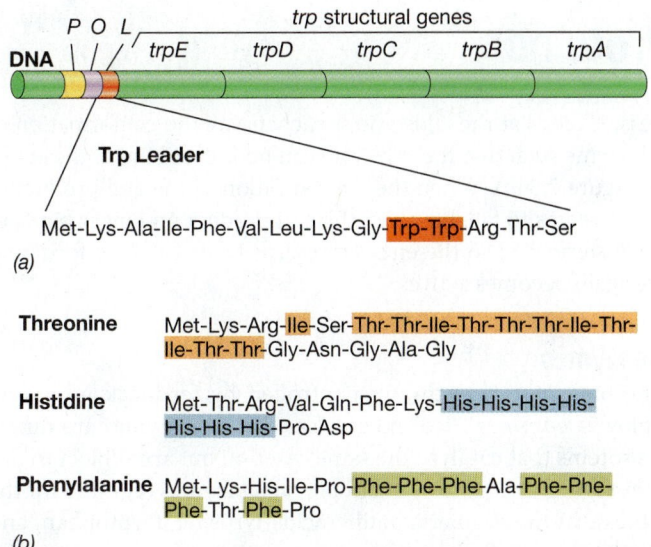

(a)

(b)

Figure 7.32 Attenuation and leader peptides in *Escherichia coli*. Structure of the tryptophan (*trp*) operon and of the tryptophan leader peptide and other leader peptides in *E. coli*. *(a)* Arrangement of the *trp* operon. Note that the leader (*L*) encodes a short peptide containing two tryptophan residues near its terminus (there is a stop codon following the Ser codon). The promoter is labeled *P*, and the operator is labeled *O*. The genes labeled *trpE* through *trpA* encode the enzymes needed for tryptophan synthesis. *(b)* Amino acid sequences of leader peptides of some other amino acid biosynthesis operons. Because isoleucine is made from threonine, it is an important constituent of the threonine leader peptide.

transcription of the remainder of the *trp* operon, which includes the structural genes for the biosynthetic enzymes. On the other hand, if tryptophan is scarce, the tryptophan-rich leader peptide will not be synthesized. If synthesis of the leader peptide is halted by a lack of tryptophan, the rest of the operon is transcribed.

Mechanism of Attenuation

How does translation of the leader peptide regulate transcription of the tryptophan genes downstream? Consider that in prokaryotic cells transcription and translation are simultaneous processes; as mRNA is released from the DNA, the ribosome binds to it and translation begins (⮌ Section 4.13). That is, while *transcription* of downstream DNA sequences is still proceeding, *translation* of already transcribed sequences is under way (**Figure 7.33**).

Transcription is attenuated because a portion of the newly formed mRNA folds into a unique stem–loop that inhibits RNA polymerase activity. The stem–loop structure forms in the mRNA because two stretches of nucleotides near each other are complementary and can thus base-pair. If tryptophan is plentiful, the ribosome translates the leader sequence until it comes to the leader stop codon. The remainder of the leader sequence then forms a stem–loop, a transcription pause site, which is followed by a uracil-rich sequence that actually causes termination (Figure 7.33*a*).

If tryptophan is limiting, transcription of genes encoding tryptophan biosynthetic enzymes is obviously desirable. During transcription of the leader, the ribosome pauses at a tryptophan codon because of a shortage of charged tryptophan tRNAs. The presence of the stalled ribosome at this position allows a stem–loop to

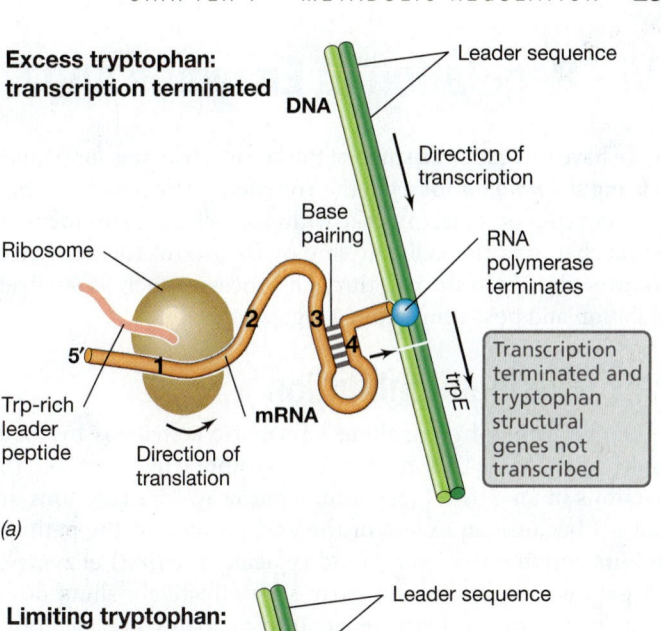

(a)

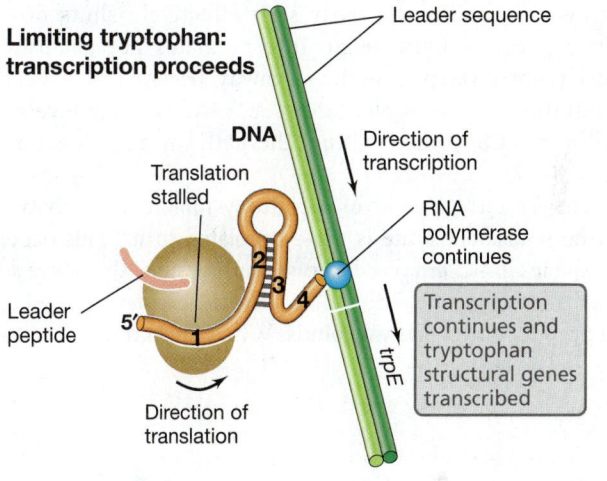

(b)

Figure 7.33 Mechanism of attenuation. Control of transcription of tryptophan (*trp*) operon structural genes by attenuation in *Escherichia coli*. The leader peptide is encoded by regions 1 and 2 of the mRNA. Two regions of the growing mRNA chain are able to form double-stranded loops, shown as 3:4 and 2:3. *(a)* When there is excess tryptophan, the ribosome translates the complete leader peptide, and so region 2 cannot pair with region 3. Regions 3 and 4 then pair to form a loop that terminates transcription. *(b)* If translation is stalled because of tryptophan starvation, a loop forms by pairing of region 2 with region 3, loop 3:4 does not form, and transcription proceeds past the leader sequence.

form (sites 2 and 3 in Figure 7.33*b*) that differs from the terminator stem–loop. This alternative stem–loop is not a transcription termination signal. Instead, it prevents the terminator stem–loop (sites 3 and 4 in Figure 7.33*a*) from forming. This allows RNA polymerase to move past the termination site and begin transcription of tryptophan structural genes. Thus, in attenuation control, the rate of transcription is influenced by the rate of translation.

MINIQUIZ

- Why does attenuation control not occur in eukaryotes?
- Explain how the formation of one stem–loop in the RNA can block the formation of another.

VI • Regulation of Enzymes and Other Proteins

We have just explored some of the key mechanisms for regulating the *amount* (or even the complete presence or absence) of an enzyme or other protein within a cell. Here we focus on the mechanisms the cell can employ to control the *activity* of enzymes already in the cell through processes such as feedback inhibition and post-translational regulation.

7.17 Feedback Inhibition

A major means of controlling enzymatic activity is by **feedback inhibition**. This mechanism temporarily shuts off the reactions in an entire biosynthetic pathway. The reactions are shut off because an excess of the end product of the pathway inhibits activity of an early (and typically the *first*) enzyme of the pathway. Inhibiting an early step effectively shuts down the entire pathway because no intermediates are generated for subsequent enzymes in the pathway (**Figure 7.34a**). Feedback inhibition is reversible, however, because once levels of the end product become limiting, the pathway again becomes functional.

How can the end product of a pathway inhibit the activity of an enzyme whose substrate is quite unrelated to it? This occurs because the inhibited enzyme has two binding sites, the *active site* (where substrate binds, ⮌ Section 3.5), and the *allosteric site*, where the end product of the pathway binds. When the end product is in excess, it binds at the allosteric site, changing the conformation of the enzyme such that the substrate can no longer bind at the active site (Figure 7.34b). When the concentration of the end product in the cell begins to fall, however, the end product no longer binds to the allosteric site, so the enzyme returns to its catalytic form and once again becomes active.

Isoenzymes

Some biosynthetic pathways controlled by feedback inhibition employ *isoenzymes* ("iso" means "same"). Isoenzymes are different proteins that catalyze the same reaction but are subject to different regulatory controls. Examples are enzymes required for the synthesis of the aromatic amino acids tyrosine, tryptophan, and phenylalanine in *Escherichia coli* (Figure 7.34c).

The enzyme 3-deoxy-D-arabino-heptulosonate 7-phosphate (DAHP) synthase plays a central role in aromatic amino acid biosynthesis. In *E. coli*, three DAHP synthase isoenzymes catalyze the first reaction in this pathway, each regulated independently by a different one of the end product amino acids. However, unlike the example of feedback inhibition where an end product completely inhibits enzyme activity, enzyme activity is diminished incrementally; enzyme activity falls to zero only when *all three* end products are present in excess (Figure 7.34c).

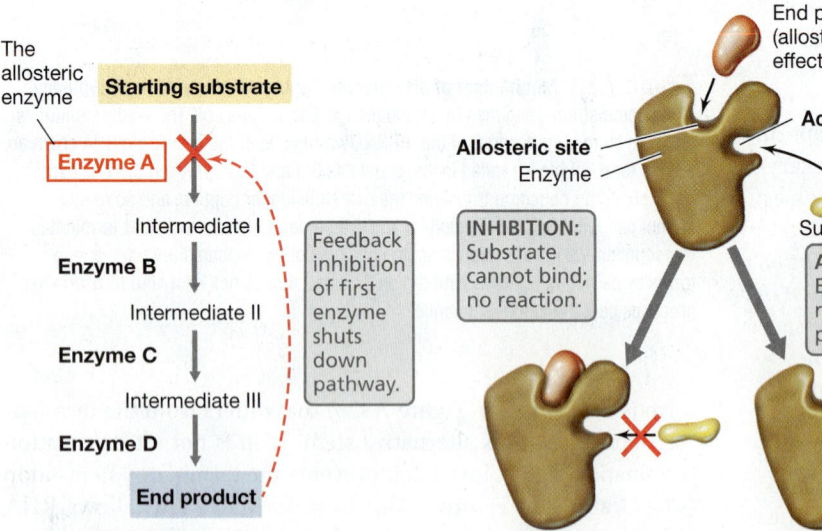

(a) Feedback inhibition

(b) Allosteric inhibition

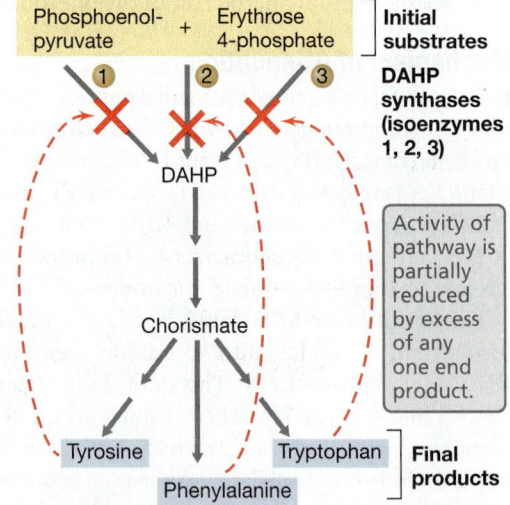

(c) Isoenzyme inhibition

Figure 7.34 Inhibition of enzyme activity.
(a) In feedback inhibition, the activity of the first enzyme of the pathway is inhibited by the end product, thus shutting off the production of the three intermediates and the end product. *(b)* The mechanism of allosteric inhibition by the end product of a pathway. When the end product binds at the allosteric site, the conformation of the enzyme is so altered that the substrate can no longer bind to the active site. However, inhibition is reversible, and end product limitation will once again activate the enzyme. *(c)* Inhibition by isoenzymes. In *Escherichia coli*, the pathway leading to the synthesis of the aromatic amino acids contains three isoenzymes of DAHP synthase. Each of these enzymes is feedback-inhibited by one of the aromatic amino acids. However, note how an excess of all three amino acids is required to completely shut off the synthesis of DAHP. In addition to feedback inhibition at the DAHP site, each amino acid feedback inhibits its further metabolism at the chorismate step.

MINIQUIZ

• What is feedback inhibition?

• What is the difference between an allosteric site and an active site?

7.18 Post-Translational Regulation

Some enzymes are regulated by covalent modification, typically the attachment or removal of some small molecule to the protein that subsequently affects enzyme activity. We have already discussed *phosphorylation*, a very common mechanism for regulating a protein post-translationally, when we considered two-component regulatory systems (Section 7.7). Biosynthetic enzymes can also be regulated by the attachment of other small molecules, such as the nucleotides adenosine monophosphate (AMP) and adenosine diphosphate (ADP), or by methylation. We consider here a single example of the well-studied case of glutamine synthetase, a key enzyme in ammonia (NH_3) assimilation (↻ Section 3.15), whose activity is modulated by AMP in a process called *adenylylation*.

Regulation of Glutamine Synthetase Activity

Each molecule of glutamine synthetase (GS) is composed of 12 identical subunits, and each subunit can be adenylylated. When GS is fully adenylylated (that is, each molecule of the enzyme contains 12 AMP groups), it is catalytically inactive. When it is partially adenylylated, it is partially active. As the glutamine pool in the cell increases, GS becomes more adenylylated, and its activity diminishes. As glutamine levels diminish, GS becomes less adenylylated and its activity increases (**Figure 7.35**). Other enzymes in the cell add and remove the AMP groups from GS, and these enzymes are themselves controlled, ultimately by levels of NH_3 in the cell.

Why should there be all of this elaborate regulation surrounding the enzyme GS? The activity of GS requires ATP, and nitrogen assimilation is a major biosynthetic process in the cell. However, when NH_3 is present at high levels in the cell, it can be assimilated into amino acids by enzymes that do not consume ATP; under these conditions, GS remains inactive. When NH_3 levels are low, however, GS becomes catalytically active. By having GS active only when NH_3 is limiting, the cell conserves ATP that would be used unnecessarily if GS were active when NH_3 was present in excess.

The modulation of GS activity in this very precise way stands in contrast to enzymes subject to feedback inhibition (Figure 7.34) whose activity is either "on" or "off" depending on the concentration of the effector molecule. This finer type of control for GS allows GS to remain partially active until NH_3 is at such high levels that NH_3-assimilating systems that have a lower affinity for NH_3 and do not require ATP are fully active.

Other Examples of Post-Translational Regulation

Throughout this chapter we have encountered other avenues in which the cell regulates the activity of proteins. One such mechanism is protein–protein interactions. In Section 7.10 we described how the σ factor RpoH is inactivated by DnaK under normal temperature conditions in the heat shock response (Figure 7.24). In addition, regulation by protein–protein interaction occurs during sporulation in *Bacillus* when the anti-sigma factor SpoIIAB binds

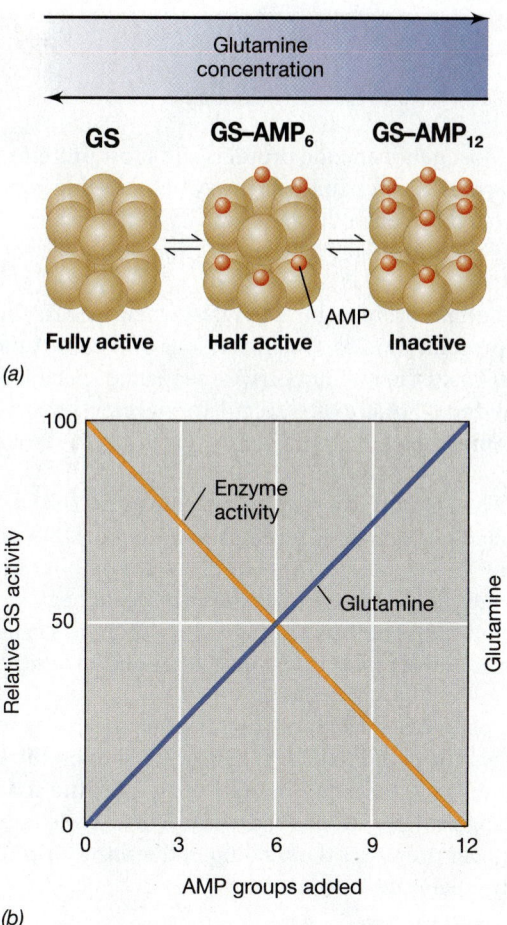

(a)

(b)

Figure 7.35 Regulation of glutamine synthetase by covalent modification. *(a)* When cells are grown with excess ammonia (NH_3), glutamine synthetase (GS) is covalently modified by adenylylation; as many as 12 AMP groups can be added. When cells are NH_3-limited, the groups are removed and form ADP. *(b)* Adenylylated GS subunits are catalytically inactive, so the overall GS activity decreases progressively as more subunits are adenylylated.

to σF, thereby preventing its association with RNA polymerase (Section 7.11 and Figure 7.25). Heterocyst formation is also controlled in part by post-translational regulation. In vegetative cells, the PatS peptide prevents HetR from activating transcription of heterocyst-formation genes (Section 7.13 and Figure 7.28).

Protease enzymes can also be employed to quickly remove damaged proteins from the cellular pool while molecular chaperones do just the opposite by refolding denatured proteins. However, regardless of the mechanism, in the final analysis it should be clear that regulating the synthesis and activities of a cell's proteins is (1) very important to its biology, (2) possible in many different ways, and (3) a major genetic investment. But the costs are worth it. At every turn in a highly competitive world, the very survival of a microorganism may well depend on its ability to conserve resources and energy output.

MINIQUIZ

• What is feedback inhibition?

• What does adenylylation do to the enzyme activity of glutamine synthetase?

BIG IDEAS

7.1 • Most genes encode proteins and most proteins are enzymes. Expression of an enzyme-encoding gene is regulated by controlling the activity of the enzyme or controlling the amount of enzyme produced.

7.2 • Certain proteins bind to DNA when specific domains of the proteins bind to specific regions of the DNA molecule. In most cases the interactions are sequence-specific. Proteins that bind to DNA are often regulatory proteins that affect gene expression.

7.3 • The amount of a specific enzyme in the cell can be controlled by regulatory proteins that bind to DNA and increase (induce) or decrease (repress) the amount of messenger RNA that encodes the enzyme. In negative control of transcription, the regulatory protein is called a repressor and it functions by inhibiting mRNA synthesis.

7.4 • Positive regulators of transcription are called activator proteins. They bind to activator-binding sites on the DNA and stimulate transcription. Inducers modify the activity of activating proteins. In positive control of enzyme induction, the inducer promotes the binding of the activator protein and thus stimulates transcription.

7.5 • Global control systems regulate the expression of many genes simultaneously. Catabolite repression is a global control system that helps cells make the most efficient use of available carbon sources. The *lac* operon is under the control of catabolite repression as well as its own specific negative regulatory system.

7.6 • *Archaea* resemble *Bacteria* in using DNA-binding activator and repressor proteins to regulate gene expression at the level of transcription.

7.7 • Signal transduction systems transmit environmental signals to the cell. In prokaryotes, signal transduction is typically carried out by a two-component regulatory system that includes a membrane-integrated sensor kinase and a cytoplasmic response regulator. The activity of the response regulator depends on its state of phosphorylation.

7.8 • Chemotactic behavior responds in a complex manner to attractants and repellents. The regulation of chemotaxis affects the activity of proteins rather than their synthesis. Adaptation by methylation allows the system to reset itself to the continued presence of a signal.

7.9 • Quorum sensing allows cells to monitor their environment for cells of their own kind. Quorum sensing depends on the sharing of specific small molecules known as autoinducers. Once a sufficient concentration of the autoinducer is present, specific gene expression is triggered.

7.10 • Cells can control sets of genes by employing alternative sigma factors. These recognize only certain promoters and thus allow transcription of a select category of genes that is appropriate under certain environmental conditions. Cells respond to both heat and cold by expressing sets of genes whose products help the cell overcome stress.

7.11 • Sporulation in *Bacillus* during adverse conditions is triggered via a complex phosphotransfer relay system that monitors multiple aspects of the environment. The sporulation factor Spo0A then sets in motion a cascade of regulatory responses under the control of several alternative sigma factors.

7.12 • Differentiation in *Caulobacter* consists of the alternation between motile cells and those that are attached to surfaces. Three major regulatory proteins—CtrA, GcrA, and DnaA—act in succession to control the three phases of the cell cycle. Each in turn controls many other genes needed at specific times in the cell cycle.

7.13 • Heterocyst formation requires expression of the major regulatory protein HetR in the protoheterocysts. However, the protein must be inactivated in vegetative cells by diffusion of the PatS peptide along the filament.

7.14 • Cells can control genes in several ways by employing regulatory RNA molecules. One way is to take advantage of base pairing and use sRNA to promote or prevent translation of mRNAs.

7.15 • Riboswitches are RNA domains at the 5′ ends of mRNA that recognize small molecules and respond by changing their three-dimensional structure to affect translation or transcriptional termination of the mRNA. Riboswitches are mostly used to control biosynthetic pathways for amino acids, purines, and a few other metabolites.

7.16 • Attenuation is a mechanism whereby transcription is controlled after initiation of mRNA synthesis. Attenuation mechanisms depend upon alternative stem–loop structures in the mRNA.

7.17 • In feedback inhibition, an excess of the final product of a biosynthetic pathway inhibits an allosteric enzyme at the beginning of the pathway. Enzyme activity can also be modulated by isoenzymes.

7.18 • Protein activity can be regulated after translation. Reversible covalent modification or interactions with other proteins can modulate protein activity.

REVIEW OF KEY TERMS

Activator protein a regulatory protein that binds to specific sites on DNA and stimulates transcription; involved in positive control

Allosteric protein a protein containing an active site for binding substrate and an allosteric site for binding an effector molecule such as the end product of a biochemical pathway.

Attenuation a mechanism for controlling gene expression that terminates transcription after initiation but before a full-length messenger RNA is produced

Autoinducer a small signal molecule that takes part in quorum sensing

Catabolite repression the suppression of alternative catabolic pathways by a preferred source of carbon and energy

Cyclic AMP a regulatory nucleotide that participates in catabolite repression

Domains regions of a protein with specific structure and function

Feedback inhibition a process in which an excess of the end product of a multistep pathway inhibits activity of the first enzyme in the pathway

Gene expression transcription of a gene followed by translation of the resulting mRNA into protein

Green fluorescent protein (GFP) a protein that fluoresces green and is widely used in genetic analysis

Heat shock proteins proteins induced by high temperature (or certain other stresses) that protect against high temperature, especially by refolding partially denatured proteins or by degrading them

Heat shock response response to high temperature that includes the synthesis of heat shock proteins together with other changes in gene expression

Induction production of an enzyme in response to a signal (often the presence of the substrate for the enzyme)

Negative control a mechanism for regulating gene expression in which a repressor protein prevents transcription of genes

Noncoding RNA (ncRNA) RNA that is not translated into protein; examples include ribosomal RNA, transfer RNA, and small regulatory RNAs

Operon one or more genes transcribed into a single RNA and under the control of a single regulatory site

Positive control a mechanism for regulating gene expression in which an activator protein functions to promote transcription of genes

Quorum sensing a regulatory system that monitors the population level and controls gene expression based on cell density

Regulatory nucleotide a nucleotide that functions as a signal rather than being incorporated into RNA or DNA

Regulon a series of operons controlled as a unit

Repression prevention of the synthesis of an enzyme in response to a signal

Repressor protein a regulatory protein that binds to specific sites on DNA and blocks transcription; involved in negative control

Response regulator protein one of the members of a two-component regulatory system; a protein that is phosphorylated by a sensor kinase and then acts as a regulator, often by binding to DNA

Riboswitch an RNA domain, usually in a messenger RNA molecule, that can bind a specific small molecule and alter its secondary structure; this, in turn, controls translation of the mRNA

Sensor kinase protein one of the members of a two-component regulatory system; a protein that phosphorylates itself in response to an external signal and then transfers the phosphoryl group to a response regulator protein

Signal transduction see two-component regulatory system

Two-component regulatory system a regulatory system consisting of two proteins: a sensor kinase and a response regulator

REVIEW QUESTIONS

1. What are the two points at which the amount of protein synthesis can be regulated? (Section 7.1)

2. Describe why a protein that binds to a specific sequence of double-stranded DNA is unlikely to bind to the same sequence if the DNA is single-stranded. (Section 7.2)

3. Most biosynthetic operons need only be under negative control for effective regulation, whereas most catabolic operons need to be under both negative and positive control. Why? (Sections 7.3 and 7.4)

4. What is the difference between an operon and a regulon? (Section 7.4)

5. Describe the mechanism by which cAMP receptor protein (CRP), the regulatory protein for catabolite repression, functions. Use the lactose operon as an example. (Section 7.5)

6. What are the two mechanisms used by archaeal repressor proteins to repress transcription? (Section 7.6)

7. What are the two components that give their name to a signal transduction system in prokaryotes? What is the function of each of the components? (Section 7.7)

8. Adaptation allows the mechanism controlling flagellar rotation to be reset. How is this achieved? (Section 7.8)

9. How can quorum sensing be considered a regulatory mechanism for conserving cell resources? (Section 7.9)

10. Describe the proteins produced when cells of *Escherichia coli* experience a heat shock. Of what value are they to the cell? (Section 7.10)

11. Explain how alternative sigma factors control sporulation in *Bacillus*. (Section 7.11)

12. What role does the DnaA protein play in differentiation in *Caulobacter*? (Section 7.12)

13. What molecule produced by heterocysts prevents differentiation in vegetative cells and how does the inhibitor reach the vegetative cells? (Section 7.13)

14. How does regulation by sRNA differ from that of riboswitches? (Sections 7.14 and 7.15)

15. Describe how transcriptional attenuation works. What is actually being "attenuated"? (Section 7.16)

16. Contrast regulation of DAHP synthase and glutamine synthetase. (Section 7.17)

17. What is the most common covalent modification to affect protein activity? (Section 7.18)

APPLICATION QUESTIONS

1. What would happen to regulation from a promoter under negative control if the region where the regulatory protein binds was deleted? What if the promoter was under positive control?

2. Promoters from *Escherichia coli* under positive control are not close matches to the promoter consensus sequence for *E. coli* (↺ Section 4.7). Why?

3. The attenuation control of the tryptophan biosynthetic pathway genes in *Escherichia coli* involves coupled transcription and translation. Can you describe why this mechanism of regulation would not be appropriate for regulating the genes involved in the utilization of lactose?

4. Most of the regulatory systems described in this chapter employ regulatory proteins. However, regulatory RNA is also important. Describe how one could achieve negative control of the *lac* operon using either of two different types of regulatory RNA.

5. Many amino acid biosynthetic operons under attenuation control are also under negative control. Considering that the environment of a bacterium can be highly dynamic, what advantage could be conferred by having attenuation as a second layer of control?

6. How would you design a regulatory system to make *Escherichia coli* use succinic acid in preference to glucose? How could you modify it so that *E. coli* prefers to use succinic acid in the light but glucose in the dark?

8 · Viruses and Virology

microbiology**now**

Where Did Viruses Come From?

Despite the fact that we know much about the molecular properties of viruses, how they co-opt cellular activities to their own advantage, and the diseases that many of them cause, we know little about viral origins. But this secret of the viral world may be beginning to show some cracks.[1]

It now seems likely that viruses arose before the appearance of the last universal common ancestor of cellular life (LUCA). This hypothesis has gained support from at least two sources. First, structural studies have shown that certain proteins in the viral shells (capsids) that surround viral genomes show significant structural homology across a wide variety of both RNA (photo, coronavirus) and DNA viruses. This suggests that despite their genomic diversity, viruses have distinct "lineages" that predate the origin of LUCA.

Second, it is clear that mutation rate and genome size are inversely related and that small RNA viruses have the highest of all known mutation rates. Moreover, single-stranded DNA viruses have lower mutation rates than single-stranded RNA viruses, and double-stranded DNA viruses have lower rates than single-stranded DNA viruses. The lowest rates of all are seen in cells, whose mutation rates are several orders of magnitude lower than those of viruses.

Perhaps Earth's first self-replicating entities resembled single-stranded RNA viruses that mutated early and often, gradually resulting in increased fitness. From these, mutants were selected that had evolved more stable (DNA-based) genomes, and from these, the first cells. Although still a hypothesis, such a process of increasing genomic stability might have been a major catalyst for the evolution of cells. If that is true, the study of viral capsid proteins and genomic mutation rates may have inadvertently provided important clues for scientists striving to understand how life began.

[1]Holmes, E.C. 2011. What does viral evolution tell us about virus origins? *J. Virol. 85:* 5247–5251.

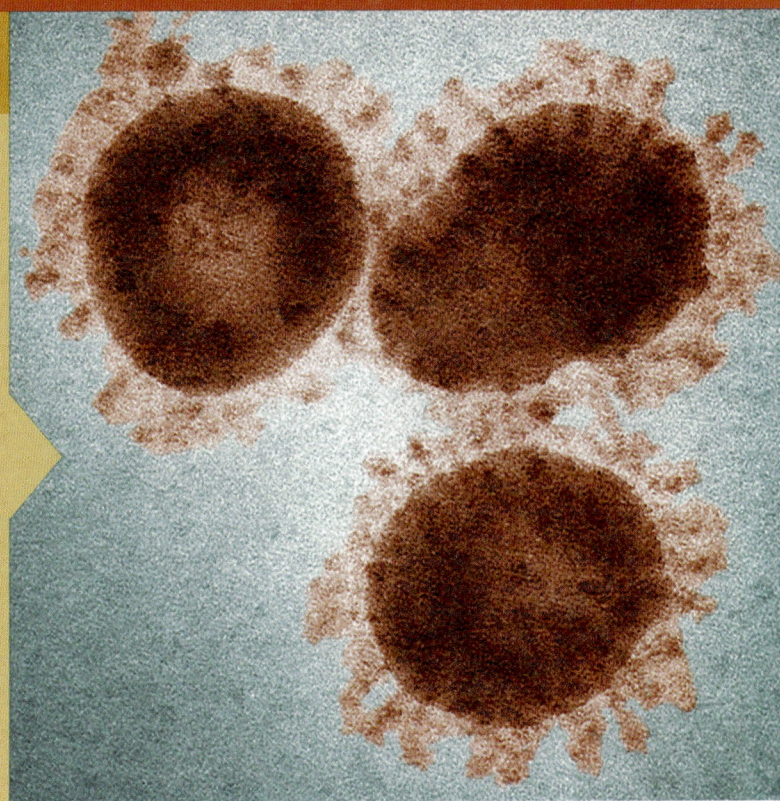

A **virus** is a genetic element that can replicate only inside a living cell, called the **host cell**. Viruses possess their own genomes and in that sense are independent of the host cell's genome. However, viruses rely on the host cell for energy, metabolic intermediates, and protein synthesis. Viruses are therefore *obligate intracellular parasites*.

Viruses infect both prokaryotes and eukaryotes and are responsible for many infectious diseases of humans and other organisms. The study of viruses is called *virology*, and this chapter covers the basic principles of the science. In Chapter 9 we consider the genomic and diversity aspects of viruses in more detail.

I • The Nature of Viruses

8.1 What Is a Virus?

Although viruses are not cells they nonetheless possess a nucleic acid genome that encodes those functions needed to replicate and an extracellular form, called the **virion**, that allows the virus to travel from one host cell to another. Viruses cannot replicate unless the virion itself (or its genome, in the case of bacterial viruses) has gained entry into a suitable host cell, a process called *infection*.

Viral Structure and Activities

The virion of a virus consists of a protein shell, the **capsid**, which contains the virus genome. Most bacterial viruses are *naked*, with no further layers, whereas many animal viruses contain an outer layer consisting of protein plus lipid called the *envelope* (**Figure 8.1**). In enveloped viruses, the inner structure of nucleic acid plus capsid protein is called the **nucleocapsid**. The virion protects the viral genome when the virus is outside the host cell, and proteins on the virion surface are important in attaching it to its host cell. The virion may also contain one or more virus-specific enzymes that play a role during infection and replication, as discussed later.

Once inside the host cell, a viral genome can orchestrate one of two quite different events. The virus may replicate and destroy the host in a **virulent** (**lytic**) infection. In a lytic infection, the virus redirects host metabolism to support virus replication and the assembly of new virions. Eventually, new virions are released, and the process can repeat itself on new host cells. Alternatively, some viruses can undergo a *lysogenic* infection; in this case, the host cell is not destroyed but is genetically altered because the viral genome becomes part of the host genome.

Viral Genomes

All cells contain double-stranded DNA genomes. By contrast, viral genomes consist of either DNA *or* RNA and are further subdivided based on whether the genome is *single-stranded* or *double-stranded*. A very few highly unusual viruses use both DNA and RNA as genetic material, but at different stages of their life cycle (**Figure 8.2**).

Viral genomes can be either linear or circular, and single-stranded viral genomes may be of either the *plus sense* or *minus sense* in terms of their base sequence. Viral genomes of the plus configuration have the *exact same* base sequence as that of the viral mRNA that will be translated to form viral proteins. By contrast, viral genomes of the minus configuration are *complementary* in base sequence to viral mRNA. This interesting feature of viral genomes requires special genetic information flow processes, and we reserve our discussion of the details of these processes to Chapter 9.

Viral genomes are typically smaller than those of cells. The smallest bacterial genome known is about 145 kilobase pairs, encoding about 170 genes. Most viral genomes encode from a few up to about 350 genes. The smallest viral genomes are those of some small RNA viruses that infect animals. The genomes of these tiny viruses contain fewer than 2000 nucleotides and only two genes. A few very large viral genomes are known, such as the 1.25-Mbp DNA genome of a marine virus called *Megavirus*, which infects protozoans. RNA viruses typically have the smallest genomes and only DNA viruses have genomes encoding more than 40 genes.

Viruses can be classified on the basis of the hosts they infect as well as by their genome structure. Thus, we have bacterial viruses, archaeal viruses, animal viruses, plant viruses, protozoan viruses, and so on. Bacterial viruses are called **bacteriophages** (or simply *phage* for short) and have been intensively studied as model systems for the molecular biology and genetics of virus replication. In this chapter we will use bacteriophages many times to illustrate simple viral principles. Indeed, many of the basic tenets of virology were first discovered with bacteriophages and subsequently applied to viruses of higher organisms. Because of their frequent medical importance, animal viruses have been extensively studied, whereas plant viruses, although of enormous importance to modern agriculture, have been less well studied.

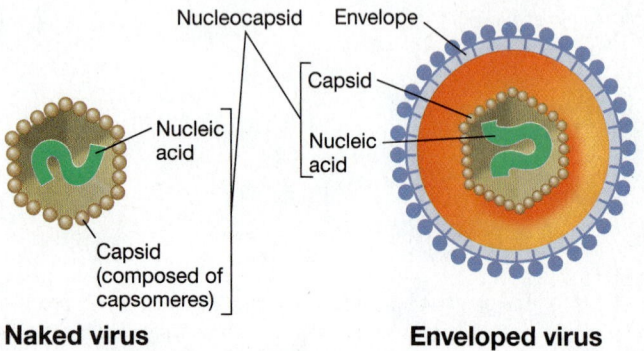

Figure 8.1 Comparison of naked and enveloped virus particles. The envelope originates from host cytoplasmic membrane.

MINIQUIZ
- How does a virus differ from a cell?
- Why does a virus need a host cell?
- Compared with cells, what is unusual about viral genomes?

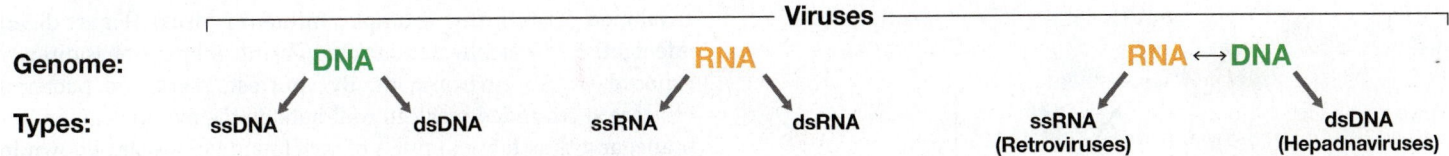

Figure 8.2 Viral genomes. The genomes of viruses can be either DNA or RNA, and some use both at different stages in their replication cycle. However, only one type of genomic nucleic acid is found in the virion of any particular type of virus. Viral genomes can be single-stranded (ss) or double-stranded (ds) and circular or linear.

UNIT 2

8.2 Structure of the Virion

Virions come in many shapes and sizes. Most viruses are smaller than prokaryotic cells, ranging in size from 0.02 to 0.3 μm (20–300 nanometers, nm). Smallpox virus, one of the larger viruses, is about 200 nm in diameter, which is about the size of the smallest known bacterial cells. Poliovirus, one of the smallest viruses, is only 28 nm in diameter, which is about the size of a ribosome, the cell's protein-synthesizing machine.

Virion Structure

The structures of virions are quite diverse, varying widely in size, shape, and chemical composition (see Figures 8.19 and 8.21). The nucleic acid of a virion is always surrounded by its capsid (Figure 8.1). The capsid is composed of a number of individual protein molecules called **capsomeres** that are arranged in a precise and highly repetitive pattern around the nucleic acid.

The small size of most viral genomes restricts the number of distinct viral proteins that can be encoded. As a consequence, a few viruses have only a single kind of protein in their capsid. An example is the well-studied tobacco mosaic virus (TMV), which causes disease in tobacco, tomato, and related plants. TMV is a single-stranded RNA virus in which the 2130 copies of the simple capsomere protein are arranged in a helix with dimensions of 18 × 300 nm (**Figure 8.3**).

The information required for the proper folding and assembly of viral proteins into capsomeres and subsequently into capsids is often embedded within the amino acid sequence of the viral proteins themselves. When this is the case, virion assembly is a spontaneous process and is called *self-assembly*. However, some virus proteins and structures require assistance from host cell folding proteins for proper folding and assembly. For example, the capsid protein of bacteriophage lambda (Section 8.8) requires assistance from the *Escherichia coli* chaperonin GroE (⮌ Section 4.14) in order to fold into its active conformation.

Virus Symmetry

Viruses are highly symmetric. When a symmetric structure is rotated around an axis, the same form is seen again after a certain number of degrees of rotation. Two kinds of symmetry are recognized in viruses, which correspond to the two primary viral shapes, rod and spherical. Rod-shaped viruses have *helical* symmetry while spherical viruses have *icosahedral* symmetry. A typical virus with helical symmetry is TMV (Figure 8.3). The lengths of helical viruses are determined by the length of the nucleic acid, and the width of the helical virion is determined by the size and packaging of the capsomeres.

Viruses with icosahedral symmetry contain 20 triangular faces and 12 vertices and are roughly spherical in shape (**Figure 8.4a**). Axes of symmetry divide the icosahedron into 5, 3, or 2 segments of identical size and shape (Figure 8.4b). Icosahedral symmetry is the most efficient arrangement of subunits in a closed shell because it requires the smallest number of capsomeres to build the shell. The simplest arrangement of capsomeres is 3 per triangular face, for a total of 60 capsomeres per virion. However, most viruses have more nucleic acid than can be packed into a shell made from 60 capsomeres and so viruses with 180, 240, or 360 capsomeres are more common. The capsid of the human papillomavirus virus (Figure 8.4c), for example, consists of 360 capsomeres, with the capsomeres arranged into 72 clusters of 5 each (Figure 8.4d).

The structure of some viruses is extremely complex, with the virion consisting of several parts each displaying its own shape and symmetry. The most complex of all viruses are the head-plus-tail

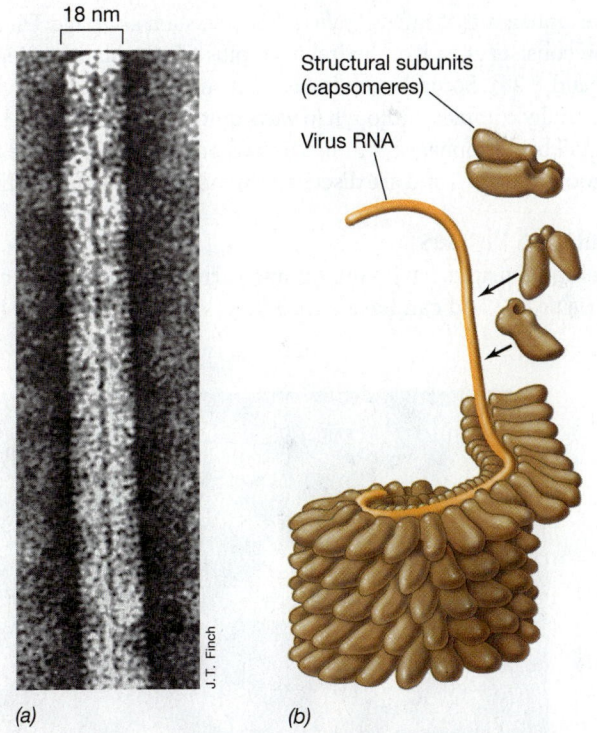

18 nm

Structural subunits (capsomeres)

Virus RNA

J.T. Finch

(a)   *(b)*

Figure 8.3 The arrangement of RNA and protein coat in a simple virus, tobacco mosaic virus. *(a)* A high-resolution electron micrograph of a portion of the tobacco mosaic virus particle. *(b)* Cutaway showing structure of the virion. The RNA forms a helix surrounded by the protein subunits (capsomeres). The center of the virus particle is hollow.

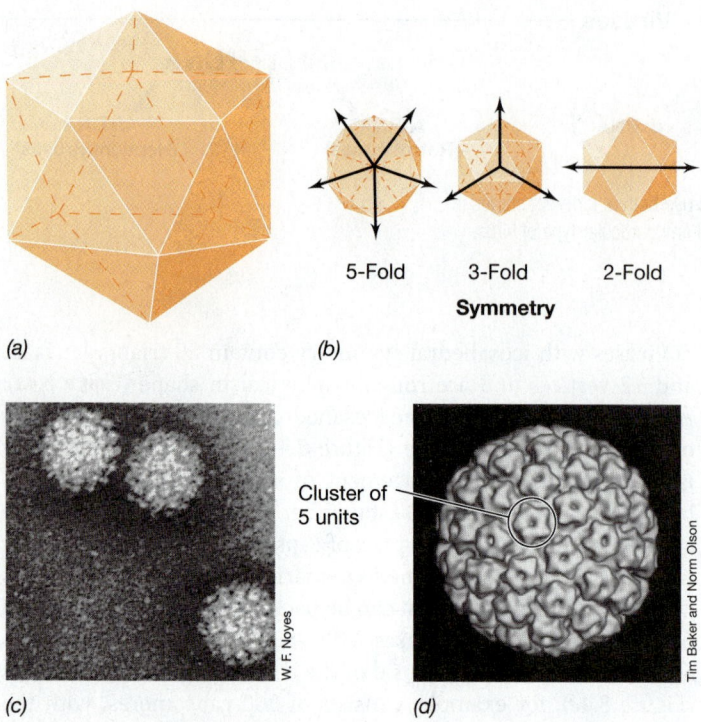

(a)

(b)

5-Fold   3-Fold   2-Fold

Symmetry

Cluster of
5 units

W. F. Noyes

Tim Baker and Norm Olson

(c)   (d)

Figure 8.4 **Icosahedral symmetry.** *(a)* Model of an icosahedron. *(b)* Three views of an icosahedron showing 5-, 3-, or 2-fold symmetry. *(c)* Electron micrograph of human papillomavirus, a virus with icosahedral symmetry. The virion is about 55 nm in diameter. *(d)* Three-dimensional reconstruction of human papillomavirus; virion contains 360 units arranged in 72 clusters of 5 each.

bacteriophages that infect *Escherichia coli*, such as phage T4. A T4 virion consists of an icosahedral head plus a helical tail (see Figures 8.19 and 8.20). Some large viruses that infect eukaryotes are also structurally complex, although in ways quite distinct from the head-plus-tail bacteriophages. Mimivirus and pox virus (see Figure 8.5*b*) are good examples and are discussed in more detail in Chapter 9.

Enveloped Viruses

Enveloped viruses have a membrane surrounding the nucleocapsid (**Figure 8.5**) and can have either RNA or DNA genomes. Most enveloped viruses (for example, influenza virus) (Figure 8.5*a*) infect animal cells in which the cytoplasmic membrane is directly exposed to the environment. By contrast, plant and bacterial cells are surrounded by a cell wall outside the cytoplasmic membrane, and thus few examples of enveloped viruses are known in these organisms. Typically, the entire virion enters an animal cell during infection, with the envelope, if present, assisting in the infection process by fusing with the host membrane. Enveloped viruses also exit more easily from animal cells. As they pass out of the host cell, they are draped in membrane material. The viral envelope consists primarily of host cytoplasmic membrane, but some viral surface proteins become embedded in the envelope as the virus passes out of the cell.

The viral envelope is important in infection as it is the component of the virion that makes contact with the host cell. The specificity of enveloped virus infection and some aspects of their penetration are thus controlled in part by the chemistry of their envelopes. The virus-specific envelope proteins are critical for both attachment of the virion to the host cell during infection and for release of the virion from the host cell after replication.

Enzymes Inside Virions

Viruses do not carry out metabolic processes and are thus metabolically inert. Nonetheless, some viruses carry enzymes in their virions that play important roles in infection. For example, some bacteriophages contain an enzyme that resembles lysozyme (↩ Section 2.10), which is used to make a small hole in the bacterium's peptidoglycan to allow nucleic acid from the virion to get into the host cytoplasm. A similar protein is produced in the later stages of infection to lyse the host cell and release new virions. Some animal viruses also contain enzymes that aid in their release from the host. For example, influenza virus (Figure 8.5*a*) has envelope proteins called *neuraminidases* that destroy glycoproteins and glycolipids of animal cell connective tissue, thus liberating the virions (↩ Section 9.9).

RNA viruses carry their own nucleic acid polymerases (called *RNA replicases*) that function to replicate the viral RNA genome and produce viral-specific mRNA. Such enzymes are necessary because cells cannot make RNA from an RNA template.

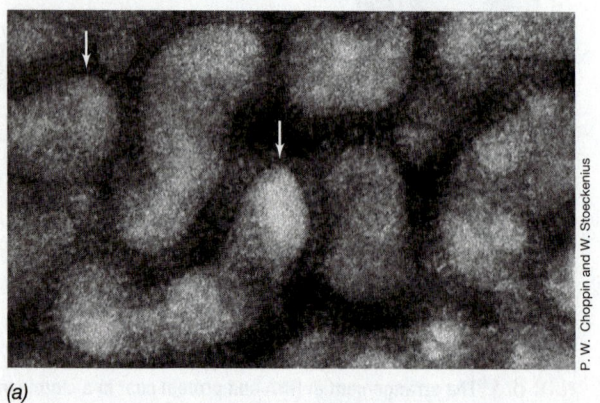

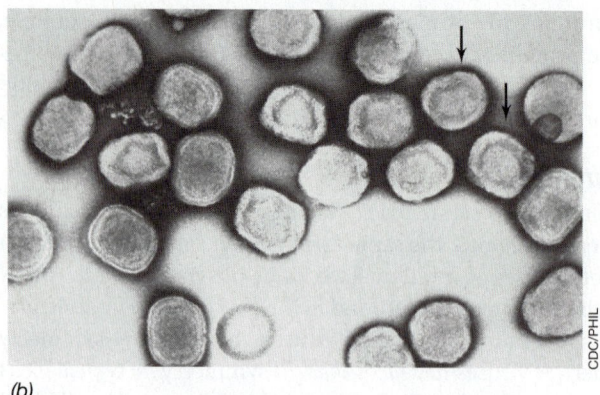

P. W. Choppin and W. Stoeckenius

CDC/PHIL

(a)   (b)

Figure 8.5 **Enveloped viruses.** *(a)* Electron micrograph of influenza virus. The virions are about 80 nm in diameter, and can have many shapes. *(b)* Electron micrograph of vaccinia virus, an enveloped icosahedral pox virus about 350 nm wide. The arrows in both micrographs point to the envelopes surrounding the nucleocapsids.

Retroviruses are unusual RNA animal viruses that replicate via DNA intermediates. Because making DNA from an RNA template is another process cells cannot do, retroviral virions contain an RNA-dependent DNA polymerase called *reverse transcriptase* (Section 8.10). So, although most viruses do not need to carry special enzymes in their virions, those that do absolutely require them for successful infection and replication.

MINIQUIZ

- Distinguish between a capsid and a capsomere. What is a common symmetry for spherical viruses?

- What is the difference between a naked virus and an enveloped virus?

- What kinds of enzymes can be found within the virions of RNA viruses? Why are they there?

8.3 Overview of the Virus Life Cycle

For a virus to replicate, it must induce a living host cell to synthesize all the essential components needed to make new virions. Because of these biosynthetic and energy requirements, dead host cells will not replicate viruses. During an active infection, viral components are assembled into new virions that are released from the cell. We use replication of a bacterial virus here as a simple example of a viral life cycle.

A cell that supports the complete replication cycle of a virus is said to be *permissive* for that virus. In a permissive host, the viral replication cycle can be divided into five steps (**Figure 8.6**).

1. *Attachment* (adsorption) of the virion to the host cell

2. *Penetration* (entry, injection) of the virion nucleic acid into the host cell

3. *Synthesis* of virus nucleic acid and protein by host cell machinery as redirected by the virus

4. *Assembly* of capsids and *packaging* of viral genomes into new virions

5. *Release* of new virions from the cell

The growth response during virus replication is illustrated in **Figure 8.7**. The response takes the form of a *one-step growth curve*, so named since a time course of virion numbers in the culture medium shows essentially no increase during the replication

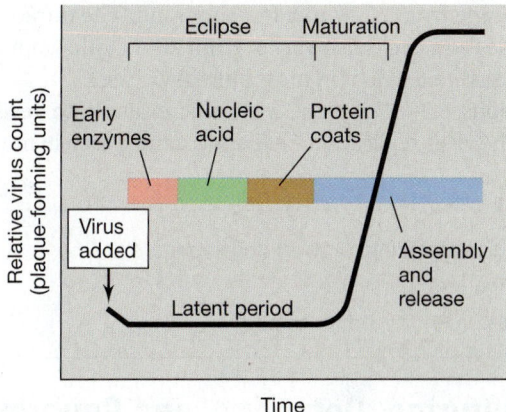

Figure 8.7 **One-step growth curve of virus replication.** Following adsorption, infectious virions cannot be detected in the growth medium, a phenomenon called *eclipse*. During the latent period, which includes the eclipse and early maturation phases, viral nucleic acid replicates and protein synthesis occurs. During the maturation period, virus nucleic acid and protein are assembled into mature virions and then released.

cycle until cells burst and release their newly synthesized virions. In the first few minutes after infection the virus enters the *eclipse* phase. Once attached to a permissive host cell, a virion is no longer available to infect another cell. This is followed by the entry of viral nucleic acid into the host cell (Figure 8.6). If the infected cell breaks open at this point, the virion no longer exists as an infectious entity since the viral genome is no longer inside its capsid.

The *maturation* phase (Figure 8.7) begins as newly synthesized viral nucleic acid molecules become packaged inside their capsids. During the maturation phase, the number of infectious virions inside the host cell rises dramatically. However, the new virions still cannot be detected in the culture medium unless the cells are artificially lysed to release them. Because newly assembled virions are not yet present outside the cell, the eclipse and maturation periods together comprise the *latent period* of viral infection (Figure 8.7).

At the end of maturation, mature virions are released, either as a result of cell lysis or by budding or excretion, depending on the virus. The number of virions released per cell, called the *burst size*, varies with the particular virus and the particular host cell,

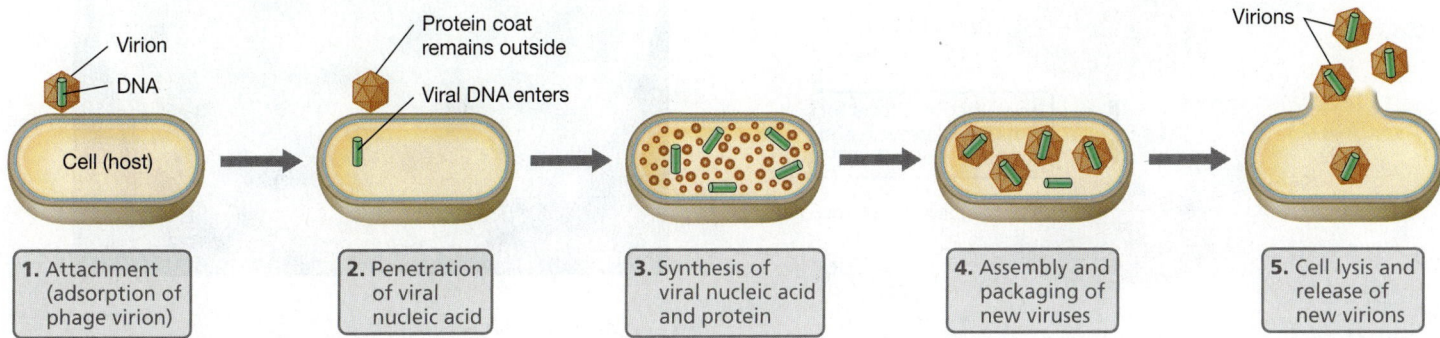

Figure 8.6 **The replication cycle of a bacterial virus.** The virions and cells are not drawn to scale. The burst size can be a hundred or more virions per host cell.

and can range from a few to a few thousand. The duration of the virus replication cycle also varies, from 20–60 min (in many bacterial viruses) to 8–40 h (in most animal viruses).

In Sections 8.5–8.7 we use a specific example to revisit these stages of the virus replication cycle and examine each in more detail.

MINIQUIZ

• What is packaged into capsids during maturation?
• Explain the term burst size.
• Why is the latent period so named?

8.4 Culturing, Detecting, and Counting Viruses

Host cells need to be grown in order for viruses to replicate in them. Pure cultures of bacterial hosts are grown either in liquid or as "lawns" on the surface of agar plates and then inoculated with a virus suspension. Animal viruses are cultivated in *tissue cultures*, which are cells obtained from an animal organ and grown in sterile glass or plastic vessels containing an appropriate culture medium (see Figure 8.9). Tissue culture media are often highly complex, containing a wide assortment of nutrients including blood serum and antimicrobial agents to prevent bacterial contamination.

Detecting and Counting Viruses: The Plaque Assay

A viral suspension can be quantified to determine the number of infectious virions present per volume of fluid, a quantity called the **titer**. This is typically done using a *plaque assay*. When a virus infects host cells growing on a flat surface, a zone of cell lysis called a **plaque** forms and appears as a clear area in the

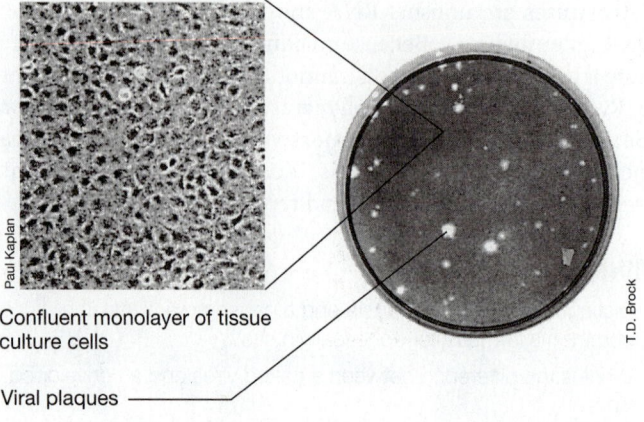

Confluent monolayer of tissue culture cells

Viral plaques

Figure 8.9 Animal cell cultures and viral plaques. The animal cells support replication of the virus, and lysed cells result in plaques.

lawn of host cells. With bacteriophages, plaques may be obtained when virions are mixed into a small volume of molten agar containing host bacteria that is spread on the surface of an agar medium (**Figure 8.8a**). During incubation, the bacteria grow and form a turbid layer (lawn) that is visible to the naked eye. However, wherever a successful viral infection has occurred, cells are lysed, forming a plaque (Figure 8.8b). By counting the number of plaques, one can calculate the titer of the virus sample (often expressed as "plaque-forming units" per milliliter). For replicating animal viruses, a tissue culture is grown and a diluted virus suspension overlaid upon it. As for bacterial viruses, plaques are revealed as cleared zones in the tissue culture cell layer, and from the number of plaques produced, an estimate of the virus titer can be made (**Figure 8.9**).

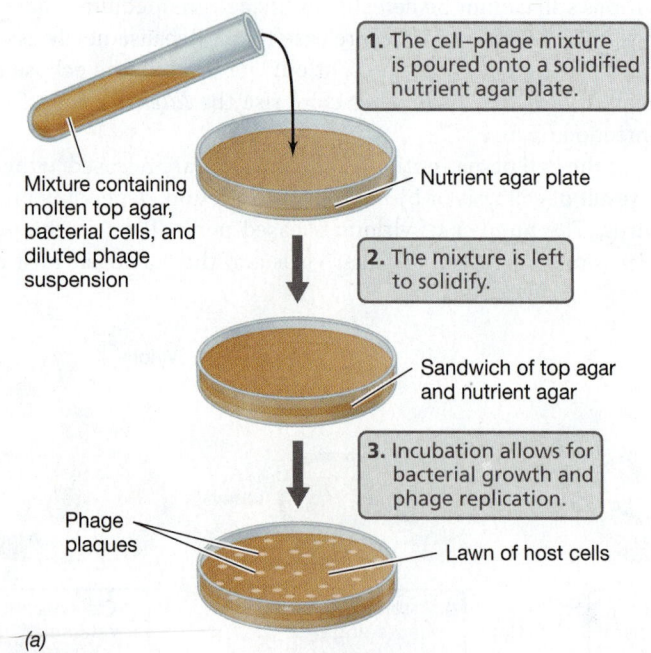

Mixture containing molten top agar, bacterial cells, and diluted phage suspension

1. The cell–phage mixture is poured onto a solidified nutrient agar plate.

Nutrient agar plate

2. The mixture is left to solidify.

Sandwich of top agar and nutrient agar

3. Incubation allows for bacterial growth and phage replication.

Phage plaques

Lawn of host cells

(a)

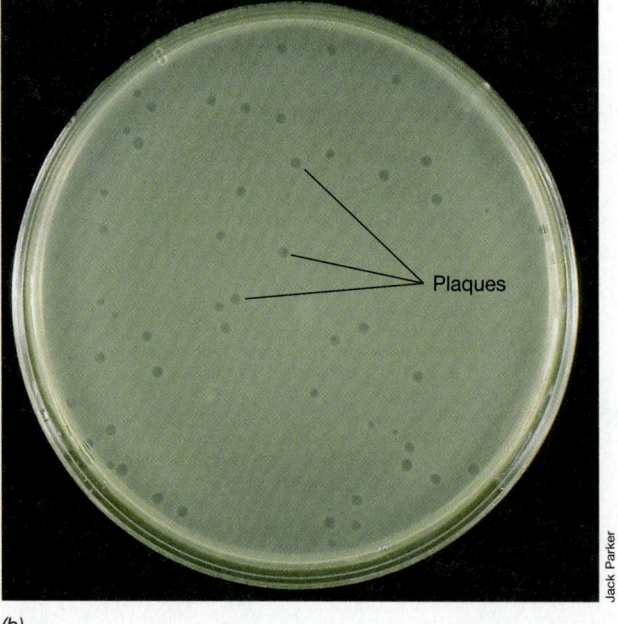

Plaques

(b)

Figure 8.8 Quantification of bacterial virus by plaque assay. (a) "Top agar" containing a dilution of virions mixed with permissive host bacteria is poured over a plate of "bottom agar." Infected cells are lysed, forming plaques in the lawn. (b) Plaques (about 1–2 mm in diameter) formed by bacteriophage T4.

The concept of *plating efficiency* is important in quantitative virology, whether of bacterial or animal viruses. In any given viral preparation, the number of plaque-forming units is always lower than actual counts of viral particles made microscopically (using an electron microscope). This is because the efficiency with which virions infect host cells is rarely 100% and may often be considerably less. Virions that fail to infect may have assembled incompletely during the maturation process or may contain defective genomes. Alternatively, a low plating efficiency may mean that viral growth conditions were not optimal or that some virions were damaged by handling or storage conditions. Although with bacterial viruses

the efficiency of plating is often higher than 50%, with many animal viruses it may be much lower, 0.1% or 1%. Knowledge of plating efficiency is useful in cultivating viruses because it allows one to estimate what a titer needs to be to yield a certain number of plaques.

MINIQUIZ --------------------------------

- What is meant by a viral titer?
- What is a plaque-forming unit?
- What is meant by the term plating efficiency?

II • Bacteriophage Life Cycles

Much of our understanding of lytic virus replication comes from the study of bacteriophages infecting *Escherichia coli*. Many RNA as well as DNA bacteriophages replicate in *E. coli* (Table 8.1). Here we choose one, bacteriophage T4, as our model for reviewing the individual stages of the virus life cycle (Figure 8.6) in more detail.

8.5 Attachment and Entry of Bacteriophage T4

The early steps in the life cycle of any bacteriophage are attachment to the surface of its host cell followed by penetration of the host cell outer layer(s) and entry of the viral genome into the cell.

Attachment

A major factor in host specificity of a virus is *attachment*. The virion itself has one or more proteins on its external surface that interact with specific host cell surface components called *receptors*. In the absence of its specific receptor, the virus cannot attach to the cell and hence cannot infect. Moreover, if the receptor is altered, for example by mutation, the host may become resistant to virus infection. The host range of a given virus is thus to a major extent determined by the presence of a suitable receptor that the virus can recognize and attach to.

Viral receptors are surface components of the host, such as proteins, carbohydrates, glycoproteins, lipids, or lipoproteins, or

cell structures made from these macromolecules (**Figure 8.10**). The receptors carry out normal functions for the cell; for example, the receptor for phage T1 is an iron-uptake protein (Figure 8.10) and that for bacteriophage lambda functions in maltose uptake. Carbohydrates in the lipopolysaccharide (LPS) outer membrane of gram-negative bacteria are the receptors recognized by bacteriophage T4, a phage that binds to the LPS of *Escherichia coli* (Figure 8.10). Appendages that project from the cell surface, such as flagella and pili, are also common receptors for bacterial viruses. Small icosahedral viruses often bind to the side of these structures, whereas filamentous bacteriophages typically bind at the tip, such as on the pilus (Figure 8.10). Regardless of the receptor used, however, once attachment has occurred, the stage is set for viral infection.

Penetration

Attachment of a virus to its host cell causes changes to both the virus and the host cell surface that result in penetration. Bacteriophages abandon the capsid outside the cell and only the viral genome reaches the cytoplasm. However, entry of the viral genome into a host cell only results in virus replication if the viral genome can be read. Consequently, for the replication of some viruses, for example RNA viruses, specific viral proteins must also enter the host cell along with the viral genome (Section 8.2).

Table 8.1 Some bacteriophages of *Escherichia coli*				
Bacteriophage	Virion structure	Genome composition[a]	Genome structure	Size of genome[b]
MS2	Icosahedral	ssRNA	Linear	3,600
φX174	Icosahedral	ssDNA	Circular	5,400
M13, f1, and fd	Filamentous	ssDNA	Circular	6,400
Lambda	Head & tail	dsDNA	Linear	48,500
T7 and T3	Head & tail	dsDNA	Linear	40,000
T4	Head & tail	dsDNA	Linear	169,000
Mu	Head & tail	dsDNA	Linear	39,000

[a]SS, single-stranded; ds, double-stranded.
[b]In bases (ss genomes) or base pairs (ds genomes). These viral genomes have been sequenced and thus their lengths are known precisely. However, the sequence and length often vary slightly among different isolates of the same virus. Hence, the genome sizes listed here have been rounded off in all cases.

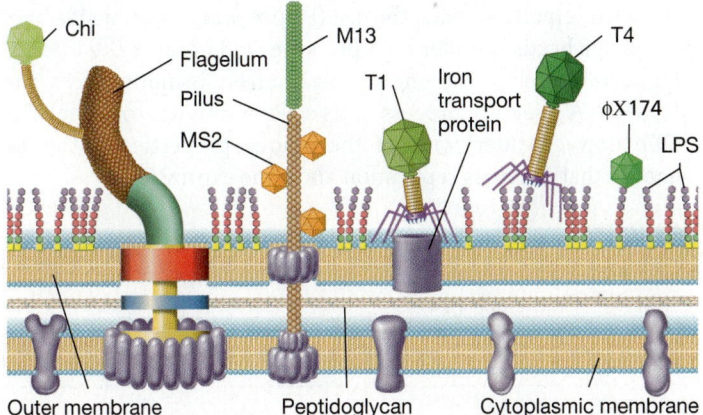

Figure 8.10 Bacteriophage receptors. Examples of the cell receptor sites used by different bacteriophages that infect *Escherichia coli*. All phages depicted except for MS2 are DNA phages.

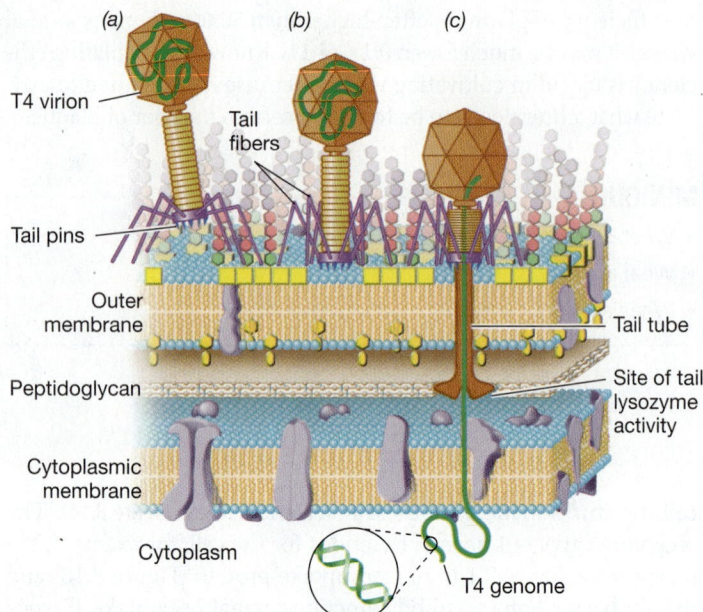

Figure 8.11 Attachment of bacteriophage T4 to an *Escherichia coli* cell. *(a)* Initial attachment of a T4 virion to the cell outer membrane by tail fiber interactions with lipopolysaccharide (LPS). *(b)* Contact of the cell wall by the tail pins. *(c)* Contraction of the tail sheath and injection of the T4 genome. The tail tube penetrates the outer membrane, and T4 lysozyme digests a small opening through the peptidoglycan layer.

The most complex viral penetration mechanisms exist with the tailed bacteriophages. Bacteriophage T4 consists of an icosahedral head, within which the viral linear double-stranded DNA is folded, and a long, complex tail, which ends in a series of tail fibers and tail pins that contact the cell surface. Phage T4 virions first attach to *Escherichia coli* cells using their tail fibers (**Figure 8.11**). The ends of the tail fibers interact specifically with polysaccharides in the cell's LPS layer and then the tail fibers retract, allowing the tail itself to contact the cell wall via the tail pins. The activity of T4 lysozyme then forms a small pore in the peptidoglycan layer and the tail sheath contracts. When this occurs, T4 DNA enters the cytoplasm of the *E. coli* cell through a tail tube in a fashion resembling that of injection by a syringe. By contrast, the T4 capsid remains outside the cell (Figure 8.11). DNA inside bacteriophage heads is under high pressure, and because the interior of a bacterial cell is also under pressure from osmotic forces, the phage DNA injection process takes several minutes to complete.

We now consider some of the unique properties of the T4 genome that affect its replication and gene expression.

MINIQUIZ -
- How does attachment contribute to virus–host specificity?
- Why does T4 need a lysozyme-like protein in order to infect its host?
- What part of phage T4 enters the host cytoplasm?

8.6 The T4 Genome

Once a permissive host cell has been infected by a virus, the earliest events surround the synthesis of new copies of the viral

genome. Because there are many types of viral genomes (Figure 8.2), there are many different schemes for virus genome replication (⇄ Section 9.1). In small DNA viruses, replication of the viral genome is done by the cell's DNA polymerase. However, in more complex DNA viruses such as bacteriophage T4, the virus encodes its own DNA polymerase. Other proteins that function in viral DNA replication such as primases and helicases (⇄ Sections 4.4–4.5) are also encoded by the T4 genome. In fact, T4 produces its own eight-protein DNA replisome complex (⇄ Section 4.6) to facilitate phage-specific genome synthesis.

Genome Replication and Circular Permutation

Chromosomes of higher organisms and genomes of bacteria contain the same genes in the same order in cells of different individuals of the same species. This is also true of many virus genomes, but not all. Sometimes a population of virions of a single virus contain genomes with the same set of genes but arranged in a different order. This is a phenomenon called *circular permutation* and is a hallmark of the T4 genome. The term circular permutation is derived from the fact that DNA molecules that are circularly permuted appear to have been linearized by opening identical circular genomes at different locations. Circularly permuted genomes are also *terminally redundant*, meaning that some DNA sequences are duplicated on both ends of the DNA molecule as a result of the mechanism that generated them.

The T4 genome is first replicated as a unit and then several genomic units are recombined end to end to form a long DNA molecule called a **concatemer** (**Figure 8.12a**). When the T4 DNA is packaged into capsids, the concatemer is not cut at a specific sequence but instead, linear segments of DNA just long enough to fill a phage head are generated. This is called *headful packaging*, and is common among bacteriophages. However, because the T4 head holds slightly more than a genome length, the headful mechanism generates terminal repeats of about 3–6 kbp at each end of the DNA molecule (Figure 8.12a).

Restriction and Modification

Although they lack the immune systems of animals, bacteria possess several weapons against viral attack. An antiviral system called CRISPR (⇄ Section 10.12) is one of these, but in addition, bacteria can destroy double-stranded viral DNA by the activity of *restriction endonucleases*, bacterial enzymes that cleave foreign DNA at specific sites (⇄ Section 11.1). This process is called *restriction* and is a general host mechanism to prevent invasion by viral (or any other foreign) DNA. For such a system to be effective, however, the host must protect its own DNA from restriction enzyme attack. The host accomplishes this by *modification* of its DNA, typically by methylation of nucleotides at the sites where the restriction enzymes cut.

Restriction enzymes are specific for double-stranded DNA and thus single-stranded DNA viruses and all RNA viruses are unaffected by restriction enzymes. Although host restriction systems confer significant protection from viral attack, some double-stranded DNA viruses have overcome host restriction by modifying their own DNA so it is no longer subject to restriction enzyme attack. Many protective mechanisms are known, but in phage T4 this is accomplished by substituting the base *5-hydroxymethylcytosine* in place of cytosine in viral DNA. The hydroxy group of this

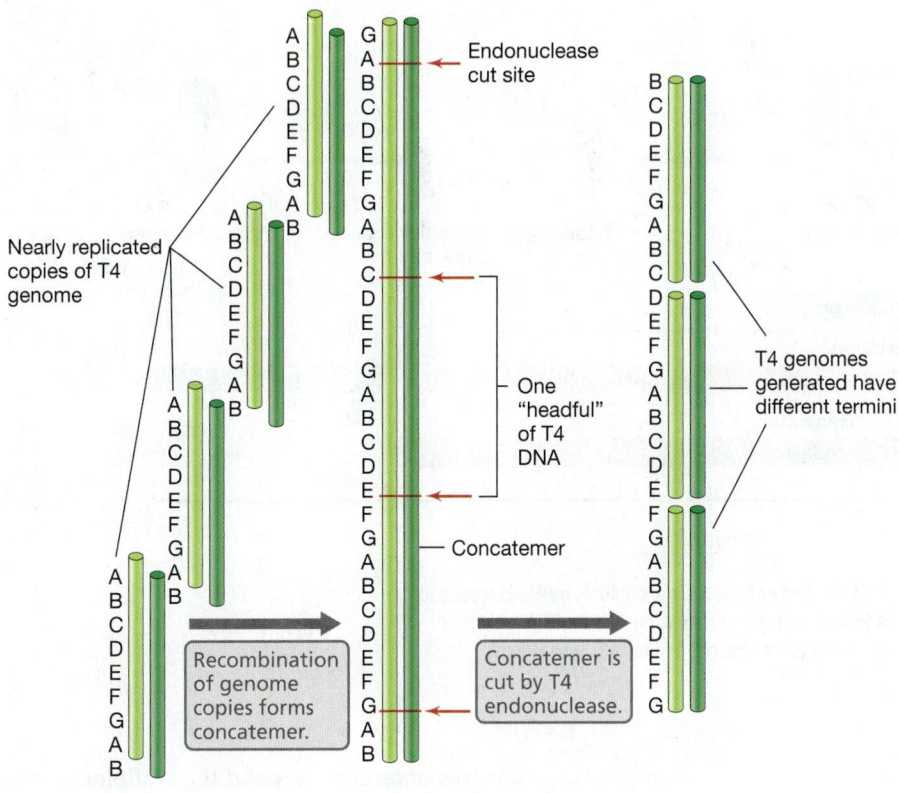

(a) Circularly permuted T4 DNA

Nearly replicated copies of T4 genome

Endonuclease cut site

Recombination of genome copies forms concatemer.

One "headful" of T4 DNA

Concatemer

Concatemer is cut by T4 endonuclease.

T4 genomes generated have different termini

Cytosine   **5-hydroxymethyl-cytosine**

Site of glucosylation

(b) An unusual base in T4 DNA

Figure 8.12 Circular permutation and the unique DNA of bacteriophage T4. *(a)* Generation of virus-length T4 DNA molecules with permuted sequences by an endonuclease that cuts off constant lengths of DNA from a concatemer regardless of their sequence. *(b)* The unique base 5-hydroxymethylcytosine in the DNA of bacteriophage T4. Once this base is glucosylated, the T4 DNA is resistant to restriction enzyme attack.

modified base is glucosylated, meaning that a molecule of glucose is added (Figure 8.12*b*), and DNA with this modification is resistant to restriction enzyme attack. By virtue of this viral protection mechanism, copies of the T4 genome are preserved until they are packaged later in the phage replication cycle.

MINIQUIZ

- What is characteristic of a circularly permuted and terminally redundant genome?
- What is a concatemer?
- How does *Escherichia coli* try to protect itself from phage attack, and how does T4 protect itself from these weapons?

8.7 Replication of Bacteriophage T4

We now examine the steps in the replication cycle of bacteriophage T4, drawing on what we already know about attachment and penetration of T4 and properties of the T4 genome from the previous two sections.

Transcription and Translation

Shortly after infection, T4 DNA is transcribed and translated, and the process of new virion synthesis begins. In less than half an hour, the process culminates in the release of new virions from the lysed cell. The major events are summarized in **Figure 8.13**.

Within a minute after T4 DNA enters the host cytoplasm, the synthesis of host DNA and RNA ceases and transcription of specific phage genes begins. Translation of viral mRNA also begins quickly, and within 4 min of infection, phage DNA replication has already begun. The T4 genome encodes three major sets of proteins called **early proteins**, **middle proteins**, and **late proteins**, the terms referring to the general order of their appearance in the cell. Early proteins include enzymes for the synthesis and glucosylation of the unusual T4 base 5-hydroxymethylcytosine (Figure 8.12*b*), enzymes that function in the T4 replisome to produce copies of the phage-specific genome, and early proteins that modify host RNA polymerase. By contrast, middle and late proteins include additional RNA polymerase–modifying proteins, and virion structural and release proteins. These include, in particular, viral head and tail proteins and the enzymes required to liberate new virions from the cell (Figure 8.13).

The T4 genome does not encode its own RNA polymerase; instead, T4-specific proteins modify the specificity of the host RNA polymerase so that it recognizes only phage promoters (recall that promoters are the regions upstream of a structural gene where RNA polymerase binds to initiate transcription, ⮌ Section 4.7). These modification proteins are encoded by T4 early genes and are transcribed by the host RNA polymerase. Host transcription is shut down shortly after this by a phage-encoded anti-sigma factor that binds to the host RNA polymerase sigma factor and prevents it from recognizing promoters on host genes. This effectively switches the activity of host RNA polymerase from transcribing host genes to transcribing T4 genes. Later in the infection process other phage proteins modify the host RNA polymerase so it now recognizes T4 middle gene promoters. Finally, transcription of T4 late genes begins, and this requires a new T4-encoded sigma factor that directs host RNA polymerase to promoters for these genes only. At this point, viral assembly can begin.

UNIT 2

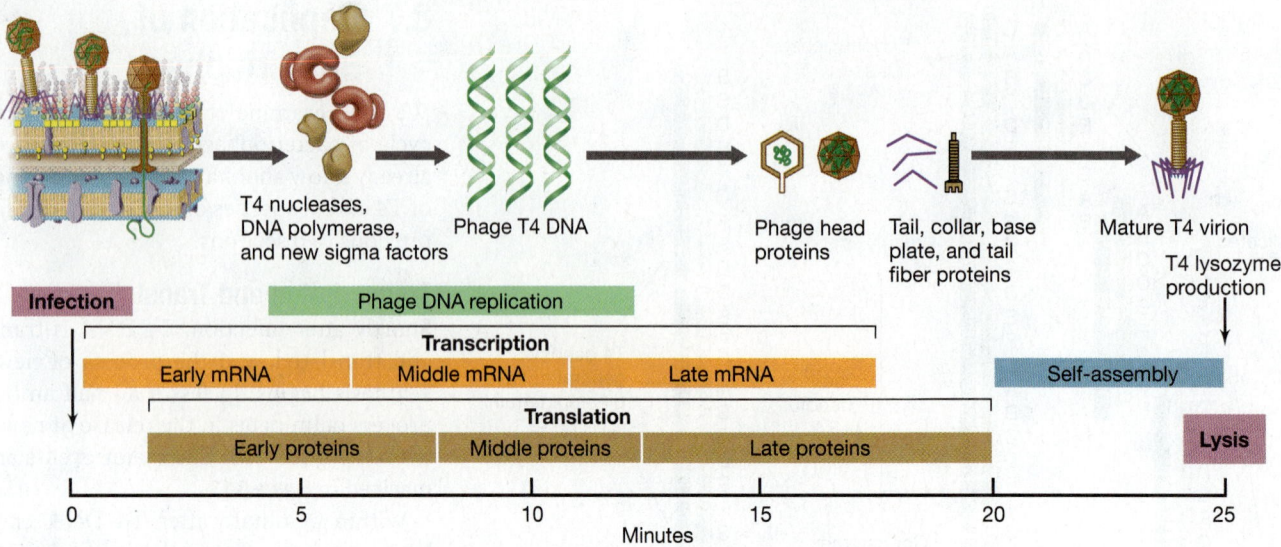

Figure 8.13 **Time course of events in phage T4 infection.** Following injection of DNA, early and middle mRNA is produced that encodes nucleases, T4 DNA polymerase, new phage-specific sigma factors, and other proteins needed for DNA replication. Late mRNA encodes virion structural proteins and T4 lysozyme, which is needed to lyse the cell and release new virions.

Packaging the T4 Genome and Virion Assembly and Release

The bacteriophage T4 DNA genome is forcibly pumped into a preassembled capsid using an energy-linked packaging motor. The motor components are encoded by viral genes, but host cell metabolism is needed to produce the proteins and supply the ATP required for the pumping process. The packaging process can be divided into three stages (Figure 8.14). First, precursors of the bacteriophage head called *proheads* are assembled but remain empty. Proheads contain temporary "scaffolding proteins" as well as head structural proteins. Second, a packaging motor is assembled at the opening to the prohead. The double-stranded linear T4 DNA genome (Figure 8.12) is then pumped into the prohead under pressure using ATP as the driving force. The prohead expands when pressurized by the entering DNA and the scaffolding proteins are simultaneously discarded. Third, the packaging motor itself is discarded and the capsid head is sealed.

After the head has been filled, the T4 tail, tail fibers, and the other components of the virion are added, primarily by self-assembly (Figures 8.13 and 8.14). The phage genome encodes a pair of very late enzymes that combine to breach the two major barriers to virion release: the host cytoplasmic membrane and peptidoglycan layer. Once these structures are compromised, the cell breaks open by osmotic lysis and the newly synthesized virions are released. After each replication cycle, which takes only about 25 min (Figure 8.13), over 100 new virions are released from each host cell (the *burst size*, Section 8.3), and these are now free to infect neighboring host cells.

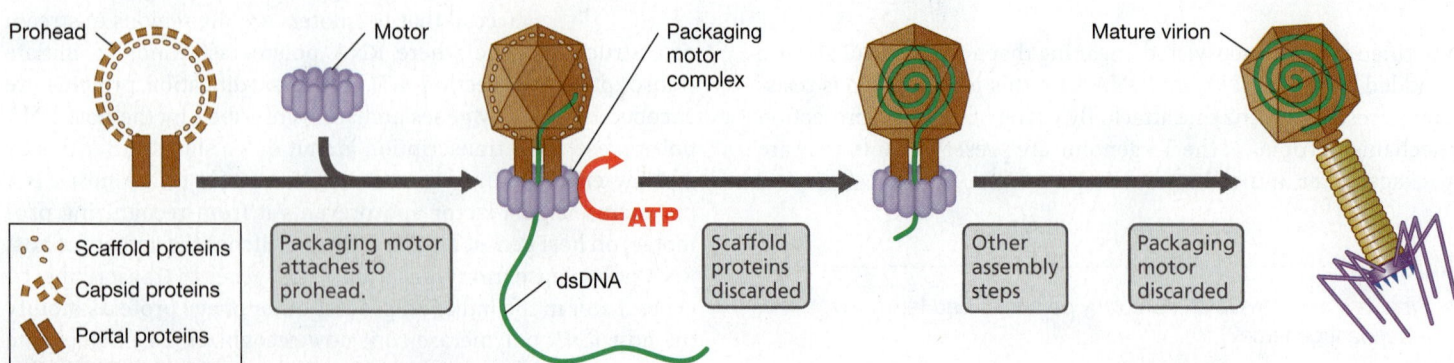

Figure 8.14 **Packaging of DNA into a T4 phage head.** Proheads are assembled from capsid and portal proteins, both of which remain in the mature virion. As the head fills with DNA, it expands and becomes more angular. Once the head is filled, the packaging motor detaches and the tail components are added.

UNIT 2

MINIQUIZ
- Give one example each of T4 early, middle, and late proteins.
- How does T4 direct host RNA polymerase to phage-specific genes?
- What is required to package the T4 genome into its phage head?

8.8 Temperate Bacteriophages and Lysogeny

Bacteriophage T4 is a virulent virus and once infection begins, it always kills its host. However, some double-stranded DNA bacterial viruses, although able to catalyze a virulent cycle, can also infect their host and establish a long-term stable relationship. These viruses are called **temperate viruses**.

Temperate viruses can enter into a state called **lysogeny**. In this state, most virus genes are not transcribed and instead, the virus genome is replicated in synchrony with the host chromosome and passed to daughter cells at cell division. The lysogenic state may confer new genetic properties on the bacterial host cell—a condition called *lysogenic conversion*—and we will see several examples

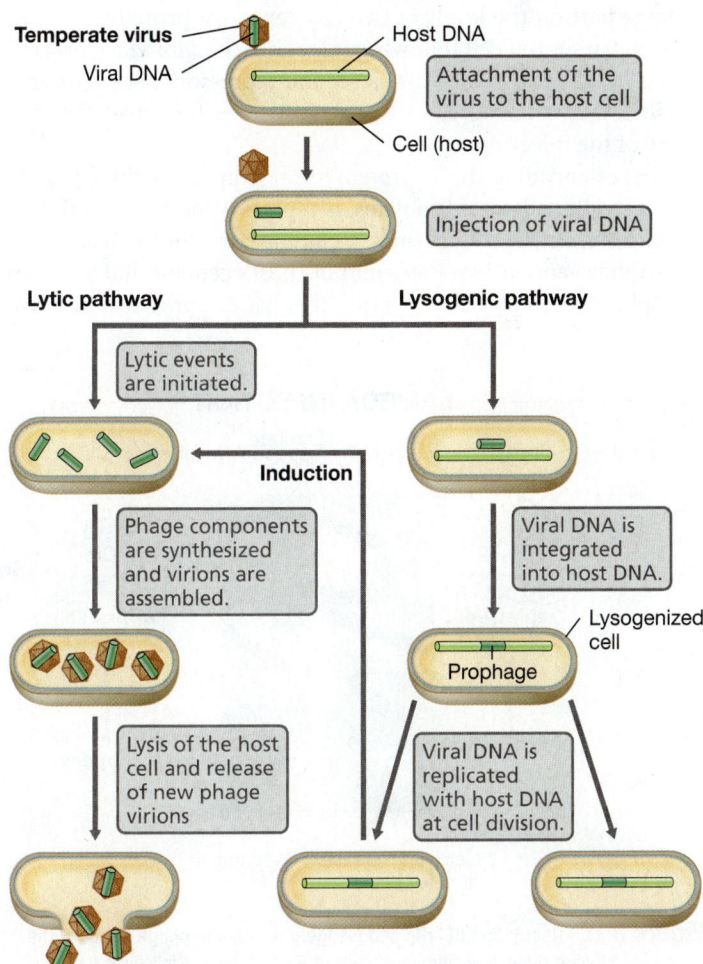

Figure 8.15 Consequences of infection by a temperate bacteriophage. The alternatives upon infection are replication and release of mature virus (lysis) or lysogeny, often by integration of the virus DNA into the host DNA, as shown here. The lysogen can be induced to produce mature virus and lyse.

in later chapters of pathogenic bacteria whose virulence (ability to cause disease) is linked to a lysogenic bacteriophage. A cell that harbors a temperate virus is called a **lysogen**.

The Replication Cycle of a Temperate Phage

Two well-characterized temperate bacteriophages are lambda and P1. The life cycle of a temperate bacteriophage is shown in **Figure 8.15**. During lysogeny, the temperate virus genome is either integrated into the bacterial chromosome (lambda) or can exist in the cytoplasm as a plasmid (P1). In either case, the viral DNA, now called a **prophage**, replicates along with the host cell as long as the genes that activate the phage virulent pathway are repressed.

Maintenance of the lysogenic state is due to a phage-encoded *repressor protein*. Normally, low-level transcription of repressor genes and their subsequent translation maintains the repressor at a low level in the cell. However, if the phage repressor is inactivated or if its synthesis is in some way prevented, the prophage can be induced into the lytic stage. If induction occurs while the viral DNA is incorporated into the bacterial chromosome, the viral DNA is excised and phage genes are transcribed and translated; new virions are then produced, and the host cell is lysed (Figure 8.15). Various cell stress conditions, especially damage to host cell DNA, can induce a prophage to enter the lytic pathway. In contrast to this process, the viral "decision" to proceed to lysogeny or the lytic pathway upon initial viral infection is another matter altogether, and has been particularly well-studied in bacteriophage lambda. We explore this story now.

Bacteriophage Lambda

Bacteriophage lambda, which infects *Escherichia coli*, is a double-stranded DNA virus with a head and tail (**Figure 8.16**). At the 5′ end of each DNA strand of the linear lambda genome is a single-stranded region 12 nucleotides long. These single-stranded "cohesive" ends are complementary in base sequence; when lambda DNA enters the host cell, they base-pair to form the *cos* site and cyclize (circularize) the genome (**Figure 8.17a**).

If lambda enters the lytic pathway, long, linear concatemers of genomic DNA are synthesized by a mechanism called **rolling circle replication**. In this process, one strand in the circular

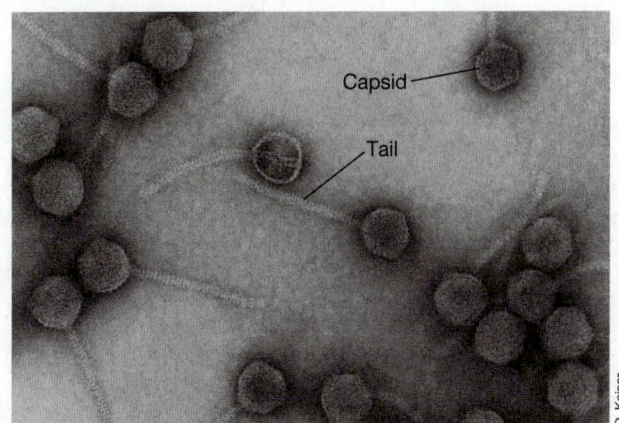

Figure 8.16 Bacteriophage lambda. Transmission electron micrograph of phage lambda virions. The head of each virion is about 65 nm in diameter and contains linear dsDNA.

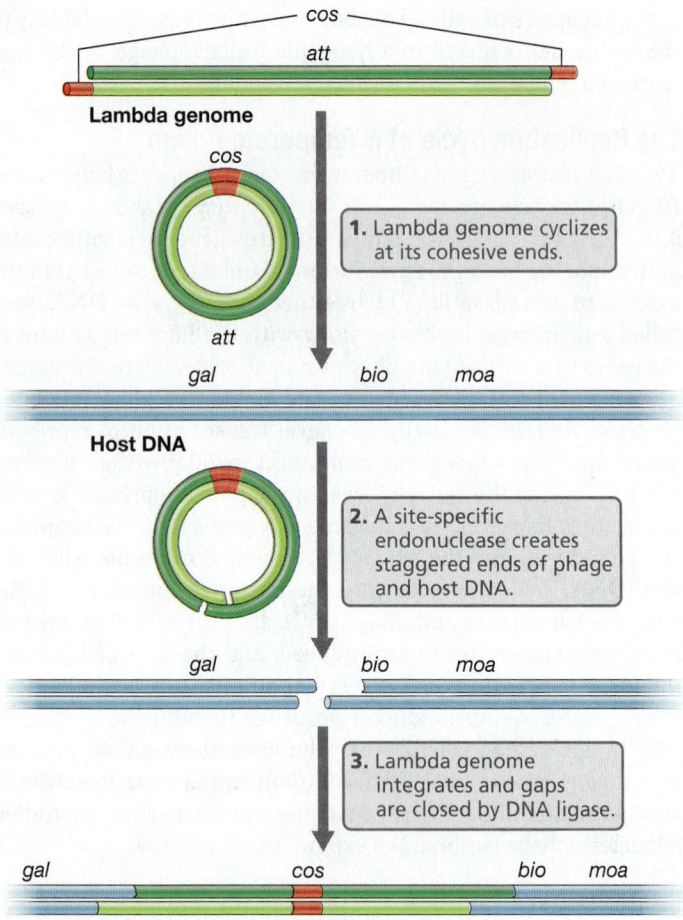

(a) Integration of lambda DNA into the host

1. Lambda genome cyclizes at its cohesive ends.

2. A site-specific endonuclease creates staggered ends of phage and host DNA.

3. Lambda genome integrates and gaps are closed by DNA ligase.

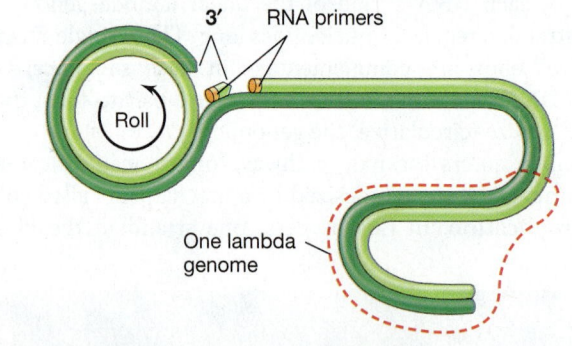

(b) Rolling circle replication of lambda genome

Figure 8.17 Integration of lambda DNA and rolling circle replication. *(a)* Lambda DNA integrates at specific attachment (*att*) sites on both the host and phage genomes. Host genes near *att* include *gal*, galactose utilization; *bio*, biotin synthesis; and *moa*, molybdenum cofactor synthesis. Lambda integrase is required, and specific pairing of the complementary ends results in integration of lambda DNA. *(b)* During rolling circle replication, as one strand (dark green) rolls out, it is both replicated at its opposite end and serves as a template for synthesis of the complementary strand.

lambda genome is nicked and is "rolled out" as a template for synthesis of the complementary strand (Figure 8.17*b*). The double-stranded concatemer is then cut into genome-sized lengths at the *cos* sites and the resulting genomes packaged into lambda phage heads. Once the tail has been added and mature

lambda virions have been assembled (Figure 8.16), cell lysis occurs and the virions are released. In its role as a lytic phage, lambda can also package a few chromosomal genes from its lysed host in newly synthesized virions and then transfer these to a second host cell, a process called *transduction*. Transduction is an important means of horizontal gene transfer in nature (Section 8.11) and is also an important tool in bacterial genetics (🔗 Section 10.7).

Instead of the lytic pathway, if lambda takes the lysogenic route, its genome integrates into the *E. coli* chromosome. This requires a protein called *lambda integrase*, a phage-encoded enzyme that recognizes the phage and bacterial genome attachment sites (*att* in Figure 8.17*a*) and facilitates integration of the lambda genome. From this relatively stable state, certain events such as host DNA damage can initiate the lytic cycle once again. After such a trigger, a lambda excision protein excises the lambda genome from the host chromosome, transcription of lambda DNA begins, and lytic events unfold.

We now consider how these opposing processes of lysis and lysogeny are controlled upon initial infection of an *E. coli* cell by a phage lambda virion.

Lysis or Lysogeny?

Whether lysis or lysogeny occurs in a lambda infection depends in large part on the levels of two key repressor proteins that can accumulate in the cell following infection: the *lambda repressor*, also called the *cI protein*, and a second repressor called *Cro*. In a nutshell, the first repressor to accumulate will control the outcome of the infection.

If genes encoding the cI protein are rapidly transcribed following infection and cI accumulates, it represses the transcription of all other lambda-encoded genes, including *cro*. When this happens, the lambda genome integrates into the host's genome and becomes a prophage (**Figure 8.18**). Cro, on the other hand, represses expression

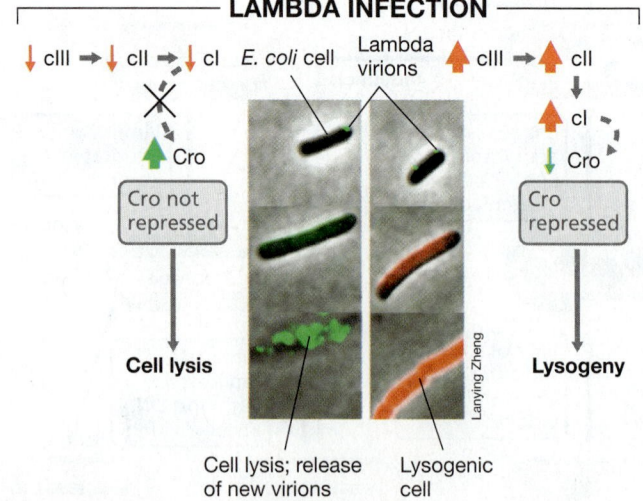

Figure 8.18 Regulation of lytic and lysogenic events in phage lambda. The photomicrographs show time courses of cells of *Escherichia coli* following a course of lytic (left panel, green) or lysogenic (right panel, red) events, as controlled by various repressors. The colors originate from genetically engineered lambda phage that trigger the production of specific fluorescent proteins when either lytic genes (green) or lysogenic genes (red) are expressed. Lytic cells are killed while *E. coli* lysogens continue to grow and divide.

of a protein called cII whose function is to activate the synthesis of cI. Hence, following infection, if cI is present at insufficient levels to repress expression of phage-specific genes, Cro can accumulate in the cell; if this happens, lambda travels the lytic pathway.

Control of these alternative lifestyles—lysis or lysogeny—of lambda has been likened to a "genetic switch," where a defined series of events must occur for one pathway to be favored over the other. Although infection of an *E. coli* cell by a lambda virion typically results in the lytic cycle, as we have said, lytic events can be switched off if sufficient concentrations of cII are present to ensure adequate levels of cI (Figure 8.18). But how does this come about? Levels of protein cII are controlled by the relative activity of a protease in the cell that slowly degrades cII and on levels of yet another protein, cIII, whose function is to stabilize cII and

prevent it from protease attack. We thus have a cascade of regulatory events here: cIII controls cII, which in turn controls cI. But even this is not the end of the story. Several other proteins not described here also play a role in the lambda lytic/lysogenic "decision," and hence the progress of a lambda infection is a highly complex series of events. Indeed, this tiny bacteriophage employs some of the most complex regulatory systems known in virology.

MINIQUIZ -
- What is a lysogen and what is a prophage?
- How does DNA replication in lambda differ from that of its host?
- What commits lambda to the lytic versus the lysogenic pathway?

III · Viral Diversity and Ecology

8.9 An Overview of Bacterial Viruses

The most common bacteriophages are head-and-tail phages containing double-stranded DNA genomes (Sections 8.5–8.7). However, many other bacteriophages are known, including a wide variety that contain single-stranded genomes. Examples of the various classes of bacteriophages, based on the properties of their genomes (Figure 8.2), are shown schematically in **Figure 8.19**.

Bacteriophages with Single-Stranded Genomes

Bacteriophages φX174, M13, and MS2 (Figure 8.19) are three well-characterized *Escherichia coli* phages that contain single-stranded genomes. Phage φX174 contains a circular DNA genome inside an icosahedral virion only 25 nm in diameter. Such small

DNA viruses possess only a few genes and rely entirely upon the DNA replication machinery of the host cell. M13 is a filamentous bacteriophage that also contains a circular DNA genome. The single-stranded genomes of bacteriophages φX174 and M13 are of the plus sense (Section 8.1), and before replication occurs, they are converted into a double-stranded **replicative form**. From these, single-stranded genome copies are derived and transcription of viral genes takes place.

Bacteriophage MS2 is a small icosahedral virus (Figure 8.19) whose single-stranded RNA genome encodes only four proteins. A key one of these is **RNA replicase**, the enzyme needed to replicate the viral RNA genome. Such an enzyme is needed because bacterial and animal cells do not possess enzymes that synthesize RNA from an RNA template. Because the RNA genome of phage MS2 is of the plus sense (Section 8.1), the genome is also mRNA and can therefore be translated directly upon entry into the cell by the host's translational machinery.

An interesting feature of many small DNA and RNA bacteriophages is their **overlapping genes**. The genomes of these tiny viruses typically contain too few genes to encode all of the proteins they need. To solve this problem, some of their open reading frames overlap, allowing the virus to produce more than a single polypeptide from a given gene. We discuss examples of overlapping genes and some of the other interesting features of small DNA and RNA bacteriophages in more detail in Chapter 9.

Head-and-Tail Bacteriophages

Head-and-tail bacteriophages with double-stranded DNA genomes have been used as models for virus replication and have revealed many of the fundamental principles of molecular biology and genetics. The first series of tailed phages to be studied were designated T1, T2, and so on, up to T7, with the T referring to the tail. T4 has a much larger genome than that of other T phages and, along with phage lambda (Section 8.8), is probably the best studied of all bacteriophages. The steps in a T phage infection were detailed in Sections 8.5–8.7.

The structure of the T4 virion is perhaps the most complicated of all viruses. The virion consists of an elongated icosahedral head

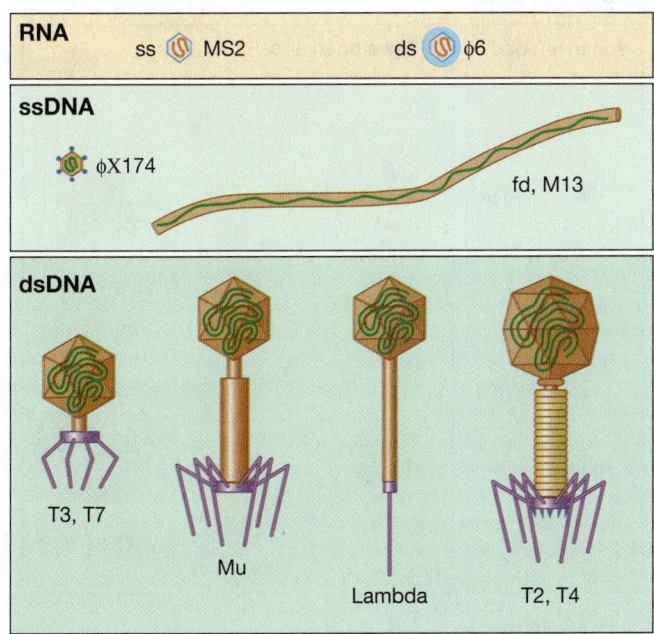

Figure 8.19 Schematic representations of the main types of bacterial viruses. Sizes are to approximate scale. The nucleocapsid of bacteriophage φ6 is surrounded by a membrane (blue).

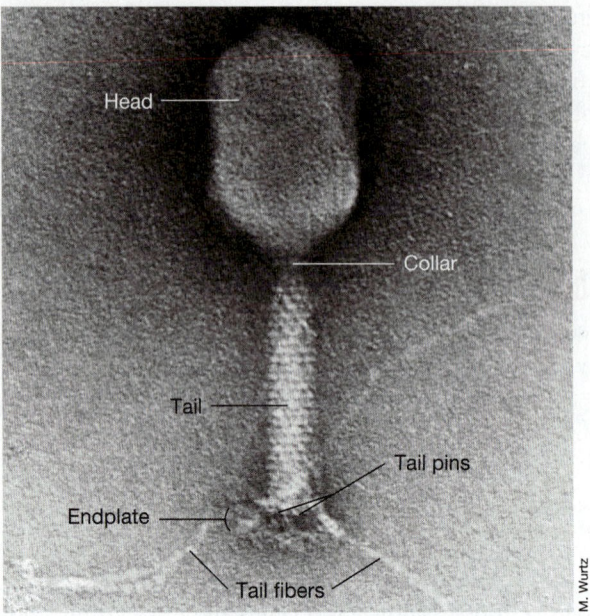

Figure 8.20 Structure of T4, a complex bacteriophage. Transmission electron micrograph of bacteriophage T4 of *Escherichia coli*. The tail components function in attachment of the virion to the host and injection of the nucleic acid (see Figure 8.11). The T4 head is about 85 nm in diameter.

whose overall dimensions are 85 × 110 nm (**Figure 8.20**). Attached to the head is a complex tail consisting of 20 different proteins that form a helical tube surrounded by a sheath. At one end the tail is joined to the head by a "neck" with a "collar" and at the other end is an end plate carrying long, jointed tail fibers (Figure 8.20). Following attachment of a T4 virion to a host cell (by way of its tail fibers, Figure 8.11), the tail contracts to make small incisions in both the host peptidoglycan and cytoplasmic membrane and inject the T4 genome into the cell. The genome of T4 is a linear DNA molecule of about 170 kilobase pairs that encodes nearly 300 proteins, including many needed for phage DNA replication (Section 8.7). Although no virus encodes its own translational apparatus, the T4 genome does encode eight of its own tRNAs. These likely assist in reading certain T4 codons, since T4 codon bias (⟳ Section 4.11) differs significantly from that of *E. coli*.

MINIQUIZ -
- What type of nucleic acid is most common in bacteriophage genomes?
- What is the function of RNA replicase?
- What does the T in T4 stand for?

8.10 An Overview of Animal Viruses

Viruses that infect plants and animals share many properties with bacterial viruses but differ in some key respects. The major tenets of virology—presence of a capsid to carry the viral DNA or RNA genome, infection and takeover of host metabolic processes, and assembly and release from the cell—are universal, regardless of the nature of the host. However, two key differences between

bacterial and animal viruses are that (1) the entire virion of animal viruses (rather than just the nucleic acid) enters the host cell, and (2) eukaryotic cells contain a nucleus, where many animal viruses replicate. We explore some aspects of animal viruses here.

Classification of Animal Viruses
Various types of animal viruses are illustrated in **Figure 8.21**. As for bacterial viruses, animal viruses are classified by the structure of their genomes (Figure 8.2). Animal viruses are known in all genomic categories, and most animal viruses that have been studied in detail are those that can replicate in cell cultures (Section 8.4 and Figure 8.9.)

The majority of important human viral diseases are caused by RNA viruses, and some examples are listed in **Table 8.2**. Most of these

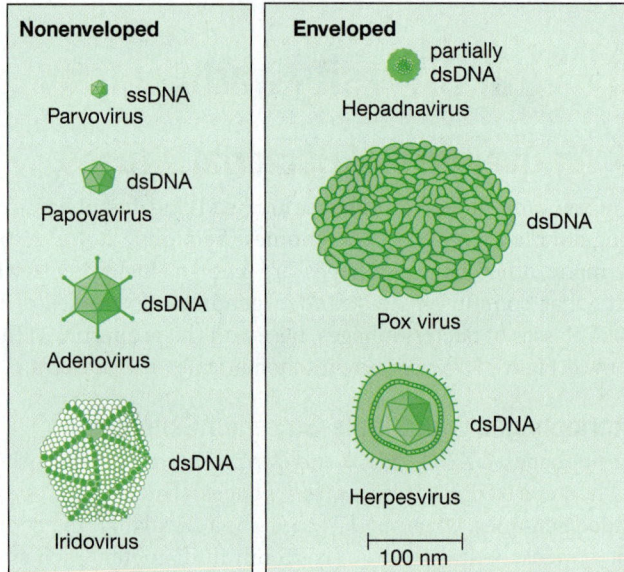

(a) **DNA viruses**

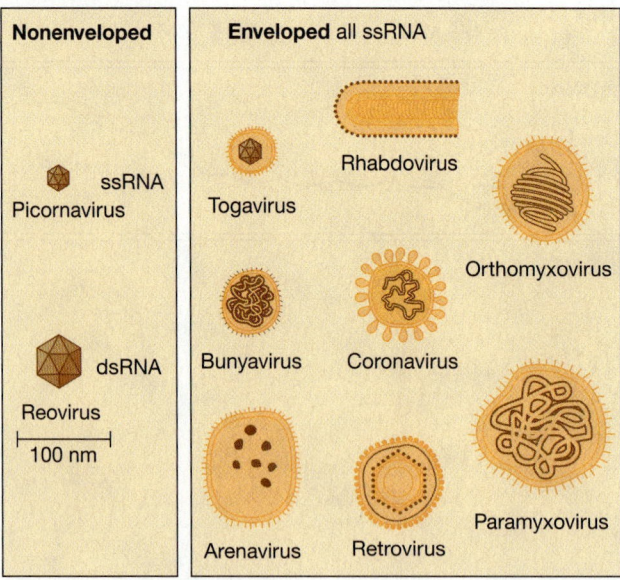

(b) **RNA viruses**

Figure 8.21 Diversity of animal viruses. The shapes and relative sizes of the major groups of vertebrate viruses. The hepadnavirus genome has one complete DNA strand and part of the complementary strand.

Table 8.2 Representative viral diseases of humans

Disease	Virus	Genome DNA or RNA[a]	Size[b]
Cold sores/genital herpes	Herpes simplex	dsDNA	152,000
Smallpox	Variola major	dsDNA	190,000
Polio	Poliovirus	ssRNA (+)	7,500
Rabies	Rabies virus	ssRNA (−)	12,000
Influenza	Influenza A virus	ssRNA (−)	13,600
Measles	Measles virus	ssRNA (−)	15,900
Ebola hemorrhagic fever	Ebola virus	ssRNA (−)	19,000
Severe acute respiratory syndrome (SARS)	SARS virus	ssRNA (+)	29,800
Infant diarrhea	Rotavirus	dsRNA	18,600
Acquired immunodeficiency syndrome (AIDS)	Human immunodeficiency virus (HIV)	ssRNA/dsDNA (a retrovirus) (+)	9,700

[a]SS, single-stranded; ds, double-stranded. +, plus-strand virus; −, negative-strand virus (Section 8.1).

[b]In bases (ss genomes) or base pairs (ds genomes). These viral genomes have been sequenced and thus their lengths are known precisely. However, the sequence and length often varies slightly among different isolates of the same virus. Hence, the genome sizes listed here have been rounded off in all cases.

RNA viruses have single-stranded genomes, the only exception being the reoviruses whose genomes consist of double-stranded RNA. As can be seen, the RNA viruses in general have relatively small genomes in contrast to the two listed DNA viruses, smallpox and herpesviruses (Table 8.2).

Unlike a bacteriophage infection, in which one of only two outcomes—lysis or lysogeny—is possible depending on the virus, other events are possible in an animal virus infection. We explore these possibilities now.

Consequences of Virus Infection in Animal Cells

Different animal viruses can catalyze at least four different outcomes (**Figure 8.22**). *Virulent infection* results in lysis of the host cell; this is the most common outcome. By contrast, in a *latent infection*, viral DNA does not replicate and the host cells are unharmed. With some enveloped animal viruses, release of virions, which occurs by a kind of budding process, may be slow, and the host cell may not be lysed. Such infections are called *persistent infections*. Finally, certain animal viruses can convert a normal cell into a tumor cell, a process called *transformation*.

Animal virus receptors are typically cell surface macromolecules used in cell–cell contact or that function in the immune system. For example, the receptors for poliovirus and for HIV (the causative agent of AIDS) are normally used in intercellular communication between human cells. In multicellular organisms, cells in different tissues or organs often express different proteins on their cell surfaces. Consequently, viruses that infect animals often infect only certain tissues. For example, viruses that cause the common cold infect only cells of the upper respiratory tract.

Animal viruses must eventually lose their outer coat to expose the viral genome. Some enveloped animal viruses are uncoated at the host cytoplasmic membrane, releasing the nucleocapsid into the cytoplasm. However, the entire virion of naked animal viruses and many enveloped animal viruses enters the cell via endocytosis. In these cases, the virion is uncoated in the host cytoplasm and the genome passes through the nuclear membrane to the nucleus, where viral nucleic acid replication occurs. Many animal viruses are enveloped, and when these exit the cell, they may pick up part of the cell's cytoplasmic membrane and use it as part of the viral envelope.

Of all the viruses listed in Figure 8.2 and Table 8.2, one group stands out as having an absolutely unique mode of replication. These are the retroviruses. We explore them next as an example of a complex and highly unusual animal virus with significant medical implications.

Retroviruses and Reverse Transcriptase

Retroviruses contain an RNA genome. However, the genome is replicated inside the host cell via a DNA intermediate. The prefix *retro* means "backward," and the term *retrovirus* refers to the

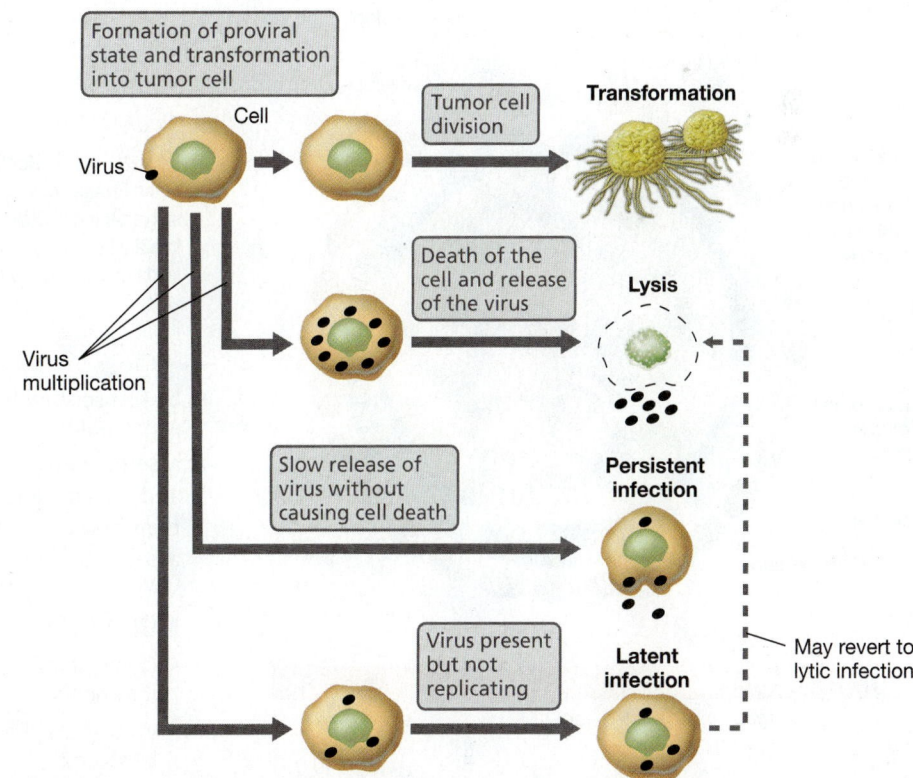

Figure 8.22 **Possible effects that animal viruses may have on cells they infect.** Most animal viruses are lytic, and only a very few are known to cause cells to transform and become cancerous.

fact that these viruses transfer information from RNA to DNA (in contrast to genetic information flow in cells, which is just the opposite). Retroviruses use the enzyme **reverse transcriptase** to carry out this unusual process. Retroviruses were the first viruses shown to cause cancer, and the human immunodeficiency virus (HIV) is a retrovirus that causes acquired immunodeficiency syndrome (AIDS).

Retroviruses are enveloped viruses that carry several enzymes in the virion (**Figure 8.23a**). These include *reverse transcriptase*, *integrase*, and a retroviral-specific *protease*. The genome of the retrovirus is unique and consists of two identical single-stranded RNAs of the plus sense (Section 8.1). The genome contains the genes *gag* (structural proteins), *pol* (reverse transcriptase and integrase), and *env* (envelope proteins) (Figure 8.23b). At each end of the retrovirus genome are repeated sequences that are essential for viral replication.

The replication of a retrovirus begins with the virion entering the host cell where the envelope is removed and reverse transcription begins in the nucleocapsid (**Figure 8.24**). A single strand of DNA is produced and then reverse transcriptase uses this as a template to make a complementary strand; double-stranded DNA is the final product. The latter is released from the nucleocapsid, enters the host nucleus along with the integrase protein, and the integrase facilitates the incorporation of the retroviral DNA into the host genome. The retroviral DNA is now a **provirus**. The latter remains in the host genome indefinitely and proviral DNA can

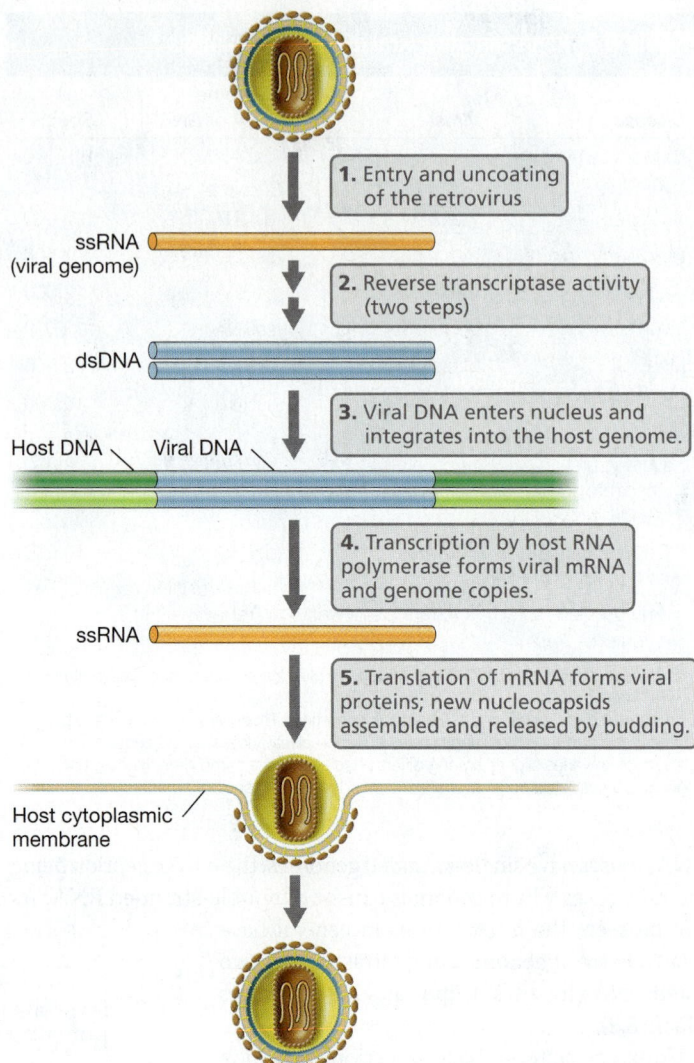

1. Entry and uncoating of the retrovirus

ssRNA (viral genome)

2. Reverse transcriptase activity (two steps)

dsDNA

3. Viral DNA enters nucleus and integrates into the host genome.

Host DNA   Viral DNA

4. Transcription by host RNA polymerase forms viral mRNA and genome copies.

ssRNA

5. Translation of mRNA forms viral proteins; new nucleocapsids assembled and released by budding.

Host cytoplasmic membrane

Figure 8.24 Replication of a retrovirus. The virion carries two identical copies of the RNA genome (orange). Reverse transcriptase, carried in the virion, makes single-stranded DNA from viral RNA and then double-stranded DNA that integrates into the host genome as a provirus. Transcription and translation of proviral genes leads to the production of new virions that are then released by budding.

be transcribed by the host RNA polymerase to form copies of the retroviral RNA genome and mRNA. Eventually, nucleocapsids are assembled that contain two copies of the retroviral RNA genome and are enveloped as they bud through the host cell cytoplasmic membrane (Figure 8.24). From here, the mature retrovirus virions are free to infect neighboring cells.

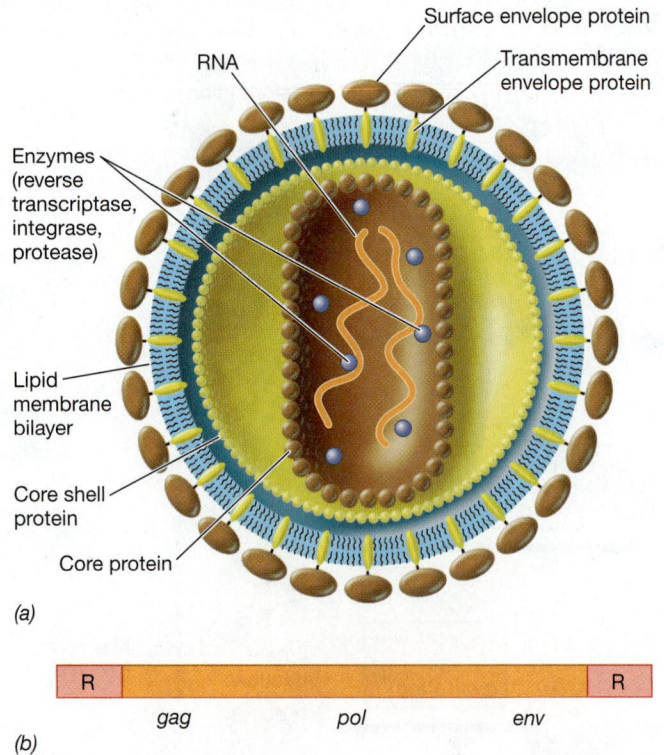

Surface envelope protein

RNA

Transmembrane envelope protein

Enzymes (reverse transcriptase, integrase, protease)

Lipid membrane bilayer

Core shell protein

Core protein

(a)

R   gag   pol   env   R

(b)

Figure 8.23 Retrovirus structure and function. *(a)* Structure of a retrovirus. *(b)* Genetic map of a typical retrovirus genome. Each end of the genomic RNA contains direct repeats (R).

MINIQUIZ -

• Contrast the ways in which animal and bacterial viruses enter their hosts.

• What is the difference between a persistent and a latent viral infection?

• Why are retroviruses so named? What is required to carry out this process?

8.11 The Virosphere and Viral Ecology

Viruses are present in every environment on Earth that contains cells and are present in enormous numbers. The number of prokaryotic cells on Earth is far greater than the total number of eukaryotic cells; estimates of total prokaryotic cell numbers are on the order of 10^{30} (⟳ Table 1.2). However, the number of viruses is even greater than this, an estimated 10^{31}. The best estimates of both cell and virus numbers in nature have come from studies done on seawater.

Viruses of Prokaryotes

There are about 10^6 prokaryotes/ml of seawater and approximately ten times as many viruses. It has been estimated that at least 5% and as many as 50% of the prokaryotes in seawater are killed by bacteriophages each day, and most of the others are eaten by protozoa. Although viruses account for most of the total microorganisms present in seawater in terms of numbers, due to their small size they constitute only about 5% of the total biomass (**Figure 8.25**).

By far the most common type of bacteriophages, at least in the oceans, are head-and-tail phages containing double-stranded DNA. By contrast, RNA-containing bacteriophages are comparatively rare. As we have seen, lysogenic bacteriophages can integrate into the genomes of their bacterial hosts (Section 8.8), and when they do, they can confer new properties on the cell. Moreover, some lytic phages facilitate the transfer of bacterial genes from one cell to another by transduction, a major means of horizontal gene transfer between prokaryotes (⟳ Section 10.7). As agents of transduction, bacteriophages are thought to have a major influence on bacterial evolution. For example, transferred genes may confer new metabolic or other beneficial properties on the recipient cells and allow them to colonize and be successful in new habitats.

Many of the prokaryotes in the marine environment are *Archaea*. In particular, a major group of marine *Archaea* of great ecological relevance are the *Thaumarchaeota*. These ammonia-oxidizing species are capable of consuming the vanishingly low levels of ammonia present in planktonic (open ocean) waters. Although lytic archaeal viruses have yet to be demonstrated for this group, at least one species of *Thaumarchaeota* has been shown to harbor a viral genome within its own genome (that is, it contains a provirus). It is thus likely that at least some, and perhaps even many, of the viruses in seawater infect marine *Archaea* instead of marine *Bacteria*. This is bolstered by the observation that virtually all known archaeal viruses are double-stranded DNA viruses, the most commonly observed group in the oceans.

Survival Strategies and Diversity of Viruses in Nature

When hosts are plentiful in nature, it is thought that bacteriophages adopt the lytic lifestyle and thus large numbers of host

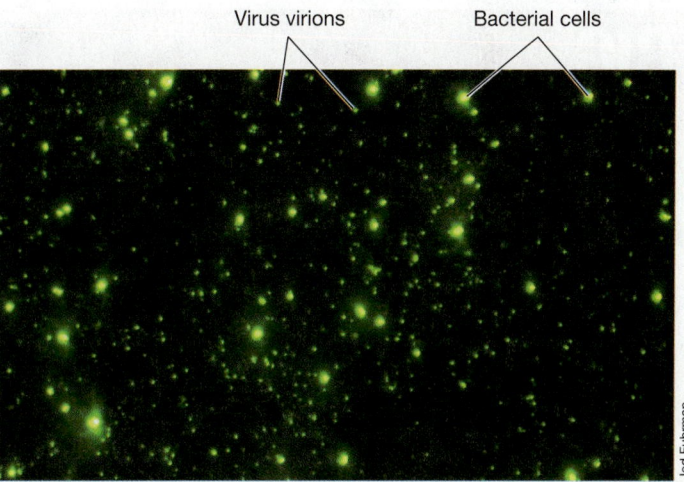

Virus virions   Bacterial cells

Jed Fuhrman

Figure 8.25 **Viruses and bacteria in seawater.** A fluorescence photomicrograph of seawater stained with the dye SYBR Green to reveal prokaryotic cells and viruses. Although viruses are too small to be seen with the light microscope, fluorescence from a stained virus is visible.

cells are killed. By contrast, when host numbers are low, it may be difficult for viruses to find a new host cell, and under such circumstances, lysogeny would be favored if the virus is lysogenic. Under these conditions, the virus would survive as a prophage until host numbers rebounded. Supporting this is the observation that in the depths of the ocean where bacterial numbers are lower than in surface waters, around half the bacteria examined have been found to contain one or more lysogenic viruses. As far as is known, single-stranded DNA viruses and all RNA viruses cannot enter a lysogenic state, and so how these viruses might survive periods of low host numbers is unknown.

Most of the genetic diversity on Earth resides in viruses, mostly bacteriophages. The *viral metagenome* is the sum total of all the virus genes in a particular environment. Several viral metagenomic studies have been undertaken, and they invariably show that immense viral diversity exists on Earth. For example, approximately 75% of the gene sequences found in viral metagenomic studies show no similarity to any other genes cataloged in viral or cellular gene databases. By comparison, surveys of bacterial metagenomes typically reveal approximately 10% unknown genes. Thus, most viruses still await discovery and most viral genes have unknown functions. This makes the study of viral diversity one of the most exciting areas of microbiology today.

MINIQUIZ

- What type of bacteriophages are most common in the oceans?
- How can bacteriophages affect bacterial evolution?
- What does the viral metagenome suggest?

BIG IDEAS

8.1 • A virus is an obligate intracellular parasite that requires a suitable host cell for replication. A virion is the extracellular form of a virus and contains either an RNA or a DNA genome inside a protein shell. Once inside the cell, either the virion or its nucleic acid redirects host metabolism to support virus replication. Viruses are classified by the characteristics of their genome and hosts. Bacteriophages infect bacterial cells.

8.2 • In the virion of a naked virus, only nucleic acid and protein are present; the entire unit is called the nucleocapsid. Enveloped viruses have one or more lipoprotein layers surrounding the nucleocapsid. The nucleocapsid is arranged in a symmetric fashion, with the icosahedron being a common morphology. Although virus particles are metabolically inert, one or more key enzymes are present within the virion in some viruses.

8.3 • The virus replication cycle can be divided into five major stages: attachment (adsorption), penetration (uptake of the entire virion or injection of the nucleic acid only), protein and nucleic acid synthesis, assembly and packaging, and virion release.

8.4 • Viruses can replicate only in their correct host cells. Bacterial viruses have proved useful as model systems because their host cells are easy to grow and manipulate in culture. Many animal viruses can be grown in cultured animal cells. Viruses can be quantified (titered) by a plaque assay. Plaques are clearings that develop on lawns of host cells, and in analogy to bacterial colonies, arise from the viral infection of a single cell.

8.5 • The attachment of a virion to a host cell is a highly specific process. Recognition proteins on the virus recognize specific receptors on the host cell. Sometimes the entire virion enters the host cell, whereas in other cases, as with most bacteriophages, only the viral genome enters.

8.6 • Bacteriophage T4 contains a double-stranded DNA genome that is both circularly permuted and terminally redundant. T4 encodes its own DNA polymerase and several other replication proteins. Cells employ restriction enzymes in attempts to destroy viral and other foreign DNA, but T4 has chemically modified its DNA to make it resistant to such attack. Cells also modify their own DNA to protect it from their own restriction enzymes.

8.7 • After a T4 virion penetrates a host cell, viral genes are expressed and regulated so as to redirect the host synthetic machinery to make viral nucleic acid and protein. Early viral genes encode viral genome replication events; middle and late viral genes encode structural proteins and capsid assembly. Once T4 components have been synthesized, new virions are made, primarily by self-assembly, and the virions released after lysis of the host cell.

8.8 • Some bacteriophages are temperate, meaning that they can initiate lytic events or integrate into the host genome as a prophage. This initiates a state called lysogeny in which the virus does not destroy the cell. A well-studied lysogenic virus of *Escherichia coli* is phage lambda; this phage uses a complex regulatory system to govern whether the lytic or lysogenic state is initiated following infection.

8.9 • The most common viruses on Earth are the complex bacteriophages with heads and tails, such as T4 and lambda. The double-stranded DNA genomes of these phages encode hundreds of proteins. These viruses have been used as model systems not only for virus replication but also for molecular biology and genetics.

8.10 • There are animal viruses with all known modes of viral genome replication. Many animal viruses are enveloped, picking up portions of host membrane as they leave the cell. Viral infection of animal host cells can result in cell lysis, but latent or persistent infections are also common, and a few animal viruses can cause cancer. Retroviruses like the AIDS virus are RNA viruses that employ the enzyme reverse transcriptase to replicate their RNA genome through a DNA intermediate. The DNA can integrate into the host chromosome where it can later be transcribed to yield viral mRNA and genomic RNA.

8.11 • The number of viruses on Earth is greater than the number of cells by 10-fold. Most of the genetic diversity on Earth resides in virus genomes, most of which are still to be investigated. Viruses affect their host cells by either culling the host population or by carrying out horizontal gene transfer from one bacterial cell to another. In the oceans, both *Bacteria* and *Archaea* are likely to be infected with viruses.

REVIEW OF KEY TERMS

Bacteriophage a virus that infects bacterial cells

Capsid the protein shell that surrounds the genome of a virus particle

Capsomere the subunit of a capsid

Concatemer two or more linear nucleic acid molecules joined covalently in tandem

Early protein a protein synthesized soon after virus infection and before replication of the virus genome

Enveloped in reference to a virus, having a lipoprotein membrane surrounding the virion

Host cell a cell inside which a virus replicates

Late protein a protein, typically a structural protein, synthesized late in virus infection

Lysogen a bacterium containing a prophage

Lysogeny a state in which the viral genome is replicated in step with the genome of the host

Lytic pathway the type of virus infection that leads to virus replication and destruction of the host cell

Middle protein a protein with either a structural or catalytic function synthesized after the early proteins in a virus infection

Nucleocapsid the complex of nucleic acid and proteins of a virus

Overlapping genes two or more genes in which part or all of one gene is embedded in the other

Plaque a zone of lysis or growth inhibition caused by virus infection of a lawn of sensitive host cells

Prophage the lysogenic form of a bacteriophage (See *provirus*)

Provirus the genome of a temperate or latent animal virus when it is replicating in step with the host chromosome

Replicative form a double-stranded DNA molecule that is an intermediate in the replication of viruses with single-stranded DNA genomes

Retrovirus a virus whose RNA genome is replicated via a DNA intermediate

Reverse transcriptase the retroviral enzyme that can produce DNA from an RNA template

RNA replicase an enzyme that can produce RNA from an RNA template

Rolling circle replication a DNA replication mechanism in which one strand is nicked and unrolled for use as a template to synthesize a complementary strand

Temperate virus a virus whose genome can replicate along with that of its host without causing cell death, in a state called lysogeny (bacterial viruses) or latency (animal viruses)

Titer the number of infectious virions in a viral suspension

Virion the infectious virus particle; the viral genome surrounded by a protein coat and sometimes other layers

Virulent virus a virus that lyses or kills the host cell after infection

Virus a genetic element containing either RNA or DNA surrounded by a protein capsid and that replicates only inside host cells

REVIEW QUESTIONS

1. Define virus. What are the minimal features needed to fit your definition? (Section 8.1)

2. In what ways do viral genomes differ from those of cells? (Section 8.1)

3. What are the major components of a virus particle? (Section 8.2)

4. Why does a one-step growth curve differ in shape from that of a bacterial growth curve? (Section 8.3)

5. Describe the events that occur on an agar plate containing a bacterial lawn when a single bacteriophage particle causes the formation of a bacteriophage plaque. (Section 8.4)

6. How is a viral suspension quantified and what is meant by the word "titer"? (Section 8.4)

7. What is required for a bacteriophage T4 virion to attach to an *Escherichia coli* cell? (Section 8.5)

8. What is different about the penetration process of bacteriophages versus animal viruses? (Sections 8.5 and 8.10)

9. In terms of its structure, how does the genome of bacteriophage T4 resemble and how does it differ from that of *Escherichia coli*? (Section 8.6)

10. Bacteriophage T4 has "early genes" and "late genes." What is meant by these classifications, and what types of proteins are encoded by each? (Section 8.7)

11. What is a temperate bacteriophage? Name a well-studied temperate phage that infects cells of *Escherichia coli*. (Section 8.8)

12. Describe different shapes of bacterial viruses. Which is the most common in nature? (Section 8.9)

13. Describe the types of genomes found in bacterial viruses. Give an example of one virus for each type of genome. (Section 8.9)

14. Describe the types of genomes found in animal viruses. (Section 8.10)

15. Why can it be said that the retrovirus genome is unique in all of biology? (Section 8.10)

16. How do viral numbers compare to those of bacteria in seawater? (Section 8.11)

17. Explain how viruses can affect *Bacteria* and *Archaea* in nature in both positive and negative ways. (Section 8.11)

APPLICATION QUESTIONS

1. What causes the viral plaques that appear on a bacterial lawn to stop growing larger?

2. The promoters on genes encoding early proteins in viruses like T4 have a different sequence than the promoters on genes encoding late proteins in the same virus. Explain how this benefits the virus.

3. Under some conditions, it is possible to obtain nucleic acid–free protein coats (capsids) of certain viruses. Under the electron microscope, these capsids look very similar to complete virions. What does this tell you about the role of the virus nucleic acid in the virus assembly process? Would you expect such particles to be infectious?

4. Contrast the enzyme(s) present in the virions of a retrovirus and a positive-strand RNA bacteriophage. Why do they differ if each has plus configuration single-stranded RNA as its genome?

CHAPTER

9 · Viral Genomes and Diversity

microbiology**now**

Ever-Expanding Viral Diversity

Viruses infect all organisms, including *Bacteria* and *Archaea,* and, collectively, viruses represent the greatest repository of genetic diversity on the planet. Many bacterial viruses (bacteriophages) and archaeal viruses have been isolated and characterized thus far. For *Bacteria*, these include both DNA and RNA phages, some with single-stranded and others with double-stranded genomes. For *Archaea,* however, no RNA viruses are known. Is this because such viruses do not exist?

All known archaeal viruses have DNA genomes, and with rare exception, double-stranded circular DNA genomes. In the past decade, researchers exploring viral diversity in hot springs in Yellowstone National Park (photo) have discovered a large number of these unusually shaped and structurally tough archaeal parasites (photo inset), but often wondered why evidence for RNA archaeal viruses never emerged in their studies. Well, now it has.

Using the powerful tools of metagenomics, researchers studying highly acidic *Archaea*-dominated Yellowstone hot springs have detected viral RNA genome segments highly diverged from those of RNA viruses of eukaryotes and even more distant from the genomes of RNA bacteriophages.[1] The viral RNA pieces were assembled into several different intact genomes that were all single-stranded and of the plus sense. Sequence analyses confirmed that each genome encoded an RNA replicase—a hallmark of RNA viruses—and that some of the archaeal viruses likely replicated by way of polyprotein formation, a replication mechanism employed by some eukaryotic plus-sense RNA viruses, such as poliovirus.

This metagenomics approach to viral diversity revealed that RNA archaeal viruses indeed exist. When further work complements this with the actual isolation of RNA virions that replicate in cultures of *Archaea,* virology will have a new window for exploring the amazing diversity of the viral world.

[1]Bolduc, B., et al. 2012. Identification of novel positive-strand RNA viruses by metagenomic analysis of *Archaea*-dominated Yellowstone hot springs. *J. Virol. 86*: 5562–5573.

Viruses have DNA or RNA genomes that can be either single-stranded or double-stranded. Compared with cells, viral genomes can create some unusual challenges for genetic information flow. In this chapter we will explore viral diversity from a genomic perspective. Our coverage will group viruses by genome structure rather than by the type of host cell they infect, since viruses with the same genome structure face common problems in genetic information flow.

I • Viral Genomes and Evolution

9.1 Size and Structure of Viral Genomes

Viral genomes vary almost a thousandfold in size from smallest to largest. DNA viruses exist along this entire gradient from the tiny circovirus, whose 1.75-*kilobase* single-stranded genome pales in comparison to that of the 1.25-*megabase*-pair double-stranded DNA genome of *Megavirus* (**Figure 9.1**). RNA genomes, whether single- or double-stranded, are typically smaller than DNA viruses. Although some viral genomes are larger than those of some prokaryotes, genomes of prokaryotes are typically much larger than those of viruses, and genomes of eukaryotes much larger than those of prokaryotes (Figure 9.1).

Whether a viral genome is large or small, once a virus has infected its host, transcription of viral genes must occur and new copies of the viral genome must be made. Only later, once viral proteins begin to appear from the translation of viral transcripts, can viral assembly begin. For certain RNA viruses, the genome is also the mRNA. For most viruses, however, viral mRNA must first be made by transcription off of the DNA or RNA genome, and we consider the variations on how this occurs now.

Viral Genome Structure: The Baltimore Scheme

The virologist David Baltimore, who along with Howard Temin and Renato Dulbecco shared the Nobel Prize for Physiology or Medicine in 1975 for the discovery of retroviruses and reverse transcriptase, developed a classification scheme for viruses. The scheme is based on the relationship of the viral genome to its mRNA and recognizes seven classes of viruses (**Figure 9.2**). By convention in virology, viral mRNA is always considered to be of the *plus* configuration. Thus, to understand the molecular biology of a particular class of virus, one must know the nature of the viral genome and what steps are necessary to produce plus complementarity mRNA (Figure 9.2).

Double-stranded DNA viruses are in Baltimore class I. The mechanism of mRNA production and genome replication of class I viruses is the same as that used by the host cell, and we saw this with bacteriophage T4, a typical class I virus (↻ Section 8.7). A virus containing a single-stranded genome may be either a **positive-strand** virus (also called a "plus-strand virus") or a **negative-strand** virus (also called a "minus-strand virus"). Class II viruses contain single-stranded plus-strand DNA genomes. Transcription of such a genome would yield a message of the minus sense. Therefore, before mRNA can be produced from class II viruses, a complementary DNA strand must first be made to form a double-stranded DNA intermediate; this is called the **replicative form**. The latter is used for transcription and as the source of new genome copies, the plus strand becoming the genome while the minus strand is discarded (Figure 9.2). With only one known exception, all single-stranded DNA viruses are positive-strand viruses.

The production of mRNA and genome replication will obviously be different for RNA viruses than for DNA viruses. Cellular RNA polymerases do not catalyze the formation of RNA from an RNA template, but instead require a DNA template. Therefore, depending on the virus, RNA viruses must either carry in their virions or encode in their genomes an RNA-dependent RNA polymerase called *RNA replicase* (↻ Section 8.2). With positive-strand RNA viruses (class IV), the genome is also mRNA. But for negative-strand RNA viruses (class V), RNA replicase must synthesize a plus strand of RNA off of the negative strand template, and the plus strand is then used as mRNA. The latter is also used as a template to make more negative-strand genomes (Figure 9.2). RNA viruses of class III face a similar problem but start with double-stranded (+/−) RNA instead of only a positive or negative strand.

Retroviruses are animal viruses whose genomes consist of single-stranded RNA of the plus configuration but which replicate through a double-stranded DNA intermediate (class VI). The

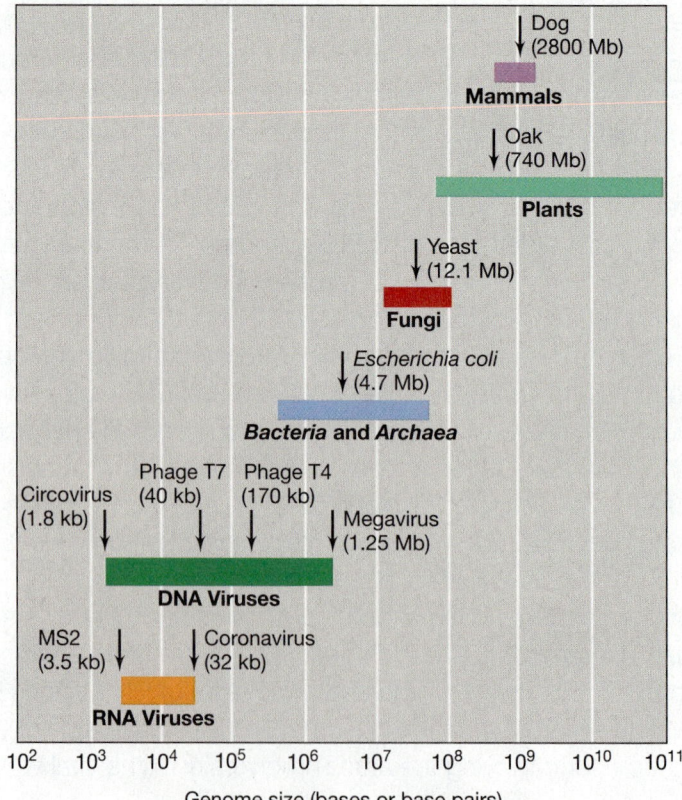

Figure 9.1 Comparative genomics. A size comparison of viral genomes and those of major groups of living organisms.

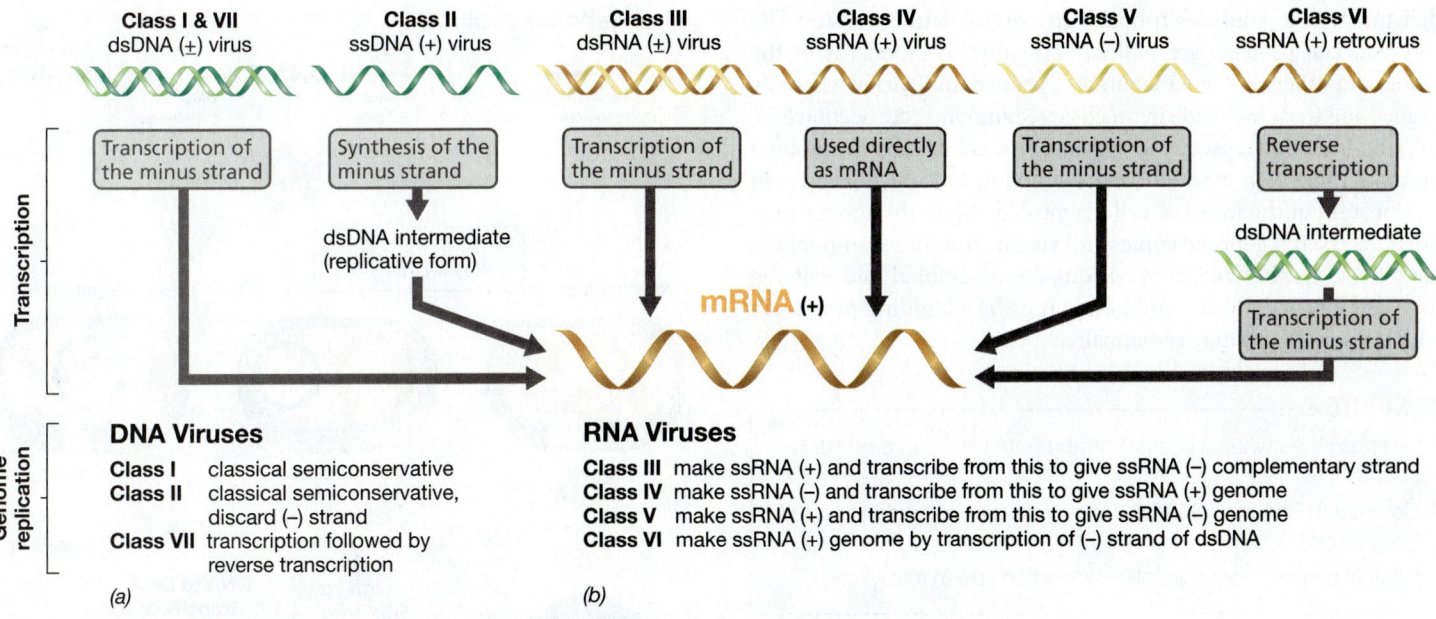

Figure 9.2 **The Baltimore classification of viral genomes.** Seven classes of viral genomes are known. The genomes can be either *(a)* DNA or *(b)* RNA, and either single-stranded (ss) or double-stranded (ds). The path each viral genome takes to form its mRNA and the strategy each uses for replication is shown.

process of copying the information found in RNA into DNA is called **reverse transcription** and is catalyzed by an enzyme called *reverse transcriptase*. Finally, class VII viruses are those highly unusual viruses whose genomes consist of double-stranded DNA but which replicate through an RNA intermediate. As we will see, these viruses also use reverse transcriptase.

Table 9.1 lists a few example viruses in each Baltimore class, and we explore the unique molecular biology of each class as we travel through this chapter.

Viral Protein Synthesis

Once viral mRNA is made (Figure 9.2), viral proteins can be synthesized. In all viruses, these proteins can be grouped into two broad categories: (1) proteins synthesized soon after infection, called *early proteins*, and (2) proteins synthesized later in the infection, called *late proteins*. Both the timing and amount of viral protein synthesis is highly regulated. Early proteins are typically enzymes that act catalytically and are therefore synthesized in relatively small amounts. These include not only nucleic acid polymerases but also proteins

Table 9.1 Some types of viral genomes

Virus	Host	DNA or RNA	Single- or double-stranded	Structure	Number of molecules	Size (bases or base pairs)[a]
H-1 parvovirus	Animals	DNA	Single-stranded	Linear	1	5,176
φX174	*Bacteria*	DNA	Single-stranded	Circular	1	5,386
Simian virus 40 (SV40)	Animals	DNA	Double-stranded	Circular	1	5,243
Poliovirus	Animals	RNA	Single-stranded	Linear	1	7,433
Cauliflower mosaic virus	Plants	DNA	Double-stranded	Circular	1	8,025
Cowpea mosaic virus	Plants	RNA	Single-stranded	Linear	2 different	9,370 (total)
Reovirus type 3	Animals	RNA	Double-stranded	Linear	10 different	23,549 (total)
Bacteriophage lambda	*Bacteria*	DNA	Double-stranded	Linear	1	48,514
Herpes simplex virus type 1	Animals	DNA	Double-stranded	Linear	1	152,260
Bacteriophage T4	*Bacteria*	DNA	Double-stranded	Linear	1	168,903
Human cytomegalovirus	Animals	DNA	Double-stranded	Linear	1	229,351

[a]The size is in bases or base pairs depending on whether the virus is single- or double-stranded. The sizes of the viral genomes chosen for this table are known accurately because they have been sequenced. However, this accuracy can be misleading because only a particular strain or isolate of a virus was sequenced. Therefore, the sequence and exact number of bases for other isolates may be slightly different. No attempt has been made to choose the largest and smallest viruses known, but rather to give a fairly representative sampling of the sizes and structures of the genomes of viruses containing both single- and double-stranded RNA and DNA.

that function to shut down host transcription and translation. By contrast, late proteins are typically structural components of the virion and other proteins that are not needed until virion assembly begins, and these are made in much larger amounts (↩ Section 8.7).

Virus infection upsets the regulatory mechanisms of the host because there is a marked overproduction of viral nucleic acid and protein in the infected cell. Eventually, when the proper proportions of viral genome copies and virion structural components have been synthesized, new virions are assembled and exit the host cell by either lysing and killing it or by a budding process in which the host cell may remain alive.

MINIQUIZ

- Distinguish between a positive-strand RNA virus and a negative-strand RNA virus.
- Contrast mRNA production in the two classes of single-stranded RNA viruses.
- What is unusual about genetic information flow in retroviruses?

9.2 Viral Evolution

When did viruses first appear on Earth and what is their relationship to cells? All known viruses require a host cell for their replication, and this leads to the natural conclusion that viruses evolved at some time *after* cells first appeared on Earth, nearly 4 billion years ago. Following this line of reasoning, viruses are most likely remnant cell components that evolved an ability to replicate with assistance from the cell. However, other hypotheses for the origin of viruses have been proposed including that viruses are relics of the "RNA world," a period in evolution where RNA is hypothesized to have been the sole carrier of genetic information (↩ Section 12.1 and see Figure 9.3), or that viruses were once cells that for some reason, perhaps to economize their genomes, discarded so many genes that they became dependent on a host for most of their replication functions.

Although *how* viruses appeared is an unanswered question, so is the question of *why* viruses appeared. One likely driver of viral evolution was as a mechanism for cells to quickly move genes about in nature. Because viruses have an extracellular form that protects the nucleic acid inside them, they could have been selected as a means of enriching the genetic diversity (and thus fitness) of cells by facilitating gene transfers between them. This function seems especially relevant for prokaryotic cells, where horizontal gene exchange is known to be a major factor in their rapid evolution (↩ Sections 6.12 and 12.7). Although many viruses kill their host cell, latent viruses do not, and it is possible that the earliest viruses were primarily latent and evolved lytic capacities only later to more rapidly access new hosts.

Viruses and the Transition from an RNA to a DNA World

Besides a likely role in facilitating genetic diversity, it is possible that viruses were the first DNA-containing entities. The hypothesis of the RNA world proposes that RNA was the genetic material of the earliest cells and that DNA eventually replaced RNA in this role because DNA is the more stable of the two molecules. Interestingly, a novel hypothesis that accounts for the RNA to DNA transition puts viruses at the center of the story (**Figure 9.3**).

Viral genome evolution

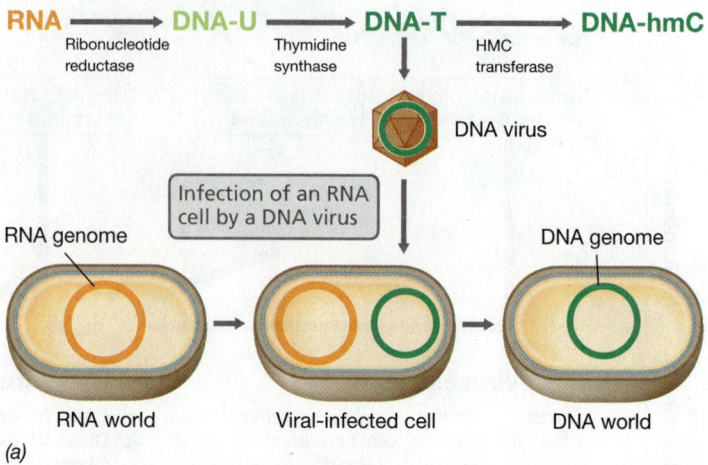

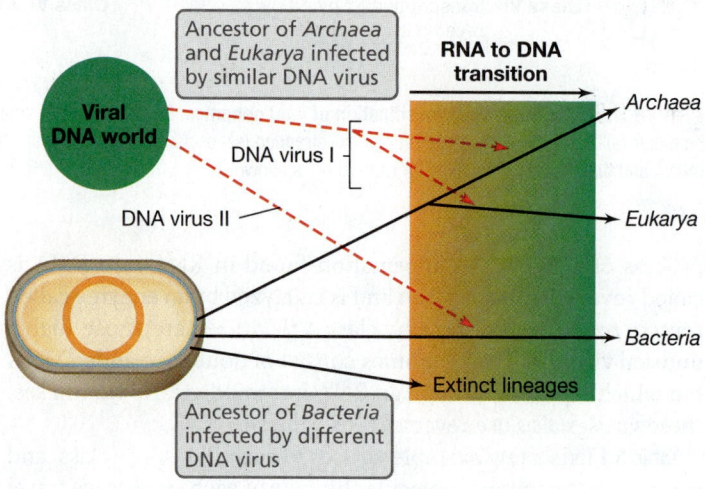

Figure 9.3 Hypothesis of viral origin of DNA. *(a)* The evolution of DNA-specific enzymes would have allowed RNA viruses to become DNA viruses; DNA-U, DNA with uracil; DNA-T, DNA with thymine; DNA-hmC, DNA with 5-hydroxycytosine. These DNA variants are known from one virus or another. *(b)* Infection of an RNA cell by a DNA virus could have exposed cells to the more stable chemistry of DNA over RNA. Infection of the ancestors of *Bacteria* with a considerably different DNA virus than those infecting *Archaea* and *Eukarya* could explain why the DNA machinery of *Archaea* and *Eukarya* differs from that of *Bacteria*.

The scenario posits that RNA viruses first evolved DNA as a modification mechanism to protect their genomes from cellular ribonucleases that could destroy them. Because DNA is not RNA and cells in the RNA world would have contained RNA genomes, DNA viruses would have had to evolve their own DNA replication machinery to replicate their genomes. It is further hypothesized that DNA viruses infected the ancestors of the three cellular domains. Gradually, by genetic exchange with the DNA viral genomes, each group of cells obtained the machinery necessary to replicate DNA and eventually converted their genomes from RNA-based to DNA-based chemistry. Furthermore, cells with RNA genomes that were not infected by DNA viruses never evolved DNA genomes, and Darwinian selection would have eventually driven these less fit cells extinct (Figure 9.3b). It is conceivable that an enzyme like reverse transcriptase was a key to the

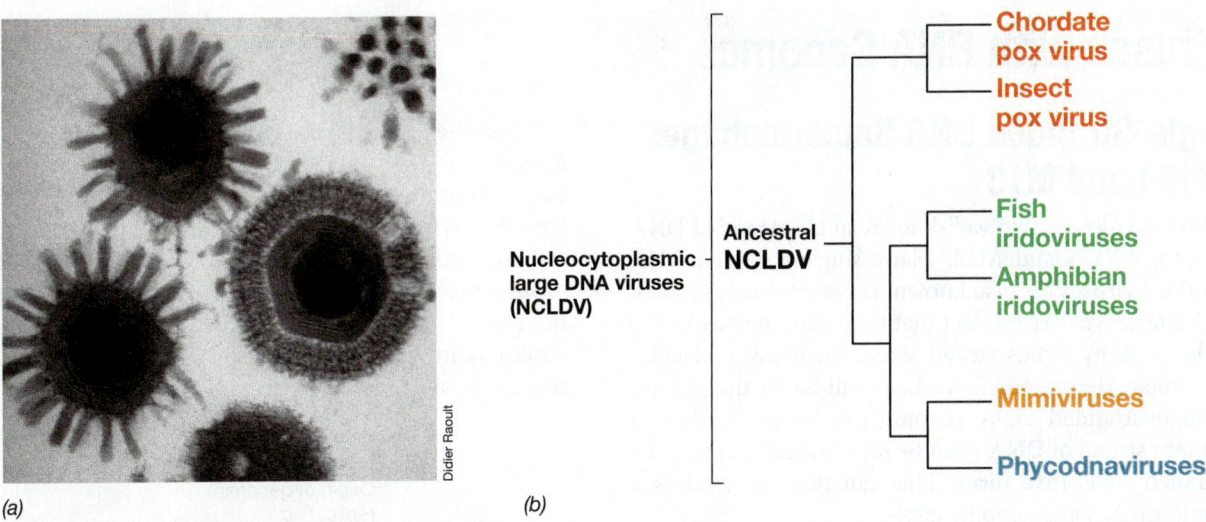

Figure 9.4 Phylogeny of nucleocytoplasmic large DNA viruses (NCLDV). *(a)* Transmission electron micrograph of Mimivirus, a member of the NCLDV group. A virion is about 0.75 μm in diameter. *(b)* Phylogeny of major groups of NCLDV based on comparative sequences of several proteins of DNA metabolism.

conversion of RNA into DNA, just as it is in retroviruses today (Section 9.11).

Why is this hypothesis appealing? The three-domain theory shows the *Archaea* and *Eukarya* to be more closely related than either is to the *Bacteria* (🗗 Figure 1.6). Although molecular analyses of the machinery required for transcription and translation support the three-domain hypothesis well, analyses of the molecular machinery for DNA replication, recombination, and repair do not. A few DNA-specific processes are similar in *Bacteria* and *Archaea*, whereas most are more similar in *Archaea* and *Eukarya*. The viral DNA scenario accounts for this discrepancy by proposing that although transcription and translation were established processes before the three domains became distinct, DNA-centered events were not. Instead, DNA biochemistry was the result of viral infections. The complement of enzymes of DNA metabolism that we see in cells today is then explained by hypothesizing that the ancestors of the *Archaea* and *Eukarya* were infected by a similar DNA virus that was distinct from the DNA virus that infected the ancestor of *Bacteria* (Figure 9.3).

The viral DNA scenario also explains how DNA originated in cells in the first place and provides a mechanism for how RNA genomes could have gradually been replaced by DNA. If this hypothesis is true, it has one ironic feature. In attempts to stay one step ahead of their hosts, one of the "avoidance maneuvers" of viruses was the evolution of DNA. But in this case, the chemistry of this molecule was such an improvement over that of RNA that cells co-opted the strategy for their own benefit.

Viral Phylogeny

Because the sequence diversity of viral genomes is so enormous—most viral genes retrieved from nature are of unknown function—it has not proven possible to construct an insightful universal phylogenetic tree of viruses like that for cells (🗗 Figure 1.6). Only in a few groups of viruses has it been possible to reliably trace phylogenies, and in these cases, trees have been assembled from sequences of a few select genes or proteins shared in

common among the group. One such example is Mimivirus and its relatives, the largest of all known viruses (**Figure 9.4**).

Mimivirus capsids are multilayered and icosahedral. The virion is surrounded by spikes and is nearly 0.75 μm in diameter, larger than some prokaryotic cells (Figure 9.4a). Mimivirus contains a 1.2-megabase-pair genome consisting of double-stranded DNA. This is over twice that of the next largest known virus, and is larger than the genomes of several prokaryotes (🗗 Table 6.1). Mimivirus infects the protozoan *Acanthamoeba* and belongs to a group of giant viruses with large genomes called *nucleocytoplasmic large DNA viruses* (NCLDV) (Figure 9.4b). The NCLDV comprise several virus families, including pox viruses (Section 9.6), iridoviruses, and certain plant viruses. These viruses share a set of highly homologous proteins, most of which function in DNA metabolism. A phylogenetic tree of these viruses constructed from DNA sequences encoding these proteins shows how they have diverged from a common ancestor (Figure 9.4b).

It is thus possible to track the phylogeny of a viral group in some cases. But to do so, one needs to begin with a group that is already known to share a number of properties in common. Other attempts to trace viral phylogeny using the comparative structural biology of capsid proteins have also proven useful with some viral groups (see page 245). Although phylogenetic trees based on cellular genomes have been constructed and support the three-domain hypothesis quite well, evolution has so scrambled viral genomes that it is unlikely that an insightful universal phylogenetic tree of viruses based on comparisons of full viral genome sequences is possible, at least not using currently available computational tools.

MINIQUIZ -

- How could viruses have accelerated the evolution of cells?
- Explain how viruses could have "invented" the genetic material found in all cells.
- Give two reasons why a universal phylogenetic tree of viruses may prove hard to construct.

II • Viruses with DNA Genomes

9.3 Single-Stranded DNA Bacteriophages: φX174 and M13

In this section we discuss two well-known single-stranded DNA bacteriophages, φX174 and M13. Many single-stranded DNA plant and animal viruses are also known. However, because these share with bacterial viruses the fact that their genomes are of the plus complementarity ("plus-strand viruses"), many molecular events are similar. Hence, our focus here will be on the phages. Before a single-stranded DNA genome can be transcribed, a complementary strand of DNA must be synthesized, forming the double-stranded replicative form. This can then be used as a source of both mRNA and genome copies.

Phage φX174

Bacteriophage φX174 contains a circular genome of 5386 nucleotides inside a tiny icosahedral virion, about 25 nm in diameter. Phage φX174 has only a few genes and shows the phenomenon of **overlapping genes**, a condition in which there is insufficient DNA to encode all viral-specific proteins unless parts of the genome are read more than once in different reading frames. For example, in the φX174 genome, gene B resides within gene A and gene K resides within both genes A and C (**Figure 9.5**). Genes D and E also overlap, gene E being contained completely within gene D. Also, the termination codon of gene D overlaps the initiation codon of gene J (Figure 9.5a).

The distinct gene products from overlapping genes are made by reinitiating transcription *in a different reading frame* within a gene to yield a second (and distinct) transcript. In addition to overlapping genes, a small protein in φX174 called A*, which functions to shut down host DNA synthesis, is synthesized by the reinitiation of *translation* (not transcription) within the mRNA for gene A. The A* protein is read from the same mRNA reading frame as A protein but has a different in-frame start codon and is thus a shorter protein.

Upon infection of an *Escherichia coli* cell by φX174, the viral DNA is separated from the protein coat and the genome is converted into a double-stranded replicative form by host enzymes. From this, several copies are made by semiconservative replication, and phage-specific transcripts are made by transcription off of the negative strand of the replicative form (Figure 9.5b). The replicative form is also the starting point for making copies of the phage genome by a mechanism we have already seen used in phage lambda (↩ Section 8.8): **rolling circle replication** (**Figure 9.6**).

In the synthesis of the φX174 genome, the rolling circle facilitates the continuous production of positive strands from the replicative form. To do this, the positive strand of the latter is nicked and the 3′ end of the exposed DNA is used to prime synthesis of a new strand (Figure 9.6). Cutting of the plus strand is accomplished by the A protein (Figure 9.5a). Continued rotation of the circle leads to the synthesis of a linear φX174 genome. Note that synthesis differs from semiconservative replication because only the negative strand serves as a template.

When the growing viral strand reaches unit length (5386 residues for φX174), the A protein cleaves it and then ligates the two ends of the newly synthesized single strand to give a single-stranded DNA circle. Ultimately, assembly of mature φX174 virions occurs and cell lysis follows. The E protein (Figure 9.5a) promotes cell lysis by inhibiting the activity of an enzyme in peptidoglycan synthesis (↩ Section 5.4). Because of the resulting weakness in newly synthesized cell wall material, the cell ruptures, releasing the phage virions.

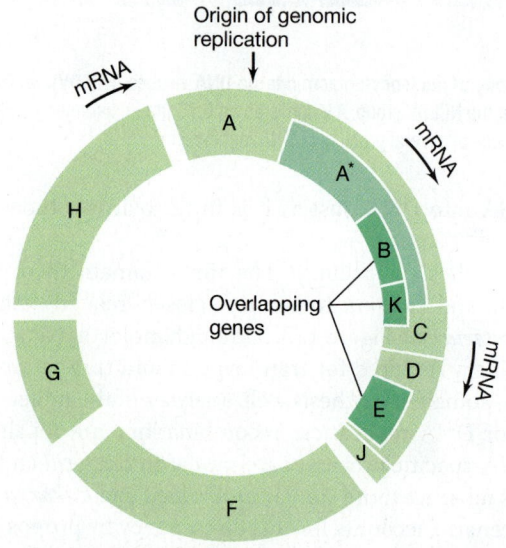

A	Replicative form DNA synthesis	E	Host cell lysis
A*	Shutoff of host DNA synthesis	F	Major capsid protein
B	Formation of capsid precursors	G	Major spike protein
C	DNA maturation	H	Minor spike protein
D	Capsid assembly	J	DNA packaging protein
		K	Function unknown

(a) **Genetic map of φX 174**

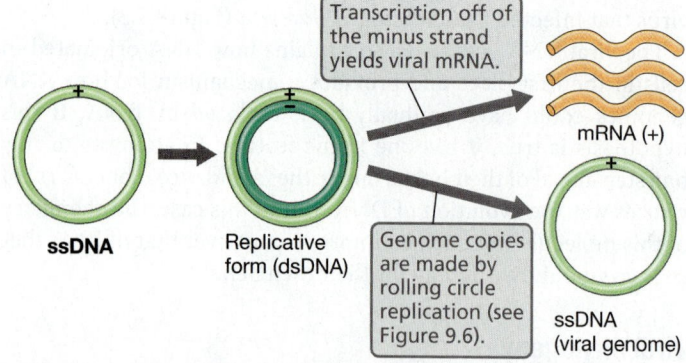

(b) **Flow of events during φX 174 replication**

Figure 9.5 **Bacteriophage φX174, a single-stranded DNA phage.** *(a)* Genetic map. Note regions of gene overlap. Protein A* is formed using only part of the coding sequence of gene A by reinitiation of translation. The key indicates the functions of the proteins encoded by each gene. *(b)* Genetic information flow in φX174. Progeny single-stranded DNA is produced from the replicative form by rolling circle replication (see Figure 9.6).

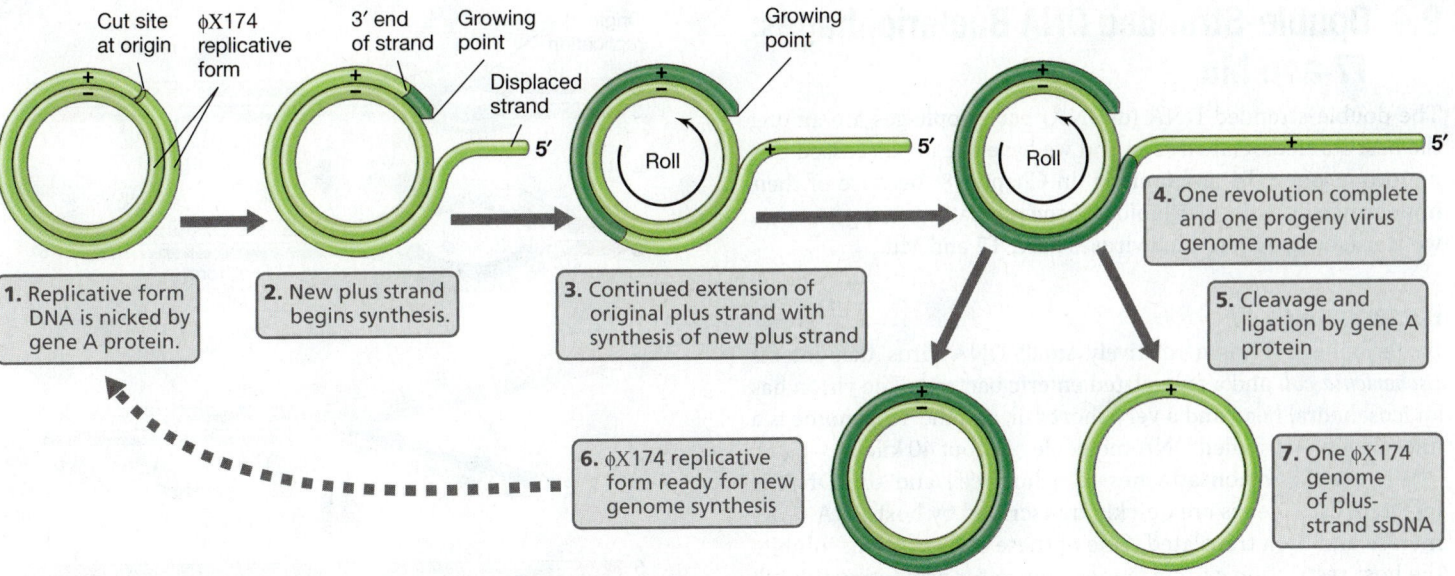

Figure 9.6 Rolling circle replication in phage φX174. Replication begins at the origin of the double-stranded replicative form with the cutting of the plus strand of DNA by gene A protein (both strands of DNA are shown in light green here for simplification). After one new progeny strand has been synthesized (one revolution of the circle), the gene A protein cleaves the new strand and ligates its two ends.

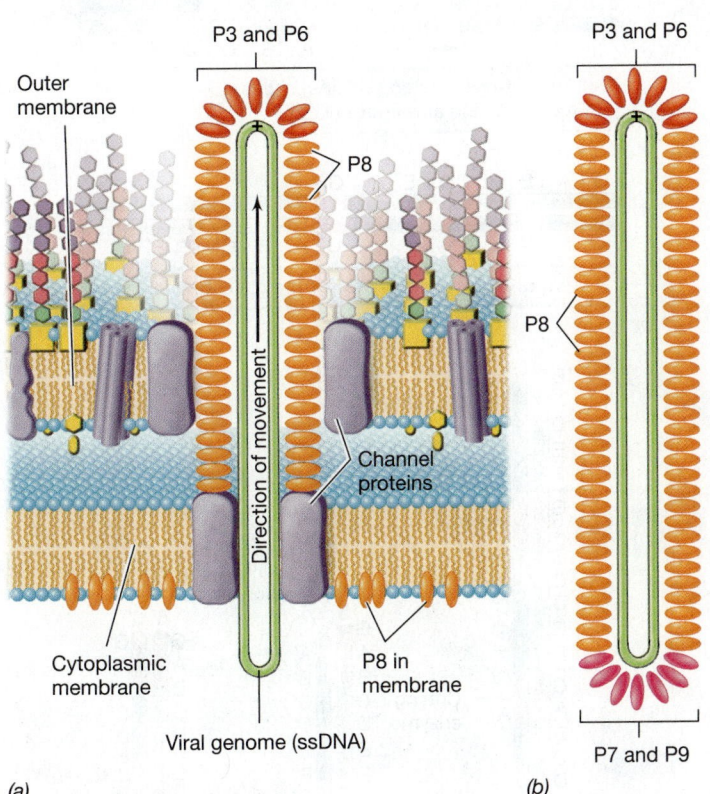

(a)   (b)

Figure 9.7 Release of phage M13. The virions of phage M13 exit infected cells without lysis. *(a)* Budding. The virus DNA crosses the cell envelope through a channel constructed from virus-encoded proteins. As this occurs, the DNA is coated with phage proteins that have been embedded in the cytoplasmic membrane. *(b)* Complete virion. The two ends of the virion are covered with small numbers of the minor coat proteins P3 and P6 (front end) or P7 and P9 (rear end).

Bacteriophage M13

Bacteriophage M13 is a filamentous phage with helical symmetry that has found extensive use in past years as a cloning and DNA-sequencing vector in genetic engineering. The virion of phage M13 is long and thin and attaches to the pilus of its host cell (⮌ Section 8.5). Filamentous phages such as M13 have the unusual property of being released without lysing the host cell; infected cells continue to grow and typical plaques are not observed. To facilitate this, M13 DNA is covered with coat proteins as it crosses the cell envelope. Four minor coat proteins cover the tips of the virion while the major coat protein covers the sides (**Figure 9.7**). Thus with M13, there is no intracellular accumulation of virions as occurs with typical lytic bacteriophages.

Several features of phage M13 make it useful as a cloning and DNA sequencing vehicle. For example, many aspects of DNA replication in M13 are similar to those of φX174 and the genome is very small; this facilitates sequencing efforts. Second, a double-stranded form of genomic DNA essential for cloning purposes is produced naturally when M13 produces its replicative form. Third, as long as infected cells are kept growing, they can be maintained indefinitely, so a continuous source of the cloned DNA is made. And finally, as in phage lambda (⮌ Section 8.8), there is a region of the M13 genome that does not encode proteins and can be replaced by foreign DNA. Consequently, phage M13 is an important component of the biotechnologist's toolbox.

MINIQUIZ

- Why is formation of the replicative form of φX174 necessary in order to make phage-specific mRNA?

- In the φX174 genome, describe the difference between how the gene B and gene A* proteins are made.

- How can M13 virions be released without killing the infected host cell?

9.4 Double-Stranded DNA Bacteriophages: T7 and Mu

The double-stranded DNA (dsDNA) bacteriophages are among the best studied of all viruses, and we have already discussed two important ones, T4 and lambda, in Chapter 8. Because of their importance in molecular biology, gene regulation, and genomics, we consider two more such viruses here, T7 and Mu.

Bacteriophage T7

Bacteriophage T7 is a relatively small DNA virus that infects *Escherichia coli* and a few related enteric bacteria. The virion has an icosahedral head and a very short tail, and the T7 genome is a linear double-stranded DNA molecule of about 40 kilobase pairs.

When a T7 virion attaches to a host cell and the DNA is injected, early genes are quickly transcribed by host RNA polymerase and then translated. One of these early proteins inhibits the host restriction system, a mechanism for protecting the cell from foreign DNA (♂⊋ Section 8.6). This occurs very rapidly, as the T7 anti-restriction protein is made and becomes active before the entire T7 genome has entered the cell. Other early proteins include a T7 RNA polymerase and proteins that inhibit host RNA polymerase activity. T7 RNA polymerase recognizes only T7 gene promoters distributed along the T7 genome. This transcriptional strategy differs from that of phage T4 because T4 uses the host RNA polymerase throughout its replication cycle but modifies it such that it recognizes phage genes only (♂⊋ Section 8.7).

Genome replication in T7 begins at an origin of replication within the molecule and proceeds bidirectionally from this point (**Figure 9.8a**). Phage T7 uses its own DNA polymerase, which is a composite protein including one polypeptide encoded by the phage and one by the host. As in phage T4, T7 DNA contains terminal repeats at both ends of the molecule and these are eventually used to form *concatemers* (Figure 9.8b). Continued replication and recombination leads to concatemers of considerable length, but ultimately a phage-encoded endonuclease cuts each concatemer at a specific site, resulting in the formation of linear DNA molecules with terminal repeats that are packaged into phage heads (Figure 9.8c). However, because T7 endonuclease cuts the concatemer at specific sequences, the DNA sequence in each T7 virion is identical. This differs from the situation in phage T4, where DNA concatemers are processed using a "headful mechanism" that generates circularly permuted genomes (♂⊋ Section 8.6).

Bacteriophage Mu

Bacteriophage Mu is a temperate phage, like lambda (♂⊋ Section 8.8), but has the unusual property of replicating by *transposition*. Transposable elements are sequences of DNA that can move

Figure 9.8 Replication of the bacteriophage T7 genome. *(a)* The linear, double-stranded DNA undergoes bidirectional replication, giving rise to intermediate "eye" and "Y" forms (for simplicity, both template strands are shown in light green and both newly synthesized strands in dark green). *(b)* Formation of concatemers by joining DNA molecules at their unreplicated terminal ends. *(c)* Production of mature viral DNA molecules from T7 concatemers by activity of the cutting enzyme, an endonuclease.

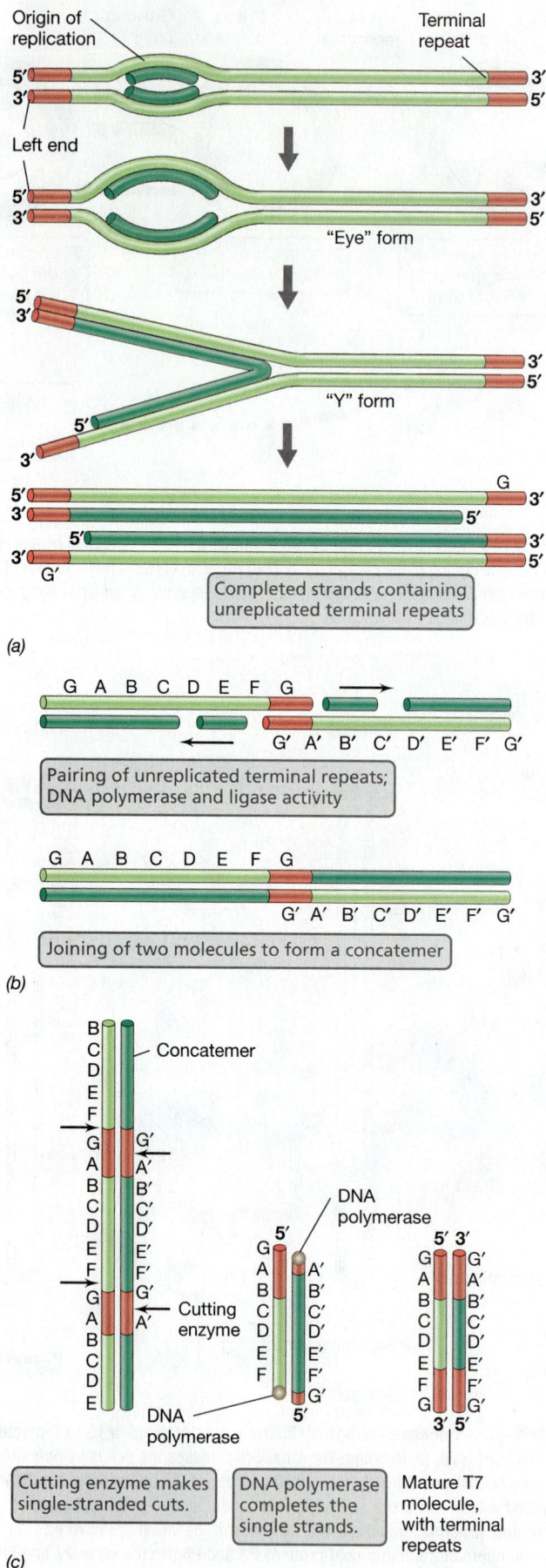

on their host genome from one location to another as discrete genetic units (⟳ Section 10.11); such movements are facilitated by an enzyme called **transposase**. Mu was so named because it generates *mu*tations when it integrates into the host cell chromosome, and thus it has been useful in bacterial genetics because it can generate mutants easily.

Bacteriophage Mu has an icosahedral head, a helical tail, and several tail fibers (**Figure 9.9a**). The genome of Mu consists of linear double-stranded DNA, and most Mu genes encode head and tail proteins, other replication factors such as the Mu transposase, and factors that affect host range. Host range is controlled by the kind of tail fibers that are made, with one type allowing only infection of *Escherichia coli* while the other type allows the phage to infect several other enteric bacteria as well.

Phage Mu replicates in a completely different manner from all other bacteriophages because its genome is replicated as part of a larger DNA molecule (Figure 9.9b). Thus, integration of Mu DNA into the host genome is essential for both lytic and lysogenic development. Integration requires the activity of Mu transposase,

and a 5-base-pair fragment of host DNA is duplicated at the target site where Mu DNA is integrated. This host DNA duplication arises because staggered cuts are made at the point in the host genome where Mu DNA is inserted. The resulting single-stranded segments are converted to the double-stranded form as part of the Mu integration process (Figure 9.9b).

Mu can enter the lytic pathway upon initial infection if the Mu repressor is not made, or it can form a lysogen if the repressor is made. In either case, Mu DNA is replicated by repeated transposition of Mu to multiple sites on the host genome. If the lytic cycle pathway is triggered, only the early genes of Mu are initially transcribed. Then, following expression of a Mu transcriptional activating protein, Mu head and tail proteins are synthesized. Following self-assembly, the cell is lysed and mature Mu virions are released. The lysogenic state in Mu requires that sufficient Mu repressor protein be present to prevent transcription of integrated Mu DNA.

MINIQUIZ --

- In what major way does transcription of phage DNA differ in phages T4 and T7?

- What is unusual about the replication mechanism of the Mu genome?

9.5 Viruses of *Archaea*

Several DNA viruses have been discovered whose hosts are species of *Archaea*, including representatives of both the *Euryarchaeota* and *Crenarchaeota* phyla (Chapter 16). Most viruses that infect species of *Euryarchaeota*, including both methanogenic and halophilic *Archaea*, are of the "head and tail" type, resembling phages that infect enteric bacteria, such as phage T4. One novel archaeal virus infects a halophile and is unusual because it is both enveloped and contains a single-stranded DNA genome. By contrast, all other characterized archaeal DNA viruses contain double-stranded and typically circular DNA genomes.

RNA archaeal viruses have been detected in thermal environments inhabited by *Crenarchaeota*. These are single-stranded plus-sense RNA viruses (positive-strand viruses, Section 9.8), but little else is known about them as they await detailed characterization and laboratory culture. However, as for *Bacteria* and *Eukarya*, it is clear that at least some *Archaea* are infected by viruses with RNA genomes (see page 265).

The most distinctive archaeal viruses infect hyperthermophilic *Crenarchaeota*. For example, the sulfur chemolithotroph *Sulfolobus* is host to several structurally unusual viruses. One such virus, called *SSV*, forms spindle-shaped virions that often cluster in rosettes (**Figure 9.10a**). Such viruses are widespread in acidic hot springs worldwide. Virions of SSV contain a circular DNA genome of about 15 kilobase pairs. A second morphological type of *Sulfolobus* virus forms a rigid, helical rod-shaped structure (Figure 9.10b). Viruses in this class, nicknamed SIFV, contain linear DNA genomes about twice the size of that of SSV. Many variations on the spindle- and rod-shaped patterns have been seen in archaeal viral isolation studies.

A spindle-shaped virus that infects the hyperthermophile *Acidianus* displays a novel behavior. The virion, called ATV, contains

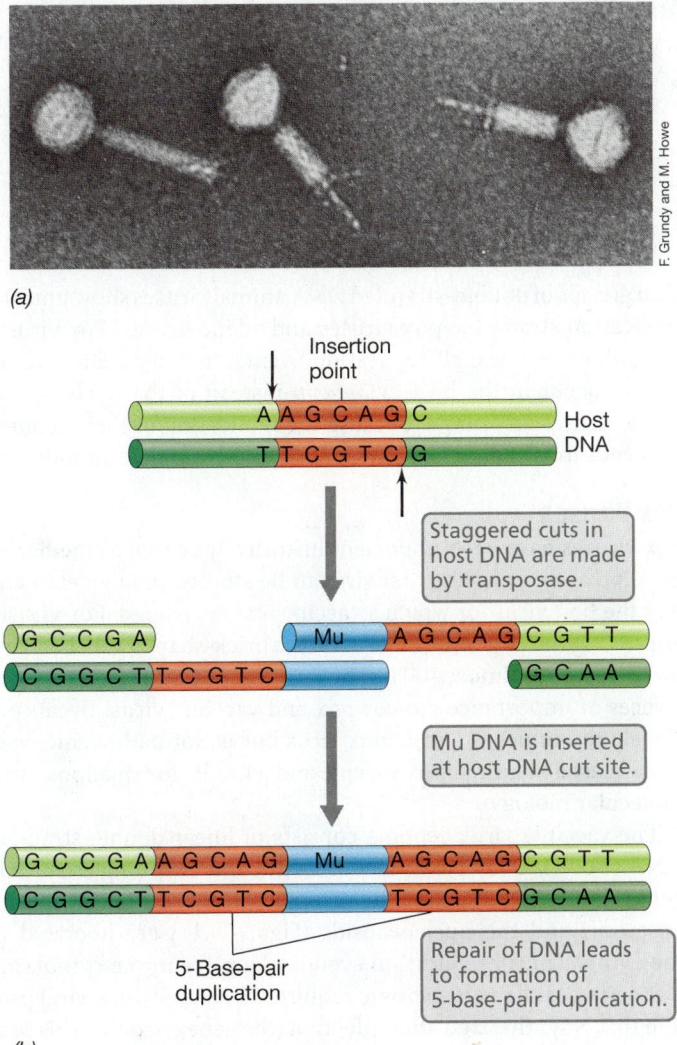

(a)

F. Grundy and M. Howe

(b)

Figure 9.9 Bacteriophage Mu. *(a)* Electron micrograph of virions of the double-stranded DNA phage Mu. *(b)* Integration of Mu into the host DNA, showing the generation of a 5-base-pair duplication of host DNA.

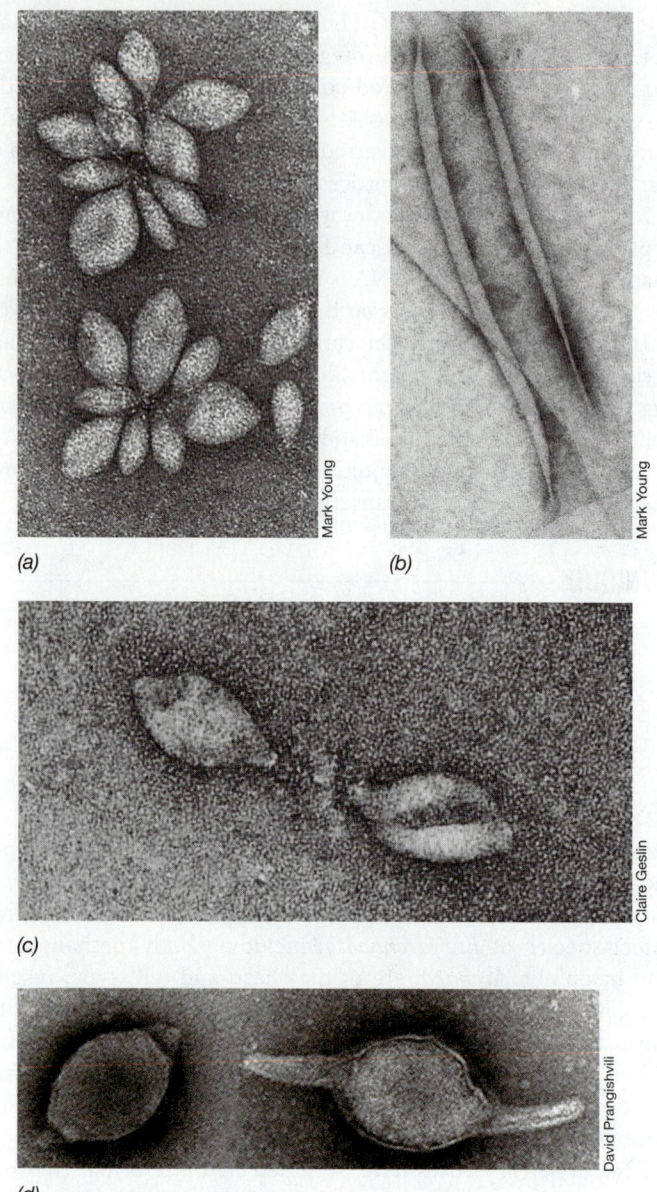

(a)

Mark Young

(b)

Mark Young

(c)

Claire Geslin

(d)

David Prangishvili

Figure 9.10 Archaeal viruses. Electron micrographs of viruses of *Crenarchaeota* (parts a, b, d), and a virus of a euryarchaeote (part c). *(a)* Spindle-shaped virus SSV1 that infects *Sulfolobus solfataricus* (virions are 40 × 80 nm). *(b)* Filamentous virus SIFV that infects *S. solfataricus* (virions are 50 × 900–1500 nm). *(c)* Spindle-shaped virus PAV1 that infects *Pyrococcus abyssi* (virions are 80 × 120 nm). *(d)* ATV, the virus that infects the hyperthermophile *Acidianus convivator*. When released from the cell the virions are lemon-shaped (left), but they proceed to grow appendages on both ends (right). ATV virions are about 100 nm in diameter.

a circular genome of about 68 kilobase pairs and is lemon-shaped when first released from the host cells. However, shortly after release from its lysed host cell, the virion produces long thin tails, one at each end (Figure 9.10d). The tails are actually tubes, and as they form, the virion becomes thinner and its volume is reduced. Remarkably, this is the first example of virus development in the complete absence of host cell contact. It is thought that the extended tails of ATV help the virus in some way survive in its hot (85°C), acidic (pH 1.5) environment. This unusually shaped virus is also lysogenic, a property rarely seen in other archaeal viruses.

A spindle-shaped virus also infects *Pyrococcus* (*Euryarchaeota*). This virus, named PAV1, resembles SSV but is larger and contains a very short tail (Figure 9.10c). PAV1 has a small circular DNA genome and is released from host cells without cell lysis, probably by a budding mechanism similar to that of the *Escherichia coli* bacteriophage M13 (Section 9.3). *Pyrococcus* has a growth temperature optimum of 100°C, meaning that PAV1 virions must be especially heat-stable. Despite their similar morphologies, genomic comparisons of PAV1 and SSV-type viruses show little sequence similarity, indicating that the two types of viruses do not have common evolutionary roots.

Replication events in the life cycles of archaeal viruses are not yet clear. However, considering that the genomes of most of these viruses are double-stranded DNA, it is unlikely that any major novel modes of replication will be uncovered. However, important molecular details, such as the extent to which viral rather than host polymerases and other enzymes are used in the replication process, await further work on these remarkably tough viruses.

MINIQUIZ -
- What type of genome is seen in most archaeal viruses?
- Compared with other archaeal viruses, what are two unusual features of the virus that infects *Acidianus*?

9.6 Uniquely Replicating DNA Animal Viruses

Two groups of double-stranded DNA animal viruses show unusual replication strategies: pox viruses and adenoviruses. Pox viruses are unique because all replication events, including DNA replication, occur in the host *cytoplasm* instead of the nucleus, and adenoviruses are unique because the replication of their genome proceeds in a leading fashion on *both* DNA template strands.

Pox Viruses

Pox viruses have been important historically as well as medically. Smallpox virus was the first virus to be studied in any detail and was the first virus for which a vaccine was developed. Pox viruses are among the largest of all viruses, the brick-shaped vaccinia virions measuring almost 400 nm in diameter (**Figure 9.11**). Other pox viruses of importance are cowpox and vaccinia virus. Because it closely resembles the smallpox virus but is not pathogenic, vaccinia is used as a smallpox vaccine and a model for smallpox virus molecular biology.

The vaccinia virus genome consists of linear double-stranded DNA about 190 kilobase pairs in length and encoding about 250 genes. Following attachment, vaccinia virions are taken up into host cells and the nucleocapsids (Figure 9.11) are liberated in the cytoplasm; all replication events take place in the cytoplasm. Uncoating of the viral genome requires the activity of a viral protein that is synthesized after infection (the gene encoding this protein is transcribed by a viral RNA polymerase contained within the virion). In addition to this uncoating gene, a number of other viral genes are transcribed, including genes that encode a DNA polymerase that synthesizes copies of the viral genome. These are

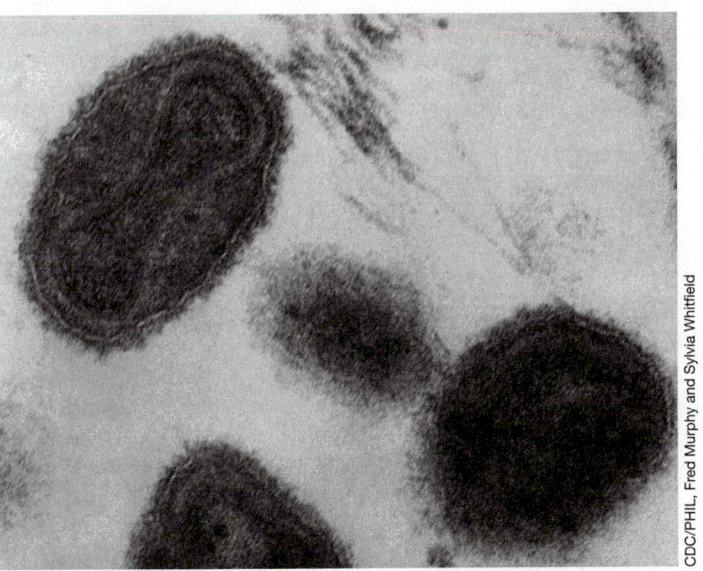

CDC/PHIL, Fred Murphy and Sylvia Whitfield

Figure 9.11 Smallpox virus. Transmission electron micrograph of a negatively stained thin section of smallpox virus virions. The virions are approximately 350 nm (0.35 μm) long. The dumbbell-shaped structure inside the virion is the nucleocapsid, which contains the double-stranded DNA genome. All pox virus replication functions occur in the host cytoplasm.

then incorporated into virions that accumulate in the cytoplasm, and the virions are released when the infected cell lyses.

Vaccinia virus has been genetically engineered to contain certain proteins from other viruses for use in vaccines (⮌ Section 11.14). A vaccine is a substance capable of eliciting an immune response in an animal that protects the animal from future infection with the same agent. Vaccinia virus causes no serious health effects in humans but elicits a strong immune response. Therefore, as a carrier of proteins from pathogenic viruses, vaccinia virus is a relatively safe and effective tool for stimulating an immune response against these pathogens. Success has been obtained with vaccinia virus vaccines against the viruses that cause influenza, rabies, herpes simplex type 1, and hepatitis B.

Adenoviruses

Adenoviruses are a group of small and naked icosahedral viruses (**Figure 9.12a**) that contain linear double-stranded DNA genomes. Adenoviruses are of minor health importance, causing mild respiratory infections in humans, but they have a unique stature in virology because of the mechanism they employ to replicate their genomes. Attached to the 5′ end of adenoviral genomic DNA is a protein called the adenoviral *terminal protein*, essential for replication of the DNA. The complementary DNA strands also have inverted terminal repeats that play a role in the replication process (Figure 9.12b).

Following infection, the adenoviral nucleocapsid is released into the host cell nucleus, and transcription of the early genes proceeds by activity of the host RNA polymerase. Most early transcripts encode important replication proteins such as the terminal protein and a viral DNA polymerase. Replication of the adenoviral genome begins at either end of the linear DNA genome and the terminal protein facilitates this process

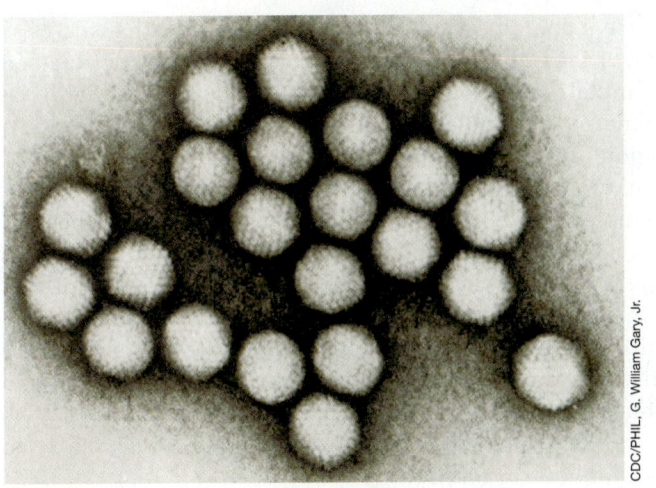

(a)

CDC/PHIL, G. William Gary, Jr.

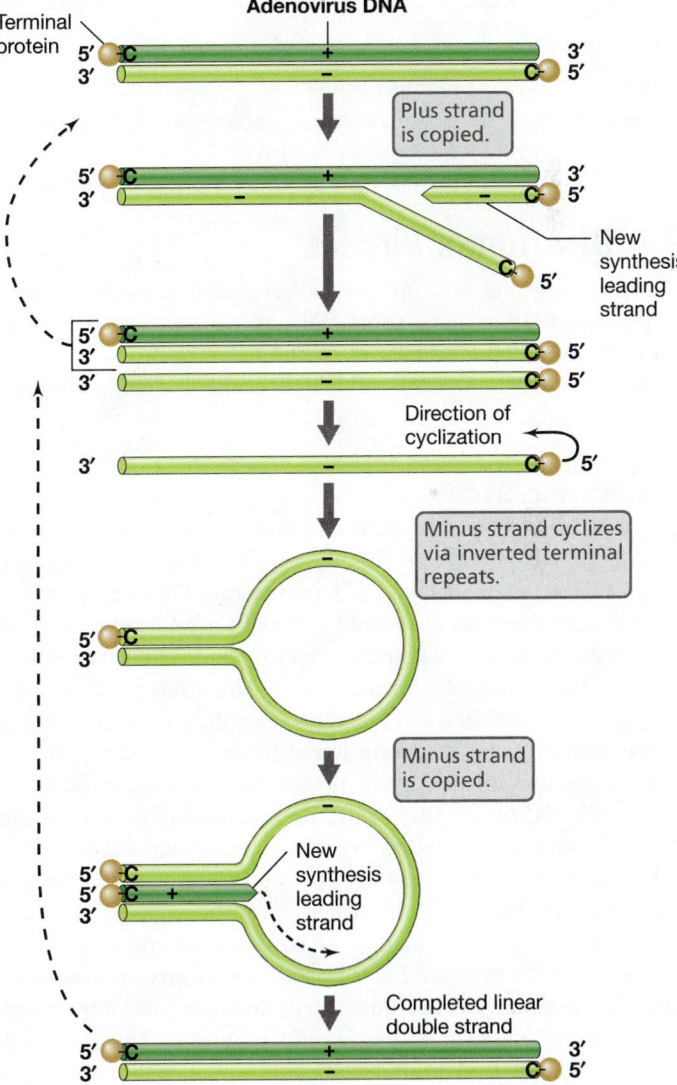

(b)

Figure 9.12 Adenoviruses. *(a)* Transmission electron micrograph of adenoviral virions. Note the icosahedral structure. *(b)* Adenoviral genome replication. Because of loop formation (cyclization), there is no lagging strand; DNA synthesis is leading on both strands. A cytosine (**C**) is attached to the terminal protein.

because it contains a covalently bound cytosine that serves as a primer for DNA polymerase (Figure 9.12*b*). The products of this initial replication are a completed double-stranded viral genome and a single-stranded minus-sense DNA molecule. At this point, a unique replication event occurs. The single DNA strand cyclizes by means of its inverted terminal repeats, and a complementary (plus-sense) DNA strand is synthesized beginning from its 5′ end (Figure 9.12*b*). This mechanism is unique because double-stranded DNA is replicated without the formation of a lagging strand, as occurs in conventional semiconservative DNA replication (⟳ Section 4.5). Once sufficient copies of the adenoviral genome have formed and virion structural components accumulate in the host cell, mature adenoviral virions are assembled and released from the cell following lysis.

MINIQUIZ

- What is unusual about genome replication in pox viruses?
- What is unusual about genome replication in adenoviruses?
- Why is the adenovirus terminal protein essential for replicating its genome?

9.7 DNA Tumor Viruses

Besides catalyzing lytic events or becoming integrated into a genome in a latent state, some DNA animal viruses can induce cancers. These include viruses of the polyomavirus family and some herpesviruses, both of which contain double-stranded DNA genomes.

Polyomavirus SV40

Polyomavirus SV40 is a naked icosahedral virus whose double-stranded DNA genome is circular (**Figure 9.13*a***). The genome is too small to encode its own DNA polymerase (Table 9.1), so host DNA polymerases are used and SV40 DNA is replicated in a bidirectional fashion from a single origin of replication. Because of the small genomes of polyomaviruses, the strategy of overlapping genes, typical of many small bacteriophages (Section 9.3), is also employed here. Transcription of the viral genome occurs in the nucleus and mRNAs are exported to the cytoplasm for protein synthesis. Eventually SV40 virion assembly occurs (in the nucleus) and the cell is lysed to release the new virions.

When SV40 infects a host cell, one of two outcomes can occur, depending on the host cell. In *permissive* hosts, virus infection results in the usual formation of new virions and the lysis of the host cell. In *nonpermissive* hosts, lytic events do not occur; instead, the viral DNA becomes integrated into host DNA, genetically altering the cells in the process (Figure 9.13*b*). Such cells can show loss of growth inhibition and become malignant, a process called *transformation* (⟳ Figure 8.22). As in certain tumor-causing retroviruses (Section 9.11), expression of specific SV40 genes is required to convert the cell to the transformed state. These tumor-inducing proteins bind to and inactivate host cell proteins that control cell division, and in this way, they promote uncontrolled cell development.

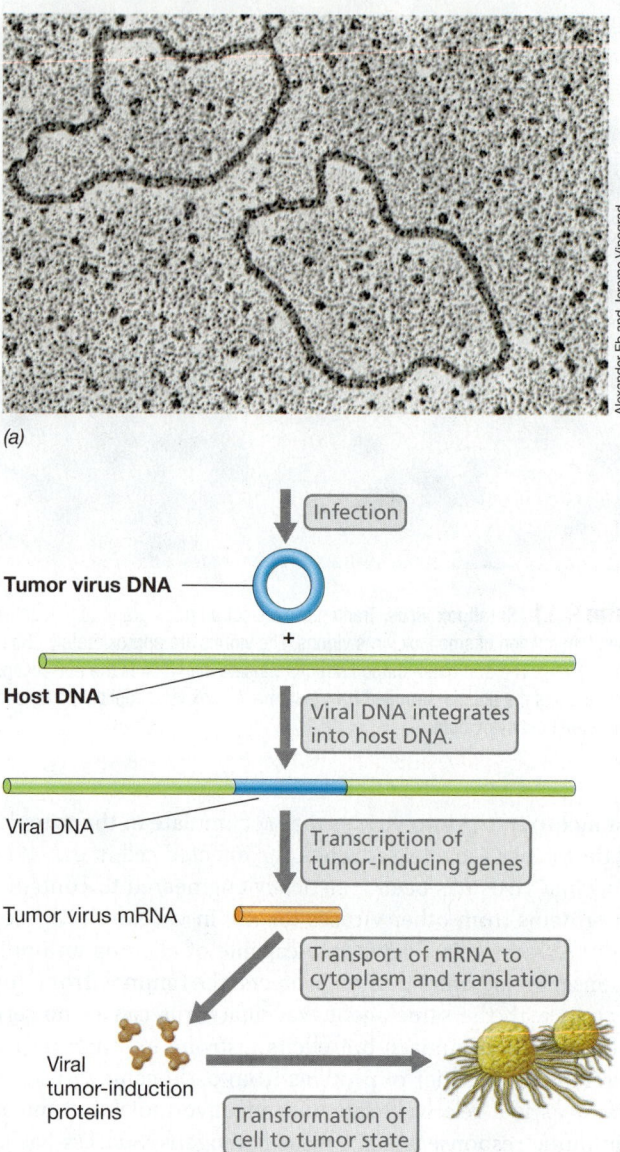

(a)

(b)

Figure 9.13 Polyomaviruses and tumor induction. *(a)* Transmission electron micrograph of relaxed (nonsupercoiled) circular DNA from a tumor virus. The contour length of each circle is about 1.5 μm. *(b)* Events in cell transformation by a polyomavirus such as SV40. Viral DNA becomes incorporated into the host genome. From there, viral genes encoding cell transformation events are transcribed and transported to the cytoplasm for translation.

Herpesviruses

Herpesviruses are a large group of double-stranded DNA viruses that cause a variety of human diseases, including fever blisters (cold sores), venereal herpes, chicken pox, shingles, and infectious mononucleosis. An important group of herpesviruses cause cancer. For example, Epstein–Barr virus causes Burkitt's lymphoma, a tumor endemic in children of central Africa and New Guinea. A widespread herpesvirus is cytomegalovirus (CMV), present in nearly three-quarters of all adults in the United States over 40 years of age. For healthy individuals, infection with CMV comes with no apparent symptoms or long-term health consequences. However, CMV can cause pneumonia, retinitis (an eye condition), and certain gastrointestinal disorders, as well as serious disease or even death in immune-compromised individuals.

Herpesviruses can remain latent in the body for long periods of time and become active under conditions of stress or when the immune system is compromised. Herpesvirus virions are enveloped and can have many distinct structural layers over the icosahedral nucleocapsid (**Figure 9.14**). Following viral attachment, the host cytoplasmic membrane fuses with the virus envelope, and this releases the nucleocapsid into the cell. The nucleocapsid is transported to the nucleus, where the viral DNA is uncoated and three classes of mRNA are produced: *immediate early*, *delayed early*, and *late* (Figure 9.14). Immediate early mRNA encodes certain regulatory proteins that stimulate the synthesis of the delayed early proteins. Among the key proteins synthesized during the delayed early stage is a viral-specific DNA polymerase and a DNA-binding protein, both of which are needed for viral DNA replication. As for other viruses, late proteins are primarily viral structural proteins.

Herpesvirus DNA replication takes place in the nucleus. After infection, the herpesvirus genome circularizes and replicates by a rolling circle mechanism. Long concatemers are formed that become processed into virus-length genomic DNA during the assembly process (Figure 9.14). Viral nucleocapsids are assembled in the nucleus, and the viral envelope is added via a budding process through the *nuclear* membrane. Mature virions are subsequently released through the endoplasmic reticulum to the outside of the cell. The assembly of herpesvirus virions thus differs from that of other enveloped viruses, which typically receive their envelope from the cytoplasmic membrane.

MINIQUIZ --

- How can the outcome of an SV40 viral infection differ in permissive versus nonpermissive hosts?
- Name two common diseases caused by herpesviruses.
- What is unusual about the envelope of a herpesvirus?

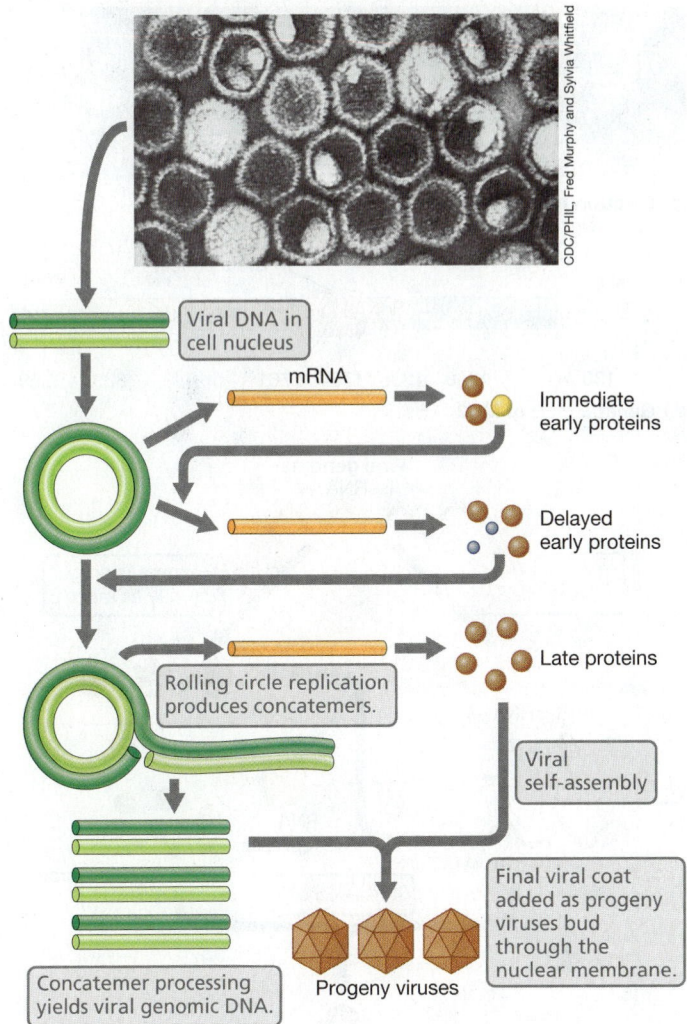

Figure 9.14 **Herpesvirus.** Flow of events in replication of herpes simplex virus starting from a transmission electron micrograph of herpes simplex virus (diameter about 150 nm). Although the viral genome is linear within the virion, it circularizes once inside the host.

III • Viruses with RNA Genomes

9.8 Positive-Strand RNA Viruses

Many viruses contain single-stranded RNA genomes of the plus sense and are therefore *positive-strand RNA viruses*. In these viruses, the sequence of the genome and the mRNA are the same (Figure 9.2). Many positive-strand animal and bacterial viruses are known, so we restrict our coverage here to just a few well-studied cases. We begin with the tiny bacteriophage MS2.

Phage MS2

Bacteriophage MS2 is about 25 nm in diameter and has an icosahedral capsid. The virus infects cells of *Escherichia coli* by attaching to the cell's pilus (**Figure 9.15*a***), a structure that normally functions in a form of horizontal gene exchange (conjugation) in bacteria. How MS2 RNA actually gets inside the *E. coli* cell from the pilus is unknown, but once it has, MS2 replication events begin quickly; the genetic map and major activities of this virus are shown in Figure 9.15*b* and *c*.

The MS2 genome encodes only four proteins, including the maturation protein, coat protein, lysis protein, and one subunit of **RNA replicase**, the enzyme that replicates the viral RNA. MS2 RNA replicase is a composite protein, with some subunits encoded by the host genome and one subunit by the viral genome. The gene encoding the MS2 lysis protein overlaps that encoding the coat protein and replicase subunit (Figure 9.15*b*). We have seen this phenomenon of *overlapping genes* before (Section 9.3) as a strategy for making small genomes more efficient.

Because the genome of phage MS2 is plus-sense RNA, it is translated directly upon entry into the cell by the host RNA polymerase. When RNA replicase is made, it begins synthesis

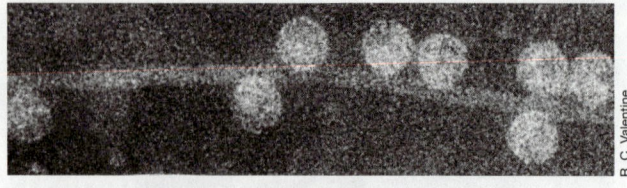

(a) **Electron micrograph of phage MS2**

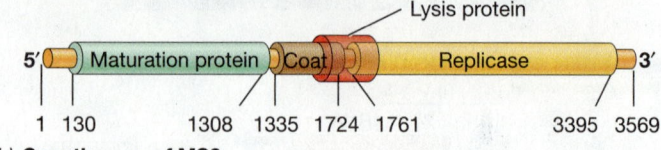

(b) **Genetic map of MS2**

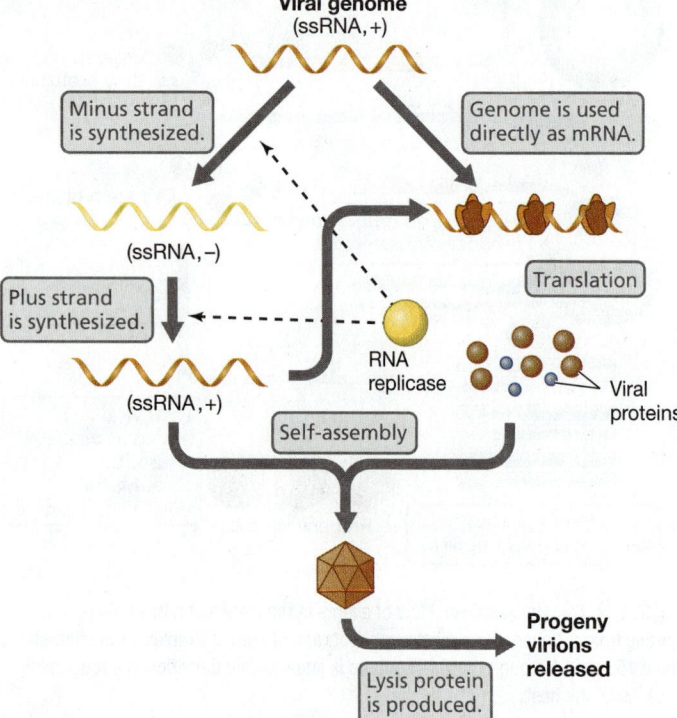

(c) **Flow of events during viral multiplication**

Figure 9.15 **A small RNA bacteriophage, MS2.** (a) Transmission electron micrograph of the pilus of a donor bacterial cell of *Escherichia coli* showing virions of phage MS2 attached. (b) Genetic map of MS2. Note how the lysis protein gene overlaps with both the coat protein and replicase genes. The numbers refer to the nucleotide positions on the RNA. (c) Flow of events during MS2 replication.

of minus-sense RNA using plus strands as templates. As minus-sense RNA copies accumulate, more plus-sense RNA is made using the minus-sense strands as templates, and some of these are translated for continued synthesis of viral structural proteins.

Phage MS2 regulates synthesis of its proteins by controlling access of host ribosomes to translational start sites on its RNA. MS2 genomic RNA is folded into a complex secondary structure. Of the four AUG translational start sites (⟲ Section 4.11) on the MS2 RNA, the most accessible to the cell's translation machinery is that for the coat protein and replicase. Hence, translation begins at these sites very early following infection. However, as

coat protein molecules accumulate, they bind to the RNA around the AUG start site for the replicase protein, effectively turning off synthesis of replicase. Although the gene for the maturation protein is at the 5′ end of the RNA, the extensive folding of the RNA limits access to the maturation protein translational start site, and consequently, only a few copies are synthesized. In this way, all MS2 proteins are made in the relative amounts needed for virus assembly. Ultimately, self-assembly of MS2 virions takes place, and virions are released from the cell as a result of cell lysis.

Poliovirus

Several positive-strand RNA animal viruses cause disease in humans and other animals. These include poliovirus, the rhinoviruses that cause many cases of the common cold, the coronaviruses that cause respiratory syndromes, including severe acute respiratory syndrome (SARS), and the hepatitis A virus. We focus here on poliovirus and coronaviruses, both of which have linear RNA genomes.

Poliovirus is one of the smallest of all viruses with an icosahedral structure containing the minimum 60 morphological units per virion (**Figure 9.16a, b**). At the 5′ terminus of the viral RNA is a protein, called the *VPg protein*, that is attached covalently to the genomic RNA, and at the 3′ terminus is a poly(A) tail (Figure 9.16c), a common feature of eukaryotic cell transcripts. The poliovirus genome is also the mRNA, and the VPg protein facilitates binding of the RNA to host ribosomes. Translation yields a **polyprotein**, a single protein that self-cleaves into several smaller proteins including virion structural proteins. Other proteins generated from the polyprotein include the VPg protein, an RNA replicase responsible for synthesis of both minus-strand and plus-strand RNA, and a virus-encoded protease, which carries out the polyprotein cleavage (Figure 9.16c). This mechanism is called *post-translational cleavage* and is common in many animal viruses as well as animal cells.

Poliovirus replication occurs in the host cell cytoplasm. To initiate infection, the poliovirus virion attaches to a specific receptor on the surface of a sensitive cell and enters the cell. Once inside the cell, the virion is uncoated, and the genomic RNA is attached to ribosomes and translated to yield the polyprotein. Replication of viral RNA begins shortly after infection by the poliovirus RNA replicase. Both the positive and negative strands that are made pick up the VPg protein, which also functions as a primer for RNA synthesis (Figure 9.16c). Once poliovirus replication begins, host events are inhibited, and about 5 h postinfection, cell lysis occurs with the release of new poliovirus virions.

Coronaviruses

Coronaviruses are single-stranded plus RNA viruses that, like poliovirus, replicate in the cytoplasm, but differ from poliovirus in their larger size and details of replication. Coronaviruses cause respiratory infections in humans and other animals, including about 15% of common colds and SARS, an occasionally fatal infection of the lower respiratory tract in humans (⟲ Section 28.3).

Coronavirus virions are enveloped and contain club-shaped glycoprotein spikes on their surfaces (**Figure 9.17a**). These give the virus the appearance of having a "crown" (*corona* is Latin for crown). Coronavirus genomes are noteworthy because they

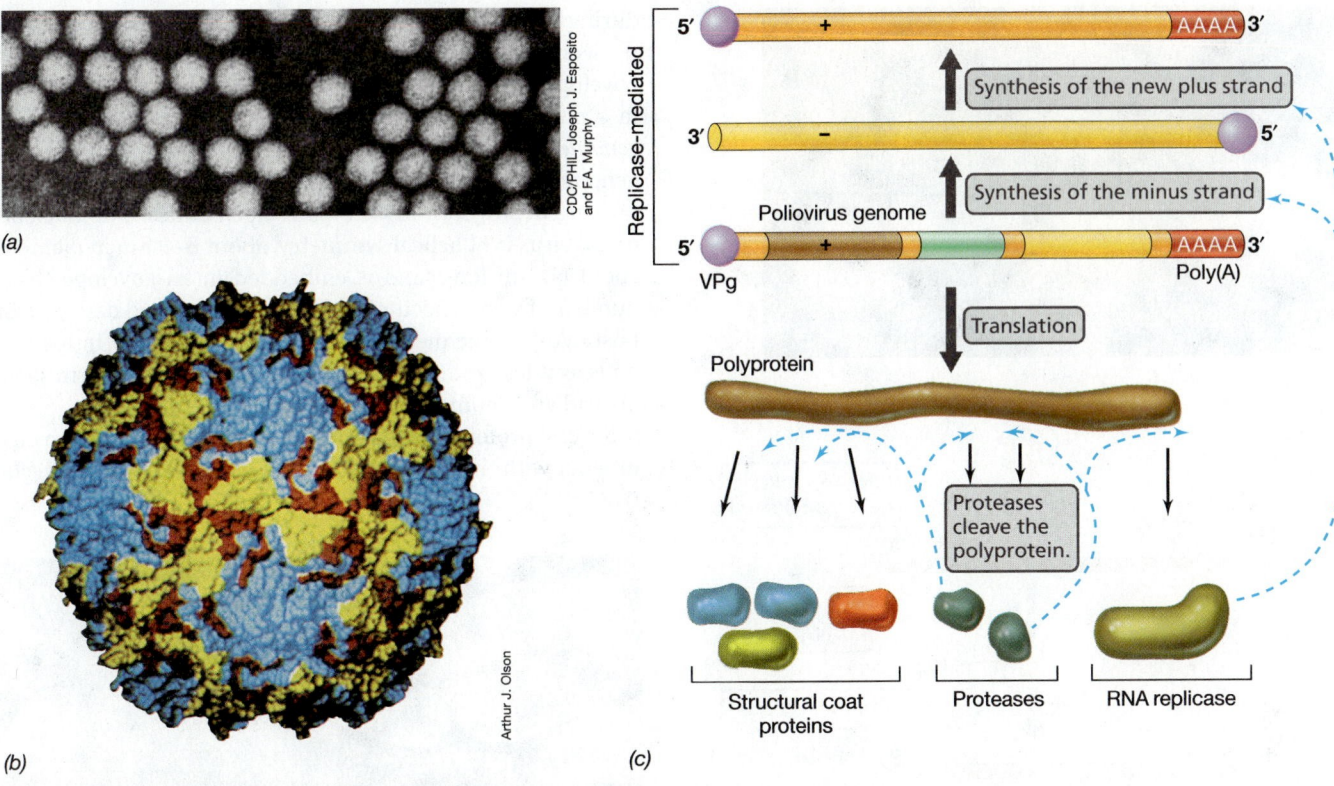

Figure 9.16 Poliovirus. *(a)* Transmission electron micrograph of poliovirus virions. *(b)* A computer model of poliovirus virions. The various structural proteins are shown in distinct colors. *(c)* The replication and translation of poliovirus. Note the importance of the RNA replicase.

are the largest of any known RNA viruses, about 30 kilobases. Because it is of the plus sense, the coronavirus genome can function directly in the cell as mRNA. Nevertheless, most coronaviral proteins are not made by translating genomic RNA. Instead, upon infection, only a portion of the genome is translated, in particular that which yields the RNA replicase (Figure 9.17*b*). This enzyme then uses the genomic RNA as a template to produce complementary negative strands from which several mRNAs are produced, and these mRNAs are translated to produce coronaviral proteins (Figure 9.17*b*). Full-length genomic RNA is also made off of the negative strands. New coronaviral virions are assembled within the Golgi complex, a major secretory organelle in eukaryotic cells (Section 2.22), and the fully assembled virions are released later from the cell surface.

Coronavirus differs from poliovirus in terms of virion and genome size, lack of the VPg protein, and absence of polyprotein formation and cleavage. Nevertheless, their single-stranded plus-sense RNA genomes dictate that many other molecular events must occur in a similar way.

MINIQUIZ --

- How can poliovirus RNA be synthesized in the cytoplasm whereas host RNA must be made in the nucleus?

- What is present in the poliovirus polyprotein?

- How are protein synthesis and genomic replication similar or different in poliovirus and the SARS virus?

9.9 Negative-Strand RNA Animal Viruses

In contrast to viruses considered in the last section, a number of animal viruses have minus-sense RNA genomes, and thus their genomes are complementary in base sequence to the mRNA. These are the *negative-strand RNA viruses*. We discuss here two important examples: rabies virus and influenza virus. There are no known negative-strand RNA bacteriophages or archaeal viruses.

Rabies Virus

Rabies virus, which causes the typically fatal disease rabies (Section 30.1), is a rhabdovirus, which refers to the characteristic shape of the virion. Rhabdoviruses are commonly bullet-shaped (**Figure 9.18***a*) and have an extensive and complex lipid envelope surrounding the helically symmetrical nucleocapsid. A rhabdovirus virion contains several enzymes that are essential for the infection process including an RNA replicase. Unlike positive-strand viruses, a rhabdovirus genome cannot be directly translated but must first be transcribed by the replicase. This occurs in the cytoplasm and generates two classes of RNAs. The first is a series of mRNAs encoding each of the viral proteins, and the second is a complementary copy of the entire viral genome; the latter serves as a template for the synthesis of genomic RNA copies (Figure 9.18*b*).

Assembly of a rhabdovirus virion is complex. Two different coat proteins are present, nucleocapsid and envelope. The nucleocapsid is formed first by assembly of nucleocapsid protein molecules during the 1918 influenza pandemic that killed millions of people worldwide (Sections 28.11 and 29.8). Influenza virus is an enveloped virus in which the viral genome is present in the virion in a number of separate pieces, a condition called a *segmented genome*. In the case of influenza A virus, a common strain, the genome is segmented into eight linear single-stranded molecules ranging in size from 890 to 2341 nucleotides. The nucleocapsid of the virus is of helical symmetry, about 6–9 nm in diameter and about 60 nm long, and is embedded in an envelope that has a number of virus-specific proteins as well as lipid derived from the host cytoplasmic membrane. Because of the way influenza virus buds as it leaves the cell, virions do not have a uniform shape and instead are pleomorphic (**Figure 9.19a**).

Several proteins on the outside of the influenza virion envelope interact with the host cell surface. One of these is *hemagglutinin*.

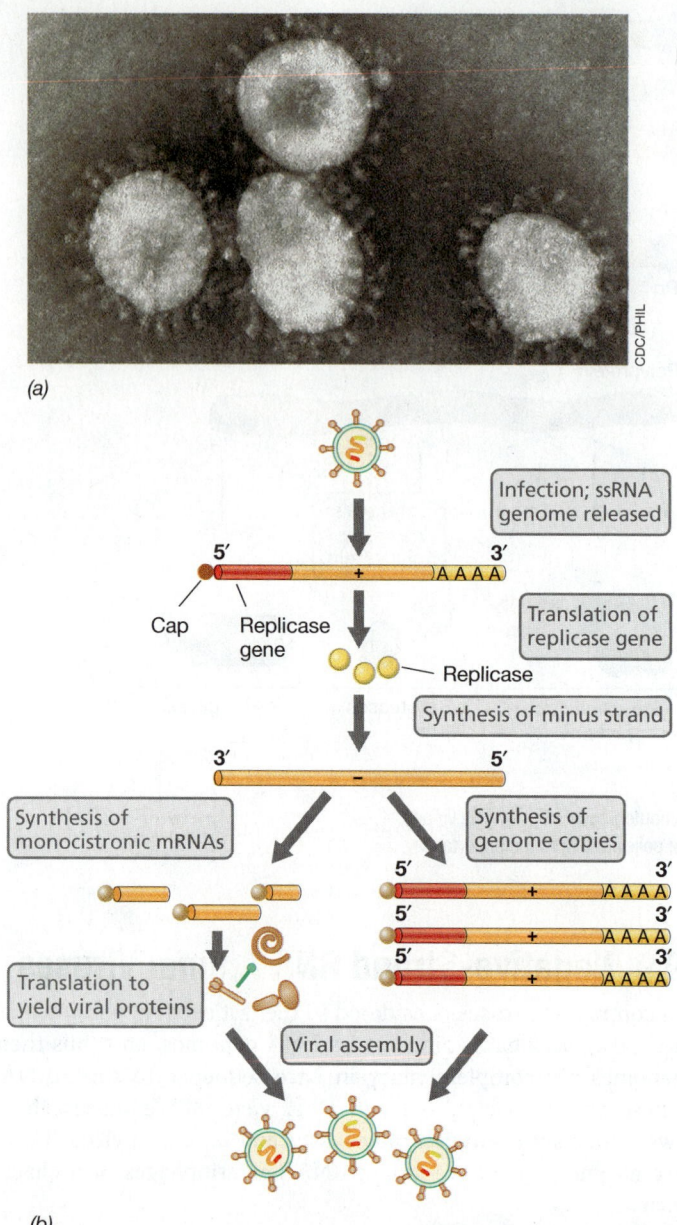

(a)

(b)

Figure 9.17 Coronaviruses. *(a)* Electron micrograph of a coronavirus; a virion is about 150 nm in diameter. *(b)* Steps in coronavirus replication. The mRNA encoding viral proteins is transcribed from the negative strand made by the RNA replicase using the viral genome as a template.

around the viral RNA genome. The envelope proteins are glycoproteins and they migrate to the cytoplasmic membrane where they are inserted into the membrane. Nucleocapsids then migrate to areas on the cytoplasmic membrane where these virus-specific glycoproteins are embedded and bud through them, becoming coated by the glycoprotein-enriched cytoplasmic membrane in the process. The final result is the release of new virions that can infect neighboring cells.

Influenza Virus

Another group of negative-strand RNA viruses contains the important human pathogen *influenza virus*. Influenza virus has been well studied over many years, beginning with early work

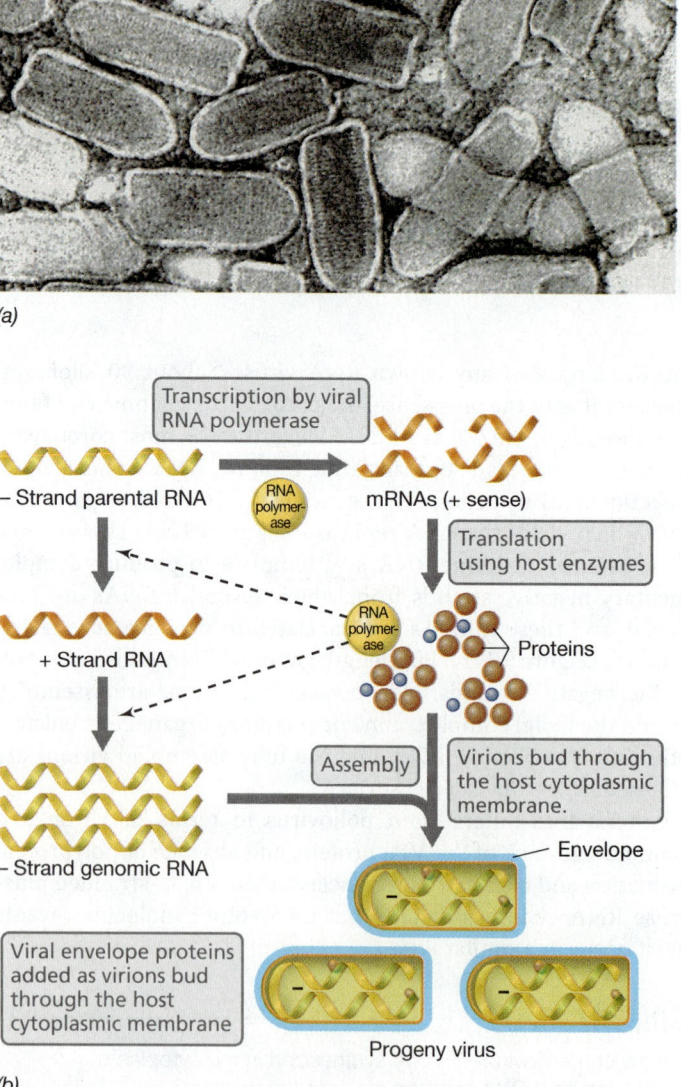

(a)

(b)

Figure 9.18 Negative-strand RNA viruses: Rhabdoviruses. *(a)* Transmission electron micrograph of vesicular stomatitis virus virions. A virion is about 65 nm in diameter. *(b)* Flow of events during replication of a negative-strand RNA virus. Note the critical importance of the viral-encoded RNA replicase.

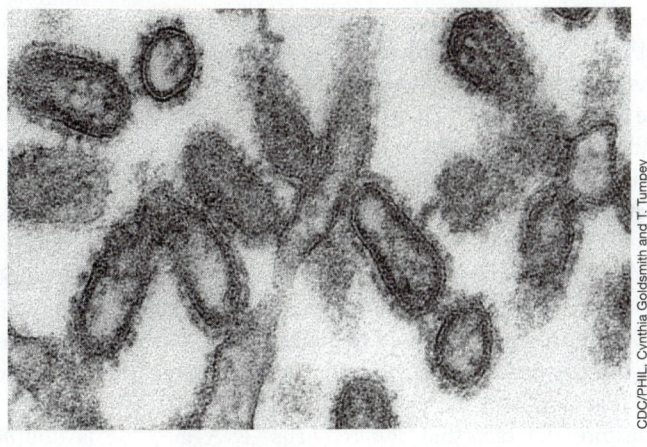

(a)

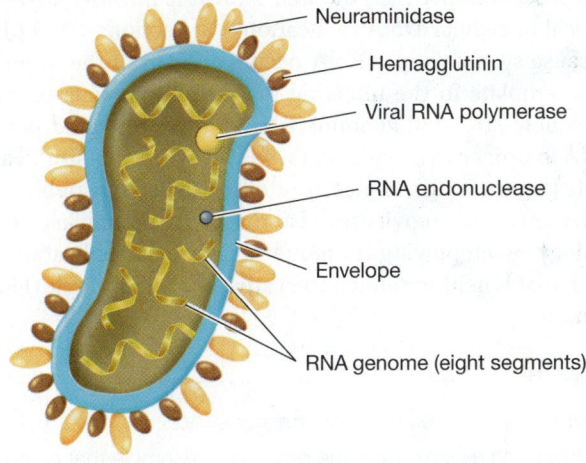

Neuraminidase

Hemagglutinin

Viral RNA polymerase

RNA endonuclease

Envelope

RNA genome (eight segments)

(b)

Figure 9.19 Influenza virus. *(a)* Transmission electron micrograph of thin sections of human influenza virus virions. *(b)* Some of the major components of the influenza virus, including the segmented genome.

Hemagglutinin is highly immunogenic and antibodies against it prevent the virus from infecting a cell. This is the mechanism by which immunity to influenza is brought about by immunization (⮌ Section 29.8). A second important influenza virus surface protein is the enzyme *neuraminidase* (Figure 9.19*b*). Neuraminidase breaks down sialic acid (a derivative of neuraminic acid) in the host cytoplasmic membrane. Neuraminidase functions primarily in virus assembly, destroying host membrane sialic acid that would otherwise block assembly or become incorporated into the virion. In addition to hemagglutinin and neuraminidase, influenza virions possess two other key enzymes. These include an RNA replicase, which converts the minus-strand genome into a plus strand, and an RNA endonuclease, which cuts the cap from host mRNAs (⮌ Section 4.9) and uses them to cap viral mRNAs so they can be translated by the host translational machinery.

After the influenza virion enters the cell, the nucleocapsid separates from the envelope and migrates to the nucleus. Uncoating activates the virus RNA replicase and transcription begins. Ten proteins are encoded by the eight segments of the influenza virus genome. The mRNAs transcribed from six segments each encode a single protein, and the other two segments encode two proteins

each. Some of the viral proteins are needed for influenza virus RNA replication, whereas others are structural proteins of the virion. The overall pattern of genomic RNA synthesis resembles that of the rhabdoviruses (Figure 9.18*b*), with full-length positive-strand RNA used as a template for making negative-strand genomic RNA. The complete enveloped virion forms by budding, as for the rhabdoviruses.

The segmented genome of the influenza virus has important practical consequences. Influenza virus exhibits a phenomenon called **antigenic shift** in which segments of the RNA genome from two different strains of the virus infecting the same cell are reassorted. This generates hybrid influenza virions that express unique surface proteins unrecognized by the immune system. Antigenic shift is thought to trigger major outbreaks of influenza because immunity to the new forms of the virus is absent from the population. We discuss antigenic shift, and a related phenomenon called *antigenic drift*, in Section 29.8.

MINIQUIZ

- Why is it essential that negative-strand viruses carry an enzyme in their virions?
- What is a segmented genome?
- In influenza virus, what is antigenic shift and how does it occur?

9.10 Double-Stranded RNA Viruses

Viruses with double-stranded RNA genomes infect animals, plants, fungi, and a few bacteria. *Reoviruses* are an important family of animal viruses with double-stranded RNA genomes, 18–30 kilobase pairs in size, and we focus on them here.

Rotavirus is a typical reovirus and is the most common cause of diarrhea in infants 6 to 24 months of age. Other reoviruses cause respiratory infections and some infect plants. Reovirus virions consist of a nucleocapsid 60–80 nm in diameter, surrounded by a double shell of icosahedral symmetry (**Figure 9.20*a, b***). As we have seen with single-stranded RNA viruses, the virions of double-stranded RNA viruses must carry their own enzyme to synthesize their mRNA and replicate their RNA genomes. Like the influenza virus genome, the reoviral genome is segmented, in this case into 10–12 molecules of linear double-stranded RNA.

To initiate infection, a reovirus virion binds to a cellular receptor protein. The attached virus then enters the cell and is transported into lysosomes, where normally it would be destroyed (⮌ Section 2.22). However, within the lysosome only the outer coats of the virion are removed by proteolytic enzymes. This reveals the nucleocapsid, which is released into the cytoplasm. This uncoating process activates the viral RNA replicase and initiates virus replication (Figure 9.20*c*).

Reovirus replication events occur exclusively in the host cytoplasm but *within* the nucleocapsid itself (Figure 9.20*c*) because the host has enzymes that recognize double-stranded RNA as foreign and would destroy it. The plus strand of the reoviral genome is inactive as mRNA, and thus the first step in replication is the synthesis of plus-sense mRNA by the viral-encoded RNA replicase, using minus-strand RNA as a template. The nucleotide triphosphates necessary for RNA synthesis are supplied by the host

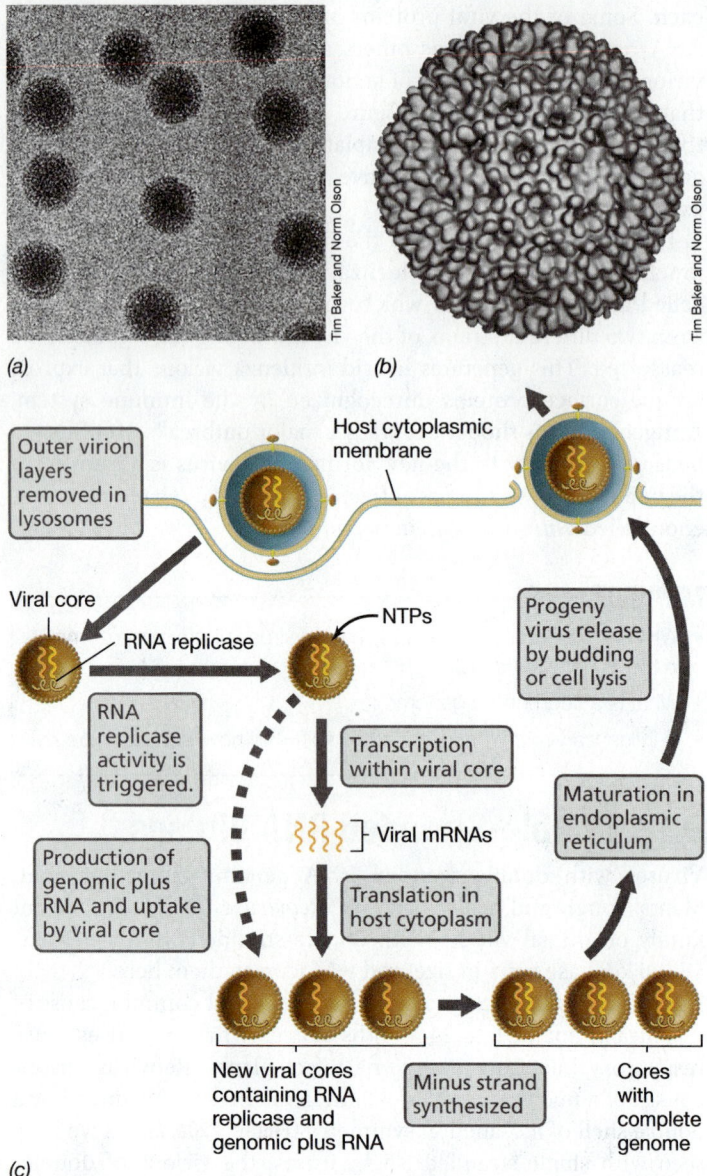

(a)   (b)

Tim Baker and Norm Olson

(c)

Figure 9.20 Double-stranded RNA viruses: The reoviruses. *(a)* Transmission electron micrograph showing reovirus virions (diameter, about 70 nm). *(b)* Three-dimensional computer reconstruction of a reovirus virion calculated from electron micrographs of frozen-hydrated virions. *(c)* The reovirus life cycle. All replication and transcription steps occur inside the nucleocapsids. NTPs, nucleotide triphosphates.

(Figure 9.20*c*). The mRNAs are then capped and methylated (as is typical of eukaryotic mRNAs, ⮌ Section 4.9) by viral enzymes and exported from the nucleocapsid into the cytoplasm and translated by host ribosomes.

Most RNAs in the reovirus genome encode a single protein, although in a few cases the protein formed is cleaved to yield the final products. However, one of the reovirus mRNAs encodes two proteins but the RNA does not have to be processed in order to translate both of these. Instead, a ribosome occasionally "misses" the start codon for the first gene in this mRNA and travels on to the start codon of the second gene to begin translation. When this occurs, the second protein, needed in small amounts, is made but the first protein is not. This "molecular mistake" can be

viewed as a primitive form of translational control that ensures that viral proteins are made in their proper amounts.

As viral proteins are formed in the host cytoplasm, they aggregate to form new nucleocapsids, trapping copies of RNA replicase inside as they form (Figure 9.20*c*). Newly formed nucleocapsids then take up the correct complement of genomic (plus-strand) RNA fragments—probably by recognition of specific sequences on each fragment—and as each single-stranded RNA enters a newly formed nucleocapsid, a double-stranded form is produced from it by RNA replicase. Once genomic synthesis is complete, viral coat proteins are added in the host's endoplasmic reticulum, and the mature reoviral virions are released by budding or cell lysis (Figure 9.20*c*).

Despite the fact that the reovirus RNA genome is double-stranded, RNA replication in these viruses is actually a *conservative* process rather than the well-known *semiconservative* process typical of cellular DNA replication (⮌ Sections 4.4–4.6). This is because synthesis of mRNA occurs *only* off of the minus strand as a template in the infecting nucleocapsids, whereas synthesis of double-stranded genomic RNA from assimilated plus-strand RNA in progeny virions occurs *only* off of the plus strand as a template (Figure 9.20*c*). Hence, in addition to having double-stranded RNA genomes, reoviruses also display their unusual molecular biology by employing a unique nucleic acid replication mechanism that is neither semiconservative nor rolling circle (Figure 9.6) in nature.

MINIQUIZ

- What does the reovirus genome consist of?
- How does reovirus genome replication resemble that of influenza virus, and how does it differ?
- Why must reoviral replication events occur within the nucleocapsid?

9.11 Viruses That Use Reverse Transcriptase

Two different classes of viruses use *reverse transcriptase*, and they differ in the type of nucleic acid in their genomes. The retroviruses have *RNA* genomes, whereas the hepadnaviruses have *DNA* genomes. Besides their unique biological properties, both classes of viruses include important human pathogens, including HIV (a retrovirus) and hepatitis B (a hepadnavirus).

Retroviruses

Retroviruses have enveloped virions that contain two identical copies of the RNA genome (⮌ Figure 8.23*a*). The virion also contains several enzymes, including reverse transcriptase, and also a specific viral tRNA. Enzymes for retrovirus replication must be carried in the virion because although the retroviral genome is of the plus sense, it is not used directly as mRNA. Instead, the genome is converted to DNA by reverse transcriptase and integrated into the host genome. The DNA formed is a linear double-stranded molecule and is synthesized within the virion and then released to the cytoplasm. An outline of the steps in reverse transcription is given in **Figure 9.21**.

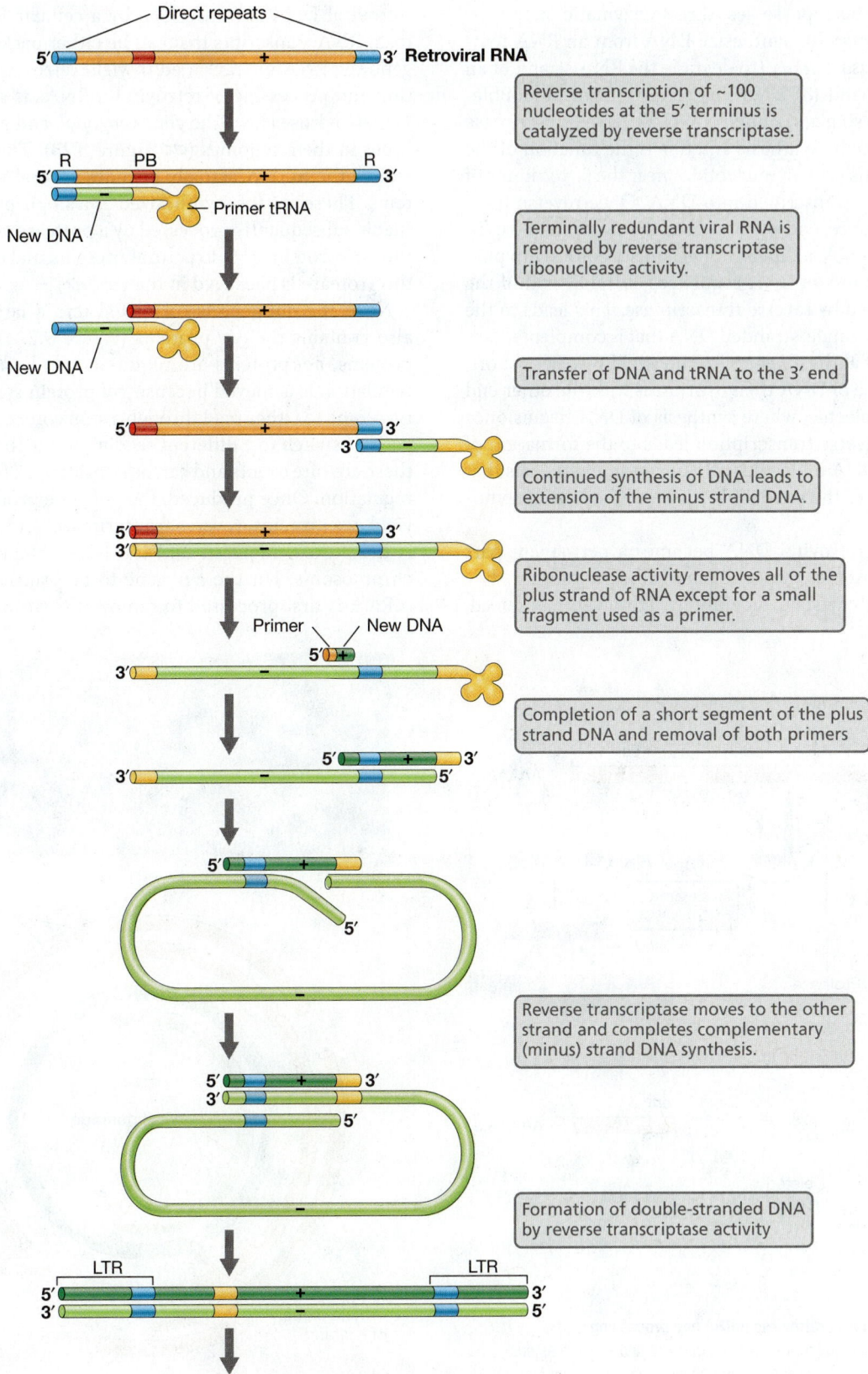

Figure 9.21 Formation of double-stranded DNA from retrovirus single-stranded genomic RNA. The sequences labeled R on the RNA are direct repeats found at either end. The sequence labeled PB is where the primer (tRNA) binds. Note that DNA synthesis yields longer direct repeats on the DNA than were originally on the RNA. These are called long terminal repeats (LTRs).

Reverse transcriptase possesses three enzymatic activities: (1) *reverse transcription* (to synthesize DNA from an RNA template), (2) *ribonuclease activity* (to degrade the RNA strand of an RNA:DNA hybrid), and (3) *DNA polymerase* (to make double-stranded DNA from single-stranded DNA). Reverse transcriptase needs a primer for DNA synthesis and this is the function of the viral tRNA. Using this primer, nucleotides near the 5′ terminus of the RNA are reverse-transcribed into DNA. Once reverse transcription reaches the 5′ end of the RNA, the process stops. To copy the remaining RNA, a different mechanism comes into play. First, terminally redundant RNA sequences at the 5′ end of the molecule are removed by reverse transcriptase. This leads to the formation of a small, single-stranded DNA that is complementary to the RNA segment at the *other end* of the viral RNA. This short, single-stranded piece of DNA then hybridizes with the other end of the viral RNA molecule, where synthesis of DNA begins once again. Continued reverse transcription leads to the formation of a double-stranded DNA molecule with long terminal repeats that assist in integration of the retroviral DNA into the host chromosome (Figure 9.21).

Once integrated, retroviral DNA becomes a permanent part of the host chromosome; the genes may be expressed or they may remain in a latent state indefinitely. However, if induced, retroviral DNA is transcribed by a cellular RNA polymerase to form RNA transcripts that can be either packaged into virions as genomic RNA or translated to yield retroviral proteins. Translation and processing of retroviral mRNAs is shown in **Figure 9.22**. All retroviruses have the genes *gag, pol,* and *env,* arranged in that order in their genomes (↺ Figure 8.23). The *gag* gene at the 5′ end of the mRNA actually encodes several viral structural proteins. These are first synthesized as a single protein (polyprotein) that is subsequently processed by a protease which itself is part of the polyprotein. The structural proteins make up the capsid, and the protease is packaged in the virion.

Next, the *pol* gene is translated into a large polyprotein that also contains the *gag* proteins (Figure 9.22a). Compared to *gag* proteins, *pol* proteins are required in only small amounts. This regulation is achieved because *pol* protein synthesis requires the ribosome to either read through a stop codon at the end of the *gag* gene or switch to a different reading frame in this region. Both of these are rare events and can be considered a form of translational regulation. Once produced, the *pol* gene product is processed to yield *gag* proteins, reverse transcriptase, and integrase; the latter is the protein required for viral DNA integration into the host chromosome. For the *env* gene to be translated, the full-length mRNA is first processed to remove the *gag* and *pol* regions, and

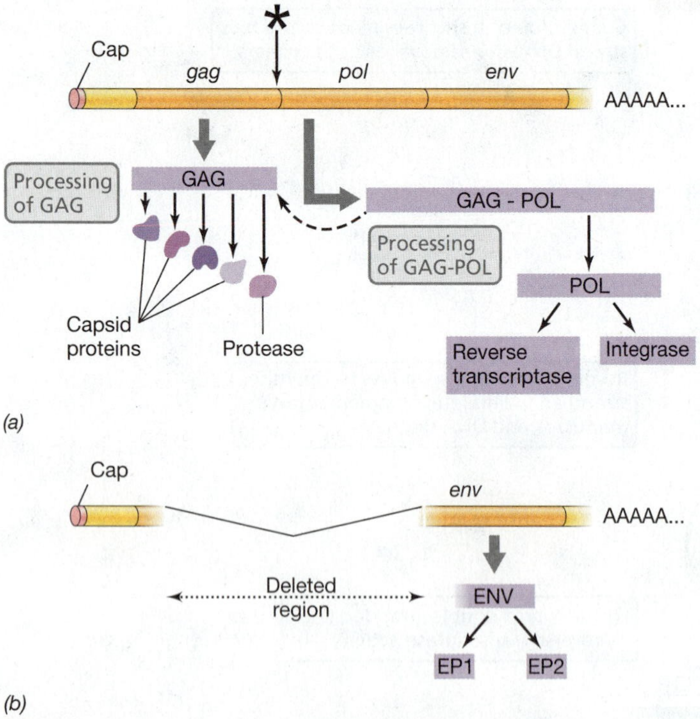

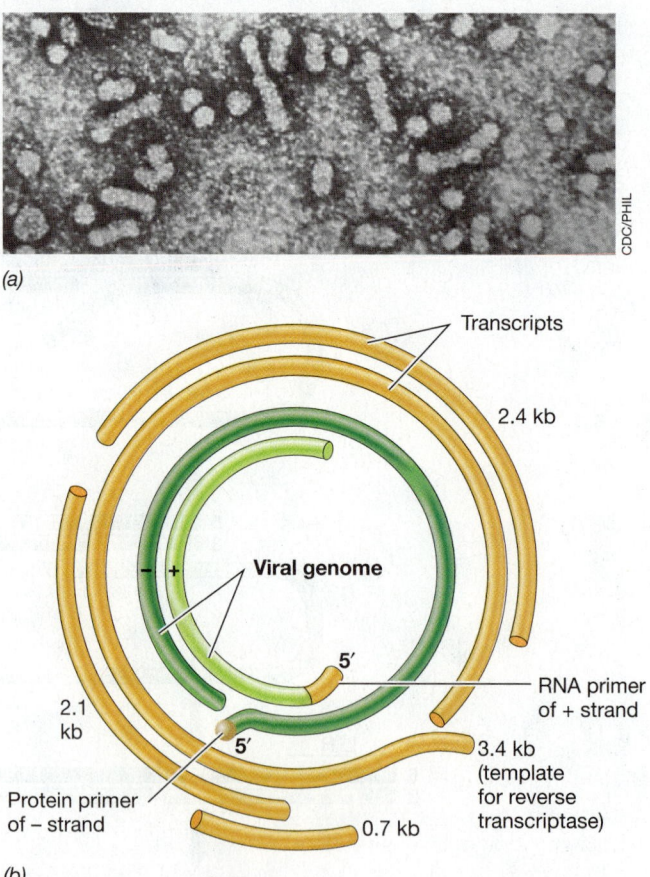

Figure 9.22 Translation of retrovirus mRNA and processing of the proteins. *(a)* Full-length retroviral mRNA encodes *gag, pol,* and *env.* The asterisk shows the site where a ribosome must read through a stop codon or do a precise shift of reading frame to synthesize the GAG-POL polyprotein. The thick gray arrows indicate translation, and the black arrows indicate protein-processing events. One of the *gag* gene products is a protease. The POL product is processed to give reverse transcriptase (RT) and integrase (IN). *(b)* The mRNA has been processed to remove most of the *gag-pol* region. This shortened message is translated to give the ENV polyprotein, which is cleaved into two envelope proteins (EP), EP1 and EP2.

Figure 9.23 Hepadnaviruses. *(a)* Electron micrograph of hepatitis B virions. *(b)* Hepatitis B genome. The partially double-stranded genome is shown in green. The sizes of the transcripts are also shown; all of the genes in the hepatitis B virus overlap. Reverse transcriptase produces the DNA genome from a single genome-length mRNA made by host RNA polymerase.

then the *env* product is made and immediately processed into two distinct envelope proteins by the viral-encoded protease (Figure 9.22*b*). Retroviral assembly occurs on the inner side of the host cytoplasmic membrane and virions are released across the membrane by budding (↻ Figure 8.24).

Hepadnaviruses

The life cycles of viruses show a variety of unusual genome structures and replication schemes, but none is more unusual than that of the **hepadnaviruses**, such as human hepatitis B virus (**Figure 9.23*a***). The tiny DNA genomes of hepadnaviruses are unusual because they are *partially* double-stranded. Despite their small size (3–4 kilobase pairs), the hepadnavirus genomes encode several proteins by employing overlapping genes, a common strategy in very small viruses (Section 9.3).

Like the retroviruses, hepadnaviruses use reverse transcriptase in their replication cycle. Besides the usual activities of this enzyme, hepadnaviral reverse transcriptase also functions as a protein primer for synthesis of one of its own DNA strands. In terms of its role in replication events, however, reverse transcriptase plays different roles in retroviral and hepadnaviral genome replication. In hepadnaviruses, the *DNA* genome is replicated through an *RNA* intermediate, whereas in retroviruses, the *RNA* genome is replicated through a *DNA* intermediate (Figures 9.21 and 9.23).

Upon infection, the hepadnavirus nucleocapsid enters the host nucleus where the partial genomic DNA strand is completed to form a complete double-stranded molecule. Transcription by host RNA polymerase yields four size classes of viral mRNAs (Figure 9.23*b*), which are subsequently translated to yield the hepadnaviral proteins. The largest of these transcripts is slightly larger than the viral genome and together with reverse transcriptase, associates with viral proteins in the host cytoplasm to form new virions. Reverse transcriptase then forms single-stranded DNA off of this large transcript inside the virion to form the DNA genome minus-sense strand and uses this as a template to form a partial plus-sense strand, yielding the incomplete double-stranded genome characteristic of hepadnaviruses (Figure 9.23*b*). Once mature virions are produced, these associate with membranes in the endoplasmic reticulum and Golgi complex, from which they are exported across the cytoplasmic membrane by budding.

MINIQUIZ
- Why are protease inhibitors an effective treatment for human AIDS?
- Contrast the genomes of HIV and hepatitis B virus.
- How does the role of reverse transcriptase in the replication cycles of retroviruses and hepadnaviruses differ?

IV • Subviral Agents

We conclude our genomic tour of the viral world by considering two *subviral* agents: the viroids and the prions. These are infectious agents that resemble viruses but which lack either nucleic acid or protein and are thus not viruses.

9.12 Viroids

Viroids are infectious RNA molecules that differ from viruses in that they lack protein. Viroids are small, circular, single-stranded RNA molecules that are the smallest known pathogens. They range in size from 246 to 399 nucleotides and show a considerable degree of sequence homology to each other, suggesting that they have common evolutionary roots. Viroids cause a number of important plant diseases and can have a severe agricultural impact (**Figure 9.24**). A few well-studied viroids include coconut cadang-cadang viroid (246 nucleotides) and potato spindle tuber viroid (359 nucleotides). No viroids are known that infect animals or microorganisms.

Viroid Structure and Function

The extracellular form of a viroid is naked RNA; there is no protein capsid of any kind. Although the viroid RNA is a single-stranded, covalently closed circle, its extensive secondary structure makes it resemble a hairpin-shaped double-stranded molecule with closed ends (**Figure 9.25**). This apparently makes the viroid sufficiently stable to exist outside the host cell. Because it lacks a capsid, a viroid does not use a receptor to enter the host cell. Instead, the viroid enters a plant cell through a wound, as from insect or other

Figure 9.24 Viroids and plant diseases. Photograph of healthy tomato plant (left) and one infected with potato spindle tuber viroid (PSTV) (right). The host range of most viroids is quite restricted. However, PSTV infects tomatoes as well as potatoes, causing growth stunting, a flat top, and premature plant death.

mechanical damage. Once inside, viroids move from cell to cell via plasmodesmata, which are the thin strands of cytoplasm that link plant cells (**Figure 9.26**).

Viroid RNA does not encode proteins and thus the viroid is totally dependent on its host for replication. Plants have several RNA polymerases, one of which has RNA replicase activity, and this is the enzyme that replicates the viroid. The replication

Figure 9.25 Viroid structure. Viroids consist of single-stranded circular RNA that forms a seemingly double-stranded structure by intra-strand base pairing.

mechanism itself resembles the rolling circle mechanism used for genome synthesis by some small viruses (Sections 9.3 and 9.7). The result is a large RNA molecule containing many viroid units joined end to end. The viroid has ribozyme (catalytic RNA) activity and this is used for self-cleavage of the large RNA molecule, releasing individual viroids.

Viroid Disease

Viroid-infected plants can be symptomless or develop symptoms that range from mild to lethal, depending on the viroid (Figure 9.24). Most disease symptoms are growth related, and it is believed that viroids mimic or in some way interfere with plant small regulatory RNAs. In fact, viroids could themselves be derived from regulatory RNAs that have evolved away from carrying out beneficial roles in the cell to inducing destructive events. Viroids are known to yield small interfering RNAs (siRNAs) as a side product during replication. It has been proposed that these siRNAs may then function by way of the RNA interference silencing pathway to suppress the expression of plant genes that show some homology to the viroid RNA, and in this way induce disease symptoms.

MINIQUIZ -

- If viroids are circular molecules, why are they depicted as hairpins?
- How might viroids cause disease in plants?

- -

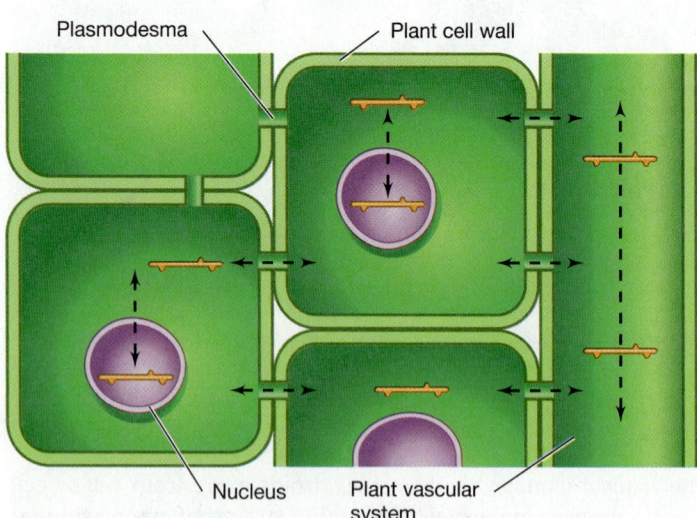

Figure 9.26 Viroid movement inside plants. After entry into a plant cell, viroids (orange) replicate either in the nucleus or the chloroplast. Viroids can move between plant cells via the plasmodesmata (thin threads of cytoplasm that penetrate the cell walls and connect plant cells). In addition, on a larger scale, viroids can move around the plant via the plant vascular system.

9.13 Prions

Prions represent the opposite extreme from that of viroids. Prions are infectious agents whose extracellular form consists entirely of protein. That is, *a prion lacks both DNA and RNA*. Prions cause several neurological diseases such as scrapie in sheep, bovine spongiform encephalopathy (BSE or "mad cow disease") in cattle, chronic wasting disease in deer and elk, and kuru and variant Creutzfeldt–Jakob disease in humans. No prion diseases of plants are known, although prions have been found in yeast. Collectively, animal prion diseases are known as *transmissible spongiform encephalopathies*.

Prion Proteins and the Prion Infectious Cycle

If prions lack nucleic acid, how is prion protein encoded? The answer to this conundrum is that the host cell itself encodes the

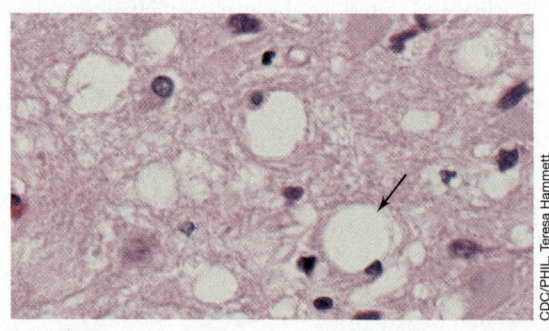

(a)

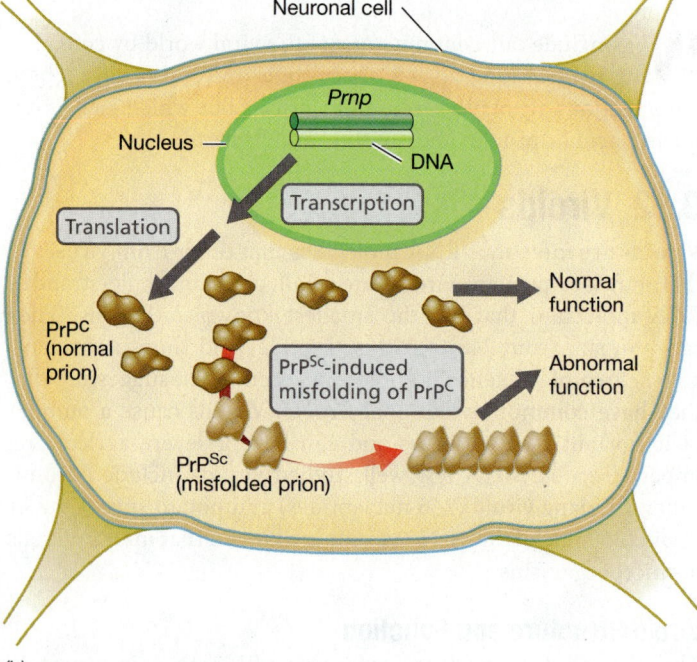

(b)

Figure 9.27 Prions. *(a)* Section through brain tissue of a human with variant Creutzfeldt–Jakob disease. Note the spongy nature of the tissue (clearings, arrow) where neural tissue has been lost. *(b)* Mechanism of prion misfolding. Neuronal cells produce the native form of the prion protein. The pathogenic form catalyzes the refolding of native prions into the pathogenic form. The pathogenic form is protease resistant, insoluble, and forms aggregates in neural cells. This eventually leads to destruction of neural tissues (see part a) and neurological symptoms.

prion. The host contains a gene, *Prnp* ("*Pri*on *p*rotein"), which encodes the native form of the prion, known as *PrP*^{C} (*Pr*ion *P*rotein *C*ellular). This is primarily found in the neurons of healthy animals, especially in the brain (**Figure 9.27a**). The pathogenic form of the prion protein is designated *PrP*^{Sc} (prion protein *Sc*rapie), because the first prion disease to be discovered was that of scrapie in sheep. PrP^{Sc} is identical in amino acid sequence to PrP^C from the same species, but has a different conformation. For example, native prion proteins are largely α-helical, whereas the pathogenic forms contain less α-helix and more β-sheet secondary structure. Prion proteins from different species of mammals are similar but not identical in amino acid sequence, and host range is linked in some way to protein sequence. For example, PrP^{Sc} from BSE-diseased cattle can infect humans, whereas PrP^{Sc} from scrapie-infected sheep apparently cannot.

When the PrP^{Sc} form enters a host cell that is expressing PrP^C, it promotes the conversion of PrP^C into the pathogenic form (Figure 9.27b). That is, the pathogenic prion "replicates" by converting preexisting native prions into the pathogenic form. As the pathogenic prions accumulate, they form insoluble aggregates in the neural cells (Figure 9.27a). This leads to disease symptoms including the destruction of brain and other nervous tissue. PrP^C functions in the cell as a cytoplasmic membrane glycoprotein, and it has been shown that membrane attachment of pathogenic prions is necessary before disease symptoms commence. Mutant versions of PrP^{Sc} that can no longer attach to nerve cell cytoplasmic membranes still aggregate but no longer cause disease.

Nonmammalian Prions

Many vertebrates, including amphibians and fish, contain genes homologous to the *Prnp* gene of mammals that are also expressed in nervous tissue. However, the proteins encoded by these genes do not have misfolded pathogenic versions and are therefore not prions. Certain fungi have proteins that fit the prion definition of an inherited self-perpetuating change in protein structure, although these proteins do not cause noticeable disease. Instead they adapt the fungal cells to altered nutritional conditions. In yeast, for example, the [URE3] prion is a protein that regulates the transcription of genes encoding certain nitrogen metabolism functions. The normal, soluble form of this protein represses genes encoding proteins that metabolize certain nitrogen sources. However, when the [URE3] prion accumulates, it forms insoluble aggregates, just as for mammalian prion protein. When this occurs, transcription of the genes for the normally repressed form of nitrogen metabolism is derepressed and the expression of these genes commences.

MINIQUIZ
- On what basis can prions be differentiated from all other infectious agents?
- What is the difference between the native and pathogenic forms of the prion protein?
- How does a prion differ from a viroid?

BIG IDEAS

9.1 • Viral genomes can be single-stranded or double-stranded DNA or RNA and vary from a few to hundreds of kilobases in size. Viral mRNA is always of the plus configuration by definition, but single-stranded genomes can be of the plus or minus configuration. Viruses with RNA genomes must either carry an RNA replicase in their virions or encode this enzyme in their genomes in order to synthesize RNA from an RNA template.

9.2 • Viruses may have evolved as agents of gene transfer in cells, or they may be degenerate cells that became dependent on a host cell for replication. Viruses may have been the first microorganisms on Earth with DNA genomes and could have conveyed this property to cells during a transition from an RNA to a DNA world. Universal viral phylogenies are not yet possible, but phylogenetic trees of various groups can be constructed.

9.3 • Single-stranded DNA viruses contain DNA of the plus configuration, and a double-stranded replicative form is necessary for transcription and genome replication. The genome of the virus φX174 is so small that some of its genes overlap, and the genome replicates by a rolling circle mechanism. Some related viruses, such as M13, have filamentous virions that are released from the host cell without lysis.

9.4 • The head-and-tail bacteriophage T7 contains a double-stranded DNA genome that encodes both early genes, transcribed by the host RNA polymerase, and late genes, transcribed by a virus-encoded RNA polymerase. Replication of the T7 genome employs T7 DNA polymerase and involves terminal repeats and concatemers. Bacteriophage Mu is a temperate virus that is also a transposable element. Mu replicates by transposition in the host chromosome.

9.5 • Several double-stranded DNA viruses infect cells of *Archaea*, most of which inhabit extreme environments. Many of these genomes are circular in contrast to the linear double-stranded DNA genomes of bacteriophages. Although head-and-tail-type viruses are known, many archaeal viruses have an unusual spindle-shaped morphology.

9.6 • Pox viruses are large double-stranded DNA viruses that replicate entirely in the cytoplasm and are responsible for several human diseases, including smallpox. Adenoviruses are double-stranded DNA viruses whose genome replication employs protein primers and a mechanism that occurs without lagging-strand synthesis.

9.7 • Some double-stranded DNA viruses cause cancer in humans. SV40 is such a tumor virus and has a tiny genome containing overlapping genes. The virus can trigger cell transformation (tumor induction) from the activity of certain genes. Some herpesviruses also cause cancer but most cause various human infectious diseases. Herpesviruses can maintain themselves in a latent state in the host indefinitely, initiating viral replication periodically.

9.8 • In single-stranded plus-sense RNA viruses, the genome is also the mRNA, and a negative strand is synthesized to produce more mRNA and genome copies. The tiny bacteriophage MS2 contains only four genes, one of which encodes a subunit of its RNA replicase. In poliovirus, the viral RNA is translated directly, producing a polyprotein that is cleaved into several small viral proteins. Coronaviruses are large RNA viruses that resemble poliovirus in some but not all of their replication features.

9.9 • In negative-strand viruses, the virus RNA is not mRNA but must first be copied to form mRNA by RNA replicase present in the virion. The positive strand is the template for production of genome copies. Important pathogenic negative-strand viruses include rabies virus and influenza virus.

9.10 • Reoviruses contain segmented linear double-stranded RNA genomes. Like negative-strand RNA viruses, reoviruses contain an RNA-dependent RNA polymerase within the virion. All replication events occur within newly forming virions.

9.11 • Some viruses employ reverse transcriptase, including retroviruses (HIV) and hepadnaviruses (hepatitis B). Retroviruses have single-stranded RNA genomes and use reverse transcriptase to make a DNA copy. Hepadnaviruses contain partially complete DNA genomes and use reverse transcriptase to make a single strand of genomic DNA from a full-length complementary strand of RNA.

9.12 • Viroids are circular single-stranded RNA molecules that do not encode proteins and are dependent on host-encoded enzymes for replication. Unlike viruses, viroid RNA is not enclosed within a capsid, and all known viroids are plant pathogens.

9.13 • Prions consist of protein but have no nucleic acid of any kind. Prions exist in two conformations, the native cellular form and the pathogenic form, which takes on a different protein structure. The pathogenic form "replicates" itself by converting native prion proteins, encoded by the host cell, into the pathogenic conformation.

REVIEW OF KEY TERMS

Antigenic shift in influenza virus, major changes in viral proteins (antigens) due to gene reassortment

Hepadnavirus a virus whose DNA genome replicates by way of an RNA intermediate

Negative strand a nucleic acid strand that has the opposite sense to (is complementary to) the mRNA

Overlapping genes two or more genes in which part or all of one gene is embedded in the other

Polyprotein a large protein expressed from a single gene and subsequently cleaved to form several individual proteins

Positive strand a nucleic acid strand that has the same sense as the mRNA

Prion an infectious protein whose extracellular form lacks nucleic acid

Replicative form a double-stranded molecule that is an intermediate in the replication of viruses with single-stranded genomes

Retrovirus a virus whose RNA genome has a DNA intermediate as part of its replication cycle

Reverse transcription the process of copying genetic information found in RNA into DNA

RNA replicase an enzyme that can produce RNA from an RNA template

Rolling circle replication a mechanism, used by some plasmids and viruses, of replicating circular DNA, which starts by nicking and unrolling one strand. For a single-stranded genome, the still-circular strand is used as a template for DNA synthesis; for a double-stranded genome, the unrolled strand is used as a template for DNA synthesis

Transposase an enzyme that catalyzes the insertion of DNA segments into other DNA molecules

Viroid an infectious RNA whose extracellular form lacks protein

REVIEW QUESTIONS

1. How much larger is the *Escherichia coli* genome than the bacteriophage T4 or T7 genomes? (Section 9.1)

2. Describe the classes of viruses based on their genomic characteristics. For each class, describe how viral mRNA is made and how the viral genome is replicated. (Section 9.1)

3. How might viruses help explain the differences observed in the DNA replication machinery of cells of the three domains? (Section 9.2)

4. What are overlapping genes? Give examples of viruses that have overlapping genes. (Sections 9.3 and 9.8)

5. Describe how the genome of bacteriophage ϕX174 is transcribed and translated. (Section 9.3)

6. Why can it be said that transcription of the bacteriophage T7 genome requires two enzymes? (Section 9.4)

7. Why is bacteriophage Mu mutagenic? What features are necessary for Mu to insert into DNA? (Section 9.4)

8. List three unusual features of the archaeal virus that infects *Acidianus* that distinguish it from bacteriophage T7. (Section 9.5)

9. Of all the double-stranded DNA animal viruses, pox viruses stand out concerning one unique aspect of their DNA replication process. What is this unique aspect and how can this be accomplished without special enzymes being packaged in the virion? (Section 9.6)

10. Explain why it can be said that adenoviruses are unique in biology. (Section 9.6)

11. List two common infectious diseases and one very serious rare disease cause by a herpesvirus. (Section 9.7)

12. If the MS2 or poliovirus virions did not contain a specific enzyme, explain why these viruses could not replicate. (Section 9.8)

13. What is the function of the VPg protein of poliovirus, and how can coronaviruses replicate without a VPg protein? (Section 9.8)

14. Describe two ways in which the genomes of poliovirus and influenza virus differ. (Sections 9.8 and 9.9)

15. Rabies virus and poliovirus both have single-stranded RNA genomes, but only in poliovirus can the genome be translated directly. Explain. (Sections 9.8 and 9.9)

16. Compare the reovirus genome to those of influenza virus and bacteriophage MS2. (Section 9.10)

17. Why do both hepadnaviruses and retroviruses require reverse transcriptase when their genomes are double-stranded DNA and single-stranded RNA, respectively? (Section 9.11)

18. What are the similarities and differences between viruses and viroids? (Section 9.12)

19. What are the similarities and differences between prions and viruses? (Section 9.13)

20. What are the similarities and differences between viroids and prions? (Sections 9.12 and 9.13)

APPLICATION QUESTIONS

1. Not all proteins are made from the RNA genome of bacteriophage MS2 in the same amounts. Can you explain why? One of the proteins functions very much like a repressor, but it functions at the translational level. Which protein is it and how does it function?

2. Replication of both strands of DNA in adenoviruses occurs in a continuous (leading) fashion. How can this happen without violating the rule that DNA synthesis always occurs in a $5' \rightarrow 3'$ direction?

3. Imagine that you are a researcher at a pharmaceutical company charged with developing new drugs against human RNA viral pathogens. Describe at least two types of drugs you might pursue, what class of virus they would affect, and why you feel that the drugs would not harm the patient.

4. Reoviruses contain genomes that are unique in all of biology. Why? Why can't reovirus replication occur in the host cytoplasm? Contrast reovirus genomic replication events with those of a cell. Why can it be said that reovirus genome replication is not semiconservative even though the reovirus genome consists of complementary strands?

10 · Genetics of *Bacteria* and *Archaea*

microbiology**now**

Defunct Viruses or Secret Agents of Gene Transfer?

How do *Bacteria* and *Archaea* acquire new and exciting characteristics that translate into the vast diversity of the microbial world? In contrast to eukaryotic organisms, several mechanisms of horizontal gene exchange occur in prokaryotes. This genetic exchange is the foundation for niche adaptation and plays a powerful role in evolution.

One example of this genetic exchange is through *gene transfer agents (GTAs)*—products of an unusual host–virus interaction. GTAs are the result of microbial cells hijacking defective viruses and using them specifically for DNA exchange. GTAs resemble tiny tailed bacteriophages (photo inset) and contain random small pieces of host DNA. They are not considered true viruses because they do not contain genes encoding their own production and do not produce characteristic viral plaques.

GTAs have been isolated from a myriad of prokaryotes including sulfate-reducing *Bacteria* and methanogenic *Archaea* and are particularly prevalent in marine prokaryotes. It is thus likely that GTAs are fairly widespread in nature. Microbial geneticists have determined that a subset of cells of the phototrophic bacterium *Rhodobacter capsulatus* produce and release GTAs during stationary phase and nutrient fluctuations.[1] This was documented by linking the promoter of a gene essential for GTA production to a reporter gene encoding a red fluorescent protein; cells making GTA turned red (photo).

While bacteriophages are considered the most abundant entities on Earth, the number of these that might actually be GTAs instead of viruses is unknown. GTAs may help explain the robust nature of DNA transfer between prokaryotes, especially those inhabiting the oceans. It also posits another question: Do GTAs play a role in the prevalence of other common genetic phenomena, such as bacterial antibiotic resistance?

[1]Fogg, P.C., et al. 2012. One for all or all for one: Heterogeneous expression and host cell lysis are key to gene transfer agent activity in *Rhodobacter capsulatus*. *PLOS One* 7: e43772.

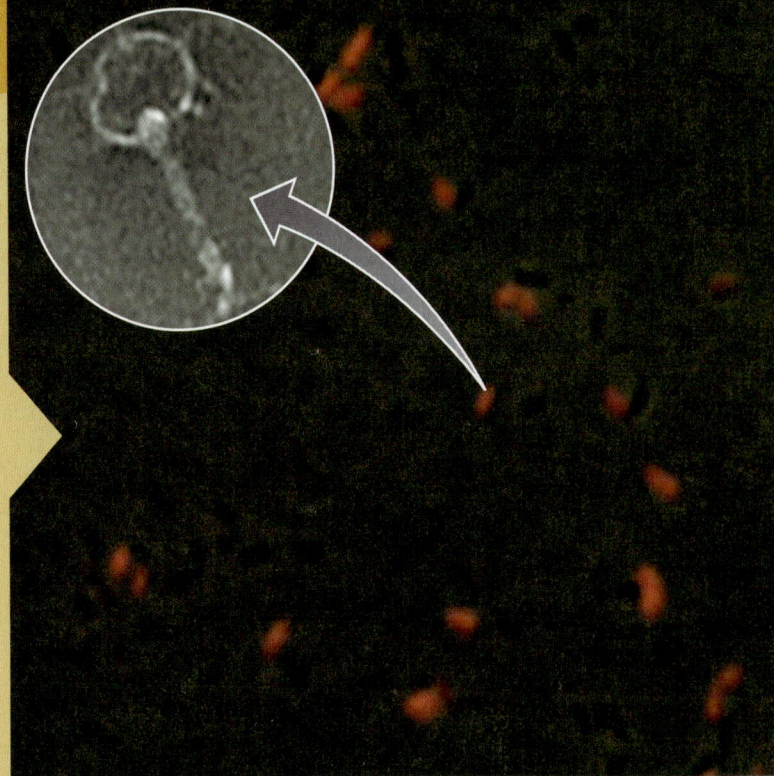

Numerous examples of microbial diversity are described throughout this book. How does this diversity arise? While prokaryotes reproduce asexually, they also possess mechanisms for exchanging genes. This gene exchange along with genetic innovations that arise from random changes in the genetic blueprint can confer an advantage that ultimately drives genetic diversity.

In this chapter we discuss the mechanisms by which *Bacteria* and *Archaea* can change their genome. We first describe how alterations arise in the genome, and then we consider how genes can be transferred from one microorganism to another by *horizontal gene transfer*. While bacterial genetics is key for microbial diversity and habitat adaptation, microorganisms also possess mechanisms to maintain genomic stability, and we finish up the chapter by considering these. Taken together, both genomic change and genomic stability are important to the evolution of an organism and its competitive success in nature.

I • Mutation

All organisms contain a specific sequence of nucleotide bases in their genome, their genetic blueprint. A **mutation** is a *heritable* change in the base sequence of that genome, that is, a change that is passed from the mother cell to progeny cells. Mutations can lead to changes—some good, some bad, but mostly neutral in effect—in the properties of an organism. Although the rate of spontaneous mutation is low (Section 10.3), the speed at which many prokaryotes divide and their exponential growth ensure that mutations accumulate surprisingly fast. Whereas mutation usually brings about only a very small amount of genetic change in a cell, genetic recombination typically generates much larger changes. Taken together, mutation and recombination fuel the evolutionary process.

We begin by considering the molecular mechanism of mutation and the properties of mutant microorganisms.

10.1 Mutations and Mutants

In all cells, the genome consists of double-stranded DNA. In viruses, by contrast, the genome may consist of single- or double-stranded DNA or RNA. A strain of any cell or virus carrying a change in nucleotide sequence is called a **mutant**. A mutant by definition differs from its parental strain in its **genotype**, the nucleotide sequence of the genome. In addition, the observable properties of the mutant—its **phenotype**—may also be altered relative to its parent. This altered phenotype is called a *mutant phenotype*. It is common to refer to a strain isolated from nature as a **wild-type strain**. The term "wild-type" may be used to refer to a whole organism or just to the status of a particular gene that is under investigation. Mutant derivatives can be obtained either directly from wild-type strains or from other strains previously derived from the wild type, for example, another mutant.

Genotype versus Phenotype

Depending on the mutation, a mutant strain may or may not differ in phenotype from its parent. By convention in bacterial genetics, the *genotype* of an organism is designated by three lowercase letters followed by a capital letter (all in italics) indicating a particular gene. For example, the *hisC* gene of *Escherichia coli* encodes a protein called HisC that functions in biosynthesis of the amino acid histidine. Mutations in the *hisC* gene would be designated as *hisC1*, *hisC2*, and so on, the numbers referring to the order of isolation of the mutant strains. Each *hisC* mutation would be different, and each *hisC* mutation might affect the HisC protein in different ways.

The *phenotype* of an organism is designated by a capital letter followed by two lowercase letters, with either a plus or minus superscript to indicate the presence or absence of that property. For example, a His$^+$ strain of *E. coli* is capable of making its own histidine, whereas a His$^-$ strain is not. The His$^-$ strain would require a histidine supplement for growth. A mutation in the *hisC* gene will lead to a His$^-$ phenotype if it eliminates the function of the HisC protein.

Isolation of Mutants: Screening versus Selection

Virtually any characteristic of an organism can be changed by mutation. Some mutations are *selectable*, conferring some type of advantage on organisms possessing them, whereas others are nonselectable, even though they may lead to a very clear change in the phenotype of an organism. A selectable mutation confers a clear advantage on the mutant strain under certain environmental conditions, so the progeny of the mutant cell are able to outgrow and replace the parent. A good example of a selectable mutation is drug resistance: An antibiotic-resistant mutant can grow in the presence of an antibiotic that inhibits or kills the parent (**Figure 10.1***a*) and is thus selected under these conditions. It is relatively easy to detect and isolate selectable mutants by choosing the appropriate environmental conditions. **Selection** is therefore an extremely powerful genetic tool, allowing the isolation of a single mutant from a population containing millions or even billions of parental cells.

An example of a nonselectable mutation is color loss in a pigmented organism (Figure 10.1*b*, *c*). Nonpigmented cells usually have neither an advantage nor a disadvantage over the pigmented parent cells when grown on agar plates, although pigmented organisms may have a selective advantage in nature. We can detect such mutations only by examining large numbers of colonies and looking for the "different" ones, a process called **screening**.

Isolation of Nutritional Auxotrophs

Although screening is more tedious than selection, methods are available for screening large numbers of colonies for certain types of mutations. For instance, nutritionally defective mutants can be detected by the technique of *replica plating* (**Figure 10.2**). A colony from a master plate can be transferred onto an agar plate lacking the nutrient by using a sterile loop, toothpick, or even a robotic arm. Parental colonies will grow normally, whereas those of the mutant will not. Thus, the inability of a colony to grow on medium lacking the nutrient signals that it is a mutant. The

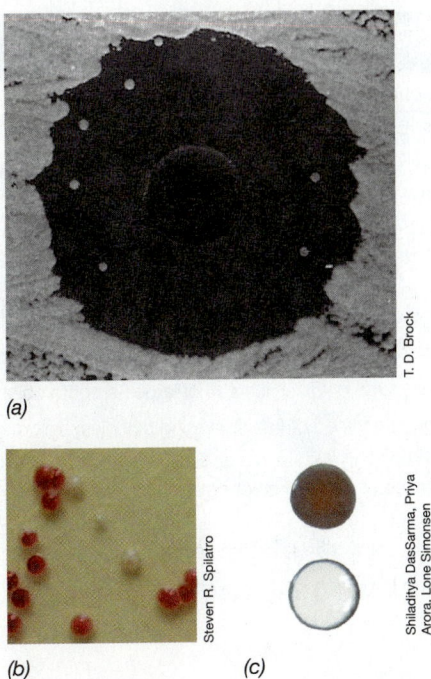

(a)

(b)   (c)

T. D. Brock

Steven R. Spilatro

Shiladitya DasSarma, Priya Arora, Lone Simonsen

Figure 10.1   Selectable and nonselectable mutations. *(a)* Development of antibiotic-resistant mutants, a type of easily selectable mutation, within the inhibition zone of an antibiotic assay disc. *(b)* Nonselectable mutations. UV-radiation-induced nonpigmented mutants of *Serratia marcescens*. The wild type has a dark red pigment. The white or colorless mutants make no pigment. *(c)* Colonies of mutants of a species of *Halobacterium*, a member of the *Archaea*. The wild-type colonies are white. The orangish-brown colonies are mutants that lack gas vesicles (🔗 Section 2.15). The gas vesicles scatter light and mask the color of the colony.

colony on the master plate corresponding to the vacant spot on the replica plate can then be picked, purified, and characterized. A mutant with a nutritional requirement for growth is called an **auxotroph**, and the parent from which it was derived is called a *prototroph*. (A prototroph may or may not be the wild type. An auxotroph may be derived from the wild type or from a mutant derivative of the wild type.) For instance, mutants of *E. coli* with a His⁻ phenotype are histidine auxotrophs. Examples of common classes of mutants and the means by which they are detected are listed in **Table 10.1**.

MINIQUIZ
- Distinguish between a mutation and a mutant.
- Distinguish between screening and selection.

10.2 Molecular Basis of Mutation

Mutations can be either spontaneous or induced. **Induced mutations** are those that are due to agents in the environment and include mutations made deliberately by humans. They can result from exposure to natural radiation (cosmic rays, and so on) that alters the structure of bases in the DNA. In addition, a variety of chemicals, including oxygen radicals (🔗 Section 5.16), can chemically modify DNA. For example, oxygen radicals can convert guanine into 8-hydroxyguanine, and this causes mutations. **Spontaneous mutations** are those that occur without external intervention. The bulk of spontaneous mutations result from occasional errors in the pairing of bases by DNA polymerase during DNA replication.

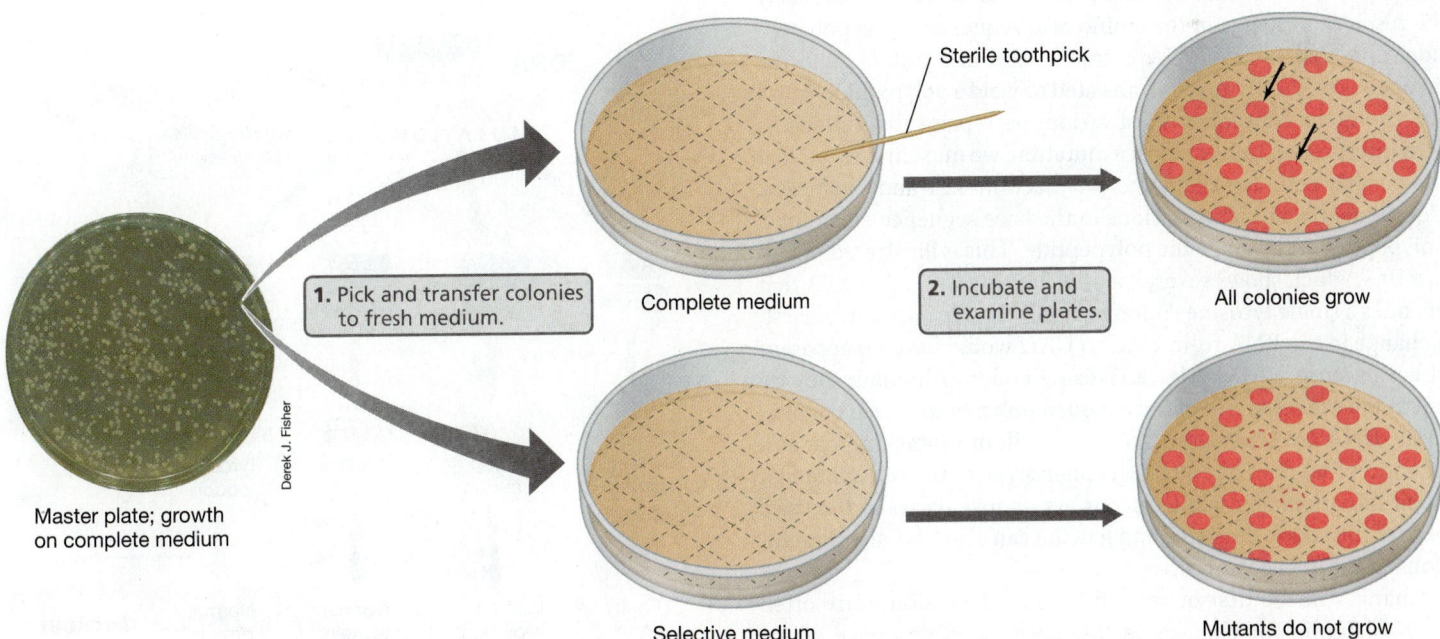

Sterile toothpick

1. Pick and transfer colonies to fresh medium.

Complete medium

2. Incubate and examine plates.

All colonies grow

Master plate; growth on complete medium

Derek J. Fisher

Selective medium

Mutants do not grow

Figure 10.2   Screening for nutritional auxotrophs. The replica-plating method can be used for the detection of nutritional mutants. Colonies from the master plate are transferred to a gridded plate containing different media for selection. The colonies not appearing on the selective medium are indicated with arrows. The selective medium lacked one nutrient (leucine) present in the master plate. Therefore, the colonies indicated with arrows on the master plate are leucine auxotrophs.

UNIT 2

Table 10.1 Kinds of mutants

Phenotype	Nature of change	Detection of mutant
Auxotroph	Loss of enzyme in biosynthetic pathway	Inability to grow on medium lacking the nutrient
Temperature-sensitive	Alteration of an essential protein so it is more heat-sensitive	Inability to grow at a high temperature that normally supports growth
Cold-sensitive	Alteration of an essential protein so it is inactivated at low temperature	Inability to grow at a low temperature that normally supports growth
Drug-resistant	Detoxification of drug or alteration of drug target or permeability to drug	Growth on medium containing a normally inhibitory concentration of the drug
Rough colony	Loss or change in lipopolysaccharide layer	Granular, irregular colonies instead of smooth, glistening colonies
Nonencapsulated	Loss or modification of surface capsule	Small, rough colonies instead of larger, smooth colonies
Nonmotile	Loss of flagella or nonfunctional flagella	Compact instead of flat, spreading colonies; lack of motility by microscopy
Pigmentless	Loss of enzyme in biosynthetic pathway leading to loss of one or more pigments	Presence of different color or lack of color
Sugar fermentation	Loss of enzyme in degradative pathway	Lack of color change on agar containing sugar and a pH indicator
Virus-resistant	Loss of virus receptor	Growth in presence of large amounts of virus

Mutations that change only one base pair are called **point mutations**. Point mutations are caused by base-pair substitutions in the DNA or by the loss or gain of a single base pair. Most point mutations do not actually cause any phenotypic change, as discussed below. However, as for all mutations, any phenotypic change that results from a point mutation depends on exactly where the mutation occurs and what type of nucleotide change occurs.

Base-Pair Substitutions

If a point mutation is within the region of a gene that encodes a polypeptide, any change in the phenotype of the cell is most likely the result of a change in the amino acid sequence of the polypeptide. The error in the DNA is transcribed into mRNA, and the erroneous mRNA in turn is translated to yield a polypeptide. **Figure 10.3** shows the consequences of various base-pair substitutions.

In interpreting the results of a mutation, we must first recall that the genetic code is degenerate (co Section 4.11 and Table 4.5). Consequently, not all mutations in the base sequence encoding a polypeptide will change the polypeptide. This is illustrated in Figure 10.3, which shows several possible results when the DNA that encodes a single tyrosine codon in a polypeptide is mutated. First, a change in the RNA from UAC to UAU would have no apparent effect because UAU is also a tyrosine codon. Although they do not affect the sequence of the encoded polypeptide, such changes in the DNA are considered one type of **silent mutation**; that is, a mutation that does not affect the phenotype of the cell. Note that silent mutations in coding regions are almost always in the third base of the codon (arginine and leucine can also have silent mutations in the first position).

Changes in the first or second base of the codon more often lead to significant changes in the amino acid sequence of the polypeptide. For instance, a single base change from UAC to AAC (Figure 10.3) results in an amino acid change within the polypeptide from tyrosine to asparagine at a specific site. This is called a **missense mutation** because the informational "sense" (precise sequence of amino acids) in the polypeptide

has changed. If the change is at a critical location in the polypeptide chain, the protein could be inactive or have reduced activity. However, not all missense mutations necessarily lead to nonfunctional proteins. The outcome depends on where the substitution lies in the polypeptide chain and on how it affects protein folding and activity. For example, mutations in the active site of an enzyme are more likely to destroy activity than mutations in other regions of the protein.

Another possible outcome of a base-pair substitution is the formation of a nonsense (stop) codon. This results in premature

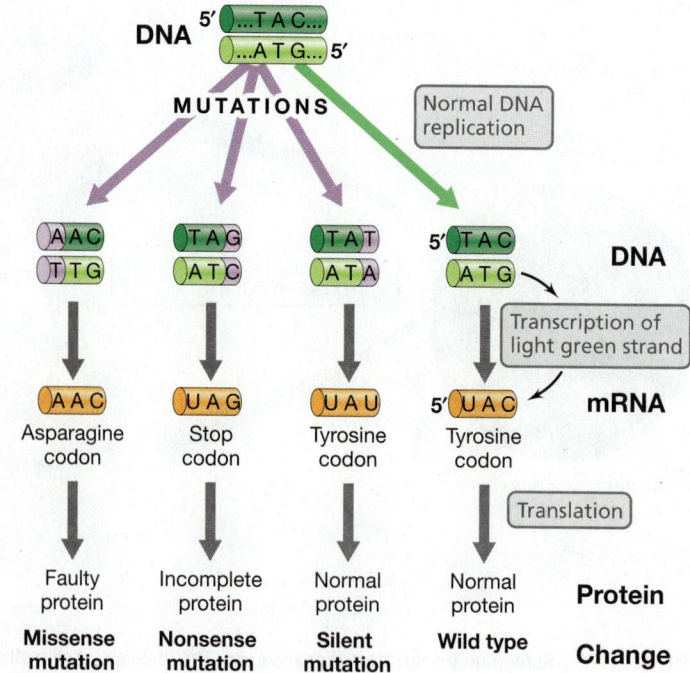

Figure 10.3 Possible effects of base-pair substitution in a gene encoding a protein. Three different protein products are possible from changes in the DNA for a single codon.

termination of translation, leading to an incomplete polypeptide (Figure 10.3). Mutations of this type are called **nonsense mutations** because the change is from a sense (coding) codon to a nonsense codon (⟳ Table 4.5). Unless the nonsense mutation is very near the end of the gene, the product is considered *truncated* or incomplete. Truncated proteins are completely inactive or, at the very least, lack normal activity.

Other terms are common in microbial genetics to describe the type of base substitution in a point mutation. **Transitions** are mutations in which one purine base (A or G) is substituted for another purine, or one pyrimidine base (C or T) is substituted for another pyrimidine. **Transversions** are point mutations in which a purine base is substituted for a pyrimidine base, or vice versa.

Frameshifts and Other Insertions or Deletions

Because the genetic code is read from one end of the nucleic acid in consecutive blocks of three bases (codons), any deletion or insertion of a single base pair results in a shift in the reading frame. These **frameshift mutations** often have serious consequences. Single base insertions or deletions change the primary sequence of the encoded polypeptide, typically in a major way (**Figure 10.4**). Such microinsertions or microdeletions can result from replication errors. Insertion or deletion of two base pairs also causes a frameshift; however, insertion or deletion of three base pairs adds or removes a whole codon. This results in addition or deletion of a single amino acid in the polypeptide sequence. Although this may well be deleterious to protein function, it is usually not as bad as a frameshift, which scrambles the entire polypeptide sequence after the mutation point.

Insertions or deletions can also result in the gain or loss of hundreds or even thousands of base pairs. Such changes inevitably result in complete loss of gene function. Some deletions are so large that they may include several genes. If any of the deleted genes are essential, the mutation will be lethal. Such deletions cannot be restored through further mutations, but only through genetic recombination. Larger insertions and deletions may arise

as a result of errors during genetic recombination. In addition, many large insertion mutations are due to the insertion of specific identifiable DNA sequences called *transposable elements* (Section 10.11). The effect of transposable elements on the evolution of bacterial genomes is discussed further in Section 6.12.

MINIQUIZ

- Do missense mutations occur in genes encoding tRNA? Why or why not?
- Why do frameshift mutations generally have more serious consequences than missense mutations?

10.3 Reversions and Mutation Rates

The rates at which different kinds of mutations occur vary widely. Some types of mutations occur so rarely that they are almost impossible to detect, whereas others occur so frequently that they present difficulties for an experimenter trying to maintain a genetically stable stock culture. Sometimes a second mutation can reverse the effect of an initial mutation. Furthermore, all organisms possess a variety of systems for DNA repair. Consequently, the observed mutation rate depends not only on the frequency of DNA changes but also on the efficiency of DNA repair.

Reversions (Back Mutations)

Point mutations are typically reversible, a process known as **reversion**. A revertant is a strain in which the original phenotype that was changed in the mutant is restored by a second mutation. Revertants can be of two types, same site or second site. In *same-site revertants*, the mutation that restores activity is at the same site as the original mutation. If the back mutation is not only at the same site but also restores the original sequence, it is called a *true revertant*.

In *second-site revertants*, the mutation is at a different site in the DNA. Second-site mutations can restore a wild-type phenotype if they function as *suppressor mutations*—mutations that compensate for the effect of the original mutation. Several classes of suppressor mutations are known. These include (1) a mutation somewhere else in the same gene that restores enzyme function, such as a second frameshift mutation near the first that restores the original reading frame; (2) a mutation in another gene that restores the function of the original mutated gene; and (3) a mutation in another gene that results in the production of an enzyme that can replace the mutated one.

An interesting class of suppressor mutations are those that result in base sequence changes in tRNAs. Nonsense mutations can be suppressed by changing the anticodon sequence of a tRNA molecule so that it now recognizes a stop codon (**Figure 10.5**). Such an altered tRNA is called a *suppressor tRNA* and will insert the amino acid it carries at the stop codon that it now reads. Suppressor tRNA mutations would be lethal unless a cell has more than one tRNA for a particular codon. One tRNA may then be mutated into a suppressor, while the other performs the original function. Most cells have multiple tRNAs and so suppressor mutations are reasonably common, at least in microorganisms. Sometimes the amino acid inserted by the suppressor tRNA is identical to

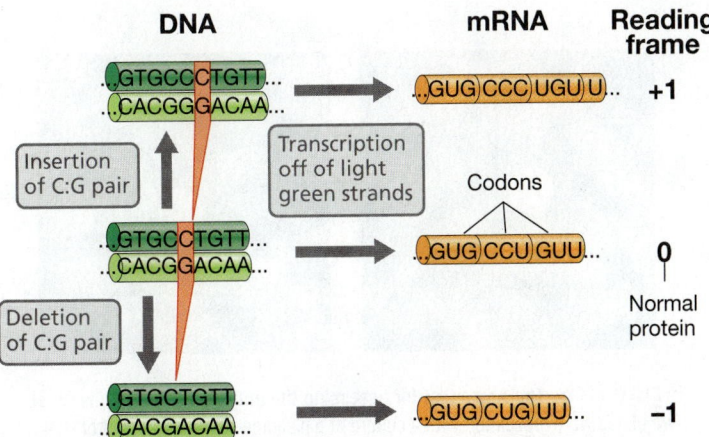

Figure 10.4 **Shifts in the reading frame of mRNA caused by insertions or deletions.** The reading frame in mRNA is established by the ribosome, which begins at the 5′ end (toward the left in the figure) and proceeds by units of three bases (codons). The normal reading frame is referred to as the 0 frame, that missing a base the −1 frame, and that with an extra base the +1 frame.

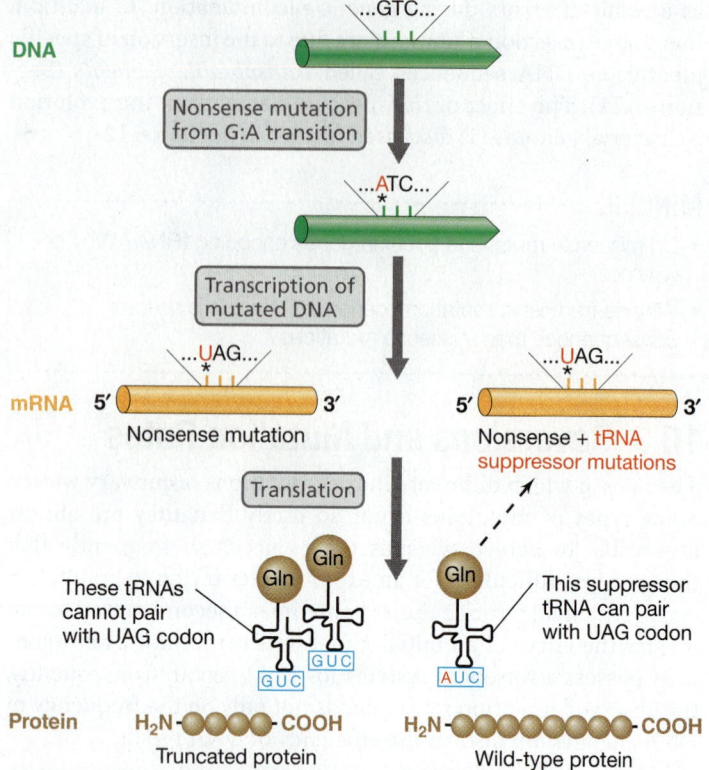

Figure 10.5 Suppression of nonsense mutations. Introduction of a nonsense mutation in a gene encoding a protein results in the incorporation of a stop codon (indicated by the *) in the corresponding mRNA. This single mutation leads to the production of a truncated polypeptide. The mutation is suppressed if a second mutation occurs in the anticodon of a tRNA, a tRNA charged with glutamine in this example, which allows the mutated tRNA or suppressor tRNA to bind to the nonsense codon.

the original amino acid and the protein is fully restored. In other cases, a different amino acid is inserted and a partially active protein may be produced.

The Ames Test

The *Ames test* (named for the biochemist Bruce Ames who developed the test) makes practical use of detecting revertants in large populations of mutant bacteria to test the mutagenicity of potentially hazardous chemicals. A standard way to test chemicals for mutagenesis in the Ames test is to look for an increased rate of back mutations (reversion) in auxotrophic strains of bacteria (**Figure 10.6**). The Ames test assays for *back* mutations instead of *forward* mutations (generating auxotrophs from the wild type) because revertants can be more easily selected.

It is important that the auxotrophic strain used in the Ames test carry a point mutation so that the reversion rate is measurable. Cells of such an auxotroph do not grow on a medium lacking the required nutrient (for example, an amino acid), and even very large populations of cells can be spread on the plate without the formation of visible colonies. However, if revertants (back mutants) are present, those cells form colonies. Thus, if 10^8 cells are spread on the surface of a single plate, even as few as 10–20 revertants can be detected by the 10–20 colonies they form (Figure 10.6, left photo). However, if the reversion rate is *increased*

by the presence of a chemical mutagen, the number of revertant colonies is even greater. After overnight incubation, the mutagenicity of the compound can be detected by looking for a halo of back mutations in the area around the paper disc (Figure 10.6).

A wide variety of chemicals have been subjected to the Ames test, and it has become one of the most useful screens for determining the potential mutagenicity of a compound. Since some mutagens can cause cancer in animals, the Ames test doubles as a means of screening for potential carcinogens as well.

Mutation Rates

For most microorganisms, errors in DNA replication occur at a frequency of 10^{-6} to 10^{-7} per thousand bases during a single round of replication. A typical gene has about 1000 base pairs. Therefore, the frequency of a mutation *in a given gene* is also in the range of 10^{-6} to 10^{-7} per round of replication. For instance, in a bacterial culture having 10^8 cells/ml, there are likely to be a number of different mutants for any given gene in each milliliter of culture. Eukaryotes with very large genomes tend to have replication error rates about 10-fold lower than typical bacteria, whereas DNA viruses, especially those with very small genomes, may have error rates 100-fold to 1000-fold higher than those of cellular organisms. RNA viruses have even higher error rates due to less proofreading (↩ Section 4.6) and the lack of RNA repair mechanisms.

Single base errors during DNA replication are more likely to lead to missense mutations than to nonsense mutations because most single base substitutions yield codons that encode other amino acids (↩ Table 4.5). The next most frequent type of codon change caused by a single base change leads to a silent mutation. This is because for the most part alternate codons for a given amino acid differ from each other by a single base change in the "silent" third position. A given codon can be changed to any of 27 other codons by a single base substitution, and on average, about two of these will be silent mutations, about one a nonsense mutation, and the rest will be missense mutations. There are also some DNA sequences, typically sequences containing short repeats,

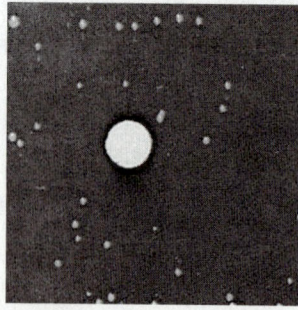
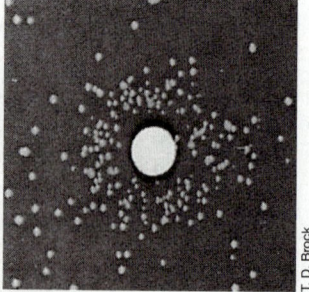

Figure 10.6 The Ames test for assessing the mutagenicity of a chemical. Two plates were inoculated with a culture of a histidine-requiring mutant of *Salmonella enterica*. The medium does not contain histidine, so only cells that revert back to wild type can grow. Spontaneous revertants appear on both plates, but the chemical on the filter-paper disc in the test plate (right) has caused an increase in the mutation rate, as shown by the large number of colonies surrounding the disc. Revertants are not seen very close to the test disc because the concentration of mutagen is lethally high there. The plate on the left was the negative control; its filter-paper disc had only water added.

that are hot spots for mutations because the error frequency of DNA polymerase is relatively high there. The error rate at hot spots is affected by the base sequence in the vicinity.

Unless a mutation can be selected for, its experimental detection is difficult, and much of the skill of the microbial geneticist requires increasing the efficiency of mutation detection. This can most readily be done by increasing the pool of mutations. As we see in the next section, it is possible to greatly increase the mutation rate by treatment with mutagenic agents. In addition, the mutation rate may change in certain situations, such as when cells are placed under high-stress conditions.

MINIQUIZ -

- Why does the Ames test measure the rate of reversion rather than the rate of forward mutation?
- Which class of mutation, missense or nonsense, is more common, and why?

10.4 Mutagenesis

The spontaneous rate of mutation is very low, but a variety of chemical, physical, and biological agents can increase the mutation rate and are therefore said to induce mutations. These agents are called **mutagens**. We discuss some of the major categories of mutagens and their activities here.

Chemical Mutagens

An overview of some of the major chemical mutagens and their modes of action is given in **Table 10.2**. Several classes of chemical mutagens exist. The *nucleotide base analogs* are molecules that resemble the purine and pyrimidine bases of DNA in structure yet display faulty pairing properties (**Figure 10.7**). If a base analog is incorporated into DNA in place of the natural base, the DNA may replicate normally most of the time. However, DNA replication errors occur at higher frequencies at these sites due to incorrect base pairing. The result is the incorporation of a mismatched base into the new strand of DNA and thus introduction of a mutation. During subsequent segregation of this strand in cell division, the mutation is revealed.

Other chemical mutagens induce *chemical modifications* in one base or another, resulting in faulty base pairing or related changes (Table 10.2). For example, alkylating agents (chemicals that react with amino, carboxyl, and hydroxyl groups by substituting them with alkyl groups) such as nitrosoguanidine are powerful mutagens and generally induce mutations at higher frequency than base analogs. Unlike base analogs, which have an effect only when incorporated during DNA replication, alkylating agents can introduce changes even in nonreplicating DNA. Both base analogs and alkylating agents tend to induce base-pair substitutions (Section 10.2).

Another group of chemical mutagens, the acridines, are planar molecules that function as *intercalating agents*. These mutagens become inserted between two DNA base pairs and push them apart. During replication, this abnormal conformation can lead to single base insertions or deletions in acridine-containing DNA. Thus, acridines typically induce frameshift mutations (Section 10.2).

Table 10.2 Chemical and physical mutagens and their modes of action

Agent	Action	Result
Base analogs		
5-Bromouracil	Incorporated like T; occasional faulty pairing with G	AT → GC and occasionally GC → AT
2-Aminopurine	Incorporated like A; faulty pairing with C	AT → GC and occasionally GC → AT
Chemicals reacting with DNA		
Nitrous acid (HNO₂)	Deaminates A and C	AT → GC and GC → AT
Hydroxylamine (NH₂OH)	Reacts with C	GC → AT
Alkylating agents		
Monofunctional (for example, ethyl methanesulfonate)	Puts methyl on G; faulty pairing with T	GC → AT
Bifunctional (for example, mitomycin, nitrogen mustards, nitrosoguanidine)	Cross-links DNA strands; faulty region excised by DNase	Both point mutations and deletions
Intercalating agents		
Acridines, ethidium bromide	Inserts between two base pairs	Microinsertions and microdeletions
Radiation		
Ultraviolet (UV)	Pyrimidine dimer formation	Repair may lead to error or deletion
Ionizing radiation (for example, X-rays)	Free-radical attack on DNA, breaking chain	Repair may lead to error or deletion

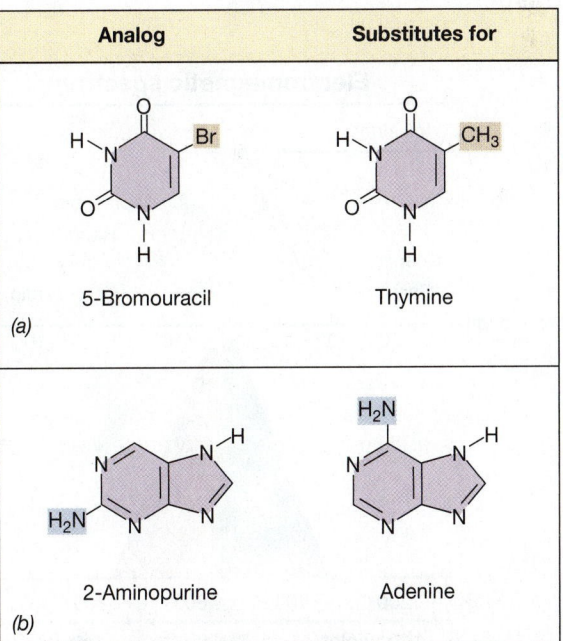

Figure 10.7 Nucleotide base analogs. Structure of two common nucleotide base analogs used to induce mutations and the normal nucleic acid bases for which they substitute. *(a)* 5-Bromouracil can base-pair with guanine, causing AT to GC substitutions. *(b)* 2-Aminopurine can base-pair with cytosine, causing AT to GC substitutions.

Ethidium bromide, which is often used to detect DNA in gel electrophoresis, is also an intercalating agent and therefore a mutagen.

Radiation

Several forms of radiation are highly mutagenic. We can divide mutagenic electromagnetic radiation into two main categories, *nonionizing* and *ionizing* (Figure 10.8). Although both kinds of radiation are used to generate mutations, nonionizing radiation such as ultraviolet (UV) radiation has the widest use.

The purine and pyrimidine bases of nucleic acids absorb UV radiation strongly, and the absorption maximum for DNA and RNA is at 260 nm. Killing of cells by UV radiation is due primarily to its effect on DNA. Although several effects are known, one well-established effect is the production of *pyrimidine dimers*, in which two adjacent pyrimidine bases (cytosine or thymine) on the same strand of DNA become covalently bonded to one another. This either greatly impedes DNA polymerase activity or greatly increases the probability of DNA polymerase misreading the sequence at this point.

Ionizing radiation is more powerful than UV radiation and includes short-wavelength radiation such as X-rays, cosmic rays, and gamma rays (Figure 10.8). These rays cause water and other substances to ionize, resulting in the formation of free radicals such as the hydroxyl radical, OH · (⊄ Section 5.16). Free radicals react with and damage macromolecules in the cell, including DNA. This causes double-stranded and single-stranded breaks that may lead to rearrangements or large deletions. At low doses of ionizing radiation only a few "hits" on DNA occur, but at higher doses, multiple hits cause fragmentation of DNA that sometimes cannot be repaired and thus leads to the death of the cell.

DNA Repair and the SOS System

By definition, a mutation is a *heritable* change in the genetic material. Therefore, if damaged DNA can be corrected before the cell divides, no mutation will occur. Cells have a variety of different DNA repair processes to correct mistakes or repair damage. While most of these DNA repair systems are virtually error-free, some are error-prone and the repair process itself introduces the mutation. Some types of DNA damage, especially large-scale damage from highly mutagenic chemicals or large doses of radiation, may cause lesions that interfere with replication. If such lesions are not removed before replication occurs, DNA replication will stall and lethal breaks in the chromosome will result. Stalled replication as well as certain types of major DNA damage activate the **SOS repair system**. The SOS system initiates a number of DNA repair processes, some of which are error-free. However, the SOS system also allows DNA repair to occur without a template, that is, with random incorporation of dNTPs. As might be expected, this results in many errors and hence many mutations. However, mutations may be less detrimental to cell survival than breaks in the chromosome, as mutations can often be corrected while chromosome breaks usually cannot.

In *Escherichia coli* the SOS repair system regulates the transcription of approximately 40 genes located throughout the chromosome that participate in DNA damage tolerance and DNA repair. In DNA damage tolerance, DNA lesions remain in the DNA, but are bypassed by specialized DNA polymerases that can move past DNA damage—a process known as translesion synthesis. Even if no template is available to allow insertion of the correct bases, it is less dangerous to cell survival to fill the gap than to let it remain. Consequently, translesion synthesis generates many errors. In *E. coli*, in which the process of mutagenesis has been studied in great detail, the two error-prone repair polymerases are DNA polymerase V, an enzyme encoded by the *umuCD* genes (Figure 10.9), and DNA polymerase IV, encoded by *dinB*. Both are induced as part of the SOS repair system.

The SOS system is regulated by two proteins, LexA and RecA. LexA is a repressor that normally prevents expression of the SOS system. The RecA protein, which normally functions in genetic recombination (Section 10.5), is activated by the presence of DNA damage, in particular by the single-stranded DNA that results when replication stalls (Figure 10.9). The activated form of RecA stimulates LexA to inactivate itself by self-cleavage. This leads to derepression of the SOS system and results in the coordinate expression of a number of proteins that take part in DNA repair. Because some of the DNA repair mechanisms of the SOS system are inherently error-prone, many mutations arise. Once the original DNA damage has been repaired, the SOS regulon is repressed and further mutagenesis ceases.

Changes in Mutation Rate and Their Evolutionary Consequence

High fidelity (low error frequency) in DNA replication is essential if organisms are to remain genetically stable. On the other hand, perfect fidelity is counterproductive because it would prevent evolution. The fact that organisms as phylogenetically distant as *Archaea* and *Bacteria* have about the same mutation rate might suggest that evolutionary pressure has selected organisms with the lowest possible mutation rates. However, this is not so. For example, mutants of some organisms that are hyperaccurate in DNA replication and repair have been selected in the laboratory.

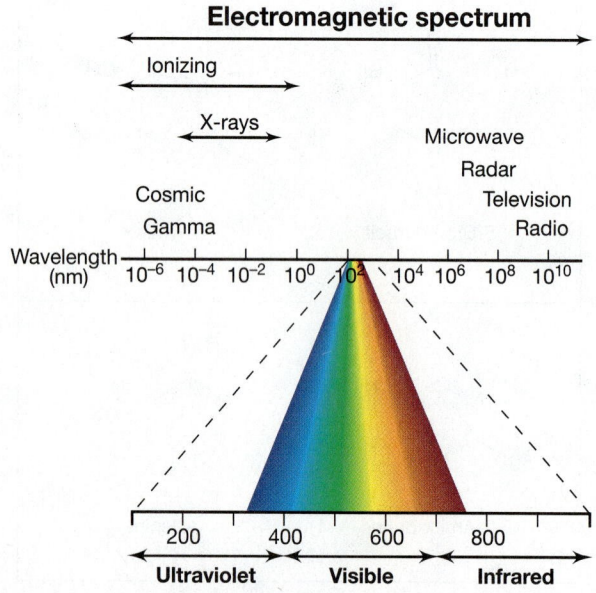

Figure 10.8 Wavelengths of radiation. Ultraviolet radiation consists of wavelengths just shorter than visible light. For any electromagnetic radiation, the shorter the wavelength, the higher the energy. DNA absorbs strongly at 260 nm.

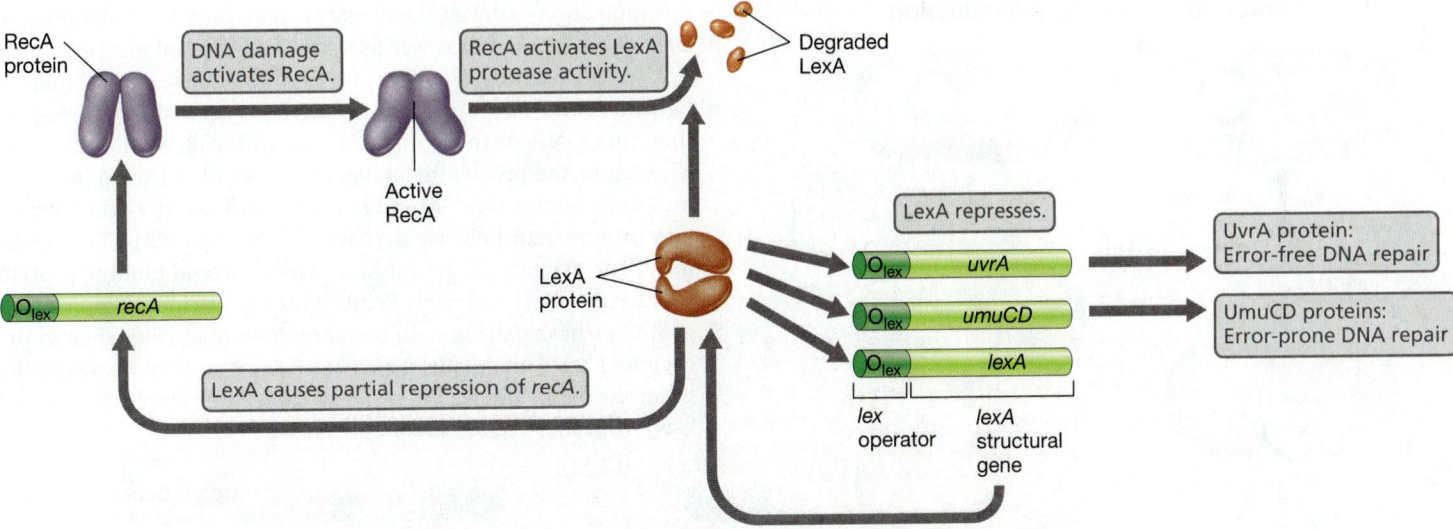

Figure 10.9 Mechanism of the SOS response. DNA damage activates RecA protein, which in turn activates the protease activity of LexA. The LexA protein then cleaves itself. LexA protein normally represses the activities of the *recA* gene and the DNA repair genes *uvrA* and *umuCD* (the UmuCD proteins are part of DNA polymerase V). However, repression is not complete. Some RecA protein is produced even in the presence of LexA protein. With LexA inactivated, these genes become highly active.

However, in these strains, the improved proofreading and repair mechanisms result in slower growth; thus, hyperaccurate mutants might well be at a disadvantage in the natural environment.

In contrast to hyperaccuracy, some organisms actually benefit from an increased mutation rate. DNA repair systems are themselves genetically encoded and thus subject to mutation. For example, the protein subunit of DNA polymerase III required for proofreading (⮌ Section 4.6) is encoded by *dnaQ*. Certain mutations in *dnaQ* lead to mutants that are still viable but have an increased rate of mutation. These are known as hypermutable or **mutator strains**. Mutations leading to a mutator phenotype are known in several other DNA repair systems as well. The mutator phenotype is apparently selected for in complex and changing environments because strains of bacteria with mutator phenotypes appear to be more abundant under these conditions. Presumably, whatever disadvantage an increased mutation rate may have in such environments is offset by the ability to generate greater numbers of useful mutations. These mutations ultimately increase evolutionary fitness of the population and make the organism more successful in its ecological niche.

As indicated earlier, a mutator phenotype may be induced in wild-type strains by stressful situations. For instance, the SOS repair system includes error-prone repair. Therefore, when the SOS repair system is activated, the mutation rate increases. In some cases this is merely an inevitable by-product of DNA repair, but in other cases, the increased mutation rate may itself be of selective value to the organism for survival purposes.

MINIQUIZ

- How do mutagens work?
- Why might a mutator phenotype be successful in an environment experiencing rapid changes?
- What is meant by "error-prone" DNA repair?

II • Gene Transfer in *Bacteria*

Comparative genomic analyses of closely related microorganisms that exhibit different phenotypes have revealed distinct genome differences. Often these idiosyncratic differences result from *horizontal gene transfer*, the movement of genes between cells that are not direct descendants of one another (⮌ Section 6.12). Horizontal gene transfer allows cells to quickly acquire new characteristics and drives metabolic diversity.

Three mechanisms of genetic exchange are known in prokaryotes: (1) *transformation*, in which free DNA released from one cell is taken up by another (Section 10.6); (2) *transduction*, in which DNA transfer is mediated by a virus (Section 10.7); and (3) *conjugation*, in which DNA transfer requires cell-to-cell contact and a conjugative plasmid in the donor cell (Sections 10.8 and 10.9). These processes are contrasted in **Figure 10.10**, and it should be noted that DNA transfer typically occurs in only one direction, from donor to recipient.

Before discussing the mechanisms of transfer, we consider the fate of transferred DNA. Whether it is transferred by transformation, transduction, or conjugation, DNA that enters the cell by horizontal gene transfer faces three possible fates: (1) It may be degraded by restriction enzymes; (2) it may replicate by itself (but only if it possesses its own origin of replication, such as a plasmid or phage genome); or (3) it may recombine with the host chromosome.

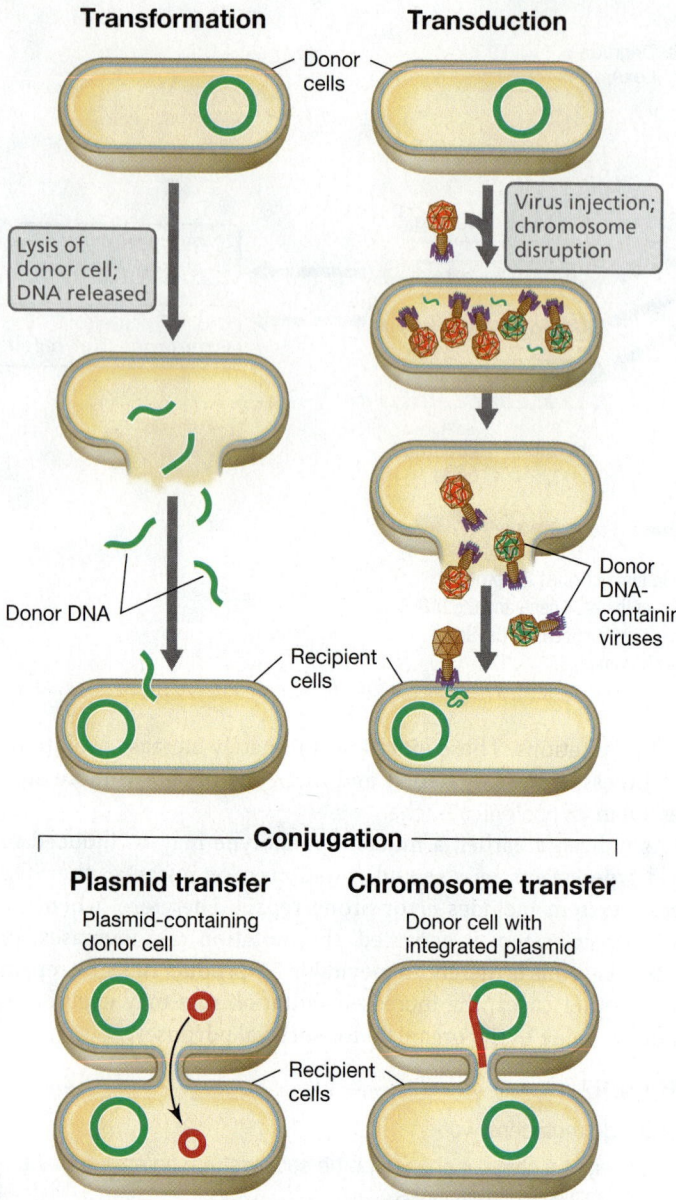

Figure 10.10 **Processes by which DNA is transferred from donor to recipient bacterial cell.** Just the initial steps in transfer are shown.

10.5 Genetic Recombination

Recombination is the physical exchange of DNA between *genetic elements* (structures that carry genetic information). In this section we focus on *homologous* recombination, a process that results in genetic exchange between homologous DNA sequences from two different sources. Homologous DNA sequences are those that have nearly the same sequence; therefore, bases can pair over an extended length of the two DNA molecules. This type of recombination drives the process of "crossing over" in classical genetics.

Molecular Events in Homologous Recombination

The RecA protein, previously mentioned in regard to the SOS repair system (Section 10.4 and Figure 10.9), is the key to homologous recombination. RecA is essential in nearly every homologous

recombination pathway. RecA-like proteins have been identified in all bacteria examined, as well as in the *Archaea* and most *Eukarya*.

A molecular mechanism for homologous recombination between two DNA molecules is shown in **Figure 10.11**. An enzyme that cuts DNA in the middle of a strand, called an *endonuclease*, begins the process by nicking one strand of the donor DNA molecule. This nicked strand is separated from the other strand by proteins with helicase activity (⮂ Section 4.5). The resulting single-stranded segment binds single-strand binding protein (⮂ Section 4.5) and then RecA. This results in a complex that promotes base pairing with the complementary sequence in the recipient DNA molecule. This base pairing in turn displaces the other strand of the recipient DNA molecule (Figure 10.11) and is appropriately called *strand invasion*.

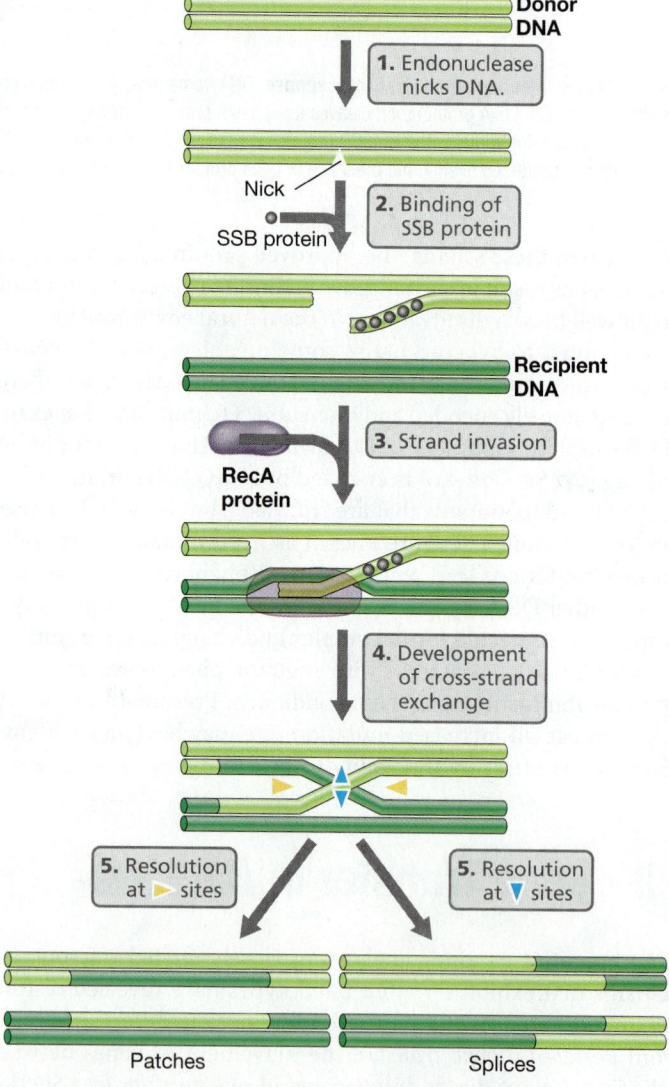

Figure 10.11 **A simplified version of homologous recombination.** Homologous DNA molecules pair and exchange DNA segments. The mechanism requires breakage and reunion of paired segments. Two of the participating proteins, single-strand binding (SSB) protein and the RecA protein, are shown. The other participating proteins are not shown. The diagram is not to scale: Pairing may occur over hundreds or thousands of bases. Resolution occurs by cutting and rejoining the cross-linked DNA molecules. Note that there are two possible outcomes, patches or splices, depending on where strands are cut during the resolution process.

The base pairing of one strand from each of the two DNA molecules over long stretches generates recombination intermediates containing long **heteroduplex** regions, where each strand has originated from a different chromosome. Finally the linked molecules are separated (resolved) by resolvase enzymes that cut and rejoin the previously unbroken strands of both original DNA molecules. Depending on the orientation of the junction during resolution, two types of products, referred to as "patches" or "splices," are formed that differ in the conformation of the heteroduplex regions remaining after resolution (Figure 10.11).

Effect of Homologous Recombination on Genotype

For homologous recombination to generate new genotypes, the two homologous sequences must be related but genetically distinct. This is obviously the case in a diploid eukaryotic cell, which has two sets of chromosomes, one from each parent. In prokaryotes, genetically distinct but homologous DNA molecules are brought together in different ways. Genetic recombination in prokaryotes occurs after fragments of homologous DNA from a donor chromosome are transferred to a recipient cell by transformation, transduction, or conjugation. It is only after the transfer event, when the DNA fragment from the donor is in the recipient cell, that homologous recombination occurs. In prokaryotes, only part of a chromosome is transferred; therefore, if recombination does not occur, the DNA fragment will be lost because it cannot replicate independently. Thus, in prokaryotes, transfer is just the first step in generating recombinant organisms.

To detect physical exchange of DNA segments, the cells resulting from recombination must be phenotypically different from both parents (**Figure 10.12**). Genetic crosses in bacteria usually depend on using recipient strains that lack some selectable character that the recombinants will gain. For instance, the recipient may be unable to grow on a particular medium, and genetic recombinants are selected that can. Various kinds of selectable markers, such as drug resistance and nutritional requirements,

were discussed in Section 10.1. The exceedingly great sensitivity of the selection process allows even a few recombinant cells to be detected in a large population of nonrecombinant cells and thus selection is an important tool for the microbial geneticist.

Complementation

In all three methods of bacterial gene transfer, only a portion of the donor chromosome enters the recipient cell. Therefore, unless recombination takes place with the recipient chromosome, the donor DNA will be lost because it cannot replicate independently in the recipient. Nonetheless, it is possible to stably maintain a state of partial diploidy for use in bacterial genetic analysis. A bacterial strain that carries *two copies* of any particular chromosomal segment is known as a partial diploid, or *merodiploid*. In general, one copy is present on the chromosome itself and the second copy on another genetic element, such as a plasmid or a bacteriophage.

Consequently, if the chromosomal copy of a gene is defective due to a mutation, it is possible to supply a functional (wild-type) copy of the gene on a plasmid or phage. For example, if one of the genes for biosynthesis of the amino acid tryptophan has a mutation resulting in a nonfunctional enzyme, this will give a Trp⁻ phenotype. That is, the mutant strain will be a tryptophan auxotroph and will require tryptophan for growth. However, if a copy of the wild-type gene is introduced into the same cell on a plasmid or viral genome, this gene will encode the necessary protein and restore the wild-type phenotype. This process is called *complementation* because the wild-type gene is said to *complement* the mutation, in this case converting the Trp⁻ cell into Trp⁺ (Figure 10.12).

MINIQUIZ

- Which protein, found in all prokaryotes, facilitates the pairing required for homologous recombination?
- Explain the fate of transferred chromosomal DNA if recombination does not occur in prokaryotes.
- What is a merodiploid?

10.6 Transformation

Transformation is a genetic transfer process by which *free DNA* is incorporated into a recipient cell and brings about genetic change. Several prokaryotes are naturally transformable, including certain species of both gram-negative and gram-positive *Bacteria* and also some species of *Archaea* (Section 10.10). Because the DNA of prokaryotes is present in the cell as a large single molecule, when a cell is gently lysed, its DNA pours out. Bacterial chromosomes break easily because of their extreme length (1700 μm in *Bacillus subtilis*, for example). Even after gentle extraction, the *B. subtilis* chromosome of 4.2 megabase pairs is converted to fragments of about 10 kilobase pairs each. Because the DNA that corresponds to an average gene is about 1000 nucleotides, each of the fragments of *B. subtilis* DNA contains about ten genes. This is a typical transformable size. A single cell usually incorporates only one or a few DNA fragments, so only a small proportion of the genes of one cell can be transferred to another in a single transformation event.

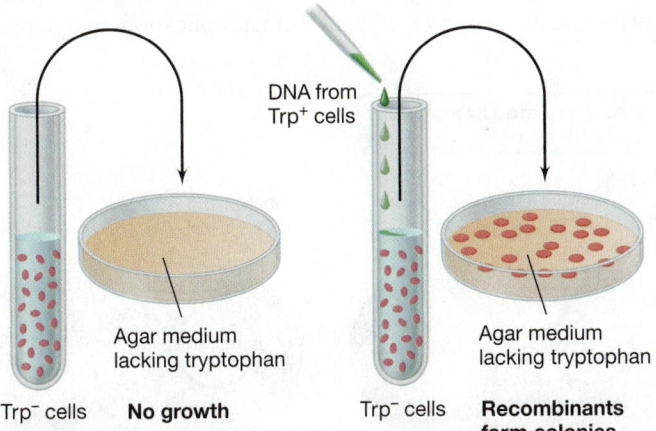

Figure 10.12 Using a selective medium to detect rare genetic recombinants. On the selective medium only the rare recombinants form colonies even though a very large population of bacteria was plated. Procedures such as this, which offer high resolution for genetic analyses, can ordinarily be used only with microorganisms. The type of genetic exchange being illustrated is transformation, but a similar outcome could result from any of the other forms of horizontal gene transfer.

DNA from Trp⁺ cells

Agar medium lacking tryptophan

Agar medium lacking tryptophan

Trp⁻ cells   **No growth**

Trp⁻ cells   **Recombinants form colonies**

Competence in Transformation

A cell that is able to take up DNA and be transformed is said to be *competent*, and this capacity is genetically determined. Competence in most naturally transformable bacteria is regulated, and special proteins play a role in the uptake and processing of DNA. These competence-specific proteins include a membrane-associated DNA-binding protein, a cell wall autolysin, and various nucleases. One pathway of natural competence in *B. subtilis*—an easily transformed species—is regulated by quorum sensing, a regulatory system that responds to cell density (⟜ Section 7.9). Cells produce and excrete a small peptide during growth, and the accumulation of this peptide to high concentrations induces the cells to become competent. But not all cells become competent. In *Bacillus*, roughly 20% of the cells in a culture become competent and stay that way for several hours. However, in *Streptococcus,* 100% of the cells can become competent, but only for a brief period during the growth cycle.

High-efficiency, natural transformation is rare among *Bacteria*. For example, *Acinetobacter, Bacillus, Streptococcus, Haemophilus, Neisseria*, and *Thermus* are naturally competent and easy to transform. By contrast, many *Bacteria* are poorly transformed, if at all, under natural conditions. For example, *Escherichia coli* and many other gram-negative bacteria fall into this category. However, if cells of *E. coli* are treated with high concentrations of Ca^{2+} and then chilled, they become adequately competent. Cells treated in this manner take up double-stranded DNA, and therefore transformation of *E. coli* by plasmid DNA can be relatively efficient. This is important because getting DNA into *E. coli*—the workhorse of genetic engineering—is critical for biotechnology, as we will see in Chapter 11.

Electroporation is a physical technique that is used to get DNA into organisms that are difficult to transform, especially those with thick cell walls. In electroporation, cells are mixed with DNA and then exposed to brief high-voltage electrical pulses. This makes the cell envelope permeable and allows entry of the DNA. Electroporation is a quick process and works for most types of cells, including *E. coli*, most other *Bacteria*, some members of the *Archaea*, and even yeast and certain plant cells.

Uptake of DNA in Transformation

During natural transformation, competent bacteria reversibly bind DNA. Soon, however, the binding becomes irreversible. Competent cells bind much more DNA than do noncompetent cells—as much as 1000 times more. As noted earlier, the sizes of the transforming fragments are much smaller than that of the whole genome, and the fragments are further degraded during the uptake process. In *Streptococcus pneumoniae* each cell can bind only about ten molecules of double-stranded DNA of 10–15 kbp each. However, as these fragments are taken up, they are converted into single-stranded pieces of about 8 kb, with the complementary strand being degraded. The DNA fragments in the mixture compete with each other for uptake and thus the probability of a transformant taking up DNA that confers an advantage or a selectable marker decreases.

Interestingly, transformation in *Haemophilus influenzae* requires the DNA fragment to have a particular 11-bp sequence for irreversible binding and uptake to occur. This sequence is found at an unexpectedly high frequency in the *Haemophilus* genome. Evidence such as this, and the fact that certain bacteria become competent in their natural environment, suggests that transformation is not a laboratory artifact but plays an important role in horizontal gene transfer in nature. By promoting new combinations of genes, naturally transformable bacteria increase the diversity and fitness of the microbial community as a whole.

Integration of Transforming DNA

During transformation, transforming DNA is bound at the cell surface by a DNA-binding protein (**Figure 10.13**). Next, either the entire double-stranded fragment is taken up, or a nuclease degrades one strand and the remaining strand is taken up, depending on the organism. After uptake, the DNA is bound by a competence-specific protein. This protects the DNA from nuclease attack until it reaches the chromosome, where the RecA protein takes over. The DNA is integrated into the genome of the recipient by recombination (Figures 10.13 and 10.11). If single-stranded DNA is integrated, a heteroduplex DNA is formed. During the next round of chromosomal replication, one parental

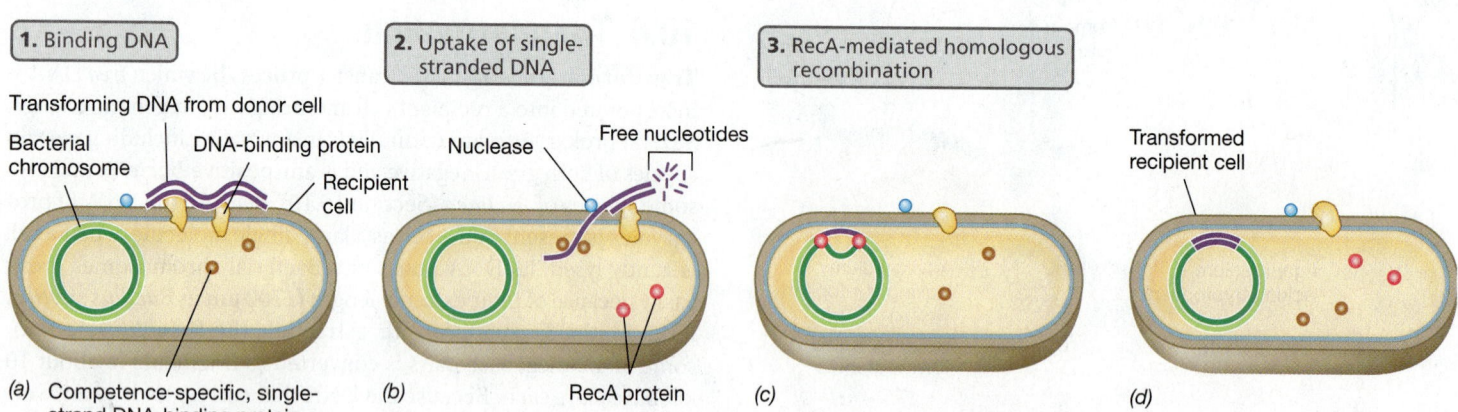

1. Binding DNA

2. Uptake of single-stranded DNA

3. RecA-mediated homologous recombination

Transforming DNA from donor cell

Bacterial chromosome

DNA-binding protein

Recipient cell

Nuclease

Free nucleotides

Transformed recipient cell

(a) Competence-specific, single-strand DNA-binding protein

(b)

RecA protein

(c)

(d)

Figure 10.13 Mechanism of transformation in a gram-positive bacterium. *(a)* Binding of double-stranded DNA by a membrane-bound DNA-binding protein. *(b)* Passage of one of the two strands into the cell while nuclease activity degrades the other strand. *(c)* The single strand in the cell is bound by specific proteins, and recombination with homologous regions of the bacterial chromosome is mediated by RecA protein. *(d)* Transformed cell.

and one recombinant DNA molecule are generated. On segregation at cell division, the recombinant molecule is present in the transformed cell, which is now genetically altered compared to its parent. The preceding applies only to small pieces of linear DNA. Many naturally transformable *Bacteria* are transformed only poorly by plasmid DNA because the plasmid must remain double-stranded and circular in order to replicate.

MINIQUIZ
- During transformation a cell usually incorporates only one or a few fragments of DNA. Explain.
- Even in naturally transformable cells, competence is usually inducible. What does this mean?

10.7 Transduction

In **transduction**, a bacterial virus (bacteriophage) transfers DNA from one cell to another. Viruses can transfer host genes in two ways. In the first, called *generalized transduction*, DNA derived from virtually any portion of the host genome is packaged inside the mature virion in place of the virus genome. In the second, called *specialized transduction*, DNA from a specific region of the host chromosome is integrated directly into the virus genome—usually replacing some of the virus genes. This occurs only with certain temperate viruses such as phage lambda ( Section 8.8).

In generalized transduction, the bacterial donor genes cannot replicate independently and are not part of a viral genome. Thus, unless the donor genes recombine with the recipient bacterial chromosome, they will be lost. In specialized transduction, homologous recombination may also occur. However, since the donor bacterial DNA is actually a part of a temperate phage genome, it may be integrated into the host chromosome during lysogeny ( Section 8.8).

Transduction occurs in a variety of *Bacteria*, including the genera *Desulfovibrio*, *Escherichia*, *Pseudomonas*, *Rhodococcus*,

Rhodobacter, *Salmonella*, *Staphylococcus*, and *Xanthobacter*, as well as *Methanothermobacter thermautotrophicus*, a species of *Archaea*. Not all phages can transduce, and not all bacteria are transducible, but with bacteriophage abundance estimated to outnumber prokaryotic cells by 10-fold in nature, the phenomenon plays an important role in gene transfer in the environment. Examples of genes transferred by transducing bacteriophages include multiple antibiotic resistance genes among strains of *Salmonella enterica* serovar Typhimurium, Shiga-like toxin genes in *Escherichia coli*, virulence factors in *Vibrio cholerae*, and genes encoding photosynthetic proteins in cyanobacteria.

While transduction plays a role in horizontal transfer of DNA in nature, microbial geneticists use both generalized and specialized transducing bacteriophages to deliver DNA to target bacterial cells. Transduction can be used to deliver DNA to strains in which transformation and conjugation are not efficient. Bacteriophages can also be used to deliver large pieces of DNA to host cells. A typical tailed phage that contains double-stranded DNA can package up to 40 kilobase pairs of DNA. The bacteriophages used for transduction in the laboratory are usually nonlytic because bacterial genes have replaced all or some necessary viral genes. To select for a transduction event, a transducing phage must infect a donor host that has a selectable marker.

Generalized Transduction

In generalized transduction, virtually any gene on the donor chromosome can be transferred to the recipient, forming a *transductant*. Generalized transduction was first discovered and extensively studied in the bacterium *Salmonella enterica* with phage P22 and has also been studied with phage P1 in *Escherichia coli*. An example of how transducing particles are formed is given in **Figure 10.14**. When a bacterial cell is infected with a phage, the lytic cycle may occur. However, during lytic infection, the enzymes responsible for packaging viral DNA into the bacteriophage sometimes package host DNA accidentally. The result is called a *transducing particle*. These cannot lead to a viral infection because they contain no viral

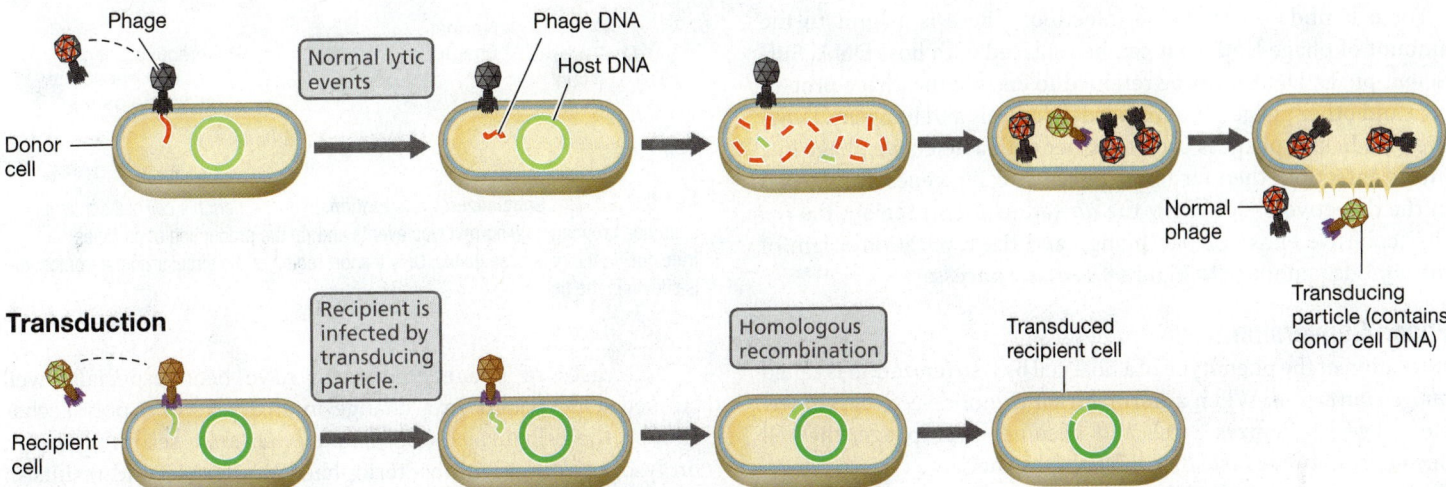

Figure 10.14 Generalized transduction. Note that "normal" virions contain phage genes, whereas a transducing particle contains host genes.

DNA, and are said to be *defective*. On lysis of the cell, the transducing particles are released along with normal virions that contain the virus genome. Consequently, the lysate contains a mixture of normal virions and transducing particles.

When this lysate is used to infect a population of recipient cells, most of the cells are infected with normal virus. However, a small proportion of the population receives transducing particles that inject the DNA they packaged from the previous host bacterium. Although this DNA cannot replicate, it can recombine with the DNA (Section 10.5) of the new host. Because only a small proportion of the particles in the lysate are defective, and each of these contains only a small fragment of donor DNA, the probability of a given transducing particle containing a particular gene is quite low. Typically, only about 1 cell in 10^6 to 10^8 is transduced for a given gene.

Lysogeny and Specialized Transduction

Generalized transduction allows the transfer of any gene from one bacterium to another, but at a low frequency. In contrast, specialized transduction allows extremely efficient transfer, but is selective and transfers only a small region of the bacterial chromosome. In the first case of specialized transduction to be discovered, genes for galactose catabolism were transduced by the temperate phage lambda of *E. coli*.

When lambda lysogenizes a host cell, the phage genome is integrated into the *E. coli* chromosome at a specific site (♻ Section 8.8). This site is next to the cluster of genes that encode the enzymes for galactose utilization. After insertion, viral DNA replication is under control of the bacterial host chromosome. Upon induction, the viral DNA separates from the host DNA by a process that is the reverse of integration (**Figure 10.15**). Usually the lambda DNA is excised precisely, but occasionally the phage genome is excised incorrectly. Some of the adjacent bacterial genes to one side of the prophage (for example, the galactose operon) are excised along with phage DNA. At the same time, some phage genes are left behind (Figure 10.15*b*). This transducing particle can subsequently transfer genes for galactose utilization to a recipient cell. This transfer can only be detected if a galactose-negative (Gal⁻) strain is infected with such a transducing particle and Gal⁺ transductants are selected.

For a lambda virion to be infectious, there is a limit to the amount of phage DNA that can be replaced with host DNA. Sufficient phage DNA must be retained to encode the phage protein coat and other phage proteins needed for lysis and lysogeny. However, if a helper phage is used together with a defective phage in a mixed infection, then far fewer phage-specific genes are needed in the defective phage. Only the *att* (attachment) region, the *cos* site (cohesive ends, for packaging), and the replication origin of the lambda genome (♻ Figure 8.17*a*) are necessary.

Phage Conversion

Alteration of the phenotype of a host cell by lysogenization is called *phage conversion*. When a normal (that is, nondefective) temperate phage lysogenizes a cell and becomes a prophage, the cell becomes immune to further infection by the same type of phage. Such immunity may itself be regarded as a change in phenotype. However, other phenotypic changes unrelated to phage immunity are often observed in phage conversion of lysogenized cells.

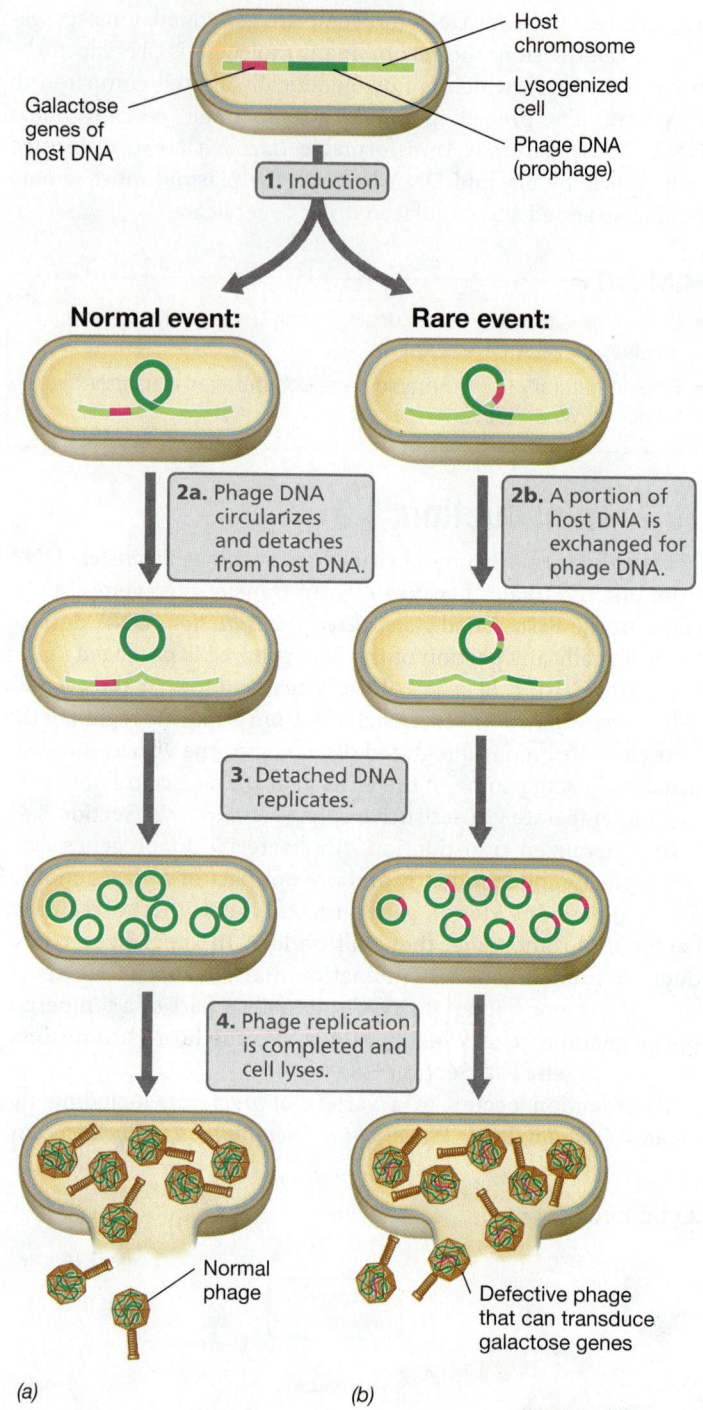

Figure 10.15 **Specialized transduction.** In an *Escherichia coli* cell containing a lambda prophage, *(a)* normal lytic events and *(b)* the production of particles transducing the galactose genes. Only a short region of the circular host chromosome is shown in the figure.

Two cases of phage conversion have been especially well studied. One results in a change in structure of a polysaccharide on the cell surface of *Salmonella enterica* serovar Anatum on lysogenization with bacteriophage ε[15]. The second results in the conversion of non-toxin-producing strains of *Corynebacterium diphtheriae* (the bacterium that causes the disease diphtheria) to toxin-producing (pathogenic) strains following lysogeny with bacteriophage β (⟳ Section 29.3). In both cases, the genes responsible for the changes are an integral part of the phage genome and hence are automatically transferred to the cell upon infection by the phage and lysogenization.

Lysogeny likely carries strong selective value for the host cell because it confers resistance to infection by viruses of the same type. Phage conversion may also be of considerable evolutionary significance because it results in genetic alteration of host cells. Many bacteria isolated from nature are natural lysogens, and thus it is likely that lysogeny is essential for survival of many host cells in nature.

MINIQUIZ

- How does a transducing particle differ from an infectious bacteriophage?
- What is the major difference between generalized transduction and transformation?
- Why is phage conversion considered beneficial to host cells?

10.8 Conjugation

Bacterial **conjugation** (mating) is a form of horizontal gene transfer that requires cell-to-cell contact. Conjugation is a plasmid-encoded mechanism that can mediate DNA transfer between unrelated cells, even between different genera. Conjugative plasmids use this mechanism to transfer copies of themselves and the genes they encode, such as those for antibiotic resistance, to new host cells.

The process of conjugation requires a *donor* cell, which contains the conjugative plasmid, and a *recipient* cell, which does not. In addition, genetic elements that cannot transfer themselves can sometimes be *mobilized* or transferred during conjugation. These other genetic elements can be other plasmids or the host chromosome itself. Indeed, conjugation was discovered because the F plasmid of *Escherichia coli* can mobilize the host chromosome (see Figure 10.21). Transfer mechanisms may differ depending on the participating plasmid, but most plasmids in gram-negative *Bacteria* employ a mechanism similar to that used by the F plasmid.

F Plasmid

The F plasmid (F stands for "fertility") is a circular DNA molecule of 99,159 bp. **Figure 10.16** shows a genetic map of the F plasmid. One region of the plasmid contains genes that regulate DNA replication. It also contains a number of transposable elements (Section 10.11) that allow the plasmid to integrate into the host chromosome. In addition, the F plasmid has a large region of DNA, the *tra* region, containing genes that encode transfer functions. Many genes in the *tra* region participate in mating pair formation, and most of these have to do with the synthesis of a surface structure, the sex pilus (⟳ Section 2.13). Only donor cells produce these pili. Different conjugative plasmids may have slightly different *tra* regions, and the pili may vary somewhat in structure. The F plasmid and its relatives encode F pili.

Pili allow specific pairing to take place between the donor and recipient cells. All conjugation in gram-negative *Bacteria*

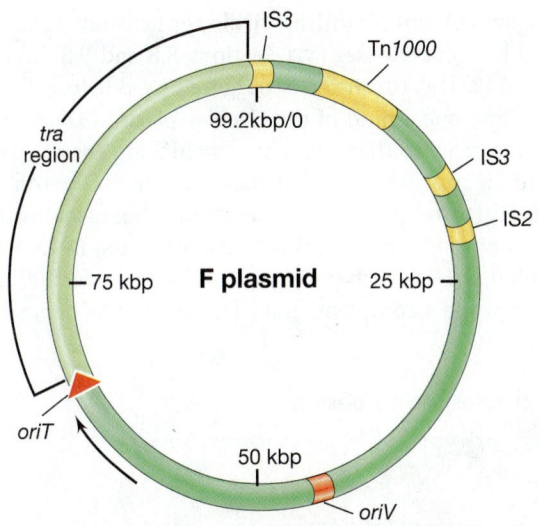

Figure 10.16 Genetic map of the F (fertility) plasmid of *Escherichia coli*. The numbers on the interior show the size in kilobase pairs (the exact size is 99,159 bp). The region in dark green at the bottom of the map contains genes primarily responsible for the replication and segregation of the F plasmid. The origin of vegetative replication is *oriV*. The light green *tra* region contains the genes needed for conjugative transfer. The origin of transfer during conjugation is *oriT*. The arrow indicates the direction of transfer (the *tra* region is transferred last). Insertion sequences are shown in yellow. These may recombine with identical elements on the bacterial chromosome, which leads to integration and the formation of different Hfr strains.

is thought to depend on cell pairing brought about by pili. The pilus makes specific contact with a receptor on the recipient cell and then is retracted by disassembling its subunits. This pulls the two cells together (**Figure 10.17**). Following this process, donor and recipient cells remain in contact by binding proteins located in the outer membrane of each cell. DNA is then transferred from donor to recipient cell through this conjugation junction.

Mechanism of DNA Transfer during Conjugation

DNA synthesis is necessary for DNA transfer by conjugation. This DNA is synthesized not by normal bidirectional replication

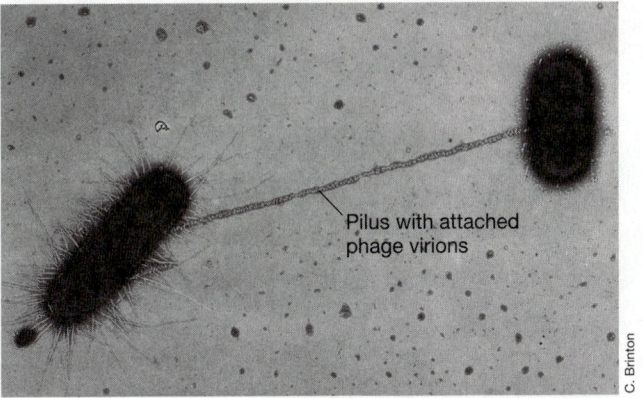

Figure 10.17 Formation of a mating pair. Direct contact between two conjugating bacteria is first made via a pilus. The cells are then drawn together to form a mating pair by retraction of the pilus, which is achieved by depolymerization. Certain small phages (F-specific bacteriophages) use the sex pilus as a receptor and can be seen here attached to the pilus.

(⮌ Section 4.6), but by **rolling circle replication**, a mechanism also used by some viruses (⮌ Sections 8.8 and 9.3) and shown in **Figure 10.18**. DNA transfer is triggered by cell-to-cell contact, at which time one strand of the circular plasmid DNA is nicked and is transferred to the recipient. The nicking enzyme required to initiate the process, TraI, is encoded by the *tra* operon of the F plasmid. TraI also has helicase activity and thus also unwinds the strand to be transferred. As this transfer occurs, DNA synthesis by the rolling circle mechanism replaces the transferred strand in the donor, while a complementary DNA strand is being made in

the recipient. Therefore, at the end of the process, both donor and recipient possess complete plasmids. For transfer of the F plasmid, if an F-containing donor cell, which is designated F^+, mates with a recipient cell lacking the plasmid (F^-), the result is two F^+ cells (Figure 10.18).

Transfer of plasmid DNA is efficient and rapid; under favorable conditions virtually every recipient cell that pairs with a donor acquires a plasmid. Transfer of the F plasmid, comprising approximately 100 kilobase pairs of DNA, takes about 5 minutes. If the plasmid genes can be expressed in the recipient, the recipient itself

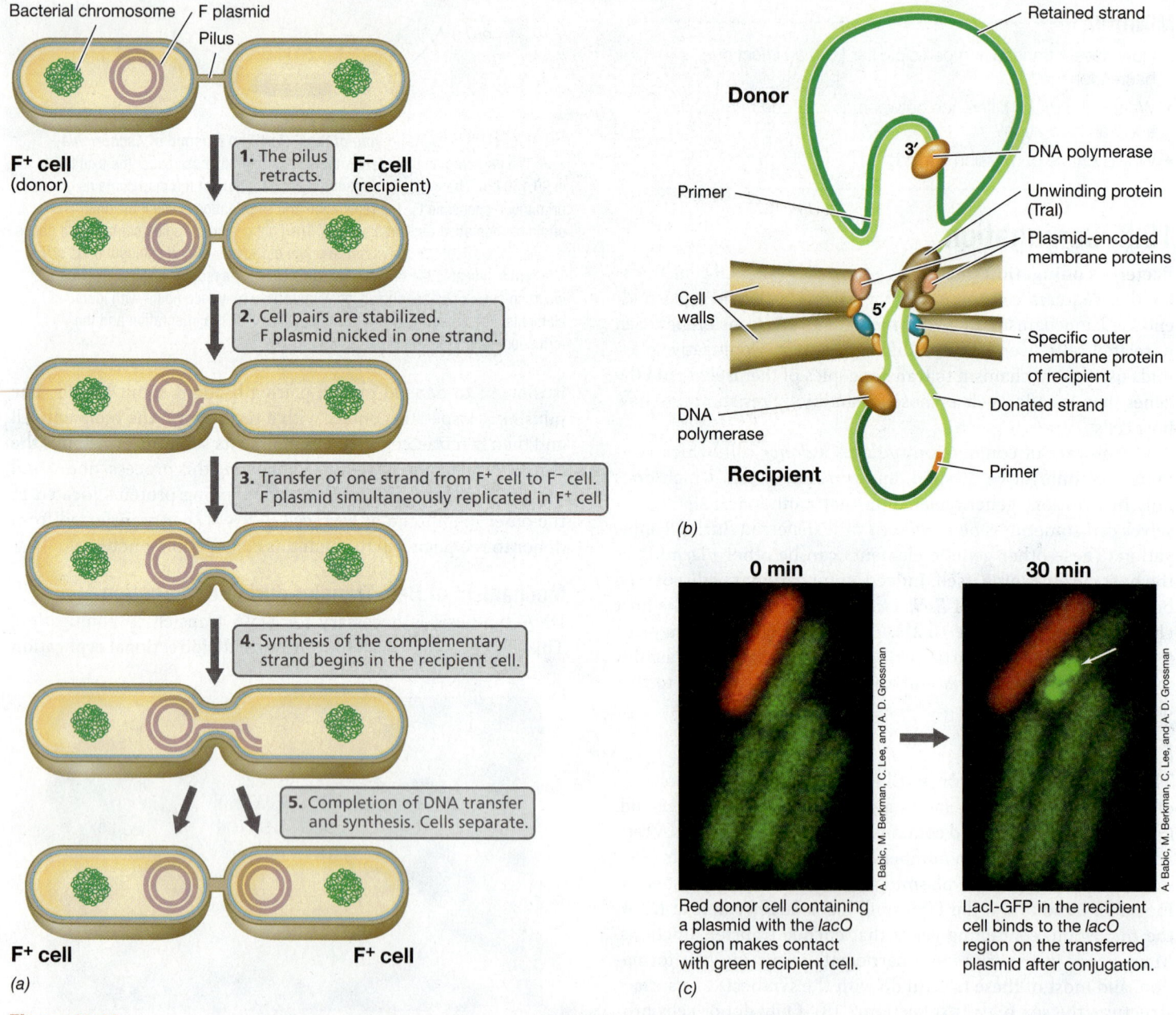

(a)

(b)

(c)

Red donor cell containing a plasmid with the *lacO* region makes contact with green recipient cell.

LacI-GFP in the recipient cell binds to the *lacO* region on the transferred plasmid after conjugation.

A. Babic, M. Berkman, C. Lee, and A. D. Grossman

Figure 10.18 Transfer of plasmid DNA by conjugation. *(a)* The transfer of the F plasmid converts an F⁻ recipient cell into an F⁺ cell. Note the mechanism of rolling circle replication. *(b)* Details of the replication and transfer process. *(c)* Visualization of DNA transfer by conjugation in *Bacillus subtilis* using fluorescent microscopy. The donor cell constitutively expresses a red fluorescent protein, while the recipient cells fluoresce green due to green fluorescent protein (GFP) fused to LacI (⮌ Figure 7.15). The DNA transferred from the donor contains a *lacO* operator region that binds LacI-GFP. Arrows indicate focal points in the recipient cell where LacI-GFP is bound to the *lacO* region obtained from conjugation.

becomes a donor and can transfer the plasmid to other recipients. In this fashion, conjugative plasmids can spread rapidly among bacterial populations, behaving much like infectious agents. This is of major ecological significance because conjugative plasmids have been found in many *Bacteria* and some *Archaea* (Section 10.10), and a few plasmid-containing cells introduced into a population of recipients can convert the entire population into plasmid-bearing (and thus donating) cells in a short time.

MINIQUIZ --

- In conjugation, how are donor and recipient cells brought into contact with each other?
- Explain how rolling circle DNA replication allows both donor and recipient to end up with a complete copy of plasmids transferred by conjugation.
- Why does F have two different origins of replication?

10.9 The Formation of Hfr Strains and Chromosome Mobilization

Chromosomal genes can be transferred by plasmid-mediated conjugation. As mentioned above, the F plasmid of *Escherichia coli* can, under certain circumstances, mobilize the chromosome for transfer during cell-to-cell contact. The F plasmid is an *episome*, a plasmid that can integrate into the host chromosome. When the F plasmid is integrated, chromosomal genes can be transferred along with the plasmid. Following genetic recombination between donor and recipient DNA, horizontal transfer of chromosomal genes by this mechanism can be very extensive.

Cells possessing a nonintegrated F plasmid are called F$^+$. Those with an F plasmid integrated into the chromosome are called **Hfr cells** (for *h*igh *f*requency of *r*ecombination). This term refers to the high rates of genetic recombination between genes on the donor and recipient chromosomes. Both F$^+$ and Hfr cells are donors, but unlike conjugation between an F$^+$ and an F$^-$, conjugation between an Hfr donor and an F$^-$ leads to transfer of genes from the host chromosome. This is because the chromosome and plasmid now form a single molecule of DNA. Consequently, when rolling circle replication is initiated by the F plasmid, replication continues on into the chromosome. Thus, the chromosome is also replicated and transferred. Hence, integration of a conjugative plasmid provides a mechanism for mobilizing a cell's genome.

Overall, the presence of the F plasmid results in three distinct alterations in a cell: (1) the ability to synthesize the F pilus (Figure 10.17), (2) the mobilization of DNA for transfer to another cell, and (3) the alteration of surface receptors so the cell can no longer act as a recipient in conjugation and is unable to take up a second copy of the F plasmid or genetically related plasmids.

Integration of F and Chromosome Mobilization

The F plasmid and the chromosome of *E. coli* both carry several copies of mobile elements called *insertion sequences* (IS; Section 10.11). These provide regions of sequence homology between chromosomal and F plasmid DNA. Consequently, homologous recombination between an IS on the F plasmid and a corresponding IS on the chromosome results in integration of the F plasmid

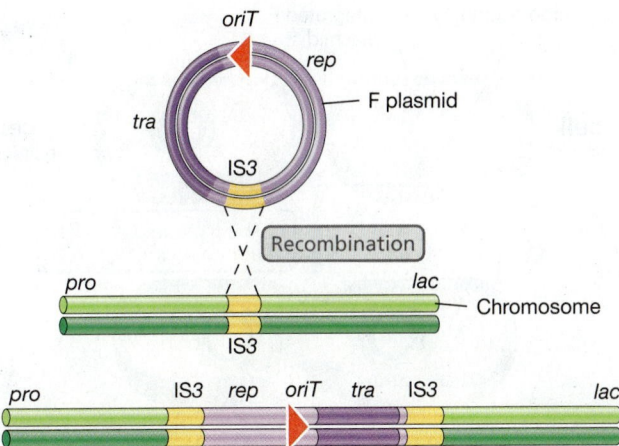

Figure 10.19 The formation of an Hfr strain. Integration of the F plasmid into the chromosome may occur at a variety of specific sites where IS elements are located. The example shown here is an IS*3* located between the chromosomal genes *pro* and *lac*. Some of the genes on the F plasmid are shown. The arrow indicates the origin of transfer, *oriT*, with the arrow as the leading end. Thus, in this Hfr *pro* would be the first chromosomal gene to be transferred and *lac* would be among the last.

into the host chromosome, as shown in **Figure 10.19**. Once integrated, the plasmid no longer replicates independently, but the *tra* operon still functions normally and the strain synthesizes pili. When a recipient is encountered, conjugation is triggered just as in an F$^+$ cell, and DNA transfer is initiated at the *oriT* (origin of transfer) site. However, because the plasmid is now part of the chromosome, after part of the plasmid DNA is transferred, chromosomal genes begin to be transferred (**Figure 10.20**). As in the case of conjugation with just the F plasmid itself (Figure 10.18), chromosomal DNA transfer also requires replication.

Because the DNA strand typically breaks during transfer, only part of the donor chromosome is transferred. Consequently, the recipient does not become Hfr (or F$^+$) because only part of the

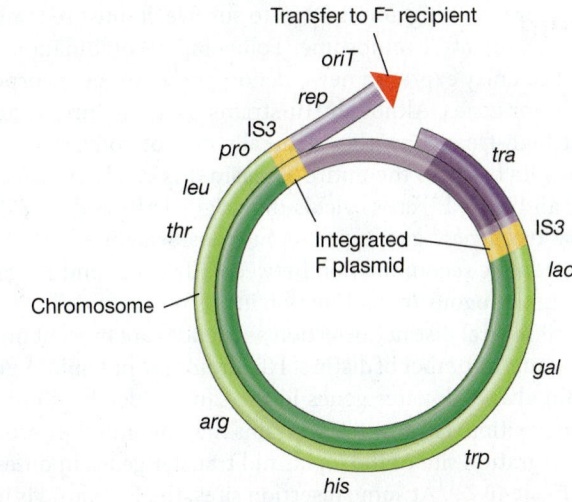

Figure 10.20 Transfer of chromosomal genes by an Hfr strain. The Hfr chromosome breaks at the origin of transfer within the integrated F plasmid. The transfer of DNA to the recipient begins at this point. DNA replicates during transfer as for a free F plasmid (Figure 10.18). This figure is not to scale; the inserted F plasmid is actually less than 3% of the size of the *Escherichia coli* chromosome.

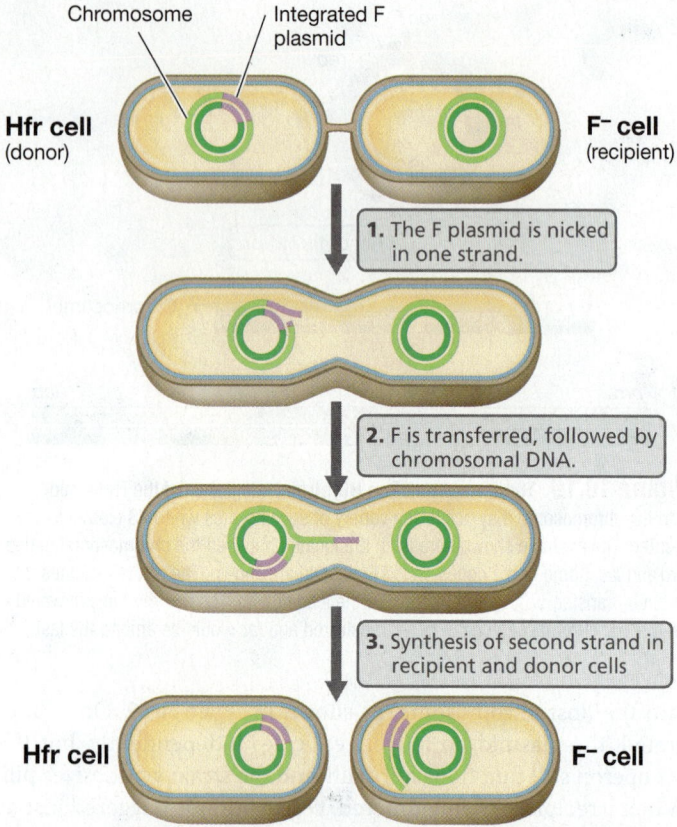

Figure 10.21 Transfer of chromosomal DNA by conjugation. Transfer of the integrated F plasmid from an Hfr strain results in the cotransfer of chromosomal DNA because this is linked to the plasmid. The steps in transfer are similar to those in Figure 10.18a. However, the recipient remains F⁻ and receives a linear fragment of donor chromosome attached to part of the F plasmid. For donor DNA to survive, it must be recombined into the recipient chromosome after transfer (not shown).

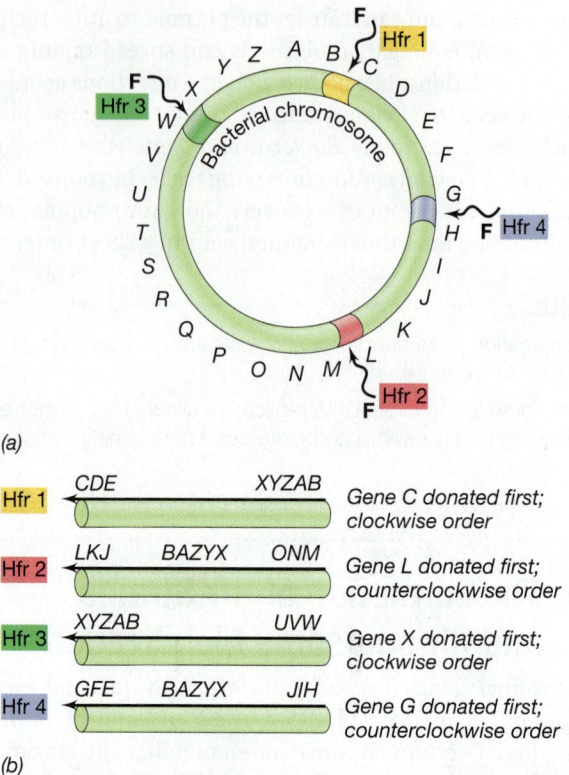

Figure 10.22 Formation of different Hfr strains. Different Hfr strains donate genes in different orders and from different origins. (a) F plasmids can be inserted into various insertion sequences on the bacterial chromosome, forming different Hfr strains. (b) Order of gene transfer for different Hfr strains.

integrated F plasmid is transferred (**Figure 10.21**). However, after transfer, the Hfr strain remains Hfr because it retains a copy of the integrated F plasmid. Because a partial chromosome cannot replicate, for incoming donor DNA to survive, it must recombine with the recipient chromosome. Following recombination, the recipient cell may express a new phenotype due to the incorporation of donor genes. Although Hfr strains transmit chromosomal genes at high frequency, they generally do not convert F⁻ cells to F⁺ or Hfr because the entire F plasmid is rarely transferred. Instead, an Hfr × F⁻ cross yields the original Hfr and an F⁻ cell that now has a new genotype. As in transformation and transduction, genetic recombination between Hfr genes and F⁻ genes involves homologous recombination in the recipient cell.

Because several distinct insertion sequences are present on the chromosome, a number of distinct Hfr strains are possible. A given Hfr strain always donates genes in the same order, beginning at the same position. However, Hfr strains that differ in the chromosomal integration site of the F plasmid transfer genes in different orders (**Figure 10.22**). At some insertion sites, the F plasmid is integrated with its origin pointing in one direction, whereas at other sites the origin points in the opposite direction. The orientation of the F plasmid determines which chromosomal genes enter the recipient first and illustrates how genes acquired from horizontal transfer and recombined into the chromosome can be transferred

to new recipient cells (Figure 10.22). By using various Hfr strains in mating experiments, it was possible to determine the arrangement and orientation of most of the genes in the *E. coli* chromosome (⤴ Section 4.3) long before it was sequenced.

Transfer of Chromosomal Genes to the F Plasmid

Occasionally, integrated F plasmids may be excised from the chromosome. During excision, chromosomal genes may sometimes be incorporated into the liberated F plasmid. This can happen because both the F plasmid and the chromosome contain multiple identical insertion sequences where recombination can occur (Figure 10.20). F plasmids containing chromosomal genes are called *F′ plasmids*. When F′ plasmids promote conjugation, they transfer the chromosomal genes they carry at high frequency to the recipients. F′-mediated transfer resembles specialized transduction (Section 10.7) in that only a restricted group of chromosomal genes is transferred by any given F′ plasmid. Transferring a known F′ into a recipient allows one to establish diploids (two copies of each gene) for a limited region of the chromosome. Such partial diploids (merodiploids) are important for complementation tests (Section 10.5).

MINIQUIZ

- In conjugation involving the F plasmid of *Escherichia coli*, how is the host chromosome mobilized?
- Why does an Hfr × F⁻ mating not yield two Hfr cells?
- At which sites in the chromosome can the F plasmid integrate?

III • Gene Transfer in *Archaea* and Other Genetic Events

Although far less studied than the genetics of *Bacteria*, the genetics of *Archaea* are beginning to emerge along with archaeal versions of the genetic tools necessary for more detailed analyses. In addition, other genetic events in *Bacteria* are important general concepts but do not deal with horizontal gene flow itself. We cover all of these topics here.

10.10 Horizontal Gene Transfer in *Archaea*

Although *Archaea* contain a single circular chromosome like most *Bacteria* (**Figure 10.23**) and genome analysis indicates that horizontal transfer of archaeal DNA also occurs in nature, the development of laboratory-based gene transfer systems lags far behind that for *Bacteria*. Practical problems here include the fact that most well-studied *Archaea* are extremophiles, capable of growth only under extreme conditions of high salt or high temperature (Chapter 16). The temperatures necessary to culture some hyperthermophiles, for example, will melt agar, and alternative materials are required to form solid media and obtain colonies.

Another problem is that most common antibiotics do not affect *Archaea*. For example, penicillins do not affect *Archaea* because their cell walls lack peptidoglycan. The choice of selectable markers for genetic crosses is therefore often limited. However, novobiocin (a DNA gyrase inhibitor) and mevinolin (an inhibitor of isoprenoid biosynthesis) are used to inhibit growth of extreme halophiles, and puromycin and neomycin (both protein synthesis inhibitors) inhibit methanogens. Auxotrophic strains of a few *Archaea* have also been isolated for genetic selection purposes.

Examples of Archaeal Genetics

No single species of *Archaea* has become a model organism for archaeal genetics, although more genetic work has been done on select species of extreme halophiles (*Halobacterium*, *Haloferax*, ⮯ Section 16.1) than on any other *Archaea*. Instead, individual mechanisms for gene transfer have been found scattered among a range of *Archaea*. In addition, several plasmids have been isolated from *Archaea* and some have been used to construct cloning vectors, allowing genetic analysis through cloning and sequencing rather than traditional genetic crosses. Transposon mutagenesis (Section 10.11) has been well developed in certain methanogen species including *Methanococcus* and *Methanosarcina*, and other tools such as shuttle vectors and other in vitro methods of genetic analysis have been developed for study of the highly unusual biochemistry of the methanogens (⮯ Sections 13.20 and 16.2).

Transformation works reasonably well in several *Archaea* although details and conditions vary from organism to organism. One approach requires removal of divalent metal ions, which in turn results in the disassembly of the glycoprotein cell wall layer surrounding many archaeal cells and hence allows access by transforming DNA. However, *Archaea* with rigid cell walls have proven difficult to transform, although electroporation sometimes works. One exception is in *Methanosarcina* species, organisms with a thick cell wall, for which high-efficiency transformation systems have been developed that employ DNA-loaded lipid preparations (liposomes) to deliver DNA into the cell.

Although viruses that infect *Archaea* are plentiful, transduction is extremely rare. Only one archaeal virus, which infects the thermophilic methanogen *Methanothermobacter thermautotrophicus*, has been shown to transduce the genes of its host. Unfortunately the low burst size (about six phages liberated per cell) makes using this system for gene transfer impractical.

Conjugation in *Archaea*

Two types of conjugation have been detected in *Archaea*. Some strains of *Sulfolobus solfataricus* (⮯ Section 16.10) contain plasmids that promote conjugation between two cells in a manner similar to that seen in *Bacteria*. In this process, cell pairing is independent of pili formation and DNA transfer is unidirectional. However, most of the genes encoding these functions seem to have little similarity to those in gram-negative *Bacteria*. The exception is a gene similar to *traG* from the F plasmid, whose protein product participates in stabilizing mating pairs. It thus seems likely that the actual mechanism of conjugation in *Archaea* is quite different from that in *Bacteria*.

Some halobacteria, in contrast, perform a novel form of conjugation. No fertility plasmids are required, and DNA transfer is bidirectional. Cytoplasmic bridges form between the mating cells and probably facilitate intercell DNA transfer. Neither type of conjugation has been developed to the point of being used for routine gene transfer or genetic analysis. However, these genetic resources will likely be useful for developing more routine genetic transfer systems for these organisms in the future.

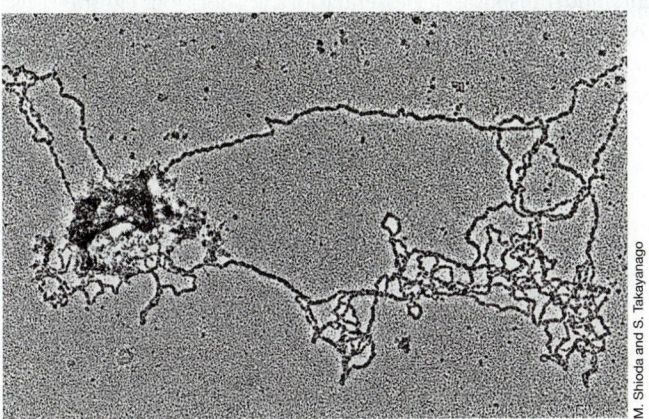

Figure 10.23 **An archaeal chromosome, as shown in the electron microscope.** The circular chromosome is from the hyperthermophile *Sulfolobus*, a member of the *Archaea*.

MINIQUIZ

- Why is it usually more difficult to select recombinants with *Archaea* than with *Bacteria*?
- Why do penicillins not kill species of *Archaea*?

10.11 Mobile DNA: Transposable Elements

As we have seen, molecules of DNA may move from one cell to another, but to a geneticist, "mobile DNA" has a specialized meaning. Mobile DNA refers to discrete segments of DNA that move as units from one location to another *within* other DNA molecules.

Although the DNA of certain viruses can be inserted into and excised from the genome of the host cell, most mobile DNA consists of **transposable elements**. These are stretches of DNA that can move from one site to another. However, transposable elements are always found inserted into another DNA molecule such as a plasmid, a chromosome, or a viral genome. Transposable elements do not possess their own origin of replication. Instead, they are replicated when the host DNA molecule into which they are inserted is replicated.

Transposable elements move by a process called *transposition* that is important both in genome rearrangement and in genetic analysis. The frequency of transposition is extremely variable, and ranges from 1 in 10^3 to 1 in 10^7 per transposable element per cell generation, depending on both the transposable element and the organism. Transposable elements are abundant and widespread in nature and can be found in the genomes of all three domains of life as well as in many viruses and plasmids, which suggests that the elements offer a selective advantage by accelerating genome rearrangement.

The two major types of transposable elements in *Bacteria* are *insertion sequences* (IS) and *transposons*. Both elements have two important features in common: They carry genes encoding *transposase*, the enzyme necessary for transposition, and they have short inverted terminal repeats at their ends that are also needed for transposition (the ends of transposable elements are not free but are continuous with the host DNA molecule into which the transposable element has inserted). **Figure 10.24** shows genetic maps of the insertion element IS2 and of the transposon Tn5.

Insertion Sequences and Transposons

Insertion sequences are the simplest type of transposable element. They are short DNA segments, about 1000 nucleotides long, and typically contain inverted repeats of 10–50 base pairs. Each different IS has a specific number of base pairs in its terminal repeats, and the only protein encoded is the transposase. Several hundred distinct IS elements have been characterized. IS elements are found in the chromosomes and plasmids of both *Bacteria* and *Archaea*, as well as in certain bacteriophages. Individual strains of the same bacterial species vary in the number and location of the IS elements they harbor. For instance, the genome of one strain of *Escherichia coli* has five copies of IS2 and five copies of IS3. Many plasmids, such as the F plasmid, also carry IS elements. Indeed, integration of the F plasmid into the *E. coli* chromosome (Figure 10.19) is due to recombination between identical IS elements on the F plasmid and the chromosome (Section 10.9).

Transposons are larger than IS elements, but have the same two essential components: inverted repeats at both ends and a gene that encodes transposase (Figure 10.24*b*). The transposase recognizes the inverted repeats and moves the segment of DNA flanked by them from one site to another. Consequently, any DNA that lies between the two inverted repeats is moved and is, in effect, part of the transposon. Genes included inside transposons vary widely. Some of these genes, such as antibiotic resistance genes, confer important new properties on the organism harboring the transposon. Because antibiotic resistance is both important and easy to detect, most highly investigated transposons have antibiotic resistance genes as selectable markers. Examples include transposon Tn5, which encodes kanamycin resistance (Figure 10.24*b*) and Tn10, which encodes tetracycline resistance.

Because any genes lying between the inverted repeats become part of a transposon, it is possible to get hybrid transposons that display complex behavior. For example, conjugative transposons contain *tra* genes and can move between bacterial species by conjugation as well as transpose from place to place within a single bacterial genome. Even more complex is bacteriophage Mu, which is both a virus and a transposon (⮌ Section 9.4). In this case a complete virus genome is contained within a transposon. Other composite genetic elements consist of a segment of DNA lying between two identical IS elements. This whole structure can move as a unit and is called a *composite transposon*. The behavior of composite transposons indicates that novel transposons likely arise periodically in cells that contain IS elements located close to one another.

Mechanisms of Transposition

Both the inverted repeats (located at the ends of transposable elements) and transposase are essential for transposition. The transposase recognizes, cuts, and ligates the DNA during transposition. When a

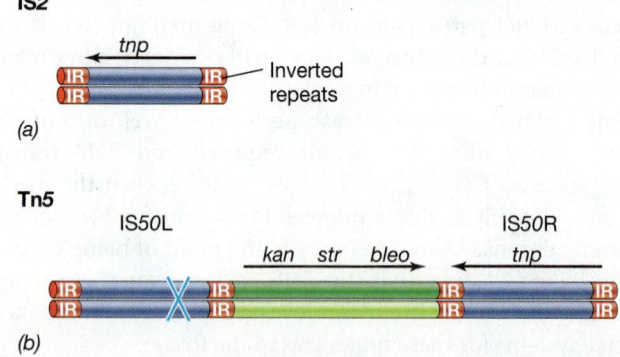

IS2

tnp

Inverted repeats

(a)

Tn5

IS50L   IS50R

kan *str* *bleo*   *tnp*

(b)

Figure 10.24 Maps of the transposable elements IS2 and Tn5. The arrows above the maps show the direction of transcription of any genes on the elements. The gene encoding the transposase is *tnp*. (a) IS2 is an insertion sequence of 1327 bp with inverted repeats of 41 bp at its ends. (b) Tn5 is a composite transposon of 5.7 kbp containing the insertion sequences IS50L and IS50R at its left and right ends, respectively. IS50L is not capable of independent transposition because there is a nonsense mutation, marked by a blue cross, in its transposase gene. The genes *kan*, *str*, and *bleo* confer resistance to the antibiotics kanamycin (and neomycin), streptomycin, and bleomycin.

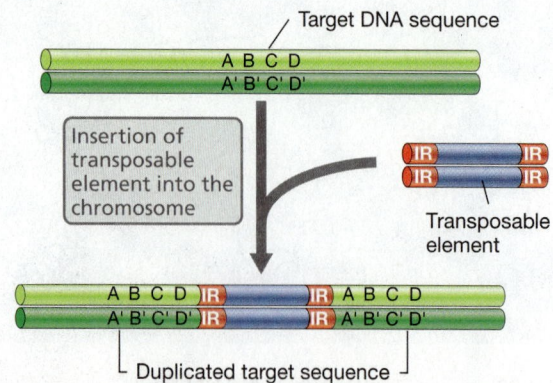

Target DNA sequence

A B C D
A' B' C' D'

Insertion of transposable element into the chromosome

Transposable element

A B C D   A B C D
A' B' C' D'   A' B' C' D'

Duplicated target sequence

Figure 10.25 Transposition. Insertion of a transposable element generates a duplication of the target sequence. Note the presence of inverted repeats (IR) at the ends of the transposable element.

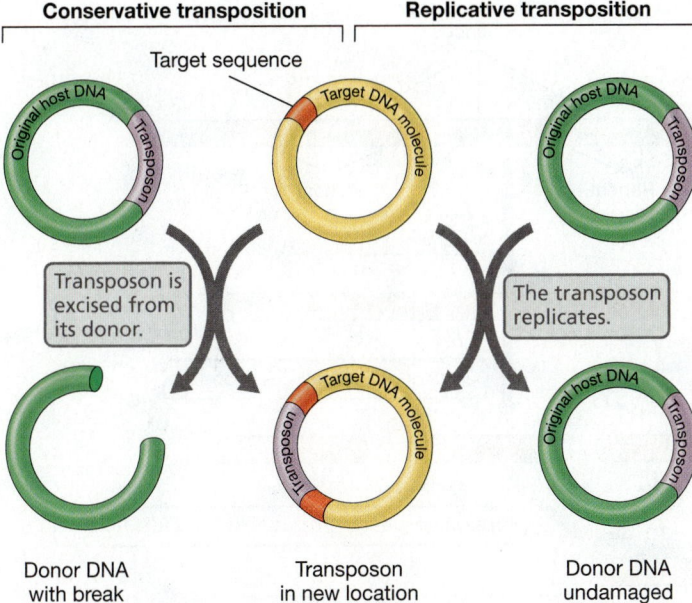

Figure 10.26 Two mechanisms of transposition. Donor DNA (carrying the transposon) is shown in green, and recipient DNA carrying the target sequence is shown in yellow. In both conservative and replicative transposition the transposase inserts the transposon (purple) into the target site (red) on the recipient DNA. During this process, the target site is duplicated. In conservative transposition, the donor DNA is left with a double-stranded break at the previous location of the transposon. In contrast, after replicative transposition, both donor and recipient DNA possess a copy of the transposon.

transposable element is inserted into target DNA, a short sequence in the target DNA at the site of integration is duplicated during the insertion process (**Figure 10.25**). The duplication arises because single-stranded DNA breaks are made by the transposase. The transposable element is then attached to the single-stranded ends that have been generated. Finally, enzymes of the host cell repair the single-strand portions, which results in the duplication.

Two mechanisms of transposition are known: *conservative* and *replicative* (**Figure 10.26**). In conservative transposition, as occurs with the transposon Tn5, the transposon is excised from one location and is reinserted at a second location. The copy number of a conservative transposon therefore remains at one. By contrast, during replicative transposition, a new copy of the transposon is produced and is inserted at the second location. Thus, after a replicative transposition event, one copy of the transposon remains at the original site, while a second copy is incorporated at the new site.

Mutagenesis with Transposons

When a transposon inserts itself within a gene, a mutation occurs in that particular gene (**Figure 10.27**). Mutations due to transposon insertion do occur naturally. However, deliberate use of transpo-

sons is a convenient way to create bacterial mutants in the laboratory. Typically, transposons carrying antibiotic resistance genes are used. The transposon is introduced into the target cell on a phage or plasmid that cannot replicate in that particular host. Consequently, antibiotic-resistant colonies will mostly be due to insertion of the transposon into the bacterial genome.

Because bacterial genomes contain relatively little noncoding DNA, most transposon insertions will occur in genes that encode proteins. This technique can be used to determine the function of a novel gene (**Figure 10.27**). If a transposon inserts into a gene encoding a product of interest such as a new antibiotic synthetase, the transposon mutant will no longer produce the antibiotic. Further analyses can be performed to reveal which gene the transposon has disrupted.

Two transposons widely used for mutagenesis of *Escherichia coli* and related bacteria are Tn5 (**Figure 10.24***b*), which confers neomycin and kanamycin resistance, and Tn*10*, which confers tetracycline resistance. Many *Bacteria*, a few *Archaea*, and the yeast *Saccharomyces cerevisiae* have all been mutagenized using engineered transposons. More recently, transposons have even been used to isolate mutations in animals, including mice.

10.12 Preserving Genome Integrity: CRISPR Interference

Bacteria and *Archaea* not only produce restriction endonucleases (⮌ Sections 8.6 and 11.1) that function to destroy incoming foreign DNA, they also have an RNA-based defense program to destroy invading DNA from viral infection and sometimes conjugation. This type of prokaryotic "immune system" helps preserve genome stability and is called the CRISPR system, which stands for *c*lustered *r*egularly *i*nterspaced *s*hort *p*alindromic *r*epeats.

The CRISPR region on the bacterial chromosome is essentially a memory bank of incoming nucleic acid sequences used for surveillance against foreign DNA. It consists of many different segments of foreign DNA called *spacers* alternating with identical repeated sequences (**Figure 10.28**). The spacer sequences correspond to pieces of foreign DNA that have previously invaded the cell. Once the spacers are recombined into the CRISPR region, the system provides resistance to any incoming DNA (and sometimes RNA) that contains the same or very closely related sequences to

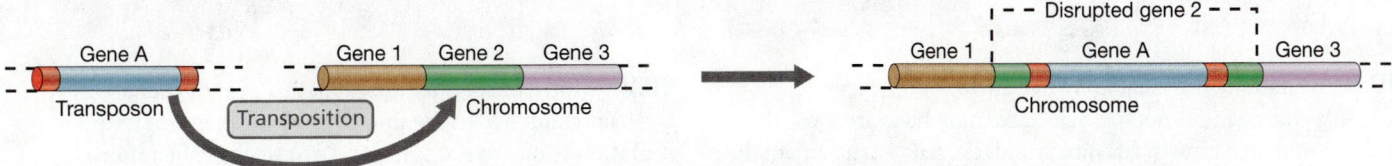

Figure 10.27 Transposon mutagenesis. The transposon moves into the middle of gene 2. Gene 2 is now disrupted by the transposon and is inactivated. Gene A from the transposon is now expressed from the chromosome.

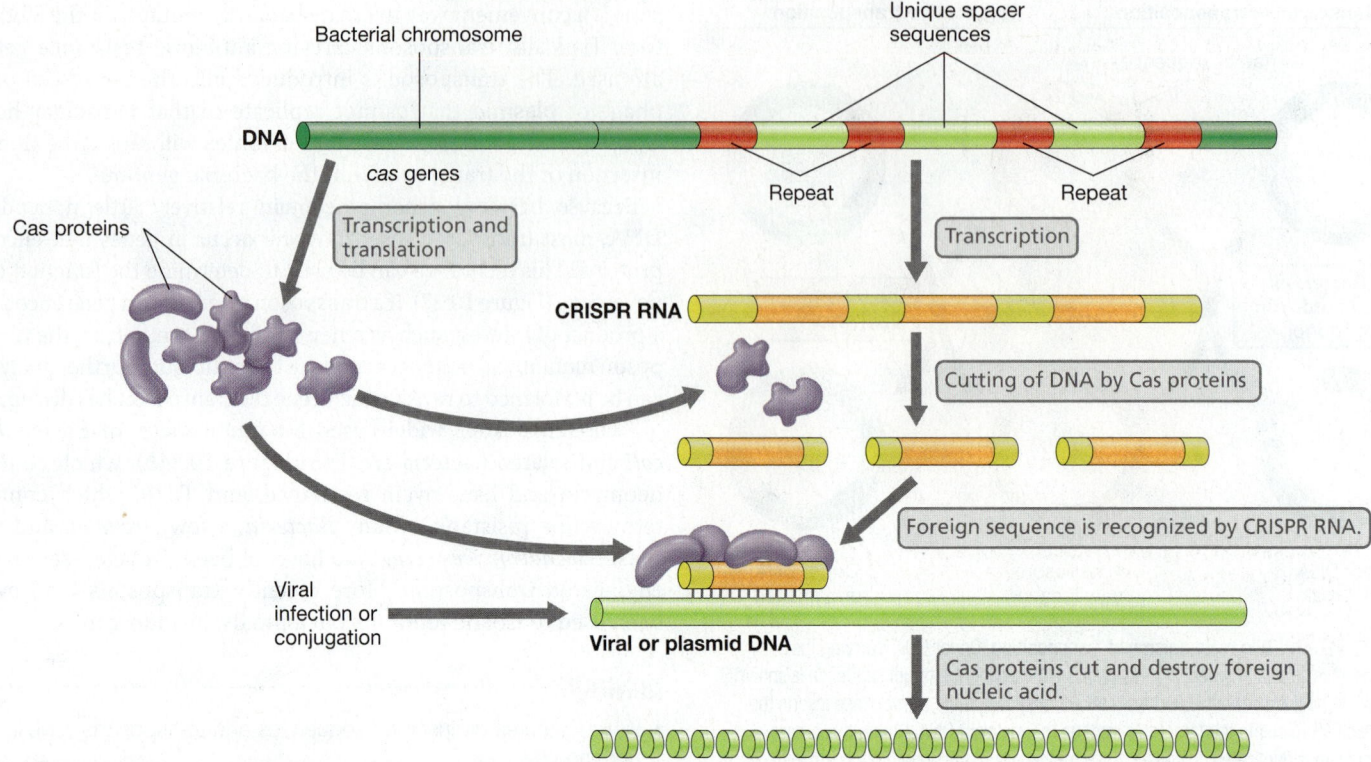

Figure 10.28 **Operation of the CRISPR system.** The CRISPR region on the bacterial chromosome is transcribed into a long RNA molecule that is then processed into segments by some of the Cas proteins. Each spacer segment corresponds to previous encounters with incoming foreign nucleic acid. If one of these short CRISPR RNA molecules (corresponding to a spacer) recognizes and base-pairs with incoming nucleic acid from transduction or conjugation, other Cas proteins destroy the foreign nucleic acid.

individual spacer regions. Proteins of the CRISPR system carry out essential functions of this RNA-based "immunity."

The proteins of the CRISPR system (CRISPR-associated proteins, or Cas proteins) perform two roles. Some participate in obtaining and storing segments of foreign DNA as spacers by recognizing specific nucleotide sequences associated with the spacers. Others use the stored sequence information to recognize intruding DNA and destroy it. The Cas proteins themselves are encoded by genes that lie upstream of the CRISPR DNA sequences (Figure 10.28).

The CRISPR region is transcribed as a whole into a long RNA molecule that is then cleaved in the middle of each of the repeated sequences by the nuclease activity of Cas proteins. This converts the long RNA molecule into spacer segments of small RNAs called *CRISPR RNAs* (*crRNAs*). If one of these crRNAs base-pairs with invading nucleic acid, then the foreign DNA or RNA is destroyed by the nuclease activity of other Cas proteins.

The CRISPR system is widely distributed in both *Archaea* and *Bacteria*. Approximately 90% of the sequenced genomes of

Archaea and 70% of those of *Bacteria* possess the CRISPR system. The utility of the system was first demonstrated in the dairy industry where starter cultures used for milk fermentation are susceptible to rampant bacteriophage infection. However, a strain of *Streptococcus thermophilus* was found to be resistant to virulent bacteriophage. The difference between this *S. thermophilus* strain and those susceptible to viral infection was its spacers within the CRISPR region. While it is unknown why some viruses are not targeted by the CRISPR system, laboratory experiments have shown that bacteriophages can overcome recognition by the Cas proteins and crRNAs by modifying their genome through mutation.

MINIQUIZ

- Why is the CRISPR system considered a prokaryotic "immune system"?
- What do the spacers within the CRISPR region correspond to?

BIG IDEAS

10.1 • Mutation is a heritable change in DNA sequence and may lead to a change in phenotype. Selectable mutations are those that give the mutant a growth advantage under certain environmental conditions and are especially useful in genetic research. If selection is not possible, mutants must be identified by screening.

10.2 • Mutations, either spontaneous or induced, are in the base sequence of the nucleic acid in a genome. A point mutation is due to a single base-pair change. In a nonsense mutation, the codon becomes a stop codon and an incomplete polypeptide is made. Deletions and insertions cause more dramatic changes in the DNA, including frameshift mutations that often result in complete loss of gene function.

10.3 • Different types of mutations occur at different frequencies. For a typical bacterium, mutation rates of 10^{-6} to 10^{-7} per kilobase pair are generally seen. Although RNA and DNA polymerases make errors at about the same rate, RNA genomes typically accumulate mutations at much higher frequencies than DNA genomes.

10.4 • Mutagens are chemical, physical, or biological agents that increase the mutation rate. Mutagens can alter DNA in many different ways. However, alterations in DNA are not mutations unless they are inherited. Some DNA damage can lead to cell death if not repaired, and both error-prone and high-fidelity DNA repair systems exist.

10.5 • Homologous recombination occurs when closely related DNA sequences from two distinct genetic elements are combined together in a single element. Recombination is an important evolutionary process, and cells have specific mechanisms for ensuring that recombination takes place.

10.6 • Certain prokaryotes exhibit competence, a state in which cells are able to take up free DNA released by other bacteria. Incorporation of donor DNA into a recipient cell requires the activity of single-strand binding protein, RecA protein, and several other enzymes. Only competent cells are transformable.

10.7 • Transduction is the transfer of host genes from one bacterium to another by a bacterial virus. In generalized transduction, defective virus particles randomly incorporate fragments of the cell's chromosomal DNA, but the transducing efficiency is low. In specialized transduction, the DNA of a temperate virus excises incorrectly and takes adjacent host genes along with it; the transducing efficiency here may be very high.

10.8 • Conjugation is a mechanism of DNA transfer in prokaryotes that requires cell-to-cell contact. Conjugation is controlled by genes carried by certain plasmids (such as the F plasmid) and requires transfer of the plasmid from a donor cell to a recipient cell. Plasmid DNA transfer requires replication using the rolling circle mechanism.

10.9 • The donor cell chromosome can be mobilized for transfer to a recipient cell. This requires an F plasmid to integrate into the chromosome to form the Hfr phenotype. Because transfer of the host chromosome is rarely complete, recipient cells rarely become F^+. F′ plasmids are previously integrated F plasmids that have excised and captured some chromosomal genes.

10.10 • Archaeal research lags behind bacterial research in the development of systems for gene transfer. Many antibiotics are ineffective against *Archaea*, making it difficult to select recombinants effectively. The unusual growth conditions needed by many *Archaea* also make genetic experimentation difficult. Nevertheless, the genetic transfer systems of *Bacteria*—transformation, transduction, and conjugation—are all known in *Archaea*.

10.11 • Transposons and insertion sequences are genetic elements that can move from one location on a host DNA molecule to another by transposition. Transposition can be either replicative or conservative. Transposons often carry genes encoding antibiotic resistance and can be used as biological mutagens.

10.12 • The clustered regularly interspaced short palindromic repeat (CRISPR) system is an RNA-based mechanism of protecting the prokaryotic genome from invading DNA resulting from infection and conjugation. If small RNA molecules resulting from the spacer regions of the CRISPR region bind to incoming complementary DNA, Cas proteins destroy the nucleic acid duplex.

MasteringMicrobiology®

Review what you know and challenge what you have learned with MasteringMicrobiology! Access study materials, chapter quizzes, animations, and microbiology lab tutorials in the Study Area to ensure that you have mastered this chapter's content.

REVIEW OF KEY TERMS

Auxotroph an organism that has developed a nutritional requirement, often as a result of mutation

Conjugation the transfer of genes from one prokaryotic cell to another by a mechanism requiring cell-to-cell contact

Frameshift mutation a mutation in which insertion or deletion of nucleotides changes the groups of three bases in which the genetic code is read within an mRNA, usually resulting in a faulty product

Genotype the complete genetic makeup of an organism; the complete description of a cell's genetic information

Heteroduplex a DNA double helix composed of single strands from two different DNA molecules

Hfr cell a cell with the F plasmid integrated into the chromosome

Induced mutation a mutation caused by external agents such as mutagenic chemicals or radiation

Insertion sequence (IS) the simplest type of transposable element, which carries only genes that participate in transposition

Missense mutation a mutation in which a single codon is altered so that one amino acid in a protein is replaced with a different amino acid

Mutagen an agent that causes mutation

Mutant an organism whose genome carries a mutation

Mutation a heritable change in the base sequence of the genome of an organism

Mutator strain a mutant strain in which the rate of mutation is increased

Nonsense mutation a mutation in which the codon for an amino acid is changed to a stop codon

Phenotype the observable characteristics of an organism

Point mutation a mutation that involves a single base pair

Recombination a resorting or rearrangement of DNA fragments resulting in a new sequence combination

Reversion an alteration in DNA that reverses the effects of a prior mutation

Rolling circle replication a mechanism of replicating double-stranded circular DNA that starts by nicking and unrolling one strand and using the other

(still circular) strand as a template for DNA synthesis

Screening a procedure that permits the identification of organisms by phenotype or genotype, but does not inhibit or enhance the growth of particular phenotypes or genotypes

Selection placing organisms under conditions that favor or inhibit the growth of those with a particular phenotype or genotype

Silent mutation a change in DNA sequence that has no effect on the phenotype

SOS repair a DNA repair system activated by DNA damage

Spontaneous mutation a mutation that occurs "naturally" without the help of mutagenic chemicals or radiation

Transduction the transfer of host cell genes from one cell to another by a virus

Transformation the transfer of bacterial genes involving free DNA

Transition a mutation in which a pyrimidine base is replaced by another pyrimidine or a purine is replaced by another purine

Transposable element a genetic element able to move (transpose) from one site to another on host DNA molecules

Transposon a type of transposable element that carries genes in addition to those required for transposition

Transversion a mutation in which a pyrimidine base is replaced by a purine or vice versa

Wild-type strain a bacterial strain isolated from nature or one used as a parent in a genetics investigation

REVIEW QUESTIONS

1. Write a one-sentence definition of the term "genotype." Do the same for "phenotype." Does the phenotype of an organism automatically change when a change in genotype occurs? Why or why not? Can phenotype change without a change in genotype? In both cases, give examples to support your answer. (Section 10.1)

2. Explain why an *Escherichia coli* strain that is His⁻ is an auxotroph and one that is Lac⁻ is not. (*Hint*: Think about how *E. coli* metabolizes histidine and lactose and what each compound is used for.) (Section 10.1)

3. What are silent mutations? From your knowledge of the genetic code, why do you think most silent mutations affect the third position in a codon? (Section 10.2)

4. What is the average rate of mutation in a cell? Can this rate change? (Section 10.3)

5. Give an example of one biological, one chemical, and one physical mutagen and describe the mechanism by which each causes a mutation. (Section 10.4)

6. What are heteroduplex regions of DNA and what process leads to their formation? (Section 10.5)

7. Explain why recipient cells do not successfully take up plasmids during natural transformation. (Section 10.6)

8. Explain how a generalized transducing particle differs from a specialized transducing particle. (Section 10.7)

9. What is a sex pilus and which cell type, F⁻ or F⁺, would produce this structure? (Section 10.8)

10. What does an F⁺ cell need to do before it can transfer chromosomal genes? (Section 10.9)

11. Explain why performing genetic selection is difficult when studying *Archaea*. Give examples of some selective agents that work well with *Archaea*. (Section 10.10)

12. What are the major differences between insertion sequences and transposons? (Section 10.11)

13. Explain why incoming DNA recognized by a short RNA molecule expressed from the CRISPR region cannot be completely foreign to the cell. (Section 10.12)

APPLICATION QUESTIONS

1. A constitutive mutant is a strain that continuously makes a protein that is inducible in the wild type. Describe two ways in which a change in a DNA molecule could lead to the emergence of a constitutive mutant. How could these two types of constitutive mutants be distinguished genetically?

2. Although a large number of mutagenic chemicals are known, none is known that induces mutations in only a single gene (gene-specific mutagenesis). From what you know about mutagens, explain why it is unlikely that a gene-specific

chemical mutagen will be found. How then is site-specific mutagenesis accomplished?

3. Why is it difficult in a single experiment to transfer a large number of genes to a recipient cell using transformation or transduction?

4. Transposable elements cause mutations when inserted within a gene. These elements disrupt the continuity of a gene. Introns also disrupt the continuity of a gene, yet the gene is still functional. Explain why the presence of an intron in a gene does not inactivate that gene but insertion of a transposable element does.

11 · Genetic Engineering and Biotechnology

microbiology**now**

From Pathogen to Tumor Assassin

Not only are cutting-edge advances in biotechnology providing critical information regarding the basic biology of life, they are also the cornerstones to improving naturally occurring products. Molecular techniques have been used to engineer the production of biofuels, drought-resistant crops, and hormones such as insulin. But what about human diseases such as cancer; is there hope here, too?

Pancreatic cancer is one of the leading causes of cancer deaths. Current radiation and chemotherapy treatments are ineffective, with dismal patient survival rates. While anticancer drugs are available, they only increase survival time for patients with advanced stages of the disease. Alternative treatments are desperately needed, and biotechnologists are coming to the rescue.

Listeria monocytogenes is a pathogen that causes listeriosis, a serious foodborne illness. *L. monocytogenes* possesses an intracellular lifestyle that allows it to evade the human immune system. However, scientists have discovered that a weakly pathogenic recombinant strain can be cleared by the immune system of healthy cells but not tumor cells. This led to an exciting idea–could this strain of *L. monocytogenes* be used to deliver anticancer agents such as therapeutic radionuclides to tumor cells only?

Radionuclides can physically destroy cancer cells, but delivering these molecules specifically to tumor cells has been problematic. Using an ingenious scheme, scientists coupled the radionuclide ^{188}rhenium to the weakly pathogenic strain of *L. monocytogenes*.[1] This "tumor-killing" strain of *Listeria* (pink in photo) was not only able to infect and multiply in pancreatic tumor cells of mice (blue), but also reduced the incidence of metastases without harming normal pancreatic cells.

This research illustrates how microbiology and biotechnology can team up to tame a virulent bacterium and convert it into a therapeutic superhero!

[1]Quispe-Tintaya, W., et al. 2013. Nontoxic radioactive *Listeria*at is a highly effective therapy against metastatic pancreatic cancer. *Proc. Natl. Acad. Sci.* 110: 8668–8673.

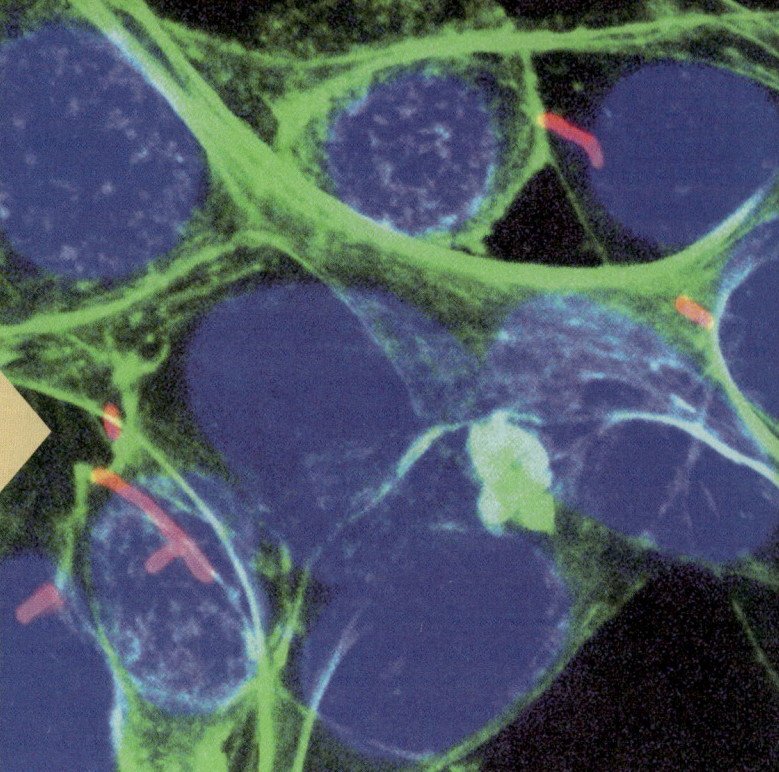

In this chapter we discuss the basic techniques of genetic engineering, in particular those used to clone, alter, and express genes efficiently in host organisms. Performing genetics *in vivo* (in living organisms) has many limitations that can be overcome by manipulating DNA *in vitro* (in a test tube). These molecular techniques are the foundation of biotechnology. Toward the end of the chapter, we will give some examples of how genetically modified organisms can be used for industrial, medical, and agricultural applications.

I • Methods for Manipulating DNA

Genetic engineering refers to the use of in vitro techniques to alter genes in the laboratory. Such altered genes may be reinserted into the original source organism or into some other host organism. Genetic engineering requires that DNA be isolated in specific fragments and purified for further manipulation. We begin by considering some of the basic tools of the genetic engineer including restriction enzymes, the separation of nucleic acids by electrophoresis, nucleic acid hybridization, amplification of DNA, and molecular cloning.

11.1 Restriction Enzymes and Nucleic Acid Separation

All cells contain enzymes that can chemically modify DNA. One major class of such enzymes is the *restriction endonucleases*, or **restriction enzymes** for short. Restriction enzymes recognize specific base sequences (recognition sequences) within DNA and cut the phosphodiester backbone, resulting in double-stranded breaks. Although restriction enzymes are widespread among both *Bacteria* and *Archaea*, they are very rare in eukaryotes. Restriction enzymes undoubtedly protect prokaryotes from hostile foreign DNA, such as virus genomes. However, restriction enzymes are also used for in vitro DNA manipulation and are a major tool of genetic engineering.

Mechanism of Restriction Enzymes

Restriction endonucleases are divided into three major classes. Type I and III restriction enzymes bind to the DNA at their recognition sequences but cut the DNA at some distance away. In contrast, the type II restriction enzymes cleave the DNA within their recognition sequences, making this class of enzymes much more useful for the specific manipulation of DNA.

Most of the DNA sequences recognized by type II restriction enzymes are short inverted repeats of 4 to 8 base pairs (bp). Figure 11.1 shows two 6-bp sequences recognized and cleaved by two different type II restriction enzymes from *Escherichia coli*; arrows indicate the cleavage sites. Note that the two strands of the recognition sequences have the same sequence if one is read from the left and the other from the right (that is, both are read 5′ → 3′). Such inverted repeat sequences are called *palindromes*. The endonuclease activity of *Eco*RI (this acronym stands for E*scherichia* coli, strain RY13, restriction enzyme *I*) makes staggered cuts, leaving short, single-stranded overhangs known as "sticky" ends at the ends of the two fragments. Other restriction enzymes such as *Eco*RV cut both strands of the DNA directly opposite each other, resulting in blunt ends (Figure 11.1*a*). As explained below, fragments with sticky ends are beneficial for molecular cloning of DNA (see Figure 11.7).

Consider again the enzymes *Eco*RI and *Eco*RV, which recognize a specific 6-bp sequence (Figure 11.1). Any specific 6-base sequence should appear in a strand of DNA about once every 4096 nucleotides on average ($4096 = 4^6$; there are 4 possible bases at each of 6 positions). This assumes that all base pairs may occur at any given position with equal probability and that the DNA consists of 50% GC. Thus, several *Eco*RI and *Eco*RV cut sites should be present in any lengthy DNA molecule. Several thousand restriction enzymes with different specificities are known, many of which are commercially available. **Table 11.1** lists a few of these enzymes and their recognition sequences.

Modification: Protection from Restriction

The natural role of restriction enzymes is to protect the cell from invasion by foreign DNA, especially viral DNA. If foreign DNA

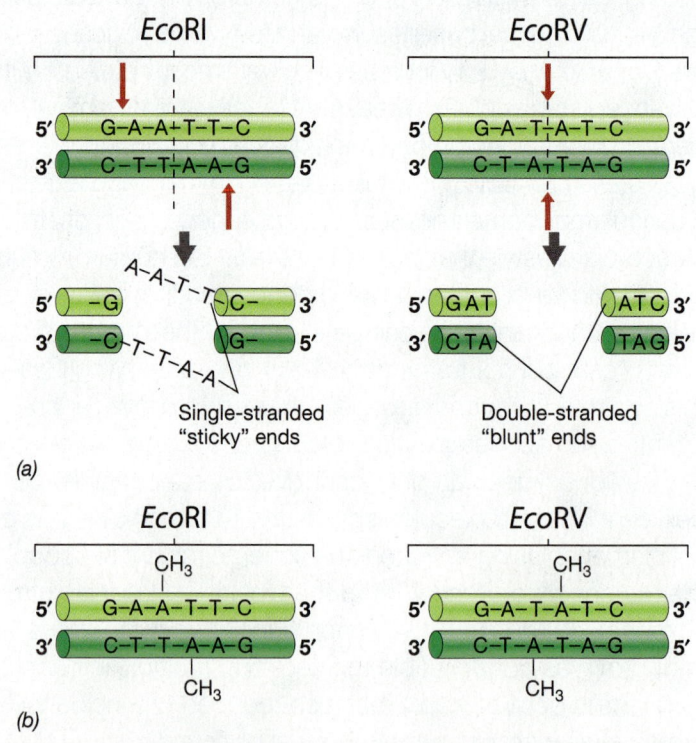

Figure 11.1 Restriction and modification of DNA. *(a)* (Top panel) Sequences of DNA recognized by the restriction endonucleases *Eco*RI and *Eco*RV. The red arrows indicate the bonds cleaved by the enzyme. The dashed line indicates the axis of symmetry of the sequence. (Bottom panel) Appearance of DNA after cutting with the restriction enzymes. Note the single-stranded "sticky" ends generated by *Eco*RI and the "blunt" ends generated by *Eco*RV. *(b)* The same sequences after modification by the corresponding methylases. The methyl groups added by these enzymes are shown, and protect the restriction site from cutting by *Eco*RI and *Eco*RV.

Table 11.1 Recognition sequences of a few restriction endonucleases

Organism	Enzyme designation[a]	Recognition sequence[b]
Bacillus globigii	BglII	A↓GATCT
Brevibacterium albidum	BalI	TGG↓C*CA
Escherichia coli	EcoRI	G↓AA*TTC[c]
Escherichia coli	EcoRV	GAT↓A*TC[c]
Haemophilus haemolyticus	HhaI	GC*G↓C
Haemophilus influenzae	HindIII	A↓AGCTT
Klebsiella pneumoniae	KpnI	GGTAC↓C
Nocardia otitidiscaviarum	NotI	GC↓GGC*CGC
Proteus vulgaris	PvuI	CGAT↓CG
Serratia marcescens	SmaI	CCC↓GGG
Thermus aquaticus	TaqI	T↓CGA*

[a]Nomenclature: The first letter of the three-letter abbreviation of a restriction endonuclease designates the genus from which the enzyme originates; the second two letters, the species. The roman numeral designates the order of discovery of enzymes in that particular organism, and any additional letters are strain designations.
[b]Arrows indicate the sites of enzymatic attack. Asterisks indicate the site of methylation (modification). G, guanine; C, cytosine; A, adenine; T, thymine. Only the 5′ → 3′ sequence is shown.
[c]See Figure 11.1a.

enters the cell, the restriction enzymes will destroy it (⇄ Section 8.6). However, a cell must protect its own DNA from inadvertent destruction by its own restriction enzymes. Such protection is conferred by **modification enzymes**. Each restriction enzyme is partnered with a corresponding modification enzyme that shares the same recognition sequence. The modification enzymes chemically modify specific nucleotides in the restriction recognition sequences of the cell's own DNA. These modified sequences can no longer be cut by the corresponding restriction enzymes. Typically, modification consists of methylating specific bases within the recognition sequence, which prevents the restriction endonuclease from binding. For example, the sequences recognized by the *Eco*RI and *Eco*RV restriction enzymes (Figure 11.1a) can be modified by methylation of the two most interior adenines (Figure 11.1b). The enzymes that perform this modification are called *methylases*. If even a single strand is modified, the recognition sequence is no longer a substrate for the corresponding restriction enzyme.

Gel Electrophoresis: Separation of DNA Molecules

In vitro manipulation of nucleic acid often requires separation of molecules based on size. For example, many restriction enzymes cut DNA molecules into segments that range in length from a few hundred to a few thousand base pairs. After the DNA is cleaved, the fragments generated can be separated from each other by **gel electrophoresis** and analyzed. Gel electrophoresis is also used to verify that amplification of a nucleic acid was successful (Section 11.3).

Electrophoresis is a procedure that separates charged molecules by migration in an electrical field. The rate of migration is determined by the charge on the molecule and by its size and shape. In gel electrophoresis (**Figure 11.2a**) the molecules are separated in a

porous gel. Gels made of *agarose*, a polysaccharide, are used for separating DNA fragments. When an electrical current is applied, nucleic acids move through the gel toward the positive electrode due to their negatively charged phosphate groups. The presence of the gel meshwork hinders the progress of the DNA, and small or compact molecules migrate more rapidly than large molecules. The higher the concentration of agarose in the gel, the greater is the resistance to movement for larger molecules. Consequently, gels of different concentrations are used to separate molecules of different size ranges.

After the gel has been run for sufficient time to separate the DNA molecules, the gel can be stained with a compound that binds to DNA, such as *ethidium bromide*, and the DNA will then

(a)

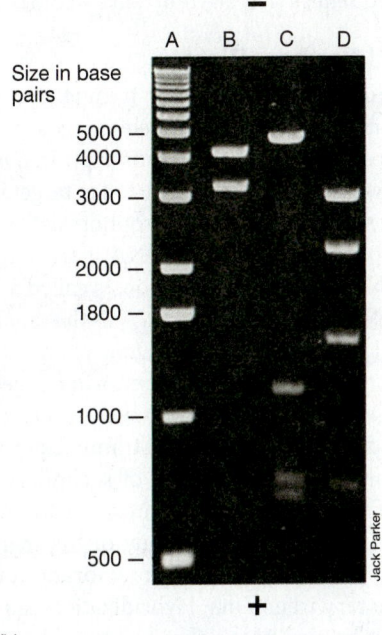

(b)

Figure 11.2 **Agarose gel electrophoresis of DNA.** *(a)* DNA samples are loaded into wells in a submerged agarose gel. *(b)* A photograph of a stained agarose gel. The DNA was loaded into wells toward the top of the gel (negative pole) as shown, and the positive electrode is at the bottom. The standard sample in lane A (DNA ladder) has fragments of known size that may be used to determine the sizes of the fragments in the other lanes. Bands stain less intensely at the bottom of the gel because the fragments are smaller, and thus there is less DNA to stain.

fluoresce under ultraviolet light (Figure 11.2*b*). To determine the size of the DNA of interest, the migration can be compared to a standard sample consisting of DNA fragments of known sizes, called a *DNA ladder*. After electrophoresis, DNA fragments can be purified from gels and used for a variety of purposes.

11.2 Nucleic Acid Hybridization

When DNA is denatured (that is, the two strands are separated), the single strands can form hybrid double-stranded molecules with other single-stranded DNA (or RNA) molecules by complementary (or almost complementary) base pairing (⟳ Section 4.2). This is called *nucleic acid hybridization,* or **hybridization** for short, and is used in detecting, characterizing, and identifying segments of DNA and RNA.

Segments of single-stranded nucleic acids whose identity is already known and that are used in hybridization are called **nucleic acid probes** or, simply, *probes*. To allow detection, probes can be made radioactive or labeled with chemicals that are colored or yield fluorescent products (⟳ Section 18.4). By varying the hybridization conditions, it is possible to adjust the "stringency" of the hybridization such that complementary base pairing must be nearly exact; this helps to avoid nonspecific pairing between nucleic acid sequences that are only partly complementary.

Southern and Northern Blots

Hybridization can be very useful for finding related sequences in different genomes or other genetic elements and to determine if a gene is expressed into an RNA transcript. In *Southern blotting*, probes of known sequence are hybridized to target DNA fragments that have been separated by gel electrophoresis (Section 11.1). The hybridization procedure in which DNA is the target sequence in the gel, and RNA or DNA is the probe, is called a **Southern blot**. By contrast, a **Northern blot** uses RNA as the target sequence and DNA or RNA as the probe to detect gene expression.

In a Southern blot the DNA fragments in the gel are first denatured to yield single strands and then transferred to a synthetic membrane. Although RNA is single-stranded, denaturant is added to the gel to prevent the formation of secondary structures (⟳ Section 4.7). The membrane is then exposed to a labeled probe. If the probe is complementary to any of the fragments, hybrids form, and the probe attaches to the membrane at the locations of the complementary fragments. Hybridization can be detected by monitoring the labeled probe that has bound to the membrane. **Figure 11.3***a* shows how a Southern blot can be used to identify fragments of DNA containing sequences that hybridize to the probe.

The procedure for Northern blots is analogous except that molecules of RNA instead of DNA are separated on a gel and transferred to a synthetic membrane where they are probed. Northern blotting is often used to identify messenger RNA (mRNA) derived from specific genes. The intensity of a Northern blot gives a rough estimate of mRNA abundance from the target gene and may therefore be used to monitor transcription (Figure 11.3*b*).

Other Hybridization Methods

Hybridization is often used to detect the presence of specific genes in genomes that have not yet been sequenced, as well as the movement of genetic elements such as a transposon (⟳ Section 10.11). To find the specific genome location of a gene of interest, total genomic DNA can be cloned (Section 11.4). Hybridization on the resulting colonies using a nucleic acid probe can detect recombinant DNA in colonies, as shown in Figure 11.8*a*. This procedure uses *replica plating* to produce a duplicate of the master plate on a

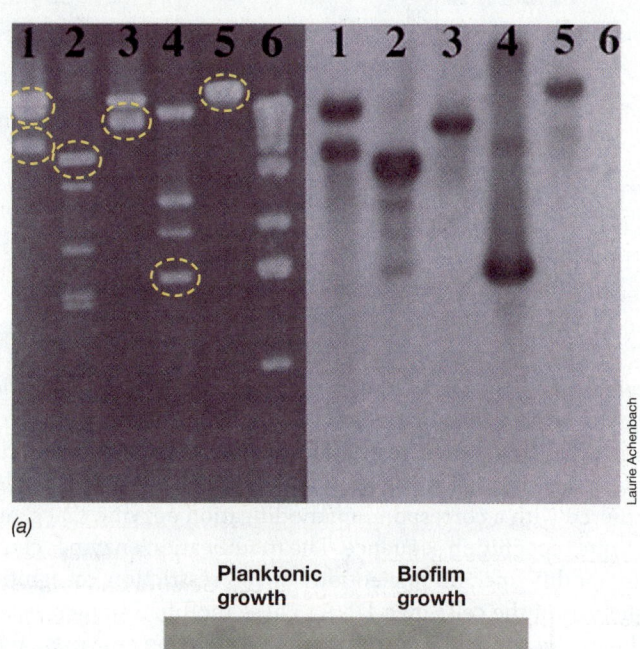

(a)

Laurie Achenbach

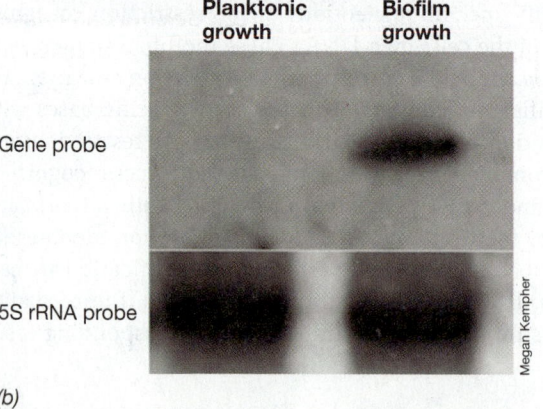

(b)

Figure 11.3 Nucleic acid hybridization. *(a)* Southern blotting. (Left panel) Purified molecules of DNA from several different plasmids were treated with restriction enzymes and then subjected to agarose gel electrophoresis. (Right panel) Blot of the DNA gel shown to the left. After blotting, DNA in the gel was hybridized to a radioactive probe. The positions of the bands were visualized by X-ray autoradiography. Note that only some of the DNA fragments (circled in yellow) have sequences complementary to the labeled probe. Lane 6 contained DNA used as a size marker and none of the bands hybridized to the probe. *(b)* Northern blotting. (Top panel) Hybridization and detection of a radioactive gene-specific probe to a blot of total RNA. The probe only bound to RNA from biofilm-grown cells, indicating that the target gene is not expressed during planktonic (suspended) growth. (Bottom panel) Hybridization and detection of a radioactive probe corresponding to the 5S rRNA to the same blot. The signal intensity indicates that equal amounts of RNA from each sample were loaded into the gel.

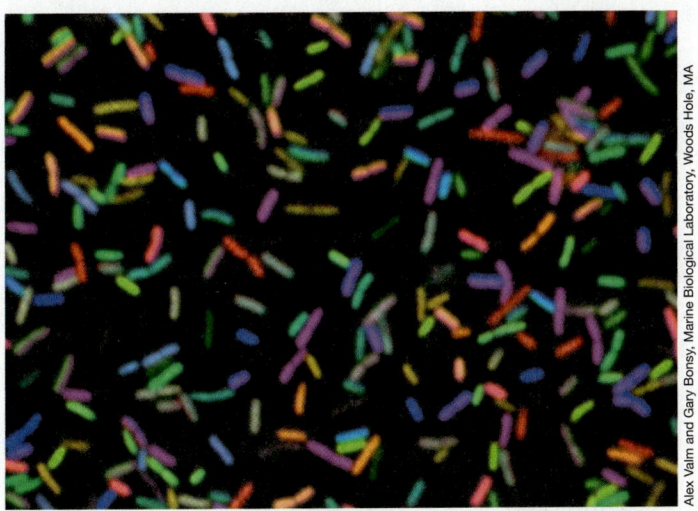

Figure 11.4 **Fluorescence spectral image of 28 differently labeled strains of *Escherichia coli*.** Cells were labeled with combinations of fluorophore-conjugated oligonucleotides that are complementary to *E. coli* 16S rRNA.

membrane filter. The cells on the filter are lysed in place to release their DNA, and the filter is treated to separate the DNA into single strands and fix them to the filter. This filter is then exposed to a labeled nucleic acid probe to allow hybridization, and unbound probe is washed away. The filter is then overlaid with X-ray film if a radioactive probe was used. After development, the X-ray film is examined for spots. Colonies corresponding to these spots are then chosen and studied further.

Hybridization is also the basis of fluorescence in situ hybridization (FISH) (⟳ Section 18.4, **Figure 11.4**). Using this technique, a range of different fluorescent signals can be covalently linked to *oligonucleotide* (short single-stranded DNA or RNA molecules) probes to target specific DNA sequences. Such probes can be used to identify particular species or strains of bacteria by hybridizing to characteristic sequences in the genes for their 16S ribosomal RNA or to ribosomal RNA directly. This approach allows the identification of pathogens in clinical samples or bacteria of interest in environmental samples. Figure 11.4 demonstrates the simultaneous use of eight different oligonucleotide probes in combinations to distinguish between 28 different strains of *Escherichia coli* whose 16S rRNA sequences varied slightly from strain to strain. The variations in color give a visual indication of the specificity and power of nucleic acid probes.

MINIQUIZ

- What are some applications of nucleic acid hybridization in molecular biology?
- What is the difference between a Southern blot and a Northern blot?

11.3 Polymerase Chain Reaction (PCR)

The **polymerase chain reaction (PCR)** is essentially DNA replication in vitro. The PCR can copy segments of DNA by up to a billionfold in the test tube, a process called *amplification*. This yields large amounts of specific genes or other DNA segments

that may be used for a range of applications in molecular biology. PCR uses the enzyme DNA polymerase, which naturally copies DNA molecules (⟳ Section 4.4). Artificially synthesized oligonucleotide primers (Section 11.5) are used to initiate DNA synthesis, but are made of DNA (rather than RNA like the primers used by cells). PCR does not actually copy whole DNA molecules but amplifies stretches of up to a few thousand base pairs (the *target*) from within a larger DNA molecule (the *template*).

The steps in PCR amplification of DNA are as follows (**Figure 11.5**):

1. Template DNA is denatured by heating.
2. Two artificial DNA oligonucleotide primers flanking the target DNA on each strand are added in excess. This ensures that

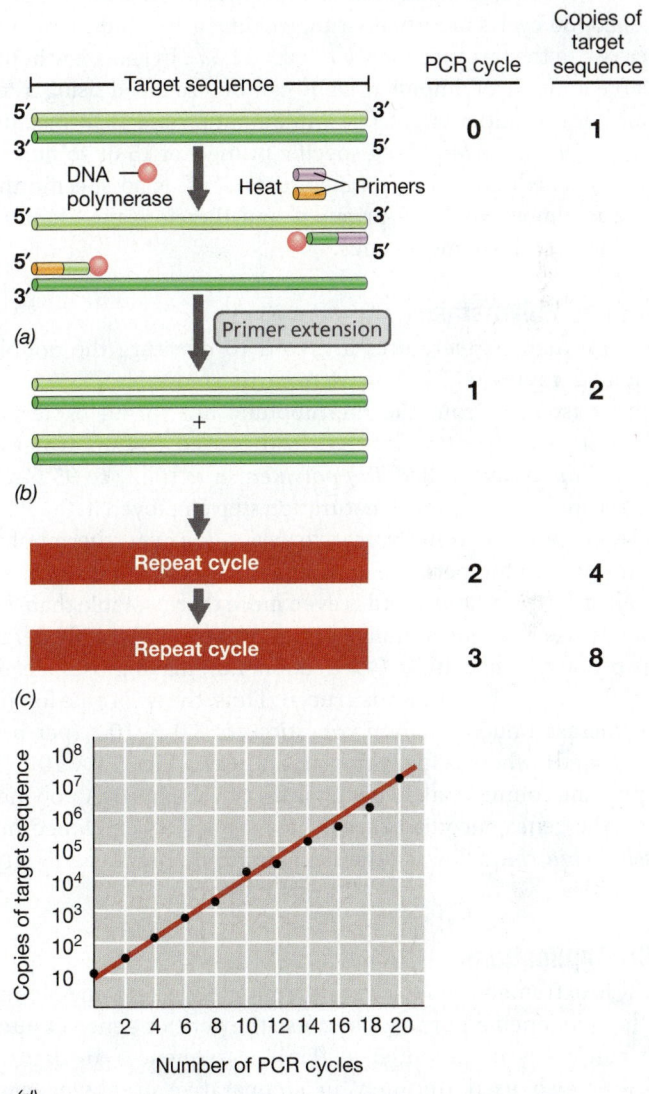

(d)

Figure 11.5 **The polymerase chain reaction (PCR).** The PCR amplifies specific DNA sequences. *(a)* Target DNA is heated to separate the strands, and a large excess of two oligonucleotide primers, one complementary to each strand, is added along with DNA polymerase. *(b)* Following primer annealing, primer extension yields a copy of the original double-stranded DNA. *(c)* Two additional PCR cycles yield four and eight copies, respectively, of the original DNA sequence. *(d)* Effect of running 20 PCR cycles on a DNA preparation originally containing ten copies of a target gene. Note that the plot is semilogarithmic.

most template strands anneal to a primer, and not to each other, as the mixture cools (Figure 11.5a).

3. DNA polymerase then extends the primers using the original DNA as the template (Figure 11.5b).

4. After an appropriate incubation period, the mixture is heated again to separate the strands, but now the target gene is present in twice the original amount. The mixture is then cooled to allow the primers to hybridize with complementary regions of newly synthesized DNA, and the whole process is repeated (Figure 11.5c).

PCR is a powerful tool and has revolutionized all of biology. It is easy to perform, extremely sensitive, specific, and highly efficient. During each round of amplification the amount of product doubles, leading to an exponential increase in the DNA. In practice, 20–30 cycles are usually run, yielding a 10^6-fold to 10^9-fold increase in the target sequence (Figure 11.5d). In just a few hours, a large amount of amplified DNA can be produced using a few molecules of initial target DNA in an automated PCR machine called a *thermocycler*. Using specific primers of 15 or so nucleotides and high annealing temperatures, PCR is so specific that there is almost no "false priming" and therefore the amplified DNA is virtually homogeneous.

PCR and Polymerases

Because high temperatures are used to denature the double-stranded copies of DNA in vitro, a thermostable DNA polymerase isolated from the thermophilic hot spring bacterium *Thermus aquaticus* (↻ Section 15.20) is used. DNA polymerase from *T. aquaticus*, called *Taq polymerase*, is stable to 95°C and thus is unaffected by the denaturation step employed in the PCR. DNA polymerase from *Pyrococcus furiosus*, a hyperthermophile with a growth temperature optimum of 100°C (↻ Section 16.4) is called *Pfu polymerase* and is even more thermostable than *Taq* polymerase. Moreover, unlike *Taq* polymerase, *Pfu* polymerase has proofreading activity (↻ Section 4.6), making it especially useful when high accuracy is crucial. Thus, the error rate for *Taq* polymerase under standard conditions is 8.0×10^{-6} (per base duplicated), whereas for *Pfu* polymerase it is only 1.3×10^{-6}. To supply the commercial demand for thermostable DNA polymerases, the genes encoding these enzymes have been cloned into *Escherichia coli*, allowing the enzymes to be produced in large quantities.

PCR Applications

PCR is extremely valuable for obtaining DNA for gene cloning or for sequencing purposes because the gene or genes of interest can easily be amplified if flanking sequences are known. PCR is also used routinely in comparative or phylogenetic studies to amplify genes from various sources. In these cases the primers are made commercially to regions of the gene that are conserved in sequence across a wide variety of organisms. Because 16S rRNA, a molecule used for phylogenetic analyses, has both highly conserved and highly variable regions (↻ Section 12.5), primers specific for the 16S rRNA gene from various taxonomic groups can be synthesized and used to survey different habitats for specific groups of organisms. This technique

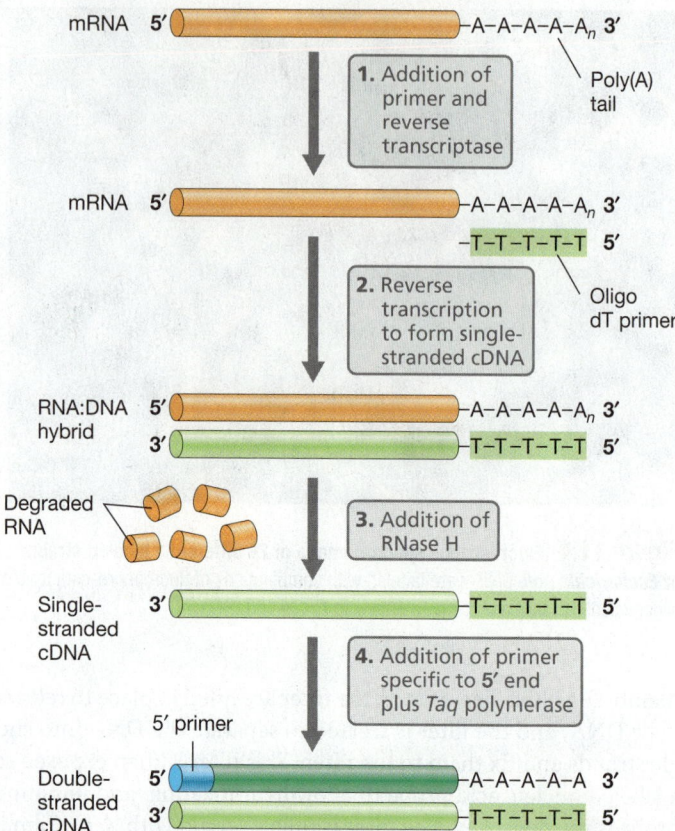

Figure 11.6 Reverse transcription PCR. Steps in the synthesis of cDNA from a eukaryotic mRNA. Reverse transcriptase synthesizes a hybrid molecule containing both RNA and DNA using the mRNA as a template and oligo-T primer as a substrate. Next, the enzyme RNaseH hydrolyzes the RNA portion of the hybrid molecule yielding a single-stranded molecule of complementary DNA (cDNA). Following the addition of a primer complementary to the 5′ end of the cDNA, *Taq* polymerase produces a double-stranded cDNA.

is widely used in microbial ecology and has revealed the enormous diversity of the microbial world, much of it as yet cultured (↻ Section 18.5).

Because it is so sensitive, PCR can be used to amplify very small quantities of DNA. For example, PCR has been used to amplify and clone DNA from sources as varied as mummified human remains and fossilized plants and animals. The ability of PCR to amplify and analyze DNA from cell mixtures has also made it a common tool of diagnostic microbiology (↻ Section 27.10). PCR has also been used in forensics to identify humans from very small samples of their DNA.

Variations in the standard PCR procedure have been developed depending on the molecular objective. **Reverse transcription** PCR (RT-PCR) can be used to make DNA from an mRNA template (**Figure 11.6**). This procedure can be used to detect if a gene is expressed or to produce an intron-free eukaryotic gene for expression in bacteria as described for the hormones insulin and somatotropin in Section 11.11. RT-PCR uses the retroviral enzyme *reverse transcriptase* to convert RNA into *complementary DNA* (cDNA) (↻ Section 9.11). To quantify the amount of initial target DNA or RNA in a sample, a procedure called *quantitative PCR (qPCR)* can also be used. This technique uses fluorescent probes to monitor the amplification process (🔗 Figures 27.18 and 27.19).

Figure 11.6 illustrates how reverse transcriptase makes a single strand of cDNA using RNA as a template. When making DNA using RNA as a template, a primer complementary to the 3′ end of the target transcript is used by the enzyme reverse transcriptase to initiate RNA synthesis. If the template is eukaryotic mRNA, a primer complementary to the poly(A) tail (🔗 Section 4.9) of the mRNA can be used. The activity of reverse transcriptase results in a hybrid nucleic acid molecule containing both DNA and RNA. RNaseH, a ribonuclease specific for the hybrid molecule, hydrolyzes the RNA, leaving the cDNA as template for standard PCR using an additional primer complementary to the 5′ end. Modifications to this procedure can be made if the 5′ end of the mRNA is not known.

MINIQUIZ

- Why is a primer needed at each end of the DNA segment being amplified by PCR?
- From which organisms are thermostable DNA polymerases obtained?
- How does RT-PCR differ from traditional PCR?

11.4 Essentials of Molecular Cloning

In **molecular cloning** a fragment of DNA is isolated and replicated. The basic strategy of molecular cloning is to isolate the desired gene (or other segment of DNA) from its original location and move it to a small, simple, and manipulable genetic element, such as a plasmid or virus, which is called a **vector** (**Figure 11.7**). Molecular cloning results in **recombinant DNA**, a DNA molecule that contains DNA from two or more sources. When the recombinant vector replicates, the cloned DNA that it contains is also replicated. Once cloned, the gene of interest can be manipulated in various ways and may eventually be inserted back into a living cell. This approach provides the foundation for much of genetic engineering and has greatly helped the detailed analysis of genomes.

The first objective of gene cloning is to isolate copies of specific genes in pure form. Consider the problem. For a genetically "simple" organism such as *Escherichia coli*, an average gene is encoded by 1–2 kbp of DNA out of a genome of over 4600 kbp (🔗 Section 4.3). An average *E. coli* gene is thus less than 0.05% of the total DNA in the cell. In human DNA the problem is even greater because the coding regions of average genes are not much larger than in *E. coli* but genes are typically split into pieces and the genome is almost 1000 times larger! Nonetheless, our knowledge of DNA chemistry and enzymology allows us to routinely break, rejoin, and replicate DNA molecules in vitro. Restriction enzymes, DNA ligase, the polymerase chain reaction (PCR), and synthetic DNA are the key tools for molecular cloning.

Steps in Gene Cloning: A Summary

The following list outlines the sequence of events in gene cloning:

1. **Isolation and fragmentation of the source DNA.** The source DNA can be total genomic DNA from an organism of interest,

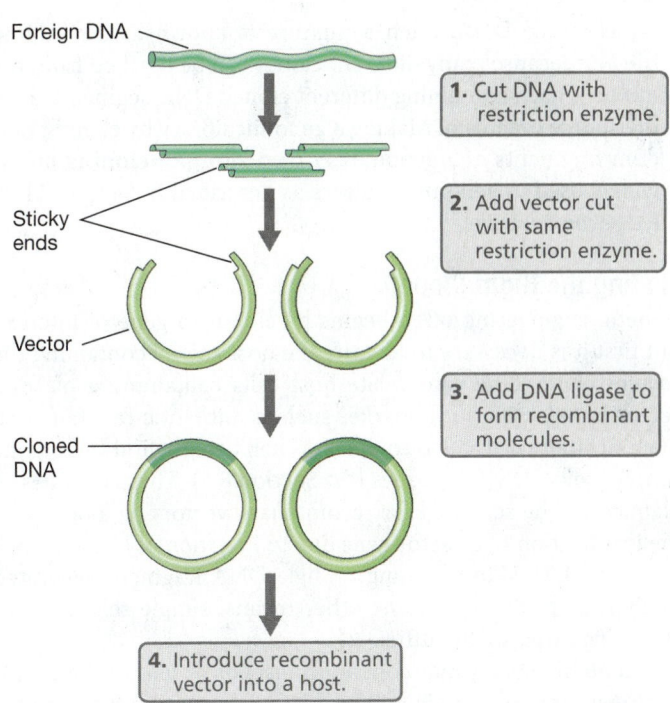

Figure 11.7 Major steps in gene cloning. The vector can be a plasmid or a viral genome. By cutting the foreign DNA and the vector DNA with the same restriction enzyme, complementary sticky ends are generated that allow foreign DNA to be inserted into the vector.

DNA synthesized from an RNA template by reverse transcriptase (Section 11.3), a gene or genes amplified by the polymerase chain reaction (Section 11.3), or even completely synthetic DNA made in vitro (Section 11.5). If genomic DNA is the source, it is first cut with restriction enzymes (Section 11.1) to give a mixture of fragments of manageable size (Figure 11.7).

2. **Inserting the DNA fragment into a cloning vector.** Cloning vectors are small, independently replicating genetic elements used to carry and replicate cloned DNA segments. Most vectors are plasmids or viruses. Cloning vectors are typically designed to allow insertion of foreign DNA at a restriction site that cuts the vector without affecting its replication (Figure 11.7). If the source DNA and the vector are both cut with the *same* restriction enzyme that yields sticky ends, joining the two molecules is greatly assisted by annealing of the complementary ends. Blunt ends generated by some restriction enzymes can be joined by direct ligation or by using synthetic DNA linkers or adapters. In either case, the strands are joined by *DNA ligase*, an enzyme that covalently links both strands of the vector and the inserted DNA. If the source DNA is PCR-generated, DNA ligase is used to join the amplified DNA to specialized vectors (see Figure 11.15).

3. **Introduction of the cloned DNA into a host organism.** Recombinant DNA molecules made in vitro are introduced into suitable host organisms where they can replicate. Transformation (🔗 Section 10.6) is often used to get recombinant DNA into cells. In practice this often yields a mixture of recombinant constructs. Some cells contain the desired cloned gene, whereas other cells may contain other cloned genes from the

same source DNA. Such a mixture is known as a **genomic library** because many different clones can be purified from the mixture, each containing different cloned DNA segments from the source organism. Making a genomic library by cloning random fragments of a genome is called **shotgun cloning** and is widely used in genomic analyses as described in Section 11.15 for *genome mining*.

Finding the Right Clone

Genetic engineering often begins by cloning a gene of interest. But first it is necessary to identify the host colony containing the correct clone. One can isolate host cells containing a plasmid vector by selecting for a marker such as antibiotic resistance, so that only these cells form colonies. When using a viral vector, one simply looks for viral plaques (Section 8.4). These colonies or plaques can be screened for recombinant vectors by looking for the inactivation of a vector gene due to insertion of foreign DNA (Section 11.7). When cloning a single DNA fragment generated by PCR or purified by some other means, simple selections or screenings are usually sufficient.

A genomic library may contain thousands or tens of thousands of clones, and often only one or a few may contain the genes of interest. Hence, identifying cells carrying cloned DNA is only the first step. The biggest challenge remains finding the clone carrying the gene of interest. One must examine colonies of bacteria or plaques from viral-infected cells growing on agar plates and detect those few that contain the gene of interest. This can be done by DNA sequencing or by restriction digests performed on plasmids extracted from a large number of colonies. Another approach is to use hybridization as described in Section 11.2 and as depicted in **Figure 11.8a**.

Detecting Proteins Expressed in the Cloning Host

If the foreign gene is expressed in the cloning host, the encoded protein can be screened for. For this to work, the host itself must not produce the protein being studied. Selection of cells containing cloned genes is relatively simple provided that the encoded protein can be assayed conveniently.

Antibodies can be used to detect a protein of interest. Antibodies are proteins of the immune system that bind in a highly specific way to a target molecule, the antigen (Section 24.4). In this case the protein encoded by the cloned gene is the antigen. Because the antibody combines specifically with the antigen, observing the binding of the antibody can identify colonies that contain the antigen. Because very little of the antigen is present in each colony, only a small amount of antibody is bound, and so a highly sensitive procedure for detecting bound antibody must be used. In practice, radioisotopes, fluorescent chemicals, or enzymes are used. Techniques for detecting antigens are discussed in Chapter 27.

The procedure using antibody detection is outlined in Figure 11.8b. *Replica plating* is used to duplicate the master plate onto a synthetic membrane filter, and all further manipulations are done with this filter. The duplicate colonies are lysed to release the antigen of interest. The antibody is then added and it binds the antigen. Unbound antibody is washed off and a radioactive reagent is added that binds specifically to the antibody. A sheet of X-ray

film is placed over the filter and exposed, and radioactive colonies appear as spots on the X-ray film after it is developed (Figure 11.8b). The location of such spots corresponds to the location of a colony on the master plate that produces the protein. This colony is picked from the master plate and subcultured.

A major limitation of this procedure is that a specific antibody must be made in an experimental animal against the antigen in question. To be successful, however, the injected protein must be pure; otherwise, antibodies against multiple antigens will be formed, leading to false-positive reactions. As one might expect, such reactions make selection of the right clones very difficult.

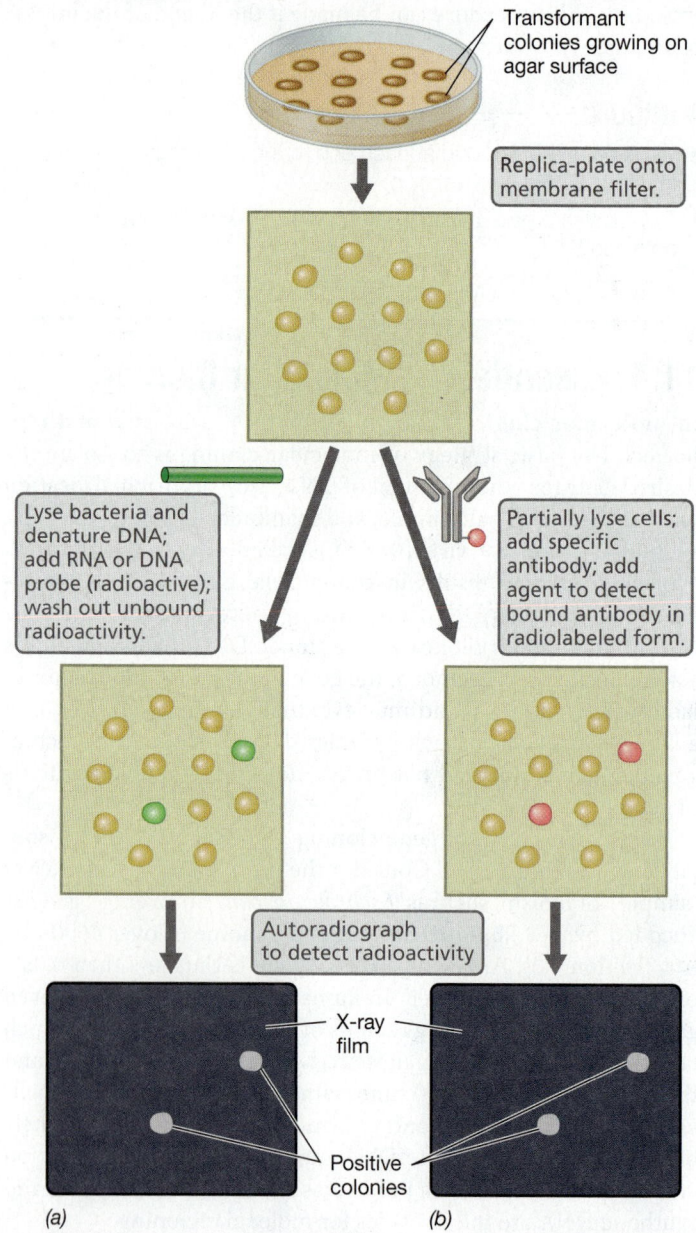

Figure 11.8 Finding the right clone. *(a)* Method for detecting recombinant clones by colony hybridization with a radioactive nucleic acid probe. Formation of a DNA duplex binds the DNA probe to a particular spot on the membrane. *(b)* Method for detecting production of protein by using a specific antibody containing a radioactive or fluorescent label.

11.5 Molecular Methods for Mutagenesis

As we have seen, conventional mutagens introduce mutations *at random* in the intact organism (⮌ Section 10.4). In contrast, in vitro mutagenesis, better known as **site-directed mutagenesis**, uses synthetic DNA plus DNA cloning techniques to introduce mutations into genes *at precisely determined sites.* In addition to changing one or just a few bases, mutations may also be engineered by inserting large segments of DNA at precisely determined locations.

Synthesizing DNA

Segments of DNA may be artificially synthesized and used as primers or probes for polymerase chain reaction, hybridization, or to provide altered versions of parts of genes or regulatory regions. Oligonucleotides of 12–40 bases are commercially available and oligonucleotides of over 100 bases in length can be made if necessary. It is also possible to synthesize entire genes if they encode small proteins (less than 600 bp) such as the subunits of insulin (Section 11.11).

DNA is synthesized in vitro in an automated solid-phase procedure in which the first nucleotide in the chain is fastened to an insoluble support, such as tiny porous glass beads. Several steps are needed for the addition of each nucleotide, and the required chemistry is complex. After each step is completed, the reaction mixture is flushed out of the solid support and the series of reactions repeated for the addition of the next nucleotide. Once the oligonucleotide is the desired length, it is cleaved from the solid-phase support by a specific reagent and purified to eliminate by-products and contaminants.

Site-Directed Mutagenesis

Site-directed mutagenesis is a powerful tool, as it allows for a change to any base pair in a specific gene and thus has many uses in genetics. By altering gene sequences to produce amino acid sequence changes, site-directed mutagenesis is used to manipulate protein characteristics such as enzyme activity or protein-binding affinity (Section 11.12). The basic procedure is to synthesize a short DNA oligonucleotide primer containing the desired base change (mutation) and to allow this to base-pair with single-stranded DNA containing the target gene. Pairing will be complete except for the region of mismatch. Then the synthetic oligonucleotide is extended using DNA polymerase, thus copying the rest of the gene, and the double-stranded molecule is inserted into a host cell by transformation. Mutants are often selected by positive selection, such as antibiotic resistance; in this case, the mutated DNA would also carry a nearby antibiotic resistance marker.

One procedure for site-directed mutagenesis is illustrated in **Figure 11.9**. The process starts with cloning the target gene into a plasmid vector. The double-stranded vector is then denatured to yield single-stranded DNA and allow for the mutagenized oligonucleotide to bind by base-pairing with the target gene. After extension by DNA polymerase, the DNA molecule contains one strand with the mismatch. After transformation into host cells, vector DNA replication, and cell division, both daughter molecules will be fully base paired but one daughter molecule will carry the mutation and the other will be wild type. Progeny bacteria are then screened for those with the mutation.

Site-directed mutagenesis may also be carried out using PCR. In this case, the short DNA oligonucleotide with the required mutation is used as a PCR primer. The mutation-carrying primer is designed to anneal to the target with the mismatch in the middle and must have enough matching nucleotides on both sides for binding to be stable during the PCR reaction. The mutant primer is then paired with a normal primer, and when the PCR reaction amplifies the target DNA, it incorporates the mutation(s) into the final amplified product.

Applications of Site-Directed Mutagenesis

Site-directed mutagenesis can be used to investigate the activity of proteins with known amino acid substitutions. For example,

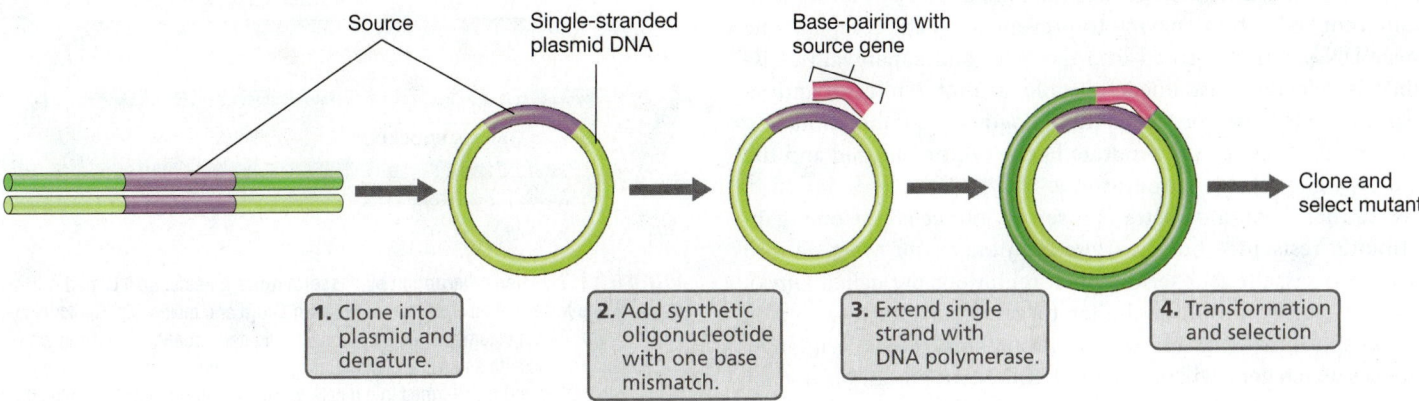

Figure 11.9 Site-directed mutagenesis using synthetic DNA. Short synthetic oligonucleotides may be used to generate mutations. Cloning the source DNA into a plasmid followed by denaturation yields the single-stranded DNA needed for site-directed mutagenesis to work.

suppose one was studying the active site of an enzyme. Site-directed mutagenesis could be used to change a specific amino acid in the active site, and the modified enzyme would then be assayed and compared to the wild-type enzyme. In such experiments, the vector encoding the mutant enzyme is inserted into a mutant host strain unable to make the original enzyme. Consequently, the activity measured is due to the mutant version of the enzyme alone.

Using site-directed mutagenesis, enzymologists can link virtually any aspect of an enzyme's activity, such as catalysis, resistance, or susceptibility to chemical or physical agents, or interactions with other proteins, to specific amino acids in the protein. In a genetic engineering example, site-directed mutagenesis has allowed scientists to alter the receptor-binding affinity of the bovine growth hormone somatotropin so that it only stimulates growth and not milk production in humans (Section 11.12).

Cassette Mutagenesis and Gene Disruption

To make more than a few base-pair changes or replace sections of a gene of interest, synthetic fragments called **DNA cassettes** (or cartridges) can be used to mutate DNA in a process known as **cassette mutagenesis**. These cassettes can be synthesized using the polymerase chain reaction or by direct DNA synthesis. The cassette can then replace sections of the DNA of interest using restriction sites. However, if sites for the appropriate restriction enzyme are not present at the required location, they can be inserted by site-directed mutagenesis as shown in Figure 11.9. Cassettes used to replace sections of genes are typically the same size as the wild-type DNA fragments they replace.

Another type of cassette mutagenesis is called **gene disruption**. In this technique, cassettes are inserted into the middle of a gene, thus disrupting the coding sequence. Cassettes used for making insertion mutations can be almost any size and can even carry an entire gene. To facilitate selection, cassettes that encode antibiotic resistance are commonly used. The process of gene disruption is illustrated in **Figure 11.10**. In this case, a DNA cassette carrying a gene conferring kanamycin resistance (the Kan cassette) is inserted at a restriction site in a cloned gene. The vector carrying the disrupted gene is then linearized by cutting with a different restriction enzyme to prevent replication. Finally, the linear DNA is transformed into the host, and kanamycin resistance is selected. The linear plasmid cannot replicate, and so resistant cells arise mostly by homologous recombination (↶ Section 10.5) between the mutated gene on the plasmid and the wild-type gene on the chromosome.

Note that when a cassette is inserted, the cells not only gain antibiotic resistance but also *lose the function of the gene* into which the cassette is inserted. Such mutations are called *knockout mutations*. These are similar to insertion mutations made by transposons (↶ Section 10.11), but here the experimenter chooses which gene will be mutated. Knockout mutations in haploid organisms (such as prokaryotes) yield viable cells only if the disrupted gene is nonessential. Indeed, the generation of gene knockouts is frequently used to investigate whether a given gene is essential.

MINIQUIZ
- Why is a solid support used during chemical synthesis of DNA?
- How can site-directed mutagenesis be useful to enzymologists?
- What are knockout mutations?

11.6 Gene Fusions and Reporter Genes

DNA manipulation has revolutionized the study of gene regulation. A coding sequence from one source (the *reporter*) may be fused to a regulatory region from another source to form a hybrid gene. Such gene fusions are often used in studying gene regulation (↶ Section 7.1), especially where assaying the levels of the natural gene product is difficult, expensive, or time consuming. Fusions may also be used to increase expression of a desired gene product.

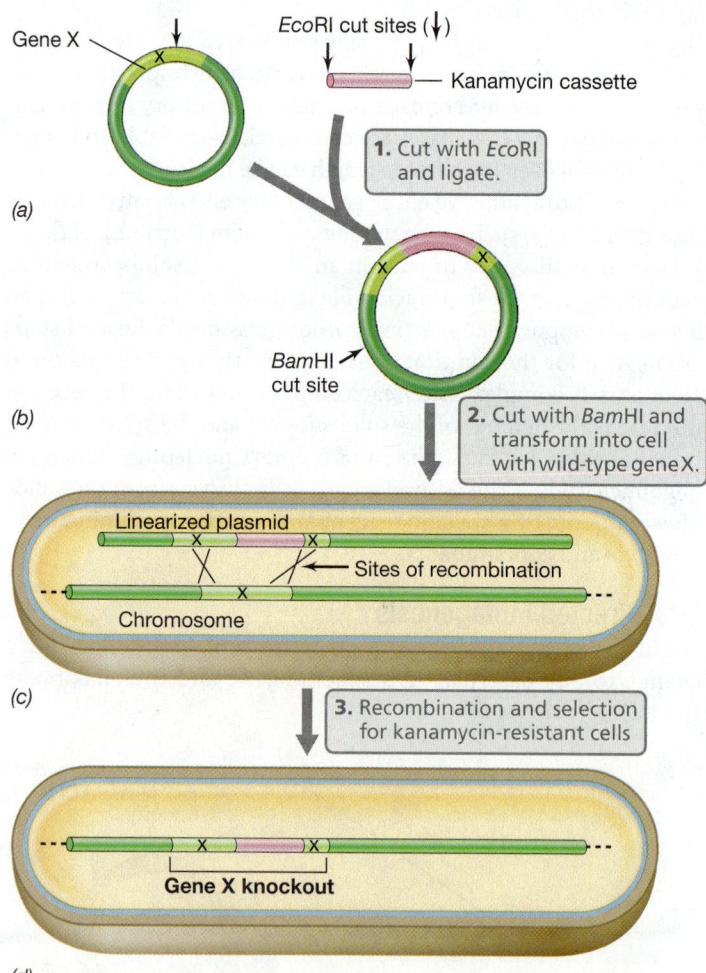

Figure 11.10   Gene disruption by cassette mutagenesis. *(a)* A cloned wild-type copy of gene X, carried on a plasmid, is cut with *Eco*RI and mixed with the kanamycin cassette. *(b)* The cut plasmid and the cassette are ligated, creating a plasmid with the kanamycin cassette as an insertion mutation within gene X. This new plasmid is cut with *Bam*HI and transformed into a cell. *(c)* The transformed cell contains the linearized plasmid with a disrupted gene X and its own chromosome with a wild-type copy of the gene. *(d)* In some cells, homologous recombination occurs between the wild-type and mutant forms of gene X. Cells that can grow in the presence of kanamycin have only a single, disrupted copy of gene X.

Reporter Genes

The key property of a **reporter gene** is that it encodes a protein that is easy to detect and assay. Reporter genes are used for a variety of purposes. They may be used to report the presence or absence of a particular genetic element (such as a plasmid) or DNA inserted within a vector. They can also be fused to other genes or to the promoter of other genes so that gene expression can be studied.

The first gene to be used widely as a reporter was the *Escherichia coli* gene *lacZ*, which encodes the enzyme β-galactosidase, required for lactose catabolism (Section 7.3). Cells expressing β-galactosidase can be detected easily by the color of their colonies on indicator plates that contain the artificial substrate Xgal (5-bromo-4-chloro-3-indolyl-β-D-galactopyranoside); Xgal is cleaved by β-galactosidase to yield a blue color (see Figure 11.14). The **green fluorescent protein (GFP)** is widely used as a reporter (**Figure 11.11**). Although the gene for GFP was originally cloned from the jellyfish *Aequorea victoria,* GFP may be expressed in most cells as it is stable and causes little or no disruption of host cell metabolism. If expression of a cloned gene is linked to the expression of GFP, the latter signals (reports) that the cloned gene has also been expressed (Figure 11.11).

Gene Fusions

It is possible to engineer constructs that consist of segments from two different genes. Such constructs are called **gene fusions**. If the promoter that controls a coding sequence is removed, the coding sequence can be fused to a different regulatory region to place the gene under the control of a different promoter. Alternatively, the promoter region can be fused to a gene whose product is easy to assay. There are two different types of gene fusions. In **operon fusions**, a coding sequence that retains its own translational start site and signals is fused to the transcriptional signals of another gene. In **protein fusions**, genes that encode two different proteins

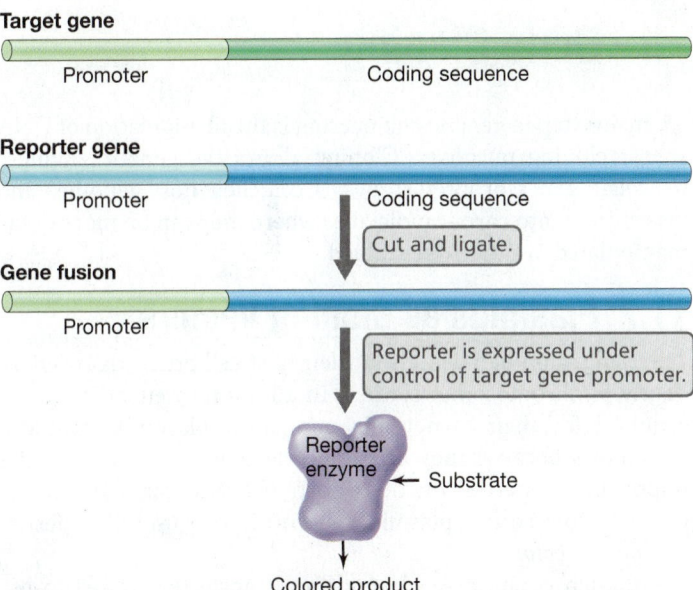

Figure 11.12 Construction and use of gene fusions. The promoter of the target gene is fused to the reporter coding sequence. Consequently, the reporter gene is expressed under those conditions where the target gene would normally be expressed. The reporter shown here is an enzyme (such as β-galactosidase) that converts a substrate to a colored product that is easy to detect. This approach greatly facilitates the investigation of regulatory mechanisms.

are fused together so that they share the same transcriptional and translational start and stop signals. Following translation, protein fusions yield a single hybrid polypeptide.

Gene fusions are often used in studying gene regulation, especially if measuring the levels of the natural gene product is difficult, expensive, or time consuming. The regulatory region of the gene of interest is fused to the coding sequence for a reporter gene, such as that for β-galactosidase or GFP. The reporter is then made under the conditions that would trigger expression of the target gene (**Figure 11.12**). The expression of the reporter is assayed under a variety of conditions to determine how the gene of interest is regulated (Section 7.1). *Transcriptional control* is assayed by fusing the transcriptional start signals of the gene of interest to a reporter gene, whereas *translational control* is assayed by fusing translational start signals of a gene of interest to a reporter gene under the control of a known promoter.

Gene fusions may also be used to test for the effects of regulatory genes. Mutations that affect regulatory genes are introduced into cells carrying gene fusions, and expression is measured and compared to cells lacking the regulatory mutations. This allows the rapid screening of multiple regulatory genes that are suspected of controlling the target gene. Besides using fusions to monitor for the presence or expression of a gene, proteins that are easily purified can also be fused to proteins of interest to aid in purification (Section 11.11).

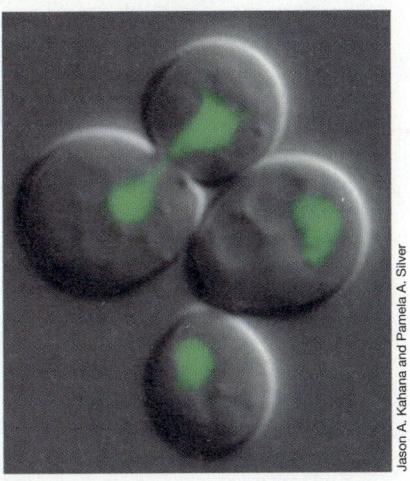

Figure 11.11 Green fluorescent protein (GFP). GFP can be used as a tag for protein localization in vivo. In this example, the gene encoding Pho2, a DNA-binding protein from the yeast *Saccharomyces cerevisiae*, was fused to the gene encoding GFP and photographed by fluorescence microscopy. The recombinant gene was transformed into budding yeast cells. These expressed the fluorescent fusion protein localized in the nucleus.

MINIQUIZ

- What is a reporter gene?
- Why are gene fusions useful in studying gene regulation?

II • Gene Cloning

A major step in genetic engineering is the manipulation of DNA for cloning purposes. Cloning allows the genetic engineer to isolate genes of interest away from their host genomes and insert them into carrier molecules where they can be more easily manipulated or otherwise studied.

11.7 Plasmids as Cloning Vectors

The replication of plasmids in their host cell proceeds independently of chromosomal control. In addition to encoding genes required for their own replication, most plasmids are natural vectors because they often encode other genes that confer important properties on their hosts (⟳ Section 4.3). As discussed below, certain plasmids have other very useful properties as *cloning vectors*.

Although conjugative plasmids are transferred by cell-to-cell contact in nature (⟳ Section 10.8), most plasmid cloning vectors have been genetically modified to abolish conjugative transfer. This prevents unwanted movement of the vector into other organisms. However, vector transfer in the laboratory can be accomplished by chemically mediated transformation or electroporation (⟳ Section 10.6). Depending on the host–plasmid system, replication of the plasmid may be under tight control, in which case only a few copies are made, or under relaxed control, in which case a large number of copies are made. Obtaining a high copy number is often important in gene cloning, and by proper selection of the host–plasmid system and manipulation of cellular macromolecule synthesis, plasmid copy numbers of several thousand per cell can be obtained.

An Example of a Cloning Vector: The Plasmid pUC19

While the first plasmid cloning vectors used were natural isolates, vectors currently used in molecular biology have been genetically engineered to show specific properties. For example, widely used cloning vectors are derivatives of the plasmid pUC19 (**Figure 11.13**). This plasmid was derived in several steps from a ColE1 toxin-encoding plasmid (⟳ Section 4.3) by removing the colicin genes and inserting genes for ampicillin resistance and for a blue–white color-screening system (see below). A short segment of artificial DNA containing cut sites for many restriction enzymes, called a *multiple cloning site* (*MCS*), was also inserted into the *lacZ* gene in pUC19; *lacZ* encodes the lactose-degrading enzyme β-galactosidase (⟳ Section 7.3). The presence of the short MCS does not inactivate *lacZ*, and cut sites for restriction enzymes present in the MCS are absent from the rest of the vector. Consequently, treatment with each of these restriction enzymes opens the vector at a unique location but does not cut the vector into multiple pieces.

Plasmid pUC19 displays a number of other characteristics that make it ideal as a cloning vehicle:

1. It is relatively small, only 2686 base pairs, which makes the DNA easier to isolate and manipulate.

2. It is stably maintained in its host (*E. coli*) in relatively high copy number, about 50 copies per cell.

3. It can be amplified to a very high number (1000–3000 copies per cell, about 40% of the cellular DNA) if necessary by inhibiting protein synthesis with the antibiotic chloramphenicol.

4. It is easy to isolate in the supercoiled form.

5. Moderate amounts of foreign DNA can be inserted, although inserts of more than 10 kilobase pairs lead to plasmid instability.

6. The complete base sequence of the plasmid is known, allowing identification of all restriction enzyme cut sites.

7. The MCS contains single cut sites for over a dozen restriction enzymes, thus increasing the versatility of the vector.

8. It has a gene conferring ampicillin resistance. This permits ready selection of host cells containing the plasmid because such hosts gain resistance to the antibiotic.

9. It can be inserted into cells easily by transformation.

10. Insertion of foreign DNA into the MCS can be detected by blue–white screening (see below) because of *lacZ*.

Cloning Genes into Plasmid Vectors

The use of plasmid vectors such as pUC19 in gene cloning is shown in **Figure 11.14**. A suitable restriction enzyme with a cut site within the MCS is chosen. Both the vector and the foreign DNA to be cloned are cut with this enzyme. The vector is linearized, and segments of the foreign DNA are inserted into the open cut site and ligated into position with the enzyme DNA ligase. This

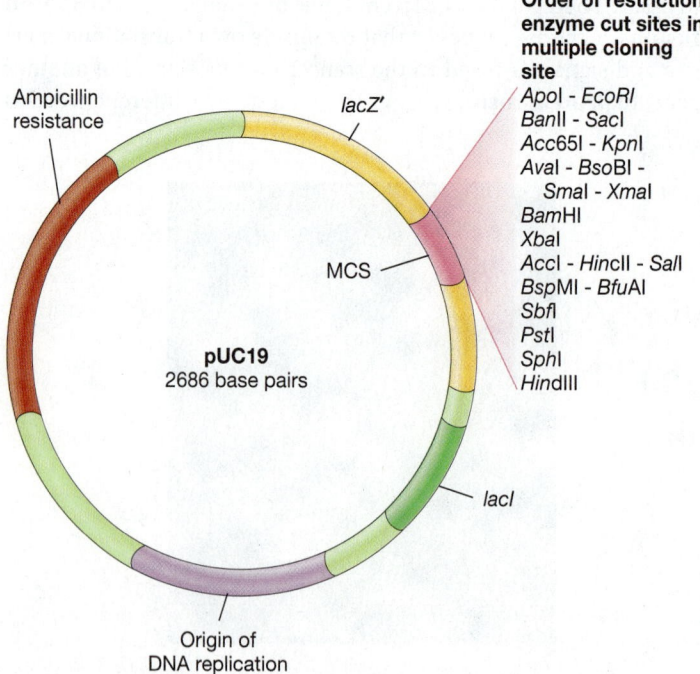

Figure 11.13 Cloning vector plasmid pUC19. Essential features include an ampicillin resistance marker and the multiple cloning site (MCS) with multiple restriction enzyme cut sites. Insertion of cloned DNA within the MCS inactivates the truncated *lacZ'* gene that encodes part of β-galactosidase and allows for easy identification of transformants by blue–white screening.

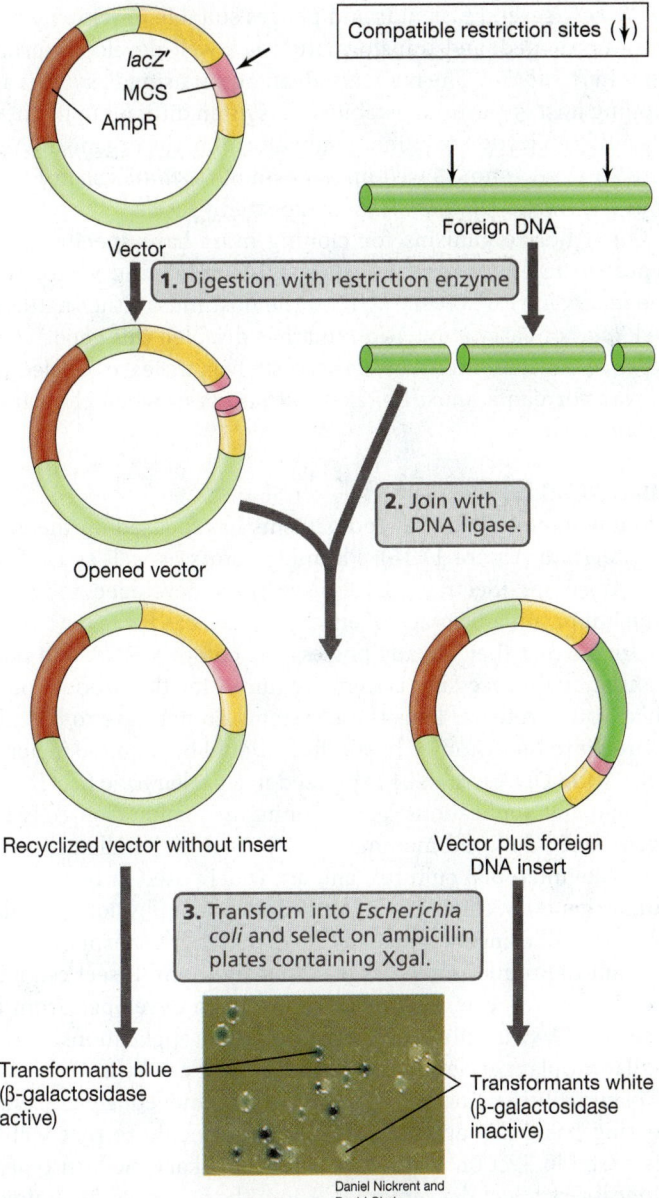

Figure 11.14 **Cloning into the plasmid vector pUC19.** The cloning vector and foreign DNA are cut with compatible restriction enzymes at positions indicated by the arrows. Insertion of DNA inactivates β-galactosidase, allowing blue-white screening for the presence of the insert. The photo on the bottom shows colonies of *Escherichia coli* on an Xgal plate. The enzyme β-galactosidase can cleave the normally colorless Xgal to form a blue product. AmpR, gene encoding ampicillin resistance.

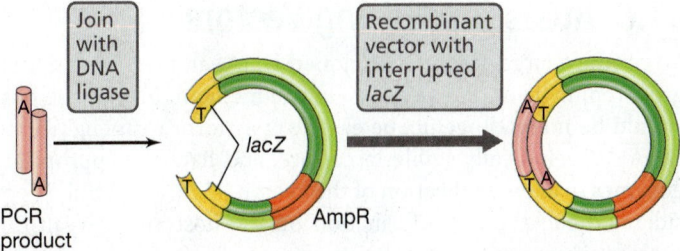

Figure 11.15 **PCR vector.** The linearized cloning vector contains overhanging thymine residues that base-pair with the adenine residues present on the 3′ ends of *Taq*-polymerase-generated PCR. Ligation of the two pieces of DNA yields a circular plasmid containing an interrupted *lacZ*. AmpR, gene encoding ampicillin resistance.

and Xgal, to test for β-galactosidase activity. Those colonies that are *blue* contain the plasmid without any inserted foreign DNA (i.e., the plasmid merely recyclized without picking up foreign DNA), whereas those colonies that are *white* contain plasmid with inserted foreign DNA and are picked for further analyses (see Figure 11.20*b* for a related example of the blue–white selection system).

Other Plasmid Vectors

Many subsequent plasmid vectors have been developed that include features similar to those of pUC19 listed above but also show other desirable features. For example, some vectors have been developed specifically for cloning DNA products synthesized by *Taq* polymerase in a polymerase chain reaction (PCR) (Section 11.3). The enzymatic activity of *Taq* polymerase adds a template-independent adenine residue to the 3′ ends of its products. Commercial linearized vectors are available that contain overhanging thymidine residues that allow for base-pairing with the PCR product and subsequent ligation using DNA ligase (**Figure 11.15**).

Other vectors have been designed to directly select for recombinant vectors through cell viability rather than by screening. For example, a vector with an MCS containing a toxin gene can be designed. Without insertion of foreign DNA into the MCS that interrupts the toxin gene, the vector encodes a protein that is lethal to the host cell. Therefore, only cells containing a recombinant plasmid in which the toxin gene has been inactivated can grow.

Cloning using plasmid vectors is versatile and widely used in genetic engineering, particularly when the fragment to be cloned is fairly small. Also, plasmids are often used as cloning vectors if expression of the cloned gene is desired, since regulatory genes can be engineered into the plasmid to obtain expression of the cloned genes under specific conditions (Section 11.9).

disrupts the *lacZ* gene, a phenomenon called *insertional inactivation*, and is used to detect the presence of foreign DNA within the vector or recombinant vector. When the colorless reagent Xgal is added to the medium, β-galactosidase cleaves it, generating a blue product. Thus, cells containing the vector *without* cloned DNA form blue colonies, whereas cells containing the vector *with* an insert of cloned DNA do not form β-galactosidase and are therefore white.

After DNA ligation, the resulting plasmids are transformed into cells of *E. coli*. The colonies are selected on media containing both ampicillin, to select for the presence of the plasmid,

MINIQUIZ

- Explain why it is necessary in cloning to use a restriction enzyme that cuts the vector in only one location.
- What is a multiple cloning site?
- What is insertional inactivation?

11.8 Hosts for Cloning Vectors

To produce large amounts of cloned DNA, an ideal host should grow rapidly in an inexpensive culture medium. Ideally, the host should be nonpathogenic, be easy to transform with engineered DNA, be genetically stable in culture, and have the appropriate enzymes to allow replication of the vector. It is also helpful if considerable background information on the host and a wealth of tools for its genetic manipulation exist.

The most useful hosts for cloning are microorganisms that are easily grown and for which we have much information. These include the bacteria *Escherichia coli* and *Bacillus subtilis*, and the yeast *Saccharomyces cerevisiae* (**Figure 11.16**). Complete genome sequences are available for all of these organisms, and they are widely used as cloning hosts. However, in some cases other hosts and specialized vectors may be necessary to get the DNA properly cloned and expressed.

Prokaryotic Hosts

Although most molecular cloning has been done in *E. coli* (Figure 11.16), this host has a few disadvantages. *E. coli* is an excellent choice for initial cloning work, but it is problematic as an expression host because it is found in the human intestine and some wild-type strains are potentially harmful (⟳ Section 31.12). However, several modified *E. coli* strains have been developed specifically for cloning purposes, and thus *E. coli* remains the organism of choice for most molecular cloning. A major problem with using any bacterial host, including *E. coli*, is the lack of systems to correctly modify eukaryotic proteins; this problem may be solved by using eukaryotic host cells, as discussed below.

Another problem with using *E. coli* is that, like all gram-negative bacteria, it has an outer membrane that hinders protein secretion. This issue may be resolved by using the gram-positive organism *B. subtilis* as a cloning host (Figure 11.16). Although the technology for cloning in *B. subtilis* is less advanced than for *E. coli*, several plasmids and phages suitable for cloning have been developed, and transformation is a well-developed procedure in *B. subtilis*. The main disadvantage of using *B. subtilis* as a cloning host is plasmid instability. It is often difficult to maintain plasmid replication over many subcultures of the organism. Also, foreign DNA is not as well maintained in *B. subtilis* as in *E. coli*; thus the cloned DNA is often unexpectedly lost.

Often host organisms for cloning must have specific genotypes to be effective. For instance, if the cloning vector uses the *lacZ* gene for screening, then the host must either naturally lack *lacZ* or carry a mutation that has disabled this gene. These types of considerations and others, such as the ease of selection of transformants, must be taken into account when choosing a cloning host.

Eukaryotic Hosts

Cloning in eukaryotic microorganisms has focused on the yeast *S. cerevisiae* (Figure 11.16). Plasmid vectors as well as artificial chromosomes (Section 11.10) have been developed for yeast. One important advantage of eukaryotic cells as hosts for cloning vectors is that they already possess the complex RNA and post-translational processing systems required for the production of eukaryotic proteins. Thus these systems do not have to be engineered into the vector or host cells as would be required if cloned eukaryotic DNA was to be expressed in a prokaryotic host.

For many applications, gene cloning in mammalian cells has been done. Cultured mammalian cells can be handled in some ways like microbial cultures, and are widely used in research on human genetics, cancer, infectious disease, and physiology. A disadvantage of using mammalian cells is that they are expensive and difficult to produce under large-scale conditions. Insect cell lines are simpler to grow, and vectors have been developed from an insect DNA virus, the baculovirus. For some applications, in particular for plant agriculture, the cloning host can be a culture line of plant cell tissue or even an entire plant. Indeed, genetic engineering has many applications in plant agriculture that will be discussed in Section 11.13. Regardless of eukaryotic host type, it is necessary to get the vector DNA into the host cells. Techniques to transfer DNA into eukaryotic cells are not discussed here, but they include transfection (see Figure 11.28), microinjection, and electroporation.

MINIQUIZ -
- Why does molecular cloning require a host?
- When would it be beneficial to use a eukaryotic host for molecular cloning?

11.9 Shuttle Vectors and Expression Vectors

Cloned genes are used for a variety of purposes. To facilitate some of the most common uses, specialized vectors have been engineered for moving cloned genes between organisms of different species or to optimize expression of cloned genes in a given host; these are called *shuttle* vectors and *expression* vectors, respectively.

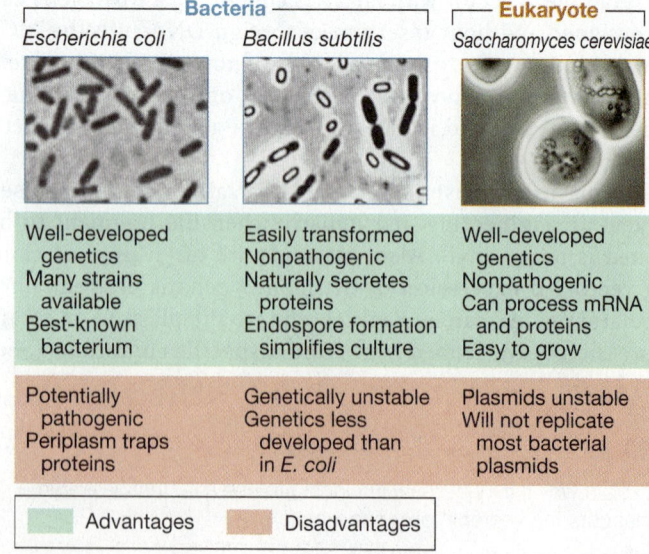

Figure 11.16 Hosts for molecular cloning. A summary of the advantages and disadvantages of some common cloning hosts.

Shuttle Vectors

Vectors that can replicate and are stably maintained in two (or more) unrelated host organisms are called **shuttle vectors**. Genes carried by a shuttle vector can thus be moved between unrelated organisms. Shuttle vectors have been developed that replicate in both *Escherichia coli* and *Bacillus subtilis*, *E. coli* and yeast, and *E. coli* and mammalian cells, as well as in many other pairs of organisms. The importance of a shuttle vector is that it allows for DNA to be easily cloned in a genetically manipulable organism and yields a recombinant vector that can replicate in a second host without any vector modifications.

Many shuttle vectors have been designed to move genes between *E. coli* and yeast. Bacterial plasmid vectors were the starting point and were modified to function in yeast as well. Because bacterial origins of replication do not function in eukaryotes, it is necessary to provide a yeast replication origin. One bonus is that DNA sequences of replication origins are similar in different eukaryotes, so the yeast origin functions in higher organisms as well. When eukaryotic cells divide, the duplicated chromosomes are pulled apart by microtubules ("spindle fibers") attached to their centromeres (♻ Section 2.20). Consequently, shuttle vectors for eukaryotes must contain a segment of DNA from the centromere in order to be properly distributed at cell division (**Figure 11.17**). Luckily, the yeast centromere recognition sequence, the CEN sequence, is relatively short and easy to insert into shuttle vectors.

Another requirement is a convenient marker to select for the plasmid in yeast. Unfortunately, yeast is not susceptible to most antibiotics that are effective against bacteria. In practice, yeast host strains are used that are defective in making a particular amino acid or purine or pyrimidine base. A functional copy of the biosynthetic gene that is defective in the host is inserted into the shuttle vector. For example, if the *URA3* gene, needed for synthesis of uracil, is used, the yeast will not grow in the absence of uracil unless it gains a copy of the shuttle vector.

Expression Vectors

Organisms have complex regulatory systems, and cloned genes are often expressed poorly or not at all in a foreign host cell. This problem is tackled by using **expression vectors** that are designed to allow the experimenter to control the expression (transcription and translation) of cloned genes. Generally, the objective is to obtain high levels of expression, especially in biotechnological applications. However, when dealing with potentially toxic gene products, a low but strictly controlled level may be appropriate.

Expression vectors contain regulatory sequences that allow manipulation of gene expression. Usually the control is transcriptional because for high levels of expression it is essential to produce high levels of mRNA. In practice, high levels of transcription require strong promoters that bind RNA polymerase efficiently (♻ Section 4.7). However, the native promoter of a cloned gene may work poorly in the new host. For example, promoters from eukaryotes or even from other bacteria function poorly or not at all in *E. coli*. Indeed, even some *E. coli* promoters function at low levels in *E. coli* because their sequences match the promoter consensus poorly and bind RNA polymerase inefficiently (♻ Section 4.7).

For this reason, expression vectors must contain a promoter that functions efficiently in the host and one that is correctly positioned to control transcription of the cloned gene. Promoters from *E. coli* that are used in expression vectors include *lac* (the *lac* operon promoter), *trp* (the *trp* operon promoter), and *tac* and *trc* (synthetic hybrids of the *trp* and *lac* promoters). These are all "strong" promoters in *E. coli*, and in addition they can be specifically regulated.

Regulation of Transcription from Expression Vectors

Although producing very high levels of mRNA and having this translated into large amounts of protein is often desirable, massive overproduction of foreign proteins can damage the host cell. Therefore, it is important to regulate the expression of cloned genes. Often, in order to avoid damaging the host cells, the culture containing the expression vector is grown without expression of the foreign gene. Then, once a large population of healthy cells is obtained, expression of the cloned gene(s) is triggered by a *genetic switch*.

Regulating transcription by a repressor protein (♻ Section 7.3) is a useful way to control a cloned gene. A strong repressor can completely block the synthesis of the proteins under its control by binding to the operator. When gene expression is required, the inducer is added. The repressor binds the inducer and is released from the DNA, thus allowing transcription of the regulated genes. The expression vector is designed such that the cloned gene is inserted just downstream from the chosen promoter and operator region. A strong ribosome-binding site is often included between the promoter and the cloned gene to allow for efficient translation. The overall result is control of the cloned gene by the chosen promoter together with efficient transcription and translation.

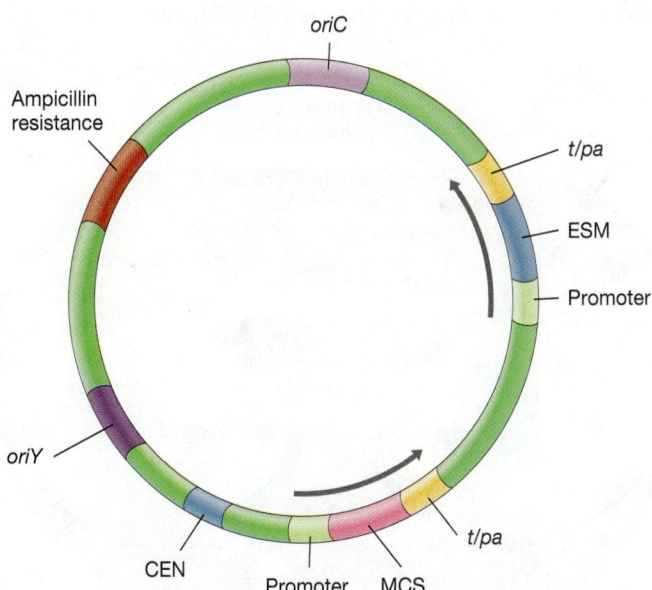

Figure 11.17 Genetic map of a shuttle vector used in yeast. The vector contains components that allow it to shuttle between *Escherichia coli* and yeast and be selected in each organism: *oriC*, origin of replication in *E. coli*; *oriY*, origin of replication in yeast; MCS, multiple cloning site; ESM, eukaryotic selectable marker; CEN, yeast centromere sequence; promoter; *t/pa*, transcription termination/polyadenylation signals. Arrows indicate the direction of transcription.

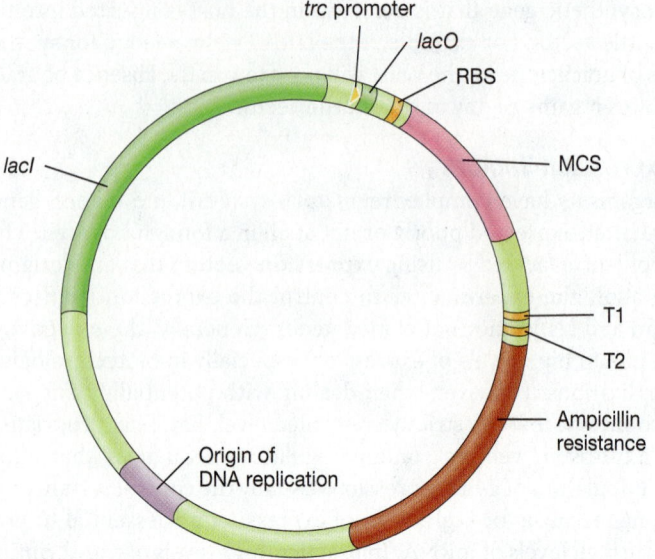

Figure 11.18 Genetic map of the expression vector pSE420. This vector was developed by Life Technologies Corp., a biotechnology company. The MCS (multiple cloning site) contains many different restriction enzyme recognition sequences to facilitate cloning. This region, plus the inserted cloned gene, are transcribed by the *trc* promoter, which is immediately upstream of the *lac* operator (*lacO*). Immediately upstream of the MCS is a sequence that encodes a ribosome-binding site (RBS) on the resulting mRNA. Downstream of the MCS are two transcription terminators (T1 and T2). The plasmid also contains the *lacI* gene, which encodes the *lac* repressor, and a gene conferring resistance to the antibiotic ampicillin. These two genes are under the control of their own promoters, which are not shown.

Figure 11.18 shows an expression vector controlled by *trc* (*trp* promoter and *lac* operator). This plasmid also contains a copy of the *lacI* gene that encodes the *lac* repressor. The level of repressor in a cell containing this plasmid is sufficient to prevent transcription from the *trc* promoter until inducer is added. Addition of lactose or related *lac* inducers triggers transcription of the cloned DNA. In addition to a strong and easily regulated promoter, most expression vectors contain an effective transcription terminator (↩ Section 4.7). This prevents transcription from the strong cloning promoter continuing on into other genes on the vector, which would interfere with vector stability. The expression vector shown in Figure 11.18 has strong transcription terminators in order to halt transcription immediately downstream from the cloned gene.

Regulating Expression with Bacteriophage T7 Control Elements

In some cases the transcriptional control system may not be a normal part of the host at all. An example of this is the use of the bacteriophage T7 promoter and T7 RNA polymerase to regulate gene expression. When T7 infects *E. coli*, it encodes its own RNA polymerase that recognizes only T7 promoters (↩ Section 9.4). In T7 expression vectors, cloned genes are placed under control of the T7 promoter. To achieve this, the gene for T7 RNA polymerase must also be present in the cell under the control of an easily regulated system, such as *lac* (**Figure 11.19**). This is usually done by integrating the gene for

T7 RNA polymerase with a *lac* promoter into the chromosome of a specialized host strain.

The BL21 series of *E. coli* host strains are especially designed to work with the pET series of T7 expression vectors. The cloned genes are expressed shortly after T7 RNA polymerase transcription has been switched on by a *lac* inducer, such as IPTG. Because it recognizes only T7 promoters, the T7 RNA polymerase transcribes only the cloned genes. The T7 RNA polymerase is so highly active that it uses most of the RNA precursors, thereby limiting transcription to the cloned genes. Consequently, host genes that require host RNA polymerase are for the most part not transcribed and thus the cells stop growing; translation in such cells then yields primarily the protein of interest. The T7 control system is thus very effective for generating large amounts of a particular protein.

Translation of the Cloned Gene

Expression vectors must also be designed to ensure that the mRNA produced is efficiently translated. To synthesize protein from an mRNA molecule, it is essential for the ribosomes to bind at the correct site and begin reading in the correct frame. In bacteria this is accomplished by having a ribosome-binding site (RBS, ↩ Section 4.11) and a nearby start codon on the mRNA. Bacterial RBSs are not found in eukaryotic genes and must be engineered into the vector if high levels of expression of the eukaryotic gene are to be obtained. The vector in Figure 11.18 shows such a site.

Other adjustments to a cloned gene may be necessary to ensure high-efficiency translation. For example, *codon usage* can be an obstacle (↩ Section 6.3 and Table 6.3). Codon usage is related to the concentration of the appropriate tRNA in the cell. Because of the redundancy of the genetic code, more than one tRNA exists for most amino acids (↩ Section 4.11). Therefore, if a cloned

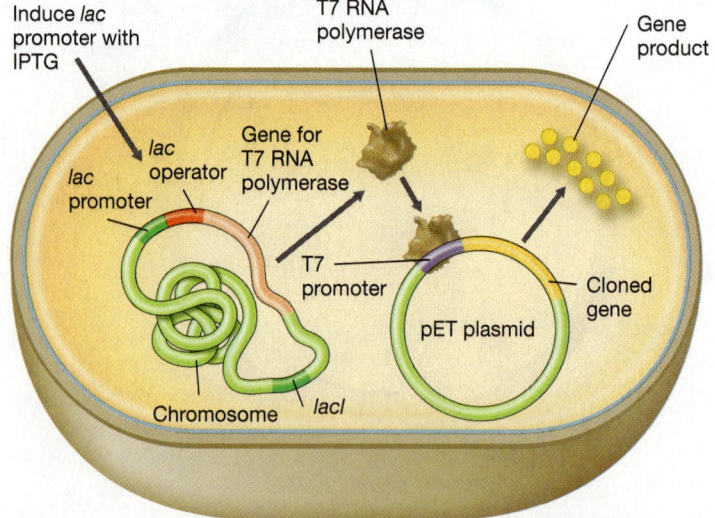

Figure 11.19 The T7 expression system. The gene for T7 RNA polymerase is in a gene fusion under control of the *lac* promoter and is inserted into the chromosome of a special host strain of *Escherichia coli*. Addition of IPTG induces the *lac* promoter, causing expression of T7 RNA polymerase. This transcribes the cloned gene, which is under control of the T7 promoter and is carried by the pET plasmid.

gene has a codon usage pattern distinct from that of its expression host, it will probably be translated inefficiently in that host. Site-directed mutagenesis (Section 11.5) can then be used to change selected codons in the gene, making it more amenable to the codon usage pattern of the host.

Finally, if the cloned gene contains introns, as eukaryotic genes typically do (⮌ Section 4.9), the correct protein product will not be made if the expression host is a prokaryote. This problem can also be corrected by using synthetic DNA. However, the usual method to create an intron-free gene is to obtain the mRNA (from which the introns have already been removed) and use reverse transcriptase to generate a complementary DNA (cDNA) copy (see Figures 11.6 and 11.23).

MINIQUIZ --
- Describe the components needed for an efficient shuttle vector.
- Describe some of the components of an expression vector that improve expression of the cloned gene.

11.10 Other Cloning Vectors

Typical plasmid vectors used for molecular cloning are limited in the amount of DNA that can be inserted, with 10 kilobase pairs (kbp) being the maximum. For applications that require cloning large genomic regions such as operons and eukaryotic genes, vectors based on bacteriophages, cosmids, and artificial chromosomes have been developed. While not discussed in detail here, derivatives of viruses are commonly used as vectors in multicellular eukaryotes. In particular, retroviruses can be used to introduce genes into mammalian cells because these viruses replicate via a DNA form that is directly integrated into the host chromosome (⮌ Section 9.11).

Cloning with Bacteriophage Lambda

Bacteriophage lambda (⮌ Section 8.8) is a useful cloning vector because its biology is well understood, it can hold larger amounts of DNA than most plasmids, and DNA can be efficiently packaged into phage particles in vitro. Also recall that during the lytic phase of phage development, the *Escherichia coli* host cell is reprogrammed to replicate large amounts of lambda DNA (⮌ Section 10.7).

Phage lambda has a large number of genes; however, a third of the lambda genome is not essential for infectivity and can be replaced with foreign DNA. This allows relatively large DNA fragments, up to about 20 kbp, to be cloned into lambda. This is twice the cloning capacity of typical small plasmid vectors. To facilitate the use of lambda as a molecular cloning vector, many of its restriction enzyme sites have been altered and a multiple cloning site (MCS) containing the gene for β-galactosidase has been added to select for recombinant vectors.

Figure 11.20a illustrates cloning with lambda vectors. The initial process is similar to cloning DNA into plasmid vectors in that restriction enzymes and DNA ligase are used. Once the DNA of interest has been inserted into the lambda DNA, the vector is *packaged* by adding extracts from lambda-lysed cells that contain the head and tail proteins and allowing the formation of mature

phage virions to occur spontaneously. These virions can be used to infect suitable host cells, and such infection is much more efficient than transformation. Individual clones are then isolated by plating on an *E. coli* host strain and picking plaques. Recombinant phage can be selected by screening for interruption of the β-galactosidase gene using a color indicator agar (Section 11.7). Plaques containing phage that do not produce β-galactosidase can be readily detected as colorless plaques among a background of colored plaques (Figure 11.20b). Nucleic acid hybridization procedures (Section 11.2) and DNA sequencing can be used to determine if recombinant lambda DNA contains the desired foreign DNA sequence.

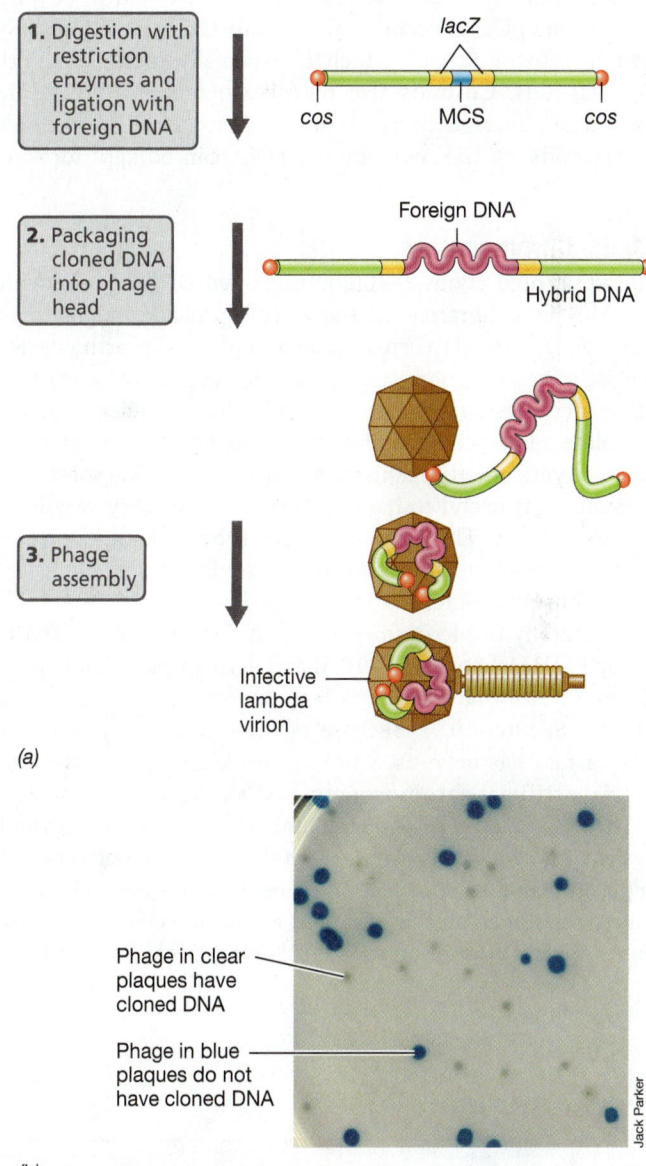

(a)

(b)

Figure 11.20 Bacteriophage lambda as a cloning vector. *(a)* Insertion of foreign DNA into lambda DNA modified to contain an MCS within a *lacZ* gene and subsequent packaging of a recombinant infective lambda virion. The maximum size of inserted DNA is about 20 kbp. *(b)* Portion of an Xgal-containing agar plate showing white plaques formed by lambda phage containing cloned DNA and blue plaques formed by phage lacking cloned DNA.

Cosmid Vectors

Cosmid vectors employ specific lambda genes and are packaged into lambda virions. *Cosmids* are plasmid vectors containing the *cos* site from the lambda genome, which yields cohesive ends when cut (⟲ Section 8.8). The *cos* site is required for packaging DNA into lambda virions. Cosmids are constructed from plasmids by ligating the lambda *cos* region to the plasmid DNA; foreign DNA is then ligated into the vector. The modified plasmid, plus cloned DNA, can then be packaged into lambda virions in vitro as described previously and the phage particles used to transduce *E. coli*.

One major advantage of cosmids is that they can be used to clone large fragments of DNA, with inserts as large as 50 kbp possible. With big inserts, fewer clones are needed to cover a whole genome. Using cosmids also avoids the necessity of having to transform *E. coli*, which is especially inefficient with larger plasmids. Cosmids also permit storage of the DNA in phage virions instead of as plasmids. Virions are more stable than plasmids, so the recombinant DNA can be kept for long periods of time.

Artificial Chromosomes

Vectors that hold about 2–10 kbp of cloned DNA are adequate for making gene libraries for sequencing prokaryotic genomes. Bacteriophage lambda vectors, which hold 20 kbp or more, are also widely used in genomics projects. However, as the size of the genome increases, so does the number of clones needed to obtain a complete sequence. Therefore, for making libraries of DNA from eukaryotic microorganisms or from higher eukaryotes such as humans, it is useful to have vectors that can carry very large segments of DNA. This allows the size of the initial library to be manageable. Such vectors have been developed and are called **artificial chromosomes**.

The naturally occurring F plasmid of *E. coli* is composed of 99.2 kbp of DNA (⟲ Section 10.8) and derivatives called F′ plasmids are known that may carry large amounts of chromosomal DNA (⟲ Section 10.9). Because of these desirable properties, the F plasmid has been used to engineer cloning vectors called **bacterial artificial chromosomes (BACs)**. **Figure 11.21** shows the structure of a 6.7-kbp BAC that contains only a few genes from F necessary for replication and to keep the copy number very low. The plasmid also contains the *cat* gene, which confers chloramphenicol resistance on the host, and a multiple cloning site that includes several restriction sites for cloning DNA. Foreign DNA

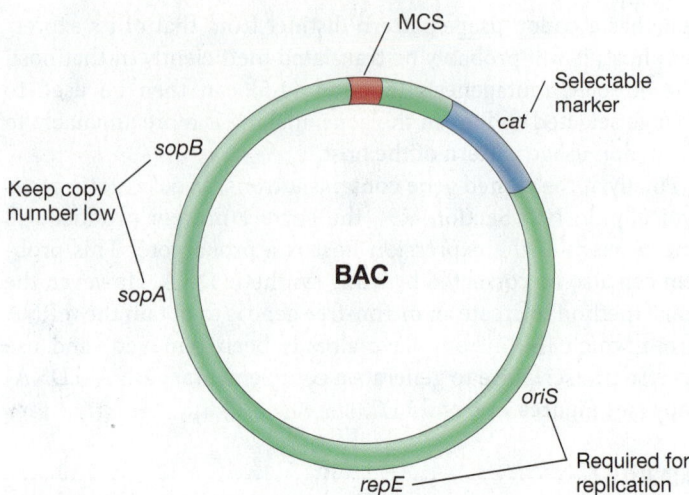

Figure 11.21 Genetic map of a bacterial artificial chromosome. The BAC shown is 6.7 kbp. The cloning region has several unique restriction enzyme sites. This BAC contains only a small fraction of the 99.2-kbp F plasmid.

of more than 300 kbp can be inserted and stably maintained in a BAC vector such as this.

Yeast artificial chromosomes (YACs) (**Figure 11.22**) are vectors that replicate in yeast like normal chromosomes, but have sites where very large fragments of DNA can be inserted. To function like normal eukaryotic chromosomes, YACs must have (1) an origin of DNA replication, (2) telomeres for replicating DNA at the ends of the chromosome, and (3) a centromere for segregation during mitosis. They must also contain a cloning site and a gene for selection following transformation into the host, which is typically the yeast *Saccharomyces cerevisiae*. Figure 11.22 shows a YAC vector into which foreign DNA has been cloned. Remarkably, YAC vectors are themselves only about 10 kbp in size but can have 200–800 kbp of cloned DNA inserted.

MINIQUIZ

- Why is the ability to package recombinant DNA in phage particles in a test tube useful?
- What do the acronyms BAC and YAC stand for?
- The yeast artificial chromosome behaves like a chromosome in a yeast cell. What makes this possible?

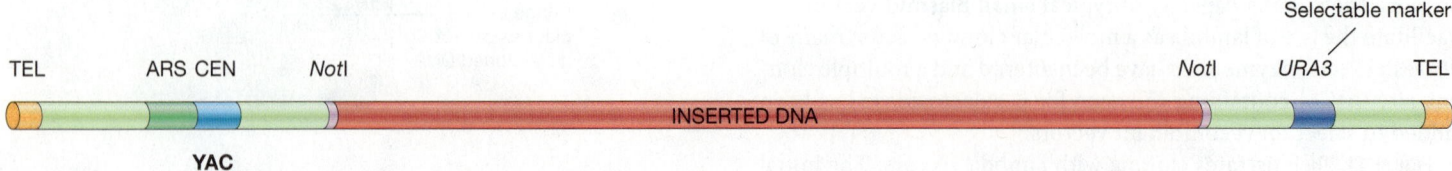

Figure 11.22 A yeast artificial chromosome containing foreign DNA. The foreign DNA was cloned into the vector at a *Not*I restriction site. The telomeres are labeled TEL and the centromere CEN. The origin of replication is labeled ARS (for autonomous replication sequence). The *URA3* gene is used for selection. The host into which the clone is transformed has a mutation in *URA3* and requires uracil for growth (Ura⁻). Host cells containing this YAC become Ura⁺. The diagram is not to scale; vector DNA is only 10 kbp whereas cloned DNA can be up to 800 kbp.

III • Products from Genetically Engineered Microorganisms

Genetic engineering has been used to transform microorganisms into tiny factories for the production of valuable products such as fuels, chemicals, drugs, and human hormones, such as insulin. Up to this point we have discussed the techniques used for manipulating, cloning, and expressing DNA. We now consider how these techniques can be directly applied to **biotechnology**, including some of the major challenges that exist with expressing eukaryotic genes in bacteria and purifying the subsequent protein products. We also discuss the genetic engineering of modified plants, animals, vaccines, and metabolic pathways.

11.11 Expressing Mammalian Genes in Bacteria

Some common problems of expression vectors were mentioned in Section 11.9, and there are other obstacles to cloning a mammalian gene in bacteria. The problems include the realities that (1) the eukaryotic genes must be placed under control of a bacterial promoter (Section 11.9); (2) any introns (⮌ Section 4.9) must be removed; (3) codon bias (⮌ Section 4.11) may require edits to gene sequences; and (4) many mammalian proteins require modification after translation to yield the active form, and bacteria cannot perform most such modifications. Here we consider solutions to these challenges.

Cloning the Gene via mRNA

The usual way to obtain an intron-free eukaryotic gene is to clone its messenger RNA (mRNA). Because introns are removed during the processing of mRNA, the mature mRNA carries an uninterrupted coding sequence (⮌ Section 4.9 and Figure 4.29). Tissues expressing the gene of interest often contain large amounts of the corresponding mRNA, although other mRNAs are also present. In certain situations, however, a single mRNA dominates in a tissue type, and extraction of bulk mRNA from that tissue provides a useful starting point for gene cloning.

In a typical mammalian cell, about 80–85% of the RNA is ribosomal RNA, 10–15% is transfer RNA, and only 1–5% is mRNA. However, eukaryotic mRNA is unique because of the poly(A) tails found at the 3′ end (⮌ Section 4.9), and this makes it easy to isolate, even though it is of low abundance. If a cell extract is passed over a chromatographic column containing strands of poly(T) linked to a cellulose support, most of the mRNA separates from other RNAs by sticking to the support by the specific pairing of A and T bases. The RNA is then released from the column by a low-salt buffer, which gives a preparation greatly enriched in mRNA.

Once mRNA has been isolated, the genetic information is converted into complementary DNA (cDNA) by RT-PCR as illustrated in Figure 11.6. This double-stranded cDNA contains the coding sequence but lacks introns (**Figure 11.23**), and thus it can be inserted into a plasmid or other vector for cloning. However, because the cDNA contains only the DNA sequences encoding the protein, it lacks a promoter and other upstream regulatory sequences necessary for expression. Special expression vectors

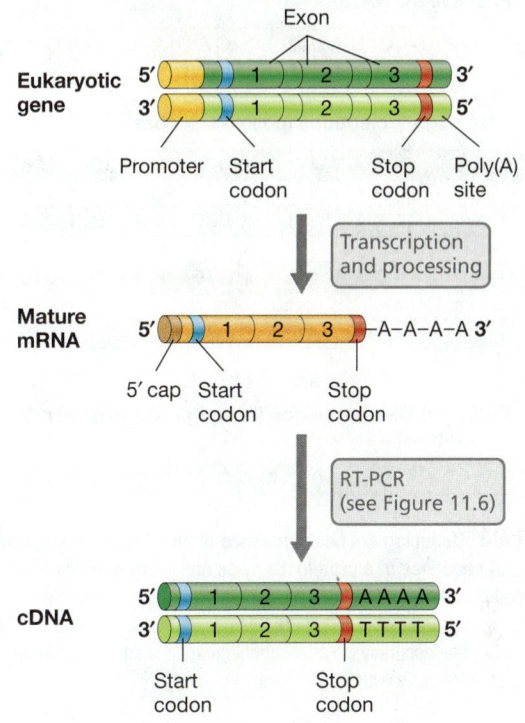

Figure 11.23 Complementary DNA (cDNA). Steps illustrating the synthesis of an intron-lacking cDNA corresponding to a eukaryotic gene using reverse transcription PCR (RT-PCR).

with bacterial promoters and ribosome-binding sites are used to obtain high-level expression of genes cloned in this way (⮌ Section 11.9).

Finding the Gene via the Protein

Knowing the sequence of an mRNA allows for the production of cDNA for cloning. In some cases, however, only the amino acid sequence of a desired protein is known. This amino acid sequence can be used to design and synthesize an oligonucleotide probe that encodes it. This process is illustrated in **Figure 11.24**. Unfortunately, degeneracy of the genetic code complicates this approach. Most amino acids are encoded by more than one codon (⮌ Table 4.5), and codon usage varies from organism to organism. Thus, the best region of a gene to synthesize as a probe is one that encodes part of the protein rich in amino acids specified by only a single codon (for example, methionine, AUG; tryptophan, UGG) or at most two codons (for example, phenylalanine, UUU, UUC; tyrosine, UAU, UAC; histidine, CAU, CAC). This strategy increases the chances that the probe will be nearly complementary to the mRNA or gene of interest. If the complete amino acid sequence of the protein is unknown, partial sequence data may be sufficient.

For very small proteins it may be feasible to artificially synthesize the entire gene (Section 11.5). Many mammalian proteins (including high-value peptide hormones) are made by protease cleavage of larger precursors. Thus, to produce a short peptide

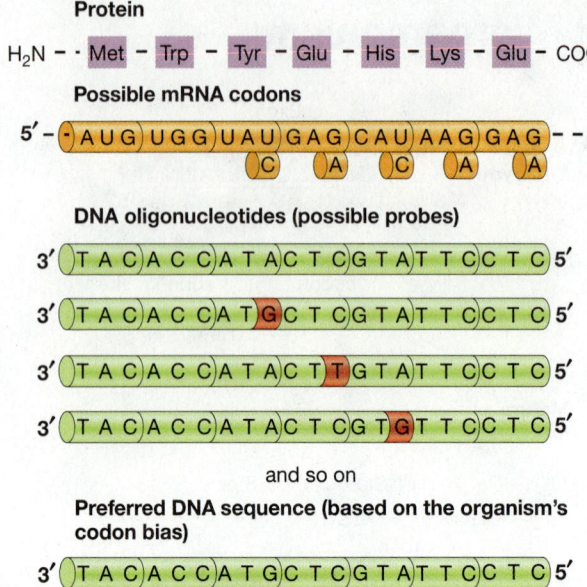

Figure 11.24 **Deducing the best sequence of an oligonucleotide probe from the amino acid sequence of a protein.** Because many amino acids are encoded by multiple codons, many nucleic acid probes are possible for a given polypeptide sequence. If codon usage by the target organism is known, a preferred sequence can be selected. Complete accuracy is not essential because a small amount of mismatch can be tolerated, especially with long probes.

hormone such as insulin, it may be more efficient to construct an artificial gene that encodes just the final hormone rather than the larger precursor protein from which it is derived naturally. Chemical synthesis also allows synthesis of modified genes that may encode useful new proteins. The synthetic approach was first used in a major way for production of the human hormone insulin in bacteria. Moreover, constructed genes are free of introns and thus the mRNA does not need processing. Also, promoters and other regulatory sequences can easily be built into the gene upstream of the coding sequences, and codon bias (⇄ Sections 4.11 and 6.3) can be accounted for.

With these techniques, many human and viral proteins have been expressed at high yield under the control of bacterial regulatory systems. These include insulin, somatostatin, viral capsid proteins, and interferon.

Protein Folding and Stability

The synthesis of a protein in a new host may spawn additional problems. For example, some proteins are susceptible to degradation by intracellular proteases and may be destroyed before they can be isolated. Moreover, some eukaryotic proteins are toxic to prokaryotic hosts, and the host cell may be killed before a sufficient amount of the product is synthesized. Further engineering of either the host or the vector may eliminate these problems.

Sometimes when foreign proteins are massively overproduced, they form inclusion bodies inside the host. Inclusion bodies consist of aggregated insoluble protein that is often misfolded or partly denatured, and they are often toxic to the host cell. Although inclusion bodies are relatively easy to purify because of

their size, the protein they contain is often difficult to solubilize and may be inactive. One possible solution to this problem is to use a host that overproduces molecular chaperones that aid in folding (⇄ Section 4.14).

Fusion Proteins for Improved Purification

Protein purification can often be made much simpler if the target protein is made as a *fusion protein* along with a carrier protein encoded by the vector. To do this, the two genes are fused to yield a single coding sequence. A short segment that is recognized and cleaved by a commercially available protease is included between them. After transcription and translation, a single protein is made. This is purified by methods designed for the carrier protein. The fusion protein is then cleaved by the protease to release the target protein from the carrier protein. Fusion proteins simplify purification of the target protein because the carrier protein can be chosen to have ideal properties for purification.

Several fusion vectors are available to generate fusion proteins. **Figure 11.25** shows an example of a fusion vector that is also an expression vector. In this example, the carrier protein is the *Escherichia coli* maltose-binding protein, and the fusion protein is easily purified by methods based on its affinity for maltose. Once purified, the two portions of the fusion protein are separated by a specific protease. In some cases the target protein is released from the carrier protein by specific chemical treatment, rather than by protease cleavage.

Fusion systems are also used for other purposes. One advantage of making a fusion protein is that the carrier protein can be

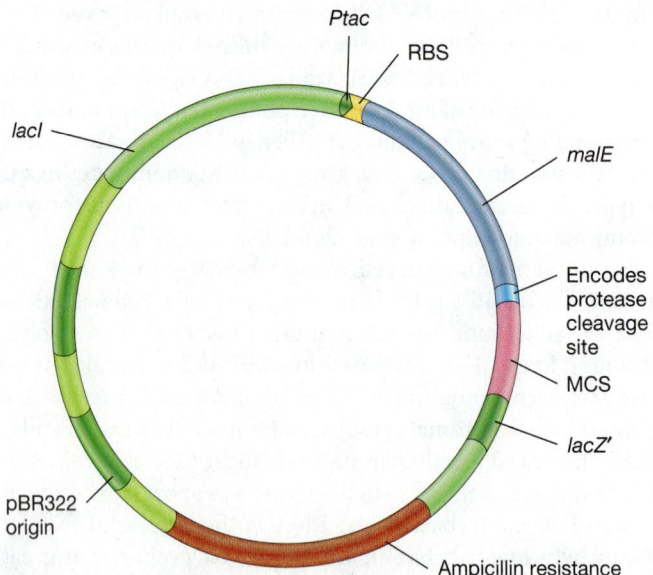

Figure 11.25 **An expression vector for fusions.** The gene to be cloned is inserted into the multiple cloning site (MCS) so it is in frame with the *malE* gene, which encodes maltose-binding protein. The insertion inactivates the gene for the alpha fragment of *lacZ*, which encodes β-galactosidase. The fused gene is under control of the hybrid *tac* promoter (Ptac) and an *Escherichia coli* ribosome-binding site (RBS). The plasmid also contains the *lacI* gene, which encodes the *lac* repressor. Therefore, an inducer must be added to turn on the *tac* promoter. The plasmid contains a gene conferring ampicillin resistance on its host. This vector was developed by New England Biolabs.

chosen to contain the bacterial *signal sequence*, a peptide rich in hydrophobic amino acids that enables transport of the protein across the cytoplasmic membrane (⟲ Section 4.14). This makes possible a bacterial expression system that not only makes mammalian proteins, but also secretes them. When the right strains and vectors are employed, the desired protein can constitute as much as 40% of the protein molecules in a cell.

MINIQUIZ

- What major advantage does cloning mammalian genes from mRNA or using synthetic genes have over PCR amplification and cloning of the native gene?
- How is a fusion protein made?

11.12 Somatotropin and Other Mammalian Proteins

One of the most economically profitable areas of biotechnology has been the production of human proteins. Many mammalian proteins have high pharmaceutical value but are typically present in very low amounts in normal tissue, and it is therefore extremely costly to purify them. Even if the protein can be produced in cell culture, this is much more expensive and difficult than growing microbial cultures that produce the protein in high yield. Therefore, the biotechnology industry has developed genetically engineered microorganisms to produce many different mammalian proteins.

Somatotropin

Although insulin was the first human protein to be produced by bacteria, the procedure had several unusual complications because insulin consists of two short polypeptides held together by disulfide bonds. A more typical example is *somatotropin* (growth hormone), and we focus on this here.

Somatotropin consists of a single polypeptide encoded by a single gene, and a shortage of somatotropin in the body results in hereditary dwarfism. Because the human somatotropin gene was successfully cloned and expressed in bacteria, children showing stunted growth can be treated with *recombinant human somatotropin*. However, dwarfism may also be caused by lack of the somatotropin receptor, and in such cases, administration of somatotropin has no effect.

The somatotropin gene was cloned as complementary DNA (cDNA) from mRNA as described in Section 11.11 (**Figure 11.26**). The cDNA was then expressed in a bacterial expression vector. The main problem with producing relatively short polypeptide hormones such as somatotropin is their susceptibility to protease digestion. This problem can be countered by using bacterial host strains defective for several proteases.

Recombinant bovine somatotropin (rBST) is used in the dairy industry (Figure 11.26). Injection of rBST into cows does not make them grow larger but instead stimulates milk production. The reason for this is that somatotropin has two binding sites. One binds to the somatotropin receptor and stimulates growth; the other binds to the prolactin receptor and promotes milk production. When human somatotropin is used to remedy

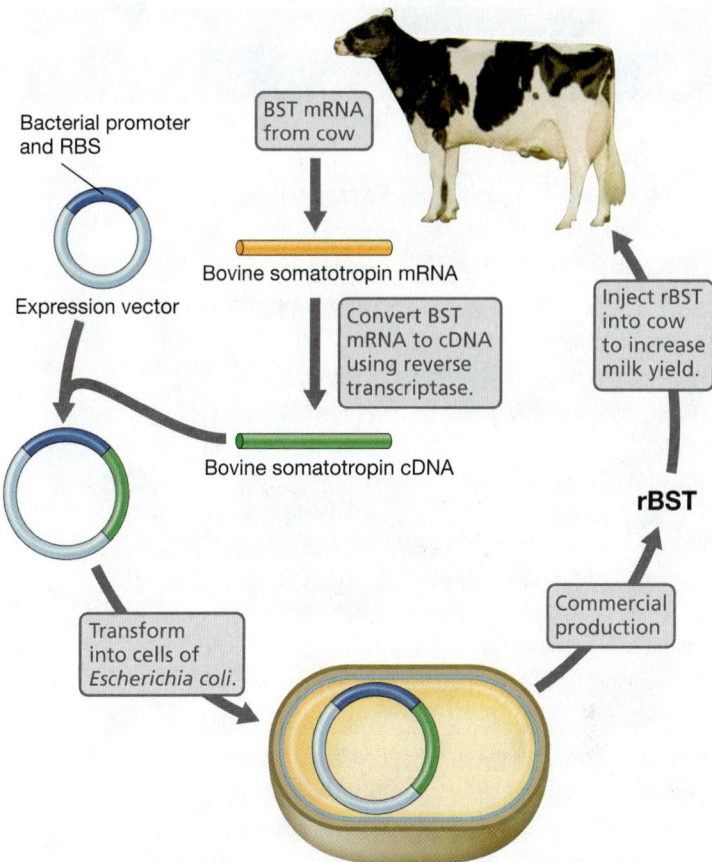

Figure 11.26 Cloning and expression of bovine somatotropin. The mRNA for bovine somatotropin (BST) is obtained from an animal. The mRNA is converted to cDNA by reverse transcriptase. The cDNA version of the somatotropin gene is then cloned into a bacterial expression vector that has a bacterial promoter and ribosome-binding site (RBS). The construct is transformed into cells of *Escherichia coli*, and recombinant bovine somatotropin (rBST) is produced. Milk production increases in cows treated with rBST.

human growth defects, it is desirable to avoid side effects from the hormone's prolactin activity (prolactin stimulates lactation). To alleviate this problem, site-directed mutagenesis (Section 11.5) of the human somatotropin gene was used to alter those amino acids of somatotropin that bind to the prolactin receptor. This shows that it is not only possible to make genuine human hormones, but also to alter their specificity and activity to make them better pharmaceuticals.

Other Mammalian Proteins

Many other mammalian proteins are produced by genetic engineering (**Table 11.2**). These include, in particular, an assortment of hormones and proteins for blood clotting and other blood processes. For example, *tissue plasminogen activator* (TPA) is a protein that dissolves blood clots in the bloodstream that may form in the final stages of the healing process. TPA is primarily used in heart patients or others suffering from poor circulation to prevent the development of clots that can be life-threatening. Heart disease is a leading cause of death in many developed countries, especially in the United States, so microbially produced TPA is in high demand.

Table 11.2 A few human medical products made by genetic engineering

Product	Function
Blood proteins	
Erythropoietin	Treats certain types of anemia
Factors VII, VIII, IX	Promotes blood clotting
Tissue plasminogen activator	Dissolves blood clots
Urokinase	Promotes blood clotting
Human hormones	
Epidermal growth factor	Wound healing
Follicle-stimulating hormone	Treatment of reproductive disorders
Insulin	Treatment of diabetes
Nerve growth factor	Treatment of degenerative neurological disorders and stroke
Relaxin	Facilitates childbirth
Somatotropin (growth hormone)	Treatment of some growth abnormalities
Immune modulators	
α-Interferon	Antiviral, antitumor agent
β-Interferon	Treatment of multiple sclerosis
Colony-stimulating factor	Treatment of infections and cancer
Interleukin-2	Treatment of certain cancers
Lysozyme	Anti-inflammatory
Tumor necrosis factor	Antitumor agent, potential treatment of arthritis
Replacement enzymes	
β-Glucocerebrosidase	Treatment of Gaucher disease, an inherited neurological disease
Therapeutic enzymes	
Human DNase I	Treatment of cystic fibrosis
Alginate lyase	Treatment of cystic fibrosis

In contrast to TPA, the blood clotting factors VII, VIII, and IX are critically important for the *formation* of blood clots. Hemophiliacs suffer from a deficiency of one or more clotting factors and can therefore be treated with microbially produced clotting factors. In the past hemophiliacs have been treated with clotting factor extracts from pooled human blood, some of which was contaminated with viruses such as HIV and hepatitis C, putting hemophiliacs at high risk for contracting these diseases. Recombinant clotting factors have eliminated this problem.

Some mammalian proteins made by genetic engineering are enzymes rather than hormones (Table 11.2). For instance, *human DNase I* is used to treat the buildup of DNA-containing mucus in the lungs of patients with cystic fibrosis. The mucus forms because cystic fibrosis is often accompanied by life-threatening lung infections by the bacterium *Pseudomonas aeruginosa*. The bacterial cells form biofilms (↺ Sections 7.9 and 19.4) within the lungs that make drug treatment difficult. DNA is released when the bacteria lyse, and this fuels mucus formation, making it harder to breathe. DNase digests the DNA and greatly decreases the viscosity of the mucus.

11.13 Transgenic Organisms in Agriculture and Aquaculture

Genetic improvement of plants and animals by traditional selection and breeding has a long history, but recombinant DNA technology has led to revolutionary changes. On the one hand, the genetic engineering of higher organisms is not truly microbiology. On the other hand, much of the DNA manipulation is carried out using bacteria and their plasmids and genes (see herbicide- and insect-resistant plants below) long before the engineered genes are finally inserted into the plant or animal. Therefore we consider the genetic manipulation of plants and animals here with a focus on the microbiology that supported it.

Because genetically engineered plants or animals contain a gene from another organism—called a *transgene*—they are **transgenic organisms**. The public knows these as **genetically modified organisms (GMOs)**. Strictly speaking, the term *genetically modified* refers to genetically engineered organisms whether or not they contain foreign DNA. In this section we discuss how foreign genes are inserted into plant and fish genomes and how transgenic organisms may be used.

The Ti Plasmid and Transgenic Plants

While recombinant DNA can be transformed into plant cells by electroporation or transfection (see Figure 11.28), the **Ti plasmid** from the gram-negative plant pathogen *Agrobacterium tumefaciens* can be used to transfer DNA directly into the cells of certain plants. This plasmid is responsible for *A. tumefaciens* virulence and encodes genes that mobilize DNA for transfer to the plant, which as a result contracts crown gall disease (↺ Section 22.4). The segment of the Ti plasmid DNA that is actually transferred to the plant is called **T-DNA**. The sequences at the ends of the T-DNA are essential for transfer, and the DNA to be transferred must be included between these ends.

One common Ti-vector system that has been used for the transfer of genes to plants is a two-plasmid system called a *binary vector*, which consists of a cloning vector plus a helper plasmid. The cloning vector contains the two ends of the T-DNA flanking a multiple cloning site, two origins of replication so that it can replicate in both *Escherichia coli* (the host for cloning) and *A. tumefaciens*, and two antibiotic resistance markers, one for selection in plants and the other for selection in bacteria. The foreign DNA is inserted into the vector, which is transformed into *E. coli* and then moved to *A. tumefaciens* by conjugation (**Figure 11.27**).

This cloning vector lacks the genes needed to transfer T-DNA to a plant. However, when placed in an *A. tumefaciens* cell that contains a suitable helper plasmid, the T-DNA can be transferred to a plant. The "disarmed" helper plasmid, called *D-Ti*, contains

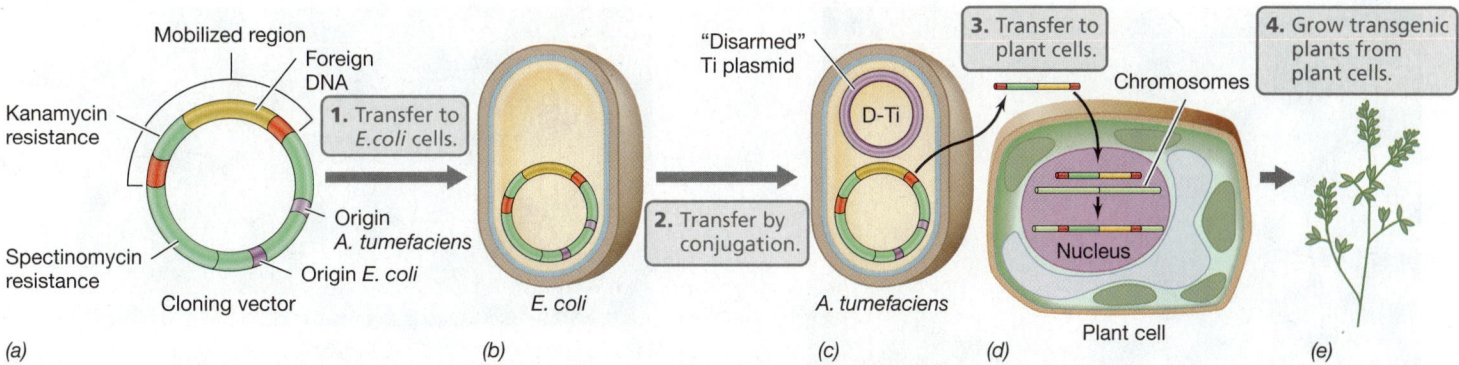

Figure 11.27 **Production of transgenic plants using a binary vector system in *Agrobacterium tumefaciens*.** *(a)* Plant cloning vector containing ends of T-DNA (red), foreign DNA, origins of replication, and resistance markers. *(b)* The vector is put into cells of *Escherichia coli* for cloning and then *(c)* transferred to *A. tumefaciens* by conjugation. The resident Ti plasmid (D-Ti) has been genetically engineered to remove key pathogenesis genes. *(d)* D-Ti can still mobilize the T-DNA region of the vector for transfer to plant cells grown in tissue culture. *(e)* From the recombinant plant cell, a whole plant can be grown. Details of Ti plasmid transfer from bacterium to plant are shown in Figure 22.21.

the virulence (*vir*) region of the Ti plasmid but lacks the T-DNA. It also lacks the genes that initiate disease but supplies all the functions needed to transfer the T-DNA from the cloning vector. The cloned DNA and the kanamycin resistance marker of the vector are mobilized by D-Ti and transferred into a plant cell where they enter the nucleus (Figure 11.27*d*). Following integration into a plant chromosome, the foreign DNA can be expressed and confer new properties on the plant.

A number of transgenic plants have been produced using the Ti plasmid of *A. tumefaciens*. The Ti system works well with broadleaf plants (dicots), including crops such as tomato, potato, tobacco, soybean, alfalfa, and cotton. It has also been used to produce transgenic trees, such as walnut and apple. The Ti system does not work with plants from the grass family (monocots, including the important crop plant, corn), but other methods of introducing DNA, such as transfection by microprojectile bombardment with a particle gun (**Figure 11.28**), have been used successfully for them.

Herbicide- and Insect-Resistant Plants

Major areas targeted for genetic improvement in plants include herbicide, insect, and microbial disease resistance, as well as improved product quality. The main genetically modified (GM) crops today are soybeans, corn, cotton, and canola. Almost all the GM soybeans and canola planted were herbicide resistant, whereas the corn and cotton were herbicide resistant or insect resistant, or both.

Herbicide resistance is genetically engineered into a crop plant to protect it from herbicides applied to kill weeds. Many herbicides inhibit a key plant enzyme or protein necessary for growth. For example, the herbicide *glyphosate* (Roundup™) kills plants by inhibiting an enzyme necessary for making aromatic amino acids. Some bacteria contain an equivalent enzyme and are also killed by glyphosate. However, mutant bacteria were selected that were resistant to glyphosate and contained a resistant form of the enzyme. The gene encoding this resistant enzyme from *A. tumefaciens* was cloned, modified for expression in plants, and transferred into important crop plants, such as soybeans. When

sprayed with glyphosate, plants containing the bacterial gene are not killed (**Figure 11.29**). Thus glyphosate can be used to kill weeds that compete for water and nutrients with the growing crop plants. Herbicide-resistant soybeans are now widely planted in the United States.

Insect Resistance: Bt Toxin

Transgenic plants resistant to damage by certain insects have been produced by genetic engineering (**Figure 11.30**). One widely used approach is based on introducing genes encoding the toxic proteins of the gram-positive bacterium *Bacillus thuringiensis* into plants. *B. thuringiensis* produces a crystalline

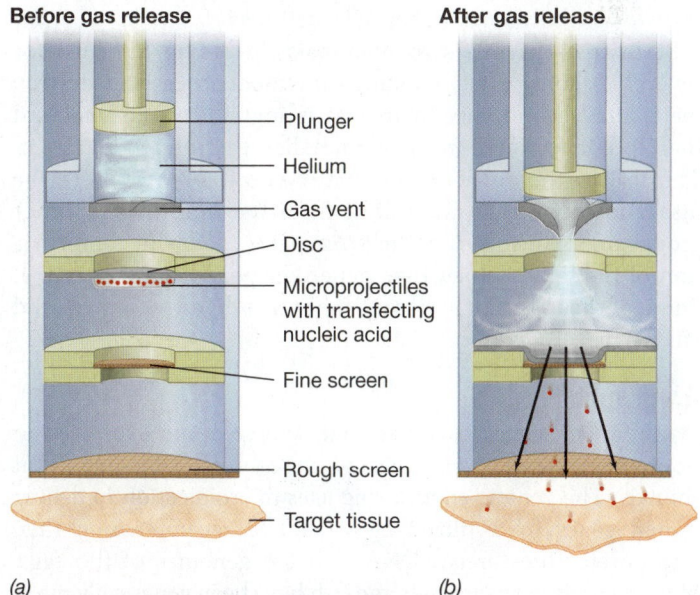

Figure 11.28 **DNA gun for transfection of eukaryotic cells.** The inner workings of the gun show how metal pellets coated with nucleic acids (microprojectiles) are projected at target cells. *(a)* Before firing and *(b)* after firing. A shock wave due to gas release throws the disc carrying the microprojectiles against the fine screen. The microprojectiles continue on into the target tissue.

Stephen R. Padgette, Monsanto Company

Figure 11.29 **Transgenic plants: herbicide resistance.** The photograph shows a portion of a field of soybeans that has been treated with Roundup™, a glyphosate-based herbicide manufactured by Monsanto Company (St. Louis, Missouri, USA). The plants on the right are normal soybeans; those on the left have been genetically engineered to be glyphosate resistant.

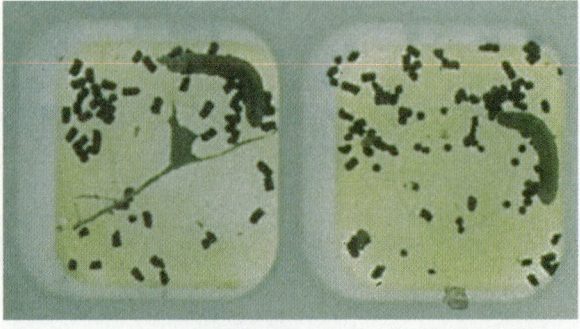

(a)

Kevin McBride, Calgene, Inc.

(b)

Figure 11.30 **Transgenic plants: insect resistance.** (a) The results of two different assays to determine the effect of beet armyworm larvae on tobacco leaves from normal plants. (b) The results of similar assays using tobacco leaves from transgenic plants that express Bt toxin in their chloroplasts.

protein called *Bt toxin* (↻ Section 15.8) that is toxic to moth and butterfly larvae. Many variants of Bt toxin exist that are specific for different insects. Certain strains of *B. thuringiensis* produce additional proteins toxic to beetle and fly larvae and mosquitoes.

The Bt transgene is normally inserted directly into the plant genome. For example, a natural Bt toxin gene was cloned into a plasmid vector under control of a chloroplast ribosomal RNA promoter and then transfected into tobacco plant chloroplasts by microprojectile bombardment (Figure 11.28). This yielded transgenic plants that expressed Bt toxin at levels that were extremely toxic to larvae from a number of insect species.

Bt toxin is harmless to mammals, including humans, for several reasons. First, cooking and food processing destroy the toxin. Second, any toxin that is ingested is digested and therefore inactivated in the mammalian gastrointestinal tract. Third, Bt toxin works by binding to specific receptors in the insect intestine that are absent from the intestines of other groups of organisms. Binding promotes a change in conformation of the toxin, which then generates pores in the intestinal lining of the insect that disrupt the insect digestive system and kill the insect.

Transgenic Fish

Many foreign genes have been incorporated and expressed in laboratory research animals and in commercially important animals. The genetic engineering uses microinjection to deliver cloned genes to fertilized eggs; genetic recombination then incorporates the foreign DNA into the genomes of the eggs. More recently, farm animals and fish have been genetically modified to improve yields.

An interesting practical example of a transgenic animal is the "fast-growing salmon" (**Figure 11.31**). These transgenic salmon do not grow to be larger than normal salmon but simply reach market size much faster. The gene for growth hormone in natural

salmon is activated by light. Consequently, salmon grow rapidly only during the summer months. In the genetically engineered salmon, the promoter for the growth hormone gene was replaced with the promoter from another fish that grows at a more or less constant rate all year round. The result was salmon that make growth hormone constantly and thus grow faster. Such salmon can be grown commercially in aquaculture operations and the transgenic fish harvested more quickly than with non-GMO farm-raised salmon.

Aqua Bounty Technologies

Figure 11.31 **Fast-growing transgenic salmon.** The *AquAdvantage*™ Salmon (top) was engineered by Aqua Bounty Technologies (St. Johns, Newfoundland, Canada). The transgenic and the control fish are 18 months old and weigh 4.5 kg and 1.2 kg, respectively.

MINIQUIZ

- What is a transgenic plant?
- Give an example of a genetically modified plant and describe how its modification benefits agriculture.
- How have transgenic salmon been engineered to reach market size faster?

11.14 Genetically Engineered Vaccines

Vaccines are substances that elicit immunity to a particular disease when injected into an animal (⇌ Section 24.6). Typically, vaccines are suspensions of killed or modified pathogenic microorganisms or viruses (or parts isolated from them). Often the part that elicits the immune response is a surface protein, for instance, a viral coat protein. Genetic engineering can be applied in many different ways to the production of vaccines.

Recombinant Vaccines

Genetic engineering can be used in vaccine development by modifying the pathogen itself. For instance, one can delete genes from a pathogenic bacterium or virus that encode virulence factors but leave those whose products elicit an immune response. This yields a recombinant and infective (but attenuated) vaccine. Conversely, one can add genes from a pathogenic virus to another, relatively harmless virus, called a *carrier virus*. Such vaccines are called **vector vaccines** and induce immunity to the pathogenic virus. Indeed, one can even combine the two approaches. For example, a recombinant vaccine is used to protect poultry against both fowlpox (a disease that reduces weight gain and egg production) and Newcastle disease (a viral disease that is often fatal). The fowlpox virus is a typical pox virus (⇌ Section 9.6) and was first modified by deleting virulence genes but not those that elicit immunity. Then immunity-inducing genes from the Newcastle virus were inserted into the modified fowlpox virus. This resulted in a **polyvalent vaccine**, a vaccine that immunizes against two different diseases at the same time.

Vaccinia virus (⇌ Section 9.6) is widely used to prepare recombinant vaccines for human use. Vaccinia virus itself is generally not pathogenic for humans and has been used for over 100 years as a vaccine against the related smallpox virus. However, cloning genes into vaccinia virus requires a selective marker, which is provided by the gene encoding the enzyme thymidine kinase. Vaccinia is unusual for a virus in carrying its own thymidine kinase, an enzyme that converts thymidine into thymidine triphosphate. However, this enzyme also converts the base analog 5-bromodeoxyuridine to a nucleotide that is incorporated into DNA, causing a lethal reaction. Therefore, cells that express thymidine kinase (whether from the host cell genome or from a virus genome) are killed by 5-bromodeoxyuridine.

Genes to be put into vaccinia virus are first inserted into an *Escherichia coli* plasmid that contains part of the vaccinia thymidine kinase (*tdk*) gene (**Figure 11.32**). The foreign DNA is inserted into the *tdk* gene, which is therefore disrupted. This recombinant plasmid is then transformed into animal cells whose own *tdk* genes have been inactivated. These cells are also infected with wild-type vaccinia virus. The two versions of the *tdk* gene—one on the plasmid and the other on the virus—then recombine. Some viruses gain a disrupted *tdk* gene plus its foreign insert (Figure 11.32). Cells infected by wild-type vaccinia virus (with an active thymidine kinase) are killed by 5-bromodeoxyuridine. By contrast, cells infected by *recombinant* vaccinia virus (with a disrupted *tdk* gene) grow long enough to yield a new generation of virions (Figure 11.32). The protocol thus selects for viruses whose *tdk* gene contains a cloned insert of foreign DNA.

Vaccinia virus does not actually need thymidine kinase to replicate. Consequently, recombinant vaccinia viruses can still infect human cells and express any foreign genes they carry. Vaccinia viruses can even be engineered to carry genes from multiple

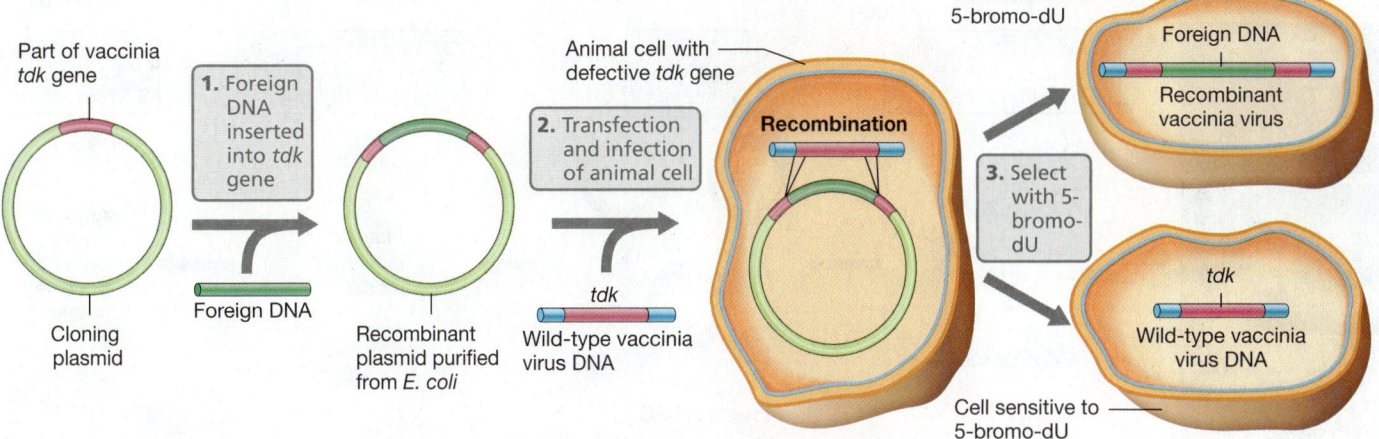

Figure 11.32 Production of recombinant vaccinia virus. Foreign DNA is inserted into a short segment of the thymidine kinase gene (*tdk*) from vaccinia virus carried on a plasmid. Following replication of this plasmid in *Escherichia coli*, both the recombinant plasmid and wild-type vaccinia virus are put into the same animal host cell to promote recombination. The animal cells are treated with 5-bromodeoxyuridine (5-bromo-dU), which kills only cells with an active thymidine kinase. Only recombinant vaccinia viruses whose *tdk* gene is inactivated by insertion of foreign DNA survive.

viruses. From an immunological standpoint, such viruses are *polyvalent vaccines*. Currently, several vaccinia vector vaccines have been developed and licensed for veterinary use, including one for rabies, while many other vaccinia vaccines are at the clinical trial stage. Vaccinia vaccines are relatively benign, yet highly immunogenic in humans, and because of this, their use will likely increase in the coming years.

Subunit Vaccines

Recombinant vaccines need not include every protein from the pathogenic organism. *Subunit vaccines* may contain only a specific protein or two from a pathogenic organism. For a pathogenic virus this is often the coat protein because coat proteins are typically highly immunogenic. The coat proteins are purified and used in high dosage to elicit a rapid and high level of immunity. Subunit vaccines are currently very popular because they can be used to produce large amounts of immunogenic proteins without the possibility that the purified products may contain the intact pathogen. Sometimes only certain portions of the viral protein are expressed rather than the entire protein, because immune cells and antibodies typically react with only small portions of the protein.

Subunit vaccines are sometimes poorly immunogenic and fail to protect in experimental tests of infection. When this occurs, it is usually because the recombinant proteins produced in bacteria are nonglycosylated, and glycosylation is necessary for the proteins to be immunologically active. Glycosylation occurs in the intact virus during the course of infection in the host animal. To solve this problem in a subunit vaccine, a eukaryotic cloning host may be necessary. For example, the first recombinant subunit vaccine approved for use in humans (against hepatitis B virus) was produced in yeast. The gene encoding a surface protein from hepatitis B virus was cloned and expressed in yeast. The protein produced was glycosylated and formed aggregates very similar to those found in patients infected with the virus. These aggregates were purified and used to effectively vaccinate humans against infection by hepatitis B virus.

11.15 Mining Genomes

Just as the total gene content of an organism is its *genome*, the collective genomes of an environment are its *metagenome* (⊘ Sections 6.10 and 18.7). Complex environments, such as fertile soil, contain vast numbers of uncultured bacteria and other microorganisms together with the viruses that prey on them (⊘ Section 6.10). Taken together, these contain correspondingly vast numbers of novel genes. Indeed, most of the genetic information on Earth exists in microorganisms and their viruses that have not been cultured. How can genetic engineering tap this resource?

Environmental Gene Mining

Gene mining is the process of isolating potentially useful novel genes from the environment without first culturing the organisms that carry them. In gene mining, DNA (or RNA) is isolated directly from environmental samples and cloned into suitable vectors to construct a metagenomic library (**Figure 11.33**). The nucleic acid includes genes from viable organisms as well as DNA from dead organisms that has been released but not yet been degraded. If RNA is isolated, it must be converted to a DNA copy by reverse transcriptase (Figure 11.6). However, isolating RNA is more time consuming and limits the metagenomic library to only those genes that have been transcribed in the environment sampled.

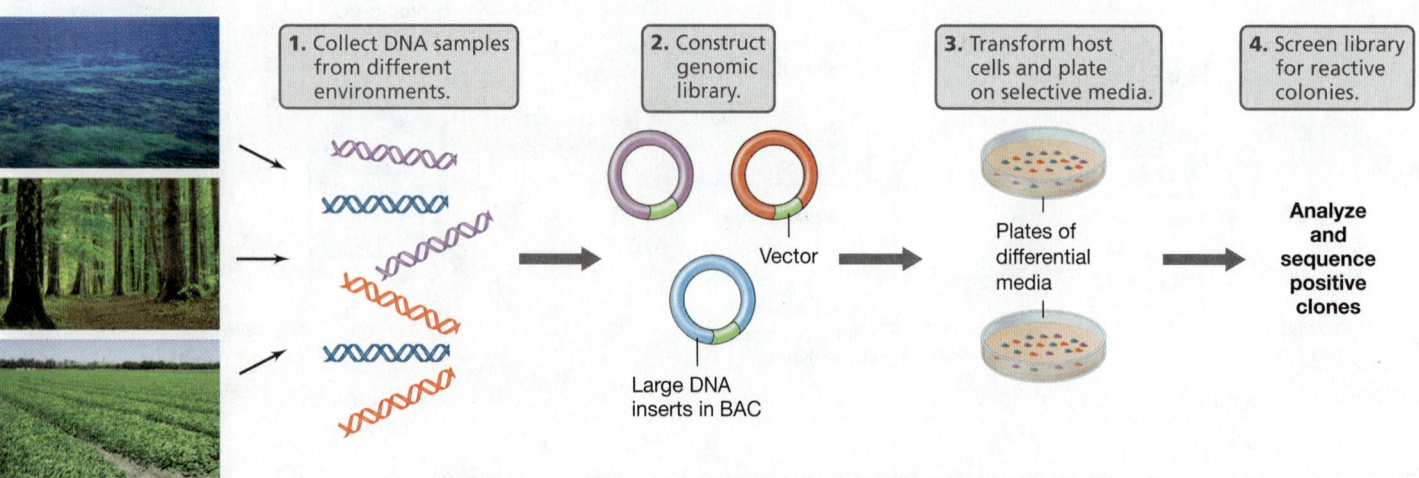

Figure 11.33 Metagenomic search for useful genes in the environment. DNA samples are obtained from different sites, such as seawater, forest soil, and agricultural soil. A clone library is constructed using bacterial artificial chromosomes (BACs) and screened for genes of interest. Possibly useful clones are analyzed further.

A metagenomic library is screened by the same techniques as any other clone library (Figure 11.8). Metagenomics has identified novel genes encoding enzymes that degrade pollutants and enzymes that make novel antibiotics. So far several lipases (see below), chitinases, esterases, and other degradative enzymes with novel substrate ranges and other properties have been isolated by this approach. Such enzymes frequently have industrial applications. Enzymes with improved resistance to industrial production conditions, such as high temperature, high or low pH, and oxidizing conditions, are especially valuable and desirable.

Retrieval of genes encoding entire metabolic pathways, such as for antibiotic synthesis, as opposed to single genes, requires vectors such as bacterial artificial chromosomes (BACs) that can carry large inserts of DNA (Section 11.10). BACs are especially useful for screening samples from rich environments, such as soil, where vast numbers of unknown genomes are present and correspondingly large numbers of genes are available to screen.

Targeted Gene Mining

Metagenomics can screen directly for enzymes with certain properties. Suppose one needed an enzyme or entire pathway capable of degrading a certain pollutant at a high temperature. The first step would be to find a hot environment polluted with the target compound. Assuming that microorganisms capable of degradation were present in the environment, a reasonable hypothesis, DNA from the environment would then be isolated and cloned. Host bacteria containing the clones would be screened for growth on the target compound. For convenience, this step is usually done in an *Escherichia coli* host, on the assumption that thermostable enzymes will still show some activity at 40°C (this is typically the case). Once suspects have been identified in the usual way, cell extracts can be tested in vitro at high temperatures for the enzyme of interest.

A gene mining strategy was used to isolate a thermostable lipase enzyme for commercial applications. Lipases catalyze the hydrolysis of triglycerides (fats), and because of this, they are sometimes incorporated into cosmetics and pharmaceuticals. But the industrial production of enzymes often requires them to retain activity at high temperatures. Hence, using DNA isolated from microorganisms in a hot spring, molecular biologists created a metagenomic library; the library was subsequently transformed into cells of *Escherichia coli*, and recombinant colonies expressing lipase activity were selected using a special medium. Analysis of enzyme extracts from these lipase-producing isolates indicated enzyme activity at temperatures up to 90°C. The gene encoding this thermostable lipase was then identified by analysis of recombinant vector DNA from the lipase-producing isolates, and subsequent cloning steps to produce an expression vector for commercial production were performed.

11.16 Engineering Metabolic Pathways

Although proteins are large molecules, expressing large amounts of a single protein that is encoded by a single gene is relatively simple. By contrast, small metabolites are typically made in biochemical pathways employing several enzymes. In these cases, not only are multiple genes needed, but their expression must be regulated in a coordinated manner as well.

Pathway engineering is the process of assembling a new or improved biochemical pathway using genes from one or more organisms. Most efforts so far have modified and improved existing pathways rather than creating entirely new ones. Because genetic engineering of bacteria is simpler than that of higher organisms, most pathway engineering has been done with bacteria. Engineered microorganisms are used to make products, including alcohols, solvents, food additives, dyes, and antibiotics. They may also be used to degrade agricultural waste, pollutants, herbicides, and other toxic or undesirable materials.

An example of pathway engineering is the production of indigo by *Escherichia coli* (Figure 11.34). Indigo is an important dye used for treating wool and cotton. Blue jeans, for example, are made of cotton dyed with indigo. In ancient times indigo and related dyes were extracted from sea snails. More recently, indigo was extracted from plants, but today it is synthesized chemically. However, the demand for indigo by the textile industry has spawned new approaches for its synthesis, including a biotechnological one.

Because the structure of indigo is very similar to that of the multi-ringed hydrocarbon naphthalene, enzymes that oxygenate

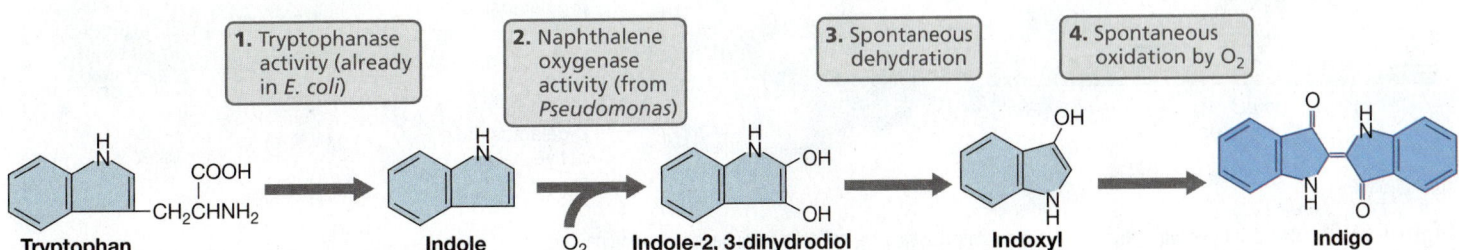

Figure 11.34 **Engineered pathway for production of the dye indigo.** *Escherichia coli* naturally expresses tryptophanase, which converts tryptophan into indole. Naphthalene oxygenase (originally from *Pseudomonas*) converts indole to dihydroxy-indole, which spontaneously dehydrates to indoxyl. Upon exposure to air, indoxyl dimerizes to form indigo, which is blue.

naphthalene also oxidize indole to its dihydroxy derivative; the latter oxidizes spontaneously in air to yield indigo, a bright blue pigment. Enzymes for oxygenating naphthalene are encoded by several plasmids found in *Pseudomonas* and other soil bacteria. When genes from such plasmids were cloned into *E. coli*, the cells turned blue due to production of indigo; the blue cells had incorporated the genes encoding the enzyme naphthalene oxygenase.

Although only the gene for naphthalene oxygenase was cloned during indigo pathway engineering, the indigo pathway consists of four steps, two enzymatic and two spontaneous (Figure 11.34). *E. coli* synthesizes the enzyme tryptophanase that carries out the first of these steps, the conversion of tryptophan to indole. For indigo production, tryptophan must be supplied to the recombinant *E. coli* cells. For commercial application this was accomplished by affixing the cells to a solid support in a bioreactor and trickling over the cells a tryptophan solution from waste protein sources. Recirculating the material over the cells several times, as is typically done in these types of immobilized cell industrial processes, steadily increases indigo levels until the dye can be harvested.

MINIQUIZ

- Explain why pathway engineering is more difficult than cloning and expressing a human hormone.
- How was *Escherichia coli* modified to produce indigo?

11.17 Synthetic Biology

Throughout this chapter we have discussed the use of genetic engineering to modify genes and organisms. But biology today can go much farther. The term *synthetic biology* refers to the use of genetic engineering to create novel biological systems out of available biological parts, often from several different organisms. These biological parts (promoters, enhancers, operators, riboswitches, regulatory proteins, enzyme domains, signal receivers, etc.) have been termed *biobricks*. Synthetic biology links these biobricks together in various combinations forming modules capable of generating complex behaviors.

One example of synthetic biology is the assembly of the world's first self-replicating synthetic bacterium by synthetic biologists at the J. Craig Venter Institute in California (USA). This feat was accomplished by artificially synthesizing a 1.08-million-base-pair (Mbp) genome based on the genome sequence of the bacterium *Mycoplasma mycoides* and then displacing the native chromosome of a *Mycoplasma capricolum* cell with the newly synthesized genome. When this was done, the *M. capricolum* cell showed all of the properties of the original *M. mycoides* cell.

A fascinating example of synthetic biology on a smaller scale is the use of genetically modified *Escherichia coli* to produce photographs. The engineered bacteria are grown as a lawn on agar plates. When an image is projected onto the lawn, bacteria in the dark make a dark pigment whereas bacteria in the light do not. The result is a primitive photograph of the projected image (**Figure 11.35**).

Construction of the photographic *E. coli* required the synthetic biology of three genetic modules: (1) a light detector and signaling module; (2) a pathway to convert heme (already present in *E. coli*) into the photoreceptor pigment phycocyanobilin; and (3) an enzyme encoded by a gene whose transcription can be switched on and off to make the dark pigment (Figure 11.35*a*). The light detector is a fusion protein. The outer half is the light-detecting part of the phytochrome protein from the cyanobacterium *Synechocystis*. This requires phycocyanobilin (an accessory light-harvesting pigment of cyanobacteria, ↩ Section 13.2), which is

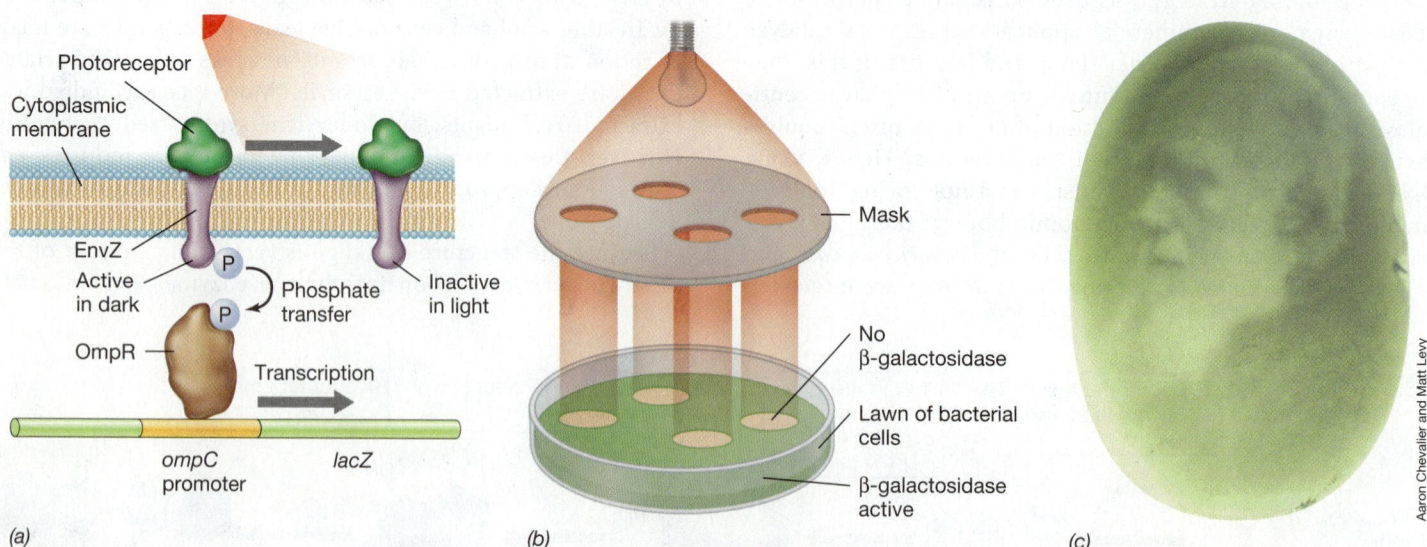

Figure 11.35 Bacterial photography. *(a)* Light-detecting *Escherichia coli* cells were genetically engineered using components from cyanobacteria and *E. coli* itself. Red light inhibits phosphate (P) transfer to the DNA-binding protein OmpR; phosphorylated OmpR is required to activate *lacZ* transcription (*lacZ* encodes β-galactosidase). *(b)* Setup for making a bacterial photograph. The opaque portions of the mask correspond to zones where β-galactosidase is active and thus to the dark regions of the final image. *(c)* A bacterial photograph of a portrait of Charles Darwin.

not made by *E. coli*, hence the need to install the pathway to make phycocyanobilin.

The inner half of the light detector is the signal transmission domain of the EnvZ sensor protein from *E. coli*. EnvZ is part of a two-component regulatory system, its partner being OmpR (↻ Section 7.7). Normally, EnvZ activates the DNA-binding protein OmpR. Activated OmpR in turn activates target genes by binding to the promoter. In the present case, the hybrid protein was designed to activate OmpR in the dark but not in the light. This is because phosphorylation of OmpR is required for activation, and red light converts the sensor to a state in which phosphorylation is inhibited. Consequently the target gene is off in the light and on in the dark. When a mask is placed over the Petri plate containing a lawn of the engineered *E. coli* cells (Figure 11.35*b*), cells in the dark make a pigment that cells in the light do not, and in this way a "photograph" of the masked image develops (Figure 11.35*c*).

The pigment made by the *E. coli* cells results from the activity of the lactose-degrading enzyme β-galactosidase, naturally present in *E. coli*. The target gene, *lacZ*, encodes this enzyme. In the dark, *lacZ* is expressed and β-galactosidase is made. The enzyme cleaves the lactose analog Xgal (Section 11.6) present in the growth medium to release galactose and a black dye. In the light, the *lacZ* gene is not expressed, no β-galactosidase is made, and so no dye is released. Contrast in the photograph is controlled by how much light cells see, which is governed by the nature of the mask that is used (Figure 11.35*c*).

Although the "synthetic" *M. capricolum* was not a cell in which all of its components—cytoplasm, membranes, ribosomes, and the like—were made from scratch, and cultures of *E. coli* will never replace digital photography, the knowledge gained in each case from assembling the necessary parts by synthetic biology helps build our understanding of how bioengineered components function in vivo. This in turn will allow for even more complex synthetic biology and may someday lead to applications of this science to solve urgent problems in medicine, agriculture, and the environment.

MINIQUIZ

- What are biobricks?
- How was *Escherichia coli* modified to produce a photograph?

BIG IDEAS

11.1 • Restriction enzymes recognize specific short sequences in DNA and make cuts in the DNA. The products of restriction enzyme digestion can be separated using gel electrophoresis.

11.2 • Complementary nucleic acid sequences may be detected by hybridization. Probes composed of single-stranded DNA or RNA and labeled with radioactivity or a fluorescent dye are hybridized to target DNA or RNA sequences.

11.3 • The polymerase chain reaction is a procedure for amplifying DNA in vitro and employs heat-stable DNA polymerases. Heat is used to denature the DNA into two single-stranded molecules, each of which is copied by the polymerase. After each cycle, the newly formed DNA is denatured and a new round of copying proceeds. After each cycle, the amount of target DNA doubles.

11.4 • The isolation of a specific gene or region of a chromosome by molecular cloning is done using a plasmid or virus as the cloning vector. Restriction enzymes and DNA ligase are used in vitro to produce a chimeric DNA molecule composed of DNA from two or more sources. Once introduced into a suitable host, the cloned DNA can be produced in large amounts under the control of the cloning vector. Identification of cloned genes is performed by a range of molecular techniques.

11.5 • Synthetic DNA molecules of desired sequence can be made in vitro and used to construct a mutated gene directly or to change specific base pairs within a gene by site-directed mutagenesis. Also, genes can be disrupted by inserting DNA fragments, called cassettes, into them, generating knockout mutants.

11.6 • Reporter genes are genes whose products, such as β-galactosidase or GFP, are easy to assay or detect. They are used to simplify and increase the speed of genetic analysis. In gene fusions, segments from two different genes, one of which is usually a reporter gene, are spliced together.

11.7 • Plasmids are useful cloning vectors because they are easy to isolate and purify and are often able to multiply to high copy numbers in bacterial cells. Antibiotic resistance genes on the plasmid are used to select bacterial cells containing the plasmid, and color-screening systems are used to identify colonies containing cloned DNA.

11.8 • The choice of a cloning host depends on the final application. In many cases the host can be a prokaryote, but in others, it is essential that the host be a eukaryote. Any host must be able to take up DNA, and there are a variety of techniques by which this can be accomplished, both natural and artificial.

11.9 • Many cloned genes are not expressed efficiently in a foreign host. Expression vectors have been developed for prokaryotic and eukaryotic hosts that contain genes or regulatory sequences that both increase transcription of the cloned gene and control the level of transcription. Signals to improve the efficiency of translation may also be present in the expression vector.

11.10 • Specialized cloning vectors such as bacteriophages, cosmids, and artificial chromosomes have been constructed for cloning very large fragments of DNA. Recombinant bacteriophages can be packaged in vitro for efficient transfer to a host cell, while cosmids are plasmid vectors that contain lambda *cos* sites. Artificial chromosomes can be used to clone DNA fragments approaching a megabase in size.

11.11 • To achieve very high levels of expression of eukaryotic genes in prokaryotes, the expressed gene must be free of introns. This can be accomplished by synthesizing cDNA from the mature mRNA encoding the protein of interest or by making an entirely synthetic gene. Protein fusions are often used to stabilize or solubilize the cloned protein.

11.12 • The first human protein made commercially using engineered bacteria was human insulin. Recombinant bovine somatotropin is widely used in the United States to increase milk yield in dairy cows.

11.13 • Genetic engineering can make plants resistant to disease and improve product quality. The Ti plasmid of the bacterium *Agrobacterium tumefaciens* can transfer DNA into plant cells. Genetically engineered commercial plants are called genetically modified organisms (GMOs).

11.14 • Many recombinant vaccines have been produced or are under development. These include live recombinant, vector, and subunit vaccines.

11.15 • Genes for useful products may be cloned directly from DNA or RNA in environmental samples without first isolating the organisms that carry them.

11.16 • In pathway engineering, genes that encode the enzymes for a metabolic pathway are assembled. These genes may come from one or more organisms, but the engineering must achieve regulation of the coordinated sequence of expression required in the pathway.

11.17 • Instead of modifying or improving a single existing pathway, synthetic biology focuses on engineering novel biological systems by linking known biological components together in various combinations.

MasteringMicrobiology® **Review what you know and challenge what you have learned with MasteringMicrobiology!** Access study materials, chapter quizzes, animations, and microbiology lab tutorials in the Study Area to ensure that you have mastered this chapter's content.

REVIEW OF KEY TERMS

Artificial chromosome a single copy vector that can carry extremely long inserts of DNA and is widely used for cloning segments of large genomes

Bacterial artificial chromosome (BAC) a circular artificial chromosome with bacterial origin of replication

Biotechnology the use of organisms, typically genetically altered, in industrial, medical, or agricultural applications

Cassette mutagenesis creating mutations by the insertion of a DNA cassette

DNA cassette an artificially designed segment of DNA that usually carries a gene for resistance to an antibiotic or some other convenient marker and is flanked by convenient restriction sites

Expression vector a cloning vector that contains the necessary regulatory sequences to allow transcription and translation of cloned genes

Gel electrophoresis a technique for separation of nucleic acid molecules by passing an electric current through a gel made of agarose or polyacrylamide

Gene disruption (also called gene knockout) the inactivation of a gene by insertion of a DNA fragment that interrupts the coding sequence

Gene fusion a structure created by joining together segments of two separate genes, in particular when the regulatory region of one gene is joined to the coding region of a reporter gene

Genetically modified organism (GMO) an organism whose genome has been altered using genetic engineering; the abbreviation GM is also used in terms such as GM crops and GM foods

Genetic engineering the use of in vitro techniques in the isolation, alteration, and expression of DNA or RNA and in the development of genetically modified organisms

Genomic library (DNA library) a collection of cloned DNA segments that is big enough to contain at least one copy of every gene from a particular organism

Green fluorescent protein (GFP) a protein that glows green and is widely used in genetic analysis

Hybridization the formation of a double helix by the base pairing of single strands of DNA or RNA from two different (but related) sources

Modification enzyme an enzyme that chemically alters bases within a restriction enzyme recognition site and thus prevents the site from being cut

Molecular cloning the isolation and incorporation of a fragment of DNA into a vector where it can be replicated

Northern blot a hybridization procedure where RNA is the target and DNA or RNA is the probe

Nucleic acid probe a strand of nucleic acid that can be labeled and used to hybridize to a complementary molecule from a mixture of other nucleic acids

Operon fusion a gene fusion in which a coding sequence that retains its own translational signals is fused to the transcriptional signals of another gene

Pathway engineering the assembly of a new or improved biochemical pathway using genes from one or more organisms

Polymerase chain reaction (PCR) artificial amplification of a DNA sequence by repeated cycles of strand separation and replication

Polyvalent vaccine a vaccine that immunizes against more than one disease

Protein fusion a gene fusion in which two coding sequences are fused so that they share the same transcriptional and translational start sites

Recombinant DNA a DNA molecule containing DNA originating from two or more sources

Reporter gene a gene used in genetic analysis because the product it encodes is easy to detect

Restriction enzyme an enzyme that recognizes a specific DNA sequence and then cuts the DNA; also known as a restriction endonuclease

Reverse transcription the conversion of an RNA sequence into the corresponding DNA sequence

Shotgun cloning making a genomic library by random cloning of DNA fragments

Shuttle vector a cloning vector that can replicate in two or more dissimilar hosts

Site-directed mutagenesis construction in vitro of a gene with a specific mutation

Southern blot a hybridization procedure where DNA is the target and RNA or DNA is the probe

T-DNA the segment of the *Agrobacterium tumefaciens* Ti plasmid that is transferred into plant cells

Ti plasmid a plasmid in *Agrobacterium tumefaciens* capable of transferring genes from bacteria to plants

Transgenic organism a plant or an animal with foreign DNA inserted into its genome

Vector (as in cloning vector) a self-replicating DNA molecule that is used to carry cloned genes or other DNA segments for genetic engineering

Vector vaccine a vaccine made by inserting genes from a pathogenic virus into a relatively harmless carrier virus

Yeast artificial chromosome (YAC) an artificial chromosome with yeast origin of replication and a centromere sequence

REVIEW QUESTIONS

1. What are restriction enzymes? Why does the presence of a restriction enzyme in a cell not cause the degradation of that cell's DNA? (Section 11.1)

2. How could you detect a colony containing a cloned gene if you already knew the sequence of the gene? (Section 11.2)

3. Describe the basic principles of gene amplification using the polymerase chain reaction (PCR). How have thermophilic and hyperthermophilic prokaryotes simplified the use of PCR? (Section 11.3)

4. Genetic engineering depends on vectors. Describe the properties needed in a well-designed plasmid cloning vector. (Section 11.4)

5. How could you detect a colony containing a cloned gene if you did not know the gene sequence but had available purified protein encoded by the gene? (Section 11.4)

6. What are the major uses for artificially synthesized DNA? (Section 11.5)

7. What does site-directed mutagenesis allow you to do that normal mutagenesis does not? (Section 11.5)

8. What is a reporter gene? Describe two widely used reporter genes. (Section 11.6)

9. How are gene fusions used to investigate gene regulation? (Section 11.6)

10. How does the insertional inactivation of β-galactosidase allow the presence of foreign DNA in a plasmid vector such as pUC19 to be detected? (Section 11.7)

11. Describe two prokaryotic cloning hosts and the beneficial and detrimental features of each. (Section 11.8)

12. Describe the similarities and differences between expression vectors and shuttle vectors. (Section 11.9)

13. How has bacteriophage T7 been used in expressing foreign genes in *Escherichia coli,* and what desirable features does this regulatory system possess? (Section 11.9)

14. What advantages are there to using a lambda-based cloning vector rather than a plasmid vector? (Section 11.10)

15. What are the essential characteristics of an artificial chromosome? What is the difference between a BAC and a YAC? (Section 11.10)

16. What is the significance of reverse transcriptase in the cloning of animal genes for expression in bacteria? (Section 11.11)

17. What classes of mammalian proteins are produced by biotechnology? How are the genes for such proteins obtained? (Section 11.12)

18. What is the Ti plasmid and how has it been of use in genetic engineering? (Section 11.13)

19. What is a subunit vaccine and why are subunit vaccines considered a safer way of conferring immunity to viral pathogens than attenuated virus vaccines? (Section 11.14)

20. How has metagenomics been used to find novel useful products? (Section 11.15)

21. What is pathway engineering? Why is it more difficult to produce an antibiotic than to produce a single enzyme via genetic engineering? (Section 11.16)

22. How does synthetic biology differ from pathway engineering? (Section 11.17)

APPLICATION QUESTIONS

1. Suppose you are given the task of constructing a plasmid expression vector suitable for molecular cloning in an organism of industrial interest. List the characteristics such a plasmid should have. List the steps you would use to create such a plasmid.

2. Suppose you have just determined the DNA base sequence for an especially strong promoter in *Escherichia coli* and you are interested in incorporating this sequence into an expression vector. Describe the steps you would use. What precautions are necessary to be sure that this promoter actually works as expected in its new location?

3. Many genetic systems use the *lacZ* gene encoding β-galactosidase as a reporter. What advantages or problems would there be if (a) luciferase or (b) green fluorescent protein were used instead of β-galactosidase as reporters?

4. You have just discovered a protein in mice that may be an effective cure for cancer, but it is present only in tiny amounts. Describe the steps you would use to produce this protein in therapeutic amounts. Which host would you want to clone the gene into and why? Which host would you use to express the protein in and why?

CHAPTER

12 · Microbial Evolution and Systematics

microbiology**now**

Gene Exchange and the Evolution of Marine *Vibrio*

Vibrio cholerae is a bacterium present in costal marine habitats around the world, and some strains cause cholera, a devastating diarrheal disease. Cholera is caused by strains of *V. cholerae* that have acquired genes encoding cholera toxin through horizontal gene transfer. Horizontal gene transfer has powerful effects on microbial evolution. In fact, many traits of bacteria, including their pathogenicity, can be altered by the acquisition of genes from other species.

Microbiologists are still struggling to understand the effects of horizontal gene transfer on microbial evolution. In *V. cholerae,* patterns of gene transfer vary with respect to the core genome and the dispensable genome. Housekeeping genes, which carry out important cellular processes and are a component of the core genome, are exchanged among *strains* of *V. cholerae* but are not exchanged between species. In contrast, integron genes, a component of the dispensable genome, can be exchanged across *species* boundaries and are readily exchanged between the species *V. cholerae* and *V. metecus* (photo inset).

It turns out that the exchange of integron genes is governed more by geographic co-occurrence than by species boundaries.[1] For example, *V. cholerae* strains isolated from a salt pond (photo) in Falmouth, Massachusetts, share more integron genes with *Vibrio metecus* strains isolated from the same pond than with other *V. cholerae* strains isolated from a geographically remote location in Bangladesh.

When it comes to gene exchange in the dispensable genome it may be all about (as a real estate agent would put it) "location, location, location." Integron genes can alter certain aspects of metabolism and cell surface characteristics. It is likely that these can help all *Vibrio* species adapt to their local environments and can impact the evolution of pathogenicity.

[1]Boucher, Y., et al. 2011. Local mobile gene pools rapidly cross species boundaries to create endemicity within global *Vibrio cholerae* populations. *mBio*. 2:e00335–10.

Evolution is the unifying theme that underlies all of biology. Charles Darwin was the first to observe that living things change over time, and he proposed that this process of evolution resulted from natural selection acting on random variations between offspring. Today we know that these random variations occur as a result of mutation and recombination in DNA sequence. The history of evolution is written in our genetic code and DNA sequences provide a record of evolution that spans billions of years. We now know that microorganisms have dominated much of the history of life on Earth. Microorganisms were teeming in the seas for billions of years before the appearance of the first plants and animals, and their activities have shaped our biosphere.

This chapter focuses on the evolution of microbial life. We will explore the evolutionary history of life and the manner in which DNA sequences can be used to classify microorganisms and to discern their evolutionary relationships. Overall, the goal of this chapter is to provide an evolutionary and systematic framework with which to comprehend the diversity of microbial life that we will explore in the next four chapters.

I • Early Earth and the Origin and Diversification of Life

In these first sections, we consider the possible conditions under which life arose, the earliest evidence for cellular life, and its divergence into three evolutionary lineages, *Bacteria, Archaea,* and *Eukarya.* Although much about these events and processes remains speculative, geological and molecular evidence has combined to build a plausible scenario for the earliest events in the evolution of life and for the fundamental impacts that microbes have had on the history of our Earth.

12.1 Formation and Early History of Earth

The Earth of 4 billion years ago would be foreign and inhospitable to human eyes, but this sterile wasteland of blasted rock and boiling seas was the incubator from which all life sprang. The story of life begins not long after the dawn of our solar system with the formation of Earth itself.

Origin of Earth

Earth formed about 4.5 billion years ago, based on analyses of slowly decaying radioactive isotopes (**Figure 12.1**). Our planet and the other planets of our solar system arose from materials making up a disc-shaped nebular cloud of dust and gases released by the supernova of a massive old star. As a new star—our sun—formed within this cloud, it began to compact, undergo nuclear fusion, and release large amounts of energy in the form of heat and light. Materials left in the nebular cloud began to clump and fuse due to collisions and gravitational attractions, forming tiny accretions that gradually grew larger to form clumps that eventually coalesced into planets. Energy released in this process heated the emerging Earth as it formed, as did energy released by radioactive decay within the condensing materials, forming a planet Earth of fiery hot magma. As Earth cooled over time, a metallic core, rocky mantle, and a lower-density, thin surface crust formed.

The inhospitable conditions of early Earth, characterized by a molten surface under intense bombardment by asteroids and other objects from space, are thought to have persisted for over 500 million years. Water on Earth originated from innumerable collisions with icy comets and asteroids and from volcanic outgassing of the planet's interior. Given Earth's heat at the time, water would have been present only as water vapor. No rocks dating to the origin of planet Earth have yet been discovered, presumably because they have undergone geological metamorphosis. Ancient crystals of the

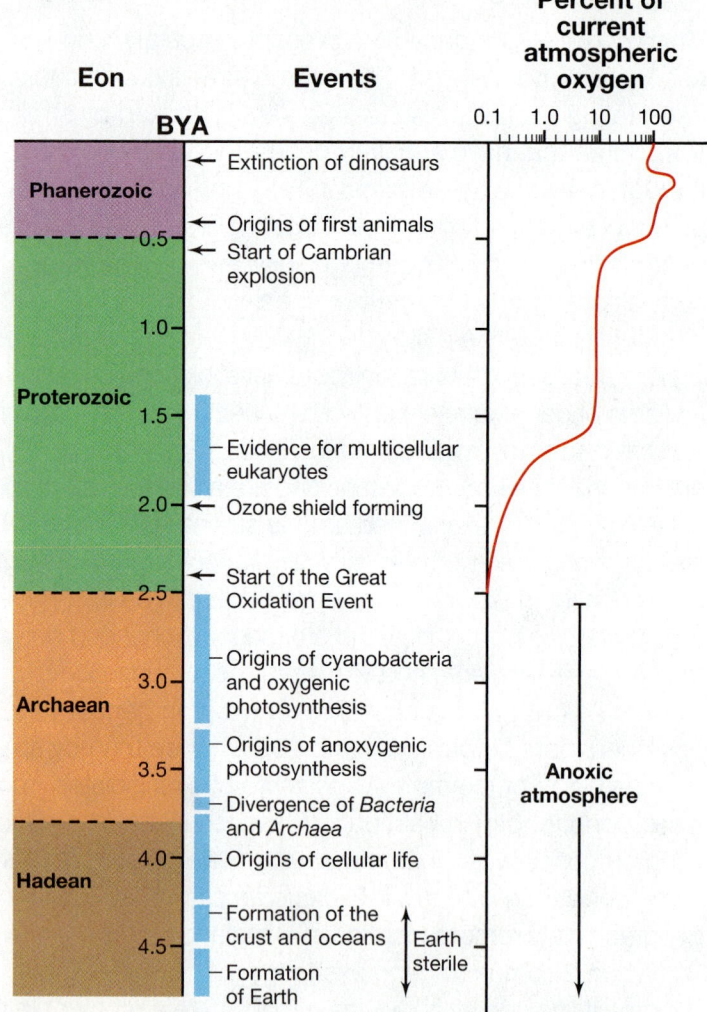

Figure 12.1 Major landmarks in biological evolution, Earth's changing geochemistry, and microbial metabolic diversification. The oldest date for the origin of life is fixed by the time of Earth's origin, and the minimum time for the origin of oxygenic photosynthesis is fixed by the Great Oxidation Event, about 2.4 billion years ago (BYA). Note how the oxygenation of the atmosphere from cyanobacterial metabolism was a gradual process, occurring over a period of about 2 billion years. Compare this figure with the introduction to the antiquity of life on Earth shown in Figure 1.4.

mineral zircon ($ZrSiO_4$) have been discovered, however, and these materials give us a glimpse of conditions on Earth at this time. Impurities trapped in the crystals and isotopic ratios of oxygen in the minerals (↩ Section 18.9) indicate that solid crust began forming and water began condensing into oceans perhaps as early as 4.3 billion years ago. The presence of liquid water implies that conditions could have been compatible with life within a couple of hundred million years after Earth was formed.

Some of the oldest sedimentary rocks discovered thus far are in southwestern Greenland; these rocks date to about 3.86 billion years ago. The sedimentary composition of these rocks indicates that oceans were present at this time. The fossilized remains of what appear to be cells (**Figure 12.2**) and the isotopically "light" carbon abundant in these rocks provide the earliest evidence for microbial life (we discuss the use of isotopic analyses of carbon and sulfur as indications of living processes in Section 18.9).

Origin of Cellular Life

The origin of life on Earth remains the greatest of mysteries, obscured by the depths of time. There are few rocks that survive unaltered to testify about this period of Earth's history. Experimental evidence indicates that organic precursors to living cells can form spontaneously under certain conditions, providing the preconditions needed for the first living systems. However, conditions on Earth's surface more than 4 billion years ago, in particular the extremely hot temperatures and levels of ultraviolet radiation, were likely hostile to the formation of life as we know it. One hypothesis holds that life may have originated well below Earth's surface at hydrothermal springs on the ocean floor (**Figure 12.3**). Deep on the ocean floor, conditions would have been less hostile and more stable than on Earth's surface. A steady and abundant supply of energy in the form of reduced inorganic compounds— hydrogen (H_2) and hydrogen sulfide (H_2S), for example—would have been available at these hydrothermal springs. The unique geochemistry of these sites may have allowed for the formation of molecules critical for the emergence of life and the formation of compartmentalized structures necessary for conserving energy.

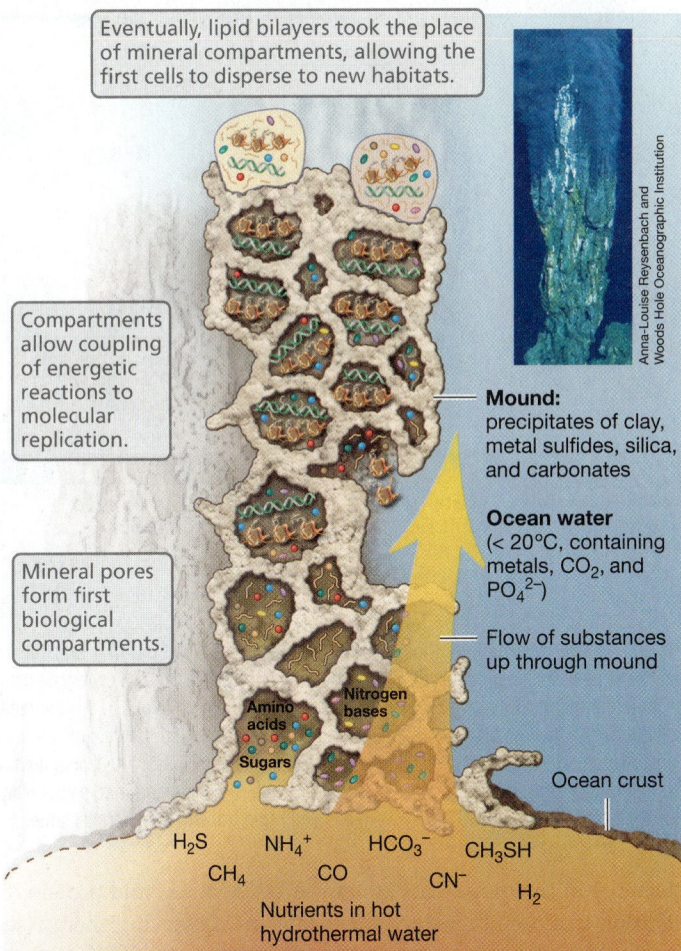

Figure 12.3 Submarine mounds and their possible link to the origin of life. Model of the interior of a hydrothermal mound with hypothesized transitions from prebiotic chemistry to cellular life depicted. Inset: photo of an actual hydrothermal mound. Hot mineral-rich hydrothermal fluid mixes with cooler, more oxidized, ocean water, forming precipitates of Fe and S compounds, clays, silicates, and carbonates. Mineral precipitates form pores that could have served as energy-rich compartments that facilitated the evolution of precellular forms of life.

Labels within figure: Eventually, lipid bilayers took the place of mineral compartments, allowing the first cells to disperse to new habitats. / Compartments allow coupling of energetic reactions to molecular replication. / Mineral pores form first biological compartments. / **Mound:** precipitates of clay, metal sulfides, silica, and carbonates / **Ocean water** (< 20°C, containing metals, CO_2, and PO_4^{2-}) / Flow of substances up through mound / Amino acids / Nitrogen bases / Sugars / Ocean crust / H_2S NH_4^+ HCO_3^- CH_3SH CH_4 CO CN^- H_2 / Nutrients in hot hydrothermal water / Anna-Louise Reysenbach and Woods Hole Oceanographic Institution

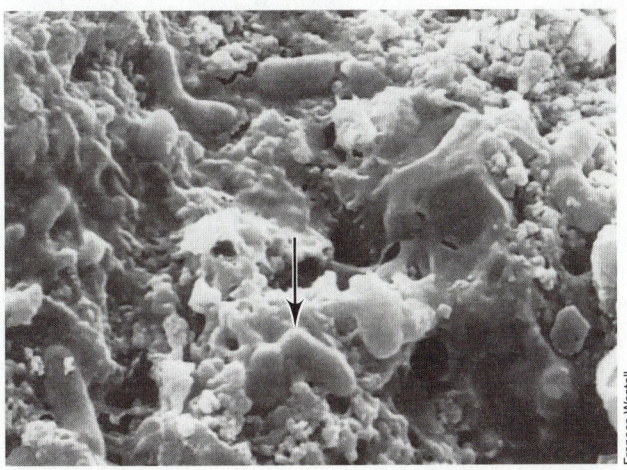

Figure 12.2 Ancient microbial life. Scanning electron micrograph of microfossil bacteria from 3.45-billion-year-old rocks of the Barberton Greenstone Belt, South Africa. Note the rod-shaped bacteria (arrow) attached to particles of mineral matter. The cells are about 0.7 μm in diameter. — Frances Westall

Whether on the seafloor or elsewhere, some form of prebiotic chemistry must have facilitated the development of the first self-replicating systems, the precursors to cellular life.

Molecules of RNA were likely a central component of the first self-replicating systems and it is possible that life began in an *RNA world* (**Figure 12.4**). RNA is a component of certain essential cofactors and molecules found in all cells (such as ATP, NADH, and coenzyme A); it can bind small molecules (such as ATP, amino acids, and other nucleotides); and it can have catalytic activity, as RNA is known to catalyze protein synthesis through the activities of rRNA, tRNA, and mRNA (↩ Section 4.13). It is possible that certain RNA molecules might once have had the ability to catalyze their own synthesis. Later, as different types of proteins emerged, some with catalytic abilities, proteins began to take over the catalytic role of RNAs. Eventually, DNA, a molecule that is inherently more stable than RNA and therefore a better repository of genetic (coding) information, arose and assumed the template role for RNA synthesis (Figure 12.4). The earliest cellular forms of life likely possessed

UNIT 3

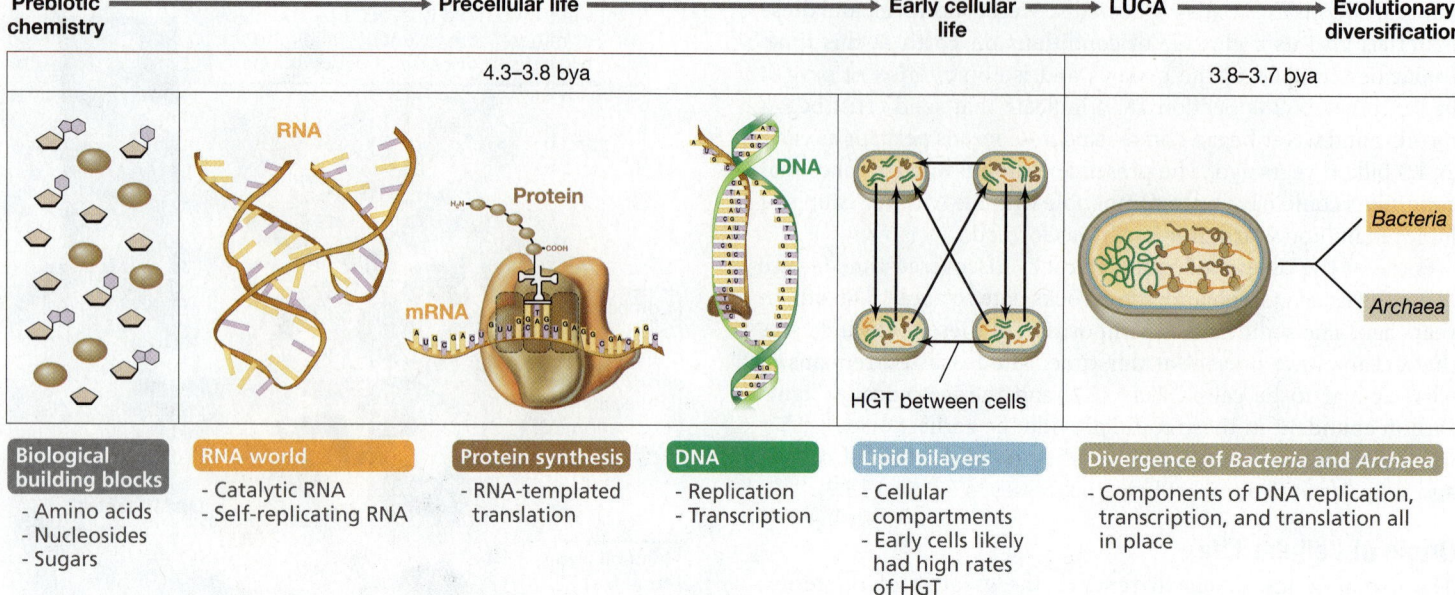

Figure 12.4 Events hypothesized to precede the origin of cellular life. The earliest self-replicating biological systems may have been based on catalytic RNA. At some point RNA enzymes evolved the capability to synthesize proteins, and proteins became the main catalytic molecules. Conversion from RNA- to DNA-based genomes required the evolution of DNA and RNA polymerases. The lipid bilayer is the site of electron transport and the evolution of this structure was likely important for energy conservation, in addition to containing and protecting biomolecules. The last universal common ancestor (LUCA), which preceded the divergence of *Bacteria* and *Archaea*, was a cellular organism that had a lipid bilayer and used DNA, RNA, and protein. Horizontal gene transfer (HGT) may have allowed rapid transfer of beneficial genes among early forms of life.

elements of this three-part system of DNA, RNA, and protein, in addition to a membrane system capable of conserving energy (see Figure 12.5). The *last universal common ancestor (LUCA)* must have existed at 3.8–3.7 billion years ago, the point at which *Bacteria* and *Archaea* diverged and life began to diversify into the forms we recognize today. One can envision a time of intensive biochemical innovation and experimentation in which much of the structural and functional machinery of these earliest self-replicating systems evolved and was refined by natural selection.

Metabolic Diversification: Consequences for Earth's Biosphere

Following the origin of cells, microbial life experienced a long period of metabolic diversification, exploiting the various resources available on Earth. The earth and all of its oceans were anoxic for much of its history. Molecular oxygen (O_2) was absent in any significant quantities until oxygenic photosynthesis by cyanobacteria evolved. Thus, the energy-generating metabolism of primitive cells would have been exclusively anaerobic and would likely have had to be heat-stable because of the temperature of early Earth.

During this era CO_2 may well have been the major source of carbon for cells (autotrophy, ↩ Section 13.5) because abiotic sources of organic carbon would quickly have become limiting. It is widely thought that H_2 was a major fuel for energy metabolism of early cells. This hypothesis is also supported by the tree of life (see Figure 12.13), in that virtually all of the earliest branching organisms in the *Bacteria* and *Archaea* use H_2 as an electron donor in energy metabolism and are autotrophs. Elemental sulfur (S^0) may have been one of the earliest electron acceptors, as the reduction of S^0

to yield H_2S is exergonic and would likely have required relatively few enzymes (**Figure 12.5**). Moreover, because of the abundance of H_2 and sulfur compounds on early Earth, this scheme would have provided cells with a nearly limitless supply of energy.

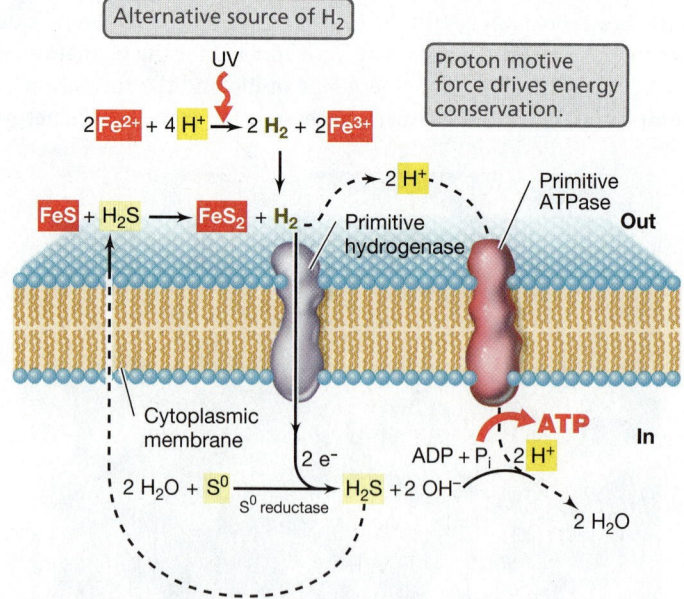

Figure 12.5 A possible energy-generating scheme for primitive cells. Formation of pyrite leads to H_2 production and S^0 reduction, which fuels a primitive ATPase. Note how H_2S plays only a catalytic role; the net substrates would be FeS and S^0. Also note how few different proteins would be required. $\Delta G^{0\prime} = -42$ kJ for the reaction $FeS + H_2S \rightarrow FeS_2 + H_2$.

Evidence suggests that ancestors of modern-day *Bacteria* and *Archaea* had already diverged by about 3.7 billion years ago (♻ Figure 1.4*b*). Early *Bacteria* may have used H_2 and CO_2 to produce acetate (♻ Section 13.19). At the same time, early *Archaea* developed the ability to use H_2 and CO_2, or possibly acetate as it accumulated, as substrates for methanogenesis (♻ Section 13.20). These early forms of chemolithotrophic metabolism driven by H_2 would likely have supported the production of large amounts of organic compounds from autotrophic CO_2 fixation. Over time, these organic materials would have accumulated and could have provided the conditions needed for the evolution of new chemoorganotrophic bacteria with diverse metabolic strategies to conserve energy from the oxidation of organic compounds.

MINIQUIZ

- What characteristics would have made the surface of Earth inhospitable to the formation of life 4.5 billion years ago?

- How do we know when oceans were first present on Earth? Why is the presence of oceans significant to the origins and diversification of life?

- What lines of reasoning support the hypothesis that the first self-replicating systems were based on RNA molecules?

12.2 Photosynthesis and the Oxidation of Earth

The evolution of photosynthesis revolutionized the chemistry of Earth. Phototrophic organisms use energy from the sun to oxidize molecules such as H_2S, S^0, or H_2O and to synthesize complex organic molecules from carbon dioxide or simple organic molecules (♻ Section 13.5). Over time, the products of photosynthesis accumulated in the biosphere, stimulating the further diversification of microbial life. Earth's first phototrophs were anoxygenic (♻ Sections 13.3 and 14.4–14.7), but from these evolved the *Cyanobacteria*, the earliest oxygenic phototrophs (Figure 12.1, ♻ Section 14.3).

Fossilized microbial formations called **stromatolites** can be found in rocks that are 3.5 billion years old, providing the earliest conclusive evidence of life on Earth (**Figure 12.6a**). Stromatolites, or "layered rocks," are formed when certain kinds of microbial mats cause the deposition of carbonate or silicate minerals that promote fossilization (we discuss microbial mats in Section 19.5). Stromatolites were diverse and common on Earth between 2.8 and 1 billion years ago, but declined dramatically in abundance over the last billion years. Stromatolites are largely gone from Earth today, and yet modern examples of this ancient microbial ecosystem can still be found in certain shallow marine basins (Figure 12.6*c, e*) or in hot springs (Figure 12.6*d*; ♻ Figure 19.9). Phototrophic bacteria, such as oxygen-producing cyanobacteria (♻ Section 14.3) and the green nonsulfur bacterium *Chloroflexus* (♻ Section 14.7), play a central role in the formation of modern stromatolites. Likewise, ancient stromatolites contain microfossils that appear remarkably similar to modern species of cyanobacteria and green algae (**Figure 12.7**). Hence, the earliest phototrophic organisms may have evolved more than 3.5 billion years ago and apparently only in *Bacteria*, giving rise to the stromatolites we observe in the fossil record.

(a) Malcolm Walter
(b) Malcolm Walter
(c) Daniel H. Buckley
(d) T. D. Brock
(e) Malcolm Walter

Figure 12.6 Ancient and modern stromatolites. *(a)* The oldest known stromatolite, found in a rock about 3.5 billion years old, from the Warrawoona Group in Western Australia. Shown is a vertical section through the laminated structure preserved in the rock. Arrows point to the laminated layers. *(b)* Stromatolites of conical shape from 1.6-billion-year-old dolomite rock from northern Australia. *(c)* Modern stromatolites, Darby Island, Bahamas. The large stromatolite in the foreground is about 1 m in diameter. *(d)* Modern stromatolites composed of thermophilic cyanobacteria growing in a thermal pool in Yellowstone National Park. Each structure is about 2 cm high. *(e)* Modern stromatolites from Shark Bay, Australia. Individual structures are 0.5–1 m in diameter.

Early forms of photosynthesis were anoxygenic, using electron donors such as H_2S and generating elemental sulfur (S^0) as a waste product (♻ Section 13.3). The ability to use solar radiation as an energy source allowed phototrophs to diversify extensively. By 2.5–3.3 billion years ago, the cyanobacterial lineage evolved a photosystem capable of oxygenic photosynthesis (♻ Section 13.4) in which H_2O supplanted H_2S in the photosynthetic reduction of CO_2, thereby generating O_2 as a waste product. As we will see in the next section, the origin of oxygenic photosynthesis and the rise of O_2 in Earth's atmosphere caused the greatest change in the history of our biosphere and set the stage for the evolution of even newer forms of life that evolved to exploit the energy available from O_2 respiration.

The Rise of Oxygen: Banded Iron Formations

In the absence of O_2, all of Earth's iron would have been present in reduced forms and there would have been abundant iron dissolved in Earth's oceans, likely making them red rather than blue. Molecular and chemical evidence indicates that oxygenic photosynthesis first appeared on Earth at least 300 million years before significant levels of O_2 appeared in the atmosphere. The O_2 that

UNIT 3

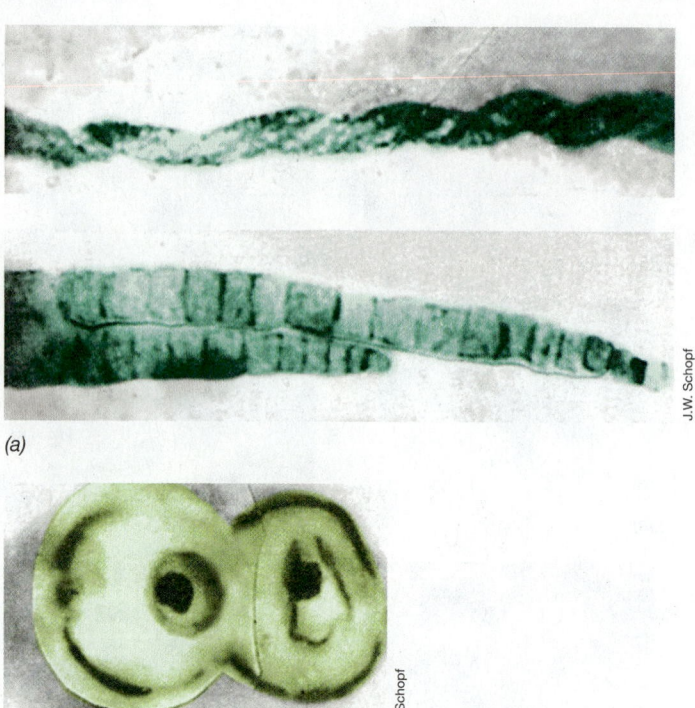

(a)

(b)

J.W. Schopf

Figure 12.7 More recent fossil bacteria and eukaryotes. *(a)* One-billion-year-old microfossils from central Australia that resemble modern filamentous cyanobacteria. Cell diameters, 5–7 μm. *(b)* Microfossils of eukaryotic cells from the same rock formation. The cellular structure is similar to that of certain modern green algae, such as *Chlorella* species. Cell diameter, about 15 μm. Color was added to make cell form more apparent.

cyanobacteria produced could not accumulate in the atmosphere because it reacted spontaneously with the reduced iron minerals in the oceans to make iron oxides. By 2.4 billion years ago, O_2 levels had risen to one part per million, a tiny amount by present-day standards, but enough to initiate what has come to be called the *Great Oxidation Event* (Figure 12.1).

The metabolism of cyanobacteria yielded O_2 that oxidized reduced minerals containing Fe^{2+} to iron oxides containing Fe^{3+}. These iron oxide minerals became a prominent marker in the geological record. Iron oxides are poorly soluble in water and would have precipitated in the oceans, raining down onto the seafloor and forming sedimentary structures known as **banded iron formations** (**Figure 12.8**), laminated sedimentary rocks formed in deposits of iron- and silica-rich materials. Much of the iron in rocks of Precambrian origin (>0.5 billion years ago, see Figure 12.1) exists in these banded iron formations, and today these minerals represent a major source of iron ore. Only after the abundant Fe^{2+} on Earth was consumed could O_2 accumulate in the atmosphere, and not until 600–900 million years ago did atmospheric O_2 reach present-day levels (~21%, Figure 12.1).

As O_2 accumulated on Earth, the atmosphere gradually changed from anoxic to oxic (Figure 12.1). Species of *Bacteria* and *Archaea* unable to adapt to this change were increasingly restricted to anoxic habitats because of the toxicity of O_2 and because it chemically oxidized the reduced substances upon which their metabolisms

John M. Hayes

Figure 12.8 Banded iron formations. An exposed cliff made of sedimentary rock about 10 m in height in Western Australia contains layers of iron oxides (arrows) interspersed with layers containing iron silicates and other silica materials. The iron oxides contain iron in the ferric (Fe^{3+}) form produced from ferrous iron (Fe^{2+}) primarily by the oxygen released by cyanobacterial photosynthesis.

depend. However, the oxic atmosphere also created conditions for the evolution of various new metabolic schemes, such as sulfide oxidation, nitrification, and the various other aerobic chemolithotrophic processes (Chapters 13, 14). Microorganisms that evolved the capacity to respire O_2 gained a tremendous energetic advantage because of the high reduction potential of the O_2/H_2O couple (⇄ Section 3.6), and with more energy at their disposal, aerobes could reproduce far more rapidly than anaerobes.

The Ozone Shield

An important consequence of O_2 for the evolution of life was the formation of *ozone* (O_3). The sun bathes Earth in intense amounts of ultraviolet (UV) radiation, which is lethal to cells and can cause severe DNA damage. When O_2 is subject to UV radiation from the sun, it is converted to ozone, which strongly absorbs UV radiation in wavelengths up to 300 nm. The conversion of O_2 to O_3 creates an ozone shield, a barrier that protects the surface of Earth from much of the UV radiation from the sun. Prior to the generation of the ozone shield, the punishing UV irradiation from the sun would have made Earth's surface fairly inhospitable for life, restricting life to environments that provided protection from UV radiation, such as in the oceans or in the subsurface. However, as Earth developed an ozone shield, organisms could range over the surface of Earth, exploiting new habitats and evolving ever-greater diversity. Figure 12.1 summarizes some landmarks in biological evolution and Earth's geochemistry as Earth transitioned from an anoxic to a highly oxic planet.

MINIQUIZ

- Why is the origin of cyanobacteria considered a critical step in evolution?
- What caused the formation of banded iron formations?
- What lines of evidence indicate that microbial life was present on Earth 3.5 billion years ago?

12.3 Endosymbiotic Origin of Eukaryotes

Until about 2 billion years ago, all cells apparently lacked a membrane-enclosed nucleus and organelles, the key characteristics of eukaryotic cells (domain *Eukarya*). Here we consider the origin of the *Eukarya* and show how eukaryotes are genetic chimeras containing genes from at least two different phylogenetic domains.

Endosymbiosis

As Earth became more oxic, organelle-containing eukaryotic microorganisms arose, and the rise in O_2 spurred their rapid evolution. While the exact origins of the eukaryotic cell remain unclear, the oldest microfossils that have recognizable nuclei are about 2 billion years old. Multicellular and increasingly complex microfossils of algae are evident from 1.9 to 1.4 billion years ago (Figure 12.7*b*). By 0.6 billion years ago, with O_2 near present-day levels, large multicellular organisms, the Ediacaran fauna, were present in the sea (Figure 12.1). In a relatively short time, multicellular eukaryotes diversified into the ancestors of modern-day algae, plants, fungi, and animals (Section 12.4).

A well-supported explanation for the origin of organelles in the eukaryotic cell is the **endosymbiotic hypothesis** (**Figure 12.9**). The hypothesis states that the mitochondria of modern-day eukaryotes arose from the stable incorporation of a respiring bacterium into other cells and that chloroplasts arose similarly from the incorporation of a cyanobacterium-like organism that performed oxygenic photosynthesis. Oxygen was almost certainly a driving force in endosymbiosis through its consumption by the ancestor of the mitochondrion and its production by the ancestor of the chloroplast. The greater energy released by aerobic respiration

undoubtedly contributed to rapid evolution of eukaryotes, as did the ability to exploit sunlight for energy.

The overall physiology and metabolism of mitochondria and chloroplasts and the sequence and structures of their genomes support the endosymbiotic hypothesis. For example, both mitochondria and chloroplasts contain ribosomes of prokaryotic size (70S), including a 16S ribosomal RNA (**16S rRNA**) molecule. The 16S RNA gene sequences (Section 12.4) of mitochondria and chloroplasts are also characteristic of *Bacteria*. Phylogenetic trees constructed from the 16S rRNA genes of mitochondria place their ancestor in the phylum *Alphaproteobacteria*, while the 16S rRNA genes of chloroplasts place their ancestor in the phylum *Cyanobacteria*. Moreover, the same antibiotics that inhibit ribosome function in free-living *Bacteria* inhibit ribosome function in these organelles. Mitochondria and chloroplasts also contain small amounts of DNA arranged in a covalently closed, circular form, which is typical of *Bacteria*, and the phylogeny of these sequences indicates a bacterial ancestry. Indeed, these and many other telltale signs of *Bacteria* are present in organelles from modern eukaryotic cells (⮂ Section 6.5).

Formation of the Eukaryotic Cell

The exact origin of the eukaryotic cell remains a major unresolved question in evolution; however, it seems clear that the modern eukaryotic cell is a genetic chimera, a cell made up of genes from both *Bacteria* and *Archaea*. There is strong support for the endosymbiotic origin of mitochondria and chloroplasts from *Bacteria* as described above, and the transfer of certain genes from these endosymbionts to the cell nucleus. Eukaryotic cells share some

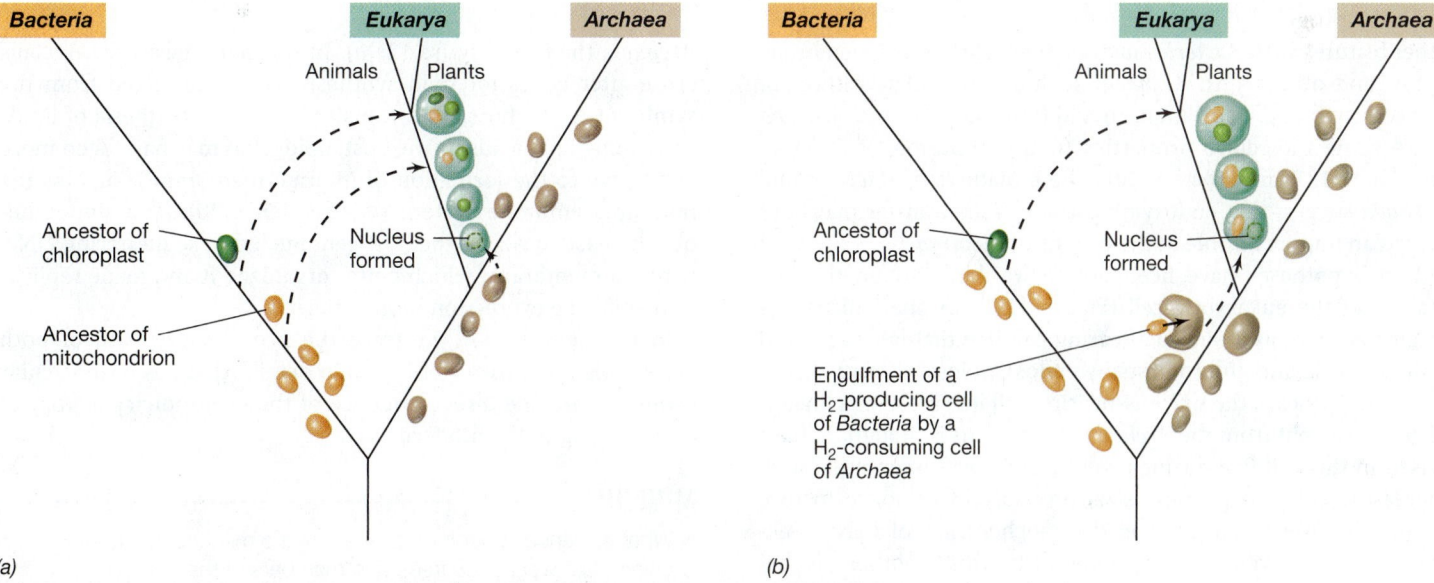

(a)   (b)

Figure 12.9 Endosymbiotic models for the origin of the eukaryotic cell. *(a)* The nucleated line diverged from the archaeal line and later acquired by endosymbiosis the bacterial ancestor of the mitochondrion and then the cyanobacterial ancestor of the chloroplast, at which point the nucleated line diverged into the lineages giving rise to plants and animals. *(b)* The hydrogen hypothesis. The bacterial ancestor of the mitochondrion was taken up endosymbiotically by a species of *Archaea* and the nucleus developed later. The cyanobacterial ancestor of the chloroplast was later taken up endosymbiotically by the ancestor of plants and algae (not shown). Note the position of the mitochondrion and plastids (chloroplasts are a type of plastid) on the universal phylogenetic tree in Figure 12.13.

UNIT 3

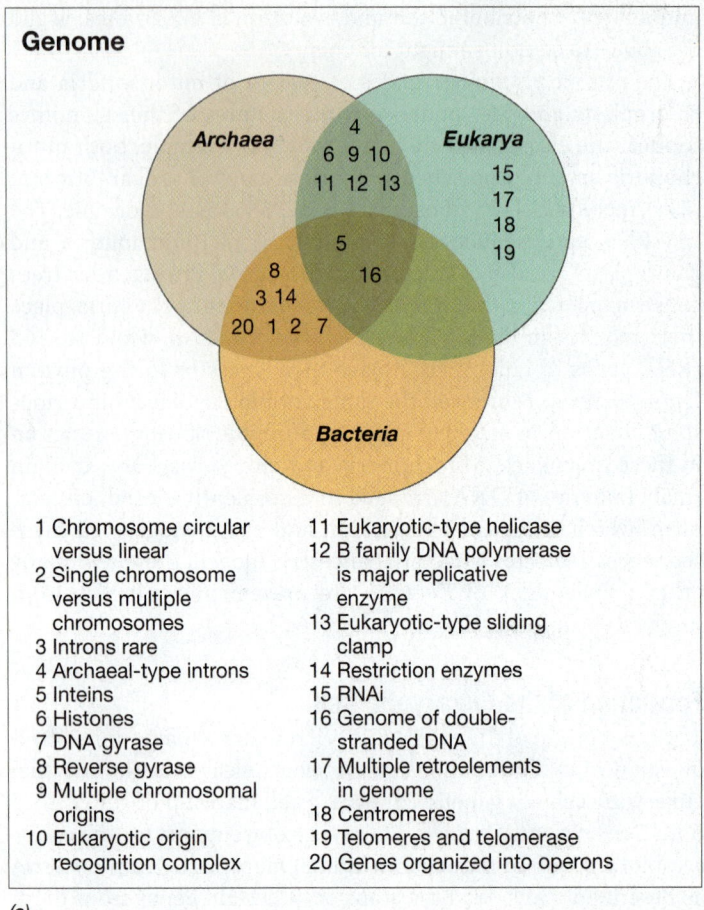

Genome

1 Chromosome circular versus linear
2 Single chromosome versus multiple chromosomes
3 Introns rare
4 Archaeal-type introns
5 Inteins
6 Histones
7 DNA gyrase
8 Reverse gyrase
9 Multiple chromosomal origins
10 Eukaryotic origin recognition complex
11 Eukaryotic-type helicase
12 B family DNA polymerase is major replicative enzyme
13 Eukaryotic-type sliding clamp
14 Restriction enzymes
15 RNAi
16 Genome of double-stranded DNA
17 Multiple retroelements in genome
18 Centromeres
19 Telomeres and telomerase
20 Genes organized into operons

(a)

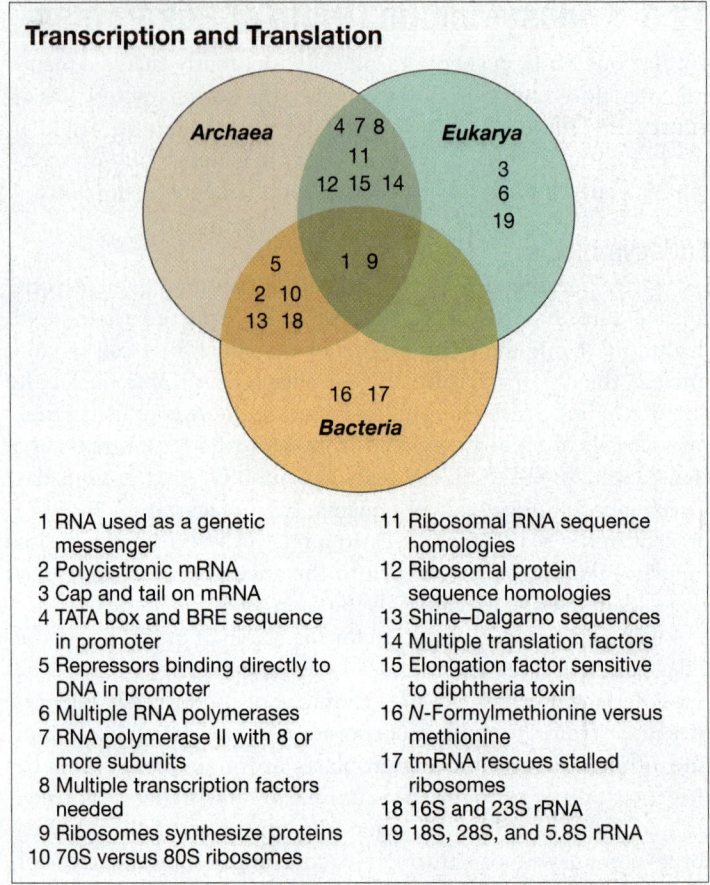

Transcription and Translation

1 RNA used as a genetic messenger
2 Polycistronic mRNA
3 Cap and tail on mRNA
4 TATA box and BRE sequence in promoter
5 Repressors binding directly to DNA in promoter
6 Multiple RNA polymerases
7 RNA polymerase II with 8 or more subunits
8 Multiple transcription factors needed
9 Ribosomes synthesize proteins
10 70S versus 80S ribosomes
11 Ribosomal RNA sequence homologies
12 Ribosomal protein sequence homologies
13 Shine–Dalgarno sequences
14 Multiple translation factors
15 Elongation factor sensitive to diphtheria toxin
16 N-Formylmethionine versus methionine
17 tmRNA rescues stalled ribosomes
18 16S and 23S rRNA
19 18S, 28S, and 5.8S rRNA

(b)

Figure 12.10 **Molecular features of the three domains.** Venn diagrams show which features are shared by the domains and which are unique. *(a)* Genomic features. *(b)* Features of transcription and translation.

other features with *Bacteria*, such as their ester-linked membrane lipids, and others with *Archaea*, such as molecular features of transcription and translation. In addition, *Bacteria* and *Archaea* share some molecular properties to the exclusion of *Eukarya* (see Table 12.1 and **Figure 12.10**). These features of *Bacteria* and *Archaea* suggest that endosymbiosis and gene transfer may have played an important role in the origins of *Eukarya*.

Two hypotheses have been put forward to explain the formation of the eukaryotic cell (Figure 12.9). In one, eukaryotes originally arose as a nucleus-bearing cell line that later acquired mitochondria and chloroplasts by endosymbiosis (Figure 12.9*a*). In this hypothesis, the nucleus-bearing cell line arose in a lineage of cells that split from the *Archaea*; the nucleus is thought to have arisen in this cell line during evolutionary experimentation with increasing cell and genome size, probably in response to oxic events that were transforming the geochemistry of Earth (Section 12.2). However, a major problem with this hypothesis is that it does not easily account for the fact that *Bacteria* and *Eukarya* have similar membrane lipids, in contrast to those of *Archaea* (⮌ Section 2.7).

The second hypothesis, called the *hydrogen hypothesis*, proposes that the eukaryotic cell arose from an association between a H_2-producing species of *Bacteria*, the symbiont, which eventually gave rise to the mitochondrion, and a species of H_2-consuming

Archaea, the host (Figure 12.9*b*). In this hypothesis, the nucleus arose after genes for lipid synthesis were transferred from the symbiont to the host. This transfer led to the synthesis of lipids containing fatty acids by the host, lipids that may have been more conducive to the formation of internal membranes, such as the nuclear membrane system (⮌ Section 2.20). The simultaneous increase in size of the host genome led to sequestering DNA within a membrane, which better organized it and made replication and gene expression more efficient.

In the next section we trace the evolutionary path of both eukaryotic and prokaryotic cells in detail. Analyses of molecular evolution provide direct evidence of the evolutionary history of cells, leading to the modern "tree of life."

MINIQUIZ

- What evidence supports the idea that the mitochondrion and chloroplast were once free-living members of the domain *Bacteria*?
- What major events in the evolution of microbial life had to occur before the origin of the acquisition of chloroplasts and mitochondria by eukaryotic cells?
- In what ways are modern eukaryotes a combination of attributes of *Bacteria* and *Archaea*?

II • Living Fossils: DNA Records the History of Life

DNA sequences provide a record of past evolutionary events and can be used to determine **phylogeny**, which is the evolutionary history of organisms. In the following sections we explore how molecular sequences can be used to build **phylogenetic trees**, diagrams that depict evolutionary history. We will consider how molecular phylogenetic analysis has changed our understanding of the history of life.

12.4 Molecular Phylogeny and the Tree of Life

The evolutionary origins of microorganisms remained a mystery until it was discovered that molecular sequences serve as a record of evolutionary history. In this section we will learn how the sequence of **ribosomal RNA (rRNA)** genes, which are found in all cells, revolutionized the understanding of microbial evolution and made it possible to construct the first **universal tree of life**.

Molecular Sequence Data Has Revolutionized Microbial Phylogeny

Following the 1859 publication of Charles Darwin's *Origin of Species*, for a hundred years and more, evolutionary history was studied primarily with the tools of paleontology, through examining fossils, and comparative biology, through comparing the traits of living organisms. These approaches led to much progress in understanding the evolution of plants and animals, but they were powerless to explain the evolution of microorganisms. The vast majority of microorganisms do not leave behind fossils, and their morphological and physiological traits provide few clues about their evolutionary history. Moreover, microorganisms do not share any morphological traits with plants and animals; thus it was impossible to create a robust evolutionary framework that included microorganisms.

The first attempt to depict the common evolutionary history of all living cells was published by Ernst Haeckel in 1866 (**Figure 12.11a**).

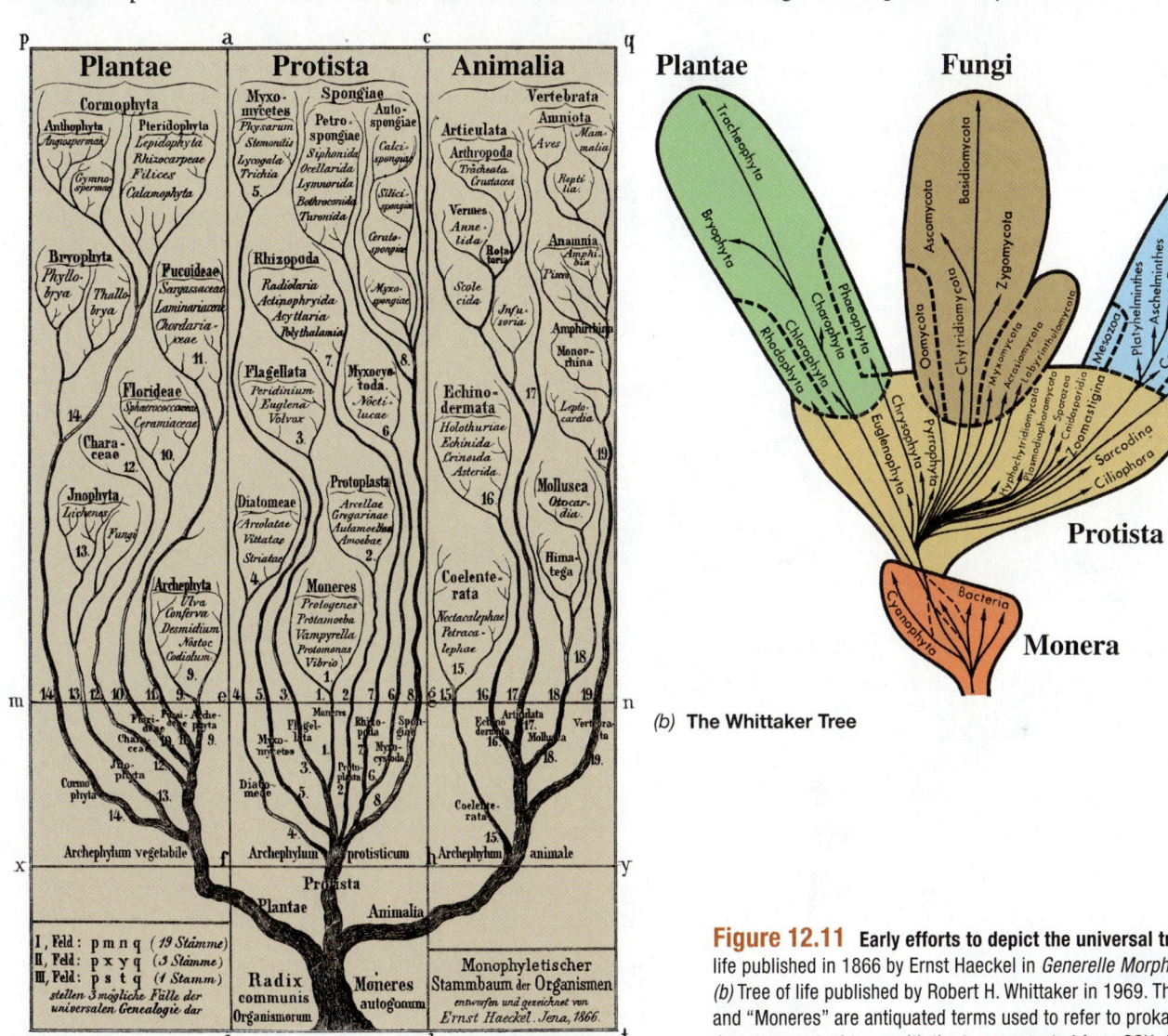

(a) **The Haeckel Tree**

(b) **The Whittaker Tree**

From *Science 163:150–160, 1967*
Reprinted with permission from AAAS

Figure 12.11   Early efforts to depict the universal tree of life. *(a)* Tree of life published in 1866 by Ernst Haeckel in *Generelle Morphologie der Organismen*. *(b)* Tree of life published by Robert H. Whittaker in 1969. The terms "Monera" and "Moneres" are antiquated terms used to refer to prokaryotic cells. Compare these conceptual trees with the tree generated from SSU rRNA gene sequences in Figure 12.13.

Haeckel correctly suggested that single-cell organisms, which he called *Monera*, were ancestral to other forms of life, but his scheme, which included plants, animals, and protists, did not attempt to resolve evolutionary relationships among microorganisms. The situation was little changed as late as 1967 when Robert Whittaker proposed a five-kingdom classification scheme (Figure 12.11*b*). Whittaker's scheme distinguished the fungi as a distinct lineage, but it was still largely impossible to resolve evolutionary relationships among most microorganisms. Hence, microbial phylogeny had made little progress since Haeckel's day.

Everything changed after the structure of DNA was discovered and it was recognized that evolutionary history is recorded in DNA sequence. Carl Woese realized in the 1970s that the sequence of rRNA molecules and their genes could be used to infer evolutionary relationships between organisms. Woese recognized that

rRNA genes are excellent candidates for phylogenetic analysis because they are (1) universally distributed, (2) functionally constant, (3) highly conserved (that is, slowly changing), and (4) of adequate length to provide a deep view of evolutionary relationships. Woese compared the sequences of **small subunit rRNA (SSU rRNA)** molecules (**Figure 12.12**) from many microorganisms and found that the sequences from methane-producing prokaryotes (methanogens) were vastly different from those of *Bacteria*. To his astonishment, he found that these sequences were as different from those of *Bacteria* as the latter were from those of *Eukarya*. He named this new group of prokaryotes the *Archaea* (originally *Archaebacteria*) and recognized them as the third **domain** of life alongside the *Bacteria* and the *Eukarya* (∞ Section 1.3 and **Figure 12.13**). More importantly, Woese demonstrated that the analysis of SSU rRNA gene sequences could be used to reveal evolutionary

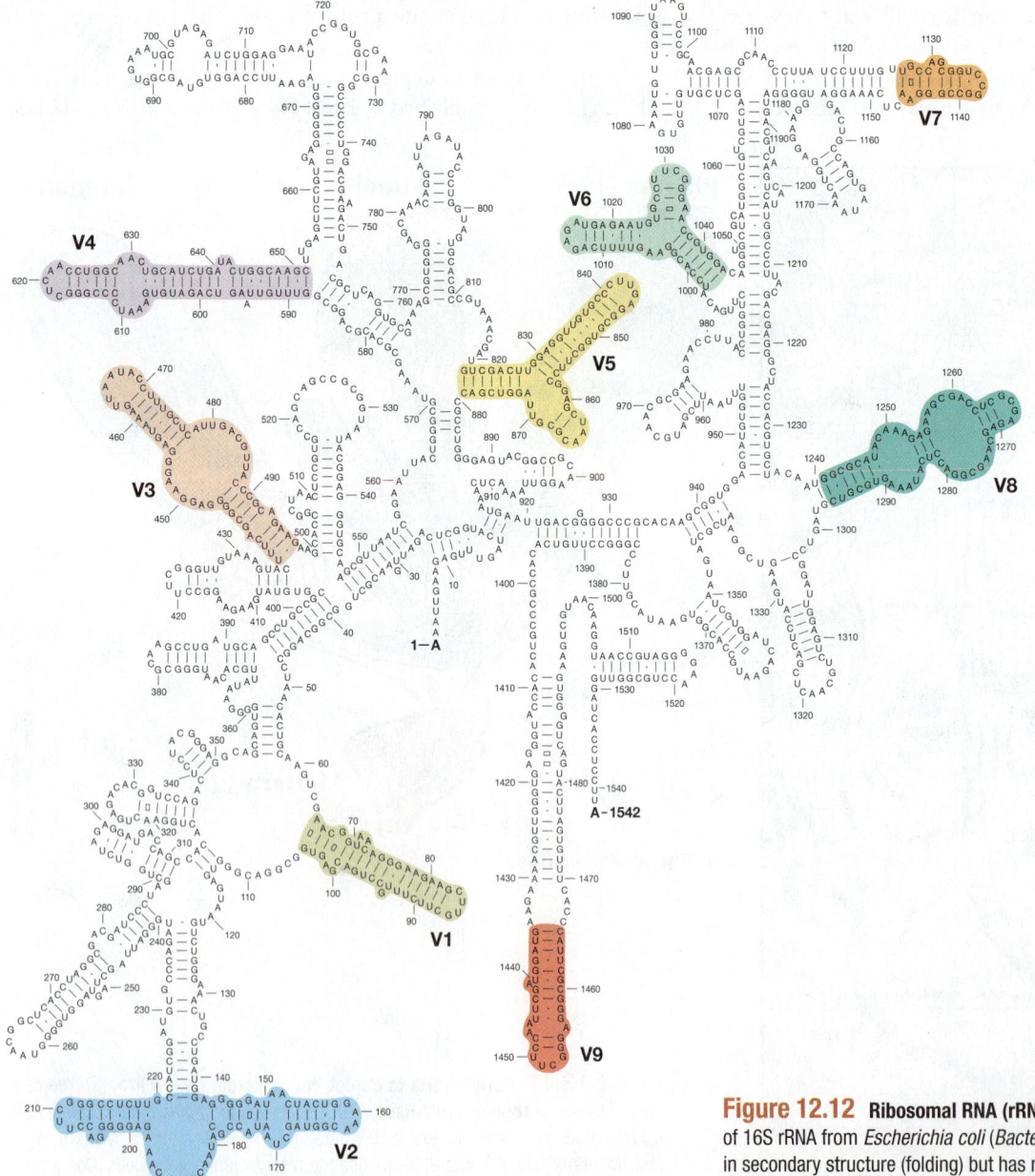

Figure 12.12 Ribosomal RNA (rRNA). Primary and secondary structure of 16S rRNA from *Escherichia coli* (*Bacteria*). The 16S rRNA from *Archaea* is similar in secondary structure (folding) but has numerous differences in primary structure (sequence). The molecule is composed of conserved and variable regions. The positions of the variable regions are indicated in color.

relationships between all cells, providing the first effective tool for the evolutionary classification of microorganisms.

Since 1977 more than 2.3 million SSU rRNA sequences have been generated and used to characterize the vast diversity of the microbial world. The Ribosomal Database Project (RDP; http://rdp.cme.msu.edu) contains an ever-growing collection of these sequences and provides computational programs for their analysis and for the construction of phylogenetic trees, a topic we will consider in Section 12.5.

The Tree of Life Based on SSU rRNA Genes

The universal phylogenetic tree of life based on SSU rRNA gene sequences (Figure 12.13) is a genealogy of all life on Earth. It depicts the evolutionary history of all cells and clearly reveals the three domains. The root of the universal tree represents a point in time when all extant life on Earth shared a common ancestor, the last universal common ancestor, LUCA (Figure 12.13 and Section 12.1). Genome analysis has revealed that the three-domain concept is supported not only by SSU rRNA sequences, but also by phylogenetic analysis of most genes that encode enzymes of transcription, translation, or DNA replication. Although there are many examples of **horizontal gene transfer** (Sections 6.12 and 12.1) between lineages within and between domains, it remains clear that the three domains represent the major evolutionary cell lineages that exist on Earth.

The manner in which the three domains were established remains a topic of debate. There are many examples of genes shared by *Bacteria*, *Archaea*, and *Eukarya*, or shared by two of the three domains (Figure 12.10). One hypothesis is that early in the history of life, before the primary domains had diverged, horizontal gene transfer was extensive, and many of the genes that encoded information-handling functions had yet to evolve. The evolution of genes that improved information handling, such as genes that encode transcription and translation proteins, would have conveyed a strong benefit and may have been transferred rapidly among early forms of life (Figure 12.4).

It is further hypothesized that over time, barriers to unrestricted horizontal gene transfer evolved. As a result, the previously promiscuous population began slowly to sort out into the

UNIT 3

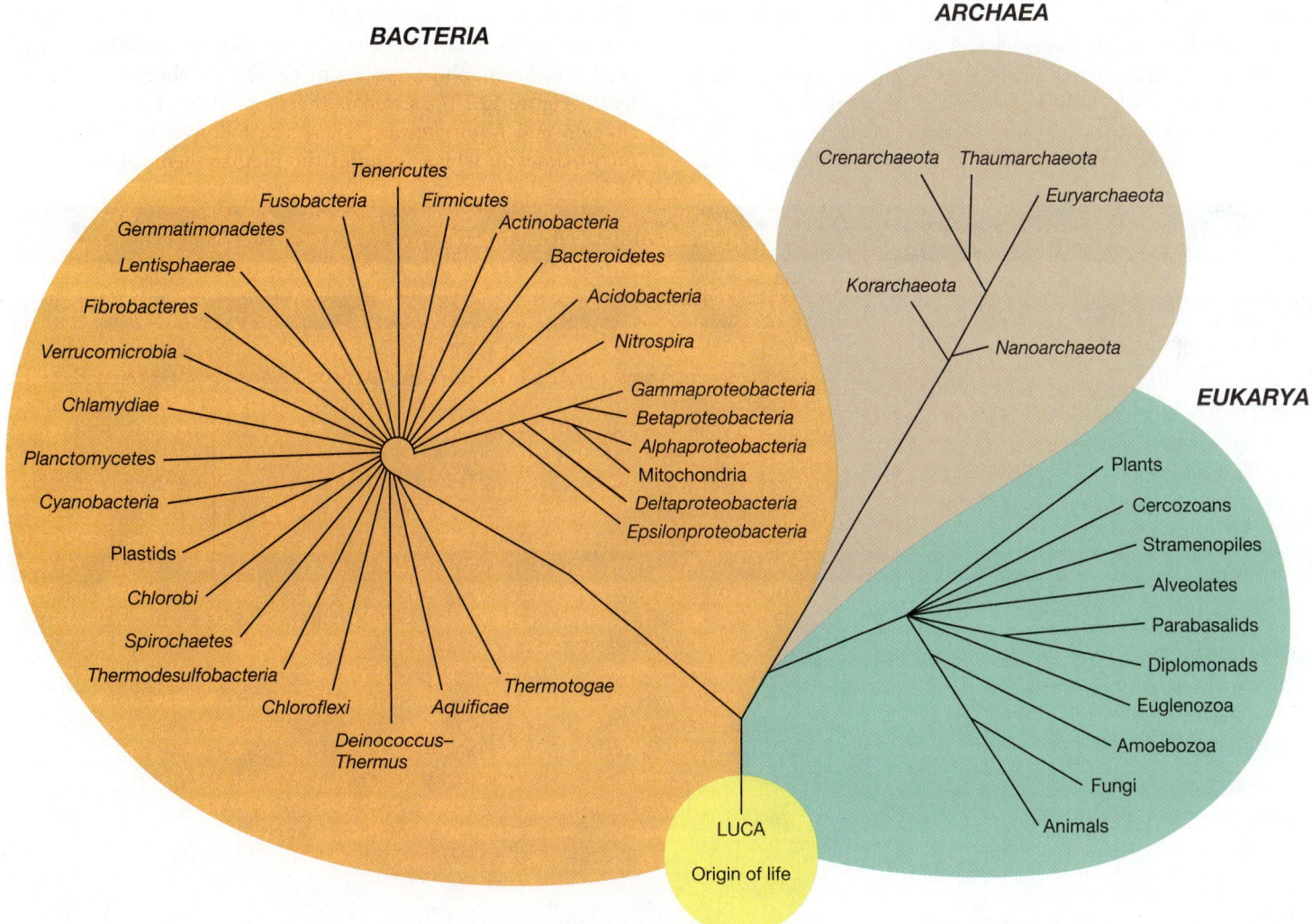

Figure 12.13 **Universal phylogenetic tree as determined from comparative SSU rRNA gene sequence analysis.** Only a few key organisms or lineages are shown in each domain. At least 84 phyla of *Bacteria* have now been identified although many of these have not yet been cultured. LUCA, last universal common ancestor.

primary lines of evolutionary descent, the *Bacteria* and *Archaea* (Figure 12.4 and Figure 12.13). There was a further bifurcation, about 2.8 billion years ago, as the *Archaea* and *Eukarya* diverged as distinct domains. As each lineage continued to evolve, certain traits became fixed within each group, giving rise to the genetic differences (Figure 12.10) and physiological and structural differences (Table 12.1) we observe between the three domains today. After nearly 4 billion years of microbial evolution, we see the grand result: three domains of cellular life that are each evolutionarily distinct and yet share certain features indicative of their common descent from a universal cellular ancestor.

Bacteria

Among *Bacteria,* at least 84 lineages (called phyla, singular **phylum,** or divisions) have been discovered thus far; only some key ones are shown in the universal tree in Figure 12.13. The *Bacteria* are discussed in detail in Chapters 14 and 15. Many lineages of *Bacteria* are known only from SSU rRNA gene sequences recovered from environmental samples (*phylotypes,* ↩ Section 18.5). Only 32 of the phyla contain species described on the basis of strains in cultivation, and more than 90% of strains in cultivation belong to one of only four phyla, the *Actinobacteria,* the *Firmicutes,* the *Proteobacteria,* and the *Bacteroidetes.* While the exact ages of these phyla are difficult to determine, it is likely that many of these phyla were established around the time at which the *Bacteria* and *Archaea* diverged.

Although the species in some of these phyla are characterized by unique phenotypic traits, such as the morphology of the spirochetes or the physiology of the cyanobacteria, most bacterial phyla contain a wide diversity of species and show tremendous physiological diversity. The *Proteobacteria* illustrate this concept well as they include organisms with a diverse array of physiological traits including aerobic respiration, fermentation, nitrification, nitrogen fixation, denitrification, sulfate reduction, sulfur and sulfide oxidation, phototrophy, dissimilatory metal reduction and oxidation, methane oxidation, and many others (Chapters 13 and 14). Species of *Proteobacteria* also possess a wide range of ecological strategies and can be found in all but the hottest and most salty environments on Earth. It is important to remember that while most phyla of plants and animals originated within the last 400 million years, bacterial phyla are billions of years old and this time has allowed for extensive experimentation and diversification.

Archaea

The domain *Archaea* consists of seven major phyla, only five of which contain species described on the basis of cultivated strains. Most described species fall within the phyla *Crenarchaeota* and *Euryarchaeota*, while only a handful of species have been described for the *Nanoarchaeota*, the *Korarchaeota*, and the *Thaumarchaeota* (Figure 12.13). We discuss *Archaea* in detail in Chapter 16. Branching close to the root of the universal tree are hyperthermophilic species of *Crenarchaeota,* such as *Pyrolobus* (Figure 12.13), as well as thermophilic species of *Nanoarchaeota* and *Korarchaeota*. These are followed by the phylum *Euryarchaeota,* which includes the methanogenic *Archaea* and

Table 12.1 Major structural and physiological characteristics of Bacteria, Archaea, and Eukarya[a]

Characteristic	Bacteria	Archaea	Eukarya
Morphological			
Prokaryotic cell structure	Yes	Yes	No
Cell wall	Peptidoglycan present	No peptidoglycan	No peptidoglycan
Membrane lipids	Ester-linked	Ether-linked	Ester-linked
Membrane-enclosed nucleus	Absent	Absent	Present
Flagella mechanism	Rotation	Rotation	Whiplike
Sensitivity to chloramphenicol, streptomycin, kanamycin, and penicillin	Yes	No	No
Physiological/special structures			
Dissimilative reduction of S^0 or SO_4^{2-} to H_2S, or Fe^{3+} to Fe^{2+}	Yes	Yes	No
Nitrification (ammonia oxidation)	Yes	Yes	No
Chlorophyll-based photosynthesis	Yes	No	Yes (in chloroplasts)
Denitrification	Yes	Yes	No
Nitrogen fixation	Yes	Yes	No
Rhodopsin-based energy metabolism	Yes	Yes	No
Chemolithotrophy (Fe^{2+}, NH_3, S^0, H_2)	Yes	Yes	No
Endospores	Yes	No	No
Gas vesicles	Yes	Yes	No
Storage granules of poly-β-hydroxyalkanoates	Yes	Yes	No
Growth above 70°C	Yes	Yes	No
Growth above 100°C	No	Yes	No

[a]Note that for many features, only particular representatives within a domain show the property.

the extreme halophiles and extreme acidophiles, such as *Thermoplasma* (Figure 12.13). The phylum *Thaumarchaeota* was first observed in the deep ocean in the 1990s but has subsequently been found in soils and marine systems all over the world. The first species of *Thaumarchaeota* were shown to be capable of ammonia oxidation. Several different species have since been isolated and all share this physiological trait (↩ Section 16.6). As for *Bacteria*, many lineages of *Archaea* are known only from SSU rRNA genes recovered from the environment and there remains great opportunity for the discovery of new lineages in the future.

Eukarya

Phylogenetic trees for *Eukarya* have been constructed from comparative sequence analysis of the 18S rRNA gene, the eukaryotic equivalent of the 16S rRNA gene in *Bacteria* and *Archaea*. As we discussed (Section 12.3), major eukaryotic organelles were clearly derived by endosymbiosis from the domain *Bacteria*, with the ancestors of mitochondria coming from within the *Proteobacteria* and those of chloroplasts from within the cyanobacteria (Figure 12.13). However, some microbial eukaryotes lack mitochondria (↩ Section 2.21). In Chapter 17 where we consider microbial eukaryotes in detail, we will see that the phylogeny of ancient eukaryotes is difficult to determine. Multigene phylogenetic trees (Section 12.9) indicate the primary eukaryotic lineages originated during a burst of evolutionary radiation about 600 million years ago that led to most lineages of microbial eukaryotes (↩ Figure 17.3). It is likely that this burst in eukaryotic evolution was triggered by the onset of oxic conditions on Earth and subsequent development of the ozone shield (Section 12.2). The latter would have greatly expanded the number of surface habitats available for colonization.

MINIQUIZ

- What kinds of evidence support the three-domain concept of life?
- How does the universal tree in Figure 12.13 support the hypothesis of endosymbiosis (Figure 12.9)?
- List three reasons that SSU rRNA genes are suitable for phylogenetic analyses.

12.5 Molecular Phylogeny: Making Sense of Molecular Sequences

All cells contain DNA as their genetic material, and DNA is passed from parent to offspring. Heritable mutations accumulate in DNA sequences over time. These mutations occur naturally and are a major cause of the random variations upon which selection acts, as described in Darwin's theory of evolution. Hence, the difference in nucleotide sequence between any two organisms will be a function of the number of mutations that have accumulated since they shared a common ancestor. As a result, differences in DNA sequences can be used to infer evolutionary relationships. In this section we will learn how DNA sequences are used in the phylogenetic analysis of microbial life.

Obtaining DNA Sequences

While analysis of microbial phylogeny relies heavily on analysis of SSU rRNA gene sequences, advances in DNA sequencing technology (↩ Section 6.2) have made genome sequencing a standard tool employed in analyses of microbial phylogeny. Obtaining gene sequence from a microorganism is relatively easy if the organism can be cultivated in isolation in the laboratory. In this case, genomic DNA is isolated and the genome sequenced directly or used to amplify one or more specific genes, using the polymerase chain reaction (PCR, ↩ Section 11.3).

PCR primers can be designed to target any region of DNA from any organism. Standard primers exist for many highly conserved genes, such as the SSU rRNA gene (Figure 12.12). Primers for the SSU rRNA gene can have different levels of phylogenetic specificity targeting discrete species, genera, and phyla, and there are even "universal" primers that will amplify the SSU rRNA gene from any organism. PCR products are visualized by agarose gel electrophoresis, excised from the gel, extracted and purified from the agarose, and then sequenced, often using the same oligonucleotides as primers for the sequencing reactions. These steps are summarized in **Figure 12.14**. Alternatively, it is also possible to amplify SSU rRNA genes from DNA that has been extracted directly from an environmental sample or to sequence directly this environmental DNA using a metagenomic approach (↩ Sections 6.10 and 18.7). These latter approaches are used widely to characterize microorganisms that are difficult to grow in laboratory culture. Once sequences are obtained, they must be aligned and analyzed, issues we turn to now.

Sequence Alignment

Phylogeny can be inferred only from genes that have **homology**, that is, genes that have been inherited from a common ancestor. Thus homology is a binary trait; sequences are either homologous or they are not. The concept of homology is often confused with that of sequence similarity. The latter is a continuous trait defined as a percentage of nucleotide positions shared between any two sequences. Sequence similarity is used to infer homology, but a similarity value can be calculated between any two sequences regardless of their function or evolutionary relationship. Thus, the terms similarity and homology are not interchangeable. Genes that have homology can be either **orthologs**, if they originate from a single ancestral gene in a common ancestor, or **paralogs**, if they are related as a result of a gene duplication (↩ Section 6.11). Phylogenetic analyses typically focus on analysis of orthologous genes that have similar function.

Phylogenetic analyses estimate evolutionary changes from the number of sequence differences across a set of homologous nucleotide positions. Some mutations introduce nucleotide insertions or deletions, and these cause gene sequences to differ in length, making it necessary to *align* nucleotide positions prior to phylogenetic analysis of gene sequences. The purpose of **sequence alignment** is to add gaps to molecular sequences in order to establish positional homology, that is, to be sure that each position in the sequence was inherited from a common ancestor of all organisms under consideration (**Figure 12.15**). Proper sequence alignment is critical to phylogenetic analysis because the assignment of mismatches and gaps caused by deletions is in effect an explicit hypothesis of how the sequences have diverged from a common ancestral sequence.

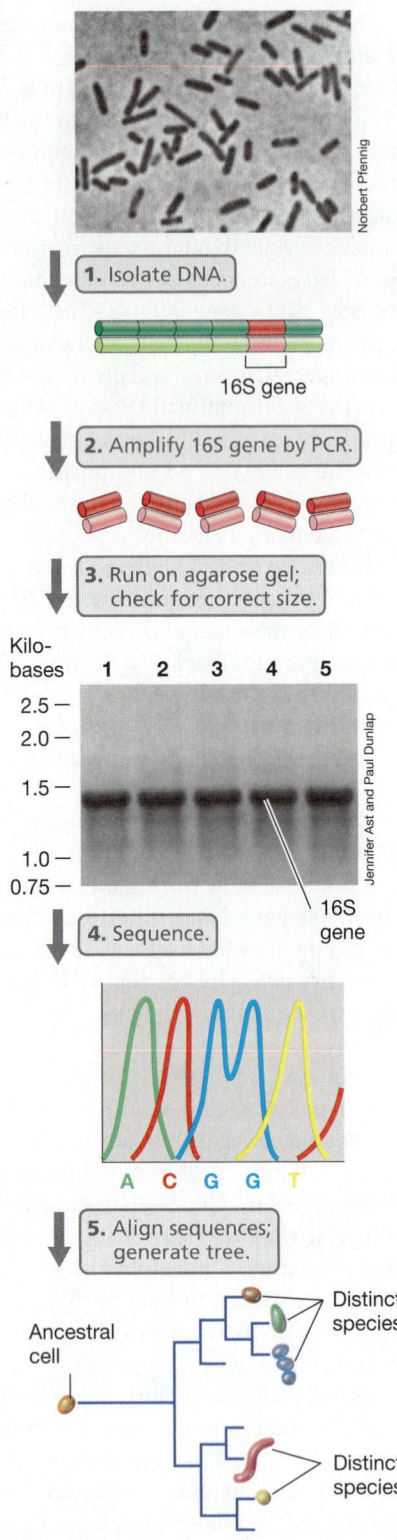

Figure 12.14 PCR amplification of the 16S rRNA gene. Following DNA isolation, primers complementary to the ends of the 16S rRNA (see Figure 12.12) are used to PCR-amplify the 16S rRNA gene from genomic DNA of five different unknown bacterial strains and the products are run on an agarose gel (photo). The bands of amplified DNA are approximately 1465 nucleotides in length. Positions of DNA kilobase size markers are indicated at the left. Excision from the gel and purification of these PCR products is followed by sequencing and analysis to identify the bacteria.

Sequences before alignment

Sequence differences

			1	2	3
1	GGA CCT AAA TTT ATA CCC	1	–	–	–
2	GGA AAA GGG CCC AAA CGC	2	11	–	–
3	GGA GGG CCT TTT ATA CCC	3	6	11	–

Sequences after alignment

			1	2	3
1	GGA – – – – – – CCT AAA TTT ATA CCC	1	–	–	–
2	GGA AAA GGG CCC – – – – – – AAA CGC	2	3	–	–
3	GGA – – – GGG CCT – – – TTT ATA CCC	3	0	3	–

(a)   (b)

Figure 12.15 Alignment of DNA sequences. *(a)* Sequences for a hypothetical region of a gene are shown for three species before alignment and after alignment. A sequence alignment should display homologous positions in vertical columns. Sequence alignment is achieved by adding gaps, indicated by hyphens, to maximize local sequence similarity between the species in the alignment. *(b)* The distance matrices show the number of sequence differences that would be inferred for each species pair both before and after alignment.

Phylogenetic Trees

A phylogenetic tree is a diagram that depicts the evolutionary history of an organism and bears some resemblance to a family tree. Most microorganisms do not leave fossils and so their ancestors are unknown, but ancestral relationships can be inferred from the DNA sequences of organisms that are alive today. Organisms that share a recent ancestor are likely to share characteristics, and thus phylogenetic trees allow us to make hypotheses about an organism's characteristics. Phylogenetic trees are also of great use in taxonomy and species identification, as we will discuss later in this chapter (Section 12.9).

A phylogenetic tree is composed of *nodes* and *branches* (**Figure 12.16**). The tips of the branches in a phylogenetic tree represent species that exist today. Phylogenetic trees can be constructed that are either *rooted trees* or *unrooted trees*. Rooted trees show the position of the ancestor of all organisms being examined. Unrooted trees depict the relative relationships among the organisms under study but do not provide evidence of the most ancestral node in the tree. The nodes represent a past stage of evolution where an ancestor diverged into two new lineages. The branch length represents the number of changes that have occurred along that branch. In a phylogenetic tree, only the position of nodes and the branch lengths are informative; rotation around nodes has no effect on the tree's topology (Figure 12.16*b*).

Tree Construction

There is only one correct phylogenetic tree that accurately depicts the evolutionary history of a group of gene sequences, but inferring the true tree from sequence data can be a challenging task. The complexity of the problem is revealed by considering the total number of trees that can represent a random set of sequences. For example, there are only three possible trees that can be drawn for any four arbitrary sequences. But if one doubles to eight the number of sequences, now 10,395 trees are possible. This complexity continues to expand exponentially such that 2×10^{182} different trees can be drawn to represent 100 arbitrary sequences.

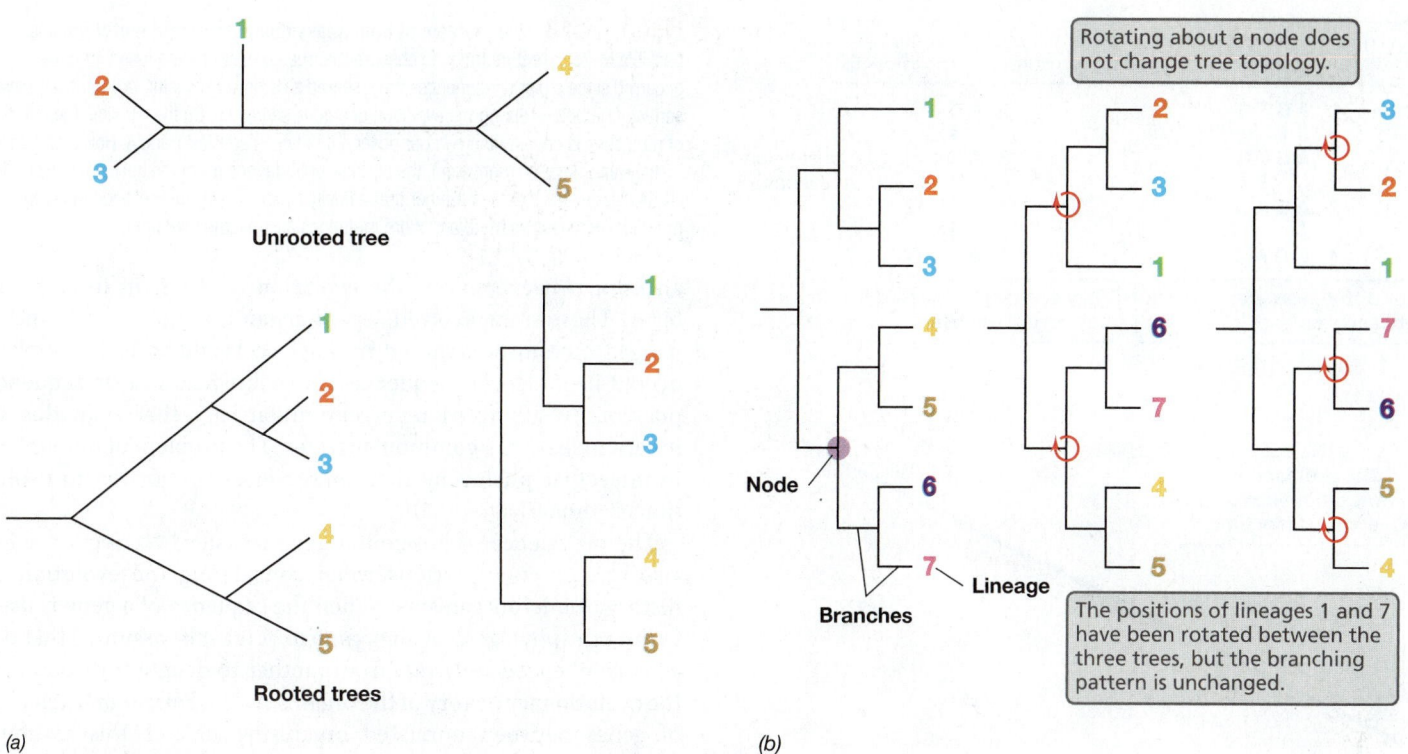

Figure 12.16 Phylogenetic trees and their interpretation. *(a)* Unrooted and rooted examples of phylogenetic trees. The tips of the branches are species (or strains) and the nodes are ancestors. Ancestral relationships are revealed by the branching order in rooted trees. *(b)* Three equivalent versions of the same phylogenetic tree are shown. The only difference between the trees is that their nodes have been rotated at the points indicated by red arrows. The vertical position of species is different between the trees but the pattern of ancestry (the nodes shared by each species) remains unchanged.

Phylogenetic analysis uses molecular sequence data in an attempt to identify the one correct tree which accurately represents the evolutionary history of a set of sequences.

A variety of methods are available for inferring phylogenetic trees from molecular sequence data. The structure of a phylogenetic tree is generally inferred by applying either an *algorithm* or some set of *optimality criteria*. An algorithm is a programmed series of steps designed to construct a single tree (**Figure 12.17**). Algorithms used to build phylogenetic trees include the *Unweighted Pair Group Method with Arithmetic Mean* (UPGMA) and *Neighbor Joining* methods. Alternatively, phylogenetic methods that employ optimality criteria include *parsimony, maximum likelihood,* and *Bayesian* analyses. These latter methods evaluate many possible trees and select the one tree that has the best optimality score, that is, they select the tree that best fits the sequence data given a discrete model of molecular evolution. Optimality scores are calculated on the basis of evolutionary models that describe how molecular sequences change over time. For example, evolutionary models can account for variation in substitution rates and base frequencies between sequence positions.

Limitations of Phylogenetic Trees

Molecular phylogeny provides powerful insights into evolutionary history, but it is important to consider the limitations of building and interpreting phylogenetic trees. For example, it can be difficult to choose the true tree based on available sequence data if several different trees fit the data equally well. *Bootstrapping*, a statistical method in which information is resampled at random, is an approach used to deal with uncertainty in phylogenetic trees. Bootstrap values indicate the percentage of the time that a given node in a phylogenetic tree is supported by the sequence data. High bootstrap values indicate that a node in the tree is likely to be correct, while low bootstrap values indicate that the placement of a node cannot be accurately determined given the data.

Homoplasy, also known as *convergent evolution*, occurs when organisms share a trait that was not inherited from a common

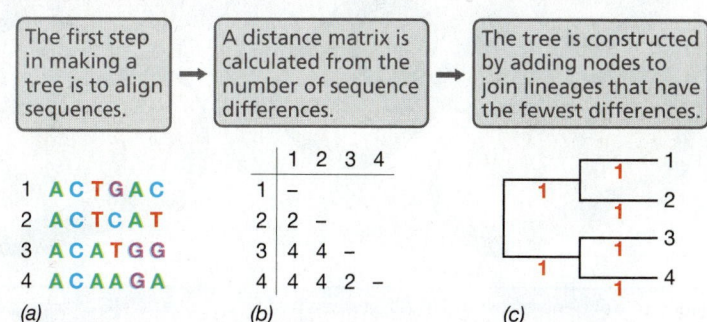

Figure 12.17 Building phylogenetic trees. The number of nucleotide differences between gene sequences can be used to build a phylogenetic tree. In the sequence alignment *(a)* we can count the number of differences between each pair of sequences to build a distance matrix *(b)*. This distance matrix can be used to build a tree *(c)* where the cumulative lengths of the horizontal branches (labeled with a red "1") between any two species in the tree are proportional to the number of nucleotide differences between these species.

Recurrent mutation can erase evolutionary information, causing sequence differences to underestimate true distances.

1	A C T G
2	A G T G
3	A G A C
4	C G A C

1	A C T G
2	A G T G
3	A A T C
4	A C T C

Substitution mutations

Four differences are observed between sequences 1 and 4.

Only one difference is observed between sequences 1 and 4.

(a)

The problem of recurrent mutation increases over evolutionary time.

Actual

Observed

Nucleotide changes

Time

(b)

Figure 12.18 The problem of homoplasy due to recurrent mutation. It is possible for recurrent mutation to obscure the true number of mutations that have occurred since a pair of sequences have shared a common ancestor. *(a)* We observe two series of mutations during the evolution of a gene sequence. On the left side, the number of mutations is equal to the number observed between species 1 and 4. However, if there is recurrent mutation (right side), the number of mutations observed between species 1 and 4 can be less than the number that actually occurred. *(b)* The likelihood of recurrent mutation increases as more and more mutations accumulate over time.

ancestor. An example is the evolution of wings in insects and birds. These traits evolved separately and do not indicate that a winged ancestor was shared among insects and birds. Homoplasy occurs in molecular sequences as well, when similar sequence positions result from recurrent mutation rather than due to inheritance from a common ancestor. The problem of homoplasy in molecular phylogeny then increases in proportion to evolutionary time (**Figure 12.18**).

The prevalence of horizontal gene transfer (⟳ Section 6.12) also creates complications when considering the evolutionary history of microorganisms. When the sequence of a gene is used to infer the phylogeny of an organism, it must be assumed that the gene is inherited *vertically* from mother to daughter throughout the evolutionary history of the organism. The *horizontal* exchange of genes between unrelated organisms violates this assumption (**Figure 12.19**). Hence, it is important to consider the difference between a *gene phylogeny*, which depicts the evolutionary

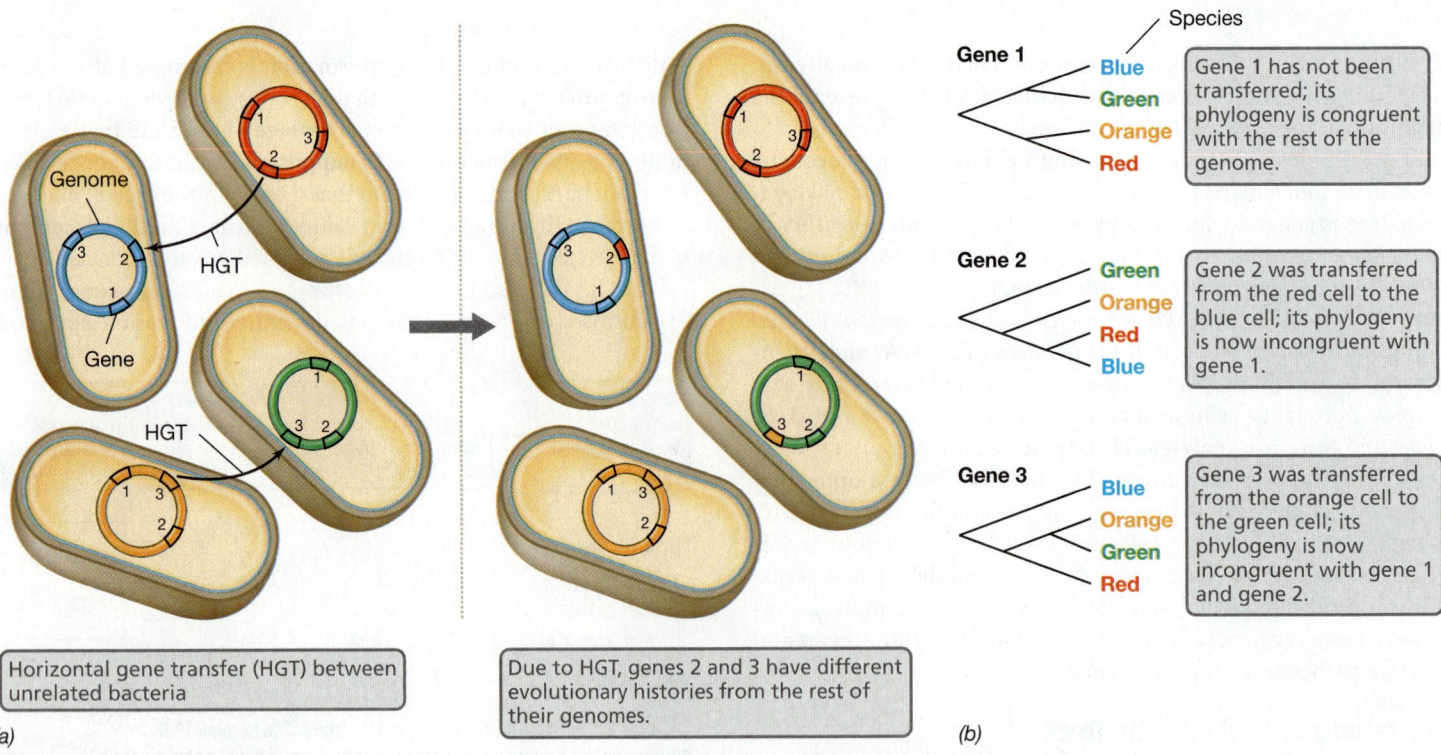

Species

Gene 1
Blue
Green
Orange
Red

Gene 1 has not been transferred; its phylogeny is congruent with the rest of the genome.

Gene 2
Green
Orange
Red
Blue

Gene 2 was transferred from the red cell to the blue cell; its phylogeny is now incongruent with gene 1.

Gene 3
Blue
Orange
Green
Red

Gene 3 was transferred from the orange cell to the green cell; its phylogeny is now incongruent with gene 1 and gene 2.

Genome
HGT
Gene
HGT

Horizontal gene transfer (HGT) between unrelated bacteria

Due to HGT, genes 2 and 3 have different evolutionary histories from the rest of their genomes.

(a)

(b)

Figure 12.19 Horizontal gene transfer. The horizontal transfer of a gene will cause it to have a different evolutionary history from the rest of the genome. *(a)* Genes are transferred horizontally between distantly related microorganisms. Colors are used to match microorganisms with their genetic material. *(b)* As a result of the horizontal transfer events in part a, we now observe different phylogenetic trees for gene 1, gene 2, and gene 3. Only the gene tree for gene 1, which was not transferred, remains congruent with the organismal phylogeny.

history of an individual gene, and an *organismal phylogeny*, which depicts the evolutionary history of the cell. In general, SSU rRNA sequences appear to be transferred horizontally at very low frequencies, and rRNA gene phylogenies agree largely with those of most genes that encode genetic informational functions in the cell. Thus, SSU rRNA gene sequences are generally considered to provide a record of organismal phylogeny. Nevertheless, many genes in microbial genomes have been acquired by horizontal gene transfer at some point in their evolutionary history and this process has important implications for microbial evolution, as we will see in the following section.

MINIQUIZ --

- How are DNA sequences obtained for phylogenetic analysis?
- What does a phylogenetic tree depict?
- Why is sequence alignment critical to phylogenetic analysis?

III • Microbial Evolution

While many of the basic principles of evolution are conserved across all domains of life, certain aspects of microbial evolution are uncommon in plants and animals. For example, *Bacteria* and *Archaea* are generally haploid and asexual, they have several mechanisms for horizontal gene transfer that result in the asymmetrical exchange of genetic material uncoupled from reproduction, and their genomes can be remarkably heterogeneous and highly dynamic. In this section we consider the processes that cause the diversification of microbial lineages and how these forces impact evolution of microbial genomes.

12.6 The Evolutionary Process

In its simplest form, evolution is a change in **allele** frequencies in a set of organisms over time. Alleles are alternate versions of a given gene. New alleles arise due to mutation and recombination, and changes in allele frequencies can occur through a variety of changes, including selection and genetic drift. How do these simple mechanisms give rise to the origin and divergence of microbial species?

Origins of Genetic Diversity

Mutations are random changes in DNA sequence that accumulate in all DNA sequences over time; they are a fundamental source of the natural variation that drives the evolutionary process. Most mutations are neutral or deleterious, though some can be beneficial. Mutations take several forms including *substitutions*, *deletions*, *insertions*, and *duplications* (Chapter 10). Duplication events produce a redundant copy of a gene that can be modified by further mutation without losing the function encoded by the original gene. Hence, duplications allow for the diversification of gene function.

Recombination is a process by which segments of DNA are broken and rejoined to create new combinations of genetic material (⮌ Section 10.5). Recombination can cause reassortment of genetic material already present in a genome and is also required for the integration into the genome of DNA acquired through horizontal gene transfer. Recombination can be broadly classified as either *homologous* or *nonhomologous*. Homologous recombination requires short segments of highly similar DNA sequence flanking the region of DNA being transferred (⮌ Section 10.5). By contrast, nonhomologous recombination is mediated by several mechanisms (⮌ Section 10.5) that share in common the fact that they do not require high levels of sequence similarity to initiate successful DNA integration.

Selection and Genetic Drift

New alleles result when mutation and recombination cause variation in gene sequences. Evolution occurs when different alleles change in frequency in a population over a span of many generations. Evolutionary biologists have described many different mechanisms that may govern this evolutionary process but chief among them are the forces of selection and genetic drift.

Selection is defined on the basis of **fitness**, the ability of an organism to produce progeny and contribute to the genetic makeup of future generations. Most mutations are *neutral* with respect to fitness and they have no effect on the cell due to the degeneracy of the genetic code (⮌ Section 4.11). These mutations generally accumulate in DNA over time. Some mutations are *deleterious*; these decrease the fitness of an organism by disrupting gene function. Deleterious mutations are generally purged from populations over time by natural selection. Some mutations can be *beneficial*, increasing the fitness of an organism, and these mutations are favored by natural selection, increasing in frequency in a population over time. An example of a beneficial mutation would be a mutation that induces antibiotic resistance in a pathogenic bacterium infecting a person undergoing antibiotic therapy. It is important to remember that all mutations occur by chance; the selective nature of the environment does not *cause* adaptive mutations but simply *selects* for the growth and reproduction of those organisms that have incurred mutations that provide a fitness advantage.

While Darwin proposed natural selection as the mechanism by which gene frequencies change over time, evolutionary change can occur through mechanisms other than selection. A chief example is **genetic drift** (**Figure 12.20**), a random process that can cause gene frequencies to change over time, resulting in evolution in the absence of natural selection. Genetic drift occurs because some members of a population will have more offspring than others simply as a result of chance; over time these chance events can result in evolutionary change in the absence of selection. Genetic drift is most powerful in small populations and in populations that experience frequent "bottleneck" events. The latter occur when a population experiences a severe reduction in population size followed by regrowth from the cells that remain. For example, genetic drift can be very important in the evolution of pathogens since each new infection is caused by a small number of cells colonizing a new host. Hence, pathogen populations can change rapidly as a result of random genetic drift as illustrated in Figure 12.20.

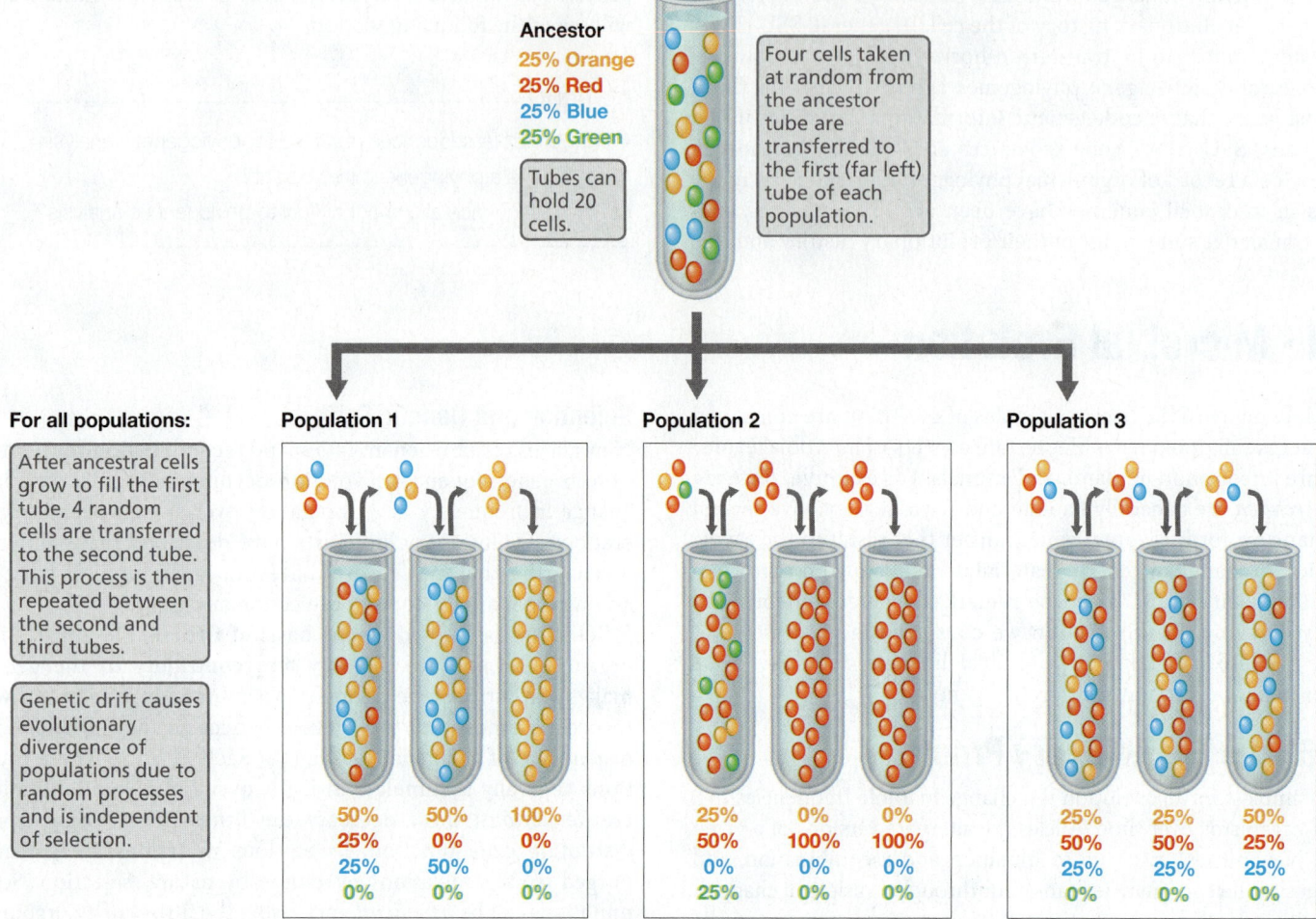

Figure 12.20 Genetic drift. Genetic drift is a random process that can cause gene frequencies in a population to change over time, causing evolution without natural selection. In this example, a population containing four different bacterial genotypes (indicated by colors), each at equal frequency, is present in the ancestor tube. Four cells at random are then transferred to each of three new tubes and the cells allowed to grow to fill each tube. There is no difference in fitness between the cells and so they grow equally. Cells taken at random are then transferred in two successive rounds. Striking differences in genotype frequencies between the populations are observed after only three rounds of transfers.

New Traits Can Evolve Quickly in Microorganisms

A change in the environment or the introduction of cells to a new environment can cause rapid evolutionary changes in microbial populations. Microorganisms typically form large populations and can reproduce quickly, producing a new generation in as few as 20 minutes for some species, and thus evolutionary events in microbial populations can often be observed in the laboratory on relatively short time scales. The heritable variation already present in a population provides the raw material upon which natural selection acts following such a change in the selective environment. Here we consider two examples of rapid evolutionary change in bacteria, one involving the rapid loss of a trait in *Rhodobacter*, and one involving the acquisition of a new trait in *Escherichia coli*.

Rhodobacter is a phototrophic purple bacterium that carries out anoxygenic photosynthesis (⌘ Section 13.3) in illuminated anoxic environments. When cultured anaerobically in either the light or the dark, the cells synthesize bacteriochlorophyll and carotenoids.

It is the absence of O_2, not the presence of light, that signals pigment synthesis in purple bacteria. In the light these pigments participate in photosynthetic reactions that lead to ATP synthesis, but in darkness, these pigments provide no benefit to the cell.

Random mutations occasionally generate *Rhodobacter* cells that produce either reduced levels of photopigments or no photopigments at all. In nature, the ability to carry out photosynthesis is an adaptive trait of significant value, and thus photosynthetic mutants are lost and wild-type cells dominate. However, in contrast to natural conditions, there is no selection against *Rhodobacter* cells that have a reduced capacity to photosynthesize if they are cultured in the laboratory in constant darkness. Mutants that produce reduced levels of photopigments arise in dark cultures just as they do in phototrophic cultures, but in the dark, these mutants are selected for and quickly take over the population (**Figure 12.21**).

Photopigments are useless in the dark, and mutants conserve energy by avoiding the metabolic cost of synthesizing them.

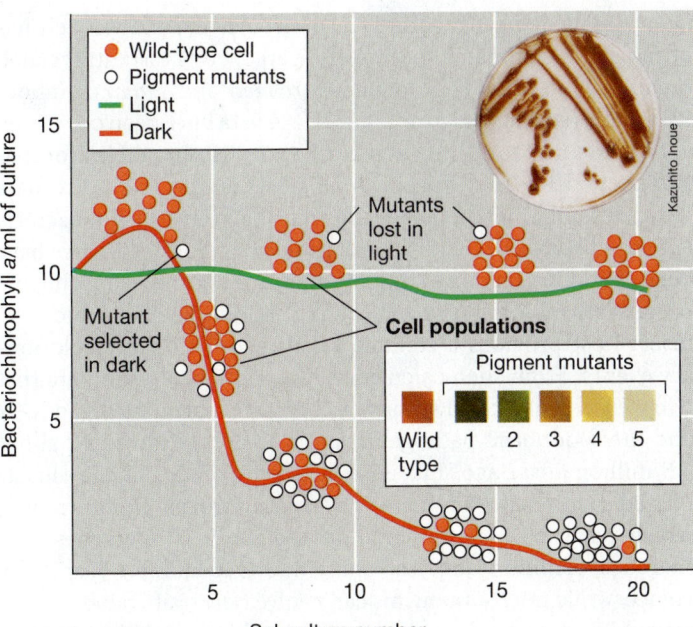

Figure 12.21 Survival of the fittest and natural selection in a population of phototrophic purple bacteria. Serial subculture of the purple bacterium *Rhodobacter capsulatus* in the dark quickly selects for nonphototrophic mutants that outcompete and grow faster than cells still making bacteriochlorophyll and carotenoids. Photos: top, plate culture showing colonies of phototrophic cells of *R. capsulatus*; bottom, close-up photos of colonies of wild type and five pigment mutants (1–5) obtained during serial dark subculture. Wild-type cells are reddish-brown from their assortment of carotenoid pigments. The color of mutant colonies reflects the absence (or reduced synthesis) of one or more carotenoids. Mutant strain 5 lacked bacteriochlorophyll and was no longer able to grow phototrophically. Mutant strains 1–4 could grow phototrophically but at reduced growth rates from the wild type. Data adapted from Madigan, M.T., et al. 1982. *J. Bacteriol. 150*: 1422–1429.

Hence, the photosynthetic mutants are able to outcompete wild-type cells that produce a full complement of photopigments. Although these mutants have reduced phototrophic capacities or in some cases have completely lost the ability to grow phototrophically (see photo inset in Figure 12.21), in permanent darkness they quickly become the fittest organisms in the population and therefore enjoy the greatest reproductive success. Mutations affecting photosynthesis occur at the same rate in the light as in the dark, but in the light the selection for phototrophy is so strong that such mutants are quickly lost from the population.

Experimental evolution is a growing field of study enabled by the rapid growth of bacterial populations and the ability to preserve bacteria indefinitely by freezing. The latter makes it possible to maintain a living "fossil record" of ancestral organisms that can be thawed later and compared to evolved strains. For example, the *Escherichia coli* long-term evolution experiment (LTEE), which has been running since 1988, has tracked the evolution of 12 parallel lines of *E. coli* through more than 50,000 generations. The *E. coli* LTEE cultures have been grown aerobically on a minimal medium with glucose as a sole source of carbon and energy. *E. coli* is typically propagated in a rich medium that contains an excess of all the nutrients cells need to grow and so the minimal glucose medium used in the LTEE represents a new adaptive environment in which *E. coli* can evolve over time.

In the LTEE, both the ancestor and the evolved lines were genetically engineered to contain a neutral marker that made their colonies either red or white. The marker made it possible to measure the fitness of evolved strains relative to the ancestor by competing them against one another (**Figure 12.22a**). Genome sequencing during the experiment revealed that mutations accumulated randomly over time in the evolved lines. However, the relative fitness of the evolved lines on minimal glucose medium increased dramatically over the first 500 generations as a result of selection acting on mutations beneficial in this new environment (Figure 12.22b). The fitness of the evolved lines continued to increase, albeit at a reduced rate, as a result of further selection over the course of the experiment. Most remarkably, after 31,500 generations, one of the evolved lines obtained the ability to use citrate as an energy source (Figure 12.22c). Citrate was present as a pH buffer in the media used in this experiment and was not considered a potential carbon source for *E. coli* because the inability to grow aerobically on citrate is a diagnostic trait for *E. coli*. However, the random accumulation of mutations in this one evolved line modified preexisting genes in such a way as to allow for the evolution of a new adaptive trait. The diverged strains can now exploit a new resource that was unavailable to the ancestral population. Since they can now use both citrate and glucose, these cells grow to a much higher cell density than the ancestor (Figure 12.22c). The fact that only one of the 12 parallel lines evolved the ability to grow on citrate demonstrates the chance nature of evolution.

The transitions shown in these experiments remind us of how quickly evolutionary pressures can shift even major properties (such as metabolic strategies) of a microbial cell population. In the case of *Rhodobacter*, a mutation that is deleterious in the wild provides a selective advantage when the organism is grown in the laboratory in a continuously dark environment. Under this new condition, evolution causes *Rhodobacter* to lose unneeded metabolic machinery. In the case of *E. coli*, the accumulation of random mutations allows for the accumulation of genetic diversity in a population. Billions of different mutations were sampled by the population over thousands of generations and some rare combination of mutations, by chance, gave the cells the ability to exploit citrate as a resource. Natural variation caused by chance mutation generated a new trait, the ability to use citrate, and since the environment in which the cells were grown happened to contain citrate, this mutation provided a selective advantage to those cells. In the absence of citrate, these mutations would still occur at the same rate. However, in the absence of a selective benefit, cells able to use citrate would likely disappear from the population over time.

Speciation of Microorganisms Can Take a Long Time

Species can possess a wide variety of individuals with different traits. As we discussed above, microorganisms can evolve new traits with remarkable speed and as a result, microbial species can be genetically and phenotypically diverse.

Sequence changes can be used as a **molecular clock**, in order to estimate the time since two lineages have diverged. Major assumptions of the molecular clock approach are that nucleotide changes accumulate in a sequence in proportion to time, that

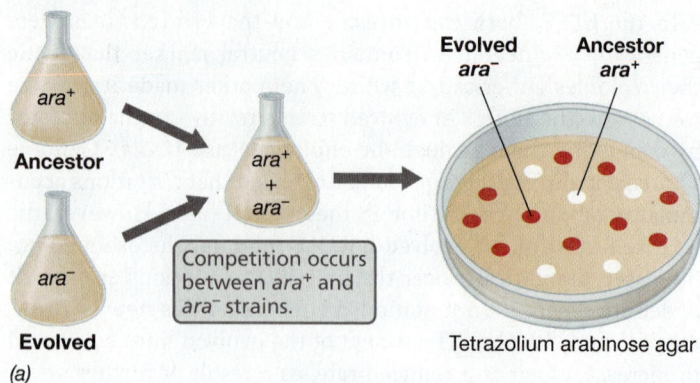

(a)

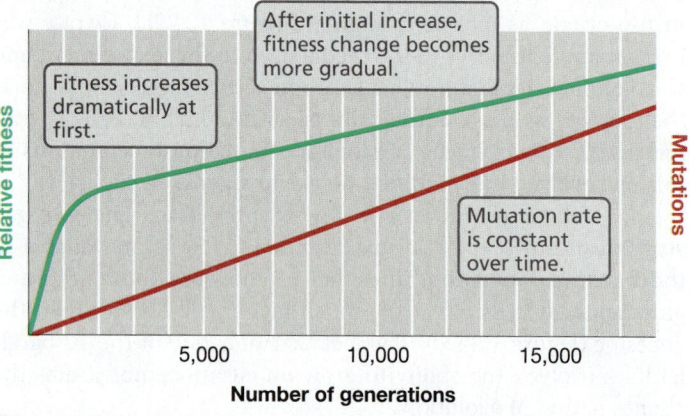

(b)

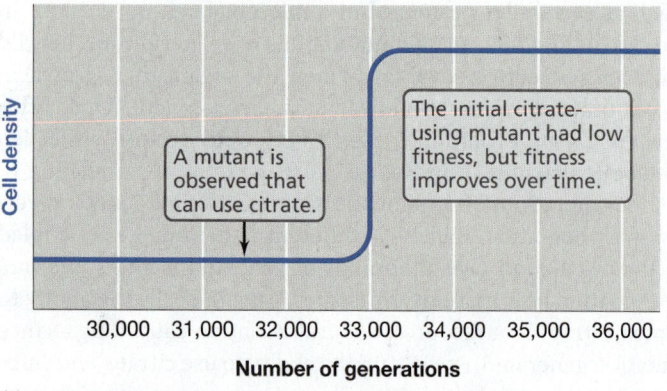

(c)

Figure 12.22 Long-term evolution of *E. coli*. *(a)* In the *Escherichia coli* long-term evolution experiment (LTEE), ancestral and derived lines differ in a mutation that affects their ability to use arabinose, allowing them to be differentiated by their colony color when grown on tetrazolium arabinose agar. *(b)* Competition experiments between evolved and ancestral strains show that relative fitness in minimal glucose media increases dramatically for evolved lines. *(c)* The ability to use citrate aerobically evolved in one of 12 LTEE lines. Cells growing on minimal glucose typically grow to low cell density, but the ability to use both glucose and citrate allowed the mutant cell line to reach significantly higher cell densities. Relative fitness is a measure of the growth rate of the evolved strain to that of the ancestral strain.

such changes are generally neutral and do not interfere with gene function, and that they are random. Molecular clock estimates are most reliable when they can be calibrated with evidence from the geological record. The molecular clock approach has been used to estimate the time of divergence of distantly related organisms, such as the domains *Archaea* and *Eukarya* (about 2.8 billion years ago, Figure 1.4*b*). These data have been combined with evidence from the geological record from stable isotopes (*⮂* Section 18.9) and specific biological markers to approximate when different metabolic patterns may have emerged in bacteria (Sections 12.1, 12.2; Figure 12.1). Molecular clock estimates have been calibrated on more contemporary time scales using obligate bacterial symbionts of insects (*⮂* Section 22.9) for which the insect host provides a suitable fossil record to date evolutionary events. From such calculations it is possible to estimate that two well-characterized strains of *E. coli*, the harmless strain K-12 and the foodborne pathogenic strain O157:H7, diverged about 4.5 million years ago. Likewise, it is estimated that the closely related *E. coli* and *Salmonella enterica* serovar Typhimurium, which have 2.8% dissimilarity in their 16S rRNA genes, last shared a common ancestor some 120–140 million years ago. Hence, while microorganisms can evolve new traits rapidly, most microbial species are ancient and microbial speciation appears to take a very long time.

MINIQUIZ

- What are the different processes that give rise to genetic variation?
- What is the difference between selection and genetic drift and how do they promote evolutionary change?
- In the experiment of Figure 12.21, why did the dark cell population lose its pigments?

12.7 The Evolution of Microbial Genomes

The dynamic nature of microbial genomes was revealed in dramatic fashion when the first genomes were sequenced from multiple strains of a single species. Genome sequencing of *Escherichia coli* strain K-12 and two pathogenic strains showed that only 39% of their genes were shared among all three genomes (**Figure 12.23**). The three genomes varied in size by more than a million base pairs in length and each contained a unique and diverse complement of genes acquired through horizontal gene transfer. Genomes of many microbial species have now been examined in this way and have revealed that genes in microbial genomes can be placed into two classes: the **core genome**, genes shared by all members of a species, and the **pan genome**, the core genome plus genes that are not shared by all members of a species and which are often acquired through horizontal gene transfer (Figure 12.19). In Chapter 6 we introduced this concept and here we consider the forces that drive these patterns of genome evolution.

The Dynamic Nature of the *Escherichia coli* Genome

More than 20 genomes have been sequenced from different strains of *E. coli*, providing further insights into the nature of the core and pan genome. *E. coli* genomes have on average 4721 genes, with individual strains having as few as 4068 or as many as

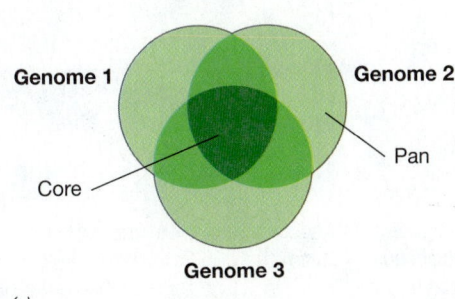

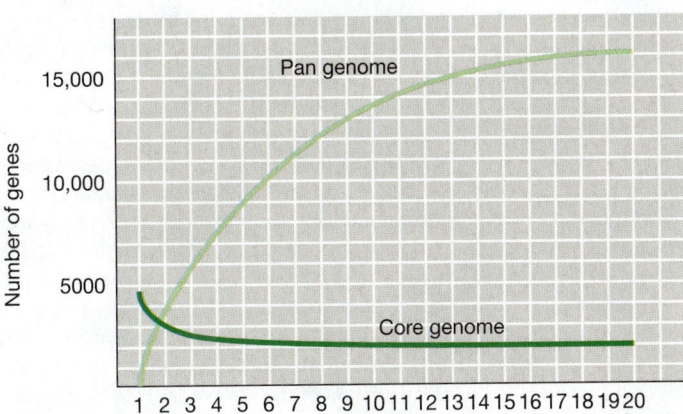

Figure 12.23 The core and pan genome concept. Microbial genomes are dynamic and heterogeneous. The first three genomes sequenced from different strains of *E. coli* were found to have only 39% of their genes in common. The core genome is considered the set of genes that are shared by all members of species (darkest green in part a), while the pan genome is the core genome plus that set of genes found unique to one strain or found in only a subset of strains (light green in part a). The size of the core and pan genome can vary between species. In *E. coli* the core genome is composed of approximately 1976 genes *(b).* The size of the pan genome in *E. coli* is not fixed, as each different strain has a unique complement of genes acquired from horizontal gene exchange. Data adapted from Touchon, M., et al. 2009. *PLoS Genetics 5*: (1) e1000344.

5379 total genes. The core genome consists of only 1976 genes present in all strains, accounting for less than half the genes present in the average *E. coli* genome. The size of the core genome can be expected to decrease as the evolutionary distance of strains increases. Taking this prediction to its extreme, only 50 to 250 genes are predicted to be universally present in all species of *Bacteria* and *Archaea*.

The number of unique genes observed continued to increase with each new *E. coli* genome sequenced such that a total of 17,838 unique genes are present in the 20 genomes (Figure 12.23*b*). Subtracting the contribution from the core genome, this indicates more than 15,862 genes not shared by all strains. A great many of these genes have clearly been inherited through horizontal gene exchange rather than through vertical patterns of inheritance. Genome analysis reveals that the core and pan genome concept is a general feature of microbial genomes, though the relative number of genes present in each pool can vary between species. The dramatic change in genome size and gene content between strains of a single species indicates that microbial genomes are highly dynamic; that is, genomes can shrink or enlarge relatively quickly over time. The existence of a pan genome suggests that *Bacteria* and *Archaea* are constantly sampling genetic information from their environment through horizontal gene transfer.

The dynamic nature of microbial genomes is a manifestation of the evolutionary mechanisms that we have already described (Section 12.6). Variations between genomes arise due to the forces of mutation and recombination, and the evolutionary dynamics of genomes are governed by selection and genetic drift. Moreover, evidence for horizontal gene transfer is widespread in microbial genomes. Patterns of gene exchange appear to be governed by phylogenetic distance, with rates of gene exchange between genomes declining as phylogenetic distance increases. In the core genome of *E. coli*, most horizontal gene transfer takes place between close relatives and occurs by replacement of homologous DNA segments of 50 to 500 base pairs in length. While many horizontal gene transfer events are due to homologous replacements, insertions resulting from nonhomologous recombination are also common in microbial genomes. Comparative analysis of *E. coli* genomes indicates that insertions average 4 genes in length but in some cases contain 10 or more genes.

Gene Deletions in Microbial Genomes

Deletions play an important role in microbial genome dynamics (see Explore the Microbial World, "The Black Queen Hypothesis"). Deletions occur with far greater frequency than insertions in microbial genomes, and this bias toward deletions is the force that maintains the small size of microbial genomes. Selection is the main force that counters the effect of deletions, preserving those genes that provide a fitness benefit to the cell. Nonessential and nonfunctional material is deleted over evolutionary time, which is why microbial genomes are tightly packed with genes and contain relatively few noncoding sequences. Most genes acquired by horizontal gene transfer, like most mutations in general, can be expected to be neutral or deleterious to the cell. Hence, it is likely that new genes are acquired from the environment and those that do not convey a fitness benefit are eroded from the genome over time due to the relentless accumulation of deletions. In addition, genetic drift (Figure 12.20) can promote the rapid accumulation of deletion events when population sizes are small or when populations pass through a bottleneck. Deletions are thought to cause the extremely small genomes found in many obligate intracellular symbionts and pathogens (⮌ Sections 6.4 and 22.9).

MINIQUIZ --

- What is the difference between the core and pan genomes of a given species?
- What kind of recombination might have the greatest impact on the core genome?
- What effects do deletions have on the evolution of microbial genomes?

I t is a common misconception that evolution inevitably causes organisms to increase in complexity over time. In reality, evolution is both a give and a take proposition. Fitness changes are completely dependent on the environment, and fitness in some environments may actually be improved by a loss, rather than a gain, of specific genes.

The *Black Queen hypothesis*[1] posits a mechanism and a rationale for this loss of function whose end result is the evolution of mutual dependence in microbial communities. The term *Black Queen* refers to the card game Hearts in which there are two winning strategies. One winning strategy is to avoid getting stuck with the queen of spades. In this strategy each player seeks to lose as many contests ("tricks") as possible so as not to be forced to collect the black queen. The second winning strategy is to "shoot the moon" by collecting all of the trump cards including the black queen. In its microbial context, the Black Queen hypothesis embraces these card game strategies by proposing that some organisms optimize fitness (that is, "win") by the selective loss of specific genes while others optimize fitness by keeping them all.

The Black Queen hypothesis proposes that certain microbial genes encode extracellular products, such as metabolites or enzymes, which can be used by all or most members of the community. If an organism remains in the community, then selection will be relaxed on genes that encode the synthesis of products that are provided by other members of the community. The presence of such shared products in the community renders genes with similar functions nonessential for some community members (**Figure 1**). The mutation bias toward deletions can then cause these genes to be lost from the genome (Section 12.7).

The fitness of organisms that lose functions and develop dependencies will actually increase in the community since these organisms no longer bear the costs of production. Such organisms will remain competitive as long as they remain within the community, but they may be unable to grow if separated from the community in which they coevolved. In this way, mutual dependencies accumulate within microbial communities over time. The Black Queen hypothesis also explains the not-uncommon observation that some microorganisms can only be grown in the laboratory in coculture with one or more other species from their environment.

In contrast to the gene loss strategy, organisms that preserve all essential functions (those that shoot the moon in the Hearts analogy) bear the costs of maintaining all gene functions, which puts them at a disadvantage to mutually dependent competitors when competing in the native community. However, cells that maintain their ability to grow independently still have a winning strategy because, unlike their mutually dependent competitors, they retain the option of dispersing to new habitats and growing outside of the native community.

Finally, in addition to describing how microbial community interdependencies might come about, the Black Queen hypothesis also reminds us of how interwoven microbial communities actually are. We will see in later chapters that several molecular tools are available to unwind this complexity and reveal both the diversity of the community and its genetic and metabolic potential.

[1]Morris J.J., R.E. Lenski, and E.R. Zinser 2012. The Black Queen hypothesis: Evolution of dependencies through adaptive gene loss. *mBio 3*: e00036-12.

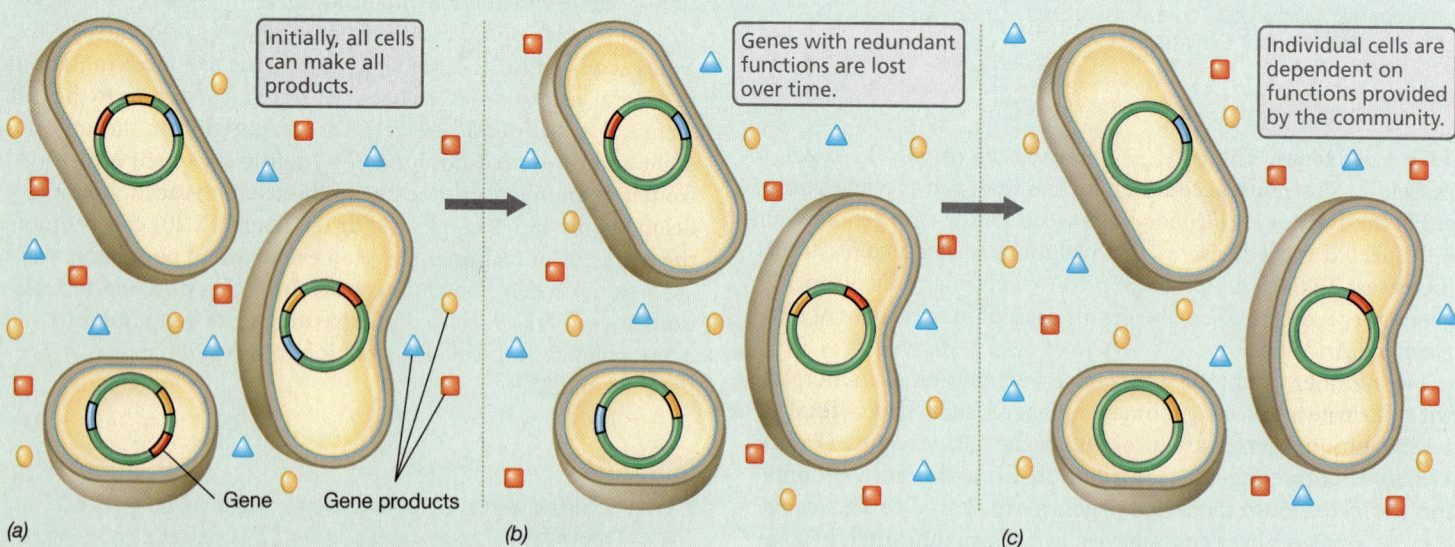

Figure 1 **The Black Queen hypothesis and the evolution of dependence in microbial communities.** *(a)* Three species in a community each have three different genes that make extracellular products that benefit the whole community (a gene and its product are shown in the same color). *(b)* Over time, random mutations cause functions to be lost from the genomes. *(c)* As long as some members of the community continue to make each product, there will be no fitness cost when a single species loses a single gene. Over time, the three species thus become mutually dependent.

IV • Microbial Systematics

Systematics is the study of the diversity of organisms and their relationships. It links phylogeny with **taxonomy**, the science in which organisms are characterized, named, and classified according to several defined criteria. Bacterial taxonomy traditionally has focused on practical aspects of identification and description, activities that rely heavily on comparisons of visible properties (phenotype). At present, the growing use of genetic information, especially DNA sequence data, is increasingly allowing taxonomy to reflect phylogenetic relationships as well.

Bacterial taxonomy has changed substantially in the past few decades, embracing a combination of methods for the identification of bacteria and description of new species. This *polyphasic approach* to taxonomy uses three kinds of methods—*phenotypic*, *genotypic*, and *phylogenetic*—for the identification and description of bacteria. Phenotypic analysis examines the morphological, metabolic, physiological, and chemical characteristics of the cell. Genotypic analysis considers characteristics of the genome. These two kinds of analysis categorize organisms based on similarities. They are complemented by phylogenetic analysis, which seeks to place organisms within an evolutionary framework using molecular sequence data (Sections 12.4, 12.5).

12.8 The Species Concept in Microbiology

At present, there is no universally accepted concept of **species** for microorganisms; this has been referred to as the "species problem" in microbiology. Species are the fundamental units of biological diversity, and how we distinguish and classify species in microbiology greatly affects our ability to explain the diversity of the microbial world. Microbial systematics combines phenotypic, genotypic, and sequence-based phylogenetic data within a framework of standards and guidelines for describing and identifying microorganisms in a taxonomic framework, but the issue of what actually constitutes a species remains controversial.

Current Definition of Microbial Species

From a taxonomic standpoint, all members of a species should be genetically and phenotypically cohesive, and their traits should be distinct from those described for other species. In addition, a species should be **monophyletic**, that is, the strains composing the species should all share a recent common ancestor to the exclusion of other species. For most of the history of microbiology it was impossible to resolve phylogenetic relationships, and thus species descriptions did not take into account the evolutionary history of a microorganism. The recognition that molecular sequences record evolutionary history created a crisis in microbial systematics as it became necessary to reconcile more classical species descriptions with insights gained through phylogenetic analyses. In addition, the discovery that microbial genomes are highly heterogeneous and contain many horizontally acquired genes poses a challenge to any definition of microbial species.

The prevailing concept for describing microbial species is best described as a *phylogenetic species concept*. The phylogenetic species concept defines a microbial species pragmatically as a group of strains that share certain diagnostic traits and that are genetically cohesive and share a unique recent common ancestor. This species concept requires that a majority of genes in the species have congruent phylogenies and share a recent common ancestor. The phylogenetic species concept is not based on an evolutionary model of speciation, and thus species described in this way do not necessarily reflect meaningful units in terms of ecological or evolutionary processes. The phylogenetic species concept was developed to facilitate taxonomy, and species justifications derived from this concept are based largely on the expert judgment of taxonomists.

Under the phylogenetic species concept for *Bacteria* and *Archaea*, species are defined operationally as a group of strains sharing a high degree of similarity in many traits and sharing a recent common ancestor for their 16S rRNA genes. Species characterization employs a polyphasic approach that considers a range of different traits in making taxonomic judgments. Traits currently considered most important for identifying species include genomic similarity based on DNA hybridization, and comparisons of small subunit rRNA sequences.

The degree of **DNA–DNA hybridization** between the genomes of two organisms (Figure 12.24) provides a measure of their genomic similarity. We discussed nucleic acid hybridization in Section 11.2. In a hybridization experiment, *probe DNA* obtained from one organism is labeled with a fluorescent or radioactive label, sheared into small pieces, and heated to separate the two DNA strands. The probe is then added to single-stranded and sheared *target DNA* from a second organism and the mixture cooled to allow the DNA strands to reanneal. The genomic similarity between the two organisms is calculated as a percentage of probe hybridized to target relative to a control (probe DNA hybridized to target DNA from the same organism).

A value of 70% or less genomic hybridization and a difference in 16S rRNA gene sequence (Sections 12.3, 12.4) of 3% or more between two organisms is taken as evidence that the two are distinct species. Experimental data suggest that these criteria are valid, reliable, and consistent in identifying new microbial species for taxonomic purposes (Figure 12.25). On the basis of the current phylogenetic species concept, over 10,000 species of *Bacteria* and *Archaea* have been formally recognized. The criteria that should be used to define a genus, the next highest taxon (see Table 12.3), is more a matter of judgment, but discrete genera typically have greater than 5% dissimilarity in their 16S rRNA gene sequences. There are no consensus criteria for defining taxonomic ranks above the level of genera.

How Many Microbial Species Are There?

The result of nearly 4 billion years of evolution is the microbial world we see today (Figure 12.13). Microbial taxonomists agree that no firm estimate of the number of species can be given at present, in part because of uncertainty about what defines a species. However, they also agree that in the final analysis, this number will be very large. Owing to the difficulty in visualizing and characterizing microorganisms, only about 10,000 species

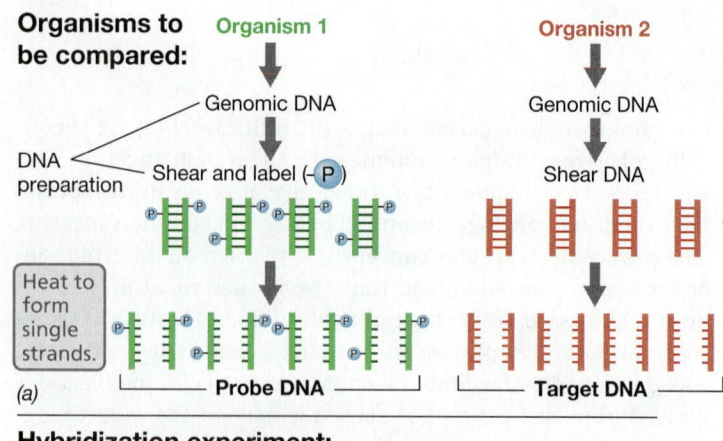

Hybridization experiment:

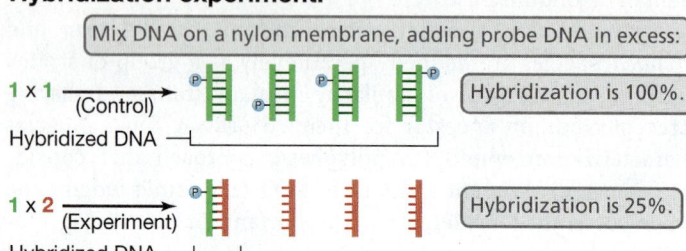

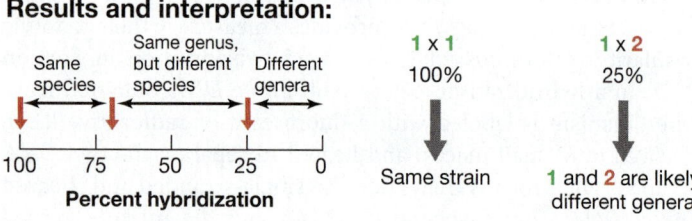

Figure 12.24 Genomic hybridization as a taxonomic tool. *(a)* Genomic DNA is isolated from the organisms to be compared and then sheared and denatured. Probe DNA is prepared from organism 1 by shearing, denaturing, and labeling the DNA (shown here as radioactive phosphate). *(b)* Sheared single-stranded target DNA from each genome is immobilized on a membrane and then hybridized with the labeled probe DNA from organism 1. Radioactivity in the hybridized DNA is measured. *(c)* Radioactivity in the control (organism 1 DNA hybridizing to itself) is taken as the 100% hybridization value.

of *Bacteria* and *Archaea* have been named using the taxonomic species concept. At this time it is impossible to accurately estimate the total number of bacterial and archaeal species on Earth, but their diversity is unquestionably higher than that of all plant and animal species combined and their total species numbers are likely several orders of magnitude higher than the 10,000 already characterized.

Every environment on Earth contains a diverse community of microorganisms. Analyses of 16S rRNA gene sequences indicate that over 10,000 different species can coexist in a single gram of soil! Nearly all plants and animals have some number of unique microorganisms associated with them as either pathogens or commensals on their surfaces or internal structures. Thus, microorganisms are not only the oldest but also the most diverse forms of life on our planet.

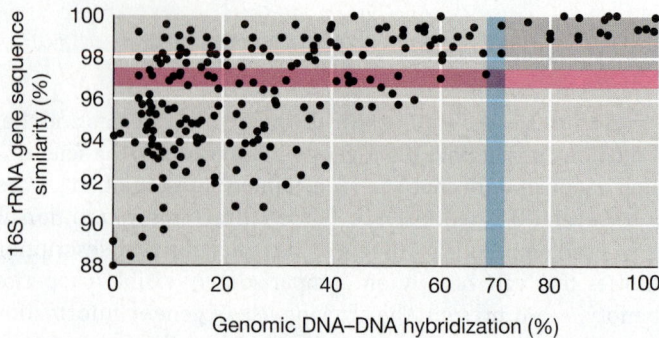

Figure 12.25 Relationship between 16S rRNA gene sequence similarity and genomic DNA–DNA hybridization for pairs of organisms. Pairs of microorganisms are compared on the basis of their 16S rRNA similarities and genomic hybridization values. Points in the upper right region represent pairs of strains that share greater than 97% 16S rRNA gene sequence similarity and 70% genomic hybridization values, and thus are likely members of the same species. Data adapted from Rosselló-Mora, R., and R. Amann. 2001. *FEMS Microbiol. Revs. 25*: 39–67, and Stackebrandt, E., and J. Ebers. 2006. *Microbiology Today. 11*: 153–155.

MINIQUIZ

- What is the difference between taxonomy and phylogeny?
- What are some key criteria from the phylogenetic species concept used to determine whether two strains belong to the same species?
- How many species of *Bacteria* and *Archaea* have been named? How many likely exist?

12.9 Taxonomic Methods in Systematics

A *polyphasic approach*, that is, an approach that uses many different methods in combination, is used to identify and name species of *Bacteria* and *Archaea* in accordance with the currently accepted taxonomic species concept. In this section we describe methods commonly used for characterizing microbial, and primarily prokaryotic, species.

Gene Sequence Analyses

As we have described, gene sequences are commonly determined from PCR-amplified fragments of DNA, and the sequences are analyzed using phylogenetic analyses (Section 12.5). Small subunit rRNA gene sequences are highly conserved, however, and while they provide valuable phylogenetic information, they are not always useful for distinguishing closely related species. By contrast, other highly conserved genes, such as *recA*, which encodes a recombinase protein, and *gyrB*, which encodes a DNA gyrase protein, can be useful for distinguishing bacteria at the species level. The DNA sequences of protein-encoding genes accumulate mutations more rapidly than rRNA genes; for this reason, sequences from such genes can distinguish bacterial species that cannot be resolved by rRNA gene sequence analyses alone (**Figure 12.26**).

Multilocus Sequence Typing

Multilocus sequence typing (MLST) is a method in which several different "housekeeping" genes from several related organisms

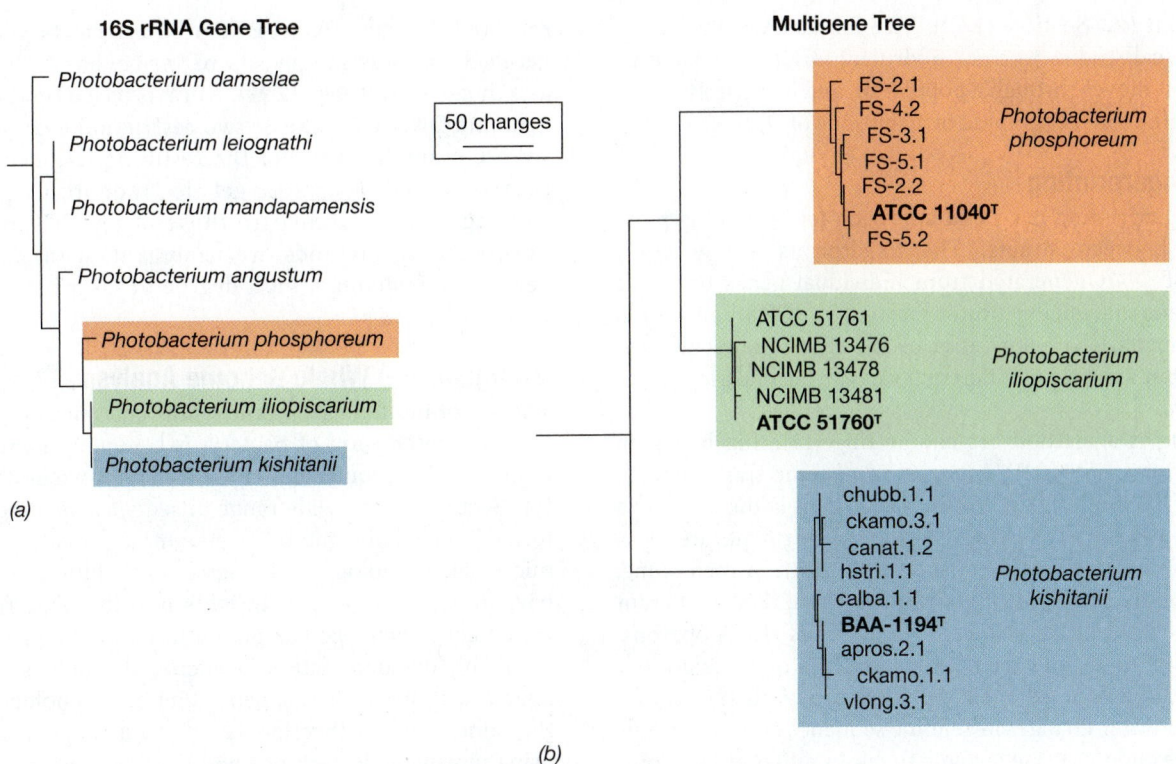

Figure 12.26 Multigene phylogenetic analysis. A phylogeny is shown for species in the genus *Photobacterium*. *(a)* 16S rRNA gene tree, showing the species to be poorly resolved. *(b)* Multigene analysis based on combined analysis of the 16S rRNA gene and *gyrB* and *luxABFE* genes in 21 isolates from three *Photobacterium* species. Multigene analysis clearly resolves the strains into three distinct phylogenetic species, *P. phosphoreum*, *P. iliopiscarium*, and *P. kishitanii*. The scale bar indicates the branch length equal to a total of 50 nucleotide changes. The type strain of each species is listed in bold. (All abbreviations are part of strain designations.) Phylogenetic analyses courtesy of Tory Hendy and Paul V. Dunlap, University of Michigan.

are sequenced and the sequences used collectively to distinguish the organisms. Housekeeping genes encode essential functions in cells and are always located on the chromosome rather than on a plasmid. For each gene, an approximately 450-base-pair sequence is amplified and then sequenced. The alleles of each gene (variants that differ by at least one nucleotide) are each assigned a number. The strain being studied is then assigned an allelic profile, or multilocus sequence type, consisting of a series of numbers representing its particular combination of alleles (**Figure 12.27**). In MLST, strains with identical sequences for a given gene have the same allele number for that gene, and two strains with identical sequences for all the genes have the same allelic profile (and would be considered identical by this method). The relatedness between each allelic profile is expressed in a dendrogram

of genetic distances that vary from 0 (strains are identical) to 1 (strains are only distantly related, if at all).

MLST has sufficient resolving power to distinguish among even very closely related strains of a given species. In practice, strains can be discriminated on the basis of a single nucleotide change in just one of the analyzed genes. However, MLST is not useful for comparing organisms above the species level; its resolution is too sensitive to yield meaningful information for grouping higher-order taxa such as genera and families.

MLST has found its greatest use in clinical microbiology, where it has been used to differentiate strains of various pathogens. This is important because some strains within a species—*Escherichia coli* K-12, for example—may be harmless, whereas others, such as strain O157:H7, can cause serious and even

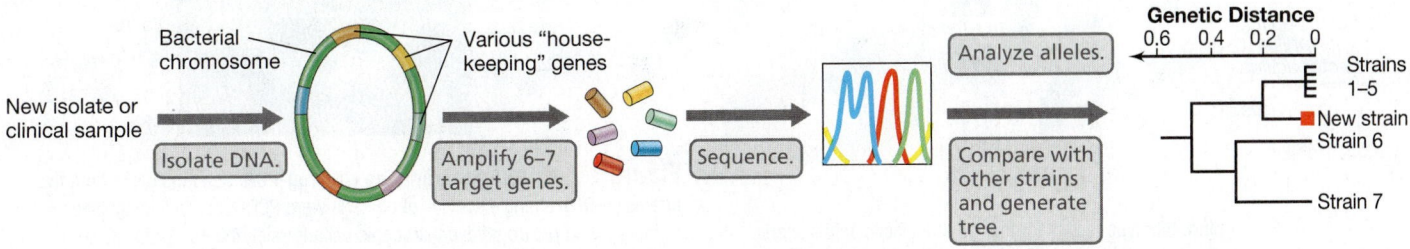

Figure 12.27 Multilocus sequence typing. Steps in MLST leading to a similarity phenogram are shown. Strains 1–5 are virtually identical, whereas strains 6 and 7 are distinct from one another and from strains 1–5.

fatal infections (∂ Section 31.12). MLST is also widely used in epidemiological studies to track a virulent strain of a bacterial pathogen as it moves through a population and in environmental studies to define the geographic distributions of strains.

Genome Fingerprinting

Genome fingerprinting is a rapid approach for evaluating polymorphisms between strains. The fingerprints are generally fragments of DNA generated from individual genes or whole genomes. Gene sequencing is often enabled by PCR amplification of gene fragments. Characterization of SSU rRNA gene sequences is common, but a variety of different genes can be of use in species classification.

Ribotyping is a method of genome fingerprinting based on the localization of SSU rRNA genes on genome fragments. In this method, genomic DNA from an organism is digested by a restriction enzyme (∂ Section 11.1) and the fragments are separated by gel electrophoresis, transferred to a nylon membrane, and labeled with an SSU rRNA gene probe (Figure 12.28). Different microbial species can have different numbers of rRNA operons, ranging from 1 to 15, and the number of rRNA operons present in a microbial genome is a conserved feature of all strains of a species. In addition, changes in genome sequence between strains can cause the endonuclease enzyme to cut in different locations, producing variation in the lengths of the restriction fragments that are visualized. Hence, the size and number of bands detected generates a specific pattern, a kind of genome fingerprint called a *ribotype*, and this pattern can be compared with patterns of reference organisms in a computer database. The ribotype of a particular organism can be unique and diagnostic, allowing rapid identification of different species and even different strains of a species. For these reasons, ribotyping has found many applications in clinical diagnostics and the microbial analyses of food, water, and beverages.

Other genome fingerprinting methods used commonly include *repetitive extragenic palindromic PCR* (*rep-PCR*) and *amplified fragment length polymorphism* (*AFLP*). The rep-PCR method is based on the presence of highly conserved repetitive DNA elements interspersed randomly around the bacterial chromosome. The number and positions of these elements differ between strains of a species. Oligonucleotide primers designed to be complementary to these elements enable PCR amplification of genomic fragments found between the

repeated elements. These PCR products can be visualized using gel electrophoresis to reveal a pattern of bands that can be used as a fingerprint (Figure 12.29). AFLP is based on the digestion of genomic DNA with one or two restriction enzymes and selective PCR amplification of the resulting fragments, which are then separated by agarose gel electrophoresis. Strain-specific banding patterns similar to those of rep-PCR or other DNA fingerprinting methods are generated, with the large number of bands giving a high degree of discrimination between strains within a species.

Multigene and Whole Genome Analyses

The use of multiple genes and whole genomes for the identification and description of bacteria is becoming increasingly common as DNA sequencing capacities improve and costs decline (∂ Section 6.2). A wide range of sequence analyses can be performed on whole microbial genomes, providing insights into microbial physiology and microbial evolution. These analyses have provided important insights into the large role that horizontal gene exchange has played in microbial evolution and on the highly dynamic nature of microbial genomes (Section 12.7). Shared orthologs, that is, genes that are homologous and share the same function (Section 12.5), can be aligned and examined using phylogenetic methods and the average nucleotide identity of these genes determined. Comparative analysis of gene content (presence or absence of genes) and *synteny*, the order of genes in the genome, and genome GC content, provide further insights into relationships between strains. Whole genome sequences can also be used for metabolic reconstruction and characterization of genetic pathways. A range of methods in comparative genomics and population genomics (Chapter 6) have been developed for use in systematic analysis.

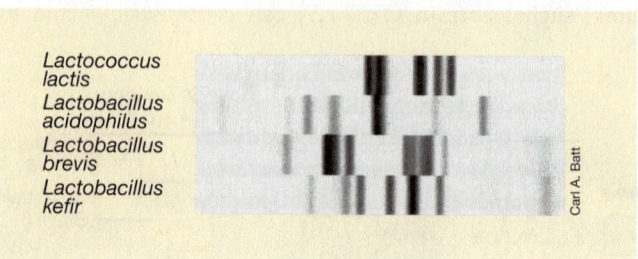

Figure 12.28 Ribotyping. Ribotype results for four different lactic acid bacteria. DNA was taken from a strain of each bacterium, digested into fragments by restriction enzymes, separated by gel electrophoresis, and then probed with a 16S rRNA gene probe. Variations in position and intensity of the bands are important in identification.

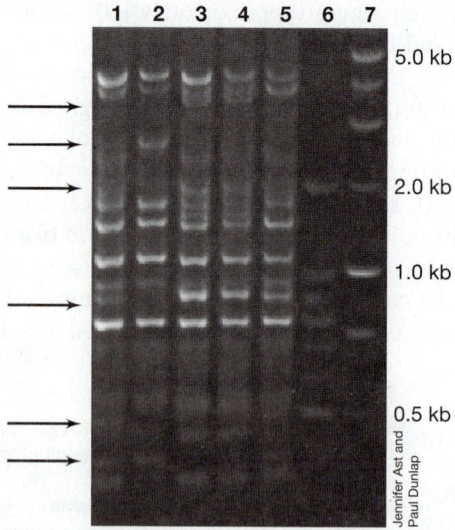

Figure 12.29 DNA fingerprinting with rep-PCR. Genomic DNAs from five strains (1–5) of a single species of bacteria were PCR-amplified using specific primers called *rep* (repetitive extragenic palindromic); the PCR products were separated in an agarose gel on the basis of size to generate DNA fingerprints. Arrows indicate some of the differing bands. Lanes 6 and 7 are 100-bp and 1-kbp DNA size markers, respectively, used for estimating sizes of the DNA fragments.

Phenotypic Analysis

The observable characteristics—the **phenotype**—of a bacterium provide many traits that can be used to differentiate species. Typically, for either describing a new species or identifying a bacterium, several of these traits are determined for the organism of interest. The results are then compared with phenotypes of known organisms, either examined in parallel with the unknowns or from published information. The specific traits used depend on the kind of organism, and which traits are chosen for testing may arise from the investigator's purpose and from substantial prior knowledge of the bacterial group to which the new organism likely belongs. For example, in applied situations, such as in clinical diagnostic microbiology, where identification may be an end in itself and time is of the essence, a well-defined subset of traits is typically used that quickly discriminates between likely possibilities. Table 12.2 lists general categories and examples of some phenotypic traits used in identifications and species descriptions, and we examine one of these traits here.

The types and proportions of fatty acids present in cytoplasmic membrane lipids and the outer membrane lipids of gram-negative bacteria are phenotypic traits often used in taxonomic analyses. The technique for identifying these fatty acids has been nicknamed **FAME**, for *fatty acid methyl ester*, and is in widespread use in clinical, public health, and food- and water-inspection laboratories where pathogens routinely must be identified. The fatty acid composition of *Bacteria* varies from species to species in chain length and in the presence or absence of double bonds, rings, branched chains, or hydroxy groups (**Figure 12.30**). For the

Table 12.2 Some phenotypic characteristics of taxonomic value

Category	Characteristics
Morphology	Colony morphology; Gram reaction; cell size and shape; pattern of flagellation; presence of spores, inclusion bodies (e.g., PHB,[a] glycogen, or polyphosphate granules, gas vesicles, magnetosomes); capsules, S-layers, or slime layers; stalks or appendages; fruiting-body formation
Motility	Nonmotile; gliding motility; swimming (flagellar) motility; swarming; motile by gas vesicles
Metabolism	Mechanism of energy conservation (phototroph, chemoorganotroph, chemolithotroph); utilization of individual carbon, nitrogen, or sulfur compounds; fermentation of sugars; nitrogen fixation; growth factor requirements
Physiology	Temperature, pH, and salt ranges for growth; response to oxygen (aerobic, facultative, anaerobic); presence of catalase or oxidase; production of extracellular enzymes
Cell lipid chemistry	Fatty acids;[b] polar lipids; respiratory quinones
Cell wall chemistry	Presence or absence of peptidoglycan; amino acid composition of cross-links; presence or absence of cross-link interbridge
Other traits	Pigments; luminescence; antibiotic sensitivity; serotype; production of unique compounds, for example, antibiotics

[a]PHB, poly-β-hydroxybutyric acid (↩ Section 2.14).
[b]Figure 12.30

Classes of Fatty Acids in *Bacteria*

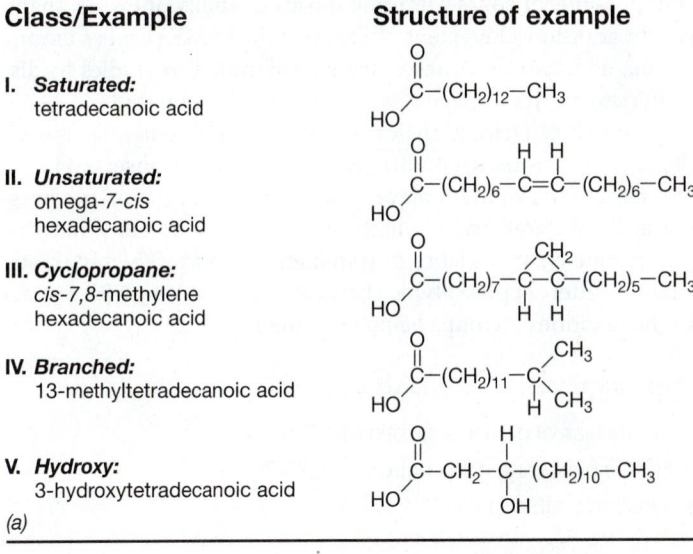

(a)

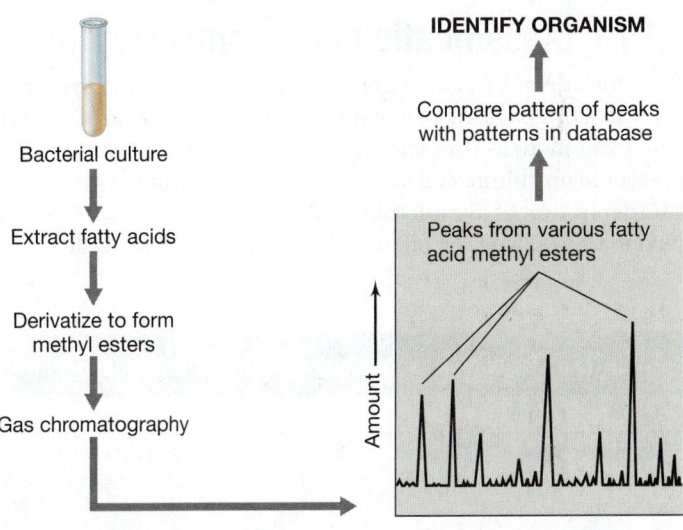

(b)

Figure 12.30 Fatty acid methyl ester (FAME) analysis in bacterial identification. *(a)* Classes of fatty acids in *Bacteria*. Only a single example is given of each class, but in fact, more than 200 structurally distinct fatty acids are known from bacterial sources. A methyl ester contains a methyl group (CH_3) in place of the proton on the carboxylic acid group (COOH) of the fatty acid. *(b)* Procedure. Each peak from the gas chromatograph is due to one particular fatty acid methyl ester, and the peak height is proportional to the amount.

analyses, fatty acids extracted from cell hydrolysates of a culture grown under standardized conditions are chemically volatilized and analyzed by gas chromatography. A chromatogram showing the types and amounts of fatty acids from the unknown bacterium is then compared with a database containing the fatty acid profiles of thousands of reference bacteria grown under the same conditions.

Fatty acid profiles of an organism, like many other phenotypic traits, can vary as a function of temperature, growth phase (exponential versus stationary), and growth medium. Hence, for valid results, it is necessary to grow the unknown organism on a specific medium and at a specific temperature. For many organisms this is impossible, of course, limiting the applicability of FAME analyses. In addition, the extent of variation in FAME profiles among strains of a species, a necessary consideration in studies to discriminate between species, is not yet well documented.

Phenotypic characteristics of strains are generally highly dependent on growth conditions, and phenotypes observed in the laboratory environment may not well represent phenotypes present in the natural environment; thus care must be taken in using phenotypic characteristics in systematic analyses. The systematic value of different phenotypic characteristics can vary with respect to the taxonomic groups being examined.

MINIQUIZ

• What class of genes is used in MLST analyses?
• How is ribotyping different from rep-PCR?
• What is FAME analysis?

12.10 Classification and Nomenclature

We conclude our treatment of microbial evolution and systematics with a brief description of how *Bacteria* and *Archaea* are classified and named—the science of *taxonomy*. Information is also presented on culture collections, which are repositories for scientific deposition of live microbial cultures; on some key taxonomic resources available for microbiology; and on the procedures for naming new microbial species. The formal description of a new microbial species and the deposition of cultures into a culture collection form an important foundation of prokaryotic systematics.

Taxonomy and Describing New Species

Classification is the organization of organisms into progressively more inclusive groups on the basis of either phenotypic similarity or evolutionary relationship. A species is made up of one to several strains, and similar species are grouped into genera (singular, genus). Similar genera are grouped into families, families into orders, orders into classes, up to the domain, the highest-level taxon based on a collection of phenotypic and genotypic information. This hierarchical scheme is illustrated in Table 12.3.

Nomenclature is the actual naming of organisms and follows the **binomial system** of nomenclature devised by the Swedish medical doctor and botanist Carl Linnaeus and used throughout biology; organisms are given genus names and species epithets. The names are Latin or Latinized Greek derivations, often descriptive of some key property of the organism, and are printed in *italics*. By classifying organisms into groups and naming them, we order the natural microbial world and make it possible to communicate effectively about all aspects of particular organisms, including their behavior, ecology, physiology, pathogenesis, and evolutionary relationships. The creation of new names must follow the rules described in the *International Code of Nomenclature of Bacteria*. This source presents the formal framework by which *Bacteria* and *Archaea* are to be officially named and the

---

Table 12.3 Taxonomic hierarchy for the purple sulfur bacterium *Allochromatium warmingii*

Taxon	Name	Properties	Confirmed by
Domain	*Bacteria*	Bacterial cells; rRNA gene sequences typical of *Bacteria*	Microscopy; 16S rRNA gene sequence analysis; presence of unique biomarkers, for example, peptidoglycan
Phylum	*Proteobacteria*	rRNA gene sequence typical of *Proteobacteria*	16S rRNA gene sequence analysis
Class	*Gammaproteobacteria*	Gram-negative bacteria; rRNA sequence typical of *Gammaproteobacteria*	Gram-staining, microscopy
Order	*Chromatiales*	Phototrophic purple bacteria	Characteristic pigments (⟳ Figures 13.2, 13.3, and 13.9)
Family	*Chromatiaceae*	Purple sulfur bacteria	Ability to oxidize H_2S and store S^0 within cells; microscopic observation of S^0 (see photo); 16S rRNA gene sequence
Genus	*Allochromatium*	Rod-shaped purple sulfur bacteria; <95% 16S gene sequence identity with all other genera	Microscopy (see photo)
Species	*warmingii*	Cells 3.5–4.0 μm × 5–11 μm; storage of sulfur mainly in poles of cell (see photo); <97% 16S gene sequence identity with all other species	Cell size measured microscopically with a micrometer; observation of polar position of S^0 globules in cells (see photo); 16S rRNA gene sequence

Sulfur (S^0) globules

Cells of *A. warmingii*

Norbert Pfennig

Table 12.4 Some national microbial culture collections

Collection	Name	Location	Web address
ATCC	American Type Culture Collection	Manassas, Virginia	http://www.atcc.org
BCCM/LMG	Belgium Coordinated Collection of Microorganisms	Ghent, Belgium	http://bccm.belspo.be
CIP	Collection de l'Institut Pasteur	Paris, France	http://www.pasteur.fr
DSMZ	Deutsche Sammlung von Mikroorganismen und Zellkulturen	Braunschweig, Germany	http://www.dsmz.de
JCM	Japan Collection of Microorganisms	Saitama, Japan	http://www.jcm.riken.go.jp
NCCB	Netherlands Culture Collection of Bacteria	Utrecht, The Netherlands	http://www.cbs.knaw.nl/nccb
NCIMB	National Collection of Industrial, Marine and Food Bacteria	Aberdeen, Scotland	http://www.ncimb.com

procedures by which existing names can be changed, for example, when new data warrant taxonomic rearrangements.

When a new microorganism is isolated from nature and thought to be unique, a decision must be made as to whether it is sufficiently different from other prokaryotes to be described as a new taxon. To achieve formal validation of taxonomic standing as a new genus or species, a detailed description of the organism's characteristics and distinguishing traits, along with its proposed name, must be published, and, as just mentioned, viable cultures of the organism must be deposited in at least two international culture collections (Table 12.4). The manuscript describing and naming a new taxon undergoes peer review before publication. A major vehicle for the description of new taxa is the *International Journal of Systematic and Evolutionary Microbiology* (*IJSEM*), the official publication of record for the taxonomy and classification of *Bacteria*, *Archaea*, and microbial eukaryotes. In each issue, the *IJSEM* also publishes an approved list of newly validated names. By providing validation of newly proposed names, publication in *IJSEM* paves the way for their inclusion in taxonomic reference sources (see later). Two websites provide listings of valid, approved bacterial names: List of Prokaryotic Names with Standing in Nomenclature (http://www.bacterio.net), and Bacterial Nomenclature Up-to-Date (http://www.dsmz.de).

The International Committee on the Systematics of Prokaryotes (ICSP) oversees the nomenclature and taxonomy of *Bacteria* and *Archaea*. The ICSP also oversees the publication of *IJSEM* and the *International Code of Nomenclature of Bacteria* and gives guidance to several subcommittees that meet to establish and revise standards for the description of new species in the different groups of prokaryotes.

Bergey's Manual and The Prokaryotes

Because taxonomy is largely a matter of scientific judgment, there is no "official" classification of *Bacteria* and *Archaea*. However, the classification system most widely accepted by microbiologists is that of *Bergey's Manual of Systematic Bacteriology*, a major taxonomic treatment of *Bacteria* and *Archaea* (see Appendix 2 for a list of higher-order taxa from the most recent edition of *Bergey's Manual*). Widely used, *Bergey's Manual* has served the community of microbiologists since 1923 and is a compendium of information on all recognized prokaryotes. Each chapter, written by experts, contains tables, figures, and other systematic information useful for identification purposes.

A second major source describing the physiology, ecology, phylogeny, enrichment, isolation, and cultivation of *Bacteria* and *Archaea* is *The Prokaryotes.* This work is available online by subscription through university libraries. Collectively, *Bergey's Manual* and *The Prokaryotes* offer microbiologists both the concepts and the details of the biology of *Bacteria* and *Archaea* as we know it today; they are the primary resources for microbiologists characterizing newly isolated organisms.

Culture Collections

National microbial culture collections (Table 12.4) are an important foundation of microbial systematics. These permanent collections catalog and store microorganisms and provide cultures upon request (for a fee) to researchers in academia, medicine, and industry. The collections play an important role in protecting microbial biodiversity, just as museums do in preserving plant and animal specimens for future study. However, unlike museums, which maintain collections of chemically preserved or dried, dead specimens, microbial culture collections store microorganisms as *viable cultures*, typically frozen or in a freeze-dried state. These storage methods maintain the cells indefinitely in a living state and prevent genetic changes that might occur if the organisms were continually subcultured.

A related and key role of culture collections is as repositories for *type strains*. When a new species of bacteria is described in a scientific journal, a strain is designated as the nomenclatural type of the taxon for future taxonomic comparison with other strains of that species (see Figure 12.26). Deposition of this type strain in the national culture collections of at least two countries—thereby making the strain publicly available—is a prerequisite for validation of the new species name. Some of the large national culture collections are listed in Table 12.4. Their websites contain searchable databases of strain holdings together with information on the environmental sources of strains and their descriptions.

MINIQUIZ

- What roles do culture collections play in microbial systematics?
- What is the *IJSEM* and what taxonomic function does it fulfill?
- Why might viable cell cultures be of more use in microbial taxonomy than preserved specimens?

BIG IDEAS

12.1 • Planet Earth is about 4.5 billion years old and the first evidence for microbial life can be found in microfossils of prokaryotic cells present in rocks 3.86 billion years old.

12.2 • In rocks 3.5 billion years old or younger, microbial formations called stromatolites are abundant and show extensive microbial diversification. The evolution of oxygenic photosynthesis caused O_2 to accumulate 2.4 billion years ago, eventually leading to the formation of banded iron formations, the ozone shield, and an oxygenated atmosphere, which set the stage for rapid diversification of metabolic types and the evolution of multicellularity.

12.3 • The eukaryotic cell developed from endosymbiotic events. The modern eukaryotic cell is a chimera with genes and characteristics from both *Bacteria* and *Archaea*. SSU rRNA sequence analyses indicate the ancestors of mitochondria are found in the phylum *Proteobacteria* and those of chloroplasts are found in the phylum *Cyanobacteria*.

12.4 • Ribosomal RNA genes have been used to construct a universal tree of life revealing that life on Earth evolved in three major directions, forming the domains *Bacteria*, *Archaea*, and *Eukarya*. The main phyla of *Bacteria* and *Archaea* diverged billions of years ago, and the best-characterized phyla of *Eukarya* diverged within the last 600 million years. The universal tree of life shows that the domains *Bacteria* and *Archaea* split from each other eons ago, and that *Eukarya* split from *Archaea* later in the history of life.

12.5 • Molecular sequences accumulate random mutations over time and molecular phylogenetic analysis is used to determine the evolutionary history of life. A phylogenetic tree is a diagram that depicts the evolutionary history of a set of genes or organisms.

12.6 • Evolution is defined as a change in allele frequencies over time. New alleles are created through the processes of mutation and recombination. Mutations occur at random and most mutations are neutral or deleterious, but some are beneficial. Natural selection and genetic drift are two mechanisms that cause allele frequencies to change over time.

12.7 • Microbial genomes are dynamic, and genome size and gene content can vary considerably between strains of a species. The core genome is defined as the set of all genes shared by a species, while the pan genome is defined as the core genome plus genes whose presence varies among strains of a species.

12.8 • At present, species in *Bacteria* and *Archaea* are defined operationally based on shared genetic and phenotypic traits. The dynamic nature of microbial genomes and the abundance of genes acquired through horizontal gene transfer have raised questions about the nature of microbial species. While many mechanisms have been proposed to describe species formation, there is no consensus on the ecological and evolutionary forces that best explain the origin of microbial species.

12.9 • Systematics is the study of the diversity and relationships of living organisms. Polyphasic taxonomy is based on phenotypic, genotypic, and phylogenetic information. Bacterial species can be distinguished genotypically on the basis of DNA–DNA hybridization, DNA fingerprinting, MLST, or multigene or whole genome analyses. Phenotypic traits useful in taxonomy include morphology, motility, metabolism, and cell chemistry, especially lipid analyses.

12.10 • Nomenclature in microbiology follows the binomial system used in all of biology. Formal recognition of a new prokaryotic species requires depositing a sample of the organism in culture collections and publishing the new species name and description. *Bergey's Manual of Systematic Bacteriology* and *The Prokaryotes* are major taxonomic compilations of *Bacteria* and *Archaea*.

MasteringMicrobiology® **Review what you know and challenge what you have learned with MasteringMicrobiology!** Access study materials, chapter quizzes, animations, and microbiology lab tutorials in the Study Area to ensure that you have mastered this chapter's content.

REVIEW OF KEY TERMS

Allele a sequence variant of a given gene

Archaea phylogenetically related prokaryotes distinct from *Bacteria*

Bacteria phylogenetically related prokaryotes distinct from *Archaea*

Banded iron formation iron oxide–rich ancient sedimentary rocks containing zones of oxidized iron (Fe^{3+}) formed by oxidation of Fe^{2+} by O_2 produced by cyanobacteria

Binomial system the system devised by the Swedish scientist Carl Linnaeus for naming living organisms in which an organism is given a genus name and a species epithet

Core genome those genes found in common in the genomes of all strains of a species

DNA–DNA hybridization the experimental determination of genomic similarity by measuring the extent of hybridization of DNA from one organism with that of another

Domain in a taxonomic sense, the highest level of biological classification

Endosymbiotic hypothesis the idea that a chemoorganotrophic bacterium and a cyanobacterium were stably incorporated into another cell type to give rise, respectively, to the mitochondria and chloroplasts of modern-day eukaryotes

Eukarya all eukaryotes: algae, protists, fungi, slime molds, plants, and animals

Evolution a change in allele frequencies over time with new alleles arising due to mutation and recombination

FAME fatty acid methyl ester; a technique for identifying microorganisms by their fatty acids

Fitness the capacity of an organism to survive and reproduce as compared to that of competing organisms

Genetic drift a process that results in a change in allele frequencies in a population as a result of random changes in the number of offspring from each individual over time

Homology having shared ancestry

Homoplasy when two organisms have the same trait as a result of recurrent mutation or convergent evolution

Horizontal gene transfer the asymmetrical and unidirectional transfer of DNA from one cell to another

Molecular clock a DNA sequence, such as a gene for rRNA, that can be used as a comparative temporal measure of evolutionary divergence

Monophyletic in phylogeny, a group descended from one ancestor

Multilocus sequence typing (MLST) a taxonomic tool for classifying organisms on the basis of gene sequence variations in several housekeeping genes

Mutation a heritable change in DNA sequence

Ortholog a gene in one organism that is similar to a gene in another organism because of descent from a common ancestor (see also *paralog*)

Pan genome the totality of the genes present in the different strains of a species

Paralog a gene whose similarity to one or more other genes in the same organism is the result of gene duplication (see also *ortholog*)

Phenotype the physical and chemical characteristics of an organism that can be observed or measured

Phylogenetic tree a diagram that depicts the evolutionary history of an organism; consists of nodes and branches

Phylogeny evolutionary history

Phylum a major lineage of cells in one of the three domains of life

Recombination resorting or rearrangement of DNA fragments resulting in a new sequence combination

Ribosomal RNA (rRNA) RNA molecules found in the small and large subunits of the ribosome

Ribotyping a means of identifying microorganisms from analysis of DNA fragments generated from restriction enzyme digestion of the genes encoding their ribosomal RNA

Selection in an evolutionary context, a process that results in a change in allele frequencies

in a population when individuals that are favored in a given environment are able to produce more offspring and make a greater contribution to the gene content of future generations

Sequence alignment the insertion of gaps into a set of molecular sequences organized in rows so that homologous positions are organized in vertical columns. Alignment is necessary prior to phylogenetic analysis because deletion and insertion mutations cause variations in the length of molecular sequences

16S rRNA a large polynucleotide (~1500 bases) that functions as part of the small subunit of the ribosome of *Bacteria* and *Archaea* and from whose gene sequence evolutionary information can be obtained; its eukaryotic counterpart is 18S rRNA

Small subunit (SSU) rRNA 16S rRNA found in the 30S ribosomal subunit of *Bacteria* and *Archaea* and its orthologous counterpart in *Eukarya*, the 18S rRNA of the 40S ribosomal subunit found in eukaryotes

Species defined in microbiology as a collection of strains that all share the same major properties and differ in one or more significant properties from other collections of strains; defined phylogenetically as a monophyletic, exclusive group based on DNA sequence

Stromatolite a laminated microbial mat, typically built from layers of filamentous *Bacteria* and other microorganisms, which can become fossilized

Systematics the study of the diversity of organisms and their relationships; includes taxonomy and phylogeny

Taxonomy the science of identification, classification, and nomenclature

Universal tree of life a phylogenetic tree that shows the positions of representatives of all domains of cellular life

REVIEW QUESTIONS

1. What is the age of planet Earth? When did the oceans form? What is the age of the earliest known microfossil? (Section 12.1)

2. What is a stromatolite and when can they be found in Earth's history? (Section 12.2)

3. Why was the evolution of cyanobacteria of such importance to the further evolution of life on Earth? What component of the geological record is used to date the evolution of cyanobacteria? (Section 12.2)

4. What is the endosymbiotic hypothesis for the origin of mitochondria and chloroplasts? What evidence supports this hypothesis? (Section 12.3)

5. What evidence supports the classification of life into three domains? (Section 12.4)

6. What major physiological and biochemical properties do *Archaea* share with *Eukarya* or with *Bacteria*? (Section 12.4)

7. Why are SSU rRNA genes good choices for phylogenetic studies, and what are their limitations? (Section 12.4)

8. Why is sequence alignment necessary prior to phylogenetic tree construction? (Section 12.5)

9. What is homoplasy and why does it create problems for molecular phylogenetic analyses? (Section 12.5)

10. What is the difference between a gene tree and an organismal tree? (Section 12.5)

11. Describe the steps for determining an SSU phylogeny of three bacteria you have isolated from nature. (Sections 12.4 and 12.5)

12. What is evolution? What processes give rise to genetic variation? What processes can cause allele frequencies to change over time? (Section 12.6)

13. What is fitness? To what degree does fitness depend on the environment in which organisms live? (Section 12.6)

14. What are some processes that influence the content of the pan genome? (Section 12.7)

15. Contrast the impacts of homologous and nonhomologous recombination on the evolution of the core genome and the pan genome. (Section 12.7)

16. What is the "species problem" and why is the concept of microbial species difficult to resolve? (Section 12.8)

17. How many bacterial species are there? Why do we not know this number more precisely? (Section 12.8)

18. What major phenotypic and genotypic properties are used to classify organisms in bacterial taxonomy? (Sections 12.8 and 12.9)

19. What is measured in FAME analyses? (Section 12.9)

20. How does 16S rRNA gene sequence analysis differ from multilocus sequence typing as an identification tool? (Section 12.9)

21. What roles do microbial culture collections play in microbial systematics? (Section 12.10)

APPLICATION QUESTIONS

1. Compare and contrast the physical and chemical conditions on Earth at the time life first arose with conditions today. From a physiological standpoint, discuss at least two reasons why *animals* could not have existed on early Earth. In what ways has microbial metabolism altered Earth's biosphere? How might life on Earth be different if oxygenic photosynthesis had not evolved?

2. For the following sequences, construct the phylogenetic tree that best depicts their evolutionary relationships.

   Taxon 1: GTTCCCTTA
   Taxon 2: GTTCGGTAT
   Taxon 3: GAAAAACCCTAT
   Taxon 4: CTTCCCTTT
   Taxon 5: GTAAAACCCGAT

3. Imagine that you have been given several bacterial strains from various countries around the world and that all the strains are thought to cause the same gastrointestinal disease and to be genetically identical. Upon carrying out a DNA fingerprint analysis of the strains, you find that four different strain types are present. What methods could you use to test whether the different strains are actually members of the same species?

4. Imagine that you have discovered a new form of microbial life, one that appears to represent a fourth domain. How would you go about characterizing the new organism and determining if it actually is evolutionarily distinct from *Bacteria*, *Archaea*, and *Eukarya*?

13 · Metabolic Diversity of Microorganisms

microbiology**now**

Unraveling Microbial Metabolisms

Once a bacterium is in laboratory culture, it is often straightforward to determine how it makes a living; that is, to determine its metabolic capacities. But how about organisms that have yet to be cultured; can we figure out what they are doing, too? We can, but by studying their *genes* rather than their *metabolisms*.

In order to probe microbial metabolic processes in an anoxic aquifer near the Colorado River (right photo), microbial ecologists isolated total DNA from the aquifer microbial community directly in the field (left photo) and from it, reconstructed and sequenced nearly 50 genomes.[1] Through sequence analyses, the scientists unraveled the metabolisms of the aquifer microbial community and examined other major aspects of the community's biology. To their surprise, they found that fermentations and H_2 metabolism rather than anaerobic respirations were the major catabolic strategies being employed. Moreover, certain metabolisms thought to be restricted to *Archaea* were also detected in aquifer *Bacteria*.

Many of the uncultivated *Bacteria* in the aquifer previously had been detected in diversity surveys of other anaerobic environments. However, because the approach in the aquifer study connected phylogeny with metabolism, the metabolic capacities of these uncultivated organisms were revealed. In fact, entire metabolic pathways could be reconstructed by examining the genetic blueprints of the microbial community. This background should assist in the design of enrichment culture experiments to cultivate the aquifer bacteria.

One of the goals of microbial ecology is to bring interesting microorganisms into laboratory culture. Meanwhile, it is still possible to probe the properties of uncultivated organisms using metagenomics. Once both the diversity and metabolic profiles of a microbial community are known, it becomes possible to predict how the ecosystem functions and how it might be disrupted by pollution or other perturbations.

[1]Wrighton, K.C., et al. 2012. Fermentation, hydrogen, and sulfur metabolism in multiple uncultivated bacterial phyla. *Science* 337: 1661–1665.

A major theme of microbiology is the great *phylogenetic* diversity of microbial life on Earth. We got a taste of this in the last chapter and will explore microbial diversity in detail in the following four chapters. In this chapter we focus on the *metabolic* diversity of microorganisms, with special emphasis on the processes and mechanisms that underlie this diversity. We will then return to the organisms themselves and unveil the phylogenetic breadth of the microbial world in the context of metabolic diversity.

I · Phototrophy

Phototrophy—the use of light energy—is widespread in the microbial world. In this unit we examine the properties and energy-conserving strategies of phototrophic microorganisms and see how these support a lifestyle based on the use of CO_2 as the sole carbon source.

13.1 Photosynthesis and Chlorophylls

The most important biological process on Earth is **photosynthesis**, the conversion of light energy to chemical energy. Organisms that carry out photosynthesis are called **phototrophs**. Phototrophic organisms are also **autotrophs**, capable of growing with CO_2 as the sole carbon source. Energy from light is used in the reduction of CO_2 to organic compounds (*photoautotrophy*). Some phototrophs can also use organic carbon as their carbon source; this lifestyle is called *photoheterotrophy*.

Photosynthesis requires light-sensitive pigments, the *chlorophylls*, found in plants, algae, and the cyanobacteria, and *bacteriochlorophylls*, in the purple and green bacteria. Cyanobacteria and purple and green bacteria are all prokaryotic phototrophs. Absorption of light energy by chlorophylls and bacteriochlorophylls begins the process of photosynthetic energy conversion, and the net result is chemical energy, ATP.

Photoautotrophy requires that two distinct sets of reactions operate in parallel: (1) ATP production and (2) CO_2 reduction to cell material. For autotrophic growth, energy is supplied from ATP, and electrons for the reduction of CO_2 come from NADH (or NADPH). The latter originates from the reduction of

$NAD(P^+)$ by electron donors in the environment. In purple and green bacteria the donor could be a reduced sulfur compound such as hydrogen sulfide (H_2S), or molecular hydrogen (H_2). By contrast, green plants, algae, and cyanobacteria derive electrons from water (H_2O).

The oxidation of H_2O produces molecular oxygen (O_2) as a by-product; because of this, the photosynthetic process in cyanobacteria is called **oxygenic photosynthesis**. However, in purple and green bacteria O_2 is *not* produced, and thus the process is called **anoxygenic photosynthesis** (Figure 13.1). Oxygen produced by cyanobacteria billions of years ago converted Earth from an anoxic to an oxic world and set the stage for an explosion of eukaryotic microbial diversity that eventually gave rise to plants and animals.

Chlorophyll and Bacteriochlorophyll

Chlorophyll and **bacteriochlorophyll** are related to tetrapyrroles that are the parent structure of the cytochromes. But unlike cytochromes, chlorophylls contain *magnesium* instead of *iron* at the center of the ring. Chlorophylls also contain specific substituents on the tetrapyrrole ring and a hydrophobic alcohol that helps anchor the chlorophyll into photosynthetic membranes. The structure of chlorophyll *a*, the principal chlorophyll of oxygenic phototrophs, is shown in Figure 13.2*a*. Chlorophyll *a* is green because it *absorbs* red and blue light and *transmits* green light; its absorption spectrum shows strong absorbance near 680 nm and 430 nm (Figure 13.2*b*). Several different chlorophylls are known, each distinguished by its unique absorption spectrum. Cyanobacteria contain chlorophyll *a*

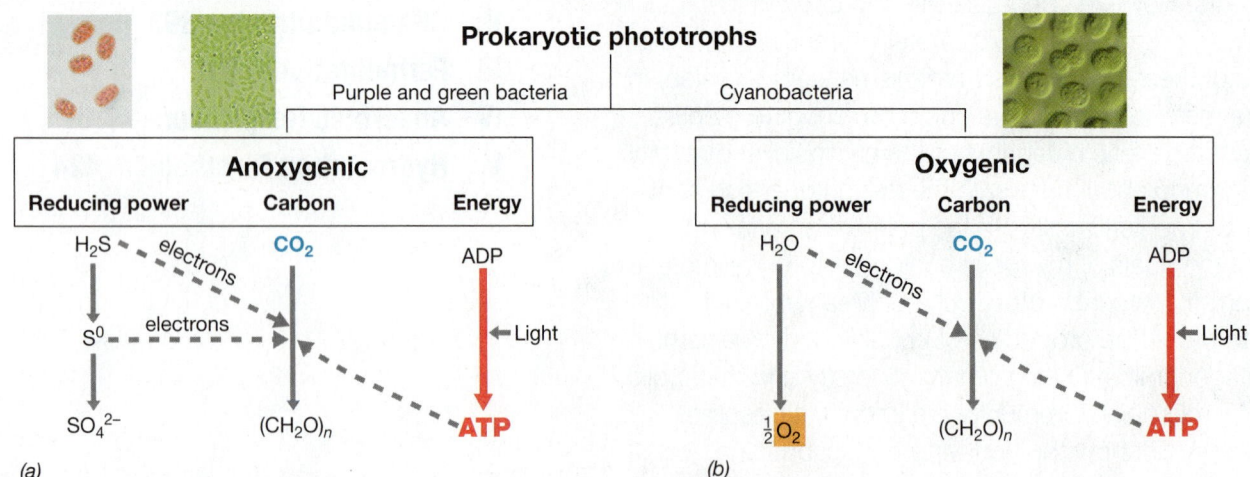

Figure 13.1 Patterns of photosynthesis. Energy and reducing power synthesis in *(a)* anoxygenic and *(b)* oxygenic phototrophs. Note that oxygenic phototrophs produce O_2, while anoxygenic phototrophs do not. Insets: Left, light photomicrographs of cells of a purple sulfur bacterium (*Chromatium*, cells 5 μm in diameter) and a green sulfur bacterium (*Chlorobium*, cells 0.9 μm in diameter). Note the sulfur globules inside or outside the cells produced from the oxidation of H_2S. Right, interference-contrast photomicrograph of cells of a coccoid-shaped cyanobacterium.

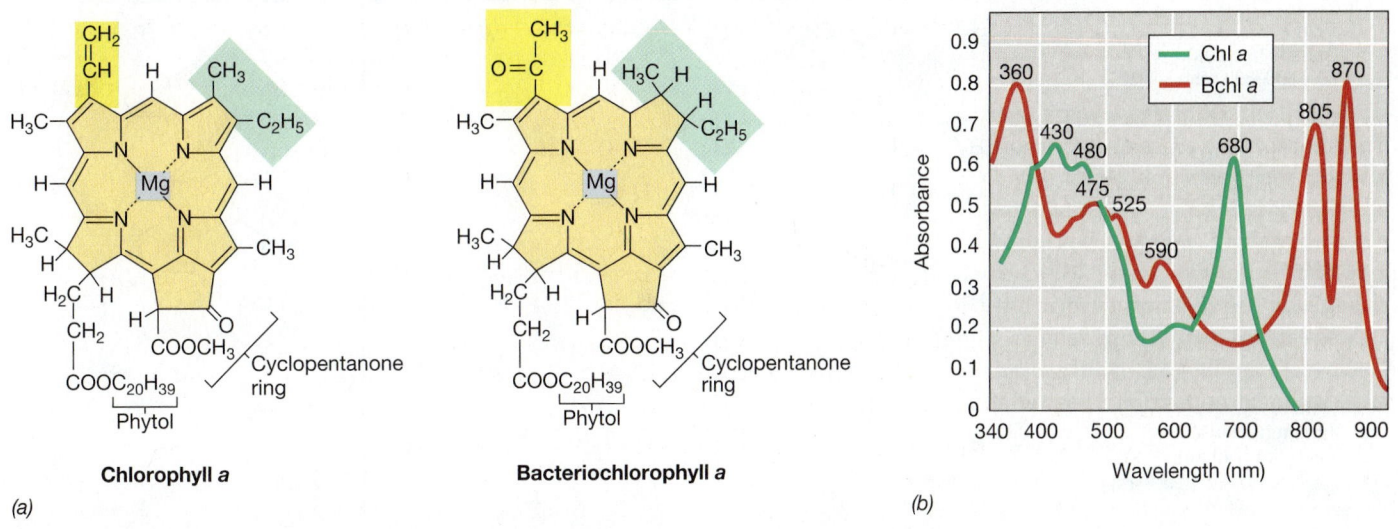

Figure 13.2 Structures and spectra of chlorophyll *a* and bacteriochlorophyll *a*. *(a)* The two molecules are identical except for those portions contrasted in yellow and green. *(b)* Absorption spectrum (green curve) of cells of the green alga *Chlamydomonas*. The peaks at 680 and 430 nm are due to chlorophyll *a*, and the peak at 480 nm is due to carotenoids. Absorption spectrum (red curve) of cells of the phototrophic purple bacterium *Rhodopseudomonas palustris*. Peaks at 870, 805, 590, and 360 nm are due to bacteriochlorophyll *a*, and peaks at 525 and 475 nm are due to carotenoids.

(a few species contain chlorophyll *d*), while their relatives the prochlorophytes produce chlorophylls *a* and *b*.

Phototrophic purple and green bacteria produce one or more bacteriochlorophylls (**Figure 13.3**). Bacteriochlorophyll *a* (Figure 13.2), present in most purple bacteria (⟳ Sections 14.4 and 14.5), absorbs maximally between 800 and 925 nm (different purple bacteria synthesize slightly different photocomplexes, and the absorption maxima of bacteriochlorophyll *a* in any given organism depend to some degree on how proteins in the photocomplexes are arranged in the photosynthetic membrane; see Figure 13.4). Other bacteriochlorophylls, whose distribution runs along phylogenetic lines, absorb in other regions of the visible and infrared spectrum (Figure 13.3).

The existence of different forms of chlorophyll or bacteriochlorophyll that absorb light of different wavelengths allows phototrophs to make better use of the available energy in the electromagnetic spectrum. By employing different pigments with distinct absorption properties, different phototrophs can coexist in the same habitat, each absorbing wavelengths of light that others are not. Thus, pigment diversity has *ecological* significance for the successful coexistence of different phototrophs in the same habitat.

Reaction Centers and Antenna Pigments

In oxygenic phototrophs and in purple anoxygenic phototrophs, chlorophyll/bacteriochlorophyll molecules do not exist freely in the cell but are attached to proteins and housed within membranes to form *photocomplexes* consisting of anywhere from 50 to 300 chlorophyll/bacteriochlorophyll molecules. Only a small number of these pigment molecules, called **reaction centers** (**Figure 13.4**), participate directly in reactions that lead to ATP synthesis. Reaction center chlorophylls/bacteriochlorophylls are surrounded by larger numbers of light-harvesting chlorophylls/ bacteriochlorophylls. These so-called **antenna pigments** (also called *light-harvesting pigments*) function to absorb light and funnel some of the energy to the reaction center (Figure 13.4). At the

low light intensities that are often found in nature, this arrangement for concentrating energy allows reaction centers to receive light energy that would otherwise be missed.

Photosynthetic Membranes, Chloroplasts, and Chlorosomes

The chlorophyll pigments and all the other components of the light-gathering apparatus exist within membranes in the cell. The location of these photosynthetic membranes differs between prokaryotic and eukaryotic microorganisms. In eukaryotic phototrophs, photosynthesis takes place in intracellular organelles, the *chloroplasts*, where the chlorophylls are attached to sheetlike membranes (**Figure 13.5**). These photosynthetic membrane systems are called **thylakoids**, and stacks of thylakoids form *grana*. The thylakoids are arranged so that the chloroplast is divided into two regions, the matrix space that surrounds the thylakoids, called the *stroma*, and the inner space within the thylakoid array. This arrangement makes possible the generation of a light-driven proton motive force that is used to synthesize ATP (Section 13.4).

Chloroplasts are absent from prokaryotic phototrophs. In purple bacteria, the photosynthetic pigments are integrated into internal membrane systems that arise from invagination of the cytoplasmic membrane. Membrane vesicles called *chromatophores* or membrane stacks called *lamellae* are common membrane arrangements in purple bacteria (**Figure 13.6**). In cyanobacteria, photosynthetic pigments reside in lamellar membranes (see Figure 13.10) also called *thylakoids* because of their resemblance to the thylakoids in the chloroplasts of algae (Figure 13.5).

The ultimate structure for capturing low light intensities is the **chlorosome** (**Figure 13.7**). Chlorosomes are present in the anoxygenic green sulfur bacteria (*Chlorobium*, Figure 13.1 and ⟳ Section 14.6) and green nonsulfur bacteria (*Chloroflexus*, ⟳ Section 14.7). Chlorosomes function as giant antenna systems, but unlike the antennae of purple bacteria or cyanobacteria, bacteriochlorophyll molecules in the chlorosome are not attached to proteins.

Pigment/Absorption maxima (in vivo)	R1	R2	R3	R4	R5	R6	R7
Bchl a (purple bacteria)/ 805, 830–890 nm	—C(=O)—CH3	—CH3[a]	—CH2—CH3	—CH3	—C(=O)—O—CH3	P/Gg[b]	—H
Bchl b (purple bacteria)/ 835–850, 1020–1040 nm	—C(=O)—CH3	—CH3[c]	=C—CH3 (with H)	—CH3	—C(=O)—O—CH3	P	—H
Bchl c (green sulfur bacteria)/745–755 nm	—C(H)(OH)—CH3	—CH3	—C2H5 / —C3H7[d] / —C4H9	—C2H5 / —CH3	—H	F	—CH3
Bchl c_s (green nonsulfur bacteria)/740 nm	—C(H)(OH)—CH3	—CH3	—C2H5	—CH3	—H	S	—CH3
Bchl d (green sulfur bacteria)/705–740 nm	—C(H)(OH)—CH3	—CH3	—C2H5 / —C3H7 / —C4H9	—C2H5 / —CH3	—H	F	—H
Bchl e (green sulfur bacteria)/719–726 nm	—C(H)(OH)—CH3	—C(H)(=O)	—C2H5 / —C3H7 / —C4H9	—C2H5 / —CH3	—H	F	—CH3
Bchl g (heliobacteria)/ 670, 788 nm	—C=CH2 (with H)	—CH3[a]	—C2H5	—CH3	—C(=O)—O—CH3	F	—H

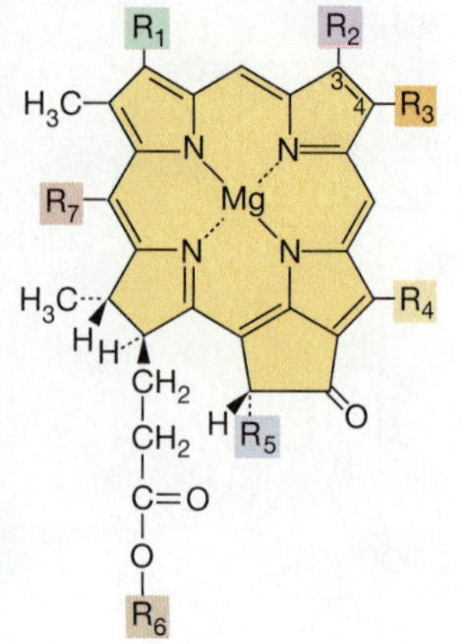

[a] No double bond between C_3 and C_4; additional H atoms are in positions C_3 and C_4.

[b] P, Phytyl ester ($C_{20}H_{39}O$—); F, farnesyl ester ($C_{15}H_{25}O$—); Gg, geranylgeraniol ester ($C_{10}H_{17}O$—); S, stearyl alcohol ($C_{18}H_{37}O$—).

[c] No double bond between C_3 and C_4; an additional H atom is in position C_3.

[d] Bacteriochlorophylls c, d, and e consist of isomeric mixtures with the different substituents on R_3 as shown.

Figure 13.3 Structure of all known bacteriochlorophylls (Bchl). The different substituents present in the positions R_1 to R_7 in the structure at the right are listed. Absorption properties can be determined by suspending intact cells of a phototroph in a viscous liquid such as 60% sucrose (this reduces light scattering and smooths out spectra) and running absorption spectra as shown in Figure 13.2b. In vivo absorption maxima are the physiologically relevant absorption peaks. The spectrum of bacteriochlorophylls extracted from cells and dissolved in organic solvents is often quite different.

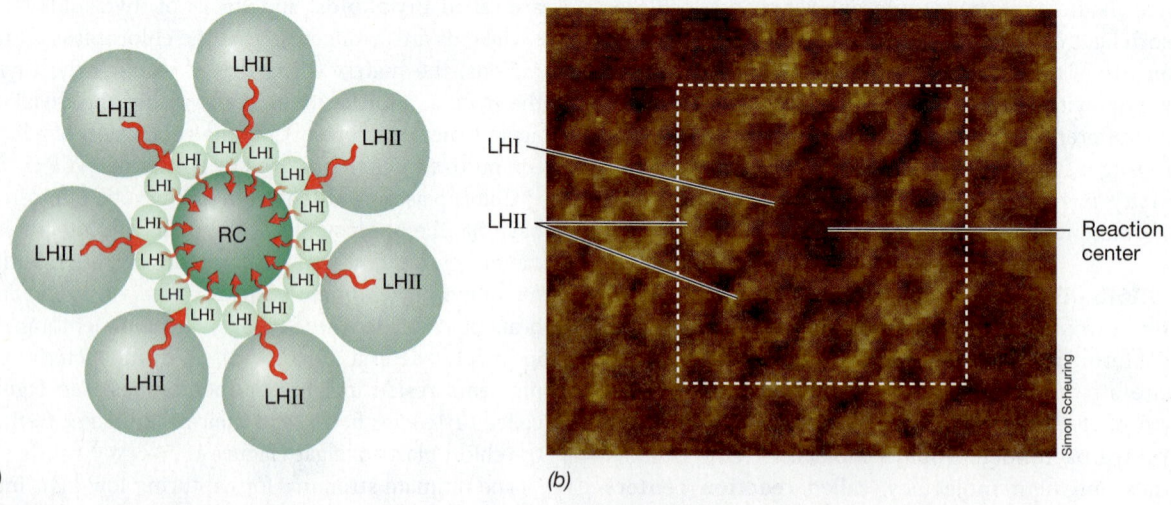

(a)

(b)

Figure 13.4 Arrangement of light-harvesting chlorophylls/bacteriochlorophylls and reaction centers within a photosynthetic membrane. (a) Light energy absorbed by light-harvesting (LH) molecules (light green) is transferred to the reaction centers (dark green, RC) where photosynthetic electron transport reactions begin. Pigment molecules are secured within the membrane by specific pigment-binding proteins. Compare this figure to Figures 13.11 and 13.12b. (b) Atomic force micrograph of photocomplexes of the purple bacterium *Phaeospirillum molischianum*. This organism has two types of light-harvesting complexes, LHI and LHII. LHII complexes transfer energy to LHI complexes, and these transfer energy to the reaction center (see Figure 13.12b).

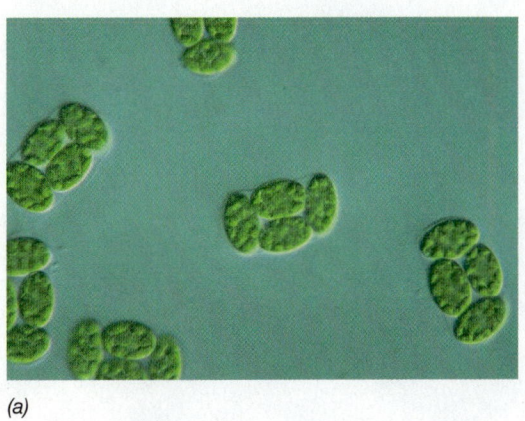

(a)

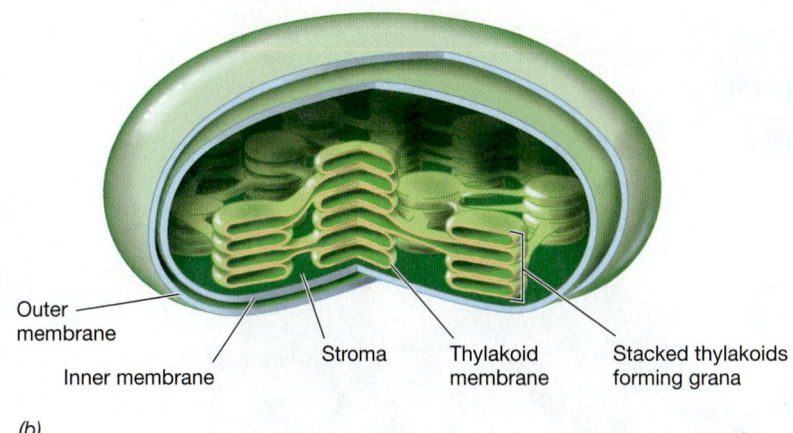

Outer membrane
Inner membrane
Stroma
Thylakoid membrane
Stacked thylakoids forming grana

(b)

Figure 13.5 **The chloroplast.** *(a)* Photomicrograph of cells of the green alga *Makinoella*. Each of the four cells in a cluster contains several chloroplasts. *(b)* Details of chloroplast structure, showing how the convolutions of the thylakoid membranes define an inner space called the stroma and form membrane stacks called grana.

UNIT 3

Chlorosomes contain bacteriochlorophyll *c, d,* or *e* (Figure 13.3) arranged in dense arrays running along the long axis of the structure. Light energy absorbed by these antenna pigments is transferred to bacteriochlorophyll *a* in the reaction center in the cytoplasmic membrane through a small protein called the *FMO protein* (Figure 13.7).

Green bacteria can grow at the lowest light intensities of all known phototrophs and are often found in the deepest waters of lakes, inland seas, and other anoxic aquatic habitats where light levels are too low to support other phototrophs. Green nonsulfur bacteria are major components of microbial mats, thick biofilms that form in hot springs and highly saline environments (∞ Section 19.5). Microbial mats experience a steep light gradient, with light levels even a few millimeters into the mat approaching darkness. Hence, chlorosomes allow green nonsulfur bacteria to grow phototrophically with only the minimal light intensities available.

MINIQUIZ -
- What is the fundamental difference between an oxygenic and an anoxygenic phototroph?
- What is the difference between the relative numbers of antenna and reaction center chlorophyll/bacteriochlorophyll molecules in a photosynthetic complex, and why?
- Why can phototrophic green bacteria grow at light intensities that will not support purple bacteria?

13.2 Carotenoids and Phycobilins

Although chlorophyll/bacteriochlorophyll is required for photosynthesis, phototrophic organisms contain other pigments as well. These pigments include, in particular, the *carotenoids* and *phycobilins.*

Carotenoids

The most widespread accessory pigments in phototrophs are the **carotenoids**. Carotenoids are hydrophobic pigments that are firmly embedded in the photosynthetic membrane. **Figure 13.8** shows the structure of a common carotenoid, *β-carotene.* Carotenoids are typically yellow, red, brown, or green and absorb light in the blue region of the spectrum (Figure 13.2).

The major carotenoids of anoxygenic phototrophs are shown in **Figure 13.9**. Because they tend to mask the color of bacteriochlorophylls, carotenoids are responsible for the brilliant colors of red, purple, pink, green, yellow, or brown that are observed in different species of anoxygenic phototrophs (∞ Figure 14.12).

Carotenoids are closely associated with chlorophyll or bacteriochlorophyll in photosynthetic complexes, and some of the energy absorbed by carotenoids can be transferred to the reaction center. However, carotenoids function primarily as photoprotective agents. Bright light can be harmful to cells because it can catalyze photooxidation reactions that can produce toxic forms of oxygen, such as singlet oxygen ($^1O_2$). Like superoxide and other forms of toxic oxygen (∞ Section 5.16), singlet oxygen can spontaneously oxidize photocomplexes, rendering them nonfunctional. Carotenoids quench toxic oxygen species by absorbing much of this harmful light and in this way prevent these dangerous photooxidations. Because phototrophic organisms by their very nature must live in the light, the photoprotection conferred by carotenoids is clearly advantageous.

Phycobiliproteins and Phycobilisomes

Cyanobacteria and the chloroplasts of red algae (which are descendants of cyanobacteria, ∞ Section 17.1) contain pigments called **phycobiliproteins**, which are the main light-harvesting systems of these phototrophs. Phycobiliproteins consist of red or blue-green linear tetrapyrroles, called *bilins,* bound to proteins, and give cyanobacteria and red algae their characteristic colors (**Figure 13.10**). The red phycobiliprotein, called *phycoerythrin,* absorbs most strongly at wavelengths around 550 nm, whereas the blue phycobiliprotein, *phycocyanin* (Figure 13.10*b*), absorbs most strongly at 620 nm. A third phycobiliprotein, called *allophycocyanin,* absorbs at about 650 nm.

Phycobiliproteins assemble into aggregates called **phycobilisomes** that attach to cyanobacterial thylakoids (Figure 13.10*c*). Phycobilisomes are arranged such that the allophycocyanin molecules are in direct contact with the photosynthetic membrane. Allophycocyanin is surrounded by phycocyanin or phycoerythrin (or both, depending on the organism). Phycocyanin and phycoerythrin absorb light of shorter wavelengths (higher energy) and transfer some energy to allophycocyanin, which is positioned

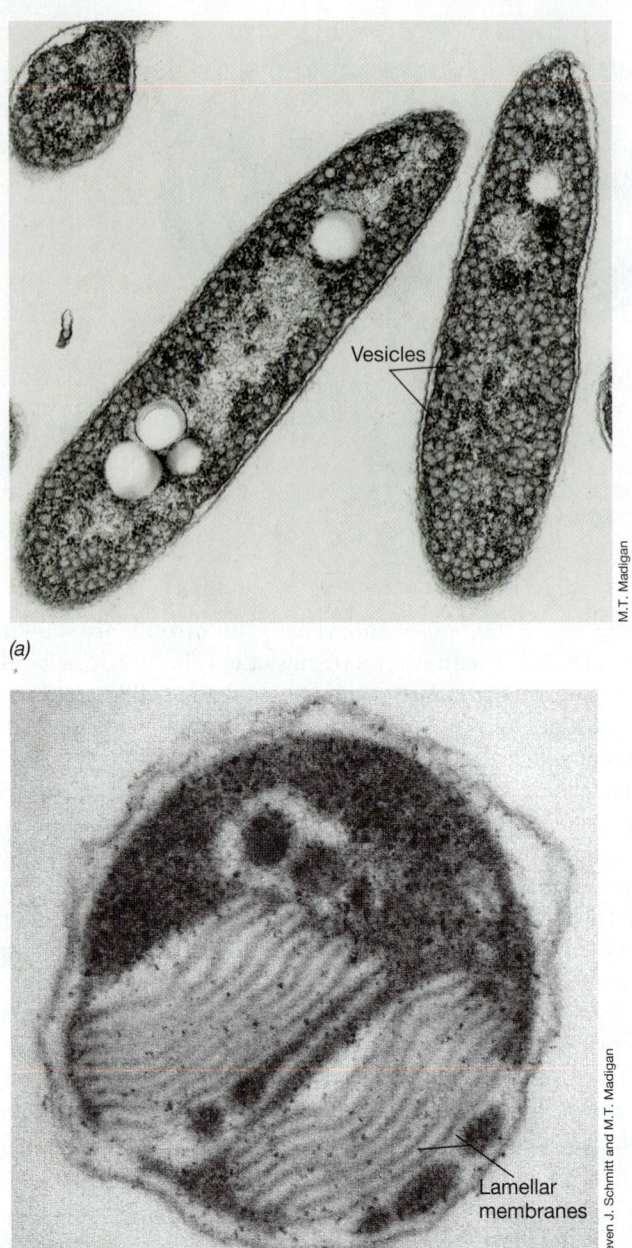

(a)

(b)

Figure 13.6 Membranes in anoxygenic phototrophs. *(a)* Chromatophores. Section through a cell of the purple bacterium *Rhodobacter* showing vesicular photosynthetic membranes. The vesicles are continuous with and arise by invagination of the cytoplasmic membrane. A cell is about 1 µm wide. *(b)* Lamellar membranes in the purple bacterium *Ectothiorhodospira*. A cell is about 1.5 µm wide. These membranes are also continuous with and arise from invagination of the cytoplasmic membrane, but instead of forming vesicles, they form membrane stacks.

closest to the reaction center chlorophyll and transfers energy to it (Figure 13.10*b*). Thus, in a fashion similar to how antenna bacteriochlorophyll systems function in anoxygenic phototrophs (Figure 13.4), energy transfer proceeds "downhill" from phycobilisomes to the reaction center. Phycobilisomes facilitate energy transfer to cyanobacterial reaction centers, allowing cyanobacteria to grow at lower light intensities than would otherwise be possible.

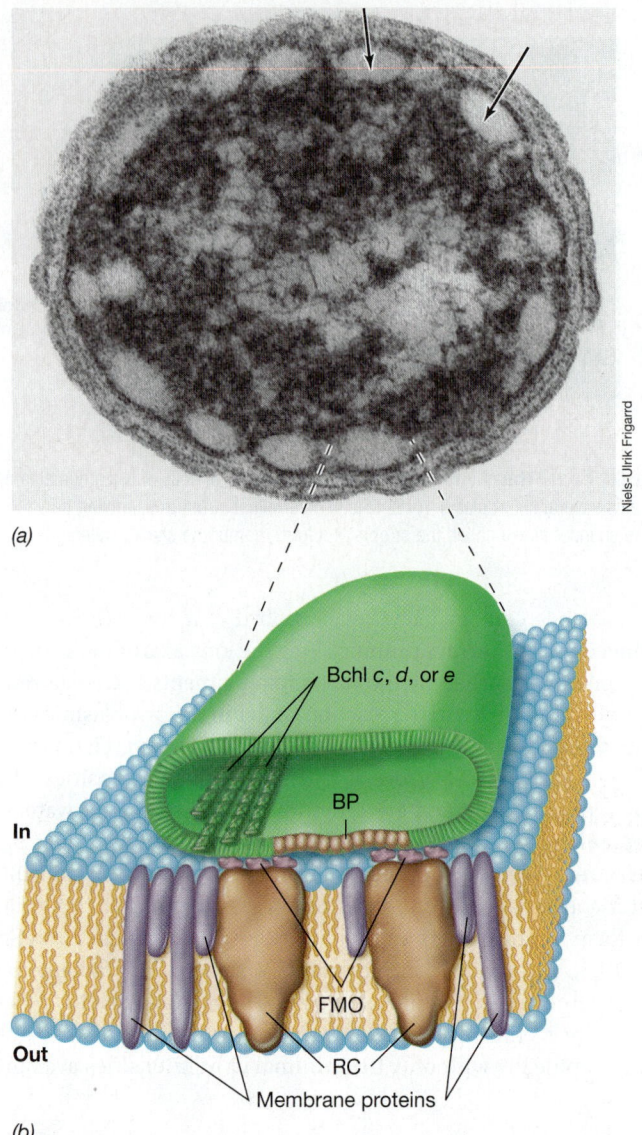

(a)

(b)

Figure 13.7 The chlorosome of green sulfur and green nonsulfur bacteria. *(a)* Transmission electron micrograph of a cross-section of a cell of the green sulfur bacterium *Chlorobaculum tepidum*. Note the chlorosomes (arrows). *(b)* Model of chlorosome structure. The chlorosome (green) lies appressed to the inside surface of the cytoplasmic membrane. Antenna bacteriochlorophyll (Bchl) molecules are arranged in tubelike arrays inside the chlorosome, and energy is transferred from these to reaction center (RC) Bchl *a* in the cytoplasmic membrane through a protein called FMO. Base plate (BP) proteins function as connectors between the chlorosome and the cytoplasmic membrane.

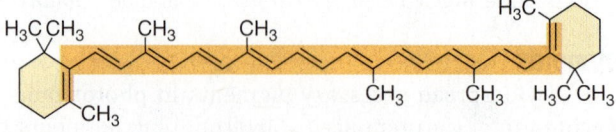

Figure 13.8 Structure of β-carotene, a typical carotenoid. The conjugated double-bond system is highlighted in orange.

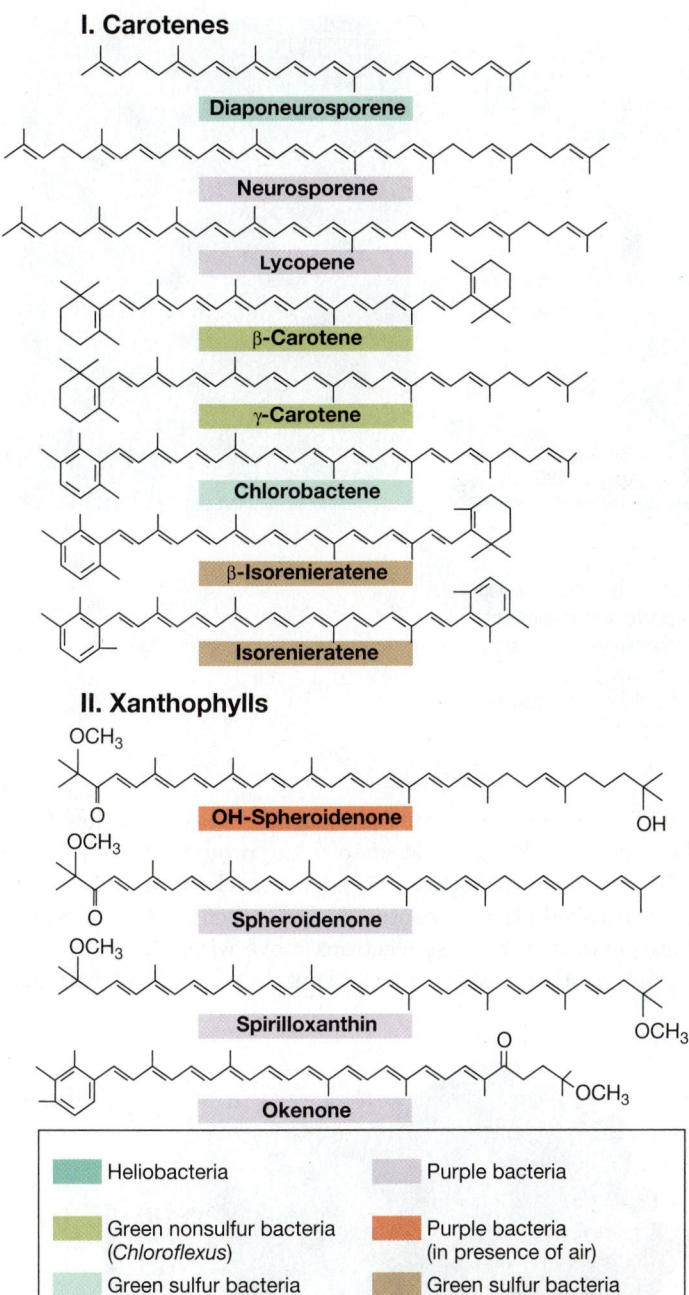

I. Carotenes

Diaponeurosporene

Neurosporene

Lycopene

β-Carotene

γ-Carotene

Chlorobactene

β-Isorenieratene

Isorenieratene

II. Xanthophylls

OH-Spheroidenone

Spheroidenone

Spirilloxanthin

Okenone

Heliobacteria	Purple bacteria
Green nonsulfur bacteria (*Chloroflexus*)	Purple bacteria (in presence of air)
Green sulfur bacteria	Green sulfur bacteria (brown-colored species)

Figure 13.9 Structures of some common carotenoids found in anoxygenic phototrophs. Carotenes are hydrocarbon carotenoids and xanthophylls are oxygenated carotenoids. Compare the structure of β-carotene shown in Figure 13.8 with how it is drawn here. For simplicity in the structures shown here, methyl (CH_3) groups are designated by bond only.

MINIQUIZ

- In which phototrophs are carotenoids found? Phycobiliproteins?
- How does the structure of a phycobilin compare with that of a chlorophyll?
- Phycocyanin is blue-green. What color of light does it absorb?

13.3 Anoxygenic Photosynthesis

In the photosynthetic light reactions, electrons traverse an electron transport chain arranged in a photosynthetic membrane in order of their increasingly more electropositive reduction potential (E_0'). This generates a proton motive force that drives ATP synthesis. Key parts of this process include photosynthetic reaction centers and photosynthetic membranes (Section 13.1).

Reaction centers of purple bacteria contain three polypeptides, designated L, M, and H. These proteins, along with a molecule of cytochrome *c*, are firmly embedded in the photosynthetic membrane (Figure 13.6) and wind through the membrane several times (**Figure 13.11**). The L, M, and H polypeptides bind two molecules of bacteriochlorophyll *a*, called the *special pair*, two additional bacteriochlorophyll *a* molecules that function in photosynthetic electron flow, two molecules of bacteriopheophytin (bacteriochlorophyll *a* minus its magnesium atom), two molecules of quinone (⮯ Section 3.10), and one carotenoid (Figure 13.11). All reaction center components are integrated in such a way that they can interact in very fast electron transfer reactions in the early stages of photosynthetic energy conversion.

Electron Flow in Purple Bacteria

Photosynthetic light reactions begin when light energy absorbed by the antenna systems is transferred to the special pair of bacteriochlorophyll *a* molecules (Figure 13.11a). This excites the special pair, converting it from a relatively weak to a very strong electron donor (very electronegative E_0', ⮯ Section 3.6). Once this strong donor has been produced, the remaining steps in photosynthetic electron flow are strongly reminiscent of those we have seen before in respiration (⮯ Section 3.10 and Figure 3.20); that is, electrons flowing through a membrane from carriers of low E_0' to those of high E_0', generating a proton motive force in the process (**Figure 13.12**).

Before excitation, the purple bacterial reaction center, which is called *P870*, has an E_0' of about +0.5 V; after excitation, it has a potential of about −1.0 V (Figure 13.12a). An excited electron within P870 proceeds to reduce a molecule of bacteriochlorophyll *a* within the reaction center (Figures 13.11a and 13.12a). This transition takes place incredibly fast, taking only about three-trillionths (3×10^{-12}) of a second. Once reduced, bacteriochlorophyll *a* reduces bacteriopheophytin *a* and the latter reduces quinone molecules within the membrane. These transitions are also very fast, taking less than one-billionth of a second (Figure 13.12). From the quinone, electrons are transported more slowly (on a millisecond scale) through a series of iron–sulfur proteins and cytochromes (Figure 13.12), eventually returning to the reaction center.

Figure 13.12b shows electron flow within the actual context of the photosynthetic membrane. Key electron transport proteins include many that also participate in respiratory electron flow (⮯ Figure 3.20)—cytochrome bc_1 and cytochrome c_2, in particular (Figure 13.12). Cytochrome c_2 is a periplasmic cytochrome (recall that the periplasm is the region between the cytoplasmic membrane and the outer membrane in gram-negative bacteria,

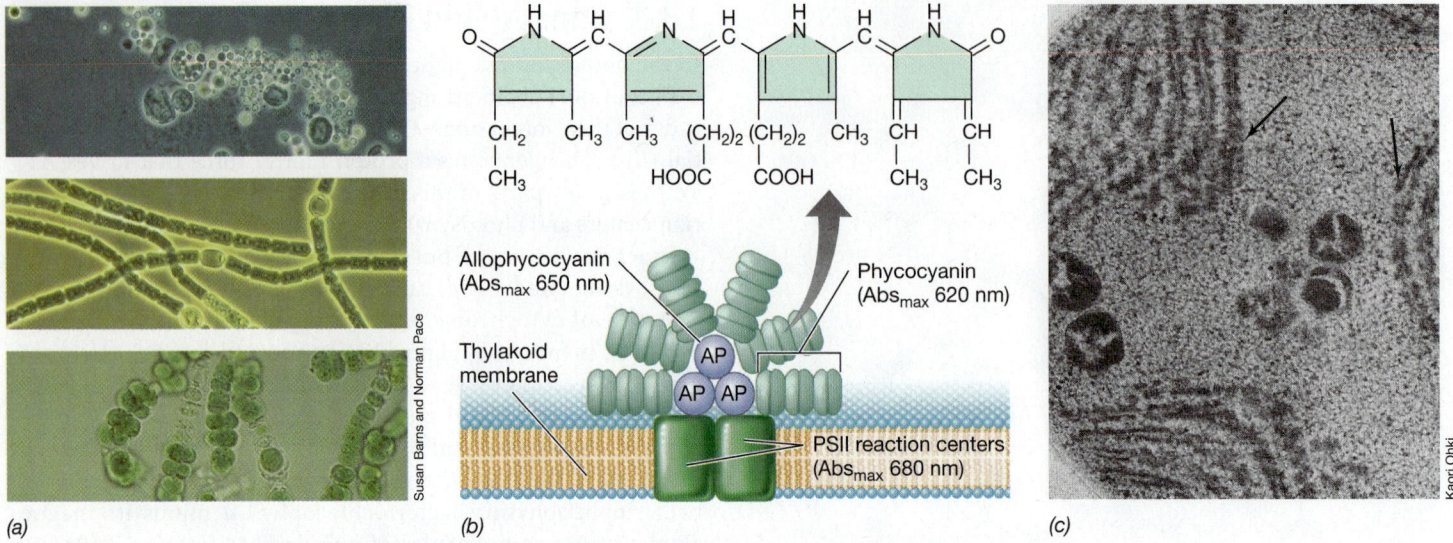

(a) Susan Barns and Norman Pace

(b)

(c) Kaori Ohki

Figure 13.10 **Phycobiliproteins and phycobilisomes.** *(a)* Light photomicrographs of cells of the cyanobacteria (top to bottom) *Dermocarpa*, *Anabaena*, and *Fischerella*, showing the typical blue-green color of cells due to phycobiliproteins. *(b)* Structure of phycocyanin (top) and a phycobilisome. Phycocyanin absorbs at higher energies (shorter wavelengths) than allophycocyanin. Chlorophyll *a* absorbs at longer wavelengths (lower energies) than allophycocyanin. Energy flow is thus phycocyanin → allophycocyanin → chlorophyll *a* of PSII. *(c)* Electron micrograph of a thin section of the cyanobacterium *Synechocystis*. Note the darkly staining ball-like phycobilisomes (arrows) attached to the lamellar membranes.

⮂ Section 2.11) that functions as an electron shuttle between the membrane-bound bc_1 complex and the reaction center (Figure 13.12*b*). Electron flow is completed when cytochrome c_2 donates an electron to the special pair to return it to its original ground-state reduction potential. The reaction center can then absorb new light energy and repeat the process.

ATP is synthesized during photosynthetic electron flow from the activity of ATPase that couples the proton motive force to ATP synthesis (⮂ Section 3.11). This mechanism of ATP synthesis is called **photophosphorylation**, specifically *cyclic photophosphorylation*, because electrons move within a closed loop. Unlike respiration, where there is a net consumption of electrons,

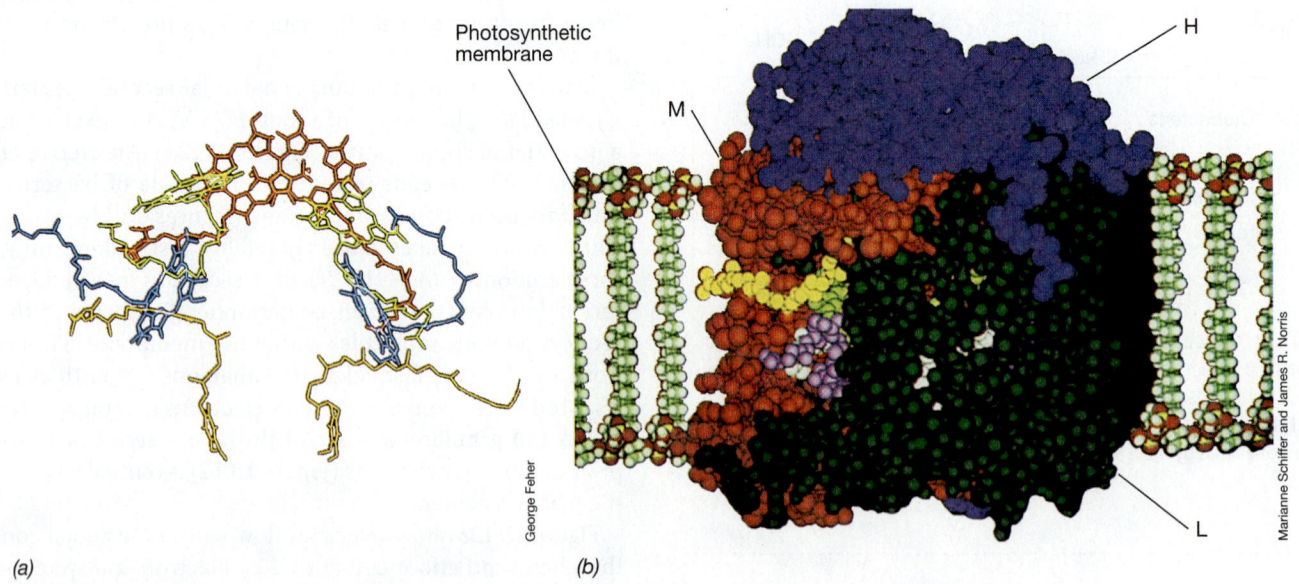

(a) George Feher

(b) Marianne Schiffer and James R. Norris

Figure 13.11 **Structure of the reaction center of a purple phototrophic bacterium.** *(a)* Arrangement of pigment molecules in the reaction center. The "special pair" of bacteriochlorophyll molecules overlap and are shown in orange at the top; quinones are in dark yellow and are at the bottom of the figure. The accessory bacteriochlorophylls are in lighter yellow near the special pair, and the bacteriopheophytin molecules are shown in blue. *(b)* Molecular model of the protein structure of the reaction center. The pigments described in part *a* are bound to membranes by protein H (blue), protein M (red), and protein L (green). The reaction center pigment–protein complex is integrated into the lipid bilayer.

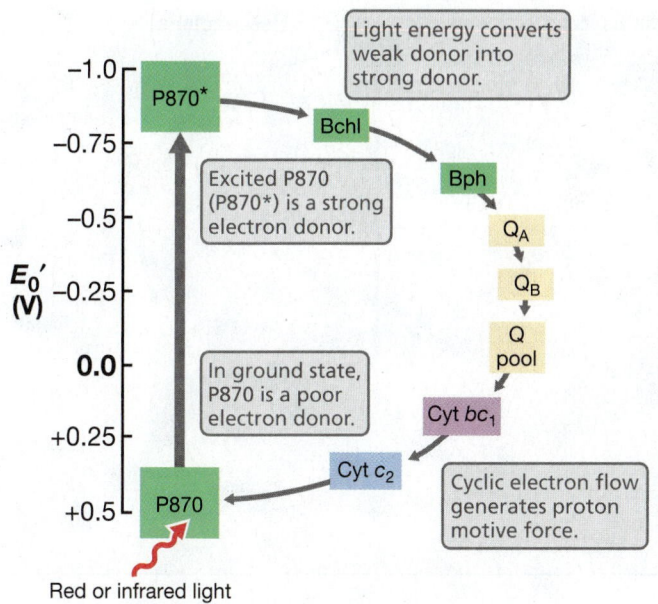

(a) **Electron flow in anoxygenic photosynthesis**

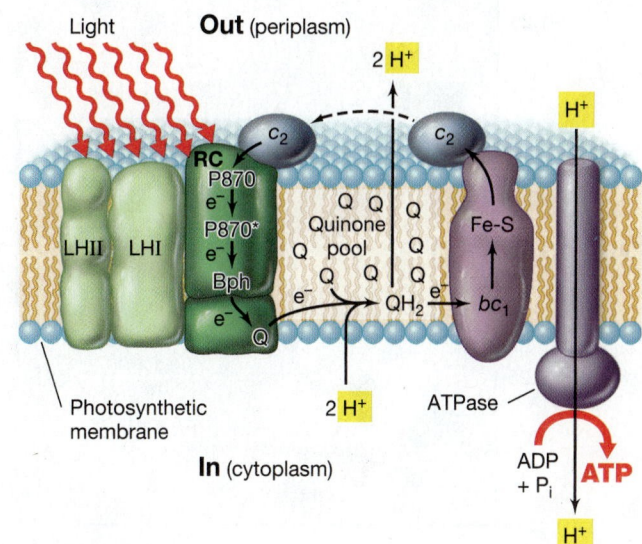

(b) **Arrangement of protein complexes in the purple bacterium reaction center**

Figure 13.12 **Electron flow in anoxygenic photosynthesis in a purple bacterium.** *(a)* Schematic of light reactions. Only a single light reaction occurs. Bph, bacteriopheophytin; Q_A, Q_B, intermediate quinones; Q pool, quinone pool in membrane; Cyt, cytochrome. *(b)* Arrangement of protein complexes in the purple bacterium reaction center leading to proton motive force (photophosphorylation) by ATPase. LH, light-harvesting bacteriochlorophyll complexes; RC, reaction center; Bph, bacteriopheophytin; Q, quinone; FeS, iron–sulfur protein; bc_1, cytochrome bc_1 complex; c_2, cytochrome c_2. For a description of ATPase function, see Section 3.11.

in cyclic photophosphorylation there is no net input or consumption of electrons; electrons simply travel a circuitous route, returning from whence they came (Figure 13.12).

Generation of Reducing Power

For a purple bacterium to grow as a photoautotroph (Section 13.1), the formation of ATP is not enough. Reducing power (NADH) is also necessary to reduce CO_2 to cell material. Reducing power for purple bacteria can come from many sources, in particular reduced sulfur compounds such as H_2S. When H_2S is the electron donor in purple sulfur bacteria, globules of S^0 are stored inside the cells (Figure 13.1). When S^0 is formed, electrons end up in the "quinone pool" (Figure 13.12). However, the E_0' of quinone (about 0 V) is insufficiently electronegative to reduce NAD^+ (-0.32 V). Hence, electrons from the quinone pool must be forced backwards (against the electrochemical gradient) to reduce NAD^+ to NADH (see Figure 13.13). This energy-requiring process, called **reverse electron transport**, is driven by the energy of the proton motive force. We will see later that reverse electron flow is also the mechanism by which chemolithotrophs obtain reducing power for CO_2 fixation; in many of these cases, the electrons come from electron donors of quite positive E_0' (Sections 13.6–13.10).

Photosynthetic Electron Flow in Other Anoxygenic Phototrophs

Thus far we have focused on electron flow in purple bacteria. Although analogous membrane-associated reactions drive

photophosphorylation in other anoxygenic phototrophs, there are significant differences in the details. The reaction centers of green nonsulfur bacteria and purple bacteria are structurally quite similar but differ from the reaction centers of green sulfur bacteria and heliobacteria, and this is reflected in differences in cyclic electron flow.

Figure 13.13 contrasts photosynthetic electron flow in purple and green bacteria and the heliobacteria. Note that in green sulfur bacteria and heliobacteria the excited state of the reaction center bacteriochlorophylls is significantly more electronegative than in purple bacteria and that actual chlorophyll *a* (green bacteria) or a structurally modified form of chlorophyll *a* (hydroxychlorophyll *a*) (heliobacteria) is present in the reaction center. Thus, unlike in purple bacteria, where the first stable acceptor molecule (quinone) has an E_0' of about 0 V (Figure 13.12*a*), the acceptors in green bacteria and heliobacteria are iron–sulfur proteins that have a much more electronegative E_0' than does NADH. Hence, reverse electron flow is unnecessary in green sulfur bacteria or heliobacteria. In green sulfur bacteria a protein called *ferredoxin* (E_0' -0.4V) is the direct electron donor for CO_2 fixation (Section 13.5). When H_2S is the source of reducing power in green sulfur or purple sulfur bacteria, globules of S^0 are produced from the oxidation of H_2S. In green bacteria the globules remain outside the cell but in purple sulfur bacteria they remain inside the cells (Figure 13.1). In both cases the S^0 eventually disappears as it is oxidized to sulfate (SO_4^{2-}) to generate additional reducing power for CO_2 fixation.

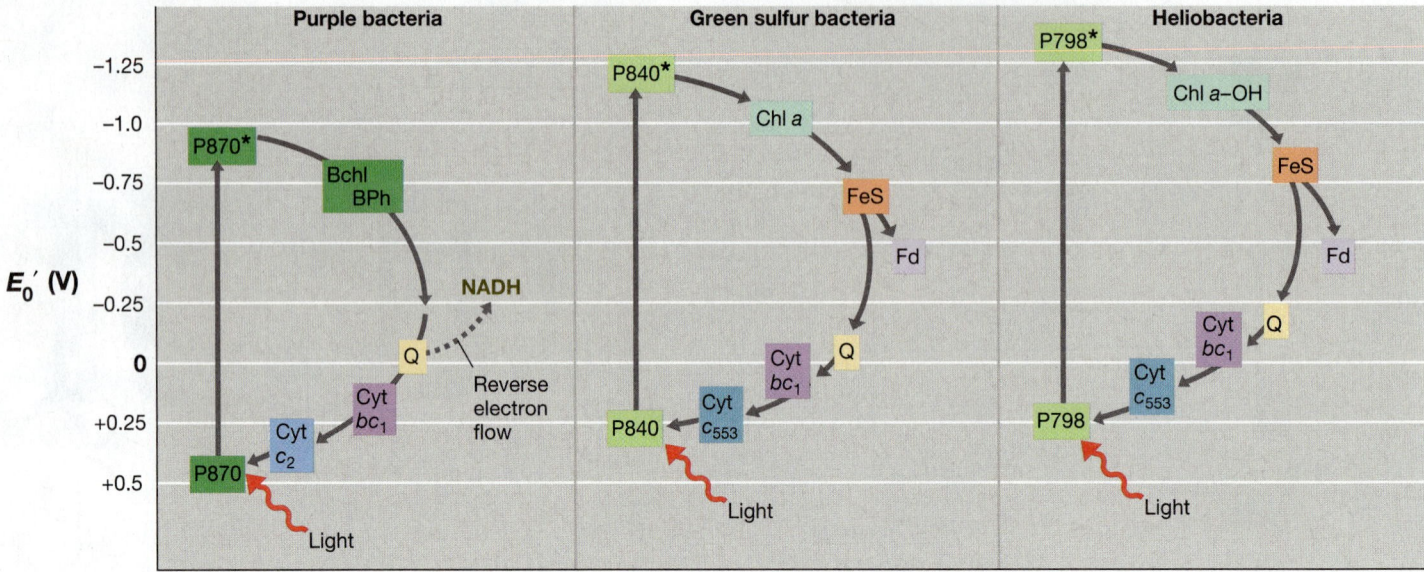

Figure 13.13 **A comparison of electron flow in purple bacteria, green sulfur bacteria, and heliobacteria.** Reverse electron flow in purple bacteria is necessary to produce NADH because the primary acceptor (quinone, Q) is more positive in potential than the NAD^+/NADH couple. In green and heliobacteria, ferredoxin (Fd), whose E_0' is more negative than that of NADH, is produced by light-driven reactions for reducing power needs. Bchl, bacteriochlorophyll; BPh, bacteriopheophytin. P870 and P840 are reaction centers of purple and green bacteria, respectively, and consist of Bchl *a*. The reaction center of heliobacteria (P798) contains Bchl *g*, and the reaction center of *Chloroflexus* is of the purple bacterial type. Note that forms of chlorophyll *a* are present in the reaction centers of green bacteria and heliobacteria.

MINIQUIZ

- What parallels exist in the processes of photophosphorylation and oxidative phosphorylation?
- What is reverse electron flow and why is it necessary? Which phototrophs need to use reverse electron flow?
- Chlorophyll *a* is not limited to oxygenic phototrophs. Explain.

13.4 Oxygenic Photosynthesis

In contrast to photosynthetic electron flow in *anoxygenic* phototrophs, in *oxygenic* phototrophs electrons flow through two distinct photosystems called *photosystem I* (*PSI*, or *P700*) and *photosystem II* (*PSII*, or *P680*). As in anoxygenic photosynthesis, the light reactions in oxygenic photosynthesis occur in photocomplexes embedded in membranes. In eukaryotic cells, the membranes are in the chloroplast (Figure 13.5), whereas in cyanobacteria, the membranes are arranged in stacks within the cytoplasm (Figure 13.10*c*).

Electron Flow and ATP Synthesis in Oxygenic Photosynthesis

PSI and PSII interact as shown in **Figure 13.14** in the "Z scheme" of photosynthesis, so named because the pathway resembles the letter "Z" turned on its side. The reduction potential of the P680 chlorophyll *a* molecule in PSII is very electropositive, even more positive than that of the O_2/H_2O couple. This is necessary to allow for the first step in electron flow, the splitting of water into oxygen and electrons (Figure 13.14). Light energy converts P680 into a strong reductant that reduces pheophytin *a* (chlorophyll *a* minus its magnesium atom), a molecule with an E_0' of about −0.5 V. An electron from H_2O is then donated to the oxidized

P680 molecule to return it to its ground-state reduction potential. From the pheophytin the electron travels through several membrane carriers of increasingly more positive E_0' including quinones, cytochromes, and a copper-containing protein called *plastocyanin*; the latter donates the electron to the PSI reaction center. The electron is accepted by P700 of PSI, which has previously absorbed light energy and donated an electron that will eventually lead to the reduction of $NADP^+$. Electrons travel through several intermediates in PSI terminating with the reduction of $NADP^+$ to NADPH (Figure 13.14).

Besides the net synthesis of NADPH, other important events take place while electrons flow from PSII to PSI. Electron transport reactions generate a proton motive force from which ATP is produced by ATPase. This mechanism for ATP synthesis is called *noncyclic photophosphorylation* because electrons do not cycle back to reduce the oxidized P680, but instead are used in the reduction of $NADP^+$. However, when reducing power is plentiful, ATP can also be produced in oxygenic phototrophs by *cyclic photophosphorylation*. This occurs when, instead of reducing $NADP^+$, electrons from PSI that would normally reduce ferredoxin are returned to travel the electron transport chain that connects PSII to PSI. In so doing, these electrons also generate a proton motive force that supports additional ATP synthesis (dashed line in Figure 13.14).

Anoxygenic Photosynthesis in Oxygenic Phototrophs

Photosystems I and II normally function in tandem in oxygenic photosynthesis. However, if PSII activity is blocked, some oxygenic phototrophs can perform photosynthesis using only PSI. Under these conditions, cyclic but not noncyclic photophosphorylation (Figure 13.14) occurs, and reducing power for CO_2 reduction

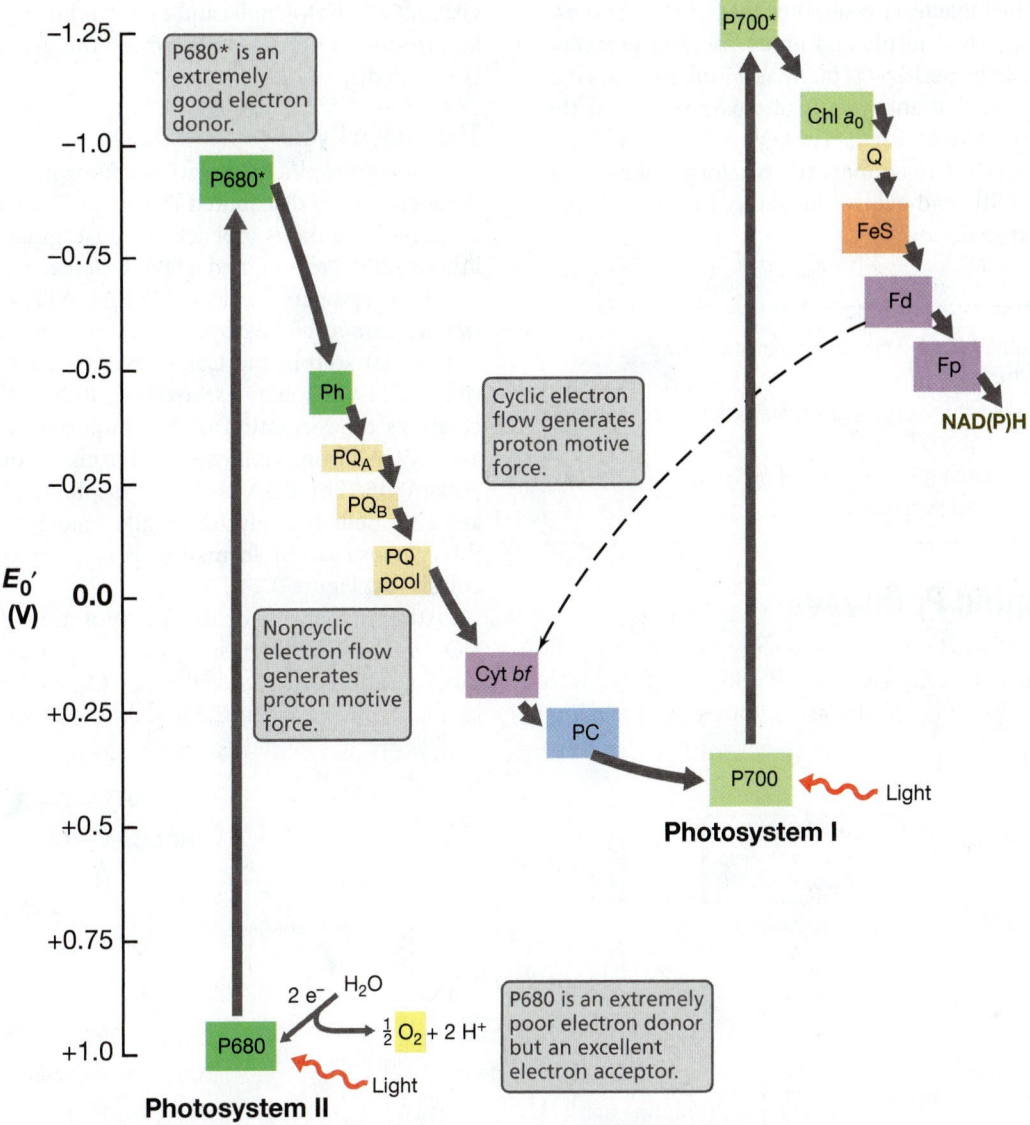

Figure 13.14 Electron flow in oxygenic photosynthesis, the "Z" scheme. Electrons flow through two photosystems, PSI and PSII. Ph, pheophytin; Q, quinone; Chl, chlorophyll; Cyt, cytochrome; PC, plastocyanin; FeS, nonheme iron–sulfur protein; Fd, ferredoxin; Fp, flavoprotein; P680 and P700 are the reaction center chlorophylls of PSII and PSI, respectively. Compare with Figure 13.12.

comes from sources other than water. In a nutshell, this is anoxygenic photosynthesis occurring in oxygenic phototrophs.

Many cyanobacteria can use H_2S as an electron donor under these conditions and many green algae can use H_2. When H_2S is used, it is oxidized to elemental sulfur (S^0), and sulfur granules similar to those produced by green sulfur bacteria (Figure 13.1) are deposited outside the cyanobacterial cells. **Figure 13.15** shows this in the filamentous cyanobacterium *Oscillatoria limnetica*. This organism lives in anoxic salt ponds where it oxidizes sulfide and carries out anoxygenic photosynthesis along with green and purple bacteria.

From an evolutionary standpoint, the process of cyclic photophosphorylation in both oxygenic and anoxygenic phototrophs is one of many indications of their close relationship. Further evidence of evolutionary relationships among phototrophs can be found in the fact that the structure of the purple bacterial and green nonsulfur photosynthetic reaction center resembles that of PSII, whereas the structure of the reaction centers of green

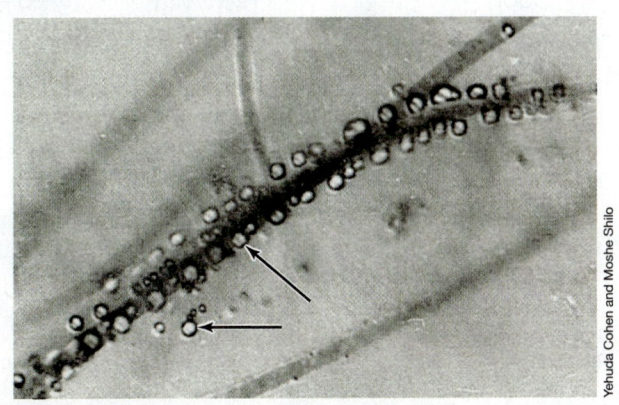

Figure 13.15 Oxidation of H_2S by *Oscillatoria limnetica*. Note the globules of S^0 (arrows), the oxidation product of H_2S, formed outside the cells. *O. limnetica* carries out oxygenic photosynthesis, but in the presence of H_2S, cells revert to the anoxygenic process.

sulfur bacteria and heliobacteria resembles that of PSI. Because the evidence is strong that purple and green bacteria preceded cyanobacteria on Earth by perhaps as many as 0.5 billion years ($\circlearrowleft$ Section 12.2), it is clear that anoxygenic photosynthesis was the first form of photosynthesis on Earth. The key evolutionary inventions of cyanobacteria were to connect the two forms of reaction centers (as PSI and PSII) and evolve the ability to use H_2O as a photosynthetic electron donor.

MINIQUIZ --

- Differentiate between cyclic and noncyclic electron flow in oxygenic photosynthesis.
- What is the key role of light energy in the initial step of the photosynthetic light reactions?
- What evidence is there that anoxygenic and oxygenic photosynthesis are related processes?

13.5 Autotrophic Pathways

Autotrophy is the process by which an energy-poor and highly oxidized form of carbon—CO_2—is reduced and assimilated into cell material. Many microorganisms are autotrophic, including virtually all phototrophs and chemolithotrophs. We focus here on autotrophy in phototrophs, where the greatest metabolic diversity is on display.

The Calvin Cycle

Several autotrophic pathways are known, but the **Calvin cycle** is the most widely distributed in nature. The Calvin cycle is present in purple bacteria, cyanobacteria, algae, green plants, most chemolithotrophic *Bacteria*, and a few *Archaea*. The cycle requires CO_2, a CO_2-acceptor molecule, NAD(P)H, ATP, and two key enzymes, *ribulose bisphosphate carboxylase* and *phosphoribulokinase*.

The first step in the Calvin cycle is catalyzed by the enzyme ribulose bisphosphate carboxylase, **RubisCO** for short. RubisCO catalyzes the formation of two molecules of 3-phosphoglyceric acid (PGA) from ribulose bisphosphate and CO_2 as shown in **Figure 13.16**. The PGA is then phosphorylated and reduced to a key intermediate of glycolysis, glyceraldehyde 3-phosphate. From this, glucose can be formed by reversal of the early steps in glycolysis ($\circlearrowleft$ Figure 3.14).

Instead of focusing on the incorporation of a single molecule of CO_2, it is easiest to consider Calvin cycle reactions based on the incorporation of 6 molecules of CO_2, as this is what is required to make one hexose ($C_6H_{12}O_6$). For RubisCO to incorporate

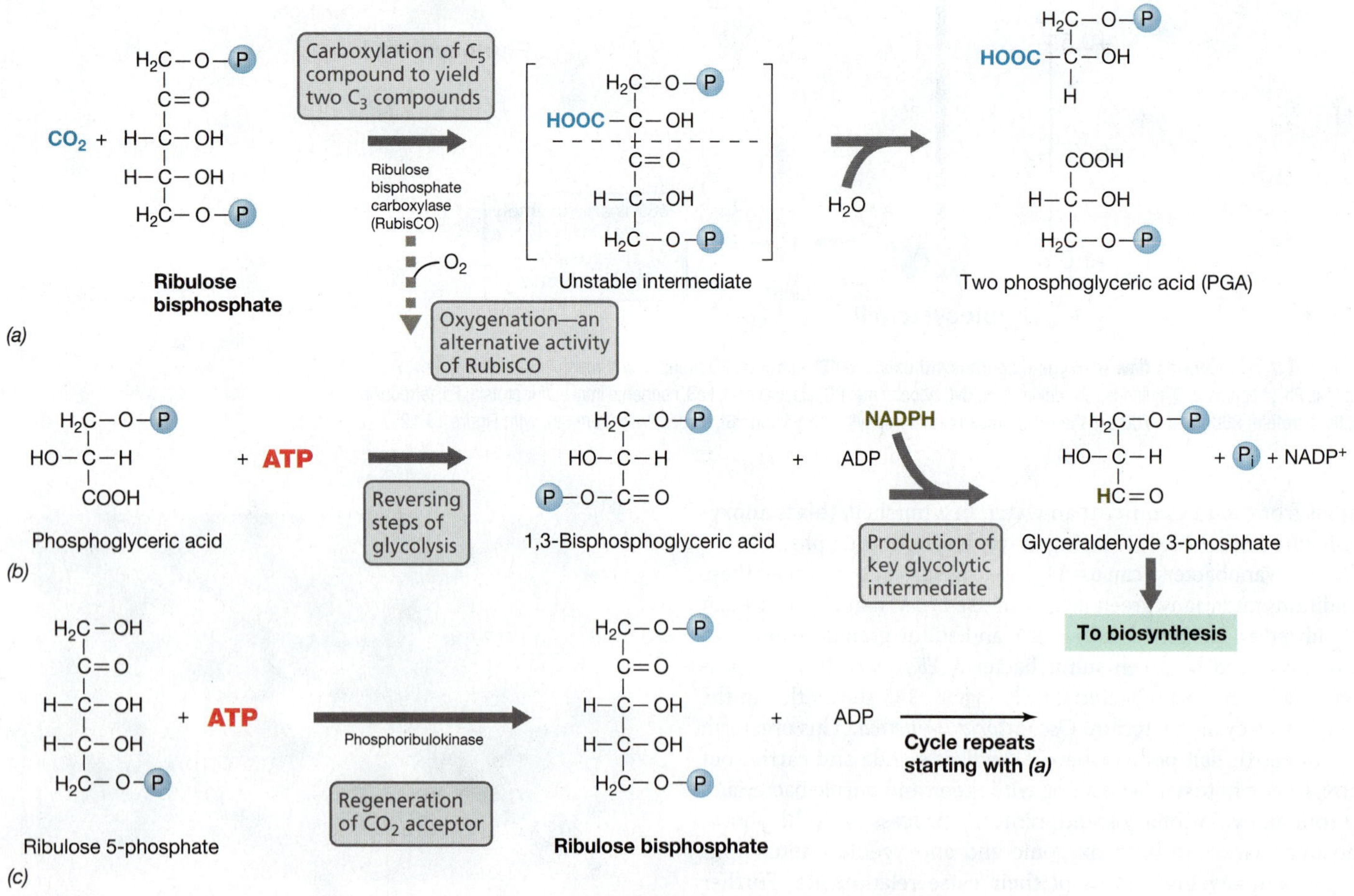

Figure 13.16 Key reactions of the Calvin cycle. *(a)* Reaction of the enzyme ribulose bisphosphate carboxylase. *(b)* Steps in the conversion of 3-phosphoglyceric acid (PGA) to glyceraldehyde 3-phosphate. Note that both ATP and NADPH are required. *(c)* Conversion of ribulose 5-phosphate to the CO_2 acceptor molecule ribulose 1,5-bisphosphate by the enzyme phosphoribulokinase.

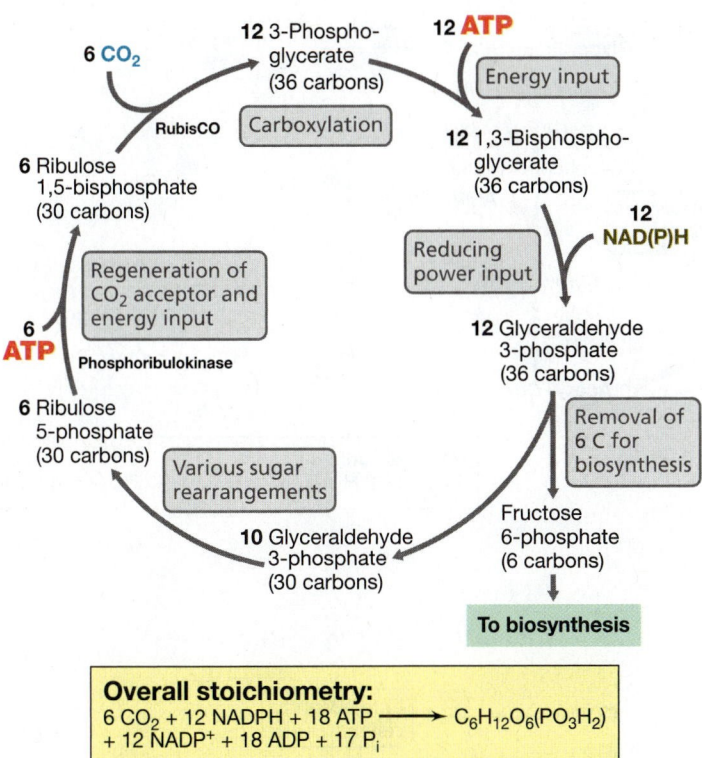

Figure 13.17 **The Calvin cycle.** Shown is the production of one hexose molecule from CO_2. For each six molecules of CO_2 incorporated, one fructose 6-phosphate is produced. In phototrophs, ATP comes from photophosphorylation and NAD(P)H from light or reverse electron flow.

Overall stoichiometry box:

Overall stoichiometry:
$$6\ CO_2 + 12\ NADPH + 18\ ATP \longrightarrow C_6H_{12}O_6(PO_3H_2)$$
$$+ 12\ NADP^+ + 18\ ADP + 17\ P_i$$

6 molecules of CO_2, 6 molecules of ribulose bisphosphate (total, 30 carbons) are required; carboxylation of these yields 12 molecules of PGA (total, 36 carbon atoms) (**Figure 13.17**). These then form the carbon skeletons for the eventual synthesis of 6 molecules of ribulose bisphosphate (total, 30 carbons) plus one hexose (6 carbons) for cell biosynthesis. A series of biochemical rearrangements between various sugars follow, resulting in 6 molecules of ribulose 5-phosphate (30 carbons). The final step in the Calvin cycle is the phosphorylation of each of these by the enzyme phosphoribulokinase (Figures 13.16*b* and 13.17) to regenerate 6 molecules of the acceptor molecule, ribulose bisphosphate. All totaled, *12 NADPH and 18 ATP* are required to synthesize one glucose from 6 CO_2 by the Calvin cycle.

Carboxysomes

Several Calvin cycle autotrophs produce polyhedral cell inclusions called **carboxysomes**. These inclusions, about 100 nm in diameter, are surrounded by a thin, protein membrane and consist of a crystalline array of RubisCO (**Figure 13.18**), with about 250 RubisCO molecules present per carboxysome.

Carboxysomes are a mechanism for concentrating CO_2 in the cell and making it readily available to RubisCO. Inorganic carbon incorporated into the cell as bicarbonate (HCO_3^-) enters the carboxysome as CO_2 through the activity of a second carboxysome enzyme, *carbonic anhydrase*. CO_2 (rather than HCO_3^-) is the actual substrate for RubisCO, and once inside the carboxysome, the CO_2 is trapped and ready for incorporation in the

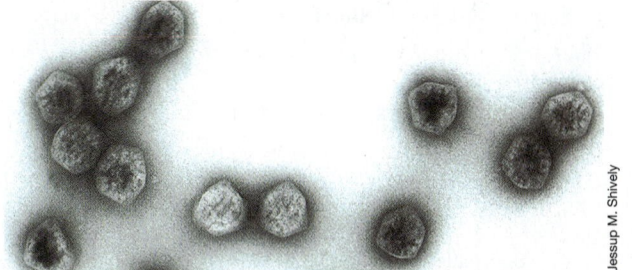

Figure 13.18 **Crystalline Calvin cycle enzymes: Carboxysomes.** Electron micrograph of carboxysomes purified from the chemolithotrophic sulfur oxidizer *Halothiobacillus neapolitanus*. The structures are about 100 nm in diameter. Carboxysomes are present in a wide variety of obligately autotrophic aerobic prokaryotes.

Jessup M. Shively

first step of the Calvin cycle. The carboxysome also functions to restrict access of RubisCO to O_2, an alternative substrate for this enzyme, and this ensures that RubisCO carboxylates rather than oxygenates ribulose bisphosphate (Figure 13.16*a*). If ribulose 1,5-bisphosphate is oxygenated, more energy and reducing power are required to incorporate it into central metabolic pathways than if it is carboxylated.

Autotrophy in Green Bacteria

Although they are autotrophs, the Calvin cycle does not operate in green sulfur and green nonsulfur bacteria. Instead, two novel autotrophic pathways are present, one in each group. Green sulfur bacteria such as *Chlorobium* (Figure 13.1) fix CO_2 by a reversal of steps in the citric acid cycle, a pathway called the **reverse citric acid cycle** (**Figure 13.19*a***). This pathway requires the activity of two ferredoxin-linked enzymes that catalyze the reductive fixation of CO_2; ferredoxin is produced in the light reactions of green sulfur bacteria (Figure 13.13).

Ferredoxin is an electron donor with a very electronegative E_0', about -0.4 V. The two ferredoxin-linked reactions catalyze (1) the carboxylation of succinyl-CoA to α-ketoglutarate, and (2) the carboxylation of acetyl-CoA to pyruvate (Figure 13.19*a*). Most of the remaining reactions of the reverse citric acid cycle are catalyzed by enzymes working in reverse of the normal oxidative direction of the cycle. One exception is *citrate lyase*, an ATP-dependent enzyme that cleaves citrate into acetyl-CoA and oxaloacetate (Figure 13.19*a*). In the oxidative direction of the cycle, citrate is produced by the enzyme *citrate synthase* (↩ Figure 3.22).

The reverse citric acid cycle operates in certain nonphototrophic autotrophs as well. For example, the hyperthermophilic *Thermoproteus* and *Sulfolobus* (Archaea; ↩ Section 16.10) and *Aquifex* (Bacteria; ↩ Section 15.19) use the reverse citric acid cycle, as do certain sulfur chemolithotrophic bacteria, such as *Thiomicrospira*. Thus, this pathway, originally discovered in green sulfur bacteria and thought to be unique to these phototrophs, is likely distributed among several groups of autotrophic prokaryotes.

Autotrophy in *Chloroflexus*

The green nonsulfur phototroph *Chloroflexus* (↩ Section 14.7) grows autotrophically with either H_2 or H_2S as electron donor.

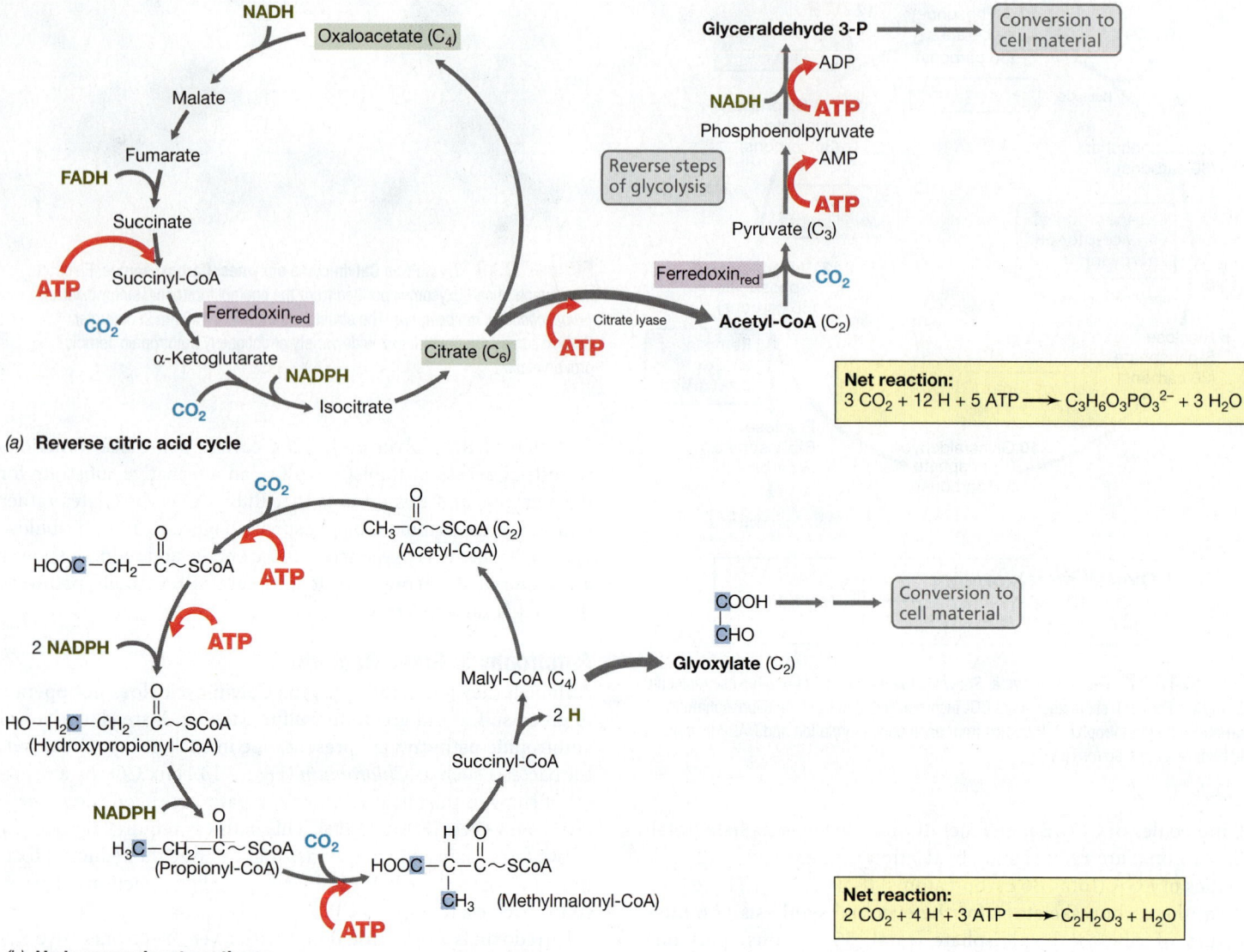

(a) Reverse citric acid cycle

Net reaction:
$$3 CO_2 + 12 H + 5 ATP \longrightarrow C_3H_6O_3PO_3{}^{2-} + 3 H_2O$$

(b) Hydroxypropionate pathway

Net reaction:
$$2 CO_2 + 4 H + 3 ATP \longrightarrow C_2H_2O_3 + H_2O$$

Figure 13.19 Unique autotrophic pathways in phototrophic green bacteria. *(a)* The reverse citric acid cycle is the mechanism of CO_2 fixation in green sulfur bacteria. Ferredoxin$_{red}$ indicates carboxylation reactions requiring reduced ferredoxin (2 H each). Starting from oxaloacetate, each turn of the cycle results in three molecules of CO_2 being incorporated and pyruvate as the product. *(b)* The hydroxypropionate pathway is the autotrophic pathway in the green nonsulfur bacterium *Chloroflexus*. Acetyl-CoA is carboxylated twice to yield methylmalonyl-CoA. This intermediate is rearranged to yield a new acetyl-CoA acceptor molecule and a molecule of glyoxylate, which is converted to cell material.

However, neither the Calvin cycle nor the reverse citric acid cycle operates in this organism. Instead, two molecules of CO_2 are reduced to glyoxylate by the **hydroxypropionate pathway**. This pathway is so named because hydroxypropionate, a three-carbon compound, is a key intermediate (Figure 13.19*b*).

In phototrophic bacteria, the hydroxypropionate pathway is found in *Chloroflexus*, thought to be one of the earliest phototrophs on Earth. This suggests that the hydroxypropionate pathway may have been one of the earliest mechanisms, if not *the* earliest, for autotrophy in anoxygenic phototrophs. In addition to *Chloroflexus*, the hydroxypropionate pathway operates in several hyperthermophilic *Archaea*, including *Metallosphaera*, *Acidianus*, and *Sulfolobus*. These are all chemolithotrophs that lie

near the base of the phylogenetic tree of *Archaea* (Chapter 16). The evolutionary roots of the hydroxypropionate pathway may thus be very deep, and it is possible that this pathway was nature's first attempt at autotrophy.

MINIQUIZ

- What reaction(s) does the enzyme RubisCO carry out?
- How much NADPH and ATP is required to make one hexose molecule by the Calvin cycle?
- What is a carboxysome and what is its function?
- Contrast autotrophy in the following phototrophs: cyanobacteria; purple and green sulfur bacteria; *Chloroflexus*.

II • Chemolithotrophy

We now turn our attention from phototrophs to the chemolithotrophs, highlighting the strategies, problems, and advantages of a lifestyle dependent on inorganic chemicals as energy sources. From an evolutionary standpoint, chemolithotrophy may have been the first form of energy conservation to evolve on Earth, as it is widespread among lineages that lie near the base of the phylogenetic trees of both *Bacteria* and *Archaea* (😞 Figures 1.6*b*, 12.13, and 16.1).

13.6 Inorganic Compounds as Electron Donors

Organisms that conserve energy from the oxidation of inorganic compounds are called **chemolithotrophs**. Most chemolithotrophic bacteria are also autotrophs. As we have noted with the phototrophs, for growth on CO_2 as the sole carbon source an organism needs (1) ATP and (2) reducing power. Some chemolithotrophs grow as **mixotrophs**, meaning that although they can conserve energy from the oxidation of an inorganic compound, they require an organic compound as their carbon source (that is, they are *not* autotrophs).

Inorganic Donors and ATP Generation

Chemolithotrophs can tap into many natural sources of inorganic electron donors, including geological, biological, and anthropogenic (the result of human activities). Volcanic activity is a major source of reduced sulfur compounds, primarily H_2S and S^0. Agricultural and mining operations add inorganic electron donors to the environment, especially reduced nitrogen and iron compounds, as does the burning of fossil fuels and the input of industrial wastes. Biological sources are also quite extensive, especially in the cases of H_2S, H_2, Fe^{2+}, and NH_3. The ecological success and metabolic diversity of chemolithotrophs is a good indication that diverse and abundant sources of inorganic electron donors are available in nature. The energy yield from the oxidation of these donors, however, varies considerably (**Table 13.1**).

In general terms, ATP generation in chemolithotrophs is similar to that in respiring chemoorganotrophs except that the electron donor is *inorganic* rather than *organic*. Just as for electrons from the oxidation of organic compounds, electrons from inorganic donors feed into electron transport chains and generate a proton motive force. Then, ATP synthesis occurs from the activity of ATPases (😞 Section 3.11). Reducing power in chemolithotrophs is obtained in either of two ways: directly from the inorganic compound (if it has a sufficiently negative reduction potential, such as H_2), or from reverse electron transport reactions (as discussed in Section 13.3 for phototrophic purple bacteria), if the inorganic electron donor is more electropositive than NADH. As we will see, with most chemolithotrophs, reverse electron transport is necessary since their electron donors are so electrochemically weak.

Energetics of Chemolithotrophy

A review of reduction potentials listed in Table 13.1 reveals that the oxidation of a number of inorganic electron donors can provide sufficient energy for ATP synthesis.

Recall from Chapter 3 that the farther apart two half reactions are in terms of the E_0' of their redox couples, the greater the amount of energy released (😞 Figure 3.9). For instance, the difference in reduction potential between the 2 H^+/H_2 couple and the $\frac{1}{2}O_2/H_2O$ couple is 1.23 V, which is equivalent to a free-energy yield of -237 kJ/mol (Appendix 1 shows how free energy values are calculated). On the other hand, the potential difference between the 2 H^+/H_2 couple and the NO_3^-/NO_2^- couple is less, 0.84 V, equivalent to a free-energy yield of -163 kJ/mol. This is still sufficient for the production of ATP (the energy-rich phosphate bond of ATP has a free energy of -31.8 kJ/mol). Thus, various inorganic electron donors and terminal electron

UNIT 3

Table 13.1 Energy yields from the oxidation of various inorganic electron donors[a]

Electron donor	Chemolithotrophic reaction	Group of chemolithotrophs	E_0' of couple (V)	$\Delta G^{0'}$ (kJ/reaction)	Number of electrons/ reaction	$\Delta G^{0'}$ (kJ/2 e^-)
Phosphite[b]	$4\,HPO_3^{2-} + SO_4^{2-} + H^+ \rightarrow 4\,HPO_4^{2-} + HS^-$	Phosphite bacteria	-0.69	-364	8	-91
Hydrogen[b]	$H_2 + \frac{1}{2}O_2 \rightarrow H_2O$	Hydrogen bacteria	-0.42	-237.2	2	-237.2
Sulfide[b]	$HS^- + H^+ + \frac{1}{2}O_2 \rightarrow S^0 + H_2O$	Sulfur bacteria	-0.27	-209.4	2	-209.4
Sulfur[b]	$S^0 + 1\frac{1}{2}O_2 + H_2O \rightarrow SO_4^{2-} + 2\,H^+$	Sulfur bacteria	-0.20	-587.1	6	-195.7
Ammonium[c]	$NH_4^+ + 1\frac{1}{2}O_2 \rightarrow NO_2^- + 2\,H^+ + H_2O$	Nitrifying bacteria	$+0.34$	-274.7	6	-91.6
Nitrite[b]	$NO_2^- + \frac{1}{2}O_2 \rightarrow NO_3^-$	Nitrifying bacteria	$+0.43$	-74.1	2	-74.1
Ferrous iron[b]	$Fe^{2+} + H^+ + \frac{1}{4}O_2 \rightarrow Fe^{3+} + \frac{1}{2}H_2O$	Iron bacteria	$+0.77$	-32.9	1	-65.8

[a]Data calculated from E_0' values in Appendix 1; values for Fe^{2+} are for pH 2, and others are for pH 7. At pH 7 the value for the Fe^{3+}/Fe^{2+} couple is about $+0.2$ V.
[b]Except for phosphite, all reactions are shown coupled to O_2 as electron acceptor. The only known phosphite oxidizer couples to SO_4^{2-} as electron acceptor. H_2 and most sulfur compounds can be oxidized anaerobically using one or more electron acceptors, and Fe^{2+} can be oxidized at neutral pH with NO_3^- as electron acceptor. For other chemolithotrophic reactions of sulfur compounds, see Table 13.2.
[c]Ammonium can also be oxidized with NO_2^- as electron acceptor (anammox, Section 13.10).

acceptors can be coupled in chemolithotrophic reactions (see Figure 13.39).

Energy calculations make it possible to predict the kinds of chemolithotrophs that should exist in nature. Because organisms must obey the laws of thermodynamics, only reactions that are exergonic are potential energy-conserving reactions for cells, and Table 13.1 lists all known classes of chemolithotrophs. We examine ecological aspects of chemolithotrophy in Chapter 20, where we will see that chemolithotrophic reactions form the heart of most nutrient cycles.

From Table 13.1 it should be clear that of the organisms listed, hydrogen bacteria obtain the most energy per two electrons oxidized while iron bacteria obtain the least. Put another way, the hydrogen bacteria oxidize a very strong electron donor while the iron bacteria oxidize a very weak donor. These differences come into play not only in terms of how much ATP can be produced per two electrons oxidized, but also in the energetic costs of making NADH for autotrophic growth. For hydrogen bacteria, this cost can be zero, while for the iron bacteria the expense is a major one, as we will see in Section 13.9.

MINIQUIZ --
- For what two purposes are inorganic compounds oxidized by chemolithotrophs?
- Why does the oxidation of H_2 yield more energy with O_2 as electron acceptor than with SO_4^{2-} as electron acceptor?

13.7 Hydrogen (H_2) Oxidation

Hydrogen (H_2) is a common product of microbial metabolism, especially of some fermentations (Sections 13.12–13.15), and a number of *aerobic* chemolithotrophs are able to use H_2 as an electron donor in energy metabolism. These are the classical "hydrogen bacteria." In addition, many *anaerobic* H_2-oxidizing *Bacteria* and *Archaea* are known, which differ in the electron acceptor they use (for example, nitrate, sulfate, ferric iron, CO_2); these organisms are discussed in Unit IV of this chapter. Here we focus on the aerobic H_2-oxidizing bacteria, organisms that couple the oxidation of H_2 to the reduction of O_2, forming water.

Energetics of H_2 Oxidation

Synthesis of ATP during H_2 oxidation by O_2 is the result of electron transport reactions that generate a proton motive force. The overall reaction

$$H_2 + \tfrac{1}{2}O_2 \rightarrow H_2O \qquad \Delta G^{0\prime} \rightarrow = -237 \text{ kJ}$$

is highly exergonic and can be coupled to the synthesis of ATP. In this reaction, which is catalyzed by the enzyme **hydrogenase**, the electrons from H_2 are initially transferred to a quinone acceptor. From there electrons travel through a series of cytochromes to generate a proton motive force and eventually reduce O_2 to water (**Figure 13.20**).

Some hydrogen bacteria synthesize two distinct hydrogenases, one cytoplasmic and one membrane-integrated. The latter enzyme participates in energetics, whereas the soluble hydrogenase has a different function. Instead of binding H_2 for use as

an electron donor in energy metabolism, the cytoplasmic hydrogenase binds H_2 and catalyzes the reduction of NAD^+ to NADH (the reduction potential of H_2 is sufficiently electronegative that reverse electron flow reactions are unnecessary). The organism *Ralstonia eutropha* has been a model for studying aerobic H_2 oxidation by species that make two hydrogenases, and we discuss some of the properties of this organism in Section 14.16. Species that synthesize only one hydrogenase make only the membrane-integrated enzyme, and it functions in both energy conservation and autotrophy in the cell.

Autotrophy in H_2 Bacteria

Although most hydrogen bacteria can also grow as chemo-organotrophs, when growing chemolithotrophically, they fix CO_2 by the Calvin cycle (Section 13.5). However, when readily usable organic compounds such as glucose are present, synthesis of Calvin cycle and hydrogenase enzymes by H_2 bacteria is repressed. Thus, H_2 bacteria are *facultative* chemolithotrophs. This undoubtedly has ecological value. In nature, H_2 levels in oxic environments are transient and low at best for at least two reasons: (1) most biological H_2 production is the result of fermentations, which are anoxic processes, and (2) H_2 can be utilized by several different anaerobic prokaryotes and thus totally consumed before it reaches oxic regions of a habitat. Hence, aerobic hydrogen bacteria must have a backup metabolism to H_2 oxidation, and in nature they likely shift between chemoorganotrophic and chemolithotrophic lifestyles as nutrients allow in their habitats. Moreover, many aerobic H_2 bacteria grow best microaerobically and are probably most competitive as H_2 bacteria in oxic–anoxic interfaces where H_2 may be in greater and more continuous supply than in fully oxic habitats.

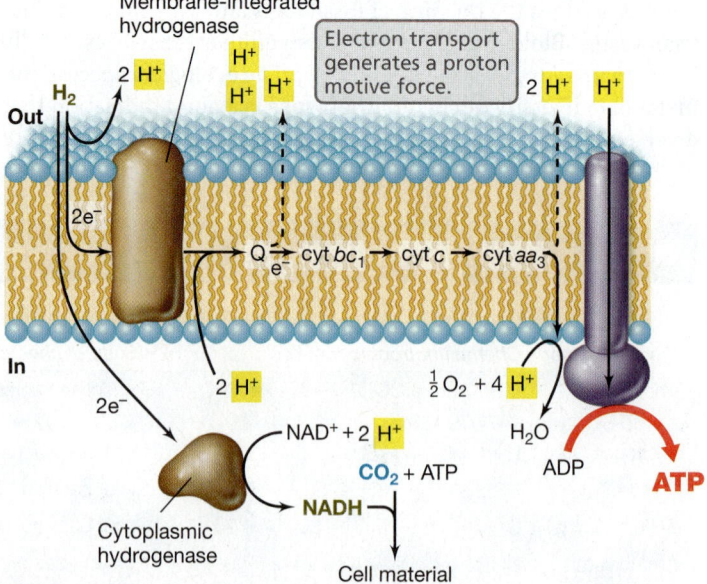

Figure 13.20 Bioenergetics and function of the two hydrogenases of aerobic hydrogen bacteria. In *Ralstonia eutropha* two hydrogenases are present; the membrane-bound hydrogenase participates in energetics, whereas the cytoplasmic hydrogenase makes NADH for the Calvin cycle. Some hydrogen bacteria have only the membrane-bound hydrogenase, and in these organisms reducing power is synthesized by reverse electron flow from Q back to NAD^+ to form NADH. Cyt, cytochrome; Q, quinone.

UNIT 3

MINIQUIZ

- What enzyme is required for hydrogen bacteria to grow as H_2 chemolithotrophs?

- Why is reverse electron flow unnecessary in H_2 bacteria that contain two hydrogenases?

13.8 Oxidation of Reduced Sulfur Compounds

Many reduced sulfur compounds can be electron donors for the colorless sulfur bacteria, called *colorless* to distinguish them from the pigmented green and purple sulfur bacteria discussed earlier in this chapter (Figure 13.1 and Section 13.3). Historically, the very concept of chemolithotrophy emerged in the late nineteenth century from studies of the sulfur bacteria by the Russian microbiologist Sergei Winogradsky (⮌ Section 1.9). This was a major new concept in microbiology at the time, and as our understanding of prokaryotic diversity has improved, it has become clear that chemolithotrophy is a major metabolic lifestyle of many *Bacteria* and *Archaea*.

Energetics of Sulfur Oxidation

The most common sulfur compounds used as electron donors are hydrogen sulfide (H_2S), elemental sulfur (S^0), and thiosulfate ($S_2O_3^{2-}$); sulfite (SO_3^{2-}) can also be oxidized (Table 13.1 and **Table 13.2**). In most cases, the final oxidation product is sulfate (SO_4^{2-}). Sulfide oxidation occurs in stages, with the first oxidation step yielding elemental sulfur, S^0. Some sulfide-oxidizing bacteria, such as *Beggiatoa*, deposit this elemental sulfur inside the cell (**Figure 13.21a**), where the sulfur exists as a potential energy (electron) reserve. When the supply of sulfide has been depleted, additional energy can then be conserved from the oxidation of sulfur to sulfate. When S^0 is present externally, the organism must attach itself to the sulfur particle because elemental sulfur is rather insoluble (Figure 13.21b). By adhering to the particle, the organism can remove sulfur atoms for oxidation to sulfate. This occurs by way of membrane or periplasmic proteins that solubilize S^0 by reducing it to HS^-; the latter is then transported into the cell and enters chemolithotrophic metabolism (see Figure 13.22).

One product of the oxidation of reduced sulfur compounds is protons (Tables 13.1 and 13.2). Consequently, one result of sulfur chemolithotrophy is acidification of the environment. Because of this, many sulfur bacteria have evolved to be acid-tolerant or even acidophilic. *Acidithiobacillus thiooxidans*, for example, grows best at a pH between 2 and 3.

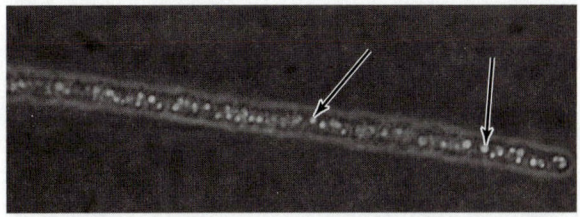

(a)

(b)

Figure 13.21 Sulfur bacteria. *(a)* Internal sulfur granules in *Beggiatoa* (arrows). *(b)* Attachment of cells of the sulfur-oxidizing archaeon *Sulfolobus acidocaldarius* to a crystal of elemental sulfur. Cells are visualized by fluorescence microscopy after being stained with the dye acridine orange. The sulfur crystal does not fluoresce.

Biochemistry of Sulfur Oxidation

The biochemical steps in the oxidation of sulfur compounds are summarized in **Figure 13.22**. At least three pathways for sulfur oxidation are known. In two of the systems, the starting substrate, HS^-, $S_2O_3^{2-}$, or S^0, is first oxidized to sulfite (SO_3^{2-}); starting from sulfide this releases six electrons. The sulfite is then oxidized to sulfate plus two electrons, and this can occur in either of two ways. The most widespread system employs the enzyme *sulfite oxidase*. This enzyme oxidizes sulfite and transfers the electrons directly to cytochrome *c*; ATP is made during subsequent electron transport reactions coupled to formation of a proton motive force (Figure 13.22b). By contrast, some sulfur chemolithotrophs oxidize SO_3^{2-} to SO_4^{2-} via a reversal of the activity of the enzyme *adenosine phosphosulfate reductase*, an enzyme essential for the metabolism of sulfate-reducing bacteria (Section 13.18 and see Figure 13.42). This reaction, run in the direction of SO_4^{2-} production by sulfur chemolithotrophs, yields one energy-rich phosphate bond when AMP is converted to ADP (Figure 13.22a). When thiosulfate is the electron donor for sulfur chemolithotrophs (Table 13.2), it is first split into S^0 and SO_3^{2-}, both of which are eventually oxidized to SO_4^{2-}.

Table 13.2 Comparison of the energetics of oxidation of some common reduced sulfur compounds

Chemolithotrophic reaction	Electrons	Stoichiometry[a]	Energetics (kJ/electron)[a]
Sulfide to sulfate	8	$H_2S + 2\,O_2 \rightarrow SO_4^{2-} + 2\,H^+$	$\Delta G^{0\prime} = -798.2$ kJ/reaction (-99.75 kJ/e$^-$)
Sulfite to sulfate	2	$SO_3^{2-} + \frac{1}{2}O_2 \rightarrow SO_4^{2-}$	$\Delta G^{0\prime} = -258$ kJ/reaction (-129 kJ/e$^-$)
Thiosulfate to sulfate	8	$S_2O_3^{2-} + H_2O + 2\,O_2 \rightarrow 2\,SO_4^{2-} + 2\,H^+$	$\Delta G^{0\prime} = -818.3$ kJ/reaction (-102 kJ/e$^-$)

[a]All reactions are balanced, both atomically and electrically. See Appendix 1 for details of calculations. For the reaction and energetics of the oxidation of sulfide to sulfur and sulfur to sulfate, see Table 13.1.

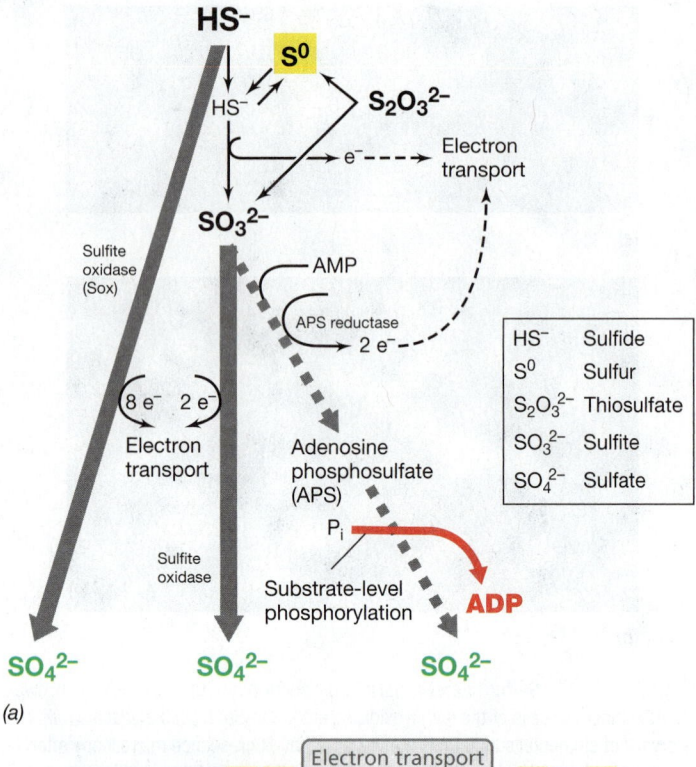

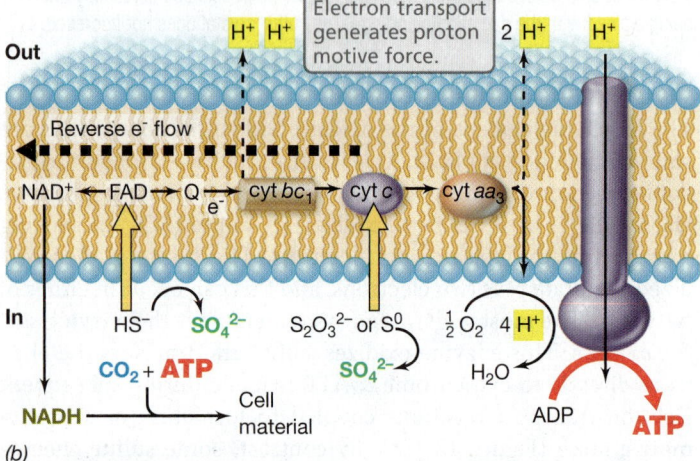

Figure 13.22 Oxidation of reduced sulfur compounds by sulfur chemolithotrophs. *(a)* Steps in the oxidation of different reduced sulfur compounds. Three different pathways are known. *(b)* Electrons from sulfur compounds feed into the electron transport chain to drive a proton motive force; electrons from $S_2O_3^{2-}$ and S^0 enter at the level of cytochrome *c*. NADH is made by reverse electron flow. Cyt, cytochrome; FAD, flavoprotein; Q, quinone. For the structure of APS, see Figure 13.42a.

A functionally distinct sulfide and thiosulfate oxidation system is present in *Paracoccus pantotrophus* and many other sulfur bacteria. This system, called the *Sox* (for *s*ulfur *ox*idation) *system*, oxidizes reduced sulfur compounds directly to sulfate without the intermediate formation of sulfite (Figure 13.22a). The Sox system contains over 15 genes encoding various cytochromes and other proteins necessary for the oxidation of reduced sulfur compounds directly to sulfate. The Sox system is present in several sulfur chemolithotrophs and also in some phototrophic sulfur bacteria that oxidize sulfide to obtain reducing power for CO_2 fixation rather than for energy conservation. The fact that

this biochemical system is distributed among prokaryotes that oxidize sulfide for very different reasons is a good indication that the genes that encode Sox have been transferred between species by horizontal gene flow (🔗 Section 6.12 and Chapter 10).

Other Aspects of Chemolithotrophic Sulfur Oxidation

Electrons from the oxidation of reduced sulfur compounds eventually reach the electron transport chain as shown in Figure 13.22b. Depending on the E_0' of the electron donor couple, electrons enter at the flavoprotein ($E_0' = -0.2$ V), quinone ($E_0' = 0$ V), or cytochrome *c* ($E_0' = +0.3$ V) levels and are transported through the chain to O_2, generating a proton motive force that forms ATP by ATPase. Electrons for autotrophic CO_2 fixation come from reverse electron flow (Section 13.3), eventually yielding NADH. Autotrophy is driven by reactions of the Calvin cycle or some other autotrophic pathway (Section 13.5). Although the sulfur chemolithotrophs are primarily an aerobic group, some species can grow anaerobically using nitrate as an electron acceptor. The sulfur bacterium *Thiobacillus denitrificans* is a classic example, reducing nitrate to dinitrogen gas (the process of denitrification, Section 13.17).

MINIQUIZ -

• How many electrons are available from the oxidation of H_2S if S^0 or SO_4^{2-} is the final product?

• In terms of intermediates, how does the Sox system differ from other sulfide-oxidizing systems?

13.9 Iron (Fe^{2+}) Oxidation

The aerobic oxidation of ferrous iron (Fe^{2+}) to ferric iron (Fe^{3+}) supports growth of the chemolithotrophic "iron bacteria" (🔗 Section 14.15). At acidic pH, only a small amount of energy is available from this reaction (Table 13.1), and for this reason the iron bacteria must couple the oxidation of large amounts of iron in order to produce only tiny amounts of cell material. The ferric iron produced spontaneously forms insoluble ferric hydroxide ($Fe^{3+} + 3 H_2O \rightarrow Fe(OH)_3 + 3 H^+$) and other iron precipitates in aquatic environments, and this drives down the pH (**Figure 13.23**). This inevitable chemical reaction probably explains why many iron-oxidizing bacteria have evolved to be strongly acidophilic.

Iron-Oxidizing Bacteria

The best-known iron bacteria, *Acidithiobacillus ferrooxidans* and *Leptospirillum ferrooxidans*, can both grow autotrophically using ferrous iron (Figure 13.23) as electron donor at pH values as low as 1; growth is optimal at pH 2–3. These bacteria are common in acid-polluted environments such as coal-mining runoff waters (Figure 13.23a). *Ferroplasma*, a species of *Archaea*, is an extremely acidophilic iron oxidizer and can grow at pH values below 0 (🔗 Section 16.3). We discuss the role of all of these organisms in acid-mine pollution and mineral oxidation in Sections 20.5, 21.1, and 21.2.

At neutral pH, Fe^{2+} spontaneously oxidizes to Fe^{3+}, so opportunities for the iron bacteria are restricted to locations where Fe^{2+} is transitioning from anoxic to oxic conditions. For example, anoxic groundwater often contains Fe^{2+}, and when it is released, as in iron-rich springwater, it becomes exposed to O_2. At such

$$Fe^{3+} + 3\,H_2O \longrightarrow Fe(OH)_3 + 3\,H^+$$

(b)

Figure 13.23 Iron-oxidizing bacteria. (a) Acid mine drainage, showing the confluence of a normal river and a creek draining a coal-mining area. At low pH values, Fe^{2+} does not oxidize spontaneously in air, but *Acidithiobacillus ferrooxidans* carries out the oxidation; insoluble $Fe(OH)_3$ and complex ferric salts precipitate. (b) Cultures of *A. ferrooxidans*. Shown is a dilution series, with no growth in the tube on the left and increasing amounts of growth from left to right. Growth is evident from the production of $Fe(OH)_3$.

interfaces, iron bacteria oxidize Fe^{2+} to Fe^{3+} before it oxidizes spontaneously. *Gallionella ferruginea*, *Sphaerotilus natans*, and *Leptothrix discophora* are examples of bacteria that live at these interfaces. They are typically seen mixed in with the characteristic ferric iron deposits they form (⮌ Figures 14.36 and 21.22).

Energy from Iron Oxidation

The bioenergetics of ferrous iron oxidation by *Acidithiobacillus ferrooxidans* and other acidophilic iron oxidizers are of considerable interest because of the very electropositive reduction potential of the Fe^{3+}/Fe^{2+} couple at acidic pH (E_0' +0.77 V at pH 2). The respiratory chain of *A. ferrooxidans* contains cytochromes of the c and aa_3 types and a periplasmic copper-containing protein called *rusticyanin* (**Figure 13.24**). There is also an iron-oxidizing protein located in the outer membrane of the cell.

Because the reduction potential of the Fe^{3+}/Fe^{2+} couple is so high, steps in electron transport to oxygen ($\frac{1}{2}O_2/H_2O$, $E_0' = +0.82$ V) can obviously be few. Iron oxidation begins in the outer membrane where the organism contacts either soluble Fe^{2+} or insoluble ferrous iron minerals. Fe^{2+} is oxidized to Fe^{3+}, a one-electron transition (Table 13.1), by an outer membrane cytochrome c that transfers electrons into the periplasm where rusticyanin ($E_0' = +0.68$ V) is the electron acceptor. This thermodynamically slightly unfavorable reaction is thought to be pulled forward by

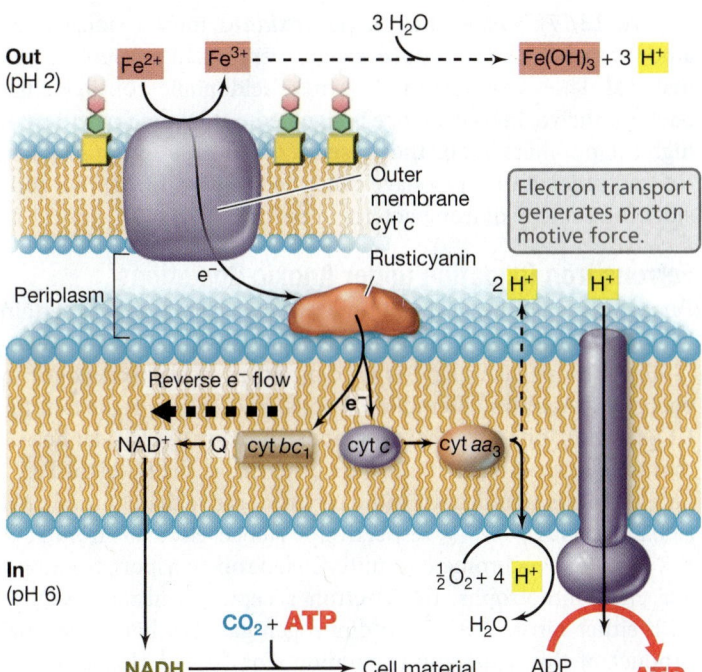

Figure 13.24 Electron flow during Fe^{2+} oxidation by the acidophile *Acidithiobacillus ferrooxidans*. The periplasmic copper-containing protein rusticyanin receives electrons from Fe^{2+} oxidized by a c-type cytochrome located in the outer membrane. From here, electrons travel a short electron transport chain, resulting in the reduction of O_2 to H_2O. Reducing power comes from reverse electron flow. Note the steep pH gradient across the membrane.

the immediate consumption of Fe^{3+} in $Fe(OH)_3$ formation (Figure 13.24). Rusticyanin then reduces a periplasmic cytochrome c, which transfers electrons to cytochrome aa_3, and it is the latter protein that reduces O_2 to H_2O; ATP is synthesized by ATPase in the usual fashion (Figure 13.24).

The nature of the proton motive force in *A. ferrooxidans* is of interest. In a highly acidic environment, a large gradient of protons already exists across the *A. ferrooxidans* membrane (the cell periplasm is pH 1–2, whereas the cytoplasm is pH 5.5–6, Figure 13.24). Although this situation might make one think that *A. ferrooxidans* can make ATP at no energetic cost, this is not the case, as the organism cannot make ATP from this natural proton motive force in the absence of an electron donor. This is because H^+ ions that enter the cytoplasm via ATPase must be consumed in order to maintain the internal pH within acceptable limits. Proton consumption occurs during the reduction of O_2 in the electron transport chain and this reaction requires electrons; the latter come from the oxidation of Fe^{2+} to Fe^{3+} (Figure 13.24).

Autotrophy in *A. ferrooxidans* is supported by the Calvin cycle (Section 13.5), and because of the high potential of the electron donor, much energy must be consumed in reverse electron flow reactions to obtain the reducing power (NADH) necessary to drive CO_2 fixation. NADH is formed by reduction of NAD^+ by electrons obtained from Fe^{2+} that are forced backwards through cytochrome bc_1 and the quinone pool at the expense of the proton motive force (Figure 13.24).

The relatively poor energetic yield from ferrous iron oxidation coupled with the large energetic demands of the Calvin cycle

(Figure 13.17) means that *A. ferrooxidans* must oxidize large amounts of Fe^{2+} to produce even a very small amount of cell material. Thus, in environments where acidophilic iron-oxidizing bacteria thrive, their presence is signaled not by the formation of high cell numbers but by the presence of the large amounts of ferric iron precipitates they have generated (Figure 13.23). We consider the ecology of iron bacteria in Sections 20.5, 21.1, and 21.2.

Ferrous Iron Oxidation under Anoxic Conditions

Ferrous iron can be oxidized under *anoxic* conditions by certain chemolithotrophs and anoxygenic phototrophic bacteria (**Figure 13.25**). In these cases, Fe^{2+} is used as either an electron donor in energy metabolism (chemolithotrophs) and/or as a reductant for CO_2 fixation (phototrophs). An important point to consider here is that at neutral pH where these organisms thrive, the E_0' of the Fe^{3+}/Fe^{2+} couple is significantly more electronegative than at acidic pH (+0.2 V versus +0.77 V, respectively). Hence, electrons from Fe^{2+} can reduce cytochrome *c* to initiate electron transport reactions. For chemolithotrophs, the electron acceptor is nitrate (NO_3^-), with either nitrite (NO_2^-) or dinitrogen gas (N_2) being the final product of this anaerobic respiration. For Fe^{2+}-oxidizing purple and green bacteria, either soluble Fe^{2+} or iron sulfide (FeS) can be used as electron donor. With FeS, both Fe^{2+} and S^{2-} are oxidized, Fe^{2+} to Fe^{3+} (one electron) and HS^- to SO_4^{2-} (eight electrons).

MINIQUIZ --

- Why is only a very small amount of energy available from the oxidation of Fe^{2+} to Fe^{3+} at acidic pH?

- What is the function of rusticyanin and where is it found in the cell?

- How can Fe^{2+} be oxidized under anoxic conditions?

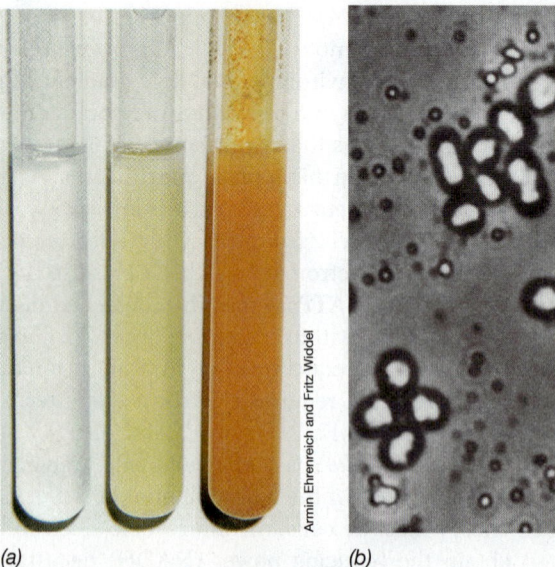

(a)   *(b)*

Figure 13.25 Fe^{2+} oxidation by anoxygenic phototrophic bacteria. *(a)* Oxidation in anoxic tube cultures. Left to right: Sterile medium, inoculated medium, a growing culture showing $Fe(OH)_3$. *(b)* Phase-contrast photomicrograph of an Fe^{2+}-oxidizing purple bacterium. The bright refractile areas within cells are gas vesicles. The granules outside the cells are iron precipitates. This organism is phylogenetically related to the purple sulfur bacterium *Chromatium*.

13.10 Nitrification and Anammox

The reduced inorganic nitrogen compounds ammonia (NH_3) and nitrite (NO_2^-) are oxidized aerobically by the chemolithotrophic *nitrifying bacteria* in the process of **nitrification** (⟳ Section 14.13). Under anoxic conditions, ammonia is also oxidized by a unique group of bacteria in the process called **anammox**.

Nitrifying and anammox bacteria are widely distributed in soils, water, wastewaters, and the oceans. Aerobically, nitrifying *Bacteria* and *Archaea* oxidize NH_3 but only to nitrite, while a distinct group of *Bacteria* oxidizes the NO_2^- to NO_3^-. The complete oxidation of NH_3 to NO_3^-, an eight-electron transfer, is thus carried out by the cooperative activities of two physiological groups of organisms—the *ammonia oxidizers* and the *nitrite oxidizers*.

Bioenergetics and Enzymology of Ammonia and Nitrite Oxidation

The bioenergetics of nitrification is based on the same principles that govern other chemolithotrophic reactions: electrons from reduced inorganic substrates (in this case, reduced nitrogen compounds) enter an electron transport chain, and electron transport reactions establish a proton motive force that drives ATP synthesis. The electron donors for the nitrifying bacteria are not particularly strong. The E_0' of the NO_2^-/NH_3 couple (the first step in the oxidation of NH_3) is +0.34 V, and the E_0' of the NO_3^-/NO_2^- couple is even more positive, about +0.43 V. By necessity, these reduction potentials force the nitrifying bacteria to donate electrons to rather high-potential electron acceptors, and this of course limits the amount of energy that can be conserved.

Several key enzymes participate in the oxidation of reduced nitrogen compounds. In ammonia-oxidizing bacteria such as *Nitrosomonas*, NH_3 is oxidized by *ammonia monooxygenase* (monooxygenases are discussed in Section 13.22), producing hydroxylamine (NH_2OH) and H_2O (**Figure 13.26**). A second key enzyme, *hydroxylamine oxidoreductase*, then oxidizes NH_2OH to NO_2^-, removing four electrons in the process. Ammonia monooxygenase is an integral membrane protein, whereas hydroxylamine oxidoreductase is periplasmic (Figure 13.26). In the reaction carried out by ammonia monooxygenase

$$NH_3 + O_2 + 2H^+ + 2e^- \rightarrow NH_2OH + H_2O$$

two electrons and protons are needed to reduce one atom of (O_2) to H_2O. These electrons originate from the oxidation of hydroxylamine and are supplied to ammonia monooxygenase from hydroxylamine oxidoreductase via cytochrome *c* and ubiquinone (Figure 13.26). Thus, for every *four* electrons generated from the oxidation of NH_3 to NO_2^-, only *two* actually reach cytochrome aa_3, the terminal oxidase that interacts with O_2 to form H_2O (Figure 13.26).

Nitrite-oxidizing bacteria such as *Nitrobacter* oxidize NO_2^- to NO_3^- by the enzyme *nitrite oxidoreductase*, with electrons traveling a very short electron transport chain (because of the high potential of the NO_3^-/NO_2^- couple) to the terminal oxidase (**Figure 13.27**). Cytochromes of the *a* and *c* types are present in the electron transport chain of nitrite oxidizers, and the activity of cytochrome aa_3 generates a proton motive force (Figure 13.27). As is the case with the iron bacteria (Section 13.9), only small

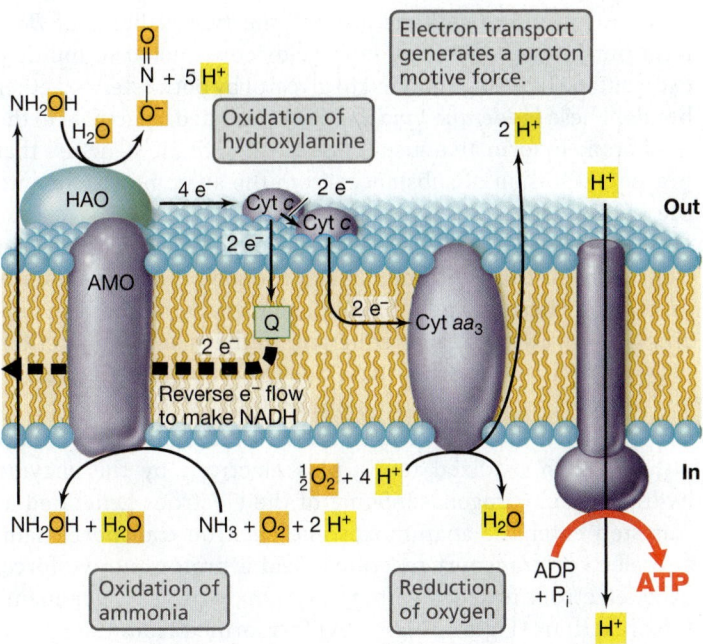

Figure 13.26 Oxidation of NH₃ and electron flow in ammonia-oxidizing bacteria. The reactants and the products of this reaction series are highlighted. The cytochrome *c* (Cyt *c*) in the periplasm is a different form of Cyt *c* than that in the membrane. AMO, ammonia monooxygenase; HAO, hydroxylamine oxidoreductase; Q, ubiquinone.

amounts of energy are available from nitrite oxidation. Hence, minimal amounts of cell material are obtained even though large amounts of nitrite may be oxidized.

Ammonia-Oxidizing *Archaea*

From a phylogenetic standpoint, *Nitrosomonas* and *Nitrobacter* are *Bacteria*. However, some *Archaea* are also ammonia oxidizers. For example, *Nitrosopumilus* is a marine ammonia-oxidizing

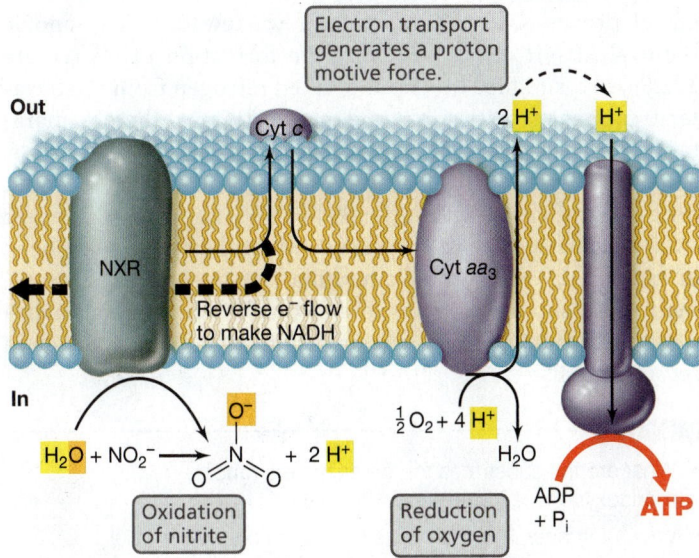

Figure 13.27 Oxidation of NO₂⁻ to NO₃⁻ by nitrifying bacteria. The reactants and products of this reaction series are highlighted to show the reaction clearly. NXR, nitrite oxidoreductase.

chemolithotroph of the archaeal phylum *Thaumarchaeota* (⊘ Section 16.6 and chapter opening page of Chapter 6). Genes in *Nitrosopumilus* that encode ammonia monooxygenase are similar to those in *Nitrosomonas*, and thus the biochemistry of ammonia oxidation is likely to have a common mechanism in both domains. One major difference, however, is that unlike ammonia-oxidizing *Bacteria*, ammonia-oxidizing *Archaea* can oxidize ammonia in the vanishingly small levels found in the open oceans. Thus, archaeal nitrifiers may dominate marine nitrification and may also play significant roles in other habitats that contain only low levels of ammonia. In this connection, the nitrifying archaeon *Nitrososphaera* inhabits soils instead of the oceans and may be a major ammonia oxidizer in terrestrial habitats. Thus far, nitrite-oxidizing *Archaea* are unknown.

Carbon Metabolism and Ecology of Nitrifying Bacteria

Like sulfur- and iron-oxidizing chemolithotrophs, aerobic nitrifying *Bacteria* employ the Calvin cycle for CO₂ fixation. The ATP and reducing power requirements of the Calvin cycle place additional burdens on an energy-generating system that already has a relatively low yield (NADH to drive the Calvin cycle in nitrifiers is formed by reverse electron flow, Figures 13.26 and 13.27). The energetic constraints are particularly severe for NO₂⁻ oxidizers, and it is perhaps for this reason that most NO₂⁻ oxidizers have alternative energy-conserving mechanisms, being able to grow chemoorganotrophically on glucose and a few other organic substrates. By contrast, species of ammonia-oxidizing bacteria are either obligate chemolithotrophs or mixotrophs. Autotrophy in ammonia-oxidizing *Archaea* is supported by a variation of the hydroxypropionate cycle (Section 13.5).

Nitrifying prokaryotes play key ecological roles in the nitrogen cycle, converting ammonia into nitrate, a key plant nutrient. Nitrifiers are also important in sewage and wastewater treatment, removing toxic amines and ammonia and releasing less toxic nitrogen compounds (⊘ Section 21.6). Nitrifiers play a similar role in the water column of lakes, where ammonia produced in the sediments from the decomposition of organic nitrogenous compounds is oxidized to nitrate, a more favorable nitrogen source for algae and cyanobacteria.

Anammox

Although the ammonia-oxidizing prokaryotes just discussed are strict *aerobes*, NH₃ can also be oxidized under anoxic conditions. This process is called *anammox* (for *an*oxic *ammo*nia *ox*idation) and is catalyzed by an unusual group of obligately anaerobic *Bacteria*.

Ammonia is oxidized in the anammox reaction using NO₂⁻ as the electron acceptor to yield N₂:

$$NH_4^+ + NO_2^- \rightarrow N_2 + 2H_2O \qquad \Delta G^{0\prime} = -357 \text{ kJ}$$

A major anammox organism, *Brocadia anammoxidans*, is a species of the *Planctomycetes* phylum of *Bacteria* (⊘ Section 15.16). *Planctomycetes* are unusual *Bacteria* in that they lack peptidoglycan and their cytoplasm contains membrane-enclosed compartments of various types (**Figure 13.28**). In cells of *B. anammoxidans* this compartment is the *anammoxosome*, and it is within this structure that the anammox reaction occurs (Figure 13.28*c*). In addition to

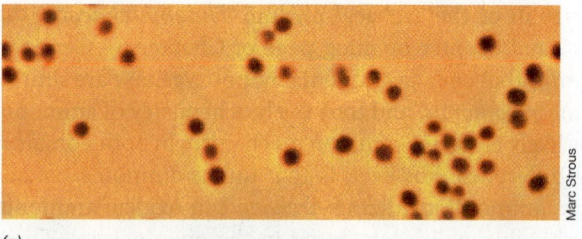

(a)

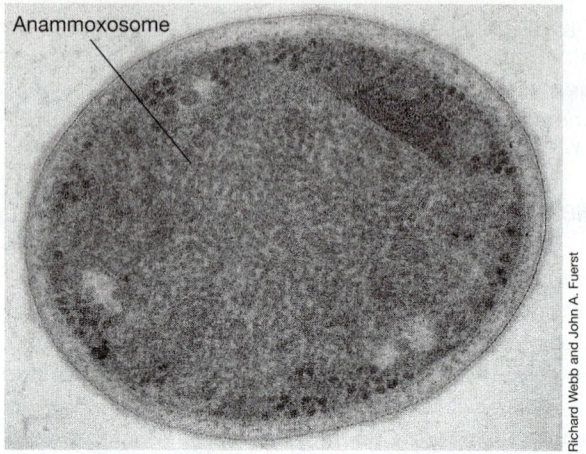

Anammoxosome

(b)

Richard Webb and John A. Fuerst

Marc Strous

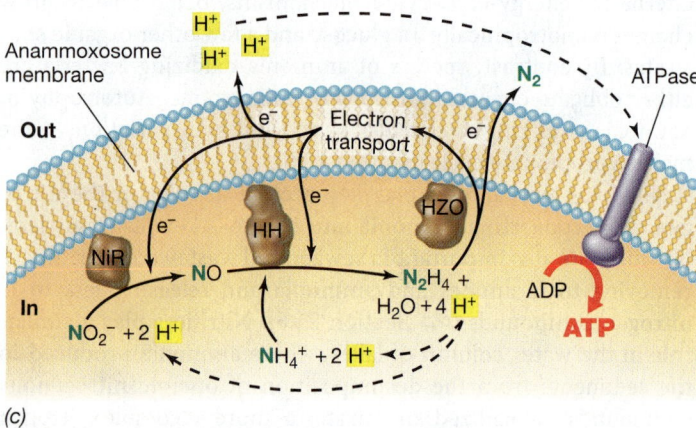

(c)

Figure 13.28 Anammox. (a) Phase-contrast photomicrograph of cells of *Brocadia anammoxidans*. A single cell is about 1 μm in diameter. (b) Transmission electron micrograph of a cell; note the membrane-enclosed compartments including the large fibrillar anammoxosome. (c) Reactions in the anammoxosome. NiR, nitrite reductase, HH, hydrazine hydrolase; HZO, hydrazine dehydrogenase.

Brocadia, several other genera of anammox bacteria are known, including *Kuenenia*, *Anammoxoglobus*, *Jettenia*, and *Scalindua*, all of which are related to *Brocadia* and also contain anammoxosomes. Like aerobic ammonia oxidizers, anammox bacteria are also autotrophs, but they do not fix CO_2 using the pathways employed by aerobic ammonia oxidizers. Instead, anammox bacteria fix CO_2 by way of the acetyl-CoA pathway, an autotrophic pathway widespread among some obligately anaerobic autotrophic *Bacteria* and *Archaea* (Section 13.19).

The anammoxosome is a unit membrane–enclosed structure (Figure 13.28b) and in this respect is technically an organelle in the eukaryotic sense of the term. Lipids that form the

anammoxosome membrane are not the typical lipids of *Bacteria* but instead consist of fatty acids constructed of multiple cyclobutane (C_4) rings bonded to glycerol by both ester and ether bonds. These *ladderane lipids*, as they are called, aggregate in the membrane to form an unusually dense membrane structure that prevents diffusion of substances from the anammoxosome into the cytoplasm.

The strong anammoxosome membrane is required to protect the cell from toxic intermediates produced during anammox reactions. These include, in particular, the compound *hydrazine* (N_2H_4), a very strong reductant. In the anammox reaction, NO_2^- is first reduced to nitric oxide (NO) by nitrite reductase, and then NO reacts with ammonium (NH_4^+) to yield N_2H_4 by activity of the enzyme hydrazine hydrolase (Figure 13.28c). N_2H_4 is then oxidized to N_2 plus electrons by the enzyme hydrazine dehydrogenase. Some of the electrons generated at this step enter the anammoxosome electron transport chain, and electron transport reactions yield a proton motive force; other electrons feed back into the system to drive earlier anammox reactions (Figure 13.28c). ATP is formed from the proton motive force by ATPases in the anammoxosome membrane (Figure 13.28c).

Ecology of Anammox

In nature the source of NO_2^- for the anammox reaction is presumably aerobic ammonia-oxidizing *Bacteria* and *Archaea*. These organisms coexist with anammox bacteria in ammonia-rich habitats such as sewage and other wastewaters. The suspended particles that form in these habitats contain both oxic and anoxic zones in which ammonia oxidizers of different physiologies can coexist in close association. In mixed laboratory cultures, high levels of oxygen inhibit anammox and favor classic nitrification, and thus it is likely that in nature, the fraction of ammonia oxidation catalyzed by anammox bacteria is governed by the concentration of O_2 in the habitat.

From an environmental standpoint, anammox is a very beneficial process in the treatment of wastewaters. The anoxic removal of NH_3 and amines by the formation of N_2 (Figure 13.28c) helps reduce the input of fixed nitrogen from wastewater treatment facilities into rivers and streams, thereby maintaining higher water quality than would otherwise be possible. The activities of marine anammox organisms likely account for the significant fraction (>50%) of NH_3 known to disappear from marine sediments, a process which has previously been unexplained. At least some ammonia-rich freshwater lake sediments also support anammox, and thus it appears that anammox can occur in any anoxic environment in which NH_3 and NO_2^- coexist.

MINIQUIZ

- What are the substrates for the enzyme ammonia monooxygenase?
- Why must anammox reactions be carried out in a special structure within the cell?
- What is the carbon source for nitrifying and anammox organisms?

III • Fermentations

We have thus far considered phototrophy and chemolithotrophy, strategies for energy conservation that do not require organic compounds as electron donors. In the next three units we focus on organic compounds as electron donors and the many ways in which chemoorganotrophs conserve energy. We begin with fermentations, major forms of anaerobic energy conservation.

13.11 Energetic and Redox Considerations

Many microbial habitats are *anoxic* (oxygen-free), and in such environments, decomposition of organic material occurs anaerobically. If sulfate (SO_4^{2-}), nitrate (NO_3^-), ferric iron (Fe^{3+}), and other electron acceptors are absent from these habitats, organic compounds are catabolized by **fermentation**. Recall from Chapter 3 that we emphasized how redox balance is achieved in fermentations by having the substrate serve as *both* electron donor and electron acceptor and that ATP is synthesized by *substrate-level phosphorylation*. We pick up on these two essential features of fermentation here (**Figure 13.29**).

Energy-Rich Compounds and Substrate-Level Phosphorylation

Energy can be conserved by substrate-level phosphorylation from many different compounds. However, central to an understanding of substrate-level phosphorylation is the concept of *energy-rich compounds*. These are organic compounds that contain an energy-rich phosphate bond or a molecule of coenzyme A. The bond is "energy-rich" because its hydrolysis is highly exergonic. **Table 13.3** lists some energy-rich compounds formed during metabolism; the hydrolysis of most of these yields sufficient free energy to be coupled to ATP synthesis ($\Delta G^{0\prime} = -31.8$ kJ/mol). If an organism can form one of these compounds during fermentative metabolism, it can likely make ATP by substrate-level phosphorylation.

Table 13.3	Energy-rich compounds that can couple to substrate-level phosphorylation[a]
Compound	Free energy of hydrolysis, $\Delta G^{0\prime}$ (kJ/mol)[b]
Acetyl-CoA	−35.7
Propionyl-CoA	−35.6
Butyryl-CoA	−35.6
Caproyl-CoA	−35.6
Succinyl-CoA	−35.1
Acetyl phosphate	−44.8
Butyryl phosphate	−44.8
1,3-Bisphosphoglycerate	−51.9
Carbamyl phosphate	−39.3
Phosphoenolpyruvate	−51.6
Adenosine phosphosulfate (APS)	−88
N^{10}-Formyltetrahydrofolate	−23.4
Energy of hydrolysis of ATP (ATP → ADP + P_i)	−31.8

[a]Data from Thauer, R.K., K. Jungermann, and K. Decker. 1977. Energy conservation in chemotrophic anaerobic bacteria. *Bacteriol. Rev. 41:* 100–180.
[b]The $\Delta G^{0\prime}$ values shown here are for "standard conditions," which are not necessarily those of cells. Including heat loss, the energy costs of making an ATP are more like 60 kJ than 32 kJ, and the energy of hydrolysis of the energy-rich compounds shown here is thus likely higher. But for simplicity and comparative purposes, the values in this table will be taken as the actual energy released per reaction.

Redox Balance and H_2 and Acetate Production

In any fermentation there must be atomic and redox balance. That is, the total number of each type of atom and electrons in the products of the reaction must balance those in the reactants (the substrates). Redox balance is achieved in fermentations by the excretion from the cell of *fermentation products*, reduced substances such as acids or alcohols that are produced as end products of the catabolism of the original fermentable substance (Figure 13.29). In several fermentations, redox balance is facilitated by the production of molecular hydrogen (H_2). The production of H_2 is associated with the activity of the iron–sulfur protein *ferredoxin*, a very low-potential electron carrier, and is catalyzed by the enzyme *hydrogenase*. H_2 can also be produced from the C_1 fatty acid formate (**Figure 13.30**). Although the H_2 can no longer be used by the fermenter and is thus excreted, H_2 is a very powerful electron donor and can be oxidized by various respiratory prokaryotes. Indeed, with its very electronegative $E_0^\prime$ (making it suitable as an electron donor for any form of respiration), H_2 is never wasted in microbial ecosystems.

Many anaerobic bacteria produce acetate or other fatty acids as a major or minor fermentation product. The production of these is energy conserving because it offers the organism the opportunity to make ATP by substrate-level phosphorylation. The key intermediate generated is the coenzyme-A derivative of each fatty acid, since these are energy-rich compounds (Table 13.3). For example, acetyl-CoA can be converted to acetyl phosphate (Figure 13.30) and the phosphate group subsequently transferred

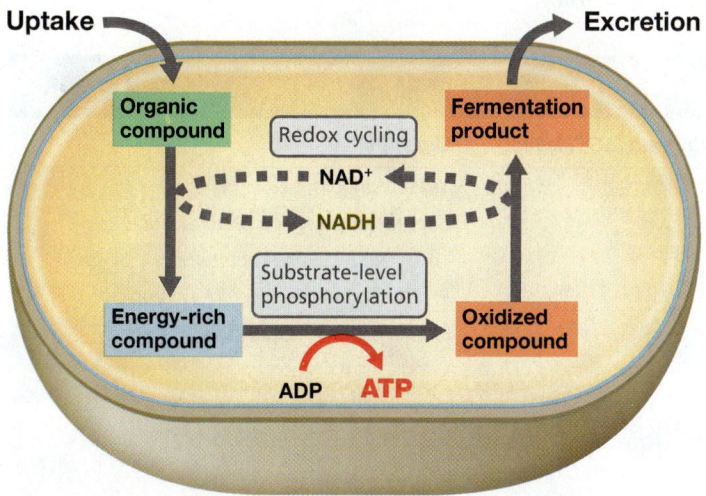

Figure 13.29 **The essentials of fermentation.** The fermentation product is excreted from the cell, and only a relatively small amount of the original organic compound is used for biosynthesis.

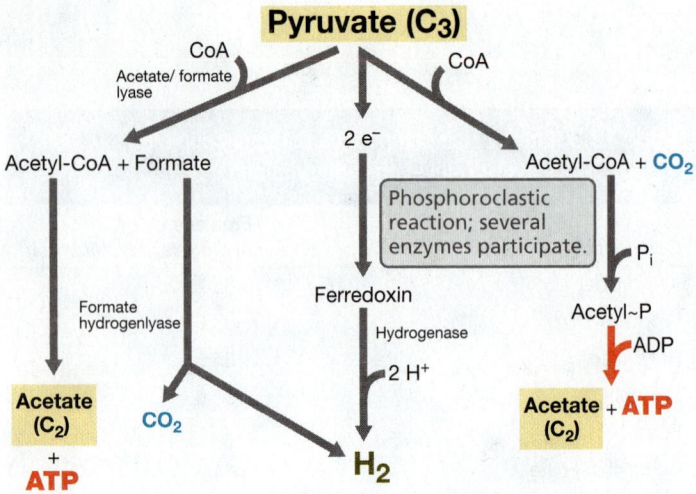

Figure 13.30 Production of H₂ and acetate from pyruvate. At least two mechanisms are known, one that produces H₂ directly and the other that makes formate as an intermediate. When acetate is produced, ATP synthesis is possible (see Table 13.3).

to ADP, yielding ATP. Fatty acid production is common in fermentations and if the fatty acid is metabolized through a Co-A intermediate, the potential for ATP synthesis by substrate-level phosphorylation is a possibility.

MINIQUIZ

- What is substrate-level phosphorylation?
- Why is acetate formation in fermentation energetically beneficial to the cell?

13.12 Lactic and Mixed-Acid Fermentations

Fermentations are classified by either the substrate fermented or the products formed. Table 13.4 lists some major fermentations classified on the basis of products formed. Note some of the broad categories, such as alcohol, lactic acid, propionic acid, mixed acid,

butyric acid, and acetogenic. Other fermentations are classified by the substrate fermented rather than the fermentation product; for instance, amino acid, purine/pyrimidine, or the succinate/oxalate fermentations. Some anaerobes even ferment aromatic compounds and other unusual substrates (Table 13.5). Clearly, a wide variety of organic compounds can be fermented, and in a few cases, only a very restricted group of anaerobes can carry out the fermentation. Many of these are metabolic specialists, having evolved the capacity to ferment a substrate not catabolized by other bacteria.

Table 13.5 Some unusual bacterial fermentations

Type	Reaction	Organisms
Acetylene	$2\ C_2H_2 + 3\ H_2O \rightarrow$ ethanol $+$ acetate$^-$ $+$ H$^+$	*Pelobacter acetylenicus*
Glycerol	4 Glycerol $+ 2\ HCO_3^- \rightarrow 7$ acetate$^-$ $+ 5$ H$^+$ $+ 4\ H_2O$	*Acetobacterium* spp.
Resorcinol (aromatic)	$2\ C_6H_4(OH)_2 + 6\ H_2O \rightarrow 4$ acetate$^-$ $+$ butyrate$^-$ $+$ 5 H$^+$	*Clostridium* spp.
Phloroglucinol (aromatic)	$C_6H_6O_3 + 3\ H_2O \rightarrow 3$ acetate$^-$ $+ 3$ H$^+$	*Pelobacter massiliensis* *Pelobacter acidigallici*
Putrescine	$10\ C_4H_{12}N_2 + 26\ H_2O \rightarrow 6$ acetate$^-$ $+ 7$ butyrate$^-$ $+$ $20\ NH_4^+ + 16\ H_2 +$ 13 H$^+$	Unclassified gram-positive nonsporulating anaerobes
Citrate	Citrate^{3-} $+ 2\ H_2O \rightarrow$ formate$^-$ $+ 2$ acetate$^-$ $+$ HCO_3^- $+$ H$^+$	*Bacteroides* spp.
Aconitate	Aconitate^{3-} $+$ H$^+$ $+ 2\ H_2O$ $\rightarrow 2\ CO_2 + 2$ acetate$^-$ $+$ H_2	*Acidaminococcus fermentans*
Glyoxylate	4 Glyoxylate$^-$ $+ 3$ H$^+$ $+$ $3\ H_2O \rightarrow 6\ CO_2 +$ $5\ H_2 +$ glycolate$^-$	Unclassified gram-negative bacterium
Benzoate	2 Benzoate$^-$ $\rightarrow$ cyclohexane carboxylate$^-$ $+ 3$ acetate$^-$ $+$ HCO_3^- $+ 3$ H$^+$	*Syntrophus aciditrophicus*

Table 13.4 Common fermentations and their energetics and example organisms

Type	Reaction	Energy yield ($\Delta G^{0\prime}$, kj/mol)	Organisms
Alcoholic	Hexose $\rightarrow 2$ ethanol $+ 2\ CO_2$	−239	Yeast, *Zymomonas*
Homolactic	Hexose $\rightarrow 2$ lactate$^-$ $+ 2$ H$^+$	−196	*Streptococcus*, some *Lactobacillus*
Heterolactic	Hexose $\rightarrow$ lactate$^-$ $+$ ethanol $+ CO_2 +$ H$^+$	−216	*Leuconostoc*, some *Lactobacillus*
Propionic acid	3 Lactate$^-$ $\rightarrow 2$ propionate$^-$ $+$ acetate$^-$ $+ CO_2 + H_2O$	−170	*Propionibacterium, Clostridium propionicum*
Mixed acid[a,b]	Hexose $\rightarrow$ ethanol $+ 2,3$-butanediol $+$ succinate^{2-} $+$ lactate$^-$ $+$ acetate$^-$ $+$ formate$^-$ $+$ H$_2 + CO_2$	—	Enteric bacteria including *Escherichia, Salmonella, Shigella, Klebsiella, Enterobacter*
Butyric acid[b]	Hexose $\rightarrow$ butyrate$^-$ $+ 2\ H_2 + 2\ CO_2 +$ H$^+$	−264	*Clostridium butyricum*
Butanol[b]	2 Hexose $\rightarrow$ butanol $+$ acetone $+ 5\ CO_2 + 4\ H_2$	−468	*Clostridium acetobutylicum*
Caproate/ Butyrate	6 Ethanol $+ 3$ acetate$^-$ $\rightarrow 3$ butyrate$^-$ $+$ caproate$^-$ $+$ $2\ H_2 + 4\ H_2O +$ H$^+$	−183	*Clostridium kluyveri*
Acetogenic	Fructose $\rightarrow 3$ acetate$^-$ $+ 3$ H$^+$	−276	*Clostridium aceticum*

[a]Not all organisms produce all products. In particular, butanediol production is limited to only certain enteric bacteria. Reaction not balanced.
[b]Stoichiometry shows major products. Other products include some acetate and a small amount of ethanol (butanol fermentation only).

We begin with two very common fermentations of sugars in which lactic acid is the sole or major product.

Lactic Acid Fermentation

Lactic acid bacteria are gram-positive nonsporulating bacteria that produce lactic acid as a major or sole fermentation product from the fermentation of sugars (⟷ Section 15.6). Two fermentative patterns are observed. One, called **homofermentative**, yields a single fermentation product, lactic acid. The other, called **heterofermentative**, yields products in addition to lactate, mainly ethanol plus CO_2.

Figure 13.31 summarizes pathways for the fermentation of glucose by homofermentative and heterofermentative lactic acid bacteria. The differences observed can be traced to the presence or absence of the enzyme *aldolase*, a key enzyme of glycolysis (⟷ Figure 3.14). Homofermentative lactic acid bacteria contain aldolase and produce *two* molecules of lactate from glucose by the glycolytic pathway (Figure 13.31*a*). Heterofermenters lack aldolase and thus cannot break down fructose bisphosphate to triose phosphate. Instead, they oxidize glucose 6-phosphate to 6-phosphogluconate and then decarboxylate this to pentose phosphate. The latter compound is then converted to triose phosphate and acetyl phosphate by the key enzyme *phosphoketolase* (Figure 13.31*b*). The early steps in catabolism by heterofermentative lactic acid bacteria are those of the pentose phosphate pathway (⟷ Figure 3.26).

In heterofermenters, triose phosphate is converted to lactic acid with the production of ATP (Figure 13.31*b*). However, to achieve redox balance the acetyl phosphate produced is used as an electron acceptor and is reduced by NADH (generated during the production of pentose phosphate) to ethanol. This occurs without ATP synthesis because the energy-rich CoA bond is lost during ethanol formation. Because of this, heterofermenters produce only *one* ATP/glucose instead of the *two* ATP/glucose produced

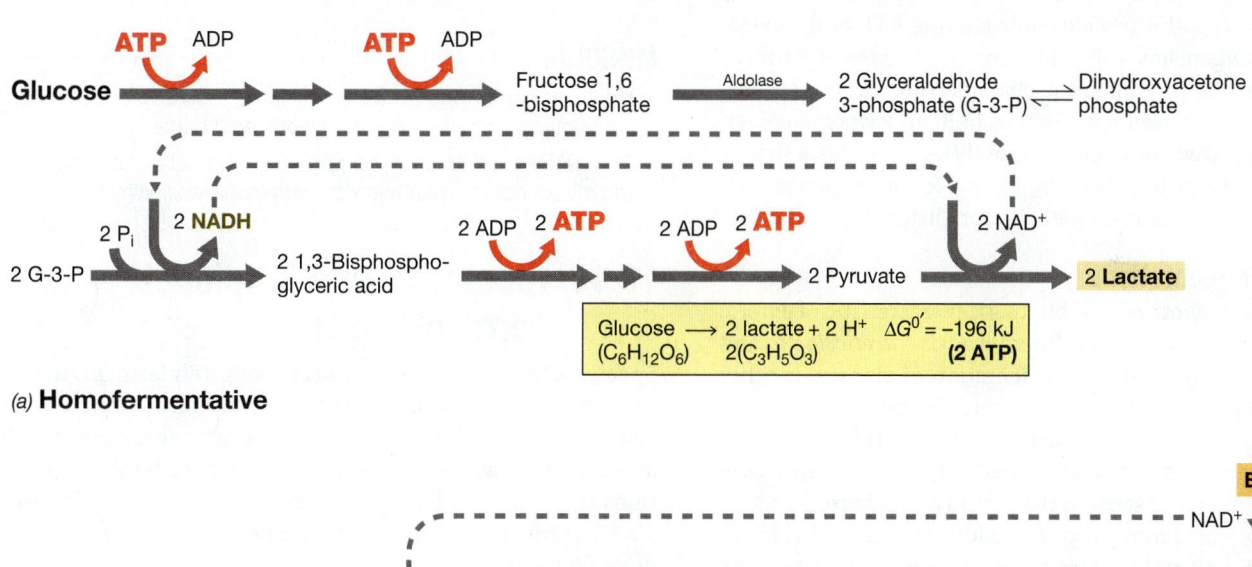

(a) **Homofermentative**

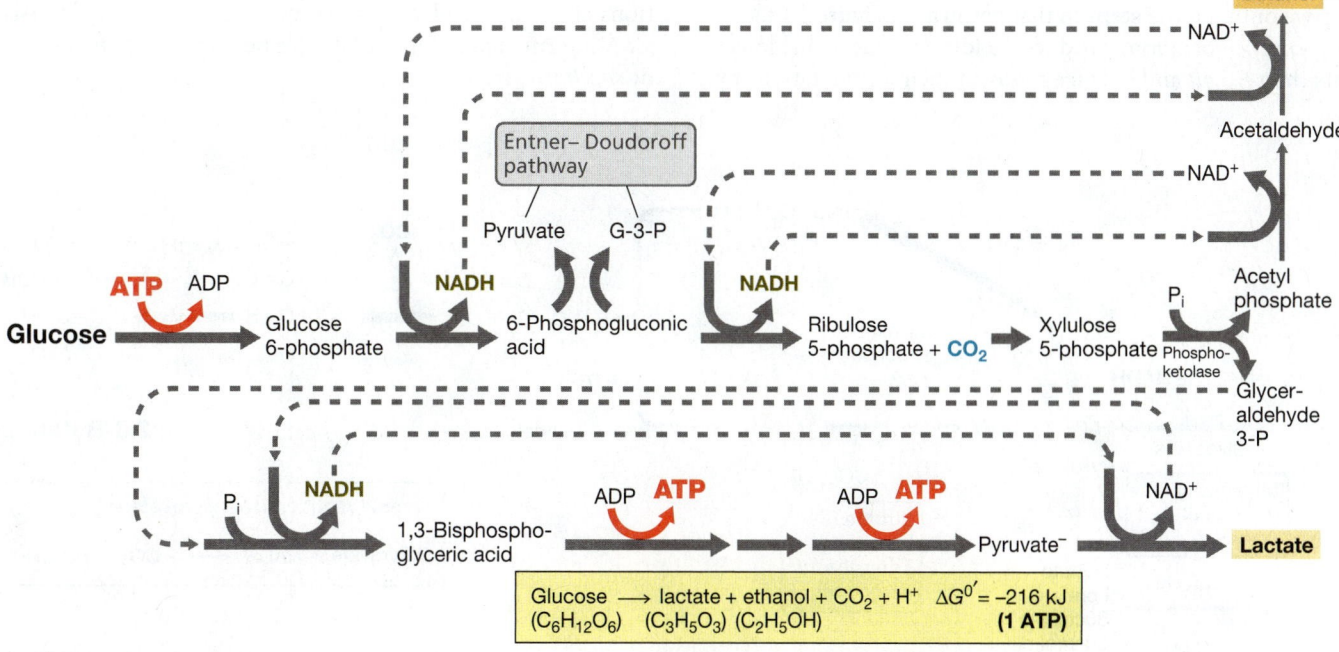

(b) **Heterofermentative**

Figure 13.31 The fermentation of glucose in *(a)* homofermentative and *(b)* heterofermentative lactic acid bacteria.
Note that no ATP is made in reactions leading to ethanol formation in heterofermentative organisms.

by homofermenters. In addition, because heterofermenters decarboxylate 6-phosphogluconate, they produce CO_2 as a fermentation product; homofermenters do not produce CO_2. Thus an easy way to differentiate a homofermenter from a heterofermenter is to observe for the production of CO_2 in laboratory cultures.

Entner–Doudoroff Pathway

A variant of the glycolytic pathway, called the *Entner–Doudoroff pathway*, is widely distributed in bacteria, especially among species of the pseudomonad group. In this pathway glucose 6-phosphate is oxidized to 6-phosphogluconic acid and NADPH; the 6-phosphogluconic acid is dehydrated and split into pyruvate and glyceraldehyde 3-phosphate (G-3-P), a key intermediate of the glycolytic pathway. G-3-P is then catabolized as in glycolysis, generating NADH and two ATP, and used as an electron acceptor to balance redox reactions (Figure 13.31*a*).

Because pyruvate is formed directly in the Entner–Doudoroff pathway and cannot yield ATP as can G-3-P (Figure 13.31), the Entner–Doudoroff pathway yields only half the ATP of the glycolytic pathway. Organisms using the Entner–Doudoroff pathway therefore share this physiological characteristic with heterofermentative lactic acid bacteria (Figure 13.31*b*). *Zymomonas*, an obligately fermentative pseudomonad, and *Pseudomonas*, a strictly respiratory bacterium (⊝⊃ Section 15.4), are major genera that employ the Entner–Doudoroff pathway for glucose catabolism.

Mixed-Acid Fermentations

In *mixed-acid fermentations* (Table 13.4), characteristic of enteric bacteria (⊝⊃ Section 15.3), three different acids—*acetic, lactic,* and *succinic*—are formed from the fermentation of glucose or other sugars that can be converted into glucose. Ethanol, CO_2, and H_2 are also typically formed as fermentation products. Glycolysis is the pathway used by mixed-acid fermenters, such as *Escherichia coli*, and we outlined the steps in that pathway in Figure 3.14.

Some enteric bacteria produce acidic products in lower amounts than *E. coli* and balance redox in their fermentations by producing larger amounts of neutral products. One key neutral product is the four-carbon alcohol *butanediol*. In this variation of the mixed-acid fermentation, butanediol, ethanol, CO_2, and H_2 are the main products observed (**Figure 13.32**). In the mixed-acid fermentation of *E. coli*, equal amounts of CO_2 and H_2 are produced, whereas in a butanediol fermentation, considerably more CO_2 than H_2 is produced. This is because mixed-acid fermenters produce CO_2 only from formic acid by means of the enzyme *formate hydrogenlyase* (Figure 13.32):

$$HCOOH \rightarrow H_2 + CO_2$$

By contrast, butanediol producers, such as *Enterobacter aerogenes*, produce CO_2 and H_2 from formic acid but also produce two additional molecules of CO_2 during the formation of each molecule of butanediol (Figure 13.32). However, because butanediol production consumes only half of the NADH generated in glycolysis (Figure 13.32), more ethanol is produced by these organisms than by non-butanediol fermenters in order to achieve redox balance.

MINIQUIZ

- How can homo- and heterofermentative metabolism be differentiated in pure cultures of lactic acid bacteria?
- Butanediol production leads to greater ethanol production than in the mixed-acid fermentation of *Escherichia coli*. Why?

13.13 Clostridial and Propionate Fermentations

Species of the genus *Clostridium* are obligately fermentative anaerobes (⊝⊃ Section 15.7). Different clostridia ferment sugars, amino acids, purines and pyrimidines, and a few other compounds. In all cases ATP synthesis is linked to substrate-level phosphorylations either in the glycolytic pathway or from the hydrolysis of a CoA intermediate (Table 13.3). We begin with sugar-fermenting, or *saccharolytic*, clostridia.

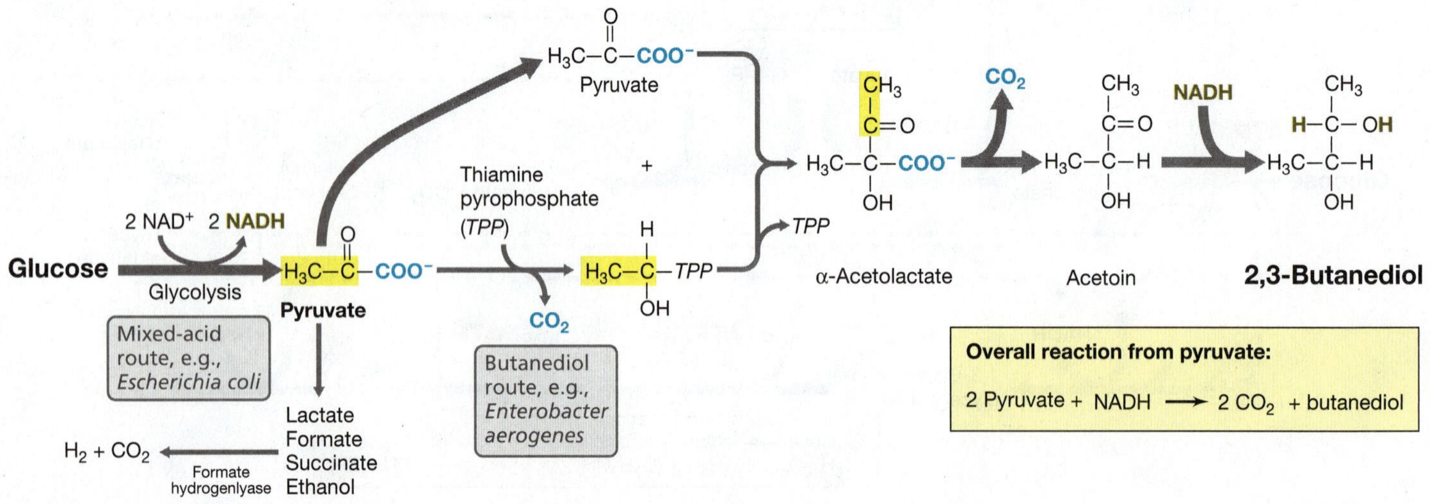

Figure 13.32 Butanediol production and mixed-acid fermentations. Note how only one NADH, but two molecules of pyruvate, are used to make one butanediol. This leads to redox imbalance and the production of more ethanol by butanediol producers than by mixed-acid fermenters.

Sugar Fermentation by *Clostridium* Species

A number of clostridia ferment sugars, producing *butyric acid* as a major fermentation product. Some species also produce the neutral products acetone and butanol; *Clostridium acetobutylicum* is a classic example of this pattern. The biochemical steps in the formation of butyric acid and neutral products from sugars are shown in **Figure 13.33**.

In saccharolytic clostridia, glucose is converted to pyruvate and NADH via the glycolytic pathway, and pyruvate is split to yield acetyl-CoA, CO_2, and H_2 (through ferredoxin) by the phosphoroclastic reaction (Figure 13.30). Most of the acetyl-CoA is then reduced to butyrate or other fermentation products using NADH derived from glycolytic reactions as electron donor. The actual products observed are influenced by the duration and the conditions of the fermentation. During the early stages of the butyric

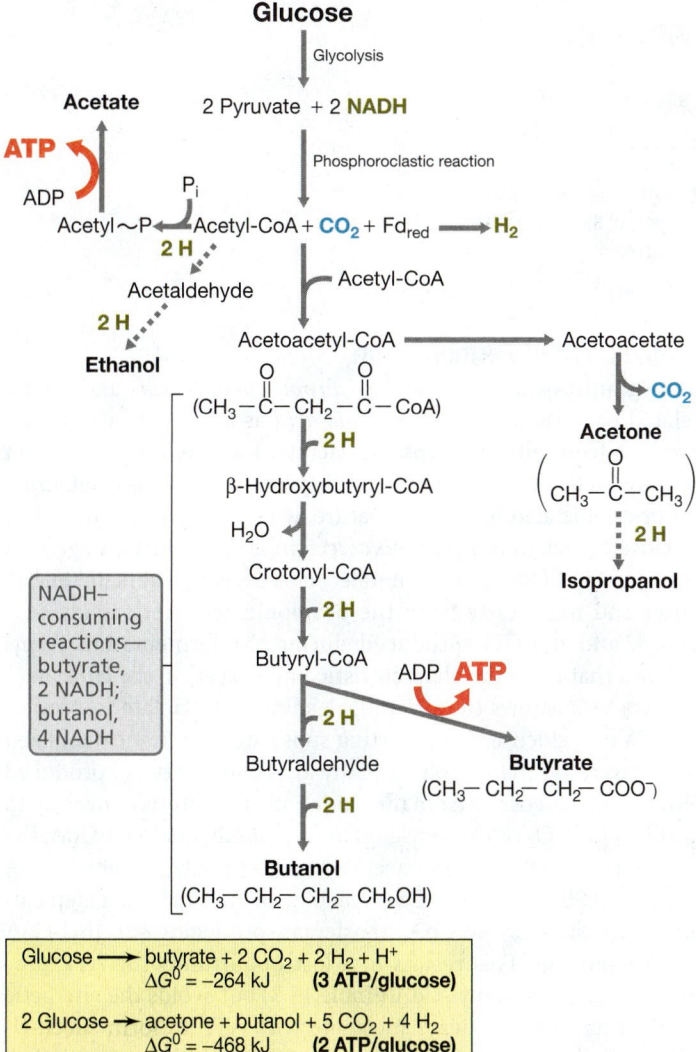

Figure 13.33 The butyric acid and butanol/acetone fermentation. All fermentation products from glucose are shown in bold (dashed lines indicate minor products). Note how the production of acetate and butyrate lead to additional ATP by substrate-level phosphorylation. By contrast, formation of butanol and acetone reduces the ATP yield because the butyryl-CoA to butyrate step is bypassed. 2 H, NADH; Fd$_{red}$, reduced ferredoxin.

fermentation, butyrate and a small amount of acetate and ethanol are produced. But as the pH of the medium drops, acid production decreases and acetone and butanol begin to appear. If the pH of the medium is kept neutral by buffering, there is very little formation of acetone and butanol; instead, butyric acid production continues, and this is for a good reason.

When *C. acetobutylicum* synthesizes butyrate, extra ATP is produced (Figure 13.33 and Table 13.3) and the organism will continue to make butyrate unless conditions become overly acidic. However, the organism is acid-sensitive, and if the pH drops below about pH 5, genes encoding enzymes that make neutral products are derepressed and the fermentation shifts to solvent production. Interestingly, the production of butanol is in part a consequence of the production of acetone. For each acetone that is made, two NADH produced during glycolysis are not reoxidized as they would be if butyrate were produced. To achieve redox balance, the cell then uses butyrate as an electron acceptor with butanol being the final fermentation product (Figure 13.33). Previously excreted butyrate can also be reincorporated by the cell and reduced to butanol and then excreted again. Although neutral product formation helps *C. acetobutylicum* keep its environment from becoming too acidic, there is an energetic price to pay for this. In producing butanol, the cell loses the opportunity to convert butyryl-CoA to butyrate and gain ATP (Figure 13.33 and Table 13.3).

Amino Acid Fermentation by *Clostridium* Species and the Stickland Reaction

Some *Clostridium* species ferment amino acids. These are the *proteolytic* clostridia, organisms that degrade proteins released from dead organisms. Some of these, such as the animal pathogen *Clostridium tetani* (tetanus), are strictly proteolytic, while other species are both saccharolytic and proteolytic.

Depending on the species, some proteolytic clostridia ferment individual amino acids, typically glutamate, glycine, alanine, cysteine, histidine, serine, or threonine. The biochemistry behind these fermentations is quite complex, but the metabolic strategy is simple. In virtually all cases, the amino acids are catabolized in such a way as to eventually yield a fatty acid–CoA derivative, typically acetyl (C_2), butyryl (C_4), or caproyl (C_6). From these, ATP is produced by substrate-level phosphorylation (Table 13.3). Other typical products of amino acid fermentation include ammonia (NH_3) and CO_2.

Some clostridia ferment only an amino acid *pair*. In this situation one amino acid functions as the electron donor and is oxidized, whereas the other amino acid is the electron acceptor and is reduced. This *coupled* amino acid fermentation is called a **Stickland reaction**, named for the scientist who discovered it. For example, *Clostridium sporogenes* ferments glycine and alanine, and in this reaction, alanine is the electron donor and glycine is the electron acceptor (**Figure 13.34**). The products of the Stickland reaction are invariably NH_3, CO_2, and a carboxylic acid with one fewer carbons than the amino acid that was oxidized (Figure 13.34).

Many of the products of amino acid fermentation by clostridia are foul-smelling substances, and the odor that results from putrefaction is mainly a result of clostridial activity. In addition

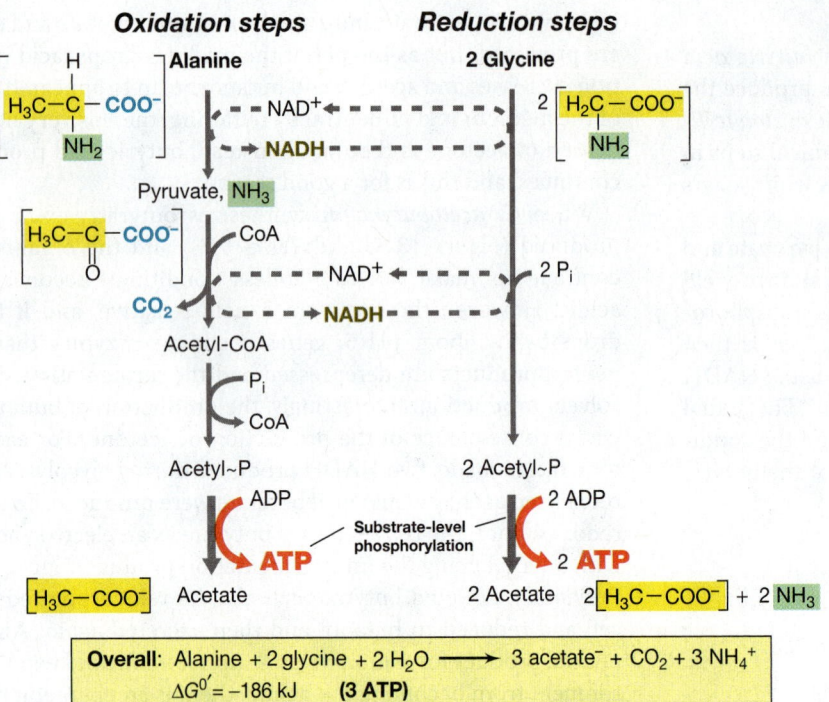

Figure 13.34 The Stickland reaction. This example shows the cocatabolism of the amino acids alanine and glycine. The structures of key substrates, intermediates, and products are shown in brackets to allow the chemistry of the reaction to be followed. Note how in the reaction shown, alanine is the electron donor and glycine is the electron acceptor.

to fatty acids, other odoriferous compounds produced include hydrogen sulfide (H_2S), methylmercaptan (CH_3SH, derived from sulfur-containing amino acids), cadaverine (from lysine), putrescine (from ornithine), and NH_3. Purines and pyrimidines, released from the degradation of nucleic acids, lead to many of the same fermentation products and yield ATP by substrate-level phosphorylation from the hydrolysis of fatty acid–CoA derivatives (Table 13.3) produced in their respective fermentative pathways.

Clostridium kluyveri Fermentation

Another species of *Clostridium* also ferments a mixture of substrates in which one is the donor and one is the acceptor, as in the Stickland reaction. However, this organism, *C. kluyveri*, ferments not amino acids but instead *ethanol plus acetate*. In this fermentation, ethanol is the electron donor and acetate is the electron acceptor. The overall reaction is shown in Table 13.4.

The ATP yield in the caproate/butyrate fermentation is low, 1 ATP/6 ethanol fermented. However, *C. kluyveri* has a selective advantage over all other fermenters in its apparently unique ability to oxidize a highly reduced fermentation product of other anaerobes (ethanol) and couple it to the reduction of another common fermentation product (acetate), reducing the latter to longer-chain fatty acids, reactions that consume NADH (see Figure 13.33). The single ATP produced per 6 ethanol oxidized comes from substrate-level phosphorylation during conversion of a fatty acid–CoA derivative formed during the fermentation. The fermentation of *C. kluyveri* is an example of a **secondary fermentation**, which can be viewed as a fermentation of fermentation products. We see another example of this next.

Propionic Acid Fermentation

The gram-positive bacterium *Propionibacterium* and some related bacteria produce *propionic acid* as a major fermentation product from either glucose or lactate. Lactate, a fermentation product of the lactic acid bacteria, is probably the major substrate for propionic acid bacteria in nature, where these two groups live in close association. *Propionibacterium* is an important agent in the ripening of Swiss (Emmentaler) cheese, which gets its unique bitter and nutty taste from the propionic and acetic acids produced, and the CO_2 produced during the fermentation forms bubbles that leave the characteristic holes (eyes) in the cheese.

Figure 13.35 shows the reactions leading from lactate to propionate. When glucose is the starting substrate, it is first catabolized to pyruvate by the glycolytic pathway. Then pyruvate, produced either from glucose or from the oxidation of lactate, is converted to acetate plus CO_2 or carboxylated to form methylmalonyl-CoA; the latter is converted into oxaloacetate and, eventually, propionyl-CoA (Figure 13.35). Propionyl-CoA reacts with succinate in a step catalyzed by the enzyme CoA transferase, producing succinyl-CoA and propionate. This results in a lost opportunity for ATP production from propionyl-CoA (Table 13.3) but avoids the energetic costs of having to activate succinate with ATP to form succinyl-CoA. The succinyl-CoA is then isomerized to methylmalonyl-CoA and the cycle is complete; propionate is formed and CO_2 regenerated (Figure 13.35).

NADH is oxidized in the steps between oxaloacetate and succinate. The reduction of fumarate to succinate (Figure 13.35) is linked to electron transport reactions and the formation of a proton motive force; this yields one ATP by oxidative phosphorylation

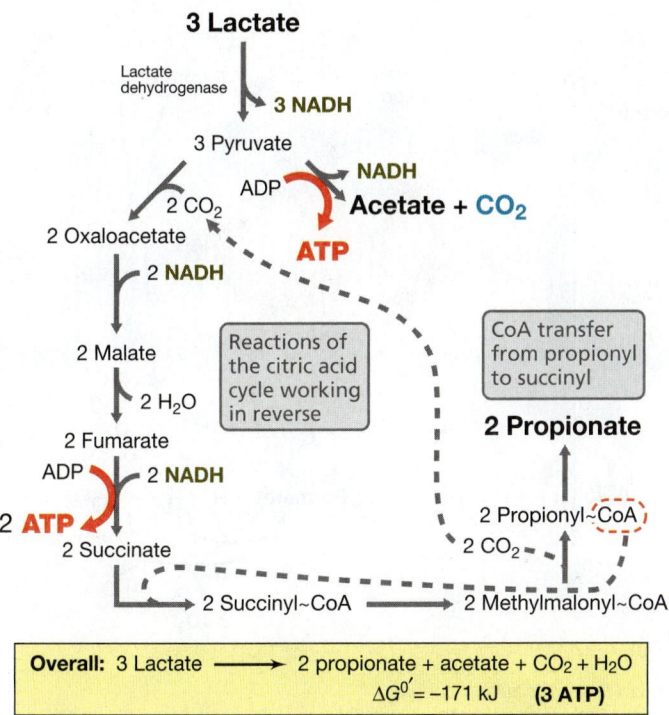

3 Lactate

Lactate dehydrogenase

3 NADH

3 Pyruvate

ADP → **NADH**

2 CO_2 → **Acetate + CO_2**

ATP

2 Oxaloacetate

2 NADH

2 Malate

2 H_2O

2 Fumarate

ADP → **2 NADH**

2 **ATP**

2 Succinate

Reactions of the citric acid cycle working in reverse

CoA transfer from propionyl to succinyl

2 Propionate

2 Propionyl~CoA
2 CO_2

2 Succinyl~CoA → 2 Methylmalonyl~CoA

Overall: 3 Lactate ⟶ 2 propionate + acetate + CO_2 + H_2O
$\Delta G^{0'} = -171$ kJ   **(3 ATP)**

Figure 13.35 **The propionic acid fermentation of *Propionibacterium*.** Products are shown in bold. The four NADH made from the oxidation of three lactate are reoxidized in the reduction of oxaloacetate and fumarate, and the CoA group from propionyl-CoA is exchanged with succinate during the formation of propionate.

(Section 13.21). The propionate pathway also converts some lactate to acetate plus CO_2, which allows for additional ATP to be made by substrate-level phosphorylation (Figure 13.35). Thus, in the propionate fermentation, both substrate-level *and* oxidative phosphorylation occur.

Propionate is also formed in the fermentation of succinate by the bacterium *Propionigenium*, but by a completely different mechanism than that described here for *Propionibacterium*. *Propionigenium*, to be considered next, is phylogenetically and ecologically unrelated to *Propionibacterium*, but aspects of its energy metabolism are of considerable interest from the standpoint of metabolic diversity and the energetic limits to life.

MINIQUIZ -
- Compare the mechanisms for energy conservation in *Clostridium acetobutylicum* and *Propionibacterium*.
- What type of substrates are fermented by saccharolytic clostridia? By proteolytic clostridia?
- What are the substrates for the *Clostridium kluyveri* fermentation? In nature, where do these come from?

13.14 Fermentations without Substrate-Level Phosphorylation

Certain fermentations yield insufficient energy to synthesize ATP by substrate-level phosphorylation (that is, less than −32 kJ, Table 13.3), yet still support anaerobic growth without added electron

acceptors. In these cases, catabolism of the compound is linked to ion pumps that establish a proton motive force or sodium motive force across the cytoplasmic membrane. Examples of these include the fermentation of succinate by *Propionigenium modestum* and the fermentation of oxalate by *Oxalobacter formigenes*.

Propionigenium modestum

Propionigenium modestum was first isolated in anoxic enrichment cultures lacking electron acceptors and fed succinate as an electron donor. *Propionigenium* inhabits marine and freshwater sediments and can also be isolated from the human oral cavity. The organism is a gram-negative short rod and, phylogenetically, is a species of *Fusobacteria* (⟳ Section 15.21). During studies of the physiology of *P. modestum*, it was shown to require sodium chloride (NaCl) for growth and to catabolize succinate under strictly anoxic conditions:

$$\text{Succinate}^{2-} + H_2O \rightarrow \text{propionate}^- + HCO_3^- \quad \Delta G^{0'} = -20.5 \text{ kJ}$$

This decarboxylation releases insufficient free energy to support ATP synthesis by substrate-level phosphorylation (Table 13.3) but sufficient free energy to pump a sodium ion (Na^+) from the cytoplasm to the periplasm across the cytoplasmic membrane. Energy conservation in *Propionigenium* is then linked to the resulting *sodium motive force*; a sodium-translocating (instead of proton-translocating) ATPase exists in the membrane of this organism that uses the sodium motive force to drive ATP synthesis (**Figure 13.36a**).

In a related decarboxylation reaction, the bacterium *Malonomonas* decarboxylates the C_3 dicarboxylic acid malonate, forming acetate plus CO_2. As for *Propionigenium*, energy metabolism in *Malonomonas* is linked to Na^+ and a sodium-driven ATPase. But the free energy available from malonate fermentation by *Malonomonas* (−17.4 kJ) is even less than that of succinate fermentation by *P. modestum*. *Sporomusa*, an endospore-forming bacterium and also an acetogen (Section 13.19), is also capable of fermenting malonate, as are a few other *Bacteria*.

Oxalobacter formigenes

Oxalobacter formigenes is a bacterium present in the intestinal tract of animals, including humans. It catabolizes the C_2 dicarboxylic acid oxalate, producing formate plus CO_2. Oxalate degradation by *O. formigenes* is thought to be important in the human colon for preventing the accumulation of oxalate, a substance that can form calcium oxalate kidney stones. *O. formigenes* is a gram-negative strict anaerobe that carries out the following reaction:

$$\text{Oxalate}^{2-} + H_2O \rightarrow \text{formate}^- + HCO_3^- \quad \Delta G^{0'} = -26.7 \text{ kJ}$$

As in the catabolism of succinate by *P. modestum*, insufficient energy is available from this reaction to drive ATP synthesis by substrate-level phosphorylation (Table 13.3). However, the reaction supports growth of the organism because the decarboxylation of oxalate is exergonic and forms formate, which is excreted from the cell. This is because the internal consumption of protons during the oxidation of oxalate and production of formate is, in effect, a proton pump; a divalent molecule (oxalate) enters the cell while a univalent molecule (formate) is excreted. The continued exchange of oxalate for formate establishes a proton motive

UNIT 3

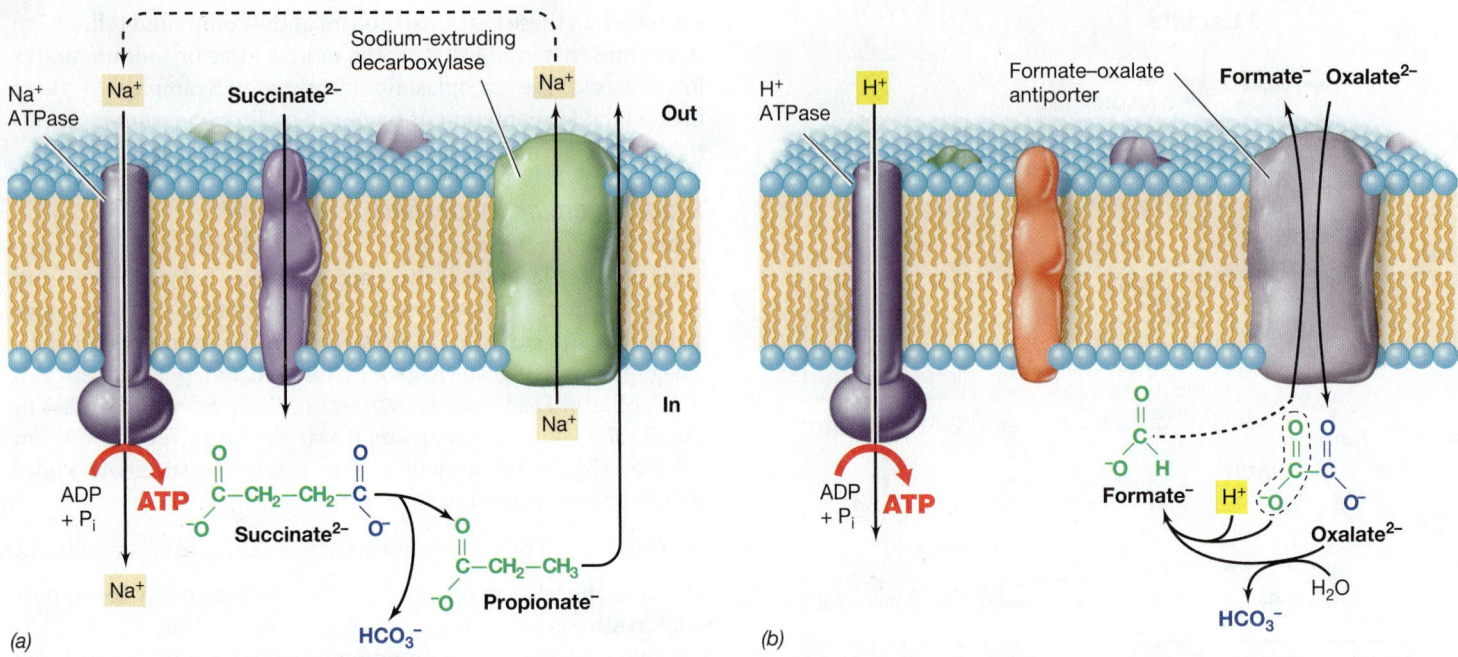

Figure 13.36 **The unique fermentations of succinate and oxalate.** *(a)* Succinate fermentation by *Propionigenium modestum*. Sodium export is linked to the energy released by succinate decarboxylation, and a sodium-translocating ATPase produces ATP. *(b)* Oxalate fermentation by *Oxalobacter formigenes*. Oxalate import and formate export by a formate–oxalate antiporter (⟲ Figure 2.21) consume cytoplasmic protons. ATP synthesis is linked to a proton-driven ATPase. All substrates and products are shown in bold.

force that is coupled to ATP synthesis by the proton-translocating ATPase in the membrane (Figure 13.36*b*).

What Can Be Learned from Decarboxylating-Type Fermentations?

The unique aspect of all of these "decarboxylation-type" fermentations is that ATP is made without substrate-level phosphorylation *or* oxidative phosphorylation driven by electron transport reactions. Instead, ATP synthesis is driven by ion pumps coupled to the small amount of energy released from the decarboxylation reaction. Organisms such as *Propionigenium*, *Malonomonas*, or *Oxalobacter* thus offer an important lesson in microbial bioenergetics: ATP synthesis from reactions that yield less than −32 kJ is still possible if the reaction is coupled to an ion pump.

At a minimum, then, an energy-conserving reaction must yield sufficient free energy to pump at least one ion. This energy requirement is estimated to be near −12 kJ. Reactions that release less free energy than this should not be able to drive ion pumps and should therefore not be potential energy-conserving reactions. However, as we will see in the next section, bacteria are known that push this theoretical limit even lower and whose energetics, consequently, are still incompletely understood. These are the syntrophs, bacteria living on the energetic margin of existence.

13.15 Syntrophy

There are many examples in microbiology of **syntrophy**, a situation in which two different organisms cooperate to degrade a substance that neither can degrade alone. Most syntrophic reactions are secondary fermentations in which organisms ferment the fermentation products of other anaerobes. We will see in Chapter 20 how syntrophy is often a key step in the anoxic catabolism that leads to the production of methane (CH_4) in nature. Here we consider the microbiology and energetic aspects of syntrophy.

Table 13.6 lists some major groups of syntrophs and the compounds they degrade. Many organic compounds can be degraded syntrophically, including even aromatic and aliphatic hydrocarbons. But the major compounds of interest in syntrophic environments are fatty acids and alcohols.

H_2 Consumption in Syntrophy: The Metabolic Link

The heart of syntrophic reactions is *interspecies H_2 transfer*—H_2 *production* by one partner, the syntroph, linked to H_2 *consumption* by the other. The H_2 consumer can be any one of a number of physiologically distinct organisms: denitrifying bacteria, ferric iron–reducing bacteria, sulfate-reducing bacteria, acetogens, or methanogens, groups we will consider in the next unit of this chapter. Consider ethanol fermentation to acetate plus H_2 by the syntroph *Pelotomaculum* coupled to the production of methane (Figure 13.37). As can be seen, the syntroph carries out a reaction whose standard free-energy change ($\Delta G^{0\prime}$) is positive. Hence, in pure culture, the organism will not grow. However, the H_2 produced by *Pelotomaculum* can be used as an electron donor by a methanogen to produce methane, an exergonic reaction. When the two reactions are summed, the overall reaction is exergonic

Table 13.6 Properties of major syntrophic bacteria[a]

Genus	Number of known species	Phylogeny[b]	Substrates fermented in coculture[c]
Syntrophobacter	4	Deltaproteobacteria	Propionate (C_3), lactate; some alcohols
Syntrophomonas	9	Firmicutes	C_4–C_{18} saturated/ unsaturated fatty acids; some alcohols
Pelotomaculum	2	Firmicutes	Propionate, lactate, several alcohols; some aromatic compounds
Syntrophus	3	Deltaproteobacteria	Benzoate and several related aromatic compounds; some fatty acids and alcohols

[a]All syntrophs are obligate anaerobes.
[b]See Chapters 14 and 15.
[c]Not all species can use all substrates listed.

(Figure 13.37), and when *Pelotomaculum* and a methanogen are cultured together (cocultured), both organisms grow luxuriously.

A second example of syntrophy is the oxidation of a fatty acid such as butyrate to acetate plus H_2 by the fatty acid–oxidizing bacterium *Syntrophomonas* (**Figure 13.38**):

$$\text{Butyrate}^- + 2\,H_2O \rightarrow 2\,\text{acetate}^- + H^+ + 2\,H_2 \qquad \Delta G^{0\prime} = +48.2\,\text{kJ}$$

The free-energy change of this reaction is even more unfavorable than that of ethanol oxidation (Figure 13.37), and in pure culture *Syntrophomonas* will obviously not grow on butyrate. However, as with ethanol fermentation by *Pelotomaculum*, if the H_2 produced by *Syntrophomonas* is consumed by a partner organism, *Syntrophomonas* will grow on butyrate in coculture with the H_2-consuming partner. How does this occur?

Energetics of H_2 Transfer

In a syntrophic relationship, the removal of H_2 by a partner organism shifts the equilibrium of the entire reaction and pulls it in the direction of product formation; this can greatly affect the energetics of the reaction. A review of the principles of free energy given in Appendix 1 indicates that the concentration of reactants and products in a reaction can have a major effect on energetics. This is usually not the case for most fermentation products because they are not consumed to extremely low levels. H_2, by contrast, can be consumed to nearly undetectable levels, and at these tiny concentrations, the energetics of reactions can be dramatically affected.

For convenience, the $\Delta G^{0\prime}$ of a reaction is calculated on the basis of *standard conditions*—one molar concentration of products and reactants (⮂ Section 3.4). By contrast, the related term ΔG is calculated on the basis of the *actual concentrations* of products and reactants present (Appendix 1 explains how to calculate

ΔG). At very low levels of H_2, the energetics of the oxidation of ethanol or fatty acids to acetate plus H_2, reactions that are endergonic under standard conditions, become exergonic. For example, if the concentration of H_2 is kept extremely low from consumption by the partner organism, ΔG for the oxidation of butyrate by *Syntrophomonas* yields -18 kJ (Figure 13.38a). As we learned in Section 13.14, this relatively low energy yield can still support growth of a bacterium.

Energetics in Syntrophs

Energy conservation in syntrophs is grounded in both substrate-level and oxidative phosphorylations. From biochemical studies of syntrophic butyrate catabolism, substrate-level phosphorylation has been shown to occur during the conversion of acetyl-CoA to acetate (Figure 13.38a) although the -18 kJ of energy released (ΔG) is in theory insufficient for this. However, the energy released is sufficient to produce *a fraction* of an ATP, so it is possible that in some way *Syntrophomonas* can couple two or more rounds of butyrate oxidation to the synthesis of one ATP by substrate-level phosphorylation.

Besides a syntrophic lifestyle, many syntrophs can also carry out anaerobic respirations (Section 13.16) in pure culture by the disproportionation of unsaturated fatty acids (disproportionation is a process in which one molecule of a substrate is oxidized while another is reduced). For example, crotonate, an intermediate in syntrophic butyrate metabolism (Figure 13.38a), supports growth of pure cultures of *Syntrophomonas*. Under these conditions some of the crotonate is oxidized to acetate and

Ethanol fermentation:

$$2\,CH_3CH_2OH + 2\,H_2O \rightarrow 4\,H_2 + 2\,CH_3COO^- + 2\,H^+$$
$$\Delta G^{0\prime} = +19.4\,\text{kJ/reaction}$$

Methanogenesis:

$$4\,H_2 + CO_2 \rightarrow CH_4 + 2\,H_2O$$
$$\Delta G^{0\prime} = -130.7\,\text{kJ/reaction}$$

Coupled reaction:

$$2\,CH_3CH_2OH + CO_2 \rightarrow CH_4 + 2\,CH_3COO^- + 2\,H^+$$
$$\Delta G^{0\prime} = -111.3\,\text{kJ/reaction}$$

(a) Reactions

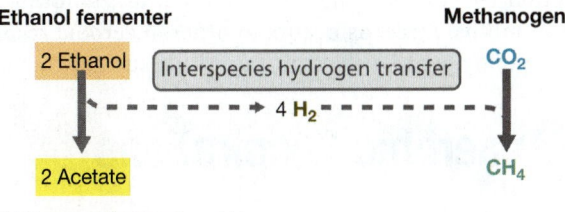

(b) Syntrophic transfer of H_2

Figure 13.37 Syntrophy: Interspecies H_2 transfer. Shown is the fermentation of ethanol to methane and acetate by syntrophic association of an ethanol-oxidizing syntroph and a H_2-consuming partner (in this case, a methanogen). *(a)* Reactions involved. The two organisms share the energy released in the coupled reaction. *(b)* Nature of the syntrophic transfer of H_2.

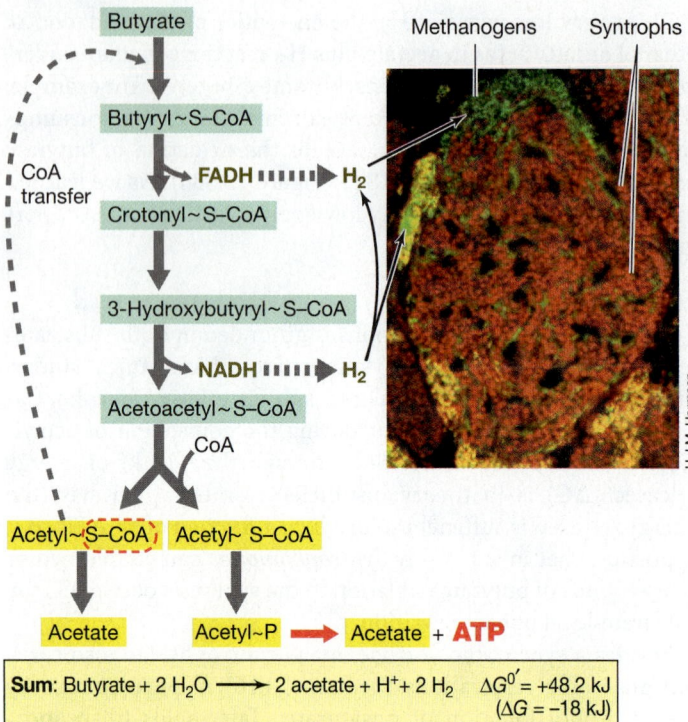

Sum: Butyrate + 2 $H_2O \longrightarrow$ 2 acetate + H^+ + 2 H_2 $\Delta G^{0'}$ = +48.2 kJ
(ΔG = −18 kJ)

(a) **Syntrophic culture**

1. Crotonate oxidation:

$CH_3HC{=}CH{-}C{\overset{O}{\underset{O^-}{\|}}}$ + 2 $H_2O \longrightarrow$ 2 acetate

+ H_2 + H^+

2. Crotonate reduction:

$CH_3HC{=}CH{-}C{\overset{O}{\underset{O^-}{\|}}}$ + $H_2 \longrightarrow$ butyrate

Proton motive force

Sum: 2 Crotonate + 2 $H_2O \longrightarrow$ 2 acetate + butyrate + H^+ $\Delta G^{0'}$ = −352 kJ

(b) **Pure culture**

Figure 13.38 **Energetics of growth of** *Syntrophomonas* **in syntrophic culture and in pure culture.** *(a)* In syntrophic culture, growth requires a H_2-consuming organism, such as a methanogen. H_2 production is driven by reverse electron flow because the E_0' values of the FADH and NADH couples are more electropositive than that of 2 H^+/H_2. *(b)* In pure culture, energy conservation is linked to anaerobic respiration with crotonate reduction to butyrate. Inset: photomicrograph of cells of a fatty acid-degrading syntrophic bacterium (red) in association with a methanogen (green-yellow).

some is reduced to butyrate (Figure 13.38*b*). Because crotonate reduction by *Syntrophomonas* is coupled to the formation of a proton motive force, as occurs in other anaerobic respirations that employ organic electron acceptors (such as fumarate

reduction to succinate, Section 13.21), it is possible that some step(s) in syntrophic metabolism generate a proton motive force as well. Pumping protons or some other ion would almost certainly be required for benzoate- and propionate-fermenting syntrophs, whose free energy yield (ΔG) is vanishingly low, only about −5 kJ per reaction.

Regardless of how ATP is made during syntrophic growth, an additional energetic problem burdens syntrophs. During syntrophic metabolism, syntrophs produce H_2 (E_0' −0.42 V) from more electropositive electron donors such as FADH (E_0' −0.22 V) and NADH (E_0' −0.32 V), generated during fatty acid oxidation reactions (Figure 13.38*a*); it is unlikely that this occurs without an energy input. Thus, some fraction of the meager ATP generated by *Syntrophomonas* during syntrophic growth is probably consumed to drive reverse electron flow reactions (Section 13.3) to produce H_2. Combining this energy drain with the inherently poor energetic yields of syntrophic reactions, it should be obvious that syntrophic bacteria thrive on a very marginal energy economy.

Ecology of Syntrophs

Ecologically, syntrophic bacteria are key links in the anoxic steps of the carbon cycle (∞ Section 20.2). Syntrophs consume highly reduced fermentation products and release a key product, H_2, for anaerobic respirations. Without syntrophs, a bottleneck would develop in anoxic environments in which electron acceptors (other than CO_2) were limiting. By contrast, when conditions are oxic or alternative electron acceptors are abundant, syntrophic relationships are unnecessary. For example, if O_2 or NO_3^- is available as an electron acceptor, the energetics of the respiration of a fatty acid or an alcohol is so favorable that syntrophic relationships are unnecessary. Thus, syntrophy is characteristic of anoxic catabolism in which primarily methanogenesis or acetogenesis are the terminal processes in the ecosystem. Methanogenesis is a major process in anoxic wastewater biodegradation, and microbiological studies of sludge granules that form in such systems have shown the close physical relationship that develops between H_2 producer and H_2 consumer in such habitats (Figure 13.38*a* inset).

MINIQUIZ

- Give an example of interspecies H_2 transfer. Why can it be said that both organisms benefit from this process?
- Why can a pure culture of *Syntrophomonas* grow on crotonate but not butyrate?

IV • Anaerobic Respirations

W e examined the process of aerobic respiration in Chapter 3. As we noted there, O_2 functions as a *terminal electron acceptor*, accepting electrons that have traversed an electron transport chain. However, we also noted that other electron acceptors can be used instead of O_2, in which case the process is called **anaerobic respiration**. Here we consider these reactions in more detail.

13.16 Principles of Anaerobic Respiration

Bacteria that carry out anaerobic respiration have electron transport chains containing the typical electron transport proteins that we have seen in aerobic respiration, photosynthesis, and chemolithotrophy—cytochromes, quinones, iron–sulfur proteins, and the like. In some organisms, such as the denitrifying bacteria, anaerobic respiration competes with aerobic respiration. In such cases, if O_2 is present, the organism will preferentially respire aerobically. Many other organisms that conserve energy through anaerobic respiration are obligate anaerobes, which are unable to respire O_2 and may even be killed by it.

Alternative Electron Acceptors and the Redox Tower

The energy released from the oxidation of an electron donor using O_2 as electron acceptor is greater than if the same compound is oxidized with an alternate electron acceptor (⮂ Figure 3.9). These energy differences are clearly apparent from the reduction potentials of each acceptor (**Figure 13.39**). Because the O_2/H_2O

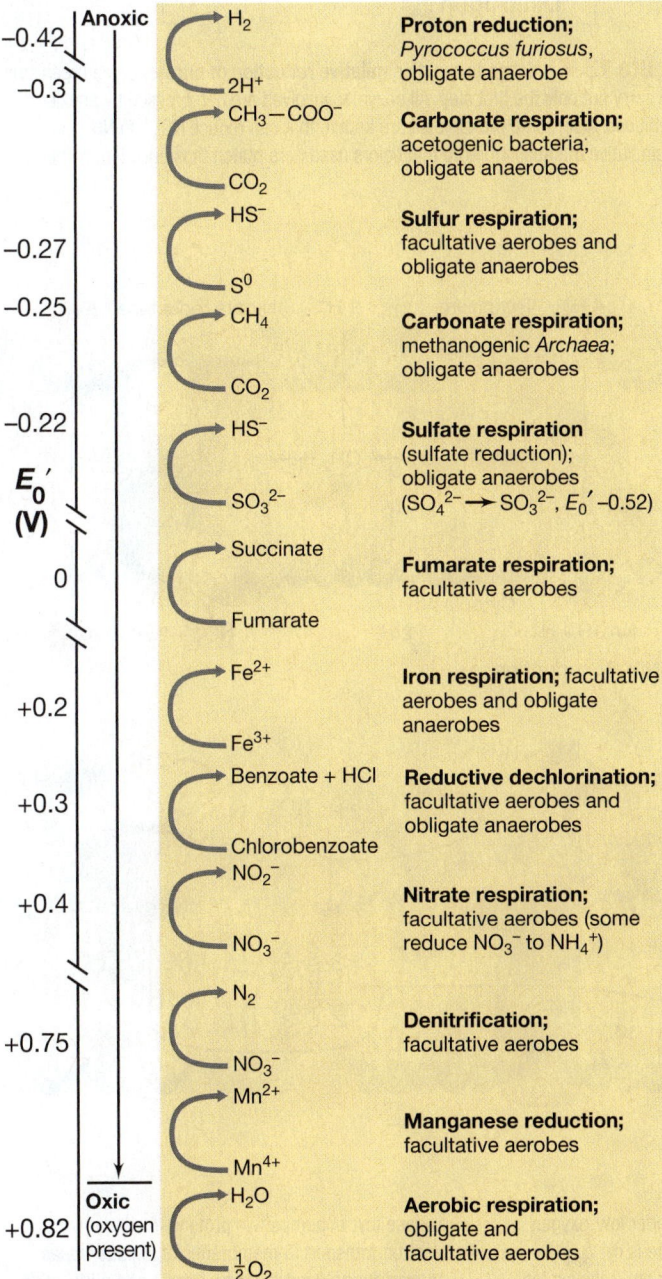

Figure 13.39 Major forms of anaerobic respiration. The redox couples are arranged in order from most electronegative E_0' (top) to most electropositive E_0' (bottom). See Figure 3.9 to compare how the energy yields of these anaerobic respirations vary.

couple is most electropositive, more energy is available when O_2 is used as a terminal electron acceptor than when any other acceptor is used. This is why aerobic respiration is the dominant process and occurs to the exclusion of anaerobic respiration in an organism in which both processes are possible. Other electron acceptors that are fairly near the O_2/H_2O couple are manganic ion (Mn^{4+}), ferric iron (Fe^{3+}), nitrate (NO_3^-), and nitrite (NO_2^-). Examples of more electronegative acceptors are sulfate (SO_4^{2-}), elemental sulfur (S^0), and carbon dioxide (CO_2); organisms that use these acceptors are typically not facultative aerobes and are thus locked into an anaerobic lifestyle. A summary of the most common types of anaerobic respiration is given in Figure 13.39.

Assimilative and Dissimilative Reductions

Inorganic compounds such as NO_3^-, SO_4^{2-}, and CO_2 are reduced by many organisms as sources of cellular nitrogen, sulfur, and carbon, respectively. The end products of such reductions are the amino groups (—NH_2) of amino acids and other nitrogenous substances, the sulfhydryl groups (—SH) of several sulfur-containing compounds in the cell, and the organic carbon found in all cell constituents, respectively. When NO_3^-, SO_4^{2-}, or CO_2 is reduced for these purposes, it is said to be *assimilated*, and the reduction process is called *assimilative* reduction. Assimilative metabolism is conceptually and physiologically quite different from the reduction of NO_3^-, SO_4^{2-}, and CO_2 during energy conservation in anaerobic metabolism. To distinguish these two kinds of reductions, use of these compounds as electron acceptors for energy purposes is called *dissimilative* reduction.

Assimilative and dissimilative metabolisms differ markedly. In assimilative metabolism, only enough of the compound (NO_3^-, SO_4^{2-}, or CO_2) is reduced to satisfy the needs for biosynthesis, and the products are eventually converted to cell material in the form of macromolecules and other biomolecules. By contrast, in dissimilative metabolism, a large amount of the electron acceptor is reduced, and the reduced product remains a small molecule (N_2, H_2S, or CH_4, for example) and is excreted from the cell.

Most organisms carry out an assortment of assimilative metabolisms, whereas a more restricted group catalyze dissimilative metabolisms. As for electron donors, virtually any organic compound that can be degraded aerobically can also be degraded under anoxic conditions by one or more forms of anaerobic respiration. Moreover, several inorganic substances can also be electron donors as long as the E_0' of their redox couple is more electronegative than that of the acceptor couple in the anaerobic respiration (Figure 13.39).

MINIQUIZ

- How does aerobic respiration differ from anaerobic respiration and why does aerobic respiration repress anaerobic respiration?
- With H_2 as an electron donor, why is the reduction of NO_3^- a more favorable reaction than the reduction of S^0?

13.17 Nitrate Reduction and Denitrification

Inorganic nitrogen compounds are some of the most common electron acceptors in anaerobic respiration. Table 13.7 summarizes the relevant forms of inorganic nitrogen with their oxidation states.

Table 13.7 Oxidation states of key nitrogen compounds

Compound	Oxidation state of N atom
Organic N (—NH₂)	−3
Ammonia (NH₃)	−3
Nitrogen gas (N₂)	0
Nitrous oxide (N₂O)	+1 (average per N)
Nitric oxide (NO)	+2
Nitrite (NO₂⁻)	+3
Nitrogen dioxide (NO₂)	+4
Nitrate (NO₃⁻)	+5

One of the most common alternative electron acceptors for dissimilative purposes is nitrate (NO_3^-), which can be reduced with two electrons to nitrite (NO_2^-), or reduced further to nitric oxide (NO), nitrous oxide (N_2O), and dinitrogen (N_2). Because NO, N_2O, and N_2 are all gases, they can be lost from the environment, and their biological production is called **denitrification** (**Figure 13.40**). For agricultural purposes denitrification is a detrimental process,

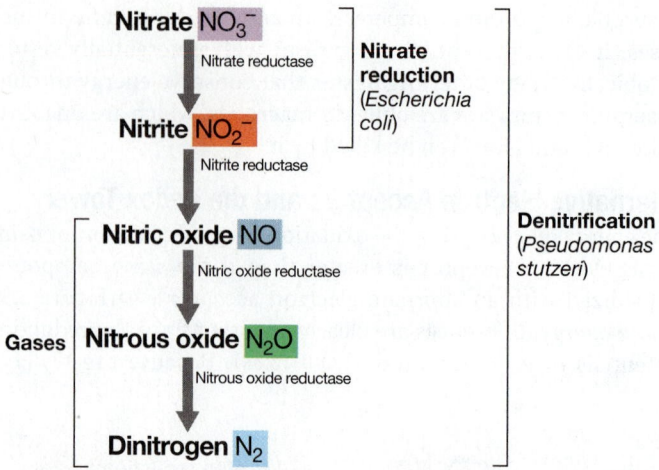

Figure 13.40 Steps in the dissimilative reduction of nitrate. Some organisms can carry out only the first step. All enzymes involved are derepressed by anoxic conditions. Also, some prokaryotes are known that can reduce NO_3^- to NH_4^+ in dissimilative metabolism. Note that colors used here match those used in Figure 13.41.

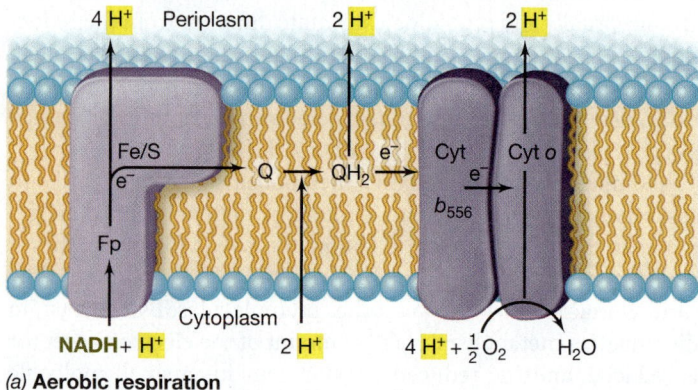

(a) **Aerobic respiration**

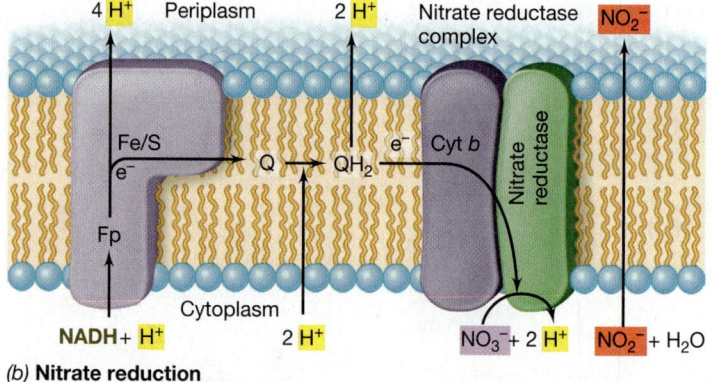

(b) **Nitrate reduction**

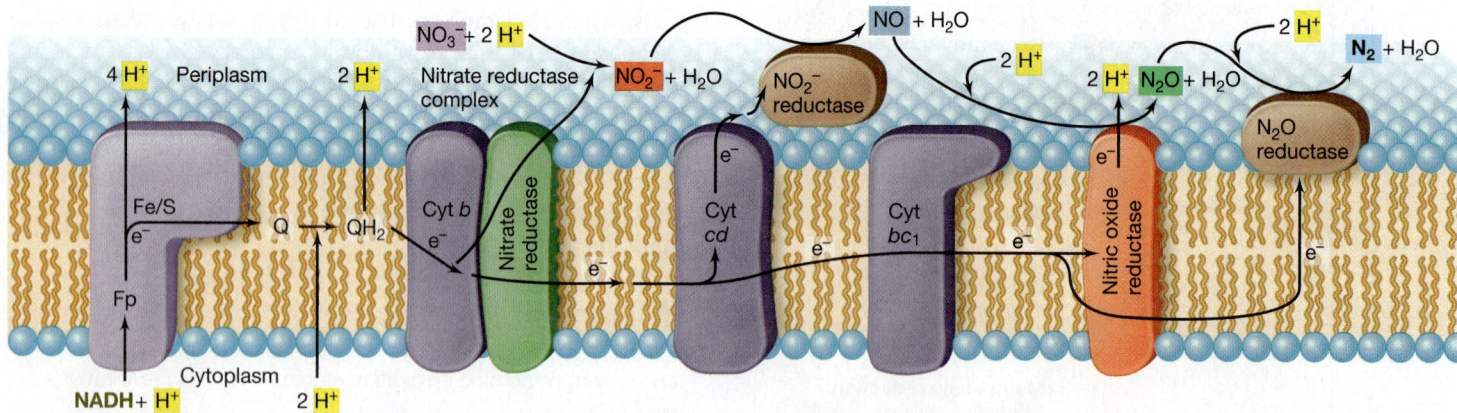

(c) **Denitrification**

Figure 13.41 **Respiration and nitrate-based anaerobic respiration.** Electron transport processes in the membrane of *Escherichia coli* when (a) O₂ or (b) NO₃⁻ is used as an electron acceptor and NADH is the electron donor. Fp, flavoprotein; Q, ubiquinone. Under high-oxygen conditions, the sequence of carriers is

cyt b_{556} → cyt o → O₂. However, under low-oxygen conditions (not shown), the sequence is cyt b_{568} → cyt d → O₂. Note how more protons are translocated per two electrons oxidized aerobically during electron transport reactions than anaerobically with NO₃⁻ as electron acceptor, because the aerobic terminal

oxidase (cyt o) pumps two protons. (c) Scheme for electron transport in membranes of *Pseudomonas stutzeri* during denitrification. Nitrate and nitric oxide reductases are integral membrane proteins, whereas nitrite and nitrous oxide reductases are periplasmic enzymes.

as it removes nitrate—often added in fertilizers—from the soil. However, for a process like sewage treatment, denitrification is beneficial because it removes fixed nitrogen, a major trigger of algal growth in rivers or lakes, from the sewage effluent (↻ Sections 19.8, 21.6, and 21.7).

Denitrifying Microorganisms

Most denitrifying *Bacteria* are phylogenetically *Proteobacteria* and are facultative aerobes physiologically. Aerobic respiration occurs when O_2 is present, even if NO_3^- is also present in the medium. Many denitrifying bacteria also reduce other electron acceptors anaerobically, such as Fe^{3+} and certain organic electron acceptors (Section 13.21), and some denitrifiers can even ferment. Thus, denitrifying bacteria are metabolically diverse in terms of alternative energy-generating mechanisms. Some species of *Archaea* can grow anaerobically by nitrate reduction to nitrite, and several can also denitrify. Interestingly, at least one eukaryote has also been shown to be a denitrifier. The protist *Globobulimina pseudospinescens*, a shelled amoeba (a foraminiferan, ↻ Section 17.7), can denitrify and likely employs this form of metabolism in its habitat, anoxic marine sediments.

Biochemistry of Dissimilative Nitrate Reduction

The electron transport pathways of aerobic respiration, nitrate respiration, and denitrification are compared in **Figure 13.41**. The enzyme that catalyzes the first step of dissimilative nitrate reduction is *nitrate reductase*, a molybdenum-containing membrane-integrated enzyme whose synthesis is repressed by O_2. All subsequent enzymes of the pathway are coordinately regulated and thus also repressed by O_2. But, in addition to anoxic conditions, nitrate must also be present before these enzymes are fully expressed.

The first product of nitrate reduction is NO_2^-, and the enzyme nitrite reductase reduces this to NO (Figure 13.41*c*). Some organisms can reduce NO_2^- to ammonia (NH_3) in a dissimilative process, but the production of gaseous products—*denitrification*—is of greatest global significance. This is because some products of denitrification, in particular N_2O and NO, are of environmental significance. N_2O can be converted to NO by sunlight, and NO reacts with and consumes ozone (O_3) in the upper atmosphere to form NO_2^-. When it rains, NO_2^- returns to Earth as nitrous acid (HNO_2) in *acid rain*.

The biochemistry of dissimilative nitrate reduction has been studied in detail in *Escherichia coli*, in which NO_3^- is reduced only to NO_2^-, and *Paracoccus denitrificans* and *Pseudomonas stutzeri*, in which denitrification occurs. The *E. coli* nitrate reductase accepts electrons from a *b*-type cytochrome, and a comparison of the electron transport chains in aerobic versus nitrate-respiring cells of *E. coli* is shown in Figure 13.41*a, b*. Because of the reduction potential of the NO_3^-/NO_2^- couple (+0.43 V), fewer protons are pumped during nitrate reduction than in aerobic respiration (O_2/H_2O, +0.82 V). In *P. denitrificans* and *P. stutzeri*, nitrogen oxides are formed from NO_2^- by the enzymes nitrite reductase, nitric oxide reductase, and nitrous oxide reductase. During these electron transport reactions, a proton motive force is established, and ATPase couples this to the synthesis of ATP (Figure 13.41*c*).

MINIQUIZ

- For *Escherichia coli*, why is more energy released in aerobic respiration than during NO_3^- reduction?

- How do the products of NO_3^- reduction differ between *E. coli* and *Pseudomonas*?

- Where is the dissimilative nitrate reductase found in the cell? What unusual metal does it contain?

13.18 Sulfate and Sulfur Reduction

Several inorganic sulfur compounds are important electron acceptors in anaerobic respiration. A summary of the oxidation states of key sulfur compounds is given in **Table 13.8**. Sulfate (SO_4^{2-}), the most oxidized form of sulfur, is reduced by the *sulfate-reducing bacteria*, a highly diverse group of obligately anaerobic bacteria widely distributed in nature. The end product of sulfate reduction is hydrogen sulfide, H_2S, an important natural product that participates in many biogeochemical processes (↻ Sections 20.4 and 21.10). Species in the genus *Desulfovibrio*, in particular *D. desulfuricans*, have been widely studied, and the general properties of sulfate-reducing bacteria are discussed in Section 14.9.

As with nitrate (Section 13.17), it is necessary to distinguish between assimilative and dissimilative sulfate metabolism. Many organisms, including plants, algae, fungi, and most prokaryotes, incorporate sulfate for biosynthetic sulfur needs; this is assimilative metabolism. By contrast, the ability to use sulfate as an electron acceptor for energy conservation requires its large-scale reduction and is restricted to the sulfate-reducing bacteria. H_2S is produced on a very large scale by these organisms and is excreted from the cell, free to be oxidized by air, used by other organisms, or combined with metals to form metal sulfides.

Table 13.8 Sulfur compounds and electron donors for sulfate reduction

Compound	Oxidation state of S atom
Oxidation states of key sulfur compounds	
Organic S (R—SH)	−2
Sulfide (H_2S)	−2
Elemental sulfur (S^0)	0
Thiosulfate ($-S-SO_3^{2-}$)	−2/+6
Sulfur dioxide (SO_2)	+4
Sulfite (SO_3^{2-})	+4
Sulfate (SO_4^{2-})	+6
Some electron donors used for sulfate reduction	
H_2	Acetate
Lactate	Propionate
Pyruvate	Butyrate
Ethanol and other alcohols	Long-chain fatty acids
Fumarate	Benzoate
Malate	Indole
Choline	Various hydrocarbons

Biochemistry and Energetics of Sulfate Reduction

As the reduction potentials in Table A1.2 and Figure 13.39 show, SO_4^{2-} is a much less favorable electron acceptor than is O_2 or NO_3^-. However, sufficient free energy to make ATP is available from sulfate reduction when an electron donor is oxidized that yields NADH or FADH. Table 13.8 lists some of the electron donors used by sulfate-reducing bacteria. Hydrogen (H_2) is used by virtually all species, whereas the use of other donors is more restricted. For example, lactate and pyruvate are widely used by species found in freshwater anoxic environments, while acetate and longer-chain fatty acids are widely used by marine sulfate-reducing bacteria. Many morphological and physiological types of sulfate reducing bacteria are known, and with the exception of *Archaeoglobus* (⮂ Section 16.5), a genus of *Archaea*, all known sulfate reducers are *Bacteria* (⮂ Section 14.9).

The reduction of SO_4^{2-} to H_2S requires eight electrons and proceeds through a number of intermediate stages. The reduction of SO_4^{2-} requires that it first be *activated* in a reaction requiring ATP. The enzyme ATP sulfurylase catalyzes the attachment of SO_4^{2-} to a phosphate of ATP, forming *adenosine phosphosulfate* (*APS*) as shown in **Figure 13.42a**. Activation raises the extremely electronegative E_0' of the SO_4^{2-}/SO_3^{2-} couple (-0.52 V) to near 0 V, making reduction of the sulfate moiety possible with electron donors such as NADH (-0.32 V).

In dissimilative sulfate reduction, the SO_4^{2-} in APS is reduced directly to sulfite (SO_3^{2-}) by the enzyme APS reductase with the release of AMP. In assimilative reduction, another phosphate is added to APS to form *phosphoadenosine phosphosulfate* (*PAPS*) (Figure 13.42a), and only then is the SO_4^{2-} reduced. However, in both cases the product of sulfate reduction is sulfite (SO_3^{2-}). Once SO_3^{2-} is formed, it is reduced to H_2S by the activity of the enzyme sulfite reductase (Figures 13.42 and 13.43).

During dissimilative sulfate reduction, electron transport reactions lead to a proton motive force and this drives ATP synthesis by ATPase. A major electron carrier in this process is *cytochrome c_3*, a periplasmic low-potential cytochrome (**Figure 13.43**). Cytochrome c_3 accepts electrons from a periplasmic hydrogenase and transfers these electrons to a membrane-associated protein complex. This complex, called *Hmc*, carries the electrons across the cytoplasmic membrane and transfers them to APS reductase and sulfite reductase, cytoplasmic enzymes that generate sulfite and sulfide, respectively (Figure 13.43).

The enzyme hydrogenase plays a central role in sulfate reduction whether *Desulfovibrio* is growing on H_2, per se, or on an organic compound such as lactate. This is because lactate is converted through pyruvate to acetate (much of the latter is either excreted or assimilated into cell material because *Desulfovibrio* cannot oxidize acetate to CO_2) with the production of H_2. This H_2 crosses the cytoplasmic membrane and is oxidized by the periplasmic hydrogenase to electrons, which are fed back into the system, and protons, which establish the proton motive force (Figure 13.43). A net of one ATP is produced for each SO_4^{2-} reduced to HS^- by H_2, and the reaction is

$$4\,H_2 + SO_4^{2-} + H^+ \rightarrow HS^- + 4\,H_2O \qquad \Delta G^{0\prime} = -152\ kJ$$

When lactate or pyruvate is the electron donor, ATP is produced not only from the proton motive force, but also by substrate-level

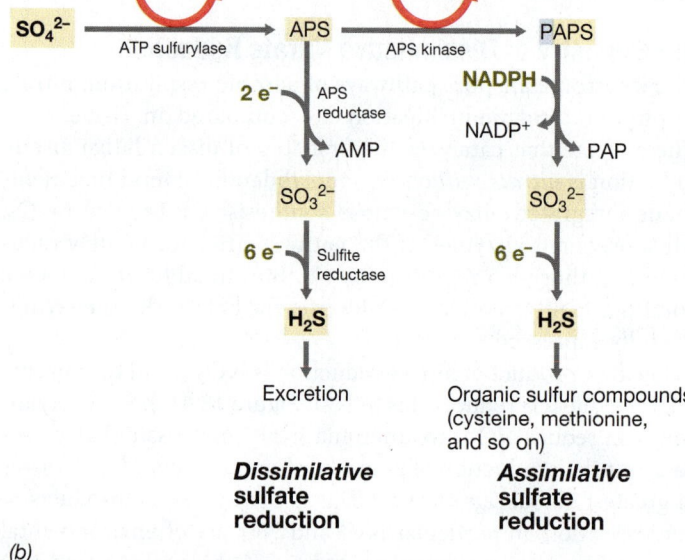

Used in *dissimilative* metabolism

APS (Adenosine 5′-phosphosulfate)

Used in *assimilative* metabolism

PAPS (Phosphoadenosine 5′-phosphosulfate)

(a)

(b)

Figure 13.42 **Biochemistry of sulfate reduction: Activated sulfate.** *(a)* Two forms of active sulfate can be made, adenosine 5′-phosphosulfate (APS) and phosphoadenosine 5′-phosphosulfate (PAPS). Both are derivatives of adenosine diphosphate (ADP), with the second phosphate of ADP being replaced by SO_4^{2-}. *(b)* Schemes of assimilative and dissimilative sulfate reduction.

phosphorylation during the oxidation of pyruvate (via acetyl-CoA and acetyl phosphate, Table 13.3) to acetate plus CO_2 (Figure 13.43).

Marine but not freshwater species of sulfate-reducing bacteria can couple the oxidation of acetate and longer-chain fatty acids to CO_2 and sulfate reduction:

$$CH_3COO^- + SO_4^{2-} + 3\,H^+ \rightarrow 2\,CO_2 + H_2S + 2\,H_2O$$

$$\Delta G^{0\prime} = -57.5\ kJ$$

The mechanism for acetate oxidation in most of these species is the *acetyl-CoA pathway*, a series of reversible reactions used by many anaerobes for acetate synthesis or acetate oxidation

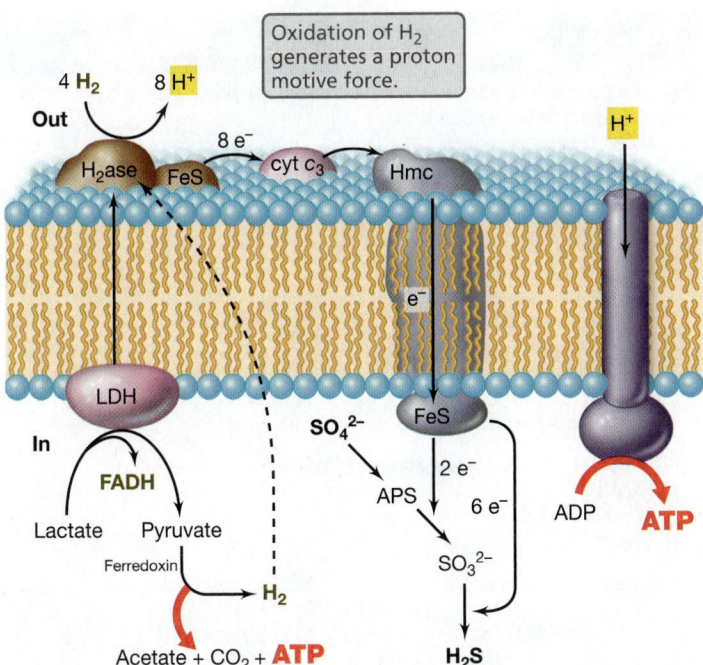

Figure 13.43 Electron transport and energy conservation in sulfate-reducing bacteria. In addition to external H_2, H_2 originating from the catabolism of organic compounds such as lactate and pyruvate can fuel hydrogenase. The enzymes hydrogenase (H_2ase), cytochrome (cyt) c_3, and a cytochrome complex (Hmc) are periplasmic proteins. A separate protein shuttles electrons across the cytoplasmic membrane from Hmc to a cytoplasmic iron–sulfur protein (FeS) that supplies electrons to APS reductase (forming SO_3^{2-}) and sulfite reductase (forming H_2S, Figure 14.14b). LDH, lactate dehydrogenase.

(Section 13.19). A few sulfate-reducing bacteria can also grow autotrophically with H_2. Under these conditions, the organisms use the acetyl-CoA pathway for making acetate as a carbon source. Such species can be cultured in a completely organic-free medium containing only mineral salts, sulfate, CO_2, and H_2.

Special Metabolisms of Sulfate-Reducing Bacteria

Certain species of sulfate-reducing bacteria can catalyze unusual reactions not characteristic of all species. These include *disproportionation*, *phosphite oxidation*, and *sulfur reduction*.

Disproportionation is a process in which one molecule of a substance is oxidized while a second molecule is reduced, ultimately forming two different products. For example, *Desulfovibrio sulfodismutans* can disproportionate thiosulfate ($S–SO_3^{2-}$) as follows:

$$S–SO_3^{2} + H_2O \rightarrow SO_4^{2-} + H_2S \qquad \Delta G^{0\prime} = -21.9 \text{ kJ/reaction}$$

Note that in this reaction, the right-hand sulfur atom of $S–SO_3^{2-}$ is oxidized (forming SO_4^{2-}), while the left-hand atom is reduced (forming H_2S). The free energy available from the oxidation of thiosulfate by *D. sulfodismutans* is insufficient to couple to substrate-level phosphorylation and so instead is coupled to a proton pump that establishes a proton motive force. Other reduced sulfur compounds such as sulfite (SO_3^{2-}) and sulfur (S^0) can also be disproportionated. These forms of sulfur metabolism allow sulfate-reducing bacteria to recover energy from sulfur intermediates produced from the oxidation of H_2S by sulfur chemolithotrophs that coexist with them in nature and also from

intermediates generated in their own metabolism during SO_4^{2-} reduction (Figure 13.42b).

At least one sulfate-reducing bacterium can couple phosphite (HPO_3^-) oxidation to SO_4^{2-} reduction. This chemolithotrophic reaction yields phosphate and sulfide:

$$4 HPO_3^- + SO_4^{2-} + H^+ \rightarrow 4 HPO_4^{2-} + HS^- \qquad \Delta G^{0\prime} = -364 \text{ kJ}$$

This bacterium, *Desulfotignum phosphitoxidans*, is an autotroph and a strict anaerobe, which by necessity it must be because phosphite spontaneously oxidizes in air. The natural sources of phosphite are likely to be organophosphorous compounds called *phosphonates* that are generated from the anoxic degradation of various organic phosphorous compounds. Along with sulfur disproportionation (also a chemolithotrophic process) and H_2 utilization, phosphite oxidation underscores the diversity of chemolithotrophic reactions carried out by sulfate-reducing bacteria.

Sulfur Reduction

Besides sulfate, most sulfate-reducing bacteria can also conserve energy from the reduction of elemental sulfur to sulfide ($S^0 + 2 H \rightarrow H_2S$). In addition, however, a variety of non-sulfate-reducing *Bacteria* and *Archaea* can also reduce sulfur in anaerobic respiration. These are the *sulfur-reducing* prokaryotes, a large group that typically coexists with sulfate-reducing bacteria in anoxic, sulfur-rich habitats.

The electrons for sulfur reduction come from H_2 or any of a number of organic compounds. For example, *Desulfuromonas acetoxidans* can oxidize acetate or ethanol to CO_2 coupled to the reduction of S^0 to H_2S. Sulfur reducers lack the capacity to activate sulfate to APS (Figure 13.42), and presumably this is what excludes them from using SO_4^{2-} as an electron acceptor. *Desulfuromonas* contains several cytochromes, including an analog of cytochrome c_3, a key electron carrier in sulfate-reducing bacteria. In culture some sulfur reducers including *Desulfuromonas* can also use Fe^{3+} as an electron acceptor in place of sulfur, but sulfur is probably the major electron acceptor used in nature. It is the reduction of oxidized sulfur compounds and the production of H_2S that connects the sulfur- and sulfate-reducing bacteria in an ecological sense.

MINIQUIZ -
- How is SO_4^{2-} converted to SO_3^{2-} during dissimilative sulfate reduction? Physiologically, how does *Desulfuromonas* differ from *Desulfovibrio*?
- Contrast the growth of *Desulfovibrio* on H_2 versus lactate as electron donors.
- Give an example of sulfur disproportionation.

13.19 Acetogenesis

Carbon dioxide (CO_2) is typically abundant in anoxic habitats because it is a major product of the energy metabolisms of anaerobic chemoorganotrophs. Two major groups of strictly anaerobic prokaryotes use CO_2 as an electron acceptor for energy conservation. One of these is the *acetogens*, and we discuss them here. The other group, the *methanogens*, are considered in the

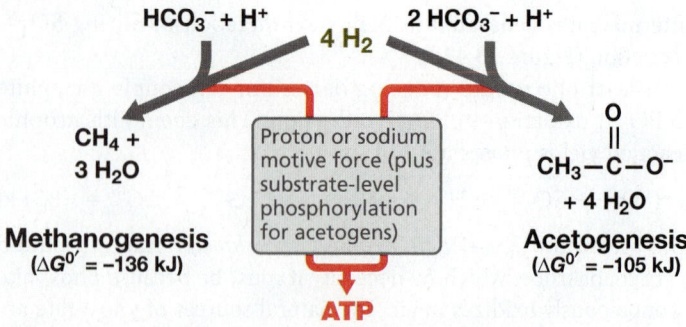

Figure 13.44 **The contrasting processes of methanogenesis and acetogenesis.** Note the difference in free energy released in the reactions.

next section. Hydrogen (H_2) is a major electron donor for both of these organisms, and an overview of their energy metabolism, **acetogenesis** and **methanogenesis**, is shown in **Figure 13.44**. Both processes are linked to ion pumps, of either protons (H^+) or sodium ions (Na^+), as the mechanism of energy conservation, and these pumps fuel ATPases in the membrane. The pathway of acetogenesis also conserves energy in a substrate-level phosphorylation reaction.

Organisms and Pathway

Acetogens carry out the reaction

$$4 H_2 + H^+ + 2 HCO_3^- \rightarrow CH_3COO^- + 4 H_2O$$

$$\Delta G^{0'} = -105 \text{ kJ}$$

In addition to H_2, electron donors for acetogenesis include various C_1 compounds such as methanol, several methoxylated aromatic compounds, sugars, organic and amino acids, alcohols, and certain nitrogen bases, depending on the organism. Many acetogens can also reduce nitrate (NO_3^-) and thiosulfate ($S_2O_3^{2-}$) in dissimilative metabolisms. However, CO_2 reduction is the major reaction of ecological relevance.

A major unifying thread among acetogens is the pathway of CO_2 reduction. Acetogens reduce CO_2 to acetate by the **acetyl-CoA pathway**, the major pathway in obligate anaerobes for the production or oxidation of acetate (see Figure 13.45). Table 13.9 lists the groups that produce or oxidize acetate via the acetyl-CoA pathway. Acetogens such as *Acetobacterium woodii* and *Clostridium aceticum* can grow either chemoorganotrophically by fermentation of sugars (reaction 1) or chemolithotrophically and autotrophically through the reduction of CO_2 to acetate with H_2 as electron donor (reaction 2). In either case, the sole product is acetate:

(1) $C_6H_{12}O_6 \rightarrow 3 CH_3COO^- + 3 H^+$

(2) $2 HCO_3^- + 4 H_2 + H^+ \rightarrow CH_3COO^- + 4 H_2O$

Acetogens catabolize glucose by way of glycolysis, converting glucose into two molecules of pyruvate and two molecules of NADH. From this point, two molecules of acetate are produced:

(3) $2 \text{ Pyruvate}^- \rightarrow 2 \text{ acetate}^- + 2 CO_2 + 2 \text{ NADH}$

The third acetate of reaction (1) comes from reaction (2), using the two molecules of CO_2 generated in reaction (3), plus the 2

Table 13.9   Organisms employing the acetyl-CoA pathway

I. *Pathway drives acetate synthesis for energy purposes*
  Acetoanaerobium noterae
  Acetobacterium woodii
  Acetobacterium wieringae
  Acetogenium kivui
  Acetitomaculum ruminis
  Clostridium aceticum
  Clostridium formicaceticum
  Clostridium ljungdahlii
  Moorella thermoacetica
  Desulfotomaculum orientis
  Sporomusa paucivorans
  Eubacterium limosum (also produces butyrate)
  Treponema primitia (from termite hindguts)

II. *Pathway drives acetate synthesis for cell biosynthesis*
  Acetogens
  Methanogens
  Sulfate-reducing bacteria

III. *Pathway drives acetate oxidation for energy purposes*
  Reaction: Acetate + 2 H_2O → 2 CO_2 + 8 H
  Group II sulfate reducers (other than *Desulfobacter*)
  Reaction: Acetate → CO_2 + CH_4
  Acetotrophic methanogens (*Methanosarcina, Methanosaeta*)

NADH generated from glycolysis and the 2 NADH generated from the oxidation of two pyruvates to two acetates [reaction (3)]. Starting from pyruvate, then, the overall production of acetate can be written as

$$2 \text{ Pyruvate}^- + 4 H \rightarrow 3 \text{ acetate}^- + H^+$$

Most acetogenic bacteria that produce acetate in energy metabolism are gram-positive *Bacteria*, and many are species of the genera *Clostridium* or *Acetobacterium* (Table 13.9). A few other gram-positive and many different gram-negative *Bacteria* and *Archaea* use the acetyl-CoA pathway for autotrophic purposes, reducing CO_2 to acetate as a source of cell carbon. The acetyl-CoA pathway supports autotrophic growth of both autotrophic sulfate-reducing bacteria and methanogens. In addition, some prokaryotes employ the acetyl-CoA pathway not to make acetate for cell carbon purposes but in the reverse direction as a means of oxidizing acetate to CO_2. These include acetate-utilizing methanogens and sulfate-reducing bacteria.

The Acetyl-CoA Pathway and Energy Conservation in Acetogenesis

Unlike other autotrophic pathways (Section 13.5), the acetyl-CoA pathway of CO_2 fixation is not a cycle. Instead, it catalyzes the reduction of CO_2 along two linear pathways, with one molecule of CO_2 being reduced to the methyl group of acetate and the other to the carbonyl group of acetate. These two C_1 units are then combined to form acetyl-CoA (**Figure 13.45**).

A key enzyme of the acetyl-CoA pathway is *carbon monoxide (CO) dehydrogenase*. CO dehydrogenase contains Ni, Zn, and Fe as cofactors and catalyzes the reaction

$$CO_2 + H_2 \rightarrow CO + H_2O$$

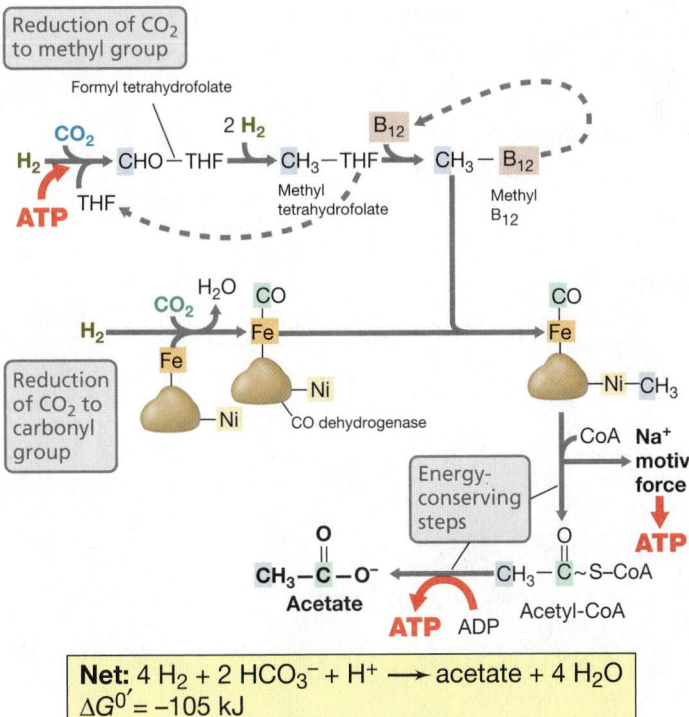

Net: $4 H_2 + 2 HCO_3^- + H^+ \longrightarrow$ acetate $+ 4 H_2O$
$\Delta G^{0\prime} = -105$ kJ

Figure 13.45 Reactions of the acetyl-CoA pathway. Carbon monoxide is bound to Fe and the CH_3 group to nickel in carbon monoxide dehydrogenase. Note that the formation of acetyl-CoA is coupled to the generation of a Na^+ motive force that drives ATP synthesis, and that ATP is also synthesized in the conversion of acetyl-CoA to acetate. THF, tetrahydrofolate; B_{12}, vitamin B_{12} in an enzyme-bound intermediate.

The CO produced by CO dehydrogenase ends up as the *carbonyl* carbon of acetate (Figure 13.45). The methyl group of acetate originates from the reduction of CO_2 by a series of reactions in which the coenzyme *tetrahydrofolate* plays a major role (Figure 13.45). The methyl group is then transferred from tetrahydrofolate to an enzyme that contains vitamin B_{12} as cofactor, and in the final step of the pathway, the methyl group is combined with CO by CO dehydrogenase to form acetyl-CoA. Conversion of acetyl-CoA to acetate plus ATP completes the reaction series (Figure 13.45).

Energy conservation in acetogenesis results from both substrate-level phosphorylation and an ion pump. One ATP is produced during the conversion of each acetyl-CoA to acetate plus ATP (Figure 13.45 and Table 13.3). In addition, however, the synthesis of acetyl-CoA from its component parts, CO and CH_3, releases free energy, some of which is conserved by a Na^+ pump that couples this to formation of a sodium motive force. This energized state of the membrane drives ATP synthesis from a Na^+-translocating ATPase. Recall that we saw a similar energetic strategy with the succinate fermenter *Propionigenium*, where succinate decarboxylation was linked to Na^+ export and a Na^+-driven ATPase (Section 13.14). Acetogens need the ATP resulting from this reaction to have a net energy gain since the single ATP made by substrate-level phosphorylation is consumed in the first step of the acetyl-CoA pathway (Figure 13.45).

MINIQUIZ --

- Draw the structure of acetate and identify the carbonyl group and the methyl group. What key enzyme of the acetyl-CoA pathway produces the carbonyl group of acetate?
- How do acetogens make ATP from the synthesis of acetate?
- If fructose catabolism by glycolysis yields only two acetates, how does *Clostridium aceticum* produce three acetates from fructose?

13.20 Methanogenesis

The biological production of methane—*methanogenesis*—is catalyzed by a group of strictly anaerobic *Archaea* called the **methanogens**. These organisms are present in freshwater sediments (**Figure 13.46**), sewage sludge digesters (⮌ Section 21.6) and other bioreactors, and the intestines of warm-blooded animals, including humans. The reduction of CO_2 by H_2 to form methane (CH_4) is a major pathway of methanogenesis and is a form of anaerobic respiration. We consider the basic properties, phylogeny, and taxonomy of the methanogens in Section 16.2. Here we focus on the bioenergetics and unique biochemistry of methanogenesis.

C_1 Carriers in Methanogenesis

Methanogenesis from CO_2 requires eight electrons, and these electrons are added two at a time. This leads to intermediary oxidation states of the carbon atom from $+4$ (CO_2) to -4 (CH_4). Several novel coenzymes participate in methanogenesis and can be divided into two classes: (1) those that carry the C_1 unit along its path of enzymatic reduction (C_1 *carriers*) and (2) those that donate electrons (*redox coenzymes*) (**Figure 13.47**). We consider the C_1 carriers first.

The coenzyme *methanofuran* is required for the first step of methanogenesis. Methanofuran contains the five-membered

Figure 13.46 Methanogenesis. Methane is collected in a funnel from swamp sediments where it was produced by methanogens and then ignited in a demonstration experiment.

I. Coenzymes that function as C₁ carriers, plus F₄₃₀

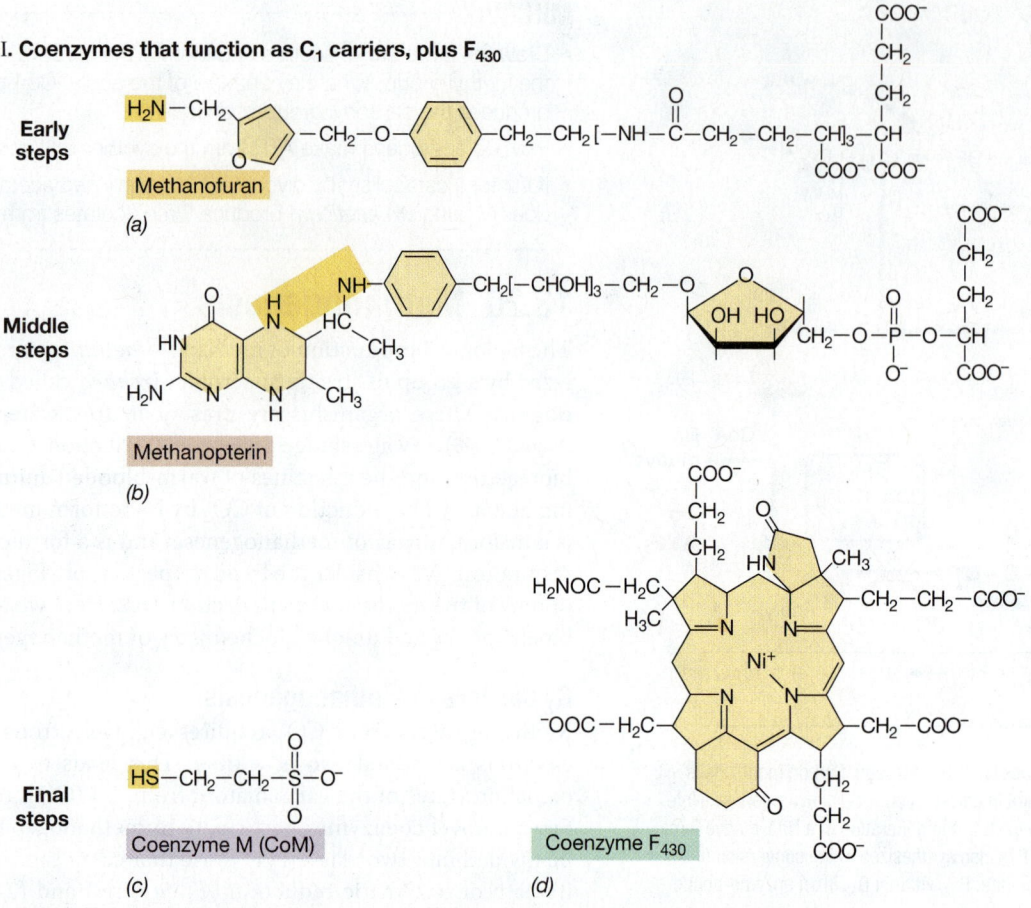

Early steps
Methanofuran
(a)

Middle steps
Methanopterin
(b)

Final steps
Coenzyme M (CoM)
(c)

Coenzyme F₄₃₀
(d)

II. Coenzymes that function as electron donors

Oxidized

−2H +2H

Reduced

Coenzyme F₄₂₀
(e)

Coenzyme B (CoB)
(f)

Figure 13.47 Coenzymes of methanogenesis. The atoms shaded in brown or yellow are the sites of oxidation–reduction reactions (brown in F_{420} and CoB) or the position to which the C_1 moiety is attached during the reduction of CO_2 to CH_4 (yellow in methanofuran, methanopterin, and coenzyme M). The colors used to highlight a particular coenzyme (CoB is orange, for example) are also in Figures 13.49 and 13.50 to follow the reactions in each figure. Coenzyme F_{430} participates in the terminal step of methanogenesis catalyzed by the enzyme methyl reductase, with the methyl group binding to Ni^+ in F_{430} prior to its reduction to CH_4.

furan ring and an amino nitrogen atom that binds CO_2 (Figure 13.47a). *Methanopterin* (Figure 13.47b) is a methanogenic coenzyme that resembles the vitamin folic acid and plays a role analogous to that of tetrahydrofolate (a coenzyme that participates in C_1 transformations; see Figure 13.45) by carrying the C_1 unit in the intermediate steps of CO_2 reduction to CH_4. *Coenzyme M* (CoM) (Figure 13.47c) is required for the terminal step of methanogenesis, the reduction of the methyl group (CH_3) to CH_4. Although not a C_1 carrier, the nickel (Ni^{2+})-containing tetrapyrrole *coenzyme F_{430}* (Figure 13.47d) also participates in the terminal step of methanogenesis as part of the methyl reductase enzyme complex (discussed later).

Redox Coenzymes

The coenzymes F_{420} and *7-mercaptoheptanoylthreonine phosphate* (also called coenzyme B, CoB) are electron donors in methanogenesis. Coenzyme F_{420} (Figure 13.47e) is a flavin derivative, structurally resembling the flavin coenzyme FMN (⮌ Figure 3.16). F_{420} plays a role in methanogenesis as the electron donor in several steps of CO_2 reduction (see Figure 13.49). Coenzyme F_{420} takes its name from the fact that its oxidized form absorbs light at 420 nm and fluoresces blue-green. Such fluorescence is useful for the microscopic identification of a methanogen (**Figure 13.48**). CoB is required for the terminal step of methanogenesis catalyzed by the *methyl reductase enzyme complex*. As shown in Figure 13.47f, the structure of CoB resembles the vitamin pantothenic acid, which is part of acetyl-CoA (⮌ Figure 3.12).

Methanogenesis from CO_2 + H_2

Electrons for the reduction of CO_2 to CH_4 typically come from H_2, but a few other substrates can also supply the electrons for CO_2 reduction in some methanogens. **Figure 13.49** shows the steps in CO_2 reduction by H_2:

1. CO_2 is activated by a methanofuran-containing enzyme and reduced to the formyl level. The immediate electron donor is ferredoxin, a strong reductant with a reduction potential (E_0') near -0.4V.

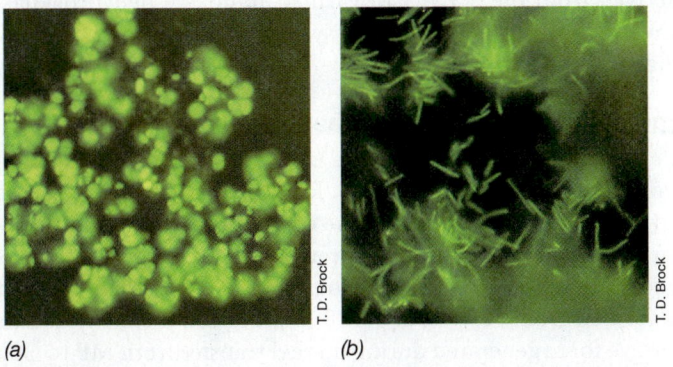

(a)   *(b)*

Figure 13.48 **Fluorescence due to the methanogenic coenzyme F_{420}.** *(a)* Autofluorescence in cells of the methanogen *Methanosarcina barkeri* due to the presence of the unique electron carrier F_{420}. A single cell is about 1.7 µm in diameter. The organisms were made visible by excitation with blue light in a fluorescence microscope. *(b)* F_{420} fluorescence in cells of the methanogen *Methanobacterium formicicum*. A single cell is about 0.6 µm in diameter.

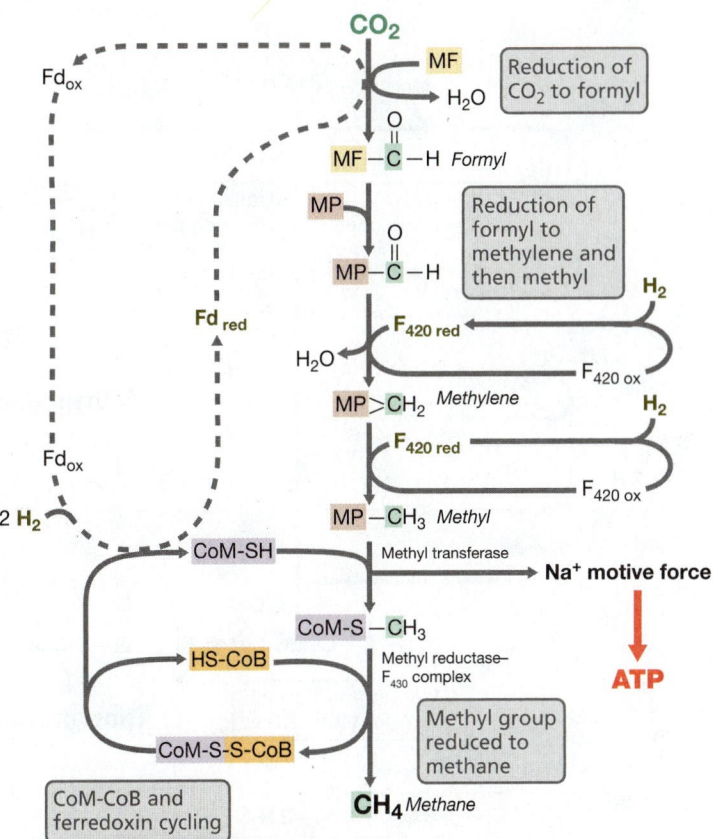

Figure 13.49 **Methanogenesis from CO_2 plus H_2.** The carbon atom reduced is highlighted in green, and the source of electrons is highlighted in brown. See Figure 13.47 for the structures of the coenzymes. MF, methanofuran; MP, methanopterin; CoM, coenzyme M; $F_{420\ red}$, reduced coenzyme F_{420}; F_{430}, coenzyme F_{430}; Fd, ferredoxin; CoB, coenzyme B.

2. The formyl group is transferred from methanofuran to an enzyme containing methanopterin (MP in Figure 13.49). It is subsequently dehydrated and reduced in two separate steps (total of 4 H) to the methylene and methyl levels. The immediate electron donor here is reduced F_{420}.

3. The methyl group is transferred from methanopterin to an enzyme containing CoM by the enzyme methyl transferase. This reaction is highly exergonic and linked to the pumping of Na^+ across the membrane from inside to outside the cell.

4. Methyl-CoM is reduced to methane by methyl reductase. In this reaction, F_{430} and CoB are required. Coenzyme F_{430} removes the CH_3 group from CH_3–CoM, forming a Ni^{2+}–CH_3 complex. This complex is reduced by CoB, generating CH_4 and a disulfide complex of CoM and CoB (CoM-S—S-CoB).

5. Free CoM and CoB are regenerated by the reduction of CoM-S—S-CoB with H_2. Simultaneously, ferredoxin is also reduced by H_2 and is thus ready for the first step of a new round of CO_2 reduction (Figure 13.49).

Methanogenesis from Methyl Compounds and Acetate

We will learn in Section 16.2 that methanogens can form CH_4 from certain methylated compounds such as methanol and acetate, as well as from H_2 + CO_2. Methanol is catabolized by donating

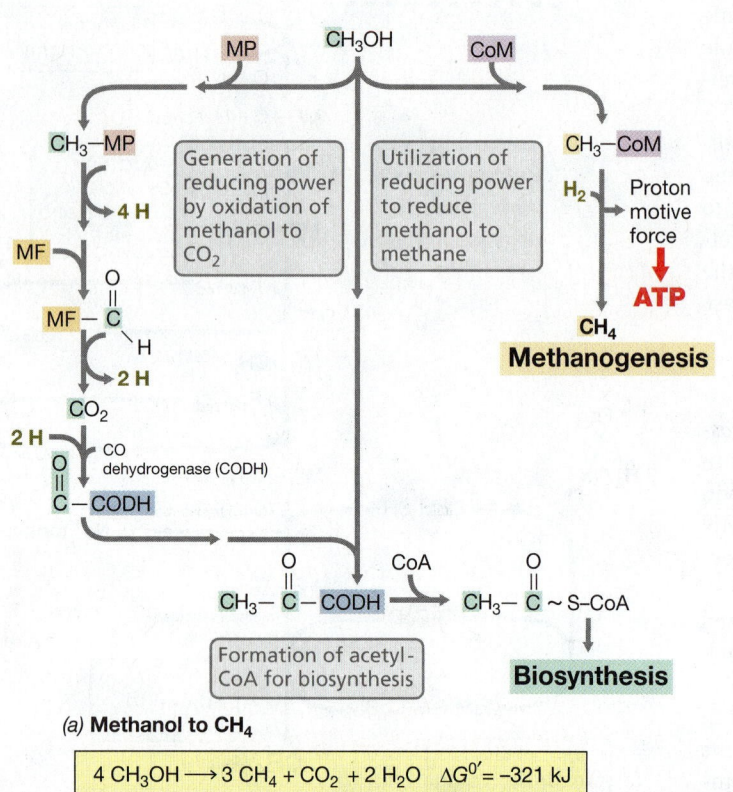

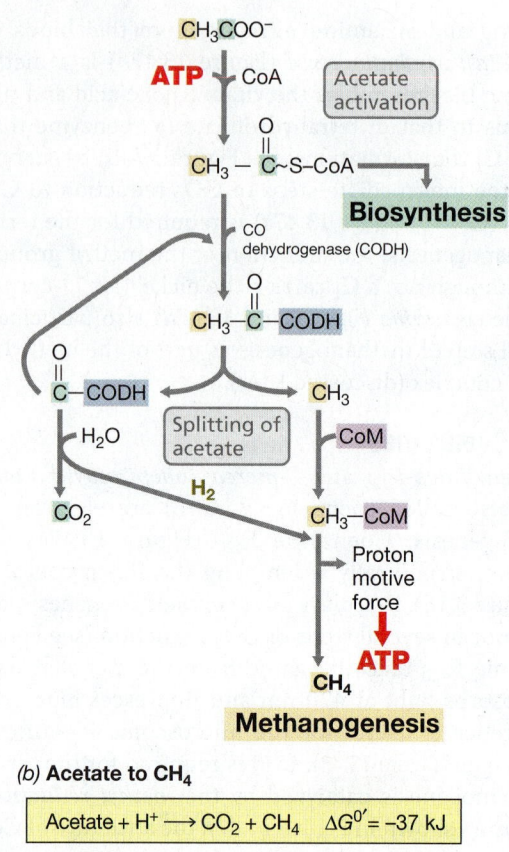

(a) Methanol to CH₄

$$4\ CH_3OH \longrightarrow 3\ CH_4 + CO_2 + 2\ H_2O \quad \Delta G^{0\prime} = -321\ kJ$$

(b) Acetate to CH₄

$$Acetate + H^+ \longrightarrow CO_2 + CH_4 \quad \Delta G^{0\prime} = -37\ kJ$$

Figure 13.50 Methanogenesis from methanol and acetate. Both reaction series contain parts of the acetyl-CoA pathway. *(a)* For growth on methanol (CH_3OH), most CH_3OH carbon is converted to CH_4, and a smaller amount is converted to either CO_2 or, via formation of acetyl-CoA, is assimilated into cell material. *(b)* Acetate is split into CH_4 and CO_2. Abbreviations and color-coding are as in Figures 13.47 and 13.49; Corr, corrinoid-containing protein; CODH, carbon monoxide dehydrogenase.

methyl groups to an enzyme containing a corrinoid coenzyme to form CH_3–corrinoid (**Figure 13.50***a*). Corrinoids are the parent structures of compounds such as vitamin B_{12} and contain a porphyrin-like ring with a central cobalt atom. The CH_3–corrinoid complex then transfers the methyl group to CoM, yielding CH_3–CoM from which methane is formed in the same way as in the terminal step of CO_2 reduction. If H_2 is unavailable to drive the terminal step, some of the methanol must be oxidized to CO_2 to yield electrons for this purpose. This occurs by reversal of steps in methanogenesis (Figures 13.49 and 13.50*a*).

When acetate is the substrate for methanogenesis, it is first activated to acetyl-CoA, which interacts with CO dehydrogenase from the acetyl-CoA pathway (Section 13.19). The methyl group of acetate is then transferred to the corrinoid enzyme to yield CH_3–corrinoid, and from there it follows the CoM-mediated terminal step of methanogenesis. Simultaneously, the CO group is oxidized to yield CO_2 and electrons (Figure 13.50*b*).

Autotrophy

Autotrophy in methanogens is supported by the acetyl-CoA pathway (Section 13.19). As we have just seen, parts of this pathway are already integrated into the catabolism of methanol and acetate by methanogens (Figure 13.50). However, methanogens lack the tetrahydrofolate-driven series of reactions of the acetyl-CoA pathway that lead to the production of a methyl

group (Figure 13.45). But this is not a problem because methanogens either derive methyl groups directly from their electron donors (Figure 13.50) or make methyl groups during methanogenesis from $H_2 + CO_2$ (Figure 13.49). Thus methanogens have access to abundant methyl groups, and the removal of some for biosynthesis is of little consequence. The carbonyl group of the acetate produced during autotrophic growth of methanogens is derived from the activity of carbon monoxide dehydrogenase, and the terminal step in acetate synthesis is as described for acetogens (Section 13.19 and Figure 13.45).

Energy Conservation in Methanogenesis

Under standard conditions, the free energy of methanogenesis from $H_2 + CO_2$ is -131 kJ/mol. Energy conservation in methanogenesis occurs at the expense of a proton or sodium motive force, depending on the substrate used; substrate-level phosphorylation (Section 13.11) does not occur. When methane is formed from $H_2 + CO_2$, ATP is produced from the sodium motive force generated during methyl transfer from MP to CoM by the enzyme methyl transferase (Figure 13.49). This energized state of the membrane then drives the synthesis of ATP, probably by way of an H^+-linked ATPase following conversion of the sodium motive force into a proton motive force by exchange of Na^+ for H^+ across the membrane. The ATP yield per CH_4 produced is about 0.5.

In some methanogens, such as *Methanosarcina*, a nutritionally versatile organism that can make methane from acetate or methanol as well as from $CO_2 + H_2$, a different mechanism of energy conservation occurs from acetate or methanol, since the methyl transferase reaction cannot be coupled to the generation of a sodium motive force under these conditions. Instead, in acetate- and methanol-grown cells energy conservation is linked to the terminal step in methanogenesis, the methyl reductase step (Figures 13.49, 13.50, and **Figure 13.51**). In this reaction, the interaction of CoB with CH_3–CoM and methyl reductase forms CH_4 and a heterodisulfide product, CoM-S—S-CoB. The latter is reduced by H_2 to regenerate CoM-SH and CoB-SH (Figure 13.49). This reduction, carried out by the enzyme *heterodisulfide reductase*, is exergonic and is coupled to the pumping of H^+ across the membrane (Figure 13.51). Electrons from H_2 flow to the heterodisulfide reductase through a membrane-associated electron carrier called *methanophenazine*. This compound is reduced by F_{420} and subsequently oxidized by a *b*-type cytochrome; the latter is the electron donor to the heterodisulfide reductase (Figure 13.51). Cytochromes and methanophenazine are absent in methanogens that can use only $H_2 + CO_2$ for methanogenesis.

In methanogens we thus see at least two mechanisms for energy conservation: (1) a proton motive force linked to the methylreductase reaction and used to drive ATP synthesis in acetate- or methanol-grown cells, and (2) a sodium motive force (that is likely converted to a proton motive force) during methanogenesis from $H_2 + CO_2$.

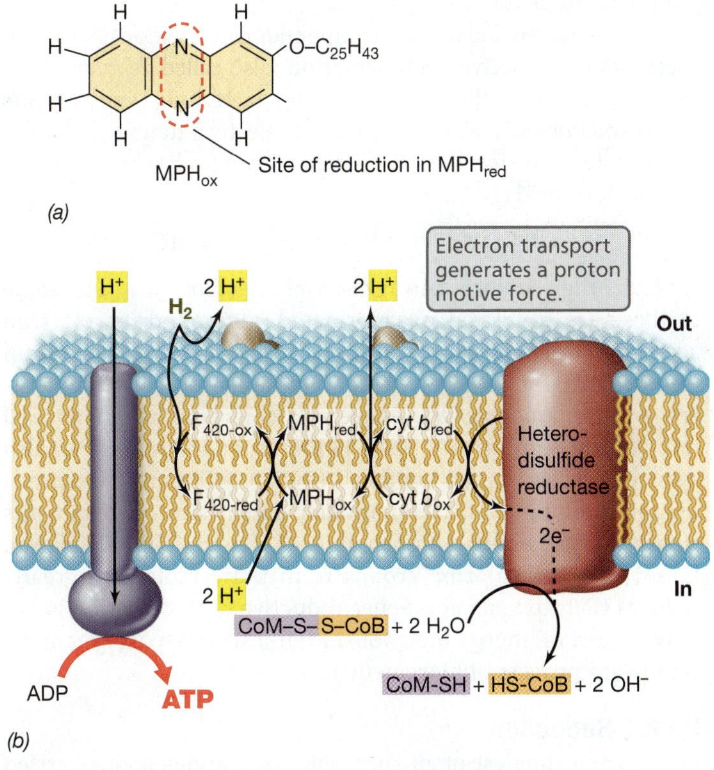

Figure 13.51 Energy conservation in methanogenesis from methanol or acetate. *(a)* Structure of methanophenazine (MPH in part b), an electron carrier in the electron transport chain leading to ATP synthesis; the central ring of the molecule can be alternately reduced and oxidized. *(b)* Steps in electron transport. Electrons originating from H_2 reduce F_{420} and then methanophenazine. The latter, through a cytochrome of the *b* type, reduces heterodisulfide reductase with the extrusion of H^+ to the outside of the membrane. In the final step, heterodisulfide reductase reduces CoM-S—S-CoB to HS-CoM and HS-CoB. See Figure 13.47 for the structures of CoM and CoB.

MINIQUIZ

- Which coenzymes function as C_1 carriers in methanogenesis? As electron donors?
- In methanogens growing on $H_2 + CO_2$, how is carbon obtained for cell biosynthesis?
- How is ATP made in methanogenesis when the substrates are $H_2 + CO_2$? Acetate?

13.21 Other Electron Acceptors

In addition to the electron acceptors for anaerobic respiration discussed thus far, several metals, metalloids, and halogenated and unhalogenated organic compounds are important electron acceptors for bacteria in nature (**Figure 13.52**). In addition to these, even protons can be used by a very few strict anaerobes. We consider these forms of anaerobic respiration here.

Metal Reduction

Several metals and metalloids can be reduced in anaerobic respirations. Ferric iron (Fe^{3+}) and manganic ion (Mn^{4+}) are the most important metals reduced. The reduction potential of

Couple	Reaction	E_0'
Fumarate/ Succinate	$^-OOC{-}CH{=}CH{-}COO^- \xrightarrow{2\,H} {}^-OOC{-}CH_2{-}CH_2{-}COO^-$	+0.03
Trimethylamine-*N*-oxide (TMAO)/ Trimethylamine (TMA)	$(CH_3)_3N{\rightarrow}O \xrightarrow{2\,H} (CH_3)_3N + H_2O$	+0.13
Arsenate/ Arsenite	$AsO_4^{3-} \xrightarrow{2\,H} AsO_3^{3-} + H_2O$	+0.14
Dimethyl sulfoxide (DMSO)/ Dimethyl sulfide (DMS)	$H_3C{-}S({\rightarrow}O){-}CH_3 \xrightarrow{2\,H} (CH_3)_2S + H_2O$	+0.16
Ferric ion/ Ferrous ion	$Fe^{3+} \xrightarrow{e^-} Fe^{2+}$	+0.20
Selenate/ Selenite	$SeO_4^{2-} \xrightarrow{2\,H} SeO_3^{2-} + H_2O$	+0.48
Manganic ion/ Manganous ion	$Mn^{4+} \xrightarrow{2\,e^-} Mn^{2+}$	+0.80
Chlorate/ Chloride	$ClO_3^- \xrightarrow{6\,H} Cl^- + 3\,H_2O$	+1.00

Figure 13.52 Some alternative electron acceptors for anaerobic respirations. Note the reaction and E_0' of each redox pair calculated for pH7.

UNIT 3

the Fe^{3+}/Fe^{2+} couple is $+0.2$ V (at pH 7), and that of the Mn^{4+}/Mn^{2+} couple is $+0.8$ V; thus, several electron donors can couple to Fe^{3+} and Mn^{4+} reduction. In these reactions, electrons typically travel from the donor through an electron transport chain that generates a proton motive force and terminates in a metal reductase system, reducing Fe^{3+} to Fe^{2+} or Mn^{4+} to Mn^{2+}. Much research on the energetics of Fe^{3+} reduction has been done with the gram-negative bacteria *Shewanella* and *Geobacter*; *Shewanella* also reduces Mn^{4+}. Each organism can oxidize several organic electron donors as well as H_2, and *Geobacter* can also oxidize acetate and the aromatic hydrocarbon toluene linked to Fe^{3+} reduction.

Other inorganic substances can function as electron acceptors for anaerobic respiration, including the metalloids selenium, tellurium, and arsenic, the transition metal vanadium, and various oxidized chlorine compounds (Figure 13.52). Most of the organisms capable of growth with these acceptors are facultative aerobes and can thus grow by aerobic respiration as well. Arsenic, selenium, and tellurium compounds are occasional pollutants in nature and can support anaerobic respirations in various bacteria. The reduction of selenate (SeO_4^{2-}) occurs to selenite (SeO_3^{2-}) and eventually to metallic selenium (Se^0), the reduction of arsenate (AsO_4^{2-}) occurs to the level of arsenite (AsO_3^{2-}), and the reduction of tellurate (TeO_4^{2-}) to tellurite (TeO_3^{2-}). Several chlorate and perchlorate-reducing bacteria have also been isolated and are likely responsible for the removal of these toxic compounds from nature; the typical end product of these reactions is chloride (Cl^-).

The sulfate-reducing bacterium *Desulfotomaculum* can reduce both AsO_4^{3-} to AsO_3^{3-} and sulfate to sulfide, and during this process, the yellow mineral orpiment (As_2S_3) precipitates spontaneously (Figure 13.53). This process is an example of *biomineralization*, the formation of a mineral by bacterial activity. As_2S_3 formation also functions as a means of detoxifying what would otherwise be a toxic compound (arsenic), and thus such microbial activities may have practical applications for the cleanup of arsenic-containing toxic wastes and groundwater.

Figure 13.53 Biomineralization during arsenate reduction by the sulfate-reducing bacterium *Desulfotomaculum auripigmentum*. Left, appearance of culture bottle after inoculation. Right, following growth for two weeks and biomineralization of arsenic trisulfide, As_2S_3. Center, synthetic sample of As_2S_3.

Organic Electron Acceptors

Several organic compounds can be electron acceptors in anaerobic respirations. Of those listed in Figure 13.52, the compound that has been most extensively studied is *fumarate*, a citric acid cycle intermediate (Figure 3.22), which is reduced to succinate. The role of fumarate as an electron acceptor for anaerobic respiration derives from the fact that the fumarate–succinate couple has a reduction potential near 0 V, which allows coupling of fumarate reduction to the oxidation of NADH, FADH, or H_2. Many facultatively aerobic bacteria can grow anaerobically on fumarate as electron acceptor, including *Escherichia coli*.

Trimethylamine oxide (TMAO) and dimethyl sulfoxide (DMSO) (Figure 13.52) are important organic electron acceptors. TMAO is a product of marine fish, and several bacteria can reduce it to trimethylamine (TMA), which has a strong odor and flavor (the odor of spoiled seafood is due primarily to TMA produced by bacterial action). Dimethyl sulfoxide (DMSO), which is reduced to dimethyl sulfide (DMS), is a common natural product and is found in both marine and freshwater environments. The reduction potentials of the TMAO/TMA and DMSO/DMS couples are the same, about $+0.15$ V, and this means that electron transport chains that terminate with TMAO or DMSO reductases must be rather short. As in fumarate reduction, in the reduction of TMAO and DMSO cytochromes of the *b* type (E_0' near 0 V) function as donors to the reductases.

Several halogenated organic compounds can function as electron acceptors in **reductive dechlorination** (also called *dehalorespiration*). For example, the sulfate-reducing bacterium *Desulfomonile* grows anaerobically with H_2 or organic electron donors and chlorobenzoate as an electron acceptor that is reduced to benzoate and hydrochloric acid (HCl):

$$C_7H_4O_2Cl^- + 2\,H \rightarrow C_7H_5O_2^- + HCl$$

Several other bacteria can reductively dechlorinate, and some of these are restricted to chlorinated compounds as electron acceptors for anaerobic respiration. For example, the bacterium *Dehalococcoides* reduces tri- and tetrachloroethylene to ethene and *Dehalobacterium* converts dichloromethane (CH_2Cl_2) into acetate and formate (**Table 13.10**). *Dehalococcoides* can also reduce polychlorinated biphenyls (PCBs). PCBs are widespread organic pollutants that contaminate freshwater environments where they accumulate in fish and other aquatic life. However, removal of the chlorine groups from these molecules greatly reduces their toxicity and hence reductive dechlorination is not only a form of energy metabolism but also an environmentally significant process of bioremediation.

Proton Reduction

Perhaps the simplest of all anaerobic respirations is one carried out by the hyperthermophile *Pyrococcus furiosus*. *P. furiosus* is a species of *Archaea* that grows optimally at 100°C (Chapter 16) on sugars and small peptides as electron donors and protons as electron acceptor. This is possible because of a unique biochemical feature of the glycolytic pathway of *P. furiosus*.

During glycolysis, the oxidation of glyceraldehyde 3-phosphate forms 1,3-bisphosphoglyceric acid, an intermediate with

Table 13.10 Characteristics of some major genera of bacteria capable of reductive dechlorination

Property	Dehalobacter	Dehalobacterium	Desulfitobacterium	Desulfomonile	Dehalococcoides
			Genus		
Electron donors	H_2	Dichloromethane (CH_2Cl_2) only	H_2, formate, pyruvate, lactate	H_2, formate, pyruvate, lactate, benzoate	H_2, lactate
Electron acceptors	Trichloroethylene, tetrachloroethylene	Dichloromethane (CH_2Cl_2) only	Ortho-, meta-, or para-chlorophenols, NO_3^-, fumarate, SO_3^{2-}, $S_2O_3^{2-}$, S^0	Metachlorobenzoates, tetrachloroethylene, SO_4^{2-}, SO_3^{2-}, $S_2O_3^{2-}$	Trichloroethylene, tetrachloroethylene
Product of reduction of tetrachloroethylene	Dichloroethylene	Not applicable	Trichloroethylene	Dichloroethylene	Ethene
Other properties[a]	Contains cytochrome b	Grows only on CH_2Cl_2 and by disproportionation as follows: $CH_2Cl_2 \rightarrow$ formate + acetate + HCl ATP is formed by substrate-level phosphorylation	Can also grow by fermentation	Sulfate-reducing bacterium; contains cytochrome c_3; requires organic carbon source; can grow by fermentation of pyruvate	Lacks peptidoglycan
Phylogeny[b]	Firmicutes	Firmicutes	Firmicutes	Deltaproteobacteria	Green nonsulfur Bacteria (Chloroflexi)

[a]All organisms are obligate anaerobes.
[b]See Chapters 14 and 15.

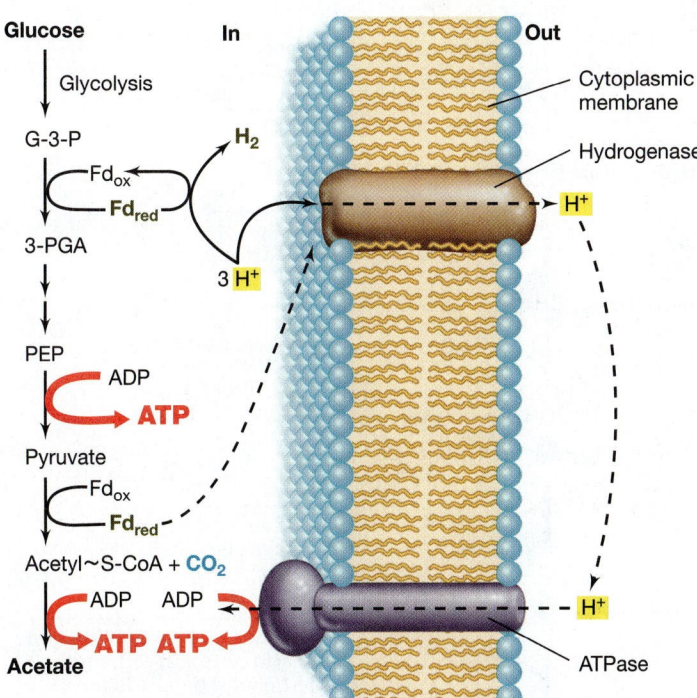

Figure 13.54 Modified glycolysis and proton reduction in anaerobic respiration in the hyperthermophile *Pyrococcus furiosus*. Hydrogen (H_2) production is linked to H^+ pumping by a hydrogenase that receives electrons from reduced ferredoxin (Fd_{red}). All intermediates from G-3-P downward in the pathway are present in two copies. Compare this figure with classical glycolysis in Figure 3.14. G-3-P, glyceraldehyde 3-phosphate; 3-PGA, 3-phosphoglycerate; PEP, phosphoenolpyruvate.

two energy-rich phosphate bonds; this compound is then converted to 3-phosphoglyceric acid plus ATP (Figure 3.14). However, in *P. furiosus* this normal glycolytic step is bypassed, and 3-phosphoglyceric acid is formed directly from glyceraldehyde 3-phosphate (**Figure 13.54**). This prevents *P. furiosus* from making ATP by substrate-level phosphorylation at this step, but this problem is compensated for by the fact that glyceraldehyde 3-phosphate oxidation is coupled to the production of *ferredoxin* rather than NADH; ferredoxin has a more negative E_0' (−0.42 V) than does NAD^+/NADH (−0.32 V). This highly negative E_0' allows for the coupling of ferredoxin oxidation to the reduction of 2 H^+ to H_2, and this reaction pumps a proton across the membrane (Figure 13.54). Proton pumping by hydrogenase is analogous to proton pumping by terminal electron carriers in other respirations (Figure 3.20 and Figure 13.41). Additional ATP is produced by *P. furiosus* by substrate-level phosphorylations in the conversions of phosphoenolpyruvate to pyruvate and acetyl-CoA to acetate (Figure 13.54).

MINIQUIZ

- With H_2 as electron donor, why is reduction of Fe^{3+} a more favorable reaction than reduction of fumarate?
- What is reductive dechlorination and why is it environmentally relevant?
- How does anaerobic glucose catabolism differ in *Lactobacillus* and *Pyrococcus furiosus*?

V • Hydrocarbon Metabolism

Hydrocarbons are widely used by microorganisms as electron donors but must first be oxygenated before they can be catabolized. Here we consider the aerobic catabolism of aliphatic and aromatic hydrocarbons, where oxygenation occurs from O_2. We then proceed to the special case of C_1 hydrocarbon catabolism and finish with a consideration of anoxic hydrocarbon metabolism, a situation where oxygenation of the hydrocarbon is still necessary, but where O_2 obviously plays no role.

13.22 Aerobic Hydrocarbon Metabolism

We previously discussed the role of molecular oxygen (O_2) as an *electron acceptor* in energy-generating reactions. By contrast, O_2 also plays an important role as a *reactant* in the catabolism of hydrocarbons, and oxygenase enzymes are key players in the process.

Oxygenases and Aliphatic Hydrocarbon Oxidation

Oxygenases are enzymes that catalyze the incorporation of O_2 into organic compounds and in some cases, inorganic compounds (Section 13.10). There are two classes of oxygenases: *dioxygenases*, which catalyze the incorporation of *both atoms* of O_2 into the molecule, and *monooxygenases*, which catalyze the incorporation of *only one* of

the two oxygen atoms of O_2 into an organic compound with the second atom of O_2 being reduced to H_2O. For most monooxygenases, the required electron donor is NADH or NADPH.

In the initial oxidation step of a saturated aliphatic hydrocarbon, one of the atoms of the O_2 is incorporated, typically at a terminal carbon atom. This reaction is catalyzed by a monooxygenase, and a typical reaction sequence is shown in **Figure 13.55a**. The end product of the reaction sequence is a fatty acid of the same length as the original hydrocarbon. The fatty acid is then oxidized by *beta-oxidation*, a series of reactions in which two carbons of the fatty acid are split off at a time (Figure 13.55b). During beta-oxidation, NADH is formed and is oxidized in the electron transport chain for energy conservation purposes. A single round of beta-oxidation releases acetyl-CoA plus a new fatty acid that is two carbon atoms shorter than the original fatty acid. The process of beta-oxidation is then repeated, and another acetyl-CoA molecule is released. The acetyl-CoA formed by beta-oxidation is either oxidized through the citric acid cycle (🔁 Figure 3.22) or used to make new cell material. With the exception of how the hydrocarbon is oxygenated, much of the biochemistry of anoxic hydrocarbon catabolism is the same as that shown for aerobic catabolism (Figure 13.55), with beta-oxidation reactions being of prime importance in both cases.

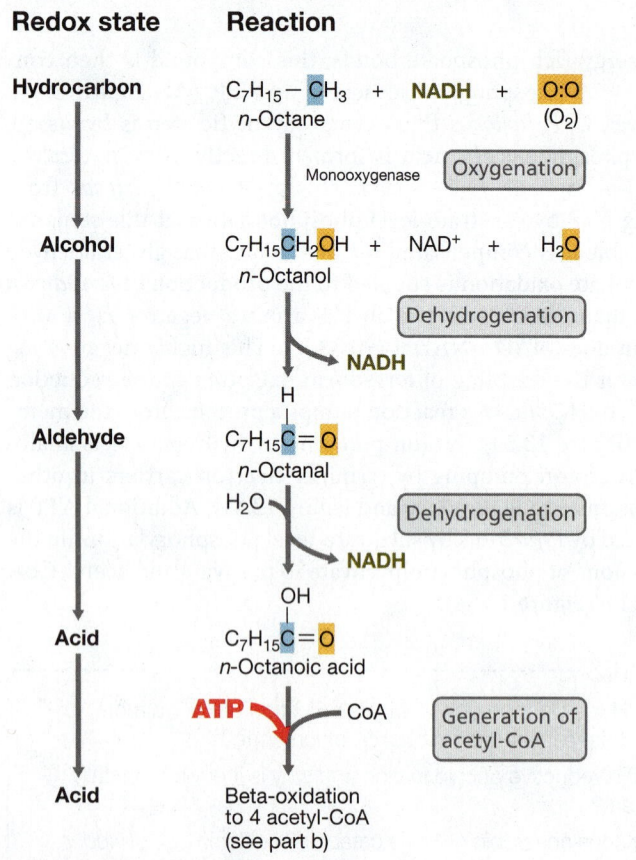

(a) Hydrocarbon oxidation

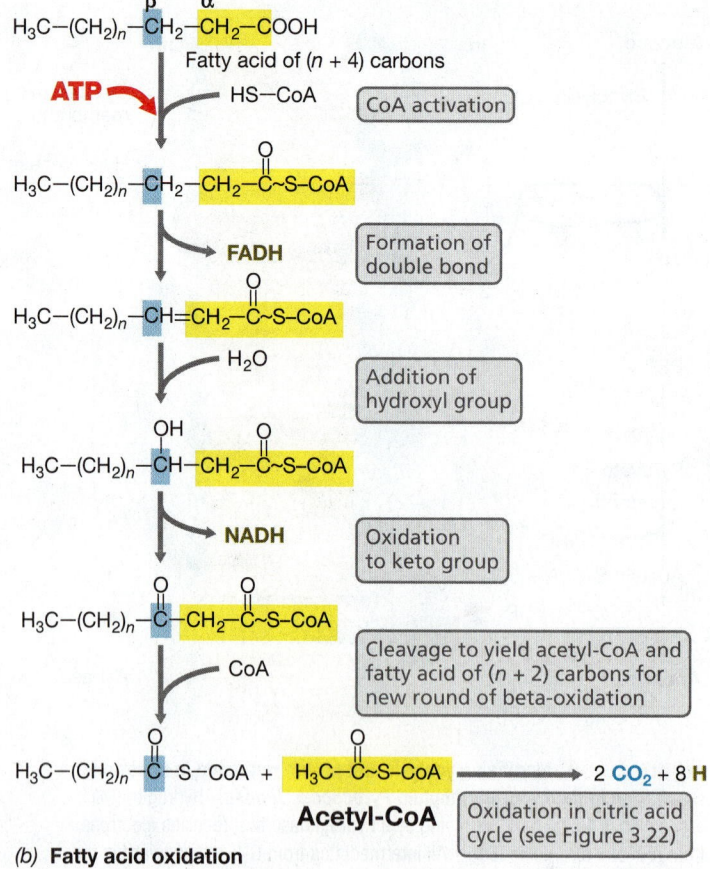

(b) Fatty acid oxidation

Figure 13.55 Monooxygenase activity and beta oxidation. *(a)* Steps in the oxidation of an aliphatic hydrocarbon, the first of which is catalyzed by a monooxygenase. *(b)* Fatty acid oxidation by beta-oxidation leads to the successive formation of acetyl-CoA.

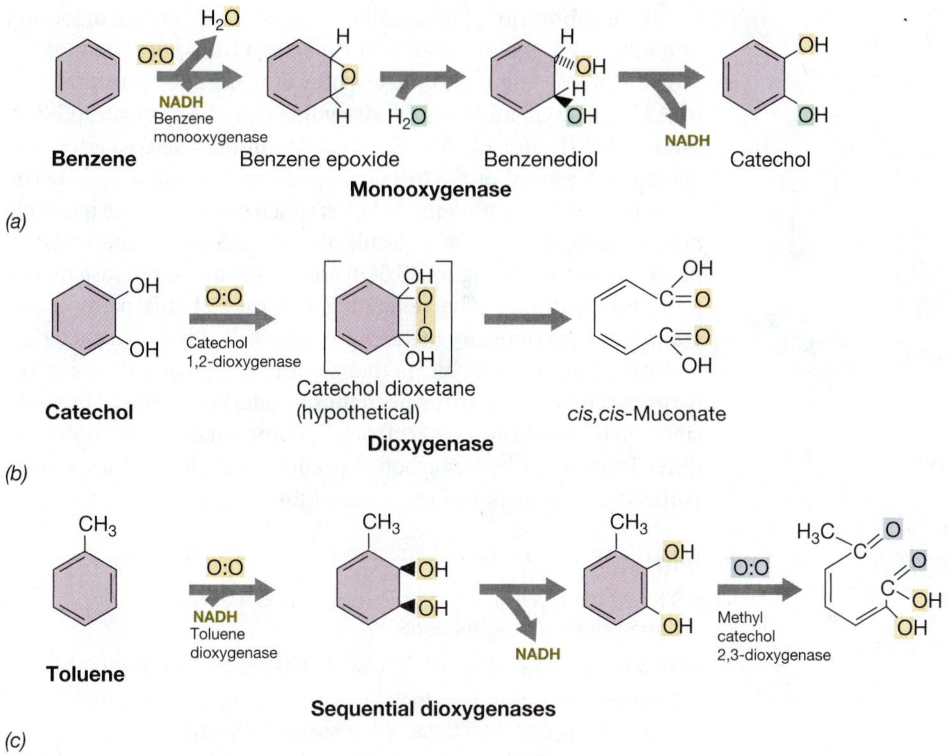

(a)

Monooxygenase

Benzene → Benzene epoxide → Benzenediol → Catechol

(b)

Dioxygenase

Catechol → Catechol dioxetane (hypothetical) → cis,cis-Muconate

(c)

Sequential dioxygenases

Toluene → → Methyl catechol →

Figure 13.56 Roles of oxygenases in catabolism of aromatic compounds. Monooxygenases introduce one atom of oxygen from O_2 into a substrate, whereas diooxygenases introduce both atoms of oxygen. *(a)* Hydroxylation of benzene to catechol by a monooxygenase in which NADH is an electron donor. *(b)* Cleavage of catechol to cis,cis-muconate by an intradiol ring-cleavage dioxygenase. *(c)* The activities of a ring-hydroxylating dioxygenase and an extradiol ring-cleavage dioxygenase in the degradation of toluene. The oxygen atoms that each enzyme introduces are distinguished by different colors. Compare aerobic toluene catabolism to anoxic toluene catabolism shown in Figure 13.59*b*.

Aromatic Hydrocarbon Oxidation

Many aromatic hydrocarbons can also be used as electron donors aerobically by microorganisms. The metabolism of these compounds, some of which contain multiple rings, such as naphthalene or biphenyls, typically has as its initial stage the formation of catechol or a structurally related compound via catalysis by oxygenase enzymes, as shown in **Figure 13.56**. Once catechol is formed it can be cleaved and further degraded into compounds that can enter the citric acid cycle, such as succinate, acetyl-CoA, and pyruvate.

Several steps in the aerobic catabolism of aromatic hydrocarbons require oxygenases. Figure 13.56*a–c* shows four different oxygenase-catalyzed reactions, one using a monooxygenase, two using a ring-cleaving dioxygenase, and one using a ring-hydroxylating dioxygenase. As in aerobic aliphatic hydrocarbon catabolism (Figure 13.55), aromatic compounds, whether single or multi-ringed, are typically oxidized completely to CO_2, with electrons entering an electron transport chain or used to make new cell material.

MINIQUIZ

- How do monooxygenases differ in function from dioxygenases?
- What is the final product of catabolism of a hydrocarbon?
- What is meant by the term "beta-oxidation"?

13.23 Aerobic Methanotrophy

Methane (CH_4) and many other C_1 compounds can be catabolized aerobically by **methylotrophs**. Methylotrophs are organisms that use organic compounds that lack C—C bonds as electron donors and carbon sources. The catabolism of compounds containing only a single carbon atom, such as the hydrocarbon methane (CH_4) and the alcohol methanol (CH_3OH), have been the best studied of these substrates. We focus here on the oxidation of CH_4 as an example of a methylotrophic lifestyle.

Methane Oxidation

The steps in CH_4 oxidation to CO_2 can be summarized as

$$CH_4 \rightarrow CH_3OH \rightarrow CH_2O \rightarrow HCOO^- \rightarrow CO_2$$

Not all methylotrophs can use methane. **Methanotrophs** are those methylotrophs that can use CH_4, and methanotrophy has been especially well studied in the gram-negative bacterium *Methylococcus capsulatus*. Methanotrophs assimilate either all or one-half of their cell carbon (depending on the pathway used) from the C_1 compound formaldehyde (CH_2O).

The initial step in the aerobic oxidation of CH_4 is catalyzed by the enzyme *methane monooxygenase* (MMO). Recall that *monooxygenases* incorporate one oxygen atom from O_2 into a carbon compound (Section 13.22 and Figure 13.55*a*). *M. capsulatus* contains two MMOs, one cytoplasmic and the other membrane-integrated; the latter has been best studied. In the MMO reaction, an atom of oxygen is introduced into CH_4, forming CH_3OH, and the second atom of O is reduced to form H_2O (**Figure 13.57**). CH_3OH is oxidized by an alcohol dehydrogenase, yielding formaldehyde (CH_2O) and NADH, and the CH_2O is either oxidized to CO_2 or used to make new cell material.

C_1 Assimilation into Cell Material

At least two distinct pathways exist for the incorporation of C_1 units into cell material in methanotrophs. The **serine pathway** is outlined in **Figure 13.58***a*. In this pathway, acetyl-CoA is synthesized from one molecule of CH_2O (produced from the oxidation of CH_3OH, Figure 13.57) and one molecule of CO_2. The serine pathway requires reducing power and energy in the form of two molecules each of NADH and ATP, respectively, for each acetyl-CoA synthesized. The serine pathway employs a number of enzymes of the citric acid cycle and one enzyme, *serine transhydroxymethylase*, unique to the pathway (Figure 13.58*a*).

An alternative pathway for C_1 incorporation is the **ribulose monophosphate pathway** (Figure 13.58*b*). This pathway is more energy efficient than the serine pathway because *all* of the carbon for cell material is derived from CH_2O. Because CH_2O is at the same oxidation level as cell material, no reducing power is needed

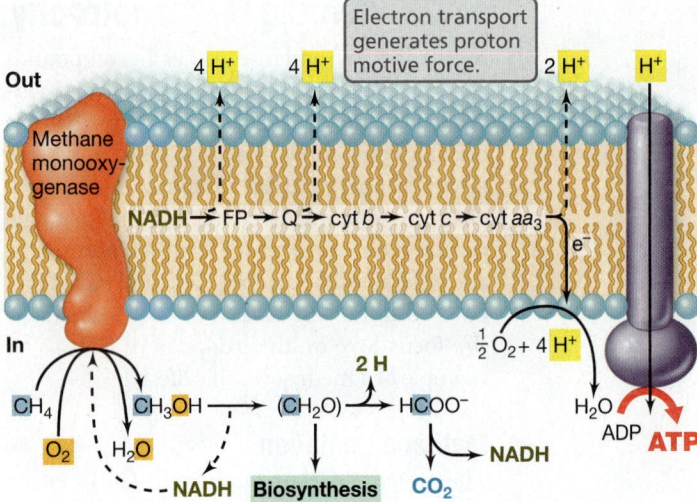

Figure 13.57 Oxidation of methane by methanotrophic bacteria. CH_4 is oxidized to CH_3OH by the membrane-integrated enzyme methane monooxygenase (MMO). A proton motive force is established from electron flow in the membrane, and this fuels ATPase. Note how carbon for biosynthesis comes from CH_2O. FP, flavoprotein; cyt, cytochrome; Q, quinone.

for its incorporation. Hence, all of the NADH from the oxidation of methane can be oxidized in the electron transport chain.

The ribulose monophosphate pathway consumes one molecule of ATP for each molecule of glyceraldehyde 3-phosphate (G-3-P) synthesized (Figure 13.58*b*); two G-3-Ps can then be converted into glucose by reversal of the glycolytic pathway (Figure 3.14). The enzymes *hexulosephosphate synthase*, which condenses one molecule of formaldehyde with one molecule of ribulose 5-phosphate, and *hexulose 6-P isomerase* (Figure 13.58*b*) are unique to the ribulose monophosphate pathway. The remaining enzymes of this pathway are enzymes of intermediary metabolism widely distributed in bacteria.

We thus see that aerobic methanotrophs share with other aerobic hydrocarbon degraders the need for oxygenase enzymes. However, since their biosyntheses start from C_1 compounds, methanotrophs differ from other hydrocarbon degraders in their need for special pathways for incorporating C_1 units into cell material.

MINIQUIZ -

- When using CH_4 as electron donor, why is *Methylococcus capsulatus* an obligate aerobe?
- Why does the oxidation of CH_4 to CH_3OH require reducing power?
- In which two ways does the ribulose monophosphate pathway save energy over reactions of the serine pathway?

- -

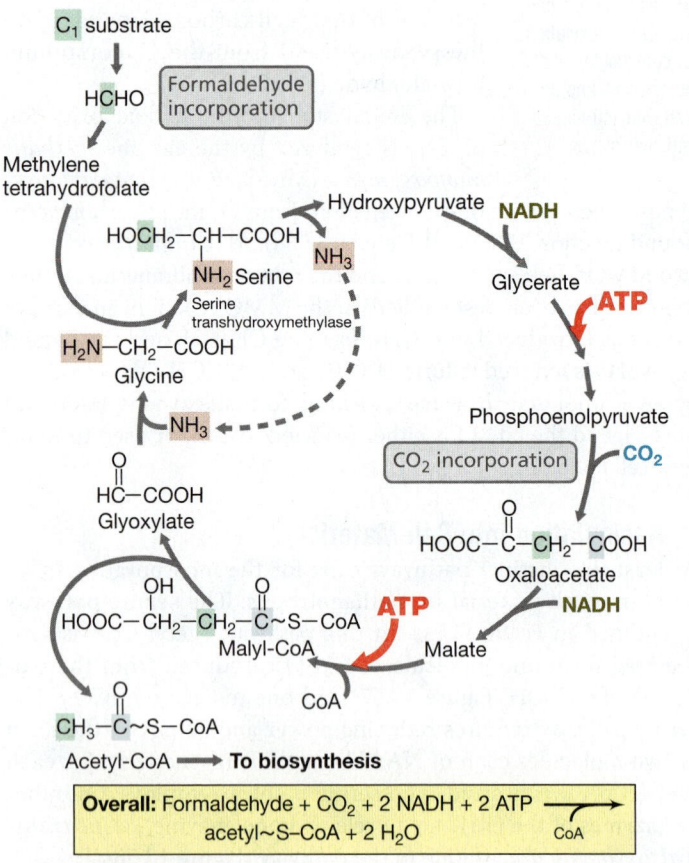

(a) The serine pathway

Overall: Formaldehyde + CO_2 + 2 NADH + 2 ATP $\longrightarrow$ acetyl~S–CoA + 2 H_2O

(b) The ribulose monophosphate pathway

Overall: 3 Formaldehyde + ATP $\longrightarrow$ glyceraldehyde-3-P

Figure 13.58 The serine and ribulose monophosphate pathways for the assimilation of C_1 units into cell material by methylotrophic bacteria. (a) Serine pathway. The product, acetyl-CoA, is used as the starting point for making new cell material. The key enzyme of the pathway is serine transhydroxymethylase. (b) Ribulose monophosphate pathway. Three molecules of CH_2O are required, with the product being glyceraldehyde 3-phosphate. The key enzyme of this pathway is hexulosephosphate synthase. The sugar rearrangements require enzymes of the pentose phosphate pathway (Figure 3.26).

13.24 Anoxic Hydrocarbon Metabolism

We saw how hydrocarbons can be catabolized *aerobically* in the preceding two sections. Now we consider their *anoxic* catabolism. Anoxic hydrocarbon oxidation can be linked to the reduction of nitrate, sulfate, or ferric iron in anaerobic respirations.

Aliphatic Hydrocarbons

Aliphatic hydrocarbons are straight-chain saturated or unsaturated organic compounds, and many are substrates for denitrifying and sulfate-reducing bacteria. Saturated aliphatic hydrocarbons as long as C_{20} support growth, although shorter-chain hydrocarbons are more soluble and readily catabolized. The mechanism of anoxic hydrocarbon degradation has been well studied for hexane (C_6H_{14}) metabolism in denitrifying bacteria (NO_3^- as electron acceptor). However,

the mechanism appears to be the same for the anoxic catabolism of longer-chain hydrocarbons and for anoxic hydrocarbon oxidation linked to other electron acceptors, and so we focus on the hexane/nitrate system here.

In anoxic hexane metabolism, hexane is modified on carbon atom 2 by attachment of a molecule of *fumarate*, a C_4 intermediate of the citric acid cycle ($\rightleftarrows$ Figure 3.22), forming the compound *1-methylpentylsuccinate* (**Figure 13.59a**). The enzymatic addition of fumarate to hexane effectively oxygenates the hexane and allows the molecule to be further catabolized anaerobically. Following the addition of coenzyme A, a series of reactions occurs that includes beta-oxidation (Figure 13.55b) and regeneration of fumarate. The electrons generated during beta-oxidation generate a proton motive force and are then consumed in nitrate or sulfate reduction.

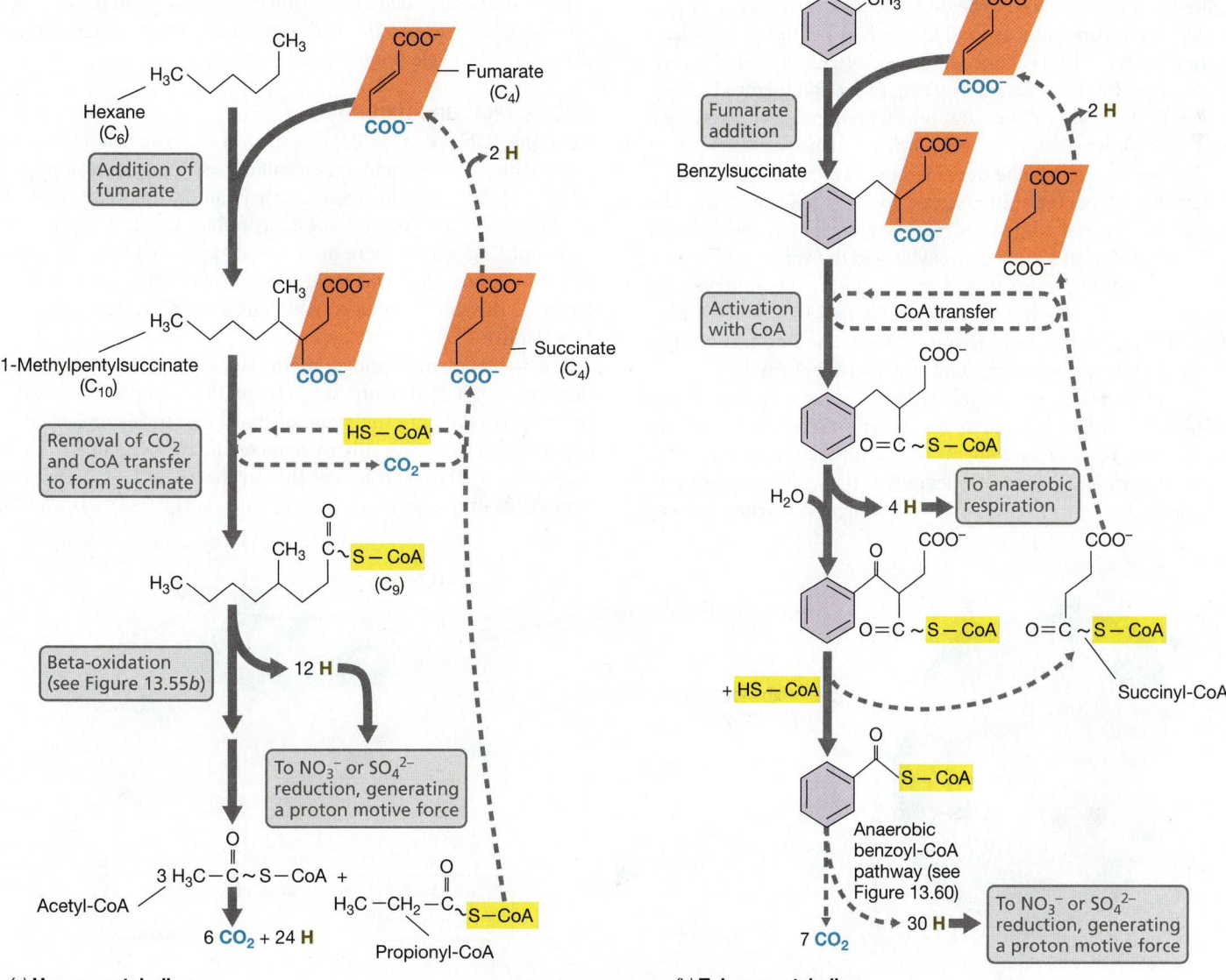

(a) Hexane catabolism

(b) Toluene catabolism

Figure 13.59 Anoxic catabolism of two hydrocarbons. *(a)* In anoxic catabolism of the aliphatic hydrocarbon hexane, the addition of fumarate provides the oxygen atoms necessary to form a fatty acid derivative that can be catabolized by beta-oxidation (see Figure 13.55) to yield acetyl-CoA. Electrons (H) generated from hexane catabolism are used to reduce sulfate or nitrate in anaerobic respirations. *(b)* Fumarate addition during the anoxic catabolism of the aromatic hydrocarbon toluene forms benzylsuccinate.

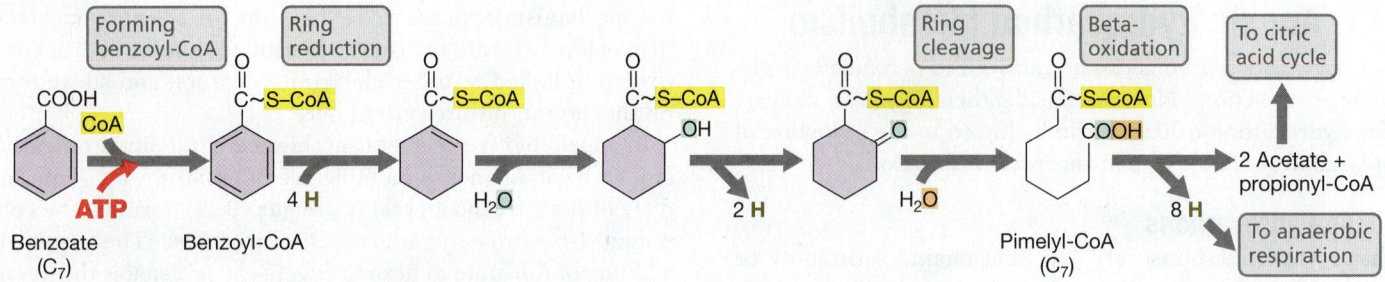

Figure 13.60 **Anoxic degradation of benzoate by the benzoyl-CoA pathway.** This pathway operates in the purple phototrophic bacterium *Rhodopseudomonas palustris* and many other facultative bacteria, both phototrophic and chemotrophic. Note that all intermediates of the pathway are bound to coenzyme A. The acetate produced is further catabolized in the citric acid cycle.

Aromatic Hydrocarbons

Aromatic hydrocarbons can be degraded anaerobically by some nitrate, ferric iron, and sulfate-reducing bacteria. For anoxic catabolism of the aromatic hydrocarbon toluene, oxygen needs to be added to the compound to begin catabolism, and this occurs by the addition of fumarate, as in aliphatic hydrocarbon catabolism (Figure 13.59a). The reaction series eventually yields benzoyl-CoA, which is further degraded by ring reduction (Figure 13.59b). Benzene (C_6H_6) can also be catabolized anaerobically, likely by a similar mechanism. Multi-ringed aromatic hydrocarbons such as naphthalene ($C_{10}H_8$) can be degraded by certain sulfate-reducing and denitrifying bacteria. In contrast to other hydrocarbons, the oxygenation of multi-ringed hydrocarbons occurs by the addition of CO_2 to the ring to form a carboxylic acid derivative rather than by fumarate addition. But this carboxylation reaction serves the same purpose as oxygenase reactions (Figures 13.55a and 13.56) or the addition of fumarate (Figure 13.59); an O atom becomes part of the hydrocarbon and facilitates its catabolism.

Many bacteria can catabolize certain aromatic hydrocarbons anaerobically, including even fermentative and phototrophic bacteria. However, except for toluene, only aromatic compounds that already contain an O atom are degraded, typically by a common mechanism. In contrast to aerobic catabolism that occurs by way of ring *oxidation* (Figure 13.56), anaerobic catabolism proceeds by ring *reduction*. Benzoate catabolism by the "benzoyl-CoA pathway" is a common theme of this biochemistry (**Figure 13.60**). Benzoate catabolism in this pathway begins by forming the coenzyme A derivative followed by ring cleavage to yield fatty or dicarboxylic acids that can be further catabolized to intermediates of the citric acid cycle (Figure 13.60).

Anoxic Methane Oxidation

Methane (CH_4) can be degraded under anoxic conditions by a consortium of two organisms, a sulfate-reducing bacterium plus a species of *Archaea* phylogenetically related to methanogens (**Figure 13.61**). The archaeal component, called ANME (*anoxic methanotroph*), of which there are several types, oxidizes CH_4 as an electron donor. Electrons from methane oxidation are then transferred to the sulfate reducer, which uses them to reduce SO_4^{2-} to H_2S (Figure 13.61b).

The mechanism of anoxic methane oxidation (AMO) remains unclear, but it is thought that the methanotroph oxidizes CH_4 to CO_2 by reversing the steps of methanogenesis and electrons are transferred to the sulfate reducer in some organic molecule (Figure 13.61b). Regardless of the mechanism, the coupled reaction yields only a small amount of free energy (−18 kJ), and how

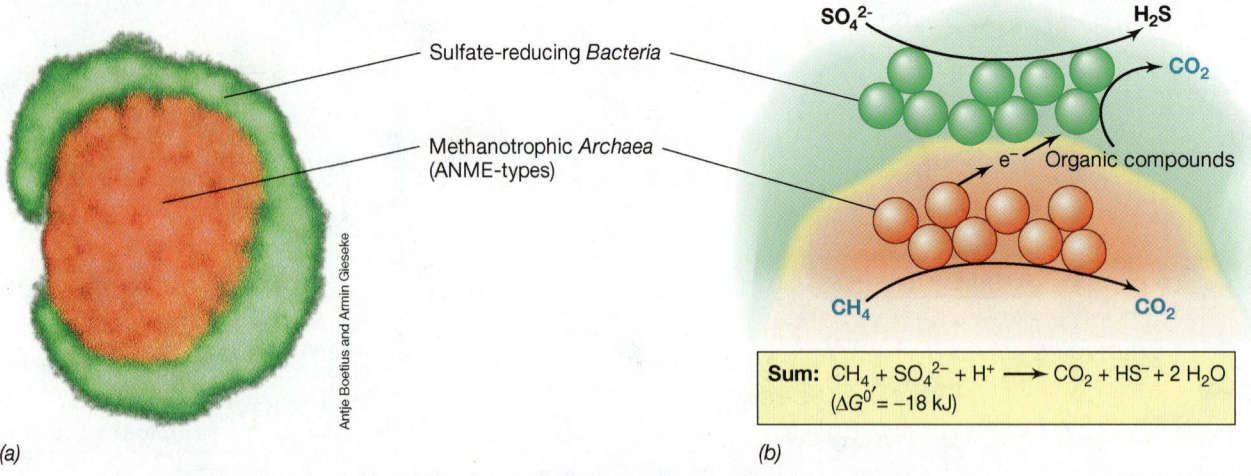

Figure 13.61 **Anoxic methane oxidation.** *(a)* Methane-oxidizing cell aggregates from marine sediments. The aggregates contain methanotrophic *Archaea* (red) surrounded by sulfate-reducing bacteria (green). Each cell type has been stained by a different FISH probe (⟲ Section 18.4). The aggregate is about 30 μm in diameter. *(b)* Mechanism for the cooperative degradation of CH_4. An organic compound or some other carrier of reducing power transfers electrons from methanotroph to sulfate reducer.

this energy is converted into ATP and split between the methanotroph and the sulfate reducer remains a major unanswered question. One possibility could be an ion pump. As we have seen in Sections 13.14 and 13.15, ion pumps can operate at very low energy yields, and thus an ion pump might play a role in the energetics of AMO as well.

AMO is not limited to sulfate-reducing bacteria consortia. Methane-oxidizing denitrifying consortia are active in anoxic environments where CH_4 and NO_3^- coexist in significant amounts, such as certain freshwater sediments. In laboratory enrichments of these consortia, some contain ANME-type methanotrophs while others are totally free of *Archaea*. AMO linked to ferric iron (Fe^{3+}) and manganic ion (Mn^{4+}) reduction are also known.

One denitrifying bacterium employs a remarkable mechanism for AMO and does not require a second organism to carry out the process. *Methylomirabilis oxyfera* oxidizes CH_4 with NO_3^- as an electron acceptor, and during CH_4 oxidation, electrons reduce NO_3^- employing most of the steps we have seen previously in denitrifying bacteria such as *Pseudomonas* (Section 13.17). These include the reduction of NO_3^- to NO_2^-, and further reduction to N_2 (Figure 13.41c). But unlike *Pseudomonas*,

in *M. oxyfera* NO_2^- is reduced to N_2 by way of nitric oxide (NO) without first producing nitrous oxide (N_2O) as an intermediate (Figure 13.41c). Instead, *M. oxyfera* splits NO into N_2 and O_2 ($2\ NO \rightarrow N_2 + O_2$) and then uses the O_2 produced as the electron acceptor for CH_4 oxidation. In other words, the organism *produces its own O_2* as an oxidant for electrons generated during the oxidation of CH_4 to CO_2 (see more on page 433).

Several metabolic strategies have evolved to catabolize methane, probably the most abundant hydrocarbon on Earth. These along with the myriad of other energy-conserving mechanisms we have seen in this chapter demonstrate the impressive breadth of microbial metabolic diversity. We now take this background and put it into the context of the organisms themselves in the next four chapters.

MINIQUIZ
- Why is toluene a hydrocarbon whereas benzoate is not?
- How is hexane oxygenated during anoxic catabolism?
- What is AMO and which organisms participate in the process?

BIG IDEAS

13.1 • In photosynthesis, ATP is generated from light and then consumed in the reduction of CO_2. Two forms of photosynthesis are known: oxygenic, where O_2 is produced (for example, cyanobacteria), and anoxygenic, where it is not (for example, purple and green bacteria). Chlorophylls and bacteriochlorophylls reside in membranes where the light reactions of photosynthesis are carried out. Antenna chlorophylls harvest light energy and transfer it to a reaction center.

13.2 • Accessory pigments including carotenoids and phycobilins absorb light and transfer the energy to reaction center chlorophyll, thus broadening the wavelengths of light usable in photosynthesis. Carotenoids also play an important photoprotective role in preventing photooxidative damage to cells.

13.3 • Electron transport reactions occur in the photosynthetic reaction center of anoxygenic phototrophs, forming a proton motive force and ATP. Reducing power for CO_2 fixation comes from substances such as H_2S, and NADH production in purple bacteria requires reverse electron transport.

13.4 • In oxygenic photosynthesis, H_2O donates electrons to drive CO_2 fixation, and O_2 is a by-product. There are two separate but interconnected photosystems in oxygenic phototrophs, PSI and PSII, whereas anoxygenic phototrophs contain a single photosystem.

13.5 • Autotrophy is supported in most phototrophic and chemolithotrophic bacteria by the Calvin cycle, in which the enzyme RubisCO plays a key role. Carboxysomes contain crystalline RubisCO and function to concentrate CO_2, the key substrate for this enzyme. The reverse citric acid and hydroxypropionate cycles are autotrophic pathways in green sulfur and green nonsulfur bacteria, respectively.

13.6 • Chemolithotrophs oxidize inorganic electron donors to conserve energy and obtain reducing power. Energy conservation occurs from respiratory processes that generate a proton motive force. Most chemolithotrophs can grow autotrophically.

13.7 • The chemolithotrophic hydrogen bacteria use H_2 as an electron donor, reducing O_2 to H_2O. The enzyme hydrogenase is required to oxidize H_2, and H_2 also supplies reducing power for the fixation of CO_2 in these autotrophs.

13.8 • Reduced sulfur compounds such as H_2S, $S_2O_3^{2-}$, and S^0 are electron donors for energy conservation in sulfur chemolithotrophs. Electrons from these substances enter electron transport chains, yielding a proton motive force. Sulfur chemolithotrophs are also autotrophs and fix CO_2 by the Calvin cycle.

13.9 • Chemolithotrophic iron bacteria oxidize Fe^{2+} as an electron donor. Most iron bacteria grow at acidic pH and are often associated with acidic pollution from mineral and coal

UNIT 3

mining. A few chemolithotrophic and phototrophic bacteria can oxidize Fe^{2+} to Fe^{3+} anaerobically.

13.10 • The ammonia-oxidizing *Bacteria* and *Archaea* produce nitrite from ammonia, which is then oxidized by nitrite-oxidizing *Bacteria* to nitrate. Anoxic ammonia oxidation (anammox) consumes both ammonia and nitrite, forming N_2. The anammox reaction occurs within a membrane-enclosed compartment called the anammoxosome.

13.11 • In the absence of external electron acceptors, organic compounds can be catabolized anaerobically only by fermentation. Most fermentations require that an energy-rich organic compound be formed that can yield ATP by substrate-level phosphorylation. Redox balance is achieved by the production of fermentation products.

13.12 • The lactic acid fermentation is carried out by homofermentative species, where lactate is the sole product, and heterofermentative species, where lactate, ethanol, and CO_2 are produced. The mixed-acid fermentation typical of enteric bacteria yields various acids plus neutral products (ethanol, butanediol), depending on the organism.

13.13 • Clostridia ferment sugars, amino acids, and other organic compounds, with butyric acid being a major product. Butyrate production allows for an additional ATP to be produced. *Propionibacterium* produces propionate, acetate, and CO_2 in a secondary fermentation of lactate where energy conservation by both substrate-level phosphorylation and anaerobic respiration occurs.

13.14 • Energy conservation in *Propionigenium*, *Oxalobacter*, and *Malonomonas* is linked to decarboxylation reactions that pump Na^+ or H^+ across the membrane; ATPases use the energy in the ion gradient to form ATP. The reactions catalyzed by these organisms yield insufficient free energy to make ATP by substrate-level phosphorylation.

13.15 • In syntrophy two organisms cooperate to degrade a compound that neither can degrade alone. In this process H_2 produced by one organism is consumed by the partner. H_2 consumption affects the energetics of the reaction carried out by the H_2 producer, allowing it to make ATP where it otherwise could not.

13.16 • Although O_2 is the most widely used electron acceptor in energy-yielding metabolism, certain other inorganic and organic compounds can be used as electron acceptors. Anaerobic respiration yields less energy than aerobic respiration but can proceed in environments where O_2 is absent.

13.17 • Nitrate is a common electron acceptor in anaerobic respiration. Nitrate reduction is catalyzed by the enzyme nitrate reductase, reducing NO_3^- to NO_2^-. Many bacteria that use NO_3^- in anaerobic respiration produce gaseous nitrogen compounds (NO, N_2O, or N_2) as final end products of reduction (denitrification).

13.18 • Sulfate-reducing bacteria are obligately anaerobic bacteria that reduce SO_4^{2-} to H_2S in a process in which SO_4^{2-} must first be activated to adenosine phosphosulfate (APS). Disproportionation is an additional energy-yielding strategy for certain species. Some organisms, such as *Desulfuromonas*, cannot reduce SO_4^{2-} but can reduce S^0 to H_2S.

13.19 • Acetogens are strict anaerobes that reduce CO_2 to acetate, usually with H_2 as electron donor. The mechanism of acetate formation is the acetyl-CoA pathway, a pathway widely distributed in obligate anaerobes for either autotrophic purposes or acetate oxidation.

13.20 • Methanogenesis is the production of CH_4 from $CO_2 + H_2$ or from acetate or methanol by strictly anaerobic methanogenic *Archaea*. Several unique coenzymes are required for methanogenesis, and energy conservation is linked to either a proton motive or sodium motive force.

13.21 • Besides inorganic nitrogen and sulfur compounds and CO_2, several other substances can function as electron acceptors for anaerobic respiration. These include Fe^{3+}, Mn^{4+}, fumarate, certain organic and chlorinated organic compounds, and even protons.

13.22 • In addition to its role as a terminal electron acceptor, O_2 can also be a substrate. In aerobic metabolism, oxygenases introduce atoms of oxygen from O_2 into hydrocarbons. Once oxygenated, aliphatic hydrocarbons can be further degraded by beta-oxidation and aromatic hydrocarbons by ring splitting and oxidation.

13.23 • Methanotrophy is the use of CH_4 as both carbon source and electron donor, and the enzyme methane monooxygenase is a key enzyme in the aerobic catabolism of methane. In methanotrophs, C_1 units are assimilated into cell material as formaldehyde or formaldehyde plus CO_2 by the ribulose monophosphate or serine pathways, respectively.

13.24 • Hydrocarbons can be oxidized under anoxic conditions following addition of the dicarboxylic acid fumarate. Aromatic compounds are catabolized anaerobically by ring reduction and cleavage to form intermediates that can be catabolized in the citric acid cycle. Methane can be oxidized under anoxic conditions by consortia containing sulfate-reducing or denitrifying *Bacteria* and methanotrophic *Archaea*.

REVIEW OF KEY TERMS

Acetogenesis energy metabolism in which acetate is produced from either H_2 plus CO_2 or from organic compounds

Acetyl-CoA pathway a pathway of autotrophic CO_2 fixation and acetate oxidation widespread in obligate anaerobes including methanogens, acetogens, and sulfate-reducing bacteria

Anaerobic respiration use of an electron acceptor other than O_2 in an electron transport–based oxidation leading to a proton motive force

Anammox anoxic ammonia oxidation

Anoxygenic photosynthesis photosynthesis in which O_2 is not produced

Antenna pigments light-harvesting chlorophylls or bacteriochlorophylls in photocomplexes that funnel energy to the reaction center

Autotroph an organism that uses CO_2 as its sole carbon source

Bacteriochlorophyll the chlorophyll pigment of anoxygenic phototrophs

Calvin cycle the biochemical pathway for CO_2 fixation in many autotrophic organisms

Carboxysomes crystalline inclusions of RubisCO

Carotenoid a hydrophobic accessory pigment present along with chlorophyll in photosynthetic membranes

Chemolithotroph a microorganism that oxidizes inorganic compounds as electron donors in energy metabolism

Chlorophyll a light-sensitive, Mg-containing porphyrin of phototrophic organisms that initiates the process of photophosphorylation

Chlorosome a cigar-shaped structure present in the periphery of cells of green sulfur and green nonsulfur bacteria and containing the antenna bacteriochlorophylls (*c*, *d*, or *e*)

Denitrification anaerobic respiration in which NO_3^- or NO_2^- is reduced to nitrogen gases, primarily N_2

Fermentation anaerobic catabolism of an organic compound in which the compound serves as both an electron donor and an electron acceptor and in which ATP is usually produced by substrate-level phosphorylation

Heterofermentative producing a mixture of products, typically lactate, ethanol, and CO_2, from the fermentation of glucose

Homofermentative producing only lactic acid from the fermentation of glucose

Hydrogenase an enzyme, widely distributed in anaerobic microorganisms, capable of oxidizing or evolving H_2

Hydroxypropionate pathway an autotrophic pathway found in *Chloroflexus* and a few *Archaea*

Methanogen a methane-producing member of the *Archaea*

Methanogenesis the biological production of CH_4

Methanotroph an organism that oxidizes CH_4

Methylotroph an organism capable of growth on compounds containing no C—C bonds; some methylotrophs are methanotrophic

Mixotroph an organism in which an inorganic compound serves as the electron donor in energy metabolism and organic compounds serve as the carbon source

Nitrification the microbial oxidation of ammonia to nitrate

Oxygenase an enzyme that catalyzes the incorporation of oxygen from O_2 into organic or inorganic compounds

Oxygenic photosynthesis photosynthesis carried out by cyanobacteria and green plants in which O_2 is evolved

Photophosphorylation the production of ATP in photosynthesis

Photosynthesis the series of reactions in which ATP is synthesized by light-driven reactions and CO_2 is fixed into cell material

Phototroph an organism that uses light as an energy source

Phycobiliprotein the antenna pigment complex in cyanobacteria that contains phycocyanin and allophycocyanin or phycoerythrin coupled to proteins

Phycobilisome an aggregate of phycobiliproteins

Reaction center a photosynthetic complex containing chlorophyll or bacteriochlorophyll and several other components, within which occur the initial electron transfer reactions of photosynthetic electron flow

Reductive dechlorination an anaerobic respiration in which a chlorinated organic compound is used as an electron acceptor, usually with the release of Cl^-

Reverse citric acid cycle a mechanism for autotrophy in green sulfur bacteria and a few other autotrophic *Bacteria*, and also in some *Archaea*

Reverse electron transport the energy-dependent movement of electrons against the thermodynamic gradient to form a strong reductant from a weaker electron donor

Ribulose monophosphate pathway a reaction series in certain methylotrophs in which formaldehyde is assimilated into cell material using ribulose monophosphate as the C_1 acceptor molecule

RubisCO the acronym for ribulose bisphosphate carboxylase, a key enzyme of the Calvin cycle

Secondary fermentation a fermentation in which the substrates are the fermentation products of other organisms

Serine pathway a reaction series in certain methylotrophs in which CH_2O plus CO_2 are assimilated into cell material by way of the amino acid serine

Stickland reaction the fermentation of an amino acid pair

Syntrophy a process whereby two or more microorganisms cooperate to degrade a substance neither can degrade alone

Thylakoids membrane stacks in cyanobacteria or in the chloroplast of eukaryotic phototrophs

REVIEW QUESTIONS

1. What are the major differences between oxygenic and anoxygenic phototrophs? (Section 13.1)

2. What are the functions of light-harvesting and reaction center chlorophylls? Why would a mutant incapable of making light-harvesting chlorophylls (such mutants can be readily isolated in the laboratory) probably not be a successful competitor in nature? (Section 13.1)

3. What accessory pigments are present in phototrophs, and what are their functions? (Section 13.2)

4. How does light result in ATP production in an anoxygenic phototroph? In what ways are photosynthetic and respiratory electron flow similar? In what ways do they differ? (Section 13.3)

5. How is reducing power for autotrophic growth obtained in a purple bacterium? In a cyanobacterium? (Sections 13.3 and 13.4)

6. How does the reduction potential (E_0') of chlorophyll a in PSI and PSII differ? Why must the reduction potential of PSII chlorophyll a be so highly electropositive? (Section 13.4)

7. What two enzymes are unique to the Calvin cycle? What reactions do these enzymes carry out? What would be the consequences if a mutant arose that lacked either of these enzymes? (Section 13.5)

8. Which organisms employ the hydroxypropionate or reverse citric acid cycles as autotrophic pathways? (Section 13.5)

9. Compare and contrast the utilization of H_2S by a purple phototrophic bacterium and by a colorless sulfur bacterium such as *Beggiatoa*. What role does H_2S play in the metabolism of each organism? (Sections 13.3, 13.6, and 13.8)

10. Which inorganic electron donors are used by the organisms *Ralstonia*, *Thiobacillus*, and *Acidithiobacillus*? (Sections 13.7–13.9)

11. Contrast classical nitrification with anammox in terms of oxygen requirements, organisms involved, and the need for monooxygenases. (Section 13.10)

12. Define the term substrate-level phosphorylation. How does it differ from oxidative phosphorylation? What compound(s) do fermentative bacteria need to synthesize in order to make ATP by substrate-level phosphorylation? (Section 13.11)

13. What are the major fermentation products of the following bacteria: *Lactobacillus*, *Clostridium*, *Propionibacterium*, *Escherichia*? (Sections 13.12 and 13.13)

14. Give an example of a fermentation that does not employ substrate-level phosphorylation. How is energy conserved in this fermentation? (Section 13.14)

15. Why is syntrophy also called "interspecies H_2 transfer"? (Section 13.15)

16. Why is NO_3^- a better electron acceptor for anaerobic respiration than is SO_4^{2-}? (Section 13.16)

17. What is the difference in nitrate respiration by *Escherichia coli* and *Pseudomonas*? (Section 13.17)

18. Why is the enzyme hydrogenase useful to *Desulfovibrio* even when it is not grown on H_2 as electron donor? (Section 13.18)

19. Compare and contrast acetogens with methanogens in terms of (1) substrates and products of their energy metabolism, (2) ability to use organic compounds as electron donors in energy metabolism, and (3) phylogeny. (Sections 13.19 and 13.20)

20. Compare and contrast ferric iron reduction with reductive dechlorination in terms of (1) product of the reduction and (2) environmental significance. (Section 13.21)

21. How do monooxygenases differ from dioxygenases in terms of the reactions they catalyze? Why are oxygenases necessary for the aerobic catabolism of hydrocarbons? (Section 13.22)

22. How does a methano*troph* differ from a methano*gen*? Which pathway for C_1 assimilation found in methanotrophs is most energetically efficient and why? (Section 13.23)

23. How do denitrifying and sulfate-reducing bacteria degrade hydrocarbons anaerobically and without oxygenases? (Section 13.24)

APPLICATION QUESTIONS

1. Where are the photosynthetic pigments located in a phototrophic purple bacterium? A cyanobacterium? A green alga? Considering the function of chlorophyll pigments, why are they not located elsewhere in the cell, for example, in the cytoplasm or in the cell wall? (Section 13.1)

2. The growth rate of the phototrophic purple bacterium *Rhodobacter* is about twice as fast when the organism is grown *phototrophically* in a medium containing malate as the carbon source as when it is grown with CO_2 as the carbon source (with H_2 as the electron donor). Discuss the reasons why this is true, and list the nutritional class in which we would place *Rhodobacter* when growing under each of the two different conditions.

3. Although physiologically distinct, chemolithotrophs and chemoorganotrophs share a number of features with respect to the production of ATP. Discuss these common features along with reasons why the growth yield (grams of cells per mole of substrate consumed) of a chemoorganotroph respiring glucose is so much higher than for a chemolithotroph respiring sulfur.

4. A fatty acid such as butyrate cannot be fermented in pure culture, although its anaerobic catabolism under other conditions occurs readily. How do these conditions differ, and why does the latter allow for butyrate catabolism? How then can butyrate be fermented in mixed culture?

5. When methane is made from CO_2 (plus H_2) or from methanol (in the absence of H_2), various steps in the metabolic pathways are shared in common. Compare and contrast methanogenesis from these two substrates, highlighting the similarities and differences in the processes.

14 · Functional Diversity of *Bacteria*

microbiology**now**

One Culture Away

Microbiologists continue to discover novel microbial processes that impact the biosphere, and the newly discovered bacterium *Methylomirabilis oxyfera* is one such example. Biological methane (CH_4) oxidation was once thought to require oxygen (O_2). Recently, however, methane oxidation has been observed in a variety of anoxic environments. Methane is a powerful greenhouse gas, and microbial consumption of methane has a major role in balancing the global methane cycle. The model for anaerobic methane oxidation was previously based on a known metabolic partnership between sulfate-reducing bacteria and methanogens able to reverse the pathway of methane production. The discovery of *M. oxyfera,* however, proves the adage that where there is energy, microorganisms will find a way to use it.

M. oxyfera emerged from an anaerobic enrichment culture begun with anoxic sediment from a canal in the Netherlands (photo). The canal received runoff from agricultural fields and contained both nitrate and methane. The enrichment culture coupled the oxidation of methane to denitrification, a process that had never before been seen. Surprisingly, however, *M. oxyfera,* though growing anaerobically, was using the standard aerobic pathway of methane oxidation! How is this possible?

M. oxyfera was found to use a novel pathway of denitrification in which two molecules of nitric oxide (NO) are used to make N_2 and O_2; the O_2 is then used immediately as the electron acceptor for the oxidation of methane.[1] Thus, O_2 production is central to the methane metabolism of *M. oxyfera* even though the bacterium inhabits anoxic environments. *M. oxyfera* also has a unique polyhedral morphology (photo inset) and belongs to a novel phylum of *Bacteria,* NC-10, from which no previous species had been isolated. The discovery of *M. oxyfera* shows that major discoveries in microbiology are often just "one culture away."

[1] Ettwig, K.F., et al. 2010. Nitrite-driven anaerobic methane oxidation by oxygenic bacteria. *Nature 464:* 543–550.

The microbial world is one of immense diversity, in both form and function. Microorganisms have been evolving for more than 3.8 billion years and have diversified to fill every available habitat on Earth. In Chapter 12 we considered the evolution of microbial life and the phylogenetic tools that have revealed it. In Chapter 13 we examined the enormous metabolic diversity of microorganisms.

In this and the next three chapters we focus on microbial diversity itself, including that of *Bacteria*, *Archaea*, and *Eukarya*.

In the present chapter we consider the *functional diversity* of particular groups of *Bacteria*. In particular we focus on organisms that share distinct physiological or ecological characteristics that do not necessarily affiliate with a single coherent phylogenetic group.

I • Functional Diversity as a Concept

Microbial diversity can be understood in terms of both phylogenetic diversity and functional diversity. In Section 14.1 we define and contrast the concepts of phylogenetic diversity and functional diversity.

14.1 Making Sense of Microbial Diversity

Phylogenetic diversity is the component of microbial diversity that deals with evolutionary relationships between microorganisms. Most fundamentally, phylogenetic diversity deals with the diversity of evolutionary lineages such as phyla, genera, and species. At its broadest, phylogenetic diversity encompasses the genetic and genomic diversity of evolutionary lineages and so can be defined on the basis of either genes or organisms (⟳ Section 12.5). Most commonly, though, phylogenetic diversity is defined on the basis of ribosomal RNA gene phylogeny, which is thought to reflect the phylogenetic history of the entire organism (⟳ Section 12.4). Phylogenetic diversity is the overarching theme of our coverage of microbial diversity in Chapters 15–17.

Functional diversity is the component of microbial diversity that deals with diversity in form and function as it relates to microbial physiology and ecology. It is useful to consider

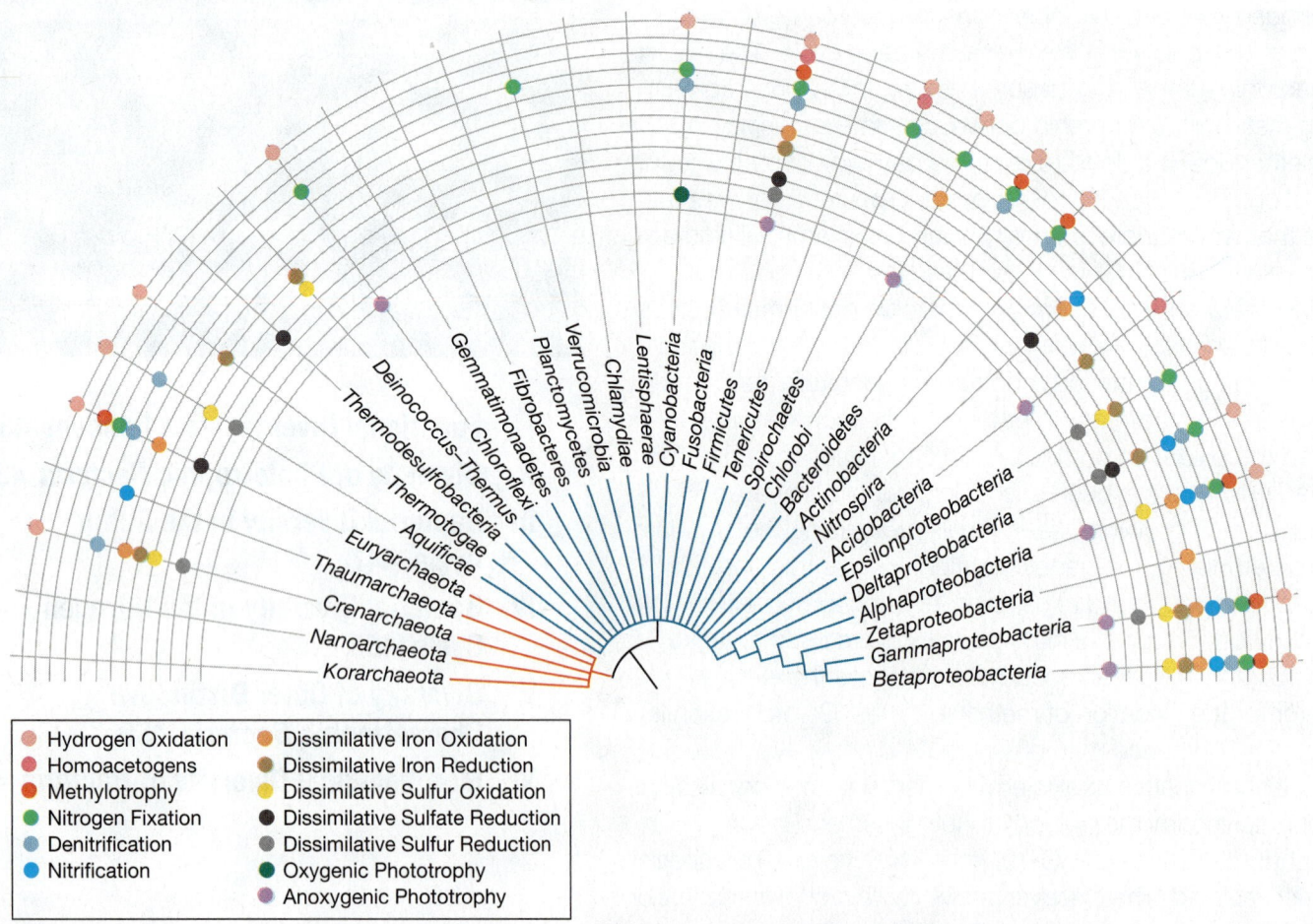

Figure 14.1 Major functional traits mapped across major phyla of *Bacteria* and *Archaea*. The dendrogram shows relationships between microbial phyla as inferred by analysis of 16S ribosomal RNA gene sequences. Blue branches are used to denote phyla of *Bacteria* and red branches phyla of *Archaea*. Colored circles indicate phyla that contain at least one species with a functional trait indicated in the color key.

microbial diversity in terms of functional groupings because organisms with common traits and common genes often share physiological characteristics and have similar ecological roles. In many cases, functional traits align with phylogenetic groups (for example, in Sections 14.3, 14.4, 14.6, 14.7, 14.20). Microbial functional diversity, however, often does not correspond with phylogenetic diversity as defined by the 16S ribosomal RNA gene. We will see many examples in this chapter where functional traits are widely distributed among the *Bacteria* and *Archaea* (**Figure 14.1**).

At least three reasons can be offered for why a functional trait is shared between divergent organisms with dissimilar 16S ribosomal RNA gene sequences. The first is *gene loss*, a situation where a trait present in the common ancestor of several lineages is subsequently lost in some lineages but retained in others which over evolutionary time became quite divergent. The second is **convergent evolution**, in which a trait has evolved independently in two or more lineages and is not encoded by homologous genes shared by these lineages. The third is **horizontal gene transfer** (⮌ Sections 6.12 and 12.5), a situation where genes that confer a particular trait are homologous and have been exchanged between distantly related lineages.

Functional diversity can be further defined in terms of physiological diversity, ecological diversity, and morphological diversity. *Physiological diversity* relates to the functions and activities of microorganisms. Physiological diversity is most commonly described in terms of microbial metabolism and cellular biochemistry (Chapter 13). *Ecological diversity* relates to relationships between organisms and their environments. Organisms with similar physiological characteristics can have different ecological strategies (Section 14.11). Causes and consequences of ecological diversity will also be considered when we consider the science of microbial ecology in Chapter 19. *Morphological diversity* relates to the outward appearance of an organism (Sections 14.20–14.24). In some cases, the morphology of a group is so distinctive that the group is essentially defined by this property, for example, with the spirochetes (Section 14.22).

The concepts of physiological, ecological, and morphological diversity are often intertwined. The examples provided in this chapter are meant to be illustrative and not exhaustive, and we will consider other organisms with important ecological functions in Chapters 15–17 and 19–22.

MINIQUIZ

- Why is it necessary to consider microbial diversity in terms of phylogenetic diversity and functional diversity?
- What are three reasons that functional traits might not correspond with distinct phylogenetic groups as defined by 16S ribosomal RNA gene sequences?

II • Diversity of Phototrophic *Bacteria*

In this section we consider the diversity of phototrophic microorganisms, those microorganisms that conserve energy from light. We will see that phototrophy is widespread within the domain *Bacteria* and that several distinct types of phototrophs can be defined on the basis of their physiological traits.

14.2 Overview of Phototrophic Bacteria

The ability to conserve energy from light evolved early in the history of life, when the Earth was anoxic (⮌ Section 12.2). Photosynthesis originated within the *Bacteria*, and the first phototrophic organisms were *anoxygenic phototrophs*, organisms that do not generate O_2 as a product of photosynthesis (⮌ Section 13.3). Instead of H_2O, these early phototrophs likely used H_2, ferrous iron, or H_2S as the electron donor for photosynthesis. Anoxygenic photosynthesis is present in five bacterial phyla: the *Proteobacteria, Chlorobi, Chloroflexi, Firmicutes,* and *Acidobacteria*. Oxygenic photosynthesis is known only within the *Cyanobacteria* (Figure 14.1). There is extensive metabolic diversity among the anoxygenic phototrophs, which are found in a wide range of habitats. It is clear that horizontal gene exchange has had a major impact on the evolution of photosynthesis and on the distribution of photosynthetic genes across the phylogenetic tree of *Bacteria*.

Phototrophic bacteria have several common features. All phototrophic bacteria use chlorophyll-like pigments and various accessory pigments to harvest energy from light and transfer this energy to a membrane-bound reaction center where it is used to drive electron transfer reactions that ultimately result in the production of ATP (⮌ Sections 13.1–13.4). Anoxygenic phototrophic bacteria have either a type I or type II photosystem. The terms "type I" and "type II" refer to the structure of the photosynthetic reaction center. Type I photosystems are most similar to photosystem I of oxygenic phototrophs while type II photosystems are most similar to photosystem II of oxygenic phototrophs. Both types of photosystems are present in cyanobacteria (⮌ Section 13.4), whereas only one type or the other is present in anoxygenic phototrophs. In some cases photosynthetic pigments are found in the cytoplasmic membrane, but often they are present in intracellular photosynthetic membrane systems that originate from invaginations of the cytoplasmic membrane. These internal membranes allow phototrophic bacteria to increase the amount of pigment they contain for better use of light of low intensities.

Many phototrophic bacteria couple light energy to carbon fixation through a variety of different mechanisms (⮌ Section 13.5), but not all phototrophs fix CO_2; some instead either prefer or require organic sources of carbon to support growth. We will see that many of the characteristics of phototrophic bacteria, including their membrane systems and photosynthetic pigments, have evolved as a result of niche adaptation for the light environment.

MINIQUIZ

- What form of photosynthesis was most likely the first to appear on Earth?

14.3 *Cyanobacteria*

Key Genera: *Prochlorococcus, Crocosphaera, Synechococcus, Trichodesmium, Oscillatoria, Anabaena*

Cyanobacteria comprise a large, morphologically and ecologically heterogeneous group of oxygenic, phototrophic *Bacteria*. As we saw in Section 12.2, these organisms were the first oxygen-evolving phototrophic organisms on Earth, and over billions of years converted the once anoxic atmosphere of Earth to the oxygenated atmosphere we see today.

Phylogeny and Classification of *Cyanobacteria*

The morphological diversity of the *Cyanobacteria* is impressive (**Figure 14.2**). Both unicellular and filamentous forms are known, and there is considerable variation within these morphological types. Cyanobacterial cells range in size from 0.5 μm in diameter to cells as large as 100 μm in diameter. *Cyanobacteria* can be divided into five morphological groups: (1) *Chroococcales* are unicellular, dividing by binary fission (Figure 14.2*a*); (2) *Pleurocapsales* are unicellular, dividing by multiple fission (colonial) (Figure 14.2*b*); (3) *Oscillatoriales* are filamentous nonheterocystous forms (Figure 14.2*c*); (4) *Nostocales* are filamentous, divide along a single axis, and are capable of cellular differentiation (Figure 14.2*d*); and (5) *Stigonematales* are morphologically similar to *Nostocales* except that cells divide in multiple planes forming branching filaments (Figure 14.2*e*). Finally, the **prochlorophytes** are a lineage of unique unicellular *Cyanobacteria* once thought to be distinct but now classified within the *Chroococcales*. **Table 14.1** lists some major genera currently recognized in each group.

Some of the major morphological classifications of *Cyanobacteria* correspond to coherent phylogenetic groups, but others do not (**Figure 14.3**). Species of *Pleurocapsales* form a coherent group within the cyanobacteria, indicating that reproduction by multiple fission arose only once in the evolutionary history of cyanobacteria (Figure 14.3). Likewise, species of the *Nostocales* and

Table 14.1 Genera and grouping of cyanobacteria

Group	Genera
Group I, *Chroococcales.* Unicellular or cell aggregates	*Gloeothece* (Figure 14.2*a*), *Gloeobacter, Synechococcus, Cyanothece, Gloeocapsa, Synechocystis, Chamaesiphon, Merismopedia, Crocosphaera* (Figure 14.7*a*), *Prochlorococcus, Prochloron*
Group II, *Pleurocapsales.* Reproduce by formation of small spherical cells called baeocytes produced through multiple fission	*Pleurocapsa* (Figure 14.2*b*), *Dermocarpa, Xenococcus, Dermocarpella, Myxosarcina, Chroococcidiopsis*
Group III, *Oscillatoriales.* Undifferentiated filamentous cells that divide by binary fission in a single plane	*Lyngbya* (Figure 14.2*c*), *Spirulina* (Figure 14.5), *Arthrospira, Oscillatoria* (Figure 14.6*a–b*), *Microcoleus, Pseudanabaena, Trichodesmium* (Figure 14.7*b*)
Group IV, *Nostocales.* Filamentous cells that produce heterocysts	*Nodularia* (Figures 14.2*d*), *Nostoc, Calothrix* (Figure 14.8*a–b*), *Anabaena* (Figure 14.6*c*), *Cylindrospermum, Scytonema, Richelia* (Figure 14.7*c*)
Group V, *Stigonematales.* Cells divide to form branches	*Fischerella* (Figures 14.2*e*, 14.8*c, d*), *Stigonema, Chlorogloeopsis, Hapalosiphon*

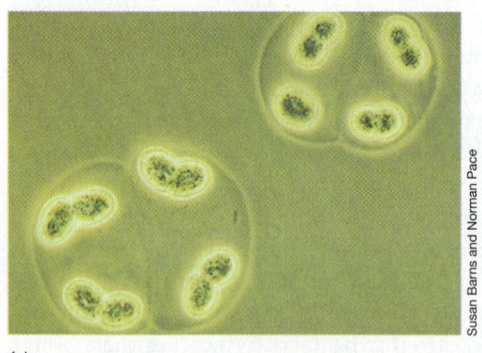

(a)

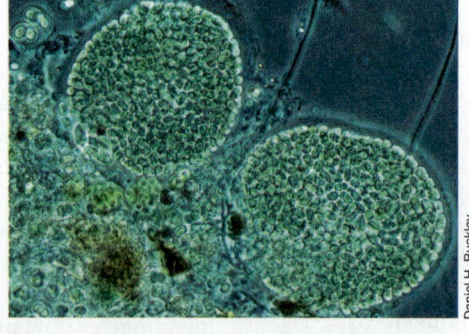

(b)

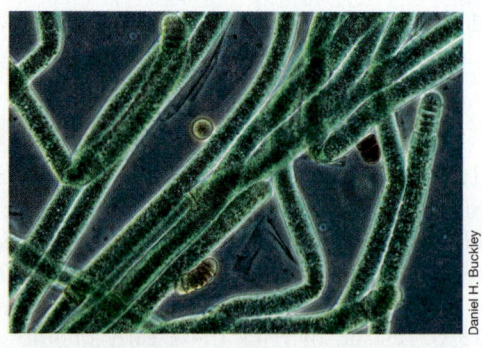

(c)

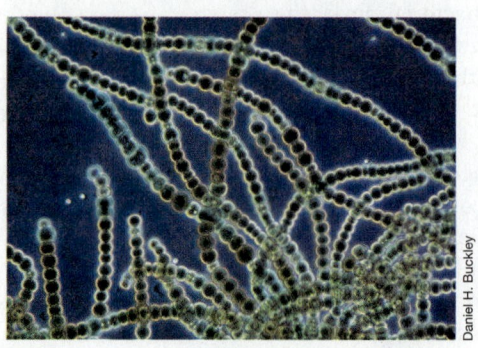

(d)

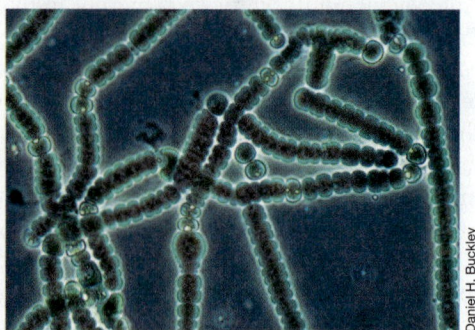

(e)

Figure 14.2 Cyanobacteria: the five major morphological types of cyanobacteria.
(a) Unicellular, *Gloeothece*; a single cell measures 5–6 μm in diameter; (b) colonial, *Pleurocapsa*; these structures contain hundreds of cells and are > 50 μm in diameter; (c) filamentous, *Lyngbya*; a single cell measures about 10 μm wide; (d) filamentous heterocystous, *Nodularia*; a single cell measures about 10 μm wide; (e) filamentous branching, *Fischerella*; a cell is about 10 μm wide. See how morphological diversity relates to phylogenetic diversity in Figure 14.3.

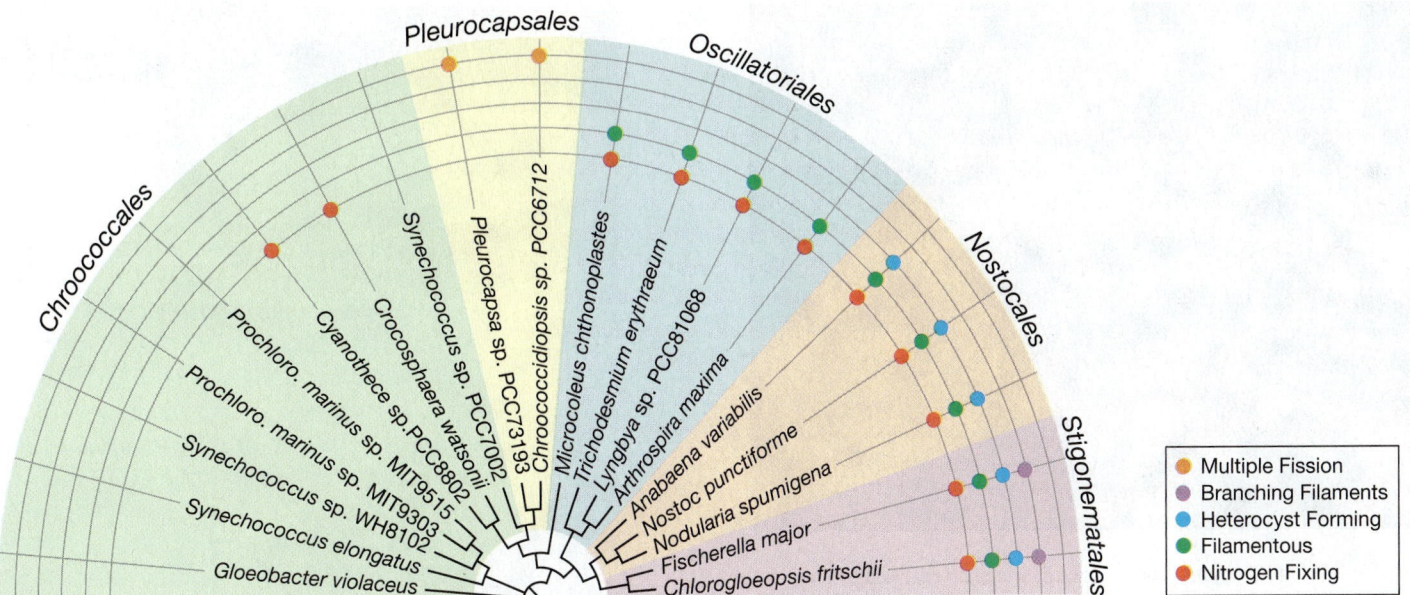

Figure 14.3 **Taxonomically informative traits mapped onto the phylogeny of *Cyanobacteria*.** The dendrogram depicts phylogenetic relationships inferred from analysis of conserved protein families in cyanobacterial genomes. Colored circles are used to indicate species traits as indicated by the key. Color shading is used to indicate taxonomic groupings. "*Prochloro.*" is used to indicate *Prochlorococcus*, which is a distinct group within the *Chroococcales*. Note that the *Chroococcales* and *Oscillatoriales* are not monophyletic in origin, meaning that these traits have arisen independently on multiple occasions in the phylogeny.

Stigonematales share a common ancestor and form a coherent phylogenetic group indicating a single origin of cellular differentiation within the *Cyanobacteria* (Figure 14.3). All *Stigonematales* share a single ancestor within the clade composed of *Nostocales* and *Stigonematales*, indicating that the capacity to form branching filaments arose only once within the lineage of *Cyanobacteria* capable of cellular differentiation (Figure 14.3). In contrast, unicellular and simple filamentous *Cyanobacteria* (*Chroococcales* and *Oscillatoriales*, respectively) are dispersed in the cyanobacterial phylogeny, and these morphological groups do not represent coherent evolutionary lineages (Figure 14.3).

Physiology and Photosynthetic Membranes

Cyanobacteria are oxygenic phototrophs and therefore have both type I and type II photosystems. All species are able to fix CO_2 by the Calvin cycle, many can fix N_2, and most can synthesize their own vitamins. Cells harvest energy from light and fix CO_2 during the day. During the night, cells generate energy by fermentation or aerobic respiration of carbon storage products such as glycogen. While CO_2 is the predominant source of carbon for most species, some cyanobacteria can assimilate simple organic compounds such as glucose and acetate if light is present, a process called *photoheterotrophy*. A few cyanobacteria, mainly filamentous species, can also grow in the dark on glucose or sucrose, using the sugar as both carbon and energy source. Finally, when sulfide concentrations are high, some cyanobacteria are able to switch from oxygenic photosynthesis to anoxygenic photosynthesis using hydrogen sulfide rather than water as electron donor for photosynthesis (↩ Figure 13.15).

Cyanobacteria have specialized membrane systems called *thylakoids* that increase the ability of cells to harvest light energy (↩ Figure 13.10). The cell wall of cyanobacteria contains peptidoglycan and is structurally similar to that of gram-negative bacteria. Photosynthesis takes place in the thylakoid membrane, a complex and multilayered photosynthetic membrane system containing photopigments and proteins that mediate photosynthesis (↩ Sections 13.1 and 13.2). In most unicellular cyanobacteria, the thylakoid membranes are arranged in regular concentric circles around the periphery of the cytoplasm (**Figure 14.4**). Cyanobacteria produce chlorophyll *a*, and most also have characteristic pigments called **phycobilins** (↩ Figure 13.10), which function as accessory pigments in photosynthesis. One class of phycobilins, *phycocyanins*, are blue and, together with the green chlorophyll *a*, are responsible for the blue-green color of most cyanobacteria. Some cyanobacteria produce *phycoerythrin*, a red phycobilin, and species producing phycoerythrin are red or brown. Photopigments are fluorescent and emit light when visualized using a

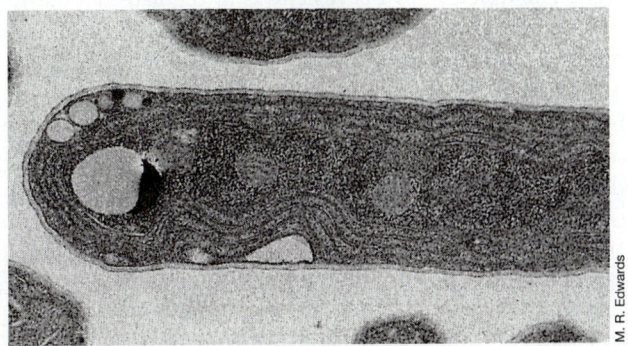

Figure 14.4 **Thylakoids in cyanobacteria.** Electron micrograph of a thin section of the cyanobacterium *Synechococcus lividus.* A cell is about 5 μm in diameter. Note thylakoid membranes running parallel to the cell wall.

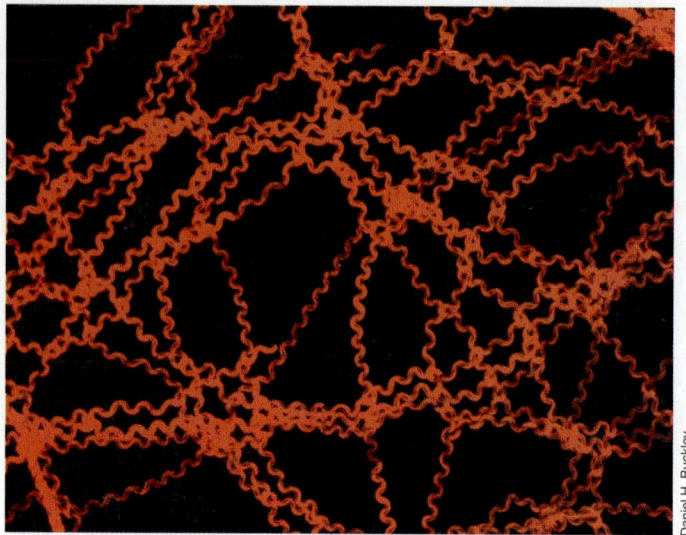

Figure 14.5 **Phycocyanin fluorescence in cyanobacteria.** Fluorescence micrograph of *Spirulina*. Filaments consist of chains of helical cells with each cell approximately 5 μm wide.

fluorescence microscope (**Figure 14.5**). Prochlorophytes, such as *Prochlorococcus* and *Prochloron*, are unique among *Cyanobacteria* in that all members of this group contain chlorophyll *a* and *b* but do not contain phycobilins.

Motility and Cellular Structures

Cyanobacteria possess several mechanisms for motility. Many cyanobacteria exhibit gliding motility (⮌ Section 2.18). Gliding occurs only when a cell or filament is in contact with a solid surface or with another cell or filament. In some cyanobacteria, gliding is not a simple translational movement but is accompanied by rotations, reversals, and flexing of filaments. Most gliding species exhibit directional movement toward light (phototaxis), and chemotaxis (⮌ Section 2.19) may occur as well. *Synechococcus* exhibits an unusual form of swimming motility that does not require flagella or any other extracellular organelle. The cell surface of *Synechococcus* has specialized proteins that provide direct thrust through a mechanism that has yet to be resolved. Gas vesicles (⮌ Section 2.15) are also found in a variety of aquatic cyanobacteria and are important in positioning cells in the water column. The function of gas vesicles is to regulate cell buoyancy such that cells can remain in a position in the water column where light intensity is optimal for photosynthesis.

Cyanobacteria are able to form a variety of structures associated with energy storage, reproduction, and survival. Many cyanobacteria produce extensive mucilaginous envelopes, or sheaths, that bind groups of cells or filaments together (Figure 14.2*a*). Some filamentous cyanobacteria can form *hormogonia* (**Figure 14.6**), short, motile filaments that break off from longer filaments to facilitate dispersal in times of stress. Some species also form resting structures called *akinetes* (Figure 14.6*c*), which protect the organism during periods of darkness, desiccation, or cold. Akinetes are cells with thickened outer walls. When conditions improve, akinetes germinate by breaking down their outer wall and initiating growth of a new vegetative filament. Many cyanobacteria also

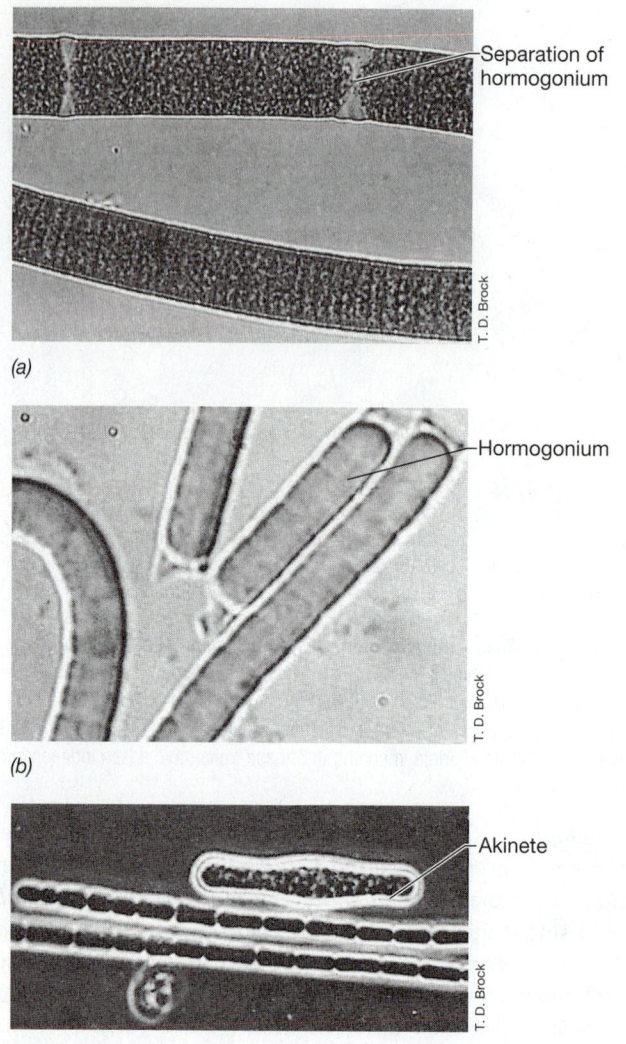

Figure 14.6 **Structural differentiation in filamentous cyanobacteria.** *(a)* Initial stage of hormogonium formation in *Oscillatoria*. Notice the empty spaces where the hormogonium is separating from the filament. *(b)* Hormogonium of a smaller *Oscillatoria* species. Notice that the cells at both ends are rounded. Cells are about 10 μm wide. Differential interference contrast microscopy. *(c)* Akinete (resting spore) of *Anabaena* in a phase-contrast micrograph, cells about 5 μm wide.

form a structure called *cyanophycin*. This structure is a copolymer of aspartic acid and arginine and is a nitrogen storage product; when nitrogen in the environment becomes deficient, cyanophycin is broken down and used as a cellular nitrogen source. Many species of the *Nostocales* and *Stigonematales* are also able to form heterocysts, as discussed next.

Heterocysts and Nitrogen Fixation

Many cyanobacteria are capable of nitrogen fixation (Figure 14.3). The nitrogenase enzyme, however, is inhibited by oxygen and thus nitrogen fixation cannot occur along with oxygenic photosynthesis (⮌ Section 3.17). Cyanobacteria have evolved several regulatory mechanisms for separating nitrogenase activity from photosynthesis (⮌ Section 7.13). For example, many unicellular cyanobacteria, such as *Cyanothece* and *Crocosphaera* (**Figure 14.7***a*), fix nitrogen only at night when photosynthesis does

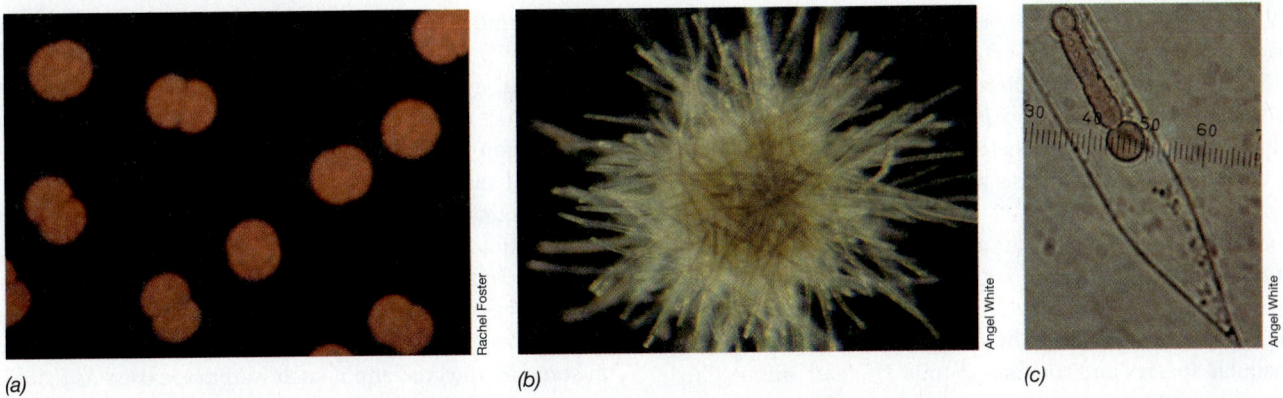

(a)   (b)   (c)

Rachel Foster   Angel White   Angel White

Figure 14.7   Marine cyanobacteria that fix N₂. *(a)* Unicellular *Crocosphaera*-like cells in the process of dividing; cells are approximately 5 µm diameter. *(b)* Colonial "tuft" of *Trichodesmium*. The tuft is composed of many attached undifferentiated unbranching filaments and has a diameter of approximately 100 µm. *(c)* A diatom containing the cyanobacterial symbiont *Richelia* (scale in micrometers). The *Richelia* symbiont is an unbranching filament with a terminal heterocyst; cells are about 5 µm wide.

not occur. In contrast, the filamentous cyanobacteria *Trichodesmium* (Figure 14.7*b*) fixes nitrogen only during the day through a mechanism that remains somewhat unclear, but appears to require transient suppression of photosynthetic activity within filaments. Finally, many filamentous cyanobacteria of the *Nostocales* and *Stigonematales* facilitate nitrogen fixation by forming specialized cells called *heterocysts,* either on the ends of filaments (**Figure 14.8*a, b***) or along the filament (Figure 14.8*c, d*).

Heterocysts arise from differentiation of vegetative cells and are the sites of nitrogen fixation in heterocystous cyanobacteria. Heterocysts are surrounded by a thickened cell wall that slows the diffusion of O₂ into the cell and permits nitrogenase activity to occur in an anoxic environment. Heterocysts lack photosystem II (Figure 14.8), the oxygen-evolving photosystem that generates reducing power from H₂O (🔗 Section 13.4). Without photosystem II, heterocysts are unable to fix CO₂ and thus lack the necessary electron donor (pyruvate) for nitrogen fixation. However, heterocysts have intercellular connections with adjacent vegetative cells that allow for mutual exchange of materials between these cells. Fixed carbon is imported by the heterocyst from adjacent vegetative cells, and this is oxidized to yield electrons for nitrogen fixation. The products of photosynthesis move from vegetative cells to heterocysts, and fixed nitrogen moves from heterocysts to vegetative cells (🔗 Section 7.13 and Figure 7.28).

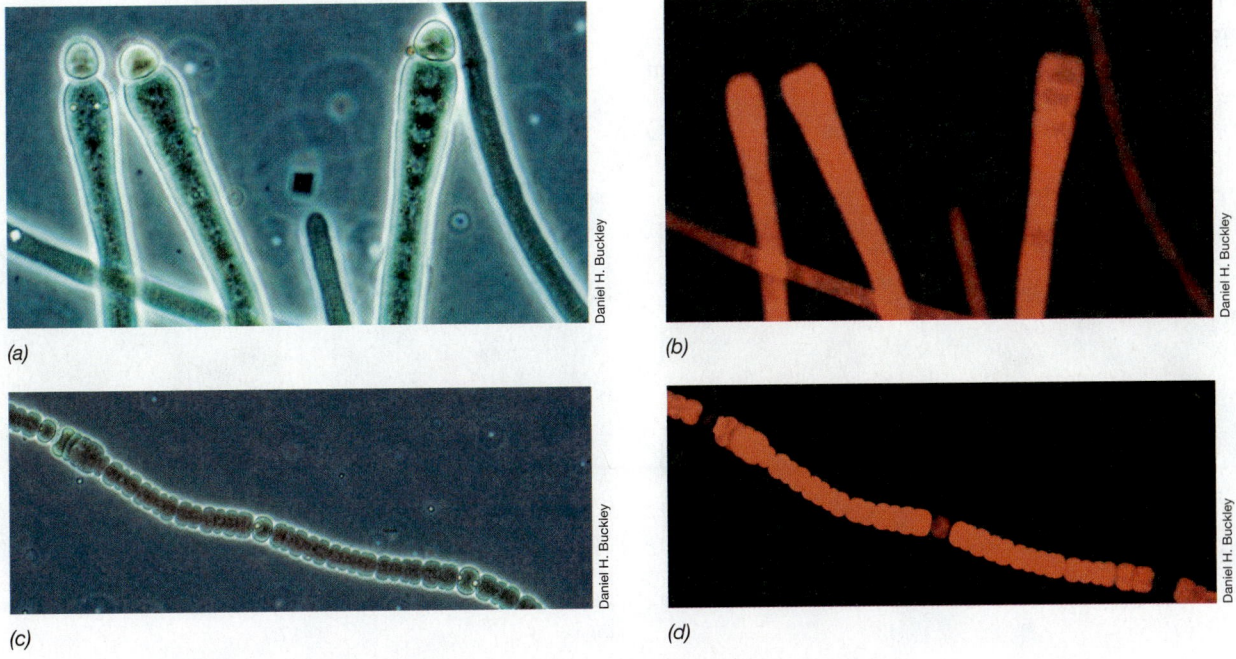

(a)   (b)
(c)   (d)

Daniel H. Buckley

Figure 14.8   Heterocysts. Differentiation of heterocysts causes the loss of photopigments and inability to carry out photosynthesis. *(a)* Phase-contrast micrograph of *Calothrix* with terminal heterocysts. *(b)* Fluorescence micrograph of the same *Calothrix* filaments; cells are about 10 µm wide. *(c)* Phase-contrast micrograph of *Fischerella.* *(d)* Fluorescence micrograph of the same *Fischerella* filaments; cells are about 10 µm wide. See how heterocyst formation is regulated at the genetic level in the well-studied cyanobacterium, *Anabaena* in Figure 7.28.

UNIT 3

Ecology of *Cyanobacteria*

Cyanobacteria are of central importance to the productivity of the oceans. Small unicellular cyanobacteria, such as *Synechococcus* and *Prochlorococcus* (Section 19.10), are the most abundant phototrophs in the oceans. Together these organisms contribute 80% of marine photosynthesis and 35% of all photosynthetic activity on Earth.

Cyanobacterial nitrogen fixation represents the dominant input of new nitrogen into vast segments of Earth's oceans, particularly in oligotrophic tropical and subtropical waters. Marine nitrogen fixation is dominated by two groups of cyanobacteria, the unicellular species and the filamentous *Trichodesmium*. Crocosphaera (Figure 14.7*a*) and relatives dominate nitrogen fixation in most of the Pacific Ocean and are widespread in tropical and subtropical habitats. *Trichodesmium* is the dominant nitrogen-fixer in the North Atlantic Ocean and parts of the Pacific where dissolved iron concentrations are elevated. *Trichodesmium* forms macroscopically visible tufts of filaments (Figure 14.7*b*) and relies on gas vesicles to remain suspended in the photic zone where it is often observed in dense masses of cells called *blooms*. In addition, other marine nitrogen-fixers including species of *Calothrix* and *Richelia* form symbiotic associations with diatoms (Figure 14.7*c*); these symbiotic associations are often observed in tropical and subtropical oceans. Finally, heterocystous cyanobacteria such as *Nodularia* (Figure 14.2*d*) and *Anabaena* can sometimes dominate nitrogen fixation in cold waters of the Northern Hemisphere and are often observed in the Baltic Sea.

Cyanobacteria are also widely found in terrestrial and freshwater environments. In general, they are more tolerant of environmental extremes, particularly extremes of desiccation, than are algae (eukaryotes). Cyanobacteria are often the dominant or sole oxygenic phototrophic organisms in hot springs, saline lakes, desert soils, and other extreme environments. In some of these environments, cyanobacterial mats of variable thickness may form (Figure 19.9). Freshwater lakes, especially those rich in inorganic nutrients, often develop blooms of cyanobacteria, especially in late summer when temperatures are warmest (Figures 19.1 and 19.17). A few cyanobacteria are symbionts of liverworts, ferns, and cycads, and a number are phototrophic components of lichens, a symbiosis between a phototroph and a fungus (Section 22.1).

Several metabolic products of cyanobacteria are of considerable practical importance. Some cyanobacteria produce potent neurotoxins, and toxic blooms may form when massive accumulations of cyanobacteria develop. Animals ingesting water containing these toxic products may be killed. Many cyanobacteria are also responsible for the production of earthy odors and flavors in some freshwater, and if such waters are used as drinking water sources, aesthetic problems may arise. The major compound produced is geosmin, a substance also produced by many actinomycetes (Section 15.12).

14.4 Purple Sulfur Bacteria

Key Genera: *Chromatium, Ectothiorhodospira*

Purple sulfur bacteria are anoxygenic phototrophs that use hydrogen sulfide (H_2S) as an electron donor for photosynthesis. Purple sulfur bacteria are a phylogenetically coherent group found within the order *Chromatiales* in the *Gammaproteobacteria*.

Purple sulfur bacteria are generally found in illuminated anoxic zones where H_2S is present. Such habitats occur commonly in lakes, marine sediments, and "sulfur springs," where H_2S produced geochemically or biologically can support the growth of purple sulfur bacteria (Figure 14.9). Purple sulfur bacteria are also commonly found in microbial mats (Section 19.5) and in salt

(a)

(b)

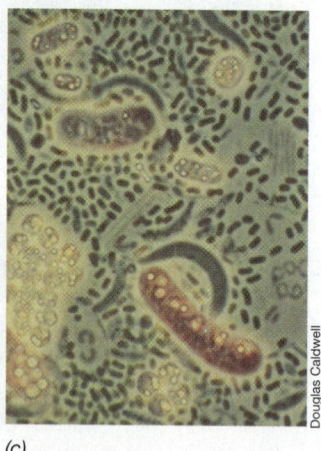
(c)

Figure 14.9 Blooms of purple sulfur bacteria. *(a) Lamprocystis roseopersicina*, in a sulfide spring. The bacteria grow near the bottom of the spring pool and float to the top (by virtue of their gas vesicles) when disturbed. The green color is from cells of the eukaryotic alga *Spirogyra*. *(b)* Sample of water from a depth of 7 m in Lake Mahoney, British Columbia; the major phototroph is *Amoebobacter purpureus*. *(c)* Phase-contrast photomicrograph of layers of purple sulfur bacteria from a small, stratified lake in Michigan. The purple sulfur bacteria include *Chromatium* species (large rods) and *Thiocystis* (small cocci).

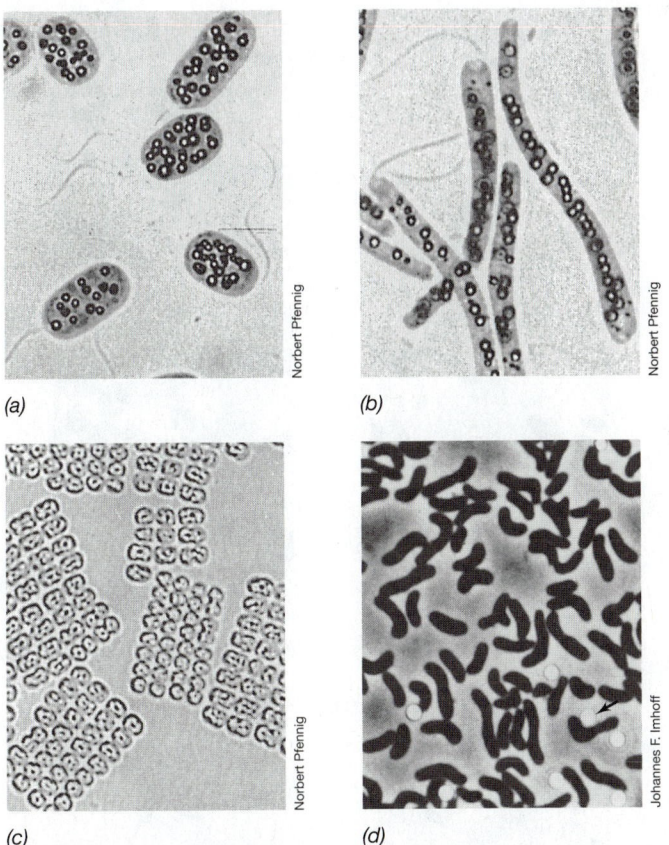

(a)

(b)

(c)

(d)

Norbert Pfennig

Norbert Pfennig

Norbert Pfennig

Johannes F. Imhoff

Figure 14.10 **Bright-field and phase-contrast photomicrographs of purple sulfur bacteria.** *(a) Chromatium okenii;* cells are about 5 μm wide. Note the globules of elemental sulfur inside the cells. *(b) Thiospirillum jenense,* a very large, polarly flagellated spiral; cells are about 30 μm long. Note the sulfur globules. *(c) Thiopedia rosea;* cells are about 1.5 μm wide. *(d)* Phase-contrast micrograph of cells of *Ectothiorhodospira mobilis;* cells are about 0.8 μm wide. Note external sulfur globules (arrow).

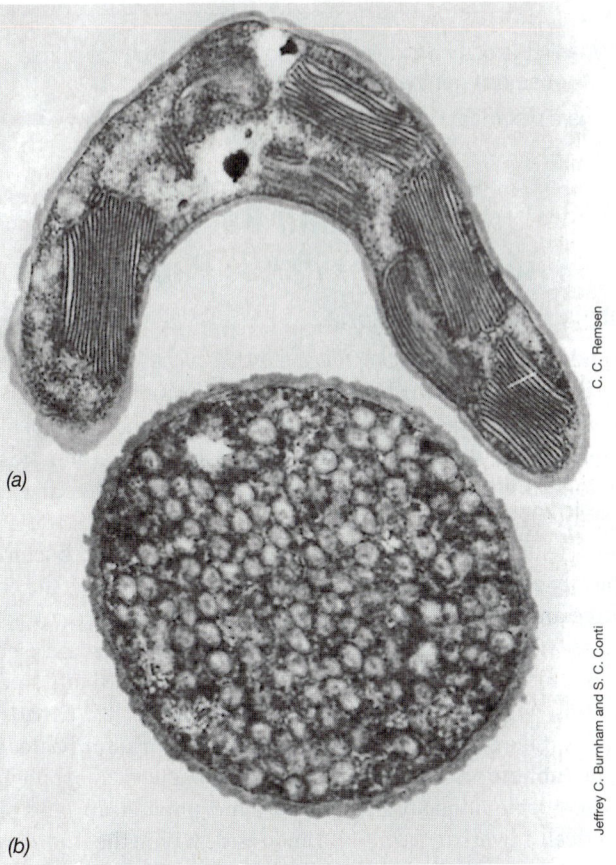

(a)

(b)

C. C. Remsen

Jeffrey C. Burnham and S. C. Conti

Figure 14.11 **Membrane systems of phototrophic purple bacteria as revealed by transmission electron microscopy.** *(a) Ectothiorhodospira mobilis,* showing the photosynthetic membranes in flat sheets (lamellae). *(b) Allochromatium vinosum,* showing the membranes as individual, spherical vesicles.

marsh sediments. The characteristic color of purple sulfur bacteria comes from their carotenoids, accessory pigments involved in light harvesting (↩ Section 13.2). These bacteria use a type II photosystem (↩ Figure 13.3), contain either bacteriochlorophyll *a* or *b*, and carry out CO_2 fixation by the Calvin cycle (↩ Section 13.5).

During autotrophic growth of purple sulfur bacteria, H_2S is oxidized to elemental sulfur (S^0), which is deposited as sulfur granules (**Figure 14.10**). When sulfide is limiting, the sulfur is used as an electron donor for photosynthesis, resulting in the oxidation of S^0 to sulfate (SO_4^{2-}). Many purple sulfur bacteria can also use other reduced sulfur compounds as photosynthetic electron donors; for example, thiosulfate ($S_2O_3^{2-}$) is commonly used to grow laboratory cultures.

The purple sulfur bacteria form two families: the *Chromatiaceae* and the *Ectothiorhodospiraceae*. Species of the two families are readily distinguished by the location of sulfur granules and by their photosynthetic membranes. *Chromatiaceae,* including the genera *Chromatium* and *Thiocapsa,* store S^0 granules *inside* their cells (in the periplasmic space) and have vesicular intracellular photosynthetic membrane systems (**Figure 14.11**). These

organisms are common in stratified lakes containing sulfide and in the anoxic sediments of salt marshes. *Ectothiorhodospiraceae,* including the two main genera *Ectothiorhodospira* and *Halorhodospira,* oxidize H_2S to S^0 that is deposited *outside* the cell (Figure 14.10*d*) and have lamellar intracellular photosynthetic membrane systems (Figure 14.11). These genera are also interesting because many species are extremely halophilic (salt-loving) or alkaliphilic (alkalinity-loving) and are among the most extreme in these characteristics of all known *Bacteria*. These organisms are typically found in saline lakes, soda lakes, and salterns, where abundant levels of SO_4^{2-} support sulfate-reducing bacteria (↩ Section 20.4 and Section 14.9), the organisms that produce H_2S.

Purple sulfur bacteria are often observed in high density in meromictic (permanently stratified) lakes. Meromictic lakes form layers because they have denser (usually saline) water on the bottom and less dense (usually freshwater) water nearer the surface. If sufficient sulfate is present to support sulfate reduction, sulfide is produced in the sediments and diffuses upward into the anoxic bottom waters. The presence of sulfide and light in the anoxic layers of the lake allow purple sulfur bacteria to form dense cell masses, usually in association with green phototrophic bacteria (Figure 14.9*b*).

UNIT 3

14.5 Purple Nonsulfur Bacteria and Aerobic Anoxygenic Phototrophs

Purple Nonsulfur Bacteria

Key Genera: *Rhodospirillum, Rhodoferax, Rhodobacter*

The **purple nonsulfur bacteria** are the most metabolically versatile of all microorganisms. Despite their name, they are not always purple; these organisms synthesize an array of carotenoids (⮌ Section 13.2) that can lend them a variety of spectacular colors (Figure 14.12). Together, these pigments give purple bacteria their colors, usually purple, red, or orange. Purple nonsulfur bacteria are typically photoheterotrophs (a condition where light is the energy source and an organic compound is the carbon source), and species are able to use a wide range of carbon sources and electron donors for photosynthesis, including organic acids, amino acids, alcohols, sugars, and even aromatic compounds like benzoate or toluene. Like purple sulfur bacteria, purple nonsulfur bacteria use a type II photosystem, and contain either bacteriochlorophyll *a* or *b*. The purple nonsulfur bacteria are morphologically and phylogenetically diverse (Figure 14.13) and reside within the *Alphaproteobacteria* (e.g., *Rhodospirillum, Rhodobacter, Rhodopseudomonas*) or *Betaproteobacteria* (e.g., *Rubrivivax, Rhodoferax*).

Purple nonsulfur bacteria are able to conserve energy through a variety of metabolic processes. For example, some species can grow photoautotrophically using H_2, low levels of H_2S, or even ferrous iron (Fe^{2+}) as the electron donor for photosynthesis with CO_2 fixation carried out by the Calvin cycle. Most species are also able to grow in darkness by using aerobic respiration of organic or even some inorganic compounds; synthesis of the photosynthetic machinery is typically repressed by O_2. Finally, some species can

Figure 14.12 **Photograph of liquid cultures of phototrophic purple bacteria showing the color of species with various carotenoid pigments.** The blue culture is a carotenoidless mutant strain of *Rhodospirillum rubrum* showing that bacteriochlorophyll *a* is actually blue. The bottle on the far right (*Rhodobacter sphaeroides* strain G) lacks one of the carotenoids of the wild type and thus is greener.

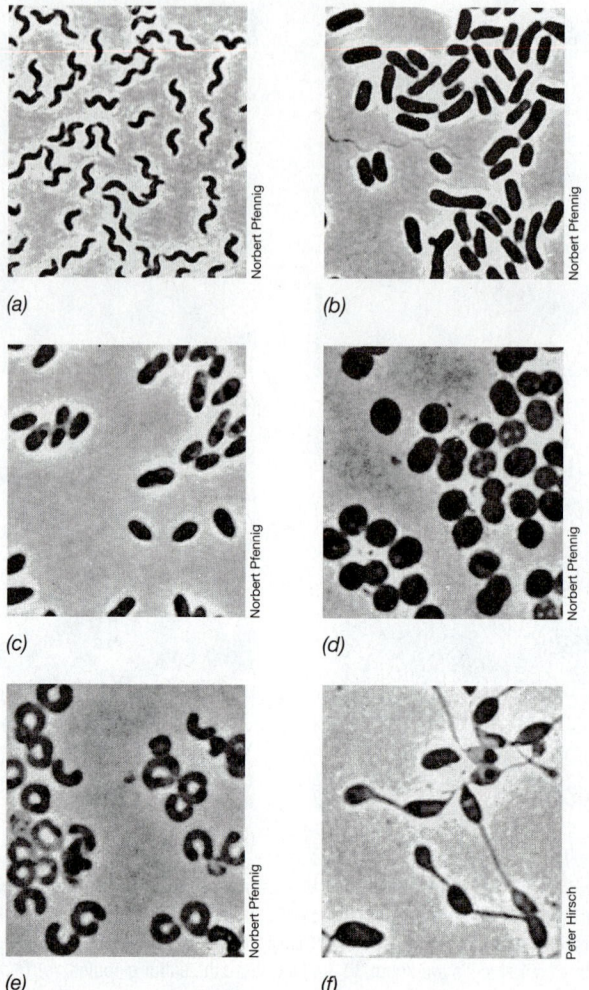

Figure 14.13 **Representatives of several genera of purple nonsulfur bacteria.** *(a) Phaeospirillum fulvum;* cells are about 3 μm long. *(b) Rhodoblastus acidophilus;* cells are about 4 μm long. *(c) Rhodobacter sphaeroides;* cells are about 1.5 μm wide. *(d) Rhodopila globiformis;* cells are about 1.6 μm wide. *(e) Rhodocyclus purpureus;* cells are about 0.7 μm in diameter. *(f) Rhodomicrobium vannielii;* cells are about 1.2 μm wide.

also grow by fermentation or anaerobic respiration using a variety of electron donors and acceptors.

Enrichment and isolation of purple nonsulfur bacteria is easy using a mineral salts medium supplemented with an organic acid as carbon source. Such media, inoculated with a mud, lake water, or sewage sample and incubated anaerobically in the light, invariably select for purple nonsulfur bacteria. Enrichment cultures can be made even more selective by omitting fixed nitrogen sources (for example, ammonia) or organic nitrogen sources (for example, yeast extract or peptone) from the medium and supplying a gaseous headspace of N_2. Virtually all purple nonsulfur bacteria can fix N_2 and will thrive under such conditions, rapidly outcompeting other bacteria.

Aerobic Anoxygenic Phototrophs

Key Genera: *Roseobacter, Erythrobacter*

The **aerobic anoxygenic phototrophs** are obligatory aerobic heterotrophs that use light as a supplemental source of energy to support growth. Like purple nonsulfur bacteria, aerobic anoxygenic phototrophs are phylogenetically diverse and are *Alphaproteo-bacteria* or *Betaproteobacteria*. The primary physiological differ-ence with the purple nonsulfur bacteria is that aerobic anoxygenic phototrophs are strict heterotrophs and employ anoxygenic pho-tosynthesis only under oxic conditions as a supplemental source of energy. Aerobic anoxygenic phototrophs contain bacteriochlo-rophyll *a* and a type II photosystem, but are unable to fix CO_2 and must rely on organic carbon for growth. Carotenoids of various types lend colors of yellow, orange, or pink to cultures.

Aerobic anoxygenic phototrophs are only able to photosynthe-size when grown on a day/night cycle. Under these conditions, bacteriochlorophyll *a* is made only in the dark and then used to conserve energy by photophosphorylation when the light returns. Aerobic anoxygenic phototrophs can account for as much as a quarter of the microbial community inhabiting coastal marine waters and 5% of gross photosynthesis in such systems (Sec-tion 19.10). Common genera found in coastal marine habitats include *Roseobacter* and *Erythrobacter*.

MINIQUIZ

- What are some similarities between purple nonsulfur bacteria and aerobic anoxygenic phototrophs? What are the differences between these two groups?
- Where would you expect to find aerobic anoxygenic phototrophs?

14.6 Green Sulfur Bacteria

Key Genera: *Chlorobium, Chlorobaculum, "Chlorochromatium"*

Green sulfur bacteria are a phylogenetically coherent group of anoxygenic phototrophs that forms the phylum *Chlorobi*. Green sulfur bacteria have little metabolic versatility and they are typi-cally nonmotile and strictly anaerobic anoxygenic phototrophic bacteria. The group is also morphologically restricted and includes primarily short to long rods (**Figure 14.14**).

Like purple sulfur bacteria, green sulfur bacteria oxidize hydro-gen sulfide (H_2S) as an electron donor for autotrophic growth, oxidizing it first to sulfur (S^0) and then to sulfate (SO_4^{2-}). But

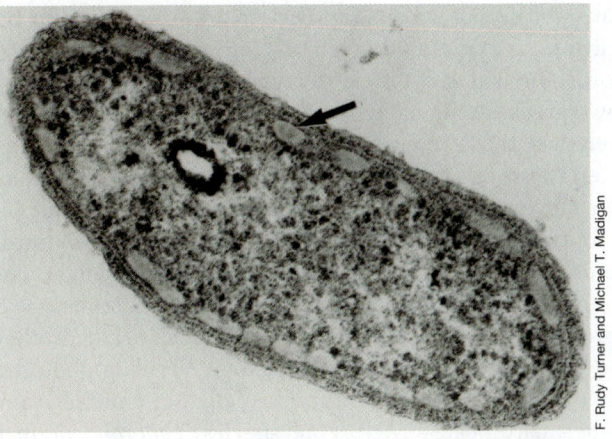

Figure 14.15 **The thermophilic green sulfur bacterium *Chlorobaculum tepidum*.** Transmission electron micrograph. Note chlorosomes (arrow) in the cell periphery. A cell is about 0.7 µm wide.

unlike most purple sulfur bacteria, the S^0 produced by green sulfur bacteria is deposited only outside the cell (Figure 14.14*a*). Autotrophy is supported not by the reactions of the Calvin cycle, as in purple bacteria, but instead by a reversal of steps in the citric acid cycle (Section 13.5 and Figure 13.19*a*), a unique means of autotrophy in phototrophic bacteria.

Pigments and Ecology

Green sulfur bacteria contain bacteriochlorophyll *c*, *d*, or *e* and house these pigments in unique structures called **chlorosomes** (**Figure 14.15**). A small amount of bacteriochlorophyll *a* is present in the reaction center and FMO protein, which connects the chloro-some to the cytoplasmic membrane (Figure 13.7*b*). Chlorosomes are oblong bacteriochlorophyll-rich bodies bounded by a thin, nonunit membrane and attached to the cytoplasmic membrane in the periphery of the cell (Figure 14.15 and Figure 13.7). Chlo-rosomes function to funnel energy into the photosystem, and this eventually leads to ATP synthesis. Unlike purple anoxygenic photo-trophs, green sulfur bacteria use a type I photosystem. Both green- and brown-colored species of green sulfur bacteria are known, the brown-colored species containing bacteriochlorophyll *e* and carot-enoids that turn dense cell suspensions brown (**Figure 14.16**).

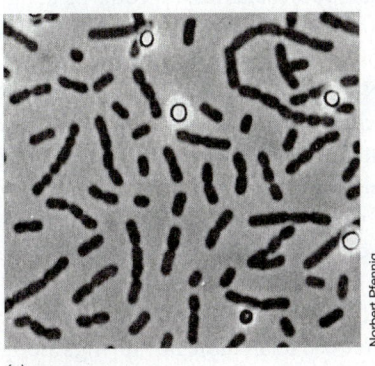
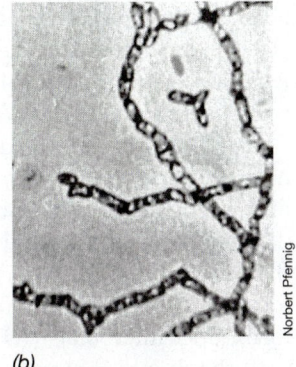

Figure 14.14 **Phototrophic green sulfur bacteria.** *(a) Chlorobium limicola*; cells are about 0.8 µm wide. Note the spherical sulfur granules deposited extracellularly. *(b) Chlorobium clathratiforme*, a bacterium forming a three-dimensional network; cells are about 0.8 µm wide.

Figure 14.16 **Green and brown chlorobia.** Tube cultures of *(a) Chlorobaculum tepidum* and *(b) Chlorobaculum phaeobacteroides*. Cells of *C. tepidum* contain bacteriochlorophyll *c* and green carotenoids, and cells of *C. phaeobacteroides* contain bacteriochlorophyll *e* and isorenieratene, a brown carotenoid.

UNIT 3

Like purple sulfur bacteria (Section 14.4), green sulfur bacteria live in anoxic, sulfidic aquatic environments. However, the chlorosome is a very efficient light-harvesting structure, which allows green sulfur bacteria to grow at light intensities much lower than those required by other phototrophs. Green sulfur bacteria also tend to have a greater tolerance of H_2S than do other anoxygenic phototrophs. As a result, green sulfur bacteria are typically found at the greatest depths of all phototrophic microorganisms in lakes or microbial mats, where light intensities are low and H_2S levels the highest. As an example, a species of green sulfur bacteria isolated from a deep-sea hydrothermal vent (e◌ Section 19.13) was found to be growing phototrophically on the weak glow of infrared radiation emitted from the geothermally heated rock. One species, *Chlorobaculum tepidum* (Figure 14.15), is thermophilic and forms dense microbial mats in high-sulfide hot springs. *C. tepidum* also grows rapidly and is amenable to genetic manipulation by both conjugation and transformation. Because of these features, *C. tepidum* has become the model organism for studying the molecular biology of green sulfur bacteria.

Green Sulfur Bacteria Consortia

Certain species of green sulfur bacteria form an intimate two-membered association, called a **consortium**, with a chemoorganotrophic bacterium. In the consortium, each organism benefits, and thus a variety of such consortia containing different phototrophic and chemotrophic components probably exist in nature. The phototrophic component, called the *epibiont*, is physically attached to the nonphototrophic central cell (**Figure 14.17**) and communicates with it in various ways (e◌ Section 22.2).

The name *"Chlorochromatium aggregatum"* (not a formal name because this is a mixed culture) has been used to describe a commonly observed green-colored consortium that is green because the epibionts are green sulfur bacteria that contain green-colored carotenoids (Figure 14.17*b*). Evidence that the epibionts are indeed green sulfur bacteria comes from pigment analyses, the presence of chlorosomes (Figure 14.17*d*), and phylogenetic staining (Figure 14.17*c*). A structurally similar consortium called *"Pelochromatium roseum"* is brown because its epibionts produce brown-colored carotenoids (e◌ Figures 22.3 and 22.4). We examine the symbiotic nature of the *Chlorochromatium* consortium in more detail in Section 22.2.

MINIQUIZ -

- Which pigments are present in the chlorosome?
- What evidence exists that the epibionts of green bacterial consortia are truly green sulfur bacteria?

14.7 Green Nonsulfur Bacteria

Key Genera: *Chloroflexus, Heliothrix, Roseiflexus*

Green nonsulfur bacteria are anoxygenic phototrophs of the phylum *Chloroflexi.* The latter contains several distinct lineages, one of which, the class *Chloroflexi,* contains green nonsulfur

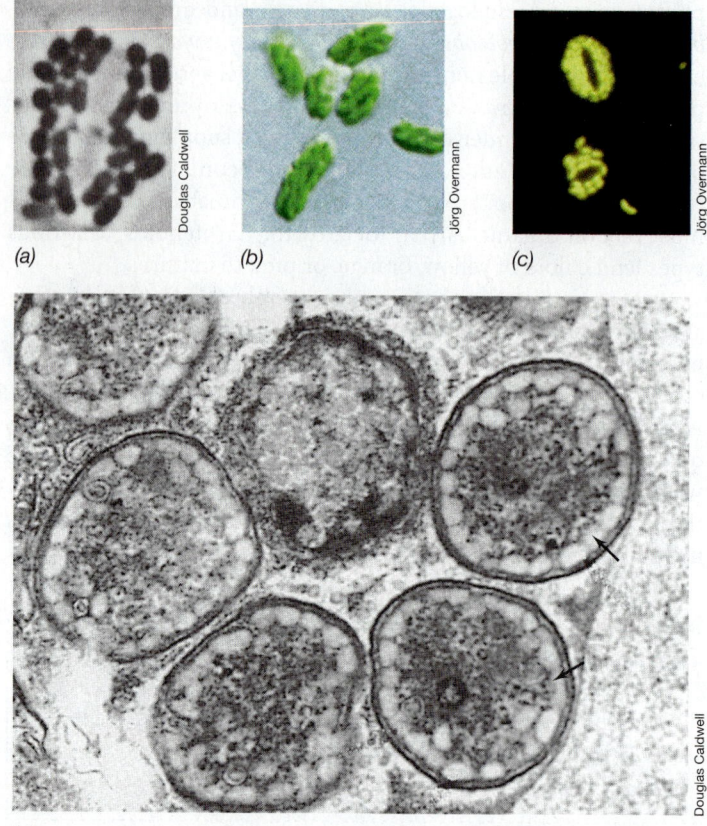

(a) (b) (c)

(d)

Figure 14.17 *"Chlorochromatium aggregatum."* Consortia of green sulfur bacteria and a chemoorganotroph. *(a)* In a phase-contrast micrograph, the nonphototrophic central organism is lighter in color than the pigmented phototrophic bacteria. *(b)* Green carotenoids lend their color to the phototrophs in a differential interference contrast micrograph. *(c)* A fluorescence micrograph shows the cells stained with a phylogenetic FISH probe specific for green sulfur bacteria. *(d)* Transmission electron micrograph of a cross section through a single consortium; note the chlorosomes (arrows) in the epibionts. The entire consortium is about 3 μm in diameter.

bacteria. The remainder of the phylum contains metabolically diverse organisms including both aerobic and anaerobic chemoorganotrophs as well as the *Dehalococcoidetes*, a group of dehalogenating bacteria that use halogenated organic compounds as electron acceptors in anaerobic respiration (e◌ Section 13.21). Analyses of 16S ribosomal RNA sequences from environmental samples (e◌ Section 18.5) indicate that species of the phylum *Chloroflexi* are widespread and that most species in the phylum have yet to be cultivated in isolation; thus the metabolic diversity of this phylum remains poorly characterized.

All cultured representatives of the green nonsulfur bacteria are filamentous bacteria that are capable of gliding motility. *Chloroflexus,* one of the most studied of the green nonsulfur bacteria, forms thick microbial mats in neutral to alkaline hot springs along with thermophilic cyanobacteria (**Figure 14.18**; e◌ Figure 19.9*b*). Green nonsulfur bacteria grow best as photoheterotrophs using simple carbon sources as electron donors for photosynthesis. However, growth also occurs photoautotrophically using H_2 or

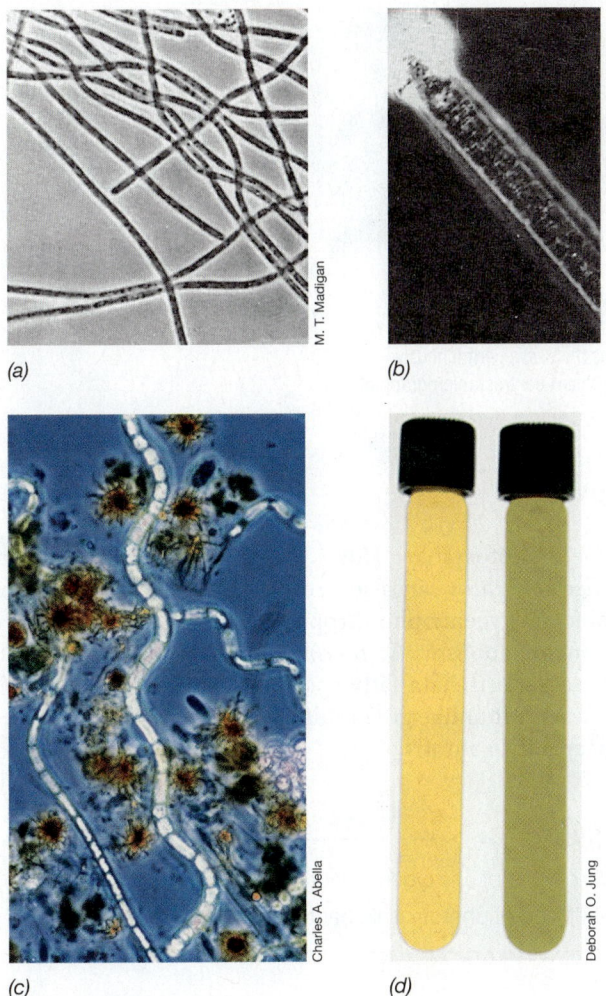

Figure 14.18 Green nonsulfur bacteria. *(a)* Phase-contrast micrograph of the anoxygenic phototroph *Chloroflexus aurantiacus*; cells are about 1 μm in diameter. *(b)* Phase-contrast micrograph of the large phototroph *Oscillochloris*; cells are about 5 μm wide. The brightly contrasting material on the top is a holdfast, used for attachment. *(c)* Phase-contrast micrograph of filaments of a *Chloronema* species; the cells are wavy filaments and about 2.5 μm in diameter. *(d)* Tube cultures of *C. aurantiacus* (right) and *Roseiflexus* (left). *Roseiflexus* is yellow because it lacks bacteriochlorophyll *c* and chlorosomes.

H₂S as electron donors for photosynthesis. The hydroxypropionate cycle, a pathway of CO_2 incorporation unique to only a few *Bacteria* and *Archaea*, supports autotrophic growth (⟳ Section 13.5). Most green nonsulfur bacteria also grow well in the dark by aerobic respiration of a wide variety of carbon sources. The photosynthetic features of the green nonsulfur bacteria are a "hybrid" between those of both green sulfur bacteria (Section 14.6) and purple phototrophic bacteria (Sections 14.4–14.5). Green nonsulfur bacteria have reaction centers that contain bacteriochlorophyll *a* and chlorosomes that contain bacteriochlorophyll *c* (Figure 14.15) and in this way are similar to green sulfur bacteria. However, in contrast to green sulfur bacteria, green nonsulfur bacteria use a type II photosystem and in this way resemble purple sulfur bacteria.

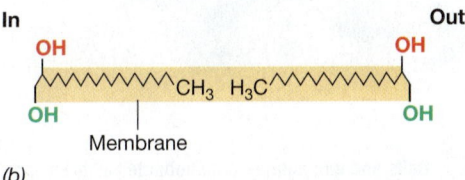

Figure 14.19 • The unusual lipids of *Thermomicrobium*. *(a)* Membrane lipids from *Thermomicrobium roseum* contain long-chain diols like the one shown here (13-methyl-1,2-nonadecanediol). Note that unlike the lipids of other *Bacteria* or of *Archaea*, neither ester- nor ether-linked side chains are present. *(b)* To form a bilayer membrane, dialcohol molecules oppose each other at the methyl groups, and the —OH groups are the inner and outer hydrophilic surfaces. Small amounts of the diols have fatty acids esterified to the secondary —OH group (shown in red), whereas the primary —OH group (shown in green) can bond a hydrophilic molecule like phosphate.

Other *Chloroflexi*

In addition to *Chloroflexus*, other phototrophic green nonsulfur bacteria include the thermophile *Heliothrix* and the large-celled mesophiles *Oscillochloris* (Figure 14.18*b*) and *Chloronema* (Figure 14.18*c*). *Oscillochloris* and *Chloronema* form rather large cells, 2–5 μm wide and up to several hundred micrometers long (Figure 14.18*c*). Species of both genera inhabit freshwater lakes containing H₂S. *Roseiflexus* and *Heliothrix* are similar to *Chloroflexus* in their filamentous morphology and thermophilic lifestyle, but differ in a major photosynthetic property. *Roseiflexus* and *Heliothrix* lack bacteriochlorophyll *c* and chlorosomes and thus more closely resemble purple phototrophic bacteria (Sections 14.4, 14.5) than *Chloroflexus*. This can be seen in cultures of *Roseiflexus* that are yellow-orange instead of green from their extensive carotenoid pigments and lack of bacteriochlorophyll *c* (Figure 14.18*d*).

Thermomicrobium is a chemotrophic genus of *Chloroflexi* and a strictly aerobic, gram-negative rod, growing optimally in complex media at 75°C. Besides its phylogenetic properties, *Thermomicrobium* is also of interest because of its membrane lipids (**Figure 14.19**). Recall that the lipids of *Bacteria* and *Eukarya* contain fatty acids esterified to *glycerol* (⟳ Section 2.7). By contrast, the lipids of *Thermomicrobium* are formed on *1,2-dialcohols* instead of glycerol, and have neither ester *nor* ether linkages (Figure 14.19). In addition, cells of *Thermomicrobium* contain only small amounts of peptidoglycan, and the cell wall is composed primarily of protein.

MINIQUIZ

- In what ways do *Chloroflexus* and *Roseiflexus* resemble *Chlorobium*? *Rhodobacter*?
- What is unique about *Thermomicrobium*?

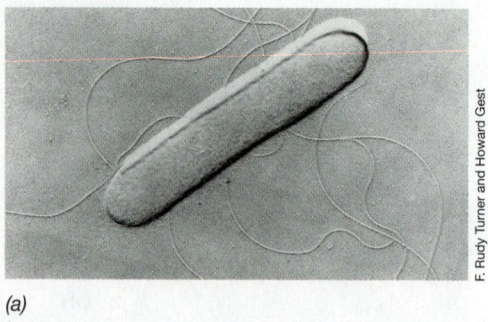
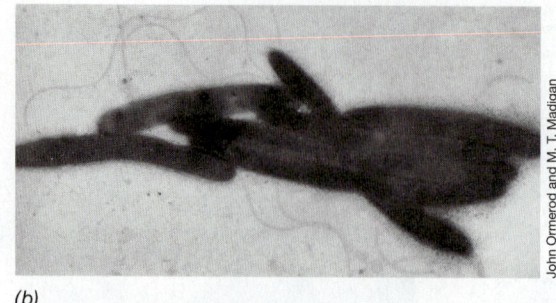
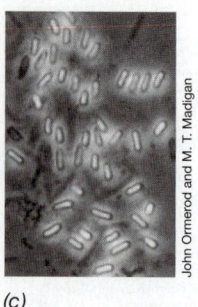

(a) (b) (c)

Figure 14.20 Cells and endospores of heliobacteria. *(a)* Electron micrograph of *Heliobacillus mobilis*, a peritrichously flagellated species. *(b)* *Heliophilum fasciatum* cell bundles as observed by electron microscopy. *(c)* Phase-contrast micrograph of endospores from *Heliobacterium gestii*. Most heliobacteria cells are about 1–2 μm in diameter.

14.8 Other Phototrophic *Bacteria*

Key Genera: *Heliobacterium, Chloracidobacterium*

Heliobacteria

Heliobacteria are a phylogenetically coherent group of phototrophic gram-positive *Bacteria* found within the phylum *Firmicutes*. The heliobacteria are anoxygenic phototrophs and produce a unique pigment, bacteriochlorophyll *g* (😊 Figure 13.3) and use a type I photosystem. Heliobacteria grow photoheterotrophically using a narrow range of organic compounds including pyruvate, lactate, acetate, or butyrate, and the group contains five genera: *Heliobacterium, Heliophilum, Heliorestis, Heliomonas,* and *Heliobacillus.* All known heliobacteria form rod-shaped or filamentous cells (**Figure 14.20**), although *Heliophilum* is unusual because its cells form into bundles (Figure 14.20*b*) that are motile as a unit.

Heliobacteria are strict anaerobes, but in addition to phototrophic growth, they can grow chemotrophically in darkness by pyruvate fermentation (as can many clostridia, close relatives of the heliobacteria). Heliobacteria produce endospores, the highly resistant structures produced by certain gram-positive bacteria (😊 Section 2.16). Like the endospores of *Bacillus* or *Clostridium* species, the endospores of heliobacteria (Figure 14.20*c*) contain elevated calcium (Ca^{2+}) levels and the signature molecule of the endospore, *dipicolinic acid.* Heliobacteria reside in soil, especially paddy (rice) field soils, where their nitrogen fixation activities may benefit rice productivity. A large diversity of heliobacteria have also been found in highly alkaline environments, such as soda lakes and surrounding alkaline soils.

Phototrophic *Acidobacteria*

A novel group of anoxygenic phototrophs has been discovered growing in photosynthetic microbial mats of a thermal spring in Yellowstone National Park. *Chloracidobacterium thermophilum* is a thermophilic oxygen-tolerant anoxygenic phototroph of the phylum *Acidobacteria* (😊 Section 15.21). Similar to green sulfur bacteria, *C. thermophilum* produces bacteriochlorophyll *a* and *c*, the latter in chlorosomes (**Figure 14.21**), and uses a type I photosystem. However, unlike green sulfur bacteria, *C. thermophilum* can also grow aerobically, as is true for the aerobic anoxygenic phototrophs (Section 14.5). In terms of its carbon metabolism, *C. thermophilum* is a photoheterotroph that uses short-chain fatty acids as electron donors for photosynthesis, but unlike green sulfur or green nonsulfur bacteria, it is incapable of autotrophy.

MINIQUIZ ---

- What types of anoxygenic phototrophs use chlorosomes?
- What kind of phototrophic bacteria make spores?

Chlorosome

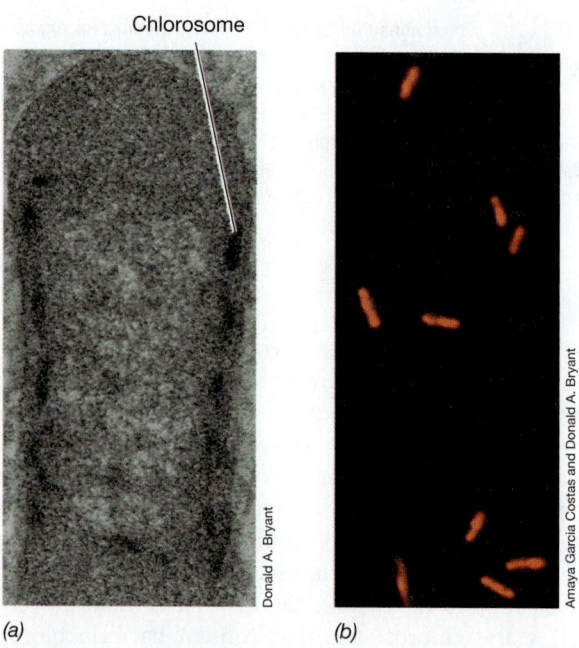

(a) (b)

Figure 14.21 Chlorosomes in *Chloracidobacterium thermophilum*, a phototrophic member of the phylum *Acidobacteria*. *(a)* Electron micrograph of *C. thermophilum* showing chlorosomes. *(b)* Fluorescence photomicrograph of *C. thermophilum.* The red color is the fluorescence of bacteriochlorophyll *c* present in chlorosomes. A cell of *C. thermophilum* is about 0.8 μm wide.

III • Bacterial Diversity in the Sulfur Cycle

Sulfur metabolism may have fueled the earliest forms of life on our planet (↩ Section 12.1), and the sulfur cycle (↩ Section 20.4) continues to support an enormous diversity of microorganisms. In this section we consider the diversity of organisms capable of *dissimilative sulfur metabolism*; that is, organisms that conserve energy through the oxidation or reduction of sulfur compounds (↩ Sections 13.8 and 13.18).

The remarkable diversity of prokaryotes capable of dissimilative sulfur metabolism is in part a function of the chemical diversity in which sulfur occurs in the biosphere. Sulfur has eight oxidation states that range from its most oxidized form, sulfate (SO_4^{2-}, oxidation state of +6), to thiosulfate ($S_2O_3^{2-}$, oxidation state of +2), to elemental sulfur (S^0, oxidation state of 0), and finally to hydrogen sulfide (H_2S, oxidation state of −2), its most reduced form. In addition, sulfur compounds can take on diverse chemical forms including inorganic sulfur compounds, organosulfur compounds, and metal sulfides.

In this section we will focus on the diversity of **dissimilative sulfate-reducers**, **dissimilative sulfur-reducers**, and **dissimilative sulfur-oxidizers**. Anoxygenic phototrophs, such as the purple and green sulfur bacteria discussed in Sections 14.4–14.6, are also important links in the sulfur cycle. However, here we restrict our focus to chemotrophic dissimilative metabolisms.

14.9 Dissimilative Sulfate-Reducing Bacteria

Key Genera: *Desulfovibrio, Desulfobacter*

Sulfate-reducing bacteria gain energy by coupling the oxidation of H_2 or organic compounds to the reduction of SO_4^{2-} (anaerobic respiration). There are more than 30 known genera of sulfate reducers found across five phyla of *Bacteria* and *Archaea* (**Figure 14.22**). Most sulfate reducers reside in the *Deltaproteobacteria*, though sulfate reducers are also found in the *Firmicutes* (e.g., *Desulfotomaculum* and *Desulfosporosinus*), *Thermodesulfobacteria* (e.g.,

Thermodesulfobacterium), and *Nitrospira* (e.g., *Thermodesulfovibrio*). Sulfate reduction also occurs in *Archaeoglobus,* a genus of the archaeal phylum *Euryarchaeota*.

Physiology of Sulfate-Reducing Bacteria

Sulfate-reducing bacteria are morphologically and biochemically diverse. The biochemistry of sulfate reduction was discussed in Section 13.18, so here we consider some of the more general physiological properties of this group. Sulfate reducers are generally obligate anaerobes, and strict anoxic techniques must be used in their cultivation (**Figure 14.23g**).

Sulfate reducers use H_2 or organic compounds as electron donors for growth, and the range of organics used is fairly broad. Lactate and pyruvate are almost universally used, and many species also oxidize short-chain alcohols (ethanol, propanol, and butanol) as electron donors. Some species, such as *Desulfosarcina* and *Desulfonema*, grow chemolithotrophically and autotrophically with H_2 as an electron donor, SO_4^{2-} as an electron acceptor, and CO_2 as the sole carbon source. A few sulfate reducers can oxidize hydrocarbons as electron donors (↩ Section 13.24).

There are two physiological types of dissimilative sulfate-reducers, the *complete oxidizer,* which can oxidize acetate and other fatty acids completely to CO_2, and the *incomplete oxidizers,* which are unable to oxidize acetate to CO_2. The latter group includes the best studied of the sulfate-reducing bacteria, *Desulfovibrio* (Figure 14.23a), along with *Desulfomonas, Desulfotomaculum,* and *Desulfobulbus* (Figure 14.23c). The acetate oxidizers include *Desulfobacter* (Figure 14.23d), *Desulfococcus, Desulfosarcina* (Figure 14.23e), and *Desulfonema* (Figure 14.23b), among many others. These bacteria specialize in the complete oxidation of fatty acids, in particular acetate, reducing SO_4^{2-} to H_2S. These two physiological groups are not phylogenetically coherent but instead are distributed widely across the phylogeny of sulfate-reducing bacteria (Figure 14.22).

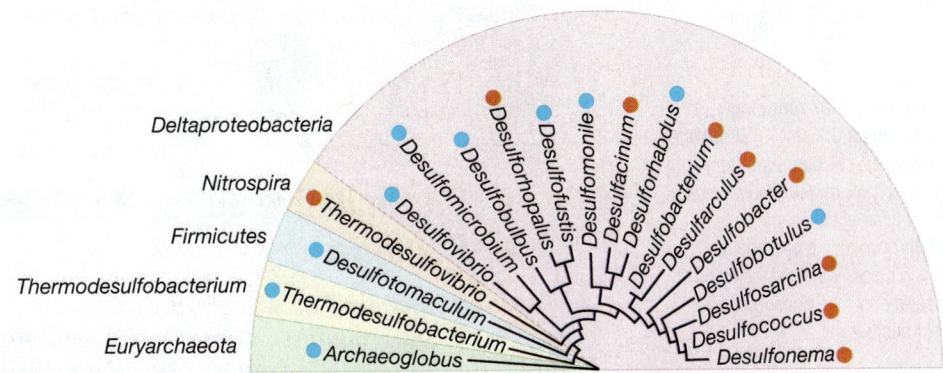

Figure 14.22 Dissimilative sulfate-reducers. The dendrogram depicts phylogenetic relationships among some genera of sulfate reducers as inferred by analysis of their 16S ribosomal RNA gene sequences. Color shading is used to differentiate the five main phyla that contain genera of sulfate reducers. Colored circles indicate whether species are complete oxidizers, which are able to oxidize acetate to CO_2, or incomplete oxidizers, which cannot oxidize acetate. The physiology of sulfate-reducing bacteria is covered in Section 13.18, and their role in the sulfur cycle in Section 20.4.

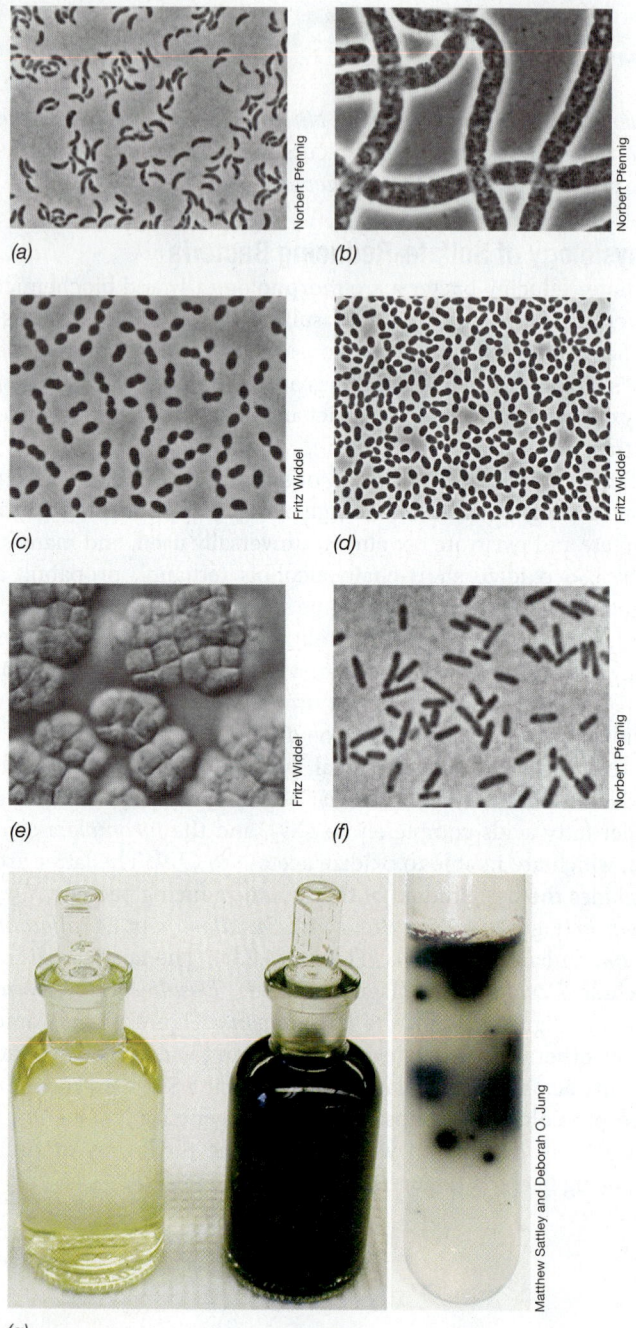

(a) *Norbert Pfennig*
(b) *Norbert Pfennig*
(c) *Fritz Widdel*
(d) *Fritz Widdel*
(e) *Fritz Widdel*
(f) *Norbert Pfennig*
(g) *Matthew Sattley and Deborah O. Jung*

Figure 14.23 **Representative sulfate-reducing and sulfur-reducing bacteria.** *(a) Desulfovibrio desulfuricans; cell diameter about 0.7 µm. (b) Desulfonema limicola; cell diameter 3 µm. (c) Desulfobulbus propionicus; cell diameter about 1.2 µm. (d) Desulfobacter postgatei; cell diameter about 1.5 µm. (e) Desulfosarcina variabilis; cell diameter about 1.25 µm. (f) Desulfuromonas acetoxidans; cell diameter about 0.6 µm. (g) Enrichment culture of sulfate-reducing bacteria. Left, sterile medium; center, a positive enrichment showing black FeS; right, colonies of sulfate-reducing bacteria in a dilution tube. Photos a–d and f are phase-contrast photomicrographs; part e is an interference contrast micrograph.*

Some sulfate-reducing bacteria can exploit alternative metabolic pathways. In addition to SO_4^{2-} or S^0, some sulfate reducers can also reduce nitrate and sulfonates (such as isethionate, HO—CH_2—CH_2—SO_3^-). Certain organic compounds can also be fermented by sulfate-reducing bacteria. The most common of these

is pyruvate, which is fermented by way of the phosphoroclastic reaction to acetate, CO_2, and H_2 (⮌ Figure 13.30). Moreover, although generally obligate anaerobes, a few sulfate-reducing bacteria are quite O_2-tolerant (primarily strains that coexist with O_2-producing cyanobacteria in microbial mats). At least one species, *Desulfovibrio oxyclinae*, can actually grow with O_2 as the electron acceptor under microaerophilic conditions.

Ecology of Sulfate-Reducing Bacteria

Sulfate reducers are widespread in aquatic and terrestrial environments that contain SO_4^{2-} and become anoxic as a result of microbial decomposition. Sulfate reducers are abundant in marine sediments, and the H_2S they generate is responsible for the pungent smell (like that of rotten eggs) often encountered near coastal ecosystems. *Desulfotomaculum*, phylogenetically a species of *Firmicutes* (gram-positive *Bacteria*), consists of endospore-forming rods found primarily in soil. Growth and reduction of SO_4^{2-} by *Desulfotomaculum* in certain canned foods leads to a type of spoilage called *sulfide stinker*. Species of *Thermodesulfobacterium*, *Thermodesulfovibrio*, and *Archaeoglobus* (an archaeon) are all thermophilic and found in geothermally heated environments such as hot springs, hydrothermal vents, and oil reserves. The remaining genera of sulfate reducers are indigenous to anoxic marine and freshwater environments and can occasionally be isolated from the mammalian gut.

The enrichment of *Desulfovibrio* species is straightforward in an anoxic lactate–sulfate medium containing ferrous iron (Fe^{2+}). A reducing agent, such as thioglycolate or ascorbate, is required to achieve a low reduction potential (E_0') in the medium. When sulfate-reducing bacteria grow, the H_2S formed from SO_4^{2-} reduction combines with the ferrous iron to form black, insoluble ferrous sulfide (Figure 14.23*g*). Purification can be accomplished by diluting the culture in molten agar tubes (⮌ Section 18.2 and Figure 18.3). Upon solidification, individual cells of sulfate-reducing bacteria become distributed throughout the agar and grow to form black colonies (Figure 14.23*g*) that can be removed aseptically to yield pure cultures.

MINIQUIZ

- What are the typical electron donors used by dissimilative sulfate-reducers?
- What bacterial phyla are known to contain dissimilative sulfate-reducers?

14.10 Dissimilative Sulfur-Reducing Bacteria

Key Genera: *Desulfuromonas, Wolinella, Sulfolobus*

Here we consider the dissimilative sulfur-reducers, microorganisms that are able to use respiratory sulfur reduction to conserve energy. Dissimilative sulfur-reducing bacteria can reduce S^0 and other oxidized forms of sulfur to H_2S but are unable to reduce SO_4^{2-}. There are more than 25 genera of dissimilative sulfur-reducers spread across five bacterial and archaeal phyla (Figure 14.1).

Most sulfur-reducing bacteria are *Proteobacteria*, primarily *Deltaproteobacteria* (e.g., *Desulfuromonas*, *Pelobacter*, *Desulfurella*, *Geobacter*), with some genera residing in the *Epsilon-* (e.g., *Wolinella* and *Sulfurospirillum*) and *Gammaproteobacteria* (e.g., *Shewanella* and *Pseudomonas mendocina*). Other sulfur-reducing bacteria are species of *Firmicutes* (e.g., *Desulfitobacterium* and *Ammonifex*), *Aquificae* (e.g., *Desulfurobacterium* and *Aquifex*), *Synergistetes* (e.g., *Dethiosulfovibrio*), or *Deferribacteres* (e.g., *Geovibrio*). The sulfur-reducing *Archaea*—of which there are many—are all genera of *Crenarchaeota* (e.g., *Acidianus*, *Sulfolobus*, *Pyrodictium*, and *Thermodiscus*).

Physiology and Ecology of Sulfur-Reducing Bacteria

The physiology of sulfur reducers is more diverse than that of sulfate reducers. Most sulfur reducers are obligate anaerobes, but facultatively aerobic species are common. Sulfur reducers are often able to reduce electron acceptors such as nitrate, ferrous iron, or thiosulfate as alternatives to S^0. Like sulfate reducers (Section 14.9), the physiology of sulfur reducers is characterized by whether they completely oxidize acetate and other fatty acids to CO_2. Species of *Desulfuromonas* (Figure 14.23*f*) are complete oxidizers that grow anaerobically by coupling the oxidation of acetate, succinate, ethanol, or propanol to the reduction of S^0. In contrast, *Sulfospirillum* and *Wolinella* are incomplete oxidizers and cannot use acetate as an electron donor. *Sulfospirillum* can reduce S^0 using either H_2 or formate as electron donor.

Dissimilative sulfur-reducing bacteria reside in many of the same habitats as dissimilative sulfate-reducing bacteria and often form associations with bacteria that oxidize H_2S to S^0, such as green sulfur bacteria (Section 14.6). The S^0 produced from H_2S oxidation is then reduced back to H_2S during metabolism of the sulfur reducer, completing an anoxic sulfur cycle (♊ Section 20.4).

MINIQUIZ
- What are the typical electron donors used by dissimilative sulfur-reducers?
- What bacterial phyla contain dissimilative sulfur-reducers?

14.11 Dissimilative Sulfur-Oxidizing Bacteria

Key Genera: *Thiobacillus, Achromatium, Beggiatoa*

Dissimilative sulfur-oxidizers are **chemolithotrophs** that oxidize reduced sulfur compounds such as H_2S, S^0, thiosulfate, or thiocyanate (^-SCN) as electron donors in energy conservation. These organisms are common in environments such as marine sediments, sulfur springs, and hydrothermal systems where H_2S produced by sulfate- or sulfur-reducing bacteria (Sections 14.9–14.10), or abiotically by geothermal reactions, is released into oxygenated waters (**Figure 14.24**). The sulfur oxidizers are found in three phyla of *Bacteria* (*Proteobacteria, Aquificae, Deinococcus–Thermus*) and one of *Archaea* (*Crenarchaeota*) (Figure 14.1). Most sulfur-oxidizing bacteria are *Beta-* (*Thiobacillus*), *Gamma-* (*Achromatium, Beggiatoa*) and *Epsilonproteobacteria* (*Thiovulum, Thiomicrospira*).

(a)

Michael F. McGlannan, Florida International University

(b)

Andreas Teske

Figure 14.24 Habitats of sulfur oxidizers. *(a)* A sulfide-containing artesian spring in Florida (USA). The outside of the spring is coated with a mat of *Thiothrix* (see Figure 14.26*b*). The mat is about 1.5 m in diameter. *(b)* Hydrothermal chimneys at Cathedral Hill in the Guaymas Basin, 2000 m depth. Sulfide-rich waters vent from the chimneys, which are covered by mats composed of orange, white, and yellow cells of *Beggiatoa*.

Physiological Diversity of Sulfur-Oxidizing Bacteria

The morphological and physiological diversity of sulfur oxidizers is significant. Cells can be less than 1 micrometer in diameter (e.g., *Thiomicrospira denitrificans*) or as large as 750 micrometers in diameter (e.g., *Thiomargarita namibiensis*). Most sulfur oxidizers are obligate aerobes; however, species of *Thiomargarita* and *Thiomicrospira* can also reduce NO_3^- in denitrification (♊ Section 13.17 and Section 14.13). Many species oxidize H_2S to elemental sulfur (S^0), which they deposit as either intracellular or extracellular granules for later use as an electron donor if H_2S becomes limiting.

Some sulfur chemolithotrophs are *obligate chemolithotrophs*, locked into a lifestyle of using inorganic instead of organic compounds as electron donors. When growing in this fashion, they are also autotrophs, converting CO_2 into cell material by reactions of the Calvin cycle. **Carboxysomes** are often present in cells of obligate chemolithotrophs (**Figure 14.25*a***). These structures contain high levels of Calvin cycle enzymes and probably increase the rate at which these organisms fix CO_2 (♊ Section 13.5).

Other sulfur chemolithotrophs are *facultative chemolithotrophs*, facultative in the sense that they can grow *either* chemolithotrophically (and thus, also as autotrophs) or chemoorganotrophically. Most species of *Beggiatoa* can obtain energy from the oxidation

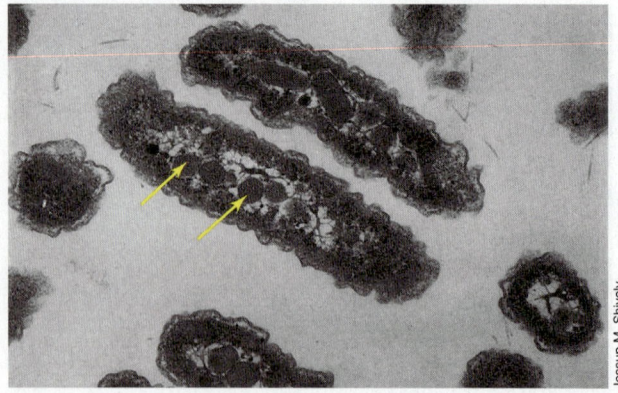

(a)

Jessup. M. Shively

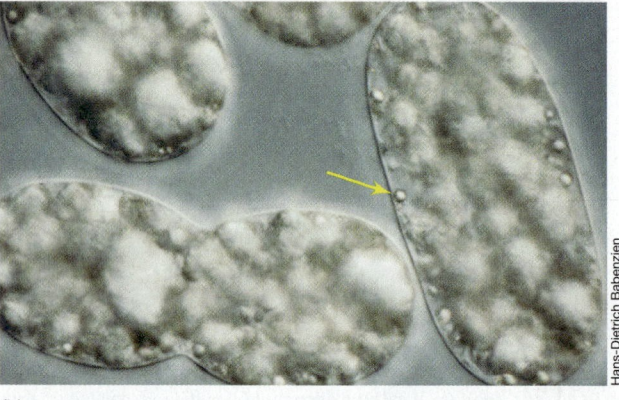

(b)

Hans-Dietrich Babenzien

Figure 14.25 **Nonfilamentous sulfur chemolithotrophs.** *(a)* Transmission electron micrograph of cells of the chemolithotrophic sulfur-oxidizer *Halothiobacillus neapolitanus.* A single cell is about 0.5 μm in diameter. Note the polyhedral bodies (carboxysomes) distributed throughout the cell (arrows) (⮎ Figure 13.18). *(b) Achromatium.* Cells photographed by differential interference contrast microscopy. The small globular structures near the periphery of the cells (arrow) are elemental sulfur, and the large granules are calcium carbonate. A single *Achromatium* cell is about 25 μm in diameter.

of inorganic sulfur compounds but lack enzymes of the Calvin cycle. They thus require organic compounds as carbon sources. Organisms with such a nutritional lifestyle are called **mixotrophs**.

Thiobacillus and *Achromatium*

The genus *Thiobacillus* and related genera contain several gram-negative, rod-shaped *Betaproteobacteria*, indistinguishable morphologically from most other gram-negative rods (Figure 14.25*a*); they are the best studied of the sulfur chemolithotrophs. The oxidation of H_2S, S^0, or thiosulfate by *Thiobacillus* generates sulfuric acid (H_2SO_4), and thus thiobacilli are often acidophilic. One highly acidophilic species, *Acidithiobacillus ferrooxidans*, can also grow chemolithotrophically by the oxidation of Fe^{2+} and is a major biological agent for the oxidation of this metal. Iron pyrite (FeS_2) is a major natural source of ferrous iron as well as of sulfide. The oxidation of FeS_2, especially in mining operations, can be both beneficial (because leaching of the ore releases the iron from the sulfide mineral) and ecologically disastrous (the environment can become acidic and contaminated with toxic metals such as aluminum, cadmium, and lead) (⮎ Section 21.2).

(a)

Verena Salman

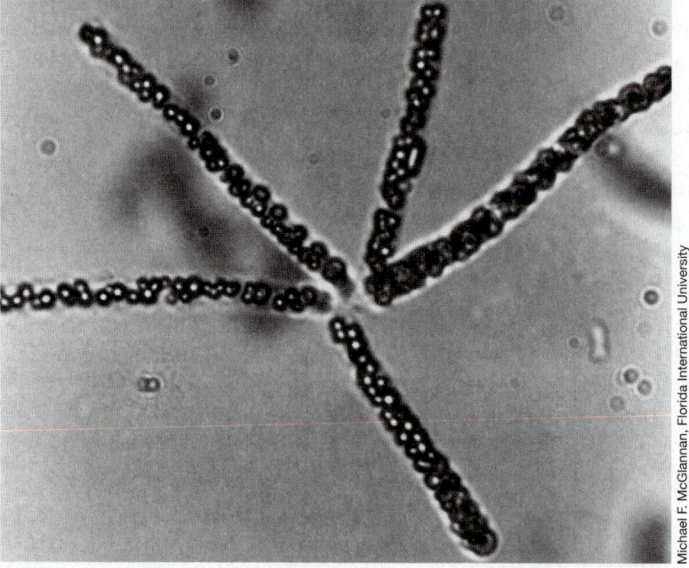

(b)

Michael F. McGlannan, Florida International University

Figure 14.26 *Thiothrix.* *(a)* Filaments of *Thiothrix* attached to plant material found in the outwash stream of a sulfidic cave in Frasassi, Italy. From the plant branch point, the longest branch is about 4 mm long. *(b)* Phase-contrast photomicrograph of a rosette of cells of *Thiothrix* isolated from the sulfide-containing artesian spring shown in Figure 14.24a. Note the internal sulfur globules produced from the oxidation of sulfide. Each filament is about 4 μm in diameter.

Achromatium is a spherical sulfur-oxidizing chemolithotroph that is common in freshwater sediments of neutral pH containing H_2S. Cells of *Achromatium* are large cocci that can have diameters of 10–100 μm (Figure 14.25*b*). *Achromatium* is a species of *Gammaproteobacteria* and is specifically related to purple sulfur bacteria, such as its phototrophic counterpart *Chromatium* (Section 14.4 and Figure 14.10*a*). Like *Chromatium*, cells of *Achromatium* store S^0 internally (Figure 14.25*b*); the granules later disappear as S^0 is oxidized to SO_4^{2-}. Cells of *Achromatium* also store large granules of calcite ($CaCO_3$) (Figure 14.25*b*), possibly as a carbon source (in the form of CO_2) for autotrophic growth. The physiology of chemolithotrophic sulfur oxidizers is discussed in Section 13.8.

Ecological Diversity and Strategies of Sulfide-Oxidizing Bacteria

Aerobic sulfide-oxidizers provide a case study that demonstrates the degree of ecological diversification that can occur among microorganisms that share the same basic metabolic features. The chemical oxidation of H_2S to H_2SO_4 is spontaneous and rapid in the presence of O_2. Hence, aerobic H_2S-oxidizers have evolved diverse ecological strategies that allow them to metabolize two molecules that otherwise react with each other spontaneously. We consider here six different strategies used by aerobic sulfide-oxidizers to cope with the chemical instability of H_2S in the presence of O_2.

1. *Thiothrix* is a filamentous sulfur chemolithotroph of the *Gammaproteobacteria* (**Figure 14.26**). *Thiothrix* forms filaments that group together at their ends by way of a holdfast to form cell arrangements called *rosettes* (Figure 14.26*b*). The ecological strategy of *Thiothrix* is to use its holdfast to position itself in high-flow environments downstream from a source of H_2S. Such environments are common near sulfur springs and in creeks draining sulfidic salt marshes where abundant H_2S is produced and carried away in waters rich with O_2 (Figure 14.26*a*). Physiologically, *Thiothrix* is an obligately aerobic mixotroph, and in this and most other respects it resembles *Beggiatoa*.

2. *Beggiatoa* are filamentous, gliding, sulfur-oxidizing *Gammaproteobacteria* that are usually large in both diameter and length, consisting of many short cells attached end to end (**Figure 14.27*a***). Filaments can flex and twist so that many filaments become intertwined to form a complex tuft. *Beggiatoa* is found primarily in microbial mats, sediments, sulfur springs, and hot springs. The ecological strategy of *Beggiatoa* is to use gliding motility to position itself at the point where H_2S and O_2 co-occur in an environment. For example, *Beggiatoa* in microbial mats can move vertically by as much as several centimeters per day in response to cyanobacterial O_2 production, moving up to obtain O_2 when photosynthesis ceases at night and down

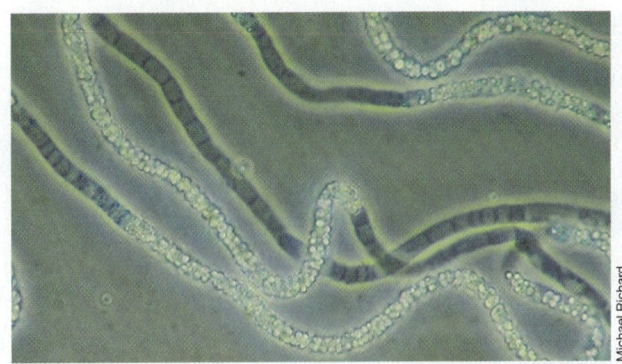

(a)

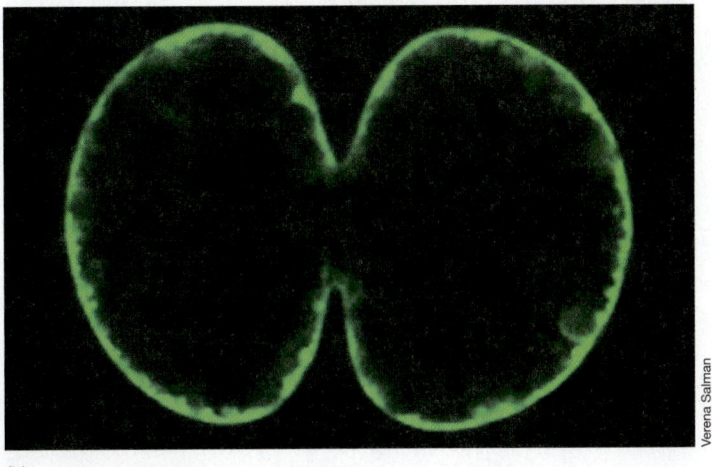

(b)

Figure 14.27   Filamentous sulfur-oxidizing bacteria. *(a)* Phase-contrast photomicrograph of a *Beggiatoa* species isolated from a sewage treatment plant. Note the abundant elemental sulfur granules in some of the cells. *(b)* Cells of a large marine *Thioploca* species. Cells contain sulfur granules (yellow) and are about 40–50 µm wide.

during the day when photosynthetic O_2 production at the mat surface causes H_2S to be found deeper in the mat.

3. The genus *Thiomargarita* contains some of the largest bacteria ever observed, with diameters that can be 0.75 mm (**Figure 14.28**).

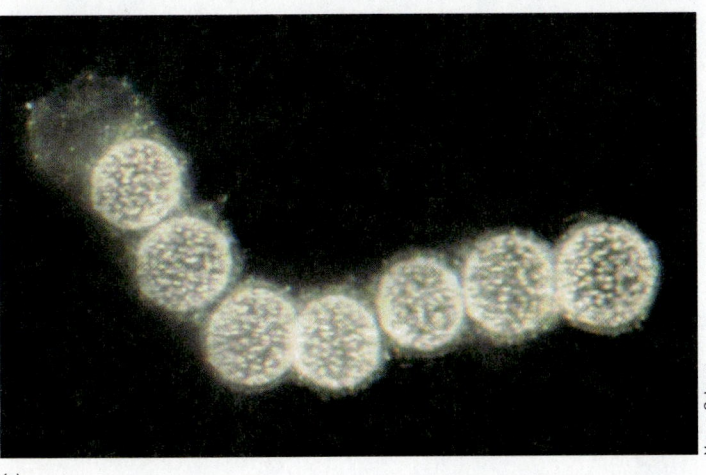

(a)

(b)

Figure 14.28   The giant sulfide-oxidizing bacterium, *Thiomargarita*. *(a) Thiomargarita namibiensis* recovered from the Namibian upwelling (off the Namibian coast, southwest Africa). Cells are about 100 µm in diameter. *(b)* Dividing cells of vacuole-containing sulfide-oxidizers recovered from the same location. Fluorescence micrograph showing ribosomes of *Thiomargarita* stained with a fluorescent nucleic acid probe. Ribosomes are found in the cytoplasm, which is present as a thin layer along the outer edge of the cells. The cytoplasm is squeezed between the cell wall and the large central vacuole, which appears dark in the image. Cells are about 50 µm wide.

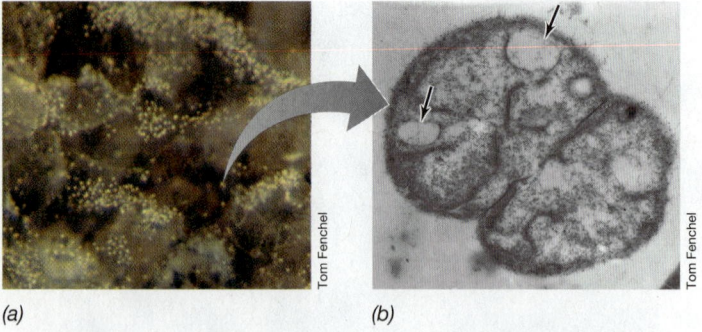

(a)   *(b)*

Figure 14.29 **The sulfur-oxidizing** *Thiovulum.* *(a)* Macrophotograph of cells of *Thiovulum* (yellow dots) that formed a thin veil in marine sand containing H_2S (large, irregular structures are sand grains). *(b)* Transmission electron micrograph of a dividing cell of *Thiovulum*. Sulfur (S^0) globules are shown with arrows. Single cells of *Thiovulum* are typically 10–20 μm in diameter.

Thiomargarita is nonmotile, and its ecological strategy is to separate in time the oxidation of H_2S from the reduction of O_2. To accomplish this, *Thiomargarita* contains a giant vacuole (Figure 14.28*b*) that it fills with high concentrations of nitrate (NO_3^-). This vacuole can fill almost the entire volume of the cell. Cells live in sulfide-rich marine sediments that are mixed occasionally with O_2-rich waters, such as that in salt marshes and in ocean upwelling zones. When buried in sediment, cells oxidize H_2S to S^0 anaerobically by reducing NO_3^- stored in the vacuole to ammonium (NH_4^+). They then store the S^0 as intracellular granules (Figure 14.28*a*). When turbulent waters mix the cells into the water column where H_2S is lacking, they switch to the aerobic oxidation of stored S^0. The energy they gain from S^0 oxidation is used to refill their vacuole with NO_3^- from the water column so they will be able to survive the next period of anoxia.

4. *Thioploca* are large filamentous bacteria that use a strategy similar to that of *Thiomargarita*. *Thioploca* also have intracellular S^0 granules and large vacuoles filled with NO_3^- (Figure 14.27*b*). However, filaments of *Thioploca* are motile by gliding and they occur in large sheaths that can be filled with many parallel filaments (Figure 14.27*b*). Sheaths are arranged vertically in the sediments and filaments glide up and down in the sheaths, going down to anaerobically respire H_2S using stored NO_3^- as electron acceptor and going up to aerobically respire S^0 and to refill their vacuoles with NO_3^- (↩ Figure 19.10).

5. *Thiovulum* are found in freshwater and marine habitats in which sulfide-rich muds interface with oxic zones (**Figure 14.29**). *Thiovulum* cells are fairly large (10–20 μm), and when motile,

they swim at exceptionally high speed, perhaps the fastest of all known bacteria (~0.6 mm/sec). The ecological strategy of *Thiovulum* is to actually control the flow of nutrients to cells. *Thiovulum* cells secrete a slime that links cells together in a veil-like structure that can be centimeters in diameter (Figure 14.29*a*). The veils, composed of many *Thiovulum* cells, are formed over a source of H_2S. Cells have long flagella that attach to the veil and to solid surfaces. Since the terminal end of the flagellum is attached and immobile, flagellar rotation causes cells to rotate along their flagellar axis. The simultaneous unidirectional rotation of all of the *Thiovulum* cells in the veil creates a flow of water through the veil, allowing the cells to generate and regulate the gradients of H_2S and O_2 they require to generate energy.

6. The final ecological strategy of sulfur chemolithotrophs is for the sulfur bacterium to form a symbiotic association with a eukaryote. There are diverse symbiotic associations in which the host provides a mechanism for regulating H_2S and O_2 levels and the sulfide-oxidizing symbiont fixes CO_2 and provides a source of carbon and energy to the host. The best example is the tubeworm *Riftia*, which contains sulfide-oxidizing endosymbionts and lives at deep-sea hydrothermal vents (↩ Section 22.12). A variety of other such symbiotic associations are present at hydrothermal vent ecosystems, including symbionts living in the gill tissue of the giant clam *Calyptogena magnifica* and on the surface of the yeti crab, which farms sulfide-oxidizing bacteria by waving its claws over sulfide-rich vent fluid. Symbioses involving invertebrates are also common in the sulfide-rich marine sediments of shallow coastal systems. For example, bivalves in the family *Solemyidae* burrow into sulfide-rich sediments and pump sulfide- and oxygen-rich water over gills that contain sulfide-oxidizing bacteria.

From these examples it should be clear how ecological diversity drives bacteria that carry out the same energy metabolism—in this case sulfide oxidation—to best exploit the different environments they inhabit. In each case, the goal of the organism is the same, to obtain the electron donor and acceptor it needs. But also in each case, the strategy to accomplish this is unique and the best fit to both the properties of the organism and the habitat it exploits.

MINIQUIZ -
- Describe the energy and carbon metabolism of *Thiobacillus* in terms of how ATP and new cell material are made.
- What are some ecological strategies that sulfur oxidizers use to compete with chemical oxidation of H_2S?

IV • Bacterial Diversity in the Nitrogen Cycle

All forms of life must assimilate nitrogen for growth and thus all organisms must catalyze certain nitrogen transformations. The *Bacteria* and *Archaea*, however, are the only domains in which representatives exist that can conserve energy from the transformation of inorganic nitrogen species. In this section we will consider the

diversity of three physiological groups of bacteria that participate in the nitrogen cycle: *diazotrophs*, *nitrifiers*, and *denitrifiers*. The physiology of these groups was considered in Sections 3.17, 13.10, and 13.17. We start our tour of microbial diversity in the nitrogen cycle by considering those bacteria that fix atmospheric nitrogen.

14.12 Diversity of Nitrogen-Fixing *Bacteria*

Key Genera: *Mesorhizobium, Desulfovibrio, Azotobacter*

Diazotrophs are microorganisms that fix dinitrogen gas (N_2) into NH_3 that can be assimilated as a source of nitrogen for cells. Nitrogen fixation is an assimilative process and requires ATP and the enzyme nitrogenase (⮌ Section 3.17). Diazotrophs typically fix N_2 only when other forms of N are absent, and nitrogenase expression is inhibited when NH_3 is available to cells (⮌ Section 3.17). Nitrogenase is irreversibly inhibited by O_2 and this is one cause of ecological diversification among diazotrophs; we will see that different organisms have evolved different solutions to protecting nitrogenase from O_2.

Nitrogen fixation is widespread among microorganisms, and it is thought that the last universal common ancestor possessed the enzyme nitrogenase. The *nifH* gene encodes the dinitrogenase reductase component of nitrogenase and is used as a measure of diazotroph diversity (⮌ Section 18.5). More than 30,000 unique *nifH* gene sequences have been described spanning nine bacterial phyla and one archaeal phylum (Figure 14.1). The phylogenetic distribution of nitrogenase in the tree of life has been influenced strongly by horizontal gene exchange. As a result, the phylogeny of *nifH* is largely inconsistent with the 16S ribosomal RNA gene phylogeny (**Figure 14.30**). We consider here the diversity of both symbiotic and free-living diazotrophs.

Symbiotic Diazotrophs

Diazotrophs form several symbiotic relationships with plants, animals, and fungi. These relationships are generally defined by the host providing a hospitable environment, including a source of carbon and energy and a system for regulating oxygen concentrations, and the microbial symbiont providing in return a supply of fixed nitrogen to the host.

The symbiosis between rhizobia and leguminous plants is one of the best-characterized nitrogen-fixing symbiotic associations (⮌ Section 22.3). Root-nodule-forming bacteria are *Alphaproteobacteria* (e.g., *Mesorhizobium, Bradyrhizobium, Sinorhizobium*), *Betaproteobacteria* (e.g., *Burkholderia*), or *Actinobacteria* (e.g., *Frankia*). Other genera of symbiotic diazotrophs are found in association with shipworms (*Teredinibacter*), termite guts (*Treponema*) (⮌ Section 22.10), endomycorrhizal fungi (*Glomeribacter*) (⮌ Section 17.12), and several fungi, algae, and plants (*Cyanobacteria*) (⮌ Sections 22.1 and 22.5). These different symbioses have evolved independently multiple times as a result of convergent evolution (Figure 14.30).

Free-Living Diazotrophs

Free-living diazotrophs need a mechanism for protecting nitrogenase from oxygen (⮌ Sections 3.17 and 7.13). The simplest solution to this problem is to grow only in anoxic environments. The origin of nitrogen fixation predates the origin of oxygenic photosynthesis and thus the first nitrogen-fixing organisms were free-living anaerobes. Obligately anaerobic free-living diazotrophs are common in anoxic environments including marine and freshwater sediments and microbial mats. Obligately anaerobic free-living diazotrophs are found in the bacterial phyla *Firmicutes* (e.g., *Clostridium*), *Chloroflexi* (e.g., *Oscillochloris*), *Chlorobi* (e.g.,

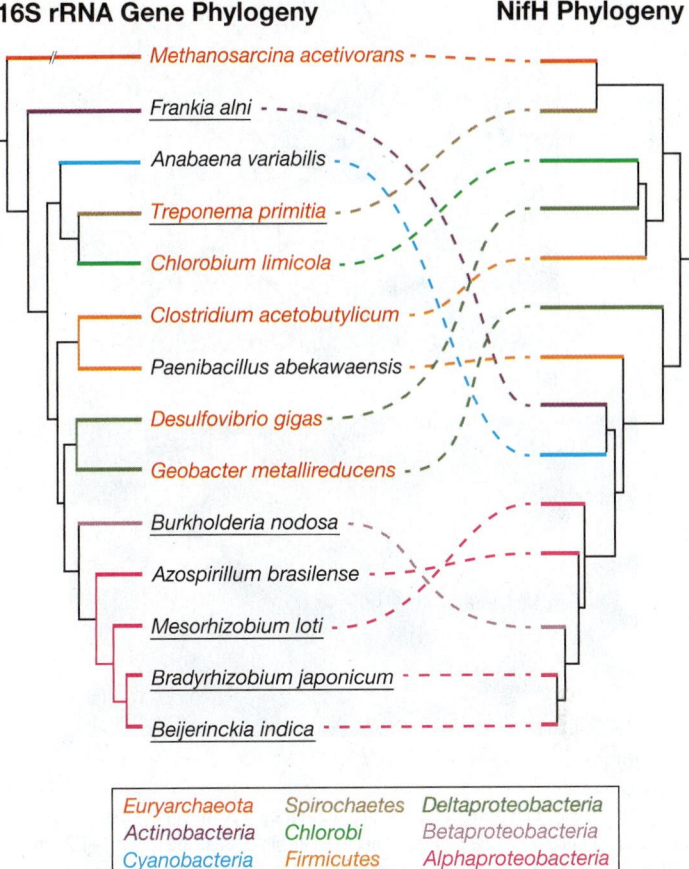

16S rRNA Gene Phylogeny | **NifH Phylogeny**

Euryarchaeota	Spirochaetes	Deltaproteobacteria
Actinobacteria	Chlorobi	Betaproteobacteria
Cyanobacteria	Firmicutes	Alphaproteobacteria

Figure 14.30 **Relationships among diazotrophic (nitrogen-fixing) bacteria as inferred from 16S ribosomal RNA gene sequences and NifH amino acid sequences.** Branches in each tree are colored to indicate phyla. The dashed lines indicate branches shared between the two trees. The incongruence between the two trees has resulted from multiple horizontal transfer events of the *nifH* gene. Red text denotes obligate anaerobes and underlined text indicates species that form symbioses with *Eukarya*.

Chlorobium), *Spirochaetes* (e.g., *Spirochaeta*), and *Proteobacteria* (e.g., *Desulfovibrio, Chromatium*) and in the archaeal phylum *Euryarchaeota* (e.g., *Methanosarcina*). *Desulfovibrio* occur in anoxic salt marsh sediments dominated by *Spartina* grass, and their N_2 fixation is an important nitrogen source to plants that live in this ecosystem.

Other simple mechanisms for protecting nitrogenase from oxygen include fixing N_2 only at times when oxygen is absent or present in low concentration. For example, facultative aerobes will often fix N_2 only while growing anaerobically (e.g., *Klebsiella*). Some aerobic nitrogen-fixers are *microaerophiles*; these organisms fix nitrogen only in environments where oxygen is present at low concentration (typically less than 2%). However, some organisms have evolved more complex mechanisms for protecting nitrogenase from oxygen and are able to grow in the presence of air.

Obligately aerobic free-living diazotrophs include the *Cyanobacteria*, which have evolved a variety of mechanisms of protecting nitrogenase from oxygen (Section 14.3), as well as a variety of unicellular free-living chemoorganotrophic bacteria. Obligately aerobic free-living diazotrophs include *Azotobacter, Azospirillum*, and *Beijerinckia*. *Azotobacter* cells are large rods or cocci

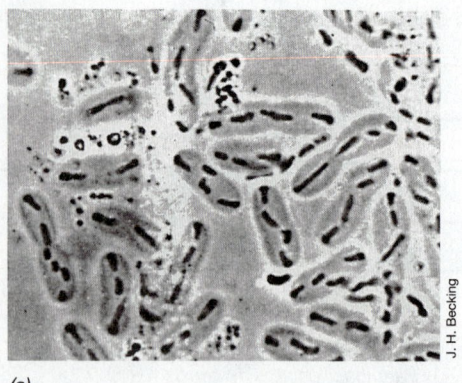

(a)

J. H. Becking

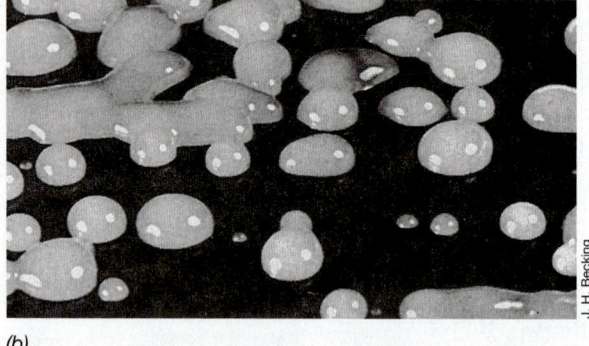

(b)

J. H. Becking

Figure 14.31 **Examples of slime production by free-living N₂-fixing bacteria.** (a) Cells of *Derxia gummosa* encased in slime. Cells are about 1–1.2 μm wide. (b) Colonies of *Beijerinckia* species growing on a carbohydrate-containing medium. Note the raised, glistening appearance of the colonies due to abundant capsular slime.

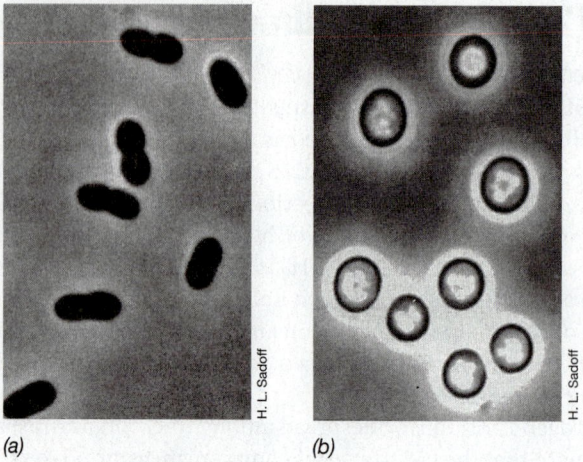

(a)   H. L. Sadoff   (b)   H. L. Sadoff

Figure 14.32 *Azotobacter vinelandii.* (a) Vegetative cells and (b) cysts visualized by phase-contrast microscopy. A cell measures about 2 μm in diameter and a cyst about 3 μm.

with diameters of 2–4 μm or more. When they are growing on N₂ as a nitrogen source, extensive capsules or slime layers are typically produced (**Figure 14.31**) (⟳ Figure 3.32*a*). It is thought that the high respiratory rate characteristic of *Azotobacter* cells and the abundant capsular slime they produce help protect nitrogenase from O₂. *Azotobacter* is able to grow on many different carbohydrates, alcohols, and organic acids, and metabolism is strictly oxidative.

Azotobacter can form resting structures called *cysts* (**Figure 14.32***b*). Like bacterial endospores, *Azotobacter* cysts show negligible endogenous respiration and are resistant to desiccation, mechanical disintegration, and ultraviolet and ionizing radiation. In contrast to endospores, however, cysts are not very heat-resistant, and they are not completely dormant because they rapidly oxidize carbon sources if supplied.

Azotobacter and Alternative Nitrogenases

We considered the important process of biological N₂ fixation in Section 3.17 and discussed the central importance of the metals molybdenum (Mo) and iron (Fe) to the enzyme nitrogenase. The species *Azotobacter chroococcum* was the first nitrogen-fixing bacterium shown capable of growth on N₂ in the absence of molybdenum. It was shown in *A. chroococcum* that either of two "alternative nitrogenases" are formed when Mo limitation prevents the normal MoFe nitrogenase from being synthesized. These nitrogenases are less efficient than the MoFe nitrogenase

and contain either vanadium (V) or Fe in place of Mo. The three different types of nitrogenase (MoFe, VFe, and FeFe) are encoded by paralogous genes, and these three related enzyme types arose as a result of gene duplication events (⟳ Section 12.6). Subsequent investigations of other nitrogen-fixing bacteria have shown that these genetically distinct "backup" nitrogenases are widely distributed among nitrogen-fixing bacteria, in particular in the *Cyanobacteria* and *Archaea*.

MINIQUIZ

• What mechanisms do free-living diazotrophs use to protect nitrogenase from oxygen?

• Where might you expect to find nitrogen-fixing bacteria?

14.13 Diversity of Nitrifying and Denitrifying *Bacteria* and *Archaea*

Microorganisms that grow by the anaerobic respiration of inorganic nitrogen (NO₃⁻, NO₂⁻) to the gaseous products NO, N₂O, and N₂ are called **denitrifiers** (⟳ Section 13.17). These organisms are typically facultative aerobes and chemoorganotrophs that use organic carbon as both carbon source and electron donor.

Microorganisms able to grow chemolithotrophically at the expense of reduced inorganic nitrogen compounds (NH₃, NO₂⁻) are called **nitrifiers** (**Figure 14.33**) (⟳ Section 13.10). These organisms are typically obligate aerobes that can also grow autotrophically; most species fix CO₂ by the Calvin cycle. A few species have also been shown to grow mixotrophically by assimilating organic carbon in addition to CO₂.

Physiology of Nitrifying *Bacteria* and *Archaea*

No chemolithotroph is known that carries out the complete oxidation of NH₃ to nitrate (NO₃⁻). Thus, nitrification results from the sequential activities of two physiological groups of organisms, the *ammonia oxidizers* (which oxidize NH₃ to

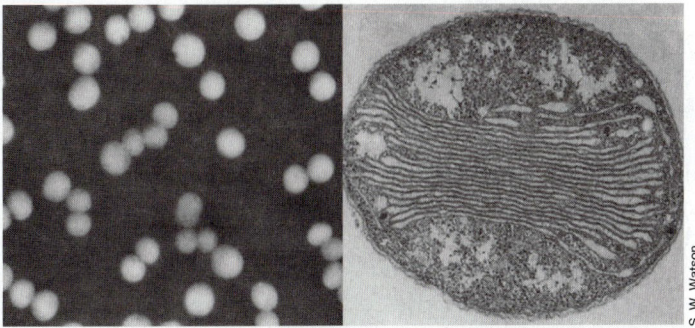

Reaction: $NH_3 + 1\frac{1}{2} O_2 \longrightarrow NO_2^- + H_2O$

(a)

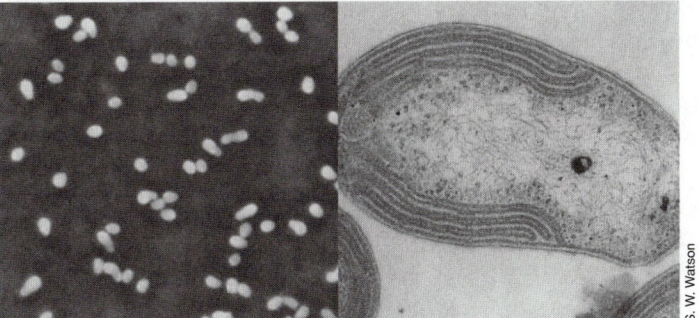

Reaction: $NO_2^- + \frac{1}{2} O_2 \longrightarrow NO_3^-$

(b)

Figure 14.33 **Nitrifying bacteria.** *(a)* Phase-contrast photomicrograph (left) and electron micrograph (right) of the ammonia-oxidizing bacterium *Nitrosococcus oceani.* A single cell is about 2 μm in diameter. *(b)* Phase-contrast photomicrograph (left) and electron micrograph (right) of the nitrite-oxidizing bacterium *Nitrobacter winogradskyi.* A cell is about 0.7 μm in diameter. Beneath each panel is the chemolithotrophic reaction that each organism catalyzes. The distinct internal membranes of each species are sites of key enzymes of nitrification.

nitrite, NO_2^-) (Figure 14.33*a*), and the *nitrite oxidizers*, the actual nitrate-producing microorganisms, which oxidize NO_2^- to NO_3^- (Figure 14.33*b*). Ammonia oxidizers typically have genus names beginning in *Nitroso-*, whereas genus names of nitrate producers begin with *Nitro-*.

Many species of nitrifiers have internal membrane stacks (Figure 14.33) that closely resemble the photosynthetic membranes found in their close phylogenetic relatives, the purple phototrophic bacteria (Section 14.4) and the methane-oxidizing (methanotrophic) bacteria (Section 14.17). The membranes are the location of key enzymes in nitrification: *ammonia monooxygenase*, which oxidizes NH_3 to hydroxylamine (NH_2OH), and *nitrite oxidoreductase*, which oxidizes NO_2^- to NO_3^- (↩ Section 13.10).

Enrichment cultures of nitrifying bacteria can be achieved using mineral salts media containing NH_3 or NO_2^- as electron donors and bicarbonate (HCO_3^-) as the sole carbon source. Because these organisms produce very little ATP from their electron donors (↩ Section 13.10), visible turbidity may not develop in cultures even after extensive nitrification has occurred. An easy means of monitoring growth is thus to assay for the production of NO_2^- (with NH_3 as electron donor) or NO_3^- (with NO_2^- as electron donor).

Nitrifying *Bacteria* and *Archaea*: Ammonia Oxidizers

Key Genera: *Nitrosomonas, Nitrosospira, Nitrosopumilus*

Ammonia oxidizers are found in the *Beta-* (e.g., *Nitrosomonas, Nitrosospira, Nitrosolobus, Nitrosovibrio*) and *Gammaproteobacteria* (*Nitrosococcus*), and in the archaeal phylum *Thaumarchaeota* (*Nitrosopumilus, Nitrosocaldus, Nitrosoarchaeum, Nitrososphaera*).

Ammonia oxidizers are widespread in soil and water. Bacterial ammonia-oxidizers are present in highest numbers in habitats where NH_3 is abundant, such as sites with extensive protein decomposition (ammonification), and also in sewage treatment facilities (↩ Section 21.6). Nitrifying bacteria develop especially well in lakes and streams that receive inputs of sewage or other wastewaters because these are frequently high in NH_3. *Nitrosomonas* is often observed in the activated sludge present in aerobic wastewater treatment facilities. Bacterial ammonia-oxidizers are also common in soils (e.g., *Nitrosospira, Nitrosovibrio*) and in the oceans (e.g., *Nitrosococcus*).

Archaeal ammonia-oxidizers (↩ Section 16.6) appear to be most common in habitats where NH_3 is present in low concentration. These organisms are thought to be the dominant ammonia oxidizers in the oceans where ammonia levels are very low (↩ Sections 19.9 and 19.11). Archaeal ammonia-oxidizers are also common in soils, and in some soils they outnumber bacterial ammonia-oxidizers by several orders of magnitude. The availability of NH_3 relative to NH_4^+ declines with pH, and thus acid soils (pH < 6.5), which are common, may favor organisms able to grow at low NH_3 concentration.

Nitrifying Bacteria: Nitrite Oxidizers

Key Genera: *Nitrospira, Nitrobacter*

Nitrite oxidizers are found in the classes *Alpha-* (*Nitrobacter*), *Beta-* (*Nitrotoga*), *Gamma-* (*Nitrococcus*), and *Deltaproteobacteria* (*Nitrospina*), as well as in the phylum *Nitrospira* (↩ Section 15.21).

Like nitrite-oxidizing *Proteobacteria, Nitrospira* oxidizes nitrite (NO_2^-) to nitrate (NO_3^-) and grows autotrophically (**Figure 14.34**). However, *Nitrospira* lacks the extensive internal membranes

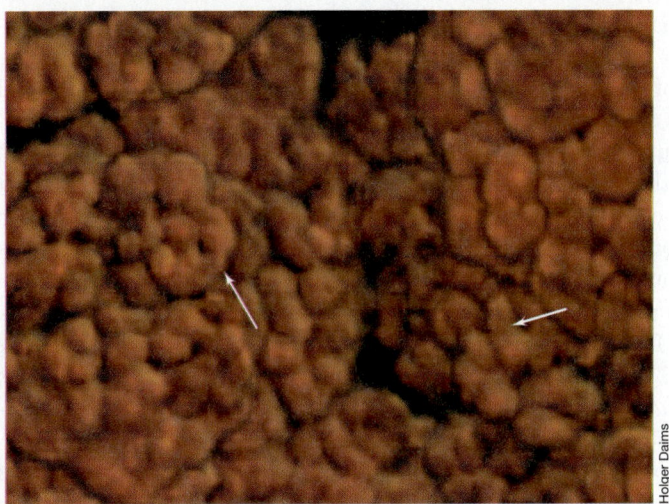

Figure 14.34 **The nitrifying bacterium *Nitrospira.*** An aggregate of *Nitrospira* cells enriched from activated sludge from a wastewater treatment facility. Individual cells are curved (arrows) and group into tetrads in the aggregate. A single cell of *Nitrospira* is about 0.3 × 1–2 μm.

UNIT 3

found in species of nitrifying *Proteobacteria*. Nevertheless, *Nitrospira* inhabits many of the same environments as nitrite-oxidizing *Proteobacteria* such as *Nitrobacter*, so it has been suggested that its capacity for NO_2^- oxidation may have been acquired by horizontal gene flow from nitrifying *Proteobacteria* (or vice versa). As we know, this mechanism for acquiring physiological traits has been widely exploited in the bacterial world (⮌ Sections 6.12 and 12.5). However, environmental surveys for the presence of nitrifying bacteria in nature have shown *Nitrospira* to be much more abundant than *Nitrobacter*; thus most of the NO_2^- oxidized in natural environments is probably due to the activities of *Nitrospira*.

Denitrifying *Bacteria* and *Archaea*

Key Genera: *Paracoccus, Pseudomonas*

Denitrifiers are capable of growth by the anaerobic respiration of NO_3^- or NO_2^- to the gaseous products NO, N_2O, and N_2 (⮌ Section 13.17). Nearly all denitrifiers are chemoorganotrophs that use organic carbon as both carbon source and electron donor. Exceptions include the denitrifying sulfur-oxidizers discussed in Section 14.11. Denitrifiers are typically facultative aerobes and in nearly all cases will grow preferentially as aerobes if O_2 is present.

Denitrifiers are of great importance in agricultural soils where they cause the loss of nitrogen fertilizers and the production of N_2O, which is a dominant component of greenhouse gases produced by agricultural soils (⮌ Section 20.8).

Denitrifiers are phylogenetically and metabolically diverse and include two archaeal phyla and six bacterial phyla, including five classes of *Proteobacteria* (Figure 14.1). One of the best-characterized denitrifiers is *Paracoccus denitrificans* (*Alphaproteobacteria*). Denitrification of NO_3^- to N_2 requires several key enzymatic steps (⮌ Section 13.17), and the genes that encode these enzymes are present throughout the tree of life. Many nitrate reducers possess only part of the denitrification pathway, and are thus unable to reduce NO_3^- completely to N_2. The phylogenetic distribution of the genes that encode denitrification has been strongly influenced by horizontal gene exchange.

MINIQUIZ ---

- Under what conditions would you expect microorganisms to grow as a result of denitrification?
- What traits are shared among ammonia and nitrite oxidizers?

V • Diversity of Other Distinctive Chemotrophic *Bacteria*

We continue our focus on functional groups whose physiological and ecological traits span different phyla as a result of convergent evolution or horizontal gene transfer. From a physiological standpoint, all groups here are chemotrophs—either chemolithotrophs or chemoorganotrophs—that contribute to specific steps in the carbon cycle or that metabolize hydrogen or metals.

14.14 Dissimilative Iron-Reducing *Bacteria*

Key Genera: *Geobacter, Shewanella*

Dissimilative iron-reducers couple the reduction of oxidized metals or metalloids to cellular growth. These organisms need to overcome the fundamental obstacle of using an insoluble solid material as an electron acceptor in respiration. A variety of microorganisms are able to enzymatically reduce metals as a consequence of either fermentation reactions or sulfur or sulfate reduction, but such organisms do not conserve energy from metal reduction. In contrast, dissimilative iron-reducers carry out metal respiration by coupling the oxidation of H_2 or organic compounds to the reduction of ferric iron (Fe^{3+}) or manganese (Mn^{6+}) (**Figure 14.35a**).

Dissimilative iron-reducers are phylogenetically diverse (Figure 14.1). Bacterial genera are found in the *Proteobacteria* (*Geobacter, Shewanella*), *Acidobacteria* (*Geothrix*), *Deferribacteres* (*Geovibrio*), *Deinococcus–Thermus* (*Thermus*), *Thermotogae* (*Thermotoga*), and *Firmicutes* (*Bacillus, Thiobacillus*), while archaeal genera are found in the *Crenarchaeota* (*Pyrobaculum*). Iron respiration likely evolved early in the history of life and its wide distribution may be due to its presence in the universal ancestor coupled with subsequent gene loss in some lineages and horizontal gene transfer to others.

Physiology

Dissimilative iron-reducers specialize in insoluble external electron acceptors, and these organisms are typically extremely versatile at anaerobic respiration. Dissimilative iron-reducers are unusual in that they possess outer membrane cytochromes that facilitate electron transfer with insoluble minerals. Most species are able to use either iron oxides or manganese oxides as electron acceptors and various species are also able to use nitrate, fumarate, inorganic sulfur compounds, cobalt, chromium, uranium, selenium, arsenic, and humic compounds (⮌ Section 13.21). Most genera of iron-reducing bacteria are obligate anaerobes, but some, such as *Shewanella* and relatives, are facultative aerobes. Electron donors are typically organic compounds such as fatty acids, alcohols, sugars, and in certain cases, even aromatic compounds. Many species are also able to use H_2 as an electron donor, but they are generally unable to grow autotrophically, requiring a source of organic carbon to support growth.

The family *Geobacteraceae* in the *Deltaproteobacteria* contains four genera of dissimilative iron-reducing bacteria (*Geobacter, Desulfuromonas, Desulfuromusa, Pelobacter*) that aptly demonstrate the physiological diversity of the obligately anaerobic metal reducers. *Geobacter, Desulfuromonas*, and *Desulfuromusa* can all use acetate as an electron donor as well as a diversity of other small organics, and they oxidize these substrates completely to CO_2. These genera typically specialize in anaerobic respiration. *Geobacter* in particular can use a wide range of electron donors and acceptors. *Geobacter* produce pili (Figure 14.35b) that contain cytochromes (Figure 14.35c), and these pili facilitate electron transfer to the surface of iron oxide minerals (for how this might occur, see the opening page of Chapter 20). *Pelobacter*, in contrast, are primarily fermentative organisms having a more limited

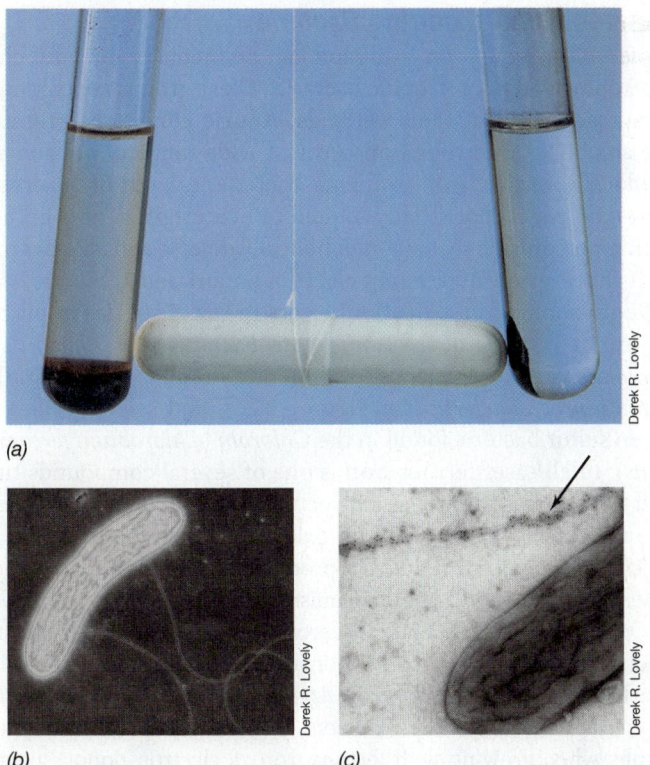

(a)

(b) · Derek R. Lovely   (c) · Derek R. Lovely

Figure 14.35 **The dissimilative iron-reducing bacterium *Geobacter*.**
(a) The uninoculated tube (left) contains an anoxic medium that includes acetate and ferrihydrite, a poorly magnetic iron oxide. Following growth of *Geobacter* (right tube) the ferrihydrite is reduced to magnetite, which is magnetic. *(b)* Transmission electron micrograph of *Geobacter sulfurreducens* showing flagella and pili. The cell is about 0.7×3.5 μm. *(c)* Transmission electron micrograph of *G. sulfurreducens* showing immunogold labeling of cytochrome OmcS on the pili (arrow).

respiratory capacity. For example, *Pelobacter carbinolicus* can only use lactate as the electron donor and can only use ferric iron or S^0 as the electron acceptors. *Pelobacter* are unable to oxidize their carbon substrates completely to CO_2.

Shewanella and its relatives *Ferrimonas* and *Aeromonas* in the *Gammaproteobacteria* are facultative aerobes and will grow aerobically when O_2 is available. *Shewanella* are able to use a wide diversity of electron donors and acceptors in addition to ferric iron and manganese. However, like *Pelobacter,* they are unable to oxidize their carbon substrates completely to CO_2 and are unable to oxidize acetate as an electron donor for anaerobic respiration.

Ecology

Dissimilative iron-reducers are common in anoxic freshwater and marine sediments. These organisms are thought to play an important role in organic matter oxidation in many anoxic habitats. Dissimilative iron-reducers are also common in the deep subsurface, found both in shallow aquifers as well in the deep subsurface environment (♻ Section 19.7). In addition, several thermophilic and hyperthermophilic iron-reducing species are known (e.g., *Thermus, Thermotoga*) and are often found in hot springs and other geothermally heated systems, including the deep subsurface.

14.15 Dissimilative Iron-Oxidizing *Bacteria*

Key Genera: *Acidithiobacillus, Gallionella*

The ability to couple the oxidation of ferrous iron (Fe^{2+}) to cell growth is widespread in the tree of life and thought to be a trait that evolved early in Earth's history. Genera capable of using ferrous iron as an electron donor to support growth are spread across five bacterial and two archaeal phyla (Figure 14.1).

Aerobic iron-oxidizer diversity and distribution are influenced strongly by pH and O_2. Ferrous iron oxidizes spontaneously to form insoluble precipitates in the presence of O_2 at neutral to alkaline pH (pH > 7) but is stable either under anoxic conditions or aerobically at acidic pH (pH < 4). Iron oxidizers can be divided into four functional groups on the basis of their basic physiology: acidophilic aerobic iron-oxidizers, neutrophilic aerobic iron-oxidizers, anaerobic chemotrophic iron-oxidizers, and anaerobic phototrophic iron-oxidizers.

Acidophilic Aerobic Iron-Oxidizing *Bacteria*

The growth of iron-oxidizing bacteria is favored in iron-rich acidic environments where soluble ferrous iron is present. Aerobic iron-oxidizers are often abundant in acid mine drainage generated from abandoned coal or iron mines or from mine tailings (♻ Sections 21.1 and 21.2). Acidophilic aerobic iron-oxidizers also inhabit iron-rich acidic springs in volcanic areas. In these environments, sulfur is often present along with ferrous iron, and many acidophilic aerobic iron-oxidizers are able to oxidize both elemental sulfur and ferrous iron. Species can be either autotrophic or heterotrophic, and commonly observed genera include *Acidithiobacillus* (*Gammaproteobacteria*), *Leptospirillum* (*Nitrospirae*), and *Ferroplasma* (*Euryarchaeota*). Other acidophilic aerobic iron-oxidizers can be found in the *Actinobacteria* and *Firmicutes*.

Neutrophilic Aerobic Iron-Oxidizing *Bacteria*

Neutrophilic aerobic iron-oxidizers are organisms adapted to a specialized niche (♻ Section 13.9). This is because ferrous iron is relatively insoluble at neutral pH and its chemical oxidation is spontaneous and rapid in the presence of air. Furthermore, at neutral pH, iron oxidation at the cell surface causes the formation of an iron oxide crust that can effectively entomb growing cells. Neutrophilic aerobic iron-oxidizers therefore thrive where iron-rich anoxic waters are exposed to air. Such habitats are common near wetlands or soils where anoxic groundwater forms a spring, but iron oxidizers also inhabit the rhizosphere of wetland plants and certain submarine hydrothermal systems.

Few genera of neutrophilic aerobic iron-oxidizers have been described and they all belong to the *Proteobacteria*. Those species found in freshwater habitats belong to a set of closely related genera in the *Betaprotobacteria*, while species found in marine habitats belong to the *Zetaproteobacteria*. The metabolism of these organisms is fairly narrow. Species are typically micro-aerophiles and obligate chemolithotrophs, though in certain cases mixotrophy has been observed. The genera *Leptothrix* and *Sphaerotilus* are an exception (Section 14.22). *Leptothrix* and *Sphaerotilus* are common in freshwater environments containing neutrophilic aerobic iron-oxidizers. They catalyze the oxidation of both iron and manganese but do not appear to conserve energy from these reactions, conserving energy instead from the oxidation of organic matter.

Characteristic species of neutrophilic aerobic iron-oxidizers are found in the genus *Gallionella* (freshwater) and the marine genus *Mariprofundus* (marine). Species of *Gallionella* and *Mariprofundus* each form a twisted stalklike structure containing $Fe(OH)_3$ from the oxidation of ferrous iron (**Figure 14.36**). The iron-encrusted stalk contains an organic matrix on which $Fe(OH)_3$ accumulates as it is excreted from the cell surface. Stalk formation is presumably an adaptation that prevents cells from becoming entombed in an iron oxide crust.

Gallionella is common in the waters draining bogs, iron springs, and other habitats where ferrous iron is present. *Mariprofundus* was first isolated from Lōʻihi Seamount, a submarine volcano found near Hawaii. *Gallionella* and *Mariprofundus* are both autotrophic chemolithotrophs containing enzymes of the Calvin cycle (⟳ Section 13.5).

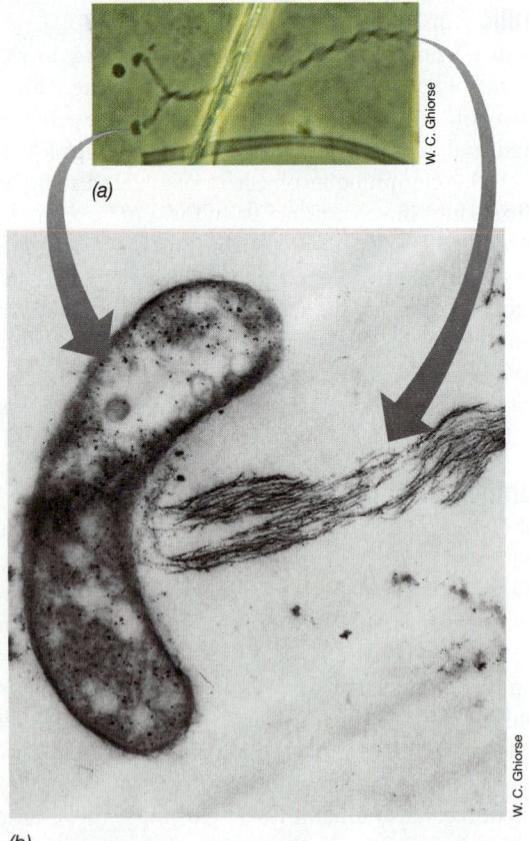

(a)

(b)

Figure 14.36 **The neutrophilic ferrous iron oxidizer, *Gallionella ferruginea*, from an iron seep near Ithaca, New York.** *(a)* Photomicrograph of two bean-shaped cells with stalks that combine to form one twisted mass. *(b)* Transmission electron micrograph of a thin section of a *Gallionella* cell with stalk. Cells are about 0.6 μm wide.

Anaerobic Iron-Oxidizing *Bacteria*

Anaerobic ferrous iron oxidation can be mediated by both chemotrophic and phototrophic bacteria. These groups are common in anoxic sediments and wetlands. Anoxic conditions promote the solubility of ferrous iron across a wide range of pH and so, unlike the aerobic iron-oxidizing bacteria, growth of anaerobic iron-oxidizers is not strictly limited to neutral pH. These groups contain organisms that are metabolically diverse and able to grow by using a variety of different electron donors and acceptors.

Phototrophic iron oxidation occurs in select species of purple nonsulfur bacteria of the *Alphaproteobacteria* (e.g., *Rhodopseudomonas palustris*), select species of purple sulfur bacteria of the *Gammaproteobacteria* (⟳ Figure 13.25), and select species of green sulfur bacteria found in the *Chlorobi* (*Chlorobium ferrooxidans*). In all cases ferrous iron is one of several compounds that these organisms can use as an electron donor in photosynthesis.

Anaerobic chemotrophic iron-oxidizers couple the oxidation of ferrous iron to nitrate reduction, producing either NO_2^- or nitrogen gases (denitrification). These organisms are *Alpha-*, *Beta-*, *Gamma-*, or *Deltaproteobacteria*, and most are also able to use various organic electron donors in nitrate reduction; many can also grow aerobically. The bacterial genera *Acidovorax*, *Aquabacterium*, and *Marinobacter* all contain anaerobic iron-oxidizers. While most species are mixotrophs when growing with ferrous iron as electron donor, species such as *Marinobacter aquaeolei* and *Thiobacillus denitrificans* are able to grow autotrophically as iron-oxidizing chemolithotrophs.

MINIQUIZ -
- What habitat characteristics govern the diversity and distribution of iron oxidizers?
- How do aerobic neutrophilic iron-oxidizers keep their cells from becoming entombed in a crust of iron?

14.16 Hydrogen-Metabolizing *Bacteria*

Key Genera: *Ralstonia, Paracoccus*

The hydrogen couple is extremely electronegative, meaning that H_2 is an excellent electron donor in energy metabolism and can couple to virtually any conceivable electron acceptor. Hence, the ability to conserve energy by H_2 oxidation is found widely in the tree of life (Figure 14.1). Examples of diverse hydrogen-oxidizing phototrophs (Sections 14.4–14.7) and anaerobic chemotrophs (Sections 14.9, 14.10 and 14.14) are discussed throughout this chapter. In addition, H_2 oxidation occurs in virtually every genus of anaerobic *Archaea* (Chapter 16). Here we consider the diversity of autotrophic and aerobic hydrogen-oxidizing chemolithotrophic *Bacteria*.

Many bacteria can grow with H_2 as the sole electron donor and O_2 as the electron acceptor in their energy metabolism:

$$H_2 + \tfrac{1}{2} \, O_2 \rightarrow H_2O \qquad \Delta G^{0'} = -237 \text{ kJ}$$

Most of these organisms, known collectively as the "hydrogen bacteria," can also grow autotrophically (using reactions of the Calvin cycle to incorporate CO_2). All hydrogen bacteria contain one or more *hydrogenase* enzymes that function to bind H_2 and use it either to produce ATP (⟳ Section 13.7) or for reducing power for autotrophic growth.

Different hydrogen-oxidizing *Proteobacteria* are scattered among the *Alpha, Beta,* and *Gamma* subclasses. These organisms should be distinguished from the many strictly anaerobic prokaryotes that oxidize H_2 in anaerobic respirations; for example, acetogens, methanogens, and sulfate-reducing bacteria (⇄ Sections 13.17–13.20). Both gram-positive and gram-negative hydrogen bacteria are known, with the best-studied representatives classified in the genera *Ralstonia* (Figure 14.37), *Pseudomonas,* and *Paracoccus. Paracoccus denitrificans* can also oxidize H_2 anaerobically by denitrification and has been particularly well studied for its bioenergetics of electron transport and generation of a proton motive force.

Physiology and Ecology of Hydrogen Bacteria

When growing chemolithotrophically on H_2, most hydrogen bacteria grow best under microaerophilic (5–10% O_2) conditions because hydrogenases are typically oxygen-sensitive enzymes. The element nickel (Ni^{2+}) must be present in the medium for chemolithotrophic growth of hydrogen bacteria because virtually all hydrogenases contain Ni^{2+} as a key metal cofactor. A few hydrogen bacteria also fix nitrogen (⇄ Section 3.17), making possible their culture in a mineral salts medium containing only gases—H_2, O_2, CO_2, and N_2—as energy, carbon, and nitrogen sources. Virtually all hydrogen bacteria are facultative chemolithotrophs, meaning that they can also grow chemoorganotrophically with organic compounds as energy sources.

CO Oxidation

Some hydrogen bacteria can grow aerobically on carbon monoxide (CO) as electron donor. CO-oxidizing bacteria, called *carboxydotrophic* bacteria, grow autotrophically using the Calvin cycle (⇄ Section 13.5) to fix CO_2 generated from the oxidation of CO. Electrons from the oxidation of CO to CO_2 by the enzyme carbon monoxide dehydrogenase travel through an electron transport chain that forms a proton motive force. Interestingly, CO

is a potent inhibitor of many cytochromes, acting as a respiratory poison. However, carboxydotrophic bacteria get around this problem by synthesizing CO-resistant cytochromes and are thus immune to any toxic effects of CO. Like the hydrogen bacteria, virtually all carboxydotrophic bacteria also grow chemoorganotrophically by oxidizing organic compounds, a likely indication that CO levels are quite variable in nature and a backup means of energy metabolism is necessary.

CO consumption by carboxydotrophic bacteria on a global basis is a significant ecological process. Although much CO is generated from human and other sources, CO levels in air have not risen significantly over many years. Because the most significant releases of CO (primarily from automobile exhaust, incomplete combustion of fossil fuels, and the catabolism of lignin, a plant product) are in oxic environments, carboxydotrophic bacteria in the upper layers of soil probably represent the most significant sink for CO in nature.

14.17 Methanotrophic and Methylotrophic *Bacteria*

Methylotrophs are organisms that grow using organic compounds lacking C—C bonds as electron donors in energy metabolism and as carbon sources. Methylotrophy occurs in the bacterial phyla *Proteobacteria, Firmicutes, Actinobacteria, Bacteroidetes, Verrucomicrobia,* and in the archaeal phylum *Euryarchaeota* (Figure 14.1). **Methanotrophs** are a subset of methylotrophs defined by their ability to use methane as a substrate for growth. We discussed the physiology of methanotrophy in Sections 13.23 and 13.24.

Aerobic methylotrophs are common in soil and aquatic environments where O_2 is present. Anaerobic methylotrophs are common in anoxic environments, particularly in marine sediments. Many anaerobic methylotrophs are methanogenic *Archaea*. In addition, a consortium of methanogenic *Archaea* and sulfate-reducing bacteria combine to oxidize methane from gas hydrates found in deep-sea sediments (⇄ Section 13.24). We consider here only the aerobic methylotrophs.

Aerobic Facultative Methylotrophs

Key Genera: *Hyphomicrobium, Methylobacterium*
Aerobic facultative methylotrophs are unable to use methane but can use many other methylated compounds. They are species of *Alpha-, Beta-,* and *Gammaproteobacteria, Actinobacteria,* and *Firmicutes.* Facultative methylotrophs are metabolically diverse and, in addition to methylated substrates, most species can grow aerobically using other organic compounds, such as organic acids, ethanol, and sugars. When growing as methylotrophs, most species can grow aerobically with methanol and some can also metabolize methylated amines, methylated sulfur compounds, and halomethanes. Most are obligate aerobes, though some species are capable of denitrification.

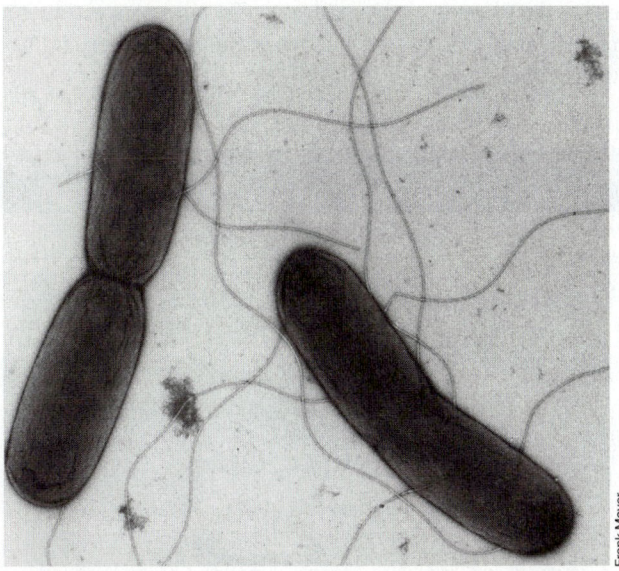

Figure 14.37 Hydrogen oxidizers. Transmission electron micrograph of negatively stained cells of the hydrogen-oxidizing chemolithotroph *Ralstonia eutropha.* A cell is about 0.6 μm in diameter and contains several flagella.

Frank Mayer

The genus *Hyphomicrobium* provides an example of the metabolic versatility of the aerobic facultative methylotrophs. Certain species of *Hyphomicrobium* can grow as aerobic methylotrophs using methanol, methylamine, or dimethylsulfide. Species of *Hyphomicrobium* can also grow as anaerobic methylotrophs using methanol as an electron donor coupled to denitrification. Finally, *Hyphomicrobium* can grow aerobically on a range of C_2 and C_4 compounds.

Aerobic Methanotrophs

Key Genera: *Methylomonas, Methylosinus*

Aerobic methanotrophs are methylotrophs that can use methane as an electron donor and typically can use it as a carbon source as well. **Table 14.2** gives a taxonomic overview of the methanotrophs. Most species of methanotrophs are *Proteobacteria* and are classified into two major groups based on their internal cell structure, phylogeny, and carbon assimilation pathway. *Type I methanotrophs* assimilate one-carbon compounds via the ribulose monophosphate cycle and are *Gammaproteobacteria*. By contrast, *type II methanotrophs* assimilate C_1 intermediates via the serine pathway and are *Alphaproteobacteria* (Table 14.2). We discussed the biochemical details of these pathways in Section 13.23. Most methanotrophs are metabolically specialized for aerobic growth on methane, through some can grow on either methane or methanol. Methanotrophs are typically obligate methylotrophs; however, the methanotrophic genus *Methylocella* contains species that can also grow using multicarbon substrates.

In addition to the proteobacterial methanotrophs described above, the phylum *Verrucomicrobia* contains the bacterium *Methylacidiphilum*. Genome analysis has shown that species of *Methylacidiphilum* lack key enzymes of both the ribulose monophosphate and serine pathways. Instead, *Methylacidiphilum* uses the Calvin cycle to assimilate carbon from CO_2.

Physiology

Methanotrophs possess a key enzyme, *methane monooxygenase*, which catalyzes the incorporation of an atom of oxygen from O_2 into CH_4, forming methanol (CH_3OH, ⮌ Section 13.23). The requirement for O_2 as a reactant in the initial oxygenation of CH_4 explains why these methanotrophs are obligate aerobes. Methane monooxygenase is located in extensive internal membrane systems that are the site of methane oxidation. Membranes in type I methanotrophs are arranged as bundles of disc-shaped vesicles distributed throughout the cell (**Figure 14.38***b*). Type II species possess paired membranes running along the periphery of the cell (Figure 14.38*a*). Verrucomicrobial methanotrophs possess membrane vesicles. Methylotrophs unable to use methane lack these internal membrane arrays.

Methanotrophs are virtually unique among bacteria in possessing relatively large amounts of sterols. Sterols are rigid planar molecules found in the cytoplasmic and other membranes of eukaryotes but are absent from most bacteria. Sterols may be an essential part of the complex internal membrane system for methane oxidation (see Figure 14.38). The only other group of bacteria in which sterols are widely distributed is the mycoplasmas, bacteria that lack cell walls and thus probably require a tougher cytoplasmic membrane (⮌ Section 15.9). Many methylotrophs contain various carotenoid pigments and high levels of cytochromes in their membranes, and these features often render colonies of aerobic methylotrophs pink.

Ecology

Aerobic methylotrophs are found in the open ocean, soils, in association with plant roots and leaf surfaces, and at the oxic interface of many anoxic environments (⮌ Section 13.23). Methanol is produced during the breakdown of plant pectin and this is likely an important substrate for methylotrophs in terrestrial ecosystems. In addition, soils contain aerobic methanotrophs that consume atmospheric methane and are an important biological sink for atmospheric methane. Aerobic methanotrophs are also common at the oxic interface of anoxic environments found in lakes, sediments, and wetlands where methanogens provide a constant source of methane. These methanotrophs play an important role in the global carbon cycle by oxidizing CH_4 and converting it into cell material and CO_2 before it reaches the atmosphere (CH_4 is a strong greenhouse gas).

Table 14.2 Some characteristics of methanotrophic bacteria

Organism	Morphology	Phylogenetic group[a]	Internal membranes[b]	Carbon assimilation pathway[c]	N_2 fixation
Methylomonas	Rod	Gamma	I	Ribulose monophosphate	No
Methylomicrobium	Rod	Gamma	I	Ribulose monophosphate	No
Methylobacter	Coccus to ellipsoid	Gamma	I	Ribulose monophosphate	No
Methylococcus	Coccus	Gamma	I	Ribulose monophosphate and Calvin cycle	Yes
Methylosinus	Rod or vibrioid	Alpha	II	Serine	Yes
Methylocystis	Rod	Alpha	II	Serine	Yes
Methylocella[d]	Rod	Alpha	II	Serine	Yes
Methylacidiphilum[d]	Rod	*Verrucomicrobiaceae*[d]	Membrane vesicles	Serine and Calvin cycle	Yes

[a]All except for *Methylacidiphilum* are *Proteobacteria*.
[b]Internal membranes: type I, bundles of disc-shaped vesicles distributed throughout the organism; type II, paired membranes running along the periphery of the cell. See Figure 14.38.
[c]See Figure 13.58.
[d]Acidophiles. For the properties of *Verrucomicrobiaceae*, see Section 15.17.

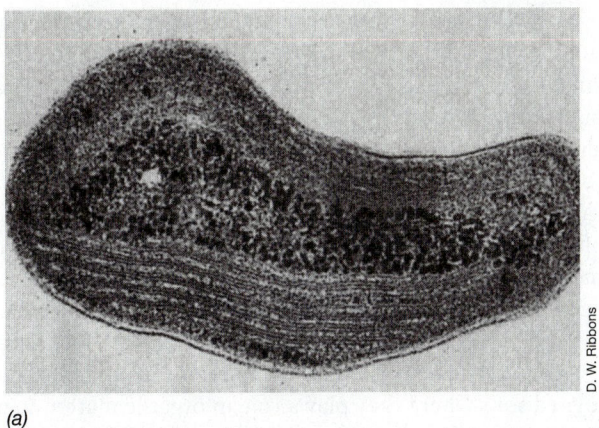

(a)

Figure 14.38 Methanotrophs. *(a)* Electron micrograph of a cell of *Methylosinus*, illustrating a type II membrane system. Cells are about 0.6 μm in diameter. *(b)* Electron micrograph of a cell of *Methylococcus capsulatus*, illustrating a type I membrane system. Cells are about 1 μm in diameter. Compare with Figure 14.33.

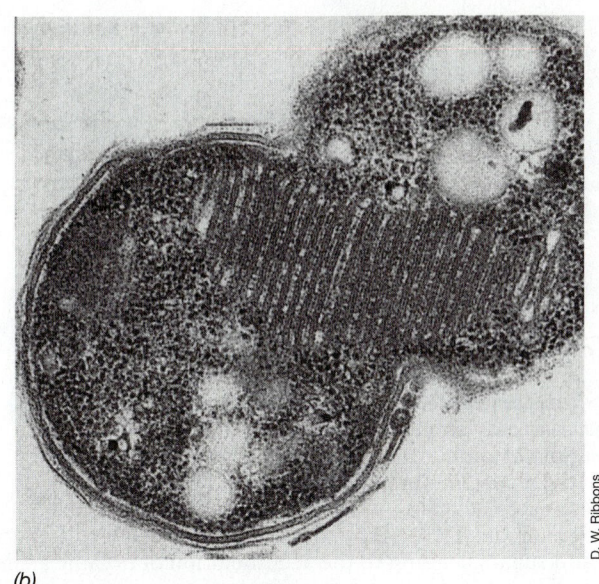

(b)

Methanotrophs also form a variety of symbioses with eukaryotic organisms. For example, some marine mussels live in the vicinity of hydrocarbon seeps on the seafloor, places where CH_4 is released in substantial amounts. Methanotrophic symbionts reside within the animal's gill tissue (**Figure 14.39**), which ensures effective gas exchange with seawater. Assimilated CH_4 is distributed throughout the animal by the excretion of organic compounds by the methanotrophs. These methanotrophic symbioses are therefore conceptually similar to those that develop between sulfide-oxidizing chemolithotrophs and hydrothermal vent tube worms and giant clams (↩ Section 22.12).

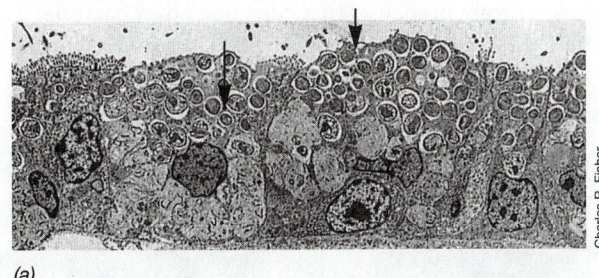

(a)

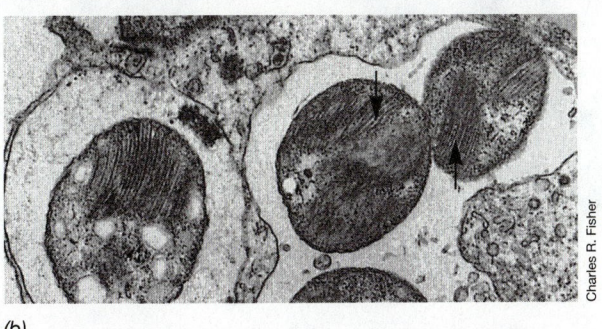

(b)

Figure 14.39 Methanotrophic symbionts of marine mussels. *(a)* Electron micrograph of a thin section at low magnification of gill tissue from a marine mussel living near hydrocarbon seeps in the Gulf of Mexico. Note the symbiotic methanotrophs (arrows) in the tissues. *(b)* High-magnification view of gill tissue showing methanotrophs with type I membrane bundles (arrows). Cells of the methanotrophs are about 1 μm in diameter. Compare with Figure 14.38*b*.

Methylomirabilis oxyfera is a methanotroph isolated from anoxic waters in the Black Sea, and was the first isolate obtained from the unique bacterial phylum NC-10 (see page 433). *M. oxyfera* is an obligate anaerobe; however, it uses the O_2-dependent enzyme of aerobic methanotrophs (methane monooxygenase) to oxidize methane to CO_2. *M. oxyfera* accomplishes this by reducing nitrite to nitric oxide (NO), which is then dismutated to N_2 and O_2 ($2\,NO \rightarrow N_2 + O_2$). The O_2 produced by this pathway is then consumed by methane monooxygenase during the oxidation of CH_4 (↩ Section 13.24). As for the methanotroph *Methylacidiphilum*, *M. oxyfera* assimilates C_1 units as CO_2, probably by the Calvin cycle.

MINIQUIZ -
- What is the difference between a methanotroph and a methylotroph?
- What is unique about the methanotroph *Methylomirabilis*?

14.18 Acetic Acid Bacteria and Acetogens

A variety of microorganisms produce *acetate* as a by-product of metabolism (e.g., fermentative microorganisms (↩ Sections 13.11–13.15), incomplete sulfate-reducers (Section 14.9), and incomplete iron-reducers (Section 14.14). Here, we consider those organisms for which acetate is the primary product of metabolism. These include the **acetic acid bacteria**, obligate aerobes used in the industrial production of acetic acid (vinegar) from sugars or alcohols. We also consider the **acetogens**, obligate anaerobes that use the acetyl-CoA pathway (↩ Section 13.19) to produce acetate from C_1 compounds. We will see that acetate production in these two groups occurs through quite different metabolic pathways and that acetic acid bacteria and acetogens differ in both their physiology and ecology.

Acetic Acid Bacteria

Key Genera: *Acetobacter, Gluconobacter*

The acetic acid bacteria comprise gram-negative, obligately aerobic, motile rods that carry out the incomplete oxidation of

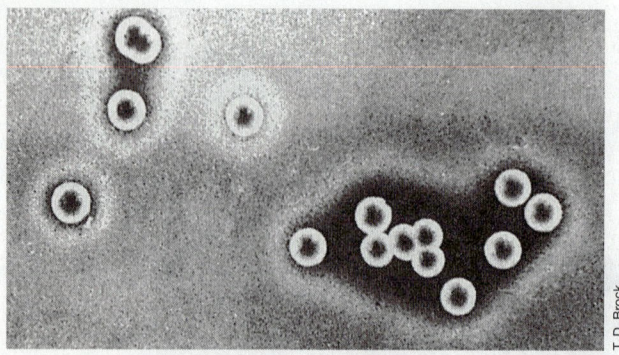

Figure 14.40 Colonies of *Acetobacter aceti* on calcium carbonate ($CaCO_3$) agar containing ethanol as electron donor. Note the clearing around the colonies due to the dissolution of $CaCO_3$ by the acetic acid produced.

alcohols and sugars, leading to the accumulation of organic acids as end products. With ethanol (C_2H_2OH) as a substrate, acetic acid ($C_2H_4O_2$) is produced, which gives the acetic acid bacteria their name. As one would expect, acetic acid bacteria are tolerant of acidic conditions; most strains can grow well at pH values lower than 5. The acetic acid bacteria are a heterogeneous assemblage of *Alphaproteobacteria*, comprising both peritrichously flagellated (*Acetobacter*) and polarly flagellated (*Gluconobacter*) organisms.

The acetic acid bacteria are commonly found in fermenting fruit juices, such as hard cider or wine, or in beer. Colonies of acetic acid bacteria can be recognized on calcium carbonate ($CaCO_3$) agar plates containing ethanol because the acetic acid they produce dissolves and causes a clearing of the otherwise insoluble $CaCO_3$ (**Figure 14.40**). Cultures of acetic acid bacteria are used in the commercial production of vinegar. In addition to ethanol, acetic acid bacteria carry out an incomplete oxidation of some higher alcohols and sugars. For example, glucose is oxidized to gluconic acid and sorbitol is oxidized to sorbose. This property of "underoxidation" is exploited in the industrial manufacture of ascorbic acid (vitamin C), where sorbose is used as a precursor for the synthesis of ascorbic acid.

Another interesting property of some acetic acid bacteria is their ability to synthesize cellulose that is chemically similar to the cellulose found in plants. Cellulose from acetic acid bacteria is formed as a matrix outside the cell wall and causes cells to become embedded in a tangled mass of cellulose microfibrils. When these species of acetic acid bacteria grow in an unshaken vessel, they form a surface pellicle of cellulose in which the bacteria develop, probably to gain access to air.

Acetogens

Key Genera: *Acetobacterium, Clostridium*

Acetogens are obligate anaerobes that produce acetate as the primary product of metabolism. The greatest diversity of acetogens is found in the phylum *Firmicutes*, but acetogens have also been described in the *Spirochaetes* and the *Acidobacteria* (Figure 14.1). True acetogens use the reductive acetyl-CoA pathway to conserve energy and assimilate carbon for growth (⇄ Section 13.19).

Many acetogens grow using simple organic compounds such as sugars, alcohols, and organic acids, by first oxidizing these molecules to acetate plus CO_2 and then conserving energy and assimilating carbon through the acetyl-CoA pathway. Use of this pathway also allows most acetogens to grow using C_1 compounds as a sole source of carbon and energy. Many acetogens can grow autotrophically using H_2, CO, or methanol as an electron donor. Acetogens are often called *homoacetogens* when acetate is the sole product of their metabolism. While acetate is generally the main product of acetogenic metabolism, some acetogens can use the acetyl-CoA pathway to produce products other than acetate (e.g., butyrate or ethanol).

Acetogens are common in anaerobic sediments and waterlogged soils where they play a role in organic matter degradation. Hydrogen-utilizing acetogens are of particular importance in gut communities where they compete for H_2 with methanogens. Acetate is readily assimilated by the gut wall and contributes to the nutrition of the host. Acetogenesis is a dominant process in the overall carbohydrate metabolism of the human gut microbiota. In addition, termites receive more than one-third of their energy from the acetate produced by acetogenic *Spirochaetes* in their hindgut communities.

MINIQUIZ

- Which industrial processes use acetic acid bacteria?
- What are the major differences between acetic acid bacteria and acetogens?

14.19 Predatory *Bacteria*

Key Genera: Bdellovibrio, Myxococcus

Some bacteria are predators that consume other bacteria. Known bacteria predators reside among several classes of *Proteobacteria*

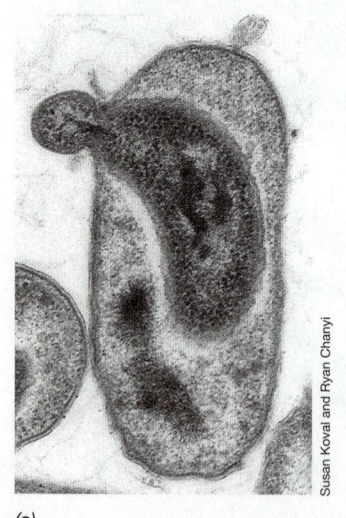
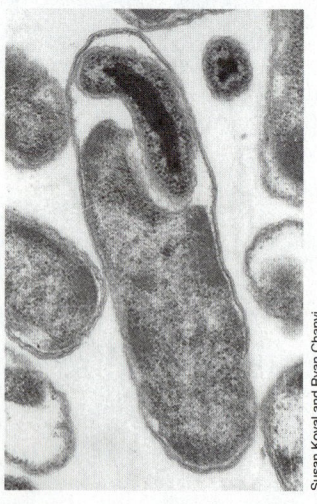

Figure 14.41 **Attack on a prey cell by *Bdellovibrio*.** Thin-section electron micrographs of *Bdellovibrio* attacking a cell of *Delftia acidovorans*. (a) Entry of the predator cell. (b) *Bdellovibrio* cell inside the host. The *Bdellovibrio* cell is enclosed in the bdelloplast and replicates in the periplasmic space. A *Bdellovibrio* cell measures about 0.3 µm in diameter.

Figure 14.42 **Developmental cycle of the bacterial predator *Bdellovibrio bacteriovorus*.** *(a)* Electron micrograph of a cell of *Bdellovibrio bacteriovorus*; note the thick flagellum A cell is 0.3 μm wide. *(b)* Events in predation. Following primary contact with a gram-negative bacterium, the highly motile *Bdellovibrio* cell attaches to and penetrates into the prey periplasmic space. Once inside the periplasmic space, *Bdellovibrio* cells elongate and within 4 h progeny cells are released. The number of progeny cells released varies with the size of the prey; 5–6 bdellovibrios are released from *Escherichia coli* and 20–30 for a larger prey cell, such as *Aquaspirillum*.

and in the *Bacteroidetes*. Several different methods of predation have been observed. Some predators, such as *Vampirococcus* (phylogeny unknown) and *Micavibrio* (*Alphaproteobacteria*), are *epibiotic predators*; they attach to the surface of their prey and acquire nutrients from its cytoplasm or periplasm. Other predators, such as *Daptobacter* (*Epsilonproteobacteria*), are *cytoplasmic predators*, as they invade their host cells and replicate in the cytoplasm, consuming their prey from the inside out. *Bdellovibrio* have a similar lifestyle as *periplasmic predators*; they invade and replicate within the periplasmic space of their prey cells. Finally, predators such as *Lysobacter* (*Gammaproteobacteria*) and *Myxococcus* (*Deltaproteobacteria*) are *social predators*. These gliding bacteria use swarming behavior to find prey, which they lyse and feed upon collectively. *Bdellovibrio* and *Myxococcus* are the most thoroughly described genera of bacterial predators.

Bdellovibrio

Bdellovibrio are small, highly motile and curved bacteria that prey on other bacteria, using the cytoplasmic constituents of their hosts as nutrients (*bdello* is a prefix meaning "leech"). After attachment of a *Bdellovibrio* cell to its prey, the predator penetrates the cell wall of the prey and replicates in the periplasmic space, eventually forming a spherical structure called a *bdelloplast*. Two stages of penetration are shown in electron micrographs in **Figure 14.41** and diagrammatically in **Figure 14.42**. A wide variety of gram-negative prey bacteria can be attacked by *Bdellovibrio*, but gram-positive cells are not attacked.

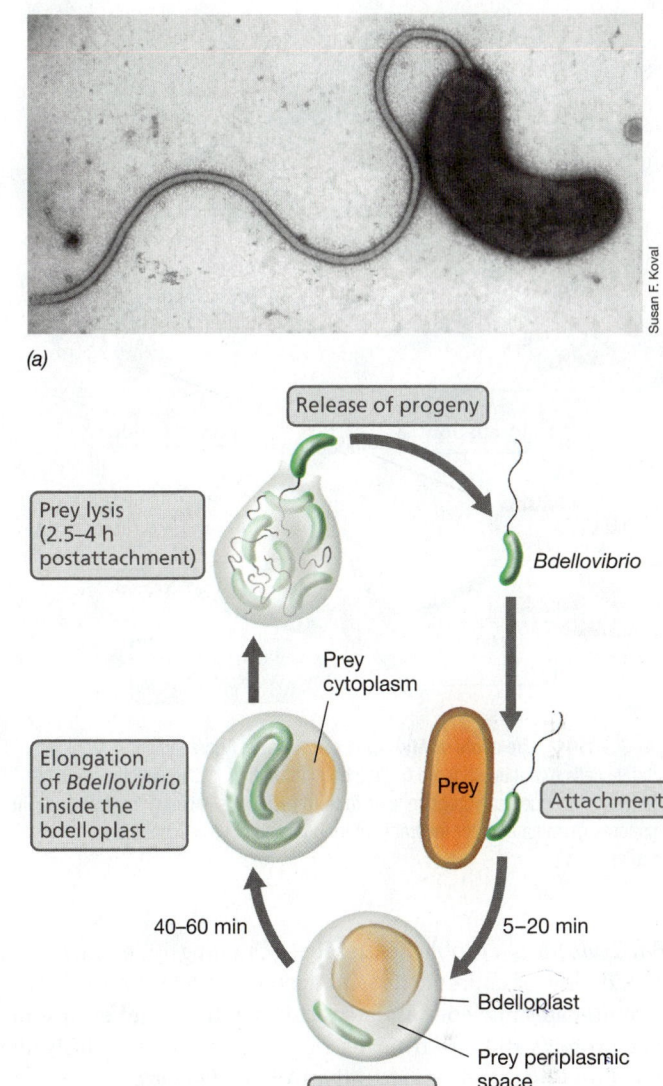

(a)

Release of progeny

Prey lysis (2.5–4 h postattachment)

Bdellovibrio

Prey cytoplasm

Elongation of *Bdellovibrio* inside the bdelloplast

Prey

Attachment

40–60 min

5–20 min

Bdelloplast

Prey periplasmic space

Penetration

(b)

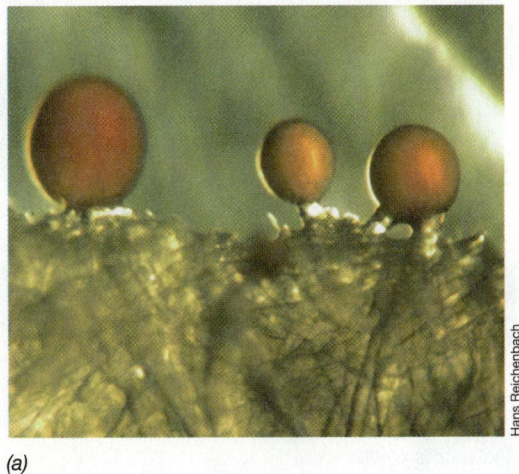

(a)

(b)

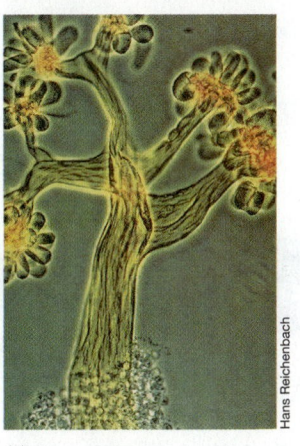

(c)

Figure 14.43 **Fruiting bodies of three species of fruiting myxobacteria.** *(a) Myxococcus fulvus* (125 μm high). *(b) Myxococcus stipitatus* (170 μm high). *(c) Chondromyces crocatus* (560 μm high).

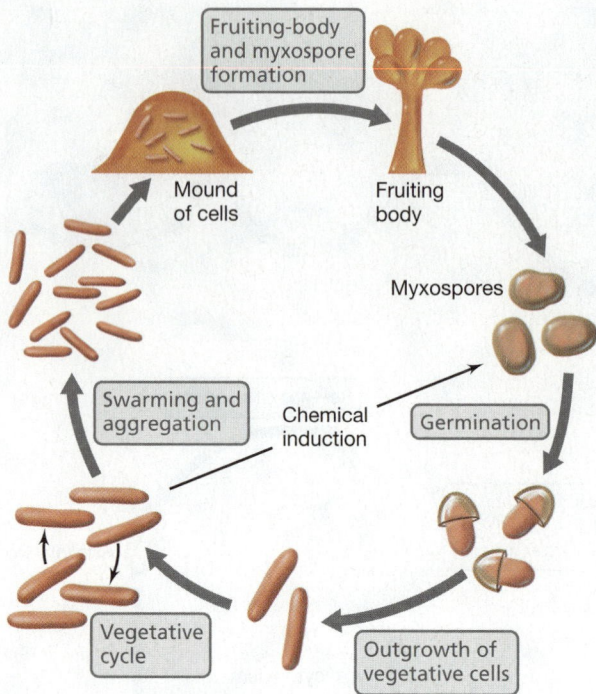

Figure 14.44 **Life cycle of *Myxococcus xanthus*.** Aggregation assembles vegetative cells that then undergo fruiting body formation, within which some vegetative cells undergo morphogenesis to form resting cells called myxospores. The myxospores germinate under favorable nutritional and physical conditions to yield vegetative cells.

Bdellovibrio is an obligate aerobe, obtaining its energy from the oxidation of amino acids and acetate. In addition, *Bdellovibrio* assimilates nucleotides, fatty acids, peptides, and even some intact proteins directly from its host without first hydrolyzing them. Prey-independent derivatives of *Bdellovibrio* can be isolated and grown on complex media, however, showing that predation is not an obligatory lifestyle.

Phylogenetically, bdellovibrios are *Deltaproteobacteria*, and they are widespread in aquatic habitats. Procedures for their

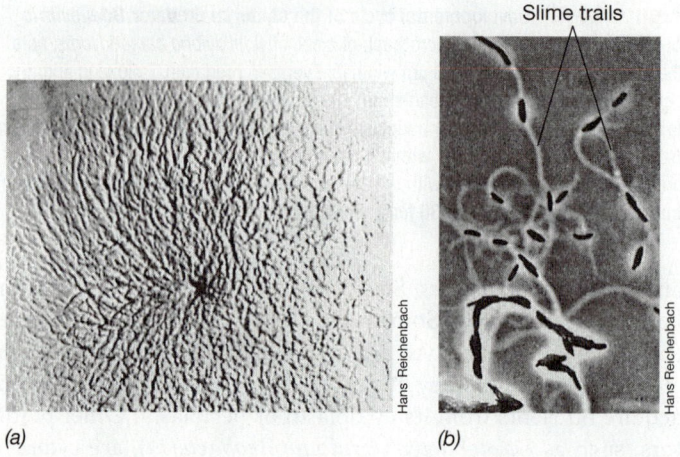

Figure 14.46 **Swarming in *Myxococcus*.** (a) Photomicrograph of a swarming colony (5-mm radius) of *Myxococcus xanthus* on agar. (b) Single cells of *Myxococcus fulvus* from an actively gliding culture, showing the characteristic slime trails on the agar. A cell of *M. fulvus* is about 0.8 µm in diameter. *M. xanthus* has been used as a model for developmental events in myxobacteria.

isolation are similar to those used to isolate bacterial viruses (⟳ Section 8.4). Prey bacteria are spread on the surface of an agar plate forming a lawn, and the surface is inoculated with a small amount of soil suspension that has been filtered through a membrane filter; the filter retains most bacteria but allows the small *Bdellovibrio* cells to pass. On incubation of the agar plate, plaques analogous to bacteriophage plaques (⟳ Figure 8.8*b*) are formed at locations where *Bdellovibrio* cells are multiplying. Pure cultures of *Bdellovibrio* can then be isolated from these plaques. *Bdellovibrio* are widely distributed, as cultures have been obtained from many soils and from sewage.

Myxobacteria

Myxobacteria exhibit the most complex behavioral patterns of all known bacteria. The life cycle of myxobacteria results in the formation of multicellular structures called *fruiting bodies*. The fruiting bodies are often strikingly colored and morphologically

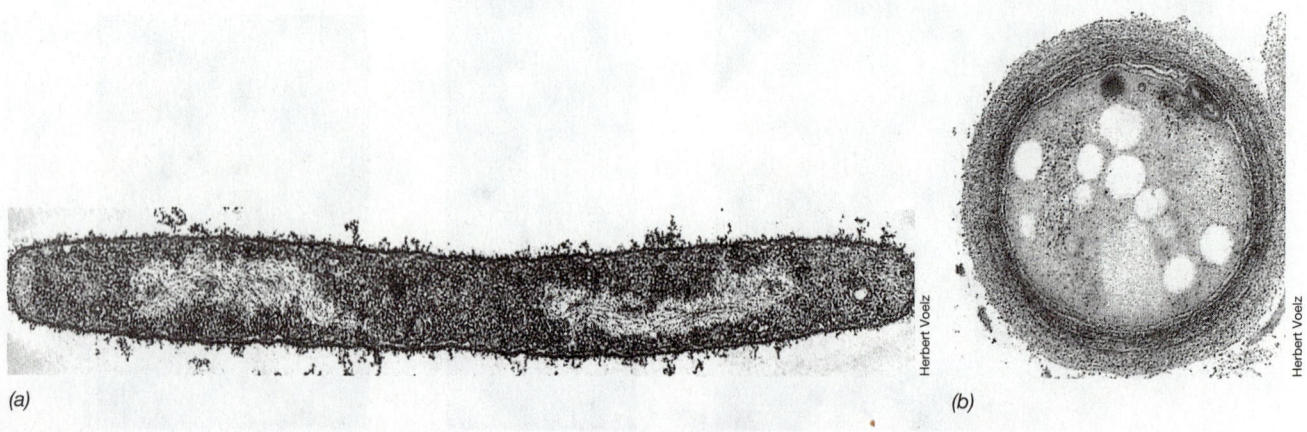

Figure 14.45 *Myxococcus.* (a) Electron micrograph of a thin section of a vegetative cell of *Myxococcus xanthus*. A cell measures about 0.75 µm wide. (b) Myxospore of *M. xanthus*, showing the multilayered outer wall. Myxospores measure about 2 µm in diameter.

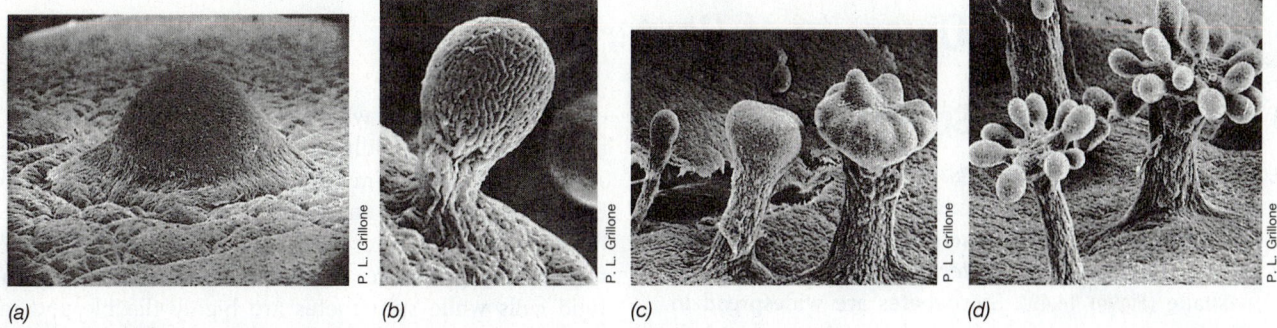

(a)   (b)   (c)   (d)

P. L. Grillone

Figure 14.47 **Scanning electron micrographs of fruiting body formation in *Chondromyces crocatus*.** *(a)* Early stage, showing aggregation and mound formation. *(b)* Initial stage of stalk formation. Slime formation in the head has not yet begun and so the cells that compose the head are still visible. *(c)* Three stages in head formation. Note that the diameter of the stalk also increases. *(d)* Mature fruiting bodies. The entire fruiting structure is about 600 μm in height (compare with Figure 14.43*c*).

elaborate (**Figure 14.43**), and these can often be seen with a hand lens on moist pieces of decaying wood or plant material. The fruiting myxobacteria are classified on morphological grounds using characteristics of the vegetative cells, the myxospores, and fruiting body structure.

The life cycle of a typical myxobacterium is shown in **Figure 14.44**. The vegetative cells of the myxobacteria are simple, non-flagellated, gram-negative rods (**Figure 14.45**) that glide across surfaces and obtain their nutrients primarily by using extracellular enzymes to lyse other bacteria and use the released nutrients. A vegetative cell excretes slime, and as it moves across a solid surface, it leaves behind a slime trail (**Figure 14.46**). The vegetative cells form a swarm that exhibits self-organizing behavior, and this allows them to behave as a single coordinated entity in response to environmental cues.

Upon nutrient exhaustion, vegetative cells of myxobacteria begin to migrate toward each other, aggregating together in mounds or heaps (**Figure 14.47**). Aggregation is likely mediated by chemotactic or quorum-sensing responses (↩ Sections 7.8 and 7.9). As the cell masses become higher, they begin to differentiate into fruiting bodies containing *myxospores*. Myxospores are specialized cells that are resistant to drying, ultraviolet radiation, and heat, but the degree of heat resistance is much less than that of the bacterial endospore (↩ Section 2.16). Fruiting bodies can be simple, consisting of masses of myxospores embedded in slime, or complex, consisting of a stalk and heads (**Figure 14.48**). The fruiting body stalk is composed of slime within which a few cells are trapped. The majority of the cells migrate to the fruiting body head, where they undergo differentiation into myxospores (Figure 14.44).

Hans Reichenbach

(a)

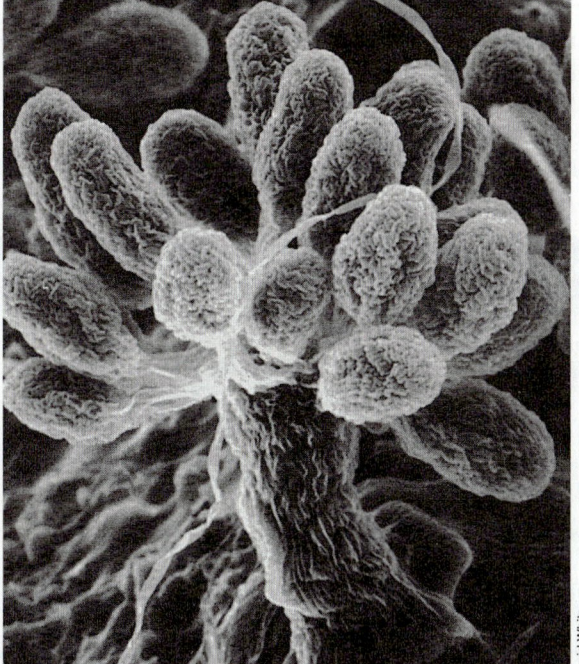

David White

(b)

Figure 14.48 ***Stigmatella aurantiaca.*** *(a)* Color photo of a single fruiting body. The color of the fruiting body is due to the production of glucosylated carotenoid pigments. The structure is about 150 μm high. *(b)* Scanning electron micrograph of a fruiting body growing on a piece of wood. Note the individual cells visible in each fruiting body.

MINIQUIZ -----------------------------------

• What environmental conditions trigger fruiting body formation in myxobacteria?

• What are the different ways in which species of *Myxococcus* and *Bdellovibrio* kill their prey?

VI • Morphological Diversity of *Bacteria*

14.20 Spirochetes and Spirilla

Key Genera: *Spirochaeta, Treponema, Cristispira, Leptospira, Borrelia*

Spirochetes are morphologically unique bacteria found only within the bacterial phylum *Spirochaetes*. Spirochetes are gram-negative, motile, tightly coiled *Bacteria*, typically slender and flexuous in shape (**Figure 14.49**). Spirochetes are widespread in aquatic sediments and in animals. Some cause diseases, including syphilis, an important human sexually transmitted disease (∞ Section 29.12). Spirochetes are classified into eight genera (**Table 14.3**) primarily on the basis of habitat, pathogenicity, phylogeny, and morphological and physiological characteristics.

Spirochetes have an unusual mode of motility conveyed by their unusual morphology. Spirochetes contain *endoflagella*, which resemble normal flagella but are found in the cell periplasm (**Figure 14.50**). The endoflagella are anchored at the cell poles and extend back along the length of the cell. Both the endoflagella and the protoplasmic cylinder are surrounded by a flexible membrane called the *outer sheath* (Figure 14.50*b*). Endoflagella rotate, as do typical bacterial flagella. However, when both endoflagella rotate in the same direction, the protoplasmic cylinder rotates in the opposite direction, placing torsion on the cell (Figure 14.50*b*). This torsion causes the spirochete cell to flex, resulting in a corkscrew-like motion that allows cells to burrow through viscous materials or tissues (Figure 14.50*b*).

Spirochetes are often confused with spirilla. **Spirilla** are helically curved rod-shaped cells, usually motile by means of polar flagella (**Figure 14.51**). The word *spirillum* refers to a type of cell shape that is widespread among *Bacteria* and *Archaea*. The number of helical turns in a single spirillum may vary from less than one complete turn (in which case the organism looks like a vibrio) to many turns. In addition, spirilla that divide terminally, such as the cyanobacterium *Spirulina* (Figure 14.5), can form long helical filaments that superficially resemble spirochetes. Spirilla, however, lack the outer sheath, endoflagella, and corkscrew-like motility of spirochetes. In addition, spirilla are typically fairly rigid cells while spirochetes are highly flexible and quite thin ($< 0.5 \, \mu m$).

Spirochaeta and *Cristispira*

The genus *Spirochaeta* includes free-living, anaerobic, and facultatively aerobic spirochetes. These organisms, of which several species are known, are common in aquatic environments such as freshwater and sediments, and also in the oceans. *Spirochaeta plicatilis* (Figure 14.49*b*) is a large spirochete found in sulfidic freshwater and marine habitats. The 20 or so endoflagella inserted at each pole of *S. plicatilis* are arranged in a bundle that winds around the coiled protoplasmic cylinder. Another species, *Spirochaeta stenostrepta* (Figure 14.49*a*), is an obligate anaerobe commonly found in H_2S-rich black muds. It ferments sugars to ethanol, acetate, lactate, CO_2, and H_2.

Cristispira (**Figure 14.52**) is a unique spirochete found in nature only in the crystalline style of certain molluscs, such as clams and oysters. The crystalline style is a flexible, semisolid rod seated in a sac and rotated against a hard surface of the digestive tract, thereby mixing and grinding the small particles of food taken in by the animal. *Cristispira* lives in both freshwater and marine molluscs, but not all species of molluscs possess this organism. Unfortunately, *Cristispira* has not been cultured, and so the physiological rationale for its restriction to this unique habitat is unknown.

Table 14.3 Genera of spirochetes and their characteristics

Genus	Dimensions (µm)	General characteristics	Number of endoflagella	Habitat	Diseases
Cristispira	30–150 × 0.5–3.0	3–10 complete coils; bundle of endoflagella visible by phase-contrast microscopy	>100	Digestive tract of molluscs; has not been cultured	None known
Spirochaeta	5–250 × 0.2–0.75	Anaerobic or facultatively aerobic; tightly or loosely coiled	2–40	Aquatic, free-living, freshwater and marine	None known
Treponema	5–15 × 0.1–0.4	Microaerophilic or anaerobic; helical or flattened coil amplitude up to 0.5 µm	2–32	Commensal or parasitic in humans, other animals	Syphilis, yaws, swine dysentery, pinta
Borrelia	8–30 × 0.2–0.5	Microaerophilic; 5–7 coils of approximately 1 µm amplitude	7–20	Humans and other mammals, arthropods	Relapsing fever, Lyme disease, ovine and bovine borreliosis
Leptospira	6–20 × 0.1	Aerobic, tightly coiled, with bent or hooked ends; requires long-chain fatty acids	2	Free-living or parasitic in humans, other mammals	Leptospirosis
Leptonema	6–20 × 0.1	Aerobic; does not require long-chain fatty acids	2	Free-living	None known
Brachyspira	7–10 × 0.35–0.45	Anaerobe	8–28	Intestine of warm-blooded animals	Causes diarrhea in chickens and swine
Brevinema	4–5 × 0.2–0.3	Microaerophile; forms deep branch in spirochete lineage as assessed by 16S rRNA sequence analysis	2	Blood and tissue of mice and shrews	Infectious for laboratory mice

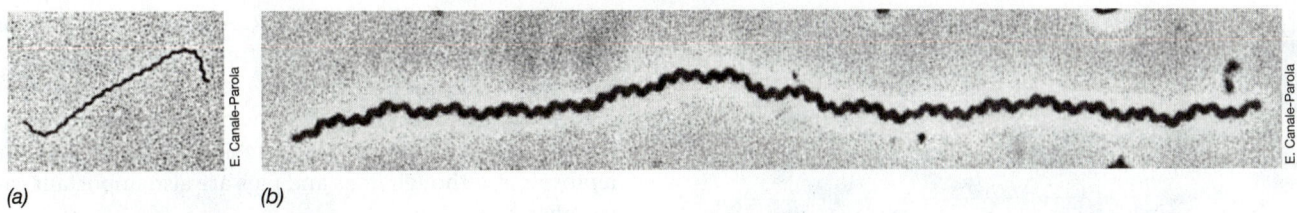

Figure 14.49   Morphology of spirochetes. Two spirochetes at the same magnification, showing the wide size range in the group. *(a) Spirochaeta stenostrepta*, by phase-contrast microscopy. A single cell is 0.25 μm in diameter. *(b) Spirochaeta plicatilis.* A single cell is 0.75 μm in diameter and can be up to 250 μm (0.25 mm) in length.

Treponema and *Borrelia*

Anaerobic or microaerophilic host-associated spirochetes that are commensals or pathogens of humans and animals reside in the genus *Treponema*. *T. pallidum*, the causal agent of syphilis (꒰꒱ Section 29.12), is the best-known species of *Treponema*. It differs in morphology from other spirochetes in that the *Treponema* cell is not helical but flat and wavy. The *T. pallidum* cell is remarkably thin, measuring only 0.2 μm in diameter. Because of this, dark-field microscopy has long been used to examine exudates from suspected syphilitic lesions (꒰꒱ Figure 29.37).

Other species of *Treponema* are also often found as commensals in humans and other animals. For example, *Treponema denticola* is common in the human oral cavity and is associated with gum disease. It ferments amino acids such as cysteine and serine, forming acetate as the major fermentation acid, as well as CO_2, NH_3, and H_2S. Spirochetes are also common in the rumen, the digestive organ of ruminant animals (꒰꒱ Section 22.7). For instance, *Treponema saccharophilum* (**Figure 14.53**) is a large, pectinolytic spirochete found in the bovine rumen where it ferments pectin, starch, inulin, and other plant polysaccharides. *Treponema primitia* can be found in the hindgut of certain termites. In the termite gut, fermentation of cellulose causes production of H_2 and CO_2. *T. primitia* is an acetogen (꒰꒱ Section 13.19) that grows on H_2 plus CO_2, forming acetate, which is an important component of the insect's nutrition. *Treponema azotonutricium* is also found in the termite hindgut and is capable of nitrogen fixation (꒰꒱ Section 3.17).

The majority of species of *Borrelia* are animal or human pathogens. *Borrelia burgdorferi* (Figure 14.53*b*) is the causative agent

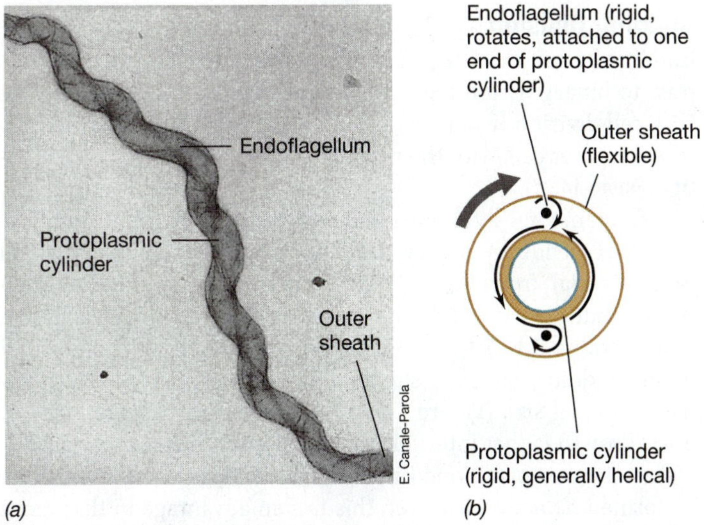

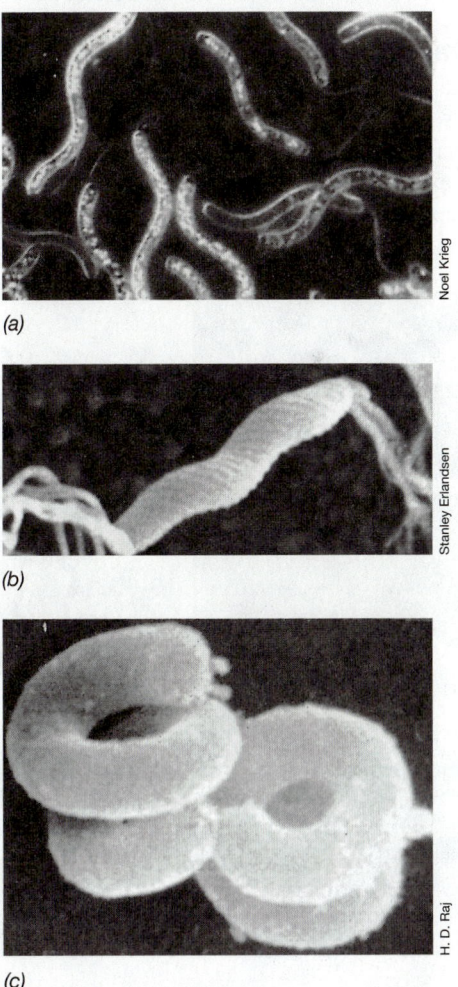

Figure 14.51   Spirilla. *(a) Spirillum volutans*, visualized by dark-field microscopy, showing flagellar bundles and volutin (polyphosphate) granules. Cells are about 1.5 × 25 μm. *(b)* Scanning electron micrograph of an intestinal spirillum. Note the polar flagellar tufts and the spiral structure of the cell surface. *(c)* Scanning electron micrograph of cells of *Ancylobacter aquaticus*. Cells are about 0.5 μm in diameter.

Figure 14.50   Motility in spirochetes. *(a)* Electron micrograph of a negatively stained cell of *Spirochaeta zuelzerae*, showing the position of the endoflagellum; the cell is about 0.3 μm in diameter. *(b)* Diagram of a spirochete cell, showing the arrangement of the protoplasmic cylinder, endoflagella, and external sheath, and how rotation of the endoflagellum generates rotation of both the protoplasmic cylinder and the external sheath.

Endoflagellum (rigid, rotates, attached to one end of protoplasmic cylinder)

Outer sheath (flexible)

Protoplasmic cylinder (rigid, generally helical)

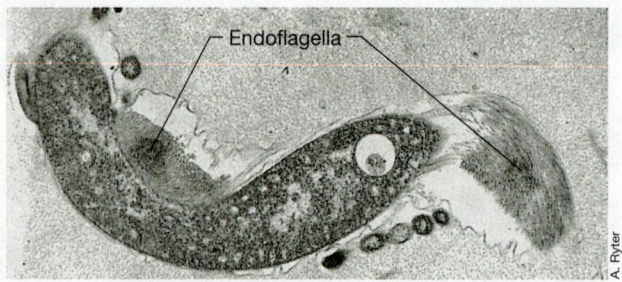

Figure 14.52 *Cristispira.* Electron micrograph of a thin section of a cell of *Cristispira.* This large spirochete is about 2 μm in diameter. Notice the numerous endoflagella.

of the tickborne *Lyme disease*, which infects humans and animals (⮐ Section 30.4). *B. burgdorferi* is also of interest because it is one of the few known bacteria that has a linear (as opposed to a circular) chromosome (⮐ Sections 4.3 and 6.4). Other species of *Borrelia* are primarily of veterinary importance, causing diseases in cattle, sheep, horses, and birds. In most cases, the bacterium is transmitted to the animal host from the bite of a tick.

Leptospira and *Leptonema*

The genera *Leptospira* and *Leptonema* contain strictly aerobic spirochetes that oxidize long-chain fatty acids (for example, the C_{18} fatty acid oleic acid) as electron donors and carbon sources. With few exceptions, these are the only substrates utilized for growth. Leptospiras are thin, finely coiled, and usually bent at

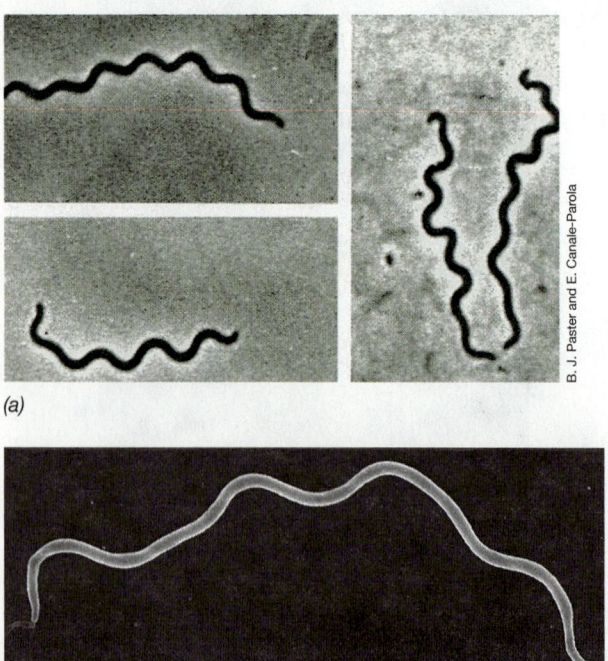

Figure 14.53 *Treponema* and *Borrelia.* *(a)* Phase-contrast micrographs of *Treponema saccharophilum*, a large pectinolytic spirochete from the bovine rumen. A cell measures about 0.4 μm in diameter. Left, regularly coiled cells; right, irregularly coiled cells. *(b)* Scanning electron micrograph of a cell of *Borrelia burgdorferi*, the causative agent of Lyme disease.

each end into a semicircular hook. At present, several species are recognized in this group, some free-living and many parasitic. Two major species of *Leptospira* are *L. interrogans* (parasitic) and *L. biflexa* (free-living). Strains of *L. interrogans* are parasitic for humans and animals. Rodents are the natural hosts of most leptospiras, although dogs and pigs are also important carriers of certain strains.

In humans the most common leptospiral syndrome is *leptospirosis*, a disorder in which the organism localizes in the kidneys and can cause renal failure or even death. Leptospiras ordinarily enter the body through the mucous membranes or through breaks in the skin during contact with an infected animal. After a transient multiplication in various parts of the body, the organism localizes in the kidneys and liver, causing nephritis and jaundice. Domestic animals such as dogs are vaccinated against leptospirosis with a killed virulent strain in the combined distemper-leptospira-hepatitis vaccine.

MINIQUIZ -
- What are the major differences between spirochaetes and spirilla?
- Name two diseases of humans caused by spirochetes.

14.21 Budding and Prosthecate/Stalked *Bacteria*

Key Genera: *Hyphomicrobium, Caulobacter*

The growth of most bacteria is coupled to cell division by the well-known process of binary fission (⮐ Section 5.1). In this section, we consider organisms that grow and divide in different ways, including budding and the formation of appendages. Budding and appendaged species often have life cycles that are distinct among bacteria.

Budding Division

Budding bacteria divide as a result of unequal cell growth. In contrast to binary fission that forms two equivalent cells (⮐ Figure 5.1), cell division in stalked and budding bacteria forms a totally new daughter cell, with the mother cell retaining its original identity (**Figure 14.54**).

A fundamental difference between budding bacteria and bacteria that divide by binary fission is the formation of new cell wall material from a single point (polar growth) rather than throughout the whole cell (intercalary growth) as in binary fission (⮐ Sections 5.1–5.4). Several genera not normally considered to be budding bacteria show polar growth without differentiation of cell size (Figure 14.54). An important consequence of polar growth is that internal structures, such as membrane complexes, are not partitioned in the cell division process and must be formed *de novo*. However, this has an advantage in that more complex internal structures can be formed in budding cells than in cells that divide by binary fission, since the latter cells would have to partition these structures between the two daughter cells. Not coincidentally, many budding bacteria, particularly phototrophic and chemolithotrophic species, contain extensive internal membrane systems.

I. **Equal products of cell division:**

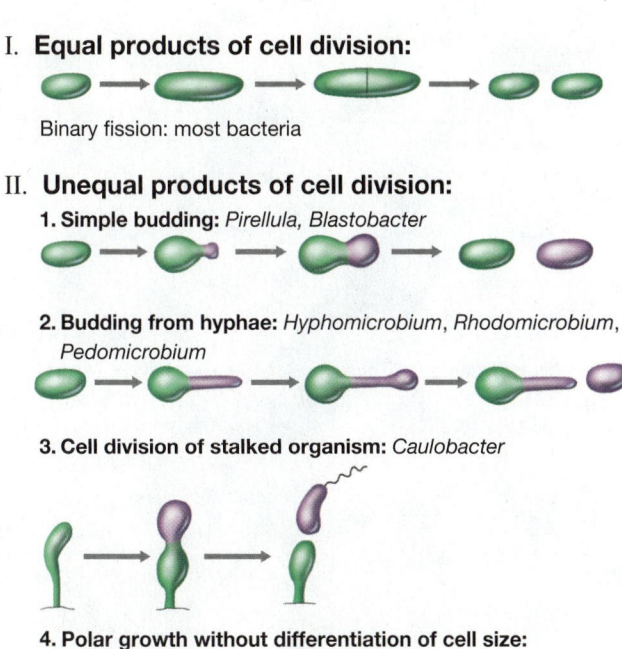

Binary fission: most bacteria

II. **Unequal products of cell division:**

1. Simple budding: *Pirellula, Blastobacter*

2. Budding from hyphae: *Hyphomicrobium, Rhodomicrobium, Pedomicrobium*

3. Cell division of stalked organism: *Caulobacter*

4. Polar growth without differentiation of cell size:

Rhodopseudomonas, Nitrobacter, Methylosinus

Figure 14.54 **Cell division in different bacteria.** Contrast between cell division in conventional bacteria and in various budding and stalked bacteria.

Budding Bacteria: *Hyphomicrobium*

Two well-studied budding bacteria are closely related *Alpha-proteobacteria*: *Hyphomicrobium*, which is chemoorganotrophic, and *Rhodomicrobium*, which is phototrophic. These organisms release buds from the ends of long, thin hyphae. The hypha is a direct cellular extension and contains cell wall, cytoplasmic membrane, and ribosomes, and can contain DNA.

Figure 14.55 shows the life cycle of *Hyphomicrobium*. The mother cell, which is often attached by its base to a solid substrate, forms a thin outgrowth that lengthens to become a hypha. At the end of the hypha, a bud forms. This bud enlarges, forms a flagellum, breaks loose from the mother cell, and swims away. Later, the daughter cell loses its flagellum and after a period of maturation forms a hypha and buds. More buds can also form at the hyphal tip of the mother cell, leading to arrays of cells connected by hyphae. In some cases, a bud begins to form directly from the mother cell without the intervening formation of a hypha, whereas in other cases a single cell forms hyphae from each end (**Figure 14.56**). Nucleoid replication events occur before the bud emerges, and then once a bud has formed, a copy of the chromosome moves down the hypha and into the bud. A cross-septum then forms, separating the still-developing bud from the hypha and mother cell (Figure 14.55).

Physiologically, *Hyphomicrobium* is a methylotrophic bacterium (Section 14.17), and it is widespread in freshwater, marine, and terrestrial habitats. Preferred carbon sources are C_1 compounds such as methanol (CH_3OH), methylamine (CH_3NH_2), formaldehyde (CH_2O), and formate ($HCOO^-$). A fairly specific enrichment procedure for *Hyphomicrobium* is to use CH_3OH as an electron donor with nitrate (NO_3^-) as an electron acceptor in a

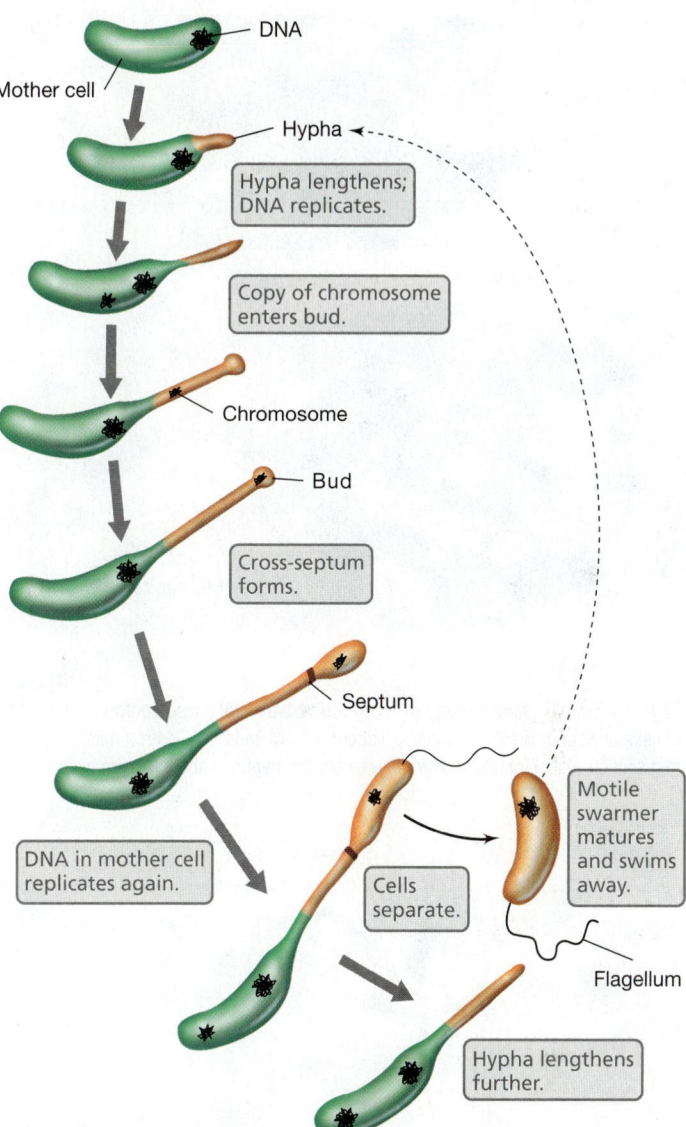

Figure 14.55 **Stages in the *Hyphomicrobium* cell cycle.** The single chromosome of *Hyphomicrobium* is circular.

dilute medium incubated under anoxic conditions. The only rapidly growing denitrifying bacterium known that uses CH_3OH as an electron donor is *Hyphomicrobium*, and so this procedure can select this organism out of a wide variety of environments.

Prosthecate and Stalked *Bacteria*

A variety of bacteria are able to produce cytoplasmic extrusions including *stalks* (**Figure 14.57**), *hyphae*, and *appendages* (**Table 14.4**). Extrusions of these kinds, which are smaller in diameter than the mature cell and contain cytoplasm and a cell wall, are collectively called **prosthecae** (**Figure 14.58**). Prosthecae allow organisms to attach to particulate matter, plant material, or other microorganisms in aquatic habitats. In addition, prosthecae can be used to increase the ratio of surface area to cell volume. Recall that the high surface-to-volume ratio of prokaryotic cells in general confers an increased ability to take up nutrients and expel wastes (⟷ Section 2.6). The unusual morphology of appendaged

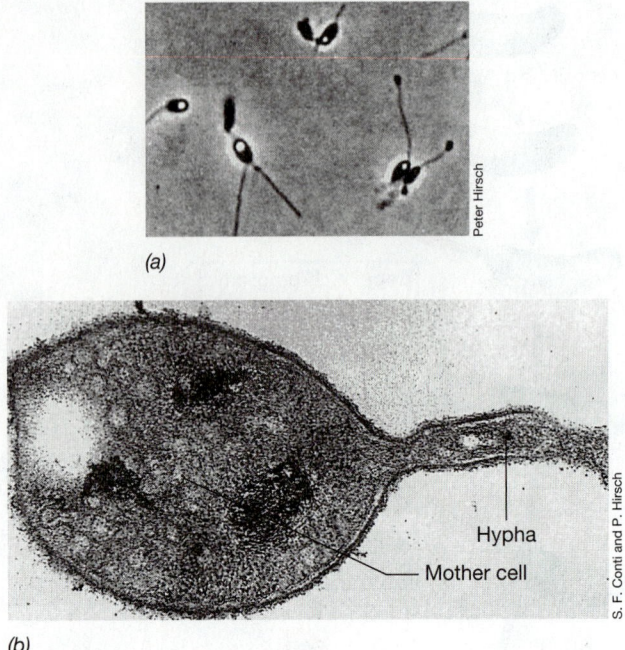

(a)

(b)

Peter Hirsch

S. F. Conti and P. Hirsch

Hypha

Mother cell

Figure 14.56 Morphology of *Hyphomicrobium*. *(a)* Phase-contrast micrograph of cells of *Hyphomicrobium*. Cells are about 0.7 μm wide. *(b)* Electron micrograph of a thin section of a single *Hyphomicrobium* cell. The hypha is about 0.2 μm wide.

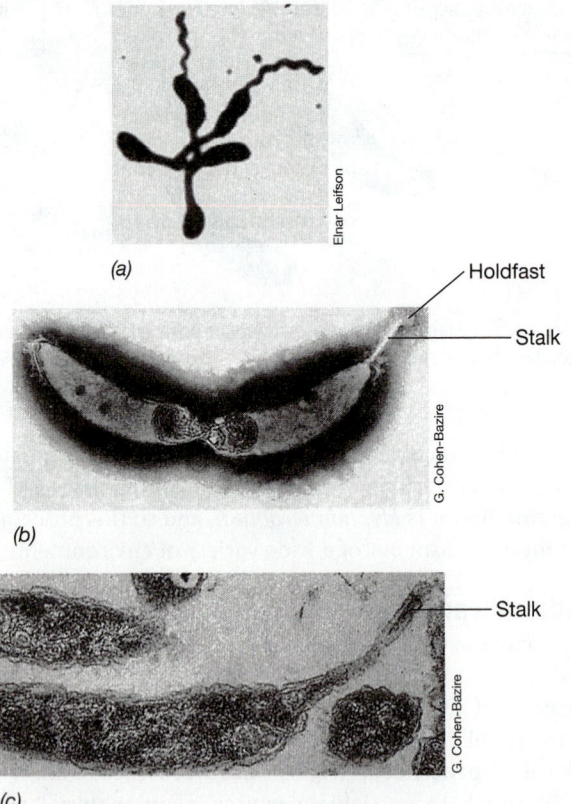

(a)

Elnar Leifson

Holdfast

Stalk

(b)

G. Cohen-Bazire

Stalk

(c)

G. Cohen-Bazire

Figure 14.57 Stalked bacteria. *(a)* A *Caulobacter* rosette. A single cell is about 0.5 μm wide. The five cells are attached by their stalks, which are also prosthecae. Two of the cells have divided, and the daughter cells have formed flagella. *(b)* Negatively stained preparation of a *Caulobacter* cell in division. *(c)* A thin section of *Caulobacter* showing that cytoplasm is present in the stalk. Parts b and c are electron micrographs.

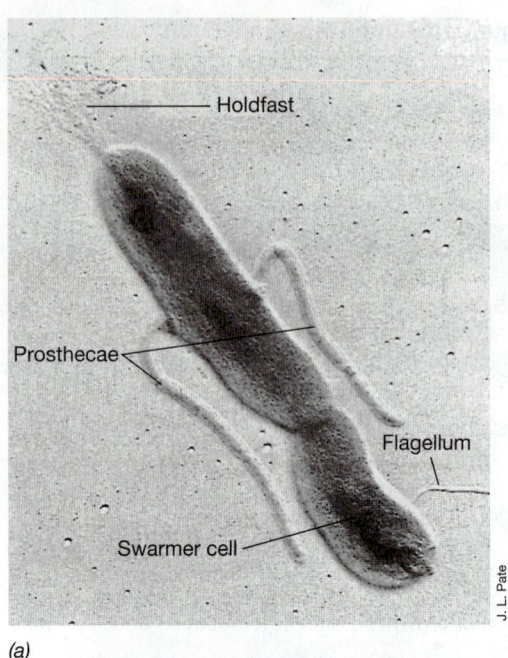

Holdfast

Prosthecae

Flagellum

Swarmer cell

J. L. Pate

(a)

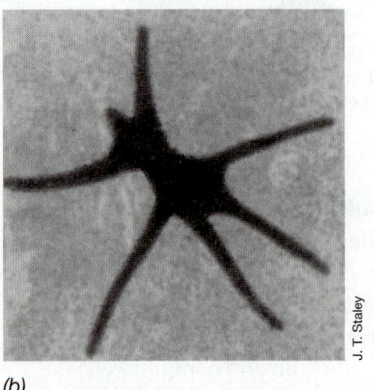

(b)

J. T. Staley

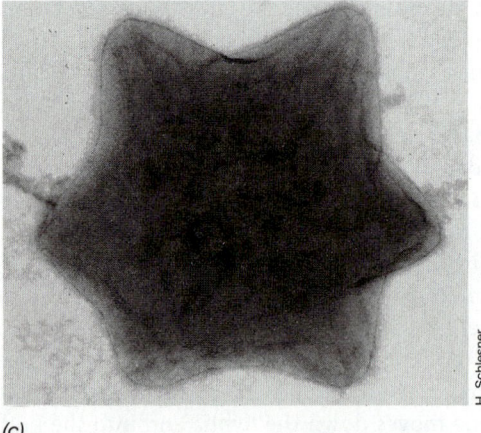

(c)

H. Schlesner

Figure 14.58 Prosthecate bacteria. *(a)* Electron micrograph of a shadow-cast preparation of *Asticcacaulis biprosthecum*, illustrating the location and arrangement of the prosthecae, the holdfast, and a swarmer cell. The swarmer cell breaks away from the mother cell and begins a new cell cycle. Cells are about 0.6 μm wide. *(b)* Negatively stained electron micrograph of a cell of *Ancalomicrobium adetum*. The prosthecae are bounded by the cell wall, contain cytoplasm, and are about 0.2 μm in diameter. *(c)* Electron micrograph of the star-shaped bacterium *Stella*. Cells are about 0.8 μm in diameter.

Table 14.4 Characteristics of major genera of stalked, appendaged (prosthecate), and budding bacteria

Characteristics	Genus	Phylogenetic group[a]
Stalked bacteria		
Stalk an extension of the cytoplasm and involved in cell division	Caulobacter	Alpha
Stalked, fusiform-shaped cells	Prosthecobacter	Verrucomicrobiaceae[b]
Stalked, but stalk is an excretory product not containing cytoplasm:		
Stalk depositing iron, cell vibrioid	Gallionella	Beta
Laterally excreted gelatinous stalk not depositing iron	Nevskia	Gamma
Appendaged (prosthecate) bacteria		
Single or double prosthecae	Asticcacaulis	Alpha
Multiple prosthecae:		
Short prosthecae, multiply by fission, some with gas vesicles	Prosthecomicrobium	Alpha
Flat, star-shaped cells, some with gas vesicles	Stella	Alpha
Long prosthecae, multiply by budding, some with gas vesicles	Ancalomicrobium	Alpha
Budding bacteria		
Phototrophic, produce hyphae	Rhodomicrobium	Alpha
Phototrophic, budding without hyphae	Rhodopseudomonas	Alpha
Chemoorganotrophic, rod-shaped cells	Blastobacter	Alpha
Chemoorganotrophic, buds on tips of slender hyphae:		
Single hypha from parent cell	Hyphomicrobium	Alpha
Multiple hyphae from parent cell	Pedomicrobium	Alpha

[a]All but *Prosthecobacter* are *Proteobacteria*.
[b]See Section 15.17.

bacteria (Figure 14.58) carries this theme to an extreme, and may be an evolutionary adaptation to life in oligotrophic (nutrient-poor) waters where these organisms are most commonly found.

Prosthecae may also function to reduce cell sinking. Because these organisms are aquatic and their metabolism is typically aerobic, prosthecae may keep cells from sinking into anoxic zones in their aquatic environments where they would be unable to respire.

Caulobacter

Two common stalked bacteria are *Caulobacter* (Figure 14.57) and *Gallionella* (Figure 14.36). The former is a chemoorganotroph that produces a cytoplasm-filled stalk, that is, a prostheca, while the latter is a chemolithotrophic iron-oxidizing bacterium whose stalk is composed of ferric hydroxide [Fe(OH)$_3$] (Section 14.15). *Caulobacter* cells are often seen on surfaces in aquatic environments with the stalks of several cells attached to form *rosettes* (Figure 14.57*a*). At the end of the stalk is a structure called a *holdfast* by which the stalk anchors the cell to a surface.

The *Caulobacter* cell division cycle (**Figure 14.59**; ↩ Section 7.12 and Figure 7.26) is unique because cells undergo unequal binary fission. A stalked cell of *Caulobacter* divides by elongation of the cell followed by binary fission, and a single flagellum forms at the pole opposite the stalk. The flagellated cell so formed, called a *swarmer,* separates from the nonflagellated mother cell and eventually attaches to a new surface, forming a new stalk at

the flagellated pole; the flagellum is then lost. Stalk formation is a necessary precursor of cell division and is coordinated with DNA synthesis (Figure 14.59). The cell division cycle in *Caulobacter* is thus more complex than simple binary fission or budding division because the stalked and swarmer cells are structurally different and the growth cycle must include both forms.

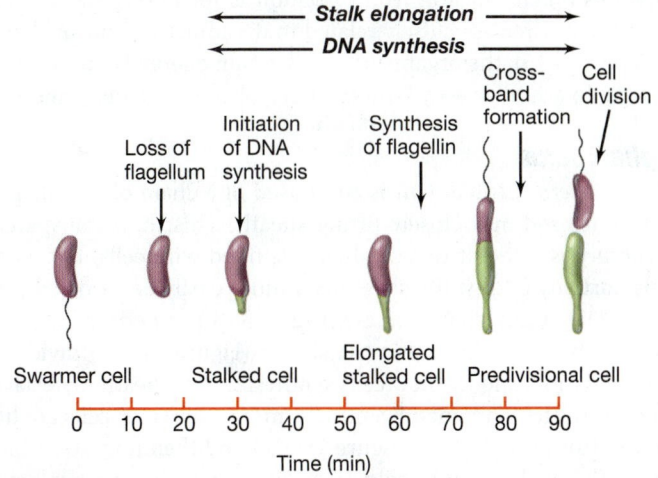

Figure 14.59 Growth of *Caulobacter*. Stages in the *Caulobacter* cell cycle, beginning with a swarmer cell. Compare with Figure 7.26.

14.22 Sheathed *Bacteria*

Key Genera: *Sphaerotilus, Leptothrix*

Bacteria in many phyla form sheaths made of polysaccharide or protein that encase one or many cells. Sheaths often function to bind cells together into long multicellular filaments (Sections 14.3, 14.11). Sheathed bacteria such as *Sphaerotilus* and *Leptothrix* are *Betaproteobacteria* that have a unique life cycle. *Sphaerotilus* and *Leptothrix* are filamentous bacteria that grow within a sheath. Under favorable conditions, the cells grow vegetatively, leading to the formation of long, cell-packed sheaths. Flagellated swarmer cells form within the sheath and under unfavorable growth conditions; the swarmer cells break out and are dispersed to new environments, leaving behind the empty sheath.

Sphaerotilus and *Leptothrix* are common in freshwater habitats that are rich in organic matter, such as wastewaters and polluted streams. Because they are typically found in flowing waters, they are also abundant in trickling filters and activated sludge digesters in sewage treatment plants (↩ Section 21.6). In habitats in which reduced iron (Fe^{2+}) or manganese (Mn^{2+}) is present, the sheaths may become coated with ferric hydroxide [$Fe(OH)_3$] or manganese oxides from the oxidation of these metals.

Leptothrix

The ability of *Sphaerotilus* and *Leptothrix* to precipitate iron oxides on their sheaths is well established, and when sheaths become iron encrusted, as occurs in iron-rich waters, they can frequently be seen microscopically (**Figure 14.60**). Iron precipitates form when ferrous iron (Fe^{2+}), chelated to organic materials such as humic or tannic acids, is oxidized. These chemoorganotrophic bacteria use the organic materials as a carbon or energy source and, when no longer chelated, the ferrous iron becomes oxidized and precipitates on the sheath. Iron oxidation is fortuitous and though these organisms are closely related to dissimilative iron-oxidizers (Section 14.15), the organism does not gain energy from iron oxidation. In a similar way, *Leptothrix* can also oxidize manganese.

Sphaerotilus

The *Sphaerotilus* filament is composed of a chain of rod-shaped cells enclosed in a closely fitting sheath. This thin, transparent structure is difficult to see when it is filled with cells, but when it is partially empty, the sheath can more easily be resolved (**Figure 14.61***a*). Individual cells are 1–2 × 3–8 μm and stain gram-negatively. The cells within the sheath (Figure 14.61*b*) divide by binary fission, and the new cells synthesize new sheath material at the tips of the filaments. Eventually, motile swarmer cells are liberated from the sheaths (Figure 14.61*c*) and then migrate, attach to a solid surface, and begin to grow, with each swarmer being the forerunner of a new filament. The sheath, which is devoid of peptidoglycan, consists of protein and polysaccharide.

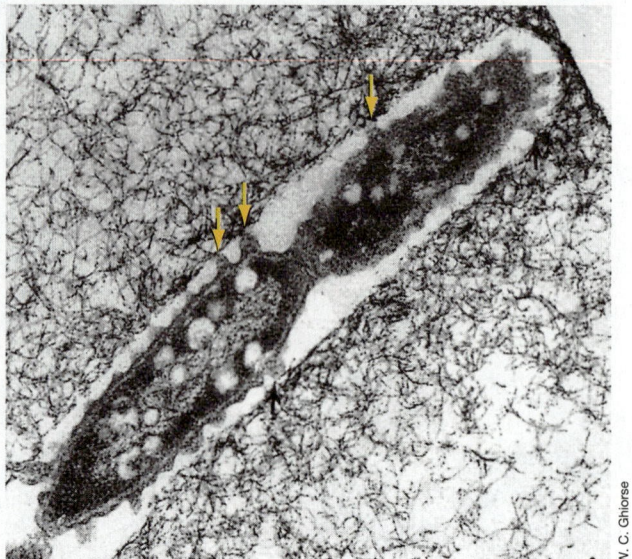

W. C. Ghiorse

Figure 14.60 *Leptothrix* **and iron precipitation.** Transmission electron micrograph of a thin section of *Leptothrix* growing in a ferromanganese film in a swamp in Ithaca, New York. A single cell measures about 0.9 μm in diameter. Note the protuberances of the cell envelope that contact the sheath (arrows).

Sphaerotilus cultures are nutritionally versatile and use simple organic compounds as carbon and energy sources. Befitting its habitat in flowing waters, *Sphaerotilus* is an obligate aerobe. Large masses (blooms) of *Sphaerotilus* often occur in the fall of the year in streams and brooks when leaf litter causes a temporary increase in the organic content of the water. In addition, its filaments are the main component of a microbial complex that wastewater engineers call "sewage fungus," a filamentous slime found on the rocks in streams receiving sewage pollution. In activated sludge of sewage treatment plants (↩ Section 21.6), *Sphaerotilus* is often responsible for a condition called *bulking*, where the tangled masses of *Sphaerotilus* filaments so increase the bulk of the sludge that it remains suspended and does not settle as it should. This has a negative effect on the oxidation of organic matter and the recycling of inorganic nutrients and leads to treatment plant discharges with high nitrogen and carbon loads.

14.23 Magnetic *Bacteria*

Key Genera: *Magnetospirillum*

Magnetic bacteria demonstrate a dramatic directed movement in a magnetic field called *magnetotaxis*. Within these cells are structures called *magnetosomes*, which consist of chains of magnetic particles made of magnetite (Fe_3O_4) or greigite (Fe_3S_4). Magnetosomes are localized within invaginations of the cell membrane that are organized in a linear conformation by a protein scaffold (↩ Section 2.14 and Figure 2.38). Magnetic bacteria orient along the north–south magnetic moment of a magnetic field, aligning

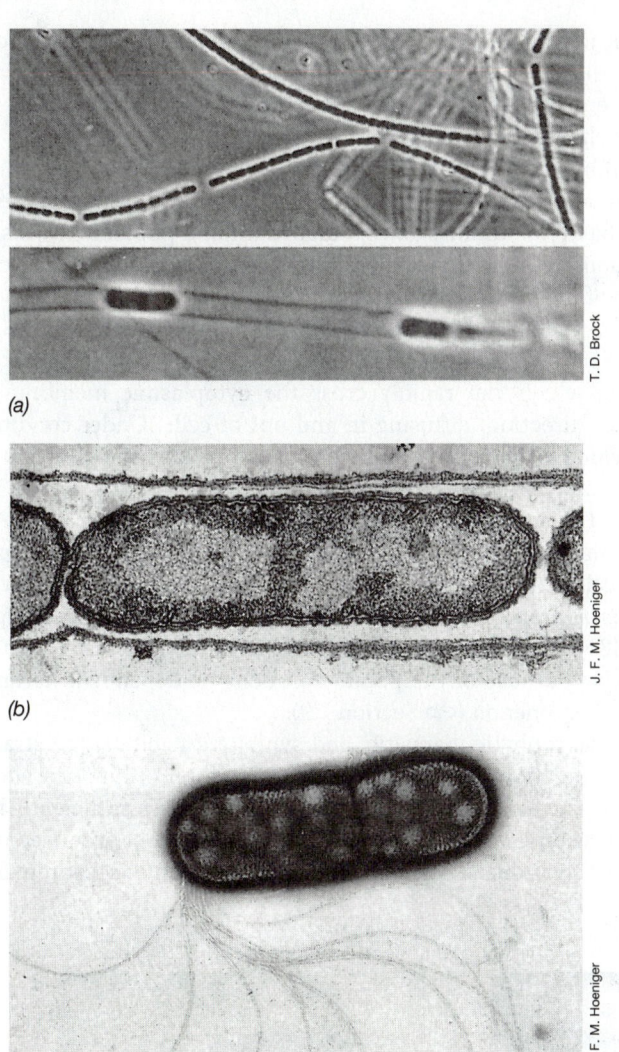

(a)

T. D. Brock

(b)

J. F. M. Hoeniger

(c)

J. F. M. Hoeniger

Figure 14.61 *Sphaerotilus natans.* A single cell is about 2 μm wide. *(a)* Phase-contrast photomicrographs of material collected from a polluted stream. Active growth stage (above) and swarmer cells leaving the sheath. *(b)* Electron micrograph of a thin section through a filament, clearly showing the sheath. *(c)* Electron micrograph of a negatively stained swarmer cell. Notice the polar flagellar tuft.

parallel to the field lines in much the same manner as a compass needle. Magnetic bacteria are typically microaerophilic or anaerobic and are most often found near the oxic-anoxic interface in sediments or stratified lakes. The magnetosomes of aerobic species typically contain the mineral magnetite while those of anaerobes contain exclusively greigite.

Although the ecological role of bacterial magnets is unclear, the ability to orient in a magnetic field may be of selective advantage in maintaining these organisms in zones of low O_2 concentration. Generally, the concentration of O_2 decreases with depth through sediments or the water column of stratified lakes. Since Earth is spherical, its magnetic field lines have a strong vertical component in the Northern and Southern Hemispheres. Thus, bacteria that orient along these field lines can preferentially swim down and away from O_2. The magnetosome functions like a compass needle to "point" the bacterium in the right direction; rotation of

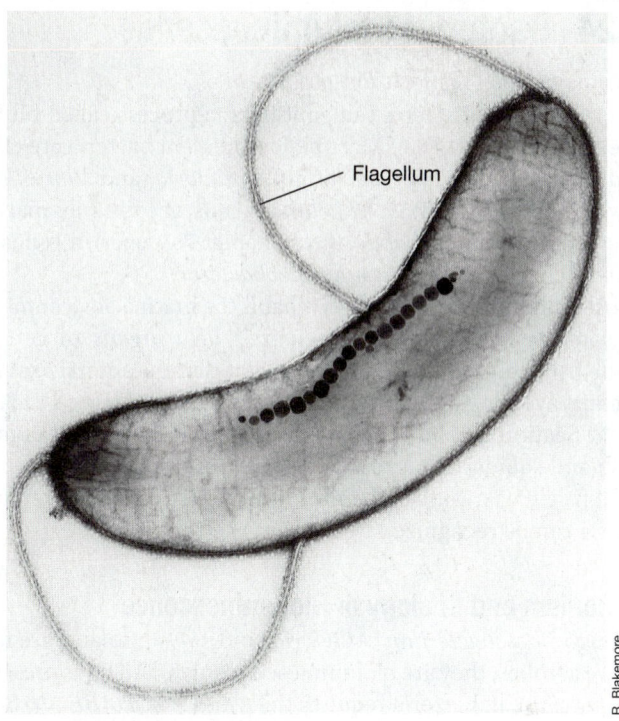

Flagellum

R. Blakemore

Figure 14.62 **A magnetotactic spirillum.** Electron micrograph of a single cell of *Magnetospirillum magnetotacticum*; a cell measures 0.3×2 μm. The cell contains particles of magnetosomes made of Fe_3O_4 arranged in a chain.

the flagellum, by contrast, is controlled by a chemotactic response to O_2 (👉 Section 2.19).

Magnetic bacteria display one of two magnetic polarities depending on the orientation of magnetosomes within the cell. Cells in the Northern Hemisphere have the north-seeking pole of their magnetosomes forward with respect to their flagella and thus move in a northward direction (which in the Northern Hemisphere is downward). Cells in the Southern Hemisphere have the opposite polarity and move southward.

Most of the magnetic bacteria that have been described are species of *Alphaproteobacteria*, but species have also been observed in the *Gammaproteobacteria*, the *Deltaproteobacteria*, and the *Nitrospira* group. The best-characterized species is *Magnetospirillum magnetotacticum* (**Figure 14.62**), which is a chemoorganotrophic microaerophile that can also grow anaerobically by reducing NO_3^- or N_2O. In contrast, the species *Desulfovibrio magneticus* is a sulfate reducer and an obligate anaerobe. In addition, magnetosomes have been observed in a few species of sulfur oxidizers and purple nonsulfur bacteria. Multicellular magnetotactic bacteria are also known. These are *Deltaproteobacteria* that form multicellular aggregates of 10–20 cells organized as a hollow sphere. While multicellular magnetotactic bacteria are obligate anaerobes, the basis of their metabolism has not yet been determined.

MINIQUIZ

- What benefit do magnetic bacteria accrue from having magnetosomes?
- Would you expect to find greigite or magnetite in the magnetosomes of *Desulfovibrio magneticus*?

14.24 Bacterial Bioluminescence

Key Genera: *Vibrio, Aliivibrio, Photobacterium*

Several species of bacteria can emit light, a process called **bioluminescence** (**Figure 14.63**). Most bioluminescent bacteria are classified in the genera *Photobacterium*, *Aliivibrio*, and *Vibrio*, but a few species reside in *Shewanella*, a genus of primarily marine bacteria, and in *Photorhabdus* (⟳ Chapter 7 opener), a genus of terrestrial bacteria (all *Gammaproteobacteria*).

Most bioluminescent bacteria inhabit the marine environment, and some species colonize specialized *light organs* of certain marine fishes and squids, producing light that the animal uses for signaling, avoiding predators, and attracting prey (Figure 14.63*c–f* and ⟳ Section 22.11). When living symbiotically in light organs of fish and squids, or saprophytically, for example on the skin of a dead fish, or parasitically in the body of a crustacean, luminous bacteria can be recognized by the light they produce.

Mechanism and Ecology of Bioluminescence

Although *Photobacterium*, *Aliivibrio*, and *Vibrio* isolates are facultative aerobes, they are bioluminescent only when O_2 is present. Luminescence in bacteria requires the genes *luxCDABE* (⟳ Section 7.9) and is catalyzed by the enzyme *luciferase*, which uses O_2, a long-chain aliphatic aldehyde (RCHO) such as tetradecanal, and reduced flavin mononucleotide ($FMNH_2$) as substrates:

$$FMNH_2 + O_2 + RCHO \xrightarrow{\text{Luciferase}} FMN + RCOOH + H_2O + light$$

The light-generating system constitutes a metabolic route for shunting electrons from $FMNH_2$ to O_2 directly, without employing other electron carriers such as quinones and cytochromes.

Luminescence in many luminous bacteria only occurs at high population density. The enzyme luciferase and other proteins of the bacterial luminescence system exhibit a population density–responsive induction, called **autoinduction**, in which transcription of the *luxCDABE* genes is controlled by a regulatory protein, LuxR, and an inducer molecule, acyl homoserine lactone (AHL, ⟳ Section 7.9 and Figure 7.20). During growth, cells produce AHL, which can rapidly cross the cytoplasmic membrane in either direction, diffusing in and out of cells. Under conditions in which a high local population density of cells of a given species is attained, as in a test tube, a colony on a plate (⟳ Figure 1.1), or in the light organ of a fish or squid (⟳ Section 22.11), AHL accumulates. Only when it reaches a certain concentration in the cell is AHL bound by LuxR, forming a complex that activates transcription of *luxCDABE*; cells then become luminous (Figure 14.58*b*). This gene regulatory mechanism is also called *quorum sensing* because of the population density–dependent nature of the phenomenon (⟳ Section 7.9).

In saprophytic, parasitic, and symbiotic habitats, the strategy for population density–responsive induction of luminescence is to ensure that luminescence develops only when sufficiently high population densities are reached to allow the light produced to be visible to animals. The bacterial light can then attract animals to

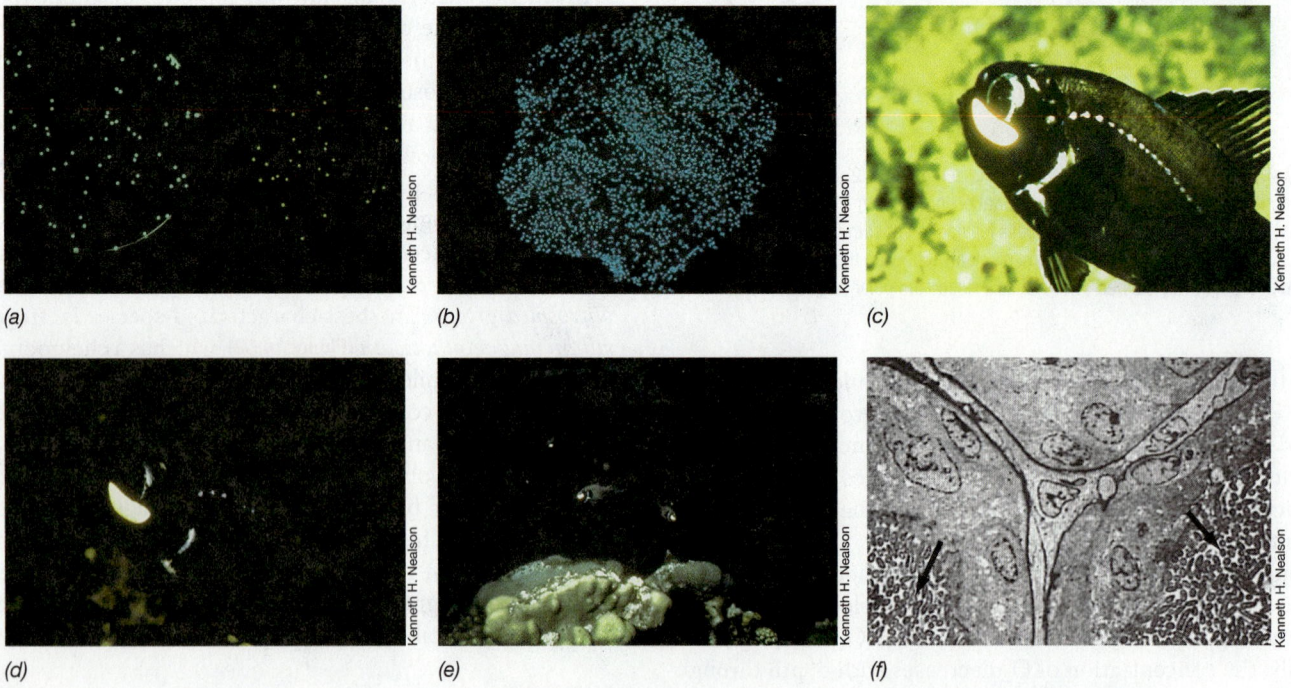

(a)   *(b)*   *(c)*

(d)   *(e)*   *(f)*

Kenneth H. Nealson

Figure 14.63 Bioluminescent bacteria and their role as light organ symbionts in the flashlight fish. *(a)* Two Petri plates of luminous bacteria photographed by their own light. Note the different colors. Left, *Aliivibrio fischeri* strain MJ-1, blue light, and right, strain Y-1, green light. *(b)* Colonies of *Photobacterium phosphoreum* photographed by their own light. *(c)* The flashlight fish *Photoblepharon palpebratus*; the bright area is the light organ containing bioluminescent bacteria. *(d)* Same fish photographed by its own light. *(e)* Underwater photograph taken at night of *P. palpebratus*. *(f)* Electron micrograph of a thin section through the light-emitting organ of *P. palpebratus* showing the dense array of bioluminescent bacteria (arrows).

feed on the luminous material, thereby bringing the bacteria into the animal's nutrient-rich gut for further growth. Alternatively, the luminous material may function as a light source in symbiotic, light organ associations.

Quorum sensing is a form of regulation that has also been found in many different nonluminous bacteria, including several animal and plant pathogens. Quorum sensing in these bacteria controls activities such as the production of extracellular enzymes and expression of virulence factors for which a high population density is beneficial if the bacteria are to have a biological effect.

MINIQUIZ

- What substrates and enzyme are required for an organism such as *Aliivibrio* to emit visible light?
- What is quorum sensing and how does it control bioluminescence?

BIG IDEAS

14.1 • Phylogenetic diversity is that component of microbial diversity that deals with evolutionary relationships between microorganisms. By contrast, functional diversity deals with diversity in form and function as it relates to microbial physiology and ecology. Incongruence between phylogeny and the functional traits of microorganisms can result from patterns of gene loss, horizontal gene transfer, and/or convergent evolution.

14.2 • Anoxygenic phototrophs, which do not produce oxygen, were the first phototrophic organisms to evolve. The evolution of photosynthesis has been impacted strongly by patterns of horizontal gene transfer.

14.3 • *Cyanobacteria* is the only bacterial phylum that contains oxygenic phototrophs. All species of cyanobacteria can fix CO_2 and many can fix N_2, making these organisms important primary producers in many ecosystems.

14.4 • Purple sulfur bacteria are anoxygenic phototrophic *Gammaproteobacteria*. Purple sulfur bacteria use H_2S and S^0 as electron donors and fix CO_2 by the Calvin cycle. These phototrophs have bacteriochlorophylls *a* or *b* and use a type II photosystem.

14.5 • Purple nonsulfur bacteria are anoxygenic phototrophic *Alpha-* and *Betaproteobacteria*. Purple nonsulfur bacteria are metabolically diverse, growing best as photoheterotrophs, and can also grow in darkness. These phototrophs have bacteriochlorophylls *a* or *b* and use a type II photosystem. Aerobic anoxygenic phototrophs have a type II photosystem but only possess bacteriochlorophyll *a*.

14.6 • Green sulfur bacteria are anoxygenic phototrophs of the phylum *Chlorobi*. Green sulfur bacteria use H_2S or S^0 as electron donors and fix CO_2 by the reverse citric acid cycle. These phototrophs contain bacteriochlorophylls *c*, *d*, or *e* (localized in their chlorosomes) as well as bacteriochlorophyll *a* in their reaction centers and use a type I photosystem.

14.7 • Green nonsulfur bacteria are anoxygenic phototrophs of the phylum *Chloroflexi* and grow best as photoheterotrophs. These phototrophs contain bacteriochlorophyll *c* in chlorosomes (as in green sulfur bacteria) and bacteriochlorophyll *a* and a type II photosystem (as in phototrophic purple bacteria).

14.8 • Heliobacteria are anoxygenic phototrophic *Firmicutes* that grow as photoheterotrophs or in darkness as chemotrophs. Heliobacteria produce bacteriochlorophyll *g* and have a type I photosystem. *Chloracidobacterium thermophilum* is an anoxygenic phototrophic acidobacterium that grows photoheterotrophically, possesses bacteriochlorophyll *a* and *c* as well as chlorosomes, and has a type I photosystem.

14.9 • Dissimilative sulfate-reducers are obligate anaerobes that grow by reducing SO_4^{2-} with H_2 or simple organic compounds as electron donors. Most sulfate reducers are *Deltaproteobacteria*. Two physiological classes of sulfate-reducing bacteria are known: complete oxidizers, which oxidize acetate to CO_2, and incomplete oxidizers, which cannot.

14.10 • Dissimilative sulfur-reducers are metabolically and phylogenetically diverse organisms that grow by reducing S^0 and other oxidized sulfur compounds as electron acceptors but are unable to reduce SO_4^{2-}.

14.11 • Sulfur chemolithotrophs, most of which are *Proteobacteria*, oxidize H_2S and other reduced sulfur compounds for energy metabolism with O_2 or NO_3^- as electron acceptors and use either CO_2 or organic compounds as carbon sources. Sulfur chemolithotrophs use a variety of ecological strategies to conserve energy from H_2S and O_2, substances that otherwise react together spontaneously.

14.12 • Diazotrophs are bacteria that assimilate N_2 through activity of the enzyme nitrogenase. Diazotrophs are metabolically and phylogenetically diverse and employ various adaptations to protect nitrogenase from oxygen inactivation.

14.13 • Nitrifying bacteria are aerobic chemolithotrophs that oxidize NH_3 to NO_2^- (prefix *Nitroso-*) or NO_2^- to NO_3^- (prefix *Nitro-*). Ammonia oxidizers are *Proteobacteria* or *Thaumarchaeota*, while nitrite oxidizers are *Proteobacteria* or *Nitrospira*. Denitrifiers are metabolically and phylogenetically diverse facultative aerobes and chemoorganotrophs that reduce NO_3^- to the gaseous products NO, N_2O, and N_2.

14.14 • Dissimilative iron-reducers reduce insoluble electron acceptors in anaerobic respirations. Most species can grow anaerobically by reducing ferric iron using H_2 or simple organic compounds as electron donor. The best-characterized genera include *Geobacter*, which contains exclusively obligate anaerobes, and *Shewanella*, which contains facultative aerobes.

14.15 • Dissimilative iron-oxidizers conserve energy from the aerobic oxidation of ferrous iron. These organisms use several ecological strategies to cope with the chemical instability of ferrous iron in oxic habitats at neutral pH. Iron oxidizers are found in four physiological groups: aerobic acidophiles, aerobic neutrophiles, anaerobic chemotrophs, and anaerobic phototrophs.

14.16 • Hydrogen bacteria oxidize H_2 with O_2 as electron acceptor and fix CO_2 by the Calvin cycle. Some hydrogen bacteria, the carboxydobacteria, oxidize carbon monoxide (CO). Most of these bacteria can also grow on organic compounds.

14.17 • Methylotrophs grow on organic compounds that lack carbon–carbon bonds. Some methylotrophs are also methanotrophs, organisms able to catabolize methane. Most methanotrophs are *Proteobacteria* that contain extensive internal membranes and incorporate carbon by either the serine or ribulose monophosphate pathways.

14.18 • The acetic acid bacteria *Acetobacter* and *Gluconobacter* are obligate aerobes that produce acetate from the oxidation of ethanol, and these acid-tolerant bacteria are often found in the fermenting fluids of alcoholic beverages. Acetogens are obligate anaerobes that use the acetyl-CoA pathway to conserve energy and produce acetate.

14.19 • Bacterial predators such as *Bdellovibrio* and *Myxococcus* consume other microorganisms. Myxobacteria have a complex developmental cycle that involves the formation of fruiting bodies that contain myxospores.

14.20 • The phylum *Spirochaetes* contains helically shaped bacteria that show a novel form of motility that allows them to "corkscrew" through viscous materials. These organisms are common in anoxic habitats and are the cause of many well-known human diseases, such as syphilis.

14.21 • Budding and prosthecate bacteria are appendaged cells that form stalks or prosthecae used for attachment or nutrient absorption, and are primarily aquatic. *Hyphomicrobium*, *Caulobacter*, and *Gallionella* are major genera and are all *Proteobacteria*.

14.22 • Sheathed bacteria are filamentous *Proteobacteria* in which individual cells form chains within an outer layer called the sheath. *Sphaerotilus* and *Leptothrix* are major genera of sheathed bacteria and can oxidize metals, such as Fe^{2+} and Mn^{2+}.

14.23 • Magnetosomes are specialized magnetic structures present in magnetotactic bacteria. Magnetosomes orient cells along the magnetic field lines of Earth, and this allows cells to use their normal chemotactic response to move vertically in a directed fashion in sediments or stratified aquatic systems.

14.24 • *Vibrio*, *Aliivibrio*, and *Photobacterium* species are marine bacteria, some of which are pathogenic and bioluminescent. Bioluminescence, catalyzed by the enzyme luciferase, is controlled by a quorum-sensing mechanism that ensures that light is not emitted until a large cell population has been attained.

MasteringMicrobiology® **Review what you know and challenge what you have learned with MasteringMicrobiology!** Access study materials, chapter quizzes, animations, and microbiology lab tutorials in the Study Area to ensure that you have mastered this chapter's content.

REVIEW OF KEY TERMS

Acetic acid bacteria obligately aerobic organisms that produce acetate from alcohols; used for making vinegar

Acetogen an obligately anaerobic organism that produces acetate by use of the reductive acetyl-CoA cycle

Aerobic anoxygenic phototroph an organism that is an aerobic heterotroph that uses anoxygenic photosynthesis as a supplemental source of energy

Autoinduction a gene regulatory mechanism involving small, diffusible signal molecules that are produced in larger amounts as population size increases

Bioluminescence the enzymatic production of visible light by living organisms

Carboxysome a polyhedral cellular inclusion of crystalline ribulose bisphosphate carboxylase (RubisCO), the key enzyme of the Calvin cycle

Chemolithotroph an organism able to oxidize inorganic compounds (such as H_2, Fe^{2+}, S^0, or NH_4^+) as energy sources (electron donors)

Chlorosome a cigar-shaped structure bounded by a nonunit membrane and containing the light-harvesting bacteriochlorophyll (*c, d,* or *e*) in green sulfur bacteria and *Chloroflexus*

Consortium a two- or more-membered association of bacteria, usually living in an intimate symbiotic fashion

Convergent evolution a circumstance where a trait or set of traits that are similar in form and/or function between two organisms are not inherited from a shared ancestor (i.e., traits that are similar but not homologous)

Cyanobacteria prokaryotic oxygenic phototrophs containing chlorophyll *a* and phycobilins

Denitrifier an organism that carries out anaerobic respiration with NO_3^-, reducing it to the gaseous products NO, N_2O, and N_2

Diazotroph an organism that can assimilate N_2 into biomass by activity of the enzyme nitrogenase

Dissimilative sulfate-reducer an anaerobic microorganism that conserves energy through the reduction of SO_4^{2-}

Dissimilative sulfur-reducer an anaerobic microorganism that conserves energy through the reduction of S^0 but cannot reduce SO_4^{2-}

Dissimilative sulfur-oxidizer a microorganism that gains energy for growth through oxidation of reduced sulfur compounds

Functional diversity the component of biological diversity that deals with the forms and functions of organisms as they relate to differences in physiology and ecology

Green nonsulfur bacteria anoxygenic phototrophs containing chlorosomes, a type II photosystem, bacteriochlorophylls *a* and *c* as light-harvesting chlorophyll, and typically growing best as photoheterotrophs

Green sulfur bacteria anoxygenic phototrophs containing chlorosomes, type I photosystem, bacteriochlorophylls *c, d,* or *e* as light-harvesting chlorophyll, and typically growing with H_2S as an electron donor

Heliobacteria anoxygenic phototrophs containing bacteriochlorophyll *g*

Horizontal gene transfer a unidirectional transfer of genes between unrelated organisms; can cause homologous genes to be dispersed in a phylogeny

Methanotroph an organism capable of oxidizing methane (CH_4) as an electron donor in energy metabolism

Methylotroph an organism capable of oxidizing organic compounds that do not contain carbon–carbon bonds; if able to oxidize CH_4, also a methanotroph

Mixotroph an organism that conserves energy from the oxidation of inorganic compounds but requires organic compounds as a carbon source

Nitrifier a chemolithotroph capable of carrying out the transformation $NH_3 \rightarrow NO_2^-$, or $NO_2^- \rightarrow NO_3^-$

Phycobilin a protein containing the pigment phycocyanin or phycoerythrin that functions as a photosynthetic accessory pigment in cyanobacteria

Prochlorophyte a bacterial oxygenic phototroph that contains chlorophylls *a* and *b* but lacks phycobilins

Prosthecae extrusion of cytoplasm, often forming distinct appendages, bounded by the cell wall

Purple nonsulfur bacteria a group of phototrophic bacteria containing bacteriochlorophyll *a* or *b* and a type II photosystem and that grow best as photoheterotrophs

Purple sulfur bacteria a group of phototrophic bacteria containing bacteriochlorophylls *a* or *b* and type II photosystem and that can oxidize H_2S to sulfur

Spirilla (singular, spirillum) spiral-shaped cells

Spirochete a slender, tightly coiled, gram-negative bacterium of the phylum *Spirochaetes* characterized by possession of endoflagella used for motility

REVIEW QUESTIONS

1. What is convergent evolution and how is it different from horizontal gene transfer? (Section 14.1)

2. Which bacterial phyla contain phototrophs? (Section 14.2)

3. How is the prochlorophyte, *Prochlorococcus*, different from other cyanobacteria? (Section 14.3)

4. Compare and contrast the metabolism, morphology, and phylogeny of purple sulfur and purple nonsulfur bacteria. (Sections 14.4–14.5)

5. Compare and contrast the metabolism of purple nonsulfur bacteria and aerobic anoxygenic phototrophs. (Section 14.5)

6. In what group of organisms would you expect to find chlorosomes? (Section 14.6)

7. What traits do green nonsulfur bacteria share with green sulfur bacteria and purple sulfur bacteria? (Section 14.7)

8. In what ways is *Chloracidobacterium thermophilum* similar to green sulfur bacteria and in what ways is it different? (Section 14.8)

9. With respect to sulfate-reducing bacteria, what is the difference between complete and incomplete oxidizers? (Section 14.9)

10. In what ways are sulfur-reducing bacteria different from sulfate-reducing bacteria and in what ways are they similar? (Sections 14.9–14.10)

11. What are some ecological strategies that aerobic sulfide-oxidizers use to compete with the chemical oxidation of H_2S by atmospheric O_2? (Section 14.11)

12. What are some ways that diazotrophs protect nitrogenase from O_2? (Section 14.12)

13. Compare and contrast the nitrogen metabolism of nitrifiers with that of denitrifiers. (Section 14.13)

14. In what ways are the dissimilative iron-reducing bacteria *Shewanella* and *Geobacter* similar and in what ways are they different? (Section 14.14)

15. Compare and contrast the metabolism of *Gallionella* and *Geobacter*. In what habitats might you find these organisms? (Sections 14.14–14.15)

16. Which group of dissimilative iron-oxidizers is the least diverse and in what way is this related to oxygen and pH? (Section 14.15)

17. Why are most aerobic hydrogen-oxidizing bacteria microaerophiles? (Section 14.16)

18. What are the differences between type I and type II methanotrophs? (Section 14.17)

19. Contrast the metabolic characteristics of acetic acid bacteria and acetogens. What characteristic do they share and in what ways do they differ? (Section 14.18)

20. Compare and contrast the life cycle of *Myxococcus* with that of *Bdellovibrio*. (Section 14.19)

21. Contrast the motility of spirochetes with that of spirilla. (Section 14.20)

22. Contrast the life cycle of *Hyphomicrobium* with that of *Caulobacter*. (Section 14.21)

23. What materials can be found in the sheath made by *Leptothrix*? (Section 14.22)

24. In what way does a magnetosome control the movement of magnetotactic bacteria? (Section 14.23)

25. Describe the manner in which cell density regulates light production in luminescent bacteria. (Section 14.24)

APPLICATION QUESTIONS

1. Describe a key physiological feature of the following *Bacteria* that would differentiate each from the others: *Acetobacter, Methylococcus, Azotobacter, Photobacterium, Desulfovibrio,* and *Spirillum.*

2. Describe the metabolism for each of the following *Bacteria* and state whether the organism is an aerobe or an anaerobe: *Thiobacillus, Nitrosomonas, Ralstonia eutropha, Methylomonas, Pseudomonas, Acetobacter,* and *Gallionella.*

15 · Diversity of *Bacteria*

microbiology**now**

Discovering Novel Microbial Phyla

Many plants and animals cannot be kept in captivity because they live in unique habitats or require complex biological interactions that cannot be replicated in a zoo or botanical garden. Many microorganisms are hard to get or keep in captivity as well, because of intimate connections to their environments. These noncultivated microorganisms were largely unknown before the development of molecular techniques for studying microbial diversity. Only 12 phyla of *Bacteria* were known in 1987, but as a result of sequencing 16S ribosomal RNA genes from environmental samples, we now know of more than 80 bacterial phyla, most of which lack cultured representatives.

A recent culturing success story occurred with phylum OP10, named after a 16S ribosomal RNA gene sequence originally recovered from Obsidian Pool, a geothermal spring in Yellowstone National Park (USA).[1] Following its discovery, 16S gene sequences from the OP10 phylum were observed in various thermal environments around the world, but a culture was never obtained. However, with persistence and improvements in culturing techniques, microbiologists were finally successful in cultivating species of OP10, now renamed as the phylum *Armatimonadetes*.[1]

One of the first cultured species of *Armatimonadetes* was the bacterium *Chthonomonas calidirosea*, isolated from geothermally heated soil at Hell's Gate, New Zealand (photo). *C. calidirosea* is a thermophile that grows best at 68°C and degrades cellulose and xylan aerobically. Strains with these characteristics often produce enzymes that are useful in industrial processes which transform plant biomass into energy or other useful products.

[1]Lee, K.C.-Y., et al. 2011. *Chthonomonas calidirosea* gen. nov., sp. nov., an aerobic, pigmented, thermophilic microorganism of a novel bacterial class, *Chthonomonadetes* classis nov., of the newly described phylum *Armatimonadetes* originally designated candidate division OP10. *Int. J. Syst. Evol. Microbiol.* 61: 2482–2490.

I   *Proteobacteria* 480

II *Firmicutes, Tenericutes*, and *Actinobacteria* 491

III *Bacteroidetes* 504

IV *Chlamydiae, Planctomycetes*, and *Verrucomicrobia* 506

V   Hyperthermophilic *Bacteria* 510

VI Other *Bacteria* 512

In the last chapter we examined microbial diversity with respect to *functional* diversity. In this and the next two chapters we shift our focus to *phylogenetic* diversity. We examine the major lineages of *Bacteria* (**Figure 15.1a**) in this chapter and the *Archaea* and microbial *Eukarya* in Chapters 16 and 17, respectively.

Including phyla of *Bacteria* known only from 16S ribosomal RNA (rRNA) gene sequences retrieved from the environment (⮑ Section 22.6), over 80 phyla can be distinguished. However, fewer than half of these contain species that have been characterized in laboratory culture (Figure 15.1b). Remarkably, more than

90% of characterized genera and species of *Bacteria* originate in only four phyla: *Proteobacteria*, *Actinobacteria*, *Firmicutes*, and *Bacteroidetes* (Figure 15.1b).

With more than ten thousand species of bacteria described, we obviously cannot consider them all. Therefore, using phylogenetic trees to focus our discussion, we will explore some of the best-known species from a broad diversity of phyla. In this chapter we will consider species from more than 20 bacterial phyla, focusing on those with the largest numbers of characterized species. We begin our tour of the *Bacteria* with the phylum *Proteobacteria*.

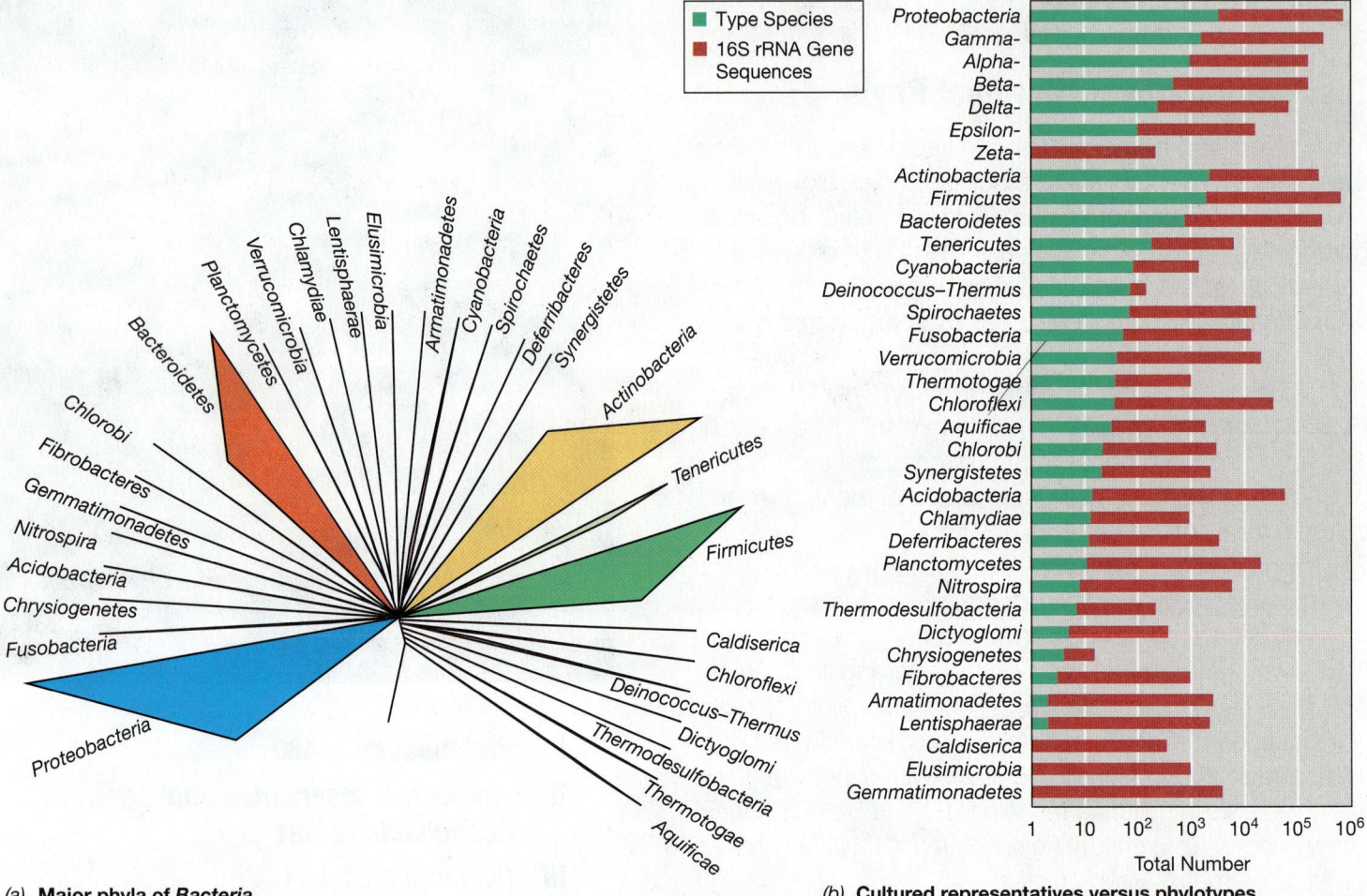

(a) **Major phyla of *Bacteria***

(b) **Cultured representatives versus phylotypes**

Figure 15.1 Some major phyla of *Bacteria* based on 16S ribosomal RNA gene sequence comparisons. (a) Depicted are the major phyla of *Bacteria* that have cultivated species. Analyses of 16S rRNA gene sequences from natural environments suggest there are more than 80 bacterial phyla. (b) Numbers of cultured and characterized species (green bars) and known 16S rRNA gene sequences (phylotypes, red bars) for each of the 29 major bacterial phyla that have at least one characterized species in pure culture. Also shown are related data for the different classes of *Proteobacteria*. Differences between the size of the red and green bars indicate the degree to which members of each group are common in natural environments but difficult to cultivate in isolation. Note that the abscissa is a log scale.

I • *Proteobacteria*

The ***Proteobacteria*** are by far the largest and most metabolically diverse phylum of *Bacteria* (**Figure 15.2**). More than a third of characterized species of *Bacteria* originate within this group (Figure 15.1b), and *Proteobacteria* constitute the majority of known bacteria of medical, industrial, and agricultural significance.

As a group, the *Proteobacteria* are all gram-negative bacteria. They show an exceptionally wide diversity of energy-generating mechanisms, with chemolithotrophic, chemoorganotrophic, and phototrophic species (Figure 15.2). Indeed, we have already seen the great diversity of energy metabolisms used by various

16S rRNA Gene Tree of *Proteobacteria*   **Proteobacterial Classes**

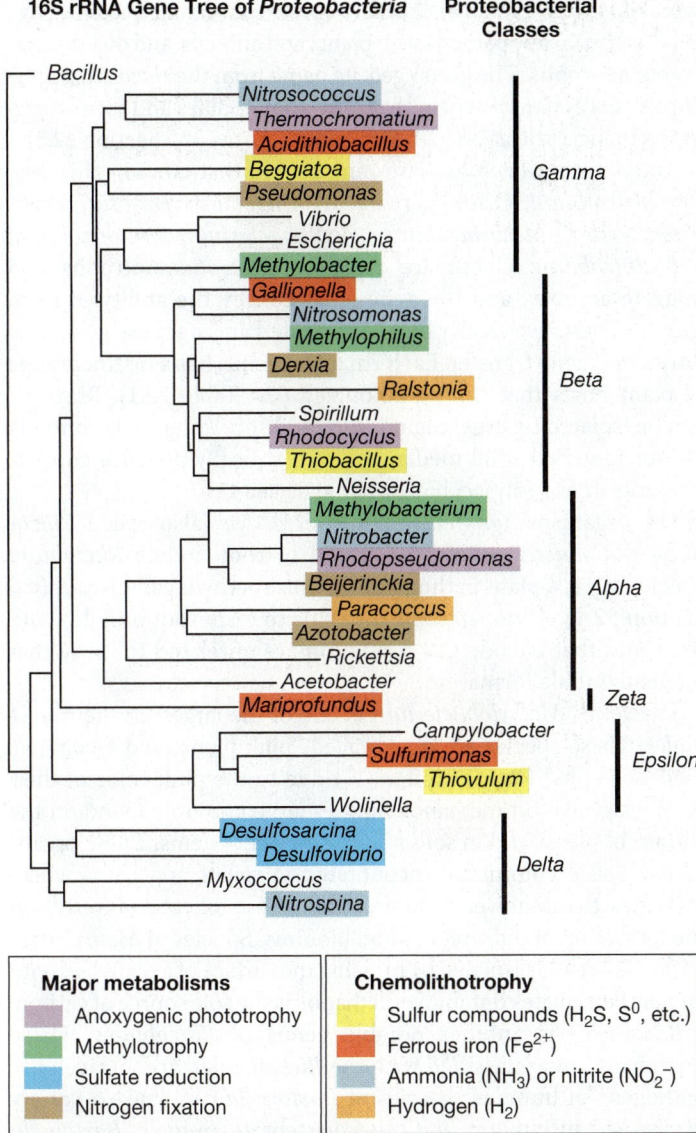

Major metabolisms

▨	Anoxygenic phototrophy
▨	Methylotrophy
▨	Sulfate reduction
▨	Nitrogen fixation

Chemolithotrophy

▨	Sulfur compounds (H_2S, S^0, etc.)
▨	Ferrous iron (Fe^{2+})
▨	Ammonia (NH_3) or nitrite (NO_2^-)
▨	Hydrogen (H_2)

Figure 15.2 Phylogenetic tree and metabolic links of some key genera of *Proteobacteria*. Phylogeny of representative genera of *Proteobacteria* as revealed by analysis of 16S rRNA gene sequences. Note how identical metabolisms are often distributed in phylogenetically distinct genera, suggesting that horizontal gene flow has been extensive in the *Proteobacteria*. Some organisms listed may have multiple properties; for example, some sulfur chemolithotrophs are also iron or hydrogen chemolithotrophs, and several of the organisms listed can fix nitrogen. Phylogenetic analyses were performed and the phylogenetic tree constructed by Marie Asao, Ohio State University.

representatives of this group in Chapter 13. The *Proteobacteria* are equally diverse in terms of their relationship to oxygen (O_2), with anaerobic, microaerophilic, and facultatively aerobic species known. Morphologically, they also exhibit a wide range of cell shapes, including straight and curved rods, cocci, spirilla, filamentous, budding, and appendaged forms.

Based on 16S rRNA gene sequences, the phylum *Proteobacteria* can be divided into six classes: *Alpha-, Beta-, Gamma-, Delta-, Epsilon-,* and *Zetaproteobacteria*. Each class contains many genera with the exception of the *Zetaproteobacteria*, which is composed of a single species, the marine iron-oxidizing

bacterium *Mariprofundus ferrooxydans* (♺ Section 14.15). Despite the phylogenetic breadth of the *Proteobacteria*, species in different classes often have similar metabolisms. For example, phototrophy and methylotrophy occur in three different classes of *Proteobacteria*, and nitrifying bacteria span four classes of *Proteobacteria* (♺ Figure 14.1). This suggests that horizontal gene flow (♺ Section 6.12) has played a major role in shaping the metabolic diversity of the *Proteobacteria*. The sharing of metabolic traits in the different classes of *Proteobacteria* is also a good reminder that phenotype and phylogeny often provide different views of prokaryotic diversity (♺ Section 14.1).

15.1 *Alphaproteobacteria*

With nearly one thousand described species, the *Alphaproteobacteria* are the second largest class of *Proteobacteria* (Figure 15.1*b*). The *Alphaproteobacteria* contain extensive functional diversity (Figure 15.2, ♺ Figure 14.1) and many genera in this group have already been considered in Chapter 14. Most species are obligate aerobes or facultative aerobes and many are **oligotrophic,** preferring to grow in environments that have low nutrient concentration. A total of 10 orders have been described within the *Alphaproteobacteria*, but the vast majority of species fall within the *Rhizobiales, Rickettsiales, Rhodobacterales, Rhodospirillales, Caulobacterales,* and *Sphingomonadales* (**Figure 15.3, Table 15.1**).

Alphaproteobacteria

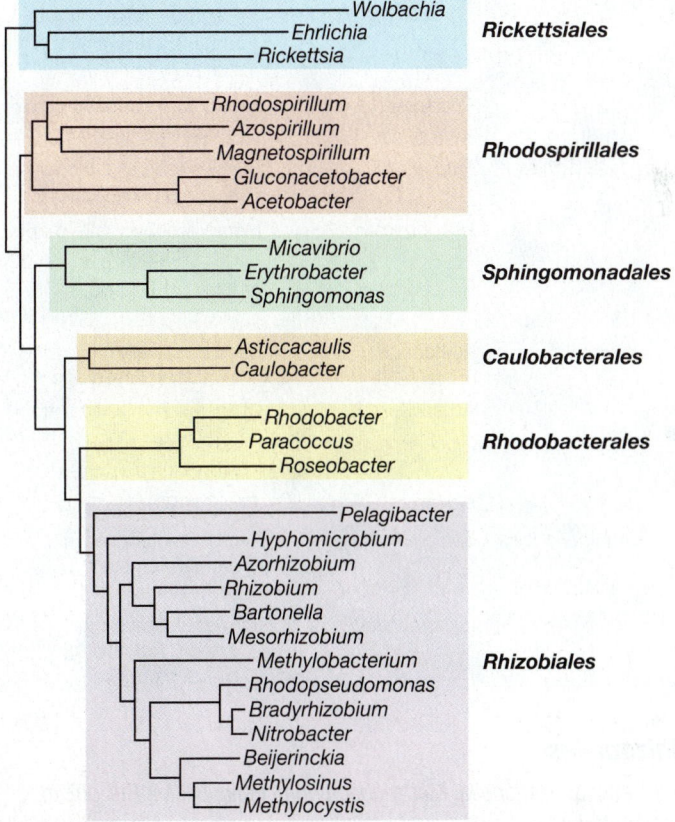

Figure 15.3 Major orders of *Proteobacteria* in the class *Alphaproteobacteria*. The phylogenetic tree was constructed using 16S rRNA gene sequences from representative genera of *Alphaproteobacteria*. Order names are shown in bold.

UNIT 3

Table 15.1 Notable genera of *Alphaproteobacteria*

Family	Genus	Notable characteristics
Caulobacterales	Caulobacter	Asymmetric cell division and formation of prosthecae
Rickettsiales	Rickettsia	Obligate intracellular parasites, transmitted by arthropods
	Wolbachia	Live within arthropods and impact their reproduction
Rhizobiales	Bartonella	Obligate intracellular parasites, transmitted by arthropods
	Bradyrhizobium	Form root nodules with soybean and other legumes
	Brucella	Facultative intracellular parasites of animals, zoonotic pathogen
	Hyphomicrobium	Stalked cells, metabolically versatile
	Mesorhizobium	Form root nodules with bird's-foot trefoil and other legumes
	Methylobacterium	Methylotroph found on plants and in soil
	Nitrobacter	Nitrifying bacterium that oxidizes NO_2^- to NO_3^-
	Pelagibacter	Oligotrophic chemoorganotroph; high abundance in ocean surface
	Rhodopseudomonas	Metabolically versatile purple nonsulfur bacterium
Rhodobacterales	Paracoccus	Species used as a model for studying denitrification
	Rhodobacter	Metabolically versatile purple nonsulfur bacteria
	Roseobacter	Aerobic anoxygenic phototroph
Rhodospirillales	Acetobacter	Used industrially for producing acetic acid
	Azospirillum	Obligately aerobic diazotroph
	Gluconobacter	Used industrially for producing acetic acid
	Magnetospirillum	Magnetotactic bacterium
Sphingomonadales	Sphingomonas	Aerobic degradation of aromatic organics, biodegradation
	Zymomonas	Ferments sugars into ethanol, potential for biofuel production

Rhizobiales

Key Genera: *Bartonella, Methylobacterium, Pelagibacter, Rhizobium, Agrobacterium*

The *Rhizobiales* (Figure 15.3) are the largest and most metabolically diverse order of *Alphaproteobacteria* and contain phototrophs (e.g., *Rhodopseudomonas*), chemolithotrophs (e.g., *Nitrobacter*),

symbionts (e.g., rhizobia), free-living nitrogen-fixing bacteria (e.g., *Beijerinckia*), a few pathogens of plants and animals, and diverse chemoorganotrophs. The group gets its name from the *rhizobia*, a *polyphyletic* collection of genera that form root nodules and fix nitrogen in symbiotic association with leguminous plants (⊂⊃ Section 22.3).

Among the *Rhizobiales* are nine genera that contain rhizobia: *Bradyrhizobium, Ochrobactrum, Azorhizobium, Devosia, Methylobacterium, Mesorhizobium, Phyllobacterium, Sinorhizobium,* and *Rhizobium*. These are typically chemoorganotrophs and obligate aerobes, and the genes that convey the ability to form root nodules have clearly been distributed among these genera by horizontal gene transfer. Each rhizobial genus has a distinct range of plant hosts that can be colonized (⊂⊃ Table 22.1). Rhizobia can be isolated by crushing nodules and spreading their contents on nutrient-rich solid media; colonies typically produce copious amounts of exopolysaccharide slime (**Figure 15.4**).

The organism *Agrobacterium tumefaciens* (also called *Rhizobium radiobacter*) is closely related to root nodule *Rhizobium* species but is a plant pathogen that causes crown gall disease (⊂⊃ Section 22.4). *A. tumefaciens* is unable to form root nodules, and the genes that encode gall formation are unrelated to those that mediate nodule formation.

The genus *Methylobacterium* is one of the largest in the *Rhizobiales*. These species are often called "pink-pigmented facultative methylotrophs" (⊂⊃ Section 14.17) due to the pink color of their colonies grown on methanol. Species are commonly found on the surface of plants and in soils and freshwater systems. These organisms are also commonly encountered in toilets and baths where their growth on shower curtains, caulk, and in toilet bowls results in the formation of pink-pigmented biofilms. Species of *Methylobacterium* are readily isolated by pressing the surface of a plant leaf onto an agar Petri plate containing methanol as the sole source of carbon.

Bartonella is another notable genus of *Rhizobiales*. These organisms, once classified with the *Rickettsiales*, are intracellular pathogens of humans. Species of *Bartonella* can cause a variety of diseases in humans and other vertebrate animals. *Bartonella quintana* is the causative agent of trench fever, a disease that decimated troops in World War I. Other species of *Bartonella* can cause bartonellosis, cat scratch disease, and a variety of inflammatory diseases. Disease transmission is mediated by arthropod vectors including fleas, lice, and sand flies. Species of *Bartonella* are fastidious and difficult to cultivate, and isolation is most commonly achieved using blood agar. When growing in tissue culture,

Figure 15.4 Colonies of *Rhizobium mongolense.* Colonies of rhizobia often produce copious exopolysaccharide slime. These colonies of *Rhizobium mongolense* were grown on a medium low in nitrogen with sucrose as carbon source.

cells of *Bartonella* grow on the outside surface of the eukaryotic host cells rather than within the cytoplasm or the nucleus.

Finally, the genus *Pelagibacter* also belongs to the *Rhizobiales*. *Pelagibacter ubique* is an oligotroph and an obligately aerobic chemoorganotroph that inhabits the photic zone of Earth's oceans. This organism can make up 25% of the bacterial cells found at the ocean's surface, and its numbers can reach 50% of cells in temperate waters in the summer; as a consequence, *Pelagibacter ubique* is likely the most abundant bacterial species on Earth (⮌ Section 19.11).

Rickettsiales

Key Genera: *Rickettsia, Wolbachia*

Rickettsiales (Figure 15.3) are all obligate intracellular parasites or mutualists of animals. Species in this order have not yet been cultivated in the absence of host cells (**Figure 15.5**) and must be grown in chicken eggs or in host cell tissue culture. Typically, *Rickettsiales* are closely associated with arthropods. Those genera that cause disease such as *Rickettsia* and *Ehrlichia* are transmitted by arthropod bites; other genera such as *Wolbachia* are obligate parasites or mutualists of insects and other arthropods.

Species of the genus *Rickettsia* are the causative agents of several human diseases, including typhus (*Rickettsia prowazekii*), and spotted fever rickettsiosis, commonly called Rocky Mountain spotted fever (*Rickettsia rickettsii*) (⮌ Section 30.3). These organisms are closely associated with arthropod vectors and can be transmitted by ticks, fleas, lice and mites. Most rickettsias are metabolically specialized, able to oxidize only the amino acids glutamate or glutamine and unable to oxidize glucose or organic acids. Rickettsias are unable to synthesize certain metabolites and must instead obtain them from host cells. Rickettsias do not survive long outside their hosts, and this may explain why they must be transmitted from animal to animal by arthropod vectors.

Electron micrographs of thin sections of rickettsial cells show a typical prokaryotic morphology including a cell wall (Figure 15.5*b*). The penetration of a host cell by a rickettsial cell is an active process, requiring both host and parasite to be viable and metabolically active. Once inside the host cell, the bacteria multiply primarily in the cytoplasm and continue replicating until the host cell is loaded with parasites (Figure 15.5; ⮌ Figure 30.6). The host cell then bursts and liberates the bacterial cells.

The genus *Wolbachia* contains intracellular parasites of many insects (**Figure 15.6**), a huge group that constitutes 70% of all known arthropod species. *Wolbachia* species can have any of several effects on their insect hosts. These include inducing parthenogenesis (development of unfertilized eggs), the killing of males, and feminization (the conversion of male insects into females).

Wolbachia pipientis is the best-studied species in the genus. Cells of *W. pipientis* colonize the insect egg (Figure 15.6), where they multiply in vacuoles of host cells surrounded by a membrane of host origin. Cells of *W. pipientis* are passed from an infected female to her offspring through this egg infection. *Wolbachia*-induced parthenogenesis occurs in a number of species of wasps. In these insects, males normally arise from unfertilized eggs (which contain only one set of chromosomes), while females arise from fertilized eggs (which contain two sets of chromosomes). However, in unfertilized eggs infected with *Wolbachia*, the organism somehow triggers a doubling of the chromosome number, thus yielding only females. Predictably, if female insects are fed antibiotics that kill *Wolbachia*, parthenogenesis ceases.

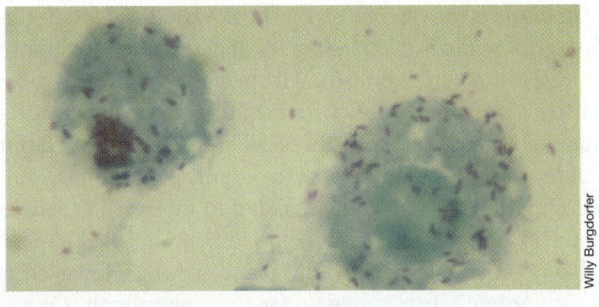

(a)

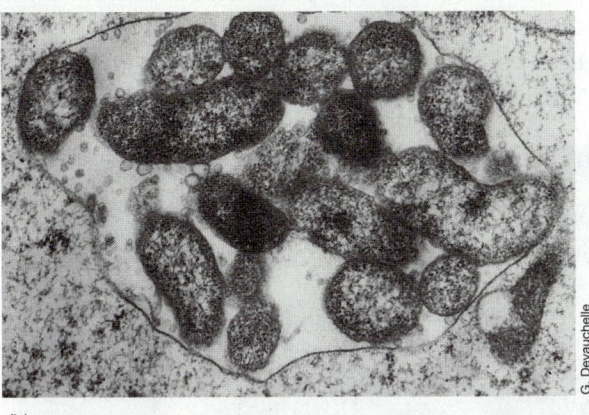

(b)

Figure 15.5 Rickettsias growing within host cells. *(a)* *Rickettsia rickettsii* in tissue culture. Cells are about 0.3 μm in diameter. *(b)* Electron micrograph of cells of *Rickettsiella popilliae* within a blood cell of its host, the beetle *Melolontha melolontha*. The bacteria grow inside a vacuole within the host cell.

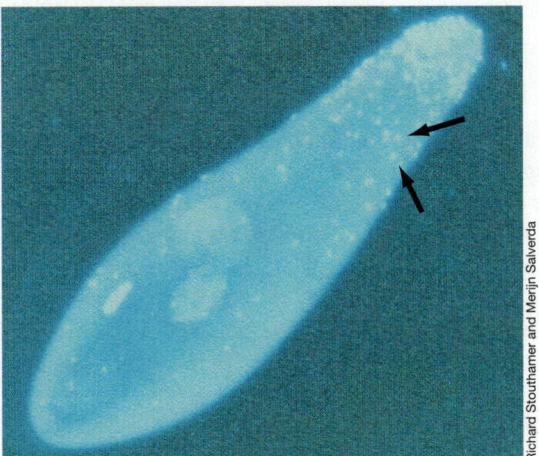

Figure 15.6 Wolbachia. Photomicrograph of a DAPI-stained egg of the parasitoid wasp *Trichogramma kaykai* infected with *Wolbachia pipientis*, which induces parthenogenesis. The *W. pipientis* cells are primarily located in the narrow end of the egg (arrows).

Other Orders of *Alphaproteobacteria*

Key Genera: *Rhodobacter, Acetobacter, Caulobacter,* and *Sphingomonas*

The families *Rhodobacterales* and *Rhodospirillales* (Figure 15.3) contain metabolically diverse organisms that have been discussed previously, including purple nonsulfur bacteria (*Rhodobacter* and *Rhodospirillum*, ↪ Section 14.5), aerobic anoxygenic phototrophs (*Roseobacter*, ↪ Section 14.5), nitrogen-fixing bacteria (*Azospirillum*, ↪ Section 14.12), denitrifiers (*Paracoccus*, ↪ Section 14.13), methylotrophs (*Methylobacterium*, ↪ Section 14.17), acetic acid bacteria (*Acetobacter* and *Gluconobacter*, ↪ Section 14.18), and magnetotactic bacteria (*Magnetospirillum*, ↪ Section 14.23), among others.

The *Caulobacterales* are typically oligotrophic and strictly aerobic chemoorganotrophs. Species typically form prosthecae or stalks (↪ Section 14.21), and many species display asymmetric forms of cell division. The characteristic genus is *Caulobacter*, which has a characteristic life cycle that we have discussed previously (↪ Sections 7.12 and 14.21).

The *Sphingomonadales* include diverse aerobic and facultatively aerobic chemoorganotrophs as well as species of aerobic anoxygenic phototrophs (*Erythrobacter*), and a few obligate anaerobes. The characteristic genus is *Sphingomonas*, which consists of obligately aerobic and nutritionally versatile species. Sphingomonads are widespread in aquatic and terrestrial environments and are notable for their ability to metabolize a wide range of organic compounds including many aromatic compounds that are common environmental contaminants (e.g., toluene, nonylphenol, dibenzo-*p*-dioxin, naphthalene, and anthracene, among others). As a consequence, sphingomonads have been widely studied as potential agents of bioremediation (↪ Section 21.5). These organisms are typically easy to cultivate and grow well on a variety of complex culture media.

MINIQUIZ

- What are some ways in which *Wolbachia* species can affect insects?
- What organisms might form the pink scum you find on the edge of a bathtub? How might you try to cultivate these organisms?

15.2 *Betaproteobacteria*

With nearly 500 described species, the *Betaproteobacteria* are the third largest class of *Proteobacteria* (**Figure 15.7**). The *Betaproteobacteria* contain an immense amount of functional diversity (Figure 15.2 and ↪ Figure 14.1), and many species in this group have already been considered in Chapter 14. A total of six orders of *Betaproteobacteria* have many characterized species: *Burkholderiales, Hydrogenophilales, Methylophilales, Neisseriales, Nitrosomonadales,* and *Rhodocyclales,* and we focus on these here.

Burkholderiales

Key Genera: *Burkholderia*

The *Burkholderiales* contain species with a wide range of metabolic and ecological characteristics. Species include strictly aerobic, facultatively aerobic, and obligately anaerobic chemoorganotrophs, anoxygenic phototrophs, obligate and facultative

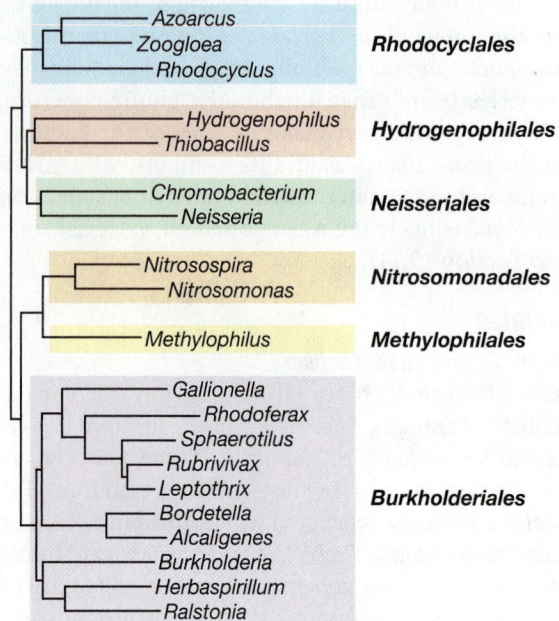

Betaproteobacteria

Figure 15.7 Major orders of *Proteobacteria* in the class *Betaproteobacteria*. The phylogenetic tree was constructed using 16S rRNA gene sequences from representative genera of *Betaproteobacteria*. Order names are shown in bold.

chemolithotrophs, free-living nitrogen fixers, and pathogens of plants, animals, and humans.

Burkholderia is the type genus for the *Burkholderiales*. The genus *Burkholderia* includes diverse species of chemoorganotrophs with strictly respiratory metabolism. All species can grow aerobically, some also grow anaerobically with nitrate as the electron acceptor, and many strains are able to fix N_2. The metabolic versatility of *Burkholderia* species with respect to organic compounds, and aromatic compounds in particular, has led to interest in their use in bioremediation (↪ Section 21.5). Certain strains of *Burkholderia* have also been shown to promote plant growth. However, many species are potentially pathogenic for plant or animals. One of the best-known of the pathogenic species is *Burkholderia cepacia*.

B. cepacia is primarily a soil bacterium but also an opportunistic pathogen (**Figure 15.8**). *B. cepacia* is often found in the rhizosphere of plants. *B. cepacia* can produce both anti-fungal and

Figure 15.8 Colonies of *Burkholderia*. Photograph of colonies of *Burkholderia cepacia* on an agar plate.

anti-nematodal compounds and thus its ability to colonize plant roots can provide disease protection and promote plant growth. However, *B. cepacia* is also known as a plant pathogen in certain circumstances, and it is a major cause of soft rot in onions. *B. cepacia* has also emerged as an opportunistic hospital-acquired infection in humans, as it is a hardy organism that is difficult to eradicate from the clinical setting. *B. cepacia* can form secondary lung infections in patients who are immunocompromised or have pneumonia or cystic fibrosis. Its ability to form biofilms in the lung and its natural resistance to many antibiotics has made this organism particularly dangerous for patients with cystic fibrosis (Section 19.4).

Rhodocyclales

Key Genera: *Rhodocyclus, Zoogloea*

Like the *Burkholderiales*, the order *Rhodocyclales* contains species with diverse metabolic and ecological characteristics. The type genus for the *Rhodocyclales* is *Rhodocyclus*, a purple non-sulfur bacterium (Section 14.5). Like most purple nonsulfur bacteria, *Rhodocyclus* species grow best as a photoheterotrophs but most can also grow as photoautotrophs with H_2 as electron acceptor. Species can also grow by respiration in darkness, but they are typically found in illuminated anoxic environments where organic matter is present.

Zoogloea is another important genus of the *Rhodocyclales*. *Zoogloea* species are aerobic chemoorganotrophs that are distinctive for producing a thick gelatinous capsule which binds cells together into a complex matrix with branching, fingerlike projections. This gelatinous matrix can cause *flocculation*, the formation of macroscopic particles that settle out of solution. *Zoogloea ramigera* is of particular importance in aerobic wastewater treatment (Section 21.6), where it degrades much of the organic carbon in the waste stream and promotes flocculation and settling, crucial steps in water purification.

Neisseriales

Key Genera: *Chromobacterium, Neisseria*

The order *Neisseriales* contains at least 29 genera of diverse chemoorganotrophs. The most well-characterized species are in the genera *Neisseria* and *Chromobacterium*. Species of *Neisseria* are commonly isolated from animals, and some of them are pathogenic. *Neisseria* species are always cocci (**Figure 15.9a**). Some

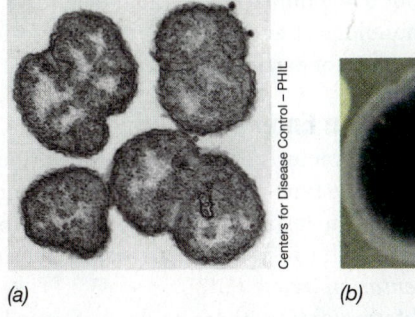
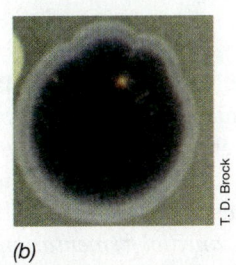

(a)   *(b)*

Figure 15.9 *Neisseria* **and** *Chromobacterium.* *(a)* Transmission electron micrograph of cells of *Neisseria gonorrhoeae* showing the typical diplococcus cell arrangements. *(b)* A large colony of *Chromobacterium violaceum.*

Neisseria are free-living saprophytes and reside in the oral cavity and other moist areas on the animal body. Others are serious pathogens, such as *Neisseria meningitidis*, which can cause a potentially fatal inflammation of the membranes lining the brain (meningitis, Section 29.5). We discuss the clinical microbiology of *Neisseria gonorrhoeae*—the causative agent of the disease gonorrhea—in Section 27.3, and the pathogenesis of gonorrhea itself in Section 29.12.

Chromobacterium is a close phylogenetic relative of *Neisseria* but is rod-shaped in morphology. The best-known *Chromobacterium* species is *C. violaceum*, a purple-pigmented organism (Figure 15.9b) found in soil and water and occasionally in pus-forming wounds of humans and other animals. *C. violaceum* and a few other chromobacteria produce the purple pigment *violacein* (Figure 15.9b), a water-insoluble pigment with both antimicrobial and antioxidant properties. *Chromobacterium* is a facultative aerobe, growing fermentatively on sugars and aerobically on various carbon sources.

Hydrogenophilales, Methylophilales, and Nitrosomonadales

Key Genera: *Hydrogenophilus, Thiobacillus, Methylophilus, Nitrosomonas*

These three orders contain organisms that have fairly specialized metabolic capabilities including chemolithotrophs and methylotrophs; most species are obligate aerobes and many are autotrophic. *Hydrogenophilus thermoluteolus* is an obligate aerobe that can grow as a chemolithotroph using H_2 as an electron donor for respiration (Sections 13.7 and 14.16) and the Calvin cycle to fix CO_2. This species is a facultative chemolithotroph, and can also grow as a chemoorganotroph on simple carbon sources. *Thiobacillus* is another important genus of *Hydrogenophilales*. Species of *Thiobacillus* can be chemoorganotrophs or chemolithotrophs. Chemolithotrophic species of *Thiobacillus* are sulfur bacteria (Sections 13.8 and 14.11) that oxidize reduced sulfur compounds as electron donors and grow by aerobic respiration or denitrification (Sections 13.17 and 14.13). Species of *Thiobacillus* can also fix CO_2 using the Calvin cycle and are commonly found in soils, sulfur springs, marine habitats, and other locales where reduced sulfur compounds are available.

The *Methylophilales* and *Nitrosomonadales* contain metabolically specialized organisms. *Methylophilus* species are obligate and facultative methylotrophs (Section 14.17) that grow on methanol and other C_1 compounds, but not on CH_4. Facultative species can grow as chemoorganotrophs through aerobic respiration of simple sugars. The order *Nitrosomonadales* contains obligately chemolithotrophic ammonia-oxidizing bacteria, the key genera being *Nitrosomonas* and *Nitrosospira* (Section 14.13).

> ### MINIQUIZ
> - List three species of *Betaproteobacteria* that are known to be human pathogens.
> - List three genera of *Betaproteobacteria* that contain chemolithotrophs.

15.3 Gammaproteobacteria— Enterobacteriales

Key Genera: *Enterobacter, Escherichia, Klebsiella, Proteus, Salmonella, Serratia, Shigella*

The *Gammaproteobacteria* are the largest and most diverse class of *Proteobacteria*, containing nearly half of the characterized species in the phylum. The class contains more than 1500 characterized species among its 15 orders (**Figure 15.10**, Figure 15.1*b*). Its species have diverse metabolic and ecological characteristics (Figure 15.2 and ↩ Figure 14.1) and include many well-known human pathogens. Species can be phototrophic (including the purple sulfur bacteria, ↩ Section 14.4), chemoorganotrophic, or chemolithotrophic, and can have either respiratory or fermentative metabolism. Members of this group often develop quickly in laboratory media and can be isolated from a wide diversity of habitats. In this section we consider the *Enterobacteriales*, one of the largest and most well-known orders within the *Gammaproteobacteria*.

The *Enterobacteriales*, commonly called the **enteric bacteria,** comprise a relatively homogeneous phylogenetic group within

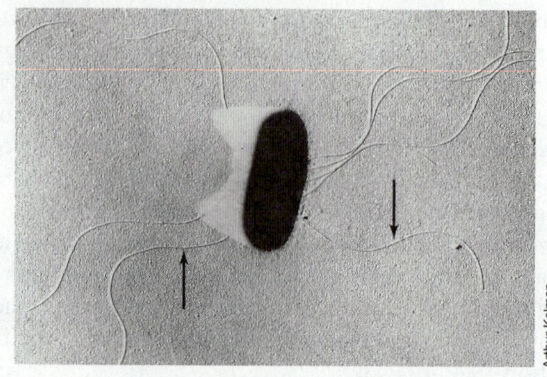

Figure 15.11 **A butanediol-producing enteric bacterium.** Electron micrograph of a shadow-cast preparation of a cell of the butanediol-producing bacterium *Erwinia carotovora*. The cell is about 0.8 μm wide. Note the peritrichously arranged flagella (arrows), typical of enteric bacteria.

the *Gammaproteobacteria* and consist of facultatively aerobic, gram-negative, nonsporulating rods that are either nonmotile or motile by peritrichous flagella (**Figure 15.11**). The *oxidase test* and the *catalase test* are common assays used to characterize bacteria (↩ Section 27.3) and these tests can be used to discriminate enteric bacteria from many other *Gammaproteobacteria*. The oxidase test is an assay for the presence of cytochrome *c* oxidase, an enzyme present in many respiring bacteria. The catalase test assays for the enzyme catalase, which detoxifies hydrogen peroxide and is commonly found in bacteria able to grow in the presence of oxygen (↩ Section 5.16 and Figure 5.31). Enteric bacteria are oxidase-negative and catalase-positive. They also produce acid from glucose and reduce nitrate but only to nitrite. Enteric bacteria have relatively simple nutritional requirements and ferment sugars to a variety of end products.

Among the enteric bacteria are many species pathogenic to humans, other animals, or plants, as well as other species of industrial importance. *Escherichia coli*, the best known of all organisms, is the classic enteric bacterium. Because of the medical importance of many enteric bacteria, an extremely large number have been characterized, and numerous genera and species have been defined, largely for ease in identification purposes in clinical microbiology. However, because enteric bacteria are genetically very closely related, their positive identification often presents considerable difficulty. In clinical laboratories, identification is typically based on the combined analysis of a large number of diagnostic tests carried out using miniaturized rapid diagnostic media kits along with immunological and genomic analyses to identify signature proteins or genes of particular species (Chapter 27).

Fermentation Patterns in Enteric Bacteria

One major taxonomic characteristic separating the various genera of enteric bacteria is the type and proportion of fermentation products generated from the fermentation of glucose. Two broad patterns are recognized, the *mixed-acid fermentation* and the *2,3-butanediol fermentation* (**Figure 15.12**).

In the mixed-acid fermentation, three acids are formed in significant amounts: acetic, lactic, and succinic. Ethanol, CO_2, and H_2 are also formed, but not butanediol. In the butanediol fermentation, smaller amounts of acids are formed, and butanediol,

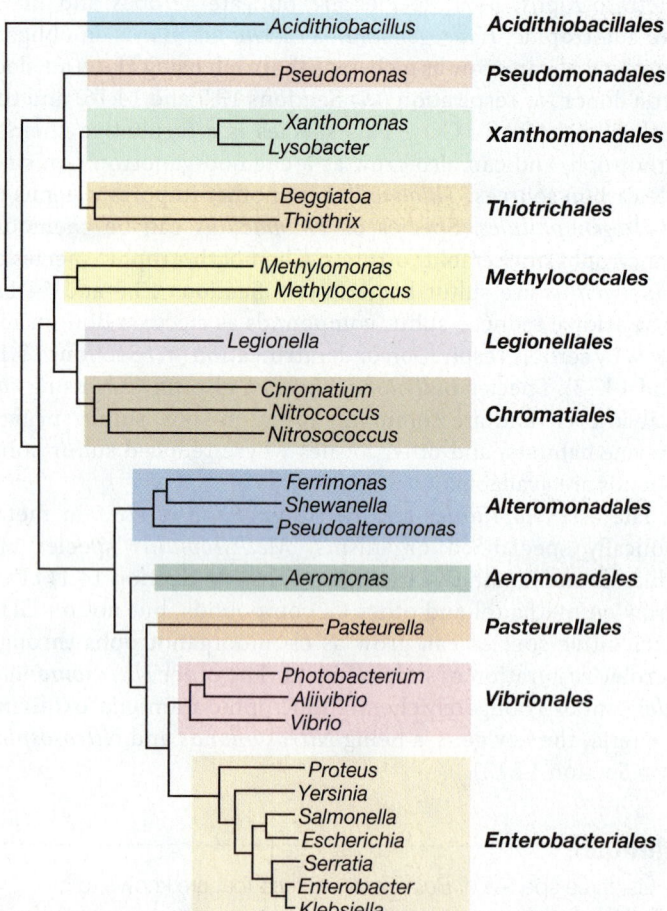

Figure 15.10 **Major orders of *Proteobacteria* in the class *Gammaproteobacteria*.** The phylogenetic tree was constructed using 16S rRNA gene sequences from representative genera of *Gammaproteobacteria*. Order names are shown in bold.

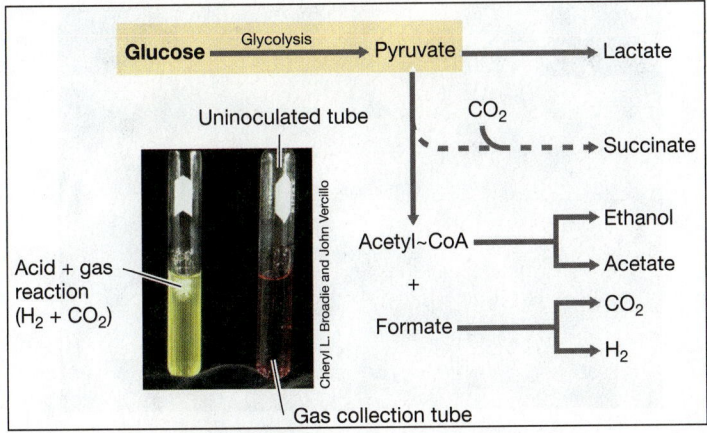

(a) **Mixed-acid fermentation** (for example, *Escherichia coli*)

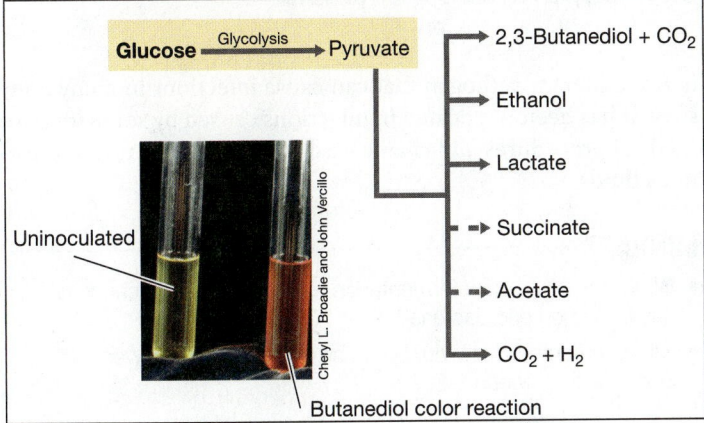

(b) **Butanediol fermentation** (for example, *Enterobacter aerogenes*)

Figure 15.12 Enteric fermentations. Distinction between (a) mixed-acid and (b) butanediol fermentation in enteric bacteria (↺ Figure 13.32). The solid arrows indicate reactions leading to major products. Dashed arrows indicate minor products. (a) The photo shows the production of acid (yellow) and gas (in the inverted Durham tube) in a culture of *Escherichia coli* carrying out a mixed-acid fermentation (purple tube was uninoculated). (b) The photo shows the pink-red color in the Voges–Proskauer (VP) test, which indicates butanediol production, following growth of *Enterobacter aerogenes*. The left (yellow) tube was not inoculated. Note that the mixed-acid fermentation produces less CO_2 but more acid products from glucose than does the butanediol fermentation.

ethanol, CO_2, and H_2 are the main products (↺ Figure 13.32). As a result of mixed-acid fermentation, equal amounts of CO_2 and H_2 are produced, whereas in the butanediol fermentation, considerably more CO_2 than H_2 is produced. This is because mixed-acid fermenters produce CO_2 only from formic acid by means of the enzyme formate hydrogenlyase:

$$HCOOH \rightarrow H_2 + CO_2$$

This reaction results in equal amounts of CO_2 and H_2. The butanediol fermenters also produce CO_2 and H_2 from formic acid, but they produce two additional molecules of CO_2 during the formation of each molecule of butanediol (Figure 15.12b). Butanediol fermentation is characteristic of *Enterobacter*, *Klebsiella*, *Erwinia*, and *Serratia*, whereas mixed-acid fermentation is observed in *Escherichia*, *Salmonella*, *Shigella*, *Citrobacter*, *Proteus*, and *Yersinia*.

Mixed-Acid Fermenters: *Escherichia, Salmonella, Shigella,* and *Proteus*

Species of *Escherichia* are almost universal inhabitants of the intestinal tract of humans and other warm-blooded animals, although they are by no means the dominant organisms in this habitat. *Escherichia* may play a nutritional role in the intestinal tract by synthesizing vitamins, particularly vitamin K. As a facultative aerobe, this organism probably also helps consume O_2, thus rendering the large intestine anoxic. Wild-type *Escherichia* strains rarely show any growth-factor requirements and are able to grow on a wide variety of carbon and energy sources such as sugars, amino acids, organic acids, and so on.

Some strains of *Escherichia* are pathogenic and have been implicated in diarrheal diseases, especially in infants; diarrheal diseases are a major public health problem in developing countries. (↺ Section 31.11). *Escherichia* is also a major cause of urinary tract infections in women. Enteropathogenic *E. coli* strains are becoming more frequently implicated in gastrointestinal infections and generalized fevers. Some strains, such as enterohemorrhagic *E. coli*, an important representative of which is strain O157:H7, can cause sporadic outbreaks of severe foodborne disease. Infection occurs primarily through consumption of contaminated foods, such as raw or undercooked ground beef, unpasteurized milk, or contaminated water. In a small percentage of cases, *E. coli* O157:H7 causes a life-threatening complication related to its production of a very potent enterotoxin.

Salmonella and *Escherichia* are quite closely related. However, in contrast to *Escherichia*, species of *Salmonella* are almost always pathogenic, either to humans or to other warm-blooded animals (*Salmonella* is also found in the intestines of cold-blooded animals, such as turtles and lizards). In humans the most common diseases caused by salmonellas are typhoid fever and gastroenteritis (↺ Sections 31.5 and 31.10). The shigellas are also genetically very closely related to *Escherichia*. Genomic analyses strongly suggest that *Shigella* and *Escherichia* have exchanged a significant number of genes by horizontal gene flow. In contrast to most *Escherichia*, however, species of *Shigella* are typically pathogenic to humans, causing a rather severe gastroenteritis called *bacillary dysentery*. *Shigella dysenteriae*, transmitted by food- and water-borne routes, is a good example of this. The bacterium, which contains endotoxin, invades intestinal epithelial cells, where it excretes a neurotoxin that causes acute gastrointestinal distress.

The genus *Proteus* typically contains highly motile cells (**Figure 15.13**) that produce the enzyme *urease*. Unlike *Salmonella* and *Shigella*, *Proteus* shows only a distant relationship to *E. coli*. *Proteus* is a frequent cause of urinary tract infections in humans and probably benefits in this regard from its ready ability to degrade urea by urease. Because of the rapid motility of *Proteus* cells, colonies growing on agar plates often exhibit a characteristic swarming phenotype (Figure 15.13b). Cells at the edge of the growing colony are more rapidly motile than those in the center of the colony. The former move a short distance away from the colony in a mass and then undergo a reduction in motility, settle down, and divide, forming a new population of motile cells that again swarm. As a result, the mature colony appears as a series of concentric rings, with higher concentrations of cells alternating with lower concentrations (Figure 15.13b).

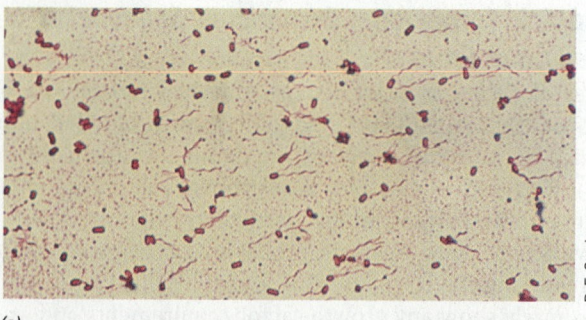

(a)

D. E. Snyder

(b)

James Shapiro

Figure 15.13 **Swarming in *Proteus*.** *(a)* Cells of *Proteus mirabilis* stained with a flagella stain; the peritrichous flagella of each cell form into a bundle to rotate in synchrony. *(b)* Photo of a swarming colony of *Proteus vulgaris*. Note the concentric rings.

Butanediol Fermenters: *Enterobacter, Klebsiella,* and *Serratia*

The butanediol fermenters are genetically more closely related to each other than to the mixed-acid fermenters, a finding that is in agreement with the observed physiological differences (Figure 15.12). *Enterobacter aerogenes* is a common species in water and sewage as well as the intestinal tract of warm-blooded animals and is an occasional cause of urinary tract infections. One species of *Klebsiella*, *K. pneumoniae*, occasionally causes pneumonia in humans, but klebsiellas are most commonly found in soil and water. Most *Klebsiella* strains also fix nitrogen (⟳ Section 3.17), a property not characteristic of other enteric bacteria.

The genus *Serratia* forms a series of red pyrrole-containing pigments called *prodigiosins* (**Figure 15.14**). Prodigiosin is produced in stationary phase as a secondary metabolite and is of interest because it contains the pyrrole ring also found in the pigments for energy transfer: porphyrins, chlorophylls, bacteriochlorophylls, and phycobilins (⟳ Sections 13.1–13.3). However, it is unclear if prodigiosin plays any role in energy transfer, and its exact function is unknown. Species of *Serratia* can be isolated from water and soil as well as from the gut of various insects and vertebrates and occasionally from the intestines of humans. *Serratia marcescens*

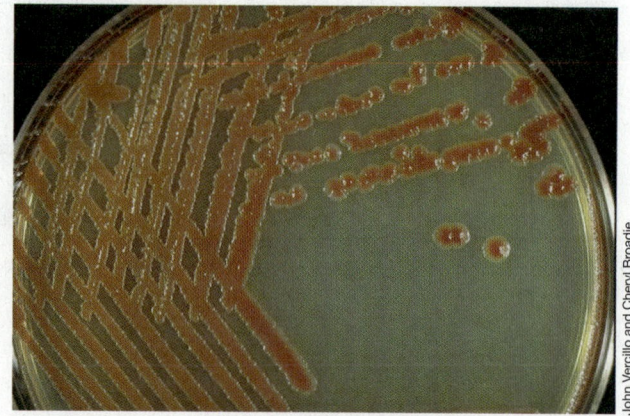

John Vercillo and Cheryl Broadie

Figure 15.14 **Colonies of *Serratia marcescens*.** The orange-red pigmentation is due to the pyrrole-containing pigment prodigiosin.

is also a human pathogen that can cause infections in many body sites. It has been implicated in infections caused by some invasive medical procedures and is an occasional contaminant in intravenous fluids.

MINIQUIZ

- What is a mixed-acid fermentation, and of what significance is this trait to enteric bacteria?
- What characteristics would you use to distinguish between *E. coli* and *K. pneumonia*?

15.4 *Gammaproteobacteria— Pseudomonadales* and *Vibrionales*

Key Genera: *Aliivibrio, Pseudomonas, Vibrio*

The phylogenetic and metabolic diversity of the *Gammaproteobacteria* makes it difficult to select the many notable species in this class of *Proteobacteria*. We focus here on the *Pseudomonadales* and *Vibrionales*, since these groups (along with the *Enterobacteriales*) represent three of the most abundant and most commonly encountered orders of *Gammaproteobacteria* (Figure 15.10).

Pseudomonadales

The *Pseudomonadales* contain exclusively chemoorganotrophs that carry out respiratory metabolisms. All species can grow as aerobes and are typically oxidase- and catalase-positive, but some are also capable of anaerobic respiration with nitrate as the electron acceptor. Most species are able to use a wide diversity of organic compounds as sources of carbon and energy for growth. These organisms are ubiquitous in soil and aquatic systems, and many species cause diseases of plants and animals, including humans. The type genus is *Pseudomonas*, and key species are defined on the basis of phylogeny and various physiological and other phenotypic characteristics, as outlined in **Table 15.2**. The term **pseudomonad** is often used to describe any gram-negative, polarly flagellated, aerobic rod that is able to use diverse carbon sources. Pseudomonads can be found in several different groups of *Proteobacteria*, but here we consider only those organisms in the order *Pseudomonadales*.

Table 15.2 Characteristics of some key *Pseudomonas* species

Species	Characteristics
Pseudomonas aeruginosa	Pyocyanin production; growth at up to 43°C; single polar flagellum; capable of denitrification; primarily a soil organism, but also common in hospital-acquired (nosocomial) infections
Pseudomonas fluorescens	Does not produce pyocyanin or grow at 43°C; tuft of polar flagella; primarily a soil organism, rarely pathogenic
Pseudomonas marginalis	Degrades pectin, causes soft rot of various plants
Pseudomonas putida	Similar to *P. fluorescens*, but does not liquefy gelatin and does grow on benzylamine
Pseudomonas syringae	Lacks arginine dihydrolase; oxidase-negative; plant pathogen causing chlorosis and necrotic lesions on leaves
Pseudomonas stutzeri	Soil saprophyte; strong denitrifier and nonfluorescent

Several species of *Pseudomonas* are pathogenic (Table 15.2). Among these, *Pseudomonas aeruginosa* (**Figure 15.15**) is frequently associated with infections of the urinary and respiratory tracts in humans. *P. aeruginosa* is not an obligate pathogen. Instead, the organism is an opportunist, initiating infections in individuals with weakened immune systems. *P. aeruginosa* is a common cause of hospital-acquired (nosocomial) infections from catheterizations, tracheostomies, lumbar punctures, and intravenous infusions, and often emerges in patients given prolonged treatment with immunosuppressive agents. *P. aeruginosa* is also a common pathogen in patients receiving treatment for severe burns or other traumatic skin damage and in people suffering from cystic fibrosis. In addition to localized infections, *P. aeruginosa* can also cause systemic infections, usually in individuals who have experienced extensive skin damage.

P. aeruginosa is naturally resistant to many widely used antibiotics, so treatment of infections is often difficult. Resistance is typically due to a resistance transfer plasmid (R plasmid) (⟲ Sections 4.3 and 27.17), which is a plasmid whose genes encode proteins that detoxify various antibiotics or pump them out of the cell. Polymyxin, an antibiotic not ordinarily used in human therapy because of its toxicity, is effective against *P. aeruginosa* and is used in critical medical situations.

Certain species of *Pseudomonas* are well-known plant pathogens (phytopathogens) (Table 15.2). Phytopathogens frequently

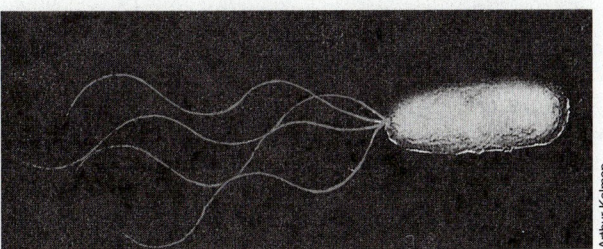

Figure 15.15 Cell morphology of pseudomonads. Shadow-cast transmission electron micrograph of a *Pseudomonas* cell. The cell measures about 1 μm in diameter.

Arthur Kelman

inhabit nonhost plants (in which disease symptoms are inapparent) and from there become transmitted to host plants and initiate infection. Disease symptoms vary considerably, depending on the particular phytopathogen and host plant. The pathogen releases plant toxins, lytic enzymes, plant growth factors, and other substances that destroy or distort plant tissue, releasing nutrients for use by the bacterium. In many cases the disease symptoms help identify the phytopathogen. Thus, *Pseudomonas syringae* is typically isolated from leaves showing chlorotic (yellowing) lesions, whereas *Pseudomonas marginalis*, a "soft-rot" pathogen, infects stems and shoots but rarely leaves.

Vibrionales

The *Vibrionales* contain facultatively aerobic rods and curved rods that employ a fermentative metabolism. One key difference between the *Vibrio* group and enteric bacteria is that *Vibrio* are oxidase-positive whereas enteric bacteria are oxidase-negative. Although *Pseudomonas* species are also oxidase-positive, they are not fermentative and so are clearly distinct from *Vibrio* species. The best-known genera in this group are *Vibrio*, *Aliivibrio*, and *Photobacterium*, which contain several species that are bioluminescent (⟲ Section 14.24).

Most vibrios and related bacteria are aquatic, found in marine, brackish, or freshwater habitats. *Vibrio cholerae* is the cause of the disease cholera in humans (⟲ Sections 28.10 and 31.3); the organism does not normally cause disease in other hosts. Cholera is one of the most common human infectious diseases in developing countries and is transmitted almost exclusively via water.

Vibrio parahaemolyticus inhabits the marine environment and is a major cause of gastroenteritis in Japan, where raw fish is widely consumed; the organism has also been implicated in outbreaks of gastroenteritis in other parts of the world, including the United States. *V. parahaemolyticus* can be isolated from seawater itself or from shellfish and crustaceans, and its primary habitat is probably marine animals, with humans being an accidental host.

MINIQUIZ ---

• What species of *Pseudomonas* is a common cause of lung infection in cystic fibrosis patients?

• What major characteristic could be used to differentiate strains of *Pseudomonas* from those of *Vibrio*?

15.5 *Deltaproteobacteria* and *Epsilonproteobacteria*

These classes of *Proteobacteria* contain fewer species and less functional diversity than we have encountered in the *Alpha-*, *Beta-*, and *Gammaproteobacteria* (Figure 15.2 and ⟲ Figure 14.1). The *Deltaproteobacteria* are primarily sulfate- and sulfur-reducing bacteria (⟲ Sections 14.9 and 14.10), dissimilative iron-reducers (⟲ Section 14.14), and bacterial predators (⟲ Section 14.19). *Epsilonproteobacteria*, by contrast, contain many species that oxidize the H_2S produced by the sulfate and sulfur reducers. The final class of *Proteobacteria*, the *Zetaproteobacteria*, contains only one characterized species (the iron oxidizer *Mariprofundus ferrooxydans*) and was considered earlier (⟲ Section 14.15).

Deltaproteobacteria

Key Genera: *Bdellovibrio, Myxococcus, Desulfovibrio, Geobacter, Syntrophobacter*

Eight orders have been characterized within the *Deltaproteobacteria* (**Figure 15.16**). The largest and most common order containing sulfate reducers is the *Desulfovibrionales*. These organisms are readily cultivated from marine sediments and nutrient-rich anoxic environments that contain sulfate. Species of *Desulfovibrionales* are typically incomplete oxidizers (↩ Section 14.9). All use sulfate as the terminal electron acceptor, and all require small organic compounds such as lactate as a source of carbon and energy for growth. Species within the orders *Desulfobacterales* and *Desulfarculales* also typically reduce sulfate; however, in contrast to the *Desulfovibrionales*, these species can be complete or incomplete acetate oxidizers (↩ Section 14.9). In addition to sulfate, some species in these three orders can also reduce sulfite, thiosulfate, or nitrate, and some are capable of certain fermentations.

The final order containing sulfate reducers is the *Syntrophobacterales*. Some but not all species of the *Syntrophobacterales* are able to reduce sulfate. In nature, however, species of *Syntrophobacterales* primarily interact with H_2-consuming bacteria in a metabolic partnership called *syntrophy* (↩ Sections 13.15 and 20.2). For example, syntrophic species such as *Syntrophobacter wolinii* oxidize propionate, producing acetate, CO_2, and H_2. However, such growth is only possible when a H_2-consuming partner is present. If sulfate is present, *S. wolinii* can grow as a sulfate reducer without the need for a partner. *S. wolinii* can also grow without a partner organism by fermenting pyruvate, fumarate, or malate.

Epsilonproteobacteria

Key Genera: *Campylobacter, Helicobacter*

The *Epsilonproteobacteria* (Figure 15.16) were initially defined by only a few pathogenic bacteria; in particular, by species of

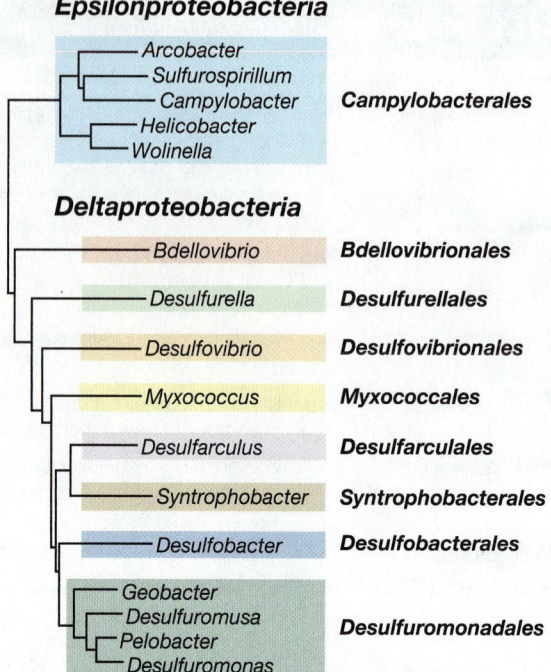

Figure 15.16 Major orders of *Proteobacteria* in the classes *Deltaproteobacteria* and *Epsilonproteobacteria*. The phylogenetic tree was constructed using 16S rRNA gene sequences from representative genera in the *Delta-* and *Epsilonproteobacteria*. Order names are shown in bold.

Campylobacter and *Helicobacter*. However, environmental studies of marine and terrestrial microbial habitats have shown that a diversity of *Epsilonproteobacteria* exist in nature, and their numbers and metabolic capabilities suggest they play important ecological roles (**Table 15.3**). Species of *Epsilonproteobacteria* are especially abundant at oxic–anoxic interfaces in sulfur-rich environments, and play major roles in the oxidation of sulfur compounds in nature.

Table 15.3 Characteristics of key genera of *Epsilonproteobacteria*

Genus	Habitat	Descriptive characters	Physiology and metabolism
Campylobacter	Reproductive organs, oral cavity, and intestinal tract of humans and other animals; pathogenic	Slender, spirally curved rods; corkscrew-like motility by single polar flagellum	Microaerophilic; chemoorganotrophic
Arcobacter	Diverse habitats (freshwater, sewage, saline environments, animal reproductive tract, plants); some species pathogenic for humans and other animals	Slender, curved rods; motile by single polar flagellum	Microaerophilic; aerotolerant or aerobic; chemoorganotrophic; oxidation of sulfide to elemental sulfur (S^0) by some species; nitrogen fixation in one species
Helicobacter	Intestinal tract and oral cavity of humans and other animals; pathogenic	Rods to tight spiral; some species with tightly coiled periplasmic fibers	Microaerophilic, chemoorganotrophic; produce high levels of urease (nitrogen assimilation)
Sulfurospirillum	Freshwater and marine habitats containing sulfur	Vibrioid to spiral-shaped cells; motile by polar flagella	Microaerophilic; reduces elemental sulfur (S^0)
Thiovulum	Freshwater and marine habitats containing sulfur; not yet in pure culture (↩ Figure 14.29)	Cells contain orthorhombic S^0 granules; rapid motility by peritrichous flagella	Microaerophilic; chemolithotrophic oxidizing H_2S
Wolinella	Bovine rumen	Rapidly motile by polar flagellum; single species known: *W. succinogenes*	Anaerobe; anaerobic respiration using fumarate, nitrate, or other compounds as terminal electron acceptor, and with H_2 or formate as electron donor

Campylobacter and *Helicobacter*

These two genera of *Epsilonproteobacteria* share a number of characteristics. *Campylobacter* and *Helicobacter* species are gram-negative, oxidase- and catalase-positive, motile spirilla, and most species are pathogenic to humans or other animals (Table 15.3). These organisms are also microaerophilic (↶ Section 5.16) and must therefore be cultured from clinical specimens at low (3–15%) O_2 and high (3–10%) CO_2.

Campylobacter species, over a dozen of which have been described, cause acute gastroenteritis that typically results in a bloody diarrhea. Pathogenesis is due to several factors, including an enterotoxin that is related to cholera toxin. *Helicobacter pylori*, also a pathogen, causes both chronic and acute gastritis, leading to the formation of peptic ulcers. We consider these diseases, including their modes of transmission and clinical symptoms, in more detail in Sections 29.10 and 31.12.

Sulfurospirillum and *Wolinella*

Species of *Sulfurospirillum*, a *Campylobacter* relative, are nonpathogenic, free-living microaerophiles found in freshwater and marine habitats (Table 15.3). These bacteria also carry out anaerobic respirations using elemental sulfur (S^0), selenate, or arsenate as electron acceptors (↶ Sections 13.18 and 13.21).

Wolinella is an anaerobic bacterium isolated from the bovine rumen (Table 15.3; ↶ Section 22.7). Unlike other *Epsilonproteobacteria*, the single known species, *W. succinogenes*, grows best as an anaerobe and can catalyze anaerobic respirations using fumarate or nitrate as electron acceptors with H_2 or formate as electron donors. Although *W. succinogenes* has thus far been found only in the rumen, its genome shows significant homologies to both the *Campylobacter* and *Helicobacter* genomes and contains additional genes that encode nitrogen fixation, extensive cell signaling mechanisms, and virtually complete metabolic pathways, absent from closely related genomes. This suggests that *Wolinella* inhabits diverse environments outside of the rumen.

Environmental *Epsilonproteobacteria*

In addition to cultured representatives of the genera mentioned above, and many additional species and genera not considered here, there are large groups within this class that are known only from 16S ribosomal RNA gene sequences obtained from the environment (↶ Section 18.5). Through environmental sequencing studies and ongoing cultivation efforts, species of *Epsilonproteobacteria* are now becoming recognized as ubiquitous in marine and terrestrial environments where sulfur-cycling activities are ongoing, particularly in deep-sea hydrothermal vent habitats where sulfide-rich and oxygenated waters mix (↶ Section 19.13). Also, living attached to the surface of animals such as the tube worm *Alvinella* and the shrimp *Rimicaris* that reside near hydrothermal vents, a large variety of uncultured *Epsilonproteobacteria* may, through their sulfur metabolism, detoxify H_2S that would otherwise be deleterious to their animal hosts, allowing the animals to thrive in a chemically hostile environment (↶ Section 22.12). Further exploration of the phylogeny, metabolic activities, and ecological roles of *Epsilonproteobacteria* will likely uncover exciting new aspects of prokaryotic diversity.

MINIQUIZ --

- What four metabolic traits are most common in species of *Deltaproteobacteria*?
- Why is *Wolinella* physiologically unusual among the *Epsilonproteobacteria*?

II • *Firmicutes*, *Tenericutes*, and *Actinobacteria*

We continue our tour of phylogenetic bacterial diversity with the gram-positive bacteria of the phyla *Actinobacteria* and *Firmicutes*, and the closely-related phylum *Tenericutes* (**Figure 15.17**). These three phyla contain nearly half of all characterized species of *Bacteria* (Figure 15.1b).

The *Actinobacteria* include the actinomycetes, a huge group of primarily filamentous soil bacteria. One distinguishing feature of the *Actinobacteria* is that they typically have genomes of high GC content and as a result are also called the **high GC gram-positive bacteria.** The *Tenericutes* include cells that lack a cell wall, and the *Firmicutes* include the endospore-forming bacteria, lactic acid bacteria, and several other groups. In contrast to the *Actinobacteria*, the genomes of *Firmicutes* generally have a low GC content, and as a result, they are also called the **low GC gram-positive bacteria.**

We begin by examining *Firmicutes* that do not form endospores.

15.6 *Firmicutes—Lactobacillales*

Key Genera: *Lactobacillus, Streptococcus*

The order *Lactobacillales* contains the **lactic acid bacteria,** fermentative organisms that produce lactic acid as a major end product of metabolism. These organisms are used widely in food production and preservation. Lactic acid bacteria are nonsporulating, oxidase- and catalase-negative rods or cocci that show an exclusively fermentative metabolism. All lactic acid bacteria produce lactic acid as a major or sole fermentation product. Members of this group lack porphyrins and cytochromes, do not carry out oxidative phosphorylation, and hence obtain energy only by substrate-level phosphorylation. Unlike many anaerobes, however, most lactic acid bacteria are not sensitive to oxygen (O_2) and can grow in its presence; thus they are called *aerotolerant anaerobes*.

Most lactic acid bacteria obtain energy only from the metabolism of sugars and therefore are usually restricted to habitats in which sugars are present. They typically have limited biosynthetic abilities, and their complex nutritional requirements include needs for amino acids, vitamins, purines, and pyrimidines (for example, ↶ Table 3.2 for *Leuconostoc mesenteroides*). One important difference between subgroups of the lactic acid bacteria lies in the pattern of products formed from the fermentation of sugars. One group, called **homofermentative,** produces a single fermentation product, *lactic acid*. The other group, called **heterofermentative,** produces other products, mainly ethanol and CO_2, as well as lactate (↶ Section 13.12 provides additional coverage of homofermentative and heterofermentative pathways).

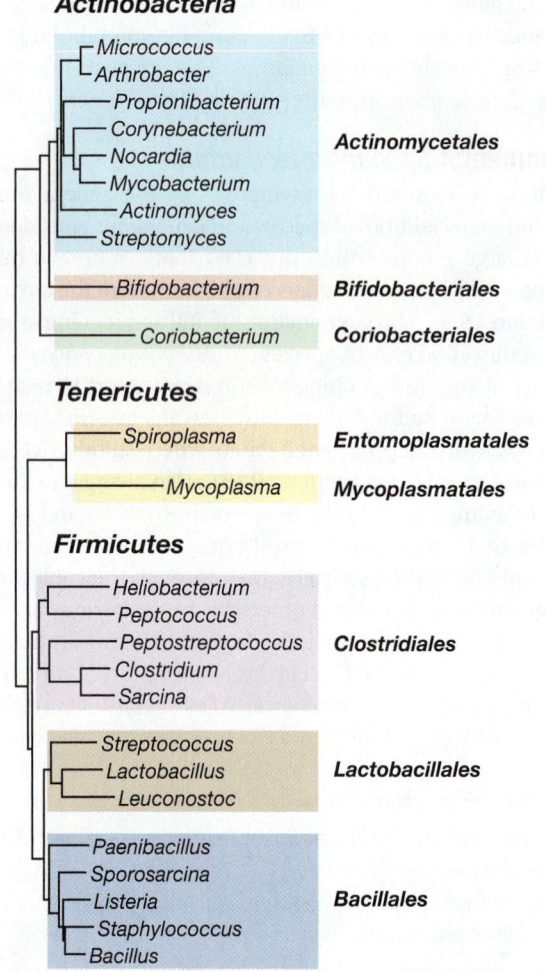

Figure 15.17 **Major orders of gram-positive bacteria and relatives.** The phylogenetic tree was constructed from 16S rRNA gene sequences of representative genera of *Actinobacteria*, *Firmicutes*, and *Tenericutes*. Order names are shown in bold.

Lactobacillus

Lactobacilli are typically rod-shaped and grow in chains, varying from long and slender to short, bent rods (**Figure 15.18**), and most are homofermentative. Lactobacilli are common in dairy products, and some strains are used in the preparation of fermented milk products. For instance, *Lactobacillus acidophilus* (Figure 15.18*a*) is used in the production of acidophilus milk; *Lactobacillus delbrueckii* (Figure 15.18*c*) is used in the preparation of yogurt; and other species are used in the production of sauerkraut, silage, and pickles (Section 31.6).

Lactobacilli are typically more resistant to acidic conditions than are other lactic acid bacteria and are able to grow well at pH values as low as 4. Because of this, they can be selectively enriched from dairy products and fermenting plant material on acidic carbohydrate-containing media. The acid resistance of the lactobacilli enables them to continue growing during natural lactic fermentations, even when the pH value has dropped too low for other lactic acid bacteria to grow. The lactobacilli are therefore typically responsible for the final stages of most lactic acid fermentations. They are rarely, if ever, pathogenic.

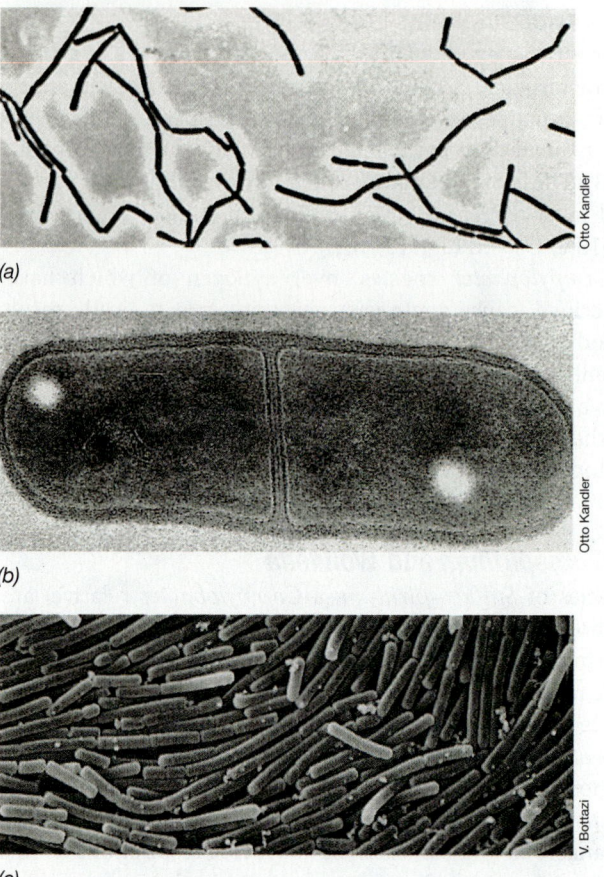

Figure 15.18 *Lactobacillus* species. *(a)* *Lactobacillus acidophilus*, phase-contrast. Cells are about 0.75 μm wide. *(b)* *Lactobacillus brevis*, transmission electron micrograph. Cells measure about 0.8 × 2 μm. *(c)* *Lactobacillus delbrueckii*, scanning electron micrograph. Cells are about 0.7 μm in diameter.

Streptococcus and Other Cocci

The genera *Lactococcus* and *Streptococcus* (**Figure 15.19**) contain homofermentative species of coccoid-shaped lactic acid bacteria with quite distinct habitats and activities that are of considerable practical importance to humans. Some species are pathogenic to humans and animals (Section 29.2). *Streptococcus* species (Figure 15.19*a*) have a characteristic cell morphology forming cocci in chains or tetrads and so are readily resolved from the rod-shaped lactobacilli. As producers of lactic acid, other streptococci play important roles in the production of buttermilk, silage, and other fermented products (Section 31.6), and certain species play a major role in the formation of dental caries (Section 23.3).

There are several other genera of homofermentative cocci. The genus *Lactococcus* (Figure 15.19*b*) contains those streptococci of dairy significance, whereas the genus *Enterococcus* includes streptococci that are primarily of fecal origin and can be human pathogens. Species of the genera *Peptococcus* and *Peptostreptococcus* are obligate anaerobes that ferment proteins rather than sugars.

Streptococci have been divided into two groups of related species: the *pyogenes subgroup*, characterized by *Streptococcus pyogenes*, the cause of strep throat (Section 29.2), and the *viridans subgroup*, characterized by *Streptococcus mutans*, the cause of dental caries (Section 23.3). Hemolysis on blood agar is

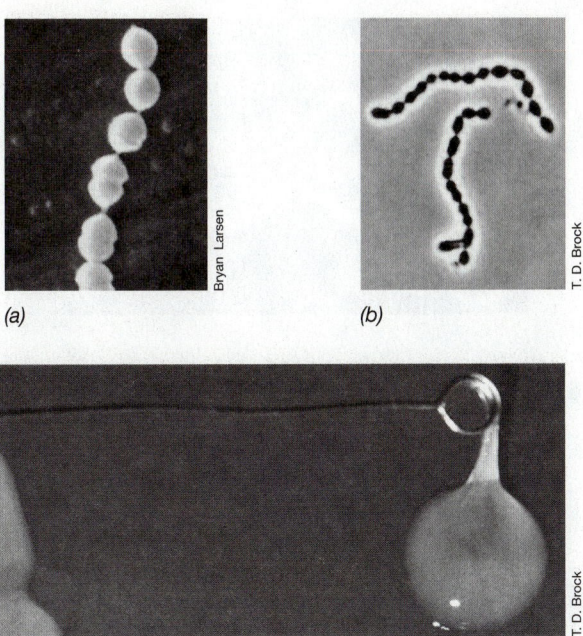

Figure 15.19 Gram-positive cocci. *(a) Streptococcus* sp., scanning electron micrograph. *(b) Lactococcus lactis*, phase-contrast micrograph. Cells in both photos are 0.5–1 μm in diameter. *(c)* Colony of *Leuconostoc mesenteroides* showing the extensive dextran slime produced by cells grown on sucrose.

of considerable importance in the subdivision of the genus into species. For example, species that produce the virulence factors streptolysin O or S form colonies surrounded by a large zone of complete red blood cell hemolysis when plated on blood agar, a condition called β-*hemolysis* (↺ Figure 23.18*a*). β-hemolysis is diagnostic for streptococci in the pyogenes subgroup. In contrast, streptococci in the viridans subgroup cause incomplete hemolysis on blood agar, a condition that leads to greening of the agar under colonies. Streptococci are also divided into immunological groups (designated by the letters A, B, C, F, G), based on the presence of specific carbohydrate antigens (antigens are substances that elicit an immune response). Those β-hemolytic streptococci found in humans usually contain the group A antigen, whereas enterococci contain the group D antigen.

Heterofermentative lactococci reside in the genus *Leuconostoc*. Strains of *Leuconostoc* also produce the flavoring ingredients diacetyl and acetoin from the catabolism of citrate; they have been used as starter cultures in dairy fermentations. Some strains of *Leuconostoc* produce large amounts of glucose or fructose polysaccharide slimes, especially when cultured on sucrose as the carbon and energy source (Figure 15.19*c*), and some of these polymers have found medical use as plasma extenders in blood transfusions.

MINIQUIZ

- How do heterofermentative and homofermentative bacteria differ physiologically?
- How can *Streptococcus pyogenes* be distinguished from *Streptococcus mutans*?

15.7 *Firmicutes*—Nonsporulating *Bacillales* and *Clostridiales*

Key Genera: *Listeria, Staphylococcus, Sarcina*

Firmicutes that form endospores reside in the orders *Bacillales* and *Clostridiales*. However, numerous *Bacillales* and *Clostridiales* are unable to form endospores, and we consider some of these here.

Listeria

The order *Bacillales* typically contains aerobic and facultatively aerobic chemoorganotrophs. Few human pathogens are found within this group, but the genus *Listeria* is a notable exception. *Listeria* are gram-positive, catalase-positive, and obligately aerobic coccobacilli that tend to form chains of three to five cells (↺ Figure 31.16). Although several species of *Listeria* are known, the species *Listeria monocytogenes* is most noteworthy because it causes a major foodborne illness, *listeriosis* (↺ Section 31.13). The organism is transmitted in contaminated, usually ready-to-eat foods, such as cheese and sausages, and can cause anything from a mild illness to a fatal form of meningitis. Species of *Listeria* often grow well at low temperatures, allowing growth in refrigerated foods.

Staphylococcus

Staphylococcus (**Figure 15.20**) is a facultative aerobe that shows a typical respiratory metabolism but can also grow fermentatively. Cells typically grow in clusters and produce acid from glucose both aerobically and anaerobically. *Staphylococcus* species are catalase-positive, and this permits their distinction from *Streptococcus* and some other genera of lactic acid bacteria. Staphylococci are relatively resistant to reduced water potential and tolerate drying and high salt (NaCl) fairly well. Their ability to grow in media containing salt provides a selective means for isolation. For example, if an appropriate inoculum such as a skin swab, dry soil, or room dust is spread on a rich-medium agar plate containing 7.5% NaCl and the plate is incubated aerobically, gram-positive cocci often form the predominant colonies. Many species are pigmented, and this provides an additional aid in selecting gram-positive cocci.

Staphylococci are common commensals and parasites of humans and animals, and they occasionally cause serious infections. In humans, there are two major species, *Staphylococcus epidermidis*, a nonpigmented, nonpathogenic organism usually found on the skin or mucous membranes, and *Staphylococcus aureus* (Figure 15.20), a yellow-pigmented species that is most commonly associated with pathological conditions including boils, pimples, pneumonia, osteomyelitis, meningitis, and arthritis. We discuss the pathogenesis of *S. aureus* in Section 23.2 and staphylococcal diseases in Sections 29.9 and 31.8.

Sarcina

The genus *Sarcina* groups obligate anaerobes that are catalase-negative within the order *Clostridiales*. *Sarcina* species divide in three perpendicular planes to yield packets of eight or more cells and are notable for this morphology (**Figure 15.21**). *Sarcina* are also extremely acid-tolerant, being able to ferment sugars and grow in environments at a pH as low as 2. Cells of one species, *Sarcina ventriculi*, contain a thick, fibrous layer of cellulose surrounding the cell wall (Figure 15.21*b*). The cellulose layers of adjacent cells

UNIT 3

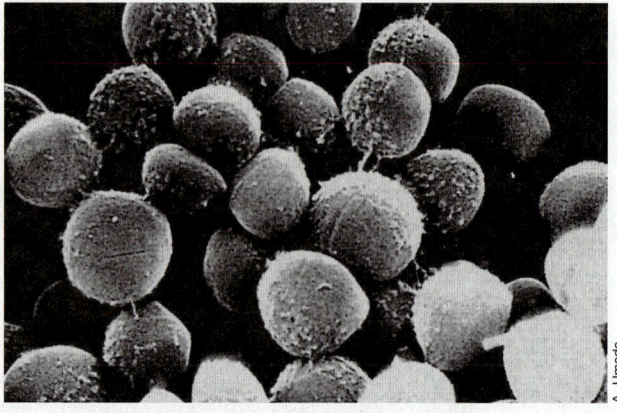

(a)

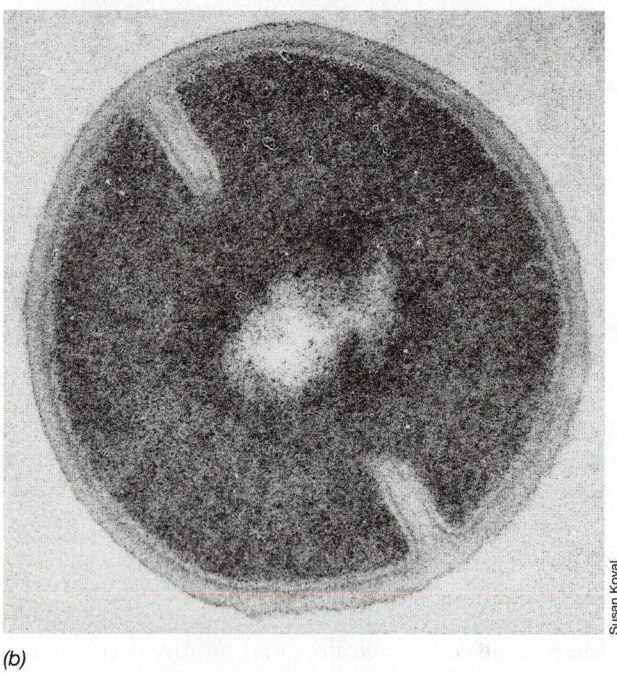

(b)

Figure 15.20 *Staphylococcus.* (a) Scanning electron micrograph of typical *Staphylococcus aureus* cells, showing the irregular arrangement of the cell clusters. Individual cells are about 0.8 μm in diameter. (b) Transmission electron micrograph of a dividing cell of *S. aureus.* Note the thick gram-positive cell wall.

become attached, and this functions as a cementing material to hold together packets of *S. ventriculi* cells.

Sarcina species can be isolated from soil, mud, feces, and stomach contents. Because of its extreme acid tolerance, *S. ventriculi* is one of only a few bacteria that can inhabit and grow in the stomach of humans and other monogastric animals. Rapid growth of *S. ventriculi* is observed in the stomach of humans suffering from certain gastrointestinal disorders, such as pyloric ulcerations. These pathological conditions retard the flow of food to the intestine and often require surgery to correct.

MINIQUIZ
- How could species of *Staphylococcus* be differentiated from *Streptococcus*?
- What characteristics differentiate *Sarcina* from *Staphylococcus*?

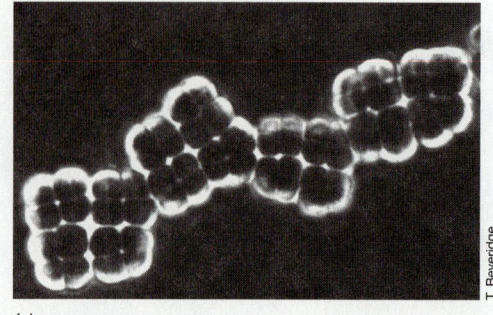

(a)

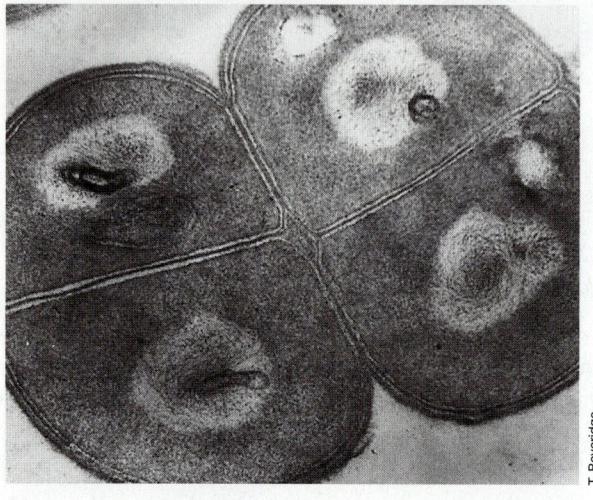

(b)

Figure 15.21 *Sarcina.* (a) Phase-contrast photomicrograph of cells of a typical gram-positive coccus *Sarcina.* A single cell is about 2 μm in diameter. (b) Electron micrograph of a thin section from *Sarcina ventriculi.* The outermost layer of the cell consists of cellulose.

15.8 *Firmicutes*—Sporulating *Bacillales* and *Clostridiales*

Key Genera: *Bacillus, Clostridium, Sporosarcina*

All endospore-forming bacteria are gram-positive species of *Bacillales* or *Clostridiales*. The ability to form endospores evolved only once in a common ancestor of the *Bacillales, Clostridiales,* and *Lactobacillales* (Figure 15.17). However, many *Bacillales* and *Clostridiales* and the entire order *Lactobacillales* are unable to form endospores. The capacity to make endospores requires many genes (🔁 Sections 2.16 and 7.11) and has not been acquired by horizontal gene transfer. It thus appears that the phylogenetic distribution of endospores has seen many cases where the capacity to form endospores has been lost during the course of evolution.

Endospore-forming bacteria (**Table 15.4**) are distinguished on the basis of cell morphology, shape and cellular position of the endospore (**Figure 15.22**), relationship to O_2, and energy metabolism. The two genera about which most is known are *Bacillus*, species of which are aerobic or facultatively aerobic, and *Clostridium*, which contains species that are obligately anaerobic and fermentative. All endospore-forming bacteria are ecologically related because they are found in nature primarily in soil. Even those species that are pathogenic to humans or other animals

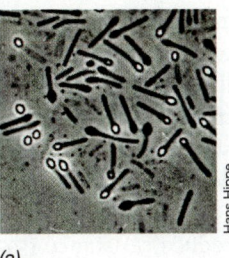 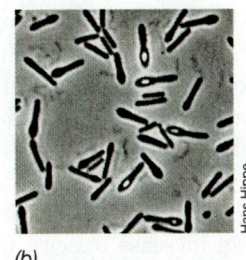 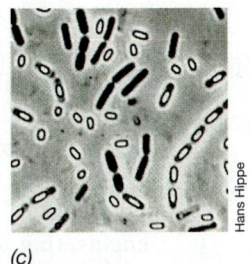

(a)   (b)   (c)

Figure 15.22 *Clostridium* **species and endospore location.** *(a) Clostridium cadaveris,* terminal endospores. Cells are about 0.9 μm wide. *(b) Clostridium sporogenes,* subterminal endospores. Cells are about 1 μm wide. *(c) Clostridium bifermentans,* central endospores. Cells are about 1.2 μm wide. All are phase-contrast micrographs.

Bacillus and Paenibacillus

A list of representatives in the *Bacillus* group is shown in Table 15.5. Species of *Bacillus* and *Paenibacillus* grow well on defined media containing any of a number of carbon sources. Many bacilli produce extracellular hydrolytic enzymes that break down complex polymers such as polysaccharides, nucleic acids, and lipids, permitting the organisms to use these products as carbon sources and electron donors. Many bacilli produce antibiotics, including bacitracin, polymyxin, tyrocidine, gramicidin, and circulin. In most cases the antibiotics are released when the culture enters the stationary phase of growth and is committed to sporulation.

Several bacilli, most notably *Paenibacillus popilliae* and *Bacillus thuringiensis,* produce toxic insecticidal proteins. *P. popilliae* causes a fatal condition called milky disease in Japanese beetle larvae and larvae of closely related beetles of the family *Scarabaeidae. B. thuringiensis* causes a fatal disease of many different groups of insects. Both of these insect pathogens form a crystalline protein during sporulation called the *parasporal body,* which

are primarily saprophytic soil organisms and infect animals only incidentally. Indeed, the ability to produce endospores should be advantageous for a soil microorganism because soil is a highly variable environment in terms of nutrient levels, temperature, and water activity.

Endospore-forming bacteria can be selectively isolated from soil, food, dust, and other materials by heating the sample to 80°C for 10 min, a treatment that effectively kills vegetative cells while any endospores present remain viable. Streaking such heat-treated samples on plates of the appropriate medium and incubating either aerobically or anaerobically selectively yields species of *Bacillus* or *Clostridium,* respectively.

Table 15.4 Major genera of endospore-forming bacteria

Characteristics	Genus
I. Rods	
Aerobic or facultatively aerobic, catalase-positive	*Bacillus*
	Paenibacillus
Microaerophilic, catalase-negative; homofermentative lactic acid producer	*Sporolactobacillus*
Anaerobic:	
Sulfate-reducing	*Desulfotomaculum*
Does not reduce sulfate, fermentative	*Clostridium*
Thermophilic, temperature optimum 65–70°C, fermentative	*Thermoanaerobacter*
Gram-negative; can grow as homoacetogen on $H_2 + CO_2$	*Sporomusa*
Halophile, isolated from the Dead Sea	*Sporohalobacter*
Produces up to five spores per cell; fixes N_2	*Anaerobacter*
Acidophile, pH optimum 3	*Alicyclobacillus*
Alkaliphile, pH optimum 9	*Amphibacillus*
Phototrophic (⇄ Section 14.8)	*Heliobacterium,* *Heliophilum,* *Heliorestis*
Syntrophic, degrades fatty acids but only in coculture with a H_2-utilizing bacterium	*Syntrophospora*
Reductively dechlorinates chlorophenols	*Desulfitobacterium*
II. Cocci	
Arranged in tetrads or packets, aerobic	*Sporosarcina* (Figure 15.24)

Table 15.5 Characteristics of representative species of bacilli

Characteristics	Species	Endospore position
I. Endospores oval or cylindrical, facultative aerobes, casein and starch hydrolyzed		
Sporangia not swollen, endospore wall thin		
Thermophiles and acidophiles	*Bacillus coagulans*	Central or terminal
	Alicyclobacillus acidocaldarius	Terminal
Mesophiles	*Bacillus licheniformis*	Central
	Bacillus cereus	Central
	Bacillus anthracis	Central
	Bacillus megaterium	Central
	Bacillus subtilis	Central
Insect pathogen	*Bacillus thuringiensis*	Central
Sporangia distinctly swollen, spore wall thick		
Thermophile	*Geobacillus stearothermophilus*	Terminal
Mesophiles	*Paenibacillus polymyxa*	Terminal
	Bacillus macerans	Terminal
	Bacillus circulans	Central or terminal
Insect pathogens	*Paenibacillus larvae*	Central or terminal
	Paenibacillus popilliae	Central
II. Endospores spherical, obligate aerobes, casein and starch not hydrolyzed		
Sporangia swollen	*Bacillus sphaericus*	Terminal
Sporangia not swollen	*Sporosarcina pasteurii*	Terminal

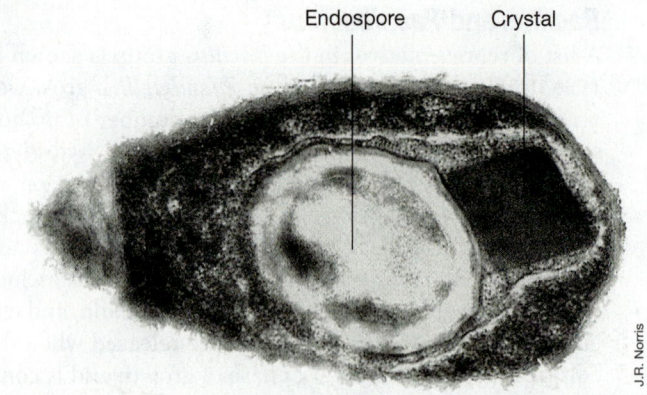

Endospore   Crystal

J.R. Norris

Figure 15.23 **The toxic parasporal crystal in the insect pathogen** *Bacillus thuringiensis.* Electron micrograph of a thin section of a sporulating cell. The crystalline protein (Bt toxin) is toxic to certain insects by causing lysis of their intestinal cells.

is deposited within the sporangium but outside the endospore proper (**Figure 15.23**). In *B. thuringiensis,* the parasporal body is a protoxin that is converted to a toxin in the insect gut. The toxin binds to specific receptors in the intestinal epithelial cells of certain insects and induces pore formation that causes leakage of the host cell cytoplasm followed by lysis. Diverse strains of *B. thuringiensis* can make different types of toxin that have specificity for different groups of insects. Endospore preparations derived from *B. thuringiensis* and *P. popilliae* are commercially available as biological insecticides.

The *cry* genes that encode crystal proteins have been isolated from several *B. thuringiensis* strains. The genes for the *B. thuringiensis* crystal protein (known commercially as "Bt toxin") have been introduced into genetically modified crops (e.g., maize, soybeans, and cotton) to render the plants resistant to insects. These genetically modified "Bt crops" are used widely around the world. Genetically altered Bt toxins have also been developed by genetic engineering to help increase toxicity and reduce resistance (⮌ Section 11.13).

Clostridium

Clostridia lack a respiratory chain, and so unlike *Bacillus* species, they obtain ATP by substrate-level phosphorylation. Many anaerobic energy-yielding mechanisms are known in the clostridia (⮌ Section 13.13). Indeed, the separation of the genus *Clostridium* into subgroups is based primarily on these properties and on the fermentable substrate used (**Table 15.6**). A number of clostridia are *saccharolytic* and ferment sugars, producing butyric acid as a major end product. Some of these also produce acetone and butanol, such as *Clostridium pasteurianum,* which is also a vigorous nitrogen-fixing bacterium.

One group of clostridia including the species *C. thermocellum, C. cellulolyticum,* and *C. cellulovorans* ferments cellulose with the formation of acids and alcohols. These species are likely the major organisms decomposing cellulose in anoxic environments such as the rumen and sediments. Cellulolytic clostridia possess *cellulosomes,* a complex multienzyme structure found on the outer surface of the

Table 15.6 Characteristics of some groups of clostridia

Key characteristics	Other characteristics	Species
Ferment carbohydrates		
Ferment cellulose	Fermentation products: acetate, lactate, succinate, ethanol, CO_2, H_2	*C. cellobioparum*[a] *C. thermocellum*
Ferment sugars, starch, and pectin; some ferment cellulose	Fermentation products: acetone, butanol, ethanol, isopropanol, butyrate, acetate, propionate, succinate, CO_2, H_2; some fix N_2	*C. butyricum* *C. cellobioparum* *C. acetobutylicum* *C. pasteurianum* *C. perfringens*
Ferment sugars primarily to acetic acid	Total synthesis of acetate from CO_2; cytochromes present in some species	*C. aceticum* *Moorella thermoacetica* *C. formicaceticum*
Ferments only pentoses or methylpentoses	Ring-shaped cells form left-handed, helical chains; fermentation products: acetate, propionate, *n*-propanol, CO_2, H_2	*C. methylpentosum*
Ferment amino acids	Fermentation products: acetate, other fatty acids, NH_3, CO_2, sometimes H_2; some also ferment sugars to butyrate and acetate; may produce exotoxins; causative agents of serious or fatal diseases	*C. sporogenes* *C. histolyticum* *C. putrefaciens* *C. tetani* *C. botulinum* *C. tetanomorphum*
	Ferments three-carbon amino acids (for example, alanine) to propionate, acetate, and CO_2	*C. propionicum*
Ferments carbohydrates or amino acids	Fermentation products from glucose: acetate, formate, small amounts of isobutyrate and isovalerate	*C. bifermentans*
Purine fermenters	Ferments uric acid and other purines, forming acetate, CO_2, NH_3	*C. acidurici*
Ethanol fermentation to fatty acids	Produces butyrate, caproate, and H_2; requires acetate as electron acceptor; does not use sugars, amino acids, or purines	*C. kluyveri*

[a]All genus names beginning with a "*C.*" are species of the genus *Clostridium*.

cell wall. The cellulosome binds insoluble cellulose and degrades it into soluble products that are transported into the cytoplasm and metabolized by the cell. This cellulosome mechanism is common to bacteria that are able to degrade cellulose anaerobically.

Another group of clostridia are *proteolytic* and conserve energy from the fermentation of amino acids. Some species ferment individual amino acids, but others ferment only amino acid pairs. The products of amino acid fermentation are typically acetate, butyrate, CO_2, and H_2. The coupled catabolism of an amino acid pair is called a *Stickland reaction*; for example, *Clostridium sporogenes* ferments glycine plus alanine. In the Stickland reaction, one amino acid functions as the electron donor and is oxidized, whereas the other is the electron acceptor and is reduced (⮢ Figure 13.34). Many of the products of amino acid fermentation by clostridia are foul-smelling substances, and the odor that results from putrefaction is mainly the result of clostridial action. In addition to butyric acid, other odoriferous compounds produced are isobutyric acid, isovaleric acid, caproic acid, hydrogen sulfide, methylmercaptan (from sulfur amino acids), cadaverine (from lysine), putrescine (from ornithine), and ammonia.

The main habitat of clostridia is the soil, where they live primarily in "pockets" made anoxic by facultative or obligately aerobic bacteria. In addition, a number of clostridia inhabit the anoxic environment of the mammalian intestinal tract. Several clostridia are capable of causing severe diseases in humans, as will be discussed in Sections 23.9 and 30.9. For example, botulism is caused by *Clostridium botulinum*, tetanus by *Clostridium tetani*, and gas gangrene by *Clostridium perfringens* and a number of other clostridia, both sugar and amino acid fermenters. These pathogenic clostridia seem in no way unusual metabolically but are distinct in that they produce specific toxins or, in those causing gas gangrene, a group of toxins. *C. perfringens* and related species can also cause gastroenteritis in humans and domestic animals (⮢ Section 31.9), and botulism outbreaks are not uncommon in birds such as ducks and a variety of other animals.

Sporosarcina

The genus *Sporosarcina* (**Figure 15.24**) is unique among endospore formers because cells are cocci instead of rods. *Sporosarcina* consists of strictly aerobic spherical to oval cells that divide

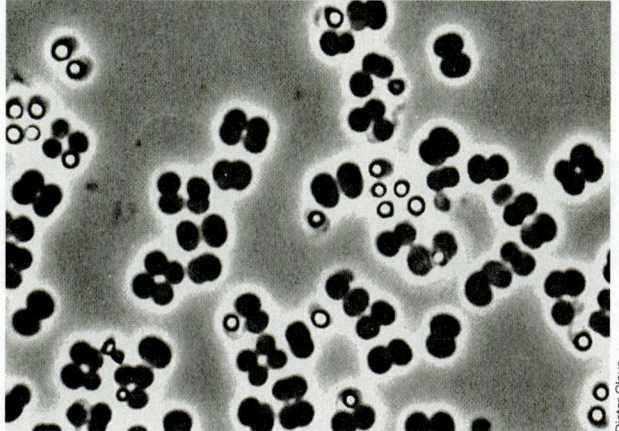

Figure 15.24 *Sporosarcina ureae.* Phase-contrast micrograph. A single cell is about 2 μm wide. Note bright refractile endospores. Most cell packets contain eight cells.

in two or three perpendicular planes to form tetrads or packets of eight or more cells. The major species is *Sporosarcina ureae*. This bacterium can be enriched from soil by plating dilutions of a pasteurized soil sample on alkaline nutrient agar supplemented with 8% urea and incubating in air. Most soil bacteria are strongly inhibited by as little as 2% urea. However, *S. ureae* tolerates this, catabolizing urea to CO_2 and ammonia (NH_3), which dramatically raises the pH. *S. ureae* is remarkably alkaline-tolerant and can be grown in media up to pH 10, and this feature can be used to advantage in its enrichment from soil.

MINIQUIZ --

- What is the major physiological distinction between *Bacillus* and *Clostridium* species?
- What is the crystalline protein made by *Bacillus thuringiensis* and what is its significance to agriculture?

15.9 *Tenericutes*: The Mycoplasmas

Key Genera: *Mycoplasma, Spiroplasma*

The *Tenericutes*, which contain the single class *Mollicutes*, are bacteria that lack cell walls (*mollis* is Latin for "soft") and are some of the smallest organisms known. This group is often called the *mycoplasmas* because *Mycoplasma*, a notable genus containing human pathogens, is the best-characterized genus in the phylum (**Table 15.7**).

Although they do not stain gram-positively (because they lack cell walls), mycoplasmas are phylogenetically related to the *Firmicutes*. Mycoplasmas typically live in close association with animal and plant hosts and this may eliminate the need for a gram-positive cell wall. These organisms also have small genomes (ranging in size from 600 to 2200 kilobase pairs), a characteristic common to obligate symbionts (⮢ Sections 6.5 and 22.9).

Properties of Mycoplasmas

The absence of cell walls in mycoplasmas has been confirmed by electron microscopy and chemical analyses, which show that peptidoglycan is absent. Mycoplasmas resemble protoplasts (bacteria treated to remove their cell walls), but they are more resistant to osmotic lysis and are able to survive conditions under which protoplasts lyse. This ability to resist osmotic lysis is at least partially determined by the presence of sterols, which make the cytoplasmic membranes of mycoplasmas more stable than that of other bacteria. Indeed, some mycoplasmas require sterols in their growth media, and this sterol requirement can aid in the classification of mycoplasmas (Table 15.7).

In addition to sterols, certain mycoplasmas contain compounds called *lipoglycans* (Table 15.7). Lipoglycans are long-chain heteropolysaccharides covalently linked to membrane lipids and embedded in the cytoplasmic membrane of many mycoplasmas. Lipoglycans in some ways resemble the lipopolysaccharides in the outer membrane of gram-negative bacteria, except that they lack the lipid A backbone (⮢ Section 2.11). Lipoglycans function to help stabilize the cytoplasmic membrane and have also been identified as facilitating attachment of mycoplasmas to cell surface receptors of animal cells.

Table 15.7 *Major characteristics of mycoplasmas*

Genus	Properties	Genome size (kilobase pairs)	Presence of lipoglycans
Require sterols			
Mycoplasma	Many pathogenic; facultative anaerobes (Figure 15.25)	600–1350	+
Anaeroplasma	May or may not require sterols; obligate anaerobes; degrade starch, producing acetic, lactic, and formic acids plus ethanol and CO_2; found in the bovine and ovine rumen	1500–1600	+
Spiroplasma	Spiral to corkscrew-shaped cells; associated with various phytopathogenic (plant disease) conditions; facultative anaerobe	940–2200	–
Ureaplasma	Coccoid cells; occasional clusters and short chains; growth optimal at pH 6; strong urease reaction; associated with certain urinary tract infections in humans; microaerophile	750	–
Entomoplasma	Facultative anaerobe; associated with insects and plants	790–1140	Unknown
Do not require sterols			
Acholeplasma	Facultative anaerobes	1500	+
Asteroleplasma	Obligate anaerobe; isolated from the bovine or ovine rumen	1500	+
Mesoplasma	Phylogenetically and ecologically related to *Entomoplasma*; facultative anaerobes	870–1100	Unknown

Growth of Mycoplasmas

Mycoplasmas can be grown in the laboratory and are small and pleomorphic cells. A single culture may exhibit small coccoid elements; larger, swollen forms; and filamentous forms, often highly branched (**Figure 15.25**). The small coccoid elements (0.2–0.3 μm in size) are among the smallest of free-living cells (Section 2.6). The mode of growth of mycoplasmas differs in liquid and agar cultures. On agar the organisms tend to grow so that they become embedded in the medium. These colonies show a characteristic "fried-egg" appearance consisting of a dense central core that penetrates downward into the agar, surrounded by a circular spreading area that is lighter in color (**Figure 15.26**). As would be expected of cells lacking cell walls, growth of *Mollicutes* is not inhibited by antibiotics that inhibit cell wall synthesis. However,

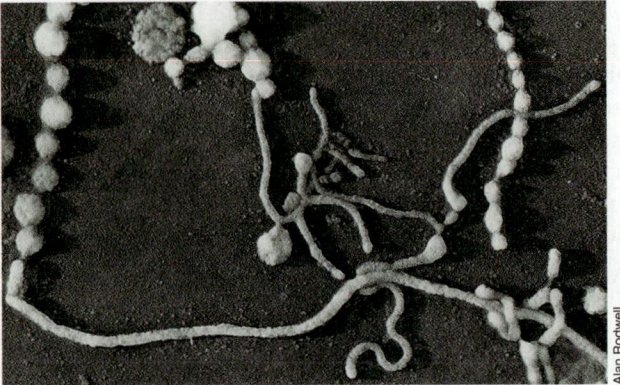

Figure 15.25 *Mycoplasma mycoides.* Metal-shadowed transmission electron micrograph. Note the coccoid and hyphalike elements. The average diameter of cells in chains is about 0.5 μm.

mycoplasmas are as sensitive as most *Bacteria* to antibiotics whose targets are other than the cell wall.

Media for the culture of mycoplasmas are typically quite complex. For many species, growth is poor or absent even in complex yeast extract–peptone–beef heart infusion media. Fresh serum or ascitic fluid (peritoneal fluid) is needed as well to provide unsaturated fatty acids and sterols. Some mycoplasmas can be cultivated on relatively simple culture media, however, and even defined media have been developed for some species. Most mycoplasmas use carbohydrates as carbon and energy sources and require vitamins, amino acids, purines, and pyrimidines as growth factors. The energy metabolism of mycoplasmas is variable; some species are strictly aerobic, whereas others are facultative aerobes or obligate anaerobes (Table 15.7).

Spiroplasma

The genus *Spiroplasma* consists of helical or spiral-shaped *Mollicutes*. Amazingly, although they lack a cell wall and flagella, spiroplasmas are motile by means of a rotary (screw) motion or a slow undulation. Intracellular fibrils that are thought to play a role in motility have been demonstrated. The organism has been isolated from ticks, the hemolymph (**Figure 15.27**) and gut of insects, vascular plant fluids and insects that feed on these fluids, and the surfaces of flowers and other plant parts. For example, *Spiroplasma citri* has been isolated from the leaves of citrus plants, where it causes a disease called citrus stubborn disease, and from corn plants suffering from corn stunt disease. A number of

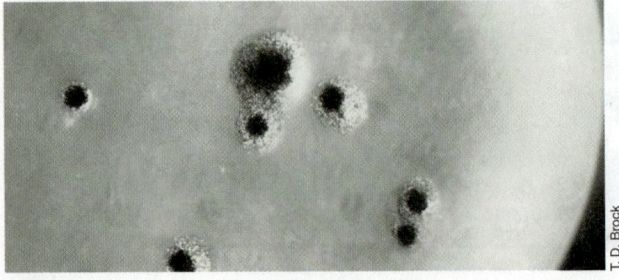

Figure 15.26 **Colonies of a *Mycoplasma* species on agar.** Note the typical "fried-egg" appearance. The colonies are about 0.5 mm in diameter.

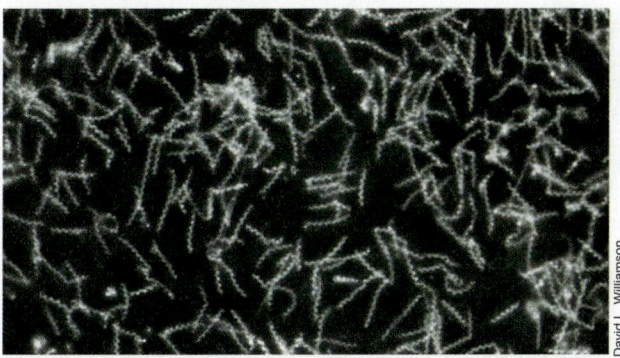

Figure 15.27 **"Sex ratio" spiroplasma from the hemolymph of the fly** *Drosophila pseudoobscura*. Dark-field micrograph. Female flies infected with the sex ratio spiroplasma bear only female progeny. Cells are about 0.15 μm in diameter.

other mycoplasma-like organisms have been detected in diseased plants by electron microscopy, which indicates that a large group of plant-associated *Mollicutes* may exist. Some species of *Spiroplasma* are known that cause insect diseases, such as honeybee spiroplasmosis and lethargy disease of the beetle *Melolontha*.

MINIQUIZ -

- Why do mycoplasmas need to have stronger cytoplasmic membranes than other bacteria?

- Motile spiroplasmas cannot contain a normal bacterial flagellum; why?

15.10 *Actinobacteria*: Coryneform and Propionic Acid Bacteria

Key Genera: *Arthrobacter, Corynebacterium, Propionibacterium*

The other major group of gram-positive bacteria is the *Actinobacteria*, which form their own phylum within the *Bacteria*. The *Actinobacteria* contain rod-shaped to filamentous and primarily aerobic bacteria that are common inhabitants of soil and plant materials. For the most part they are harmless commensals, species of *Mycobacterium* (for example, *Mycobacterium tuberculosis*) being notable exceptions. Some are of great economic value in either the production of antibiotics or certain fermented dairy products. While there are nine orders of *Actinobacteria*, the vast majority of species belong to the order *Actinomycetales* (Figure 15.17). We consider here the coryneform bacteria, species of *Actinomycetales* that have an unusual method of cell division, and the propionic acid bacteria, important agents in the ripening of Swiss cheese.

Coryneform Bacteria

Coryneform bacteria are gram-positive, aerobic, nonmotile, rod-shaped organisms that form irregular-shaped, club-shaped, or V-shaped cell arrangements during growth. V-shaped cells arise as a result of an abrupt movement that occurs just after cell division, a process called *snapping division* (**Figure 15.28**). Snapping division occurs because the cell wall consists of two layers; only the inner layer participates in cross-wall formation, and so after the cross-wall is formed, the two daughter cells remain attached by the outer layer

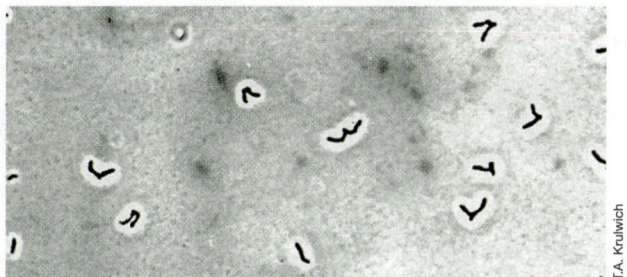

Figure 15.28 **Snapping division in *Arthrobacter*.** Phase-contrast micrograph of characteristic V-shaped cell groups in *Arthrobacter crystallopoietes* resulting from snapping division. Cells are about 0.9 μm in diameter.

of the cell wall. Localized rupture of this outer layer on only one side of the cell results in a bending of the two cells away from the ruptured side (**Figure 15.29**) and thus development of V-shaped forms.

The main genera of coryneform bacteria are *Corynebacterium* and *Arthrobacter*. The genus *Corynebacterium* consists of an extremely diverse group of bacteria, including animal and plant pathogens and saprophytes. Some species, such as *Corynebacterium diphtheriae*, are pathogenic (diphtheria, ⮑ Section 29.3). The genus *Arthrobacter*, consisting primarily of soil organisms, is distinguished from *Corynebacterium* on the basis of a developmental cycle involving conversion from rod to coccus and back to rod again (**Figure 15.30**). However, some coryneform bacteria are pleomorphic and form coccoid cells during growth, and so the distinction between the two genera on the basis of life cycle is not

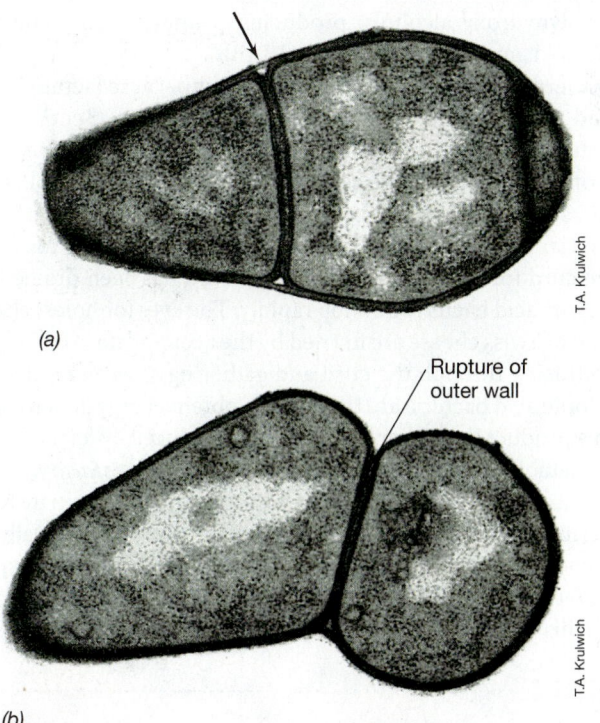

(a)

Rupture of outer wall

(b)

Figure 15.29 **Cell division in *Arthrobacter*.** Transmission electron micrograph of cell division in *Arthrobacter crystallopoietes*, illustrating how snapping division and V-shaped cell groups arise. *(a)* Before rupture of the outer cell wall layer (arrow). *(b)* After rupture of the outer layer on one side. Cells are 0.9–1 μm in diameter.

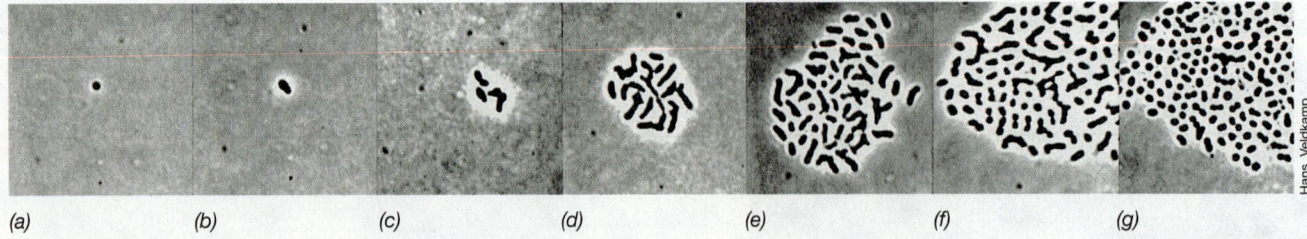

(a)   (b)   (c)   (d)   (e)   (f)   (g)

Hans Veldkamp

Figure 15.30 **Stages in the life cycle of *Arthrobacter globiformis* as observed in slide culture.** *(a)* Single coccoid element; *(b–e)* conversion to rod and growth of a microcolony consisting predominantly of rods; *(f–g)* conversion of rods to coccoid forms. Cells are about 0.9 μm in diameter.

absolute. The *Corynebacterium* cell frequently has a swollen end, so it has a club-shaped appearance, whereas *Arthrobacter* species are less commonly club-shaped.

Along with the *Acidobacteria* (Section 15.21), species of *Arthrobacter* are among the most common of all soil bacteria. They are remarkably resistant to desiccation and starvation, despite the fact that they do not form spores or other resting cells. Arthrobacters are a heterogeneous group that have considerable nutritional versatility, and strains have been isolated that decompose herbicides, caffeine, nicotine, phenols, and other unusual organic compounds.

Propionic Acid Bacteria

The **propionic acid bacteria** (genus *Propionibacterium*) were first discovered in Swiss (Emmentaler) cheese, where their fermentative production of CO_2 produces the characteristic holes and the propionic acid they produce is at least partly responsible for the unique flavor of the cheese. The bacteria in this group are gram-positive anaerobes that ferment lactic acid, carbohydrates, and polyhydroxy alcohols, producing primarily propionic acid, acetic acid, and CO_2 (⮂ Section 13.13).

The fermentation of lactate is of interest because lactate itself is an end product of fermentation for many bacteria (Section 15.6). The starter culture in Swiss cheese manufacture consists of a mixture of homofermentative streptococci and lactobacilli, plus propionic acid bacteria. The homofermentative organisms carry out the initial fermentation of lactose to lactic acid during formation of the curd (protein and fat). After the curd has been drained, the propionic acid bacteria develop rapidly. The eyes (or holes) characteristic of Swiss cheese are formed by the accumulation of CO_2, the gas diffusing through the curd and gathering at weak points. The propionic acid bacteria are thus able to obtain energy anaerobically from a product that other bacteria have produced by fermentation. This metabolic strategy is called a *secondary fermentation*.

Propionate is also formed in the fermentation of succinate by the bacterium *Propionigenium*. This organism is phylogenetically and ecologically unrelated to *Propionibacterium*, but energetic aspects of its fermentation are of considerable interest. We discussed the mechanism of the *Propionigenium* fermentation in Section 13.14.

MINIQUIZ

- What is snapping division and what organism exhibits it?
- What organism is involved in the production of Swiss cheese, and what products does it make that help to flavor the cheese and make the holes?

15.11 *Actinobacteria*: *Mycobacterium*

Key Genus: *Mycobacterium*

The genus *Mycobacterium* contains several notable human pathogens, chief among them *Mycobacterium tuberculosis*, the cause of tuberculosis (⮂ Section 29.4). Species are rod-shaped bacteria that at some stage of their growth cycle possess the distinctive staining property called **acid-fastness**. This property is due to the presence of unique lipids called *mycolic acids*, found only in species of the genus *Mycobacterium*, on the surface of the mycobacterial cell. Mycolic acids are a group of complex branched-chain hydroxylated lipids (**Figure 15.31***a*) covalently bound to peptidoglycan in the cell wall; the complex gives the cell surface a waxy, hydrophobic consistency.

Because of their waxy surface, mycobacteria do not stain well with Gram stain. A mixture of the red dye basic fuchsin and phenol is used in the acid-fast (Ziehl–Neelsen) stain. The stain is driven into the cells by slow heating, and the role of the phenol is to enhance penetration of the fuchsin into the lipids. After washing in distilled water, the preparation is decolorized with acid alcohol and counterstained with methylene blue. Cells of acid-fast organisms stain red, whereas the background and non-acid-fast organisms appear blue (⮂ Figure 29.15*a*).

Mycobacteria are somewhat pleomorphic and may undergo branching or even filamentous growth. However, in contrast to the filaments of the actinomycetes (Section 15.12), the filaments of the mycobacteria do not form a true mycelium. Mycobacteria can be separated into two major groups: slow-growing species (e.g., *M. tuberculosis*, *M. avium*, *M. bovis*, and *M. gordonae*) and fast-growing species (e.g., *M. smegmatis*, *M. phlei*, *M. chelonae*, *M. parafortuitum*). *Mycobacterium tuberculosis* is a typical slow

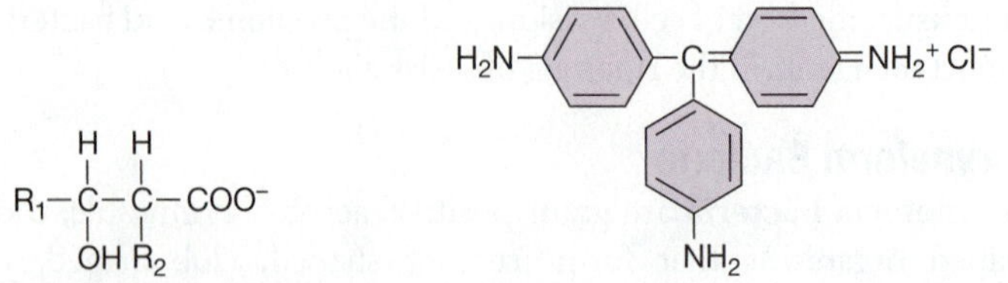

(a) **Mycolic acid; R₁ and R₂ are long-chain aliphatic hydrocarbons**

(b) **Basic fuchsin**

Figure 15.31 **Acid-fast staining.** Structure of *(a)* mycolic acid and *(b)* basic fuchsin, the dye used in the acid-fast stain. The fuchsin dye combines with mycolic acids in the cell wall via ionic bonds between COO^- and NH_2^+.

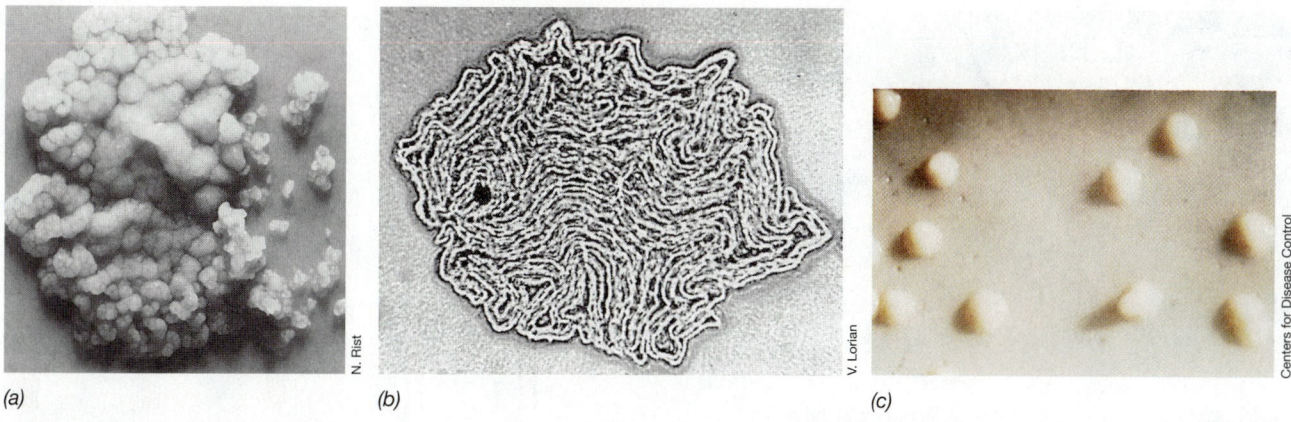

(a)   (b)   (c)

N. Rist   V. Lorian   Centers for Disease Control

Figure 15.32 **Characteristic colony morphology of mycobacteria.** *(a)* *Mycobacterium tuberculosis*, showing the compact, wrinkled appearance of the colony. The colony is about 7 mm in diameter. *(b)* A colony of virulent *M. tuberculosis* at an early stage, showing the characteristic cordlike growth. Individual cells are about 0.5 μm in diameter. (See also the historic drawings of *M. tuberculosis* cells made by Robert Koch, ↺ Figure 1.22). *(c)* Colonies of *Mycobacterium avium* from a strain of this organism isolated as an opportunistic pathogen from an AIDS patient.

grower, and visible colonies are produced from dilute inoculum only after days to weeks of incubation. When growing on solid media, mycobacteria form tight, compact, often wrinkled colonies (**Figure 15.32**). This colony morphology is probably due to the high lipid content and hydrophobic nature of the cell surface, which facilitates cells sticking together.

For the most part, mycobacteria have relatively simple nutritional requirements. Most species can grow aerobically in a simple mineral salts medium with ammonium as the nitrogen source and glycerol or acetate as the sole carbon source and electron donor. Growth of *M. tuberculosis* is more difficult and is stimulated by lipids and fatty acids. The virulence of *M. tuberculosis* cultures has been correlated with the formation of long, cordlike structures (Figure 15.32*b*) that form due to side-to-side aggregation and intertwining of long chains of bacteria. Growth in cords reflects the presence of a characteristic glycolipid, the *cord factor*, on the cell surface (**Figure 15.33**). The pathogenesis of tuberculosis, along with the related mycobacterial disease leprosy, is discussed in Section 29.4.

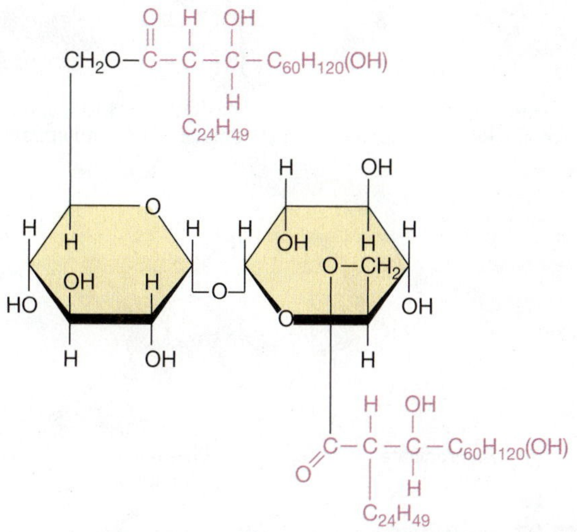

Some mycobacteria produce yellow carotenoid pigments (Figure 15.32*c*), and pigmentation can aid in identification. Mycobacteria can either be nonpigmented (e.g., *M. tuberculosis*, *M. bovis*, *M. smegmatis*, *M. chelonae*); or can form pigment only when cultured in light, a property called *photochromogenesis* (e.g., *M. parafortuitum*); or can form pigment even when cultured in the dark, a property called *scotochromogenesis* (e.g., *M. gordonae*, *M. phlei*). Photochromogenesis is triggered by the blue region of the visible spectrum and is characterized by the photoinduction of one of the early enzymes in carotenoid biosynthesis. As with other carotenoid-containing bacteria, it is likely that carotenoids protect mycobacteria against oxidative damage from singlet oxygen (↺ Section 5.16).

MINIQUIZ --

- What is mycolic acid, and what properties does this substance confer on mycobacteria?

15.12 Filamentous *Actinobacteria*: *Streptomyces* and Relatives

Key Genera: *Streptomyces, Actinomyces, Nocardia*

The **actinomycetes** are a large group of phylogenetically related, filamentous and aerobic gram-positive *Bacteria* common in soils. Many actinomycetes have a characteristic developmental cycle that culminates in the production of desiccation-resistant spores. Filaments elongate from their ends and form branching *hyphae*. Hyphal growth results in a network of filaments called a *mycelium* (**Figure 15.34**), analogous to that formed by filamentous fungi (↺ Section 17.9). When nutrients are depleted, the mycelium forms aerial hyphae that differentiate into spores that allow for survival and dispersal. We focus here on the genus *Streptomyces*, the most important genus in this group.

Streptomyces

Over 500 species of *Streptomyces* are recognized. *Streptomyces* filaments are typically 0.5–1.0 μm in diameter and of indefinite

Figure 15.33 **Structure of cord factor, a mycobacterial glycolipid: 6,6′-di-*O*-mycolyl trehalose.** The two identical long-chain dialcohol groups are shown in purple.

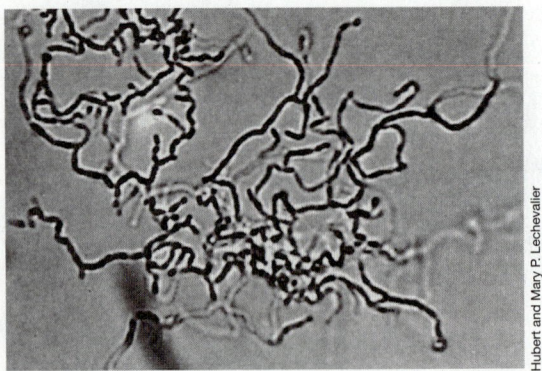

Figure 15.34 *Nocardia*. A young colony of an actinomycete of the genus *Nocardia*, showing typical filamentous cellular structure (mycelium). Each filament is about 0.8–1 µm in diameter.

length, and often lack cross-walls in the vegetative phase. *Streptomyces* grow at the tips of the filaments and may branch often. Thus, the vegetative phase consists of a complex, tightly woven matrix, resulting in a compact, convoluted mycelium and subsequent colony. As the colony ages, characteristic aerial filaments called *sporophores* are formed, which project above the surface of the colony and give rise to spores (**Figure 15.35**).

Streptomyces spores, called *conidia*, are quite distinct from the endospores of *Bacillus* and *Clostridium*. Unlike the elaborate cellular differentiation that leads to the formation of an endospore, conidia are produced by the formation of cross-walls in the multinucleate

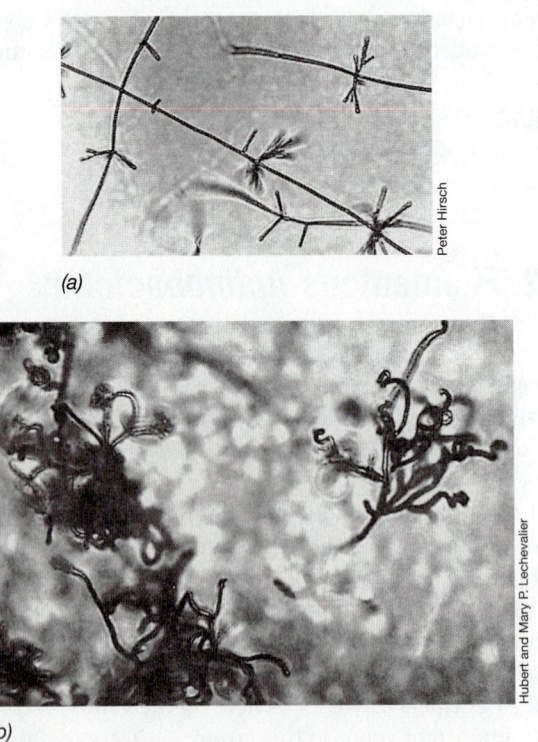

(a)

(b)

Figure 15.35 **Spore-bearing structures of actinomycetes.** Phase-contrast micrographs. Compare these photos with the art in Figure 15.37. *(a) Streptomyces*, a monoverticillate type. *(b) Streptomyces*, a closed spiral type. Filaments are about 0.8 µm wide in both types.

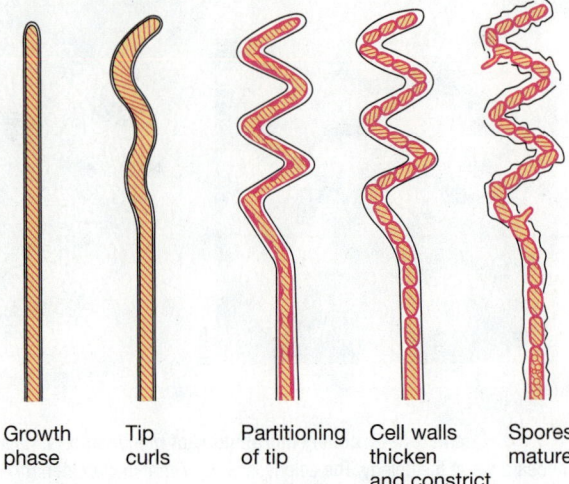

Growth phase | Tip curls | Partitioning of tip | Cell walls thicken and constrict | Spores mature

Figure 15.36 **Spore formation in *Streptomyces*.** Diagram of stages in the conversion of an aerial hypha (sporophore) into spores (conidia).

sporophores followed by separation of the individual cells directly into spores (**Figure 15.36**). Differences in the shape and arrangement of aerial filaments and spore-bearing structures of various species are among the fundamental features used in classifying the *Streptomyces* species (**Figure 15.37**). The conidia and sporophores are often pigmented and contribute a characteristic color to the mature

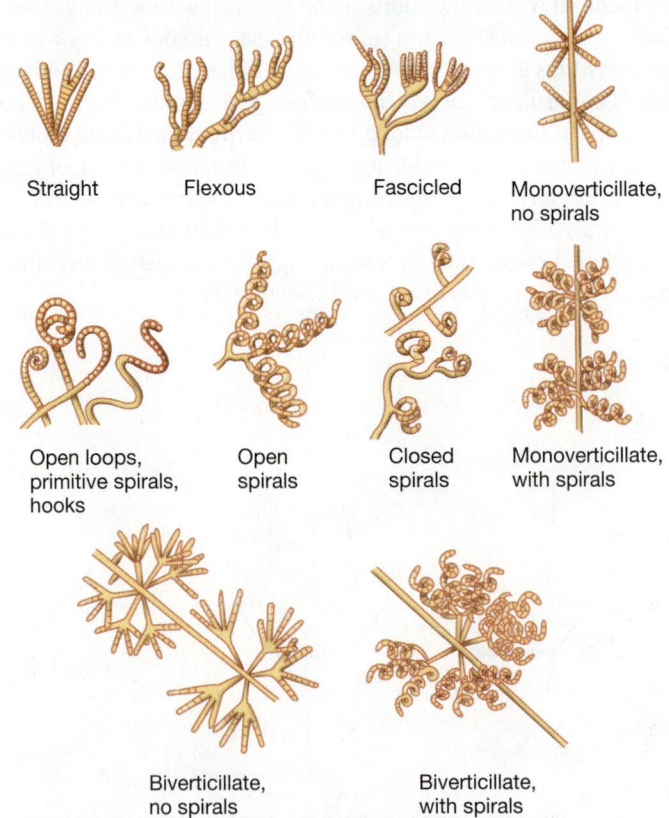

Straight | Flexous | Fascicled | Monoverticillate, no spirals

Open loops, primitive spirals, hooks | Open spirals | Closed spirals | Monoverticillate, with spirals

Biverticillate, no spirals | Biverticillate, with spirals

Figure 15.37 **Morphologies of spore-bearing structures in the streptomycetes.** A given species of *Streptomyces* produces only one morphological type of spore-bearing structure. The term "verticillate" means "whorls."

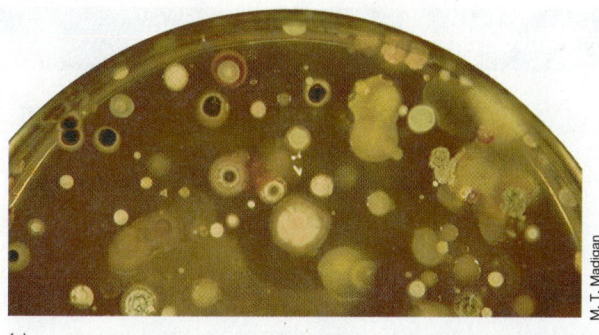

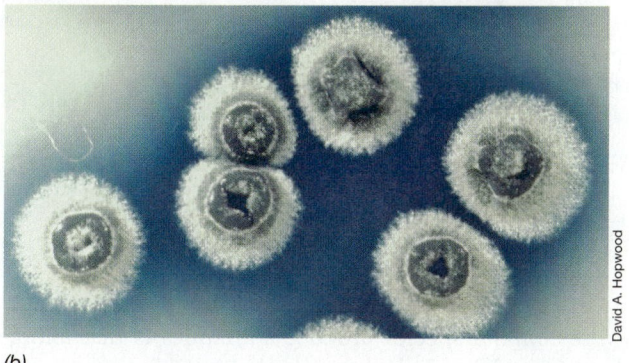

Figure 15.38 **Streptomycetes.** *(a)* Colonies of *Streptomyces* and other soil bacteria derived from spreading a soil dilution on a casein–starch agar plate. The *Streptomyces* colonies are of various colors (several black *Streptomyces* colonies are near the top of the plate) but can easily be identified by their opaque, rough, nonspreading morphology. *(b)* Close-up photo of colonies of *Streptomyces coelicolor*.

colony (**Figure 15.38**). The dusty appearance of the mature colony, its compact nature, and its color make detection of *Streptomyces* colonies on agar plates relatively easy (Figure 15.38*b*).

Ecology and Isolation of *Streptomyces*

Although a few streptomycetes are aquatic, they are primarily soil organisms. In fact, the characteristic earthy odor of soil is caused by the production by streptomycetes of a series of complex metabolites all called *geosmin*. Alkaline to neutral soils are more favorable for the development of *Streptomyces* than are acid soils. Moreover, higher numbers of *Streptomyces* are found in well-drained soils (such as sandy loams or soils covering limestone), where conditions are more likely to be aerobic, than in waterlogged soils, which quickly become anoxic.

Isolation of *Streptomyces* from soil is relatively easy: A suspension of soil in sterile water is diluted and spread on selective agar medium, and the plates are incubated aerobically at 25°C (Figure 15.38). Media selective for *Streptomyces* contain mineral salts plus polymeric substances such as starch or casein as organic nutrients. Streptomycetes typically produce extracellular hydrolytic enzymes that permit utilization of polysaccharides (starch, cellulose, and hemicellulose), proteins, and fats, and some strains can use hydrocarbons, lignin, tannin, and other polymers. After incubation for 5–7 days in air, the plates are examined for the presence of the characteristic *Streptomyces* colonies (Figure 15.38), and spores from colonies can be restreaked to isolate pure cultures.

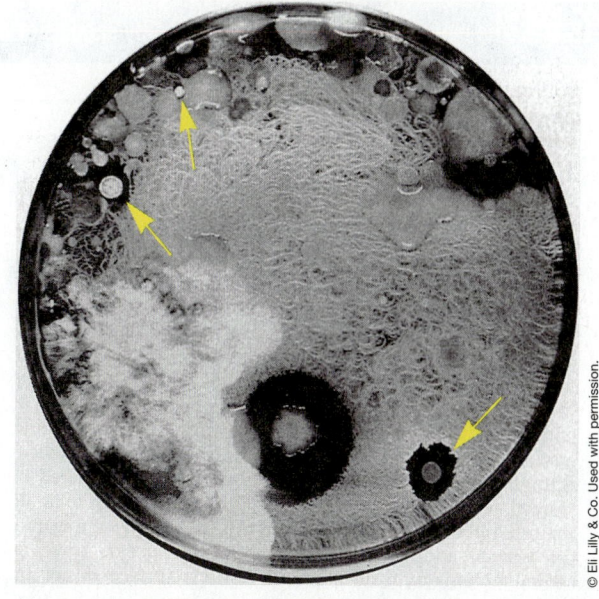

Figure 15.39 **Antibiotics from *Streptomyces.*** *(a)* Antibiotic action of soil microorganisms on a crowded plate. The smaller colonies surrounded by inhibition zones (arrows) are streptomycetes; the larger, spreading colonies are *Bacillus* species, some of which are also producing antibiotics. *(b)* The red-colored antibiotic undecylprodigiosin is being excreted by colonies of *S. coelicolor*.

Antibiotics of *Streptomyces*

Perhaps the most striking physiological property of the streptomycetes is the extent to which they produce *antibiotics* (**Table 15.8**). Evidence for antibiotic production is often seen on the agar plates used in their initial isolation: Adjacent colonies of other bacteria show zones of inhibition (**Figure 15.39*a***).

About 50% of all *Streptomyces* isolated have been found to be antibiotic producers. Over 500 distinct antibiotics are produced by streptomycetes and many more are suspected; most of these have been identified chemically (Figure 15.39*b*). Some species produce more than one antibiotic, and often the several antibiotics produced by one organism are chemically unrelated. Although an antibiotic-producing organism is resistant to its own antibiotics, it usually remains sensitive to antibiotics

Table 15.8 Some common antibiotics synthesized by species of *Streptomyces* and related *Actinobacteria*

Chemical class	Common name	Produced by	Active against[a]
Aminoglycosides	Streptomycin	S. griseus[b]	Most gram-negative *Bacteria*
	Spectinomycin	*Streptomyces* spp.	*Mycobacterium tuberculosis*, penicillinase-producing *Neisseria gonorrhoeae*
	Neomycin	S. fradiae	Broad spectrum, usually used in topical applications because of toxicity
Tetracyclines	Tetracycline	S. aureofaciens	Broad spectrum, gram-positive and gram-negative *Bacteria*, rickettsias and chlamydias, *Mycoplasma*
	Chlortetracycline	S. aureofaciens	As for tetracycline
Macrolides	Erythromycin	*Saccharopolyspora erythraea*	Most gram-positive *Bacteria*, frequently used in place of penicillin; *Legionella*
	Clindamycin	S. lincolnensis	Effective against obligate anaerobes, especially *Bacteroides fragilis*, the major cause of anaerobic peritoneal infections
Polyenes	Nystatin	S. noursei	Fungi, especially *Candida* (a yeast) infections
	Amphotericin B	S. nodosus	Fungi
None	Chloramphenicol	S. venezuelae	Broad spectrum; drug of choice for typhoid fever

[a]Most antibiotics are effective against several different *Bacteria*. The entries in this column refer to the common clinical application of a given antibiotic. The structures and mode of action of many of these antibiotics are discussed in Sections 27.11–27.14.
[b]All species names beginning with an "*S.*" are species of *Streptomyces*.

produced by other streptomycetes. Many genes are required to encode the enzymes for antibiotic synthesis, and because of this, the genomes of *Streptomyces* species are typically quite large (8 megabase pairs and larger; ↩ Table 6.1). More than 60 streptomycete antibiotics have been used in human and veterinary medicine, and some of the most commonly used are listed in Table 15.8.

Ironically, despite the extensive research done on antibiotic-producing streptomycetes by the antibiotic industry and the fact that *Streptomyces* antibiotics are a multibillion-dollar-a-year industry, the ecology of *Streptomyces* remains poorly understood. The interactions of these organisms with other bacteria and the ecological rationale for antibiotic production remains an important topic about which we know very little. One hypothesis for why *Streptomyces* species produce antibiotics is that antibiotic production, which is linked to sporulation (a process itself triggered by nutrient depletion), might be a mechanism to inhibit the growth of other organisms competing with *Streptomyces* cells for limiting nutrients. This would allow the *Streptomyces* to complete the sporulation process and form a dormant structure that would increase their chances of survival.

MINIQUIZ --

- Contrast spores and sporulation in *Streptomyces* and *Bacillus* species.
- Why might antibiotic production be of advantage to streptomycetes?

III • *Bacteroidetes*

The phylum *Bacteroidetes* contains more than 700 characterized species spread across four orders: *Bacteroidales*, *Cytophagales*, *Flavobacteriales*, and *Sphingobacteriales* (**Figure 15.40**). The *Bacteroidetes* are gram-negative nonsporulating rods; species are typically saccharolytic and can be aerobic or fermentative, including obligate aerobes, facultative aerobes, and obligate anaerobes. Gliding motility (↩ Section 2.18) is widespread in the phylum, though many species are nonmotile and a few are motile by flagella. The genus *Bacteroides* has been particularly well studied as these organisms are a major component of the microbial community in the human gut.

15.13 *Bacteroidales*

Key Genus: *Bacteroides*

The order *Bacteroidales* primarily contains obligately anaerobic fermentative species. The type genus is *Bacteroides*, which contains species that are saccharolytic, fermenting sugars or proteins

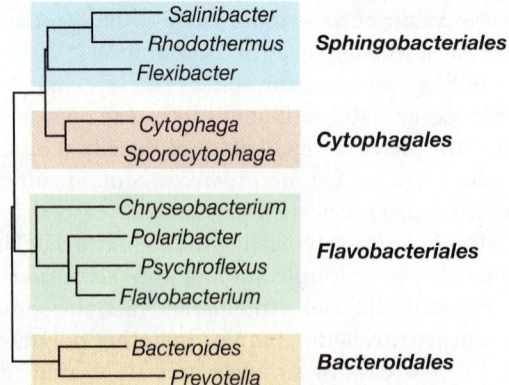

Bacteroidetes

Figure 15.40 **Major orders of *Bacteroidetes*.** The phylogenetic tree was constructed from 16S rRNA gene sequences of representative genera of *Bacteroidetes*. Order names are shown in bold.

Figure 15.41 Sphingolipids. Comparison of (a) glycerol with (b) sphingosine. In sphingolipids, characteristic of *Bacteroides* species, sphingosine is the esterifying alcohol; a fatty acid is bonded by peptide linkage through the N atom (shown in red), and the terminal —OH group (shown in green) can be any of a number of compounds including phosphatidylcholine (sphingomyelin) or various sugars (cerebrosides and gangliosides).

(depending on the species) to acetate and succinate as major fermentation products. *Bacteroides* are normally commensals, found in the intestinal tract of humans and other animals. In fact, *Bacteroides* species are the numerically dominant bacteria in the human large intestine, where measurements have shown that 10^{10}–10^{11} prokaryotic cells are present per gram of feces (⟳ Section 22.6). However, species of *Bacteroides* can occasionally be pathogens and are the most important anaerobic bacteria associated with human infections such as *bacteremia* (bacteria in the blood).

Bacteroides thetaiotaomicron is one of the most prominent species of *Bacteroides* found in the lumen of the large intestine. *B. thetaiotaomicron* specializes in the degradation of complex polysaccharides. A majority of its genome is devoted to making enzymes that degrade polysaccharides. The diversity and number of genes for carbohydrate metabolism found in its genome far exceeds those found in any other bacterial species. *B. thetaiotaomicron* produces many enzymes that are not encoded by the human genome and thus it vastly increases the diversity of plant polymers that can be degraded in the human digestive tract.

Species of *Bacteroides* are unusual in that they are one of the few groups of bacteria to synthesize a special type of lipid called *sphingolipid* (**Figure 15.41**), a collection of lipids characterized by the long-chain amino alcohol sphingosine in place of glycerol in the lipid backbone. Sphingolipids such as sphingomyelin, cerebrosides, and gangliosides are common in mammalian tissues, especially in the brain and other nervous tissues, but rare in most bacteria. The production of sphingolipids can be found in a number of other genera in the phylum *Bacteroidetes* including *Flectobacillus*, *Prevotella*, *Porphyromonas*, and *Sphingobacterium*.

MINIQUIZ -

- What is the role of *Bacteroides thetaiotaomicron* in the human gut?

15.14 *Cytophagales, Flavobacteriales, and Sphingobacteriales*

Key Genera: *Cytophaga, Flavobacterium, Flexibacter*

Cytophagales

The order *Cytophagales* (Figure 15.40) contains almost exclusively obligate aerobes, though some species have limited fermentative capabilities. Cells are typically long, slender, gram-negative rods, often containing pointed ends, and move by gliding (**Figure 15.42**).

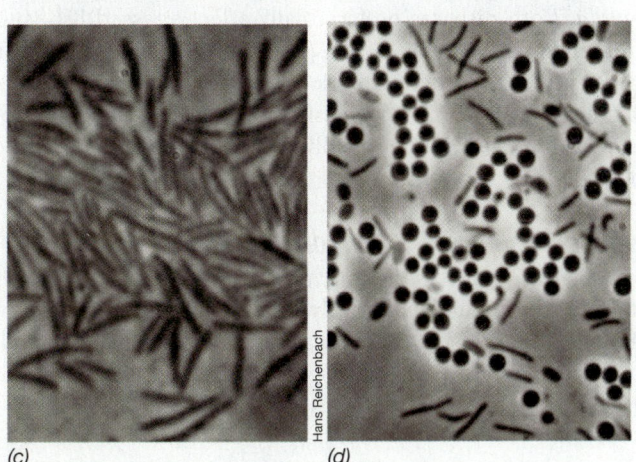

Figure 15.42 *Cytophaga* and *Sporocytophaga*. (a) Streak of an agarolytic marine *Cytophaga* hydrolyzing agar in a Petri dish. (b) Colonies of *Sporocytophaga* growing on cellulose. Note the clearing zones (arrows) where the cellulose has been degraded. (c) Phase-contrast micrograph of cells of *Cytophaga hutchinsonii* grown on cellulose filter paper (cells are about 1.5 μm in diameter). (d) Phase-contrast micrograph of the rod-shaped cells and spherical microcysts of *Sporocytophaga myxococcoides* (cells are about 0.5 μm and microcysts about 1.5 μm in diameter). Although *Sporocytophaga* microcysts are only slightly more heat-tolerant than vegetative cells, they are extremely resistant to desiccation and thus help the organism survive dry periods in soil.

Cytophagas specialize in the degradation of complex polysaccharides. They are widespread in oxic soils and freshwaters, where they probably account for much of the bacterial cellulose digestion.

Cellulose decomposers can easily be isolated by placing small crumbs of soil on pieces of cellulose filter paper laid on the surface of mineral salts agar. The bacteria attach to and digest the cellulose fibers, forming spreading colonies (Figure 15.42c).

Cellulose degradation by cytophagas can proceed by two different mechanisms. The typical mechanism is the free cellulase mechanism in which cells secrete extracellular enzymes called *exoenzymes* that degrade insoluble cellulose outside of the cell. A complex mixture of enzymes is secreted including *processive endocellulases*, which cleave *internal* β-1,4 glucosidic bonds, and *processive exocellulases*, which cleave *terminal* β-1,4 glucosidic bonds, releasing cellobiose. These *exoenzymes* degrade insoluble cellulose into soluble polysaccharides and disaccharides that can be readily assimilated by cells. *Cytophaga hutchinsonii* does not produce processive cellulases, and its degradation of cellulose likely requires physical contact of cellulose fibers with cellulase enzymes located on the outer surface of its cell wall.

The genus *Cytophaga* contains species that can degrade not only cellulose (Figure 15.42c) but also agar (Figure 15.42a) and chitin. In pure culture *Cytophaga* can be grown on agar containing embedded cellulose fibers (Figure 15.42b). The related genus *Sporocytophaga* is similar to *Cytophaga* in morphology and physiology, but the cells form resting spherical structures called *microcysts* (Figure 15.42d), similar to those produced by some fruiting myxobacteria (⟳ Section 14.19).

Several species of *Cytophaga* are fish pathogens and can cause serious problems in the cultivated fish industry. Two of the most important diseases are *columnaris disease*, caused by *Cytophaga columnaris*, and *cold-water disease*, caused by *Cytophaga psychrophila*. Both diseases preferentially affect stressed fish, such as those living in waters receiving pollutant discharges or living in high-density confinement situations such as fish hatcheries and aquaculture facilities. Infected fish show tissue destruction, frequently around the gills, probably from proteolytic activities of the *Cytophaga* pathogen.

Flavobacteriales and *Sphingobacteriales*

Flavobacteriales and *Sphingobacteriales* (Figure 15.40) typically contain aerobic and facultatively aerobic chemoorganotrophs. Like most *Bacteroidetes*, these organisms are gram-negative rods, and are saccharolytic with many species motile by gliding. Species are found widely in soils and in aquatic habitats, where they typically degrade complex polysaccharides.

Flavobacteriales can be particularly abundant in marine waters including aquatic systems in polar environments. *Flavobacterium* species are primarily found in aquatic habitats, both freshwater and marine, as well as in foods and food-processing plants. Most species are obligate aerobes, though some species are able to reduce nitrate in an anaerobic respiration. Flavobacteria frequently produce yellow pigments and are generally saccharophilic; most can also degrade starch and proteins. Flavobacteria are rarely pathogenic; however, one species, *Flavobacterium meningosepticum*, has been implicated in cases of infant meningitis, and several fish pathogens are also known.

Some *Flavobacteriales* are psychrophilic or psychrotolerant (⟳ Section 5.12). These include, in particular, the genera *Polaribacter* and *Psychroflexus*, organisms commonly isolated from cold environments, especially permanently cold environments such as polar waters and sea ice. Many related genera are also capable of good growth below 20°C and can thus be agents of food spoilage. None are pathogenic.

Sphingobacteriales are phenotypically similar to many *Flavobacteriales*. In terms of physiology, species of *Sphingobacteriales* are generally able to degrade a greater breadth of complex polysaccharides than are *Flavobacteriales*, and in this regard they resemble species of *Cytophagales*. The genus *Flexibacter* is typical of many genera of *Sphingobacteriales*. Species of *Flexibacter* differ from those of *Cytophaga* in that they usually require complex media for good growth and are unable to degrade cellulose. Cells of some *Flexibacter* species also undergo changes in cell morphology from long, gliding, threadlike filaments lacking cross-walls to short, nonmotile rods. Many flexibacteria are pigmented due to carotenoids located in their cytoplasmic membrane, or from related pigments called *flexirubins*, located in the cell's outer membrane. *Flexibacter* species are common in soil and freshwaters where they degrade polysaccharides, and none have been identified as pathogens.

MINIQUIZ --

- Describe a method for isolating *Cytophaga* species from nature.
- What characteristics are shared between the genera *Cytophaga* and *Bacteroides*, and in what ways do they differ?

IV • *Chlamydiae*, *Planctomycetes*, and *Verrucomicrobia*

The phyla *Chlamydiae*, *Planctomycetes*, and *Verrucomicrobia* share an ancestor and are more closely related to each other than to other bacterial phyla (**Figure 15.43**). These three groups contain organisms that can be found in a variety of habitats including soils, aquatic systems, and in association with eukaryotic hosts. We first consider the chlamydia, a group of small gram-negative bacteria that cause some serious human and animal diseases.

15.15 *Chlamydiae*

Key Genera: *Chlamydia, Chlamydophila, Parachlamydia*

The phylum *Chlamydiae* contains a single order, the *Chlamydiales*. The entire phylum consists of obligate intracellular parasites of eukaryotes. Though the species that are human pathogens have been characterized in most detail, the phylum contains diverse species that interact with a wide variety of eukaryotic hosts.

Planctomycetes

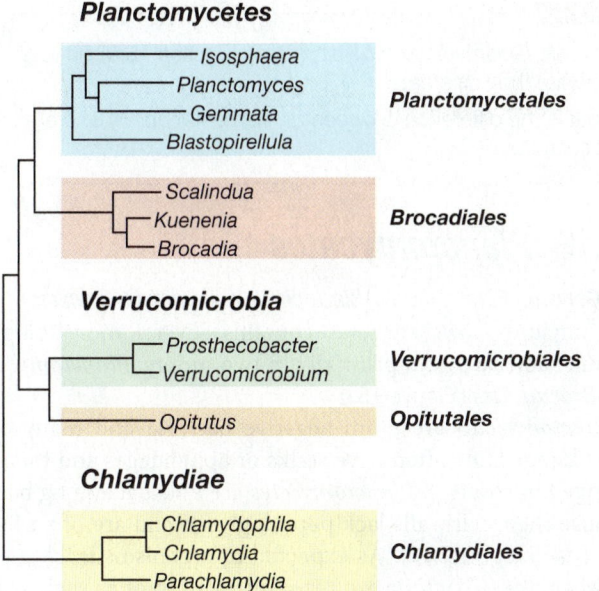

Figure 15.43 **Major orders of *Chlamydiae*, *Planctomycetes*, and *Verrucomicrobia*.** The phylogenetic tree was constructed from 16S rRNA gene sequences of representative genera of *Chlamydiae*, *Planctomycetes*, and *Verrucomicrobia*. Order names are shown in bold.

Species are typically very small cocci, approximately 0.5 μm in diameter, and display a distinctive developmental cycle. Like many obligate parasites and symbionts, the genomes of *Chlamydiae* are typically reduced, ranging in size from 550–1000 kilobase pairs (ↄↄ Section 6.4).

Life Cycle of *Chlamydiae*

All species of *Chlamydiae* demonstrate a unique chlamydial life cycle (**Figure 15.44**). Two types of cells are seen in the life cycle: (1) a small, dense cell, called an *elementary body*, which is relatively resistant to drying and is the means of dispersal, and (2) a larger, less dense cell, called a *reticulate body*, which divides by binary fission and is the vegetative form.

Elementary bodies are nonmultiplying cells specialized for infectious transmission. By contrast, reticulate bodies are noninfectious forms that function only to multiply inside host cells to form a large inoculum for transmission. Unlike the rickettsias, the chlamydias are not transmitted by arthropods but are primarily airborne invaders of the respiratory system—hence the significance of resistance to drying of the elementary bodies. A dividing reticulate body can be seen in **Figure 15.45**. After a number of cell divisions, these vegetative cells are converted into elementary bodies that are released when the host cell disintegrates (Figure 15.44*b*) and can then infect other nearby host cells. Generation times of 2–3 h have been measured for reticulate bodies, considerably faster than times found for the rickettsias (Section 15.1).

Notable Genera of *Chlamydiae*

Chlamydiae are particularly well adapted to invading and colonizing eukaryotic cells, and different species can infect a diverse array of eukaryotic hosts. The species *Parachlamydia acanthamoebae* infects free-living amoebae, particularly amoebae in the genus *Acanthamoeba*. *Parachlamydia* demonstrates the typical chlamydial life cycle during infection of amoebae (Figure 15.44). Most species of *Chlamydiae* can multiply or survive within free-living

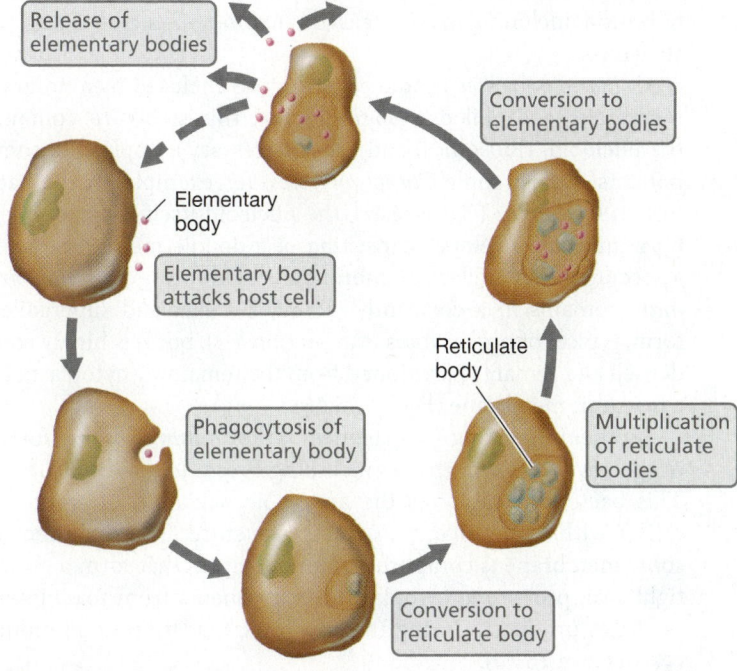

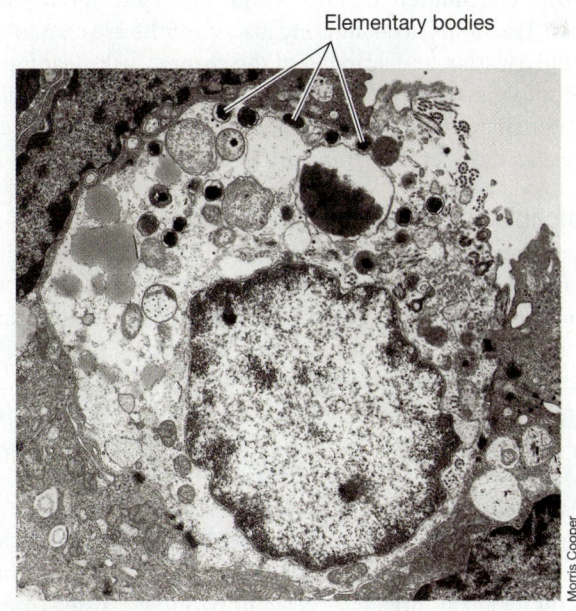

(a) (b)

Figure 15.44 **The infection cycle of a chlamydia.** *(a)* Schematic diagram of the cycle: The entire cycle takes about 48 h. *(b)* Human chlamydial infection. Elementary bodies (∼0.3 μm in diameter) are the infectious form and reticulate bodies (∼1 μm in diameter) are the multiplying form. An infected fallopian tube cell is bursting, releasing mature elementary bodies.

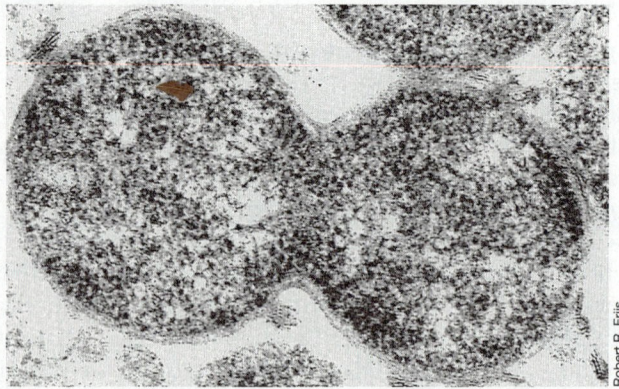

Figure 15.45 *Chlamydia.* Thin-section electron micrograph of a dividing reticulate body of *Chlamydophila psittaci* within a mouse tissue-culture cell. A single chlamydial cell is about 1 μm in diameter.

amoebae, and these hosts may be important for the survival and dispersal of *Chlamydiae* in nature. A diversity of 16S rRNA gene sequences from *Chlamydiae* can be detected in natural environments, suggesting that these organisms are widespread and that many of their natural hosts have yet to be identified. While free-living amoebae are the natural hosts for *P. acanthamoebae*, this species can also infect humans, although only weakly compared with *Chlamydiae* whose natural hosts are human.

The best-studied human pathogens are found in the genera *Chlamydia* and *Chlamydophila*. Several species are recognized within these genera: *Chlamydophila psittaci*, the causative agent of the disease psittacosis; *Chlamydia trachomatis*, the causative agent of trachoma and a variety of other human diseases; and *Chlamydophila pneumoniae*, the cause of some respiratory syndromes. Psittacosis is an epidemic disease of birds that is occasionally transmitted to humans and causes pneumonia-like symptoms. Trachoma, a debilitating disease of the eye characterized by vascularization and scarring of the cornea, is the leading cause of blindness in humans. Other strains of *C. trachomatis* infect the genitourinary tract, and chlamydial infections are currently one of the leading sexually transmitted diseases (ↄ Section 29.13).

Molecular and Metabolic Properties

The chlamydias are among the most biochemically limited of all known *Bacteria*. Indeed, their genomes, approximately 1 Mbp in size, appear to be even more biosynthetically limited than those of the rickettsias, the other group of obligate intracellular parasites known among the *Bacteria* (Section 15.1). Interestingly, the *C. trachomatis* genome lacks a gene encoding the protein FtsZ, a key protein in septum formation during cell division (ↄ Section 5.2) and thought to be indispensable for growth of all prokaryotes. The cell wall of *C. trachomatis* also appears to lack peptidoglycan, although genes for peptidoglycan biosynthesis are present in its genome. Interestingly, some genes in *C. trachomatis* are distinctly eukaryotic, indicating horizontal transfer from host to bacterium; these genes may encode functions that facilitate the pathogenic lifestyle of *C. trachomatis* (ↄ Section 29.13). In sum, the chlamydias appear to have evolved with an efficient and effective survival strategy including parasitizing the resources of the host and producing resistant cell forms for transmission.

MINIQUIZ
- How are *Chlamydia* and *Mycoplasma* (Section 15.9) similar? How are they different?
- What is the difference between an elementary body and a reticulate body?

15.16 *Planctomycetes*

Key Genera: *Planctomyces, Blastopirellula, Gemmata, Brocadia*

The phylum *Planctomycetes* contains several morphologically unique bacteria found primarily in two orders, *Planctomycetales* and *Brocadiales* (Figure 15.43).

Planctomycetes are gram-negative bacteria and many divide by budding. They often have stalks or appendages and their cells arranged in rosettes. *Planctomycetes* are unusual among bacteria because their cell walls lack peptidoglycan and are of an S-layer type (ↄ Section 2.12). As expected of organisms lacking peptidoglycan, these bacteria are resistant to antibiotics such as penicillin and cephalosporin that disrupt peptidoglycan synthesis. Another remarkable feature of *Planctomycetes* is that they often contain intracellular compartments that resemble the organelles of eukaryotes.

Compartmentalization in *Planctomycetes*

We learned in Section 1.2 of the major structural differences between prokaryotic and eukaryotic cells. In particular, eukaryotes have a membrane-enclosed nucleus whereas in prokaryotes, DNA supercoils and compacts to form the nucleoid present in the cytoplasm. However, *Planctomycetes* are unique among all known prokaryotes in that they show extensive cell compartmentalization, including in some cases a membrane-enclosed nuclear structure.

All *Planctomycetes* produce a structure enclosed by a nonunit membrane and called a *pirellulosome*; this structure contains the nucleoid, ribosomes, and other necessary cytoplasmic components. But in some *Planctomycetes*, for example, in the bacterium *Gemmata* (**Figure 15.46**), the nucleoid itself is surrounded by a "nuclear envelope" consisting of a double membrane layer as occurs in the nuclear membrane of eukaryotes. DNA in *Gemmata* remains in a covalently closed, circular, and supercoiled form, typical of prokaryotes (ↄ Section 4.3), but it is highly condensed and remains partitioned from the remaining cytoplasm by a true unit membrane (Figure 15.46).

Another interesting compartment is the *anammoxosome*, found in species of the *Brocadiales* including *Brocadia anammoxidans*. This bacterium catalyzes the anaerobic oxidation of ammonia (NH₃) within the anammoxosome structure. The anammoxosome membrane is composed of unique lipids that form a rather tight seal, protecting cytoplasmic components from toxic intermediates produced during the anaerobic oxidation of ammonia (ↄ Section 13.10).

Planctomyces

Planctomyces is the best-characterized genus in the *Planctomycetes*. In Section 14.21 we considered the stalked proteobacterium *Caulobacter*. *Planctomyces* is also a stalked bacterium (**Figure 15.47**).

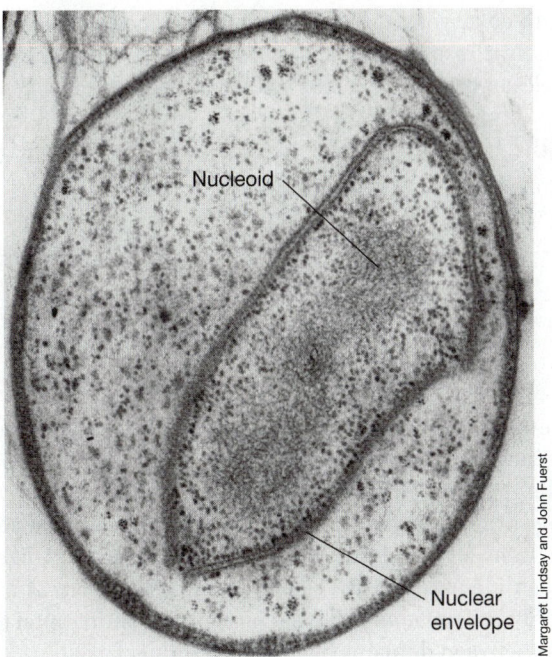

Figure 15.46 *Gemmata*: **a nucleated bacterium.** Thin-section electron micrograph of a cell of *Gemmata obscuriglobus* showing the nucleoid surrounded by a nuclear envelope. The cell is about 1.5 μm in diameter.

However, unlike *Caulobacter*, the stalk of *Planctomyces* consists of protein and does not contain a cell wall or cytoplasm (compare Figure 15.47 with Figure 14.57). The *Planctomyces* stalk presumably functions in attachment, but it is a much narrower and finer structure than the prosthecal stalk of *Caulobacter*.

Like *Caulobacter* (Figures 7.26 and 14.57), *Planctomyces* is a budding bacterium with a life cycle. Its motile swarmer cells attach to a surface, grow a stalk from the attachment point, and generate a new cell from the opposite pole by budding. This daughter cell produces a flagellum, breaks away from the attached mother cell, and begins the cycle anew. Physiologically, *Planctomyces* species are facultatively aerobic chemoorganotrophs, growing either by fermentation or respiration of sugars.

The habitat of *Planctomyces* is primarily aquatic, both freshwater and marine, and the genus *Isosphaera* is a filamentous, gliding hot spring bacterium. The isolation of *Planctomyces* and relatives, like that of *Caulobacter*, requires dilute media. In addition, because all known members of this group lack peptidoglycan, enrichments can be made even more selective by the addition of penicillin.

MINIQUIZ --

- How does the stalk of *Planctomyces* differ from the stalk of *Caulobacter*?

- What is unusual about the bacterium *Gemmata*?

15.17 *Verrucomicrobia*

Key Genera: *Verrucomicrobium, Prosthecobacter*

The phylum *Verrucomicrobia* contains at least four orders with characterized species, but most are found within the order *Verrucomicrobiales* (Figure 15.43). Species of *Verrucomicrobia* are aerobic or facultatively aerobic bacteria capable of fermenting sugars. An exception is the genus *Methylacidiphilum*, which contains aerobic methanotrophs (Section 14.17). In addition, some *Verrucomicrobia* form symbiotic associations with protists. *Verrucomicrobia* are widespread in nature, inhabiting freshwater and marine environments as well as forest and agricultural soils. The *Verrucomicrobia* can have membrane-bound intracellular structures similar to those found in the *Planctomycetes*. The *Verrucomicrobia* typically form cytoplasmic appendages called *prosthecae* (Section 14.21). *Verrucomicrobia* share with other prosthecate bacteria the presence of peptidoglycan in their cell walls and in this way are clearly distinct from *Planctomycetes*.

The genera *Verrucomicrobium* and *Prosthecobacter* produce two to several prosthecae per cell (**Figure 15.48**). Unlike cells of *Caulobacter* (Figures 7.26 and 14.57), which contain a single prostheca and produce flagellated and nonprosthecate swarmer cells, *Verrucomicrobium* and *Prosthecobacter* divide symmetrically, and both mother and daughter cells contain prosthecae at the time of cell division. The genus name *Verrucomicrobium* derives from Greek roots meaning "warty," which is an appropriate description

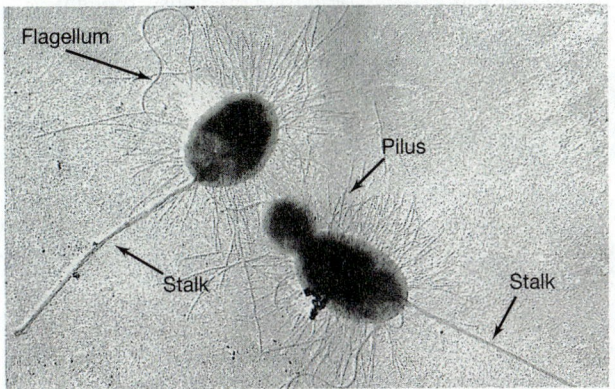

Figure 15.47 *Planctomyces maris.* Metal-shadowed transmission electron micrograph. A single cell is about 1–1.5 μm long. Note the fibrillar nature of the stalk. Pili are also abundant. Note also the flagella (curly appendages) on each cell and the bud that is developing from the nonstalked pole of one cell.

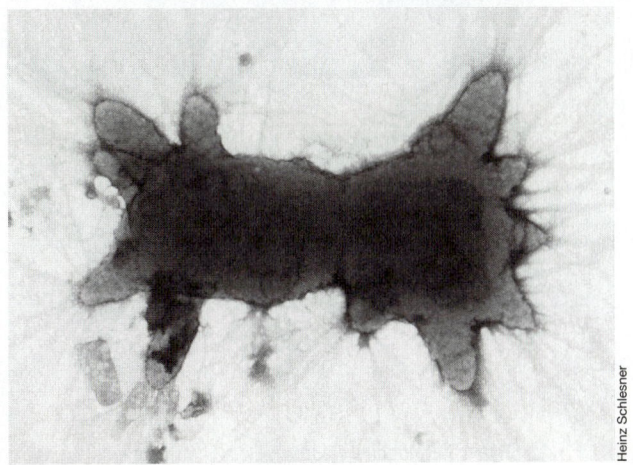

Figure 15.48 *Verrucomicrobium spinosum.* Negatively stained transmission electron micrograph. Note the wartlike prosthecae. A cell is about 1 μm in diameter.

of cells of *Verrucomicrobium spinosum* with their multiple projecting prosthecae (Figure 15.48).

Species of the genus *Prosthecobacter* contain two genes that show significant homology to the genes that encode tubulin in eukaryotic cells. Tubulin is the key protein that makes up the cytoskeleton of eukaryotic cells (↩ Section 2.22). Although the important cell division protein FtsZ (↩ Section 5.2) is also a tubulin homolog, the *Prosthecobacter* proteins are structurally more similar to eukaryotic tubulin than is FtsZ. The role of the tubulin proteins in *Prosthecobacter* is unknown since a eukaryotic-like cytoskeleton has not been observed in these organisms.

MINIQUIZ -
- Describe two ways that *Verrucomicrobia* differ from *Planctomycetes*.

V • Hyperthermophilic *Bacteria*

Three phyla of hyperthermophilic bacteria cluster deep in the phylogenetic tree of *Bacteria*, near the root (Figure 15.1). Each group consists of one or two major genera, and a key physiological feature of most species is hyperthermophily—optimal growth at temperatures above 80°C (↩ Section 5.13). We begin with *Thermotoga* and *Thermodesulfobacterium*, each representative of its own lineage.

15.18 *Thermotogae* and *Thermodesulfobacteria*

Key Genera: *Thermotoga, Thermodesulfobacterium*

Thermotoga species are rod-shaped hyperthermophiles that form a sheathlike envelope (called a *toga*; thus the genus name) (**Figure 15.49a**), stain gram-negatively, and are nonsporulating. *Thermotoga* species are fermentative anaerobes, catabolizing sugars or starch and producing lactate, acetate, CO_2, and H_2 as fermentation products. The organisms can also grow by anaerobic respiration using H_2 as an electron donor and ferric iron as an electron acceptor. Species of *Thermotoga* have been isolated from terrestrial hot springs as well as marine hydrothermal vents.

Despite being bacterial, the genome of *Thermotoga* contains many genes that show strong homology to genes from hyperthermophilic *Archaea*. In fact, over 20% of the genes of *Thermotoga* probably originated from *Archaea* by horizontal gene transfers (↩ Sections 6.12 and 12.5). Although a few archaea-like genes have been identified in the genomes of other *Bacteria* and vice versa, only in *Thermotoga* has such large-scale horizontal transfer of genes between domains been detected thus far.

Thermodesulfobacterium (**Figure 15.50**) is a thermophilic sulfate-reducing bacterium, positioned on the phylogenetic tree in a separate phylum between *Thermotoga* and *Aquifex* (Figure 15.1). *Thermodesulfobacterium* is a strict anaerobe that uses compounds such as lactate, pyruvate, and ethanol (but not acetate) as electron donors, as do sulfate-reducing bacteria such as *Desulfovibrio* (↩ Section 14.9), reducing SO_4^{2-} to H_2S.

An unusual biochemical feature of *Thermodesulfobacterium* is the production of *ether-linked lipids*. Recall that such lipids are a hallmark of the *Archaea* and that a polyisoprenoid C_{20} hydrocarbon (phytanyl) replaces fatty acids as the side chains in archaeal lipids (↩ Section 2.7). However, the ether-linked lipids in *Thermodesulfobacterium* are unusual because the glycerol side chains are not phytanyl groups, as they are in *Archaea*, but instead are composed of a unique C_{17} hydrocarbon along with some fatty acids (Figure 15.50b). Thus we see in *Thermodesulfobacterium* both a deep phylogenetic lineage (Figure 15.1) and a lipid profile that combines features of both the *Archaea* and the *Bacteria*. However, a few other *Bacteria* have also been found to contain ether-linked lipids, and thus these lipids may be more common among *Bacteria* than previously thought.

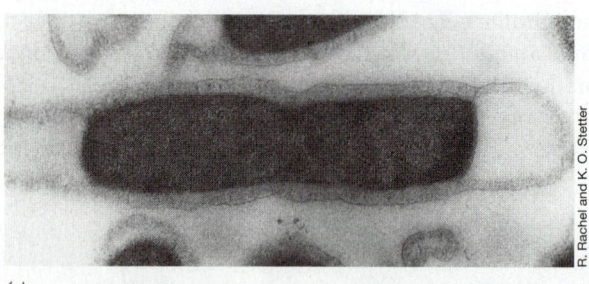

(a)

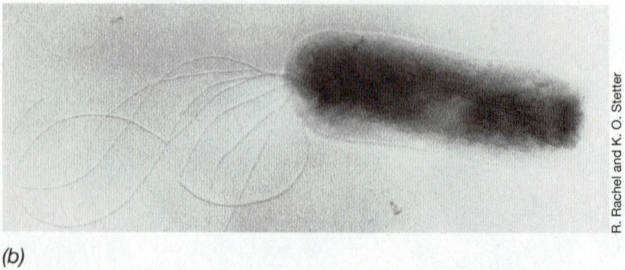

(b)

Figure 15.49 Hyperthermophilic *Bacteria*. Electron micrographs of two hyperthermophiles: (a) *Thermotoga maritima*—temperature optimum, 80°C. Note the outer covering, the toga. (b) *Aquifex pyrophilus*—temperature optimum, 85°C. Cells of *Thermotoga* measure 0.6 × 3.5 μm; cells of *Aquifex* measure 0.5 × 2.5 μm.

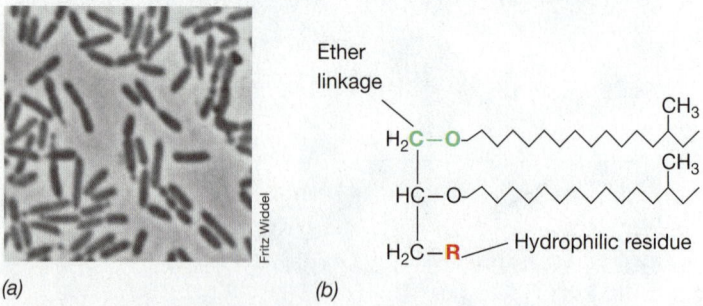

(a) (b)

Figure 15.50 *Thermodesulfobacterium*. (a) Phase-contrast micrograph of cells of *Thermodesulfobacterium thermophilum*. (b) Structure of one of the lipids of *Thermodesulfobacterium mobile*. Note that although the two hydrophobic side chains are ether-linked, they are not phytanyl units, as in *Archaea*. The designation "R" is for a hydrophilic residue, such as a phosphate group.

MINIQUIZ

• What is unique about the genome of *Thermotoga* and the lipids of *Thermodesulfobacterium*?

15.19 *Aquificae*

Key Genera: *Aquifex, Thermocrinis*

The genus *Aquifex* (Figure 15.49*b*) is an obligately chemolithotrophic and autotrophic hyperthermophile and is the most thermophilic of all known *Bacteria*. Various *Aquifex* species utilize H_2, sulfur (S^0), or thiosulfate ($S_2O_3^{2-}$) as electron donors and O_2 or nitrate (NO_3^-) as electron acceptors, and grow at temperatures up to 95°C. *Aquifex* can tolerate only very low O_2 concentrations (microaerophilic), and is unable to oxidize all tested organic compounds. *Hydrogenobacter*, a relative of *Aquifex*, shows most of the same properties as *Aquifex*, but is an obligate aerobe.

Aquifex and Autotrophy

Autotrophy in *Aquifex* occurs by way of the reverse citric acid cycle, a series of reactions previously detected only in green sulfur bacteria (⟳ Sections 13.3 and 14.6) within the domain *Bacteria*. The complete genome sequence of *Aquifex aeolicus* has been determined, and its entirely chemolithotrophic and autotrophic lifestyle is encoded by a very small genome of only 1.55 megabase pairs (one-third the size of the *Escherichia coli* genome). The discovery that so many hyperthermophilic species of *Archaea* and *Bacteria*, like *Aquifex*, are H_2 chemolithotrophs, coupled with the finding that they branch as very early lineages on their respective phylogenetic trees (Figure 15.1), suggests that H_2 was a key electron donor for energy metabolism in primitive organisms that appeared on early Earth (⟳ Sections 12.1 and 16.14).

Thermocrinis

Thermocrinis (**Figure 15.51**) is a relative of *Aquifex* and *Hydrogenobacter*. This bacterium grows optimally at 80°C as a chemolithotroph oxidizing H_2, $S_2O_3^{2-}$, or S^0 as electron donors, with O_2 as electron acceptor. *Thermocrinis ruber*, the only known species, grows in the outflow of certain hot springs in Yellowstone National Park (Figure 15.51*a*) where it forms pink "streamers" consisting of a filamentous form of the cells attached to siliceous sinter (Figure 15.51*b*). In static culture, cells of *T. ruber* grow as individual rod-shaped cells (Figure 15.51*c*). However, when cultured in a flowing system in which growth medium is trickled over a solid glass surface to which cells can attach, *Thermocrinis* assumes the streamer morphology it forms in its constantly flowing habitat in nature.

T. ruber is of historical significance in microbiology because it was one of the organisms discovered in the 1960s by Thomas Brock, a pioneer in the field of thermal microbiology. The discovery by Brock that the pink streamers (Figure 15.51*b*) contained protein and nucleic acids clearly indicated that they were living organisms and not just mineral debris. Moreover, the presence of streamers in 80–90°C hot spring outflow waters but not those of lower temperatures supported Brock's hypothesis that these organisms actually *required* heat for growth and were therefore likely to be present in even boiling or superheated

Figure 15.51 *Thermocrinis.* *(a)* Octopus Spring, Yellowstone National Park. The source water of this alkaline and siliceous hot spring is 92°C. *(b)* Cells of *Thermocrinis ruber* growing as filamentous streamers (arrow) attached to siliceous sinter in the outflow (85°C) of Octopus Spring. *(c)* Scanning electron micrograph of rod-shaped cells of *T. ruber* grown on a silicon-coated cover glass. A single cell of *T. ruber* is about 0.4 μm in diameter and from 1 to 3 μm long.

waters. Both of these conclusions were subsequently supported by the discovery by Brock and other microbiologists of dozens of genera of hyperthermophilic prokaryotes inhabiting hot springs, hydrothermal vents, and other thermal environments. More coverage of hyperthermophiles can be found in Sections 5.11, 5.13, and in Chapter 16.

MINIQUIZ

• Of what evolutionary significance is the fact that organisms in the *Aquifex* lineage are both hyperthermophilic and H_2 chemolithotrophs?

VI • Other *Bacteria*

Thus far in this chapter we have focused on phyla that have many described species (Figure 15.1). Beyond these mainstream bacterial phyla are many others that have but one or at most a handful of characterized species (Figure 15.1*b*). In addition, many more phyla are known only from community sampling of 16S rRNA genes from nature (↩ Section 18.5). We cannot cover them all. So in this final unit of the chapter we consider one phylum that has been well studied and then summarize some other phyla that are emerging into the mainstream of microbial diversity.

15.20 *Deinococcus–Thermus*

Key Genera: *Deinococcus, Thermus*

The *deinococci* group contains only a few characterized genera in two orders, the *Deinococcales* and the *Thermales*. Members of this phylum are typically aerobic chemoorganotrophs that metabolize sugars, amino and organic acids, or various complex mixtures. Though deinococci stain gram-positively, they have a gram-negative cell wall structure (**Figure 15.52**) made up of several layers, including an outer membrane, which is characteristic of gram-negative bacteria (↩ Section 2.11). However, unlike the outer membrane of bacteria such as *Escherichia coli*, the outer membrane of deinococci lack lipid A. Deinococci also contain an unusual form of peptidoglycan in which ornithine replaces diaminopimelic acid in the *N*-acetylmuramic acid cross-links (↩ Section 2.10).

Species of *Thermales* are typically thermophiles or hyperthermophiles and the type genus is *Thermus*. *Thermus aquaticus*, discovered in a Yellowstone National Park hot spring in the mid-1960s by Thomas Brock, has been a model organism for studying life at high temperatures. *T. aquaticus* has subsequently been isolated from many geothermal systems, and is the source of *Taq* DNA polymerase. Because it is so heat-stable, *Taq* polymerase allowed the polymerase chain reaction (PCR) technique for amplifying DNA to be fully automated (↩ Section 11.3), an advance that has revolutionized all of biology.

Radiation Resistance of *Deinococcus radiodurans*

Species of *Deinococcales* have the unusual property of being extremely radiation resistant, and *Deinococcus radiodurans* is the best-studied species in this regard. Most deinococci are red or pink due to carotenoids, and many are highly resistant to both radiation and desiccation. Resistance to ultraviolet (UV) radiation can be used to advantage in isolating deinococci. These remarkable organisms can be selectively isolated from soil, ground meat, dust, and filtered air following exposure of the sample to intense UV (or even gamma) radiation and plating on a rich medium containing tryptone and yeast extract. For example, *D. radiodurans* cells survive exposure to 15,000 grays (Gy) of ionizing radiation (1 Gy = 100 rad). This is sufficient to shatter the organism's chromosome into hundreds of fragments (by contrast, a human can be killed by exposure to less than 10 Gy).

In addition to impressive radiation resistance, *D. radiodurans* is resistant to the mutagenic effects of many mutagenic agents. The only chemical mutagens that seem to work on *D. radiodurans* are agents such as nitrosoguanidine, which induces deletions in DNA. Deletions are apparently not repaired as efficiently as point mutations in this organism, and mutants of *D. radiodurans* can be isolated in this way.

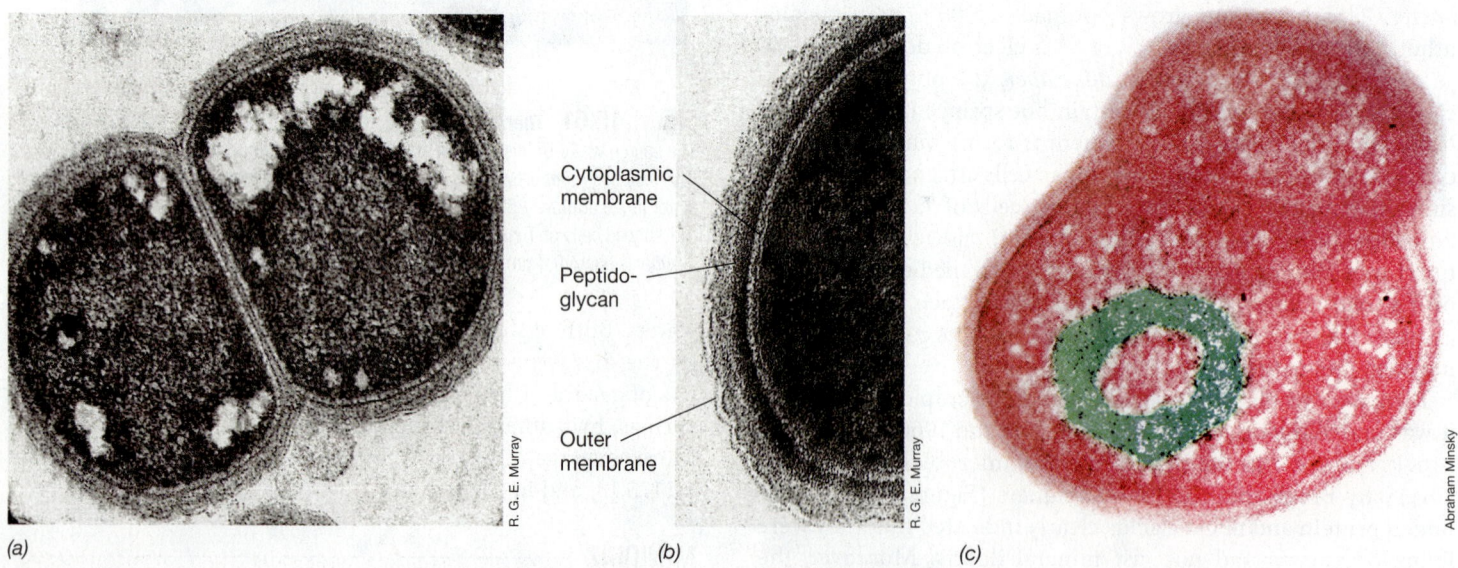

(a) *(b)* *(c)*

Figure 15.52 **The radiation-resistant coccus *Deinococcus radiodurans*.** An individual cell is about 2.5 μm in diameter. *(a)* Transmission electron micrograph of *D. radiodurans*. Note the outer membrane layer. *(b)* High-magnification micrograph of wall layer. *(c)* Transmission electron micrograph of cells of *D. radiodurans* colored to show the toroidal morphology of the nucleoid (green).

Cytoplasmic membrane

Peptidoglycan

Outer membrane

DNA Repair in *Deinococcus radiodurans*

Studies of *D. radiodurans* have shown that it is highly efficient in repairing damaged DNA. Several different DNA repair enzymes exist in *D. radiodurans*. In addition to the DNA repair enzyme RecA (⮂ Section 10.4), several RecA-independent DNA systems exist in *D. radiodurans* that can repair breaks in single- or double-stranded DNA, and excise and repair misincorporated bases. In fact, repair processes are so effective that the chromosome can even be reassembled from a fragmented state.

It is also thought that the unique arrangement of DNA in *D. radiodurans* cells plays a role in radiation resistance. Cells of *D. radiodurans* always exist as pairs or tetrads (Figure 15.52*a*). Instead of scattering DNA within the cell as in a typical nucleoid, DNA in *D. radiodurans* is ordered into a toroidal (coiled, or stack of rings) structure (Figure 15.52*c*). Repair is then facilitated by the fusion of nucleoids from adjacent compartments, because their toroidal structure provides a platform for homologous recombination. From this extensive recombination, a single repaired chromosome emerges, and the cell containing this chromosome can then grow and divide.

MINIQUIZ -

- Describe a commercial application of *Thermus aquaticus*.
- Describe an unusual biological feature of *Deinococcus radiodurans*.

15.21 Other Notable Phyla of *Bacteria*

The basic properties of seven other phyla of *Bacteria* are discussed briefly below. Although most of these have few cultured representatives (Figure 15.1*b*), many may well be of considerable ecological importance. If so, future research on their culture and ecological activities will supply the necessary proof. Until then, we paint a picture of these phyla with a broad brush to summarize their major characteristics in a general way.

Acidobacteria

Acidobacteria are widespread in the environment as revealed by analyses of 16S rRNA genes retrieved from environmental samples (Figure 15.1*b*). *Acidobacteria* are abundant in soils, particularly acid soils (pH < 6.0) where they often comprise a majority of some soil communities. *Acidobacteria* also inhabit freshwater, hot spring microbial mats, wastewater treatment reactors, and sewage sludge. There is evidence for as many as 25 major subgroups within the *Acidobacteria*, indicating substantial phylogenetic and metabolic diversity of the species in this phylum. Their abundance, widespread distribution, and likely metabolic diversity indicate they play important ecological roles, especially in soil. Unfortunately, while *Acidobacteria* are widespread in the environment, they have proven difficult to cultivate; as a result, few species have been isolated (Figure 15.1*b*) and only a handful of genera have been described.

The few species of *Acidobacteria* that have been characterized are metabolically diverse, including both chemoorganotrophs and photoheterotrophs as well as obligate aerobes and obligately fermentative anaerobes. Three species of *Acidobacteria* have been well characterized, *Acidobacterium capsulatum*, *Geothrix fermentans*, and *Holophaga foetida*; all are gram-negative chemoorganotrophs. *A. capsulatum* is an acidophilic, encapsulated, obligately aerobic bacterium isolated from acid mine drainage; it utilizes various sugars and organic acids. *G. fermentans*, a strict anaerobe, oxidizes simple organic acids (acetate, propionate, lactate, fumarate) to CO_2 coupled to the reduction of ferric iron as electron acceptor (dissimilative iron reduction, ⮂ Section 14.14), and can also ferment citrate to acetate plus succinate as products. *H. foetida* is a strictly anaerobic homoacetogen (⮂ Sections 13.19 and 14.18) that grows by degrading methylated aromatic compounds to acetate. Some *Acidobacteria* degrade polymers such as cellulose and chitin, and at least one genus, *Chloracidobacterium*, is phototrophic (⮂ Section 14.8).

Nitrospirae, Deferribacteres, and Chrysiogenetes

The phylum *Nitrospirae* is named for the genus *Nitrospira*, a chemolithotroph that oxidizes nitrite to nitrate and grows autotrophically (⮂ Section 14.13), as do species of the proteobacterium *Nitrobacter* (⮂ Section 13.10). *Nitrospira* inhabits many of the same environments as *Nitrobacter*. However, environmental surveys have shown that *Nitrospira* is much more abundant than *Nitrobacter* in nature, and thus most of the nitrite oxidized in nitrogen-rich environments such as wastewater treatment plants and ammonia-rich soils is probably due to *Nitrospira*. Other key *Nitrospirae* include *Leptospirillum*, an aerobic, acidophilic, iron-oxidizing chemolithotroph (⮂ Section 14.15) common in acid mine drainage associated with the mining of coal and iron (⮂ Section 21.1).

The phyla *Deferribacteres* and *Chrysiogenetes* (Figure 15.1) contain anaerobic chemoorganotrophs that display considerable metabolic diversity with respect to the electron acceptors used in anaerobic respirations (Chapter 13). Most, though not all, species are able to grow through anaerobic respiration of nitrate to nitrite or ammonium. The *Deferribacteres* group is named for the genus *Deferribacter*, a thermophilic dissimilative ferric iron-reducer (⮂ Sections 13.21 and 14.14) that can also reduce nitrate and metal oxides. *Geovibrio* is a related genus that can also grow using elemental sulfur (⮂ Section 14.10) as an electron acceptor. The bacterium *Chrysiogenes arsenatis* and its relatives are notable for the ability to couple the oxidation of acetate and a few other organic compounds to the reduction of arsenate as a terminal electron acceptor, reducing it to arsenite. In addition to arsenate, many species of *Chrysiogenetes* can reduce selenate, nitrite, nitrate, thiosulfate, and elemental sulfur in anaerobic respirations (⮂ Section 13.21).

Synergistetes, Fusobacteria, Fibrobacteres

The phyla *Synergistetes*, *Fusobacteria*, and *Fibrobacteres* contain relatively few characterized species (Figure 15.1*b*), but those that have been cultured employ fermentative metabolisms. Species in these groups are often associated with the gastrointestinal tracts of animals and some have been associated with human disease.

Synergistetes are gram-negative nonsporulating rods found in association with animals and in anoxic environments in terrestrial and marine systems. Described species are typically obligate anaerobes that degrade proteins and are capable of fermenting amino acids. In animals they are most often found in the gastrointestinal tract; for example, *Synergistes jonesii* inhabits the rumen. In humans, species of *Synergistetes* have been associated with certain soft tissue wounds and abscesses, dental plaque, and periodontal conditions.

Fusobacteria are gram-negative nonsporulating rods found in sediments and the gastrointestinal systems and oral cavities of animals. *Fusobacteria* are obligate anaerobes that ferment carbohydrates, peptides, and amino acids. Species of the genus *Fusobacterium* are common components of the human microbiome where they colonize mucous membranes. Different species can be found in the oral cavity, the gastrointestinal tract, and the vagina. *Fusobacterium nucleatum* is often found in gingival crevices in the human oral cavity. Some fusobacteria may be human pathogens and *F. nucleatum* is often present in patients suffering from periodontal disease.

While 16S ribosomal RNA genes from *Fibrobacteres* can be recovered from a wide range of habitats, the only characterized species have come from either the rumen or gastrointestinal tracts of animals. The genus *Fibrobacter* contains gram-negative fermentative strict anaerobes. However, unlike most *Fusobacteria* and *Synergistetes*, species of *Fibrobacter* are unable to ferment proteins or amino acids and specialize instead in the fermentation of carbohydrates, including cellulose. In the rumen, cellulose is the major source of energy, and in such environments it supports not only cellulolytic bacteria such as *Fibrobacter* but many noncellulolytic anaerobes that use glucose released during cellulose degradation.

MINIQUIZ

- What is a major habitat for many species of *Acidobacteria*?
- How do *Nitrospira* and *Deferribacter* differ in terms of lifestyle and metabolism?
- What metabolic characteristics are shared by most *Synergistetes*, *Fusobacteria*, and *Fibrobacteres*, and what disease in humans has been correlated with the presence of *Synergistetes* and *Fusobacteria*?

BIG IDEAS

15.1 • The *Alphaproteobacteria* are the second largest class of *Proteobacteria* and metabolically diverse. Key genera are: *Rhizobium*, *Rickettsia*, *Rhodobacter*, and *Caulobacter.*

15.2 • The *Betaproteobacteria* are the third largest class of *Proteobacteria* and metabolically diverse. Key genera are *Burkholderia*, *Rhodocyclus*, *Neisseria*, and *Nitrosomonas.*

15.3 • The *Gammaproteobacteria* are the largest and most diverse class of *Proteobacteria* and contain many human pathogens. The *Enterobacteriales*, or enteric bacteria, are the most heavily studied of all bacteria. Key genera are *Escherichia* and *Salmonella.*

15.4 • The *Pseudomonadales* and *Vibrionales* are among the most common *Gammaproteobacteria*. Key genera are *Pseudomonas* and *Vibrio.*

15.5 • The *Deltaproteobacteria* and *Epsilonproteobacteria* are smaller and less metabolically diverse classes of *Proteobacteria*. Key genera of *Deltaproteobacteria* are *Myxococcus*, *Desulfovibrio*, and *Geobacter*. Key genera of *Epsilonproteobacteria* are *Campylobacter* and *Helicobacter.*

15.6 • Lactic acid bacteria such as *Lactobacillus* and *Streptococcus* produce lactate as the primary end product of fermentation, and they have many roles in food production and preservation. The *Firmicutes* are one of the two main phyla of gram-positive bacteria.

15.7 • Many genera of *Firmicutes* in the orders *Bacillales* and *Clostridiales*, including *Staphylococcus*, *Listeria*, and *Sarcina*, are unable to form endospores.

15.8 • Production of endospores is a hallmark of the key genera *Bacillus* and *Clostridium* and is only found in the phylum *Firmicutes.*

15.9 • The phylum *Tenericutes* contains the mycoplasmas, organisms that lack cell walls and have very small genomes. Many species are pathogenic for humans, other animals, and plants. The key genus is *Mycoplasma.*

15.10 • *Actinobacteria* are the second major phylum of gram-positive bacteria. *Corynebacterium* and *Arthrobacter* are common gram-positive soil bacteria. *Propionibacterium* ferments lactate to propionate and is the key agent responsible for the unique flavor and texture of Swiss cheese.

15.11 • Species of *Actinobacteria* in the genus *Mycobacterium* are mainly harmless soil saprophytes, but *Mycobacterium tuberculosis* causes the disease tuberculosis.

15.12 • The streptomycetes are a large group of filamentous, gram-positive bacteria that form spores at the end of aerial filaments and are found in the phylum *Actinobacteria*. Many clinically useful antibiotics such as tetracycline and neomycin have come from *Streptomyces* species.

15.13 • The phylum *Bacteroidetes* includes gram-negative rods that do not form spores, many of which have gliding motility. Most species in the order *Bacteroidales* are obligate anaerobes that ferment carbohydrates in anoxic environments. The genus *Bacteroides* contains species that are common in the gastrointestinal tract of animals.

15.14 • The *Cytophagales* and *Flavobacteriales* are orders in the *Bacteroidetes* that include aerobic bacteria able to degrade complex polysaccharides such as cellulose. These bacteria are important in organic matter decomposition.

15.15 • The phylum *Chlamydiae* includes small obligate intracellular parasites that are adept at invading eukaryotic cells. Many species cause various diseases in humans and other animals.

15.16 • The *Planctomycetes* are a group of stalked, budding bacteria that form intracellular compartments of various types, in some cases indistinguishable from the nucleus of eukaryotic cells.

15.17 • Species of *Verrucomicrobia* are distinguished by their multiple prosthecate cells and their unique phylogeny.

15.18 • *Thermotogae* and *Thermodesulfobacteria* form two deeply branching phyla within the *Bacteria*. These hyperthermophilic bacteria have proven that extensive horizontal gene transfer has occurred from *Archaea* to *Bacteria* (*Thermotoga*) and that ether-linked lipids are not limited to the *Archaea* (*Thermodesulfobacterium*).

15.19 • The *Aquifex* phylum contains a group of hyperthermophilic, H_2-oxidizing bacteria that form the earliest branch on the tree of the domain *Bacteria*.

15.20 • *Deinococcus* and *Thermus* are the major genera in a distinct phylum of *Bacteria*. *Thermus* is the source of the key enzyme in automated PCR, whereas *Deinococcus* is the most radiation-resistant bacterium known, exceeding even endospores in this regard.

15.21 • *Acidobacteria* are widespread in many environments, especially soils, and show various physiologies. The genus *Nitrospira* includes nitrite-oxidizing bacteria, while species of *Deferribacteres* and *Chrysiogenetes* specialize in various forms of anaerobic respiration. Species of *Synergistetes*, *Fusobacteria*, and *Fibrobacteres* are fermentative anaerobes that inhabit the gastrointestinal tract and other anoxic niches in animals.

MasteringMicrobiology®

Review what you know and challenge what you have learned with MasteringMicrobiology! Access study materials, chapter quizzes, animations, and microbiology lab tutorials in the Study Area to ensure that you have mastered this chapter's content.

REVIEW OF KEY TERMS

Acid-fastness a property of *Mycobacterium* species in which cells stained with the dye basic fuchsin resist decolorization with acidic alcohol

Actinomycetes a term used to refer to aerobic filamentous bacteria in the phylum *Actinobacteria*

Coryneform bacteria gram-positive, aerobic, nonmotile, rod-shaped organisms with the characteristic of forming irregular-shaped, club-shaped, or V-shaped cell arrangements, typical of several genera of unicellular *Actinobacteria*

Enteric bacteria a large group of gram-negative rod-shaped *Bacteria* characterized by a facultatively aerobic metabolism and commonly found in the intestines of animals

Heterofermentative in reference to lactic acid bacteria, capable of making more than one fermentation product

High GC gram-positive bacteria a term that refers to bacteria in the *Actinobacteria*

Homofermentative in reference to lactic acid bacteria, producing only lactic acid as a fermentation product

Lactic acid bacteria fermentative bacteria that produce lactic acid, are found in the *Firmicutes*, and are important in the production and preservation of many foods

Low GC gram-positive bacteria a term that refers to bacteria in the *Firmicutes*

Oligotrophic a term that refers to organisms that grow best under low-nutrient conditions

Propionic acid bacteria gram-positive fermentative bacteria that generate propionate as a fermentation end product and are important in the production of cheese

Proteobacteria the largest and most metabolically diverse phylum of bacteria

Pseudomonad a term used to refer to any gram-negative, polarly flagellated, aerobic rod able to use a diverse suite of carbon sources

REVIEW QUESTIONS

1. Which four phyla of bacteria contain the most species that have been characterized? (Section 15.1)

2. Which phylum contains the bulk of known gram-negative bacteria? What subgroup of this phylum contains the bacterium *Escherichia coli*? Which contains *Pseudomonas aeruginosa*? (Sections 15.1–15.5)

3. What morphological and physiological features distinguish *Burkholderia cepacia* from *Pseudomonas aeruginosa*? In what common environments might these organisms be encountered? (Sections 15.2, 15.4)

4. What is the catalase test? What catalase reaction would you expect from an obligate aerobe? What reaction would you expect from an obligate anaerobe? (Section 15.3)

5. What morphological and physiological features distinguish *Escherichia* from *Vibrio*? (Sections 15.3, 15.4)

6. In which phyla and genera would you expect to find endospore-forming bacteria? (Section 15.8)

7. What key features could be used to differentiate the following genera of gram-positive bacteria: *Bacillus*, *Mycoplasma*, *Staphylococcus*, *Propionibacterium*, *Streptomyces*, and *Mycobacterium*? (Sections 15.6–15.12)

8. In what phylum would you expect to find a gram-negative, obligately anaerobic, nonsporulating, gliding rod-shaped bacterium? Name an environment where these bacteria may be found. (Sections 15.13–15.14)

9. What traits do the chlamydia and the rickettsias have in common? In what ways do they differ? What is the function of each of the two types of cells formed by *Chlamydia*? (Sections 15.1, 15.15)

10. What do species in the *Planctomycetes* have in common with *Archaea*? With *Eukarya*? (Section 15.16)

11. Describe a key feature that would differentiate each of the following *Bacteria*: *Streptococcus*, *Planctomyces*, *Verrucomicrobium*, and *Gemmata*. (Sections 15.6, 15.16, 15.17)

12. Describe a key physiological feature of the following *Bacteria* that would differentiate each from the others: *Lactobacillus*, *Nitrospira*, and *Geothrix*. (Sections 15.6, 15.21)

13. What major physiological property unites species of *Thermotoga*, *Aquifex*, and *Thermocrinis*? (Sections 15.18, 15.19)

14. Why might *Deinococcus* thrive in soils that have been contaminated by radioactive fallout? (Section 15.20)

15. What are four ways in which different species of *Acidobacteria* have been shown to generate energy? (Section 15.21)

16. List three different genera of anaerobic bacteria that are able to degrade cellulose. (Sections 15.8, 15.12, 15.14, 15.21)

APPLICATION QUESTIONS

1. Enteric bacteria, lactic acid bacteria, and propionic acid bacteria have distinctive metabolic traits that can be used to characterize and identify these organisms. Describe the metabolic characteristics of these organisms, name a genus that belongs to each group, and indicate in what way these organisms can be differentiated.

2. Microorganisms can have a variety of different relationships with oxygen. Describe the terms used to characterize a cell's response to oxygen, and give an example from this chapter of an organism that can be described by each of these terms.

16 · Diversity of *Archaea*

microbiology**now**

Archaea and Global Warming

Anthropogenic CO_2 emissions have significantly affected global climate. However, *Archaea* and *Bacteria* have also profoundly affected our planet, including its climate. One example comes from the Arctic, where soil is frozen as permafrost. Permafrost can be 100 meters deep and it encompasses 25% of the terrestrial surface of the Earth. Within permafrost is stored an enormous mass of organic carbon, most of which has been locked away in ice for more than 20,000 years. But this ice is starting to thaw, and the result could have global consequences.

The Intergovernmental Panel on Climate Change predicts that Arctic temperatures will increase 7°C by the year 2100. When permafrost melts, it is converted into wetlands, and these are major habitats for *Archaea* that produce methane (methanogens). Methane is a greenhouse gas with a warming potential 25 times more powerful than CO_2. Hence, if Arctic warming continues at its present pace, much permafrost carbon could be converted into methane, significantly accelerating global climate change.

At Stordalen Mire in northern Sweden, microbiologists are investigating methanogens in thawed permafrost.[1] Chambers are used to trap and measure methane produced in the wetlands that have replaced thawed permafrost (photo). The source of most of the methane was found to be a novel methanogen, *Methanoflorens stordalenmirensis*, which grows rapidly in thawed permafrost. *M. stordalenmirensis* represents a novel order of methanogens previously called "Rice Cluster II." These methanogens are present in wetlands worldwide but *M. stordalenmirensis* is the first characterized species of this new taxonomic family, the *Methanoflorentaceae*.

In addition to human impacts on climate, future control of global climate change may well depend in a major way on what is discovered about the ecology of methanogenesis by *M. stordalenmirensis*.

[1]Mondav, R., et al. 2012. Microbial dynamics in a thawing world: Linking microbial communities to increased methane flux. *Proc. 14th Int. Symp. Microbial Ecology,* Copenhagen, Denmark.

We now consider organisms in the domain *Archaea*. A phylogenetic tree of *Archaea* is shown in **Figure 16.1**. The tree, based on comparative sequences of ribosomal proteins, reveals several phyla, including the **Euryarchaeota**, **Crenarchaeota**, **Thaumarchaeota**, **Korarchaeota**, and **Nanoarchaeota**. The exact ancestry of these groups remains a contentious issue, and phylogenetic trees constructed from 16S ribosomal RNA gene sequences often conflict with those made using other genomic loci (for example, compare Figures 12.13 and 16.1).

The evolutionary history of the *Archaea* is ancient and complex, involving horizontal gene transfers within and between phyla. Common traits shared by all *Archaea* include their ether-linked lipids, their lack of peptidoglycan in cell walls (Chapter 2), and their structurally complex RNA polymerases, which resemble those of *Eukarya* (꒰꒱ Figure 4.21). But beyond this, *Archaea* show enormous phenotypic diversity.

Archaea include species that carry out chemoorganotrophic or chemolithotrophic metabolisms, and both aerobic and anaerobic species are common (summarized later in Table 16.6). Chemoorganotrophy is widespread among *Archaea*, and fermentations and anaerobic respirations are common. Chemolithotrophy is also well established in the *Archaea*, with H$_2$ being a common electron donor (Section 16.14), and with ammonia oxidation found among species of *Thaumarchaeota*. Anaerobic respiration, especially forms employing elemental sulfur (S^0) as an electron acceptor, is prevalent among the *Archaea*, especially *Crenarchaeota*. By contrast, aerobic respiration occurs widely in *Thaumarchaeota* and is common among a few groups of *Euryarchaeota* but is characteristic of only a few species of *Crenarchaeota*.

Many metabolic traits of archaeal species are also found in *Bacteria* but others are unique to *Archaea*. **Methanogens**, for example, are *Euryarchaeota* that conserve energy from the production of methane (꒰꒱ Section 13.20). *Methanogenesis* is a globally important process that is uniquely archaeal (꒰꒱ Sections 13.20, 20.1, and 20.2). *Archaea* are also well known for containing many species of **extremophiles**, including species that are **hyperthermophiles** (organisms with growth temperature optima above 80°C), *halophiles*, and *acidophiles* (Chapter 5). However, a great many species in the *Euryarchaeota* and most *Thaumarchaeota* are not extremophiles and are found in soils, sediments, oceans, lakes, in association with animals, and even in the human gut!

With this brief background and the phylogeny of *Archaea* (Figure 16.1) firmly in mind, we now consider the organismal diversity of this fascinating domain of life.

I · *Euryarchaeota*

Euryarchaeota comprise a large and physiologically diverse group of *Archaea*. This phylum includes methanogens as well as many genera of extremely halophilic (salt-loving) *Archaea*. As a study in physiological contrasts, these two groups are remarkable: Methanogens are the strictest of anaerobes while extreme halophiles are primarily obligate aerobes. Other groups of euryarchaeotes include the hyperthermophiles *Thermococcus* and *Pyrococcus*, the hyperthermophilic methanogen *Methanopyrus*,

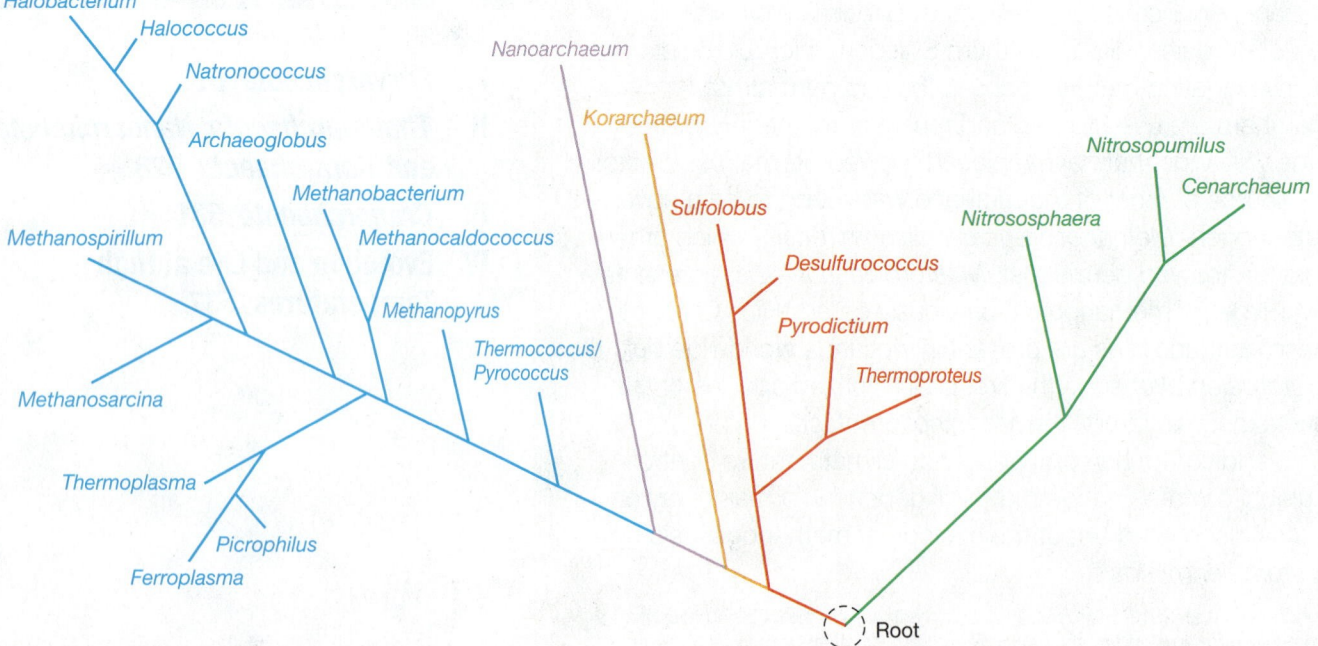

Figure 16.1 Detailed phylogenetic tree of the *Archaea* based on comparisons of ribosomal proteins from sequenced genomes. Each of the five archaeal phyla is indicated in a different color. The *Korarchaeota* and *Nanoarchaeota* are each represented by only a single known species.

and the cell wall–less *Thermoplasma*, an organism phenotypically similar to the mycoplasmas (⮌ Section 15.9). We begin our review of *Euryarchaeota* by reviewing the halophilic *Archaea*.

16.1 Extremely Halophilic *Archaea*

Key Genera: *Halobacterium, Haloferax, Natronobacterium*

Extremely halophilic *Archaea*, often called the "haloarchaea," are a diverse group that inhabits environments high in salt. These include naturally salty environments, such as solar salt evaporation ponds and salt lakes, and artificial saline habitats such as the surfaces of heavily salted foods, for example, certain fish and meats. Such salty habitats are called *hypersaline* (**Figure 16.2**). The term **extreme halophile** is used to indicate that these organisms are not only halophilic, but that their requirement for salt is very high, in some cases at levels near saturation (⮌ Figure 5.26).

An organism is considered an extreme halophile if it requires 1.5 M (about 9%) or more sodium chloride (NaCl) for growth. Most species of extreme halophiles require 2–4 M NaCl (12–23%) for optimal growth. Virtually all extreme halophiles can grow at 5.5 M NaCl (32%, the limit of saturation for NaCl), although some species grow very slowly at this salinity. Some phylogenetic relatives of extremely halophilic *Archaea*, for example species of *Haloferax* and *Natronobacterium*, are able to grow at much lower salinities, such as at or near that of seawater (about 2.5% NaCl); nevertheless, these organisms are phylogenetic relatives of other extreme halophiles.

Hypersaline Environments: Chemistry and Productivity

Hypersaline habitats are common throughout the world, but extremely hypersaline habitats are rare. Most such environments are in hot, dry areas of the world. Salt lakes can vary considerably in ionic composition. The predominant ions in a hypersaline lake depend on the surrounding topography, geology, and general climatic conditions.

Great Salt Lake in Utah (USA) (Figure 16.2a), for example, is essentially concentrated seawater. In this hypersaline lake the relative proportions of the various ions [e.g., sodium (Na^+), chloride (Cl^-), and sulfate (SO_4^{2-})] are those of seawater, although the

(a) *T. D. Brock*

(b) *NASA*

(c) *Michael T. Madigan*

(d) *Francisco Rodríguez-Valera*

Figure 16.2 Hypersaline habitats for halophilic *Archaea*. *(a)* The north arm of Great Salt Lake, Utah, a hypersaline lake in which the ratio of ions is similar to that in seawater, but in which absolute concentrations of ions are several times that of seawater. The green color is primarily from cells of cyanobacteria and green algae. *(b)* Aerial view near San Francisco Bay, California, of a series of seawater evaporating ponds where solar salt is prepared. The red-purple color is predominantly due to bacterioruberins and bacteriorhodopsin in cells of haloarchaea. *(c)* Lake Hamara, Wadi El Natroun, Egypt. A bloom of pigmented haloalkaliphiles is growing in this pH 10 soda lake. Note the deposits of trona ($NaHCO_3 \cdot Na_2CO_3 \cdot 2\ H_2O$) around the edge of the lake. *(d)* Scanning electron micrograph of halophilic bacteria including square *Archaea* present in a saltern in Spain.

overall concentration of ions is much higher. In addition, the pH of this hypersaline lake is slightly alkaline.

Soda lakes, in contrast, are highly alkaline, hypersaline environments. The water chemistry of soda lakes resembles that of hypersaline lakes such as Great Salt Lake, but because high levels of carbonate minerals are also present in the surrounding strata, the pH of soda lakes is quite high. Waters of pH 10–12 are not uncommon in these environments (Figure 16.2c). In addition, calcium (Ca^{2+}) and Mg^{2+} are virtually absent from soda lakes because they precipitate out at high pH and carbonate concentrations.

The diverse chemistries of hypersaline habitats have selected for a large diversity of halophilic microorganisms. Some organisms are unique to one environment while others are widespread. Moreover, despite their extreme conditions, salt lakes can be highly productive ecosystems (the word *productive* here means high levels of autotrophic CO_2 fixation). *Archaea* are not the only microorganisms present. The eukaryotic alga *Dunaliella* (↩ Figure 17.33a) is the major, if not the sole, oxygenic phototroph in most salt lakes. In highly alkaline soda lakes where *Dunaliella* is absent, anoxygenic phototrophic purple bacteria of the genera *Ectothiorhodospira* and *Halorhodospira* (↩ Section 14.5) predominate. Organic matter originating from primary production by oxygenic or anoxygenic phototrophs sets the stage for growth of haloarchaea, which are chemoorganotrophic organisms. In addition, a few extremely halophilic chemoorganotrophic *Bacteria*, such as *Halanaerobium*, *Halobacteroides*, and *Salinibacter*, thrive in such environments.

Marine salterns are also habitats for extreme halophiles. Marine salterns are enclosed basins filled with seawater that are left to evaporate, yielding solar sea salt (Figure 16.2b, d). As salterns approach the minimum salinity limits for haloarchaea, the waters turn a reddish purple color due to the massive growth—called a *bloom*—of cells (the red coloration apparent in Figure 16.2b and c is due to carotenoids and other pigments to be discussed later). Morphologically unusual *Archaea* are often present in salterns, including species with a square or cup-shaped morphology (Figure 16.2d). Extreme halophiles are also present in highly salted foods, such as certain types of sausages, marine fish, and salted pork.

Taxonomy and Physiology of Extremely Halophilic *Archaea*

Table 16.1 lists several of the currently recognized genera of extremely halophilic *Archaea*. Besides the term haloarchaea, these *Archaea* are sometimes called "halobacteria," because the genus *Halobacterium* (**Figure 16.3**) was the first in this group to be described (prior to the discovery of *Archaea*) and is still the best-studied representative of the group. *Natronobacterium*, *Natronomonas*, and their relatives differ from other extreme halophiles in being extremely alkaliphilic as well as halophilic. As befits their soda lake habitat (Table 16.1 and Figure 16.2c), natronobacteria grow optimally at very low Mg^{2+} concentrations and high pH (9–11).

Haloarchaea stain gram-negatively, reproduce by binary fission, and do not form resting stages or spores. Cells of the various cultured genera are rod-shaped, cocci, or cup-shaped, but even cells that form squares are known (Figure 16.2d). Cells of *Haloquadratum* are square in shape and are only about 0.1 μm thick. *Haloquadratum* also forms gas vesicles that allow it to float in its salty hypersaline habitat, probably as a means to be in contact with air since most extreme halophiles are obligate aerobes.

Table 16.1 Some genera of extremely halophilic *Archaea*

Genus	Morphology	Habitat
Extreme halophiles		
Halobacterium	Rods	Salted fish; hides; hypersaline lakes; salterns
Halorubrum	Rods	Dead Sea; salterns
Halobaculum	Rods	Dead Sea
Haloferax	Flattened discs	Dead Sea; salterns
Haloarcula	Irregular discs	Salt pools, Death Valley, CA; marine salterns
Halococcus	Cocci	Salted fish; salterns
Halogeometricum	Pleomorphic flat cells	Solar salterns
Haloterrigena	Rods, ovals	Saline soil
Haloquadratum	Flat squares	Salterns
Haloalkaliphiles		
Natronobacterium	Rods	Highly saline soda lakes
Natrinema	Rods	Salted fish; hides
Natrialba	Rods	Soda lakes; beach sand
Natronomonas	Rods	Soda lakes
Natronococcus	Cocci	Soda lakes
Natronorubrum	Flattened cells	Soda lakes

Many other extremely halophilic *Archaea* also produce gas vesicles. Most species of extreme halophiles lack flagella, but a few strains are weakly motile by flagella that rotate to propel the cell forward (↩ Section 2.17). The genomes of *Halobacterium* and *Halococcus* are unusual in that large plasmids containing up to 30% of the total cellular DNA are present and the GC base ratio of these plasmids (near 60% GC) differs significantly from that of chromosomal DNA (66–68% GC). Plasmids from extreme halophiles are among the largest naturally occurring plasmids known.

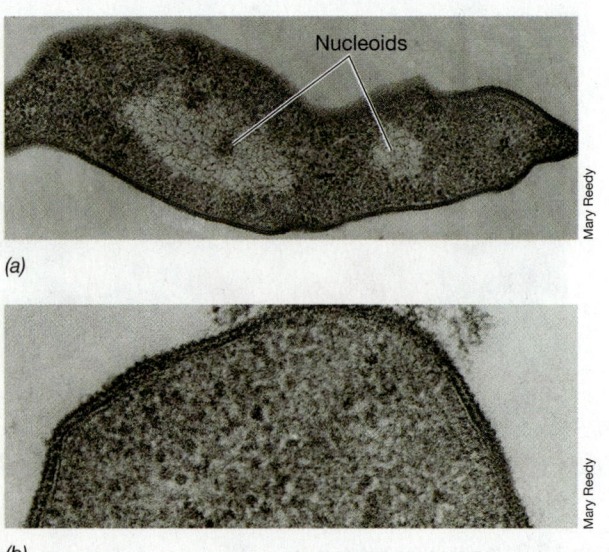

Nucleoids

(a)

(b)

Figure 16.3 **Electron micrographs of thin sections of the extreme halophile** ***Halobacterium salinarum.*** A cell is about 0.8 μm in diameter. *(a)* Longitudinal section of a dividing cell showing the nucleoids. *(b)* High-magnification electron micrograph showing the glycoprotein subunit structure of the cell wall.

Most species of extremely halophilic *Archaea* are obligate aerobes. Most haloarchaea use amino acids or organic acids as electron donors and require a number of growth factors such as vitamins for optimal growth. A few haloarchaea oxidize carbohydrates aerobically, but this capacity is rare; sugar fermentation does not occur. Electron transport chains containing cytochromes of the *a*, *b*, and *c* types are present in *Halobacterium*, and energy is conserved during aerobic growth via a proton motive force arising from electron transport. Some haloarchaea have been shown to grow anaerobically, as growth by anaerobic respiration (↻ Section 13.16) linked to the reduction of nitrate or fumarate has been demonstrated in certain species.

Water Balance in Extreme Halophiles

Extremely halophilic *Archaea* require large amounts of NaCl for growth. Detailed salinity studies of *Halobacterium* have shown that the requirement for Na^+ cannot be satisfied by any other ion, even the chemically related ion potassium (K^+). However, cells of *Halobacterium* need *both* Na^+ and K^+ for growth, because each plays an important role in maintaining osmotic balance.

As we learned in Section 5.15, microbial cells must withstand the osmotic forces that accompany life. To do so in a high-solute environment such as the salt-rich habitats of *Halobacterium*, organisms must either accumulate or synthesize solutes intracellularly. These solutes are called **compatible solutes**. These compounds counteract the tendency of the cell to become dehydrated under conditions of high osmotic strength by placing the cell in positive water balance with its surroundings. Cells of *Halobacterium*, however, do not synthesize or accumulate organic compounds but instead pump large amounts of K^+ from the environment into the cytoplasm. This ensures that the concentration of K^+ *inside* the cell is even greater than the concentration of Na^+ *outside* the cell (**Table 16.2**). This ionic condition maintains positive water balance.

The *Halobacterium* cell wall (Figure 16.3*b*) is composed of glycoprotein and is stabilized by Na^+. Sodium ions bind to the outer surface of the *Halobacterium* wall and are absolutely essential for maintaining cellular integrity. When insufficient Na^+ is present, the cell wall breaks apart and the cell lyses. This is a consequence of the exceptionally high content of the *acidic* (negatively charged) amino acids aspartate and glutamate in the glycoprotein of the *Halobacterium* cell wall. The negative charge on the carboxyl group of these amino acids is bound to Na^+; when Na^+ is diluted away, the negatively charged parts of the proteins tend to repel each other, leading to cell lysis.

Halophilic Cytoplasmic Components

Like cell wall proteins, cytoplasmic proteins of *Halobacterium* are highly acidic, but it is K^+, not Na^+, that is required for activity. This makes sense because K^+ is the predominant cation in the cytoplasm of cells of *Halobacterium* (Table 16.2). Besides having a high acidic amino acid composition, halobacterial cytoplasmic proteins typically contain lower levels of hydrophobic amino acids and lysine, a positively charged (basic) amino acid, than proteins of nonhalophiles. This is also to be expected because in a highly ionic cytoplasm, polar proteins would tend to remain in solution whereas nonpolar proteins would tend to cluster and perhaps lose activity. The ribosomes of *Halobacterium* also require high KCl levels for stability, whereas ribosomes of nonhalophiles have no KCl requirement.

Table 16.2 Concentration of ions in cells of *Halobacterium salinarum*[a]

Ion	Concentration in medium (M)	Concentration in cells (M)
Na^+	4.0	1.4
K^+	0.032	4.6
Mg^{2+}	0.13	0.12
Cl^-	4.0	3.6

[a]Data from *Biochim. Biophys. Acta* 65: 506–508 (1962).

Extremely halophilic *Archaea* are thus well adapted, both internally and externally, to life in a highly ionic environment. Cellular components exposed to the external environment require high Na^+ for stability, whereas internal components require high K^+. With the exception of a few extremely halophilic members of the *Bacteria* that also use KCl as a compatible solute, in no other group of prokaryotes do we find this unique requirement for such high amounts of specific cations.

Bacteriorhodopsin and Light-Mediated ATP Synthesis in Halobacteria

Certain species of haloarchaea can catalyze a light-driven synthesis of ATP. This occurs without chlorophyll pigments, so it is not photosynthesis. However, other light-sensitive pigments are present, including red and orange carotenoids—primarily C_{50} pigments called *bacterioruberins*—and inducible pigments involved in energy conservation; we discuss these pigments here.

Under conditions of low aeration, *Halobacterium salinarum* and some other haloarchaea synthesize a protein called **bacteriorhodopsin** and insert it into their cytoplasmic membranes. Bacteriorhodopsin is so named because of its structural and functional similarity to rhodopsin, the visual pigment of the eye. Conjugated to bacteriorhodopsin is a molecule of retinal, a carotenoid-like molecule that can absorb light energy and pump a proton across the cytoplasmic membrane. The retinal gives bacteriorhodopsin a purple hue. Thus cells of *Halobacterium* that are switched from growth under high-aeration conditions to oxygen-limiting growth conditions (a trigger of bacteriorhodopsin synthesis) gradually change color from orange-red to purple-red as they synthesize bacteriorhodopsin and insert it into their cytoplasmic membranes.

Bacteriorhodopsin absorbs green light around 570 nm. Following absorption, the retinal of bacteriorhodopsin, which normally exists in a *trans* configuration (Ret_T), becomes excited and converts to the *cis* (Ret_C) form (**Figure 16.4**). This transformation is coupled to the translocation of a proton across the cytoplasmic membrane. The retinal molecule then decays to the *trans* isomer along with the uptake of a proton from the cytoplasm, and this completes the cycle. The proton pump is then ready to repeat the cycle (Figure 16.4). As protons accumulate on the outer surface of the membrane, a proton motive force is generated that is coupled to ATP synthesis through the activity of a proton-translocating ATPase (↻ Section 3.11; Figure 16.4).

Bacteriorhodopsin-mediated ATP production in *H. salinarum* supports slow growth of this organism under anoxic conditions. The light-stimulated proton pump of *H. salinarum* also functions to pump Na^+ out of the cell by activity of a Na^+–H^+ antiport system and to drive the uptake of nutrients, including the K^+ needed for osmotic balance. Amino acid uptake by *H. salinarum* is

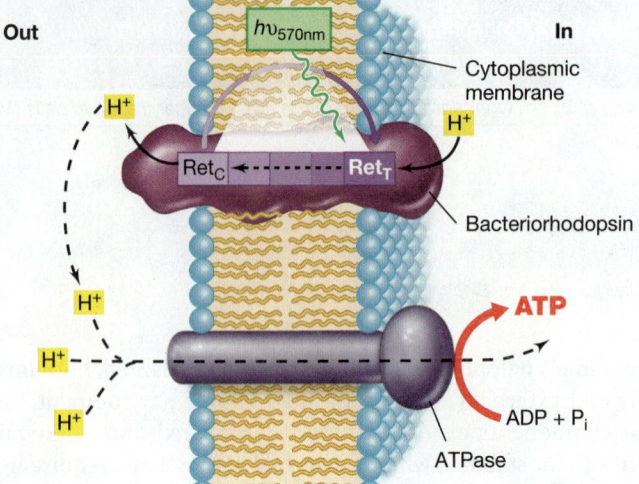

Out | **In**

H$^+$

Cytoplasmic membrane

hv_{570nm}

Ret$_C$ ← - - - - - Ret$_T$

H$^+$

Bacteriorhodopsin

H$^+$

ATP

H$^+$

H$^+$

ADP + P$_i$

ATPase

Figure 16.4 Model for the mechanism of bacteriorhodopsin. Light of 570 nm (hv_{570nm}) converts the protonated retinal of bacteriorhodopsin from the *trans* form (Ret$_T$) to the *cis* form (Ret$_C$), along with translocation of a proton to the outer surface of the cytoplasmic membrane, thus establishing a proton motive force. ATPase activity is driven by the proton motive force.

indirectly driven by light as well, because amino acids are cotransported into the cell with Na$^+$ by an amino acid–Na$^+$ symporter (⟲ Section 2.9); removal of Na$^+$ from the cell occurs by way of the light-driven Na$^+$–H$^+$ antiporter.

Other Rhodopsins

Besides bacteriorhodopsin, at least three other rhodopsins are present in the cytoplasmic membrane of cells of *H. salinarum*. **Halorhodopsin** is a light-driven chloride (Cl$^-$) pump that brings Cl$^-$ into the cell as the anion for K$^+$. The retinal of halorhodopsin binds Cl$^-$ and transports it into the cell. Two other light sensors, called *sensory rhodopsins*, are present in *H. salinarum*. These light sensors control phototaxis (movement toward light, ⟲ Section 2.19) by the organism. Through the interaction of a cascade of proteins similar to those in chemotaxis (⟲ Sections 2.19 and 7.8), sensory rhodopsins affect flagellar rotation, moving cells of *H. salinarum* toward light where bacteriorhodopsin can function to make ATP (Figure 16.4).

We will learn when we consider marine microbiology (⟲ Sections 19.9 and 19.11) that several *Proteobacteria* (domain *Bacteria*) that inhabit the upper layers of the ocean contain bacteriorhodopsin-like proteins called *proteorhodopsins*. As far

as is known, proteorhodopsin functions like bacteriorhodopsin except that different spectral forms exist, each form being tuned to the absorption of its own specific wavelengths of light. Proteorhodopsin as a mechanism for energy conservation in marine bacteria makes good ecological sense because levels of dissolved organic matter in the open oceans are typically very low, and thus a strictly chemoorganotrophic lifestyle would be difficult.

MINIQUIZ --

• If cells of *Halobacterium* require high levels of Na$^+$ for growth, why is this not true for the organism's cytoplasmic enzymes?

• What benefit does bacteriorhodopsin confer on a cell of *Halobacterium salinarum*?

16.2 Methanogenic *Archaea*

Key Genera: *Methanobacterium, Methanocaldococcus, Methanosarcina*

Many *Euryarchaeota* are methanogens, microorganisms that produce methane (CH$_4$) as an integral part of their energy metabolism (methane production is called *methanogenesis*). In Section 13.20 we considered the biochemistry of methanogenesis. Later, we will learn how methanogenesis is the terminal step in the biodegradation of organic matter in many anoxic habitats in nature (⟲ Section 20.2). Table 16.3 lists the major sources of biogenic methane in nature.

Diversity and Physiology of Methanogens

Methanogens show a variety of morphologies (**Figure 16.5** and **Table 16.4**). Their taxonomy is based on both phenotypic and phylogenetic analyses, with several taxonomic orders being recognized (in taxonomy, an order contains groups of related families, each of which contains one or more genera; ⟲ Section 12.10).

Methanogens show a diversity of cell wall chemistries. These include the pseudomurein walls of *Methanobacterium* species and relatives (**Figure 16.6a**), walls composed of methanochondroitin (so named because of its structural resemblance to chondroitin, the connective tissue polymer of vertebrate animals) in *Methanosarcina* and relatives (Figure 16.6b), the protein or glycoprotein walls of *Methanocaldococcus* (**Figure 16.7a**) and *Methanoplanus* species, respectively, and the S-layer walls of *Methanospirillum* (Figure 16.5c; ⟲ Section 2.12).

Physiologically, methanogens are obligate anaerobes, and strict anoxic techniques are necessary to culture them. Most

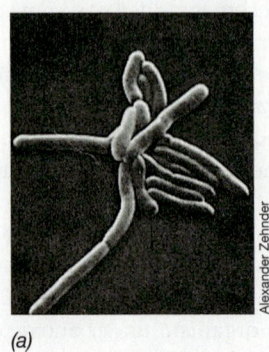
(a)

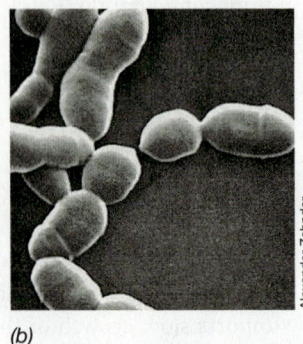
(b)

(c)

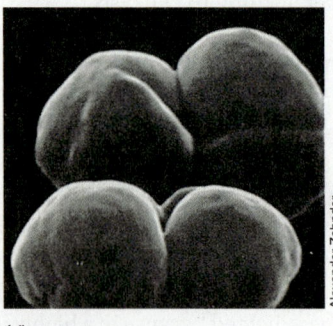
(d)

Figure 16.5 Scanning electron micrographs of cells of diverse species of methanogenic *Archaea*. *(a) Methanobrevibacter ruminantium.* A cell is about 0.7 μm in diameter. *(b) Methanobrevibacter arboriphilus.* A cell is about 1 μm in diameter. *(c) Methanospirillum hungatei.* A cell is about 0.4 μm in diameter. *(d) Methanosarcina barkeri.* A cell is about 1.7 μm wide.

Table 16.3 Habitats of methanogens

I. Anoxic sediments: marsh, swamp, and lake sediments, paddy fields, moist landfills
II. Animal digestive tracts:[a]
  A. Rumen of ruminant animals such as cattle, sheep, elk, deer, and camels
  B. Cecum of cecal animals such as horses and rabbits
  C. Large intestine of monogastric animals such as humans, swine, and dogs
  D. Hindgut of cellulolytic insects (for example, termites)
III. Geothermal sources of H_2 + CO_2: hydrothermal vents
IV. Artificial biodegradation facilities: sewage sludge digesters
V. Endosymbionts of various anaerobic protozoa

[a]See Sections 22.6 and 22.7 and Figures 22.27 and 22.28.

Table 16.5 Substrates converted to methane by various methanogenic *Archaea*

I. CO_2-type substrates
Carbon dioxide, CO_2 (with electrons derived from H_2, certain alcohols, or pyruvate)
Formate, $HCOO^-$
Carbon monoxide, CO

II. Methylated substrates
Methanol, CH_3OH
Methylamine, $CH_3NH_3^+$
Dimethylamine, $(CH_3)_2NH_2^+$
Trimethylamine, $(CH_3)_3NH^+$
Methylmercaptan, CH_3SH
Dimethylsulfide, $(CH_3)_2S$

III. Acetotrophic substrates
Acetate, CH_3COO^-
Pyruvate, CH_3COCOO^-

Table 16.4 Characteristics of some methanogenic *Archaea*[a]

Order/Genus	Morphology	Substrates for methanogenesis
Methanobacteriales		
Methanobacterium	Long rods	H_2 + CO_2, formate
Methanobrevibacter	Short rods	H_2 + CO_2, formate
Methanosphaera	Cocci	Methanol + H_2 (both needed)
Methanothermus	Rods	H_2 + CO_2
Methanothermobacter	Rods	H_2 + CO_2, formate
Methanococcales		
Methanococcus	Irregular cocci	H_2 + CO_2, pyruvate + CO_2, formate
Methanothermococcus	Cocci	H_2 + CO_2, formate
Methanocaldococcus	Cocci	H_2 + CO_2
Methanotorris	Cocci	H_2 + CO_2
Methanomicrobiales		
Methanomicrobium	Short rods	H_2 + CO_2, formate
Methanogenium	Irregular cocci	H_2 + CO_2, formate
Methanospirillum	Spirilla	H_2 + CO_2, formate
Methanoplanus	Plate-shaped cells—occurring as thin plates with sharp edges	H_2 + CO_2, formate
Methanocorpusculum	Irregular cocci	H_2 + CO_2, formate, alcohols
Methanoculleus	Irregular cocci	H_2 + CO_2, alcohols, formate
Methanofollis	Irregular cocci	H_2 + CO_2, formate
Methanolacinia	Irregular rods	H_2 + CO_2, alcohols
Methanosarcinales		
Methanosarcina	Large, irregular cocci in packets	H_2 + CO_2, methanol, methylamines, acetate
Methanolobus	Irregular cocci in aggregates	Methanol, methylamines
Methanohalobium	Irregular cocci	Methanol, methylamines
Methanococcoides	Irregular cocci	Methanol, methylamines
Methanohalophilus	Irregular cocci	Methanol, methylamines, methyl sulfides
Methanosaeta	Long rods to filaments	Acetate
Methanosalsum	Irregular cocci	Methanol, methylamines, dimethylsulfide
Methanimicrococcus	Irregular cocci	Methanol, methylamines (H_2 needed with any methanogenic substrate)
Methanopyrales		
Methanopyrus	Rods in chains	H_2 + CO_2

[a]Taxonomic orders are listed in bold. An order is a taxonomic rank that consists of several families; families consist of several genera.

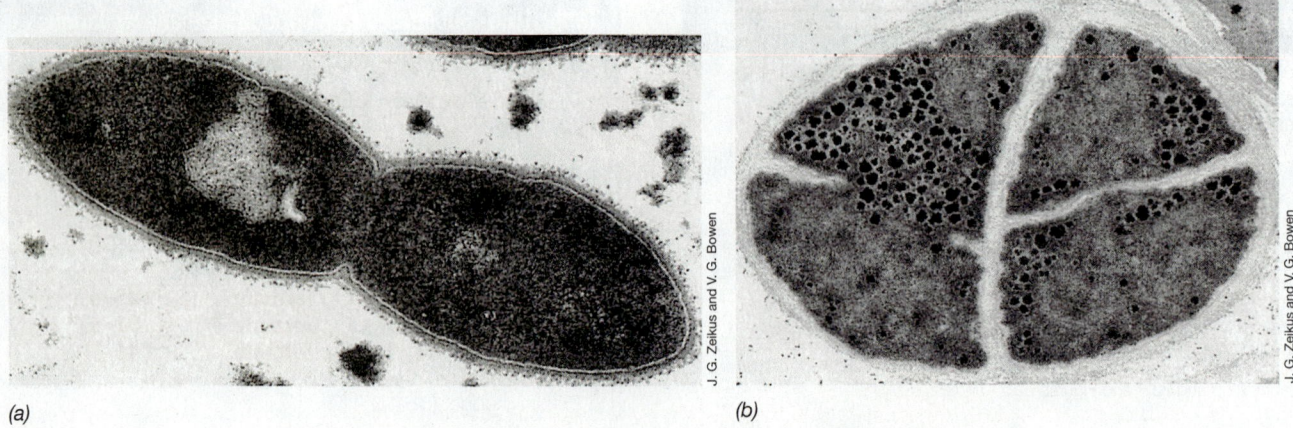

(a) — J. G. Zeikus and V. G. Bowen

(b) — J. G. Zeikus and V. G. Bowen

Figure 16.6 **Transmission electron micrographs of thin sections of methanogenic *Archaea*.** *(a) Methanobrevibacter ruminantium.* A cell is 0.7 μm in diameter. *(b) Methanosarcina barkeri*, showing the thick cell wall and the manner of cell segmentation and cross-wall formation. A cell is 1.7 μm in diameter.

methanogens are mesophilic and nonhalophilic, although species that grow optimally at very high (Figure 16.7) or very low temperatures, at very high salt concentrations, or at extremes of pH, have also been described. Several substrates can be converted to CH_4 by methanogens. Interestingly, these substrates do not include such common compounds as glucose and organic or fatty acids (other than acetate and pyruvate). Compounds such as glucose can be converted to CH_4, but only in reactions in which methanogens and other anaerobes cooperate. With the right mixture of organisms, virtually any organic compound, even hydrocarbons, can be converted to CH_4 plus CO_2 (⮌ Section 20.2).

Three classes of compounds make up the list of methanogenic substrates shown in **Table 16.5**. These are *CO_2-type* substrates, *methylated* substrates, and *acetate*. CO_2-type substrates include CO_2 itself, which is reduced to CH_4 using H_2 as the electron donor. Other substrates of this type include formate (which is $CO_2 + H_2$ in combined form) and CO, carbon monoxide. Methylated substrates include methanol (CH_3OH) and many others (Table 16.5).

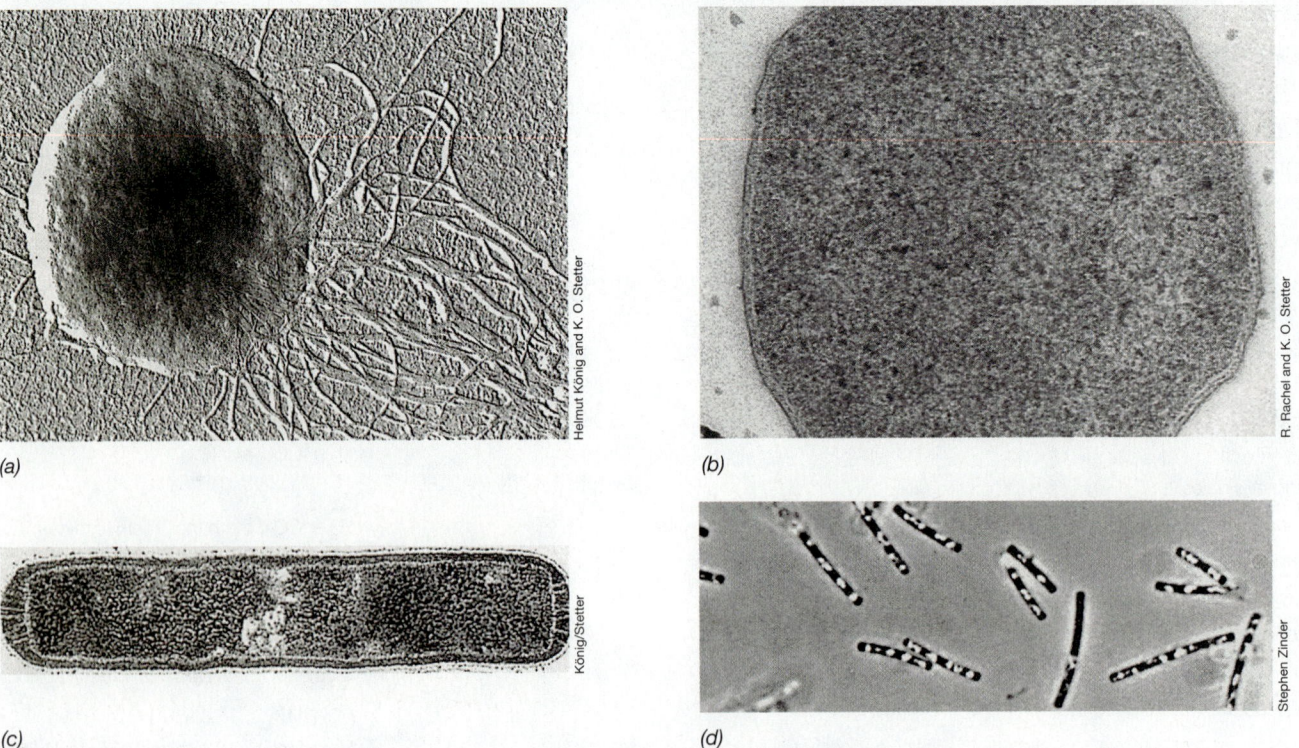

(a) — Helmut König and K. O. Stetter

(b) — R. Rachel and K. O. Stetter

(c) — König/Stetter

(d) — Stephen Zinder

Figure 16.7 **Hyperthermophilic and thermophilic methanogens.** *(a) Methanocaldococcus jannaschii* (temperature optimum, 85°C), shadowed preparation electron micrograph. A cell is about 1 μm in diameter. *(b) Methanotorris igneus* (temperature optimum, 88°C), thin section. A cell is about 1 μm in diameter. *(c) Methanothermus fervidus* (temperature optimum, 88°C), thin-sectioned electron micrograph. A cell is about 0.4 μm in diameter. *(d) Methanosaeta thermophila* (temperature optimum, 60°C), phase-contrast micrograph. A cell is about 1 μm in diameter. The refractile bodies inside the cells are gas vesicles.

Methanol (CH_3OH) can be reduced using an external electron donor such as H_2, or, alternatively, in the absence of H_2, some CH_3OH can be oxidized to CO_2 to generate the electrons needed to reduce other molecules of CH_3OH to CH_4 (⮌ Figure 13.50*a*). The final methanogenic process is the cleavage of acetate to CO_2 plus CH_4. Only a few known methanogens are acetotrophic (Tables 16.4 and 16.5), although acetate is a major source of CH_4 in nature. The biochemistry of methanogenesis from each of these classes of substrates is considered in Section 13.20 along with how CH_4 formation is linked to energy conservation.

Methanocaldococcus jannaschii as a Model Methanogen

The genome of the hyperthermophilic methanogen *Methanocaldococcus jannaschii* (Figure 16.7*a*) and many other methanogens have been sequenced. The 1.66-Mbp circular genome of *M. jannaschii*, an organism that has been used as a model for the molecular study of methanogenesis and archaeal motility (see "Archaeal Tortoise and Hare", page 25) contains about 1700 genes, and genes encoding enzymes of methanogenesis and several other key cell functions have been identified. Interestingly, the majority of *M. jannaschii* genes encoding functions such as central metabolic pathways and cell division are similar to those in *Bacteria*. By contrast, most of the *M. jannaschii* genes encoding core molecular processes such as transcription and translation more closely resemble those of eukaryotes. These findings reflect the various traits shared by organisms in the three cellular domains and are consistent with our understanding of how the three domains of cells evolved, as discussed in Chapter 12. However, analyses of the *M. jannaschii* genome also show that fully 40% of its genes have no counterparts in genes from either of the other domains. Some of these are genes that encode the enzymes needed for methanogenesis, of course, but many others likely encode novel cellular functions absent from cells in the other domains or may encode redundant functions carried out by classes of enzymes distinct from those found in *Bacteria* and *Eukarya*.

MINIQUIZ

- What are the major substrates for methanogenesis?
- What is unusual about the *Methanocaldococcus jannaschii* genome?

16.3 *Thermoplasmatales*

Key Genera: *Thermoplasma, Picrophilus, Ferroplasma*

A phylogenetically distinct line of *Archaea* contains thermophilic and extremely acidophilic genera: *Thermoplasma, Ferroplasma,* and *Picrophilus* (Figure 16.1). These prokaryotes are among the most acidophilic of all known microorganisms, with *Picrophilus* being capable of growth even below pH 0. Most are thermophilic as well. These genera also form their own taxonomic order within the *Euryarchaeota,* the *Thermoplasmatales.* We begin with a description of the mycoplasma-like organisms *Thermoplasma* and *Ferroplasma.*

Archaea Lacking Cell Walls

Thermoplasma and *Ferroplasma* lack cell walls, and in this respect they resemble the mycoplasmas (⮌ Section 15.9). *Thermoplasma* (**Figure 16.8**) is a chemoorganotroph that grows optimally at 55°C and pH 2 in complex media. Two species of *Thermoplasma* have

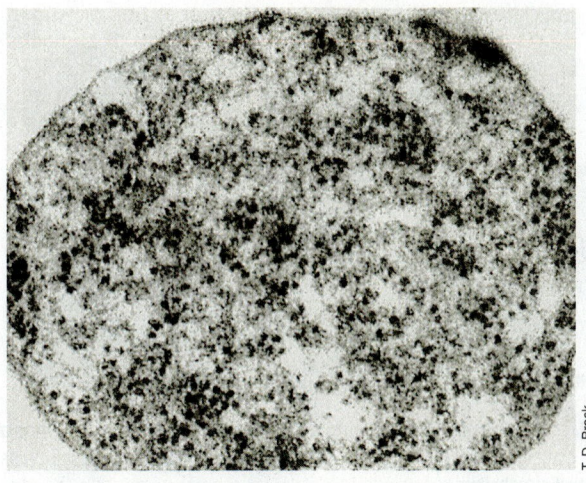

(a)

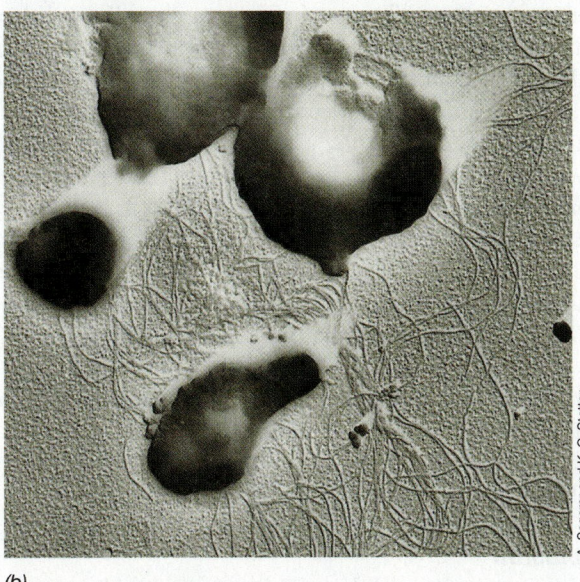

(b)

Figure 16.8 *Thermoplasma* **species.** *(a) Thermoplasma acidophilum,* an acidophilic and thermophilic mycoplasma-like archaeon; electron micrograph of a thin section. The diameter of cells varies from 0.2 to 5 μm. The cell shown is about 1 μm in diameter. *(b)* Shadowed preparation of cells of *Thermoplasma volcanium* isolated from hot springs. Cells are 1–2 μm in diameter. Notice the abundant flagella and irregular cell morphology.

been described, *Thermoplasma acidophilum* and *Thermoplasma volcanium.* Species of *Thermoplasma* are facultative aerobes, growing either aerobically or anaerobically by sulfur respiration (⮌ Section 13.18). Most strains of *T. acidophilum* have been obtained from self-heating coal refuse piles. Coal refuse contains coal fragments, pyrite (FeS_2), and other organic materials extracted from coal. When dumped into piles in surface-mining operations, coal refuse heats as a result of microbial metabolism bringing it to combustion temperature (**Figure 16.9**). This sets the stage for growth of *Thermoplasma,* which likely metabolizes organic compounds leached from the hot coal refuse. The second species, *T. volcanium,* has been isolated in hot acidic soils throughout the world and is highly motile by multiple flagella (Figure 16.8*b*).

To survive the osmotic stresses of life without a cell wall and to withstand the dual environmental extremes of low pH and high temperature, *Thermoplasma* has evolved a unique cytoplasmic

Figure 16.9 **A typical self-heating coal refuse pile, habitat of *Thermoplasma*.** The pile, containing coal debris, pyrite, and other microbial substrates, self-heats due to microbial metabolism.

membrane structure. The membrane contains a lipopolysaccharide-like material called *lipoglycan*. This substance consists of a tetraether lipid monolayer membrane with mannose and glucose (**Figure 16.10**). This molecule constitutes a major fraction of the total lipids of *Thermoplasma*. The membrane also contains glycoproteins but not sterols. These molecules render the *Thermoplasma* membrane stable to hot, acidic conditions.

Like mycoplasmas (⮌ Section 15.9), *Thermoplasma* contains a relatively small genome (1.5 Mbp). In addition, *Thermoplasma* DNA is complexed with a highly basic DNA-binding protein that organizes the DNA into globular particles resembling the nucleosomes of eukaryotic cells. This protein is homologous to the histone-like DNA-binding protein HU of *Bacteria*, which plays an important role in organization of the DNA in the cell. In contrast, several other *Euryarchaeota* contain basic proteins homologous to the DNA-binding histone proteins of eukaryotic cells.

Ferroplasma

Ferroplasma is a chemolithotrophic relative of *Thermoplasma*. *Ferroplasma* is a strong acidophile; however, it is not a thermophile, as it grows optimally at 35°C. *Ferroplasma* oxidizes ferrous iron (Fe^{2+}) to ferric iron (Fe^{3+}) to obtain energy (this reaction generates acid, see Figure 16.18*d*) and uses CO_2 as its carbon source (autotrophy). *Ferroplasma* grows in mine tailings containing pyrite (FeS_2), which is its energy source. The extreme acidophily of *Ferroplasma* allows it to drive the pH of its habitat down to extremely acidic values. After moderate acidity is generated from Fe^{2+} oxidation by acidophilic organisms such as *Acidithiobacillus ferrooxidans* and *Leptospirillum ferrooxidans* (⮌ Section 20.5), *Ferroplasma* becomes active and subsequently generates the very

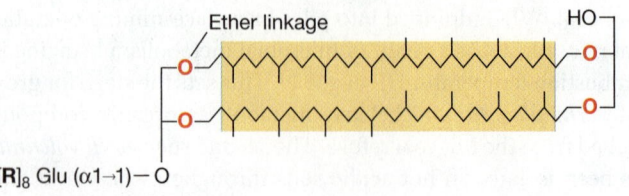

Figure 16.10 **Structure of the tetraether lipoglycan of *Thermoplasma acidophilum*.** Glu, Glucose; Man, mannose. Note the ether linkages and the fact that this lipid would form a monolayer rather than a bilayer membrane (compare the structure of lipoglycan with the membranes shown in Figure 2.17*e*).

low pH values typical of acid mine drainage. Acidic waters at pH 0 can be generated by the activities of *Ferroplasma*.

Picrophilus

A phylogenetic relative of *Thermoplasma* and *Ferroplasma* is *Picrophilus*. Although *Thermoplasma* and *Ferroplasma* are extreme acidophiles, *Picrophilus* is even more so, growing optimally at pH 0.7 and capable of growth at pH values lower than 0. *Picrophilus* also has a cell wall (an S-layer; ⮌ Section 2.12) and a much lower DNA GC base ratio than does *Thermoplasma* or *Ferroplasma*. Although phylogenetically related, *Thermoplasma*, *Ferroplasma*, and *Picrophilus* have quite distinct genomes. Two species of *Picrophilus* have been isolated from acidic Japanese solfataras, and like *Thermoplasma*, both grow heterotrophically on complex media.

The physiology of *Picrophilus* is of interest as a model for extreme acid tolerance. Studies of its cytoplasmic membrane point to an unusual arrangement of lipids that forms a highly acid-impermeable membrane at very low pH. By contrast, at moderate acidities such as pH 4, the membranes of cells of *Picrophilus* become leaky and disintegrate. Obviously, this organism has evolved to survive only in highly acidic habitats.

MINIQUIZ

- In what ways are *Thermoplasma* and *Picrophilus* similar? In what ways do they differ?
- How does *Thermoplasma* strengthen its cytoplasmic membrane to survive without a cell wall?

16.4 *Thermococcales* and *Methanopyrus*

Key Genera: *Thermococcus, Pyrococcus, Methanopyrus*

A few euryarchaeotes thrive in thermal environments and some are hyperthermophiles. We consider here three hyperthermophilic euryarchaeotes that branch very near the root of the *Euryarchaeota* (Figure 16.1). Two of these, *Thermococcus* and *Pyrococcus*, form a distinct taxonomic order: the *Thermococcales*. The third organism, *Methanopyrus*, is a methanogen that closely resembles other methanogens (Section 16.2 and Table 16.4) in its basic physiology but is unusual in its hyperthermophily, lipids, and phylogenetic position (Figure 16.1).

Thermococcus and *Pyrococcus*

Thermococcus is a spherical hyperthermophilic euryarchaeote indigenous to anoxic thermal waters in various locations throughout the world. The spherical cells contain a tuft of polar flagella and are thus highly motile (**Figure 16.11**). *Thermococcus* is an obligately anaerobic chemoorganotroph that metabolizes proteins and other complex organic mixtures (including some sugars) with elemental sulfur (S^0) as electron acceptor at temperatures from 55 to 95°C.

Pyrococcus is morphologically similar to *Thermococcus* (Figure 16.11*b*). *Pyrococcus* differs from *Thermococcus* primarily by its higher temperature requirements; *Pyrococcus* grows between 70 and 106°C with an optimum of 100°C. *Thermococcus* and *Pyrococcus* are also metabolically quite similar. Proteins, starch, or maltose are oxidized as electron donors, and S^0 is the terminal electron acceptor and is reduced to hydrogen sulfide (H_2S). Both *Thermococcus* and *Pyrococcus* form H_2S when S^0 is present, but form H_2 when S^0 is absent (see Table 16.8).

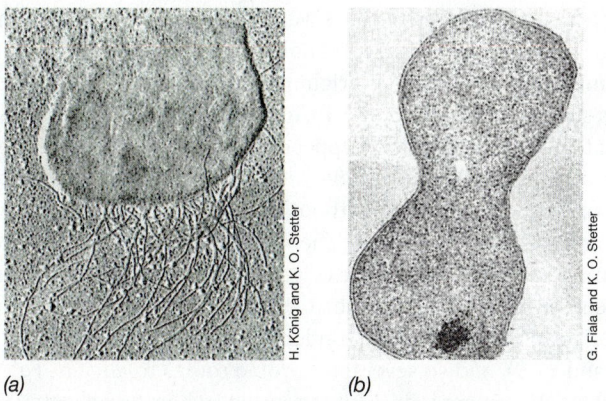

(a)   (b)

H. König and K. O. Stetter

G. Fiala and K. O. Stetter

Figure 16.11 Spherical hyperthermophilic *Euryarchaeota* from submarine volcanic areas. *(a) Thermococcus celer.* Electron micrograph of shadowed cells (note tuft of flagella). *(b)* Dividing cell of *Pyrococcus furiosus.* Electron micrograph of thin section. Cells of both organisms are about 0.8 μm in diameter.

Methanopyrus

Methanopyrus is a rod-shaped hyperthermophilic methanogen (**Figure 16.12**). *Methanopyrus* was isolated from hot sediments near submarine hydrothermal vents and from the walls of "black smoker" hydrothermal vent chimneys (Section 16.11; ⟳ Section 19.13). *Methanopyrus* shares phenotypic properties with both the hyperthermophiles and the methanogens. *Methanopyrus* produces CH_4 only from $H_2 + CO_2$ and grows rapidly for an autotrophic organism (generation time <1 h at 100°C). In special pressurized vessels, growth of one strain of *Methanopyrus* has been recorded at 122°C, the highest temperature yet shown to support microbial growth.

Methanopyrus is also unusual because it contains membrane lipids found in no other known organism. Recall that in the lipids of *Archaea*, the glycerol side chains contain **phytanyl** rather than fatty acids bonded in ether linkage to the glycerol (⟳ Section 2.7). In *Methanopyrus*, this ether-linked lipid is an *unsaturated* form of the otherwise saturated dibiphytanyl tetraethers found in all other hyperthermophilic *Archaea*. These unusual lipids may help to stabilize the cytoplasmic membrane of *Methanopyrus* at its unusually high growth temperatures.

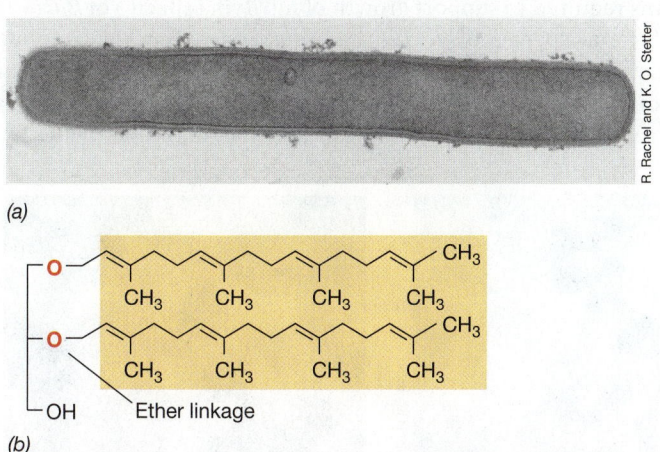

(a)

R. Rachel and K. O. Stetter

(b)   Ether linkage

Figure 16.12 *Methanopyrus.* *Methanopyrus* grows optimally at 100°C and can make CH_4 only from $CO_2 + H_2$. *(a)* Electron micrograph of a cell of *Methanopyrus kandleri*, the most thermophilic of all known organisms (upper temperature limit, 122°C). This cell measures 0.5 × 8 μm. *(b)* Structure of the novel lipid of *M. kandleri*. This is the normal ether-linked lipid of the *Archaea* except that the side chains are an unsaturated form of phytanyl (geranylgeraniol).

MINIQUIZ

- How do *Thermococcus* and *Pyrococcus* make ATP?
- In what way(s) does *Methanopyrus* display properties of key organisms in both the *Euryarchaeota* and *Crenarchaeota*, and what is unusual about its lipids?

16.5 *Archaeoglobales*

Key Genera: *Archaeoglobus, Ferroglobus*

We will see later that a number of hyperthermophilic *Crenarchaeota* catalyze anaerobic respirations in which elemental sulfur (S^0) is used as an electron acceptor, being reduced to H_2S (see Table 16.6). One hyperthermophilic euryarchaeote, *Archaeoglobus*, can reduce sulfate (SO_4^{2-}) and forms a phylogenetically distinct lineage within the *Euryarchaeota* (Figure 16.1).

Archaeoglobus

Archaeoglobus was isolated from hot marine sediments near hydrothermal vents. In its metabolism, *Archaeoglobus* couples the oxidation of H_2, lactate, pyruvate, glucose, or complex organic compounds to the reduction of SO_4^{2-} to H_2S. Cells of *Archaeoglobus* are irregular cocci (**Figure 16.13a**) and grow optimally at 83°C.

Archaeoglobus and methanogens share some characteristics. We learned in Section 13.20 about the unique biochemistry of methanogenesis. Briefly, this process requires a series of novel coenzymes,

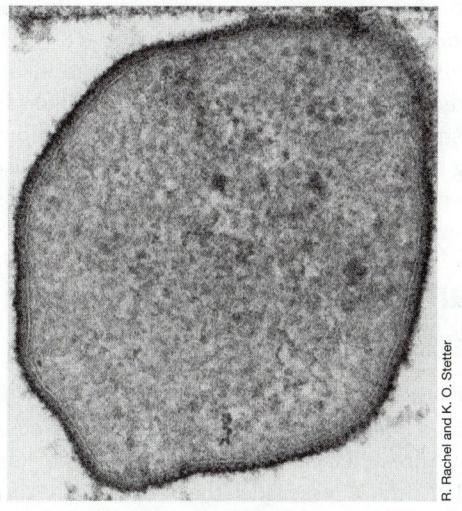

(a)

R. Rachel and K. O. Stetter

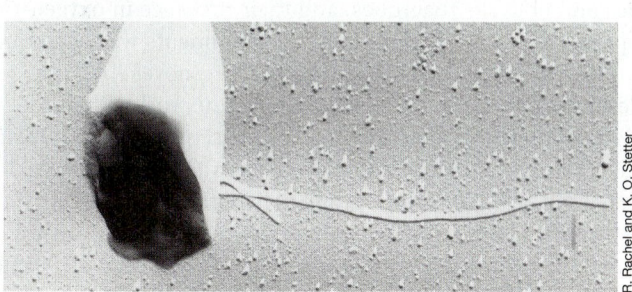

(b)

R. Rachel and K. O. Stetter

Figure 16.13 *Archaeoglobales.* *(a)* Transmission electron micrograph of the sulfate-reducing hyperthermophile *Archaeoglobus fulgidus.* The cell measures 0.7 μm in diameter. *(b)* Freeze-etched electron micrograph of *Ferroglobus placidus*, a ferrous iron–oxidizing, nitrate-reducing hyperthermophile. The cell measures about 0.8 μm in diameter.

UNIT 3

and with rare exceptions, these coenzymes have only been found in methanogens. Surprisingly, however, *Archaeoglobus* also contains many of these coenzymes and cultures of this organism actually produce small amounts of CH_4. Thus, *Archaeoglobus*, which also shows a rather close phylogenetic relationship to methanogens (Figure 16.1), may be a metabolically intermediate type of organism, bridging the energy-conserving processes of methanogenesis and other forms of respiration among *Archaea*. Not surprisingly then, the genome of *Archaeoglobus*, which contains about 2400 genes, shares a number of genes in common with methanogens (Section 16.2).

Ferroglobus

Ferroglobus (Figure 16.13b) is related to *Archaeoglobus* but is not a sulfate reducer. Instead, *Ferroglobus* is an iron-oxidizing chemolithotroph, conserving energy from the oxidation of Fe^{2+} to Fe^{3+} coupled to the reduction of nitrate (NO_3^-) to nitrite (NO_2^-) (see Table 16.6). *Ferroglobus* grows autotrophically and can also use H_2 or H_2S as electron donors in its energy metabolism. *Ferroglobus* was isolated from a shallow marine hydrothermal vent and grows optimally at 85°C.

Ferroglobus is interesting for several reasons, but especially for its ability to oxidize Fe^{2+} to Fe^{3+} under anoxic conditions. This process might help explain the origin of the abundant Fe^{3+} in ancient rocks such as the banded iron formations (⊘ Section 12.2), rocks dated to before the predicted appearance of cyanobacteria on Earth. With organisms like *Ferroglobus*, it would have been possible for Fe^{2+} oxidation to proceed without the need for molecular oxygen (O_2) as an electron acceptor. The metabolism of *Ferroglobus* thus has implications for dating the origin of cyanobacteria and the subsequent oxygenation of Earth. Certain anoxygenic phototrophic bacteria can also oxidize Fe^{2+} under anoxic conditions (⊘ Sections 13.3 and 13.9), and so several anaerobic routes to ancient Fe^{3+} are possible. This makes it difficult to estimate when cyanobacteria first appeared on Earth and to what degree nonphototrophic organisms helped trigger the Great Oxidation Event (⊘ Figure 12.1).

MINIQUIZ --
- Compare the energy-yielding metabolisms of *Archaeoglobus* and *Ferroglobus*.

II • *Thaumarchaeota*, *Nanoarchaeota*, and *Korarchaeota*

Our understanding of *Archaea* has been revolutionized by development of molecular phylogeny (⊘ Sections 12.4 and 12.5) and methods for studying microorganisms without the need for cultivation in laboratory cultures (⊘ Sections 18.5–18.7). The *Thaumarchaeota*, *Nanoarchaeota*, and *Korarchaeota* were all discovered and characterized initially with the aid of techniques for analysis of 16S ribosomal RNA genes. From these initial efforts, species representing each of these phyla were subsequently isolated or at least grown in enrichment cultures. We begin our consideration of these unusual phyla with the *Thaumarchaeota*.

16.6 *Thaumarchaeota* and Nitrification in *Archaea*

Key Genera: *Nitrosopumilus, Nitrososphaera*

Early surveys of 16S ribosomal RNA genes from open ocean microbial communities resulted in the shocking conclusion that *Archaea* were abundant and widespread in the oceans. At the time, the archaeal domain was considered to contain only extremophiles and obligate anaerobes, and their presence in oxygen-rich temperate and even polar oceanic environments was something of a mystery. Even more remarkable, these novel *Archaea* were widespread and common in soils all over the world. Phylogenetic analysis of their 16S ribosomal RNA gene sequences initially suggested that this novel group of *Archaea* was a deeply divergent lineage of the *Crenarchaeota*, a group of hyperthermophilic *Archaea* (see Section 16.9). It was only after genome sequence analysis of the marine nitrifier *Nitrosopumilus maritimus* that it became clear that the *Thaumarchaeota* are a distinct phylum of *Archaea*. Analyses of genome sequences confirm that *Thaumarchaeota* constitute a unique phylum of *Archaea* and that they diverged from the primary line of archaeal descent prior to the divergence of *Crenarchaeota* and *Euryarchaeota* (Figure 16.1).

Physiological Characteristics of *Thaumarchaeota*

The physiology of *Thaumarchaeota* remained a mystery until the isolation of *Nitrosopumilus maritimus* (**Figure 16.14**). *N. maritimus* grows chemolithotrophically by aerobically oxidizing ammonia (NH_3) to nitrite (NO_2^-), the first step in nitrification (⊘ Sections 13.10, 14.13, and 20.3). This organism uses CO_2 as its sole carbon source (autotrophy), as do nitrifying *Bacteria* (⊘ Section 14.13). However, unlike ammonia-oxidizing *Bacteria* such as *Nitrosomonas*, *N. maritimus* is adapted to life under extreme nutrient limitation, as would befit an organism indigenous to open ocean waters. *N. maritimus* can grow at NH_3 concentrations that are a hundred times lower than those required by bacterial nitrifiers, and are actually growth inhibited at the higher NH_3 concentrations required to support growth of nitrifying species of *Bacteria*.

Several species of *Thaumarchaeota* have been isolated and characterized, revealing a number of properties common to this group. Species have been isolated from habitats including the oceans, marine sediment, an estuary, soil, and hot springs. All

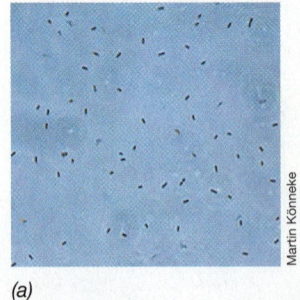
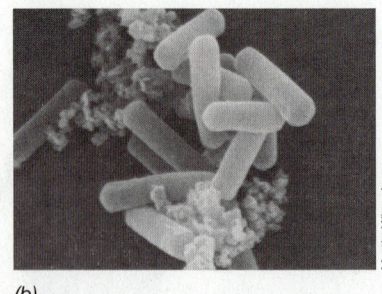

(a)   *(b)*

Figure 16.14 *Nitrosopumilus maritimus*, a nitrifying species of *Archaea*. This organism can oxidize NH_3 present at the very low amounts typical of marine environments. *(a)* Phase-contrast photomicrograph. *(b)* Scanning electron micrograph. A single cell of *N. maritimus* is about 0.2 μm in diameter.

existing isolates are chemolithotrophic ammonia-oxidizers, and most species, like *N. maritimus*, are able to grow at very low concentrations of NH_3. The membranes of all *Thaumarchaeota* also have a unique lipid called *crenarchaeol* (↩ Figure 2.17c), a compound limited to species of this phylum. In addition, autotrophy in *Thaumarchaeota* is supported by the 3-hydroxypropionate/4-hydroxybutyrate cycle, which further distinguishes archaeal nitrifiers from nitrifying *Bacteria* that employ the Calvin cycle for CO_2 fixation (↩ Section 13.5). The 3-hydroxypropionate/4-hydroxybutyrate cycle also allows for the assimilation of organic carbon, and some archaeal nitrifiers have been shown to assimilate pyruvate during mixotrophic growth. Growth temperatures of *Thaumarchaeota* vary widely, as some species thrive in polar seas while others inhabit hot spring environments up to about 75°C.

Nitrososphaera viennensis, which represents a lineage of *Thaumarchaeota* found widely in soils, can grow at a wide range of NH_3 concentrations. Like marine species of *Thaumarchaeota*, *N. viennensis* can grow at low concentrations of NH_3, but *N. viennensis* can also tolerate high levels (up to 10 mM) of ammonium at neutral pH. Hence, *N. viennensis*, and other archaeal nitrifiers, may be active in soils that have fairly high levels of ammonia and in these environments they may compete directly with bacterial nitrifiers. In addition, several species of *Thaumarchaeota*, including *N. viennensis*, possess urease activity. *N. viennensis* can grow with urea as the sole source of energy, hydrolyzing it to ammonia, which is subsequently used as an electron donor.

Environmental Distribution of *Thaumarchaeota*

We now know that *Thaumarchaeota*, ubiquitous in soils and found throughout the marine water column from the equator to the polar seas, are one of the most abundant and widespread phyla on our planet. In surveys of soil or marine samples, thaumarchaea are often found to be the dominant group of *Archaea*. By using fluorescent phylogenetic probes (FISH, ↩ Section 18.4), thaumarchaea have been detected in oxic marine waters worldwide; they thrive even in waters and sea ice near Antarctica (**Figure 16.15**). Marine species are planktonic (suspended freely or attached to suspended particles in the water column, Figure 16.15b) and

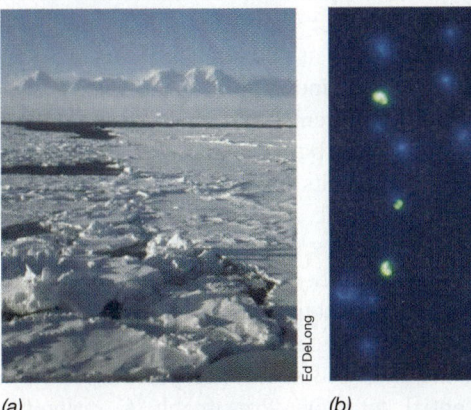

(a)   (b)

Ed DeLong

Figure 16.15 Cold-dwelling *Thaumarchaeota*. (a) Photo of the Antarctic Peninsula taken from shipboard. The frigid waters that lie under the surface ice shown here are habitats for cold-dwelling *Thaumarchaeota*. (b) Fluorescence photomicrograph of seawater treated with a FISH probe (↩ Section 18.4) specific for species of *Thaumarchaeota* (green cells). Blue cells are stained with DAPI, a fluorescent DNA stain that stains all cells.

present in significant numbers ($\sim10^4$/ml) in waters that are both nutrient-poor and very cold (0–4°C in seawater and below 0°C in sea ice). Marine thaumarchaea are found throughout the oceans and may constitute 20% of picoplankton (very small prokaryotes) worldwide. They are particularly abundant in the deep ocean where they can also account for up to 40% of picoplankton (↩ Section 19.12). The NH_3 concentration in marine waters is often at the threshold for archaeal nitrification, suggesting that *Thaumarchaeota* may play a major role in controlling the levels of NH_3 in the oceans (see "Genomics and New *Archaea*," page 183).

Thaumarchaea are also common in soils, comprising as much as 1–2% of the total ribosomal RNA in soil communities and, in some soils, outnumbering nitrifying *Bacteria* by 1000-fold. They are found in soils across a wide range of pH from 3.5 to 8.7. While widely present in soils, thaumarchaea may be particularly important in acid soils (pH < 5.5), which make up more than 30% of all soils. Nitrifiers oxidize NH_3, but at low pH, NH_4^+ predominates and is thus unavailable for nitrification. While nitrification occurs in acid soils, and often at high rates, bacterial nitrifiers have not been observed to grow below pH 6.3. In contrast, the thaumarchaeotal species *Nitrosotalea devanaterra*, isolated from acidic agricultural soil, grows optimally at pH 4–5. The ability of thaumarchaea to grow at low NH_3 concentrations explains how they can be successful in acidic soils where free NH_3 is present in very low concentration.

MINIQUIZ
- How does the organism *Nitrosopumilus maritimus* conserve energy and obtain carbon?
- In what environments might you expect to find species of *Thaumarchaeota*?

16.7 *Nanoarchaeota* and the "Hospitable Fireball"

Key Genus: *Nanoarchaeum*

The *Nanoarchaeota* are represented by a single species, the highly unusual *Nanoarchaeum equitans*. *N. equitans* is one of the smallest cellular organisms known and has the smallest genome among species of *Archaea* (0.49 Mb). The coccoid cells of *N. equitans* are very small, about 0.4 μm in diameter, and have only about 1% of the volume of an *Escherichia coli* cell. They cannot grow in pure culture and replicate only when attached to the surface of their host organism, *Ignicoccus hospitalis* (Section 16.11), a hyperthermophilic species of *Crenarchaeota* whose name means "the hospitable fireball." *N. equitans* grows to 10 or more cells per *Ignicoccus* cell and lives an apparently parasitic lifestyle (**Figure 16.16**), making it the only known archaeal symbiont.

Nanoarchaeum and Its Host

N. equitans and its host *Ignicoccus* were first isolated from a submarine hydrothermal vent (↩ Section 19.13) off the coast of Iceland. However, environmental sampling of 16S ribosomal RNA genes (↩ Section 18.5) indicates that organisms phylogenetically similar to *N. equitans* exist in other submarine hydrothermal vents and in terrestrial hot springs, so *Archaea* of this kind are probably distributed worldwide in suitable hot habitats. Like its host *Ignicoccus*, *N. equitans* grows at temperatures from 70 to 98°C and optimally at 90°C.

UNIT 3

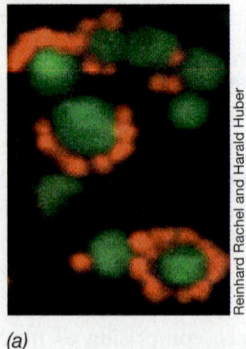
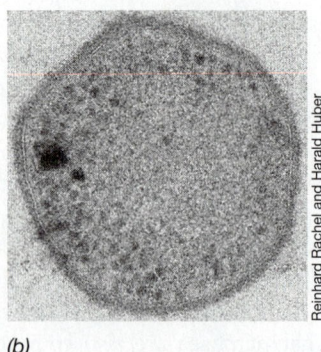

(a) (b)

Figure 16.16 *Nanoarchaeum equitans.* *(a)* Fluorescence micrograph of cells of *N. equitans* (red) attached to cells of *Ignicoccus* (green). Cells were stained by FISH (♻ Section 18.4) using specific nucleic acid probes targeted to each organism. *(b)* Transmission electron micrograph of a thin section of a cell of *N. equitans*. Note the distinct cell wall. Cells of *N. equitans* are about 0.4 μm in diameter. Note the unique phylogeny of *Nanoarchaeum* in Figure 16.1.

The metabolism of *Nanoarchaeum* is not fully understood, but it likely depends on its host for many metabolic functions. *Ignicoccus* is an autotroph that uses H_2 as an electron donor and S^0 as an electron acceptor and so probably supplies *N. equitans* with organic carbon. *N. equitans* is incapable of metabolizing H_2 and S^0 for energy, and whether it generates ATP from substances obtained from *Ignicoccus* or obtains its ATP directly from its host is unknown. The appearance of *N. equitans* cells is typical of *Archaea*, with a cell wall consisting of an S-layer (♻ Section 2.12) that overlays what appears to be a periplasmic space (Figure 16.16*b*).

Although the sequence of its 16S ribosomal RNA gene clearly places *N. equitans* in the domain *Archaea*, the sequence differs at many sites from 16S ribosomal RNA gene sequences from other *Archaea*, even in regions of the molecule that are highly conserved among *Archaea*. These differences initially led to the conclusion that *N. equitans* was an early-branching lineage of *Archaea* (♻ Figure 12.13). However, more detailed phylogenetic analyses of genes encoding ribosomal proteins suggest that the divergence of *N. equitans* occurred at about the time that the *Euryarchaeota* were formed (Figure 16.1). Some analyses even suggest that *N. equitans* may be a species of *Euryarchaeota*. However, genomic analyses show the organism to lack several genes that encode information processing and cell division in *Euryarchaeota*. Conclusive phylogenetic placement of the *Nanoarchaeota* will ultimately require the discovery of more species from this group.

The *N. equitans* Genome

The sequence of the *N. equitans* genome provides insight into this organism's obligately parasitic lifestyle. Its single, circular genome is only 490,885 nucleotides long, one of the smallest cellular genomes yet sequenced (♻ Table 6.1). Genes for several important metabolic functions are missing from the *N. equitans* genome, including those for the biosynthesis of amino acids, nucleotides, coenzymes, and lipids. Also missing are genes encoding proteins for widely distributed catabolic pathways, such as glycolysis. Presumably, all of these functions are carried out for *N. equitans* by its *Ignicoccus* host, with transfer of needed substances from *Ignicoccus* to the attached *N. equitans* cells. *N. equitans* also lacks some

of the genes necessary to encode ATPase, and this indicates that it may not synthesize a functional ATPase. If true, this would be a first for cellular organisms. If no ATPase is present and substrate-level phosphorylation does not occur (due to the lack of glycolytic enzymes), then *N. equitans* would be dependent on *Ignicoccus* for energy as well as carbon.

With so many genes missing, which genes remain in the *N. equitans* genome? *N. equitans* contains genes encoding the key enzymes for DNA replication, transcription, and translation as well as genes for DNA repair enzymes. In addition to its small size, the genome of *N. equitans* is also among the most gene dense of any organism known; over 95% of the *N. equitans* chromosome encodes proteins. It appears that *N. equitans* is living near the limits of life in terms of cell volume and genetic capacity.

MINIQUIZ
- Which aspects of the biology of *Nanoarchaeum equitans* make it especially interesting from an evolutionary point of view?
- Why can it be said that *Nanoarchaeum equitans* is both a carbon and an energy parasite?

16.8 *Korarchaeota* and the "Secret Filament"

Key Genus: *Korarchaeum*

Ribosomal RNA sequences of *Korarchaeota* have been observed in a range of geothermal habitats, both submarine and terrestrial. However, *Korarchaeum cryptofilum*, whose name means "the cryptic filament of youth," is the only characterized species in the phylum *Korarchaeota*.

First observed as a 16S ribosomal RNA gene phylotype recovered from the hot spring named Obsidian Pool in Yellowstone National Park, USA, *K. cryptofilum* has yet to be grown in pure culture, as for *N. equitans* (Section 16.7). However, its genome sequence has been determined from metagenomic analyses (see ♻ Sections 6.10 and 18.7) of an enrichment culture. *K. cryptofilum* is an obligately anaerobic chemoorganotroph and a hyperthermophile, growing at 85°C. Cells are long, thin (<0.2-μm diameter) filaments of variable length (**Figure 16.17a–c**), with most filaments being around 15 μm long but some reaching as much as 100 μm. Filaments of *K. cryptofilum* have a tough paracrystalline S-layer (Figure 16.17*d*), which maintains cell integrity in its extremely hot habitat.

Though *K. cryptofilum* cannot be grown in isolation, its genome sequence provides clues about its lifestyle. *K. cryptofilum* lacks the ability to perform anaerobic respiration (with the possible exception of proton reduction, ♻ Section 13.21) and lives a fermentative lifestyle. Similar to other archaeal hyperthermophiles, *K. cryptofilum* grows by fermentation of peptides or amino acids (see Table 16.6). *K. cryptofilum* lacks many core genes in biosynthesis including the ability to synthesize purines, coenzyme A, and several essential cofactors. Presumably *K. cryptofilum* obtains these essential components from its environment. The inability of *K. cryptofilum* to synthesize molecules essential for its own growth may be explained by the evolution of mutual dependence as described by the Black Queen hypothesis (♻ Explore the Microbial World, Chapter 12). This dependence on other

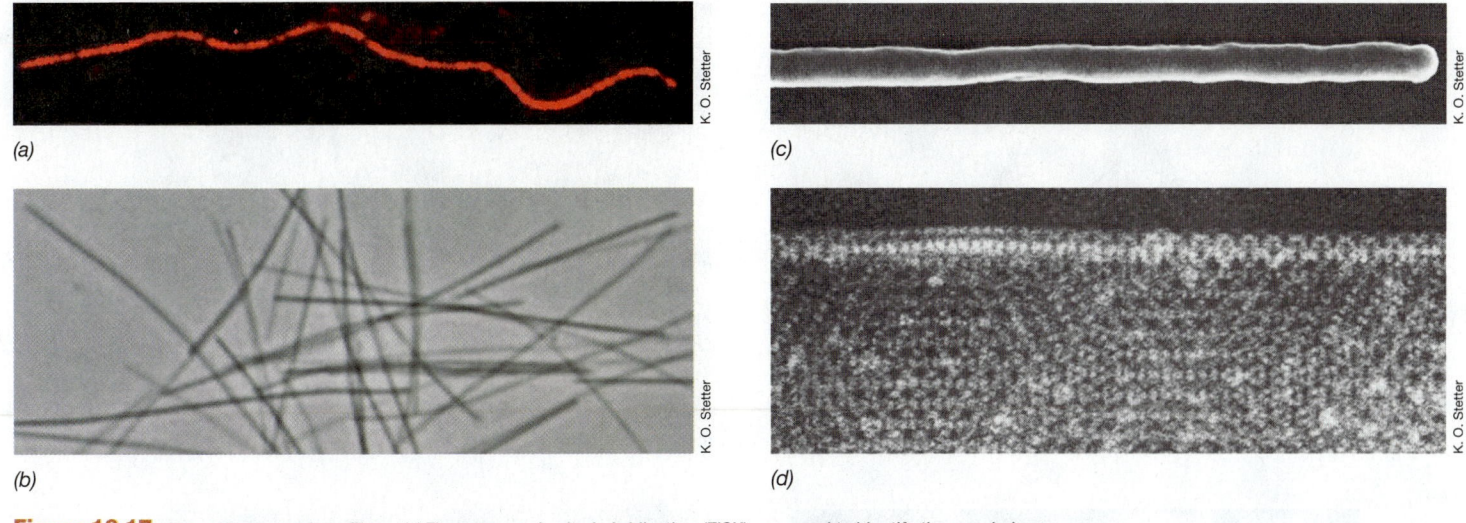

(a)

(b)

(c)

(d)

Figure 16.17 *Korarchaeum cryptofilum. (a)* Fluorescence in situ hybridization (FISH) was used to identify the morphology of *Korarchaeota* growing in an enrichment culture at 85°C. *(b)* Phase-contrast image of filaments of *K. cryptofilum. (c)* Scanning electron micrograph of a *K. cryptofilum* filament. *(d)* Transmission electron micrograph of the surface of a *K. cryptofilum* filament showing the paracrystalline S-layer (⟳ Section 2.12). Filaments of *K. cryptofilum* are about 0.17 μm wide and 15 μm long.

members of the hot spring microbial community may explain why *K. cryptofilum* has proven difficult to obtain in pure culture.

As with the *Nanoarchaeota*, there is some uncertainty about the phylogenetic position of the *Korarchaeota*. The genome of *K. cryptofilum* includes some gene families that share affinity with *Euryarchaeota* and others that share affinity with *Crenarchaeota*. For example, phylogenetic analysis of ribosomal proteins, RNA polymerase subunits, and ribosomal RNA genes indicate affinity between *Crenarchaeota* and *Korarchaeota*. In contrast, genes for cell division, tRNA maturation, and DNA replication and repair indicate affinity between *Euryarchaeota* and *Korarchaeota*. The unique genetic composition of *K. cryptofilum* supports its placement near the base of the archaeal radiation (Figure 16.1), and future work on this interesting archaeon should clarify its actual phylogenetic position.

MINIQUIZ

- What is the most likely reason that *Korarchaeum cryptofilum* has been difficult to isolate in pure culture?

III • *Crenarchaeota*

Among *Archaea* in laboratory culture, the **Crenarchaeota** are mostly hyperthermophiles and include species growing optimally above the boiling point of water. Many hyperthermophiles are chemolithotrophic autotrophs, and because no phototrophs can survive such temperatures, these organisms are the sole primary producers in these habitats.

16.9 Habitats and Energy Metabolism of *Crenarchaeota*

Most hyperthermophilic *Archaea* have been isolated from geothermally heated soils or waters containing S^0 and H_2S, and most species metabolize sulfur in one way or another. In terrestrial environments, sulfur-rich springs, boiling mud, and soils may have temperatures up to 100°C and are mildly to extremely acidic owing to the production of sulfuric acid (H_2SO_4) from the biological oxidation of H_2S and S^0 (⟳ Sections 13.8 and 20.4). Such hot, sulfur-rich environments, called **solfataras**, are found throughout the world (**Figure 16.18**), including Italy, Iceland, New Zealand, and Yellowstone National Park in Wyoming (USA). Depending on the surrounding geology, solfataras can be mildly acidic to slightly alkaline (pH 5–8) or extremely acidic, with pH values below 1. Hyperthermophilic crenarchaeotes have been obtained from all of these environments, but most inhabit neutral or weakly acidic thermal habitats.

Hyperthermophilic *Crenarchaeota* also inhabit undersea hot springs called **hydrothermal vents**. We discuss the geology and microbiology of these habitats in Section 19.13. Here we only note that submarine waters can be much hotter than surface waters because the water is under hydrostatic pressure. Indeed, all hyperthermophiles with growth temperature optima above 100°C originate from submarine sources. These sources include shallow (2–10 m depth) vents such as those off the coast of Vulcano, Italy, as well as deep (2000–4000 m depth) vents near ocean-spreading centers (see Figure 16.24). Deep hydrothermal vents are the hottest habitats so far known to yield prokaryotes.

With a few exceptions, hyperthermophilic *Crenarchaeota* are obligate anaerobes. Their energy-yielding metabolism is either chemoorganotrophic or chemolithotrophic (or both, for example, in *Sulfolobus*) and is dependent on diverse electron donors and acceptors. Fermentation is rare and most bioenergetic strategies involve anaerobic respirations (**Table 16.6**). Energy is conserved

Figure 16.18 **Terrestrial habitats of hyperthermophilic** *Archaea:* **Yellowstone National Park (Wyoming, USA).** *(a)* A typical solfatara; steam rich in H_2S rises to the surface. *(b)* Sulfur-rich hot spring, a habitat containing dense populations of *Sulfolobus*. The acidity in solfataras and sulfur springs comes from the oxidation of H_2S and S^0 to H_2SO_4 (sulfuric acid) by *Sulfolobus* and related prokaryotes. *(c)* A typical neutral pH boiling spring, Imperial Geyser. Many different species of hyperthermophilic *Archaea* may reside in such a habitat. *(d)* An acidic iron-rich geothermal spring, another *Sulfolobus* habitat; here the oxidation of Fe^{2+} to Fe^{3+} generates acidity.

during these respiratory processes by the same general mechanism widespread in *Bacteria*: electron transfer within the cytoplasmic membrane leading to the formation of a proton motive force from which ATP is made by way of proton-translocating ATPases (🔁 Section 3.11). Classical substrate-level phosphorylation, a hallmark of fermentative *Bacteria* (🔁 Section 3.7), is unknown in *Archaea*.

Many hyperthermophilic crenarchaeotes can grow chemolithotrophically under anoxic conditions, with H_2 as the electron donor and S^0 or NO_3^- as the electron acceptor; a few can also oxidize H_2 aerobically (Table 16.6). H_2 respiration with ferric iron (Fe^{3+}) as electron acceptor also occurs in several hyperthermophiles. Other chemolithotrophic lifestyles include the oxidation of S^0 and Fe^{2+} aerobically or Fe^{2+} anaerobically with NO_3^- as the acceptor (Table 16.6). Only one sulfate-reducing hyperthermophile is known (the euryarchaeote *Archaeoglobus*, Section 16.5). The only bioenergetic option apparently impossible is photosynthesis, a means of energy conservation that is apparently limited to temperatures below 74°C (see Figure 16.28).

MINIQUIZ -

• Why is it unlikely that hyperthermophiles with growth temperature optima >100°C would reside in terrestrial hot springs?

• What form of energy metabolism is widespread among hyperthermophiles?

16.10 *Crenarchaeota* from Terrestrial Volcanic Habitats

Key Genera: *Sulfolobus, Acidianus, Thermoproteus, Pyrobaculum*

Terrestrial volcanic habitats can have temperatures as high as 100°C and are thus suitable for hyperthermophilic *Archaea*. Two phylogenetically related organisms isolated from these environments are *Sulfolobus* and *Acidianus*. These genera form the heart of an order called the *Sulfolobales* (**Table 16.7**). In addition, *Sulfolobus* has been a model organism for molecular biology studies of *Archaea*.

Table 16.6 Energy-yielding reactions of hyperthermophilic *Archaea*

Nutritional class	Energy-yielding reaction	Metabolic type[a]	Example genera[b]
Chemoorganotrophic	Organic compound + $S^0 \rightarrow H_2S + CO_2$	AnR	Thermoproteus, Thermococcus, Desulfurococcus, Thermofilum, Pyrococcus
	Organic compound + $SO_4^{2-} \rightarrow H_2S + CO_2$	AnR	Archaeoglobus
	Organic compound + $O_2 \rightarrow H_2O + CO_2$	AeR	Sulfolobus
	Organic compound $\rightarrow CO_2 + H_2$ + fatty acids	AnR	Staphylothermus, Pyrodictium
	Organic compound + $Fe^{3+} \rightarrow CO_2 + Fe^{2+}$	AnR	Pyrodictium
	Organic compound + $NO_3^- \rightarrow CO_2 + N_2$	AnR	Pyrobaculum
	Pyruvate $\rightarrow CO_2 + H_2$ + acetate	AnR	Pyrococcus
	Peptides	F	Hyperthermus, Korarchaeum
Chemolithotrophic	$H_2 + S^0 \rightarrow H_2S$	AnR	Acidianus, Pyrodictium, Thermoproteus, Stygiolobus, Ignicoccus
	$H_2 + NO_3^- \rightarrow NO_2^- + H_2O$ (NO_2^- is reduced to N_2 by some species)	AnR	Pyrobaculum
	$4 H_2 + NO_3^- + H^+ \rightarrow NH_4^+ + 2 H_2O + OH^-$	AnR	Pyrolobus
	$H_2 + 2 Fe^{3+} \rightarrow 2 Fe^{2+} + 2 H^+$	AnR	Pyrobaculum, Pyrodictium, Archaeoglobus
	$2 H_2 + O_2 \rightarrow 2 H_2O$	AeR	Acidianus, Sulfolobus, Pyrobaculum
	$2 S^0 + 3 O_2 + 2 H_2O \rightarrow 2 H_2SO_4$	AeR	Sulfolobus, Acidianus
	$2 FeS_2 + 7 O_2 + 2 H_2O \rightarrow 2 FeSO_4 + 2 H_2SO_4$	AeR	Sulfolobus, Acidianus, Metallosphaera
	$2 FeCO_3 + NO_3^- + 6 H_2O \rightarrow 2 Fe(OH)_3 + NO_2^- + 2 HCO_3^- + 2 H^+ + H_2O$	AnR	Ferroglobus
	$4 H_2 + SO_4^{2-} + 2 H^+ \rightarrow 4 H_2O + H_2S$	AnR	Archaeoglobus
	$4 H_2 + CO_2 \rightarrow CH_4 + 2 H_2O$	AnR	Methanopyrus, Methanocaldococcus Methanothermus

[a]AnR, anaerobic respiration; AeR, aerobic respiration; F, fermentation.
[b]Most are *Crenarchaeota*; see Figure 16.1.

Sulfolobales

Sulfolobus grows in sulfur-rich acidic thermal areas (Figure 16.18) at temperatures up to 90°C and at pH values of 1–5. *Sulfolobus* is an aerobic chemolithotroph that oxidizes H_2S or S^0 to H_2SO_4 and fixes CO_2 as a carbon source. *Sulfolobus* can also grow chemoorganotrophically. Cells of *Sulfolobus* are more or less spherical but contain distinct lobes (**Figure 16.19a**). Cells adhere tightly to sulfur crystals, where they can be seen with a microscope after staining with fluorescent dyes (↩ Figure 13.21b). Besides the aerobic respiration of sulfur or organic compounds, *Sulfolobus* can also oxidize Fe^{2+} to Fe^{3+}, and this ability has been harnessed for the high-temperature leaching of iron and copper ores (↩ Sections 20.5 and 21.1).

A facultative aerobe resembling *Sulfolobus* also lives in acidic solfataric springs. This organism, *Acidianus* (Figure 16.19b), differs from *Sulfolobus* most clearly by its ability to grow using S^0 both anaerobically as well as aerobically. Under aerobic conditions the organism uses S^0 as an electron *donor*, oxidizing S^0 to H_2SO_4, with O_2 as an electron acceptor. Anaerobically, *Acidianus* uses S^0 as an electron *acceptor* with H_2 as an electron donor, forming H_2S as the reduced product. Thus, the metabolic fate of S^0 in cultures of *Acidianus* depends on the presence or absence of O_2. Like *Sulfolobus*, *Acidianus* is roughly spherical in shape but is not as lobed (Figure 16.19b). It grows at temperatures from 65°C up to a maximum of 95°C, with an optimum of about 90°C. As a group, then, the *Sulfolobales* contain the most thermophilic of all highly acidophilic *Archaea*.

Thermoproteales

Key genera within the *Thermoproteales* are *Thermoproteus*, *Thermofilum*, and *Pyrobaculum*. The genera *Thermoproteus* and *Thermofilum* consist of rod-shaped cells that inhabit neutral or slightly acidic hot springs. Cells of *Thermoproteus* are rigid rods about 0.5 μm in diameter and highly variable in length, ranging from short cells of 1–2 μm (**Figure 16.20a**) up to filaments 70–80 μm long. Filaments of *Thermofilum* are thinner, some as little as 0.17–0.35 μm wide, with filament lengths ranging up to 100 μm (Figure 16.20b).

Both *Thermoproteus* and *Thermofilum* are strict anaerobes that carry out an S^0-based anaerobic respiration (Table 16.6). Most *Thermoproteus* isolates can grow chemolithotrophically on H_2 or chemoorganotrophically on complex carbon substrates such as yeast extract, small peptides, starch, glucose, ethanol, malate, fumarate, or formate (Table 16.6). *Pyrobaculum* (Figure 16.20c) is a rod-shaped hyperthermophile but is physiologically distinct from other *Thermoproteales* in that some species of *Pyrobaculum* can respire aerobically. However, *Pyrobaculum* can also grow by anaerobic respiration with NO_3^-, Fe^{3+}, or S^0 as electron acceptors and H_2 as an electron donor (that is, they can grow chemolithotrophically and autotrophically). Other species of *Pyrobaculum* can grow anaerobically on organic electron donors, reducing S^0 to H_2S. The growth temperature optimum of *Pyrobaculum* is 100°C, and species of this organism have been isolated from terrestrial hot springs and from hydrothermal vents.

Table 16.7 Properties of some hyperthermophilic *Crenarchaeota*

Order/Genus[a]	Morphology	Relationship to O_2[b]	Minimum	Optimum	Maximum	Optimum pH
Sulfolobales						
Sulfolobus	Lobed coccus	Ae	55	75	87	2–3
Acidianus	Coccus	Fac	60	88	95	2
Metallosphaera	Coccus	Ae	50	75	80	2
Stygiolobus	Lobed coccus	An	57	80	89	3
Sulfurisphaera	Coccus	Fac	63	84	92	2
Sulfurococcus	Coccus	Ae	40	75	85	2.5
Thermoproteales						
Thermoproteus	Rod	An	60	88	96	6
Thermofilum	Rod	An	70	88	95	5.5
Pyrobaculum	Rod	Fac	74	100	102	6
Caldivirga	Rod	An	60	85	92	4
Thermocladium	Rod	An	60	75	80	4.2
Desulfurococcales						
Desulfurococcus	Coccus	An	70	85	95	6
Aeropyrum	Coccus	Ae	70	95	100	7
Staphylothermus	Cocci in clusters	An	65	92	98	6–7
Pyrodictium	Disc-shaped with filaments	An	82	105	110	6
Pyrolobus	Lobed coccus	Fac	90	106	113	5.5
Thermodiscus	Disc-shaped	An	75	90	98	5.5
Ignicoccus	Irregular coccus	An	65	90	103	5
Hyperthermus	Irregular coccus	An	75	102	108	7
Stetteria	Coccus	An	68	95	102	6
Sulfophobococcus	Disc-shaped	An	70	85	95	7.5
Thermosphaera	Coccus	An	67	85	90	7
Strain 121[c]	Coccus	An	85	106	121	7

[a]The group names ending in "ales" are order names.
[b]Ae, aerobe; An, anaerobe; Fac, facultative
[c]Also known by the unofficial taxonomic name of "*Geodemma barossii*."

MINIQUIZ

• What are the major differences between *Sulfolobus* and *Pyrolobus*?

• Among *Thermoproteales*, what is unusual about the metabolism of *Pyrobaculum*?

16.11 *Crenarchaeota* from Submarine Volcanic Habitats

Key Genera: *Pyrodictium, Pyrolobus, Ignicoccus, Staphylothermus*

We now consider the microbiology of submarine volcanic habitats, homes to the most thermophilic of all known *Archaea*. These habitats include both shallow-water thermal springs and deep-sea hydrothermal vents. We discuss the geology of these fascinating microbial habitats in Section 19.13 and the interesting animal communities that develop there in Section 22.12. The organisms to be described here constitute an order of *Archaea* called the *Desulfurococcales* (Table 16.7).

Pyrodictium and *Pyrolobus*

Pyrodictium and *Pyrolobus* are examples of microorganisms whose growth temperature optimum lies above 100°C; the optimum for *Pyrodictium* is 105°C and for *Pyrolobus* is 106°C. Cells of *Pyrodictium* are irregularly disc-shaped and grow in culture in a mycelium-like layer attached to crystals of S^0. The cell mass consists of a network of fibers to which individual cells are attached (Figure 16.21a). The fibers are hollow and consist of protein arranged in a fashion similar to that of bacterial flagella (↩ Section 2.17). However, the filaments do not function in motility but instead as organs of attachment. The cell walls of *Pyrodictium* are composed of glycoprotein. Physiologically, *Pyrodictium* is a strict anaerobe that grows chemolithotrophically on H_2 as an electron donor and S^0 as an electron acceptor or chemoorganotrophically on complex mixtures of organic compounds (Table 16.6).

Pyrolobus fumarii (Figure 16.21c) is one of the most thermophilic of the hyperthermophiles. Its growth temperature

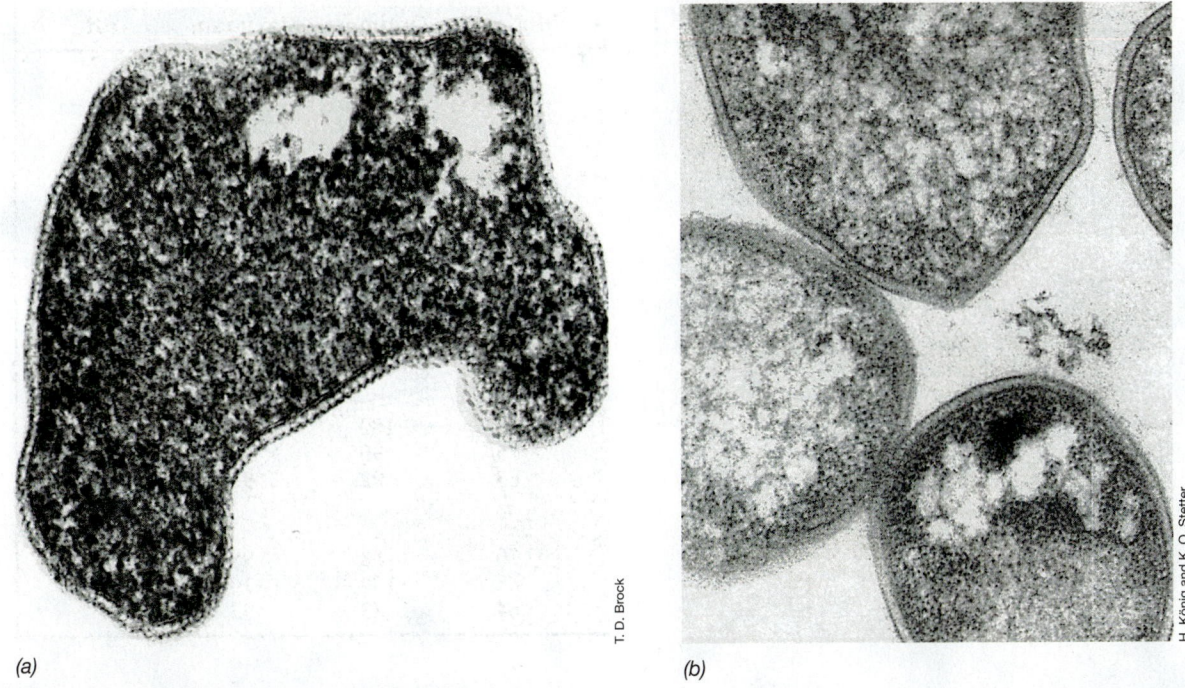

(a)

T. D. Brock

(b)

H. König and K. O. Stetter

Figure 16.19 Acidophilic hyperthermophilic *Archaea*, the *Sulfolobales*. *(a) Sulfolobus acidocaldarius.* Electron micrograph of a thin section. *(b) Acidianus infernus.* Electron micrograph of a thin section. Cells of both organisms vary from 0.8 to 2 μm in diameter. *Sulfolobales* typically show temperature optima below 90°C (Table 16.7).

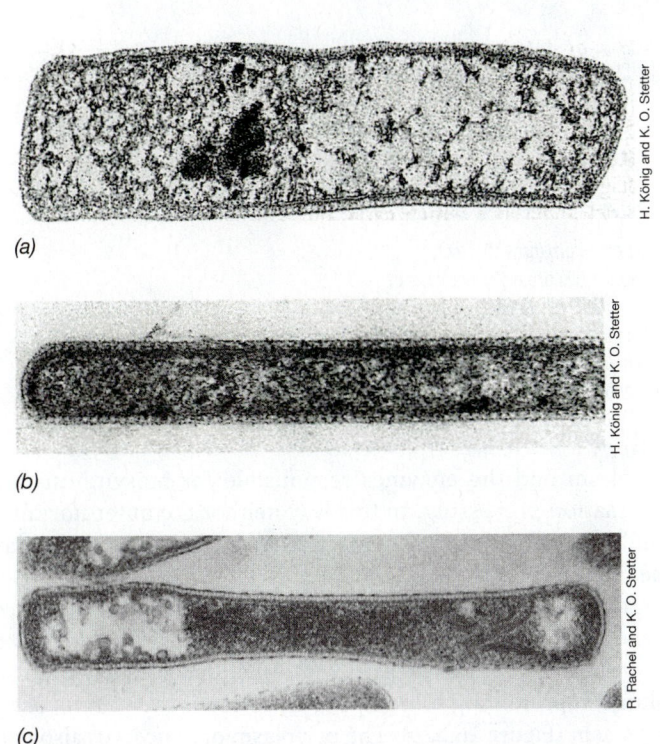

(a)

H. König and K. O. Stetter

(b)

H. König and K. O. Stetter

(c)

R. Rachel and K. O. Stetter

Figure 16.20 Rod-shaped hyperthermophilic *Archaea*, the *Thermoproteales*. *(a) Thermoproteus neutrophilus.* Electron micrograph of a thin section. A cell is about 0.5 μm in diameter. *(b) Thermofilum librum.* A cell is about 0.25 μm in diameter. *(c) Pyrobaculum aerophilum.* Transmission electron micrograph of a thin section; a cell measures 0.5 × 3.5 μm. Although the temperature optimum of *P. aerophilum* is 100°C, optima for other *Thermoproteales* are all below 90°C (Table 16.7).

maximum is 113°C (Table 16.7). *P. fumarii* lives in the walls of "black smoker" hydrothermal vent chimneys (c⊋ Section 19.13 and Figures 19.35, 19.37, and 19.38) where its autotrophic abilities contribute organic carbon to this otherwise inorganic environment. *P. fumarii* cells are coccoid-shaped (Figure 16.21*c*), and the cell wall is composed of protein. The organism is an obligate H_2 chemolithotroph, growing by the oxidation of H_2 coupled to the reduction of NO_3^- to ammonium (NH_4^+), thiosulfate ($S_2O_3^{2-}$) to H_2S, or very low concentrations of O_2 to H_2O. Besides its extremely thermophilic nature, *P. fumarii* can withstand temperatures substantially above its growth temperature maximum. For example, cultures of *P. fumarii* survive autoclaving (121°C) for 1 h, a condition that even bacterial endospores (c⊋ Section 2.16) cannot withstand.

Another organism in this group shares with *Pyrolobus* a growth temperature optimum of 106°C. However, "Strain 121," as this organism has been called, actually shows weak growth at 121°C, and cells remain viable for 2 h at 130°C. Only *Methanopyrus*, a hyperthermophilic methanogen, can grow at a higher temperature (122°C, Section 16.4). Strain 121 consists of coccoid, flagellated cells (Figure 16.21*d*); the organism is also a strict anaerobe and grows chemolithotrophically and autotrophically with Fe^{3+} as electron acceptor and formate or H_2 as electron donors. It is thus clear that the *Pyrodictium*/*Pyrolobus* group collectively contains the most hyperthermophilic examples of all known prokaryotes.

Desulfurococcus and *Ignicoccus*

Other notable members of the *Desulfurococcales* include *Desulfurococcus*, the genus for which the order is named (**Figure 16.22**), and *Ignicoccus*. *Desulfurococcus* is a strictly anaerobic S^0-reducing

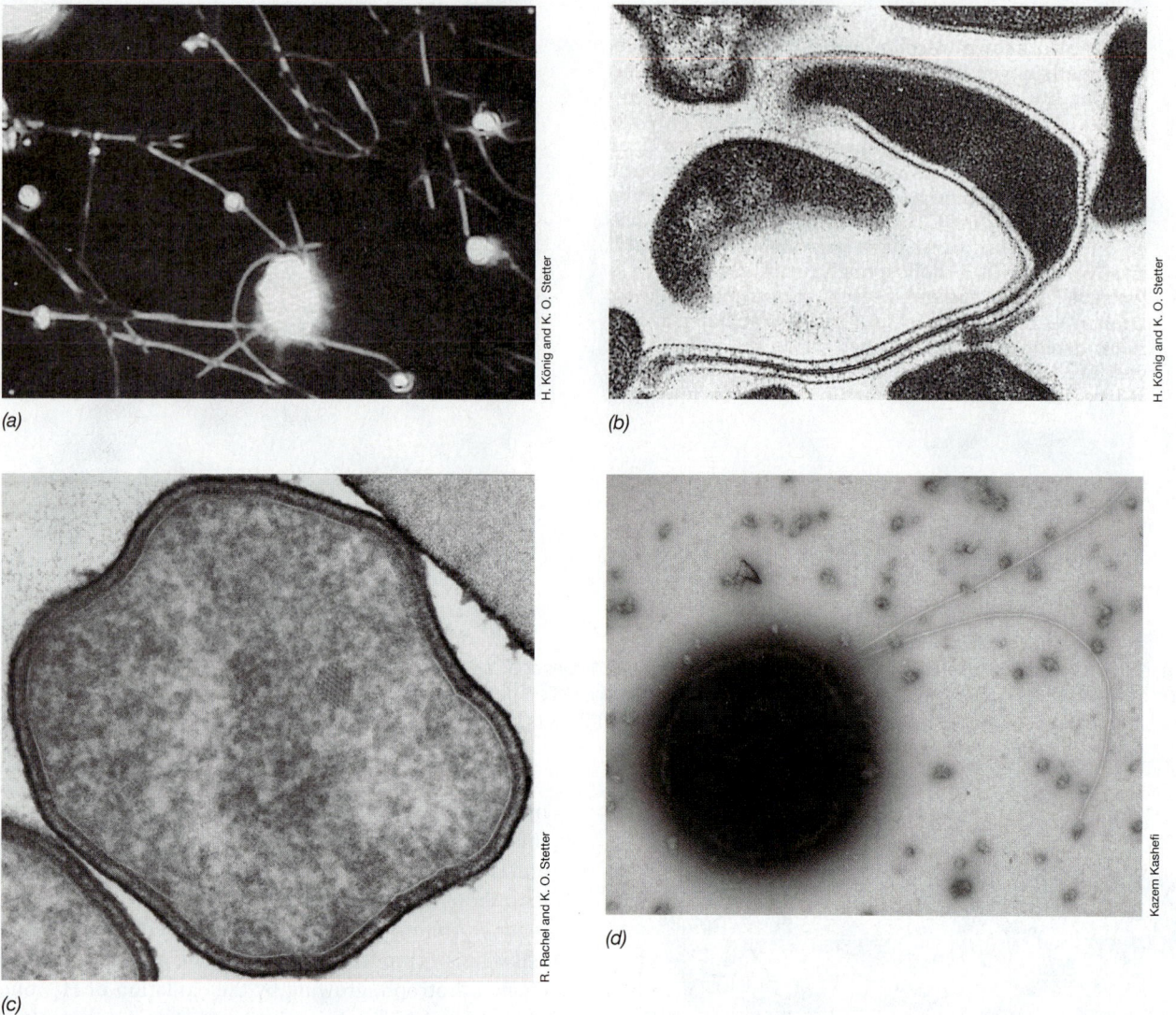

(a) — H. König and K. O. Stetter

(b) — H. König and K. O. Stetter

(c) — R. Rachel and K. O. Stetter

(d) — Kazem Kashefi

Figure 16.21 *Desulfurococcales* with growth temperature optima >100°C. *(a)* Pyrodictium occultum (growth temperature optimum, 105°C), dark-field micrograph. *(b)* Thin-section electron micrograph of *P. occultum*. Cells are highly variable in diameter from 0.3 to 2.5 μm. *(c)* Thin section of a cell of *Pyrolobus fumarii*, one of the most thermophilic of all known bacteria (growth temperature optimum, 106°C); a cell is about 1.4 μm in diameter. *(d)* Negative stain of a cell of "Strain 121," capable of growth at 121°C; a cell is about 1 μm wide. Although the *Desulfurococcales* contain the greatest number of hyperthermophiles capable of growth above 100°C, the most thermophilic of all known *Archaea* is actually a euryarchaeote, *Methanopyrus* (Section 16.4).

organism like *Pyrodictium*, but differs from this organism in its phylogeny and the fact that it is much less thermophilic, growing optimally at about 85°C.

Ignicoccus grows optimally at 90°C, and its energy metabolism is based on H₂ as an electron donor and S⁰ as an electron acceptor, as is that of so many hyperthermophilic *Archaea* (Table 16.6). Some *Ignicoccus* species are hosts to the small parasitic archaeon, *Nanoarchaeum equitans* (Section 16.7). *Ignicoccus* (Figure 16.22b) has a novel cell structure that lacks an S-layer and possesses a unique *outer cellular membrane*. This outer cellular membrane is distinct in several ways from the outer membrane of gram-negative *Bacteria* (⇆ Section 2.11). Most notably, the outer cellular membrane of *Ignicoccus* contains ATPase and is the site of energy conservation. *Ignicoccus* also contains an inner cellular membrane that contains the

cytoplasm and the enzymes responsible for biosynthesis and information processing. In this way neither the outer nor inner cellular membrane satisfies the typical definition of a cytoplasmic membrane (⇆ Section 2.7).

Between the inner and outer cellular membranes of *Ignicoccus* is a large *intermediate compartment* that is analogous to the periplasm of gram-negative *Bacteria* but is much larger in volume, representing some two to three times the volume of the cytoplasm (Figure 16.22b). The periplasm of *Ignicoccus* also contains membrane-bound vesicles (Figure 16.22b) that may function in exporting substances outside the cell. In this way, the cell structure of *Ignicoccus* resembles that of *Eukarya*. Hence, *Ignicoccus* has been proposed to be a modern descendant of the ancestral cell type that gave rise to the origin of eukaryotic cells (⇆ Figure 12.9).

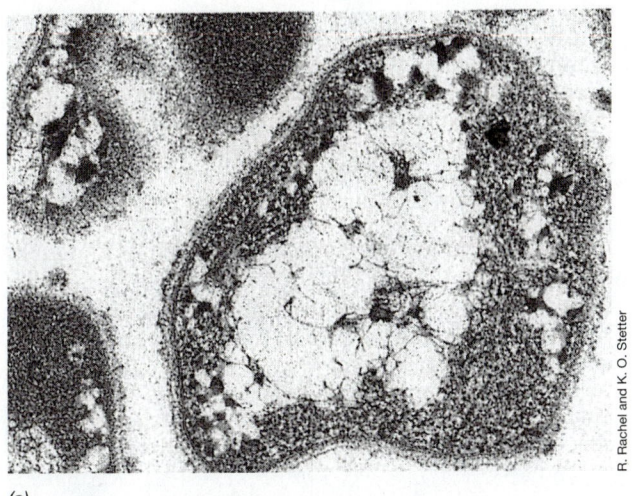

(a)

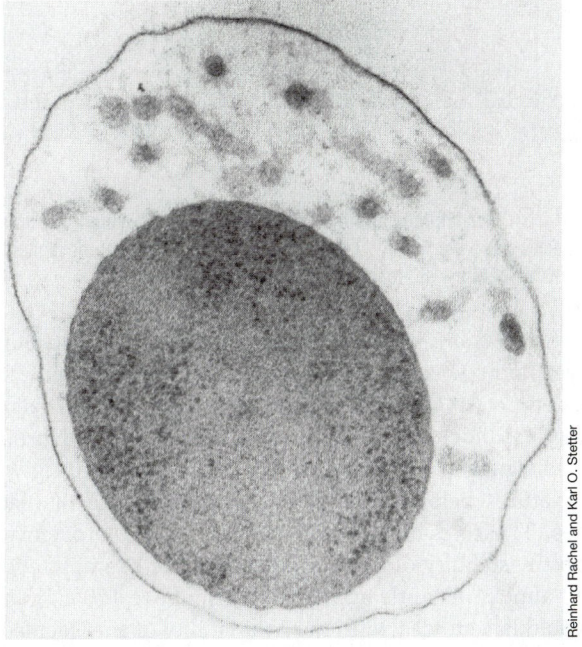

(b)

R. Rachel and K. O. Stetter

Reinhard Rachel and Karl O. Stetter

Figure 16.22 *Desulfurococcales* **with growth temperature optima <100°C.** *(a)* Thin section of a cell of *Desulfurococcus saccharovorans*; a cell is 0.7 μm in diameter. *(b)* Thin section of a cell of *Ignicoccus islandicus*. The cell proper is surrounded by an extremely large periplasm. The cell itself measures about 1 μm in diameter and the cell plus periplasm measures 1.4 μm.

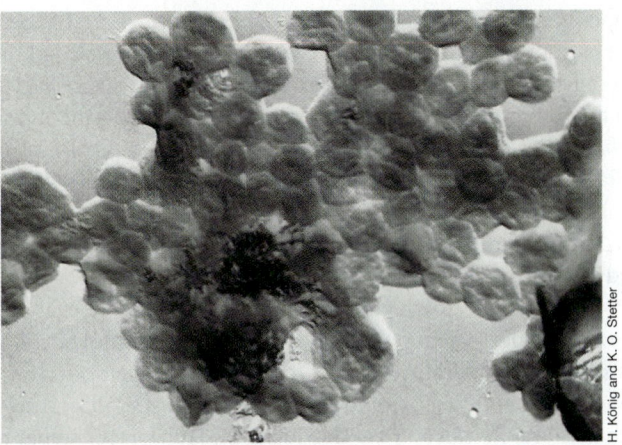

H. König and K. O. Stetter

Figure 16.23 **The hyperthermophile *Staphylothermus marinus*.** Electron micrograph of shadowed cells. A single cell is about 1 μm in diameter.

Staphylothermus

A morphologically unusual member of the order *Desulfurococcales* is the genus *Staphylothermus* (**Figure 16.23**). Cells of *Staphylothermus* are spherical, about 1 μm in diameter, and form aggregates of up to 100 cells, much like its morphological counterpart among the *Bacteria*, *Staphylococcus* (Figures 15.20 and 29.29*a*). Unlike many hyperthermophiles, *Staphylothermus* is not a chemolithotroph, but instead a chemoorganotroph, growing optimally at 92°C. Energy is obtained from the fermentation of peptides, producing the fatty acids acetate and isovalerate as fermentation products (Table 16.6).

Isolates of *Staphylothermus* have been obtained from both shallow marine hydrothermal vents and very hot black smokers (see Figure 16.24; Section 19.13). This organism is apparently widely distributed in submarine thermal areas, where it is likely to play a significant role in consuming proteins released from dead organisms.

MINIQUIZ -

- What can we conclude about the *Pyrodictium/Pyrolobus* group in terms of life at high temperature?
- What unusual structural features are present in *Ignicoccus* and *Staphylothermus*?

IV • Evolution and Life at High Temperatures

Most of the hyperthermophiles discovered so far are species of *Archaea* and some grow near to what may be the upper temperature limit for life. Here we consider the major factors that likely define the upper temperature limit for life and the biological adaptations of hyperthermophiles that permit them to exist at the exceptionally high temperatures of 100°C and higher. We end with a discussion of the importance of hydrogen (H_2) metabolism to the biology of hyperthermophiles.

16.12 An Upper Temperature Limit for Microbial Life

Habitats that contain liquid water—a prerequisite for cellular life—and that have temperatures higher than 100°C are only found where geothermally heated water flows out of vents or rifts in the ocean floor (Figures 12.3 and 19.34–19.38). The hydrostatic pressure that overlies the water keeps it from boiling, allowing it to reach temperatures of up to about 400°C in vents at several thousand meters' depth. In contrast, terrestrial hot springs can boil and therefore only attain temperatures near 100°C. It is not surprising, then, that hydrothermal vents have been rich sources of hyperthermophilic *Archaea* with growth temperature optima above 100°C (Table 16.7).

Black smokers emit hydrothermal vent fluid at 250–350°C or higher. Metallic mounds or more upright structures called *chimneys* form from the metal sulfides that precipitate out of the hot fluid as it mixes with the surrounding, much cooler seawater (**Figure 16.24**). As far as is known, the superheated vent water itself is sterile. However, hyperthermophiles thrive in mounds or smoker chimney walls where temperatures are compatible with their survival and growth (⮌ Section 19.13 and Figure 19.37). By studying structures such as these, we can address the question, "What is the upper temperature for microbial (and presumably all forms of) life?"

What Is the Upper Temperature Limit for Life?

How high a temperature can hyperthermophiles withstand? Over the past several decades, the known upper temperature limit for life has been pushed higher and higher with the isolation and characterization of new species of thermophiles and hyperthermophiles (**Figure 16.25**). For some time the record holder was *Pyrolobus fumarii* (Figure 16.21*c*), with its upper temperature limit for growth of 113°C. The current record holder, *Methanopyrus* (Section 16.4 and Figure 16.12), however, has pushed the limit somewhat higher, with the ability to grow at 122°C and to survive substantial periods at even higher temperatures. Given the trend over the past several years (Figure 16.25), one can predict that *Archaea* even more hyperthermophilic than *Methanopyrus* may inhabit hydrothermal environments but have yet to be isolated. Indeed, many experts predict that the upper temperature limit for prokaryotic life is likely to exceed 140°C, perhaps even 150°C, and that the maximum temperature allowing survival but not growth is even hotter yet.

Biochemical Problems at Supercritical Temperatures

Whatever the upper temperature limit is for life, it is likely to be defined by one or more biochemical challenges that evolution has

Figure 16.24 Hydrothermal vents. Hydrothermal mound from the Rainbow vent field, Mid-Atlantic Ridge hydrothermal system. The hydrothermal fluid emitting from the two short chimneys is >300°C.

Anna-Louise Reysenbach and Woods Hole Oceanographic Institution

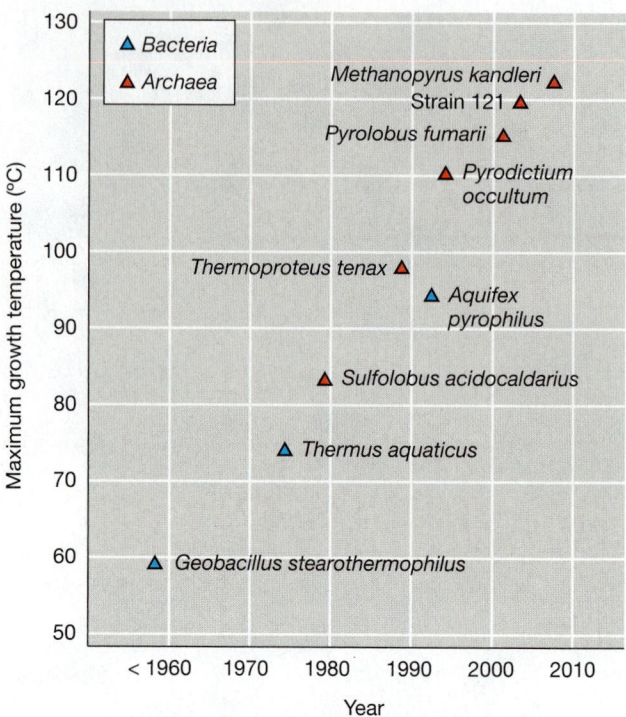

Figure 16.25 Thermophilic and hyperthermophilic prokaryotes. The graph gives the species that were, in turn, the record holders for growing at the highest temperature, from before 1960 to the present.

been unable to solve. There is obviously an upper limit, but we do not yet know what it is. Water samples taken directly from superheated (>250°C) hydrothermal vent discharges are devoid of measurable biochemical markers (DNA, RNA, and protein) that would signal life as we know it, while vents emitting water at temperatures below about 150°C yield evidence of macromolecules. These results are consistent with laboratory experiments on the stability of key biomolecules. For example, ATP is degraded almost instantly at 150°C. Thus, above 150°C, any life forms would have to deal with the heat lability of a molecule that is, as far as is known, universally distributed in cells. As a caveat, however, the stability of small molecules such as ATP may be significantly greater under cytoplasmic conditions of high levels of dissolved solutes than in pure solutions tested in the laboratory. Nevertheless, if life forms exist at temperatures above 150°C, they must be unique in many ways, either using a suite of novel small molecules absent from cells as we know them, or deploying special protection systems that maintain small molecules in a stable state such that biochemistry can proceed.

MINIQUIZ -
- Where are the hottest potential microbial habitats located on Earth?
- Why would it be impossible for organisms to grow at 200 or 300°C?

16.13 Molecular Adaptations to Life at High Temperature

Because all cellular structures and activities are affected by heat, hyperthermophiles are likely to exhibit multiple adaptations to the exceptionally high temperatures of their habitats. Here we briefly examine some adaptations employed by hyperthermophiles to protect their proteins and nucleic acids at high temperatures.

Protein Folding and Thermostability

Because most proteins denature at high temperatures, much research has been done to identify the properties of thermostable proteins. Protein thermostability derives from the folding of the molecule itself, not because of the presence of any special amino acids. Perhaps surprisingly, however, the amino acid composition of thermostable proteins is not particularly unusual except perhaps in their slight bias for increased levels of amino acids that promote alpha-helical secondary structures. In fact, many enzymes from hyperthermophiles contain the same major structural features in both primary and higher-order structure (⟲ Section 4.14) as their heat-labile counterparts from organisms that grow best at much lower temperatures.

Thermostable proteins typically do display some structural features that likely improve their thermostability. These include having highly hydrophobic cores, which decrease the tendency of the protein to unfold in an ionic environment, and more ionic interactions on the protein surfaces, which also help hold the protein together and work against unfolding. Ultimately, it is the *folding* of the protein that most affects its heat stability, and noncovalent ionic bonds called *salt bridges* on a protein's surface likely play a major role in maintaining the biologically active structure. But, as previously stated, many of these changes are possible with only minimal changes in primary structure (amino acid sequence), as seen when thermostable and heat-labile forms of the same protein are compared.

Chaperonins: Assisting Proteins to Remain in Their Native State

Earlier we discussed a class of proteins called *chaperonins* (heat shock proteins; ⟲ Section 4.14) that function to refold partially denatured proteins. Hyperthermophilic *Archaea* have special classes of chaperonins that function only at the highest growth temperatures. In cells of *Pyrodictium abyssi* (**Figure 16.26**), for example, a major chaperonin is the protein complex called the **thermosome**. This complex keeps other proteins properly folded and functional at high temperature, helping cells survive even at temperatures above their maximal growth temperature. Cells of *P. abyssi* grown near its maximum temperature (110°C) contain high levels of the thermosome. Possibly because of this, the cells can remain viable following a heat shock, such as a 1-h treatment in an autoclave (121°C). In cells experiencing such a treatment and then returned to the optimum temperature, the thermosome, which is itself quite heat-resistant, is thought to refold sufficient copies of key denatured proteins that *P. abyssi* can once again begin to grow and divide. Thus, due to chaperonin activity, the upper temperature limit at which many hyperthermophiles can *survive* is higher than the upper temperature at which they can *grow*. The "safety net" of chaperonin activity probably ensures that cells in nature that briefly experience temperatures above their growth temperature maximum are not killed by the exposure.

DNA Stability: Solutes, Reverse Gyrase, and DNA-Binding Proteins

What keeps DNA from melting at high temperatures? Various mechanisms are known to contribute. One such mechanism

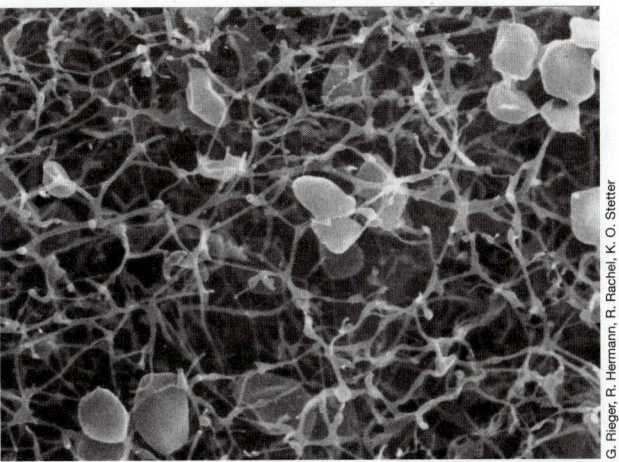

Figure 16.26 *Pyrodictium abyssi*, **scanning electron micrograph.** *Pyrodictium* has been studied as a model of macromolecular stability at high temperatures. Cells are enmeshed in a sticky glycoprotein matrix that binds them together.

increases cellular solute levels, in particular potassium (K^+) or compatible organic compounds. For example, the cytoplasm of the hyperthermophilic methanogen *Methanopyrus* (Section 16.4) contains molar levels of potassium cyclic 2,3-diphosphoglycerate. This solute prevents chemical damage to DNA, such as depurination or depyrimidization (loss of a nucleotide base through hydrolysis of the glycosidic bond) from high temperatures, events that can lead to mutation (⟲ Section 10.2). This compound and other compatible solutes, such as potassium di-*myo*-inositol phosphate, which protects against osmotic stress, and the polyamines putrescine and spermidine, which stabilize both ribosomes and nucleic acids at high temperature, help maintain key cellular macromolecules in hyperthermophiles in their active forms.

A unique protein found *only* in hyperthermophiles is responsible for DNA stability in these organisms. All hyperthermophiles produce a special DNA topoisomerase called **reverse DNA gyrase**. This enzyme introduces positive supercoils into the DNA of hyperthermophiles (in contrast to the negative supercoils introduced by DNA gyrase present in all other prokaryotes; ⟲ Section 4.4). Positive supercoiling stabilizes DNA to heat and thereby prevents the DNA helix from spontaneously unwinding. The noticeable absence of reverse DNA gyrase in prokaryotes whose growth temperature optima lie below 80°C strongly suggests a specific role for this enzyme in DNA stability at high temperatures.

Species of *Euryarchaeota* also contain highly basic (positively charged) DNA-binding proteins that are remarkably similar in amino acid sequence and folding properties to the core histones of the *Eukarya* (⟲ Figure 2.61). Archaeal histones from the hyperthermophilic methanogen *Methanothermus fervidus* (Figure 16.7c) have been particularly well studied. These proteins wind and compact DNA into nucleosome-like structures (**Figure 16.27**) and maintain the DNA in a double-stranded form at very high temperatures. Archaeal histones are found in most *Euryarchaeota*, including extremely halophilic *Archaea*, such as *Halobacterium*. However, because the extreme halophiles are not thermophiles, archaeal histones may have other functions besides DNA stability, in particular in assisting in gene expression by opening the helix to allow for transcriptional proteins to bind.

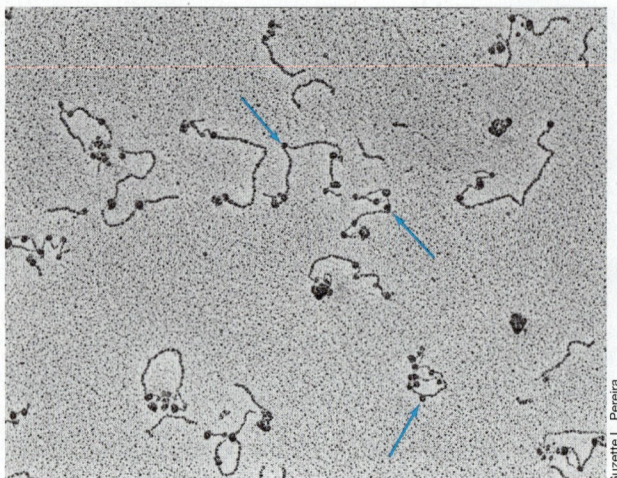

Figure 16.27   Archaeal histones and nucleosomes. Electron micrograph of linearized plasmid DNA wrapped around copies of archaeal histone Hmf (from the hyperthermophilic methanogen *Methanothermus fervidus*) to form the roughly spherical, darkly stained nucleosome structures (arrows). Compare this micrograph with an artist's depiction of the histones and nucleosomes of *Eukarya* shown in Figure 2.61*b*.

Lipid and Ribosomal RNA Stability

How have the lipids and the protein-synthesizing machinery of hyperthermophiles adjusted to high temperatures? Virtually all hyperthermophilic *Archaea* synthesize lipids of the dibiphytanyl tetraether type (↩ Section 2.7). These lipids are naturally heat-resistant because the phytanyl units forming each half of the membrane structure are covalently bonded to one another; this yields a *lipid monolayer* membrane instead of the normal lipid bilayer (↩ Figure 2.17). This structure resists the tendency of heat to pull apart a lipid bilayer constructed of fatty acid or phytanyl side chains that are not covalently bonded.

A final point on molecular adaptations to life at high temperatures is that of the base composition of ribosomal RNAs. Ribosomal RNAs are key structural and functional components of the ribosome, the cell's protein-synthesizing apparatus (↩ Section 4.13). Hyperthermophilic species of both *Bacteria* and *Archaea* show as much as a 15% greater proportion of GC base pairs in their small ribosomal subunit RNAs compared with organisms that grow at

lower temperatures. GC base pairs form three hydrogen bonds compared to the two of AU base pairs (↩ Figure 4.2), and thus the higher GC content of the ribosomal RNAs should confer greater thermal stability on the ribosomes of these organisms and this should assist protein synthesis at high temperatures. By contrast to ribosomal RNAs, the GC content of genomic DNA of hyperthermophiles is often rather low, which suggests that the thermal stability of ribosomal RNA might be an especially significant factor for life under hyperthermophilic conditions.

MINIQUIZ

- How do hyperthermophiles keep proteins and DNA from being destroyed by high heat?
- How are the lipids and ribosomes of hyperthermophiles protected from heat denaturation?

16.14 Hyperthermophilic *Archaea*, H₂, and Microbial Evolution

When cellular life first arose on Earth nearly 4 billion years ago, it is virtually certain that temperatures were far hotter than they are today. Thus, for hundreds of millions of years, Earth may have been suitable only for hyperthermophiles. Given the discussion above on the temperature limits to life, it has been hypothesized that biological molecules, biochemical processes, and the first cells arose on Earth around hydrothermal springs and vents on the seafloor as they cooled to temperatures compatible with biological molecules (↩ Section 12.1 and Figures 12.3 and 12.4). The phylogeny of modern hyperthermophiles (Figure 16.1), as well as the similarities in their habitats and metabolism to those of early cells on Earth, suggests that hyperthermophiles may be the closest remaining descendants of ancient cells and are a living window into the biology of ancient microbial life.

Hyperthermophilic Habitats and H₂ as an Energy Source

The oxidation of H_2 linked to the reduction of Fe^{3+}, S^0, NO_3^-, or, rarely, O_2 is a widespread form of energy metabolism in hyperthermophiles (Table 16.6 and **Figure 16.28**). This, coupled with the likelihood that these hyperthermophiles best characterize early Earth

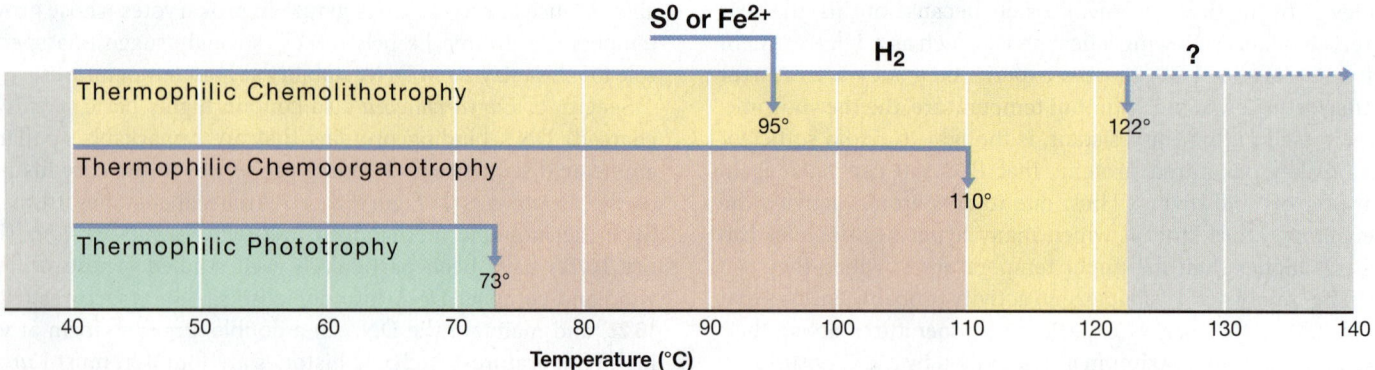

Figure 16.28   Upper temperature limits for energy metabolism. The record holder for phototrophy is *Synechococcus lividus* (*Bacteria*, cyanobacteria); for chemoorganotrophy, *Pyrodictium occultum* (*Archaea*); for chemolithotrophy with S⁰ as electron donor, *Acidianus infernus* (*Archaea*); for chemolithotrophy with Fe^{2+} as electron donor, *Ferroglobus placidus* (*Archaea*); and for chemolithotrophy with H₂ as electron donor, *Methanopyrus kandleri* (*Archaea*, 122°C).

phenotypes, points to the important role H_2 has played in the evolution of microbial life. Hydrogen metabolism may have evolved in primitive organisms because of the ready availability of H_2 and suitable inorganic electron acceptors in their primordial environments, but also because a H_2-based energy economy requires relatively few proteins (↻ Figure 12.5). As chemolithotrophs, these organisms may have obtained all of their carbon from CO_2 or might have assimilated available organic compounds directly for biosynthetic needs. Either way it is likely that the oxidation of H_2 was the energetic driving force for maintaining life processes.

If one compares microbial energy conservation mechanisms as a function of temperature from data of cultured prokaryotes, only chemolithotrophic organisms are known at the hottest temperatures (Figure 16.28). Chemoorganotrophy occurs up to at least 110°C, as this is the upper temperature limit for growth of *Pyrodictium occultum*, an organism that can conserve energy and grow by fermentation and by chemolithotrophic growth on H_2 with S^0 as electron acceptor (Table 16.6). Photosynthesis is the least heat-tolerant of all bioenergetic processes, with no hyperthermophilic representatives known and an apparent upper temperature limit of 73°C. This is consistent with the conclusion that anoxygenic photosynthesis first appeared on Earth some hundreds of millions of years after the first life forms are thought to have appeared (↻ Figure 12.1).

Comparisons of bioenergetic options as a function of temperature (Figure 16.28) point to the H_2-oxidizing hyperthermophilic *Archaea* and *Bacteria* as the most likely extant examples of Earth's earliest cellular life forms. More so than any other prokaryotes, these organisms retain the metabolic and physiological traits one would predict to be necessary for existence on a hot early Earth.

MINIQUIZ -
- What phylogenetic and physiological evidence suggests that today's hyperthermophiles are the closest living links to Earth's earliest cells?

BIG IDEAS

16.1 • Extremely halophilic *Archaea* require large amounts of NaCl for growth and accumulate large levels of KCl in their cytoplasm as a compatible solute. These salts affect cell wall stability and enzyme activity. The light-mediated proton pump bacteriorhodopsin helps extreme halophiles make ATP.

16.2 • Methanogenic *Archaea* are strict anaerobes whose metabolism is linked to the production of CH_4. Methane can be produced by CO_2 reduction by H_2, from methyl substrates such as CH_3OH, or from acetate.

16.3 • *Thermoplasma*, *Ferroplasma*, and *Picrophilus* are extremely acidophilic thermophiles that form their own phylogenetic family of *Archaea*. Cells of *Thermoplasma* and *Ferroplasma* lack cell walls, resembling the mycoplasmas in this regard.

16.4 • *Methanopyrus* is a hyperthermophilic methanogen that contains unusual lipids and can grow at 122°C, the highest temperature supporting growth of any life form.

16.5 • *Archaeoglobus* and *Ferroglobus* are related anaerobic *Archaea* that carry out different anaerobic respirations. *Archaeoglobus* is a sulfate reducer and *Ferroglobus* is a nitrate reducer that oxidizes ferrous iron.

16.6 • *Thaumarchaeota* are widespread and abundant in soils and marine environments. All cultivated species of thaumarchaea are autotrophic ammonia-oxidizers and these organisms are important in the global nitrogen cycle.

16.7 • *Nanoarchaeum equitans* is a hyperthermophile that forms its own phylum, the *Nanoarchaeota*, and is a parasite of the crenarchaeote *Ignicoccus*. *N. equitans* has a tiny, highly compact genome and depends on *Ignicoccus* for most of its cellular needs, including both carbon and energy.

16.8 • *Korarchaeum cryptofilum* forms its own phylum, the *Korarchaeota*, and is a hyperthermophile that lacks important biosynthetic pathways, obtaining key building blocks from its environment. *K. cryptofilum* has some genes that are similar to *Euryarchaeota* and other genes that are similar to *Crenarchaeota*.

16.9 • A wide variety of chemoorganotrophic and chemolithotrophic energy metabolisms have been found in hyperthermophilic *Crenarchaeota*, including fermentation and anaerobic respirations. Strictly autotrophic lifestyles are common but photosynthesis is absent.

16.10 • Hyperthermophilic *Crenarchaeota* thrive in terrestrial hot springs of various chemistries. These include in particular organisms such as *Sulfolobus*, *Acidianus*, *Thermoproteus*, and *Pyrobaculum*.

16.11 • In deep-sea hydrothermal systems, *Crenarchaeota* such as *Pyrolobus*, *Pyrodictium*, *Ignicoccus*, and *Staphylothermus* thrive. With the exception of the methanogen *Methanopyrus* (*Euryarchaeota*), species of these genera grow at the highest temperatures of all *Archaea*, in many cases well above the boiling point of water.

16.12 • Life as we know it is probably limited to temperatures below 150°C. Key small molecules, such as ATP, are quickly destroyed above this temperature although extremely heat-stable macromolecules may not be.

16.13 • Macromolecules in hyperthermophiles are protected from heat denaturation by their heat-stable folding patterns (proteins), solutes and binding proteins (DNA), unique monolayer membrane architecture (lipids), and the high GC content of their ribosomal RNAs.

16.14 • Hydrogen metabolism is likely to have been the driving force behind the energetics of the earliest cells on Earth. Chemolithotrophic metabolisms based on H_2 as an electron donor are found in the most heat-tolerant of all known prokaryotes.

REVIEW OF KEY TERMS

Bacteriorhodopsin a protein containing retinal that is found in the membranes of certain extremely halophilic *Archaea* and that is involved in light-mediated ATP synthesis

Compatible solute an organic or inorganic substance that is accumulated in the cytoplasm of a halophilic organism and maintains osmotic pressure

Crenarchaeota a phylum of *Archaea* that contains both hyperthermophilic and cold-dwelling organisms

Euryarchaeota a phylum of *Archaea* that contains primarily methanogens, extreme halophiles, *Thermoplasma*, and some marine hyperthermophiles

Extreme halophile an organism whose growth is dependent on large concentrations (generally 9% or more) of NaCl

Extremophile an organism whose growth is dependent on extremes of temperature, salinity, pH, pressure, or radiation, which are generally inhospitable to most forms of life

Halorhodopsin a light-driven chloride pump that accumulates Cl^- within the cytoplasm

Hydrothermal vent a deep-sea hot spring emitting warm (~20°C) to superheated (>300°C) water

Hyperthermophile an organism with a growth temperature optimum of 80°C or greater

Korarchaeota a phylum of *Archaea* that contains the hyperthermophile *Korarchaeum cryptophilum*

Methanogen a CH_4-producing organism

Nanoarchaeota a phylum of *Archaea* that contains the hyperthermophilic parasite *Nanoarchaeum equitans*

Phytanyl a branched-chain hydrocarbon containing 20 carbon atoms and commonly found in the lipids of *Archaea*

Reverse DNA gyrase a protein universally present in hyperthermophiles that introduces positive supercoils into circular DNA

Solfatara a hot, sulfur-rich, generally acidic environment commonly inhabited by hyperthermophilic *Archaea*

Thaumarchaeota a phylum of *Archaea* that contains widespread species capable of aerobic ammonia oxidization

Thermosome a heat shock (chaperonin) protein complex that functions to refold partially heat-denatured proteins in hyperthermophiles

REVIEW QUESTIONS

1. How can organisms such as *Halobacterium* survive in a high-salt environment, whereas an organism such as *Escherichia coli* cannot? (Section 16.1)

2. Contrast the roles of bacteriorhodopsin, halorhodopsin, and sensory rhodopsin in *Halobacterium salinarum*. (Section 16.1)

3. What is the electron donor for methanogenesis when CO_2 is reduced to CH_4? (Section 16.2)

4. What two major physiological features unify species of *Thermoplasmatales*? Why does this allow some of them to successfully colonize coal refuse piles? (Section 16.3)

5. What is physiologically unique about *Methanopyrus* compared with another methanogen such as *Methanobacterium*? (Section 16.4) What is physiologically unique about *Archaeoglobus*? (Section 16.5)

6. What is physiologically unusual about the thaumarchaeotal species *Nitrosopumilus maritimus*? (Section 16.6)

7. How is *Nanoarchaeum* similar to other *Archaea*? How does it differ? (Section 16.7)

8. Why is it difficult to determine the phylogenetic placement of *Nanoarchaeota* and *Korarchaeota*? (Sections 16.7 and 16.8)

9. What forms of energy metabolism are present in *Crenarchaeota*? What form is absent? (Section 16.9)

10. What is unusual about the metabolism of S^0 by *Acidianus*? (Section 16.10)

11. What is unusual about the organism *Pyrolobus fumarii*? (Section 16.11)

12. What organism is the current record holder for the upper temperature limit for growth? (Section 16.12)

13. What is reverse DNA gyrase and why is it important to hyperthermophiles? (Section 16.13)

14. Why might H_2 metabolism have evolved as a mechanism for energy conservation in the earliest organisms on Earth? (Section 16.14)

APPLICATION QUESTIONS

1. Using the phylogenetic tree in Figure 16.1 as a guide, discuss what indicates that bacteriorhodopsin may have been a late evolutionary invention and that anaerobic respiration with S^0 as electron acceptor might have been an early evolutionary invention.

2. Defend or refute the following statement: The upper temperature limit to life is unrelated to the stability of proteins or nucleic acids.

17 · Diversity of Eukaryotic Microorganisms

microbiology**now**

Horizontal Gene Transfer in an Extremophilic Eukaryote

Horizontal gene transfer (HGT) is a hallmark of *Bacteria* and *Archaea,* but what about microbial eukaryotes—does HGT occur in them as well? The answer is clearly yes, because researchers have now discovered a microbial eukaryote that has acquired several important genes from its bacterial and archaeal neighbors.

The red alga *Galdieria sulphuraria* inhabits hot, acidic, sulfur- and metal-rich environments (photo). The alga is remarkably tolerant of heat, acid, salt, and toxic metals, including arsenic, aluminum, cadmium, and mercury. A close inspection of the genome of *G. sulphuraria* (photo inset) found that at least 75 genes had been acquired by HGT from prokaryotes and that many of these conferred key extremophilic properties on this alga.[1] Chief among these were a family of genes encoding soluble ATPases found in hot spring *Archaea* that likely contribute to heat tolerance; such ATPases have not been found in any other eukaryotes.

Other key genes acquired by *G. sulphuraria* allow the alga to survive salt stress and metal toxicity and to synthesize a cytoplasmic membrane with low proton permeability, a necessity for thriving in highly acidic environments. *G. sulphuraria* also has the unusual ability (for an alga) to grow in darkness on over 50 different carbon sources. This diverse capacity has been linked to genes acquired by HGT, including genes encoding transporters for sugars, amino acids, fatty acids, and glycerol.

As more genomes from microbial eukaryotes are unraveled, the mosaic nature of the *G. sulphuraria* genome may turn out to be more common. But for now, this physiologically tough red alga stands out as a remarkable example of a microbial eukaryote whose genome has been bolstered by horizontal gene transfers across phylogenetic domains.

[1]Schönknecht, G., et al. 2013. Gene transfer from *Bacteria* and *Archaea* facilitated evolution of an extremophilic eukaryote. *Science 339:* 1207–1210.

I • Organelles and Phylogeny of Microbial Eukaryotes

In Chapter 2 we discussed the architecture of the microbial eukaryotic cell. There we reviewed the typical components of the eukaryotic cell: the nucleus, mitochondrion, endoplasmic reticulum, and Golgi complex (structures common to virtually all eukaryotes), and the chloroplast, present in phototrophic eukaryotes (⮌ Sections 2.20–2.22). In this chapter we consider the phylogeny and diversity of microbial eukaryotes. We begin by considering the mitochondrion and chloroplast, structures whose evolutionary history is distinct from that of the eukaryotic cell itself (⮌ Sections 2.21 and 12.3).

17.1 Endosymbioses and the Eukaryotic Cell

Biologists agree that the eukaryotic cell is a genetic chimera. The main part of the eukaryotic cell including its cytoplasm (and likely its nucleus) can be traced to the *Eukarya* domain, whereas its energy-producing organelles—the mitochondria and chloroplasts—contain their own DNA and are clearly derived from *Bacteria.*

Initial speculation on the link between organelles and bacteria goes back over a century and was based on the fact that microscopically, mitochondria and chloroplasts "looked like" bacteria. Through the years this idea slowly gathered experimental support to yield the current view that mitochondria and chloroplasts are ancestors of respiratory or phototrophic *Bacteria*, respectively, that established residence inside another cell type as a source of ATP in exchange for a safe and stable existence. This is the **endosymbiotic hypothesis** (⮌ Section 12.3) and is a major tenet of modern biology.

Support for the Endosymbiotic Hypothesis

Several lines of evidence support the endosymbiotic hypothesis:

1. **Mitochondria and chloroplasts contain DNA.** Although most mitochondrial and chloroplast proteins are encoded by nuclear DNA, a few are encoded by a small genome residing within the organelle itself. These include proteins of the respiratory chain (mitochondrion) and photosynthetic apparatus (chloroplast) as well as ribosomal RNAs and transfer RNAs. Most mitochondrial DNA and all chloroplast DNA is of a covalently closed circular form like that of most *Bacteria* (⮌ Section 1.2). Organellar DNA can be visualized in eukaryotic cells with special staining methods (**Figure 17.1**).

2. **The eukaryotic nucleus contains genes derived from *Bacteria.*** Genomic sequences of eukaryotic cells have clearly shown that several nuclear genes encode functions specific to mitochondria and chloroplasts. Moreover, because these gene sequences more closely resemble those of *Bacteria* than those of *Archaea* or *Eukarya*, it is concluded that these genes were translocated to the nucleus from ancestors of mitochondria and chloroplasts during the transition from engulfed cells to dedicated organelles.

3. **Organellar ribosomes and their phylogeny.** Ribosomes are either 80S in size, typical of the cytoplasm of eukaryotic cells, or 70S, typical of *Bacteria* and *Archaea*. Mitochondria and chloroplasts contain 70S ribosomes, and phylogenetic analyses of their ribosomal RNA gene sequences (Chapter 12) along with genomic studies of organellar DNA (⮌ Section 6.5) show unequivocally that these structures were originally *Bacteria.*

4. **Antibiotic specificity.** Several antibiotics (for example, streptomycin) kill or inhibit *Bacteria* by interrupting 70S ribosome protein synthesis functions. These same antibiotics also inhibit protein synthesis in mitochondria and chloroplasts.

5. **Hydrogenosomes.** Hydrogenosomes are membrane-enclosed organelles found in certain amitochondriate eukaryotic anaerobes that supply the cell with ATP from fermentative reactions (⮌ Figure 2.64*b*). Like mitochondria, hydrogenosomes also contain their own DNA and ribosomes, and phylogenetic analyses of hydrogenosome ribosomal RNA have revealed their connection to *Bacteria.*

Secondary Endosymbiosis

The mitochondrion, chloroplast, and hydrogenosome are structures that originated from *primary* endosymbiosis events. That is, these structures are derived from cells of *Bacteria*. Primary endosymbioses gave rise to the chloroplast in the common ancestor of green algae, red algae, and plants (**Figure 17.2** and see Figure 17.3). However, following this primary event, several unrelated groups of nonphototrophic microbial eukaryotes also acquired chloroplasts but by *secondary* rather than primary endosymbioses. The secondary events occurred when entire green algal or red algal cells were engulfed and their chloroplasts stably retained, thereby making the engulfing cell phototrophic.

Secondary endosymbioses within the green algae account for the presence of chloroplasts in euglenids and chlorarachniophytes, while alveolates (ciliates, apicomplexans, and dinoflagellates) and stramenopiles obtained their chloroplasts through

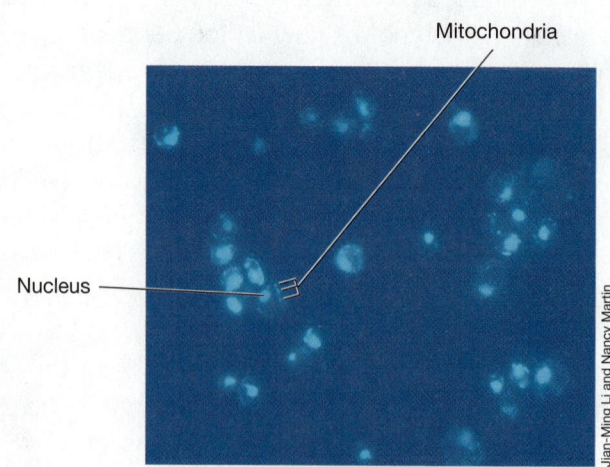

Figure 17.1 **Organellar DNA.** Cells of the yeast *Saccharomyces cerevisiae* have been stained with the fluorescent dye DAPI that binds to DNA. Each mitochondrion has two to four circular chromosomes that stain blue with the dye.

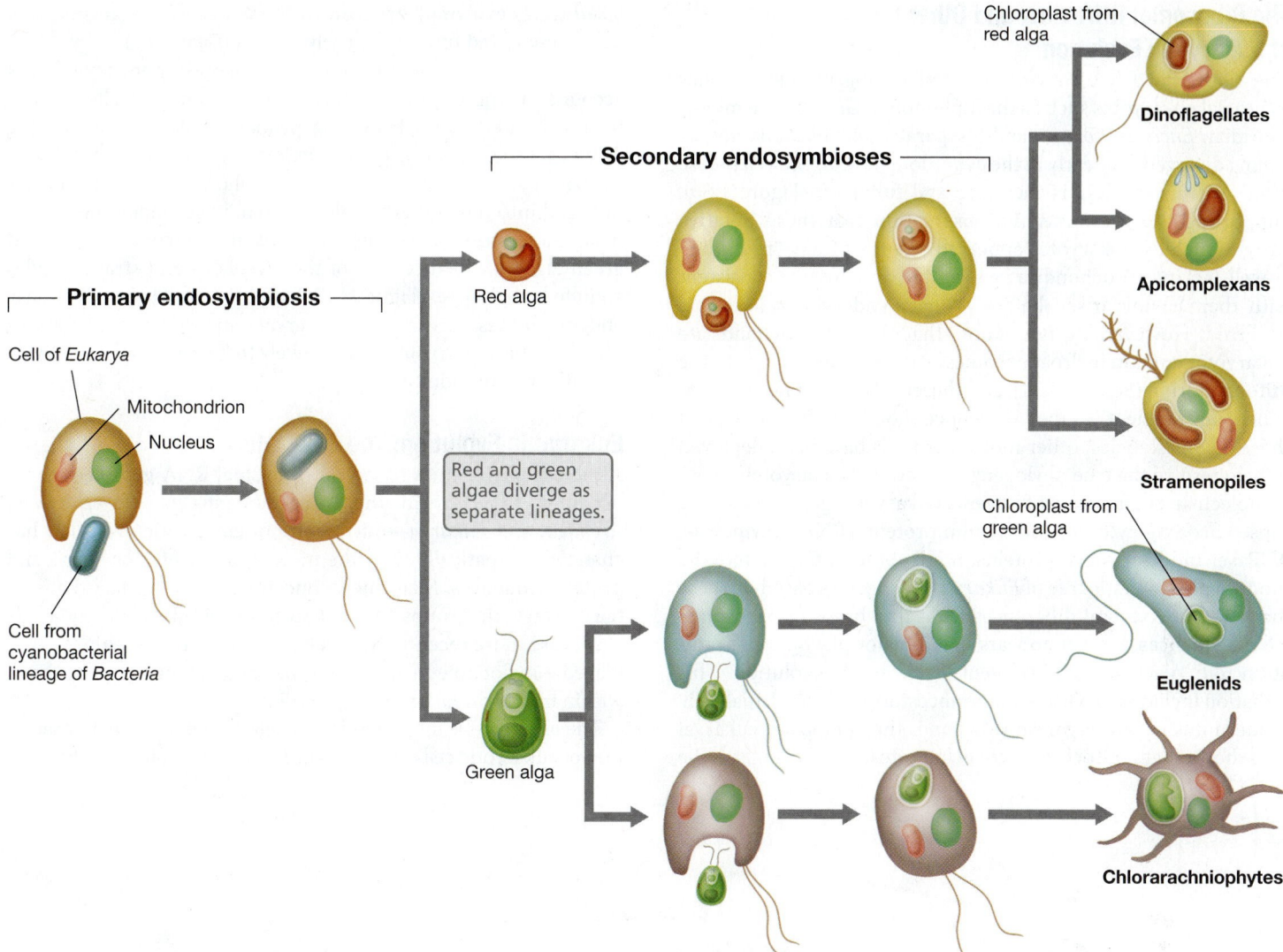

Figure 17.2 Endosymbioses. Following primary endosymbiotic association(s) leading to the mitochondrion, primary endosymbioses with phototrophic *Bacteria* led to the red and green algae. Secondary symbioses of green and red algae spread the property of photosynthesis to many independent lineages of protists.

secondary endosymbioses with red algae (Figure 17.2 and see Figure 17.3). The ancestral red algal chloroplasts were apparently lost from some lineages, such as the ciliates, or became greatly reduced in size in others, such as the apicomplexans, where only traces of chloroplasts remain. In some other organisms, such as the dinoflagellates, the red algal chloroplast was apparently replaced altogether with a chloroplast from different algae, including green algae.

These many examples of endosymbiotic events underscore the importance of endosymbiosis in the evolution and diversification of microbial eukaryotes. It is unlikely that primary endosymbiotic events occurred only once in evolutionary history—after all, trial and error is the essence of evolution—and secondary endosymbioses almost certainly occurred quite commonly (Figure 17.2). Even today there are many examples of nonphototrophic protists that engulf phototrophic protists, and the entrapped phototrophs carry out photosynthesis for extended periods (⟲ Section 22.14). Indeed, endosymbioses are apparently a common and ongoing occurrence in the eukaryotic world.

MINIQUIZ -

- What is the endosymbiotic hypothesis?
- Summarize the molecular evidence that supports the relationship of organelles to *Bacteria*.
- Distinguish between primary and secondary endosymbiosis.

17.2 Phylogenetic Lineages of *Eukarya*

From the universal phylogenetic tree of life (⟲ Figures 1.6*b* and 12.13) we learned that *Eukarya* are more closely related to *Archaea* than to *Bacteria*. The phylogeny of microbial *Eukarya* was originally inferred from sequences of ribosomal RNA (rRNA) obtained from eukaryotic cell cytoplasmic (18S) ribosomes. However, unlike the 16S rRNA-based trees of prokaryotes, certain aspects of the 18S phylogenetic tree of eukaryotes have proven unreliable, and hence the modern eukaryal tree has been deduced from a combination of comparative sequencing methods.

The Ribosomal RNA View and Other Views of Eukaryotic Evolution

The ribosomal RNA view of eukaryal phylogeny distinguishes microbial eukaryotes such as the diplomonad *Giardia*, the microsporidian *Encephalitozoon*, and the parabasalid *Trichomonas*, as having diverged very early in the evolution of *Eukarya*, well before other organisms such as the algae and fungi (↩ Figure 1.6*b*). Supporting this view was the observation that these putative early-branching eukaryotes appeared phenotypically "primitive" as well, lacking mitochondria for example, and this was consistent with their having arisen before primary endosymbiotic events occurred. However, we now know that these amitochondriate eukaryotes contain hydrogenosomes, structures analogous to the mitochondrion (Section 17.1), and therefore may not be as phylogenetically "ancient" as they were once thought to be. Because of these inconsistencies, other molecular tools have been deployed to help resolve the true phylogeny of microbial eukaryotes.

Molecular sequencing of several eukaryotic genes including those encoding cytoskeleton tubulin proteins, RNA polymerase, ATPase, and heat shock proteins, has been used to generate the modern phylogenetic tree of *Eukarya*. Phylogenies based on these markers show several differences from that based on ribosomal RNA sequences. First, it appears that a major phylogenetic radiation took place as an early event in eukaryote evolution. This radiation included evolution of the ancestors of all, or virtually all, modern eukaryotic organisms. Second, the composite eukaryal tree shows that amitochondriate eukaryotes, once thought to be basal (early evolving), are instead *highly derived* organisms, and that animals and fungi are closely related (**Figure 17.3**).

The tree of *Eukarya* also shows how secondary endosymbioses account for the origin of chloroplasts in some unicellular phototrophic eukaryotes. Following primary endosymbiosis of the cyanobacterial ancestor of chloroplasts by early mitochondrion-containing eukaryotes, these now phototrophic eukaryotes diverged into red and green algae. Then, in secondary endosymbioses, ancestors of the euglenozoans and cercozoans engulfed green algae while ancestors of the alveolates and stramenopiles engulfed red algae (Figures 17.2 and 17.3). These secondary endosymbioses account for the great phylogenetic diversity of phototrophic eukaryotes and are likely to have occurred relatively recently in evolutionary time.

Eukaryotic Evolution: The Big Picture

Although phylogenies based on ribosomal RNA gene sequences (Chapter 12) confirm the three domains of life—*Bacteria*, *Archaea*, and *Eukarya*—our picture of eukaryotic evolution has changed dramatically with the incorporation of other gene and protein sequences. Major new concepts include the fact that certain eukaryotic groups thought to have arisen very early probably arose more recently and that secondary endosymbioses have played a major role in spreading the capacity for photosynthesis within microbial *Eukarya* (Figure 17.3).

The origin of the mitochondrion likely predated this major radiation of eukaryotic cells, as all extant *Eukarya* contain mitochondria

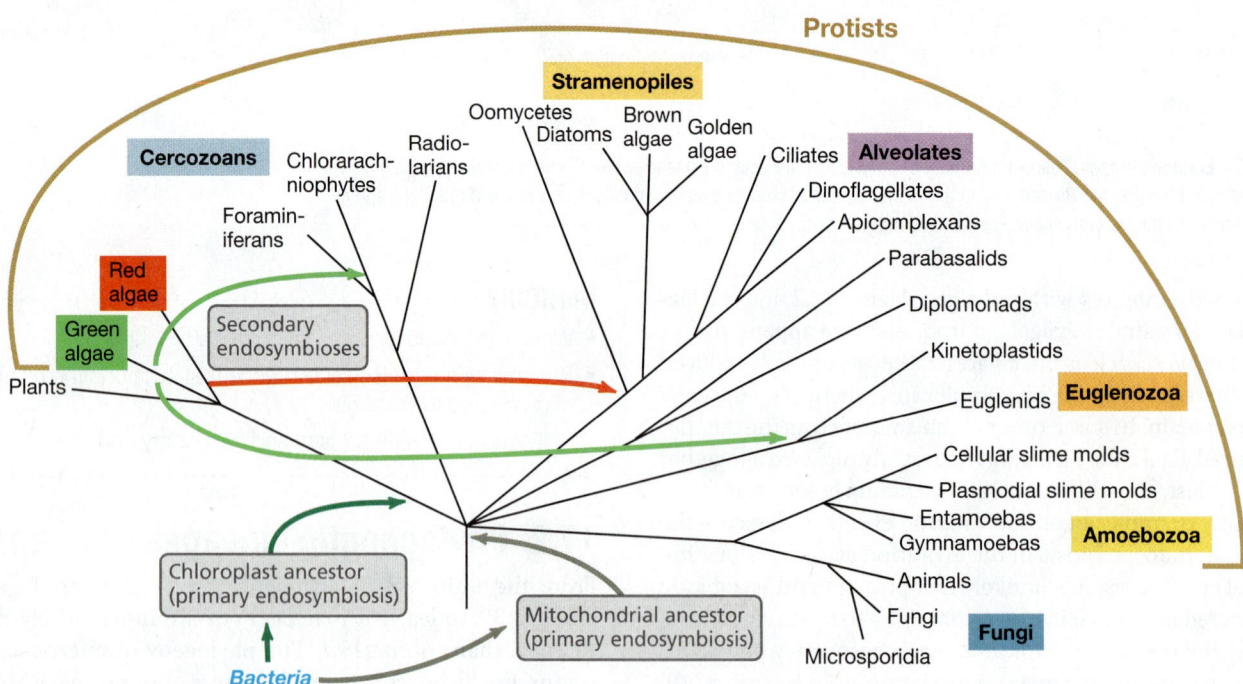

Figure 17.3 Phylogenetic tree of *Eukarya*. This composite tree is based on sequences of several genes and proteins. Dark green and red arrows indicate primary endosymbiotic events for the acquisition of the mitochondrion (red) and the chloroplast (green). Light green arrows indicate secondary endosymbiotic acquisition of chloroplasts from green algae by various protists, and the red arrow indicates secondary endosymbiotic acquisition of chloroplasts from red algae. Note that the greatest diversity in the eukaryotic world resides in the protists.

or hydrogenosomes or some genetic trace of these structures. The mitochondrion or similar structure would have provided the early eukaryotic cell with dramatic new metabolic capabilities and likely triggered the evolutionary radiation of eukaryotic microorganisms. What promoted this primary endosymbiotic event is unknown, but quite possibly it was the accumulation of O_2 in the atmosphere from cyanobacterial photosynthesis (Figure 12.1). Somewhat later in evolutionary time, the ancestor of the chloroplast was acquired in a primary endosymbiotic event, with eukaryotic phototrophic diversity unfolding later through secondary endosymbioses (Figure 17.2).

The phylogenetic tree shown in Figure 17.3 should not be considered the final word on eukaryotic evolution. As new results reveal previously unsuspected aspects of eukaryotic biology, new pictures of eukaryotic phylogeny will emerge periodically. But at this point it appears that two points are clear. First, a composite tree, rather than an 80S ribosomal RNA tree, is the scaffold upon which the eukaryotic tree of life will rest; and second, acquisition of the mitochondrion by primitive *Eukarya* was central to the evolutionary success of this domain.

MINIQUIZ

- What does the endosymbiotic hypothesis propose?
- How does the composite tree of eukaryotes differ from the ribosomal RNA–based tree?
- How does secondary endosymbiosis help explain the diversity of phototrophic eukaryotes?

II • Protists

With the big picture of eukaryotic cell phylogeny in mind, we proceed to examine the major groups of eukaryotic microorganisms. We begin with protists other than the green and red algae. **Protists** include both phototrophic and nonphototrophic microbial eukaryotes. These organisms are widely distributed in nature, exhibit a wide range of morphologies, and show great phylogenetic diversity. Indeed, protists represent much of the diversity found in the domain *Eukarya* (Figure 17.3).

17.3 Diplomonads and Parabasalids

Key Genera: *Giardia, Trichomonas*

Diplomonads and parabasalids are unicellular, flagellated protists that lack mitochondria and chloroplasts. They live in anoxic habitats, such as animal intestines, either symbiotically or as parasites, and conserve energy from fermentation. Some diplomonads cause serious and common diseases in fish, domestic animals, and humans, and one parabasalid causes a major sexually transmitted disease of humans. Both groups share a relatively recent common ancestor before they diverged to form separate phylogenetic lineages (Figure 17.3).

Diplomonads

Diplomonads (**Figure 17.4a**) characteristically contain two nuclei of equal size, and also contain mitosomes, much reduced mitochondria lacking electron transport proteins and enzymes of the citric acid cycle. The diplomonad *Giardia* has a relatively small genome for a eukaryote, about 12 megabase pairs. The genome is also quite compact, contains few introns, and lacks genes for many metabolic pathways, including the citric acid cycle (Figure 3.22). These characteristics likely account for the organism's parasitic lifestyle. *Giardia intestinalis* (Figure 17.4a), also known as *Giardia lamblia*, causes giardiasis, one of the most common waterborne diarrheal diseases in the United States. We examine the disease giardiasis in Section 32.4.

Parabasalids

Parabasalids contain a *parabasal body* that, among other functions, gives structural support to the cell's Golgi complex. These microbial eukaryotes lack mitochondria but contain hydrogenosomes for anaerobic metabolism (Section 2.21). Parabasalids live in the intestinal and urogenital tract of vertebrates and invertebrates as parasites or as commensal symbionts (Section 32.4). The parabasalid *Trichomonas vaginalis* is motile by a tuft of flagella (Figure 17.4b) and causes a widespread sexually transmitted disease in humans.

The genomes of parabasalids are unique among eukaryotes in that most of them lack introns, the noncoding sequences characteristic of eukaryotic genes (Sections 4.9 and 6.6). In addition, the genome of *T. vaginalis* is surprisingly huge for a parasitic organism, about 160 megabase pairs, and shows evidence of genes acquired from bacteria by horizontal gene transfer. Much of the genome of *T. vaginalis* contains repetitive DNA sequences and transposable

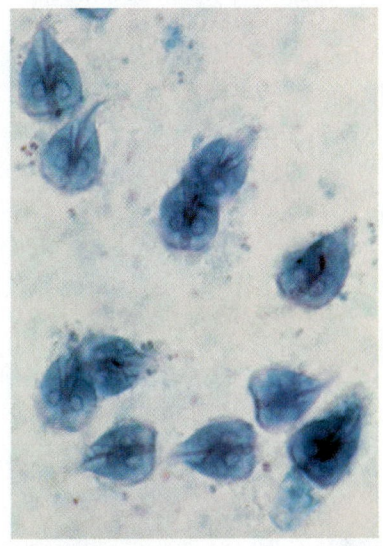
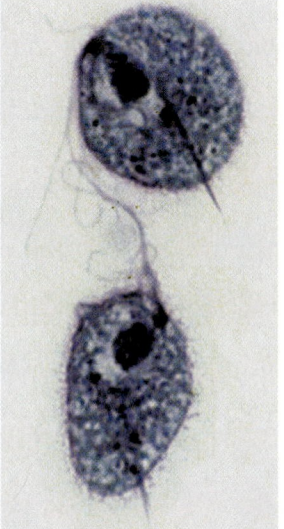

(a)

(b)

Figure 17.4 Diplomonads and parabasalids. *(a)* Light photomicrograph of cells of *Giardia intestinalis*, a typical diplomonad. Note the dual nuclei. Cells are about 10 μm wide. *(b)* Light photomicrograph of cells of the parabasalid *Trichomonas vaginalis*. Cells are about 6 μm wide. The spear-like structure (axostyle) is used to attach the cell to urogenital tissues.

elements (Section 10.11), which has made genomic analyses difficult. But *Trichomonas* is still thought to contain nearly 60,000 genes, about twice that of the human genome and near the upper limit observed thus far for eukaryotic genomes.

17.4 Euglenozoans

Key Genera: *Trypanosoma, Euglena*

Euglenozoans are a diverse assemblage of unicellular, free-living or parasitic flagellated eukaryotes that includes the kinetoplastids and euglenids. These microbial eukaryotes shared a very recent common ancestor before they diverged into separate phylogenetic lineages (Figure 17.3).

Kinetoplastids

Kinetoplastids are a well-studied group of euglenozoans and are named for the presence of the *kinetoplast*, a mass of DNA present in their single, large mitochondrion. Kinetoplastids live primarily in aquatic habitats, where they feed on bacteria. Some species, however, are parasites of animals and cause serious diseases in humans and vertebrate animals. Cells of *Trypanosoma*, a genus infecting humans, are small, about 20 μm long, thin, and crescent-shaped. Trypanosomes have a single flagellum that originates in a basal body and folds back laterally across the cell where it is enclosed by a flap of cytoplasmic membrane (**Figure 17.5**). Both the flagellum and the membrane participate in propelling the organism, making effective movement possible even in viscous liquids, such as blood, where pathogenic trypanosomes are often found.

Trypanosoma brucei (Figure 17.5) causes *African sleeping sickness*, a chronic and usually fatal human disease. The parasite lives and grows primarily in the bloodstream, but in the later stages of the disease it invades the central nervous system, causing an inflammation of the brain and spinal cord that is responsible for the characteristic neurological symptoms of the disease. The parasite is transmitted from host to host by the tsetse fly, *Glossina* spp., a bloodsucking fly found only in certain parts of Africa. After moving from the human to the fly in a blood meal, the parasite proliferates in the intestinal tract of the fly and invades the insect's salivary glands and mouthparts, from which it is transferred to a new human host by a fly bite (Section 32.6).

Other kinetoplastids that are human parasites include *Trypanosoma cruzi*, the causative agent of *Chagas' disease*, and *Leishmania* species, the causative agents of cutaneous and systemic *leishmaniasis*. Chagas' disease is spread by the bite of a blood-feeding insect called the "kissing bug." The disease is usually self-limiting, but it can become chronic and lead to a fatal infection. Leishmaniasis is a disease of tropical and subtropical regions transmitted to humans and other mammals by a bite from the sandfly. This potentially fatal disease can be localized to the skin surrounding a fly bite or can infect the spleen and liver and cause systemic infection. Both Chagas' disease and leishmaniasis are covered in more detail in Section 32.6.

Euglenids

Another well-studied group of euglenozoans are the euglenids (**Figure 17.6**). Unlike the kinetoplasts, these motile microbial eukaryotes are nonpathogenic and both chemotrophic *and* phototrophic. Most euglenids contain two flagella, dorsal and ventral, and their active motility allows the organisms to access both illuminated and dark habitats in their environment to support their alternate nutritional lifestyles.

Euglenids live exclusively in aquatic habitats, both freshwater and marine, and contain chloroplasts, which support phototrophic growth (Figure 17.6). In darkness, however, cells of *Euglena*, a typical euglenid, can lose their chloroplasts and exist as chemoorganotrophs. Many euglenids can also feed on bacterial

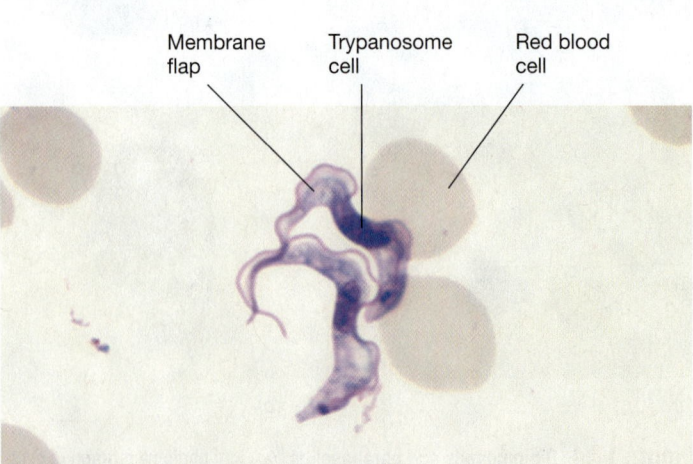

Figure 17.5 **Trypanosomes.** Photomicrograph of the flagellated euglenozoan *Trypanosoma brucei*, the causative agent of African sleeping sickness. Blood smear preparation. A cell is about 3 μm wide.

Membrane flap • Trypanosome cell • Red blood cell

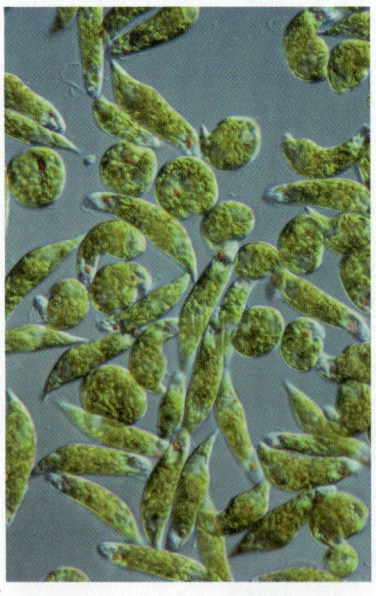
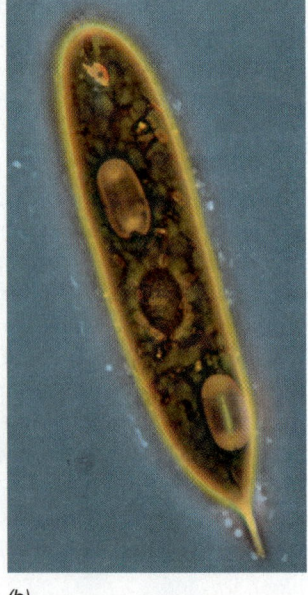

(a) (b)

Figure 17.6 *Euglena*, a euglenozoan. (a) This phototrophic protist, like other euglenids, is not pathogenic. A cell is about 15 μm wide. (b) High-magnification view.

cells via **phagocytosis,** a process of surrounding a particle with a portion of their flexible cytoplasmic membrane to engulf the particle and bring it into the cell where it is digested.

MINIQUIZ -

- Contrast the two nutritional options for *Euglena*.
- How do cells of *Trypanosoma brucei* get from one human host to another?
- What diseases are caused by *Trypanosoma cruzi* and *Leishmania* spp., respectively?

17.5 Alveolates

Key Genera: *Gonyaulax, Plasmodium, Paramecium*

The alveolates as a group are characterized by their *alveoli,* cytoplasmic sacs located just under the cytoplasmic membrane. Although the function of alveoli is unknown, they may help the cell maintain osmotic balance by controlling water influx and efflux, and in the dinoflagellates, may function as armor plates (see Figure 17.9). Three phylogenetically distinct, although related, kinds of alveolates are known: the *ciliates,* which use cilia for motility; the *dinoflagellates,* which are motile by means of a flagellum; and the *apicomplexans,* which are animal parasites (Figure 17.3).

Ciliates

Ciliates possess *cilia* (**Figure 17.7**) at some stage of their life cycle. Cilia are structures that function in motility and may cover the cell or form tufts or rows, depending on the species. Probably the best-known and most widely distributed ciliates are those of the genus *Paramecium* (Figure 17.7). Like many other ciliates, *Paramecium* uses cilia not only for motility but also to obtain food by ingesting particulate materials such as bacterial cells through a distinctive funnel-shaped oral groove. Cilia that line the oral groove move material down the groove to the cell mouth, also called the *gullet* (Figure 17.7b). There, it is enclosed in a vacuole by phagocytosis. Digestive enzymes secreted into the vacuole then break down the material as a source of nutrients.

Ciliates are unique among protists in having two kinds of nuclei, *micronuclei* and *macronuclei*. Genes in the macronucleus regulate basic cellular functions, such as growth and feeding, whereas those of the micronucleus are involved in sexual reproduction, which occurs through a partial fusion of two *Paramecium* cells and exchange of micronuclei. The genome of *Paramecium* is huge, with macronuclear genes numbering about 40,000, nearly twice that of humans (↺ Section 6.6).

Many *Paramecium* species (as well as many other protists) are hosts for endosymbiotic prokaryotes or eukaryotes, the latter usually green algae. These organisms may play a nutritional role, synthesizing vitamins or other growth factors used by the host cell. Several anaerobic ciliated protists contain endosymbiotic prokaryotes. For example, ciliated protists in the termite hindgut contain endosymbiotic methanogens (*Archaea*) that consume H_2 plus CO_2 to yield methane (CH_4). Ciliates themselves can also be symbiotic: Obligately anaerobic ciliates are present in the rumen, the forestomach of ruminant animals, and play an important role in the digestive and fermentative processes of the animal (↺ Section 22.7).

In contrast to symbioses, some ciliates are animal parasites, although this lifestyle is less common in ciliates than in some other groups of protists. The species *Balantidium coli* (**Figure 17.8**), for example, is primarily an intestinal parasite of domestic animals, but occasionally it infects the intestinal tract of humans, producing dysentery-like symptoms. Cells of *B. coli* form cysts (Figure 17.8) that promote disease transmission in infected food or water.

Dinoflagellates

Dinoflagellates are a diverse group of marine and freshwater phototrophic alveolates (**Figure 17.9**) that acquired the capacity to

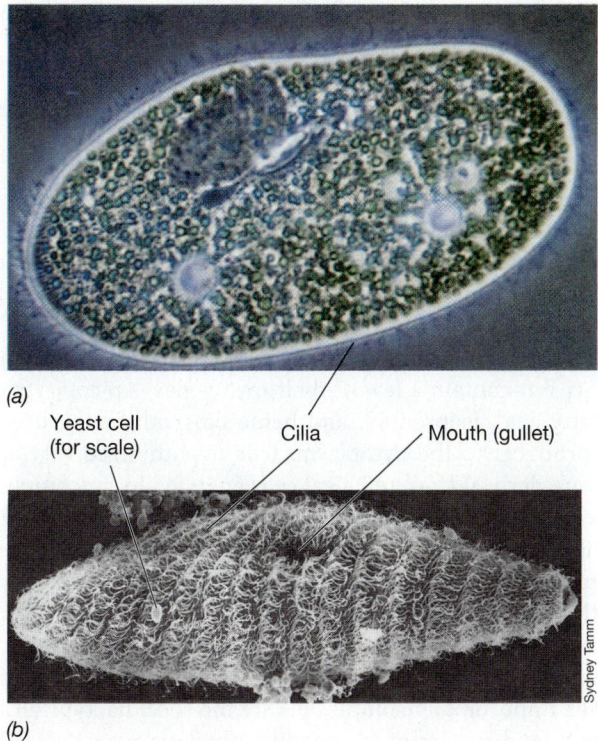

(a)

Yeast cell (for scale) Cilia Mouth (gullet)

(b)

Sydney Tamm

Figure 17.7 *Paramecium*, **a ciliated protist.** *(a)* Phase-contrast photomicrograph. *(b)* Scanning electron micrograph. Note the cilia in both micrographs. A single *Paramecium* cell is about 60 μm in diameter.

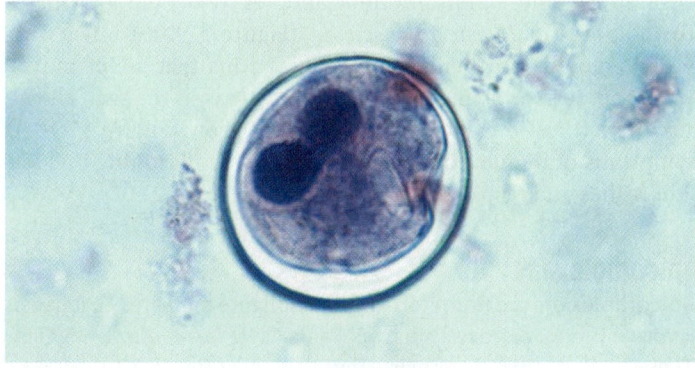

Figure 17.8 *Balantidium coli*, **a ciliated protist that causes a dysentery-like disease in humans.** The dark blue-stained structure in this *B. coli* cyst obtained from swine intestine is the macronucleus. The cell is about 50 μm wide.

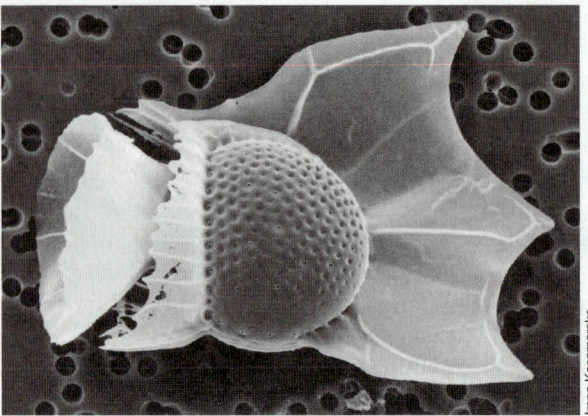

Figure 17.9 **The marine dinoflagellate *Ornithocercus magnificus* (an alveolate).** The cell proper is the globular central structure; the attached ornate structures are called *lists*. A cell is about 30 µm wide.

photosynthesize through secondary endosymbioses (Figures 17.2 and 17.3). Flagella encircling the cell impart spinning movements that give dinoflagellates their name (*dinos* is Greek for "whirling"). Dinoflagellates have two flagella of different lengths and with different points of insertion into the cell, transverse and longitudinal. The transverse flagellum is attached laterally, whereas the longitudinal flagellum originates from the lateral groove of the cell and extends lengthwise (see Figure 17.10*b*). Some dinoflagellates are free-living, whereas others live a symbiotic existence with animals that form coral reefs, obtaining a sheltered and protected habitat in exchange for supplying phototrophically fixed carbon as a food source for the reef.

Several species of dinoflagellates are toxic. For example, dense suspensions of *Gonyaulax* cells, called "red tides" (**Figure 17.10*a***) due to the red-colored pigments of this organism, can form in warm and typically polluted coastal waters. Such blooms are often associated with fish kills and poisoning in humans following consumption of mussels that have accumulated *Gonyaulax* through filter feeding. Toxicity results from a neurotoxin that can cause a condition called *paralytic shellfish poisoning* in humans and some marine animals, such as sea otters. Symptoms include numbness of the lips, dizziness, and difficulty breathing; in severe cases, death can result from respiratory failure. *Pfiesteria* is another toxic dinoflagellate. Toxic spores of *Pfiesteria piscicida* (Figure 17.10*b*) infect fish and eventually kill them due to neurotoxins that affect movement and destroy skin. Lesions form on areas of the fish, allowing opportunistic bacterial pathogens to grow (Figure 17.10*c*). Symptoms of human toxemia from *Pfiesteria* poisoning include skin rashes and respiratory problems.

Apicomplexans

Apicomplexans are nonphototrophic obligate parasites that cause severe human diseases such as malaria (*Plasmodium* species) (**Figure 17.11*a***), toxoplasmosis (*Toxoplasma*) (Figure 17.11*b*), and coccidiosis (*Eimeria*). These organisms are characterized by nonmotile adult stages, and nutrients are taken up in soluble form across the cytoplasmic membrane as in bacteria and fungi.

(a)

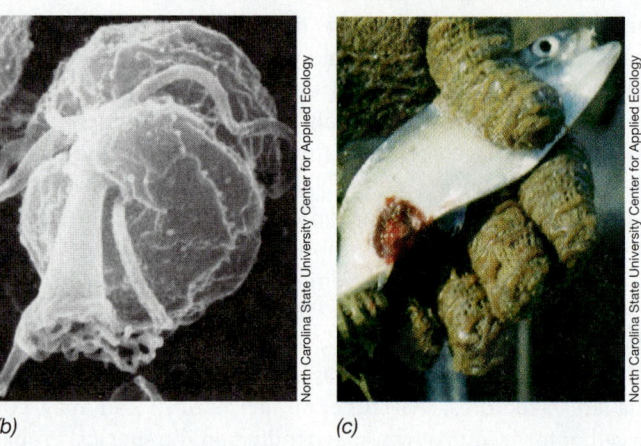

(b)   *(c)*

Figure 17.10 **Toxic dinoflagellates (alveolates).** *(a)* Photograph of a "red tide" caused by massive growth of toxin-producing dinoflagellates such as *Gonyaulax*. The toxin is excreted into the water and also accumulates in shellfish that feed on the dinoflagellates. *(b)* Scanning electron micrograph of a toxic spore of *Pfiesteria piscicida*; the structure is about 12 µm wide. *(c)* A fish killed by *P. piscicida*; note the lesions of decaying flesh.

Apicomplexans produce structures called *sporozoites* (Figure 17.11*b*), which function in transmission of the parasite to a new host, and the name apicomplexan derives from the presence at one apex of the sporozoite of a complex of organelles that penetrate host cells. Apicomplexans also contain *apicoplasts*. These are degenerate chloroplasts that lack pigments and phototrophic capacity but contain a few of their own genes. Apicoplasts catalyze fatty acid, isoprenoid, and heme biosyntheses, and export their products to the cytoplasm. It is hypothesized that apicoplasts are derived from red algal cells engulfed by apicomplexans in a secondary endosymbiosis (Figures 17.2 and 17.3). Over time, the chloroplast of the red algal cell degenerated to play a nonphototrophic role in the apicomplexan cell.

Both vertebrates and invertebrates can be hosts for apicomplexans. In some cases, an alternation of hosts takes place, with some stages of the life cycle linked to one host and some to another. Important apicomplexans are the coccidia, typically bird parasites, and species of *Plasmodium* (malaria parasites) (Figure 17.11*a*). We reserve detailed discussion of malaria—a disease that throughout history has killed more humans than any other disease—for Section 32.5.

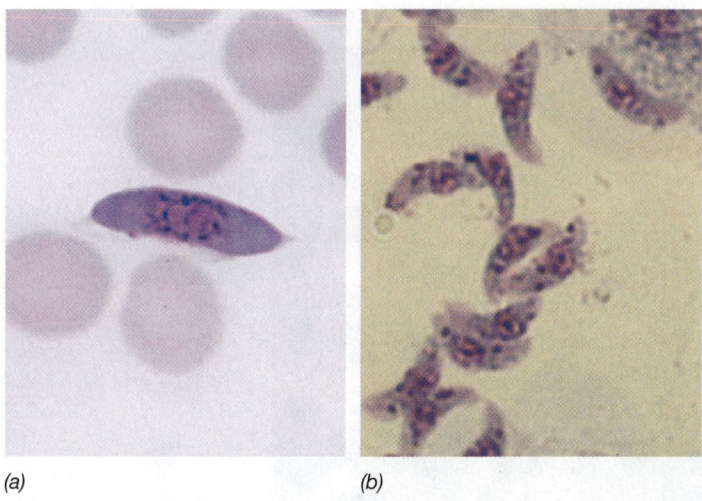

(a)   *(b)*

Figure 17.11 **Apicomplexans.** *(a)* A gametocyte of *Plasmodium falciparum* in a blood smear. The gametocyte is the stage in the malarial parasite life cycle that infects the mosquito vector. *(b)* Sporozoites of *Toxoplasma gondii.*

17.6 Stramenopiles

Key Genera: *Phytophthora, Nitzschia, Ochromonas, Macrocystis*

The *stramenopiles* include both chemoorganotrophic and phototrophic microorganisms as well as macroorganisms. Members of this group bear flagella with many short, hairlike extensions (Figure 17.2), and this morphological feature gives the group its name (from Latin *stramen* for "straw" and *pilus* for "hair"). The diatoms, oomycetes, golden algae, and brown algae are the major groups of stramenopiles (Figure 17.3).

Diatoms

Diatoms include over 200 genera of unicellular, phototrophic, microbial eukaryotes, and are major components of the planktonic (suspended) phytoplankton microbial community in marine and fresh waters. Diatoms characteristically produce a cell wall made of silica to which protein and polysaccharide are added. The wall, which protects the cell against predation, exhibits widely different shapes in different species and can be highly ornate (**Figure 17.12**). The external structure formed by this wall, called a *frustule*, often remains after the cell dies and the organic materials have disappeared. Diatom frustules typically show morphological symmetry, including *pinnate symmetry* (having similar parts arranged on opposite sides of an axis, as in the common diatom *Nitzschia*, Figure 17.12b), and *radial symmetry*, as in the marine diatoms *Thalassiosira* and *Asterolampra* (Figure 17.12c, d). Because the diatom frustules, which are composed mainly of silica, are resistant to decay, these structures can remain intact for long periods of time and often sink and remain in the sediments

for millions of years. Diatom frustules constitute some of the best unicellular eukaryotic fossils known, and from dating of frustule samples, it has been shown that diatoms first appeared on Earth relatively recently, about 200 million years ago.

Oomycetes

The oomycetes, also called *water molds*, were previously grouped with fungi based on their filamentous growth and the presence of **coenocytic** (that is, multinucleate) hyphae, morphological traits characteristic of fungi (Section 17.9). Phylogenetically, however, the oomycetes are distant from fungi and are closely related to other stramenopiles (Figure 17.3). Oomycetes differ from fungi in other fundamental ways, as well. For example, the cell walls of oomycetes are typically made of cellulose instead of the chitin cell walls of fungi, and the water molds have flagellated cells, which are lacking in all but a few fungi. Nonetheless, oomycetes are ecologically similar to fungi in that they grow as a mass of hyphae decomposing dead plant and animal material in aquatic habitats.

Oomycetes have had a major impact on human society, as many species are plant pathogens (phytopathogens). The oomycete *Phytophthora infestans*, which causes late blight disease of potatoes, contributed to massive famines in Ireland in the mid-nineteenth century. The famines led to the death of a million Irish and triggered great waves of Irish immigration to North America. Other major phytopathogens include *Pythium*, a common pathogen of greenhouse seedlings, and *Albugo*, which causes "white rusts" on several agricultural crops.

Golden Algae and Brown Algae

Along with the diatoms, golden and brown algae form major lineages of stramenopiles. Golden algae, also called *chrysophytes*, are primarily unicellular marine and freshwater phototrophs. Some species are chemoorganotrophs and feed by either phagocytosis or by transporting soluble organic compounds across the cytoplasmic membrane. Some golden algae, such as *Dinobryon* (**Figure 17.13a**), found in freshwater, are colonial. However, most golden algae are unicellular and motile by the activity of two flagella of unequal length.

Golden algae are so named because of their golden-brown color (Figure 17.13a, c). This is due to chloroplast pigments dominated by the brown-colored carotenoid fucoxanthin. The major chlorophyll pigment in golden algae is chlorophyll *c* rather than chlorophyll *a*, and they lack the phycobiliproteins present in red algal chloroplasts (Section 17.15). Cells of the unicellular golden alga *Ochromonas*, the best-studied genus of this group, have only one or two chloroplasts (Figure 17.13c).

Brown algae are primarily marine and are multicellular and typically macroscopic. No unicellular brown algae are known. The kelps, such as the giant kelp *Macrocystis* (Figure 17.13b), which can grow up to 50 m in length, are perhaps the most widespread of brown algae. *Fucus*, another common seaweed of intertidal regions, can grow up to 2 m. As their name implies, brown algae are brown or green-brown in color depending on how much of the carotenoid pigment fucoxanthin they produce. Most marine "seaweeds" are brown algae and their rapid growth, especially in cold marine waters, can cause nuisance problems when they wash ashore and decay.

UNIT 3

(a)

(b)   *(c)*   *(d)*

Figure 17.12 Diatom frustules. *(a)* Dark-field photomicrograph of a collage of frustules from different diatom species showing various forms of symmetry. *(b–d)* Scanning electron micrographs of diatom frustules showing pinnate (part b) or radial (parts c, d) symmetry. Diatoms vary considerably in size from very small species about 5 μm wide to larger species up to 200 μm wide.

MINIQUIZ

- What structure of diatoms accounts for their excellent fossil record?
- In what ways do oomycetes differ from and resemble fungi?
- Which chlorophyll pigment is found in golden and brown algae?

17.7 Cercozoans and Radiolarians

Cercozoans and radiolarians have diverged from one another rather recently (Figure 17.3) and are distinguished from other protists by their threadlike cytoplasmic extrusions (pseudopodia) by which they move and feed. Cercozoans were previously classified as *amoeba* because of their pseudopodia, but it is now known that many phylogenetically diverse organisms employ pseudopodia for motility and feeding purposes.

Cercozoans

Cercozoa include the chlorarachniophytes and foraminiferans as major groups. Chlorarachniophytes are freshwater and marine amoeba-like phototrophs that develop a flagellum for dispersal; their acquisition of green algal chloroplasts is a prime example of a secondary endosymbiosis (Figure 17.2) and shows how extensively this process has molded several phylogenetically distinct lineages of microbial eukaryotes (Figure 17.3).

In contrast to chlorarachniophytes, foraminifera are exclusively marine and form shell-like structures called *tests*, which have distinctive characteristics and are often quite ornate (**Figure 17.14a**). Tests are typically made of organic materials reinforced with calcium carbonate. The test is not firmly attached to the cell, and the amoeba-like cell may extend partway out of the test during feeding. However, because of the weight of the test, the cell

(a)   (b)   (c)

Figure 17.13 Golden and brown algae. *(a) Dinobryon,* a golden alga (family *Chrysophyceae*) that forms branched colonies. *(b) Macrocystis,* a marine kelp belonging to the brown algae (family *Phaeophyceae*). *(c) Ochromonas,* a unicellular chrysophyte. The golden or brown color of the chloroplasts of these algae is due to the pigment fucoxanthin.

usually sinks to the bottom of the water column, and it is thought that the organisms feed on dissolved organic matter and particulate deposits, primarily bacteria, other protists, and the remains of dead organisms near the sediments. Foraminiferan cells can also host a variety of algae that form endosymbiotic relationships with the protist and supply it with organic carbon, probably in exchange for inorganic nutrients derived from the breakdown of dead organisms. Phototrophs are found primarily in planktonic foraminifera that remain suspended in the water column to provide their endosymbionts with sufficient sunlight.

Foraminiferan tests (Figure 17.14*a*) are relatively resistant to decay and are readily fossilized. These buried and preserved tests are quite useful to geologists. Because particular taxa of foraminifera are typically associated with particular strata in the geological

record, fossilized tests in samples obtained from exploratory wells are used by oil industry paleontologists as a means to date and assess the petroleum potential of a given drill site.

Radiolarians

Radiolarians are mostly planktonic marine eukaryotes, which, like cercozoans, also form threadlike pseudopodia (Figure 17.14*b*). Radiolarians are strictly heterotrophic and mainly reside in the upper 100 m or so of ocean waters where they consume bacteria and particulate organic matter. Some species associate with algae that take on a symbiotic (but not endosymbiotic) role and supply nutrients to the radiolarian.

The name "radiolarian" comes from the radial symmetry of their tests, transparent or translucent mineral skeletons made of silica in

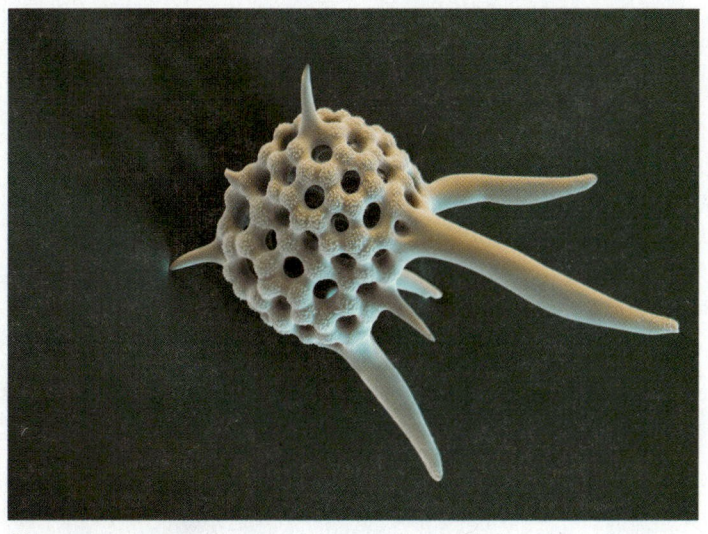

(a)   (b)

Figure 17.14 Cercozoa and radiolaria. *(a)* A foraminiferan. Note the ornate and multilobed test. The test is about 1 mm wide. *(b)* A spiked radiolarian of the *Nassellaria* group. A test is about 150 μm wide. Both *(a)* and *(b)* are colorized scanning electron micrographs.

one fused piece (Figure 17.14*b*). Along with the accumulation of lipid droplets and large cytoplasmic vacuoles, the needle-like pseudopodia of radiolarians probably help keep the organisms from sinking in their mainly open ocean (planktonic) habitats. However, when cells eventually die, their tests settle to the ocean floor and can build up over time into thick layers of slowly decaying cell material.

MINIQUIZ -

- What structure distinguishes cercozoans and radiolarians from other protists?

- How are chlorarachniophytes thought to have acquired the ability to photosynthesize?

17.8 Amoebozoa

Key Genera: *Amoeba, Entamoeba, Physarum, Dictyostelium*

The amoebozoa are a large group of terrestrial and aquatic protists that use lobe-shaped pseudopodia for movement and feeding, in contrast to the threadlike pseudopodia of cercozoans and radiolarians. The major groups of amoebozoa are the *gymnamoebas*, the *entamoebas*, and the *plasmodial* and *cellular slime molds*. Phylogenetically, the amoebozoa diverged from a lineage that eventually led to the fungi and animals (Figure 17.3).

Gymnamoebas and Entamoebas

The gymnamoebas are free-living protists that inhabit aquatic and soil environments. They use pseudopodia to move by a process called *amoeboid movement* (**Figure 17.15**) and feed by phagocytosis on bacteria, other protists, and particulate organic materials. Amoeboid movement results from streaming of the cytoplasm as it flows forward at the less contracted and viscous cell tip, taking the path of least resistance. Cytoplasmic streaming is facilitated by microfilaments (⮌ Section 2.22), which exist in a thin layer just beneath the cytoplasmic membrane. *Amoeba* (Figure 17.15) is a common organism in pond waters, with species varying in size from 15 μm in diameter—clearly microscopic—to over 750 μm—visible with the naked eye.

In contrast to gymnamoebas, the entamoebas are parasites of vertebrates and invertebrates. Their usual habitat is the oral cavity or intestinal tract of animals. *Entamoeba histolytica* is pathogenic in humans and can cause amebic dysentery, an ulceration of the intestinal tract that results in a bloody diarrhea. This parasite forms cysts that are transmitted from person to person by fecal contamination of water, food, and eating utensils. In Section 32.3 we discuss the etiology and pathogenesis of amebic dysentery, an important cause of death from intestinal parasites in humans.

Slime Molds

The **slime molds** were previously grouped with fungi since they undergo a similar life cycle and produce fruiting bodies with spores for dispersal. As protists, however, slime molds are motile and can move across a solid surface fairly quickly (see Figures 17.16–17.18). Slime molds are divided into two groups, *plasmodial slime molds* (also called *acellular slime molds*), whose vegetative forms are masses of protoplasm of indefinite size and shape called plasmodia (**Figure 17.16**), and *cellular slime molds*, whose vegetative forms are single amoebae. Slime molds live primarily on decaying plant matter, such as leaf litter, logs, and soil where they consume other microorganisms, especially bacteria, which they ingest by phagocytosis. Slime molds can maintain themselves in a vegetative state for long periods but eventually form differentiated sporelike structures that can remain dormant and then germinate later to once again generate the active amoeboid state.

Plasmodial slime molds, such as *Physarum*, exist in the vegetative phase as an expanding single mass of protoplasm called the *plasmodium* that contains many diploid nuclei (Figure 17.16). The plasmodium is actively motile by amoeboid movement, and from this phase, a sporangium containing haploid spores can be produced; when conditions are favorable, the spores germinate to yield haploid flagellated swarm cells. The fusion of two swarm cells then regenerates a diploid plasmodium.

In contrast to their plasmodial relatives, cellular slime molds are individual haploid cells and form diploids only under certain conditions. The well-studied cellular slime mold *Dictyostelium discoideum* undergoes an asexual life cycle in which vegetative cells

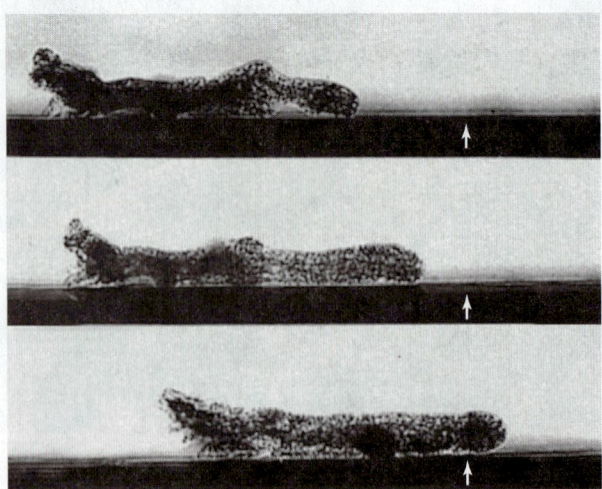

M. Haberey

Figure 17.15 Time-lapse view of the amoebozoan *Amoeba proteus*. The time interval from top to bottom is about 6 sec. The arrows point to a fixed spot on the surface. A single cell is about 80 μm wide.

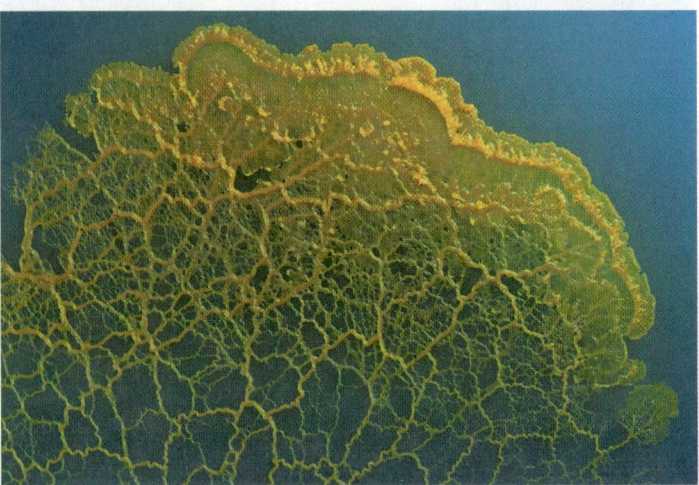

Figure 17.16 Slime mold. The plasmodial slime mold *Physarum* growing on an agar surface. The plasmodium is about 5 cm long and 3.5 cm wide.

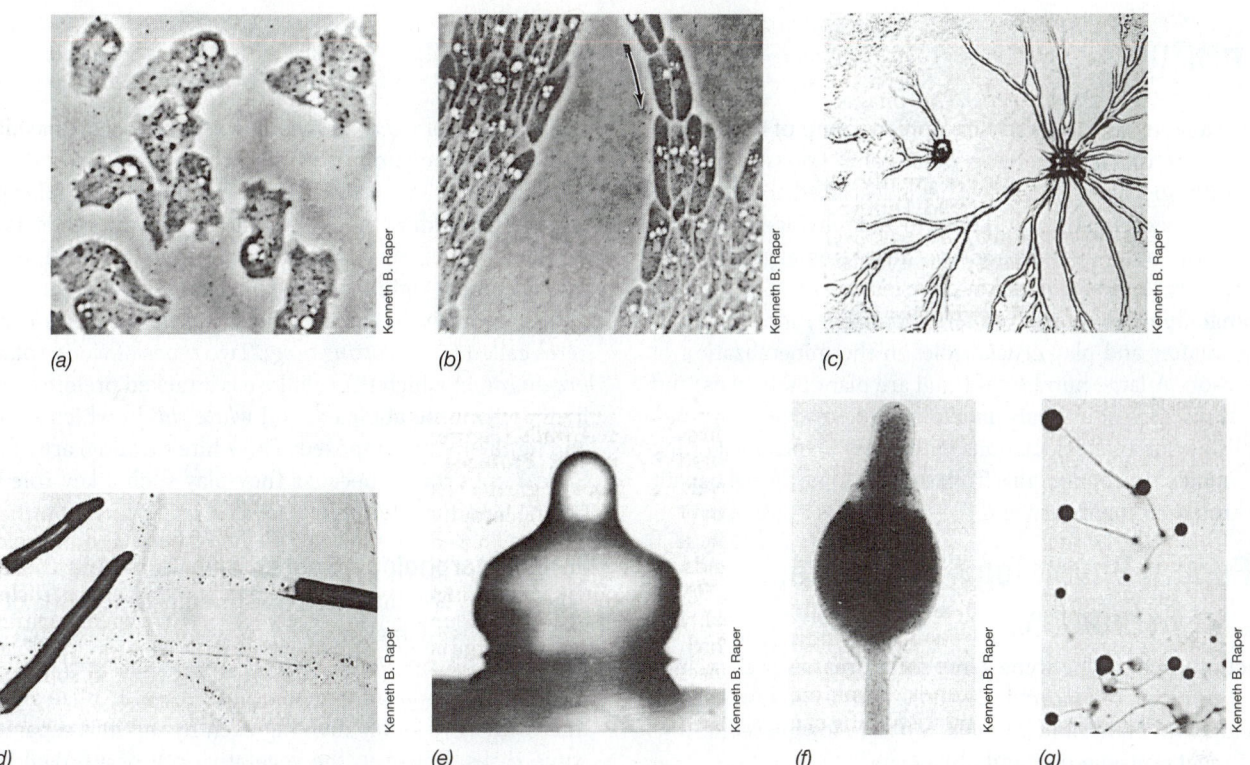

Figure 17.17 Photomicrographs of various stages in the life cycle of the cellular slime mold *Dictyostelium discoideum*. *(a)* Amoebae in preaggregation stage. *(b)* Aggregating amoebae. Amoebae are about 300 μm in diameter. *(c)* Low-power view of aggregating amoebae. *(d)* Migrating pseudoplasmodia (slugs) moving on an agar surface and leaving trails of slime behind. *(e, f)* Early stage of fruiting body. *(g)* Mature fruiting bodies. Figure 17.18 shows the sizes of these structures.

aggregate, migrate as a cell mass, and eventually produce fruiting bodies in which cells differentiate and form spores (**Figures 17.17** and **17.18**). When cells of *Dictyostelium* are starved, they aggregate and form a pseudoplasmodium; in this stage cells lose their individuality, but do not fuse. Aggregation is triggered by the production of cyclic adenosine monophosphate (cAMP). The first cells of *Dictyostelium* that produce this compound attract neighboring cells and eventually aggregate into motile masses of cells called *slugs*. Fruiting body formation begins when the slug becomes stationary and vertically oriented. The emerging structure differentiates into a stalk and a head, with stalk cells forming cellulose, which provides the rigidity of the stalk, and the head cells differentiating into spores. Eventually, spores are released and dispersed, with each spore forming a new amoeba (Figures 17.17 and 17.18).

In addition to this asexual process, *Dictyostelium* can produce sexual spores. These form when two amoebae in an aggregate fuse to form a single giant amoeba. A thick cellulose wall develops around this cell to form a structure called the *macrocyst*, and this can remain dormant for long periods. Eventually, the diploid nucleus undergoes meiosis to form haploid nuclei that become integrated into new amoebae that can once again initiate the asexual cycle.

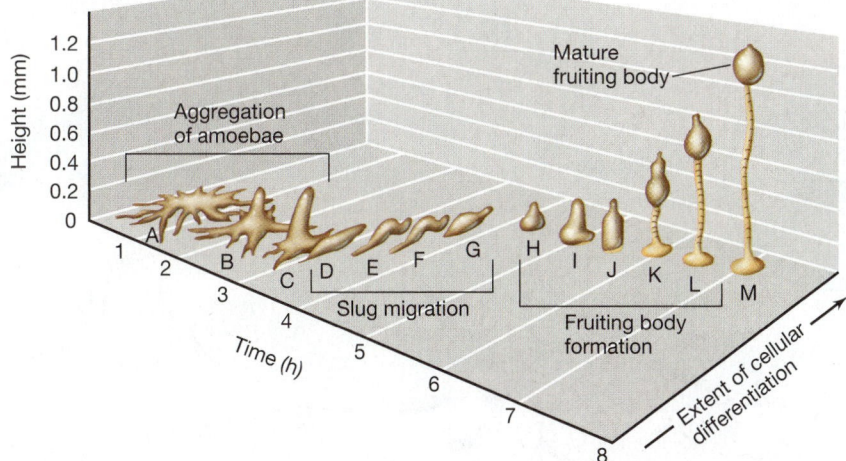

Figure 17.18 Stages in fruiting body formation in the cellular slime mold *Dictyostelium discoideum*. (A–C) Aggregation of amoebae. (D–G) Migration of the slug formed from aggregated amoebae. (H–I) Culmination of migration and formation of the fruiting body. (M) Mature fruiting body composed of stalk and head. Cells from the rear of the slug form the head and become spores. *Dictyostelium* also undergoes sexual reproduction (not shown) when two amoebae fuse to form a macrocyst; the fused nuclei in the macrocyst return to the haploid stage when meiosis forms new vegetative amoebae.

MINIQUIZ

- How can amoebozoans be distinguished from cercozoans and radiolarians?
- Compare and contrast the lifestyles of gymnamoebas and entamoebas.
- Describe the major steps in the life cycle of *Dictyostelium discoideum*.

III • Fungi

Fungi are a large, diverse, and widespread group of organisms, consisting of the *molds*, *mushrooms*, and *yeasts*. Approximately 100,000 species of fungi have been described, and as many as 1.5 million species may exist. Fungi form a phylogenetic cluster distinct from other protists and are the microbial group most closely related to animals (Figure 17.3).

Most fungi are microscopic and terrestrial. They inhabit soil or dead plant matter and play crucial roles in the mineralization of organic carbon. A large number of fungi are plant pathogens, and a few cause diseases of animals, including humans. Certain fungi also establish symbiotic associations with many plants, facilitating the plant's acquisition of minerals from soil, and many fungi benefit humans through fermentation and the synthesis of antibiotics.

17.9 Fungal Physiology, Structure, and Symbioses

In this section we describe some general features of fungi, including their physiology, cell structure, and the symbiotic associations they develop with plants and animals. In the following section we examine fungal reproduction and phylogeny.

Nutrition and Physiology

Fungi are chemoorganotrophs—typically displaying simple nutritional requirements—and most are aerobic. Fungi feed by secreting extracellular enzymes that digest polymeric materials, such as polysaccharides or proteins, into monomers that are assimilated as sources of carbon and energy. As decomposers, fungi digest dead animal and plant materials. As parasites of plants or animals, fungi use the same mode of nutrition but take up nutrients from the living cells of the plants and animals they invade rather than from dead organic materials.

A major ecological activity of fungi, especially basidiomycetes, is the decomposition of wood, paper, cloth, and other products derived from these natural sources. Lignin, a complex polymer in which the building blocks are phenolic compounds, is an important constituent of woody plants, and in association with cellulose it confers rigidity on them. Lignin is decomposed in nature almost exclusively through the activities of certain basidiomycetes called *wood-rotting fungi*. Two types of wood rot are known: *brown rot*, in which the cellulose is attacked preferentially and the lignin left unmetabolized, and *white rot*, in which both cellulose and lignin are decomposed. The white rot fungi are of major ecological importance because they play such a key role in decomposing woody materials in forests.

Fungal Morphology, Spores, and Cell Walls

Most fungi are multicellular, forming a network of filaments called *hyphae* (singular, hypha) from which asexual spores are produced (**Figure 17.19**). Hyphae are tubular cell walls that surround the cytoplasmic membrane. Fungal hyphae are often septate, with cross-walls dividing each hypha into separate cells. In some cases, however, the vegetative cell of a fungal hypha contains more than one nucleus, and hundreds of nuclei can form due to repeated nuclear divisions without the formation of cross-walls, a condition called *coenocytic*. Each hyphal filament grows mainly at the tip by extension of the terminal cell (Figure 17.19).

Hyphae typically grow together across and above a surface to form a compact, macroscopically visible tuft called a *mycelium* (**Figure 17.20a**). From the mycelium, aerial hyphae reach up into the air above the surface, and spores called **conidia** are formed on their tips (Figure 17.20b). Conidia are asexual spores and they are often pigmented black, green, red, yellow, or brown (Figure 17.20). Conidia give the mycelium a dusty appearance (Figure

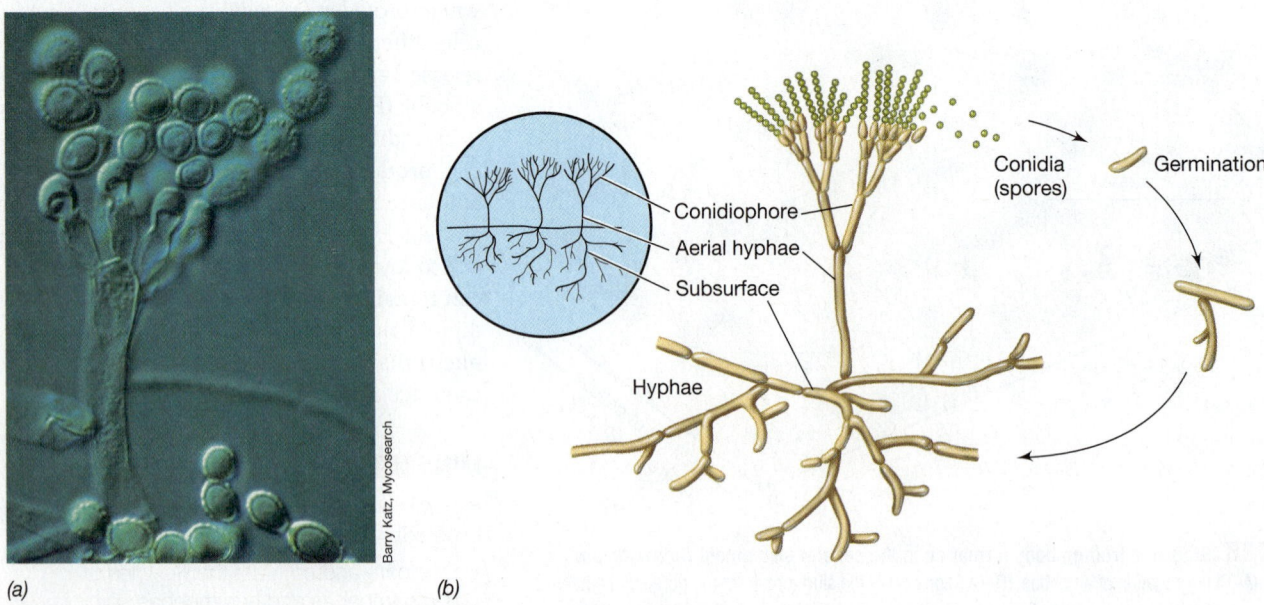

(a)   *(b)*

Barry Katz, Mycosearch

Figure 17.19 Fungal structure and growth. *(a)* Photomicrograph of a typical mold fungus. Spherical structures at the ends of aerial hyphae are asexual spores (conidia). *(b)* Diagram of a mold life cycle. The conidia can be dispersed by either wind or animals and are about 2 μm wide.

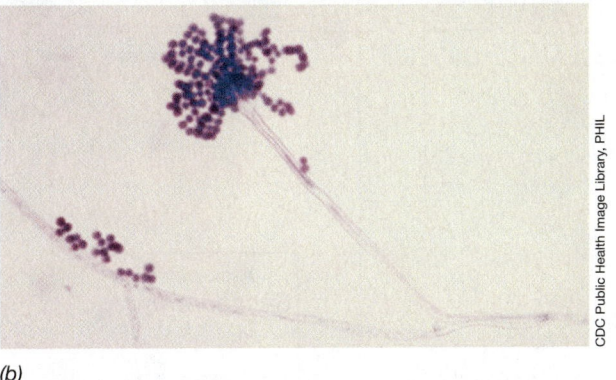

Figure 17.20   Hyphal fungi (molds). *(a)* Colonies of an *Aspergillus* species (ascomycete), growing on an agar plate. Note the masses of filamentous cells (mycelia) and asexual spores that give the colonies a dusty, matted appearance. *(b)* Conidiophore and conidia of *Aspergillus fumigatus* (see Figure 17.19). The conidiophore is about 300 μm long and the conidia about 3 μm wide.

17.20*a*) and function to disperse the fungus to new habitats. Some fungi form macroscopic reproductive structures called *fruiting bodies* (**mushrooms** or puff balls, for example), in which millions of spores are produced that can be dispersed by wind, water, or animals (**Figure 17.21**). In contrast to mycelial fungi, some fungi grow as single cells; these are the **yeasts.**

Most fungal cell walls consist of **chitin,** a polymer of *N*-acetyl-glucosamine. Chitin is arranged in the walls in microfibrillar bundles, as is cellulose in plant cell walls, to form a thick, tough wall structure. Other polysaccharides such as mannans and galactosans, or even cellulose itself, replace or supplement chitin in some fungal cell walls. Fungal cell walls are typically 80–90% polysaccharide, with only small amounts of proteins, lipids, polyphosphates, and inorganic ions making up the wall-cementing matrix.

Symbioses and Pathogenesis

Most plants are dependent on certain fungi to facilitate their uptake of minerals from soil. These fungi form symbiotic associations with the plant roots called *mycorrhizae* (the word means, literally, "fungus roots"). Mycorrhizal fungi establish close physical contact with the roots and help the plant obtain phosphate and other minerals and also water from the soil. In return, the fungi obtain nutrients such as sugars from the plant root (⟳ Figure 22.24). There are two kinds of mycorrhizal associations. One, *ectomycorrhizae*, typically forms between basidiomycetes (Section 17.14) and the roots of woody plants, while the second,

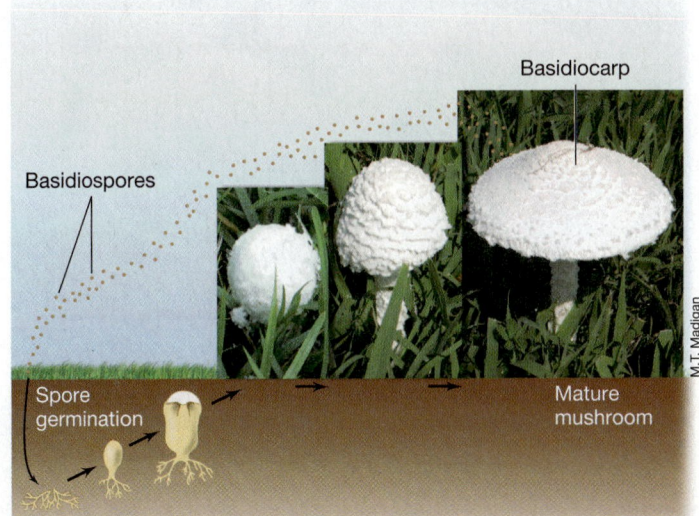

Figure 17.21   Mushroom life cycle. Mushrooms typically develop underground and then emerge on the surface rather suddenly (usually overnight), triggered by an influx of moisture. Photos of stages in formation of a common lawn mushroom (see also Section 17.14).

endomycorrhizae, forms between glomeromycete fungi (Section 17.12) and many nonwoody plants. Some fungi also form associations with cyanobacteria or green algae. These are the *lichens*, the colorful and crusty growths often seen on the surfaces of trees and rocks. We explore the biology of lichens and mycorrhizae in more detail in Sections 22.1 and 22.5, respectively.

Fungi can invade and cause disease in plants and animals. Fungal plant pathogens cause widespread crop and plant damage worldwide, and fruit and grain crops in particular suffer significant yearly losses due to fungal infection. Human fungal diseases, called *mycoses*, range from relatively minor and easily cured conditions, such as athlete's foot, to serious, life-threatening systemic mycoses, such as histoplasmosis. Section 32.2 describes the major human diseases caused by fungi.

MINIQUIZ -

- What are conidia? How does a conidium differ from a hypha? A mycelium?
- What is chitin and where is it present in fungi?
- Distinguish between mycorrhizae and lichens.

17.10 Fungal Reproduction and Phylogeny

Fungi reproduce by *asexual* means in one of three ways: (1) by the growth and spread of hyphal filaments; (2) by the asexual production of spores (conidia; Figures 17.20 and 17.21); or (3) by simple cell division, as in budding yeasts (**Figure 17.22**). Most fungi also form sexual spores, typically as part of an elaborate life cycle. Some fungi, such as the well-known mold *Penicillium* (the source of the antibiotic penicillin), were long thought to lack a sexual stage and reproduce only by way of conidia. But it has now been shown that *Penicillium* (and probably all fungi of its taxonomic class, the *Deuteromycetes*) go through a sexual stage in their life cycles.

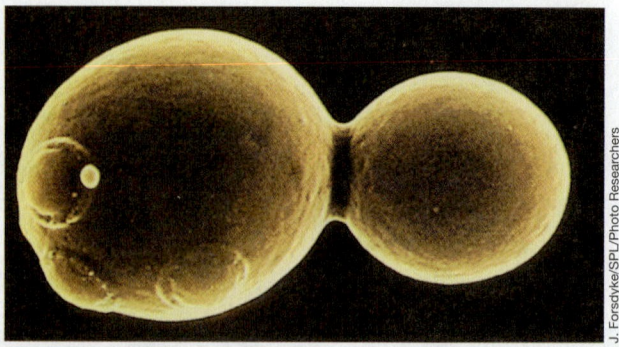

J. Forsdyke/SPL/Photo Researchers

Figure 17.22 **The common baker's and brewer's yeast *Saccharomyces cerevisiae* (ascomycetes).** In this colorized scanning electron micrograph, note the budding division and scars from previous buds. A single cell is about 6 μm in diameter.

Sexual Spores of Fungi

Some fungi produce spores as a result of sexual reproduction. The spores develop from the fusion of either unicellular gametes or specialized hyphae called *gametangia*. Alternatively, sexual spores can originate from the fusion of two haploid cells to yield a diploid cell; this then undergoes meiosis and mitosis to yield individual haploid spores. Depending on the group, different types of sexual spores are produced. Spores formed within an enclosed sac (ascus) are called *ascospores*. Many yeasts produce ascospores, and we consider sporulation in the common baker's yeast *Saccharomyces cerevisiae* in Section 17.13. Sexual spores produced on the ends of a club-shaped structure (basidium) are *basidiospores* (Figure 17.21 and see Figure 17.30*c*). *Zygospores*, produced by zygomycetous fungi such as the common bread mold *Rhizopus* (Section 17.12), are macroscopically visible structures that result from the fusion of hyphae and genetic exchange. Eventually the zygospore matures and produces asexual spores that are dispersed by air and germinate to form new fungal mycelia. Chytrid fungi produce motile sexual spores called *zoospores*.

Sexual spores of fungi are typically resistant to drying, heating, freezing, and some chemical agents. However, neither sexual nor asexual spores of fungi are as resistant to heat as bacterial endospores (⟲ Section 2.16). Both asexual and sexual spores of fungi can germinate and develop into a new hypha.

The Phylogeny of Fungi

Fungi share a more recent common ancestor with animals than does any other group of eukaryotic organisms (Figure 17.3). Fungi and animals are thought to have diverged approximately 1.5 billion years ago. The earliest fungal lineage is thought to be the chytridiomycetes, an unusual group of motile fungi in which cells produce flagellated spores (zoospores, Section 17.11). Thus the lack of flagella in most fungi indicates that motility is a characteristic that has been lost at various times in different fungal lineages.

A detailed picture of fungal phylogeny is shown in the evolutionary tree in **Figure 17.23**. The phylogeny shown in this figure, based on comparative sequencing of 18S ribosomal RNA (which can be used to resolve fairly close, but not distant, relationships among eukaryotes; see Section 17.2), defines several distinct

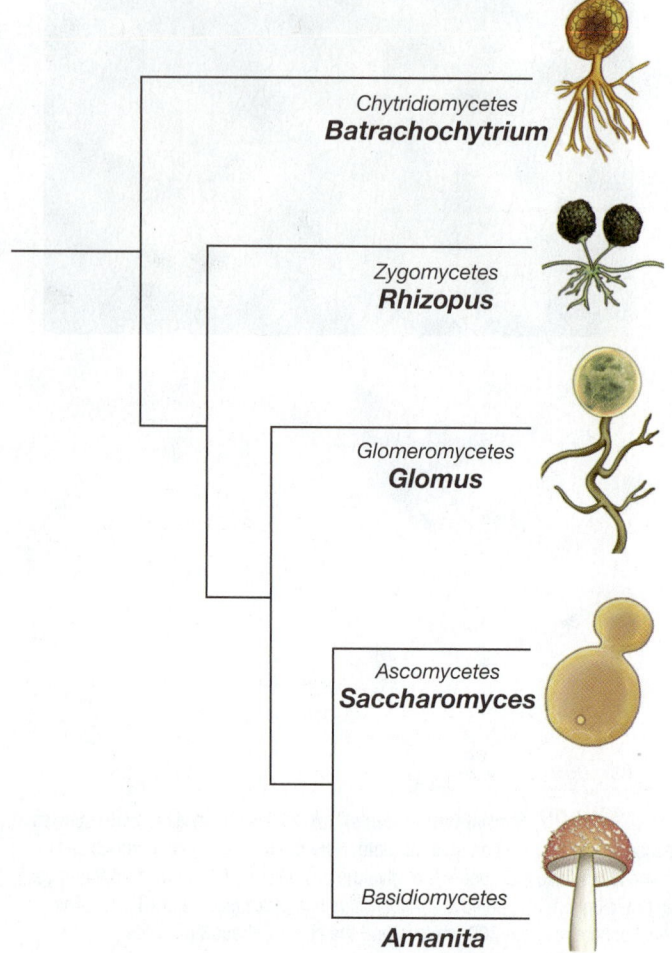

Chytridiomycetes
Batrachochytrium

Zygomycetes
Rhizopus

Glomeromycetes
Glomus

Ascomycetes
Saccharomyces

Basidiomycetes
Amanita

Figure 17.23 **Phylogeny of fungi.** This generalized phylogenetic tree based on 18S ribosomal RNA gene sequences depicts the relationships among the major groups (phyla) of fungi. A typical genus is listed for each group and depicted in the tree.

fungal groups: the chytridiomycetes, zygomycetes, glomeromycetes, ascomycetes, and basidiomycetes. Figure 17.23 also supports the idea that the chytridiomycetes lie phylogenetically basal to all other fungal groups and that the most derived groups of fungi are the basidiomycetes, which includes the mushrooms (Figure 17.21 and see Figure 17.30), and the ascomycetes, which includes the yeasts such as *Saccharomyces* (Figure 17.22) and molds such as *Aspergillus* (Figure 17.20).

MINIQUIZ

- Why is the mold *Penicillium* economically important?
- What are the major differences between ascospores and conidia?
- To what major group of macroorganisms are fungi most closely related?

17.11 Chytridiomycetes

Key Genera: *Allomyces, Batrachochytrium*

Chytridiomycetes, or *chytrids*, are the earliest diverging lineage of fungi (Figure 17.23), and their name refers to the structure of

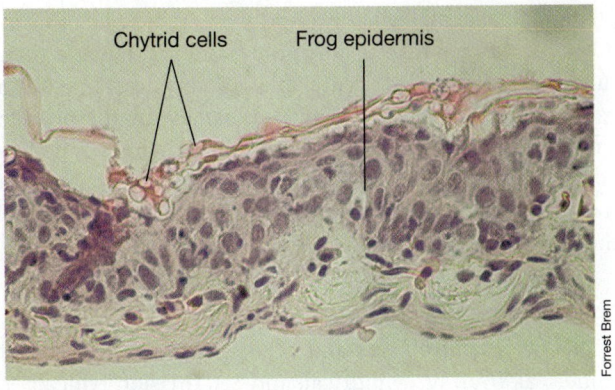

Chytrid cells   Frog epidermis

Forrest Brem

Figure 17.24 Chytridiomycetes. Cells of the chytrid *Batrachochytrium dendrobatidis* stained pink growing on the surface of the epidermis of a frog.

the fruiting body, which contains their sexual spores (*zoospores*). These spores are unusual among fungal spores in being flagellated and motile, and are ideal for dispersal of these organisms in the aquatic environments, mostly freshwater and moist soils, where they are commonly found.

Many species of chytrids are known and some exist as single cells, whereas others form colonies with hyphae. They include both free-living forms that degrade organic material, such as *Allomyces*, and parasites of animals, plants, and protists. The chytrid *Batrachochytrium dendrobatidis* causes chytridiomycosis of frogs (**Figure 17.24**), a condition in which the organism infects the frog's epidermis and interferes with the ability of the frog to respire across the skin. Chytrids have been implicated in the massive die-off of frogs and some other amphibians worldwide, probably in response to increases in global temperatures that have stimulated chytrid proliferation and to increased animal susceptibility due to habitat loss and aquatic pollution.

Some chytrids are obligately anaerobic, a highly unusual property for eukaryotic cells, and inhabit the rumen of ruminant animals. The rumen is that part of the ruminant digestive system where the breakdown of cellulose and related polysaccharides occurs (⟳ Section 22.7). The chytrid *Neocallimastix*, for example, inhabits the rumen and conserves energy from the fermentation of sugars to acids, alcohol, and H_2. Cells of *Neocallimastix* lack mitochondria and instead contain hydrogenosomes that assist in their fermentative lifestyle by degrading pyruvate to acetate plus CO_2 plus H_2 (⟳ Figure 2.64).

Unresolved aspects of the phylogeny of chytrids suggest that this group is not monophyletic. That is, some organisms currently classified as chytrids may actually be more closely related to species of other fungal groups, such as the zygomycetes. As is true for protists, much about the evolution of this and other groups of fungi remains to be learned.

MINIQUIZ

- What animal group has been most affected by chytrids?
- What is one feature of chytrids that distinguishes them from other fungi?
- What is physiologically unusual about the chytrid *Neocallimastix*?

17.12 Zygomycetes and Glomeromycetes

Key Genera: *Rhizopus, Encephalitozoon, Glomus*

We consider two groups of fungi here, the zygomycetes, known primarily for their role in food spoilage, and the glomeromycetes, important fungi in certain mycorrhizal associations. Zygomycetes are commonly found in soil and on decaying plant material, whereas glomeromycetes form symbiotic relationships with plant roots. All of these fungi are coenocytic (multinucleate), and a unifying feature is the formation of sexual spores called *zygospores* (Section 17.10).

Rhizopus, the Common Bread Mold

The black bread mold *Rhizopus* (**Figure 17.25a**) is a common zygomycete. This organism undergoes a complex life cycle that includes both asexual and sexual reproduction. In the asexual phase the mycelia form sporangia within which haploid spores are produced. Once released, spores disperse and eventually germinate, giving rise to vegetatively growing mycelia. In the sexual phase, mycelial gametangia of different mating types (analogous to male and female, see Section 17.13) fuse to yield a cell with two nuclei called a *zygosporangium*, which can remain dormant and resist dryness and other unfavorable conditions. When conditions are favorable, the different haploid nuclei fuse to form a diploid nucleus followed by meiosis to yield haploid spores. As in the asexual phase, the release of the spores, in this case genetically nonidentical spores, disperses the organism for vegetative hyphal growth.

Microsporidia and Glomeromycetes

Microsporidia are tiny (2–5 μm) and unicellular parasites of animals and protists. Based on 18S ribosomal RNA gene sequencing and their lack of mitochondria, microsporidia were once thought to form a very early-branching lineage of *Eukarya*. However, composite gene and protein sequencing has shown the microsporidia to be closely related to the zygomycetes (Figure 17.3).

Microsporidia have adapted to a parasitic lifestyle through the elimination or loss of many key aspects of eukaryotic biology; they are even more structurally stripped down than other amitochondriate eukaryotes. The microsporidium *Encephalitozoon* (Figure 17.25b), for example, lacks mitochondria and hydrogenosomes,

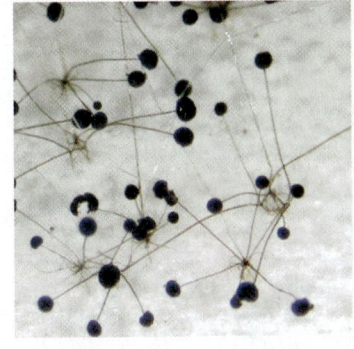
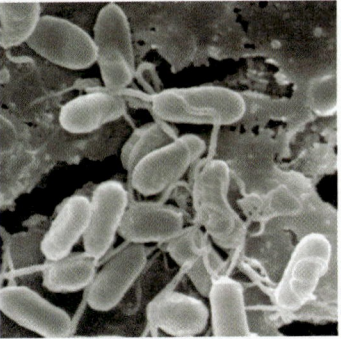

(a)   *(b)*

Figure 17.25 Zygomycetes and microsporidia. *(a)* Stained mycelium of the mold *Rhizopus* showing the spherical sporangia that contain asexual spores. *(b)* Scanning electron micrograph of cells of *Encephalitozoon intestinalis*.

and even lacks the Golgi complex (♻ Figure 2.60). Moreover, the organism contains a very small genome of only 2.9 megabase pairs and contains only about 2000 genes (this is 1.5 megabase pairs and 2600 genes smaller than that of the bacterium *Escherichia coli*!). The *Encephalitozoon* genome lacks genes for major metabolic pathways, such as the citric acid cycle, meaning that this pathogen must depend on its host for even the most basic of metabolic processes. *Encephalitozoon* causes chronic debilitating diseases of the intestine, lung, eye, muscle, and some internal organs but is uncommon among healthy adults with normal immune systems. However, microsporidial diseases have appeared with increasing frequency in immune-compromised individuals, such as those with AIDS or those on long-term administration of immune-suppressing drugs.

The glomeromycetes are a relatively small and unique group of obligately symbiotic fungi in which all known species form associations with plants called *endomycorrhizae* (Section 17.9 and ♻ Section 22.5). As many as 80% or more of land plant species form these associations in which the fungal hyphae enters the plant cell and aids the plant's acquisition of minerals from the soil in return for fixed carbon from the plant. As plant symbionts, glomeromycetes are thought to have played a pivotal role in the ability of early vascular plants to colonize land. As far as is known, glomeromycetes reproduce only asexually and are mostly coenocytic in their hyphal morphology. Spores of *Glomus* (Figure 17.23), a major genus of endomycorrhizae, are collected from the roots of cultivated plants and used as an agricultural inoculant to ensure vigorous symbiotic associations.

MINIQUIZ --

- Contrast the habitats of zygomycetes and glomeromycetes.
- What is unusual about the genome of microsporidia?
- How does the fungus *Glomus* aid the acquisition of nutrients by plants?

17.13 Ascomycetes

Key Genera: *Saccharomyces, Candida, Aspergillus*

The ascomycetes are a large and highly diverse group of fungi that range from single-celled species, such as the baker's yeast *Saccharomyces* (**Figure 17.26** and Figure 17.22), to species that grow as filaments, such as the common mold *Aspergillus* (Figure 17.20). The group ascomycetes, species of which are found in aquatic and terrestrial environments, takes its name from the production of

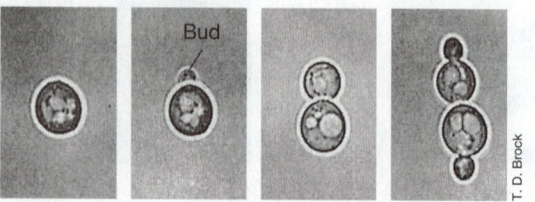

Figure 17.26 Growth by budding division in *Saccharomyces cerevisiae*. A time-lapse series of phase-contrast micrographs shows the budding division process starting from a single cell. Note the pronounced nucleus. A single cell of *S. cerevisiae* is about 6 μm in diameter.

asci (singular, ascus), cells in which two haploid nuclei from different mating types fuse to form a diploid nucleus that eventually undergoes meiosis to form haploid ascospores. In addition to ascospores, ascomycetes reproduce asexually by the production of conidia that form at the tips of specialized hyphae called *conidiophores* (Figure 17.20). Both saprophytic and pathogenic yeasts, such as *Candida albicans*, are common in nature. We focus here on the yeast *Saccharomyces* as a model ascomycete.

Saccharomyces cerevisiae

The cells of *Saccharomyces* and other single-celled ascomycetes are spherical, oval, or cylindrical, and cell division typically takes place by budding. In the budding process, a new cell forms as a small outgrowth of the old cell; the bud gradually enlarges and then separates from the parent cell (Figures 17.22 and 17.26).

Yeast cells are typically much larger than bacterial cells and can be distinguished from bacteria microscopically by their larger size and by the obvious presence of internal cell structures, such as the nucleus or cytoplasmic vacuoles (Figure 17.26). Yeasts flourish in sugar-rich habitats such as fruits, flowers, and the bark of trees. Yeasts are typically facultative aerobes, growing aerobically as well as by fermentation. Several yeasts live symbiotically with animals, especially insects, and a few species are pathogenic for animals and humans (♻ Section 32.2). The most important commercial yeasts are the baker's and brewer's yeasts, which are species of *Saccharomyces*. The yeast *S. cerevisiae* has been studied as a model eukaryote for many years and was the first eukaryote to have its genome completely sequenced (♻ Section 6.6).

Mating Types and Sexual Reproduction in *Saccharomyces*

Saccharomyces can reproduce by sexual means in which two cells fuse. Within the fused cell, called a *zygote*, meiosis occurs and ascospores are eventually formed. The life cycle of *S. cerevisiae* is described in **Figure 17.27**. Cells of *S. cerevisiae* can grow vegetatively in either a haploid or diploid stage. *S. cerevisiae* forms two different types of haploid cells called *mating types*. These are designated α (alpha) and *a* (encoded by genes α and *a*) and are analogous to male and female gametes. The α and *a* genes regulate the production of the peptide hormones α factor or *a* factor, which are excreted by yeast cells during mating. The hormones bind to cells of the opposite mating type and bring about changes in their cell surfaces that enable the cells to fuse; once mating has occurred, the nuclei fuse, forming a diploid zygote (**Figure 17.28**). The zygote grows vegetatively by budding, but under starvation conditions it undergoes meiosis and generates ascospores (Figure 17.27).

Haploid strains of *S. cerevisiae* are genetically predisposed to be either *a* or α but are able to switch their mating type. This switch occurs when the active mating-type gene is replaced with one of two otherwise "silent" genes, as shown in **Figure 17.29**. There is a single location on one of the *S. cerevisiae* chromosomes called the *MAT* (for *mating type*) locus, at which either gene *a* or gene α can be inserted. At this locus, the *MAT* promoter controls transcription of whichever gene is present. If gene *a* is at that locus, then the cell is mating-type *a*, whereas if gene α is at that locus, the cell is mating-type α. Elsewhere in the yeast genome are copies of genes *a* and α that are not expressed, and these are the source of the

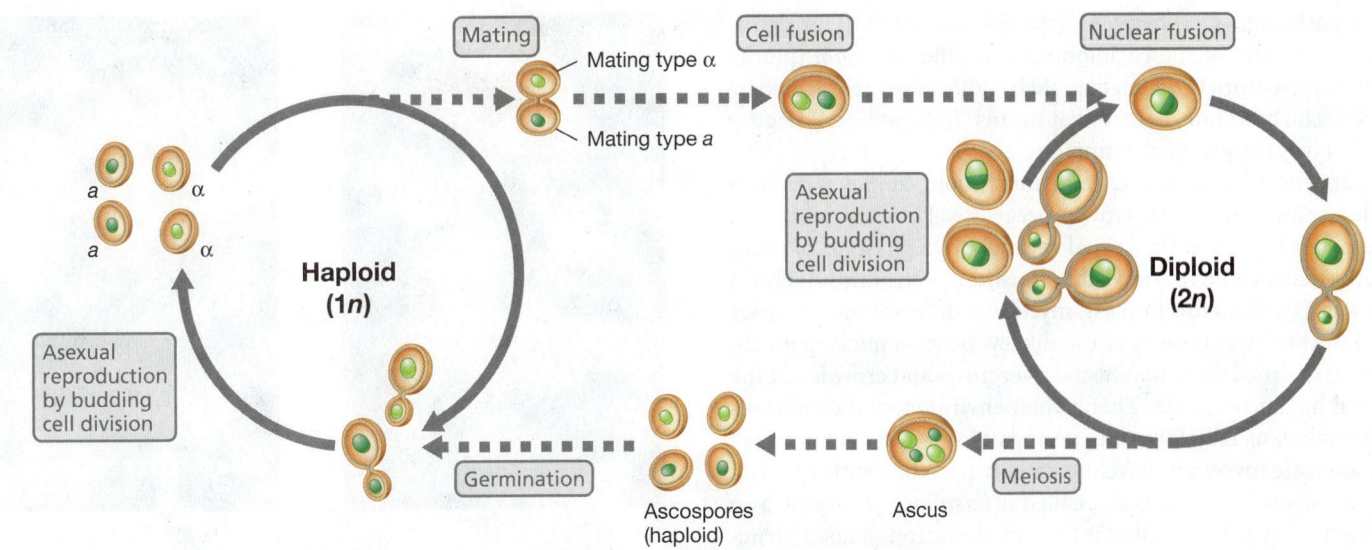

Figure 17.27 Life cycle of a typical ascomycete yeast, *Saccharomyces cerevisiae*. Cells can grow vegetatively for long periods as haploid cells or as diploid cells before life cycle events (dashed lines) generate the alternate genetic form.

inserted gene. In the switch (Figure 17.29), the appropriate gene, *a* or α, is copied from its silent site and inserted into the *MAT* location, replacing the gene already present. The old mating-type gene is excised and discarded, and the new gene is inserted. Whichever gene is inserted in the *MAT* locus is the one that will govern the mating type of the strain. It is thus possible for cells from a pure culture of *S. cerevisiae* derived from a single cell to mate, following a mating-type switch in one or more cells in the culture.

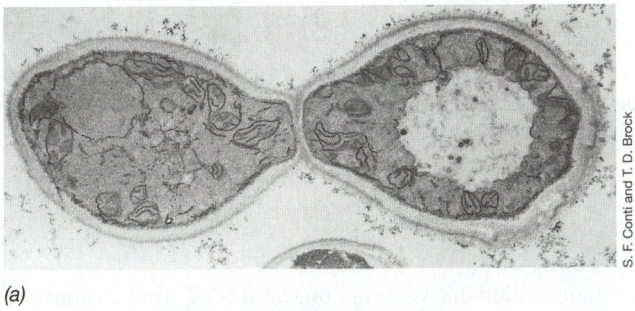

(a)

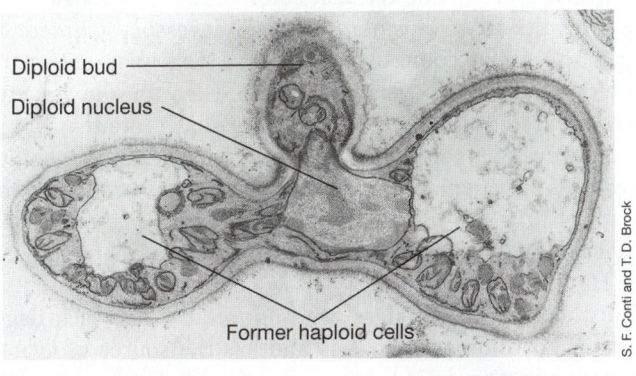

Diploid bud

Diploid nucleus

Former haploid cells

(b)

Figure 17.28 Electron micrographs of mating in the ascomycete yeast *Hansenula wingei*. (a) Two cells have fused at the point of contact. (b) Late stage of mating. The nuclei of the two cells have fused, and a diploid bud has formed at a right angle to the mating cells. This bud becomes the progenitor of a diploid cell line. A cell of *Hansenula* is about 10 μm in diameter.

MINIQUIZ

- Are ascospores haploid or diploid cells?
- Explain how a *single* haploid cell of *Saccharomyces* can eventually yield a diploid cell.

17.14 Mushrooms and Other Basidiomycetes

Key Genera: *Agaricus, Amanita*

Basidiomycetes are a large group of fungi, with over 30,000 species described. Many are the commonly recognized mushrooms and toadstools, some of which are edible, such as the commercially grown mushroom *Agaricus*. Others, such as the mushroom *Amanita* (Figure 17.30*a*), are highly poisonous. Other basidiomycetes include puffballs, smuts, rusts, and an important human

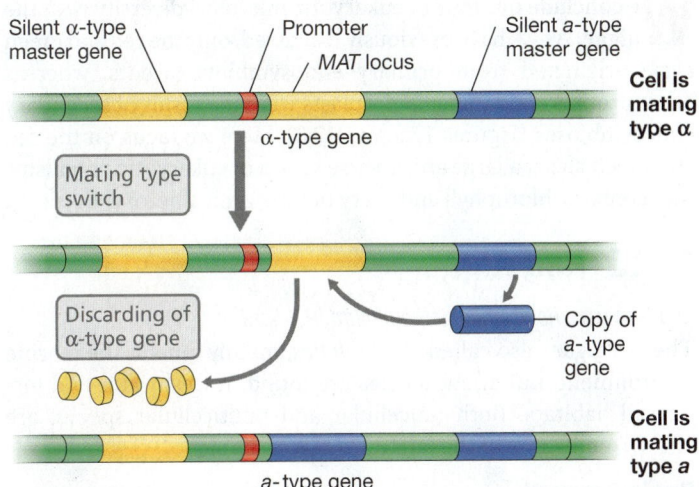

Figure 17.29 The cassette mechanism that switches an ascomycete yeast from mating type α to *a*. The cassette inserted at the *MAT* locus determines the mating type. The process shown is reversible, so type *a* can also revert to type α.

fungal pathogen, *Cryptococcus* (⊂⊃ Section 32.2). The defining characteristic of the basidiomycetes is the *basidium* (plural, basidia), a structure in which haploid basidiospores are formed by meiosis. The basidium, a word that means "little pedestal" (Figure 17.30*c*), gives the group its name.

During most of its existence, a mushroom fungus lives as a simple haploid mycelium, growing vegetatively in soil, leaf litter, or decaying logs. It is the sexual reproductive phase of basidiomycetes that produces the visible mushroom structure (Figures 17.21 and 17.30). In this process, mycelia of different mating types fuse, and the faster growth of the dikaryotic (two nuclei per cell) mycelium formed from that fusion overgrows and crowds out the parental haploid mycelia. Then, when environmental conditions are favorable, usually following periods of wet and cool weather, the dikaryotic mycelium develops rapidly into the fruiting body.

A mushroom fruiting body, called a *basidiocarp*, begins as a mycelium that differentiates into a small button-shaped structure underground that then expands into the full-grown basidiocarp that we see aboveground, the mushroom (Figures 17.21 and 17.30). The dikaryotic basidia are borne on the underside of the basidiocarp on flat plates called *gills*, which are attached to the cap of the mushroom (Figure 17.30*b*, *c*). The basidia then undergo a fusion of the two nuclei, forming basidia with diploid nuclei. The two rounds of meiotic division generate four haploid nuclei in the basidia, and each of the nuclei becomes a basidiospore. The genetically distinct basidiospores can then be dispersed by wind to new habitats to begin the cycle again, germinating under favorable conditions and growing as haploid mycelia (Figure 17.21).

MINIQUIZ ---

- Which basidiomycetes are edible?
- Are basidiospores haploid or diploid?

(a)   (b)

Gills

(c)

Figure 17.30   Mushrooms. *(a) Amanita*, a highly poisonous mushroom. *(b)* Gills on the underside of the mushroom fruiting body contain the spore-bearing basidia. *(c)* Light micrograph of basidia and basidiospores from the mushroom *Coprinus*.

IV • Red and Green Algae

We conclude our tour of eukaryotic microbial diversity with the algae. As we have previously discussed, only the red and green algae originated from primary endosymbiotic events, whereas other protists containing chloroplasts were the result of secondary endosymbioses (Figures 17.2 and 17.3). Here we focus on the red and green algae, a large and diverse group of eukaryotic organisms that contain chlorophyll and carry out oxygenic photosynthesis.

17.15 Red Algae

Key Genera: *Polysiphonia, Cyanidium, Galdiera*

The red algae, also called *rhodophytes*, mainly inhabit the marine environment, but a few species are found in freshwater and terrestrial habitats. Both unicellular and multicellular species are known, and some of the latter are macroscopic.

Basic Properties

Red algae are phototrophic and contain chlorophyll *a*; their chloroplasts lack chlorophyll *b* but contain phycobiliproteins,

the major light-harvesting pigments of the cyanobacteria (⊂⊃ Section 13.2). The reddish color of many red algae (**Figure 17.31**) results from phycoerythrin, an accessory pigment that masks the green color of chlorophyll. This pigment is present along with phycocyanin and allophycocyanin in structures called *phycobilisomes*, the light-harvesting (antenna) components of cyanobacteria. At greater depths in aquatic habitats, where less light penetrates, cells compensate by producing more phycoerythrin and are a darker red, whereas shallow-dwelling species often have less phycoerythrin and can be green in color (see Figure 17.32).

Most species of red algae are multicellular and lack flagella. Some are considered seaweeds and are the source of agar, the solidifying agent used in bacteriological media, and carrageenans, thickening and stabilizing agents used in the food industry. Other species of red algae, such as the genus *Porphyra*, are harvested and dried and used in making sushi. Different species of red algae are filamentous, leafy, or, if they deposit calcium carbonate, *coralline*

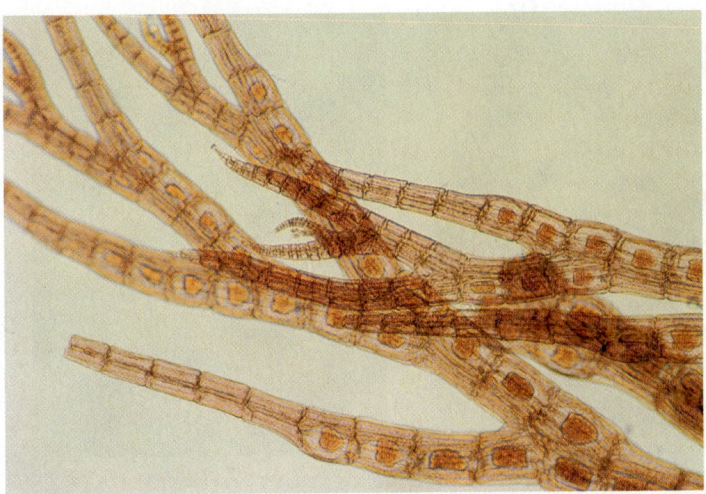

Figure 17.31 *Polysiphonia*, **a filamentous marine red alga.** Light micrograph. *Polysiphonia* grows attached to the surfaces of marine plants. Cells are about 150 μm wide.

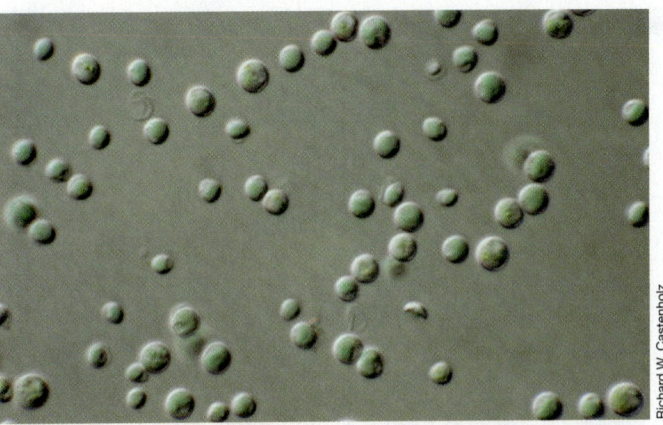

Richard W. Castenholz

Figure 17.32 *Galdieria*, **a unicellular red alga.** This alga grows at low pH and high temperature in hot springs. The cells are about 25 μm in diameter and are more blue-green than red in color because *Galdieria* contains mainly phycocyanin rather than phycoerythrin as its phycobilin. See page 543 for interesting features of the genome of *Galdieria*.

(coral-like) in morphology. Coralline red algae play an important role in the development of coral reefs and help strengthen reefs from wave damage (Section 22.14).

Cyanidium and Relatives

Unicellular species of red algae are also known. One such group, members of the *Cyanidiales* that includes the genera *Cyanidium*, *Cyanidioschyzon*, and *Galdieria* (**Figure 17.32**), live in acidic hot springs at temperatures from 30 to 60°C and at pH values from 0.5 to 4.0; under these extreme conditions, no other phototrophic microorganisms (including anoxygenic phototrophs) can exist. The unicellular red algae are unusual in other ways as well. For example, cells of *Cyanidioschyzon merolae* are unusually small (1–2 μm in diameter) for eukaryotes, and the genome of this species, approximately 16.5 Mbp, is one of the smallest genomes known for a phototrophic eukaryote.

MINIQUIZ

- What traits link cyanobacteria and red algae?
- What physiological properties would be necessary for *Galdieria* to live in its habitat?

17.16 Green Algae

Key Genera: *Chlamydomonas, Volvox*

The green algae, also called *chlorophytes*, have chloroplasts containing chlorophylls *a* and *b*, which give them their characteristic green color, but they lack phycobiliproteins and so do not develop the red or blue-green colors of red algae (Figures 17.31 and 17.32). In the composition of their photosynthetic pigments, green algae are similar to plants and are closely related to plants phylogenetically. There are two main groups of green algae, the chlorophytes, examples of which are the microscopic *Chlamydomonas* and *Dunaliella* (**Figure 17.33a**), and the charophyceans such as *Chara* (Figure 17.33b), macroscopic organisms that often resemble land plants and are actually most closely related to land plants.

Most green algae inhabit freshwater while others are found in moist soil or growing in snow, to which they impart a pink color (Figure 5.22). Other green algae live as symbionts in lichens (Section 22.1). The morphology of chlorophytes ranges from unicellular (Figure 17.33*a*, *c*) to filamentous, with individual cells arranged end to end (Figure 17.33*e*), to **colonial**, as aggregates of cells (Figure 17.33*f*). Even multicellular species exist, an example of which is the seaweed *Ulva*. Most green algae have a complex life cycle, with both sexual and asexual reproductive stages.

Very Small Green Algae and Colonial Green Algae

One of the smallest eukaryotes known is the green alga *Ostreococcus tauri*, a common unicellular species of marine phytoplankton (Section 19.10). Cells of *O. tauri* have a diameter of approximately 2 μm, and the organism contains the smallest genome of any known phototrophic eukaryote, approximately 12.6 Mbp. *Ostreococcus* has thus provided a model organism for research into the evolution of genome reduction and specialization in eukaryotes.

At the colonial level of organization in green algae is *Volvox* (Figure 17.33*f*). This alga forms colonies composed of several hundred flagellated cells, some of which are motile and primarily carry out photosynthesis, while others specialize in reproduction. Cells in a *Volvox* colony are interconnected by thin strands of cytoplasm that allow the entire colony to swim in a coordinated fashion. *Volvox* has been a long-term model for research on the genetic mechanisms controlling multicellularity and the distribution of functions among cells in multicellular organisms.

Some colonial green algae have potential as sources of biofuels. For example, the colonial green alga *Botryococcus braunii* excretes long-chain (C_{30}–C_{36}) hydrocarbons that have the consistency of crude oil (Figure 17.33*g*). About 30% of the *B. braunii* cell dry weight consists of this petroleum, and there has been heightened interest in using this and other oil-producing algae as renewable sources of petroleum. Evidence from biomarker studies have shown that some known petroleum reserves originated from green algae such as *B. braunii* that settled in lakebeds in

UNIT 3

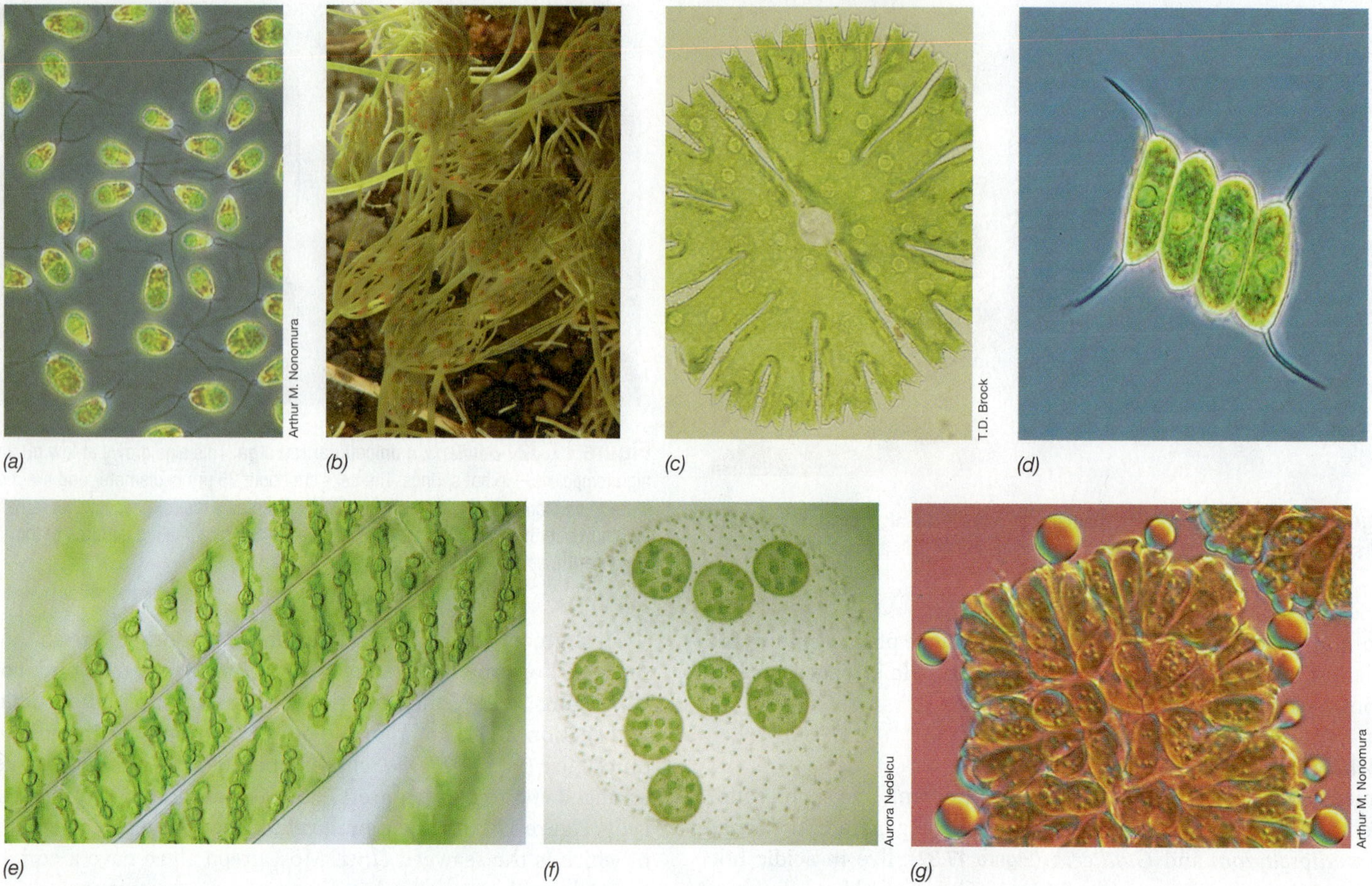

(a)   (b)   (c)   (d)

(e)   (f)   (g)

Figure 17.33 Green algae. *(a)* A single-celled, flagellated green alga, *Dunaliella*. A cell is about 5 μm wide. *(b)* The plantlike green alga *Chara*. *(c)* *Micrasterias*. This single multilobed cell is about 100 μm wide. *(d)* *Scenedesmus*, showing packets of four cells each. *(e)* *Spirogyra*, a filamentous alga with cells about 20 μm wide. Note the green spiral-shaped chloroplasts. *(f)* *Volvox carteri* colony with eight daughter colonies. *(g)* The petroleum-producing green alga, *Botryococcus braunii*. Note the excreted oil droplets surrounding the cell.

ancient times. Hence, if the scale-up challenges for commercial algal petroleum production could ever be met, it is possible that some fraction of the world's oil supply could someday come from photosynthesis by green algae.

Endolithic Phototrophs

Some green algae grow inside rocks. These *endolithic* (*endo* means "inside") phototrophs inhabit porous rocks, such as those containing quartz, and are typically found in layers near the rock surface (**Figure 17.34**). Endolithic phototrophic communities are most common in dry environments such as deserts or cold dry environments such as Antarctica. For example, in the McMurdo Dry Valleys of Antarctica, where temperatures and humidity are extremely low (⟳ Figure 5.21*d*), life within a rock has its advantages. Rocks in these harsh environments are heated by the sun, and water from snowmelt can be absorbed and retained for relatively long periods, supplying moisture needed for growth.

(a)

(b)

Figure 17.34 Endolithic phototrophs. *(a)* Photograph of a limestone rock from the Dry Valleys region of Antarctica broken open to show the layer of endolithic green algae. *(b)* Light micrograph of cells of the green alga *Trebouxia*, a widespread endolithic alga in Antarctica.

Moreover, water absorbed by a porous rock makes the rock more transparent, thus funneling more light to the algal layers.

A wide variety of phototrophs can form endolithic communities, including cyanobacteria and various green algae (Figure 17.34). In addition to being free-living phototrophs, green algae and cyanobacteria coexist with fungi in endolithic lichen communities (🔗 Section 22.1 for discussion of the lichen symbiosis). Metabolism and growth of these internal rock microbial communities slowly weathers the rock, allowing gaps to develop where water can enter, freeze and thaw, and eventually crack the rock, producing new habitats for microbial colonization. The decomposing rock also forms a crude soil that can support development of plant and animal communities in environments where conditions (temperature, moisture, and so on) permit.

MINIQUIZ

- What phototrophic properties link green algae and plants?
- What is unusual about the green algae *Ostreococcus*, *Volvox*, and *Botryococcus*?
- What are endolithic phototrophs?

BIG IDEAS

17.1 • Key metabolic organelles of eukaryotes are the chloroplast, which functions in photosynthesis, and the mitochondrion or hydrogenosome, which function in respiration or fermentation. These organelles were originally *Bacteria* that established permanent residence inside other cells (endosymbiosis).

17.2 • Ribosomal RNA gene sequences do not yield as reliable a phylogenetic tree of the *Eukarya* as do other genes and proteins. The modern, multigene tree of eukaryotes shows a major radiation of eukaryotic diversity emerging at some time following symbiotic events that led to the mitochondrion.

17.3 • Diplomonads such as *Giardia* are unicellular, flagellated, nonphototrophic protists. Parabasalids such as *Trichomonas* are human pathogens and contain huge genomes that lack introns.

17.4 • Euglenozoans are unicellular, flagellated protists. Some are phototrophic. This group includes some important human pathogens, such as *Trypanosoma*, and some well-studied nonpathogens, such as *Euglena*.

17.5 • Three groups make up the alveolates: ciliates, dinoflagellates, and apicomplexans. Most ciliates and dinoflagellates are free-living organisms, whereas apicomplexans are obligate parasites of animals.

17.6 • Stramenopiles are protists that bear a flagellum with fine, hairlike extensions. They include oomycetes, diatoms, and brown and golden algae.

17.7 • Cercozoans and radiolarians are two related groups of protists. The cercozoans include the phototrophic chlorarachniophytes and foraminiferans, whereas the radiolarians are chemoorganotrophs.

17.8 • Amoebozoa are protists that use pseudopodia for movement and feeding. Within amoebozoa are gymnamoebas, entamoebas, and slime molds. Plasmodial slime molds form masses of motile protoplasm, whereas cellular slime molds are individual cells that aggregate to form fruiting bodies from which spores are released.

17.9 • Fungi include the molds, mushrooms, and yeasts. Other than phylogeny, fungi primarily differ from protists by their rigid cell wall, production of spores, and lack of motility.

17.10 • A variety of sexual spores are produced by fungi, including ascospores, basidiospores, and zygospores. From a phylogenetic standpoint, fungi are the closest relatives of animals, and chytridiomycetes are the earliest lineage of fungi.

17.11 • Chytrids are primarily aquatic fungi and are thought to be the most ancient of fungi, lying basal to all other known fungal groups on the 18S rRNA gene tree. Some chytrids are amphibian pathogens.

17.12 • Zygomycetes form coenocytic hyphae and undergo both asexual and sexual reproduction, and the common bread mold *Rhizopus* is a good example. Microsporidia, once thought to be an early lineage of *Eukarya*, are closely related to the zygomycetes. Glomeromycetes are fungi that form endomycorrhizal associations with plants.

17.13 • The ascomycetes are a large and diverse group of saprophytic fungi. Some, such as *Candida albicans*, can be pathogenic in humans. There are two mating types in the yeast *Saccharomyces cerevisiae*, and yeast cells can convert from one type to the other by a genetic switch mechanism.

17.14 • Basidiomycetes include the mushrooms, puffballs, smuts, and rusts. Basidiomycetes undergo both vegetative reproduction as haploid mycelia and sexual reproduction via fusion of mating types and formation of haploid basidiospores.

17.15 • Red algae are mostly marine and range from unicellular to multicellular. Their reddish color is due to the pigment phycoerythrin, a key cyanobacterial pigment, present in their chloroplast.

17.16 • Green algae are common in aquatic environments and can be unicellular, filamentous, colonial, or multicellular. A unicellular green alga, *Ostreococcus*, has the smallest genome known for a phototrophic eukaryote, while the green alga *Volvox* is a model colonial phototroph.

MasteringMicrobiology® **Review what you know and challenge what you have learned with MasteringMicrobiology!** Access study materials, chapter quizzes, animations, and microbiology lab tutorials in the Study Area to ensure that you have mastered this chapter's content.

REVIEW OF KEY TERMS

Algae phototrophic eukaryotes, both microorganisms and macroorganisms

Chitin a polymer of *N*-acetylglucosamine commonly found in the cell walls of fungi

Ciliate any protist characterized in part by rapid motility driven by numerous short appendages called cilia

Coenocytic the presence of multiple nuclei in fungal hyphae without septa

Colonial the growth form of certain protists and green algae in which several cells live together and cooperate for feeding, motility, or reproduction; an early form of multicellularity

Conidia the asexual spores of fungi

Endosymbiotic hypothesis the idea that a respiratory bacterium and a cyanobacterium were stably incorporated into another cell type to yield the mitochondria and chloroplasts, respectively, of eukaryotic cells

Fungi nonphototrophic eukaryotic microorganisms with rigid cell walls

Mushroom the aboveground fruiting body, or basidiocarp, of basidiomycete fungi

Phagocytosis a mechanism for ingesting particulate material in which a portion of the cytoplasmic membrane surrounds the particle and brings it into the cell

Protist a unicellular eukaryotic microorganism; may be flagellate or aflagellate, phototrophic or nonphototrophic, and most lack cell walls; includes algae and protozoa

Secondary endosymbiosis the acquisition by a mitochondrion-containing eukaryotic cell of a red or green algal cell

Slime mold a nonphototrophic protist that lacks cell walls and that aggregates to form fruiting structures (cellular slime molds) or masses of protoplasm (acellular slime molds)

Yeast the single-celled growth form of various fungi

REVIEW QUESTIONS

1. If streptomycin blocks protein synthesis in organelles, what does this tell you about the organelles' relationship to *Bacteria*? (Section 17.1)

2. Distinguish between a primary and a secondary endosymbiosis. Which groups of protists are derived from which form of endosymbiosis? (Section 17.1)

3. Examine the phylogenetic tree of life in Figure 1.6b. In what ways and why does it differ from the tree in Figure 17.3? (Section 17.2)

4. In what ways do diplomonads and parabasalids differ from each other? (Section 17.3)

5. What morphological feature unites kinetoplastids and euglenids? (Section 17.4)

6. What organism causes "red tides" and why is this organism toxic? (Section 17.5)

7. In terms of their photosynthetic pigments, how are brown and golden algae similar? (Section 17.6)

8. What morphological trait links cercozoans and radiolarians and distinguishes them from other protists? (Section 17.7)

9. Although both are slime molds, *Dictyostelium* and *Physarum* differ in major ways. Explain. (Section 17.8)

10. What is the major difference between a mold and a yeast? (Section 17.9)

11. List the different types of sexual spores of fungi. Are conidia sexual or asexual spores? (Section 17.10)

12. In what way do chytrids differ from other fungi? (Section 17.11)

13. What is the major feature of the ecology of glomeromycetes? (Section 17.12)

14. How is the mating type of a yeast cell determined? (Section 17.13)

15. What morphological feature unites the basidiomycetes, and where is this feature found? (Section 17.14)

16. In what kinds of habitats would one likely find red algae? (Section 17.15)

17. What traits link green algae and plants? (Section 17.16)

APPLICATION QUESTIONS

1. Explain why the process of endosymbiosis can be viewed as both an ancient event and a more recent event. What advantages could endosymbiosis give to both the endosymbiont and the host?

2. Summarize the evidence for endosymbiosis. How could the endosymbiotic hypothesis have originated before the era of molecular biology? How has molecular biology supported the theory?

microbiology**now**

Stitching Together Genomes

Thus far only a tiny fraction of microbial diversity has been brought into laboratory culture. Although we discuss in this chapter advanced methods that are helping to culture the previously uncultured, sequencing of DNA recovered directly from an environmental sample is an alternative way to evaluate the metabolic capacities of a natural microbial community.

This approach, called *metagenomics,* relates gene sequences to specific biochemical functions, revealing the metabolic and other capabilities of the microbial community. However, microbial communities are not simply a collection of genes; rather, they are a system of interacting organisms, each of which contains a specific complement of genes that determine the properties of that organism. To gain a true understanding of ecosystem function, the assembly of individual genomes is an essential step.

The ability to assemble complete genomes from many hundreds of millions of short bits of sequence (typically 50–200 nucleotides) obtained in a metagenomic analysis has recently become possible using advanced computational methods. The figure here shows a "connection graph" that depicts an assembly of genomes from a coastal marine water sample. A total of 58.5 *billion* nucleotides in the metagenome were used to stitch together these complete and near-complete genomes.[1] The strands are colored by differences in the percentage of guanine plus cytosine content of their genomic DNAs. Long strands correspond to prokaryotic genomes and the small circular strands are most likely from viruses or plasmids.

This massive metagenomic study paid off handsomely, as it revealed the physiological capacities of an abundant but not yet cultured group of marine *Euryarchaeota* (*Archaea*). The genomes showed these organisms to be motile photoheterotrophs that degrade protein and lipids as carbon sources using light as their energy source.

[1]Iverson, V., et al. 2012. Untangling genomes from metagenomes: Revealing an uncultured class of marine *Euryarchaeota*. *Science 335:* 587–590.

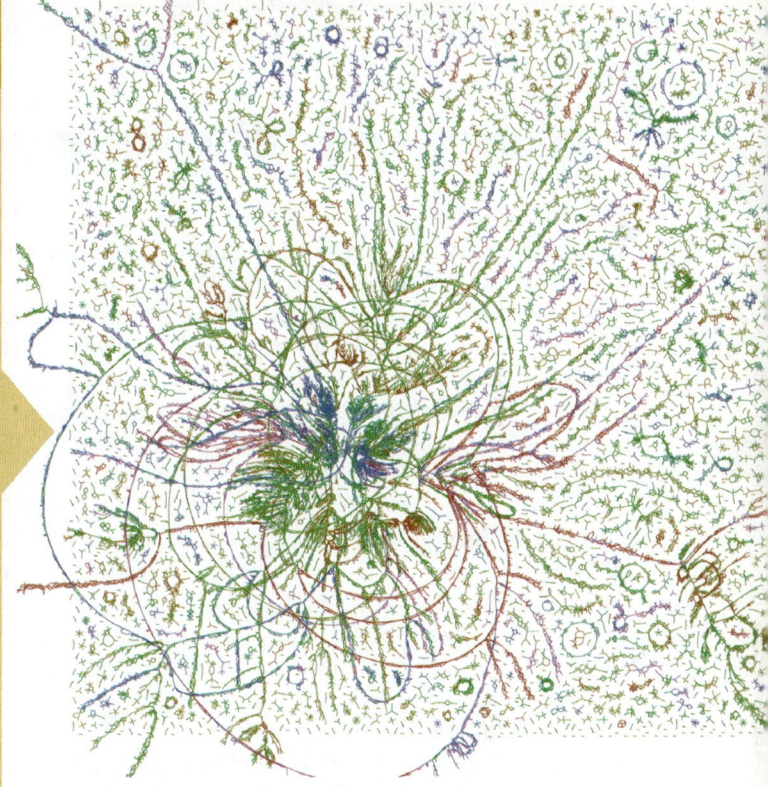

We now begin a new unit devoted to microorganisms in their natural habitats. We learned in Chapter 1 that *microbial communities* consist of cell populations living in association with other populations in nature. The science of **microbial ecology** is focused on how microbial populations assemble to form communities and how these communities interact with each other and their environments.

The major components of microbial ecology are *biodiversity* and *microbial activity*. To study biodiversity, microbial ecologists must identify and quantify microorganisms in their habitats. Knowing how to do this is often helpful for isolating organisms of interest as well, another goal of microbial ecology. To study

microbial activity, microbial ecologists must measure the metabolic processes that microorganisms carry out in their habitats. In this chapter we consider modern methods for assessing microbial diversity and activity. Chapter 19 will outline the basic principles of microbial ecology and examine the types of environments that microorganisms inhabit. Chapters 20, 21, and 22 will complete our coverage of microbial ecology with a consideration of nutrient cycles, applied microbiology, and the role that microorganisms play in symbiotic associations with higher life forms.

We begin with the microbial ecologist's toolbox, which includes a collection of powerful tools for dissecting the structure and function of microbial communities in relation to their natural habitats.

I • Culture-Dependent Analyses of Microbial Communities

The vast majority of microorganisms, more than 99% of all species, have never been grown in laboratory cultures. Recognition of this fact, based on molecular diversity surveys (Sections 18.3–18.7) of microbial habitats, has stimulated the development of new methods for separating out particular microbial species—that is, *isolating* them—to establish pure cultures. Culturing a microorganism remains the only way to fully characterize its properties and predict its impact on an environment.

In the first part of this chapter we cover the enrichment approach, a time-honored and useful method for isolating microorganisms from nature but one with limitations. Enrichment is based on culturing in a selective growth medium, and the tools and methods used in this approach are referred to, collectively, as *culture-dependent* analyses. As we will see, considerable progress has been made in culturing the more elusive microorganisms in natural populations by using robotics to set up large numbers of enrichment cultures that can be monitored using molecular tools. In the second and third parts of this chapter we consider *culture-independent* analyses, techniques that can tell us much about the structure and function of microbial communities in the absence of actual laboratory cultures. In the final part of this chapter, we consider methods for measuring microbial activities in nature and linking them to specific organisms.

18.1 Enrichment

For an **enrichment culture**, a medium and a set of incubation conditions are established that are *selective* for the desired organism and counterselective for undesired organisms. Effective enrichment cultures duplicate as closely as possible the resources and conditions of a particular ecological niche. Literally hundreds of different enrichment strategies have been devised, and **Tables 18.1** and **18.2** provide an overview of some successful ones.

Inocula

Successful enrichment requires an appropriate inoculum containing the organism of interest. Thus, the making of an enrichment culture begins with collecting a sample from the appropriate habitat to serve as the inoculum (Tables 18.1 and 18.2). Enrichment cultures are established by placing the inoculum into selective

media and incubating under specific conditions. In this way, many common prokaryotes can be isolated. For example, the great Dutch microbiologist Martinus Beijerinck, who conceptualized the enrichment culture technique (⟳ Section 1.9), used enrichment cultures to isolate for the first time the nitrogen-fixing bacterium *Azotobacter* (**Figure 18.1**). Because *Azotobacter* is a rapidly growing bacterium capable of N_2 fixation in air

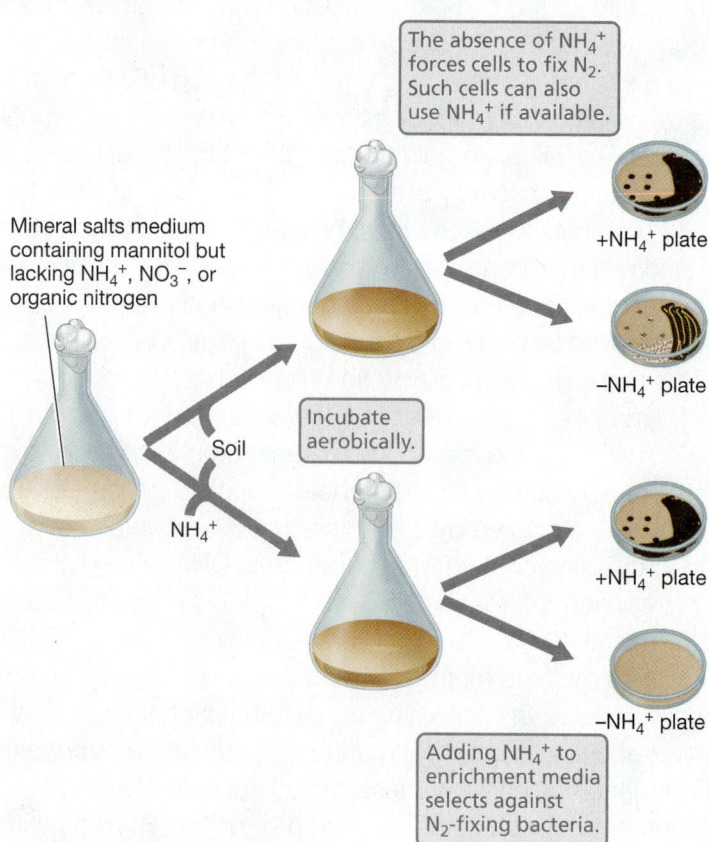

Figure 18.1 **The isolation of *Azotobacter*.** Selection for aerobic nitrogen-fixing bacteria usually results in the isolation of *Azotobacter* or its relatives. See Section 1.9 and Figure 1.23 for more on the historical importance of *Azotobacter*.

Table 18.1 Some enrichment culture methods for phototrophic and chemolithotrophic bacteria

Light-phototrophic bacteria: main C source, CO_2

Incubation condition	Organisms enriched	Inoculum
Incubation in air		
N_2 as nitrogen source	Cyanobacteria	Pond or lake water; sulfide-rich muds; stagnant water; raw sewage; moist, decomposing leaf litter; moist soil exposed to light
NO_3^- as nitrogen source, 55°C	Thermophilic cyanobacteria	Hot spring microbial mat
Anoxic incubation		
H_2 or organic acids; N_2 as sole nitrogen source	Purple nonsulfur bacteria, heliobacteria	Same as above plus hypolimnetic lake water (⮀ Section 19.8); pasteurized soil (heliobacteria); microbial mats for thermophilic species
H_2S as electron donor	Purple and green sulfur bacteria	
Fe^{2+}, NO_2^- as electron donor	Purple bacteria	

Dark-chemolithotrophic bacteria: main C source, CO_2 (medium must lack organic C)

Electron donor	Electron acceptor	Organisms enriched	Inoculum
Incubation in air: aerobic respiration			
NH_4^+	O_2	Ammonia-oxidizing *Bacteria* (*Nitrosomonas*) or *Archaea* (*Nitrosopumilus*)	Soil, mud, sewage effluent, seawater
NO_2^-	O_2	Nitrite-oxidizing bacteria (*Nitrobacter, Nitrospira*)	
H_2	O_2	Hydrogen bacteria (various genera)	
H_2S, S^0, $S_2O_3^{2-}$	O_2	*Thiobacillus* spp.	
Fe^{2+}, low pH	O_2	*Acidithiobacillus ferrooxidans*	
Anoxic incubation			
S^0, $S_2O_3^{2-}$	NO_3^-	*Thiobacillus denitrificans*	Mud, lake sediments, soil
H_2	NO_3^-	*Paracoccus denitrificans*	
Fe^{2+}, neutral pH	NO_3^-	*Acidovorax* and various other gram-negative autotrophic bacteria	

(⮀ Section 3.17), enrichment using media devoid of fixed nitrogen, such as ammonia or nitrate, and incubation in air selects strongly for this bacterium and its close relatives. Non-nitrogen-fixing bacteria and anaerobic nitrogen-fixing bacteria are counterselected in this technique.

Enrichment Culture Outcomes

For success with enrichment cultures, attention to both the culture medium and the incubation conditions is important. That is, *resources* (nutrients) and *conditions* (temperature, pH, osmotic considerations, and the like) must mimic those of the habitat to give the best chance of obtaining the organism of interest (⮀ Table 19.1). Some enrichment cultures yield nothing. This may be because the organism capable of growing under the enrichment conditions specified is absent from the habitat. Alternatively, even though the organism of interest exists in the habitat sampled, the resources and conditions of the laboratory culture may be insufficient for its growth. Thus enrichment cultures can yield a firm positive conclusion (that an organism with certain capacities exists in a particular environment) but never a firm negative conclusion (that such an organism is not present). Moreover, the isolation of the desired organism from an enrichment culture says nothing about the ecological importance or abundance of the organism in its habitat; a positive enrichment proves only that the organism was present in the sample, and in theory, this requires only a single viable cell in the enrichment culture inoculum.

The Winogradsky Column

The **Winogradsky column** is an artificial microbial ecosystem and a long-term source of various bacteria for enrichment cultures. Winogradsky columns have been used to isolate phototrophic purple and green bacteria, sulfate-reducing bacteria, and many other anaerobes. Named for the famous Russian microbiologist Sergei Winogradsky (⮀ Section 1.9), the column was first used by Winogradsky in the late nineteenth century in his classic studies of soil microorganisms.

A Winogradsky column is prepared by filling a glass cylinder about half full with organically rich, preferably sulfide-containing mud into which carbon substrates have been mixed. The substrates determine which organisms are enriched. Fermentative substrates, such as glucose, that can lead to acidic conditions and excessive gas formation (which can create gas pockets that

Table 18.2 Some enrichment culture methods for chemoorganotrophic and strictly anaerobic bacteria[a]

Electron donor (and nitrogen source)	Electron acceptor	Typical organisms enriched	Inoculum
Incubation in air: aerobic respiration			
Lactate + NH_4^+	O_2	*Pseudomonas fluorescens*	Soil, mud; lake sediments; decaying vegetation; pasteurize inoculum (80°C for 15 min) for all *Bacillus* enrichments
Benzoate + NH_4^+	O_2	*Pseudomonas fluorescens*	
Starch + NH_4^+	O_2	*Bacillus polymyxa*, other *Bacillus* spp.	
Ethanol (4%) + 1% yeast extract, pH 6.0	O_2	*Acetobacter, Gluconobacter*	
Urea (5%) + 1% yeast extract	O_2	*Sporosarcina ureae*	
Hydrocarbons (e.g., mineral oil, gasoline, toluene) + NH_4^+	O_2	*Mycobacterium, Nocardia, Pseudomonas*	
Cellulose + NH_4^+	O_2	*Cytophaga, Sporocytophaga*	
Mannitol or benzoate, N_2 as N source	O_2	*Azotobacter*	
CH_4 + NO_3^-	O_2	*Methylobacter, Methylomicrobium*	Lake sediments, thermocline (⮌ Section 19.8) of stratified lake
Anoxic incubation: anaerobic respiration			
Organic acids	NO_3^-	*Pseudomonas* (denitrifying species)	Soil, mud; lake sediments
Yeast extract	NO_3^-	*Bacillus* (denitrifying species)	
Organic acids	SO_4^{2-}	*Desulfovibrio, Desulfotomaculum*	
Acetate, propionate, butyrate	SO_4^{2-}	Fatty acid–oxidizing sulfate reducers	As above; or sewage digester sludge; rumen contents; marine sediments
Acetate, ethanol	S^0	*Desulfuromonas*	
Acetate	Fe^{3+}	*Geobacter, Geospirillum*	
Acetate	ClO_3^-	Various chlorate-reducing bacteria	
H_2	CO_2	Methanogens (chemolithotrophic species only), homoacetogens	Mud, sediments, sewage sludge
CH_3OH	CO_2	*Methanosarcina barkeri*	
CH_3NH_2 or CH_3OH	NO_3^-	*Hyphomicrobium*	
Hydrocarbons	SO_4^{2-} or NO_3^-	Anoxic hydrocarbon-degrading bacteria	Freshwater or marine sediments
Acetate + H_2 + NH_4^+	Tetrachloroethene (PCE)	*Dehalococcoides* spp.	PCE-polluted groundwater
Anoxic incubation: fermentation			
Glutamate or histidine	No exogenous electron acceptors added	*Clostridium tetanomorphum* or other proteolytic *Clostridium* species	Mud, lake sediments; rotting plant or animal material; dairy products (lactic and propionic acid bacteria); rumen or intestinal contents (enteric bacteria); sewage sludge; soil; pasteurize inoculum for *Clostridium* enrichments
Starch + NH_4^+	None	*Clostridium* spp.	
Starch + N_2 as N source	None	*Clostridium pasteurianum*	
Lactate + yeast extract	None	*Veillonella* spp.	
Glucose or lactose + NH_4^+	None	*Escherichia, Enterobacter*, other fermentative organisms	
Glucose + yeast extract (pH 5)	None	Lactic acid bacteria (*Lactobacillus*)	
Lactate + yeast extract	None	Propionic acid bacteria	
Succinate + NaCl	None	*Propionigenium*	
Oxalate	None	*Oxalobacter*	
Acetylene	None	*Pelobacter* and other acetylene fermenters	

[a]All media must contain an assortment of mineral salts including N, P, S, Mg^{2+}, Mn^{2+}, Fe^{2+}, Ca^{2+}, and other trace elements (⮌ Sections 3.1–3.2). Certain organisms may have requirements for vitamins or other growth factors. This table is meant as an overview of enrichment methods and does not speak to the effect incubation temperature might have in isolating thermophilic (high temperature), hyperthermophilic (very high temperature), and psychrophilic (low temperature) species, or the effect that extremes of pH or salinity might have, assuming an appropriate inoculum was available. Some enrichment substrates are naturally more specific than others. For example, glucose is quite nonspecific as an enrichment substrate compared with benzoate or methanol.

disrupt the enrichment and let in air) are avoided. The mud is supplemented with small amounts of calcium carbonate ($CaCO_3$) as a buffer and gypsum ($CaSO_4$) as a source of sulfate. The mud is packed tightly in the cylinder, taking care to avoid trapping air, and then covered with lake, pond, or ditch water (or seawater if it is a marine column). The top of the cylinder is covered to prevent evaporation, and the container is placed near a window that receives diffuse sunlight for a period of months.

In a typical Winogradsky column a diverse community of organisms develops (**Figure 18.2a**). Algae and cyanobacteria develop quickly in the upper portions of the water column; by producing O_2 these organisms help to keep this zone of the column oxic. Decomposition processes in the mud lead to the production of organic acids, alcohols, and H_2, suitable substrates for sulfate-reducing bacteria (⇄ Section 13.18). Hydrogen sulfide (H_2S) from the sulfate reducers triggers the development of purple and green sulfur bacteria (anoxygenic phototrophs, ⇄ Section 13.3) that use sulfide as a photosynthetic electron donor. These organisms typically grow in patches in the mud on the sides of the column but may bloom in the water itself if oxygenic phototrophs are scarce (Figure 18.2b). The pigmented cells of the anoxygenic phototrophs can be sampled with a pipette for microscopy, isolation, and characterization (Table 18.1).

Winogradsky columns have been used to enrich both aerobic and anaerobic prokaryotes. Besides supplying a ready source of inocula for enrichment cultures, columns can also be supplemented with a specific compound to enrich an organism in the inoculum that can degrade it. Once a crude enrichment has been established in the column, culture media can

be inoculated for the isolation of pure cultures, as discussed in the next section.

Enrichment Bias

Although the enrichment culture technique is powerful, there exists a bias, and sometimes a very severe bias, in the outcome of enrichments. This bias is typically most profound in liquid enrichment cultures where the most rapidly growing organism(s) for the chosen set of conditions dominate. However, using molecular techniques to be described later, we now know that the most rapidly growing organisms in laboratory cultures are often only minor components of the microbial community rather than the most abundant and ecologically relevant organisms carrying out the process of interest. This could be for several reasons including the fact that the levels of resources available in laboratory cultures are typically much higher than those in nature, and the conditions in the natural habitat, including both the types and proportions of different organisms present as well as the physical and chemical conditions, are nearly impossible to reproduce and maintain for long periods in laboratory cultures.

This problem of **enrichment bias** can be demonstrated by comparing the results obtained in dilution cultures (Section 18.2) with classical liquid enrichment. Dilution of an inoculum followed by liquid enrichment or plating often yields different organisms than liquid enrichments established with the same but undiluted inocula. It is thought that dilution of the inoculum eliminates quantitatively insignificant but rapidly growing "weed" species, allowing development of organisms that are more abundant in the community but slower growing. Dilution of the

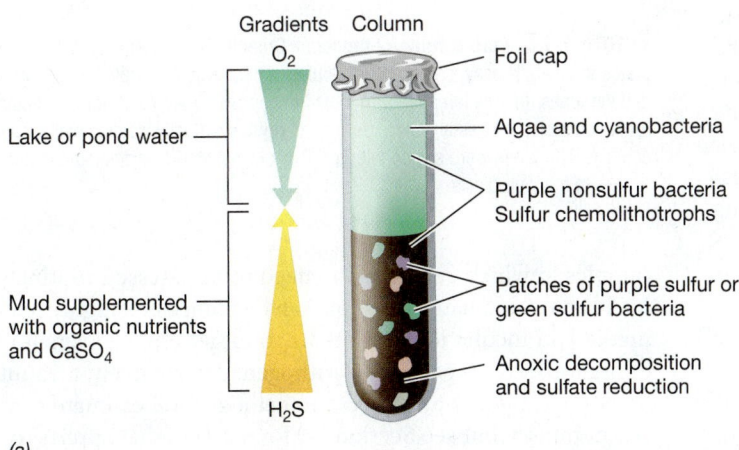

(a)

(b)

Figure 18.2 The Winogradsky column. *(a)* Schematic view of a typical column used to enrich phototrophic bacteria. The column is incubated in a location that receives subdued sunlight. Anoxic decomposition leading to SO_4^{2-} reduction creates the gradient of H_2S. *(b)* Photo of Winogradsky columns that have remained anoxic up to the top; each column had a bloom of a different phototrophic bacterium. Left to right: *Thiospirillum jenense, Chromatium okenii*, both of which are purple sulfur bacteria, and *Chlorobium limicola* (green sulfur bacterium).

inoculum is thus a common practice in enrichment culture microbiology today. As discussed below, the problem of overgrowth by "weed" species can also be circumvented by physical isolation of the desired organism before introducing it into a growth medium. This is partly accomplished by use of the dilution method. However, more recently, sophisticated methods have been developed to physically isolate single cells of interest (or single cell types) and place them in a growth medium that is free of undesired cells. We consider these techniques in the next section.

MINIQUIZ

- Describe the enrichment strategy behind Beijerinck's isolation of *Azotobacter*.
- Why is sulfate (SO_4^{2-}) added to a Winogradsky column?
- What is enrichment bias? How does dilution reduce enrichment bias?

18.2 Isolation

A *pure culture*—one containing a single kind of microorganism—can be isolated from an enrichment culture in many ways. Common isolation procedures include the streak plate, the agar dilution, and liquid dilution. For organisms that form colonies on agar plates, the streak plate is quick, easy, and the method of choice (**Figure 18.3**); if a well-isolated colony is selected and restreaked several successive times, a pure culture can usually be obtained. With proper incubation facilities (for example, anoxic jars or anoxic chambers for anaerobes, ⟳ Section 5.16), it is possible to purify both aerobes and anaerobes on agar plates by the streak plate method.

Agar Dilution Tubes and the Most-Probable-Number Technique

In the agar dilution tube method, a mixed culture is diluted in tubes of molten agar medium, resulting in colonies embedded in the agar. This method is useful for purifying anaerobic organisms such as phototrophic sulfur bacteria and sulfate-reducing bacteria from samples taken from Winogradsky columns or other sources. A culture is purified by successive dilutions of cell suspensions in tubes of molten agar medium (Figure 18.3). Repeating this procedure using a colony from the highest-dilution tube as inoculum for a new set of dilutions usually yields pure cultures. A related procedure called the *roll tube method* uses tubes containing a thin layer of agar on their inner surface. The agar can then be streaked for isolated colonies. Because the tubes can be flushed with an oxygen-free gas during streaking, the roll tube method is primarily used for the isolation of anaerobic prokaryotes.

Another purification procedure is the serial dilution of an inoculum in a liquid medium until the final tube in the series shows no growth. When a 10-fold serial dilution is used, for example, the last tube showing growth should have originated from ten or fewer cells. Besides being a method for obtaining pure cultures, serial dilution techniques are widely used to estimate viable cell numbers in the **most-probable-number (MPN) technique** (**Figure 18.4**). MPN methods have been used for estimating the numbers of microorganisms in foods, wastewater, and other

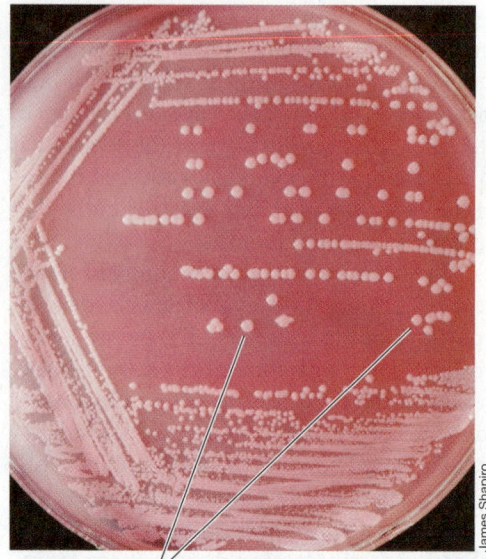
(a)

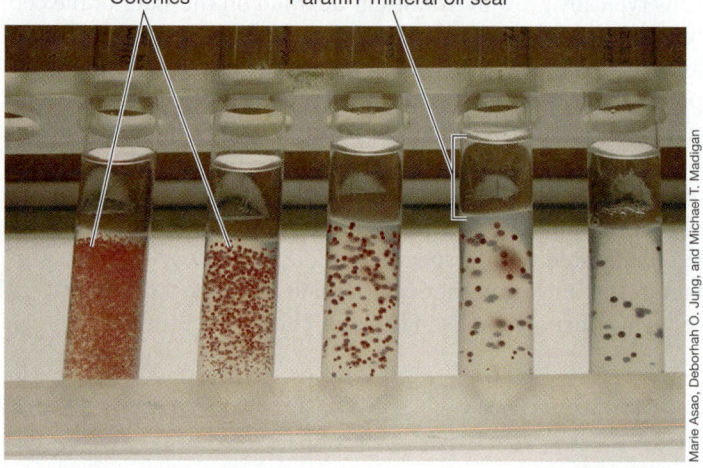
Colonies   Paraffin–mineral oil seal

(b)

Figure 18.3 Pure culture methods. *(a)* Organisms that form distinct colonies on plates are usually easy to purify. *(b)* Colonies of phototrophic purple bacteria in agar dilution tubes; the molten agar was cooled to approximately 45°C before inoculation. A dilution series was established from left to right, eventually yielding well-isolated colonies. The tubes were sealed with a 1:1 mixture of sterile paraffin and mineral oil to maintain anaerobiosis.

samples in which cell numbers need to be assessed routinely. An MPN count of a natural sample can be done using highly selective media and incubation conditions to target one or a small group of organisms or a particular pathogen. Alternatively, a count can be done using complex media to get a general estimate of viable cell numbers (but see Section 5.9 for a caveat that applies to such estimates).

Use of several replicate tubes at each dilution improves accuracy of the final MPN obtained.

Criteria for Purity

Regardless of the methods used to purify a culture, once a putative pure culture has been obtained, it is essential to verify its purity. This is typically done through a combination of (1) microscopy, (2) observation of colony characteristics on plates or in dilution

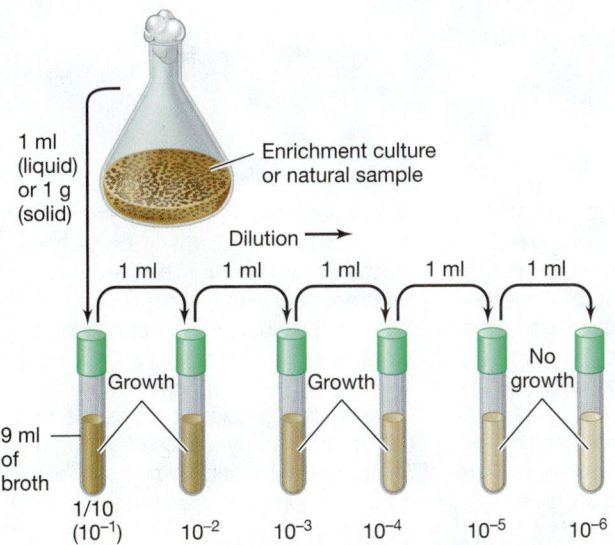

Figure 18.4 Procedure for a most-probable-number (MPN) analysis. Growth in the 10^{-4} but not the 10^{-5} dilution means that cell numbers were at least 10^{4} cells/ml in the sample used for inoculation. Since particle-attached microorganisms can skew numbers significantly, gentle methods to disassociate microorganisms from particles are often used prior to dilution.

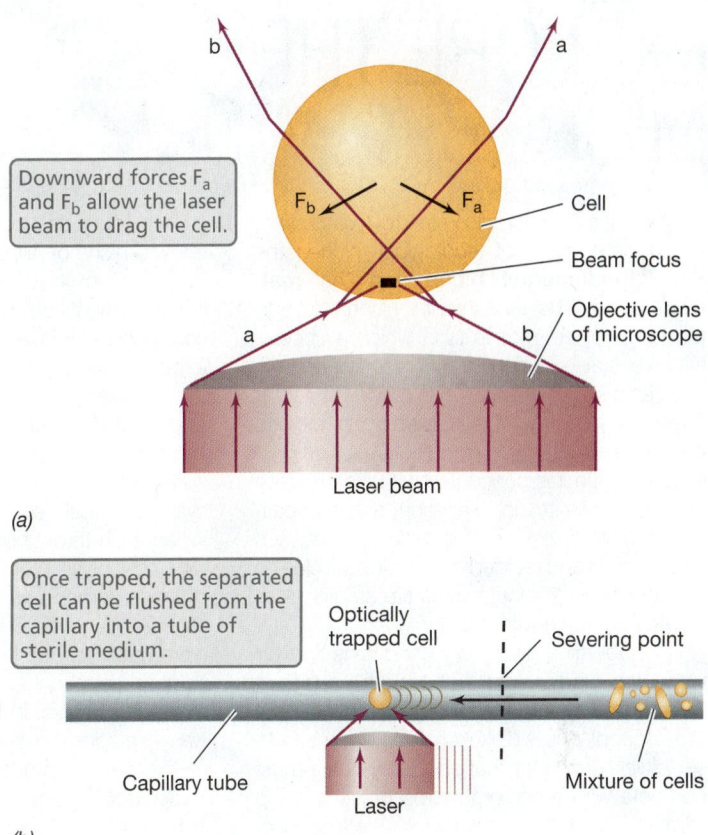

Figure 18.5 The laser tweezers for the isolation of single cells. A technique for physically isolating individual cells for subsequent growth in pure culture.

tubes, and (3) tests of the culture for growth in other media. In the latter, it is important to test the culture for growth in media in which it is predicted that the desired organism will grow poorly or not at all but that contaminants will grow vigorously. In the final analysis, the microscopic observation of a single morphological type of cell that displays uniform staining characteristics (for example, in a Gram stain) coupled with uniform colony characteristics and the absence of contamination in growth tests with various culture media is good evidence that a culture is pure (also called *axenic*).

The molecular methods described in the following sections for characterizing environmental populations can also be applied to verification of culture purity. However, these techniques are generally complementary and do not substitute for the more basic observations of culture characteristics and cellular morphology.

Selective Single-Cell Isolation: The Laser Tweezers and Flow Cytometry

In addition to the methods just described, other more technologically demanding tools for obtaining pure cultures are available, including laser tweezers and flow cytometry. These methods are especially useful for isolating slow-growing microorganisms that would otherwise be overgrown by "weed" species in enrichment cultures or to isolate organisms present in such low numbers that they would be missed using dilution-based enrichment methods.

Laser tweezers consist of an inverted light microscope equipped with a strongly focused infrared laser and a micromanipulation device. Trapping a single cell is possible because the laser beam creates a force that pushes down on a microbial cell (or other small object) and holds it in place (**Figure 18.5a**). Then when the laser beam is moved, the trapped cell moves along with it. If a mixed sample is in a capillary tube, a single cell can be optically trapped and moved away from contaminating organisms (Figure 18.5b). The cell can

then be isolated by breaking the tube at a point between the cell and the contaminants and flushing the cell into a small tube of sterile medium. Laser tweezers, when coupled with staining techniques that identify particular organisms (Sections 18.3 and 18.4), can be used to select organisms of interest from a mixture for purification and further laboratory study.

A second method for selective isolation of single cells employs **flow cytometry**, a technique for counting and examining microscopic particles by suspending them in a stream of fluid and passing them through an electronic detector. Flow cytometers assess selected criteria (including size, shape, or fluorescent properties) of single cells as they pass through a detector at rates of many thousands of cells per second, and can also sort out single cells based on measured criteria (see Section 18.10 and Figure 18.29). The latter capacity of the flow cytometer can be used to enrich a particular cell type from a mixture of many types.

Single-cell sorting capacity can arrange single cells on the surface of a solid growth medium or deposit them into individual wells of a multiwelled (microtiter) plate, where each well contains the same growth medium or slightly different growth media. Because the growth requirements of some organisms include organic compounds and metabolites produced by other organisms that share their environment, addition of filter-sterilized source water (for aquatic organisms) or soil water extract (for soil organisms) has been successful for bringing some previously uncultured organisms into laboratory culture.

EXPLORE THE MICROBIAL WORLD

General ecologists refer to two distinct categories of ecological niche, the **fundamental niche** and the **realized niche**. The fundamental niche refers to the range of environments in which a species will be sustained when it is not resource-limited, such as may result from competition with other species. By contrast, the realized niche refers to the range of natural environments supporting a species when it is confronted with factors such as resource limitation, predation, and competition from other species. This distinction between fundamental and realized niches yields a clue as to why some microorganisms are so difficult to isolate from the environment.

Establishing laboratory conditions that fall within the fundamental niche will be sufficient to maintain an organism in pure culture, but those conditions will likely not be adequate to selectively enrich for that species when it must compete with other organisms present in the sample inoculum. Since the realized niche of most microorganisms is unknown, there has been an increasing emphasis on developing high-throughput methods for cultivation, using robotics to set up many cultivation attempts in parallel. High-throughput methods allow for simultaneous investigation of many alternative

growth conditions in an attempt to replicate the realized niche or, alternatively, allow the organism to occupy its fundamental niche by relieving it from competition. The latter approach is more commonly employed, since fewer media need to be prepared to sustain growth.

Methods for collecting and depositing a single cell in nutrient media are well developed, including the established dilution methods, as well as the more recent applications of flow cytometric cell sorting and laser tweezers. When high-throughput cultivation is coupled with molecular probes to screen for growth of novel organisms identified previously by culture-independent methods (Section 18.5), there has been increasing success in isolating some interesting bacteria from the uncultured majority in nature. In fact, this general method was used for the successful isolation of one of the most abundant organisms on Earth, *Pelagibacter ubique* (**Figure 1**).[1] As discussed in Section 19.10, this bacterium is a species of highly successful marine bacterioplankton (suspended bacterial cells) that thrive on the very dilute pool of dissolved organic matter present in the open oceans.

Dedicated patience is needed in any cultivation effort, as the discovery of slow-growing

or dormant organisms may require months of incubation. Also, many or perhaps even most microorganisms in nature are adapted to extremely low nutrient concentrations, being inhibited by nutrient concentrations used to grow organisms commonly studied in the laboratory. In addition, others may depend on complex interspecies relationships that cannot be replaced by a defined growth medium. Thus, high-throughput methods are ideally suited to surveying various combinations of resources to find the set that best supports the laboratory culture of the most interesting and ecologically relevant organisms from nature.

Culture-independent methods have identified over 50 major divisions (phyla) within the domain *Bacteria* alone. Remarkably, only 12 phyla were known in 1987! Among the major bacterial divisions so far identified, only about half now have cultured representatives. Additionally, even when cultures are available for a division, they are generally few in number and therefore do not fully encompass the phylogenetic diversity within the group. Thus, the challenge of obtaining representative cultures is one of both *breadth* of coverage (obtaining at least one member of each division) and *depth* of coverage (developing a culture collection that spans the phylogenetic diversity of each division). The relatively poor representation of the natural diversity of *Bacteria* in culture collections today (↪ Figure 15.1) also applies to microbial eukaryotes and the *Archaea*. However, this situation is not so much a problem as it is a remarkable opportunity for a new generation of microbiologists interested in microbial diversity. We now have the understanding and technology necessary to more fully explore the remarkable diversity of microbial life through culture-based analyses.[1]

[1]Rappé, M.S., S.A.Connon, K.L. Vergin, & S.J. Giovannoni, 2002. Cultivation of the ubiquitous SAR11 marine bacterioplankton clade. *Nature 418:* 630–633.

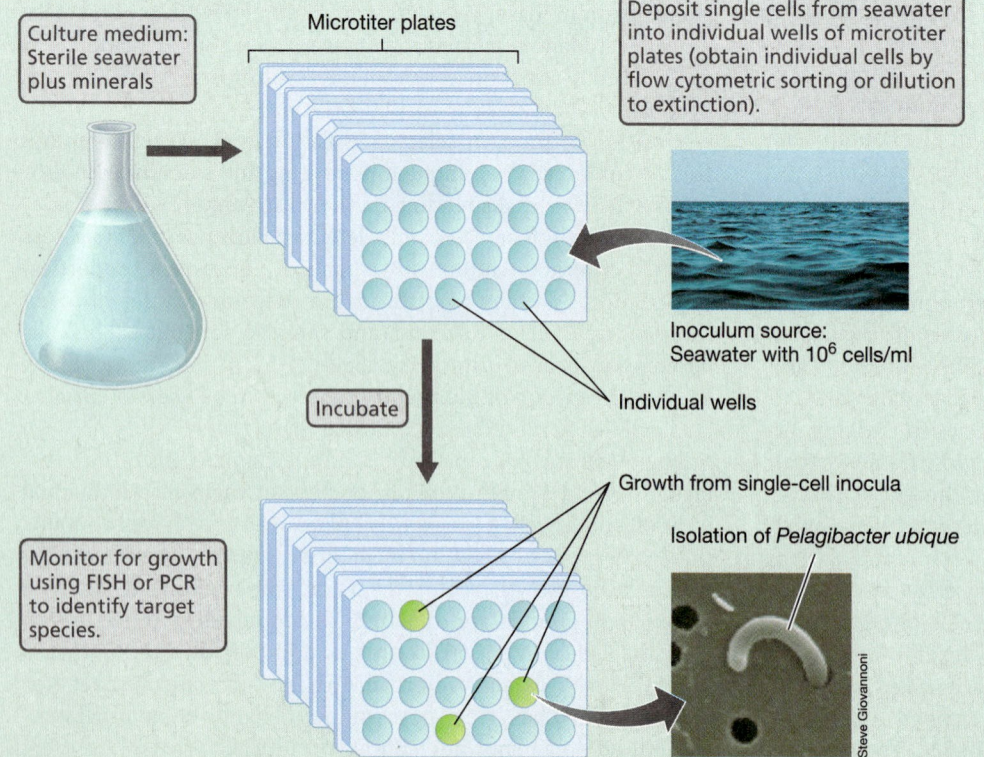

Culture medium: Sterile seawater plus minerals

Microtiter plates

Deposit single cells from seawater into individual wells of microtiter plates (obtain individual cells by flow cytometric sorting or dilution to extinction).

Inoculum source: Seawater with 10^6 cells/ml

Individual wells

Incubate

Growth from single-cell inocula

Isolation of *Pelagibacter ubique*

Monitor for growth using FISH or PCR to identify target species.

Steve Giovannoni

Figure 1 Methodological pipeline for high-throughput cultivation of previously uncultured microorganisms. The method shown here was used to isolate *Pelagibacter ubique*, one of the most abundant bacteria on Earth. Following the addition of filter-sterilized seawater and low nutrient concentrations to the individual wells, pure cultures of *Pelagibacter* and other novel marine *Bacteria* were obtained.

Continuing developments of these and related methods for culture screening have spawned a new field of *high-throughput technology* for culturing previously uncultured microorganisms (see Explore the Microbial World, "Culturing the Uncultured"). High-throughput methods also include the use of robotic systems to quickly test hundreds of combinations of nutrients for growth or to assay hundreds or thousands of different wells for DNA sequences that will identify the organisms being enriched, all run simultaneously to yield rapid results (that is, at high throughput).

MINIQUIZ

- How does the agar dilution method differ from streaking to obtain isolated colonies?

- How might you isolate a morphologically unique bacterium present in an enrichment culture in relatively low numbers?

- What is meant by "high-throughput" in culturing microorganisms? How has it benefited microbiology?

II • Culture-Independent Microscopic Analyses of Microbial Communities

Microbial ecologists quantify cells in a microbial habitat to estimate relative abundances of different species. Cell stains are necessary to obtain these types of data, and we detail these methods here. Organisms in natural environments can also be detected by assaying their genes. Genes encoding either ribosomal RNA (rRNA, ↩ Section 12.4) or enzymes that support a specific physiology are the usual targets in these studies. *Environmental genomics* (Section 18.7) is a method for assessing the entire gene complement of a habitat, revealing both the biodiversity and metabolic capabilities of the microbial community at the same time.

18.3 General Staining Methods

Several staining methods are suitable for quantifying microorganisms in natural samples. Although these methods do not reveal the physiology or phylogeny of the cells, they are nonetheless reliable and widely used by microbial ecologists for measuring total cell numbers. One method to be described also allows for an assessment of cell viability.

Fluorescent Staining with Dyes That Bind Nucleic Acids

Fluorescent dyes can be used to stain microorganisms from virtually any microbial habitat. **DAPI** (4′,6-diamidino-2-phenylindole) is a popular stain for this purpose, as is the dye **acridine orange**.

There is also increasing use of *SYBR Green I*, a dye that confers very bright fluorescence to all microorganisms, including viruses. These stains bind to DNA and are strongly fluorescent when exposed to ultraviolet (UV) radiation (DAPI absorption maximum, 400 nm; acridine orange absorption maximum, 500 nm; SYBR Green I absorption maximum, 497 nm), making the microbial cells in the sample readily visible and easy to enumerate. Cells stained with DAPI fluoresce blue, cells stained with acridine orange fluoresce orange or greenish-orange, and cells stained with SYBR Green I fluoresce green (**Figure 18.6**).

Dyes that stain DNA are widely used for the enumeration of microorganisms in environmental, food, and clinical samples. Depending on the sample, background staining is occasionally a problem with fluorescent stains, but because these dyes specifically stain nucleic acids, they are for the most part nonreactive with inert matter. Thus, for many samples, from soil as well as aquatic sources, they can give a reasonable estimate of the cell numbers present. Staining with the brightly fluorescent SYBR Green I also provides excellent enumeration of aquatic virus populations (↩ Section 19.11). For dilute aquatic samples, cells can be stained following collection on a membrane surface by filtration.

DNA staining is a nonspecific process; *all* microorganisms in a sample are stained. Although this may at first seem desirable, it is not necessarily so. For example, DAPI and acridine orange fail

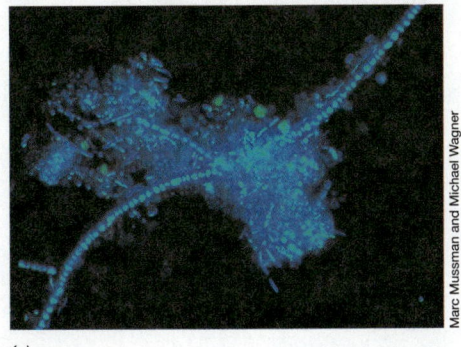

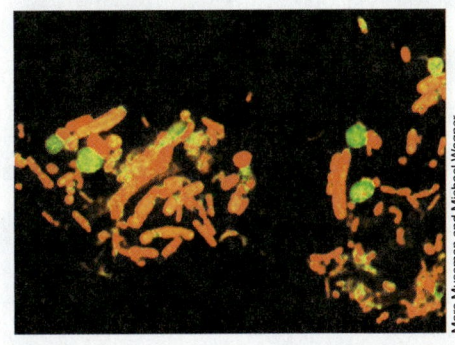

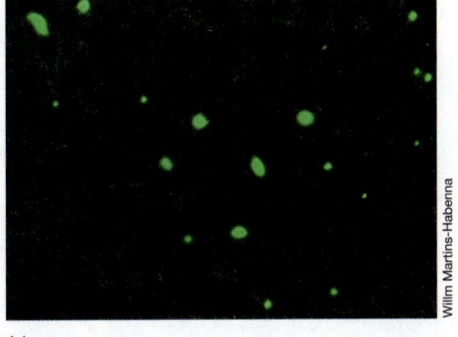

Figure 18.6 Nonspecific fluorescent stains. *(a)* DAPI and *(b)* acridine orange staining showing microbial communities inhabiting activated sludge in a municipal wastewater treatment plant. With acridine orange, cells containing low RNA levels stain green. *(c)* SYBR Green–stained sample of Puget Sound (Washington, USA) surface water showing green-fluorescing bacterial cells. The large cells near the center of the field are 0.8–1.0 μm in diameter.

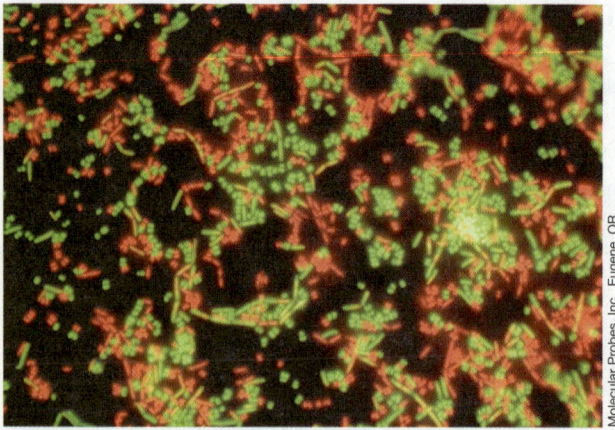

Figure 18.7 Viability staining. Live (green) and dead (red) cells of *Micrococcus luteus* (cocci) and *Bacillus cereus* (rods) stained by the LIVE/DEAD BacLight Bacterial Viability Stain.

to differentiate between living and dead cells or between different species of microorganisms, so they cannot be used to assess cell viability or to track species of microorganisms in an environment.

Viability Staining

Viability staining differentiates live cells from dead ones. Hence, viability stains yield both abundance and viability data at the same time. The basis of differentiating between live and dead cells lies with whether a cell's cytoplasmic membrane is intact. Two dyes that fluoresce green and red are added to a sample; the green-fluorescing dye penetrates all cells, viable or not, whereas the red dye, which contains the chemical propidium iodide, penetrates only those cells whose cytoplasmic membrane is no longer intact and that are therefore dead. Thus, when viewed microscopically, green cells are scored as alive and red cells as dead, yielding an instant assessment of both abundance and viability (**Figure 18.7**).

Although useful for research that uses laboratory cultures, the live/dead staining method is not suitable for use in the direct microscopic examination of samples from many natural habitats because of problems with nonspecific staining of background materials. However, procedures have been developed to overcome this problem in analyses of aquatic environments; a water sample is filtered and the filters are stained with the live/dead stain and examined microscopically. Thus in aquatic microbiology, live/dead staining is often used to measure the viability of cell populations in the water column of lakes or oceans, or in the flowing waters of streams, rivers, and other aquatic environments.

Fluorescent Proteins as Cell Tags and Reporter Genes

Bacterial cells can be altered by genetic engineering to make them autofluorescent. As discussed earlier (⮌ Section 11.6), a gene encoding the green **fluorescent protein** (GFP) can be inserted into the genome of virtually any cultured bacterium. When the GFP gene (*gfp*) is expressed, cells fluoresce green when observed with ultraviolet microscopy (**Figure 18.8**). Although GFP is not useful for the study of natural populations of microorganisms (because these cells lack the GFP gene), GFP-tagged cells can be introduced into an environment, such as plant roots, and then tracked over time by microscopy. Using this method, microbial ecologists can study competition between the native microflora and a GFP-tagged introduced strain and can assess the effect of perturbations of an environment on the survivability of the introduced strain. GFP tagging is also used extensively in the study of microbial symbiotic associations with plants and animals (Chapter 22). However, GFP requires O_2 to become fluorescent, and thus the GFP method is not suitable for tracking cells introduced into strictly anoxic habitats. The photophysical properties of GFP and other fluorescent proteins isolated from different marine invertebrates (jellyfish, corals, anemones) have since been altered

Excite (nm)	Emit (nm)	FP
399	456	FP1
433	475	FP2
466	507	FP3
467	509	FP4
485	510	FP5 (GFP)
515	528	FP6
516	529	FP7
554	581	FP8
568	592	FP9
587	610	FP10
588	633	FP11
600	650	FP12

(a)

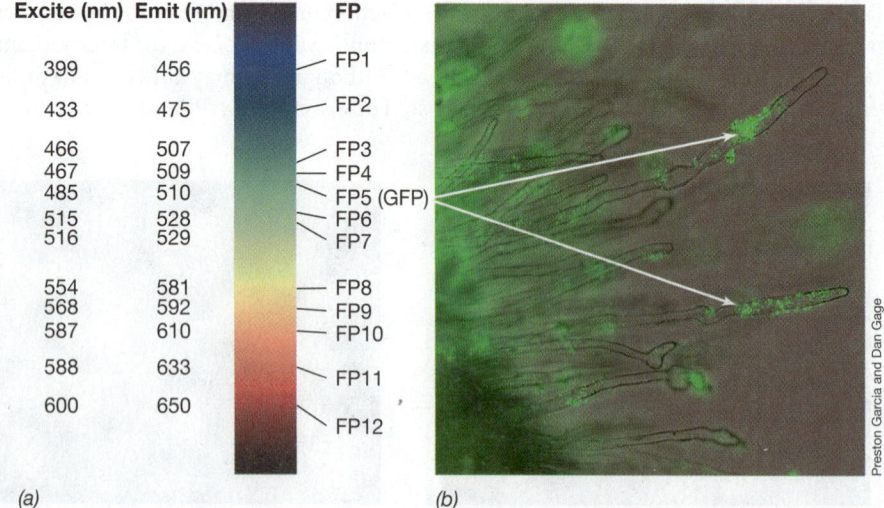

(b)

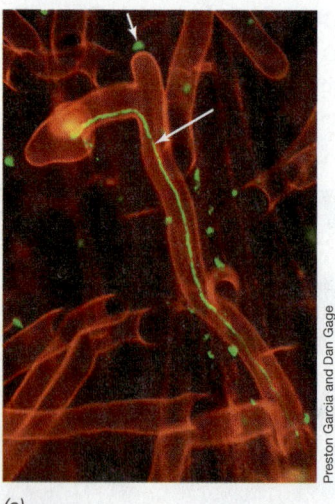

(c)

Figure 18.8 Fluorescent protein reporters. (a) Twelve different fluorescent proteins (FP1–FP12) are known that have distinct excitation (Excite) and emission (Emit) properties. (b) Cells of *Sinorhizobium meliloti* (arrows) carrying a plasmid with an alpha-galactoside inducible promoter fused to the GFP (FP5); the cells are on clover seedling roots. Green fluorescence indicates that alpha-galactosides are released and available to support the growth of this bacterium. (c) *S. meliloti* cells (arrow) carrying a plasmid with a succinate-inducible promoter fused to GFP; green fluorescence indicates that succinate or other C_4 dicarboxylic acids have been secreted by the plant root hairs.

through mutation to yield a broad palette of fluorescent proteins of varying spectral properties (Figure 18.8*a*), offering the experimental basis for simultaneous monitoring of multiple species.

The gene *gfp* and those encoding other fluorescent proteins have also been used extensively in laboratory cultures of various bacteria and in controlled environments as a *reporter gene*. When this gene is fused with an operon under the control of a specific regulatory protein, transcription can be studied by using fluorescence as the indicator (a "reporter") of activity. That is, when genes containing the fused fluorescent protein gene are transcribed and translated, both the protein of interest and the fluorescent protein are made, and cells fluoresce the characteristic color (⮂ Section 11.6 and Figures 11.11). For example, expression of *gfp* was used to demonstrate that colonization of alfalfa roots by *Sinorhizobium meliloti* (legume–root nodule symbiosis, ⮂ Section 22.3) is promoted by sugars and dicarboxylic acids released by the plant (Figure 18.8*b, c*).

Limitations of Microscopy

The microscope is an essential tool for exploring microbial diversity and for enumerating and identifying microorganisms in natural samples. However, microscopy alone does not suffice for the study of microbial diversity. Prokaryotes vary greatly in size (⮂ Section 2.6 and Table 2.1). Very small cells can be a major problem and can go totally unnoticed, and some cells are near the limits of resolution of the light microscope. Such cells can easily be overlooked in the examination of natural samples, especially if the sample contains high levels of particulate matter or high numbers of large cells. Also, it is often difficult to differentiate live cells from dead cells or cells in general from certain inert materials in natural samples. However, the biggest limitation with the microscopic methods we have discussed thus far is that none of them reveal the phylogenetic diversity of the microorganisms in the habitat under study.

We will see in the next section and get a preview here (**Figure 18.9**) of powerful staining methods that can reveal the *phylogeny* of organisms observed in a natural sample. These methods

have revolutionized microbial ecology and have helped microbiologists overcome the major limitation of the light microscope in microbial ecology: identifying from a phylogenetic perspective cells observed in a microscopic field. These methods have also taught microbial ecologists an important lesson—when observing unstained or nonspecifically stained natural populations of microorganisms under the microscope, one must remember that the sample almost certainly contains a genetically diverse community, even if many cells "look" the same (Figure 18.9). The simple shapes of bacteria conceal their remarkable diversity.

MINIQUIZ -
- How does viability staining differ from stains like DAPI?
- What is a reporter gene?
- Why is it incorrect to say that the GFP is a "staining" method?

18.4 Fluorescence In Situ Hybridization (FISH)

Because of their great specificity, nucleic acid probes are powerful tools for identifying and quantifying microorganisms. Recall that a **nucleic acid probe** is a DNA or RNA oligonucleotide complementary to a sequence in a target gene or RNA; when the probe and the target come together, they hybridize (⮂ Section 11.2). Nucleic acid probes can be made fluorescent by attaching fluorescent dyes to them. The fluorescent probes can often be used to identify organisms that contain a nucleic acid sequence complementary to the probe. This technique is called **fluorescence in situ hybridization (FISH)**, and different applications are described here, including methods that target phylogeny (**Figure 18.10**) or gene expression (see Figure 18.12).

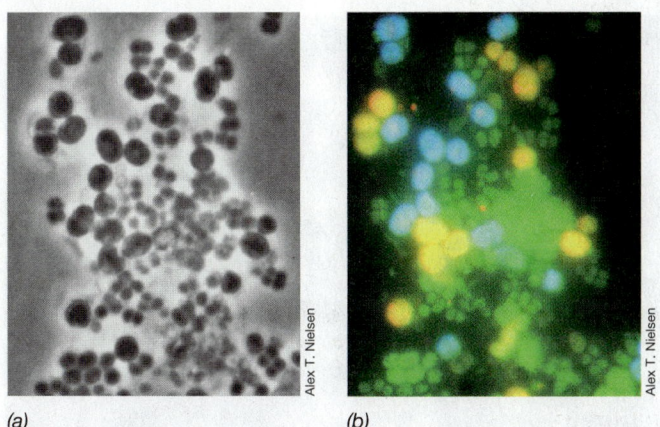

(a)   (b)

Alex T. Nielsen

Figure 18.9 Morphology and genetic diversity. The photomicrographs shown here, produced by *(a)* phase-contrast and *(b)* a technique called phylogenetic FISH (Section 18.4), are of the same field of cells. Although the large oval cells are of a rather unusual size for prokaryotic cells and all look similar by phase-contrast microscopy, the phylogenetic stains reveal that there are two genetically distinct types (one stains yellow and one stains blue). Both cell types are about 2.25 μm in diameter. The green cells in pairs or clusters are about 1 μm in diameter.

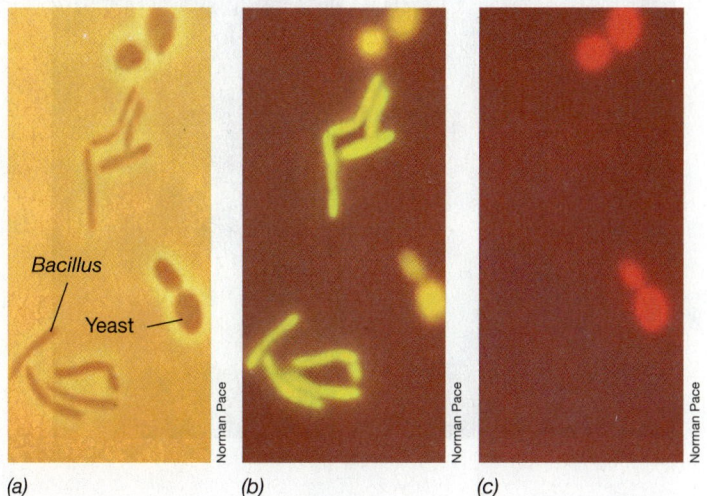

Bacillus

Yeast

Norman Pace

(a)   (b)   (c)

Figure 18.10 Fluorescently labeled rRNA probes: Phylogenetic stains. *(a)* Phase-contrast photomicrograph of cells of *Bacillus megaterium* (rod, *Bacteria*) and the yeast *Saccharomyces cerevisiae* (oval cells, *Eukarya*). *(b)* Same field; cells stained with a yellow-green universal rRNA probe (this probe hybridizes with rRNA from organisms of any phylogenetic domain). *(c)* Same field; cells stained with a eukaryal probe (only cells of *S. cerevisiae* react). Cells of *B. megaterium* are about 1.5 μm in diameter and cells of *S. cerevisiae* are about 6 μm in diameter.

Phylogenetic Staining Using FISH

Phylogenetic FISH stains are fluorescing oligonucleotides complementary in base sequence to sequences in ribosomal RNA (16S or 23S rRNA in prokaryotes or 18S or 28S rRNA in eukaryotes, ⮌ Section 12.4). Phylogenetic stains penetrate cells without lysing them and hybridize with rRNA directly in the ribosomes. The number of fluorescent probes bound to a cell reflects the number of its ribosomes. As single microbial cells can contain tens of thousands of ribosomes, strong signals can be achieved. Because ribosomes are scattered throughout the cell in most prokaryotes, the entire cell becomes fluorescent (Figures 18.9b and 18.10).

By targeting sites in the rRNA that are variable between different organisms, phylogenetic stains can be designed to be very specific and react with only one species or a handful of related microbial species. Alternatively, by targeting conserved stretches in the rRNA they can be made more general and react with, for example, all cells of a given phylogenetic domain. Using FISH, an investigator can identify or track an organism of interest or a domain of interest in a natural sample. For example, if one wishes to determine the percentage of a given microbial population that are *Archaea*, an archaeal-specific phylogenetic stain may be used in combination with DAPI (Section 18.3) to assess *Archaea* and total numbers, respectively, and a percentage could then be derived by calculation.

FISH technology can also employ multiple phylogenetic probes. With a suite of probes, each designed to react with a particular organism or group and each containing its own fluorescent dye, FISH can image multiple taxa in a habitat in a single experiment (**Figure 18.11**). If FISH is combined with confocal microscopy (⮌ Section 2.3), it is possible to explore microbial populations with depth, as, for example, in a biofilm (⮌ Section 19.4). In addition to microbial ecology, FISH is also an important tool in the food industry and in clinical diagnostics for the microscopic detection of specific pathogens in food products or clinical specimens.

CARD-FISH

Besides characterizing the abundance of different taxa in a habitat, FISH can be used to measure *gene expression* in organisms in a natural sample. Because the target in this case is messenger RNA (mRNA), a form of RNA that is much less abundant in the cells than is rRNA, standard FISH techniques cannot be applied. Instead, the signal (fluorescence) must be amplified. A FISH method that enhances the signal is called *catalyzed reporter deposition FISH* (*CARD-FISH*).

In CARD-FISH the specific nucleic acid probe contains a molecule of the enzyme peroxidase conjugated to it instead of a fluorescent dye. After there has been time for hybridization, the preparation is treated with a fluorescently labeled soluble compound called *tyramide*, which is a substrate for peroxidase. Within cells containing the nucleic acid probe, the tyramide is converted by the activity of peroxidase into a very reactive intermediate that covalently binds to adjacent proteins; this amplifies the signal sufficiently to be detected by fluorescence microscopy (**Figure 18.12**). Each molecule of peroxidase activates many molecules of tyramide so that even mRNAs present at very low abundance can be visualized.

Besides detecting mRNA, CARD-FISH is also useful in phylogenetic studies of prokaryotes that may be growing very slowly, for example organisms inhabiting the open oceans where cold temperatures and low nutrient concentrations limit growth rates (Figure 18.12). Because such cells have few ribosomes compared with more actively growing cells, standard FISH often yields only a weak signal.

MINIQUIZ --

- What structure in the cell is the target for fluorescent probes in phylogenetic FISH?
- FISH and CARD-FISH can be used to reveal different things about cells in nature. Explain.

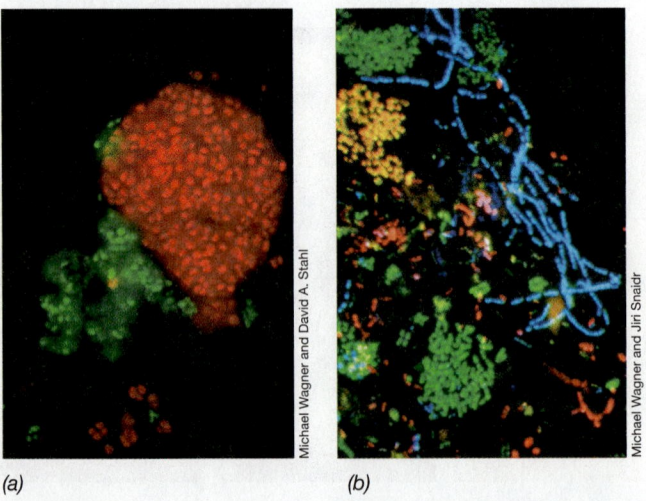

Figure 18.11 **FISH analysis of activated sludge from a wastewater treatment plant.** *(a)* Nitrifying bacteria. Red, ammonia-oxidizing bacteria; green, nitrite-oxidizing bacteria. *(b)* Confocal laser scanning micrograph of a sewage sludge sample treated with three phylogenetic FISH probes, each containing a fluorescent dye (green, red, or blue) that identifies a particular group of *Proteobacteria*. Green-, red-, or blue-stained cells reacted with only a single probe; other cells reacted with two (turquoise, yellow, purple) or three (white) probes.

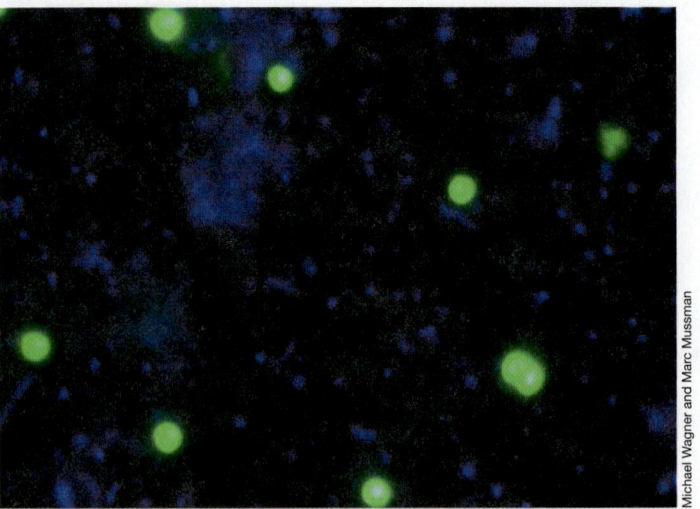

Figure 18.12 **Catalyzed reporter deposition FISH (CARD-FISH) labeling of *Archaea*.** Archaeal cells in this preparation fluoresce intensely (green) relative to DAPI-stained cells (blue).

III • Culture-Independent Genetic Analyses of Microbial Communities

Microbial biodiversity studies can forgo isolating organisms or even quantifying or identifying them microscopically using the stains described in the previous sections. Instead, *specific genes* can be used as measures of biodiversity and metabolic capacity. Some genes are unique to particular organisms. Detection of such a gene in an environmental sample implies that the organism is present. The major techniques employed in this type of microbial community analysis are the polymerase chain reaction (PCR), DNA fragment analysis by gel electrophoresis (DGGE, T-RFLP, ARISA) or molecular cloning, and DNA sequencing and analysis. In addition, as we will see in Section 18.7, entire genomes from the cells present in an environmental sample can also be analyzed as a measure of the biodiversity of microbial communities.

18.5 PCR Methods of Microbial Community Analysis

PCR and Microbial Community Analysis

We discussed the principle of PCR in Section 11.3. Recall the major steps in PCR: (1) Two nucleic acid primers are hybridized to a complementary sequence in a target gene; (2) DNA polymerase copies the target gene; and (3) multiple copies of the target gene are made by repeated melting of complementary strands, hybridization of primers, and new synthesis (ↄↄ Figure 11.5). From a single copy of a gene, several million copies can be made.

Which genes are suitable as target genes for microbial community analyses? Because genes encoding the small subunit ribosomal (SSU) rRNAs are phylogenetically informative and techniques for their analysis well developed (ↄↄ Sections 12.4 and 12.5), they are widely used in community analyses. Moreover, because rRNA genes are universal and contain several regions of high sequence conservation, it is possible to amplify them from all organisms using only a few different PCR primers, even though the organisms may be phylogenetically distantly related. In addition to rRNA genes, genes that encode enzymes for metabolic functions unique to a specific organism or group of related organisms can be the target genes (**Table 18.3**).

Genes such as those encoding rRNAs that have changed in sequence over time as species have diverged are called *orthologs* (ↄↄ Sections 6.11 and 12.5). Organisms that share the same or very closely related orthologous genes are called a **phylotype**. In microbial ecology, the phylotype concept is primarily used to provide a natural (phylogenetic) framework for describing the microbial diversity of a given habitat, regardless of whether the identified phylotypes are cultured organisms or not. Thus, the word *phylotype* is widely used to describe the microbial diversity of a habitat based solely on nucleic acid sequences. It is only when additional physiological and genetic information becomes available, typically after the organism is brought into laboratory culture, that proposing a genus and species name for a phylotype becomes possible.

In a typical community analysis experiment, total DNA is isolated from a microbial habitat (**Figure 18.13**). Commercially available

Table 18.3 Genes commonly used for evaluating specific microbial processes in the environment using PCR

Metabolic process[a]	Target gene	Encoded enzyme
Denitrification	narG	Nitrate reductase
	nirK, nirS	Nitrite reductase
	norB	Nitric oxide reductase
	nosZ	Nitrous oxide reductase
Nitrogen fixation	nifH	Nitrogenase
Nitrification	amoA	Ammonia monooxygenase
Methane oxidation	pmoA	Methane monooxygenase
Sulfate reduction	apsA	Adenosine phosphosulfate reductase
	dsrAB	Sulfite reductase
Methane production	mcrA	Methyl coenzyme M reductase
Degradation of petroleum compounds	nahA	Naphthalene dioxygenase
	alkB	Alkane hydroxylase
Anoxygenic photosynthesis	pufM	M subunit of photosynthetic reaction center

[a]All of these metabolic processes are discussed in Chapter 13 and Section 3.17.

kits that yield high-purity DNA from soil and other complex habitats are available for this purpose. The DNA obtained is a mixture of genomic DNA from all of the microorganisms that were in the sample from the habitat (Figure 18.13). From this mixture, PCR is used to amplify the target gene and make multiple copies of each variant (phylotype) of the target gene. If RNA is isolated instead of DNA (to detect those genes being transcribed), the RNA can be converted into complementary DNA (cDNA) by the enzyme reverse transcriptase (ↄↄ Sections 9.11 and 27.10) and the cDNA subjected to PCR as for isolated DNA. However, regardless of whether DNA or RNA is originally isolated, the different phylotypes need to be sorted out following the PCR step before they can be sequenced. Sorting can be accomplished using one of three different methods: (1) physical separation by gel electrophoresis, (2) clone library construction, and (3) next-generation sequencing technology. We consider these methods now.

Denaturing Gradient Gel Electrophoresis: Separating Very Similar Genes

One method to resolve phylotypes is **denaturing gradient gel electrophoresis (DGGE)**, which separates genes of the same *size* that differ in their melting (denaturing) profile because of differences in their *base sequence* (**Figure 18.14a, b**). DGGE employs a gradient of a DNA denaturant, typically a mixture of urea and formamide. When a double-stranded DNA fragment moving through the gel reaches a region containing sufficient denaturant, the strands begin to "melt"; at this point, their migration stops (Figures 18.13 and 18.14b). Differences in base sequence cause differences in the melting properties of DNA. Thus, the different

UNIT 4

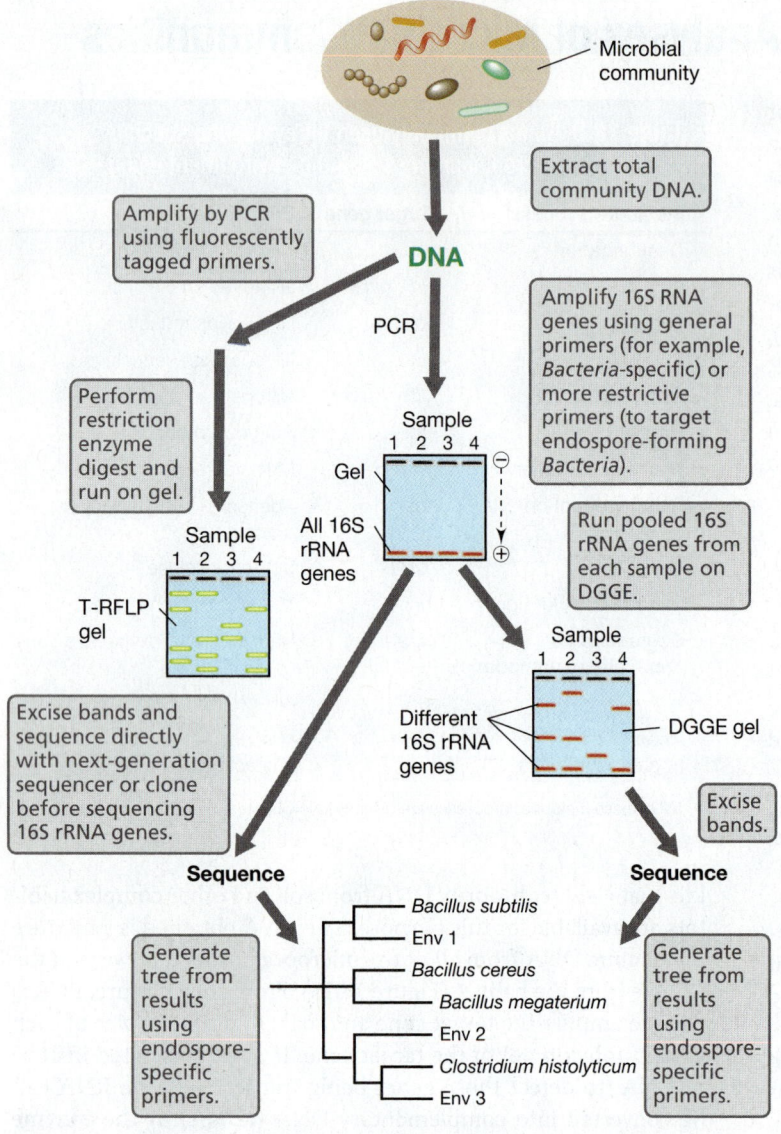

Figure 18.13 Steps in single-gene biodiversity analysis of a microbial community. From total community DNA, 16S rRNA genes are amplified using, in the DGGE example, primers that target only *Firmicutes*, a group of gram-positive *Bacteria* that includes the endospore-forming genera *Bacillus* and *Clostridium*. The PCR bands are excised and the different 16S rRNA genes separated by either cloning or DGGE. Following sequencing, a phylogenetic tree is generated. "Env" indicates an environmental sequence (phylotype). In T-RFLP analyses, the number of bands indicates the number of phylotypes.

bands observed in a DGGE gel are phylotypes that can differ in base sequence significantly or by as little as a single base change.

Once DGGE has been performed, the individual bands are excised and sequenced (Figure 18.13). With 16S rRNA as the target gene, for example, the DGGE pattern immediately reveals the number of phylotypes (distinct 16S rRNA genes) present in a habitat (Figure 18.14c). The method provides an excellent mechanism to quickly evaluate temporal and spatial shifts in microbial community structure (Figure 18.14c). Following the sequencing of each DGGE band, the actual species present in the community can be determined by phylogenetic analyses (⟳ Sections 12.4 and 12.5; Figure 18.13). If PCR primers specific for genes other than 16S rRNA are used, such as a metabolic gene (Table 18.3),

the variants of this specific gene that exist in the sample can also be assessed. Thus, although the number of bands on a DGGE gel is an overview of the biodiversity in a habitat (Figure 18.14c), sequence analysis is still required for identification and to infer phylogenetic relationship.

T-RFLP and ARISA

Another method for rapid microbial community analysis is *terminal restriction fragment length polymorphism* (*T-RFLP*). In this method a target gene (usually an rRNA gene) is amplified by PCR from community DNA using a primer set in which one of the primers is end-labeled with a fluorescent dye. The PCR products are then treated with a restriction enzyme (⟳ Section 11.1) that cuts the DNA at specific sequences. Restriction enzymes with recognition sites of only four base pairs are commonly used because they cut frequently within a relatively short PCR product. This generates a series of DNA fragments of varying length, the number of which depends on how many restriction cut sites are present in the DNA. The fluorescently labeled terminal fragments are then separated by gel electrophoresis and the digestion products further separated and sized on an automated DNA sequencer that detects fragments based on fluorescence. Therefore, only the terminal dye-labeled fragments are detected. The pattern obtained shows the rRNA sequence variation in the microbial community sampled (Figure 18.13).

DGGE and T-RFLP both measure single-gene diversity, but in different ways. The pattern of bands on a DGGE gel reflects the number of same-length sequence variants of a single gene (Figure 18.14), whereas the pattern of bands on a T-RFLP gel reflects variants differing in DNA sequence of a single gene as measured by differences in restriction enzyme cut sites. The information obtained from a T-RFLP analysis, in addition to providing insight into the diversity and population abundances of a microbial community, can also be used to infer phylogeny. Diagnostic information for each fragment includes knowledge of sequences near both ends (primer sequence and restriction enzyme cut site), knowledge that a second restriction site does not exist within the fragment, and fragment length. Using specialized software, this information can be used to search for matching 16S rRNA sequences in public databases. Although this is of some predictive value, many closely related sequences are often not differentiated by these criteria. Thus, T-RFLP generally underestimates the diversity within a microbial community.

A technique related to T-RFLP that provides more detailed analysis of microbial communities is *automated ribosomal intergenic spacer analysis* (*ARISA*), which exploits the proximity of the 16S rRNA and 23S rRNA genes in prokaryotes. The DNA separating these two genes, called the *internal transcribed spacer* (*ITS*) region, differs in length among species and often also differs in length among the multiple rRNA operons of a single species (**Figure 18.15a**). The PCR primers for ARISA are complementary to conserved sequences in the 16S and 23S rRNA genes that flank the spacer region. Amplification (Figure 18.15b) and analysis (Figure 18.15c) are conducted as described for T-RFLP, resulting in a complex

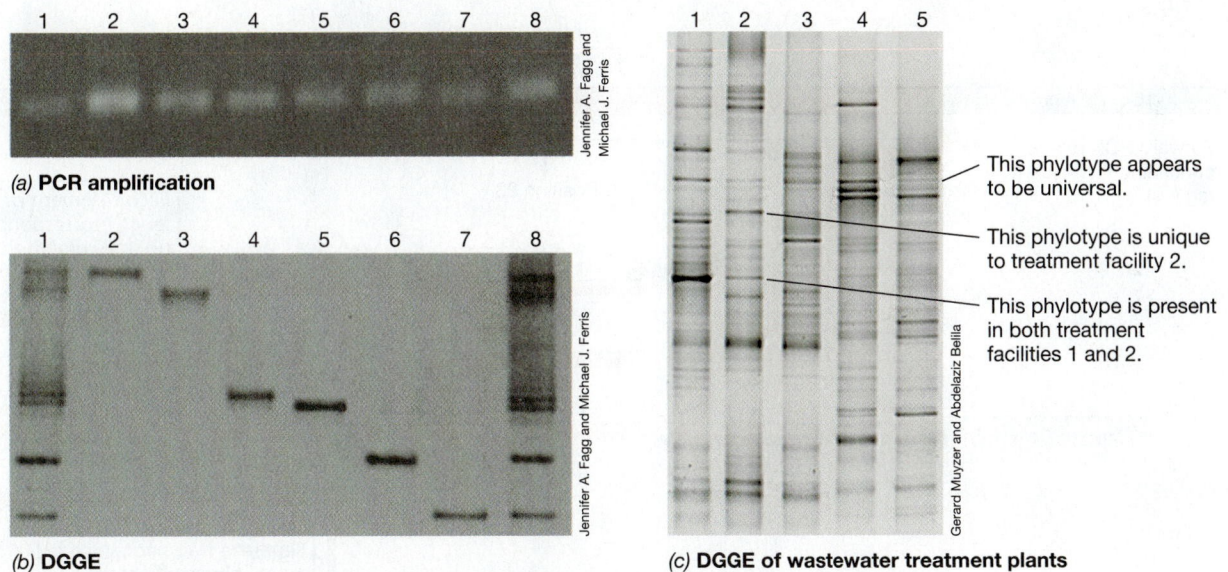

(a) PCR amplification

Jennifer A. Fagg and Michael J. Ferris

(b) DGGE

Jennifer A. Fagg and Michael J. Ferris

This phylotype appears to be universal.

This phylotype is unique to treatment facility 2.

This phylotype is present in both treatment facilities 1 and 2.

Gerard Muyzer and Abdelaziz Belila

(c) DGGE of wastewater treatment plants

Figure 18.14   PCR and DGGE gels. Bulk DNA was isolated from a microbial community and amplified by PCR using primers for 16S rRNA genes of *Bacteria* (*a*, lanes 1 and 8). Six bands later resolved by DGGE (*b*, lanes 2–7) were excised and reamplified and each gave a single band at the same location on the PCR gel (*a*, lanes 2–7). However, by DGGE analysis, each band migrated to a different location on the gel (*b*, lanes 2–7). Note that all bands migrate to the same location in the nondenaturing PCR gel because they are all of the same size, but migrate to different locations on the DGGE gel because they have different sequences. (*c*) DGGE profiles of microbial communities from different wastewater treatment facilities amplified using primers for the 16S rRNA genes of *Bacteria*.

pattern of bands that can be used for community analysis. However, ARISA differs from T-RFLP in that ARISA does not require a restriction enzyme digestion following PCR amplification. The word "automated" in the ARISA acronym refers to the use of a DNA sequencer that automatically identifies and assigns sizes to each dye-labeled fragment (Figure 18.15*c*), as can also be done in T-RFLP analyses. ARISA has received greatest application in the study of microbial community dynamics by monitoring, for example, changes in the presence and relative abundance of a specific community member through time and space.

Diversity Studies Using Clone Libraries or Next-Generation Sequencing

Most earlier molecular microbial diversity research relied on the construction of clone libraries to separate individual amplified DNA molecules (*amplicons*); each clone in the library contained a unique sequence that was then used as a template for sequence determination (c⊋ Section 6.2). Figure 18.14*a* shows that a 16S rRNA gene amplicon mixture appears as a single band when selected on a nondenaturing gel. However, because the amplified target gene came from a mixture of different cells, the phylotypes need to be sorted out before they are sequenced. This can be accomplished by either DGGE (Figure 18.14*b*, *c*), molecular cloning (Figure 18.13, c⊋ Section 11.4), or by high-throughput sequencing systems (c⊋ Section 6.2) that do not require cloning for sequence determination.

Clone library construction and sequencing remains a standard method for analysis of microbial community phylogenetic diversity and for assessing functional potential (Table 18.3). However, since next-generation sequencers do not require a cloning step,

individual DNA fragments are separated and amplified on the sequencing device itself; thus, the PCR products can be used directly for sequencing. Since hundreds of thousands of amplification reactions are conducted simultaneously on next-generation sequencers, the total number of sequencing reads vastly exceeds what is possible by sequencing individual clones obtained in a clone library on a one-by-one basis (**Figure 18.16**). The tremendous volume of sequence generated by the new sequencing technology provides for extremely *deep sequence analyses*, meaning that minor phylotypes that were possibly missed by the more limited clone library method can now be revealed (Figure 18.16*b*). For example, imagine that a particular phylotype is present at only 0.01% in a library of cloned sequences. It would then be necessary to sequence well over a thousand clones, on a one-by-one basis, to have any reasonable chance of observing that specific phylotype. By contrast, the power of next-generation sequencing would detect this low-abundance phylotype along with its more abundant neighbors. The collection of minor phylotypes, which represent a substantial fraction of total diversity but only a minor component of total organism abundance in most environments, has been referred to as the *rare biosphere* (Figure 18.16).

Results of PCR Phylogenetic Analyses

Phylogenetic analyses of microbial communities have yielded surprising results. For example, using the gene encoding 16S rRNA as the target, analyses of natural microbial communities typically show that many phylogenetically distinct prokaryotes (phylotypes) are present whose rRNA gene sequences differ from those of all known laboratory cultures (Figure 18.13). Moreover, using additional methods that allow a quantitative assessment of

UNIT 4

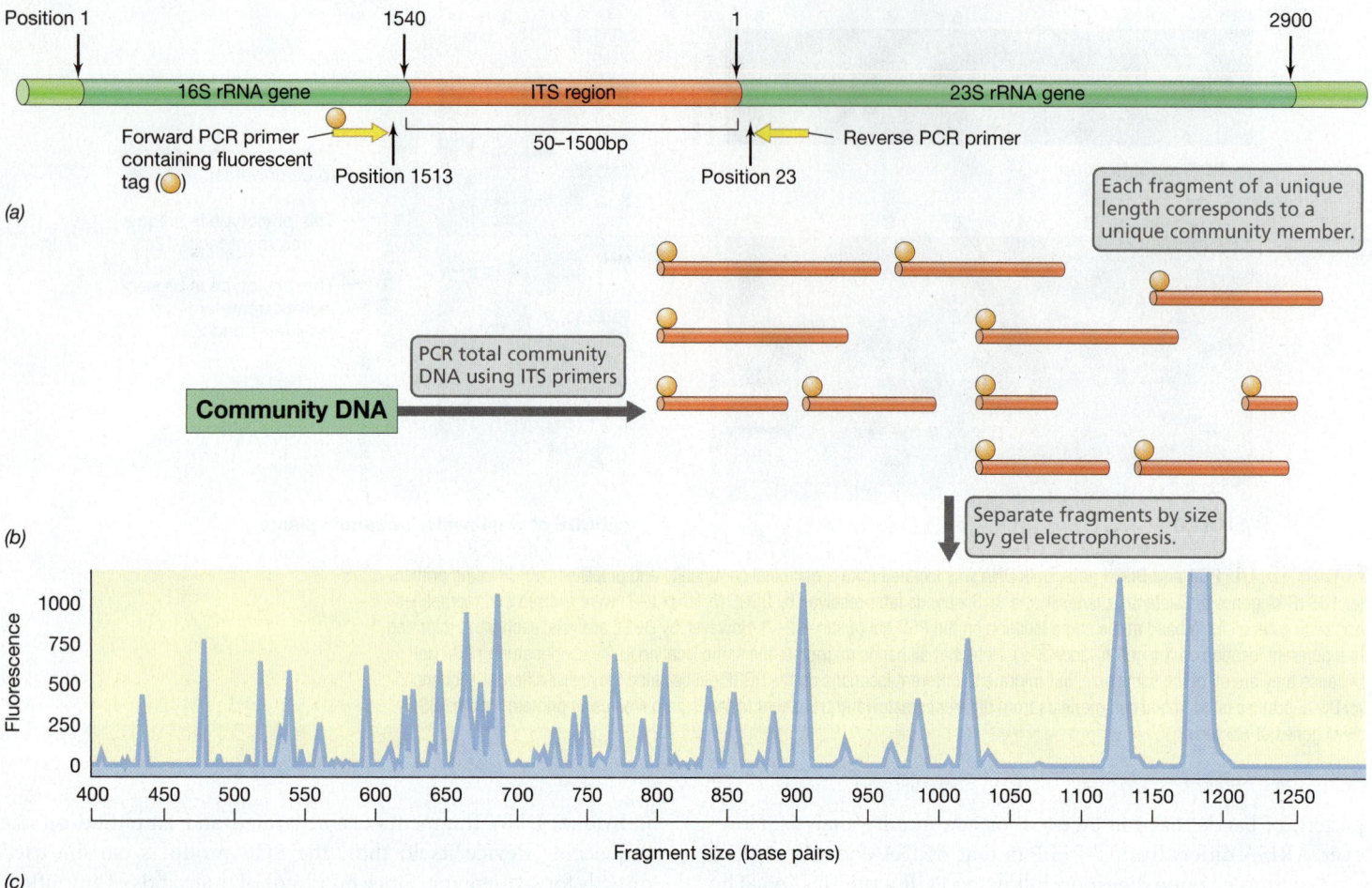

Figure 18.15 **Automated ribosomal intergenic spacer analysis (ARISA).** *(a)* Structure of rRNA operon spanning the 16S rRNA gene (positions 1–1540), an internal transcribed spacer (ITS) region of variable length, and the 23S rRNA gene (positions 1–2900). The PCR primers, one labeled with a fluorescent dye, are complementary to conserved sequences near the ITS region. *(b)* Amplified DNA fragments of different lengths, each corresponding to a community member. *(c)* Fragment analysis determined by an automated DNA sequencer. Peaks, corresponding to different ITS regions, can be identified by cloning and sequencing the amplified products.

each phylotype, it has been discovered that with few exceptions, the most abundant phylotypes in a natural microbial community are ones that have thus far defied laboratory culture. These sobering results make it clear that our knowledge of microbial diversity from enrichment cultures is very incomplete and that enrichment bias (Section 18.1) is a serious problem in culture-dependent biodiversity studies. In fact, microbial ecologists estimate that less than 0.1% of the phylotypes revealed by molecular community analyses have ever been grown in laboratory cultures. Clearly there is much work in store for microbiologists who seek to understand microbial diversity.

MINIQUIZ -

- What could you conclude from PCR/DGGE analysis of a sample that yielded one band by PCR and one band by DGGE? One band by PCR and four bands by DGGE?

- What surprising finding has come out of many molecular studies of natural habitats using 16S rRNA as the target gene?

18.6 Microarrays for Analysis of Microbial Phylogenetic and Functional Diversity

We previously considered the use of DNA chips, a type of *microarray*, for assessing overall gene expression in microorganisms (⟳ Section 6.7). More general microarrays can be constructed for rapid analyses of biodiversity and the functional potential of natural communities. Microarrays designed for biodiversity studies, called *phylochips*, have been developed for screening microbial communities for specific groups of prokaryotes. Another type of microarray has been designed to detect genes encoding functions of biogeochemical significance, such as genes encoding proteins required for sulfate respiration, ammonia oxidation, denitrification, or nitrogen fixation (Table 18.3). Because genes encoding functionally comparable enzymes can vary significantly in their primary sequence, arrays of function-specific genes, sometimes referred to as *functional gene microarrays*, must contain many thousands of probes in order to achieve reasonable coverage of natural diversity. Even then, these arrays may only sample a small fraction of natural functional diversity.

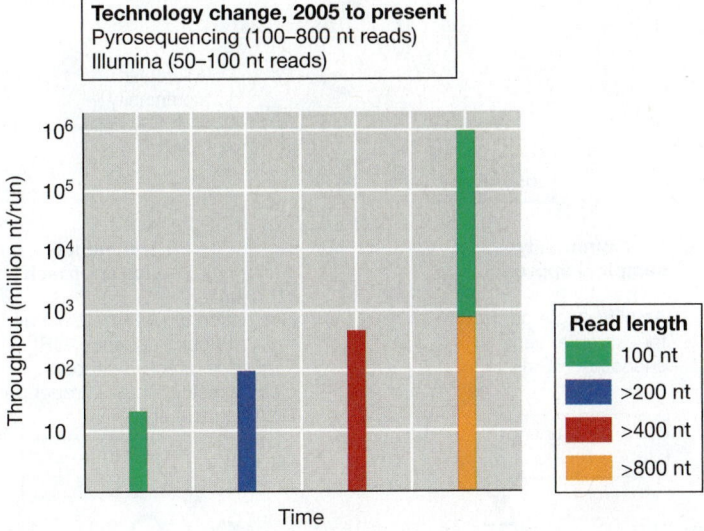

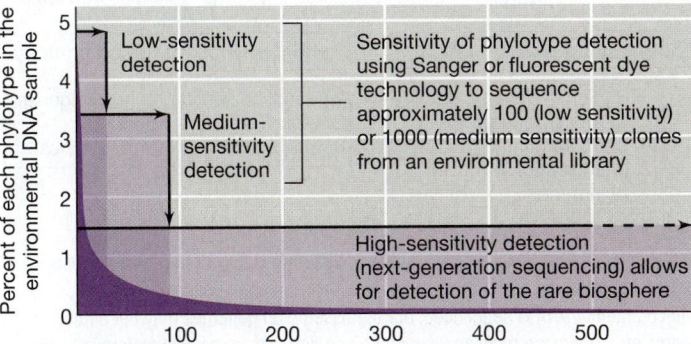

Figure 18.16 Community diversity analyses using next-generation sequencing technology. *(a)* Current sequencing platforms (⟳ Section 6.2) have the capacity to generate 10^{12} nucleotides (nt) of sequence in a single sequencing run (requiring a week or less), with individual read lengths varying from 100 to 800 nucleotides. *(b)* This enormous sequencing capacity has revealed many unique phylotypes that were not detected using DGGE or clone library sequencing. Fewer than 100 unique phylotypes would be detected by Sanger sequencing of 1000 clones in a library of 16S rRNA gene PCR amplicons. Jed Fuhrman is acknowledged for input to part b.

Phylochips are constructed by affixing rRNA probes or rRNA gene–targeted oligonucleotide probes to the chip surface in a known pattern. Each phylochip can be made as specific or general as required for the study by adjusting the specificity of the probes, and several thousand different probes can be added to a single phylochip. As an example, consider a phylochip designed to assess the diversity of sulfate-reducing bacteria (⟳ Sections 13.18 and 14.9) in a sulfidic environment, such as a marine sediment. Arranged in a known pattern on the phylochip are oligonucleotides complementary to specific sequences in the 16S rRNA genes of all known sulfate-reducing bacteria (over 100 species). Then, following the isolation of total community DNA from the sediment and PCR amplification and fluorescence labeling of the 16S rRNA genes, the environmental DNA is hybridized with the probes on the phylochip. The species that are present are determined by assessing which probes hybridized sample DNA (**Figure 18.17**). Alternatively, rRNA might be extracted directly from the microbial community, labeled with a fluorescent dye, and hybridized directly to the phylochip without an amplification step. Much more general and inclusive phylochips have also been developed. For example, one such phylochip contains 500,000 rRNA gene-targeted oligonucleotides with coverage of over 8000 individual microbial taxa.

A functional gene microarray called the *GeoChip* contains about 50,000 gene sequences from more than 290 gene categories. The categories encompass very broad metabolic capacities, including the production and consumption of methane, alternative respiratory systems (e.g., dissimilative metal reduction, halorespiration), heavy metal resistance, degradation of recalcitrant chlorinated pollutants, and common oxidative and reductive steps in the nitrogen, carbon, and sulfur cycles (Chapter 20).

Phylochips and functional gene microarrays like the GeoChip circumvent many of the time-consuming steps—PCR, DGGE, cloning, and sequencing—that are done in the microbial community analyses considered earlier (Figure 18.13). An important advantage of these methods compared to sequencing methods is reproducibility, especially for low-abundance taxa. However, an important caveat to interpretation of any gene microarray is the possibility of nonspecific hybridization. That is, gene variants that are closely related in sequence may not be resolved because of overlapping hybridization patterns. Moreover, totally unrelated genes may yield false positive results if they are sufficiently complementary to the probe to cause hybridization. Nevertheless, phylochips and functional gene arrays comprise another important tool for the culture-independent assessment of microbial biodiversity and potential metabolic activities.

MINIQUIZ
- What is a phylochip and what can it tell you?
- What are the advantages and disadvantages of microarray technology compared to sequencing PCR products?
- Why does T-RFLP analysis generally not completely capture the diversity of phylotypes in an environmental sample?

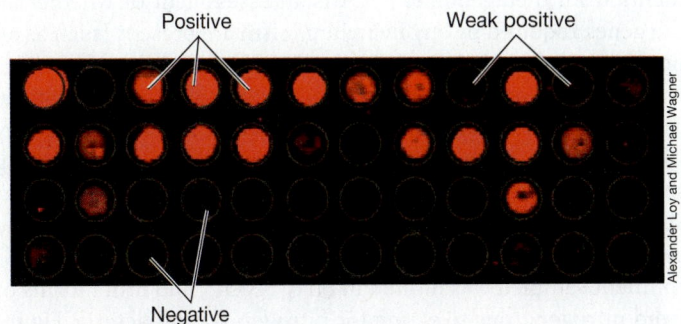

Figure 18.17 Phylochip analysis of sulfate-reducing bacteria diversity. Each spot on the microarray shown has an oligonucleotide complementary to a sequence in the 16S rRNA of a different species of sulfate-reducing bacteria. After the microarray is hybridized with 16S rRNA genes PCR-amplified from a microbial community and then fluorescently labeled, the presence or absence of each species is signaled by fluorescence (positive or weak positive) or nonfluorescence (negative), respectively.

UNIT 4

18.7 Environmental Genomics and Related Methods

A more encompassing approach to the molecular study of microbial communities is **environmental genomics**, also called **metagenomics**. These methods employ the sequencing and analysis of all microbial genomes in a particular environment as a means of characterizing the entire genetic content of that environment. Metagenomics initially focused on capturing random fragments of environmental DNA in small- or large-insert plasmids, which were used to create clone libraries of environmental DNA for sequencing; alternatively, the libraries could be screened for novel genes, such as those that encode antibiotic production. However, the introduction of high-throughput DNA sequencing technology (∂ Sections 6.2 and 18.5) rapidly accelerated this technique and removed the need for cloning DNA. Instead, DNA could now be sequenced directly from total DNA.

Before the metagenomics era, microbial community analyses typically focused on the diversity of a *single* gene in an environmental sample. By contrast, in environmental genomics, *all* genes in a given microbial community can be sampled, and if done with proper experimental design, the information obtained can support a much deeper understanding of the structure and function of the community than can single-gene analyses.

It is not the immediate goal of environmental genomics to generate complete and finished genome sequences, as has been done for many cultured microorganisms (Chapter 6). Instead, the idea is to detect as many genes as possible encoding recognizable proteins and then, if possible, to determine the phylogeny of the organism(s) to which the genes belong. However, this limitation is now being reduced by the increased coverage possible using the latest high-throughput DNA sequencing technology (Figure 18.16) and improved algorithms used for assembly of metagenomic sequence data. These advances have enabled genomes to be routinely reconstructed from community DNA (see the opening page of this chapter).

A problem with genomes assembled from a mixture of environmental DNA sequence reads, however, is that they are unlikely to be clonal, instead being composed of fragments of DNA from closely related strains of a species (**Figure 18.18**). Of importance to "reassembly" of genomes or near-complete genomic fragments from metagenomic DNA is an assessment of whether all the genes required by any living organism are present (such as all necessary stable RNAs—tRNAs and rRNAs) and therefore diagnostic of a complete genome. In addition, an assessment of the relative abundance of genes encoding specific functions is equally valuable, since abundance changes suggest interactions among species or a common response to a particular environmental variable. For example, if a high number of genes were recovered in the pathway for nitrogen fixation, this would suggest that the environment sampled was limited in NH_4^+, NO_3^-, and other forms of fixed nitrogen, thus selecting for nitrogen-fixing bacteria. Figure 18.18 contrasts the environmental genomic approach with single-gene analysis of microbial communities.

New Metagenomic Technologies

An early metagenomic study of prokaryotes in the Sargasso Sea (a low-nutrient region of the Atlantic Ocean near Bermuda)

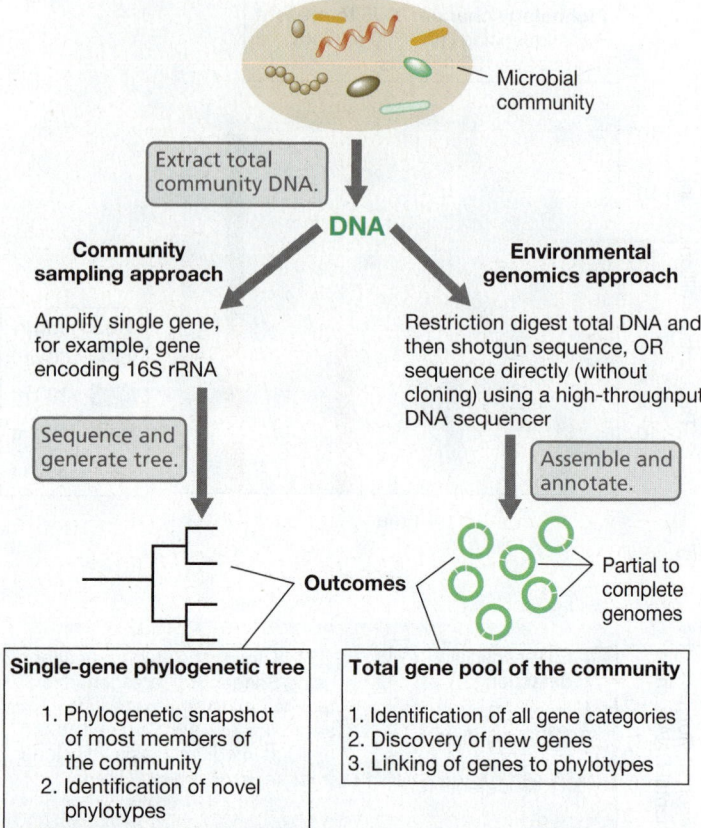

Figure 18.18 Single-gene versus environmental genomic approaches to microbial community analysis. In the environmental genomic approach, all community DNA is sequenced, but the assembled genomes may not all be complete. Total gene recovery is variable and depends on several factors including the complexity of the habitat and the amount of sequence determined. Recovery is typically better when diversity is low and sequence redundancy is high.

revealed remarkable diversity. This study was based on analysis of about one billion bp of sequence data from a random shotgun plasmid library of DNA (∂ Section 6.2) obtained from surface water. The results suggested that at least 1800 bacterial and archaeal species were present, including 148 previously unknown phylotypes and many novel genes. Many of these species had previously been missed by rRNA-based community analyses. This is because the low-sensitivity detection generally afforded by sequencing clone libraries often misses minor species (Figure 18.16) and also because not all of the 16S rRNA genes that were present in the Sargasso Sea microbial community could be amplified with the primers used for PCR amplification. Genes that fail to amplify, of course, remain undetected in community analyses. Metagenomics sidesteps this problem by sequencing DNA *without first amplifying it* by gene-specific PCR (Table 18.3). Thus, genes are sequenced whether they can be amplified by PCR or not.

Although 1 billion base pairs of sequence is an enormous single data set that cost more than US$1 million with then available technology, it was insufficient to fully describe the microbial species diversity in the Sargasso Sea sample! Indeed, 1 milliliter of seawater contains approximately 5 trillion bp of bacterial genomic DNA (given an average genome size of 5 million bp, and a density of 1 million cells), and would therefore require 5,000 times that sequencing effort just to cover each base pair once on average. Even with current technology, which can generate over 800 billion bp of sequence in 10 days (Figure 18.16), no one environment has yet been sequenced completely. Moreover, this astonishing sequencing capacity that allows for metagenomic analyses of both the abundant as well as the rare components in a given habitat is also placing unprecedented demands on the computational capacity required for the sequence analyses. Indeed, major leaps in computational efficiency and storage capacities will be needed to keep pace with the volume of metagenomic data of the future.

Some Examples of Environmental Genomics

Environmental genomics can detect both new genes in known organisms and known genes in new organisms. In addition, many of the metagenomic sequence fragments show DNA motifs that suggest they encode proteins, but these proteins have no known homologs in the existing public databases and share no apparent phylogenetic relationship with any known species. These are known as "orphan" or "ORFan" genes, which is a play on the abbreviation for open reading frame (ORF). In the Sargasso Sea study mentioned previously, genes encoding proteins that function in known metabolisms were occasionally found embedded within the genomes of organisms not previously known to carry out such metabolisms. For instance, the discovery of genes related to those encoding ammonia monooxygenase, a key enzyme of ammonia-oxidizing *Bacteria* (Table 18.3; ⇄ Sections 13.10, 14.13, and 16.6), on a DNA fragment that also contained archaeal genes suggested the possible existence of ammonia-oxidizing *Archaea*. This was later established when microbiologists were successful in isolating nitrifying *Archaea* from the marine environment (*Nitrosopumilus maritimus*, ⇄ Sections 13.10 and 16.6).

In a second example from the Sargasso Sea study, genes encoding *proteorhodopsin*, the light-mediated proton pump present in certain *Proteobacteria* and related to bacteriorhodopsin of extreme halophiles (⇄ Section 16.1), were found within the genomes of several new phylogenetic lineages of *Bacteria*. The gene for proteorhodopsin had been discovered earlier in an uncultured group of marine *Gammaproteobacteria* by cloning and sequencing large DNA fragments isolated from ocean water. Ongoing metagenomic analyses have since revealed that proteorhodopsin is broadly distributed, including in marine *Archaea* and freshwater *Bacteria*. These discoveries pointed to the importance of light to the physiology and ecology of these organisms and suggested new strategies for how to enrich and isolate them in laboratory culture. Proteorhodopsin has now been identified in a number of cultured microorganisms (including *Alpha-Beta-*, and *Gammaproteobacteria*, species of *Bacteroidetes*, and eukaryotic marine dinoflagellates) and is associated primarily with bioenergetic functions.

Genomic approaches have also revealed variations in genes associated with a single phylotype; that is, in strains that contain identical, or nearly identical, rRNA genes. For example, in studies of *Prochlorococcus*, the most abundant cyanobacterium (oxygenic phototroph) in the ocean (⇄ Section 14.3), comparison of the genome sequences of cultured strains with *Prochlorococcus* genes obtained from metagenomic analyses of ocean water identified extensive regions shared between the cultured and environmental populations (**Figure 18.19**). This high level of gene conservation confirms that the organisms in culture are typical of environmental populations. However, these analyses also identified several highly variable regions in which the genomes of cultured strains differed significantly from environmental populations. These variable regions were clustered in the genome as *genomic islands*, also called *chromosomal islands* (⇄ Section 6.13), and likely encode functions that control the growth response of particular *Prochlorococcus* populations to environmental variables such as temperature or light quality and intensity.

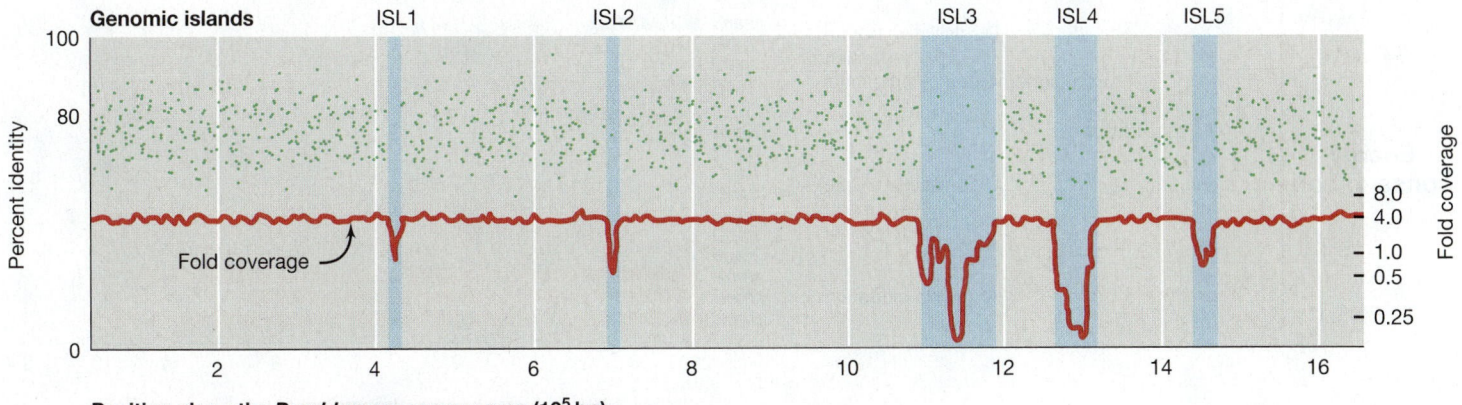

Figure 18.19 Metagenomic analysis. Sequences (represented as green dots) from the Sargasso Sea metagenome that align to the genome sequence of a cultured *Prochlorococcus*, showing regions where the cultured strain has genes of high similarity (high % identity) with sequences in the metagenome, and other regions (shaded) where it lacks genes in common (genomic islands, ISL1–ISL5). Since the DNA sequence contained within the genomic islands is thought to encode niche-specific functions, the cultured strain would likely not exhibit the same environmental distribution as strains containing all the island genes. Fold coverage is a measure of how completely the various regions in the *Prochlorococcus* genome are accounted for by similar sequences in the metagenome.

Metatranscriptomics and Metaproteomics

The application of genomic methods has spawned two related techniques, *metatranscriptomics* and *metaproteomics*. **Metatranscriptomics** is analogous to metagenomics but analyzes the sequences of community *RNA*, rather than *DNA*. The isolated RNA is converted into cDNA by reverse transcription (⟳ Sections 9.11 and 27.10) before sequencing. Although metagenomics describes the functional capacities of the community (for example, the relative abundance of specific genes), metatranscriptomics reveals which genes in the community are actually being expressed, and the relative level of that expression, at a specific time and place. Because the expression of most genes in prokaryotes is controlled at the level of transcription (⟳ Section 7.1), mRNA abundance can be considered a census of individual gene expression levels. Thus, gene transcript abundance determined for an entire community can be used to infer the operation of major metabolic processes catalyzed by that community at the time of sampling (**Figure 18.20**).

Metaproteomics, the measure of the diversity and abundance of different *proteins* in a community, is an even more direct measure of cell function than is metatranscriptomics. This is because different mRNAs have different half-lives and efficiencies of translation, and thus will not all yield the same number of protein copies. However, metaproteomics is much more of a technical challenge than is either metagenomics or metatranscriptomics (⟳ Section 6.8). Protein identification, usually by mass spectrometric characterization of peptides released from enzymatic digestion of the total protein pool using a protease that cleaves at arginine or lysine residues, relies on naturally available material since it is not possible to amplify protein sequences as one does using PCR to amplify nucleic acids for sequencing. Protein identification also requires at least partial physical separation of the individual

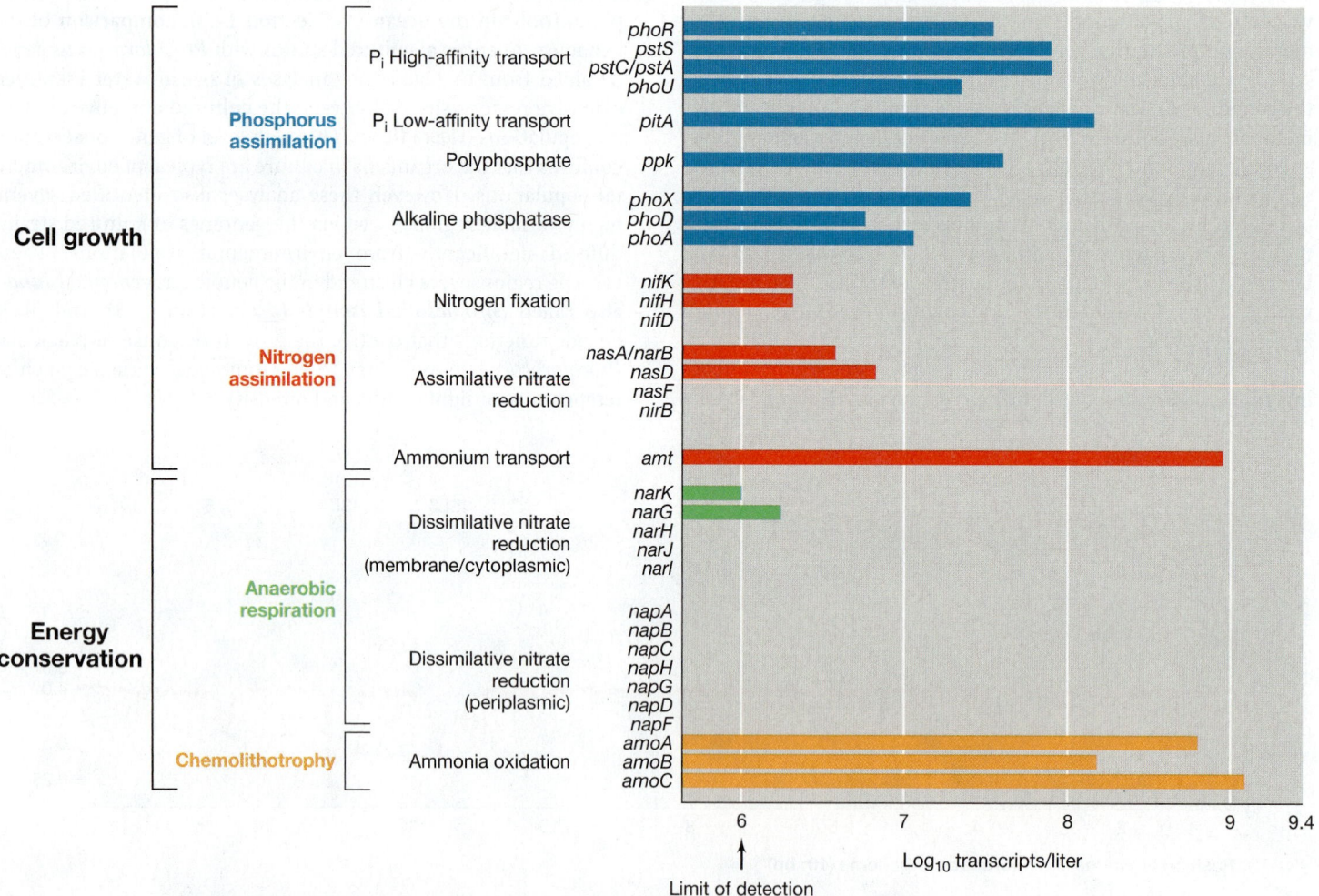

Figure 18.20 **Metatranscriptomic analysis of coastal marine surface waters.** Expression of genes for key steps in the N and P cycle in a seawater sample determined by sequencing environmental mRNA. These data showed that the microbial community was using both inorganic (high expression of P transporters) and organic (alkaline phosphatase) forms of phosphate (PO_4^{3-}). Low levels of transcripts for genes required for NO_3^- assimilation contrasted with the high expression of genes for NH_3 transport and chemolithotrophic NH_3 oxidation. Also, as expected for oxic marine surface waters, there was little expression of genes for NO_3^- respiration. Data courtesy of Mary Ann Moran.

peptides in order to reduce the complexity of samples analyzed by mass spectrometry. A final complication is the variable recovery of membrane-bound and cytoplasmic proteins. As a consequence, metaproteomics has thus far been restricted mostly to the qualitative characterization of rather simple microbial communities, such as those in some extreme environments, or to the characterization of only very abundant proteins in more complex communities. We discussed how proteins are identified in proteomic analyses and other aspects of metaproteomics in Section 6.8.

MINIQUIZ --

- What is a metagenome?

- How do environmental genomic approaches differ from environmental single-gene analyses, such as that based on 16S rRNA gene analysis for microbial community characterization?

- How can the most metabolically active cell populations in a community be identified using environmental genomic methods?

IV • Measuring Microbial Activities in Nature

So far in this chapter our discussion has focused on measuring microbial *diversity*. We now turn to how microbial ecologists measure microbial *activity*; that is, what microorganisms are actually *doing* in their environment. The techniques we consider include the use of radioisotopes, microsensors, stable isotopes, and several genomic methods.

Activity measurements in a natural sample are *collective* estimates of the physiological reactions occurring in the entire microbial community, although several techniques to be discussed later (see Sections 18.10 and 18.11) allow for a more targeted assessment of physiological activity. Activity measurements reveal both the types and rates of major metabolic reactions in a habitat, and the various techniques can be used alone or in combination in microbial community analyses. In conjunction with biodiversity estimates and gene expression analyses, these help define the structure and function of the microbial ecosystem, the ultimate goal of microbial ecology. Activity measurements can also provide valuable information for the design of enrichment cultures.

18.8 Chemical Assays, Radioisotopic Methods, and Microsensors

In many studies, direct chemical measurements of microbial reactions are sufficient for assessing microbial activity in an environment. For example, the fate of lactate oxidation by sulfate-reducing bacteria in a sediment sample can be tracked easily. If sulfate-reducing bacteria are present and active in a sediment sample, then lactate added to the sediment will be consumed and SO_4^{2-} will be reduced to H_2S. Since lactate, SO_4^{2-}, and S^{2-} can all be measured with fairly high sensitivity using simple chemical assays, the transformations of these substances relative to one another in a sample can easily be followed (**Figure 18.21a**).

Radioisotopes

When very high sensitivity is required, or turnover rates need to be determined, or the fate of portions of a molecule needs to be followed, *radioisotopes* are more useful than strictly chemical assays. For instance, if measuring photoautotrophy is the goal, the light-dependent uptake of radioactive carbon dioxide ($^{14}CO_2$) into microbial cells can be measured (Figure 18.21b). If SO_4^{2-} reduction is of interest, the rate of conversion of $^{35}SO_4^{2-}$ to $H_2{}^{35}S$ can be assessed (Figure 18.21c). Heterotrophic activities can

be measured by tracking the release of $^{14}CO_2$ from ^{14}C-labeled organic compounds (Figure 18.21d), and so on.

Both isotopic and chemical methods are widely used in microbial ecology. To be valid, however, these must employ proper controls because some isotopic transformations might be due to abiotic processes. The *killed cell control* is the key control in such experiments. That is, it is essential to show that the transformation being measured stops when chemical agents or heat treatments that kill microorganisms are applied to the sample. Formalin at a final concentration of 4% is commonly used as a chemical sterilant in microbial ecology studies. This kills all cells, and transformations of radiolabeled materials in the presence of 4% formalin can be ascribed to abiotic processes (Figure 18.21).

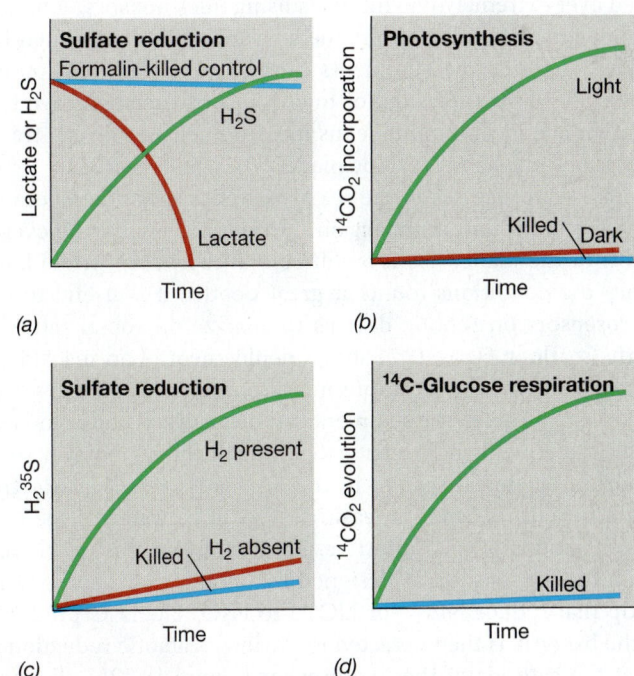

Figure 18.21 Microbial activity measurements. *(a)* Chemical measurements of lactate and H_2S transformations during SO_4^{2-} reduction. Radioisotopic measurements: *(b)* photosynthesis measured with $^{14}CO_2$; *(c)* SO_4^{2-} reduction measured with $^{35}SO_4^{2-}$; *(d)* production of $^{14}CO_2$ from ^{14}C-glucose.

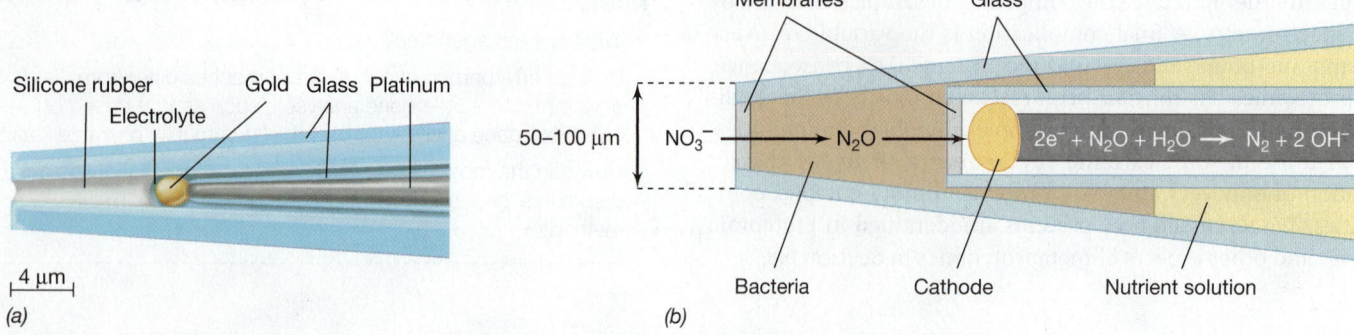

Figure 18.22 Microsensors. *(a)* Schematic drawing of an oxygen (O_2) microsensor. Oxygen diffuses through the silicone membrane in the microsensor tip and reacts with electrons on the gold surface of the cathode, forming hydroxide ions (OH^-); the latter generates a current proportional to the O_2 concentration in the sample. Note the scale of the electrode. *(b)* Biological microsensor for the detection of nitrate (NO_3^-). Bacteria immobilized at the sensor tip denitrify NO_3^- or NO_2^- to N_2O, which is detected by electrochemical reduction to N_2 at the cathode. Based on drawings by Niels Peter Revsbech.

Microsensors

Microsensors in the form of glass needles containing a sensing mechanism at the tip have been used to study the activity of microorganisms in nature. Microsensors have been constructed that measure many chemical species including pH, O_2, NO_2^-, NO_3^-, nitrous oxide (N_2O), CO_2, H_2, and H_2S. As the name *microsensor* implies, these devices are very small, their tips ranging in diameter from 2 to 100 μm (**Figure 18.22**). The sensors are carefully inserted into the habitat in small increments to follow microbial activities over very short distances.

Microsensors have many applications. For example, O_2 concentrations in microbial mats (⮂ Figure 19.19*c*), aquatic sediments, or soil particles (⮂ Figure 19.3) can be very accurately measured over extremely fine intervals using microsensors. A micromanipulator is used to insert the sensors gradually through the sample such that measurements can be taken every 50–100 μm (**Figure 18.23**). Using a bank of microsensors, each sensitive to a different chemical, simultaneous measurements of several transformations in a habitat can be made.

Microbial processes in the sea are extensively studied because they have a profound impact on nutrient cycles and the overall health of the planet. As it is difficult to reproduce in the laboratory the conditions found at great depths, it is useful to use microsensors on robotic devices to analyze microbial activities on the seafloor. **Figure 18.24** shows deployment of an instrument "lander" equipped with various microsensors so that the distribution of chemicals in the sediment can be analyzed and compared with that in overlying ocean water. One of the biologically most important chemical species in the oceans is NO_3^-, but electrochemical sensors cannot measure NO_3^- in seawater, as the high concentrations of salts interfere. To circumvent this problem, a "living" microsensor was designed that contains bacteria within its tip that reduce NO_3^- (or NO_2^-) to N_2O. The N_2O produced by the bacteria is then detected following its abiotic reduction to N_2 at the cathode of the microsensor (Figure 18.22*b*); this provides an electrical impulse signaling the presence of NO_3^-. In the oxic layer of marine sediments, NO_3^- is produced from the oxidation of NH_4^+ (nitrification, ⮂ Section 13.10), so there is often a peak of NO_3^- in the sediment surface layer (Figure 18.23). In the deeper, anoxic layers of the sediment, NO_3^- is consumed by denitrification and dissimilative nitrate reduction to ammonia (DRNA) (⮂ Section 13.17), and NO_3^- therefore disappears a few millimeters below the oxic–anoxic interface (Figure 18.23).

MINIQUIZ -
- Why are radioisotopes so useful in measuring microbial activities?
- If a large pulse of organic matter entered the sediment, how would that change the profiles of NO_3^- and O_2 shown in Figure 18.23?

- -

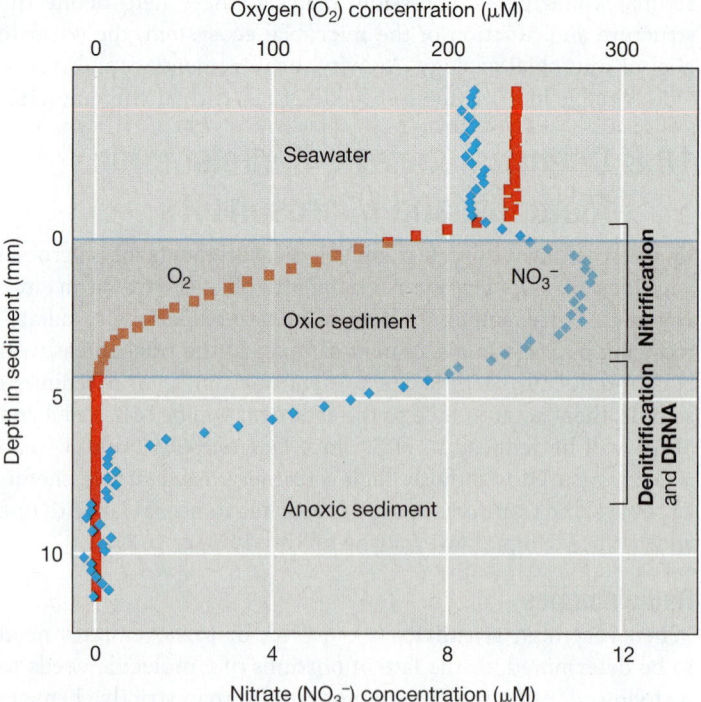

Figure 18.23 Depth profiles of O_2 and NO_3^-. Data obtained using the lander (see Figure 18.24) equipped with microelectrode sensors for remote chemical characterization of deep-sea sediments. Note the zones of nitrification and denitrification. DRNA, dissimilative reduction of NO_3^- to NH_4^+. Based on data and drawings by Niels Peter Revsbech.

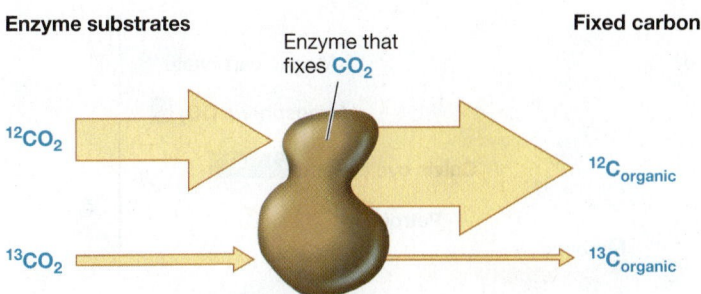

Enzyme substrates

Enzyme that fixes CO_2

Fixed carbon

$^{12}CO_2$

$^{13}CO_2$

$^{12}C_{organic}$

$^{13}C_{organic}$

Figure 18.25 Mechanism of isotopic fractionation with C as an example. Enzymes that fix CO_2 preferentially fix the lighter isotope (^{12}C). This results in fixed carbon being enriched in ^{12}C and depleted in ^{13}C relative to substrate CO_2. The size of the arrows indicates the relative abundance of each isotope of carbon.

is discriminated against when both are metabolized by an enzyme (**Figure 18.25**). For example, when CO_2 is fixed into cell material by an autotrophic organism, the cellular C becomes *enriched* in ^{12}C and *depleted* in ^{13}C, relative to an inorganic carbon standard of known isotopic composition. Likewise, the S atom in H_2S produced from the bacterial reduction of SO_4^{2-} is isotopically lighter than H_2S that has formed geochemically. These discriminations are called **isotopic fractionations** (Figure 18.25) and are typically the result of biological activities. Thus this technique can be used as a measure of whether or not a particular transformation has been catalyzed by microorganisms.

The isotopic fractionation of C in a sample is calculated as the extent of ^{13}C depletion relative to a standard having an isotopic composition of geological origin. The standard for C isotope analysis is rocks from a Cretaceous (65- to 150-million-year-old) limestone formation (the Pee Dee belemnite). Because the magnitude of fractionation is usually very small, depletion is calculated as "per mil" (‰, or parts per thousand) and reported as the $\delta^{13}C$ (pronounced "delta C 13") of a sample using the following formula:

$$\delta^{13}C = \frac{(^{13}C/^{12}C \text{ sample}) - (^{13}C/^{12}C \text{ standard})}{(^{13}C/^{12}C \text{ standard})} \times 1000‰$$

The same formula is used to calculate the fractionation of S isotopes, in this case using iron sulfide (FeS) mineral from the Canyon Diablo meteorite as the standard:

$$\delta^{34}S = \frac{(^{34}S/^{32}S \text{ sample}) - (^{34}S/^{32}S \text{ standard})}{(^{34}S/^{32}S \text{ standard})} \times 1000‰$$

Use of Isotopic Fractionation in Microbial Ecology

The isotopic composition of a material can reveal its biological or geological past. For example, plant material and petroleum (which is derived from plant material) have similar isotopic compositions (**Figure 18.26**). Carbon from both plants and petroleum is isotopically lighter than the CO_2 from which it was formed because the biochemical pathway used to fix CO_2 discriminated against $^{13}CO_2$ (Figures 18.25 and 18.26). Moreover, methane (CH_4) produced by methanogenic *Archaea* (↺ Section 16.2) is isotopically extremely light, indicating that methanogens discriminate strongly against $^{13}CO_2$ when they reduce CO_2 to CH_4 (↺ Section 13.20). By contrast, carbon in isotopically heavier marine carbonates is clearly of geological origin (Figure 18.26).

Figure 18.24 Deployment of a deep-sea lander. The lander is equipped with a bank of microsensors (arrow) to measure distribution of chemicals in marine sediments.

Niels Peter Revsbech

18.9 Stable Isotopes

For many of the chemical elements different isotopes exist, varying in their number of neutrons. Certain isotopes are unstable and break down as a result of radioactive decay. Others, called *stable isotopes*, are not radioactive, but are metabolized differently by microorganisms and can be used to study microbial transformations in nature. There are two methods in which stable isotopes can yield information on microbial activities. We describe isotopic fractionation in this section and stable isotope probing in Section 18.11.

Isotopic Fractionation

The two elements most useful for stable isotope studies in microbial ecology are carbon (C) and sulfur (S), although the heavy isotope of nitrogen, ^{15}N, is also widely used. Carbon (C) exists in nature primarily as ^{12}C, but about 5% exists as ^{13}C. Likewise, S with its four stable isotopes exists primarily as ^{32}S. Some S is found as ^{34}S and very small amounts as ^{33}S and ^{36}S. The relative abundance of these isotopes changes when C or S is metabolized by microorganisms because enzymes typically favor the *lighter* isotope. That is, relative to the lighter isotope, the heavier isotope

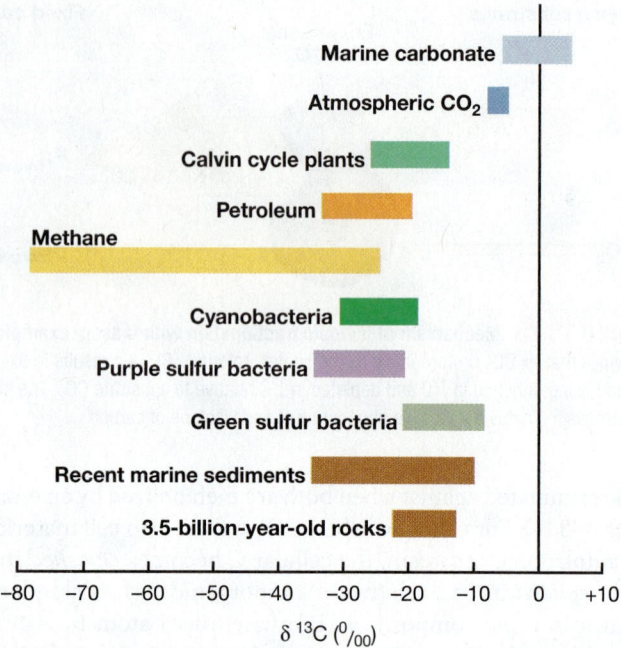

Figure 18.26 **Isotopic geochemistry of ^{13}C and ^{12}C.** Note that C fixed by autotrophic organisms is enriched in ^{12}C and depleted in ^{13}C. Methane formed from the reduction of CO_2 with H_2 by methanogenic *Archaea* shows extreme isotopic fractionation.

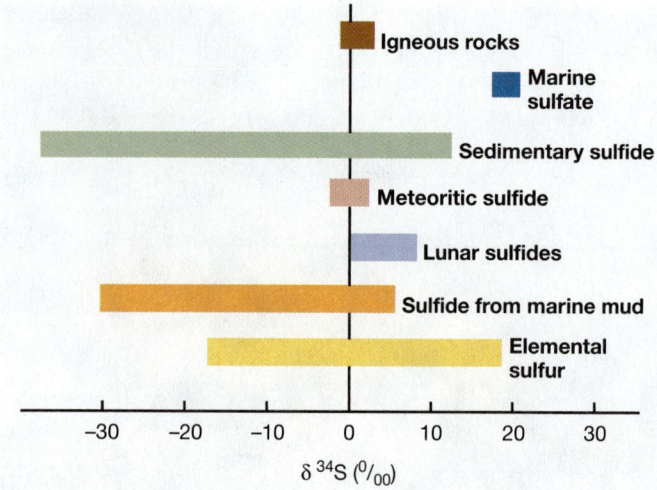

Figure 18.27 **Isotopic geochemistry of ^{34}S and ^{32}S.** Note that H_2S and S^0 of biogenic origin are enriched in ^{32}S and depleted in ^{34}S.

Because of the differences in the proportion of ^{12}C and ^{13}C in carbon of biological versus geological origin, the ^{13}C/^{12}C ratio of rocks of different ages has been used as evidence for or against past biological activity in Earth's ancient environments. Organic C in rocks as old as 3.5 billion years shows evidence of isotopic fractionation (Figure 18.26), supporting the idea that autotrophic life existed at this time. Indeed, we now believe that the first life on Earth appeared somewhat before this, about 3.8–3.9 billion years ago (⮌ Sections 1.3 and 12.1).

The activity of sulfate-reducing bacteria is easy to recognize from their fractionation of stable S isotopes in sulfides (**Figure 18.27**). As compared with an H_2S standard, sedimentary H_2S is highly enriched in ^{32}S (depleted in ^{34}S, Figure 18.27). Fractionation during sulfate reduction allows one to identify biologically produced S and has been widely used to trace the activities of sulfur-cycling prokaryotes through geological time. Sulfur isotopic analyses have also been used as evidence for the lack of life on the Moon. For example, the data in Figure 18.27 show that the isotopic composition of sulfides in lunar rocks closely approximates that of the H_2S standard, which represents primordial Earth, and differs from that of microbially produced H_2S.

MINIQUIZ

- How can the ^{13}C/^{12}C composition of a substance reveal its biological or geological origin?
- What is the simplest explanation for why lunar sulfides are isotopically similar to those of the primordial Earth?
- What is the expected isotopic composition of carbon in methanotrophs (bacteria that consume CH_4)?

18.10 Linking Genes and Functions to Specific Organisms: SIMS, Flow Cytometry, and MAR-FISH

The isotopic methods described thus far used samples containing large numbers of cells to infer that specific processes such as autotrophy or sulfate reduction were occurring within a community. These methods give an overview of community activities but do not reveal the contribution of individual cells. To do this, new isotopic methods have been developed that can measure the activity and the elemental and isotopic composition of single cells. Coupled with advanced DNA sequencing methods that can determine a genome sequence from the DNA contained in a single cell (⮌ Section 6.10 and Chapter 6 Explore the Microbial World, "Genomics, One Cell at a Time"), these techniques are at the cutting edge of microbial ecology today.

Imaging Metabolic Activities of Single Cells by Secondary Ion Mass Spectrometry (SIMS)

Secondary ion mass spectrometry (SIMS) is based on the detection of ions released from a sample placed under a focused high-energy primary ion beam, for example, of cesium (Cs$^+$); from the data generated, the elemental and isotopic composition of released materials can be obtained. When the primary ion beam impacts the sample, most chemical bonds are broken and atoms or polyatomic fragments are ejected from a very thin layer (1–2 nm) of the surface as either neutral or charged particles (secondary ions), a process called *sputtering*. These secondary ions are directed to a mass spectrometer, an instrument that can determine their mass-to-charge ratio.

NanoSIMS instruments are SIMS devices designed to yield information on single cells. The instrument is equipped with Cs$^+$ and O_2 primary beam sources with a resolution of 50 nm for the Cs$^+$ ion beam and 200 nm for the O_2 beam. The O_2 beam generates positive secondary ions and is used to analyze metals (e.g., Fe, Na, Mg) while the Cs$^+$ beam generates negative secondary ions for the analysis of major cellular elements (C, N, P, S, O, H) and halogens. The NanoSIMS instrument also records where on the specimen the ion beam is directed such that a two-dimensional image of the distribution of specific ions on the sample surface is obtained. In addition, by focusing the ion beam on the same spot during repeated cycles of sputtering, material can be slowly burned away to expose deeper regions of the sample. This high-resolution SIMS analysis is where the term *NanoSIMS* got its name. NanoSIMS instruments have multiple detectors that provide for the simultaneous analysis of ions of different mass-to-charge ratios originating from the same sample location (**Figure 18.28**).

When combined with FISH (Section 18.4), NanoSIMS can be used to track the incorporation of different elements, natural isotopes, or isotope-labeled substrates into individual cells of specific cell populations. An initial application of the pairing of these two technologies (FISH-SIMS) was to characterize the composition of C isotopes in structured aggregates of anaerobic methane-oxidizing prokaryotes. A form of anaerobic methane oxidation widespread in marine sediments is the result of a syntrophic association between sulfate-reducing *Bacteria* and methane-oxidizing (methanotrophic) *Archaea* that form aggregates, with the metabolically coupled sulfate reducers surrounding an interior core of the archaeal methanotrophs (⟲ Sections 13.15, 13.24, and 20.1). Because biogenic CH_4 is highly depleted in ^{13}C, NanoSIMS technology could be used to confirm incorporation of the lighter (^{12}C) methane carbon into archaeal methanotrophs.

A variation on the FISH-SIMS method that greatly simplifies the identification of cells scanned by NanoSIMS uses probe-conferred deposition of a halide (Br, Fl, I), either through direct incorporation of the halide into an oligonucleotide probe (SIMS-in situ hybridization or SIMSISH) or by using a halide-containing tyramide substrate (see CARD-FISH, Section 18.4). This method is referred to as "elemental FISH" (EL-FISH) and also as "halogen *in situ* hybridization-SIMS" (HISH-SIMS). Halogens possess a high ionization yield compared with other elements and are thus easy to detect, and are typically of low natural abundance. Thus, one of the NanoSIMS detectors is dedicated to identifying cells to which the probe has hybridized (Figure 18.28*d*) by halogen ionization while the remaining detectors are used for assessing elemental composition (Figure 18.28*c*).

Having excellent spatial resolution, NanoSIMS are increasingly used to examine metabolite transfer among single cells of interacting microorganisms. For example, labeling with $^{15}N_2$ followed by NanoSIMS was used to demonstrate the transfer of N_2 fixed by methanotrophic *Archaea* to the surrounding sulfate-reducing bacteria in the aggregates mentioned previously, and to demonstrate transfer of N_2 fixed by filamentous cyanobacteria to attached heterotrophic bacteria (Figure 18.28). Labeling with $^{15}NH_4$ and ^{13}C-labeled CO_2 or organic substrates is also being used to explore the assimilation of key nutrients and the transfer of metabolites among microbial species in both aquatic and soil environments.

Flow Cytometry and Multiparametric Analyses

Because of the large population sizes of natural microbial communities—typically well in excess of 10^6 cells per milliliter of water or per gram of soil—methods that rely on microscopy can examine only a very small part of a whole community. Although image analysis software can help automate the process, most microscopic analyses still rely on the practiced eye of the investigator. It is particularly difficult to assess cell numbers by counting cells microscopically, and this problem is compounded if populations are present in low numbers. However, a technique called flow cytometry offers an alternative to more labor-intensive microscopic methods. Flow cytometers can examine specific cell parameters such as size, shape, or fluorescent properties as the cells pass through a detector at rates of many thousands of cells per second (**Figure 18.29**). Fluorescence may be intrinsic (for example, chlorophyll fluorescence of phototrophic microorganisms; or it may be conferred by DNA staining, or by differential staining of live versus dead cells (vital stains), or by fluorescent DNA probes (FISH), all methods discussed in this chapter. A major advantage of flow cytometry is the ability to carry

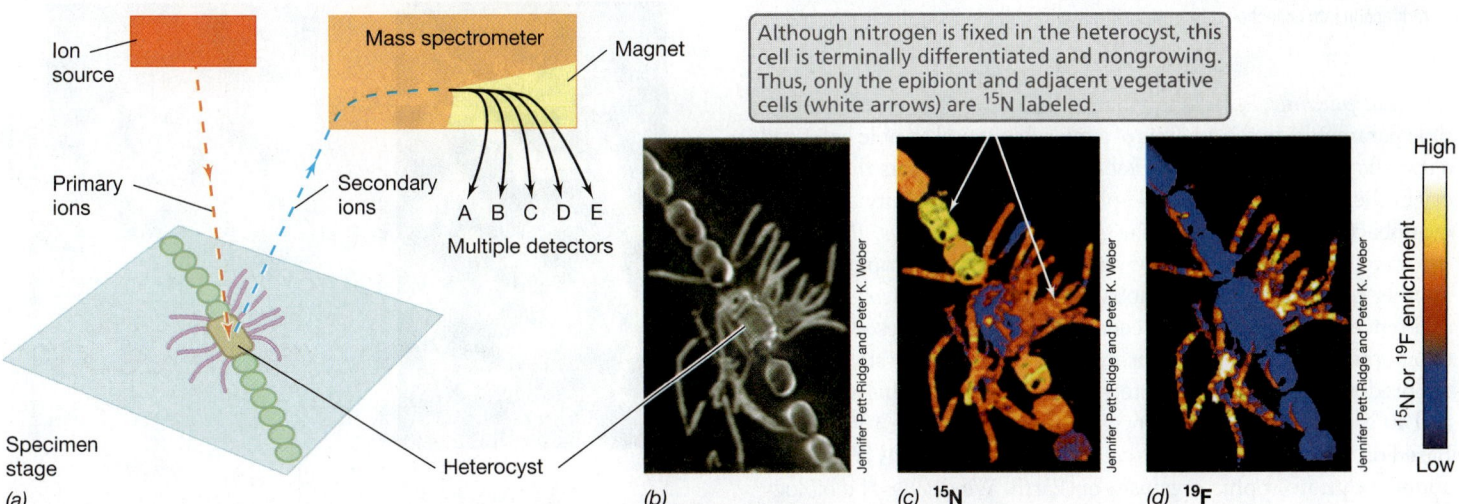

Figure 18.28 NanoSIMS technology. *(a)* Schematic of NanoSIMS operation showing the beams of primary (red) and secondary (blue) ions and five different detectors, each of which identifies ions of a different mass-to-charge ratio. *(b–d)* Demonstration of interspecies nutrient transfer from a filamentous cyanobacterium (*Anabaena*) to a *Rhizobium* species attached to the cyanobacterial heterocyst. The coculture was incubated with $^{15}N_2$, and the transfer of ^{15}N-labeled compounds from *Anabaena* to *Rhizobium* was imaged using a combination of EL-FISH and NanoSIMS. *(b)* Total ^{12}C abundance. *(c)* ^{15}N enrichment. *(d)* ^{19}F abundance conferred by a probe that hybridizes only to the attached rhizobial cells (EL-FISH).

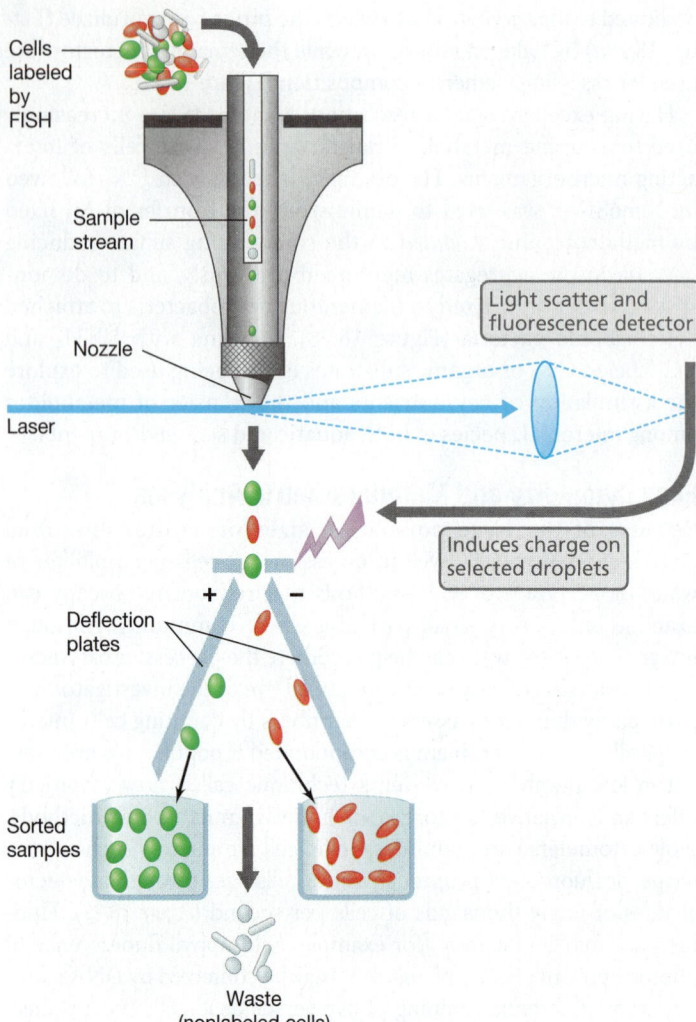

Figure 18.29 Flow cytometric cell sorting. As the fluid stream exits the nozzle, it is broken into droplets containing no more than a single cell. Droplets containing desired cell types (detected by fluorescence or light scatter) are charged and collected by redirection into collection tubes by positively or negatively charged deflection plates.

out *multiparametric analyses*, that is, the capacity to combine multiple parameters in the analysis of a microbiological sample or to find a specific population. A remarkable example of this was the discovery in the late 1980s of a novel and abundant community of marine cyanobacteria, all species of the genus *Prochlorococcus*. *Prochlorococcus* cells are smaller and have different fluorescent properties than another common marine cyanobacterium, *Synechococcus*. Based on differences in size and fluorescence, flow cytometry resolved these two populations and *Prochlorococcus* was subsequently shown to be the predominant oxygenic phototroph in ocean waters between 40°S and 40°N latitude, reaching concentrations greater than 10^5 cells/ml. Based on this finding, it can be said that *Prochlorococcus* is the most abundant phototrophic organism on Earth. We discuss the biology of *Prochlorococcus* in more detail in Section 19.10 and Figure 18.19.

Radioisotopes in Combination with FISH: Microautoradiography-FISH

Radioisotopes are used as measures of microbial activity in a microscopic technique called **microautoradiography (MAR)**. In this method, cells from a microbial community are exposed to a substrate containing a radioisotope, such as an organic compound or CO_2. Heterotrophs take up the radioactive organic compounds and autotrophs take up the radioactive CO_2. Following incubation in the substrate, cells are affixed to a slide and the slide is dipped in photographic emulsion. While the slide is left in darkness for a period, radioactive decay from the incorporated substrate induces formation of silver grains in the emulsion; these appear as black dots above and around the cells. **Figure 18.30a** shows a MAR experiment in which an autotrophic cell has taken up $^{14}CO_2$.

Microautoradiography can be done simultaneously with FISH (Section 18.4) in **MAR-FISH**, a powerful technique that combines identification with activity measurements. MAR-FISH allows a microbial ecologist to determine (by MAR) which organisms in a natural sample are metabolizing a particular radiolabeled substance while at the same time identifying these organisms (by FISH) (Figure 18.30). MAR-FISH thus goes a step beyond phylogenetic identification by revealing physiological information on the organisms, as is also true of NanoSIMS. Such data are useful not only for understanding the activity of the microbial ecosystem but also for guiding enrichment cultures. For example, knowledge of the phylogeny and morphology of an organism metabolizing a particular

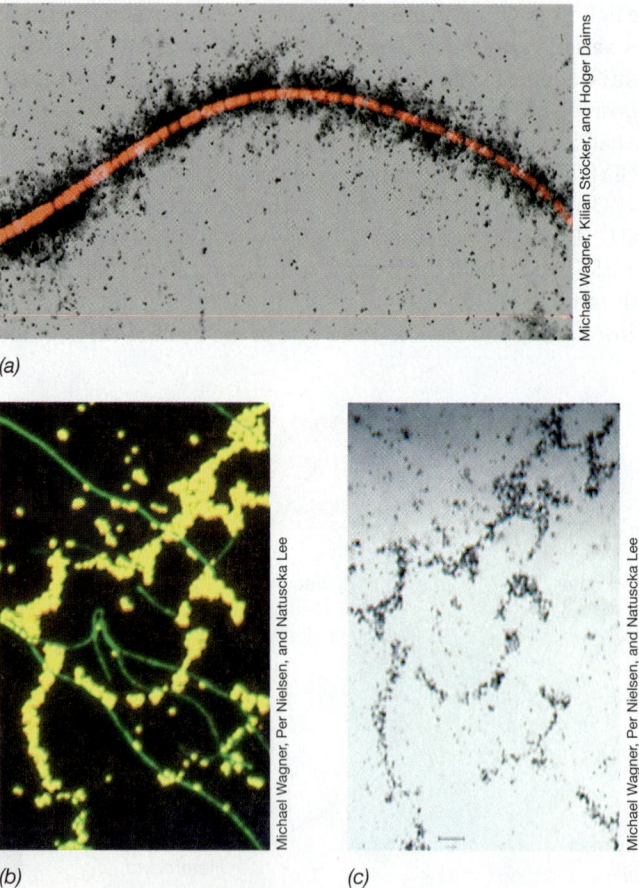

Figure 18.30 MAR-FISH. Fluorescence in situ hybridization (FISH) combined with microautoradiography (MAR). *(a)* An uncultured filamentous cell belonging to the *Gammaproteobacteria* (as revealed by FISH) is shown to be an autotroph (as revealed by MAR-measured uptake of $^{14}CO_2$). *(b)* Uptake of ^{14}C-glucose by a mixed culture of *Escherichia coli* (yellow cells) and *Herpetosiphon aurantiacus* (filamentous green cells). *(c)* MAR of the same field of cells shown in part *b*. Incorporated radioactivity exposes the film and shows that glucose was assimilated mainly by cells of *E. coli*.

substrate in a natural sample can be used to design an enrichment protocol to isolate the organism. In addition, MAR-FISH results can be quantified by counting the silver grains as a measure of the amount of substrate consumed by single cells, allowing the activity distribution in a community to be described. The technique is limited only by the availability of suitable radioactive isotopes. For example, although C-labeled substrates work well, it is not feasible to track N incorporation using MAR-FISH, since the radioactive isotope ^{13}N has a very short half-life. However, it is feasible to track N incorporation using the nonradioactive ^{15}N with NanoSIMS, as we saw earlier (Figure 18.28).

MINIQUIZ
- How could NanoSIMS be used to identify a nitrogen-fixing bacterium?
- Compared with microscopy, what are the advantages and disadvantages of flow cytometry for characterizing a microbial community?
- How does MAR-FISH link microbial diversity and activity?

18.11 Linking Genes and Functions to Specific Organisms: Stable Isotope Probing and Single-Cell Genomics

We have seen in the previous section how the combination of FISH with MAR or FISH with NanoSIMS allows for analyses of both microbial diversity and activity. These are powerful methods for connecting specific microbial populations with a specific activity or ecological niche, but in both cases, the phylogeny of the organisms of interest must be known for the FISH probe to be developed (Section 18.4). An alternative method of coupling diversity to activity is **stable isotope probing (SIP)**, a method that employs stable isotopes such as ^{13}C or ^{15}N or even ^{18}O to label DNA of organisms in a community. In addition to SIP, improvements in DNA sequencing technology allow genomics to be performed on single cells obtained from the environment. We look at the power of both of these methods here.

Stable Isotope Probing

How is a SIP experiment done? Let's say the goal of a research project was to characterize organisms capable of catabolizing aromatic compounds in lake sediment. Using benzoate as a model aromatic compound, ^{13}C-enriched benzoate would be added to a sediment sample, the sample incubated for an appropriate period, and then total DNA extracted from the sample (**Figure 18.31**). As shown in Figure 18.13, such DNA originates from *all* of the organisms in the microbial community. However, organisms that incorporate ^{13}C-benzoate will synthesize DNA containing ^{13}C. ^{13}C-DNA is heavier, albeit only slightly heavier, than ^{12}C-DNA, but the difference is sufficient to separate the heavier DNA from the lighter DNA by a special type of centrifugation technique (Figure 18.31). Once the ^{13}C-DNA is isolated, it can be analyzed using various genomics techniques for genes of interest.

If the goal of the benzoate study was to characterize the phylogeny of the organism(s) catabolizing the benzoate, PCR amplification and analysis of 16S rRNA genes from the ^{13}C-DNA could be used to do so (Figures 18.13 and 18.14). However, in addition to phylogenetic analyses, functional genes could also be targeted once the ^{13}C-DNA was obtained. For example, SIP has been employed in studies of the phylogeny and metabolism of methylotrophs, organisms that specialize in the catabolism of C_1 compounds (Section 13.23), in natural environments. In these studies, ^{13}CH$_4$ or ^{13}C-labeled methanol (^{13}CH$_3$OH) was used to label the methylotrophs' DNA followed by PCR amplification of 16S rRNA genes and genes encoding specific methane oxidation functions (Table 18.3) from the ^{13}C-DNA. Whole genome analyses are also possible using SIP. For example, in a second methylotroph study, SIP was used in combination with metagenomic analyses (Section 18.7), and the results pointed to a previously unsuspected methylotroph as being a key species for the catabolism of C_1 compounds in that particular environment.

SIP can also be done with labeled N. In this case, the isotopically heavy isotope of N, ^{15}N, competes with the more abundant and lighter isotope, ^{14}N. To study nitrogen fixation, for example, a sample would be supplied with ^{15}N$_2$, and those organisms that can fix N$_2$ (Section 3.17) will incorporate some of the ^{15}N$_2$. Some of the ^{15}N will end up in the organism's DNA, making it isotopically "heavy"; such DNA can be separated from isotopically lighter (^{14}N) DNA by ultracentrifugation (Figure 18.31) and then analyzed for specific genes.

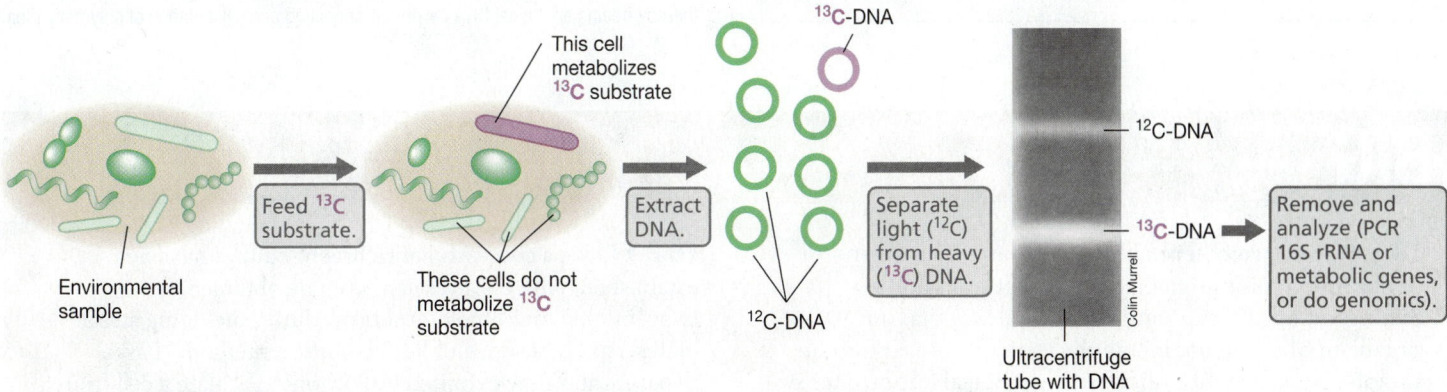

Figure 18.31 Stable isotope probing. The microbial community in an environmental sample is fed a ^{13}C-substrate. Organisms that can metabolize the substrate produce ^{13}C-DNA as they grow and divide; ^{13}C-DNA can be separated from lighter ^{12}C-DNA by density gradient centrifugation (photo). The isolated DNA is then subjected to specific gene analysis or entire genomic analysis.

Single-Cell Genomics

A major stumbling block in PCR-based gene recovery method is the requirement that a specific gene that will react with the primers used in the amplification be identified prior to analysis. Newer methods of DNA amplification now provide an alternative method for associating specific genes with a specific organism without the problems and biases associated with PCR. These methods employ *single-cell genomics* (⟳ Section 6.10 and Chapter 6 Explore the Microbial World, "Genomics, One Cell at a Time"), one of the most recent tools to enter the microbial ecologist's toolbox. **Multiple displacement amplification (MDA)** (**Figure 18.32**) is key to single-cell genomics and is used to amplify chromosomal DNA from a single cell isolated from a natural environment using a cell sorting technique, such as flow cytometry (Figure 18.29).

MDA uses a DNA polymerase from a specific bacteriophage to initiate replication of cell DNA at random points in the chromosome, displacing the complementary strand as each polymerase molecule synthesizes new DNA. This polymerase has strong strand displacement activity, resulting in the synthesis of numerous high-molecular-weight DNA products. The number of genome copies produced by amplification is sufficient to determine the complete, or nearly complete, genome sequence using next-generation sequencing platforms. In this way, both phylogenetic and metabolic functions can be inferred from the genome sequence and PCR is not required.

MDA requires stringent control over purity to eliminate contaminating DNA, but when combined with high-throughput DNA sequencing methods, MDA provides a powerful tool for linking specific metabolic functions to individual cells that have never been grown in laboratory culture. Information about the metabolic capacities of these uncultured organisms can then be used to develop strategies to recover them by enrichment culture methods and get them growing in laboratory culture.

MINIQUIZ

- How can stable isotope probing reveal the identity of an organism that carries out a particular process?
- What key method is required to do genomics on a single cell?
- When grown on $^{15}N_2$, would the DNA of a nitrogen-fixing bacterium be lighter or heavier than that of a bacterium unable to fix nitrogen?

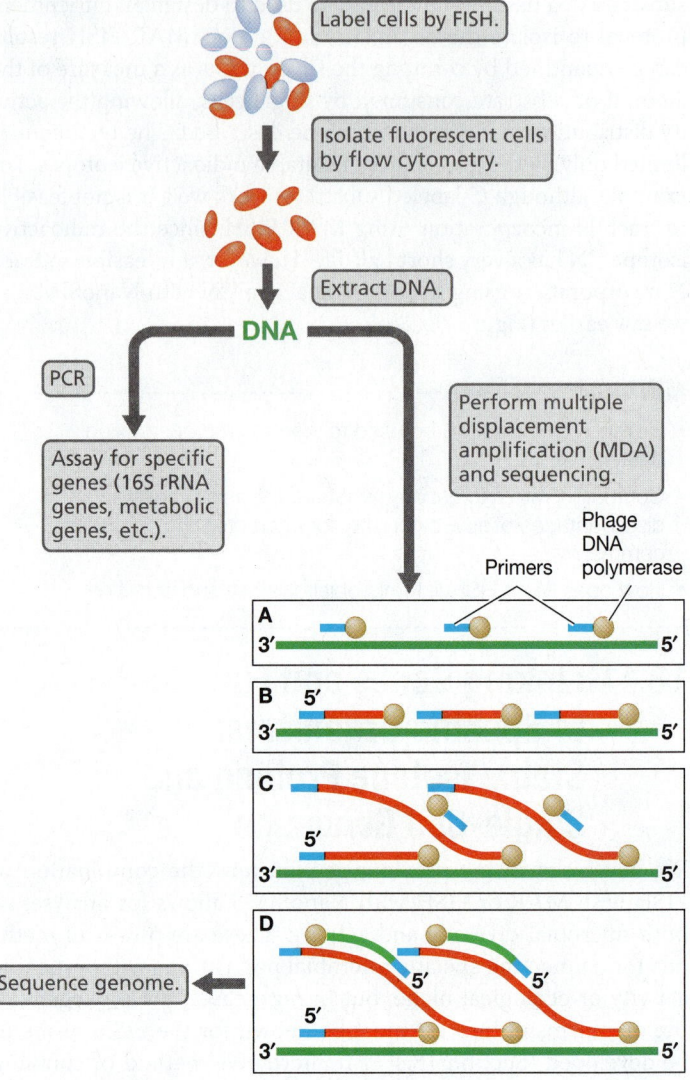

Figure 18.32 Genetic analysis of sorted cells. DNA is recovered from a specific population of cells following FISH labeling and flow cytometric sorting (Figure 18.29). DNA is characterized by PCR amplification and sequencing of specific genes, or by amplification of the entire genome by MDA followed by sequencing. For MDA, an amount of DNA sufficient for full genome sequence determination is produced using short DNAs of random sequence as primers (A) to initiate genome replication by a bacteriophage DNA polymerase. The bacteriophage polymerase copies DNA from multiple points in the genome and also displaces newly synthesized DNA (B, C), thereby freeing additional DNA for primer annealing and (D) initiation of polymerization.

BIG IDEAS

18.1 • The enrichment culture technique is a means of obtaining microorganisms from natural samples. Successful enrichment and isolation prove that an organism of a specific metabolic type was present in the sample, but does not indicate its ecological importance or abundance.

18.2 • Once a successful enrichment culture has been established, pure cultures can often be obtained by conventional microbiological procedures, including streak plates, agar dilution, and liquid dilution methods. Laser tweezers and flow cytometry allow one to isolate a cell from a microscope field and move it away from contaminants.

18.3 • DAPI, acridine orange, and SYBR Green are general stains for quantifying microorganisms in natural samples. Some stains can differentiate live versus dead cells. The GFP makes cells autofluorescent and is a means for tracking cells introduced into the environment and reporting gene expression. In natural samples, morphologically identical cells may actually be genetically distinct.

18.4 • FISH methods have combined the power of nucleic acid probes with fluorescent dyes and are thus highly specific in their staining properties. FISH methods include phylogenetic stains and CARD-FISH.

18.5 • PCR can be used to amplify specific target genes such as rRNA genes or key metabolic genes. DGGE can identify the different variants of these genes present in different species in a community.

18.6 • Phylochips combine microarray and phylogenetic technologies and are used to screen microbial communities for specific groups of prokaryotes.

18.7 • Environmental genomics (metagenomics) is based on cloning, sequencing, and analysis of the collective genomes of the organisms present in a microbial community. Metatranscriptomics and metaproteomics are offshoots of metagenomics whose focus is mRNA and proteins, respectively.

18.8 • The activity of microorganisms in natural samples can be assessed very sensitively using radioisotopes or microsensors, or both. The measurements obtained give the net activity of the microbial community.

18.9 • Isotopic compositions can reveal the biological origin and/or biochemical mechanisms involved in the formation of various substances. Isotopic fractionation is a result of the activity of enzymes that discriminate against the heavier form of an element when binding their substrates.

18.10 • A variety of advanced technologies such as NanoSIMS and MAR-FISH make it possible to examine metabolic activity, gene content, and gene expression in natural microbial communities. NanoSIMS employs secondary ion mass spectrometry technology while MAR-FISH combines the uptake of radiolabeled substrates (MAR) along with phylogenetic identification (FISH).

18.11 • SIP employs substrates labeled with heavy isotopes to generate "heavy" DNA that can be separated from the remaining ("light") DNA. Genomic analyses of the heavy DNA allow processes to be linked to specific organisms. Single-cell genomics incorporates methods for analyzing the genome of individual cells isolated from a natural microbial community.

REVIEW OF KEY TERMS

Acridine orange a nonspecific fluorescent dye used to stain DNA in microbial cells in a natural sample

DAPI a nonspecific fluorescent dye that stains DNA in microbial cells; used to obtain total cell numbers in natural samples

Denaturing gradient gel electrophoresis (DGGE) an electrophoretic technique capable of separating nucleic acid fragments of the same size that differ in base sequence

Enrichment bias a problem with enrichment cultures in which "weed" species tend to dominate in the enrichment, often to the exclusion of the most abundant or ecologically significant organisms in the inoculum

Enrichment culture highly selective laboratory culture methods for obtaining microorganisms from natural samples

Environmental genomics (metagenomics) the use of genomic methods (sequencing and analyzing genomes) to characterize natural microbial communities

Flow cytometry a technique for counting and examining microscopic particles by suspending them in a stream of fluid and passing them by an electronic detection device

Fluorescence in situ hybridization (FISH) a method employing a fluorescent dye covalently bonded to a specific nucleic acid probe for identifying or tracking organisms in the environment

Fluorescent protein any of a large group of proteins that fluoresce different colors, including the green fluorescent protein, for tracking genetically modified organisms and determining conditions that induce the expression of specific genes

Fundamental niche the range of environments in which a species will be sustained when it is not resource-limited, such as may result from competition with other species

Isotopic fractionation the discrimination by enzymes against the heavier isotope of the various isotopes of C or S, leading to enrichment of the lighter isotopes

Laser tweezers a device for obtaining pure cultures by optically trapping a single cell with a laser beam and moving it away from surrounding cells into sterile growth medium

MAR-FISH a technique that combines identification of microorganisms with measurement of metabolic activities

Metatranscriptomics the measurement of whole-community gene expression using RNA sequencing

Metaproteomics the measurement of whole-community protein expression using mass spectrometry to assign peptides to the amino acid sequences coded by unique genes

Microautoradiography (MAR) the measurement of the uptake of radioactive substrates by visually observing the cells in an exposed photographic emulsion

Microbial ecology the study of the interaction of microorganisms with each other and their environment

Microsensor a small glass sensor or electrode for measuring pH or specific compounds such as O_2, H_2S, or NO_3^- that can be immersed into a microbial habitat at microscale intervals

Most-probable-number (MPN) technique the serial dilution of a natural sample to

determine the highest dilution yielding growth

Multiple displacement amplification (MDA) a method to generate multiple copies of chromosomal DNA from a single organism

Nucleic acid probe an oligonucleotide, usually 10–20 bases in length, complementary in base sequence to a nucleic acid sequence in a target gene or RNA

Phylotype one or more organisms with the same or related sequences of a phylogenetic marker gene

Realized niche the range of natural environments supporting a species when

that organism is confronted with factors such as resource limitation, predation, and competition from other species

Stable isotope probing (SIP) a method for characterizing an organism that incorporates a particular substrate by supplying the substrate in ^{13}C or ^{15}N form and then isolating heavy isotope–enriched DNA and analyzing the genes

Winogradsky column a glass column packed with mud and overlaid with water to mimic an aquatic environment, in which various bacteria develop over a period of months

REVIEW QUESTIONS

1. What is the basis of the enrichment culture technique? Why is an enrichment medium usually suitable for the enrichment of only a certain group or groups of organisms? (Section 18.1)

2. What is the principle of the Winogradsky column, and what types of organisms does it serve to enrich? How might a Winogradsky column be used to enrich organisms present in an extreme environment, like a hot spring microbial mat? (Section 18.1)

3. Describe the principle of MPN for enumerating bacteria from a natural sample. (Section 18.2)

4. Why would the laser tweezers be a method superior to dilution and liquid enrichment for obtaining an organism present in a sample in low numbers? (Section 18.2)

5. What is the GFP? In what ways does a green fluorescing cell differ from a cell fluorescing from, for example, phylogenetic staining? (Sections 18.3 and 18.4)

6. Compare and contrast the use of FISH and phylochips for enumerating microbial cells in natural environments. What advantages and limitations do each of these methods have? (Sections 18.3 and 18.6)

7. Can nucleic acid probes in microbial ecology be as sensitive as culturing methods? What advantages do nucleic acid methods have compared with culture methods? What disadvantages? (Sections 18.4 and 18.6)

8. How can a phylogenetic picture of a microbial community be obtained without culturing its inhabitants? (Section 18.5)

9. After PCR amplification of total community DNA using a specific primer set, why is it necessary to either clone or run DGGE on the products before sequencing them? (Section 18.5)

10. Why is a microarray not suitable for characterizing community-wide transcription? (Sections 18.6 and 18.7)

11. Give an example of how environmental genomics has discovered a known metabolism in a new organism. (Section 18.7)

12. Why is environmental proteomics limited by natural abundance of microbial populations, whereas environmental genomics and metatranscriptomics are not so limited? (Section 18.7)

13. What are the major advantages of radioisotopic methods in the study of microbial ecology? What type of controls (discuss at least two) would you include in a radioisotopic experiment to show $^{14}CO_2$ incorporation by phototrophic bacteria or to show $^{35}SO_4^{2-}$ reduction by sulfate-reducing bacteria? (Section 18.8)

14. Will autotrophic organisms contain more or less ^{12}C in their organic compounds than was present in the CO_2 that fed them? (Section 18.9)

15. What can MAR-FISH tell you that FISH alone cannot? (Section 18.10)

16. What is the advantage of having multiple detectors on a NanoSIMS instrument? (Section 18.10)

17. How might you combine SIP and NanoSIMS to identify novel methane-consuming cells in a natural community? (Sections 18.10 and 18.11)

APPLICATION QUESTIONS

1. Design an experiment for measuring the activity of sulfur-oxidizing bacteria in soil. If only certain species of the sulfur oxidizers present were metabolically active, how could you tell this? How would you prove that your activity measurement was due to biological activity?

2. You wish to know whether *Archaea* exist in a lake water sample but are unsuccessful in culturing any. Using techniques described in this chapter, how could you determine whether *Archaea* existed in the sample, and if they did, what proportion of the cells in the lake water were *Archaea*?

3. Design an experiment to solve the following problem. Determine the rate of methanogenesis ($CO_2 + 4 H_2 \rightarrow CH_4 + 2H_2O$) in

anoxic lake sediments and whether or not it is H_2-limited. Also, determine the morphology of the dominant methanogen (recall that these are *Archaea*, ↺ Section 16.2). Finally, calculate what percentage the dominant methanogen is of the total archaeal and total prokaryotic populations in the sediments. Remember to specify necessary controls.

4. Design a SIP experiment that would allow you to determine which organisms in a lake water sample were capable of oxidizing the hydrocarbon hexane (C_6H_{14}). Assume that four different species could do this. How would you combine SIP with other molecular analyses to identify these four species?

microbiology**now**

Living in a World of Extreme Energy Limitation

Cold anoxic marine sediments comprise the largest organic carbon sink on Earth and are one of the most significant of all microbial habitats, harboring an estimated 3×10^{29} cells; this is similar to the total number of microbial cells in seawater.

Organic matter enters the sediments when cells and organic particles sink through the water column. Easily degraded components are removed by microbial processes in the water column or in surface sediments, leaving behind a dilute pool of deeply buried organic matter that is not readily degradable. Because of this poor nutrient quality, most of the deep-sediment microbial community exists in a state of metabolic slow motion, with populations doubling every one hundred to one thousand years.

Archaea are abundant in this energy-starved environment, and it is thought they may be better adapted than *Bacteria* to energy limitation.[1] However, since the major sediment populations also represent novel archaeal lineages that have not yet been cultured, insight into their physiology has only recently emerged from the application of molecular and analytical tools suited to the analysis of single cells.

Several partial genome sequences have been determined from single cells of *Archaea* plucked from sediments collected by coring the seafloor (photo). Genomic analyses confirmed their affiliation with the novel archaeal groups and revealed a physiological capacity to degrade and assimilate protein. Thus, these organisms apparently specialize in the degradation and assimilation of proteins buried by sedimentation or released when other sediment microorganisms die.

Since sediment microorganisms largely control the fate of carbon in this vast subsurface reservoir of organic matter, the discovery of these archaeal protein eaters living at the thermodynamic edge of life has given microbial ecologists a new perspective on carbon cycling in marine sediments.

[1]Lloyd, K.G., et al. 2013. Predominant archaea in marine sediments degrade detrital proteins. *Nature 496:* 215–218.

Microorganisms do not live alone in nature but instead interact with other organisms and with their environment. In so doing, microorganisms carry out many essential activities that support all life on Earth. In this chapter we explore some of the major habitats of microorganisms; these include soil, freshwater, and the oceans. In addition to these, microorganisms have also established more specific, and often very intimate, associations with plants and animals. We examine a few examples of such microbial partnerships and symbioses in Chapter 22.

I • Microbial Ecology

We begin with a broad overview of the science of microbial ecology, including ways that organisms interact with each other and their environments and the difference between species *diversity* and species *abundance*. These basic ecological concepts pervade this and the next two chapters.

19.1 General Ecological Concepts

The distribution of microorganisms in nature resembles that of macroorganisms in the sense that a given species resides in certain places but not others; that is, everything is not everywhere. Also, environments differ in their abilities to support diverse microbial populations. We examine these concepts here.

Ecosystems and Habitats

An **ecosystem** is a dynamic complex of plant, animal, and microbial communities and their abiotic surroundings, all of which interact as a functional unit. An ecosystem contains many different **habitats**, parts of the ecosystem best suited to one or a few populations. Although microorganisms are present in any habitat containing plants and animals, many microbial habitats are unsuitable for plants and animals. For example, microorganisms are ubiquitous on Earth's surface and even deep within it; they inhabit boiling hot springs and solid ice, acidic environments near pH 0, saturated brines, environments contaminated with radionuclides and heavy metals, and the interior of porous rocks that contain only traces of water. So some ecosystems are mostly or even exclusively microbial.

Collectively, microorganisms show great metabolic diversity and are the primary catalysts of nutrient cycles in nature (Chapter 20). The *types* of microbial activities possible in an ecosystem are a function of the species present, their population sizes, and the physiological state of the microorganisms in each habitat. By contrast, the *rates* of microbial activities in an ecosystem are controlled by the nutrients and growth conditions that prevail. Depending on several factors, microbial activities in an ecosystem can have minimal or profound impacts and can diminish or enhance the activities of both the microorganisms themselves and the macroorganisms that may coexist with them.

Species Diversity in Microbial Habitats

A group of microorganisms of the same species that reside in the same place at the same time constitutes a microbial **population**. A microbial population may be descended from a single cell. As noted in earlier chapters, a microbial **community** consists of populations of one species living in association with populations of one or more other species. The mix of species that are found in a certain habitat are those best able to grow with the nutrients and conditions that prevail there.

The diversity of microbial species in a community can be expressed in two ways (**Figure 19.1**). One is **species richness**, the total *number* of different species present. Identifying cells is, of course, basic to determining microbial species richness, but this need not require their isolation and culture. Species richness may also be expressed in molecular terms by the diversity of phylotypes (for example ribosomal RNA genes, ↺ Section 18.5) observed in a given community. **Species abundance**, by contrast, is the *proportion* of each species in the community. Species richness and

(a)

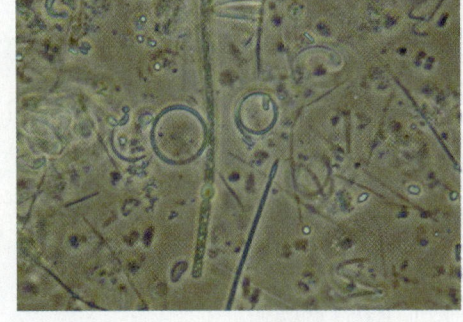
(b)

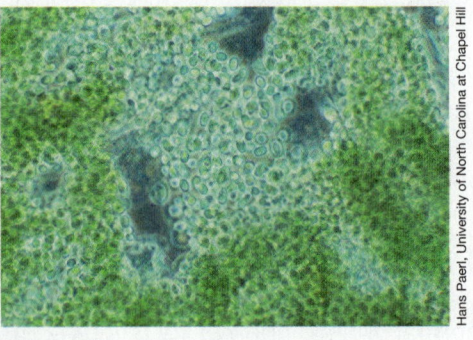
Hans Paerl, University of North Carolina at Chapel Hill
(c)

Figure 19.1 Microbial species diversity: Richness versus abundance. *(a)* Collecting samples from Lake Taihu, China, following a bloom of the cyanobacterium *Microcystis*. *(b)* High species richness in St. John's River, Florida, shown by microscopy of planktonic microorganisms including cyanobacteria, diatoms, green algae, flagellates, and bacteria. *(c)* Shift of St. John's River community to low richness but high abundance following a bloom of the cyanobacterium *Microcystis*.

Table 19.1 Resources and conditions that govern microbial growth in nature

Resources
Carbon (organic, CO_2)
Nitrogen (organic, inorganic)
Other macronutrients (S, P, K, Mg)
Micronutrients (Fe, Mn, Co, Cu, Zn, Mn, Ni)
O_2 and other electron acceptors (NO_3^-, SO_4^{2-}, Fe^{3+})
Inorganic electron donors (H_2, H_2S, Fe^{2+}, NH_4^+, NO_2^-)

Conditions
Temperature: cold → warm → hot
Water potential: dry → moist → wet
pH: 0 → 7 → 14
O_2: oxic → microoxic → anoxic
Light: bright light → dim light → dark
Osmotic conditions: freshwater → marine → hypersaline

abundance can change quickly over a short time as shown in Figure 19.1. One goal of microbial ecology is to understand species richness and abundance in microbial communities along with the community's associated activities and the abiotic environment. Once all of these factors are known, microbial ecologists can model the ecosystem by perturbing it in some way and observing whether predicted changes match experimental results.

The microbial species richness and abundance of a community are functions of the conditions that prevail and the kinds and amounts of nutrients available in the habitat. **Table 19.1** lists common nutrients and conditions relevant to microbial growth. In some microbial habitats, such as undisturbed organic-rich soils, high species richness is common (see Figure 19.14), with most species present at only moderate abundance. Nutrients in such a habitat are of many different types, and this helps select for high species richness. In other habitats, such as some extreme environments, species richness is often very low and abundance of one or a few species very high. This is because the conditions in the environment exclude all but a handful of species, and key nutrients are present at such high levels that the highly adapted species can grow to high cell densities. Bacteria that catalyze acid mine runoff from the oxidation of iron are a good example. These organisms thrive in highly acidic, iron-rich but organic-poor waters, where the acidic conditions and the dearth of organic carbon limit species richness. However, the elevated levels of ferrous iron (Fe^{2+}) present, which is oxidized to Fe^{3+} in energy-yielding reactions (♻ Section 13.9), fuel high species abundance. We examine the activities of acidophilic iron-oxidizing microorganisms in Sections 20.5 and 21.1.

MINIQUIZ -
- What is the difference between species richness and species abundance?
- How does an ecosystem differ from a habitat?
- What are the characteristics of a microbial population?

19.2 Ecosystem Service: Biogeochemistry and Nutrient Cycles

In any ecosystem whose resources and growth conditions are suitable, microorganisms will grow to form populations. Metabolically similar microbial populations that exploit the same resources in a similar way are called **guilds**. A habitat that is shared by a guild and supplies the resources and conditions the cells require for growth is called a **niche**. Sets of guilds form microbial communities (**Figure 19.2**). Microbial communities interact with macroorganisms and abiotic factors in the ecosystem in a way that defines the workings of that ecosystem.

Energy Inputs to the Ecosystem

Energy enters ecosystems as sunlight, organic carbon, and reduced inorganic substances. Light is used by phototrophs to make ATP and synthesize new organic matter (Figure 19.2). In addition to carbon (C), new organic matter contains nitrogen (N), sulfur (S), phosphorus (P), iron (Fe), and the other elements of life (♻ Section 3.1). This newly synthesized organic material along with organic matter that enters the ecosystem from the outside (called *allochthonous* organic matter) fuels the catabolic activities of chemoorganotrophic organisms. These activities oxidize the organic matter to CO_2 by respiration or ferment it to various reduced substances. If chemolithotrophs are present and metabolically active in the ecosystem, they can conserve energy from

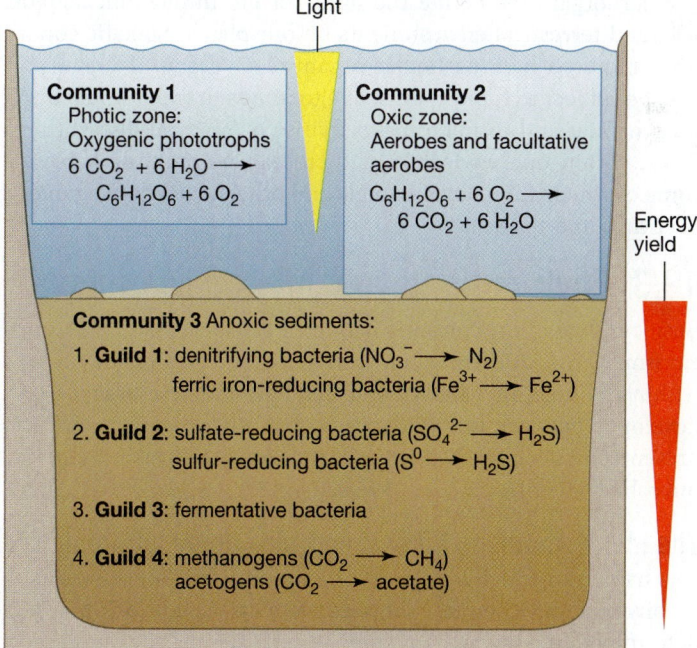

Figure 19.2 Populations, guilds, and communities. Microbial communities consist of populations of cells of different species. A freshwater lake ecosystem would likely have the communities shown here. The reduction of NO_3^-, Fe^{3+}, SO_4^{2-}, S^0, and CO_2 are examples of anaerobic respirations. The region of greatest activity for each of the different respiratory processes would differ with depth in the sediment. As more energetically favorable electron acceptors are depleted by microbial activity near the surface, less favorable reactions occur deeper in the sediment.

the oxidation of inorganic electron donors, such as H_2, Fe^{2+}, S^0, or NH_3 (Chapters 13 and 14), and contribute new organic matter through their autotrophic activities (Figure 19.2).

Biogeochemical Cycling

Microorganisms play an essential role in cycling elements, in particular C, N, S, and Fe, between their different chemical forms. The study of these transformations is part of **biogeochemistry**, an interdisciplinary science that includes biology, geology, and chemistry. Figure 19.2 shows how the activities of different guilds of microorganisms influence the chemistry of one environment, a lake ecosystem. The sequence of changing chemistry with increasing depth in the sediments corresponds to the layers of different microbial guilds. The location of each guild in the ecosystem is primarily determined by the availability of electron donors and acceptors, both of which tend to decrease with increasing depth in the sediments.

A *biogeochemical cycle* defines the transformations of an element that are catalyzed by either biological or chemical means (or both). Many different microorganisms participate in biogeochemical cycling reactions, and in many cases, microorganisms are the *only* biological agents capable of regenerating forms of the elements needed by other organisms, particularly plants. Thus, biogeochemical cycles are often also *nutrient cycles*, reactions that generate important nutrients for other organisms.

Most biogeochemical cycles proceed by oxidation–reduction reactions as the element moves through the ecosystem and are often tightly *coupled*, with transformations in one cycle impacting one or more other cycles. For example, hydrogen sulfide (H_2S) is oxidized by phototrophic and chemolithotrophic microorganisms to sulfur (S^0) and sulfate (SO_4^{2-}), the latter being a key nutrient for plants. Phototrophs and chemolithotrophs are autotrophic organisms, and thus impact the carbon cycle by producing new organic carbon from CO_2. However, SO_4^{2-} can be reduced to H_2S by activities of the sulfate-reducing bacteria, organisms that consume organic carbon, and this reduction closes the biogeochemical sulfur cycle while regenerating CO_2. The cycling of nitrogen is also a microbial process and is key to the regeneration of forms of nitrogen usable by plants and other organisms. The nitrogen cycle is driven by both chemolithotrophic and chemoorganotrophic bacteria, organisms that produce and consume organic carbon, respectively. We consider the theme of biogeochemical cycles and their coupled nature in more detail in Chapters 13, 14, and 20.

MINIQUIZ

- How does a microbial guild differ from a microbial community?
- What is a biogeochemical cycle? Give an example based on sulfur. Why are biogeochemical cycles also called nutrient cycles?

II • The Microbial Environment

Microorganisms define the limits of life throughout aquatic and terrestrial environments on our planet. Specific conditions required by a particular organism or group of organisms may be subject to rapid change due to inputs to and outputs from their habitat and to microbial activities or physical disturbance. Thus, within one environment there can be multiple habitats, some of which are relatively stable and others that change rapidly over time and space.

19.3 Environments and Microenvironments

Besides living in the common habitats of soil and water, microorganisms thrive in extreme environments and also reside on and within the cells of other organisms. The intimate associations developed between microorganisms and other organisms will be presented in Chapter 22. Here we focus on terrestrial and aquatic microbial habitats.

The Microorganism, Niches, and the Microenvironment

The habitat in which a microbial community resides is governed by physical and chemical (physicochemical) conditions that are determined in part by the metabolic activities of the community. For example, the organic material used by one species may have been a metabolic by-product of a second species. Another example is oxygen (O_2), which can become limiting if biological consumption exceeds the rate at which it is supplied.

Because microorganisms are very small, they directly experience only a tiny local environment; this small space is called their **microenvironment**. For example, for a typical 3-μm rod-shaped

bacterium, a distance of 3 mm is equivalent to that which a human would experience over a distance of 2 km! As a consequence of the smallness of microorganisms, the variable metabolic activities of nearby microorganisms, and the changes in physicochemical conditions over short intervals of time and distance, numerous microenvironments can exist within a given habitat. The conditions supporting growth within a microenvironment correspond to the general requirements for growth we considered in Chapter 5.

Ecological theory states that for every organism there exists at least one niche, the *realized niche* (also called the *prime niche*), where it will be most successful. The organism dominates the realized niche but may also inhabit other niches; in other niches it is less ecologically successful than in its realized niche but it may still be able to compete. The full range of environmental conditions under which an organism can exist is called its *fundamental niche*. The word "niche" should not be confused with the word "microenvironment" because the microenvironment describes conditions at a specific location and can change rapidly. In other words, the general conditions that describe a specific niche may be transient at many places in a microenvironment.

Another important consequence for microorganisms of being small is that diffusion often determines the availability of resources. Consider, for example, the distribution of an important microbial nutrient such as O_2 in a soil particle. Microsensors (↩ Section 18.8) can be used to measure oxygen concentrations throughout small soil particles. As shown in the data from an actual microsensor experiment (**Figure 19.3**), soil particles are not homogeneous in terms of their O_2 content but instead contain many adjacent

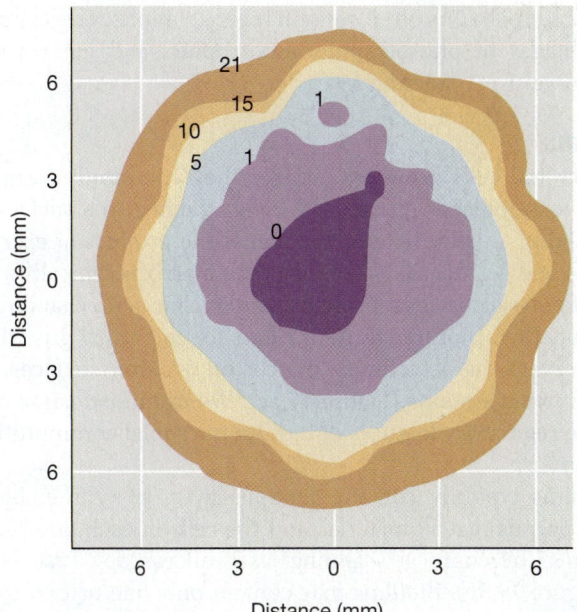

Figure 19.3 Oxygen microenvironments. Contour map of O_2 concentrations in a small soil particle as determined by a microsensor ($\rightleftharpoons$ Section 18.8). The axes show the dimensions of the particle. The numbers on the contours are percentages of O_2 concentration (air is 21% O_2). Each zone can be considered a different microenvironment.

microenvironments. The outer layer of the soil particle may be fully oxic (21% O_2) while the center, only a very short distance away (in human terms, but of course a great distance from a microbial standpoint), may be anoxic (O_2-free). The microorganisms near the outer edges consume all of the O_2 before it can diffuse to the center of the particle. Thus, anaerobic organisms could thrive near the center of the particle, microaerophiles (aerobes that require very low oxygen levels) farther out, and obligately aerobic organisms in the outermost region of the particle. Facultatively aerobic bacteria (organisms that can grow either aerobically or anaerobically) could be distributed throughout the particle ($\rightleftharpoons$ Section 5.16). Nutrient transfer is particularly important in thick assemblages of cells, such as biofilms and microbial mats, and we explore this in the next section.

Physicochemical conditions in a microenvironment are subject to rapid change in both time and space. For example, the O_2 concentrations shown in the soil particle in Figure 19.3 represent "instantaneous" values. Measurements taken in the same particle following a period of intense microbial respiration or disturbance due to wind, rain, or disruption by soil animals could differ dramatically from those shown. During such events certain populations may temporarily dominate the activities in the soil particle and grow to high numbers, while others remain dormant or nearly so. However, if the microenvironments shown in Figure 19.3 are eventually reestablished, the various microbial activities characteristic of different regions of the soil particle will eventually return as well.

Nutrient Levels and Growth Rates

Resources (Table 19.1) typically enter an ecosystem intermittently. A large pulse of nutrients—for example, an input of leaf litter or the carcass of a dead animal—may be followed by a period of nutrient deprivation. Because of this, microorganisms in nature often face a "feast-or-famine" existence. It is thus common for them to produce storage polymers as reserve materials when resources are abundant and draw upon these reserves in periods of starvation. Examples of storage materials are poly-β-hydroxyalkanoates, polysaccharides, and polyphosphate ($\rightleftharpoons$ Section 2.14).

Extended periods of exponential microbial growth in nature are probably rare. Microorganisms typically grow in spurts, linked closely to the availability and nature of resources. Because all relevant physicochemical conditions in nature are rarely optimal for microbial growth at the same time, growth rates of microorganisms in nature are usually well below the maximum growth rates recorded in the laboratory. For instance, the generation time of *Escherichia coli* in the intestinal tract of a healthy adult eating at regular intervals is about 12 h (two doublings per day), whereas in pure culture it can grow much faster, with a minimum generation time of about 20 min under optimal conditions. In addition, research-based estimates indicate that most cultured soil bacteria typically grow in nature at less than 1% of the maximal growth rate measured in the laboratory.

These slower growth rates in nature than in laboratory culture reflect the facts that (1) resources and growth conditions (Table 19.1) are frequently suboptimal; (2) the distribution of nutrients throughout the microbial habitat is not uniform; and (3) except in rare instances, microorganisms in nature grow in mixed populations rather than pure culture. An organism that grows rapidly in pure culture may grow much slower in a natural environment where it must compete with other organisms that may be better suited to the resources and growth conditions available.

Microbial Competition and Cooperation

Competition among microorganisms for resources in a habitat may be intense, with the outcome dependent on several factors, including rates of nutrient uptake, inherent metabolic rates, and ultimately, growth rates. A typical habitat contains a mixture of different species (Figures 19.1 and 19.2), with the density of each population dependent on how closely its niche resembles its realized niche.

Some microorganisms work together to carry out transformations that neither can accomplish alone—a process called *syntrophy*—and these microbial partnerships are particularly important for anoxic carbon cycling ($\rightleftharpoons$ Sections 13.15 and 20.2). Metabolic cooperation can also be seen in the activities of organisms that carry out complementary metabolisms. For example, we have previously considered metabolic transformations that are carried out by two distinct groups of organisms, such as those of the nitrifying prokaryotes ($\rightleftharpoons$ Sections 13.10, 14.13, and 16.6). Together, the nitrifying prokaryotes oxidize ammonia (NH_3) to nitrate (NO_3^-), although neither the ammonia oxidizers (having both archaeal and bacterial representatives) nor the nitrite-oxidizing bacteria are capable of doing this alone. Because nitrite (NO_2^-), the product of the ammonia-oxidizing microorganisms, is the substrate for the nitrite-oxidizing bacteria, the two groups of organisms often live in nature in tight association within their habitats ($\rightleftharpoons$ Figure 18.11).

UNIT 4

19.4 Surfaces and Biofilms

Surfaces are important microbial habitats, typically offering greater access to nutrients, protection from predation and physicochemical disturbances, and a means for cells to remain in a favorable habitat and not be washed away. Moreover, flow across a colonized surface increases transport of nutrients to the surface, providing more resources than are available to planktonic cells (cells that live a floating existence) in the same environment. A surface may also be provided by another organism or by a nutrient such as a particle of organic matter. For example, plant roots become heavily colonized by soil bacteria living on organic exudates from the plant, as revealed when fluorescent stains are used (**Figure 19.4***a*).

Virtually any natural or artificial surface exposed to microorganisms will be colonized. For example, microscope slides have been used as experimental surfaces to which organisms can attach and grow. A slide can be immersed in a microbial habitat, left for a period of time, and then retrieved and examined microscopically (Figure 19.4*b*). Clusters of a few cells that develop from a single colonizing cell, called *microcolonies*, form readily on such surfaces, much as they do on natural surfaces in nature. In fact, periodic microscopic examination of immersed microscope slides has been used to measure growth rates of attached organisms in nature. Surface colonization may be sparse, consisting only of microcolonies and not visible to the eye, or may consist of so many cells that microbial accumulation becomes visible as, for example, in a stagnant toilet bowl. Surface growth can be particularly problematic in the hospital setting where microbial colonization of indwelling devices such as catheters and intravenous lines can cause serious infection. In a few extreme environments that lack small animal grazers (for example, hot springs), microbial accumulation on a surface can be many centimeters in thickness. Called **microbial mats**,

such accumulations often contain highly complex yet very stable assemblages of phototrophic, autotrophic, and heterotrophic microorganisms (Section 19.5).

Biofilms

As bacterial cells grow on surfaces they commonly form **biofilms**—assemblages of bacterial cells attached to a surface and enclosed in an adhesive matrix that is the product of excretion by cells and cell death (**Figure 19.5**). The matrix is typically a mixture of polysaccharides, proteins, and nucleic acids that bind the cells together. Biofilms trap nutrients for microbial growth and help prevent the detachment of cells on dynamic surfaces, such as in flowing systems (Figure 19.5*c*). We examined some of the genetic regulatory features of biofilm microbial communities in Section 7.9.

Biofilms typically contain multiple layers of cells embedded in the porous matrix material, and the cells in each layer can be examined by confocal scanning laser microscopy ($\Rightarrow$ Section 2.3; Figure 19.5*b*). Biofilms may contain only one or two species or, more commonly, many species of bacteria. The biofilms that form on tooth and soft surfaces of the mouth, for example, contain between 100 and 200 different phylotypes ($\Rightarrow$ Section 18.5), including species of both *Bacteria* and *Archaea*; in total, the human mouth is a habitat for approximately 700 phylotypes. Biofilms are

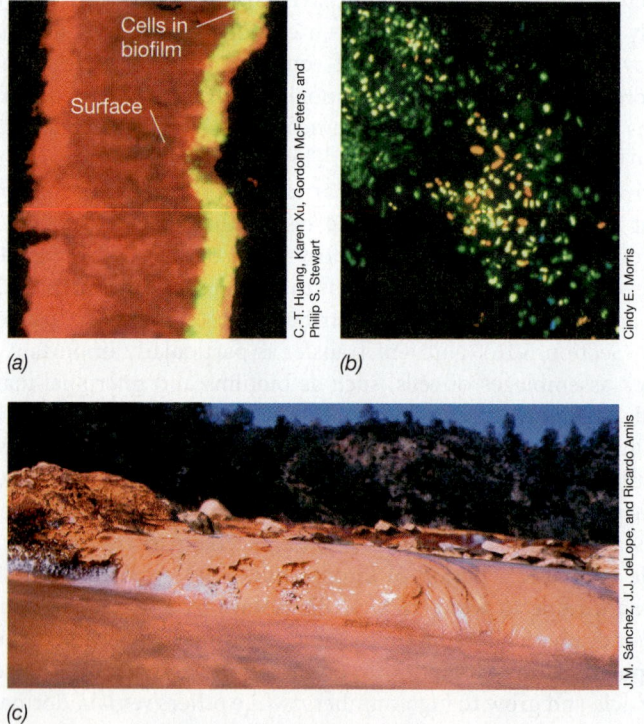

(a)

(b)

(c)

Figure 19.5 Examples of microbial biofilms. *(a)* A cross-sectional view of an experimental biofilm made up of cells of *Pseudomonas aeruginosa*. The yellow layer (about 15 μm in depth) contains cells and is stained by a reaction showing activity of the enzyme alkaline phosphatase. *(b)* Confocal scanning laser microscopy of a natural biofilm (top view) on a leaf surface. The color of the cells indicates their depth in the biofilm: red, surface; green, 9-μm depth; blue, 18-μm depth. *(c)* A biofilm of iron-oxidizing bacteria attached to rocks in the Rio Tinto, Spain. As Fe^{2+}-rich water passes over and through the biofilm, the organisms oxidize Fe^{2+} to Fe^{3+}.

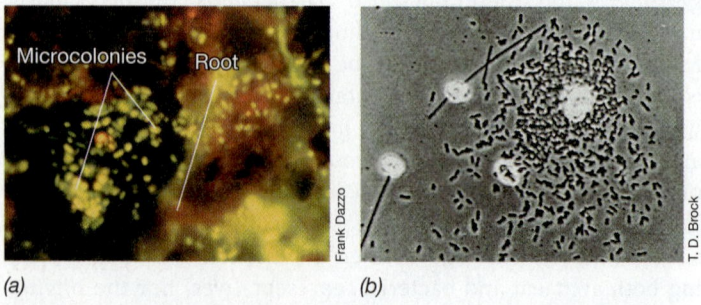

(a)

(b)

Figure 19.4 Microorganisms on surfaces. *(a)* Fluorescence photomicrograph of a natural microbial community living on plant roots in soil and stained with acridine orange. Note microcolony development. *(b)* Bacterial microcolonies developing on a microscope slide that was immersed in a river. The bright particles are mineral matter. The short, rod-shaped cells are about 3 μm long.

thus functional and growing microbial communities and not just cells trapped in a sticky matrix. We contrasted microbial growth in biofilms with that of planktonic (suspended) growth in Chapter 5 Explore the Microbial World, "Stick or Swim."

Wherever submerged surfaces are present in natural environments, biofilm growth is almost always more extensive and diverse than the planktonic growth in the liquid that surrounds the surface. Biofilms differ from planktonic communities in supporting critical transport and transfer processes, which generally control growth in biofilm environments. For example, if consumption of O_2 by populations near the surface exceeds diffusion of O_2 into deeper regions of the biofilm, the deeper regions will become anoxic, opening up new niches for colonization by obligate anaerobes or facultative aerobes. This is similar to the depletion of O_2 in the interior of a soil particle that was depicted in Figure 19.3.

One of the most clinically and industrially relevant properties of biofilm microbial communities is their inherent tolerance to antibiotics and other antimicrobial stressors. A given species growing in a biofilm can be up to 1000 times more tolerant of an antimicrobial substance than planktonic cells of the same species. Reasons for the greater tolerance include slower growth rates in biofilms, reduced penetration of antimicrobial substances through the extracellular matrix, and the expression of genes that increase tolerance to stress. Their tolerance of antimicrobial substances may explain why biofilms are responsible for many untreatable or difficult-to-treat chronic infections and are also hard to eradicate in industrial systems where surface growth (fouling) by microorganisms may impair important processes.

Biofilm Formation

How do biofilms form? Random collision of cells with a surface accounts for initial cell attachment, with adhesion promoted by interaction between one or more cellular structures and the surface. Cellular structures promoting attachment include protein appendages (pili, flagella) and cell surface proteins (for example, the large protein LapA exposed on the cell surface is required for attachment and biofilm formation by *Pseudomonas fluorescens*, as we will see later in this section).

Attachment of a cell to a surface is a signal for the expression of biofilm-specific genes. These include genes encoding proteins that synthesize intercellular signaling molecules and production of extracellular polysaccharides that initiate matrix formation (**Figure 19.6a**). Once committed to biofilm formation, a previously planktonic cell typically loses its flagella and becomes nonmotile. However, biofilms are not static entities and cells may also be released from the biofilm matrix through an active process of dispersal (Figure 19.6a).

Although the mechanism remains to be discovered, bacteria somehow "sense" a suitable surface and this coordinates events that lead to the biofilm growth mode. How surface sensing takes place is an area of active research, but the actual switch from planktonic to biofilm growth is triggered by the production of the chemical *cyclic di-guanosine monophosphate* (*c-di-GMP*) (**Figure 19.7**). We examined this second messenger as a regulator of biofilm formation in Section 7.9, showing that c-di-GMP signaling operates at multiple levels to modulate gene expression and enzyme activity, such as by binding to transcriptional regulators,

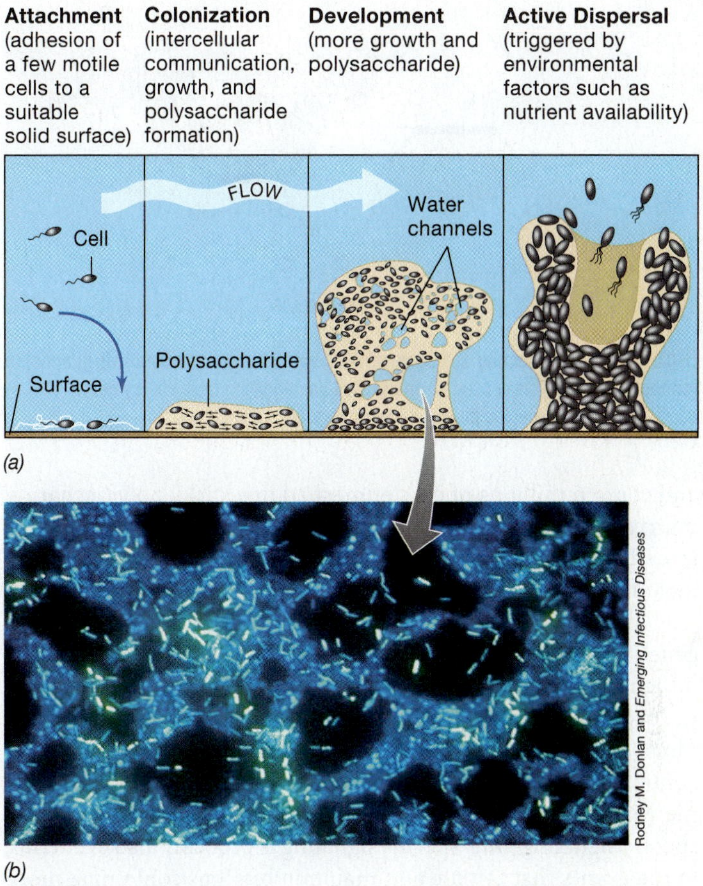

(a)

(b)

Rodney M. Donlan and *Emerging Infectious Diseases*

Figure 19.6 Biofilm formation. *(a)* Biofilms begin with the attachment of a few cells that then grow and communicate with other cells. The matrix is formed and becomes more extensive as the biofilm grows, eventually releasing cells. *(b)* Photomicrograph of a DAPI-stained biofilm that developed on a stainless steel pipe. Note the water channels.

mRNA (riboswitches, ⟳ Section 7.15), and specific proteins to alter enzyme activity. For example, c-di-GMP binds to proteins that reduce the activity of the flagellar motor, regulates cell surface proteins required for attachment, and mediates the biosynthesis of extracellular matrix polysaccharides of the biofilm.

Pseudomonas aeruginosa and Biofilms

Besides the intracellular activities triggered by c-di-GMP, intercellular communication is required for the development and maintenance of bacterial biofilms. For example, in *Pseudomonas aeruginosa*, a notorious biofilm former (**Figure 19.8**), the major intercellular signaling molecules are *acyl homoserine lactones*. As these molecules accumulate, they signal adjacent *P. aeruginosa* cells that the population of this species is enlarging (that is, they are a signal in quorum sensing, ⟳ Section 7.9). The signaling lactones then control expression of genes that contribute to biofilm formation. Genes turned on at this time include, in particular, those that increase the intracellular levels of c-di-GMP.

Elevated c-di-GMP levels initiate the production of extracellular polysaccharide and decrease flagellar function. Over time, in nutrient-rich conditions, *P. aeruginosa* cells can develop mushroom-shaped microcolonies that can be over 0.1 mm high

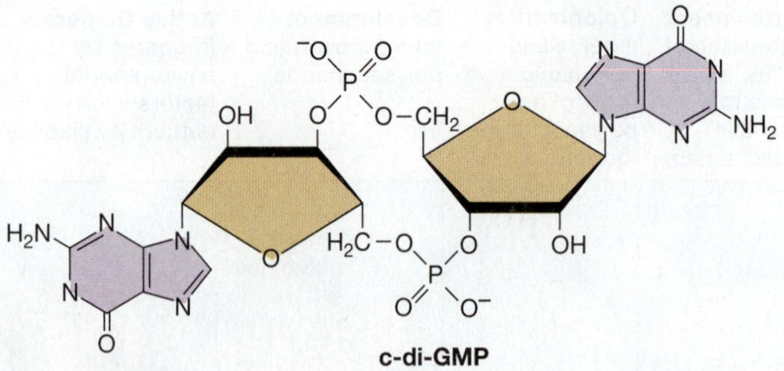

c-di-GMP

Figure 19.7 Molecular structure of the second messenger cyclic di-guanosine monophosphate. This is used as an intracellular signaling molecule by many bacteria to control specific physiological processes.

and contain millions of cells enmeshed in a sticky polysaccharide matrix (Figure 19.8). The final architecture of the biofilm is determined by multiple factors in addition to signaling molecules, including nutritional factors and local flow environment.

P. aeruginosa biofilms form in human lungs in patients with the genetic disease *cystic fibrosis*. In the biofilm state, *P. aeruginosa* is difficult to treat with antibiotics and the biofilm appears to help the bacteria persist in individuals with this disease. Like most biofilms, that which develops in the lungs of cystic fibrosis patients contains more than one bacterial species (🔁 Chapter 5 Explore the Microbial World, "Stick or Swim"). So, in addition to *intra*species signaling, *inter*species signaling is probably also occurring in the events that initiate and maintain biofilms containing more than one species.

In *P. fluorescens*, a related biofilm-forming organism, increases in c-di-GMP also promote biofilm formation. However, the biofilm machinery regulated by c-di-GMP in this species is very different from that of *P. aeruginosa*. In *P. fluorescens*, changes in c-di-GMP affect secretion and cell surface localization of a large adhesion protein called LapA that helps stick the cell to surfaces. For example, in response to low extracellular phosphate, *P. fluorescens* cells maintain a low c-di-GMP level that prevents localization of LapA to the outer membrane, thereby disabling the attachment mechanism required to initiate biofilm formation. If phosphate levels continue to fall within the biofilm, the associated reduction in c-di-GMP levels also results in the activation of a protease that cleaves LapA; this releases already attached cells and promotes their dispersal to explore nearby habitats for nutrients

(Figure 19.6). Other environmental cues for *P. fluorescens* dispersal include carbon or oxygen depletion and changes in temperature or iron availability.

Although quorum sensing–based signaling has been primarily associated with biofilm formation, the quorum-sensing system of *Staphylococcus aureus* plays a role in dispersal. Thus, no single developmental program for biofilm formation is shared among species. This lack of unity has greatly complicated the development of biofilm control strategies, as discussed at the end of this section.

Why Bacteria Form Biofilms

At least four reasons have been proposed for the formation of biofilms. First, biofilms are a means of microbial self-defense that increase survival. Biofilms resist physical forces that could otherwise remove cells only weakly attached to a surface. Biofilms also resist phagocytosis by protozoa and cells of the immune system, and retard the penetration of toxic molecules such as antibiotics. These advantages improve the chances for survival of cells in the biofilm. Second, biofilm formation allows cells to remain in a favorable niche. Biofilms attached to nutrient-rich surfaces, such as animal tissues, or to surfaces in flowing systems (Figure 19.5c) fix bacterial cells in locations where nutrients may be more abundant or are constantly being replenished. Third, biofilms form because they allow bacterial cells to live in close association with each other. As we have already seen for *P. aeruginosa* and the biofilm that forms in cystic fibrosis patients, this facilitates cell-to-cell communication and increases chances for survival. Moreover, when cells are in close proximity to one another, there are more opportunities for nutrient and genetic exchange. Finally, biofilms seem to be the typical way bacterial cells grow in nature. The biofilm may be the "default" mode of growth for prokaryotes in natural environments, which differ dramatically in nutrient levels from the rich liquid culture media used in the laboratory. Planktonic growth may be the norm only for those bacteria adapted to life at extremely low nutrient concentrations (discussed in Sections 19.9 and 19.11).

Biofilm Control

Biofilms have significant implications in human medicine and commerce. In the body, bacterial cells within a biofilm are protected from attack by the immune system, and antibiotics and other antimicrobial agents sometimes fail to penetrate the biofilm. Besides

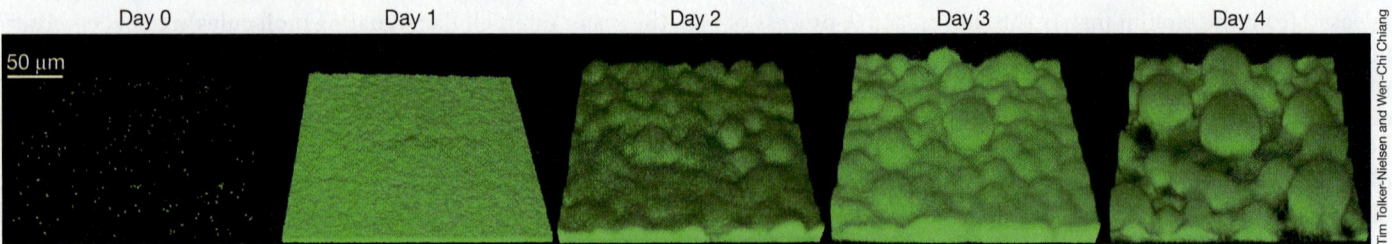

Figure 19.8 *Pseudomonas aeruginosa* biofilm development. Confocal scanning laser micrographs of a developing *Pseudomonas aeruginosa* biofilm in a flow-cell continuously irrigated with nutrient-rich medium. *P. aeruginosa* cells first attach to the glass surface (day 0), then rapidly grow and move on the surface to cover the entire surface (day 1); by day 4 mushroom-shaped microcolonies over 0.1 mm high have developed.

cystic fibrosis, biofilms have been implicated in several medical and dental conditions, including periodontal disease, chronic wounds, kidney stones, tuberculosis, Legionnaires' disease, and *Staphylococcus* infections. Medical implants are ideal surfaces for biofilm development. These include both short-term devices, such as a urinary catheter, as well as long-term implants, such as artificial joints. It is estimated that 10 million people a year in the United States experience biofilm infections from implants or intrusive medical procedures. Biofilms explain why routine oral hygiene is so important for maintaining dental health. Dental plaque is a typical biofilm and contains acid-producing bacteria responsible for dental caries (⮌ Section 23.3).

In industrial situations biofilms can slow the flow of water, oil, or other liquids through pipelines and can accelerate corrosion of the pipes themselves. Biofilms also initiate the degradation of submerged objects, such as structural components of offshore oil platforms, boats, and shoreline installations. The safety of drinking water may be compromised by biofilms that develop in water distribution pipes (Figure 19.6*b*), many of which in the United States are nearly 100 years old (⮌ Section 21.9). Water-pipe biofilms mostly contain harmless microorganisms, but if pathogens successfully colonize a biofilm, standard chlorination practices may fail to kill them. Periodic releases of pathogenic cells can then lead to outbreaks of disease. For example, there is concern that *Vibrio cholerae*, the causative agent of cholera (⮌ Section 31.3), may be propagated in this manner.

Biofilm control is big business, and thus far, only a limited number of tools exist to fight biofilms. Collectively, industries commit huge financial resources to treating pipes and other surfaces to keep them free of biofilms. New antimicrobial agents that can penetrate biofilms, as well as drugs that eliminate biofilm formation by interfering with intercellular communication, are being developed. A class of chemicals called *furanones*, for example, has shown promise as biofilm preventives on abiotic surfaces.

MINIQUIZ

- Why might a biofilm be a good habitat for bacterial cells living in a flowing system?

- Give an example of a medically relevant biofilm that forms in virtually all healthy humans.

- How do different intercellular and intracellular signaling molecules modulate biofilm formation and dispersal?

19.5 Microbial Mats

Microbial mats are among the most visibly conspicuous of microbial communities and can be considered extremely thick biofilms. Built by phototrophic or chemolithotrophic bacteria, these layered microbial communities can be several centimeters thick (**Figure 19.9**). The layers are composed of species of different microbial guilds whose activities are governed by light availability and other resources (Table 19.1). The combination of microbial metabolism and nutrient transport controlled by diffusion results in steep concentration gradients of different microbial nutrients and metabolites, creating unique niches at different depth intervals in the mats. The most abundant and versatile phototrophic mat builders are filamentous cyanobacteria, oxygenic phototrophs many of which grow under extreme environmental conditions. For example, some species of cyanobacteria grow in waters as hot as 73°C or as cold as 0°C, and others tolerate salinities in excess of 12% and pH values as high as 10.

Cyanobacterial Mats

Cyanobacterial mats are complete microbial ecosystems, containing large numbers of **primary producers** (cyanobacteria and other phototrophic bacteria) that use light energy to synthesize new organic material from CO_2. These along with populations of

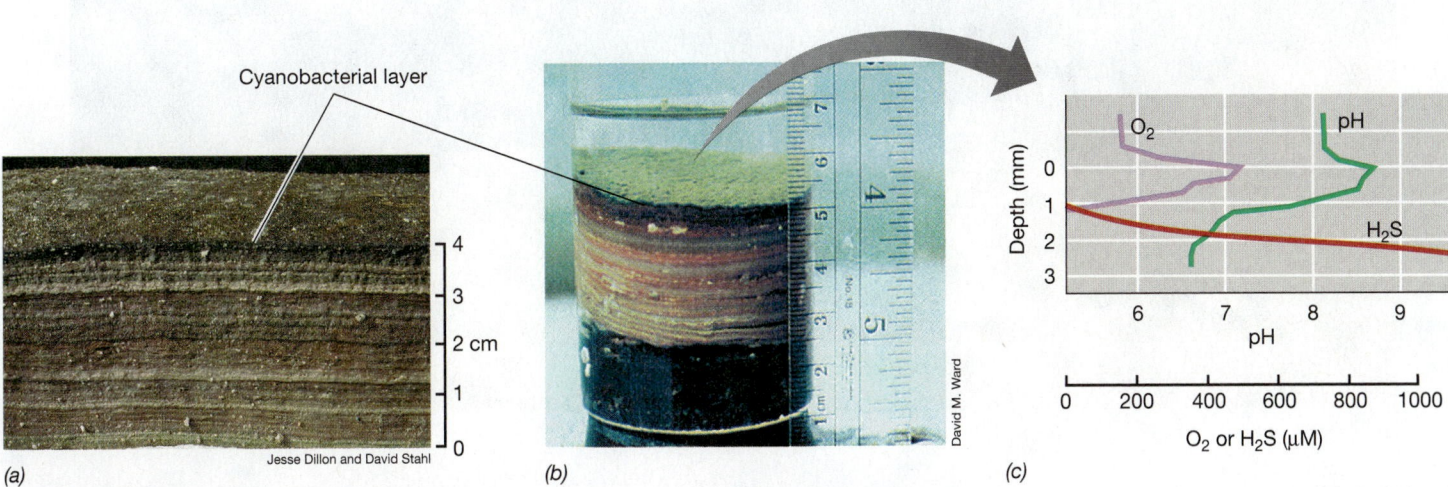

Figure 19.9 **Microbial mats.** *(a)* Mat specimen collected from the bottom of a hypersaline pond at Guerrero Negro, Baja California (Mexico). Most of the bottom of this shallow pond is covered with mats built by the major primary producer, the filamentous cyanobacterium *Microcoleus chthonoplastes*. *(b)* Microbial mat core from an alkaline Yellowstone National Park (USA) hot spring. The upper (green) layer contains mainly cyanobacteria, while the reddish layers contain anoxygenic phototrophic bacteria. *(c)* Oxygen (O_2), H_2S, and pH profiles through a hot spring mat core such as that shown in part b.

consumers in the mat community mediate all key nutrient cycles. Although microbial mat ecosystems have existed for over 3.5 billion years, the evolution of metazoan grazers and competition with macrophytes (aquatic plants) triggered their decline about a billion years ago.

Microbial mats develop today only in aquatic environments where specific environmental stresses restrict grazing and competition, conditions most commonly found in hypersaline or geothermal habitats. Well-studied microbial mats are found in hypersaline solar evaporation basins, either formed naturally, such as Solar Lake (Sinai, Egypt), or those constructed for the recovery of sea salt (Figure 19.9a). Because microbial mats are restricted to extreme environments, most are found in remote locations and many are not readily accessible to study. In contrast, however, the cyanobacterial mats that colonize the outflow channels of hot springs in Yellowstone National Park (USA) and many other thermal regions in the world are easily accessible to scientific research (Figure 19.9b, c).

The chemical and biological structure of a microbial mat can change dramatically during a 24-h period (called a *diel cycle*) as a consequence of changing light intensity. Using microsensors (⮌ Section 18.8) it is possible to measure pH, H_2S, and O_2 repeatedly over a diel cycle in zones in the mat separated vertically by only a few micrometers. During the day, there is intense oxygen production in the photic surface layer of microbial mats and active sulfate reduction throughout the lower regions. Near the zone where O_2 and H_2S begin to mix, intense metabolic activity by phototrophic and chemolithotrophic sulfur bacteria may consume these substrates rapidly over very short vertical distances. Detecting the rate of these changes reveals the zones of greatest microbial activity (Figure 19.9c). These gradients disappear at night when the entire mat turns anoxic and H_2S accumulates. Some mat organisms rely on motility to follow the shifting chemical gradients. For example, sulfur-oxidizing filamentous phototrophic bacteria such as *Chloroflexus* and *Roseiflexus* (⮌ Section 14.7) follow the up-and-down movement of the O_2–H_2S interface on a diel basis.

(a)

(b)

(c)

Andreas Teske and Markus Huettel

Figure 19.10 *Thioploca* mats. *(a, c)* Filaments of the large sulfur-oxidizing chemolithotroph *Thioploca* extend into the water above the sediment (87 m depth) in the Bay of Concepción off the Chilean coast. *(b)* From 10 to 20 filaments (trichomes) are bundled together by a gelatinous sheath, each bundle approximately 1.5 mm in diameter and 10–15 cm in length. Two species of *Thioploca* commonly inhabit the same bundle: *T. chileae*, about 20 μm in diameter, and *T. araucae*, about 40 μm in diameter. Individual trichomes glide independently within the sheaths and can extend up to 3 cm into the water. See page 631 for another example of filamentous sulfide-oxidizing bacteria and their unique metabolic strategy.

Chemolithotrophic Mats

The most common types of chemolithotrophic mats are composed of filamentous sulfur-oxidizing bacteria, such as *Beggiatoa* and *Thioploca* species, which grow on marine sediment surfaces at the interface between O_2 supplied from the overlying water and H_2S produced by sulfate-reducing bacteria living in the sediment. In these habitats the bacteria oxidize H_2S to support energy conservation and autotrophic reactions (⮌ Sections 13.8 and 14.11).

Chemolithotrophic mats composed of sulfur-oxidizing *Thioploca* species on sediments of the Chilean and Peruvian continental shelf are thought to be the most extensive microbial mats on Earth (**Figure 19.10**). *Thioploca* has developed a remarkable strategy to bridge spatially separated resources. These chemolithotrophic mat organisms contain large internal vacuoles that store high concentrations of nitrate (NO_3^-) as an electron acceptor to support anaerobic respiration. Much like a scuba diver filling tanks with oxygen to dive into the water, cells of *Thioploca* migrate up to the sediment surface to charge internal vacuoles

with NO_3^- from the water column (Figure 19.10*a, b*). They then return ("dive") to the anoxic depths of the sediment (gliding at speeds of 3–5 mm per hour) to use their stored NO_3^- as an electron acceptor for H_2S oxidation.

The physical and biological structures of both biofilms and microbial mats are determined by metabolic interactions among microorganisms and the diffusion of nutrients. Thus, as biofilms form on a surface they become increasingly more complex, and in so doing generate new niches for organisms of differing physiologies. This diversity reaches its maximum in mature microbial mats (Figure 19.9), as these structures have been shown to be among the most complex microbial communities characterized thus far by molecular community sampling (⮌ Section 18.5).

MINIQUIZ
- What is a microbial mat?
- How would motile aerobic bacteria in a microbial mat respond to changing O_2 concentrations over a diel cycle?

III • Terrestrial Environments

Extensive microbial habitats on Earth are in two terrestrial environments that are similar in lacking sunlight, being periodically or permanently anoxic, and having other physicochemical conditions in common. The two terrestrial environments are soils and water enclosed in soils and bedrock. In each of the next two sections we cover these microbial habitats, and in each case we begin with the abiotic part of the environment and conclude with a discussion of the microbial communities that live there.

19.6 Soils

The word *soil* refers to the loose outer material of Earth's surface, a layer distinct from the bedrock that lies underneath (**Figure 19.11**). Soil develops over long periods of time through complex interactions among the parent geological materials (rock, sand, glacial drift materials, and so on), the topography, climate, and the presence and activities of living organisms.

Soils can be divided into two broad groups: *mineral soils* are derived from the weathering of rock and other inorganic materials, and *organic soils* are derived from sedimentation in bogs and marshes. Most soils are a mixture of these two basic types. Although mineral soils, which are the primary focus of this section, predominate in most terrestrial environments, there is increasing interest in the role that organic soils play in carbon storage. A detailed understanding of carbon storage (sinks) and sources (such as release of CO_2) is of great relevance to the science of climate change. The carbon cycle is considered in Chapter 20.

Soil Composition and Formation

Vegetated soils have at least four components. These include (1) inorganic mineral matter, typically 40% or so of the soil volume; (2) organic matter, usually about 5%; (3) air and water, roughly

50%; and (4) microorganisms and macroorganisms, about 5%. Particles of various sizes are present in soil. Soil scientists classify soil particles on the basis of size: Those in the range of 0.1–2 mm in diameter are called *sand*, those between 0.002 and 0.1 mm *silt*, and those less than 0.002 mm *clay*. Different textural classes of soil are then given names such as "sandy clay" or "silty clay" based on the percentages of sand, silt, and clay they contain. A soil in which no one particle size dominates is called a *loam*.

Physical, chemical, and biological processes all contribute to the formation of soil. An examination of almost any exposed rock reveals the presence of algae, lichens, or mosses. These organisms are phototrophic and produce organic matter, which supports the growth of chemoorganotrophic bacteria and fungi. More complex chemoorganotrophic communities composed of *Bacteria*, *Archaea*, and eukaryotes then develop as the extent of the earlier colonizing organisms increases. Carbon dioxide produced during respiration becomes dissolved in water to form carbonic acid (H_2CO_3), which slowly dissolves the rock, especially rocks containing limestone ($CaCO_3$). In addition, many chemoorganotrophs excrete organic acids, which also promote the dissolution of rock into smaller particles.

Freezing, thawing, and other physical processes assist in soil formation by forming cracks in the rocks. As the particles generated combine with organic matter, a crude soil forms in these crevices, providing sites needed for pioneering plants to become established. The plant roots penetrate farther into the crevices, further fragmenting the rock; the excretions of the roots promote development in the **rhizosphere** (the soil that surrounds plant roots and receives plant secretions) of high microbial cell abundance (Figure 19.4*a*). When the plants die, their remains are added to the soil and become nutrients for more extensive microbial development. Minerals are rendered soluble, and as water percolates, it carries some of these substances deeper into the soil.

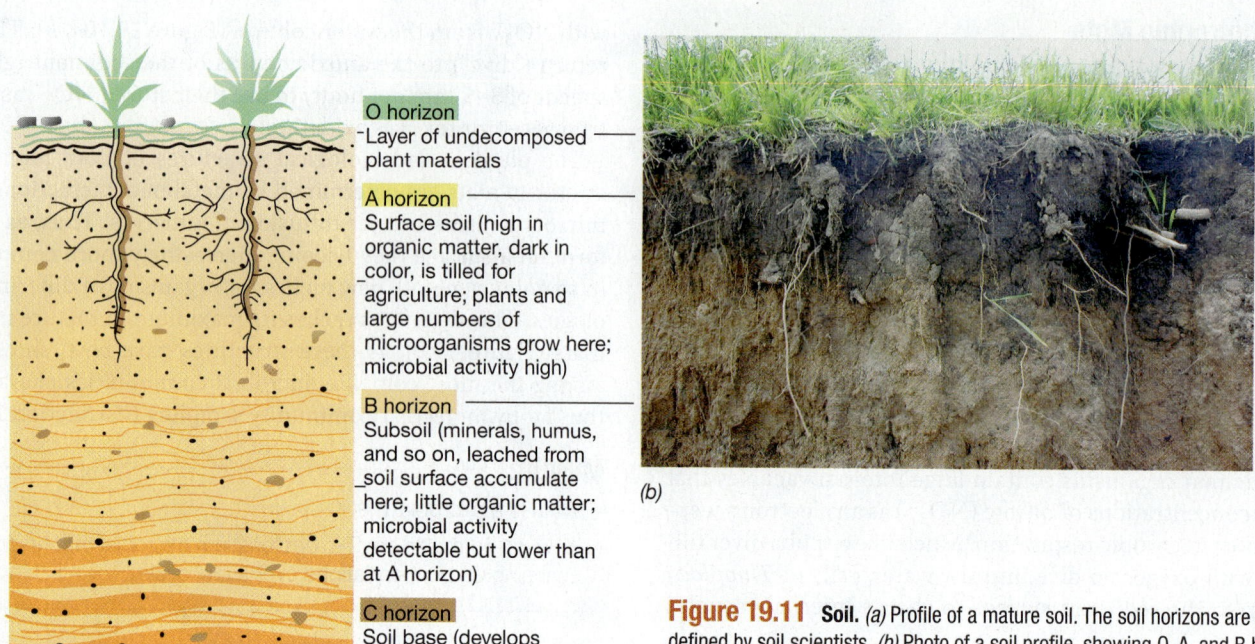

O horizon
Layer of undecomposed plant materials

A horizon
Surface soil (high in organic matter, dark in color, is tilled for agriculture; plants and large numbers of microorganisms grow here; microbial activity high)

B horizon
Subsoil (minerals, humus, and so on, leached from soil surface accumulate here; little organic matter; microbial activity detectable but lower than at A horizon)

C horizon
Soil base (develops directly from underlying bedrock; microbial activity generally very low)

(a)

(b)

Figure 19.11 Soil. *(a)* Profile of a mature soil. The soil horizons are zones defined by soil scientists. *(b)* Photo of a soil profile, showing O, A, and B horizons. This soil from Carbondale, Illinois (USA), is rich in clay and very compact. Such soils are not as well drained as those rich in sand. Note the clear color delineation between the organic-rich A horizon and the less-organic-rich B horizon.

As weathering proceeds, the soil increases in depth and becomes able to support the development of larger plants and small trees. Soil animals such as earthworms colonize the soil and play an important role in keeping the upper layers of the soil mixed and aerated. Eventually, the movement of materials downward results in the formation of soil layers, called a *soil profile* (Figure 19.11). The rate of development of a typical soil profile depends on climatic and other factors, but it can take hundreds to thousands of years.

Water Availability: Vegetated and Dryland Soils as Microbial Habitats

The limiting nutrients in soils are often inorganic nutrients such as phosphorus and nitrogen, key components of several classes of macromolecules. Another major factor affecting microbial activity in soil is the availability of water, and we have previously emphasized the importance of water for microbial growth (⟲ Section 5.15).

Water is a highly variable component of soil, and a soil's water content depends on soil composition, rainfall, drainage, and plant cover. Water is held in the soil in two ways—by adsorption onto surfaces or as free water in thin sheets or films between soil particles (**Figure 19.12**). The water present in soils has materials dissolved in it, and the mixture is called the *soil solution*. In well-drained soils, air penetrates readily, and the oxygen concentration of the soil solution can be high, similar to that of the soil surface. In waterlogged soils, however, the only oxygen present is that dissolved in water, and this can be rapidly consumed by the resident microflora. Such soils then become anoxic, and, as described for freshwater environments (Section 19.8), show profound changes in their biological activities.

There is also water in the larger channels in soil, where bulk flow is important for rapid transport of microorganisms and their substrates and products.

Arid Soils

The greatest microbial activity in soils is in the organic-rich surface layers in and around the rhizosphere (Figure 19.4a). However, some soils are so dry that plant coverage is greatly limited and only special microbial communities can thrive.

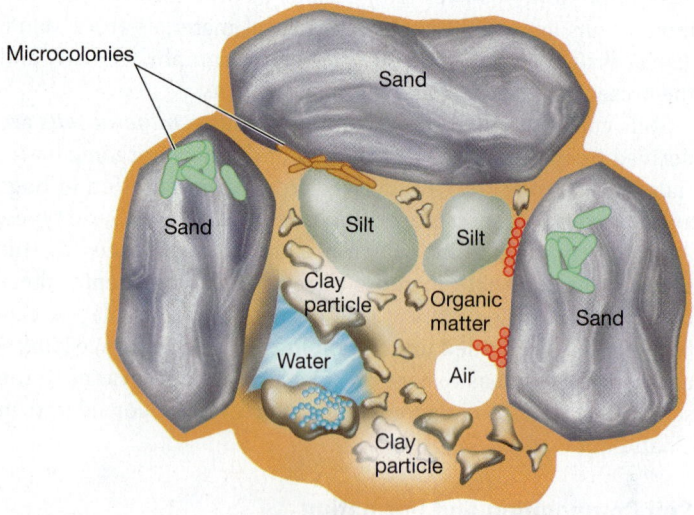

Microcolonies

Sand

Sand

Silt

Silt

Clay particle

Organic matter

Sand

Water

Air

Clay particle

Figure 19.12 A soil microbial habitat. Very few microorganisms are free in the soil solution; most of them reside in microcolonies attached to the soil particles. Note the relative size differences among sand, clay, and silt particles.

These are *arid soils*, and approximately 35% of Earth's landmass is permanently or seasonally arid. Aridity can be defined by the *aridity index*, expressed as the ratio of precipitation to potential evapotranspiration (P/PET). Evapotranspiration is the sum of water loss through evaporation and plant transpiration. A region is classified as arid when there is a P/PET of less than 1; that is, water entering through precipitation (and fog and dew) is less than that lost through evapotranspiration.

Arid soils are among the most extreme environments on Earth, with temperature highs in excess of 60°C and lows of −24°C, high insolation (exposure to solar rays), and low water activity. Although arid regions are typically nearly devoid of leafy plants, they sustain important microbial communities that assemble in and stabilize the near-soil surface and reside within and on the surfaces of rocks. The dominant microorganisms present in these carbon-limited environments are cyanobacteria, with lesser numbers of green algae, fungi, heterotrophic bacteria, lichens, and mosses.

Dryland microbial habitats include *biological soil crusts* (BSCs) (Figure 19.13), ventral surfaces of translucent stones (*hypolithic* colonists), exposed rock surfaces (*epilithic* colonists), and the interior pore spaces, cracks, and fissures of rocks (*endolithic* colonists). The soil crusts are dominated by cyanobacterial *Microcoleus* species (Figure 19.13*b, c*), whereas coccoid *Chroococcidiopsis* species are the predominant endolithic population. The rock colonists play an important role in weathering and soil formation as described above; here we primarily consider the BSC communities.

The BSC serves a critical function in soil stabilization of desert ecosystems. Stabilization is critical because of the very slow rate of desert soil formation (<1 cm per 1000 years). Here, the filamentous cyanobacteria (*Microcoleus*) and fungi provide soil cohesion, which is further stabilized aboveground by lichens and mosses when present. Importantly, this microbial network functions to eliminate soil erosion from wind and water. The BSCs are major determinants of water infiltration and influence local hydrological cycles and water availability to vegetation. Remarkably, when moisture and temperature conditions are optimal, the photosynthetic rates of BSC are comparable to those of vascular plant leaves. Cyanobacteria and other nitrogen-fixing bacteria (⮂ Sections 3.17, 7.13, and 14.3) provide nitrogen, and much of the fixed nitrogen is released immediately and made available to other soil biota.

The disruption of BSCs is a major contributor to *desertification*, a process exacerbated by climate change and human activities. Dust storms resulting from BSC destruction reduce soil fertility, and when heavy dust is deposited on nearby snowfields it accelerates melt and evapotranspiration rates, thereby reducing freshwater inputs to rivers. Once compromised, soil crusts have recovery times varying from 15 to 50 years. Given the expansive terrestrial presence of BSCs, their importance to human and ecosystem function, and the projected increase in aridity associated with climate change, a better understanding of BSC formation and the rehabilitation of compromised BSCs is important for a healthy planet Earth.

A Phylogenetic Snapshot of Soil Prokaryotic Diversity

As we saw in Figure 19.3, even a single soil particle can contain many different microenvironments and can thus support the growth of several physiological types of microorganisms. To examine soil particles directly for microorganisms, fluorescence microscopes are often used, the organisms in the soil having been

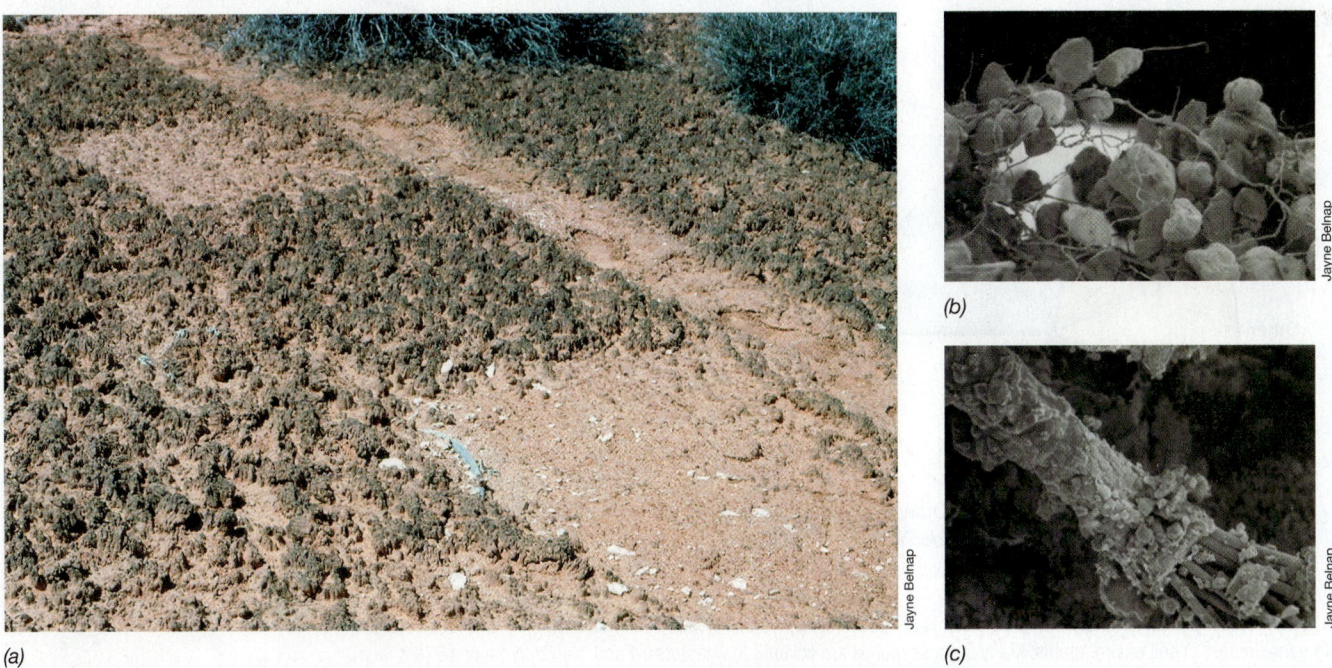

(a)

(b)

(c)

Figure 19.13   Biological soil crust (BSC). *(a)* BSC on the Colorado Plateau shown adjacent to lighter disturbed soils.
(b, c) Scanning electron micrographs of filamentous cyanobacteria (*Microcoleus* species) that bind sand grains together with their sheath material. The sand grains in part b are about 100 μm in diameter and the filaments in part c about 5 μm in diameter.

previously stained with a fluorescent dye. To visualize a specific microorganism in a soil particle, fluorescent antibody staining or gene probes (⥲ Sections 18.3, 18.4) can also be used. Microorganisms can also be observed on soil surfaces directly by scanning electron microscopy (Figure 19.13*b, c*).

We learned in Chapter 18 that sequence analyses of 16S ribosomal RNA (rRNA) genes obtained from the environment can be used as a measure of prokaryotic diversity (⥲ Section 18.5). As yet, no natural communities have been so thoroughly characterized by these techniques that all resident species have been

identified. However, within limits, the method is widely considered to be a valid measure of microbial diversity and avoids the more serious problems of enrichment bias that plague culture-dependent diversity studies (⥲ Section 18.1). Here and in later sections of this chapter we present a "phylogenetic snapshot" of major microbial habitats, with the goal of emphasizing trends and patterns rather than absolute details.

Molecular community sampling of a typical vegetated surface has shown typically *thousands* of different species of *Bacteria* and *Archaea* in a single gram of soil, likely reflecting the numerous

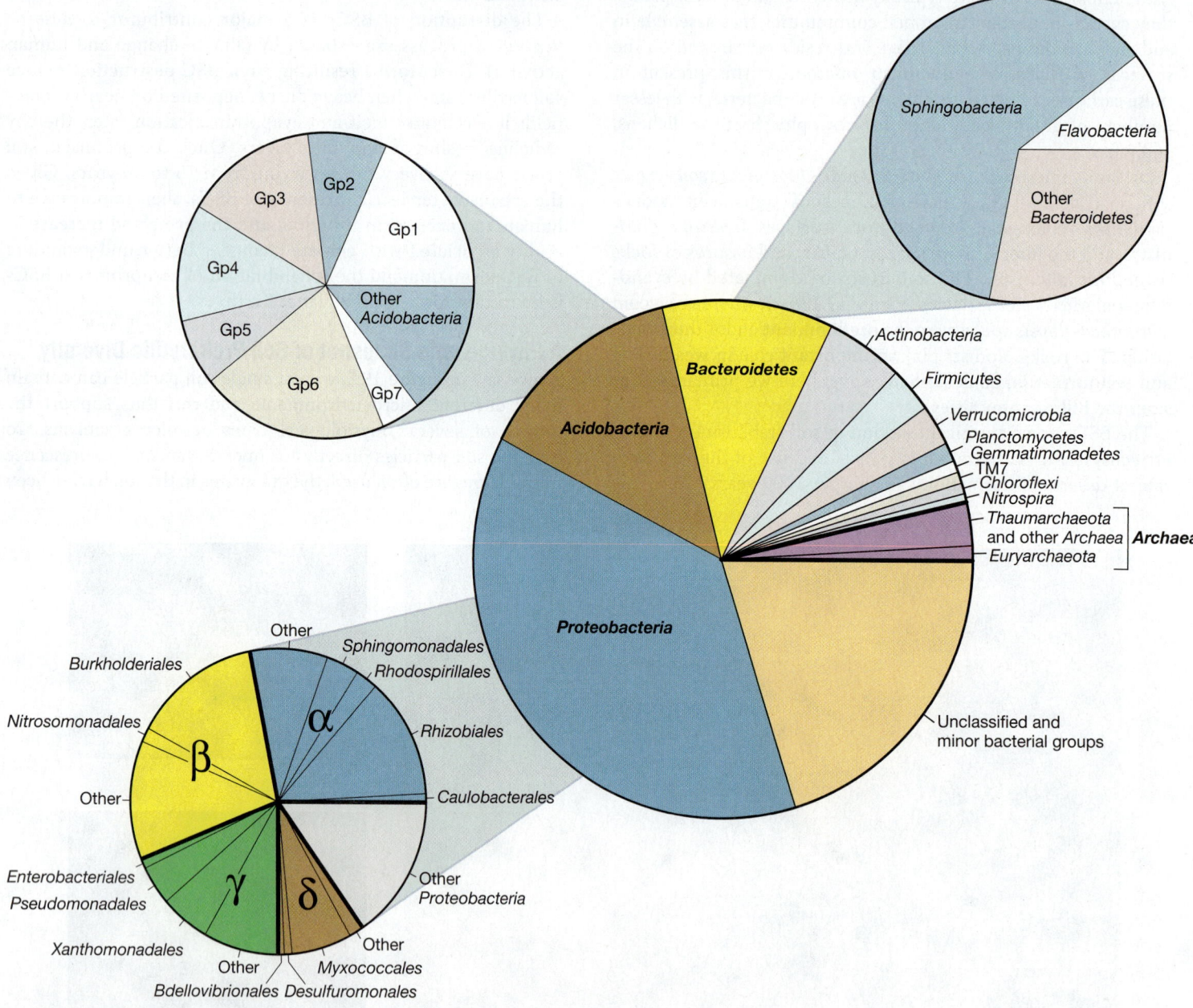

Figure 19.14   Soil prokaryotic diversity. The results are pooled analyses of 287,933 sequences from several studies of the 16S rRNA gene content of soil environments. Many of these groups are covered in Chapters 14 and 15 (*Bacteria*) or 16 (*Archaea*). For *Proteobacteria*, *Acidobacteria*, and *Bacteroidetes*, major subgroups are indicated (Gp, group). Note high species richness as indicated by the large proportion of the total community composed of unclassified and minor bacterial groups. Also note the relatively low proportion of the total prokaryotic soil community that consists of *Archaea* and that many soil *Archaea* are not close relatives of known *Euryarchaeota* or *Crenarchaeota*. Data assembled and analyzed by Nicolas Pinel.

microenvironments present there. A "species" is operationally defined here as a 16S rRNA gene sequence obtained from a microbial community that differs from all other sequences by more than 3% (⮌ Section 12.8). Such an environmental sequence is called a *phylotype*. Besides very large species numbers, soil microbial diversity studies have also showed that diversity varies with soil type and geographical location. For example, analysis of an Alaska forest soil, an Oklahoma prairie soil, and a Minnesota farm soil (all sites in the USA) revealed approximately 5000, 3700, and 2000 different phylotypes, respectively. The Alaska and Minnesota soils showed similar distributions at the phylum level of taxonomy (for example, *Proteobacteria*, *Acidobacteria*, *Bacteroidetes*, *Actinobacteria*, *Verrucomicrobia*, and *Planctomycetes*) but shared only about 20% of their species in common. This indicates that although the *proportions* of the dominant phyla in different soils are relatively constant, the *actual species present* within a phylum may vary considerably between different soils. In addition, lower bacterial diversity was observed in the farm soil than the Alaska soil, probably because modern intensive agricultural practices rely heavily on fertilization, low plant diversity, and the chemical suppression of unwanted plants and animals.

Figure 19.14 shows the general composition of prokaryotic soil communities based on pooled 16S rRNA sequence data taken from several soils. As can be seen, *Proteobacteria* (Chapters 14 and 15) make up nearly half of the total phylotypes recovered, with all major subgroups except for *Epsilonproteobacteria* well represented. *Acidobacteria* and *Bacteroidetes* are also abundant groups; *Actinobacteria* and *Firmicutes* are less so. In addition to these, a major proportion of soil phylotypes are unclassified species or members of minor bacterial groups. This underscores the high bacterial diversity typical of soil ecosystems. In contrast to *Bacteria*, the diversity of *Archaea* in soil is minimal, with relatively few sequences within each major phylum of *Archaea* (*Euryarchaeota*, *Thaumarchaeota*, and *Crenarchaeota*) represented. However, since there have been fewer selective surveys of archaeal diversity in soils, their diversity may be greater than now recognized.

A similar study to that shown in Figure 19.14 but performed on hydrocarbon-polluted soil showed that the general taxonomic makeup of polluted and unpolluted soils is similar: *Proteobacteria* comprise the largest fraction in both soil types, followed by significant representation of *Acidobacteria*, *Bacteroidetes*, *Actinobacteria*, and *Firmicutes*. However, there was a significant shift in fractional representation of these taxa in the two soils. Polluted soils are enriched in *Actinobacteria* and *Euryarchaeota* but diminished in *Bacteroidetes*, *Acidobacteria*, and unclassified *Bacteria* relative to nonpolluted soils. Notably, *Crenarchaeota* are absent from all surveys of hydrocarbon-polluted soils, suggesting that hydrocarbon pollutants suppress the *Crenarchaeota* and the ammonia-oxidizing *Thaumarchaeota* (*Archaea*, ⮌ Sections 16.6, 16.9). The impact of hydrocarbon pollution on *Bacteria* was reflected by a greater proportion of *Gammaproteobacteria* and a single dominant *Bacteroidetes* phylotype. By contrast, unpolluted soils contained several phylotypes of *Bacteroidetes* (Figure 19.14). The diversity of *Acidobacteria*, a major group in unpolluted soils, is also significantly reduced in polluted soils (Figure 19.14).

Although the *functional* significance of the observed diversity of microbial communities in polluted versus unpolluted soils is unknown, the shifts observed signal that the two soils will likely differ in their capacity to process carbon and nitrogen and to carry out other important nutrient cycling events. However, despite this lack of a functional connection, different 16S rRNA gene surveys of soils agree on two things: (1) undisturbed, unpolluted soils support very high prokaryotic diversity, and (2) perturbations in a soil trigger measurable shifts in community composition—presumably toward species that are more competitive in the disturbed soil environment—and an overall reduction in prokaryotic diversity.

MINIQUIZ --

- Which phylum of *Bacteria* dominates bacterial diversity in vegetated soil?
- What factors govern the extent and type of microbial activity in soils?
- Which region of soil is the most microbially active?

19.7 The Subsurface

In the soils and rocks of Earth's subsurface, there is water. This underground water, called *groundwater*, is a vast but little-explored microbial habitat (see MicrobiologyNow, page 379). As recently as three decades ago most microbiologists were of the opinion that significant microbial numbers were limited to the top 100 m or so of Earth's crust. However, from research made possible by the development of improved drilling and aseptic sampling technology, it is now known that microbial life extends down at least 3 *kilometers* into the Earth in regions containing trapped water. The microbiology of relatively shallow groundwater is quite similar to the microbiology of soils. However, microorganisms in deep subsurface waters exist at temperatures that can exceed 50°C and in anoxic and nutrient-depleted surroundings. What do we know about these organisms?

Microbiology of the Deep Subsurface

Subsurface microbiology initially focused on relatively shallow and easily accessible aquifer systems, revealing diverse populations of *Archaea* and *Bacteria* and a limited presence of protozoa and fungi. An *aquifer* is an underground layer of water-bearing permeable material, such as fractured rock or gravel. Microorganisms in aquifers are metabolically active and greatly influence the chemistry of groundwater. For example, the presence of ferrous iron (Fe^{2+}) in groundwater is largely attributable to the activity of microorganisms such as *Geobacter* that reduce ferric iron (Fe^{3+}) as an electron acceptor (⮌ Section 13.21).

The period of time a water mass remains within a region of the subsurface varies from weeks to millions of years, depending on its proximity to the surface and the rate of recharge (movement of surface water into groundwater). The long-term isolation of microorganisms in deep subsurface groundwater that is not recharged has been suggested as a mechanism for *allopatric* speciation (the emergence of new microbial species as a consequence of geographic isolation). However, the microbial diversity discovered in the subsurface thus far using culture-independent

(a)

(b)

Esta van Heerden

Terry Hazen

Figure 19.15 **Sampling the deep subsurface.** *(a)* Sampling hot (55°C) fissure water from a depth of 3000 m in the Tau Tona South African gold mine. *(b)* Drilling to 600 m in Allendale, SC (USA), for the U.S. Department of Energy (DOE) Deep Subsurface Microbiology Program.

techniques (Chapter 18) has for the most part been unremarkable; the organisms closely resemble surface or near-surface species.

Research on the deep microbial biosphere has been facilitated by mining and drilling operations that expose water in fractured rock at great depths. For example, samples collected from a nearly 3-km-deep gold-mining operation in South Africa (**Figure 19.15**) revealed chemolithotrophic and autotrophic *Bacteria* and *Archaea*. DNA extracted from fissure water showed that an H_2-oxidizing, sulfate-reducing bacterium was virtually the only organism present. Genome analysis of the organism, as yet uncultured but given the provisional name *Desulforudis audaxviator*, indicated that it should be thermophilic and should be capable of autotrophic growth using H_2 as the electron donor for respiration and CO_2 fixation. In addition, the organism contained genes encoding nitrogen fixation proteins (↩ Section 3.17), meaning that it could live on a diet of a few minerals, CO_2, SO_4^{2-}, N_2, and H_2. Such an organism would be well suited to long-term isolation in the deep subsurface and could be a model for the types of physiologies one would expect in such a nutrient-deficient environment.

Possible sources of H_2 for chemolithotrophs in the deep subsurface include the radiolysis of water by uranium, thorium, and other radioactive elements, and geochemical processes such as the release of H_2 from the oxidation of iron silicate minerals in aquifers. As an electron donor, H_2 can satisfy the needs of bacteria that carry out many different anaerobic respirations, including sulfate reduction, methanogenesis, acetogenesis, and ferric iron reduction (Chapter 13), and examples of all these physiologies have been identified in various subsurface microbiology research

projects. Thus the current consensus is that these types of chemolithotrophs likely dominate the deep subsurface.

Growth Rates and the Future of Subsurface Microbiology

Bacterial numbers in uncontaminated groundwater vary by several orders of magnitude (10^2–10^8 per ml), reflecting primarily nutrient availability, mostly in the form of dissolved organic carbon. Measured and estimated generation times for deep subsurface bacteria vary by many orders of magnitude, from days to centuries, as determined by the physicochemical environment, the physiology of the resident populations, and nutrient availability. However, relevant data are scarce and our knowledge about the microbial ecology of the subsurface will be greatly advanced by emerging technologies such as single-cell genomics for characterizing individual cells in their natural environment (↩ Chapter 6 Explore the Microbial World, "Genomics, One Cell at a Time," and Section 18.11). For example, microorganisms appear to be attached to surfaces or within biofilms in the nutrient-depleted subsurface, but it is unknown whether these are genetically or physiologically distinct from microorganisms in planktonic populations.

These many unanswered questions in subsurface microbiology have supported the establishment of permanent science laboratories at great depths in the Earth. For example, the Sanford Underground Research Facility in Lead, South Dakota (USA), is being constructed at a depth of 2400 m in the former Homestake gold mine with government and private support for research in physics, geology, and microbiology. The Integrated Ocean Drilling Program, an international effort, has probed for microbial populations at great depths below the seafloor. Results thus far have shown *Archaea* and *Bacteria* as far down as 1600 m below the seafloor in rocks more than 100 million years old. Although this may sound ancient, such ages are relatively young compared with viable bacteria that have been recovered from salt crystals nearly a half billion years old. Obviously, prokaryotic cells can remain viable for enormously long periods of time.

MINIQUIZ
- Why could allopatric speciation be possible in the deep subsurface?
- What environmental factors determine the abundance and type of cells in the deep subsurface?

IV • Aquatic Environments

Freshwater and marine environments differ in many ways including salinity, average temperature, depth, and nutrient content, but both provide many excellent habitats for microorganisms. In this unit we focus first on freshwater microbial habitats. We then consider two marine environments: (1) coastal and ocean waters, and (2) the deep sea. Much new information is emerging about marine microorganisms from studies using the molecular tools of microbial ecology, especially genetic stains, microbial community sampling, and metagenomics (Chapter 18).

19.8 Freshwaters

Freshwater environments are highly variable in the resources and conditions (Table 19.1) available for microbial growth. Both oxygen-producing and oxygen-consuming organisms are present in aquatic environments, and the balance between photosynthesis and respiration (Figure 19.2) controls the natural cycles of oxygen, carbon, and other nutrients (nitrogen, phosphorus, metals).

Among microorganisms, oxygenic phototrophs include the algae and cyanobacteria. These can either be *planktonic* (floating) and distributed throughout the water columns of lakes, sometimes accumulating in large numbers at a particular depth, or *benthic*, meaning they are attached to the bottom or sides of a lake or stream. Oxygenic phototrophs, which obtain their energy from light and use water as an electron donor to reduce CO_2 to organic matter (Chapter 13), are the main primary producers in freshwater aquatic ecosystems.

The activity and diversity of chemoorganotrophic aquatic microbial communities depend to a major extent on primary production, in particular its rates and temporal and spatial distributions. Oxygenic phototrophs produce new organic material as well as O_2. If primary production rates are very high, the resultant excessive organic matter production can lead to bottom-water O_2 depletion from respiration and the development of anoxic conditions. This in turn stimulates anaerobic metabolisms, including anaerobic respirations and fermentations (Chapter 13). Like oxygenic phototrophs, anoxygenic phototrophs can also fix CO_2 into organic material. But these organisms use reduced substances other than water, such as H_2S or H_2, as electron donors in photosynthesis (⮁ Section 13.3). Organic matter produced by anoxygenic phototrophs can also support and enhance respiration, accelerating the spread of anoxia.

Oxygen Relationships in Freshwater Environments

The biological and nutrient structure of lakes is greatly influenced by seasonal changes in physical gradients of temperature and salinity. In many lakes in temperate climates the water column becomes stratified, separated into layers of differing physical and chemical characteristics that constitute a **stratified water column**. During the summer, warmer and less dense surface layers, called the **epilimnion**, are separated from the colder and denser bottom layers (the **hypolimnion**). The *thermocline* is the transition zone from epilimnion to hypolimnion (**Figure 19.16**).

In the late fall and early winter, lake surface waters become colder and thus more dense than the bottom layers. This, combined with wind-driven mixing, causes the cooled surface water to sink and the lake to "turn over," mixing surface and bottom waters and their nutrients. The separation of a relatively well-mixed surface layer from a relatively static bottom layer limits the transfer of nutrients between layers until fall turnover once again mixes the water column.

During periods of stratification, transfer between surface and bottom waters is controlled not by mixing but by the much slower process of diffusion. As a result, bottom waters can experience seasonal periods of either low or no dissolved O_2. Although O_2 is one of the most plentiful gases in the atmosphere (21% of air), it has relatively limited solubility in water, and in a large body of water its exchange with the atmosphere is slow. Whether a body of water actually becomes O_2-depleted depends on several factors, including the amount of organic matter present and the

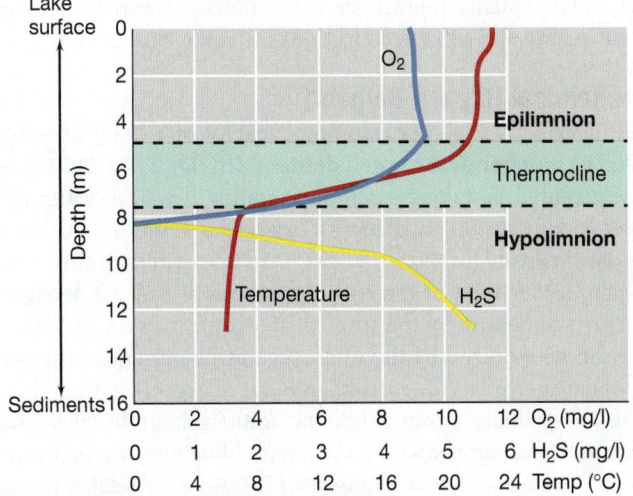

Figure 19.16 Development of anoxic conditions in a temperate lake due to summer stratification. The colder bottom waters are more dense and contain H_2S from bacterial sulfate reduction. The thermocline is the zone of rapid temperature change. As surface waters cool in the fall and early winter, they reach the temperature and density of hypolimnetic waters and sink, displacing bottom waters and effecting lake turnover. Data from a small freshwater lake in northern Wisconsin (USA).

degree of mixing of the water column. Organic matter that is not consumed in surface layers sinks to the depths and is decomposed by anaerobes (Figure 19.2). Lakes may contain high levels of dissolved organic matter because inorganic nutrients that run off the surrounding land can trigger algal and cyanobacterial blooms; these organisms typically excrete various organic compounds and also release complex organic compounds when they die and decay. The combination of water body stratification during early summer, high organic loading, and limited O_2 transfer results in O_2 depletion of the bottom waters (Figure 19.16), making them unsuitable for aerobic organisms such as plants and animals.

The annual turnover cycle allows the bottom waters of a lake to pass from oxic to anoxic and back to oxic. Microbial activity and community composition is altered with these changes in oxygen content, but other factors that accompany fall turnover of the water column, especially changes in temperature and nutrient levels, govern microbial diversity and activity as well. If organic matter is sparse, as it is in pristine lakes or in the open ocean, there may be insufficient substrate available for chemoorganotrophs to consume all the oxygen. The microorganisms that dominate such environments are typically **oligotrophs**, organisms adapted to growth under very dilute conditions (Section 19.11). Alternatively, where currents are strong or there is turbulence because of wind mixing, the water column may be well mixed, and consequently oxygen may be transferred to the deeper layers.

Oxygen levels in rivers and streams are also of interest, especially those that receive inputs of organic matter from urban, agricultural, or industrial pollution. Even in a river well mixed by rapid water flow and turbulence, large organic inputs can lead to a marked oxygen deficit from bacterial respiration (**Figure 19.17a**). As the water moves away from a point source input, for example, from an input of sewage, organic matter is gradually consumed, and the oxygen content returns to previous levels. As in lakes, nutrient inputs to rivers and streams from sewage or other pollutants can trigger massive blooms of cyanobacteria and algae (Figure 19.1) and aquatic plants (Figure 19.17b), thereby diminishing overall water quality and growth conditions for aquatic animals.

Biochemical Oxygen Demand

The microbial oxygen-consuming capacity of a body of water is called its **biochemical oxygen demand (BOD)**. The BOD of water is determined by taking a sample, aerating it well to saturate the water with dissolved O_2, placing it in a sealed bottle, incubating it in the dark (usually for 5 days at 20°C), and determining the residual oxygen in the water at the end of incubation. A BOD determination gives a measure of the amount of organic material in the water that can be oxidized by the microorganisms present in the water. As a lake or river recovers from an input of organic matter or from excessive primary production, the initially high BOD becomes lower and is accompanied by a corresponding increase in dissolved oxygen in the ecosystem (Figure 19.17a). Another related measure of the organic material in a body of water is the *chemical oxygen demand* (*COD*). This determination uses a strong oxidizing agent, such as acidic potassium dichromate, to oxidize the organic matter to CO_2; the amount of organic matter present is proportional to the amount of dichromate consumed. COD is often used as a rapid measure of water quality and of its potential BOD.

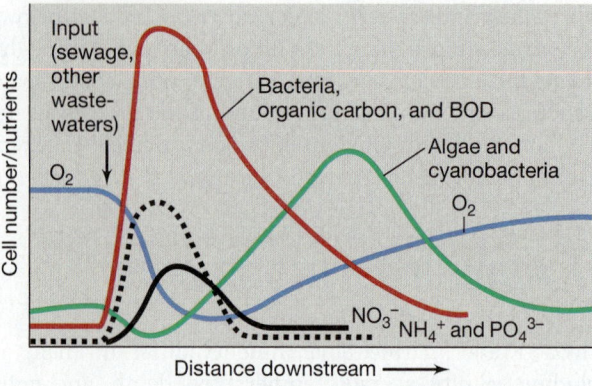

(b)

Figure 19.17 **Effect of the input of organic-rich wastewaters into aquatic systems.** *(a)* In a river, bacterial numbers increase and O_2 levels decrease with a spike of organic matter. The rise in algae and cyanobacteria is a response to inorganic nutrients, especially PO_4^{3-}. *(b)* Photo of a eutrophic (nutrient-rich) lake, Lake Mendota, Madison, Wisconsin (USA), showing algae, cyanobacteria, and aquatic plants that bloom in response to nutrients from agricultural runoff. (See also Figure 19.1.)

We thus see that in freshwaters the oxygen and carbon cycles are linked, with the levels of organic carbon and oxygen being inversely related. Although oxygenic photosynthesis produces O_2, the corresponding production of organic matter leads to O_2 deficiencies. Anoxic aquatic environments, which are typically rich in organic material, are the end result of respiratory processes that remove dissolved oxygen from the ecosystem, leaving the remaining organic material to be mineralized by organisms employing the anaerobic energy metabolisms we discussed in Chapter 13. It is also important to recognize the importance of storms, floods, and droughts in determining delivery, transport, and cycling of organic matter and inorganic nutrients in freshwater systems, including streams, rivers, lakes, and reservoirs. These less predictable changes also affect microbial productivity, diversity, distribution, and interactions in freshwater systems.

A Phylogenetic Snapshot of Freshwater Prokaryotic Diversity

The importance of prokaryotes in lakes, streams, and rivers to the production, regeneration, and mobilization of nutrients is well recognized. However, only more recently have molecular methods been used to identify the participating microbial populations, their interactions, and seasonal patterns. As we saw for studies of soil diversity (Section 19.6), 16S ribosomal RNA gene sequencing is used as a culture-independent method to identify and quantify microbial phylotypes (↩ Section 18.5). Since most molecular studies of freshwater systems have focused on lakes, the emerging picture of lake community structure is examined here.

Figure 19.18 shows the major prokaryotic groups that inhabit lake surface samples (the epilimnion). Five major bacterial groups, or phyla, are routinely observed in the following approximate order of decreasing representation: *Proteobacteria, Actinobacteria, Bacteroidetes, Cyanobacteria,* and *Verrucomicrobia. Archaea* affiliated with *Euryarchaeota, Crenarchaeota,* and *Thaumarchaeota* are also present. This phylum-level composition shares features in common with the ocean, where *Proteobacteria* and *Bacteroidetes* also comprise the greater part of diversity (see Figure 19.27). However, unlike the high diversity of *Betaproteobacteria* in lakes, in the oceans *Gammaproteobacteria* and *Alphaproteobacteria* are the more diverse subgroups of the *Proteobacteria* (see Figure 19.27).

A functional interpretation of lake prokaryotic community structure is constrained by the limited availability of cultured representatives. Freshwater *Thaumarchaeota* affiliate with known ammonia-oxidizing species, but the metabolism of freshwater *Euryarchaeota* is not yet known. *Actinobacteria* are chemoorganotrophic bacteria that in lakes may be responsible for the breakdown of nucleic acid and proteins. In addition, metagenomic analyses (↩ Section 18.7) have shown that at least some *Actinobacteria* contain genes related to those that encode bacteriorhodopsin, a membrane-integrated protein that converts light energy into ATP (Section 19.11 and ↩ Section 16.1) in extremely halophilic *Archaea* (the *Actinobacteria* analog is called *actinorhodopsin*). Hence some *Actinobacteria* may be able to harvest light as an energy source.

Bacteroidetes are well represented in lake ecosystems. These organisms are known for their significant metabolic diversity and are likely to be important in lakes in the degradation of various biopolymers and humic materials. The abundant *Betaproteobacteria* tend to be fast-growing species that can respond quickly to pulses of organic nutrients, whereas the *Alphaproteobacteria* are more competitive under conditions of low availability of organic nutrients; this likely accounts for their greater abundance in the oligotrophic open ocean (see Figure 19.27).

Taken as a whole, high prokaryotic diversity in freshwater lakes (Figure 19.18) reflects the dynamic character of these habitats. Lakes typically receive seasonally variable inputs of endogenous and exogenous nutrients, a pattern that sustains a phylogenetically and metabolically complex prokaryotic community.

MINIQUIZ

- What is a primary producer? In a freshwater lake, would primary producers more likely reside in the epilimnion or the hypolimnion, and why?

- Will addition of organic matter to a water sample increase or decrease its BOD?

- What factors might account for the prokaryotic diversity of freshwater lakes?

19.9 The Marine Environment: Phototrophs and Oxygen Relationships

With the exception of oxygen, nutrient levels in the open ocean (the *pelagic zone*) are often very low compared with many freshwater environments. This is especially true of key inorganic nutrients for phototrophic organisms, such as nitrogen, phosphorus, and iron. In addition, water temperatures in the oceans are cooler and more constant seasonally than those of most freshwater lakes. The activity of marine phototrophs is limited by these factors, and thus total microbial cell numbers are typically about 10-fold lower in the oceans than in freshwater environments ($\sim 10^6$/ml versus 10^7/ml, respectively). These are average numbers, and studies of marine prokaryotic diversity are just beginning to reveal recurrent temporal patterns of diversity and abundance.

The Bermuda Atlantic Time-Series Study (BATS) has a history of continuous biogeochemical monitoring of ocean waters since the mid-1950s, and is now incorporating molecular analyses of microbial population structure. BATS has revealed three seasonal microbial communities in ocean waters: (1) the community corresponding to the spring surface water bloom (featuring small eukaryotic algae, marine actinobacteria, and two groups of *Alphaproteobacteria*); (2) the summertime community in the upper water column associated with water column stratification (featuring *Pelagibacter, Puniceispirillum,* and two groups of *Gammaproteobacteria*); and (3) the deeper, more stable community (featuring *Nitrosopumilus,* representatives of the SAR11 group with which the genus *Pelagibacter* affiliates, a group of *Deltaproteobacteria,* and species of two additional groups related to the *Chloroflexi* and *Fibrobacter*). Thus, a complex and poorly understood interplay of seasonal changes in physicochemical and biotic conditions controls the structure of this oceanic microbial community in recurring annual cycles.

Many different prokaryotes and eukaryotes inhabit ocean waters, but most are very small cells, a typical characteristic of organisms living in nutrient-poor environments. Smallness is an adaptive feature for nutrient-limited microorganisms in that it requires less energy for cellular maintenance. The trade-off is

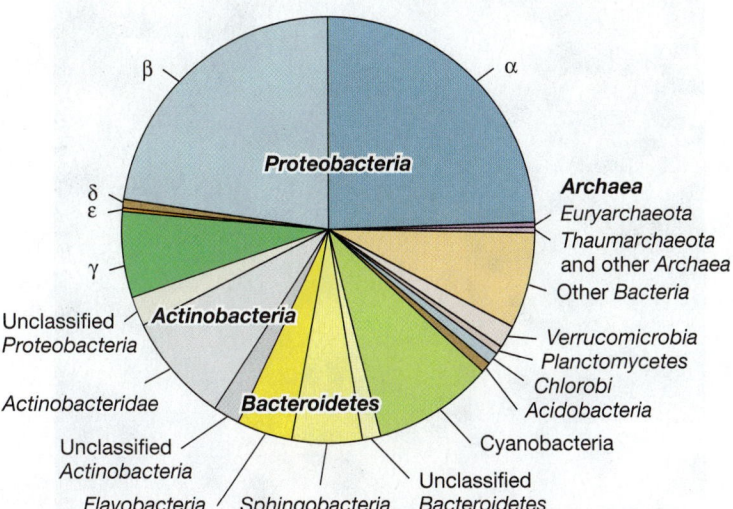

Figure 19.18 Freshwater lake prokaryotic diversity. Distribution of 16S ribosomal RNA gene sequences by phylum determined from analysis of a collective dataset of 16S genes detected in the epilimnion of several freshwater lakes. Data assembled and analyzed by Nicolas Pinel.

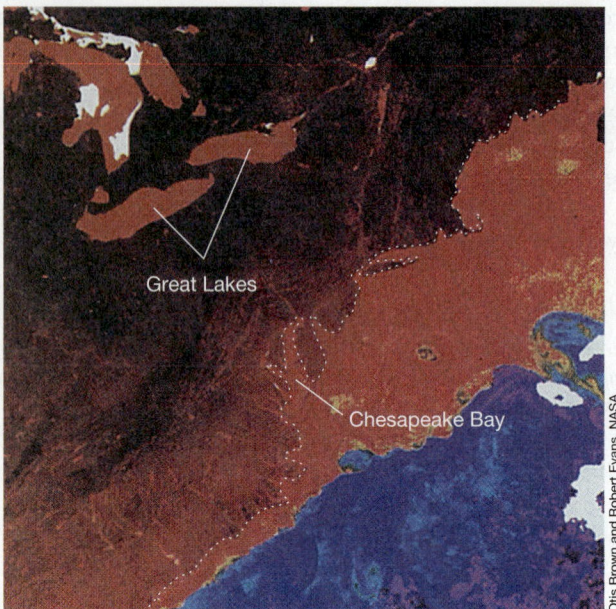

Figure 19.19 **Distribution of chlorophyll in the western North Atlantic Ocean as recorded by satellite.** The east coast of the United States from the Carolinas to northern Maine is shown in dotted outline. Areas rich in phototrophic plankton are shown in red (>1 mg chlorophyll/m³); blue and purple areas have lower chlorophyll concentrations (<0.01 mg/m³). Note the high primary productivity of coastal areas and the Great Lakes.

that a greater number of transport enzymes relative to cell volume are needed for organisms to acquire nutrients from very dilute (oligotrophic) than from nutrient-rich (eutrophic) aquatic environments. For example, ammonia-oxidizing *Archaea* (*Nitrosopumilus*, ⮎ Section 16.6) are the dominant chemolithotrophs in pelagic waters and have very high-affinity transport systems for acquiring the ammonia they need as an electron donor in energy metabolism.

In pelagic waters there is a lower return of nutrients from the bottom waters than in freshwater lakes, and thus lower average primary productivity. However, because the oceans are so large, the collective carbon dioxide sequestration and oxygen production from oxygenic photosynthesis in the oceans are major factors in Earth's carbon balance. Salinity is more or less constant in the pelagic zone but is more variable in coastal areas. Terrestrial inputs, retention of nutrients, and upwelling of nutrient-rich waters combine to support higher populations of phototrophic microorganisms in near-shore waters than in pelagic waters (**Figure 19.19**); the more productive near-shore waters in turn support higher densities of chemoorganotrophic bacteria and aquatic animals, such as fish and shellfish.

In shallow marine waters such as marine bays and inlets, nutrient inputs can actually lead to the waters becoming intermittently anoxic from the removal of O_2 by respiration and the production of H_2S by sulfate-reducing bacteria. An extensive region (6000–7000 square miles) of oxygen depletion in the Gulf of Mexico is associated with high loads of nitrogen and phosphorus carried in by the Mississippi River from agricultural runoff in the Mississippi Valley. This region, called the *Gulf of Mexico Dead Zone*, contributes to the loss and impairment of fish and benthic sea life that sustain major seafood industries in this region. The Gulf of Mexico experiences other ecological problems as well, as we examine now.

The Deepwater Horizon Catastrophe

In addition to the chronic degradation of the Gulf of Mexico ecology through agricultural runoff, increased offshore oil drilling also poses significant environmental risk. A major catastrophe for the Gulf of Mexico was the April 2010 explosion and sinking of the Deepwater Horizon offshore drilling platform; failure to control well pressure resulted in the rupture of the wellhead at a depth of 1.5 km and the release of over 4 million barrels of oil before the well was capped three months later (**Figure 19.20**). This

(a)

(b)

Figure 19.20 **The Deepwater Horizon oil spill in the Gulf of Mexico.** *(a)* Inferno resulting from the wellhead blowout. *(b)* NASA Terra satellite image taken on May 24, 2010, of the Gulf of Mexico near New Orleans, Louisiana. A large plume of oil was released at about 1500 m depth, some of which reached the surface where sunlight reflects off of the oil slick (arrows).

largest marine oil spill ever was unique in that most of the oil was released as a plume at great depths in the water column. Typically, marine oil spills contaminate primarily the surface waters, resulting in rapid volatilization and loss to the atmosphere of low-molecular-weight oil components (such as naphthalene, ethylbenzene, toluene, and xylene). By contrast, the Deepwater Horizon spill released both low-molecular-weight components *and* natural gas (methane, ethane, propane) deep into the water column. These components comprised about 35% of the hydrocarbon plume that extended many miles into the Gulf from the surface to depths greater than 800 m (Figure 19.20*b*).

The microbial response to hydrocarbon contamination was tracked over several months using both culture-based and molecular methods, including 16S ribosomal RNA gene and metagenomic sequencing, and phylochip microarray analyses (⮌ Sections 18.5–18.7). These methods showed that the initial microbial response to the spill (May and June 2010) was a bloom of hydrocarbon-degrading *Gammaproteobacteria* species related to genera in the *Oceanospirillales* group, and of the genera *Colwellia* and *Cycloclasticus*. Increased numbers of *Colwellia* and *Oceanospirillales* species were attributed to their use of gaseous hydrocarbons, since both grew rapidly when ethane or propane was added to enrichment cultures (⮌ Section 18.1). *Colwellia* species also contributed to the degradation of a variety of other hydrocarbons, as indicated by their growth in crude oil enrichment cultures lacking natural gas and by stable isotope probing experiments (⮌ Section 18.11) showing their incorporation of ^{13}C benzene. Although there remains considerable uncertainty about the fate of all the hydrocarbons released during the Deepwater Horizon spill, it appears that the early stimulation of a bloom of hydrocarbon-degrading bacteria by the more easily degraded, soluble, low-molecular-weight components helped reduce the environmental impact of this immense oil spill.

Oxygen Minimum Zones

Another feature of the marine water column is **oxygen minimum zones (OMZs)**, regions of oxygen-depleted waters at intermediate depths, typically in waters between 100 and 1000 m, that extend over wide expanses of the open and coastal ocean. These oxygen-depleted regions arise when the respiratory demand for oxygen exceeds oxygen availability, and they are associated with nutrient-rich, highly productive regions. In this way they are similar to the depletion of oxygen caused by agricultural runoff in coastal zones, such as contributes to the Gulf of Mexico Dead Zone. However, OMZs predate human activity and originate naturally in regions of high surface production and little mixing with oxygen-rich water.

The oxygen saturation values of the largest of the OMZs in the eastern Pacific off the coast of Peru are less than 10% of that at the surface. Oxygen levels at certain depth intervals in the OMZs of the Bay of Bengal and Arabian Sea approach zero. Because of this, OMZs have been recognized as significant sinks for the loss of fixed nitrogen through denitrification (⮌ Section 13.17) and anammox processes (⮌ Section 13.10). In addition to contributing to a significant fraction of the 50% loss of fixed nitrogen from the oceans, these regions are also a source of nitrous oxide (N_2O), a potent greenhouse gas of which approximately one-third is emitted from the oceans.

Ongoing studies of OMZs have shown that these regions of oxygen depletion are expanding, and that their recent expansion is almost certainly associated with global warming. As the oceans absorb more heat, warming of the surface waters increases stratification of near-surface waters and reduces oxygen transfer through mixing to deeper regions. Expansion of the OMZs will favor anaerobic microbial processes at the expense of the aerobic processes that sustain critical oceanic food webs. These changes may further affect atmospheric chemistry by increasing the release of N_2O and will negatively impact marine food webs by reducing levels of fixed nitrogen. Ultimately, these changes are expected to impact commercial fisheries.

MINIQUIZ

- What did the Deepwater Horizon spill tell us about how mixed hydrocarbons are degraded in nature?
- Will addition of organic matter to a water sample increase or decrease its BOD?
- What is an oxygen minimum zone and why is expansion of these zones a problem for marine and global ecology?

19.10 Major Marine Phototrophs

The oceans contain large numbers of phototrophic microorganisms, including both prokaryotic and eukaryotic oxygenic phototrophs as well as significant numbers of a special group of purple (anoxygenic) phototrophs. We consider these organisms here as a prelude to exploring the marine prokaryotic world in general in the next section.

Primary Productivity: *Prochlorococcus*

Much of the primary productivity in the open oceans, even at significant depths, comes from photosynthesis by prochlorophytes, tiny prokaryotic phototrophs that are phylogenetically related to cyanobacteria (⮌ Section 14.3); **prochlorophytes** contain chlorophylls *a* and *b* but do not contain phycobilins. The organism *Prochlorococcus* is a particularly important primary producer in the marine environment (**Figure 19.21**). Because *Prochlorococcus* lacks phycobilins, the accessory pigments of the cyanobacteria (⮌ Section 13.2), dense suspensions of *Prochlorococcus* cells are olive green (as are green algae) rather than the blue-green color of cyanobacteria (compare Figures 19.1*c* and 19.21).

Prochlorococcus accounts for up to half of the photosynthetic biomass and primary production in the tropical and subtropical regions of the world's oceans, reaching cell densities of 10^5/ml. At least four strains of *Prochlorococcus* have been identified, and each inhabits its own depth range in pelagic waters. The different *Prochlorococcus* strains are considered distinct *ecotypes*, genetic variants of a species that differ physiologically and therefore occupy slightly different niches. The different *Prochlorococcus* ecotypes photosynthesize at different light intensities and use different inorganic and organic nitrogen and phosphorus sources. *Prochlorococcus* is thus distributed in both surface waters and deeper waters to depths of 200 m, and when an oxygen minimum zone (Section 19.9) is present, it extends into the upper regions of this zone. This is near the bottom of the photic zone where light

Figure 19.21 *Prochlorococcus*, the most abundant oxygenic phototroph in the oceans. A bottle of *Prochlorococcus* showing the olive green color of the cells containing chlorophyll *a* and *b*. Inset: FISH-stained cells of *Prochlorococcus* in a marine water sample (↩ Section 18.4 and Figure 18.19).

intensities are very low (see Figure 19.24). Genome sequences of a dozen or so *Prochlorococcus* ecotypes have been determined and comparisons have revealed that although each ecotype contains about 2000 genes, only about 1100 genes are shared by all ecotypes. Each ecotype contains approximately 200 unique genes, which presumably have adaptive significance for growth in its realized niche. This was illustrated in Chapter 18 where we compared the genome of a single cultured *Prochlorococcus* ecotype to metagenome sequences obtained from pelagic waters (↩ Section 18.7 and Figure 18.19).

Other Pelagic Oxygenic Phototrophs

In tropical and subtropical oceans, the planktonic filamentous marine cyanobacterium *Trichodesmium* (**Figure 19.22a**) is a widespread and occasionally abundant phototroph. Cells of *Trichodesmium* form puffs (colonies) of filaments. Each puff can contain many hundreds of individual filaments, each filament composed of 20–200 cells. In the Caribbean Sea, colonies of *Trichodesmium* can approach 100/m^3. *Trichodesmium* is a nitrogen-fixing cyanobacterium, and the production of fixed nitrogen by this organism is thought to be an important link in the marine nitrogen cycle. *Trichodesmium* contains phycobilins, absent from prochlorophytes, and thus differs from these organisms in its absorption properties (↩ Section 13.2).

Very small phototrophic eukaryotes also inhabit coastal and pelagic waters, and some of these are among the smallest eukaryotic cells known. Three common genera—*Bathycoccus*, *Micromonas*, and *Ostreococcus*—contain only one mitochondrion and one chloroplast per cell. These genera are now assigned to the *Prasinophyceae*, a family of green algae that diverged early from other lineages of green algae (↩ Section 17.16). Cells of *Ostreococcus* are cocci that measure only about

0.7 μm in diameter (Figure 19.22*b*), which is even smaller than a cell of *Escherichia coli*.

Although cells of *Ostreococcus* and *Prochlorococcus* are of roughly the same dimensions and they are both oxygenic phototrophs, their genomes are distinct. The genome of *Ostreococcus* is 12.6 Mbp (distributed over 20 chromosomes), which is more than seven times the size of the *Prochlorococcus* genome. Even though this is large relative to cyanobacteria, the *Ostreococcus* genome is very gene dense, containing about 8000 genes, and thought to be near the minimum genome size of a free-living photosynthetic eukaryote. As a reference, the genome of a common plant, Japanese rice (*Oryza sativa* subsp. *japonica*), is 420 Mbp and contains about 50,000 genes.

In many marine waters, other small eukaryotic cells are present at about 10^4/ml. Although many of these are *Ostreococcus* or relatives, some are chemoorganotrophs and some are phototrophs unrelated

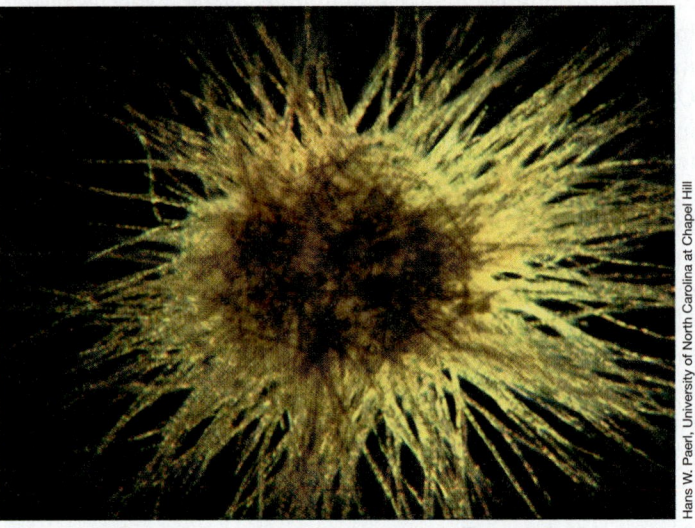

(a)

(b)

Figure 19.22 *Trichodesmium* and *Ostreococcus*. *(a)* Light photomicrograph of a puff of cells of the nitrogen-fixing cyanobacterium *Trichodesmium*. The filaments in the puff are chains of cells, each of which is about 6 μm in diameter. *(b)* Transmission electron micrograph of a cell of *Ostreococcus*, a small green alga (eukaryote) found primarily in marine coastal waters. The arrow points to the chloroplast. An *Ostreococcus* cell is about 0.7 μm in diameter.

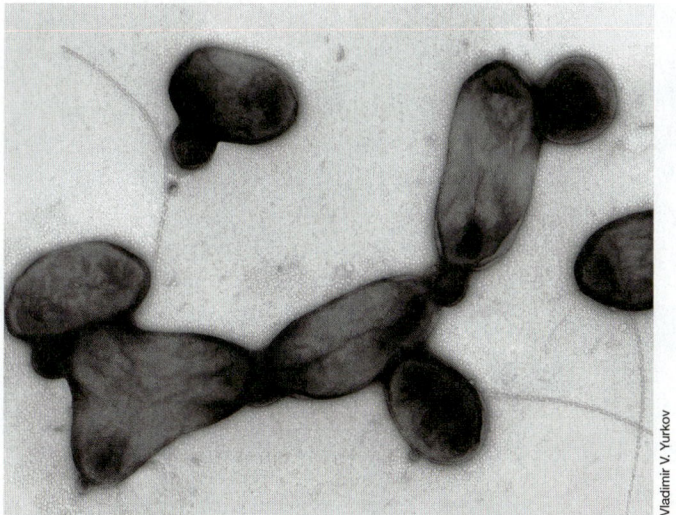

Figure 19.23 Aerobic anoxygenic phototrophic bacteria. Transmission electron micrograph of negatively stained cells of *Citromicrobium*. Cells of this marine, aerobic anoxygenic phototroph produce bacteriochlorophyll *a* only under oxic conditions and divide by both budding and binary fission, yielding morphologically unusual and irregular-shaped cells.

to *Ostreococcus* that incorporate small amounts of organic matter to supplement their primarily phototrophic lifestyle.

Aerobic Anoxygenic Phototrophs

Besides *oxygenic* phototrophs, *anoxygenic* phototrophs are also present in coastal and pelagic marine waters. Like purple anoxygenic phototrophs, these organisms contain bacteriochlorophyll *a* (⮌ Sections 13.1, 13.3, 14.4, and 14.5). However, unlike classical purple bacteria that carry out photosynthesis only under *anoxic* conditions, these anoxygenic phototrophs carry out photosynthetic light reactions only under *oxic* conditions.

Aerobic anoxygenic phototrophs include bacteria such as *Erythrobacter*, *Roseobacter*, and *Citromicrobium* (Figure 19.23), all genera of *Alphaproteobacteria*. Aerobic anoxygenic phototrophs synthesize ATP by photophosphorylation when oxygen is present (which is all of the time in oxic pelagic waters), but they are unable to grow autotrophically and thus rely on organic carbon for their carbon sources (a nutritional condition called *photoheterotrophy*). These organisms thus use the ATP produced by photophosphorylation to supplement their otherwise chemoorganotrophic metabolism.

Surveys have shown that a great diversity of aerobic anoxygenic phototrophs exist in marine waters, especially near-shore waters. Oligotrophic and highly oxic freshwater lakes are also habitats for these interesting phototrophic bacteria. The physiology of aerobic anoxygenic phototrophs is thus ideal for their illuminated and highly oxic habitats.

MINIQUIZ

- How does *Ostreococcus* differ from *Prochlorococcus*? What do they have in common?
- How does the organism *Prochlorococcus* contribute to both the carbon and oxygen cycles in the oceans?
- How does *Roseobacter* differ from *Prochlorococcus*?

19.11 Pelagic *Bacteria*, *Archaea*, and Viruses

Despite vanishingly low nutrient levels, significant numbers of prokaryotes live a planktonic existence in pelagic waters. These include species of both *Bacteria* and *Archaea*, and one organism in particular has garnered significant attention, a bacterium named *Pelagibacter*.

Distribution and Activity of *Archaea* and *Bacteria* in Pelagic Waters

The abundance of prokaryotic cells in the open oceans decreases with depth. In surface waters, cell numbers average about 10^6/ml. Below 1000 m, however, total cell numbers fall to between 10^3 and 10^5/ml. The distribution of *Bacteria* and *Archaea* with depth has been tracked in pelagic waters using fluorescence in situ hybridization (FISH) technology (⮌ Section 18.4).

Species of *Bacteria* tend to predominate in waters above 1000 m, although cells of *Bacteria* and *Archaea* are found in near-equal abundance in deeper waters (Figure 19.24). Deep-water *Archaea* are almost exclusively species of *Thaumarchaeota* (⮌ Section 16.6), and many or perhaps even most are ammonia-oxidizing chemolithotrophs (⮌ Sections 13.10 and 16.6); these organisms play an important role in coupling the marine carbon and nitrogen cycles (Chapter 20). Extrapolating from the data in Figure 19.24, it is

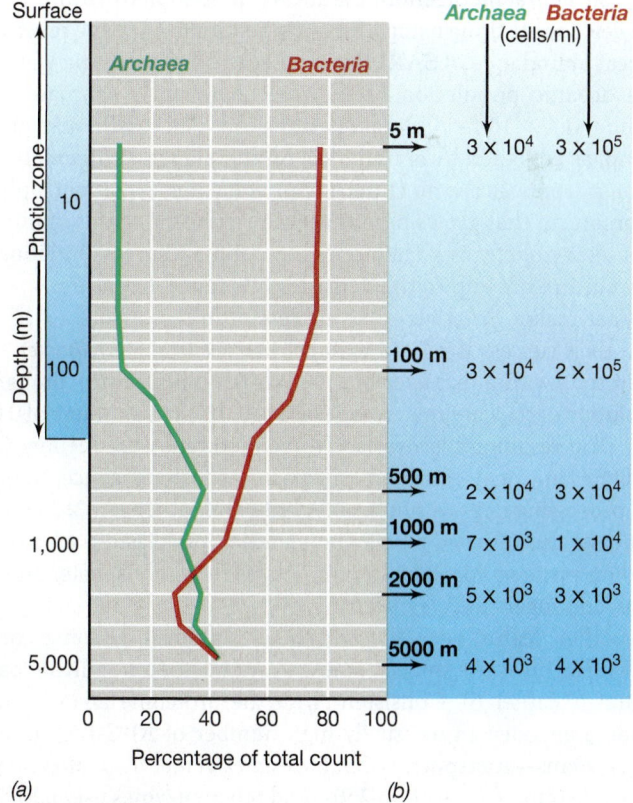

Figure 19.24 Percentage of total prokaryotes belonging to *Archaea* and *Bacteria* in North Pacific Ocean water. *(a)* Distribution of *Archaea* and *Bacteria* with depth. *(b)* Absolute numbers per milliliter of *Archaea* and *Bacteria* with depth in the open ocean.

estimated that 1.3×10^{28} and 3.1×10^{28} cells of *Archaea* and *Bacteria*, respectively, exist in the world's oceans. This means that the oceans contain the largest microbial biomass on Earth's surface.

Pelagic *Bacteria* and *Archaea* are ecologically important because they consume dissolved organic carbon in the oceans, one of the largest pools of usable organic carbon on Earth. These small and free-living planktonic prokaryotes consume about half the total oceanic organic carbon produced from photosynthesis and are responsible for about half of all marine respiration and nutrient regeneration. Planktonic marine prokaryotes thus return organic matter to the marine food web that would otherwise be lost because of the inability of larger marine organisms to take up such diluted organic nutrients. This so-called "secondary production" is balanced by cell losses from bacterial grazing protists and from virus attack (see Figure 19.26), leading to a near-steady state in which bacterial abundance in the open ocean remains roughly constant over time. But importantly, secondary production both recycles nutrients and allows some of the dissolved organic carbon in seawater to reach larger organisms, including fish, because protists are passed up the food web by the feeding activities of larger organisms.

Pelagibacter

Very small planktonic chemoorganotrophic bacteria inhabit pelagic marine waters in numbers of 10^5–10^6 cells/ml. The most abundant of these are members of the SAR11 group within the *Alphaproteobacteria*, which includes the genus *Pelagibacter*. Environmental metagenomic studies (Sections 6.10 and 18.7) and cell counts done using FISH (Section 18.4) have revealed a great abundance of SAR11 group organisms in pelagic waters. The oceanic population of the SAR11 group is estimated to number 2.4×10^{28} cells, making it the most successful microbial group, as reflected by abundance, on the planet. *Pelagibacter* is an oligotroph, as are most pelagic prokaryotes. An oligotroph is an organism that grows best at very low concentrations of nutrients. *Pelagibacter* is a chemoorganotroph and grows in laboratory culture only up to the densities it is found in nature.

What makes *Pelagibacter* so successful in the open oceans? In part, their success is related to small size. Cells of *Pelagibacter* are small rods with a diameter of only 0.2–0.5 μm, near the limits of resolution of the light microscope (Figure 19.25), and a volume of 0.01 μm³. The resulting high surface-to-volume ratio (Section 2.6) facilitates nutrient transport, increasing substrate concentration and processing rates within the cell. Proteomic analyses (Section 6.8) have also revealed a high abundance of periplasmic substrate-binding proteins for soluble nutrients such as phosphate, amino acids, and sugars in *Pelagibacter*.

Another feature of *Pelagibacter* is its small genome, only 1.3 Mbp. This is the smallest known genome for a free-living bacterium (Chapter 6). Consistent with the proteome analysis, the genome encodes an unusually high number of ABC-type transport systems—transporters that have an extremely high affinity for their substrates (Section 2.9)—and other enzymes useful for an oligotrophic organism. The highly reduced genome is also highly "streamlined," having the shortest intergenic spacing (an average of only 3 base pairs) of any sequenced genome. Having a highly compact and efficient genome reduces the cost of replication.

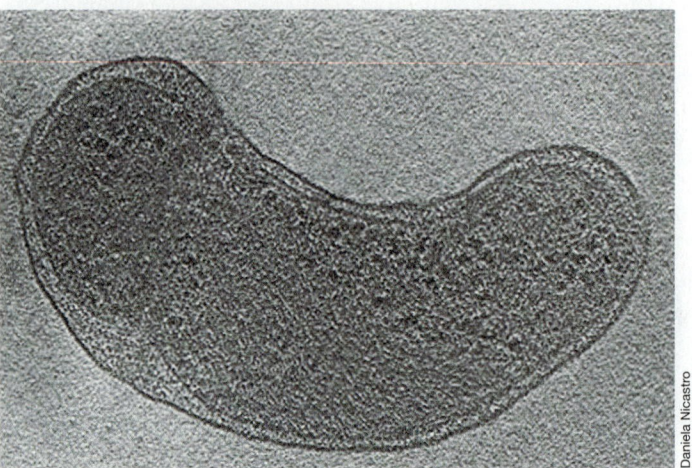

Figure 19.25 *Pelagibacter*, the most abundant prokaryote in the ocean. Electron micrograph taken by electron tomography, a technique that introduces a three-dimensional effect. A single cell of *Pelagibacter* is about 0.2 μm in diameter.

In addition to its small size and compact genome, *Pelagibacter* has genes that encode a form of the visual pigment rhodopsin that can convert light energy into ATP. In Section 16.1 we discussed the now well-studied molecule *bacteriorhodopsin*, a light-activated protein complex present in the extreme halophile *Halobacterium* (*Archaea*); bacteriorhodopsin functions in ATP synthesis as a simple light-driven proton pump (Figure 16.4). The form of rhodopsin in *Pelagibacter* and other pelagic prokaryotes is structurally similar to bacteriorhodopsin and has been called **proteorhodopsin** ("proteo" referring to *Proteobacteria*). Although proteorhodopsin was first discovered in species of *Proteobacteria*, it is actually fairly widely distributed in *Bacteria*, including many *Gamma-* and *Alphaproteobacteria*, *Bacteroidetes*, and *Actinobacteria*, and has also been found in nonhalophilic species of *Archaea*, such as species of the marine *Euryarchaeota*. The different variants of proteorhodopsins in marine microorganisms have absorption properties that reflect changing spectral properties of light at increasing depths in the water column, with near-surface variants absorbing green light and those at greater depths absorbing blue light.

Studies of various cultured proteorhodopsin-containing marine bacteria have now conclusively shown these organisms survive starvation better in the light than in the dark. That is, energy-starved cells use light-mediated ATP production to compensate for energy unavailable from carbon respiration due to the low levels of organic carbon present. Remarkably, proteorhodopsins have been estimated to exist in 80% of marine bacteria in some waters. Proteorhodopsin is thus a widespread strategy to supplement the energy metabolism of marine prokaryotes such that they need not rely solely on scarce organic carbon for their energy needs.

Marine Viruses

In the oceans, viruses are more abundant than cellular microorganisms, often numbering over 10^7 virions/ml in typical seawater (Section 8.11). In coastal waters, where bacterial cell numbers are higher than in the oceans, viral numbers are also higher, as many as 10^8 virions/ml. Most of these viruses are

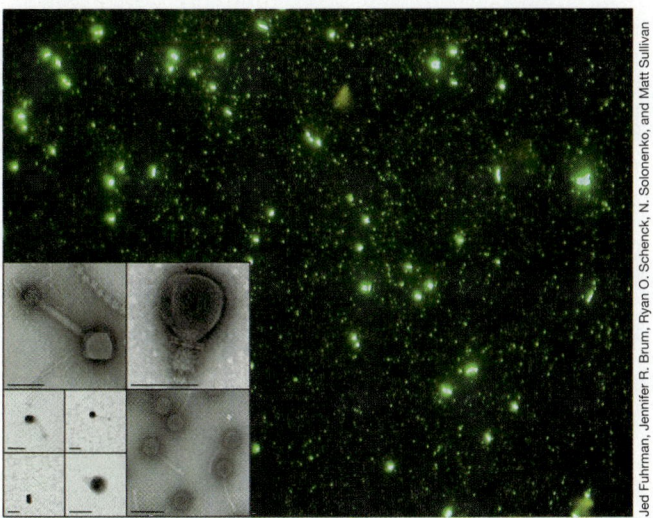

Figure 19.26 **Viruses in seawater.** A water sample collected on a 0.02-μm filter is stained with SYBR Green and viewed by epifluorescence microscopy. The tiny green dots are viruses while the larger, brighter dots are prokaryotes about 0.5 μm in diameter. Viruses are typically 10 times more abundant than prokaryotes in seawater. Inset transmission electron micrographs show various marine bacterial viruses (scale bars, 100 nm in all images).

bacteriophages, which infect species of *Bacteria*, and archaeal viruses, which infect species of *Archaea*. The number of virions in seawater is about 10-fold greater than average prokaryotic cell numbers, suggesting that viruses are actively infecting their hosts, replicating, and being released into seawater (**Figure 19.26**). Only a small fraction of released viruses (an average of one per burst) successfully infects a new host, and most are inactivated or destroyed by sunlight and hydrolytic enzymes. In these ways, the entire viral population is replaced in periods of only a few days or weeks. We considered bacterial and archaeal viral diversity in Chapter 9.

Along with feeding by protists, marine virus infections probably help to maintain prokaryotic numbers at the levels that are observed, but viruses may also have other important ecosystem functions. These include facilitating genetic exchange between prokaryotic cells and allowing for lysogeny, the state in which a virus genome integrates within the cellular genome; lysogeny can confer new genetic properties on the cell (ợợ Sections 8.8 and 10.7). For example, the discovery that some of the viruses that infect *Prochlorococcus*, the most abundant oxygenic phototroph in the oceans (Figure 19.21 and Section 19.10), contain genes that encode proteins for photosynthesis indicates that even key metabolic properties may be encoded by viruses. Although the genetic diversity of marine viruses is just now being recognized, it is thought that the diversity of marine viral genomes could surpass even that of all prokaryotic cells, making the oceans a hotbed of genetic diversity.

A Phylogenetic Snapshot of Marine Prokaryotic Diversity

Several studies have attempted to characterize the diversity of planktonic marine prokaryotes by analysis of 16S ribosomal RNA genes obtained from seawater. The existence of abundant alpha-proteobacterial populations to which *Pelagibacter* is affiliated was first revealed by such 16S rRNA sequence analysis. Mesophilic *Archaea* related to *Nitrosopumilus maritimus* (ợợ Section 16.6) were discovered using similar methods.

Major bacterial groups now recognized as abundant in the open ocean include *Alpha-* and *Gammaproteobacteria*, cyanobacteria, *Bacteroidetes*, and to a lesser extent, *Betaproteobacteria* and *Actinobacteria*; *Firmicutes* are only minor components (**Figure 19.27**). As for soil, a large proportion of unclassified and minor bacterial groups are also present in seawater. A major group of marine *Gammaproteobacteria* is the yet to be cultured "SAR86 group," which accounts for approximately 10% of the total prokaryotic community in the ocean surface layer. Representing the *Archaea* in pelagic waters is a rather restricted diversity of *Euryarchaeota*,

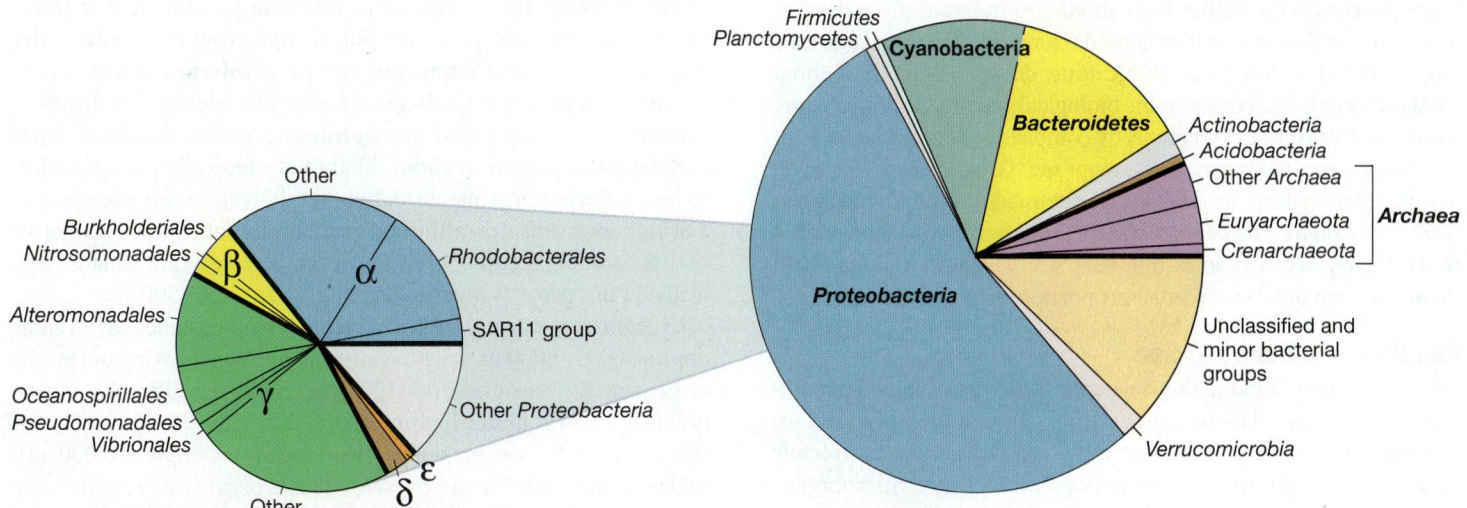

Figure 19.27 **Ocean prokaryotic diversity.** The results are pooled analyses of 25,975 16S ribosomal RNA sequences from several studies of pelagic ocean waters. Many of these groups are covered in Chapters 14 and 15 (*Bacteria*) or 16 (*Archaea*). For *Proteobacteria*, major subgroups are indicated. Note the high proportion of cyanobacterial and *Gammaproteobacteria* sequences. Data assembled and analyzed by Nicolas Pinel. Compare the prokaryotic diversity of seawater with that of freshwater shown in Figure 19.18.

Crenarchaeota, and *Thaumarchaeota*, most of which have not yet been brought into laboratory culture.

With the exception of the cyanobacteria, most marine *Bacteria* are thought to be chemoorganotrophs adapted to extremely low nutrient availability, some augmenting energy conservation through proteorhodopsin or aerobic anoxygenic phototrophy (Section 19.10). The discovery of the chemolithotroph *Nitrosopumilus* suggested the possibility that many marine *Archaea* specialize in ammonia oxidation, although heterotrophic species likely exist as well. "Dilution culture" methods employing very dilute culture media have been successful in bringing some pelagic prokaryotes into culture (⮌ Section 18.2). It appears that most of these organisms have evolved to grow only at very low nutrient concentrations, so it is either difficult or impossible to culture them to higher cell densities. Cell densities of marine oligotrophs in laboratory cultures are similar to those in their natural environments (10^5–10^6/ml), which renders many of the common tools for measuring cell growth (turbidity, microscopic counts) useless on samples that are not first concentrated. Nevertheless, there have been notable successes with dilution culturing of marine bacteria and the aforementioned *Pelagibacter* is a good example (⮌ Chapter 18 Explore the Microbial World, "Culturing the Uncultured").

MINIQUIZ

- What is proteorhodopsin and why is it so named? Why might proteorhodopsin make a bacterium such as *Pelagibacter* more competitive in its habitat?

- How do numbers of pelagic prokaryotes and viruses compare?

- Which phylum and subgroups of *Bacteria* dominate pelagic marine waters?

19.12 The Deep Sea and Deep-Sea Sediments

Light penetrates no farther than about 300 m in pelagic waters; as has been mentioned, this illuminated region is called the *photic zone* (Figure 19.24). Beneath the photic zone, down to a depth of about 1000 m, there is still considerable biological activity. However, water at depths greater than 1000 m is, by comparison, much less biologically active and is known as the *deep sea*. Greater than 75% of all ocean water is deep-sea water, lying primarily at depths between 1000 and 6000 m. The deepest waters in the oceans lie below 10,000 m. However, because holes this deep are very rare, the waters in them make up only a very small proportion of all pelagic waters.

Conditions in the Deep Sea

Organisms that inhabit the deep sea face three major environmental extremes: (1) low temperature, (2) high pressure, and (3) low nutrient levels. In addition, deep-sea waters are completely dark such that photosynthesis is impossible. Thus, microorganisms that inhabit the deep sea must be chemotrophic and able to grow under high pressure and oligotrophic conditions in the cold.

Below depths of about 100 m, ocean water temperatures stay constant at 2–3°C. We discussed the responses of microorganisms to changes in temperature in Sections 5.11 and 5.12. As would be

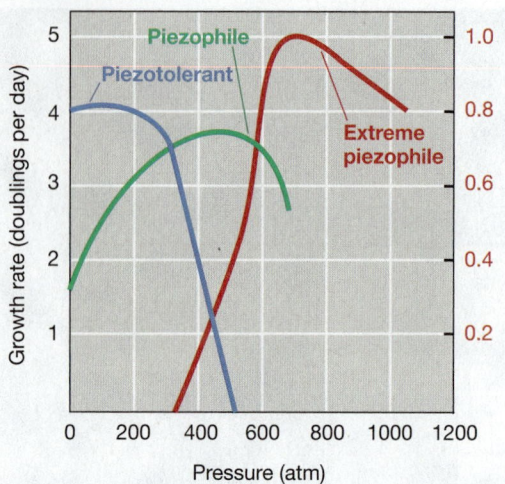

Figure 19.28 **Growth of piezotolerant, piezophilic, and extremely piezophilic bacteria.** The extreme piezophile (*Moritella*) was isolated from the Mariana Trench, off the Philippines, Pacific Ocean (Figure 19.30). Compare the slower growth rate of the extreme piezophile (right ordinate) with the growth rate of the piezotolerant and piezophilic bacteria (left ordinate), and note the inability of the extreme piezophile to grow at low pressures.

expected, bacteria isolated from marine waters below 100 m are psychrophilic (cold-loving) or at least psychrotolerant. Deep-sea microorganisms must also be able to withstand the enormous hydrostatic pressures associated with great depths. Pressure increases by 1 atm for every 10 m of depth in a water column. Thus, an organism growing at a depth of 5000 m must be able to withstand pressures of 500 atm. We will see that microorganisms are remarkably tolerant of high hydrostatic pressures; many species can withstand pressures of 500 atm, and some species can withstand far more than this.

Piezotolerant and Piezophilic *Bacteria* and *Archaea*

Different physiological responses to pressure are observed in different deep-sea microorganisms. Some organisms simply tolerate high hydrostatic pressure, but do not grow optimally under such pressure; these organisms are **piezotolerant** (**Figure 19.28**). By contrast, others actually *grow best* under elevated hydrostatic pressure; these are called **piezophiles**. Organisms isolated from surface waters down to about 3000 m are typically piezotolerant. In piezotolerant organisms, higher metabolic rates are observed at 1 atm than at 300 atm, although growth rates at the two pressures may be similar (Figure 19.28). However, piezotolerant isolates typically do not grow at pressures greater than about 500 atm.

By contrast, cultures derived from samples taken at greater depths, 4000–6000 m, are typically piezophilic, growing optimally at pressures of around 300–400 atm. However, although piezophiles grow best under high pressure, they can still grow at 1 atm (Figure 19.28). In even deeper waters (for example, 10,000 m), **extreme piezophiles** are present. These organisms require very high pressure for growth (**Figure 19.29**). For example, the extreme piezophile *Moritella*, isolated from the Mariana Trench (Pacific Ocean, >10,000-m depth) (**Figure 19.30**), grows optimally at a pressure of 700–800 atm and grows nearly as well at 1035 atm, the pressure it experiences in its natural habitat.

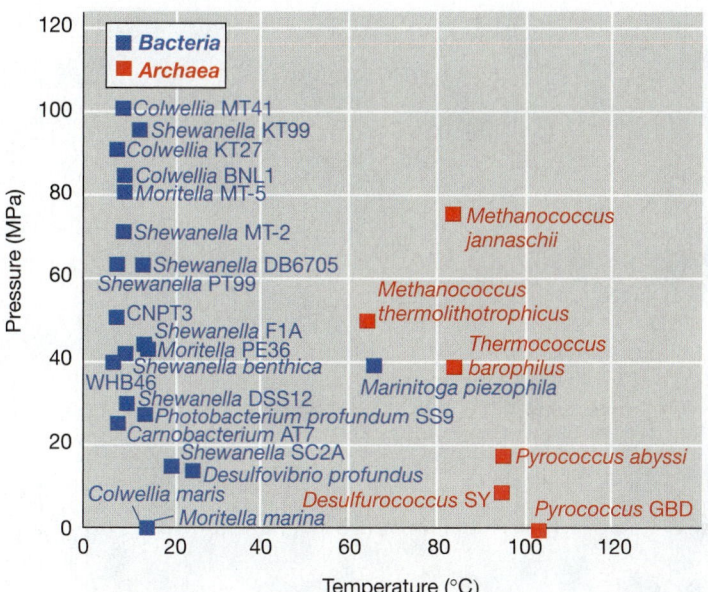

Figure 19.29 **Pressure and temperature optima for cultured bacterial and archaeal piezophiles.** Pressure is in pascals (Pa), the SI units for pressure. One megapascal (MPa) corresponds to approximately 10 atm. Note that different species of the same genus can have vastly different pressure optima. Data assembled by Doug Bartlett.

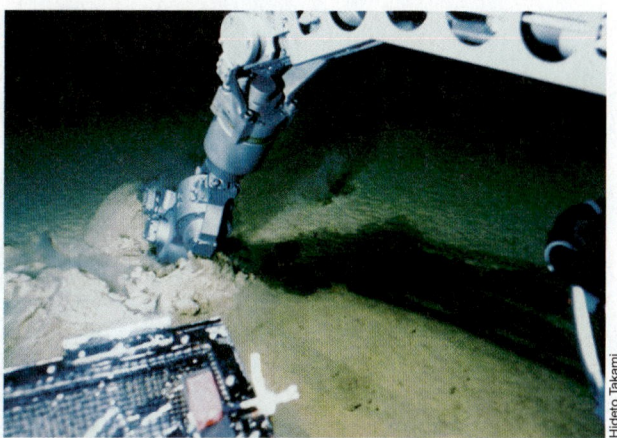

Hideto Takami

Figure 19.30 **Sampling the deep sea.** The unmanned submersible *Kaiko* collecting a sediment sample on the seafloor of the Mariana Trench at a depth of 10,897 m. The tubes of sediment are used for enrichment and isolation of piezophilic bacteria.

Molecular Effects of High Pressure

High pressure affects cellular physiology and biochemistry in many ways. In general, pressure decreases the ability of the subunits of multi-subunit proteins to interact. Thus, large protein complexes in extreme piezophiles must interact in such a way as to minimize pressure-related effects. Protein synthesis, DNA synthesis, and nutrient transport are sensitive to high pressure. Piezophilic bacteria grown under high pressure have a higher proportion of unsaturated fatty acids in their cytoplasmic membranes than when grown at 1 atm. Unsaturated fatty acids allow membranes to remain functional and keep from gelling at high pressures or at low temperatures. The rather slow growth rates of extreme piezophiles such as *Moritella* compared with other marine bacteria (Figure 19.28) are likely due to the combined effects of pressure and low temperature; low temperature slows down the reaction rates of enzymes, directly affecting cell growth (⇄ Sections 5.11 and 5.12).

Studies of gene expression and adaptive features contributing to growth at high pressure have required special pressurized incubation devices (**Figure 19.31**). These studies have shown that when a gram-negative piezophile is grown under high pressure, a specific outer membrane protein called OmpH (*o*uter *m*embrane *p*rotein H) is present that is absent from cells grown at 1 atm. OmpH is a type of porin. Porins are proteins that form channels through which molecules diffuse into the periplasm (⇄ Section 2.11). Presumably, the porin made by cells grown at 1 atm cannot function properly at high pressure and thus a different porin must be synthesized. Interestingly, pressure controls transcription of *ompH*, the gene encoding OmpH. In this piezophile a pressure-sensitive membrane protein complex is present that monitors pressure and triggers transcription of *ompH* only when conditions of high pressure warrant it. Transcriptomic analyses (⇄ Section 6.7) indicate that even relatively modest changes

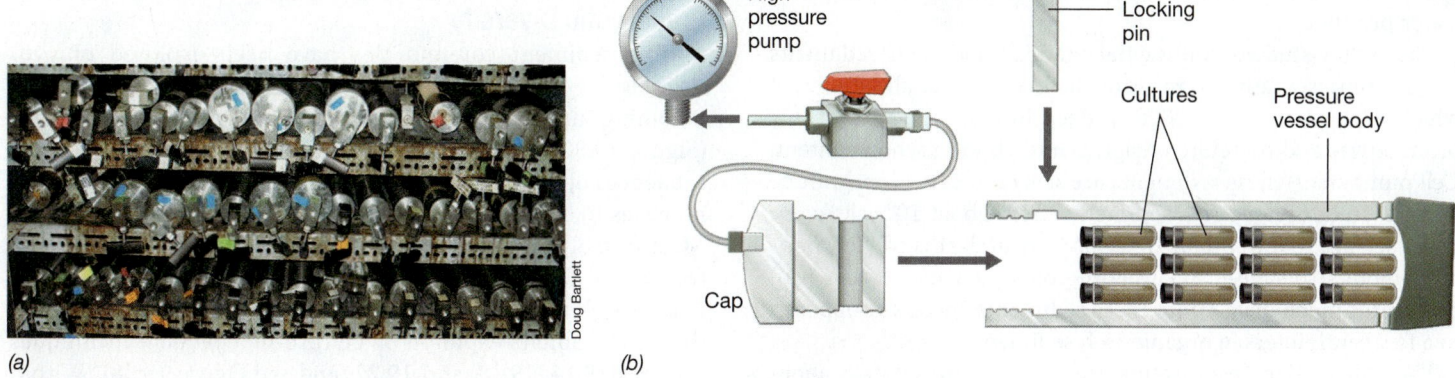

Doug Bartlett

Figure 19.31 **Pressure cells for growing piezophiles under elevated pressure.** *(a)* Photo of several pressure cells incubating in a cold room (4°C). *(b)* Schematic design of a pressure cell. These vessels are designed to maintain pressures of 1000 atm. Illustration based on drawing by Doug Bartlett.

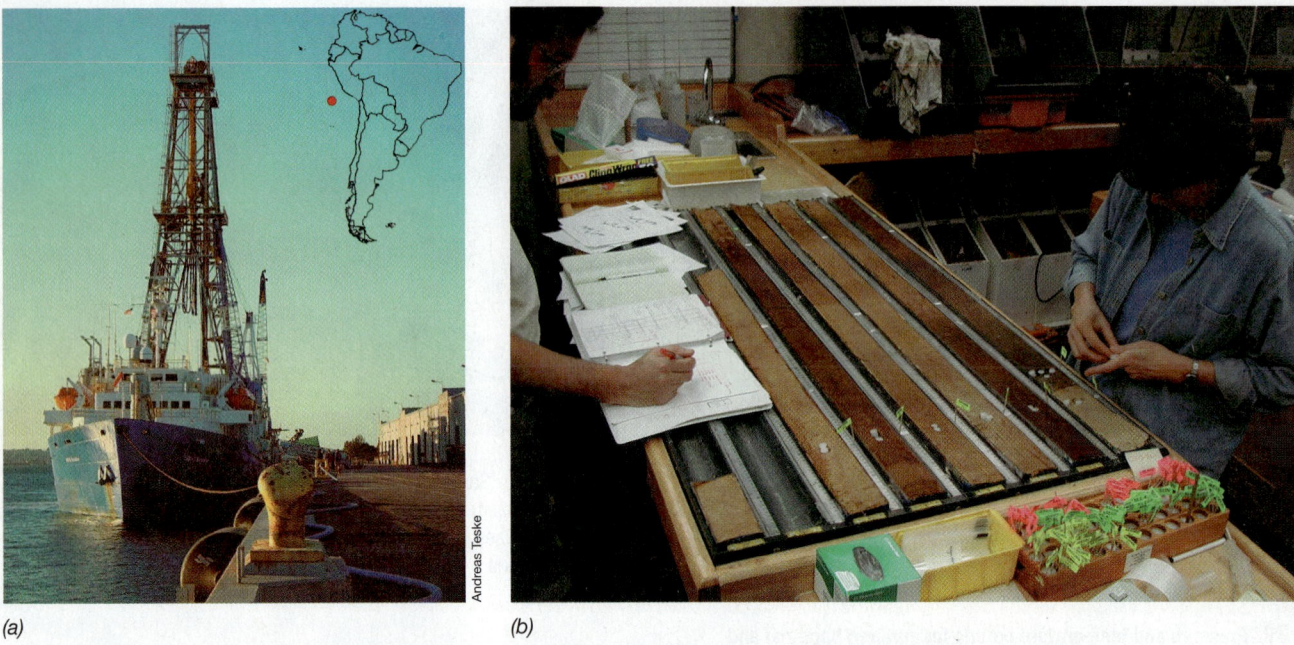

(a)

(b)

Andreas Teske

Figure 19.32 Drilling deep-sea sediments. *(a)* Deep-sea drilling vessel the JOIDES *Resolution.* Inset: red dot indicates the location of sediment sampling in the Peru Basin. *(b)* Sediment cores recovered from the Peru Basin at 4800 m depth. Cores were split lengthwise to allow subsampling for molecular characterization. See Section 19.5 and Figure 19.10 for discussion of sulfide-oxidizing microbial mats that grow on the sediment surface off the Chilean and Peruvian coasts.

in hydrostatic pressure alter the expression of a large number of genes in piezophiles, so it is likely that many other pressure-monitoring proteins exist in these organisms.

Deep-Sea Sediments

Another vast and mostly unexplored microbial ecosystem exists deep below the seafloor. Deep drilling expeditions to explore the depths below the ocean seafloor have revealed both archaeal and bacterial populations as deep as 1600 m (**Figure 19.32**). Most studies thus far have focused on relatively organic-rich deep-subsurface sediments at continental margins. Here, cell numbers typically decrease from about 10^9 cells/g of surface sediment to about 10^6 cells/g at depths as great as 1000 m below the seafloor. The decrease in cell numbers with depth is correlated with the amount of organic carbon entering the sediment system, primarily through past particle transport from surface water production.

The better-studied continental margins and shelf sediments are not representative of most of the ocean floor, about 90% of which is at greater than 2000 m deep in marine waters of low productivity and therefore of significantly lower carbon content. Cell numbers in these sediments are several orders of magnitude lower than in organic-rich sediments, from about 10^6 cells/gram at the surface to fewer than 10^3 cells/gram at depths of a few hundred meters. Because of lower microbial activity, oxygen (O_2) penetrates much deeper into these sediments, meters as opposed to a few centimeters in organic-rich sediments.

The sub-seafloor ecosystems are estimated to contain about 4 petagrams (1 petagram is 10^{15} g) of microbial cellular carbon, approximately 0.6% of Earth's total living biomass. Sequencing

of 16S ribosomal RNA genes selectively amplified by PCR (↩ Section 18.5) using DNA extracted from drilling cores, as well as more limited metagenomic surveys, has identified relatively few sequences related to the classical sulfate-reducing bacteria (↩ Section 14.9) or methanogenic and methane-oxidizing *Archaea* (↩ Sections 13.20, 13.24, and 16.2) common in surface sediments. Most of the *Archaea* in the subsurface, identified only by their 16S rRNA sequences, affiliate with novel phylum-level clades of *Archaea* having no presently cultured representatives. In turn, distinct lineages of novel *Archaea* preferentially inhabit the organic-rich coastal margin and the organic-poor sediments comprising most of the ocean floor, possibly reflecting varying electron donor and electron acceptor availability in these two types of sediment.

A Phylogenetic Snapshot of Marine Sediment Prokaryotic Diversity

Marine sediment communities have been explored only to a limited extent, given the great difficulty and expense of obtaining uncontaminated drilling cores from great depth (Figure 19.32). Analyses of available 16S ribosomal RNA gene sequences obtained from deep coring samples show these communities to be distinct from open-ocean and soil communities. Most notably, *Archaea* of unknown affiliation make up a large fraction of the diversity (**Figure 19.33**). By contrast, in shallow marine sediments *Proteobacteria* dominate, as they do in all of the other habitats explored by culture-independent techniques (Figures 19.14, 19.18, and 19.27, and see Figure 19.38). Within marine sediment *Proteobacteria*, phylotypes associated with sulfate-reducing bacteria such as the *Desulfobacterales* are quite

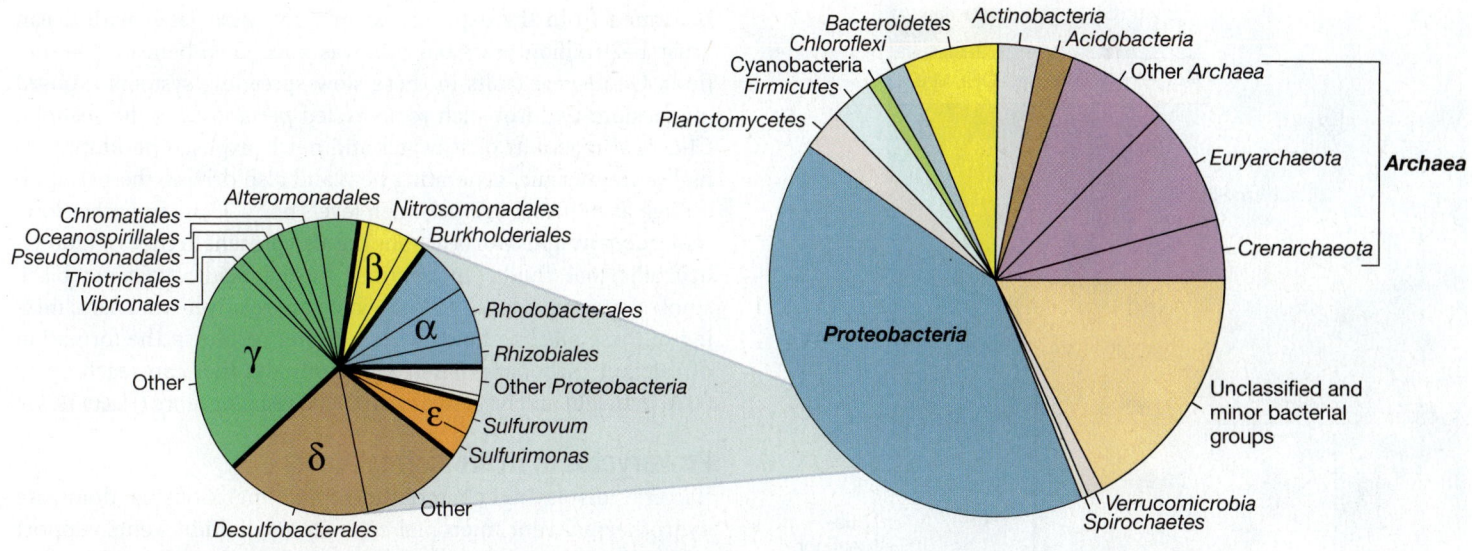

Figure 19.33 **Marine sediment prokaryotic diversity.** The results are pooled analyses of 13,360 16S ribosomal RNA gene sequences from several studies of shallow and deep marine sediments. Many of the groups indicated are covered in Chapters 14 and 15 (*Bacteria*) or 16 (*Archaea*). For *Proteobacteria*, major subgroups are indicated. Note the high proportion of archaeal sequences and of *Gamma-*, *Delta-*, and *Epsilonproteobacteria*. Data assembled and analyzed by Nicolas Pinel. Compare the prokaryotic diversity of marine sediments with that of open ocean water shown in Figure 19.27.

common (Figure 19.33); sulfate reduction is the major form of anaerobic respiration in marine sediments (↺ Sections 13.18 and 14.9). *Bacteroidetes* and the unclassified/minor groups are also well represented in shallow marine sediments.

Although major players in marine waters, cyanobacteria make up just a tiny proportion of the total cell population in the permanently dark and anoxic sediments and probably represent cells that have reached the sediments after attaching to a particle or dead animal that eventually sank. How organisms in deep marine sediments survive in the nutrient-depleted depths far below the seafloor is unclear, but it will not be surprising if it turns out they employ many of the strategies we have seen in pelagic prokaryotes, including small cell size and small, compact genomes.

MINIQUIZ --

- How does pressure change with depth in a water column?
- What molecular adaptations are found in piezophiles that allow them to grow optimally under high pressure?
- Why are sulfate-reducing bacteria common in marine sediments?

19.13 Hydrothermal Vents

Although we have thus far described the deep sea as a remote, low-temperature, high-pressure environment suitable only for slow-growing piezotolerant and piezophilic microorganisms, there are some amazing exceptions. Thriving animal and microbial communities are found clustered in and around thermal springs in deep-sea waters throughout the world. These hot springs are located at depths from less than 1000 m to greater

than 4000 m from the ocean surface in regions of the seafloor where volcanic magma and hot rock have caused the floor to rift apart at crustal spreading centers (**Figure 19.34**), or where iron and magnesium minerals associated with ancient rocks react with seawater and generate heat. Seawater seeping into these dynamic cracking regions of the crust reacts with hot rock, resulting in hot springs saturated with inorganic chemicals and dissolved gases. Collectively, these types of underwater hot springs are called **hydrothermal vents**. We discuss several remarkable symbiotic associations between hydrothermal vent–associated animals and microorganisms in Chapter 22. Here we consider the vent environment as a habitat for free-living microorganisms.

Types of Vents

Volcanic hydrothermal systems are typically either warm (~5 to >50°C), diffuse vents or very hot vents that emit hydrothermal fluids at 270 to >400°C. The gently flowing, warm, diffuse fluids are emitted from cracks in the seafloor and the exterior walls of hydrothermal chimneys. The fluids originate from the mixing of cold seawater with hot hydrothermal fluids in subsurface regions of the sediments. Hot vents, called *black smokers*, form upright sulfide edifices called *chimneys* that can be less than 1 m to over 30 m in height. Chimneys form when acidic hydrothermal fluids rich in dissolved metals and magmatic gases are suddenly mixed with cold, oxygenated seawater. The rapid mixing causes fine-grained metal sulfide minerals such as pyrite and sphalerite to precipitate out, forming dark, buoyant plumes that rise above the seafloor (**Figure 19.35**).

A quite different type of hydrothermal vent environment is the "Lost City" formation located in the mid-Atlantic Ocean. Lost City

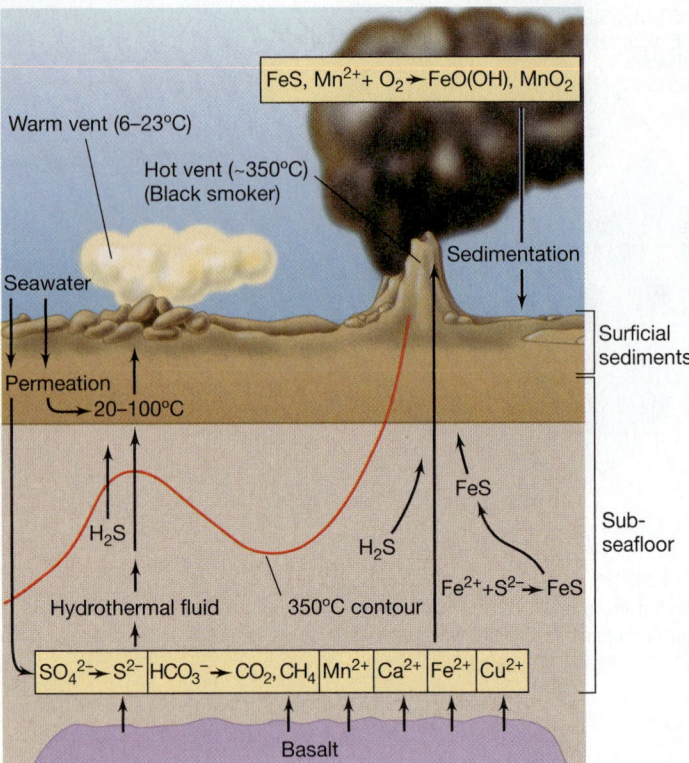

$$FeS, Mn^{2+} + O_2 \rightarrow FeO(OH), MnO_2$$

Warm vent (6–23°C)

Hot vent (~350°C) (Black smoker)

Sedimentation

Seawater

Surficial sediments

Permeation

20–100°C

Hydrothermal fluid

H_2S

H_2S

FeS

350°C contour

$Fe^{2+} + S^{2-} \rightarrow FeS$

Sub-seafloor

$$SO_4^{2-} \rightarrow S^{2-} \mid HCO_3^- \rightarrow CO_2, CH_4 \mid Mn^{2+} \mid Ca^{2+} \mid Fe^{2+} \mid Cu^{2+}$$

Basalt

Figure 19.34 Hydrothermal vents. Schematic showing geological formations and major inorganic chemicals and minerals that are emitted from warm vents and black smokers. In warm vents, the hot hydrothermal fluid is cooled by cold 2–3°C seawater permeating the sediments. In black smokers, hot hydrothermal fluid near 350°C reaches the seafloor directly. Surficial is a geological term pertaining to Earth's surface.

is formed from the exposure of minerals associated with ocean crust 1–2 million years old that was once deep beneath the seafloor. Geological faults in these slow-spreading systems exposed magnesium and iron-rich rocks called *peridotites* at the seafloor. Chemical reactions of seawater and newly exposed peridotite are highly exothermic, generating heat and also driving the pH up to as high as pH 11. Extremely high levels of H_2, CH_4, and other low-molecular-weight hydrocarbons are also present in the hot (200°C) hydrothermal fluids. In contrast to the acidic volcanic black smoker systems (Figure 19.34), which are relatively transient, mixing of these alkaline fluids with seawater results in the formation of calcium carbonate (limestone) chimneys that can reach up to 60 m in height and be active for 100,000 years or more (Figure 19.36).

Prokaryotes in Hydrothermal Vents

Bacteria displaying chemolithotrophic metabolisms dominate hydrothermal vent microbial ecosystems. Sulfidic vents support sulfur bacteria, whereas vents that emit other inorganic electron donors support nitrifying, hydrogen-oxidizing, iron- and manganese-oxidizing, or methylotrophic bacteria, the latter presumably growing on the CH_4 and carbon monoxide (CO) emitted from the vents. Table 19.2 summarizes the inorganic electron donors and electron acceptors that are thought to play a role in chemolithotrophic metabolisms at hydrothermal vents. All of these metabolisms were discussed in Chapter 13.

Figure 19.35 A hydrothermal vent black smoker emitting sulfide- and mineral-rich water at temperatures of 350°C. The walls of the black smoker chimneys display a steep temperature gradient and contain several types of prokaryotes.

Figure 19.36 Massive carbonate chimney formation at Lost City peridotite-hosted vent system. Microbial colonization of freshly exposed mineral surfaces was studied by placing sterile mineral fragments in the green-topped device placed over an actively venting area of the chimney. The diameter of the cylindrical collection device is approximately 10 cm.

Table 19.2 Chemolithotrophic prokaryotes present near deep-sea hydrothermal vents[a]

Chemolithotroph	Electron donor	Electron acceptor	Product from donor
Sulfur-oxidizing	HS^-, S^0, $S_2O_3^{2-}$	O_2, NO_3^-	S^0, SO_4^{2-}
Nitrifying	NH_4^+, NO_2^-	O_2	NO_2^-, NO_3^-
Sulfate-reducing	H_2	S^0, SO_4^{2-}	H_2S
Methanogenic	H_2	CO_2	CH_4
Hydrogen-oxidizing	H_2	O_2, NO_3^-	H_2O
Iron- and manganese-oxidizing	Fe^{2+}, Mn^{2+}	O_2	Fe^{3+}, Mn^{4+}
Methylotrophic	CH_4, CO	O_2	CO_2

[a]See Chapter 13 for detailed discussions of these metabolisms and Chapters 14–16 for further coverage of each group of organisms.

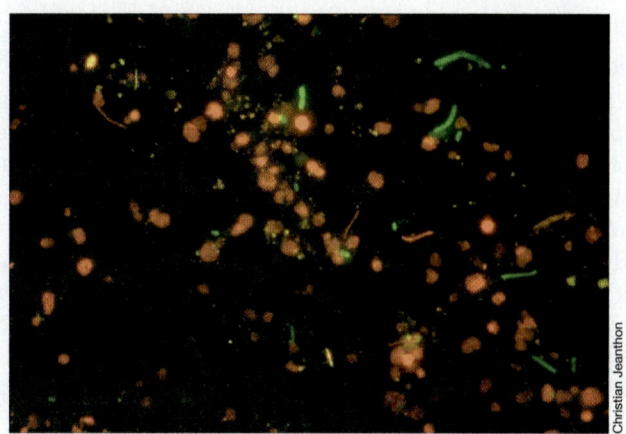

Figure 19.37 **Phylogenetic FISH staining of black smoker chimney material.** Taken from the Snake Pit vent field in the Mid-Atlantic Ridge, depth of 3500 m. A green fluorescing dye was conjugated to a probe that reacts with the 16S rRNA of all *Bacteria* and a red dye to a 16S rRNA probe for *Archaea*. The hydrothermal fluid going through the center of this chimney was at 300°C.

Although prokaryotes cannot survive in the superheated hydrothermal fluids of black smokers, thermophilic and hyperthermophilic organisms do thrive in the *gradients* that form as the superheated water mixes with cold seawater. For example, the walls of smoker chimneys are teeming with hyperthermophiles such as *Methanopyrus*, a species of *Archaea* that oxidizes H_2 and makes CH_4 (⇄ Section 16.4). Phylogenetic FISH staining (⇄ Section 18.4) has detected cells of both *Bacteria* and *Archaea* in smoker chimney walls (**Figure 19.37**). The most thermophilic of all known sulfur-reducing prokaryotes, species of *Pyrolobus* and *Pyrodictium* (Chapter 16), were isolated from black smoker chimney walls. In contrast to the significant microbial diversity in volcanic vent chimney walls, the carbonate

chimney walls of the Lost City vents are comprised primarily of methanogens of the genus *Methanosarcina*. These organisms are presumably nourished by the H_2-rich fluids that permeate the porous chimney walls.

When smokers plug up from mineral debris, hyperthermophiles presumably drift away to colonize active smokers and somehow become integrated into the growing chimney wall. Surprisingly, although they require very high temperatures for growth, hyperthermophiles are remarkably tolerant of cold temperatures and oxygen. Thus, transport of cells from one vent site to another in cold oxic seawater apparently is not a problem.

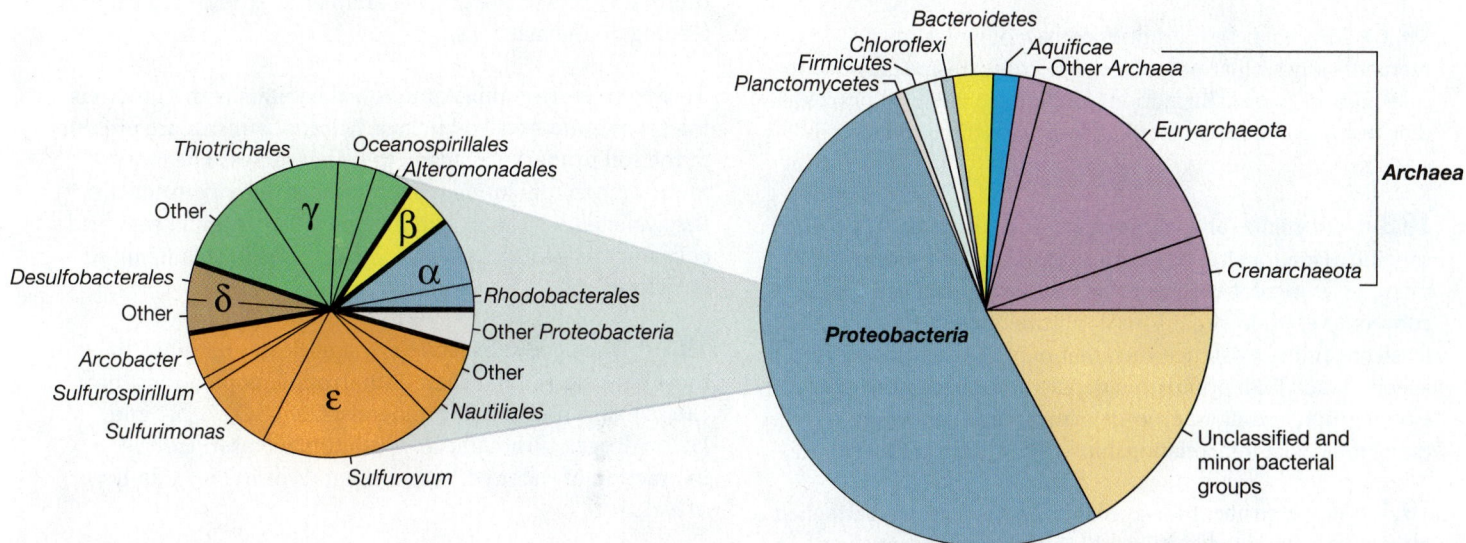

Figure 19.38 **Hydrothermal vent prokaryotic diversity.** The results are pooled analyses of 14,293 sequences from several studies of the 16S rRNA gene content of warm and hot hydrothermal vents. Many of these groups are covered in Chapters 14 and 15 (*Bacteria*) or 16 (*Archaea*). For *Proteobacteria*, major subgroups are indicated. Note the high proportion of *Archaea* and of *Epsilonproteobacteria*. The physiology of many of these organisms is summarized in Table 19.2. Data assembled and analyzed by Nicolas Pinel.

A Phylogenetic Snapshot of Hydrothermal Vent Prokaryotic Diversity

Using the powerful tools developed for microbial community sampling (🔗 Section 18.5), studies of prokaryotic diversity near volcanic hydrothermal vents have revealed an enormous diversity of *Bacteria*. These 16S rRNA gene sequence surveys include both warm and hot vents. Hydrothermal vent microbial communities are dominated by *Proteobacteria*, in particular *Epsilonproteobacteria* (🔗 Section 15.5; **Figure 19.38**). *Alpha-*, *Delta-*, and *Gammaproteobacteria* are also abundant, whereas *Betaproteobacteria* are much less so. Many *Epsilon-* and *Gammaproteobacteria* oxidize sulfide and sulfur as electron donors with either O_2 or nitrate (NO_3^-) as electron acceptors.

As shown in the blowup diagram of *Proteobacteria* in Figure 19.38, vent *Epsilonproteobacteria* phylotypes most closely match those of chemolithotrophic sulfur bacteria such as *Sulfurimonas*, *Arcobacter*, *Sulfurovum*, and *Sulfurospirillum*. These bacteria oxidize reduced sulfur compounds as electron donors (🔗 Sections 13.8 and 14.11), and such a physiology is consistent with their presence near vent fluids charged with sulfur and sulfide. In addition, most *Deltaproteobacteria* specialize in anaerobic metabolisms using oxidized sulfur compounds as electron acceptors.

In contrast to *Bacteria*, the diversity of volcanic hydrothermal vent *Archaea* is quite limited. Estimates of the number of unique phylotypes indicate that the diversity of *Bacteria* near hydrothermal vents is about 10 times that of *Archaea*. However, *Archaea* are prevalent in samples recovered from the walls of hot vent chimneys (Figure 19.37). Most of the *Archaea* detected near hydrothermal vents are either methanogens (🔗 Section 16.2) or species of marine *Crenarchaeota* and *Euryarchaeota* (🔗 Figure 16.1). With the exception of the ammonia-oxidizing thaumarchaeote *Nitrosopumilus* (🔗 Section 16.6), organisms in these groups remain uncultured and their physiologies poorly understood.

MINIQUIZ
- How does a warm hydrothermal vent differ from a black smoker, both chemically and physically?
- Why is 350°C water emitted from a black smoker not boiling?
- Which phylum of *Bacteria* and which subgroups of this phylum dominate hydrothermal vent ecosystems, and why?

BIG IDEAS

19.1 • Ecosystems consist of organisms, their environments, and all of the interactions among the organisms and environments. The organisms are members of populations and communities and are adapted to habitats. Species richness and abundance are aspects of species diversity in a community and an ecosystem.

19.2 • Microbial communities consist of guilds of metabolically similar organisms. Microorganisms play major roles in energy transformations and biogeochemical processes that result in the recycling of elements essential to living systems.

19.3 • The niche for a microorganism consists of the specific assortment of biotic and abiotic factors within a microenvironment in which that microorganism can be competitive. Microorganisms in nature often live a feast-or-famine existence such that only the best-adapted species reach high population density in a given niche. Cooperation among microorganisms is also important in many microbial interrelationships.

19.4 • When surfaces are available, bacteria grow in attached masses of cells called biofilms. Biofilm formation involves both intra- as well as intercellular communication and confers several protective advantages on cells. Biofilms can have significant medical and economic impacts on humans when unwanted biofilms develop on inert as well as living surfaces.

19.5 • Microbial mats are extremely thick biofilms consisting of microbial cells and trapped particulate materials. Microbial mats are widespread in hypersaline or thermal waters where grazing animals are prevented from feeding on the mat cells.

19.6 • Soils are complex microbial habitats with numerous microenvironments and niches. Microorganisms are present in the soil primarily attached to soil particles. The most important factors influencing microbial activity in soil are the availability of water and nutrients. However, in very arid soils microorganisms play important roles in stabilizing soil structure.

19.7 • The deep subsurface is a significant microbial habitat, most likely sustaining chemolithotrophic populations that can live on a diet of a few minerals, CO_2, SO_4^{2-}, N_2, and H_2. Hydrogen is thought to be continually produced by interaction of water with iron minerals or by the radiolysis of water.

19.8 • In freshwater aquatic ecosystems, phototrophic microorganisms are the main primary producers.

Most of the organic matter produced is consumed by bacteria, which can lead to depletion of oxygen in the environment. The BOD of a body of water indicates its relative content of organic matter that can be biologically oxidized.

19.9 • Pelagic marine waters are more nutrient deficient than most freshwaters, yet substantial numbers of prokaryotes inhabit the oceans. However, in some highly productive and expansive oceanic regions, oxygen can be drawn down to low levels at depths between 100 and 1000 m called oxygen minimum zones.

19.10 • The major microbial oxygenic phototrophs in the open oceans include the prokaryote *Prochlorococcus* and eukaryote *Ostreococcus*; both of these phototrophs are small microorganisms. Marine anoxygenic phototrophs include *Roseobacter* and its relatives, the aerobic phototrophic purple bacteria.

19.11 • Species of *Bacteria* tend to predominate in marine surface waters, whereas in deeper waters *Archaea* comprise a larger fraction of the microbial community. Many pelagic *Bacteria* use light to make ATP by rhodopsin-driven proton pumps. Viruses outnumber prokaryotes by several orders of magnitude in marine waters.

19.12 • The deep sea is a cold, dark habitat where hydrostatic pressure is high and nutrient levels are low. Piezophiles grow best under pressure but do not require pressure, whereas extreme piezophiles require high pressure, typically several hundred atmospheres, for growth.

19.13 • Hydrothermal vents are deep-sea hot springs where either volcanic activity or unusual chemistry generates fluids containing large amounts of inorganic electron donors that can be used by chemolithotrophic bacteria.

MasteringMicrobiology® **Review what you know and challenge what you have learned with MasteringMicrobiology!** Access study materials, chapter quizzes, animations, and microbiology lab tutorials in the Study Area to ensure that you have mastered this chapter's content.

REVIEW OF KEY TERMS

Biochemical oxygen demand (BOD) the microbial oxygen-consuming properties of a water sample

Biofilm colonies of microbial cells encased in a porous organic matrix and attached to a surface

Biogeochemistry the study of biologically mediated chemical transformations in the environment

Community two or more cell populations coexisting in a certain area at a given time

Ecosystem a dynamic complex of organisms and their physical environment interacting as a functional unit

Epilimnion the warmer and less dense surface waters of a stratified lake

Extreme piezophile an organism requiring several hundred atmospheres of pressure for growth

Guild metabolically similar microbial populations that exploit the same resources in a similar way

Habitat an environment within an ecosystem where a microbial community could reside

Hydrothermal vents warm or hot water–emitting springs associated with crustal spreading centers on the seafloor

Hypolimnion the colder, denser, and often anoxic bottom waters of a stratified lake

Microbial mat a thick, layered, diverse community nourished either by light in a hypersaline or an extremely hot aquatic environment, in which cyanobacteria are essential; or by chemolithotrophs growing on the surface of sulfide-rich marine sediments

Microenvironment a micrometer-scale space surrounding a microbial cell or group of cells

Niche in ecological theory, the biotic and abiotic characteristics of the microenvironment that contribute to an organism's competitive success

Oligotroph an organism that grows only or grows best at very low levels of nutrients

Oxygen minimum zone (OMZ) an oxygen-depleted region of intermediate depth in the marine water column

Piezophile an organism that grows best under a hydrostatic pressure greater than 1 atm

Piezotolerant able to grow under elevated hydrostatic pressures but growing best at 1 atm

Population a group of organisms of the same species in the same place at the same time

Primary producer an organism that synthesizes new organic material from CO_2 and obtains energy from light or from oxidation of inorganic compounds

Prochlorophyte a prokaryotic oxygenic phototroph that contains chlorophylls *a* and *b* and lacks phycobiliproteins

Proteorhodopsin a light-sensitive protein present in some pelagic *Bacteria* that fuels a proton pump that yields ATP

Rhizosphere the region immediately adjacent to plant roots

Species abundance the proportion of each species in a community

Species richness the total number of different species present in a community

Stratified water column a body of water separated into layers having distinct physical and chemical characteristics

UNIT 4

REVIEW QUESTIONS

1. List some of the key resources and conditions that microorganisms need to thrive in their habitats. (Section 19.1)

2. In what forms does potential energy enter a microbial ecosystem? Which energy classes of microorganisms can exploit each? (Section 19.2)

3. Explain why both obligately anaerobic and obligately aerobic bacteria can often be isolated from the same soil sample. (Section 19.3)

4. The surface of a rock in a flowing stream will often contain a biofilm. What advantages could be conferred on bacteria growing in a biofilm compared with growth within the flowing stream? (Section 19.4)

5. How can biofilms complicate treatment of infectious diseases? (Section 19.4)

6. How do microbial mats compare with biofilms in terms of dimensions and microbial diversity? (Section 19.5)

7. In what soil horizon are microbial numbers and activities the highest, and why? (Section 19.6)

8. What are biological soil crusts and what functions do they provide in arid regions? (Section 19.6)

9. How are nutrients for microbial growth replenished in the deep subsurface as opposed to the near subsurface? (Section 19.7)

10. How and in what way does an input of organic matter, such as sewage, affect the oxygen content of a river or stream? (Section 19.8)

11. Why are OMZs of harm to macrobiological marine life? (Section 19.9)

12. Which microorganisms are the major phototrophs in the oceans? (Section 19.10)

13. Many pelagic prokaryotes can use light energy but are not considered "phototrophs" in the same sense as cyanobacteria or purple bacteria. Explain. (Section 19.11)

14. What is the difference between piezotolerant and piezophilic bacteria? Between these two groups and extreme piezophiles? What properties do piezotolerant, piezophilic, and extremely piezophilic microorganisms have in common? (Section 19.12)

15. Why are chemolithotrophic bacteria so prevalent at hydrothermal vents? (Section 19.13)

APPLICATION QUESTIONS

1. Imagine a sewage plant that is releasing sewage containing high levels of ammonia and phosphate and very low levels of organic carbon. Which types of microbial blooms might be triggered by this sewage? How would the graphs of oxygen near and beyond the plant's release point differ from the graph shown in Figure 19.17a?

2. Keeping in mind that the open-ocean waters are highly oxic, predict the possible metabolic lifestyles of open-ocean *Archaea* and *Bacteria*. Why might rhodopsin-like pigments be more abundant in one group of organisms than in the other?

3. Global warming has been suggested to result in reduced transfer of oxygen to deeper waters in the ocean (Section 19.9). How might global warming also result in reduced nutrient availability to planktonic species in marine surface waters?

microbiology**now**

Microbial Power Lines

An exciting new area of microbial research surrounds how insoluble electron acceptors such as iron and manganese oxides are reduced in anoxic habitats. Although it is widely believed that electrically conductive structures exist on the bacterial cell surface that deliver electrons to the insoluble electron acceptor, molecular details have remained elusive. However, a recent discovery made with a soluble electron acceptor system may have shed light on this mystery.

A morphologically distinctive filamentous bacterium related to sulfate-reducing bacteria was found to oxidize hydrogen sulfide (H_2S) in marine sediments using oxygen (O_2) as the electron acceptor.[1] However, the sulfide and oxygen were separated from each other by *more than a centimeter,* leading to the obvious question of how the donor and acceptor reactions were coupled.

Microscopic and microsensor analyses revealed that the filamentous bacteria extended vertically as single filaments connecting the oxic surface and the anoxic sulfide-rich zone. The capacity of the filaments to couple electron transfer over long distances was established by using a very thin wire to slice between the oxic and anoxic zones; this severed the filaments and stopped sulfide oxidation.

This remarkable system of electron transfer was associated with cable-like structures that formed a ring around each filament and extended the entire length of each filament (photos). These are thought to function as "living electrical cables," coupling the transfer of electrons derived from sulfide oxidized at one end of the filament to the reduction of oxygen at the other end.

The mechanism of this remarkable metabolic feat, and its relationship to systems used by bacteria that reduce metal oxides, remains to be determined. However, it is clear that bacterially mediated electron transfers can occur over rather long distances in order to couple donor and acceptor events.

[1]Pfeffer, C., et al. 2012. Filamentous bacteria transport electrons over centimetre distances. *Nature 491:* 218–221.

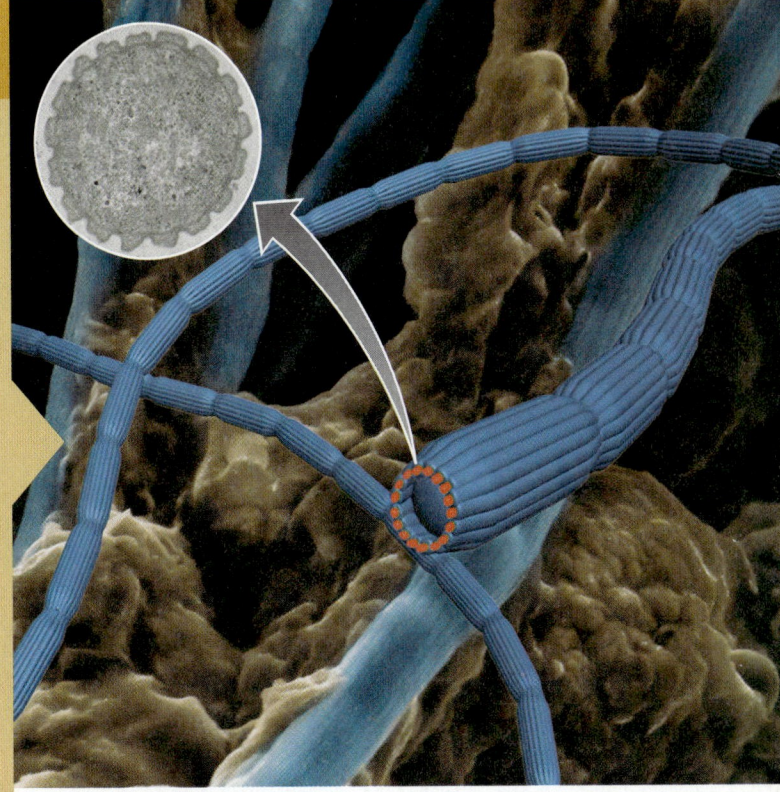

In the previous chapter we examined a variety of microbial habitats in order to set the stage for the consideration of some major microbial activities in this chapter. Here we explore the concept of *nutrient cycles* and the impact humans are having on those cycles. In particular we focus on the biogeochemical activities of microorganisms and see how these activities interrelate.

I • Carbon, Nitrogen, and Sulfur Cycles

The key nutrients for life are cycled by both microorganisms and macroorganisms, but for any given nutrient, it is microbial activities that dominate. Understanding how microbial nutrient cycles work is important because the cycles and their many feedback loops are essential for plant agriculture and the overall health of sustainable plant life.

We begin our coverage of nutrient cycles with the carbon cycle. Major areas of interest here are the magnitude of carbon reservoirs on Earth, the rates of carbon cycling within and between reservoirs, and the coupling of the carbon cycle to other nutrient cycles. We emphasize the gases *carbon dioxide* (CO_2) and *methane* (CH_4) as major components of the carbon cycle and of human impacts on the global ecosystem.

20.1 The Carbon Cycle

On a global basis, carbon (C) cycles as CO_2 through all of Earth's major carbon reservoirs: the atmosphere, the land, the oceans, freshwaters, sediments and rocks, and biomass (Figure 20.1). As we have already seen for freshwater environments, the carbon and oxygen cycles are intimately linked (♻ Section 19.8). All nutrient cycles link in some way to the carbon cycle, but the nitrogen (N) cycle links particularly strongly because, other than water (H_2O), C and N make up the bulk of living organisms (♻ Section 3.1 and see Figure 20.4).

Carbon Reservoirs

By far the largest C reservoir on Earth is the sediments and rocks of Earth's crust (Figure 20.1), but the rate at which sediments and rocks decompose and carbon is removed as CO_2 is so slow that flux out of this reservoir is insignificant on a human time scale. A large amount of C is found in land plants. This is the organic C of forests, grasslands, and agricultural crops—the major sites of phototrophic CO_2 fixation. However, more C is present in dead organic material, called **humus**, than in living organisms. Humus is a complex mixture of organic materials that have resisted rapid decomposition and is derived primarily from dead plants and microorganisms. Some humic substances are quite recalcitrant, with a decomposition time of several decades, but certain other humic components decompose much more rapidly.

The most rapid means of transfer of C is via the atmosphere. Carbon dioxide is removed from the atmosphere primarily by photosynthesis of land plants and marine microorganisms and is returned to the atmosphere by respiration of animals and chemoorganotrophic microorganisms (Figure 20.1). The single most important contribution of CO_2 to the atmosphere is by microbial decomposition of dead organic material. However, since the Industrial Revolution, human activities have increased atmospheric CO_2 levels by nearly 40%, primarily from the combustion of fossil fuels. This rise in CO_2, a major *greenhouse gas*, has triggered a period of steadily increasing global temperatures called

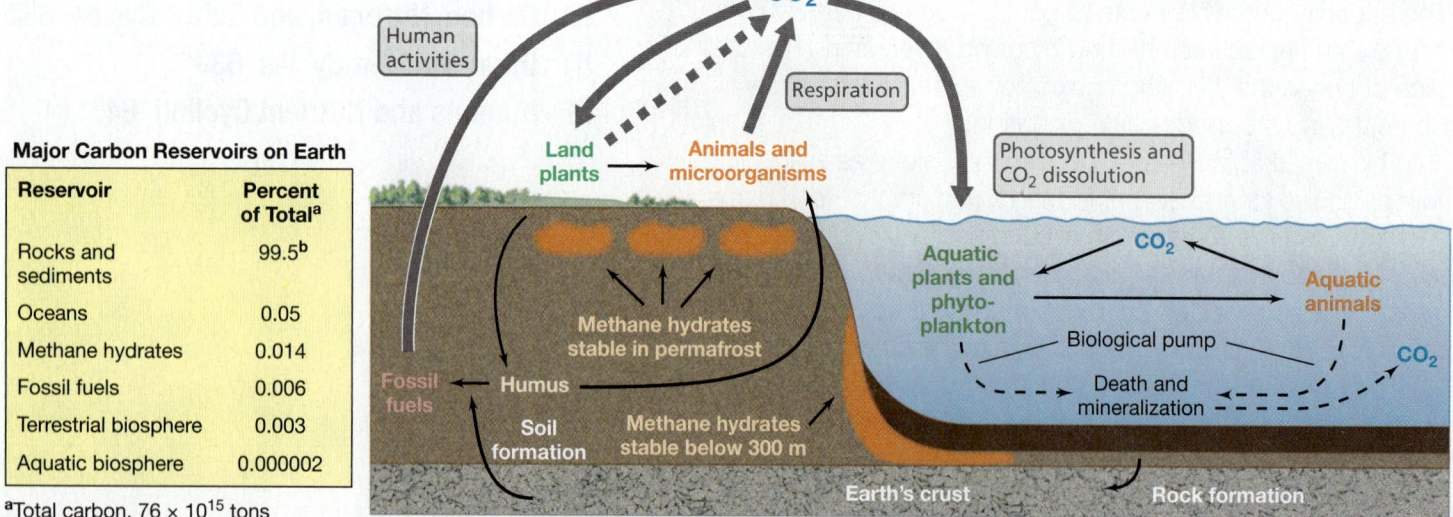

Major Carbon Reservoirs on Earth

Reservoir	Percent of Total[a]
Rocks and sediments	99.5[b]
Oceans	0.05
Methane hydrates	0.014
Fossil fuels	0.006
Terrestrial biosphere	0.003
Aquatic biosphere	0.000002

[a]Total carbon, 76×10^{15} tons
[b]80% inorganic

Figure 20.1 The carbon cycle. The carbon and oxygen cycles are closely connected, as oxygenic photosynthesis both removes CO_2 and produces O_2, and respiration both produces CO_2 and removes O_2. As the accompanying table shows, by far the greatest reservoir of carbon on Earth is in rocks and sediments, and most of this is in inorganic form as carbonates.

global warming (see Figure 20.18). Although the consequences of global warming on microbial nutrient cycling are currently unpredictable, everything we know about the biology of microorganisms tells us that microbial activities in nature will change in response to higher temperatures. Whether these responses will be favorable or unfavorable to higher organisms including people is today a major area of active research (Section 20.8).

Photosynthesis and Decomposition

New organic compounds are biologically synthesized on Earth only by CO_2 fixation by phototrophs and chemolithotrophs. Most organic compounds originate in photosynthesis and thus phototrophic organisms are the foundation of the carbon cycle (Figure 20.1). However, phototrophic organisms are abundant in nature only in habitats where light is available. The deep sea, deep terrestrial subsurface, and other permanently dark habitats are devoid of indigenous phototrophs. There are two groups of oxygenic phototrophic organisms: *plants* and *microorganisms*. Plants are the dominant phototrophic organisms of terrestrial environments, whereas phototrophic microorganisms dominate in aquatic environments.

The redox cycle for C (**Figure 20.2**) begins with photosynthetic CO_2 fixation, driven by the energy of light:

$$CO_2 + H_2O \rightarrow (CH_2O) + O_2$$

CH_2O represents organic matter at the oxidation–reduction level of cell material. Phototrophic organisms also carry out respiration,

both in the light and the dark. The overall equation for respiration is the reverse of oxygenic photosynthesis:

$$(CH_2O) + O_2 \rightarrow CO_2 + H_2O$$

For organic compounds to accumulate, the rate of photosynthesis must exceed the rate of respiration. In this way, autotrophic organisms build biomass from CO_2, and then this biomass in one way or another supplies the C heterotrophic organisms need. Anoxygenic phototrophs and chemolithotrophs also produce excess organic compounds, but in most environments the contributions of these organisms to the accumulation of organic matter are minor compared to the inputs of oxygenic phototrophs. This is because the reductant used by oxygenic phototrophs, H_2O, is in virtually unlimited supply.

Organic compounds are degraded biologically to CH_4 and CO_2 (Figure 20.2). Carbon dioxide, most of which is of microbial origin, is produced by aerobic and anaerobic respirations (⮌ Section 13.16). Methane is produced in anoxic environments by *methanogens* from the reduction of CO_2 with hydrogen (H_2) or from the splitting of acetate into CH_4 and CO_2. However, any naturally occurring organic compound can eventually be converted to CH_4 from the cooperative activities of methanogens and various fermentative bacteria, as we will see in the next section. Methane produced in anoxic habitats is insoluble and diffuses to oxic environments, where it is either released to the atmosphere or oxidized to CO_2 by *methanotrophs* (Figure 20.2). Hence, most of the C in organic compounds eventually returns to CO_2, and the links in the carbon cycle are closed.

Methane Hydrates

Although present in the atmosphere at levels lower than even CO_2, CH_4 is a potent greenhouse gas that is over 20 times more effective in trapping heat than is CO_2. Some CH_4 enters the atmosphere from methanogenic production, but not all biologically produced CH_4 is immediately consumed or released to the atmosphere. Huge amounts of CH_4 derived primarily from microbial activities are trapped underground or under marine sediments as *methane hydrates*, molecules of frozen CH_4. Methane hydrates form when sufficient CH_4 is present in environments of high pressure and low temperature such as beneath the permafrost in the Arctic and in marine sediments (Figure 20.1). These deposits can be up to several hundred meters thick and are estimated to contain 700–10,000 petagrams (1 petagram = 10^{15} g) of CH_4. This exceeds other known CH_4 reserves on Earth by several orders of magnitude.

Methane hydrates are highly dynamic, absorbing and releasing CH_4 in response to changes in pressure, temperature (**Figure 20.3**), and fluid movement. Methane hydrates also fuel deep-water ecosystems, called *cold seeps*. Here, the slow release of CH_4 from seafloor hydrates nourishes not only anaerobic methane-oxidizing *Archaea* (⮌ Section 13.24), but also animal communities that contain aerobic methane-oxidizing endosymbionts that oxidize CH_4 and release organic matter to the animals (⮌ Section 22.12). Anaerobic oxidation of CH_4 is coupled to the reduction of sulfate (SO_4^{2-}), nitrate (NO_3^-), and oxides of iron and manganese [e.g., FeO(OH)], and climate scientists now fear that global warming could catalyze a catastrophic release of CH_4 from methane hydrates, an event that would rapidly affect Earth's climate.

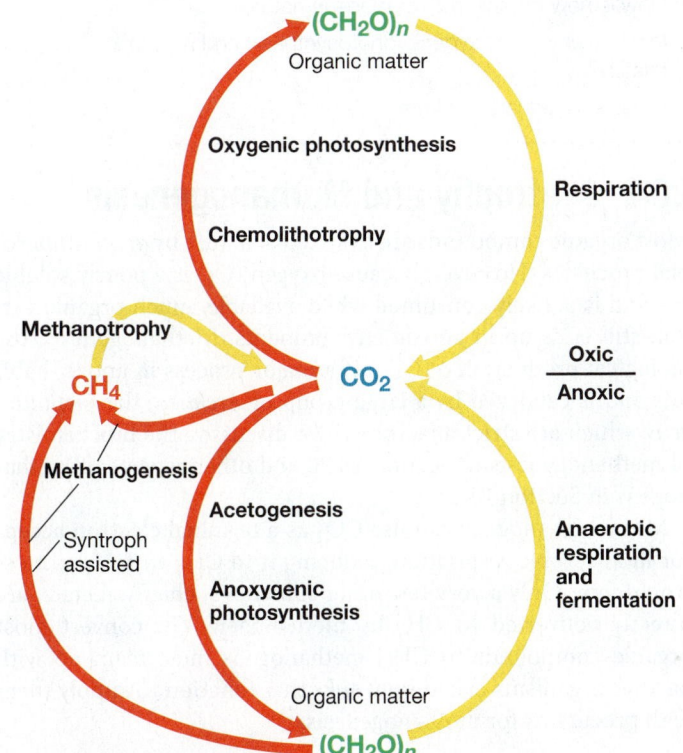

Figure 20.2 Redox cycle for carbon. The diagram contrasts autotrophic processes ($CO_2 \rightarrow$ organic compounds) and heterotrophic processes (organic compounds $\rightarrow CO_2$). Yellow arrows indicate oxidations; red arrows indicate reductions.

Figure 20.3 **Burning methane hydrate.** Frozen methane ice retrieved from marine sediments is ignited.

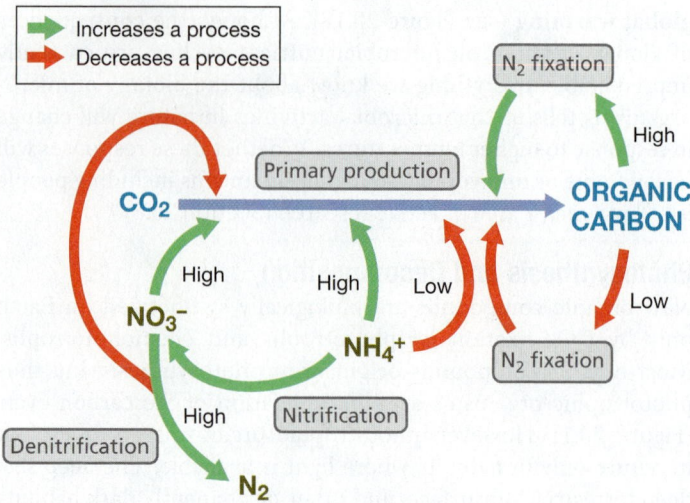

Figure 20.4 **Coupled cycles.** All nutrient cycles are interconnected, but the carbon and nitrogen cycles are intimately coupled. In the carbon cycle, CO_2 supplies the C for carbon compounds. The N cycle, shown in more detail in Figure 20.7, supplies N for many of the compounds.

In fact, the sudden release of large amounts of CH_4 from methane hydrates may have triggered the Permian–Triassic extinctions some 250 million years ago. These extinctions, the worst in Earth's history, wiped out virtually all marine animals and over 70% of all terrestrial plant and animal species. In addition to the release of methane hydrates, as permafrost melts, its huge reserve of organic matter could trigger the formation of additional methane (see *Archaea* and Global Warming, page 517).

Carbon Balances and Coupled Cycles

Although it is convenient to consider carbon cycling as a series of reactions separate from those in other nutrient cycles, awareness of how the various nutrient cycles feed back upon one another and are interconnected is extremely important. In reality, all nutrient cycles are *coupled cycles*; major changes in one cycle affect the functioning of others. But certain cycles, such as the carbon and nitrogen cycles (**Figure 20.4**), are very closely coupled and experience severe human impacts with untoward consequences for health of the planet (see Section 20.8). The rate of primary productivity (CO_2 fixation) is controlled by several factors, in particular by the magnitude of photosynthetic biomass and by available N, often a limiting nutrient. Thus, large-scale reductions in biomass, for instance by widespread deforestation, reduce rates of primary productivity and increase levels of CO_2. High levels of organic C stimulate nitrogen fixation ($N_2 \rightarrow NH_3$) and this in turn adds more fixed N to the pool for primary producers; low levels of organic C have just the opposite effect (Figure 20.4). High levels of ammonia (NH_3) stimulate primary production and nitrification, but inhibit nitrogen fixation. High levels of nitrate (NO_3^-), an excellent N source for plants and aquatic phototrophs, stimulate primary production but also increase the rate of denitrification; the latter removes fixed forms of N from the environment and feeds back in a negative way on primary production (Figure 20.4).

This simple example illustrates how nutrient cycles are anything but isolated entities; they are coupled systems that maintain a delicate balance of inputs and outputs. Thus, one could expect these cycles to respond to large inputs in specific links (for example,

through inputs of CO_2 or nitrogen fertilizers) in ways that are not always beneficial to the biosphere (Section 20.8). This is particularly true of the C and N cycles because next to H_2O, C and N are the most abundant elements in living organisms and their cycles interact with each other in such major ways.

MINIQUIZ

• How is new organic matter made in nature?

• In what ways are oxygenic photosynthesis and respiration related?

• What is a methane hydrate?

20.2 Syntrophy and Methanogenesis

Most organic compounds are oxidized in nature by *aerobic* microbial processes. However, because oxygen (O_2) is a poorly soluble gas and is actively consumed when available, much organic carbon still ends up in anoxic environments. Methanogenesis, the biological production of CH_4, is a major process in anoxic habitats and is catalyzed by a large group of *Archaea*, the *methanogens*, which are strict anaerobes. We discussed the biochemistry of methanogenesis in Section 13.20 and other aspects of methanogens in Section 16.2.

Most methanogens can use CO_2 as a terminal electron acceptor in anaerobic respiration, reducing it to CH_4 with H_2 as electron donor. Only a very few other substrates, chiefly acetate, are directly converted to CH_4 by methanogens. To convert most organic compounds to CH_4, methanogens must team up with partner organisms called *syntrophs* that function to supply them with precursors for methanogenesis.

Anoxic Decomposition and Syntrophy

In Section 13.15 we discussed the biochemistry of **syntrophy**, a process in which two or more organisms cooperate in the anaerobic degradation of organic compounds. Here we consider the

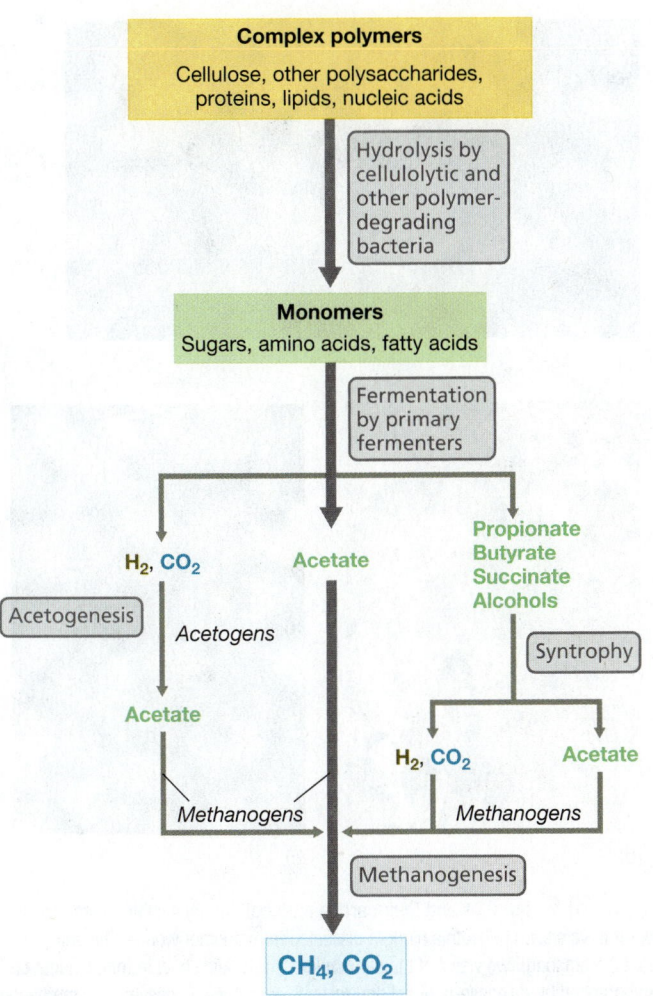

Figure 20.5 Anoxic decomposition. In anoxic decomposition various groups of fermentative anaerobes cooperate in the conversion of complex organic materials to CH_4 and CO_2. This pattern holds for environments in which sulfate-reducing bacteria play only a minor role; for example, in freshwater lake sediments, sewage sludge bioreactors, or the rumen.

interactions of syntrophic bacteria with their partner organisms and their significance for the carbon cycle. Our focus will be anoxic freshwater sediments and anoxic wastewater treatment, both of which are major sources of CH_4.

Polysaccharides, proteins, lipids, and nucleic acids from dead organisms find their way into anoxic habitats, where they are catabolized. The monomers released by hydrolysis of these polymers become major electron donors for energy metabolism. For the breakdown of a typical polysaccharide such as cellulose, the process begins with *cellulolytic* bacteria (**Figure 20.5**). These organisms hydrolyze cellulose into glucose, which is catabolized by fermentative organisms to short-chain fatty acids (acetate, propionate, and butyrate), alcohols such as ethanol and butanol, and the gases H_2 and CO_2. Hydrogen (H_2) and acetate are consumed by methanogens directly, but the bulk of the carbon remains in the form of fatty acids and alcohols; these cannot be directly catabolized by methanogens and require the activities of syntrophic bacteria (Section 13.15; Figure 20.5).

Syntrophic bacteria are *secondary* fermenters because they ferment the products of the primary fermenters, yielding H_2, CO_2, and acetate as products. For example, *Syntrophomonas wolfei* oxidizes C_4 to C_8 fatty acids, yielding acetate, CO_2 (if the fatty acid was C_5 or C_7), and H_2 (**Table 20.1** and Figure 20.5). Other species of *Syntrophomonas* use fatty acids up to C_{18} in length, including some unsaturated fatty acids. *Syntrophobacter wolinii* specializes in propionate (C_3) fermentation, generating acetate, CO_2, and H_2, and *Syntrophus gentianae* degrades aromatic compounds such as benzoate to acetate, H_2, and CO_2 (Table 20.1). Despite rather extensive metabolic diversity, syntrophs are unable to carry out any of these reactions in pure culture. Instead, they depend on a H_2-consuming partner organism because of the unusual bioenergetics linked to the syntrophic process.

As described in Section 13.15, H_2 consumption by a partner organism is absolutely essential for growth of syntrophic bacteria (in the absence of other electron acceptors), and the association of H_2 producer and H_2 consumer can be very intimate. In fact, it

UNIT 4

Table 20.1 Major reactions in the anoxic conversion of organic compounds to methane[a]

Reaction type	Reaction	$\Delta G^{0\prime}$ [b]	ΔG [c]
		Free energy change (kJ/reaction)	
Fermentation of glucose to acetate, H_2, and CO_2	Glucose + $4H_2O \rightarrow 2$ acetate$^-$ + $2HCO_3^-$ + $4H^+$ + $4H_2$	−207	−319
Fermentation of glucose to butyrate, CO_2, and H_2	Glucose + $2H_2O \rightarrow$ butyrate$^-$ + $2HCO_3^-$ + $2H_2$ + $3H^+$	−135	−284
Fermentation of butyrate to acetate and H_2	Butyrate$^-$ + $2H_2O \rightarrow 2$ acetate$^-$ + H^+ + $2H_2$	+48.2	−17.6
Fermentation of propionate to acetate, CO_2, and H_2	Propionate$^-$ + $3H_2O \rightarrow$ acetate$^-$ + HCO_3^- + H^+ + H_2	+76.2	−5.5
Fermentation of ethanol to acetate and H_2	2 Ethanol + $2H_2O \rightarrow 2$ acetate$^-$ + $4H_2$ + $2H^+$	+19.4	−37
Fermentation of benzoate to acetate, CO_2, and H_2	Benzoate$^-$ + $7H_2O \rightarrow 3$ acetate$^-$ + $3H^+$ + HCO_3^- + $3H_2$	+70.1	−18
Methanogenesis from H_2 + CO_2	$4H_2$ + HCO_3^- + $H^+ \rightarrow CH_4$ + $3H_2O$	−136	−3.2
Methanogenesis from acetate	Acetate$^-$ + $H_2O \rightarrow CH_4$ + HCO_3^-	−31	−24.7
Acetogenesis from H_2 + CO_2	$4H_2$ + $2HCO_3^-$ + $H^+ \rightarrow$ acetate$^-$ + $4H_2O$	−105	−7.1

[a]Data adapted from Zinder, S. 1984. Microbiology of anaerobic conversion of organic wastes to methane: Recent developments. *Am. Soc. Microbiol.* 50:294–298.
[b]Standard conditions: solutes, 1 M; gases, 1 atm; 25°C.
[c]Concentrations of reactants in typical anoxic freshwater ecosystems: fatty acids, 1 mM; HCO_3^-, 20 mM; glucose, 10 μM; CH_4, 0.6 atm; H_2, 10^{-4} atm. For calculating ΔG from $\Delta G^{0\prime}$, refer to Appendix 1.

is thought that H_2 transfer in some syntrophic associations may be through *direct conduction*, where electrons are transferred between species using electrically conductive wirelike structures (see Explore the Microbial World later in this chapter). But no matter how the transfer occurs, it is the transfer of H_2 itself that makes the syntrophic association work. How is this so?

When the reactions listed in Table 20.1 for the fermentation of butyrate, propionate, ethanol, or benzoate are written with all reactants at standard conditions (solutes, 1 M; gases, 1 atm, 25°C), the reactions yield free-energy changes ($\Delta G^{0\prime}$, ᴄᴈ Section 3.4) that are positive in arithmetic sign; that is, the reactions *require* rather than *release* energy. But the consumption of H_2 dramatically affects the energetics, making the reaction favorable and allowing energy to be conserved. This can be seen in Table 20.1, where the ΔG values (free-energy change measured under actual conditions in the habitat) are negative in arithmetic sign if H_2 concentrations are kept near zero by a H_2-consuming partner organism. This allows the syntrophic bacterium to conserve a small amount of energy that is used to produce ATP.

The final products of the syntrophic partnership are CO_2 and CH_4 (Figure 20.5), and any naturally occurring organic compound that enters a methanogenic habitat will eventually be converted to these products. This includes even complex aromatic and aliphatic hydrocarbons. Additional organisms other than those shown in Figure 20.5 may participate in such degradations, but eventually fatty acids and alcohols will be generated and they will be converted to methanogenic substrates by syntrophs. Acetate produced by syntrophs (as well as by the activities of acetogenic bacteria, ᴄᴈ Section 13.19) is a direct methanogenic substrate and is converted to CO_2 and CH_4 by various methanogens.

Methanogenic Symbionts and Acetogens in Termites

A variety of anaerobic protists that thrive under strictly anoxic conditions, including ciliates and flagellates, play a major role in the carbon cycle. Methanogenic *Archaea* live within some of these protist cells as H_2-consuming endosymbionts. For example, methanogens are present *within* cells of trichomonal protists inhabiting the termite hindgut (**Figure 20.6**) where methanogenesis and acetogenesis are major metabolic processes. Methanogenic symbionts of protists are species of the genera *Methanobacterium* or *Methanobrevibacter* (ᴄᴈ Section 16.2).

In the termite hindgut, endosymbiotic methanogens along with acetogenic bacteria are thought to benefit their protist hosts by consuming H_2 generated from glucose fermentation by cellulolytic protists. The acetogens are not endosymbionts but instead reside in the termite hindgut itself, consuming H_2 from primary fermenters and reducing CO_2 to make acetate. Unlike methanogens, acetogens can ferment glucose directly to acetate. Acetogens can also ferment methoxylated aromatic compounds to acetate. This is especially important in the termite hindgut because termites live on wood, which contains lignin, a complex polymer of methoxylated aromatic compounds. The acetate produced by acetogens in the termite hindgut is consumed by the insect as its primary energy source. Microbial symbioses in the termite hindgut are discussed in more detail in Section 22.10.

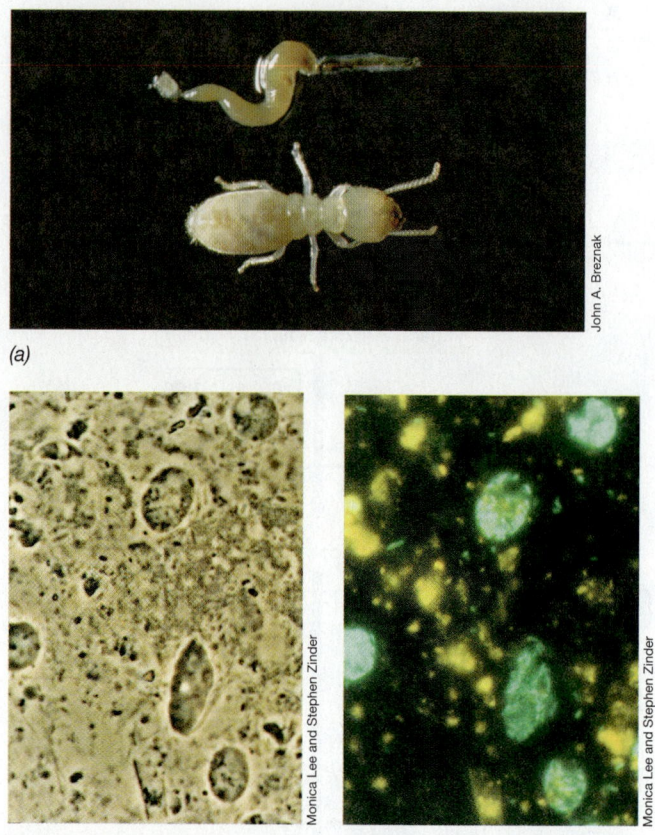

(a)

(b)　　　　　(c)

Figure 20.6 Termites and their carbon metabolism. *(a)* A subterranean termite worker larva shown beneath a hindgut dissected from another worker. The animal is about 0.5 cm long. Two views of the same microscopic field show termite hindgut protists photographed by *(b)* phase-contrast and *(c)* epifluorescence. Endosymbiotic methanogens in the protist cells fluoresce blue-green due to the methanogenic coenzyme F_{420} (compare with Figure 13.48). The average diameter of the protist cells is 15–20 μm.

MINIQUIZ

- Why does *Syntrophomonas* need a partner organism to ferment fatty acids or alcohols?
- What kinds of organisms are used in coculture with *Syntrophomonas*?
- What is the final product of acetogenesis?

20.3 The Nitrogen Cycle

Nitrogen is an essential element for life (ᴄᴈ Section 3.1) and exists in a number of oxidation states. We have discussed four major microbial N transformations thus far: nitrification, denitrification, anammox, and nitrogen fixation (Chapter 13). These and other key N transformations are summarized in the redox cycle shown in **Figure 20.7**.

Nitrogen Fixation and Denitrification

Nitrogen gas (N_2) is the most stable form of N and is a major reservoir for N on Earth. However, only a relatively small number of prokaryotes are able to use N_2 as a cellular N source by the process of *nitrogen fixation* ($N_2 + 8H \rightarrow 2NH_3 + H_2$) (ᴄᴈ Section 3.17). The N recycled on Earth is mostly already "fixed N"; that is, N in

Key Processes and Prokaryotes in the Nitrogen Cycle

Processes	Example organisms
Nitrification ($NH_4^+ \longrightarrow NO_3^-$)	
$NH_4^+ \longrightarrow NO_2^-$	*Nitrosomonas, Nitrosopumilus (Archaea)*
$NO_2^- \longrightarrow NO_3^-$	*Nitrobacter*
Denitrification ($NO_3^- \longrightarrow N_2$)	*Bacillus, Paracoccus, Pseudomonas*
N_2 Fixation ($N_2 + 8\,H \longrightarrow NH_3 + H_2$)	
Free-living	
Aerobic	*Azotobacter*
	Cyanobacteria
Anaerobic	*Clostridium*, purple and green phototrophic bacteria
	Methanobacterium (Archaea)
Symbiotic	*Rhizobium*
	Bradyrhizobium
	Frankia
Ammonification (organic-N $\longrightarrow NH_4^+$)	
	Many organisms can do this
Anammox ($NO_2^- + NH_3 \longrightarrow 2\,N_2$)	*Brocadia*

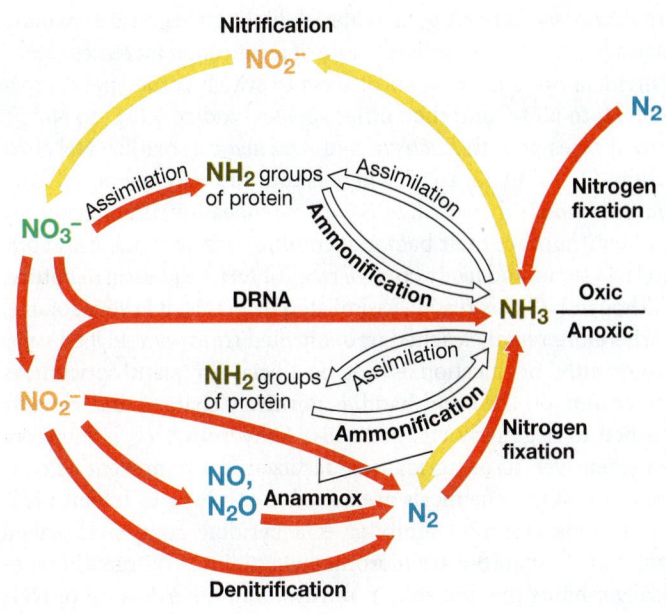

Figure 20.7 Redox cycle for nitrogen. Oxidation reactions are shown by yellow arrows and reductions by red arrows. Reactions without redox change are in white. The actual anammox reaction is $NH_4^+ + NO_2^- \rightarrow N_2 + 2H_2O$ (⟳ Figure 13.28). DRNA, dissimilative reduction of nitrate to ammonia.

combination with other elements, such as in ammonia (NH_3) or nitrate (NO_3^-). In many environments, however, the short supply of fixed N puts a premium on biological nitrogen fixation, and in these habitats, nitrogen-fixing bacteria flourish.

We discussed the role of NO_3^- as an alternative electron acceptor in anaerobic respiration in Section 13.17. Under most conditions, the end product of NO_3^- reduction is N_2, nitric oxide (NO), or nitrous oxide (N_2O). The reduction of NO_3^- to these gaseous N compounds, called **denitrification** (Figure 20.7), is the main means by which N_2 and N_2O are formed biologically. On the one hand, denitrification is a detrimental process. For example, if agricultural fields fertilized with NO_3^- fertilizer become waterlogged following heavy rains, anoxic conditions can develop and denitrification can be extensive; this removes fixed N from the soil. On the other hand, denitrification can aid in wastewater treatment (⟳ Sections 21.6 and 21.7). By converting NO_3^- to volatile forms of N, denitrification minimizes fixed N and thus algal growth when the treated sewage is discharged into lakes and streams.

The production of N_2O and NO by denitrification can have other environmental consequences. Nitrous oxide can be photochemically oxidized to NO in the atmosphere. Nitric oxide reacts with ozone (O_3) in the upper atmosphere to form nitrite (NO_2^-), and this returns to Earth as nitric acid (HNO_2). In addition, N_2O is a very potent greenhouse gas. Although N_2O molecules persist on average only about 100 years because of their reactivity, on a per weight basis, the contribution of N_2O to warming is about 300 times that of CO_2. Thus, denitrification contributes to global warming; to O_3 destruction, which increases passage of ultraviolet radiation to the surface of Earth; and to acid rain, which increases acidity of soils. Increases in soil acidity can change microbial community

structure and function and, ultimately, soil fertility, impacting both plant diversity and agricultural yields of crop plants.

Ammonification and Ammonia Fluxes

Ammonia is released during the decomposition of organic N compounds such as amino acids and nucleotides, a process called *ammonification* (Figure 20.7). Another process contributing to the generation of NH_3 is the respiratory reduction of NO_3^- to NH_3, called *dissimilative reduction of nitrate to ammonia* (DRNA, Figure 20.7). DRNA dominates NO_3^- and nitrite (NO_2^-) reduction in reductant-rich anoxic environments, such as highly organic marine sediments and the human gastrointestinal tract. It is thought that nitrate-reducing bacteria exploit this pathway primarily when NO_3^- is limiting because DRNA consumes eight electrons compared with the four and five electrons consumed when NO_3^- is reduced to N_2O or N_2, respectively.

At neutral pH, NH_3 exists as ammonium (NH_4^+). Much of the NH_4^+ released by aerobic decomposition in soils is rapidly recycled and converted to amino acids in plants and microorganisms. However, because NH_3 is volatile, some of it can be lost from alkaline soils by vaporization, and there are major losses of NH_3 to the atmosphere in areas with dense animal populations (for example, cattle feedlots). On a global basis, however, NH_3 constitutes only about 15% of the N released to the atmosphere, the rest being primarily N_2 or N_2O from denitrification.

Nitrification and Anammox

Nitrification, the oxidation of NH_3 to NO_3^-, is a major process in well-drained oxic soils at neutral pH, and is carried out by the nitrifying prokaryotes (Figure 20.7). Whereas denitrification *consumes* NO_3^-, nitrification *produces* NO_3^-. If materials high in NH_3, such as manure or sewage, are added to soils, the rate of nitrification increases.

Nitrification is a two-step process in which some species oxidize NH_3 to NO_2^- and then other species oxidize NO_2^- to NO_3^-. Many species of both *Bacteria* and *Archaea* can oxidize NH_3 (⮂ Sections 13.10, 14.13, 16.6), whereas thus far, only species of *Bacteria* are known that oxidize NO_2^-. Archaeal nitrifiers generally greatly outnumber their bacterial counterparts in marine and terrestrial systems and likely control rates of NH_3 oxidation in nature.

Although NO_3^- is readily assimilated by plants, it is very soluble, and therefore rapidly leached or denitrified from waterlogged soils; consequently, nitrification is not beneficial for plant agriculture. Ammonium, on the other hand, is positively charged and strongly adsorbed to negatively charged soils. Anhydrous NH_3 is therefore used extensively as an agricultural fertilizer, but to prevent its conversion to NO_3^-, chemicals are added to the NH_3 to inhibit nitrification. One common inhibitor is a pyridine compound called *nitrapyrin* (2-chloro-6-trichloromethylpyridine). Nitrapyrin specifically inhibits the *first* step in nitrification, the oxidation of NH_3 to NO_2^-. However, this effectively inhibits both steps in nitrification because the second step, $NO_2^- \rightarrow NO_3^-$, depends on the first (⮂ Section 13.10). The addition of nitrapyrin to anhydrous NH_3 has greatly increased the efficiency of crop fertilization and has helped prevent pollution of waterways by NO_3^- leached from nitrified soils.

Ammonia can be oxidized under anoxic conditions by the bacterium *Brocadia* in the process called *anammox*. In this reaction, NH_3 is oxidized anaerobically with NO_2^- as the electron acceptor, forming N_2 as the final product (Figure 20.7), which is released to the atmosphere. Although a major process in sewage and in anoxic marine basins and sediments, anammox is not significant in well-drained (oxic) soils. The microbiology and biochemistry of anammox was discussed in Section 13.10.

MINIQUIZ
- What is nitrogen fixation and why is it important to the nitrogen cycle?
- How do the processes of nitrification and denitrification differ? How do nitrification and anammox differ?
- How does the compound nitrapyrin benefit both agriculture and the environment?

20.4 The Sulfur Cycle

Microbial transformations of sulfur (S) are even more complex than those of N because of the large number of oxidation states of S and the fact that several transformations of S also occur spontaneously (abiotically). Chemolithotrophic S oxidation and sulfate (SO_4^{2-}) reduction were covered in Sections 13.8, 13.18, 14.9, and 14.11. The redox cycle for microbial S transformations is shown in **Figure 20.8**.

Although a number of oxidation states of S are possible, only three are significant in nature, -2 (sulfhydryl, R–SH, and sulfide, HS^-), 0 (elemental sulfur, S^0), and $+6$ (sulfate, SO_4^{2-}). The bulk of Earth's S is in sediments and rocks in the form of sulfate minerals, primarily gypsum ($CaSO_4$) and sulfide minerals (pyrite, FeS_2), but the oceans constitute the most significant reservoir of SO_4^{2-} in the biosphere. A significant amount of S, in particular sulfur dioxide (SO_2, a gas), enters the S cycle from human activities, primarily the burning of fossil fuels.

Hydrogen Sulfide and Sulfate Reduction

A major volatile S gas is hydrogen sulfide (H_2S). Hydrogen sulfide is produced from bacterial sulfate reduction ($SO_4^{2-} + 4H_2 \rightarrow H_2S + 2H_2O + 2OH^-$) (Figure 20.8) or is emitted from sulfide springs and volcanoes. Although H_2S is volatile, different forms

Key Processes and Prokaryotes in the Sulfur Cycle

Process	Organisms
Sulfide/sulfur oxidation ($H_2S \rightarrow S^0 \rightarrow SO_4^{2-}$)	
Aerobic	Sulfur chemolithotrophs (*Thiobacillus*, *Beggiatoa*, many others)
Anaerobic	Purple and green phototrophic bacteria, some chemolithotrophs
Sulfate reduction (anaerobic) ($SO_4^{2-} \rightarrow H_2S$)	*Desulfovibrio*, *Desulfobacter* *Archaeoglobus* (*Archaea*)
Sulfur reduction (anaerobic) ($S^0 \rightarrow H_2S$)	*Desulfuromonas*, many hyperthermophilic *Archaea*
Sulfur disproportionation ($S_2O_3^{2-} \rightarrow H_2S + SO_4^{2-}$)	*Desulfovibrio*, and others
Organic sulfur compound oxidation or reduction ($CH_3SH \rightarrow CO_2 + H_2S$) ($DMSO \rightarrow DMS$)	Many organisms can do this
Desulfurylation (organic–S $\rightarrow H_2S$)	Many organisms can do this

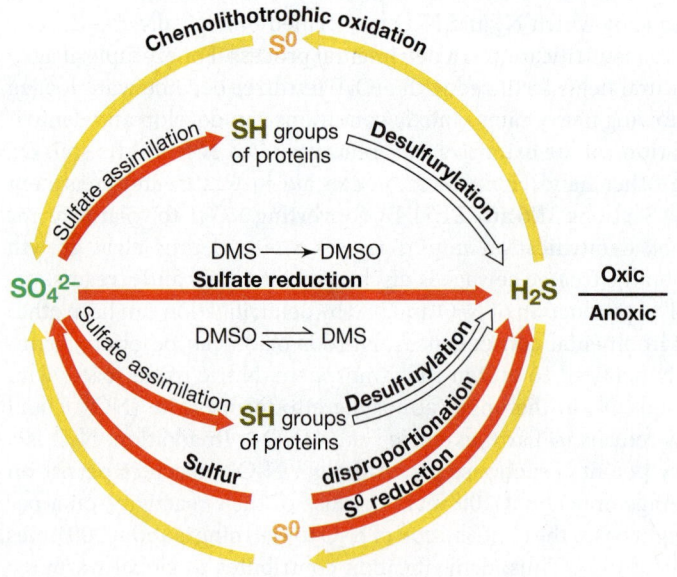

Figure 20.8 Redox cycle for sulfur. Oxidations are indicated by yellow arrows and reductions by red arrows. Reactions without redox changes are in white. DMS, dimethyl sulfide; DMSO, dimethyl sulfoxide.

exist depending on pH: H_2S predominates below pH 7 and the nonvolatile HS^- and S^{2-} predominate above pH 7. Collectively, H_2S, HS^-, and S^{2-} are referred to as "sulfide."

Sulfate-reducing bacteria are a large and highly diverse group (🔗 Sections 13.18 and 14.9) and are widespread in nature. However, in anoxic habitats such as freshwater sediments and many soils, sulfate reduction is limited by SO_4^{2-} availability. Moreover, because organic electron donors (or H_2, which is a product of the fermentation of organic compounds) are needed to support sulfate reduction, it only occurs where significant amounts of organic material are present.

In marine sediments, the rate of sulfate reduction is typically carbon-limited and can be greatly increased by an influx of organic matter. This is important because the disposal of sewage or garbage in the oceans or coastal regions can trigger sulfate reduction. Hydrogen sulfide is toxic to many plants and animals and therefore its formation is potentially detrimental (sulfide is toxic because it combines with the iron of cytochromes and blocks respiration). Sulfide is commonly detoxified in nature by combination with iron, forming the insoluble minerals FeS (pyrrhotite) and FeS_2 (pyrite). The black color of sulfidic sediments or sulfate-reducing bacterial cultures is due to these metal sulfide minerals (🔗 Figure 14.23g).

Sulfide and Elemental Sulfur Oxidation–Reduction

Under oxic conditions, sulfide rapidly oxidizes spontaneously at neutral pH. Sulfur-oxidizing chemolithotrophic bacteria, most of which are aerobes (🔗 Sections 13.8 and 14.11), can catalyze the oxidation of sulfide. However, because of the rather rapid spontaneous reaction, microbial sulfide oxidation is significant only in areas where H_2S emerging from anoxic environments meets air. Where light is available, there can be anoxic oxidation of sulfide, catalyzed by the phototrophic purple and green sulfur bacteria (🔗 Sections 13.3, 14.4, and 14.6).

Elemental sulfur is chemically stable but is readily oxidized by sulfur-oxidizing chemolithotrophic bacteria such as *Thiobacillus* and *Acidithiobacillus*. Because S^0 is insoluble, the bacteria that oxidize it must attach to the S^0 crystals to obtain their substrate (🔗 Figure 13.21). The oxidation of S^0 forms sulfuric acid (H_2SO_4), and thus S^0 oxidation characteristically lowers the pH in the environment, sometimes drastically. For this reason, S^0 is sometimes added to alkaline soils as an inexpensive and natural way to lower the pH, relying on the ubiquitous sulfur chemolithotrophs to carry out the acidification process.

Elemental sulfur can be reduced as well as oxidized. The reduction of S^0 to sulfide (a form of anaerobic respiration) is a major ecological process of some *Bacteria* and hyperthermophilic *Archaea* (🔗 Section 14.10 and Chapter 16). Although sulfate-reducing bacteria can also reduce S^0, in sulfidic habitats most S^0 is reduced by the physiologically specialized sulfur reducers, organisms that are incapable of SO_4^{2-} reduction (🔗 Section 14.10). The habitats of the sulfur reducers are generally those of the sulfate reducers, so from an ecological standpoint, the two groups form a metabolic guild unified by their formation of H_2S.

Organic Sulfur Compounds

In addition to *inorganic* forms of S, several *organic* S compounds are also cycled in nature. Many of these foul-smelling compounds are highly volatile and can thus enter the atmosphere. The most abundant organic S compound in nature is *dimethyl sulfide* (CH_3—S—CH_3); it is produced primarily in marine environments as a degradation product of dimethylsulfoniopropionate, a major osmoregulatory solute in marine algae (🔗 Section 5.15). This compound can be used as a carbon source and electron donor by microorganisms and is catabolized to CH_3—S—CH_3 and acrylate (CH_2=$CHCOO^-$). The latter, a derivative of the fatty acid propionate, is used to support growth.

Dimethyl sulfide released to the atmosphere undergoes photochemical oxidation to methanesulfonate (CH_3SO_3), SO_2, and SO_4^{2-}. By contrast, CH_3—S—CH_3 produced in anoxic habitats can be microbially transformed in at least three ways: (1) by methanogenesis (yielding CH_4 and H_2S), (2) as an electron donor for photosynthetic CO_2 fixation in phototrophic purple bacteria (yielding dimethyl sulfoxide, DMSO), and (3) as an electron donor in energy metabolism in certain chemoorganotrophs and chemolithotrophs (also yielding DMSO). DMSO can be an electron acceptor for anaerobic respiration (🔗 Section 13.21), producing CH_3—S—CH_3. Many other organic S compounds affect the global sulfur cycle, including methanethiol (CH_3SH), dimethyl disulfide (H_3C—S—S—CH_3), and carbon disulfide (CS_2), but on a global basis, CH_3—S—CH_3 is the most significant.

MINIQUIZ -

- Is H_2S a substrate or a product of the sulfate-reducing bacteria? Of the chemolithotrophic sulfur bacteria?
- Why does the bacterial oxidation of sulfur result in a pH drop?
- What organic sulfur compound is most abundant in nature?

II • Other Nutrient Cycles

In this unit we explore the interactions of microorganisms with metals—in particular iron and manganese—and with some nonmetals whose microbial transformations are of major global significance.

20.5 The Iron and Manganese Cycles

Iron (Fe) is one of the most abundant elements in Earth's crust. On the surface of Earth, Fe exists naturally in two oxidation states, ferrous [Fe^{2+}, also Fe(II)] and ferric [Fe^{3+}, also Fe(III)]. A third oxidation state,

Fe^0, is abundant in Earth's core and is also a major product of human activities from the smelting of iron ores to form cast iron.

In nature, iron cycles primarily between the Fe^{2+} and Fe^{3+} forms, and these redox transitions are one-electron oxidations and reductions. Ferric iron is reduced both chemically and as a form of anaerobic respiration, and Fe^{2+} is oxidized both chemically and as a form of chemolithotrophic metabolism (**Figure 20.9**). Manganese (Mn), although present at 5- to 10-fold lesser abundance than Fe

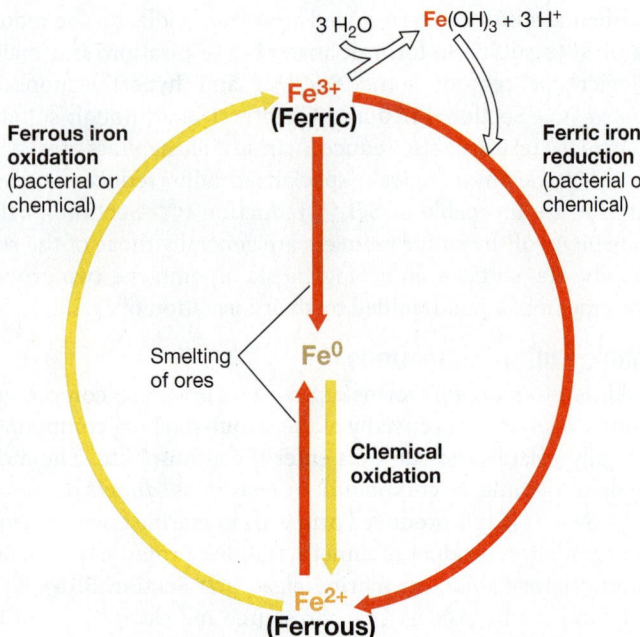

Figure 20.9 Redox cycle for iron. The major forms of iron in nature are Fe^{2+} and Fe^{3+}. Fe^0 is primarily a product of smelting of iron ores. Oxidations are shown by yellow arrows and reductions by red arrows. Fe^{3+} forms various minerals such as ferric hydroxide, $Fe(OH)_3$.

in the near surface environment, is another redox-active metal of microbiological significance, existing primarily in two oxidation states (Mn^{2+} and Mn^{4+}, see Figure 20.10).

A key feature of the iron and manganese cycles is the different solubilities of these metals in their oxidized versus reduced forms. Reduced iron (Fe^{2+}) and manganese (Mn^{2+}) are soluble. In contrast, oxidized minerals of iron such as iron oxide-hydroxides [e.g., $Fe(OH)_3$, $FeOOH$, and Fe_2O_3] and manganese oxide (MnO_2) are insoluble and tend to settle out in aquatic environments. As a consequence, these strong oxidants can comprise several percent by weight of marine and freshwater sediments, making them among the most abundant of potential electron acceptors in many anoxic systems (see Figure 20.10).

Bacterial Reduction of Iron and Manganese Oxides

Some *Bacteria* and *Archaea* can use Fe^{3+} as an electron acceptor in anaerobic respiration (Section 13.21). These organisms also commonly have the capacity to use Mn^{4+} as an electron acceptor, and some have the capacity to reduce oxidized uranium (Section 21.3).

Ferric iron and manganese oxide reduction is common in waterlogged soils, bogs, and anoxic lake sediments (**Figure 20.10**). When soluble reduced iron and manganese reach oxic regions, for example, through diffusion from anoxic regions of sediments, they are oxidized chemically [e.g., $Fe^{2+} + \frac{1}{4}O_2 + 2\frac{1}{2}H_2O \rightarrow Fe(OH)_3 + 2H^+$] or microbiologically (Figure 20.9). The chemical oxidation of Fe^{2+} is very rapid at near-neutral pH. Although the spontaneous oxidation of Mn^{2+} is very slow at neutral pH, the rate of oxidation can be increased up to five orders of magnitude by a variety of manganese-oxidizing bacteria and even fungi. The oxidized metal

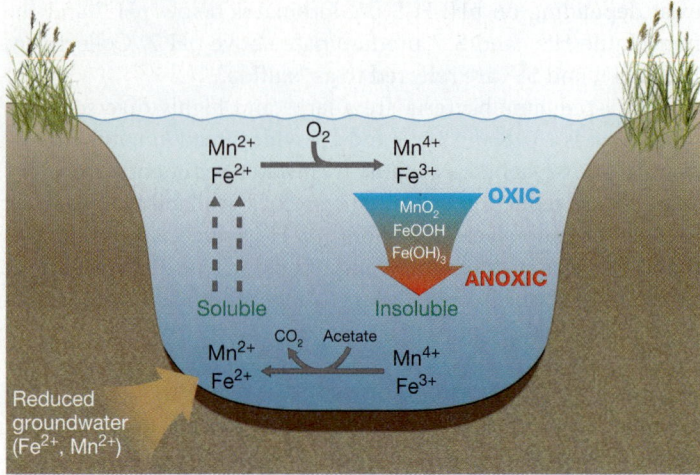

Figure 20.10 Iron and manganese redox cycling in a typical freshwater system. Iron and manganese oxides in sediments are used as electron acceptors by metal-reducing bacteria. The resulting reduced forms are soluble and diffuse into the oxic regions of the sediment or water column, where they are oxidized microbially or chemically. Precipitation of the insoluble oxidized metals then returns the metals to the sediments, completing the redox cycle.

oxides and hydroxides then precipitate, returning the oxidized metals to the sediments where they can again serve as electron acceptors, completing the cycle.

Oxidized iron (Fe^{3+}) and manganese (Mn^{4+}) are chemically very reactive. Phosphate is trapped as insoluble ferric phosphate precipitates. Chemical oxidation of refractory organic compounds by Mn^{4+} oxide may yield more available sources of carbon for microbial growth. Other metals [e.g., copper (Cu), cadmium (Cd), cobalt (Co), lead (Pb), arsenic (As)] form insoluble complexes with the iron and manganese oxides. When these oxides are subsequently reduced, the bound phosphate and metals are also liberated along with the soluble forms of these metals.

In recent years it has been recognized that the surfaces and appendages of cells of bacteria that interact with iron and manganese oxides, such as *Geobacter*, are electrically conductive,

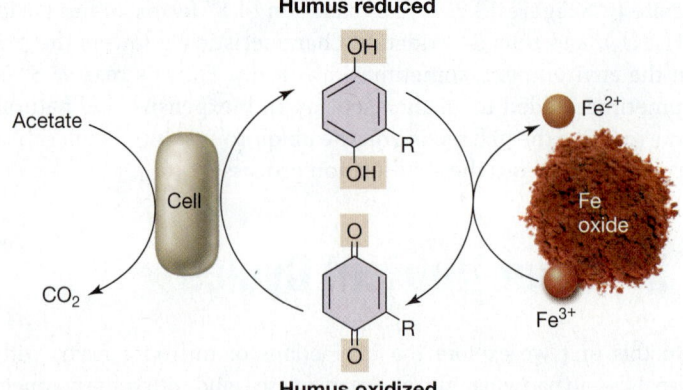

Figure 20.11 Role of humic substances in humus as an electron shuttle in microbial metal reduction. Quinone-like functional groups in humus are reduced by acetate-oxidizing bacteria. The reduced humus then donates electrons to metal oxides, releasing reduced soluble iron (Fe^{2+}) and oxidized humus. The cycle continues as oxidized humus is again reduced by the bacteria.

Regardless of the electron acceptor they use, when bacteria respire, they carry out oxidations and reductions that generate electricity. They do this when they oxidize an organic or inorganic electron donor and separate electrons from protons during electron transport. The electrons eventually reduce some electron acceptor and the protons generate the proton motive force.

In any form of respiration, electron disposal is necessary for energy conservation. When the electron acceptor is oxygen (O_2), nitrate (NO_3^-), or many of the other soluble substances used by bacteria as electron acceptors (⤺ Sections 13.16–13.21), the final product diffuses away from the cell. Many bacteria reduce ferric iron (Fe^{3+}) as an electron acceptor under anoxic conditions, including the bacterium *Geobacter sulfurreducens* (**Figure 1**). However, in contrast to soluble electron acceptors, Fe^{3+} is typically present in nature as an insoluble mineral, such as an iron oxide (Figure 20.11), and thus the reduction of Fe^{3+} occurs *outside* the cell. Under such conditions, the ferric iron functions as an electrical anode, and the bacterial cell facilitates transfer of electrons from the electron donor to the anode.[1]

Research has shown that *Geobacter* forms direct electrical connections with insoluble materials that can either accept or donate electrons. Electron transfer involves cytochromes localized along the length of pili that are generally 10–20 micrometers long (Figure 14.35c). These electrically conductive structures function as electrical nanowires, much as copper wire does in a household electrical circuit. Being conductive structures, nanowires can form direct electrical connections with insoluble materials that either accept or donate electrons, or alternatively, nanowires can form connections between cells. In this way, electrons obtained by *Geobacter* from

the oxidation of organic electron donors or from H_2 can be shuttled to a suitable electron acceptor. Although the cytochromes are required for the ultimate transfer of electrons, for example, to the reduction of the iron oxide electron acceptor, studies of the process suggest that the organic material comprising the pili is itself electrically conductive.

Surprisingly, bacterial electron shuttling can occur over rather large spatial distances, much larger than the cell itself. In studies of hydrogen sulfide (H_2S) oxidation in anoxic marine sediments (sulfide is the product of sulfate-reducing bacteria), the oxidation of H_2S deep in the sediments released electrons that reduced O_2 at the sediment water interface, some 20 cm away (see "Microbial Power Lines" on page 631).[2] The electrical conductors in the sediment are filamentous bacteria affiliated with the *Desulfobulbaceae* family of sulfate-reducing bacteria (⤺ Section 14.9). Although phylogenetically affiliated with *sulfate reducers*, the filamentous bacteria actually function as *sulfide oxidizers*, using O_2 as the terminal electron acceptor.

The surface of the filamentous bacterial cells has ridges running along its entire length. Microscopically these ridges appear much like cables, each microbial filament surrounded by 15–17 structures 400–700 nm in width that run continuously along the length of the filament. These structures are implicated in electron transfer from the sulfide oxidized at one end of the filament to the reduction of O_2 near the sediment surface at the other end of the filament. Although reminiscent of *Geobacter* nanowires, the mechanism for electron transfer over such large distances is unknown.

In nature, electrical communication between bacterial cells may be a major way by which electrons generated from microbial metabolism in anoxic habitats are shuttled to oxic

Figure 1 Cells of *Geobacter* attached to ferric iron precipitates (arrows) reduce Fe^{3+} to Fe^{2+}.

regions. Moreover, research on the microbiology of the process indicates that this microbial electricity could be harnessed in the form of microbial "fuel cells" that could oxidize toxic and waste carbon compounds in anoxic environments, with the resulting electrons coupled to power generation. In such a scheme, bacteria would be exploited to function as catalysts for diverting electrons from electron donors directly to artificial anodes, with the resulting electrical current being siphoned off to supply a portion of human power needs.

[1]Lovley, D.R. 2006. Bug juice: Harvesting electricity with microorganisms. *Nat. Rev. Microbiol. 4:* 497–508.

[2]Pfeffer, C., S. Larsen, J. Song, M.D. Dong, F. Besenbacher, R.L. Meyer, K.U. Kjeldsen, L. Schreiber, Y.A. Gorby, M.Y. El-Naggar, K.M. Leung, A. Schramm, N. Risgaard-Petersen, and L.P. Nielsen. 2012. Filamentous bacteria transport electrons over centimetre distances. *Nature 491:* 218–221.

---

functioning as "nanowires" to move electrons around in microbial habitats. This movement of electrons is a form of electricity, and the process may eventually have commercial applications for power generation (see Explore the Microbial World, "Microbially Wired"). Humic substances (Section 20.1) can also facilitate microbial metal reduction. Since some constituents of humics can alternate between oxidized and reduced forms, they can function to shuttle electrons from the bacterium to the reduction of the iron or manganese oxides (**Figure 20.11**).

Microbial Oxidation of Reduced Iron and Manganese

At neutral pH, ferrous iron (Fe^{2+}) is rapidly oxidized abiotically in oxic environments. In contrast, at *acidic* pH (pH < 4) Fe^{2+} is not oxidized spontaneously. Thus, much of the research on microbial oxidation of iron is focused on acidic, iron-rich habitats, where acidophilic chemolithotrophs such as *Acidithiobacillus ferrooxidans* and *Leptospirillum ferrooxidans* oxidize Fe^{2+} to Fe^{3+} (**Figure 20.12**).

The oxidation of Fe^{2+} to Fe^{3+} yields a single electron; consequently very little energy can be conserved (⤺ Sections 13.9

Ricardo Amils

Figure 20.12 Oxidation of ferrous iron (Fe^{2+}). A microbial mat growing in the Rio Tinto, Spain. The mat consists of acidophilic green algae (eukaryotes) and various iron-oxidizing chemolithotrophic prokaryotes. The Rio Tinto has a pH of about 2 and contains high levels of dissolved metals, in particular Fe^{2+}. The red-brown precipitates consist of Fe(OH)$_3$ and other ferric minerals.

and 14.15) and so these bacteria must oxidize large amounts of Fe^{2+} in order to grow. In such environments, even a relatively small population of cells can precipitate a large amount of iron minerals. Although O$_2$ is the most environmentally significant electron acceptor, Fe^{2+} oxidation can also be coupled to NO$_3^-$ reduction by some organisms (Sections 13.9 and 14.15) and functions as an electron donor in photosynthesis for some anoxygenic phototrophs (Sections 13.9, 14.2, and 14.5). Even though the oxidation of Mn^{2+} to Mn^{4+} is also potentially energetically favorable for growth, and many microorganisms catalyze Mn^{2+} oxidation, as yet no organism has been conclusively shown to derive energy from the oxidation of reduced manganese.

Since the abiotic oxidation of reduced iron is rapid at near-neutral pH, iron-oxidizing bacteria that inhabit nonacidic environments are restricted to a very narrow redox region in which ferrous iron–rich water impinges on oxygenated water (Figure 20.10). These microoxic habitats include freshwater and coastal sediments, slow-moving streams, ferrous iron–rich waters from springs, and hydrothermal vents (**Figure 20.13**). For example, when ferrous iron–rich groundwaters are exposed to air, Fe^{2+} is oxidized at the interface of these two zones by iron-oxidizing bacteria such as *Gallionella* and *Leptothrix* (Figure 20.13*b, c, d*; Sections 13.9 and 14.15). Thus, their physiology dictates that they maintain a position within a narrow environment of low levels of O$_2$ and high levels of reduced metals. How these organisms secure and maintain a position within such a narrow range of abiotic conditions is not well understood, but the sheath and stalk structures typically found in iron oxidizers may assist their proper positioning (Figure 20.13*b, d, f*; Figures 14.36 and 14.60).

As we have seen, organisms that reduce insoluble metal oxides can use extracellular conductors for electron transfer, such as electrically conductive pili or cell-surface-associated cytochromes. However, a similar problem exists for organisms that oxidize metals: Insoluble metal oxides are the product of metal oxidation, and the organism must ensure that these insoluble oxides are deposited external to the cell. Thus, organisms that oxidize Fe^{2+} or Mn^{2+} use surface-associated electron transfer proteins to ensure that metals are oxidized *outside* the cytoplasm. Cytochromes participate in both iron reduction and iron oxidation, and genomes of metal-oxidizing *Gallionella* and *Sideroxydans* species contain genes encoding surface-associated cytochromes (e.g., MtrA) that resemble those encoding proteins known to reduce metal oxides in *Shewanella*, suggesting that mechanistically similar electron transfer pathways are likely used for both the reduction and oxidation of extracellular metals.

Although possibly sharing similar electron transfer mechanisms, metal-oxidizing bacteria are confronted with another problem—their metabolism could soon encase the cell in an iron oxide shell. To prevent this, metal oxidizers produce extracellular organic material that captures metal oxides and deposits it some distance away from the cell. Some metal oxidizers, such as *Gallionella*, produce extended organic stalks that become encrusted with metal oxides away from the cell (Figure 20.13*d*; see also Figure 14.36). An alternative strategy is used by *Leptothrix* species. These bacteria produce an organic sheath surrounding the cells that becomes encrusted with metal oxides (Figure 20.13*b* and Figure 14.60). In this case the cells can move out of the sheath, leaving the metal oxide crust behind.

Although not all metal oxidizers produce such morphologically conspicuous structures, it is thought that most if not all metal oxidizers are forced to produce some form of extracellular organic material in order to sequester the insoluble product of their energy metabolism. In addition, the incorporation of this organic matter into the metal oxides likely alters the physical and chemical properties of the minerals themselves.

MINIQUIZ -
- In what oxidation state is Fe in Fe(OH)$_3$? In FeS? How is Fe(OH)$_3$ formed?
- Why does biological Fe^{2+} oxidation under oxic conditions occur mainly at acidic pH?
- Why is excreted organic matter important to many iron oxidizers?

20.6 The Phosphorus, Calcium, and Silica Cycles

Many other chemical elements undergo microbial cycling and we focus on three key ones here—phosphorus (P), calcium (Ca), and silica (Si). The cycling of these elements is important in aquatic environments, particularly in the oceans, which are major reservoirs of Ca and Si. In the oceans, huge amounts of Ca and Si are incorporated into the exoskeletons of certain microorganisms. However, unlike the C, N, and S cycles, in the P, Ca, and Si cycles there are no redox changes or gaseous forms that can escape and alter Earth's atmospheric chemistry. Nevertheless, as we will see, keeping these cycles in balance—especially that of Ca—is important for maintaining sustainable life on Earth.

Phosphorus

Phosphorus exists in nature primarily as organic and inorganic phosphates. Phosphorus reservoirs include phosphate-containing

(a) David Emerson
(b) David Emerson
(c) David Emerson
(d) David Emerson
(e) Woods Hole Oceanographic Institution
(f) Clara Chan
(g) David Emerson

Figure 20.13 **Fe-oxidizing microbial mats.**
(a) Freshwater microbial mat in a slow-moving stream where Fe^{2+}-enriched groundwater is mixing with oxygenated surface water, triggering growth of Fe^{2+}-oxidizing bacteria and precipitation of iron oxides. *(b, c)* Phase-contrast and epifluoresence photomicrographs of the sheath-forming Fe-oxidizer *Leptothrix ochracea*

(the sheath is approximately 2 µm wide). *(d)* The stalk-forming Fe^{2+}-oxidizer *Gallionella ferruginea* showing bean-shaped cells in the process of cell division at the end of the iron oxide–encrusted stalk (each bean-shaped cell is about 2 µm long). *(e)* An iron-oxidizing mat at a deep-sea hydrothermal vent (1000-meter depth) at Lōʻihi Seamount. *(f)* TEM image of biogenic oxides

produced at Lōʻihi; note the variety of helical stalks and tubular sheathlike filaments (the filaments vary from 2 to 4 µm wide). *(g)* Phase-contrast photomicrograph of marine Fe^{2+}-oxidizers growing at the ends of iron oxide filaments (cells denoted by arrows) from an experimental incubation at Lōʻihi (the filaments are approximately 2 µm wide).

minerals in rocks, dissolved phosphates in freshwaters and marine waters, and the nucleic acids and phospholipids of living organisms. Although P has multiple oxidation states, most environmental phosphates are at the +5 oxidation state (for example, inorganic phosphate, HPO_4^-). In nature P cycles through living organisms (as cellular P), waters and soils (as inorganic and organic P), and Earth's crust (as inorganic P). P is typically the limiting nutrient for photosynthesis in freshwaters, which receive it from the weathering of rocks.

In the oceans, a fraction of dissolved P is organic, in the form of phosphate esters and *phosphonates*. Phosphonates are organophosphate compounds that contain a direct bond between the P and C atoms. Phosphonates are produced by certain microorganisms and comprise about a quarter of the organic P pool in nature; however, for many organisms phosphonates are a less available source of P than is HPO_4^- because of the enzymes required to degrade phosphonates. Organisms lacking these enzymes can be phosphorus-limited even when

sufficient P is present as phosphonates. Moreover, the degradation of methylphosphonate (CH_5O_3P) by some marine microorganisms—a process that liberates CH_4—may explain the previously puzzling observation that relatively high levels of CH_4 are present in the oxygenated surface waters of the ocean (methanogenic *Archaea* are strict anaerobes; ↩ Section 16.2).

Calcium

The major global reservoirs of Ca are calcareous rocks and the oceans. In the oceans, where dissolved Ca exists as Ca^{2+}, calcium cycling is a highly dynamic process although the concentration of Ca^{2+} in seawater remains constant at about 10 mM. Several marine eukaryotic phototrophic microorganisms take up Ca^{2+} to form their calcareous exoskeletons; these include in particular the *coccolithophores* and *foraminifera* (**Figure 20.14**; ↩ Section 17.7). The calcium-cycling activities of these planktonic phototrophs are also tightly coupled with inorganic components of the carbon cycle.

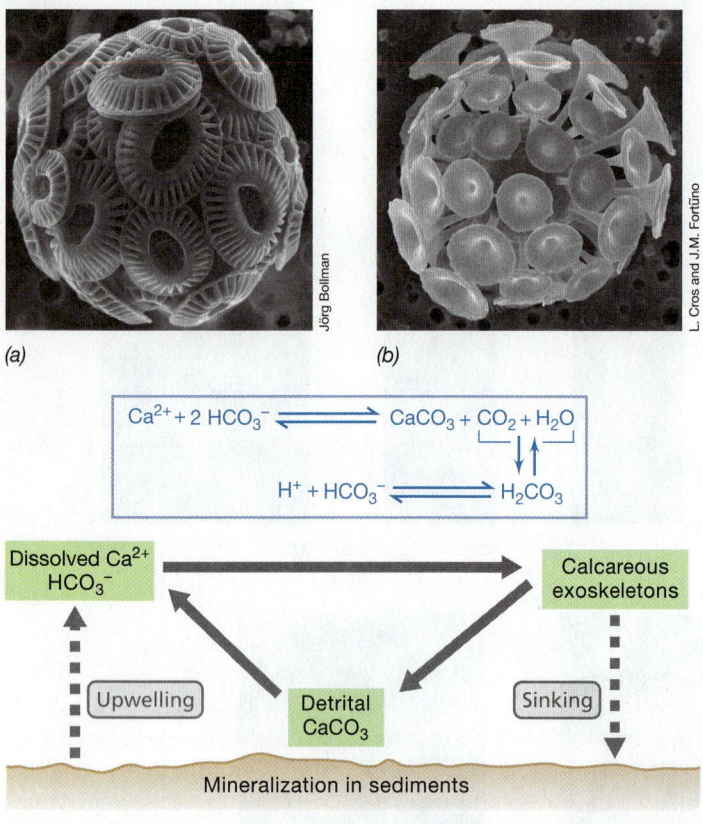

(a)   (b)

$$Ca^{2+} + 2\,HCO_3^- \rightleftharpoons CaCO_3 + CO_2 + H_2O$$

$$H^+ + HCO_3^- \rightleftharpoons H_2CO_3$$

Dissolved Ca^{2+} HCO_3^- → Calcareous exoskeletons

Upwelling — Detrital $CaCO_3$ — Sinking

Mineralization in sediments

(c)

Figure 20.14 The marine calcium (Ca) cycle. Scanning electron micrographs of cells of the calcareous phytoplankton *(a) Emiliania huxleyi* and *(b) Discosphaera tubifera.* The exoskeletons of these coccolithophores are made of calcium carbonate ($CaCO_3$). A cell of *Emiliana* is about 8 μm wide and a cell of *Discosphaera* is about 12 μm wide. *(c)* The marine calcium cycle; dynamic pools of Ca^{2+} are shaded in green. Detrital $CaCO_3$ is that in fecal pellets and other organic matter from dead organisms. Note how H_2CO_3 formation decreases ocean pH when it dissolves to form H^+ and HCO_3^-.

The precipitation of calcium carbonate ($CaCO_3$) to form the shells of calcareous phytoplankton controls both CO_2 flux into ocean surface water and inorganic C transport into deep ocean water and the sediments. Moreover, the formation of $CaCO_3$ both depletes surface dissolved bicarbonate (HCO_3^-) and increases the level of dissolved CO_2 (Figure 20.14*c*). The latter reduces the influx of atmospheric CO_2 into surface ocean waters and this helps maintain the slightly alkaline pH of the oceans. When these calcareous organisms die and sink toward the sediments, inorganic and organic C and Ca^{2+} are transported to the deep ocean from which they are only slowly released over long periods.

The formation of $CaCO_3$ exoskeletons brings into play a delicate balance between Ca^{2+} and C and is a process sensitive to changes in atmospheric CO_2 levels. This is because increased levels of atmospheric CO_2 increase the formation of carbonic acid (H_2CO_3), and as this dissociates to form HCO_3^- and H^+, $CaCO_3$ dissolves and seawater pH decreases (Figure 20.14*c*). The more acidic oceans that will result from rising atmospheric CO_2 are predicted to reduce the rate of formation of calcareous shells, which will likely have effects on other microbial nutrient cycles and plant and animal communities (Section 20.8).

Silica

The marine Si cycle is controlled primarily by unicellular eukaryotes (diatoms, silicoflagellates, and radiolarians) that build ornate external cell skeletons called *frustules* (**Figure 20.15*a***) (⮂ Sections 17.6 and 17.7). These structures are not constructed of $CaCO_3$ as in the coccolithophores, but of opal (SiO_2), whose formation begins with the uptake by the cell of dissolved silicic acid (Figure 20.15*b*).

Diatoms are rapidly growing phototrophic eukaryotes and often dominate blooms of phytoplankton in coastal and open ocean waters. However, unlike other major phytoplankton groups, diatoms require Si and can become silica-limited when blooms develop. Also, because of their large size, diatom cells tend to sink faster than other organic particles, and in this way, they contribute significantly to the return of Si and C to deeper ocean waters. The transport of organic material produced through primary production in near-surface waters to deeper ocean waters, primarily by sinking particles, is called the *biological pump* and is an important aspect of the carbon cycle in terms of carbon burial and mineralization in marine environments (Figure 20.1).

In addition to the major nutrient requirements of any phototrophic organism (CO_2, N, P, Fe), diatoms require sufficient dissolved Si, and in nature this originates primarily from Si released from the skeletons of dead diatoms (Figure 20.15*b*). Although Si is released fairly rapidly following cell death, during periods of high

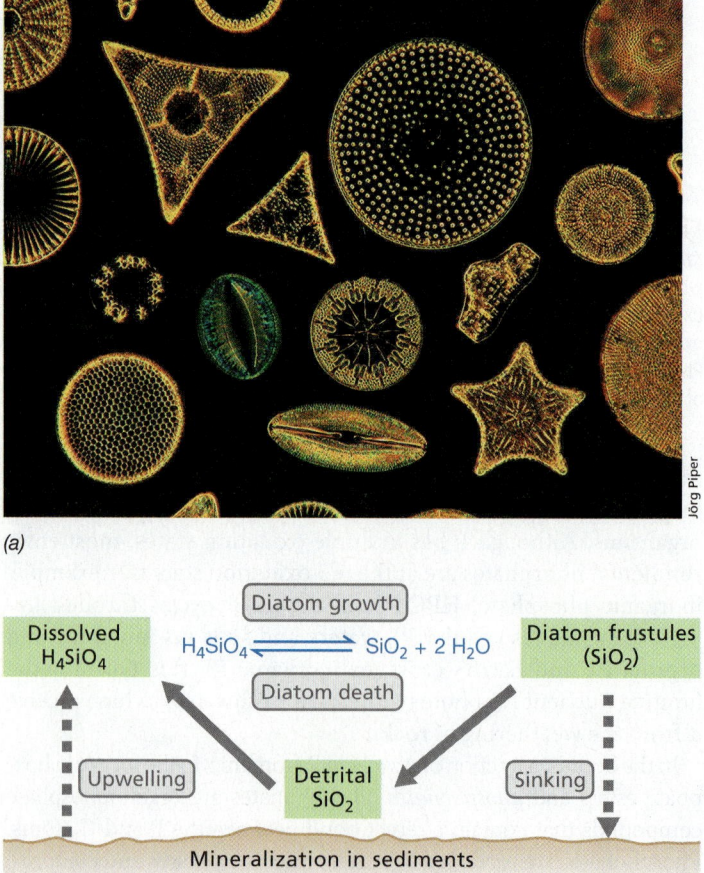

(a)

Dissolved H_4SiO_4 Diatom growth $H_4SiO_4 \rightleftharpoons SiO_2 + 2\,H_2O$ Diatom death   Diatom frustules (SiO_2)

Upwelling — Detrital SiO_2 — Sinking

Mineralization in sediments

(b)

Figure 20.15 The marine silica cycle. *(a)* Dark-field photomicrograph of a collection of diatom shells (frustules). The frustules are made of SiO_2. *(b)* The marine silica cycle; dynamic pools of Si are shaded in green.

diatom production in relatively shallow waters, a significant fraction of dissolved Si can be buried in sediments and remain there for millions of years. This has consequences for continued diatom growth and their phototrophic consumption of dissolved CO_2 from ocean waters. The flux of CO_2 into and out of ocean water affects its pH (Figure 20.14c), and through this link, the Si and C cycles are coupled in similar fashion to what we have seen with the Ca and C cycles.

III • Humans and Nutrient Cycling

Humans have a profound impact on microbial nutrient cycles by adding and removing components of the cycles in large amounts. Here we consider human inputs of three major species: mercury (Hg), CO_2 and other atmospheric gases, and various fixed N compounds. These compounds either cause toxicity problems (Hg) or affect our planet in globally significant ways (gases and N compounds). We begin with the very toxic metal Hg, which is transformed by bacteria in many different ways.

20.7 Mercury Transformations

Mercury is not a biological nutrient but microbial transformations of various mercuric compounds help to detoxify some of its most toxic forms. Mercury is a widely used industrial product, especially in the electronics industry. Mercury is also an active ingredient in many pesticides, a pollutant from the chemical and mining industries and from the combustion of fossil fuels and municipal wastes, and a common contaminant of aquatic ecosystems and wetlands. Because of its propensity to concentrate in living tissues, Hg is of considerable environmental importance. The major form of Hg in the atmosphere is elemental mercury (Hg^0), which is volatile and is oxidized to mercuric ion (Hg^{2+}) photochemically. Most Hg thus enters aquatic environments as Hg^{2+} (**Figure 20.16**).

Microbial Redox Cycle for Mercury

Mercuric ion readily adsorbs to particulate matter and can be metabolized from there by microorganisms. Microbial activity

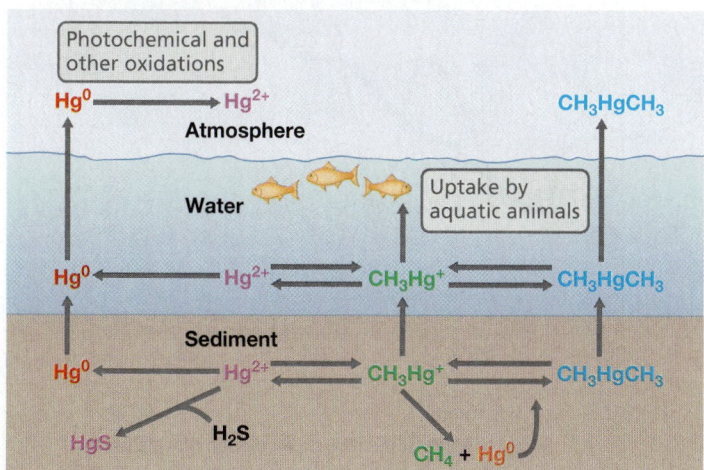

Figure 20.16 Biogeochemical cycling of mercury. The major reservoirs of mercury are water and sediments. Mercury in water can be concentrated in animal tissues; it can be precipitated as HgS from sediments. The forms of mercury commonly found in aquatic environments are each shown in a different color.

methylates Hg, yielding *methylmercury* (CH_3Hg^+) (Figure 20.16). Methylmercury is extremely toxic to animals because it can be readily absorbed through the skin and is a potent neurotoxin. But in addition, CH_3Hg^+ is soluble and can be concentrated in the food chain, primarily in fish, or can be further methylated by microorganisms to yield the volatile compound *dimethylmercury* (CH_3—Hg—CH_3). Both methyl- and dimethylmercury accumulate in animals, especially in muscle tissues. Methylmercury is about 100 times more toxic than Hg^0 or Hg^{2+}, and its accumulation in the aquatic food chain seems to be particularly acute in freshwater lakes and marine coastal waters where enhanced levels of CH_3Hg^+ have been detected in fish caught for human consumption. Mercuric compounds can cause liver and kidney damage in humans and other animals.

Several other microbial Hg transformations occur, including reactions catalyzed by sulfate-reducing bacteria ($H_2S + Hg^{2+} \rightarrow HgS$) and methanogens ($CH_3Hg^+ \rightarrow CH_4 + Hg^0$) (Figure 20.16). The solubility of mercuric sulfide (HgS) is very low, so in sulfidic sediments, most Hg exists as HgS. But upon aeration, HgS can be oxidized to Hg^{2+} and SO_4^{2-} by metal-oxidizing bacteria (Section 20.5), and the Hg^{2+} is eventually converted to CH_3Hg^+. Note, however, that it is not the Hg in HgS that is oxidized here, but instead the *sulfide*, probably by organisms related to *Acidithiobacillus* (Section 14.11).

Mercury Resistance

At sufficiently high concentrations, Hg^{2+} and CH_3Hg^+ can be toxic to microorganisms as well as to macroorganisms. However, several gram-positive and gram-negative bacteria convert toxic forms of Hg to nontoxic or less toxic forms. These mercury-resistant bacteria employ the enzyme *organomercury lyase* to degrade the highly toxic CH_3Hg^+ to Hg^{2+} and methane (CH_4), and the NADPH (or NADH)-linked enzyme *mercuric reductase* to reduce Hg^{2+} to Hg^0, which is volatile and thus mobile (**Figure 20.17**).

In many mercury-resistant bacteria, genes encoding Hg resistance reside on plasmids or transposons (Sections 4.3 and 10.11). These *mer* genes are arranged in an operon under control of the regulatory protein MerR, which can function as either a repressor or an activator of transcription (Sections 7.3 and 7.4), depending on Hg availability. In the absence of Hg^{2+}, MerR functions as a *repressor* and binds to the operator region of the *mer* operon, thus preventing transcription of the structural genes, *merTPABD*. However, when Hg^{2+} is present, it forms a complex with MerR, which then binds to the *mer* operon and functions as an *activator* of transcription of *mer* structural genes (Figure 20.17).

The protein MerP is a periplasmic mercuric ion–binding protein. MerP binds Hg^{2+} and transfers it to the membrane transport protein MerT, which interacts with mercuric reductase (MerA) to

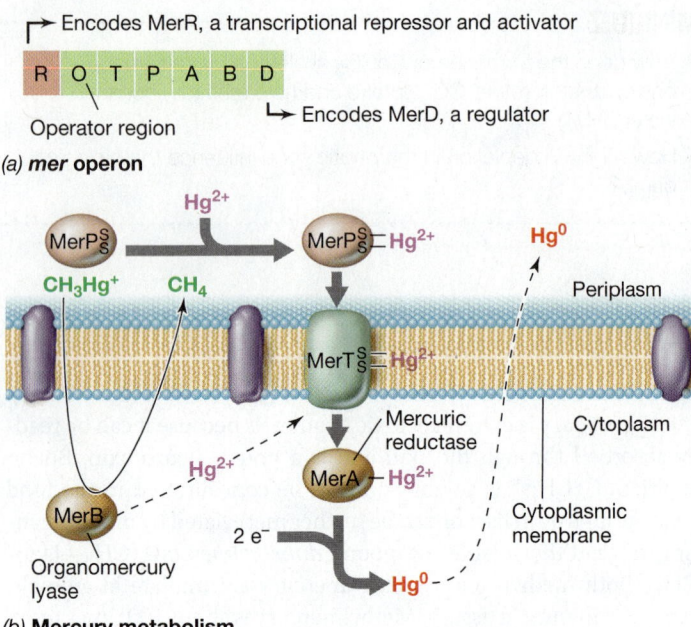

Figure 20.17 Mechanism of mercury transformations and resistance.
(a) The *mer* operon. MerR can function as either a repressor (in the absence of Hg^{2+}) or a transcriptional activator (in the presence of Hg^{2+}). *(b)* Transport and reduction of Hg^{2+} and CH_3Hg^+; the Hg^{2+} is bound by cysteine residues in the MerP and MerT proteins. MerA is the enzyme mercuric reductase and MerB is organomercury lyase.

reduce Hg^{2+} to Hg^0 (Figure 20.17*b*). Thus, Hg^{2+} is not released into the cytoplasm and the final result is the release of Hg^0 from the cell. Mercuric ion produced from the activity of MerB is trapped by MerT and reduced by MerA, again releasing Hg^0 (Figure 20.17*b*). In this way, Hg^{2+} and CH_3Hg^+ are converted to the relatively nontoxic Hg^0.

MINIQUIZ

- What forms of mercury are most toxic to organisms?
- How is mercury detoxified by bacteria?

20.8 Human Impacts on the Carbon and Nitrogen Cycles

Human activities are having major impacts on the carbon and nitrogen cycles, and these impacts have significance for the health of our planet in general. The period of marked human influence on these nutrient cycles began with the Industrial Revolution and is informally termed the *Anthropocene*, a new geological epoch. Although the greatest human impacts have been on the release of CO_2 through the burning of fossil fuels (oil, gas, and coal) and from extensive and ongoing deforestation, human activity has also profoundly affected the nitrogen cycle. We discussed earlier the close coupling of the carbon and nitrogen cycles (Section 20.1), and here we consider some of the projected biogeochemical consequences of human alteration of these two critical nutrient cycles.

CO_2 and Global Warming

Atmospheric CO_2 levels have increased approximately 40% since the beginning of the Industrial Revolution in the 1800s, and are now higher than at any time in the last 800,000 years. Carbon dioxide is one of several *trace gases* (primarily water vapor, CO_2, CH_4, and N_2O) which comprise less than 0.5% of the atmosphere but contribute significantly to terrestrial and atmospheric warming due to the *greenhouse effect*, the ability of these gases to trap the infrared radiation emitted by the Earth. The increase in atmospheric CO_2 concentration, measured across a global network of sampling stations (**Figure 20.18**), is currently about 2 parts per million per year. This increase would be much more rapid were it not for the high solubility of CO_2 in water, which produces carbonic acid; much anthropogenic CO_2 thus dissolves in the oceans (Figures 20.1 and 20.14).

The surface waters of the oceans have taken up an estimated 500 billion tons of CO_2 from the atmosphere out of a total of 1300 billion tons of total anthropogenic emissions, thus modulating the greenhouse effect. The increase in average Earth air temperature (estimated to have increased 0.75°C in the twentieth century and projected to increase by anywhere from 1.1 to 6.4°C in the twenty-first century) would also have been more rapid without the buffering influence of the oceans. Since three orders of magnitude more energy is required to raise the temperature of a cubic meter of water than a cubic meter of air, over 80% of the heat retained on Earth due to the greenhouse effect has actually entered the ocean.

Although there is considerable uncertainty about the consequences of ocean warming and CO_2 consumption on Earth's biological systems, there is agreement on how these changes will affect biogeochemistry. Warmer ocean surface waters are more buoyant (because of their lower density) than are deeper waters. Thus, as occurs seasonally in lakes (⮌ Section 19.8), the oceans will become more stratified with future global warming. Stratification tends to slow the transfer of nutrients from deeper waters that are needed to nourish phytoplankton production at the base of the food web in surface waters. This in turn reduces ocean productivity and export of a portion of that production to the deeper ocean through sedimentation

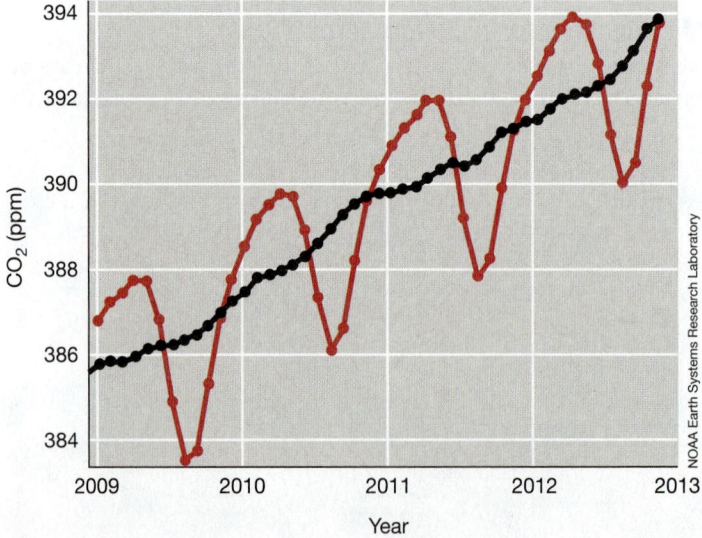

Figure 20.18 Global average of monthly mean carbon dioxide (CO_2) concentrations in air above marine surface sites. These data are continuously collected by the Global Monitoring Division of NOAA/Earth System Research Laboratory. The red curve shows variations in monthly mean values associated with annual fluctuations of temperature and precipitation affecting photosynthesis and respiration on land. The black curve shows the average monthly increase in CO_2 after correcting for seasonal cycle influence.

(the biological pump, Figure 20.1). The biological pump is important to the long-term removal of carbon from the atmosphere. Melting of polar sea ice by global warming might, however, mitigate this effect somewhat by opening up new waters for phytoplankton production.

Ocean warming is also contributing to the expansion of oxygen minimum zones (OMZs), regions of naturally occurring low O_2 concentration in subsurface waters between 100 and 1000 m in depth (Section 19.9). OMZs are a consequence of both the reduced solubility of O_2 in warmer water and the increasing stratification associated with surface warming, which reduces mixing of surface and subsurface waters. Animals will be excluded from the expanding OMZs whereas anaerobic microbial processes, such as denitrification and anammox that directly influence the nitrogen cycle and production of the greenhouse gas N_2O, will be enhanced.

Acidification of the ocean resulting from the ongoing uptake of anthropogenic CO_2 has reduced ocean pH by 0.1 pH units since the beginning of the Industrial Revolution, and may further reduce the pH by 0.3–0.4 units by the year 2100. The ongoing reduction in carbonate (CO_3^{2-}) concentration, a consequence of increasing acidification, is expected to be detrimental to marine calcifiers (organisms synthesizing $CaCO_3$ shells or skeletons, Figure 20.14). Since the concentration of Ca in seawater is relatively constant, continued reduction in CO_3^{2-} will ultimately reach a point where the dissolution of existing $CaCO_3$ is chemically favored, ultimately releasing more dissolved CO_2 (Figure 20.14), which reduces the capacity of the oceans to absorb more atmospheric CO_2.

Although the biological response to ocean acidification is unknown, it is likely that coral reef ecosystems, a major component of the marine biosphere (Section 22.14), will cease to occur naturally on Earth if CO_2 emissions continue at their present rate (Figure 20.18). Calcification in foraminifera (Section 17.7) will likely be impaired significantly by ocean acidification, as will calcification in coccolithophores (Figure 20.14). Over periods of a century or so, the invasion of anthropogenic CO_2 into the deep ocean will ultimately result in a significant reduction in the levels of $CaCO_3$ sequestered there, and this is predicted to disrupt the carbon cycle in major ways.

Anthropogenic Effects on the Nitrogen Cycle

Anthropogenic impacts on the nitrogen cycle are as profound as those on the carbon cycle (Figure 20.1). The yearly industrial production of nitrogenous fertilizers through the Haber–Bosch process, which combines $N_2 + H_2$ to form NH_3 under high temperature and pressure, is now comparable to the amount of fixed nitrogen entering the biosphere through biological nitrogen fixation, a key link in the nitrogen cycle (Section 20.3). This includes nitrogen fixation by both microorganisms living alone and those which live in symbiotic associations with plants or algae. Most of the industrially produced N is applied to farmland, but a significant fraction runs off to the oceans and contributes to coastal eutrophication (Section 19.9). Large amounts are also lost as gaseous nitrogen compounds (N_2, N_2O, and NO) from nitrification of NH_3 and denitrification of NO_3^- (Section 20.3).

Transport of N from industrial and agricultural centers through the atmosphere fertilizes both terrestrial and marine systems. Atmospheric deposition of fixed N_2 to the oceans is now about the same as that which enters through biological nitrogen fixation. The ecological consequences of this fertilization are a major unknown. On the one hand, if deposition suppresses microbial nitrogen fixation, this would to some degree mitigate the fertilization effect. On the other hand, a greater supply of both CO_2 and iron (caused by greater deposition of dust from areas of increasing desertification, Section 19.6) along with increased N depositions could enhance primary production, since iron is also often a limiting nutrient. Either way, major effects on the carbon cycle should be expected from human inputs in the nitrogen cycle.

Although changes in Earth's biosphere from human intervention in microbial nutrient cycles are a certainty, precisely what these changes will be is less clear. However, because major nutrient cycles are closely coupled (Section 20.1 and Figure 20.4), it is likely that any significant changes in the carbon and nitrogen cycles will have feedback effects on other cycles as well. Collectively, these events could upset the interrelationships of the nutrient cycles we have explored in this chapter and have significant (and likely negative) consequences for Earth's higher organisms.

MINIQUIZ --

- What is the greenhouse effect and what causes it?
- What is the fate of most nitrogen used in agricultural applications?
- Why are the OMZs expanding and what are the likely impacts on nutrient cycles?

BIG IDEAS

20.1 • The oxygen and carbon cycles are interconnected through the complementary activities of autotrophic and heterotrophic organisms. Microbial decomposition is the single largest source of CO_2 released to the atmosphere.

20.2 • Under anoxic conditions, organic matter is degraded to CH_4 and CO_2. Methane is formed primarily from the reduction of CO_2 by H_2 and from acetate, both supplied by syntrophic bacteria; these organisms depend on H_2 consumption as the basis of their energetics. On a global basis, biogenic CH_4 is a much larger source than abiogenic CH_4.

20.3 • The principal form of nitrogen on Earth is N_2, which can be used as a N source only by nitrogen-fixing bacteria. Ammonia produced by nitrogen fixation or by ammonification can be assimilated into organic matter or oxidized to NO_3^-. Denitrification and anammox cause major losses of fixed nitrogen from the biosphere.

20.4 • Bacteria play major roles in both the oxidative and reductive sides of the sulfur cycle. Sulfur- and sulfide-oxidizing bacteria produce SO_4^{2-}, whereas sulfate-reducing bacteria consume SO_4^{2-}, producing H_2S. Because sulfide is toxic and reacts with various metals, SO_4^{2-} reduction is an important biogeochemical process. Dimethyl sulfide is the major organic sulfur compound of ecological significance in nature.

20.5 • Iron and manganese exist naturally in two oxidation states, Fe^{2+}/Fe^{3+} and Mn^{2+}/Mn^{4+}. Bacteria reduce the oxidized metals in anoxic environments and oxidize the reduced forms primarily in oxic environments. At neutral pH, bacteria compete with abiotic oxidation in the presence of O_2. Ferrous iron oxidation is common in coal-mining regions, where it causes a type of pollution called acid mine drainage.

20.6 • P, Ca, and Si are elements cycled by microbial activities, primarily in aquatic environments. Calcium and silica play important roles in the biogeochemistry

of the oceans as components of the exoskeletons of coccolithophores and diatoms, respectively.

20.7 • A major toxic form of Hg in nature is CH_3Hg^+, which can yield Hg^{2+}, which is reduced by bacteria to Hg^0. Genes conferring resistance to the toxicity of Hg, such as those that encode enzymes that can detoxify or pump out the metal, often reside on plasmids or transposons.

20.8 • Anthropogenic inputs of CO_2 and reactive nitrogen are impacting major nutrient cycles. Although some consequences are reasonably well understood, including expansion of OMZs and impaired growth of calcareous organisms, the long-term changes to the nutrient cycles that sustain Earth's biosphere are not well understood.

MasteringMicrobiology®   **Review what you know and challenge what you have learned with MasteringMicrobiology!** Access study materials, chapter quizzes, animations, and microbiology lab tutorials in the Study Area to ensure that you have mastered this chapter's content.

REVIEW OF KEY TERMS

Denitrification the biological reduction of nitrate (NO_3^-) to gaseous N compounds

Global warming the predicted and ongoing warming of the atmosphere and oceans attributed to anthropogenic release of

greenhouse gases, primarily CO_2, that trap infrared radiation emitted by Earth

Humus dead organic matter, some of which functions as electron shuttles for the microbial reduction of metal oxides

Syntrophy the cooperation of two or more microorganisms to degrade anaerobically a substance neither can degrade alone

REVIEW QUESTIONS

1. Why can it be said that the carbon and nitrogen cycles are "coupled"? (Section 20.1)

2. How can organisms such as *Syntrophobacter* and *Syntrophomonas* grow when their metabolism is based on thermodynamically unfavorable reactions? How does coculture of these syntrophs with certain other bacteria allow them to grow? (Section 20.2)

3. Compare and contrast the processes of nitrification and denitrification in terms of the organisms involved, the environmental conditions that favor each process, and the changes in nutrient availability that accompany each process. (Section 20.3)

4. Which group of bacteria cycle sulfur compounds under anoxic conditions? If sulfur chemolithotrophs had never evolved, would there

be a problem in the microbial cycling of sulfur compounds? Which organic sulfur compounds are most abundant in nature? (Section 20.4)

5. Why are most iron-oxidizing chemolithotrophs obligate aerobes and why are the better-studied iron oxidizers acidophilic? (Section 20.5)

6. In what ways are Ca and Si cycling in ocean waters similar, and in what ways do they differ? How do the calcium and silica cycles couple to the carbon cycle? (Section 20.6)

7. How are Hg^{2+} and CH_3Hg^+ detoxified by the *mer* system? (Section 20.7)

8. What negative effects on the oceans do rising CO_2 levels have? (Section 20.8)

APPLICATION QUESTIONS

1. Compare and contrast the carbon, sulfur, and nitrogen cycles in terms of the physiologies of the organisms that participate in the cycle. Which physiologies are part of one cycle but not another?

2. ^{14}C-labeled cellulose is added to a vial containing some anoxic freshwater lake sediments and sealed under anoxic conditions.

A few hours later, $^{14}CH_4$ appears in the vial. Discuss what has happened to yield such a result.

3. Carbon can be sequestered in the ocean in a variety of forms. Discuss the different forms, their biological sources, and how global warming will influence them.

21 · Microbiology of the Built Environment

microbiology**now**

Your Subway System: What's in the Air?

Does the congregation of people in buildings and transportation systems influence the microflora of these "built environments"? A recent study of air quality in a metropolitan transit system posed this very question using culture-independent methods to characterize the diversity and abundance of the indoor airborne microflora for its possible impact on public health.[1]

The New York City municipal subway system moved a total of 1.6 billion passengers in 2011 and was the site of the first intensive molecular survey of aerosol microbiology in heavily trafficked enclosed structures. Investigators placed air samplers designed to efficiently capture bacteria-sized particles at various locations in boarding areas of the subway system (photos). Following collection of the microflora from several cubic meters of air, DNA was extracted for phylogenetic analyses of ribosomal RNA genes as a metric of microbial diversity.

Surprisingly, phylogenetic analyses showed no evidence of pathogens. Instead, the microflora was mostly composed of organisms associated with outdoor air, along with a much smaller component of microorganisms typically found on humans. For example, about 5% of the subway air microflora were organisms found on human skin. Since the normal microflora of the human skin has been well characterized and shown to vary by body location, these investigators could also conclude that the skin-derived microorganisms in New York subway air came mainly from the feet, hands, arms, and heads of riders—the generally more exposed areas of the human body.

This survey is reassuring in its discovery that serious microbial pathogens are not floating around the subway air system. But the study also served as a prototype for how the microbial composition of the built environment can be monitored in a scientifically sound and objective way.

[1]Robertson, C.E., et al. 2012. Culture-independent analysis of aerosol microbiology in a metropolitan subway system. *Appl. Environ. Microbiol. 79:* 3485–3493.

This chapter addresses the microbiology of built systems. These include the infrastructure for drinking water and wastewater distribution and treatment, gas and oil transmission, building materials, and environments modified for mineral extraction or for the cleanup of pollutants. Built systems create new microbial habitats, promoting both desired and undesired microbial activities. Examples of systems designed to select for desirable microbial activities include the construction of biological reactors for the treatment of wastewater and the stimulation of microbial activity in aquifers to clean up environmental pollutants.

A notable example of an unwanted activity is microbially influenced corrosion of the pipelines used for transmission of wastewater, drinking water, and oil. These are natural processes in which microorganisms simply exploit the resources provided to them in the built environment. Essential infrastructure costing several billion dollars is lost every year to microbially influenced corrosion. For example, the American Association of Civil Engineers estimates that in the next 30 years, about 30% of the drinking water distribution system in the United States will need to be replaced at an annual cost of $11 billion.

I • Mineral Recovery and Acid Mine Drainage

The biogeochemical capacities of microorganisms seem almost limitless, and it is often said that microorganisms are "Earth's greatest chemists." The activities of these great little chemists have been exploited in many ways. Here we consider how microbial activities help extract valuable metals from low-grade ores.

21.1 Mining with Microorganisms

One of the most common forms of iron in nature is **pyrite** (FeS_2), which is often present in bituminous coals and in metal ores. Sulfide (HS^-) also forms insoluble minerals with many metals, and many ores mined as sources of these metals are sulfide ores. If the concentration of metal in the ore is low, it may be economically feasible to mine the ore only if the metals of interest are first concentrated by **microbial leaching** (Figure 21.1). The promotion of acid production and dissolution of FeS_2 by acidophilic bacteria such as *Acidithiobacillus ferrooxidans* is used to leach the metal ores in large-scale mining operations. Leaching is especially useful for copper ores because copper sulfate ($CuSO_4$), formed during the oxidation of copper sulfide ores, is very water-soluble. Indeed, approximately a quarter of all copper mined worldwide is obtained by microbial leaching.

The Leaching Process

The susceptibility to oxidation varies among minerals, and those minerals that are most readily oxidized are most amenable to microbial leaching. Thus, iron and copper sulfide ores such as

(a)   *(b)*   *(c)*

Figure 21.1 The leaching of low-grade copper ores using iron-oxidizing bacteria. *(a)* A typical leaching dump. The low-grade ore has been crushed and dumped in such a way that the surface area exposed is as high as possible. Pipes distribute the acidic leach water over the surface of the pile. The acidic water slowly percolates through the pile and exits at the bottom. *(b)* Effluent from a copper leaching dump. The acidic water is very rich in Cu^{2+}. *(c)* Recovery of copper as metallic copper (Cu^0) by passage of the Cu^{2+}-rich water over metallic iron in a long flume. *(d)* A small pile of metallic copper removed from the flume, ready for further purification.

(d)

pyrrhotite (FeS) and covellite (CuS) are readily leached, whereas lead and molybdenum ores are much less so. In microbial leaching, low-grade ore is dumped in a large pile called the *leach dump* and a dilute sulfuric acid solution at pH 2 is percolated down through the pile (Figure 21.1). The liquid emerging from the bottom of the pile (Figure 21.1*b*) is rich in dissolved metals and is transported to a precipitation plant (Figure 21.1*c*) where the desired metal is precipitated and purified (Figure 21.1*d*). The liquid is then pumped back to the top of the pile and the cycle repeated. As needed, acid is added to maintain an acidic pH.

We illustrate microbial leaching of copper with the common copper ore CuS, in which copper exists as Cu^{2+}. *A. ferrooxidans* oxidizes the sulfide in CuS to SO_4^{2-}, releasing Cu^{2+} as shown in **Figure 21.2**. However, this reaction can also occur spontaneously. Indeed, the key reaction in copper leaching is actually not the bacterial oxidation of sulfide in CuS but the spontaneous oxidation of sulfide by ferric iron (Fe^{3+}) generated from the bacterial oxidation of ferrous iron (Fe^{2+}) (Figure 21.2). In any copper ore, FeS_2 is also present, and its oxidation by bacteria leads to the formation of Fe^{3+} (Figure 21.2). The spontaneous reaction of CuS with Fe^{3+} proceeds in the absence of O_2 and forms Cu^{2+} plus Fe^{2+}; importantly for efficiency of the leaching process, this reaction can take place deep in the leach dump where conditions are anoxic.

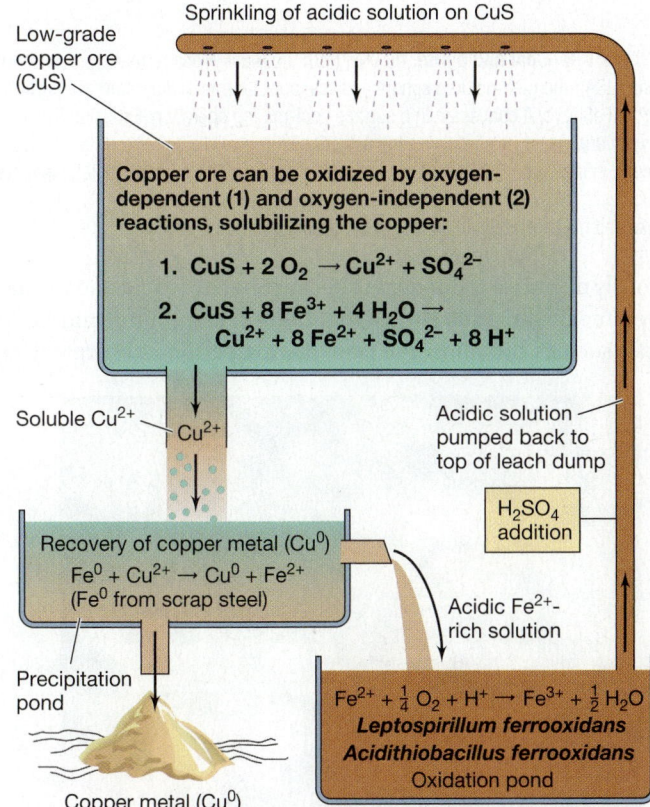

Sprinkling of acidic solution on CuS

Low-grade copper ore (CuS)

Copper ore can be oxidized by oxygen-dependent (1) and oxygen-independent (2) reactions, solubilizing the copper:

1. $CuS + 2 O_2 \rightarrow Cu^{2+} + SO_4^{2-}$

2. $CuS + 8 Fe^{3+} + 4 H_2O \rightarrow$
   $Cu^{2+} + 8 Fe^{2+} + SO_4^{2-} + 8 H^+$

Soluble Cu^{2+}

Cu^{2+}

Acidic solution pumped back to top of leach dump

H_2SO_4 addition

Recovery of copper metal (Cu^0)
$Fe^0 + Cu^{2+} \rightarrow Cu^0 + Fe^{2+}$
(Fe^0 from scrap steel)

Acidic Fe^{2+}-rich solution

Precipitation pond

$Fe^{2+} + \frac{1}{4} O_2 + H^+ \rightarrow Fe^{3+} + \frac{1}{2} H_2O$
Leptospirillum ferrooxidans
Acidithiobacillus ferrooxidans
Oxidation pond

Copper metal (Cu^0)

Figure 21.2 **Arrangement of a leaching pile and reactions in the microbial leaching of copper sulfide minerals to yield metallic copper.** Reaction 1 occurs both biologically and chemically. Reaction 2 is strictly chemical and is the most important reaction in copper-leaching processes. For reaction 2 to proceed, it is essential that the Fe^{2+} produced from the oxidation of sulfide in CuS to sulfate be oxidized back to Fe^{3+} by iron chemolithotrophs (see chemistry in the oxidation pond).

Metal Recovery

The precipitation plant is where the Cu^{2+} from the leaching solution is recovered (Figure 21.1*c, d*). Shredded scrap iron (a source of elemental iron, Fe^0) is added to the precipitation pond to recover copper from the leach liquid by the chemical reaction shown in the lower part of Figure 21.2. This results in a Fe^{2+}-rich liquid that is pumped to a shallow oxidation pond where iron-oxidizing chemolithotrophs oxidize the Fe^{2+} to Fe^{3+}. This now ferric iron–rich acidic liquid is pumped to the top of the pile and the Fe^{3+} is used to oxidize more CuS (Figure 21.1). The entire CuS leaching operation is thus driven by the oxidation of Fe^{2+} to Fe^{3+} by iron-oxidizing bacteria.

Temperatures rise in a leaching dump and this leads to shifts in the iron-oxidizing microbial community. *A. ferrooxidans* is a mesophile, and when heat generated by microbial activities raises temperatures above about 30°C inside a leach dump, this bacterium is outcompeted by mildly thermophilic iron-oxidizing chemolithotrophic *Bacteria* such as *Leptospirillum ferrooxidans* and *Sulfobacillus*. At even higher temperatures (60–80°C), hyperthermophilic *Archaea* such as *Sulfolobus* (⟳ Section 16.10) predominate in the leach dump.

Other Microbial Leaching Processes: Uranium and Gold

Bacteria are also used in the leaching of uranium (U) and gold (Au) ores. In uranium leaching, *A. ferrooxidans* oxidizes U^{4+} to U^{6+} with O_2 as an electron acceptor. However, U leaching depends more on the abiotic oxidation of U^{4+} by Fe^{3+} with *A. ferrooxidans* contributing to the process mainly through the reoxidation of Fe^{2+} to Fe^{3+}, as in copper leaching (Figure 21.2). The reaction observed is as follows:

$$UO_2 + Fe_2(SO_4)_3 \rightarrow UO_2SO_4 + 2 FeSO_4$$

$$(U^{4+})\ (Fe^{3+}) \qquad (U^{6+}) \qquad (Fe^{2+})$$

Unlike UO_2, the uranyl sulfate (UO_2SO_4) formed is highly soluble and is concentrated by other processes.

Gold is typically present in nature in deposits associated with minerals containing arsenic (As) and FeS_2. *A. ferrooxidans* and related bacteria can leach the arsenopyrite minerals, releasing the trapped Au:

$$2 FeAsS[Au] + 7 O_2 + 2 H_2O + H_2SO_4 \rightarrow Fe_2(SO_4)_3$$
$$+ 2 H_3AsO_4 + [Au]$$

The Au is then complexed with cyanide (CN^-) by traditional gold-mining methods. Unlike copper leaching, which is done in a huge dump (Figure 21.1*a*), gold leaching is done in small bioreactor tanks (**Figure 21.3**), where more than 95% of the trapped Au can be released. Moreover, the potentially toxic As and CN^- residues from the mining process are removed in the gold-leaching bioreactor. Arsenic is removed as a ferric precipitate, and CN^- is removed by its bacterial oxidation to CO_2 plus urea in later stages of the Au recovery process. Small-scale microbial-bioreactor leaching has thus become popular as an alternative to the environmentally devastating gold-mining techniques that leave a toxic trail of As and CN^- at the extraction site. Pilot processes are also being developed for bioreactor leaching of zinc, lead, and nickel ores.

UNIT 4

Figure 21.3 Gold bioleaching. Gold leaching tanks in Ghana (Africa). Within the tanks, a mixture of *Acidithiobacillus ferrooxidans*, *Acidithiobacillus thiooxidans*, and *Leptospirillum ferrooxidans* solubilizes the pyrite/arsenic mineral containing trapped gold, which releases the gold.

MINIQUIZ

• What is required to oxidize CuS under anaerobic conditions?
• What key role does *Acidithiobacillus ferrooxidans* play in the copper leaching process?

21.2 Acid Mine Drainage

Although microbial leaching has tremendous value in mining operations, the same process has contributed to extensive environmental destruction where mining operations improperly handle or dispose of pyrite containing coal and mineral deposits. Bacterial and spontaneous oxidation of sulfide minerals is the major cause of **acid mine drainage**, an environmental problem worldwide caused by surface mining operations. As described for the oxidation of copper sulfides promoted in microbial mining (Section 21.1), the oxidation of FeS_2 is a combination of chemically and bacterially catalyzed reactions, and two electron acceptors participate in the process: O_2 and Fe^{3+}. When FeS_2 is first exposed in a mining operation (**Figure 21.4b**), a slow chemical reaction with O_2 begins (Figure 21.4c). This reaction, called the *initiator reaction*, leads to the oxidation of HS^- to SO_4^{2-} and the development of acidic conditions as Fe^{2+} is released. *A. ferrooxidans* and *L. ferrooxidans* then oxidize Fe^{2+} to Fe^{3+}, and the Fe^{3+} formed under these acidic conditions, being soluble, reacts spontaneously with more FeS_2 and oxidizes the HS^- to sulfuric acid (H_2SO_4), which immediately dissociates into SO_4^{2-} and H^+:

$$FeS_2 + 14\,Fe^{3+} + 8\,H_2O \rightarrow 15\,Fe^{2+} + 2\,SO_4^{2-} + 16\,H^+$$

Again, the bacteria oxidize Fe^{2+} to Fe^{3+}, and this Fe^{3+} reacts with more FeS_2. Thus, there is a progressive, rapidly increasing rate at which FeS_2 is oxidized, called the *propagation cycle* (Figure 21.4c). Under natural conditions some of the Fe^{2+} generated by the bacteria leaches away and is subsequently carried by anoxic groundwater into surrounding streams. However, bacterial or spontaneous oxidation of Fe^{2+} then takes place in the aerated streams, and because O_2 is present, the insoluble $Fe(OH)_3$ is formed.

As we have seen (Figure 21.4c), the breakdown of FeS_2 ultimately leads to the formation of H_2SO_4 and Fe^{2+}; in waters in which these products have formed, pH values can be lower than 1. Mixing of acidic mine waters into rivers (**Figure 21.5**) and lakes

(a)

(b)

Propagation cycle

Initiator reaction

$$FeS_2 + 3\tfrac{1}{2}\,O_2 + H_2O \longrightarrow Fe^{2+} + 2\,SO_4^{2-} + H^+$$

Fe^{3+}

Acidification

Spontaneous (Fe^{3+} is oxidant for propagation cycle)

Bacteria or spontaneous

(c)

Figure 21.4 Coal and pyrite. *(a)* Coal from the Black Mesa formation in northern Arizona (USA); the gold-colored spherical discs (about 1 mm in diameter) are particles of pyrite (FeS_2). *(b)* A coal seam in a surface coal-mining operation. Exposing the coal to oxygen and moisture stimulates the activities of iron-oxidizing bacteria growing on the pyrite in the coal. *(c)* Reactions in pyrite degradation. The primarily abiotic initiator reaction sets the stage for the primarily bacterial oxidation of Fe^{2+} to Fe^{3+}. The Fe^{3+} attacks and oxidizes FeS_2 abiotically in the propagation cycle.

seriously degrades water quality because both the acid and the dissolved metals (in addition to iron, there is aluminum, and heavy metals such as cadmium and lead) are toxic to aquatic organisms.

Figure 21.5 Acid mine drainage from a surface coal-mining operation. The yellowish-red color is due to the precipitated iron oxides in the drainage (see Figure 21.4c for the reactions in acid mine drainage).

The O_2 requirement for the oxidation of Fe^{2+} to Fe^{3+} explains how acid mine drainage develops. As long as the pyritic material is not mined, FeS_2 cannot be oxidized because O_2, water, and the bacteria cannot reach it. However, when a mineral or coal seam is exposed (Figure 21.4b), O_2 and water are introduced, making both spontaneous and bacterial oxidation of FeS_2 possible. The acid formed can then leach into surrounding aquatic systems (Figure 21.5).

Where acid mine drainage is extensive and Fe^{2+} levels high, a strongly acidophilic species of *Archaea*, *Ferroplasma*, is often present. This aerobic iron-oxidizing organism is capable of growth at pH 0 and at temperatures up to 50°C. Cells of *Ferroplasma* lack a cell wall and are phylogenetically related to *Thermoplasma*, also a cell-wall-lacking and strongly acidophilic (but chemoorganotrophic) member of the *Archaea* (⮂ Section 16.3).

MINIQUIZ --

- In what oxidation state is iron in the mineral $Fe(OH)_3$? In FeS? How is $Fe(OH)_3$ formed?

- Natural pyritic deposits, such as underground coal seams, do not contribute to acid mine drainage; why not?

II • Bioremediation

The term **bioremediation** refers to the microbial cleanup of oil, toxic chemicals, or other environmental pollutants, usually by stimulating the activities of indigenous microorganisms in some way. These pollutants include both natural materials, such as petroleum products, and **xenobiotic** chemicals, synthetic chemicals not produced by organisms in nature.

Although bioremediation of many toxic substances has been proposed, most successes have been in cleaning up spills of crude oil or the leakage of hydrocarbons from bulk storage tanks. More recently, the targeted destruction of chlorinated environmental pollutants, including commonly used solvents and pesticides, has become more amenable to bioremediation as a result of better understanding of associated microbiology. There has also been increasing success in the bioremediation of uranium-contaminated environments, many of which are the legacy of poorly regulated past mining of uranium for nuclear fuel and weapons. We begin here with a consideration of this very toxic pollutant.

21.3 Bioremediation of Uranium-Contaminated Environments

Major classes of inorganic pollutants are metals and radionuclides that cannot be destroyed, but only altered in chemical form. Often the extent of environmental pollution is so great that physical removal of the contaminated material is impossible. Thus, *containment* is the only real option, and a common goal in the bioremediation of inorganic pollutants is to change their mobility, making them less likely to move with groundwater and so contaminate surrounding environments. Here we consider how the radioactive element uranium can be contained by the activities of bacteria.

Bioremediation of Uranium

Uranium contamination of groundwater has occurred at sites in the United States and elsewhere where uranium ores have been processed or stored (**Figure 21.6**), and the movement of radioactive materials offsite via groundwater is a threat to environmental and human health. Because the contamination is often widespread, making mechanical methods of recovery very expensive, microbiologists have joined forces with engineers to develop biological treatments that exploit the ability of some bacteria to reduce U^{6+} to U^{4+}. Uranium as U^{6+} is soluble, whereas U^{4+} forms an immobile uranium mineral called *uraninite*, thus limiting the movement of U into groundwater and potential contact with humans and other animals.

Bacterial Transformations of Uranium

The major strategy for immobilizing uranium has been to use bacteria to change the oxidation state of U in major uranium contaminants to a form that will stabilize

Figure 21.6 Uranium bioremediation. An experimental plot at a United States Department of Energy uranium-contaminated site. Organic carbon (acetate) is being infused into the site (see inset photo) and travels in groundwater in the direction of the arrow shown in the main photo. Acetate is an electron donor for reduction of U^{6+} to U^{4+}, which immobilizes the uranium.

the element. In this regard, *Bacteria*, including metal-reducing *Shewanella* and *Geobacter* species (Section 14.14) and sulfate-reducing *Desulfovibrio* species (Section 14.9), couple the oxidation of organic matter and H_2 to the reduction of U^{6+} to U^{4+}.

Field studies in which organic electron donors have been injected into uranium-contaminated aquifers to stimulate U^{6+} reduction have shown that this approach can lower U levels to below the U.S. Environmental Protection Agency's drinking water standard of 0.126 μM. However, even though uraninite is stable under reducing conditions, if conditions become oxic, it reoxidizes. Thus, much ongoing uranium bioremediation research is focused on questions of whether microbially reduced uranium is stable if the composition of the microbial community changes or if oxidants, such as O_2, NO_3^-, and Fe^{3+}, are introduced via groundwater. This is obviously an important question because uraninite stability must be targeted for the long term in order to account for the long half-life of nuclear decay of uranium.

MINIQUIZ --

- Which reaction, oxidation or reduction, is key to uranium bioremediation?

- Why is immobilization a good strategy for dealing with uranium pollution?

21.4 Bioremediation of Organic Pollutants: Hydrocarbons

Organic pollutants, unlike inorganic pollutants, can generally be completely degraded by microorganisms, eventually to CO_2. This is true of petroleum released in oil spills (**Figure 21.7**), which can be attacked by many different microorganisms. These organisms have been exposed to complex mixtures of hydrocarbons through natural oil seeps for millennia and thus have evolved the catabolic machinery necessary to degrade this naturally occurring pollutant. In contrast, xenobiotic pollutants tend to be more persistent and are degraded by more specialized groups of microorganisms. In this section we focus on hydrocarbons and in the next section on xenobiotics.

Petroleum and Hydrocarbon Bioremediation

Petroleum is a rich source of organic matter, and because of this, microorganisms readily attack hydrocarbons when petroleum is pumped to Earth's surface and comes into contact with air and moisture. Under some circumstances, such as in bulk petroleum storage tanks, microbial growth is undesirable. However, in oil spills, biodegradation is desirable and can be promoted by the addition of inorganic nutrients to balance the huge influx of organic carbon from the oil (Figure 21.7).

The biochemistry of hydrocarbon catabolism was covered in Sections 13.22–13.24. Both oxic and anoxic biodegradation is possible. We emphasized that under oxic conditions oxygenase enzymes play an important role in introducing oxygen atoms into the hydrocarbon. Our discussion here will focus on *aerobic* processes, because it is only when O_2 is present that oxygenase enzymes can function and hydrocarbon bioremediation can be effective in a relatively short time.

Diverse bacteria, fungi, and a few green algae can oxidize petroleum products aerobically. Small-scale oil pollution of aquatic and terrestrial ecosystems from human as well as natural activities is common. Oil-oxidizing microorganisms develop rapidly on oil films and slicks, and hydrocarbon oxidation is most extensive if the temperature is warm enough and supplies of inorganic nutrients (primarily N and P) are sufficient.

Because oil is insoluble in water and is less dense, it floats to the surface and forms slicks. There, hydrocarbon-degrading bacteria attach to the oil droplets (**Figure 21.8**) and eventually decompose the oil and disperse the slick. Certain oil-degrading bacteria are specialist species; for example, the bacterium *Alcanivorax borkumensis* grows only on hydrocarbons, fatty acids, or pyruvate. This organism produces surfactant chemicals that help break up the oil and solubilize it. Once solubilized, the oil can be incorporated more readily and catabolized as an electron donor and carbon source.

In large surface oil spills such as those shown in Figure 21.7, volatile hydrocarbons, both aliphatic and aromatic, evaporate quickly without bioremediation, leaving nonvolatile components for cleanup crews and microorganisms to tackle. Microorganisms

(a)

(b)

(c)

Figure 21.7 Environmental consequences of large oil spills and the effect of bioremediation. *(a)* A contaminated beach along the coast of Alaska containing oil from the *Exxon Valdez* spill of 1989. *(b)* The rectangular plot (arrow) was treated with inorganic nutrients to stimulate bioremediation of spilled oil by microorganisms, whereas areas above and to the left were untreated. *(c)* Oil spilled into the Mediterranean Sea from the Jiyeh (Lebanon) power plant that flowed to the port of Byblos during the 2006 war in Lebanon.

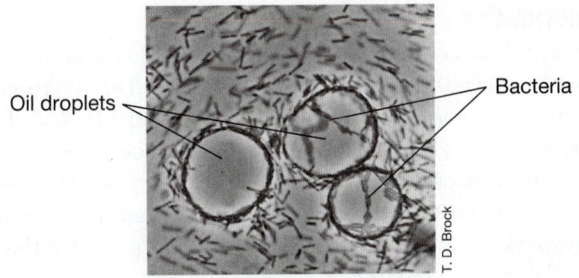

Figure 21.8 **Hydrocarbon-oxidizing bacteria in association with oil droplets.** The bacteria are concentrated in large numbers at the oil–water interface, but are actually not within the droplet itself.

consume oil by oxidizing it to CO_2. When bioremediation activities are promoted by inorganic nutrient application, oil-oxidizing bacteria typically develop quickly after an oil spill (Figure 21.7b), and under ideal conditions, 80% or more of the nonvolatile oil components can be oxidized within one year. However, certain oil fractions, such as those containing branched-chain and polycyclic hydrocarbons, are not preferred microbial substrates and remain in the environment much longer. Spilled oil that finds its way into sediments is even more slowly degraded and can have a significant long-term impact on fisheries that depend on unpolluted waters for productive yields.

A notable exception to the more common surface spill of oil was the 2010 sinking of the Deepwater Horizon offshore drilling platform in the Gulf of Mexico, resulting in the rupture of the wellhead at a depth of 1.5 km and release of over 4 million barrels of oil into the deep ocean (⟲ Section 19.9 and Figure 19.20). About 35% of the resulting hydrocarbon plume was comprised of low-molecular-weight components and natural gas (methane, ethane, propane). The availability of these more easily degraded oil components is thought to have accelerated the natural degradation process by stimulating the development of a large bloom of bacteria having the capacity to oxidize both the easily degraded and more recalcitrant hydrocarbon components. It remains uncertain whether the industry decision to promote dispersal of the oil (which was intended to increase the oil's surface area and bioavailability) by injecting thousands of gallons of chemical dispersants directly into the plume actually accelerated microbial degradation. Regardless, although some legacy of this major oil spill remains, much of the oil did disappear from a combination of volatilization and microbial activities.

Degradation of Stored Hydrocarbons

Interfaces where oil and water meet often form on a large scale. Besides water that separates from crude petroleum during storage and transport, moisture can condense inside bulk fuel storage tanks (**Figure 21.9**) where there are leaks. This water eventually accumulates in a layer beneath the petroleum. Gasoline and crude oil storage tanks are thus potential habitats for hydrocarbon-oxidizing microorganisms. If sufficient sulfate (SO_4^{2-}) is present in the oil, as it often is in crude oils, sulfate-reducing bacteria can grow in the tanks, consuming hydrocarbons under anoxic conditions (⟲ Sections 13.24 and 14.9). The sulfide (H_2S) produced is highly corrosive and causes pitting and subsequent leakage of

Figure 21.9 **Bulk petroleum storage tanks.** Fuel tanks often support microbial growth at oil–water interfaces.

the tanks along with souring of the fuel. Aerobic degradation of stored fuel components is less of a problem because the storage tanks are sealed and the fuel itself contains little dissolved O_2.

MINIQUIZ
- Why do petroleum-degrading bacteria need to attach to the *surface* of oil droplets?
- What is unique about the physiology of the bacterium *Alcanivorax*?

21.5 Bioremediation of Organic Pollutants: Pesticides and Plastics

Unlike hydrocarbons, many chemicals that humans put into the environment have never been there before. These are the xenobiotics, and we consider their microbial degradation here.

Pesticide Catabolism

Xenobiotics include pesticides, polychlorinated biphenyls (PCBs), munitions, dyes, and chlorinated solvents, among many other chemicals. Some xenobiotics differ chemically in such major ways from anything organisms have experienced in nature that they biodegrade extremely slowly, if at all. Other xenobiotics are structurally related to one or more natural compounds and can sometimes be degraded slowly by enzymes that normally degrade the structurally related natural compounds. We focus here on pesticide bioremediation.

Over 1000 pesticides have been marketed worldwide for pest control purposes. Pesticides include *herbicides*, *insecticides*, and *fungicides*. Pesticides display a wide variety of chemistries, and include chlorinated, aromatic, and nitrogen- and phosphorus-containing compounds (**Figure 21.10**). Some of these substances can be used as carbon and energy sources by microorganisms, whereas others are utilized only poorly or not at all. Highly chlorinated compounds are typically the pesticides most resistant to microbial attack. However, related compounds may differ

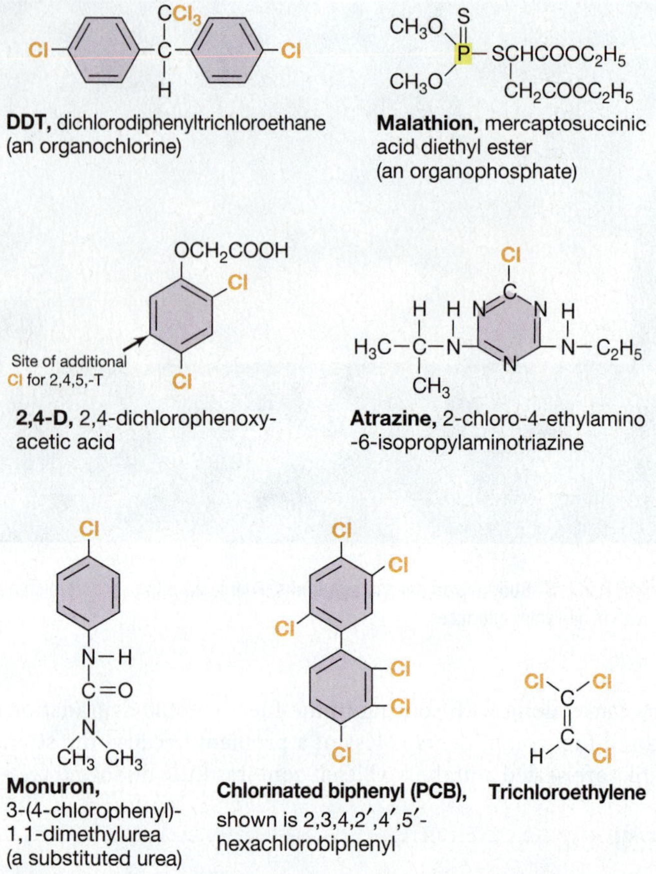

DDT, dichlorodiphenyltrichloroethane (an organochlorine)

Malathion, mercaptosuccinic acid diethyl ester (an organophosphate)

2,4-D, 2,4-dichlorophenoxy-acetic acid

Site of additional Cl for 2,4,5,-T

Atrazine, 2-chloro-4-ethylamino-6-isopropylaminotriazine

Monuron, 3-(4-chlorophenyl)-1,1-dimethylurea (a substituted urea)

Chlorinated biphenyl (PCB), shown is 2,3,4,2′,4′,5′-hexachlorobiphenyl

Trichloroethylene

Figure 21.10 Examples of xenobiotic compounds. Although none of these compounds exist naturally, microorganisms exist that can break them down.

remarkably in their degradability. For example, chlorinated compounds such as DDT persist relatively unaltered for years in soils, whereas chlorinated compounds such as 2,4-D are significantly degraded in just a few weeks.

Environmental factors, such as temperature, pH, aeration, and organic content of the soil, influence the rate of pesticide decomposition, and some pesticides can disappear from soils nonbiologically by volatilization, leaching, or spontaneous chemical breakdown. In addition, some pesticides are degraded only when other organic material is present that can be used as the primary energy source, a phenomenon called *cometabolism*. In most cases, pesticides that are cometabolized are only partially degraded, generating new xenobiotic compounds that may be even more toxic or difficult to degrade than the original compound. Thus, from an environmental standpoint, cometabolism of a pesticide is not always good.

Dechlorination

Many xenobiotics are chlorinated compounds and their degradation proceeds through *dechlorination*. For example, the bacterium *Burkholderia* dechlorinates the pesticide 2,4,5-T aerobically, releasing chloride ion (Cl^-) in the process (**Figure 21.11**); this reaction is catalyzed by oxygenase enzymes (⇆ Section 13.22). Following dechlorination, a dioxygenase enzyme breaks the aromatic ring to yield compounds that can enter the citric acid cycle and yield energy.

Although the aerobic breakdown of chlorinated xenobiotics is undoubtedly ecologically important, **reductive dechlorination** may be even more so because of the rapidity with which anoxic conditions develop in polluted microbial habitats. We previously described reductive dechlorination as a form of anaerobic respiration in which chlorinated organic compounds such as chlorobenzoate ($C_7H_4O_2Cl^-$) are terminal electron acceptors and when reduced, release chloride (Cl^-), a nontoxic substance (⇆ Section 13.21).

Many compounds can be reductively dechlorinated including dichloro-, trichloro-, and tetrachloro- (perchloro-) ethylene, chloroform, dichloromethane, and polychlorinated biphenyls (Figure 21.10). In addition, several brominated and fluorinated organic compounds can be dehalogenated in analogous fashion. Many of these chlorinated or halogenated compounds are highly toxic and some have been linked to cancer (particularly trichloroethylene). Some of these compounds, such as PCBs, have been widely used as insulators in electrical transformers and enter anoxic environments from slow leakage of the transformer or from leaking storage containers. Eventually these compounds end up in groundwater or sediment, where they are among the most common contaminants detected in the United States. There is therefore great interest in reductive dechlorination as a bioremediation strategy for their removal from anoxic environments.

Plastics

Plastics are classic examples of xenobiotics, and the plastics industry worldwide produces over 40 million tons of plastic per year, almost half of which are discarded rather than recycled. Plastics are polymers of various chemistries (**Figure 21.12a**). Many plastics remain essentially unaltered for long periods in landfills, refuse dumps, and as litter in the environment. This problem has fueled the search for biodegradable alternatives called **microbial plastics** as replacements for some synthetic plastics.

Polyhydroxyalkanoates (PHAs) are a common bacterial storage polymer (⇆ Section 2.14), and these readily biodegradable polymers have many of the desirable properties of xenobiotic plastics. PHAs can be biosynthesized in various chemical forms, each with its own unique physical properties (stiffness, shear and impact

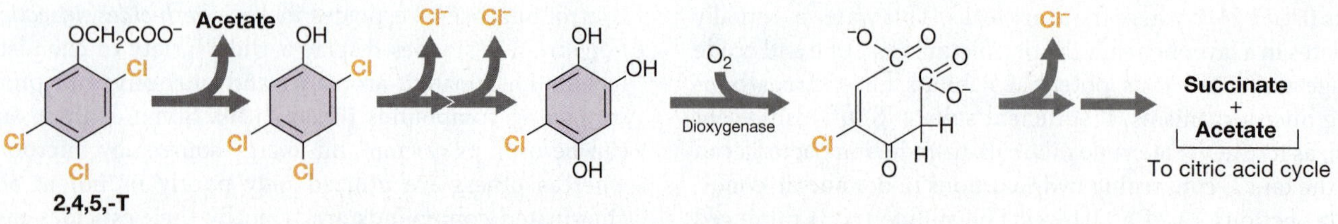

Figure 21.11 Biodegradation of the herbicide 2,4,5-T. Pathway of aerobic 2,4,5-T biodegradation; note the importance of a dioxygenase enzyme (⇆ Section 13.22) in the degradation process.

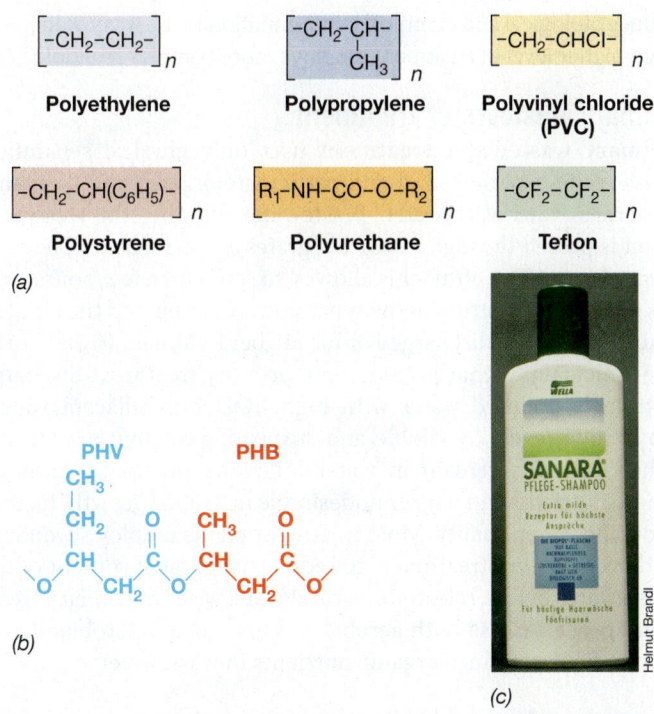

Figure 21.12 **Synthetic and microbial plastics.** *(a)* The monomeric structure of several synthetic plastics. *(b)* Structure of the copolymer of poly-β-hydroxybutyrate (PHB) and poly-β-hydroxyvalerate (PHV). *(c)* A brand of shampoo previously marketed in Germany and packaged in a bottle made of the PHB/PHV copolymer.

strength, and the like). A PHA *copolymer* containing equal amounts of poly-β-hydroxybutyrate and poly-β-hydroxyvalerate (Figure 21.12*b*) has been marketed in Europe as a container for personal care products and has had the greatest success as a plastic substitute thus far (Figure 21.12*c*). However, because synthetic plastics are currently less expensive than microbial plastics, synthetic petroleum-based plastics make up virtually the entire plastics market today.

The bacterium *Ralstonia eutropha* has been used as a model organism for the commercial production of PHAs. This genetically manipulable and metabolically diverse bacterium produces PHAs in high yield, and specific copolymers can be obtained by simple nutritional modifications. Nevertheless, the microbial plastics industry is burdened by the reality that the best substrates for PHA biosyntheses are glucose and related organic compounds, substances obtained from corn or other crops. And even at today's prices for oil, plant products cannot compete with oil as feedstocks for the plastics industry.

MINIQUIZ

- Why might the addition of inorganic nutrients stimulate oil degradation whereas the addition of glucose would not?
- What is reductive dechlorination and how does it differ from the reactions shown in Figure 21.11?
- What main advantage do microbial plastics have over synthetic plastics?

III • Wastewater and Drinking Water Treatment

Water is the most important potential common source of infectious diseases and a potential source for chemically induced intoxications. This is because a single water source often serves large numbers of people, as, for example, in large cities. Everyone in these cities must use the available water, and contaminated water has the potential to spread disease to all exposed individuals. Similarly, appropriate treatment of wastewater is essential for maintaining environmental quality and for reducing the spread of disease.

The outbreak of cholera in Haiti following the 2010 earthquake is a reminder of the importance of well-maintained waste and drinking water treatment systems in securing public health. Here we examine systems built for the chemical and biological treatment of water and the transmission systems used for delivering treated water to consumers. We also examine the human health significance of the microbial ecology that develops within the pipes of municipal water distribution systems and premise plumbing.

21.6 Primary and Secondary Wastewater Treatment

Wastewater is domestic sewage or liquid industrial waste that cannot be discarded in untreated form into lakes or streams due to public health, economic, environmental, and aesthetic considerations. Wastewater treatment employs physical and chemical methods as well as industrial-scale use of microorganisms. Wastewater enters a treatment plant and, following treatment, the **effluent water**—treated wastewater discharged from the wastewater treatment facility—is suitable for release into surface waters such as lakes and streams or to drinking water purification facilities (**Figure 21.13**).

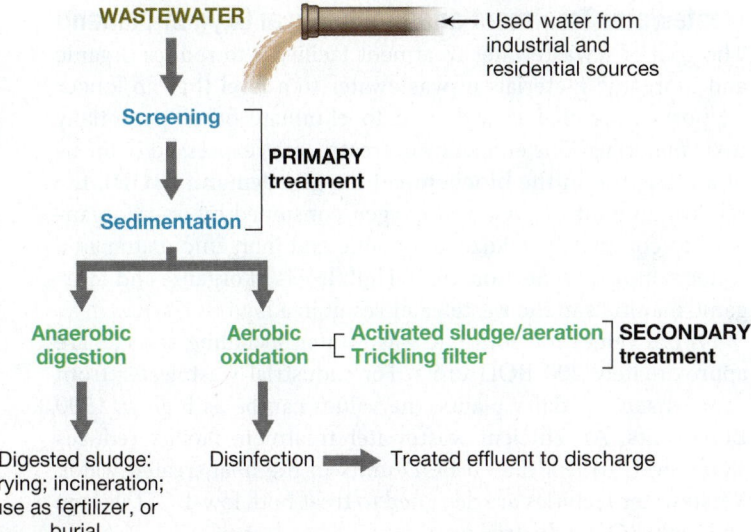

Figure 21.13 **Wastewater treatment processes.** Effective water treatment plants use the primary and secondary treatment methods shown here. Tertiary treatment may also be used to reduce biochemical oxygen demand (BOD) levels in effluent water to undetectable levels.

Wastewater and Sewage

Wastewater from domestic sewage or industrial sources cannot be discarded in untreated form into lakes or streams. **Sewage** is liquid effluent contaminated with human or animal fecal materials. Wastewater may also contain potentially harmful inorganic and organic compounds as well as pathogenic microorganisms. Wastewater treatment can use physical, chemical, and biological (microbiological) processes to remove or neutralize contaminants.

On average, each person in the United States uses 100–200 gallons of water every day for washing, cooking, drinking, and sanitation. Wastewater collected from these activities must be treated to remove contaminants before it can be released into surface waters. About 16,000 publicly owned treatment works (POTW) operate in the United States. Most POTWs are fairly small, treating 1 million gallons (3.8 million liters) or less of wastewater per day. Collectively, however, these plants treat about 32 billion gallons of wastewater daily. Wastewater plants are usually constructed to handle both domestic and industrial wastes. Domestic wastewater is made up of sewage, "gray water" (the water resulting from washing, bathing, and cooking), and wastewater from small-scale food processing in homes and restaurants.

Industrial wastewater includes liquid discharged from the petrochemical, pesticide, food and dairy, plastics, pharmaceutical, and metallurgical industries. Industrial wastewater may contain toxic substances; some manufacturing and processing plants are required by the U.S. Environmental Protection Agency (EPA) to pretreat toxic or heavily contaminated discharges before they enter POTWs. Pretreatment may involve mechanical processes in which large debris is removed. Some wastewaters are pretreated biologically or chemically to remove highly toxic substances such as cyanide; heavy metals such as arsenic, lead, and mercury; or organic materials such as acrylamide, atrazine (a herbicide), and benzene. These substances are converted to less toxic forms by treatment with chemicals or microorganisms capable of neutralizing, oxidizing, precipitating, or volatilizing these wastes. The pretreated wastewater can then be released to the POTW.

Wastewater Treatment and Biochemical Oxygen Demand

The goal of a wastewater treatment facility is to reduce organic and inorganic materials in wastewater to a level that no longer supports microbial growth and to eliminate other potentially toxic materials. The efficiency of treatment is expressed in terms of a reduction in the **biochemical oxygen demand (BOD)**, the relative amount of dissolved oxygen consumed by microorganisms to completely oxidize all organic and inorganic matter in a water sample (🔗 Section 19.8). High levels of organic and inorganic materials in the wastewater result in a high BOD.

Typical values for domestic wastewater, including sewage, are approximately 200 BOD units. For industrial wastewater from sources such as dairy plants, the values can be as high as 1500 BOD units. An efficient wastewater treatment facility reduces BOD levels to less than 5 BOD units in the final treated water. Wastewater facilities are designed to treat both low-BOD sewage and high-BOD industrial wastes.

Treatment is a multistep operation employing a number of independent physical and biological processes (Figure 21.13). *Primary, secondary,* and sometimes additional treatments are employed to reduce biological and chemical contamination in the wastewater, and each higher level of treatment employs more complex technologies.

Primary Wastewater Treatment

Primary wastewater treatment uses only physical separation methods to separate solid and particulate organic and inorganic materials from wastewater. Wastewater entering the treatment plant is passed through a series of grates and screens that remove large objects. The effluent is allowed to settle for a few hours. Solids settle to the bottom of the separation reservoir and the effluent is drawn off to be discharged or for further treatment (**Figure 21.14**).

Municipalities that provide only primary treatment discharge extremely polluted water with high BOD into adjacent waterways; high levels of soluble and suspended organic matter and other nutrients remain in water following primary treatment. These nutrients can trigger undesirable microbial growth, further reducing water quality. Most treatment plants employ secondary and even *tertiary* treatments to reduce the organic content of the wastewater before release to natural waterways. Secondary treatment processes use both aerobic and anaerobic microbial digestion to further reduce organic nutrients in wastewater.

Secondary Anaerobic Wastewater Treatment

Secondary anaerobic wastewater treatment involves a series of degradative and fermentative reactions carried out by various prokaryotes under anoxic conditions. Anaerobic treatment is typically used to treat wastewater containing large quantities of insoluble organic matter (and therefore having a very high BOD) such as fiber and cellulose waste from food and dairy plants. The anaerobic degradation process is carried out in large, enclosed tanks called *sludge digesters* or *bioreactors* (**Figure 21.15**). The process requires the collective activities of many different types of prokaryotes. The major reactions are summarized in Figure 21.15c.

First, anaerobes use polysaccharidases, proteases, and lipases to digest suspended solids and large macromolecules into soluble components. These soluble components are then fermented to yield a mixture of fatty acids, H_2, and CO_2; the fatty acids are

Figure 21.14 Primary treatment of wastewater. Wastewater is pumped into the reservoir (left) where solids settle. As the water level rises, the water spills through the grates to successively lower levels. Water at the lowest level, now virtually free of solids, enters the spillway (arrow) and is pumped to a secondary treatment facility.

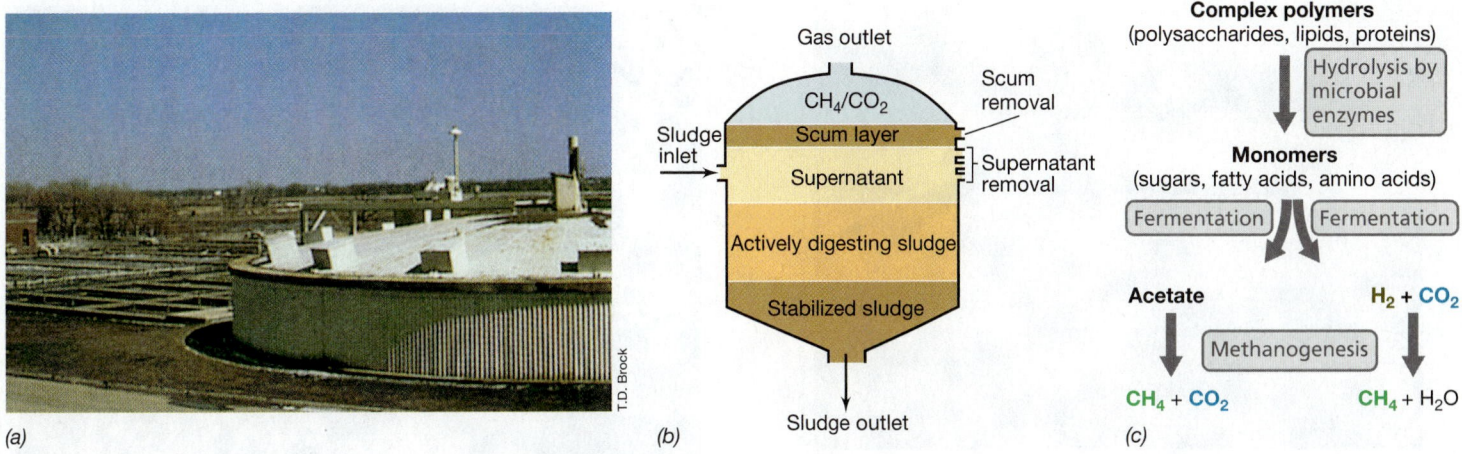

Figure 21.15 Secondary anaerobic wastewater treatment. *(a)* Anaerobic sludge digester. Only the top of the tank is shown; the remainder is underground. *(b)* Inner workings of a sludge digester. *(c)* Major microbial processes in anaerobic sludge digestion. Methane (CH_4) and carbon dioxide (CO_2) are the major products of anaerobic biodegradation.

further fermented by the cooperative actions of syntrophic bacteria (⊂⊃ Section 13.15) to produce acetate, CO_2, and H_2. These products are then used as substrates by methanogenic *Archaea* (⊂⊃ Section 16.2), fermenting acetate to produce methane (CH_4) and CO_2, the major products of anoxic sewage treatment (Figure 21.15*c*). The CH_4 is burned off or used as fuel to heat and power the wastewater treatment plant.

Secondary Aerobic Wastewater Treatment

Secondary aerobic wastewater treatment uses oxidative degradation reactions carried out by microorganisms under aerobic conditions to treat wastewater containing low levels of organic materials (**Figure 21.16a, b**). In general, wastewaters that originate from residential sources can be treated efficiently using only aerobic treatment. Several aerobic degradative processes can be used for wastewater treatment; *activated sludge* methods are the most common (Figure 21.16*a, b*). Here, wastewater is continuously mixed and aerated in large tanks. Slime-forming aerobic bacteria, including *Zoogloea ramigera* and others, grow and form aggregated masses called flocs (**Figure 21.17**). The biology of *Zoogloea* is discussed in Section 15.2. Protists, small animals, filamentous bacteria, and fungi attach to the flocs. Oxidation of organic matter occurs on the floc as it is agitated and exposed to air. The aerated effluent containing the flocs is pumped into a holding tank or clarifier where the flocs settle. Some of the floc material (called activated sludge) is then returned to the aerator as inoculum for new wastewater, and the rest is pumped to the anaerobic sludge digester (Figure 21.15) or is removed, dried, and burned, or is used for fertilizer.

Wastewater normally stays in an activated sludge tank for 5–10 hours, a time too short for complete oxidation of all organic matter. However, during this time much of the soluble organic matter is adsorbed to the floc and incorporated by the microbial cells. The BOD of the liquid effluent is considerably reduced (up to 95%) when compared to the incoming wastewater; most of the material with high BOD is now in the settled flocs. The flocs can then be transferred to the anoxic sludge digester for conversion to CO_2 and CH_4.

The *trickling filter* method is also commonly used for secondary aerobic treatment (Figure 21.16*c*). A trickling filter is a bed of crushed rocks, about 2 m thick. Wastewater is sprayed on top of the rocks and slowly passes through the bed. The organic material in the wastewater adsorbs to the rocks, and microorganisms grow on the large, exposed rock surfaces. The complete mineralization of organic matter to CO_2, ammonia, nitrate, sulfate, and phosphate takes place in the extensive microbial biofilm that develops on the rocks.

Most treatment plants chlorinate the effluent after secondary treatment to further reduce the possibility of biological contamination. The treated effluent can then be discharged into streams or lakes. In the eastern United States, many wastewater treatment facilities use UV radiation to disinfect effluent water. Ozone (O_3), a strong oxidizing agent that is an effective bacteriocide and viricide, is also used for wastewater disinfection in some treatment plants in the United States.

MINIQUIZ

- What is biochemical oxygen demand (BOD), and why is its reduction important in wastewater treatment?
- How do primary and secondary wastewater treatment methods differ?
- Other than treated water, what are the final products of wastewater treatment? How might these end products be used?

21.7 Advanced Wastewater Treatment

Advanced wastewater treatment is any process designed to produce an effluent of higher quality than normally achieved by secondary treatment. It includes tertiary treatment, physical-chemical treatment, or combined biological-physical treatment. **Tertiary wastewater treatment** is defined as any treatment process in which unit operations are added for the further processing of the secondary treatment effluent.

Typical goals of advanced treatment include additional removal of organic matter and suspended solids, removal of key inorganic nutrients required for microbial growth (including ammonia, nitrate, nitrite, phosphorus, or dissolved organic carbon), and

UNIT 4

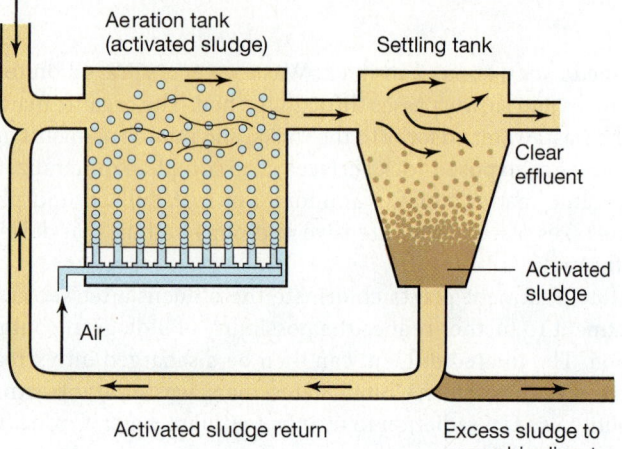

Wastewater from primary treatment

Aeration tank (activated sludge)

Settling tank

Clear effluent

Activated sludge

Air

Activated sludge return

Excess sludge to anaerobic digester

(b)

(c)

Figure 21.16 Secondary aerobic wastewater treatment processes. Parts a and b show the activated sludge method. (a) Aeration tank of an activated sludge installation in a metropolitan wastewater treatment plant. The tank is 30 m long, 10 m wide, and 5 m deep. (b) Wastewater flow through an activated sludge installation. Recirculation of activated sludge to the aeration tank introduces microorganisms responsible for oxidative degradation of the organic components of the wastewater. (c) Trickling filter method. The booms rotate, distributing wastewater slowly and evenly on the rock bed. The rocks are 10–15 cm in diameter and the bed is 2 m deep.

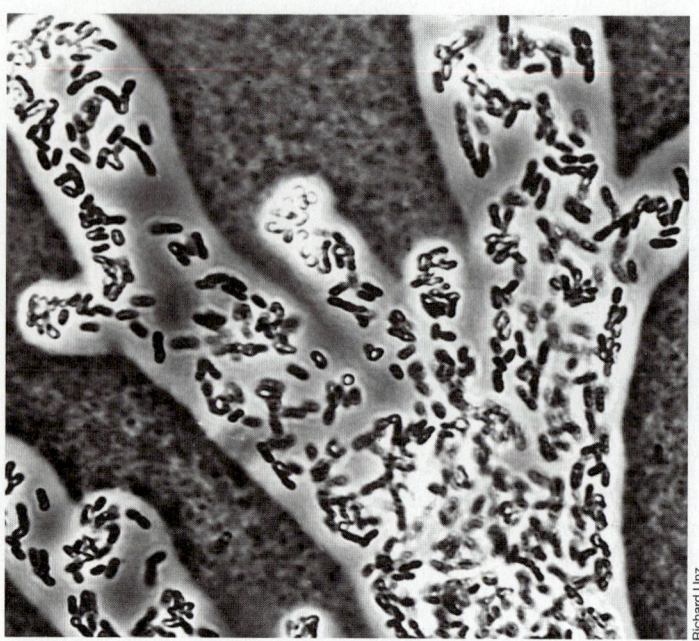

Figure 21.17 A wastewater floc formed by the bacterium *Zoogloea ramigera*. Floc formed in the activated sludge process consists of a large number of small, rod-shaped cells of *Z. ramigera* surrounded by a polysaccharide slime layer, arranged in characteristic fingerlike projections in this negative stain with India ink.

degradation of any potentially toxic materials. Advanced water treatment is the most complete method of treating sewage but has not been widely adopted due to the costs associated with such complete nutrient removal. Here we examine biological removal of phosphorus and trace contaminants, two areas of advanced treatment of increasing importance to wastewater treatment.

Biological Phosphorus Removal

Conventional secondary biological treatment removes only about 20% of phosphorus from wastewater, necessitating additional chemical or biological treatment. Chemical precipitation is the most commonly used process, removing up to 90% of the influent phosphorus. Removal is accomplished by the addition of either Fe or Al as chloride or sulfate salts, with Fe^{2+} or Fe^{3+} salts more commonly used. At near-neutral pH, the Fe^{3+} forms insoluble ferric phosphate ($FePO_4$) or ferric hydroxide-phosphate complexes. These then precipitate and are removed as sludge.

The chemical precipitation process results in up to 95% more sludge, contributing to additional disposal problems. As an alternative, tertiary treatment that encourages the growth of phosphorus-accumulating bacteria can also remove up to 90% of phosphorus, a process called *enhanced biological phosphorus removal* (*EBPR*). Here the waste stream is processed by sequential passage through anaerobic and aerobic bioreactors (**Figure 21.18**). In the anaerobic reactor, *phosphorus-accumulating organisms* (*PAOs*) use energy available from stored polyphosphate to assimilate short-chain fatty acids, and produce intracellular polyhydroxyalkanoates (PHAs) (Figure 21.18*a*; ↩ Section 2.14); as this occurs, soluble orthophosphate (PO_4^{3-}) is released. During the following aerobic phase of treatment, the stored PHA is metabolized, providing energy and carbon for new cell growth. The energy is used to form intracellular polyphosphate,

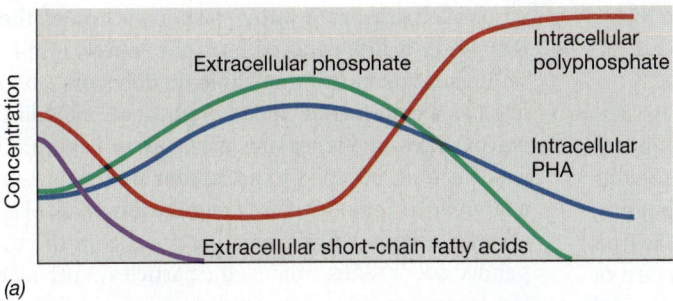

(a)

(b)

Figure 21.18 Enhanced biological phosphorus removal process. During passage through the reactor system the microbial community transitions from anaerobic to aerobic growth. In the anaerobic zone, short-chain fatty acids are taken up and internal stores of polyphosphate (polyP) are released as extracellular orthophosphate. In the aerobic zone, the extracellular phosphate is reassimilated as polyP and the intracellular stores of polyhydroxyalkanoates (PHAs) are metabolized. High-phosphorus sludge is harvested for disposal.

removing orthophosphate from solution (Figure 21.18a). The new biomass (sludge) with high polyphosphate content is then collected for phosphorus removal (Figure 21.18b).

The EBPR process sometimes fails due to the overgrowth of competing microbial populations, commonly microorganisms that accumulate glycogen as opposed to phosphorus, thus rendering the process less efficient. Hence, better control of the process will require improved understanding of the ecology and physiology of the PAOs. Recent progress in this area has been made with identification of one of the principal PAOs, *Accumulibacter phosphatis*. *A. phosphatis* is part of a clade of related phosphorus-accumulating *Betaproteobacteria* (⟳ Section 15.2) that have been identified in different EBPR systems. Although no pure cultures are yet available, laboratory reactor systems enriched in these organisms are now providing insight into operating conditions necessary for stable operation of the EBPR.

Contaminants of Emerging Concern

Until recently, studies of the environmental fate of chemicals have focused primarily on priority pollutants, including heavily used agricultural products and chemicals that demonstrate acute toxicity or carcinogenicity. However, it is now clear that new bioactive pollutants are entering the environment and will likely pose new challenges for microbial bioremediation. These pollutants include pharmaceuticals, active ingredients in personal care products, fragrances, household products, sunscreens, and many other unusual or xenobiotic molecules.

Unlike pesticides, these "new" pollutants are more or less continuously discharged to the environment primarily through release of treated or untreated sewage, and because of this, they do not need to persist to have environmental effects. For example, it is known that synthetic estrogen compounds, excreted in the urine of women taking birth control pills and eventually discharged from wastewater treatment plants, can activate estrogen response genes in aquatic animals such as fish and contribute to the feminization of males.

Wastewater treatment plants were originally designed to handle natural materials, primarily human and industrial wastes, but now there is a growing interest in carefully researching the design of future treatment facilities to stimulate bioremediation of these emerging contaminants. Because these contaminants are often present in very low concentrations and are often new classes of xenobiotic chemicals, they may not actually support microbial growth but be degraded only by cometabolism or by highly specialized species. We can therefore expect that the bioremediation of emerging contaminants will be an active area of microbiological research and public policy in coming years.

MINIQUIZ -

- What are the advantages of EBPR relative to traditional chemical removal of phosphorus? Are there any disadvantages?

- Give an example of an "emerging" contaminant.

UNIT 4

21.8 Drinking Water Purification and Stabilization

Wastewater treated by secondary methods can usually be discharged into rivers and streams. However, such water is not **potable** (safe for human consumption). The production of potable water requires further treatment to remove potential pathogens, eliminate taste and odor, reduce nuisance chemicals such as iron and manganese, and decrease **turbidity**, which is a measure of suspended solids. **Suspended solids** are small particles of solid pollutants that resist separation by ordinary physical means.

Intestinal infections due to waterborne pathogens are still common, even in developed countries (⇌ Section 31.1), and some estimates indicate that waterborne diseases impact the health of several million people each year in the United States alone. Water treatment practices, however, have significantly improved access to safe water, starting with public works projects coupled with the application and development of water microbiology in the early twentieth century.

A century ago, water purification in the United States was limited to *filtration* to reduce turbidity, and this resulted in high rates of waterborne disease. Although filtration significantly decreased the microbial load of water, many microorganisms still passed through the filters. However, around 1913, **chlorination** using Cl_2 came into use as a disinfectant for large water supplies. Chlorine gas was an effective and inexpensive general disinfectant for drinking water, and its use quickly reduced the incidence of waterborne disease (⇌ Section 28.5). Major improvements in public health in the United States were largely due to the adoption of water filtration and disinfection treatment procedures. Public works engineering and microbiology working hand in hand were thus the major contributors to the dramatic advances in public health in the United States and other developed countries in the twentieth century.

Physical and Chemical Purification

A typical city installation for drinking water treatment is shown in **Figure 21.19a**. Figure 21.19*b* shows the process that purifies **raw water** (also called **untreated water**) that flows through the treatment plant. Raw water is first pumped from the source, in this case a river, to a sedimentation basin where anionic polymers, alum (aluminum sulfate), and chlorine are added. **Sediment**, including soil, sand, mineral particles, and other large particles, settles out. The sediment-free water is then pumped to a **clarifier** or coagulation basin, which is a large holding tank where **coagulation** takes place. The alum and anionic polymers form large particles from the much smaller suspended solids. After mixing, the particles continue to interact, forming large, aggregated masses, a process called **flocculation**. The large, aggregated particles (floc) settle out by gravity, trapping microorganisms and adsorbing suspended organic matter and sediment.

After coagulation, flocculation, and sedimentation, the clarified water undergoes **filtration** through a series of filters designed to remove organic and inorganic solutes, as well as remaining suspended particles and microorganisms. The filters typically consist of thick layers of sand, activated charcoal, and ion exchangers. When combined with previous purification steps, the filtered water is free of particulate matter, most organic and inorganic chemicals, and nearly all microorganisms.

Disinfection

Clarified, filtered water must be disinfected before it is released to the supply system as pure, potable **finished water**. **Primary disinfection** is the introduction of sufficient disinfectant into clarified, filtered water to kill existing microorganisms and inhibit further microbial growth. Chlorination is the most common method of primary disinfection. In sufficient doses, chlorine kills most microorganisms within 30 minutes. A few pathogenic protists such as *Cryptosporidium*, however, are not easily killed by chlorine treatment (⇌ Sections 28.7 and 32.4). In addition to killing microorganisms, chlorine oxidizes and effectively neutralizes many organic compounds. Since most taste- and odor-producing chemicals are organic compounds, chlorination improves water taste and smell. Chlorine is added to water either from a concentrated solution of sodium hypochlorite or calcium hypochlorite, or as chlorine gas

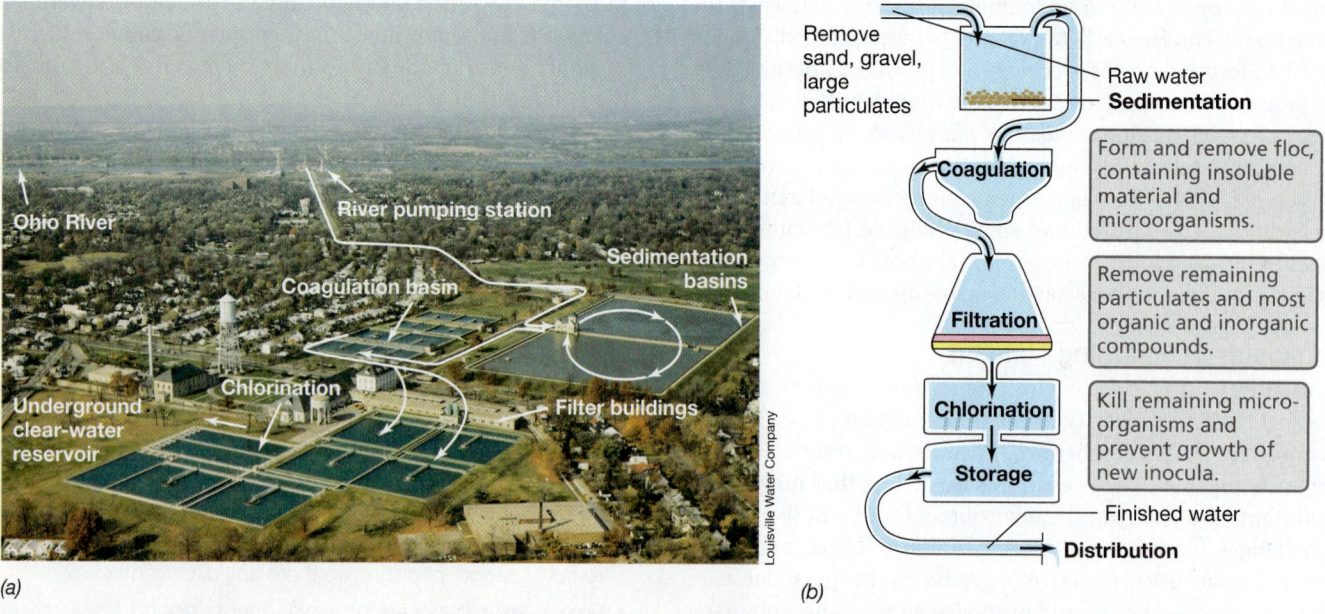

(a)

(b)

Figure 21.19 Water purification plant. *(a)* Aerial view of a water treatment plant in Louisville, Kentucky, USA. The arrows indicate direction of flow of water through the plant. *(b)* Schematic overview of a typical community water purification system.

from pressurized tanks. Chlorine gas is commonly used in large water treatment plants because it is most amenable to automatic control. When dissolved in water, chlorine gas is extremely volatile and dissipates within hours from treated water. To maintain adequate levels of chlorine for primary disinfection, many municipal water treatment plants introduce ammonia gas with the chlorine to form the more stable, nonvolatile chlorine-containing compound **chloramine**, $HOCl + NH_3 \rightarrow NH_2Cl + H_2O$.

Chlorine is consumed when it reacts with organic materials. Therefore, sufficient quantities of chlorine must be added to finished water containing organic materials so that a small amount, called the *chlorine residual*, remains. The chlorine residual reacts to kill any remaining microorganisms. The water plant operator performs chlorine analyses on the treated water to determine the level of chlorine to be added for **secondary disinfection**, the maintenance of sufficient chlorine residual or other disinfectant residual in the water distribution system to inhibit microbial growth. A chlorine residual level of 0.2–0.6 mg/liter is suitable for most water supplies. After chlorine treatment, the now potable water is pumped to storage tanks from which it flows by gravity or pumps through a **distribution system** of storage tanks and supply lines to the consumer. Residual chlorine levels inhibit growth of bacteria in the finished water prior to reaching the consumer. It does not protect against catastrophic system failures such as a broken pipe in the distribution system.

Ultraviolet (UV) radiation is also used as an effective means of disinfection. As we discussed in Section 5.18, UV radiation is used to treat secondarily treated effluent from water treatment plants. In Europe, UV irradiation is commonly used for drinking water applications, and it is increasingly used in the United States. For disinfection, UV light is generated from mercury vapor lamps. Their major energy output is at 253.7 nm, a wavelength that is bacteriocidal and may also kill cysts and oocysts of protists such as *Giardia* and *Cryptosporidium*, important eukaryotic pathogens in water (Section 32.4). Viruses, however, are more resistant.

UV radiation has several advantages over chemical disinfection procedures like chlorination. First, UV irradiation is a physical process that introduces no chemicals into the water. Second, UV radiation–generating equipment can be used in existing flow systems. Third, no disinfection by-products are formed with UV disinfection. Especially in smaller systems where finished water is not pumped long distances or held for long periods (reducing the need for residual chlorine), UV disinfection may be preferable to reduce dependence on chlorination.

MINIQUIZ

- What specific purposes do sedimentation, coagulation, filtration, and disinfection accomplish in the drinking water treatment process?
- What general procedures are used to reduce microbial numbers (microbial load) in water supplies?
- What are the advantages of UV disinfection versus, or as a complement to, chemical disinfection with chlorine?

21.9 Municipal and Premise Water Distribution Systems

Once drinking water leaves the treatment facility, the water often travels through many miles of municipal and premise distribution

pipes from the facility to the consumer (Figure 21.20). In addition to taste and odor problems often associated with source water, the long transit and residence times may also contribute to undesirable taste and odors from biological and chemical processes. Although undesirable, taste and odor alone usually do not signal a health threat. However, water distribution systems may also promote the growth of obligate or opportunistic pathogens, sequester and protect pathogens, or select for more pathogenic and resistant forms of microorganisms. Even though drinking water–associated disease often goes unreported, in the United States alone between 2007–2008, 36 disease outbreaks associated with drinking water affected at least 4000 persons and were linked to three deaths.

The Microbiology of Municipal Water Distribution Systems

Microbial growth in drinking water distribution systems can be eliminated only through complete nutrient removal (elimination of growth substrates originating from the source water and from distribution system structural materials) or by maintaining appropriate residual chlorine levels throughout the distribution system. In reality, neither of these is attainable. Growth is unavoidable as a consequence of reduction in chlorine concentration with increasing distance from the point of production together with the tendency for microorganisms to form biofilms on the pipe walls. Microorganisms in biofilms are more resistant to disinfection (Section 19.4) and significant

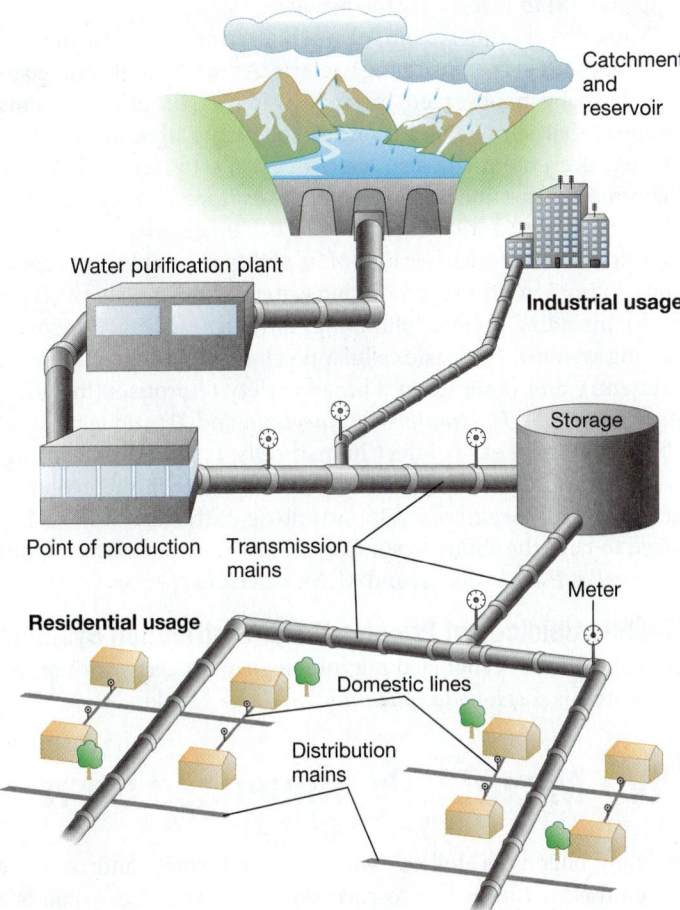

Figure 21.20 Drinking water distribution system. A municipal distribution system includes a surface reservoir, water purification plant, distribution mains, and domestic lines that encompass many miles of pipes in a typical community.

microbial accumulation is found in all distribution systems, over 90% of which is in the form of biofilms that coat the pipe walls.

Only recently have culture-independent molecular techniques, including 16S rRNA sequence analysis (⮌ Section 18.5), begun to fully resolve the species that commonly colonize water distribution pipes. Although these studies are showing that pathogenic species are rare, some opportunistic pathogens (⮌ Section 23.6) are present and can infect susceptible humans, including infants and the elderly or individuals with compromised immune systems. Opportunistic pathogens that have been found in water distribution systems include (1) nontuberculous mycobacteria (including *Mycobacterium avium*, *M. intracellulare*, *M. kansasii*, and *M. fortuitum*) associated with many thousands of clinical cases each year in the United States; (2) *Legionella pneumophila* (the causative agent of Legionnaires' disease, ⮌ Section 31.4); (3) *Pseudomonas aeruginosa* (which can infect the eyes, ears, skin, and lungs); and (4) opportunistic protozoan pathogens such as *Naegleria* and *Acanthamoeba* (⮌ Section 32.3) that can cause keratitis and encephalitis.

Because infection by these and other opportunistic pathogens is often of unclear origin and much waterborne disease goes unreported, the significance of water distribution systems as a source (or reservoir) for pathogenic microorganisms is unclear. However, because of the potential large-scale health risk, the issue of pathogens in drinking water has been receiving much greater attention in recent years, including the use of molecular microbial ecology (Chapter 18) to investigate the problem.

Water distribution systems also support numerous grazing protists that subsist by consuming bacteria. As many as 300 amoebae/ cm^2 have been observed in some water distribution systems. Bacteria that survive and replicate following ingestion by these protists are potentially also less susceptible to clearance by the mammalian immune system. The best example of this is *Legionella*, an opportunistic pathogen that has emerged as a relatively new public health risk because of its ability to establish residence and replicate in protists inhabiting water handling systems (**Figure 21.21**), including premise plumbing, shower heads, and air conditioning systems. The basic cellular mechanisms *Legionella* uses to gain entry and replicate in a broad variety of protists (including *Acanthamoeba*, *Hartmannella*, *Naegleria* and *Tetrahymena*) also allow it to more easily infect human cells. It has even been suggested that protists have been the driving force in the evolution of pathogenic *Legionella*. Opportunistic pathogens now recognized to have the ability to survive or grow within protists include *Legionella*, *Pseudomonas*, and *Mycobacterium* species.

The Microbiology of Premise Water Distribution Systems

One of the best-recognized microbiological concerns with premise water is *Legionella pneumophila* (⮌ Section 31.4). This

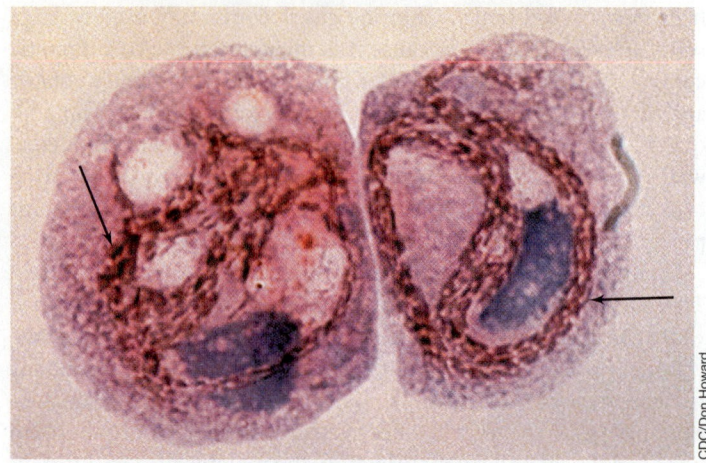

Figure 21.21 Protists as reservoirs of *Legionella*. Two cells of the protist *Tetrahymena* contain chains of the rod-shaped pathogen *Legionella pneumophila* (arrows). In premise water systems, protists can persist and be reservoirs of bacterial pathogens.

pathogen multiplies in premise water systems at temperatures between 20 and 46°C. It survives for months in drinking water, and survival is augmented by the presence of other bacteria and protozoa—in which intracellular growth is possible (Figure 21.21)—and also through sequestration in biofilms. Temperatures greater than 50°C lead to a decrease in numbers, and temperatures greater than 60°C result in rapid elimination (cell death). Thus, to prevent growth of *L. pneumophila*, premise water must be kept below 20°C or above 50°C from storage units to the tap.

Nontuberculous mycobacteria (including *Mycobacterium avium*, *M. intracellulare*, *M. kansasii*, and *M. fortuitum*) are also more resistant to chlorine disinfection and protozoal grazing, and are now known to be enriched in showerheads receiving municipal water that still shows a chlorine residual. As yet the significance of showers as a reservoir of opportunistic pathogens is unknown. However, the increasing frequency of showering as opposed to bathing, and possible aerosolization of opportunistic pathogens through showering, has prompted additional research in this area. The general picture that is emerging is that changes in treatment processes and the architecture of water distribution systems, coupled with the aging condition of some systems, can compromise human health.

MINIQUIZ -
- Trace the treatment of water through a drinking water treatment plant, from the inlet to the final distribution point (faucet).
- What features of a premise water distribution system might encourage the growth of *Legionella*? Suppress growth?

IV • Microbially Influenced Corrosion

Many billions of dollars' worth of metal, stone, and concrete infrastructure is lost to corrosion every year. Corrosion is a complex process that may be influenced and accelerated by microbial activity. Microorganisms accelerate corrosion through alteration of pH or redox, production of corrosive metabolites, and creation of corrosive microenvironments in biofilms. However, since the balance between chemistry and biology is often not well resolved, corrosion in which microorganisms are implicated is referred to as **microbially influenced corrosion (MIC)**.

In this section we examine a few cases in which the microbial contribution to corrosion is relatively well understood. However, it is likely that the general models for MIC will be modified as we learn more about how microorganisms interact with and modify different structural materials.

21.10 Microbially Influenced Corrosion of Metals

Iron is the most commonly used metal in the built environment. On a global basis, millions of miles of water, gas, and oil distribution pipelines are made of metal, and their corrosion contributes to the greatest loss of infrastructure in the built environment. Corrosion of iron by oxygen in air is thought to be solely an electrochemical process. However, much critical infrastructure is buried or submerged, restricting exposure to oxygen. At near-neutral pH, in the absence of oxygen, corrosion of iron and steel is significantly accelerated by microbial activity. Microbial groups implicated in MIC include sulfate reducers (⟳ Sections 13.18 and 14.9), ferric-iron-reducing bacteria (⟳ Sections 14.14 and 20.5), ferrous-iron-oxidizing bacteria (⟳ Sections 13.9, 14.15, and 20.5), and methanogens (⟳ Sections 13.20, 16.2, and 20.2).

Metal Corrosion by Sulfate-Reducing Bacteria

Metal structures submerged in the marine environment and pipelines used for transmission of low-grade oil are particularly subject to MIC through the activities of sulfate-reducing microorganisms. Corrosion by sulfate-reducing bacteria is partly attributable to the chemically corrosive nature of hydrogen sulfide (H_2S), the product of their metabolism (⟳ Section 14.9). Crude oils containing more than about 0.5% sulfur by weight are called "sour" and may be naturally corrosive because of the H_2S that is present. In oilfields near the ocean, such as the Middle East and Alaska, seawater is injected to maintain reservoir pressure and force oil into the producing well. Since seawater contains nearly 30 mM sulfate, an undesirable consequence of injection is further souring by stimulating the growth of sulfate-reducing microorganisms.

A strategy now used by the petroleum industry to control souring is inclusion of nitrate (NO_3^-) in the injection water, stimulating the growth of nitrate-reducing microorganisms. Since nitrate respiration is energetically more favorable than sulfate respiration (⟳ Sections 13.17 and 19.2), the nitrate reducers outcompete sulfate reducers for usable organic electron donors in the oil. Nitrate also stimulates the growth of sulfide-oxidizing, nitrate-reducing microorganisms (⟳ Sections 13.8 and 14.11), thereby reversing the souring by removing the sulfide.

Mechanisms of Metal Corrosion

The mechanisms by which sulfate reducers contribute directly to corrosion are more controversial, as now represented by two major competing models. One model is based on the "cathodic depolarization theory," in which hydrogen consumption by the sulfate reducer accelerates electrochemical pitting of the iron surface (**Figure 21.22a**). This model is based on the capacity of many sulfate reducers to use hydrogen (H_2) as an electron donor, thereby accelerating the energetically favorable but kinetically slow H_2 production originating from the chemical oxidation of iron ($Fe + 2H^+ \rightarrow Fe^{2+} + H_2$), with the following overall stoichiometry, whose free energy change ($\Delta G^{0\prime}$) is highly favorable:

$$4 Fe^0 + SO_4^{2-} + 3 HCO_3^- + 5 H^+ \rightarrow FeS + 3 FeCO_3 + 4 H_2O$$
$$(\Delta G^{0\prime} = -925 \text{ kJ})$$

However, this model has been questioned because H_2 formation from the iron surface at neutral pH is an intrinsic bottleneck, controlled by the limited availability of protons required for the reactions generating H_2.

Detailed electrochemical studies have since shown that some sulfate-reducing bacteria, such as *Desulfopila corrodens*, have the capacity to take up electrons directly from the metal (Fe^0, Figure 21.22b). In this model, the sulfate reducers attached to the metal surface engage in direct (cathodic) electron uptake from the metal through an electroconductive sulfidic corrosion layer (Figure 21.22b). A similar ability to take up electrons directly from Fe^0 has been observed for a *Methanobacterium*-like methanogen that produces methane (CH_4) rather than sulfide from growth on elemental iron. The direct electron uptake model also suggests that associated with the cell surface are redox-active proteins, or other conductive structures, that conduct electrons from the corrosion layer to the cell. This represents yet another of a growing number

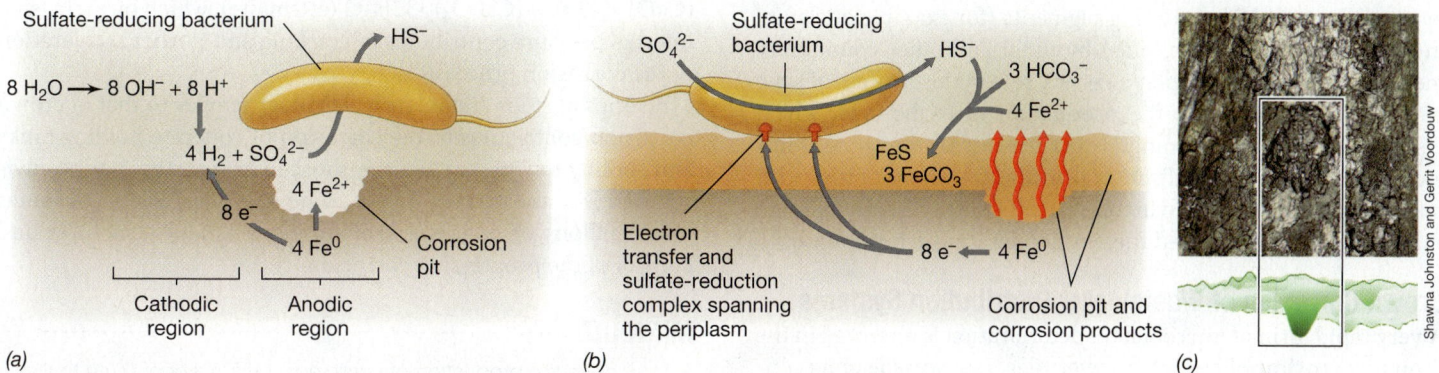

Figure 21.22 Corrosion of iron by sulfate-reducing bacteria. Two models for the activities of sulfate-reducing bacteria in metal corrosion. *(a)* Accelerating oxidation of metallic iron by consuming H_2 produced abiotically by proton reduction at the metal surface. *(b)* Direct electron transfer from the metal using electron-conductive outer cell wall structures connecting to an electron transfer system spanning the periplasm. *(c)* Top: photo of a model iron surface undergoing sulfidic corrosion. Bottom: scan of a side view of the metal surface in the photo revealing the areas where corrosion and pitting of the metal surface has occurred.

of examples of the microbial use of conductive cellular structures for the oxidation or reduction of insoluble electron acceptors or electron donors, respectively (⇄ Sections 14.14 and 20.5).

MINIQUIZ -

- How does a nitrate addition prevent sulfide souring of crude oil?
- Why is accelerated microbial corrosion of iron metal thought to require a direct interaction between the sulfate reducers and the metal surface?

21.11 Biodeterioration of Stone and Concrete

In the same way that microorganisms contribute to soil formation through the dissolution of mineral surfaces through combined physical and metabolic activities (⇄ Section 19.6), building materials composed of natural stone or concrete are also subject to microbial colonization that may contribute to a slow loss of structural integrity through their metabolic activities. This process is referred to as *biodeterioration*.

Biodeterioration of Stone Building Materials

Microbial colonization of natural and structural stone building material is ubiquitous. Microorganisms can colonize the surface and penetrate several millimeters into rocky material depending on its physical characteristics (e.g., surface roughness, porosity, light penetration). Organisms can also grow on and within the facades of buildings constructed of limestone, sandstone, granite, basalt, and soapstone. These "within stone," or *endolithic*, communities (⇄ Section 17.16) are phylogenetically diverse, comprised of chemoorganotrophic and chemolithotrophic *Bacteria*, *Archaea*, fungi, algae, and cyanobacteria. The cyanobacteria and algae primarily nourish the community, living in close or symbiotic association with other microbial members. For example, endolithic fungi have been observed to enclose the phototrophs in lichen-like associations.

Although not generally included in discussions of "extreme environments," life on and within stone building materials requires adaptation to multiple extreme conditions, including intense solar radiation, desiccation, temperature and moisture fluctuations, and lack of nutrients. Protection from solar radiation is conferred by production of UV-absorbing pigments (for example, melanin, mycosporines, and carotenoids) by fungi and other community members. The fungi also play a central role in this process of slow biodeterioration through the production of oxalic acid, which dissolves and mobilizes mineral constituents of the stone. Mineral dissolution and mobilization provides the communities with nutrients and is believed to increase habitability by enlarging pore spaces within the stone and thereby accelerating deterioration.

Crown Corrosion of Wastewater Distribution Systems

A very rapid form of microbial biodeterioration is observed in the **crown corrosion** of concrete sewer tiles, a process leading ultimately to the collapse of the pipe. Corrosion is a consequence of a complex ecology between sulfate-reducing bacteria (⇄ Sections 13.18 and 14.9) and chemolithotrophic sulfur-oxidizing bacteria (⇄ Sections 13.8 and 14.11) in these underground wastewater transmission systems (**Figure 21.23**).

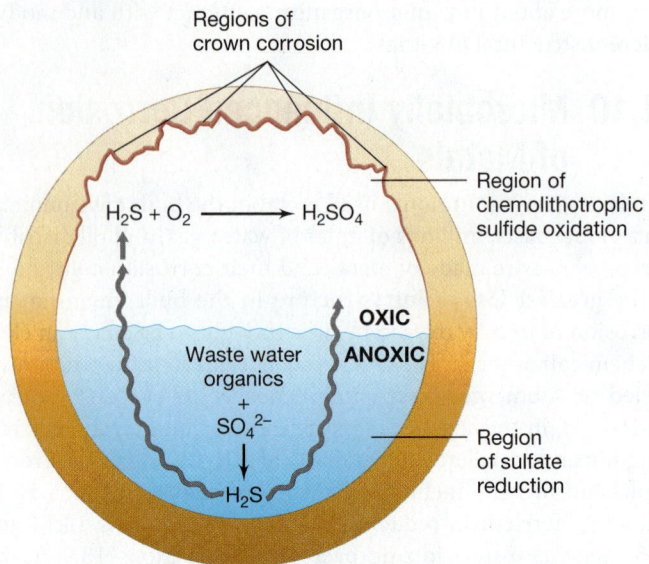

Figure 21.23 Crown corrosion of concrete sewer pipes. Corrosion is the result of a microbial sulfur cycle that develops within the transmission pipe. Sulfate-reducing bacteria consume organic material in the anoxic wastewater, producing H_2S. The latter is oxidized by sulfur-oxidizing chemolithotrophic bacteria that attach to the oxic upper (crown) pipe surface, accelerating corrosion from the production of H_2SO_4 (sulfuric acid).

The first step in crown corrosion is the reduction of sulfate in the sewage to H_2S by sulfate-reducing bacteria, using primarily organic electron donors available in the waste stream water for sulfate reduction. The H_2S then flows into the headspace of the pipe where conditions are oxic. The sulfide, or partially oxidized intermediates such as thiosulfate or sulfur, is then oxidized by neutrophilic thiobacilli such as *Thiobacillus thioparus* (⇄ Section 14.11). As the pH drops to 4–5 with continued microbial production of sulfuric acid, acidophilic sulfur-oxidizing species such as *Acidithiobacillus thiooxidans* displace the neutrophilic species. Destruction and ultimate structural failure of the concrete results from reaction of sulfuric acid with the free lime in the concrete, producing $CaSO_4\cdot 2H_2O$ (gypsum) that penetrates into the concrete. The gypsum then reacts with calcium aluminate in the concrete, leading to the production of $(CaO)_3\cdot(Al_2O_3)\cdot(CaSO_4)_3\cdot 32H_2O$ (ettringite), which by increasing internal pressure contributes to cracking and further acceleration of the corrosion process.

A series of steps and microbial ecology similar to that of crown corrosion contributes to the corrosion of concrete holding tanks and cooling towers, particularly those in or near the marine environment. In the United States alone such corrosion consumes many billions of dollars a year for replacement structures and control of the progressing corrosion.

MINIQUIZ -

- How does the production of oxalic acid by fungi contribute to the deterioration of stone building materials?
- Prior to better regulatory control of metal release into domestic wastewater systems, crown corrosion of sewer tiles was less of a problem. Why?

BIG IDEAS

21.1 • The capacity of bacteria to oxidize Fe^{2+} aerobically at acidic pH is used to mine metals, principally copper-, uranium-, and gold-containing low-grade ores, through a process called microbial leaching. Bacterial oxidation of Fe^{2+} to Fe^{3+} is the key reaction in most microbial leaching processes because Fe^{3+} can oxidize extractable metals in the ores under either oxic or anoxic conditions.

21.2 • Spontaneous microbial oxidation of ferrous iron in pyritic ore or coal that has been exposed to air and water, such as occurs during some coal-mining operations, causes a type of pollution called acid mine drainage.

21.3 • Although an inorganic pollutant such as uranium cannot be destroyed, containment is possible by reducing its mobility. For example, metal-reducing microorganisms in a region of uranium contamination can be stimulated to reduce U^{6+} to U^{4+}, forming an immobile uranium mineral, *uraninite*, that does not move into the groundwater.

21.4 • Hydrocarbons are excellent carbon sources and electron donors for bacteria and are readily oxidized when O_2 is available. Hydrocarbon-oxidizing bacteria bioremediate spilled oil, and their activities can be assisted by addition of inorganic nutrients.

21.5 • Some xenobiotics (chemicals new to nature) persist, whereas others are readily degraded, depending on their chemistries. Dechlorination is a major means of detoxifying xenobiotics that reach anoxic environments. With the exception of readily degradable microbial plastics, recalcitrant synthetic plastics are major environmental concerns.

21.6 • Sewage and industrial wastewater treatment reduces the BOD of wastewater. Primary, secondary, and tertiary wastewater treatment employs physical, biological, and physicochemical processes. After secondary or tertiary treatment, effluent water has significantly reduced BOD and is suitable for release into the environment.

21.7 • Advanced wastewater treatment, such as enhanced biological phosphorus removal, is used to improve the quality of the treated wastewater. Of increasing concern are pharmaceuticals and ingredients in personal care products that are not degraded by conventional treatment systems and can have adverse environmental effects even at very low concentrations.

21.8 • Drinking water purification plants employ industrial-scale physical and chemical systems that remove or neutralize biological, inorganic, and organic contaminants from natural, community, and industrial sources. Water purification plants employ clarification, filtration, and chlorination processes to produce potable water.

21.9 • The many miles of pipes for municipal drinking water distribution systems and premise plumbing have created new microbial habitats. Most microorganisms are associated with the pipe walls as biofilms, resulting in a community that is more resistant to chlorine and that can sustain or sequester opportunistic pathogenic bacteria, such as *Mycobacterium*, *Legionella*, and *Pseudomonas*. The ability of some of these to grow within protist cells may increase their pathogenicity.

21.10 • Corrosion of metal structures exposed to the environment can be accelerated by microbial activity during microbially influenced corrosion. Structures in or near seawater are particularly prone to corrosion as a consequence of the direct and indirect activities of sulfate-reducing bacteria.

21.11 • Microbial contribution to the structural degradation of stone and concrete is called biodeterioration. Complex microbial communities colonize the stone and produce substances that dissolve and mobilize its mineral constituents. Crown corrosion of concrete sewer lines results from the concerted activities of sulfate-reducing and sulfur-oxidizing bacteria growing within the wastewater and the headspace of sewer pipes, respectively. Sulfuric acid is primarily responsible for the destruction of the concrete.

MasteringMicrobiology®

Review what you know and challenge what you have learned with MasteringMicrobiology! Access study materials, chapter quizzes, animations, and microbiology lab tutorials in the Study Area to ensure that you have mastered this chapter's content.

REVIEW OF KEY TERMS

Acid mine drainage acidic water containing H_2SO_4 derived from the microbial and spontaneous oxidation of iron sulfide minerals released by coal mining

Biochemical oxygen demand (BOD) the relative amount of dissolved oxygen consumed by microorganisms for complete oxidation of organic and inorganic material in a water sample

Bioremediation the cleanup of oil, toxic chemicals, and other pollutants by microorganisms

Chloramine a disinfectant chemical manufactured on-site by combining chlorine and ammonia at precise ratios

Chlorination disinfecting water with Cl_2 at a sufficiently high concentration that a

residual level is maintained throughout the distribution system

Clarifier a reservoir in which suspended solids in raw water are coagulated and removed through precipitation

Coagulation the formation of large insoluble particles from much smaller, colloidal particles by the addition of aluminum sulfate and anionic polymers

Crown corrosion the destruction of the upper half, or crown, of concrete wastewater pipes by sulfuric acid produced through the concerted activities of sulfate-reducing and sulfur-oxidizing bacteria.

Distribution system water pipes, storage reservoirs, tanks, and other equipment used to deliver drinking water to consumers or store it before delivery

Effluent water treated wastewater discharged from a wastewater treatment facility

Filtration the removal of suspended particles from water by passing it through one or more permeable membranes or media (e.g., sand, anthracite, or diatomaceous earth)

Finished water water delivered to the distribution system after treatment

Flocculation the water treatment process after coagulation that uses gentle stirring to cause suspended particles to form larger, aggregated masses (flocs)

Microbial leaching the extraction of valuable metals such as copper from sulfide ores by microbial activities

Microbial plastics polymers consisting of microbially produced (and thus biodegradable) substances, such as polyhydroxyalkanoates

Microbially influenced corrosion (MIC) the contribution of microbial metabolic activities to accelerating the corrosion of metal and concrete structures

Potable drinkable; safe for human consumption

Primary disinfection the introduction of sufficient chlorine or other disinfectant into clarified, filtered water to kill existing microorganisms and inhibit further microbial growth

Primary wastewater treatment physical separation of wastewater contaminants, usually by separation and settling

Pyrite a common iron-containing ore, FeS_2

Raw water surface water or groundwater that has not been treated in any way (also called untreated water)

Reductive dechlorination an anaerobic respiration in which a chlorinated organic compound is used as an electron acceptor, usually with the release of Cl^-

Secondary aerobic wastewater treatment oxidative reactions carried out by microorganisms under aerobic conditions to treat wastewater containing low levels of organic materials

Secondary anaerobic wastewater treatment degradative and fermentative reactions carried out by microorganisms under anoxic conditions to treat wastewater containing high levels of insoluble organic materials

Secondary disinfection the maintenance of sufficient chlorine or other disinfectant residual in the water distribution system to inhibit microbial growth

Sediment soil, sand, minerals, and other large particles found in raw water

Sewage liquid effluents contaminated with human or animal fecal material

Suspended solid a small particle of solid pollutant that resists separation by ordinary physical means

Tertiary wastewater treatment the physicochemical or biological processing of wastewater to reduce levels of inorganic nutrients

Turbidity a measurement of suspended solids in water

Untreated water surface water or groundwater that has not been treated in any way (also called raw water)

Wastewater liquid derived from domestic sewage or industrial sources, which cannot be discarded in untreated form into lakes or streams

Xenobiotic a synthetic compound not produced by organisms in nature

REVIEW QUESTIONS

1. How is *Acidithiobacillus ferrooxidans* useful in the mining of copper ores? Which crucial step in the indirect oxidation of copper ores is carried out by *A. ferrooxidans*? How is copper recovered from copper solutions produced by leaching? (Section 21.1)

2. Which *Bacteria* and *Archaea* play a major role in acid mine drainage? Why do they carry out the reactions that they do? Why is air necessary for this process? (Section 21.2)

3. Describe a strategy for bioremediating a site that contains buried nuclear weapons that are leaking uranium. What could thwart your bioremediation efforts? (Section 21.3)

4. What physical and chemical conditions are necessary for the rapid microbial degradation of oil in aquatic environments? Design an experiment that would allow you to test which conditions optimized the oil oxidation process. (Section 21.4)

5. What are xenobiotic compounds and why might microorganisms have difficulty catabolizing them? (Section 21.5)

6. Trace the treatment of wastewater in a typical plant from incoming water to release. What is the overall reduction in the BOD for typical household wastewater? What is the overall reduction in the BOD for typical industrial wastewater? (Section 21.6)

7. Why is advanced wastewater treatment desirable from an environmental point of view? (Section 21.7)

8. Identify (stepwise) the process of purifying drinking water. What important contaminants are targeted by each step in the process? (Section 21.8)

9. Discuss the microbial habitats that develop within drinking water distribution and premise plumbing systems. How do microorganisms persist in the presence of chlorine? What features of the distribution system habitat might contribute to a microbial health hazard? (Section 21.9)

10. Why is the presence or absence of sulfate (SO_4^{2-}) so important in the extent of metal corrosion? (Section 21.10)

11. What types of structures are subject to crown corrosion? Why is sulfate important here as well as in metal corrosion? (Section 21.11)

APPLICATION QUESTIONS

1. Acid mine drainage is in part a chemical process and in part a biological process. Discuss the chemistry and microbiology that lead up to acid mine drainage and point out the key reactions that are biological. What ways can you think of to prevent acid mine drainage?

2. Why is reduction of BOD in wastewater a primary goal of wastewater treatment? What are the consequences of releasing wastewater with a high BOD into local water sources such as lakes or streams?

3. Discuss the microbial ecology contributing to crown corrosion of concrete sewer lines. In consideration of this ecology, what intervention strategies might be useful in reducing or eliminating corrosion?

22 · Microbial Symbioses

microbiology**now**

A Symbiotic Trio Sustains Seagrass Ecosystems

Seagrass meadows function as keystone habitats for various animals that live in coastal areas, including fish, migrating corals, water birds, and turtles. Seagrasses (top photo) also protect coastlines from erosion and function as important carbon and nutrient sinks. However, the remarkable success of seagrasses in temperate to tropical waters is enigmatic; how do these plants avoid being poisoned by the highly toxic hydrogen sulfide produced by sulfate-reducing bacteria that are especially active in coastal sediments? A research study has now solved this mystery.

The success of seagrass ecosystems is linked to the activities of a small clam (bottom photos), one of a variety of lucinid bivalve species that live near the root systems of the seagrass.[1] The clam functions as a sulfide sink, using oxygen released from roots of the seagrass to oxidize the sulfide. The ability of the clam to oxidize sulfide is in turn the result of its symbiotic association with sulfide-oxidizing bacteria residing in its gill tissue. These chemolithotrophic bacteria synthesize sugars that sustain both themselves and the clams. Experiments in which the seagrass was grown in isolation or in association with the clam confirmed the key role of the clams in sulfide depletion, thereby enhancing the growth of the plant.

Field surveys of seagrass beds confirmed a global association between lucinids and seagrass, and paleo-records show this association has existed ever since seagrasses evolved around 100 million years ago. The ecological success of seagrass can thus be attributed to a long and remarkably successful partnership between plant, animal, and bacterium—a symbiotic trio. An understanding of the importance of this unique partnership is essential for the restoration of seagrass meadows in coastal areas where these critical habitats are in decline from pollution.

[1]van der Heide, T., et al. 2012. A three-stage symbiosis forms the foundation of seagrass ecosystems. *Science 336:* 1432–1434.

In this chapter we consider relationships of microorganisms with other microorganisms or with macroorganisms—prolonged and intimate relationships of a type called **symbiosis**, a word that means "living together." Microorganisms that live within or on plants and animals can be grouped according to how they affect their hosts. *Parasites* are microorganisms that benefit at some expense to the host, *pathogens* actually cause a disease in the host, *commensals* have no discernible effect on the host, and *mutualists* are beneficial to the host. One way or another, all microbial symbioses benefit the microorganism.

Pathogenic and parasitic associations will be addressed in Chapter 23 and in following chapters covering specific infectious diseases. In this chapter we focus on **mutualisms**—relationships in which both partners benefit. We view the microorganisms as intimate evolutionary partners that influence both the evolution and physiology of their hosts. Many mutualistic symbioses of microorganisms with plants and animals have their origins many millions of years ago. A mutualism that persists over evolutionary time beneficially modifies the physiology of both partners. This process of reciprocal change between a microorganism and its host is called **coevolution**, and over time, the changes may be so extensive that the symbiosis becomes obligate—either the microorganism or the host (or both) cannot survive independent of the other.

I • Symbioses between Microorganisms

Many microbial species—both prokaryotes and eukaryotes—have intimate and beneficial associations with other microbial species. Direct microscopic observations of natural samples show that many microorganisms are not solitary entities but are associated with other microorganisms on surfaces or as suspended aggregates of cells. In most cases, the advantages conferred by an association are unknown. Because microbial ecologists have recognized that *communities* of interacting microbial populations—not individual organisms—control critical environmental processes, research to discover the advantages of strictly microbial symbioses has increased. We present in Part I two types of microbial mutualisms where the advantages to both partners are clear.

22.1 Lichens

Lichens are readily visible leafy or encrusting microbial symbioses often found growing on bare rocks, tree trunks, house roofs, and bare soils—surfaces where other organisms typically do not grow (**Figure 22.1**). A lichen is a mutualistic association between two microorganisms, a fungus and either an alga or a cyanobacterium. The alga or cyanobacterium is the phototrophic partner and produces organic matter, which then feeds the fungus. The fungus, unable to carry out photosynthesis, provides a firm anchor within which the phototrophic partner can grow, protected from erosion by rain or wind. Cells of the phototroph are embedded in defined layers or clumps among cells of the fungus (**Figure 22.2**). The morphology of a lichen is primarily determined by the fungus, and many fungi are able to form lichen associations. Diversity among the phototrophs is much lower, and many different kinds of lichens can have the same phototrophic partner. Many cyanobacteria that partner in lichens are nitrogen-fixing species, organisms such as *Anabaena* or *Nostoc* (🔗 Sections 3.17 and 14.3).

The fungus clearly benefits from associating with the phototroph in the lichen symbiosis, but how does the phototroph

(a)

(b)

Figure 22.1 **Lichens.** *(a)* A lichen growing on a branch of a dead tree. *(b)* Lichens coating the surface of a large rock.

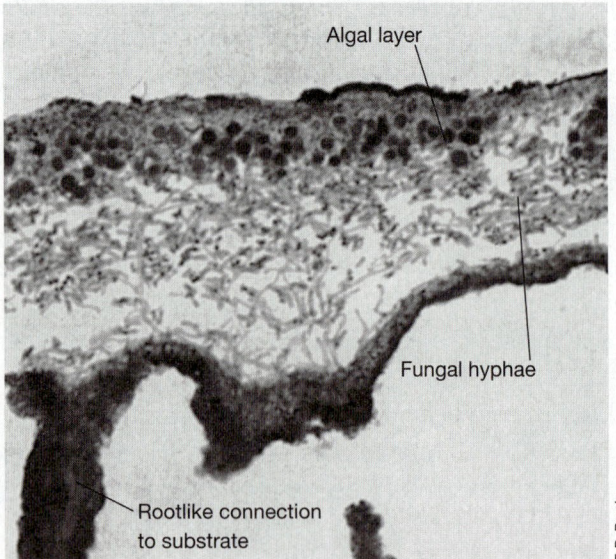

Figure 22.2 **Lichen structure.** Photomicrograph of a cross section through a lichen. The algal layer is positioned within the lichen structure near the top so as to receive the most sunlight.

benefit? *Lichen acids*, complex organic compounds secreted by the fungus, promote the dissolution and chelation of inorganic nutrients from the rock or other surface that are needed by the phototroph. Another role of the fungus is to protect the phototroph from drying; most of the habitats in which lichens live are dry, and fungi are, in general, better able to tolerate dry conditions than are the phototrophs. The fungus actually facilitates the uptake of water and sequesters some for the phototroph.

Lichens typically grow quite slowly. For example, a lichen 2 cm in diameter growing on the surface of a rock may be several years old. Lichen growth varies from 1 mm or less per year to over 3 cm per year, depending on the organisms composing the symbiosis and the amount of rainfall and sunlight received.

MINIQUIZ -

- What are the benefits to both partners in the lichen mutualism?
- Besides organic compounds, of what benefit to the fungus is a mutualism with *Anabaena*?

22.2 "*Chlorochromatium aggregatum*"

In freshwater environments there are microbial mutualisms called **consortia**. A common consortium develops between nonmotile, phototrophic, green sulfur bacteria, which may be colored either green or brown (↩ Section 14.6), and motile, nonphototrophic bacteria. These consortia are found worldwide in stratified sulfidic freshwater lakes, and can account for up to 90% of the green sulfur bacteria and nearly 70% of the bacterial biomass in these lakes. The basis of the mutualism of these consortia is in the phototrophic production of organic matter by the green sulfur bacterium and the motility of the partner organism. Each consortium has been given a genus and species name, but since these names do not denote true species (because they are not a single organism), the names are enclosed in quotation marks. We examined the general biology of these consortia in Section 14.6.

Nature of the Consortium

The morphology of a green sulfur bacterial consortium depends upon the species composition. The consortium generally consists of 13–69 green sulfur bacteria, called *epibionts*, surrounding and attached to a central, colorless, flagellated, rod-shaped bacterium (**Figure 22.3**). Several distinct motile phototrophic consortia have been recognized based on the color, morphology, and presence or absence of gas vesicles (↩ Section 2.15) of the epibionts. For example, in "*Chlorochromatium aggregatum*" the central bacterium is surrounded by rod-shaped green bacteria. In "*Pelochromatium roseum*" the epibiont is brown. The consortium "*Chlorochromatium glebulum*" is bent and includes green epibionts that contain gas vesicles (Figure 22.3).

Green sulfur bacteria are obligately anaerobic phototrophs that form a distinct phylum (*Chlorobi*, ↩ Section 14.6). The green and brown species differ in the types of bacteriochlorophyll and carotenoids they contain. Both green and brown species are found in stratified lakes where light penetrates to depths at which the water contains hydrogen sulfide (H_2S). In the stratified lakes, the motile consortia reposition rapidly to remain where conditions

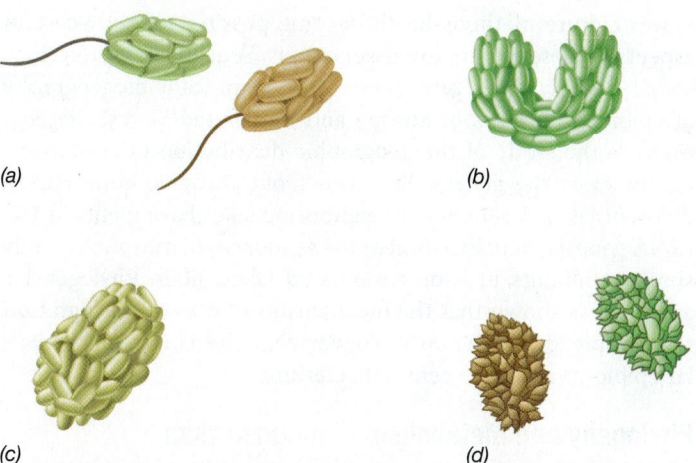

Figure 22.3 Drawings of some motile phototrophic consortia found in freshwater lakes. Green epibionts: *(a)* "*Chlorochromatium aggregatum*," *(b)* "*C. glebulum*," *(c)* "*C. magnum*," *(d)* "*C. lunatum*." Brown epibionts: *(a)* "*Pelochromatium roseum*," *(d)* "*P. selenoides*." The epibionts are 0.5–0.6 μm in diameter. Adapted from Overmann, J., and H. van Gemerden. 2000. *FEMS Microbiol. Rev. 24*: 591–599.

are favorable for photosynthesis in the constantly changing gradients of light, oxygen, and sulfide. Water samples collected from depths where these conditions are favorable are enriched in this morphologically conspicuous consortium (**Figure 22.4**). The consortia show dark aversion (scotophobotaxis, ↩ Section 2.19) and positive chemotaxis toward sulfide. Some free-living green sulfur bacteria, such as *Pelodictyon phaeoclathratiforme*, have gas vesicles that regulate buoyancy and vertical position in the water column. However, the time they require for repositioning is from one to several days, which is not fast enough for tracking the more rapidly changing gradients. By contrast, motile consortia move up and down in the water column fast enough to follow the gradients of light and sulfide as they change on a diel basis.

Although green bacterial consortia were discovered almost a century ago, only with the advent of molecular techniques and

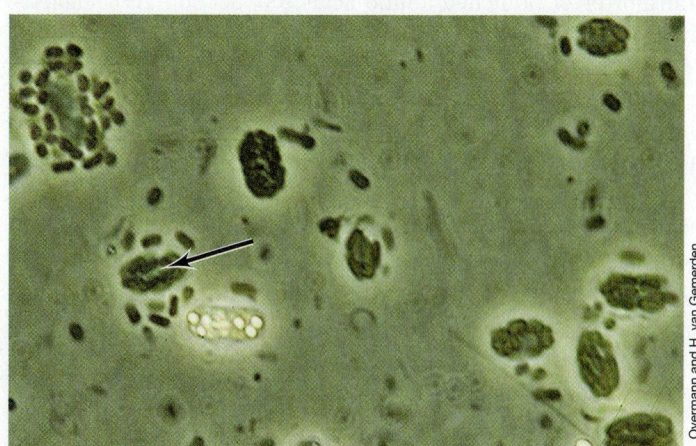

Figure 22.4 Phase-contrast micrograph of "*Pelochromatium roseum*" from Lake Dagow (Brandenburg, Germany). The preparation was compressed between a coverslip and microscope slide to reveal the central rod-shaped bacterium (arrow). A single consortium is about 3.5 μm in diameter. Used with permission from J. Overmann and H. van Gemerden. 2000. *FEMS Microbiol. Rev. 24*: 591–599.

UNIT 4

newer culture methods has it become possible to study certain aspects of these remarkable associations. Sequencing of 16S ribosomal RNA (rRNA) genes revealed a significant biogeography of epibionts in lakes of Europe and the United States. *Biogeography* is the study of the geographic distribution of organisms; in this case, the genetically distinct phototrophic consortia in different lakes. Epibionts in neighboring lakes have identical 16S rRNA gene sequences, whereas the sequences of morphologically similar epibionts in widely separated lakes differ. Phylogenetic analysis has shown that the mechanisms of cell–cell recognition responsible for stable morphology have evolved between particular epibionts and their central bacterium.

Phylogeny and Metabolism of a Consortium

The epibiont of "*Chlorochromatium aggregatum*" has been isolated and grown in pure culture. Although this green sulfur bacterium, named *Chlorobium chlorochromatii*, can be grown in pure culture, no naturally free-living variant has been observed, supporting the view that in nature, a symbiotic lifestyle is obligate for epibionts. The central bacterium of "*Chlorochromatium aggregatum*" belongs to the *Betaproteobacteria* (꜀꜂ Section 15.2). Interestingly, this bacterium requires α-ketoglutarate, an intermediate of the citric acid cycle (꜀꜂ Section 3.12), and this is presumably supplied to it by the epibiont. However, the central cell only assimilates fixed carbon in the presence of light and sulfide—conditions in which the epibionts are active and can transfer nutrients to the central bacterium.

Recent studies comparing the transcriptome and proteome (꜀꜂ Sections 6.7 and 6.8) of *C. chlorochromatii* growing alone or in association with the central rod bacterium have identified some features specifically related to the symbiosis. Approximately 50 proteins are unique to the symbiotic state. Most of approximately 350 differentially regulated genes are repressed when the organism is symbiotically associated, whereas only 19 genes are more highly expressed. Many of these up-regulated genes encode proteins of amino acid metabolism and nitrogen regulation. These include the enzyme glutamate synthase and an ABC transporter of branched amino acids (꜀꜂ Section 2.9), suggesting that the metabolic coupling between the epibiont and central rod bacterium involves the exchange of amino acids. Although it is not yet known whether the central bacterium transfers any organic compounds to the epibiont, this hypothesis can be tested now that the genome sequence of the central bacterium is known.

Scanning electron microscopy of the consortium (**Figure 22.5**) has revealed that tubular extensions of the central bacterium's periplasm (꜀꜂ Section 2.11) cover much of its surface and appear to fuse with the periplasm of the epibiont. If the two bacterial

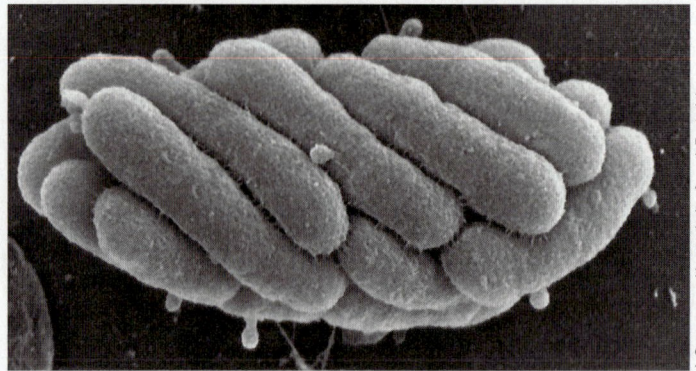

(a)

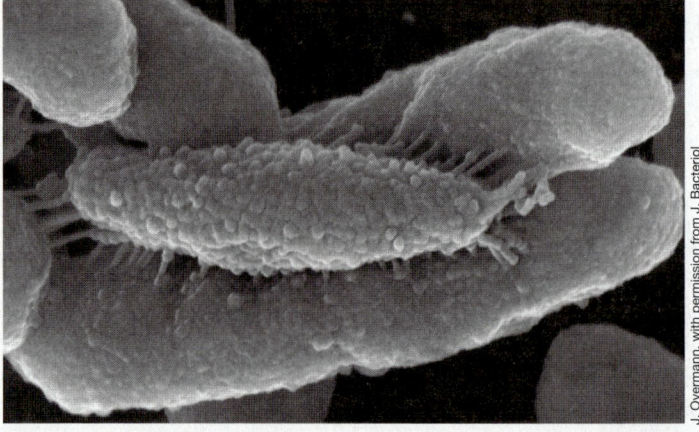

(b)

Figure 22.5 Scanning electron micrographs of "*Chlorochromatium aggregatum.*" *(a) Chlorobium chlorochromatii* epibionts tightly clustered around a flagellated central bacterium. *(b)* The central bacterium exhibits numerous protrusions of its outer membrane that make intimate contact with the epibionts, possibly fusing the periplasms of the two gram-negative organisms. Cells of the epibiont are about 0.6 μm in diameter. Used with permission from G. Wanner et al. 2008. *J. Bacteriol. 190:* 3721–3730.

partners actually share a common periplasmic space, this would facilitate the transfer of nutrients from phototroph to chemotroph.

MINIQUIZ
- What is the evidence that "*Chlorochromatium aggregatum*" is a stable product of evolution?
- What advantage does motility offer a phototrophic consortium?
- How might nutrients be shuttled between phototroph and chemotroph in the consortium?

II • Plants as Microbial Habitats

Plants interact closely with microorganisms through their roots and leaf surfaces and even more intimately within their vascular tissue and cells. Most mutualisms between plants and microorganisms increase nutrient availability to the plants or defend them against pathogens. We consider three examples in the following sections: (1) a mutualism (root nodules, Section 22.3), (2) a symbiosis that is harmful to the plant (crown gall disease, Section 22.4), and (3) a mutualism in which plants expand and interconnect their root system through association with a fungus (mycorrhizae, Section 22.5).

22.3 The Legume–Root Nodule Symbiosis

A plant–bacterial mutualism of great importance to humans is that of leguminous plants and nitrogen-fixing bacteria. *Legumes* are plants that bear their seeds in pods. This third largest family of flowering plants includes such agriculturally important plants as soybeans, clover, alfalfa, beans, and peas. These plants are key commodities for the food and agricultural industries, and the ability of legumes to grow without nitrogen fertilizer saves farmers millions of dollars in fertilizer costs yearly and reduces the polluting effects of fertilizer runoff.

The partners in a symbiosis are called *symbionts*, and most nitrogen-fixing bacterial symbionts of plants are collectively called *rhizobia*, derived from the name of a major genus, *Rhizobium*. Rhizobia are species of *Alpha-* or *Betaproteobacteria* (⮂ Sections 15.1 and 15.2) (**Figure 22.6**) that can grow freely in soil or infect leguminous plants and establish a symbiotic relationship. The same genus (or even species) can contain both rhizobial and nonrhizobial strains. Infection of legume roots by rhizobia leads to the formation of **root nodules** (**Figure 22.7**) in which the bacteria fix gaseous nitrogen (N_2) (⮂ Section 3.17). Nitrogen fixation in root nodules accounts for a fourth of the N_2 fixed annually on Earth and is of enormous agricultural importance, as it increases the fixed nitrogen content of soil. Nodulated legumes can grow well on unfertilized bare soils that are nitrogen deficient, while other plants grow only poorly on them (**Figure 22.8**).

Leghemoglobin and Cross-Inoculation Groups

In the absence of its bacterial symbiont, a legume cannot fix N_2. Rhizobia, on the other hand, can fix N_2 when grown in pure culture under microaerophilic conditions (a low-oxygen environment is necessary because nitrogenases are inactivated by high levels of O_2, ⮂ Section 3.17). In the nodule, O_2 levels are precisely controlled by the O_2-binding protein **leghemoglobin**. Production of this iron-containing protein in healthy N_2-fixing nodules (**Figure 22.9**) is induced through the interaction of the plant and bacterial partners. Leghemoglobin functions as an "oxygen buffer," cycling between the oxidized (Fe^{3+}) and reduced (Fe^{2+}) forms of iron to keep unbound O_2 within the nodule low.

Figure 22.7 **Soybean root nodules.** The nodules developed from infection by *Bradyrhizobium japonicum*. The main stem of this soybean plant is about 0.5 cm in diameter.

The ratio of leghemoglobin-bound O_2 to free O_2 in the root nodule is thus maintained on the order of 10,000:1.

There is a marked specificity between the species of legume and rhizobium that can establish a symbiosis. A particular rhizobial species is able to infect certain species of legumes but not others. A group of related legumes that can be infected by a particular rhizobial species is called a *cross-inoculation group*—there are, for example, a clover group, a bean group, an alfalfa group, and so on (**Table 22.1**). If legumes are inoculated with the correct rhizobial strain, leghemoglobin-rich, N_2-fixing nodules develop on their roots (Figures 22.7–22.9).

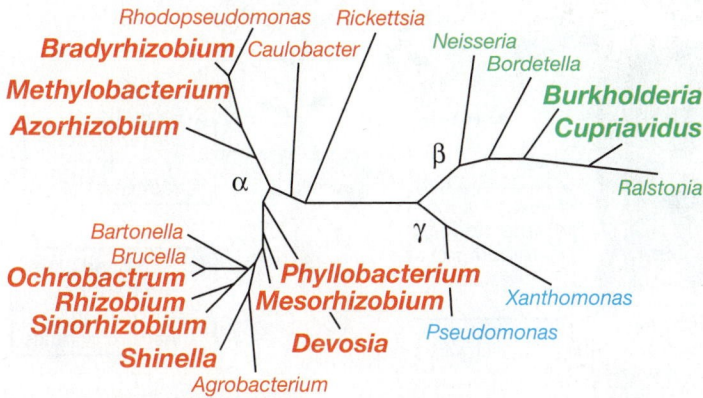

Figure 22.6 **Phylogeny of rhizobial (names in boldface) and related genera inferred from analysis of 16S rRNA gene sequences.** There are rhizobia in 12 genera and more than 70 species of *Alpha-* and *Betaproteobacteria*.

Figure 22.8 **Effect of nodulation on plant growth.** A field of unnodulated (left) and nodulated (right) soybean plants growing in nitrogen-poor soil. The yellow color is typical of chlorosis, the result of nitrogen starvation.

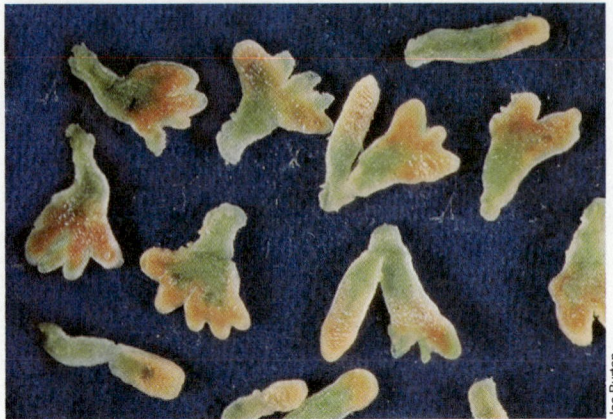

Figure 22.9 Root nodule structure. Sections of root nodules from the legume *Coronilla varia*, showing the reddish pigment leghemoglobin.

Table 22.1 Major cross-inoculation groups of leguminous plants	
Host plant	*Nodulated by*
Pea	*Rhizobium leguminosarum* biovar *viciae*[a]
Bean	*Rhizobium leguminosarum* biovar *phaseoli*[a]
Bean	*Rhizobium tropici*
Lotus	*Mesorhizobium loti*
Clover	*Rhizobium leguminosarum* biovar *trifolii*[a]
Alfalfa	*Sinorhizobium meliloti*
Soybean	*Bradyrhizobium japonicum*
Soybean	*Bradyrhizobium elkanii*
Soybean	*Sinorhizobium fredii*
Sesbania rostrata (a tropical legume)	*Azorhizobium caulinodans*

[a]Several varieties (biovars) of *Rhizobium leguminosarum* exist, each capable of nodulating a different legume.

Steps in Root Nodule Formation

How root nodules form is well understood for most rhizobia (**Figure 22.10**). The steps are as follows:

1. Recognition of the correct partner by both plant and bacterium and attachment of the bacterium to the root hairs
2. Secretion of oligosaccharide signaling molecules (Nod factors) by the bacterium
3. Bacterial invasion of the root hair
4. Movement of bacteria to the main root by way of the infection thread
5. Formation of modified bacterial cells (bacteroids) within the plant cells and development of the N_2-fixing state
6. Continued plant and bacterial cell division, forming the mature root nodule

Another mechanism of nodule formation that does not require Nod factors is used by some species of phototrophic rhizobia. This mechanism has yet to be fully elucidated, but appears to require the bacterial production of *cytokinins*. Cytokinins are plant hormones, derived from adenine or phenylurea, necessary for cell growth and differentiation.

Attachment and Infection

The roots of leguminous plants secrete organic compounds that stimulate the growth of a diverse rhizosphere microbial community. If rhizobia of the correct cross-inoculation group are in the soil, they will form large populations and eventually attach to the root hairs that extend from the roots of the plant (Figure 22.10). An adhesion protein called *rhicadhesin* is present on the cell surfaces of rhizobia. Other substances, such as carbohydrate-containing proteins called *lectins* and specific receptors in the plant cytoplasmic membrane, also play roles in plant–bacterium attachment.

After attaching, a rhizobial cell penetrates into the root hair, which curls in response to substances secreted by the bacterium.

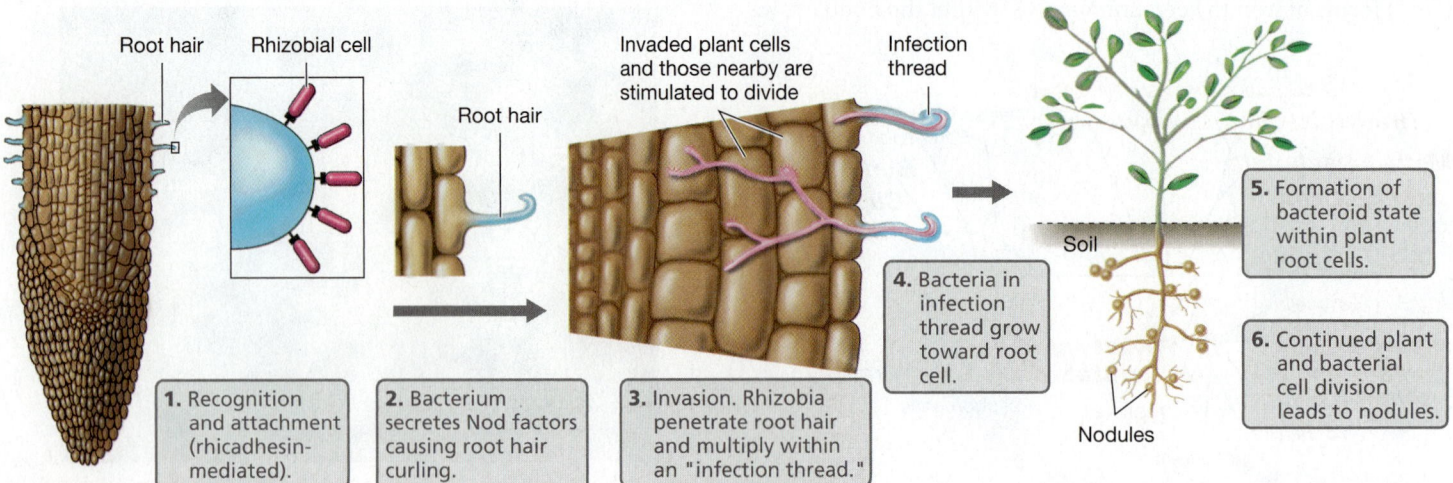

Figure 22.10 Steps in the formation of a root nodule in a legume infected by *Rhizobium*. Formation of the bacteroid state is a prerequisite for nitrogen fixation. See Figure 22.15 for physiological activities in the nodule.

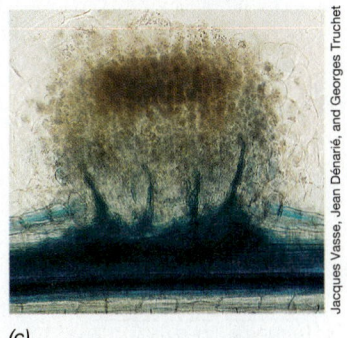
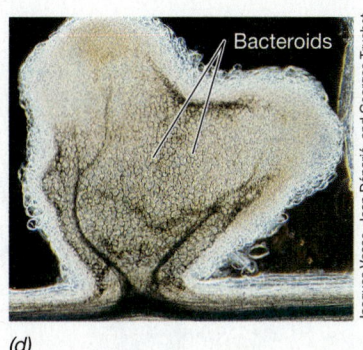

(a)   (b)   (c)   (d)

Figure 22.11 The infection thread and formation of root nodules. *(a)* An infection thread induced by cells of *Rhizobium leguminosarum* biovar *trifolii* in a root hair of white clover (*Trifolium repens*). The infection thread consists of a cellulosic tube through which bacteria move to root cells. *(b–d)* Nodules from alfalfa roots infected with cells of *Sinorhizobium meliloti* shown at different stages of development. Cells of both *R. leguminosarum* biovar *trifolii* and *S. meliloti* are about 2 mm long. The time course of nodulation events from infection to effective nodule is about 1 month in both soybean and alfalfa. Bacteroids are about 2 µm long. Photos *b–d* reprinted with permission from *Nature 351:* 670–673 (1991), © Macmillan Magazines Ltd.

The bacterium then induces formation by the plant of a cellulosic tube, called the **infection thread** (Figure 22.11*a*), which spreads down the root hair. Root cells adjacent to the root hairs subsequently become infected by rhizobia, and plant cells divide. Continued plant cell division forms the tumorlike nodule (Figure 22.11*b–d*). A different mechanism of infection is used by some rhizobia adapted to aquatic or semiaquatic tropical legumes (see Figure 22.16). These rhizobia enter the plant at the loose cellular junctions of roots emerging perpendicular from an established root (*lateral roots*). Following entry into the plant, some of the rhizobia develop infection threads, whereas others do not.

Bacteroids

The rhizobia multiply rapidly within the plant cells and become transformed into swollen, misshapen, and branched cells called **bacteroids**. A microcolony of bacteroids becomes surrounded by portions of the plant cytoplasmic membrane to form a structure called the *symbiosome* (Figure 22.11*d*), and only after the symbiosome forms does N₂ fixation begin. Nitrogen-fixing nodules can be detected experimentally by the reduction of acetylene to ethylene (⮌ Section 3.17). When the plant dies, the nodule deteriorates, releasing bacteroids into the soil. Although bacteroids are incapable of division, a small number of dormant rhizobial cells are always present in the nodule. These now proliferate, using some of the products of the deteriorating nodule as nutrients. The bacteria can then initiate infection the next growing season or maintain a free-living existence in the soil.

Nodule Formation: *nod* Genes, Nod Proteins, and Nod Factors

Rhizobial genes that direct the steps in nodulation of a legume are called *nod genes*. It is thought that the ability to form nodules has independently emerged multiple times through the horizontal transfer of such genes as *nod* and *nif* that are located on plasmids or transferable regions of chromosomal DNA. In *Rhizobium leguminosarum* biovar *viciae*, which nodulates peas, ten *nod* genes have been identified. The *nodABC* genes encode proteins that

produce oligosaccharides called **Nod factors**; these induce root hair curling and trigger cell division in the pea plant, eventually leading to formation of the nodule (see Figure 22.15 for a description of root nodule biochemistry).

Nod factors are lipochitin oligosaccharides to which various substituents are bonded (Figure 22.12) that function as primary rhizobial signal molecules triggering legumes to develop new

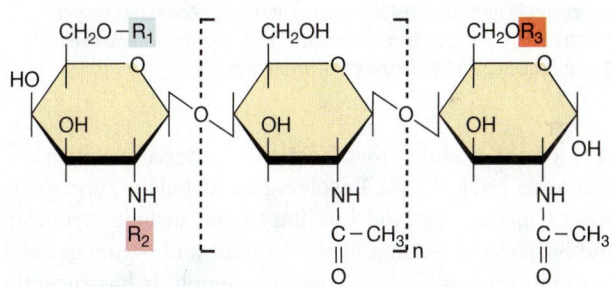

(a)

Rhizobial or AM fungus species	R₁	R₂	R₃
Sinorhizobium meliloti (alfalfa)	Ac	C16:2 or C16:3	SO₃H
Rhizobium leguminosarum biovar *viciae* (pea)	Ac	C18:1 or C18:4	H or Ac
Glomus intraradices (many agricultural crops)	H	C16 or C16:1 or C16:2 or C18 or C18:1Δ9Z	H or SO₃H

(b)

Figure 22.12 Nod and Myc factors. *(a)* General structure of Nod factors produced by rhizobia species (*Sinorhizobium meliloti* and *Rhizobium leguminosarum* biovar *viciae*) and the Myc factor produced by the *Glomus intraradices*, an arbuscular mycorrhizal (AM) fungus (Section 22.5). *(b)* Table of the structural differences (R₁, R₂, R₃) that define the precise signaling factors of each species. The central hexose unit can repeat up to three times for different Nod factors, and repeat either two or three times for the different Myc factors. C16:1, C16:2, and C16:3, palmitic acid with either one, two, or three double bonds, respectively; C18:1, oleic acid with one double bond; C18:1Δ9Z, the *trans* isomer of oleic acid with one double bond at the 9th C–C bond; C18:4, oleic acid with four double bonds; Ac, acetyl.

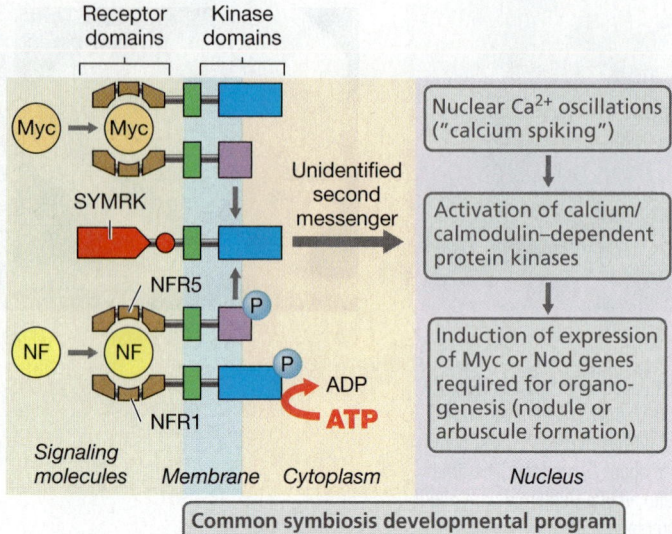

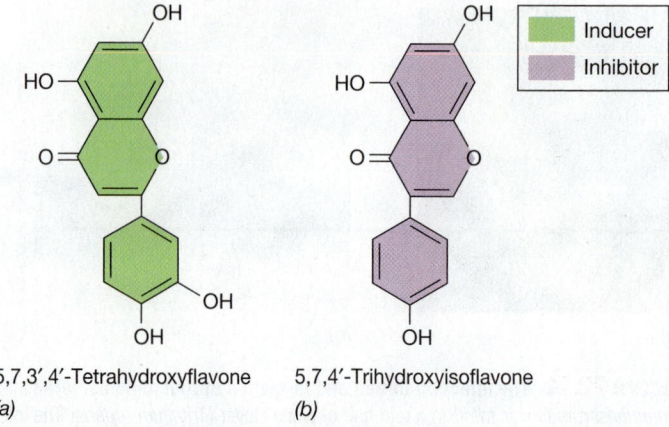

5,7,3′,4′-Tetrahydroxyflavone
(a)

5,7,4′-Trihydroxyisoflavone
(b)

Figure 22.14 **Plant flavonoids and nodulation.** Structures of flavonoid molecules that are *(a)* an inducer of *nod* gene expression and *(b)* an inhibitor of *nod* gene expression in *Rhizobium leguminosarum* biovar *viciae*. Note the similarities in the structures of the two molecules. The common name of the structure shown in part a is *luteolin*, and it is a flavone derivative. The structure in part b is called *genistein*, and it is an isoflavone derivative.

Figure 22.13 **Nod and Myc signaling pathways in root nodule and arbuscule formation.** Nod factor (NF) signaling involves at least three membrane-associated receptors (NFR1, NFR5, and SYMRK) that together initiate nodulation development via protein phosphorylation. NFR1 and SYMRK have active kinase domains (blue), whereas NFR5 kinase is inactive. The direct binding of NF to a complex of NFR1 and NFR5 at the plant cell cytoplasmic membrane initiates signal transduction by activation of the NFR1 kinase. The resulting auto- or trans-phosphorylation of the NFR cytoplasmic domains triggers events leading to formation of the infection thread. Signal transduction to the SYMRK by the NFR1-NFR5-Nod factor complex (or by an unidentified receptor for the Myc factor) is part of a conserved symbiosis program, in which induction of calcium signaling in the plant cell nucleoplasm triggers gene expression changes and production of plant growth hormones (cytokinins) required for nodule or arbuscule formation. Neither the identity of the second messenger nor participation of NFR-like receptors in the Myc signaling pathway have been established. See Section 22.5 for discussion of mycorrhizae.

plant organs: root nodules that host the bacteria as nitrogen-fixing bacteroids (**Figure 22.13**). Resolving the details of the signaling pathway triggered by Nod binding to cell surface receptors (NFR1 and NFR2) and leading to the induction of organogenesis (nodule formation) is an active area of research. It has recently been shown that many elements of the signaling pathway leading to nodulation are also used by the mycorrhizal fungi for infection of plant roots (Figure 22.13 and Section 22.5).

Which plants a given rhizobial species can infect is in part determined by the structure of the Nod factor it produces. Besides the *nodABC* genes, which are universal and whose products synthesize the Nod backbone, each cross-inoculation group contains *nod* genes that encode proteins that chemically modify the Nod factor backbone to form its species-specific molecule (Figure 22.12). In *R. leguminosarum* biovar *viciae*, *nodD* encodes the regulatory protein NodD, which controls transcription of other *nod* genes. After interacting with inducer molecules, NodD promotes transcription and is thus a positive regulatory protein (♂ Section 7.4). NodD inducers are plant flavonoids, organic molecules that are widely secreted by plants. Some flavonoids that are structurally very closely related to *nodD* inducers in *R. leguminosarum* biovar *viciae* inhibit *nod* gene expression in other rhizobial species (**Figure 22.14**). This indicates that part of the specificity observed between plant and bacterium in the rhizobia–legume

symbioses lies in the chemistry of the flavonoids secreted by each species of legume.

Biochemistry of Root Nodules

As discussed in Section 3.17, N_2 fixation requires the enzyme *nitrogenase*. Nitrogenase from bacteroids shows the same biochemical properties as the enzyme from free-living N_2-fixing bacteria, including O_2 sensitivity and the ability to reduce acetylene as well as N_2. Bacteroids are dependent on the plant for the electron donor for N_2 fixation. The major organic compounds transported across the symbiosome membrane and into the bacteroid proper are citric-acid-cycle intermediates—in particular, the C_4 organic acids *succinate, malate,* and *fumarate* (**Figure 22.15**). These are used as electron donors for ATP production and, following conversion to pyruvate, as the ultimate source of electrons for the reduction of N_2.

The product of N_2 fixation is ammonia (NH_3), and the plant assimilates most of this NH_3 by forming organic nitrogen compounds. The NH_3-assimilating enzyme glutamine synthetase is present in high levels in the plant cell cytoplasm and can convert glutamate and NH_3 into glutamine (♂ Section 3.15). This and a few other organic nitrogen compounds transport bacterially fixed nitrogen throughout the plant.

Stem-Nodulating Rhizobia

Although most leguminous plants form N_2-fixing nodules on their *roots*, a few legume species bear nodules on their *stems*. Stem-nodulated leguminous plants are widespread in tropical regions where soils are often nitrogen deficient because of leaching and intense biological activity. The best-studied system is the tropical aquatic legume *Sesbania*, which is nodulated by the bacterium *Azorhizobium caulinodans* (**Figure 22.16**). Stem nodules typically form in the submerged portion of the stems or just above the water level. The general sequence of events by which stem nodules form in *Sesbania* resembles that of root nodules: attachment, formation of an infection thread, and bacteroid formation.

Plant cytoplasm

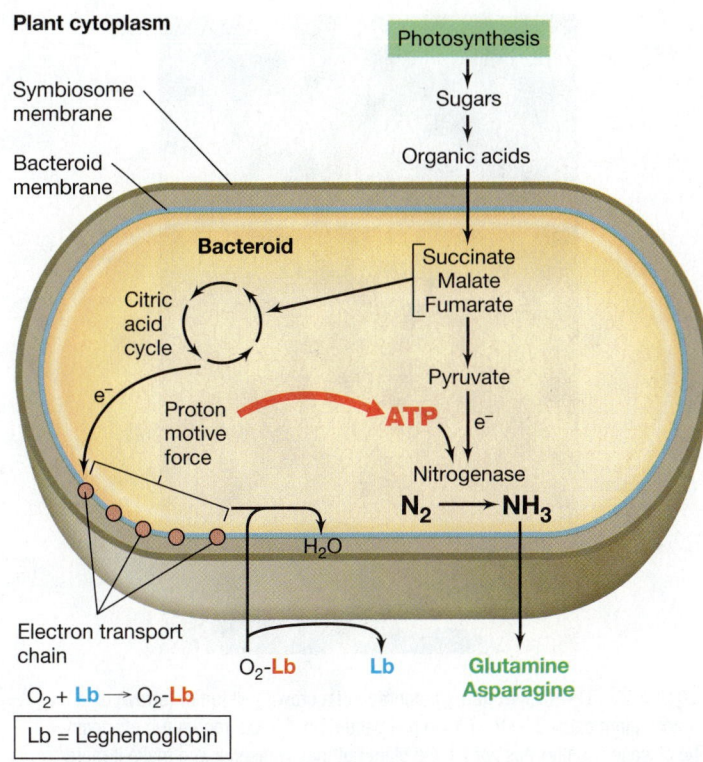

Figure 22.15 **The root nodule bacteroid.** Schematic diagram of major metabolic reactions and nutrient exchanges in the bacteroid. The symbiosome is a collection of bacteroids surrounded by a membrane originating from the plant.

Some stem-nodulating rhizobia produce bacteriochlorophyll *a* and thus have the potential to carry out anoxygenic photosynthesis (⇄ Section 13.3). Bacteriochlorophyll-containing rhizobia, called photosynthetic *Bradyrhizobium*, are widespread in nature, particularly in association with tropical legumes. In these species, light energy converted to chemical energy (ATP) in photosynthesis is likely to be at least part of the energy source needed by the bacterium to support N_2 fixation.

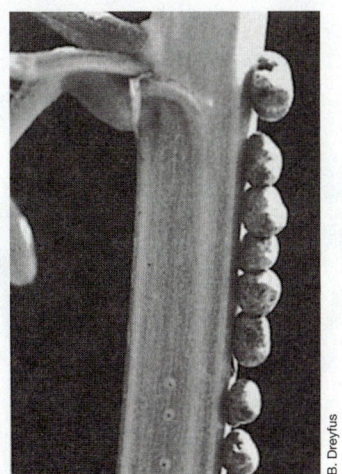

Figure 22.16 **Stem nodules formed by stem-nodulating *Azorhizobium*.** The right side of this stem of the tropical legume *Sesbania rostrata* was inoculated with *Azorhizobium caulinodans*, but the left side was not.

Figure 22.17 *Azolla–Anabaena* **symbiosis.** (a) Intact association showing a single plant of *Azolla pinnata*. The diameter of the plant is approximately 1 cm. (b) Cyanobacterial symbiont *Anabaena azollae* as observed in crushed leaves of *A. pinnata*. Single cells of *A. azollae* are about 5 μm wide. Vegetative cells are oblong; the spherical heterocysts (lighter color, arrows) are differentiated for nitrogen fixation.

Nonlegume N_2-Fixing Symbioses: *Azolla–Anabaena* and *Alnus–Frankia*

Various nonleguminous plants form N_2-fixing symbioses with bacteria other than rhizobia. For example, the water fern *Azolla* harbors within small pores of its fronds a species of heterocystous N_2-fixing cyanobacteria (⇄ Section 14.3) called *Anabaena azollae* (**Figure 22.17**). *Azolla* has been used for centuries to enrich rice paddies with fixed nitrogen. Before planting rice, the farmer allows the surface of the rice paddy to become densely covered with *Azolla*. As the rice plants grow, they eventually crowd out the *Azolla*, causing its death and the release of its nitrogen, which is assimilated by the rice plants. By repeating this process each growing season, rice farmers can obtain high yields of rice without applying nitrogenous fertilizers.

The alder tree (genus *Alnus*) has N_2-fixing root nodules (**Figure 22.18a**) that harbor filamentous, N_2-fixing actinomycetes of

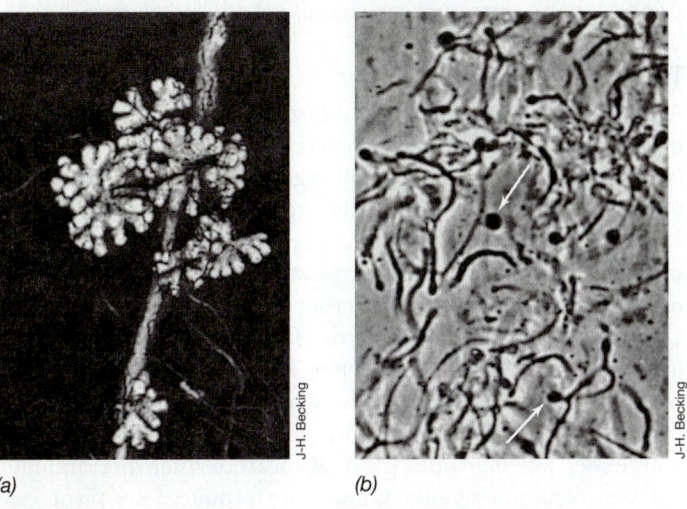

Figure 22.18 *Frankia* **nodules and *Frankia* cells.** (a) Root nodules of the common alder *Alnus glutinosa*. (b) *Frankia* culture purified from nodules of *Comptonia peregrina*. Note vesicles (arrows) on the tips of hyphal filaments.

UNIT 4

the genus *Frankia*. When assayed in cell extracts the nitrogenase of *Frankia* is sensitive to O_2, but cells of *Frankia* fix N_2 at full oxygen tensions. This is because *Frankia* protects its nitrogenase from O_2 by localizing the enzyme in terminal swellings on the cells called *vesicles* (Figure 22.18*b*). The vesicles contain thick walls that retard O_2 diffusion, thus maintaining the O_2 tension within vesicles at levels compatible with nitrogenase activity. In this regard, *Frankia* vesicles resemble the heterocysts produced by some filamentous cyanobacteria as localized sites of N_2 fixation (⇄ Section 14.3).

Alder is a characteristic pioneer tree able to colonize nutrient-poor soils, probably because of its ability to enter into a symbiotic N_2-fixing relationship with *Frankia*. A number of other small or bushy, woody plants are nodulated by *Frankia*. However, unlike the rhizobial symbionts of legumes, a single strain of *Frankia* can form nodules on several different species of plants, suggesting that the *Frankia*–root nodule symbiosis is less specific than that of leguminous plants.

Figure 22.19 Crown gall. Photograph of a crown gall tumor (arrow) on a tobacco plant caused by the crown gall bacterium *Agrobacterium tumefaciens*. The disease usually does not kill the plant but may weaken it and make it more susceptible to drought and diseases.

MINIQUIZ

- How do rhizobial root nodules benefit a plant?
- What are Nod factors and what do they do?
- What is a bacteroid and what occurs within it? What is the function of leghemoglobin?
- What are the major similarities and differences between rhizobia and *Frankia*?

22.4 *Agrobacterium* and Crown Gall Disease

Some microorganisms develop parasitic symbioses with plants. The genus *Agrobacterium*, a relative of the root nodule bacterium *Rhizobium* (Figure 22.6), is such an organism, causing the formation of tumorous growths on diverse plants. The two species of *Agrobacterium* most widely studied are *Agrobacterium tumefaciens* (also called *Rhizobium radiobacter*), which causes *crown gall disease*, and *Agrobacterium rhizogenes*, which causes *hairy root disease*.

The Ti Plasmid

Although wounded plants often form a benign accumulation of tissue called a *callus*, the growth in crown gall disease (**Figure 22.19**) is different in that it is uncontrolled growth, resembling an animal tumor. *A. tumefaciens* cells induce tumor formation only if they contain a large plasmid called the **Ti plasmid** (Ti for *tumor induc-ing*). In *A. rhizogenes*, a similar plasmid called the *Ri plasmid* is necessary for induction of hairy root disease. Following infection, a part of the Ti plasmid called the *transferred DNA* (T-DNA) is integrated into the plant's genome. T-DNA carries the genes for tumor formation and also for the synthesis of a number of modified amino acids called *opines*. Octopine [N^2-(1,3-dicarboxyethyl)-L-arginine] and *nopaline* [N^2-(1,3-dicarboxypropyl)-L-arginine] are two common opines. Opines are produced by plant cells transformed by T-DNA and are a source of carbon and nitrogen, and sometimes phosphate, for the parasitic *A. tumefaciens* cells. These nutrients are the benefits for the bacterial symbiont.

Recognition and T-DNA Transfer

To initiate the tumorous state, *A. tumefaciens* cells attach to a wound site on the plant. Following attachment, the synthesis of cellulose microfibrils by the bacteria helps anchor them to the wound site, and bacterial aggregates form on the plant cell surface. This sets the stage for plasmid transfer from bacterium to plant.

The general structure of the Ti plasmid is shown in **Figure 22.20**. Only the T-DNA is actually transferred to the plant. The T-DNA contains genes that induce tumorigenesis. The *vir* genes on the Ti plasmid encode proteins that are essential for T-DNA transfer.

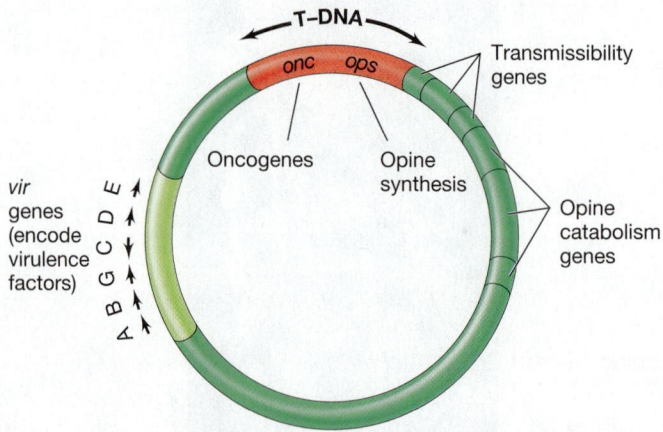

Figure 22.20 Structure of the Ti plasmid of *Agrobacterium tumefaciens*. T-DNA is the region transferred to the plant. Arrows indicate the direction of transcription of each gene. The entire Ti plasmid is about 200 kilobase pairs of DNA and the T-DNA is about 20 kilobase pairs.

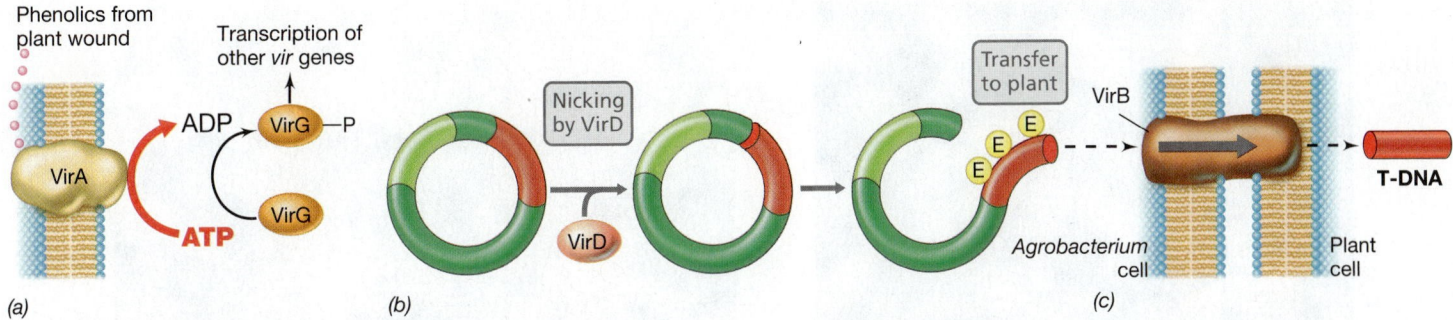

Figure 22.21 Mechanism of transfer of T-DNA to the plant cell by *Agrobacterium tumefaciens*. *(a)* VirA activates VirG by phosphorylation and VirG activates transcription of other *vir* genes. *(b)* VirD is an endonuclease that nicks the Ti plasmid, exposing the T-DNA. *(c)* VirB functions as a conjugation bridge between the *A. tumefaciens* cell and the plant cell, and VirE is a single-strand binding protein that assists in T-DNA transfer. Plant DNA polymerase produces the complementary strand to the transferred single strand of T-DNA.

Transcription of *vir* is induced by metabolites synthesized by wounded plant tissues. Examples of inducers include the phenolic compounds acetosyringone and ferulate. The transmissibility genes on the Ti plasmid (Figure 22.20) allow the plasmid to be transferred by conjugation from one bacterial cell to another.

The *vir* genes are the key to T-DNA transfer. The *virA* gene encodes a protein kinase (VirA) that interacts with inducer molecules and then phosphorylates the product of the *virG* gene (Figure 22.21). VirG is activated by phosphorylation and functions to activate other *vir* genes. The product of the *virD* gene (VirD) has endonuclease activity and nicks DNA in the Ti plasmid in a region adjacent to the T-DNA. The product of the *virE* gene is a DNA-binding protein that binds the single strand of T-DNA in the plant cell to protect it from destruction by nucleases. The *virB* operon encodes 11 different proteins that form a type IV secretion system (⮂ Section 4.14) for single-strand T-DNA and protein transfer between bacterium and plant (Figure 22.21) and thus resembles bacterial conjugation (⮂ Section 10.8). Laboratory studies of *A. tumefaciens* have shown that it can transfer T-DNA into many types of eukaryotic cells, including fungi, algae, protists, and even human cell lines.

Once inside the plant cell, T-DNA then becomes inserted into the genome of the plant. Tumorigenesis (*onc*) genes on the Ti plasmid (Figure 22.20) encode enzymes for plant hormone production and at least one key enzyme of opine biosynthesis. Expression of these genes leads to tumor formation and opine production. The Ri plasmid responsible for hairy root disease also contains *onc* genes. However, in this case the genes confer increased auxin responsiveness to the plant, and this promotes overproduction of root tissue and the symptoms of the disease. The Ri plasmid also encodes several opine biosynthetic enzymes.

Genetic Engineering with the Ti Plasmid

From the standpoint of microbiology and plant pathology, crown gall disease and hairy root disease both require intimate interactions that lead to genetic exchange from bacterium to plant. In other words, tumor induction in these diseases is the result of a natural plant-transformation system. Thus, in recent years interest in the Ti–crown gall system has shifted away from the disease itself toward applications of this natural genetic exchange process in plant biotechnology.

Several modified Ti plasmids that lack disease genes but that can still transfer DNA to plants have been developed by genetic engineering. These have been used for the construction of genetically modified (transgenic) plants. Many transgenic plants have been constructed thus far, including crop plants carrying genes for resistance to herbicides, insect attack, and drought. We discuss the use of the Ti plasmid as a vector in plant biotechnology in Section 11.13.

MINIQUIZ
- What are opines and whom do they benefit?
- How do the *vir* genes differ from T-DNA in the Ti plasmid?
- How has an understanding of crown gall disease benefited plant agriculture?

22.5 Mycorrhizae

Mycorrhizae are mutualisms between plant roots and fungi in which nutrients are transferred in both directions. The fungus transfers nutrients—in particular, phosphorus and nitrogen—from the soil to the plant, and the plant in turn transfers carbohydrates to the fungus. These mutualisms are harnessed in agricultural applications. From fungal spores produced in culture or from root scrapings of infected plants, soil inoculants are produced that enhance plant growth.

Classes of Mycorrhizae

There are two classes of mycorrhizae. In *ectomycorrhizae*, fungal cells form an extensive sheath around the outside of the root with only a slight penetration into the root tissue itself (Figure 22.22). In *endomycorrhizae*, a part of the fungus becomes deeply embedded within the root tissue. Ectomycorrhizae are found mainly on the roots of forest trees, especially conifers, beeches, and oaks, and are most highly developed in boreal and temperate forests. In such forests, almost every root of every tree is mycorrhizal. The root system of a mycorrhizal tree such as a pine (genus *Pinus*) is composed of both long and short roots. The short roots, which

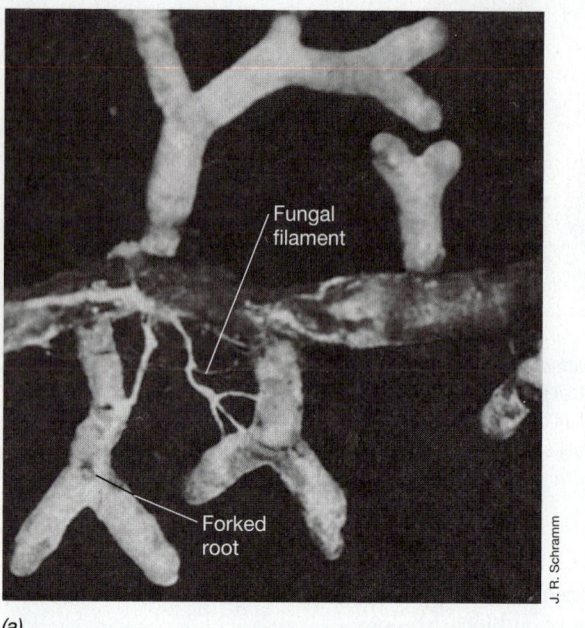

(a)

(b)

Figure 22.22 Mycorrhizae. *(a)* Typical ectomycorrhizal root of the pine *Pinus rigida* with filaments of the fungus *Thelephora terrestris*. *(b)* Seedling of *Pinus contorta* (lodgepole pine), showing extensive development of the absorptive mycelium of its fungal associate *Suillus bovinus*. This grows in a fanlike formation from the ectomycorrhizal roots to capture nutrients from the soil. The seedling is about 12 cm high.

are characteristically dichotomously branched in *Pinus* (Figure 22.22*a*), show typical fungal colonization, and long roots are also frequently colonized. Most mycorrhizal fungi do not catabolize cellulose and other leaf litter polymers. Instead, they catabolize simple carbohydrates and typically have one or more vitamin requirements. They obtain their carbon from root secretions and obtain inorganic minerals from the soil. Mycorrhizal fungi are rarely found in nature except in association with roots, and many are probably obligate symbionts.

Despite the close symbiotic association between fungus and root, a single species of tree can form multiple mycorrhizal associations. One pine species can associate with over 40 species of fungi. This relative lack of host specificity allows ectomycorrhizal mycelia to interconnect trees, providing linkages for transfer of carbon and other nutrients between trees of the same or different species. Nutrient transfer from well-illuminated overstory plants to shaded trees is thought to help equalize resource availability, subsidizing young trees and increasing biodiversity by promoting the coexistence of different species.

Arbuscular Mycorrhizae

Although ectomycorrhizal fungi play a significant role in the ecology of forests, there is a greater diversity of endomycorrhizae. Most are *arbuscular mycorrhizae* (*AM*) that comprise a phylogenetically distinct fungal division, the *Glomeromycota* (⮌ Section 17.12), of which all or most species are obligate plant mutualists (the word "arbuscular" means "little tree"). AM colonize 70–90% of all terrestrial plants, including most grassland species and many crop species. The association between plants and the *Glomeromycota* is thought to be the ancestral type of mycorrhizae, established 400–460 million years ago and an important evolutionary step in the successful invasion of dry land by terrestrial plants.

It is now known that AM fungi produce lipochitin oligosaccharide signaling factors (**Myc factors**) very closely related to Nod factors (Section 22.3), and these initiate formation of the mycorrhizal state (Figures 22.12 and 22.13). Root colonization by an AM fungus begins with germination of a soil-borne spore, producing a short germination mycelium that recognizes the host plant through reciprocal chemical signaling; the fungus then forms a contact structure called the *hyphopodium* with root epidermal cells (**Figure 22.23**). Penetrating hyphae extend into the

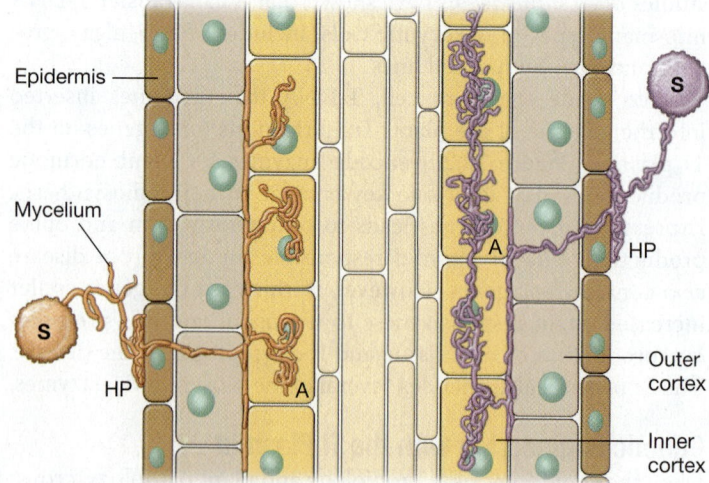

Epidermis

Mycelium

HP

A

S

A

HP

Outer cortex

Inner cortex

Figure 22.23 Arbuscular mycorrhizae root colonization. A spore (S) near a tree root generates a short mycelium that is attracted to the root by chemical signaling, forming an attachment structure called the hyphopodium (HP). The mycelium then enters the inner cortex region of the root by penetrating epidermal cells and cells of the outer cortex. Arbuscules (dichotomously branched invaginations, A) are formed by mycelia spreading either intercellularly (left) or intracellularly (right).

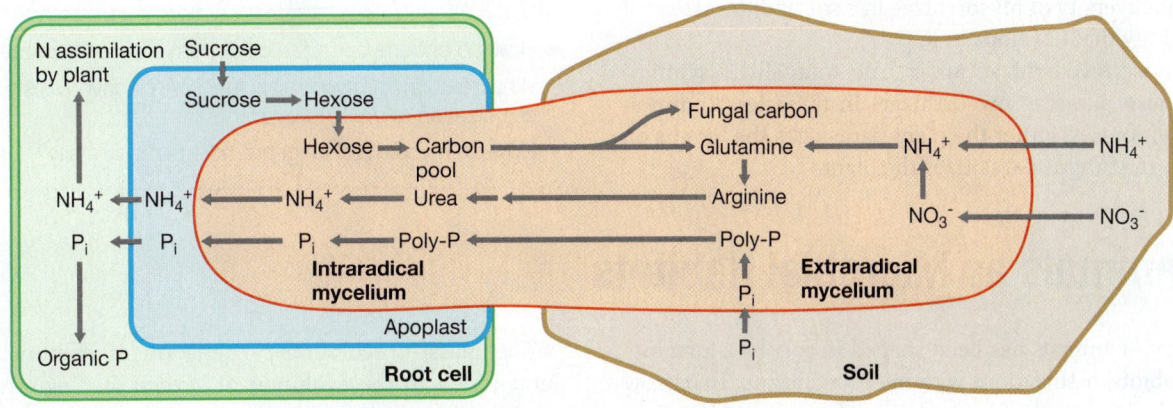

Figure 22.24 Pathways of N, P, and C exchange between plant and AM fungi. Inorganic nitrogen (NH_4^+ and NO_3^-) and phosphorus (P_i) mined from the soil by the extraradical (soil-associated) mycelia is translocated to the plant as arginine and polyphosphate (poly-P) through the mycelial network and delivered to the plant at the intraradical (plant cell–associated) mycelium. Ammonia and phosphate are regenerated in the intraradical mycelium for transfer to the plant cell. In exchange for the N and P, the plant provides organic carbon to the fungus.

plant from each hyphopodium, usually taking an intracellular path through epidermal and outer cortical cell layers of the root before forming dichotomously branched or coiled hyphal structures called **arbuscules** within cells of the plant's inner cortex, near to vascular tissues. However, the arbuscular hyphae remain separated from plant protoplasm by an extensive plant cytoplasmic membrane that forms a region called the *apoplast* (**Figure 22.24**), which functions to increase the surface area of contact between plant and fungus. Inorganic nitrogen and phosphorus are "mined" from the soil by the fungi, converted to arginine and polyphosphate, and then translocated through the hyphae to the plant (Figure 22.24).

Myc factors are very similar to the rhizobial Nod factors and only relatively minor modifications of the chitin backbone structure confer specificity (Figure 22.12). It is now suspected that the basic signaling and developmental systems used in the legume–root nodule symbiosis (Section 22.3), which arose about 60 million years ago, first evolved in the much more ancient AM fungi–plant symbiosis. Apparently the AM fungal system was recruited and adapted for the legume–root nodule symbiosis (Figure 22.13).

Although the arbuscular mycorrhizae are a much more ancient and widely distributed microorganism–plant symbiosis, understanding of their signaling and developmental program has been much slower to develop because the AM fungi cannot be maintained in pure culture. AM fungi are obligately *biotropic* (meaning that they obtain their nutrients only from living cells of their symbiotic partner), and have no supporting genetic system as has been exploited to help unravel the complex developmental steps leading to legume–root nodule formation.

Benefits for the Plant

The beneficial effect of the mycorrhizal fungus on the plant is best observed in poor soils where plants that are mycorrhizal thrive, but nonmycorrhizal ones do not. For example, if trees planted in prairie soils, which ordinarily lack a suitable fungal inoculum, are artificially inoculated at the time of planting, they grow much more rapidly than uninoculated trees (**Figure 22.25**).

The mycorrhizal plant can absorb nutrients from its environment more efficiently and thus has a competitive advantage (Figure 22.24). This improved nutrient absorption is due to the greater surface area provided by the fungal mycelium. For example, in the pine seedling shown in Figure 22.22*b*, the ectomycorrhizal fungal mycelium makes up the overwhelming part of the absorptive capacity of the plant root system. The mycorrhizal plant is better able to function physiologically and compete successfully in a species-rich plant community, and the fungus benefits from a steady supply of organic nutrients.

In addition to helping plants absorb nutrients, mycorrhizae also play a significant role in supporting plant diversity. Field experiments have clearly shown a positive correlation between the

Figure 22.25 Effect of mycorrhizal fungi on plant growth. Six-month-old seedlings of Monterey pine (*Pinus radiata*) growing in pots containing prairie soil: left, nonmycorrhizal; right, mycorrhizal.

abundance and diversity of mycorrhizae in a soil and the extent of the plant diversity that develops in it.

Although most mycorrhizae are a true mutualistic symbiosis, there are also parasitic mycorrhizae. In these less common mycorrhizal symbioses, either the plant parasitizes the fungus or, in some cases, the fungus parasitizes the plant.

MINIQUIZ
- How do endomycorrhizae differ from ectomycorrhizae?
- What features of mycorrhizal fungi might have assisted in colonization of dry land by plants?
- How do mycorrhizal fungi promote plant diversity?

III • Mammals as Microbial Habitats

The evolution of animals has been shaped in part by a long history of symbiotic associations with microorganisms. To narrow our focus and look in depth at some details of these symbioses, we consider only mammals here. Microorganisms inhabit all sites on mammalian bodies, but the greatest diversity and density of microorganisms are found in the mammalian gut, and we center our discussion there. And finally, of the many mammals on Earth, we restrict our attention to ruminants and humans, the best-studied animals in terms of their gut microflora.

22.6 The Mammalian Gut

Some mammals are *herbivores*, consuming only plant materials, whereas others are *carnivores*, eating primarily the flesh of other animals. *Omnivores* eat both plants and animals. As **Figure 22.26** indicates, closely related mammals have evolved adaptations for differing diets. Notice that mammals of different lineages independently evolved the herbivorous lifestyle, mostly during the Jurassic period, an era in Earth's history of roughly 60 million years beginning about 200 million years ago.

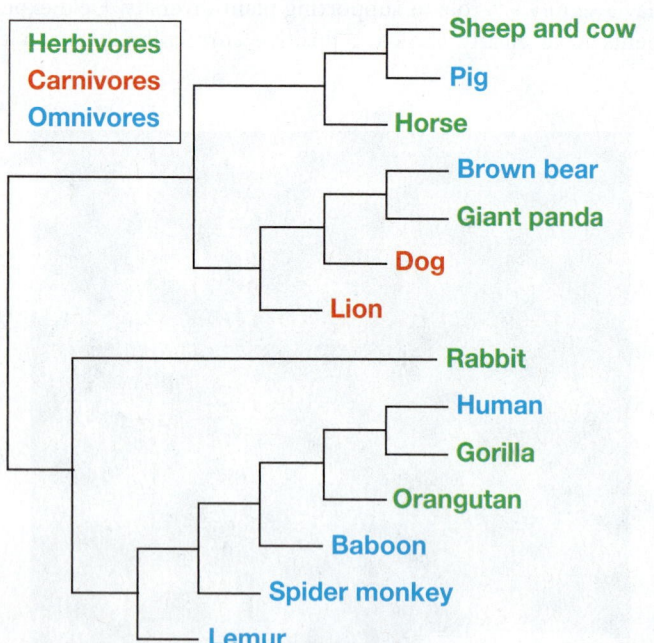

Figure 22.26 Phylogenetic tree showing multiple origins of herbivory among mammals. Some of the herbivores listed are foregut fermenters, while others are hindgut fermenters (Figure 22.27). Instead of animal flesh, some mammalian carnivores eat only insects (the insectivores, such as bats), or fish (the piscivores, such as the river otter).

The massive evolutionary radiation of mammals during the Jurassic led to the evolution of several feeding strategies. Most mammalian species evolved gut structures that foster mutualistic associations with microorganisms. As anatomical differences evolved, microbial fermentation remained important or essential in mammalian digestion. *Monogastric* mammals, such as humans, have a single compartment, the stomach, positioned before the intestine. Such animals may get a substantial part of their energy requirement from microbial fermentation of otherwise indigestible foods, but herbivores are totally dependent on such fermentations.

Plant Substrates

Microbial associations with various mammalian species led to the capacity to catabolize plant fiber, the structural component of plant cell walls. Fiber is composed primarily of insoluble polysaccharides of which cellulose is the most abundant component. Mammals—and indeed almost all animals—lack the enzymes necessary to digest cellulose and certain other plant polysaccharides. Only microorganisms have genes encoding the glycoside hydrolases and polysaccharide lyases required to decompose these polysaccharides.

As the most abundant organic compound on Earth and one composed exclusively of glucose, cellulose offers a rich source of carbon and energy for animals that can digest it. The two primary traits that evolved to support herbivory are (1) an enlarged anoxic fermentation chamber for holding ingested plant material and (2) an extended retention time—the time that ingested material remains in the gut. A longer retention time allows for a longer association of microorganisms with the ingested material and thus a more complete degradation of the plant polymers.

Foregut versus Hindgut Fermenters

Two digestive patterns have evolved in herbivorous mammals. In herbivores with a *foregut* fermentation, the microbial fermentation chamber *precedes* the small intestine. This gut architecture originated independently in ruminants, colobine monkeys, sloths, and macropod marsupials (**Figure 22.27**). These all share the common feature that ingested nutrients are degraded by the gut microbiota *before* reaching the acidic stomach and small intestine. We examine the digestive processes of ruminants, as examples of foregut fermenters, in the next section.

Horses and rabbits are herbivorous mammals, but they are not foregut fermenters. Instead, these animals are *hindgut* fermenters. They have only one stomach, but use an organ called the *cecum*, a digestive organ located between the small and large intestines, as their fermentation vessel. The cecum contains fiber- and cellulose-digesting (cellulolytic) microorganisms. Mammals, such as

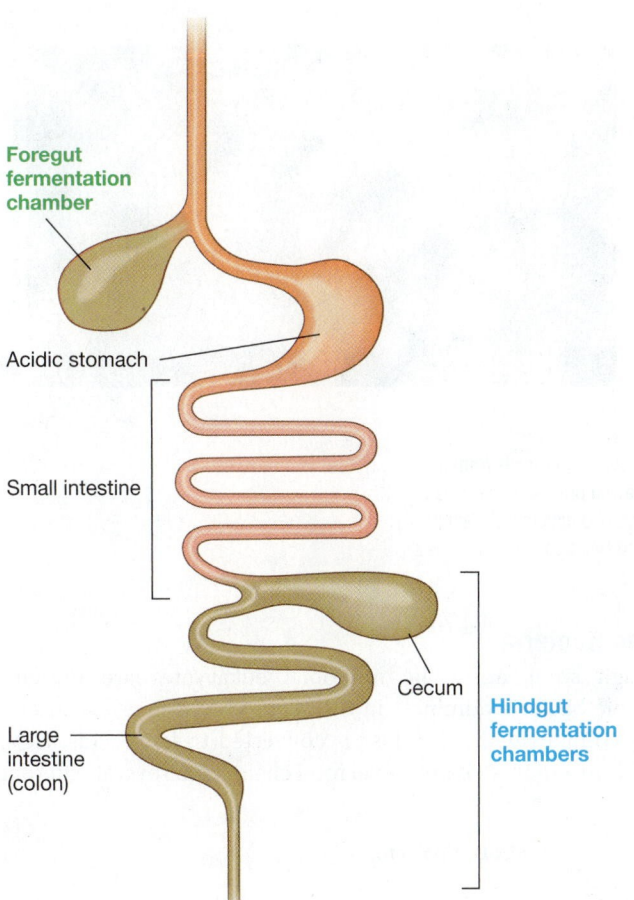

Foregut fermentation chamber

Acidic stomach

Small intestine

Cecum

Large intestine (colon)

Hindgut fermentation chambers

Foregut fermenters Examples: Ruminants (photo 1), colobine monkeys, macropod marsupials, hoatzin (photo 2)

1.

2.

Bernard Swain

Hindgut fermenters Examples: Cecal animals (photos 3 and 4), primates, some rodents, some reptiles

3.

Nancy L. Spear

4.

UNIT 4

Figure 22.27 **Variations on vertebrate gut architecture.** All vertebrates have a small intestine, but vary in other gut structures. Most host absorption of dietary nutrients occurs in the small intestine, whereas microbial fermentation can occur in the forestomach, cecum, or large intestine (colon). Foregut fermentation is found in four major clades of mammals and one avian species (the hoatzin). Hindgut fermentation, either in the cecum or large intestine/colon, is common to many clades of mammals (including humans), birds, and reptiles. Compare with Figure 22.26.

the rabbit, that rely primarily on microbial breakdown of plant fiber in the cecum are called *cecal fermenters*. In other hindgut fermenters, both the cecum and colon are major sites of fiber breakdown by microorganisms.

Anatomical differences among monogastric mammals, foregut fermenters, and hindgut fermenters are summarized in Figure 22.27. Nutritionally, foregut fermenters have an advantage over hindgut fermenters in that the cellulolytic microbial community of the foregut eventually passes through an acidic stomach. As this occurs, most microbial cells are killed by the acidity and become a protein source for the animal. By contrast, in animals such as horses and rabbits, the remains of the cellulolytic community pass out of the animal in the feces because of its position posterior to the acidic stomach.

MINIQUIZ --

• How do animals with foregut and hindgut fermentation differ in recovery of nutrients from plants?

• How does retention time affect microbial digestion of food in a gut compartment?

22.7 The Rumen and Ruminant Animals

A very successful group of foregut fermenters are *ruminants*, herbivorous mammals that possess a special digestive organ, the **rumen**, within which cellulose and other plant polysaccharides are digested by microorganisms. Some of the most important domesticated animals—cows, sheep, and goats—are ruminants. Camels, buffalo, deer, reindeer, caribou, and elk are also ruminants. Indeed, ruminants are Earth's dominant herbivores. Because the human food economy depends to a great extent on ruminant animals, rumen microbiology is of considerable economic significance and importance.

Rumen Anatomy and Activity

Unique features of the rumen as a site of cellulose digestion are its relatively large size (capable of holding 100–150 liters in a cow, 6 liters in a sheep) and its position in the gastrointestinal system *before* the acidic stomach. The rumen's warm and constant temperature (39°C), narrow pH range (5.5–7, depending on when the animal was last fed), and anoxic environment are also important factors in overall rumen function.

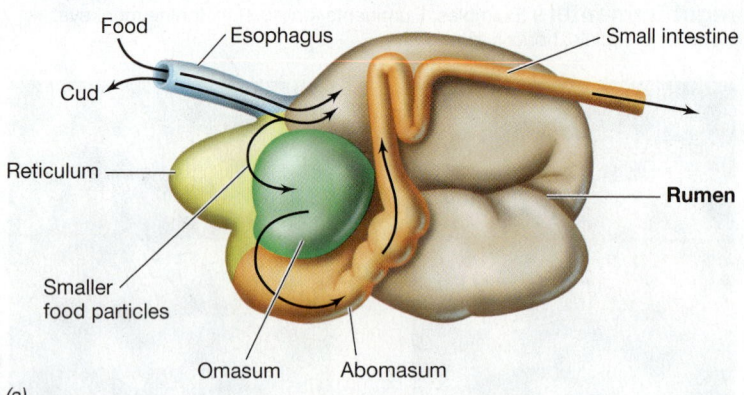

(a)

(b)

Figure 22.28 **The rumen.** (a) Schematic diagram of the rumen and gastrointestinal system of a cow. Food travels from the esophagus into the reticulo-rumen, consisting of the reticulum and rumen. Cud is regurgitated and chewed until food particles are small enough to pass from the reticulum into the omasum, abomasum, and intestines, in that order. The abomasum is an acidic vessel, analogous to the stomach of monogastric animals like pigs and humans. (b) Photo of a fistulated Holstein cow. The fistula, shown unplugged, is a sampling port that allows access to the rumen.

Figure 22.28a shows the relationship of the rumen to other parts of the ruminant digestive system. The digestive processes and microbiology of the rumen have been well studied, in part because it is possible to implant a sampling port, called a *fistula*, into the rumen of a cow (Figure 22.28b) or a sheep and remove samples for analysis.

After a cow swallows its food, the food enters the first chamber of the four-compartment stomach, the reticulum. Partially digested plant materials flow freely between the rumen and reticulum, sometimes referred to together as the *reticulo-rumen*. The main function of the reticulum is to collect smaller food particles and move them to the omasum. Larger food particles (called cud) are regurgitated, chewed, mixed with saliva containing bicarbonate, and returned to the reticulo-rumen, where they are digested by rumen bacteria. Solids may remain in the rumen for more than a day during digestion. Eventually, small and more thoroughly digested food particles are passed to the omasum and from there to the abomasum, an organ similar to a true, acidic stomach. In the abomasum, chemical digestive processes begin that continue in the small and large intestine.

Microbial Fermentation in the Rumen

Food remains in the rumen for 20–50 h depending on the feeding schedule and other factors. During this relatively long retention time, cellulolytic microorganisms hydrolyze cellulose, which frees glucose. The glucose then undergoes bacterial fermentation with the production of **volatile fatty acids (VFAs)**, primarily *acetic*, *propionic*, and *butyric* acids, and the gases carbon dioxide (CO_2) and methane (CH_4) (**Figure 22.29**). The VFAs pass through the rumen wall into the bloodstream and are oxidized by the animal as its main source of energy. The gaseous fermentation products CO_2 and CH_4 are released by eructation (belching).

The rumen contains enormous numbers of bacteria (10^{10}–10^{11} cells/g of rumen contents). Most of the bacteria adhere tightly to food particles. These particles proceed through the gastrointestinal tract of the animal where they undergo further digestive processes similar to those of nonruminant animals. Bacterial cells that digested plant fiber in the rumen are themselves digested in the acidic abomasum. Because bacteria living in the rumen biosynthesize amino acids and vitamins, the digested bacterial cells are a major source of protein and vitamins for the animal.

Rumen Bacteria

Although some anaerobic microbial eukaryotes are present, anaerobic bacteria dominate in the rumen because it is a strictly anoxic compartment. Cellulose is converted to fatty acids, CO_2, and CH_4 in a multistep microbial food chain, with several different

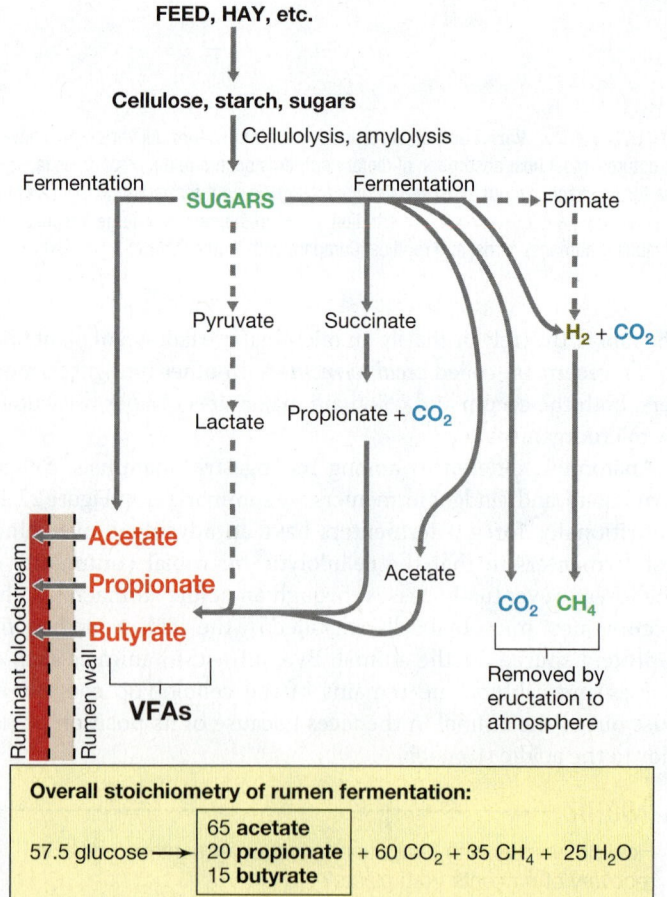

Overall stoichiometry of rumen fermentation:

57.5 glucose → | 65 **acetate** / 20 **propionate** / 15 **butyrate** | + 60 CO_2 + 35 CH_4 + 25 H_2O

Figure 22.29 **Biochemical reactions in the rumen.** The major pathways are solid lines; dashed lines indicate minor pathways. Approximate steady-state rumen levels of volatile fatty acids (VFAs) are acetate, 60 mM; propionate, 20 mM; butyrate, 10 mM.

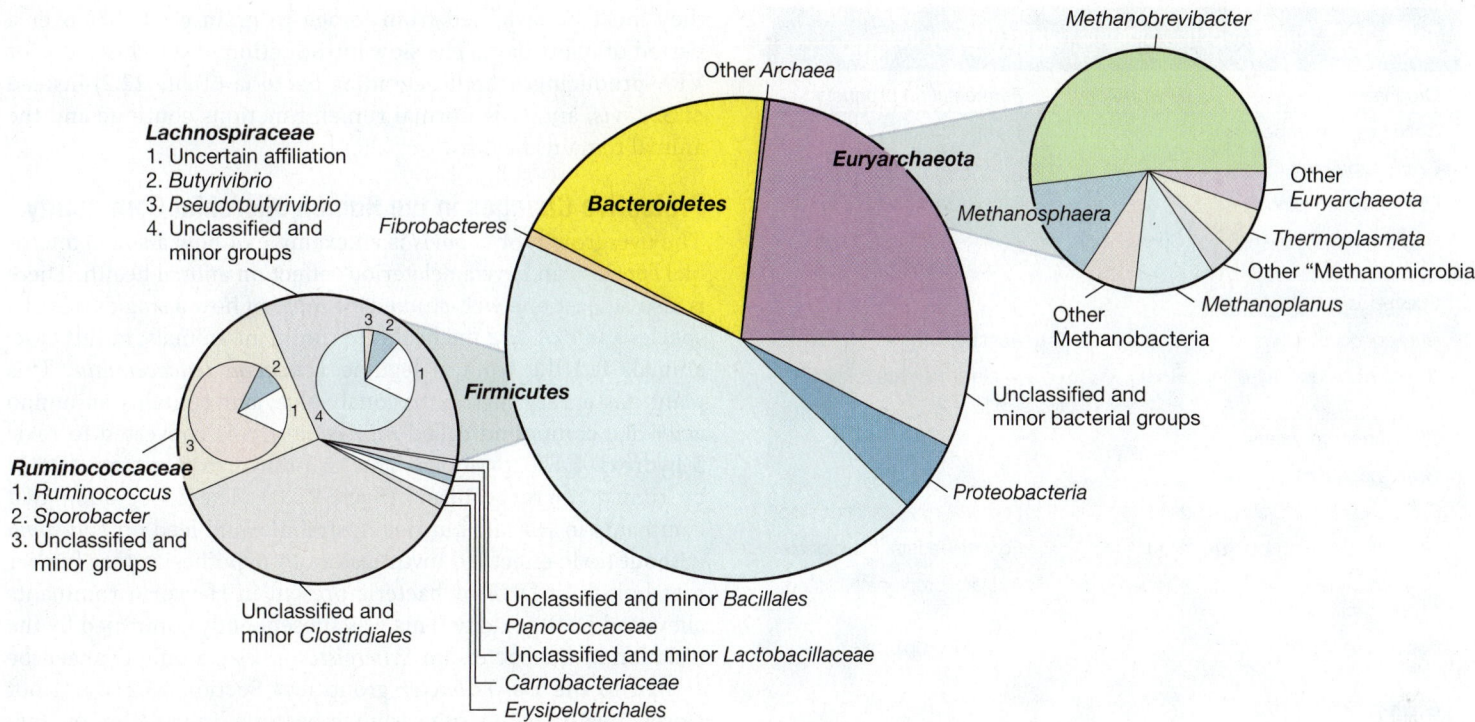

Figure 22.30 Ruminal microbial community inferred from 16S rRNA gene sequences. The results are pooled analyses of 14,817 sequences from several studies of ruminant animals, including cow, sheep, goat, and deer. They provide information primarily on diversity, not relative abundance. Data assembled and analyzed by Nicolas Pinel.

anaerobes participating in the process. Recent estimates of rumen microbial diversity from analysis of 16S rRNA gene sequences suggest that the typical rumen contains 300–400 bacterial "species" (defined as "operational taxonomic units" sharing less than 97% sequence identity, ↺ Section 12.8) (**Figure 22.30**). This is more than 10 times higher than culture-based diversity estimates. Molecular surveys show that species of *Firmicutes* and *Bacteroidetes* dominate the *Bacteria* in the rumen, while methanogens make up virtually the entire archaeal population (Figure 22.30).

A number of rumen anaerobes have been cultured and their physiology characterized (**Table 22.2**). Several different rumen bacteria hydrolyze cellulose to sugars and ferment the sugars to VFAs. *Fibrobacter succinogenes* and *Ruminococcus albus* are the two most abundant cellulolytic rumen anaerobes. Although both organisms produce cellulases, *Fibrobacter*, a gram-negative bacterium, produces enzymes localized to the outer membrane. *Ruminococcus*, a gram-positive bacterium which therefore lacks an outer membrane, produces a cellulose-degrading protein complex stabilized by scaffold proteins and bound to the cell wall. Both organisms therefore need to bind to cellulose particles in order to degrade them.

If a ruminant is gradually switched from cellulose to a diet high in starch (grain, for instance), the starch-digesting bacteria *Ruminobacter amylophilus* and *Succinomonas amylolytica* grow to high numbers in the rumen. On a low-starch diet these organisms are typically minor constituents. If an animal is fed legume hay, which is high in pectin, a complex polysaccharide containing both hexose and pentose sugars, then the pectin-digesting bacterium *Lachnospira multipara* (Table 22.2) becomes an abundant member of the rumen microbial community. Some

of the fermentation products of these rumen bacteria are used as energy sources by secondary fermenters in the rumen. For example, succinate is fermented to propionate plus CO_2 (Figure 22.29) by the bacterium *Schwartzia*, and lactate is fermented to acetate and other fatty acids by *Selenomonas* and *Megasphaera* (Table 22.2). Hydrogen (H_2) produced in the rumen by fermentative processes never accumulates because it is quickly consumed by methanogens for the reduction of CO_2 to CH_4. H_2 removal facilitates greater fermentative activity since H_2 accumulation negatively affects the energetics of fermentative reactions that produce H_2 (↺ Section 13.15).

Dangerous Changes in the Rumen Microbial Community

Significant changes in the microbial composition of the rumen can cause illness or even death of the animal. For example, if a cow is changed abruptly from forage to a grain diet, the gram-positive bacterium *Streptococcus bovis* grows rapidly in the rumen. The normal level of *S. bovis*, about 10^7 cells/g, is an insignificant fraction of total rumen bacterial numbers. But if large amounts of grain are fed abruptly, numbers of *S. bovis* can quickly rise to dominate the rumen microbial community to over 10^{10} cells/g. This occurs because grasses contain mainly cellulose, which does not support growth of *S. bovis*, while grain contains high levels of starch, on which *S. bovis* grows rapidly.

Because *S. bovis* is a lactic acid bacterium (↺ Sections 13.12 and 15.6), large populations are capable of producing large amounts of lactic acid. Lactic acid is a much stronger acid than the VFAs produced during normal rumen function. Lactate production thus acidifies the rumen below its lower functional limit of about pH 5.5, thereby disrupting the activities of normal rumen bacteria. Rumen

Table 22.2 Characteristics of some rumen prokaryotes

Organism[a]	Morphology	Fermentation products
Cellulose decomposers		
Gram-negative		
Fibrobacter succinogenes[b]	Rod	Succinate, acetate, formate
Butyrivibrio fibrisolvens[c]	Curved rod	Acetate, formate, lactate, butyrate, H_2, CO_2
Gram-positive		
Ruminococcus albus[c]	Coccus	Acetate, formate, H_2, CO_2
"*Clostridium lochheadii*"	Rod (endospores)	Acetate, formate, butyrate, H_2, CO_2
Starch decomposers		
Gram-negative		
Prevotella ruminicola[d]	Rod	Formate, acetate, succinate
Ruminobacter amylophilus	Rod	Formate, acetate, succinate
Selenomonas ruminantium	Curved rod	Acetate, propionate, lactate
Succinomonas amylolytica	Oval	Acetate, propionate, succinate
Gram-positive		
Streptococcus bovis	Coccus	Lactate
Lactate decomposers		
Gram-negative		
Selenomonas ruminantium subsp. *lactilytica*	Curved rod	Acetate, succinate
Megasphaera elsdenii	Coccus	Acetate, propionate, butyrate, valerate, caproate, H_2, CO_2
Succinate decomposer		
Gram-negative		
Schwartzia succinovorans	Rod	Propionate, CO_2
Pectin decomposer		
Gram-positive		
Lachnospira multipara	Curved rod	Acetate, formate, lactate, H_2, CO_2
Methanogens		
Methanobrevibacter ruminantium	Rod	CH_4 (from H_2 + CO_2 or formate)
Methanomicrobium mobile	Rod	CH_4 (from H_2 + CO_2 or formate)

[a]Except for the methanogens, which are *Archaea*, all organisms listed are species of *Bacteria*.
[b]These species also degrade xylan, a major plant cell wall polysaccharide.
[c]Also degrades starch.
[d]Also ferments amino acids, producing NH_3. Several other rumen bacteria ferment amino acids as well, including *Peptostreptococcus anaerobius* and *Clostridium sticklandii*.

acidification, a condition called *acidosis,* causes inflammation of the rumen epithelium, and severe acidosis can cause hemorrhaging in the rumen, acidification of the blood, and death of the animal.

Despite the activities of *S. bovis,* ruminants such as cattle can be fed a diet exclusively of grain. However, to avoid acidosis, they must be switched from forage to grain *gradually* over a period of many days. The slow introduction of starch selects for VFA-producing, starch-degrading bacteria (Table 22.2) instead of *S. bovis,* and thus normal rumen functions continue and the animal remains healthy.

Protective Changes in the Rumen Microbial Community

The overgrowth of *S. bovis* is an example of how a single microbial species can have a deleterious effect on animal health. There is also at least one well-studied example of how a single bacterial species can *enhance* the health of ruminant animals; in this case, animals fed the tropical legume *Leucaena leucocephala*. This plant has a very high nutritional value, but contains an amino acid–like compound called *mimosine* that is converted to toxic 3-hydroxy-4(1H)-pyridone and 2,3-dihydroxypyridine (DHP) by rumen microorganisms (Figure 22.31). The observation that ruminants in Hawaii, but not Australia, could feed on *Leucaena* without toxic effect led investigators to hypothesize that further metabolism of DHP by bacteria present in Hawaiian ruminants alleviated DHP toxicity. This was subsequently confirmed by the isolation of the bacterium *Synergistes jonesii*, a unique anaerobe related to the *Deferribacter* group (↩ Section 15.21) and not closely related to any other rumen bacteria. Inoculation of Australian ruminants with cells of *S. jonesii* conferred resistance to mimosine by-products, allowing the animals to feed on *Leucaena* without ill effect.

The success of this single-organism modification of the rumen microbial community has encouraged further studies of this sort, including genetic engineering of bacteria to improve their ability to utilize available nutrients or to detoxify toxic substances. A notable success has been inoculation of the rumen of sheep with genetically engineered cells of *Butyrivibrio fibrisolvens* (Table 22.2) containing a gene encoding the enzyme fluoroacetate dehalogenase; this successfully prevented fluoroacetate poisoning of sheep fed plants containing high levels of this highly toxic inhibitor of the citric acid cycle.

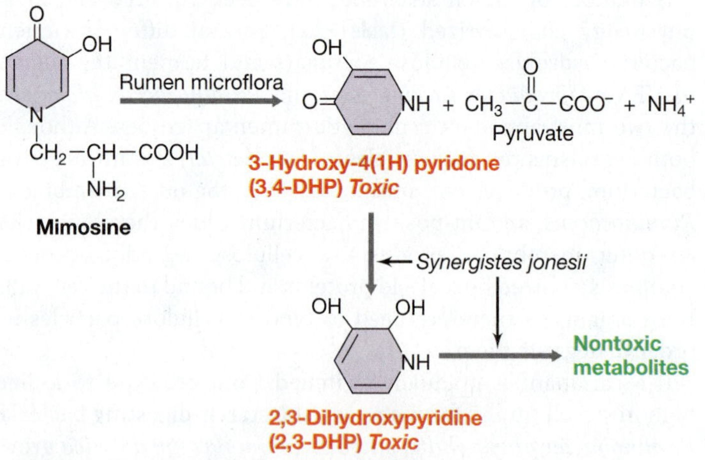

Figure 22.31 Conversion of mimosine to toxic pyridine and pyridone metabolites by ruminal microorganisms. Mimosine is converted to toxic 3,4-DHP by normal ruminal microbiota. *Synergistes jonesii* converts 3,4-DHP to nontoxic metabolites through a 2,3-DHP intermediate, preventing buildup of toxic metabolites of mimosine.

Rumen Protists and Fungi

In addition to huge populations of prokaryotes, the rumen has characteristic populations of ciliated protists (Chapter 17) present at a density of about 10^6 cells/ml. Many of these protists are obligate anaerobes, a property that is rare among eukaryotes. Although these protists are not essential for rumen fermentation, they contribute to the overall process. In fact, some protists are able to hydrolyze cellulose and starch and ferment glucose with the production of the same VFAs formed by cellulose-fermenting bacteria (Figure 22.29 and Table 22.2). Rumen protists also consume rumen bacteria and smaller rumen protists and are likely to play a role in controlling bacterial densities in the rumen. An interesting commensal interaction has been observed between rumen protists that produce VFAs and H_2 as products and methanogenic bacteria that consume the H_2, producing CH_4. Because their cells autofluoresce (Section 13.20), methanogens are easily observed in rumen fluid bound to the surface of H_2-producing protists.

Anaerobic fungi also inhabit the rumen and play a role in its digestive processes. Rumen fungi are typically species that alternate between a flagellated and a thallus form, and studies with pure cultures have shown that they can ferment cellulose to VFAs. *Neocallimastix*, for example, is an obligately anaerobic fungus that ferments glucose to formate, acetate, lactate, ethanol, CO_2, and H_2. Although a eukaryote, this fungus lacks mitochondria and cytochromes and thus lives an obligately fermentative existence. However, *Neocallimastix* cells contain a redox organelle called the *hydrogenosome*; this mitochondrial analog evolves H_2 and has otherwise been found only in certain anaerobic protists (Section 2.21).

Rumen fungi play an important role in the degradation of polysaccharides other than cellulose, including a partial solubilization of lignin (the strengthening agent in the cell walls of woody plants), hemicellulose (a derivative of cellulose that contains pentoses and other sugars), and pectin.

MINIQUIZ

- What physical and chemical conditions prevail in the rumen?
- What are VFAs and of what value are they to the ruminant?
- Why is the metabolism of *Streptococcus bovis* of special concern to ruminant nutrition?

22.8 The Human Microbiome

The human microbiome encompasses all sites of the human body inhabited by microorganisms. These sites include the mouth, nasal cavities, throat, stomach, intestines, urogenital tracts, and skin (Sections 23.1–23.5). It is estimated that the number of microorganisms in the human microbiome is approximately 10^{14}, which is ten times more than the total number of human cells in a single person.

Importance to Human Health

The gut microbial community in the healthy human was once considered to consist of microorganisms that were merely commensals, but we now know that this community is important in early development and overall health and predisposition to disease.

Recognition that the gut microflora function as mutualists that play a central role in human health has spawned two major international research programs. One, based in the United States, is called the *Human Microbiome Project* (HMP), and a second major project, supported by the European Commission, is called the *Metagenomics of the Human Intestinal Tract* (MetaHIT).

To date, the HMP has examined the microbial diversity in 250 healthy volunteers in two American cities by sequencing over 5000 samples taken from each subject one to three times and from 15 to 18 body sites (nine oral, four skin, one nasal, one stool, three vaginal). These and other ongoing studies of the human microbiome—including its relationship to disease, ethnicity, and diet—are coordinated under the International Human Genome Consortium. Some of the major questions posed by these integrated projects include: (1) Do individuals share a core human microbiome? (2) Is there a correlation between microbial population structure and host genotype? (3) Do differences in the human microbiome correlate with differences in human health? (4) Are differences in the relative abundance of different bacteria important?

Human microbiome studies based on surveys of the human gut microflora using 16S rRNA gene sequencing and metagenomic analyses have shown that the diversity between different persons is so great that no one microbial species is found at high abundance in all individuals. Similarities among individuals are more evident at higher bacterial taxonomic levels (such as phyla) and in the distribution of genes of similar function in the gut community. The possible benefits of these analyses for clinical medicine include the development of biomarkers for predicting predisposition to specific diseases, the design of drugs targeting selected members of the intestinal microbial community, personalized drug therapies, and tailor-made probiotics (Section 23.4).

The Human Gut Microbial Community

Humans are monogastric and omnivorous animals (Figure 22.27). In the human duodenum, ingested food passed down from the stomach is blended with bile, bicarbonate, and digestive enzymes. About 1–4 h after ingestion, food reaches the gut (large intestine) and by this time it is near neutral pH, and bacterial numbers have increased from about 10^4/g to about 10^8/g (**Figure 22.32**). Both the host and its gut microorganisms share the easily digestible nutrients. The large intestine is the most heavily colonized area of the gastrointestinal tract and contains 10^{11}–10^{12} bacterial cells per g.

Colonization of an initially sterile gut begins immediately after birth; a succession of microbial populations replaces each other in turn until a stable, adult microbial community is established. The source of early colonizers is not clear, although some species are clearly transmitted from mother to infant. The infant gut community is dominated by bifidobacteria, which are fermentative species of the bacterial class *Actinobacteria* (Section 15.10) and does not reach an adultlike composition until about age 3. Recent studies have correlated frailty in the elderly population with two major factors: (1) a general decrease in gut diversity, and (2) reduced levels of *Firmicutes* and increased *Bacteroides*. This second factor also correlates with the lower amounts of free glutarate and the anti-inflammatory short-chain fatty acid butyrate in the elderly.

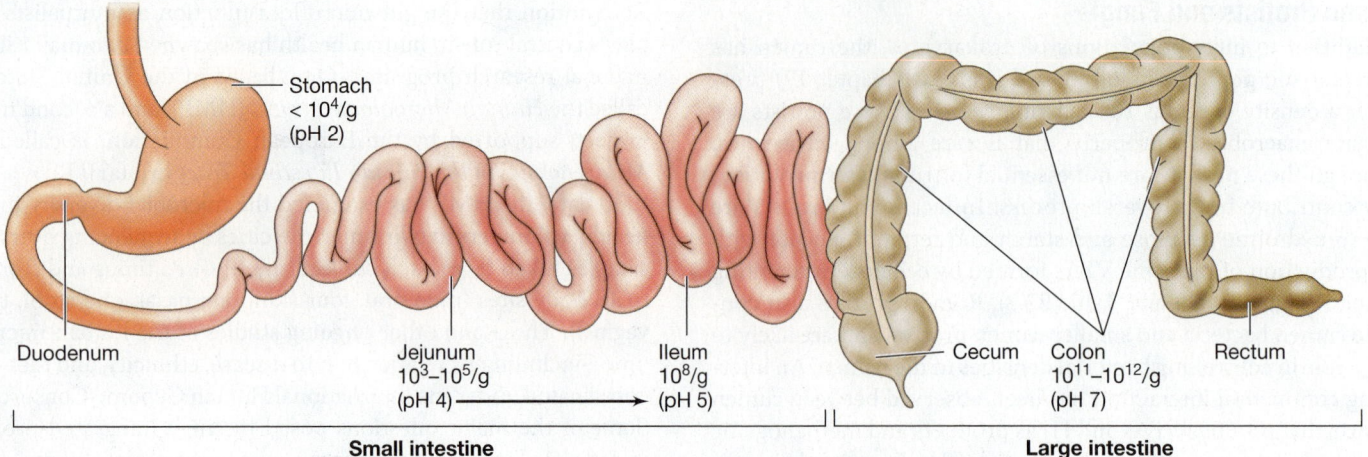

Figure 22.32 **Numbers of bacteria in the monogastric human gastrointestinal tract.** The small intestine is composed of the duodenum, jejunum, and ileum. Numbers in the individual sections are estimates of bacteria per gram of intestinal contents in healthy humans.

As is now recognized for most microbial communities, early descriptions of diversity based on culturing microorganisms greatly underestimated true diversity. For example, although we think of *Escherichia coli* as a significant gut bacterium, the entire phylum *Gammaproteobacteria* (to which *E. coli* belongs; ↩ Section 15.3) makes up less than 1% of all gut bacteria. *E. coli* simply grows extremely well in laboratory culture and can thus be readily detected even when present in low numbers.

Somewhat surprisingly, mammalian gut communities are composed of only a few major phyla and show a species composition distinct from that of any free-living microbial communities (Chapter 19). The vast majority (~98%) of all human gut phylo-types fall into one of four major bacterial phyla: *Firmicutes, Bacteroidetes, Proteobacteria,* and *Actinobacteria* (**Figure 22.33**). The *Bacteroidetes* and *Firmicutes* phyla dominate, but vary tremendously in relative abundance among individuals—differences in individual abundance range from >90% *Bacteroidetes* to >90% *Firmicutes*. In contrast to the limited phylum-level diversity, the species diversity in the mammalian gut is enormous. The most recent census of diversity in human fecal samples, based on millions of 16S rRNA sequences, has identified between 3,500 and 35,000 "species." This large difference is mainly due to whether the 16S rRNA similarity threshold for defining a species (↩ Section 12.8) is set at the 97% sequence identity cut-off or is more stringent

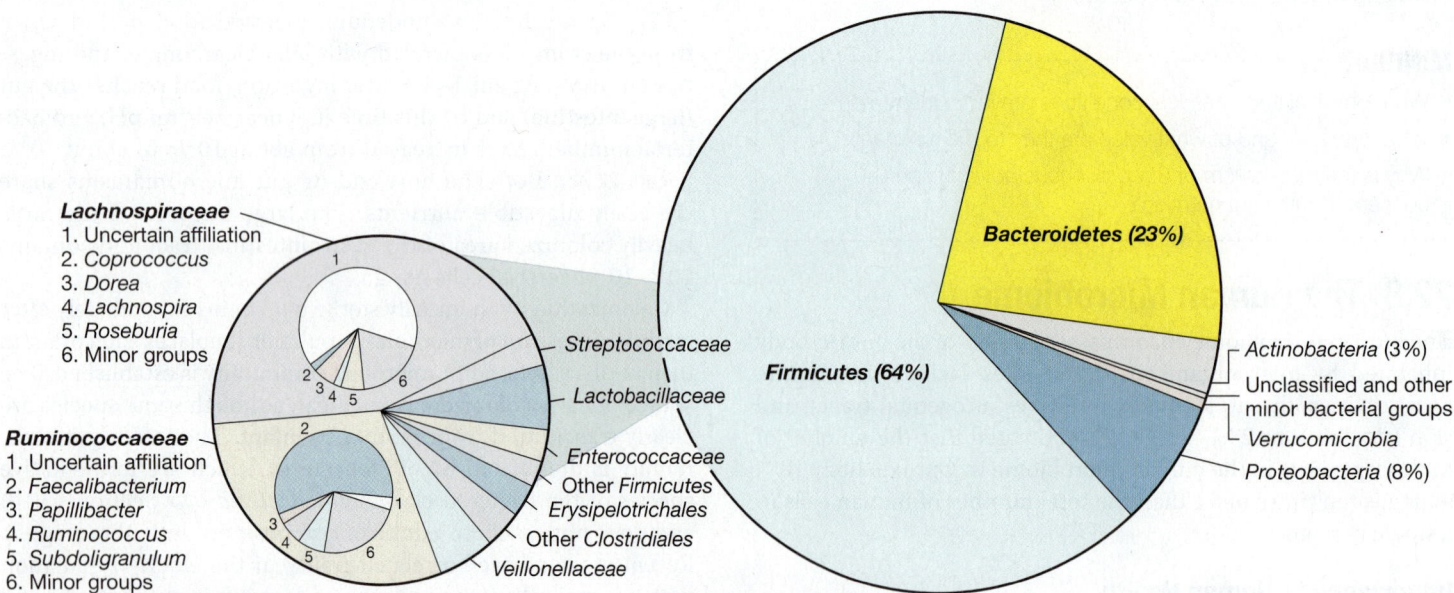

Figure 22.33 **Microbial composition of the human colon inferred from 16S rRNA gene sequences.** The results are pooled analyses of 17,242 sequences mostly obtained from the distal colon (fecal samples) of several individuals. The data provide information primarily of diversity, not relative abundance. Studies of abundance patterns have shown that the *Firmicutes–Bacteroidetes* ratio is highly variable among individuals. Data assembled and analyzed by Nicolas Pinel.

(~98–99% identity). *Archaea* (represented by a phylotype closely related to the methanogen *Methanobrevibacter smithii*), yeasts, fungi, and protists make up only a minor part of the human gut community (compare this with the rumen, Section 22.7).

Comparative studies have also shown that humans share more genera with each other than with other species of mammals. This suggests that the mammalian gut microflora may be "fine-tuned" to each mammalian species. Interestingly, although there is high variability from person to person in gut community composition, an individual's community is relatively stable over long periods. Also, ongoing metagenomic sequencing studies hint at the existence of a limited number of distinct well-balanced general types of gut communities. Three such gut communities, called *enterotypes*, have been proposed. The association of an individual with any one enterotype transcends national borders, nutritional background, and ethnicity. Metabolic pathway reconstructions based on annotation of metagenomic gene sequences suggests that enterotypes are functionally distinct, for example, differing in their capacity for vitamin production. The existence of alternative symbiotic states, reflected by enterotypes, suggests that an individual's enterotype might influence that person's response to diet and drug therapy, or health status in general. If true, this could bring exciting new concepts and practices into the field of clinical medicine.

Contribution of Gut Microorganisms to Human Metabolism

Human gut microorganisms synthesize a large variety of enzymes that allow for the processing of complex dietary carbohydrates into monosaccharides and the production of VFAs. The genomes of *Bacteroides* species common in human adults encode enzymes that catabolize polysaccharides, consistent with these bacteria being adapted to a gut environment rich in polysaccharides. Even though the *Bacteroidetes–Firmicutes* ratio varies tremendously among individuals, the gut community metagenome (⮌ Section 6.10) presents a very similar set of genes related to complex carbohydrate degradation. Gut microorganisms also function in nitrogen metabolism. Of the 20 amino acids that humans require, 10 are said to be essential nutrients because we cannot synthesize them in adequate amounts. Although we obtain essential amino acids, such as lysine, from food, these nutrients may also be produced and secreted by certain gut microorganisms. For example, the infant microbiome has a higher level of folate-producing enzymes than that of adults, possibly associated with the ability of adults to obtain folate (an essential vitamin) from a more complex diet.

Gut microorganisms are also known to contribute to the "maturing" of the gastrointestinal tract. This includes triggering the expression of genes whose products catalyze nutrient uptake and metabolism in gut epithelial cells, priming the immune system early in life to recognize the normal gut microflora as nonforeign, and the development of a mucosal barrier to prevent colonization by foreign bacteria. Studies of experimental colonization of germ-free mice with individual microbial species or microbial communities have demonstrated that colonization triggers the expression of genes for glucose uptake and lipid absorption and transport in the ileum. This also indicates that there may be a link between gut microbial composition and the ability of the host to harvest energy from its diet, contributing to nutritional abnormalities such as obesity, and we focus on this now.

Role of Gut Microorganisms in Obesity

Obesity is a significant health risk that contributes to high blood pressure, cardiovascular disease, and diabetes. Gut microorganisms likely play a part in human obesity, although mechanisms remain hypothetical. Initial evidence relating gut microorganisms to host fat accumulation came from studies using germ-free mice. In these experiments, normal mice had 40% more total body fat than those raised under germ-free conditions, although both mouse populations were fed the same rations. After germ-free mice were inoculated with cecal material from a normal mouse, they developed a gut microflora and their total body fat increased although there had been no changes in food intake or energy expenditure.

Mice that are genetically obese have microbial gut communities that differ from those of normal mice, with 50% fewer *Bacteroidetes*, a proportional increase in *Firmicutes*, and a greater number of methanogenic *Archaea* (**Figure 22.34**). Methanogens are thought to increase the efficiency of microbial conversion of fermentable substrates by removing hydrogen (H_2), as mentioned for fermentation in the rumen (Section 22.7). Hydrogen removal should stimulate fermentation, making more nutrients available for absorption by the host and thus contributing to obesity.

Animal model inferences have been more difficult to demonstrate with human subjects, since strict control of diet and host genotype is not feasible, and gut flora manipulation is much more difficult to achieve. Nevertheless, studies of humans, while not strictly confirming the *Bacteroidetes–Firmicutes* relationship established in mice, have shown that obese individuals are more likely to harbor species of *Prevotella* (a genus of *Bacteroidetes*) and methanogenic *Archaea*. Thus, the general model in humans appears to be the same as that for the mouse (Figure 22.34). That is, the methanogens are proposed to remove H_2 produced by *Prevotella*, facilitating fermentation by *Prevotella* and increasing short-chain fatty acids' availability to the host. This general model is also supported by the study of mice co-colonized with *Bacteroides thetaiotaomicron* (having a metabolism similar to *Prevotella*) and the methanogen *Methanobrevibacter smithii*. Relative to monocolonized controls, these mice have a higher number of total gut bacteria, higher acetate levels in the intestinal lumen and blood, and greater body fat. The discovery that gut microflora can affect obesity offers at least one nongenetic explanation for why obesity often "runs in families."

Remarkably, the increase in body fat associated with pregnancy may also be influenced by the gut microbiota. The period between the first and third trimester of pregnancy is associated with a decrease in gut microbial diversity and enrichment in the gut community of species of *Proteobacteria* and *Actinobacteria*. These changes are associated with the increased body fat and insulin insensitivity that develop later in gestation. A simple interpretation of these findings is that a pregnant woman's body manipulates her gut microbiome as part of its preparations for a greater demand on stored energy reserves.

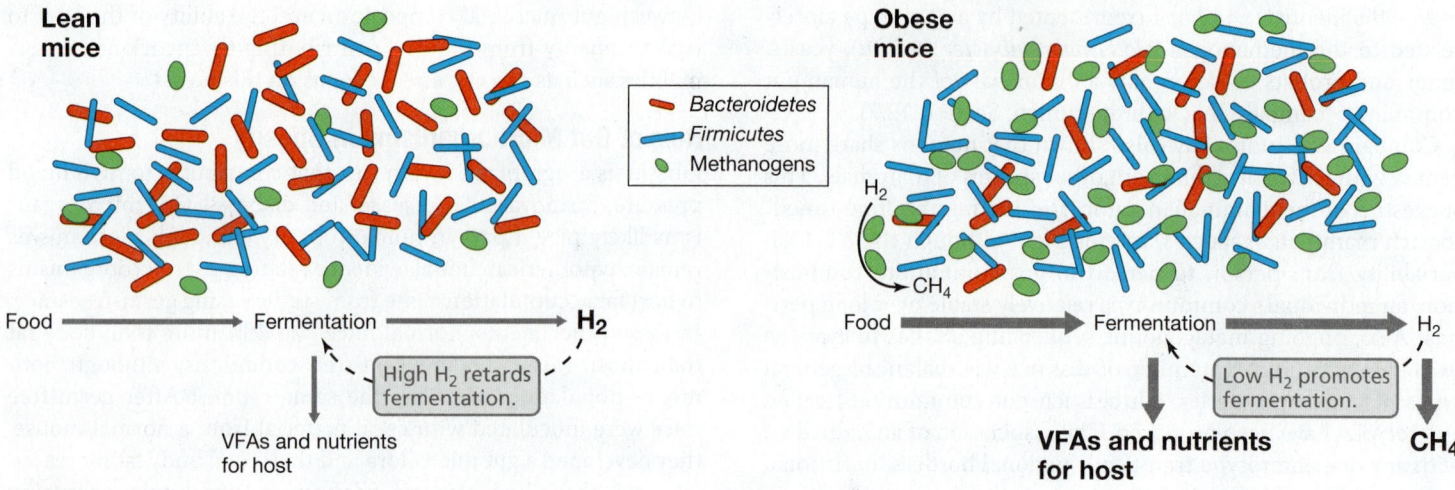

Figure 22.34 Differences in gut microbial communities between lean and obese mice. Obese mice have more methanogens, a 50% reduction in *Bacteroidetes*, and a proportional phylum-wide increase in *Firmicutes*. Nutrient production from fermentation is higher in obese mice due to removal of H₂ by methanogens.

Microbial Communities in the Human Mouth and Skin

Besides the gut, the mouth and skin are also sites heavily colonized by microorganisms. The oral microbiome is essentially as diverse as the gut, but individuals share greater proportions of common taxa for the mouth than for the gut. Abundant genera included *Streptococcus*, *Haemophilus*, *Veillonella*, *Actinomyces*, and *Fusobacterium*. As for all microbial communities reexamined by molecular methods, 16S rRNA-based sequence surveys of the oral cavity have shown that culture-based methods provided a very incomplete census of diversity. At least 750 species of aerobic and anaerobic microorganisms, including a minor representation of methanogenic *Archaea* and yeast, are known to reside in the oral cavity, distributed among teeth, tissue surfaces, and saliva. Because of the high species diversity, current research is focused on those genera having the largest representation in healthy adults.

The oral cavity provides a variety of habitats, each colonized by species that are present primarily as biofilms (⮌ Section 19.4). The primary colonizers of clean tooth surfaces are species of *Streptococcus*; obligate anaerobes such as *Veillonella* and *Fusobacterium* colonize habitats below the gum line. Most of these organisms contribute to the health of the host by keeping pathogenic species in check and not allowing them to adhere to mucosal surfaces. Tooth decay, gum inflammation, and periodontal disease are among the most visible manifestations of a breakdown in these generally stable mutualisms. We discuss the normal microbial community of the oral cavity in more detail in Section 23.3.

The skin is a critical human organ functioning primarily to prevent loss of moisture and restrict the entry of pathogens. Skin is also part of the human microbiome. Although total microbial numbers are typically low relative to the oral and gut communities, molecular analyses have shown that the skin harbors a rich and diverse microbial community of bacteria and fungi (primarily yeast) that vary significantly with location on the body. A 16S rRNA sequencing comparison of twenty diverse skin sites categorized as moist, dry, or oily, revealed tremendous diversity and variation among sites and individuals, but also showed some

common patterns. All together, nearly 20 bacterial phyla were detected, but most sequences affiliated with four groups: *Actinobacteria* (52%), *Firmicutes* (24%), *Proteobacteria* (16%), and *Bacteroidetes* (6%). More specific coverage of the normal flora of human skin can be found in Section 23.2 and Figure 23.2.

Changes in the Human Microbiome Associated with Disease

Changes in the human microbiome have long been associated with inflammatory bowel disease (IBD), the chronic inflammation of all or part of the digestive tract. It is widely accepted that IBD is not caused by a specific pathogen, but rather an imbalance between the immune system and the normal gut microbiota. This type of disruption of the homeostasis between microbiota and the host is called **dysbiosis**.

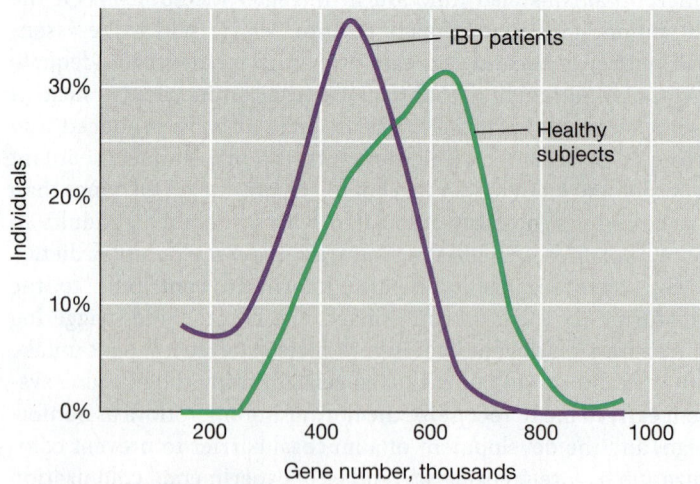

Figure 22.35 Reduced functional capacity of the gut microbiome of patients with inflammatory bowel disease. Metagenomic analysis of human gut microbiota in healthy subjects and patients with inflammatory bowel disease (IBD) revealed a tendency toward fewer nonredundant bacterial genes in patients with IBD.

Mice models point to a complex but transmissible etiology of IBD. Fostering or co-caging healthy mice with IBD-predisposed mice was sufficient to cause IBD development in the healthy mice and was correlated with the transfer of *Enterobacteriaceae* species *Klebsiella pneumoniae* and *Proteus mirabilis* from the IBD mice to the healthy mice. However, as for the relationship between the gut microbiome and obesity, IBD causality in humans is less well understood. Metagenomic analyses of healthy subjects and patients with IBD showed that the gut microbiota of IBD patients shared fewer genes in common with healthy subjects, relative to the number of genes shared among healthy subjects. The microbial community of IBD patients also tended to have significantly reduced functional capacity, as reflected by a reduction in the number of nonredundant genes relative to healthy subjects (**Figure 22.35**).

Other conditions known to be associated with changes in the human microbiome include type 2 (non-insulin-dependent) diabetes, asthma, atopic dermatitis, colorectal cancer, kidney stones, periodontitis, and psoriasis. As we learn more about the relationship between the human microbiome and health and disease, therapeutic intervention may well be possible. This might include promoting the growth of protective symbiotic bacteria or inhibiting the growth of individual microorganisms (or assemblages of microorganisms) that compromise health, and transplanting microbial communities from healthy to diseased individuals.

MINIQUIZ

- Which major phyla of *Bacteria* dominate the human gut?
- How might increased numbers of methanogens in the gut contribute to obesity?
- List some practical outcomes of characterizing the human microbiome.

IV • Insects as Microbial Habitats

Insects are the most abundant class of animals living today, with over one million species known. As many as 20% of all insects are thought to support symbiotic microorganisms in a mutually beneficial way. The symbioses contribute to the insects' ecological success by providing them either nutritional advantages or protection. Some symbionts are found on insects' outer surfaces or in their digestive tracts. *Endosymbionts* are intracellular bacteria and are typically localized to specialized organs within the insect.

22.9 Heritable Symbionts of Insects

How symbionts are transferred from one generation to the next determines how a mutualism functions and how stable it is. Microbial symbionts can either be acquired by a host from an environmental reservoir (horizontal transmission) or be transferred directly from the parent to the next generation (*heritable* or *vertical* transmission). The mode of symbiont transmission is related to the specificity and persistence of an association. In general, less specificity is associated with horizontal transmission. In this section we focus only on mutualisms in which the microbial symbiont has no free-living form; that is, the symbionts are transmitted in a vertical fashion.

Types of Heritable Symbionts

All known heritable symbionts of insects lack a free-living replicative stage. Thus, they are *obligate* symbionts. However, although these bacteria require the host for replication, not all hosts are dependent upon the symbiont. Relative to host dependence, heritable symbionts are either *primary symbionts* or *secondary symbionts*. Primary symbionts are required for host reproduction. They are restricted to a specialized region called the **bacteriome** present in several insect groups; within the bacteriome the bacterial cells reside in specialized cells called **bacteriocytes**. Secondary symbionts are not required for host reproduction. Unlike primary symbionts, secondary symbionts are not always present in every individual of a species and are not restricted to particular host tissues.

Secondary symbionts are broadly distributed among insect groups. Like pathogens, they invade different cell types and may live extracellularly within the insect's *hemolymph* (the fluid bathing the body cavity). In insects with bacteriomes, secondary symbionts can invade the bacteriocytes, co-residing with or sometimes displacing the primary symbionts (**Figure 22.36**). However, in order to persist in the insect host, the secondary symbiont must confer some advantage, such as a nutritional advantage or protection from environmental stresses such as heat. For example, whiteflies infected with *Rickettsia* bacteria (⮯ Section 15.1) produce offspring at about twice the rate of uninfected flies, and more offspring survive to adulthood. Secondary symbionts may also provide protection against invasion by pathogens or predators. A *Spiroplasma* bacterium (⮯ Section 15.9), which was first observed in *Drosophila neotestacea* in the 1980s, provides protection against a parasitic nematode worm. In

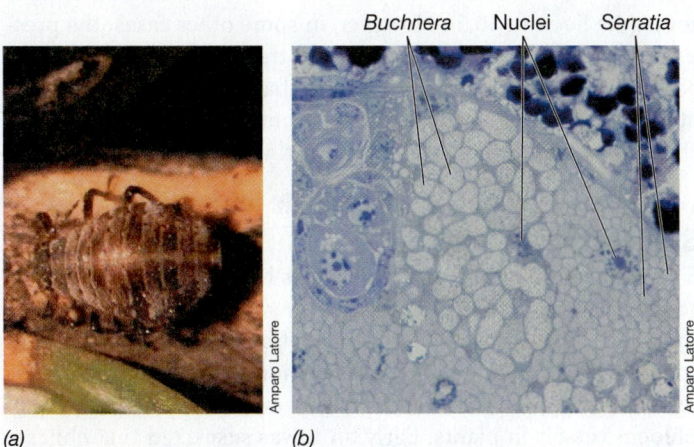

Figure 22.36 Primary and secondary symbionts of an aphid. *(a)* The cedar aphid *Cinara cedri*, a model organism for studies of symbioses. *(b)* Transmission electron micrograph of the bacteriome of *C. cedri* showing two bacteriocytes. Packed within each bacteriocyte are cells of *Buchnera aphidicola* (the primary symbiont) or *Serratia symbiotica*, the smaller, secondary symbiont. Arrows identify the nucleus of each bacteriocyte. The bacteriocyte containing *Buchnera* cells is about 40 μm wide.

most cases the basis for increased fitness or protection is unknown, but in one case a toxin encoded by a lysogenic bacteriophage (⊘ Section 8.8) carried by the symbiont is known to confer protection on the insect from infection by a parasitic wasp.

There are heritable parasitic symbionts that manipulate the host's reproductive system, increasing the frequency of female progeny (sex ratio skewing, ⊘ Figure 15.27). Because most heritable symbionts are transmitted maternally, the suppression of male progeny serves to expand the number of infected individuals and increase the rate of spread through an insect population. Since symbiont-conferred functions can spread rapidly within a population, acquisition of symbiont-encoded traits provides a mechanism for much more rapid adaptation than is possible through mutations in insect genes. *Rickettsia* infection of the whitefly population provides one example of how rapidly symbiont-conferred traits can spread through a population. Only 1% of whiteflies were infected with *Rickettsia* in 2000. In 2006, 97% of flies were infected. In another example, a strain of *Wolbachia* (⊘ Section 15.1) swept through populations of *Drosophila simulans* in California in only 3 years.

An important applied benefit of improved basic understanding of insect symbionts is the increased use of symbionts in insect pest management and the control of vectorborne diseases, such as malaria and filariasis in humans (⊘ Sections 32.5 and 32.7). For example, symbiotic *Wolbachia*, which are reproductive manipulators, are widely distributed among insect species (possibly infecting as many as 60–70% of all insect species). The sperm of *Wolbachia*-infected males can sterilize uninfected females. Although the mechanism for sterilization is not fully understood, the phenomenon is being tested as a means to suppress disease transmission. Release of a large number of *Wolbachia*-infected male *Culex quinquefasciatus* mosquitoes, the vector of the filarial nematode causing elephantiasis (⊘ Section 32.7), in Myanmar (Burma) effectively eliminated the local mosquito population.

In some cases, the presence of the symbiont decreases insect transmission of disease. *Aedes aegypti* mosquitoes infected with *Wolbachia* are less likely to transmit the virus causing dengue fever (⊘ Section 30.5). However, in some other cases, the presence of the symbiont *increases* disease transmission. For example, whiteflies infected with *Hamiltonella* bacteria (a symbiont affiliated with the *Enterobacteriaceae*) are more likely than uninfected flies to transmit tomato yellow leaf curl virus.

Nutritional Significance of Obligate Intracellular Symbionts of Insects

The association of bacteria and insects has allowed many insects to use food resources that are rich in some nutrients, but poor in others. To achieve adequate nutrition, some insects exploit the metabolic potential of their symbionts. For instance, aphids feed on the carbohydrate-rich but otherwise nutrient-poor sap of phloem vessels in plants. Early on it was suspected that obligate symbionts might benefit the insect by providing nutrients not provided by their primary diet, and this is now known to be true.

Molecular analyses have shown that most families of aphids harbor the bacterium *Buchnera* in their bacteriomes (⊘ Section 6.5). The role of *Buchnera* in host nutrition was first indicated by experiments using defined diets to examine the nutrient requirements of aphids. Compared with infected controls, symbiont-free aphids required a diet containing all amino acids that are either lacking or rare in phloem sap. Subsequent genomic studies documented the presence in *Buchnera* of genes encoding the biosynthesis of nine amino acids missing from the sap. There are also examples of synergy between host and symbiont where the synthesis of certain amino acids becomes a joint venture. For example, *Buchnera* lacks the enzyme needed for the last step in leucine biosynthesis, but the necessary gene is present in the aphid's genome. Presumably, this enzyme is made by the aphid and participates in the leucine biosynthetic pathway along with the bacterial enzymes.

A secondary symbiont can also contribute to a joint venture. For example, the *Buchnera* symbiont of the cedar aphid is unable to supply tryptophan to the aphid. Two genes in the tryptophan biosynthetic pathway are present in *Buchnera*, but the remaining genes for the pathway are located on the chromosome of a secondary endosymbiont (Figure 22.36). Thus, different parts of a required metabolic pathway can be encoded by different endosymbionts present in the same insect. The fungus-cultivating ants provide yet another example of a complex symbiosis that has formed between an insect and multiple microorganisms (see Explore the Microbial World, "The Multiple Microbial Symbionts of Fungus-Cultivating Ants").

Mealybugs (*Planococcus citri*) present one of the most unusual examples of a partnership between two symbionts infecting the same insect. Mealybugs have two stable bacterial symbionts, "*Candidatus* Tremblaya princeps" (a *Betaproteobacterium*) and "*Candidatus* Moranella endobia" (a *Gammaproteobacterium*) (the term "*Candidatus*" means that these organisms are not yet in pure culture). These symbionts cooperate in providing essential amino acids to the host missing in its diet, as is true for the symbionts of many sap-feeding insects. However, the Moranella bacterium lives inside of Tremblaya! This is the only known example of a bacterium-within-a-bacterium symbiosis. The highly reduced Tremblaya genome has lost all genes for tRNA synthetases, an essential function either supplied by the host or by the Moranella residing within its cytoplasm.

Genome Reduction and Gene Transfer Events

Common features of primary symbionts are extreme genome reduction (⊘ Table 6.1), high adenine plus thymine content, and accelerated rates of mutation. Genomes of insect symbionts fall within a range from 0.14 to 0.80 Mbp and 16.5 to 33% G+C (Table 22.3). The 0.14-Mb genome of "*Candidatus* Tremblaya princeps" is the smallest genome known for any cell. In contrast, the genomes of related free-living bacteria range from 2 to 8 Mbp with a base composition closer to 50% G+C. Two common types of spontaneous mutation, cytosine deamination and the oxidation of guanosine, if not repaired, change a GC pair to an AT pair (⊘ Section 10.2). Symbionts with reduced genomes have fewer DNA repair enzymes (⊘ Section 10.4) and this likely facilitates a shift over time to genomes of lower G+C content.

The streamlined genomes of insect symbionts have lost genes from most functional categories (Chapter 6) and tend to retain only genes required for host fitness and essential molecular

EXPLORE THE MICROBIAL WORLD

The Multiple Microbial Symbionts of Fungus-Cultivating Ants

The attine ants are an example of an elaborate symbiotic association between multiple microbial species and insect. These ants have established an obligate mutualism with a fungus they cultivate in fungal gardens for food, using small leaf fragments to mulch these gardens. A close symbiotic relationship between ant and fungus was first indicated by the observation that one specific fungus was cultivated by each ant lineage. The ants and their mutualistic fungi can be divided into five agricultural systems, each requiring distinct lineages of ants and fungi. Ants grouped in the "lower attine agriculture" system form associations with specific groups of fungi they capture from the environment. By contrast, the "higher attine agriculture" group cultivates fungi that apparently are no longer capable of existing apart from the ant mutualism.

In addition to the close mutualistic relationship between ant species and the specific fungus they cultivate, this symbiosis is now known to include four other microbial symbionts: a small fungus that is parasitic on the garden fungus, nitrogen-fixing bacteria (⇄ Section 3.17) associated with the garden fungus, an actinobacterium that antagonizes the parasitic fungus, and a black yeast that interferes with the actinobacterium.

The fungus is vertically transmitted between ant generations by colony-founding queens. The queen collects a pellet of fungus prior to her mating flight, storing it in a pouch in the oral cavity. After mating, she uses the fungus pellet to establish a new nest and fungus garden (**Figure 1a**). Nitrogen-fixing *Klebsiella* and *Pantoea* species associated with the fungus enrich the nutritional quality of the garden by adding new nitrogen to the nitrogen-poor leaf growth substrate. A single leaf-cutter ant

(a)   (b)

Figure 1 Attine ants. *(a)* Queen and worker ants in their fungal gardens. *(b)* Mutualism with *Actinobacteria* can cover much of the exoskeleton of workers (white areas).

colony may contribute as much as 1.8 kg of fixed nitrogen per year. This new nitrogen benefits the ant colony and also results in higher overall plant diversity near leaf-cutter colonies.

However, the garden is at risk of being destroyed by a parasitic fungus of the genus *Escovopsis*. To repel the parasitic microfungus, the ant has formed another symbiotic association with an actinobacterium (genus *Pseudonocardia*) that appears as a "waxy bloom" growing on the cuticle of the ant (**Figure 1b**). These bacteria, housed in specialized cuticular modifications on the ant's body, secrete secondary metabolites that inhibit the growth of *Escovopsis*. The *Pseudonocardia* likely receive nourishment from the ant from glandular secretions through pores localized in regions of cuticular modification. Comparative genomic sequencing has revealed

good congruence between the phylogenies of the ants, fungal cultivars, *Escovopsis*, and *Pseudonocardia*, pointing to very specific interactions among microorganisms and ants in this complex symbiosis.

The fourth and final microorganism identified in this symbiosis is a yeast that grows in the same cuticular regions colonized by the bacterium *Pseudonocardia*. This black-pigmented yeast interferes with chemical protection of the garden by stealing nutrients from the *Pseudonocardia*, thereby indirectly reducing its ability to suppress *Escovopsis* growth. The attine ant–microbial symbiosis is thus a complex maze of interactions between ant, fungi, and bacteria. Another example of a symbiotic trio—in this case a plant, animal, and bacterium—was described on page 669.

processes, such as translation, replication, and transcription. Genome reduction implies that the symbionts are reliant on the host for many functions no longer encoded in the symbiont genome (⇄ Section 6.5). For example, in many cases genes needed for the biosynthesis of cell wall components are missing, including lipid A and peptidoglycan, suggesting that the host supplies these functions or that the structures are not required to form stable cells within the bacteriocyte.

There is an interesting genomic contrast between primary symbionts and typical disease-causing bacteria (pathogens). While primary symbionts tend to lose genes encoding proteins

required in *catabolic* pathways, pathogenic bacteria typically retain these, but lose genes for *anabolic* pathways. This reflects their differing relationships with their hosts; the insect symbiont provides the host with essential biosynthetic nutrients while the pathogen obtains important biosynthetic nutrients from the host.

Because genome sequences for a large number of insects and their symbionts are now appearing, microbiologists can begin to evaluate the frequency of gene transfer between them. Horizontal gene transfer is the movement of genetic information across normal mating barriers (Chapters 10 and 12). Although early

693

Table 22.3 Genome features of some endosymbionts of animals[a]

Host	Symbiont (genus)	Genome size (Mbp)	G+C (%)	Genes
Aphid	Heterotroph (*Buchnera*)	0.42–0.62	20–26	362–574
Tsetse fly	Heterotroph (*Wigglesworthia*)	0.70	22	617
Carpenter ant	Heterotroph (*Blochmannia*)	0.71–0.79	27–30	583–610
Sharpshooter	Heterotroph (*Sulcia*)	0.25	22	227
Mealybug	Heterotroph ("*Candidatus* Moranella endobia" Gammaproteobacteria)	0.54	43.5	406
Mealybug	Heterotroph ("*Candidatus* Tremblaya princeps" Betaproteobacteria)	0.14	58.8	121
Clam (*Calyptogena okutanii*)	Sulfur oxidizer (unnamed)	1.0	32	975
Clam (*Calyptogena magnifica*)	Sulfur oxidizer (*Ruthia*)	1.2	34	1248
Tube worm (*Riftia pachyptila*)	Sulfur oxidizer (unnamed)	3.3[b]	NA	NA

[a]All listed symbionts are obligately associated with their hosts, with the exception of the symbiont of *Riftia*, which also has a free-living stage.
[b]The free-living sulfur-oxidizing bacterium *Thiomicrospira crunogena* has a genome significantly smaller (2.4 Mb) than this symbiont.

research demonstrated that DNA of *Wolbachia* bacteria has been transferred to the nuclear genomes of their insect and nematode hosts, inspection of other insect mutualisms for which both host and symbiont genome sequences have become available (e.g., aphid and body louse), indicate that DNA transfer is very rare. This suggests that horizontal transfer is highly variable for reasons yet to be determined.

MINIQUIZ

- What factors stabilize the presence of a secondary insect symbiont?
- What are the consequences of symbiont genome reduction?
- How could you determine if a symbiont and host have experienced a long period of coevolution?

22.10 Termites

Microorganisms are primarily responsible for the degradation of wood and cellulose in natural environments. However, the activities of free-living microbial species have been exploited by certain groups of insects that have established symbiotic associations with protists and bacteria that can digest lignocellulosic materials. Like the rumen of herbivorous animals, the insect gut provides a protective niche for microbial symbionts, and in return, the insect gains access to nutrients derived from an otherwise indigestible carbon source. Termites are among the most abundant representatives of this type of symbiotic alliance.

Termite Natural History and Biochemistry

Microbial symbionts in termites decompose the greater part of cellulose (74–99%) and hemicellulose (65–87%) in the plant material they ingest. In contrast to the insect examples discussed in the previous section, most termites do not harbor *intracellular* bacteria. Instead, the symbiotic bacteria are present in digestive organs (guts) as in the case of mammals.

Termite diets include lignocellulosic plant materials (either intact or at various stages of decay), dung, and soil organic matter (humus). About two-thirds of the terrestrial environment supports one or more termite species, with the greatest representation in tropical and subtropical regions, where termites may constitute as much as 10% of all animal biomass and 95% of soil insect biomass. In savannas, their numbers sometimes exceed $4000/m^2$, and their biomass density $(1–10 \text{ g}/m^2)$ may be higher than that of grazing mammalian herbivores.

Termites are categorized as higher or lower based on their phylogeny, and this classification correlates with different symbiotic strategies. The posterior alimentary tract of *higher* termites (family *Termitidae*, comprising about three-fourths of termite species) contains a dense and diverse community of mostly anaerobic bacteria, including cellulolytic species. In contrast, the *lower* termites harbor diverse populations of both anaerobic bacteria and cellulolytic protists. Bacteria of lower termites participate little or not at all in cellulose digestion; only the protists phagocytize and degrade the wood particles ingested by the termites. The termite itself produces cellulases in the salivary glands or the midgut epithelium, but the relative contributions of microbial and termite enzymes to lignocellulosic breakdown is unknown.

The termite gut consists of a foregut (including the crop and muscular gizzard), a tubular midgut (site of secretion of digestive enzymes and absorption of soluble nutrients), and a relatively large hindgut of about 1 microliter volume (**Figure 22.37**). In lower termites, the hindgut consists primarily of a single chamber, the *paunch* (Figure 22.37*a*). The hindgut of most higher termites is more complex, being divided into several compartments (Figure 22.37*b*). For both higher and lower termites, the hindgut harbors a dense and diverse microbial community and is a major site of nutrient absorption. Acetate and other organic acids are produced during microbial fermentation of carbohydrate in the hindgut, and these products are primary carbon and energy sources for the termite (Figure 22.37*c*). High O_2 consumption by bacteria near the gut wall keeps the interior of the hindgut anoxic. However, microsensor measurements (⇄ Section 18.8) have shown that O_2 can penetrate up to 200 μm into the gut before it is completely removed by microbial respiratory activity. Thus, this tiny gut compartment offers distinct microbial niches with respect to O_2 and can support diverse microbial activities.

Bacterial Diversity and Lignocellulose Digestion in Higher Termites

In termites of different genera, the microbial gut communities differ significantly. Analysis of 16S rRNA gene sequences from hindgut contents of species of the higher termite genus *Nasutitermes* revealed a high diversity of microbial species from 12 phyla of *Bacteria*, but few *Archaea* (**Figure 22.38**). Spirochetes of

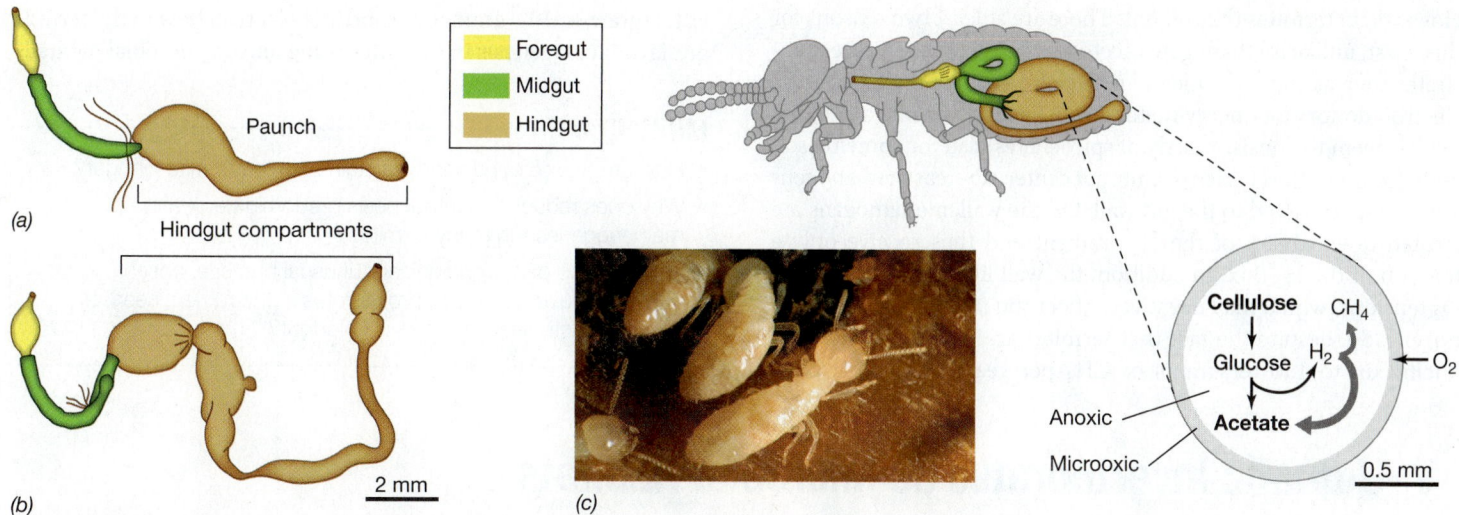

Figure 22.37 Termite gut anatomy and function. Gut architecture of lower *(a)* and higher *(b)* termites, showing the foregut, midgut, and differing complexity of the hindgut compartments. *(c)* Photo of workers, gut architecture, and biochemical activities of the lower termite *Coptotermes formosanus*. Acetate and other products of microbial fermentations are assimilated by the termite. Hydrogen produced by fermentation is consumed primarily by CO_2-reducing acetogens, with a smaller amount going to hydrogenotrophic methanogens. Methanogenesis and acetogenesis are discussed in Sections 13.20 and 13.19, respectively.

the genus *Treponema* (↩ Section 14.20) dominated, with a lesser contribution from thus far uncultured organisms distantly related to the phylum *Fibrobacteres* (↩ Section 15.21), a group also present in the rumen (Figure 22.30). Metagenomic analysis (↩ Section 6.10) of the *Nasutitermes* hindgut microbial community has revealed bacterial genes encoding glycosyl hydrolases that hydrolyze cellulose and hemicelluloses. These metagenomic data clearly implicate spirochetes and *Fibrobacteres* in the digestion of lignocellulose, although the corresponding cellulolytic bacteria have not yet been isolated from the higher termites (Figure 22.38). At every molting of an individual termite, gut symbionts are lost, yet there is good conservation of the gut community within each termite species. Stable horizontal transmission of gut symbionts likely occurs due to the intimate social behavior and close contact characteristic of termites.

Acetogenesis and Nitrogen Fixation in the Termite Gut

Genes encoding enzymes of the acetyl-CoA pathway are highly represented in the spirochetes of the *Nasutitermes* hindgut, consistent with their function as the major CO_2-reducing acetogens (↩ Section 13.19). The termite gut microbial communities have long been recognized as important to host nitrogen metabolism, providing new fixed nitrogen through nitrogen fixation (↩ Section 3.17) and helping to conserve nitrogen by recycling excretory nitrogen back to the insect for biosynthesis. Consistent with this, metagenomic analyses reveal that many bacteria, including *Fibrobacteres* and treponeme spirochetes, contain genes encoding nitrogenase, the enzyme required to fix N_2.

From a simple energetic viewpoint, methanogenesis from H_2 and CO_2 is more favorable than acetogenesis from the same substrates (-34 kJ/mol of H_2 versus -26 kJ/mol of H_2, respectively), and thus methanogens should have a competitive advantage in all habitats in which the two processes compete (↩ Sections 13.19–13.20).

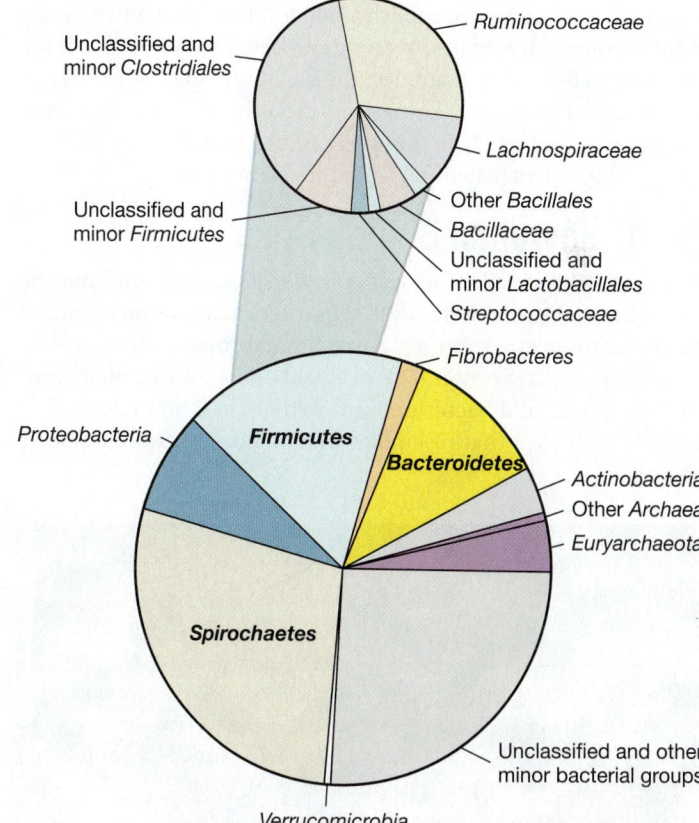

Figure 22.38 Microbial composition of termite hindgut inferred from 16S rRNA sequences. The results are pooled analyses of 5075 sequences from amplified or metagenomic sequencing studies of three genera of wood-feeding higher termites, *Nasutitermes*, *Reticulitermes*, and *Microcerotermes*. The data provide information primarily of diversity, not relative abundance. Data assembled and analyzed by Nicolas Pinel.

However, in termites they do not. There are at least two reasons for this. First, unlike methanogens, acetogens are able to use other substrates such as sugars or methyl groups from lignin degradation as electron donors for energy metabolism. Second, termite acetogens (which seem to consist mostly of spirochetes) can for some reason better colonize the H_2-rich termite gut center, whereas methanogens are largely restricted to the gut wall. On the wall, methanogens are located downstream of the H_2 gradient and thus receive only a fraction of the H_2 flux. In addition, the wall likely contains higher O_2 tensions, which may negatively affect the physiology of methanogens. So, despite the fact that termites are methanogenic, producing up to 150 teragrams of CH_4 per year on a global basis (1 teragram = 10^{12} grams), carbon and electron flow in the termite gut favor acetogenesis in this interesting anoxic microbial habitat.

V · Aquatic Invertebrates as Microbial Habitats

Thus far in this chapter we have discussed how certain macroorganisms that live in terrestrial environments provide habitats for microbial symbionts. Aquatic environments—especially marine environments—impose different constraints on symbioses and offer different opportunities and challenges for the evolution of symbioses between macroorganisms and microorganisms. Nevertheless, microbial symbioses with marine animals, especially with invertebrates, are common. By finding habitats in marine invertebrates, microorganisms establish a safe residence in a nutritionally rich environment. And the invertebrates benefit, too, as we will see with two well-studied examples: the squid and the hydrothermal vent animal symbioses. These, and the other microbial–animal associations discussed in this section, are thus true symbioses, with both partners benefiting from the relationship.

22.11 Hawaiian Bobtail Squid

The Hawaiian bobtail squid, *Euprymna scolopes*, is a small marine invertebrate (**Figure 22.39a**) that sequesters large populations of the bioluminescent gram-negative gammaproteobacterium *Aliivibrio fischeri* (Section 15.4) in a light organ located on its ventral side. Squid and bacterium are partners in a mutualism. The bacteria emit light that resembles moonlight penetrating marine waters, and this is thought to camouflage the squid from predators that strike from beneath. Several other species of *Euprymna* inhabit marine waters near Japan and Australia and in the Mediterranean, and these contain *Aliivibrio* symbionts as well.

The Squid–*Aliivibrio* System as a Model Symbiosis

Many features of the *E. scolopes–A. fischeri* symbiosis have made it an important model for studies of animal–bacterial symbioses. These include the facts that the animals can be grown in the laboratory and that there is only a single bacterial species in the symbiosis in contrast to the huge number in symbioses such as those of the rumen (Figure 22.30) or the mammalian large intestine (Figure 22.33). In addition, the symbiosis is not an essential one; both the squid and its bacterial partner can be cultured apart from each other in the laboratory. This allows juvenile squid to be grown without bacterial symbionts and then experimentally colonized. Experiments can be done to study specificity in the symbiosis, the number of bacterial cells needed to initiate an infection, the capacity of genetically defined mutants of *A. fischeri* to initiate infection of the squid, and many other aspects of the relationship. Moreover, because the genome of *A. fischeri* has been sequenced, the powerful techniques of microbial genomics may be employed.

(a)

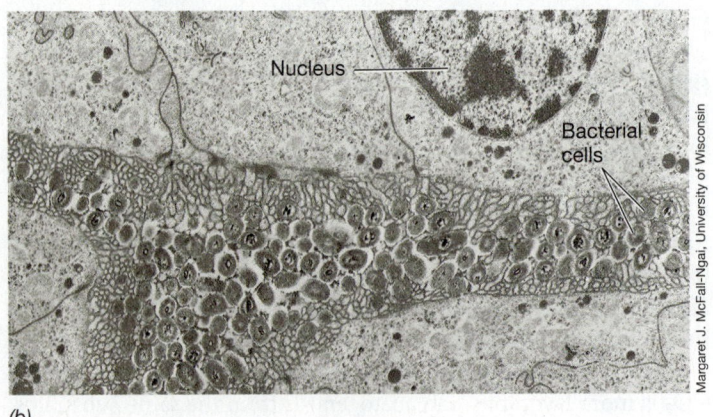

Nucleus

Bacterial cells

(b)

Figure 22.39 Squid–*Aliivibrio* symbiosis. *(a)* An adult Hawaiian bobtail squid, *Euprymna scolopes*, is about 4 cm long. *(b)* Thin-sectioned transmission electron micrograph through the *E. scolopes* light organ shows a dense population of bioluminescent *Aliivibrio fischeri* cells.

Establishing the Squid–*Aliivibrio* Symbiosis

Juvenile squid just hatched from eggs do not contain cells of *A. fischeri*. Thus, transmission of bacterial cells to juvenile squid is a horizontal (environmental) rather than a vertical (parent to offspring) event. Almost immediately after juveniles emerge from eggs, cells of *A. fischeri* in surrounding seawater begin to colonize them, entering through ciliated ducts that end in the immature light organ. Amazingly, the light organ becomes colonized specifically with *A. fischeri* and not with any of the many other species of gram-negative bacteria present in the seawater. Even if large numbers of other species of bioluminescent bacteria are offered to juvenile squid along with low numbers of *A. fischeri*, only *A. fischeri* establishes residence in the light organ. This implies that the animal in some way recognizes and accepts *A. fischeri* cells and excludes those of other species.

The squid–*Aliivibrio* symbiosis develops in several stages. Contact of the squid with any bacterial cells triggers recognition in a very general way. Upon contact with peptidoglycan (a component of the cell wall of *Bacteria*, ⮂ Section 2.10), the young squid secretes mucus from its developing light organ. The mucus is the first layer of specificity in the symbiosis, as it makes gram-negative but not gram-positive bacteria aggregate. Within the aggregates of gram-negative cells that may contain only low numbers of *A. fischeri*, this bacterium somehow outcompetes the other gram-negative bacteria to form a monoculture. The monoculture is established within 2 h of a juvenile's hatching from an egg. The highly motile *A. fischeri* cells present in the aggregate migrate up the ducts and into the light organ tissues. Once there, they lose their flagella, become nonmotile, divide to form dense populations (Figure 22.39b), and trigger developmental events that lead to maturation of the host light organ. The light organ in a mature *E. scolopes* contains between 10^8 and 10^9 *A. fischeri* cells.

Colonization of *A. fischeri* by the squid is assisted by the gas nitric oxide (NO). Nitric oxide is a well-known defense response of animal cells to attack by bacterial pathogens; the gas is a strong oxidant and causes sufficient oxidative damage to bacterial cells to kill them (⮂ Section 25.1). Nitric oxide produced by the squid is incorporated into the mucus aggregates and is present in the light organ itself. As *A. fischeri* colonizes the light organ, NO levels diminish rapidly. It appears that cells of *A. fischeri* can tolerate exposure to NO and consume it through the activity of NO-inactivating enzymes. The inability of other gram-negative bacteria in the mucus aggregates to detoxify NO helps explain the sudden enrichment of *A. fischeri* in the ducts even before the actual colonization of the light organ. Then, after establishment, continued production of NO in the light organ prevents colonization by other bacterial species.

Propagating the Symbiosis

The squid matures into an adult in about two months and then lives a strictly nocturnal existence in which it feeds mostly on small crustaceans. During the day, the animal buries itself and remains quiescent in the sand. Each morning the squid nearly empties its light organ of *A. fischeri* cells and begins to grow a new population of the bacterium. The bacterial cells grow rapidly in the light organ; by midafternoon, the structure contains the dense populations of *A. fischeri* cells required for the production of visible light. The actual emission of light requires a certain density of cells and is controlled by the regulatory mechanism called *quorum sensing* (⮂ Section 7.9). The daily expulsion of bacterial cells is thought to be

a mechanism for seeding the environment with cells of the bacterial symbionts. This, of course, increases the chances that the next generation of juvenile squid will be colonized.

A. fischeri grows much faster in the light organ than in the open ocean, presumably because it is supplied with nutrients by the squid. Thus *A. fischeri* benefits from the symbiosis by having an alternative habitat to seawater in which rapid growth and dense populations are possible. Isolation studies have shown that *A. fischeri* is not a particularly abundant marine bacterium. Daily expulsion of *A. fischeri* cells from the light organ increases the bacterium's numbers in the microbial community. Thus, the symbiotic relationship of the bacterium with the squid probably helps maintain larger *A. fischeri* populations than would exist if all cells were free-living. Because the competitive success of a microbial species is to some degree a function of population size (⮂ Section 19.1), this boost in cell numbers may confer an important ecological advantage on *A. fischeri* in its marine habitat.

22.12 Marine Invertebrates at Hydrothermal Vents and Gas Seeps

Diverse invertebrate communities develop near undersea hot springs called *hydrothermal vents*. We covered the geochemistry and microbiology of hydrothermal vents in Section 19.13. Here we focus on hydrothermal vent animals and their microbial symbionts.

Macroinvertebrates, including tube worms over 2 m in length and large clams and mussels, are present near these vents (**Figure 22.40**). Photosynthesis cannot support these invertebrate communities because they exist below the photic zone. However, hydrothermal fluids contain large amounts of reduced inorganic materials, including H_2S, Mn^{2+}, H_2, and CO (carbon monoxide), and some vents contain high levels of ammonium (NH_4^+) instead of H_2S. All of these are good electron donors for chemolithotrophs, *Bacteria* and *Archaea* that use inorganic compounds as electron donors and fix CO_2 as their carbon source (Chapter 13). Thus, these hydrothermal vent invertebrates can exist in permanent darkness because they are nourished through a symbiotic association with these autotrophic bacteria.

Tube Worms, Mussels, and Giant Clams

Hydrothermal vent–associated animals either feed directly on free-living chemolithotrophs or have formed tight symbiotic associations with them. Mutualistic chemolithotrophs are either tightly attached to the animal surface (that is, as *epibionts*) or actually live within the animal tissues, supplying organic compounds to the animals in exchange for a safe residence and ready access to the electron donors needed for their energy metabolism. For example, the 2-m-long tube worms (Figure 22.40a) lack a mouth, gut, and anus, but contain an organ consisting primarily of spongy tissue called the *trophosome*. This structure, which constitutes half the worm's weight, is filled with sulfur granules and large populations of spherical sulfur-oxidizing

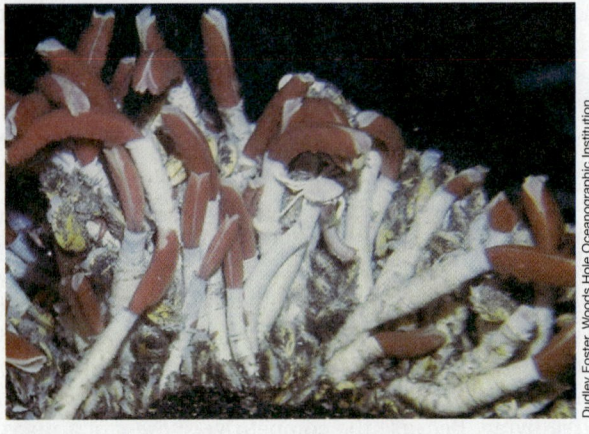

(a)

(b)

Figure 22.40 **Invertebrates living near deep-sea thermal vents.** *(a)* Tube worms (phylum Annelida), showing the sheath (white) and plume (red) of the worm bodies. *(b)* Mussel bed in vicinity of a warm vent. Note yellow deposition of elemental sulfur from the oxidation of H_2S emitted from the vents.

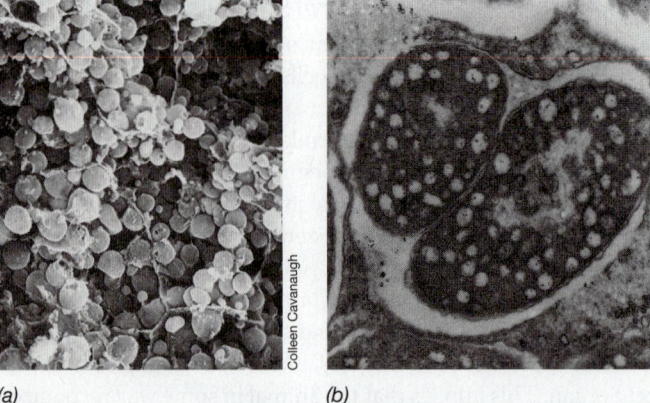

(a)   *(b)*

Figure 22.41 **Chemolithotrophic sulfur-oxidizing bacteria associated with the trophosome tissue of tube worms from hydrothermal vents.** *(a)* Scanning electron micrograph of trophosome tissue showing spherical chemolithotrophic sulfur-oxidizing bacteria. Cells are 3–5 μm in diameter. *(b)* Transmission electron micrograph of bacteria in sectioned trophosome tissue. The cells are frequently enclosed in pairs by an outer membrane of unknown origin. Reprinted with permission from *Science 213*: 340–342 (1981), © AAAS.

bacteria (**Figure 22.41**). Bacterial cells taken from trophosome tissue show activity of enzymes of the Calvin cycle, a major pathway for autotrophy (⟳ Section 13.5), but interestingly, they also contain enzymes of the reverse citric acid cycle, a second autotrophic pathway (⟳ Section 13.5). In addition, they show a suite of sulfur-oxidizing enzymes necessary to obtain energy from reduced sulfur compounds (⟳ Sections 13.8 and 14.11). The tube worms are thus nourished by organic compounds produced from CO_2 and secreted by the sulfur chemolithotrophs.

Along with tube worms, giant clams and mussels (Figure 22.40*b*) are also common near hydrothermal vents, and sulfur-oxidizing bacterial symbionts have been found in the gill tissues of these animals. Phylogenetic analyses have shown that each individual animal harbors one or more different strains of bacterial symbiont and that a variety of species of bacterial symbionts inhabit different species of vent animal. With the exception of the symbiont of *Riftia*, which also has a free-living stage (Table 22.3), none of the bacterial symbionts of hydrothermal vent animals have yet been obtained in laboratory culture, even though they are fairly closely related to free-living sulfur chemolithotrophs (⟳ Sections 13.8, 14.11, and 15.5).

The red plume of the tube worm (Figure 22.40*a*) is rich in blood vessels and is used to trap and transport inorganic substrates to the bacterial symbionts. Tube worms contain unusual hemoglobins that bind H_2S and O_2; these are then transported to the trophosome where they are released to the bacterial symbionts. The CO_2 content of tube-worm blood is also high, about 25 mM, and presumably this is released in the trophosome as a carbon source for the symbionts. In addition, stable isotope analyses (⟳ Section 18.9) of elemental sulfur from the trophosome have shown that its $^{34}S/^{32}S$ composition is the same as that of the sulfide emitted from the vent. This ratio is distinct from that of seawater sulfate and is further proof that geothermal sulfide is actually entering the worm in large amounts.

Other marine invertebrates have coevolved bacterial symbioses that supply their nutrition as well (**Table 22.4**). For example, methanotrophic (CH_4-consuming) symbionts are present in giant clams that live near natural gas seeps at relatively shallow depths in the Gulf of Mexico. Although not autotrophs (CH_4 is an organic compound), the methanotrophs do provide nutrition to the clams; the methanotrophs use CH_4 as their electron donor and carbon source and secrete organic carbon to the clams. Molecular hydrogen (H_2) was recently shown to be used as an electron donor by the mussel *Bathymodiolus puteoserpentis*, the most abundant macrofauna in vent fields associated with the peridotite-hosted vent systems of the Mid-Atlantic Ridge (⟳ Section 19.13). These systems release extremely high levels of H_2 and CH_4, with measured H_2 concentrations as high as 19 mM. This mussel was previously shown to live in a dual symbiosis with methane-oxidizing bacteria and chemolithotrophic sulfur-oxidizing bacteria localized to the gill tissue. Remarkably, the sulfur-oxidizing symbiont of *B. puteoserpentis* also has the capacity to use H_2 as an energy source, making this mussel one of the most versatile of vent macrofauna.

Genomics and Hydrothermal Vent Symbioses

Genome sequencing is revealing additional features of the metabolic interaction and coevolution of marine invertebrates and their bacterial symbionts. The genome sequence of the gill endosymbiont of the giant vent clam *Calyptogena magnifica* offers direct evidence for carbon fixation via the Calvin cycle; the genome encodes

Table 22.4 Marine animals with chemolithotrophic or methanotrophic endosymbiotic bacteria

Host (genus or order)	Common name	Habitat	Symbiont metabolic type
Porifera (Demospongiae)	Sponge	Seeps	Methanotrophs
Platyhelminthes (Catenulida)	Flatworm	Shallow water	Sulfur chemolithotrophs
Nematoda (Monhysterida)	Mouthless nematode	Shallow water	Sulfur chemolithotrophs
Mollusca (Solemya, Lucina)	Clam	Vents, seeps, shallow water	Sulfur chemolithotrophs
Mollusca (Calyptogena)	Clam	Vents, seeps, whale falls[a]	Sulfur chemolithotrophs
Mollusca (Bathymodiolus)	Mussel	Vents, seeps, whale and wood falls[a]	Sulfur and H_2 chemolithotrophs, methanotrophs
Mollusca (Alviniconcha)	Snail	Vents	Sulfur chemolithotrophs
Annelida (Riftia)	Tube worm	Vents, seeps, whale and wood falls[a]	Sulfur chemolithotrophs

[a]Whale and wood falls are sunken whale carcasses and wood, respectively.

the key enzymes of the Calvin cycle, ribulose bisphosphate carboxylase (RubisCO) and phosphoribulokinase (Section 13.5), and genes encoding key sulfur oxidation processes. The genome of this symbiont also encodes the biosynthesis of most vitamins and cofactors and all 20 amino acids needed to support the host. However, because few substrate-specific transporters are encoded by the symbiont genome, it is suspected that the clam actually digests symbiont cells for nutrition, as do mussels (Table 22.4).

Like the obligate symbionts of insects, most symbionts of marine invertebrates have small genomes (Table 22.3), indicating reduced function and an obligate association with their host. The bacterial symbiont of the giant tube worm *Riftia pachyptila* is an exception, having a genome larger than some free-living sulfur-oxidizing chemolithotrophs (Table 22.3). The *R. pachyptila* symbiont is acquired by uninfected juvenile animals from the environment (horizontal transmission), and its larger genome is likely important for survival as a free-living bacterium.

MINIQUIZ

- How do giant tube worms receive their nutrition?
- What are the similarities of the obligate symbioses of insects and hydrothermal vent invertebrates?
- What factors determine the genome size of the symbionts of marine invertebrates?

22.13 Leeches

Leeches are parasitic annelids (segmented worms). Leeches are related to earthworms and share several properties with them. Some leeches live in marine environments, but our example here, the medicinal leech *Hirudo verbana* (**Figure 22.42a**), lives in freshwater.

Parasitic Lifestyle of Leeches

Like many animals that depend on a microbial partner, medicinal leeches have a restricted diet. They feed exclusively on vertebrate blood and secrete powerful anticoagulants and vasodilators that stimulate blood flow. In a single feeding, *H. verbana* can consume over five times its body weight in blood. The blood meal is stored in the *crop*, which is the largest compartment of the leech digestive tract (Figure 22.42b). During feeding, water and salts are absorbed from the crop content until most water is removed and the fluid is in osmotic balance with the leech hemolymph. Excess water and nitrogenous waste are secreted through several pairs of bladders. Both the digestive tract and the bladder house microbial communities. It is thought that one function of the symbionts is to provide essential nutrients, such as vitamin B_{12}, absent or in low amounts in the blood meal.

This amazing ability of medicinal leeches to remove blood and secrete pharmacologically active compounds has been used for ages for the medical practice of *bloodletting*, and in recent times most commonly in plastic and reconstructive surgery. A challenge for medical replants and transplants is the connection of the veins. If, after transplant surgery, the number of functional veins exiting from the surgically introduced tissue is insufficient, the flow of fresh oxygenated blood into the tissue is stopped. The lack of oxygen can result in failure of the transplant. Leeches applied to the area remove blood, letting fresh blood enter the introduced tissue, and this procedure increases the transplant success rate.

The Leech Microbial Community

The leech digestive tract has two major compartments that house microbial communities, the digestive tract (the large crop and the

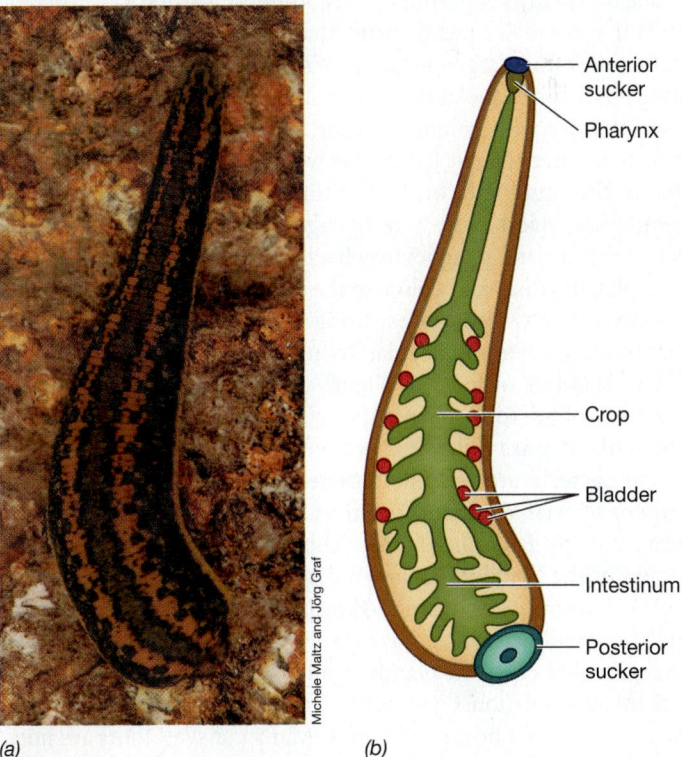

Figure 22.42 Medicinal leech *Hirudo verbana*. (a) An animal of about 6 cm in length. (b) Anatomy of *H. verbana*, showing the crop, intestinum, and bladder pairs.

smaller intestinum), where the digestion of the erythrocytes and absorption of nutrients are thought to occur, and the bladders (Figure 22.42*b*). The microbial community of the crop is surprisingly simple. Culture-independent studies using a combination of 16S rRNA gene analyses and fluorescence in situ hybridization (FISH, ⊘ Section 18.4) revealed that the microbial community inside the crop is dominated by two species, *Aeromonas veronii* (*Gammaproteobacteria*) and a *Rikenella*-like (*Bacteroidetes*) bacterium. Farther along the alimentary canal toward the intestinum (Figure 22.42*b*) the complexity of the microbial community increases. In the intestinum various *Alpha-* and *Gammaproteobacteria*, along with *Bacteroidetes* and *Firmicutes*, prevail.

The functional significance of the simple microbial community of the crop has been explored through a combination of metagenomic and metatranscriptomic analyses (Chapter 6), together suggesting that the *Rikenella*-like symbiont has the capacity to forage on sulfated- and sialated-mucin polysaccharides (glycans) that line the surface of the crop epithelium. Mucins comprise a family of heavily glycosylated proteins that form gels functioning in lubrication or as protective barriers on the surface of epithelial tissue. The sequence-based insight prompted the development of a mucin-based cultivation medium subsequently used for the successful selective enrichment and isolation of the *Rikenella*-like symbiont, confirming its ability to ferment host-derived polysaccharides to acetate. Since acetate is likely utilized by *Aeromonas veronii*, this discovery also provided a possible physiological link between the two major crop symbionts. Also, the ability of the *Rikenella*-like symbiont to utilize the leech-derived mucin may explain the ability of the symbionts to persist within the leech gut for up to 6 months between blood feedings.

The unusually simple microbial community inside the crop suggests that there are mechanisms that prevent other microorganisms from colonizing. Specificity of symbiotic associations can be affected by the mode of transmission and molecular mechanisms that interfere with colonization or maintenance of microorganisms that enter the gut habitat, as we saw for *Aliivibrio* colonization of the squid light organ (Section 22.11). For example, leech hemocytes, invertebrate macrophage-like cells (⊘ Section 24.1), patrol the gut and phagocytose bacteria. *A. veronii* is able to prevent phagocytosis and colonize the leech gut by injecting toxins directly into the hemocytes, using a bacterial secretion system that functions like a molecular syringe (⊘ Section 4.14).

The bladders of leeches (Figure 22.42*b*) house an interesting ensemble of microorganisms. The epithelial cells lining the lumen of the bladder are tightly packed with an *Ochrobactrum* species. These bacteria are related to beneficial and pathogenic alphaproteobacterial symbionts, such as *Sinorhizobium meliloti* (Section 22.3) and *Brucella abortus*. The microbial community in the lumen of the bladder displays a distinct stratification: Two species of *Bacteroidetes* colonize the epithelial side and two species of *Betaproteobacteria* colonize the luminal side of the biofilm-like structure that coats the bladder wall (**Figure 22.43**).

Symbiotic relationships require transmission of the microbial partners between host generations. Many gut symbionts are horizontally (environmentally) transmitted, but *A. veronii* appears to be vertically transmitted from the parent to the offspring through the "cocoons" in which the embryos develop (juvenile leeches removed from cocoons are already infected with cells of *A. veronii*).

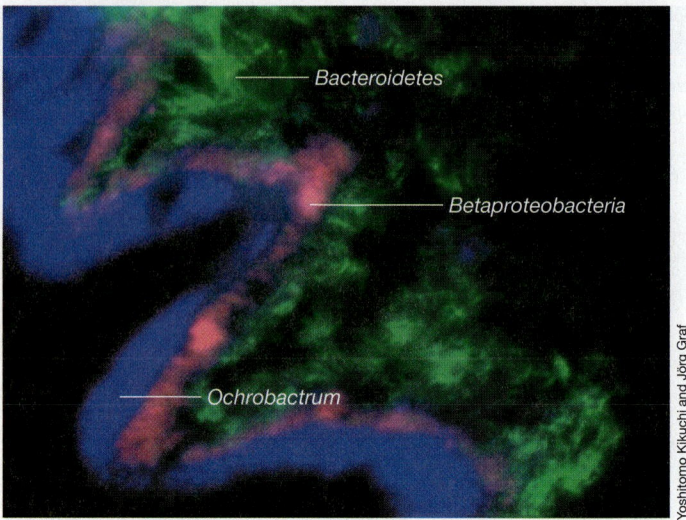

Figure 22.43 **Micrograph of a FISH-stained microbial community in the bladder of *Hirudo verbana*.** A probe (red) targeted at the 16S rRNA of *Betaproteobacteria* and a probe (green) targeted at the 16S rRNA of *Bacteroidetes* reveal distinct layers of different bacteria in the lumen of the bladder. Staining with DAPI (blue), which binds to DNA, reveals the intracellular alphaproteobacterium *Ochrobactrum* and host nuclei.

Similarly, most of the bladder symbionts have been detected in juveniles taken from cocoons. Such vertical transmission of the symbionts ensures their safe transfer to the next host generation.

MINIQUIZ

- How do leeches transmit symbionts to their progeny?
- In what way does the *Aeromonas veronii* symbiont of the leech resemble a pathogenic bacterium?
- What is the suggested metabolic interaction between the *A. veronii* and *Rikenella*-like symbionts?

22.14 Reef-Building Corals

Coral reef ecosystems are the products of mutualistic associations between algae and simple marine animals. The extensive ecosystems associated with the worldwide distribution of these mutualisms support tens of thousands of species.

Phototrophic Symbioses with Animals

We saw in the beginning of this chapter that a lichen is a mutualism between a fungus and a phototrophic partner—an alga or cyanobacterium. Like the fungi, some animals establish mutualistic associations with photosynthetic algae or cyanobacteria (Table 22.5). The animals in most of these associations are in phyla that display very simple body plans; for example, the Porifera (sponges) and Cnidaria (corals, sea anemones, and hydroids). These mutualistic animal–bacterial associations live in clear tropical waters where nutrients for the animals are scarce, and the animal body typically has a large surface area relative to its volume and is thus well suited for capturing light.

The coral skeleton is an extremely efficient light-gathering structure that greatly enhances light harvesting. There are only a few instances of algae forming associations with more complex animals, such as those in the phyla Platyhelminthes (flatworms), Mollusca (snails and clams), and Urochordata (sea squirts).

Table 22.5 Symbioses between animals and phototrophic symbionts

Host	Common name	Symbionts
Porifera	Sponge	Cyanobacteria, *Chlorella*, *Symbiodinium*
Cnidaria	Coral, sea anemone	*Symbiodinium*, *Chlorella*
Platyhelminthes	Flatworm	Diatoms, primitive chlorophytes
Mollusca	Snail, clam	*Symbiodinium*, *Chlorella*
Ascidia	Sea squirt	Cyanobacteria

In these cases either the animal has a suitable surface-to-volume ratio or has evolved specific light-gathering surfaces. The unicellular phototrophic symbionts are phylogenetically diverse and include cyanobacteria (⇄ Section 14.3), and red and green algae, diatoms, and dinoflagellates (Chapter 17). Most common are the green algae *Chlorella* (associating with sponges and freshwater hydras), cyanobacteria (associating with marine sponges), and species of the dinoflagellate genus *Symbiodinium*.

The most spectacular and ecologically significant of these mutualisms is between the cnidarian stony corals (order Scleractinia) and the dinoflagellate *Symbiodinium* (**Figure 22.44**). Together the corals and dinoflagellates form the trophic and structural foundation of the coral reef ecosystem. The cnidarians possess a very simple two-tissue-layer body plan (ectoderm and gastroderm) and harbor the dinoflagellate symbiont intracellularly in vacuoles called *symbiosomes* within cells of the inner (gastrodermal) tissue layer (Figure 22.44c). The algae receive key inorganic nutrients from host metabolism and pass photosynthetically produced organic compounds to the corals. This mutualism has allowed coral reefs to develop in large expanses of nutrient-poor ocean waters. Dinoflagellates and other alveolates comprise eight genera and around 2000 extant species (⇄ Section 17.5). Although dinoflagellate mutualisms are common, most are between species of *Symbiodinium* and marine invertebrates or protists (Figure 22.44). We focus here on the symbiotic association between *Symbiodinium* and the stony coral cnidarians.

Transmission, Specificity, and Benefits of the *Symbiodinium*–Coral Association

Reef-building corals reproduce sexually by releasing gametes into the seawater (broadcast spawning). A male and a female gamete fuse to form a free-swimming larva that later settles on a surface, where it may initiate a new coral colony. Algal symbionts are typically present in the egg before it is released from the parent (vertical transmission), although free-living *Symbiodinium* cells can also be ingested by juvenile corals (horizontal transmission). A developing coral that ingests dinoflagellates digests all of them except the particular *Symbiodinium* of its mutualism. After establishing an association, the coral controls the growth of *Symbiodinium* via chemical signaling and, following each cell division, each *Symbiodinium* daughter cell is allocated to a new symbiosome.

Both partners in the cnidarian–dinoflagellate mutualism have evolved adaptations for nutritional exchange. The dinoflagellates donate most of their photosynthetically fixed carbon (in the form of small molecules such as sugars, glycerol, and amino acids) to the cnidarian in exchange for inorganic nitrogen, phosphorus, and inorganic carbon from the host. Moreover, in addition to providing protection and inorganic nutrients, the calcium carbonate skeleton of corals is one of the most efficient collectors of solar radiation in nature, amplifying the incident light field for the symbionts by as much as fivefold; this benefits the symbiont in carrying out photosynthesis under a light-absorbing water column.

Coral Bleaching—The Risk of Harboring a Phototrophic Symbiont in a Changing World

Many of the extensive coral reef systems in the oceans worldwide are now threatened with extinction, primarily as a consequence of human activities. Ongoing loss of these beautiful and productive ecosystems is thought to be the result of elevated atmospheric CO_2; namely, increased sea surface temperature, rising sea levels, and ocean acidification (⇄ Sections 20.6 and 20.8). Coastal development also threatens reef systems, contributing to pollution from sewage discharge, eutrophication from nutrient runoff, and overfishing. These environmental changes are contributing to high mortality through disease, loss of coral structure from reduced calcification caused by acidification, and bleaching. Healthy corals harbor millions of cells of *Symbiodinium* per square centimeter of tissue. Coral bleaching is the loss of color from host tissues caused by the lysis of these symbionts, revealing the underlying white limestone skeleton (**Figure 22.45**).

Coral reefs live close to their optimum temperature and it is the synergistic effect of increased sea surface temperature and irradiance

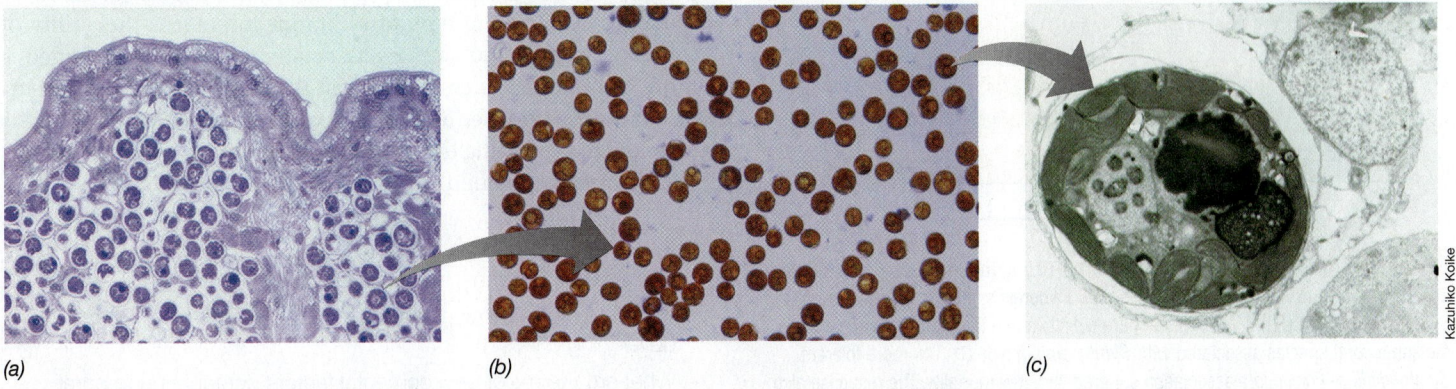

(a)   *(b)*   *(c)*

Figure 22.44 ***Symbiodinium* symbiont of marine invertebrates.** *(a)* Thin-section micrograph of *Symbiodinium* in the mantle tissue of a giant clam. *(b)* *Symbiodinium* cells recovered from a soft coral. *(c)* Transmission electron micrograph of a *Symbiodinium* cell within a vacuole of a cell of the stony coral *Ctenactis echinata*. The *Symbiodinium* cell is about 10 µm in diameter.

(a)

(b)

Figure 22.45 Coral bleaching. *(a)* Two colonies of the brain coral *Colpophyllia natans*. The coral on the left is a healthy brown color, whereas the coral on the right is fully bleached. *(b)* A large colony of partially bleached *Montastraea faveolata*.

that causes massive bleaching. Elevated temperature and high irradiance impair the photosynthetic apparatus of the dinoflagellates, resulting in the production of reactive oxygen species (for example, singlet oxygen and superoxide (⇨ Section 5.16) that cause damage to both host and symbiont. Bleaching is thought to be caused by a protective immune response of the host that destroys compromised symbionts. Increases in sea surface temperatures as small as 0.5–1.5°C above the local maximum, if sustained for several weeks, can induce rapid coral bleaching. A significant decrease in temperature below the optimum range for coral growth can have a similar impact. Thermal stress, accentuated by seasonal increases in electromagnetic radiation of ultraviolet and some visible wavelengths, has resulted in bleaching of huge expanses of coral reefs.

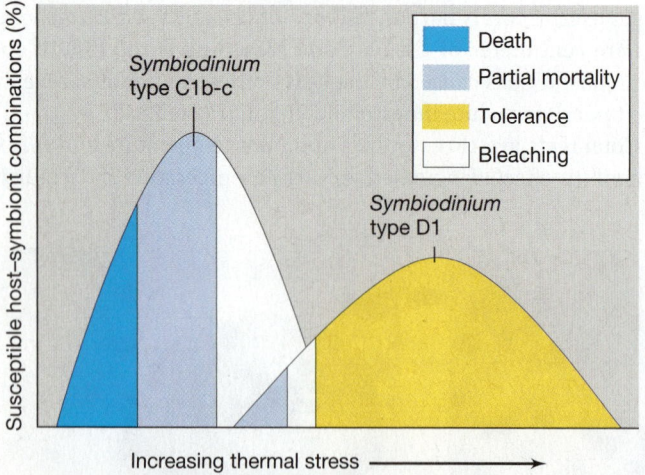

Figure 22.46 Differential stress tolerance of a coral species associated with different *Symbiodinium* phylotypes. *Pocillopora* corals symbiotically associated with *Symbiodinium* type C1b-c were much more sensitive to a thermal stress event than the same coral species associated with *Symbiodinium* type D1. The more tolerant *Symbiodinium*–*Pocillopora* association suffered very low mortality. The response also suggested additional genetic variation within each *Symbiodinium* type, since the two mutualisms displayed a range of sensitivity to increasing thermal stress.

Although coral reefs are clearly threatened, there is much uncertainty in projecting their future. The more ominous projections, based on projected increases in sea temperature, point to a collapse of Indian Ocean coral reef systems within only a few years and a possible global collapse of coral reefs by the middle of this century. However, these projections still lack basic knowledge about the vulnerability of individual coral species and the adaptive capacity of individual coral–symbiont mutualisms. For example, thermal tolerance is in part conferred by the species or strain of *Symbiodinium*, and following a bleaching event the mutualism can shift to a more thermally tolerant symbiont (**Figure 22.46**).

Molecular results have indicated that there are over 150 different *Symbiodinium* phylotypes, each possibly representing a distinct species with different stress tolerance. Both symbiont swapping and symbiont switching have been proposed as the underlying mechanism for shifting between symbionts. In switching, the symbiont is taken up from a water column population. In swapping, the shift results from differential growth of a genetic variant already associated with the coral, but in very low numbers, thereby swapping places with the previously dominant mutualist following the bleaching event. Most studies indicate that swapping is the more common adaptive mechanism, but uncertainty remains. Because the type of symbiont influences the ability of the coral to adapt to stresses associated with climate change, a more complete understanding of the alternative mechanisms of adaptive response, including possible symbiont switching, is essential to predicting the future health of corals, their symbionts, and the reefs they build.

MINIQUIZ -

• What gives corals their spectacular colors?

• What are the two mechanisms of *Symbiodinium* transfer to developing corals?

• What are the major environmental factors contributing to coral bleaching?

BIG IDEAS

22.1 • Lichens are a mutualistic association between a fungus and an oxygenic phototroph.

22.2 • The consortium "*Chlorochromatium aggregatum*" is a mutualism between a phototrophic green sulfur bacterium and a motile heterotroph. Mutual benefit is based on the phototroph supplying organic matter to the heterotroph in exchange for motility that permits rapid repositioning in stratified lakes to obtain optimal light and nutrients.

22.3 • One of the most agriculturally important plant–microbial symbioses is that between legumes and nitrogen-fixing bacteria. The bacteria induce the formation of root nodules within which nitrogen fixation occurs. The plant provides the energy needed by the root nodule bacteria, and the bacteria provide fixed nitrogen for the plant.

22.4 • The crown gall bacterium *Agrobacterium* enters into a unique relationship with plants. Part of the Ti plasmid in the bacterium can be transferred into the genome of the plant, initiating crown gall disease. The Ti plasmid has also been used for the genetic engineering of crop plants.

22.5 • Mycorrhizae are mutualistic associations between fungi and the roots of plants that allow the plant to extend its root system via intimate interaction with an extensive network of fungal mycelia. Both ectomycorrhizae and endomycorrhizae are known. The mycelia network provides the plant with essential inorganic nutrients, and the plant, in turn, supplies organic compounds to the fungus.

22.6 • Microbial fermentation is important for digestion in all mammals. Several microbial mutualisms have evolved in different mammals that allow for the digestion of different types of food. Herbivores derive almost all of their carbon and energy from plant fiber.

22.7 • The rumen, the digestive organ of ruminant animals, specializes in cellulose digestion, which is carried out by microorganisms. Bacteria, protists, and fungi in the rumen produce volatile fatty acids that provide energy for the ruminant. Rumen microorganisms synthesize vitamins and amino acids and are also a major source of protein—all used by the ruminant.

22.8 • The human microbiome encompasses all sites of the human body inhabited by microorganisms. The microorganisms are critical to early development, health, and predisposition to disease. The human gut microbial community is unique when compared with that of other mammals. The gut microflora affects energy recovery from food, and a shift in gut community structure may be a major factor in obesity.

22.9 • A large proportion of insects have established obligate mutualisms with bacteria, the basis of the mutualism often being bacterial biosynthesis of nutrients such as amino acids that are absent from the food the insect feeds on. Long-established obligate mutualisms are marked by extreme genome reduction of the symbiont, with retention of only those genes essential for the mutualism.

22.10 • Termites associate symbiotically with bacteria and protists capable of digesting plant cell walls. The unique termite gut configuration and the hindgut microbial community composed largely of cellulolytic bacteria and protists and acetogenic bacteria result in high levels of acetate, the primary source of carbon and energy for the termite.

22.11 • A light-emitting organ on the underside of the Hawaiian bobtail squid provides a habitat for bioluminescent cells of *Aliivibrio fischeri*. From the mutualism in the light organ, the squid gains protection from predators while the bacterium benefits from a habitat in which it grows quickly and contributes cells to its free-living population.

22.12 • Most invertebrates living on the seafloor near regions receiving hydrothermal fluids have established obligate mutualisms with chemolithotrophic bacteria. These mutualisms are nutritional, allowing the invertebrates to thrive in an environment enriched in reduced inorganic materials, such as H_2S, that are abundant in vent fluids. The invertebrates provide the symbionts an ideal nutritional environment in exchange for organic nutrients.

22.13 • Leeches and particular bacterial species form symbioses in regions of the host body that are important for host nutrition and nitrogen retention. The existence of mechanisms for vertical transmission of the symbionts indicates that these mutualisms are highly evolved and functionally important.

22.14 • The mutualism between the dinoflagellate *Symbiodinium* and the stony corals produces the extensive worldwide coral reef ecosystems that sustain a tremendous diversity of marine life. Coral bleaching caused by climate change threatens these ecosystems.

REVIEW OF KEY TERMS

Arbuscule branched or coiled hyphal structure within cells of the inner cortex of plants with a mycorrhizal infection

Bacteriocyte a specialized insect cell in which bacterial symbionts reside

Bacteriome a specialized region in several insect groups that contains insect bacteriocyte cells packed with bacterial symbionts

Bacteroid the misshapen cells of rhizobia inside a leguminous plant root nodule; can fix N_2

Coevolution evolution that proceeds jointly in a pair of intimately associated species owing to the effects each has on the other

Consortium a mutualism between bacteria, for example, a phototrophic green sulfur bacterium and a motile nonphototrophic bacterium

Dysbiosis an alteration or imbalance of an individual's microbiome relative to the normal, healthy state, primarily observed in the microbiota of the digestive tract or the skin.

Infection thread in the formation of root nodules, a cellulosic tube through which *Rhizobium* cells can travel to reach and infect root cells

Leghemoglobin an O_2-binding protein found in root nodules

Lichen a fungus and an alga (or cyanobacterium) living in symbiotic association

Mutualism a symbiosis in which both partners benefit

Myc factors lipochitin oligosaccharides produced by mycorrhizal fungi to initiate symbiosis with a plant

Mycorrhizae a symbiotic association between a fungus and the roots of a plant

Nod factors lipochitin oligosaccharides produced by root nodule bacteria that help initiate the plant–bacterial symbiosis

Root nodule a tumorlike growth on plant roots that contains symbiotic nitrogen-fixing bacteria

Rumen the first vessel in the multichambered stomach of ruminant animals in which cellulose digestion occurs

Symbiosis an intimate relationship between two organisms, often developed through prolonged association and coevolution

Ti plasmid a conjugative plasmid in the bacterium *Agrobacterium tumefaciens* that can transfer genes into plants

Volatile fatty acids (VFAs) the major fatty acids (acetate, propionate, and butyrate) produced during fermentation in the rumen

MasteringMicrobiology® **Review what you know and challenge what you have learned with MasteringMicrobiology!** Access study materials, chapter quizzes, animations, and microbiology lab tutorials in the Study Area to ensure that you have mastered this chapter's content.

REVIEW QUESTIONS

1. Describe the similarities and differences between the lichen and coral symbioses. (Sections 22.1, 22.14)

2. In the "*Chlorochromatium*" symbiosis, how does each partner benefit? (Section 22.2)

3. Describe the steps in the development of root nodules on a leguminous plant. What is the nature of the recognition between plant and bacterium and how do Nod factors help control this? How does this compare with recognition in the *Agrobacterium*–plant system? (Sections 22.3 and 22.4)

4. Compare and contrast the production of a plant tumor by *Agrobacterium tumefaciens* and a root nodule by a *Rhizobium* species. In what ways are these structures similar? In what ways are they different? Of what importance are plasmids to the development of both structures? (Sections 22.3 and 22.4)

5. How do mycorrhizae improve the growth of trees? In what way(s) are the root nodule and mycorrhizal symbioses similar? (Section 22.5)

6. What is a rumen and how do the digestive processes operate in the ruminant digestive tract? What are the major benefits and the disadvantages of a rumen system? How does a cecal animal compare with a ruminant? (Sections 22.6 and 22.7)

7. What is an example of a single microbial species contributing to herbivore health? What is an example of a single microbial species contributing to herbivore pathology? (Section 22.7)

8. What is a possible mechanism by which the microbial community of the human gut increases energy recovery, thereby contributing to obesity? (Section 22.8)

9. Why was *Escherichia coli* long thought to be a dominant member of the human gut microbial community? (Section 22.8)

10. How is it possible for aphids to feed only on the carbohydrate-rich but nutrient-poor sap of phloem vessels in plants? (Section 22.9)

11. Why do symbionts that are transmitted horizontally show less genome reduction, as opposed to the significant genome reduction observed in heritable symbionts? (Section 22.9)

12. How do the microbial communities of guts of higher and lower termites differ in composition and degradation of cellulose? (Section 22.10)

13. How is the correct bacterial symbiont selected in the squid–*Aliivibrio* symbiosis? (Section 22.11)

14. How does a tube worm obtain nutrients if it lacks a mouth, gut, and anus? (Section 22.12)

15. Compare the microbial communities in the medicinal leech crop, intestinum, and bladder. (Section 22.13)

16. How does the body plan of corals influence their ability to symbiotically associate with *Symbiodinium*? (Section 22.14)

APPLICATION QUESTION

1. Imagine that you have discovered a new animal that consumes only grass in its diet. You suspect it to be a ruminant and have available a specimen for anatomical inspection. If this animal is a ruminant, describe the position and basic components of the digestive tract you would expect to find and any key microorganisms and substances you might look for. What metabolic types of microorganisms or specific genes would you predict would be present?

23 · Microbial Interactions with Humans

microbiology**now**

The Fungal Microbiome of the Skin

A major challenge to developing an accurate picture of the human microbiome is its incredible diversity and the issues inherent with capturing data that accurately reflect this diversity. Besides bacteria, other microorganisms inhabit the human body, in particular, fungi, and it is important to include these organisms when deriving a census of the human microbiome. A recent study does just that.[1]

Ten healthy human volunteers were sampled at 14 different body sites and the samples processed for both fungal and bacterial diversity. Using genomic methods that analyzed a key phylogenetic gene in fungi, the researchers generated more than 5 million sequences to compare with 16S rRNA gene sequences derived from bacteria obtained from the same samples. By analyzing and comparing the diversity in their samples, they compiled an inclusive list of the human skin microbiome, which includes over 200 genera each of fungi and bacteria.

Analysis of the data showed that the fungus *Malassezia* (photos) was present at all body sites and was the predominant genus in all but sites on the foot. The latter showed high fungal diversity but relatively low bacterial diversity. In general, fungal diversity correlated with body site, with the greatest diversity on the foot. Bacterial diversity, on the other hand, depended more on the physiology of the skin sites (moist, dry, or oily), as we will explore further in this chapter.

Our understanding of the role of the skin in health and disease depends on our understanding of the normal skin microflora, including the fungi. The role of the normal microflora in protection against disease and the mechanism(s) by which the immune system learns to tolerate the normal microflora are important questions for the future of this field.

[1]Findley, K., et al. 2013. Topographical diversity of fungal and bacterial communities in human skin. *Nature* doi:10.11038/nature12171.

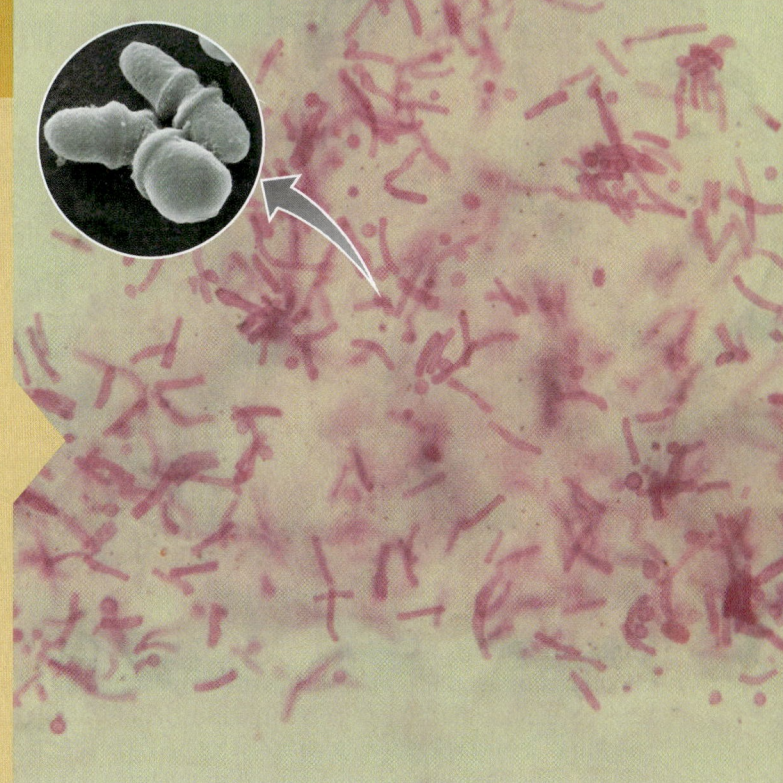

Humans have an extensive population of microorganisms, including large populations of bacteria and fungi, on their skin and the mucous membranes lining the mouth, gut, and excretory and reproductive systems. The human body is composed of about 10^{13} cells, but about ten times as many microorganisms live on or in the body that are beneficial and even necessary to maintain good health.

Microorganisms called *pathogens* can invade, infect and damage the human body. Pathogens use attachment structures, growth factors, enzymes, and toxins to gain access and damage host tissues. We begin our coverage here with microorganisms normally found in and on the human body. We then look at selected pathogens and some of their disease-producing strategies. We conclude by introducing the nonspecific defense mechanisms our bodies use to suppress or destroy most pathogens.

I • Normal Human–Microbial Interactions

Through normal everyday activities, the human body is exposed to microorganisms in the environment. Hundreds of species and countless individual microbial cells, collectively called the **normal microflora**, grow on or in the human body. This is the *human microbiome*, the sum total of all microorganisms that live on or in the human body.

23.1 Beneficial Human–Microbial Interactions

The normal microflora have developed a symbiotic relationship with the mammalian host. They contribute to the health and well-being of the host by producing beneficial microbial products and inhibiting growth of dangerous microorganisms. In turn, the host contributes various microenvironments that support microbial growth. The normal microflora are first introduced at birth.

Colonization

Mammals develop in a microbially sterile environment in utero and have no exposure to microorganisms. **Colonization**, the growth of a microorganism after it has gained access to host tissues, begins as mammals are exposed to microorganisms in the birth process. The skin surfaces are readily colonized by many species. Likewise, the mucous membranes of the oral cavity and gastrointestinal tract acquire microorganisms through feeding and exposure to the mother's body, which, along with other environmental sources, initiates colonization of the skin, oral cavity, upper respiratory tract, and gastrointestinal tract (**Figure 23.1**).

Different populations of microorganisms colonize individuals in different localities and at different times. Moreover, the normal microflora are highly diverse and may differ significantly between individuals, even in a given population. As we go along we will point out patterns of colonization by particular groups of microorganisms that inhabit specific niches, presumably because of their ability to access nutritional and metabolic support at particular body sites.

Mammalian hosts are rich in the organic nutrients and growth factors required by bacteria and provide conditions of controlled pH, osmotic pressure, and temperature that are favorable for the growth of microorganisms. Bodies, however, are not uniform environments. Each body region such as the skin, respiratory tract, and gastrointestinal tract differs chemically and physically from others, providing a selective environment that favors the growth of certain microorganisms and prevents the growth of others. Thus, these different environments each support the growth of a diverse and regionally unique microflora. For example, the relatively dry environment of the skin favors the growth of dehydration-resistant species such as gram-positive streptococci and staphylococci (↶ Sections 15.6 and 15.7), while the anoxic environment of the large intestine supports growth of obligately anaerobic bacteria such as *Bacteroides* (↶ Section 15.13).

Table 23.1 shows some of the major types of microorganisms normally found in association with various body surfaces in humans. Normal microflora do not colonize the internal organs, blood, lymph, or the nervous systems. The growth of microorganisms in these normally microbially sterile environments indicates serious infectious disease.

Colonization Sites

Colonization typically begins at sites in the **mucous membranes** (Figure 23.1). Mucous membranes consist of *epithelial cells*, tightly packed cells that interface with the external environment. They are found throughout the body, lining the urogenital, respiratory, and gastrointestinal tracts. The epithelial cells in mucous membranes secrete **mucus**, a thick liquid secretion that contains water-soluble proteins and glycoproteins. Mucus retains moisture and inhibits microbial attachment; invaders are usually swept away by physical processes like swallowing or sneezing, but some microorganisms adhere to the epithelial surface and colonize.

Microorganisms are also found on surfaces of the body that are not mucous membranes but are exposed to the environment, especially the skin. As we shall see next, the skin is actually a complex organ consisting of several distinct microenvironments, each with its own characteristic microflora.

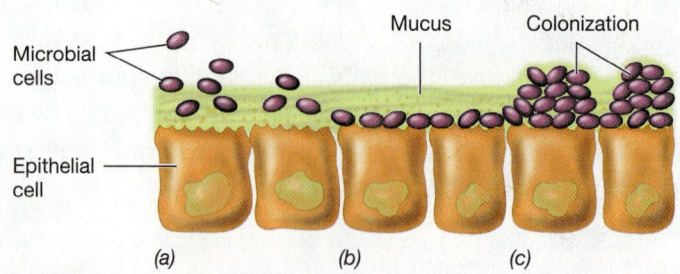

Figure 23.1 **Bacterial interactions with mucous membranes.** *(a)* Loose association. *(b)* Adhesion. *(c)* Colonization.

Table 23.1 Representative normal microflora of humans

Anatomical site	Most prevalent taxa[a]
Skin	*Acinetobacter, Corynebacterium, Enterobacter, Klebsiella, Malassezia* (f), *Micrococcus, Propionibacterium, Proteus, Pseudomonas, Staphylococcus, Streptococcus*
Mouth	*Streptococcus, Lactobacillus, Fusobacterium, Veillonella, Corynebacterium, Neisseria, Actinomyces, Geotrichum* (f), *Candida* (f), *Capnocytophaga, Eikenella, Prevotella,* spirochetes (several genera)
Respiratory tract	*Streptococcus, Staphylococcus, Corynebacterium, Neisseria, Haemophilus*
Gastrointestinal tract[b]	*Lactobacillus, Streptococcus, Bacteroides, Bifidobacterium, Eubacterium, Peptococcus, Peptostreptococcus, Ruminococcus, Clostridium, Escherichia, Klebsiella, Proteus, Enterococcus, Staphylococcus, Methanobrevibacter,* gram-positive bacteria, *Proteobacteria, Actinobacteria, Fusobacteria*
Urogenital tract	*Escherichia, Klebsiella, Proteus, Neisseria, Lactobacillus, Corynebacterium, Staphylococcus, Candida* (f), *Prevotella, Clostridium, Peptostreptococcus, Ureaplasma, Mycoplasma, Mycobacterium, Streptococcus, Torulopsis* (f)

[a]This list is not meant to be exhaustive, and not all of these organisms are found in every individual. Distribution may vary with age (adults vs. children) and sex. Many of these organisms are opportunistic pathogens under certain conditions. Some taxa are found at more than one body area. (f), fungi.
[b]For a molecular picture of the prokaryotic diversity of the human large intestine, see ↩ Section 22.8.

MINIQUIZ

- Identify factors necessary to support colonization of body surfaces by normal microflora.
- At what body sites are microflora found?

23.2 Microflora of the Skin

An average adult human has about two square meters (2 m^2) of skin surface that varies greatly in chemical composition and moisture content. One distinct microenvironment includes moist skin areas such as the inside of the nostril, the armpit, and the umbilicus. Moist skin is separated by only a few centimeters from dry microenvironments such as the forearms and the palms of the hands. A third microenvironment consists of areas with high concentrations of sebaceous glands that produce an oily substance called *sebum*. Sebaceous areas are those by the side of the nose, the back of the scalp, and the upper chest and back.

The skin microflora has been examined by molecular ecology methods employing comparative rRNA gene sequencing (↩ Section 18.5). In one study, nineteen different bacterial phyla were detected, but four phyla predominated: *Actinobacteria, Firmicutes, Proteobacteria,* and *Bacteroidetes* (**Figure 23.2a**). Over 200 different genera were identified, but members of three genera, *Corynebacteria* (*Actinobacteria*), *Propionibacteria* (*Actinobacteria*), and *Staphylococcus* (*Firmicutes*) comprised more than 60% of the sequences (Figure 23.2b). Each microenvironment showed a unique microbiota. Moist sites were dominated by corynebacteria and staphylococci while drier sites supported a mixed population dominated by *Betaproteobacteria,* corynebacteria, and *Flavobacteriales.* Sebaceous areas had predominantly propionibacteria and staphylococci (Figure 23.2b).

Analysis of these data provides an overview of the human normal microflora, but individuals showed variations from the composite patterns, meaning that there is no single normal microflora pattern. Instead, a predictable group of microorganisms is likely

UNIT 5

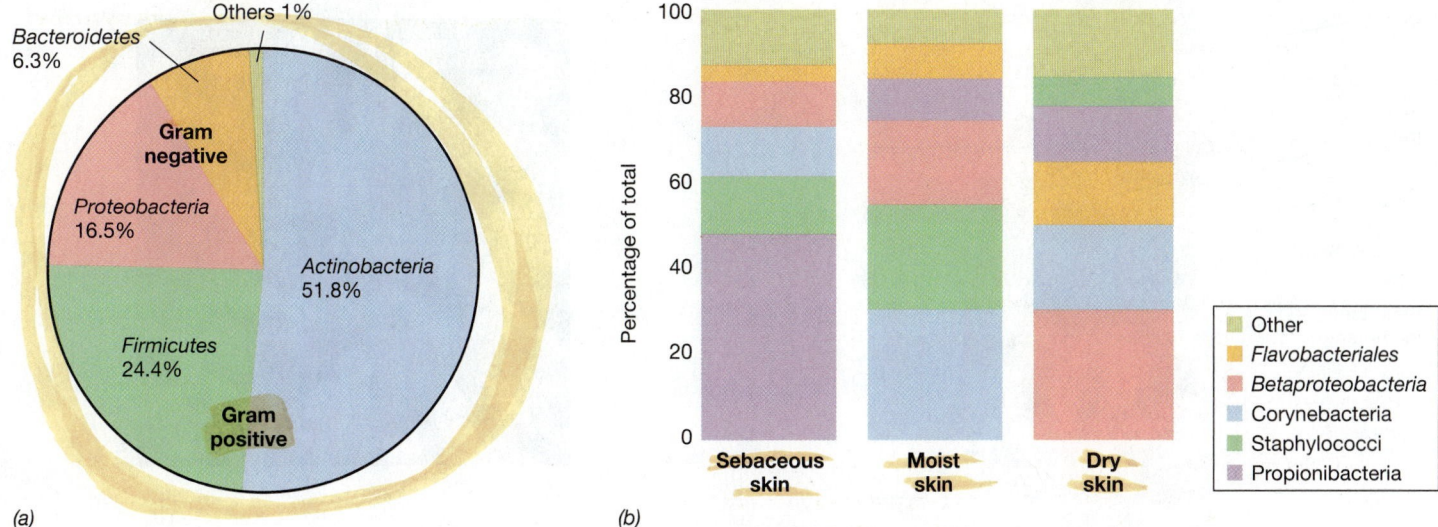

Figure 23.2 Normal skin microflora. *(a)* Analysis of the skin microbiome from 10 healthy human volunteers detected 19 bacterial phyla. Four phyla were predominant. *(b)* Composite populations of *Bacteria* from the same volunteers, divided according to sebaceous, moist, and dry skin microenvironments. Data are adapted from Grice et al., 2009, *Science 324:* 1190.

to comprise the normal microflora of a given individual. As is the case for the microbiome of the gut (⮂ Section 22.8), metagenomic analyses cannot define the total number of organisms present, but only estimate the number of different species present; thus, such analyses are estimates of *diversity* rather than *abundance*.

Eukaryotic microorganisms are also present on the skin. The chapter opening vignette (page 705) highlights microbiome studies that defined the genera and location of common fungi. *Malassezia* species are the most common fungi found on the skin, and at least five different species of this yeast are typically found in healthy individuals. In the absence of host resistance, as in patients with HIV/AIDS or whose normal microflora has been compromised, *Candida* and other fungi can also colonize the skin and cause serious infections.

Environmental and host factors influence the composition of the normal skin microflora. For example, the *weather* may cause an increase in skin temperature and moisture, which increases the density of the skin microflora. The *age* of the host also has an effect; young children have a more varied microflora and carry more potentially pathogenic gram-negative *Bacteria* than do adults. *Personal hygiene* influences the resident microflora; individuals with poor hygiene typically have higher microbial population densities on their skin. And finally, many microorganisms that would otherwise colonize skin cannot survive there simply because of its low moisture content and acidic pH.

MINIQUIZ

- Compare the populations of microorganisms in the three major skin microenvironments.
- Describe the properties of microorganisms that grow well on the skin.

23.3 Microflora of the Oral Cavity

The oral cavity is a complex, heterogeneous microbial habitat. Several different microenvironments exist in the oral cavity that can support a broad microbial diversity.

The Oral Microenvironment

Saliva contains microbial nutrients, but it is not a good growth medium because the nutrients are present in low concentration and saliva contains antibacterial substances. In particular, saliva contains *lysozyme*, an enzyme that cleaves glycosidic linkages in peptidoglycan of the bacterial cell wall, weakening the wall and causing cell lysis (⮂ Section 2.10). Another enzyme, *lactoperoxidase*, found in both milk and saliva, kills bacteria by a reaction in which singlet oxygen is generated (⮂ Section 5.16). Despite the activity of these antibacterial substances, food particles and cell debris provide high concentrations of nutrients near surfaces such as teeth and gums, creating favorable conditions for extensive local microbial growth, tissue damage, and disease.

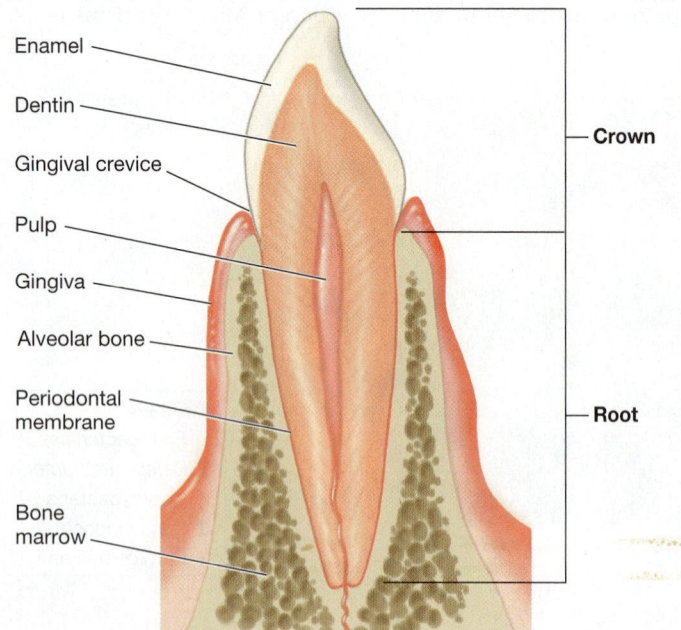

Figure 23.3 Section through a tooth. The diagram shows the tooth architecture and the surrounding tissues that anchor the tooth in the gum.

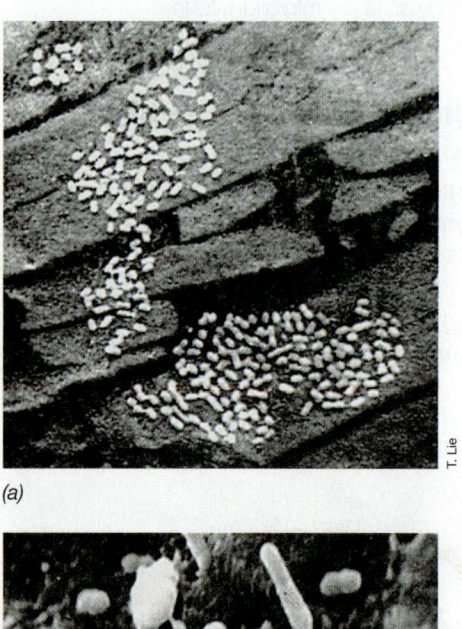

(a)

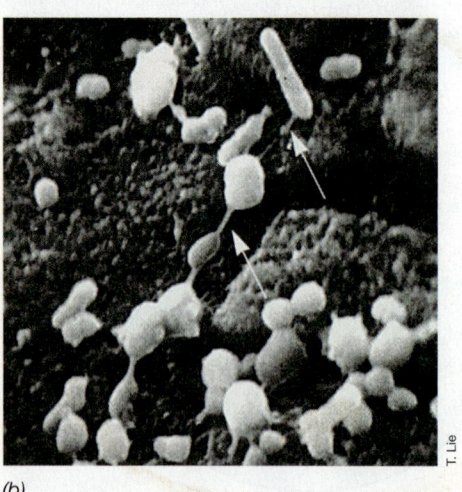

(b)

Figure 23.4 Colonization of tooth surfaces. *(a)* The colonies are growing on a model tooth surface inserted into the mouth for 6 h. *(b)* Higher magnification of the preparation in part a. Note the diverse morphology of the organisms present and the slime layer (arrows) holding the organisms together.

The tooth consists of a mineral matrix of calcium phosphate crystals (enamel) surrounding living tooth tissue (dentin and pulp) (**Figure 23.3**). Bacteria found in the mouth during the first year of life (when teeth are absent) are predominantly aerotolerant anaerobes such as streptococci and lactobacilli, and a few aerobes. When the teeth appear, the newly created surfaces are rapidly colonized by anaerobes that are specifically adapted to growth in biofilms on the surfaces of the teeth and in the gingival crevices (**Figure 23.4**).

Oral Microflora

Metagenomic analysis of the human oral microflora show a complex microbial community. Samples acquired from a number of subjects show over 600 taxa (**Table 23.2**). Most of these microorganisms have facultatively aerobic metabolisms, but some, such as *Bacteroidetes*, are obligately anaerobic and some have aerobic metabolisms, such as the *Neisseria*, *Acinetobacter*, and *Moraxella* genera in the *Proteobacteria* phylum. The most abundant genera are among the *Firmicutes*; *Veillonella parvula*, an obligate anaerobe, is the most abundant single species and *Streptococcus* is the most abundant genus in the mouth, comprising about 25% of bacteria found in some individuals. The related *Firmicutes* genera *Abiotrophia*, *Gemella*, and *Granulicatella* are also extremely common; species from these genera were among the 10 taxa most frequently detected. Most are present in much lower numbers, with only 17 taxa each contributing more than 1% of the oral microbiome. As is the case for the skin microbiome (Section 23.2), not all taxa are present or similarly distributed in all individuals.

MINIQUIZ

- Compare the microbial microenvironments in the oral cavity in newborns and adults.
- Identify the microorganisms that predominate in the adult oral cavity by taxa and metabolic requirements.

23.4 Microflora of the Gastrointestinal Tract

The human gastrointestinal tract consists of the stomach, small intestine, and large intestine (**Figure 23.5**). The gastrointestinal tract is responsible for digestion of food and absorption of nutrients, and many important nutrients are produced by the indigenous microflora. Starting with the stomach, the digestive tract is a column of nutrients mixed with microorganisms, primarily *Bacteria*. The nutrients move through this column and as they do, they encounter ever-changing microbial communities. Here we examine the organisms as well as their functions and special properties throughout the entire gastrointestinal tract. In Section 22.8 we examined the microbial diversity of the human large intestine, emphasizing the symbiotic nature of the microbial community and its host.

The gastrointestinal tract has about 400 m^2 of surface area and is home to about 10^{14} microbial cells. Our current view of the diversity and numbers of microorganisms that reside here has come from a combination of culture-dependent methods and culture-independent molecular methods (Chapter 18).

Table 23.2 Predominant microbial phyla and taxa in the oral cavity[a]

Domain and phylum	Number of taxa (percent)
Bacteria	
Firmicutes	227 (36.7)
Bacteroidetes	107 (17.3)
Proteobacteria	106 (17.1)
Actinobacteria	72 (11.6)
Spirochaetes	49 (7.9)
Fusobacteria	32 (5.2)
TM7	12 (1.9)
Synergistetes	10 (1.6)
Chlamydiae	1 (0.2)
Chloroflexi	1 (0.2)
SR1	1 (0.2)
Archaea	
Euryarchaeota	1 (0.2)
Total	619 (100)

[a]Data from Dewhirst, F.E., et al., 2010. *J. Bacteriol.* 192: 5002–5017. *Bacteria* are discussed in Chapters 14 and 15 and *Archaea* in Chapter 16.

The Stomach

Because stomach fluids are highly acidic (about pH 2), the stomach is a chemical barrier to the entry of microorganisms into the gastrointestinal tract. However, microorganisms do populate this seemingly hostile environment. The stomach microbial population consists of several different bacterial taxa. Each person has a unique population, but all contain several species of gram-positive bacteria as well as *Proteobacteria*, *Bacteroidetes*, *Actinobacteria*, and *Fusobacteria* (Figure 23.5). *Helicobacter pylori*, the most common single organism found, colonizes the stomach wall in many, but not all, individuals and can cause ulcers in susceptible hosts (⟲ Section 29.10). Some of the bacteria that populate the stomach consist of organisms found in the oral cavity, introduced with the passage of food.

Distal to the stomach, the intestinal tract consists of the *small* intestine and the *large* intestine, each of which is divided into different anatomical segments. The composition of the intestinal microflora in humans varies considerably and is somewhat dependent on diet. For example, persons on a meat-based diet show higher numbers of *Bacteroides* and lower numbers of coliforms and lactic acid bacteria than do individuals with a vegetarian diet. Representative microorganisms found in the gastrointestinal tract are shown in Figure 23.5.

The Small Intestine

The small intestine has two distinct environments in the *duodenum* and the *ileum*, which are connected by the *jejunum*. The duodenum, adjacent to the stomach, is fairly acidic and its normal microflora resembles that of the stomach. From the duodenum to the ileum, the pH gradually becomes less acidic and bacterial numbers increase. In the lower ileum, cell numbers of

UNIT 5

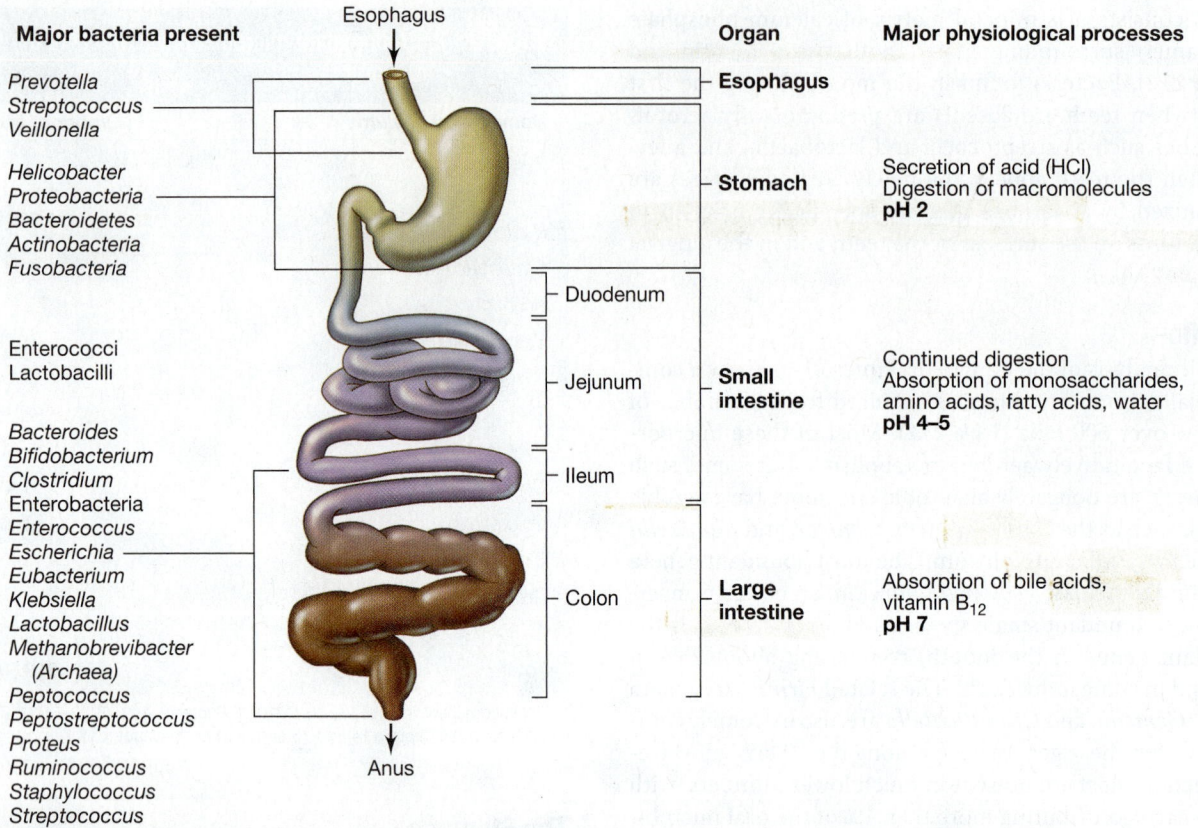

Major bacteria present

Esophagus

Prevotella
Streptococcus
Veillonella

Helicobacter
Proteobacteria
Bacteroidetes
Actinobacteria
Fusobacteria

Enterococci
Lactobacilli

Bacteroides
Bifidobacterium
Clostridium
Enterobacteria
Enterococcus
Escherichia
Eubacterium
Klebsiella
Lactobacillus
Methanobrevibacter
(Archaea)
Peptococcus
Peptostreptococcus
Proteus
Ruminococcus
Staphylococcus
Streptococcus

Anus

Organ	Major physiological processes
Esophagus	
Stomach	Secretion of acid (HCl) Digestion of macromolecules **pH 2**
Duodenum Jejunum Ileum — **Small intestine**	Continued digestion Absorption of monosaccharides, amino acids, fatty acids, water **pH 4–5**
Colon — **Large intestine**	Absorption of bile acids, vitamin B$_{12}$ **pH 7**

Figure 23.5 Microflora of the human gastrointestinal tract. These taxa are representative of microorganisms found in healthy adults. Not every individual harbors all of these microorganisms.

10^5–10^7/gram of intestinal contents are common, even though the environment becomes progressively more anoxic. Fusiform anaerobic bacteria are typically present, attached to the intestinal wall at one end (**Figure 23.6**, ↩ Figure 1.10*b*).

The Large Intestine

The ileum empties into the *cecum*, the connecting portion of the large intestine. The *colon* makes up the rest of the large intestine. In the colon, prokaryotes are present in enormous numbers. The colon is essentially an *in vivo* fermentation vessel; many bacteria live here, using nutrients derived from the digestion of food (Figure 23.5). Facultative aerobes such as *Escherichia coli* are present but in smaller numbers than other bacteria; total counts of facultative aerobes are less than 10^7/gram of intestinal contents. The facultative aerobes consume any remaining oxygen, making the large intestine strictly anoxic. Anoxia promotes growth of obligate anaerobes such as *Clostridium* and *Bacteroides* species.

The total number of obligate anaerobes in the colon is enormous. Bacterial counts of 10^{10} to 10^{11} cells/gram in distal gut and fecal contents are normal, with *Bacteroidetes* and gram-positive species accounting for greater than 99% of all *Bacteria*. The archaeal methanogen *Methanobrevibacter smithii* (↩ Section 16.2) may also be present in significant numbers. Protists are not found in the gastrointestinal tract of healthy humans but they can cause gastrointestinal infections if ingested in contaminated food or water (Chapter 31). Section 22.8 gives a molecular snapshot of bacterial diversity in the human large intestine.

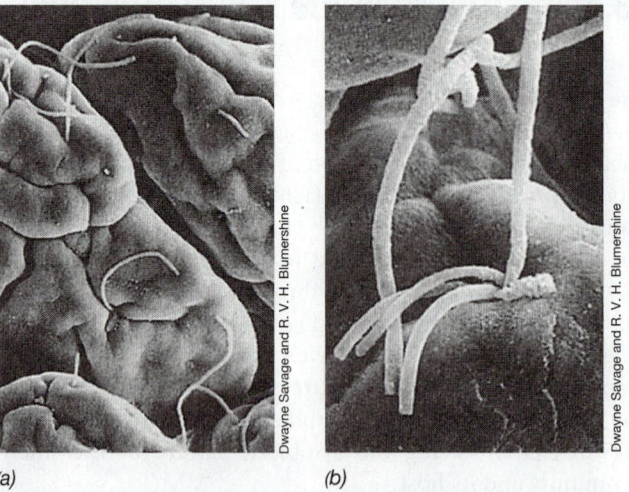

(a)　(b)

Dwayne Savage and R. V. H. Blumershine

Figure 23.6 Microflora in the small intestine. Scanning electron micrographs of the microbial community on epithelial cells in the mouse ileum. *(a)* An overview at low magnification shows long, filamentous fusiform bacteria on epithelial cells in the mouse ileum. *(b)* Higher magnification shows several filaments attached at a single depression. The attachment is at the end of the filaments. Individual cells are 10–15 μm long.

EXPLORE THE
MICROBIAL WORLD

Microorganisms that are part of the normal microflora grow on and in the body and are essential to the well-being of all higher organisms. The microorganisms we acquire and retain compete at various sites in the body with pathogens, inhibiting colonization by these organisms. Commensals that reside in the gut are active participants in the digestion of food and manufacture of essential nutrients.

It follows that humans could manipulate their commensal bacteria, perhaps altering, regulating, or enhancing the normal flora to enhance the positive benefits of certain selected bacteria. In theory, the ingestion of selected microorganisms might be used to change or reestablish our gastrointestinal microflora to promote health, especially in individuals who experience major changes in their normal microflora due to disease, surgery, or other medical treatments, or whose normal microflora is altered for other reasons, such as poor diet. Intentionally ingested microorganisms used for this purpose are called *probiotics* (**Figure 1**). Probiotics are suspensions of live microorganisms that, when administered in adequate amounts, confer a perceived health benefit on the host.

Are probiotics really useful? Few studies conclusively show that the alteration of commensal microbial populations in normal healthy adults has major, long-lasting, positive health effects. Most probiotic products (Figure 1) are directed toward replacing or reconstituting the intestinal microflora of humans by ingesting live, concentrated microbial cultures; the products are directed at preventing or correcting digestive problems. Probiotic products may thus confer short-term benefits, but there is little evidence for long-lasting establishment or reestablishment of an altered microflora without the continued consumption of the probiotic.

Probiotics are routinely used in production farm animals as part of their normal diet to prevent digestive problems. These same probiotic treatments may also be beneficial by reducing antibiotic use and preventing the development of antibiotic-resistant foodborne pathogens. The focus here is on the preventative nature of the treatments; animal probiotics are not used as a cure for disease. Strains of *Saccharomyces* (yeast), *Lactobacillus, Bacillus,* and *Propionibacterium* have been used for these purposes. Conceivably, similar treatments would also be beneficial for humans.

A number of human ailments respond positively to probiotics administration, although the mechanisms by which this occurs are unclear. For example, the watery diarrhea children experience from rotavirus infection can be shortened by administration of several probiotics preparations. *Saccharomcyes* may reduce the recurrence of diarrhea and shorten infections due to *Clostridium difficile*. Probiotic lactobacilli have also been used to treat urogenital infections in humans.

The composition of the gut microflora can change rapidly when probiotics are administered. In many cases, the makers of probiotics recommend that the microbial supplements should be consumed on a regular basis over a long period of time to achieve the intended result; if consumption is stopped, the gut microflora returns to its original state, indicating that the effects of probiotics are likely only short term. Thus, while probiotics may offer several benefits, especially for reestablishing the gut's normal microflora following catastrophic events such as severe diarrheal disease, evidence for positive and lasting benefits is not well established. Carefully designed and scientifically controlled studies must be conducted to document the outcomes of probiotic treatment. The studies must use standardized preparations of probiotics containing known organisms and administered in precise doses to test efficacy.

Sources: Walker, R., and M. Buckley. 2006. *Probiotic Microbes: The Scientific Basis*. American Academy of Microbiology. Goldin, B.R., and Gorbach, S.L. 2008. *Clinical Indications for Probiotics: An Overview. Clinical Infectious Diseases 46*: S96–100.

UNIT 5

Deborah O. Jung and John Martinko

Figure 1 Probiotics. Probiotic foods and supplements widely available in the United States.

Products of Intestinal Microflora

Intestinal microorganisms carry out a variety of metabolic reactions that produce several important compounds (Table 23.3). The composition of the intestinal microflora and the diet influence the type and amount of compounds produced. Among these products are vitamins B_{12} and K. These essential vitamins are not synthesized by humans (and vitamin B_{12} is not present in plants) but are made by the intestinal microflora and absorbed from the colon. In addition, steroids, produced in the liver and released into the intestine from the gallbladder as bile acids, are modified in the intestine by the microflora; the modified bioactive steroid compounds are then absorbed from the gut.

Other products generated by the activities of fermentative bacteria and methanogens include gas and several odoriferous substances (Table 23.3). Normal adults expel several hundred milliliters of gas from the intestines each day (flatus), of which about half is nitrogen (N_2) from swallowed air. Some foods metabolized by fermentative bacteria in the intestines result in the production of hydrogen (H_2) and carbon dioxide (CO_2). Methanogens, found in the intestines of many but not all adults, convert H_2 and CO_2 produced by fermentative bacteria to methane (CH_4). The methanogens in the rumen of cattle (Section 22.7) produce huge amounts of methane, up to a quarter of total global production.

During the passage of food through the gastrointestinal tract, water is absorbed from the digested material, which gradually becomes more concentrated and is converted to feces. Bacteria compose about one-third of the weight of fecal matter. Organisms living in the lumen of the large intestine are continuously displaced downward by the flow of material, and bacteria that are lost are continuously replaced by new growth, similar to an *in vitro* continuous culture system (Section 5.7). The time needed for passage of material through the human gastrointestinal tract is about 24 h, and the growth rate of bacteria in the lumen is one to two doublings per day. In humans, about 10^{13} bacterial cells are shed each day in feces.

Changing the Normal Microflora

When an antibiotic is taken orally it inhibits the growth of the normal flora as well as the targeted pathogen(s), leading to the loss of antibiotic-susceptible bacteria in the intestinal tract; this alteration of the intestinal microbial community is often signaled by the voiding of loose feces or diarrhea. In the absence of a full complement of normal flora, opportunistic pathogens such as antibiotic-resistant *Staphylococcus, Proteus, Clostridium difficile*, or the yeast *Candida albicans* can become established and can affect digestive functions or actually cause disease. For example, antibiotic treatment allows bacteria such as *C. difficile* that are less susceptible to antibiotics to grow without competition from the normal flora, causing infection and colitis.

When antibiotic therapy ends, the normal intestinal flora are quickly reestablished in adults. To speed this process, recolonization of the gut by desired species can be accomplished by the administration of **probiotics**, live cultures of intestinal bacteria that, when administered to a host, may confer a health benefit. Rapid recolonization of the gut may reestablish a competitive local flora that can outcompete pathogens and provide desirable microbial metabolic products (see Explore the Microbial World, "Probiotics").

MINIQUIZ

- Why might the small intestine be more suitable for growth of facultative aerobes than the large intestine?
- Identify several essential compounds made by indigenous intestinal microorganisms.

23.5 Microflora of Mucosal Tissues

Mucous membranes support the growth of a normal microflora that prevents infection by pathogenic microorganisms. Here we discuss two mucosal environments and their resident microorganisms.

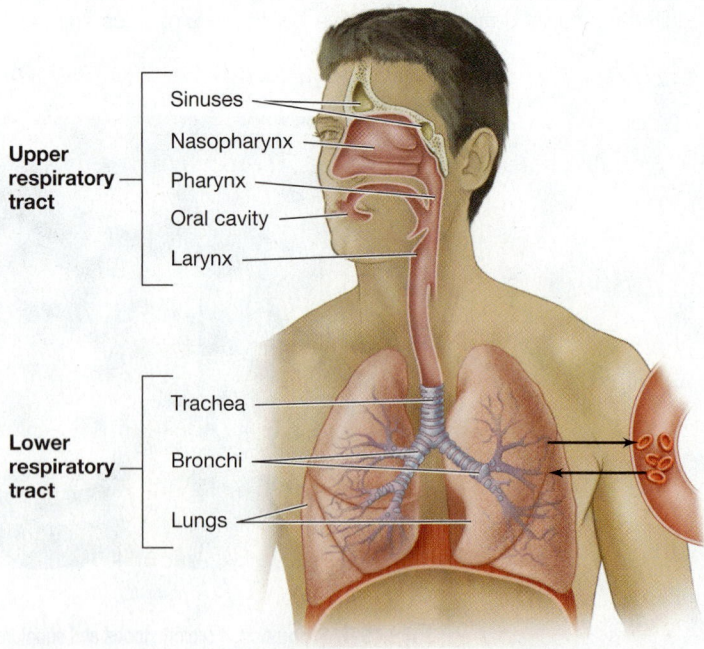

Figure 23.7 **The respiratory tract.** In healthy individuals the upper respiratory tract has a large variety and number of microorganisms. By contrast, the lower respiratory tract in a healthy person has few if any microorganisms.

Table 23.3 Biochemical/metabolic contributions of intestinal microorganisms	
Process	**Product**
Vitamin synthesis	Thiamine, riboflavin, pyridoxine, B_{12}, K
Gas production	CO_2, CH_4, H_2
Odor production	H_2S, NH_3, amines, indole, skatole, butyric acid
Organic acid production	Acetic, propionic, butyric acids
Glycosidase reactions	β-Glucuronidase, β-galactosidase, β-glucosidase, α-glucosidase, α-galactosidase
Steroid metabolism (bile acids)	Esterified, dehydroxylated, oxidized, or reduced steroids

Respiratory Tract

The anatomy of the respiratory tract is shown in **Figure 23.7**. In the **upper respiratory tract** (nasopharynx, oral cavity, larynx, and pharynx), microorganisms live in areas bathed with secretions from the mucous membranes. Bacteria continually enter the upper respiratory tract from the air during breathing, but most are trapped in the mucus of the nasal and oral passages and expelled with nasal secretions, or swallowed. A few microorganisms, however, colonize respiratory mucosal surfaces in all individuals. The microorganisms most commonly present are staphylococci, streptococci, diphtheroid bacilli, and gram-negative cocci.

Occasionally, potential pathogens such as *Staphylococcus aureus* and *Streptococcus pneumoniae* are part of the normal flora in the nasopharynx of healthy individuals (Table 23.1). These individuals are *carriers* of the pathogens but do not normally develop disease, presumably because the other resident microorganisms compete successfully for nutritional and metabolic resources and limit pathogen activities. The innate immune system (⮌ Section 24.2) and components of the adaptive immune system such as secreted antibodies (⮌ Section 25.7) are particularly active at mucosal surfaces and inhibit growth and invasion by potential pathogens.

The **lower respiratory tract** (trachea, bronchi, and lungs) has no resident microflora in healthy adults, despite the large number of organisms potentially able to reach this region during normal breathing. Dust particles, which are fairly large, settle in the upper respiratory tract. As the air passes into the lower respiratory tract, the flow rate decreases, and organisms settle onto the walls of the respiratory passages. The walls of the entire respiratory tract are lined with ciliated epithelial cells, and the cilia, beating upward, push bacteria and other particulate matter toward the upper respiratory tract where they are then expelled in saliva and nasal secretions or are swallowed. Only particles smaller than about 10 μm in diameter reach the lungs. Nevertheless, some pathogens reach these locations and cause disease, most notably pneumonia caused by certain bacteria or viruses (⮌ Sections 29.2 and 29.6).

Urogenital Tract

In healthy male and female urogenital tracts (**Figure 23.8**), the kidneys and bladder are sterile; however, epithelial cells lining the distal urethra are colonized by facultatively aerobic gram-negative *Bacteria* (Table 23.1). Potential pathogens such as *Escherichia coli* and *Proteus mirabilis*, normally present in small numbers in the body or in the local environment, can multiply in the urethra

Female

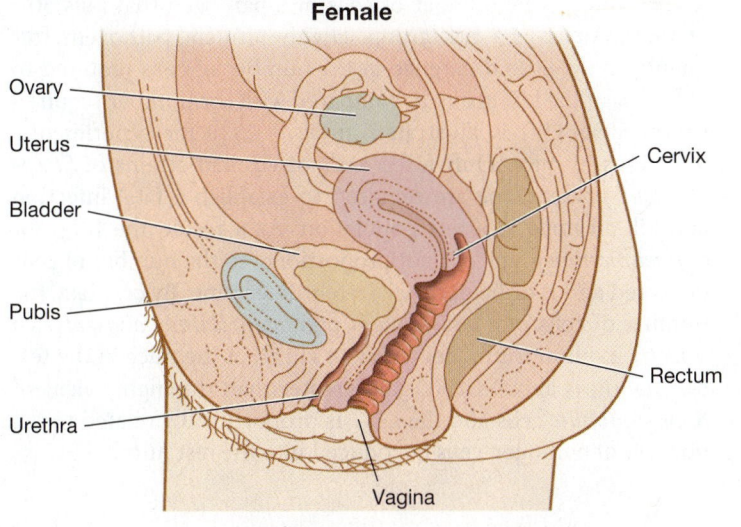

Ovary
Uterus
Bladder
Pubis
Urethra
Cervix
Rectum
Vagina

(a)

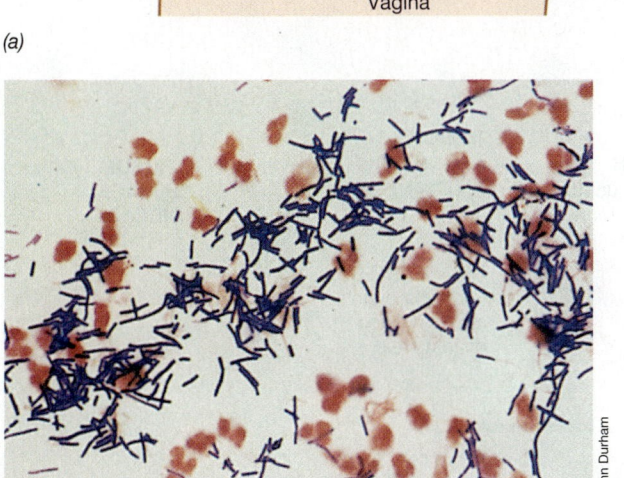

(b)

John Durham

Male

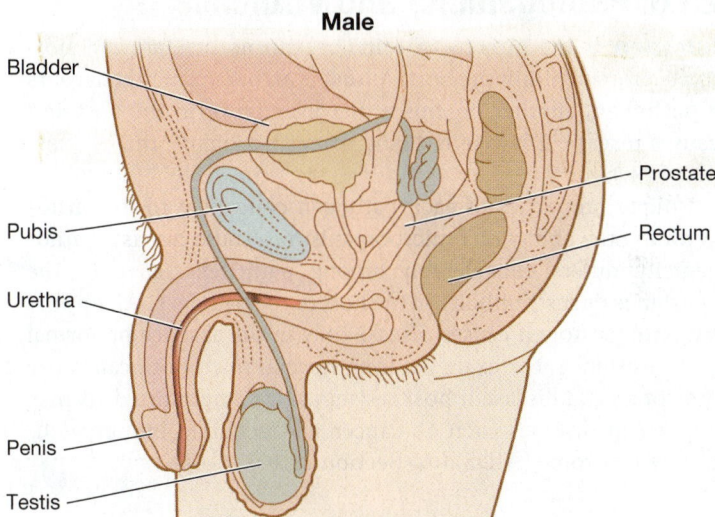

Bladder
Pubis
Urethra
Penis
Testis
Prostate
Rectum

Figure 23.8 Microbial growth in the genitourinary tract. *(a)* The genitourinary tracts of the human female and male, showing regions (red) where microorganisms often grow. The upper regions of the genitourinary tracts of both males and females are sterile in healthy individuals. *(b)* Gram stain of *Lactobacillus acidophilus*, the predominant organism in the vagina of women between the onset of puberty and the end of menopause. The individual gram-positive rods are 3–4 μm long.

UNIT 5

and cause disease if conditions such as changes in pH occur. Such organisms are a frequent cause of urinary tract infections, especially in females.

The vagina of the adult female is weakly acidic (pH <5) and contains significant amounts of glycogen. *Lactobacillus acidophilus*, a resident organism in the vagina, ferments the polysaccharide glycogen, producing lactic acid that maintains a local acidic environment (Figure 23.8*b*). Other organisms, such as species of the yeasts *Torulopsis* and *Candida*, various streptococci, and *E. coli*, may also be present. Before puberty, *L. acidophilus* is absent, the female vagina is neutral and does not produce glycogen, and

the flora consist predominantly of staphylococci, streptococci, diphtheroids, and *E. coli*. After menopause, glycogen production ceases, the pH rises, and the flora again resembles that found before puberty.

II • Pathogenesis

Microbial *pathogenesis* is the process by which microorganisms cause disease. Microbial pathogenesis begins with exposure and adherence of microorganisms to host cells (Figure 23.1), followed by invasion, infection, and ultimately, disease (**Figure 23.9**). We begin with some important terms commonly used in medical microbiology.

23.6 Pathogenicity and Virulence

Infection is the growth of microorganisms that are not normally present within the host. A **host** is an organism that harbors a **pathogen**, another organism that lives on or in the host and causes disease. **Disease** is tissue damage or injury that impairs host function.

Unique properties of each pathogen contribute to its **pathogenicity**, the ability of a microorganism to cause disease. Pathogenicity differs considerably among pathogens, as does the resistance or susceptibility of the host to the pathogen. An **opportunistic pathogen** causes disease only in the absence of normal host resistance. For example, even normal microflora can cause infections and disease if host resistance is compromised, as may happen in diseases such as cancer and acquired immunodeficiency syndrome (AIDS) (⟲ Section 29.14).

Virulence

The measure of pathogenicity is called **virulence**, the relative ability of a pathogen to cause disease. Virulence is the outcome of host–pathogen interactions, a dynamic relationship between the two organisms, influenced by ever-changing conditions in the pathogen, the host, and the environment. Neither the virulence of the pathogen nor the relative resistance of the host is a constant factor.

Virulence can be measured in experimental studies of the LD_{50} (lethal dose$_{50}$), the number of cells of a pathogen that kills 50% of the animals in a test group. Highly virulent pathogens frequently show little difference in the number of cells required to kill 100% of the test group as compared with the number required to kill 50%. This is illustrated in **Figure 23.10** for experimental infections in mice. Only a few cells of virulent strains of *Streptococcus pneumoniae* are required to establish a fatal infection and kill all mice in a test population. As a result, the LD_{50} for *S. pneumoniae* in mice is not proportional to the number of cells delivered and is difficult to accurately determine. By contrast, the number of cells of a less virulent pathogen, *Salmonella enterica* serovar Typhimurium, necessary to kill all of the mice in the test population is about 10,000-fold greater than the highly virulent *S. pneumoniae* cells and the LD_{50} is proportionally related to the number of pathogen cells introduced into the test mice.

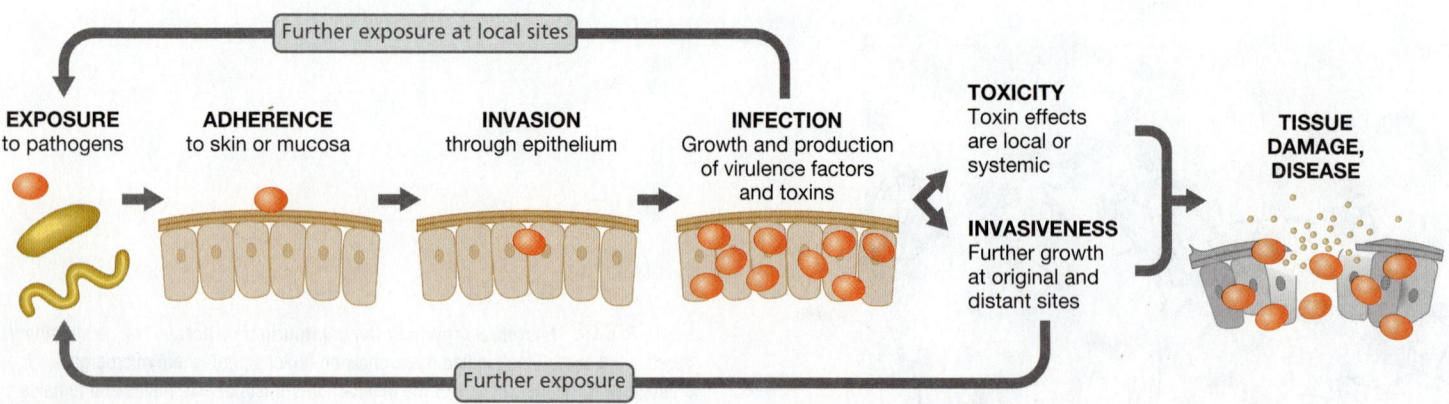

Figure 23.9 Microbial pathogenesis. Following exposure to a pathogenic microorganism, pathogen-directed events can result in disease.

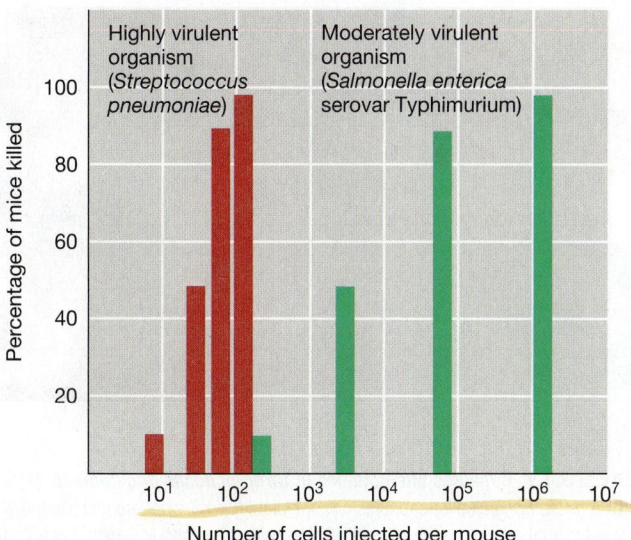

Figure 23.10 Microbial virulence. Differences in microbial virulence demonstrated by the number of cells of *Streptococcus pneumoniae* and *Salmonella enterica* serovar Typhimurium required to kill mice.

Attenuation

Attenuation is the decrease or loss of virulence of a pathogen. When pathogens are kept in laboratory culture rather than isolated from diseased animals, their virulence often decreases, or is even completely lost. Strains that have either a reduced virulence or are no longer virulent are said to be *attenuated*. Attenuation probably occurs because nonvirulent or weakly virulent mutants grow faster than virulent strains in laboratory media, where virulence has no selective advantage. After successive transfers in fresh media, such mutants are therefore selectively favored. If an attenuated culture is reinoculated into an animal, the organism may regain its original virulence, especially with continued in vivo passage. But in many cases, a loss of virulence is permanent. Nevertheless, attenuated strains can be very valuable because they are often used for the production of vaccines, especially viral vaccines. For example, measles, mumps, and rubella vaccines, and rabies vaccines for animals other than humans, employ attenuated strains of each virus.

MINIQUIZ -

- How can the LD_{50} test be used to define virulence of a pathogen?

- What circumstances can contribute to attenuation of a pathogen?

- -

23.7 Adherence

Bacteria or viruses able to initiate infection often adhere to epithelial cells through specific interactions between molecules on the pathogen and molecules on the host tissues. In addition, pathogens often adhere to each other, forming biofilms (⮌ Section 19.4). In medical microbiology, **adherence** is the enhanced ability of a microorganism to attach to a cell or surface.

Noncovalent Adherence Factors

Some macromolecules responsible for bacterial adherence are not covalently attached to the bacteria. These surface molecules are collectively known as a **glycocalyx**, a polymer secreted by a bacterium that coats the surface of the bacterium. These are usually macromolecular polysaccharide polymers. For example, in the case of *Bacillus anthracis* (the bacterium that causes anthrax), the gycocalyx is a capsule composed of a polymer of D-glutamic acid. This capsule can be readily visualized, and the encapsulated cells of *B. anthracis* typically form smooth slimy colonies (**Figure 23.11**).

Cell surface structures were considered in Section 2.13. A loose network of polymers extending outward from a cell is called a **slime layer** (Figure 23.4*b*). A coat consisting of a dense, well-defined polymer layer surrounding the cell is called a **capsule** (**Figure 23.12***a*). Slime layers and capsules are important for adherence to other bacteria as well as to host tissues, but many pathogens, for example *Vibrio cholerae*, the causative agent of the disease cholera, do not require slime layers or capsules for attachment (Figure 23.12*b*).

Capsules are particularly important for protecting pathogenic bacteria from host defense mechanisms. For example, the only known virulence factor for *Streptococcus pneumoniae* is

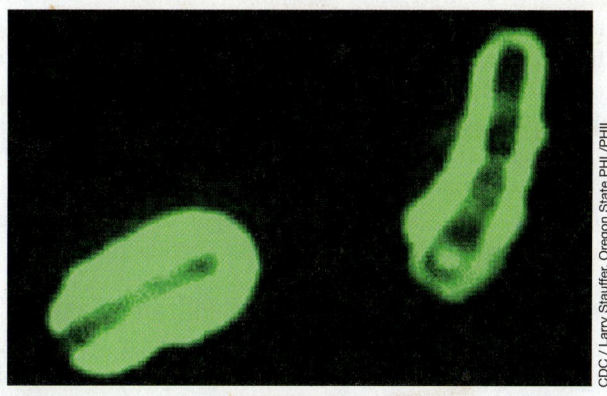

(a)

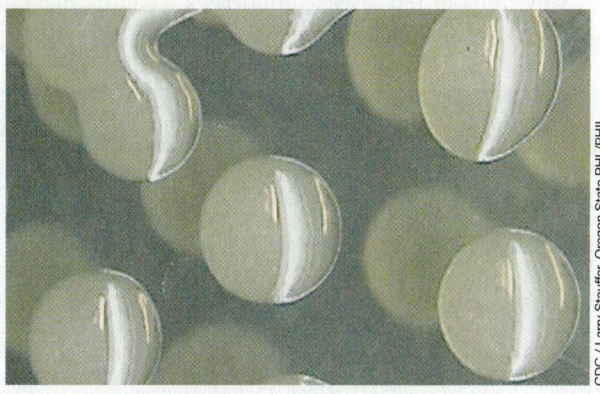

(b)

Figure 23.11 *Bacillus anthracis* **and capsules.** (a) Fluorescent stain of *B. anthracis* capsules. Antibodies for the *B. anthracis* capsule are prepared with a fluorescent dye attached that stains the capsule green; the capsule extends up to 1 μm from the cell. (b) *B. anthracis* growing on an agar plate. Colonies of encapsulated cells are about 0.5 cm in diameter and mucoid in appearance.

UNIT 5

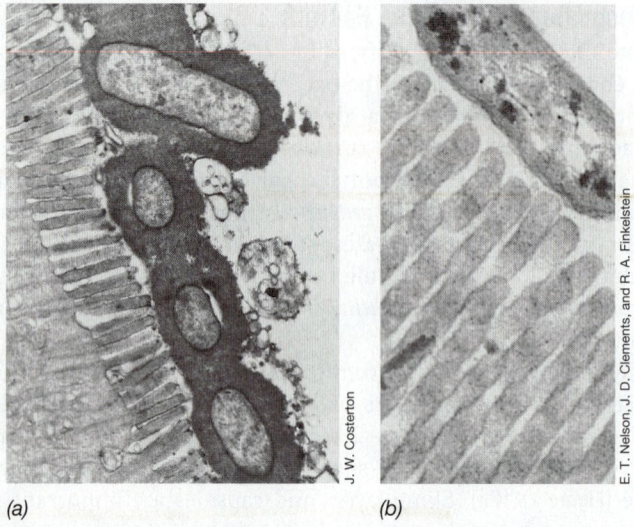

(a)   (b)

Figure 23.12 **Adherence of pathogens to tissues.** *(a)* Enteropathogenic *Escherichia coli* is attached to the brush border of intestinal microvilli through a distinct capsule. The *E. coli* cells are about 0.5 μm in diameter. *(b)* Transmission electron micrograph of a thin section of *Vibrio cholerae* adhering to the brush border of microvilli in the intestine. This organism has no capsule.

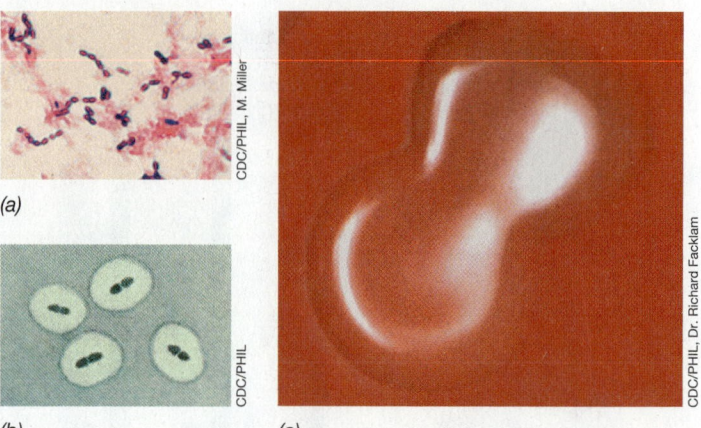

(a)

(b)   (c)

Figure 23.13 **Capsules and colonies in *Streptococcus pneumoniae*.** *(a)* Gram stain of *S. pneumoniae* cells; capsules are not visible. *(b)* *S. pneumoniae* treated with anticapsular antibodies (Quellung reaction) that make the capsule visible. *(c)* Colonies of encapsulated *S. pneumoniae* cells grown on blood agar show a mucoid morphology with a sunken center. The colonies are about 2–3 mm in diameter.

its polysaccharide capsule (**Figure 23.13**). Encapsulated strains of *S. pneumoniae* grow in lung tissues in enormous numbers, where they initiate host responses that lead to pneumonia, interfere with lung function, and cause extensive host damage or even death (⮌ Section 29.2). By contrast, nonencapsulated strains are quickly and efficiently ingested and destroyed by phagocytes, white blood cells that ingest and kill bacteria by a process called *phagocytosis*. Thus, *S. pneumoniae* capsules (Figure 23.13*b*) are essential for pathogenicity; the capsules defeat a major defense mechanism used by the host to prevent invasion (⮌ Section 24.2).

Other Adherence Factors: Fimbriae, Pili, and Flagella

Many pathogens selectively adhere to particular types of cells through cell surface structures other than capsules or slime layers. For example, *Neisseria gonorrhoeae*, the pathogen that causes the sexually transmitted disease gonorrhea, adheres specifically

to mucosal epithelial cells in the genitourinary tract, eye, rectum, and throat; other tissues are not infected. *N. gonorrhoeae* has a cell surface protein called Opa (*o*pacity *a*ssociated *p*rotein) that binds specifically to a host protein called CD66 found only on the surface of these cells, allowing adherence of the pathogen to host cells (**Table 23.4**). Likewise, influenza virus targets lung mucosal cells and attaches specifically to lung epithelial cells by way of the protein hemagglutinin present on the virus surface (⮌ Section 29.8).

Fimbriae and pili are bacterial cell surface protein structures (⮌ Section 2.13) that function in the attachment process. For instance, the pili of *Neisseria gonorrhoeae* play a key role in attachment to urogenital epithelia, and fimbriated strains of *Escherichia coli* (**Figure 23.14**) are more frequent causes of urinary tract infections than strains lacking fimbriae. Among the best-characterized fimbriae are the *type I fimbriae* of enteric bacteria (*Escherichia, Klebsiella, Salmonella,* and *Shigella*). Type I fimbriae are uniformly distributed on the surface of cells. Pili are typically longer

Table 23.4 Major adherence factors used to facilitate attachment of microbial pathogens to host tissues[a]

Factor	Example
Capsule/slime layer (Figures 23.4, 23.11, 23.12, 23.13, 23.15)	Pathogenic *Escherichia coli*—capsule promotes adherence to the brush border of intestinal microvilli
	Streptococcus mutans—dextran slime layer promotes binding to tooth surfaces
Adherence proteins	*Streptococcus pyogenes*—M protein on the cell binds to receptors on respiratory mucosa
	Neisseria gonorrhoeae—Opa protein on the cell binds to CD66 receptors on epithelium
Lipoteichoic acid (⮌ Figure 2.27)	*Streptococcus pyogenes*—lipoteichoic acid facilitates binding to respiratory mucosal receptor (along with M protein)
Fimbriae (pili) (Figure 23.14)	*Neisseria gonorrhoeae*—pili facilitate binding to epithelium
	Salmonella species—type I fimbriae facilitate binding to epithelium of small intestine
	Pathogenic *Escherichia coli*—fimbrial colonization factor antigens (CFAs) facilitate binding to epithelium of small intestine

[a]Most receptor sites on host tissues are glycoproteins or complex lipids such as gangliosides or globosides.

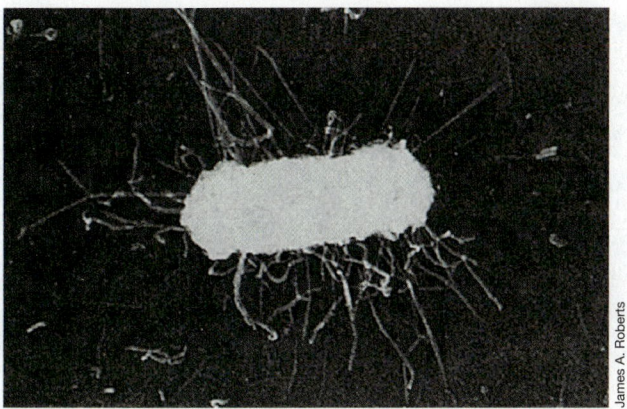

Figure 23.14 Fimbriae. Shadow-cast electron micrograph of the bacterium *Escherichia coli* showing type P fimbriae, which resemble type I fimbriae but are somewhat longer. The cell is about 0.5 μm in diameter.

James A. Roberts

than fimbriae, with fewer pili found on the cell surface. Both pili and fimbriae function by binding host cell surface glycoproteins, initiating adherence. Flagella can also increase adherence to host cells (see Figure 23.17).

Studies of diarrhea caused by some pathogenic strains of *E. coli* indicate specific interactions between the mucosal epithelium and pathogens. Most strains of *E. coli* are nonpathogenic inhabitants of the cecum and the colon (Figure 23.5). Several nonpathogenic strains of *E. coli* are usually present in the body at the same time, and large numbers routinely pass through the body and are voided in feces. However, pathogenic enterotoxigenic strains of *E. coli* contain genes encoding fimbrial colonization factor antigens; these proteins adhere specifically to cells in the host small intestine. From here, they multiply and produce enterotoxins that cause diarrhea as well as other illnesses (⤴ Section 31.11).

Another notorious pathogen, *Streptococcus pyogenes*, the causative agent of strep throat and scarlet and rheumatic fevers (⤴ Section 29.2), utilizes fimbriae-associated lipoteichoic acid, along with two specific proteins, F and M, to facilitate attachment to host cells (Table 23.4). M protein is also responsible for resistance to phagocytosis by neutrophils, immune cells important in antibacterial resistance (⤴ Sections 24.2 and 29.2).

MINIQUIZ --

- Describe glycocalyx, slime layers, and capsules.
- How do Opa proteins on *Neisseria gonorrhoeae* and fimbrial colonization factor antigens on *Escherichia coli* and *Salmonella* cells influence adherence to mucosal tissues?

23.8 Invasion, Infection, and Virulence Factors

The initial inoculum of a pathogen is usually insufficient to cause host damage, even if a pathogen gains access to tissues. The pathogen must therefore multiply and colonize the tissue (Figure 23.9). To do so the pathogen must find appropriate nutrients and environmental conditions to grow and cause infection in the host.

Invasion

Following colonization a pathogen must usually invade tissues to initiate disease. **Invasion** is the ability of a pathogen to enter into host cells or tissues, spread, and cause disease. In most cases, microbial infections begin at breaks or wounds in the skin or on the mucous membranes of the respiratory, digestive, or genitourinary tract, surfaces that are normally microbial barriers. In some cases, growth may also begin on intact mucosal surfaces, especially if the normal flora has been altered or eliminated, for example, by antibiotic therapy.

Some pathogens remain localized after initial entry, multiplying and invading at a single focus of infection such as the boil that may arise from *Staphylococcus* skin infections (⤴ Section 29.9). If bacterial growth occurs, organisms may cause **bacteremia**, the presence of bacteria in the bloodstream, from where they can travel to distant parts of the body. Spread of the pathogen through the blood and lymph systems can also result in a bloodborne systemic infection called **septicemia**, and the organism may spread to other tissues. Septicemia may lead to massive inflammation, culminating in septic shock and rapid death, as we discuss in Section 24.5. Bacteremia and septicemia usually start as an infection in a specific organ such as the intestine, kidney, or lung.

Infection and Disease

Infection refers to any situation in which a microorganism not a member of the local flora is established and growing in a host, whether or not the host is harmed. Infection is not synonymous with disease because growth of a microorganism, even a pathogen, on a host does not always cause host damage. The normal microflora also grow on the host and are usually harmless, but may infect the host as opportunistic pathogens and cause disease if host resistance is compromised, for instance by cancer or acquired immunodeficiency syndrome (AIDS) (⤴ Section 29.14).

Infection requires growth of microorganisms after they have attached to surfaces. This requires that the host environment supply adequate nutrients to support growth. However, not all growth factors, both organic as well as trace metals, are in adequate supply in all tissues at all times, even in a vertebrate host. For example, iron is a major growth-limiting micronutrient that influences microbial growth. Host proteins called *transferrin* and *lactoferrin* are very high-affinity iron-binders and function to sequester iron in the host, which can limit infection. But many pathogens produce iron-chelating compounds called *siderophores* to counter host iron sequestration, and some of these are so effective that they can actually remove iron from host tissues.

The attachment and infection process has been well studied with the formation of biofilms (⤴ Section 19.4) on tooth surfaces by the oral microflora. Even on a freshly cleaned tooth surface, acidic glycoproteins from the saliva form a thin organic film several micrometers thick; this film provides an attachment site for bacterial cells. Streptococci quickly colonize the glycoprotein film. These include, in particular, the two species implicated in tooth decay, *Streptococcus sobrinus* and *S. mutans* (**Figure 23.15**). *S. sobrinus* has an affinity for salivary glycoproteins secreted onto smooth tooth surfaces (Figure 23.4). *S. mutans*, by contrast, resides in crevices and small fissures where it produces dextran, a strongly adhesive polysaccharide that it uses to attach to tooth

UNIT 5

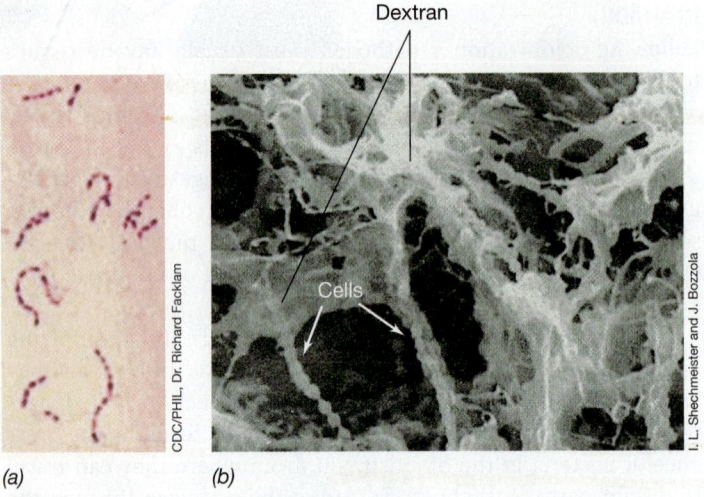

(a) *(b)*

Figure 23.15 **The cariogenic bacterium *Streptococcus mutans*.** *(a)* Stained cells in a light micrograph show the characteristic cell chains of streptococci. *(b)* Scanning electron micrograph of the sticky dextran material that holds cells together in filaments. Individual cells of *S. mutans* are about 1 μm in diameter.

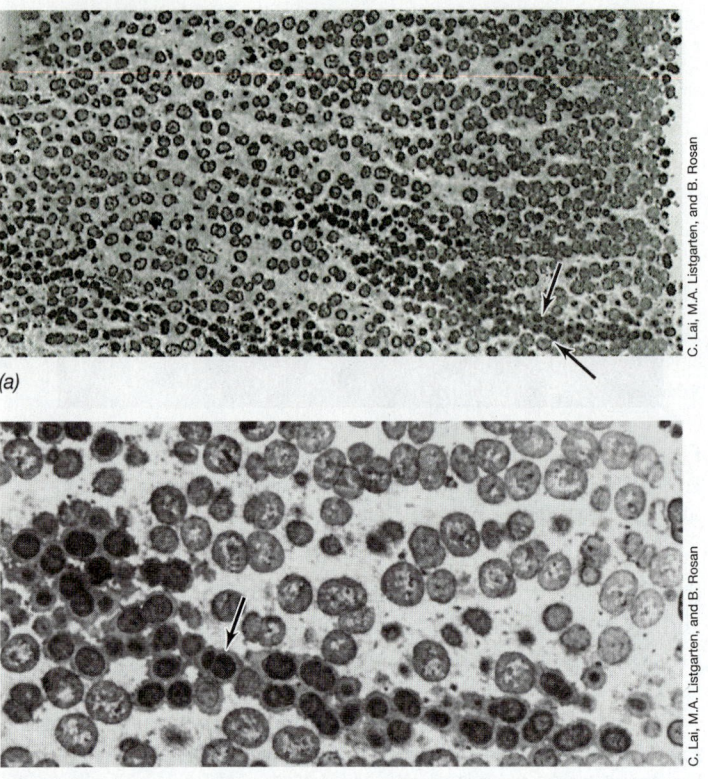

(b)

Figure 23.16 **Dental plaque.** The left side of the transmission electron micrograph is the base of the plaque; the right side is the portion exposed to the oral cavity. *(a)* The low-magnification micrograph shows predominantly streptococci. *Streptococcus sobrinus*, labeled by an antibody-microchemical technique, appears darker than the rest. *S. sobrinus* cells are seen as two distinct chains (arrows). *(b)* Higher-magnification micrograph showing the region with *S. sobrinus* cells (dark, arrow). Note the extensive slime layer surrounding the *S. sobrinus* cells. Individual cells are about 1 μm in diameter. The genus *Streptococcus* resides in the *Firmicutes* and is discussed in Section 15.6.

surfaces (see Figure 23.16). Both *S. sobrinus* and *S. mutans* are lactic acid bacteria that ferment glucose to lactic acid, the agent that destroys tooth enamel.

Extensive bacterial growth results in a thick oral biofilm called **dental plaque** (Figure 23.16). As plaque continues to form, filamentous anaerobes such as *Fusobacterium* species appear. The filamentous bacteria embed in the matrix formed by the streptococci and extend perpendicular to the tooth surface, making an ever-thicker biofilm. Associated with oral filamentous bacteria are spirochetes such as *Borrelia* species, gram-positive rods, and gram-negative cocci. In heavy plaque, filamentous obligately anaerobic organisms such as *Actinomyces* may predominate. Dental plaque is thus a mixed-culture biofilm composed of several different genera and their accumulated products.

The microbial populations within dental plaque exist in a microenvironment of their own making and maintain themselves in the face of wide variations in the macroenvironmental conditions of the oral cavity. As dental plaque accumulates, the microflora produce locally high concentrations of organic acids, in particular lactic acid, that cause decalcification of the tooth enamel resulting in **dental caries** (tooth decay). Tooth enamel is calcified tissue, and the ability of microorganisms to invade this tissue plays a major role in the extent of dental caries. Thus, dental caries is an infectious disease.

Virulence Factors

Many pathogens produce *virulence factors* that indirectly or directly enhance invasiveness by promoting pathogen infection. Many of these virulence factors are enzymes. For example, streptococci, staphylococci, and certain clostridia produce *hyaluronidase* (Table 23.5), an enzyme that promotes spreading of organisms in tissues by breaking down the polysaccharide hyaluronic acid. The latter is an intercellular cement in animals, and its breakdown allows the pathogens to spread from an initial infection site. Similarly, the clostridia that cause gas gangrene (⇌ Section 30.9)

produce *collagenase*, which destroys collagen and enables these organisms to spread through the body. Many pathogenic streptococci and staphylococci also produce proteases, nucleases, and lipases that degrade host proteins, nucleic acids, and lipids, respectively (Table 23.5).

Two virulence factors are enzymes that affect fibrin, the insoluble blood protein that forms clots. The clotting mechanism, triggered by tissue injury, isolates the pathogens, limiting infection to a local region. Some pathogens counter this process by producing fibrinolytic enzymes that dissolve the fibrin clots and make further invasion possible. One fibrinolytic substance produced by *Streptococcus pyogenes* is called *streptokinase* (Table 23.5).

By contrast to the destructive activity of streptokinase, some pathogens produce enzymes that actually promote the formation of fibrin clots. These clots protect the pathogen from host responses. For example, *coagulase* (Table 23.5), produced by pathogenic *Staphylococcus aureus*, causes insoluble fibrin to be deposited on *S. aureus* cells, protecting them from attack by host cells. The fibrin matrix produced as a result of coagulase activity may account for the localized nature of many staphylococcal

Table 23.5 Exotoxins and other extracellular virulence factors produced by human pathogens

Organism	Disease	Toxin or factor[a]	Action/Enzyme type
Bacillus anthracis	Anthrax	Lethal factor (LF) Edema factor (EF) Protective antigen (PA) (AB)	PA is the cell-binding B component, EF causes edema, LF causes cell death
Bacillus cereus	Food poisoning	Enterotoxin complex	Induces fluid loss from intestinal cells
Bordetella pertussis	Whooping cough	Pertussis toxin (AB)	Blocks G protein signal transduction, kills cells
Clostridium botulinum	Botulism	Neurotoxin (AB)	Flaccid paralysis (Figure 23.21)
Clostridium tetani	Tetanus	Neurotoxin (AB)	Spastic paralysis (Figure 23.22)
Clostridium perfringens	Gas gangrene, food poisoning	α-Toxin (CT)	Hemolysis (lecithinase, Figure 23.18b)
		β-Toxin (CT)	Hemolysis
		γ-Toxin (CT)	Hemolysis
		δ-Toxin (CT)	Hemolysis (cardiotoxin)
		κ-Toxin (E)	Collagenase
		λ-Toxin (E)	Protease
		Enterotoxin (CT)	Alters permeability of intestinal epithelium
Corynebacterium diphtheriae	Diphtheria	Diphtheria toxin (AB)	Inhibits protein synthesis in eukaryotes (Figure 23.20)
Escherichia coli (enterotoxigenic strains only)	Gastroenteritis	Enterotoxin (Shiga-like toxin) (AB)	Inhibits protein synthesis, induces bloody diarrhea and hemolytic uremic syndrome
Haemophilus ducreyi	Chancroid	Cytolethal distending toxin (AB)	Genotoxin (DNA lesions cause apoptosis in host cells)
Pseudomonas aeruginosa	P. aeruginosa infections	Exotoxin A (AB)	Inhibits protein synthesis
Salmonella spp.	Salmonellosis, typhoid fever, paratyphoid fever	Enterotoxin (AB)	Inhibits protein synthesis, lyses host cells
		Cytotoxin (CT)	Induces fluid loss from intestinal cells
		Injectisome	Mechanism for injecting toxins into host cells (⮂ Figure 4.43)
Shigella dysenteriae	Bacterial dysentery	Shiga toxin (AB)	Inhibits protein synthesis, induces bloody diarrhea and hemolytic uremic syndrome
Staphylococcus aureus	Pyogenic (pus-forming) infections (boils and so on), respiratory infections, food poisoning, toxic shock syndrome, scalded skin syndrome	α-Toxin (CT)	Hemolysis
		Toxic shock syndrome toxin (SA)	Systemic shock
		Exfoliating toxin A and B (SA)	Peeling of skin, shock
		Leukocidin (CT)	Destroys leukocytes
		β-Toxin (CT)	Hemolysis
		γ-Toxin (CT)	Kills cells
		δ-Toxin (CT)	Hemolysis, leukolysis
		Enterotoxin A, B, C, D, and E (SA)	Induce vomiting, diarrhea, shock
		Coagulase (E)	Induces fibrin clotting
Streptococcus pyogenes	Pyogenic infections, tonsillitis, scarlet fever	Streptolysin O (CT)	Hemolysis
		Streptolysin S (CT)	Hemolysis (Figure 23.18a)
		Erythrogenic toxin (SA)	Causes scarlet fever
		Streptokinase (E)	Dissolves fibrin clots
		Hyaluronidase (E)	Dissolves hyaluronic acid in connective tissue
Vibrio cholerae	Cholera	Enterotoxin (AB)	Induces fluid loss from intestinal cells (Figure 23.23)

[a]AB, AB toxin; CT, cytolytic toxin; E, enzymatic virulence factor; SA, superantigen toxin, see Section 24.9.
[b]Cytolethal distending toxin is found in other gram-negative pathogens including Campylobacter jejuni, Escherichia coli, Helicobacter spp., Salmonella enterica serovar Typhi, and Shigella dysenteriae.

UNIT 5

infections, as in boils and pimples (👉 Section 29.9). Coagulase-positive *S. aureus* strains are usually more virulent than coagulase-negative strains.

Virulence in *Salmonella*: Pathogenicity Islands and Plasmids

Salmonella species infect humans, leading to various gastrointestinal illnesses. *Salmonella* species encode a large number of virulence factors that are important in adherence and infection (Figure 23.17). Several genes that direct the invasion process in *Salmonella* and related gram-negative pathogens are found clustered together on the chromosome as *pathogenicity islands* (👉 Section 6.13). For example, *Salmonella* pathogenicity island 1 (SPI1) is a collection of virulence genes of *Salmonella* that encode at least 10 different proteins that promote invasion. One is *invH*, which encodes a surface adhesion protein. Other *inv* genes encode proteins important for trafficking of virulence proteins. The InvJ regulator protein controls assembly of structural proteins InvG, PrgH, PrgI, PrgJ, and PrgK that form a type III secretion system called the *injectisome*, an organelle in the bacterial envelope that allows direct transfer of virulence proteins into host cells through a needle-like assembly (Figure 23.17, 👉 Figure 4.43).

Another *Salmonella* pathogenicity island, SPI2, contains genes that are responsible for causing more systemic disease and resistance to host cell defenses. In addition, several plasmid-borne virulence factors such as antibiotic resistance genes encoded on R plasmids can be spread between most *Salmonella* species as well as other enteric bacteria (👉 Section 4.3). Pathogenicity islands and R plasmids allow for the facile and rapid transfer of virulence factors. It is thus not uncommon for genes encoding factors in one pathogen to be very similar if not identical to those in another because of transfer of parts or all of the islands between species by horizontal gene exchange.

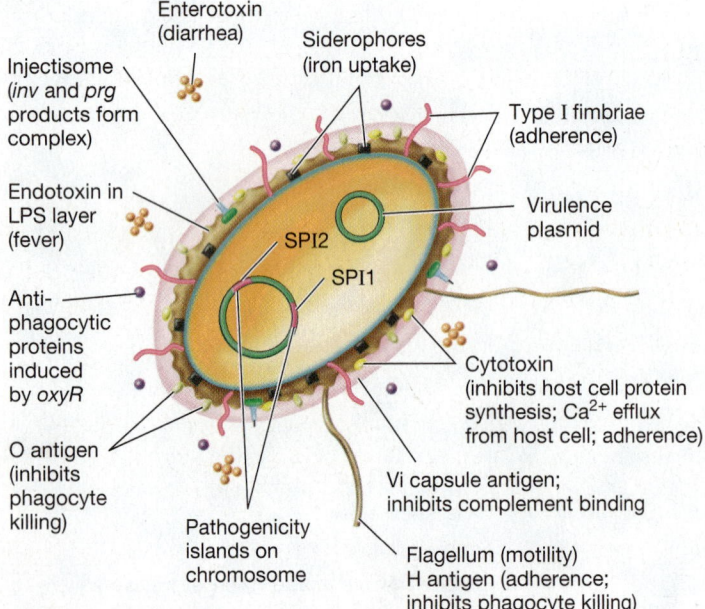

Figure 23.17 Virulence factors in *Salmonella*. Factors important for virulence and the development of pathogenesis in this gram-negative enteric pathogen are shown.

23.9 Exotoxins

Toxicity is the ability of an organism to cause disease by means of a preformed toxin that inhibits host cell function or kills host cells. **Exotoxins** are toxic *proteins* released from the pathogen as it grows. These toxins travel from a site of infection and cause damage at distant sites. Table 23.5 provides a summary of the properties and actions of some of the known bacterial exotoxins as well as other extracellular virulence factors released from growing bacteria.

Exotoxins fall into three categories in terms of mechanisms: the *cytolytic toxins*, the *AB toxins*, and the *superantigen toxins*. Cytolytic toxins work by degrading cytoplasmic membrane integrity, causing lysis. AB toxins consist of two subunits, A and B. The B component binds to a host cell surface molecule, facilitating the transfer of the A subunit across the cytoplasmic membrane, where it damages the cell. Superantigen toxin work by stimulating large numbers of immune cells, resulting in extensive inflammation and tissue damage (👉 Section 24.9).

A subset of the exotoxins is the **enterotoxins**, exotoxins whose activity affects the small intestine, generally causing secretion of fluid into the intestinal lumen resulting in vomiting and diarrhea. Usually acquired by ingestion of contaminated food or water, enterotoxins are produced by a variety of bacteria, including the food-poisoning organisms *Staphylococcus aureus*, *Clostridium perfringens*, and *Bacillus cereus* and the intestinal pathogens *Vibrio cholerae*, *Escherichia coli*, and *Salmonella enterica* serovar Typhimurium (Chapter 31). As with the other exotoxins, enterotoxins may be cytolytic toxins, AB toxins, or superantigen toxin.

Cytotoxins

Cytotoxins, also called cytolytic toxins, are secreted, soluble, extracellular proteins produced by a variety of pathogens. Cytolytic toxins damage the host cytoplasmic membrane, causing cell lysis and death. Because the lytic activity of these toxins is most easily observed in assays that use red blood cells (erythrocytes), the toxins are often called *hemolysins* (Table 23.5). However, hemolysins also lyse cells other than erythrocytes. The production of hemolysins is demonstrated in the laboratory by streaking the pathogen on a blood agar plate (a rich medium containing 5% sterile blood). During growth of the colonies, hemolysin is released and lyses the surrounding red blood cells, releasing hemoglobin and creating a clear area, called a zone of *hemolysis*, around the growing colonies (Figure 23.18).

Some hemolysins attack the phospholipid of the host cytoplasmic membrane. Because the phospholipid lecithin (phosphatidylcholine) is often used as a substrate, these enzymes are called *lecithinases* or *phospholipases*. An example is the α-toxin of *Clostridium perfringens*, a lecithinase that dissolves membrane lipids, resulting in cell lysis (Table 23.5, Figure 23.18*b*). Because

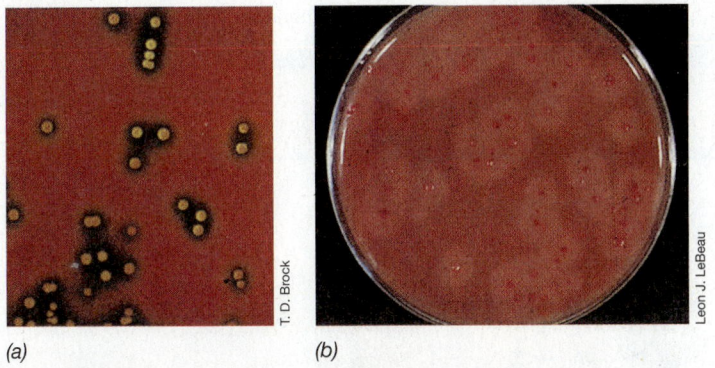

(a)   (b)

T. D. Brock

Leon J. LeBeau

Figure 23.18 Hemolysis. (a) Zones of hemolysis around colonies of *Streptococcus pyogenes* growing on a blood agar plate. (b) Action of lecithinase, a phospholipase, around colonies of *Clostridium perfringens* growing on an agar medium containing egg yolk, a source of lecithin. Lecithinase dissolves the cytoplasmic membranes of red blood cells, producing cloudy zones of hemolysis around each colony.

the cytoplasmic membranes of all organisms contain phospholipids, phospholipases can destroy bacterial as well as animal cell cytoplasmic membranes.

Some hemolysins, however, are not phospholipases. Streptolysin O, a hemolysin produced by streptococci, affects the sterols of the host cytoplasmic membrane. *Leukocidins* (Table 23.5) lyse white blood cells and may decrease the host immune response. Staphylococcal α-toxin (**Figure 23.19** and Table 23.5) kills nucleated cells and lyses erythrocytes. Toxin subunits first bind to the phospholipid bilayer. The subunits then oligomerize into nonlytic heptamers, now associated with the membrane. Following oligomerization, each heptamer undergoes conformational changes to produce a membrane-spanning pore. The pore releases the cytoplasmic contents and allows the influx of extracellular material, thus killing the cell (Figure 23.19).

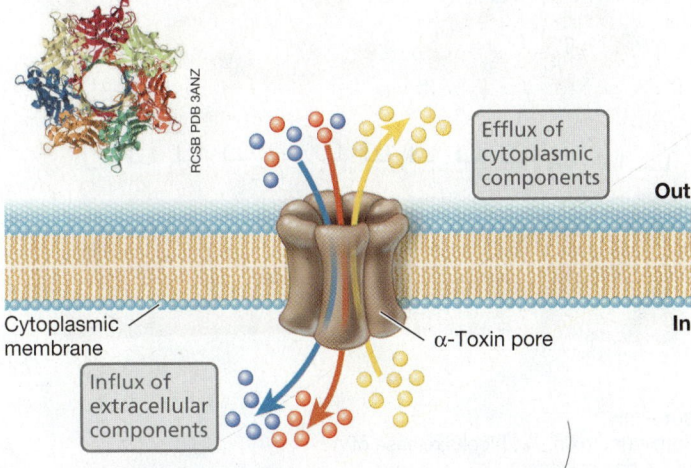

RCSB PDB 3ANZ

Efflux of cytoplasmic components

Out

Cytoplasmic membrane

α-Toxin pore

In

Influx of extracellular components

Figure 23.19 Staphylococcal α-toxin. Staphylococcal α-toxin is a pore-forming cytotoxin that is produced by growing *Staphylococcus* cells. Released as monomers, seven identical protein subunits oligomerize in the cytoplasmic membrane of target cells. The oligomer forms a pore, releasing the contents of the cell. In red blood cells, hemolysis occurs, visually indicating cell lysis. The thumbnail photo on top shows the structure of α-toxin looking down through the pore. Each of the seven identical subunits is shown in a different color.

Diphtheria Toxin

The diphtheria toxin produced by *Corynebacterium diphtheriae* is an AB toxin and an important virulence factor of the pathogen (⟳ Section 29.3). Diphtheria toxin inhibits protein synthesis in eukaryotes. Rats and mice are relatively resistant to diphtheria toxin, but humans and many other animals are very susceptible, with only a single molecule of toxin sufficient to kill a cell.

Diphtheria toxin is secreted by *C. diphtheriae* as a single polypeptide. One component, subunit B, specifically binds to a host cell protein on eukaryotic cells, the heparin-binding epidermal growth factor (**Figure 23.20**). After binding, proteolytic cleavage between subunit B and the remaining portion of the protein, subunit A, allows subunit A to move across the host cytoplasmic membrane into the cytoplasm. Here subunit A disrupts protein synthesis by blocking transfer of an amino acid from tRNA to growing polypeptide chains. Diphtheria toxin specifically inactivates elongation factor 2 (EF-2), a protein that functions in growth of the polypeptide chain, by catalyzing the attachment of adenosine diphosphate (ADP) ribose from NAD$^+$. Following ADP-ribosylation, the activity of the modified EF-2 decreases dramatically and protein synthesis stops.

Diphtheria toxin is not encoded by the bacterium but instead by a viral gene called *tox* present in the genome of the lysogenic bacteriophage β. Lysogenic phages are those whose genomes have become integrated into their host's chromosome (⟳ Section 8.8). Toxigenic, pathogenic strains of *C. diphtheriae* are infected with phage β and hence produce the toxin. Nontoxigenic, nonpathogenic strains of *C. diphtheriae* can be converted to pathogenic strains by infection with phage β, a process called *phage conversion* (⟳ Section 10.7).

Exotoxin A of *Pseudomonas aeruginosa* functions similarly to diphtheria toxin, also modifying EF-2 by ADP-ribosylation (Table 23.5). The enterotoxin produced by *Shigella dysenteriae*, called *Shiga toxin*, and the Shiga-like toxin produced by enteropathogenic *E. coli* O157:H7 (⟳ Section 31.12), are also AB toxins (Table 23.5). The Shiga-like toxins target cells of the small intestine near where the pathogen colonized, shutting down protein synthesis and leading to bloody diarrhea and hemolytic uremic syndrome, a kidney disease that may result in kidney failure, especially in children.

Botulinum and Tetanus Toxins

Clostridium botulinum and *Clostridium tetani* are endospore-forming bacteria commonly found in soil. These organisms occasionally cause disease in animals from highly poisonous AB exotoxins that function as *neurotoxins* (⟳ Sections 30.9 and 31.9). Neither *C. botulinum* nor *C. tetani* is very invasive, and therefore pathogenicity is almost exclusively due to neurotoxicity. Botulinum toxin and tetanus toxin both block release of neurotransmitters involved in muscle control, but the mode of action and disease symptoms are quite distinct (**Figure 23.21** and **Figure 23.22**).

C. botulinum sometimes grows directly in the body, causing infant or wound botulism. Frequently, however, *C. botulinum* grows and produces toxin in improperly preserved foods. As a result, infection and growth in the body are unnecessary, and ingestion of preformed toxin is the common method of acquiring

UNIT 5

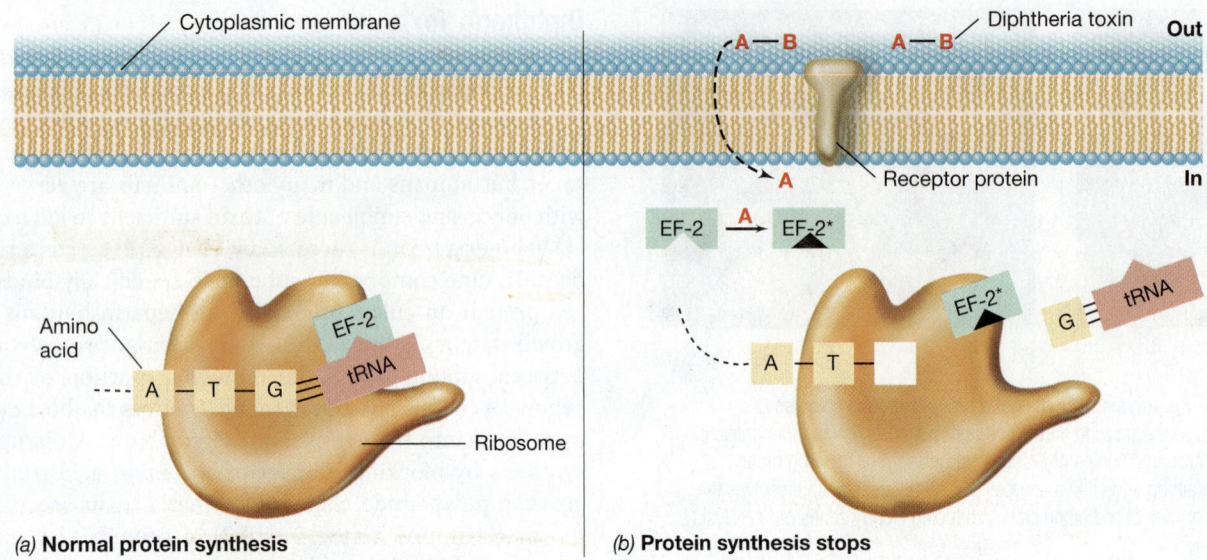

(a) **Normal protein synthesis**

(b) **Protein synthesis stops**

Figure 23.20 Diphtheria toxin activity. Diphtheria toxin is an AB toxin produced by *Corynebacterium diphtheriae.* *(a)* In eukaryotic cells, elongation factor 2 (EF-2) normally binds to the ribosome and facilitates attachment of an amino acid–charged tRNA to the ribosome, resulting in protein elongation. *(b)* Diphtheria toxin binds to the cytoplasmic membrane by way of its B subunit. Cleavage of the toxin allows the A subunit to enter the cell where it catalyzes ADP-ribosylation of elongation factor 2 (EF-2 → EF-2*). The modified elongation factor no longer binds to the ribosome, resulting in the cessation of protein synthesis and cell death.

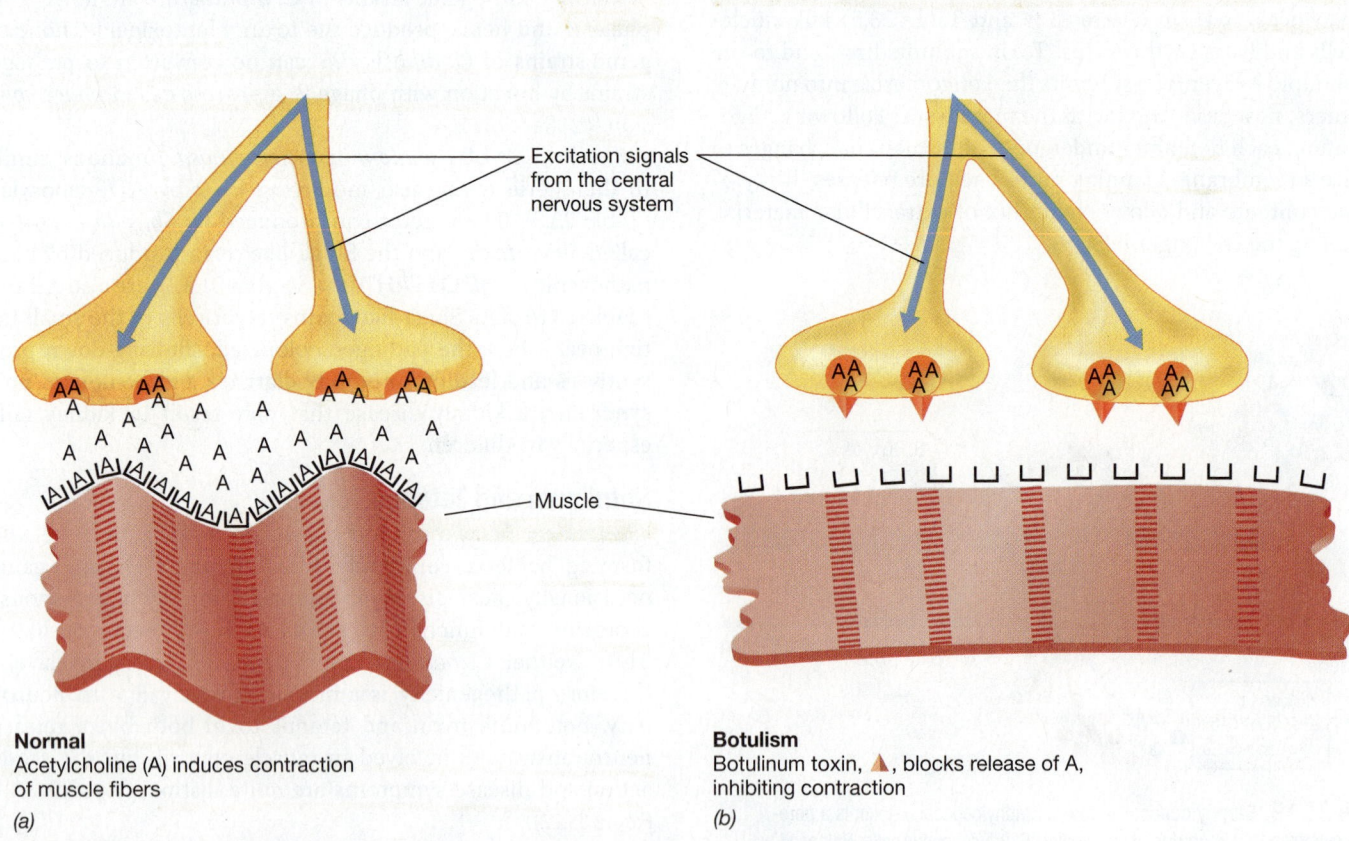

Normal
Acetylcholine (A) induces contraction of muscle fibers
(a)

Botulism
Botulinum toxin, ▲, blocks release of A, inhibiting contraction
(b)

Figure 23.21 The activity of botulinum toxin from *Clostridium botulinum.* *(a)* Upon stimulation of peripheral and cranial nerves, acetylcholine (A) is normally released from vesicles at the neural side of the motor end plate. Acetylcholine then binds to specific receptors on the muscle, inducing contraction. *(b)* Botulinum toxin acts at the motor end plate to prevent release of acetylcholine (A) from vesicles, resulting in a lack of stimulus to the muscle fibers, irreversible relaxation of the muscles, and flaccid paralysis.

botulism. Botulinum toxins, the most potent biological toxins known, are seven related AB toxins. One nanogram (10^{-9} g) of botulinum toxin is sufficient to kill a guinea pig. Of the seven distinct botulinum toxins known, at least two are encoded on lysogenic bacteriophages specific for *C. botulinum*. The major toxin is a protein that forms complexes with nontoxic botulinum proteins to yield a bioactive protein complex. The complex then binds to presynaptic membranes on the termini of the stimulatory motor neurons at the neuromuscular junction, blocking the release of acetylcholine. Normal transmission of a nerve impulse to a muscle cell requires acetylcholine interaction with a muscle receptor; botulinum toxin prevents the poisoned muscle from receiving the excitatory acetylcholine signal (Figure 23.21). This prevents muscle contraction and leads to a flaccid paralysis and death by suffocation, the fatal outcome of botulism.

In contrast to *C. botulinum*, *C. tetani* grows in the body in deep wounds that become anoxic, such as punctures. *C. tetani* cells rarely leave the wound where they were first introduced, growing relatively slowly at the wound site. The toxin spreads systemically via the neural cells and causes spastic paralysis, the hallmark of tetanus (⮂ Section 30.9 and Figure 30.22*b*). On contact with the central nervous system, tetanus toxin is transported through the motor neurons to the spinal cord, where it binds specifically to ganglioside lipids at the termini of the inhibitory interneurons.

The inhibitory interneurons normally work by releasing an inhibitory neurotransmitter, typically the amino acid glycine, which binds to receptors on the motor neurons. Glycine from the inhibitory interneurons then stops the release of acetylcholine by the motor neurons and inhibits muscle contraction, allowing relaxation of the muscle fibers. However, if tetanus toxin blocks glycine release, the motor neurons cannot be inhibited, resulting in the continual release of acetylcholine and uncontrolled contraction of the poisoned muscles (Figure 23.22). The outcome is a spastic, twitching paralysis, as affected muscles are constantly contracted. If the muscles of the mouth are involved, the prolonged contractions restrict the mouth's movement, resulting in a condition called *lockjaw*. If respiratory muscles are involved, prolonged contraction may result in death due to asphyxiation.

Cholera Toxin

Cholera toxin, an AB-type enterotoxin produced by *V. cholerae*, causes cholera (⮂ Section 31.3). Cholera is characterized by massive fluid loss from the intestines, resulting in severe diarrhea, life-threatening dehydration, and electrolyte depletion (**Figure 23.23**). The disease starts by ingestion of *V. cholerae* cells from contaminated food or water. The organism travels to the intestine, where it colonizes and secretes cholera toxin (Figure 23.23). In the gut, the B subunit of the toxin, consisting of five identical

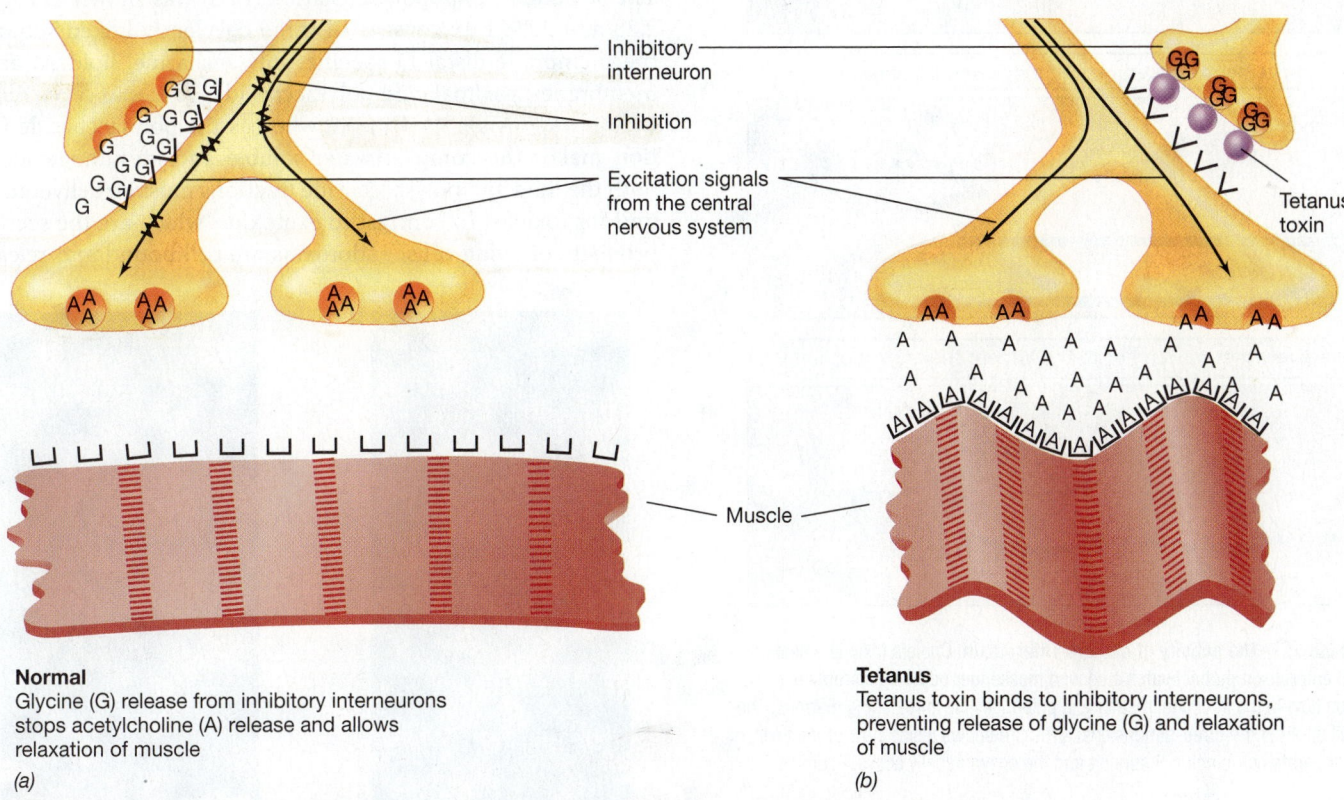

Normal
Glycine (G) release from inhibitory interneurons stops acetylcholine (A) release and allows relaxation of muscle

(a)

Tetanus
Tetanus toxin binds to inhibitory interneurons, preventing release of glycine (G) and relaxation of muscle

(b)

Figure 23.22   The activity of tetanus toxin from *Clostridium tetani*. *(a)* Muscle relaxation is normally induced by glycine (G) release from inhibitory interneurons. Glycine acts on the motor neurons to block excitation and release of acetylcholine (A) at the motor end plate. *(b)* Tetanus toxin binds to the interneuron to prevent release of glycine from vesicles, resulting in a lack of inhibitory signals to the motor neurons, constant release of acetylcholine to the muscle fibers, irreversible contraction of the muscles, and spastic paralysis. For purposes of illustration, the inhibitory interneuron is shown near the motor end plate, but it is actually in the spinal cord.

UNIT 5

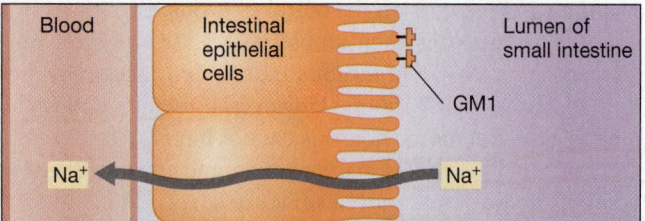

1. Normal ion movement, Na⁺ from lumen to blood, no net Cl⁻ movement

Blood | Intestinal epithelial cells | Lumen of small intestine

GM1

Na⁺ ← Na⁺

2. Infection and toxin production by *V. cholerae*

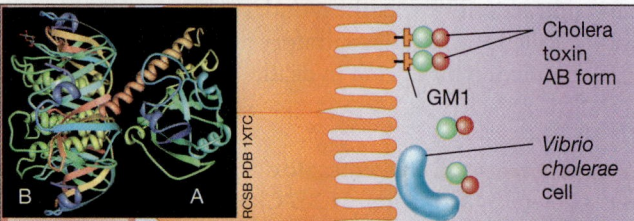

RCSB PDB 1XTC

B   A

Cholera toxin AB form

GM1

Vibrio cholerae cell

3. Activation of epithelial adenylate cyclase by cholera toxin

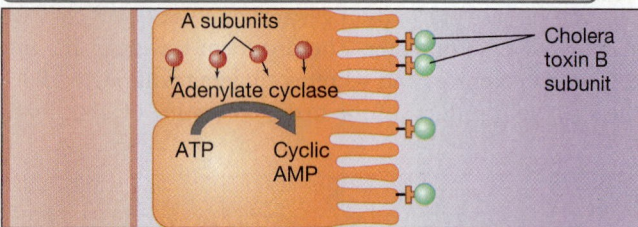

A subunits

Adenylate cyclase

ATP → Cyclic AMP

Cholera toxin B subunit

4. Na⁺ movement blocked, net Cl⁻ movement to lumen

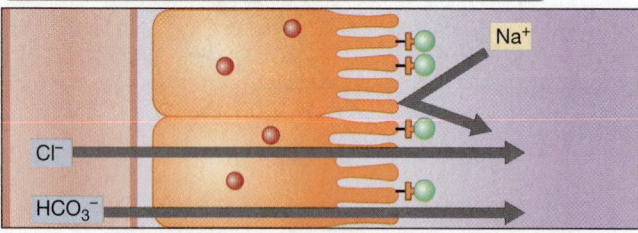

Na⁺

Cl⁻

HCO₃⁻

5. Massive water movement to the lumen; cholera symptoms

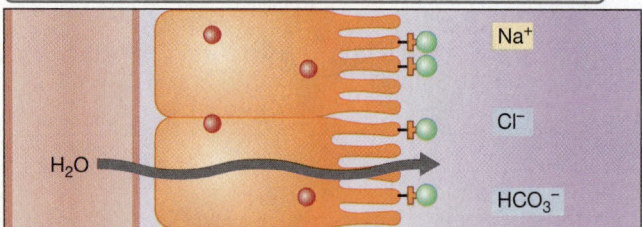

Na⁺

Cl⁻

H₂O

HCO₃⁻

Figure 23.23 The activity of cholera enterotoxin. Cholera toxin is a heat-stable AB enterotoxin that activates a second messenger pathway, disrupting normal ion flow in the intestine, resulting in potentially life-threatening diarrhea. The thumbnail photo of the three-dimensional structure shows a side view of the toxin, with the separate cell-binding B subunit and the enzymatically active A subunit.

monomers, binds specifically to GM1 ganglioside, a complex glycolipid found in the cytoplasmic membrane of intestinal epithelial cells.

The B subunit targets the toxin specifically to the intestinal epithelium but has no toxicity itself; toxicity is a function of the

A subunit, which crosses the cytoplasmic membrane and activates adenylate cyclase, the enzyme that converts ATP to cyclic adenosine monophosphate (cAMP). This molecule is a cyclic nucleotide (⊘ Figure 7.14) that mediates several regulatory systems in cells, including ion balance. The increased cAMP induced by cholera enterotoxin induces secretion of chloride and bicarbonate (HCO_3^-) from small intestine epithelial cells into the intestinal lumen. This change in ion concentrations leads to the secretion of large amounts of water; the rate of water loss into the small intestine is greater than the possible reabsorption of water by the large intestine, resulting in a large net fluid loss and watery diarrhea.

MINIQUIZ --

- What key features are shared by all AB exotoxins?
- Are bacterial growth and infection in the host necessary for the production of toxins? Explain and cite examples for your answer.

23.10 Endotoxins

Endotoxins are the toxic lipopolysaccharides found in most gram-negative *Bacteria*. They are structural components of the gram-negative outer membrane (⊘ Section 2.11) and thus unlike exotoxins are not soluble products of growing bacteria.

Endotoxin Structure and Biology

The structure of lipopolysaccharide (LPS) was shown in Figures 2.28 and 2.29. LPS consists of three covalently linked subunits: the membrane-distal O-specific polysaccharide, lipid A, and a membrane-proximal core polysaccharide. The lipid A portion of LPS is responsible for toxicity, whereas the polysaccharide fraction makes the complex water-soluble and immunogenic, but both the lipid and polysaccharide fractions must be delivered as a unit for toxicity. In contrast to exotoxins, which are the secreted products of living cells, endotoxins are cell bound and released

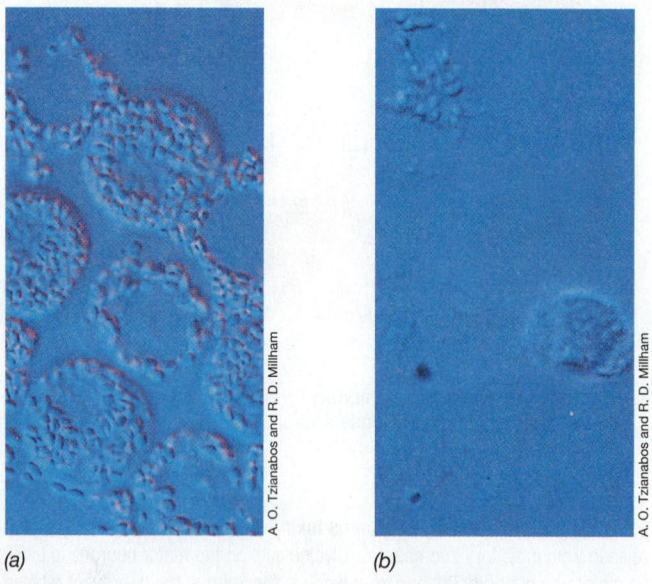

(a)   (b)

A. O. Tzianabos and R. D. Millham

Figure 23.24 *Limulus* amoebocytes. (a) Normal amoebocytes from the horseshoe crab, *Limulus polyphemus*. (b) Amoebocytes following exposure to bacterial lipopolysaccharide (LPS). LPS induces degranulation and lysis of the cells.

Table 23.6 Properties of exotoxins and endotoxins

Property	Exotoxins	Endotoxins
Chemical properties	Proteins, excreted by certain gram-positive or gram-negative *Bacteria*; generally heat-labile	Lipopolysaccharide–lipoprotein complexes, released on cell lysis as part of the outer membrane of gram-negative *Bacteria*; extremely heat-stable
Mode of action; symptoms	Specific; usually binds to specific cell receptors or structures; either cytotoxin, enterotoxin, or neurotoxin with defined, specific action on cells or tissues	General; fever, diarrhea, vomiting
Toxicity	Often highly toxic in picogram to microgram quantities, sometimes fatal	Moderately toxic in tens to hundreds of microgram amounts, rarely fatal
Immunogenicity response	Highly immunogenic; stimulate the production of neutralizing antibody (antitoxin)	Relatively poor immunogen; immune response not sufficient to neutralize toxin
Toxoid potential	Heat or chemical treatment may destroy toxicity, but treated toxin (toxoid) remains immunogenic	None
Fever potential	Nonpyrogenic; does not produce fever in the host	Pyrogenic; often induces fever in the host
Genetic origin	Often encoded on extrachromosomal elements	Chromosomal genes

in toxic amounts only when the cells lyse. Endotoxins have been studied in *Escherichia*, *Shigella*, and especially *Salmonella*, where they are another of the many virulence factors that contribute to pathogenesis (Figure 23.17). The properties of exotoxins and endotoxins are compared in Table 23.6.

Endotoxins cause a variety of physiological effects. Fever is an almost universal result of endotoxin exposure because endotoxin stimulates host cells to release cytokines, soluble proteins secreted by phagocytes and other cells that function as *endogenous pyrogens*, proteins that affect the temperature-controlling center of the brain, causing fever. Cytokines released due to endotoxin exposure can also cause diarrhea, a rapid decrease in the numbers of lymphocytes and platelets, and generalized inflammation (↩ Section 24.5). Large doses of endotoxin can cause death from hemorrhagic shock and tissue necrosis. Endotoxins are, however, generally less toxic than most exotoxins. For instance, in mice the LD_{50} for endotoxin is 200–400 micrograms per animal, whereas the LD_{50} for botulinum toxin is about 25 picograms, about 10 million times less.

Limulus Amoebocyte Lysate Assay for Endotoxin

Because endotoxins induce fever, pharmaceuticals such as antibiotics and intravenous solutions must be free of endotoxin. An endotoxin assay of very high sensitivity has been developed using lysates of amoebocytes from the horseshoe crab, *Limulus*

polyphemus. Endotoxin specifically causes lysis of the amoebocytes (Figure 23.24). In the *Limulus* amoebocyte lysate (LAL) assay, *Limulus* amoebocyte extracts are mixed with the solution to be tested. If endotoxin is present, the amoebocyte extract forms a gel and precipitates, causing a change in turbidity. This reaction is measured quantitatively with a spectrophotometer and can detect as little as 10 pg/ml of LPS.

The LAL assay is used to detect endotoxin in clinical samples such as serum or cerebrospinal fluid. A positive test is presumptive evidence for infection by gram-negative bacteria. Drinking water, water used for formulation of injectable drugs, and injectable aqueous solutions are routinely tested using the LAL to identify and eliminate endotoxin contamination from gram-negative organisms. A commercially available assay uses horseshoe crab factor C made by recombinant DNA techniques (factor C is the key protein activated by endotoxin in the LAL assay). This allows for a more standardized assay protocol and has the advantage of being free of animal products.

MINIQUIZ

- Why do gram-positive bacteria not produce endotoxins?
- Why is it necessary to test for endotoxin in water used for injectable drug preparations?

III • Host Factors in Infection and Disease

Humans have certain innate resistance factors that keep them from contracting most infectious diseases. These include a number of chemical and physical barriers to microbial infection, the usual prerequisite for disease. In addition, the condition of the host can play a role in whether or not the diseased state can become established. We conclude this chapter with a discussion of these factors, which are often the tipping point between health and disease.

23.11 Innate Resistance to Infection

The presence of the normal microflora is a critically important mechanism for resisting pathogen infection, especially on the skin and in the gut. Pathogens do not easily infect tissues on which normal microflora are already colonized because the normal microflora limit available microbial nutrients and sites for infection. Here we introduce several other resistance factors common to vertebrate hosts. These factors nonspecifically inhibit infection by most pathogens (**Figure 23.25**).

Natural Host Resistance

The ability of a particular pathogen to cause disease in an individual animal species is highly variable. In the disease rabies, for instance, certain animal species are much more susceptible than others. For example, raccoons and skunks are extremely susceptible to rabies as compared with opossums, which rarely develop the disease. Anthrax infects many species of animals, causing disease symptoms varying from fatal blood poisoning in cattle to the mild pustules of human cutaneous anthrax (⮌ Section 30.8). Introduction of the same pathogen by other routes, however, may challenge the resistance of the host. For example, pulmonary or airborne anthrax, such as that induced by weaponized strains used for bioterrorism (⮌ Section 28.8), is typically fatal in humans. Anthrax causes a localized infection when acquired through the skin, but a lethal, systemic infection when acquired through the mucous membranes of the lungs.

As another example of innate host resistance, diseases of warm-blooded animals are rarely transmitted to cold-blooded species, and vice versa. Presumably, the anatomical and metabolic features of one group are not compatible with pathogens that infect the other group.

Infection Site and Tissue Specificity

Most pathogens must adhere and infect at the site of exposure to initiate infection. Even if pathogens adhere to an exposure site, the organisms cannot infect the host if the site is not compatible with the pathogen's nutritional and metabolic needs. For example, if cells of *Clostridium tetani* were ingested, tetanus would not normally result because the pathogen is either killed by the acidity of the stomach or cannot compete with the well-developed normal intestinal microflora. If, on the other hand, *C. tetani* cells or endospores were introduced into a deep wound, the organism would grow and produce tetanus toxin in the anoxic zones created by local tissue death. Conversely, enteric bacteria such as *Salmonella* and *Shigella* do not normally cause wound infections but often infect the intestinal tract and cause disease.

In some cases, pathogens interact exclusively with members of a few closely related host species because the hosts share tissue-specific receptors. Human immunodeficiency virus (HIV), for instance, infects only higher primates including the great apes and humans. This is because a protein called CXCR4 found on human T lymphocytes (cells in the immune system) and a protein called CCR5 found on human macrophages (a phagocyte found in many human tissues) are also expressed in great apes. These proteins are the only cell surface receptors for HIV and specifically bind the gp120 protein of HIV (⮌ Section 29.14). Other animals, even most primates, lack the CXCR4 and CCR5 proteins, cannot bind HIV, and are therefore not susceptible to HIV infection. Likewise, influenza viruses are specific for their species of origin, normally infecting only birds or humans and occasionally swine (⮌ Section 29.8). **Table 23.7** presents some examples of pathogen specificity.

Physical and Chemical Barriers

The structural integrity of tissue surfaces poses a barrier to penetration by microorganisms. The tight junctions between epithelial cells in all body tissues inhibit invasion and infection. In the skin and mucosal tissues, potential pathogens must first adhere to tissue surfaces and then grow at these sites before traveling elsewhere in the body. Mucosal surfaces are bathed in mucus. Epithelial cells have cilia that expel suspended pathogens and keep them from adhering to tissues. In addition to the normal microflora growing at potential infection sites, resistance to infection and invasion is enhanced by antibacterial peptides called *defensins*, produced in the skin, lungs, and gut.

Sebaceous glands in the skin secrete fatty acids and lactic acid, lowering the acidity of the skin to pH 5 and inhibiting colonization of many pathogenic bacteria (blood and internal organs are about pH 7.4). Microorganisms inhaled through the nose or mouth are removed by ciliated epithelial cells on the mucosal surfaces of the nasopharynx and trachea. Potential pathogens ingested in food or water must survive the strong acidity (pH 2) and digestive enzymes such as pepsin in the stomach. The pH changes rapidly

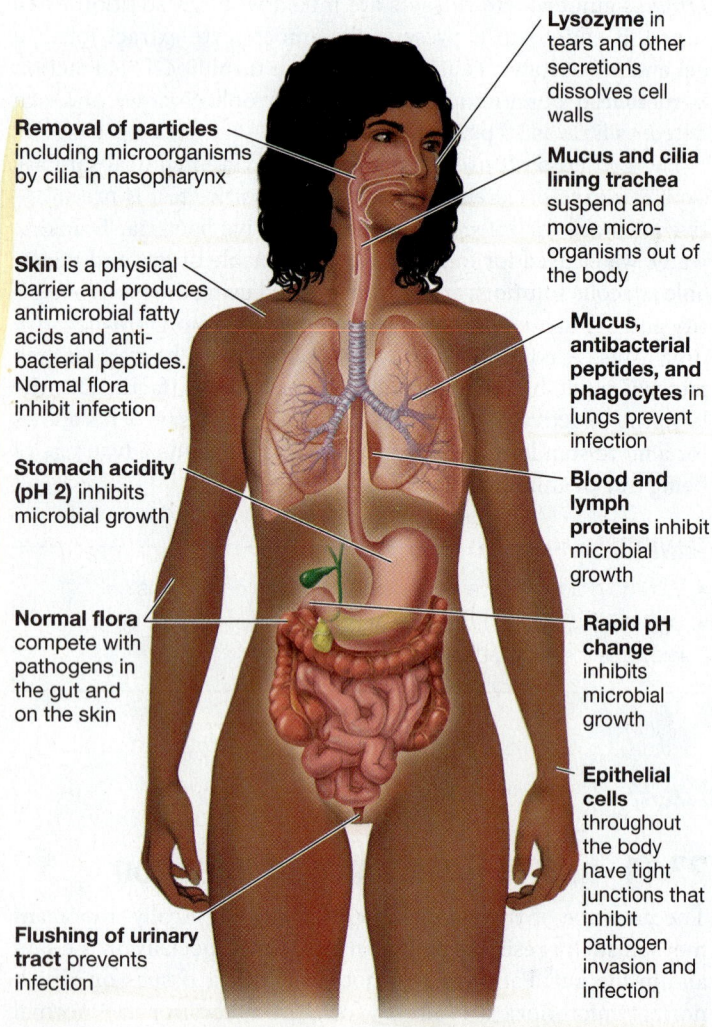

Lysozyme in tears and other secretions dissolves cell walls

Mucus and cilia lining trachea suspend and move microorganisms out of the body

Mucus, antibacterial peptides, and phagocytes in lungs prevent infection

Blood and lymph proteins inhibit microbial growth

Rapid pH change inhibits microbial growth

Epithelial cells throughout the body have tight junctions that inhibit pathogen invasion and infection

Removal of particles including microorganisms by cilia in nasopharynx

Skin is a physical barrier and produces antimicrobial fatty acids and antibacterial peptides. Normal flora inhibit infection

Stomach acidity (pH 2) inhibits microbial growth

Normal flora compete with pathogens in the gut and on the skin

Flushing of urinary tract prevents infection

Figure 23.25 Physical, chemical, and anatomical barriers to infection. These barriers provide natural resistance to colonization and infection by pathogens.

Table 23.7 Tissue specificity in infectious disease

Disease	Tissue infected	Organism
Acquired immunodeficiency syndrome (AIDS)	T helper lymphocytes	Human immunodeficiency virus (HIV)
Botulism	Motor end plate	Clostridium botulinum
Cholera	Small intestine epithelium	Vibrio cholerae
Dental caries	Oral epithelium	Streptococcus mutans, S. sobrinus, S. mitis
Diphtheria	Throat epithelium	Corynebacterium diphtheriae
Gonorrhea	Mucosal epithelium	Neisseria gonorrhoeae
Influenza	Respiratory epithelium	Influenza A and influenza B virus
Malaria	Blood (erythrocytes)	Plasmodium spp.
Pyelonephritis	Kidney medulla	Proteus spp.
Spontaneous abortion (cattle)	Placenta	Brucella abortus
Tetanus	Inhibitory interneuron	Clostridium tetani

in the lower intestinal tract and, if pathogens survive, they must then compete with the abundant resident microflora present in the small and large intestines (Figure 23.5). The lumen of the kidney, the eye, the respiratory system, and the cervical mucosa are constantly bathed with secretions such as tears and mucus containing lysozyme, an enzyme that can kill bacteria by digesting the cell wall. Altogether the body is protected by numerous physical structures, chemicals, and secretions, all aimed at suppressing pathogen invasion and infection.

MINIQUIZ

- Describe host tissue specificity for pathogens.
- Identify physical and chemical barriers to pathogens. How might these barriers be compromised?

23.12 Risk Factors for Infection

In addition to the barriers to infection discussed in the previous section, a number of innate factors contribute to the susceptibility of the host to infection and disease. We conclude our consideration of microbial interactions with the host by considering how these factors may facilitate invasion by pathogens and lead to infectious disease.

Age, Stress, and Diet

Age of the host is an important factor for determining susceptibility to infectious disease. In general, infectious diseases are more common in the very young and in the very old. In the infant, for example, an intestinal microflora develops quickly, but the normal flora of an infant is not the same as that of an adult. Before the development of an adult flora, and especially in the days immediately following birth, pathogens thus have a greater opportunity to become established and produce disease. Thus, infants under one year of age often acquire diarrhea caused by enteropathogenic strains of *Escherichia coli* or viruses such as rotavirus.

In adults over 65 years of age, infectious diseases are much more common than in younger adults. For example, the elderly are much more susceptible to respiratory infections such as those caused by influenza virus, probably because of a declining ability to generate an effective immune response to respiratory pathogens. Anatomical changes associated with age may also encourage infection. For example, enlargement of the prostate gland, a common condition in men over the age of 50, frequently leads to a decreased urinary flow rate, allowing pathogens to infect the male urinary tract more readily and cause infections (Figure 23.8).

Stress can predispose a healthy individual to disease. In studies with animals, physiological stressors such as fatigue, exertion, poor diet, dehydration, or drastic climate changes increase the incidence and severity of infectious diseases. For example, rats subjected to intense physical activity for long periods show a higher mortality rate from experimental *Salmonella* infections compared with rested control animals. Hormones that are produced under stress inhibit immune responses and may play a role in stress-mediated disease. For example, cortisol, a hormone produced at high levels in the body in times of stress, is an anti-inflammatory agent that inhibits the activation of phagocytes and the immune response.

Diet plays a role in host susceptibility to infection. Inadequate diets low in protein and calories alter the normal flora, allowing opportunistic pathogens a better chance of multiplying and increasing the overall susceptibility of the host to known pathogens. For example, the number of *Vibrio cholerae* cells necessary to produce cholera in an individual is drastically reduced if the individual is malnourished. Moreover, the consumption of pathogen-contaminated food is an obvious way to acquire infections, and ingestion of pathogens with food can sometimes enhance the ability of the pathogen to cause disease. The number of organisms necessary to induce cholera, for example, is greatly reduced when the *V. cholerae* is ingested in food, presumably because the food neutralizes stomach acids that would normally destroy the pathogen on its way to colonizing the small intestine.

The presence of a particular substance in the diet can also be the trigger for disease. We discussed dental caries in Section 23.8. Perhaps the most important factor in the development of dental caries is a diet high in sucrose (table sugar). A common lactic acid bacterium present in oral plaque, *Streptococcus mutans*, produces dextran through the activity of the enzyme dextransucrase but only in the presence of sucrose, the substrate for this enzyme:

$$n \text{ sucrose} \xrightarrow{\text{dextransucrase}} \text{dextran } (n \text{ glucose}) + n \text{ fructose}$$

The sticky dextran provides a scaffold for further growth of *S. mutans* and allows coinfection by *S. sobrinus*, the organism that along with *S. mutans* causes dental caries (Figures 23.15 and 23.16). Without dietary sucrose, the highly cariogenic *S. mutans* is unable to synthesize the dextran layer needed to keep the bacterial cells attached to the teeth.

The Compromised Host

A *compromised host* is one in whom one or more resistance mechanisms are inactive and in whom the probability of infection is

UNIT 5

therefore increased. Many hospital patients with noninfectious diseases (for example, cancer and heart disease) acquire microbial infections more readily because they are compromised hosts (⟲ Section 27.2). Such *healthcare-associated infections,* also called *nosocomial infections,* affect up to 2 million individuals each year in the United States, causing up to 100,000 deaths. Invasive medical procedures such as catheterization, hypodermic injection, spinal puncture, biopsy, and surgery may unintentionally introduce microorganisms into the patient. The stress of surgery and the anti-inflammatory drugs given to reduce pain and swelling can also reduce host resistance. For example, organ transplant patients are treated with immunosuppressive drugs to prevent immune rejection of the transplant, but suppressed immunity also reduces the ability of the patient to resist infection.

Some factors can compromise host resistance even outside the hospital. Smoking, excess consumption of alcohol, intravenous drug use, lack of sleep, poor nutrition, and acute or chronic infection with another agent are conditions that reduce host resistance. For example, infection with the human immunodeficiency virus (HIV) predisposes a patient to infections from microorganisms that are not pathogens in uninfected individuals. HIV causes acquired immunodeficiency (AIDS) by destroying one type of immune cell, the CD4 T lymphocytes, involved in the immune response. The reduction in CD4 T cells reduces immunity, and an opportunistic pathogen, one that does not cause disease in a healthy, uninfected host, can then cause serious disease or even death (⟲ Sections 28.9 and 29.14).

Finally, certain genetic conditions compromise the host. For example, genetic diseases that eliminate important parts of the immune system predispose individuals to infections. Individuals with such conditions frequently die at an early age, not from the genetic condition itself, but from microbial infection.

MINIQUIZ

- Identify factors that influence susceptibility to infectious disease in infants, adults, and the aged.
- Identify factors that influence susceptibility to infection and can be controlled by the host.

BIG IDEAS

23.1 • The animal body is a favorable environment for the growth of microorganisms, most of which are beneficial. The number of microorganisms on and in our body is greater than the number of body cells.

23.2 • The skin has at least three different microenvironments, sebaceous, moist, and dry, that harbor distinctly different populations of microorganisms. Environmental and host factors influence the quantity and makeup of the normal skin microflora.

23.3 • The microflora in the oral cavity is extremely complex, with no predominant taxa. The microflora develops in the variable microenvironments associated with the teeth and gums.

23.4 • The gastrointestinal tract supports a diverse population of microorganisms in a variety of nutritional and environmental conditions. The populations of microorganisms are influenced by the diet of the individual and by the unique physical conditions in each distinct site along the gastrointestinal tract.

23.5 • A robust population of normal nonpathogenic microorganisms in the respiratory and urogenital tracts is essential for optimal organ function in normal individuals. The normal microflora help prevent the colonization of pathogens by competing for nutrients and sites for attachment.

23.6 • Virulence is a quantitative measure of pathogenicity. Pathogens use a wide variety of mechanisms and virulence factors to establish an infection.

23.7 • Pathogens gain access to host tissues by adherence at mucosal surfaces through interactions between pathogen and host macromolecules on tissue surfaces.

23.8 • Pathogen invasion starts at the site of adherence and may spread throughout the host via the circulatory or lymphatic systems. A pathogen must gain access to nutrients and appropriate growth conditions before it can infect host tissue. Virulence factors such as various enzymes or cell capsules aid in establishing infection.

23.9 • Exotoxins contribute to the virulence of pathogens. Cytotoxins such as hemolysins, and AB toxins such as diphtheria and cholera toxins are potent exotoxins. Each exotoxin affects a specific host cell function. Enterotoxins are exotoxins that affect the small intestine. Bacterial exotoxins such as botulinum and tetanus toxins are some of the most poisonous substances known.

23.10 • Endotoxins are lipopolysaccharides derived from the outer membrane of gram-negative bacteria such as *Salmonella*. Both the lipid and the polysaccharide components of endotoxin are necessary for toxicity.

23.11 • Innate resistance factors, as well as physical, anatomical, and chemical barriers, prevent colonization of the host by most pathogens. Breakdown of these passive defenses may result in susceptibility to infection and disease.

23.12 • Age, general health, genetic makeup, lifestyle factors such as stress and diet, and chronic diseases can all contribute to susceptibility of the host to infectious disease.

MasteringMicrobiology®

Review what you know and challenge what you have learned with MasteringMicrobiology! Access study materials, chapter quizzes, animations, and microbiology lab tutorials in the Study Area to ensure that you have mastered this chapter's content.

REVIEW OF KEY TERMS

Adherence the enhanced ability of a microorganism to attach to a cell or surface

Attenuation a decrease or loss of virulence

Bacteremia the presence of microorganisms in the blood

Capsule a dense, well-defined polysaccharide or protein layer closely surrounding a cell

Colonization the growth of a microorganism after it has gained access to host tissues

Dental caries tooth decay resulting from bacterial infection

Dental plaque bacterial cells encased in a matrix of extracellular polymers and salivary products, found on the teeth

Disease an injury to a host organism, caused by a pathogen or other factor, that affects host function

Endotoxin the lipopolysaccharide portion of the cell envelope of most gram-negative *Bacteria*, which is a toxin when solubilized

Enterotoxin a protein released extracellularly by a microorganism as it grows that produces immediate damage to the small intestine of the host

Exotoxin a protein released extracellularly by a microorganism as it grows that produces immediate host cell damage

Glycocalyx polymer secreted by a microorganism that coats the surface of the microorganism

Host an organism that can harbor a pathogen

Infection a microorganism not a member of the local flora is established and grows in a host, whether or not the host is harmed

Invasion the ability of a pathogen to enter into host cells or tissues, spread, and cause infection

Lower respiratory tract the trachea, bronchi, and lungs

Mucous membrane layer of mucus-covered epithelial cells that interact with the external environment

Mucus a liquid secretion that contains water-soluble glycoproteins and proteins that retain moisture and aid in resistance to microbial invasion on mucosal surfaces

Normal microflora microorganisms that are usually found associated with healthy body tissue

Opportunistic pathogen an organism that causes disease in the absence of normal host resistance

Pathogen an organism, usually a microorganism, that grows in or on a host and causes disease

Pathogenicity the ability of a pathogen to cause disease

Probiotic a live microorganism that, when administered to a host, may confer a health benefit

Septicemia a bloodborne systemic infection

Slime layer a diffuse layer of polymer fibers, typically polysaccharides, that forms an outer surface layer on the cell

Toxicity the ability of an organism to cause disease by means of a preformed toxin that inhibits host cell function or kills host cells

Upper respiratory tract the nasopharynx, oral cavity, and throat

Virulence the relative ability of a pathogen to cause disease

REVIEW QUESTIONS

1. Identify organs in the human body that are normally colonized by microorganisms. Which organs are normally devoid of microorganisms? What do the organs in each set have in common? (Section 23.1)

2. Identify the most common resident microorganisms on the skin. How were these resident microorganisms identified experimentally? (Section 23.2)

3. Describe the microenvironments in the oral cavity. How can anaerobic bacteria survive in the mouth? (Section 23.3)

4. How do pH and oxygen affect the microbial communities that grow in each different region of the gastrointestinal tract? (Section 23.4)

5. Describe the relationship between *Lactobacillus acidophilus* and glycogen in the vagina. What factors influence the differences between the normal vaginal flora of adult females as compared to that of prepubescent juvenile females? (Section 23.5)

6. Define virulence and identify parameters to distinguish between highly virulent and moderately virulent pathogens. (Section 23.6)

7. Identify the role of the capsule and the fimbriae of bacteria in microbial adherence. (Section 23.7)

8. Explain the role of the availability of nutritional factors in infection by microorganisms in the body. (Section 23.8)

9. Distinguish between the mechanism of cytotoxins and AB toxins. Provide at least one example of each. (Section 23.9)

10. Identify the structural features, origins, and major effects of endotoxins. (Section 23.10)

11. Identify at least four mechanisms by which a healthy host resists infection. (Section 23.11)

12. Identify common factors that can lead to host compromise. Which factors are controllable by the host? Which factors are not controllable by the host? (Section 23.12)

UNIT 5

APPLICATION QUESTIONS

1. Mucous membranes are barriers against colonization and growth of microorganisms. However, mucous membranes, for example in the mouth and gut, are colonized with a variety of different microorganisms, some of which are potential pathogens. Explain how these potential pathogens are controlled under normal circumstances. Then describe at least one set of circumstances that might encourage pathogenicity.

2. Coagulase is a virulence factor for *Staphylococcus aureus* that acts by causing clot formation at the site of *S. aureus* growth.

Streptokinase is a virulence factor for *Streptococcus pyogenes* that acts by dissolving clots at the site of *S. pyogenes* growth. Reconcile these opposing strategies for enhancing pathogenicity.

3. Although mutants incapable of producing exotoxins are relatively easy to isolate, mutants incapable of producing endotoxins are much harder to isolate. From what you know of the structure and function of these types of toxins, explain the differences in mutant recovery.

microbiology**now**

A Cure for Peanut Allergies?

Can allergies be cured? Some allergies are curable by *desensitization,* a process where the foreign substance, the *allergen,* is injected into the patient. This works well for allergens such as ragweed pollen, but not for food allergens such as peanuts. Allergies to peanuts and tree nuts are the most common causes of severe and sometimes fatal food-induced allergic reactions, shown in the photo of the child. These serious reactions, called *anaphylaxis,* are becoming more common, especially in developed countries. Now, however, one research team has developed a method to desensitize peanut-allergic patients.

The researchers reasoned that if desensitization techniques using injected allergens did not work for peanut allergies, then perhaps the route of desensitization would change the outcome. Maybe a small dose of the allergen given to the immune cells in the oral cavity would induce tolerance (nonresponsiveness) to the allergen. To test this, they delivered the allergen *sublingually*—under the tongue—and measured the response.

The results from "feeding" peanut allergens were astonishing. When given very small sublingual doses of allergen over long periods of time, almost all patients could then eat peanuts. And the longer that patients took the sublingual desensitizing doses, the more peanuts they could eat without developing anaphylaxis. The key was to administer the desensitizing dose by the *same route* that the allergen would normally contact the immune system—in this case, by mouth. When administered in this way, the immune system became tolerant of peanut allergens and no longer responded to the high doses from actually eating peanuts.

It is hoped that this approach will work for other food allergies such as those to shellfish and certain fruits. If so, this would be a major benefit to those that suffer from these harmful immune responses.

Fleischer, D.M., et al. 2013. Sublingual immunotherapy for peanut allergy: A randomized, double-blind, placebo-controlled multicenter trial. *J. Allergy Clin. Immunol. 131:* 119–127.

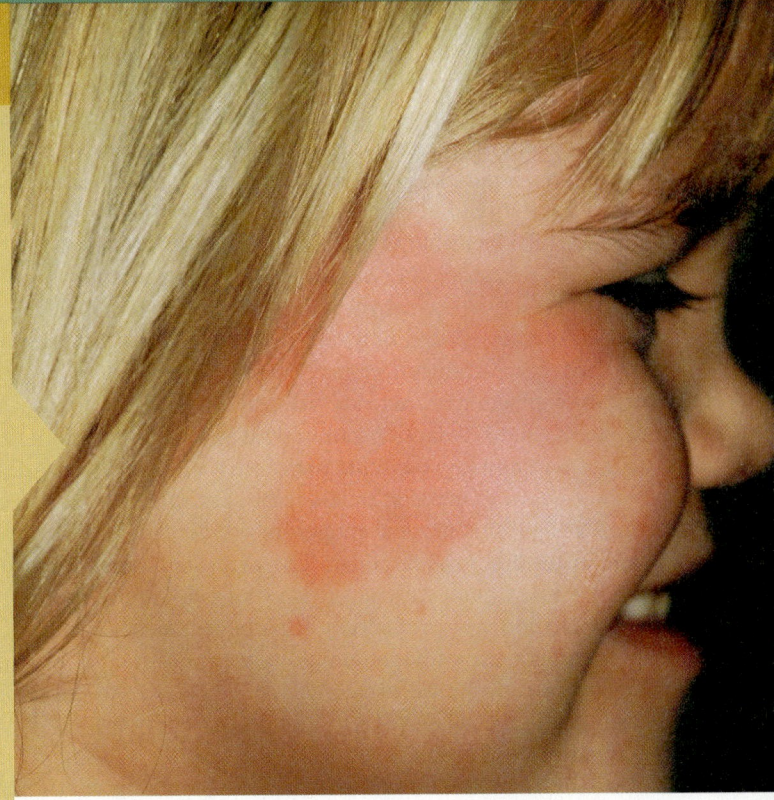

We discussed passive protection against pathogen invasion, infection, and disease in Chapter 23. In the next three chapters, we focus on active mechanisms evolved by vertebrates to resist pathogens and their diseases. In this chapter we present an overview of **immunity**, the active ability to resist disease. Multicellular organisms have a built-in, or innate, immune system that targets common pathogens regardless of their identity. A second tier of immunity, the adaptive immune system, targets specific pathogens to minimize their harmful effects.

We first consider the basic features of the immune response to pathogens. We then discuss vaccinations, a practical tool used to recruit the adaptive immune response for protection against future pathogen challenges. We conclude by describing immune system reactions such as allergies that cause disease.

I • Immunity

Innate immunity, the body's built-in ability to recognize and destroy pathogens or their products, is largely a function of **phagocytes**, cells that can ingest, kill, and digest most microbial pathogens. Innate immune responses develop within hours after contact and infection with a pathogen. Structural features such as the cell wall constituents shared by many pathogens interact with universal receptors on the phagocytes. The interactive phagocytes then activate genes that lead to pathogen destruction.

Some pathogens are, however, so virulent that innate responses are not completely effective. When this happens, the innate response phagocytes activate **adaptive immunity** to deal with these infections. Adaptive immunity is the acquired ability to recognize and destroy a specific pathogen or its products. Adaptive responses are directed at unique pathogen molecules called **antigens**. Phagocytes present antigen molecules to lymphocytes, key cells in the adaptive response. The antigens bind specific receptors on the lymphocyte, triggering genes that promote lymphocyte multiplication and production of pathogen-specific proteins that interact with the pathogen, marking it for destruction. A protective adaptive response usually takes several days to develop; the strength of the adaptive response increases as the numbers of antigen-reactive lymphocytes increase.

We begin with the cells and organs common to the entire immune system and then consider the cells and mechanisms active in innate and adaptive immunity.

24.1 Cells and Organs of the Immune System

The cells active in innate and adaptive immunity develop from common precursors called **stem cells**. Immunity results from the actions of cells that circulate throughout the body, primarily through the blood and **lymph**, a fluid similar to blood that contains nucleated cells and proteins but lacks red blood cells. Blood and lymph interact directly or indirectly with every major organ system.

Stem Cells, Blood, and Lymph

Multipotent stem cells are precursor cells that can differentiate into any blood cell (**Figure 24.1**). Stem cells grow in the bone marrow where they differentiate into a variety of mature cells under the influence of soluble **cytokines** and **chemokines**, proteins that influence many aspects of immune cell differentiation. The differentiated cells travel through the blood and lymph to other parts of the body.

Blood consists of cellular and noncellular components, including many cells and molecules active in the immune response. The most numerous cells in human blood are *erythrocytes* (red blood cells), nonnucleated cells that function to carry oxygen from the lungs to the tissues (**Table 24.1**). About 0.1% of the cells in blood, however, are nucleated cells called **leukocytes** or *white blood cells*. Leukocytes include the phagocytes of the innate immune system and **lymphocytes**, the cells active in the adaptive response.

Whole blood is composed of suspended cells and **plasma**, a liquid containing proteins and other solutes. Outside the body, whole blood or plasma quickly forms an insoluble fibrin clot, remaining liquid only when an anticoagulant such as potassium citrate or heparin is added. When blood clots, the insoluble proteins trap the cells in a large, insoluble mass. The remaining fluid, called **serum**, contains no cells or clotting proteins. Serum does, however, contain a high concentration of other proteins, including soluble immune proteins called *antibodies* that bind pathogen antigens.

Blood and Lymph Circulation

Blood is pumped by the heart through arteries and capillaries throughout the body and is returned through the veins (**Figure 24.2a, b**). In the capillary beds, leukocytes and solutes pass to and from the blood into the lymphatic system, a separate circulatory system containing lymph (Figure 24.2a–c).

Lymph drains from extravascular tissues into lymphatic capillaries, lymph ducts, and then into lymph nodes throughout the lymphatic system (Figure 24.2d). **Lymph nodes** are organs that contain lymphocytes and phagocytes arranged to encounter microorganisms and antigens as they travel through the lymphatic circulation. The **mucosa-associated lymphoid tissue (MALT)**, another part of the lymphatic system, interacts with antigens and microorganisms that enter the body through mucous membranes, including those of the gut, the genitourinary tract, and bronchial tissues. The MALT also contains phagocytes and lymphocytes. Lymph fluid with antibodies and immune cells empties into the blood circulatory system via the thoracic lymph duct.

The spleen consists of red pulp, which is rich in red blood cells, and white pulp, which consists of organized lymphocytes and phagocytes arranged to filter the blood, analogous to lymph nodes and MALT in the lymphatic system. Collectively, the lymph nodes, MALT, and spleen are called **secondary lymphoid organs**. The secondary lymphoid organs are the sites where antigens interact with antigen-presenting phagocytes and lymphocytes to generate an adaptive immune response (Figure 24.2a).

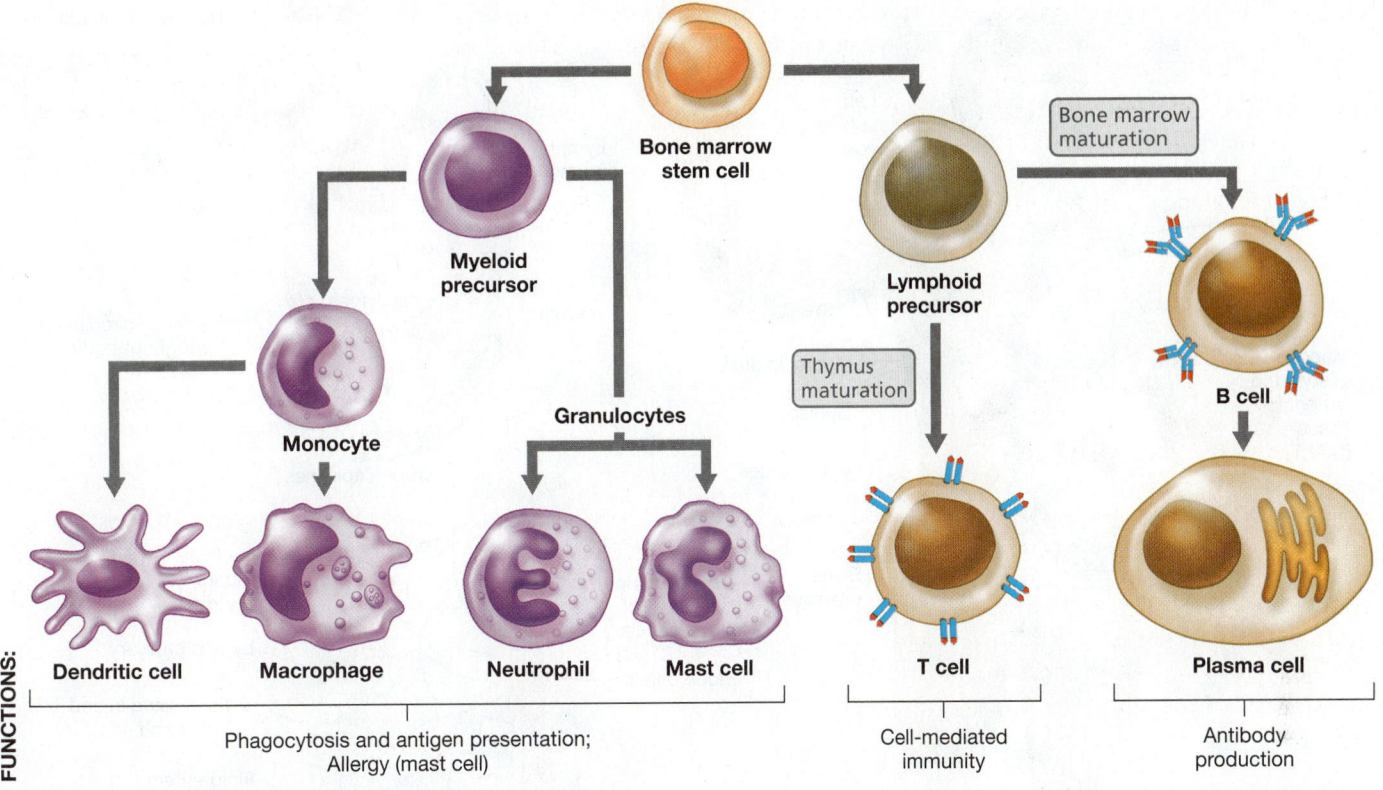

Figure 24.1 Immune response cells. Immune cells develop from multipotent stem cells in the bone marrow into either myeloid precursors or lymphoid precursors. These precursors, in turn, differentiate into end cells that have various immune functions.

Table 24.1 Major cell types found in normal human blood

Cell type	Cells per milliliter
Erythrocytes	$4.2–6.2 \times 10^9$
Leukocytes[a]	$4.5–11 \times 10^6$
Lymphocytes	$1.0–4.8 \times 10^6$
Granulocytes and monocytes	Up to 7.0×10^6

(a) Red blood cells (erythrocytes)

(b) Lymphocyte

(c) Neutrophil (a granulocyte)

(d) Monocyte

J. Martinko and M. T. Madigan

[a]Leukocytes include all nucleated blood cells. They include lymphocytes and cells derived from myeloid stem cells, the monocytes and granulocytes, such as neutrophils.

Leukocytes

Leukocytes are nucleated white blood cells found in the blood and the lymph. Several distinct leukocytes (Table 24.1 and Figure 24.1) participate in innate and adaptive immunity.

Myeloid cells, active in innate immunity, are derived from myeloid precursor cells. Mature myeloid cells can be divided into two lineages, the monocytes and the granulocytes (Figure 24.1). The monocyte lineage develops into specialized phagocytic cells, the **antigen-presenting cells (APCs)**. These cells, in addition to the B cells we discuss below, engulf, process, and present antigens to lymphocytes. APCs include *macrophages* and *dendritic cells*. Immature cells called monocytes are circulating precursors of macrophages and dendritic cells. **Macrophages** are generally the first defense cells that interact with a pathogen. They are abundant in many tissues, especially spleen, lymph nodes, and MALT. **Dendritic cells** are phagocytes that specialize in presenting antigen to lymphocytes.

Granulocytes are the second lineage of cells derived from myeloid precursors. Granulocytes contain cytoplasmic inclusions, or granules, that can be visualized by staining. These granules contain toxins or enzymes that are released to kill target cells. The phagocytic activity of one granulocyte, the **neutrophil**, also called a *polymorphonuclear leukocyte* (*PMN*), is central to innate immunity. Release of granules, a process called *degranulation*, from a granulocyte called a mast cell can cause allergy symptoms and inflammation.

Lymphocytes are specialized leukocytes involved exclusively in the adaptive immune response. Mature lymphocytes circulate

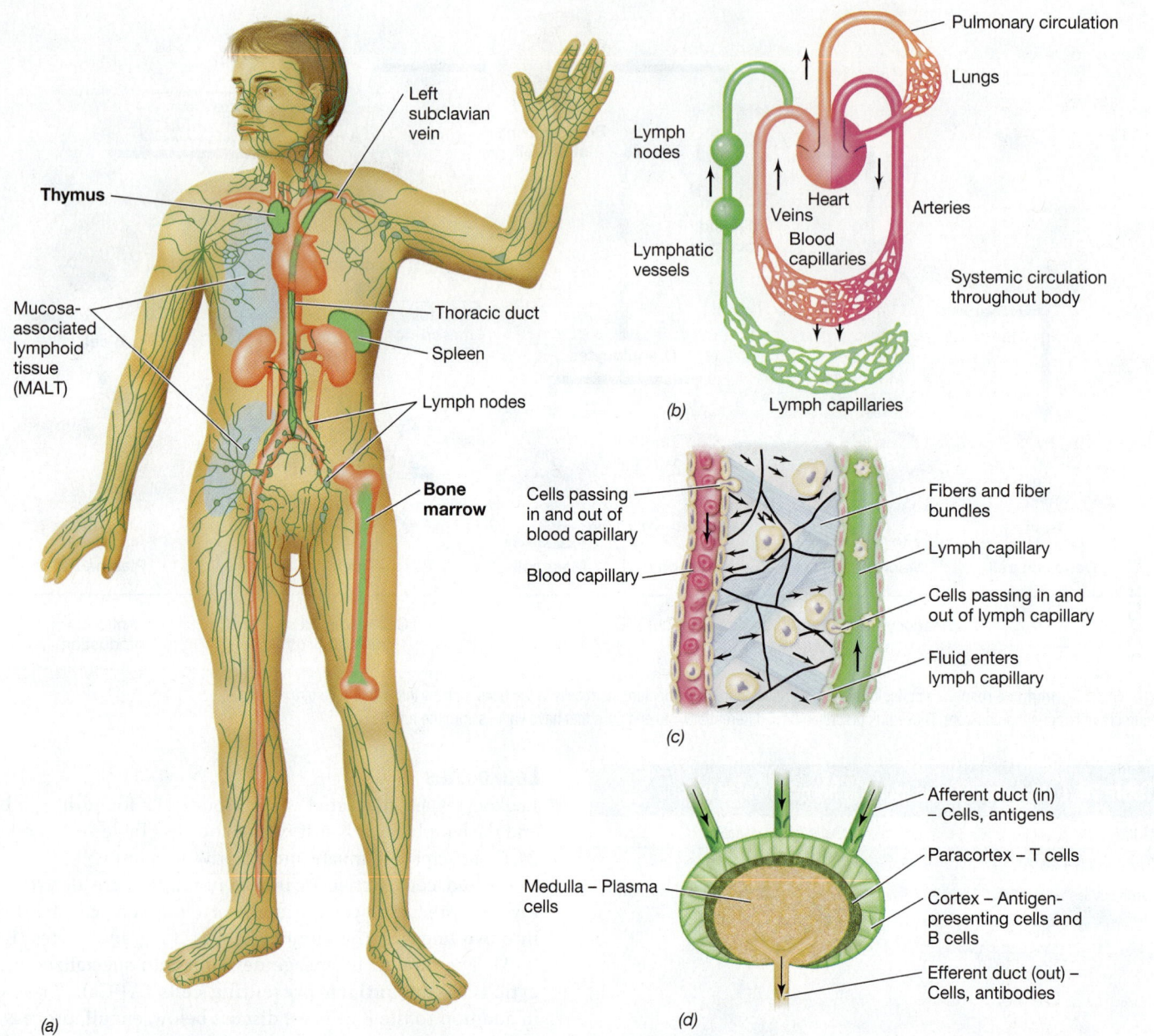

Figure 24.2 **The blood and lymphatic systems.**
(a) The blood and lymphatic circulatory systems. Major blood vessels and associated organs are shown in red. Major lymphatic organs and vessels are shown in green. The primary lymphoid organs are the bone marrow and thymus. The secondary lymphoid organs are the lymph nodes, spleen, and MALT. *(b)* Connections between the lymphatic and blood systems. Blood flows from the veins to the heart, to the lungs, and then through the arteries to capillaries in tissues. Exchange of solutes and cells occurs between blood and lymphatic capillaries. Lymph drains from the thoracic duct into the left subclavian vein of the blood circulatory system. *(c)* The exchange of cells between the blood and lymphatic systems is shown in detail. Both blood and lymphatic capillaries are closed vessels, but cells pass from blood capillaries to lymphatic capillaries and back by a process known as extravasation. *(d)* A secondary lymphoid organ, the lymph node, showing the major anatomical areas and the immune cells concentrated in each area. The anatomy of the MALT and the spleen is analogous to that of the lymph nodes.

through the blood and lymph system, but are concentrated in the lymph nodes and spleen where they interact with antigens. There are two types of lymphocytes, B cells (B lymphocytes) and T cells (T lymphocytes) (Figure 24.1). **B cells** originate and mature in the **bone marrow**. They are specialized APCs and the precursors of antibody-producing **plasma cells**. **Antibodies**, also called **immunoglobulins (Igs)**, are soluble proteins produced by B cells and plasma cells. Antibodies interact with particular antigens. **T cells**,

which interact with antigen, begin their development in the bone marrow but travel to the **thymus** to mature. The bone marrow and thymus in mammals are called **primary lymphoid organs** because they are the sites where the lymphoid stem cells develop into functional antigen-reactive lymphocytes (Figure 24.2*a*).

All leukocytes move throughout the body and pass from blood to interstitial spaces, then to lymphatic vessels, and back to the blood circulatory system, a process called *extravasation* (Figure 24.2*c*).

MINIQUIZ

- Trace the development of B cells, T cells, and macrophages from the common stem cell.
- Describe the circulation of a leukocyte from the blood to the lymph and back to the blood.

24.2 Innate Immunity

Innate immunity is the noninducible, preexisting ability to recognize and destroy a pathogen or its products. Innate immunity does not require previous exposure to a pathogen or its products and is mediated by phagocytes. Eukaryotes from the lowest plants through the most evolved vertebrates have functionally similar phagocytic recognition mechanisms that lead to rapid and effective host defense. We know that the innate system receptors present in vertebrates, for example, have structural and evolutionary homologs in phylogenetic groups as distant as the insect *Drosophila* (fruit fly).

Pathogen-Associated Molecular Patterns

The macromolecules inside and on the surface of pathogens display **pathogen-associated molecular patterns (PAMPs)**, consisting of repeating subunits. An example of a PAMP is the lipopolysaccharide (LPS) common to all gram-negative bacterial outer membranes (⮐ Section 2.11). Other PAMPs include bacterial flagellin, the double-stranded RNA (dsRNA) of certain viruses, and the lipoteichoic acids of gram-positive bacteria (⮐ Section 2.10). These macromolecules consist of repeating structural elements shared among broadly related pathogens.

Pattern Recognition Receptors

Phagocytes such as macrophages and neutrophils are the first line of defense against pathogens that the body has never before encountered. Phagocytes can interact quickly and effectively with pathogens because they have evolved specialized molecules that interact directly with PAMPs. These specialized molecules are called **pattern recognition receptors (PRRs)** (**Figure 24.3**). Each PRR interacts with a particular PAMP to activate the phagocyte. One PRR found on all phagocytes, for example, interacts with the LPS on gram-negative bacteria, including all pathogenic strains of *Salmonella* spp., *Escherichia coli*, and *Shigella* spp. Another phagocyte PRR interacts with the peptidoglycan on gram-positive cells. Still other PRRs interact with conserved pathogen features such as the dsRNA found in some viruses and flagellin on certain motile bacteria. The interaction of a PAMP with a PRR activates the phagocyte to ingest and destroy the targeted pathogen by phagocytosis. All phagocytes have PRRs that are instantly available to interact with invasive pathogens.

MINIQUIZ

- Identify a pathogen-associated molecular pattern shared by a group of microorganisms.
- Identify the cell types that use pattern recognition receptors to provide innate immunity to pathogens.

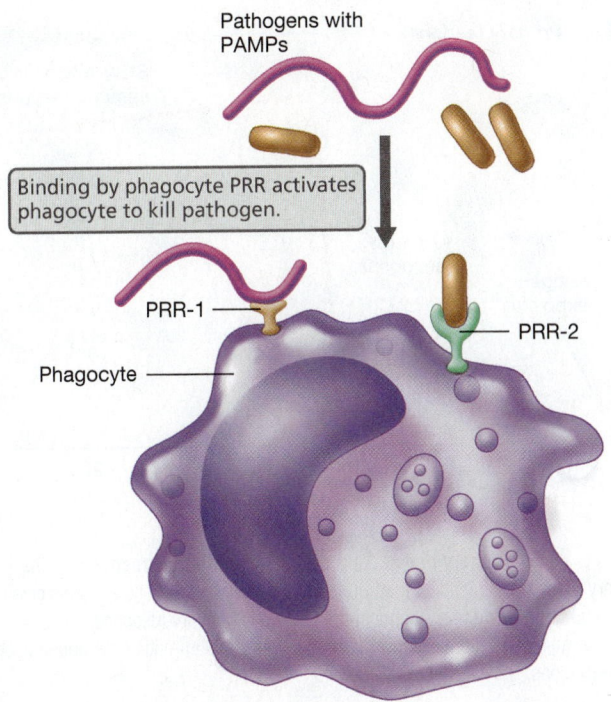

Pathogens with PAMPs

Binding by phagocyte PRR activates phagocyte to kill pathogen.

PRR-1

PRR-2

Phagocyte

Figure 24.3 Innate immunity. Phagocytes interact with pathogens by recognizing pathogen-associated molecular patterns (PAMPs) with preformed pattern recognition receptors (PRRs). Binding of a PAMP by a phagocyte PRR stimulates the phagocyte to destroy the pathogen and activate other phagocytes.

24.3 Adaptive Immunity

The phagocytes responsible for innate immunity often trigger *adaptive* immunity in vertebrate animals. As compared with the common targets that trigger innate immunity, adaptive immunity is directed toward a molecular component of the pathogen called an antigen. In adaptive immunity, pathogen-specific receptors are produced in large numbers only after exposure to the pathogen or its products.

The first exposure to an antigen generates a **primary adaptive immune response**: The antigen contact stimulates growth and multiplication of antigen-reactive cells, creating **clones**, large numbers of identical antigen-reactive cells. These clones persist for years and confer long-term specific immunity.

Antigen-reactive lymphocytes are divided into T cells and B cells. Each cell type produces a unique protein that interacts with a single antigen. These proteins thus have **specificity** for that antigen. The antigen-binding proteins of T cells are the **T cell receptors (TCRs)**; cell-surface antibodies on a B cell are the **B cell receptors (BCRs)**.

As compared to innate immunity, the adaptive response is inducible only when triggered by a unique antigen on a pathogen. For example, polysaccharide antigen from the LPS of a particular gram-negative bacterium is unique for a genus and sometimes even for a species within the genus. For example, an individual lymphocyte clone that interacts with an LPS polysaccharide on *Salmonella* will not interact with the LPS on other bacteria. The terminal sugars that constitute the polysaccharides of *Salmonella* spp. are unique identifying antigens for the genus and are not

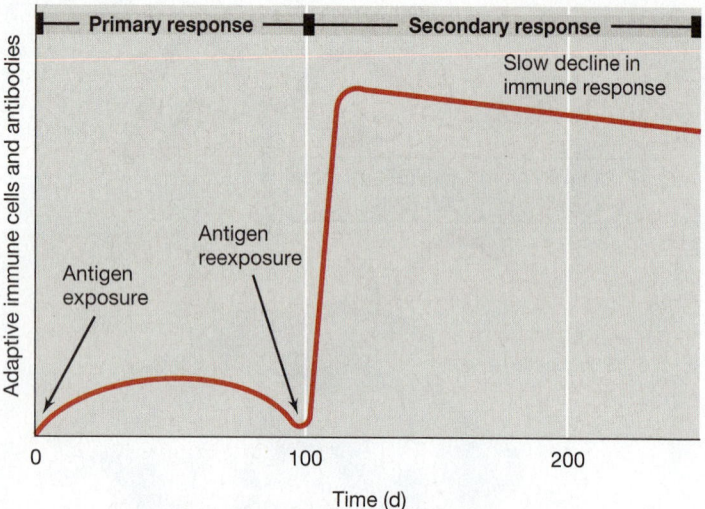

Figure 24.4 Primary and secondary adaptive immune responses. The primary response induces both immune cells and antibodies. The antigens given at day 0 and day 100 must be identical to induce a secondary response. The secondary response may generate more than 10-fold increases in immune cells and antibody concentrations.

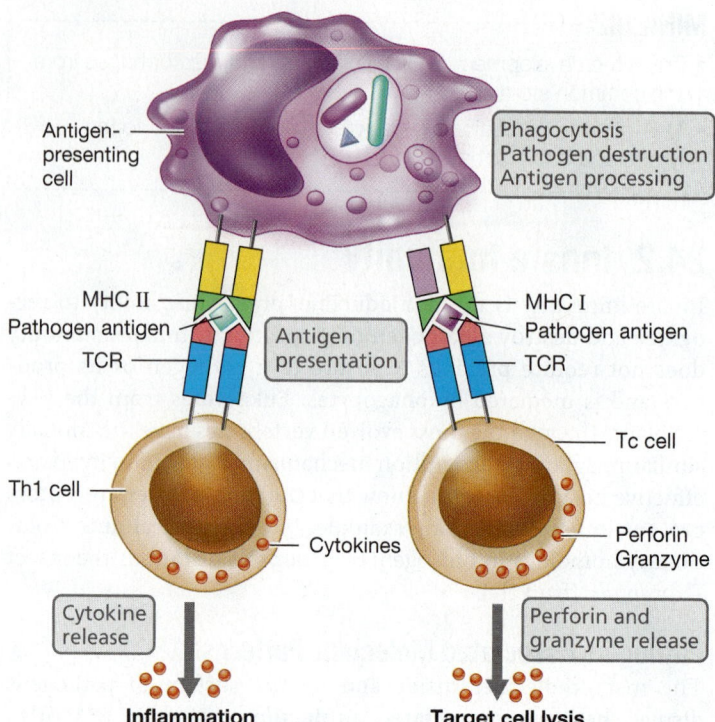

Figure 24.5 T cell immunity. Antigen-presenting cells such as the phagocytes in innate immunity ingest, degrade, and process antigens. They then present antigens to T cells that secrete protein cytokines that activate the adaptive immune response. Antigen-reactive T cells include inflammatory T-helper (Th1) cells that make cytokines that activate other cells, causing inflammation. T-cytotoxic (Tc) cells produce perforins and granzymes, proteins that enter and lyse nearby target cells.

shared by other bacteria, even other gram-negative enteric bacteria such as *Escherichia coli* or *Shigella* spp.

A second exposure to the same antigen activates the clones of antigen-reactive cells and generates a faster, stronger **secondary adaptive immune response** that peaks within several days (**Figure 24.4**). The products of this secondary immune response quickly target the pathogen for destruction. This rapid increase in adaptive immunity after a second antigen exposure is called **memory**. Finally, the adaptive immune system exhibits **tolerance**, the acquired *inability* to generate an immune response against self antigens. Tolerance ensures that adaptive immunity is directed to outside agents that pose genuine threats to the host, and not to host proteins.

T Cells and Antigen Presentation

Adaptive immunity begins with the interactions of immune T lymphocytes with peptide antigens on infected cells. The infected cells that are first recognized by T cells may include the same phagocytes that participate in the innate immune response. The T cell, with its TCR, can recognize peptide only when the peptides are complexed with self proteins called **major histocompatibility complex (MHC)** proteins, found on host cell surfaces (**Figure 24.5**). All host cells display MHC I proteins that present peptides from viruses and other intracellular pathogens for immune recognition; APCs (macrophages, dendritic cells, and B cells) also display an additional antigen-presenting protein, MHC II. Macrophages are found in all organs of the body, but the other APCs are localized in the secondary lymphoid organs—spleen, lymph nodes, and MALT. These secondary lymphoid organs are the sites where the adaptive immune response begins. APCs ingest bacteria, viruses, and other antigenic material by phagocytosis (in macrophages and dendritic cells) or through internalization of molecular antigen bound to a BCR. After ingestion, the APCs degrade the antigens to small peptides. The MHC proteins inside

the APC bind the peptides derived from the digested pathogens. The MHC-embedded peptides are then transported to the phagocyte surface where the complex is displayed, a process called *antigen presentation*. For example, a phagocyte infected with influenza virus will display MHC I and II proteins embedded with influenza peptides. These influenza peptide–MHC complexes are the targets for T cells.

T Lymphocyte Subsets

T cells interact with the peptide–MHC complex using the TCR. Each T cell expresses a TCR that is specific for a single peptide–MHC complex. Antigen-specific T cells are found closely associated with the APCs in the spleen, lymph nodes, and MALT. T cells constantly sample surrounding APCs for peptide–MHC complexes; peptide–MHC complexes that interact with the TCR signal the T cell to grow and divide, producing antigen-reactive clones that interact directly to kill cells displaying the same foreign peptide or help other cells to eliminate cells with the same peptide.

Antigen-reactive T cells consist of three basic T cell subsets, based on their functional properties. These T cell subsets interact with other cells to initiate immune reactions. *T-helper (Th) cells* interact with peptide–MHC II complexes on the surface of APCs. This interaction causes the Th cells to differentiate into one of two subsets that indirectly mediate immune reactions. Antigen-activated T cell subsets, termed Th1 and Th2, respond by proliferating and producing soluble cytokines. The cytokines interact

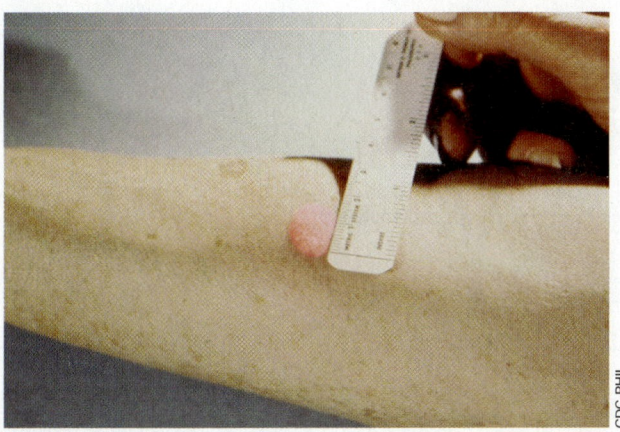

Figure 24.6 **Th1 cells and macrophage activation.** This tuberculin test shows a positive reaction. Macrophages activated by antigen-specific Th1 cells caused the localized reaction to a tuberculosis antigen, tuberculin, at the site of injection. The raised area of inflammation on the forearm is about 1.5 cm in diameter.

with receptors on other cells, activating them to initiate an immune response. Antigen-specific Th1 cells that interact with peptide–MHC II complexes on the surface of macrophages stimulate the Th1 cell to produce cytokines that activate the macrophages, enhancing phagocytosis of cells displaying the target peptide and causing inflammatory reactions that limit the spread of infections (Figure 24.5, **Figure 24.6**). For example, *Mycobacterium tuberculosis* infects macrophages and other cells in the lung, causing tuberculosis. Activated macrophages kill *M. tuberculosis* inside the cell, stopping the spread of infection to other cells. An inflammatory reaction associated with *M. tuberculosis* is termed the *tuberculin reaction* and is used as a diagnostic test for

M. tuberculosis exposure. This test uses *tuberculin*, an extract from *M. tuberculosis*, to attract immune Th1 cells that then produce cytokines, activating macrophages and causing a localized red, hot, hardened, and swollen area that typifies inflammation and effective immunity. Differentiated Th2 cells, the other T-helper subset, use cytokines to stimulate ("help") antigen-reactive B cells to produce antibodies, as we discuss below. *T-cytotoxic (Tc) cells* recognize the peptide–MHC I complex on an infected cell. When Tc cells interact with the infected cell, they secrete proteins that kill the peptide-bearing infected cell (Figure 24.5).

MINIQUIZ --

- Explain the process of antigen presentation to T cells.
- Define the role of Tc and Th1 cells in adaptive immunity.

24.4 Antibodies

Antibodies are soluble proteins made by B cells and plasma cells (Figure 24.1) in response to exposure to antigens. Each antibody binds specifically to a single antigen. Antibody-mediated immunity controls the spread of infection by recognizing pathogens and their products in extracellular environments such as blood and mucus secretions.

B Cell Functions

B cells are specialized lymphocytes that have antibodies on their surface; each B cell displays multiple copies of a single antibody that is specific for a single antigen. To make antibodies, B cells must first bind antigens through interactions with the BCR (**Figure 24.7**). The surface antibody–antigen interaction induces the B cell

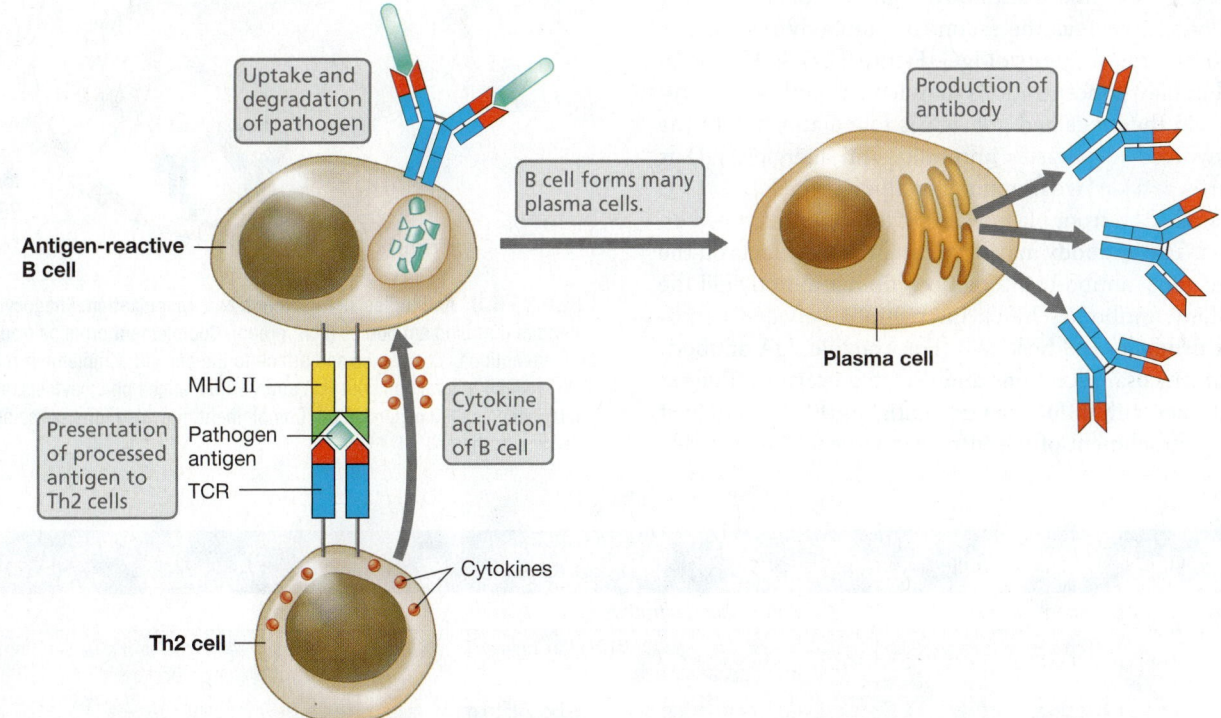

Figure 24.7 **Antibody-mediated immunity.** Antibody on a B cell binds antigen and, through collaboration with a Th2 cell, the B cell forms clones which can then differentiate into antibody-producing plasma cells.

to ingest the antigen-containing pathogen by phagocytosis. The B cell then kills and digests the pathogen, producing a battery of pathogen-derived peptide antigens. These peptides are then complexed with MHC II and displayed, or *presented*, on the surface of the B cell to the antigen-specific Th2 cell.

The Th2 cell does not interact directly with the pathogen but stimulates ("helps") other cells, in this case, the antigen-presenting B cell on which it recognized the MHC–peptide. Th2 cells produce cytokines that stimulate antigen-reactive B cells, which in turn respond by growing and dividing, establishing clones of the original antigen-reactive B cell. Many of these activated B cells then differentiate into plasma cells that produce soluble antibodies (Figure 24.7). This primary antibody response is detectable within about five days after antigen exposure, and antibodies reach peak quantities within several weeks. Some of the activated B cells from the clone remain in circulation in the immune system as memory B cells. Subsequent exposure to the same antigen, for example, by reinfection with the same pathogen, stimulates the antigen-reactive memory B cells, producing a secondary antibody response, characterized by a faster development of higher quantities of antibodies (Figure 24.4). The secondary response, also called *immune memory*, is the basis for vaccination, as we discuss later. The antibody response is highly specific for the eliciting antigen; antigen binding by antibodies can trigger neutralization or destruction of pathogens or their products through several mechanisms (**Figures 24.8** and **24.9**).

Antibody Classes and Functions

Several different classes of antibodies are distinguished from one another by their primary amino acid sequence. Each antibody class has a defined general function (**Table 24.2**). IgM and IgG are found in blood. The primary antibody response consists mostly of IgM antibodies, whereas the secondary antibody response is characterized by large amounts of IgG (Figure 24.4). IgA is found in blood and in high concentrations in mucous membrane secretions, such as in the lungs and gut. IgE is found attached to the mast cells involved in parasite immunity and allergies. IgD is found primarily as a surface immunoglobulin on B cells.

Antibodies released from plasma cells interact with antigen on the pathogens. The antibody may have one or more effects on the pathogen, but most antibody interactions do not directly kill the pathogen. Many antibodies block interactions between pathogens or their products and host cells. For example, IgA antibodies present in mucosal secretions and directed against influenza virus may interact with influenza virus antigens that bind to host cells, blocking attachment of the influenza virus to the host cell.

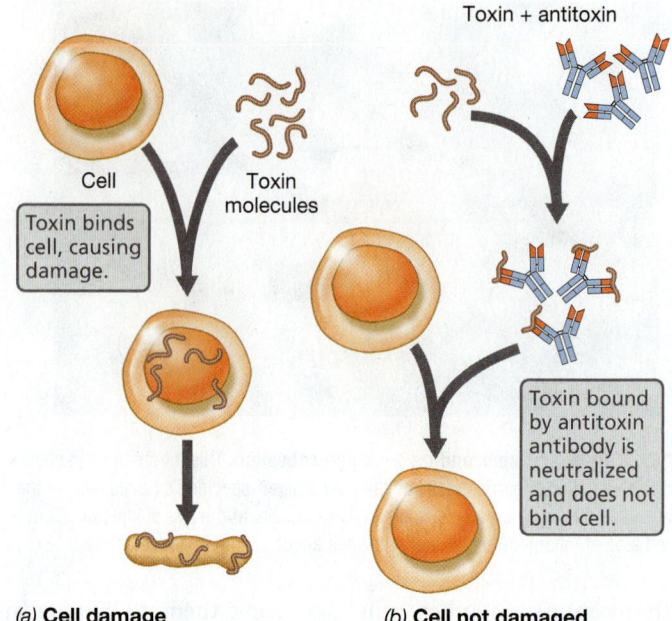

Figure 24.8 **Neutralization of an exotoxin by an antitoxin antibody.** *(a)* Toxin results in cell destruction. *(b)* Antitoxin prevents cell destruction.

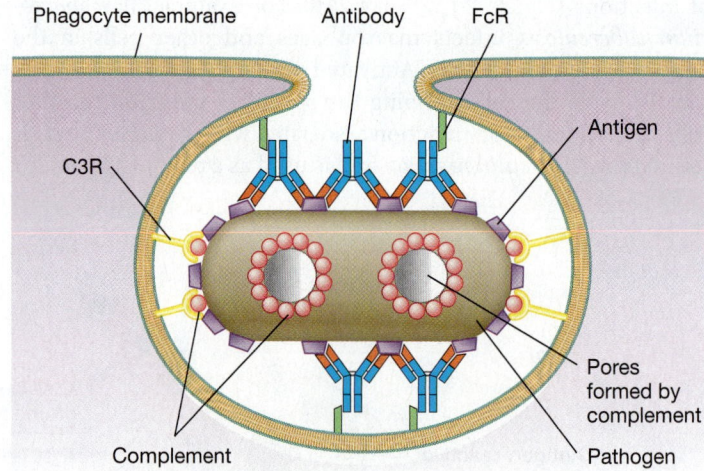

Figure 24.9 **Antibodies, complement, and opsonization.** Phagocytes have receptors that bind antibodies (FcR; green). Complement proteins (red) bind to antigen–antibody complexes and attach to the cell via complement receptors (C3R; yellow). Interaction with FcR and C3R enhances phagocytosis, a phenomenon called opsonization. Complement may also form pores and directly lyse the cell.

Table 24.2 Major classes of antibodies

Antibody class	Location	Serum Concentration	Functions
IgA	Serum and mucus secretions	2.1 mg/ml in serum and high local amounts at mucosal surfaces	Secondary response and mucosal immunity for extracellular pathogens
IgE	Mast cells	Low amounts in serum; all bound to mast cells	Parasite immunity, allergies
IgG	Serum	13.5 mg/ml (highest in serum)	Secondary response for extracellular pathogens
IgM	Serum	1.5 mg/ml	Primary response for extracellular pathogens

Specific serum antibodies can also bind toxins such as tetanus toxin, again blocking the binding of toxin to host cell receptors. This process is called *neutralization* (Figure 24.8). In many cases, antibodies mark the pathogen for destruction by phagocytosis. Phagocytes have general antibody receptors called *Fc receptors* (*FcR*) that bind to any antibody attached to an antigen. This interaction results in enhanced phagocytosis of the antibody-coated cells, a process known as *opsonization* (Figure 24.9).

Complement

Antibody-mediated destruction of pathogens may also involve a group of proteins known collectively as *complement* (Figure 24.9). The complement proteins attach to pathogen surfaces, attracted by IgM or IgG antibodies bound to the pathogen. The complement proteins bound to the cell surface by the antibody have two possible effects on the pathogen. First, complement proteins can form a pore in the pathogen cytoplasmic membrane, directly lysing the pathogen cell. This complement–antibody interaction affects only those pathogen cells with bound antibodies. For example, antibodies specific for cell surface proteins of *Salmonella* interact only with *Salmonella*. Complement causes

lysis of only the antibody-coated *Salmonella* cell, but not of a nearby *Escherichia coli* cell that is not coated with antibodies. Many pathogens, such as the thick-walled gram-positive *Streptococcus* species, are relatively resistant to complement-mediated lysis because the cell wall makes the cytoplasmic membrane less accessible to complement proteins. Antibodies to the external cell wall components, however, still attract complement proteins to the pathogen surface. The second effect of complement binding is the stimulation of phagocytosis. Pathogen-bound complement proteins are recognized by complement receptors called *C3 receptors* (*C3R*) found on the surface of phagocytes such as neutrophils and macrophages. This interaction results in opsonization and phagocytosis of the cells sensitized by antibody and complement (Figure 24.9).

MINIQUIZ --------------------------------------

- Summarize antibody production starting with pathogen interaction with a B cell.
- Define the role of antibody and complement in pathogen destruction.

II • Host Defense

The immune system protects the host through several mechanisms, including inflammation and active and passive adaptive immune responses that rid the body of pathogens. We can also generate protective immunity without infection through vaccination with antigens derived from pathogens.

24.5 Inflammation

Inflammation is a general, nonspecific reaction to noxious stimuli such as toxins and pathogens. Inflammation is characterized by redness (erythema), swelling (edema), pain, and heat, usually localized at the site of infection (Figure 24.6 and **Figure 24.10**). The

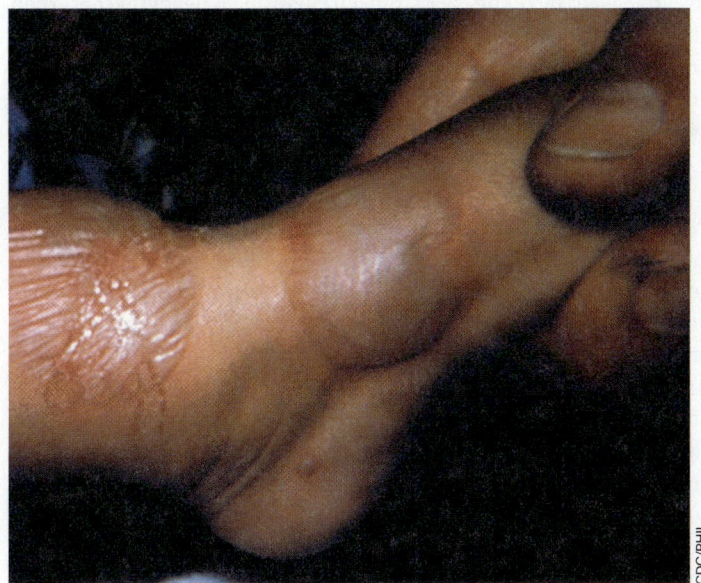

(a) **Photograph of a child's foot showing swelling due to infection with vaccinia virus; fluid accumulation results from inflammatory activities.**

Figure 24.10 Inflammation.

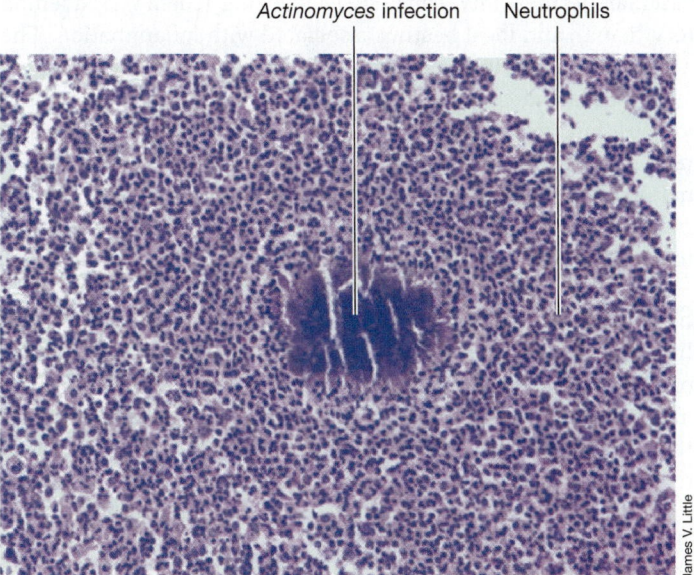

Actinomyces infection   Neutrophils

(b) **The dark mass in the center of the photomicrograph resulted from infection by *Actinomyces*, a filamentous bacterium. The darkly stained cells surrounding the mass are neutrophils, indicating acute inflammation.**

molecular activators of inflammation are a group of cell activators and chemoattractants, including cytokines and chemokines. These activators are produced by various cells. The most important chemokines and cytokines are termed *proinflammatory* because of their inflammation-inducing abilities; they are produced in high concentrations by phagocytes and lymphocytes.

Both innate and adaptive immune responses to infection can cause inflammation; both immune recognition systems induce the activators that recruit and activate effector cells such as neutrophils. An immune response normally activates inflammation to isolate and limit tissue damage by destroying pathogen invaders and removing damaged cells. In some cases, however, inflammation results in considerable damage to healthy host tissue.

Inflammatory Cells and Local Inflammation

Immune-mediated inflammation is an acute condition that begins at the site of pathogen entry into the body. The innate PRRs on macrophages and other tissue cells at the site of infection engage the pathogen PAMPs (Figure 24.3). This action activates the local cells to produce and release mediators including cytokines and chemokines that interact with cytokine and chemokine receptors on other cells such as neutrophils. For example, local tissue macrophages that are activated by PAMP–PRR interaction secrete a chemokine called CXCL8. The CXCL8, through a CXCL8 receptor, activates neutrophils that then migrate along the chemokine gradient toward the source of the CXCL8, where they begin to ingest and kill the pathogen. The neutrophils, in turn, secrete even more CXCL8, attracting more neutrophils and amplifying the response, eventually destroying the pathogens (Figure 24.10*b*).

The chemokine and cytokine mediators released by injured cells and phagocytes contribute to inflammation. For example, the macrophage and other cells at the site of infection produce proinflammatory cytokines including interleukin-1 (IL-1), IL-6, and tumor necrosis factor α (TNF-α). These cytokines increase vascular permeability, causing the swelling (edema), reddening (erythema), and local heating associated with inflammation. The edema stimulates local neurons, causing pain (**Figure 24.11***a*).

The usual outcome of the inflammatory response is a rapid localization and destruction of the pathogen by macrophages and recruited neutrophils. As the pathogens are removed, the inflammatory cells are no longer stimulated, their numbers at the site are reduced, cytokine production decreases, attraction of phagocytes toward the cytokine producers stops, and inflammation subsides.

Systemic Inflammation and Septic Shock

In some cases, the inflammatory response fails to localize the pathogens and the reaction spreads throughout the body. An uncontrolled systemic inflammation can be more dangerous than the original infection, with inflammatory cells and mediators contributing to large-scale inflammation. An inflammatory response that spreads inflammatory cells and mediators through the entire circulatory and lymphatic systems can lead to septic shock, a life-threatening condition.

A common cause of septic shock is systemic infection by enteric bacteria such as *Salmonella* or *Escherichia coli* caused by a ruptured or leaking bowel that releases the gram-negative organisms into the intraperitoneal cavity or the bloodstream. The primary

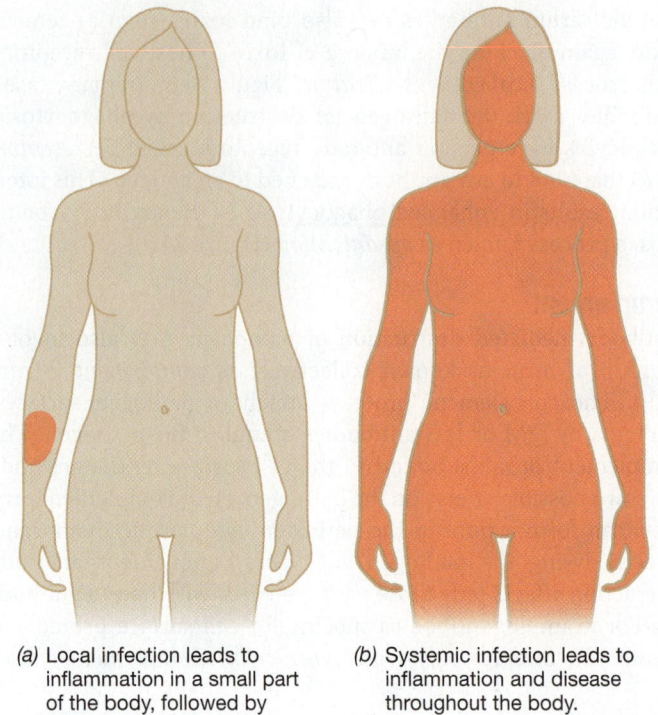

(a) Local infection leads to inflammation in a small part of the body, followed by healing.

(b) Systemic infection leads to inflammation and disease throughout the body.

Figure 24.11 Local and systemic inflammation. *(a)* Local infection, mediated by proinflammatory cytokines from local macrophages, results in inflammation that subsides as the infection is cleared. *(b)* Systemic infection causes systemic release of proinflammatory cytokines, resulting in widespread systemic inflammatory symptoms including severe edema, fever, and septic shock, even if the infection is controlled.

infection is often cleared by the phagocytes or is treated successfully with antibiotics. However, the endotoxic outer membrane LPS from these organisms interacts with a PRR on phagocytes, stimulating production of proinflammatory cytokines, which are released into the systemic circulation. The cytokines then induce systemic responses that parallel the localized inflammatory response, but on a larger scale involving many organ systems, ultimately leading to an extensive whole-body inflammatory event with potentially life-threatening consequences (Figure 24.11*b*).

The proinflammatory cytokines IL-1, IL-6, and TNF-α, for example, are *endogenous pyrogens*. They stimulate the brain to produce prostaglandins, chemical signals that raise body temperature, causing a fever. By contrast, the same cytokines released in small amounts at local sites of infection produce local heating; this increases blood flow and promotes healing. The large quantities of endogenous pyrogens released into the general circulation as a result of a systemic infection induce organism-wide heating characterized by uncontrollable high fever. In addition, the large amounts of inflammatory mediators released systemically, instead of causing local edema due to vasodilation and increased vascular permeability, cause the same reactions on a system-wide scale. The result is a massive efflux of fluids from the central vascular tissue with loss of systemic blood pressure due to reduced blood volume, and severe edema due to influx of fluids from vascular tissues into extravascular spaces.

In short, the proinflammatory cytokines, produced in high quantities to combat systemic infections, induce high fever, extremely low blood pressure, and severe edema. This condition, termed septic shock, causes death in up to 30% of affected individuals.

MINIQUIZ
- Identify the molecular mediators of inflammation and define their individual roles.
- Identify the major symptoms of localized inflammation and of septic shock.

24.6 Immunity and Immunization

Both the innate and adaptive immune responses protect the host from infections by pathogens, and both innate and adaptive immunity are essential for survival (**Figure 24.12**). For example, individuals that lack innate immunity because they fail to produce phagocytes develop recurrent infections from bacteria, viruses, and fungi and die at an early age without extraordinary intervention. Individuals who lack adaptive immunity have the same outcome, but may survive longer if the innate immune system is functional.

Immunity may be naturally occurring or artificially induced by exposure to antigens (vaccination). We acquire immunity either *actively*, for example when we generate an immune response through exposure to antigen, or *passively*, for example when we receive antibodies or immune cells from an immune individual. Active and passive immunity are illustrated in **Figure 24.13** and contrasted in **Table 24.3**.

Natural Immunity

Animals normally develop *natural active immunity* by acquiring a natural infection that initiates an adaptive immune response (Figure 24.13). Natural active immunity is the outcome of exposure to antigens through infection and usually results in protective immunity from antibodies and T cells. For example, virtually all adults have acquired natural active immunity to many strains of influenza and cold viruses through immune responses to infections. Natural

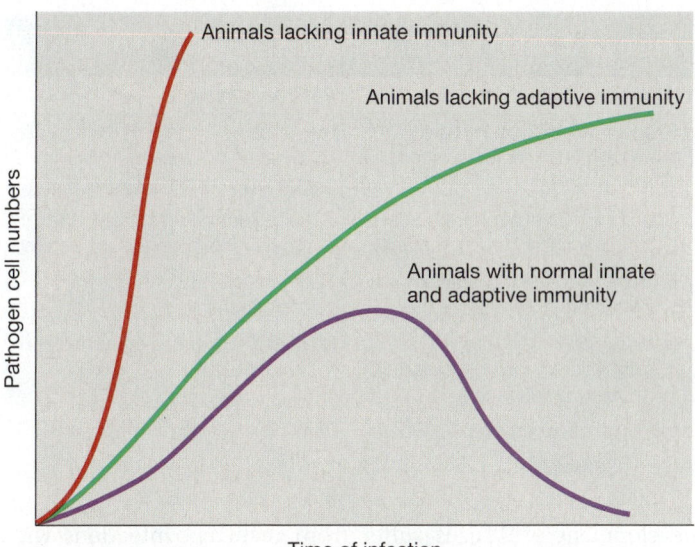

Figure 24.12 Infection and pathogen clearance in normal and immune-deficient animals. Animals with genetic defects that prevent development of the phagocytes critical for innate immunity have recurrent, incurable, lethal infections. Genetic defects that prevent development of mature, antigen-reactive B and T cells critical for adaptive immunity also allow recurrent infections, but the innate response controls these infections for a longer time and these animals live longer than the animals lacking innate immunity. Animals with normal innate and adaptive immunity rapidly clear most infections.

immunity, however, requires that an infection has occurred, which in some cases could involve potentially dangerous pathogens.

Active immunity is critical for infectious disease resistance. We know this because of problems caused by genetic defects and diseases that affect the immune system. For example, patients who cannot produce antibodies because of genetic defects in their B cells acquire serious infections from extracellular pathogens, especially bacteria. Individuals with genetic defects that prevent

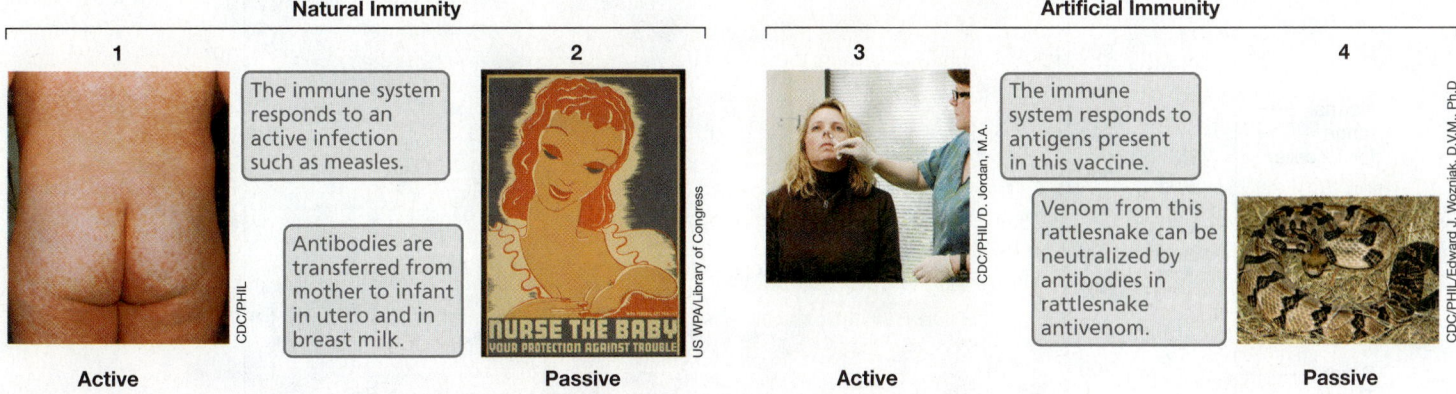

Figure 24.13 Natural and passive immunity. Photos, left to right: (1) Childhood measles showing typical systemic measles rash. Natural active immunity requires infection with a pathogen to activate the adaptive immune response. (2) A 1934 United States government poster promoting breast-feeding. Natural passive immunity occurs when immunity is transferred from one individual to another by natural means, such as the transfer of antibodies in breast milk. (3) Vaccination by nasal inhalation of antigen. Artificial active immunity occurs from exposure to particular antigens in a vaccine. (4) Timber rattlesnakes produce highly poisonous venom. An antivenom consisting of purified timber rattlesnake antivenom antibodies can be prepared in horses and artificial passive immunity conferred on a snakebite victim by injecting the victim with the antivenom.

Table 24.3 Active and passive immunity

Active immunity	Passive immunity
Exposure to antigen; immunity achieved by injecting antigen or through infection	No exposure to antigen; immunity achieved by injecting antibodies or antigen-reactive T cells
Specific immune response made by individual achieving immunity	Specific immune response made by the donor of antibodies or T cells
Immunity activated by antigen; immune memory in effect	No immune system activation; no immune memory
Immunity can be maintained via stimulation of memory cells (i.e., booster immunization)	Immunity cannot be maintained and decays rapidly
Immunity develops over a period of weeks	Immunity develops immediately

development of T cells suffer from recurrent infections with viruses and other intracellular pathogens. Individuals with *severe combined immune deficiency syndrome* (*SCID*) have a genetic defect that prevents proper formation of either B or T cells; they have no effective adaptive immunity and die from multiple recurrent infections (Figure 24.12) unless they receive supportive therapy such as a bone-marrow transplant and antibiotics. Immunodeficiency can also be caused by toxic reactions to drugs, environmental contaminants, or infections. For example, the loss of the adaptive immune response is the defining characteristic of *acquired immunodeficiency syndrome* (*AIDS*). In AIDS patients, infection with the human immunodeficiency virus (HIV) causes depletion of Th cells, resulting in a lack of effective immunity (**Figure 24.14**). In cases of immunodeficiency the result is similar; patients suffer from recurrent life-threatening infections.

Natural passive immunity (Figure 24.13) occurs when a nonimmune person acquires preformed immune cells or antibodies through natural transfer of cells or antibodies from an immune person. For example, for several months after birth, newborns have maternal IgG antibodies in their blood; these protective antibodies were transferred across the placenta before birth. Similarly, IgA antibodies are transferred to newborns in breast milk. In both cases, the antibodies that are protective for the infant were made in the mother and are passively acquired by natural means. These preformed antibodies provide disease protection while the immune system of the newborn matures.

Artificial Passive Immunity

In *artificial passive immunity*, the individual receiving the antibodies played no active part in antibody production; he or she receives preformed antibodies through injection of an *antiserum* (serum containing antibodies from the blood of an immune individual) or purified antibodies (immunoglobulin) derived from an immune individual. These antibodies gradually disappear from the body and a later exposure to the antigen cannot elicit a secondary response.

Artificial passive immunity is used to prevent or treat acute infectious diseases such as tetanus and in the treatment of bites by venomous animals (Figure 24.13). Cells or antibodies from an immune individual are transferred to a nonimmune individual. For example, tetanus antiserum may be administered to passively immunize an individual suspected of being exposed to *Clostridium tetani* due to a traumatic injury such as a car accident. The antibody injection provides immediate immune protection against tetanus toxin. The antibody-containing preparation is known as an *antiserum*, an *antitoxin* if the antibodies are directed against a toxin, or an *antivenom* if the antibodies are directed toward a venom. Antisera are obtained from immunized animals such as horses, or from humans with high levels of antibody. The immunoglobulin from serum pooled from a number of individuals is also used for passive immunization to prevent certain viral diseases since pooled sera contain antibodies to various common pathogens such as hepatitis A.

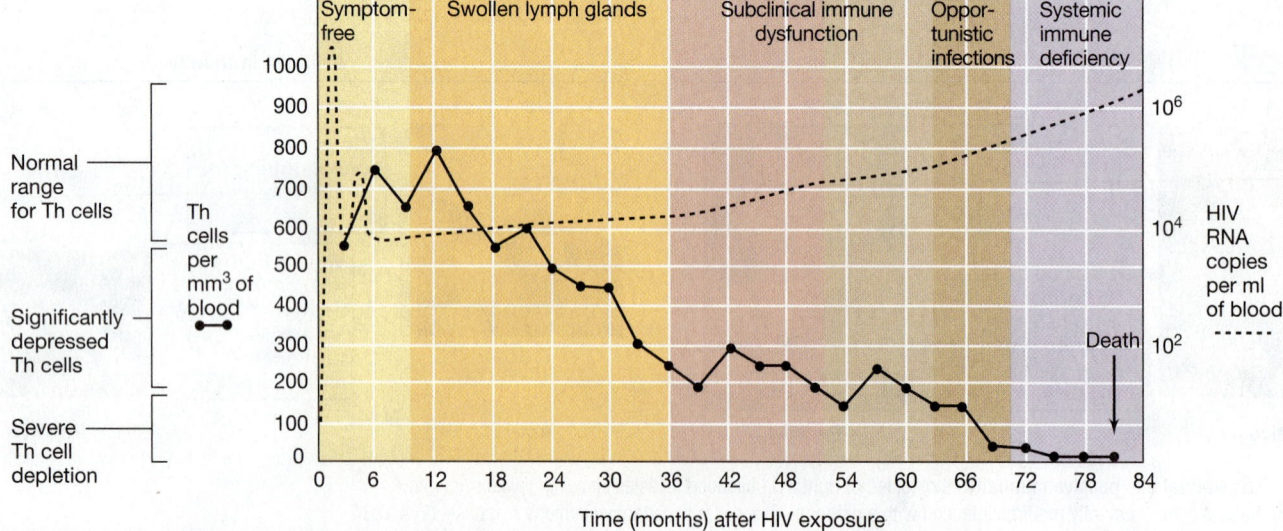

Figure 24.14 Decline of T-helper (Th) lymphocytes and progress of HIV/AIDS. An untreated HIV infection progresses to AIDS. The number and functional ability of the Th cells gradually declines, while the viral load, measured as HIV-specific RNA copies per milliliter of blood, increases after an initial decline. The lack of an effective immune response leads to an increase in life-threatening infections and death.

Artificial Active Immunity: Vaccination

Vaccination or **immunization** is the purposeful artificial induction of active immunity, and is a major weapon for the prevention and treatment of many infectious diseases. Artificial immunity can be induced actively by injecting an individual with a pathogen or its products (Figure 24.13). The goal is to produce a protective immune response to the pathogen. In active immunity, introduction of antigen induces changes in the host: The immune system produces antibodies in a primary immune response and, more importantly, produces a large number of immune memory cells. A second ("booster") dose of the same antigen results in a faster response yielding much higher levels of antibodies and immune T cells due to this memory, or secondary, immune response. Active immunity may remain throughout life as a result of immune memory.

Artificial active immunity is used to protect a person against future attack by a pathogen, but takes several days to weeks to induce protective immunity. For example, immunization with tetanus toxoid, a nontoxic version of *C. tetani* (tetanus) exotoxin, protects individuals against future encounters with exotoxin, but it is not an effective therapy for the trauma victim in the car accident above because effective adaptive immunity takes a week or more to develop.

Vaccines

The antigen or antigen mixture used to induce artificial active immunity is known as a **vaccine** or an *immunogen*. A summary of diseases for which vaccines are available for human use is given in **Table 24.4**. Immunization with a vaccine designed to produce artificial active immunity may introduce risks of infection and other adverse reactions. To reduce risks, pathogens or their products are inactivated or otherwise made harmless. For example, many vaccines consist of pathogens killed by chemical agents such as phenol or formaldehyde, or physical agents such as heat. Formaldehyde is also used to inactivate viruses for vaccines, such as in the inactivated (Salk) polio vaccine. Likewise, the active form of an exotoxin cannot be used as an immunogen because of the toxic effects. Many exotoxins, however, can be modified chemically so they retain their antigenicity but are no longer toxic. A modified exotoxin is called a **toxoid**. Toxoids such as the vaccine for *C. tetani* exotoxin can be given safely. The toxoid vaccine induces long-term protective immunity against the exotoxin. In other cases, antigenic components are extracted from pathogens grown in vitro, purified, and injected as a vaccine. Such is the case for some pneumococcal vaccines; they consist of a mixture of pneumococcal polysaccharide capsule antigens derived from the most common pathogenic strains.

Immunization with live cells or virus is usually more effective than immunization with dead or inactivated material. It is often possible to isolate a mutant strain of a pathogen that has lost its virulence but still retains the immunizing antigens; strains of this type are called *attenuated strains* (⤺ Section 27.6). However, because attenuated strains of pathogens are still viable, some individuals, especially those who are immunocompromised, may acquire active disease. Serious cases of disease in immunocompromised individuals have occurred, for example, from attenuated poliovirus and smallpox vaccines.

Table 24.4 Vaccines for infectious diseases in humans

Disease	Type of vaccine used
Bacterial diseases	
Anthrax	Toxoid
Cholera	Killed cells or cell extract (*Vibrio cholerae*)
Diphtheria	Toxoid
Haemophilus influenzae type b meningitis	Conjugated vaccine (polysaccharide of *Haemophilus influenzae* type b conjugated to protein)
Meningitis	Purified polysaccharide from *Neisseria meningitidis*
Paratyphoid fever	Killed bacteria (*Salmonella enterica* serovar Paratyphi)
Pertussis	Killed bacteria (*Bordetella pertussis*) or acellular proteins
Plague	Killed cells or cell extract (*Yersinia pestis*)
Pneumonia (bacterial)	Purified polysaccharide from *Streptococcus pneumoniae* or polysaccharide–toxoid conjugate
Tetanus	Toxoid
Tuberculosis	Attenuated strain of *Mycobacterium tuberculosis* (BCG)
Typhoid fever	Killed bacteria (*Salmonella enterica* serovar Typhi)
Typhus	Killed bacteria (*Rickettsia prowazekii*)
Viral diseases	
Hepatitis A	Recombinant DNA vaccine
Hepatitis B	Recombinant DNA vaccine or inactivated virus
Human papillomavirus (HPV)	Recombinant DNA vaccine
Influenza (seasonal)	Inactivated or attenuated virus
Influenza (H1N1)	Inactivated or attenuated virus
Japanese encephalitis	Inactivated virus
Measles	Attenuated virus
Monkeypox	Cross-reacting virus (vaccinia)
Mumps	Attenuated virus
Polio	Attenuated virus
Rabies	Attenuated virus (Sabin) or inactivated virus (Salk)
Rotavirus	Inactivated virus (human) or attenuated virus (dogs and other animals)
Rubella	Attenuated virus
Smallpox	Cross-reacting virus (vaccinia)
Varicella (chicken pox/shingles)	Attenuated virus
Yellow fever	Attenuated virus

Many effective viral vaccines are live attenuated vaccines. Attenuated vaccines tend to provide long-lasting T cell–mediated immunity, as well as vigorous antibody immunity and a strong secondary response upon reimmunization or infection with the targeted pathogen. However, attenuated vaccine strains are difficult

UNIT 5

to select, standardize, and maintain. Live, attenuated vaccines, such as most measles vaccines, have a limited shelf life and require refrigeration for storage. Killed virus vaccines, on the other hand, tend to provide short-lived immune responses with less long-term memory, but they are relatively easy to store and maintain their potency for long periods of time.

Most bacterial vaccines are provided as antigens in an inactivated form, such as the toxoids that protect against tetanus and diphtheria. One current acellular pertussis (whooping cough) vaccine consists of the inactivated pertussis toxin (PT) and filamentous hemagglutinin (FHA) antigens. Inactivated bacterial vaccines induce antibody-mediated protection without exposing recipients to the risk of infection, but primary and secondary responses are somewhat variable with each vaccine and individual, requiring periodic reimmunization to establish and maintain immunity.

Immunization Practices

Infants acquire natural passive immunity from maternal antibodies transferred across the placenta or in breast milk. As a result, infants are immune to many common infectious diseases during the first 6 months of life. However, infants should be immunized to prevent key infectious diseases as soon as possible so that their own active immunity can replace the maternal passive immunity. As discussed in Section 24.3, a single exposure to antigen does not lead to a high antibody *titer*, or antibody quantity. After an initial immunization, a series of secondary or "booster" reimmunizations are given to produce a secondary response and a high antibody titer. A summary of current vaccine recommendations in the United States is shown in **Figure 24.15**.

The importance of immunization in controlling infectious diseases is well established. For example, introduction of an effective vaccine into a population has reduced the incidence of formerly epidemic childhood diseases such as measles, mumps, and rubella (⮌ Section 29.6) and has eliminated smallpox altogether (⮌ Section 28.8). The degree of immunity obtained by vaccination, however, varies greatly with the individual as well as with the quality and quantity of the vaccine. Lifelong immunity is rarely achieved by means of a single injection, or even a series of injections, and the immune cells and antibodies induced by immunization gradually disappear from the body. On the other hand, natural infections may stimulate immune memory. In the complete absence of antigenic stimulation, the length of effective immunity varies considerably with different vaccines. For example, protective immunity to tetanus from toxoid immunization may last many years, but it is not lifelong. Current recommendations call for reimmunization in adults every 10 years to maintain protective immunity. Immunity induced by inactivated influenza virus vaccine disappears within a year or two without reimmunization through active infection or reimmunization.

Immunizations benefit individuals by preventing acute disease, and they also benefit the entire population. *Herd immunity* is the

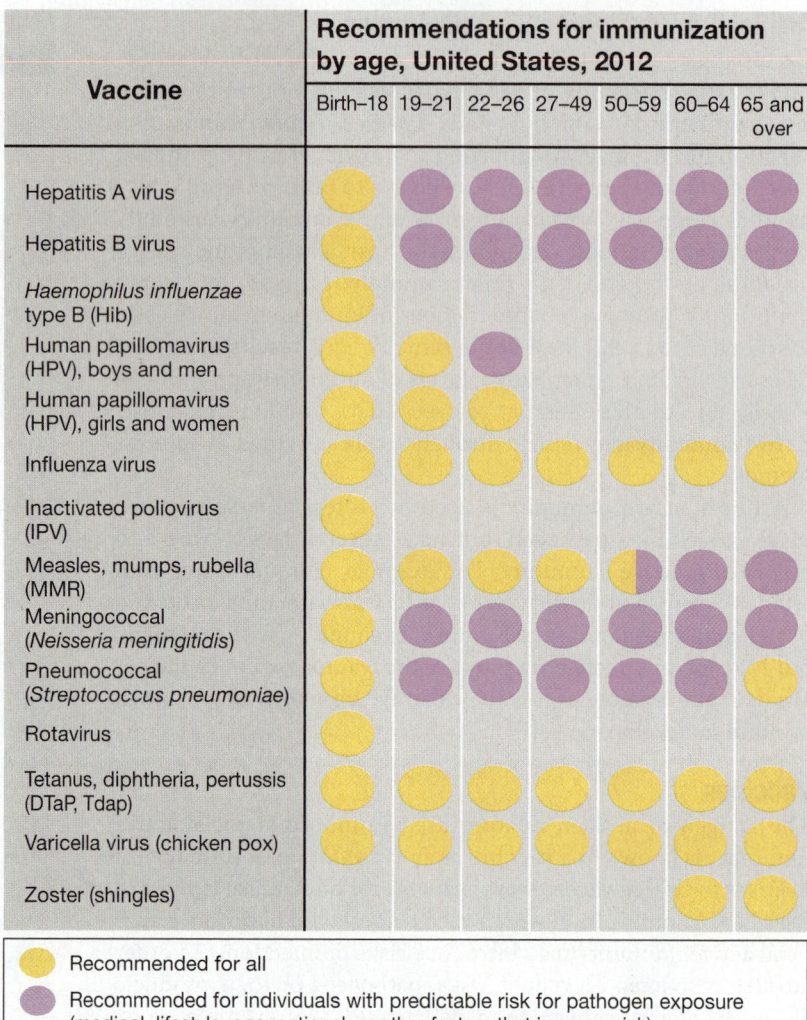

Vaccine	Recommendations for immunization by age, United States, 2012						
	Birth–18	19–21	22–26	27–49	50–59	60–64	65 and over
Hepatitis A virus	🟡	🟣	🟣	🟣	🟣	🟣	🟣
Hepatitis B virus	🟡	🟣	🟣	🟣	🟣	🟣	🟣
Haemophilus influenzae type B (Hib)	🟡						
Human papillomavirus (HPV), boys and men	🟡	🟡	🟣				
Human papillomavirus (HPV), girls and women	🟡	🟡	🟡				
Influenza virus	🟡	🟡	🟡	🟡	🟡	🟡	🟡
Inactivated poliovirus (IPV)	🟡						
Measles, mumps, rubella (MMR)	🟡	🟡	🟡	🟡	🟡🟣	🟣	🟣
Meningococcal (*Neisseria meningitidis*)	🟡	🟣	🟣	🟣	🟣	🟣	🟣
Pneumococcal (*Streptococcus pneumoniae*)	🟡	🟣	🟣	🟣	🟣	🟣	🟡
Rotavirus	🟡						
Tetanus, diphtheria, pertussis (DTaP, Tdap)	🟡	🟡	🟡	🟡	🟡	🟡	🟡
Varicella virus (chicken pox)	🟡	🟡	🟡	🟡	🟡	🟡	🟡
Zoster (shingles)						🟡	🟡

🟡 Recommended for all

🟣 Recommended for individuals with predictable risk for pathogen exposure (medical, lifestyle, occupational, or other factors that increase risk)

Figure 24.15 Immunization recommendations for children and adults in the United States. These immunizations are recommended by the U.S. Centers for Disease Control and Prevention, Atlanta, Georgia, as of 2012. The CDC Vaccines & Immunization website (http://www .cdc.gov) has specific recommendations for timing and dose of immunizations for all age groups and recommendations for special populations such as international travelers, women of childbearing age, and persons with medical conditions such as immunodeficiencies and chronic diseases.

concept that infections spread poorly in populations with a large proportion of immune individuals. Immunization is therefore a major tool for infectious disease control programs.

MINIQUIZ

- Distinguish between natural and artificial active immunity and natural and artificial passive immunity.
- Review the immunization recommendations for individuals in your age group. How do these artificial active immunizations benefit the immunized individual? The population?

24.7 Immunization Strategies

Historically, many vaccines have been derived from attenuated or inactivated whole pathogens. Today genetic engineering methods can generate pathogen-derived antigens for use as vaccines, altogether avoiding exposure to whole pathogens either living or dead.

Synthetic and Genetically Engineered Vaccines

In Chapter 11 we discussed how genetic engineering tools can be used to synthesize new genes and proteins. The simplest alternate approach to vaccine development is to make use of these methods to produce *synthetic peptides*. To make a vaccine, a genetic engineer can synthesize a peptide that corresponds to an antigen of an infectious agent. For example, the structure of the toxin from the foot-and-mouth disease virus, an important animal pathogen, is known. Because the whole protein is toxic, it cannot be used as a vaccine. However, a peptide of 20 amino acids is an important antigen in the protein, but the peptide is too small to be an effective vaccine by itself. Genetic engineers attached a synthetic version of the small peptide to a large, innocuous protein that acts as a carrier molecule. The synthetic *conjugate vaccine* produces a protective response to foot-and-mouth disease virus infection. This strategy has great promise for creating vaccines to a number of pathogens, but the entire sequence of the disease-causing protein must be known and a protective antigen recognized by lymphocytes must be identified before an effective vaccine can be engineered. However, the entire genomic sequences of many pathogens are now known, providing the information necessary to identify the most antigenic part of each.

Two widely available conjugate vaccines couple extracted bacterial polysaccharide to a protein toxoid, provoking a more robust immune response with better immune memory than injection of the polysaccharide antigen alone. One pneumococcal vaccine uses pneumococcal polysaccharide coupled to diphtheria toxoid

(Figure 24.16). Likewise, the vaccine for *Haemophilus influenzae* type b (Hib) uses Hib polysaccharide coupled to tetanus toxoid. Polysaccharide antigens are not processed efficiently by B cells and generally provide only a primary response with little immune memory, but the conjugated protein toxoids are processed and activate Th2 cells, resulting in a primary response followed by a strong secondary response and immune memory.

Genomic information is particularly useful for making viral vaccines. For example, genes that encode antigens from virtually any virus can be cloned into the vaccinia virus genome and expressed. Inoculation with the antigen-producing vaccinia virus can induce immunity to the product of the cloned gene. Such a preparation is called a *recombinant-vector vaccine*. This method depends on the identification and cloning of the gene that encodes the antigen and also on the ability of the vaccinia virus to express the cloned gene as an antigenic protein. An effective recombinant vaccinia–rabies vaccine has been developed for use in animals. (Recombinant DNA methods for vaccine development were discussed in Section 11.14.)

Another immunization strategy involves the use of proteins made from cloned DNA as immunogens (see Explore the Microbial World: Vaccines and Public Health). First, a pathogen gene must be cloned in a suitable microbial host that expresses the protein encoded by the cloned gene. The pathogen protein can then be harvested and used as a vaccine; such a vaccine is called a *recombinant-antigen vaccine* (↩ Section 11.14). For example, the current hepatitis B virus vaccine is a major hepatitis surface protein antigen (HbsAg) expressed by genetically modified yeast cells. A vaccine that is effective against human

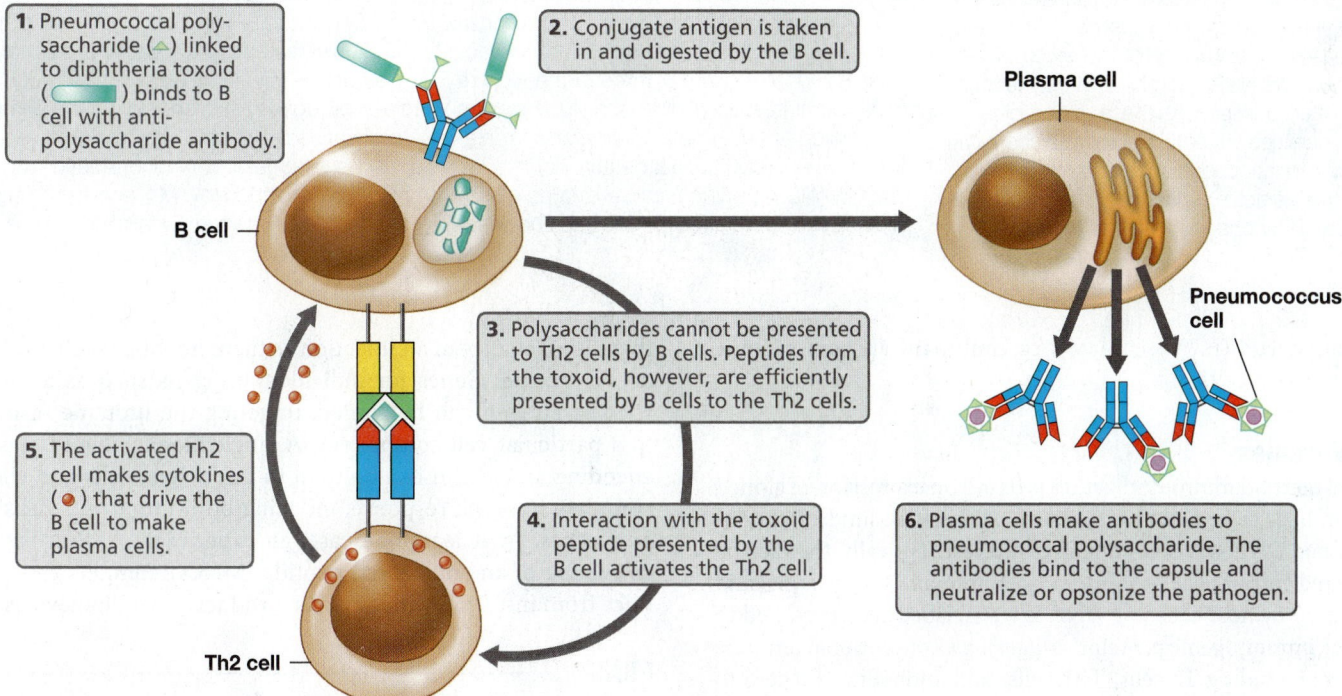

Figure 24.16 Conjugate vaccines. Conjugate vaccines such as *Streptococcus pneumoniae* (pneumococcus) polysaccharide covalently linked to diphtheria toxoid and *Haemophilus influenzae* type b (Hib) polysaccharide coupled to tetanus toxoid provide effective immunity to polysaccharide antigens. Polysaccharides are poor immunogens in the absence of protein carriers such as toxoids.

Various strains of human papillomavirus (HPV) infect up to 75% of sexually active people, causing genital warts and vulvar, vaginal, and cervical cancers in women (**Figure 1**) and genital warts and anal cancers in men. In the United States, about 12,000 women develop cervical cancer each year, and about 4,000 die. Worldwide, nearly half a million women develop cervical cancer annually and about 250,000 die from it. Vaccines directed to HPV antigens are highly protective against viral infection, and prevent genital warts and the cancers caused by the targeted HPVs.

Two HPV vaccines are approved for use. One of these vaccines, *Gardasil*, targets HPV types 6, 11, 16, and 18. Another vaccine, *Cervarix*, is effective against HPV 16 and 18, two strains that account for 70% of cervical cancers and 90% of genital warts (Figure 1). *Gardasil* has been approved for use in females and males 9–26 years of age, and *Cervarix* is approved for use in females ages 9–25. Immunization is recommended for females and males ages 9–26.

The vaccines are a preparation of virus-like particles of the major L1 capsid proteins from the HPV strains. The L1 proteins have been genetically engineered to be expressed by the yeast *Saccharomyces cerevisiae* and are released by disruption of the recombinant yeast cells as self-assembled virus-like particles. After purification, these recombinant-antigen particles are adsorbed onto a chemical adjuvant, an innocuous substance that immobilizes the particles, enhancing their ability to be taken up by phagocytes after injection.

Because HPV is responsible for most cervical cancers, immunization of susceptible individuals with strain-specific HPV vaccines will prevent many cancers in the immunized individuals. Perhaps as important, however, herd immunity resulting from immunization of a large proportion of the population can significantly reduce the spread of these viruses, providing protection even for individuals who are not immunized (⮌ Section 28.2). Models suggest that herd immunity can be raised to meaningful levels by immunizing about 80% of young women. The recommended immunization of boys and men might provide the further herd immunity necessary to stop transmission of the targeted HPV strains, and is also effective for prevention of genital warts and anal cancers in men. HPV vaccines are the only ones effective against any sexually transmitted infection in humans.

The HPV vaccine is a recent example of effective vaccine development and implementation, but a number of important infectious diseases still cannot be prevented by vaccination. For example, no vaccines exist for most diarrheal diseases (2.2 million annual deaths worldwide) or respiratory diseases (about 4 million annual deaths worldwide), with the notable exceptions of influenza and pneumococcal vaccines. Effective vaccines still do not exist for three of the world's most lethal infectious diseases: tuberculosis (1.5 million annual deaths worldwide), malaria (about 900,000 annual deaths worldwide), and HIV/AIDS (2 million annual deaths worldwide). Even some effective vaccines have se-

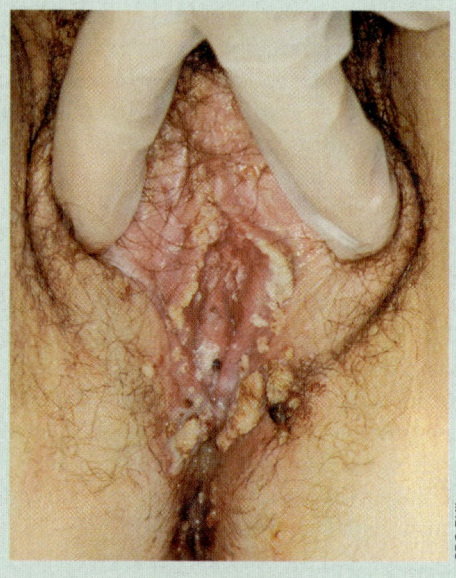

Figure 1 Genital warts in a female caused by infection with human papillomavirus.

rious limitations. For example, influenza vaccines are only effective for one year; they are reformulated each year to target the strain-specific H and N antigens currently circulating (⮌ Sections 28.2 and 28.11). Developments of a universal influenza vaccine that targets a common influenza virus antigen, M1, has been proposed, as it should induce immunity to all influenza strains with a single vaccine. The immunogenicity and protection against influenza provided by M1 and other shared antigens are, however, unproven.

papillomavirus (HPV) is also a recombinant-antigen vaccine made in yeast cells.

DNA Vaccines

A novel method for immunization is based on expression of cloned genes in host cells. *DNA vaccines* are bacterial plasmids that contain cloned DNA with the antigen of interest. Typically, the vaccine is injected intramuscularly into a host animal. Once the plasmid is taken up by host cells, the DNA is transcribed and translated to produce immunogenic proteins, triggering a conventional immune response including Tc cells, Th1 cells, and antibodies directed to the protein encoded by the cloned DNA.

DNA vaccine strategies provide considerable advantages over conventional immunization methods. For instance, because only a single pathogen gene is cloned into the plasmid and injected, there is no chance of an infection as there might be with an attenuated vaccine. Genes for individual antigens such as a tumor-specific antigen can be cloned, targeting the immune response to a particular cell component. A single bioengineered plasmid encoding an antigen can be used to infect host cells and elicit a complete immune response, inducing both immune T cells and antibodies. In at least one case, an experimental DNA vaccine consisting of an engineered peptide–MHC I complex protected mice from infection with a cancer-producing papillomavirus.

MINIQUIZ
- Identify immunization strategies used for approved vaccines.
- Identify the advantages of alternative immunization strategies as compared to traditional immunization procedures.

III • Immune Diseases

Immune reactions can cause host cell damage and disease. **Hypersensitivity** is an inappropriate immune response that results in host damage. Hypersensitivity diseases are categorized according to the antigens and the mechanisms that produce disease. Here we discuss these diseases, and ones produced by superantigen proteins produced by certain bacteria and viruses. Superantigens cause host damage by activating massive immune inflammatory responses.

24.8 Allergy, Hypersensitivity, and Autoimmunity

Antibody-mediated **immediate hypersensitivity** is commonly called *allergy*. Cell-mediated hypersensitivities also cause allergy-like diseases but, because of the delayed onset of symptoms, cell-mediated reactions are termed **delayed-type hypersensitivity (DTH)**. **Autoimmunity** is a harmful immune reaction directed against self antigens. These diseases are categorized as type I, II, III, or IV hypersensitivities based on immune effectors, antigens, and symptoms (Table 24.5).

Immediate Hypersensitivity

Immediate hypersensitivity, or type I hypersensitivity, is caused by the release of vasoactive products from mast cells coated with IgE (Figure 24.17). Immediate hypersensitivity reactions occur within minutes after exposure to an *allergen*, the antigen that caused the type I hypersensitivity. Depending on the individual and the allergen, immediate hypersensitivity reactions can be very mild or can cause a life-threatening reaction called *anaphylaxis*.

About 20% of the population suffers from immediate hypersensitivity allergies to pollens, molds, animal dander, certain foods (strawberries, nuts, and shellfish), insect venoms, mites in house dust, and other agents. Most allergens enter the body at the surface of mucous membranes such as the lungs or the gut. Initial exposure to allergens stimulates mucosa-associated Th2 cells to produce cytokines that induce B cells to make IgE antibodies. Rather than circulating like IgG or IgM, the allergen-specific IgE antibodies bind to IgE receptors on mast cells (Figure 24.17). Mast cells are nonmotile granulocytes (Section 24.1) associated with the connective tissue adjacent to capillaries throughout the body. With any subsequent exposure to the immunizing allergen, the mast cell–bound IgE molecules bind the antigen. Cross-linking of IgEs by an antigen triggers the release of soluble allergic mediators from the mast cells, a process called *degranulation*. These mediators cause allergic symptoms within minutes of antigen exposure. After initial sensitization by an allergen, the allergic individual responds to each subsequent reexposure to the allergen.

The principal chemical mediators released from mast cells are histamine and serotonin, modified amino acids that cause rapid dilation of blood vessels and contraction of smooth muscle, initiating symptoms ranging from mild local discomfort to systemic *anaphylactic shock*. Local symptoms typically include mucus production; rash; sneezing; itchy, watery eyes; and hives (Figure 24.18). Anaphylactic shock symptoms may include vasodilation (causing a sharp drop in blood pressure) and asthma due to smooth muscle constriction in the lungs. Severe anaphylaxis is treated immediately with epinephrine to counter smooth muscle contraction, increase blood pressure, and promote breathing. Less serious allergic symptoms can be treated

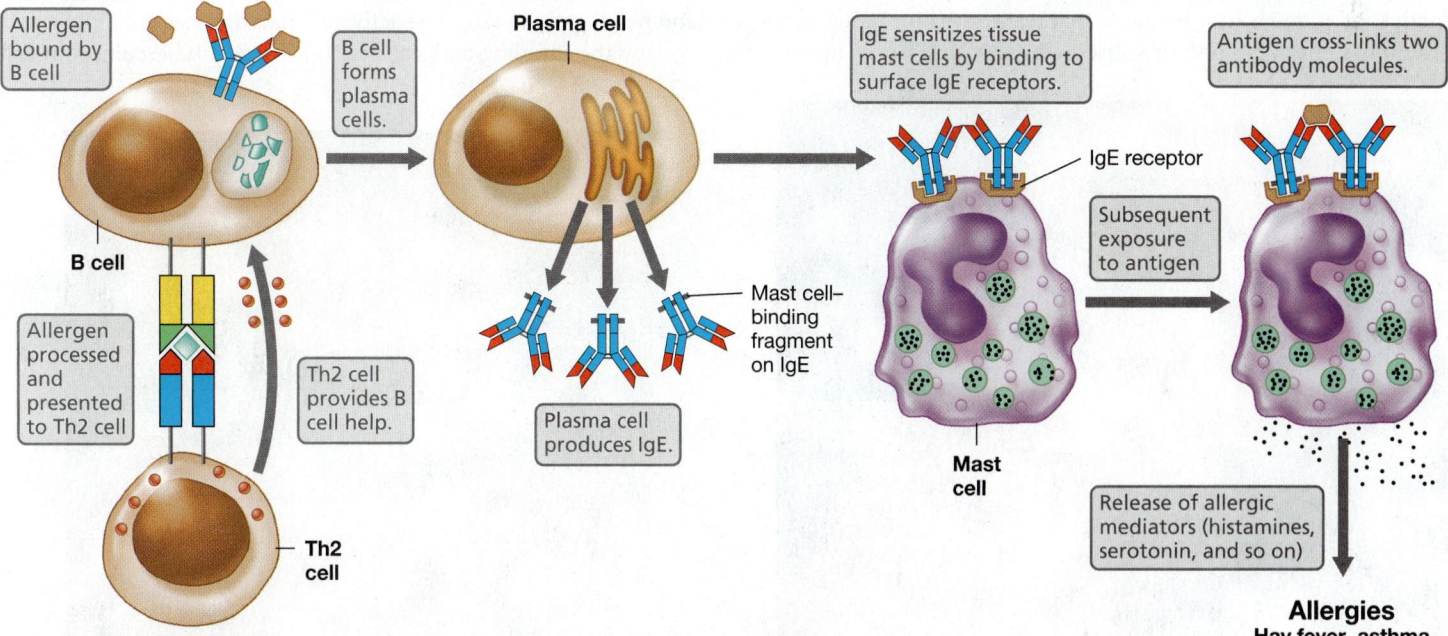

Figure 24.17 Immediate hypersensitivity. Certain antigens such as pollens stimulate IgE production. IgE binds to mast cells by means of a high-affinity surface receptor and arms the mast cell. Antigen cross-links surface IgE, causing release of soluble mediators such as histamine. These mediators produce symptoms ranging from mild allergic symptoms to life-threatening anaphylaxis.

UNIT 5

Table 24.5 Hypersensitivity

Classification	Description	Immune mechanism	Time of latency	Examples
Type I	Immediate	IgE sensitization of mast cells	Minutes	Reaction to bee venom (sting) Hay fever
Type II	Cytotoxic[a]	IgG interaction with cell surface antigen	Hours	Drug reactions (penicillin)
Type III	Immune complex	IgG interaction with soluble or circulating antigen	Hours	Systemic lupus erythematosus (SLE)
Type IV	Delayed type	Th1 inflammatory cell activation of macrophages	Days (24–48 h)	Poison ivy Tuberculin test

[a]Autoimmune diseases may be caused by type II, type III, or type IV reactions.

with *antihistamines*, drugs that neutralize histamine. Treatment for symptoms may also include anti-inflammatory steroids. Finally, immunization with increasing doses of the allergen may shift antibody production from IgE to IgG and IgA. The IgG and IgA interact with the allergens, blocking binding to the IgE on sensitized mast cells, thus stopping allergic symptoms and inhibiting production of more IgE. This procedure is called *desensitization*.

Delayed-Type Hypersensitivity

Delayed-type hypersensitivity (DTH), or type IV hypersensitivity, is cell-mediated hypersensitivity characterized by tissue damage due to inflammation produced by Th1 cells (Table 24.5). Delayed-type hypersensitivity symptoms appear several hours after secondary exposure to the eliciting antigen, with a maximal response usually occurring in 24 to 48 hours. Typical DTH antigens include chemicals that are not normally antigens but become so when they covalently bind to skin proteins, which can create new antigens and elicit a DTH response. Hypersensitivity to these newly created antigens is known as *contact dermatitis* and results in, for example, skin reactions to poison ivy (**Figure 24.19**), jewelry, cosmetics, latex, and other chemicals that react with host tissues. Several hours after a second or subsequent exposure to the antigen,

the skin feels itchy at the site of contact. Reddening and swelling appear, often with localized tissue destruction in the form of blistering, and reach a maximum in several days. The delayed onset and the progress of the inflammatory response are the hallmarks of the DTH reaction. As discussed below, certain self antigens may also elicit DTH responses, resulting in autoimmune disease.

Another example of delayed-type hypersensitivity is the development of protective immunity to the causal agent of tuberculosis, *Mycobacterium tuberculosis* (Section 29.4). This cellular immune response was discovered by Robert Koch during his classic studies on tuberculosis (Section 1.8). When antigens derived from the bacterium are injected subcutaneously into a person previously infected with *M. tuberculosis*, a skin reaction called the *tuberculin reaction* develops. A positive tuberculin reaction develops fully after 24–48 hours (Figure 24.6), in contrast to IgE-mediated immediate hypersensitivity that develops within minutes after antigen exposure. Th1 cells stimulated by the mycobacterial antigens release cytokines in the region of the introduced antigen that attract and activate large numbers of macrophages, which in turn produce a characteristic local inflammation, including induration, edema, erythema, pain, and heating of the skin. The activated macrophages then ingest and destroy the invading antigen. The DTH-based tuberculin skin test

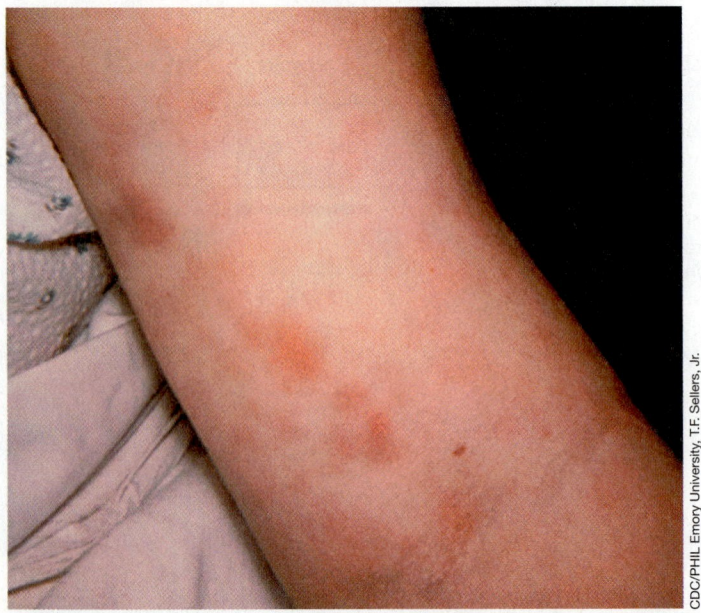

Figure 24.18 Hives due to immediate hypersensitivity. The raised, red areas are typical symptoms after contact with allergens that cause immediate hypersensitivity.

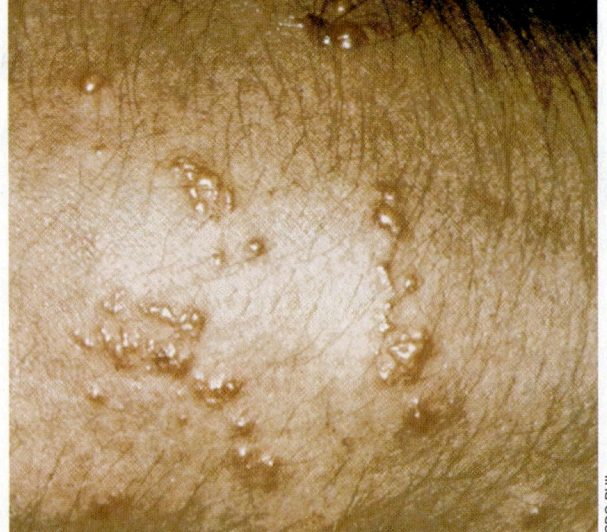

Figure 24.19 Delayed-type hypersensitivity. Poison ivy blisters on an arm. The raised rash appears 24–48 hours after exposure to plants of the genus *Rhus* due to macrophage activation by Th1 cells sensitized to *Rhus* antigens.

determines current or previous infection with *M. tuberculosis* or immunization with the BCG-attenuated tuberculosis vaccine.

A number of other infectious diseases due to intracellular pathogens elicit DTH reactions. These include bacterial diseases such as leprosy, brucellosis, psittacosis; viral diseases such as mumps; and fungal diseases such as coccidioidomycosis, histoplasmosis, and blastomycosis. Visible antigen-specific skin responses resembling the tuberculin reaction occur after injection of antigens derived from the pathogens, indicating pathogen exposure and Th1-mediated immunity.

Autoimmunity

As lymphocytes develop, T and B cells that can react with self antigens are normally eliminated. Autoimmune diseases result when these cells are instead activated to produce immune reactions against self proteins (Table 24.6). For example, Th1-mediated DTH can cause autoimmune responses directed against self antigens, as in the case of the Th1-mediated response to brain-derived antigens in allergic encephalitis. In type 1 (juvenile) diabetes mellitus, Th1 cells directed to antigens on pancreatic cells cause reactions that destroy the insulin-producing beta cells in the pancreas. Many autoimmune diseases, however, are caused by **autoantibodies**, antibodies that interact with self antigens.

In many cases, autoantibodies interact with organ-specific antigens. For example, in *Hashimoto's disease*, autoantibodies are made against thyroglobulin, a product of the thyroid gland. In this case, antibodies to thyroglobulin bind complement proteins, leading to local inflammation and destruction of the thyroid cells and function, the hallmarks of a type II hypersensitivity disease (Table 24.6).

Systemic lupus erythematosus (*SLE*) is an example of a disease caused by type III hypersensitivity. This disease and others like it are caused by autoantibodies directed against soluble, circulating self antigens. In SLE, the antigens include nucleoproteins and DNA. Antibodies bind to soluble proteins, producing insoluble immune complexes. Disease results when circulating antigen–antibody complexes deposit in different body tissues such as the kidney,

lungs, and spleen. Here the antibodies bind complement, resulting in inflammation and local, often severe, cell damage. Thus, type III hypersensitivity is an immune complex disorder (Table 24.5).

Organ-specific autoimmune diseases are sometimes more easily controlled clinically than diseases that affect multiple organs. For example, the product of organ function, such as thyroxine in autoimmune hypothyroidism or insulin in juvenile diabetes, can often be supplied in pure form from another source. SLE, rheumatoid arthritis, and other autoimmune diseases that affect multiple organs and sites can often be controlled only by general immunosuppressive therapy, such as the use of steroid drugs. General immunosuppression, however, significantly increases chances of opportunistic infections.

Heredity influences the incidence, type, and severity of autoimmune diseases. Many autoimmune diseases correlate strongly with the presence of certain MHC proteins (⊂⊃ Section 25.4). Studies of model autoimmune diseases in mice support such a genetic link, but the precise conditions necessary for developing autoimmunity may also depend on factors such as prior infections, gender, age, and health status. Women, for example, are about 20 times more likely to develop SLE than are men.

MINIQUIZ -
- Discriminate between immediate hypersensitivity and delayed-type hypersensitivity with respect to antigens and immune effectors.
- Provide examples and mechanisms for an antibody-mediated autoimmune disease directed against a specific organ and one involving circulating immune complexes.

24.9 Superantigens: Overactivation of T Cells

We discussed the mechanisms of action for several different categories of bacterial toxins in Chapter 23. Most toxins interact directly with host cells to cause tissue damage. Endotoxins, for

Table 24.6 Autoimmune diseases of humans		
Disease	*Organ, cell, or molecule affected*	*Mechanism (hypersensitivity type)*[a]
Juvenile diabetes (insulin-dependent diabetes mellitus)	Pancreas	Cell-mediated immunity and autoantibodies against surface and cytoplasmic antigens of beta cells of pancreatic islets (II and IV)
Myasthenia gravis	Skeletal muscle	Autoantibodies against acetylcholine receptors on skeletal muscle (II)
Goodpasture's syndrome	Kidney	Autoantibodies against basement membrane of kidney glomeruli (II)
Rheumatoid arthritis	Cartilage	Autoantibodies against self IgG antibodies, which form complexes deposited in joint tissue, causing inflammation and cartilage destruction (III)
Hashimoto's disease (hypothyroidism)	Thyroid	Autoantibodies to thyroid surface antigens (II)
Male infertility (some cases)	Sperm cells	Autoantibodies agglutinate host sperm cells (II)
Pernicious anemia	Intrinsic factor	Autoantibodies prevent absorption of vitamin B_{12} (III)
Systemic lupus erythematosus (SLE)	DNA, cardiolipin, nucleoprotein, blood clotting proteins	Autoantibody response to various cellular constituents results in immune complex formation (III)
Addison's disease	Adrenal glands	Autoantibodies to adrenal cell antigens (II)
Allergic encephalitis	Brain	Cell-mediated response against brain tissue (IV)
Multiple sclerosis	Brain	Cell-mediated and autoantibody response against central nervous system (II and IV)

[a]See Table 24.5.

example, interact directly with many cell types, causing release of endogenous pyrogens and other soluble mediators and producing fever and general inflammation (Section 24.5). Most exotoxins also interact directly with cells to cause cell damage. However, certain exotoxins, the superantigens, act indirectly on host cells, subverting the immune system so that T cells and their cytokine products extensively damage host cells (Figure 24.20).

Superantigens are proteins capable of eliciting a very strong response because they activate more T cells than a normal immune response. Superantigens interact directly with TCRs and MHC proteins (Figure 24.21). They are produced by many viruses and bacteria. Streptococci and staphylococci, for example, produce several different and very potent superantigens (⮌ Sections 29.2 and 29.9).

Superantigen interaction with TCRs differs from the conventional antigen–TCR binding shown in Figure 24.5. Conventional foreign antigens, presented by an MHC protein, bind to a TCR at a defined antigen-binding site. However, superantigens bind to a site on the TCR that is outside the antigen-specific TCR binding site. A superantigen binds to all TCRs with a shared common structure, and many different TCRs share the same structure outside the antigen-binding site. In some cases, superantigens can bind 5–25% of all T cells, whereas less than 0.01% of all available T cells interact with a conventional foreign antigen in a typical immune response. Superantigens also bind MHC II on APCs, again at a

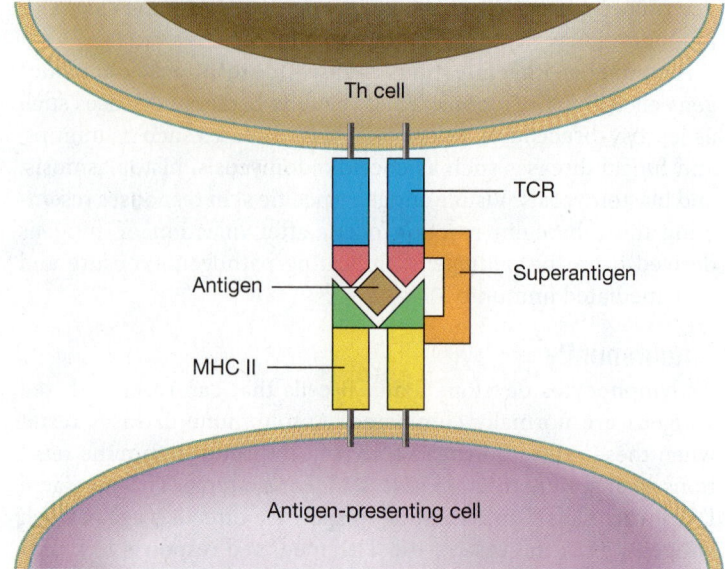

Figure 24.21 Superantigens. Superantigens bind to conserved regions of both the MHC and TCR proteins at positions outside the normal binding site. Superantigens interact with large numbers of T cells, causing large-scale T cell activation, cytokine release, and systemic inflammation.

site outside the normal peptide-binding site. These interactions mimic conventional antigen presentation and stimulate large numbers of T cells to grow and divide. As in normal responses, the activated T cells produce cytokines that stimulate other cells, such as macrophages and other phagocytes. The extensive cytokine production by the large proportion of superantigen-activated T cells triggers a widespread cell-mediated response characterized by systemic inflammatory reactions. The resulting fever, diarrhea, vomiting, mucus production, and even systemic shock may be fatal in extreme cases. Clinically, superantigen shock is indistinguishable from septic shock (Section 24.5).

A very common superantigen disease is *Staphylococcus aureus* food poisoning, characterized by fever, vomiting, and diarrhea, and caused by one of several superantigen staphylococcal enterotoxins. *S. aureus* also produces the superantigen responsible for *toxic shock syndrome* (Figure 24.20). *Streptococcus pyogenes* produces erythrogenic toxin, the superantigen responsible for scarlet fever (⮌ Section 29.2).

MINIQUIZ -
- Discriminate between normal and superantigen activation of T cells.
- Identify the binding site for superantigens on T cells and APCs.

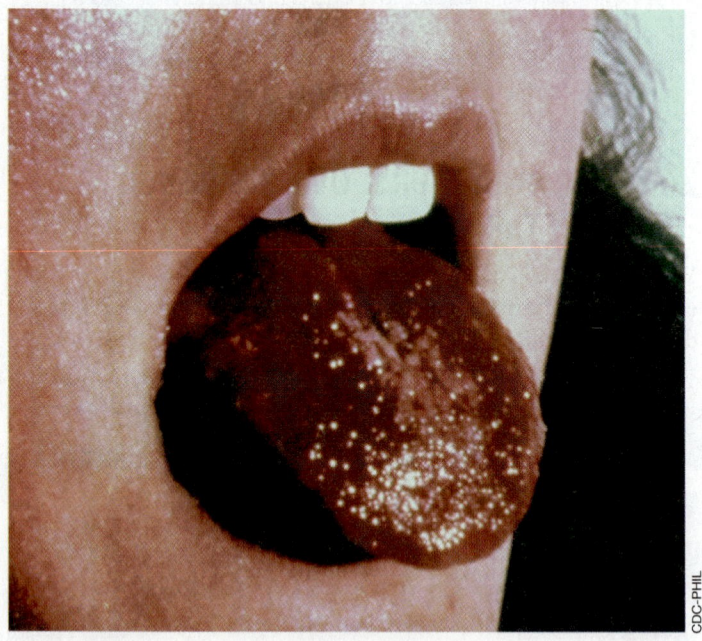

Figure 24.20 Toxic shock syndrome. This individual exhibits "strawberry tongue," a symptom of toxic shock syndrome caused by a *Staphylococcus aureus* superantigen.

BIG IDEAS

24.1 • Cells involved in innate and adaptive immunity originate from bone-marrow stem cells. The blood and lymph systems circulate cells and proteins that are important components of the immune response. Several different leukocytes participate in immune responses in all parts of the body.

24.2 • Innate immunity is a natural protective response to infection characterized by recognition of common pathogen-associated molecular patterns on pathogens. Phagocytes recognize these patterns through preformed pathogen recognition receptors, and the recognition and interaction process stimulates phagocytes to destroy the pathogens.

24.3 • Adaptive immunity is triggered by the specific interactions of T cells with antigens presented on APCs. Peptide antigens embedded in MHC proteins are presented to T cells. Tc cells kill antigen-bearing target cells directly. Th cells act through cytokines to promote immune reactions. Th1 cells initiate inflammation and immunity by activating macrophages.

24.4 • Th2 cells stimulate B cells that have been exposed to antigen to differentiate into plasma cells, which then produce antibodies. Antibodies are soluble, antigen-specific proteins that interact with antigens. Antibodies provide targets for interaction with proteins of the complement system, resulting in destruction of antigens through lysis or opsonization.

24.5 • Inflammation, characterized by pain, swelling (edema), redness (erythema), and heat, is a normal and generally desirable outcome due to activation of nonspecific immune response effectors. Uncontrolled systemic inflammation, called septic shock, can lead to serious illness and death.

24.6 • Adaptive immunity develops naturally and actively through immune responses to infections, or naturally and passively through antibody transfer from the placenta or breast milk. The inability to generate an innate or adaptive immune response results in recurrent, uncontrollable infections. Artificial passive immunity occurs when antibodies or immune cells are transferred from an immune individual to a nonimmune individual. Immunization induces artificial active immunity and is widely used to prevent infectious diseases. Vaccines are either attenuated or inactivated pathogens or pathogen products or are genetically engineered antigens.

24.7 • Immunization strategies using bioengineered molecules eliminate exposure to microorganisms and, in some cases, even to protein antigen. Application of these strategies is providing safer vaccines targeted to individual pathogen antigens.

24.8 • Hypersensitivity is the induction by foreign antigens of antibody-mediated or cell-mediated immune responses that damage host tissue. In autoimmunity, the immune response is directed against self antigens. Damage to host tissue is caused by the inflammation produced by immune mechanisms.

24.9 • Superantigens are components of certain bacterial and viral pathogens that bind and activate large numbers of T cells. Superantigen-activated T cells may produce diseases characterized by systemic inflammatory reactions.

MasteringMicrobiology® **Review what you know and challenge what you have learned with MasteringMicrobiology!** Access study materials, chapter quizzes, animations, and microbiology lab tutorials in the Study Area to ensure that you have mastered this chapter's content.

REVIEW OF KEY TERMS

Adaptive immunity the acquired ability to recognize and destroy a particular pathogen or its products, dependent on previous exposure to the pathogen or its products; also called specific immunity and antigen-specific immunity

Antibody a soluble protein produced by B cells and plasma cells that interacts with antigen; also called immunoglobulin

Antigen a molecule that interacts with specific components of the immune system

Antigen-presenting cell (APC) a macrophage, dendritic cell, or B cell that takes up and processes antigen and presents it to T-helper cells

Autoantibody an antibody that reacts to self antigens

Autoimmunity a harmful immune reaction directed against self antigens

B cell a lymphocyte that has immunoglobulin surface receptors, produces immunoglobulin, and may present antigens to T cells

B cell receptor (BCR) a cell-surface antibody that acts as an antigen receptor on a B cell

Bone marrow the primary lymphoid organ containing the pluripotent precursor cells for all blood and immune cells

Chemokine a soluble protein that modulates an immune response

Clone a copy of an antigen-reactive lymphocyte

Cytokine a soluble protein produced by a leukocyte that modulates an immune response

Delayed-type hypersensitivity (DTH) an inflammatory allergic response mediated by Th1 lymphocytes

Dendritic cell a phagocytic antigen-presenting cell found in various body tissues; transports antigen to secondary lymphoid organs

Hypersensitivity an immune response leading to damage to host tissues

Immediate hypersensitivity an allergic response mediated by vasoactive products released from IgE-sensitized mast cells

Immunity the ability of an organism to resist infection

Immunoglobulin (Ig) a soluble protein produced by B cells and plasma cells that interacts with antigen; also called antibody

Inflammation a nonspecific reaction to noxious stimuli such as toxins and pathogens, characterized by redness (erythema), swelling (edema), pain, and heat (fever), usually localized at the site of infection

Innate immunity the noninducible ability to recognize and destroy an individual pathogen or its products that does not rely on previous exposure to a pathogen or its products; also called nonspecific immunity

Leukocyte a nucleated cell in blood; also called a white blood cell

Lymph a fluid that circulates through the lymphatic system; like blood but lacking red blood cells

Lymph nodes organs that contain lymphocytes and phagocytes arranged to encounter microorganisms and antigens as they travel through the lymphatic circulation

Lymphocytes a subset of nucleated cells in blood involved in the adaptive immune response

Macrophage a large leukocyte found in tissues that has phagocytic and antigen-presenting capabilities

Major histocompatibility complex (MHC) a genetic region that encodes several proteins important for antigen processing and presentation. MHC I proteins are expressed on all cells. MHC II proteins are expressed only on antigen-presenting cells

Memory (immune memory) the ability to rapidly produce large quantities of specific immune cells or antibodies after subsequent exposure to a previously encountered antigen

Mucosa-associated lymphoid tissue (MALT) a part of the lymphatic system that interacts with antigens and microorganisms that enter the body through mucous membranes, including those of the gut, the genitourinary tract, and bronchial tissues

Neutrophil a leukocyte exhibiting phagocytic properties, a granular cytoplasm (granulocyte), and a multilobed nucleus; also called polymorphonuclear leukocyte or PMN

Pathogen-associated molecular pattern (PAMP) a repeating structural component of a microorganism or virus recognized by a pattern recognition receptor (PRR)

Pattern recognition receptor (PRR) a protein in a phagocyte membrane that recognizes a pathogen-associated molecular pattern (PAMP)

Phagocyte a cell that engulfs foreign particles, and can ingest, kill, and digest most pathogens

Plasma the liquid portion of the blood containing proteins and other solutes

Plasma cell a differentiated B cell that produces soluble antibodies

Primary adaptive immune response the production of antibodies or immune T cells on first exposure to antigen; the antibodies are mostly of the IgM class

Primary lymphoid organ an organ in which antigen-reactive lymphocytes develop and become functional; the bone marrow is the primary lymphoid organ for B cells; the thymus is the primary lymphoid organ for T cells

Secondary adaptive immune response the enhanced production of antibodies or immune T cells on second and subsequent exposures to antigen; the antibodies are mostly of the IgG class

Secondary lymphoid organ an organ at which antigens interact with antigen-presenting phagocytes and lymphocytes to generate an adaptive immune response; these include lymph nodes, spleen, and mucosa-associated lymphoid tissue

Serum the liquid portion of the blood with clotting proteins removed

Specificity the ability of the immune response to interact with particular antigens

Stem cell a progenitor cell that can develop into other cell types

Superantigen a pathogen product capable of eliciting an inappropriately strong inflammatory immune response by stimulating greater than normal numbers of T cells

T cell a lymphocyte that interacts with antigens through a T cell receptor for antigen; T cells are divided into functional subsets including Tc (T-cytotoxic) cells and Th (T-helper) cells. Th cells are further subdivided into Th1 (inflammatory) cells and Th2 cells, which aid B cells in antibody formation

T cell receptor (TCR) an antigen-specific receptor protein on the surface of T cells

Thymus the primary lymphoid organ in which T cells develop

Tolerance the acquired inability to produce an immune response to particular antigens

Toxoid a form of a toxin that retains antigenicity but has lost toxicity

Vaccination (immunization) the inoculation of a host with inactive or weakened pathogens or pathogen products to stimulate protective active immunity

Vaccine an inactivated or attenuated pathogen, or a harmless pathogen product used to induce artificial active immunity

REVIEW QUESTIONS

1. What is the origin of the phagocytes and lymphocytes active in the immune response? Track the maturation of B cells and T cells. (Section 24.1)

2. Identify the cells that express pattern recognition receptors (PRRs). How do PRRs associate with pathogen-associated molecular patterns (PAMPs) to promote innate immunity? (Section 24.2)

3. Identify the lymphocytes and the antigen-specific receptors involved in cell-mediated adaptive immunity. (Section 24.3)

4. Identify the lymphocytes and the antigen-specific receptors involved in antibody-mediated adaptive immunity. (Section 24.4)

5. Identify the cells that initiate inflammation and the cells that are activated by inflammatory signals. (Section 24.5)

6. List the immunizations recommended for adults in the United States. (Section 24.6)

7. List the diseases for which you have been immunized. List the diseases for which you may have acquired immunity naturally. (Sections 24.6 and 24.7)

8. The acellular pertussis vaccine uses a biotechnology-based immunization strategy that has been adapted for this approved vaccine. What advantages does the biotechnology-based vaccine have over a conventional vaccine? What disadvantages, if any? (Section 24.7)

9. Define the differences between immediate and delayed-type hypersensitivity in terms of immune effectors, target tissues, antigens, and clinical outcome. (Section 24.8)

10. Describe the general mechanism used by superantigens to activate T cells. How does superantigen activation differ from T cell activation by conventional antigens? (Section 24.9)

APPLICATION QUESTIONS

1. Describe the relative importance of innate immunity compared to adaptive immunity. Is one more important than the other? Can we survive in a normal environment without immunity?

2. Inflammation is the hallmark of an active immune response. Explain how inflammation is triggered by both innate and adaptive immune mechanisms. Are the inflammatory cells the same for both methods of activation? Why does inflammation subside as an infection is controlled?

3. Many infectious diseases have no effective vaccines. Pick several of these diseases (for example, AIDS, malaria, the common cold) and explain why current vaccine strategies have not been effective. Prepare some alternate strategies for immunization against the diseases you have chosen.

4. Are superantigen reactions desirable for the host? Do they confer protection for the host or do they benefit the pathogen?

25 · Immune Mechanisms

microbiology**now**

Why Is Alum in Your Vaccine?

Alum is an insoluble aluminum salt that has been used for over 80 years in human vaccines. Alum is an *adjuvant,* a substance that enhances the immune response nonspecifically when administered with a vaccine. Alum is the only adjuvant that is approved for use in vaccines for humans. For example, the diphtheria–tetanus–acellular pertussis (DTaP) vaccine contains an alum adjuvant. How does alum work?

It is well known that vaccines administered without alum are ineffective and that vaccine proteins (antigens) adsorb to the surface of alum particles. It is also known that without alum, few antigen-specific T helper cells are induced and the immune response is poor. The explanation was that alum increases the half-life of antigen in the body, thus extending exposure of the antigen to immune cells. But is this the whole story?

When a vaccine is injected intramuscularly (see photo), monocytes migrate to the injection site, causing inflammation. A research study[1] found that alum accumulates into nodules that are held together by host chromatin (including DNA) acquired from the inflammatory monocytes. Chromatin DNA and the adsorbed vaccine proteins move into the cytosol of antigen-presenting dendritic cells. From there, the host DNA activates pathways that increase the ability of dendritic cells to interact with antigen-specific T helper cells.

Alum thus works as an adjuvant not simply by prolonging the time antigen stays in the body, but also by enhancing DNA signaling. This in turn promotes more vigorous antigen presentation by dendritic cells and better stimulation of T helper cells. The final result is a stronger and more effective immune response.

[1]McKee, A.M., et al. 2013. Host DNA released in response to aluminum adjuvant enhances MHC class II-mediated antigen presentation and prolongs CD4 T-cell interactions with dendritic cells. *Proc. Natl. Acad. Sci. (USA) 110*(12): E1122–1131. doi: 10.1073/pnas.1300392110.

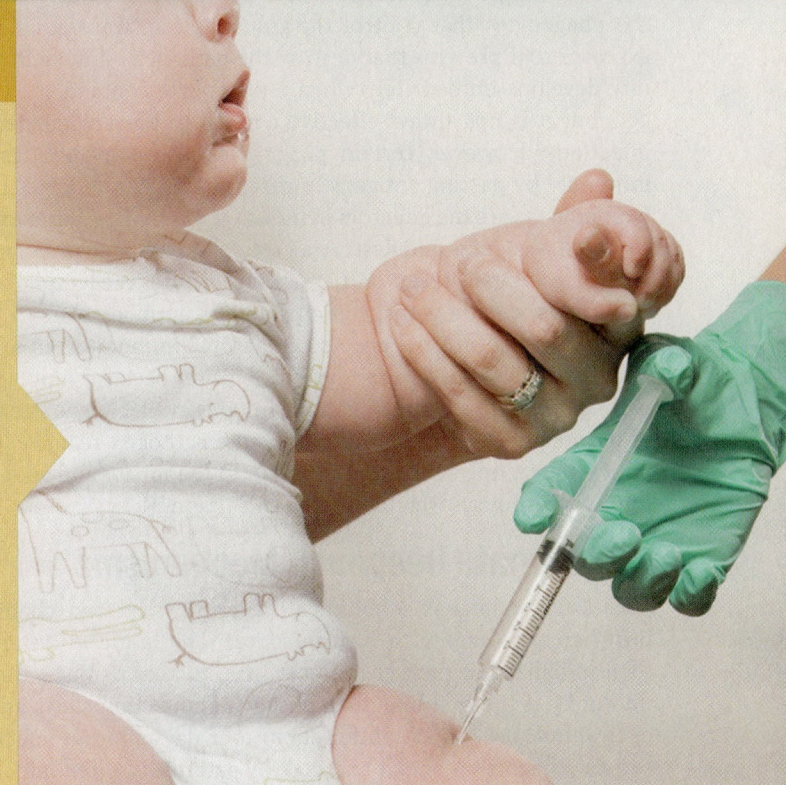

We discussed the key features of innate and adaptive immunity in terms of protection against pathogen infection and disease in Chapter 24. Here we focus on the *mechanisms* by which immunity is achieved by examining the cellular and molecular processes central to innate and adaptive immunity.

I • Basic Immune Mechanisms

Innate immunity is primarily a function of *phagocytes*. Innate responses recognize common structural features found on and in pathogens. Interactions with pathogens activate genes in the phagocytes that control the transcription, translation, and expression of proteins that destroy the pathogens. Innate immunity develops immediately when a phagocyte contacts a pathogen, but it is not always effective enough to prevent dangerous infections. However, certain phagocytes also activate adaptive immunity by passing antigens to receptors on lymphocytes. The lymphocytes are the effectors in *adaptive immunity*, the acquired ability to recognize and destroy a particular pathogen or its products. The antigen–receptor interaction activates the lymphocytes that produce pathogen-specific proteins—antibodies and T cell receptors—that are the agents of adaptive immunity. An adaptive response takes several days to develop because only a few pathogen-reactive lymphocytes are initially available; the strength of the adaptive response increases as the pathogen-reactive lymphocytes multiply. Cells that function in the innate and adaptive immune responses were described in Section 24.1.

25.1 Innate Response Mechanisms

Successful pathogens sometimes breach physical and chemical host barriers, leading to host infection. The immune system is then mobilized to protect the host from further damage. Innate immunity is the first line of defense when infection starts and is critical for host protection for about four days after an infection begins. Phagocytes engulf and destroy pathogens, often initiating complex host-mediated inflammatory reactions (⮌Section 24.5).

Phagocytes

The first cell type active in the innate response is typically a *phagocyte* (literally, a cell that eats). The primary function of a phagocyte is to engulf and destroy pathogens. A secondary function is to process the pathogen for antigens that initiate the adaptive immune response.

Phagocytes include monocytes, neutrophils, macrophages, and dendritic cells (**Figure 25.1**). Found in tissues and fluids throughout the body, most have inclusions called *lysosomes*, intracellular vacuoles containing bacteriocidal substances such as hydrogen peroxide, lysozyme, proteases, phosphatases, nucleases, and lipases. Phagocytes trap pathogens on surfaces such as blood vessel walls or fibrin clots. The membrane surrounding the pathogen pinches off and forms a *phagosome*. The phagosome, a vacuole containing the engulfed pathogen, then moves into the cytoplasm and fuses with a lysosome to form a phagolysosome. The toxic substances and enzymes inside the phagolysosome usually kill and digest the engulfed microbial cell (**Figure 25.2**).

The *neutrophil* is an actively motile phagocyte that contains lysosomes (Figure 25.1*a*). Derived from myeloid stem cells (⮌ Figure 24.1), neutrophils are found predominantly in the bloodstream and bone marrow, from where they migrate to sites of active infection in tissues. Neutrophils present in higher than normal numbers in the blood, a condition called *neutrophilia*, or at a site of inflammation indicate an active response to a current infection.

Monocytes are circulating precursors of *macrophages*, a major phagocytic cell type (Figure 25.1*a*). Macrophages are large noncirculating phagocytic cells found in almost all tissues (Figure 25.1*b*),

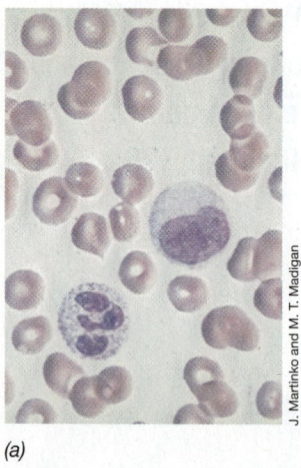
(a)
J. Martinko and M. T. Madigan

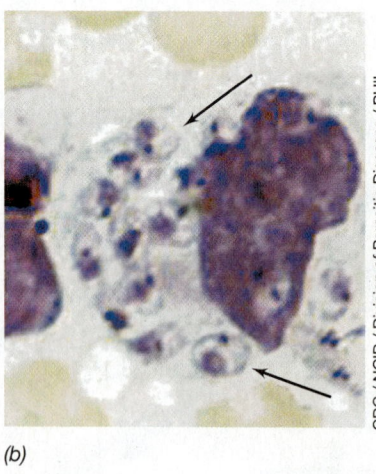
(b)
J. Martinko and M. T. Madigan

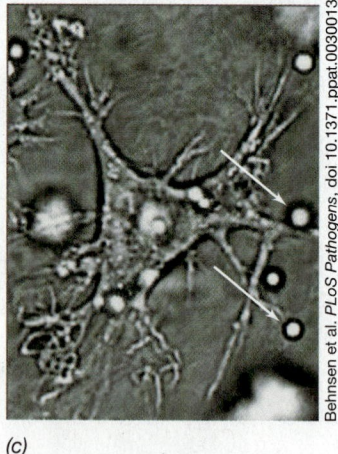
(c)
CDC / NCID / Division of Parasitic Diseases / PHIL

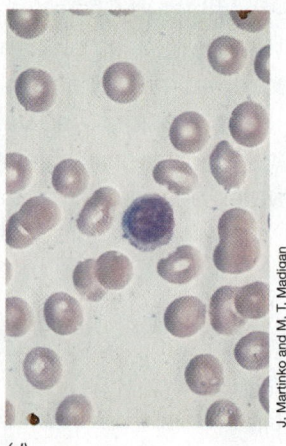
(d)
Behnsen et al. *PLoS Pathogens*, doi 10.1371.ppat.0030013
J. Martinko and M. T. Madigan

Figure 25.1 Major immune cell types. *(a)* The nucleated cell in the lower left center is a neutrophil (polymorphonuclear leukocyte, PMN), characterized by a segmented nucleus (violet stain) and granular cytoplasm. The nucleated cell to the right and slightly above the neutrophil is a monocyte. These phagocytes are 12–15 m in diameter. The nonnucleated red blood cells are about 6 μm in diameter. *(b)* A skin macrophage that has ingested numerous *Leishmania* (arrows), a protozoan. *(c)* A dendritic cell interacting with fungal conidia (arrows). *(d)* The nucleated cell is a circulating lymphocyte. The lymphocyte is about 10 μm in diameter and has almost no visible cytoplasm.

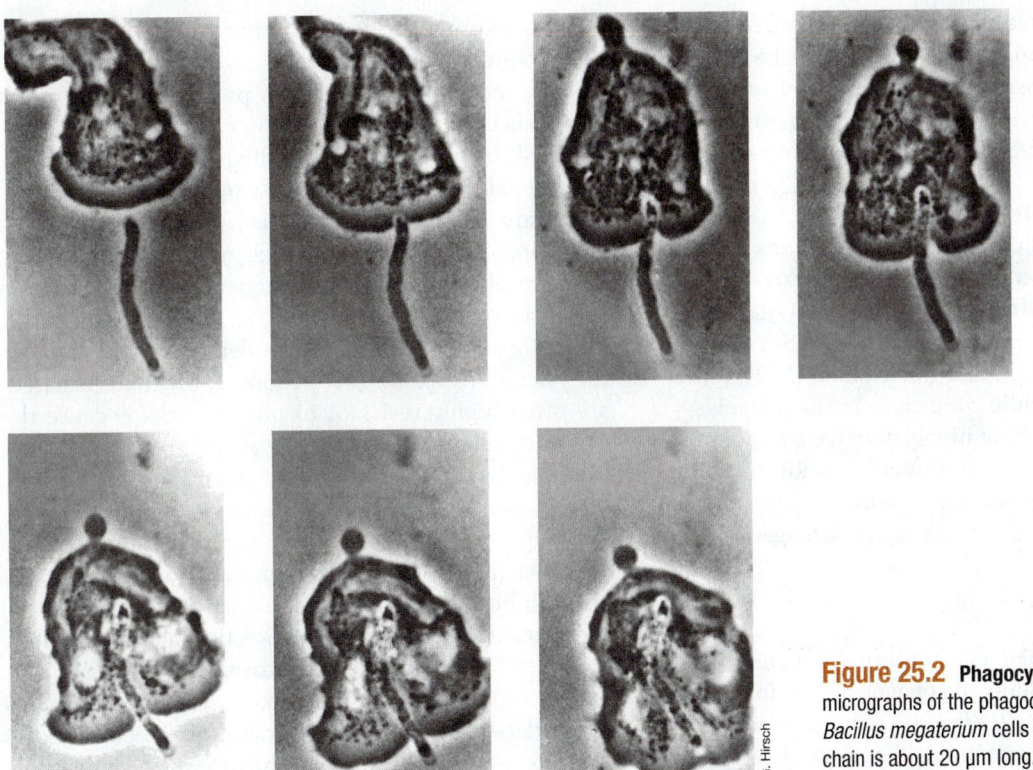

Figure 25.2 Phagocytosis. Time-lapse phase-contrast micrographs of the phagocytosis and digestion of a chain of *Bacillus megaterium* cells by a human macrophage. The bacterial chain is about 20 μm long. The macrophage is one of a group of cells that ingests and degrades pathogens and pathogen products.

J. G. Hirsch

where they may constitute up to 10–15% of the total cells. Because they ingest and destroy most pathogens and foreign molecules that invade the body, macrophages are essential to the innate response. They are also critically important for initiating adaptive immunity by presenting antigens to T lymphocytes.

Dendritic cells (Figure 25.1*c*) also have the dual function of phagocytosis and antigen presentation. Derived from the same monocyte progenitors as macrophages, immature dendritic cells are found throughout the body tissues, where they function as active phagocytes. When the dendritic cells ingest antigen, they migrate to the lymph nodes, where they present antigen to T lymphocytes. The specialized antigen-presenting properties of macrophages and dendritic cells are examined in Section 25.4.

Pathogen Recognition by Phagocytes

Phagocytes have a pathogen-recognition system that triggers a timely and appropriate response, generally leading to recognition, containment, and destruction of the pathogen. This system employs evolutionarily conserved *pattern recognition receptors* (*PRRs*). PRRs are membrane-bound phagocyte proteins that recognize a *pathogen-associated molecular pattern* (*PAMP*), a structural component on a microbial cell or virus (⮂ Figure 24.3). PRRs were first observed in phagocytes in *Drosophila*, the fruit fly, where they are called *Toll receptors*. Each **Toll-like receptor (TLR)** on a human phagocyte recognizes a specific PAMP. For example, TLR-2, a PRR on human phagocytes, interacts with peptidoglycan, a PAMP present in the cell wall of nearly all bacteria (⮂ Section 2.10); the interaction activates the phagocytes, targeting gram-positive pathogens with exposed peptidoglycan (**Figure 25.3**). Access to the peptidoglycan of gram-negative

cell walls is blocked by the surface lipopolysaccharides. Other TLRs recognize PAMPs such as the unmethylated CpG oligonucleotides, the lipopolysaccharide of gram-negative bacteria, or the double-stranded RNA of certain viruses. Several soluble host molecules function similarly to these phagocyte-associated PRRs. Later in this chapter we discuss the soluble PRRs in the

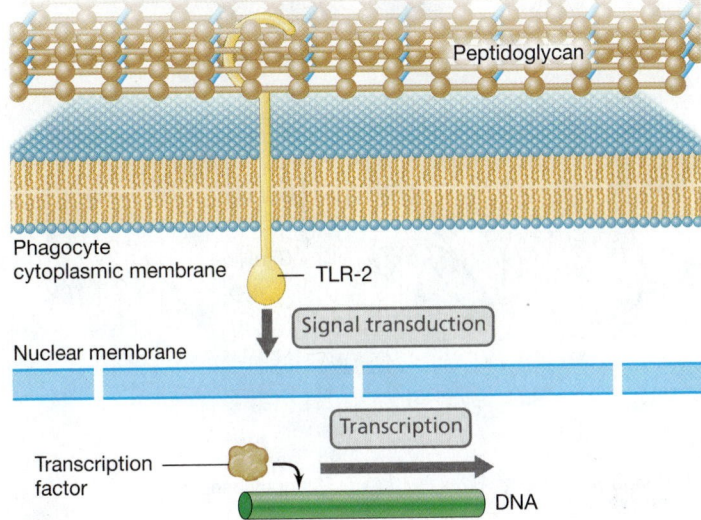

Figure 25.3 A Toll-like receptor. Membrane-spanning TLR-2 interacts with peptidoglycan from gram-positive pathogens. This interaction stimulates signal transduction, activating transcription factors in the nucleus. The result is translation of proteins that induce inflammation and other phagocyte activities. All Toll-like receptors have analogous mechanisms for activating innate immunity.

UNIT 5

context of their ability to activate proteins that enhance phagocytosis and destruction of pathogens (Section 25.9). The PAMP-PRR interaction triggers a transmembrane signal that results in production of important defense proteins, including some that produce toxic oxygen compounds that can cause pathogen death.

Oxygen-Dependent Killing in Phagocytes

Genes that control the production of oxygen compounds toxic to pathogens are highly transcribed in activated phagocytes. These toxic compounds include hydrogen peroxide (H_2O_2), superoxide anions (O_2^-), hydroxyl radicals ($OH \cdot$), singlet oxygen ($^1O_2$), hypochlorous acid (HOCl), and nitric oxide (NO) (**Figure 25.4**) ($\rightleftarrows$ Section 5.16). The acidic conditions in the phagolysosome favor the production of these highly reactive compounds. Phagocytic cells use toxic oxygen compounds to kill ingested bacterial cells by oxidizing key cellular constituents. The reactions occur within the phagocyte, which is not damaged by the toxic oxygen products.

Inhibiting Phagocytes

Some pathogens have developed mechanisms for neutralizing toxic phagocyte products, for killing the phagocytes, or for avoiding phagocytosis. For example, *Staphylococcus aureus* produces pigmented compounds called carotenoids that neutralize singlet oxygen and prevent killing ($\rightleftarrows$ Section 29.9). Intracellular pathogens such as *Mycobacterium tuberculosis* (the cause of tuberculosis) grow and persist within phagocytic cells ($\rightleftarrows$ Section 29.4). *M. tuberculosis* uses its cell wall glycolipids to absorb hydroxyl

radicals and superoxide anions, the most lethal toxic oxygen species produced by phagocytes.

Some intracellular pathogens produce phagocyte-killing proteins called *leukocidins*. In such cases, the pathogen is ingested as usual, but the leukocidin kills the phagocyte, releasing the pathogen. Dead phagocytes make up much of the material of *pus*; organisms such as *Streptococcus pyogenes* and *S. aureus*, major leukocidin producers, are called *pyogenic* (pus-forming) pathogens. Localized infections by pyogenic bacteria can form boils or abscesses.

Another important pathogen defense against phagocytosis is the bacterial capsule ($\rightleftarrows$ Section 2.13). Encapsulated bacteria are often highly resistant to phagocytosis because the capsule prevents adherence of the phagocyte to the bacterial cell. For example, fewer than ten cells of an encapsulated strain of *Streptococcus pneumoniae* can kill a mouse within a few days after exposure ($\rightleftarrows$ Figure 23.10). Nonencapsulated strains are completely avirulent. Surface components other than capsules can also inhibit phagocytosis. For instance, pathogenic *S. pyogenes* produces M protein, a substance that alters the surface of the pathogen and inhibits phagocytosis.

Antibodies or soluble PRRs that interact with capsules or other cell surface molecules can reverse the protective effect of bacterial defense mechanisms and enhance phagocytosis, the process of opsonization. As an example, the effective vaccine directed against *Streptococcus pneumoniae*, an organism that causes serious bacterial pneumonia, uses capsule polysaccharides to induce protective antibodies ($\rightleftarrows$ Section 24.7).

MINIQUIZ

- Describe the cellular location and molecular specificity of PAMPs and PRRs.
- Identify the mechanism used by phagocytes to induce pathogen killing.

25.2 Adaptive Response Properties

Adaptive immunity is the acquired ability to recognize and destroy a particular pathogen or its products. In contrast to innate immunity, effective adaptive immunity requires activation by pathogen exposure. B lymphocytes produce antibodies, which protect against extracellular antigens. T lymphocytes, through their antigen-specific T cell receptor (TCR), protect against intracellular pathogens such as viruses and certain bacteria. Adaptive immunity is characterized by the properties of *specificity, memory*, and *tolerance*. None of these properties is found in the innate response.

Specificity

The immune response is incredibly specific, but the innate and the adaptive systems differ in this regard. Innate immunity is directed against features common to pathogens such as the peptidoglycan of all gram-positive bacteria or the lipopolysaccharide of all gram-negative bacteria. By contrast, adaptive immunity is directed to interactions with particular pathogen-specific macromolecules such as the M-protein antigen on a

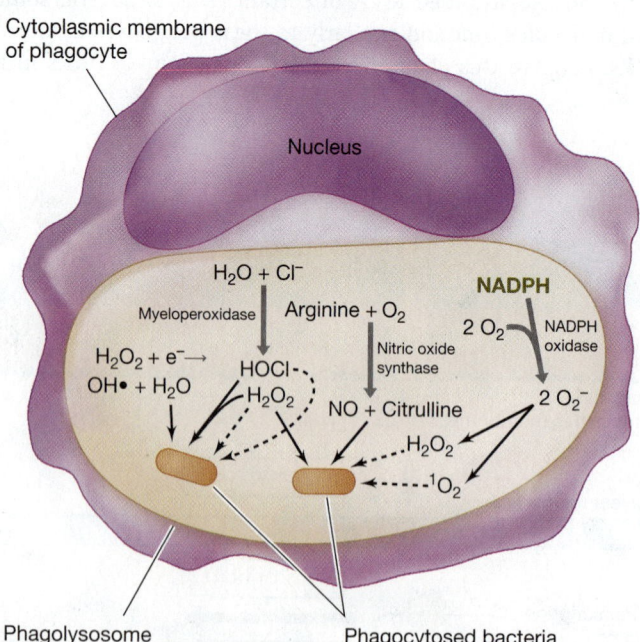

Figure 25.4 Activity of phagocyte enzymes in generating toxic oxygen compounds. These compounds include hydrogen peroxide (H_2O_2), the hydroxyl radical ($OH \cdot$), hypochlorous acid (HOCl), the superoxide anion (O_2^-), singlet oxygen ($^1O_2$), and nitric oxide (NO). Formation of these toxic compounds requires a substantial increase in the uptake and utilization of molecular oxygen, O_2. This increase in oxygen uptake and consumption by activated phagocytes is known as the respiratory burst.

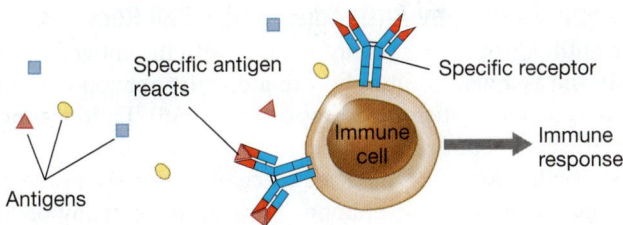

Specificity: Immune cells have surface receptors that interact with individual antigens.

(a)

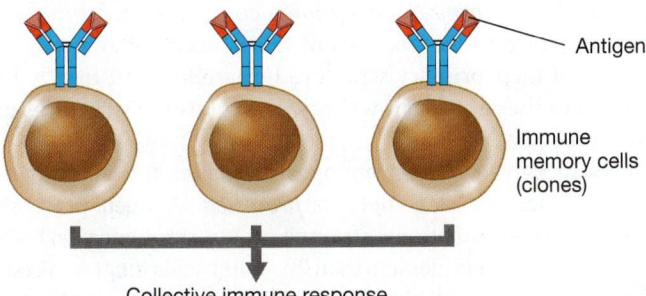

Memory: The first antigen exposure induces multiplication of antigen-reactive cells, resulting in multiple copies, or *clones*. After a subsequent exposure to the same antigen, the immune response is faster and stronger due to the large number of responding cells.

(b)

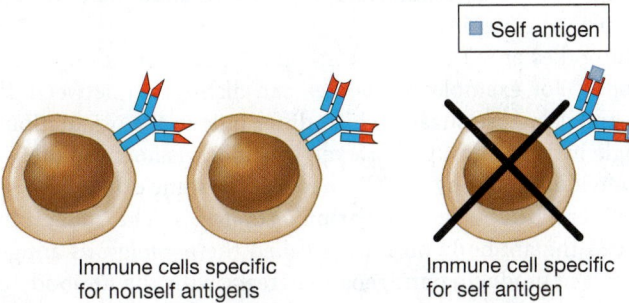

Tolerance: Immune cells are not able to react with self antigen. Self-reactive cells are destroyed during development of the immune response.

(c)

Figure 25.5 The adaptive immune response. Key features of antibody-mediated and cell-mediated immunity are *(a)* specificity, *(b)* memory, and *(c)* tolerance.

single strain of *Streptococcus pyogenes* (🔗 Section 29.2). The *specificity* of the antigen–antibody or antigen–TCR interaction is dependent on the capacity of the lymphocyte cell receptor to interact with particular antigens. Thus, in contrast to the innate system, the adaptive immune response, triggered by antigen contact, is exclusively and specifically directed to the unique antigen (**Figure 25.5***a*).

Memory and Tolerance

The immune system must encounter antigen to stimulate production of detectable and effective antigen-activated antibodies or TCRs. A subsequent exposure to the same antigen stimulates rapid production of large quantities of the same T cells or antibodies. This capacity to respond more quickly and vigorously to subsequent exposures to the eliciting antigen is *immune memory* (Figure 25.5*b*). Immune memory provides the host with immediate resistance to previously encountered pathogens. Clinical medicine takes advantage of immune memory by vaccinating susceptible individuals with dead or weakened pathogens or their products to artificially stimulate and enhance immunity for a number of dangerous pathogens (🔗 Section 24.7).

Besides the important aspects of specificity and memory, *tolerance* is important to the immune system and is defined as the acquired *inability* to make an adaptive immune response directed to *self* antigens. Because macromolecules in the host are also antigens, the host immune system must avoid recognizing host macromolecules as they could be damaged if they were recognized as antigens by antibodies or T cells. Thus, the adaptive immune response must develop the capacity to discriminate between *foreign* (nonself and dangerous) antigens and *host* (self and not dangerous) antigens (Figure 25.5*c*).

- - -

MINIQUIZ -

- What controls specificity of an immune cell?
- Distinguish between the terms immune memory and immune tolerance.
- If immune specificity, memory, or tolerance were impaired, what would be the outcome for the host?

- - -

II • Antigens and Antigen Presentation

The adaptive immune response recognizes a broad range of pathogen-derived macromolecules. The macromolecules are degraded and processed in host cells to produce antigens that are in turn presented to T cells. We first discuss antigens and then focus on the mechanisms of antigen processing and presentation to T cells.

25.3 Immunogens and Antigens

Antigens are substances that react with antibodies or TCRs. Most, but not all, antigens are **immunogens**, substances that induce an immune response. Here we examine the features of

effective immunogens and then define the features of antigens that promote interactions with antibodies and TCRs.

Intrinsic Properties of Immunogens

Immunogens share several intrinsic properties that enable them to induce an adaptive immune response. First, *molecular size* is an important property of immunogenicity; for a molecule to be immunogenic, it must be sufficiently large. For example, low-molecular-weight compounds called *haptens* cannot induce an immune response but can be bound by antibodies. However, because haptens are bound by antibodies, they are antigens even though they are

Table 25.1 Properties of immunogens

Properties intrinsic to the immunogen	
Size	>10,000 molecular weight
Complexity	Polymers > monomers
Form	Aggregated > soluble
Properties extrinsic to the immunogen	
Dose	10 µg to 1 g
Route	Intravenous, intradermal, or subcutaneous > oral or topical
Foreignness	Nonself >> self

not immunogens. Haptens such as sugars, amino acids, and other low-molecular-weight organic compounds become effective immunogens only when they are coupled to a larger protein, generally with a molecular weight of 10,000 or greater. Thus, sufficient molecular size is a prerequisite for immunogenicity; this property and the other key properties discussed next are summarized in **Table 25.1**.

Complex, nonrepeating polymers such as proteins are effective immunogens. Complex carbohydrates can also be very good immunogens. In contrast, nucleic acids, simple polysaccharides with repeating subunits, and lipids, because they are composed of chains of identical or nearly identical monomers, tend to be poor immunogens. Thus, *sufficient molecular complexity* is another property of immunogenicity.

Large, complex macromolecules in insoluble or aggregated form (for example, proteins precipitated by heating) are usually excellent immunogens. The insoluble material is readily taken up by a phagocyte, leading to an adaptive immune response. By contrast, the soluble form of the same molecule may be a very poor immunogen because the soluble molecule is not ingested efficiently by phagocytes. Thus, *appropriate physical form* is another property of immunogenicity.

Extrinsic Properties of Immunogens

Although many substances are intrinsically immunogenic, extrinsic factors also influence immunogenicity. Three extrinsic factors important for immunogens include the *dose*, the *route* of administration, and the *foreign nature* of the immunogen to the host.

The dose of an immunogen administered to a host can be important for an effective immune response, but a broad range of doses ordinarily provides satisfactory immunity. In general, doses of 10 µg to 1 g are effective in most mammals. Doses of immunogen higher than 1 g or lower than 10 µg may not stimulate an immune response but may actually suppress a specific immune response and instead induce tolerance.

The route of administration of an immunogen is also important. Immunizations given by parenteral (outside of the gastrointestinal tract) routes, usually by injection, are normally more effective than those given topically or orally. When given by oral or topical routes, antigens may be significantly degraded before contacting a phagocyte.

Finally, an effective immunogen must be foreign with respect to the host. The adaptive immune system recognizes and eliminates only foreign antigens. Self antigens are not recognized because individuals are tolerant of their own molecules (Section 25.2).

Antigen Binding by Antibodies and T Cell Receptors

The antibody or TCR does not interact with the antigenic macromolecule as a whole but only with a distinct portion of the molecule called an antigenic **epitope** (**Figure 25.6**). Epitopes include sugars, short peptides, and other organic molecules.

Antibodies interact with a sequence of four to six amino acids, the optimal size for an epitope. Thus proteins, many of which consist of hundreds or even thousands of amino acids, are arrays of overlapping epitopes. The surface of a bacterial cell or virus consists of a mosaic of proteins, polysaccharides, and other macromolecules, all with individual epitopes. In many cases, antibodies recognize *conformational epitopes* composed of, for example, amino acids from two portions of the molecule that are distant in terms of their primary structure but are brought together by folding into the secondary, tertiary, or quaternary structures of a macromolecule.

TCRs recognize protein epitopes only after the immunogens have been partially degraded, or *processed*. Antigen processing destroys the conformational structure of a macromolecule, breaking proteins into peptides less than 20 amino acids long. As a result, TCRs recognize only *linear epitopes* in the primary protein structure. Processed antigens are presented to T cells on the surface of specialized APCs or target cells, as we will discuss in Section 25.4.

Antibodies, on the other hand, can recognize conformational epitopes on proteins or polysaccharides expressed in native conformations on macromolecular surfaces. In addition, antibodies recognize linear epitopes.

Antibodies and TCRs can distinguish between closely related epitopes. For example, antibodies can distinguish between glucose and galactose sugars, which differ only in the orientation of a single hydroxyl group. However, specificity is not absolute, and an individual antibody or TCR may react to some extent with several different but structurally similar epitopes. The antigen that induced the antibody or TCR is called the *homologous* antigen, and any noninducing antigens that react with the antibody are called *heterologous* antigens. An interaction between an antibody or TCR and a heterologous antigen is called a *cross-reaction*.

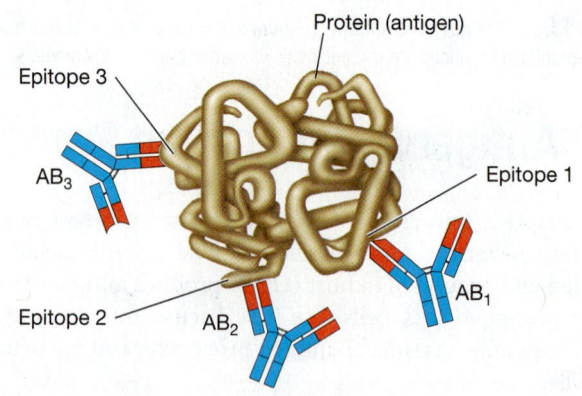

Figure 25.6 Antigens and epitopes for antibodies. Antigens may contain several different epitopes, each capable of reacting with a different antibody (AB). The epitope 1 recognized by AB_1 is a conformational epitope. Epitope 1 consists of two nonlinear parts of the folded polypeptide; the folding brings two distant portions of the polypeptide together to make a single epitope.

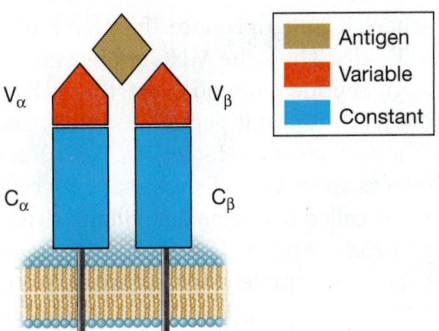

Figure 25.7 The T cell receptor. The V domains of the α chain and β chain combine to form the peptide antigen-binding site.

25.4 Antigen Presentation to T Cells

T cells are thymus-derived lymphocytes that interact with antigens and activate the adaptive immune response through TCRs (⟳ Section 24.3). Here we examine how the TCR interacts with antigens on an antigen-presenting phagocyte cell or on an infected target cell.

The T Cell Receptor

The TCR is a membrane-spanning protein that extends from the T cell surface into the extracellular environment. Each T cell has thousands of copies of the same TCR on its surface. A functional TCR consists of two polypeptides, an α chain and a β chain. Each polypeptide consists of several **domains**, regions of the protein that have defined structural and functional properties. Each chain has a variable (V) domain and a constant (C) domain (**Figure 25.7**). The V_α and V_β domains interact cooperatively to form an antigen-binding site. As we will see in Section 26.7, the adaptive immune response can generate TCRs that will bind nearly every known peptide antigen. TCRs recognize only MHC–peptide; other antigens, such as complex polysaccharides, are not recognized by TCRs, but these may be bound by the immunoglobulin receptors on B cells. TCRs recognize and bind a peptide antigen only when it is bound to a *self* protein, the major histocompatibility complex protein.

Major Histocompatibility Complex Proteins

A linked set of genes called the *major histocompatibility complex (MHC)* is found in all vertebrates. The MHC encodes a group of proteins important in antigen presentation. The MHC proteins in humans, called *human leukocyte antigens* or *HLAs*, were first identified as the major antigens responsible for immune-mediated organ transplant rejection. We now know that MHC proteins function primarily as antigen-presenting molecules, binding pathogen-derived peptides and displaying these peptides for interaction with TCRs.

There are two classes of MHC proteins called class I and class II. **MHC class I proteins** are found on the surfaces of all nucleated cells. **MHC class II proteins** are found only on the surface of B lymphocytes, macrophages, and dendritic cells, all of which are APCs. This differential cellular distribution relates to the functions of the class I and class II proteins.

An MHC class I protein consists of two polypeptides, a membrane-embedded alpha chain encoded in the MHC gene region and a smaller protein called *beta-2 microglobulin* ($\beta_2 m$) encoded by a non-MHC gene (**Figure 25.8a**). An MHC class II protein consists of two noncovalently linked polypeptides called α and β. Like MHC class I α chains, these polypeptides are embedded in the cytoplasmic membrane and project outward from the cell surface (Figure 25.8b).

In different individuals of the same species, MHC proteins are not structurally identical. Different individuals usually have subtle differences in the amino acid sequence of homologous MHC proteins. These genetically encoded MHC variants, of which there are over 3500 in humans, are called *polymorphisms*. Polymorphisms in MHC proteins are the major antigenic barriers for tissue transplantation from one individual to another; tissue transplants not matched for MHC identity are recognized as nonself by the immune system and are rejected. We present the detailed molecular structure and genetic organization of the MHC genes and proteins in Chapter 26.

Antigen Processing and Presentation

The MHC proteins cannot be expressed on the cell surface unless they are complexed with peptide. These MHC–peptide complexes reflect the composition of the proteins inside the cell. For example, a cell that contains no pathogens or foreign antigens displays MHC proteins complexed with self peptides derived from the normal degradation of self proteins during cell growth. On the other hand, host cells that have ingested foreign proteins or pathogens, and cells infected with viruses, produce peptides that also interact with MHC proteins. The host cells then degrade (process) the antigens to form small peptides. In this case, the MHC proteins expressed on the cell surface are complexed with foreign peptides. These MHC proteins with embedded peptides provide the foreign-antigen targets for T cells. T cells, through their TCRs, continually sample the surface of other cells to identify cells carrying foreign antigens; the TCR

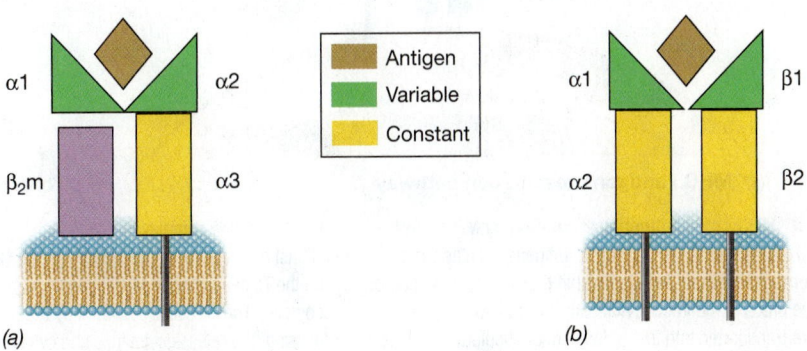

Figure 25.8 The MHC proteins. *(a)* MHC class I protein. The α1 and α2 domains interact to form the peptide antigen-binding site. *(b)* MHC class II protein. The α1 and β1 domains combine to form the peptide antigen-binding site.

UNIT 5

interacts with the foreign antigen presented on an MHC protein. No T cells can react with the MHC–peptide complexes on uninfected cells because self-reactive T cells have been eliminated during the development of tolerance in the immune system. Two distinct antigen-processing schemes are at work, one for MHC I antigen presentation and one for MHC II antigen presentation (**Figure 25.9**).

MHC I proteins present peptide epitopes derived from pathogen proteins in the cytoplasm of cells that have been infected by viruses and other intracellular pathogens; such infected cells are called *target cells* (Figure 25.9*a*). Proteins derived from infecting viruses, for example, are taken up and digested in the cytoplasm in a structure called the *proteasome*. Peptides about ten amino acids long are transported into the endoplasmic reticulum through a pore formed by two proteins, called the *transporters associated with antigen processing* (*TAP*). Once the peptides have entered the ER, they are bound by the MHC I protein, which has been assembled in the ER and held in place near the TAP site by a group of *chaperone proteins* until a peptide is bound. The MHC I–peptide complex is then released from the chaperones and moves to the

cell surface, where it integrates into the membrane and can be recognized by T cells. Thus, the MHC proteins are a platform to which the foreign peptide is bound. Next, the TCR on the surface of a T cell interacts with both peptide (nonself) and MHC protein (self) on the surface of the target cell. This T cell–target cell interaction induces specialized T-cytotoxic (Tc) cells to produce cytotoxic proteins called perforins that kill the virus-infected target cell (Section 25.5). Any nucleated cell can act as a target cell for T cells recognizing peptide–MHC I complexes.

MHC II proteins are the antigen-presenting proteins in a second pathway (Figure 25.9*b*). MHC II proteins are expressed exclusively on the phagocytic APCs, where they function to present peptides from engulfed pathogens such as bacteria. MHC class II proteins are initially assembled in the endoplasmic reticulum much as MHC I proteins are assembled. However, a difference from the assembly pathway of MHC I proteins is that a chaperone protein called Ii, or invariant chain, binds to the MHC II protein, blocking peptide loading inside the endoplasmic reticulum. These MHC II–Ii complexes are transported from the endoplasmic

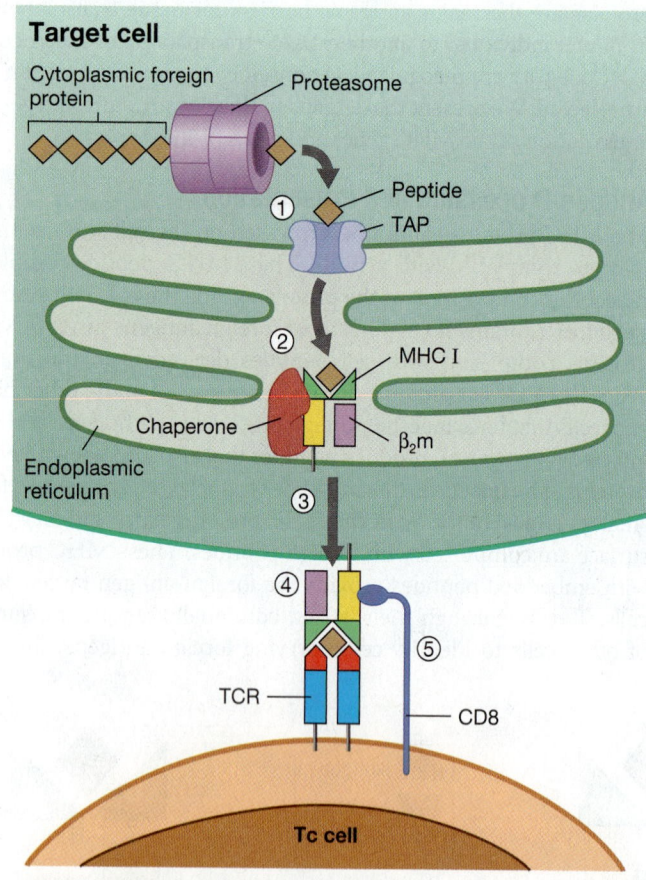

(a) **MHC I antigen presentation pathway**

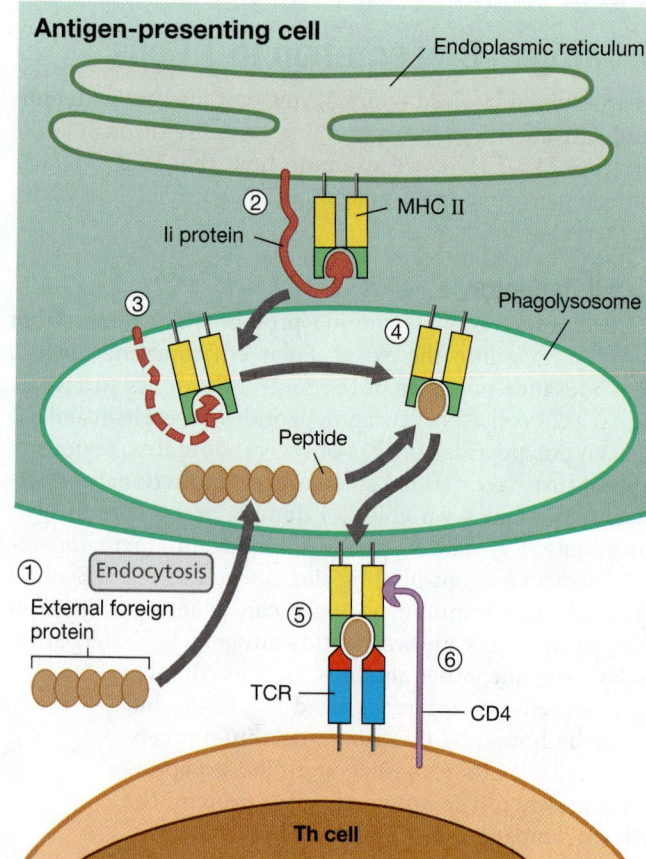

(b) **MHC II antigen presentation pathway**

Figure 25.9 **Antigen presentation by MHC I and MHC II proteins.** *(a)* ① Protein antigens such as virus components manufactured within the cell are degraded by the proteasome in the cytoplasm. The peptide fragments are transported into the endoplasmic reticulum (ER) through a pore formed by the TAP proteins. ② MHC I proteins in the ER are stabilized by chaperones until peptide fragments are bound. ③ When peptide fragments are bound by MHC I, the complex is transported to the cell surface. ④ The

MHC I–peptide complex interacts with T cell receptors (TCRs) on the surface of Tc cells. ⑤ The CD8 coreceptor on the Tc cell engages MHC I, resulting in a stronger complex. The Tc cell is activated by the binding events, causing it to release cytokines and cytolytic toxins and kill the target cell. *(b)* ① Foreign proteins, imported from outside the cell by endocytosis, are digested into peptide fragments in phagosomes. ② MHC II proteins in the ER are assembled with Ii, a blocking protein that prevents MHC

II from binding with peptides in the ER. ③ The MHC II–Ii assembly is transported to the lysosome, where it remains until the lysosome fuses with the phagosome, forming a phagolysosome where Ii is degraded, ④ freeing the MHC II protein to bind the foreign peptide fragments. ⑤ The MHC II–peptide complex is transported to the cell surface, where it interacts with TCRs and ⑥ the CD4 coreceptor on Th cells. The Th cells then release cytokines that interact with other cells to promote an immune response.

reticulum to lysosomes. After phagocytosis of a pathogen, the phagosome containing the foreign antigen fuses with the lysosome to form a phagolysosome. In the latter, the foreign antigens and the Ii peptide are digested by lysosomal enzymes. The foreign peptides, generally about 11–15 amino acids long (slightly larger than MHC I–binding peptides), bind to the newly opened MHC II antigen-binding site. The complex is transported to the cytoplasmic membrane, where it is displayed on the cell surface to specialized **T-helper (Th) cells**. The Th cells, through the TCR, recognize the MHC II–peptide complex. This interaction activates the Th cells to secrete cytokines, stimulating antibody production by B cells or causing inflammation.

CD4 and CD8 Coreceptors

In addition to the TCR, each T cell expresses a unique cell surface protein that functions as a coreceptor. Th cells express a **CD4 coreceptor** protein, and Tc cells express a **CD8 coreceptor**

protein (Figure 25.9). When the TCR binds to the peptide–MHC complex, the coreceptor on the T cell also binds to the MHC protein on the antigen-presenting cell, strengthening the molecular interactions between the cells and enhancing activation of the T cell. CD4 binds only to the MHC II protein, strengthening Th cell interaction with APCs that express MHC II protein. Likewise, CD8 binds only to the MHC I protein, enhancing the binding of Tc cells to MHC I–bearing target cells. The CD4 and CD8 proteins are also used for in vitro tests as T cell markers to differentiate Th (CD4) cells from Tc (CD8) cells.

MINIQUIZ --

• Identify the cells that display MHC I and MHC II proteins on their surface.

• Define the sequence of events for processing and presenting antigens from both intracellular and extracellular pathogens.

III • T Lymphocytes and Immunity

Antigen presentation activates precursor T lymphocytes to differentiate into T cells that carry out antigen-specific cell-mediated immunity. These functions include cell-mediated killing, inflammatory responses, and "help" for antibody-producing B cells. In the absence of antigen-activated T cells, there is little antigen-specific immunity and no immune memory.

25.5 T-Cytotoxic Cells and Natural Killer Cells

In the previous section we introduced two subsets of T cells, the T-cytotoxic cells and the T-helper cells. Here we examine the antigen-specific cell-killing function of the T-cytotoxic cells in detail. We also introduce the natural killer (NK) cell, a lymphocyte-like cell that uses another mechanism to recognize and kill cells infected with intracellular pathogens.

T-Cytotoxic Cells

T-cytotoxic (Tc) cells, also called cytotoxic T lymphocytes (CTLs), are CD8 T cells that directly kill cells that display foreign surface antigens. As we discussed in Section 25.4, Tc cells recognize foreign peptides embedded in MHC I proteins. Cells displaying the foreign antigen are killed by the Tc cells. For example, a viral peptide embedded in MHC I, displayed on a virus-infected cell, marks the cell for interaction and killing by a Tc cell.

Contact between a Tc cell and the target cell is required for cell death (**Figure 25.10**). The point of initial contact is between the TCR and the peptide–MHC I complex. The CD8 protein on the Tc cell then binds the MHC I protein, strengthening the interaction. On contact with the target cell, granules in the Tc cell migrate to the contact site, where the contents of the granules are released (degranulation). The granules contain perforin and proteases called *granzymes*. Perforin enters the membrane of the target cell and forms a pore, through which granzymes enter the target cell. Granzymes are cytotoxins that cause *apoptosis*, or

programmed cell death, characterized by death and degradation of the cell from within. The Tc cells, however, remain unaffected; their membranes are not damaged by perforin. Tc cells kill only those cells displaying the foreign antigen because the granules are released only at the contact surface between the Tc and the target cell bearing peptide–MHC I. Cells lacking the peptide recognized by the Tc cells do not make contact and are not killed.

Natural Killer Cells

Natural killer cells (NK cells) are cytotoxic lymphocytes that are distinct from T cells and B cells. Nevertheless, NK cells resemble Tc cells in their ability to destroy cancer cells and cells infected with intracellular pathogens. NK cells also use perforin and granzymes to kill their targets, destroying cancer cells and virus-infected cells without prior exposure or contact with the foreign

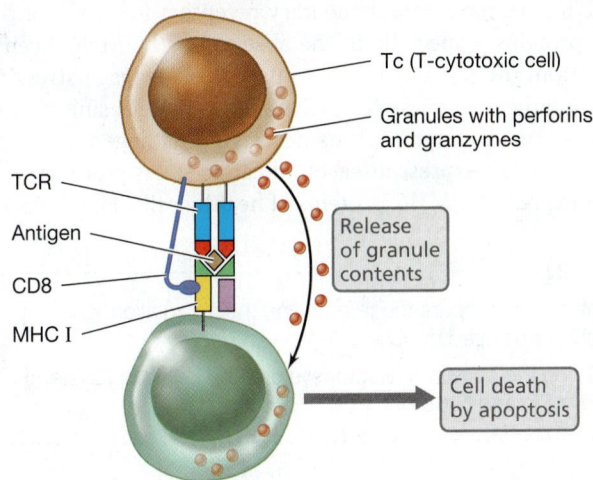

Figure 25.10 T-cytotoxic cells. When the TCR on a Tc cell binds MHC I–peptide complexes on any cell, the Tc cell releases granules that contain perforin and granzymes, cytotoxins that perforate the target cell and cause apoptosis, respectively.

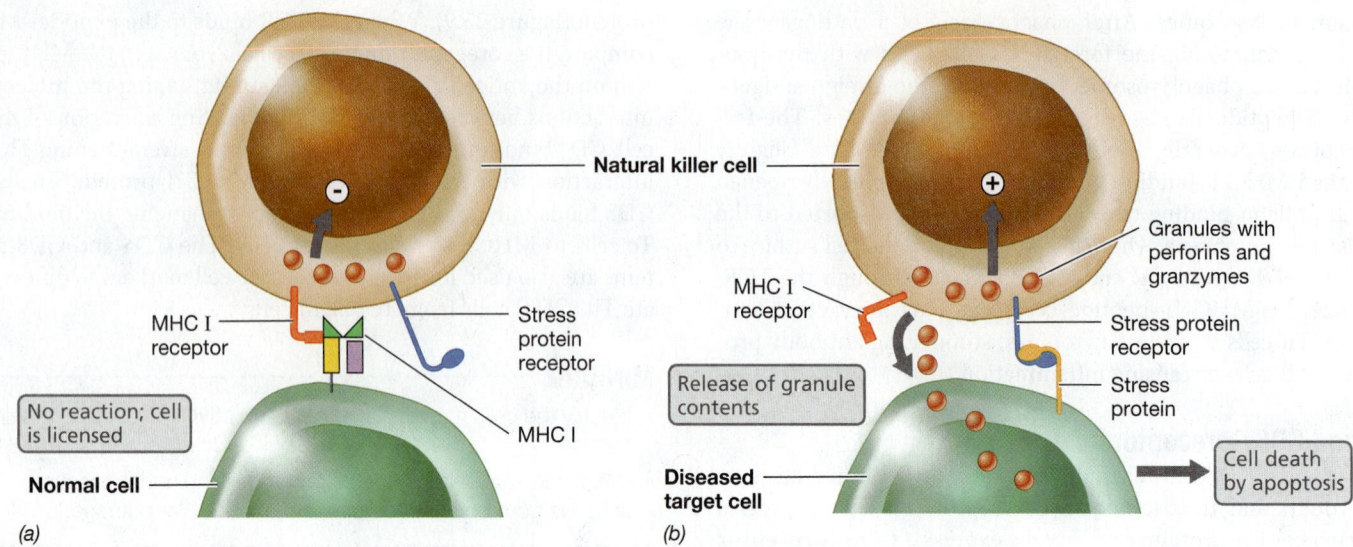

Figure 25.11 Natural killer cells. Natural killer (NK) cells have two receptors: one interacts with MHC I on healthy cells; the second one interacts with cell stress proteins, found only on tumor cells or pathogen-infected cells. *(a)* MHC I recognition licenses the healthy cells, preventing the NK cell from releasing its contents. *(b)* Pathogen-infected cells or tumor cells express stress proteins and often reduce MHC I expression. In the absence of MHC I recognition, the NK cell interacts with the stress protein and releases perforins and granzymes, killing the diseased cell.

cells (**Figure 25.11**). This is important because many tumor cells and virus-infected cells reduce or eliminate normal MHC I expression to evade the antigen-specific Tc response. In this process, the number of NK cells does not increase, nor do they exhibit memory after interaction with target cells.

NK cells recognize and destroy tumor cells or pathogen-infected cells using a two-receptor system. The molecular targets of NK cells are proteins on the surface of other cells (Figure 25.11a). As NK cells circulate and interact with the cells in the body, they use special MHC I receptors to recognize MHC I proteins on normal, healthy cells. Binding of the NK receptors to MHC I deactivates the NK cell, turning off the perforin and granzyme killing mechanisms, a process called *licensing*. Tumor cells or pathogen-infected cells, however, may express stress proteins on their surface; NK cells have complementary receptors for many of these stress proteins. Especially in the absence of the MHC licensing interaction, the stress receptors on NK cells engage stress proteins on target cells. The NK cell responds by releasing cytotoxic perforins and granzymes, thus destroying pathogen-infected or tumor cells that express disease-indicating stress proteins and no longer express the MHC proteins of healthy cells (Figure 25.11b).

MINIQUIZ

- Identify and compare the targets and the recognition mechanisms used by Tc and NK cells.
- Describe the common effector system (the cell-killing system) used by Tc and NK cells.

25.6 T-Helper Cells

Interactions with APCs drive CD4 Th cells to differentiate into several subsets, each producing unique combinations of cytokines that recruit effector cells such as macrophages, antibody-producing B cells, and neutrophils. Cells of Th1 and Th2 subsets play a role in adaptive immunity, promoting inflammation and antibody production respectively. Th17 cells amplify the innate immune response, and Treg cells suppress immunity when it is not needed (**Table 25.2**).

Th1 Cells and Macrophage Activation

Macrophages play a central role as APCs in cell-mediated immunity. As illustrated in **Figure 25.12**, macrophages engulf, process, and present antigen to **Th1** cells. Th1 cells produce IL-2 (*inter-leukin-2*), a cytokine that promotes growth and activation of other T cells, and activate macrophages through other cytokines. As a general rule, even inactive macrophages phagocytose and kill most bacteria, but a few pathogens survive and multiply within the phagolysosome. These intracellular pathogens include *Mycobacterium tuberculosis*, *Mycobacterium leprae*, and *Listeria monocytogenes*, the bacteria that cause tuberculosis, leprosy, and listeriosis, respectively. Animals inoculated with a moderate dose of *M. tuberculosis* are able to overcome the infection and develop a protective T cell–mediated immune response mediated by T-inflammatory cells, the Th1 subset. Th1 cells activate the macrophages through the cytokines interferon gamma (IFN-γ), tumor necrosis factor alpha (TNF-α), and granulocyte–monocyte colony-stimulating factor (GM-CSF) (Figure 25.12). The Th1-activated macrophages take up and kill foreign cells more efficiently than resting macrophages. Surprisingly, the activated macrophages also phagocytose and kill cells infected with unrelated intracellular pathogens such as *Listeria*. Specificity is at the level of the Th1 Cells and Macrophage Activation effector macrophages act nonspecifically; activated macrophages kill intracellular bacteria that normally multiply in nonactivated macrophages or other cell types.

Th1-activated macrophages not only kill pathogen-infected cells but also help destroy tumor cells because tumor cells often produce

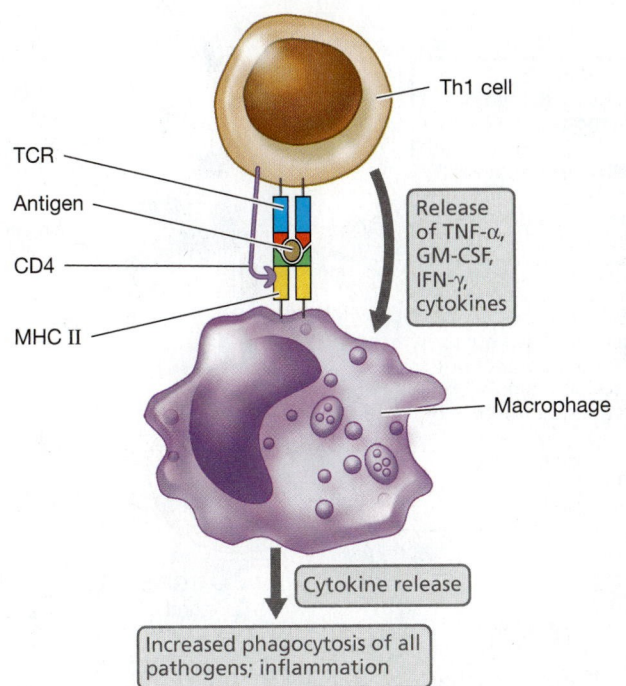

Figure 25.12 **Th1 cells.** Th1 cells (T-inflammatory cells) are activated by antigens presented on macrophages in the context of MHC II protein. Activated Th1 cells produce cytokines that stimulate the macrophages to increase phagocyte activity and promote inflammation.

Table 25.2 T-helper cell subsets

	Th subset			
	Th1	*Th2*	*Th17*	*Treg*
Antigen-presenting cell	Macrophage	B cell	Activated dendritic cell	Nonactivated dendritic cell
Major cytokines produced	IL-2, IFN-γ, TNF-α	IL-4, IL-5	IL-17, IL-6	IL-10, TGF-β
Cellular effects	Activation of T cells (IL-2) and macrophages	Activation of B cells	Activation and recruitment of neutrophils	Suppression of adaptive immune cells
Systemic effects	Cell-mediated immunity	Antibody-mediated immunity	Amplification of innate immunity	Control of Th immunity

tumor-specific antigens not found on normal cells. Tumor cells can be destroyed by macrophages activated by the Th1 cells that react with the tumor-specific antigen. Transplantation rejection, a major problem encountered after organs or tissues are transplanted from one person to another, is also mediated by Th1-activated macrophages. In this case, Th1 cells recognize the nonself MHC proteins of the transplant, triggering macrophage activation and transplant destruction.

Th2 Cells

Th2 cells play a pivotal role in B cell activation and antibody production. As discussed in Section 24.4, B cells make antibodies. Differentiated B cells are coated with antibodies that are antigen receptors. Antigen binds to the B cell antigen receptors but the B cell does not immediately produce soluble antibodies (**Figure 25.13**). The antibody-bound antigen is first taken into the B cell by endocytosis and degraded. Peptides from the degraded antigen are then presented on the B cell's MHC II protein. In this way the B cell serves a dual role, first as an APC, and second as an antibody producer. As an APC, the B cell takes up and processes antigen into peptides and loads them into MHC II. The B cell then presents the MHC II–peptide to a Th2 cell. The Th2 cell responds by producing IL-4 and IL-5, cytokines that activate the B cell. The activated B cell differentiates into

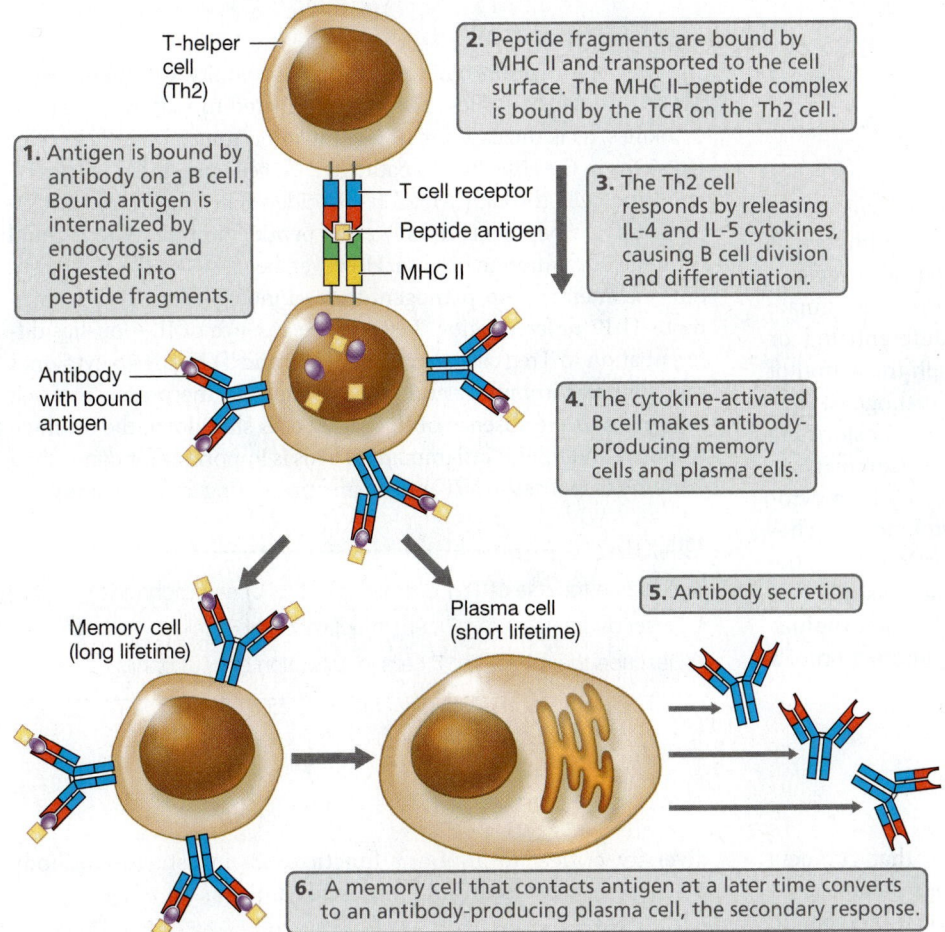

1. Antigen is bound by antibody on a B cell. Bound antigen is internalized by endocytosis and digested into peptide fragments.

2. Peptide fragments are bound by MHC II and transported to the cell surface. The MHC II–peptide complex is bound by the TCR on the Th2 cell.

3. The Th2 cell responds by releasing IL-4 and IL-5 cytokines, causing B cell division and differentiation.

4. The cytokine-activated B cell makes antibody-producing memory cells and plasma cells.

5. Antibody secretion

6. A memory cell that contacts antigen at a later time converts to an antibody-producing plasma cell, the secondary response.

Figure 25.13 **T cell–B cell interaction and antibody production.** B cells interact with antigen and Th2 cells to produce antibodies. B cells initially function as antigen-presenting cells using the antigen-specific Ig receptor to trap antigen. Antigen processing by the B cell and presentation to the Th2 cell induces the Th2 cell to make cytokines that activate the B cell to become an antibody-producing plasma cell or a memory cell.

UNIT 5

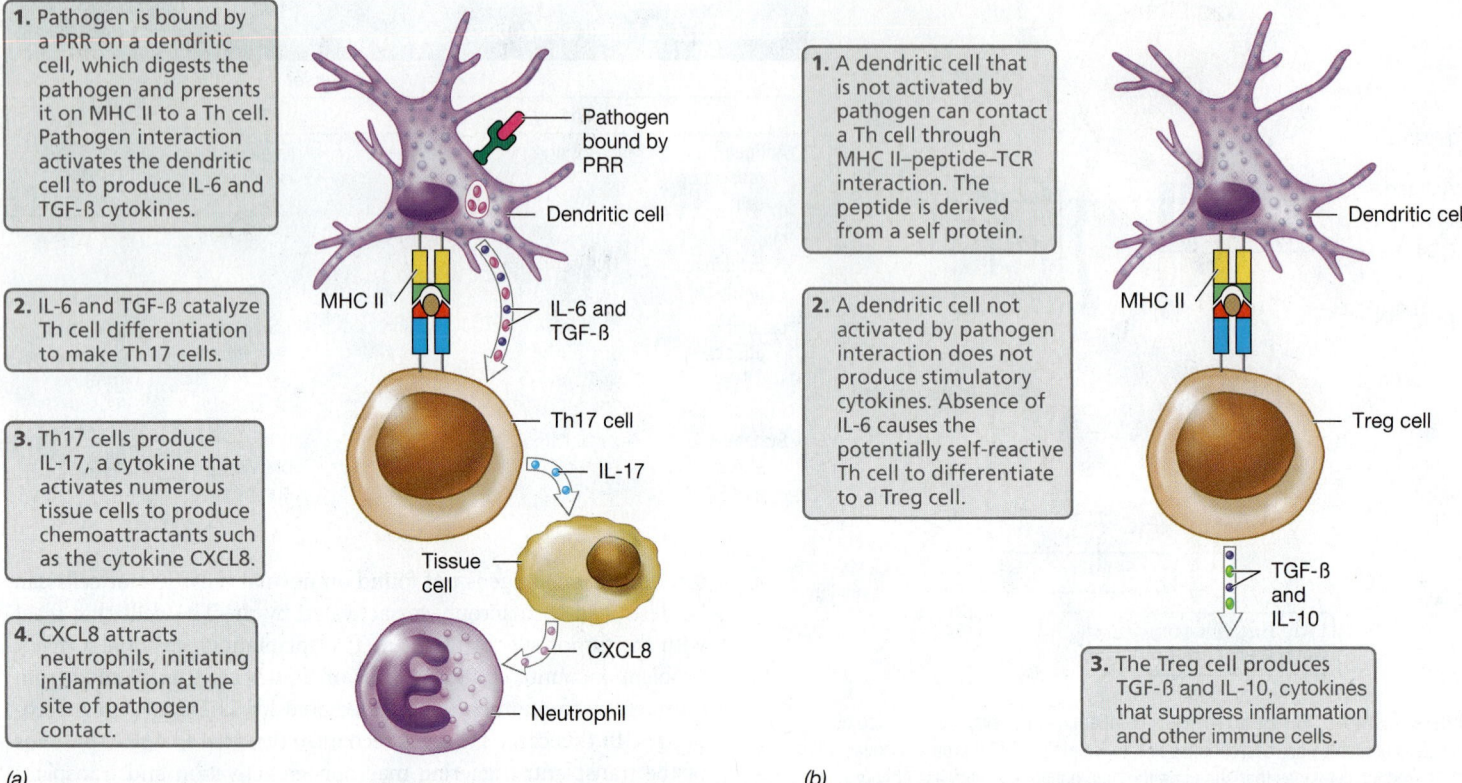

1. Pathogen is bound by a PRR on a dendritic cell, which digests the pathogen and presents it on MHC II to a Th cell. Pathogen interaction activates the dendritic cell to produce IL-6 and TGF-ß cytokines.

2. IL-6 and TGF-ß catalyze Th cell differentiation to make Th17 cells.

3. Th17 cells produce IL-17, a cytokine that activates numerous tissue cells to produce chemoattractants such as the cytokine CXCL8.

4. CXCL8 attracts neutrophils, initiating inflammation at the site of pathogen contact.

Pathogen bound by PRR
Dendritic cell
MHC II
IL-6 and TGF-ß
Th17 cell
IL-17
Tissue cell
CXCL8
Neutrophil

(a)

1. A dendritic cell that is not activated by pathogen can contact a Th cell through MHC II–peptide–TCR interaction. The peptide is derived from a self protein.

2. A dendritic cell not activated by pathogen interaction does not produce stimulatory cytokines. Absence of IL-6 causes the potentially self-reactive Th cell to differentiate to a Treg cell.

3. The Treg cell produces TGF-ß and IL-10, cytokines that suppress inflammation and other immune cells.

Dendritic cell
MHC II
Treg cell
TGF-ß and IL-10

(b)

Figure 25.14 Th17 and Treg cells. *(a)* Th17 cells interact with pathogen-stimulated dendritic cells to draw neutrophils to the pathogen, leading to inflammation and pathogen control. *(b)* Treg cells interact with nonactivated dendritic cells and respond by making immunosuppressive cytokines that control reactions to self antigens.

a plasma cell that produces and secretes antibodies, as we will discuss in Section 25.8.

Th17 and Treg Cells

Antigen presentation by dendritic cells plays a critical role in the development of Th17 and Treg cells. **Th17** cells are important in the first stages of the adaptive immune response; triggered by antigen interactions, Th17 cells recruit neutrophils, innate immune cells important in inflammation. Undifferentiated or *naive* Th cells differentiate into Th17 cells through the action of dendritic cells. When dendritic cells encounter pathogens, they present antigen and secrete two cytokines, IL-6 and transforming growth factor-β (TGF-β), which catalyze differentiation of Th to Th17 cells (**Figure 25.14***a*). Th17 cells then produce IL-17, a cytokine that activates other tissue cells to produce cytokines and chemokines that attract neutrophils to the site of infection. Thus, the function of Th17 cells is to produce IL-17, starting a cascade that draws neutrophils to infection sites. By recruiting neutrophils, Th17 cells amplify innate immunity triggered by interaction of a pathogen with the dendritic cell.

Treg cells are important in control of immunity. Undifferentiated Th cells remain so unless they are stimulated to mature by certain cytokines, as is the case for IL-6 stimulation to produce Th17 cells. However, in the absence of a pathogen, Th cells can still interact with dendritic cells through MHC II–peptide–TCR (Figure 25.14*b*). In this case, the peptide is usually a self peptide; immunity to it could cause an autoimmune disease. However, because the dendritic cells did not interact with pathogen, they cannot produce IL-6 to promote Th17 differentiation. Instead, the absence of IL-6 pushes differentiation to Treg cells that make IL-10 and TGF-β, two cytokines that suppress immunity and inflammation. In the presence of self antigens and the absence of IL-6, Treg cells shut down the immune response and inhibit inflammation. This is important for controlling immune responses to self antigen and preventing autoimmunity.

MINIQUIZ
- Describe the role of Th1 cells in activation of macrophages.
- Describe the role of Th2 cells in activation of B cells.
- Describe the role of Th17 cells in activation of neutrophils.

IV • Antibodies and Immunity

Antibodies provide antigen-specific immunity that protects against extracellular pathogens and dangerous soluble proteins such as toxins. After considering the structure and genetic diversity critical to antibody function, we investigate antibody mechanisms that neutralize or destroy antigens.

25.7 Antibody Structure

Antibodies, also called *immunoglobulins* (*Ig*), are proteins that interact specifically with antigenic epitopes. They are found in the serum and other body fluids such as mucosal secretions and milk. Serum containing antigen-specific antibodies is called *antiserum* (⟳ Section 24.6). Immunoglobulins are separated into five major classes based on their physical, chemical, and immunological properties: *IgG, IgA, IgM, IgD,* and *IgE* (Table 25.3).

Immunoglobulin G Structure

IgG is the most common circulating antibody, comprising about 80% of the serum immunoglobulins. IgG is composed of four polypeptide chains (Figure 25.15). Disulfide bridges (S—S bonds) connect the individual chains. In each IgG protein, two identical light chains are paired with two identical heavy chains. Each light chain has about 220 amino acids, and each heavy chain has about 440 amino acids. Each heavy chain interacts with a light chain to form a functional antigen-binding site. An IgG antibody, therefore, is *bivalent* because it has two identical sites that bind two identical epitopes.

Heavy and light chains consist of a series of distinct protein domains of about 110 amino acids (Figure 25.15). A heavy-chain variable domain is connected to three constant domains about 110 amino acids long. The amino acid sequence in the variable domain differs in each different antibody. The variable domain binds antigen. The three constant domains of each heavy chain are identical in all Ig molecules of a given class. Each light chain consists of one variable and one constant domain. The variable domain of a light chain interacts with the variable domain of a heavy chain to bind antigen. The amino acid sequence in the constant domain is the same in light chains of the same type.

The Antigen-Binding Site

The antigen-binding site of IgG and all other antibodies form by cooperative interaction between the variable domains of heavy and light

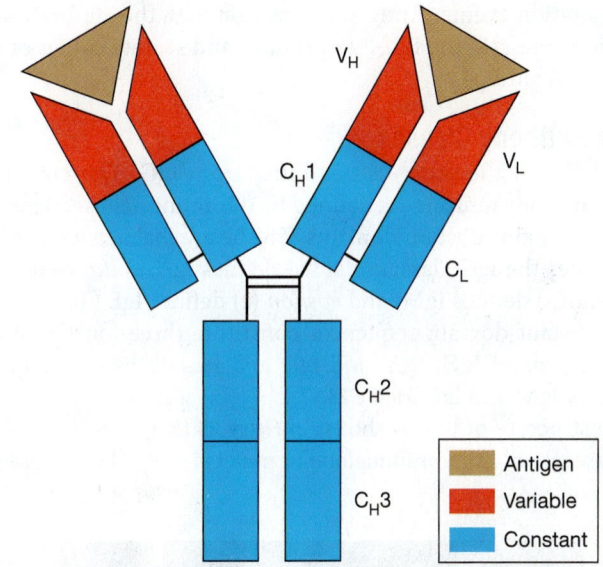

Figure 25.15 Immunoglobulin G structure. IgG consists of two heavy chains (50,000 molecular weight) and two light chains (25,000 molecular weight), with a total molecular weight of 150,000. One heavy and one light chain interact to form an antigen-binding unit. The variable domains of the heavy and light chains (V_H and V_L) bind antigen. The constant domains (C_H1, C_H2, C_H3, C_L) are identical in all IgG proteins. The chains are covalently joined with disulfide bonds.

chains (Figure 25.16). The variable domains of the two chains interact, forming a receptor that binds antigen strongly but noncovalently. The measurable strength of binding of antibody to antigen is called *binding affinity*. A high-affinity antibody binds tightly to antigen.

Each individual's immune system has the capacity to recognize, or bind, countless antigens, and each antigen is bound by a unique antigen-binding site. To accommodate all possible antigens, each individual can produce billions of different antigen-binding sites, each encoded by a gene formed through somatic recombination

---

Table 25.3 Properties of human immunoglobulins

Class/H chain[a]	Molecular weight/ formula[b]	Serum (mg/ml)	Antigen-binding sites	Properties	Distribution
IgG γ	150,000 2(H + L)	13.5	2	Major circulating antibody; four subclasses: IgG1, IgG2, IgG3, IgG4; IgG1 and IgG3 activate complement	Extracellular fluid; blood and lymph; crosses placenta
IgM μ	970,000 (pentamer) 5[2(H + L)] + J	1.5	10	First antibody to appear after immunization; strong complement activator	Blood and lymph; monomer is B cell surface receptor
	175,000 (monomer) 2(H + L)	0	2		
IgA α	150,000 2(H + L)	3.5	2	Important circulating antibody	Secretions (saliva, colostrum, cellular and blood fluids); monomer in blood and dimer in secretions
	385,000 (secreted dimer) 2[2(H + L)] + J + SC	0.05	4	Major secretory antibody	
IgD δ	180,000 2(H + L)	0.03	2	Minor circulating antibody	Blood and lymph; B lymphocyte surfaces
IgE ε	190,000 2(H + L)	0.00005	2	Involved in allergic reactions and parasite immunity	Blood and lymph; C_H4 binds to mast cells and eosinophils

[a]All immunoglobulins may have either λ or κ light chain types, but not both.
[b]Based on the number and arrangement of heavy (H) and light (L) chains in each functional molecule. J is joining protein present in serum IgM and secretory IgA. SC is the secretory component found in secreted IgA.

and mutation events. Antigen interaction with the antibodies on a B cell stimulates the B cell to produce and secrete copies of the preformed antibody.

Other Antibody Classes

Antibodies of the other classes differ from IgG. The class of a given antibody molecule is defined by the amino acid sequence of its heavy-chain constant domains. The heavy chain called *gamma* (γ) defines the IgG class; *alpha* (α) defines IgA; *delta* (δ) defines IgD; *mu* (μ) defines IgM; and *epsilon* (ε) defines IgE (Table 25.3). The constant domain sequences constitute three-fourths of the heavy chains of IgG, IgA, and IgD and four-fifths of the heavy chains of IgM and IgE (**Figure 25.17**).

The structure of IgM is shown in **Figure 25.18**. IgM usually forms an aggregate of five immunoglobulin molecules attached by at least

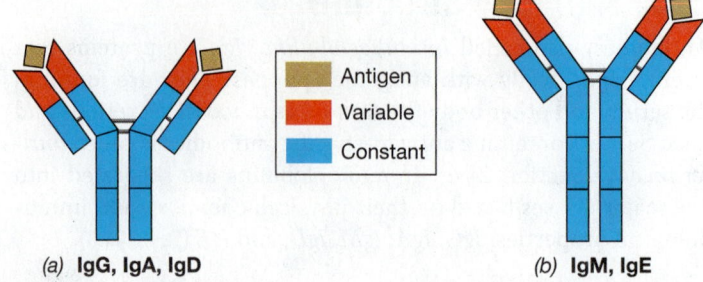

(a) **IgG, IgA, IgD**   (b) **IgM, IgE**

Figure 25.17 **Immunoglobulin classes.** All classes of Igs have V_H and V_L that bind antigen. *(a)* IgG, IgA, and IgD have three constant domains. *(b)* The heavy chains of IgM and IgE have a fourth constant domain.

one J (joining) chain. IgM is the first class of Ig made in a typical immune response to a bacterial infection, but IgMs generally have low *affinity* (binding strength) for antigen. Overall antigen-binding strength is enhanced, however, by the high *valence* of the pentameric IgM molecule; ten binding sites are available for interaction with antigen (Table 25.3 and Figure 25.18). The combined strength of antigen binding by the multiple antigen-binding sites on IgM is called *avidity*. Thus, IgM has *low* affinity but *high* avidity for antigen. Up to 10% of serum antibodies are IgM. IgM monomers are the class of antibodies found on the surface of B cells.

Dimers of IgA are present in body fluids such as saliva, tears, breastmilk colostrum, and mucosal secretions from the gastrointestinal, respiratory, and genitourinary tracts. These mucosal surfaces are associated with mucosa-associated lymphoid tissue (MALT) that produces IgA. In an average adult, the mucosal

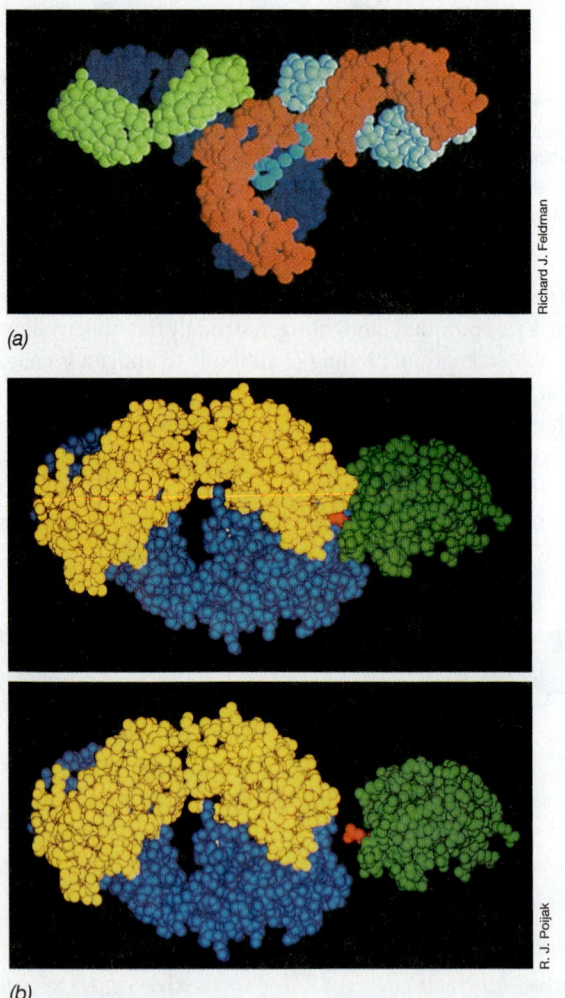

(a)

(b)

Richard J. Feldman

R. J. Poljak

Figure 25.16 **Immunoglobulin structure and the antigen-binding site.** *(a)* Space-filling model of an IgG molecule. The heavy chains are red and dark blue. The light chains are green and light blue. *(b)* Space-filling model of the binding interactions between an antigen and an immunoglobulin. The antigen (lysozyme) is green. The variable domain of the Ig heavy chain is blue; the light-chain variable domain is yellow. The amino acid in red is a glutamine residue in the enzyme lysozyme. The glutamine fits into a pocket on the Ig molecule, but overall antigen–antibody interaction involves contacts between many other amino acids on the surfaces of both the Ig and the antigen. Reprinted with permission from *Science* 233:747 (1986) ©AAAS.

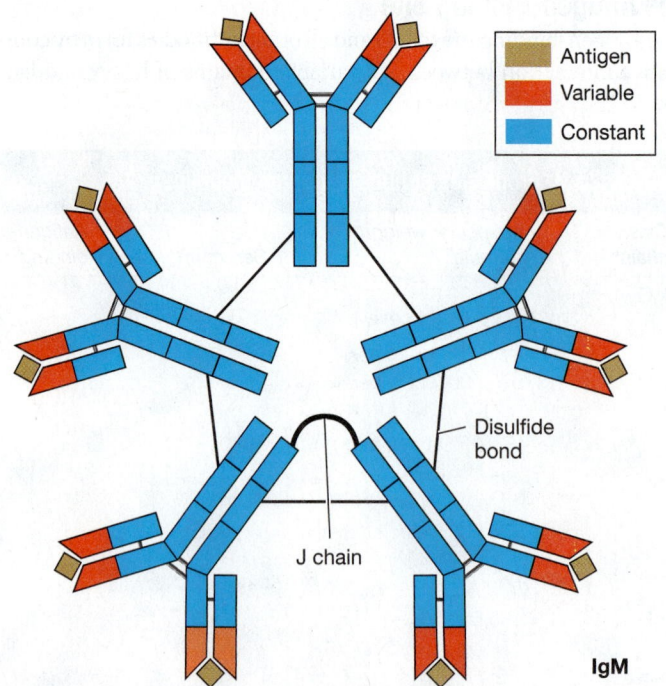

IgM

Figure 25.18 **Immunoglobulin M.** IgM is found in serum as a pentameric protein consisting of five IgM proteins covalently linked to one another via disulfide bonds and a J chain protein. Because it is a pentamer, IgM can bind up to ten antigens, as shown.

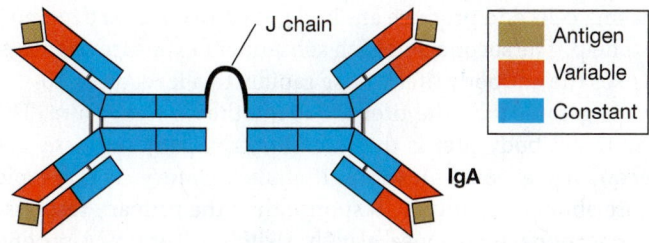

Figure 25.19 **Immunoglobulin A.** Secretory IgA (sIgA) is often found in body secretions as a dimer consisting of two IgA proteins covalently linked to one another by a joining (J) chain protein. A secretory component, not shown, aids in transport of IgA across mucosal membranes.

surfaces total about $400 \, m^2$ (skin has about $6 \, m^2$), and large amounts of secretory IgA are produced—about 10 g per day. By contrast, the serum IgG produced in an individual is about 5 g per day, only half as much as IgA. Secretory IgA has two IgA molecules covalently linked by a J chain peptide and a protein called the *secretory component* that aids in transport of IgA across membranes (**Figure 25.19**). IgA is also present in serum as a monomer (Table 25.3).

IgE is found in extremely small amounts in serum (about 1 of every 50,000 serum Ig molecules is IgE). Most IgE is bound to cells. For example, IgE antibody, through its constant region, binds eosinophils, arming these granulocytes to target eukaryotic parasites like schistosomes and other worms. IgE also binds to tissue mast cells. Binding of antigen to the variable antigen-binding portions of IgE on mast cells causes release of the mast cell contents in a process called degranulation. Degranulation of mast cells triggers immediate-type hypersensitivities (allergies, ↺ Section 24.8). The molecular weight of IgE is significantly higher than most other Igs (Table 25.3) because, like IgM, IgE has a fourth constant domain (Figure 25.17). The additional constant domain of IgE binds to eosinophils and mast cells, a critical step for activating the protective and allergic reactions associated with these cell types (↺ Figure 24.17).

IgD (Figure 25.17), present in serum in low concentrations, has no known unique immune protective function. However, IgD, like IgM, is abundant on the surfaces of B cells, especially memory B cells.

MINIQUIZ --

- Identify the antibody heavy- and light-chain domains that bind antigen.
- Differentiate among antibody classes using structural characteristics, expression patterns, and functional roles.

25.8 Antibody Production

A predictable sequence of events leads to antibody production after antigen exposure. At the genetic level, B cells and T cells use gene recombination and mutation mechanisms to generate rearranged genes encoding a limitless number of unique antigen-binding receptors. At the cellular level, complex interactions between B cells and T cells produce antibodies that provide effective antigen-specific immunity.

Generation of Antigen Receptor Diversity

Each individual is capable of producing billions of different antibodies and TCRs, each aimed to interact with one of the countless antigens in our environment. How does the immune system produce all of these antigen-specific proteins? Immune receptor diversity is generated by a mechanism found only in B and T cells. Antibody production starts with stepwise rearrangements of the Ig-encoding genes. During development of B cells in the bone marrow, both heavy-chain and light-chain genes rearrange. The genes are recombined—individual gene pieces are mixed and matched in various combinations—by gene splicing and rearrangements in the differentiating B cells, a process called *somatic recombination*.

Figure 25.20 shows how rearrangement of a single human light-chain kappa gene results in formation of one light-chain protein. The heavy-chain gene complex has even more gene segments, allowing for more recombinations and potential heavy chains (↺ Figure 26.8). Only one rearrangement, however, is produced in any given B cell. The final result is a single functional heavy-chain gene and a single functional light-chain gene. Each of these rearranged genes is transcribed and translated to make an antibody consisting of two heavy-chain proteins and two light-chain proteins. The unique antibody is then expressed on the surface of the B cell.

Antigen exposure is necessary to stimulate the B cell to produce soluble antibodies and differentiate to plasma cells that will produce more soluble antibody copies. In addition, antigen exposure induces mutations at an accelerated rate in antibody genes, a phenomenon called *somatic hypermutation*. This process further modifies and diversifies the antibodies produced.

The large number of possible gene rearrangements coupled with the somatic hypermutation events after antigen exposure ensure almost unlimited antibody diversity. Similar rearrangements also occur during

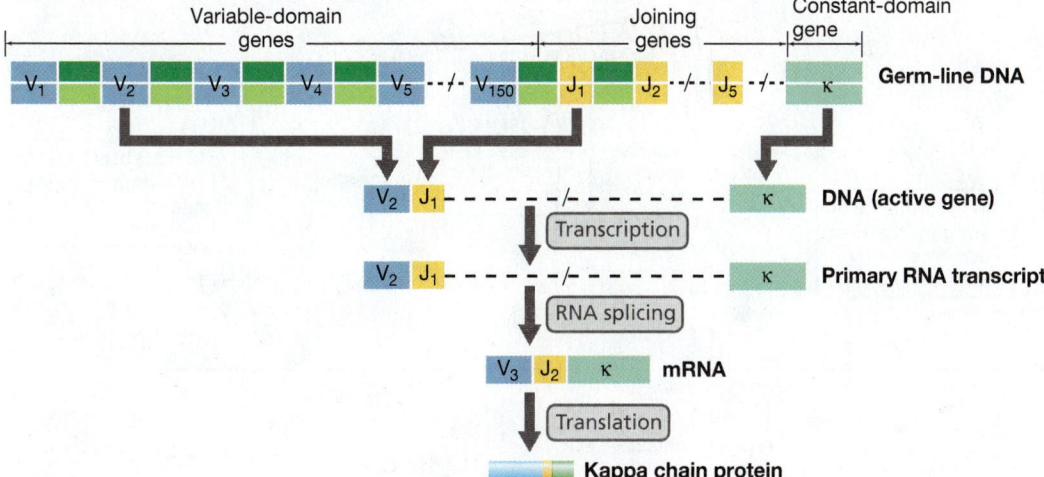

Figure 25.20 **Immunoglobulin kappa chain gene rearrangement in human B cells.** The gene segments are arranged in tandem in the kappa (κ) light-chain genes on chromosome 2. DNA rearrangements are completed in the maturing B cell. Any one of the 150 V (variable) sequences may combine with any one of the 5 J sequences. Thus, 750 (150 × 5) recombinations are possible, encoding 750 distinct kappa variable domains, but only one productive rearrangement occurs in each cell.

T cell development, resulting in the generation of considerable diversity in TCRs. T cells, however, do not use hypermutation to expand diversity.

Antibody Production and Immune Memory

Starting with a B cell, antibody production begins with antigen exposure and culminates with the production and secretion of an antigen-specific antibody according to a defined sequence:

1. Antigens are introduced via the lymphatic and blood circulatory systems to secondary lymphoid organs such as lymph nodes, spleen, and MALT (⮌ Section 24.1 and Figure 24.2). The route of antigen exposure influences the class of the antibodies produced. Antigen introduced into the bloodstream by injection or infection travels to the spleen, where IgM, IgG, and serum IgA antibodies are formed. Antigen introduced subcutaneously, intradermally, topically, or intraperitoneally is carried by the lymphatic system to the nearest lymph nodes, again stimulating production of IgM, IgG, and serum IgA. Antigen introduced to mucosal surfaces is delivered to the nearest MALT. For example, antigen delivered orally is delivered to the MALT in the intestinal tract, preferentially stimulating production of secretory IgA in the gut.

2. Following the initial antigen contact, each antigen-stimulated B cell multiplies and differentiates to form antibody-secreting plasma cells and memory cells (Figure 25.13). *Plasma cells* are relatively short-lived (less than 1 week), but they produce and secrete large amounts of mostly IgM antibody in the *primary antibody response* (**Figure 25.21**). After a short latent period, antibody appears in the blood and a gradual increase in *antibody titer* (antibody quantity) occurs. As antigen disappears, the primary antibody response decreases slowly.

3. The **memory B cells** generated by the initial exposure to antigen may live for years. A reexposure to the same antigen immediately stimulates memory B cells to differentiate to plasma cells and produce antibody; memory B cells need no T cell help. The second and each subsequent exposure to antigen causes the antibody titer to rise rapidly to a level often 10–100 times greater than the titer following the first exposure. This rise in antibody titer is the *secondary antibody response*. The secondary response illustrates immune memory: a more rapid, more abundant antibody response than the primary response. The secondary response usually switches from IgM production to another antibody class. In serum, the most common antibody class switches are from IgM to IgG and IgA. This phenomenon is called *class switching* (Figure 25.21).

4. The titer slowly decreases over time, but subsequent exposures to the same antigen can cause another memory response. The rapid and strong memory response is the basis for the immunization procedure known as a "booster shot" (for example, the yearly rabies shot given to domestic animals). Periodic reimmunization maintains high levels of memory B cells and circulating antibody specific for a certain antigen, providing long-term active protection against individual infectious diseases.

MINIQUIZ --

- Explain how a limited numbers of immunoglobulin genes can be rearranged to form almost limitless numbers of antibody proteins.
- Explain the rationale for periodic revaccination in children and adults.

25.9 Antibodies, Complement, and Pathogen Destruction

Complement is a group of sequentially interacting proteins, many with enzymatic activity, important in both innate and adaptive immunity. Complement activity can be triggered by innate or adaptive mechanisms.

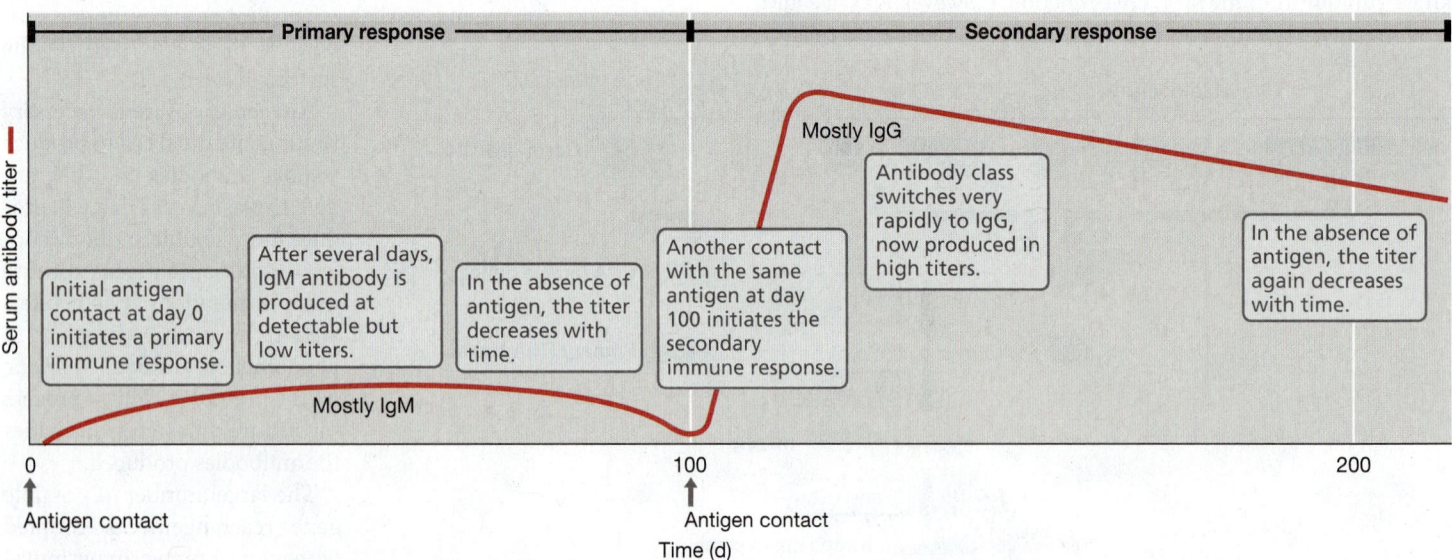

Figure 25.21 Primary and secondary antibody responses in serum. The antigen injected at day 0 and day 100 must be identical to induce a secondary response. The secondary response, also called a booster response, may be more than 10-fold greater than the primary response. Note the class switch from IgM production in the primary response to IgG production in the secondary response.

Classical Complement Activation and Cell Damage

Complement proteins react in a sequence after exposure to antigen–antibody complexes on a target cell. Complement activation may result in membrane damage and lysis of the target cell or enhanced phagocytosis of the target cell, a process called *opsonization*. Serum contains complement, and most antigen-bound IgG or IgM antibodies can bind complement (Table 25.3).

The individual proteins of complement are designated C1, C2, C3, and so on. Classical activation of complement occurs when IgG or IgM antibodies bind antigens, especially on cell surfaces. The antibodies are said to *fix* (bind) the ever-present complement proteins. The complement proteins react in a defined sequence, with activation of one complement protein leading to activation of the next, and so on. The key steps, shown in

Figure 25.22, start with binding of antibody to antigen (initiation) and binding of C1 components (C1q, C1r, and C1s) to the antibody–antigen complex, leading to C4-C2 deposition at an adjacent membrane site. This complex is called a *C3 convertase*, an enzyme that cleaves C3 to C3a and C3b. C3b then binds to the convertase, forming a complex that initiates a C5-C6-C7 interaction at a second membrane site. C8 and C9 are then deposited with the C5-C6-C7. The membrane-bound C5–9 components, called the *membrane attack complex* (*MAC*), insert through the membrane to form a pore and allow the contents to leak out, lysing the cell (**Figure 25.23**).

By-products of complement activation include chemoattractants called *anaphylatoxins*; these molecules cause inflammatory reactions at the site of complement deposition. For example, when C3 is cleaved to C3a and C3b, C3b fixes to the target cell,

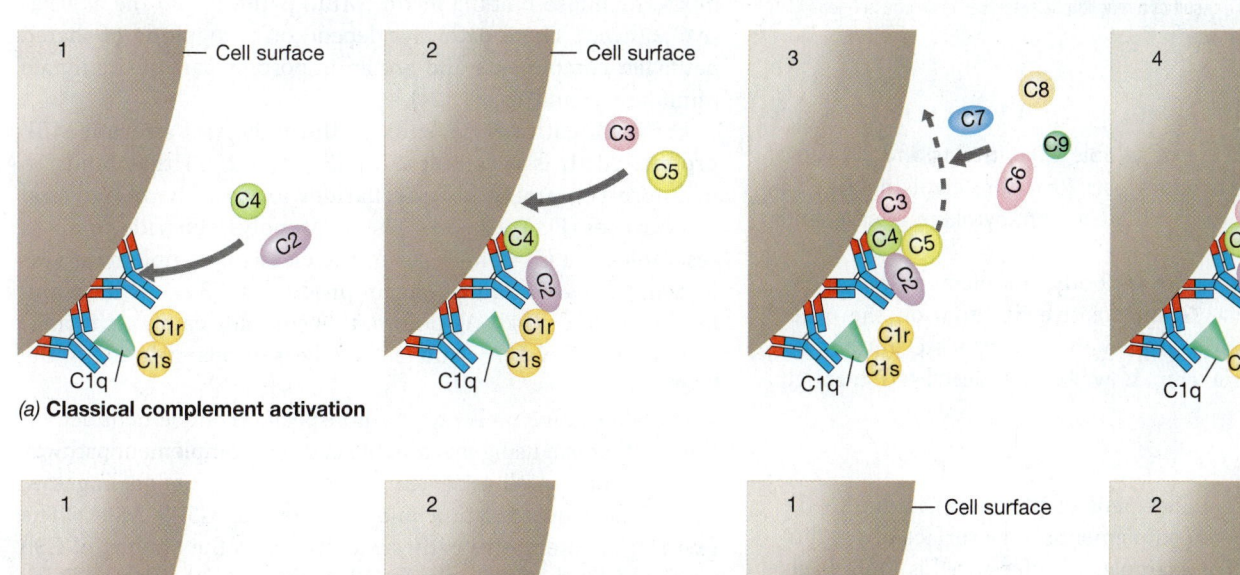

(a) **Classical complement activation**

(b) **Mannose-binding lectin (MBL) pathway activation**

(c) **Alternative pathway activation**

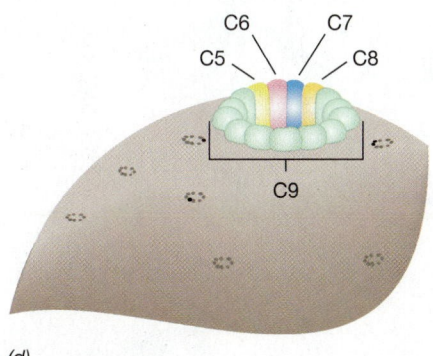

(d)

Figure 25.22 Complement activation. *(a)* The sequence, orientation, and activity of the components of the classical complement pathway as they interact to lyse a cell. 1. Binding of the antibody and the C1 protein complex (C1q, C1r, and C1s). 2. The C42 complex interacts with C3. 3. The C423 complex activates C5, which then binds an adjacent membrane site. 4. Sequential binding of C6, C7, C8, and C9 to C5 produces a pore in the membrane. C5–C9 is the membrane attack complex (MAC). *(b)* The mannose-binding lectin (MBL) pathway. MBL binds to the mannose on the bacterial membrane, and anchors formation of C423. The membrane-bound C3 activates C5, as in 3 above, and initiates formation of the MAC (4 above). *(c)* The alternative pathway. C3 bound to the cell attracts protein B and D, which activates C3. The C3B complex is further stabilized on the membrane by factor P (properdin). C3B then acts on C3 in the blood, causing more C3 to bind to the membrane. Bound C3 then activates C5, as in step 3 of the classical activation pathway above, and initiates formation of the MAC (4 above). *(d)* A schematic view of the pore formed by complement components C5 through C9.

UNIT 5

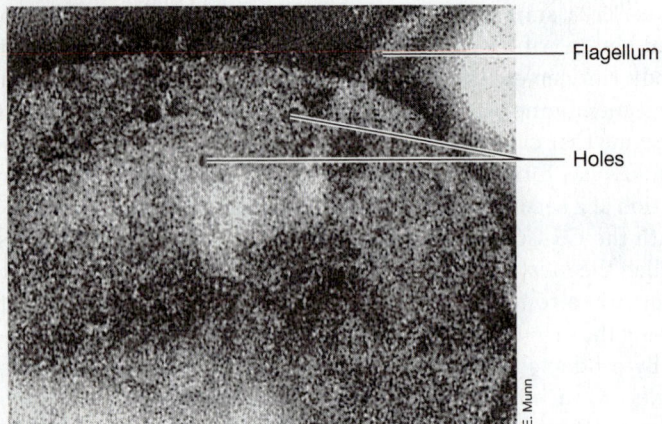

Flagellum

Holes

E. Munn

Figure 25.23 Complement activity on bacterial cells. This electron micrograph of *Salmonella enterica* serovar Paratyphi shows holes that formed in the bacterial cell envelope as a result of a reaction between cell envelope antigens, specific antibody, and complement.

as outlined above. Release of soluble C3a attracts and activates phagocytes, increasing phagocytosis. Reactions involving the C5a cleavage product lead to T cell attraction, cytokine release, and inflammation.

When activated by specific antibody, complement lyses many gram-negative bacteria. Gram-positive bacteria, on the other hand, are not lysed by complement and specific antibodies. Gram-positive bacteria can, however, be destroyed through opsonization.

Opsonization

Opsonization is the enhancement of phagocytosis due to the deposition of antibody or complement on the surface of a pathogen or other antigen. For example, a bacterial cell is more likely to be phagocytosed when antibody binds antigen on its surface. If complement binds to an antibody–antigen complex on the cell surface, the cell is even more likely to be ingested. This is because most phagocytes, including neutrophils, macrophages, and B cells, have antibody receptors (FcR) as well as C3 receptors (C3R). These receptors bind the antibody constant domain and C3 complement protein, respectively. Normal phagocytic processes are enhanced about 10-fold by antibody–FcR interactions and amplified another 10-fold by C3–C3R interactions. Antibodies bound to surface antigens on gram-positive *Bacteria* use the classical complement pathway to promote opsonization, leading to enhanced phagocytosis and pathogen destruction.

Mannose-Binding Lectin and Alternative Pathways

In addition to complement activation by the classical pathway, C3 can be deposited on membranes and the MAC can be activated by the mannose-binding lectin (MBL) pathway and the alternative pathway. These pathways depend on recognition of shared pathogen components and are an important part of the innate immune system (**Figure 25.24**).

The MBL pathway depends on the activity of a serum MBL protein. MBL is a soluble PRR (Section 25.1) that binds to mannose-containing polysaccharides found only on bacterial cell surfaces (Figure 25.24). The MBL–polysaccharide complex resembles the C1 complexes of the classical complement system and fixes C4 and C2, again producing C3 convertase and binding C3b to C42. As before, this complex catalyzes formation of the C5–9 MAC and leads to lysis or opsonization of the bacterial cell.

The alternative pathway is a nonspecific complement activation mechanism using many of the classical complement pathway components as well as several unique serum proteins. Together they induce opsonization and activate the C5–9 MAC. The first step in alternative pathway activation is the binding of C3b produced by the classical or MBL pathway to the bacterial cell

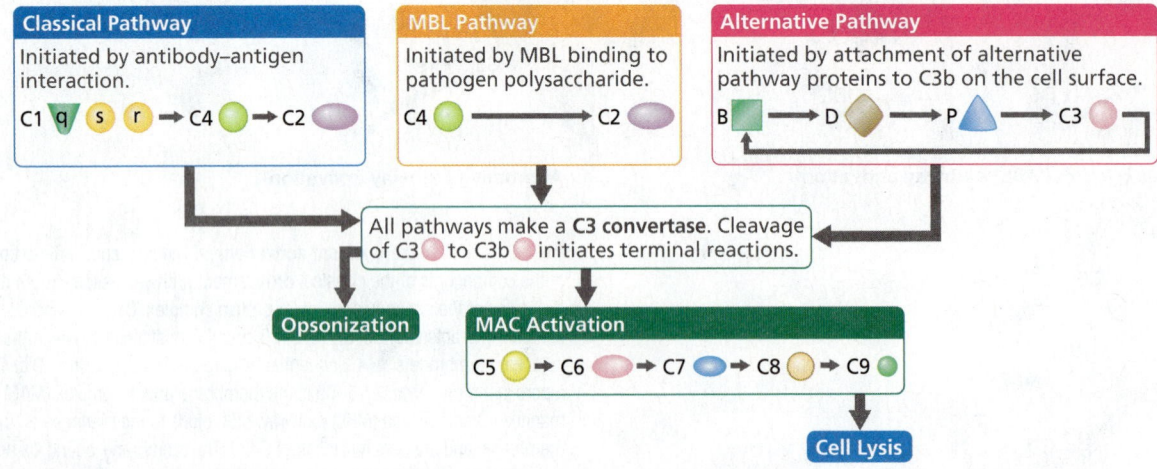

Figure 25.24 Activation of the complement system. Proteins that activate the complement system in the classical, mannose-binding lectin (MBL), and alternative pathways are shown. The proteins interact in an ordered sequence as shown by the arrows, with each pathway independently leading to production of a C3 convertase that cleaves C3 to produce C3b, a protein necessary for initiation of the terminal complement-mediated outcomes of opsonization or cell lysis.

surface. C3b on the membrane can then bind the alternative pathway serum protein factor B, which is cleaved by factor D to give soluble Ba and Bb. The C3bBb complex is another C3 convertase. Factor P, or properdin, also binds to bacterial cell walls. P may join C3bBb to form C3bBbP. This is a very stable C3 convertase because it is fixed on the cell, as are the C3 convertases produced by the classical and MBL pathways (Figure 25.24). C3bBbP then attracts more C3, which is deposited on the membrane, initiating the same reaction as the membrane-bound C423 complex of the classical complement pathway. The result is formation of the C5–9 MAC and cell destruction or enhanced opsonization via phagocyte C3 receptors.

Both the alternative pathway and the MBL pathway nonspecifically target bacterial invaders and lead to activation of the membrane attack complex and enhanced opsonization via formation of stable C3 convertases. MBL, factors B, D, and P, and classical complement proteins are part of the innate immune response. Neither the alternative pathway nor the MBL pathway requires prior antigen exposure or the presence of antibodies for activation. Through the alternative and MBL pathways, C3 convertase triggers formation of the C5–9 MAC or enhances opsonization via C3 receptors on phagocytes.

MINIQUIZ --

- Which antibody classes bind complement?
- What is opsonization and how does opsonization help prevent bacterial disease?
- Why are mannose-binding lectin and alternative pathways considered part of the innate immune system?

BIG IDEAS

25.1 • Phagocytes use their membrane-bound pattern recognition receptors (PRRs) to recognize pathogen-associated molecular patterns (PAMPs). PRR–PAMP interactions activate phagocyte killing of pathogens and cause inflammation. Many pathogens have developed mechanisms to inhibit phagocytes.

25.2 • The adaptive immune response is characterized by *specificity* for the antigen, the ability to respond more vigorously when reexposed to the same antigen (*memory*), and the acquired inability to interact with self antigens (*tolerance*).

25.3 • Immunogens are foreign macromolecules that induce an immune response. Immunogens initiate an immune response when introduced into a host. Antigens are molecules recognized by antibodies or TCRs. Antibodies recognize linear and conformational epitopes; TCRs recognize linear peptide epitopes.

25.4 • T cells interact with antigen-bearing cells including dedicated APCs and pathogen-infected cells. TCRs bind peptide antigens presented by MHC proteins on infected cells or APCs. These interactions activate T cells to kill antigen-bearing cells or to promote inflammation or antibody production.

25.5 • T-cytotoxic (Tc) cells recognize antigens on virus-infected host cells and tumor cells through antigen-specific TCRs. Antigen-specific recognition triggers killing via perforins and granzymes. NK cells do not require antigen, but respond to stress proteins on virus-infected cells and tumor cells, using perforins and granzymes to kill the targets.

25.6 • Through the action of cytokines, Th1 inflammatory cells activate macrophage effector cells; Th2 helper cells activate B cells. Th17 cells are activated by pathogen-activated dendritic cells and recruit neutrophils. Treg cells produce cytokines that suppress adaptive immunity.

25.7 • Each immunoglobulin (antibody) protein consists of two heavy and two light chains. The antigen-binding site is formed by the interaction of the variable regions of one heavy and one light chain. Each antibody class has different structural characteristics, expression patterns, and functional roles.

25.8 • Antibody production is initiated when an antigen contacts an antigen-specific B cell. The antigen-reactive B cell processes the antigen and presents it to an antigen-specific Th2 cell. The Th2 cell becomes activated, producing cytokines that signal the antigen-specific B cell to clonally expand and differentiate to produce antibodies. Activated B cells live for years as memory cells and can rapidly expand and differentiate to produce high titers of antibodies after reexposure to antigen.

25.9 • The complement system catalyzes bacterial opsonization and cell destruction. Complement is triggered by antibody interactions or by interactions with nonspecific activators such as mannose-binding lectin. Complement activation may be a product of either innate or adaptive immunity. Complement may enhance phagocytosis or cause target cell lysis.

REVIEW OF KEY TERMS

Antigen a molecule capable of interacting with specific components of the immune system

CD4 coreceptor a protein found exclusively on Th cells that interacts with MHC II on an antigen-presenting cell

CD8 coreceptor a protein found exclusively on Tc cells that interacts with MHC I on a target cell

Complement a series of proteins that react in a sequential manner with antibody–antigen complexes, mannose-binding lectin, or alternative activation pathway proteins to potentiate or amplify target cell destruction

Domain a region of a protein having a defined structure and function

Epitope the portion of an antigen that reacts with a specific antibody or T cell receptor

Immunogen a molecule capable of eliciting an immune response

MHC class I protein an antigen-presenting molecule found on all nucleated vertebrate cells

MHC class II protein an antigen-presenting molecule found on macrophages, B cells, and dendritic cells

Memory B cell a long-lived cell responsive to a specific antigen

Natural killer (NK) cell a specialized lymphocyte that recognizes and destroys foreign cells or infected host cells in a nonspecific manner

Opsonization the enhancement of phagocytosis due to the deposition of

antibody or complement on the surface of a pathogen or other antigen

T-cytotoxic cell (Tc) a lymphocyte that interacts with MHC I–peptide complexes through its T cell receptor and produces cytotoxins that kill the interacting target cell

T-helper (Th) cell a lymphocyte that interacts with MHC II–peptide complexes through its T cell receptor and produces cytokines that act on other cells. Th subsets include **Th1** cells that activate macrophages; **Th2** cells that activate B cells; **Th17** cells that activate neutrophils; and **Treg** cells that suppress adaptive immunity

Toll-like receptor (TLR) a pattern recognition receptor on phagocytes that interacts with a pathogen-associated molecular pattern

REVIEW QUESTIONS

1. Identify some pathogen-associated molecular patterns (PAMPs) that are recognized by pattern recognition receptors (PRRs). What is the significance of the interactions between these molecules? (Section 25.1)

2. Explain how phagocytes engulf and kill microorganisms, with particular attention to oxygen-dependent mechanisms. (Section 25.1)

3. Identify the three defining characteristics of the adaptive immune response. (Section 25.2)

4. What molecules induce immune responses? What properties are necessary for a molecule to induce an immune response? (Section 25.3)

5. Describe the basic structure of class I and class II major histocompatibility complex (MHC) proteins. In what functional ways do they differ? (Section 25.4)

6. Differentiate between Tc cells and NK cells. What is the activation signal for each cell type? (Section 25.5)

7. How do Th cells differ from Tc cells? Differentiate between the functional roles of Th1, Th2, Th17, and Treg cells. (Section 25.6)

8. Describe the structural and functional differences among the five major classes of antibodies. (Section 25.7)

9. Identify the cell interactions in production of antibodies by B cells. (Section 25.8)

10. Describe the complement system. Is the order of protein interactions important? Why or why not? Identify the components of the mannose-binding lectin pathway for complement activation. Identify the components of the alternative pathway for complement activation. (Section 25.9)

APPLICATION QUESTIONS

1. Describe the potential problems that would arise if a person had an acquired inability to phagocytose pathogens. Could the person survive in a normal environment such as a college campus? What defects in the phagocyte might cause lack of phagocytosis? Explain.

2. Specificity and tolerance are necessary qualities for an adaptive immune response. However, memory seems to be less critical, at least at first glance. Define the role of immune memory and explain how the production and maintenance of memory cells might benefit the host in the long term. Is memory a desirable trait for innate immunity? Explain.

3. What problems would arise if a person had a hereditary deficiency that resulted in an inability to present antigens to Tc cells? What

would the problems be if the person had a deficiency in presenting antigen to Th1 cells? To Th2 cells? To all T cells? What molecules might be deficient in each situation? Could a person having any one of these deficiencies survive in a normal environment? Explain for each.

4. Antibodies of the IgA class are probably more prevalent than those of the IgG class. Explain this and define the benefits this may have for the host.

5. Do you agree with the following statement? Complement is a critical component of antibody-mediated defense. Explain your answer. What might happen to persons who lack complement component C3? C5? Factor B (alternative pathway)? Mannose-binding lectin (MBL)?

26 · Molecular Immunology

microbiology**now**

Ancient Hominids Helped Shape Modern Immunity

Genes encoding the human immune system class I major histocompatibility proteins (also called the human leukocyte antigens, or HLAs) are highly polymorphic, with several hundred alleles at some loci. HLA gene products are critically important for presenting antigens to T cells in early steps of the immune response. We can trace an individual's ancestry by inspecting that person's spectrum of HLA genes, which is also one method of proving (or disproving) paternity. But in exciting new research on human origins, HLA genes are being used to track the ancestry and migration of ancient humans.[1]

The research study suggests that some HLA genes in modern humans were acquired by early human interbreeding with at least two related hominid groups that later became extinct, the Neanderthals and Denisovans. The Denisovans, a lineage recently discovered in Siberia, were a group of hominids closely related to the Neanderthals (photo) that populated large parts of Eurasia.

Denisovan HLA genes are found today in contemporary humans from South Pacific islands and in the Aboriginal peoples of Australia. The implication is that the Denisovans lived in Eurasia and interbred with humans as they migrated out of Africa and eventually populated the South Pacific islands and beyond. However, since other modern Asians do not have Denisovan HLAs, a second migration must have occurred, one that did not interbreed with Denisovans, presumably because they were extinct.

Denisovan HLA genes are not found in African peoples. Thus, our human ancestors that migrated out of Africa must have acquired Denisovan genes only after the migrations had occurred. The Denisovan HLA genes provided modern humans with additional genetic diversity and likely enhanced the overall immune response and improved survival.

[1]Abi-Rached, et al. 2011. The shaping of modern human immune systems by multiregional admixture with archaic humans. *Science* 334: 89–94.

In this chapter we examine molecules of the innate immune response and the adaptive response, and the mechanisms that affect cell differentiation and activation in immunity. First we examine the innate immune response proteins that interact with shared molecular targets on pathogens. We then discuss the structure and development of receptors of the adaptive immune response. We conclude by presenting the mechanisms that activate the adaptive immune response.

I • Receptors and Immunity

26.1 Innate Immunity and Pattern Recognition

An *innate* recognition system for recognizing and controlling pathogens is widely distributed in multicellular organisms from primitive plants to vertebrate animals. Many invertebrates have genes homologous to the pattern recognition receptors found in higher animals.

Pathogen-Associated Molecular Patterns and Pattern Recognition Receptors

Pathogen-associated molecular patterns (*PAMPs*) are structural components common to a particular group of infectious agents. PAMPs are often macromolecules and include polysaccharides, proteins, nucleic acids, or even lipids. The lipopolysaccharide (LPS) of the gram-negative bacterial cell wall (↪ Section 2.11) is an excellent example of a PAMP.

Pattern recognition receptors (*PRRs*) are a group of soluble and membrane-bound host proteins that interact with PAMPs (Table 26.1). An example of a soluble extracellular PRR is mannose-binding lectin (MBL) (↪ Section 25.9). The PAMP recognized by MBL is the sugar mannose, found as a repeating subunit in bacterial and fungal polysaccharides (mannose on mammalian cells is inaccessible to MBL). C-reactive protein, an extracellular soluble PRR, is an *acute phase protein* produced by the liver in response to inflammation. C-reactive protein interacts with the phosphocholine macromolecules of gram-positive bacteria cell walls. Both of these PRRs target pathogen surface PAMPs, and both bind complement proteins, leading to lysis or opsonization of the targeted cell.

Table 26.1 Receptors and targets in the innate immune response

Pattern recognition receptors (PRRs)	Pathogen-associated molecular patterns (PAMPs) and targets	Result of interaction
Soluble extracellular PRRs		
Mannose-binding lectin[a] (soluble)	Mannose-containing cell surface microbial components, as in gram-negative bacteria	Complement activation
C-reactive protein (soluble)	Components of gram-positive cell walls	
Plasma membrane–associated PRRs		
TLR-1[b] (Toll-like receptor 1)	Lipoproteins in mycobacteria	Signal transduction, phagocyte activation, and inflammation[c]
TLR-2	Peptidoglycan on gram-positive bacteria; zymosan in fungi	
TLR-4	LPS (lipopolysaccharide) in gram-negative bacteria	
TLR-5	Flagellin in bacteria	
TLR-6	Lipoproteins in mycobacteria; zymosan in fungi	
Endosomal membrane–associated PRRs		
TLR-3	Double-stranded viral RNA	Signal transduction, phagocyte activation, and inflammation[c]
TLR-7	Single-stranded viral RNA	
TLR-8	Single-stranded viral RNA	
TLR-9	Unmethylated CpG oligonucleotides in bacteria	
Cytoplasmic PRRs		
NLRs (NOD-like receptors)		
NOD1	Peptidoglycan on gram-negative bacteria	Stimulate production of antimicrobial peptides and proinflammatory cytokines
NOD2	Peptidoglycan on gram-positive bacteria	
NLRP3	Inflammasome component	Triggers release of proinflammatory cytokines, increasing inflammation

[a]The extracellular soluble PRRs are produced by liver cells in response to inflammatory cytokines.
[b]Toll-like receptors are membrane-integrated PRRs expressed in phagocytes. TLR-1, -2, -4, -5, and -6 are in the cytoplasmic membrane. TLR-3, -7, -8, and -9 are found in intracellular organelle membranes such as in lysosomes.
[c]Toll-like receptors are involved in phagocyte activation via signal transduction.

Multicellular organisms such as invertebrates and plants lack adaptive immunity, but they have a well-developed innate response to a wide variety of pathogens. Virtually all of these respond by recognizing molecules found on the pathogen cell or virus. These molecules contain conserved, repetitive structures called pathogen-associated molecular patterns (PAMPs) that include such things as the LPS and flagellin of gram-negative bacteria, the peptidoglycan of gram-positive bacteria, and the double-stranded RNA molecules unique to certain viruses, among others. By recognizing features shared by many pathogens, the innate immune mechanism has evolved to provide protection against most common pathogens.

Responses to pathogens by the fruit fly, *Drosophila melanogaster* (**Figure 1**), have provided insight into innate immune mechanisms in many other groups of organisms. Several proteins required for fruit fly development are also important receptors for recognizing invading bacteria, functioning as pattern recognition receptors (PRRs) that interact with PAMPs on the macromolecules produced by the pathogen. The best example of a PRR is *Drosophila* Toll, a transmembrane protein essential for dorsoventral axis formation as well as in the innate immune response of the fly.

Toll immune signaling is initiated by the interaction of a pathogen or its components with the Toll protein displayed on the surface of phagocytes. *Drosophila* Toll, however, does not interact directly with the pathogen. Signal transduction events start with the binding of a PAMP such as the lipopolysaccharide (LPS) of gram-negative bacteria (⮌ Section 2.11) by one or more accessory proteins (Figure 26.1 shows the analogous TLR-4 system in humans). The LPS–accessory protein complex then binds to Toll. The membrane-integrated Toll protein initiates a signal transduction cascade, activating a nuclear transcription factor and inducing transcription of several genes that encode antimicrobial peptides. Toll-associated transcription factors induce expression of antimicrobial peptides, including drosomycin, active against fungi; diptericin, active against gram-negative bacteria; and defensin, active against gram-positive bacteria. The peptides, produced in the liver-like fat body of *Drosophila*, are released into the fly's circulatory system where they interact with the target organism and cause cell lysis.

Structurally, the Toll proteins are related to lectins, a group of proteins found in all multicellular organisms including invertebrates and plants. Lectins interact specifically with certain oligosaccharide monomers. In humans, Toll-like receptors (TLRs) react with a wide variety of PAMPs. As with *Drosophila* Toll, human TLR-4 provides innate immunity against gram-negative bacteria through indirect interactions with LPS, initiating a kinase signal cascade and activating the nuclear transcription factor

Figure 1 ***Drosophila melanogaster,* the common fruit fly.** The Toll protein, a homolog of the Toll-like receptors of higher vertebrates, was first discovered in the fruit fly.

NFκB. NFκB activates transcription of cytokines and other phagocyte proteins involved in the host responses (Figure 26.1).

Drosophila Toll is an evolutionary, structural, and functional relative of the Toll-like receptors in higher vertebrates, including humans. Toll and its homologs are evolutionarily ancient, highly conserved components of the innate immune system in animals and have even been found in plants.

Jarmo Holopainen

PRRs were first recognized as evolutionarily conserved membrane-associated receptors in the invertebrate *Drosophila* (the fruit fly), where they were called Toll receptors (see Explore the Microbial World, "*Drosophila* Toll Receptors—An Ancient Response to Infections"). Structural, functional, and evolutionary homologs of the Toll receptors, called *Toll-like receptors* (*TLRs*), are widely expressed on mammalian innate immune cells. TLRs are found associated with membranes on the surface or in intracellular vesicles in macrophages, monocytes, dendritic cells, and neutrophils, all of which are phagocytes that have the ability to engulf and destroy pathogens. At least nine TLRs in humans interact with a variety of cell surface and soluble PAMPs from viruses, bacteria, and fungi.

The *NOD-like receptors* (*NLRs*) are a family of PRRs having a nucleotide-binding oligomerization domain (NOD). NLRs are soluble PRRs found in the cytoplasm. NOD1 and NOD2 interact with peptidoglycan components of gram-negative and gram-positive bacterial cell walls, respectively, stimulating production of antimicrobial peptides and inflammatory cytokines (Table 26.1). NOD-like receptor pyrin 3 (NLRP3) interacts with other proteins to form a structure called an *inflammasome*. The cytoplasmic inflammasome senses cellular stress indicators such as the loss of potassium (K^+) ions from damaged cells, triggering the production of proinflammatory cytokines and initiating inflammation.

Several TLRs interact with more than one PAMP. For example, TLR-4 is part of the innate immune response to bacterial LPS and also responds to *heat shock proteins* produced by damaged host cells. Both LPS and heat shock protein interact with TLR-4 via receptor proteins that, in turn, interact with TLR-4. In other cases, the TLR binds directly to the PAMP without the interactions of receptor proteins, as is the case for TLR-5 and its target, flagellin.

Signal Transduction in Phagocytes

Interaction of a PAMP with a PRR triggers transmembrane *signal transduction*, initiating gene transcription and translation

UNIT 5

of host-response proteins similarly to the membrane signal transduction mechanisms in prokaryotes (∞ Section 7.7). Signal transduction initiated by PAMP–PRR interaction results in enhanced phagocytosis, killing of pathogens, inflammation, and tissue healing (∞ Section 24.5).

For example, a signal transduction pathway may be activated by the binding of LPS (a PAMP) to TLR-4 (a PRR) (**Figure 26.1**). TLR-4 then binds proteins in the cytosol, starting a cascade of reactions that activates transcription factors such as NFκB (nuclear factor kappa B), a protein that binds to specific regulatory sites on DNA, initiating transcription of downstream genes. Many of the NFκB-regulated genes encode host-response proteins such as the cytokines that activate cells and initiate inflammation.

TLR-4 consists of three distinct protein domains, each with a separate function. The external domain of TLR-4 contains a binding site for LPS complexed with a cell surface protein called CD14 (Figure 26.1). A transmembrane domain in TLR-4 connects the external domain to a cytoplasmic domain. Binding of the CD14–LPS complex by the TLR-4 external domain causes a change in the conformation of a third TLR-4 domain extending into the cytoplasm, exposing a site that interacts with an adaptor protein, MyD88. MyD88, in turn, is conformationally altered and binds a protein tyrosine kinase (PTK), IRAK4. PTKs transfer energy-rich phosphates from ATP to target-protein tyrosines that are exposed when binding alters conformation. In this case, binding of MyD88 by IRAK4 activates the kinase activity of IRAK4, initiating a *kinase* cascade through ATP-mediated phosphorylation of TRAF6, another kinase. TRAF6 then phosphorylates IκK (inhibitor of kappa kinase). IκK then phosphorylates the IκB (inhibitor of kappa B) protein, causing it to dissociate from NFκB. The uninhibited NFκB then diffuses across the nuclear membrane, binds to NFκB-binding motifs on DNA, and initiates transcription of downstream genes.

As this example shows, signal transduction pathways initiate activation of transcription through ligand–receptor binding on the cell surface. The ligand–receptor interaction outside the cell induces the binding, recruitment, and concentration of the adaptor proteins and kinase enzymes inside the cell. A single kinase can phosphorylate many signal cascade proteins, amplifying the effect of a single ligand–receptor interaction. Signal transduction leading to activation of shared transcription factors and protein production is also the activation mechanism for lymphocytes in adaptive immunity, as we will discuss below.

MINIQUIZ

- Identify the steps leading to recognition of LPS by TLR-4.
- Outline the general features of a signal transduction pathway starting with binding of a PAMP by a membrane-associated PRR.

26.2 Adaptive Immunity and the Immunoglobulin Superfamily

The **immunoglobulin gene superfamily** includes genes and their protein products that share structural, evolutionary, and functional features with immunoglobulin genes and proteins.

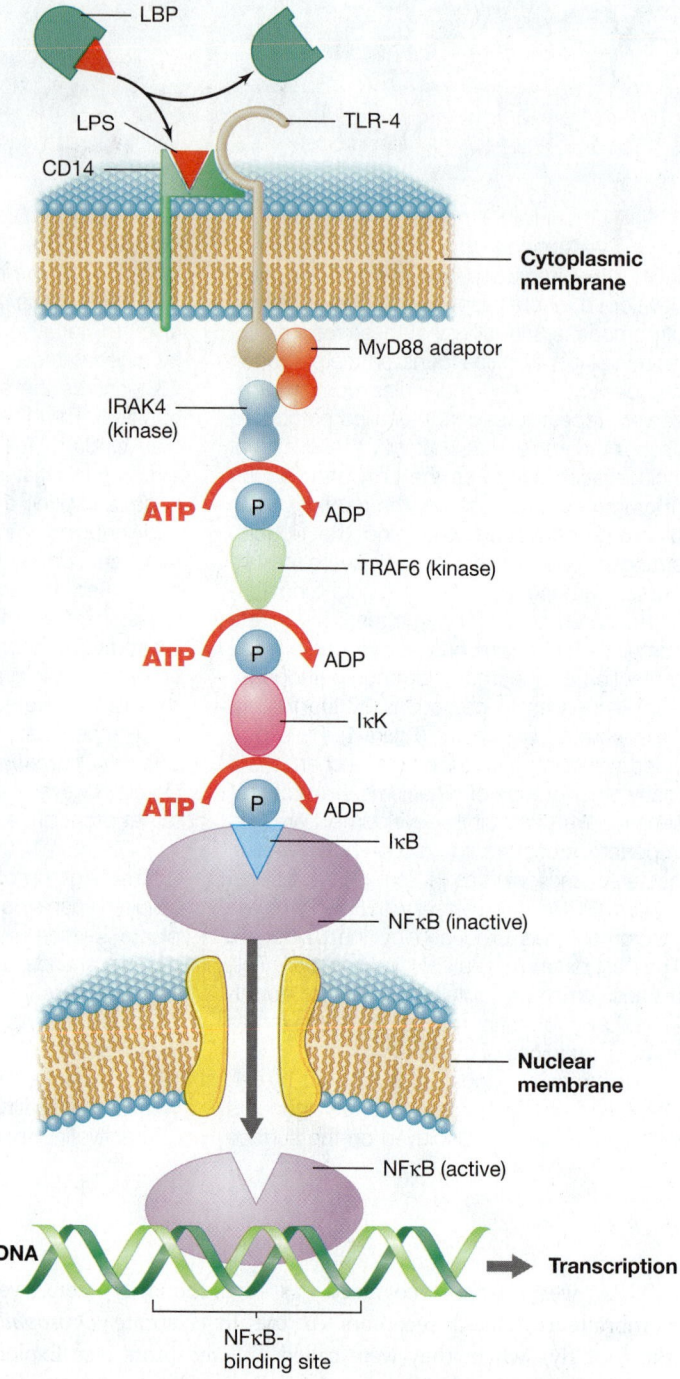

Figure 26.1 Signal transduction in innate immunity. Signal transduction is initiated when LPS, a PAMP, is bound by LBP (lipopolysaccharide-binding protein), which then transfers LPS to CD14 on the surface of a phagocyte. The LPS–CD14 complex then binds to the transmembrane TLR-4 receptor. The binding of TLR-4 initiates a series of reactions involving adaptor proteins and kinases, resulting in activation of the transcription factor NFκB. NFκB then diffuses across the nuclear membrane, binds to DNA, and initiates transcription of proteins essential for innate immunity.

The antigen-binding proteins of the adaptive immune response are part of this extended gene family.

As we discussed in Chapters 24 and 25, three different cell surface proteins interact directly with antigens during the adaptive immune response. These are the *immunoglobulins* (*Igs* or *antibodies*) produced by B cells that interact with antigens; the antigen-binding *T cell receptors* (*TCRs*) on the surface of T cells; and the proteins of the *major histocompatibility complex* (*MHC*) that present processed antigen. Each of these three antigen-binding proteins has a different location, structure, and function. Igs, found on the surface of B lymphocytes and in serum and mucosal secretions, interact directly with extracellular antigens (⊂⊃ Section 25.7). TCRs, found exclusively on T cells, also bind antigen, but only when it is presented by MHC proteins (⊂⊃ Section 25.4).

Structure and Evolution of Antigen-Binding Proteins

Ig, TCR, and MHC proteins share structural features and have evolved by duplication and selection of genes encoding primordial antigen receptors. Some important Ig superfamily proteins are shown in **Figure 26.2**. The proteins consist of a number of discrete domains.

Each protein has at least one domain with a highly conserved amino acid sequence called a *constant* (*C*) *domain*. The C domain typically has about 100 amino acids with an intrachain disulfide bond spanning 50–70 amino acids. C domains provide structural integrity for the antigen-binding molecules, attach the antigen-binding V domains to the cytoplasmic membrane, and give each protein its characteristic shape. C domains can also provide recognition sites for accessory molecules. For example, C domains of most IgG and all IgM proteins are bound by the C1q component of complement, a critical first step in initiating the complement activation sequence (⊂⊃ Section 25.9). Likewise, MHC class I C domains bind to the accessory CD8 protein on T-cytotoxic (Tc) lymphocytes, and homologous MHC class II C domains bind CD4 on T-helper (Th) cells. MHC I–CD8 and MHC II–CD4 interactions are critical steps for T cell activation and immune response development (⊂⊃ Section 25.4).

The *variable* (*V*) *domains* of TCR and Ig are about the same size as the constant domains, but V-domain structures can be considerably different from one another and from the C domains. Ig and TCR V domains have evolved to interact with a wide variety of antigens. By contrast, the V domains of MHC proteins have evolved independently of Ig and TCR V domains; they interact with nonself peptides, resulting in the MHC–peptide complex recognized by a TCR.

TCR, Ig, and MHC proteins each consist of two nonidentical polypeptides. The TCR consists of an alpha (α) and a beta (β) chain. MHC proteins also consist of two different polypeptide chains, again designated α and β (⊂⊃ Section 25.4). Igs have a separate heavy and light chain (⊂⊃ Section 25.7). These heterodimers are expressed on a cell surface and bind antigens. However, the specific function of each of these molecules is quite different. Igs can be anchored on B cell surfaces where they bind to pathogens and their products such as toxins. Igs are also produced in large quantities as soluble serum and mucosal proteins. TCRs are found exclusively on T lymphocytes and interact with antigenic peptides derived from processed pathogen proteins. These peptides are presented by the MHC proteins on target cells or specialized antigen-presenting cells, or APCs (as we discussed in Section 25.4, APCs include macrophages, dendritic cells, and B lymphocytes).

Signal Transduction in Antigen-Reactive Lymphocytes

B cells and T cells interact with antigen through their Ig and TCR antigen receptors, respectively. As with the membrane-integrated PRRs in the innate immune response, the antigen-specific Igs and TCRs transmit a signal from receptor binding across the cytoplasmic membrane to enhance transcription and activate the cell. B and T lymphocytes use the antigen-binding Ig and TCR proteins to transfer signals across the membrane by connecting

UNIT 5

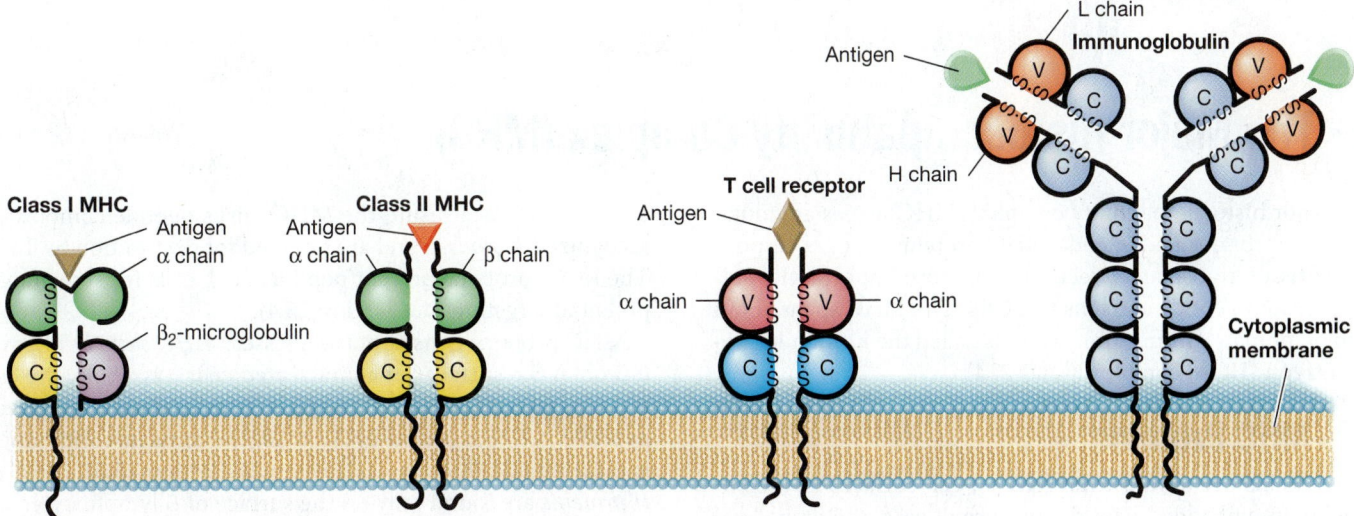

Figure 26.2 Immunoglobulin gene superfamily proteins. Constant domains have homologous amino acid sequences and higher-order structures. The Ig-like C domains in each protein chain indicate evolutionary relationships that identify the proteins as members of the Ig gene superfamily. The V domains of Igs and TCRs are also Ig domains, but the peptide-binding domains of MHC class I and class II proteins are not because their structures vary considerably from the basic features of the Ig domain.

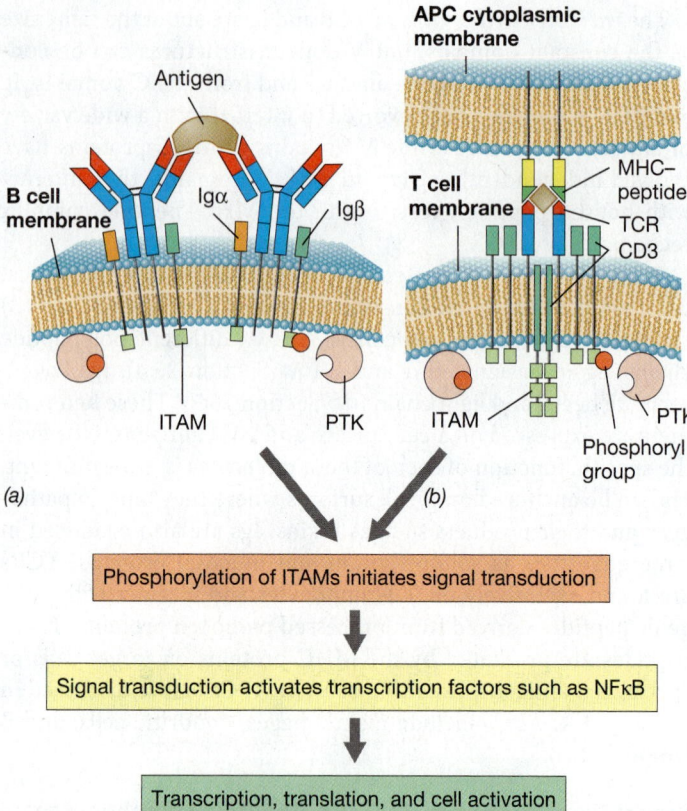

(a) (b)

Phosphorylation of ITAMs initiates signal transduction

Signal transduction activates transcription factors such as NFκB

Transcription, translation, and cell activation

Figure 26.3 Signal transduction in adaptive immunity. *(a)* The surface Ig on B cells associates with two adaptor proteins, Igα and Igβ, in the cytoplasmic membrane. The adaptor proteins contain multiple ITAMs (immune-based tyrosine-activation motifs) that are exposed when antigen binds and cross-links Ig. Phosphorylation of the ITAMs by kinases initiates signal cascades and induces activation of transcription factors, including NFκB, which initiate transcription. *(b)* The TCR also associates with adaptor proteins, collectively called CD3. TCR interaction with the MHC–peptide complex exposes the ITAMs on the CD3 components, leading to phosphorylation, signal transduction, activation of transcription factors, and translation of T cell–specific proteins.

to the common signal transduction pathways inside the cell. The antigen receptors, however, cannot directly connect to the signal transduction pathways because Igs and TCRs have very small cytoplasmic domains. These domains do not interact directly with the adaptor proteins common to signal transduction pathways. In addition, the cytoplasmic domains of neither Igs nor TCRs have cytoplasmic tyrosines that can be phosphorylated (**Figure 26.3**).

To get around this problem, both receptors associate with adaptor molecules that have *immune-based tyrosine-activation motifs* (*ITAMs*), possessing tyrosine residues that can be phosphorylated. These adaptor proteins are Igα and Igβ for immunoglobulins (Figure 26.3*a*) and the CD3 complex for TCRs (Figure 26.3*b*). The adaptors associate noncovalently with their respective antigen receptors in the membrane. Antigen binding by the Ig or TCR provokes conformational changes in the adaptor proteins. These changes expose the adaptor proteins' cytoplasmic ITAMs, which are then phosphorylated by a family of protein–tyrosine kinases (PTKs), as shown in Figure 26.3*b*. As in signal transduction in the innate response, the kinase reaction initiates a cascade, culminating in the activation of NFκB and other transcription factors, and initiating transcription of downstream genes.

These transcription factors induce the transcription of downstream genes such as the heavy and light chain immunoglobulin genes in B cells, leading to translation and immunoglobulin production. In T cells, transcription factors may cause transcription of TCR genes and also cytokine genes, leading directly to translation and expression of cytokines and activation of more antigen-reactive cells.

MINIQUIZ

- Describe and compare the structural features of Ig-family constant and variable domains.
- How is signal transduction and activation of transcription factors influenced by antigen binding to Igs and TCRs?

II • The Major Histocompatibility Complex (MHC)

The major histocompatibility complex (MHC) genes are found in all vertebrates and encode MHC proteins that are responsible for tissue transplant rejection and, more importantly, present processed peptide antigens to TCRs. The MHC spans about 4 Mbp on human chromosome 6 and is called the **human leukocyte antigen (HLA)** complex (**Figure 26.4**).

26.3 MHC Proteins

MHC proteins are always expressed on cell surfaces as a complex with an embedded peptide. In normal cells, the embedded peptide is derived from breakdown products of cell metabolism. Thus, the MHC proteins hold embedded *self* peptides. In cells infected with a virus, however, many of the embedded peptides are derived from the virus. These viral peptides complexed with the MHC protein look much like the variant MHC proteins on

a transplant. As a result, the MHC–virus peptide complexes are recognized as *nonself* and are targeted for destruction by Tc cells. The MHC proteins present peptides to T cells for screening and potential targeting (⊃ Section 25.4).

MHC proteins consist of two classes. *MHC class I proteins* are found on the surfaces of all nucleated cells. As a rule, the class I proteins present peptide antigens to Tc cells. If class I–embedded peptides are recognized by Tc cells, the antigen-containing cell is targeted and directly destroyed (⊃ Section 25.5). *MHC class II proteins* are found only on the surface of B lymphocytes, macrophages, and dendritic cells, the professional APCs (⊃ Section 25.4). Through the class II proteins, the APCs present antigens to the Th cells, stimulating cytokine production that leads to antibody-mediated immunity or inflammatory responses (⊃ Section 25.6).

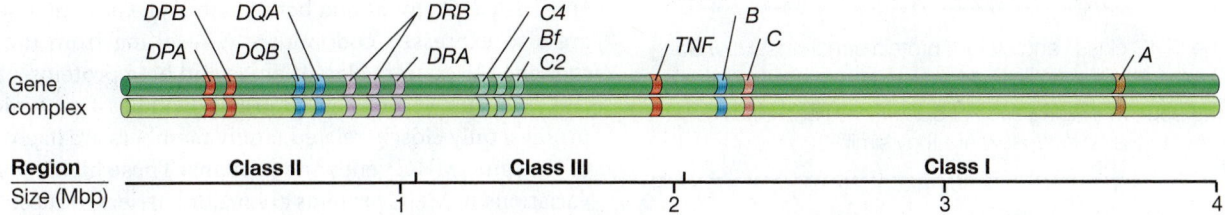

Figure 26.4 **The human leukocyte antigen (HLA) gene map.** The HLA complex, located on chromosome 6, is more than 4 million bases in length. Class II genes *DPA* and *DPB* encode class II proteins DPα and DPβ; *DQA* and *DQB* encode DQα and DQβ; *DRA* and two *DRB* loci encode DRα and DRβ proteins. The class I MHC proteins HLA-B, HLA-C, and HLA-A are encoded by genes *B*, *C*, and *A*. The class II and class I loci are highly polymorphic and encode peptide-binding proteins. Class III MHC genes encode proteins associated with immune-related functions such as complement proteins C4 and C2, and the cytokine TNF (tumor necrosis factor).

Class I MHC Proteins

An MHC class I protein consists of two polypeptides (**Figure 26.5***a*). The gene for the membrane-integrated alpha (α) chain is in the MHC gene region on chromosome 6. The other class I polypeptide is the noncovalently associated beta-2 microglobulin (β₂m). The three-dimensional structure of class I MHC protein interacts with the antigen peptide and the TCR simultaneously. The class I α chain folds to form a groove that is closed on both ends. The groove lies between two α-helices that straddle a β-sheet. In the endoplasmic reticulum, the MHC I groove is loaded with peptides of about 8 to 11 amino acids in length, derived from degraded *endogenous* proteins (Figure 26.5*b*). For example, viral proteins produced inside the cell are degraded into peptides and loaded into class I MHC proteins. The MHC–peptide complex then moves to the cell surface to be recognized by TCRs on Tc cells (⟳ Section 25.4).

Class II MHC Proteins

An MHC class II protein consists of two noncovalently associated, membrane-integrated polypeptides, α and β, found only on APCs. One α and one β polypeptide, expressed together, form a functional heterodimer (Figure 26.5*c*). The heterodimers may be arranged in pairs or trimers that enhance their stability. The α1 and β1 domains of the class II protein interact to form a binding site for TCR–peptide similar to the class I binding site for TCR–peptide. However, the ends of the groove are open, permitting the class II protein to bind and display peptides that may be significantly longer than 8–11 amino acids. Class II–binding peptides, generally 10 to 20 amino acids long, are proteolytic fragments derived from *exogenous* pathogens internalized and processed by the APCs (⟳ Section 25.4). The APCs use the class II–peptide complex to interact with TCRs on Th cells, leading to Th activation (⟳ Section 25.6).

UNIT 5

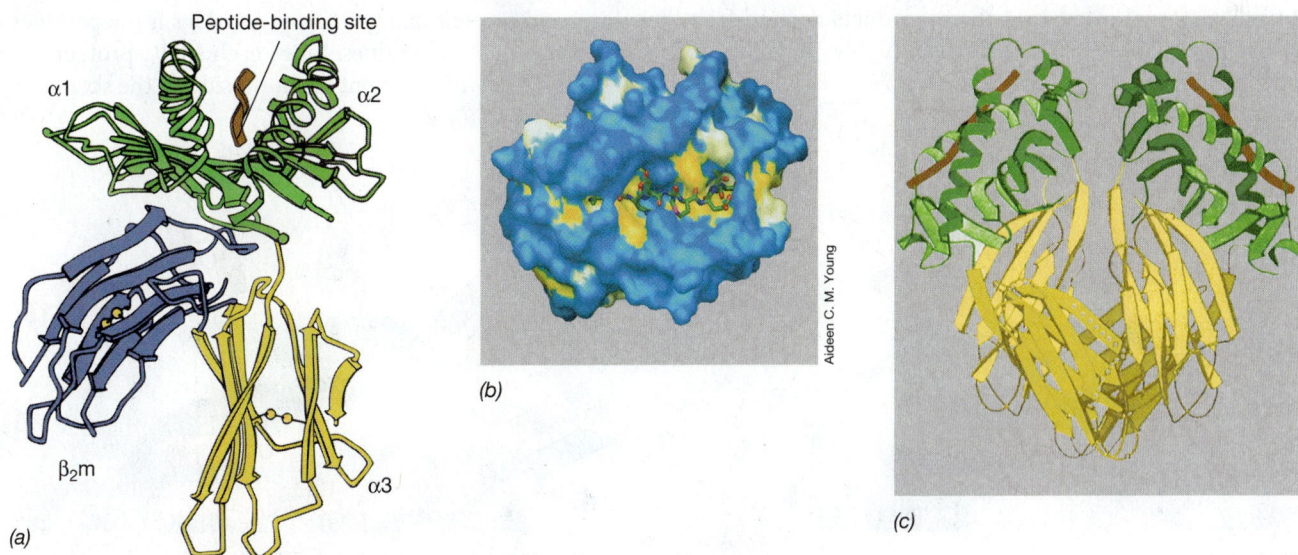

Figure 26.5 **MHC protein structure.** *(a)* The MHC class I protein. Beta-2 microglobulin (β₂m) binds noncovalently to the α chain. The antigen peptide (brown) is bound cooperatively by the α1 and α2 domains. A helical shape indicates α-helix protein structure, and a flat shape indicates a β-sheet. *(b)* An MHC I protein with a bound peptide, as seen from above. A peptide of nine amino acids is shown as a carbon backbone structure, embedded in a space-filling model of a mouse MHC I protein. *(c)* A class II protein dimer. The peptides (brown) are shown in their positions in the binding sites of the MHC II proteins.

26.4 MHC Polymorphism, Polygeny, and Peptide Antigen Binding

The human MHC class I and class II genes encode peptide-binding proteins that bind antigen peptides for presentation to T cells. The many variations of these proteins collectively bind all known peptides.

Polymorphism and Polygeny

Polymorphism is the occurrence in a population of multiple alleles (alternate forms of a gene) at a specific locus (the location of the gene on the chromosome) in frequencies that cannot be explained by recent random mutations. For example, the MHC class I locus *HLA-A* has 2013 known alleles. These genetic variations encode 2013 distinct HLA-A proteins. Each person, however, has only two of the *HLA-A* alleles; one allele is of paternal origin and one is of maternal origin; the polymorphisms reside in the entire human population. The two allelic protein products are expressed *codominantly* (equally) (**Figure 26.6a**).

In addition, gene duplication events during evolution resulted in two additional genetically, structurally, and functionally related polymorphic loci, *HLA-B* and *HLA-C*. The occurrence of multiple copies of evolutionarily, genetically, structurally, and functionally related genes is called **polygeny**. *HLA-B* has 2605 known allelic variants; *HLA-C* has 1551 known alleles. Thus, an individual usually displays six structurally distinct proteins derived from the three polymorphic class I loci (three products derived from maternal origin and three products derived from paternal origin) (Figure 26.6b). Likewise, highly polymorphic alleles encode MHC class II proteins at the *HLA-DR, HLA-DP,*

and *HLA-DQ* alpha- and beta-chain loci. Class II gene products are also expressed codominantly, resulting from the 12 alleles that encode distinct class II alpha and beta proteins.

As a result of polygeny, most individuals have unique MHC profiles; only closely related family members are likely to have all of the same MHC genes and proteins. These highly polymorphic variations in MHC proteins are major barriers to successful tissue transplants because the MHC proteins on the donor tissue (graft) are recognized as foreign antigens by the recipient's immune system. An immune response directed against the graft MHC proteins causes graft death and rejection. Tissue graft rejection, however, is minimized by matching MHC alleles between donors and recipients. Control of rejection can also be accomplished through drugs that suppress the immune system.

Peptide Binding

Most of the allelic variations in MHC proteins occur as amino acid changes concentrated in the antigen-binding groove, and each polymorphic variation of the MHC protein binds a different set of peptide antigens. The peptides bound by a single MHC protein share a common structural pattern, or peptide **motif**, and each different MHC protein binds a different motif. For example, for a certain class I protein, the bound peptides contain eight amino acids with a phenylalanine (F) at position 5 and a leucine (L) at position 8. All other positions in the peptide can be occupied by any amino acid (X). Thus, all peptides sharing the sequence X-X-X-X-**F**-X-X-**L** would bind that MHC protein. Another MHC class I protein encoded by a different MHC allele binds a different motif, with nine amino acids and invariant amino acids tyrosine (Y) at position 2 and isoleucine (I) at position 9 (X-**Y**-X-X-X-X-X-X-**I**).

The invariant amino acids in each motif are *anchor residues*: They bind directly and specifically within an individual MHC–peptide binding groove. Thus, an individual MHC protein can bind and present many different peptides if the peptides contain the same anchor residues. Since each MHC protein binds a different motif with different anchor residues, the six possible MHC

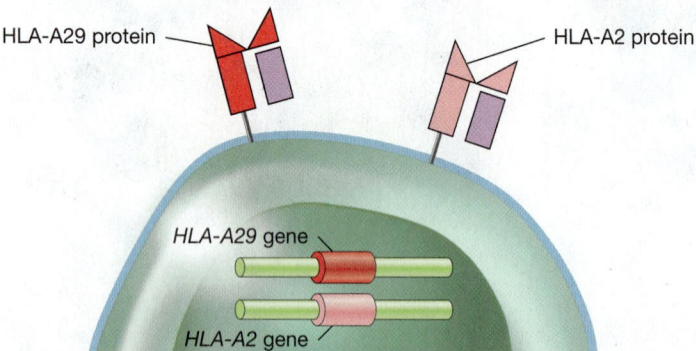

(a) **Polymorphism**

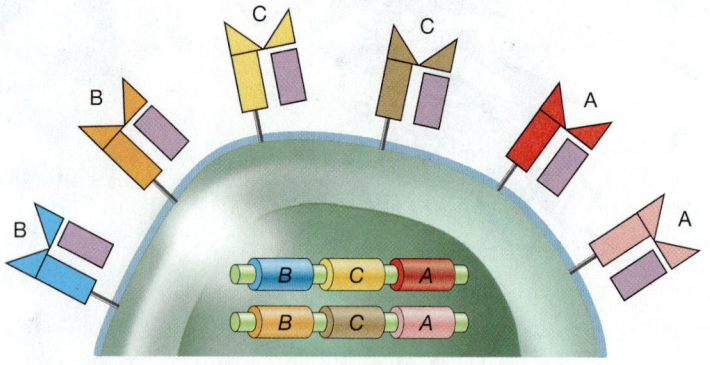

(b) **Polygeny**

Figure 26.6 Polymorphism and polygeny in MHC genes and proteins. *(a)* Polymorphism in *HLA-A* loci results in codominant expression of proteins encoded by both alleles. There are over 2000 HLA-A alleles in the human population, but only two (one at each locus) are found in each individual. *HLA-B* and *HLA-C* exhibit similar levels of polymorphism. *(b)* Polygeny in MHC results in duplicated polymorphic *HLA-A, HLA-B,* and *HLA-C* genes that potentially encode three pairs of different MHC proteins. The colors represent alternate alleles of each gene and their respective protein.

I proteins in an individual bind six different motifs. In this way, each individual can present a large number of different peptide antigens using the limited number of MHC I molecules available. MHC II proteins bind peptides in an analogous manner. As a result, within the human species, at least a few peptide antigens from each pathogen will display a motif that will be bound and presented by the MHC proteins. This system is very different from the mechanisms employed by Igs and TCRs that also bind antigens. Each Ig or TCR interacts very specifically with only a *single* antigen. As we shall see, these proteins employ unique genetic mechanisms to generate virtually unlimited diversity (Sections 26.6 and 26.7).

MINIQUIZ
- Define polymorphism and polygeny as they apply to MHC genes.
- How does a single MHC protein present many different peptides to T cells?

III • Antibodies and T Cell Receptors

Here we look at the structure, antigen-binding function, genetic organization, and generation of diversity in the infinitely variable immunoglobulins and T cell receptors.

26.5 Antibody Proteins and Antigen Binding

Antibodies consist of four polypeptides, two heavy chains (H) and two light chains (L) (Figure 26.2), arranged as a pair of heterodimers. Each heterodimer consists of a light-chain–heavy-chain pair and is a complete antigen-binding unit. The heavy and light chains are further divided into C (constant) and V (variable) domains. The C domains are responsible for common functions such as complement binding (⟲ Section 25.9). The V domains of one H and one L chain interact to form an antigen-binding site (**Figure 26.7**). Here we examine the structural features of the V domains and the antigen-binding site.

Variable Domains

Amino acid sequences are considerably different in the V domains of different Igs (Figure 26.7). Amino acid variability is especially apparent in several **complementarity-determining regions (CDR)**. The three CDRs in each of the V domains provide most of the molecular contacts with antigen. CDR1 and CDR2 differ somewhat between different immunoglobulins, but the CDR3s differ dramatically from one another. The CDR3 of the heavy chain has a particularly complex structure, encoded within three distinct gene segments, as we shall see shortly. The CDR3 consists of the carboxy-terminal portion of the V domain, followed by a short "diversity" (D) segment of about three amino acids, and a longer "joining" (J) segment about 13–15 amino acids long. The light-chain CDR3 is similar, but lacks the D segment. The heavy- and light-chain CDRs cooperate in antigen binding.

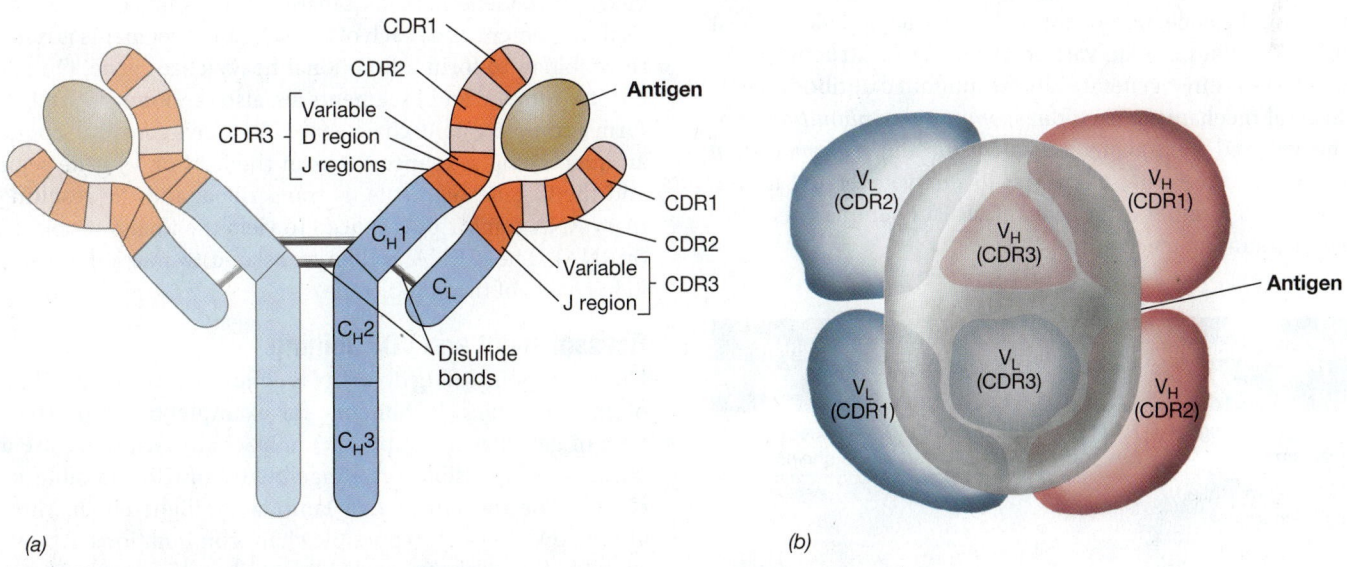

(a)   *(b)*

Figure 26.7 **Antigen binding by immunoglobulin light and heavy chains.** *(a)* An Ig is shown schematically, with bound antigen. The V domains on the H and L chains are shown in red, with the antigen-binding CDR1, CDR2, and CDR3. C_H1, C_H2, and C_H3 are constant domains in the H chain, and C_L is the constant domain in the L chain. *(b)* Complementarity-determining regions (CDRs) from both H (red) and L (blue) chains, shown from above, are conformed to make a single antigen-binding site on the Ig. The highly variable CDR3s from both H and L chains cooperate at the center of the site. Antigen, shown in gray, may contact all CDRs. The shape of the site may be a shallow groove or a deep pocket, depending on the antibody–antigen pair involved.

UNIT 5

Antigen Binding

The Ig three-dimensional structure was shown in Figure 25.16. Each antigen–antibody reaction requires the specific combination of the antigen with the cooperating variable domains of the heavy and light chains. The antigen-binding site of an antibody molecule measures about 2×3 nm, large enough to accommodate a small portion of the antigen, called an *epitope*, about 10 to 15 amino acids long. Antigen binding is ultimately a function of the Ig folding pattern of the heavy and light polypeptide chains. The Ig folds of the V region bring all six CDRs (CDR1, 2, and 3 from both heavy and light chains) together. The result is a unique and specific antigen-binding site (🔗 Figure 25.16 and Figure 26.7). In the next section, we examine the genetic mechanisms that generate the tremendous diversity found in the Ig proteins.

Each antibody binds each antigen with a characteristic affinity (binding strength). The affinity of an antibody is typically highest for the antigen for which it was selected, and antibodies usually do not bind other antigens. However, some antibodies will interact, usually weakly, with antigens other than the selecting antigen. This phenomenon is called a *cross reaction*.

MINIQUIZ

- Draw a complete Ig molecule and identify antigen-binding sites on the antibody.
- Describe antigen binding to the CDR1, 2, and 3 regions of the heavy-chain and light-chain variable domains.

26.6 Antibody Genes and Diversity

For most proteins, one gene encodes one protein. However, this is not the case with the heavy and light chains of immunoglobulins. Because the collection of antibodies in each individual must recognize and bind a wide variety of molecular structures, the immune system must generate almost unlimited antibody variation. Several mechanisms including *somatic recombination*, random heavy- and light-chain *reassortment*, and *hypermutation* all contribute to the almost limitless diversity generated from a relatively small number of Ig genes. **Table 26.2** summarizes these diversity-generating mechanisms.

Table 26.2 Generation of antigen-binding receptor diversity in B cells and T cells		
Diversity-generating mechanism	*B cell Ig receptors, heavy and light chains*	*T cell receptors, α and β chains*
Somatic recombination of tandem genes	Yes	Yes
Random reassortment	Yes	Yes
Imprecise V-D-J or V-J joining	Yes	Yes
Nucleotide additions at V-D-J or V-J junctions	Yes	Yes
D's in 3 reading frames	No	Yes
Somatic hypermutation	Yes	No

Immunoglobulin Genes

The gene encoding each immunoglobulin H or L chain is constructed from several gene segments. In each B cell, Ig gene segments undergo a series of somatic, random rearrangements (recombination followed by deletion of intervening sequences) to produce a single functional antibody gene derived from the pool of antibody genes. Molecular studies have verified this "genes in pieces" hypothesis by demonstrating that the V, D, and J gene segments encoding heavy-chain V domains, as well as the genes encoding C domains, are separated from one another in the genome. As the B cell matures, the gene segments are brought together (somatically recombined) to form a single Ig heavy-chain gene (**Figure 26.8**). A single V gene encodes CDR1 and CDR2, whereas CDR3 is encoded by a mosaic of the 3′ end of the V gene, followed by the D and J genes.

In each B cell, only one protein-producing rearrangement occurs in the heavy- and light-chain genes. Called *allelic exclusion*, this mechanism ensures that *each B cell produces only one Ig*. Finally, the class-defining constant domains of Igs are encoded by separate C genes. Thus, four different gene segments, V, D, J, and C, recombine to form one functional heavy-chain gene. Similarly, light chains are encoded by recombination products of light-chain V, J, and C genes.

The gene segments required for all Igs exist in all cells but undergo recombination only in developing B lymphocytes. As shown in Figure 26.8, each B cell contains multiple kappa (κ) and a corresponding set of lambda (λ) light-chain V and J genes arranged in tandem. Each B cell also contains tandem heavy-chain V genes, D genes, and J genes. In addition, the heavy-chain constant-domain (C_H) genes and the light-chain constant-domain genes (C_L) are present. The V, D, J, and C genes are separated by noncoding sequences (introns) typical of gene arrangements in eukaryotes. Genetic recombination occurs in each B cell during its development. One each of the V, D, and J segments is randomly recombined to form a functional heavy-chain gene. On another chromosome, V and J segments are also randomly recombined to form a complete light-chain gene. The active gene, still containing an intervening sequence between the VDJ or VJ gene segments and the C gene segments, is transcribed, and the resulting primary RNA transcript is spliced to yield the final messenger RNA (mRNA). The mRNA is then translated to make the heavy and light chains of the Ig molecule.

Reassortment and VDJ Joining

Up to this point, all Ig diversity is generated from recombination of existing genes. In humans, for example, based on the numbers of genes at the kappa (κ) light-chain loci, there are about $40\ V \times 5\ J$ possible rearrangements, or 200 possible κ light chains. For the alternative lambda (λ) light chain, there are about $30\ V \times 4\ J = 120$ possible chain combinations. About 6000 possible heavy chains can be formed by the rearrangement of about $40\ V \times 25\ D \times 6\ J$ genes. Each heavy and light chain gene has a theoretically equal possibility of being expressed in each B cell, randomly reassorting all heavy and light chains. The final light chain and heavy chain produced by a given B cell result from assembly of the single translated heavy- and light-chain genes (Figure 26.8c). Assuming that each heavy chain and light

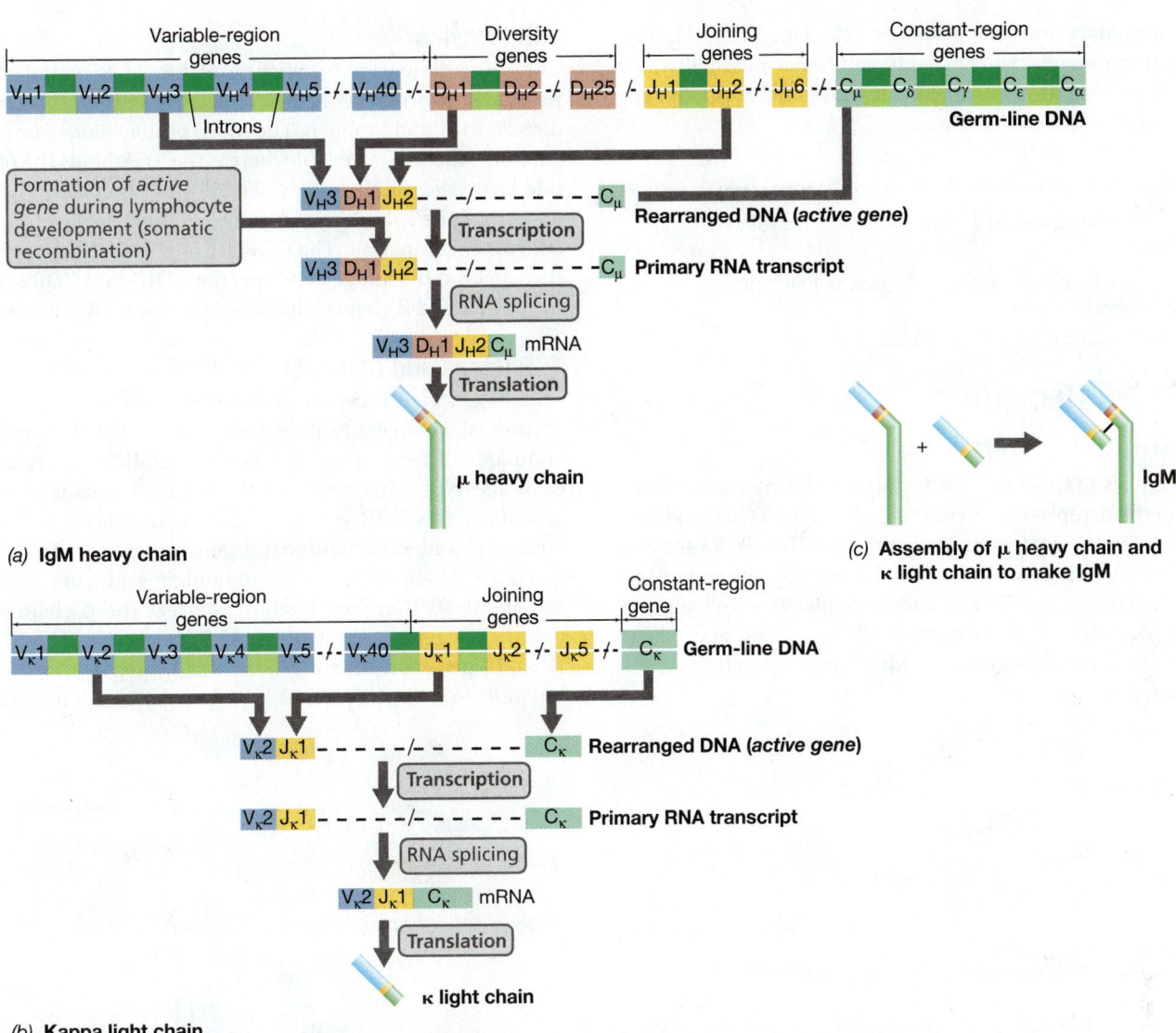

Figure 26.8 Immunoglobulin gene rearrangement in human B cells. Ig genes are arranged in tandem on three different chromosomes. *(a)* The H chain gene complex on chromosome 14. The filled boxes represent Ig coding genes. The broken lines indicate intervening sequences and are not shown to scale. *(b)* The κ light-chain complex on chromosome 2. The λ light-chain genes are in a similar complex on chromosome 22. *(c)* Assembly of one-half of an antibody molecule.

chain has an equal chance to be expressed, there are 6000 × 200 = 1,200,000 possible immunoglobulins with κ light chains and 6000 × 120 = 720,000 possible immunoglobulins with λ chains. Adding these possibilities together, at least 1,920,000 possible antibodies can be expressed!

Additional diversity is generated in the CDR3 of both heavy and light chains by several unique mechanisms. First, the DNA-joining mechanism responsible for construction of V-D or D-J segments in the heavy chain or the V-J gene segments in the light chain is imprecise; the final sequence at these coding joints frequently varies by a few nucleotides from the genomic sequence. Even more diversity is generated at V-D and D-J coding joints in the heavy-chain genes and at V-J coding joints in light-chain genes by nucleotide additions. Either random (N) or template-specific (P) nucleotides may be added. Since these coding joints are contained within the sequences responsible for encoding CDR3

on both heavy and light chains, N and P diversity at V-domain coding joints changes or adds amino acids in the CDR3 of both heavy and light chains.

Hypermutation

Finally, antibody diversity is expanded even more in B cells by **somatic hypermutation**, the mutation of Ig genes at much higher rates than the mutation rates observed in other genes. Somatic hypermutation of Ig genes is typically evident after a second exposure to an immunizing antigen. Somatic hypermutation occurs only in the V regions of rearranged heavy- and light-chain genes, creating B cells with mutated Ig receptors. These mutated B cells then compete for available antigen. This process selects B cells with receptors having higher antigen-binding strength (affinity) than the original B cell receptor. This *affinity maturation* process is one of the factors responsible for a dramatically stronger secondary immune response ($\hookrightarrow$ Figure 25.21). The affinity maturation mechanism adds additional possibilities to the generation of Ig diversity, making the potential antibody repertoire almost limitless.

26.7 T Cell Receptors: Proteins, Genes, and Diversity

T cell receptors (TCRs) are cell surface antigen receptors integrated into the cytoplasmic membrane of T cells. TCRs bind foreign peptides in the context of MHC proteins. The TCRs accomplish this dual binding function through a binding site composed of the V domains of the TCR α and β chains. The α-chain and β-chain V domains of TCRs contain CDR1, CDR2, and CDR3 segments that bind directly to the MHC–peptide antigen complex ($\hookrightarrow$ Section 25.4).

TCR Proteins

The three-dimensional structure of the TCR bound to MHC–peptide is shown in **Figure 26.9**. Both TCR and MHC proteins bind directly to peptide antigen. The MHC protein binds one face of the peptide—the MHC motif—whereas the TCR binds the other peptide face—the T cell epitope. The CDR regions of the TCR bind directly to the MHC–peptide complex, and each CDR has a specific binding function. The CDR3 regions of the TCR α chain and β chain bind the antigen epitope; the CDR1 and CDR2 regions of the TCR α and β chains bind mainly to the MHC proteins.

TCR Genes and Diversity

T cells generate receptor diversity in ways similar to the generation of Ig diversity in B cells. Table 26.2 summarizes and compares the receptor diversity-generating mechanisms for each cell type. Analogous to the H and L chains of immunoglobulins, the TCR α and β chains are encoded by distinct constant- and variable-domain gene segments. TCR V-region genes are arranged as a series of tandem segments. The α chain has about 80 V and 61 J genes, whereas the β chain has 52 V genes, 2 D genes, and 13 J genes (**Figure 26.10**). The β-chain V, D, and J genes and the α-chain V and J genes undergo recombination to form functional V-region genes. As in Igs, somatic

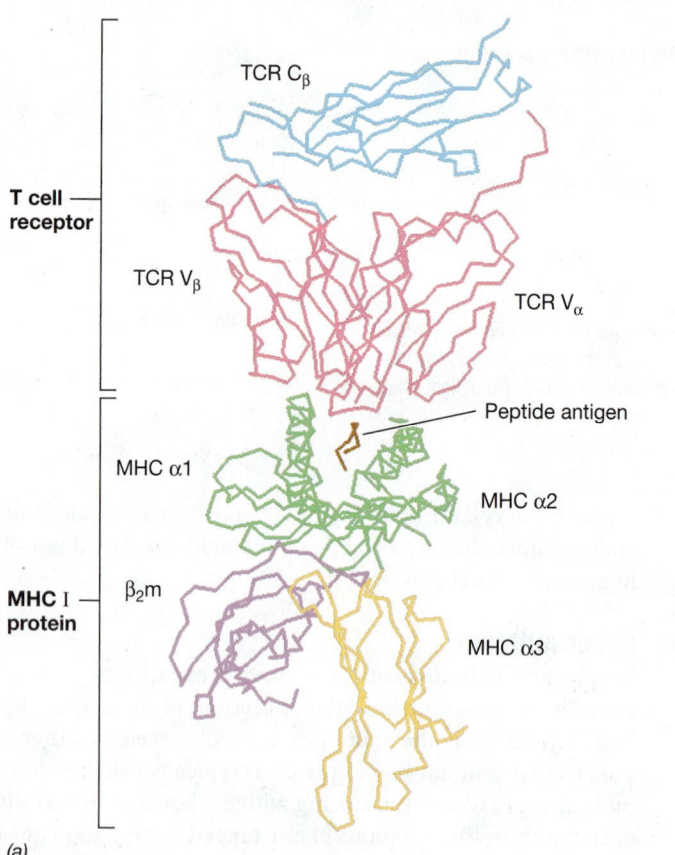

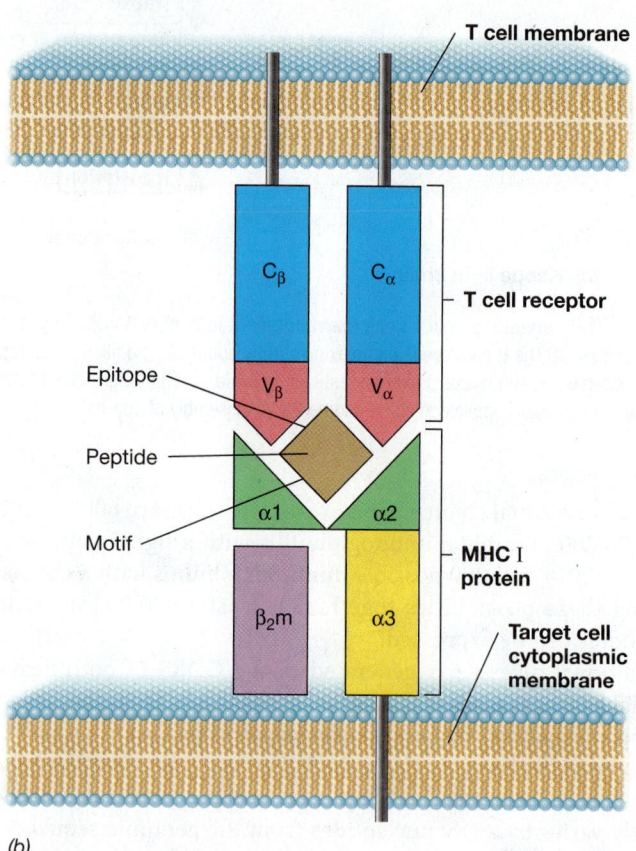

(a)

(b)

Figure 26.9 The TCR:MHC I–peptide complex. *(a)* A three-dimensional structure showing the orientation of TCR, peptide (brown), and MHC. This structure was derived from data deposited in the Protein Data Bank. *(b)* A diagram of the TCR:MHC–peptide structure. Note that the peptide is bound by both MHC and TCR proteins and has a distinct surface structure that interacts with each.

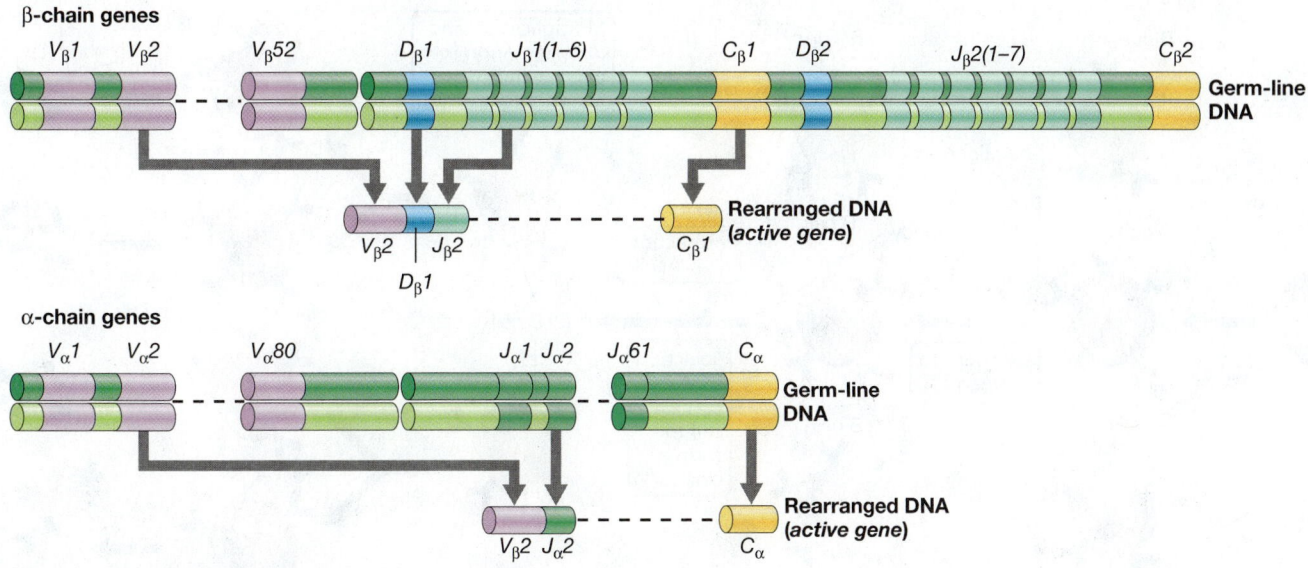

Figure 26.10 Organization of the human TCR α- and β-chain genes. The α-chain genes are located on chromosome 14 and the β-chain genes are on chromosome 6.

mutations result from N and P diversity at V-D and D-J coding joints in the β chain and at the V-J coding joint in the α chain. Finally, the D region of the β chain can be transcribed in all three reading frames, leading to production of three separate transcripts from each D-region gene and creating greater diversity than would be expected from the D gene segments alone. As for assembly of Ig H and L chains, individual α and β chains are produced by each T cell at random and assembled to form a complete α:β heterodimer. The somatic hypermutation mechanisms responsible for increased receptor diversity in Ig genes do not operate in T cells and do not generate additional TCR diversity. Potential TCR diversity, however, is still enormous, and on the order of 10^{15} different TCRs can be generated.

MINIQUIZ
- Distinguish among the functions of the CDR1, CDR2, and CDR3 segments of the T cell receptor.
- Identify diversity-generating mechanisms unique to TCRs as compared to diversity-generating mechanisms in Igs.

IV • Molecular Switches in Immunity

In addition to antigen interaction, several factors control immune activity. Clonal selection selects antigen-reactive cells that respond to foreign antigens while ignoring self antigens. A set of unique cell-surface molecular signals are required for activating T cells or B cells. Finally, cytokines and chemokines produced by activated cells recruit other cells in the immune response.

26.8 Clonal Selection and Tolerance

T cells must be able to discriminate between the potentially dangerous nonself antigens and the harmless self antigens that compose our body tissues. Thus, T cells acquire *tolerance*, or specific unresponsiveness to self antigens. Mature immune lymphocytes interact only with nonself antigens.

Clonal Selection

The **clonal selection** theory states that each antigen-reactive B cell or T cell has a cell surface receptor for a single antigen epitope.

When stimulated by interaction with that antigen, each cell can replicate, and antigen-stimulated B and T cells grow and differentiate, producing a pool of cells that expresses the same antigen-specific receptors. A *clone* comprises the identical progeny of the initial antigen-reactive cell (**Figure 26.11**). Cells that have not interacted with antigen do not proliferate.

To respond to the seemingly infinite variety of antigens, a nearly infinite number of antigen-reactive cells are needed in the body. As we have discussed, the immune system generates a nearly limitless number of antigen-specific B and T cell receptors. Inevitably, some of these receptors will have the potential to react with self antigens in the host; the immune system must eliminate or suppress these self-reactive cells while at the same time selecting cells that may be useful against nonself antigens.

T Cell Selection and Tolerance

T cells undergo immune selection *for* potential antigen-reactive cells and selection *against* those cells that react strongly with self

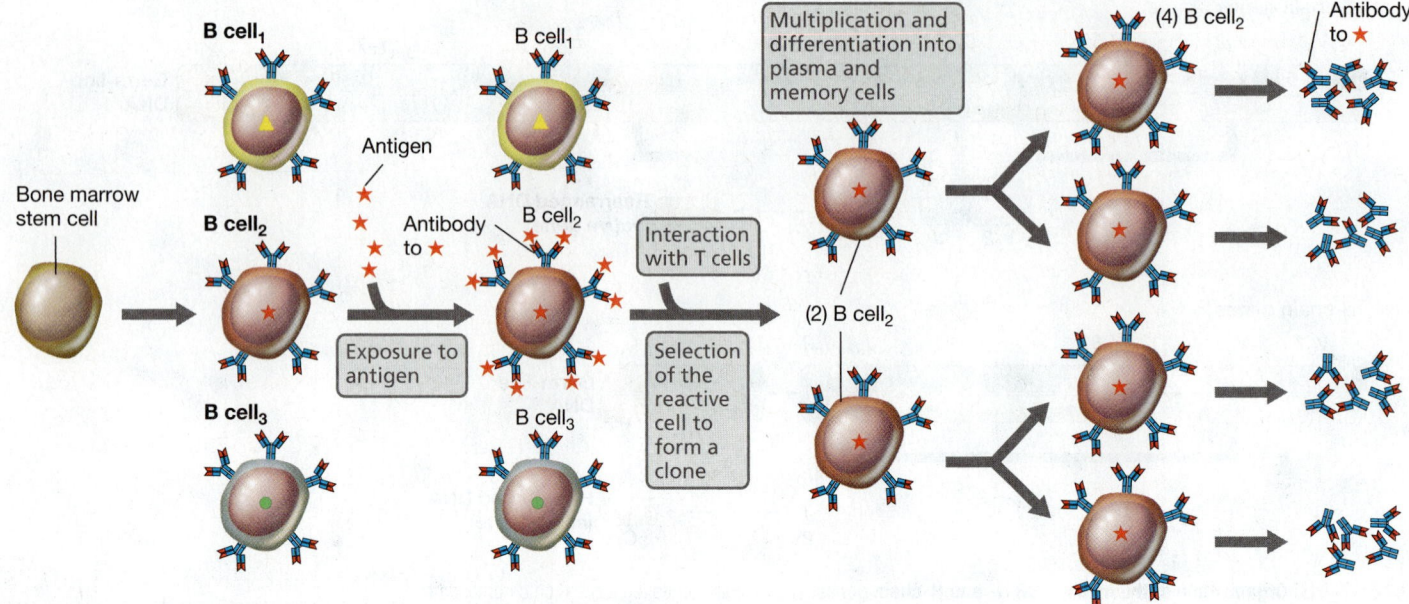

Figure 26.11 Clonal selection. Individual B cells, specific for a single antigen, proliferate and expand to form a clone after interaction with the specific antigen. The antigen drives selection and then proliferation of the individual antigen-specific B cell. Clonal copies of the original antigen-reactive cell have the same antigen-specific surface antibody. Continued exposure to antigen results in continued expansion of the clone.

antigens. Selection against self-reactive cells results in the development of tolerance. The failure to develop tolerance may result in dangerous reactions to self antigens, a condition called *autoimmunity* (↪ Section 24.8).

T lymphocyte precursors leave the bone marrow and enter the thymus, a primary lymphoid organ, via the bloodstream (**Figure 26.12**). During the process of T cell maturation in the thymus, immature T cells undergo a two-step selection process

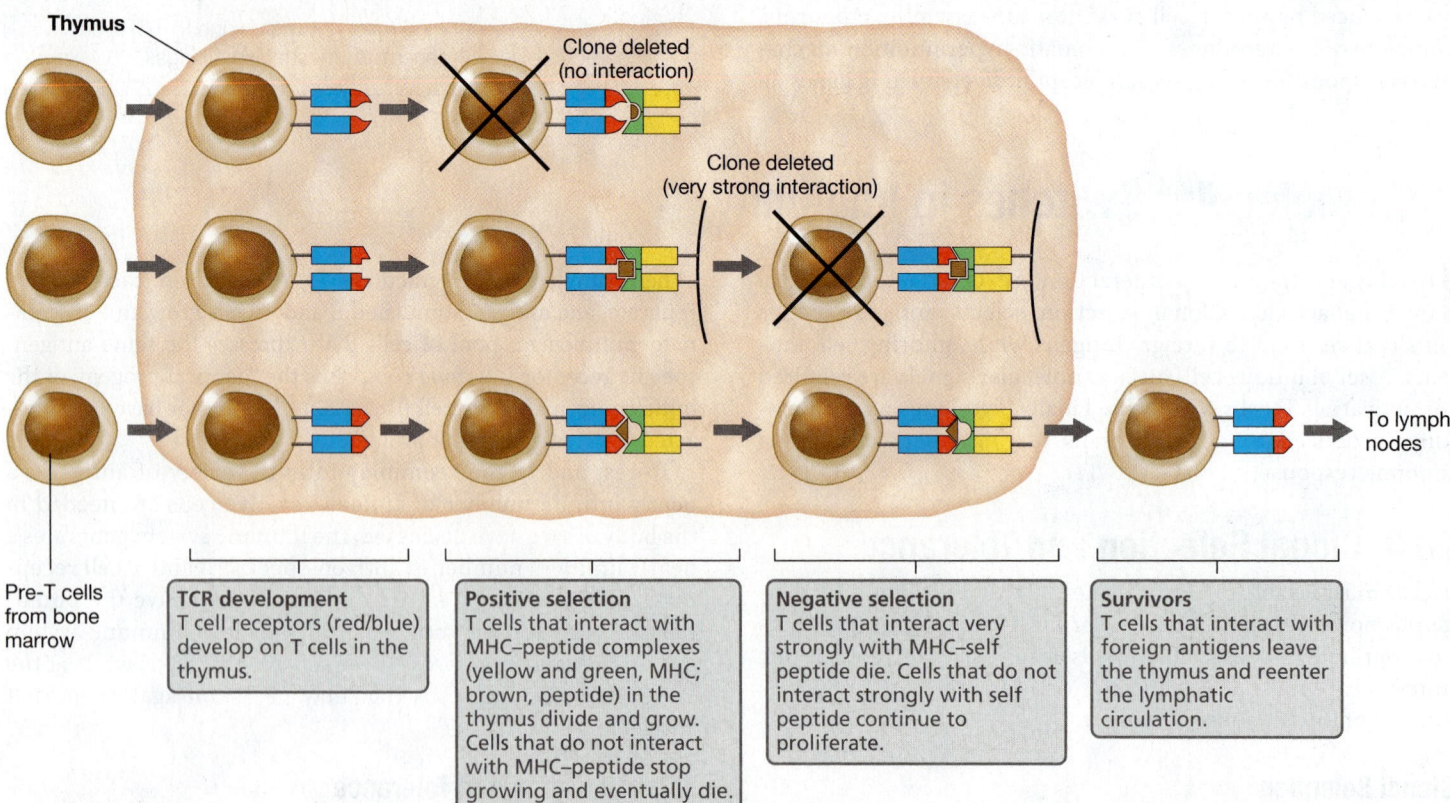

TCR development
T cell receptors (red/blue) develop on T cells in the thymus.

Positive selection
T cells that interact with MHC–peptide complexes (yellow and green, MHC; brown, peptide) in the thymus divide and grow. Cells that do not interact with MHC–peptide stop growing and eventually die.

Negative selection
T cells that interact very strongly with MHC–self peptide die. Cells that do not interact strongly with self peptide continue to proliferate.

Survivors
T cells that interact with foreign antigens leave the thymus and reenter the lymphatic circulation.

Figure 26.12 T cell selection and clonal deletion. T cells undergo selection for recognition of dangerous nonself antigens in the thymus.

to (1) select potential antigen-reactive cells (positive selection) and (2) eliminate cells that react with self antigens (negative selection). **Positive selection** requires the interaction of new T cells in the thymus with the thymic self antigens; the peptide antigens in the thymus are of self origin. Using their TCRs, some T cells bind to MHC–peptide complexes on the thymic tissue. The T cells that do not bind MHC–peptide complexes undergo *apoptosis*, or programmed cell death, and are permanently eliminated. By contrast, those T cells that bind thymic MHC proteins receive survival signals and continue to divide and grow. Positive selection retains T cells that recognize MHC–peptide and deletes T cells that do not recognize MHC–peptide and would therefore be unable to recognize MHC–peptide outside the thymus.

The second stage of T cell maturation is **negative selection**. Here the positively selected T cells continue to interact with thymic MHC–peptide. T cells that react with thymic self antigens are potentially dangerous if they react very strongly with these antigens (autoimmunity). The very strongly self-reactive T cells bind tightly to thymus cells; these self-reactive T cells cannot divide and eventually die. TCRs that react less strongly with self MHC–peptide survive this selection and live. This two-stage thymic selection process for selecting self-tolerant, antigen-reactive T cells results in **clonal deletion**. Precursors of T cell clones that are either useless (do not bind) or harmful (bind too tightly) die in the thymus; more than 95% of all T cell precursors that enter the thymus do not survive the selection process.

The remaining selected T cells are destined to interact very strongly with nonself antigens. They are not destroyed in the thymus because their weak binding interactions with thymic self antigens signal them to proliferate. The selected and growing T cells leave the thymus and migrate to the spleen, mucosa-associated lymphoid tissue, and lymph nodes, where they can contact foreign antigens presented by B lymphocytes and other APCs.

B Cell Tolerance

The development of tolerance in B cells is also necessary because antibodies produced by self-reactive B cells (autoantibodies) may cause autoimmunity and damage to host tissue (⇄ Section 24.8). B cells also undergo a process of clonal deletion. Many self-reactive B cells are eliminated during development in the bone marrow, the primary lymphoid organ responsible for B cell development in mammals.

In addition to clonal deletion, **clonal anergy** (clonal unresponsiveness) also plays a role in final selection of the B cell repertoire. Some immature B cells are reactive to self antigens, but do not become activated even when exposed to high concentrations of self antigens. This is because B cell activation requires a second signal from Th cells, as we shall now see. If no second signal is generated because the available Th cells have been rendered tolerant to the antigen in the thymus, the B cell remains unresponsive.

MINIQUIZ

- Distinguish between positive and negative T cell selection. How do positive and negative selection control the development of tolerance in T cells?
- Distinguish between clonal deletion and clonal anergy in B cells.

26.9 T Cell and B Cell Activation

T and B cells require additional molecular signals for activation in addition to antigen interactions through Igs or TCRs. Lack of these signals results in unresponsive cells, even if they are exposed to antigen. This mechanism helps to prevent autoimmunity.

T Cell Activation

As we have described, T cells that react strongly with self-antigens are deleted in the thymus. However, many self-antigens are not expressed in the thymus. As a result, many T cell clones responsive to nonthymus antigens avoid clonal deletion in the thymus. These self-reactive T cells become anergic, but may persist as unresponsive T cells. The key to maintaining clonal anergy in these potentially dangerous self-reactive T cells is the signal mechanism used to activate T cells after they leave the thymus.

When positively and negatively selected T cells leave the thymus, they migrate to the secondary lymphoid organs (lymph nodes, spleen, and mucosa-associated lymphoid tissue; ⇄ Section 24.1). These antigen-reactive T cells have not yet encountered specific antigen; they are naive or uncommitted T cells. Uncommitted T cells must be activated by an APC to become competent effector cells.

The first step in activation of uncommitted T cells is binding of the MHC–foreign peptide complex on the APC by the TCR (**Figure 26.13**). This first signal is absolutely required for activation. Without TCR interacting with MHC–peptide, a Tc cell cannot be activated. The next step requires the interaction of two more proteins, one found on the APC, called B7, and one found only on T cells, called CD28. The binding of B7 to CD28, a second signal, activates the Tc cell, making it an effector cell. In the absence of a B7–CD28 interaction, the T cell is not activated (Figure 26.13). A Tc cell that is activated will kill any target cell that displays the MHC–foreign peptide complex, even those cells that do not display B7. After a T cell is activated, only the first signal (TCR binding to MHC–peptide) is necessary to induce killer activity. An analogous situation occurs with Th cells.

T Cell Anergy

The requirement for a second activation signal has major implications for establishing and maintaining clonal anergy. Self antigens that are not found in the thymus are present on many other cells in the body. An uncommitted Tc cell that interacts with a self antigen found on a cell that is not an APC will receive only an MHC–peptide signal because non-APCs do not display the B7 protein necessary to complete the second signal. In the absence of the B7–CD28 interaction, a Tc cell that engages MHC–peptide is permanently anergized and can never be activated (Figure 26.13); the B7–CD28 second signal is absolutely required for activation. Uncommitted Th lymphocytes are activated in the same way, also using the B7–CD28 coreceptor second signal.

B Cell Activation

The B cell also has independent signals other than antigen interaction for activation and antibody production. However, different signals activate B cells than those that activate T cells. As we have seen, B cells are responsible for antigen uptake, processing,

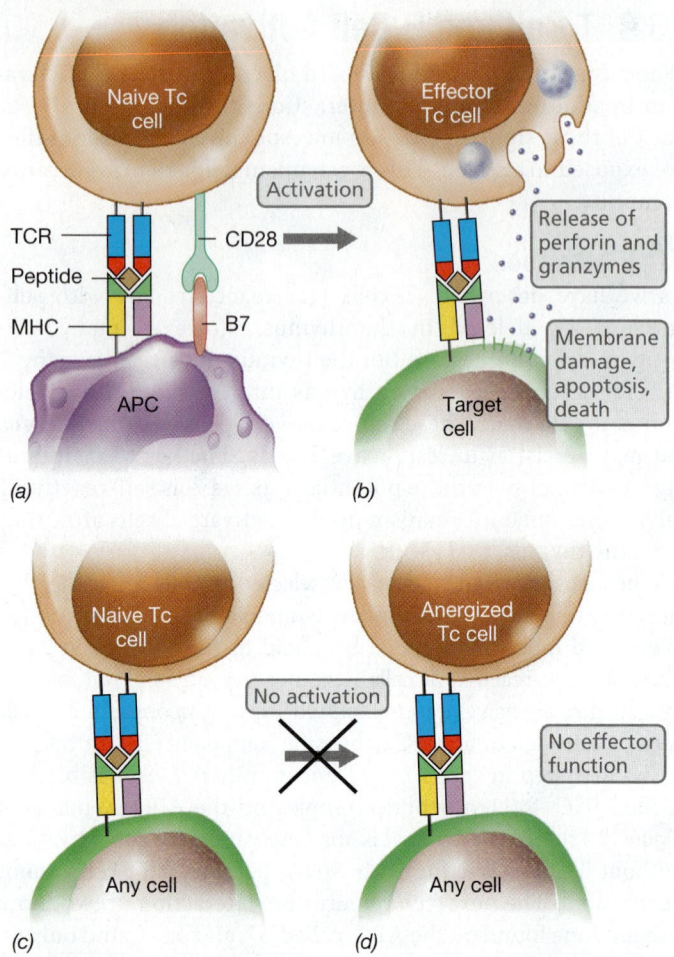

(a)

(b)

(c)

(d)

Figure 26.13   T cell activation. *(a)* A naive Tc cell interacts via TCR with the MHC–peptide complex on an APC, the first required activation signal. CD28 on the Tc cell interacts with B7 on the APC, the second required interaction necessary to activate the naive T cell. *(b)* The activated Tc cell can kill any target cell with the same MHC–peptide. *(c)* A naive Tc cell interacts via the TCR with the MHC–peptide complex on any cell. Conditions for the first signal (interactions via TCR with the MHC–peptide complex) are met, but the second signal cannot be generated because only APCs display the B7 protein. *(d)* In the absence of the second signal, the Tc cell becomes permanently unresponsive, or *anergized*.

and presentation as well as the production of specific antibodies (**Figure 26.14** and ⮌ Section 25.8). The first signal for the B cell is antigen binding and cross-linking of surface immunoglobulin.

The second signal for B cell activation involves several molecules. The antigen–Ig interaction first signal generates a transmembrane signal that stimulates the B cell to express CD40 on its surface. Meanwhile, the B cell ingests the antigen bound on the Igs, processes the ingested antigen to peptides, and presents peptide antigen embedded in MHC II to neighboring Th cells (both Th1 and Th2 cells can be involved in this process). Interaction via the TCR:MHC II–peptide complex stimulates expression of CD40L (CD40 ligand) by the Th cells, which in turn binds to the B cell CD40. The CD40L–CD40 interaction initiates signal transduction in the Th cell, leading to transcription of a number of T cell proteins, including IL-4 and other soluble cytokines. The cytokines, secreted by the T cell, interact with cytokine receptors

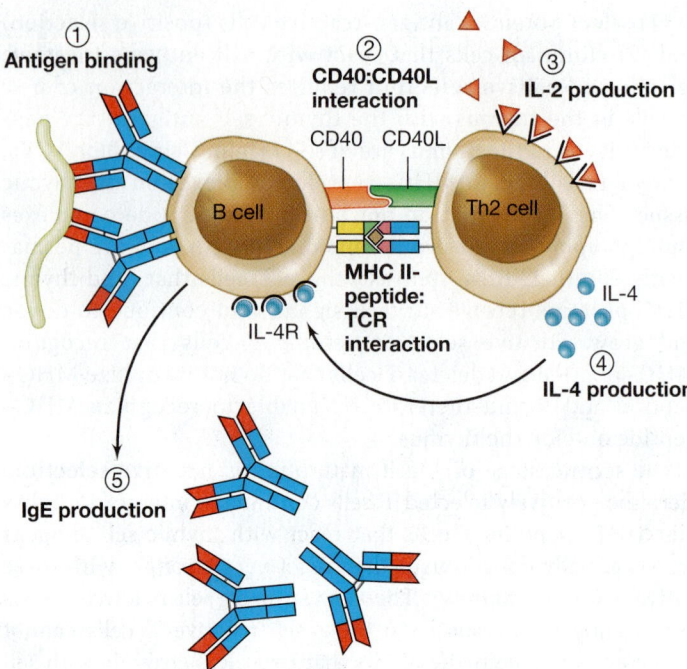

Figure 26.14   B cell activation. ①Antigen binds and cross-links the Ig receptors on a naive B cell, stimulating the B cell to produce CD40 and express it on the cell surface. The B cell processes the antigen and presents it to a Th2 cell via MHC II. ②The Th2 cell TCR interacts with the MHC II–peptide. The Th2 CD40L then interacts with the B cell CD40. ③These interactions stimulate the Th2 cell to produce IL-2, which stimulates the same Th2 cell (autocrine function). ④The stimulated Th2 cell can make several cytokines, one of which is IL-4. IL-4 is a final activation signal for this B cell. ⑤The IL-4–stimulated B cell then produces IgE. Th2 cytokines stimulate both T and B lymphocytes.

on the B cell, completing the second signal for the B cells and stimulating antibody production. Thus, the complete second signal for B cell activation requires two distinct interactions: (1) the interaction of CD40 on the B cell with CD40L on the Th cell; and (2) the interaction of a cytokine produced by the Th cell with the cytokine receptor on the B cell.

After a B cell is activated, it no longer needs T cell interactions or cytokines to make antibody; antigen interaction alone can then stimulate antibody production. Some activated B cells will be transformed into plasma cells that will secrete large amounts of antibody in the primary immune response. Others remain as *memory B cells* and play a major role in the secondary immune response during subsequent exposure to an antigen (⮌ Section 25.8).

MINIQUIZ -
- Define the activation signals for an uncommitted T cell.
- Define the activation signals for an uncommitted B cell.

26.10 Cytokines and Chemokines

Intercellular communication in the immune system is accomplished in many cases through a heterogeneous family of *cytokines,* soluble proteins produced by leukocytes and other cells.

Cytokines regulate cellular functions in immune cells and activate various cell types. The cytokines produced by lymphocytes are often called *lymphokines* or *interleukins* (*ILs*).

Cytokines bind specific receptors. Some cytokines bind to receptors on the cell that produced them. Thus, these cytokines have autocrine (self-stimulatory) abilities. Other cytokines bind to receptors on other cells. Cytokine–receptor binding generally activates a signal transduction pathway that controls transcription and protein synthesis. These signals ultimately result in cell differentiation and clonal proliferation.

Chemokines are a group of small proteins produced by macrophages, lymphocytes, and other cells in response to bacterial products, viruses, and other cell-damaging agents. Chemokines attract phagocytes and T cells to the site of injury, stimulating an inflammatory response and recruiting the cells necessary to initiate a specific immune response.

Table 26.3 lists some important immune cytokines and chemokines, the cells that produce them, their most common target cells, and their most important biological effects. Over 50 cytokines are known, most of which are produced by either T cells or monocytes and macrophages. About 40 chemokines are known. We first examine the activity of cytokines required for the induction of an antigen-specific, antibody-mediated immune response. We then look at cytokines produced by Th1 cells that activate macrophages to produce the cytokines and chemokines that trigger macrophage-mediated inflammation.

Cytokines and Antibody Production

B cells are responsible for antigen uptake, processing, and presentation as well as the production of specific antibodies. As we discussed in the previous section, B cells require two independent signals for activation and antibody production. B cells are activated by antigen binding to surface immunoglobulin (signal 1) followed by interaction between the B cell CD40 and CD40L on the T cell (Figure 26.14). The activated Th cell responds by

producing IL-2, which is secreted and bound by the IL-2R on the surface of the Th cells. Thus, IL-2 can activate the same cell that secreted it. Under the influence of IL-2, the cell divides, making clonal copies. In the process, the Th cell also makes other cytokines such as IL-4 and IL-5.

IL-4 then binds to the IL-4R on the original antigen-presenting B cell. The IL-4:IL-4R interaction stimulates the B cell to differentiate into a plasma cell, which ultimately produces antibodies (⮌ Section 25.8). The IL-4 generated by the responding T cell is the second signal (signal 2) necessary for initiation of antibody production. In addition, the IL-4 interaction signals an immunoglobulin class switch. For example, IL-4:IL-4R interaction can switch antibody production from IgM to IgE or IgG1.

Alternatively, IL-5 produced by Th cells can bind an IL-5R on the antigen-presenting B cell. Parallel to the situation with IL-4:IL-4R interaction, the IL-5:IL-5R also stimulates the B cell to differentiate into a plasma cell, which ultimately produces antibodies, and induces a class switch, but this time to IgA.

As these two examples show, IL-2, IL-4, and IL-5 cytokines are soluble mediators and activators for both T and B lymphocytes. They interact to induce the antibody-mediated immune response; Th cells in different locations produce different B cell–activating cytokines to focus the antibody response to that environment. For example, Th cells near the skin produce more IL-4, inducing production of IgE, while Th cells in the gut produce IL-5, inducing production of secretory IgA. Thus, IL-4 and IL-5 not only control activation of the B cell, but also control the quality of the antibody response, directing the class switch from IgM to IgE or IgA, respectively, thereby focusing the antibody production for a particular environment.

Th1 Activation of Macrophages

Table 26.3 shows the activity of several cytokines produced by some Th1 cells. These cytokines are important in the activation of macrophages. The cytokines IFN-γ (gamma interferon), GM-CSF

UNIT 5

Table 26.3 Major immune cytokines and chemokines			
Cytokine (chemokine)	Major producer cells	Major target cells	Major effect
IL-4[a]	Th2	B cells	Activation, proliferation, differentiation, IgG1 and IgE synthesis
IL-5	Th2	B cells	Activation, proliferation, differentiation, IgA synthesis
IL-2	Naive T cells, Th1, and Tc	T cells	Proliferation (often autocrine)
IFN-γ[b]	Th1	Macrophages	Activation
GM-CSF[c]	Th1	Macrophages	Growth and differentiation
TNF-α[d]	Th1	Macrophages	Activation, production of proinflammatory cytokines
	Macrophages	Vascular epithelium	Activation, inflammation
IL-1β	Macrophages	Vascular epithelium, lymphocytes	Activation, inflammation
IL-6	Macrophages, dendritic cells	Lymphocytes	Activation
IL-12	Macrophages, endothelial cells	NK cells, naive T cells	Activation, enhances differentiation to Th1
IL-17	Th17	Neutrophils	Activation
CXCL8 (chemokine)	Macrophages	Neutrophils, basophils, T cells	Chemotactic factor
CCL2 (MCP-1[e]) (chemokine)	Macrophages	Macrophages, T cells	Chemotactic factor, activator

[a]IL, interleukin; [b]IFN, interferon; [c]GM-CSF, granulocyte–monocyte colony-stimulating factor; [d]TNF, tumor necrosis factor, [e]MCP, macrophage chemoattractant protein.

(granulocyte–monocyte colony-stimulating factor), and TNF-α (tumor necrosis factor alpha) are produced by antigen-activated Th1 cells. These cytokines stimulate macrophage differentiation and activation.

Stimulated macrophages produce a number of cytokines and chemokines, many of which play a role in initiating inflammation. Some of the most important macrophage-produced *proinflammatory* cytokines are IL-1β, TNF-α, IL-6, and IL-12. IL-1β and TNF-α induce activation of vascular endothelium. IL-6 activates lymphocytes, and all except IL-12 induce fever at the systemic level. IL-12 acts to stimulate natural killer (NK) cells and to induce naive T cells to differentiate to Th1 cells (Table 26.3).

Chemokines produced by activated macrophages include CXCL8 and CCL2, also called MCP-1. CXCL8, also called IL-8, is produced by monocytes, macrophages, and other cells. CXCL8 is secreted by the activated cells and binds to receptors on T cells and neutrophils, where it acts as a chemoattractant. This results in a neutrophil-mediated inflammatory response followed by a specific immune response by the attracted T cells. As is the case for the cytokine receptors, engaged chemokine receptors on the target cells act through signal transduction pathways to induce activation of effector cells such as neutrophils or T cells.

CCL2 is produced by macrophages and other cells. CCL2 attracts basophils, eosinophils, monocytes, dendritic cells, natural killer cells, and T cells, stimulating production of more inflammatory mediators and potentially organizing an antigen-specific immune response.

MINIQUIZ

- Identify the major cytokines and chemokines produced by Th1 cells, Th2 cells, and macrophages.
- Identify the proinflammatory cytokines, the cells that produce them, and their effects on other cells.

BIG IDEAS

26.1 • PRRs interact with PAMPs shared by various pathogens, activating complement and phagocytes to target and destroy pathogens. PRR–PAMP interactions initiate signal transduction cascades that activate effector cells.

26.2 • The Ig gene superfamily encodes proteins that are evolutionarily, structurally, and functionally related to immunoglobulins. The antigen-binding Igs, TCRs, and MHC proteins are members of this family. Antigen binding to Ig or TCR facilitates signal transduction through adaptor molecules containing ITAMs.

26.3 • Class I MHC proteins are expressed on all nucleated cells and function to present endogenous antigenic peptides to TCRs on Tc cells. Class II MHC proteins are expressed only on APCs. They function to present exogenously derived peptide antigens to TCRs on Th cells.

26.4 • MHC genes encode proteins used to present peptide antigens to T cells. Class I and class II MHC genes are highly polymorphic. MHC class I and class II alleles encode proteins that bind and present peptides with conserved structural motifs.

26.5 • The antigen-binding site of Ig is composed of the V (variable) domains of one heavy chain and one light chain. Each V region contains three complementarity-determining regions, or CDRs, that are folded together to form the antigen-binding site.

26.6 • Immunoglobulin diversity is generated by several mechanisms. Somatic recombination of gene segments allows shuffling of the various Ig gene segments. Random reassortment of the heavy- and light-chain genes, imprecise joining of VDJ and VJ gene segments, and hypermutation mechanisms contribute to nearly unlimited immunoglobulin diversity.

26.7 • T cell receptors bind to peptide antigens presented by MHC proteins. The CDR3 regions of both the α chain and the β chain bind to the antigen epitope; the CDR1 and CDR2 regions bind to the MHC protein. The β-chain V domain is encoded by VDJ gene segments. The α-chain V domain is encoded by VJ gene segments. TCR diversity, generated by a variety of mechanisms, is nearly unlimited.

26.8 • The thymus is a primary lymphoid organ that provides an environment for the maturation of antigen-reactive T cells. Immature T cells that do not interact with MHC–peptide (positive selection) or that react strongly with self antigens (negative selection) are eliminated by clonal deletion in the thymus. T cells that survive positive and negative selection leave the thymus and can participate in the immune response. B cell reactivity to self antigens is controlled through clonal deletion and anergy.

26.9 • Uncommitted T cells are activated in the secondary lymphoid organs by first binding MHC–peptide with their TCRs (signal 1), followed by binding of the B7 APC protein to the CD28 T cell protein (signal 2). B cell activation is initiated by antigen interaction with surface immunoglobulin (signal 1), followed by interaction between the B cell CD40 protein and CD40L on the T cell to generate cytokine production (signal 2).

26.10 • Cytokines produced by leukocytes and other cells are soluble mediators that regulate interactions between cells. Several cytokines, such as IL-2 and IL-4, affect lymphocytes and are critical components in the generation of specific immune responses. Other cytokines, such as IFN-γ and TNF-α, affect a wide variety of cell types. Chemokines produced by various cells are released in response to injury and are strong attractants for nonspecific inflammatory cells and T cells.

MasteringMicrobiology®   **Review what you know and challenge what you have learned with MasteringMicrobiology!** Access study materials, chapter quizzes, animations, and microbiology lab tutorials in the Study Area to ensure that you have mastered this chapter's content.

REVIEW OF KEY TERMS

Clonal anergy the inability to produce an immune response to specific antigens due to the neutralization of effector cells

Clonal deletion for T cell selection in the thymus, the killing of useless or self-reactive clones

Clonal selection the production by a B or T cell of copies of itself after antigen interaction

Complementarity-determining region (CDR) a varying amino acid sequence within the variable domains of immunoglobulins or T cell receptors where contacts with antigen are made

Human leukocyte antigen (HLA) an antigen-presenting protein encoded by a major histocompatibility complex gene in humans

Immunoglobulin gene superfamily a family of genes that are evolutionarily, structurally, and functionally related to immunoglobulins

Motif in antigen presentation, a conserved amino acid sequence found in all peptides that bind to a given MHC protein

Negative selection in T cell selection, the deletion of T cells that interact with self antigens in the thymus (*see also* clonal deletion)

Polygeny the occurrence of multiple copies of evolutionarily, genetically, structurally, and functionally related genes

Polymorphism in a population, the occurrence of multiple alleles for a gene locus at a higher frequency than can be explained by recent random mutations

Positive selection in T cell selection, the growth and development of T cells that interact with self MHC–peptide in the thymus

Somatic hypermutation the mutation of immunoglobulin genes at rates higher than those observed in other genes

REVIEW QUESTIONS

1. Identify at least one soluble pattern recognition receptor (PRR), its interacting pathogen-associated molecular pattern (PAMP), and the resulting host response. (Section 26.1)

2. Define the criteria used to assign a gene and its encoded protein to the Ig gene superfamily. (Section 26.2)

3. Identify the major structural features of class I and class II MHC proteins. (Section 26.3)

4. Polymorphism implies that each different MHC protein binds a different peptide motif. For the MHC class I polymorphisms, how many different MHC proteins are expressed in an individual? By the entire human population? (Section 26.4)

5. Which Ig chains are used to construct a complete antigen-binding site? Which domains? Which CDRs? (Section 26.5)

6. Calculate the total number of germ-line-encoded V_H and V_L domains that can be constructed from the available Ig genes.

Compare this to the numbers of germ-line-encoded V_β and V_α domains of the TCR. How is diversity expanded in each pair of antigen-receptor proteins? (Section 26.6)

7. In TCRs, diversity can be generated by recombination and reassortment events as in Igs. As is the case in Igs, additional diversity is generated with somatic events such as N nucleotide additions and reading of the D segment in all three reading frames. Explain these diversity-generating mechanisms. (Section 26.7)

8. Explain positive and negative selection of T cells. (Section 26.8)

9. What molecular interactions are necessary for activation of uncommitted T cells? For activation of uncommitted B cells? (Section 26.9)

10. What are the major cytokines and their effects in an antibody-mediated response? In a Th1-mediated response? (Section 26.10)

APPLICATION QUESTIONS

1. Identify the consequences of a genetic mutation that eliminates a PRR by predicting the outcome for the host. Do this for at least one soluble PRR and one membrane-bound PRR.

2. Polymorphism implies that each different MHC protein binds a different peptide motif. However, for the MHC class I proteins, only 6 peptide motifs can be recognized in an individual, whereas over 6000 motifs can be recognized by the entire human population. What advantage does recognition of multiple motifs have for the individual? What potential advantage does recognition of the extremely large number of motifs have for the population? Can everyone process and present the same antigens?

3. Although genetic recombination events are important for generating significant diversity in the antigen-binding site of Igs, post-recombination somatic events may be even more important in achieving overall Ig diversity. Do you agree or disagree with this statement? Explain.

4. What would be the result of activation of all T cells that contact antigen? How does the multiple signal mechanism prevent this from happening?

27 · Diagnostic Microbiology

microbiology**now**

Antibiotics and Honeybees

Honeybees (*Apis mellifera*, photo) produce honey and are key agricultural pollinators. In the 1950s, the broad-spectrum antibiotic oxytetracycline was introduced into honeybee hives in the United States to prevent bee diseases caused by bacterial pathogens. After over 50 years of routine antibiotic use, a study now shows that the gut microflora of most honeybees in the United States contain a set of eight tetracycline resistance genes in high copy number.[1]

The composition of the honeybee gut bacterial community is surprisingly conserved and is dominated by eight species that contribute to defense against parasitic infections and aid in nutrition. The resistance genes are carried between and among these species on plasmids and other mobile genetic elements. Comparisons of American honeybees with those from Europe and New Zealand, where antibiotic use in honeybees is not allowed, show that the bacterial community from bees in untreated hives has not acquired resistance genes. This is also true of honeybees in the United States that have not been exposed to antibiotics for over 25 years and for bumblebees, which share a similar gut microflora with honeybees but are not domesticated and therefore not exposed to oxytetracycline.

Do we really need to treat honeybees with antibiotics, and can we accurately foresee the consequences of long-term antibiotic use? There is no evidence that tetracycline-resistant honeybee bacteria have been transferred to or caused disease in humans, but these antibiotic-resistant bacteria comprise a reservoir of resistance genes that can be transferred to honeybee pathogens and other bacteria. More important, prolonged exposure to a single antibiotic may alter the honeybee microflora in unknown ways that could affect the insects' long-term health and disease resistance.

[1]Tian, B., et al. 2012. Long-term exposure to antibiotics has caused accumulation of resistance determinants in the gut microbiota of honeybees. *mBio 3(6):* e00377–12. doi:10.1128/mBi.oo377–12.

Clinical microbiology laboratories must identify pathogens safely, rapidly, and efficiently. The clinical microbiologist examines patient samples using direct observation, culture, immunological assays, and molecular tools to identify pathogens. Identification of pathogens guides infection control by targeting antimicrobial drugs to specific pathogens.

I • The Clinical Environment

27.1 Safety in the Microbiology Laboratory

Safety in the clinical laboratory prevents the spread of infections to laboratory workers. Standard laboratory practices for handling clinical samples have been established to prevent accidental laboratory infections.

Laboratory Safety

The clinical laboratory has potential biohazards for all personnel and is especially dangerous for untrained personnel or those who do not employ the necessary precautions. All laboratories that handle human or primate tissue must have an occupational exposure control plan for handling bloodborne pathogens, designed specifically to protect workers from infection by hepatitis B virus (HBV, the cause of infectious hepatitis, ↩ Section 29.11) and human immunodeficiency virus (HIV, the cause of acquired immunodeficiency syndrome [AIDS], ↩ Section 29.14). The occupational exposure plan limits infection by all pathogens.

The two most common causes of laboratory infections are ignorance and carelessness. Training and enforcement of established safety procedures, however, can prevent most accidental infections. Most laboratory-acquired infections do not result from identifiable exposures like culture spills but from routine handling of patient specimens. Infectious aerosols generated during microbiological procedures are the most common causes of laboratory infections. Clinical laboratories follow the general safety rules outlined in **Table 27.1** to minimize laboratory infections. Adherence to safety rules ensures a safe laboratory environment that is in compliance with governmental regulations.

These are general standards in all laboratories that handle potential infectious agents and are the basis for all aspects of healthcare infection control. Laboratories that handle particularly dangerous or transmissible agents may have additional rules and procedures in addition to these to ensure a safe work environment, as discussed below. In the final analysis, safety in the workplace is the responsibility of laboratory personnel.

Biological Containment and Biosafety Levels

The level of containment used to prevent accidental infections or accidental environmental contamination (escape) in clinical, research, and teaching laboratories must be adjusted to counter the biohazard potential of the organisms handled in the laboratory. Laboratories are classified according to their containment capabilities from least to greatest by their *biosafety level (BSL)*, designated as *BSL-1, BSL-2, BSL-3,* and *BSL-4* (**Figure 27.1**). Personnel in laboratories working at all biosafety levels must follow good laboratory practices that ensure basic cleanliness and limit contamination, as we outlined in Table 27.1. **Table 27.2** shows requirements for each biosafety level. At each level, the precautions, equipment, and costs increase.

Most universities have BSL-1 and BSL-2 facilities for teaching and research. Standard clinical laboratories operate at BSL-2. The

Table 27.1 Microbiology laboratory safety standards	
Rule	*Implementation*
Restrict access	Only laboratory workers and trained support personnel have access.
Practice good personal hygiene	Eating, drinking, and application of cosmetics are forbidden in the lab. Handwashing prevents spread of pathogens.
Use personal protection equipment	Lab coats, gloves, eye protection, and respirators are recommended or required depending on the pathogens being handled.
Vaccinate	Personnel must be vaccinated for agents they may contact.
Handle specimens safely	Consider clinical specimens to be infectious and handle appropriately.
Decontaminate	After use or exposure, decontaminate specimens, surfaces, and materials by disinfecting, autoclaving, or incinerating.

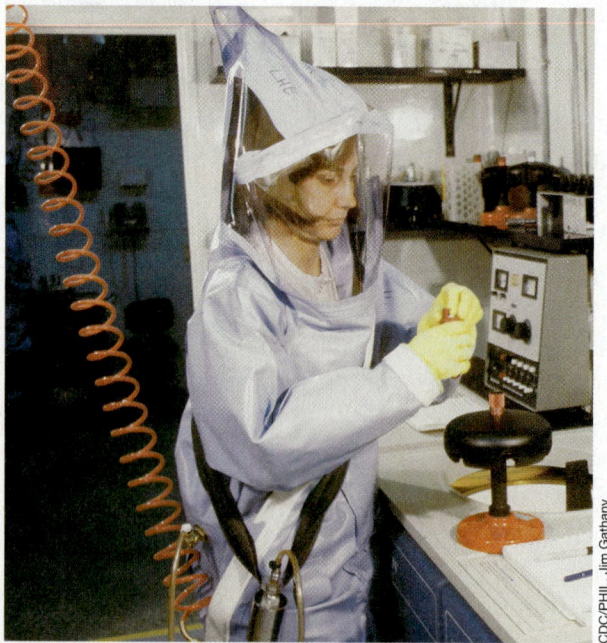

CDC/PHIL, Jim Gathany

Figure 27.1 A worker in a BSL-4 (biosafety level 4) laboratory. BSL-4 is the highest level of biological control, affording maximum worker protection and pathogen containment. The worker has a whole-body sealed suit with an outside air supply and ventilation system. Air locks control all access to the laboratory. All material leaving the laboratory is autoclaved or chemically decontaminated.

Table 27.2 Biosafety levels and guidelines for microbiology laboratories

Biosafety level	Access	Precautions/specialized equipment [a]	Example and purpose	Microorganism examples
BSL-1	Should be limited	Barrier protection (coats, gloves) *should* be used.	Teaching laboratory that does not work with known pathogens	*Bacillus subtilis*
BSL-2	Must be limited	Barrier protection *must* be used. Manipulations that may generate aerosols must be performed in a biological safety cabinet.	Laboratory that works with moderate-risk pathogens	*Streptococcus pyogenes*, *Escherichia coli*
BSL-3	Must be limited; separated from public corridors	Barrier protection *must* be used. The laboratory is negatively pressurized and fitted with filters to prevent pathogen escape. Manipulations must be performed in a biological safety cabinet.	Laboratory that works with emerging and high-risk pathogens	*Mycobacterium tuberculosis*, human immunodeficiency virus (HIV)
BSL-4	Must be limited; separated from public corridors	Barrier protection *must* be used. Manipulations must be performed in a sealed biological safety cabinet or by personnel wearing positive pressure suits with an air supply, in addition to BSL-3 requirements (Figure 27.1).	Laboratory that works with emerging and high-risk pathogens, especially those spread by aerosols, or for which there is no treatment, cure, or vaccine	Ebola virus, drug-resistant *Mycobacterium tuberculosis*

specialized physical requirements for BSL-3 facilities limit them to major clinical centers and research settings. Because BSL-4 facilities must ensure total isolation and physical containment of pathogens, fewer than fifty BSL-4 laboratories are operational worldwide. Most BSL-4 laboratories are associated with government facilities such as the Centers for Disease Control and Prevention (Atlanta, Georgia, USA) and the U.S. Army Medical Research Institute of Infectious Diseases (USAMRIID; Fort Detrick, Maryland, USA).

MINIQUIZ
- What are the major causes of laboratory infections?
- Identify the biological hazard containment features of BSL-1, BSL-2, BSL-3, and BSL-4 laboratories.

27.2 Healthcare-Associated Infections

A **healthcare-associated infection (HAI)** is a local or systemic infection acquired by a patient in a healthcare facility, particularly during a stay in the facility. HAIs cause significant morbidity and mortality. About 1 in 20 patients admitted to healthcare facilities acquire HAIs, also called *nosocomial infections* (*nosocomium* is the Latin word for "hospital"). Nearly 1.7 million HAIs occur annually in the United States, leading directly or indirectly to almost 100,000 deaths.

Mechanisms of Transfer of Healthcare-Associated Infections

Some HAIs are acquired from patients with communicable diseases, but others are caused by pathogens that are selected and maintained within the hospital environment, spread by cross-infection from patient to patient or from healthcare personnel. Healthcare-associated pathogens are often present as normal flora in either patients or healthcare staff.

Healthcare facilities are high-risk environments for the spread of infectious disease because these facilities concentrate individuals who have infectious disease or are at risk for acquiring infectious disease because of their underlying health conditions. Some of the common risk factors for acquiring infectious diseases in healthcare settings are summarized in Table 27.3.

The most common sites of HAIs are shown in Figure 27.2. Of the nearly 100,000 estimated annual deaths caused by HAIs in the United States, about 36,000 are from pneumonia, 31,000 from bloodstream infections, 13,000 from urinary tract infections, 8,000 from surgical site infections, and 11,000 from other sites. A relatively small number of pathogens cause most HAIs (Table 27.4), but a number of other infectious agents can also cause HAIs.

Common Pathogens in Healthcare-Associated Infections

Staphylococcus aureus is one of the most important and widespread HAI pathogens (⟳ Section 29.9). It is the most common cause of pneumonia, the third most common cause of blood

Table 27.3 Risk factors for hospital-acquired infections

Risk factor	Rationale
Patients	Patients are already ill or compromised
Newborn infants and the elderly	Not fully immune competent
Infectious disease patients	Pathogen reservoirs
Patient proximity	Increases cross-infection
Healthcare personnel	Can transfer pathogens between and among patients
Medical procedures (blood draws, etc.)	Breaching the skin barrier can introduce pathogens
Surgery	Exposes internal organs, may introduce pathogens, and stress lowers resistance to infection
Anti-inflammatory drugs	Lower resistance to infection
Antibiotic treatment	May select for resistant and opportunistic pathogens

Table 27.4 Healthcare-associated pathogens

Pathogen	Common infection sites and diseases	Micrographs[b]
[a]*Acinetobacter*	Wound/surgical site, bloodstream, pneumonia, urinary	*Acinetobacter*
Burkholderia cepacia	Pneumonia	*B. cepacia*
Clostridium difficile, *C. sordellii*	Gastrointestinal Pneumonia, endocarditis, arthritis, peritonitis, myonecrosis	*C. difficile*
[a]*Enterobacteriaceae,* carbapenem-resistant, especially *Escherichia* coli and *Klebsiella*	Pneumonia, wound/surgical site, bloodstream, meningitis	*E. coli*
[a]Vancomycin-resistant *Enterococcus* (VRE)	Wound/surgical site, bloodstream, urinary	*Klebsiella*
Hepatitis	Chronic liver infection	*Enterococcus*
Human immunodeficiency virus (HIV)	Immunodeficiency	Hepatitis B virus
Influenza virus	Pneumonia	HIV
Mycobacterium abscessus	Skin and soft tissue infections	Influenza virus
[a]*M. tuberculosis*	Chronic lung infection (tuberculosis)	*M. tuberculosis*
Norovirus	Gastroenteritis	Norovirus
[a]*Staphylococcus aureus* Methicillin-resistant (MRSA) Vancomycin-intermediate and resistant (VISA, VRSA)	Bloodstream, pneumonia, endocarditis, osteomyelitis	*S. aureus*

[a]Antibiotic-resistant organisms that exhibit multiple drug resistance.
[b]All inset micrographs are colorized scanning or transmission electron micrographs and all are from CDC/PHIL. Additional micrograph credits (numbers run top to bottom): 1–5, 10, and 12, Janice Haney Carr; 6, Peta Wardell; 7, Erskine Palmer; 8, A. Harrison and P. Feorino; 9, Frederick Murphy; 11, Charles D. Humphrey.

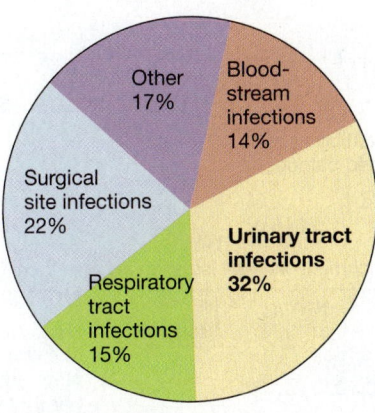

Figure 27.2 Healthcare-associated infections. About 1.7 million healthcare-associated infections occur annually in the United States. Data are from Klevens et al., *Public Health Reports 122:* 160–166, 2007.

infections, and is particularly problematic in nurseries. Many hospital strains of *S. aureus* are unusually virulent and are resistant to common antibiotics, making treatment very difficult. The staphylococci are the most common cause of bloodborne HAIs and are also very prevalent in wound infections.

Staphylococcus and *Enterococcus*, and *Escherichia coli, Klebsiella pneumonia*, and other *Enterobacteriaceae* all have the potential for causing HAIs, but they may also be members of the normal flora in some individuals, making it difficult to eliminate them in healthcare settings. In addition, these organisms can acquire drug resistance. Pathogens that are not part of the normal flora such as *Acinetobacter* and *Mycobacterium* spp. can be eliminated from the healthcare environment. These pathogens are carried into the healthcare environment by infected individuals or, in the case of some mycobacteria, as environmental contaminants that enter in dust and air.

Prevention of HAIs involves cooperation between the healthcare facility infection-control team and the facility staff, including direct healthcare workers and supporting staff such as housekeeping. Infection control starts with management of incoming patients at the point of entry to the healthcare facility; incoming patients should be assessed for possible infections and isolated as necessary to prevent spread of infections to other staff and patients. From this point, the healthcare facility staff employs standard procedures that limit infection, applying the same general precautions outlined for laboratory workers in Table 27.1.

MINIQUIZ ---

• Why are patients in healthcare facilities more susceptible than normal individuals to pathogens?

• How can the spread of HAIs be controlled?

II • Microbiological Identification of Pathogens

The observation and growth of pathogens from patient specimens are important strategies for identification of the causative agent of an infectious disease. Identification leads to antimicrobial drug susceptibility testing and development of a specific treatment plan. We begin by looking at methods for observing, growing, and isolating pathogens, followed by methods for identification and drug susceptibility testing.

27.3 Direct Detection of Pathogens

If clinically relevant organisms are to be isolated and identified, the specimen must be obtained and handled properly to ensure that the pathogen survives. First, the specimen must be obtained from the actual site of the infection; the sample must be taken aseptically to avoid contamination with irrelevant microorganisms. Next, the sample size must be large enough to ensure an inoculum sufficient for growth. Third, the metabolic requirements for organism survival must be maintained during sampling, storage, and transport. Finally, the sample must be processed as quickly as possible to avoid degradation. For example, samples obtained from anoxic sites must be obtained, stored, and transported under anoxic conditions to ensure the survival of potential anaerobic pathogens.

Samples of tissues or fluids are collected for microbiological, immunological, and molecular biological analyses if a healthcare provider suspects a disease is caused by an infectious agent (**Figure 27.3**). Typical samples include blood, urine, feces, sputum, cerebrospinal fluid, or pus from a wound. Sterile swabs are often used to obtain samples from suspected infected areas such as wounds, skin, nares, or throat (**Figure 27.4**). The swab is then used to inoculate the surface of an agar plate or a tube of liquid culture medium. In some cases, a small piece of tissue (biopsy) may be obtained for culture or microscopic examination. Blood and other fluids are initially analyzed using automated microbiology methods.

Most pathogens are detected by direct means using one or more of several diagnostic tests. The reliability of any diagnostic test depends on both the *specificity* and the *sensitivity* of the test. **Specificity** is the ability of the test to recognize a single pathogen. Optimal specificity implies that the test is specific for a single pathogen, and will not identify any other pathogen. High specificity reduces false-positive results. For example, for the detection of *Neisseria gonorrhoeae*, the organism that causes gonorrhea, the specificity of Gram-stained smears of urethral exudates from men is about 99% and about 95% for endocervical exudates from women; false-positive tests for gonorrhea are therefore rare.

Sensitivity defines the lowest numbers of a pathogen or the lowest amount of a pathogen product that can be detected. The highest level of sensitivity requires that the test be capable of identifying a single organism or molecule. High sensitivity prevents false-negative reactions. For example, for the detection of *N. gonorrhoeae*, the sensitivity of Gram-stained smears of urethral exudates from men is about 90%, and about 50% for endocervical exudates from women. Thus, the test is a sensitive indicator for gonorrhea in men, but is much less sensitive for women.

UNIT 5

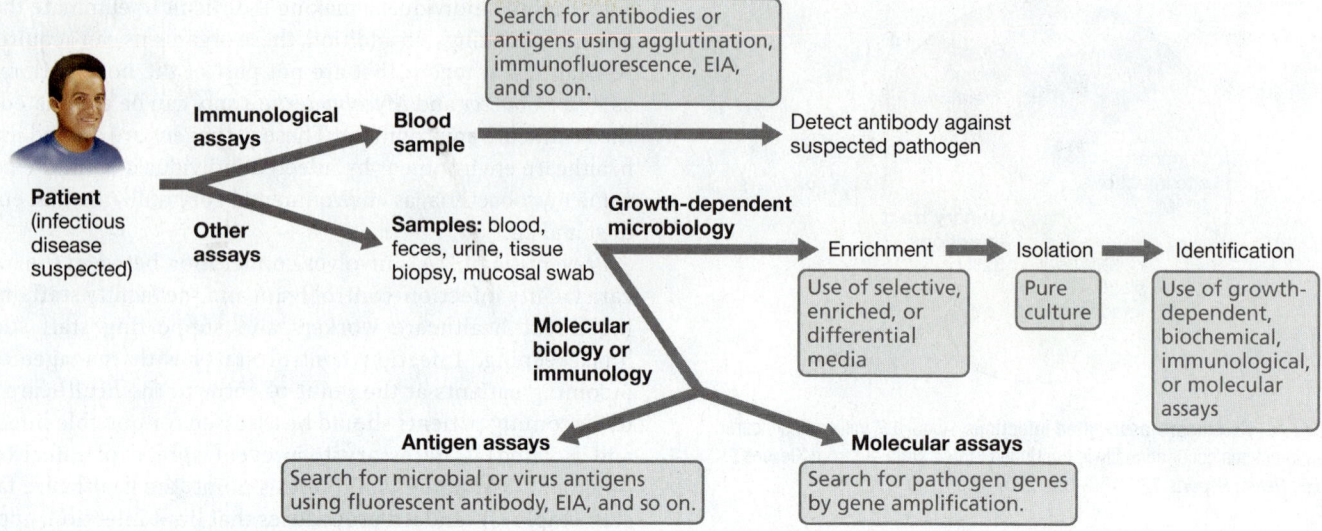

Figure 27.3 Laboratory identification of microbial pathogens. The flowchart shows alternative paths for identifying pathogens or pathogen exposure in the clinical laboratory.

In suspected cases of gonorrhea in females, false-negative Gram stains are relatively common, and thus females must be examined by more sensitive methods, including culture techniques.

Direct Observation and Culture

Direct observation of pathogens in clinical specimens is an important tool for diagnosis of many infectious diseases. Examples include the observation of acid-fast stained *Mycobacterium tuberculosis* in sputum from patients, proving infection with *M. tuberculosis* (➾ Section 29.4 and Figure 29.16a). Likewise, infection with *N. gonorrhoeae* can be diagnosed by direct observation of Gram-stained smears of a patient sample such as urethral exudates from men. The presence of gram-negative diplococci in clumps and in inclusions in neutrophils is diagnostic for the disease (**Figure 27.5a**). In women, however, cervical smears often do not reveal the presence of the infecting organism; cultures or molecular techniques are used to establish or confirm a diagnosis of gonorrhea (➾ Section 29.12 and Figure 27.5b).

Most pathogens can be readily grown in laboratory cultures. **Enrichment culture**, the use of selected culture media and incubation conditions to isolate microorganisms from samples (➾ Section 18.1), is thus an important tool in the clinical laboratory.

Most microorganisms of clinical importance can be grown, isolated, and identified using specialized growth media. Clinical samples are inoculated on **general-purpose media** such as *blood agar* (**Figure 27.6a**) and *chocolate agar* (Figure 27.6b) that support the growth of many aerobic and facultatively aerobic pathogens. Chocolate agar, named because of its deep brown appearance, contains heat-lysed blood. The heated blood interacts with the other media components, absorbing compounds that are toxic for fastidious microorganisms such as *N. gonorrhoeae*. More specialized media may also be used.

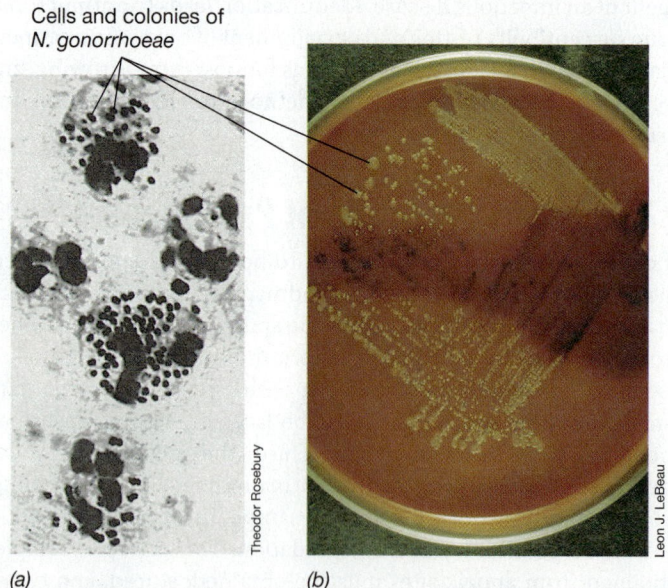

Cells and colonies of *N. gonorrhoeae*

(a)   (b)

Theodor Rosebury   Leon J. LeBeau

Figure 27.5 Identification of *Neisseria gonorrhoeae*. (a) *N. gonorrhoeae* cells within human polymorphonuclear leukocytes from a urethral exudate. Note the paired diplococci (leader). (b) *N. gonorrhoeae* growing on Thayer–Martin agar. The plate has been stained in the middle with a reagent that turns colonies blue if cells contain cytochrome *c* (the oxidase test). *N. gonorrhoeae* colonies in contact with the reagent are blue, indicating that they are oxidase-positive.

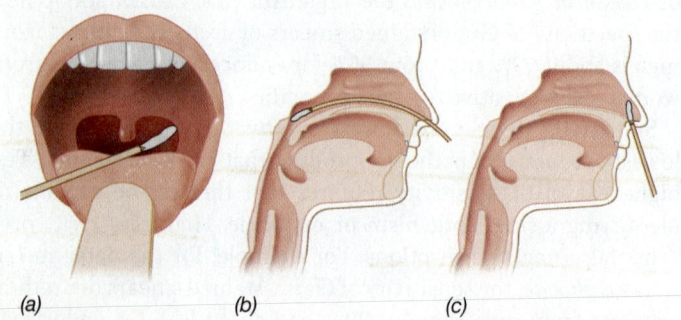

(a)   (b)   (c)

Figure 27.4 Specimens from the upper respiratory tract. (a) Throat swab. (b) Nasopharyngeal swab passed through the nose. (c) Swabbing the inside of the nose.

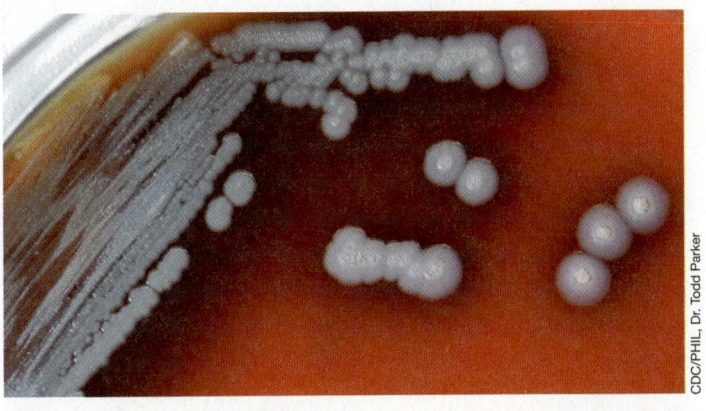

(a)

(b)

(c)

CDC/PHIL, Dr. Todd Parker

CDC/PHIL, Dr. Todd Parker

John Martinko and Cheryl Broadie

Figure 27.6 Enriched media. *(a) Burkholderia* growing on sheep blood agar (SBA); the red color is from blood suspended in a rich medium such as trypticase soy agar. *(b) Francisella tularensis* growing on chocolate agar; the brown color is due to heat-lysed blood in a rich medium such as trypticase soy agar. *(c) Escherichia coli*, a lactose fermenter (left), and *Pseudomonas aeruginosa*, a non–lactose fermenter (right) growing on an eosin–methylene blue (EMB) agar. The green metallic sheen of the colonies identifies *E. coli* as a lactose fermenter.

The next step in identification makes use of *enriched media* to culture selected pathogens. **Enriched media** contain specific growth factors that enhance the growth of selected pathogens. For example, the Thayer–Martin medium enhances the growth of bacteria such as *N. gonorrhoeae* (Figure 27.5*b*). **Selective media** allow for some organisms to grow while inhibiting the growth of others due to the presence of inhibitory agents. Finally,

differential media are specialized media that allow identification of organisms based on their growth, color, and appearance on the medium. Eosin–methylene blue (EMB) agar is a selective medium that inhibits growth of Gram-positive organisms while supporting growth of Gram-negative organisms. In addition, EMB agar is a differential medium because it distinguishes lactose fermenters such as *Escherichia coli* from non-lactose-fermenting Gram-negative organisms such as *Pseudomonas aeruginosa* (Figure 27.6*c*). Table 27.5 indicates the specimens usually inoculated onto each of the most commonly used general-purpose and enriched culture media.

Blood and Liquid Specimens

Pathogens in large-volume liquid samples like blood and cerebrospinal fluid are routinely detected using automated culture systems, followed by microscopic examination and subculture.

The standard blood culture procedure is to draw 10 to 20 ml of blood aseptically from a vein and inject it into two culture bottles containing both an anticoagulant and a general-purpose culture medium. One bottle is incubated aerobically and one is incubated under anoxic conditions, and both are kept at 35°C for up to 5 days. Automated culture systems detect growth by monitoring gas consumption (in oxic conditions) or production (carbon dioxide in anoxic conditions) as often as every 10 minutes. Some systems also measure turbidity. Most clinically significant bacteria are recovered within 2 days, but growth of organisms such as mycobacteria or fungi may take 3 to 5 days or longer. Bottles that contain growth are examined first by Gram stain (꘍ Section 2.2) and then inoculated onto enrichment and differential media for further isolation and identification.

Bacteremia is the presence of bacteria in the blood. Bacteremia is extremely uncommon in healthy individuals, normally occurring only transiently due to tooth brushing, dental surgery, or trauma. The prolonged presence of bacteria in the blood is generally

Table 27.5 Enriched and selective media for primary isolation of pathogens

Specimen	Media[a]		
	Blood agar	CA	Enteric agar
Fluids from chest, abdomen, pericardium, joint	+	+	+
Feces: rectal or enteric transport swabs	+	+	+
Surgical tissue biopsies	+	−	+
Throat, sputum, tonsil, nasopharynx, lung, lymph nodes	+	+	+
Urethra, vagina, cervix	+	+	+
Urine	+	−	+
Blood[b]	+	+	+
Wounds, abscesses, exudates	+	+	+

[a]Blood agar, 5% whole sheep blood in trypticase soy agar; CA, chocolate agar (heated blood in trypticase soy agar); enteric agar, for example eosin–methylene blue (EMB) agar.
[b]Blood is cultured initially in broth. Depending on the Gram stain characteristics of isolates, subculturing is done on enteric agar (gram-negative) or chocolate agar (gram-positive).

UNIT 5

indicative of systemic infection. **Septicemia**, or **sepsis**, is a blood infection by a virulent organism that enters the blood from a focus of infection, multiplies, and travels to various body tissues to initiate new infections. Sepsis is extremely serious and can be fatal.

The most common pathogens found in blood include gram-positive *Staphylococcus* spp. and *Enterococcus* spp., but about 2–3% of blood cultures are contaminated by microorganisms such as *Staphylococcus epidermidis*, coryneform bacteria, or propionibacteria introduced from the skin during blood sampling. However, these organisms can also infect the heart (subacute bacterial endocarditis) or colonize artificial heart valves. Thus, the results of a positive blood culture must be reconciled with clinical observations for an accurate diagnosis.

Urinary Tract and Fecal Cultures

Urinary tract infections are common, especially in women. Interpretation of microbiological findings from urine cultures can be confusing because the disease-causing agents are often members of the normal flora (for example, *E. coli*). In most cases, the urinary tract becomes infected by organisms that ascend into the bladder from the urethra. Urinary tract infections are, however, the most common healthcare-associated infections, often introduced through catheters.

Direct microscopic examination of urine may indicate *bacteriuria*, the presence of abnormal numbers of bacteria in the urine, but nearly all urine contains some bacterial growth. A Gram stain may be done directly on urine samples to identify the morphology of potential urinary tract pathogens. This method can be used to putatively identify gram-negative rods including the enteric bacteria, gram-negative cocci such as *Neisseria*, and gram-positive cocci such as *Enterococcus*. The Gram stain and other direct staining methods are also useful for direct detection of bacteria in other body fluids such as sputum and wound exudates.

A significant urinary tract infection typically results in bacterial counts of 10^5 or more organisms per milliliter of urine. The most common urinary tract pathogens are enteric bacteria, with *E. coli* accounting for about 90% of the cases. Two media types are normally used to culture urinary tract pathogens. Blood agar can be used for initial isolation. Selective and differential enteric media such as eosin–methylene blue (EMB) agar permit the initial differentiation of lactose fermenters from non–lactose fermenters and inhibit the growth of possible contaminants such as gram-positive *Staphylococcus* species (Figure 27.6c). Additional differential and selective media may be used to differentiate between and among potential gram-negative pathogens (**Figure 27.7**).

Urine cultures can be done quantitatively by counting colonies on blood agar or a selective agar medium. A calibrated loop is used to deliver a specified amount of urine, usually 1 μl, as the inoculum for a plate. If no bacterial growth is obtained despite persistent urinary tract infection symptoms, a clinician may request additional cultures for more nutritionally demanding organisms such as *N. gonorrhoeae* and *Chlamydia trachomatis*.

Proper collection and preservation of feces is important for the isolation of intestinal pathogens. During storage, feces become more acidic, so delay between sampling and processing must be minimized. This is especially critical for the isolation of acid-sensitive pathogens such as *Shigella* and *Salmonella*.

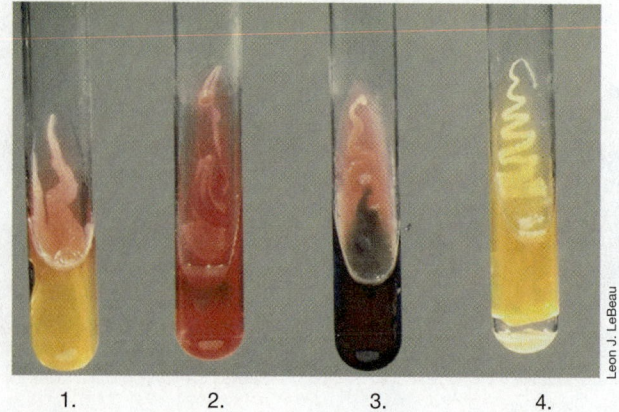

Leon J. LeBeau

Figure 27.7 **A growth-dependent diagnostic test for gram-negative pathogens.** Four different enteric bacteria are shown growing on triple-sugar iron (TSI) agar. The medium contains glucose, lactose, and sucrose. Organisms able to ferment only glucose cause acid formation only in the butt, whereas lactose- or sucrose-fermenting organisms cause acid formation throughout the slant. Gas formation is indicated by the breaking up of the agar in the butt. Hydrogen sulfide formation (either from protein degradation or from reduction of thiosulfate in the medium) is indicated by blackening due to reaction with ferrous iron in the medium. The medium is inoculated both on the surface of the slant and by stabbing into the solid agar butt. From left to right: 1. Fermentation of glucose only, typical of *Shigella*. 2. Growth but no fermentation, typical of *Pseudomonas*, a nonfermenting obligate aerobe. 3. Hydrogen sulfide formation, typical of *Salmonella*. 4. Fermentation of sugars and hydrogen gas production, typical of *Escherichia coli*.

Freshly collected fecal samples are placed in a sterile sealed container for transport to the laboratory. Bloody or pus-containing stools as well as stools from patients with suspected foodborne or waterborne infections are inoculated into selective media for isolation of individual bacteria. Intestinal eukaryotic pathogens are identified by direct microscopic observation of cysts in the stool sample or through antigen-detection assays rather than by culture methods. Many laboratories also use selective and differential media to identify *E. coli* O157:H7 and *Campylobacter*, two important intestinal pathogens typically acquired from contaminated food or water (∞ Sections 31.11 and 31.12).

Wounds and Abscesses

Infections associated with traumatic injuries such as animal bites, burns, or cuts are sampled to recover the relevant pathogen. The results must be interpreted carefully to differentiate between infection and contamination. Wound infections and abscesses are frequently contaminated with normal flora, and swab samples from such lesions are frequently misleading. For abscesses and other purulent lesions, pus is aspirated with a sterile syringe and needle following disinfection of the skin surface. Internal purulent lesions are sampled by biopsy or from tissues removed in surgery.

Pathogens commonly associated with wound infections are *Staphylococcus aureus*, enteric bacteria, *Pseudomonas aeruginosa*, and anaerobes such as *Bacteroides* and *Clostridium*. Because of the varied oxygen requirements of these bacteria, samples must be obtained, transported, and cultured under anoxic as well as oxic conditions. The major isolation media are blood agar, selective media for enteric bacteria, and enrichment media containing additional supplements and reducing agents for obligate anaerobes. Gram stains from such specimens are examined directly by microscopy.

Genital Specimens and Culture for Gonorrhea

In males, a purulent urethral discharge indicates a sexually transmitted infection (STI). These STIs are classified as nongonococcal or gonococcal urethritis. Nongonococcal urethritis is usually caused by *Chlamydia trachomatis, Ureaplasma urealyticum,* or *Trichomonas vaginalis* (⮁ Section 29.13). Gonococcal urethritis is caused by *Neisseria gonorrhoeae* (⮁ Section 29.12).

N. gonorrhoeae is usually found as gram-negative diplococci. No similar microorganisms are normally found in the urogenital tract and so a Gram stain of a urethral, vaginal, or cervical smear showing gram-negative diplococci is diagnostic for gonorrhea. Microscopic examination of purulent discharges usually reveals gram-negative diplococci in neutrophils (Figure 27.5*a*). Chocolate agar, a nonselective enriched medium, is often used for specimens suspected to contain *N. gonorrhoeae*. A selective medium used for primary isolation is modified Thayer–Martin (MTM) agar (Figure 27.5*b*). This medium incorporates the antibiotics vancomycin, nystatin, trimethoprim, and colistin to suppress the growth of normal flora. These antibiotics have no effect on *N. gonorrhoeae* or *Neisseria meningitidis*, a cause of bacterial meningitis (⮁ Section 29.5).

Inoculated plates are incubated in a humid environment in an atmosphere containing 3–7% CO_2 for 24 and 48 hours and tested for their oxidase reaction because *Neisseria* are oxidase-positive (Figure 27.5*b*). Oxidase-positive, gram-negative diplococci growing on chocolate agar or selective media are presumed to be gonococci if the inoculum was derived from genitourinary sources. Definitive identification of *N. gonorrhoeae* requires determination of carbohydrate utilization patterns and immunological or nucleic acid probe tests. Laboratory testing of urogenital samples for *N. gonorrhoeae* (and the often-associated *C. trachomatis*) is often done using DNA amplification via polymerase chain reaction (PCR) or other molecular methods.

Culture of Anaerobic Microorganisms

Obligately anaerobic bacteria are common causes of infection, and their identification requires special isolation and culture methods. In general, media for anaerobes do not differ greatly from those used for aerobes, except that they are (1) usually richer in organic constituents, (2) contain reducing agents (usually cysteine or thioglycolate) to remove oxygen, and (3) contain a redox indicator to indicate that conditions are anoxic. Collection, handling, and processing of specimens must exclude oxygen contamination because oxygen is toxic to obligately anaerobic organisms.

Several habitats in the body, such as portions of the oral cavity and the lower intestinal tract, are anoxic and support the growth of anaerobic normal flora. Other parts of the body, however, can also become anoxic as a result of tissue injury or trauma that reduces blood supply and oxygen perfusion to the injured site. These anoxic sites can then be colonized by obligate anaerobes. In general, potentially pathogenic anaerobic bacteria are part of the normal flora but are kept in check by competition from other members of the normal flora. Under certain conditions, however, these normally benign bacteria may become opportunistic

pathogens. For example, *Clostridium difficile* is usually a harmless member of the normal flora in the lower intestinal tract, but commonly emerges as a healthcare-associated pathogen after intensive antibiotic therapy destroys the competing normal microbial flora (Table 27.4).

Isolation, growth, and identification of anaerobic pathogens are complicated by specimen contamination as well as the constant challenge of maintaining an anoxic environment during collection, transport, and culture. Samples collected by syringe aspiration or biopsy must be immediately placed in a tube containing oxygen-free gas, usually with a dilute salt solution containing a reducing agent such as thioglycolate and a redox indicator such as resazurin (⮁ Figure 5.27) to monitor oxygen contamination.

For anoxic incubation, specimens are inoculated into anaerobic culture bottles in an automated culture system or incubated under anoxic conditions after inoculation of culture media containing reducing agents, usually in an anoxic "glove box" filled with an oxygen-free gas such as nitrogen or hydrogen (⮁ Figure 5.28*b*).

MINIQUIZ --

- Identify culture methods and conditions used for blood, wound, urine, fecal, and genital specimens.
- Describe the methods used to maintain optimum conditions for the isolation of anaerobic pathogens.

27.4 Growth-Dependent Identification Methods

If growth is detected after inoculation of a specimen on a general-purpose medium, the clinical microbiologist must identify the organism or organisms present. Many microorganisms recovered from clinical samples can be identified using growth-dependent assays.

Based on its growth characteristics on the enriched media used for primary isolation, a presumptive pathogen is subcultured onto specialized media designed to measure one of many different biochemical reactions. Here we give examples of a few standard laboratory biochemical tests. Each test is designed to differentiate bacterial growth and metabolism patterns, with the goal of identifying an individual pathogen based on its unique pattern of growth on the selected media.

The media employed are selective, differential, or both. Eosin–methylene blue (EMB) agar, for example, is a widely used selective and differential medium for the isolation and differentiation of enteric bacteria. Methylene blue is selective because it inhibits the growth of gram-positive bacteria, and thus only gram-negative organisms can grow. EMB agar has an initial pH of 7.2 and contains lactose and sucrose, but not glucose, as energy sources. Acidification changes eosin, the differential media component, from colorless to red or black. Strong lactose-fermenting bacteria such as *Escherichia coli* acidify the medium and the colonies appear black with a greenish sheen. Enteric bacteria such as *Klebsiella* or *Enterobacter* produce less acid, and colonies on EMB are pink to red. Colonies of non–lactose fermenters, such as *Salmonella*, *Shigella*, and *Pseudomonas*, are translucent or pink (Figure 27.6c). Thus, EMB is selective for the growth of gram-negative bacteria and also differentiates among common enteric bacteria.

Many differential media incorporate biochemical tests to measure the presence or absence of enzymes involved in catabolism of a specific substrate or substrates. Hundreds of differential biochemical tests are known, but only about 20 are used routinely. One example is the differential triple-sugar iron (TSI) agar test used to differentiate enteric pathogens. The fermentation and gas production patterns differentiate these bacteria at the genus and sometimes species levels (Figure 27.7). Another testing method uses chromogenic substrates that alter the color of colonies of targeted organisms. For instance, CHROMagar, a proprietary selective and differential media, inhibits the growth of most microorganisms. Methicillin-resistant *Staphylococcus aureus* (MRSA), however, produces distinctive pink colonies; the fluorogenic medium contains compounds that fluoresce when metabolized by MRSA.

The biochemical reaction patterns for pathogens are stored in a databank. Using the results of differential tests on an unknown pathogen, the computer matches the characteristics of the unknown organism to metabolic patterns of known pathogens, allowing identification. As few as three or four key tests are sufficient to make an unambiguous identification of many pathogens. However, in some cases more sophisticated identification procedures are required. The clinical microbiologist decides which diagnostic tests to use based on the origin of the clinical specimen, the characteristics of a pure culture of the specimen grown on general-purpose media (for example, morphology and Gram stain), and previous experience with similar cases.

MINIQUIZ -
- Distinguish between general-purpose, selective, and differential media. Give an example of a medium used for each purpose.
- Identify the selective and differential components of EMB agar and explain how each works.

27.5 Antimicrobial Drug Susceptibility Testing

Pathogens isolated from clinical specimens are identified to confirm medical diagnoses and to guide antimicrobial therapy. For many pathogens, appropriate and effective antimicrobial treatment is based on current experience and practices. For a select group of pathogens, however, decisions about appropriate antimicrobial therapy must be made on a case-by-case basis. Such pathogens include those for which antimicrobial drug resistance is common (for example, gram-negative enteric bacteria), those that cause life-threatening disease (for example, meningitis caused by *Neisseria meningitidis*), and those that require bacteriocidal rather than bacteriostatic drugs (⮌ Section 5.19) to prevent disease progression and tissue damage. Bacteriocidal agents are indicated, for example, for organisms that cause bacterial endocarditis, where total and rapid killing of the pathogen is critical for patient survival.

Minimum Inhibitory Concentration

Antimicrobial susceptibility is measured by determining the smallest amount of agent needed to completely inhibit the growth of the tested organism in vitro, a value called the **minimum inhibitory concentration (MIC)**. To determine the MIC for a given agent against a given organism, a series of culture tubes is prepared and inoculated with the same number of microorganisms. Each tube contains medium with an increasing concentration of the antimicrobial agent. After incubation, the tubes are checked for visible growth (turbidity). In practice, this is done in an automated format using microliter amounts of media and reagents. The MIC is the lowest concentration of an agent that completely inhibits the growth of the test organism (⮌ Figure 5.40). A miniaturized version of this test shows a standard microtiter method for determining MIC using twofold dilutions of several antibiotics in medium inoculated with a standard amount of the test bacteria (**Figure 27.8a**). In a clinical microbiology laboratory, the tests for routine MIC determinations are automated.

Measuring Antimicrobial Susceptibility

The standard assay for antimicrobial activity is the *disc diffusion test* (Figure 27.8b–f). A Petri plate containing an agar medium is inoculated by evenly spreading a suspension of a pure culture of the suspected pathogen on the agar surface. Known amounts of an antimicrobial agent on filter-paper discs are then placed on the surface of the agar. During incubation, the agent diffuses from the disc into the agar during a specified period of incubation, establishing a gradient; the farther the chemical diffuses away from the filter paper, the lower is the concentration of the agent. At some distance from the disc, the effective MIC is reached. Beyond this point the microorganism grows, but closer to the disc, growth is absent. A *zone of inhibition* forms with a diameter proportional to the amount of antimicrobial agent added to the disc, the solubility of the agent, the diffusion coefficient, and the overall effectiveness of the agent.

Figure 27.8g demonstrates antibiotic susceptibility using a preformed and predefined gradient of an antimicrobial agent immobilized on a plastic strip. The concentration gradient covers an MIC range across 15 twofold dilutions. When applied to the surface of an inoculated agar plate, the gradient transfers from the strip to the agar and remains stable for a period that covers the wide range of critical times associated with the growth characteristics of different pathogenic bacteria. After a suitable incubation period, an elliptical zone of inhibition centered along the axis of the strip develops. The MIC value (in micrograms per milliliter) can be read at the point where the ellipse edge intersects the precalibrated test strip, providing a precise MIC.

The MIC is not a constant for a given agent; it varies with the test organism, the inoculum size, the composition of the culture medium, the incubation time, and the conditions of incubation, such as temperature, pH, and aeration. When culture conditions are standardized, however, different antimicrobial agents can be compared to determine which is most effective against the isolated pathogen. Standards for antimicrobial agents are constantly updated by the Clinical and Laboratory Standards Institute, an organization that develops and establishes voluntary consensus standards for antimicrobial testing (http://www.clsi.org). The United States Food and Drug Administration sets standards for automated instruments used for susceptibility testing in the United States.

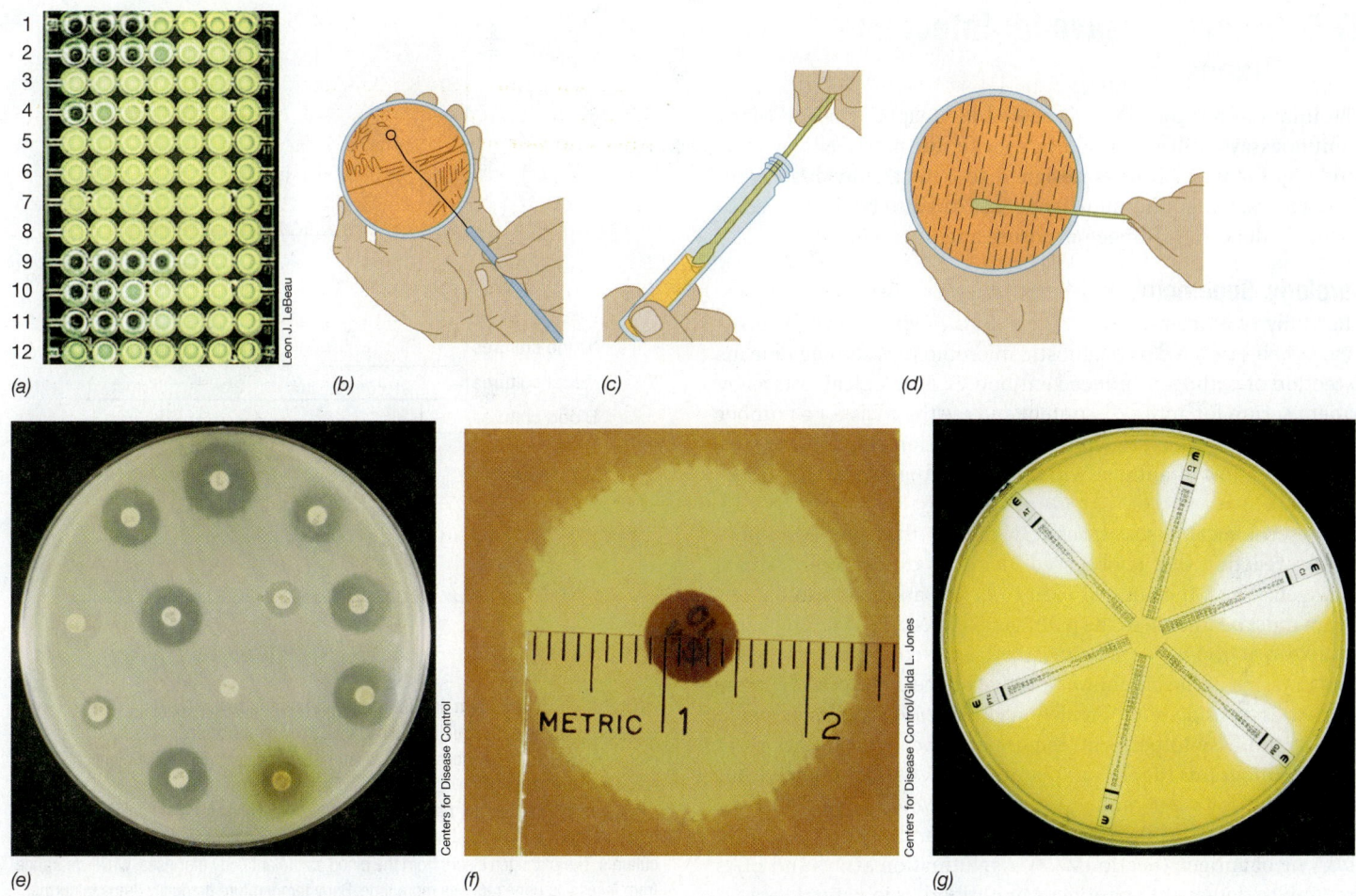

(a) (b) (c) (d)

(e) (f) (g)

Figure 27.8 **Antibiotic susceptibility testing.**
Methods for determining the susceptibility of an organism to antibiotics. *(a)* Antibiotic susceptibility as determined by the broth dilution method in a microtiter plate. The organism is *Pseudomonas aeruginosa*. Each row has a different antibiotic. The end point is the well with the lowest concentration of antibiotic that shows no visible bacterial growth. The highest concentration of antibiotic is in the well at the left; serial twofold dilutions are made in the wells to the right. In rows 1 and 2, the end point is the third well.

In row 3, the antibiotic is ineffective at the concentrations tested, since there is bacterial growth in all the wells. In row 4, the end point is in the first well. *(b)* For the disc diffusion test, isolated pure colonies are homogenized in liquid medium to standard turbidity. *(c, d)* A sterile cotton swab is dipped into the bacterial suspension and streaked evenly over the surface of an appropriate agar medium. *(e)* Discs containing known amounts of different antibiotics are placed on the inoculated agar surface. *(f)* After incubation, inhibition zones are observed and measured; the

susceptibility of the organism is determined by reference to an interpretive chart of zone sizes. *(g)* Antibiotic susceptibility determined by the Etest (AB BIODISK, Solna, Sweden) for different antibiotics. Each strip, laid on an inoculated plate before incubation, is calibrated in µg/ml starting with the lowest concentration from the center of the plate. The lowest concentration of antibiotic that inhibits bacterial growth is the MIC value. For example, the MIC for cefotaxime (CT) is 16 µg/ml. This organism is resistant to imipenem (IP); MIC > 27 µg/ml.

Hospital infection-control microbiologists generate and examine susceptibility data to generate periodic reports called **antibiograms**. These reports define the susceptibility of clinically isolated organisms to the antibiotics in current use. Antibiograms are used to monitor control of known pathogens, to track the emergence of new pathogens, and to identify the emergence of antibiotic resistance at the local level.

MINIQUIZ

- Describe the disc diffusion test for antimicrobial susceptibility. For an individual organism and antimicrobial agent, what do the results indicate?

- What is the value of antimicrobial drug susceptibility testing for the microbiologist, the physician, and the patient?

III • Growth-Independent Diagnostic Methods

Immunological assays are used in clinical, reference, and research laboratories to detect specific pathogens or pathogen products. When culture methods for pathogens are not routinely available or are prohibitively difficult to perform, as is the case with many viral infections or some bacterial pathogens, immunoassays or pathogen-specific nucleic acid amplification assays using the polymerase chain reaction (PCR) provide means to identify individual pathogens or host exposure to pathogens.

27.6 Immunoassays for Infectious Diseases

The immune response was discussed in Chapters 24–26. Many immunoassays utilize antibodies specific for pathogens or their products for in vitro tests designed to detect individual infectious agents. Patient immune responses can also be monitored to obtain evidence of pathogen exposure and infection.

Serology, Specificity, and Sensitivity

The study of antigen–antibody reactions in vitro is called **serology**. When extended to diagnostic microbiology, serology means detection of pathogen-induced antibodies. Serological tests assay patient serum for antibody content and are the basis for a number of diagnostic tests. Antigen–antibody reactions rely on the specific interaction of antigen with an antibody molecule (Section 25.7).

For serological tests, *specificity* means that the antibody–antigen reaction that is observed identifies exposure to a single pathogen. Thus, the antigen used to detect antibodies in patient serum must be unique to the pathogen in question, avoiding false-positive reactions. The *sensitivity* of some common serological tests in terms of the amount of antibody necessary to detect antigen varies considerably. Passive *agglutination* reactions, which are fast and easy to perform, require antibody concentrations of up to 6 nanograms (ng, 10^{-9}g) per ml, while the very sensitive but more technically demanding *enzyme immunoassay* (*EIA*) tests require as little as 0.1 ng of antibody per ml and can detect as little as 0.1 ng of antigen (Section 27.9). Agglutination assays and EIAs are used to detect either antigens or antibodies in patient sera.

Antibody Titers

An alternative approach to isolation and culture of a pathogen that provides strong indirect evidence for infection is to measure antibody *titer* (quantity) directed to antigens produced by the suspected pathogen. If an individual is infected with a suspected pathogen, the immune response—in this case, the antibody titer—to that pathogen should become elevated. Serial dilutions of patient serum are assayed by methods we will discuss in Sections 27.7–27.9. The **titer** is defined as the highest dilution (lowest concentration) of serum at which an antigen–antibody reaction is observed (**Figure 27.9**).

A positive antibody titer indicates previous infection or exposure to a pathogen. For pathogens rarely found in a population, a single positive test for a pathogen-specific antibody without a follow-up test may indicate ongoing, active infection. For example, this is the case for hantavirus diseases (Section 30.2). In most cases, however, the mere presence of antibody does not indicate active infection. Antibody titers typically remain detectable for long periods after a previous infection has been resolved. To link an acute illness to a particular pathogen, it is essential to show a *rise* in antibody titer in serum samples taken from a patient during the acute disease and later during the convalescent phase of the disease. Frequently, the antibody titer is low during the acute stage of the infection and rises during convalescence (Figure 27.9). A rise in antibody titer is strong circumstantial evidence that the illness is due to the suspected pathogen.

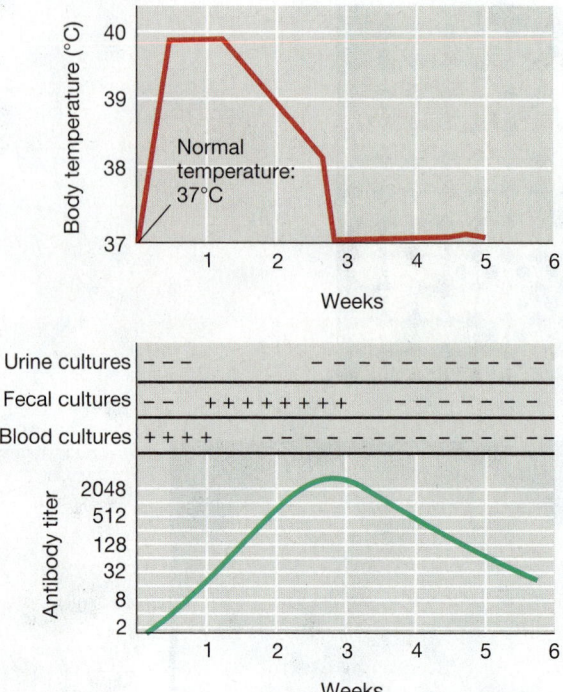

Figure 27.9 Infection and immunity in typhoid fever. The data are a composite pattern seen in untreated patients. Body temperature indicates acute disease progression over time. Antibody titer was measured by determining the highest twofold serum dilution causing agglutination of *Salmonella enterica* serovar Typhi (Section 31.5). Titer is shown as the *reciprocal* of the highest dilution showing an agglutination reaction. Presence of viable bacteria in blood, feces, and urine was determined from cultures. The pathogen clears from the blood as the antibody titer rises, while clearance from feces and urine requires more time. Body temperature gradually drops to normal as the antibody titer rises. Typhoid used to be a major public health threat in the United States before drinking water was routinely filtered and chlorinated (Section 28.5).

Skin Tests

A number of pathogens induce a delayed-type hypersensitivity (DTH) response mediated by Th1 cells (Sections 24.3 and 24.8). For these pathogens, skin testing may be useful for determining exposure. As an example, a commonly used skin test is the *tuberculin test*, which consists of an intradermal injection of a soluble extract from cells of *Mycobacterium tuberculosis*. A positive inflammatory reaction at the site of injection within 48 hours indicates current infection or previous exposure to *M. tuberculosis*. This test identifies responses caused by pathogen-specific inflammatory Th1 cells (Figure 24.6). Skin tests are routinely used for aiding in diagnosis of tuberculosis, Hansen's disease (leprosy), and some fungal diseases because the antibody responses for intracellular and fungal infections are often undetectable or are very weak.

If a pathogen is extremely localized, there may be little induction of a systemic immune response and no rise in antibody titer or skin test reactivity, even if the pathogen is proliferating profusely at the site of infection. A good example is gonorrhea, caused by infection of mucosal surfaces with *Neisseria gonorrhoeae*. Gonorrhea does not elicit a systemic or protective immune response, there is no serum antibody titer or skin test reactivity, and reinfection of individuals is common (Section 29.12).

MINIQUIZ

- Explain the reasons for changes in antibody titer for a single infectious agent, from the acute phase through the convalescent phase of the infection.

- Describe the method, time frame, and rationale for the tuberculin skin test. What component of the immune response does this test detect?

27.7 Agglutination

Agglutination is a reaction between antibody and particle-bound antigen resulting in visible clumping of the particles. Agglutination tests can be done in test tubes or in small-volume microtiter plates, or can be done by mixing reagents on glass slides. Agglutination tests are widely used in clinical and diagnostic laboratories; they are simple to perform, highly specific, inexpensive, rapid, and reasonably sensitive. Standardized agglutination tests are used for the identification of blood group (red blood cell) antigens (**Figure 27.10a**) as well as pathogens and pathogen products. To determine blood groups, blood samples are mixed with either anti-A antisera or anti-B antisera and the agglutination of red blood cells is assessed (Figure 27.10).

Direct agglutination results when soluble antibody causes clumping due to interaction with an antigen that is an integral part of the surface of a cell or other insoluble particle such as a red blood cell (erythrocyte). Agglutination of red blood cells is called *hemagglutination* and is the basis for human blood typing (Figure 27.10). *Passive agglutination* is the agglutination of soluble antigens or antibodies that have been adsorbed or chemically coupled to insoluble particles such as latex beads or charcoal particles.

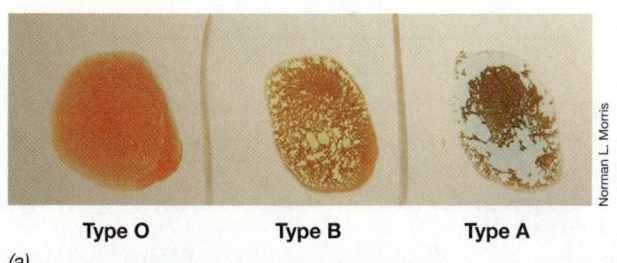

Blood type	Caucasian	African American	Hispanic	Asian
O	45%	51%	57%	40%
A	40%	26%	31%	28%
B	11%	19%	10%	25%
AB	4%	4%	2%	7%

(b)

Figure 27.10 Direct agglutination of human red blood cells: ABO blood typing. *(a)* A drop of whole blood was mixed with antigen-specific antisera for each reaction. The reaction on the left shows no agglutination with antibody, typical of blood group O. The reaction in the center shows the diffuse agglutination pattern indicating a positive reaction for the B blood group. The reaction on the right shows the strong agglutination pattern with large, clumped agglutinates typical for the A blood group. *(b)* ABO blood types by race and ethnicity in the U.S. population. Data are from the American Red Cross.

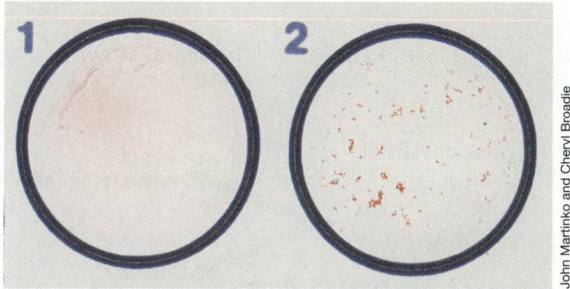

Figure 27.11 Latex bead agglutination test for *Staphylococcus aureus*. Panel 1 shows a negative control. Note the uniform pink color of the suspended latex beads coated with antibodies to protein A and clumping factor, two antigens found exclusively on the surface of *S. aureus* cells. Panel 2 shows the same suspension after a loopful of material from a bacterial colony was mixed into the suspension. The bright red clumps indicate a positive agglutination, identifying the colony as *S. aureus*.

The insoluble antigen or antibody can then be detected by agglutination reactions. The cell or particle serves as an inert carrier. Passive agglutination is up to five times more sensitive than direct agglutination.

The agglutination of antigen-coated or antibody-coated latex beads by complementary antibody or antigen from a patient is a typical rapid assay. Small (0.8-μm diameter) latex beads coated with a specific antigen are mixed with patient serum on a microscope slide and incubated for a short period. If patient antibody binds the antigen on the bead surface, the milky white latex suspension will become visibly clumped, indicating a positive agglutination reaction and exposure to the pathogen.

Latex agglutination is also used to detect bacterial surface antigens by mixing a small amount of a bacterial colony with antibody-coated latex beads. For example, a commercially available suspension of latex beads coated with antibodies to protein A and clumping factor, two proteins found exclusively on the surface of *Staphylococcus aureus* cells, is specific for identification of clinical isolates of *S. aureus*. Unlike traditional growth-dependent tests for *S. aureus*, the latex bead assay takes less than a minute and can be used directly on a clinical sample, such as the material from a purulent infection possibly caused by *S. aureus* (**Figure 27.11**). Latex bead agglutination assays have also been developed to identify other common pathogens such as *Streptococcus pyogenes*, *Neisseria gonorrhoeae*, *Escherichia coli* O157:H7, the fungus *Candida albicans*, and many others.

Passive agglutination assays require no expensive equipment or particular expertise, and can be highly specific and reasonably sensitive. The cost-effective nature of the assays makes them suitable for large-scale screening programs. These tests are therefore widely used in clinical applications.

MINIQUIZ

- Distinguish between direct and passive agglutination. Which tests are more sensitive?

- What advantages does agglutination have over other immunoassays? What disadvantages?

UNIT 5

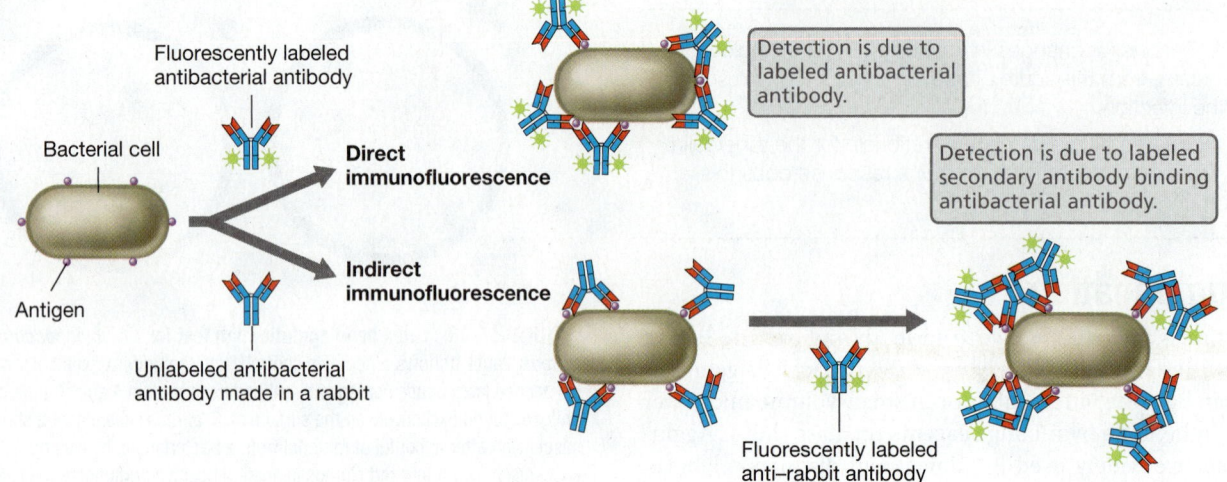

Figure 27.12 Fluorescent antibody methods for detection of microbial surface antigens. Note how indirect immunofluorescence requires a labeled secondary antibody that binds to the primary antibody.

27.8 Immunofluorescence

Antibodies containing conjugated fluorescent dyes can be used to detect antigens on intact cells. Such **fluorescent antibodies** are widely used for diagnostic and research applications.

Fluorescent Methods

Fluorescent antibody–staining methods can be either direct or indirect (**Figure 27.12**). In the *direct method*, the antibody that interacts with the surface antigen is itself covalently linked to the fluorescent dye. In the *indirect method*, the presence of a nonfluorescent antibody on the surface of a cell is detected by the use of a fluorescent antibody directed against the nonfluorescent antibody.

Antibodies can be covalently modified by fluorescent dyes such as rhodamine B, which fluoresces red, or fluorescein isothiocyanate, which fluoresces yellow-green. The attached dyes do not alter the specificity of the antibody but make it possible to detect the complex by use of a fluorescence microscope once it has bound to cell or tissue surface antigens. The cell-bound fluorescent antibodies emit their characteristic fluorescent color when excited with light of particular wavelengths. Fluorescent antibodies can be used to identify a microorganism directly in a patient specimen (in situ), bypassing the need for the isolation and culture of the organism.

Applications

In a typical test using fluorescent antibodies, a specimen containing a suspected pathogen is mixed with a specific fluorescent antibody and observed with a fluorescence microscope. If the pathogen contains surface antigens reactive with the antibody, the pathogen cells fluoresce (**Figure 27.13**). Fluorescent antibodies can be applied directly to infected host tissues, permitting diagnosis long before primary isolation techniques yield a suspected pathogen. For example, for diagnosing legionellosis (or Legionnaires' disease), a form of infectious pneumonia (Section 31.4), a positive identification can be made by staining biopsied lung tissue directly with fluorescent antibodies specific for cell wall antigens

of *Legionella pneumophila* (Figure 27.13*b*), the causative agent of the disease. Likewise, a direct fluorescent antibody test detecting the capsule of *Bacillus anthracis* can be used to confirm a diagnosis for anthrax (Figure 23.11*a*).

Direct or indirect fluorescent antibody tests are also used to help diagnose viral infections with pathogens such as human B lymphotropic virus (HBLV), respiratory syncytial virus (RSV), and Epstein–Barr virus (EBV) (**Figure 27.14**). The common respiratory pathogens influenza A and B, parainfluenza, and adenovirus can also be identified using immunofluorescence of tissue-culture cells infected with virus from respiratory tract specimens.

MINIQUIZ

- Explain and compare direct and indirect fluorescent antibody assays, including the advantages and disadvantages of each.
- Why can fluorescent antibodies be used to identify specific cells in complex mixtures such as blood?

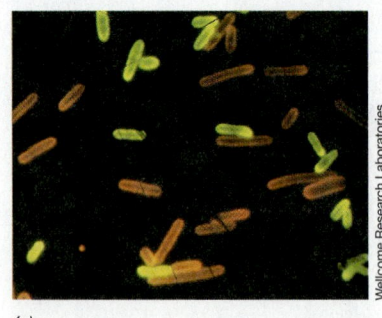
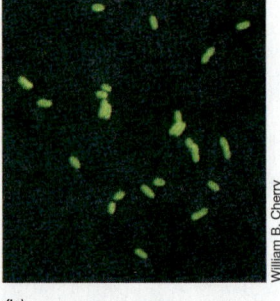

Figure 27.13 Fluorescent antibody identification of bacteria. *(a)* Cells of *Clostridium septicum* were stained with antibody conjugated with fluorescein isothiocyanate, which fluoresces yellow-green. Cells of *Clostridium chauvoei* were stained with antibody conjugated with rhodamine B, which fluoresces red-orange. *(b)* Immunofluorescent stained cells of *Legionella pneumophila*, the cause of legionellosis. The specimen was taken from biopsied lung tissue. The individual organisms are 2–5 μm in length. The cells were stained green with antibodies coupled to fluorescein isothiocyanate.

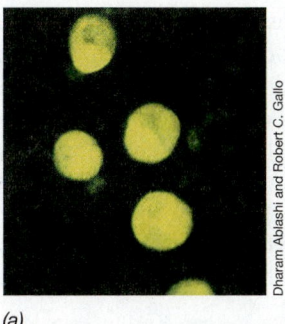
(a)

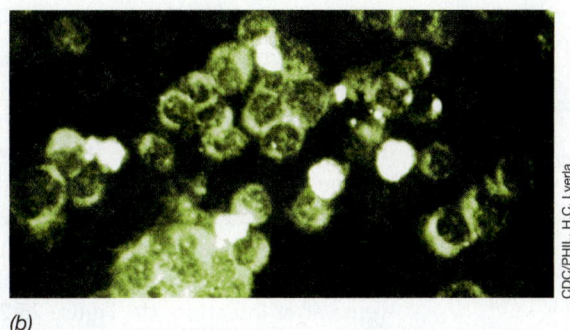
(b)

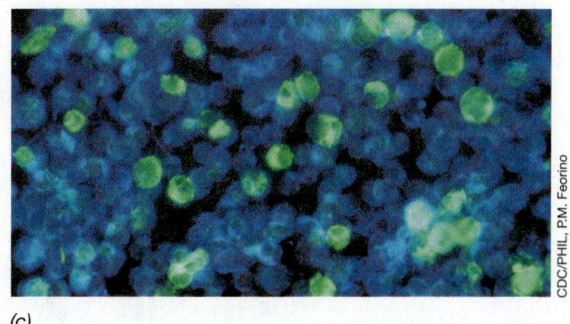
(c)

Figure 27.14 Fluorescent antibody identification of virus pathogens. *(a)* Detection of virus-infected cells by immunofluorescence. Human herpesvirus 6 (HHV-6)-infected spleen cells were incubated with serum containing antibodies to HHV-6. Cells were then treated with fluorescein isothiocyanate–conjugated anti–human IgG antibodies. HHV-6-infected cells fluoresce bright yellow. Unstained cells did not react with the serum. Individual cells are about 10 μm in diameter. *(b)* Detection of respiratory syncytial virus (RSV)-infected cells using indirect immunofluorescence. The brightly stained cells are infected. *(c)* Detection of Epstein–Barr virus (EBV)-infected cells using indirect immunofluorescence. EBV causes mononucleosis and lymphoma. The green-stained cells are infected.

27.9 Enzyme Immunoassays, Rapid Tests, and Immunoblots

Enzyme immunoassays (EIAs), or *enzyme-linked immunosorbent assays* (*ELISAs*), are very sensitive immunological assays widely used in clinical and research applications. An EIA can detect as little as 0.01 nanograms of antigen or antibody. The speed (the assay typically can be done in several hours), low cost, lack of hazardous waste, long shelf life, high specificity, and high sensitivity of EIA tests make them particularly useful immunodiagnostic tools. *Rapid tests* are similar to EIAs except that results can often be reported within minutes. Many rapid tests are designed to be given as point-of-care tests but are generally not as specific or sensitive as EIAs. The comparatively complex and time-consuming **immunoblot (Western blot)** uses immobilized pathogen proteins to bind antibodies from patient specimens such as serum, providing highly specific evidence for pathogen exposure. The immunoblot is often used to confirm results obtained from other serological tests such as rapid tests or EIAs.

EIA

In EIA tests, an enzyme is covalently attached to an antigen or antibody molecule, creating an immunological tool with high specificity and high sensitivity. The enzyme's catalytic properties and the specific binding properties of the labeled antigen or antibody are unaltered. Enzymes typically bound to antigen or antibody include peroxidase, alkaline phosphatase, and β-galactosidase, all of which interact with enzyme-specific substrates to form colored reaction products that can be detected in very low amounts.

Four EIA formats are commonly used for evaluation of specimens for infectious disease, one for detecting antigen (*direct EIA*), another for detecting antibodies (*indirect EIA*), one that detects antibody using a sandwich technique (*antigen sandwich EIA*), and one that detects both antigen and antibody (*combination EIA*). The principal features of each platform are illustrated in **Figure 27.15**.

Direct EIAs are designed to detect antigens such as virus particles in a blood or fecal sample (Figure 27.15a). Antibodies directed to a pathogen antigen are coated onto a matrix such as a plastic microtiter plate. The patient sample such as serum is then added. After antigen in the sample binds to the antibody, a second antibody is added. This antibody is also specific for the antigen, and coupled to an enzyme. Finally, enzyme substrate is added and the enzyme converts the substrate to its colored product in proportion to the amount of patient antigen bound by the enzyme–antibody complex. Direct EIAs are useful for detecting antigens including cholera toxin, enteropathogenic *Escherichia coli* toxin, and *Staphylococcus aureus* enterotoxin. Viruses currently detected using direct EIA techniques include influenza, rotavirus, hepatitis viruses, rubella virus, bunyavirus, measles virus, mumps virus, and parainfluenza virus.

Indirect EIAs are used to detect antibodies to pathogens in body fluids (Figure 27.15b). The indirect test starts with pathogen antigen immobilized on the matrix. Patient serum is added and, if antibodies are present, they bind to the antigen. Next, an antibody–enzyme complex specific for the patient antibodies is added. Finally, the addition of the enzyme substrate results in the development of a colored product that is proportional to the concentration of patient antibody on the sample.

Indirect EIAs have been developed for detecting serum antibodies to *Salmonella* (gastrointestinal diseases), *Yersinia* (plague), rickettsias (Rocky Mountain spotted fever, typhus, Q fever), *Vibrio cholerae* (cholera), *Mycobacterium tuberculosis* (tuberculosis), *Legionella pneumophila* (legionellosis), *Borrelia burgdorferi* (Lyme disease), and *Treponema pallidum* (syphilis), among others (Chapters 29, 30, and 31). EIAs have also been developed for detecting antibodies to *Candida* (yeast) and other eukaryotic pathogens, including those causing amebiasis, Chagas' disease, schistosomiasis, and malaria (Chapter 32).

The antigen sandwich EIA also detects antibodies to pathogens in body fluids (Figure 27.15c). The sandwich test starts with pathogen antigen immobilized on the matrix. Patient serum is added and, if antibodies are present, they bind to the antigen. Next, the same antigen coupled to enzyme is added. Finally, the addition of the enzyme substrate results in the development of colored product that is proportional to the concentration of patient antibody on the sample.

UNIT 5

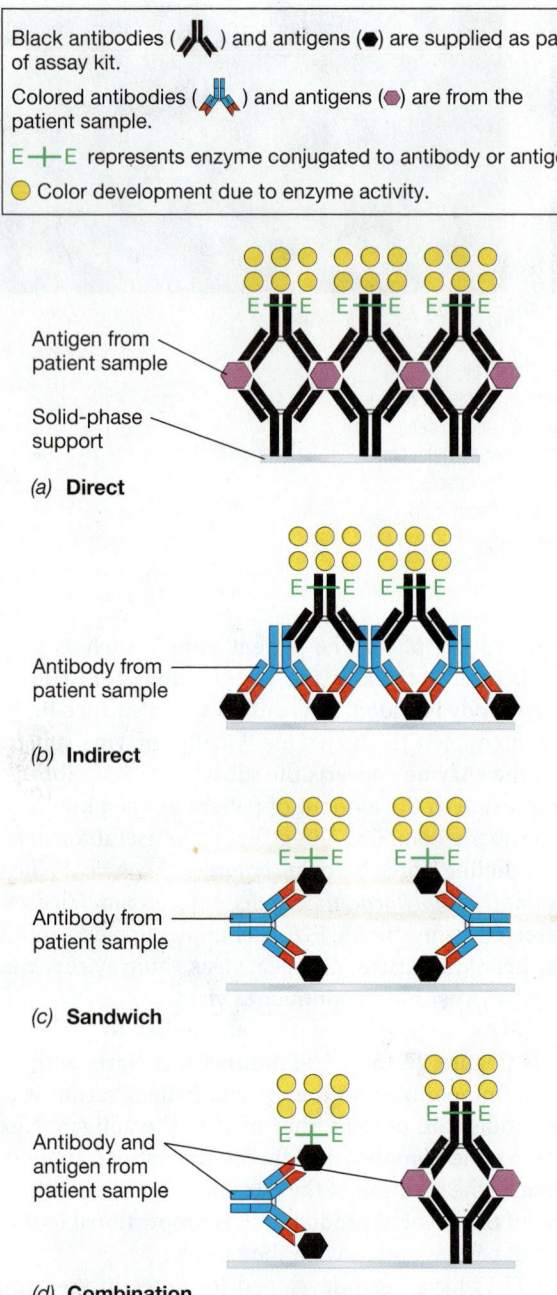

Black antibodies ( ) and antigens ( ) are supplied as part of assay kit.

Colored antibodies ( ) and antigens ( ) are from the patient sample.

E—E represents enzyme conjugated to antibody or antigen.

Color development due to enzyme activity.

Antigen from patient sample

Solid-phase support

(a) Direct

Antibody from patient sample

(b) Indirect

Antibody from patient sample

(c) Sandwich

Antibody and antigen from patient sample

(d) Combination

Figure 27.15 Enzyme immunoassays (EIAs). Patient samples are in color. Assay reagents are shown in black. All assays are fixed to a solid-phase support (light blue). Enzymes bound to antigen or antibody convert substrate to a colored product, shown in yellow. In each assay, the amount of colored product produced is proportional to the amount of pathogen-specific antibody or antigen derived from a patient sample. *(a)* The *direct EIA* uses immobilized pathogen-specific antibody and enzyme-labeled pathogen-specific antibody to detect pathogen antigen in patient samples such as blood. *(b)* The *indirect EIA* uses immobilized pathogen antigen and enzyme-labeled antibody directed to immunoglobulin to detect pathogen-specific antibodies in patient samples such as blood. *(c)* The *sandwich EIA* uses immobilized pathogen antigen and enzyme-labeled pathogen antigen to detect pathogen-specific antibodies in patient samples such as blood. The sandwich EIA is more sensitive than the direct or indirect EIA methods. *(d)* The *combination EIA* uses both the sandwich and direct assays in one platform to identify antibody and antigen in patient samples, maximizing sensitivity. Although a less sensitive assay, the immunoblot (see Figure 27.17) has greater specificity than EIAs.

The antigen sandwich method is considered the most sensitive antibody screening method because it detects pathogen-specific antibody irrespective of antibody isotype. This method is used for HIV screening (third-generation HIV test) and is preferable to indirect EIA tests because the sandwich test can detect the IgM isotype, produced in the primary immune response to HIV, about four weeks after infection. Most indirect tests use anti-IgG as the enzyme-conjugated antibody, delaying observation of antibodies to HIV to the onset of a secondary antibody response, about five weeks after infection (Figure 25.21).

The combination EIA, shown in Figure 27.15*d*, makes use of a direct EIA to detect pathogen antigen and the sandwich method to detect pathogen-specific antibodies, both on a single matrix. This method is used for a fourth-generation HIV test. This fourth-generation test is more sensitive than the third-generation sandwich test; antigen can be detected as little as 2.5 weeks after HIV infection, reducing the time to treatment.

Rapid Tests

Rapid immunoassay procedures use reagents adsorbed to a fixed support material such as paper strips, nitrocellulose, or plastic membranes. These tests cause a color change on the strip in a very short time. These "point-of-care" tests are rapid diagnostic aids for infectious diseases such as HIV/AIDS (Section 29.14), "strep throat" (pharyngeal infection with *Streptococcus pyogenes*, Section 29.2), influenza (Section 29.8), and gonorrhea (Section 29.12). Results can often be reported in minutes; however, the sensitivity and specificity of rapid tests are usually not as good as EIA tests for the same pathogens.

In most of these tests, a body fluid, generally urine, blood, saliva, or sputum, is applied to the reagent–support matrix (**Figure 27.16**). To detect patient antibody positive for a given antigen, as in rapid HIV tests, the matrix contains soluble antigen conjugated to a colored molecule called a *chromophore*. As the liquid sample diffuses through the matrix, antibodies bind the labeled antigen. Incorporated into the matrix is a single line of antigen fixed to the matrix. The labeled antigen–antibody complex migrates through the matrix, where it contacts and binds the fixed antigen, immobilizing the labeled antigen–antibody complex. As the concentration of labeled complex builds up, the chromophore becomes visible as a colored line on the line of fixed antigen, indicating a positive test for the antibody.

Similarly, to detect antigen in patient samples, for example for identification of infection by *S. pyogenes* (strep throat), the same system is used, but the chromophore is now bonded to antibody, and the putative antigen sample is applied to the matrix where it binds the soluble antibody–chromophore complex. As the liquid sample diffuses through the matrix, the patient antigen binds the labeled antibody. Fixed to the matrix is a single line of antibody. The now-labeled antigen–antibody complex migrates through the matrix, where it contacts the fixed antibody. The labeled antibody–antigen complex binds the fixed antibody, immobilizing the entire complex. As the concentration of labeled complex builds up, the chromophore becomes visible on the line of fixed antibody, indicating a positive test for the antigen.

These tests are valuable for point-of-care analysis at the site where the sample is collected (for instance, away from a clinical

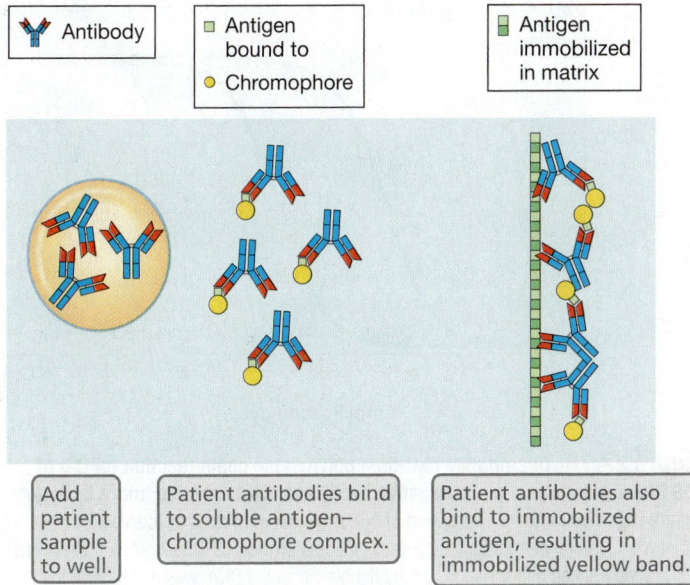

Figure 27.16 Rapid tests. A body fluid such as urine, blood, saliva, or sputum is applied to the reagent–support matrix containing soluble antigen conjugated to a chromophore. The liquid sample diffuses through the matrix, and antibodies bind the chromophore-labeled antigen. Incorporated into the matrix is a single line of antigen fixed to the matrix. When the labeled antigen–antibody complex contacts and binds the fixed antigen, the concentration of labeled complex builds and the chromophore becomes visible as a colored line on the line of fixed antigen, indicating a positive test for the antibody.

laboratory) and for rapid tests in the laboratory. Results can be reported almost immediately, avoiding the need for delays in patient care or for follow-up visits to obtain test results. The drawback to these tests, however, is that they are often less specific or less sensitive than more elaborate tests. As a result, rapid tests are often confirmed by EIA or other tests.

Immunoblots

Immunoblot method employs three techniques: (1) the *separation* of proteins on polyacrylamide gels, (2) the *transfer* (blotting) of proteins from gels to a nitrocellulose or nylon membrane, and (3) the *identification* of the proteins by specific antibodies. The HIV immunoblot (Figure 27.17) is widely used for confirmation of HIV infection. HIV immunoblots are generally less sensitive, more laborious, more time consuming, and more costly than the HIV EIA, so they are not used as HIV exposure screening tools. Immunoblots are used for confirmation of HIV infection, however, because the HIV EIA, while very sensitive, occasionally yields false-positive results. The more specific immunoblot is used to confirm positive EIA results. The HIV immunoblot procedures are similar to those used for immunoblots to diagnose infection by other pathogens. In general, immunoblot procedures are used to detect pathogen-specific antibody in patient samples.

To perform the HIV immunoblot, membrane strips with fixed HIV proteins are incubated with the patient serum sample. If the sample is HIV-positive, patient antibodies will bind to the HIV proteins on the membrane. To detect whether antibodies from the serum sample have bound to HIV antigens, a detecting antibody, anti–human IgG conjugated to an enzyme, is added to the strips. If the detecting antibody binds, the activity of the conjugated enzyme, after addition of substrate, will form a colored band on the strip at the site of antibody binding. The patient is HIV-positive if the position of the bands resolved by exposure to the patient serum and a positive control serum are identical; a control negative serum is also analyzed in parallel and must show no bands (Figure 27.17). Although the intensity of the bands obtained in the HIV immunoblot varies somewhat from sample to sample, the interpretation of an immunoblot is generally unequivocal; the test is most commonly used to confirm positive EIA results for HIV and eliminate false positives.

MINIQUIZ

- Compare indirect EIA, direct EIA, and combination EIA with respect to their intended ability to identify HIV infection.
- Compare the advantages and disadvantages of EIA, rapid tests, and immunoblots with respect to speed, sensitivity, and specificity.

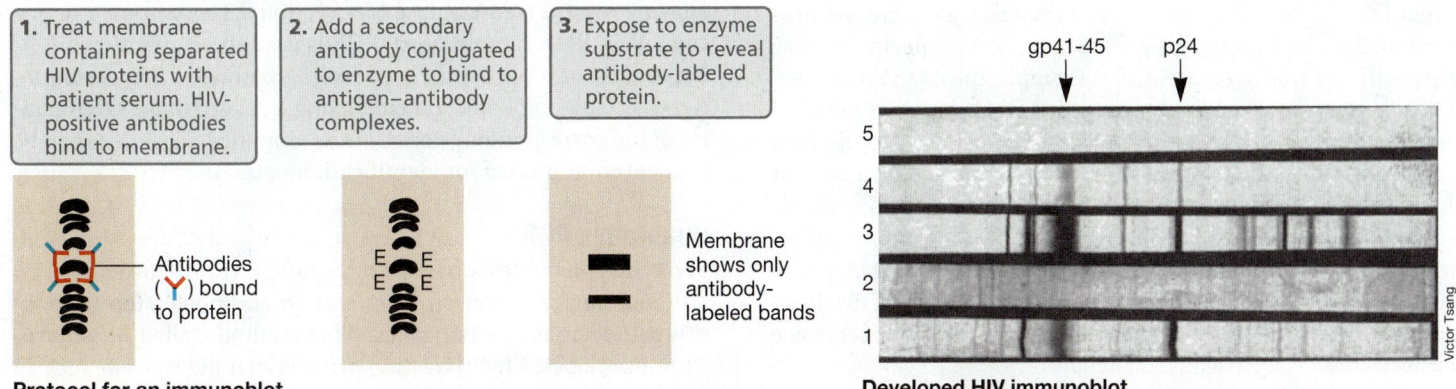

Protocol for an immunoblot

Developed HIV immunoblot

Figure 27.17 The immunoblot (Western blot) and its use in the diagnosis of human immunodeficiency virus (HIV) infection. The molecules p24 (capsid protein) and gp41 (envelope glycoprotein) are diagnostic for HIV. Lane 1, positive control serum (from known AIDS patients); lane 2, negative control serum (from healthy volunteer); lane 3, strong positive from patient sample; lane 4, weak positive from patient sample; lane 5, reagent blank to check for background binding.

UNIT 5

27.10 Nucleic Acid Amplification

We discussed in Section 11.3 how the polymerase chain reaction (PCR) amplifies nucleic acids, forming multiple copies of target sequences. PCR techniques use primers for a pathogen-specific gene to examine DNA derived from suspected infected tissue, even in the absence of an observable, culturable pathogen. As a result, PCR-based tests are widely used for identification of a number of individual pathogens and are particularly useful for identifying viral and intracellular infections, where culturing the responsible agents may be very difficult or even impossible. Extremely sensitive methods based on nucleic acid analyses are widely used in clinical microbiology to detect pathogens.

These methods do not depend on pathogen isolation or growth, or on the detection of an immune response to the pathogen. Instead, it is species-specific nucleic acid sequences that are detected in the assays.

PCR Testing and Analysis

PCR-based tests include three basic components. First, DNA or RNA must be extracted from the sample to be tested. Second, the nucleic acid must be amplified using appropriate gene-specific nucleic acid primers. Short oligonucleotides (typically 15–27 nucleotides in length) are used as primers for PCR amplification of a specific gene or genes characteristic for a specific pathogen. This requires knowledge of the sequence of the pathogen target genes. Third, the amplified nucleic acid product (the *amplicon*) must be visualized, a procedure that can involve gel electrophoresis or other methods. The presence of the appropriate amplified gene segment confirms the presence of the pathogen. PCR diagnostic methods are used for identification of nearly all pathogens.

Quantitative PCR

Many PCR tests employ *quantitative real-time PCR (qPCR)*. The qPCR test uses fluorescently labeled PCR amplicons that yield an almost immediate result, avoiding the need for postamplification nucleic acid purification and visualization. The accumulation of target DNA is monitored during the qPCR process. This is achieved by adding fluorescent probes to the PCR reaction mixture. Probe fluorescence increases upon binding to DNA. As the target DNA is amplified, the level of fluorescence increases proportionally. The fluorescent probes may be nonspecific or may be specific for the target DNA. For example, the dye SYBR Green binds *nonspecifically* to double-stranded DNA, but does not bind to single-stranded DNA or RNA; SYBR Green added to the PCR mixture becomes fluorescent only when bound, indicating that double-stranded DNA is present, in this case due to the amplification process (**Figure 27.18**). *Gene-specific* fluorescent probes are made by attaching a fluorescent dye to a short DNA probe that matches the target sequence being amplified: The dye fluoresces only when double-stranded DNA of the correct sequence accumulates.

Because qPCR amplification can be monitored continuously, visualization by gel electrophoresis or other detection methods is not necessary to confirm amplification; detection of a gene diagnostic for a particular pathogen in a clinical sample may be performed in several hours instead of the usual overnight processing

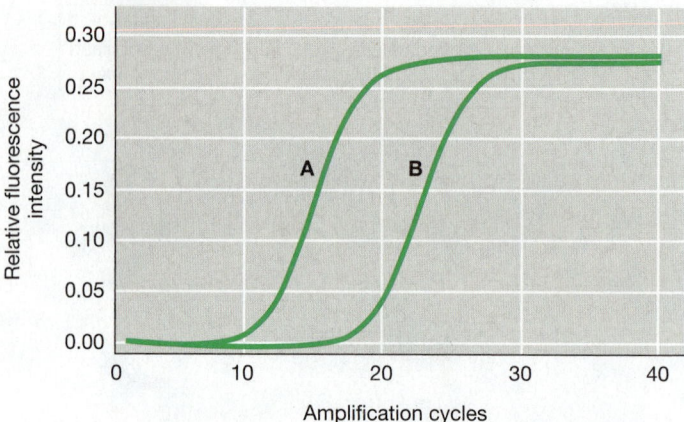

Figure 27.18 **Quantitative real-time polymerase chain reaction (qPCR) of 16S RNA genes from a gram-negative bacterium.** DNA extracted from a laboratory culture was monitored for expression of 16S RNA (curve A) and *npt* (curve B), a kanamycin resistance marker, using gene-specific primers. SYBR Green, a fluorescent dye that fluoresces only when bound by double-stranded DNA, was mixed with the PCR mixture and used to visualize amplified DNA as it formed. The curve on the left (A) had 0.15 fluorescence units after 15 cycles, while the curve on the right (B) had 0.15 fluorescence units after 22 cycles, indicating that the 16S RNA had a higher template abundance in this strain as compared to the abundance of the template for *npt*.

required by standard PCR. Moreover, by monitoring the *rate* of fluorescence increase in the PCR reaction, it is possible to accurately determine the *amount* of target DNA present in the original sample; qPCR can be used to assess the abundance of a pathogen in a sample by quantifying a gene characteristic for that particular organism.

Reverse Transcription PCR

PCR has been extended by the development of *reverse transcription PCR (RT-PCR)*. This technique is applied to the analysis of environmental samples in addition to clinical samples for identification of pathogens.

RT-PCR uses pathogen-specific RNA to produce complementary DNA (cDNA) directly from patient samples, and can be used for detection of RNA retroviruses such as HIV and other RNA viruses (♂♀ Figure 11.6). The first step in RT-PCR is to use the enzyme reverse transcriptase (♂♀ Section 9.11) to make a cDNA copy of an RNA sample. Next, PCR is used to amplify the cDNA. Expression of a particular gene from a pathogen may be monitored by isolating RNA and employing qPCR to make DNA copies of the corresponding gene(s). The amplified DNA can then be sequenced or probed for identification.

Qualitative PCR

Some diagnostic tests based on the qPCR format use a slightly different amplification protocol and an additional step to identify pathogen-associated genes. This method, called *qualitative PCR*, uses labeled hybridization primers that incorporate into an amplicon product of a qPCR reaction (**Figure 27.19**). In the example shown, the hybridization probes hybridize to the DNA *pol* gene of both herpes simplex 1 and 2 viruses (HSV-1 and HSV-2).

The amplicon is detected by fluorescence using a pair of specific hybridization probes. The probes are two different

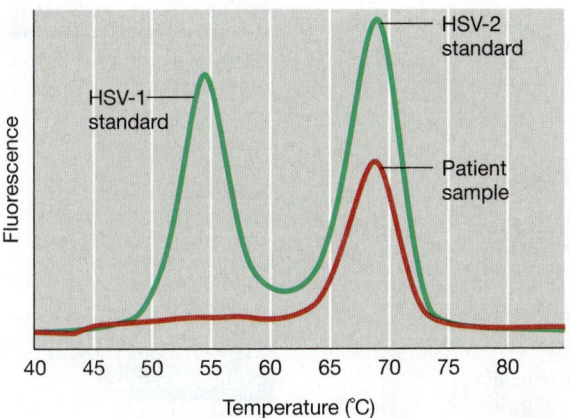

Figure 27.19 **Qualitative real-time polymerase chain reaction for the** *pol* **gene of HSV-1 (herpes simplex 1 virus) and HSV-2.** DNA from a patient sample was assayed for the *pol* gene of both HSV-1 and HSV-2 following quantitative PCR (qPCR). The assay uses two hybridization probes that incorporate into the qPCR amplicon product. The probes are labeled with fluorescent dyes that hybridize with an internal sequence of the amplified fragment of each viral genome during the PCR cycle. After hybridization to the template DNA, the probes are excited by a light source and their fluorescence is measured. After the PCR cycle, the DNAs are melted and each virus shows a distinct melting curve. The melting point in the patient sample (red) corresponds to the HSV-2 standard (green), indicating that the patient is infected with HSV-2.

oligonucleotides, labeled with fluorescent dyes. The probes hybridize to an internal sequence of the amplified fragment during the annealing phase of the PCR cycle. After hybridization to the template DNA, the probes are excited by a light source in the PCR instrument. The emitted fluorescence is then measured and, after the PCR cycle, a melting curve analysis is performed to differentiate between samples positive for HSV-1 and HSV-2. The melting curve will show a type-specific melting behavior due to nucleotide polymorphisms in the targeted hybridization sequences of HSV-1 and HSV-2. Because of these polymorphisms, the melting points for HSV-1 and HSV-2 are significantly different and allow clear determination of the HSV subtype.

This method provides results within hours. Results are compared to internal assay control reactions, avoiding the need for postamplification nucleic acid purification and visualization.

MINIQUIZ

- What advantage does nucleic acid amplification have over standard culture methods for identification of microorganisms? What disadvantages?
- How do qPCR and qualitative PCR differ?

IV • Antimicrobial Drugs

27.11 Synthetic Antimicrobial Drugs

Synthetic antimicrobial drugs include *growth factor analogs* that interfere with microbial metabolism and *quinolones* that interfere with DNA packaging in bacteria.

Growth Factor Analogs

Growth factors are specific chemical substances required by an organism that is unable to synthesize them. As a result, an outside source is essential for survival (ᴄᴐ Section 3.1). A **growth factor analog** is a synthetic compound that is structurally similar to a growth factor, but subtle structural differences between the analog and the growth factor prevent the analog from functioning in the cell, thereby disrupting cell metabolism. Analogs are known for many important biomolecules such as certain vitamins, amino acids, and nucleosides. Some growth factor analogs are antibacterial. Those effective for the treatment of viral and fungal infections will be discussed in Sections 27.15 and 27.16.

Antimicrobial drugs are compounds that kill or control the growth of microorganisms in the host (in vivo). Effective antimicrobial drugs exhibit **selective toxicity**; they inhibit or kill pathogens without adversely affecting the host. Antimicrobial drugs are classified based on their molecular structure, mechanism of action (**Figure 27.20**), and spectrum of antimicrobial activity (**Figure 27.21**). Worldwide, 10,000 metric tons or more of various antimicrobial drugs are manufactured and used annually (**Figure 27.22**). Antimicrobial agents fall into two broad categories, *synthetic antimicrobial drugs* and *antibiotics*.

Sulfa drugs were the first widely used growth factor analogs that specifically inhibited the growth of bacteria. The discovery of the first sulfa drug resulted from the large-scale screening of chemicals for activity against streptococcal infections in experimental animals. Sulfanilamide, the simplest sulfa drug, is an analog of *p*-aminobenzoic acid, which is a part of the vitamin folic acid, a nucleic acid precursor (**Figure 27.23**).

Sulfanilamide blocks the synthesis of folic acid, thereby inhibiting nucleic acid synthesis. Sulfanilamide is selectively toxic in bacteria because bacteria synthesize their own folic acid, whereas humans and most animals obtain folic acid from their diet. Initially, sulfa drugs were widely used for treatment of streptococcal infections. However, resistance to sulfonamides increased because many formerly susceptible pathogens developed an ability to take up folic acid from their environment. Antimicrobial therapy with sulfamethoxazole (a sulfa drug) plus trimethoprim, a related folic acid synthesis competitor, is still effective in many instances because the drug combination blocks two sequential steps in the folic acid synthesis pathway; resistance to this drug combination requires two mutations in genes of the same pathway, a relatively rare event.

Isoniazid (ᴄᴐ Figure 29.17) is an important growth factor analog with a very narrow spectrum of activity (Figure 27.21). Effective only against *Mycobacterium*, isoniazid interferes with the synthesis of mycolic acid, a mycobacterial cell wall component. A nicotinamide (vitamin) analog, isoniazid is the most effective single drug used for control and treatment of tuberculosis (ᴄᴐ Section 29.4).

Quinolones

The **quinolones** are synthetic antibacterial compounds that disrupt bacterial metabolism by interfering with bacterial DNA

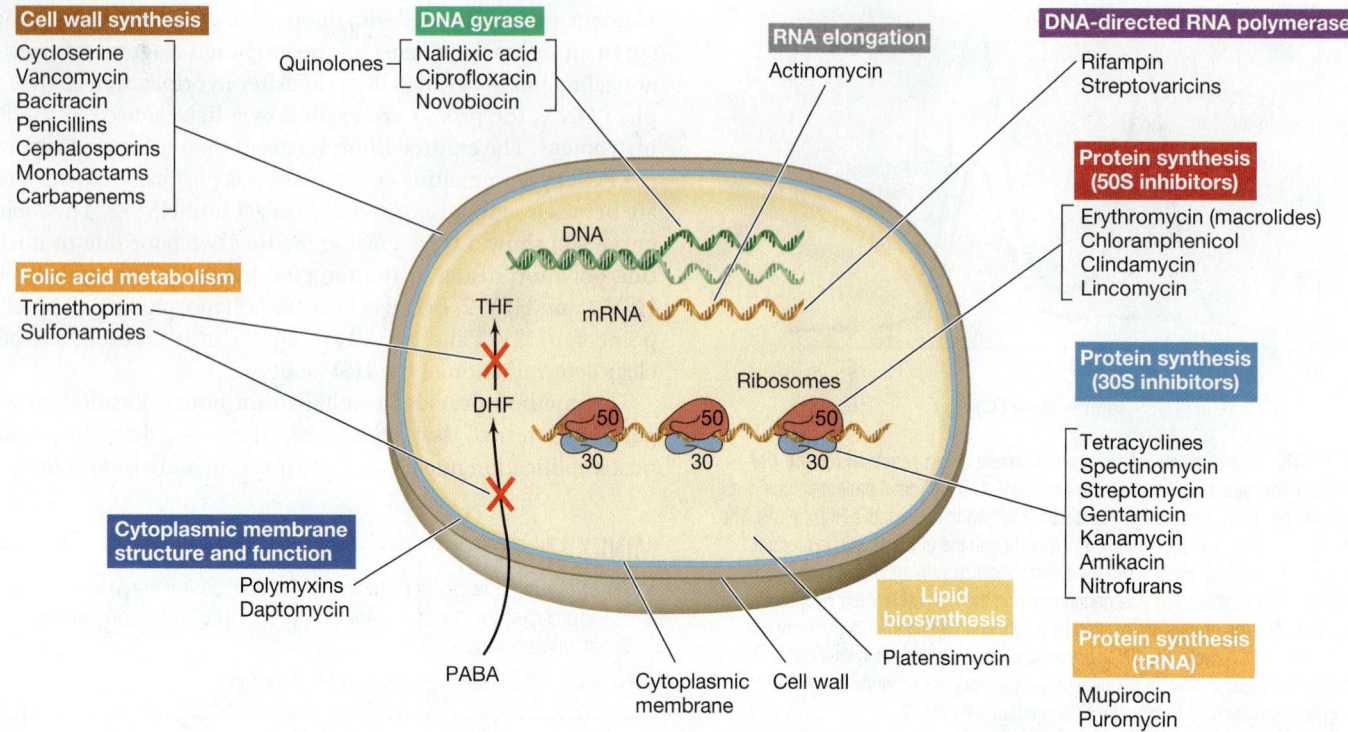

Figure 27.20 **Targets of major antibacterial agents.** Agents are classified according to their target structures in the bacterial cell. THF, tetrahydrofolate; DHF, dihydrofolate; mRNA, messenger RNA.

gyrase, preventing the supercoiling of DNA, a required step for packaging DNA in the bacterial cell ($\Leftrightarrow$ Section 4.3). Because DNA gyrase is found in all *Bacteria*, the fluoroquinolones are effective for treating both gram-positive and gram-negative bacterial infections (Figure 27.21). Fluoroquinolones such as ciprofloxacin (**Figure 27.24a**) are routinely used to treat urinary tract infections in humans and have been widely used in the beef and poultry industries for prevention and treatment of respiratory

diseases in animals. Ciprofloxacin is also the drug of choice for treating anthrax because some strains of *Bacillus anthracis*, the causative agent of anthrax ($\Leftrightarrow$ Section 30.8), are resistant to penicillin. Moxifloxacin (Figure 27.24b) is one of only a few drugs proven effective for treatment of tuberculosis. In combination with other anti-tuberculosis drugs ($\Leftrightarrow$ Section 29.4), moxifloxacin may significantly reduce treatment time, a major problem with isoniazid-based treatments.

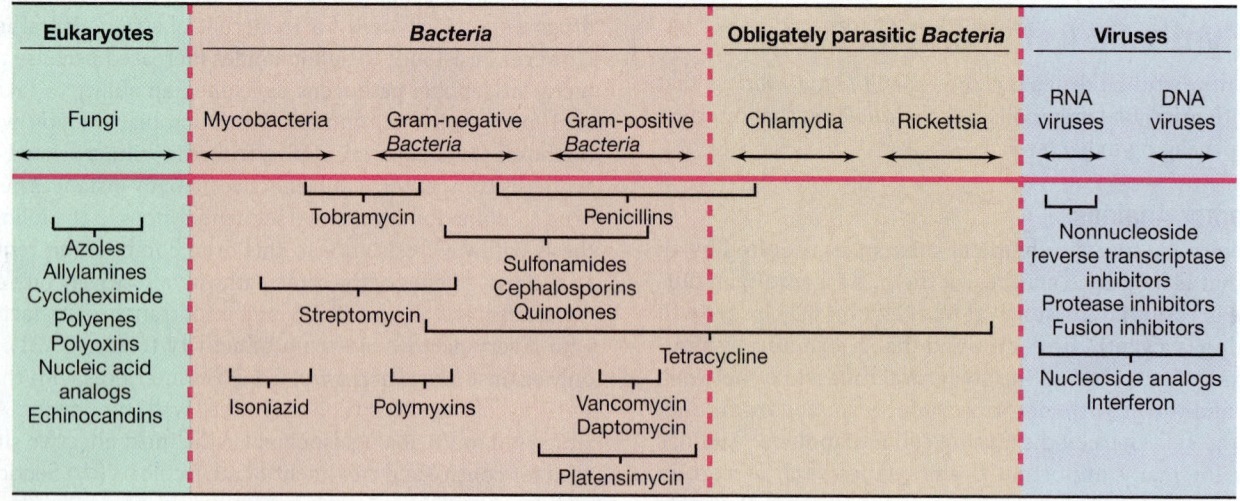

Figure 27.21 **Antimicrobial spectrum of activity.** Each antimicrobial agent affects a limited and well-defined group of microorganisms. A few agents are very specific and affect the growth of only a single genus. For example, isoniazid affects only organisms in the genus *Mycobacterium*.

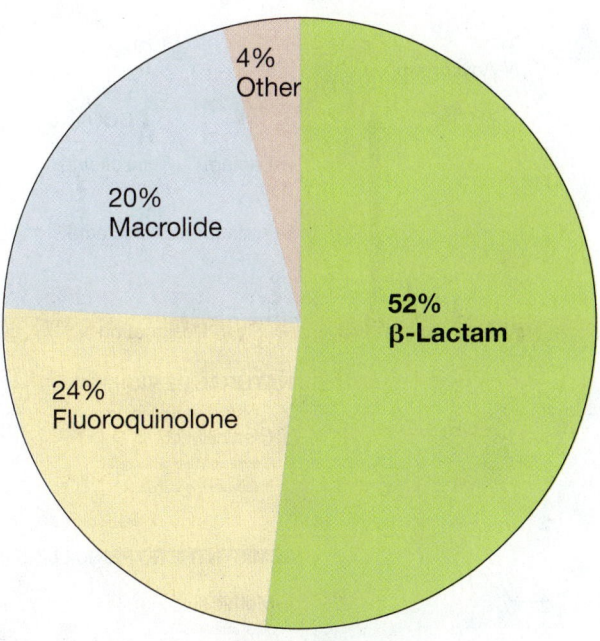

Figure 27.22 Annual worldwide production and use of antibiotics. An estimated 10,000 metric tons of antimicrobial agents are manufactured worldwide per year. The β-lactam antibiotics include cephahalosporins (30%), penicillins (7%), and other β-lactams (15%). "Other" includes tetracyclines, aminoglycosides, and all other antimicrobial drugs.

MINIQUIZ

- Explain the concept of selective toxicity in terms of antimicrobial therapy.
- Describe the action of any one of the synthetic antimicrobial drugs.

27.12 Natural Antimicrobial Drugs: Antibiotics

Antibiotics are antimicrobial agents produced by microorganisms. Antibiotics are produced by a variety of *Bacteria* and fungi (*Eukarya*) and apparently function in nature as they do clinically:

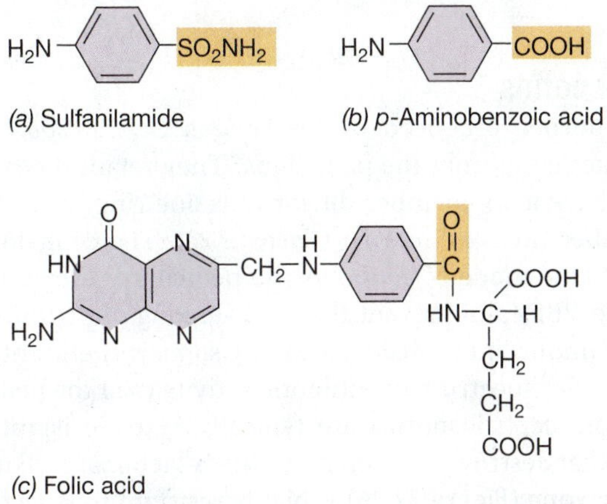

(a) Sulfanilamide

(b) *p*-Aminobenzoic acid

(c) Folic acid

Figure 27.23 Sulfa drugs. *(a)* The simplest sulfa drug, sulfanilamide. *(b)* Sulfanilamide is an analog of *p*-aminobenzoic acid, a precursor of *(c)* folic acid, a growth factor.

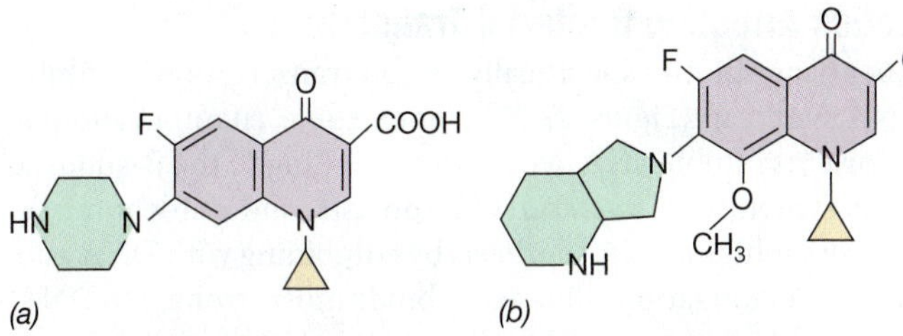

(a)

(b)

Figure 27.24 Quinolones. *(a)* Ciprofloxacin, a fluorinated derivative of nalidixic acid with broad-spectrum activity, is more soluble than the parent compound, allowing it to reach therapeutic levels in blood and tissues. *(b)* Moxifloxacin, a fluoroquinolone approved for treatment of *Mycobacterium* infections.

to inhibit or kill other microorganisms. Although thousands of antibiotics are known, less than 1% are clinically useful, often because of problems with host toxicity or lack of uptake by host cells. However, the clinically useful antibiotics have had a dramatic impact on the treatment of infectious diseases. Natural antibiotics can often be artificially modified to enhance their efficacy. These are said to be *semisynthetic* antibiotics.

Antibiotics and Selective Antimicrobial Toxicity

The susceptibility of individual microorganisms to individual antimicrobial agents varies significantly (Figure 27.21). For example, gram-positive *Bacteria* and gram-negative *Bacteria* differ in their susceptibility to an individual antibiotic such as penicillin; gram-positive *Bacteria* are generally affected, whereas most gram-negative *Bacteria* are naturally resistant. Certain **broad-spectrum antibiotics** such as tetracycline, however, are effective against both groups. As a result, a broad-spectrum antibiotic finds wider medical use than a narrow-spectrum antibiotic. An antibiotic with a limited spectrum of activity may, however, be quite valuable for the control of pathogens that fail to respond to other antibiotics. A good example is vancomycin, a narrow-spectrum antibiotic that is an effective bacteriocidal agent for gram-positive, penicillin-resistant enterococci as well as *Staphylococcus*, *Bacillus*, and *Clostridium* (Figure 27.21). Ribosomes, the cell wall, the cytoplasmic membrane, lipid biosynthesis enzymes, and DNA replication and transcription elements are important targets of antibiotics in *Bacteria* (Figure 27.20).

Antibiotics Affecting Translation

Some antibiotics inhibit protein synthesis by disrupting translation through interactions with the ribosome, often involving binding to ribosomal RNA (rRNA) (Figure 27.20). Many, such as chloramphenicol, streptomycin, and the tetracyclines, target ribosomes from only one phylogenetic domain, *Bacteria*, but have no effect on the cytoplasmic ribosomes of *Eukarya*. However, since mitochondria and chloroplast ribosomes in *Eukarya* have their evolutionary origins from bacterial ribosomes (that is, they are 70S ribosomes), antibiotics that inhibit protein synthesis in *Bacteria* also inhibit protein synthesis in these organelles (🔗 Section 17.1). For example, tetracycline antibiotics inhibit 70S ribosomes, but are still medically useful because the eukaryotic 70S mitochondrial ribosomes are affected only at higher concentrations than are used for antimicrobial therapy.

UNIT 5

Antibiotics Affecting Bacterial Transcription

A number of antibiotics specifically inhibit transcription by inhibiting RNA synthesis (Figure 27.20). For example, rifampin and the streptovaricins inhibit RNA synthesis by binding to the β-subunit of RNA polymerase in *Bacteria*, chloroplasts, and mitochondria. Actinomycin inhibits RNA synthesis by combining with DNA and blocking RNA elongation. This agent binds most strongly to DNA at guanine–cytosine base pairs, fitting into the major groove in the double strand where RNA is synthesized.

MINIQUIZ --

• Distinguish differences in the mode of action of antibiotics and growth factor analogs.

• What is a broad-spectrum antibiotic?

27.13 β-Lactam Antibiotics: Penicillins and Cephalosporins

β-lactam antibiotics are cell-wall synthesis inhibitors that include the medically important penicillins and cephalosporins. These antibiotics share a characteristic structural component, the *β-lactam ring* (**Figure 27.25**). Together, the β-lactam antibiotics account for over one-half of all antibiotics produced and used worldwide (Figure 27.22).

Penicillins

In 1929 Alexander Fleming characterized the first antibiotic, an antibacterial compound called **penicillin** (Figure 27.25) isolated from the fungus *Penicillium chrysogenum*. This new β-lactam antibiotic was dramatically effective for controlling staphylococcal and pneumococcal infections and was more effective for treating streptococcal infections than sulfa drugs. Penicillin revolutionized the treatment of infectious diseases when it came into general use after World War II.

The β-lactam antibiotics are inhibitors of bacterial cell wall synthesis. An important feature of bacterial cell wall synthesis is *transpeptidation*, the reaction that results in the cross-linking of two glycan-linked peptide chains (⮌ Section 5.4). Because the cell wall and its synthesis mechanisms are unique to *Bacteria*, the β-lactam antibiotics are highly selective and nontoxic to host cells.

Penicillin G (Figure 27.25) is active primarily against gram-positive *Bacteria* because gram-negative *Bacteria* are impermeable to the antibiotic, but chemical modification of the penicillin G structure significantly changes the resulting antibiotic; many chemically modified semisynthetic penicillins are effective against gram-negative *Bacteria*. For example, ampicillin and carbenicillin are semisynthetic penicillins effective against some gram-negative *Bacteria*. The structural differences in these semisynthetic penicillins allow them to be transported inside the gram-negative outer membrane (⮌ Section 2.11), where they inhibit cell wall synthesis. Penicillin G is also sensitive to β-lactamase, an enzyme produced by a number of penicillin-resistant *Bacteria* (see Section 27.17). Oxacillin and methicillin are widely used β-lactamase-resistant semisynthetic penicillins.

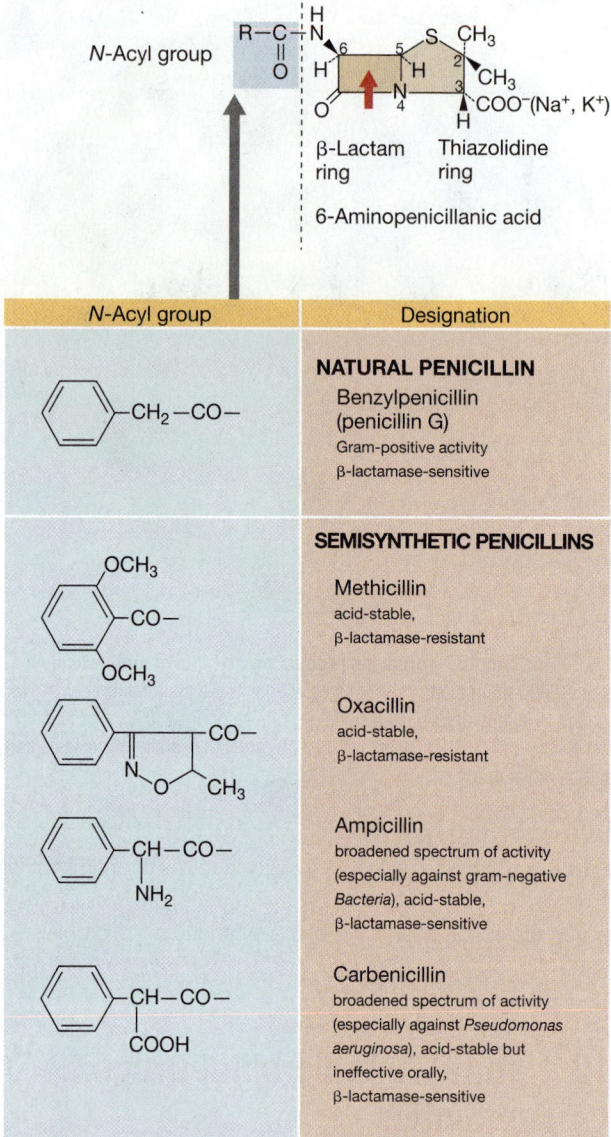

Figure 27.25 **Penicillins.** The red arrow (top panel) is the site of activity of most β-lactamases, enzymes that destroy penicillin and other β-lactam antibiotics. Most acid-stable penicillins can be administered orally.

Cephalosporins

Cephalosporins, produced by the fungus *Cephalosporium* sp., differ structurally from the penicillins. They retain the β-lactam ring but have a six-member dihydrothiazine ring instead of the five-member thiazolidine ring (**Figure 27.26**). The cephalosporins have the same mode of action as the penicillins; they bind irreversibly to PBPs and prevent the cross-linking of peptidoglycan. Clinically important cephalosporins are semisynthetic antibiotics with a broader spectrum of antibiotic activity than the penicillins. In addition, cephalosporins are typically more resistant to the enzymes that destroy β-lactam rings, the β-lactamases. For example, ceftriaxone (Figure 27.26) is highly resistant to β-lactamases and has replaced penicillin for treatment of *Neisseria gonorrhoeae* (gonorrhea) infections because many *N. gonorrhoeae* strains are now resistant to penicillin (⮌ Section 29.12).

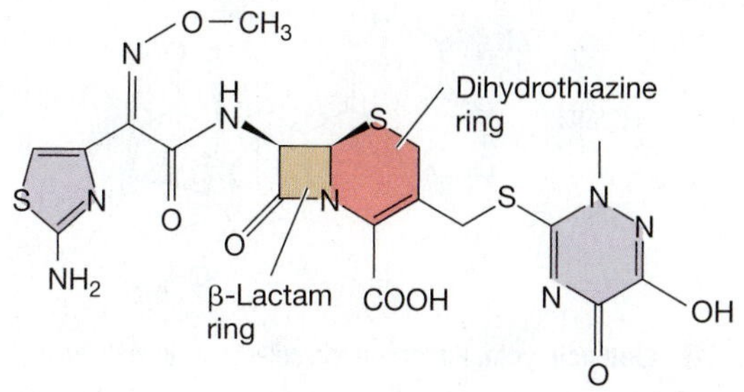

Figure 27.26 Ceftriaxone. Ceftriaxone is a β-lactam antibiotic that is resistant to most β-lactamases due to the adjacent six-member dihydrothiazine ring. Compare this structure to the five-member thiazolidine ring of the β-lactamase-sensitive penicillins (Figure 27.25).

MINIQUIZ

- How do the β-lactam antibiotics function?
- Of what clinical value are semisynthetic penicillins over natural penicillin?

27.14 Antibiotics from *Bacteria*

Many antibiotics active against *Bacteria* are also produced by *Bacteria*. These include many antibiotics that have major clinical applications.

Aminoglycosides

Antibiotics that contain amino sugars bonded by glycosidic linkage are called **aminoglycosides**. Clinically useful aminoglycosides include streptomycin (produced by *Streptomyces griseus*) and its relatives, kanamycin (**Figure 27.27**), neomycin, and gentamicin. The aminoglycosides target the 30S subunit of the ribosome,

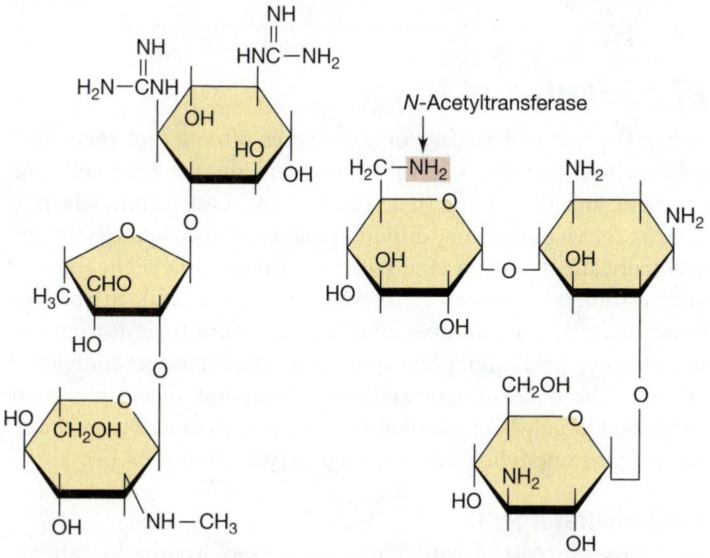

Figure 27.27 Aminoglycoside antibiotics: streptomycin and kanamycin. The amino sugars are in yellow. At the position indicated, kanamycin can be modified by a resistance plasmid that encodes *N*-acetyltransferase. Following acetylation, the antibiotic is inactive. Both kanamycin and streptomycin are synthesized by *Streptomyces* species.

inhibiting protein synthesis (Figure 27.20), and are useful for treatment of infections by gram-negative *Bacteria* (Figure 27.21).

The use of aminoglycosides for treatment of gram-negative infections has decreased since the development of the semisynthetic penicillins (Section 27.13) and tetracyclines (discussed later in this section). Aminoglycoside antibiotics are now used primarily when other antibiotics fail.

Macrolides

Macrolide antibiotics contain lactone rings bonded to sugars (**Figure 27.28**). Variations in both the lactone ring and the sugars result in a large number of macrolide antibiotics. Common macrolides are erythromycin (produced by *Streptomyces erythreus*), clarithromycin, azithromycin, and telithromycin. The macrolides account for about 20% of the total world production and use of antibiotics (Figure 27.22). Erythromycin is a broad-spectrum antibiotic that targets the 50S subunit of the bacterial ribosome, partially inhibiting protein synthesis (Figure 27.20). The partial inhibition of protein synthesis leads to preferential translation of some proteins and restricts translation of others, resulting in an imbalance in the proteome, which may disrupt metabolic functions at all levels. Often used clinically in patients allergic to penicillin or other β-lactam antibiotics, erythromycin is particularly useful for treating legionellosis (⊃⊅ Section 31.4).

Tetracyclines

The **tetracyclines**, produced by several species of *Streptomyces*, are broad-spectrum antibiotics that inhibit almost all gram-positive and gram-negative *Bacteria* (Figure 27.21). The basic structure of the tetracyclines consists of a naphthacene ring system (**Figure 27.29**). Substitutions to the basic naphthacene ring occur naturally and form new tetracycline analogs. Semisynthetic tetracyclines having substitutions in the naphthacene ring system have also been developed. Like erythromycin and the aminoglycoside antibiotics, tetracycline is a protein synthesis inhibitor, interfering with bacterial 30S ribosome subunit function (Figure 27.20).

The tetracyclines and the β-lactam antibiotics comprise the two most important groups of antibiotics in the medical field. The tetracyclines are also widely used in veterinary medicine and in some countries are used as nutritional supplements for poultry and swine. Because extensive nonmedical uses of medically important antibiotics can contribute to antibiotic resistance among pathogens, this use is now discouraged.

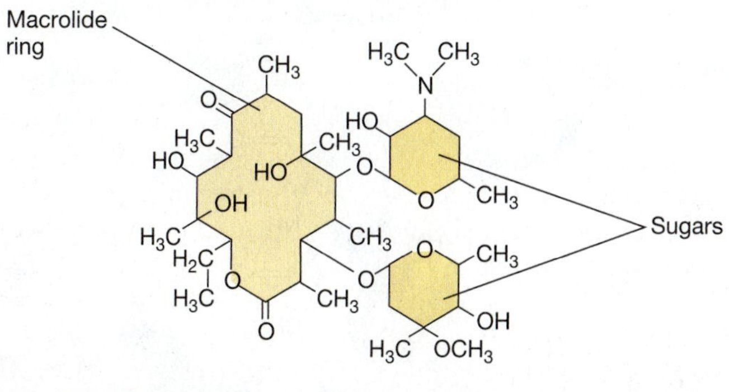

Figure 27.28 Erythromycin, a macrolide antibiotic. Erythromycin is a widely used broad-spectrum antibiotic.

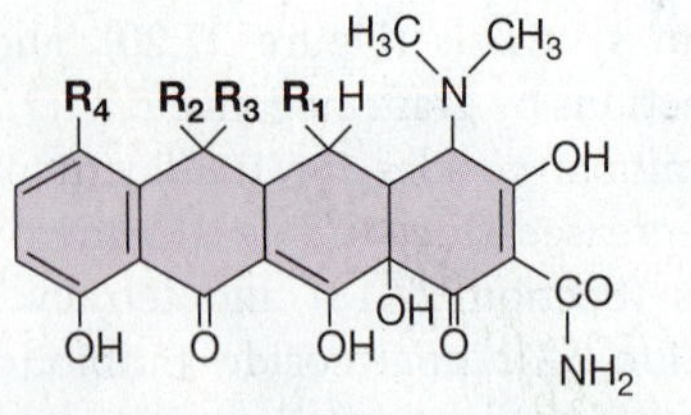

Figure 27.29 **Tetracycline.** Tetracycline is a broad-spectrum antibiotic with many active semisynthetic analogs.

Tetracycline analog	R_1	R_2	R_3	R_4
Tetracycline	H	OH	CH_3	H
7-Chlortetracycline (aureomycin)	H	OH	CH_3	Cl
5-Oxytetracycline (terramycin)	OH	OH	CH_3	H

Novel Antibiotics: Daptomycin and Platensimycin

Some antibiotics have novel structures or targets. For example, the antibiotic daptomycin is produced by a species of *Streptomyces* and is a cyclic lipopeptide (**Figure 27.30**) with a unique mode of action. Used mainly to treat infections by gram-positive *Bacteria*

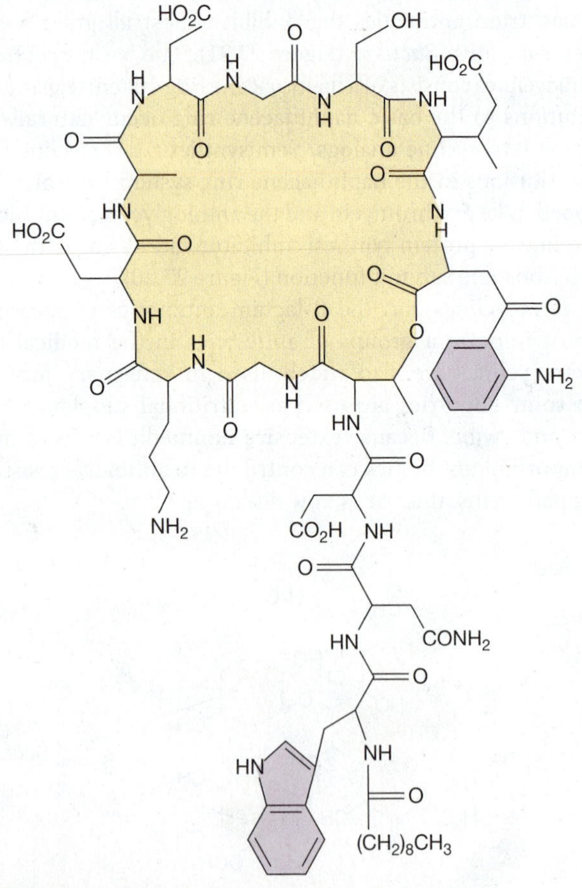

Figure 27.30 **Daptomycin.** Daptomycin is a cyclic lipopeptide that depolarizes cytoplasmic membranes in gram-positive *Bacteria*.

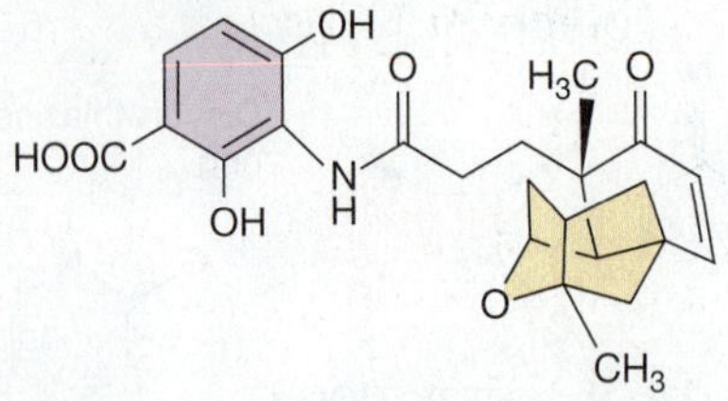

Figure 27.31 **Platensimycin.** Platensimycin selectively inhibits lipid biosynthesis in *Bacteria*.

such as the pathogenic streptococci and staphylococci (cp Sections 29.2 and 29.9), daptomycin binds specifically to bacterial cytoplasmic membranes, forms a pore, and induces rapid depolarization of the membrane. The depolarized cell quickly loses its ability to synthesize macromolecules such as nucleic acids and proteins, resulting in cell death. Alterations in cytoplasmic membrane structure, however, may lead to resistance.

Platensimycin (**Figure 27.31**), produced by *Streptomyces platensis*, is the first member of a new class of antibiotics that inhibits fatty acid biosynthesis, thus disrupting lipid biosynthesis. Platensimycin is effective against a broad range of gram-positive *Bacteria*, including nearly untreatable infections caused by methicillin-resistant *Staphylococcus aureus* and vancomycin-resistant enterococci. Platensimycin has a unique mode of action, shows no host toxicity, and there is no known potential for development of resistance by pathogens. We discuss the discovery of platensimycin in Section 27.18.

MINIQUIZ

- What are the sources of aminoglycosides, tetracyclines, macrolides, daptomycin, and platensimycin?
- How does the activity of each antibiotic class lead to control of affected pathogens?

27.15 Antiviral Drugs

Drugs that control viruses of eukaryotes often affect their host cells as well because the viruses depend on the host cell biosynthetic machinery for their replication. As a result, selective toxicity for viruses is very difficult to attain; only agents that preferentially affect unique viral pathways involved in replication or viral assembly of structural components are useful. In spite of these limitations, a number of drugs are more toxic for viruses than for the host, and a few agents specifically target individual viruses. The development and use of antiviral agents has been successful largely because of efforts to control infections with the human immunodeficiency virus (HIV) (cp Section 29.14).

HIV Antiviral Agents

The most successful and commonly used agents for antiviral chemotherapy are the nucleoside analogs (**Table 27.6**). The first compound to be widely used in this category was zidovudine, also called *azidothymidine* (*AZT*) (cp Figure 29.48a). AZT inhibits multiplication of HIV and other retroviruses by blocking reverse transcription and production of the virally encoded DNA

Table 27.6 Antiviral compounds

Category/ examples	Mechanism of action	Virus affected
Fusion inhibitor		
Enfuvirtide	Blocks HIV–T lymphocyte membrane fusion	HIV (human immunodeficiency virus)
Interferons		
α, β, γ-Interferon	Induces proteins that inhibit viral replication	Broad spectrum (host-specific)
Neuraminidase inhibitors		
Oseltamivir (Tamiflu®) and zanamivir (Relenza®)	Block active site of influenza neuraminidase	Influenza A and B
Nonnucleoside reverse transcriptase inhibitor (NNRTI)		
Nevirapine	Reverse transcriptase inhibitor	HIV
Nucleoside analogs		
Acyclovir (↩ Figure 29.42)	Viral polymerase inhibitor	Herpes viruses, Varicella zoster
Zidovudine (AZT) (↩ Figure 29.48a)	Reverse transcriptase inhibitor	HIV
Ribavirin	Blocks capping of viral RNA	Respiratory syncytial virus, influenza A and B, Lassa fever
Nucleotide analogs		
Cidofovir	Viral polymerase inhibitor	Cytomegalovirus, herpesviruses
Tenofovir (TDF)	Reverse transcriptase inhibitor	HIV
Protease inhibitors		
Indinavir, saquinavir (see Figure 27.37)	Viral protease inhibitors	HIV

intermediate (↩ Section 29.14); it is chemically related to thymidine but is a dideoxy derivative, lacking the 3′-hydroxyl group. A number of other nucleoside analogs such as acyclovir (Table 27.6 and ↩ Figure 29.42) have analogous mechanisms and were developed for the treatment of HIV and other viruses.

Nearly all nucleoside analogs are **nucleoside reverse transcriptase inhibitors (NRTIs)** and work by inhibiting elongation of the viral nucleic acid chain by a nucleic acid polymerase. Because the normal cell function of nucleic acid replication is targeted, these drugs usually induce some host toxicity. Many NRTIs also lose their antiviral potency due to the emergence of drug-resistant viruses over time (↩ Section 29.14).

Several antiviral agents target the key enzyme of retroviruses, reverse transcriptase. Nevirapine, a **nonnucleoside reverse transcriptase inhibitor (NNRTI)**, binds directly to reverse transcriptase and inhibits reverse transcription. Phosphonoformic acid, an analog of inorganic pyrophosphate, inhibits normal internucleotide linkages, preventing synthesis of viral nucleic acids. As with the NRTIs, the NNRTIs generally induce some level of host toxicity because their action also affects normal host cell nucleic acid synthesis.

Protease inhibitors are another class of antiviral drugs that are effective for treatment of HIV (Table 27.6 and see Figure 27.37). These drugs prevent viral replication by binding the active site of HIV protease, inhibiting this enzyme from processing large viral proteins into individual viral components, thus preventing virus maturation (↩ Section 9.11).

Another category of anti-HIV drugs is represented by a single drug, enfuvirtide, a **fusion inhibitor** composed of a 36-amino acid synthetic peptide that binds to the gp41 membrane protein of HIV (Table 27.6 and ↩ Section 29.14). Binding of the gp41 protein by enfuvirtide stops the conformational changes necessary for the fusion of HIV and T lymphocyte membranes, thus preventing infection of cells by HIV.

Other Antiviral Agents

A single category of drugs effectively limits influenza infection. The neuraminidase inhibitors oseltamivir (brand name Tamiflu) and zanamivir (Relenza) block the active site of neuraminidase in influenza A and B viruses, inhibiting virus release from infected cells. Zanamivir is used only for treatment of influenza, whereas oseltamivir is used for both treatment and prophylaxis (Table 27.6).

Virus interference is a phenomenon in which infection with one virus interferes with subsequent infection by another virus. Several small proteins called interferons are the cause of interference. **Interferons** are small proteins in the cytokine family (↩ Section 26.10) that prevent viral replication by stimulating the production of antiviral proteins in uninfected cells. Interferons are formed in response to live virus, inactivated virus, and viral nucleic acids. Interferon is produced in large amounts by cells infected with viruses of low virulence, but highly virulent viruses inhibit host protein synthesis before interferon can be produced, significantly reducing interferon production. Interferons are also induced by double-stranded RNA (dsRNA). In nature, dsRNA exists only in virus-infected cells as the replicative form of RNA viruses such as rhinoviruses (cold viruses) (↩ Section 29.7); the viral dsRNA signals the animal cell to produce interferon.

Interferon activity is *host*-specific rather than *virus*-specific. That is, interferon produced by one species activates receptors only on cells from the same species. As a result, interferon produced by cells of an animal in response to, for example, a rhinovirus, could also inhibit multiplication of, for example, influenza viruses in cells within the same species, but has no effect on the multiplication of any virus, including rhinovirus, in other species.

Clinical utility of interferon depends on our ability to deliver it to local areas in the host through injections or aerosols to stimulate the production of antiviral proteins in uninfected host cells. Alternatively, appropriate interferon stimulators such as viral nucleotides, nonvirulent viruses, or even synthetic nucleotides, if given to host cells prior to infection, may stimulate natural production of interferon.

MINIQUIZ --

- Why are there relatively few effective antiviral agents? Such agents are not used to treat common viral illnesses such as colds; why not?

- What steps in the viral maturation process are inhibited by nucleoside analogs? By protease inhibitors? By interferons?

27.16 Antifungal Drugs

Like viruses, fungi pose special problems for the development of effective drugs. Because fungi are *Eukarya*, antifungal agents that act on metabolic pathways in fungi often affect corresponding pathways in host cells, making the drugs toxic. As a result, many antifungal drugs can be used only for topical (surface) applications. However, a few drugs are selectively toxic for fungi because they target unique fungal structures or metabolic processes. Fungus-specific drugs are becoming increasingly important as fungus infections in immunocompromised individuals become more prevalent (ᏘᏘ Sections 32.1 and 32.2).

Ergosterol inhibitors encompass two types of antifungal compounds that work by interacting with ergosterol or inhibiting its synthesis (**Table 27.7**). Ergosterol is present in fungal cytoplasmic membranes in place of the cholesterol found in animal cell cytoplasmic membranes. Ergosterol inhibitors include the *polyene* antibiotics, a group of antifungal drugs produced by species of *Streptomyces*. Polyenes bind specifically to ergosterol, causing membrane permeability and eventually cell death (**Figure 27.32**).

Azoles and *allylamines* constitute a second functional group of drugs that affect ergosterol. They work by selectively inhibiting ergosterol biosynthesis and therefore have broad antifungal activity. Treatment with azoles results in abnormal fungus cytoplasmic membranes, leading to membrane damage and alteration of critical membrane transport activities. Allylamines also inhibit ergosterol biosynthesis, but are restricted to topical use because they are not readily taken up by animal tissues.

Echinocandins are cell wall inhibitors that block the activity of 1,3-β-D-glucan synthase, the enzyme that forms β-glucan polymers in the fungal cell wall (Figure 27.32 and Table 27.7). Because mammalian cells do not have 1,3-β-D-glucan synthase (or cell walls), the action of these agents is specific, resulting in selective fungal cell death. Echinocandins are used to treat infections with fungi such as *Candida* and some fungi that are resistant to other agents (ᏘᏘ Section 32.1).

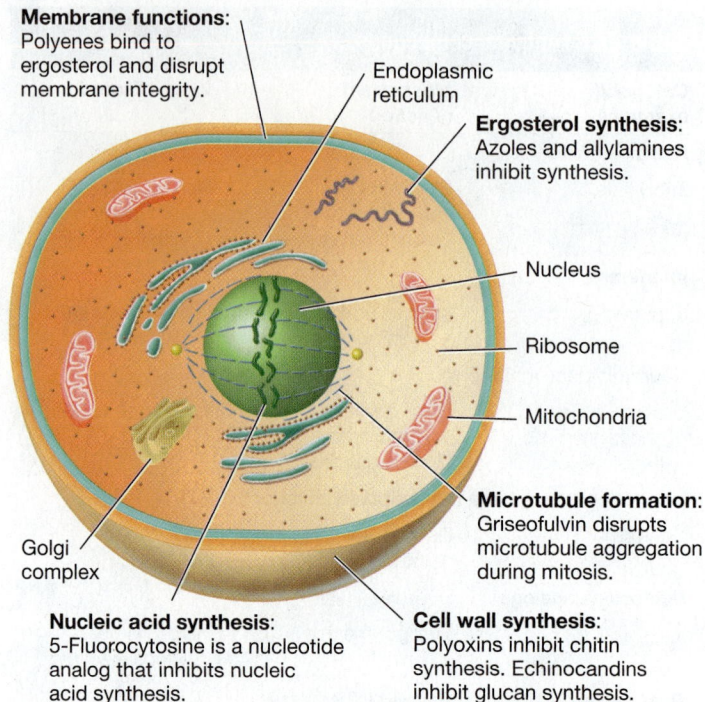

Membrane functions: Polyenes bind to ergosterol and disrupt membrane integrity.

Ergosterol synthesis: Azoles and allylamines inhibit synthesis.

Endoplasmic reticulum

Nucleus

Ribosome

Mitochondria

Microtubule formation: Griseofulvin disrupts microtubule aggregation during mitosis.

Golgi complex

Nucleic acid synthesis: 5-Fluorocytosine is a nucleotide analog that inhibits nucleic acid synthesis.

Cell wall synthesis: Polyoxins inhibit chitin synthesis. Echinocandins inhibit glucan synthesis.

Figure 27.32 Targets of some antifungal agents. Traditional antibacterial agents are generally ineffective because fungi are eukaryotic cells. The cytoplasmic membrane and cell wall targets shown here are unique structures not present in vertebrate host cells.

Fungal cell walls also contain chitin, a polymer of *N*-acetylglucosamine found only in fungi and insects. Several polyoxins inhibit cell wall synthesis by interfering with chitin biosynthesis. Polyoxins are not used clinically, but are widely used as agricultural fungicides.

Other antifungal drugs inhibit folate biosynthesis, interfere with DNA topology during replication, or, in the case of drugs such as griseofulvin, disrupt microtubule aggregation during

Table 27.7 Antifungal agents			
Category	*Target*	*Examples*	*Use*
Allylamines	Ergosterol synthesis	Terbinafine	Oral, topical
Aromatic antibiotic	Mitosis inhibitor	Griseofulvin	Oral
Azoles	Ergosterol synthesis	Clotrimazole	Topical
		Fluconazole	Oral
		Itraconazole	Oral
		Ketoconazole	Oral
		Miconazole	Topical
		Posaconazole	Experimental
Chitin synthesis inhibitor	Chitin synthesis	Nikkomycin Z	Experimental
Echinocandins	Cell wall synthesis	Caspofungin	Intravenous
Nucleic acid analogs	DNA synthesis	5-Fluorocytosine	Oral
Polyenes	Ergosterol synthesis	Amphotericin B	Oral, intravenous
		Nystatin	Oral, topical
Polyoxins	Chitin synthesis	Polyoxin A	Agricultural
		Polyoxin B	Agricultural

mitosis. Moreover, the nucleic acid analog 5-fluorocytosine (flucytosine) is an effective nucleic acid synthesis inhibitor in fungi. Some very effective antifungal drugs also have other applications. For example, vincristine and vinblastine are effective antifungal agents and also have anticancer properties.

Predictably, the use of antifungal drugs has resulted in the emergence of populations of resistant fungi and the emergence of opportunistic fungal pathogens. For example, *Candida* species, which are normally not pathogenic, now produce disease in immunocompromised individuals. Drug-resistant *Candida*

infections now occur in individuals who have been treated with antifungal drugs, and some *Candida* are resistant to multiple antifungal agents (see Figure 27.35).

MINIQUIZ

- Why are there very few clinically effective antifungal agents?
- What factors contribute to an increased incidence of fungal infections?

V • Antimicrobial Drug Resistance

Antimicrobial drug resistance is the acquired ability of a microorganism to resist the effects of an antimicrobial agent to which it is normally susceptible. No single antimicrobial agent inhibits all microorganisms, and some form of antimicrobial drug resistance is an inherent property of virtually all microorganisms.

27.17 Resistance Mechanisms and Spread

As we have discussed, antibiotic producers are microorganisms and we know that genes encoding antibiotic resistance are present in virtually every organism that makes an antibiotic. Widespread antimicrobial drug resistance is routinely passed by horizontal gene transfer between and among microorganisms. Thus, for any of several different reasons, some microorganisms are resistant to certain antibiotics.

Resistance Mechanisms

Some specific examples of bacterial resistance to antibiotics are shown in **Table 27.8**. Sites of enzymatic modification of selected antibiotics are shown in **Figure 27.33**. Antibiotic resistance can be genetically encoded by the microorganism on either the bacterial chromosome or on a plasmid called an *R* (for *resistance*) *plasmid* (᠅ Section 4.3) (Table 27.8). Because of widespread existing antibiotic resistance and continual emergence of new resistance, pathogens isolated from clinical specimens must be tested for antibiotic susceptibility to ensure appropriate treatment of an infection (Section 27.5).

Most drug-resistant bacteria isolated from patients contain drug-resistance genes located on horizontally transmitted R plasmids rather than on the chromosome. The R plasmid genes encode enzymes that modify and inactivate the drug (Figure 27.33) or genes that encode enzymes that prevent uptake of the drug or actively pump it out. For example, *Bacteria* carrying R plasmids that encode resistance for the aminoglycoside streptomycin may make enzymes that phosphorylate, acetylate, or adenylate the drug. The modified drug then lacks antibiotic activity.

For the penicillins, R plasmids encode *β-lactamase*, an enzyme that splits the β-lactam ring, inactivating the antibiotic (Figure 27.33). Chloramphenicol resistance is due to an R plasmid–encoded enzyme that acetylates the antibiotic. R plasmids may contain multiple resistance genes and confer antibiotic resistance on a cell previously sensitive to each antibiotic.

R plasmids and resistance genes predate the widespread use of antibiotics. A strain of *Escherichia coli* that was freeze-dried in 1946 contained a plasmid with genes conferring resistance to both tetracycline and streptomycin, even though neither of these antibiotics was used clinically until several years later. Similarly, R plasmid genes for resistance to semisynthetic penicillins existed before the semisynthetic penicillins had been synthesized. Another study examined 30,000-year-old permafrost and found bacterial genes for β-lactamases as well as tetracycline and vancomycin resistance genes. The inevitable conclusion is that human and veterinary antibiotic use does not produce resistance but rather selects for microorganisms with preexisting resistance mechanisms.

Of perhaps even more ecological significance, R plasmids with antibiotic resistance genes are found in many nonpathogenic gram-negative soil bacteria. While the ecological role of antibiotics and R plasmids in nature is open to speculation—concentrations of

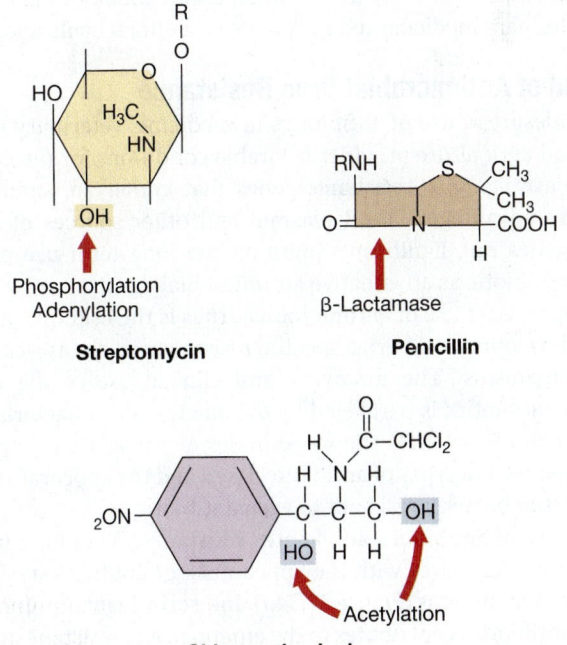

Figure 27.33 Sites at which antibiotics are modified by enzymes encoded by R plasmid genes. Antibiotics may be selectively inactivated by chemical modification or cleavage. For the complete structure of streptomycin, see Figure 27.27, and for penicillin, Figure 27.25.

Table 27.8 Bacterial resistance to antibiotics

Resistance mechanism	Antibiotic example	Genetic basis of resistance	Mechanism present in:
Reduced permeability	Penicillins	Chromosomal	*Pseudomonas aeruginosa*
			Enteric bacteria
Inactivation of antibiotic Examples: β-lactamases; modifying enzymes such as methylases, acetylases, phosphorylases, and others	Penicillins	Plasmid and chromosomal	*Staphylococcus aureus*
			Enteric *Bacteria*
	Chloramphenicol	Plasmid and chromosomal	*Neisseria gonorrhoeae*
			Staphylococcus aureus
	Aminoglycosides	Plasmid	Enteric bacteria
Alteration of target Examples: RNA polymerase, rifamycin; ribosome, erythromycin, and streptomycin; DNA gyrase, quinolones)	Erythromycin	Chromosomal	*Staphylococcus aureus*
	Rifamycin		Enteric bacteria
	Streptomycin		Enteric bacteria
	Norfloxacin		Enteric bacteria
			Staphylococcus aureus
Development of resistant biochemical pathway	Sulfonamides	Chromosomal	Enteric bacteria
			Staphylococcus aureus
Efflux (pumping out of cell)	Tetracyclines	Plasmid	Enteric bacteria
	Chloramphenicol	Chromosomal	*Staphylococcus aureus*
			Bacillus subtilis
	Erythromycin	Chromosomal	*Staphylococcus*

antibiotics in natural environments are so low that they pose no threat to neighboring bacteria—some suggest that antibiotics function as signaling molecules between neighboring cells. Regardless of their true function, it is clear that antibiotic resistance genes arose in bacteria long before antibiotics were ever used in clinical medicine. These naturally occurring genes have been strongly selected for in the past 80 years or so by the common use of antibiotics in human and veterinary medicine and in certain agricultural applications.

Spread of Antimicrobial Drug Resistance

The widespread use of antibiotics in medicine, veterinary medicine, and agriculture provides favorable conditions for the spread of R plasmids with resistance genes that confer an immediate selective advantage. The R plasmid and other sources of resistance genes put significant limits on the long-term use of any single antibiotic as an effective antimicrobial agent.

Inappropriate use of antimicrobial drugs is the leading cause of rapid development of drug-specific resistance in disease-causing microorganisms. The discovery and clinical use of the many known antibiotics is paralleled by the emergence of bacteria that resist them. **Figure 27.34** shows examples of correlations between the amounts and types of antibiotics used and the appearance and numbers of bacteria resistant to each antibiotic.

Overuse of antibiotics accelerates resistance. Antibiotic use in agriculture correlates with the appearance of antibiotic-resistant infections in humans (Figure 27.34*a*). Indiscriminant, nonmedical use of antibiotics contributes to the emergence of resistant strains. In addition to their traditional use as a treatment for infections, antibiotics are used in agriculture as supplements to animal feeds both as growth-promoting substances and as prophylactic additives to prevent the occurrence of disease. Worldwide, about 50% of all antibiotics produced are used in farm animal applications.

In one study in China, manure from swine farms that used antibiotics was tested for resistance genes using PCR arrays; the tests detected 149 unique resistance genes. The most prevalent genes were enriched as much as 28,000-fold as compared to antibiotic-free farms or soil. In addition, the manure itself still contained residual antibiotics, further enhancing potential selection for the resistance genes outside the host animal.

The broad-spectrum fluoroquinolones such as ciprofloxacin have been used extensively for over 20 years as growth-promoting and prophylactic agents in agriculture. Correlating with this use, fluoroquinolone-resistant *Campylobacter jejuni* emerged as a foodborne pathogen in poultry (Section 31.12), presumably because of the routine treatment of poultry flocks with fluoroquinolones to prevent respiratory diseases. Voluntary guidelines used by both poultry and drug producers are in place to monitor and reduce the use of fluoroquinolones with the goal of preventing resistance to new fluoroquinolone antibiotics.

Increasingly, the antimicrobial agent prescribed for treatment of a particular infection must be changed because of increased resistance of the microorganism causing the disease. A classic example is the development of resistance to penicillin and other antimicrobial drugs in *Neisseria gonorrhoeae*, the bacterium that causes the sexually transmitted disease gonorrhea (Figure 27.34*b*). Penicillin, widely used until after 1980, was replaced by ciprofloxacin, but its effective use to treat gonorrhea in selected populations lasted only about ten years, prompting a change in treatment recommendations to ceftriaxone, a penicillinase-resistant β-lactam antibiotic (Figure 27.34*c*). Treatment guidelines are updated nearly every year to cope with continually emerging drug resistance in *N. gonorrhoeae* (Section 29.12).

Antibiotics are used in clinical practice far more often than necessary. Antibiotic treatment is warranted in about 20% of

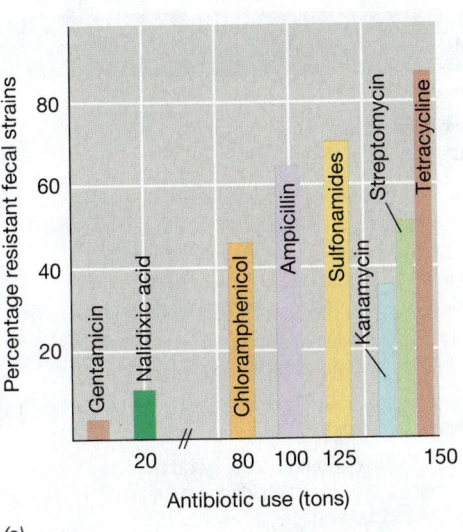

(a)

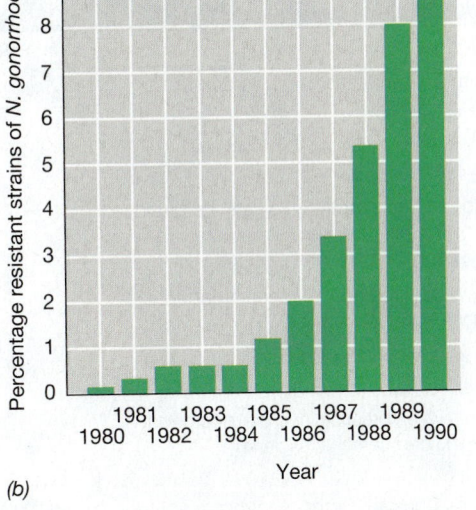

(b)

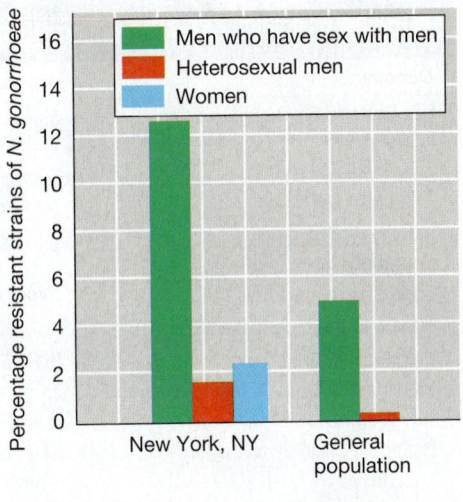

(c)

Figure 27.34 Patterns of drug resistance in pathogens. *(a)* The relationship between antibiotic use and the percentage of antibiotic-resistant bacteria isolated from diarrheal patients. Those agents that have been used in the largest amounts, as indicated by the amount produced commercially, are those for which drug-resistant strains are most frequent. *(b)* Percentage of reported cases of gonorrhea caused by drug-resistant strains. The actual number of reported drug-resistant cases in 1985 was 9,000. This number rose to 59,000 by 1990. Greater than 95% of reported drug-resistant cases were due to penicillinase-producing strains of *Neisseria gonorrhoeae*. Since 1990, penicillin has not been recommended for treatment of gonorrhea because of the emerging drug resistance. *(c)* The prevalence of fluoroquinolone-resistant *N. gonorrhoeae* in certain populations in the United States in 2003. Ciprofloxacin, a fluoroquinolone, is no longer recommended as a primary choice for treatment of *N. gonorrhoeae* infections.

individuals who seek treatment, but antibiotics are prescribed up to 80% of the time. Furthermore, in up to 50% of cases, prescribed doses or duration of treatments are not correct. This is compounded by patient noncompliance: Many patients stop taking medications, particularly antibiotics, as soon as they feel better. For example, the emergence of isoniazid-resistant tuberculosis correlates with a patient's failure to take the oral medication daily for the full course of 6–9 months (⮑ Section 29.4). Exposure of virulent pathogens to sublethal doses of antibiotics for inadequate periods of time selects for drug-resistant strains. Other recent studies, however, indicate that this trend is changing in the United States. Physicians prescribe about one-third fewer antibiotics for treatment of childhood infections than they did in the year 2000. This reduction is largely due to better education of physicians, healthcare providers, and patients concerning the proper role of antibiotic therapy.

Antibiotic-Resistant Pathogens

Largely as a result of failures to properly use antibiotics and monitor resistance, many pathogens have developed resistance to some antimicrobial agents since widespread use of antimicrobial drugs began in the 1950s (**Figure 27.35**). Penicillin and sulfa drugs, the first widely used antimicrobial agents, are not used as extensively today because so many pathogens have acquired resistance. Even the organisms that are still uniformly sensitive to penicillin, such as *Streptococcus pyogenes* (the cause of strep throat, scarlet fever, and rheumatic fever, ⮑ Section 29.2), now require larger doses of penicillin for successful treatment.

A few pathogens have developed resistance to all known antimicrobial agents (Figure 27.35). Among these are several isolates of methicillin-resistant *Staphylococcus aureus* (MRSA; methicillin is a semisynthetic penicillin) (⮑ Section 29.9). Although MRSA infections most often occur in healthcare settings, MRSA also causes a significant number of community-associated infections. An increasing number of independently derived MRSA strains have developed reduced susceptibility to even vancomycin and are termed "vancomycin intermediate *Staphylococcus aureus*" (VISA) strains.

Vancomycin-resistant *Enterococcus faecium* (VRE) and some isolates of *Mycobacterium tuberculosis* and *Candida albicans* have

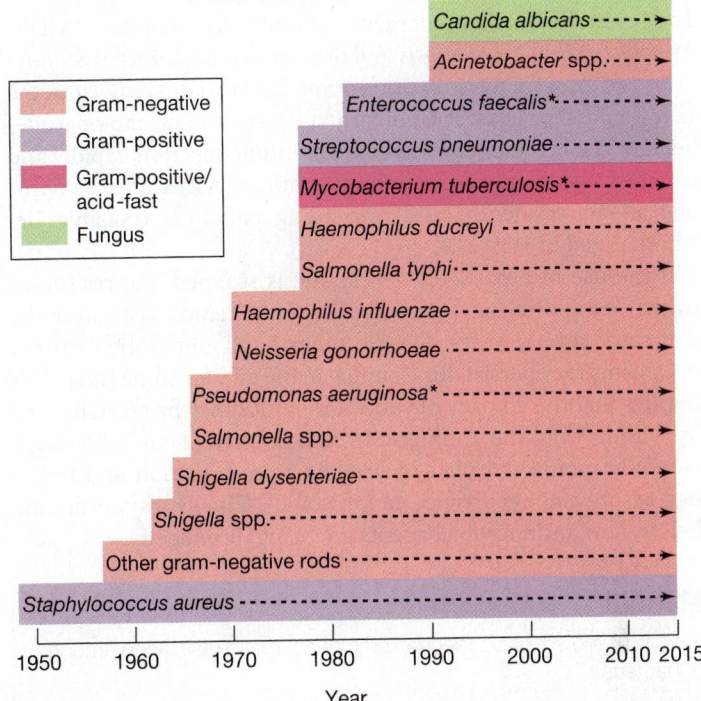

Figure 27.35 The appearance of antimicrobial drug resistance in some human pathogens. The asterisks indicate that some strains of these pathogens are now untreatable with known antimicrobial drugs.

Table 27.9 Guidelines for prevention of antimicrobial drug resistance

Directive	Action/examples	Rationale
Vaccinate to prevent common diseases	Immunize with DPT and other required and recommended vaccines.	Immunization prevents diseases that may require antimicrobial drug treatment.
Avoid unnecessary invasive procedures	Avoid catheters, biopsies, and so on unless absolutely necessary.	Parenteral procedures increase the risk of pathogen exposure.
Identify and target the pathogen	Use the antibiotic that selectively targets the pathogen of concern. For example, treat streptococcal sore throat with penicillin instead of erythromycin.	Exposure to broader-spectrum antibiotics is unnecessary and may damage local microflora, leading to infection by opportunistic pathogens.
Treat with the oldest effective antimicrobial drug	For example, treat streptococcal sore throat with penicillin instead of vancomycin.	Treatment with the latest antimicrobials enhances potential for resistance selection.
Monitor antimicrobial use	Discontinue treatment after the prescribed course.	Unnecessary treatment enhances potential for antibiotic resistance selection.
Break the chain of contagion	Isolate patients when practical and practice good housekeeping and personal hygiene.	Reducing contact and practicing good hygiene limits potential for cross-contamination of healthcare providers and clients.
Access experts	Consult with healthcare infection-control teams.	Local information can best inform selection of antibiotic therapy.

also developed resistance to all known antimicrobial drugs. Antibiotic resistance can be minimized if drugs are used only for treatment of susceptible diseases and are given in sufficiently high doses and for sufficient lengths of time to eradicate the microbial population in the host before resistant mutants can propagate. Combining two unrelated antimicrobial agents may also reduce resistance; it is less likely that a mutant strain resistant to one antibiotic will also be resistant to the second antibiotic. However, certain common R plasmids confer multiple drug resistance and make multiple antibiotic therapy less useful as a clinical treatment strategy.

Preventing Antimicrobial Drug Resistance

To prevent the further emergence of multi-drug-resistant (MDR) *Mycobacterium tuberculosis* and drug-resistant strains of *Staphylococcus aureus*, *Enterococcus faecium*, and *Candida albicans*, the U.S. Centers for Disease Control and Prevention promotes guidelines stressing the importance of preventing infection, rapidly and conclusively diagnosing and treating infections, using antimicrobial agents prudently, and preventing pathogen transmission, summarized in **Table 27.9**.

If the use of a particular antibiotic is stopped, the resistance to that antibiotic can be reversed, at least temporarily, over the course of several years. On the other hand, antibiotic-resistant organisms may persist, for example in the gut, for some time. This implies that the efficacy of some antibiotics may be reestablished by withdrawing the antibiotic from use, but only by following a carefully monitored plan of prudent reintroduction and future use to prevent recurrence of resistance. Finally, as we discuss below, new antimicrobial agents are being developed.

MINIQUIZ

- Identify the six basic mechanisms of antibiotic resistance among bacteria.
- What practices encourage the development of antibiotic-resistant pathogens?

27.18 New Antimicrobial Drugs

Resistance will eventually develop to any antimicrobial drug. Conservative, appropriate use of antibiotics can prolong or even resurrect the effective clinical use of available drugs, but the long-term solution to antimicrobial drug resistance requires continuous development of new antimicrobial drugs through design or discovery.

Candidate antimicrobial drugs must then be tested for efficacy and toxicity in animals and finally in clinical trials in humans. This entire process from discovery through laboratory and clinical trials typically takes 10–25 years before a drug is approved for clinical use. Each year, the pharmaceutical industry spends up to $4 billion on new antimicrobial drug development and each new drug approved for human use costs the industry over $500 million in initial investment.

Antimicrobial Drug Design

New analogs of existing antimicrobial compounds are often effective because the new compounds have the same proven mechanism of action. The analog may actually be more effective than the parent compound and, because resistance is based on structural recognition, analogs may not be recognized by resistance factors. For example, Figure 27.29 shows the structure of tetracycline. Using natural tetracycline as the lead compound, systematic chemical substitutions at the four R group sites can generate an almost endless series of tetracycline analogs. Using this basic strategy, semisynthetic analogs of tetracycline, β-lactam antibiotics (Section 27.13), and vancomycin (**Figure 27.36**) have been synthesized.

Novel antimicrobial compounds are much more difficult to identify than analogs of existing drugs because new antimicrobial compounds must work at unique sites in metabolism or be structurally dissimilar to existing compounds to avoid known resistance mechanisms. Computer technology and structural biology methods make it possible to model and design novel compounds, maximizing binding and efficacy in the computer environment, all at relatively low cost.

Vancomycin

Figure 27.36 Vancomycin. Intermediate drug resistance to the parent structure of vancomycin has developed in recent years. However, modification at the position shown in red by substitution of a methylene (=CH₂) group for the carbonyl oxygen restores much of the lost activity. Like penicillin, vancomycin functions by preventing cross-linking of peptidoglycan and is most effective against gram-positive pathogens.

One of the most dramatic successes in computer-directed drug design is the development of *saquinavir*, a protease inhibitor that is used to slow the growth of the human immunodeficiency virus (HIV) in infected individuals (**Figure 27.37**). Saquinavir binds the active site of the HIV protease enzyme. The structure of saquinavir was modeled on the known three-dimensional structure of the protease–substrate complex. The HIV protease normally cleaves a virus-encoded precursor protein to produce the mature viral core and activate the reverse transcriptase enzyme necessary for replication (Sections 8.10 and 9.11). Saquinavir is a high-affinity peptide analog of the HIV precursor protein that displaces the authentic protein substrate, inhibiting virus maturation. A number of other computer-designed protease inhibitors are in use as antiviral drugs for the treatment of HIV/AIDS.

Novel Targets for Antibiotics

Traditional candidate antimicrobial drugs were biomolecules isolated from natural sources such as *Streptomyces* or *Penicillium* and systematically screened for antimicrobial activity. These sources and their semisynthetic derivatives isolated through the combinatorial chemistry and computer design methods discussed above have been largely exhausted.

One key to overcoming this problem is to select for antibiotics that interact with targets that are relatively unexploited. A second key is to increase assay sensitivity to select antibiotics that are produced in smaller amounts than those recognized by traditional antibiotic susceptibility methods. Platensimycin (Figure 27.31) was discovered using these principles. First, platensimycin is the first antimicrobial drug targeted to disrupt bacterial lipid biosynthesis. It is especially active against gram-positive pathogens, including drug-resistant staphylococci and enterococci. To select

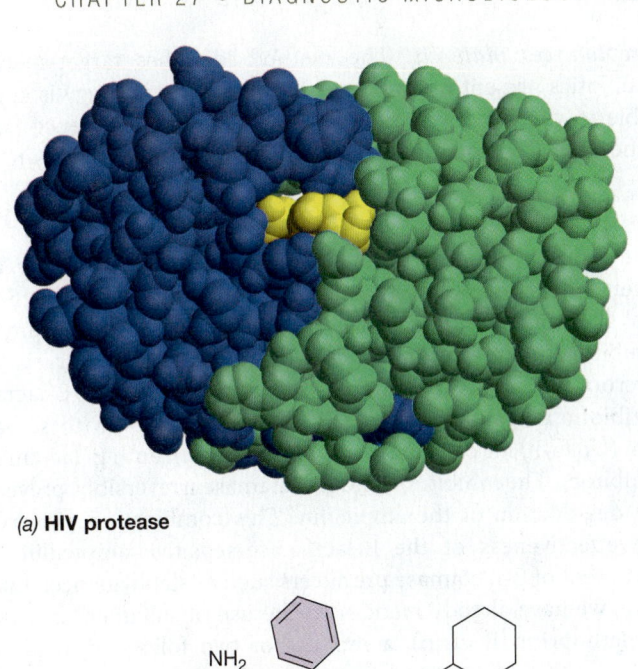

(a) **HIV protease**

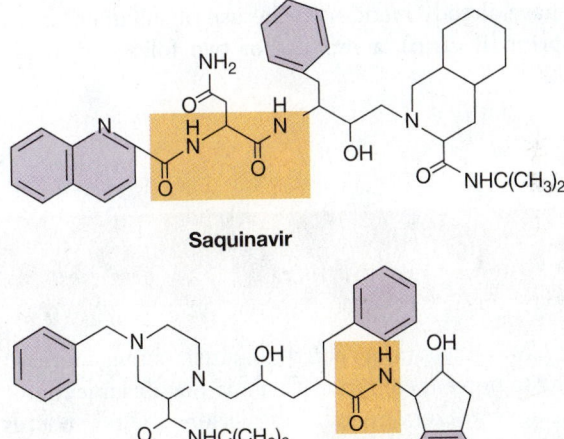

Saquinavir

Indinavir

(b)

Figure 27.37 Computer-generated anti-HIV drugs. *(a)* The HIV protease homodimer. Individual polypeptide chains are shown in green and blue. A peptide (yellow) is bound in the active site. HIV protease cleaves an HIV precursor protein, a necessary step in virus maturation. Blocking of the protease site by the bound peptide inhibits precursor processing and HIV maturation. This structure is derived from information in the Protein Data Bank. *(b)* These anti-HIV drugs are peptide analogs called protease inhibitors that were designed by computer to block the active site of HIV protease. The areas highlighted in orange show the regions analogous to peptide bonds in proteins.

an agent for a defined target, in this case an enzyme in the lipid synthesis pathway of gram-positive bacteria, scientists reduced the amount of the target lipid molecule by introducing a defect in the lipid synthase gene FabF in *Staphylococcus aureus*. They achieved this by using a strain expressing antisense FabF RNA (Section 7.14). The gene-specific antisense RNA decreased expression of FabF, reducing fatty acid synthesis and increasing the sensitivity of the crippled *S. aureus* strain to antibiotics that inhibit fatty acid synthesis. After screening natural products from 83,000 strains of potential antibiotic producers, the scientists identified and isolated platensimycin from a soil microorganism,

UNIT 5

Streptomyces platensis. The method identifies target-specific antibiotics present in low concentrations. This strategy is applicable to virtually any target for which the gene sequence (and, hence, the corresponding antisense RNA sequence) is known.

Drug Combinations

The efficacy of some antibiotics can be retained if they are given with compounds that inhibit antibiotic resistance. Several β-lactam antibiotics can be combined with β-lactamase inhibitors to preserve antibiotic activity in β-lactamase-resistant microorganisms. For example, the broad-spectrum β-lactam antibiotic ampicillin (Figure 27.25) can be mixed with sulbactam (Unasyn), or clavulanic acid (Augmentin), both β-lactamase inhibitors. The inhibitors bind β-lactamase irreversibly, preventing degradation of the ampicillin. This combination preserves the effectiveness of the β-lactamase-sensitive ampicillin for treatment of β-lactamase producers such as staphylococci. Likewise, we have already mentioned the use of sulfamethoxazole–trimethoprim (Bactrin), a mixture of two folic acid synthesis inhibitors, to prevent the loss of efficacy through mutation and selection for resistance (Section 27.11).

Drug combination therapy approaches have revolutionized treatment of HIV infections. Currently, a combination therapy consisting of nucleoside analogs and a protease inhibitor is recommended. This drug treatment protocol is termed HAART, for highly active anti-retroviral therapy. As with antimicrobial combination regimens, HAART is designed to target two independent viral functions; nucleoside analogs target virus replication and the protease inhibitors target virus maturation. Because the probability of a single virus developing resistance to multiple drugs is less than the probability of developing resistance to a single drug, HAART-resistant strains are relatively uncommon (Section 29.14).

MINIQUIZ

- Explain the advantages and disadvantages of developing new drugs based on existing drug analogs.
- Identify other methods of developing new drugs.

BIG IDEAS

27.1 • Clinical laboratory safety requires training and planning to prevent contamination and possible infection of laboratory workers. Specific precautions and procedures proportional to the risk of infection by a given agent, designated by biosafety levels (BSL), must be in place to handle contaminated materials and patient specimens.

27.2 • Patients in healthcare facilities are unusually susceptible to infectious disease because of their compromised health and potential exposure to various pathogens in the facilities. Many healthcare-associated infections are drug resistant.

27.3 • Appropriate sampling, observation, and culture techniques are necessary to isolate and identify potential pathogens. The selection of techniques requires knowledge of the ecology, physiology, and metabolism of suspected pathogens.

27.4 • Most pathogens exhibit unique metabolic patterns when grown on specialized selective and differential media. Growth-dependent patterns provide information necessary for accurate pathogen identification.

27.5 • Pathogens isolated from clinical samples are often tested for antibiotic susceptibility to ensure appropriate antibiotic therapy. Testing is based on the minimum inhibitory concentration of an agent necessary to completely inhibit growth of a pathogen.

27.6 • An immune response is often a natural outcome of infection. Specific immune responses involving a rise in antibody titers and positive T cell–mediated skin tests can be used to provide evidence for infections and to monitor convalescence.

27.7 • Direct agglutination tests are used for determination of blood types. Passive agglutination tests are available for identification of a variety of pathogens and pathogen-related products. Agglutination tests are rapid, relatively sensitive, highly specific, simple to perform, and inexpensive.

27.8 • Fluorescent antibodies are used for quick, accurate identification of pathogens and other antigenic substances in tissue samples, blood, and other complex mixtures. Fluorescent antibody–based methods can be used for identification of a variety of prokaryotic and eukaryotic cell types.

27.9 • Enzyme immunoassays, rapid tests, and immunoblots are sensitive and specific immunological assays. These tests can be engineered to detect either antibody or antigen for diagnosis of infections by a large number of pathogens.

27.10 • Nucleic acid amplification (PCR) methods are applied as extremely specific diagnostic tools used for a large number of pathogens. qPCR and qualitative PCR techniques provide quantification and identification of pathogens.

27.11 • Synthetic antimicrobial agents are selectively toxic for bacteria, viruses, and fungi. Synthetic growth factor analogs are metabolic inhibitors. Quinolones inhibit the action of DNA gyrase in *Bacteria*.

27.12 • Antibiotics are chemically diverse antimicrobial compounds produced by microorganisms. Although many antibiotics are known, only a few are clinically effective. Each antibiotic works by inhibiting a specific cellular process in the target microorganisms.

27.13 • The β-lactam antibiotics including penicillins and cephalosporins are the most important class of clinical antibiotics. These antibiotics and their semisynthetic derivatives target cell wall synthesis in *Bacteria*. They have low host toxicity and collectively have a broad spectrum of activity.

27.14 • The aminoglycosides, macrolides, and tetracycline antibiotics selectively interfere with protein synthesis in *Bacteria*. Daptomycin and platensimycin are structurally novel antibiotics that target cytoplasmic membrane functions and lipid biosynthesis, respectively. These antibiotics are structurally complex molecules produced by *Bacteria* and are active against other *Bacteria*.

27.15 • Antiviral agents selectively target virus-specific enzymes and processes. Useful agents include analogs and compounds that inhibit nucleic acid polymerases and viral genome replication. Protease inhibitors interfere with viral maturation steps. Host cells also produce antiviral interferon proteins that stop viral replication.

27.16 • Antifungal agents exhibiting selective toxicity are hard to find because fungi are *Eukarya*, but some effective antifungal agents are available. Treatment of fungal infections is an emerging human health issue.

27.17 • The use of antimicrobial drugs inevitably leads to resistance in the targeted microorganisms. The development of resistance can be accelerated by the indiscriminate use of the drugs. Many pathogens have developed resistance to common antimicrobial drugs.

27.18 • New antimicrobial compounds are constantly being discovered and developed to deal with drug-resistant pathogens and to enhance our ability to treat infectious diseases.

MasteringMicrobiology®

Review what you know and challenge what you have learned with MasteringMicrobiology! Access study materials, chapter quizzes, animations, and microbiology lab tutorials in the Study Area to ensure that you have mastered this chapter's content.

UNIT 5

REVIEW OF KEY TERMS

Agglutination a reaction between antibody and particle-bound antigen resulting in visible clumping of the particles

Aminoglycoside an antibiotic such as streptomycin, containing amino sugars linked by glycosidic bonds

Antibiogram a report indicating the susceptibility of clinically isolated microorganisms to the antibiotics in current use

Antibiotic a chemical substance produced by a microorganism that kills or inhibits the growth of another microorganism

Antimicrobial drug resistance the acquired ability of a microorganism to resist the effects of an antimicrobial agent to which it is normally susceptible

Bacteremia the presence of bacteria in the blood

β-lactam antibiotic penicillin or a related antibiotic that contains the four-membered heterocyclic β-lactam ring

Broad-spectrum antibiotic an antibiotic that acts on both gram-positive and gram-negative *Bacteria*

Differential media growth media that allow identification of microorganisms based on phenotypic properties

Enriched media media that allow metabolically fastidious microorganisms to grow because of the addition of specific growth factors

Enrichment culture the use of selective culture media and incubation conditions to isolate specific microorganisms from natural samples

Enzyme immunoassay (EIA) a test that uses antibodies linked to enzymes to detect antigens or antibodies in body fluids

Fluorescent antibody an antibody molecule covalently modified with a fluorescent dye that makes the antibody visible under fluorescent light

Fusion inhibitor a peptide that blocks the fusion of viral and target cytoplasmic membranes

General-purpose media growth media that support the growth of most aerobic and facultatively aerobic organisms

Growth factor analog a chemical agent that is related to and blocks the uptake or utilization of a growth factor

Healthcare-associated infection (HAI) a local or systemic infection acquired by a patient in a healthcare facility, particularly during a stay in the facility. Also called *nosocomial infection*.

Immunoblot (Western blot) the use of labeled antibodies to detect specific proteins after separation by electrophoresis and transfer to a membrane

Interferon a cytokine protein produced by virus-infected cells that induces signal transduction in nearby cells, resulting in transcription of antiviral genes and expression of antiviral proteins

Minimum inhibitory concentration (MIC) the smallest amount of an agent needed to completely inhibit the growth of an organism in vitro

Nonnucleoside reverse transcriptase inhibitor (NNRTI) a nonnucleoside analog used to inhibit viral reverse transcriptase

Nucleoside reverse transcriptase inhibitor (NRTI) a nucleoside analog used to inhibit viral reverse transcriptase

Penicillin a class of antibiotics that inhibit bacterial cell wall synthesis, characterized by a β-lactam ring

Protease inhibitor an inhibitor of a viral protease

Quinolone a synthetic antibacterial compound that interacts with DNA gyrase and prevents supercoiling of bacterial DNA

Selective media media that enhance the growth of certain organisms while retarding the growth of others due to an added media component

Selective toxicity the ability of a compound to inhibit or kill pathogenic microorganisms without adversely affecting the host

Sensitivity the lowest amount of antigen that can be detected by a diagnostic test

Septicemia (sepsis) a blood infection

Serology the study of antigen–antibody reactions in vitro

Specificity the ability of an antibody or a lymphocyte to recognize a single antigen, or of a diagnostic test to identify a specific pathogen

Tetracycline an antibiotic characterized by the four-ring naphthacene structure

Titer the quantity of antibody present in a solution

REVIEW QUESTIONS

1. How are most laboratory-associated infections contracted? What action can be taken to prevent laboratory infections? (Section 27.1)

2. Healthcare environments are conducive to the spread of infectious diseases. Review the reasons for the enhanced spread of infection in healthcare facilities. What are the sources of most healthcare-associated infections? (Section 27.2)

3. Describe the standard procedure for obtaining and culturing a throat culture and a blood sample. What special precautions must be taken while obtaining the blood sample? (Section 27.3)

4. Why is it important to process clinical specimens as rapidly as possible? What special procedures and precautions are necessary for the isolation and culture of anaerobes? (Section 27.3)

5. Differentiate between selective and differential media. Is eosin–methylene blue agar a selective medium or a differential medium? How and why is it used in a clinical laboratory? (Section 27.4)

6. Describe the disc diffusion test for antibiotic susceptibility. Why should potential pathogens from patient isolates be tested for antibiotic susceptibility? (Section 27.5)

7. Why does the antibody titer rise after infection? Is a high antibody titer indicative of an ongoing infection? Explain. Why is it necessary to obtain an acute and a convalescent blood sample to monitor infections? (Section 27.6)

8. Agglutination tests are widely used for clinical diagnostic purposes. Why is this so? (Section 27.7)

9. How are fluorescent antibodies used for the diagnosis of viral diseases? What advantages do fluorescent antibodies have over other tests for viral infections? (Section 27.8)

10. Enzyme immunoassays (EIAs) are extremely sensitive, as compared with agglutination. Why is this so? (Section 27.9)

11. Why is the immunoblot (Western blot) procedure used to confirm screening tests that are positive for human immunodeficiency virus (HIV)? (Section 27.9)

12. Distinguish between quantitative and qualitative PCR. (Section 27.10)

13. Growth factor analogs are distinguished from antibiotics by a single important criterion. Explain. (Section 27.11)

14. Identify common sources for naturally occurring antimicrobial drugs. (Section 27.12)

15. Describe the mode of action of a β-lactam antibiotic. Why are these antibiotics generally more effective against gram-positive bacteria than against gram-negative bacteria? (Section 27.13)

16. Distinguish between the modes of action of three of the protein synthesis–inhibiting antibiotics. (Section 27.14)

17. Why do antiviral drugs generally exhibit host toxicity? (Section 27.15)

18. Identify the targets that allow the selective toxicity of antifungal agents. (Section 27.16)

19. Identify six mechanisms responsible for antibiotic resistance. (Section 27.17)

20. Explain how selection for metabolic mutants can extend traditional methods of natural product selection for antibiotic discovery. (Section 27.18)

APPLICATION QUESTIONS

1. A blood culture is positive for *Staphylococcus epidermidis*. Explain the finding. Is it likely that the patient has *S. epidermidis* bacteremia? Prepare a list of possibilities and questions for a discussion with the physician in charge. What additional information will be needed to confirm or rule out a bloodstream infection due to *S. epidermidis*?

2. Define the procedures you would use to isolate and identify a new pathogen. Keep in mind Koch's postulates (Section 1.8) as you form your answer. Be sure to include growth-dependent assays, immunoassays, and molecular assays. Where would you report your findings? Which of your assays could be adapted to be used as a routine, high-throughput test for rapid clinical diagnosis?

3. Viruses and fungi present special problems for drug therapy. Explain the issues inherent in drug treatment of both groups and explain whether or not you agree with the preceding statement. Give specific examples and suggest at least one group of agents that might target both types of infectious agents.

4. Explain the genetic basis of acquired resistance to β-lactam antibiotics in *Staphylococcus aureus*. Design laboratory experiments to reverse resistance to the β-lactam antibiotics. Can your experiment be applied "in the field" to promote deselection of antibiotic-resistant organisms?

microbiology**now**

MERS-CoV: An Emerging Disease

New emerging diseases often appear suddenly. For example, Middle East respiratory syndrome coronavirus (MERS-CoV) was first identified as the cause of serious cases of pneumonia in 2012. The first outbreak occurred in Saudi Arabia and then spread by way of travelers to other Arabic countries and Europe. From September 2012 through June 2013, 58 cases were confirmed, with 33 deaths.[1]

An investigative team traveled to Saudi Arabia to observe an ongoing healthcare-associated outbreak of MERS-CoV and reported that the virus spread easily from patient to patient in three different healthcare facilities. Two healthcare workers were also infected, with 15 fatalities in 23 cases. The incubation time for infection is about 5 days, with rapid spread to the next host. Genomic sequencing data confirmed that a single group of closely related viruses (photo) was responsible for the outbreak.

The confirmed cases and two possible cases were linked to contact with infected individuals in the healthcare facilities. MERS-CoV is highly virulent and spreads easily in healthcare settings, probably by way of airborne droplets or indirect contacts. Thus healthcare facilities that care for MERS-CoV patients are on high alert to protect both workers and patients from further spread of this disease.

Another coronavirus, SARS-CoV, discussed later in this chapter, causes a similar illness to MERS-CoV with correspondingly high mortality rates. Like SARS-CoV, MERS-CoV probably originated in bats. All of its reservoirs are unknown but include camels. Is it spread from bats and camels via respiratory droplets and close contact? Because MERS-CoV spreads so easily from person to person and has such high mortality, epidemiologists need to answer these questions if they are to break the chain of infection and prevent a possible MERS-CoV pandemic.

[1]Assiri, A., et al. 2013. Hospital outbreak of Middle East respiratory syndrome coronavirus. *New England Journal of Medicine, 369:* 407–416.

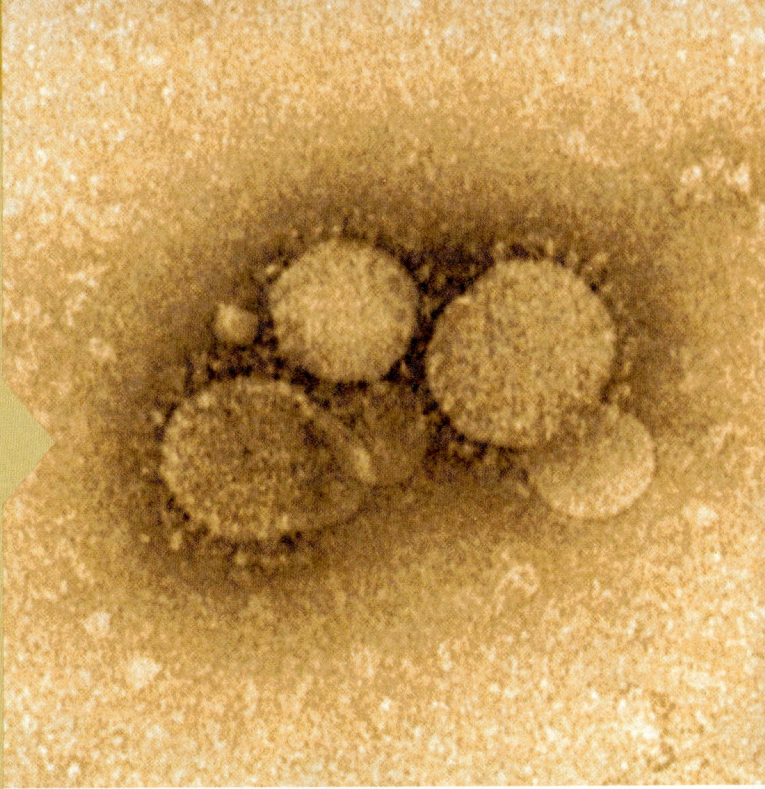

Epidemiology is the study of the occurrence, distribution, and determinants of health and disease in a population; it deals with **public health**, the health of the population as a whole. Here we consider the epidemiology of infectious diseases and the public health methods used to control them.

In Chapter 1 we compared the current common causes of death in the United States with those at the beginning of the twentieth century (ↄ Figure 1.8). In the United States and other developed countries, infectious diseases no longer kill as they once did, but globally they still account for almost one-quarter of annual deaths. Even developed countries are affected by the antibiotic-resistant pathogens and pandemic influenzas that cause new illnesses and new infectious diseases to continue to emerge throughout the world (**Figure 28.1**). Identifying and solving problems associated with infectious disease is the goal of the epidemiologist.

I • Principles of Epidemiology

Here we consider the principles of epidemiology of infectious diseases and define key terms in the language of the epidemiologist.

28.1 Epidemiology Basics

The epidemiologist traces the spread of a disease to identify its origin and mode of transmission. Epidemiological data are obtained by collecting disease information in a *population*. Data are gathered from disease-reporting surveillance networks, clinical records, and patient interviews with the goal of defining common factors for an illness. This is in contrast to individual patient treatment and diagnosis in the clinic or laboratory. The epidemiologist couples host–pathogen population dynamics to clinical problems to formulate effective public health measures for disease control.

Epidemiology Vocabulary

A well-adapted pathogen lives in balance with its host, taking what it needs for existence and causing only a minimum of harm. Such pathogens may cause **chronic infections** (long-term infections) in the host. When there is a balance between host and pathogen, both host and pathogen survive. On the other hand, a host whose resistance is compromised because of factors such as poor diet, age, and other stressors can be harmed (ↄ Section 23.12). In addition, new pathogens occasionally emerge to which the individual host, specific populations, or even an entire species has not developed resistance. Such emerging pathogens often cause **acute infections**, characterized by rapid and dramatic disease onset.

Certain terms have specialized meanings in epidemiology. A disease is an **epidemic** when it simultaneously infects an unusually high number of individuals in a population; a **pandemic** is a widespread, usually worldwide, epidemic. By contrast, an **endemic disease** is one that is constantly present, usually in low numbers, in a population (**Figure 28.2**). An endemic disease implies that the pathogen may not be highly virulent or that the majority of individuals in the population may be immune, resulting in low but persistent numbers of cases. Individuals infected with an

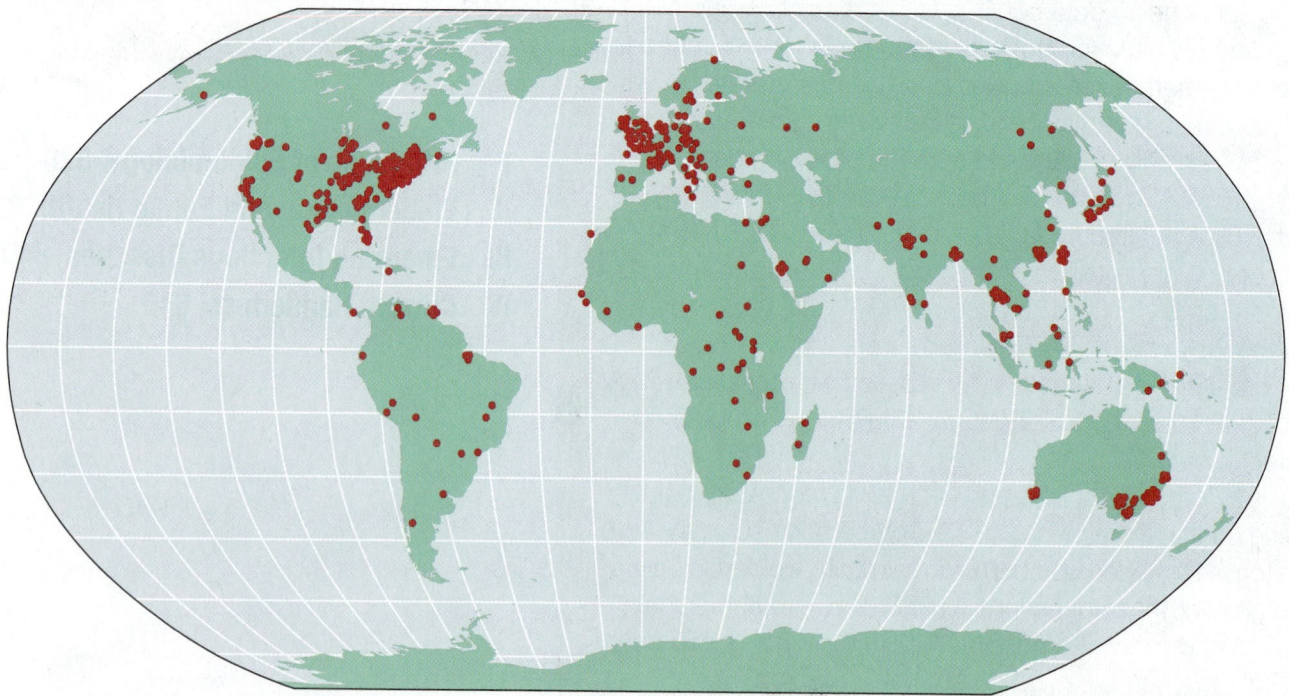

Figure 28.1 Worldwide emerging infectious diseases, 1940–2004. The number of emerging and reemerging infectious diseases is significant in spite of extensive sanitation, vaccination, and antimicrobial drug therapy programs that have reduced mortality from infectious diseases in the last century in the developed world. Each red dot indicates an emerging infectious disease. Adapted from Jones et al., *Nature 451*: 990–993, 2008.

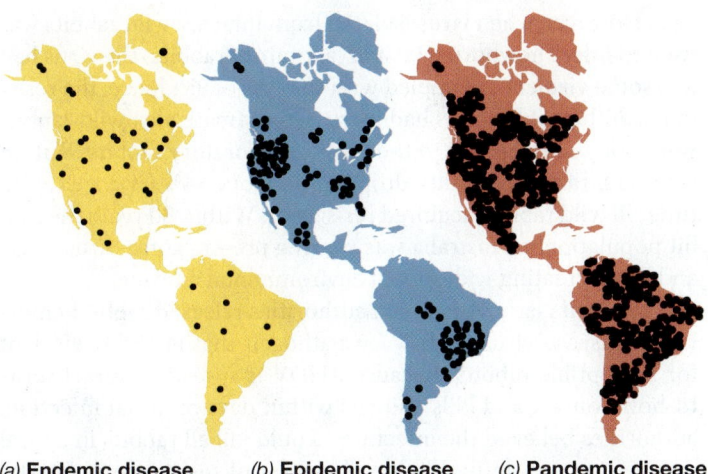

(a) Endemic disease **(b) Epidemic disease** **(c) Pandemic disease**

Figure 28.2 **Endemic, epidemic, and pandemic disease.** Each dot represents a disease outbreak. *(a)* Endemic diseases are present in the population in particular limited geographical areas. *(b)* Epidemic diseases show high incidence in a wider area, usually developing from an endemic focus. *(c)* Pandemic diseases are distributed worldwide.

organism that causes an endemic disease are **reservoirs** of infection, a source of infectious agents from which susceptible individuals may be infected.

The **incidence** of a particular disease is the *number of new cases* in a population in a given time period. For example, in 2010 there were 47,500 new cases of HIV infection in the United States, for an incidence of 15.5 new cases per 100,000 people per year. The **prevalence** of a given disease is the *total number of new and existing disease cases* in a population in a given time period. For example, within the United States there were 803,771 persons living with HIV/AIDS at the end of 2010. Expressed another way, the prevalence of HIV/AIDS in the United States was about 262 cases per 100,000 persons in 2010. Thus, *incidence* provides a record of new cases of a disease, whereas *prevalence* indicates the total disease burden in a population. The incidence and prevalence of disease are indicators of the public health of a particular group such as the total global population or the population of a localized region, such as a city, state, or country.

Sporadic cases of a disease occur one at a time in geographically separated areas, suggesting that the cases are not related. A disease **outbreak**, on the other hand, is the appearance of a large number of cases in a short time in an area previously experiencing only sporadic cases. Diseased individuals who show no symptoms or only mild symptoms are said to have *subclinical infections*. Subclinically infected individuals are frequently **carriers** of the particular pathogen, with the pathogen reproducing within them and being shed into the environment. Finally, the term **virulence** is often used in epidemiological parlance, and is a measure of the relative ability of a pathogen to cause disease.

Public health conditions and concerns vary with location and time. That is, assessment of public health at a given time provides only a snapshot of a dynamic situation. Public health policies and laws are evaluated by examining public health statistics over long time periods, with the goal of reducing the incidence and prevalence of disease.

Mortality and Morbidity

Mortality is the incidence of *death* in a population. Infectious diseases were the major causes of death in 1900 in all countries and geographic regions, but they are now less prevalent in developed countries. Noninfectious "lifestyle" diseases such as heart disease and cancer are now much more prevalent and cause higher mortality than do infectious diseases (∞ Figure 1.8). However, this could change rapidly if public health measures were to break down. Worldwide, and especially in developing countries, infectious diseases are still major causes of mortality (Table 28.1 and Section 28.6).

Morbidity is the incidence of *disease* in a population and includes both fatal and nonfatal diseases. Morbidity statistics indicate the public health of a population more precisely than mortality statistics because many diseases have relatively low mortality. Put another way, the major causes of *illness* are quite different from the major causes of *death*. For example, high-morbidity infectious diseases include acute respiratory diseases such as the common cold and acute digestive disorders. However, seldom do these diseases cause death in developed countries. Thus, both of these diseases have high morbidity, but low mortality. On the other hand, Ebola virus infects only several hundred people worldwide every year, but the mortality in some outbreaks approaches 70%. Thus, Ebola has low morbidity, but high mortality.

Disease Progression

The progression of clinical symptoms for a typical acute infectious disease can be divided into stages:

1. *Infection:* The organism invades, colonizes, and grows in the host.
2. *Incubation period:* Some time always passes between infection and the appearance of disease signs and symptoms. Some

Table 28.1 Worldwide deaths due to infectious diseases, 2004[a]

Disease	Deaths	Causative agent(s)
Respiratory infections[b]	4,259,000	Bacteria, viruses, fungi
Diarrheal diseases	2,163,000	Bacteria, viruses
Acquired immunodeficiency syndrome (AIDS)	2,040,000	Virus
Tuberculosis[c]	1,464,000	Bacterium
Malaria	889,000	Protist
Measles[c]	424,000	Virus
Meningitis, bacterial[c]	340,000	Bacterium
Pertussis (whooping cough)[c]	254,000	Bacterium
Tetanus[c]	163,000	Bacterium
Hepatitis (all types)[d]	159,000	Viruses
Other communicable diseases	1,645,000	Various agents

[a]Data show the ten leading causes of death due to infectious diseases. Globally, there were about 58.7 million deaths from all causes in 2004. About 13.8 million deaths, or 23.5%, were from infectious diseases, nearly all in developing countries. Data are from the World Health Organization (WHO), Geneva, Switzerland.
[b]For some acute respiratory agents such as influenza and *Streptococcus pneumoniae* there are effective vaccines; for others, such as colds, there are no vaccines.
[c]Diseases for which effective vaccines are available.
[d]Vaccines are available for hepatitis A virus and hepatitis B virus. There are no vaccines for other hepatitis agents.

UNIT 6

diseases, like influenza, have very short incubation periods, measured in days; others, like AIDS, have longer ones, sometimes extending for years. The incubation period for a given disease is determined by inoculum size, the virulence and life cycle of the pathogen, and resistance of the host. At the end of the incubation period, the first signs and symptoms, for example in the case of an ensuing cold, a mild cough and a feeling of general fatigue, usually appear.

3. *Acute period:* The disease is at its height, with overt symptoms and signs such as fever and chills.

4. *Decline period:* Disease signs and symptoms subside. Any fever subsides, usually following a period of intense sweating, and a feeling of well-being develops. The decline period may be rapid (within one day), in which case decline occurs by *crisis*, or it may be slower, extending over several days, in which case decline occurs by *lysis*.

5. *Convalescent period:* The patient regains strength and returns to normal.

After the acute period, the immune mechanisms of the host become increasingly important for complete recovery from the disease.

MINIQUIZ --

- Why do epidemiologists acquire population-based data about infectious diseases?

- Distinguish between an endemic disease, an epidemic disease, and a pandemic disease.

28.2 The Host Community

The colonization of a susceptible host population by a pathogen may lead to explosive infections, transmission to uninfected hosts, and an epidemic. As the host population develops resistance, however, the spread of the pathogen is checked, and eventually a balance is reached in which host and pathogen populations reach an equilibrium. In an extreme case, failure to reach equilibrium could result in death and eventual extinction of the host species. If the pathogen has no other host, then the extinction of the host also results in extinction of the pathogen. Thus, the evolutionary success of a pathogen may depend on its ability to establish an equilibrium with the host population rather than destroy the host population. In most cases, the evolution of the host and the pathogen affect one another; that is, the host and pathogen *coevolve*.

Coevolution of a Host and a Pathogen

A striking example of host and pathogen coevolution is a case where myxoma virus was intentionally introduced in Australia to control wild rabbits that were causing massive crop and vegetation damage. The virus, spread by the bite of mosquitoes, is extremely virulent and causes fatal infections in susceptible animals. Within several months, the virus infection had spread over a large area, rising to peak incidence in the summer when the mosquito vectors were present, and then declining in the winter as mosquitoes disappeared. Over 95% of the infected rabbits died during the first

year. However, when virus isolated from infected wild rabbits was used to infect newborn wild and laboratory rabbits, the virus had lost some virulence. Coupled with the loss of virulence, the resistance of the wild rabbits had increased dramatically; wild rabbits were not as susceptible to the virus as laboratory rabbits. Within six years, rabbit mortality dropped to about 84% (**Figure 28.3**). In time, all wild rabbits acquired resistance. Within 30 years the rabbit population in Australia was nearing pre–myxoma virus levels and again creating widespread environmental damage.

A few years later, Australian authorities released rabbit hemorrhagic disease virus (RHDV), a pathogen that is highly virulent for susceptible rabbits. Because RHDV is spread by direct host-to-host contact and kills animals within days of initial infection, authorities believed the infections would kill all rabbits in a local population, preventing the development of resistance to RHDV. At first RHDV was very effective at reducing local rabbit populations. However, natural infection of some rabbits by an indigenous but nonlethal hemorrhagic fever virus conferred immune cross-resistance to the introduced RHDV. This unanticipated immunity reduced RHDV virulence in some areas. As with the myxoma virus, the host developed resistance to the RHDV control agent, moving the host–pathogen balance toward equilibrium.

For pathogens that do not rely on host-to-host transmission such as *Clostridium tetani*, a common soil bacterium that causes tetanus, there is no selection for decreased virulence to support mutual coexistence. Vectorborne pathogens usually transmitted by the bite of ticks or other arthropods are also under no evolutionary pressure to spare the human host. As long as a vector can obtain its blood meal after infection and before the host dies, the pathogen can maintain a high level of virulence and kill its human host. For example, malaria parasites (*Plasmodium* spp.) show antigenic variations in coat proteins that aid in avoiding the immune response of the host. This ability to avoid the host

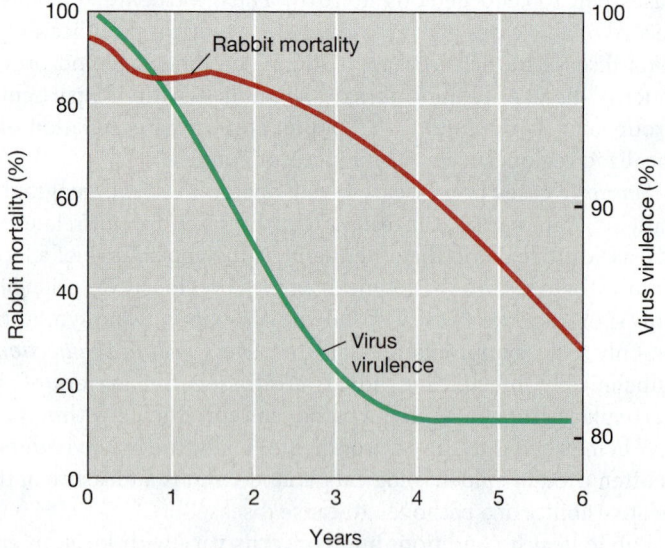

Figure 28.3 Myxoma virus and host coevolution. Myxoma virus was introduced into Australia to control the wild rabbit population. Virus virulence is shown as the average mortality in laboratory rabbits for virus recovered from the field each year. Rabbit mortality was determined by removing young wild rabbits from dens and infecting them with a viral strain that killed 90–95% of control laboratory rabbits.

immunity increases pathogen virulence without regard to the susceptibility of the host.

Other evidence for the phenomenon of unchecked pathogen virulence comes from studies of diarrhea in newborns. In hospital nurseries, *Escherichia coli* can cause severe diarrhea and even death; virulence seems to increase with each passage of the pathogen through a patient. The pathogens replicate in one host and are then inadvertently transferred to another through healthcare providers or objects such as soiled bedding and furniture. Extraordinary efforts such as completely disinfecting the nursery and furniture, coupled with transferring healthcare staff to other services, are sometimes necessary to interrupt the cycle of these highly virulent infections.

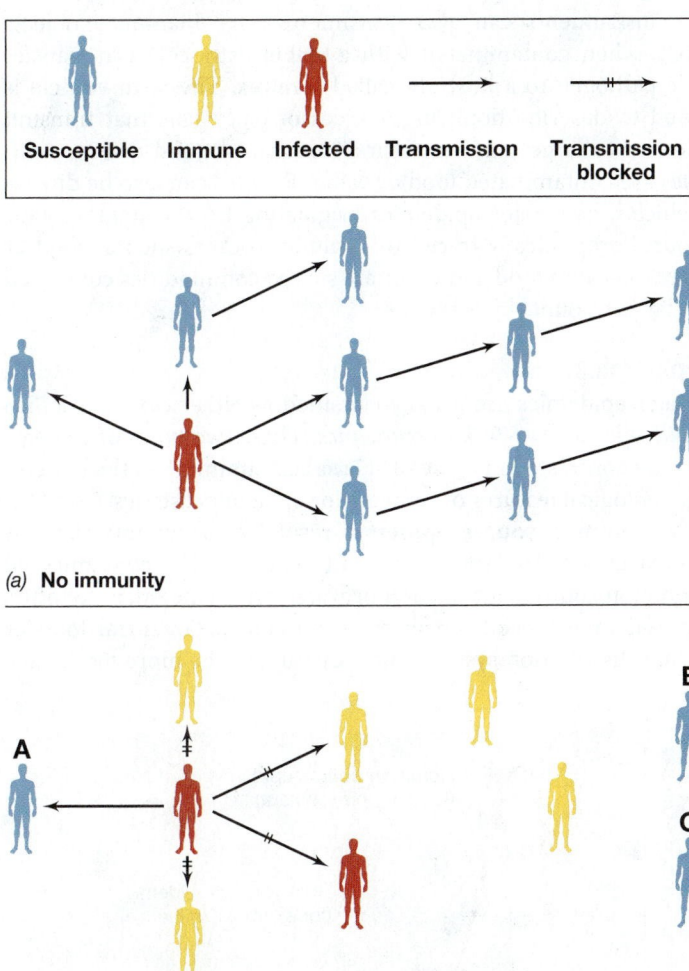

(a) **No immunity**

(b) **Herd immunity**

Figure 28.4 Herd immunity and transmission of infection. Immunity in some individuals protects individuals without immunity from infection. *(a)* In a population with no immunity, transfer of a pathogen from one infected individual can ultimately infect (arrows) all the individuals as newly infected individuals in turn transfer the pathogen to other individuals. *(b)* In a population that is only moderately dense and that has some immunity against a moderately transmissible pathogen such as influenza, an infected individual cannot transfer the pathogen to all susceptible individuals because resistant individuals, immune from previous exposure or immunization, break the cycle of pathogen transmission: Susceptible individual A becomes infected, but susceptible individuals B and C are protected. The proportion of a population that must be immune for herd immunity to be effective also varies with the disease; highly infectious diseases require a higher proportion of immune individuals for herd immunity to prevent transmission.

Herd Immunity

If a high proportion of the individuals in a group are immune to a pathogen, then the whole population will be protected; this resistance to infection is called **herd immunity** (**Figure 28.4**). Assessment of herd immunity is important for understanding the development of epidemics. The more highly infectious a pathogen, or the longer its period of infectivity, the greater the proportion of immune individuals necessary to prevent disease spread. For a highly infectious disease such as measles, 90–95% of the population must be immune to confer herd immunity. By contrast, a lower proportion of immune individuals can prevent an epidemic by a less infectious agent or one with a brief period of infectivity. Mumps virus, which is less infectious than measles virus, exhibits this pattern. In the absence of immunity, even poorly infectious agents can be transmitted person-to-person if susceptible hosts have repeated or constant contact with an infected individual. This is the case for the transmission of H5N1 avian influenza among humans (Section 28.11).

MINIQUIZ --

• Explain coevolution of host and pathogen. Cite a specific example.

• How does herd immunity prevent a nonimmune individual from acquiring a disease? Give an example.

28.3 Infectious Disease Transmission

Epidemiologists follow the transmission of a disease by correlating geographic, climatic, social, and demographic data with disease incidence. These correlations are used to identify possible modes of transmission and disease patterns. A disease limited to a restricted geographic location, for example, may suggest a particular vector. This is the case with malaria, a disease of tropical regions that is only transmitted by mosquito species restricted to tropical regions (⮌ Section 32.5).

A marked seasonality or periodicity of a disease often indicates certain modes of transmission. For example, influenza occurs in an annual cyclic pattern, causing epidemics propagated among school children and other populations of susceptible individuals. Influenza infectivity is high in crowded environments such as schools because the virus is transmitted by a respiratory route. Epidemic strains of influenza virus change almost every year, and as a result, most children are susceptible to infection. On the introduction of virus into a school, an explosive propagated epidemic results. Nearly every individual becomes infected and then becomes immune. As the immune population increases, the epidemic subsides, but introduction of a new influenza virus, usually in the following year, will trigger another epidemic.

The mode of pathogen transmission is usually related to the preferred habitat of the pathogen in the body. Respiratory pathogens, for example, are typically transmitted by an airborne route, whereas intestinal pathogens are spread through fecal contamination of food or water. In some cases, environmental factors such as weather patterns may influence the survival of the pathogen. For example, California encephalitis virus and other encephalitis viruses cause disease during the summer and fall months and

UNIT 6

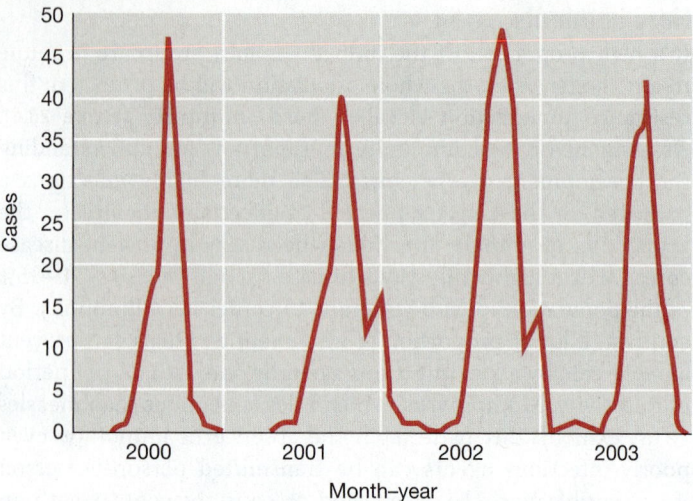

Figure 28.5 **California encephalitis in the United States.** Disease incidence shows a sharp rise in late summer, followed by a complete decline in winter. The disease cycle follows the yearly cycle of the mosquito vector, which dies out in the winter months. Data are from the CDC, Atlanta, Georgia, USA.

disappear every winter in a cyclical pattern (**Figure 28.5**). The virus is transmitted from mosquito vectors that die during the winter months, causing the disease to disappear until the insect vector reappears and again transmits the virus in the summer months.

Pathogens can be classified by their mechanism of transmission, but all mechanisms have three stages in common: (1) escape from the host, (2) travel, and (3) entry into a new host. Pathogen transmission can be by direct or indirect mechanisms.

Direct Host-to-Host Transmission

Host-to-host transmission occurs when an infected host transmits a disease directly to a susceptible host without the assistance of an intermediate host or inanimate object. Upper respiratory infections such as the common cold and influenza are most often transmitted host to host by droplets resulting from sneezing or coughing (↩ Figure 29.1). Many of these droplets, however, do not remain airborne for long. Transmission, therefore, requires close, although not necessarily intimate, person-to-person contact.

Some pathogens are extremely sensitive to environmental factors such as drying and heat and are unable to survive for significant periods of time away from the host. These pathogens, transmitted only by intimate person-to-person contact such as exchange of body fluids in sexual intercourse, include those responsible for sexually transmitted diseases including syphilis (*Treponema pallidum*), gonorrhea (*Neisseria gonorrhoeae*), and HIV/AIDS (HIV, human immunodeficiency virus; AIDS, acquired immunodeficiency syndrome).

Direct contact also transmits skin pathogens such as staphylococci (boils and pimples) and fungi (ringworm). These pathogens often spread by indirect means as well because they are relatively resistant to environmental conditions such as drying.

Indirect Host-to-Host Transmission

Indirect transmission of an infectious agent can be facilitated by either living or inanimate carriers. Living carriers that transmit

pathogens are called **vectors**. Commonly, arthropods (mites, ticks, or fleas) or vertebrates (dogs, cats, or rodents) act as vectors. Arthropod vectors may not be hosts for the pathogen, but may carry the agent from one host to another. Many arthropods obtain their nourishment by biting and sucking blood, and if the pathogen is present in the blood, the arthropod vector may ingest the pathogen and transmit it when biting another individual. In some cases viral pathogens replicate in the arthropod vector, which is then considered an *alternate host*. Such is the case for West Nile virus (↩ Section 30.6). These replications lead to greater pathogen numbers, increasing the probability that a subsequent bite will lead to infection.

Inanimate agents such as bedding, toys, books, and surgical instruments can also transmit disease. Inanimate objects that, when contaminated with a viable pathogen, can transfer the pathogen to a host are called **fomites**. The term **vehicle** is used to describe nonliving sources of pathogens that transmit disease to large numbers of individuals; common disease vehicles are contaminated food or water. Fomites can also be disease vehicles, but major epidemics originating from a single-vehicle source are typically traced to common sources such as food or water because food and water are shared commodities consumed in large amounts by everyone.

Epidemics

Major epidemics are usually classified as either *common-source epidemics* or *host-to-host epidemics*. These two types of epidemics are contrasted in **Figure 28.6**. **Table 28.2** summarizes the key epidemiological features of some major epidemic diseases.

A **common-source epidemic** results from an infection (or intoxication) of a large number of people from a contaminated common source such as food or water. Such epidemics are often caused by a breakdown in the sanitation of a central food or water distribution system, but they can also be more local, such

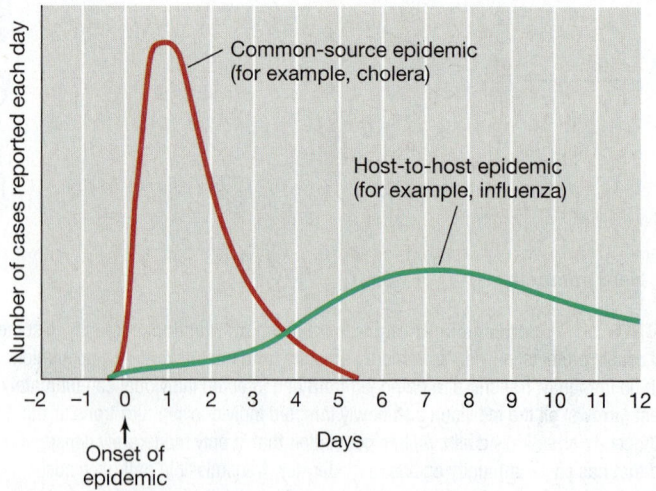

Figure 28.6 **Types of epidemics.** The shape of the curve that plots incidence of an epidemic disease against time identifies the likely type of the epidemic. For a common-source epidemic such as one resulting from contaminated food or water shared by the people who become infected, the curve rises sharply to a peak and then declines rapidly. Host-to-host infectious disease incidence rises relatively slowly as new cases accumulate.

Table 28.2 Selected epidemic/pandemic diseases

Disease	Causative agent	Infection sources	Reservoirs	Control measures
Common-source epidemic				
Cholera	*Vibrio cholerae* (B[a])	Fecal contamination of food and water	Humans	Decontamination of public water sources; immunization
Host-to-host epidemics *Respiratory diseases*				
Tuberculosis	*Mycobacterium tuberculosis* (B)	Sputum from human cases; infected milk	Humans, cattle	Treatment with antimicrobial drugs; pasteurization of milk
Influenza	Influenza virus (V)	Aerosol and fomite spread from human cases	Humans, animals	Immunization
Host-to-host epidemics *Sexually transmitted diseases*				
HIV/AIDS (human immunodeficiency virus/acquired immunodeficiency syndrome)	Human immunodeficiency virus (HIV)	Infected body fluids, especially blood and semen	Humans	Treatment with metabolic inhibitors (not curative)
Vectorborne disease				
Malaria	*Plasmodium* spp. (P)	Bite from *Anopheles* mosquito	Humans, mosquito	Control mosquito population; treat and prevent human infections with antimalarial drugs

[a]B, Bacteria; V, virus; P, protist.

as contaminated food in a particular restaurant. Foodborne and waterborne common-source epidemics are primarily intestinal diseases; the pathogen leaves the body in fecal material, contaminates food or water supplies due to improper sanitary procedures, and then enters the intestinal tract of the recipient during ingestion of the food or water. Waterborne and foodborne diseases are generally controlled by public health measures, which we discuss further in Chapter 31. A classic common-source epidemic is cholera. In 1855 the British physician John Snow correlated cholera incidence with fecal contamination of the water distribution systems in London. Snow clearly showed that the infectious agent, the bacterium *Vibrio cholerae*, was transmitted through consumption of the contaminated common-source vehicle, water (Section 28.10 and ⮌ Section 31.3).

The disease incidence for a common-source outbreak is characterized by a rapid rise to a peak incidence because a large number of individuals become ill within a relatively brief period of time (Figure 28.6). Assuming that the pathogen-contaminated common source is discovered and sanitized, the incidence of a common-source illness also declines rapidly, although the decline is less rapid than the rise. Cases continue to be reported for a period of time approximately equal to the duration of one incubation period for the disease.

In a **host-to-host epidemic**, the disease incidence shows a relatively slow, progressive rise (Figure 28.6) and a gradual decline. Cases continue to be reported over a period of time equivalent to several incubation periods of the disease. A host-to-host epidemic can be initiated by the introduction of a single infected individual into a susceptible population, with this individual infecting one or more people. The pathogen then replicates in susceptible individuals, reaches a communicable stage, is transferred to other susceptible individuals, and again replicates and becomes communicable. Influenza and chicken pox are examples of diseases

that are typically spread in host-to-host epidemics. Chapter 29 discusses these and a number of other diseases propagated by host-to-host transmission.

Basic Reproduction Number (R_0)

The infectivity of a pathogen can be predicted using mathematical models that estimate the **basic reproduction number (R_0)**, defined as the number of expected secondary transmissions from each single case of a disease in an entirely susceptible population. **Table 28.3** shows the R_0 values for selected infectious diseases.

Table 28.3 Basic reproduction number (R_0) and herd immunity necessary for community protection from selected infectious diseases

Disease	[a]R_0	Herd immunity
Diphtheria	7	85%
Ebola	1.8	
Influenza[b]	1.6	29%
Measles	18	94%
Mumps	7	86%
Pertussis	17	94%
Polio	7	86%
Rubella	7	85%
SARS-CoV	3.6	
Smallpox	7	85%

[a]R_0 and herd immunity values are the highest estimates for each disease. Herd immunity values are shown only for those diseases for which vaccines are available.
[b]Values shown are for the pandemic (H1N1) 2009 influenza. Each influenza epidemic has a different R_0 and herd immunity value. Herd immunity values assume a 100% effective vaccine. Vaccine efficacy for influenza is about 60% and observed herd immunity values are 40% or greater depending on the susceptible host populations.

Handling of the *severe acute respiratory syndrome* (SARS) epidemic early in this decade is an excellent example of the successful application of the principles of epidemiology. Like many other rapidly emerging diseases, SARS was viral and originated in animals. Such characteristics have the potential to trigger explosive disease in humans when the infectious agents cross host species barriers.

The SARS epidemic began in late 2002 in Guangdong Province, China. By the following February, the virus had spread to 28 countries. Global travel provided the major vehicle for SARS dissemination. The cause of SARS was quickly traced to a coronavirus derived from an animal source. The coronavirus entered the human food chain through an exotic food animal such as a civet (a small, catlike, nocturnal animal). The SARS coronavirus (SARS-CoV) (**Figure 1**) most probably originated in bats. Civets apparently acquired the virus by consuming fruit contaminated by bat feces. SARS-CoV likely evolved over an extended period of time in bats and developed, quite by accident, the ability to infect civets and then humans.

Much like common cold viruses, SARS-CoV is a relatively hardy, very infectious RNA virus that is difficult to contain (R_0 of 3.6). Once in humans, SARS-CoV quickly spreads from person to person by sneezing and coughing or by contact with contaminated fomites or feces. Ordinarily, a new coldlike virus would be of little concern, but SARS-CoV caused infections with significant mortality. From the 8500 known SARS-CoV infections, there were over 800 deaths, for an overall mortality rate approaching 10%. In persons over 65 years of age, the mortality rate approached 50%, attesting to the virulence of SARS-CoV as a human pathogen.

About 20% of all SARS cases were in healthcare workers, demonstrating the high infectivity of the virus. Standard containment

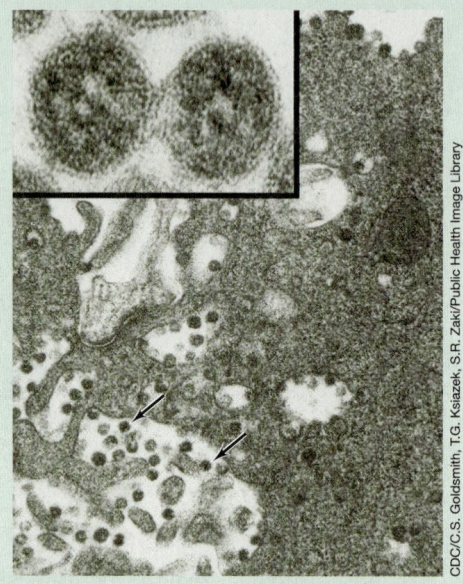

Figure 1 Severe acute respiratory syndrome coronavirus (SARS-CoV). The upper left panel shows isolated SARS-CoV virions. An individual virion is 128 nm in diameter. The large panel shows coronaviruses within cytoplasmic membrane–bound vacuoles and in the rough endoplasmic reticulum of host cells. The virus replicates in the cytoplasm and exits the cell through the cytoplasmic vacuoles.

CDC/C.S. Goldsmith, T.G. Ksiazek, S.R. Zaki/Public Health Image Library

and infection control methods practiced by healthcare personnel were not effective in controlling spread of the disease. As a result, SARS patients were confined for the course of the disease in strict isolation in negative-pressure rooms. To prevent infection, healthcare staff caring for SARS patients wore respirators when working with patients or when handling fomites (bed linens, eating utensils, and so on).

The recognition and containment of the SARS outbreak was the start of an international response involving clinicians, scientists, and public officials. Almost immediately, travel to and from the endemic area was restricted,

limiting further outbreaks. SARS-CoV was quickly isolated and cultured, and its genome was rapidly sequenced. This information was used to develop PCR tests to detect the virus in samples. As laboratory work progressed, epidemiologists traced the virus back to the civet food source in China and stopped further transmission to humans by restricting the sale of civets and other foods from wild sources. These actions collectively stopped the outbreak.

SARS is an example of a serious infection that emerged very rapidly from a unique source. However, rapid identification and characterization of the SARS pathogen, nearly instant development of worldwide notification procedures and diagnostic tests, and a concerted effort to understand the biology and genetics of this novel pathogen controlled the disease; there have been no cases of SARS since early 2004. The rapid emergence of SARS, and the equally rapid and successful international effort to identify and control the outbreak, provide a model for the control of emerging epidemics.

As international travel and trade expand, the chances for propagation and rapid dissemination of new exotic diseases will increase. For example, in early 2013, a SARS-like viral illness rapidly emerged in Saudi Arabia and nearby countries that caused severe (and in some cases fatal) symptoms similar to those of SARS. This virus, nicknamed MERS-CoV (Middle East respiratory syndrome—coronavirus) (see opening page of this chapter) is a novel coronavirus that suddenly appeared in healthcare settings and can be rapidly transmitted from person to person. Because of these challenges, epidemiologists and healthcare workers must prepare for the emergence of other serious infectious diseases, including pandemic influenza in particular. However, it is likely that the lessons learned from the SARS epidemic will pay dividends when other emerging diseases appear.

The R_0 value predicts the risk of disease spread in a population. An R_0 of 1 means that each infected person will pass the disease to one susceptible person, maintaining the disease in the population. An R_0 that is greater than 1 indicates that each infected person will pass the disease to *more than one* susceptible person, propagating the outbreak and leading to a possible epidemic or even pandemic. In contrast, an R_0 that is less than 1 indicates that each infected person will pass the disease to *fewer than one* susceptible person, and the disease will die out under these circumstances.

R_0 directly correlates with the herd immunity necessary to prevent spread of infection; the higher the R_0 value, the greater the herd immunity required to stop infection. For example, under ideal circumstances, to stop the spread of the highly infectious measles virus ($R_0 = 18$), 94% of the population must be immune, while only 29% of the population must be immune to stop the spread of influenza ($R_0 = 1.6$) (Table 28.3). Unfortunately, conditions are not always ideal and the mathematical models that predict R_0 may not take into account such factors as numbers of recovered individuals, population density (close contact), length of contact time, populations of high-risk individuals, and other variables that may affect disease spread. As a result, R_0 can only estimate theoretical infectivity, but it is still useful as a comparative gauge of the relative infectivity of a pathogen and helps to establish goals for immunization coverage to prevent disease spread.

The *observed reproduction number*, *R*, calculated from studies of actual disease spread, is a more useful term because it takes into account observed transmissions from infected to susceptible individuals. For most disease outbreaks, *R* cannot be reliably obtained because precise information about disease spread is not available; too many individuals are involved to accurately track the source of each infection. However, for the SARS outbreak of 2003 (see Explore the Microbial World, "SARS—A Model of Epidemiological Success"), the observed *R* was 3.6. Public health officials, recognizing the potential for a serious epidemic, instituted infection controls such as isolation of infected individuals and strict barrier protection for healthcare personnel. These measures reduced the SARS *R* value to 0.7, ending the threat of further disease spread. Likewise in one Ebola filovirus outbreak (Table 28.5), the reproduction number was reduced from a theoretical R_0 of 1.8 to an *R* of 0.7, again by using strict infection control measures that stopped the spread of the infection, thus averting a potential epidemic.

MINIQUIZ

- Distinguish between direct and indirect transmission of disease. Cite at least one example of each.
- Define the basic reproduction number for a pathogen.

28.4 Disease Reservoirs and Epidemics

Reservoirs are sites at which infectious agents remain viable and from which individuals may become infected. Reservoirs may be either animate or inanimate. Some pathogens whose reservoirs are not in animals only incidentally infect humans and cause disease. For example, organisms in the genus *Clostridium*, common soil bacteria, occasionally infect humans, causing life-threatening diseases such as tetanus, botulism, gangrene, and certain gastrointestinal diseases. Host damage and even death cause no harm to those pathogen populations because they are normal inhabitants of the soil; the pathogen is not dependent on the host for survival, so host–pathogen balance is not required for pathogen survival. Such a pathogen can cause devastatingly acute disease with no consequences for the pathogen.

For many pathogens, however, living organisms are the only reservoirs. In these cases, the reservoir host is essential for the life cycle of the infectious agent; maintenance of human pathogens of this kind requires person-to-person transmission. Many viral and bacterial respiratory pathogens and sexually transmitted pathogens require human hosts; certain staphylococci and streptococci are examples of human-restricted pathogens, as are the agents that cause diphtheria, gonorrhea, and mumps. As we shall see, many pathogens that live their entire life cycle dependent on a single host species, especially humans, can be eradicated or at least controlled. Table 28.2 lists some human infectious diseases with epidemic potential and their reservoirs.

Zoonosis

Some infectious diseases are caused by pathogens that reproduce in both humans and animals. A disease that primarily infects animals but is occasionally transmitted to humans is called a **zoonosis**. Animal-to-animal transmission of veterinary diseases can be high because public health measures are less developed, especially for wild animal populations. Occasionally, transmission of a zoonotic disease is from animal to human; person-to-person transfer of these pathogens is rare, but does occur. Factors leading to the emergence of zoonotic disease include the existence and propagation of the infectious agent in an animal host, the proper environment for propagation and transfer of the agent, and the presence of the new susceptible host species. When there is animal-to-human transmission, a new and highly virulent infectious disease may suddenly emerge in the exposed human population. This was the case, for example, with SARS (see the Explore the Microbial World feature).

Control of a zoonotic disease in the human population does not usually eliminate the disease as a potential public health problem. Eradication of the human form of a zoonotic disease can generally be achieved only through elimination of the disease in the animal reservoir. This is because the essential maintenance of the pathogen depends on animal-to-animal transfer; humans are incidental, nonessential hosts. As one example, plague is primarily a disease of rodents. Effective control of human cases of plague is achieved by control of the infected rodent population and the insect (flea) vector that carries the pathogen to humans. Control of the zoonotic reservoir and vector are more effective in preventing plague transmission than interventions such as vaccines in the incidental human host (⮌ Section 30.7). As another example, zoonotic bovine tuberculosis is easily spread to humans and is clinically indistinguishable from human tuberculosis. The disease was brought under control in developed countries primarily by identifying and destroying infected animals. Pasteurization of milk was also important because milk was the main vehicle for transmission of bovine tuberculosis to humans (⮌ Section 29.4).

Certain infectious diseases are caused by organisms such as protists that have more complex life cycles including an obligate transfer from a nonhuman host to a human host, followed by an obligate transfer back to the nonhuman host (for example, malaria, ⮌ Section 32.5). In such cases, the disease may potentially be controlled in either humans or the nonhuman host.

Carriers

A living *carrier* is a pathogen-infected individual who has a subclinical infection and shows no symptoms or only mild symptoms of clinical disease. Carriers are potential sources of infection for others. Carriers may be in the incubation period of the disease, in which case the carrier state precedes the development of actual symptoms. Respiratory infections such as colds and influenza, for example, are often spread via carriers who are unaware of their infection and so are not taking any precautions against infecting others. The carrier state lasts only a short time for carriers who develop acute disease. However, chronic carriers may spread disease for extended periods of time. Chronic carriers usually appear healthy. They may have recovered from a clinical disease but still harbor viable pathogens, or their infections may not be apparent.

Carriers can be identified using diagnostic techniques; culture or immunoassay surveys are conducted in populations to identify carriers. For example, skin testing with *Mycobacterium tuberculosis* antigens tests for delayed hypersensitivity. The test reveals previous or current infection and is widely used to identify carriers of *M. tuberculosis* (ↄ Section 24.3). Other diseases in which carriers contribute to the spread of infection include hepatitis, typhoid fever, and HIV/AIDS. Culture or immunoassay surveys of food handlers and healthcare workers are sometimes used to identify individuals who are carriers and pose a risk as sources of infection.

A famous example of a chronic carrier was the woman known as Typhoid Mary, a cook in New York City in the early part of the twentieth century. Typhoid Mary (her real name was Mary Mallon) was employed as a cook during a typhoid fever epidemic in 1906. Investigations revealed that Mary was associated with a number of the typhoid outbreaks. She was the likely source of infection because her feces contained large numbers of the typhoid bacterium, *Salmonella enterica* serovar Typhi. She remained a carrier throughout her life, probably because her gallbladder was infected and continuously secreted the pathogen into her intestine. She refused to have her gallbladder removed and was imprisoned. Released on the pledge that she would not cook or handle food for others, Mary disappeared, changed her name, and continued to cook in restaurants and public institutions, leaving behind epidemic outbreaks of typhoid fever. After several years, she was again arrested and imprisoned and remained in custody until her death in 1938.

MINIQUIZ -
- What is a zoonotic disease? A disease reservoir?
- Distinguish between acute and chronic carriers. Provide an example of each.

II • Epidemiology and Public Health

Here we identify some of the methods used to identify, track, contain, and eradicate infectious diseases within populations. We also identify some important current and potential future threats from infectious diseases.

28.5 Public Health and Infectious Disease

Public health refers to the health of the general population and to the activities of public health authorities in the control of disease. The incidence and prevalence of many infectious diseases dropped dramatically during the twentieth century, especially in developed countries, because of universal improvements in public health from advances in basic living conditions. Access to safe water and food, improved public sewage treatment, less crowded living conditions, and lighter workloads have contributed immeasurably to disease control, primarily by reducing exposure to infectious agents. Several historically important diseases, including smallpox, typhoid fever, diphtheria, brucellosis, and poliomyelitis, have been controlled and in some cases eliminated (for example, smallpox) by active, disease-specific public health measures such as quarantine and vaccination.

Controls Directed against Common Vehicles

Common vehicles for pathogen dispersal include food, water and air. The transmission of pathogens in food or water can be eliminated by preventing contamination (Chapter 31). For example, water purification methods have dramatically reduced the incidence of typhoid fever (**Figure 28.7**). Laws controlling food purity and preparation have greatly decreased the probability of transmission of foodborne pathogens to humans. As we mentioned previously, the destruction of infected cattle and pasteurization

of milk have virtually eliminated the spread of bovine tuberculosis to humans.

Transmission of respiratory pathogens carried in the air is difficult to prevent. Attempts at chemical disinfection of air have been unsuccessful. Air filtration is a viable method but is limited to small, enclosed areas. In Japan, many people wear face masks when they have upper respiratory infections to prevent transmission to others, but such methods, although effective, are voluntary and are difficult to institute as public health measures.

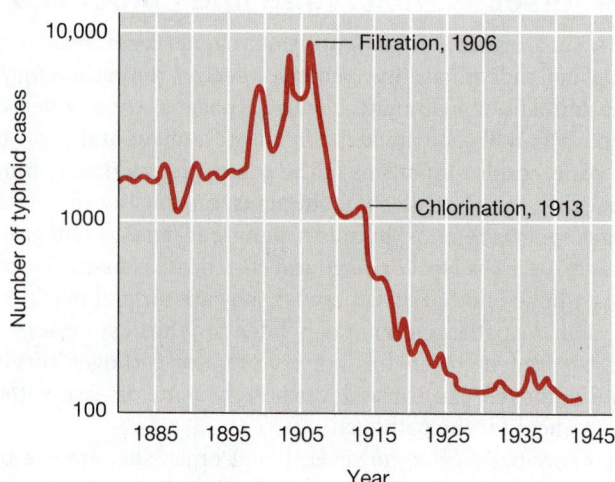

Figure 28.7 Typhoid fever in Philadelphia. The introduction of filtration and chlorination eliminated typhoid fever in Philadelphia and other cities with well-regulated water supplies.

Controls Directed against the Reservoir

When the disease reservoir is primarily domestic animals, the infection of humans can be prevented if the disease is eliminated from the infected animal population. Immunization or destruction of infected animals may eliminate the disease in animals and, consequently, in humans. These procedures have nearly eliminated brucellosis and bovine tuberculosis in humans and have controlled bovine spongiform encephalitis (mad cow disease), caused by a prion (⇄ Sections 9.13 and 31.14) in cattle in the United Kingdom, Canada, and the United States. In the process, the health of the domestic animal population is also improved.

When the disease reservoir is a wild animal, eradication is much more difficult. Rabies, for example, is a disease of both wild and domestic animals that is transmitted to domestic animals primarily from wild animals. Thus, control of rabies in domestic animals and in humans can be achieved by immunization of domestic animals. However, because the majority of rabies cases in the United States are in wild animals (⇄ Section 30.1), eradication of rabies would require the immunization or destruction of all wild animal reservoirs, including such diverse species as raccoons, bats, skunks, and foxes. Although oral rabies immunization is practical and recommended for rabies control in restricted wild animal populations, its efficacy is untested in large, diverse animal populations such as the wild animal reservoir in the United States.

When insects such as the mosquito vectors that transmit malaria are also required hosts, effective control of the disease can be accomplished by eliminating the vector–host with insecticides or other agents. The use of toxic or carcinogenic chemicals, however, must be balanced with environmental concerns, because in some cases, the elimination of one public health problem only creates another. For example, the insecticide dichlorodiphenyltrichloroethane (DDT) is very effective against mosquitoes and is credited with eradicating yellow fever and malaria in North America. DDT use, however, was banned in the United States in 1972 because of serious environmental and wildlife concerns. DDT is still used in many developing countries to control mosquito-borne diseases, but its use is declining worldwide.

When humans are the disease reservoir (as, for example, in HIV/AIDS), control and eradication can be difficult, especially if there are asymptomatic carriers. By contrast, certain diseases that are limited to humans have no asymptomatic phase. If these can be prevented through immunization or treatment with antimicrobial drugs, the disease can be eradicated if those who have contracted the disease and all possible contacts are strictly quarantined, immunized, and treated. Such a strategy was successfully employed by the World Health Organization to eradicate smallpox and is currently being used to eradicate polio, as we discuss later.

Immunization

Smallpox, diphtheria, tetanus, pertussis (whooping cough), measles, mumps, rubella, and poliomyelitis have been controlled primarily by immunization. Diphtheria, for example, is no longer considered an endemic disease in the United States. Vaccines are available for a number of other infectious diseases (⇄ Table 24.4). As we discussed in Section 28.2, 100% immunization is not necessary for disease control in a population because of herd immunity,

although the percentage needed to ensure disease control varies with the infectivity and virulence of the pathogen and with the living conditions of the population (for example, crowding).

Measles epidemics offer an example of the effects of herd immunity. The occasional resurgence of the highly contagious measles virus ($R_0 = 18$, Table 28.3) emphasizes the importance of maintaining appropriate immunization levels for a given pathogen. Until 1963, the year an effective measles vaccine was licensed, nearly every child in the United States acquired measles through natural infections, resulting in over 300,000 annual cases. After introduction of the vaccine, the number of annual measles infections decreased rapidly (**Figure 28.8**). Case numbers reached a low of 1497 by 1983. However, by 1990, the percentage of children immunized against measles fell to 70%, and the number of new cases rose to 27,786. A concerted effort to increase measles immunization levels to above 90% virtually eliminated indigenous measles transmission in the United States, and a total of only 312 measles cases were reported in 1993. Currently, about 100 cases of measles are reported each year in the United States, most due to infections imported by visitors from other countries.

Most children in the United States are now adequately immunized, but up to 80% of adults lack effective immunity to important infectious diseases because immunity from childhood vaccinations declines with time. When childhood diseases occur in adults, they can have devastating effects. For example, if a woman contracts rubella (a vaccine-preventable viral disease) (⇄ Section 29.6) during pregnancy, the fetus may develop serious developmental and neurological disorders. Measles, mumps, and chicken pox are also more serious diseases in adults than in children.

All adults are advised to review their immunization status and check their medical records to ascertain dates of immunizations. This is particularly true for individuals who are traveling abroad. Tetanus immunizations, for example, must be renewed at least every 10 years to provide effective immunity. Surveys of adult populations indicate that more than 10% of adults under the

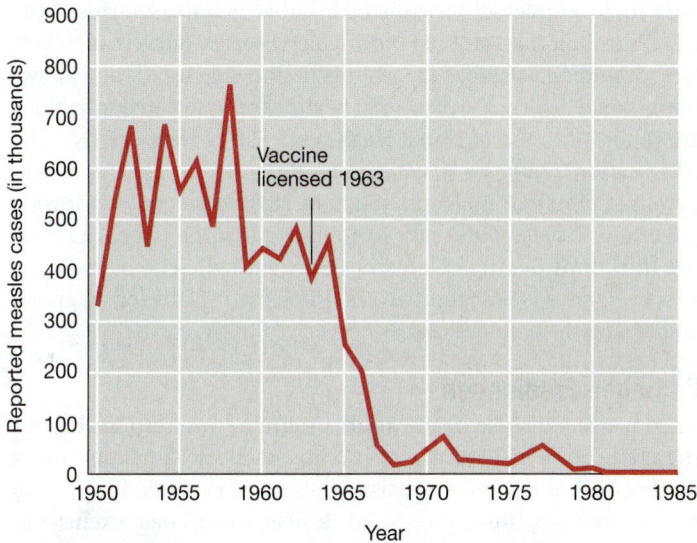

Figure 28.8 Measles immunization in the United States. The introduction of a measles vaccine eliminated measles as a common childhood infection within 20 years.

age of 40 and more than 50% of those over 60 are not adequately immunized. General recommendations for immunization were discussed in Section 24.6 and those for specific infections will be discussed in Chapters 29 through 32.

Isolation, Quarantine, and Surveillance

Isolation is the separation of persons who have an infectious disease from those who are healthy. **Quarantine** is the separation and restriction of well persons who may have been exposed to an infectious disease to see if they develop the disease. The length of isolation or quarantine for a given disease is the longest period of communicability for that disease. To be effective, these measures must prevent infected or potentially infected individuals from contacting uninfected susceptible individuals.

By international agreement, six diseases require isolation and quarantine: *smallpox, cholera, plague, yellow fever, typhoid fever,* and *relapsing fever.* Each is a very serious, particularly communicable disease. Spread of certain other highly contagious diseases such as Ebola hemorrhagic fever, SARS, H5N1 influenza, and meningitis may also be controlled by quarantine or isolation as outbreaks occur.

Surveillance is the observation, recognition, and reporting of diseases as they occur. **Table 28.4** lists the diseases currently under surveillance in the United States. Some of the epidemic diseases (listed in Table 28.2) and emerging diseases (see Table 28.5) are not on the surveillance list. However, many of these diseases such as seasonal influenza are surveyed through regional laboratories that identify *index cases*—those cases of disease that exhibit unusually high incidence, new syndromes or characteristics, or are linked to new or evolving pathogens that have high potential for causing new epidemics.

The **Centers for Disease Control and Prevention (CDC)** in the United States is the agency of the United States Public Health Service that tracks disease trends, provides disease information to the public and to healthcare professionals, and forms public policy regarding disease prevention and intervention. The CDC operates a number of infectious disease surveillance programs. For a few examples, diseases listed in Table 28.4 are reported to the CDC through the National Notifiable Diseases Surveillance System, providing a database that spots nationwide trends and allows for national healthcare planning; healthcare-associated infections are reported to the National Healthcare Safety Network (NHSN) (ᒐᗺ Section 27.2); the Select Agents Program handles reports of potential bioterror incidents (Section 28.8); emerging worldwide infectious disease outbreaks are reported to the Global Emerging Infections Sentinel Network (GeoSentinel). The overall goal of surveillance is to formulate and implement plans for diagnosis and treatment of infections.

Pathogen Eradication

A concerted disease eradication program was responsible for the eradication of naturally occurring smallpox. Smallpox was a disease with a reservoir consisting solely of the individuals with acute smallpox infections, and transmission was exclusively person-to-person. Infected individuals transmitted the disease through direct contact with previously unexposed individuals. Although smallpox, a viral disease, cannot be treated once

Table 28.4 Reportable infectious agents and diseases in the United States, 2013

Diseases caused by *Bacteria*

Anthrax	Q fever
Botulism	Salmonellosis
Brucellosis	Shiga toxin–producing *Escherichia coli* (STEC)
Chancroid	
Chlamydia trachomatis infection	Shigellosis
	Spotted fever rickettsiosis
Cholera	Streptococcal toxic shock syndrome
Diphtheria	
Ehrlichiosis/Anaplasmosis	*Streptococcus pneumoniae,* invasive disease
Gonorrhea	
Haemophilus influenzae, invasive disease	Syphilis, all stages
	Tetanus
Hansen's disease (leprosy)	Toxic shock syndrome (staphylococcal)
Hemolytic uremic syndrome	
Legionellosis	Tuberculosis
Listeriosis	Tularemia
Lyme disease	Typhoid fever
Meningococcal disease (*Neisseria meningitidis*)	Vancomycin-intermediate *Staphylococcus aureus* (VISA)
Pertussis	Vancomycin-resistant *Staphylococcus aureus* (VRSA)
Plague	Vibriosis (non-cholera *Vibrio* infections)
Psittacosis	

Diseases caused by viruses

Arboviruses (encephalitis and non-neuroinvasive disease)
Dengue
Hantavirus pulmonary syndrome
Hepatitis A, B, C
HIV infection/AIDS
Novel influenza A
Measles
Mumps
Polio
Rabies
Rubella
Severe acute respiratory syndrome (SARS-CoV)
Smallpox
Varicella (chicken pox)
Viral hemorrhagic fevers
West Nile virus
Yellow fever

Diseases caused by protists

Babesiosis
Cryptosporidiosis
Cyclosporiasis
Malaria
Giardiasis

Disease caused by a helminth

Trichinellosis (trichinosis)

Disease caused by a fungus

Coccidioidomycosis

acquired, immunization practices were very effective; vaccination with the related vaccinia virus conferred complete immunity.

The World Health Organization (WHO) implemented a smallpox eradication plan in 1967. Because of the success of vaccination programs worldwide, endemic smallpox had already been confined to Africa, the Middle East, and the Indian subcontinent. WHO workers then vaccinated everyone in remaining endemic areas, with the goal of providing direct and herd immunity to all potential contacts. Each subsequent outbreak or suspected outbreak was targeted by WHO teams that traveled to the outbreak site, quarantined individuals with active disease, and vaccinated all contacts. To break the chain of possible infection, they then immunized everyone who had contact with the contacts. This aggressive policy to provide vaccinations eliminated the active natural disease within a decade, and WHO announced the eradication of smallpox in 1980.

Poliomyelitis, another viral disease with a human-only reservoir, is also preventable with an effective vaccine and is targeted for eradication (endemic polio has been eradicated from the Western Hemisphere). Using much the same strategy to target polio as was used for smallpox, WHO undertook a widespread immunization program in 1988, concentrating efforts in remaining endemic areas. In all, over 2 billion individuals, mostly children, have been immunized, preventing an estimated 5 million cases of paralytic polio. By 2012 endemic polio was restricted to Nigeria, Pakistan, and Afghanistan. Only 650 polio cases were reported in 2011. Individual outbreaks are treated by immunization of all susceptible persons in the region of the outbreak.

Hansen's disease (leprosy), another disease restricted to humans, is also targeted for eradication. Active cases of Hansen's disease can now be effectively treated with a multidrug therapy that cures the patient and also prevents spread of *Mycobacterium leprae*, the causal agent (⤿ Section 29.4).

Other communicable diseases are candidates for eradication. These include Chagas' disease (by treating active cases and destroying the insect vector of the *Trypanosoma cruzi* parasite in the American tropics) (⤿ Section 32.6) and dracunculiasis (by treating drinking water in Africa, Saudi Arabia, Pakistan, and other places in Asia to prevent transmission of *Dracunculus medinensis*, the Guinea helminth parasite). Eradication of syphilis may be possible because the disease is found only in humans and is treatable. Diphtheria, caused by *Corynebacterium diphtheriae*, is no longer endemic in North America. The disease could be globally eradicated by application of the strict immunization protocols that have virtually eliminated it from North America.

MINIQUIZ

- Compare public measures for controlling infectious disease caused by insect vectors and human carriers.
- Outline the steps taken to eradicate smallpox and polio.

28.6 Global Health Comparisons

The World Health Organization has divided the world into six geographic regions for the purpose of collecting and reporting health information such as causes of morbidity and mortality. These geographic regions are Africa, the Americas (North America, the Caribbean, Central America, and South America), the eastern Mediterranean, Europe, Southeast Asia, and the Western Pacific. Here we compare mortality data from a relatively developed region, the Americas, to those from a developing region, Africa.

Infectious Disease in the Americas and Africa: A Comparison

Mortality statistics in developed and developing countries are significantly different, as illustrated by a comparison of data from the Americas and from Africa in 2008 when the worldwide population was nearly 6.9 billion. Worldwide, 60.8 million individuals died, giving a mortality rate of 8.8 deaths per 1000 inhabitants per year. About 15.8 million (26%) of these deaths were attributable to infectious diseases. There were 924 million people in the Americas in 2008 and there were 5.6 million deaths, or 6.1 deaths per 1000 inhabitants per year. In Africa, there were 837 million people in 2008 and 14.1 million deaths, or 16.8 deaths per 1000 inhabitants per year. These statistics clearly show differences in overall mortality between developed and developing countries, but a comparative examination of the *causes* of mortality is even more instructive.

Figure 28.9 indicates that infectious diseases caused the most deaths in Africa, whereas in the Americas, noninfectious diseases such as cancer, cardiovascular disease, and diabetes were the leading causes of mortality. In Africa, there were about 6.6 million

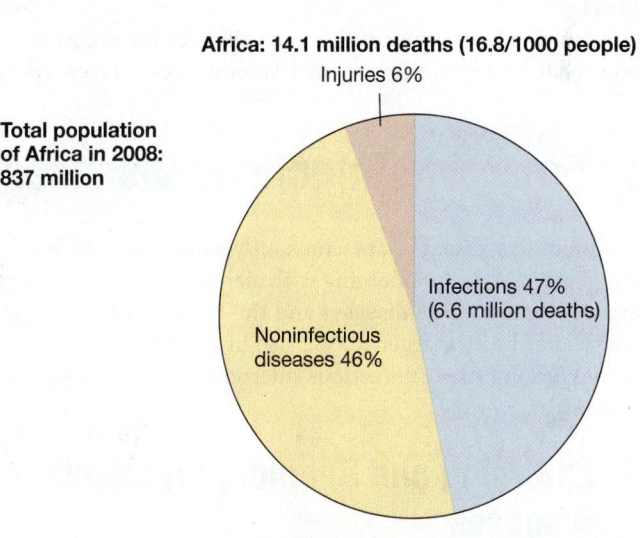

Africa: 14.1 million deaths (16.8/1000 people)

Injuries 6%

Total population of Africa in 2008: 837 million

Infections 47% (6.6 million deaths)

Noninfectious diseases 46%

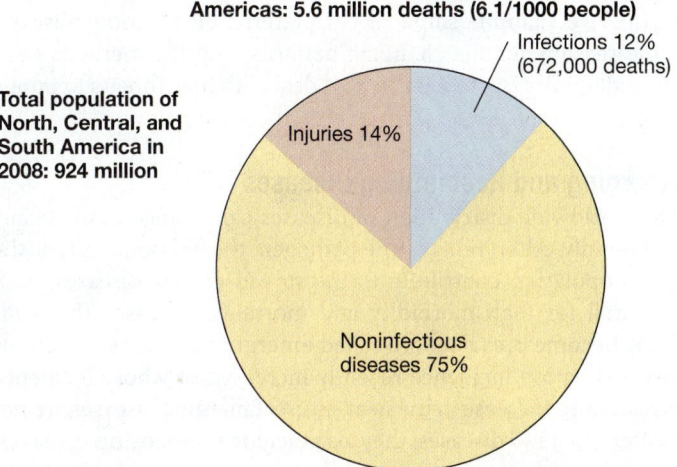

Americas: 5.6 million deaths (6.1/1000 people)

Infections 12% (672,000 deaths)

Total population of North, Central, and South America in 2008: 924 million

Injuries 14%

Noninfectious diseases 75%

Figure 28.9 Causes of death in Africa and the Americas, 2008. Noninfectious diseases include cancer, cardiovascular diseases, and diabetes. Injuries include accidents, murder, suicide, and war. Data are from the World Health Organization.

UNIT 6

deaths due to infectious diseases and the life expectancy was 54 years of age. The African death toll due to infectious diseases was 10% of the total deaths in the world. In stark contrast, only 672,000 died of infectious disease in the Americas and the life expectancy was 76 years of age. In developed countries, the increased life expectancy is a direct consequence of the reduction in death rates from infection over the last century (↩ Figure 1.8). Most of these gains are due to the advances in public health. By contrast, lack of resources in developing countries limits access to adequate sanitation, safe food and water, immunizations, healthcare, and medicines, leading to increases in infectious diseases and, by extension, to dramatically shorter life expectancy.

Travel to Endemic Areas

The high incidence of disease in many parts of the world is a concern for people traveling to such areas. However, travelers can be immunized against many of the diseases that are endemic in foreign countries. Specific recommendations for immunization for those traveling abroad are updated biannually and published by the Centers for Disease Control and Prevention (http://www.cdc.gov/).

For many countries, immunization certificates for yellow fever are required for entry from areas with endemic yellow fever. These areas include much of equatorial South America and Africa. Most other nonstandard immunizations such as those for rabies and plague are recommended only for people who are expected to be at high risk, such as veterinary healthcare providers. The CDC summarizes current information for the potential for infectious disease transmission throughout the world, including diseases for which there are no effective immunizations (for example, HIV/AIDS, malaria, Ebola hemorrhagic fever, dengue fever, amebiasis, encephalitis, and typhus). Travelers should take precautions such as avoiding unprotected sex, avoiding insect and animal bites, drinking only water that has been properly treated to kill all microorganisms, eating properly stored and prepared food, and undergoing antibiotic and chemotherapeutic programs for prophylaxis or for suspected exposure. Although these precautions do not guarantee that one will remain disease-free, adhering to them greatly reduces the risk of infection.

MINIQUIZ
- Contrast mortality due to infectious diseases in Africa and the Americas.
- List infectious diseases for which you have not been immunized and with which you could come into contact next year.

III • Emerging Infectious Diseases

New infectious diseases are constantly emerging and established diseases are reemerging with alarming frequency. Here we discuss some of these diseases and the reasons for their sudden emergence or reemergence. We also investigate the potential for the purposeful use of infectious microorganisms as agents of war and terror.

28.7 Emerging and Reemerging Infectious Diseases

Infectious diseases are global, dynamic health problems. In this section we examine some recent patterns of infectious disease, some reasons for the changing patterns, and the methods used by epidemiologists to identify and deal with new threats to public health.

Emerging and Reemerging Diseases

The worldwide distribution of diseases can change dramatically and rapidly. Alterations in the pathogen, the environment, or the host population contribute to the spread of new diseases, with potential for high morbidity and mortality. Diseases that suddenly become prevalent are called **emerging diseases**, infectious diseases whose incidence recently increased or whose incidence threatens to increase in the near future. Emerging diseases are not limited to "new" diseases; they also include **reemerging diseases**, diseases that were previously under control but suddenly appear as a new epidemic.

Examples of global emerging and reemerging disease were shown in Figure 28.1, and recent examples are shown in **Figure 28.10**.

Select diseases with high potential for emergence or reemergence are described in **Table 28.5**. The epidemic diseases listed in Table 28.2 also have the potential to emerge or reemerge in epidemic or pandemic forms.

Emerging epidemic diseases are not a new phenomenon. Among the diseases that rapidly and sometimes catastrophically emerged in the past are syphilis (caused by *Treponema pallidum*) and plague (caused by *Yersinia pestis*). In the Middle Ages, up to one-third of all humans were killed by the plague epidemics that swept Europe, Asia, and Africa (↩ Chapter 1 Explore the Microbial World, "The Black Death Decoded" and Section 30.7). Influenza caused a devastating worldwide pandemic in 1918–1919, claiming up to 100 million lives. In the 1980s, HIV/AIDS and Lyme disease emerged as new diseases. Important emerging pathogens in the last decade include West Nile virus (↩ Section 30.6) and pandemic (H1N1) 2009 influenza, the strain that emerged in spring 2009 in Mexico and rapidly spread throughout the world (Section 28.11). Health officials worldwide are particularly concerned about the potential for rapid emergence of pandemic influenza developing from H5N1 avian influenza (↩ Section 29.8).

Emergence Factors: Demographics, Land Use, and Transportation

Some factors responsible for the emergence of new pathogens, some current examples, and some epidemics attributable at least in part to these emergence factors are outlined in **Table 28.6**. Human demographics and behavior can affect disease spread. In 1800, less than 2% of the world's population lived in urban areas.

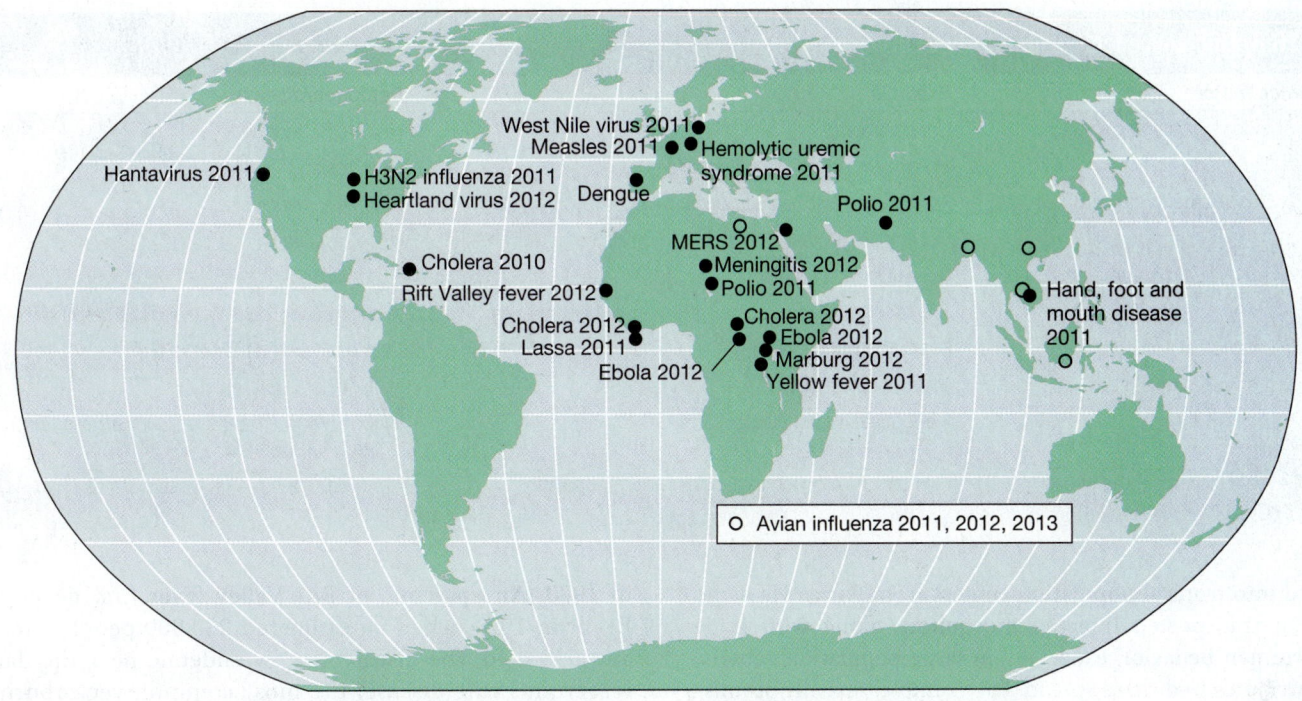

Figure 28.10 **Recent outbreaks of emerging and reemerging infectious diseases.** The diseases shown are local outbreaks capable of producing widespread epidemics and pandemics. Not shown are established pandemic diseases such as HIV/AIDS and cholera, and predictable annual epidemic diseases such as seasonal epidemic human influenza.

Today nearly one-half of the world's population lives in cities. The high density of human hosts in cities can facilitate disease transmission. For example, dengue fever (Table 28.4) is a serious disease now concentrated in tropical and subtropical urban areas, with up to 100 million infections due to the spread of dengue virus in the mosquito *Aedes aegypti* (**Figure 28.11**). Prior to 1981, dengue fever was unknown in the Americas as the mosquito vector *Aedes aegypti* was not present (⮂ Section 30.5). However, by 2003 both the mosquito and dengue virus were present in all Central American and tropical South American countries, possibly

Table 28.5 Emerging and reemerging epidemic infectious diseases

Agent	Disease and symptoms	Mode of transmission	Cause of emergence
Bacteria, rickettsias, and chlamydias			
Borrelia burgdorferi	Lyme disease: rash, fever, neurological and cardiac abnormalities, arthritis	Bite of infective *Ixodes* tick	Increase in deer and human populations in wooded areas
Mycobacterium tuberculosis	Tuberculosis: cough, weight loss, lung lesions	Sputum droplets (exhaled through a cough or sneeze) from a person with active disease	Antimicrobial drug resistance as multidrug resistant (MDR) and extensively drug resistant (XDR) tuberculosis
Vibrio cholerae	Cholera: severe diarrhea, rapid dehydration	Water contaminated with the feces of infected persons; food exposed to contaminated water	Poor sanitation and hygiene; carried to non-endemic areas via infected travelers and commerce
Viruses			
Dengue	Hemorrhagic fever	Bite of an infected mosquito (primarily *Aedes aegypti*)	Poor mosquito control; increased urbanization in tropics; increased travel and shipping
Filoviruses (Marburg, Ebola)	Fulminant, high mortality, hemorrhagic fevers	Direct contact with infected blood, organs, secretions, and semen	Contact with vertebrate reservoirs
Influenza H5N1 (avian influenza)	Fever, headache, cough, pneumonia, high mortality	Direct contact with infected animals or humans, not easily spread via respiratory aerosols	Danger of animal–human virus reassortment; antigenic shift
Fungi			
Candida	Candidiasis: fungal infections of the gastrointestinal tract, vagina, and oral cavity	Endogenous flora becomes an opportunistic pathogen; contact with secretions or excretions from infected persons	Immunosuppression; medical devices (catheters); antibiotic use

UNIT 6

Table 28.6 Emergence factors for infectious disease

Emergence factor	Example	Pathogen/disease
Human demographics and behavior	Urbanization	HIV/AIDS, dengue fever
Technology and industry	Healthcare-associated infections	Drug-resistant pathogens
Economic development and land use	Aswan High Dam construction	Rift Valley fever
	Changes in recreational and housing patterns	Lyme disease
International travel and commerce	Central food distribution	*E. coli* O157:H7 food infections
	Shipment of animals across international boundaries	Marburg and Ebola-Reston filovirus outbreaks
Pathogen adaptation and change	RNA virus mutations	Pandemic (H1N1) 2009 influenza
Breakdown of public health	Cholera epidemics	Haiti cholera outbreak 2010
	Inadequate immunization	Whooping cough outbreaks in Eastern Europe and the United States
Unusual events that upset the usual host–pathogen balance	Mild climate allowing expanded rodent population	Hantavirus outbreaks, United States

imported into tropical population centers from Africa via cargo ships that transported infected mosquitoes along with trade goods. Human behavior, especially in large population centers, also contributes to disease spread. For example, sexually promiscuous practices in population centers contribute to the spread of hepatitis and HIV/AIDS.

Technological advances and industrial development have a generally positive impact on living standards worldwide, but in some cases these advances have contributed to the spread of diseases. For example, the number and kind of healthcare-associated infections have increased dramatically in recent years (�❧ Section 27.2). Antibiotic resistance in microorganisms is another negative outcome of modern healthcare practices. For instance, vancomycin-resistant enterococci and staphylococci and drug-resistant *Streptococcus pneumoniae* and *Mycobacterium tuberculosis* are emerging pathogens in developed countries.

Economic development and changes in land also promote disease spread. For example, Rift Valley fever, another mosquito-borne viral infection, has been on the increase since the completion of the Aswan High Dam on the Nile River in Egypt

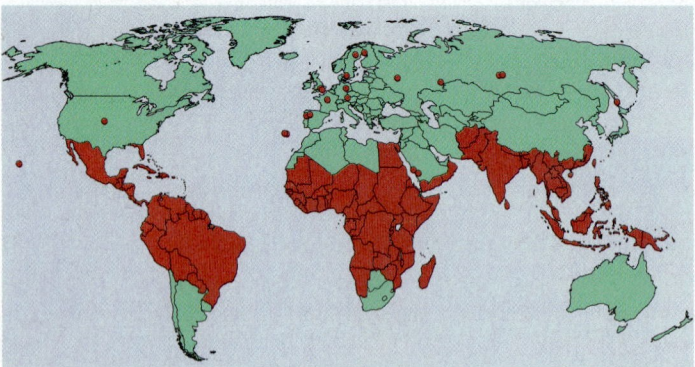

Figure 28.11 Dengue virus 2013. Dengue virus is now found in all tropical and subtropical countries due to the spread of its *Aedes aegypti* mosquito vector. The red areas are now endemic for the virus and mosquito vector. The red dots indicate outbreaks outside the known endemic areas. Prior to 1981, dengue virus was unknown in the Americas. Data are from the CDC, Atlanta, Georgia, USA.

in 1970. An epidemic of Rift Valley fever first developed in Egypt in 1977, when an estimated 200,000 people became ill and 598 died; the disease is now endemic near the dammed reservoir. Lyme disease, the most common vectorborne disease in the United States, is on the rise largely due to changes in land use patterns (⌐❧ Section 30.4). Reforestation and the resulting increase in populations of deer and mice (the natural reservoirs for the disease-producing *Borrelia burgdorferi*) have resulted in greater numbers of the arthropod tick vector. New homes and recreational areas in and near forests increase contact between the infected ticks and humans, consequently increasing disease incidence.

Transportation, bulk processing, and central distribution methods have become increasingly important for quality assurance and economy in the food industry. However, these same factors can increase the potential for common-source epidemics when sanitation measures fail. For example, a single U.S. meat-processing plant spread *Escherichia coli* O157:H7 to people in eight states in 2009. The contaminated food source, ground beef, was recalled and the epidemic was eventually stopped, but not before several people died.

International travel and commerce also affect the spread of pathogens. For example, filoviruses (*Filoviridae*), a group of RNA viruses, cause fevers culminating in hemorrhagic disease in infected hosts (⌐❧ Explore the Microbial World, "Handling Viral Hemorrhagic Fever Viruses," in Chapter 30). These untreatable viral diseases typically have a mortality rate above 20%. Most outbreaks have been restricted to equatorial central Africa, where the natural primate hosts and other vectors live (Figure 28.10), but travel of potential hosts to or from endemic areas can result in disease transmission. For example, one of the filoviruses was imported into Marburg, Germany, in 1967 with a shipment of African green monkeys used for laboratory work. The virus spread from the primate host to some of the human handlers. Seven of 31 infected people died in this outbreak of what became known as the *Marburg virus*. In 1989, another shipment of laboratory monkeys brought a different filovirus to Reston, Virginia, in the United States. The virus, now called Ebola-Reston, did not cause overt disease in any humans but, having an effective respiratory transmission mode, Ebola-Reston infected and killed most of the monkeys at the Reston facility within days. The high mortality potential of filovirus infections, especially one with respiratory transmission, could devastate population centers worldwide in a matter of weeks.

Emergence Factors: Pathogen Variation, Public Health Breakdowns, and Climate Change

Pathogen adaptation and change can contribute to disease emergence. For example, nearly all RNA viruses, including influenza (⇌ Section 29.8), HIV, and the hemorrhagic fever viruses, mutate rapidly. Because RNA viruses lack correction mechanisms for mutations during RNA replication, they incorporate mutations at an extremely high rate compared with most DNA viruses. These mutant RNA viruses present major epidemiological problems because their altered genomes may affect their antigens, making immunity to old viruses ineffective for neutralizing the mutants (see an example with influenza in Figure 29.27). Bacterial genetic mechanisms are capable of enhancing virulence and promoting emergence of new epidemics. Virulence-enhancing factors are often carried by bacteriophages, plasmids, and transposons as mobile genetic elements that can be transferred between and among members of the same species, and sometimes to other species and genera. Table 28.7 lists some mobile virulence factors that can contribute to pathogen emergence.

A breakdown of public health measures is sometimes responsible for the emergence or reemergence of diseases. For instance, cholera (caused by *Vibrio cholerae*) can be adequately controlled, even in endemic areas, by providing proper sewage disposal and water treatment. In 2010 contaminated water supplies probably caused by inadequate sanitation measures by United Nations peacekeeping forces brought cholera to Haiti for the first time in over 100 years (Section 28.10). In 1993, the municipal water supply of Milwaukee, Wisconsin, was contaminated with the chlorine-resistant protist *Cryptosporidium*, resulting in over 400,000 cases of intestinal disease.

Inadequate public vaccination programs can lead to the resurgence of previously controlled diseases. For example, pertussis, a vaccine-preventable childhood respiratory disease, has increased recently in Eastern Europe and in the United States partly due to inadequate immunization among adults and children.

Finally, abnormal natural occurrences sometimes upset the usual host–pathogen balance. For example, hantavirus is a human pathogen that is endemic in some rodent populations (⇌ Section 30.2). An abnormally high number of cases of human hantavirus infections leading to several deaths were reported in 1993 in the American Southwest and were linked to exposure to wild mice and their urine and droppings. The likelihood of exposure to mice and their waste products was increased due to a larger than normal wild mouse population resulting from near-record rainfall, a long growing season, and a mild winter, leading to a dramatic increase in the zoonotic host and pathogen density and enhanced exposure for susceptible human hosts. Similar factors led to a hantavirus outbreak in Yosemite National Park (California, USA) in 2011 (Figure 28.10). As global warming progresses, we are likely to see permanent alterations in climate that significantly alter the range of hosts, vectors, and pathogens.

Addressing Emerging Diseases

Emerging diseases have, at least at first, low incidence and are usually absent from the official notifiable disease list for the United States (Table 28.4). The keys for addressing emerging diseases are recognition of the disease and intervention to prevent pathogen transmission.

Emerging diseases are first recognized from their unique epidemic incidence, clusterings and other epidemiological patterns, and clinical symptoms unrelated to known pathogens. These diseases warrant intensive public health surveillance, followed by specific interventions designed to control further outbreaks. Methods such as isolation, quarantine, immunization, and drug treatment can be applied to contain outbreaks. For vectorborne and zoonotic diseases, the nonhuman host or vector must be identified to intervene in the life cycle of the pathogen and stop human infection.

International public health surveillance and intervention programs were instrumental in controlling the emergence of severe

Table 28.7 Virulence factors encoded by bacteriophages, plasmids, and transposons

Genetic element	Organism	Virulence factors
Bacteriophage	*Streptococcus pyogenes*	Erythrogenic toxin
	Escherichia coli	Shiga-like toxin
	Staphylococcus aureus	Enterotoxins A, D, E, staphylokinase, toxic shock syndrome toxin-1 (TSST-1)
	Clostridium botulinum	Neurotoxins C, D, E
	Corynebacterium diphtheriae	Diphtheria toxin
Plasmid	*Escherichia coli*	Enterotoxins, pili colonization factor, hemolysin, urease, serum resistance factor, adherence factors, cell invasion factors
	Bacillus anthracis	Edema factor, lethal factor, protective antigen, poly-D-glutamic acid capsule
	Yersinia pestis	Coagulase, fibrinolysin, murine toxin
Transposon	*Escherichia coli*	Heat-stable enterotoxins, aerobactin siderophores, hemolysin and pili operons
	Shigella dysenteriae	Shiga toxin
	Vibrio cholerae	Cholera toxin

UNIT 6

acute respiratory syndrome (SARS), a disease that emerged rapidly, explosively, and unpredictably from a zoonotic source. On the other hand, even a rapid and focused response was unsuccessful in containing the spread of pandemic (H1N1) 2009 influenza (Section 28.11).

MINIQUIZ ---

• What factors are important in the emergence or reemergence of potential pathogens?

• Indicate general and specific methods that would be useful for identifying emerging infectious diseases.

28.8 Biological Warfare and Biological Weapons

Biological warfare is the use of biological agents to incapacitate or kill a military or civilian population in an act of war or terrorism. The use and development of biological weapons are forbidden by international law. Nevertheless, biological weapons have been used against targets in the United States, and biological weapon-making facilities may be in the hands of several governments as well as avowed terrorist groups.

Characteristics of Biological Weapons

Effective biological weapons are pathogens or toxins that are (1) easy to produce and deliver, (2) safe for use by the offensive forces, and (3) able to incapacitate or kill people in a systematic and consistent manner. Although biological weapons are potentially useful in the hands of conventional military forces, the greatest likelihood of biological weapons use is probably by terrorists. This is in part due to the availability and low cost of producing and propagating many of the organisms useful for biological warfare.

Virtually all pathogenic bacteria or viruses are potentially useful for biological warfare, and several of the most likely candidate organisms are relatively simple to grow and disseminate. *Select agents* that have significant potential for use as biological weapons are listed in **Table 28.8**. The most frequently mentioned candidates for use as biological weapons are smallpox virus and *Bacillus anthracis*, the cause of anthrax.

Select agents are classified according to their potential as bioweapons into three categories by the Centers for Disease Control and Prevention. The highest level of threat comes from Category A agents. These can be easily disseminated by, for example, aerosols, or are easily transmissible from person to person. These agents characteristically cause high mortality and consequently have high impact on public health. Preparations for attacks by such agents require a specific plan for each agent. Category B agents are moderately easy to spread, result in moderate morbidity and low mortality, and require specialized diagnostic and surveillance capabilities. Category C agents are emerging pathogens for which no specific control and containment plans can be anticipated.

Smallpox

Smallpox virus has intimidating potential as a Category A biological warfare agent because it can be easily spread by contact or aerosol spray and it has a mortality rate of 30% or more. Its

Table 28.8 Select agents and diseases by bioweapons threat category[a]
Category A
Highest-priority agents that pose a risk to national security. These agents are easily disseminated or transmitted and result in high mortality rates. They require special action for public health preparedness.
Disease/Pathogen
Anthrax (*Bacillus anthracis*)
Botulism (*Clostridium botulinum* toxin)
Plague (*Yersinia pestis*)
Smallpox (*Variola major*)
Tularemia (*Francisella tularensis*)
Viral hemorrhagic fevers (filoviruses [e.g., Ebola, Marburg] and arenaviruses [e.g., Lassa, Machupo])
Category B
Second-highest-priority agents. These agents are moderately easy to disseminate, result in moderate morbidity and low mortality, and require specific enhancements of public health diagnostic capacity and disease surveillance.
Disease/Pathogen
Brucellosis (*Brucella* species)
Epsilon toxin of *Clostridium perfringens*
Food safety threats (e.g., *Salmonella* spp., *Escherichia coli* O157:H7, *Shigella*)
Glanders (*Burkholderia mallei*)
Melioidosis (*Burkholderia pseudomallei*)
Psittacosis (*Chlamydophila psittaci*)
Q fever (*Coxiella burnetii*)
Ricin toxin from *Ricinus communis* (castor beans)
Staphylococcal enterotoxin B (*Staphylococcus aureus*)
Typhus fever (*Rickettsia prowazekii*)
Viral encephalitis (alphaviruses such as Venezuelan equine encephalitis, eastern equine encephalitis, western equine encephalitis)
Water safety threats (*Vibrio cholerae*, *Cryptosporidium parvum*, and others)
Category C
Third-highest-priority agents are emerging pathogens that are available, easily produced and disseminated, with high potential for high morbidity and mortality.
Pathogens
Emerging infectious diseases such as hantavirus

[a]Source: The Centers for Disease Control and Prevention, Atlanta, Georgia, USA.

potential for use as a military biological weapon is considered low, however, because the only known stocks of smallpox virus are in guarded repositories in the United States and Russia and military personnel are routinely vaccinated. A possibility remains, however, for terrorist groups or military forces to gain access to the smallpox virus and spread it among the public at large. Because of this, the United States government has made provisions to immunize frontline healthcare and public safety personnel for smallpox.

Although there is an extremely effective smallpox vaccine that uses the closely related vaccinia virus as the immunogen, this vaccine has not been in general use for almost 40 years because wild smallpox was eradicated worldwide over 35 years ago. Vaccinia immunization is very effective, but it carries significant risk; about one to two people per million who receive the vaccine will likely die from a vaccinia virus complication. Consequently the vaccine is no longer recommended and over 90% of the world population is inadequately immune and susceptible to the disease. Preparations for a potential smallpox attack in the United States have included recommendations for immunization of certain individuals: persons having close contact with smallpox patients; workers evaluating, caring for, or transporting smallpox patients; laboratory personnel handling clinical specimens from smallpox patients; and other persons such as housekeeping personnel who might come into contact with infectious materials from smallpox patients.

Anthrax

Bacillus anthracis, a ubiquitous saprophytic soil bacterium, is a Category A biowarfare and bioterrorism agent. Its unique properties make it particularly useful as a bioweapon. It grows as an aerobic gram-positive rod and produces endospores resistant to heat and drying, and it forms distinctive colonies on blood agar plates (**Figure 28.12**). Endospore formation enhances the ability to disseminate *B. anthracis* in aerosols.

B. anthracis endospores are the normal means of acquiring anthrax; we discuss the pathology of anthrax in detail in Section 30.8. Animals and humans occasionally acquire the spores by contact with contaminated plants or soil. There are three forms of the disease. *Cutaneous anthrax* (**Figure 28.13a**) is contracted when abraded skin is contaminated by *B. anthracis* endospores produced by a toxigenic poly-D-glutamic acid–encapsulated strain. Cutaneous anthrax cases in humans are rare in the United States. *Gastrointestinal anthrax* is contracted from consumption of endospore-contaminated plants or meat from animals infected with anthrax. Human gastrointestinal anthrax is also rare in the United States. *Inhalation anthrax* (also called pulmonary anthrax) is contracted when the endospores are inhaled. Inhalation of the endospores or the live bacteria results in pulmonary infections characterized by pulmonary and cerebral hemorrhage

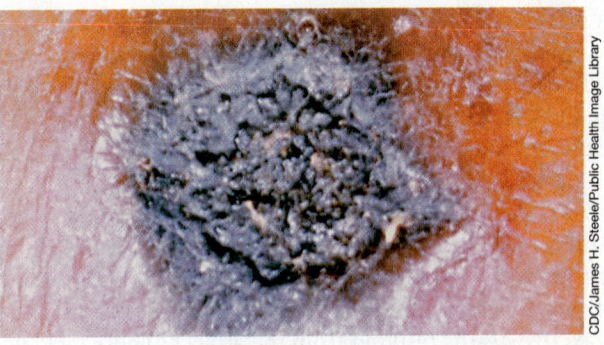

(b)

Figure 28.13 **Anthrax.** (a) Cutaneous anthrax. The blackened lesion on the forearm of a patient, about 2 cm in diameter, results from tissue necrosis. (b) Inhalation anthrax can cause cerebral hemorrhage, as shown by the dark coloration in this fixed and sectioned human brain. See also Figure 30.21.

(Figure 28.13b). Untreated inhalation anthrax has a mortality rate of nearly 100%. Fortunately, inhalation anthrax cases, even in agricultural workers, are extremely rare. The most recent naturally acquired inhalation anthrax infection in the United States occurred in 1976. However, several cases of inhalation anthrax identified in 2001 in the United States were caused by a bioterrorism attack.

Pathogenesis results from inhalation of endospores from toxigenic *B. anthracis*. Pathogenic strains produce three key virulence proteins—*protective antigen* (PA), *lethal factor* (LF), and *edema factor* (EF). PA and LF form *lethal toxin*. PA and EF form *edema toxin* (⮌ Table 23.5). Growth of *B. anthracis* in the lymph nodes and lymphatic tissues leads to sore throat, fever, and muscle aches. Symptoms from the growing toxemia escalate in several days to include difficulty in breathing and systemic shock. Fatality rates can approach 90% even when exposure is recognized and treatment is started, and is nearly 100% in cases for which treatment is not started until after the onset of symptoms.

Weaponized anthrax strains are usually endospore preparations that have properties that enhance dissemination. Such strains and preparations were developed in several countries in the post–World War II era, but development of new biological weapons was halted by international treaty in 1972. The physical characteristics of weaponized anthrax preparations typically include a

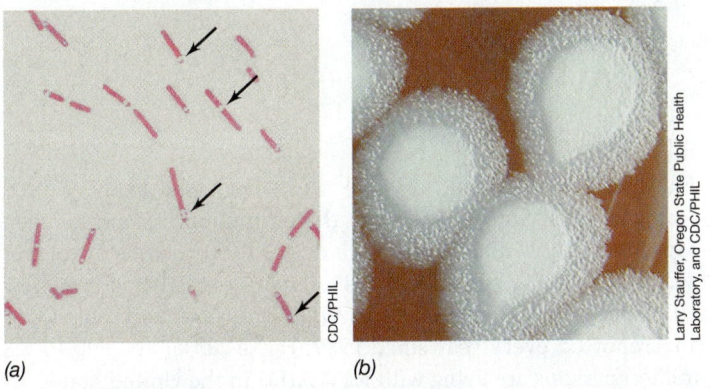

(a)   (b)

Figure 28.12 *Bacillus anthracis.* (a) *Bacillus anthracis* is a gram-positive endospore-forming rod approximately 1 μm in diameter and 3–4 μm in length. Note the developing endospores (arrows). (b) Characteristic "ground glass" appearance of colonies of *B. anthracis* on blood agar plates.

UNIT 6

small particle size interspersed with a very fine particulate agent such as talc. This powdered form ensures that the endospores will spread easily by air currents. Thus, opening an envelope containing endospores or releasing the powder–endospore mixture into a ventilation system or other air current can contaminate surrounding areas and personnel.

In 2001 anthrax attacks were carried out in the United States by mailing envelopes or packages containing weaponized anthrax spores. In all, the attacks were responsible for 22 cases of anthrax, 11 of which were cutaneous anthrax, and 11 were inhalation anthrax; five deaths resulted from the attacks. These incidents in the United States were not the first or the most serious anthrax biological weapons infections. In 1979, *B. anthracis* spores were inadvertently released into the atmosphere from a biological weapons facility in Sverdlovsk, Russia. Less than 1 g of endospores was released, and everyone in the area surrounding the facility was immunized and given prophylactic antibiotic therapy as soon as the first anthrax case was diagnosed. However, even with these quick reactive measures, 77 persons outside the facility contracted inhalation anthrax and 66 died.

Vaccination for anthrax is restricted to individuals who are considered at risk. This includes agricultural animal workers and military personnel.

Delivery of Biological Weapons

As for anthrax, organisms most suitable for use as biological weapons can be spread as an aerosol, providing simple, rapid, widespread dissemination leading to infection. Examples of several other aerosol exposures are instructive.

In 1962, one of the last outbreaks of smallpox in a developed country occurred in Germany. A German worker developed smallpox after returning from Pakistan, where smallpox was still endemic. The patient, who had a cough, was immediately hospitalized and quarantined; the cough aerosolized the virus and caused 19 cases of smallpox and one death, even though every individual who became infected had been immunized.

There have been planned bioterrorist attacks in the United States and other countries before the anthrax attacks of 2001. In 1984 in The Dalles, Oregon (USA), terrorists sprayed salad bars in ten restaurants with a culture of *Salmonella enterica* serovar Typhimurium, causing 751 cases of foodborne salmonellosis in a region that usually has fewer than 10 cases per year. In 1995, a terrorist group released sarin nerve gas, a chemical weapon, into a Tokyo subway, killing several people and injuring many more. This incident is relevant to a discussion of biological weapons because this group also possessed anthrax cultures, bacteriological media, drone airplanes, and spray tanks. They attempted at least one biological attack, but were unsuccessful.

Delivery of preformed bacterial toxins such as botulinum toxin or staphylococcal enterotoxin to large populations may be impractical because most exotoxins are proteins that lose effectiveness as they are diluted or denatured, and are destroyed in common sources such as drinking water. However, delivery of toxins could be aimed at selected individuals and small groups, or delivered randomly to instigate panic.

Prevention and Response to Biological Weapons

Proactive measures against the deployment of biological weapons have already begun with periodic updating of the international agreements of the 1972 Biological and Toxic Weapons Convention. The fifth and most recent update was in 2002. At the practical level, governments are now supporting the large-scale production and distribution of vaccines along with the development of strategic and tactical plans to prevent and contain biological weapons.

The United States government, through the Centers for Disease Control and Prevention (CDC), has developed the Select Agent Program surveillance system to monitor possession and use of potential bioterrorism agents. The CDC Laboratory Response Network and the Health Alert Network have been upgraded to enhance their diagnostic capabilities and increase the reporting abilities of local and regional healthcare centers to rapidly identify bioterrorism events as well as emerging diseases (Table 28.5).

MINIQUIZ --

- What characteristics make a pathogen or its products particularly useful as a biological weapon?
- Indicate the steps you would take to identify and treat infections from smallpox virus or anthrax in a bioterror attack.

IV • Current Pandemics

Here we examine data collected by national and worldwide disease-surveillance programs that provide a picture of current and emerging disease patterns for three major pandemic diseases: HIV/AIDS, cholera, and influenza.

28.9 The HIV/AIDS Pandemic

HIV/AIDS is a continuum of disease, starting with the infection of an individual with the human immunodeficiency virus (HIV), leading to the clinical disease, acquired immunodeficiency syndrome (AIDS), a disease that attacks the immune system (⮌ Section 29.14).

HIV/AIDS Numbers

The first reported cases of AIDS were diagnosed in the United States in 1981. Since then, more than 1 million cases have been reported in the United States with over 500,000 deaths. A total of 36,870 persons in 2009 and 35,741 persons in 2010 acquired new HIV infections. Over 33,000 new AIDS cases have been diagnosed and reported every year since 1989 (**Figure 28.14**); 800,000 to 1.3 million persons are living with HIV/AIDS in the United States.

Worldwide, from 1981 through 2010, over 80 million people have been infected with HIV. About 46 million people have already died from AIDS, and about 34 million are currently infected with HIV (**Table 28.9**), with most—23.5 million—living in sub-Saharan Africa.

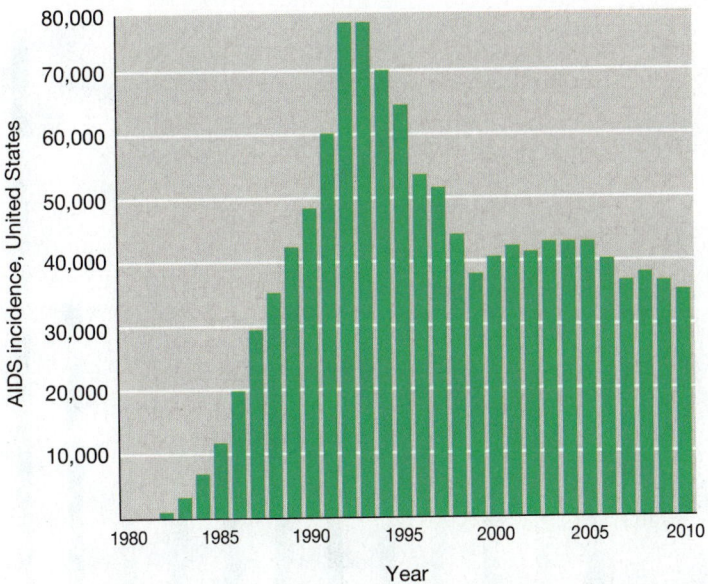

Figure 28.14 **Annual new cases of human immunodeficiency virus/acquired immunodeficiency syndrome (HIV/AIDS) in the United States.** Cumulatively, there were about 1.1 million cases of HIV/AIDS through 2010. In 2009, the HIV/AIDS case definition changed to include all new HIV infections and AIDS diagnoses. Data are from the HIV/AIDS Surveillance Report and Division of HIV/AIDS Prevention—Surveillance and Epidemiology, CDC, Atlanta, Georgia, USA.

Globally, 2.5 million individuals are newly infected with HIV and about 1.7 million deaths occur each year, with 1.2 million deaths occurring in sub-Saharan Africa.

HIV/AIDS Epidemiology

Case studies in the United States in the 1980s initially suggested a high AIDS prevalence among men who have sex with men and intravenous drug abusers. This indicated a transmissible agent, presumably transferred during sexual activity or by blood-contaminated needles. Individuals receiving blood or blood products were also at high risk: Hemophiliacs who required infusions of blood products, usually pooled from multiple donors, acquired AIDS, as did a small number of individuals who received blood transfusions or tissue transplants before 1982 (when blood-screening procedures were implemented). Today the incidence of HIV in hemophiliacs and in blood transfusion and organ

Table 28.9 HIV/AIDS infections worldwide, 2011[a]	
Location	*HIV/AIDS infections*
The Americas	3.0 million
Western and Central Europe	0.9 million
Eastern Europe and Central Asia	1.4 million
Africa	23.8 million
East Asia and Pacific	0.9 million
South and Southeast Asia	4.0 million
Oceania	53,000

[a]The total number of individuals living with HIV/AIDS is estimated to be about 34 million. About 1.7 million people died from AIDS in 2011. Data are from the World Health Organization.

recipients has been virtually eliminated through rigorous screening of blood and biological products.

Soon after the discovery of HIV, laboratory immunosorbent assays and immunoblot tests (↺ Section 27.9) were developed to detect antibodies to the virus in a blood sample. Extensive surveys of HIV incidence and prevalence defined the spread of HIV and ensured that new cases would not be transmitted by blood transfusions. The pattern illustrated in **Figure 28.15** is typical of an agent transmissible by blood or other body fluids. The identification of defined high-risk groups implied that HIV was not transmitted from person to person by casual contact, such as the respiratory route, or by contaminated food or water. Instead, body fluids, primarily blood and semen, were identified as the vehicles for transmission of HIV.

Figure 28.15 shows that in the United States the number of AIDS cases is disproportionately high in men who have sex with men, but the patterns in women and in certain racial and ethnic groups indicate that male-to-male sex is not the only risk factor for acquiring AIDS. Among women, for example, heterosexuals are the largest risk group, whereas in African American and Hispanic men, intravenous drug use as well as sexual activity is linked to HIV infection.

Racial differences in HIV incidence rates in the United States indicate that social and economic factors also contribute to infection risk. African Americans acquired 44% of all new HIV infections in 2010, but made up only about 14% of the U.S. population. Black men accounted for 70% of these infections; the incidence of new HIV infections for black men was seven times as high as that for white men.

The study of individuals who are at high risk for acquiring AIDS indicates that virtually all who acquire HIV today share two specific behavior patterns. First, they engage in activities (sex or drug use) in which body fluids, usually semen or blood, are transferred. Second, they exchange body fluids with multiple partners through sexual activity, needle-sharing drug activity, or both. With each

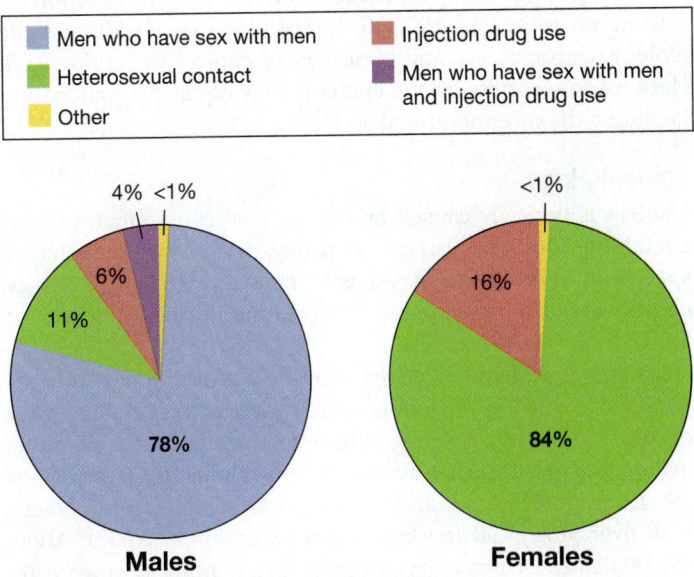

Figure 28.15 **Distribution of AIDS cases by risk group and sex in adolescents and adults in the United States, 2010.** Information was obtained from 38,000 males and 9,500 females diagnosed with HIV/AIDS in 2010. Data are from the CDC, Atlanta, Georgia, USA.

UNIT 6

encounter they have a probability of receiving body fluids from an HIV-infected individual and therefore a chance of being infected with HIV.

HIV can be transmitted to the fetus by infected mothers and also in mothers' milk; in 2010, there were 162 new cases of perinatal HIV infection in the United States. Infants born to HIV-infected mothers have maternally derived antibodies to HIV in their blood. However, a positive diagnosis of HIV infection in infants must wait a year or more after birth because about 70% of infants showing maternal HIV antibodies at birth show no signs of being infected later on.

Heterosexual transmission of HIV is the norm in Africa. In some regions, fewer men than women are infected with HIV. The identification of high-risk groups such as prostitutes has led to the development of health education campaigns that inform the public of HIV transmission methods and define high-risk behaviors. Because no cure or proven immunization for AIDS is available, public health education remains the most effective approach to the control of HIV/AIDS. We discuss the pathology and therapy of HIV/AIDS in Section 29.14.

MINIQUIZ -

- Describe the major risk factors for acquiring HIV infection. Tailor your answer to your country of origin.

- Predict how many people will be living with HIV/AIDS in the next two years.

28.10 Cholera Pandemics

Cholera is a severe diarrheal disease that is now largely restricted to the developing world. Cholera is an example of a major waterborne disease that can be controlled by application of appropriate public health measures for water treatment. Global incidence estimates range from 3 to 5 million cases per year, mostly unreported, causing an estimated 100,000–120,000 deaths. We discuss the biology, pathogenesis, and treatment of cholera in Section 31.3. Here we concentrate on the epidemiology of cholera pandemics, including the recent outbreak in Haiti.

Epidemiology

Cholera is typically caused by ingestion of contaminated water containing *Vibrio cholerae*, a gram-negative, curved rod–shaped species of *Proteobacteria*. As with many waterborne diseases, cholera can also be acquired by consumption of contaminated food (⇄ Section 31.3).

Cholera is endemic in Africa, Southeast Asia, the Indian subcontinent, and Central and South America. Epidemic cholera occurs frequently in areas where sewage treatment is either inadequate or altogether absent. Worldwide in 2008, there were 190,130 reported cases and 5143 reported deaths from cholera, with over 98% of all reported cases occurring in Africa. About 100,000 cases or more have been reported annually since 2000, with a low of 95,560 cases in 2004, and a high of 589,854 cases in 2011 (**Figure 28.16**).

The World Health Organization estimates that only 5–10% of cholera cases are reported, so the total incidence of cholera exceeds 1 million cases per year. Even in developed countries, the

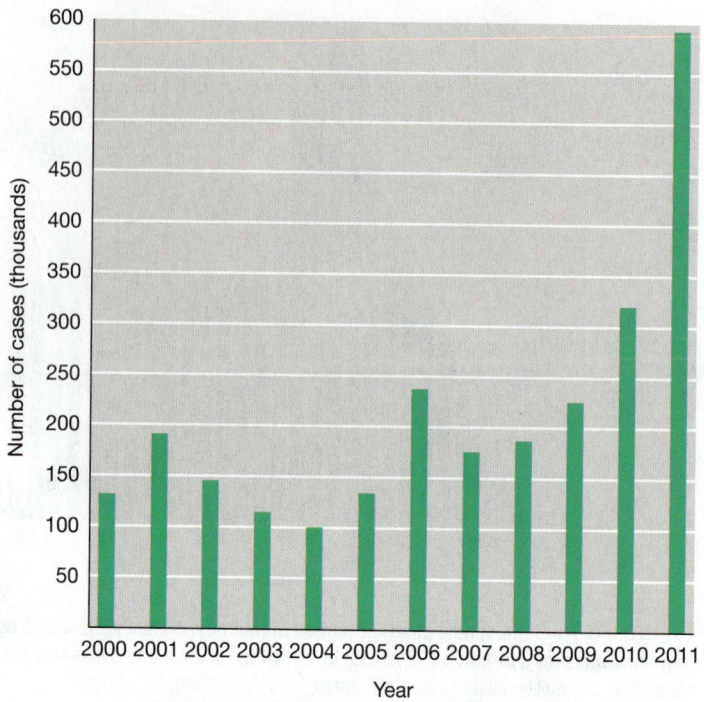

Figure 28.16 **Cholera cases.** The reported cholera cases from 2000 to 2011 show an increasing trend indicating a continuation of the seventh pandemic (or the beginning of an eighth pandemic). Up to 95% of cholera cases are unreported. Data were provided by the World Health Organization.

disease is a threat. A handful of cases are reported each year in the United States, but rarely from drinking water sources. Most of these cases are imported, often in food. A few cases are possibly from endemic sources; raw shellfish seems to be the most common vehicle, presumably because *V. cholerae* can be free-living in coastal waters in endemic areas where the pathogen adheres to marine microflora ingested by the shellfish (⇄ Section 31.3).

Epidemic cholera may develop into pandemics when travelers from endemic areas carry the pathogen to new locations with susceptible populations. Since 1817, cholera has swept the world in seven major, and nearly contiguous, pandemics (**Figure 28.17**). All cholera pandemics have originated from the Indian subcontinent, where cholera is endemic. Two distinct pandemic strains of *V. cholerae* are recognized, known as the *classic* and the *El Tor* biotypes. The *V. cholerae* O1 El Tor biotype started the seventh pandemic in Indonesia in 1961, and its spread continues to the present. This pandemic has caused over 5 million cases of cholera and at least 250,000 deaths and continues to be a major cause of morbidity and mortality, especially in developing countries. In 1992, a genetic variant of the El Tor biotype known as *V. cholerae* O139 Bengal arose in Bangladesh and caused an extensive epidemic. *V. cholerae* O139 Bengal has continued to spread since 1992, causing several major epidemics, and may be the agent of an eighth pandemic.

Haitian Outbreak

In October 2010 Haiti experienced its first cholera in over 100 years. Through July 2012, Haiti experienced 581,952 cases and 7455 deaths due to cholera, with no end in sight. The outbreak began in the aftermath of a catastrophic 2010 earthquake. There were two possible triggers of the cholera outbreak, the first being

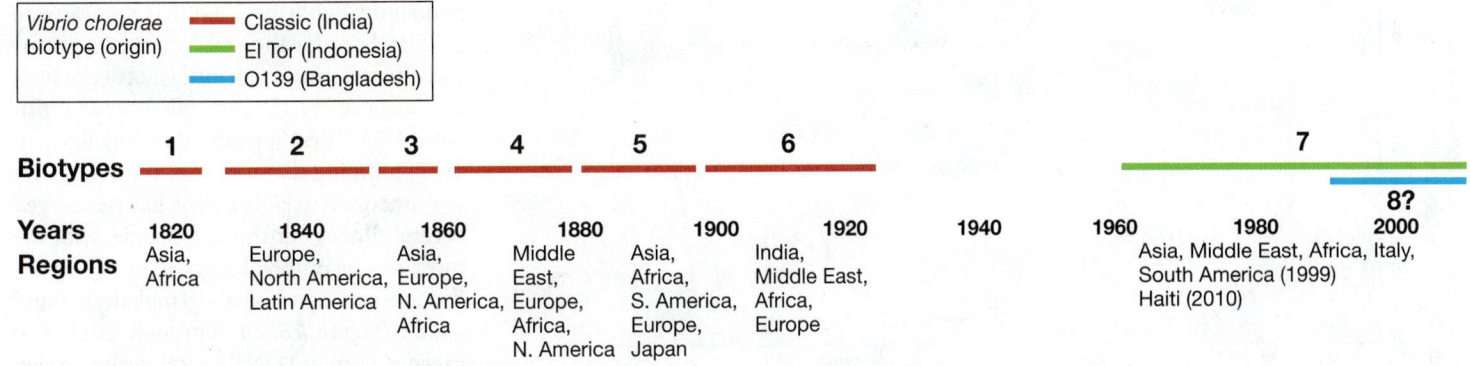

Figure 28.17 Cholera pandemic timeline. Seven cholera pandemics have been nearly contiguous for over 200 years. The seventh pandemic started in 1961 and is ongoing. The O139 strain that appeared in 1991 is endemic to Bangladesh and the Bay of Bengal and is causing epidemics that may be the prelude to an eighth pandemic.

a classic scenario of poor sanitation and the second an accidental importation from an outside source.

Based on climatic data and the possibility of the earthquake disturbing coastal waters where *Vibrio cholerae* is endemic in local microflora, some scientists hypothesized that the vibrios had grown explosively offshore and been washed into coastal freshwater, where they contaminated drinking water sources. The poor sanitary conditions following the earthquake included the breakdown of water and sewage treatment infrastructure that, they reasoned, provided the perfect scenario for an outbreak.

The second possibility was that the outbreak was the result of poor sanitation and disposal of waste from a battalion of United Nations (UN) peacekeeping troops. The troops had arrived from Nepal where there had been a recent cholera outbreak. Sewage from their camp ran into the Artibonite River, a major water source for much of Haiti. The first cases were from towns near the Artibonite River and the UN camp, inland from the coastal areas that would be the likely initial sites of an outbreak due to an offshore source. Whole genome sequence typing confirmed that the strain responsible for the Haiti outbreak was nearly identical to *V. cholerae* serogroup O1, serotype Ogawa, the strain that had caused a recent outbreak in Nepal. This serotype had never before been seen in the Western Hemisphere.

Though scientists were not permitted to take samples from the Nepalese soldiers, the molecular and classical epidemiology data lead to the almost certain conclusion that the Nepalese troops had introduced cholera to Haiti. Cholera has since spread to the Dominican Republic, Haiti's sister country on the island of Hispaniola, to other areas of the Caribbean, and to Mexico.

MINIQUIZ -
- Identify the most likely means for acquiring cholera.
- Why are cholera pandemics still occurring?

28.11 Influenza Pandemics

The influenza pandemic of 2009–2010 began with an outbreak in Mexico. Influenza pandemics occur every 10 to 40 years due to major genetic changes in the influenza A virus genome (⮌ Table 29.2). We discuss the disease influenza in detail in Section 29.8.

Swine Flu—Pandemic (H1N1) 2009 Influenza

An influenza pandemic began in March 2009 with the outbreak of epidemics in Mexico. Typical year-to-year antigenic variation, called *antigenic drift*, is caused by point mutations in the RNA of the influenza genome. These mutations seldom cause pandemics, but do cause annual influenza outbreaks. Pandemic influenza strains arise from a much larger change in the viral genome termed *antigenic shift*. In the "swine flu" pandemic—officially, pandemic (H1N1) 2009—swine in Mexico were simultaneously infected with swine influenza, bird influenza, and human influenza (⮌ Figure 29.27). During viral maturation, viral genomic RNA segments, in this case from three sources, are mixed together and packaged to form genetically unique viruses, a process called *reassortment*. Such *reassortant viruses* cause new pandemics as new strains are mixed in susceptible animals and spread to susceptible human populations (⮌ Section 29.8).

A virus that results from antigenic shift has the potential to contain antigens to which no human has had prior exposure. This means that immunity to a new virus is nonexistent; the only way humans can obtain immunity to a new strain is to become infected (or artificially immunized) and produce an immune response. For the virus of pandemic (H1N1) 2009, almost no one less than 50 years old had any immunity because they were never exposed to similar H1N1 viral strains. As a result, many of the deaths in this pandemic were of people younger than 50 who were healthy until they were infected by the virus. However, pandemic (H1N1) 2009 is related to the 1957 influenza pandemic nicknamed the "Asian flu," and, farther back, to the 1918 influenza pandemic that killed over 2 million people worldwide. Hence, most people 50 or older had probably been infected with a strain of influenza virus related to the pandemic (H1N1) 2009 virus and had immune cells and antibodies (immune memory) that responded to control this pandemic virus. For younger people, unfortunately, immune memory to pandemic (H1N1) 2009 was nonexistent.

Within six months of its emergence, pandemic (H1N1) 2009 had spread to almost every country in the world, causing significant mortality in most of those. The pattern of spread was similar to that of seasonal influenza, but had one major difference. The virus began its spread from an initial focus of infection in Mexico and the southwestern United States in March, at the very end of the traditional winter flu season. Instead of dying out, as

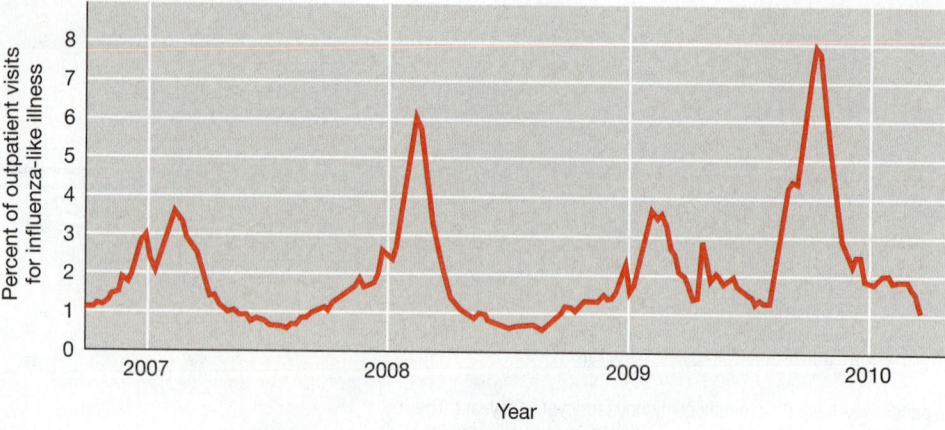

Figure 28.18 **Influenza incidence in the United States.** Pandemic (H1N1) 2009 influenza virus caused a higher than normal incidence of influenza from the middle of 2009 through 2010. In the 2009–2010 flu season the peak incidence of disease was higher than in the three previous seasons and occurred 3–4 months earlier than usual. Data are adapted from the CDC, Atlanta, Georgia, USA.

most seasonal flu outbreaks do at the end of the winter season, pandemic (H1N1) 2009 continued to spread through the summer months in the United States, especially in susceptible populations such as children at youth group camps. The 2009–2010 flu season in the United States thus differed from the typical seasonal influenza infection pattern; the highest incidence occurred in October and November and tapered off during the usual peak flu season in January through March (**Figure 28.18**). In the Southern Hemisphere, where the flu season runs from about April to September, the pandemic strain spread with all of the characteristics of seasonal flu.

Future Influenza Pandemics

Perhaps the greatest threat to biosecurity worldwide is another influenza pandemic that has the virulence and infectivity of the 1918 pandemic. Public health officials worldwide are watching for the emergence of a potentially devastating avian influenza designated influenza A H5N1, also called *avian influenza*. H5N1 first appeared in Hong Kong in 1997, jumping directly from the avian host to humans. The H5N1 virus has reemerged several times over the last decade, with the most recent outbreaks occurring in Egypt, Indonesia, Cambodia, Bangladesh, and China (Figure 28.10). Through 2012, 610 cases of human H5N1 infections have been confirmed, resulting in 360 deaths, for a mortality rate of almost 60%.

The H5N1 virus is spread directly from birds such as domestic chickens or ducks to humans through prolonged contact or consumption of infected meat; avian influenza is spread human to human only after prolonged close contact, but some reports indicate that H5N1 has infected swine. If further reassortment with human influenza strains occurs, a new reassortant virus for which there is no immunity in humans could trigger an influenza pandemic with the potential for unprecedented mortality. Plans are in place nationally and internationally to provide appropriate vaccines and support for potential pandemics initiated by this and other emergent influenza strains. A recombinant vaccine for the H5N1 virus is available on a limited basis.

MINIQUIZ

- Identify the features that distinguished pandemic (H1N1) 2009 influenza from a seasonal influenza epidemic.
- Why is H5N1 avian influenza considered a major health threat?

BIG IDEAS

28.1 • Epidemiology is the study of the occurrence, distribution, and determinants of health and disease in populations. An endemic disease is continually present at low incidence in a population, whereas an epidemic disease is one that has increased to unusually high incidence in a population. Incidence is a record of new cases of a disease, whereas prevalence is a record of total cases of a disease in a population. Infectious diseases cause morbidity (illness) and may cause mortality (death). An infectious disease follows a predictable clinical pattern in the host.

28.2 • Effects on both populations and individuals must be studied to understand infectious disease. The interactions of pathogens with hosts can be dynamic, affecting the long-term evolution and survival of all species involved. Herd immunity provides disease protection for uninfected or unimmunized hosts.

28.3 • Infectious diseases can be transmitted directly from one host to another host, indirectly from living vectors or inanimate objects (fomites), or from common-source vehicles such as food and water. Epidemics may be of host-to-host origin or originate from a common source.

28.4 • Many pathogens exist only in humans and are maintained only by transmission from person to person. Many other human pathogens, however, have reservoirs in soil, water, or animals. An understanding of disease reservoirs, carriers, and pathogen life cycles is critical for controlling disease epidemics.

28.5 • Food and water purity regulations, vector control, immunization, quarantine, isolation, and disease surveillance are public health measures that reduce the incidence of communicable diseases.

28.6 • Infectious diseases account for almost 25% of all mortality worldwide. Most cases of infectious diseases are in developing countries. Control of infectious diseases can be accomplished by public health measures.

28.7 • Changes in host, vector, or pathogen conditions, whether natural or artificial, can encourage the explosive emergence or reemergence of infectious diseases. Global surveillance and intervention programs must be in place to prevent new epidemics and pandemics.

28.8 • Bioterrorism is a threat in a world of rapid international travel and easily accessible technical information. Biological agents can be used as weapons by military forces or by terrorist groups. Aerosols or common sources such as food and water are the most likely modes of delivery. Prevention and containment measures rely on a well-prepared public health infrastructure.

28.9 • HIV/AIDS is a major worldwide public health problem that affects individuals who exchange bodily fluids. Those who engage in promiscuous unprotected sex or intravenous drug use are particularly at risk.

28.10 • Cholera pandemics have occurred nearly constantly over the last 200 years. Cholera control can be achieved by maintaining adequate clean water and waste sanitation measures.

28.11 • Influenza pandemics occur cyclically. New pandemic influenza strains resulting from bird–swine–human influenza reassortments present the biggest predictable infectious disease threat worldwide.

MasteringMicrobiology®   **Review what you know and challenge what you have learned with MasteringMicrobiology!** Access study materials, chapter quizzes, animations, and microbiology lab tutorials in the Study Area to ensure that you have mastered this chapter's content.

REVIEW OF KEY TERMS

Acute infection a short-term infection, usually characterized by dramatic onset

Basic reproduction number (R_0) the number of expected secondary transmissions from each single case of a disease in an entirely susceptible population

Biological warfare the use of biological agents to incapacitate or kill a military or civilian population in an act of war or terrorism

Carrier a subclinically infected individual who may spread a disease

Centers for Disease Control and Prevention (CDC) the agency of the United States Public Health Service that tracks disease trends, provides disease information to the public and to healthcare professionals, and forms public policy regarding disease prevention and intervention

Chronic infection a long-term infection

Common-source epidemic an infection (or intoxication) of a large number of people from a contaminated common source such as food or water

Emerging disease an infectious disease whose incidence recently increased or whose incidence threatens to increase in the near future

Endemic disease a disease that is constantly present, usually in low numbers, in a population

Epidemic the occurrence of a disease in unusually high numbers in a localized population

Epidemiology the study of the occurrence, distribution, and determinants of health and disease in a population

Fomite an inanimate object that when contaminated with a viable pathogen can transfer the pathogen to a host

Herd immunity the resistance of a population to a pathogen as a result of the immunity of a large portion of the population

Host-to-host epidemic an epidemic resulting from person-to-person contact, characterized by a gradual rise and fall in number of new cases

Incidence the number of new disease cases reported in a population in a given time period

Isolation in the context of infectious disease, the separation of persons who have an infectious disease from those who are healthy

Morbidity the incidence of disease in a population

Mortality the incidence of death in a population

Outbreak the occurrence of a large number of cases of a disease in a short period of time

Pandemic a worldwide epidemic

Prevalence the total number of new and existing disease cases reported in a population in a given time period

Public health the health of the population as a whole

Quarantine the separation and restriction of well persons who may have been exposed to an infectious disease to see if they develop the disease

Reemerging disease an infectious disease previously under control but that produces a new epidemic

Reservoir a source of infectious agents from which susceptible individuals may be infected

Surveillance the observation, recognition, and reporting of diseases as they occur

Vector a living agent that transfers a pathogen (differs from genetic vector, discussed in Chapter 11)

Vehicle a nonliving source of pathogens that transmits the pathogens to large numbers of individuals; common vehicles are food and water

Virulence the relative ability of a pathogen to cause disease

Zoonosis any disease that occurs primarily in animals but can be transmitted to humans

UNIT 6

REVIEW QUESTIONS

1. Distinguish between *acute* and *chronic*, *mortality* and *morbidity*, *prevalence* and *incidence*, and *epidemic* and *pandemic*, as these terms relate to infectious disease. (Section 28.1)

2. How does herd immunity protect the nonimmune members of the population from acquiring a disease? Will herd immunity work for diseases that have a common source, such as water? Why or why not? (Section 28.2)

3. Give examples of host-to-host transmission of disease via direct contact. Also give examples of indirect host-to-host transmission of disease via vector agents and fomites. (Section 28.3)

4. Identify the disease reservoirs for the diseases botulism, gonorrhea, and plague. How do the reservoirs influence our ability to control or eradicate the disease? (Section 28.4)

5. Describe the major medical and public health measures developed in the twentieth century that were instrumental in controlling the spread of infectious diseases in developed countries. (Section 28.5)

6. Compare the contribution of infectious diseases to mortality in developed and developing countries. (Section 28.6)

7. Review the major reasons for the emergence of new infectious diseases. What methods are available for identifying and controlling the emergence of new infectious diseases? (Section 28.7)

8. Describe the general properties of an effective biological warfare agent. How do smallpox and *Bacillus anthracis* meet these criteria? Identify other organisms that meet the basic requirements for a bioweapon. (Section 28.8)

9. Identify the major risk factors for acquiring human immunodeficiency virus (HIV) infection in the United States. Does this pattern hold for other geographic regions? (Section 28.9)

10. Review the current cholera pandemic. Where and when did the pandemic start? (Section 28.10)

11. Why is influenza H5N1 considered a very important biological threat? (Section 28.11)

APPLICATION QUESTIONS

1. Smallpox, a disease that was limited to humans, was eradicated. Plague, a disease with a zoonotic reservoir in rodents (⮂ Section 30.7), can never be eradicated. Explain this statement and why you agree or disagree with the possibility of eradicating plague on a global scale. Devise a plan to eradicate plague in a limited environment such as a town or city. Be sure to use methods that involve the reservoir, the pathogen, and the host.

2. Identify a pathogen that would be a suitable agent for biological warfare and is not listed as a Category A or B agent (Table 28.8). Describe the properties of the pathogen in the context of its use as a biological weapon. Describe equipment and other resources necessary for growing large amounts of the pathogen. Identify a suitable delivery method. As you will propagate, weaponize, and deliver the pathogen, describe the precautions you will take to protect yourself. Now reverse your role. As a public health official in your community, describe how you would recognize and diagnose the disease caused by the agent. Indicate the measures you would take to treat the illnesses caused by the agent. How could you best limit the damage? Would quarantine and isolation methods be useful? What about immunization and antibiotics?

3. Human immunodeficiency virus/acquired immunodeficiency syndrome (HIV/AIDS) is a disease that can be eliminated because it is propagated by person-to-person contact and there are no known animal reservoirs. Do you agree or disagree with this statement? Explain your answer. Design a program for eliminating HIV/AIDS in a developed country and in a developing country. How would these programs differ? What factors would work against the success of your program, both in terms of human behavior and in terms of HIV/AIDS disease progression? Why are the numbers of HIV/AIDS patients continuing to grow, especially in developing countries? HIV/AIDS incidence (new cases) in developed countries has been virtually unchanged in this century (Figure 28.14). The numbers of individuals living with HIV/AIDS, however, is increasing. Explain this contradiction.

4. H5N1 avian influenza has high potential for causing an influenza pandemic under certain circumstances. Discuss conditions in which avian influenza may spread in humans as an epidemic. If such a human–avian strain were to evolve in Asia, from the perspective of a national public health official, what measures would you employ to stop the spread of the new influenza to the United States? If you fail to contain the new virus, based on the known mortality figures for the current avian influenza strain, and assuming that 10% of the population became infected, predict the overall number of deaths in the United States.

29 · Person-to-Person Bacterial and Viral Diseases

microbiology**now**

Is Another Influenza Pandemic on the Way?

In 1918 a pandemic (worldwide epidemic) of influenza swept across the world infecting 500 million people and killing nearly a fifth of them. The pandemic strain of influenza virus was particularly virulent, killing many otherwise healthy people. In 2005, scientists revived the 1918 killer strain of influenza (photo) from tissues of one of its victims in order to determine why this strain was so virulent.

Health officials today are concerned that new highly virulent influenza viruses may be formed from the exchange of genes between different strains. It is well known that influenza viruses that infect humans can also infect birds and swine. When an animal is infected by more than one strain, their genes can mix—a process called reassortment—to form viruses with new properties. A major worry today is that influenza virus strain H5N1, which has caused large influenza outbreaks in poultry and wild birds but does not transmit well to or between people, could reassort and trigger a new human influenza pandemic.

Scientists are focused on how H5N1 gains transmissibility in a new host. In a major study,[1] researchers were surprised to find that either one of two different genes transferred to the H5N1 virus from a recent human influenza strain (H1N1) allowed H5N1 to be spread by the airborne route between guinea pigs, a new phenomenon for this virus. Changes in the transmission pattern of influenza virus can thus occur very quickly and from a minimal amount of genetic exchange.

Influenza viruses are particularly dangerous pathogens because they are easily spread in infectious droplets and their genetics allow them to rapidly become transmissible in new hosts. Is a new human influenza pandemic in the making? Stay tuned.

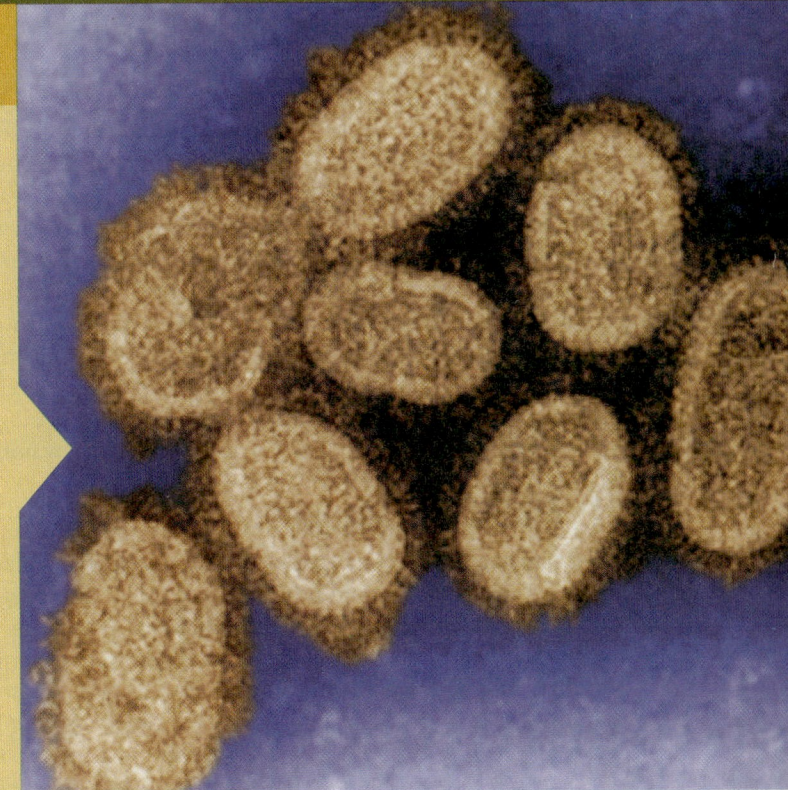

[1]Zhang, Y., et al. 2013. H5N1 hybrid viruses bearing 2009/H1N1 virus genes transmit in guinea pigs by respiratory droplet. *Science* 340: 1459–1463.

As many as several million species of microorganisms may exist in nature, but only a few hundred of them cause disease. In this and the next three chapters we focus on this vitally important subset of the microbial world. We investigate the biology of the pathogens as well as the diseases they cause, including disease diagnosis, treatment, and prevention.

Our infectious disease coverage is organized around each pathogen's *mode of transmission*. In this chapter we explore diseases transmitted from person to person, whether through the air, by direct contact, or through intimate contact. Using this approach, we will establish the ecological connections between biologically diverse pathogens. In Chapters 30 and 31 we focus on diseases transmitted by animal and arthropod vectors and diseases from common sources such as water or food, respectively. In Chapter 32 we examine fungal and parasitic infections, diseases caused by microbial *Eukarya*.

I • Airborne Bacterial Diseases

Worldwide, acute respiratory infections kill more than 4 million people a year, mainly in developing countries. Children and the elderly make up most of the fatalities, but in general, respiratory infections are the most common of all human diseases. Aerosols, such as those generated by a sneeze (**Figure 29.1**), as well as by coughing, talking, or breathing, are major vehicles for person-to-person transmission of respiratory diseases. Besides directly infecting a new host, infectious mucus from an aerosol can also contaminate objects, such as a door handle, and transmit infection well after the aerosol event. In these ways, respiratory diseases spread quickly, especially in congested areas, as airborne pathogens exploit a simple yet highly effective means of infecting new hosts.

29.1 Airborne Pathogens

Microorganisms found in air are derived from soil, water, plants, animals, people, surfaces, and other sources. Most microorganisms survive poorly in air. As a result, airborne pathogens are effectively transmitted between people only over short distances. Certain pathogens, however, survive drying well and can remain alive in dust or on fomites for long periods of time. For example, because of their thick, rigid cell walls, gram-positive bacteria (*Staphylococcus, Streptococcus*) are generally more resistant to drying than are gram-negative bacteria. Likewise, the waxy layer of *Mycobacterium* cell walls resists drying and promotes survival of pathogens such as *Mycobacterium tuberculosis*.

Large numbers of droplets can be expelled during a sneeze (Figure 29.1). Infectious droplets are about 10 μm in diameter and each droplet can contain one or more microbial cells or virus virions. The initial speed of the droplet movement is about 100 m/s (more than 325 km/h) in a violent sneeze and ranges from 15 to 50 m/s during coughing or shouting. The number of bacteria in a single sneeze varies from 10^4 to 10^6, and viral numbers can be much higher than this. Because of their small size, the droplets evaporate quickly in the air, leaving behind dried mucus in which the airborne pathogens remain embedded.

The human respiratory tract is divided into upper and lower regions and specific airborne pathogens tend to exploit one region or the other, or sometimes both (**Figure 29.2**). The speed at which air moves through the human respiratory tract varies, and in the lower respiratory tract the rate is quite slow. As air slows down, particles in it stop moving and settle. Large particles settle first and the smaller ones later; only particles smaller than about 3 μm travel as far as the bronchioles in the lower respiratory tract (Figure 29.2).

Upper respiratory infections such as the common cold are typically acute and non-life-threatening. By contrast, *lower* respiratory infections, such as bacterial or viral pneumonia, are often chronic and can be quite serious, especially in the elderly or an immune-compromised person. Also, although most common respiratory infections are not serious in an otherwise healthy host, they can set the stage for *secondary infections* that can be life-threatening. For example, death of an elderly person from pneumonia following a severe case of influenza is not an uncommon event.

Most human respiratory pathogens are transmitted from person to person because humans are the only reservoir for the pathogens. However, many airborne pathogens, such as *Streptococcus* spp., cold viruses, and influenza, can also be transmitted by direct contact (for example, by a handshake) or on fomites. Accurate and rapid diagnosis and treatment of respiratory infections are well developed in the clinical setting and if practiced effectively, can limit host damage. Many bacterial and viral pathogens transmitted by an airborne route can be controlled by immunization, and most respiratory bacterial pathogens respond readily

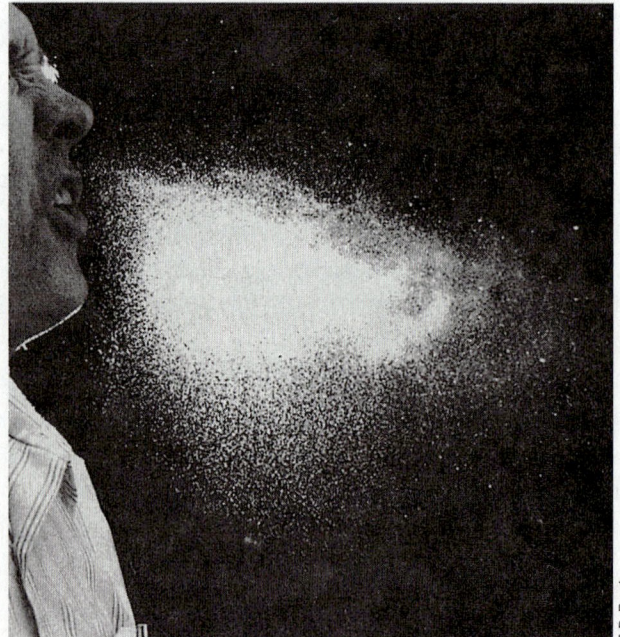

Figure 29.1 **High-speed photograph of an unstifled sneeze.** Effluent is emerging at over 325 km/s (200 miles/s).

T. D. Brock

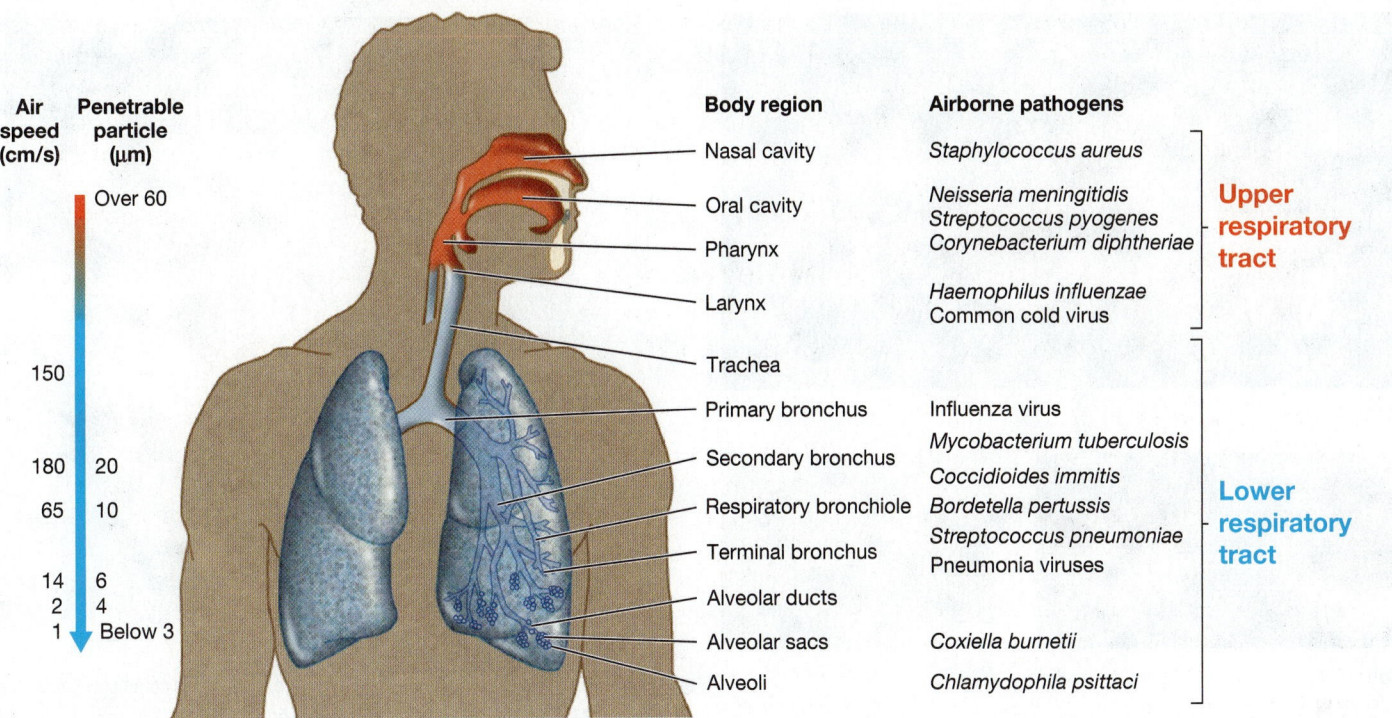

Air speed (cm/s)	Penetrable particle (μm)	Body region	Airborne pathogens	
	Over 60	Nasal cavity	*Staphylococcus aureus*	
		Oral cavity	*Neisseria meningitidis*	**Upper respiratory tract**
		Pharynx	*Streptococcus pyogenes* *Corynebacterium diphtheriae*	
		Larynx	*Haemophilus influenzae* Common cold virus	
150		Trachea		
		Primary bronchus	Influenza virus	
180	20	Secondary bronchus	*Mycobacterium tuberculosis* *Coccidioides immitis*	**Lower respiratory tract**
65	10	Respiratory bronchiole	*Bordetella pertussis*	
		Terminal bronchus	*Streptococcus pneumoniae* Pneumonia viruses	
14	6	Alveolar ducts		
2	4			
1	Below 3	Alveolar sacs	*Coxiella burnetii*	
		Alveoli	*Chlamydophila psittaci*	

Figure 29.2   The human respiratory system. The microorganisms listed typically initiate infections at the locations indicated.

to antibiotic therapy. Antiviral therapies, on the other hand, are rather limited, and recovery from viral infections is often due solely to the immune response.

MINIQUIZ

- Why can it be said that respiratory pathogens have exploited an effective means of transmission?
- Identify pathogens more commonly found in the upper respiratory tract. Identify pathogens more commonly found in the lower respiratory tract.

29.2 Streptococcal Diseases

The bacteria *Streptococcus pyogenes* (**Figure 29.3**) and *Streptococcus pneumoniae* are important human respiratory pathogens. Streptococci are nonsporulating, homofermentative but aerotolerant gram-positive cocci (♻ Section 15.6). Cells of *S. pyogenes* (Figure 29.3) typically grow in elongated chains, as do many other species of the genus. Pathogenic strains of *S. pneumoniae* typically grow in pairs or in short chains, and virulent strains produce an extensive polysaccharide capsule (see Figure 29.11). Virulent strains of *Streptococcus* can form vicious pus-forming wounds in humans and other warm-blooded animals (**Figure 29.4** and see Figure 29.10). But in addition, many other serious conditions whose symptoms are less dramatic than these are also associated with streptococcal infections.

Streptococcus pyogenes

Streptococcus pyogenes (Figure 29.3), also called *group A streptococci*, is frequently isolated from the upper respiratory tract

of healthy adults. Although numbers are typically low, if host defenses are weakened or a new, highly virulent strain is introduced, serious infections are possible.

S. pyogenes is the cause of *streptococcal pharyngitis*, better known as *strep throat* (**Figure 29.5**). Most clinical isolates of *S. pyogenes* produce an exotoxin (♻ Section 23.9) that lyses red blood cells in culture media, a condition called *β-hemolysis* (Figure 29.4b and see Figure 29.8). Streptococcal pharyngitis is characterized by a severe sore throat with enlarged tonsils and red spots on the soft palate (Figure 29.5); tender cervical lymph nodes; and a mild fever and feeling of general malaise. *S. pyogenes* can also

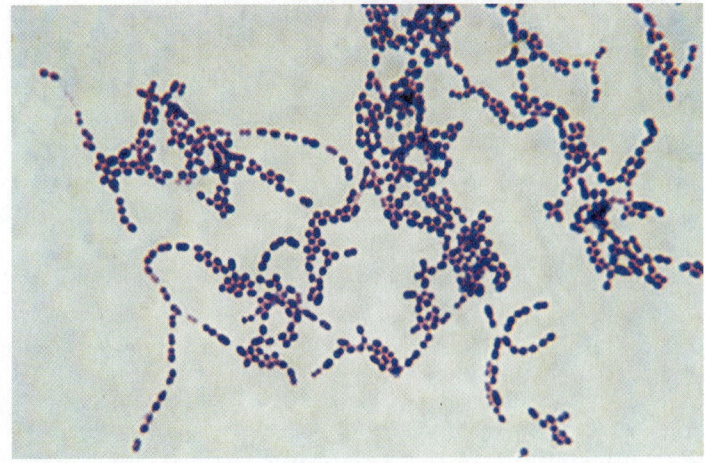

Figure 29.3   *Streptococcus pyogenes*. Cells of *Streptococcus pyogenes* grow in chains, and cells range in size from 0.6 to 1 μm in diameter.

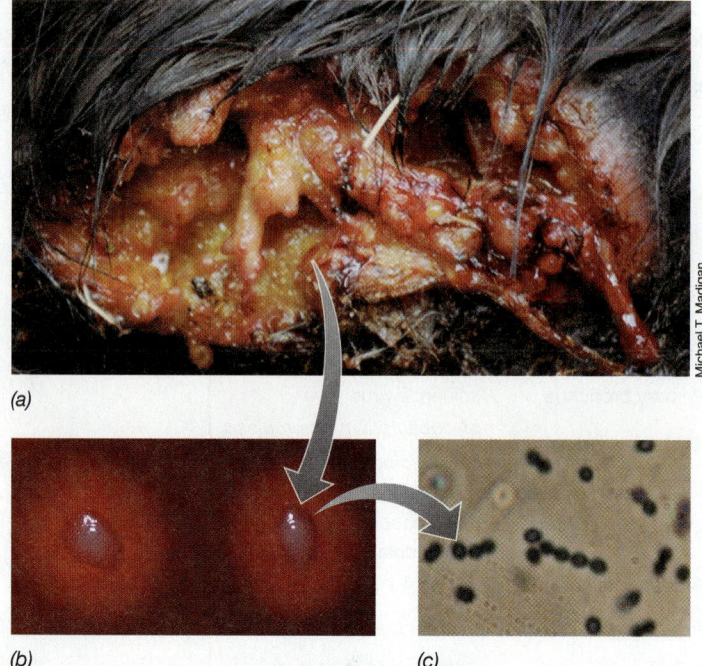

Figure 29.4 **Pus-forming wound in a horse from β-hemolytic streptococci.** *(a)* Pus and coagulated blood from a *Streptococcus equi* infection of the horse's salivary glands (the salivary glands have burst open from the infection). *(b)* Colonies of *S. equi* showing β-hemolysis on blood agar. *(c)* Phase-contrast photomicrograph of cells of *S. equi*. Cells are 1 μm in diameter.

cause infections of the middle ear (*otitis media*) and of the mammary glands (*mastitis*); infections of the superficial layers of the skin called *impetigo* (**Figure 29.6**); *erysipelas*, an acute streptococcal skin infection (**Figure 29.7**); and other conditions linked to the aftereffects of streptococcal infections.

About half of the clinical cases of severe sore throat are due to *S. pyogenes*; most others are due to viral infections. Because of this, an accurate and rapid diagnosis of a severe sore throat is important. If the sore throat is due to *S. pyogenes*, immediate treatment

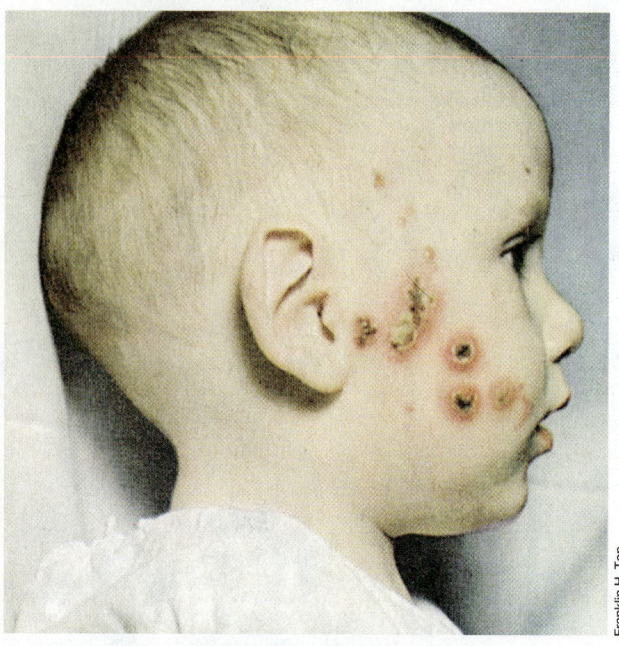

Figure 29.6 **Typical lesions of impetigo.** Impetigo is commonly caused by *Streptococcus pyogenes* or *Staphylococcus aureus*.

is important because untreated group A streptococcal infections can lead to serious secondary diseases such as scarlet fever, rheumatic fever, acute glomerulonephritis, and streptococcal toxic shock syndrome. On the other hand, if the sore throat is due to a virus, treatment with antibiotics will be useless and will only promote drug resistance on the part of the normal microflora.

Clinical tools for quickly diagnosing strep throat are widely available and in routine use in primary care clinics. These tools include, in particular, rapid antigen detection systems that contain antibodies specific for cell surface proteins of *S. pyogenes*

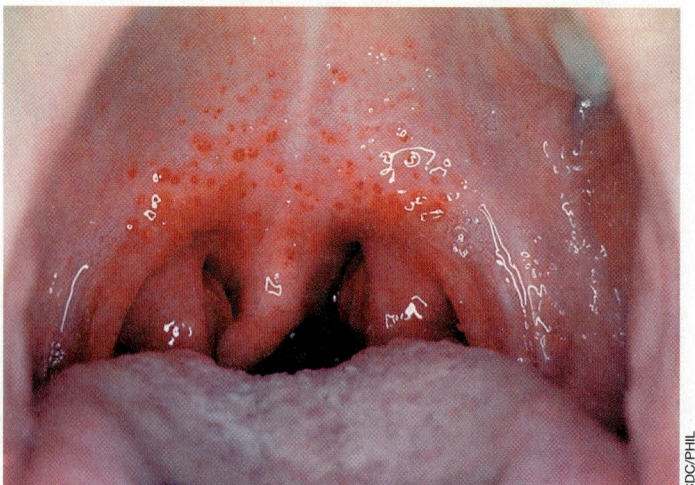

Figure 29.5 **A case of strep throat caused by *Streptococcus pyogenes*.** The back of the throat is inflamed and shows small red spots, typical of streptococcal pharyngitis.

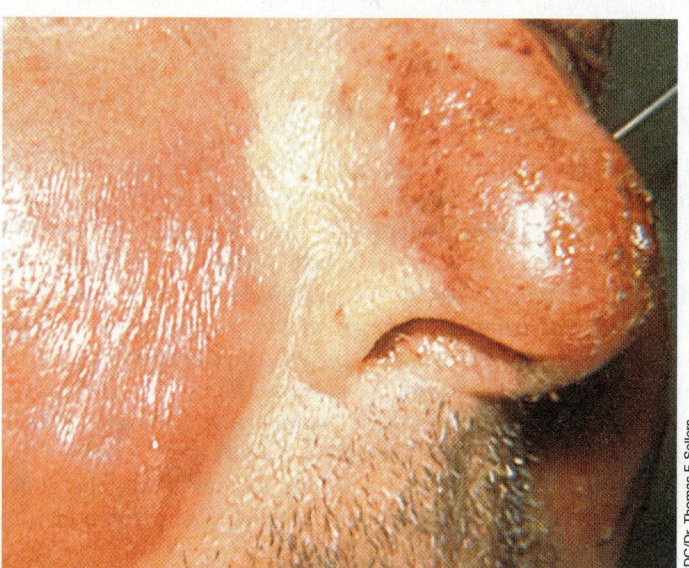

Figure 29.7 **Erysipelas.** Erysipelas is a *Streptococcus pyogenes* infection of the skin, shown here on the nose and cheeks, characterized by redness and distinct margins of infection. Other commonly infected body sites include the ears and the legs.

(⟳ Section 27.9). A more sensitive and accurate confirmation is possible by obtaining an actual culture of *S. pyogenes* from the throat or other suspected lesion on a blood agar plate (Figure 29.8). In contrast to rapid tests, however, results of a throat culture may take up to 48 h to process and such a delay in treatment can have adverse effects, as we consider now.

Scarlet and Rheumatic Fevers and Other Group A Strep Syndromes

Certain strains of group A streptococci carry a lysogenic bacteriophage that encodes streptococcal pyrogenic exotoxin A (SpeA), SpeB, SpeC, and SpeF. These exotoxins are responsible for most of the symptoms of *streptococcal toxic shock syndrome* and **scarlet fever** (Figure 29.9). Streptococcal pyrogenic exotoxins are superantigens that recruit large numbers of T cells to the infected tissues (⟳ Sections 23.9 and 23.9). Toxic shock results when the activated T cells secrete cytokines, which in turn activate large numbers of macrophages and neutrophils, causing severe inflammation and tissue destruction.

Scarlet fever, signaled by a severe sore throat, fever, and characteristic rash (Figure 29.9), is readily treatable with antibiotics or may be self-limiting. But treatment is always advisable since several undesirable conditions can emerge from a case of scarlet fever. Occasionally group A streptococcal infections cause fulminant (sudden and severe) invasive systemic infections, such as cellulitis, a skin infection in subcutaneous layers; or *necrotizing fasciitis*, a rapid and progressive disease resulting in extensive destruction of subcutaneous tissue, muscle, and fat (Figure 29.10). Necrotizing fasciitis is a clinical term for the condition caused by "flesh-eating bacteria." In these cases, SpeA, SpeB, SpeC, and SpeF exotoxins and the bacterial cell surface M protein function as superantigens; the associated inflammation results in extensive tissue destruction and can be fatal (Figure 29.10).

Untreated or insufficiently treated *S. pyogenes* infections may lead to other severe conditions 1 to 4 weeks after the onset of infection. For example, the immune response to the invading pathogen can produce antibodies that cross-react with host tissue

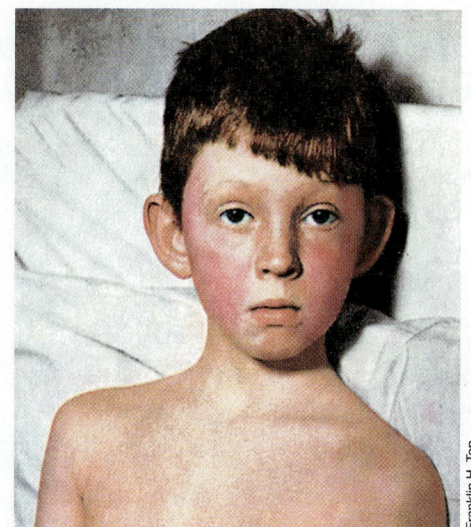

Figure 29.9 **Scarlet fever.** The typical rash of scarlet fever results from the activity of the pyrogenic exotoxins produced by *Streptococcus pyogenes*.

antigens of the heart, joints, and kidneys, resulting in damage to these tissues. The most serious of these syndromes is **rheumatic fever** caused by rheumatogenic strains of *S. pyogenes*. These strains contain cell surface antigens that are similar in structure to heart valve and joint proteins. Thus rheumatic fever is, in effect, an *autoimmune disease* (⟳ Section 24.8), because antibodies directed against streptococcal antigens cross-react with heart valve and joint antigens, causing host inflammation and tissue destruction. Damage to host tissues may be permanent, and is often exacerbated by subsequent streptococcal infections that lead to recurring bouts of rheumatic fever. Another streptococcal syndrome is *acute poststreptococcal glomerulonephritis*, a painful kidney disease. This "immune complex" disease develops transiently due to streptococcal antigen–antibody complexes in the blood that lodge in the glomeruli (filtration membranes of the kidney) and cause inflammation, a condition called *nephritis*.

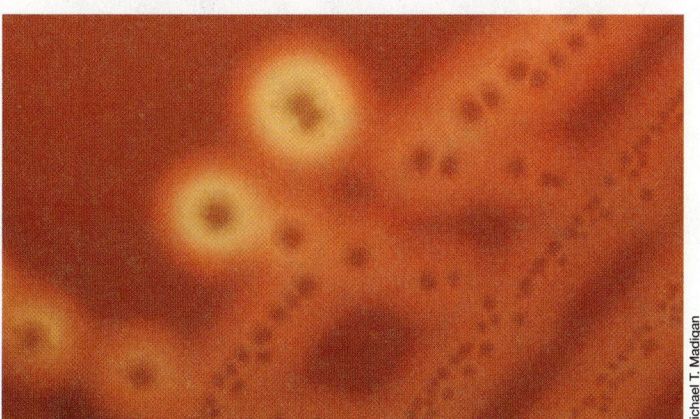

Figure 29.8 **β-Hemolysis**. The ability of a bacterium to lyse red blood cells and form a clear zone around a colony on a blood agar plate indicates secretion of the protein β-hemolysin. See also Figures 23.18*a* and 29.4*b*.

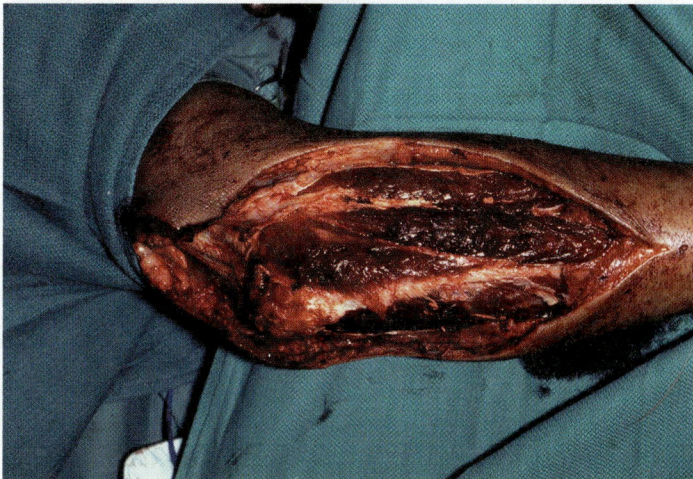

Figure 29.10 **Necrotizing fasciitis (flesh-eating bacteria).** Soft tissue infection of a human forearm by a group A *Streptococcus pyogenes*. Flesh on the arm has split open from the infection to reveal muscle fascia and infected internal tissues.

UNIT 6

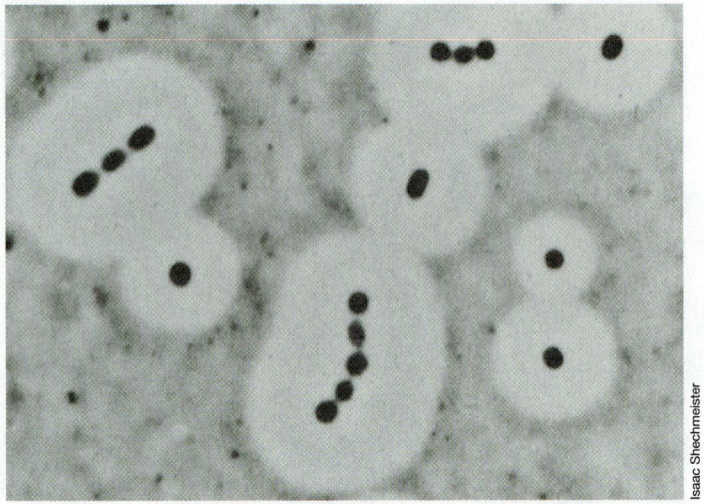

Figure 29.11 *Streptococcus pneumoniae.* India ink negative stain of cells of *Streptococcus pneumoniae.* An extensive capsule surrounds the cells, which are 1.0–1.2 μm in diameter.

Streptococcus pneumoniae

A second major human streptococcal pathogen is *Streptococcus pneumoniae* (Figure 29.11), a species that can cause invasive lung infections, typically as secondary infections to other respiratory disorders. Encapsulated strains of *S. pneumoniae* (Figure 29.11) are particularly pathogenic because they are very invasive. Cells invade the lower respiratory tract where the capsule enables the cells to resist phagocytosis yet generate a strong host inflammatory response. Reduced lung function, called *pneumonia*, results from the accumulation of recruited phagocytic cells and fluid. Cells of *S. pneumoniae* can then spread from the focus of infection as a bacteremia, sometimes infecting the bones, middle ear, and heart valves (endocarditis). *S. pneumoniae* infection is often the cause of death in elderly persons who die from "respiratory failure."

Unlike the case with *S. pyogenes*, effective vaccines are available for preventing infection by the most common strains of *S. pneumoniae*. A vaccine for adults consists of a mixture of 23 capsular polysaccharides (Figure 29.11) from the most prevalent pathogenic strains. The vaccine is recommended for those over age 60, health-care providers, individuals with compromised immunity, and any other high-risk population. *S. pneumoniae* infections typically respond quickly to penicillin therapy, but up to 30% of pathogenic isolates now exhibit resistance to this drug. Resistance to the antibiotics erythromycin and cefotaxime is also found in some strains but thus far, all strains have been found sensitive to vancomycin, an antibiotic held in reserve for treating pneumonia and several other bacterial diseases where antibiotic resistance is widespread.

MINIQUIZ ---

- How does *Streptococcus pyogenes* infection cause rheumatic fever?
- What is the primary virulence factor for *Streptococcus pneumoniae*?

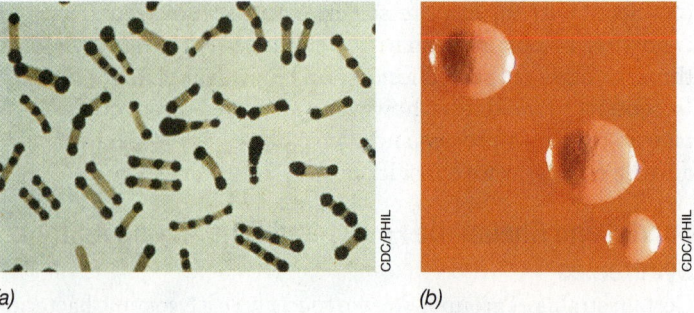

(a)   *(b)*

Figure 29.12 *Corynebacterium* **and diphtheria.** *(a)* Cells of *Corynebacterium diphtheriae* showing typical club-shaped appearance. The gram-positive cells are 0.5–1.0 μm in diameter and may be several micrometers in length. *(b)* Colonies of *C. diphtheriae* grown on a selective medium of blood agar plus tellurite.

29.3 Diphtheria and Pertussis

Diphtheria is a severe respiratory disease that typically infects young children. Diphtheria is caused by *Corynebacterium diphtheriae*, a gram-positive, nonmotile, and aerobic club-shaped bacterium that forms small, smooth colonies on blood agar plates (Figure 29.12). **Pertussis**, also known as **whooping cough**, is a serious respiratory disease caused by infection with *Bordetella pertussis*, a small, gram-negative, aerobic coccobacillus (see Figure 29.14*a*). Pertussis mostly affects children but can cause serious respiratory disease in adults as well. Both diphtheria and pertussis can be prevented by vaccination and cured with antibiotics.

Diphtheria

Cells of *C. diphtheriae* (Figure 29.12*a*) enter the host from airborne droplets, infecting the tissues of the throat and tonsils. Throat tissues respond to *C. diphtheriae* infection by forming a characteristic lesion called a *pseudomembrane* (Figure 29.13), which consists of damaged host cells and cells of *C. diphtheriae*. Pathogenic strains of *C. diphtheriae* carry a lysogenic bacteriophage whose genome

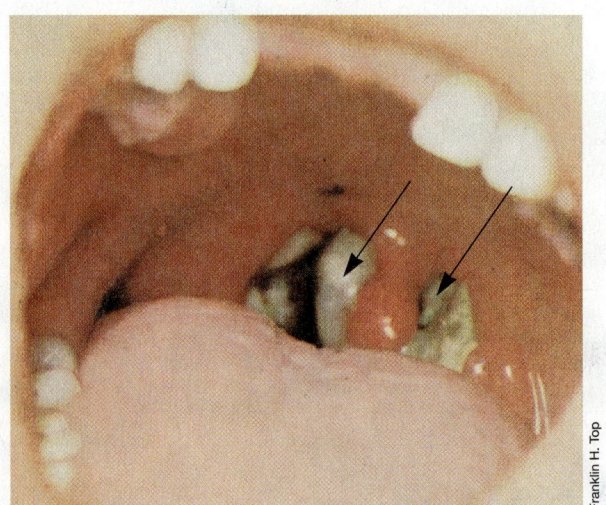

Figure 29.13 **The pseudomembrane in diphtheria.** The pseudomembrane (arrows) in an active case of diphtheria restricts airflow and swallowing and is associated with a severe sore throat.

encodes a powerful exotoxin called *diphtheria toxin*. This toxin inhibits protein synthesis in the host, leading to cell death (⮌ Figure 23.20). Death from diphtheria is typically due to a combination of partial suffocation by the pseudomembrane and tissue destruction by diphtheria exotoxin. *C. diphtheriae* isolated from the throat is diagnostic for diphtheria. Nasal or throat swabs are used to inoculate blood agar containing tellurite (Figure 29.12*b*) or Loeffler's medium, a selective medium that inhibits the growth of most other respiratory pathogens.

Prevention of diphtheria is accomplished with a highly effective toxoid vaccine, part of the *DTaP* (*d*iphtheria toxoid, *t*etanus toxoid, and *a*cellular *p*ertussis) *vaccine* (⮌ Section 24.6). Diphtheria is all but absent from developed countries where this vaccine is widely used. Penicillin, erythromycin, and gentamicin are generally effective treatments for diphtheria, but in life-threatening cases, diphtheria antitoxin (an antiserum to diphtheria toxoid produced in horses) may be administered in addition to antibiotic therapy.

Pertussis

Pertussis (whooping cough) is an acute, highly infectious respiratory disease. Infants less than 6 months old, who are too young to be effectively vaccinated, have the highest incidence of disease and also have the most severe symptoms. Cells of *B. pertussis* (**Figure 29.14a**) attach to host cells of the upper respiratory tract and excrete *pertussis exotoxin*. This potent toxin induces synthesis of cyclic adenosine monophosphate (cyclic AMP, ⮌ Figure 7.14), which is at least partially responsible for the events that lead to host tissue damage. *B. pertussis* also produces an endotoxin (⮌ Section 23.10), which may induce some of the symptoms of whooping cough. Clinically, whooping cough is characterized by a recurrent, violent cough that can last up to 6 weeks. The spasmodic coughing gives the disease its name; a whooping sound results from the patient inhaling deep breaths to obtain sufficient air.

Worldwide, up to 50 million cases and over 250,000 deaths occur each year from pertussis, most in developing countries. *B. pertussis* is endemic worldwide and pertussis remains a problem, even in developed countries, usually due to inadequate immunization.

In the United States there has been a gradual upward trend of pertussis since the 1980s, with spikes in reported cases in 2005, 2010, and 2012 (Figure 29.14*b*); many of these have been in young adults under age 20. In the United States, pertussis causes fewer than 20 deaths per year. But pertussis is a classic endemic disease; incidence rises cyclically as populations become susceptible and are exposed to the pathogen. A combination of lax vaccination protocols and the fact that pertussis is a much more common disease than diphtheria have probably fueled the overall higher incidence of pertussis in recent years.

Whooping cough can be treated with ampicillin, tetracycline, or erythromycin, although antibiotics alone do not seem to effect a complete cure, as patients continue to show symptoms and remain infectious for up to 2 weeks after beginning antibiotic therapy. This indicates that the immune response may be as important as antibiotics in ridding the pathogen from the body.

MINIQUIZ -
- Contrast the disease symptoms of diphtheria and pertussis.
- What measures can be taken to decrease the current incidence of pertussis in a population?

29.4 Tuberculosis and Leprosy

The famous pioneering microbiologist Robert Koch, the founder of the field of medical microbiology, isolated and described the causative agent of tuberculosis, *Mycobacterium tuberculosis*, in 1882 (⮌ Section 1.8). A related species, *Mycobacterium leprae*, causes leprosy (Hansen's disease). Mycobacteria are grampositive bacteria and share the property of being *acid-fast* due to the waxy mycolic acid constituent of their cell walls (⮌ Section 15.11). Mycolic acid allows these organisms to retain the red dye carbol-fuchsin after washing a mycobacterial smear on a slide in 3% hydrochloric acid in alcohol. Colonies of *M. tuberculosis* grow slowly on plates and have a characteristically wrinkled morphology (**Figure 29.15**).

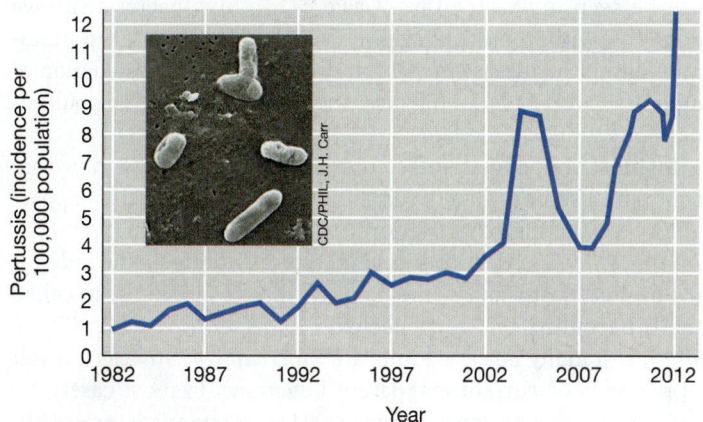

(a) **Gram-stained cells of *Bordetella pertussis*, the cause of whooping cough**

(b) **Pertussis in the United States with inset showing SEM of cells of *Bordetella* sp.**

Figure 29.14 ***Bordetella* and pertussis.** Cells of *B. pertussis* are typically coccobacilli 0.2–0.5 um in diameter and about 1 um long. Pertussis has been on the rise, with more than 41,000 cases reported in 2012, a 60-year high. Data are from the CDC.

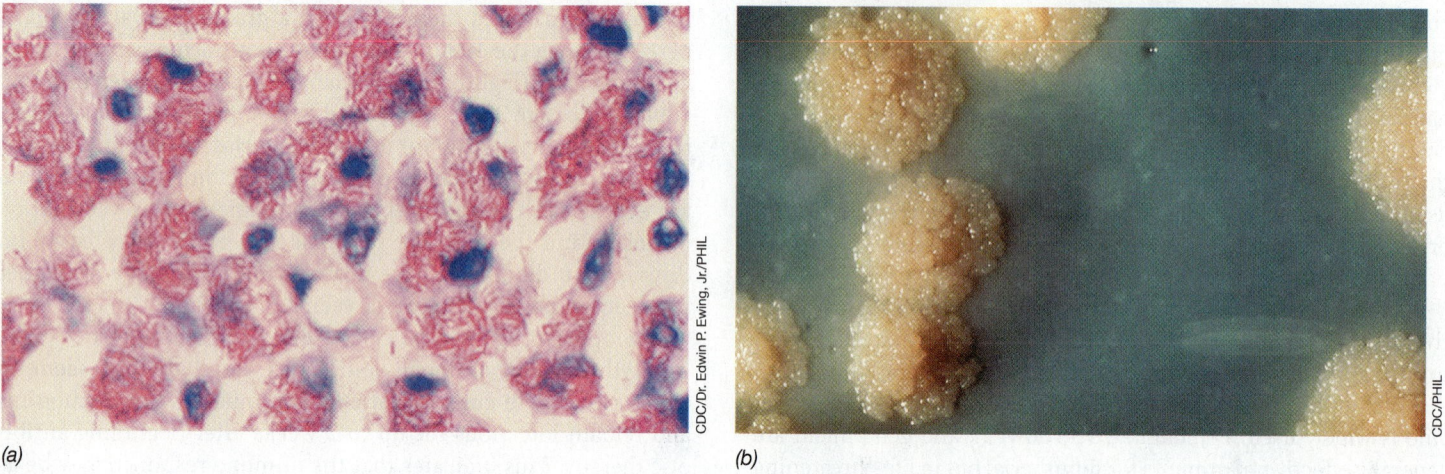

(a)   (b)

Figure 29.15 Mycobacteria. *(a)* Acid-fast stained lymph node biopsy from a patient with HIV/AIDS shows cells of *Mycobacterium avium*, a relative of *M. tuberculosis*. Multiple bacilli, stained red with carbol-fuchsin and treated with 3% hydrochloric acid, are evident inside each human cell. The individual rods are about 0.4 μm in diameter and up to 4 μm in length. *(b)* Colonies of *Mycobacterium tuberculosis*. The rough wrinkled surface is typical of mycobacterial colonies.

Tuberculosis

Tuberculosis (TB) is easily transmitted by the respiratory route, and at one time it was the most important infectious disease of humans. Worldwide, TB still accounts for nearly 1.5 million deaths per year. About one-third of the world's population has been infected with *M. tuberculosis*, though most do not show active disease because cell-mediated immunity (⊃⊂ Sections 24.3, 24.8, and 25.1) plays a critical role in the prevention of active disease after infection.

Tuberculosis can take several forms. TB can be a *primary* infection (initial infection) or *postprimary* infection (reinfection). Primary infection typically results from inhalation of droplets containing *M. tuberculosis*, after which the bacteria settle in the lungs and grow. The host mounts an immune response to *M. tuberculosis*, resulting in the formation of aggregates of activated macrophages, called *tubercles*. Bacteria are found in the sputum of individuals with active disease, and areas of destroyed tissue can be seen in chest X-rays (**Figure 29.16**). Mycobacteria survive and grow within macrophages in the tubercles, forming granulomas, and if the disease is not controlled, extensive destruction of lung tissue can occur. If the disease reaches this stage, the pulmonary infection is often fatal.

In most individuals infected with *M. tuberculosis*, however, acute disease does not occur; instead, the infection is inapparent. Nevertheless, the infection hypersensitizes the individual to *M. tuberculosis* or its products and typically protects the individual against postprimary infections. A diagnostic skin test, called the **tuberculin test**, can detect this hypersensitivity (⊃⊂ Figure 24.6), and many healthy adults are *tuberculin-positive* as a result of previous or current inapparent infections. In most cases, the cell-mediated immune response to *M. tuberculosis* is protective and lifelong. However, some tuberculin-positive patients develop postprimary tuberculosis through reinfection from bacteria that have remained dormant in lung macrophages for years. Because

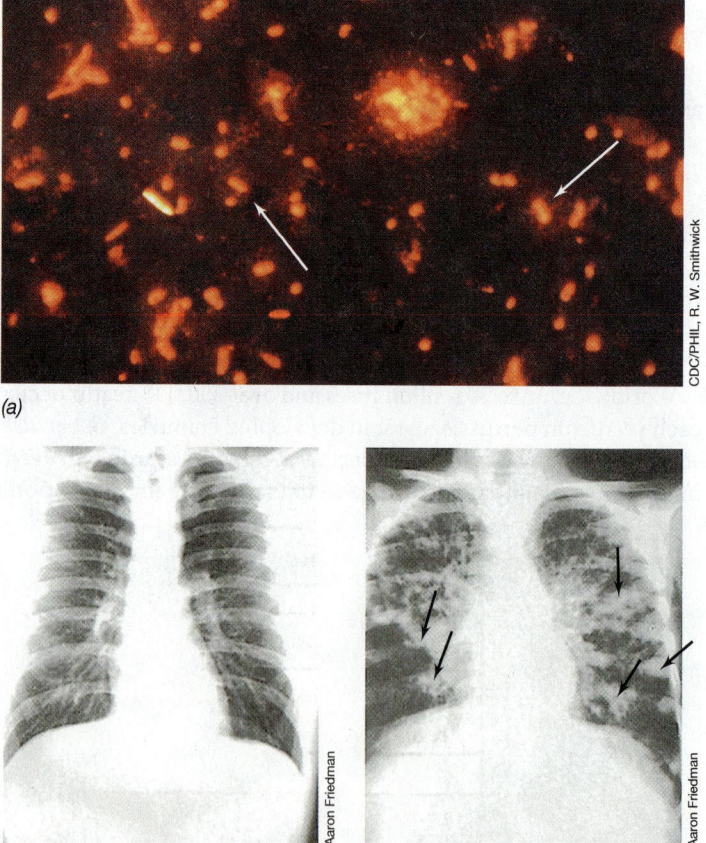

(a)

(b)   (c)

Figure 29.16 Tuberculosis symptoms. *(a)* Sputum sample from a patient with tuberculosis stained by the Smithwick acridine orange method. Cells of *M. tuberculosis* are the yellow-orange rod-shaped structures (arrows). *(b)* Normal chest X-ray. The faint white lines are arteries and other blood vessels. *(c)* Chest X-ray of an advanced case of pulmonary tuberculosis; white patches (arrows) indicate areas of tubercles that contain viable cells of *Mycobacterium tuberculosis*.

of this, individuals who have a positive tuberculin test are typically treated with anti-tuberculosis drugs for extended periods to ensure that all mycobacteria have been killed.

Antimicrobial therapy of TB has been a major means of controlling the disease. Streptomycin was the first effective anti-tuberculosis antibiotic, but the real revolution in tuberculosis treatment came with the discovery of isonicotinic acid hydrazide, called *isoniazid* (INH) (Figure 29.17). This drug is highly effective and readily absorbed when given orally. Isoniazid is a growth factor analog (Section 27.11) of the structurally related molecule nicotinamide; in mycobacteria the drug inhibits mycolic acid synthesis and this compromises cell wall integrity. Following treatment with isoniazid, mycobacteria lose their acid-fast properties, in keeping with the role of mycolic acid in this staining property.

Treatment of tuberculin-positive individuals is typically achieved with daily doses of isoniazid and the antibiotic rifampin for 2 months, followed by biweekly doses for a total of 9 months. This treatment eradicates pockets of *M. tuberculosis* cells and prevents emergence of antibiotic-resistant derivatives. Multiple drug therapy reduces the possibility that strains having resistance to more than one drug will emerge. Resistance of *M. tuberculosis* to isoniazid and other drugs, however, is increasing, especially in HIV/AIDS patients, in whom TB is a common infection (see Figure 29.45g). Treatment of these strains, called *multi-drug-resistant tuberculosis strains*, requires the use of second-line tuberculosis drugs that are generally more toxic, less effective, and more costly than rifampin and isoniazid.

Leprosy

Mycobacterium leprae, a relative of *M. tuberculosis*, causes the disease *leprosy*, more formally known as *Hansen's disease*. The most serious form of Hansen's disease is *lepromatous* leprosy, characterized by folded, bulblike lesions on the body, especially on cooler parts of the body such as the face and extremities (Figure 29.18). The lesions are due to the growth of *M. leprae* cells in the skin and contain large numbers of bacterial cells. Like cells of other mycobacteria (Figure 29.15a), cells of *M. leprae* from the lesions stain deep red with carbol-fuchsin in the acid-fast staining procedure, providing a definitive demonstration of active infection.

In severe untreated cases of leprosy, the disfiguring lesions lead to destruction of peripheral nerves; muscles then atrophy and motor function is impaired. The loss of sensation in the extremities leads to inapparent injuries, such as burns and cuts. Loss of

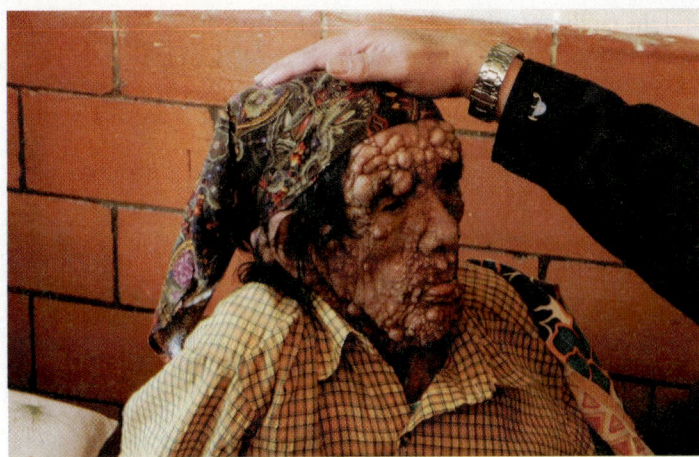

Figure 29.18 Lepromatous leprosy lesions on the skin. Lepromatous leprosy is caused by infection with *Mycobacterium leprae*. The lesions can contain up to 10^9 bacterial cells per gram of tissue, indicating an active uncontrolled infection with a poor prognosis.

bone calcium leads to a slow shrinking of the digits and their transition to clawlike forms in late-stage leprosy. Pathogenicity in the disease is due to a combination of delayed-type hypersensitivity (Section 24.8) and the highly invasive activities of *M. leprae*, which can grow within macrophages and lead to the characteristic lesions (Figure 29.18). Leprosy is transmitted by direct contact as well as by an airborne route, but is not as highly contagious as TB. Historically, leprosy has been associated with poverty, malnutrition, and poor sanitation and hygiene. Among other things, these factors undoubtedly affect an individual's ability to resist infection.

Many Hansen's disease patients exhibit less-pronounced lesions from which *M. leprae* cells cannot be obtained; these individuals have the *tuberculoid* form of the disease. Tuberculoid leprosy is characterized by a vigorous immune response and a good prognosis for spontaneous recovery. Hansen's disease of either form, and the continuum of intermediate forms, is treated using a multiple drug therapy protocol, which includes some combination of extended therapy of up to 1 year with *dapsone* (4,4'-sulfonylbis-benzeneamine, an inhibitor of folic acid synthesis), *rifampin*, a bacterial RNA polymerase inhibitor, and *clofazimine*, a drug that targets bacterial respiration and ion transport.

Nearly 250,000 new cases of leprosy were reported in 2009, with most cases occurring in Africa, the Indian subcontinent, and Brazil. In the United States only about 200 cases occur per year, mainly in immigrants. Until recently, a leprosy diagnosis relied on the identification of *M. leprae* cells from lesions. However, a quick, inexpensive, and specific blood test is now available that should greatly assist in identifying early stage leprosy, the most treatable form.

In addition to *M. tuberculosis* and *M. leprae*, several other mycobacteria are human pathogens. These include in particular *M. bovis*, a close relative of *M. tuberculosis* and a common pathogen of dairy cattle. *M. bovis* can initiate classic symptoms of TB in humans; however, a combination of the pasteurization of milk

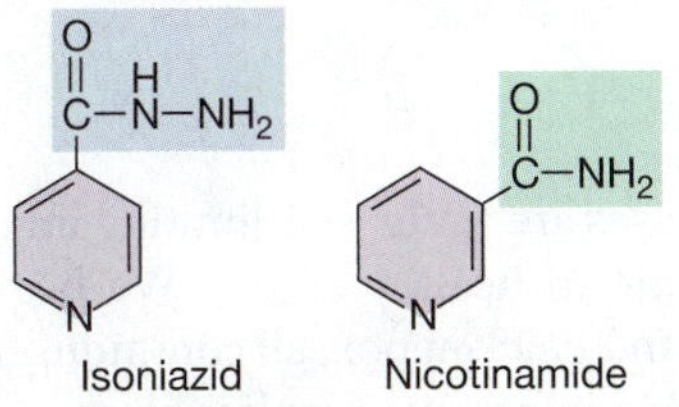

Figure 29.17 Structure of isoniazid (isonicotinic acid hydrazide). Isoniazid is an effective chemotherapeutic agent for tuberculosis. Note the structural similarity to nicotinamide.

and the culling of infected cattle has greatly reduced the incidence of bovine-to-human transmission of this form of TB.

29.5 Meningitis and Meningococcemia

Meningitis is an inflammation of the meninges, the membranes that are the protective covering of the central nervous system, that is, the spinal cord and brain. Several different microorganisms, including certain viruses, bacteria, fungi, and protists, can cause meningitis. Here we focus on the severe bacterial form of the disease called *infectious meningitis*, caused by the bacterium *Neisseria meningitidis*.

Pathogen and Disease Syndromes

Neisseria meningitidis, often called the *meningococcus*, is a gramnegative and obligately aerobic coccus about 0.6–1.0 μm in diameter (**Figure 29.19***a*); it is a relative of the bacterium that causes gonorrhea, *Neisseria gonorrhoeae*. The bacterium is transmitted to a new host, usually via the airborne route from an infected individual, and attaches to the cells of the nasopharynx. Once there, the organism quickly gains access to the bloodstream, causing widespread dissemination (bacteremia) and upper respiratory tract symptoms. Meningitis is characterized by the sudden onset of a headache accompanied by vomiting and a stiff neck, and can progress to coma and death in less than a day. Instead of or in addition to full-blown meningitis, *N. meningitidis* bacteremia sometimes leads to fulminant **meningococcemia**, a condition characterized by intravascular coagulation and tissue destruction (gangrene, Figure 29.19*b*), shock, and death in over 10% of cases.

Meningococcal meningitis often occurs in epidemics, usually in populations living in close proximity such as in military barracks or college dormitories. Anyone can get meningococcal disease, but the incidence is much higher in infants, school-age children, and young adults. Up to 30% of people carry *N. meningitidis* in their nasopharynx with no apparent harmful effects, and the trigger for conversion from the asymptomatic carrier state to the disease state is unknown.

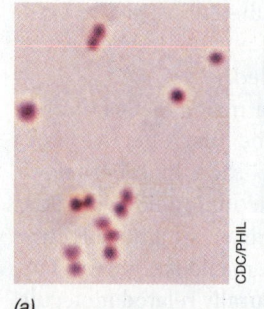
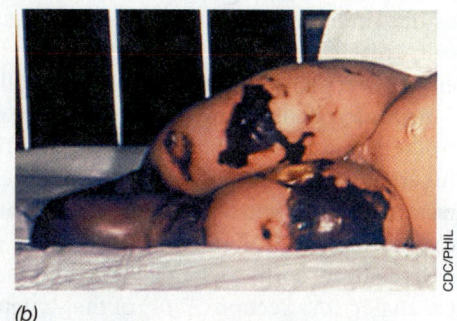

(a)　　　　　　　　　　*(b)*

Figure 29.19 *Neisseria meningitidis.* The organism causes meningitis and meningococcemia. *(a)* Gram stain of cells of *N. meningitidis*; cocci are about 0.6–1.0 μm in diameter. *(b)* Four-month old infant with gangrene on legs from meningococcemia.

Diagnosis, Treatment, and Vaccines

Meningococcal meningitis is definitively diagnosed from cultures of *N. meningitidis* isolated from nasopharyngeal swabs, blood, or cerebrospinal fluid. Thayer–Martin medium (⮌ Figure 27.5), a selective medium for the growth of pathogenic *Neisseria*, including both *N. meningitidis* and *N. gonorrhoeae*, is used to isolate *N. meningitidis*, and colonies containing gram-negative diplococci (Figure 29.19*a*) are further tested. However, due to the rapid onset of life-threatening symptoms in infectious meningitis, preliminary diagnosis is often based on clinical symptoms and treatment is started before culture tests confirm infection with *N. meningitidis*. Treatment is typically with penicillin, and intravenous application is often needed to speed antibiotic infusion.

Naturally occurring antibodies acquired by subclinical infections with *N. meningitidis* are effective for preventing infectious meningitis in most adults. Vaccines consisting of purified polysaccharides or polysaccharides from the most prevalent pathogenic strains are available to immunize certain susceptible populations such as military recruits and students living in dormitories, especially if an outbreak has already occurred. In addition, the antibiotic rifampin is often used to eradicate the carrier state and prevent meningococcal disease in close contacts of infected individuals.

II · Airborne Viral Diseases

29.6 Viruses and Respiratory Infections

The most prevalent and difficult to treat of all human infectious diseases are those caused by viruses. This is because viruses can often remain infectious for long periods in dried mucus (Figure 29.1) or on fomites, and because viruses require host cells for replication. Hence, killing the virus often means killing the cell as well.

Most viral diseases are acute, self-limiting infections, but some can be problematic in healthy adults. We begin with measles, rubella, mumps, and chicken pox, all common, endemic viral diseases transmitted in infectious droplets by an airborne route.

Measles and Rubella

Measles (*rubeola* or *7-day measles*) affects susceptible children as an acute, highly infectious, often epidemic disease. The measles

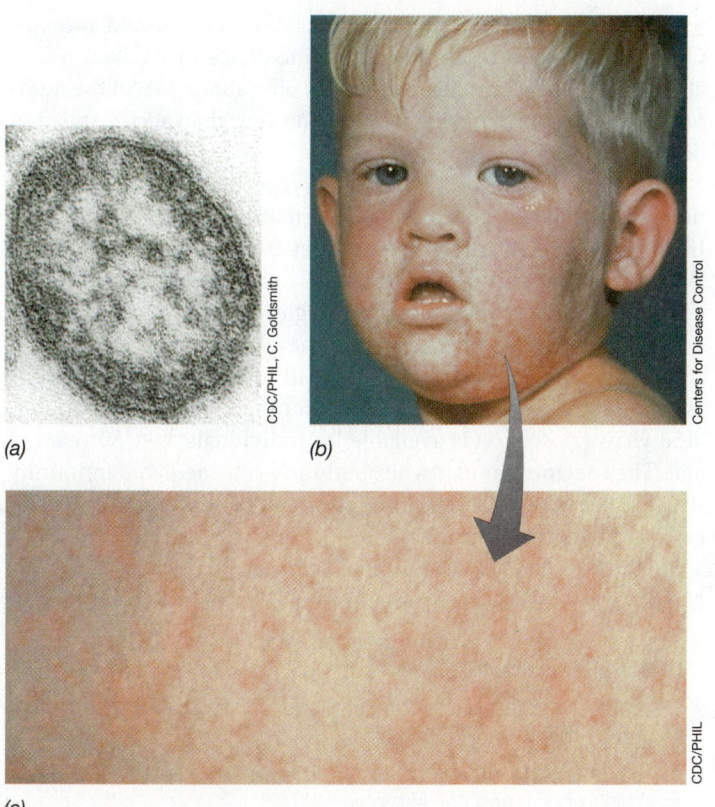

(a)

(b)

(c)

Figure 29.20 **Measles in children.** *(a)* Transmission electron micrograph of a measles virus virion; a virion is about 150 nm in diameter. *(b, c)* Measles rash. The light pink rash starts on the head and neck, and can spread to the chest, trunk, and limbs. Discrete papules coalesce into blotches as the rash progresses for several days.

virus (**Figure 29.20***a*) is a *paramyxovirus*, a single-stranded, minus-sense RNA virus (⟳ Section 9.9) that enters the nose and throat by airborne transmission, quickly leading to a systemic viremia. Symptoms start with nasal discharge and redness of the eyes. As the disease progresses, fever and cough appear and rapidly intensify, followed by a characteristic rash (Figure 29.20*b, c*).

Symptoms of measles generally persist for 7–10 days, and no drugs are available that will eliminate symptoms. However, the measles virus generates a strong immune response. Circulating antibodies to measles virus are measurable within 5 days of infection; these serum antibodies along with T-cytotoxic lymphocytes combine to eliminate the virus from the host. Possible postinfection complications include middle ear infection, pneumonia, and, in rare cases, measles encephalomyelitis.

Although once a common childhood illness, measles is limited to rare isolated outbreaks in the United States because of widespread immunization programs begun in the 1960s. Those outbreaks that have occurred have been in populations that were either not immunized or inadequately immunized. Worldwide, measles remains endemic, however, and still causes over 400,000 annual deaths, mostly in children. Active immunity to measles is conferred with an attenuated virus preparation as part of the MMR (measles, mumps, rubella) vaccine (⟳ Figure 24.15). Because the disease is highly infectious, all public school systems in the United States require proof of measles immunization

before a child can enroll. A childhood case of measles generally confers lifelong immunity.

Rubella (sometimes called *German measles* or *3-day measles*) is caused by a single-stranded, positive-sense RNA virus (⟳ Section 9.8). Symptoms of rubella resemble those of measles (**Figure 29.21**) but are often restricted to just the upper torso. Rubella is less contagious than measles, and thus a significant proportion of the population has never been infected. However, during the first three months of pregnancy, rubella virus can infect the fetus by placental transmission and cause serious fetal abnormalities including stillbirth, deafness, heart and eye defects, and brain damage, events called *congenital rubella syndrome*. Thus, women should not be immunized with the rubella vaccine or contract a rubella infection during pregnancy. Also for this reason, routine childhood immunization against rubella should be practiced. An attenuated rubella virus is administered as part of the MMR vaccine. The low incidence of rubella since 2001, coupled with the high degree of protection by the vaccine and the relatively low infectivity of the virus, combine to make rubella very rare in the United States. An active rubella vaccination program worldwide is also decreasing case numbers significantly, as total reported active cases in 2009 were fewer than 125,000 and cases of congenital rubella syndrome were 165.

Mumps

Mumps, like measles, is caused by a paramyxovirus and is also highly infectious by the airborne route. Mumps is spread by airborne droplets, and the disease is characterized by inflammation of the salivary glands, typically the parotid gland, the largest of the salivary glands, leading to swelling of the jaws and neck (**Figure 29.22**). The virus spreads through the bloodstream and may infect other organs, including the testes and pancreas, and may cause encephalitis in rare severe cases. As for measles, the immune response rather than drug treatment is what cures a case of mumps. The host immune response produces antibodies to mumps virus surface proteins, and this generally leads to a quick recovery and lifelong immunity to reinfection.

An attenuated mumps vaccine (part of the MMR) is highly effective in preventing disease. Hence, the prevalence of mumps in developed countries is low, with disease generally restricted to individuals who did not receive the vaccine. However, an outbreak of mumps in the midwestern United States in 2006 involved more than 5000 cases, which is significantly up from the typical

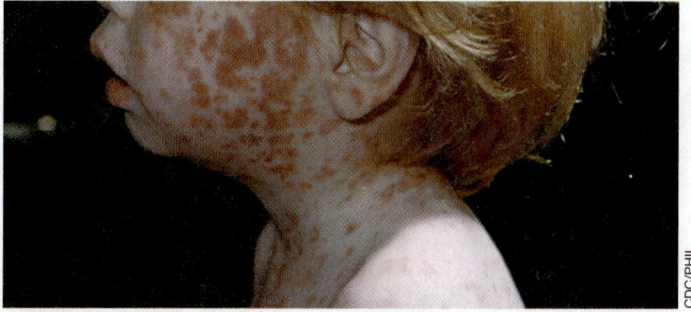

Figure 29.21 **Rubella.** The rash of rubella (German measles) on the face of a young child.

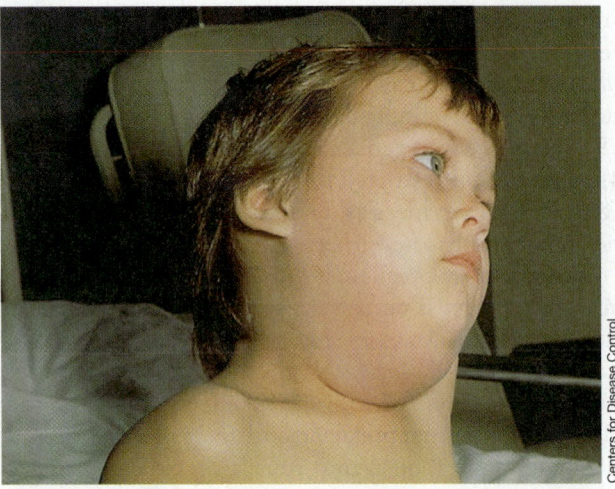

Centers for Disease Control

Figure 29.22 **Mumps.** Glandular swelling characterizes infection with the mumps virus. Mumps symptoms typically last about one week and a person is infectious both before and during the symptomatic stages.

number of fewer than 300 cases per year since 2001. The outbreak affected mainly young adults (18–34). As a result, recommendations for immunizations were revised to target school-age children, healthcare workers, and adults who had not previously had mumps.

Chicken Pox and Shingles

Chicken pox (*varicella*) is a common childhood disease caused by the varicella-zoster virus (VZV), a double-stranded DNA herpesvirus (🔗 Section 9.7). VZV is a mild but highly contagious disease and is transmitted by infectious droplets, especially when susceptible individuals are in close contact. In schoolchildren, for example, close confinement during the winter months leads to the spread of VZV through airborne droplets from infected classmates and through direct contact with chicken pox blisters of other children or contaminated fomites. The virus enters the respiratory tract, multiplies, and is quickly disseminated via the bloodstream, resulting in a systemic papular rash (**Figure 29.23**) that heals quickly without scarring. An attenuated chicken pox virus vaccine (Varivax) is used in the United States but not as

widely as the MMR vaccine for measles, rubella, and mumps. Consequently, the reported annual incidence of chicken pox is about 40,000 cases per year, but this is only about 25% of the number of cases reported prior to 1995, the year the vaccine was first licensed.

VZV establishes a lifelong latent (permanent) infection in nerve cells. The virus can remain dormant there indefinitely, but in some individuals the virus migrates from this reservoir to the skin surface, often years or decades later, causing a painful skin eruption called *shingles* (*zoster*). Shingles most commonly strikes immunosuppressed individuals or the elderly, and often causes severe blisters and a rash on the head, neck, or upper torso. A fairly effective shingles vaccine containing concentrated attenuated virus (Zostavax) is available for individuals over 50 years of age. The vaccine stimulates antibody and cell-mediated immunity to VZV, keeping VZV from migrating out of nerve ganglia to skin cells and triggering shingles.

MINIQUIZ -

- How do the genomes of the measles virus and the German measles virus differ?
- Describe the potential serious outcomes of infection by measles, mumps, rubella, and VZV viruses.
- Identify the effects of immunization on the incidence of measles, mumps, rubella, and chicken pox.

29.7 Colds

Colds are the most common of infectious diseases. Colds are typically upper respiratory tract viral infections that are transmitted via droplets spread from coughs, sneezes, and respiratory secretions. Colds are usually of short duration, lasting a week or so, and the symptoms are milder than other respiratory diseases such as influenza. **Table 29.1** contrasts the usually distinct symptoms and incidence of colds and influenza.

Cold symptoms include *rhinitis* (inflammation of the nasal region, especially the mucous membranes), nasal obstruction, watery nasal discharges, muscle aches, and a general feeling of malaise, usually without fever. *Rhinoviruses* are single-stranded

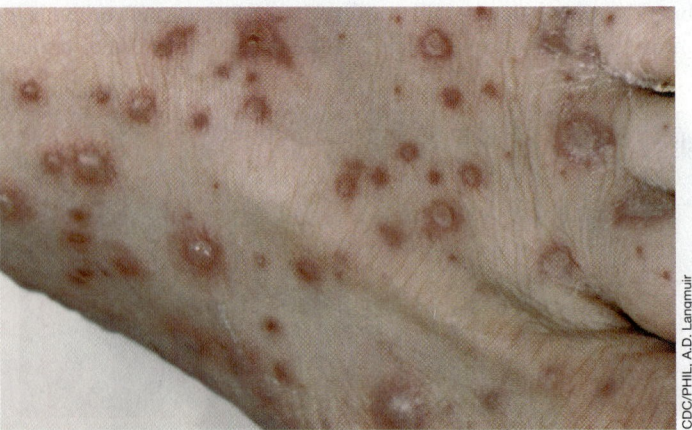

CDC/PHIL, A.D. Langmuir

Figure 29.23 **Chicken pox.** Papular rash on the foot of an adult. The papules are due to infection by varicella-zoster virus, the herpesvirus that causes chicken pox.

Table 29.1	Colds and influenza	
Symptoms	**Cold**	**Influenza**
Fever	Rare	Common (39–40°C); sudden onset
Headache	Rare	Common
General malaise	Slight	Common; often quite severe; can last several weeks
Nasal discharge	Common and abundant	Less common; usually not abundant
Sore throat	Common	Less common
Vomiting and/or diarrhea	Rare	Common in children
Incidence[a]	340	50

[a]Cases/100 people per year in the United States. Incidence of all other infectious diseases totals about 30 cases/100 people per year.

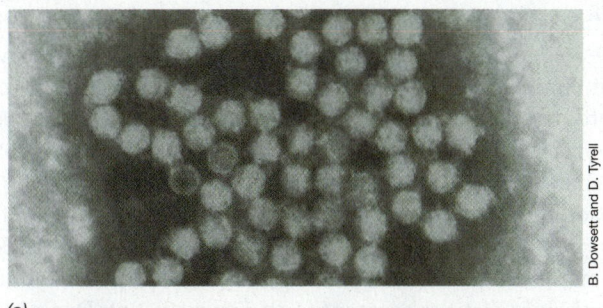

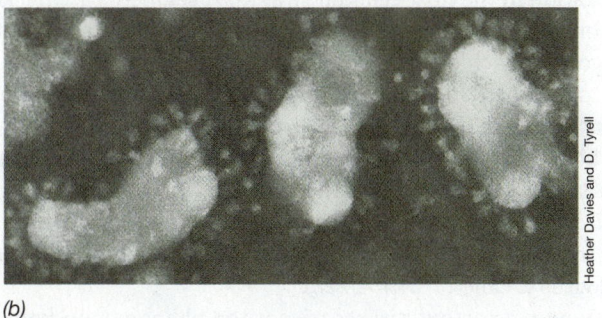

Figure 29.24 Transmission electron micrographs of common cold viruses. *(a)* Human rhinovirus; a virion is about 30 nm in diameter. *(b)* Human coronavirus; a virion is about 60 nm in diameter.

plus-sense RNA viruses of the picornavirus group (**Figure 29.24a**) (🔗 Section 9.8) and are the most common causes of colds. Over 100 different rhinoviruses have been identified. About one-quarter of all colds are due to infections with other viruses. These include in particular the *coronaviruses* (Figure 29.24b). Adenoviruses, coxsackie viruses, respiratory syncytial viruses (RSV), and orthomyxoviruses are collectively responsible for only a small percentage of common colds.

Aerosol transmission is a major means of spreading colds, although experiments with volunteers suggest that direct contact and indirect contact involving fomites are also important means of transmission, perhaps even more important than aerosols. Incidence of the common cold rises when people are indoors in the winter months, although it is possible to "catch a cold" at any time of year. Most antiviral drugs are ineffective against common cold viruses, although some have shown promise for preventing the onset of symptoms following rhinovirus exposure. Moreover, new antiviral drugs are being designed based on knowledge of the three-dimensional structure of cold viruses. For example, antirhinovirus drugs that bind to the virus and change its surface properties in such a way as to prevent it from attaching to host cells have been developed. But thus far, most "cold drugs" on the market simply help to reduce the severity of symptoms—the cough, nasal discharges, headache, and the like.

Because colds are generally self-limiting and not serious diseases, treatment is aimed at controlling symptoms, especially nasal discharges, with antihistamine and decongestant drugs. Cold viruses also induce an antibody-mediated immune response. However, the number of immunologically unique strains of each type of cold virus makes immunity to the common cold from previous exposure unlikely. Thus, the common cold is a recurrent

event, with the severity of symptoms a function of both the type of cold virus involved and the overall health and well-being of the person at the time of infection.

MINIQUIZ --
- Define the cause and symptoms of common colds.
- Discuss the possibilities for effective treatment and prevention of colds.

29.8 Influenza

Influenza is a highly infectious airborne disease of viral origin. Influenza viruses contain a single-stranded, negative-sense, segmented RNA genome surrounded by an envelope composed of protein, a lipid bilayer, and external glycoproteins (**Figure 29.25**) (🔗 Section 9.9). There are three classes of influenza viruses: influenza A, B, and C. Here we consider only influenza A because it is the most important human pathogen.

Antigenic Drift and Antigenic Shift

Each strain of influenza A virus can be identified by a unique set of surface glycoproteins. These glycoproteins are *hemagglutinin* (HA or H antigen) and *neuraminidase* (NA or N antigen). Each virus has one type of HA and one type of NA on its viral capsid. The HA antigen is important in *attaching* the influenza virus to host cells while the NA antigen is instrumental in *releasing* the virus from host cells; each antigen is composed of several individual proteins (**Figure 29.26**).

Infection or immunization with influenza virus results in the production of antibodies that react with the HA and NA glycoproteins. When these antibodies bind to HA or NA, the virus is blocked from either attaching or releasing and is effectively neutralized, stopping the infection process. However, over time,

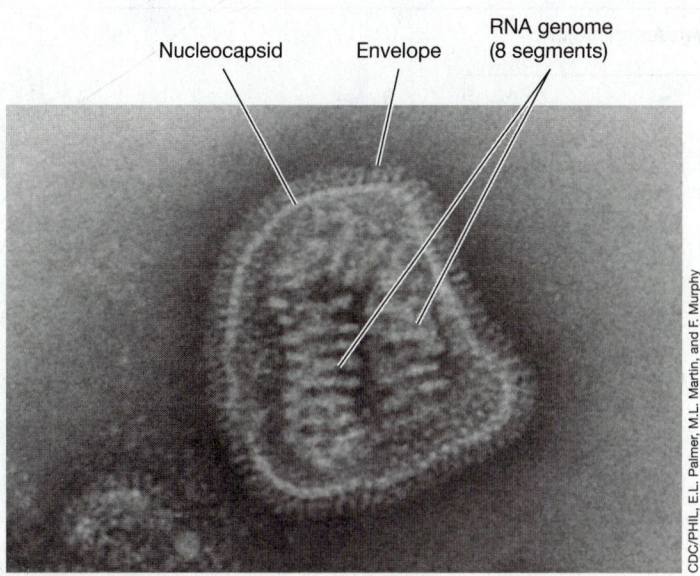

Figure 29.25 Influenza A virus. The virus contains a single-stranded negative-sense RNA genome in eight segments; a virion is about 100 nm in diameter. Major factors in the success of influenza virus as a pathogen are antigenic drift and antigenic shift (see Figure 29.27).

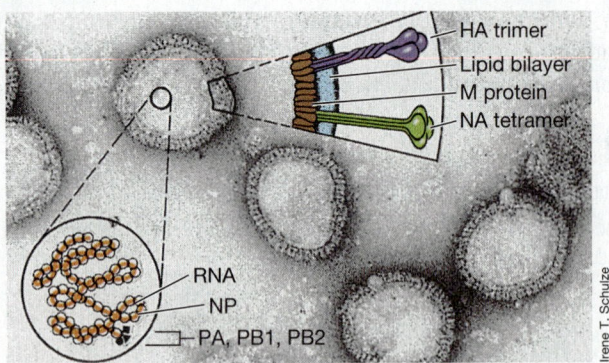

Figure 29.26 **Influenza virus structure.** Major viral coat proteins are: HA, hemagglutinin (three copies make up the HA coat spike); NA, neuraminidase (four copies make up the NA coat spike); M, coat protein; NP, nucleoprotein; PA, PB1, PB2, and other internal proteins, some of which have enzymatic functions.

the viral genes encoding the HA and NA glycoprotein antigens mutate, rendering minor changes to their amino acid sequence and hence antigenic structure. Mutations that alter as few as one amino acid in the glycoprotein can affect how an antibody binds to these antigens. This slight variation in the structure of influenza viral surface antigens is at the heart of a phenomenon in influenza biology called **antigenic drift**. As a result of these subtle yet important changes, host immunity to a given virus strain diminishes as the strain mutates, and reinfection with the mutated strain can occur. This phenomenon is why last year's influenza vaccine may work only poorly against this year's crop of influenza viruses (**Figure 29.27a**).

In addition to antigenic drift, there is a second feature of influenza virus biology that aids virulence. The single-stranded RNA genome of influenza viruses is *segmented*, with genes found on each of eight distinct segments (Figure 9.19b). During virus maturation in the host cell, the viral RNA segments are packaged randomly. To be infective, a virus must be packaged so it contains one copy of each of the eight gene segments. Occasionally, however, more than one strain of influenza virus infects a single animal at one time. In such cases, if the two strains infect the same cell, both viral genomes are replicated; when genome packaging occurs, the segments from the two strains may intermix. The result is a genetically unique virus that is now a *new virus strain*. This mixing of gene fragments between different strains of influenza virus is called *reassortment*.

Unique *reassortant viruses* trigger the phenomenon known as **antigenic shift** (Figure 29.27b), a major change in a surface antigen resulting from the total replacement of the RNA that encoded it.

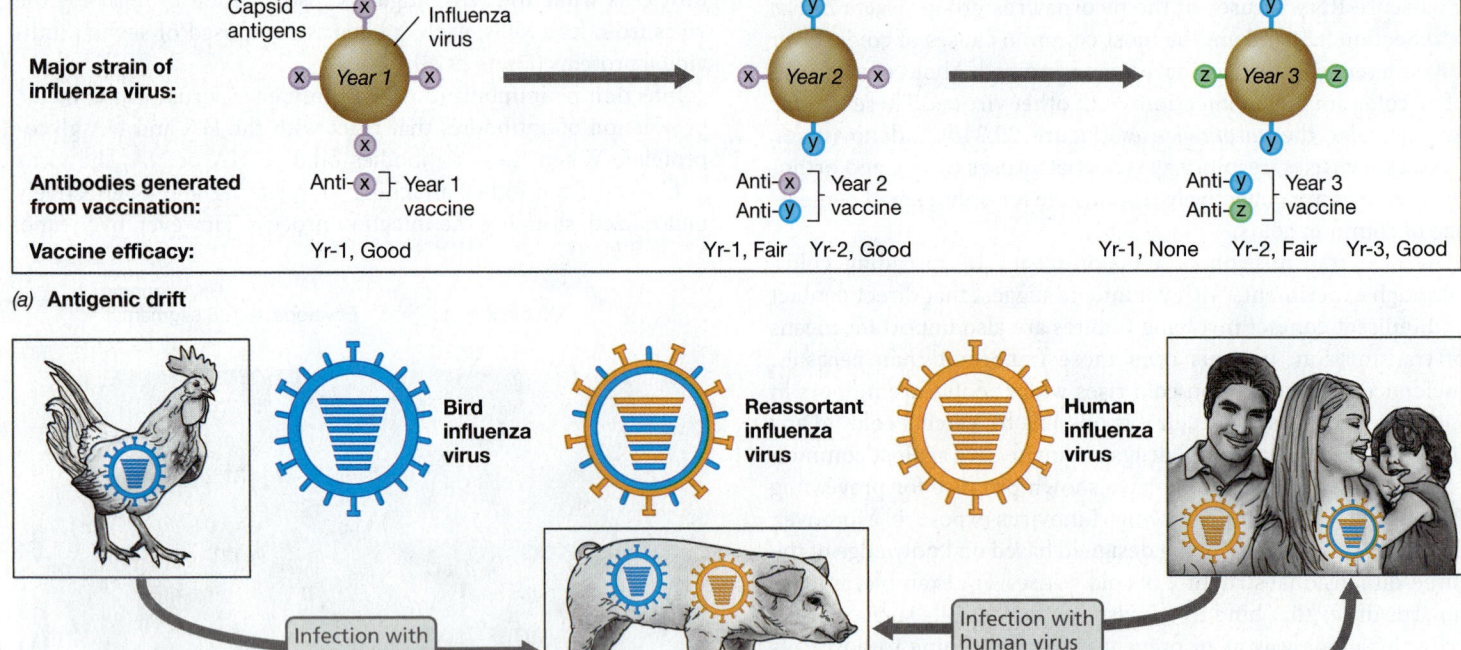

Figure 29.27 **Antigenic drift and antigenic shift in influenza virus biology.** *(a)* Antigenic drift. A new vaccine is prepared each year against the major strain of influenza circulating among the population. However, vaccine efficacy wanes with time as immunologically new surface antigens appear from mutations in genes encoding viral surface proteins. *(b)* Antigenic shift. Influenza strains that originate in birds and humans can also infect swine. If a pig becomes infected with both bird and human viruses simultaneously, the viral genomes can be mixed, forming reassortant viruses. If such viruses, which now contain several unique antigens, infect humans, influenza pandemics can be triggered (see page 853).

Antigenic shift can immediately and completely change one or both of the major HA and NA viral glycoproteins in a major way. As a result, reassortant viruses are essentially unrecognized by immune responses to previous influenza infections. Reassortant viruses also frequently display one or more unique virulence properties that help to trigger unusually strong clinical symptoms and are the usual catalysts of influenza *pandemics*, which we consider shortly.

Symptoms and Treatment of Influenza

Human influenza virus is transmitted from person to person through the air, primarily in droplets expelled during coughing and sneezing (Figure 29.1). The virus infects the mucous membranes of the upper respiratory tract and occasionally invades the lungs. Symptoms include a low-grade fever lasting up to a week, chills, fatigue, head and muscle aches, a cough and/or a sore throat, and general malaise (Table 29.1). Most of the serious consequences of seasonal influenza occur not from the disease itself but from bacterial secondary infections, especially in persons whose resistance has been lowered by the influenza infection. For example, in infants and the elderly, influenza can be followed by bacterial pneumonia (Section 29.2), sometimes in fatal form.

Most individuals develop protective immunity to the infecting strain of influenza virus, making it impossible for that strain or a very closely related strain to cause widespread infection (epidemic) until the virus encounters another susceptible population. Immunity occurs from both antibody- and cell-mediated immune responses directed at HA and NA glycoproteins. Influenza epidemics can be controlled by immunization. Developing an effective vaccine, however, is complicated by the large number of existing influenza viral strains resulting from antigenic drift and antigenic shift (Figure 29.27). Through careful worldwide surveillance, samples of the major emerging strains of influenza virus are obtained each year before the onset of seasonal epidemics and used to prepare that year's vaccine. In most years, this approach confers adequate protective immunity.

Most human influenza viruses respond to antiviral drugs. The adamantanes—*amantadine* and *rimantadine*—are synthetic amines that inhibit viral replication, and the neuraminidase inhibitors *oseltamivir* (Tamiflu) and *zanamivir* (Relenza) (⟲ Table 27.6) block release of newly replicated human influenza virions. These drugs are often used early on to shorten the course and severity of infection, especially in immune-compromised people or the elderly.

Influenza Pandemics

Influenza pandemics—worldwide epidemics—are much less frequent than outbreaks and epidemics, occurring from 10 to 40 years apart (Table 29.2). Flu pandemics result from antigenic shift, and virtually all have been due to avian and human influenza viruses reassorting in swine (Figure 29.27*b*) since swine can propagate both avian and human influenza viruses. This results in a highly virulent

Table 29.2 Influenza pandemics		
Year	**Name**	**Strain**
1889	Russian	H2N2
1900	Old Hong Kong	H3N8
1918	Spanish	H1N1
1957	Asian	H2N2
1968	Hong Kong	H3N2
2009	Swine	H1N1

influenza strain for which there is no preexisting immunity in humans.

The "Spanish flu" pandemic of 1918 was the most catastrophic in recorded history, killing somewhere between 20 and 50 million people worldwide, 2 million in the United States. The extreme virulence of the 1918 H1N1 virus is thought to have triggered host production and release of unusually large amounts of inflammatory substances, resulting in systemic inflammation and more severe symptoms than those typical of yearly flu epidemics. The 1957 Asian flu was also a memorable pandemic (Figure 29.28), beginning in China and spreading to the United States and shortly thereafter to Europe and South America. In this case, the pandemic influenza strain was a highly virulent H2N2 virus, differing antigenically from all previous strains. Immunity to this strain was thus absent, allowing the virus to spread rapidly throughout the world, killing nearly 2 million people. However, early detection of the Asian flu virus and more sophisticated vaccine manufacturing capabilities kept this influenza pandemic under better control than the 1918 pandemic where an effective vaccine was unavailable.

Pandemic influenza A (H1N1) 2009 virus, nicknamed the "swine flu," spread much more rapidly in 2009 than even the 1957

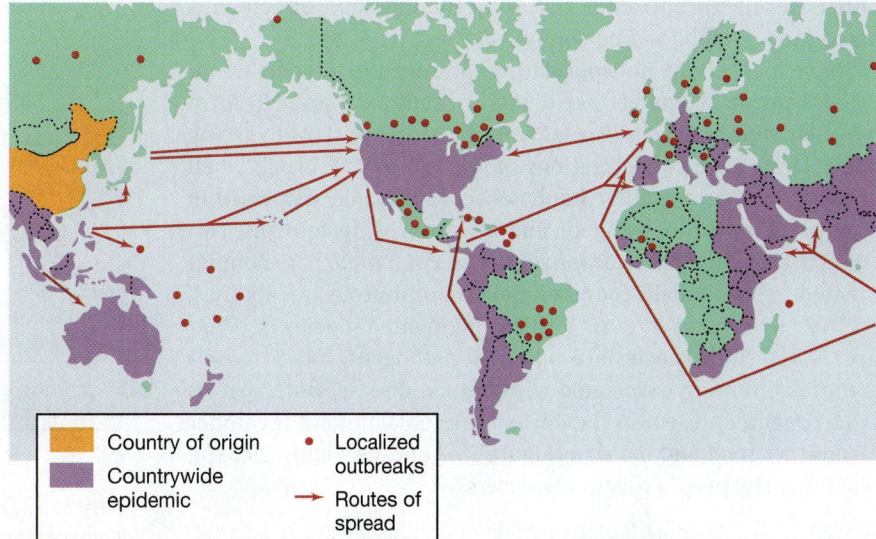

Figure 29.28 An influenza pandemic. Map of the Asian influenza pandemic of 1957. Lax agricultural practices with poultry and swine coupled with human interactions with these animals allowed the reassortment of influenza viral genomes from the three host species, producing a new strain for which there was no immune memory in humans. (see Figure 29.27*b* and page 853).

UNIT 6

Asian flu, starting in Mexico and spreading quickly to the United States, Europe, and Central and South America. H1N1 was a classic case of influenza virus genome reassortment in swine (Figure 29.27*b*), and from the swine reservoir, a highly virulent virus emerged to infect humans. Although the H1N1 virus spread quickly, the overall mortality rate was much lower than in the Spanish H1N1 pandemic, an estimated 0.1–0.2%, which is only slightly higher than seasonal influenza mortality.

Influenza A H5N1, nicknamed the "bird flu," appeared in Hong Kong in 1997, jumping directly from its avian host to humans without a reassortment step. H5N1 has now been detected in birds throughout Asia, Europe, the Middle East, and North Africa. Effective human-to-human spread of H5N1 apparently only occurs following prolonged close contact, and so this strain has not yet triggered a pandemic. However, if this virus were to infect swine and yield an easily transmissible reassortant virus that subsequently jumps to humans (Figure 29.27*b*), such a virus could initiate a major new influenza pandemic. Health organizations worldwide have recognized this potential and have plans in place to provide appropriate vaccines and support for potential pandemics initiated by this and other emergent influenza viruses (see this chapter's opening page).

MINIQUIZ --

- Distinguish between antigenic drift and antigenic shift in influenza.
- Discuss the possibilities for effective immunization programs for influenza and compare them to the possibilities for immunization for colds.

III • Direct-Contact Diseases

Some pathogens are spread primarily by direct contact with an infected person or by direct contact with blood or excretions from an infected person. Many of the respiratory diseases we have just discussed can also be spread by direct contact, but here we consider diseases spread primarily through direct contact with infected individuals rather than by an airborne route. These include staphylococcal infections, ulcers, and certain types of hepatitis.

29.9 *Staphylococcus aureus* Infections

The genus *Staphylococcus* contains pathogens of humans and other animals. Staphylococci commonly infect skin and wounds and may also cause pneumonia. Most staphylococcal infections result from the transfer of staphylococci in the normal flora of an infected, asymptomatic individual, to a susceptible individual. Others result from toxemia following the ingestion of contaminated food ("staph food poisoning," ↩ Section 31.8).

Staphylococci are nonsporulating, gram-positive cocci about 0.5–1.5 μm in diameter that divide in multiple planes to form irregular clumps of cells (**Figure 29.29**). They are resistant to drying and tolerate high concentrations of salt (up to 10% NaCl) when grown on artificial media. Staphylococci are readily dispersed in dust particles through the air and on surfaces. In humans, two species are important: *Staphylococcus epidermidis*, a nonpigmented species usually found on the skin or mucous membranes, and *Staphylococcus aureus*, a yellow-pigmented species (see Figure 29.31). Both species are potential pathogens, but *S. aureus* is more commonly associated with human disease. Both species are frequently present in the normal microbial flora of the upper respiratory tract and the skin (Figure 29.2 and see Figure 29.31*b*), making many people potential carriers.

Epidemiology and Pathogenesis

Staphylococcal diseases include acne, boils, pimples, impetigo, pneumonia, osteomyelitis, carditis, meningitis, and arthritis. Many of these diseases are *pyogenic* (pus-forming) (**Figure 29.30a**). Those strains of *S. aureus* that cause human disease produce a variety of virulence factors (↩ Table 23.5). At least four different *hemolysins* (proteins that lyse red blood cells, see Figure 29.8) have been recognized, and a single strain often produces several. A key

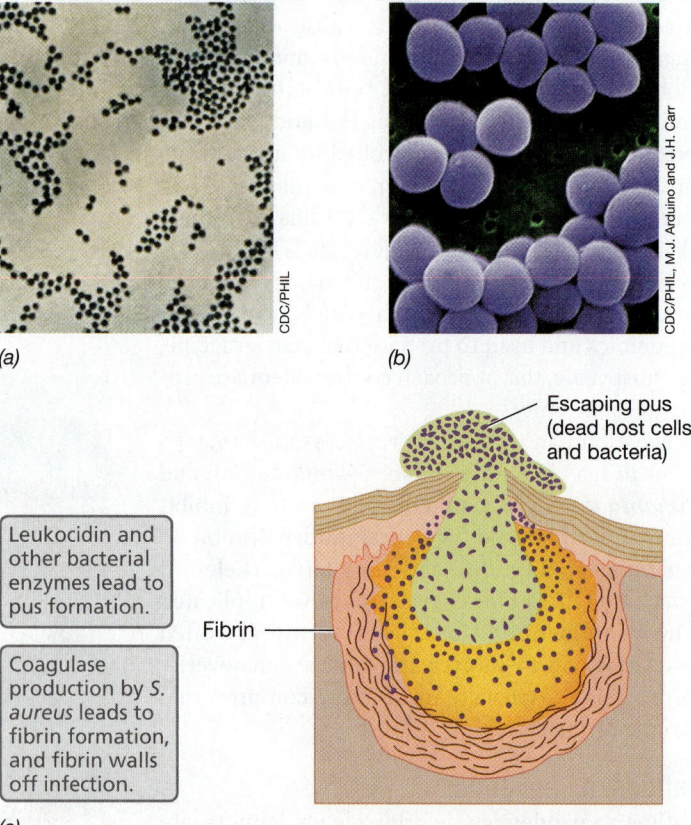

(a) CDC/PHIL
(b) CDC/PHIL, M.J. Arduino and J.H. Carr

(c)

Escaping pus (dead host cells and bacteria)

Leukocidin and other bacterial enzymes lead to pus formation.

Fibrin

Coagulase production by *S. aureus* leads to fibrin formation, and fibrin walls off infection.

Figure 29.29 *Staphylococcus aureus* and *S. aureus* infections. Cells divide in several planes, giving the appearance of a cluster of grapes. *(a)* Gram stain; an individual coccus is about 1 μm in diameter. *(b)* Scanning electron micrograph of cells. *(c)* Structure of a boil. Staphylococci initiate a localized skin infection and become walled off by coagulated blood and fibrin through the activity of the enzyme coagulase, a major virulence factor. The ruptured boil releases pus, consisting of dead host cells and bacteria. See also Figure 29.30.

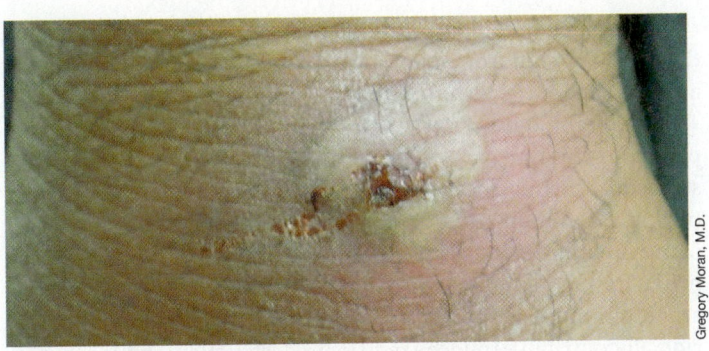

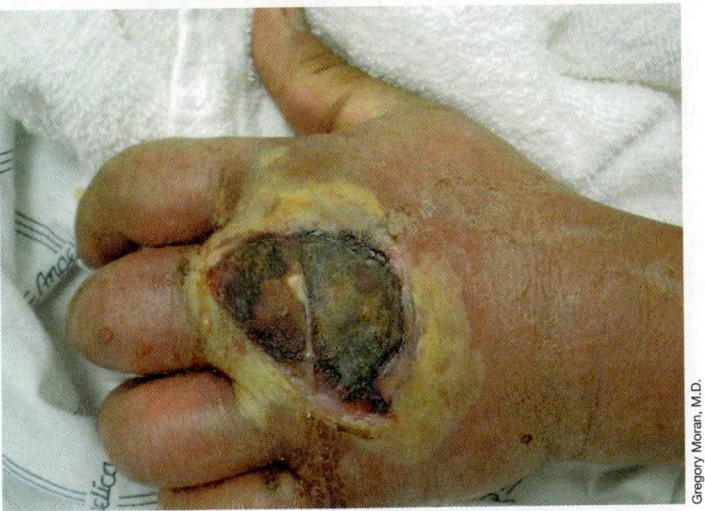

Figure 29.30 Pus-forming staphylococcal wounds. *(a)* A typical pus-forming wound on the hand. Pus lies just under the epidermal layer. *(b)* Abscess on the hand caused by a methicillin-resistant strain of *Staphylococcus aureus* (MRSA strain). If no treatment for a pus-forming wound is sought or if penicillin is administered first, MRSA infections can cause extensive tissue destruction, as shown here.

virulence factor produced by *S. aureus* is *coagulase*, an enzyme that converts fibrin to fibrinogen, forming a localized clot. Clotting induced by coagulase results in the accumulation of fibrin around the bacterial cells, making it difficult for host immune cells to contact the bacteria and initiate phagocytosis. Most *S. aureus* strains also produce *leukocidin*, a protein that destroys white blood cells. Production of leukocidin in skin lesions such as boils and pimples results in host cell destruction and is one of the factors responsible for pus (Figure 29.30). Some strains of *S. aureus* also produce other virulence proteins including *hyaluronidase*, *fibrinolysin*, *lipase*, *ribonuclease*, and *deoxyribonuclease*.

Certain strains of *S. aureus* cause **toxic shock syndrome (TSS)**, a serious outcome of staphylococcal infection, characterized by high fever, rash, vomiting, diarrhea, and death. TSS was first recognized in women and was associated with the use of highly absorbent tampons. However, TSS is now seen in both men and women and is typically initiated by staphylococcal infections following surgery. The symptoms of TSS result from an exotoxin called *toxic shock syndrome toxin-1*. This very potent toxin is a superantigen (⇄ Section 24.9) that is released during cell growth and recruits large numbers of T cells to the site of infection. These then cause a major inflammatory response that is fatal in about 70% of cases. TSS can also be due to superantigens from other pathogens, including *Streptococcus pyogenes* (Section 29.2).

Diagnosis and Treatment

To diagnosis an *S. aureus* infection through laboratory culture, a specimen, typically from a pus-forming wound (Figure 29.30*a*), is cultured on a selective and differential medium containing 7.5% NaCl, mannitol, and phenol red, a pH indicator (mannitol–salt agar, **Figure 29.31**). The salt inhibits the growth of nonhalophilic bacteria while allowing staphylococci to grow. In addition, because *S. aureus* ferments mannitol, it generates acidity and changes the medium from red to yellow; other staphylococci, such as *S. epidermidis*, do not (Figure 29.31).

In major clinical laboratories, the polymerase chain reaction (PCR) is used to amplify genes unique to *S. aureus* from DNA isolated from a clinical sample, and this speeds up the diagnosis (results from laboratory culture take 24 h). For specific identification of methicillin-resistant strains of *S. aureus* (MRSA), a special selective and differential medium is available as well as a PCR protocol to identify *mecA*, the gene that encodes methicillin resistance in MRSA strains.

Historically, *S. aureus* infections have been treated with various penicillin and cephalosporin antibiotics. However, extensive use of these antibiotics for many years has selected for resistant strains that now predominate, especially in the clinical environment. Surgical patients, for example, may acquire staphylococci from healthcare personnel who are asymptomatic carriers of drug-resistant strains. As a result, appropriate antimicrobial drug therapy for *S. aureus* infections is a major problem in healthcare environments. The antibiotics clindamycin and various tetracycline drugs are currently used to treat MRSA infections.

MRSA infections (Figure 29.30*b*) are becoming more common. Over 100,000 cases of MRSA are reported each year in the United States, but actual infections are probably closer to ten times this number. Many of these cases are hospital-acquired (nosocomial) MRSA, but many others are not. Because of the potential severity of MRSA infections, it is important to rapidly identify these

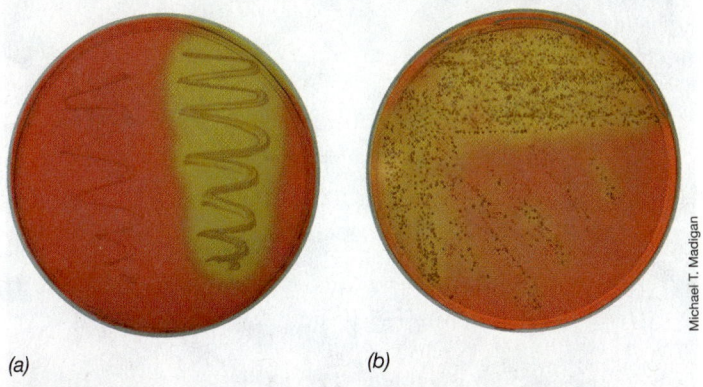

Figure 29.31 Mannitol–salt agar in the isolation of staphylococci. *(a)* Mannitol–salt agar (MSA) is both selective and differential for staphylococci. The presence of 7.5% NaCl makes MSA selective and phenol red makes it differential. Left, *Staphylococcus epidermidis*; right, *Staphylococcus aureus*. *(b)* A nasal swab of the senior author of this textbook supports the observation that most humans are carriers of *S. aureus*.

strains in clinical specimens so that an effective treatment is begun as soon as possible. Delayed treatment of a MRSA infection, whether due to hesitation to seek treatment or treatment with an ineffective antibiotic, can lead to extensive tissue damage (Figure 29.30b).

Prevention of staphylococcal infections is virtually impossible because many people are asymptomatic carriers of *S. aureus*, either on their skin or upper respiratory tract (Figure 29.31b). However, identification and treatment or isolation of MRSA-carrying healthcare providers who serve in surgical or nursery units has helped limit transmission of these very aggressive strains. As is true of many direct-contact diseases, MRSA transmission can also be greatly diminished by practicing good basic hygiene, avoiding contact with the personal items (including clothing and towels) of others, and keeping wounds covered.

MINIQUIZ --

- What is the normal habitat of *Staphylococcus aureus*? How does *S. aureus* spread from person to person?

- What is MRSA, and why is it a health problem?

29.10 *Helicobacter pylori* and Gastric Ulcers

Helicobacter pylori is a gram-negative, highly motile, spiral-shaped bacterium (**Figure 29.32**) associated with gastritis, ulcers, and gastric cancers. This bacterium colonizes the non-acid-secreting mucosa of the stomach and the upper intestinal tract. Up to 80% of gastric ulcer patients have concomitant *H. pylori* infections, and up to 50% of asymptomatic adults in developing countries are chronically infected. Although there is no known nonhuman reservoir of *H. pylori*, infection occurs in high incidence within families, suggesting person-to-person transmission.

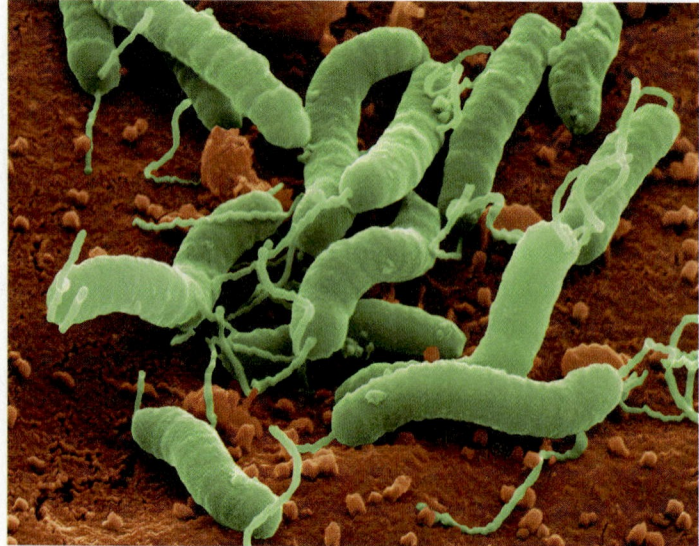

Figure 29.32 *Helicobacter pylori.* Colorized scanning electron micrograph of cells attached to the mucous lining of the stomach. Cells range in size from 3 to 5 μm in length and are about 0.5 μm in diameter. Note the flagella.

H. pylori infections also occasionally occur in clusters, suggesting that transmission from common sources such as food or water is also possible.

H. pylori is only slightly invasive and colonizes the surfaces of the gastric mucosa, where it is protected from the effects of stomach acids by the gastric mucus layer. After mucosal colonization, a combination of virulence factors and host responses cause inflammation, tissue destruction, and ulceration. Pathogen products such as the cytotoxin VacA (an exotoxin), urease, and an autoimmune response triggered by *H. pylori* lipopolysaccharide all contribute to localized tissue destruction and ulceration. Individuals who acquire *H. pylori* tend to have chronic infections unless they are treated with antibiotics. Treatment is both simple and important, as chronic inflammation of the gastroduodenum (gastritis) due to untreated *H. pylori* infection may lead to the development of gastric cancers.

Clinical signs of *H. pylori* infection include belching and stomach (epigastric) pain. Definitive diagnosis requires the isolation or observation of *H. pylori* from a gastric ulcer biopsy. A simple diagnostic test for the *H. pylori* enzyme urease is used for a noninvasive diagnosis. In this test, a small amount of ^{13}C- or ^{14}C-labeled urea ($H_2N-CO-NH_2$) is ingested; if *H. pylori* is present, the bacterium will hydrolyze the urea, forming labeled CO_2 and ammonia. Hence, the presence of labeled CO_2 in the patient's breath is highly suggestive of *H. pylori* infection.

The best evidence for a causal association between *H. pylori* and gastric ulcers comes from antibiotic treatments for the disease. Long-term treatment with antacids helps alleviate gastric ulcer symptoms temporarily, but most patients relapse within 1 year. However, by treating the *cause* rather than the *effect* of the disease, actual cures can be obtained. *H. pylori* infection is typically treated with a combination of drugs, including the antibacterial compound metronidazole, an antibiotic such as tetracycline or amoxicillin, and a bismuth-containing antacid preparation. The combination treatment, administered for 14 days, abolishes the *H. pylori* infection and provides a long-term cure.

For their contributions to unraveling the connection between *H. pylori* and peptic and duodenal ulcers, the Australian scientists Robin Warren and Barry Marshall were awarded the 2005 Nobel Prize in Physiology or Medicine.

MINIQUIZ --

- Describe infection by *Helicobacter pylori* and the resulting development of an ulcer.

- How can gastric ulcers be diagnosed? How can they be cured?

29.11 Hepatitis Viruses

Hepatitis is a liver inflammation commonly caused by an infectious agent. Hepatitis sometimes results in acute illness followed by destruction of functional liver anatomy and cells, a condition known as **cirrhosis**. Hepatitis due to infection can cause chronic or acute disease, and some forms lead to liver cancer.

Although many viruses and a few bacteria can cause hepatitis, a restricted group of viruses is often associated with liver disease. Hepatitis viruses A, B, C, D, and E are phylogenetically diverse viruses

Table 29.3 Hepatitis viruses

Disease	Virus and genome	Vaccine	Clinical illness	Transmission route
Hepatitis A	*Hepatovirus* (HAV) ssRNA	Yes	Acute	Enteric (food)
Hepatitis B	*Orthohepadnavirus* (HBV) dsDNA	Yes	Acute, chronic, oncogenic	Parenteral, sexual
Hepatitis C	*Hepacivirus* (HCV) ssRNA	No	Chronic, oncogenic	Parenteral
Hepatitis D	*Deltavirus* (HDV) ssRNA	No	Fulminant, only with HBV	Parenteral
Hepatitis E	*Caliciviridae* family (HEV) ssRNA	No	Fulminant disease in pregnant women	Enteric (water)

but share in common their ability to infect the liver (Table 29.3). Hepatitis A and E viruses, although occasionally transmitted person to person, are more commonly transmitted by food (hepatitis A virus) or water (hepatitis E virus). We cover hepatitis A viral disease in Chapter 31. Here our focus is on hepatitis viruses transmitted by direct contact.

The incidence of hepatitis A and B, the most common forms, has decreased significantly in the past 20 years because of effective vaccines (Figure 29.33) and increases in surveillance. By comparison to hepatitis A and B, hepatitis C infections are of very low incidence (Figure 29.33).

Epidemiology

Infection with *hepatitis B virus* (HBV) is often called *serum hepatitis* because it is transmitted in blood or body fluids in contact with blood. HBV is a hepadnavirus, a partially double-stranded DNA virus (⮌ Section 9.11). The mature virus particle containing the viral genome is called a *Dane particle* (Figure 29.34). HBV causes acute, often severe disease that can lead to liver failure and death. Chronic HBV infection can lead to cirrhosis and liver cancer.

HBV is transmitted by a *parenteral route*, which means "outside the gut." The main means of HBV transmission is from blood transfusions, contact with infected blood in a hypodermic needle, and from mother to child during childbirth. HBV may also be transmitted through exchanges of body fluids during sex. The number of new HBV infections has remained low and more or less constant since the year 2000 (Figure 29.33). Nevertheless, over 100,000 people worldwide and nearly 5000 people in the United States die yearly from liver failure or liver cancer caused by chronic HBV infection.

Hepatitis D virus (HDV) is a *defective virus* (⮌ Section 10.7) that lacks genes encoding its own capsid. HDV is also transmitted by parenteral routes, but because it is a defective virus, it cannot replicate and form an intact virion unless the cell is also infected with HBV. The HDV genome replicates independently but relies on HBV to produce capsid proteins (which are the same as those used by HBV) to form infectious virions. Thus, HDV infections are always coinfections with HBV.

Hepatitis C virus (HCV) is also transmitted parenterally. HCV generally produces a mild or even asymptomatic disease at first, but later on up to 85% of those infected develop chronic hepatitis, with up to 20% proceeding to chronic liver disease and cirrhosis. Chronic infection with HCV leads to hepatocarcinoma (liver cancer) in 3–5% of infected individuals. The latency period for development of cancer can be several decades after the primary infection. Only a fraction of the estimated 25,000 annual new infections with HCV are recognized and reported in the United States (Figure 29.33). Large numbers of HCV-related deaths occur annually due to chronic HCV infections that develop into liver cancer. HCV-induced liver disease accounts for up to 10,000 of the 25,000 annual deaths due to liver cancer, other chronic liver diseases, and cirrhosis.

Other Aspects of Hepatitis Syndromes

Hepatitis is an acute disease of the liver, a vital organ that plays a role in several key metabolic processes, including carbohydrate,

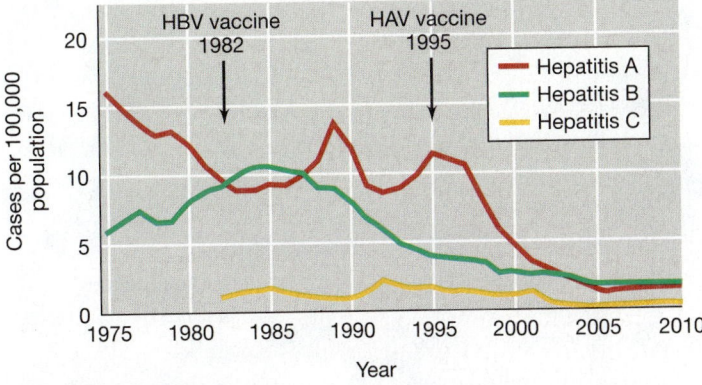

Figure 29.33 Hepatitis A, B, and C in the United States. In 2010 there were 1670 reported cases of hepatitis A, 3350 reported cases of hepatitis B, and 850 reported cases of hepatitis C. Data obtained from the CDC, Atlanta, Georgia, USA.

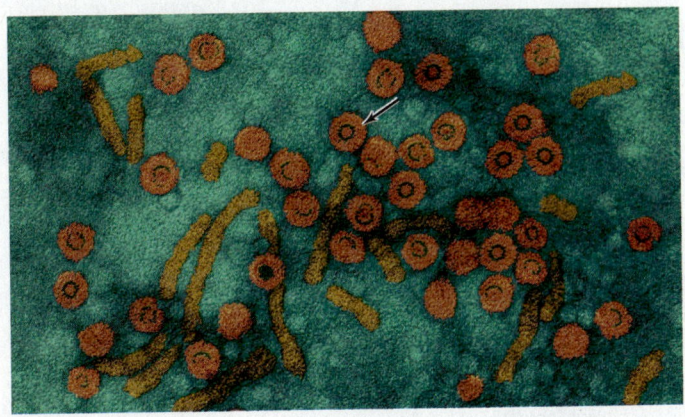

Figure 29.34 Hepatitis B virus (HBV). The arrow indicates a complete HBV virion, which is called a Dane particle. A Dane particle is about 40 nm in diameter.

UNIT 6

lipid, and protein syntheses, as well as detoxification and many other functions. Symptoms of hepatitis include fever, jaundice (yellowing of the skin and the whites of the eyes), and liver enlargement and cirrhosis. All hepatitis viruses cause similar acute symptoms and cannot be readily distinguished based on clinical findings alone. Chronic hepatitis infections, usually caused by HBV or HCV, are often asymptomatic or produce very mild symptoms, but nonetheless cause serious liver disease, even in the absence of liver cancer.

Diagnosis of hepatitis is based on a combination of clinical symptoms and laboratory tests that assess liver function, especially key liver enzymes. Cirrhosis is diagnosed by visual examination of biopsied liver tissue. Virus-specific molecular assays are typically used to confirm a diagnosis, positively identify the infectious agent, and determine a course of treatment; isolation and culture of hepatitis viruses is usually not attempted.

Many of the immunological and molecular diagnostic tools discussed in Chapter 27 are used in hepatitis diagnoses. These include enzyme immunoassays that target viral-specific proteins or antiviral antibodies in a blood sample, immunoblots (Western blots), and immunofluorescence (microscopic) methods. Polymerase chain reaction (PCR) tests are also used to detect hepatitis viral genomes in blood or in liver tissue obtained by biopsy.

Infection with HAV or HBV can be prevented with effective vaccines. HBV vaccination is recommended and in most cases is required for school-age children in the United States. No effective vaccines are available for the other hepatitis viruses. For those unvaccinated, the practice of *universal precautions* will prevent infection. The precautions prescribe a high level of vigilance and aseptic handling and containment procedures to deal with patients, body fluids, and infected waste materials (⮌ Section 27.1). Most treatment of hepatitis is supportive, providing rest and time for the immune system to attack the infection and allow liver damage to be repaired. In some cases, in particular for HBV infections, some antiviral drugs are available that offer effective treatment.

MINIQUIZ -----------------------------

- What host organ do hepatitis viruses attack? How are hepatitis A, B, and C viruses transmitted?
- Describe potential prevention and treatment methods for hepatitis A, B, and C viruses.

IV · Sexually Transmitted Infections

Sexually transmitted infections (STIs), also called *sexually transmitted diseases* (STDs), are caused by a wide variety of bacteria, viruses, protists, and even fungi (Table 29.4). Unlike respiratory pathogens that can be shed constantly in large numbers by an infected individual, sexually transmitted pathogens are typically found only in body fluids from the genitourinary tract (and blood, in the case of HIV). Because they require a protected and moist environment, sexually transmitted pathogens preferentially and sometimes exclusively colonize the genitourinary tract.

Table 29.4 Sexually transmitted infections and treatment guidelines

Disease	Causative organism(s)[a]	Recommended treatment[b]
Gonorrhea	*Neisseria gonorrhoeae* (B)	Cefixime or ceftriaxone, *and* azithromycin or doxycycline
Syphilis	*Treponema pallidum* (B)	Benzathine penicillin G
Chlamydia trachomatis infections	*Chlamydia trachomatis* (B)	Doxycycline or azithromycin
Nongonococcal urethritis	*C. trachomatis* (B)	Azithromycin or doxycycline
	or *Ureaplasma urealyticum* (B)	
	or *Mycoplasma genitalium* (B)	
	or *Trichomonas vaginalis* (P)	Metronidazole
Lymphogranuloma venereum	*C. trachomatis* (B)	Doxycycline
Chancroid	*Haemophilus ducreyi* (B)	Azithromycin
Genital herpes	Herpes simplex 2 (V)	No known cure; symptoms can be controlled by several antiviral drugs
Genital warts	Human papillomavirus (HPV) (certain strains)	No known cure; symptomatic warts can be removed surgically, chemically, or by cryotherapy
Trichomoniasis	*Trichomonas vaginalis* (P)	Metronidazole
Acquired immunodeficiency syndrome (AIDS)	Human immunodeficiency virus (HIV)	No known cure; several drugs can stop viral replication and slow disease progression
Pelvic inflammatory disease	*N. gonorrhoeae* (B) or	Cefotetan and doxycycline
	C. trachomatis (B)	
Vulvovaginal candidiasis	*Candida albicans* (F)	Butoconazole

[a]B, bacterium; V, virus; P, protist; F, fungus.
[b]Recommendations of the U.S. Department of Health and Human Services, Public Health Service.

Since the transmission of STIs is limited to sexual activity, their spread can be controlled by sexual abstinence and minimized by the use of condoms that stop the exchange of body fluids during sex. With the exception of HIV/AIDS, most STIs are curable and many can have only minor symptoms. These realities, combined with the fact that those infected are sometimes reluctant to seek treatment, make treatment of STIs an ongoing public health challenge. However, delaying or forgoing treatment of STIs only serves to maintain lines of transmission and can lead to long-term health problems such as infertility, cancer, heart disease, degenerative nerve disease, birth defects, stillbirth, or destruction of the immune system, any of which can result in death.

29.12 Gonorrhea and Syphilis

Gonorrhea and *syphilis* are ancient STIs, but because of major differences in their symptoms, the overall pattern of disease differs significantly between the two. In the United States, cases of gonorrhea peaked following the introduction of birth control pills in the mid-1960s, and gonorrhea is still quite prevalent today; cases of syphilis, on the other hand, have a much lower incidence (**Figure 29.35**). This is partly because syphilis exhibits very obvious symptoms in its primary stage and infected individuals usually seek immediate treatment.

Gonorrhea

Neisseria gonorrhoeae, often called the *gonococcus*, causes gonorrhea. *N. gonorrhoeae* is a gram-negative and obligately aerobic diplococcus related biochemically and phylogenetically to *Neisseria meningitidis* (Section 29.5). Cells of *N. gonorrhoeae* are killed rapidly by drying, sunlight, and ultraviolet radiation and thus normally do not survive away from the mucous membranes of the pharynx, conjunctiva, rectum, or genitourinary tract (**Figure 29.36**). Because of this, gonorrhea can only be transmitted by intimate person-to-person contact. We discussed the clinical microbiology and diagnosis of gonorrhea in Section 27.3.

The symptoms of gonorrhea are quite different in the male and female. In females, gonorrhea may be asymptomatic or cause a mild vaginitis that is difficult to distinguish from vaginal

infections caused by other organisms; hence, the infection may easily go unnoticed. Complications from untreated gonorrhea in females, however, can lead to a chronic condition called *pelvic inflammatory disease* (*PID*), which can cause sterility. In men, *N. gonorrhoeae* causes a painful infection of the urethral canal and typical puslike urethral discharge. Complications from untreated gonorrhea affecting both males and females include damage to heart valves and joint tissues due to inflammatory reactions from immune complexes that deposit in these areas. In addition to disease in adults, *N. gonorrhoeae* can also cause eye infections in newborns. Infants born of infected mothers may acquire eye infections during birth. Therefore, prophylactic treatment of the eyes of all newborns with an ointment containing erythromycin is generally mandatory to prevent gonococcal infection in infants.

Treatment of gonorrhea with penicillin was the method of choice until the 1980s when strains of *N. gonorrhoeae* resistant to penicillin arose. The quinolones ciprofloxacin, ofloxacin, or levofloxacin were also used, but by 2006, a significant fraction of *N. gonorrhoeae* strains isolated in the United States had developed resistance to these drugs as well. Strains resistant to penicillin and quinolones respond to alternative antibiotic therapy with a single dose of the β-lactam antibiotics cefixime or ceftriaxone.

Despite the fact that drugs are still effective in treating gonorrhea, incidence of gonorrhea remains relatively high (Figure 29.35) for at least three reasons. First, although anti-gonococcal antibodies are generated by an infection, they are strain-specific and provide no protection from infection by other strains of *N. gonorrhoeae*. As a consequence, gonorrhea reinfection is possible and quite common in high-risk populations. In addition, within a single *N. gonorrhoeae* strain, antigenic switches can thwart the immune response. For example, by mutation *N. gonorrhoeae* can alter the structure of its pilus proteins, thus creating new serotypes to challenge the immune response. Second, oral contraceptives cause a rise in vaginal pH; when this occurs, lactic acid bacteria normally found in the adult vagina (↩ Figure 23.8b) fail to develop, and this reduces competition for colonization by *N. gonorrhoeae*. And finally, and quite importantly, symptoms of gonorrhea in the female are often so mild that the disease may go unrecognized.

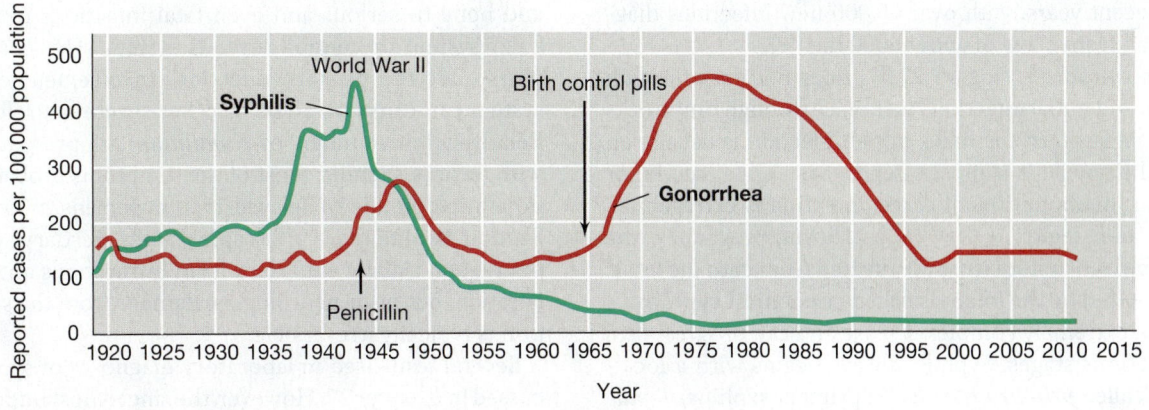

Figure 29.35 Reported cases of gonorrhea and syphilis in the United States. Note the downward trend in disease incidence after the introduction of antibiotics and the upward trend in the incidence of gonorrhea after the introduction of birth control pills. In 2011 there were 321,849 new cases of gonorrhea and 46,042 new cases of primary and secondary syphilis in the United States.

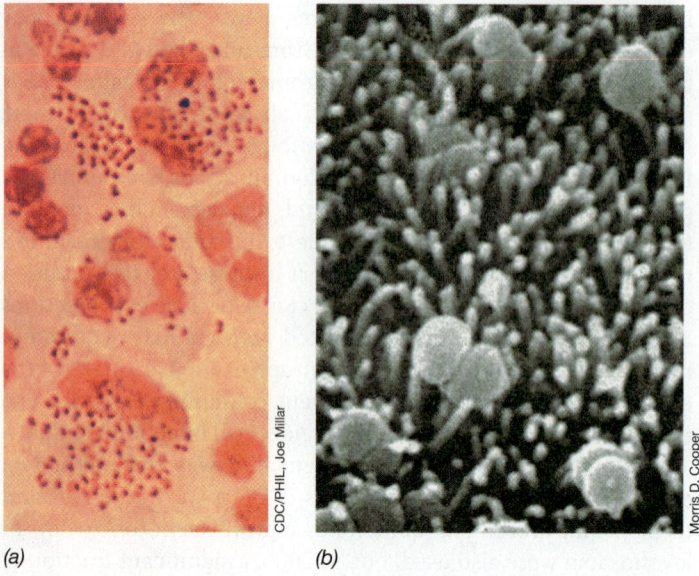

Figure 29.36 **The causative agent of gonorrhea, *Neisseria gonorrhoeae*.** *(a)* Gram stain of a urethral discharge. *(b)* Scanning electron micrograph of the microvilli of human fallopian tube mucosa with cells of *N. gonorrhoeae* attached to the surface of epithelial cells. Cells of *N. gonorrhoeae* are about 0.8 μm in diameter. *Neisseria* species are *Betaproteobacteria* (↺ Section 15.2).

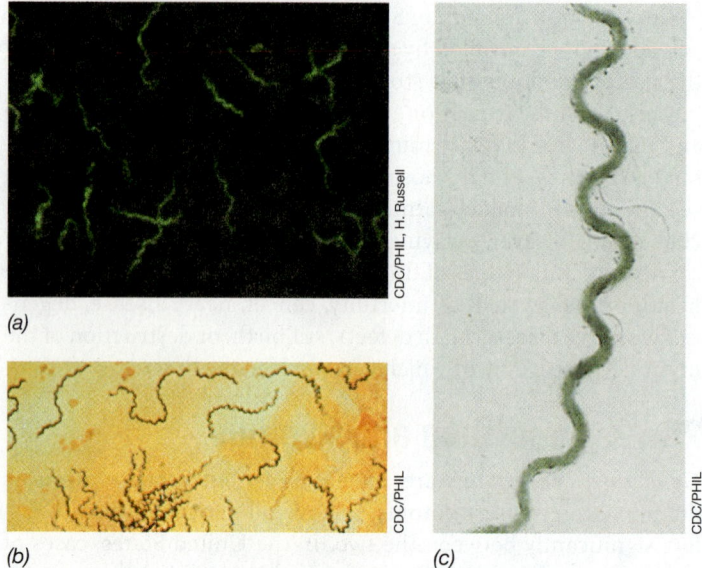

Figure 29.37 **The syphilis spirochete, *Treponema pallidum*.** *(a)* Cells stained with a fluorescent antibody measure 0.15 μm wide and 10–15 μm long. *(b)* Silver-stained (Fontana method) preparation of a specimen from a syphilitic chancre. *(c)* Shadow-cast electron micrograph of a cell of *T. pallidum*. The endoflagella are typical of spirochetes. (↺ Section 14.20).

Syphilis

Syphilis is caused by a spirochete, *Treponema pallidum,* a long and extremely thin coiled cell (Figure 29.37). Like *N. gonorrhoeae,* *T. pallidum* is very sensitive to environmental stress and drying, and thus syphilis is only transmitted through intimate sexual contact or from mother to fetus during pregnancy. The biology of the spirochetes and the genus *Treponema* is discussed in Section 14.20.

Syphilis is often transmitted along with gonorrhea as coinfections. However, syphilis is potentially the more serious disease. For example, syphilis kills about 100,000 people per year worldwide, whereas gonorrhea by itself kills only about 1000 people per year. Nevertheless, largely because of differences in the symptoms and pathobiology of the two diseases, the incidence of syphilis in the United States is much lower than the incidence of gonorrhea. The incidence of syphilis in the United States, however, has increased in recent years, with over 10,000 new infections diagnosed each year, from a low of about 6000 in 1997.

The syphilis spirochete (Figure 29.37) does not pass through unbroken skin, and initial infection takes place through tiny breaks in the epidermal layer. In the male, initial infection is usually on the penis; in the female it is most often in the vagina, cervix, or perineal region. In about 10% of cases, infection is extragenital, usually in the oral region (Figure 29.38). During pregnancy, the organism can be transmitted from an infected woman to the fetus; the disease acquired by the infant is called **congenital syphilis**.

Syphilis is an extremely complex disease and can progress into increasingly serious stages. Syphilis always begins with a localized infection called *primary syphilis.* In primary syphilis, *T. pallidum* multiplies at the initial site of entry, and a characteristic lesion called a *chancre* forms within 2 weeks to 2 months (Figure 29.38a, b). Microscopy of a syphilitic chancre exudate reveals the actively motile spirochetes (Figure 29.37). In most cases the chancre heals spontaneously and *T. pallidum* disappears from the site. In untreated cases, however, some cells spread from the initial site to various parts of the body, such as the mucous membranes, eyes, joints, bones, or central nervous system, where extensive multiplication occurs. A hypersensitivity reaction to the treponemes often takes place, revealed by the development of a generalized skin rash; this rash is the key symptom of *secondary syphilis* (Figure 29.38c).

In the absence of treatment, the subsequent course of the disease varies from case to case. About one-fourth of infected individuals undergo a spontaneous cure and are free of any further disease symptoms. Another one-fourth exhibit no further symptoms, but maintain a persistent, chronic, syphilitic infection. Roughly half of untreated patients develop *tertiary syphilis,* with symptoms ranging from relatively mild infections of the skin and bone to serious and even fatal infections of the cardiovascular system or central nervous system. This may occur many years after the primary infection. Involvement of the nervous system can cause paralysis or other severe neurological damage. Relatively low numbers of *T. pallidum* are present in individuals with tertiary syphilis; most of the symptoms probably result from inflammation due to delayed-type hypersensitivity reactions (↺ Section 24.8) to the syphilis spirochetes. Tertiary syphilis can still be treated, usually with long-term intravenous antibiotic administration, but prior neurological damage from the syphilitic infection is typically irreversible.

Several tests used in laboratory diagnoses of syphilis were discussed in Chapter 27. However, the single most important physical sign of a primary syphilis infection, the chancre (Figure 29.38a, b), is highly diagnostic for the disease. Infected individuals generally seek treatment for syphilis because of the chancre. Penicillin remains highly effective in syphilis therapy, and the primary and

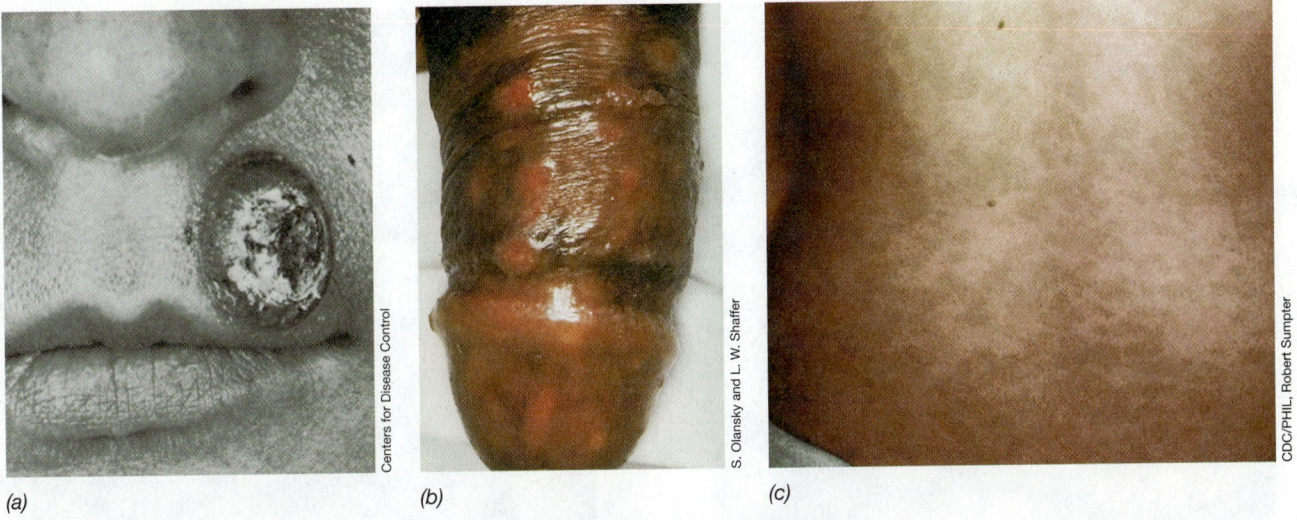

(a) Centers for Disease Control *(b)* S. Olansky and L. W. Shaffer *(c)* CDC/PHIL, Robert Sumpter

Figure 29.38 Primary and secondary syphilis. *(a)* Chancres on the lip and *(b)* the penis in cases of primary syphilis. The chancre is the characteristic lesion of primary syphilis at the site of infection by *Treponema pallidum*. *(c)* Syphilitic rash on the lower back of a patient showing secondary syphilis.

secondary stages of the disease can typically be cured by a single injection of benzathine penicillin G.

MINIQUIZ

- Explain at least one potential reason for the high incidence of gonorrhea as compared with syphilis.
- Describe the progression of untreated gonorrhea and untreated syphilis. Do treatments produce a cure for each disease?

29.13 Chlamydia, Herpes, and Human Papillomavirus

STIs caused by *Chlamydia* (a bacterium) and herpesvirus and human papillomavirus are very prevalent among sexually active adults and are often more difficult to diagnose and treat than are syphilis and gonorrhea.

Chlamydia

A number of sexually transmitted diseases can be ascribed to infection by the obligately intracellular bacterium *Chlamydia trachomatis* (**Figure 29.39**). This organism is one of a small group of parasitic bacteria that form their own phylum (the *Chlamydiae*) of *Bacteria* (⊂⊃ Section 15.15). Because *C. trachomatis* must be grown in host cells (tissue culture), its rapid isolation and identification is not as straightforward as for *Neisseria gonorrhoeae*.

The total incidence of sexually transmitted *C. trachomatis* infections probably greatly outnumbers the incidence of gonorrhea. Over 1 million chlamydial cases are now reported annually in the United States, but because of their often inapparent nature, there may be more than 4 million new sexually transmitted chlamydial infections every year. Because of this, chlamydia is the most prevalent STI and reportable communicable disease in the United States. *C. trachomatis* also causes a serious eye infection called *trachoma*, but the strains of *C. trachomatis* responsible for STIs are distinct from those causing trachoma. Chlamydial infections may also be transmitted congenitally to the newborn in the birth canal, causing newborn conjunctivitis and pneumonia.

Nongonococcal urethritis (NGU) due to *C. trachomatis* is one of the most frequently observed sexually transmitted diseases in males and females, but the infections are often inapparent. In a small percentage of cases, chlamydial NGU leads to serious acute complications, including testicular swelling and prostate inflammation in men and cervicitis, pelvic inflammatory disease, and fallopian tube damage in women. During NGU, cells of *C. trachomatis* can attach to microvilli of fallopian tube cells, enter, multiply, and eventually lyse the cells (Figure 29.39*b*). Untreated NGU in a female can thus cause infertility. Infections with the protist *Trichomonas vaginalis* can cause symptoms similar to those of chlamydial NGU, and we consider trichomoniasis along with other parasitic infections in Chapter 32.

Chlamydial NGU is frequently observed as a secondary infection following gonorrhea. Both *Neisseria gonorrhoeae* and *C. trachomatis* are often transmitted to a new host simultaneously. However, treatment of gonorrhea does not eliminate the chlamydia. Although cured of gonorrhea, these patients are still infected with chlamydia and eventually experience an apparent recurrence of gonorrhea that is instead a case of chlamydial NGU. Thus, patients treated for gonorrhea with drugs such as cefixime or ceftriaxone are also given azithromycin or doxycycline to treat a potential coinfection with *C. trachomatis*. A variety of clinical techniques including nucleic acid and immunological analyses are available for making a positive diagnosis of *C. trachomatis* infection, but drug therapy in the absence of a positive diagnosis is often prescribed.

Lymphogranuloma venereum (LGV) is a sexually transmitted disease caused by distinct strains of *C. trachomatis* (LGV 1, 2, and 3). The disease occurs most frequently in males and is characterized by infection and swelling of the lymph nodes in and about the groin. From the infected lymph nodes, chlamydial cells may travel to the rectum and cause a painful inflammation of rectal tissues called *proctitis*. LGV has the potential to cause

UNIT 6

Figure 29.39 **Cells of *Chlamydia trachomatis* (arrows) attached to human fallopian tube tissues.** *(a)* Cells attached to the microvilli of a fallopian tube. *(b)* A damaged fallopian tube containing a cell of *C. trachomatis* (arrow) in the lesion.

regional lymph node damage and the complications of proctitis. It is the only chlamydial infection that invades beyond the epithelial cell layer.

Herpes

Herpesviruses are a large group of double-stranded DNA viruses (♻ Section 9.7), many of which are human pathogens. The herpes simplex viruses are responsible for both cold sores and genital infections.

Herpes simplex 1 virus (*HSV-1*) infects the epithelial cells around the mouth and lips, causing cold sores, also known as fever blisters (**Figure 29.40**). HSV-1 is spread via direct contact with infectious lesions or through saliva. The incubation period of HSV-1 infections is short (3–5 days), and the lesions heal without treatment in 2–3 weeks. However, latent herpes infections are common, because the virus typically persists in low numbers in nerve tissue. Recurrent acute herpes infections can then occur when the virus is triggered by coinfections with other pathogens or by bodily stress. Oral herpes caused by HSV-1 is quite common and

apparently has no long-term harmful effects on the host, beyond the discomfort of the oral blisters.

Herpes simplex 2 virus (*HSV-2*) infections are associated primarily with the anogenital region, where the virus causes painful blisters on the penis of males or on the cervix, vulva, or vagina of females (**Figure 29.41**). HSV-2 infections are generally transmitted by direct sexual contact, and the disease is most easily transmitted when active blisters are present, but may also be transmitted during asymptomatic periods, even when the infection is presumably latent. HSV-2 occasionally infects other sites such as the mucous membranes of the mouth and can also be transmitted to a newborn by contact with herpetic lesions in the birth canal at

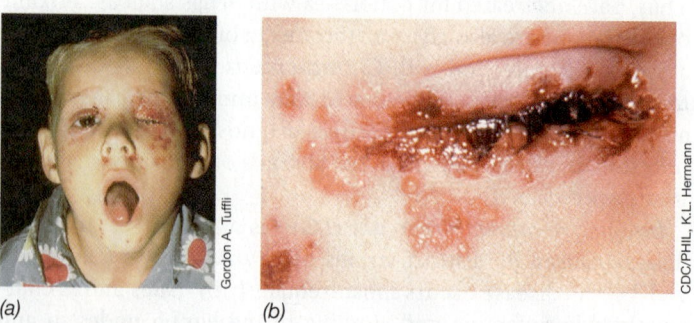

Figure 29.40 **Herpes simplex 1 virus infections.** *(a)* A severe case of herpes blisters on the face due to infection with herpes simplex 1 virus. *(b)* Close-up view of herpes blisters by the eye.

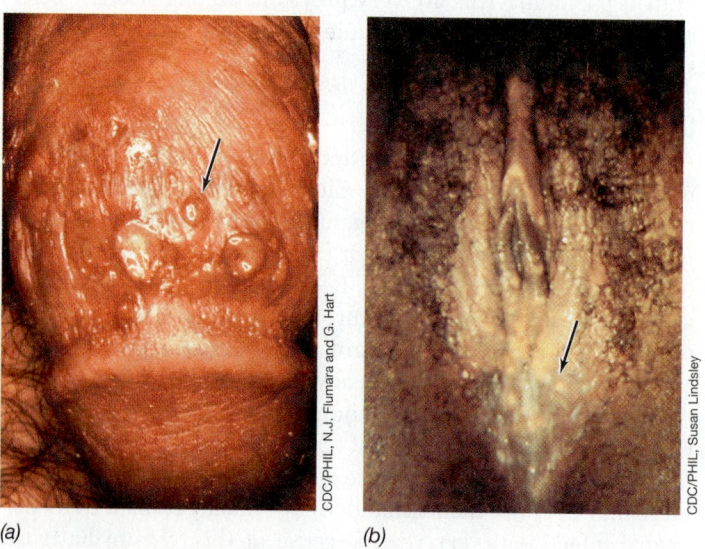

Figure 29.41 **Herpes simplex 2 virus infections.** Herpes simplex 2 virus blisters on the *(a)* penis and *(b)* vulva. As for herpes type 1, acute type 2 herpes infections can seemingly be cured only to reappear later from a persistent virus infection (♻ Figure 8.22).

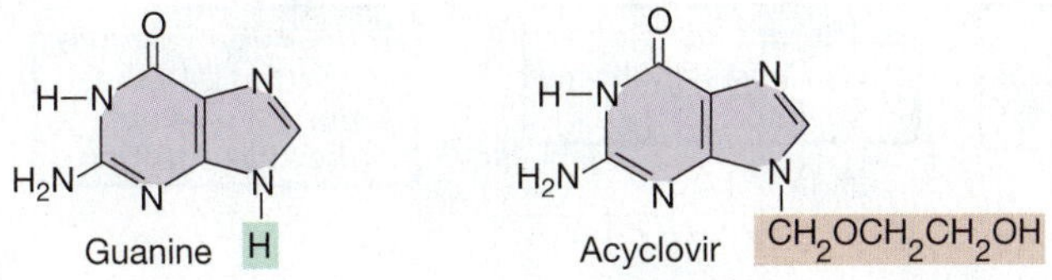

Figure 29.42 Guanine and the guanine analog acyclovir. Acyclovir has been used therapeutically to control genital herpes (HSV-2) blisters (Figure 29.41).

birth. The disease in the newborn varies from latent infections with no apparent damage to systemic disease resulting in brain damage or even death. To avoid herpes infections in newborns, delivery by cesarean section is advised for pregnant women with genital herpes infections.

The long-term effects of genital herpes infections are not fully understood. However, studies have indicated a significant correlation between genital herpes infections and cervical cancer in females. Genital herpes infections are presently incurable, although a limited number of drugs have been successful in controlling the infectious blister stages. The guanine analog acyclovir (Figure 29.42), given orally and also applied topically, is particularly effective in limiting the shed of active virus from blisters and promoting the healing of blistering lesions (Figure 29.41). Acyclovir, valacyclovir, and vidarabine are nucleoside analogs that interfere with herpesvirus DNA polymerase, inhibiting viral DNA replication. (Section 27.15).

Human Papillomavirus

As for herpesviruses, **human papillomaviruses (HPV)** comprise a family of double-stranded DNA viruses. Of more than 100 different strains, about 30 are transmitted sexually, and several of these cause genital warts and cervical cancer (Chapter 24 Explore the Microbial World "Vaccines and Public Health"). About 20 million people in the United States are infected, and up to 80% of women over age 50 have had at least one HPV infection. Over 6 million people acquire new HPV infections annually, leading to almost 10,000 cases of cervical cancer and about 3700 deaths.

Most HPV infections are asymptomatic, with some progressing to cause genital warts. Others cause cervical neoplasia (abnormalities in cells of the cervix), and a few progress to cervical cancers. Most HPV infections resolve spontaneously but, as with many viral infections, there is no adequate treatment or cure for active infections. Because human papillomaviruses are potentially oncogenic (cancer-causing), an HPV vaccine has been developed and is currently recommended for use in females 11–26 years of age. The vaccine has also been recommended for males because immunized males no longer carry HPV and thus cannot infect females, and because HPV infection in males can lead to anal and penile cancers.

MINIQUIZ

- Describe pertinent clinical features and treatment protocols for chlamydia, herpes, trichomoniasis, and human papillomavirus.

- Why are these diseases more difficult to diagnose than gonorrhea or syphilis?

29.14 HIV/AIDS

Acquired immunodeficiency syndrome (AIDS) is caused by the human immunodeficiency virus (HIV). Worldwide, more than 80 million people have been infected with HIV, and there are nearly 3 million new infections and over 2 million deaths each year. We covered the epidemiology of AIDS in Section 28.9.

HIV and a Definition of AIDS

HIV is of two types, *HIV-1* and *HIV-2*, but since more than 99% of global AIDS cases are due to HIV-1, we focus on HIV-1 here. HIV-1 is a retrovirus (Sections 8.10 and 9.11) that replicates in macrophages and T cells of the human immune system (Chapters 24 and 25). HIV infection eventually leads to the destruction of key immune system cells, greatly crippling the host immune response. Death from AIDS is the result of a secondary infection, typically caused by one or more **opportunistic pathogens**, potential pathogens that in a healthy individual would be controlled by the immune system.

The current definition of a case of HIV/AIDS is a patient who tests positive for HIV in immunological and/or nucleic acid-based tests and meets at least one of the following two criteria:

1. A CD4 T cell number of less than 200/μl of whole blood (the normal count is 600–1000/μl) or a CD4 T cell/total lymphocytes percentage of less than 14%.

2. A CD4 T cell number of more than 200/μl *and* any of the following diseases: candidiasis, coccidioidomycosis, cryptococcosis, histoplasmosis, isosporiasis, *Pneumocystis jiroveci* pneumonia, cryptosporidiosis, or toxoplasmosis of the brain (all fungal diseases) (Section 32.2); pulmonary tuberculosis or other mycobacterial infections, or recurrent *Salmonella* septicemia (bacterial diseases); cytomegalovirus infection, HIV-related encephalopathy, HIV wasting syndrome, chronic ulcers, or bronchitis due to herpes simplex (viral infections); or certain malignant diseases such as invasive cervical cancer, Kaposi's sarcoma, Burkitt's lymphoma, primary lymphoma of the brain, or immunoblastic lymphoma, or recurrent pneumonia due to any agent.

Pathogenesis of HIV/AIDS

HIV infects cells that have the CD4 cell surface protein. The two cell types most commonly infected are macrophages and a class of lymphocytes called T-helper (Th) cells, both of which are important components of the immune system (Chapter 24). Infection normally occurs first in macrophages. At the macrophage cell surface, the CD4 molecule binds to the gp120/gp41 capsid protein of HIV as the virus interacts with the macrophage receptor CCR5 (Figure 29.43). CCR5 is a coreceptor for HIV and, together with CD4, forms the docking site where the HIV envelope fuses with the host cytoplasmic membrane; this is required for the viral nucleocapsid to be inserted into the cell. Within the macrophage, HIV replicates (Figure 9.21) and makes an altered form of gp120 that recognizes a different coreceptor, CXCR4, on Th cells. HIV virions are released from macrophages and proceed to infect and replicate in Th lymphocytes; Th cells that produce HIV no longer divide and are eventually diminished by attrition.

In some HIV/AIDS patients, HIV infection does not progress immediately to killing host immune cells. HIV can exist

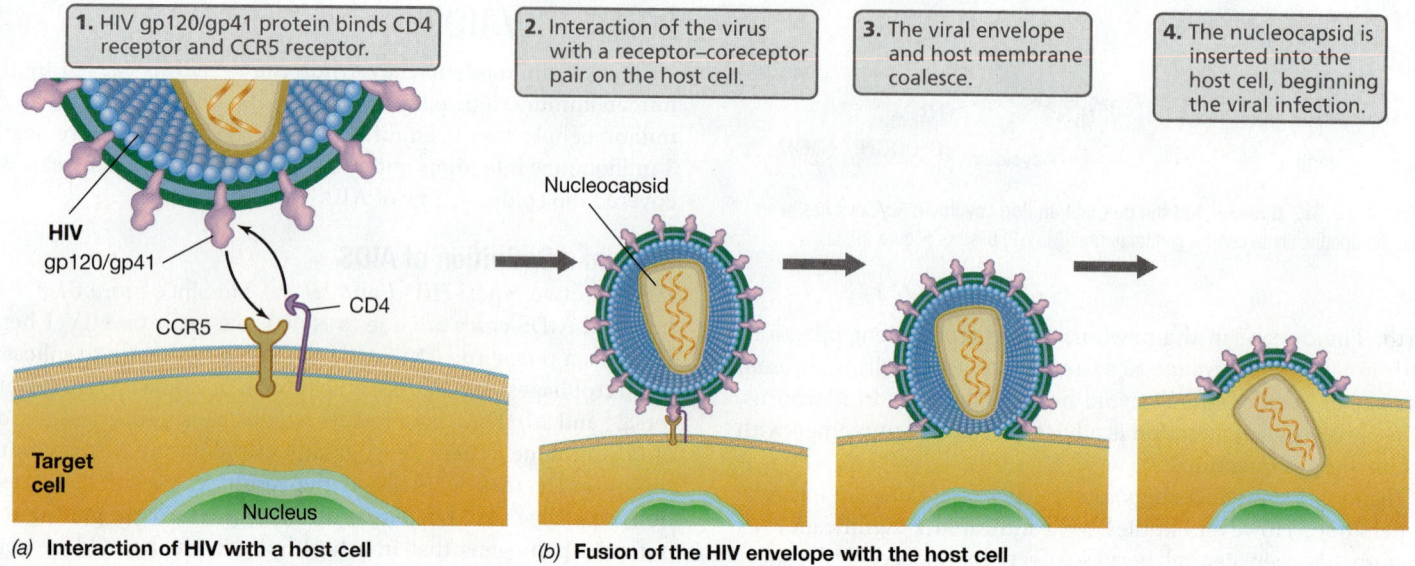

1. HIV gp120/gp41 protein binds CD4 receptor and CCR5 receptor.

2. Interaction of the virus with a receptor–coreceptor pair on the host cell.

3. The viral envelope and host membrane coalesce.

4. The nucleocapsid is inserted into the host cell, beginning the viral infection.

(a) **Interaction of HIV with a host cell**

(b) **Fusion of the HIV envelope with the host cell**

Figure 29.43 Infection of a CD4 target cell with HIV. (a) Recognition and binding of the virus by CCR5 and CD4 receptors. (b) The viral nucleocapsid eventually enters the cell. Details of the replication of the HIV genome are shown in Figures 8.24 and 9.21.

in a dormant state as a provirus; under these conditions, the reverse-transcribed HIV genome, now in the form of DNA, is integrated into host chromosomal DNA (⮂ Figure 8.24). At this point the cell may show no outward sign of infection. Indeed, HIV DNA can remain latent for long periods, replicating only as the host cell DNA replicates. However, sooner or later, HIV begins to replicate, and progeny virus are produced and released from the cell.

Symptoms of HIV/AIDS

Ongoing HIV infection results in a progressive decline in CD4 cell numbers. In a healthy human, CD4 cells constitute about 70% of the total T cell pool. In those with HIV/AIDS, CD4 numbers

steadily decrease, and by the time opportunistic infections begin to appear, CD4 cells are all but absent (**Figure 29.44**). The progression of untreated HIV infection to AIDS follows a typical pattern. First, there is an intense immune response to HIV and HIV numbers drop. But eventually, the immune response is overwhelmed and HIV levels slowly increase while CD4 T cells slowly decrease. When T cell numbers have dropped below about 200/mm³ of blood, the door is open for infections by opportunistic pathogens (Figure 29.44).

Opportunistic infections caused by normally benign protists, fungi, bacteria, and viruses occur with high prevalence in those with HIV/AIDS and are typically the actual cause of death (**Figure 29.45**). The most common opportunistic disease in

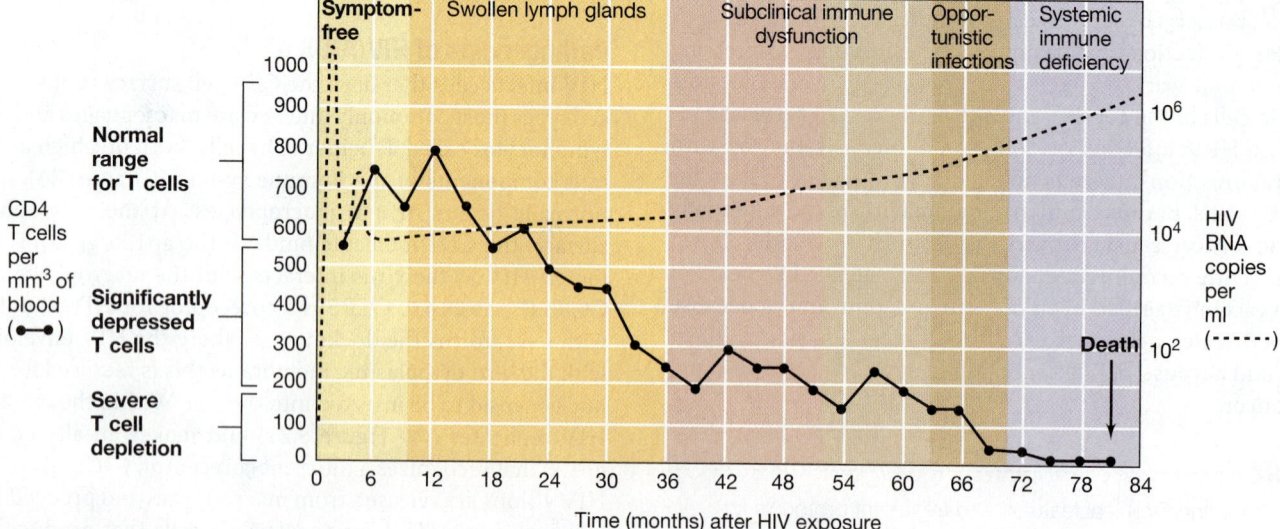

Figure 29.44 Decline of CD4 T lymphocytes and progress of HIV infection. During the typical progression of untreated AIDS, there is a gradual loss in the number and functional ability of the CD4 T cells, while the viral load, measured as HIV-specific RNA copies per milliliter of blood, gradually increases after an initial decline.

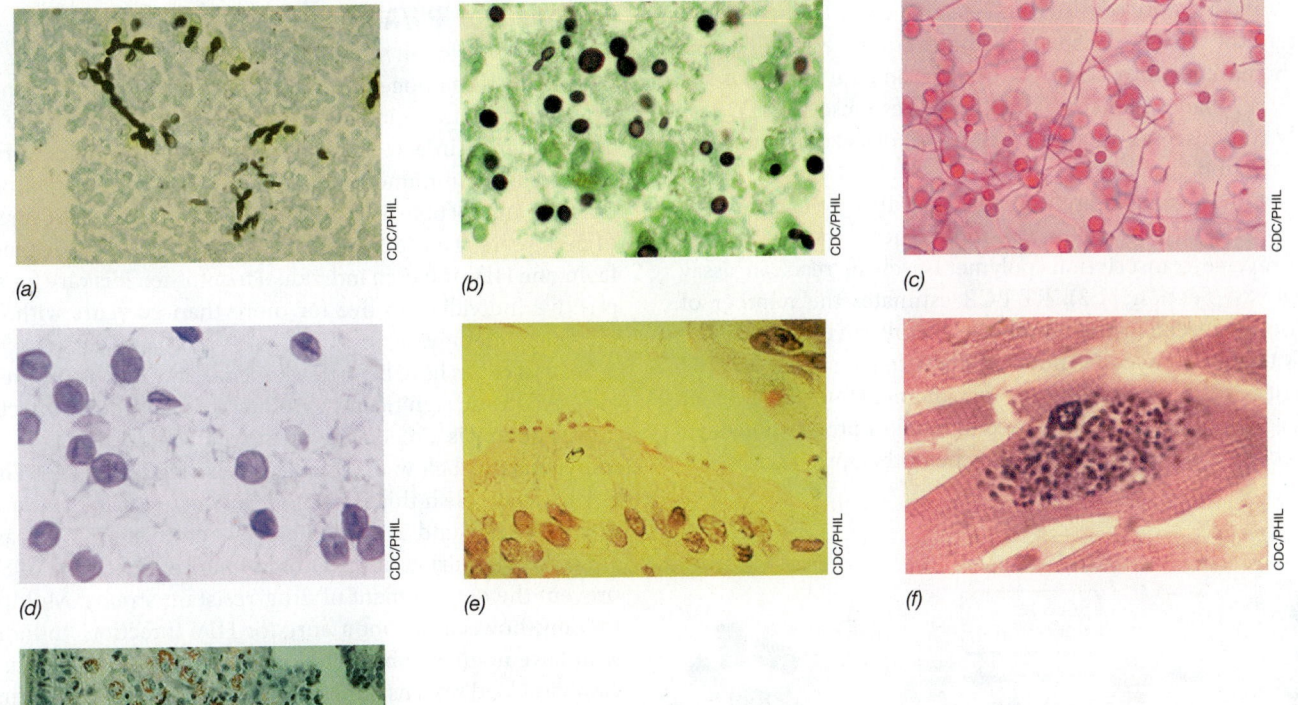

Figure 29.45 Opportunistic pathogens associated with HIV/AIDS. *(a) Candida albicans*, from heart tissue of patient with systemic *Candida* infection. *(b) Cryptococcus neoformans*, from lung tissue of an AIDS patient. *(c) Histoplasma capsulatum*, showing reproductive structures called macroconidia. *(d) Pneumocystis jiroveci*, from lungs of immune-compromised patient. *(e) Cryptosporidium* sp. from small intestine of a patient with cryptosporidiosis. *(f) Toxoplasma gondii*, from heart tissue of patient with toxoplasmosis. *(g) Mycobacterium* spp. infection of the small bowel (acid-fast stain). Coverage of many of these fungal and parasitic diseases can be found in Chapter 32.

HIV/AIDS patients is pneumonia caused by the fungus *Pneumocystis jiroveci* (Figure 29.45*d*), but infections by various molds, yeasts, protists, and bacteria are also seen (Figure 29.45). Almost all of these opportunistic pathogens are difficult to treat. For example, many of the drugs used to treat infections from fungi and protists (both *Eukarya*) have significant negative side effects on the host, and mycobacterial infections are often from drug-resistant strains (Section 29.4). A cancer frequently seen in HIV/AIDS patients is *Kaposi's sarcoma*, a cancer of the cells lining the blood vessels characterized by purple splotches on the skin, especially in the extremities (**Figure 29.46**). Kaposi's sarcoma is caused by coinfection of HIV and human herpesvirus 8 (HHV-8) and is rarely seen outside of HIV/AIDS patients.

Diagnosing HIV/AIDS

HIV infection is typically diagnosed by identifying antibodies to the pathogen in a patient blood sample. The HIV enzyme immunoassay (EIA, ↩ Figure 27.15*b, c*) is used for screening purposes, typically on a large scale, for example, of donated blood. A positive HIV EIA must be confirmed by an HIV immunoblot (Western blot, ↩ Figure 27.17) or by immunofluorescence (↩ Figure 27.14) to rule out the possibility of a false-positive screening test. Rapid and inexpensive HIV tests are available for preliminary screening of blood in clinics. One test requires only a single drop of patient blood and detects the gp41 HIV surface antigen (Figure

29.43) by producing a visible agglutination reaction. A second uses saliva as a source of anti-HIV antibodies and yields a colored product. In general, however, these rapid tests are not as sensitive or specific as the standard HIV EIA and thus positive tests should

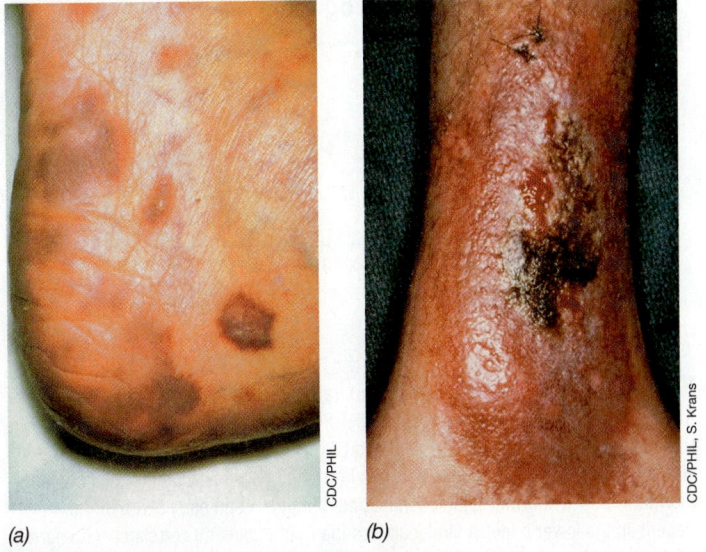

Figure 29.46 Kaposi's sarcoma. Lesions are shown as they appear on *(a)* the heel and lateral foot, and *(b)* the distal leg and ankle.

be confirmed by more sensitive and specific tests. Unfortunately, no matter how sensitive or specific, none of these tests will detect those who have recently acquired the virus and are infectious but have not yet made a detectable antibody response to HIV; this antibody response can require a period of 6 weeks or more following infection.

Diagnostic procedures are available that directly measure the number of HIV virions in a blood sample. These tests use a virus-specific reverse transcription–polymerase chain reaction assay (RT-PCR, ↺ Section 11.3). RT-PCR estimates the number of HIV virions present in the blood, the so-called **viral load**. The RT-PCR test for HIV load is not routinely used to screen for HIV because it is costly and technically demanding. However, after an initial diagnosis, the test is often used to monitor progression of an HIV infection and the effectiveness of chemotherapy (**Figure 29.47**).

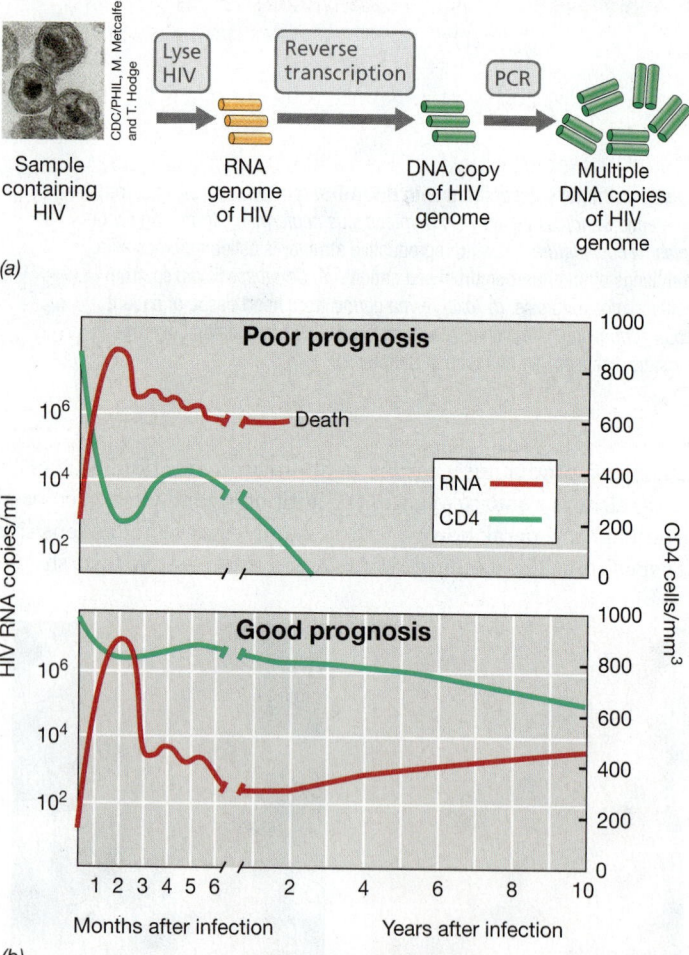

(a)

(b)

Figure 29.47 Monitoring of HIV load. *(a)* Procedure for detecting HIV by reverse transcription–polymerase chain reaction (RT-PCR) techniques. *(b)* Time course for HIV infection as monitored by HIV load and CD4 T cell counts. In the upper panel, a viral load greater than 10^4 copies/ml correlates with below normal CD4 cell numbers (normal = 600–1500/mm³), indicating a poor prognosis and early death of the patient. In the lower panel, a viral load less than 10^4 copies/ml correlates with normal CD4 cell numbers, indicating a good prognosis and extended survival of the patient. Data are adapted from the CDC, Atlanta, Georgia, USA.

Treatment of HIV/AIDS

The prognosis for an *untreated* HIV-infected individual is poor, as opportunistic pathogens or malignancies (Figures 29.45 and 29.46) eventually kill those infected. Long-term studies indicate that the average person infected with HIV progresses through several stages of decreasing immune function, with CD4 cells dropping from a normal range of 600–1000/mm³ of blood to near zero over a period of 5–7 years (Figure 29.44). Although the rate of decline varies from one HIV-infected individual to another, it is rare for an HIV-positive individual to live for more than 10 years without anti-HIV drug therapy.

Several drugs have been developed that delay the progression of HIV/AIDS and significantly prolong the life of those infected with HIV. Therapy is aimed at reducing the viral load of HIV-infected individuals to below detectable levels (Figure 29.44). The strategy to accomplish this is called *highly active anti-retroviral therapy* (HAART) and is carried out by administering at least three anti-retroviral drugs at once to inhibit the replication of HIV and prevent the development of drug-resistant strains. Multiple drug therapy, however, is not a cure for HIV infection. In individuals who have no detectable viral load after drug treatment, a significant viral load returns if therapy is interrupted or discontinued, or if multiple drug resistance develops.

Effective anti-HIV drugs fall into four categories, including two classes of *reverse transcriptase inhibitors*, various *protease inhibitors*, *fusion inhibitors*, and *integrase inhibitors*. Reverse transcriptase is the enzyme that converts the single-stranded RNA genome of HIV into cDNA and then double-stranded DNA and is essential for viral replication (↺ Sections 8.10 and 9.11). Cells lack reverse transcriptase and thus reverse transcriptase inhibitors are viral-specific. *Azidothymidine* (AZT) closely resembles the nucleoside thymidine but lacks the correct attachment site for the next base in a replicating nucleotide chain, resulting in termination of the growing DNA chain. AZT is thus a **nucleoside reverse transcriptase inhibitor** (**Figure 29.48***a*). **Nonnucleoside reverse transcriptase inhibitors**, such as *nevirapine* (Figure 29.48*b*), inhibit the activity of reverse transcriptase in a different way by interacting with the protein and altering the conformation of the catalytic site.

Another category of anti-HIV drugs is the **protease inhibitors**, such as *saquinavir* (Figure 29.48*c*). These are peptide analogs that inhibit processing of retroviral polypeptides (↺ Figure 9.22) by binding to the active site of the processing enzyme, *HIV protease*; this effectively inhibits viral maturation. **Fusion inhibitors** include *enfuvirtide*, a synthetic peptide that functions by binding to the gp41 protein on HIV capsids (Figure 29.43); this stops fusion of the viral envelope and the CD4 cell cytoplasmic membrane. Finally, there are the **integrase inhibitors**, such as *elvitegravir* and *raltegravir*. These drugs target HIV integrase, the protein that integrates the HIV genome into host cell DNA. The interference with integration of viral DNA into the host cell genome interrupts the HIV replication cycle.

All anti-HIV drugs rapidly decrease the viral load when given to HIV-infected individuals, but drug-resistant strains of HIV arise quickly if only a single drug is administered. A typical HAART protocol for treatment of an established HIV infection includes at

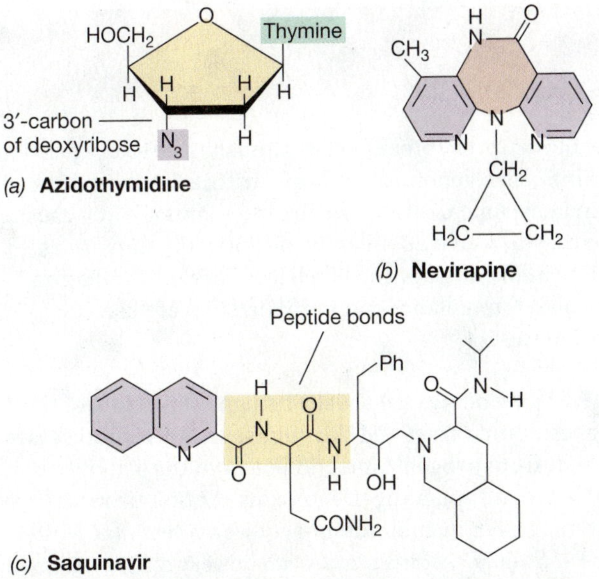

(a) **Azidothymidine**

(b) **Nevirapine**

(c) **Saquinavir**

Figure 29.48 HIV/AIDS chemotherapeutic drugs. (a) Azidothymidine (AZT), a nucleoside reverse transcriptase inhibitor. This nucleoside analog is missing the –OH group on the 3′ carbon, causing nucleotide chain elongation to terminate when the analog is incorporated, inhibiting virus replication. (b) Nevirapine, a nonnucleoside reverse transcriptase inhibitor, binds directly to the catalytic site of HIV reverse transcriptase, also inhibiting elongation of the nucleotide chain. (c) Saquinavir, a protease inhibitor, was designed by computer modeling to fit the active site of the HIV protease. Saquinavir is a peptide analog: The tan-highlighted area shows the region analogous to peptide bonds. Blocking the activity of HIV protease prevents the processing of HIV proteins and maturation of the virus.

least one protease or nonnucleoside reverse transcriptase inhibitor plus a combination of two nucleoside reverse transcriptase inhibitors. A resistant virus would, therefore, have to develop resistance to three drugs simultaneously, and the probability of this occurring is very small. A patient receiving this combination therapy is then monitored to track changes in viral load (Figures 29.44 and 29.47). An effective HAART protocol reduces viral load to nondetectable levels within several days. Drug therapy is then continued and the patient monitored for viral load indefinitely. If the viral load again reaches detectable limits, the drug cocktail is changed because an increase in viral load indicates the emergence of drug-resistant HIV.

In addition to drug resistance, some anti-retroviral drugs are toxic to the host. In many cases, nucleoside analogs are not well tolerated by patients, presumably because they interfere with host functions such as cell division. In general, the nonnucleoside reverse transcriptase inhibitors and the protease inhibitors are better tolerated because they target virus-specific functions. However, drug resistance and host toxicity are major problems in all forms of HIV therapy. Thus, new chemotherapeutic agents and drug protocols are constantly being developed and tailored to the needs of individual patients.

HIV/AIDS Prevention

Public education about how HIV/AIDS is transmitted, sexual abstinence, and avoidance of high-risk behavior remain the major tools used to prevent HIV/AIDS. HIV spread is linked to promiscuous sexual activities and other activities that involve exchange of body fluids, which include not only men who have sex with men, but also prostitution and intravenous drug use where needles are shared. In some countries, the fastest growing mode of HIV transmission is actually between heterosexual partners. Effective prevention of HIV transmission therefore requires avoiding high-risk behaviors, regardless of sexual partners.

The United States Surgeon General has issued specific recommendations for avoiding HIV infection. These include, in addition to the avoidance of intravenous drug use where needles are shared:

1. Avoiding mouth contact with penis, vagina, or rectum.

2. Avoiding all sexual activities that could cause cuts or tears in the linings of the rectum, vagina, or penis.

3. Avoiding sexual activities with individuals from high-risk groups. These include prostitutes (both male and female); those who have multiple sex partners, particularly homosexual men and bisexuals; and intravenous drug users.

4. If a person has had sex with someone in a high-risk group, a blood test should be done to determine if infection with HIV has occurred. The blood test should be repeated at intervals for a year or longer because of the lag time in the immune response. If the test is positive, the sexual partners of the HIV-positive individual must be protected by use of a condom during sexual activities.

MINIQUIZ

• Review the definition of HIV/AIDS. Which symptoms of HIV/AIDS are shared by all HIV/AIDS patients?

• What does the enzyme reverse transcriptase do and why is it a good target for anti-HIV drugs?

• What are the current prevention guidelines for HIV/AIDS infection? Are they effective?

UNIT 6

BIG IDEAS

29.1 • Bacterial and viral respiratory pathogens are transmitted in air. Most respiratory pathogens are transferred from person to person via respiratory aerosols generated by coughing, sneezing, talking, or breathing, or by direct or fomite contact. Respiratory pathogens infect either the upper or lower respiratory tracts and sometimes both.

29.2 • Streptococcal diseases include strep throat and pneumococcal pneumonia. *Streptococcus pyogenes* infections may progress into serious conditions such as scarlet and rheumatic fevers, and pneumococcal pneumonia can have high mortality. Both pathogens can be cultured and both are treatable with antimicrobial drugs including penicillin.

29.3 • Diphtheria is an acute respiratory disease caused by *Corynebacterium diphtheriae*. Early childhood immunization is effective for preventing this very serious respiratory disease. Whooping cough is an endemic disease caused by *Bordetella pertussis*. Immunization of children, adolescents, and adults can control its propagation and spread.

29.4 • Tuberculosis is one of the most prevalent and dangerous infectious diseases in the world. Its incidence is increasing in developed countries in part because of the emergence of drug-resistant strains of *Mycobacterium tuberculosis*. The pathology of tuberculosis and other mycobacterial diseases such as Hansen's disease (leprosy) is influenced by the cellular immune response.

29.5 • *Neisseria meningitidis* is a common cause of meningococcemia and meningitis in young adults and occasionally occurs in epidemics in enclosed populations. Bacterial meningitis and meningococcemia can have high mortality rates, and treatment and prevention strategies including vaccines are available.

29.6 • Viral respiratory diseases are highly infectious and may cause serious health problems, although most are controllable and not life-threatening. The measles/mumps/rubella (MMR) vaccine is highly effective in controlling these diseases.

29.7 • Colds are the most common infectious viral diseases. Usually caused by a rhinovirus, colds are generally mild and self-limiting diseases; "cold drugs" may help to moderate symptoms but are not a cure. Each infection induces specific, protective immunity, but the large number of cold viruses precludes complete protective immunity or vaccines.

29.8 • Influenza is caused by an RNA virus that contains a segmented genome and is easily transmitted by the airborne

route. Influenza outbreaks occur annually due to the plasticity of the influenza genome. Antigenic drift varies the nature of the viral envelope of influenza viruses in minor ways, causing influenza seasonal epidemics, while antigenic shift varies the virus in major ways and can trigger periodic influenza pandemics. Surveillance and immunization are used to control influenza.

29.9 • Staphylococci are usually benign inhabitants of the upper respiratory tract and skin, but several serious diseases can result from pyogenic infection or from the activity of staphylococcal superantigen exotoxins. Antibiotic resistance is common, even in community-acquired infections. MRSA strains of *Staphylococcus aureus* can be very difficult to treat and cause significant tissue damage.

29.10 • *Helicobacter pylori* infection is the common cause of gastric ulcers. Gastric ulcers are now treated with antibiotics as an infectious disease, promoting a permanent cure.

29.11 • Viral hepatitis can result in acute liver disease, which may be followed by chronic liver disease (cirrhosis). Hepatitis B and C viruses in particular are transmitted by direct contact and can cause chronic infections leading to liver cancer. Vaccines are available for hepatitis viruses A and B. Viral hepatitis is still a major public health problem because of the high infectivity of the viruses and the lack of effective treatments.

29.12 • Gonorrhea and syphilis, caused by *Neisseria gonorrhoeae* and *Treponema pallidum*, respectively, are STIs with potential serious consequences if infections are not treated. In the United States, the incidence of gonorrhea has decreased in the last several years, but the incidence of syphilis has increased.

29.13 • Chlamydia is the most prevalent of STIs, and if left untreated, can cause serious complications in both males and females. Herpes simplex viruses cause incurable infections transmitted by oral or genital contact with herpes 1 or herpes 2, respectively. Human papillomaviruses cause widespread STIs that may lead to cervical and other cancers, but effective HPV vaccines are available.

29.14 • HIV is a retrovirus that destroys the immune system, leading to AIDS, and opportunistic pathogens eventually kill the host. There is no effective cure or vaccine for HIV infection, although antiviral drugs may slow or stop the progress of AIDS. Preventing HIV infection requires education and avoidance of high-risk behaviors involving exchange of body fluids.

REVIEW OF KEY TERMS

Antigenic drift a minor change in influenza virus antigens due to gene mutation

Antigenic shift a major change in influenza virus antigen due to gene reassortment

Cirrhosis breakdown of normal liver architecture, resulting in fibrosis

Congenital syphilis syphilis contracted by an infant from its mother during pregnancy

Fusion inhibitor a synthetic polypeptide that binds to viral glycoproteins, inhibiting fusion of viral and host cell membranes

Hepatitis liver inflammation, commonly caused by an infectious agent

Human papillomavirus (HPV) a sexually transmitted virus that causes genital warts, cervical neoplasia, and cancer

Integrase inhibitor drug that interrupts the HIV replication cycle by interfering with integrase, the HIV protein that catalyzes the integration of viral dsDNA into host cell DNA

Meningitis inflammation of the meninges (brain tissue), sometimes caused by *Neisseria meningitidis* and characterized by sudden onset of headache, vomiting, and stiff neck, often progressing to coma within hours

Meningococcemia a rapidly progressing severe disease caused by *Neisseria meningitidis* and characterized by septicemia, intravascular coagulation, and shock

Nonnucleoside reverse transcriptase inhibitor a nonnucleoside compound that inhibits the action of retroviral reverse transcriptase by binding directly to the catalytic site

Nucleoside reverse transcriptase inhibitor a nucleoside analog compound that inhibits the action of retroviral reverse transcriptase by competing with nucleosides

Opportunistic pathogen an organism that causes disease in the absence of normal host resistance

Pertussis (whooping cough) a disease caused by an upper respiratory tract infection with *Bordetella pertussis*, characterized by a deep, persistent cough

Protease inhibitor a compound that inhibits the action of viral protease by binding directly to the catalytic site, preventing viral protein processing

Rheumatic fever an inflammatory autoimmune disease triggered by an immune response to infection by *Streptococcus pyogenes*

Scarlet fever characteristic reddish rash resulting from an exotoxin produced by *Streptococcus pyogenes*

Sexually transmitted infection (STI) an infection that is usually transmitted by sexual contact

Toxic shock syndrome (TSS) the acute systemic shock resulting from a host response to an exotoxin produced by *Staphylococcus aureus*

Tuberculin test a skin test for previous infection with *Mycobacterium tuberculosis*

Viral load a quantitative assessment of the amount of virus in a host organism, usually in the blood

REVIEW QUESTIONS

1. Why do gram-positive bacteria cause respiratory diseases more frequently than gram-negative bacteria? (Section 29.1)

2. What are the typical symptoms of a streptococcal respiratory infection? Why should streptococcal infections be treated promptly? (Section 29.2)

3. Describe the causal agents and the symptoms of diphtheria and pertussis. Why has diphtheria incidence declined in the United States, while pertussis incidence is higher than a decade ago? (Section 29.3)

4. Describe the process of infection by *Mycobacterium tuberculosis*. Does infection always lead to active tuberculosis? Why or why not? How is exposure to *M. tuberculosis* detected in humans? (Section 29.4)

5. Describe the symptoms of meningococcemia and meningitis. How are these diseases treated? What is the prognosis for each? (Section 29.5)

6. Compare and contrast measles, mumps, and rubella. Include a description of the pathogen, major symptoms encountered, and any potential consequences of these infections. Why is it important that women be vaccinated against rubella before puberty? (Section 29.6)

7. Why are colds such common respiratory diseases and why are vaccines not used to prevent colds? (Section 29.7)

8. Why is influenza such a common respiratory disease? How are influenza vaccines chosen? (Section 29.8)

9. Distinguish between pathogenic staphylococci and those that are part of the normal flora. (Section 29.9)

10. Describe the evidence linking *Helicobacter pylori* to gastric ulcers. How can these ulcers be cured? (Section 29.10)

11. Describe the major pathogenic hepatitis viruses. How are they related to one another? How is each spread? (Section 29.11)

12. Why did the incidence of gonorrhea rise dramatically in the mid-1960s, while the incidence of syphilis actually decreased at the same time? (Section 29.12)

13. For the sexually transmitted infections of chlamydia, herpes, and human papillomavirus, describe the organism that causes each. In each case, is treatment possible, and if so, is it an effective cure? Why or why not? (Section 29.13)

14. Describe how human immunodeficiency virus (HIV) effectively shuts down both humoral immunity and cell-mediated immunity. What is HAART therapy? (Section 29.14)

APPLICATION QUESTIONS

1. Why is it that you get a cold or two each year but if you had a case of measles, it was a one-time occurrence?

2. Why does active tuberculosis often lead to a permanent reduction in lung capacity, whereas most other respiratory diseases cause only temporary respiratory problems? Worldwide, the prevalence of tuberculosis infection is very high, but active disease is much lower. Explain.

3. Your college roommate goes home for the weekend, becomes extremely ill, and is diagnosed with bacterial meningitis at a local hospital. Because he was away, university officials are not aware of his illness. What should you do to protect yourself against meningitis? Should you notify university health officials?

4. Contrast an HIV infection with an infection by any other viral pathogen considered in this chapter, regardless of mode of transmission. Why do untreated cases of HIV infection inevitably lead to death whereas untreated cases of chicken pox, influenza, or even hepatitis typically do not?

5. Discuss the molecular biology of antigenic shift in influenza viruses and comment on the immunological consequences for the host. Why does antigenic shift prevent the production of a single universally effective vaccine for influenza control? Next, compare antigenic shift to antigenic drift. Which causes the greatest antigenic change? Which creates the biggest problems for vaccine developers? Which can lead to pandemic influenza, and why?

6. As the director of your dormitory's public health advisory group, you are charged to present information on chlamydia, herpes, trichomoniasis (see Chapter 32), and human papillomavirus infection, all STIs. Besides this textbook, where can you get reliable information about STIs and what information would you present? For each of these diseases, discuss the individual and public health issues that must be addressed.

30 · Vectorborne and Soilborne Bacterial and Viral Diseases

microbiology**now**

Vampire Bats and Rabies

Rabies is a disease of warm-blooded animals—primarily mammals—caused by an RNA virus. The rabies virus triggers an acute encephalitis and eventual death of the animal. Over 55,000 people, mainly children, die each year from rabies worldwide. Most rabies deaths are linked to bites from rabid dogs. Cases of human rabies from wild animal bites, such as from a fox, raccoon, or skunk, are rare. In the United States, most cases of human rabies, which average about three per year, are linked to bites from bats.

At least 47 species of bats inhabit the United States and Canada. Most of these have a diet of either insects or fruits. One bat not native to this part of North America is the vampire bat—a species that thrives on blood (photo). Vampire bats are New World animals and inhabit warm climates. They range from Mexico to as far south as Argentina, and until 2010, no cases of human rabies in the United States had been linked to a vampire bat bite. However, in August of that year, a migrant worker died of rabies in Louisiana. The worker had received a bat bite 2 weeks earlier in Mexico and was beginning to show some of the neurological symptoms of rabies. Less than a month later the worker was dead and samples collected at autopsy showed infection with a strain of rabies virus previously linked to vampire bats.[1]

This rabies death signals the possibility of a serious new means of rabies transmission in the United States. With global climate change pushing subtropical climates ideal for the vampire bat farther north, it is likely that these bats will begin to inhabit southern regions of the United States soon. And because vampire bats consume blood, they could become important new rabies vectors, transmitting the disease to livestock and other animals not normally associated with rabies.

[1]Balsamo, G., et al. 2011. Human rabies from exposure to a vampire bat in Mexico–Louisiana, 2010. *Morbidity and Mortality Weekly Report. 60*: 1050–1052.

In this chapter we focus on pathogenic bacteria and viruses transmitted to humans by animals, arthropods, or soil. Animal-transmitted pathogens have their origins in nonhuman vertebrates, and these infected animal populations can transmit infections to humans. Some arthropods are disease vectors, spreading pathogens to new hosts from a bite. Soilborne pathogens are transmitted to humans through either direct contact with soil or contact with infected animal fur or hides. A few of the diseases we will explore in this chapter produce only mild symptoms and are typically self-limiting. But most are highly dangerous with life-threatening symptoms and high mortality rates. These include such dreaded diseases as rabies, hantavirus syndromes, yellow fever, and plague.

I • Animal-Transmitted Viral Diseases

A zoonosis is an animal disease transmissible to humans, generally by direct contact, aerosols, or bites. Immunization and veterinary care control many infectious diseases in domesticated animals, reducing the transfer of zoonotic pathogens to humans. However, wild animals neither receive veterinary care nor are they immunized, making them a source of potential zoonoses. Diseases in animals may be **enzootic**, present endemically in certain populations, or **epizootic**, with incidence reaching epidemic proportions. In this unit we focus on two typically enzootic viral diseases, rabies and hantavirus syndromes, both of which can be transmitted to humans.

30.1 Rabies Virus and Rabies

Rabies occurs in wild animals, and the major enzootic reservoirs of rabies virus in the United States are raccoons, skunks, coyotes, foxes, and bats (see page 885). A small number of rabies cases also occur annually in domestic animals (**Figure 30.1**).

Symptoms and Pathology of Rabies

Rabies is caused by a rhabdovirus, a single-stranded minus-sense RNA virus (⇄ Section 9.9) that infects cells of the central nervous system in most warm-blooded animals, almost invariably leading to death once symptoms have developed. The virus (**Figure 30.2a**) enters the body from virus-contaminated saliva through a wound from a bite or through contamination of mucous membranes. Rabies virus multiplies at the site of inoculation and travels to the central nervous system. The incubation period before the onset of symptoms is highly variable and depends on the host, the size, location, and depth of the inoculating wound, and the titer of rabies virions transmitted in the bite. In dogs, the incubation period for rabies is less than two weeks. By contrast, in humans, 9 months or more may pass before rabies symptoms become apparent.

Rabies virus proliferates in the brain, especially in the thalamus and hypothalamus. Infection leads to fever, excitation, dilation of the pupils, excessive salivation, and anxiety. A fear of swallowing (hydrophobia, an early name for rabies) develops from uncontrollable spasms of the throat muscles, and death eventually results from respiratory paralysis. In humans, an *untreated* rabies infection in which symptoms have begun is almost always fatal. Fortunately for both domestic animals and humans, a very effective rabies vaccine exists and this keeps the incidence of rabies low in domestic animals (Figure 30.1a) and a rarity in humans.

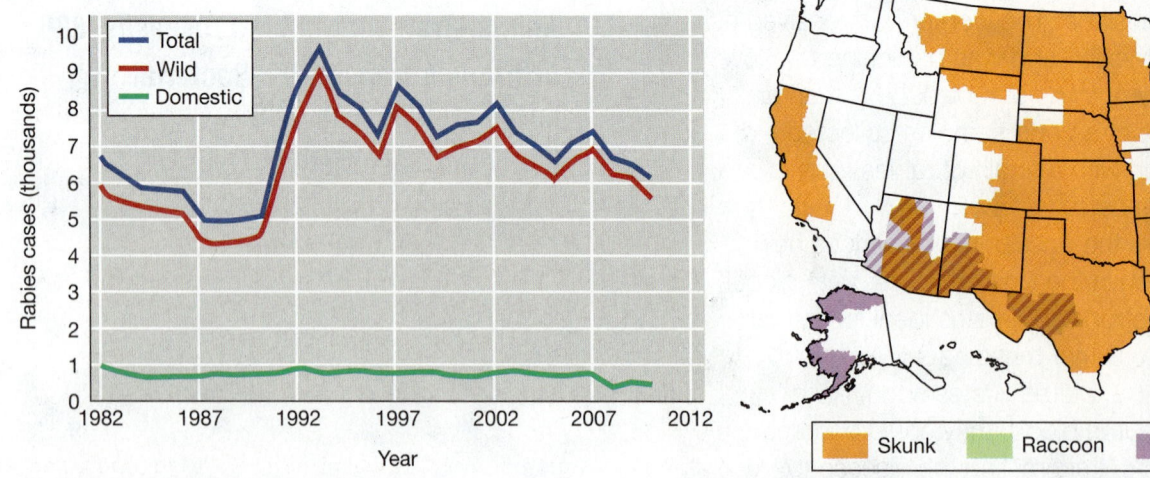

(a) **Incidence of rabies in the United States**

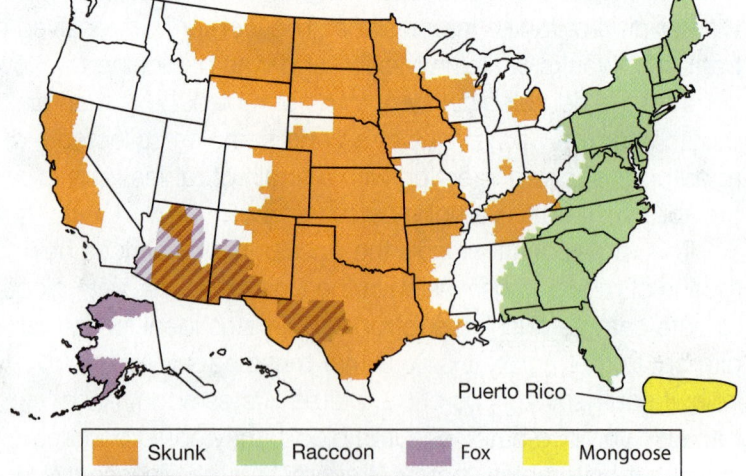

| Skunk | Raccoon | Fox | Mongoose |

(b) **Major vectors of rabies in the United States**

Figure 30.1 Rabies cases in wild and domestic animals in the United States. *(a)* Incidence of rabies by year. Human cases are fewer than 5 per year. *(b)* Major reservoirs of rabies virus. In some areas, for example, southwest Texas, both skunks and foxes are the major rabies reservoirs. Over 90% of all reported rabies cases occur in wild animals. However, actual numbers are probably significantly higher than shown in part a due to undiagnosed cases and undiscovered rabid animal carcasses. Data are from the Centers for Disease Control and Prevention, Atlanta, Georgia, USA.

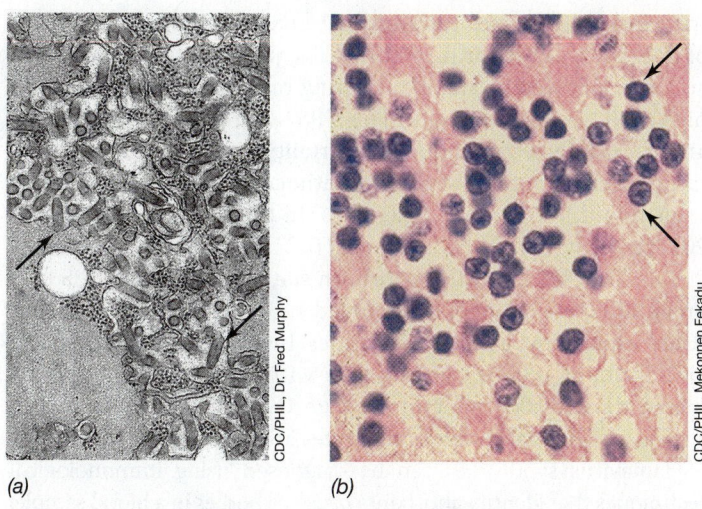

(a)   (b)

CDC/PHIL, Dr. Fred Murphy

CDC/PHIL, Mekonnen Fekadu

Figure 30.2 Rabies virus. *(a)* The bullet-shaped rabies virions (arrows) shown in this transmission electron micrograph of a tissue section from a rabid animal are about 75 × 180 nm. *(b)* Pathology of rabies in humans. In brain tissue, rabies virus causes characteristic cytoplasmic inclusions called Negri bodies (arrows), which contain rabies virus antigens. Negri bodies are about 2–10 μm in diameter.

Diagnosis, Treatment, and Prevention of Rabies

Rabies is diagnosed in the laboratory by examining tissue samples for the virus. Fluorescent antibodies that bind to rabies virus in brain tissues are used to confirm a case of rabies in a postmortem examination. Viral inclusions called *Negri bodies* are seen in the cytoplasm of nerve cells stained for light microscopy and these characteristic structures confirm rabies virus infection as well (Figure 30.2*b*).

Because rabies is such a serious disease, firm guidelines for treating possible human exposure to rabies have been established, and the details can be found in the rabies section of the World Health Organization website (http://www.WHO.int). In summary, the guidelines state that if a wild or stray animal is suspected of being rabid, it should be immediately examined for evidence of the rabies virus. If a domestic animal, generally a dog, cat, or ferret, bites a human, especially if the bite is unprovoked, the animal should be held in quarantine for 10 days to check for signs of rabies. If the animal exhibits rabies symptoms, or a definitive diagnosis of its illness cannot be made after 10 days, the human should be passively immunized with rabies immune globulin (purified human antibodies to rabies virus) injected at both the site of the bite and intramuscularly. The patient should also be actively immunized with a rabies virus vaccine. Because of the very slow progression of rabies in humans, this combination of passive and active immune therapy (⇄ Section 24.6) is nearly 100% effective, stopping the onset of the disease.

Rabies is prevented largely through immunization. An inactivated rabies vaccine is used in the United States for both humans and domestic animals. Prophylactic rabies immunization is practiced for individuals at high risk, such as veterinarians, animal control personnel, animal researchers, and individuals who work in rabies research or rabies vaccine production laboratories. The rabies problem is primarily with wild animals (Figure 30.1), where traditional means of vaccination are impossible. However, experimental

trials with an oral rabies vaccine administered in food "baits" have reduced the incidence and spread of rabies in limited geographic areas. If herd immunity (⇄ Section 28.2) could be established in some of the key carriers of rabies (Figure 30.1*b*), it might be possible to reduce incidence of the disease dramatically. Some states and countries, such as Hawaii and Great Britain, are rabies-free, and any animal imported into these areas is subject to quarantine.

Although rabies is vaccine-preventable, about 55,000 people per year die from rabies, primarily in developing countries in Asia and Africa where rabies is enzootic in domestic animals due to inadequate vaccination practices. Worldwide, nearly 14 million people receive prophylactic treatment for rabies after exposure annually, and in the United States, over 20,000 individuals receive such treatment. Fewer than three cases of human rabies are reported in the United States each year, nearly always the result of bites from wild animals. Because domestic animals often have exposure to wild animals, dogs and cats are routinely vaccinated against rabies beginning at 3 months of age. Large farm animals, especially horses, are often immunized against rabies, as well.

MINIQUIZ

- What is the procedure for treating a human bitten by an animal if the animal cannot be found?
- What major advantage does an oral vaccine have over a parenteral (injected) vaccine for rabies control in wild animals?

30.2 Hantavirus and Hantavirus Syndromes

Hantaviruses cause two severe, emerging diseases, **hantavirus pulmonary syndrome (HPS)**, an acute respiratory and cardiac disease, and **hemorrhagic fever with renal syndrome (HFRS)**, an acute disease characterized by shock and kidney failure. Both diseases are caused by hantaviruses transmitted from infected rodents. Hantavirus is named for Hantaan, Korea, the site of a hemorrhagic fever outbreak where the virus was first recognized as a human pathogen.

Symptoms and Pathology of Hantavirus Syndromes

Hantaviruses are enveloped viruses with single-stranded minus-sense RNA genomes arranged in segments (**Figure 30.3**; ⇄ Section 9.9); hantaviruses are related to other hemorrhagic fever viruses such as Lassa fever virus and Ebola virus (⇄ Section 28.7). Hantaviruses infect rodents including mice, rats, lemmings, and voles, without causing disease. The virus is transmitted from these reservoirs to humans by inhalation of virus-contaminated rodent excreta. Humans are accidental hosts and are infected only when they come into contact with rodents, their waste, or their saliva.

HPS is characterized by a sudden onset of fever, muscle pain, a reduction in the number of blood platelets along with an increase in the number of circulating leukocytes, and hemorrhaging. Death if it occurs takes several days, and is usually a result of systemic shock and cardiac complications precipitated by leakage of fluid into the lungs, causing suffocation and heart failure. These symptoms are typical of hantaviruses, but other symptoms such as kidney failure are common, depending on the strain of virus causing the disease. HFRS is characterized by intense headache, back and abdominal pain, renal dysfunction, and various hemorrhagic complications. HPS strains

UNIT 6

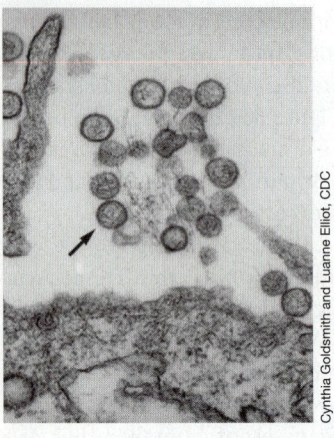
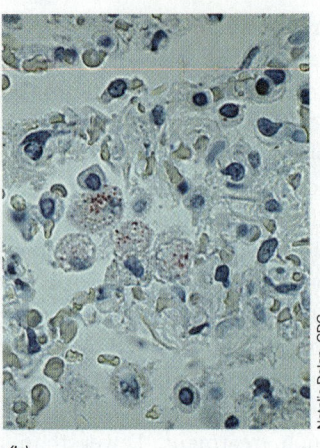

(a)   (b)

Figure 30.3 Hantavirus. *(a)* An electron micrograph of the Sin Nombre hantavirus. The arrow indicates one of several virions that are about 100 nm in diameter. *(b)* Immunostaining of Andes hantavirus antigens in alveolar macrophages. Each granular dark blue–stained area indicates cellular infection of an individual macrophage (each cell is about 15 μm in diameter).

are more prevalent in the Americas, whereas HFRS strains are more prevalent in hantavirus outbreaks in Eurasia. HPS strains typically show a significantly higher mortality rate than HFRS strains.

Hantaviruses can be cultured in the laboratory but must be handled with biosafety level 4 (BSL-4; ↩ Section 27.1) precautions. In the world of infectious diseases, hantavirus and other BSL-4 viral pathogens are considered "the worst of the worst" and are thus handled in the United States by the Special Pathogens Branch of the Centers for Disease Control and Prevention in Atlanta (Georgia, USA) (Explore the Microbial World, "Handling Viral Hemorrhagic Fever Viruses").

Epidemiology, Diagnosis, and Prevention of Hantavirus Syndromes

A significant HPS outbreak in the United States occurred near the Four Corners region of Arizona, Colorado, New Mexico, and Utah in 1993. The outbreak resulted from an enlarged population of deer mice in the spring of 1993. The previous winter was mild and was followed by abundant spring rains, triggering unusually high food levels for the mice. The HPS outbreak caused 32 deaths among 53 infected people (60% mortality), illustrating the potential danger of outbreaks due to pathogens that can be directly transmitted from animal reservoirs. In total from 1993 through 2011, there have been 587 cases of HPS in the United States, with 211 deaths (36%), mostly in western states. In 2012, an HPS outbreak affecting 10 people occurred in Yosemite National Park (California, USA) with three fatalities. Incidence of infection with HFRS strains is much higher than with HPS strains. It is estimated that 200,000 infections occur annually, chiefly in China, Korea, and Russia, but mortality rates are typically very low.

Hantavirus syndromes can be diagnosed using immunological techniques that identify anti-hantavirus antibodies in a blood sample. These include immunoassays (Figure 30.3*b* and ↩ Section 27.9) that detect both exposure to the virus and the strength of the immune response. The presence of the viral RNA genome from circulating virions can also be detected using RT-PCR (↩ Sections 11.3 and 27.10) on patient tissue or blood samples.

There is no virus-specific treatment or vaccine for hantavirus diseases. Treatment amounts to isolation, rest, rehydration, and alleviation of other symptoms. Hantavirus infection can be prevented by avoiding rodent contact and rodent habitat. Destruction of mouse habitat, restricting food supplies (for example, keeping human food in sealed containers), and aggressive rodent extermination measures are the only effective controls, since animal surveys have shown that in areas that have experienced a hantavirus outbreak, a high proportion of mice carry the virus.

MINIQUIZ -

- Why are hantaviruses considered a major public health problem in the United States?
- Describe the spread of hantaviruses to humans. What are some effective measures for preventing infection by hantaviruses?

II • Arthropod-Transmitted Bacterial and Viral Diseases

Pathogens can be spread to new hosts from the bite of an infected arthropod. In the bacterial and viral diseases we consider here—the rickettsial illnesses, yellow and dengue fevers, Lyme disease, and plague—humans are only *accidental hosts* for the pathogen. The *reservoir* of the pathogen is the arthropod vector. Nevertheless, the diseases can be devastating and often fatal.

30.3 Rickettsial Diseases

The **rickettsias** are small *Bacteria* that live an obligate intracellular existence and are associated with bloodsucking arthropods such as fleas, lice, or ticks. We discussed the biology of rickettsias in Section 15.1. Of the diseases that rickettsias can cause in humans and other vertebrates, the most important are *typhus fever, spotted fever rickettsiosis (Rocky Mountain spotted fever)*,

and *ehrlichiosis*. Rickettsias have not been cultured in artificial culture media but can be grown in laboratory animals, ticks and lice, mammalian tissue culture cells, and the yolk sac of chick embryos (see Figure 30.6*b*). In animals, growth takes place primarily in phagocytes, such as macrophages.

Rickettsias are divided into three groups, based loosely on the clinical diseases they cause. The groups are (1) the *typhus group*, such as *Rickettsia prowazekii*; (2) the *spotted fever group*, such as *Rickettsia rickettsii*; and (3) the *ehrlichiosis group*, characterized by *Ehrlichia chaffeensis*.

The Typhus Group: *Rickettsia prowazekii*

Typhus is transmitted from person to person by the common body or head louse (**Figure 30.4*a***), and humans are the only known

EXPLORE THE
MICROBIAL WORLD

Handling Viral Hemorrhagic Fever Viruses

The Special Pathogens Branch of the Centers for Disease Control and Prevention (CDC) specializes in the handling of a subgroup of dangerous viral pathogens, the *hemorrhagic fever viruses*. These viruses are the most lethal infectious agents known and cause a series of viral hemorrhagic fevers (VHFs), including hantavirus syndromes and Ebola.

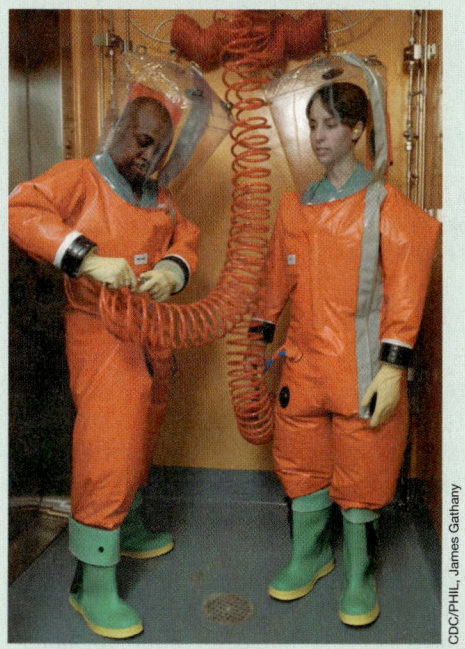

Figure 1 Biosafety level 4. Microbiologists suit up before entering the CDC BSL-4 lab. The suits supply filtered air and are positively pressurized to prevent back-flushing of laboratory air or particles.

Hemorrhagic fever viruses are handled under biosafety level 4 (BSL-4) standards (**Figure 1**). BSL-4 is the highest level of biological containment and is used only when working with samples that pose a high risk of life-threatening disease and for which no treatment or effective vaccines exist. VHF viruses are definitely in this category as they cause severe symptoms that affect multiple organ systems and show a high rate of mortality.

In the United States, the only endemic hemorrhagic fever viruses are the hantaviruses. With hantaviruses we understand the vectors and hosts and how to prevent human infections. However, with many hemorrhagic fever viruses, the vectors and mechanisms of transmission are unknown. For example, Ebola virus (**Figure 2**) is endemic in rural parts of central Africa and can spread quickly among humans. There have been seven major Ebola outbreaks in Africa since 1976, the latest in 2009, and the results have been devastating. For example, of 1748 confirmed cases of Ebola, 1162 have been fatal (67% mortality). Obviously, if Ebola were to infect a densely populated area, a major medical emergency would ensue. The primary goal of the Special Pathogens Branch is to see that something like this does not happen.

The overwhelmingly high mortality rates for VHFs makes them some of the most feared of diseases, and so research to produce vaccines and treatments is ongoing. In the meantime, the Special Pathogens Branch of the CDC remains focused on the current disease scene. The branch is charged with managing hantavirus patients in the United

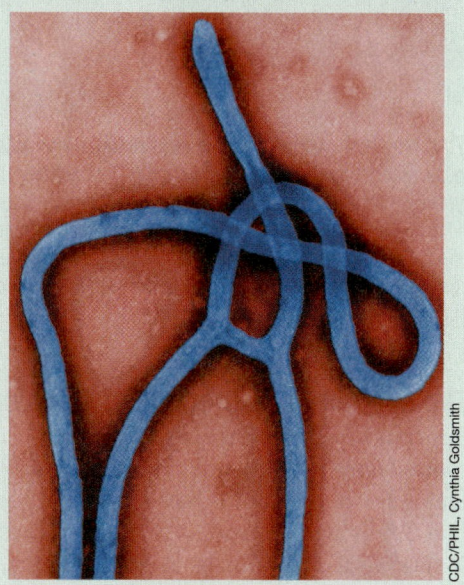

Figure 2 Ebola virus. Colorized transmission electron micrograph of a negatively stained preparation of Ebola virus virions. A single virion is about 80 nm in diameter.

States, developing diagnostic tools to identify known and emerging VHF viruses, and understanding the biology, pathology, and transmission of these dangerous pathogens. A major goal is to be able to predict outbreaks of VHFs, quickly identify them when they occur, and implement adequate measures to stop the outbreak quickly. For more information about the Special Pathogens Branch of the CDC, explore the website at http://www.cdc.gov/.

UNIT 6

mammalian host. During World War I, a typhus epidemic spread throughout Eastern Europe and caused almost 3 million deaths. Typhus has historically been a problem among troops in wartime. Because of the unsanitary, cramped conditions characteristic of wartime military operations, infected lice can spread easily among soldiers with devastating results. Up until World War II, typhus caused more military deaths than did combat.

Cells of *R. prowazekii* are introduced through the skin when a puncture caused by a louse bite becomes contaminated with louse feces that contain the rickettsial cells. During an incubation period of 1–3 weeks, the organism multiplies inside cells lining the small blood vessels. Symptoms of typhus (fever, headache,

and general body weakness) then begin to appear. Several days later, a characteristic rash is observed in the armpits and generally spreads over the body, except for the face, palms of the hands, and soles of the feet. Complications from untreated typhus include damage to the central nervous system, lungs, kidneys, and heart. Epidemic typhus has a mortality rate of as much as 30%. Tetracycline and chloramphenicol are most commonly used to control infections caused by *R. prowazekii*. *Rickettsia typhi*, the organism that causes murine typhus, is another important pathogen in the typhus group and can also infect humans. A typhus vaccine is available but is typically only administered to those traveling to typhus endemic areas.

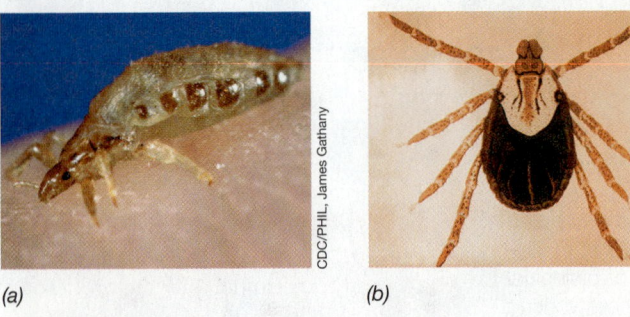

(a)

CDC/PHIL, James Gathany

(b)

CDC/PHIL

Figure 30.4 Arthropod vectors of rickettsial diseases. *(a)* The female body louse, about 3 mm long, can carry *Rickettsia prowazekii*, the agent that causes typhus. In addition, the body louse can carry *Borrelia recurrentis*, the agent of relapsing fever, and *Bartonella quintana*, the agent of trench fever. *(b)* The American dog tick that carries *Rickettsia rickettsii*, the causative agent of Rocky Mountain spotted fever, is about 5 mm long, but can expand to three times this size when engorged with blood.

The Spotted Fever Group: *Rickettsia rickettsii*

Spotted fever rickettsiosis, commonly called *Rocky Mountain spotted fever* (*RMSF*), was first recognized in the western United States about 1900 but is more prevalent today in the central and mid-South region (**Figure 30.5**). RMSF is caused by *R. rickettsii* and is transmitted to humans by various ticks, most commonly the dog tick (Figure 30.4*b*) and wood ticks. Over 2000 people acquire RMSF yearly in the United States, a significant increase since 2002 which is probably due to increased human activities in tick-infested areas. Humans acquire the pathogen from the bite of an infected tick; rickettsial cells are present in the salivary glands of the tick and in the ovaries of female ticks.

Cells of *R. rickettsii*, unlike other rickettsias, grow within the nucleus of the host cell as well as in host cell cytoplasm (**Figure 30.6***a, c*). Following an incubation period of 3–12 days, characteristic symptoms, including fever and a severe headache, occur. A few days later, a systemic rash breaks out (Figure 30.6*d*), generally accompanied by gastrointestinal problems

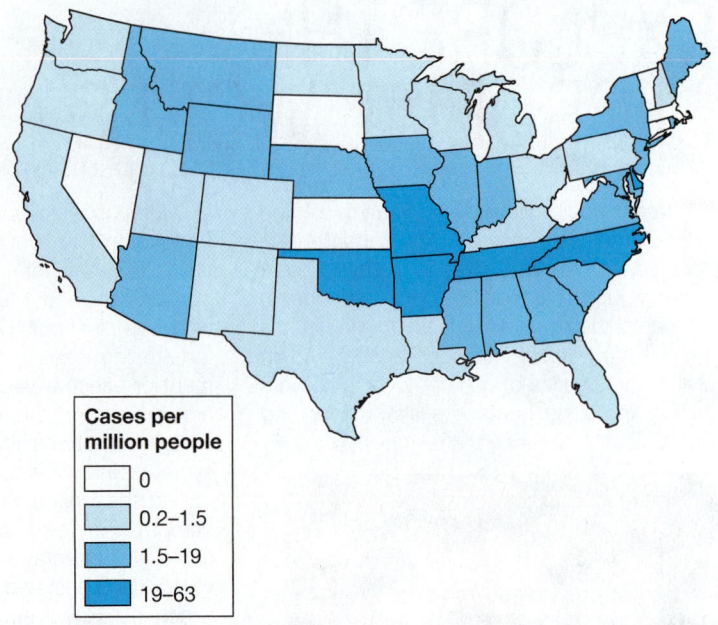

Cases per million people
- ☐ 0
- ☐ 0.2–1.5
- ☐ 1.5–19
- ☐ 19–63

Figure 30.5 Spotted fever rickettsiosis (Rocky Mountain spotted fever) in the United States, 2010. Despite the name, cases of Rocky Mountain spotted fever are currently concentrated in the eastern and mid-South states west to Oklahoma.

such as diarrhea and vomiting. The clinical symptoms of RMSF persist for over 2 weeks if the disease is untreated. Tetracycline or chloramphenicol generally promotes a prompt recovery from RMSF if administered early in the course of the infection, and treated patients have less than 1% mortality. Mortality in untreated cases resembles that of typhus, up to 30%. No effective vaccine against RMSF is currently available.

Ehrlichiosis and Tickborne Anaplasmosis

Ehrlichia and related genera (🔗 Section 15.1) are responsible for two emerging tickborne diseases in the United States, *human monocytic ehrlichiosis* (*HME*) and *human granulocytic anaplasmosis*

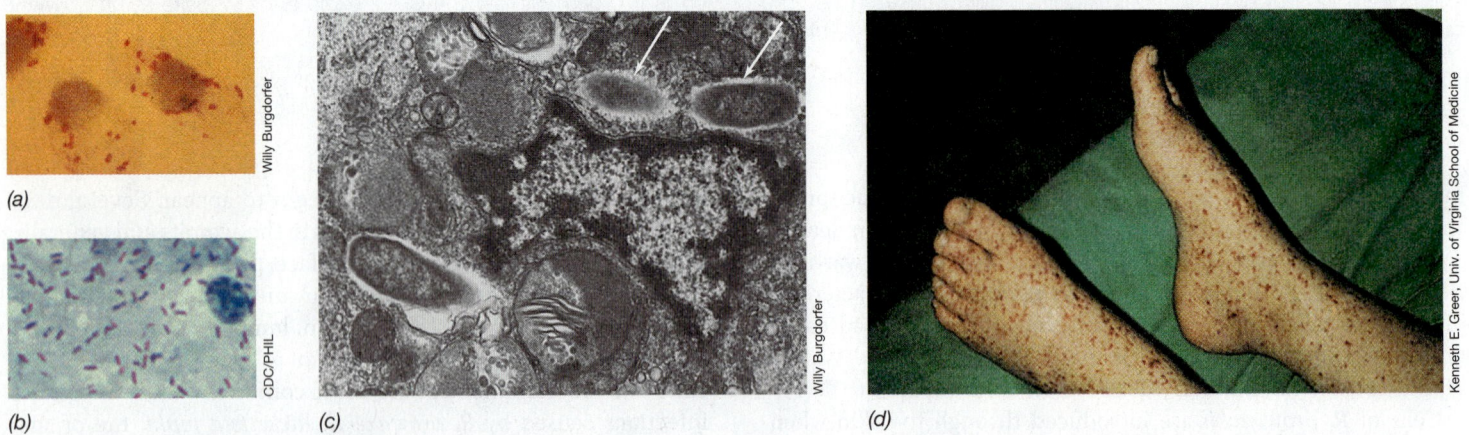

(a)

Willy Burgdorfer

(b)

CDC/PHIL

(c)

Willy Burgdorfer

(d)

Kenneth E. Greer, Univ. of Virginia School of Medicine

Figure 30.6 *Rickettsia rickettsii* and spotted fever rickettsiosis. *(a)* Cells of *R. rickettsii*, growing in the cytoplasm and nucleus of tick hemocytes and *(b)* in chicken egg yolk sacs; cells are about 0.4 μm in diameter. *(c)* Transmission electron micrograph of *R. rickettsii* (arrows) in a granular hemocyte of an infected wood tick. *(d)* Rash of spotted fever rickettsiosis on the feet. The whole-body rash is indicative of spotted fever rickettsiosis and helps distinguish it from typhus, in which the rash does not cover the whole body.

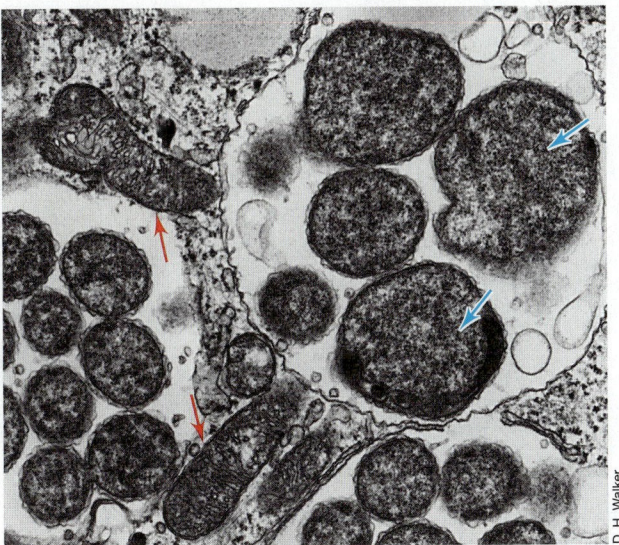

D. H. Walker

Figure 30.7 *Ehrlichia chaffeensis*, **the causative agent of human monocytic ehrlichiosis (HME).** The electron micrograph shows inclusions in a human monocyte that contains large numbers of *E. chaffeensis* cells. The blue arrows indicate two of the many bacteria in each inclusion. The *E. chaffeensis* cells are about 0.3–0.9 μm in diameter. Mitochondria are shown with red arrows.

(*HGA*). The pathogens that cause HME are *Ehrlichia chaffeensis* and *Rickettsia sennetsu*, and those that cause HGA are *Ehrlichia ewingii* and *Anaplasma phagocytophilum*.

The onset of these clinically indistinguishable rickettsial diseases is characterized by flulike symptoms that can include fever, headache, malaise, changes in liver function, and a reduction in white blood cell numbers. Peripheral blood leukocytes have visible inclusions of rickettsial cells, a diagnostic indicator for the diseases (**Figure 30.7**). The symptoms, except for the inclusions, are similar to other rickettsial infections, and can range from subclinical to fatal. Long-term complications for progressive untreated cases may include respiratory and renal insufficiency and serious neurological involvement.

HGA and HME are spread by ticks of various species, and mammalian reservoirs of the pathogens include deer, some rodents, and humans. In the United States, HGA occurs primarily in the upper Midwest and coastal New England, while HME is concentrated in the lower Midwest and the East Coast; together, almost 2000 cases are reported each year, with cases of HGA predominating. Diagnosis of rickettsial syndromes is not straightforward because the rash observed can be mistaken for other diseases, such as scarlet fever, or even measles or syphilis. Confirmation of a rickettsial disease requires immunological tests, including fluorescent antibodies or immunoassays, or PCR-based analyses that detect pathogen DNA.

Prevention of HGA and HME is best achieved either by avoiding tick habitat or by wearing tick-proof clothing and applying insect repellents containing diethyl-*m*-toluamide (DEET). It is also good practice to examine yourself carefully for ticks after hiking in tick habitat and to remove any ticks immediately, taking care to remove all tick mouthparts if the tick has already attached. Doxycycline, a tetracycline antibiotic, is the drug of choice for the treatment of HGA and HME. Vaccines are currently unavailable for the prevention of HGA and HME.

Q Fever

Q fever is a pneumonia-like infection caused by the obligate intracellular parasite *Coxiella burnetii*, a bacterium related to the rickettsias (ᕳᕲ Section 15.1). Although not transmitted to humans by an insect bite, *C. burnetii* cells are transmitted to animals such as sheep, cattle, and goats by insect bites, and from these reservoirs to humans. Domestic animals generally have inapparent infections, but may shed large quantities of *C. burnetii* cells in their urine, feces, milk, and other body fluids. Infected animals or contaminated animal products such as wool, meat, and milk are potential sources for human infection. The resulting influenza-like illness can progress to include prolonged fever, headache, chills, chest pains, pneumonia, and endocarditis (inflammation of the inner lining of the heart). In the United States, Q fever is most prevalent in rural states with large farm or ranch animal populations, and only about 150 cases are reported annually.

As for rickettsial infections, laboratory diagnosis of *C. burnetti* infection is typically made by immunological tests designed to measure host antibodies to the pathogen. Q fever responds well to tetracycline, and therapy should be started quickly in any suspected case to prevent endocarditis and heart valve damage. Q fever is also a potential biological warfare agent (ᕳᕲ Section 28.8).

MINIQUIZ --

- What are the arthropod vectors and animal hosts for typhus, spotted fever rickettsiosis, ehrlichiosis, and anaplasmosis?

- What precautions can be taken to prevent rickettsial infections?

30.4 Lyme Disease and *Borrelia*

Lyme disease is a tickborne disease that affects humans and other animals. Lyme disease was named for Old Lyme, Connecticut, where cases were first recognized, and is currently the most prevalent arthropod-borne disease in the United States. Lyme disease is caused by infection with a spirochete, *Borrelia burgdorferi* (**Figure 30.8**; ᕳᕲ Section 14.20), transmitted by a tick bite. The ticks that carry *B. burgdorferi* feed on the blood of

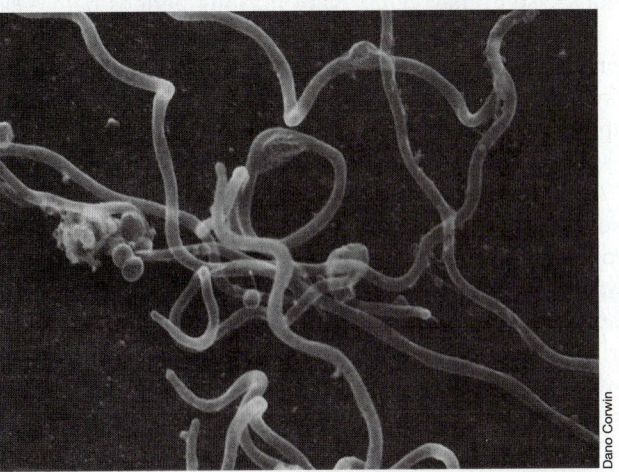

Dario Corwin

Figure 30.8 Scanning electron micrograph of the Lyme spirochete, *Borrelia burgdorferi*. A single cell is approximately 0.4 μm in diameter.

UNIT 6

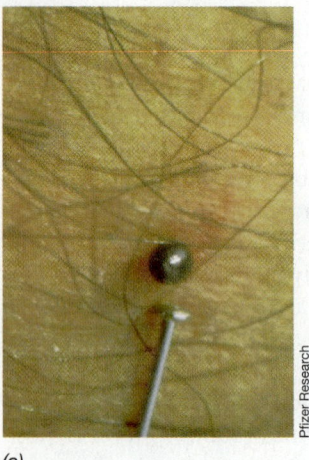

(a)

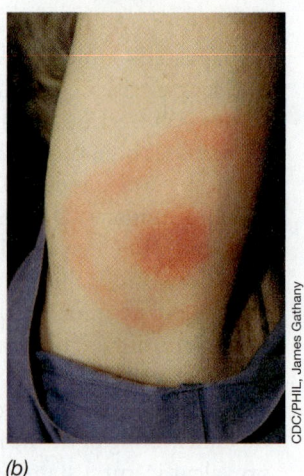

(b)

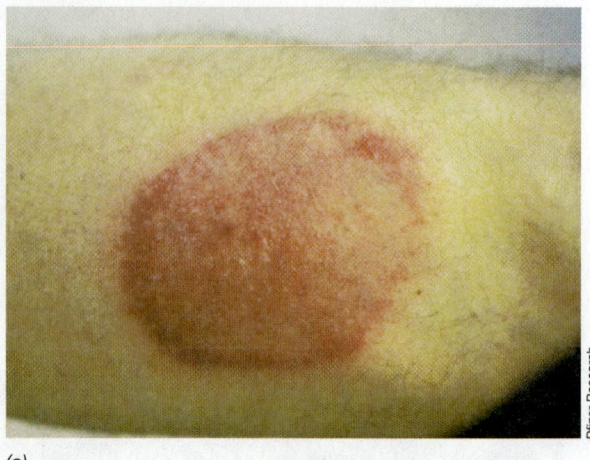

(c)

Figure 30.9 Lyme disease infection. (a) Transmission from a deer tick obtaining a blood meal from a human. (b), (c) Characteristic Lyme disease rash. The rash starts at the site of a tick bite and grows in a concentric circular fashion (b) or bull's-eye fashion (c) over a period of several days. A typical rash is about 5 cm in diameter.

birds, domesticated animals, various wild animals, and humans. *B. burgdorferi* is of interest in a nonmedical way as well, because it is one of only a handful of *Bacteria* that contain a linear (as opposed to a circular) chromosome (no linear chromosomes are known in *Archaea*) (↩ Section 4.3).

Pathology, Diagnosis, and Treatment of Lyme Disease

Cells of *B. burgdorferi* are transmitted to humans while the tick is obtaining a blood meal (**Figure 30.9a**). A systemic infection develops, leading to the acute symptoms of Lyme disease: headache, backache, chills, and fatigue. In about 75% of Lyme cases, a concentric circular or "bull's-eye" rash forms within a week at the site of the tick bite (Figure 30.9b, c). During this acute stage, Lyme disease is readily treatable with tetracycline or penicillin.

Untreated cases of Lyme disease may progress to a chronic stage weeks to months after the initial tick bite, causing arthritis in about half of those infected. Neurological problems such as palsy, weakness in the limbs, and heart damage can also occur. In untreated cases, cells of *B. burgdorferi* infecting the central nervous system may lie dormant for long periods before causing additional chronic symptoms, including problems with vision and facial muscle movements, or seizures. Interestingly, the symptoms of chronic Lyme disease, especially neurological symptoms, mimic those of chronic syphilis, caused by a different spirochete, *Treponema pallidum* (↩ Sections 14.20 and 29.12). Unlike syphilis, however, Lyme disease is not spread person to person.

No toxins or other major virulence factors have been identified in Lyme disease pathogenesis, but the pathogen triggers a strong immune response. Antibodies to *B. burgdorferi* appear 4–6 weeks after infection and can be detected by various immunological assays. However, because antibodies to *B. burgdorferi* antigens persist for years after infection, the presence of these antibodies does not necessarily indicate a recent infection. A PCR assay (↩ Section 11.3) is also in use to detect *B. burgdorferi* DNA in body fluids and tissues. In practice, however, Lyme disease is typically diagnosed from clinical symptoms and only confirmed later by

laboratory assays. If a patient has Lyme disease symptoms and other findings such as facial tics or arthritis, or has had recent tick exposure or exhibits the characteristic Lyme rash (Figure 30.9), a presumptive diagnosis of Lyme disease is made and antibiotic treatment is initiated.

Treatment of early stage Lyme disease is usually with doxycycline or amoxicillin for 20 to 30 days. For patients having neurological or cardiac symptoms, the antibiotic ceftriaxone is administered intravenously because this drug crosses the blood–brain barrier and can thus kill spirochetes in the central nervous system.

Epidemiology and Prevention of Lyme Disease

Deer and white-footed field mice are prime mammalian reservoirs of *B. burgdorferi* in the northeastern United States, a hotbed of infection (see Figure 30.11). These animals become infected from bites by the deer tick, *Ixodes scapularis* (**Figure 30.10**), although some other ticks can transmit the Lyme spirochete as well. Lyme disease has also been identified in Europe and Asia. In these countries, both the tick vector and the species of *Borrelia* differ from those in the United States, which shows that Lyme

Figure 30.10 Deer ticks, the major vector of Lyme disease. Left to right, male and female adult ticks, nymph, and larva forms. The length of an adult female is about 3 mm. Although all forms feed on humans, the female nymphal and adult ticks are principally responsible for transmitting *Borrelia burgdorferi*.

Figure 30.11 Lyme disease in the United States, 2011. Each dot represents a confirmed case. Confirmed and probable cases totaled over 36,000, with 96% of these localized to 13 states in the upper regions of the Midwest and East Coast. Lyme disease is reported through the National Notifiable Diseases Surveillance System of the Centers for Disease Control and Prevention, Atlanta, Georgia, USA.

disease has a broad geographic distribution. But in all cases, Lyme disease is caused by related species of pathogenic *Borrelia* transmitted to humans by tick vectors.

Deer ticks are typically smaller than many other ticks, making them easy to overlook. Moreover, unlike the case with ticks that carry other tickborne diseases (Figure 30.4), a very high percentage of deer ticks carry *B. burgdorferi*. Both of these factors—small vector size and high occurrence of the pathogen—undoubtedly contribute to the fact that Lyme is the most commonly reported vectorborne disease in the United States. Most cases of Lyme disease in the United States have been reported from the Northeast and upper Midwest—areas of the country where deer are abundant—but cases have been observed in nearly every state (**Figure 30.11**). The incidence of Lyme disease in the United States is significant, with over 30,000 cases reported in 2011.

As for any tickborne infection, prevention of Lyme disease begins by avoiding contact with the vector. Insect repellents containing DEET or the wearing of snug-fitting clothing is helpful, as is a thorough body exam following walks in tick-infested environments. Lyme disease vaccines are available for domestic animals, but no human Lyme disease vaccine is currently in use.

MINIQUIZ
- What are the primary symptoms of Lyme disease?
- In the United States, where is Lyme disease most prevalent?
- Outline methods for prevention of *Borrelia burgdorferi* infection.

30.5 Yellow Fever and Dengue Fever

Several arthropod-transmitted diseases are caused by flaviviruses. These are single-stranded plus-sense RNA viruses (⮌ Section 9.8) transmitted by the bite of an infected arthropod. Because of this characteristic mode of transmission, these viruses are also called *arboviruses* (*ar*thropod-*bo*rne viruses).

Many serious human diseases are caused by arboviruses including various types of encephalitis and hemorrhagic fevers. Here we consider two potentially fatal flavivirus diseases still common in developing countries, yellow fever and dengue. Both viruses are transmitted by the same vector, infected mosquitoes of the genus *Aedes* (**Figure 30.12**), and some of the disease symptoms are similar.

Yellow Fever

Yellow fever is an endemic disease of tropical and subtropical climates, especially in Latin America and Africa. Brazil, Colombia, Venezuela, and parts of Bolivia and Peru, along with most countries in sub-Saharan central Africa, experience the greatest incidence. Yellow fever is absent from the United States except in unvaccinated individuals who contract the disease through travel to an endemic area. Yellow fever virus is related to dengue virus (see later), West Nile virus (Section 30.6), and certain encephalitis viruses. Yellow fever is one of only a handful of infectious diseases for which isolation and quarantine are practiced (⮌ Section 28.5). In the case of yellow fever, although the disease is not transmitted person to person, isolation of active cases prevents local mosquitoes from taking a blood meal from the infected individual and transmitting the disease to others.

Following a bite from an infected mosquito, the yellow fever virus replicates in lymph nodes and certain immune system cells

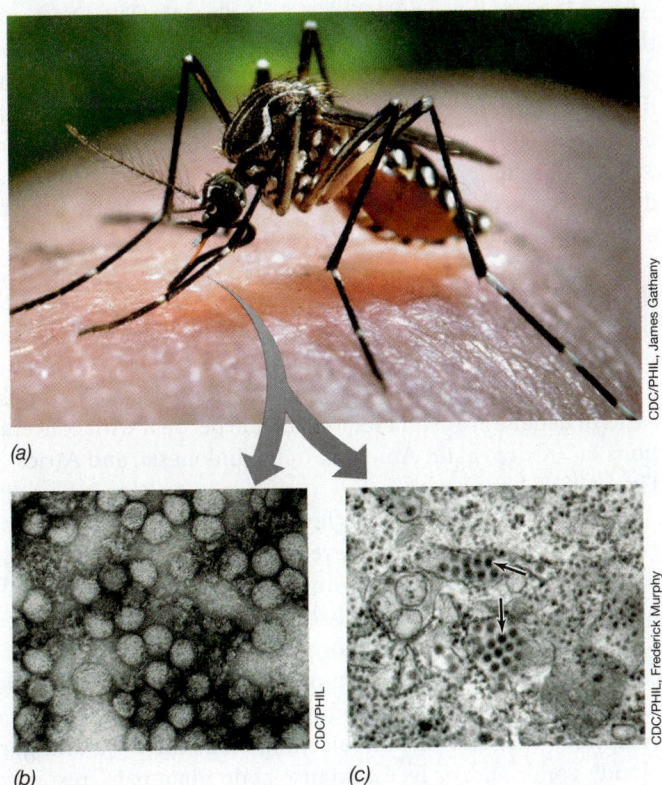

(a)

(b)   (c)

Figure 30.12 Yellow fever and dengue fever. *(a)* Yellow and dengue fever viruses are both transmitted by the bite of an infected *Aedes aegypti* mosquito. Transmission electron micrographs of *(b)* yellow fever virus and *(c)* dengue fever virus (arrows, in a tissue specimen). Both yellow and dengue fever viruses are about 50 nm in diameter and are plus-sense RNA viruses that replicate by way of polyprotein formation, as in poliovirus (⮌ Figure 9.16).

UNIT 6

and eventually travels to the liver. Once a person is infected, symptoms range anywhere from none to major organ failure and death. Most infected individuals display a mild fever with accompanying chills, a headache and back pains, nausea, and other symptoms that are not diagnostically useful. Presumably these are cases in which the immune system has the infection under control. However, in about one in five yellow fever cases, the disease enters its toxic phase, characterized by jaundice (thus the name, *yellow fever*) and by hemorrhaging from the mouth, eyes, and gastrointestinal tract. This triggers the onset of bouts of bloody vomit, and if bleeding continues, it leads to toxic shock and multiple organ failure. About 20% of cases that reach this stage are fatal. Humans and nonhuman primates are the main reservoir for the yellow fever virus.

Yellow fever is fully preventable by an effective vaccine. A yellow fever vaccine was developed in the 1930s and widely used by military and support personnel in tropical battlefields. Historically, the disease has been controlled by a combination of vaccination and elimination of both the vector (mosquito) population by chemical agents and vector breeding grounds by draining swamps and low-lying wetlands in endemic areas.

The yellow fever vaccine is highly recommended for those traveling to endemic areas, and many countries require proof of vaccination for anyone entering their country from a foreign country where yellow fever is endemic. In addition, the World Health Organization (WHO) has initiated a mass vaccination program in Africa. Despite the availability of a vaccine, the WHO estimates that each year nearly 200,000 cases of yellow fever occur, mostly unreported, and that about 15% of all cases are fatal. No treatment for yellow fever is known. However, once the disease is diagnosed, typically by detecting anti–yellow fever virus antibodies in a blood sample, the patient is isolated and prescribed rest and drugs to control symptoms. Recovery without entering the toxemia stage is due to the immune response.

Dengue Fever

Like yellow fever, dengue (pronounced deng-gay) fever is transmitted by mosquitoes of the genus *Aedes* (Figure 30.12) and is a disease of tropical and subtropical regions. Up to 100 million cases of dengue are estimated worldwide per year with concentrations in Mexico, Latin America, India, Indonesia, and Africa (⟳ Figure 28.11).

Dengue begins with a high fever and headache or joint pains and in some patients, severe eye pain and a systemic rash. Most infected individuals show self-improvement within a week and no further symptoms, presumably because of an immune response to the dengue virus. But, as for yellow fever, a dengue infection can take a more severe course of events and proceed to *dengue hemorrhagic fever*. This condition is characterized by severe symptoms that can include bleeding from the nose and gums, bloody vomit and/or feces, intense abdominal pain, respiratory distress, and a general feeling of malaise. The blood pressure of a dengue patient can drop dramatically during the hemorrhagic fever stage and a small percentage of these cases are fatal. Treatment for dengue is primarily to relieve symptoms, particularly dehydration from loss of blood and other fluids. Unlike yellow fever, no effective vaccine exists for dengue, and thus rest and symptom relief, even in cases of dengue hemorrhagic fever, is the only effective treatment.

Dengue can be controlled by eliminating either the vector or contact with the vector. Extensive chemical spraying for mosquito eradication was widely practiced in urban centers of the southern United States and kept dengue in check in the twentieth century. Now, however, spraying programs are less common and global climate change is moving tropical temperature patterns northward. In response, the *Aedes* mosquito has become entrenched in southern and central regions of the United States. Besides *Aedes aegypti* (Figure 30.12), the Asian tiger mosquito (*Aedes albopictus*), which is spreading quickly in the United States, also carries the dengue virus. Drainage of small puddles of water, such as those in discarded tires or other water traps, removes mosquito breeding grounds and greatly reduces the opportunities for a dengue outbreak. Personal protection from mosquito bites by using effective insect repellents and clothing is also a proven means of preventing infection.

The WHO estimates that nearly a half million cases of dengue hemorrhagic fever occur yearly with about 22,000 being fatal. Cases of dengue in the United States are rare, with virtually all of them being imported from endemic areas. Dengue is a prime example of an *emerging disease* (⟳ Section 28.7), since cases were geographically restricted until the mid-twentieth century when global commerce is thought to have transmitted species of the *Aedes* mosquito beyond their original range.

MINIQUIZ
- Identify the vector and reservoir for yellow fever and dengue viruses.
- Contrast the procedures for preventing infection in yellow and dengue fevers.

30.6 West Nile Fever

West Nile virus (WNV) causes **West Nile fever**, a human viral disease that is transmitted through the bite of a mosquito and thus is a seasonal disease. WNV is a flavivirus, as are the yellow fever and dengue viruses (Section 30.5), and has an enveloped capsid (**Figure 30.13b**) containing a plus-sense, single-stranded RNA genome (⟳ Section 9.8). The virus can invade the nervous system of its warm-blooded hosts, which include some species of both birds and mammals.

WNV Transmission and Pathology

WNV causes active disease in over 100 species of birds and is transferred to hosts by the bite of an infected mosquito. Over 40 species of mosquito can carry the virus including *Culex* species (Figure 30.13a), common in central and eastern states of the United States and in urban centers throughout the country. Infected birds develop a systemic viral infection (viremia) that is often fatal. Mosquitoes feeding on viremic birds are infected and can then infect other susceptible birds, renewing the cycle. In contrast to birds, humans and other animals are dead-end hosts for the virus because they do not develop the viremia necessary to infect mosquitoes.

(a)

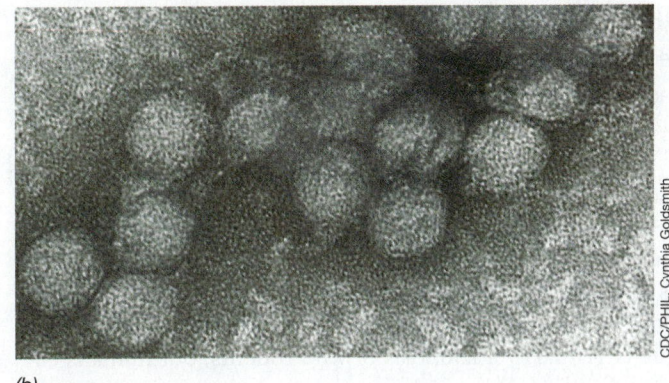

(b)

Figure 30.13 **West Nile virus.** *(a)* The mosquito *Culex quinquefasciatus*, shown here engorged with human blood, is a West Nile virus vector. *(b)* An electron micrograph of the West Nile virus. The icosahedral virion is about 40–60 nm in diameter.

Mortality rates for WNV infection are species-specific. For example, the human mortality rate from WNV is 4% while that for horses is significantly higher, near 40%. Most human infections are asymptomatic or mild and are not reported. After an incubation period of 3–14 days, about 20% of infected individuals develop West Nile fever, a mild illness lasting 3–6 days. The fever may be accompanied by headache, nausea, myalgia, rash, lymphadenopathy (swelling of lymph nodes), and malaise. Less than 1% of infected individuals develop serious neurological diseases such as *West Nile encephalitis* or *West Nile meningitis* from viral replication in neural tissues (**Figure 30.14**). These are more common in adults over age 50, and the neural effects may be permanent. About 5% of West Nile cases that progress to these forms are fatal. Diagnosis of WNV disease includes assessment of clinical symptoms followed by confirmation by immunological tests that detect WNV antibodies in blood.

Control and Epidemiology of WNV

Human WNV disease was first identified in the West Nile region of Africa in 1937 and spread from there to Egypt and Israel. In the 1990s there were WNV outbreaks in horses, birds, and humans in African and European countries. In 1999, the first cases were reported in the United States in the Northeast, around New York. Within three years this emerging disease had shifted from the East Coast to the Midwest, with a peak reported number of cases in Illinois and a nationwide case total of 4156. Shortly thereafter the disease moved even farther west and by 2012, the majority of the total of 5674 cases reported were firmly entrenched in the central and south central United States and California (**Figure 30.15**). West Nile disease is now enzootic in the bird population in the United States and in only low incidence in the human population, its accidental host.

Control of WNV illness is much the same as for other vectorborne diseases: Limit exposure to mosquitoes using insect repellents or wear tight-fitting clothing. Spraying for mosquitoes has limited effectiveness, but removal of mosquito breeding grounds, particularly sources of standing water, is very helpful in controlling mosquito populations. A veterinary WNV vaccine is widely used in horses where mortality rates demand it, but no human WNV vaccine is currently available.

UNIT 6

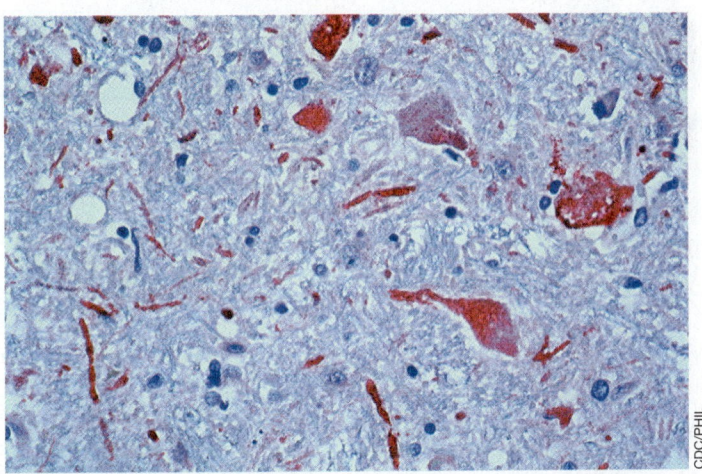

Figure 30.14 **West Nile encephalitis.** Brain section from a West Nile encephalitis victim. Red areas in the tissue are neurons containing West Nile virus as detected using an immunostaining technique.

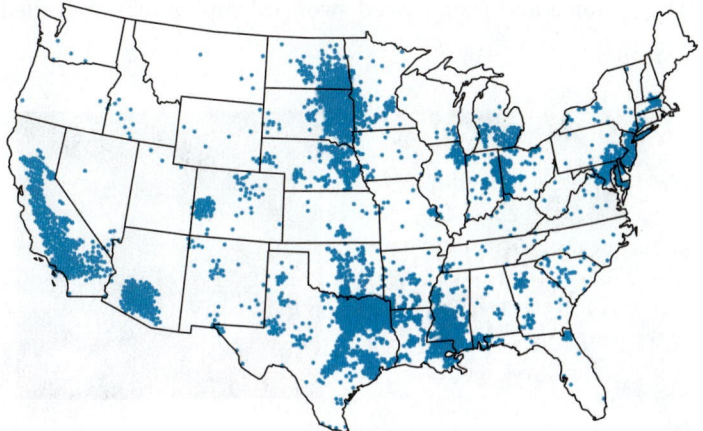

Figure 30.15 **West Nile disease in the United States, 2012.** The virus caused 5674 cases of human disease and 286 deaths. West Nile virus is now endemic in mosquitoes and birds throughout the United States. Data are from the Centers for Disease Control and Prevention, Atlanta, Georgia, USA.

MINIQUIZ

• Identify the vector and reservoir for West Nile virus.

• Trace the progress of West Nile virus in the United States since 1999.

30.7 Plague

Plague has caused more human deaths than any other infectious disease except for malaria and tuberculosis (⟲ Chapter 1 Explore the Microbial World, "The Black Death Decoded"). Plague is primarily a zoonosis of wild rodents, but humans can become accidental hosts when rodent populations experience a die-off. Plague is caused by *Yersinia pestis*, a gram-negative, facultatively aerobic, rod-shaped, and encapsulated enteric bacterium (⟲ Section 15.3) that is easily grown in laboratory culture (**Figure 30.16**).

Pathology and Treatment of Plague

The pathogenesis of plague is not clearly understood, but cells of *Y. pestis* produce several virulence factors that contribute to the disease process. The V and W antigens in *Y. pestis* cell walls are protein–lipoprotein complexes that inhibit immune cell phagocytosis. *Murine toxin*, an exotoxin that is lethal for mice, is produced by virulent strains of *Y. pestis*. Murine toxin is a respiratory inhibitor that causes systemic shock, liver damage, and respiratory distress in mice. The toxin likely plays a role in human plague as well, because these symptoms are common in plague patients. *Y. pestis* also produces a highly immunogenic endotoxin that may play a role in the disease process.

Plague can occur in several forms (see Figure 30.19). *Sylvatic plague* is enzootic among wild rodents. Plague is transmitted by several species of fleas, a main one being *Xenopsylla cheopis*, the rat flea (**Figure 30.17a**). Fleas ingest *Y. pestis* cells in a blood meal and the bacterium multiplies in the flea's intestine. From there, the infected flea transmits the disease to rodents or humans in the next bite. The most common form of plague in humans is *bubonic plague*. In this case, cells of *Y. pestis* travel to the lymph nodes, where they replicate and cause swelling. The regional and pronounced swollen lymph nodes are called

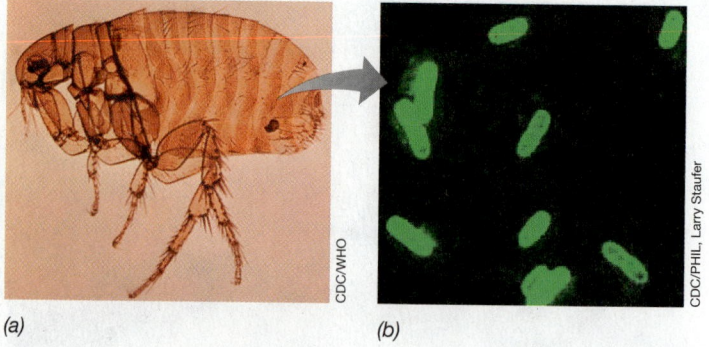

(a)   (b)

Figure 30.17 **The rat flea, a major vector of plague.** *(a)* The rat flea *Xenopsylla cheopis* carries cells of *Yersinia pestis*. The bacterium replicates in the flea gut and *(b)* cells of *Y. pestis* are transmitted to a host in a flea bite. The rat flea was the major vector for the pandemics of plague that ravaged medieval Europe in the fourteenth century. Cells in part b were stained with a fluorescent antibody prepared against *Y. pestis* cell surface antigens.

buboes, and the disease gets its name from these structures (**Figure 30.18a**). The buboes become filled with *Y. pestis* cells, and the bacterium's capsule prevents phagocytosis and destruction by cells of the immune system. Secondary buboes form in peripheral lymph nodes, and cells eventually enter the bloodstream, causing septicemia. Multiple local hemorrhages produce dark splotches on the skin, giving plague its historical name, the "Black Death" (Figure 30.18b). If the infection is not treated quickly, the symptoms of plague (lymph node swelling and pain, prostration, shock, and delirium) usually progress and cause death within 3–5 days.

Pneumonic plague occurs when cells of *Y. pestis* are either inhaled directly or reach the lungs via the blood or lymphatic circulation. Significant symptoms are usually absent until the last day or two of the disease when large amounts of bloody sputum are produced. About 90% of untreated cases of pneumonic plague die within 48 h. Moreover, pneumonic plague is highly contagious and can spread rapidly from person to person if those infected are not immediately isolated. *Septicemic plague* is the rapid spread

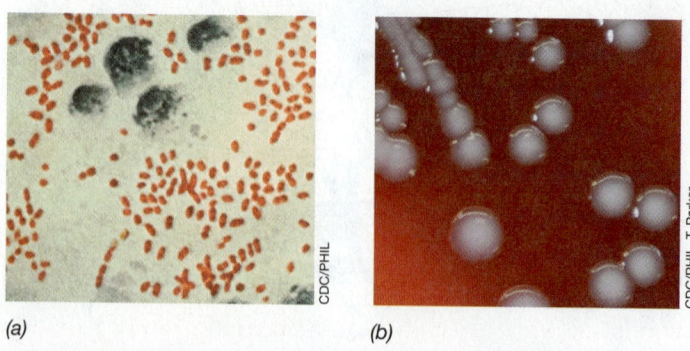

(a)   (b)

Figure 30.16 *Yersinia pestis.* *(a)* Blood smear containing cells of *Y. pestis* (red) and a few white blood cells. The bacterial cells are about 0.8 μm in diameter. *(b)* Colonies of *Y. pestis* grown on blood agar.

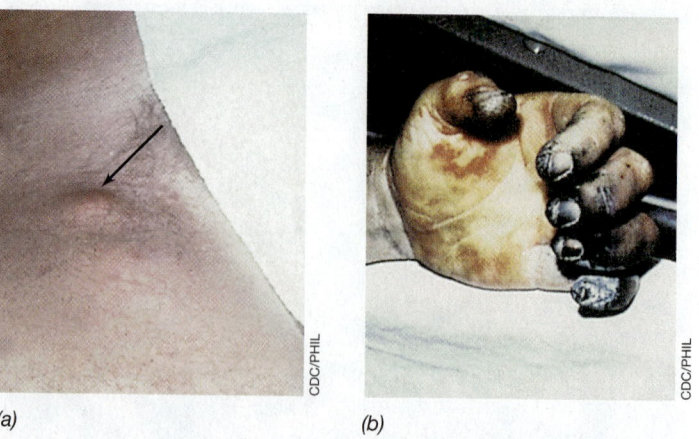

(a)   (b)

Figure 30.18 **Plague in humans.** *(a)* A bubo formed in the groin. *(b)* Gangrene and sloughing of skin in the hand of a plague victim. Human plague can manifest itself in three different forms: bubonic, pneumonic, and septicemic (see Figure 30.19).

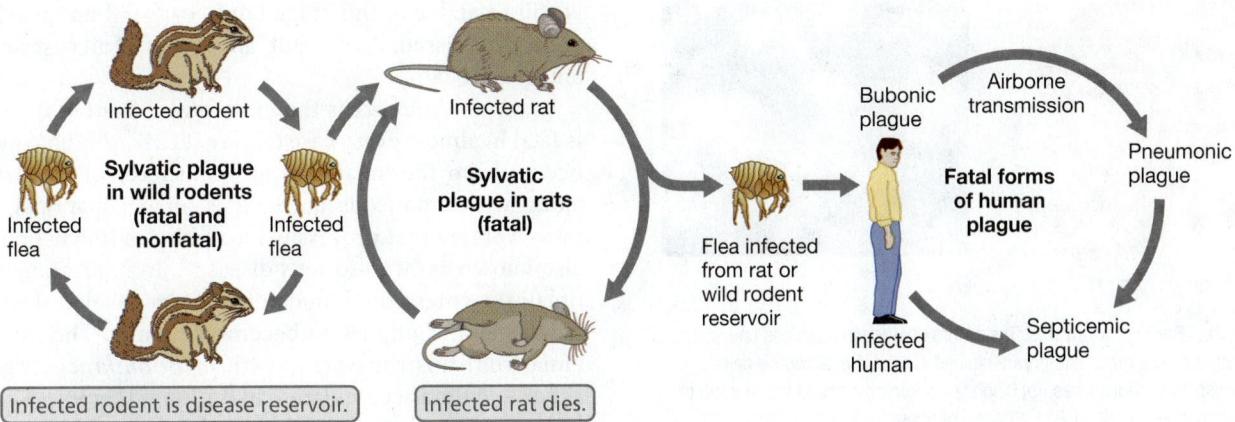

Figure 30.19 Plague epidemiology. In some wild rodents, sylvatic plague causes only a mild infection but diseased animals remain a reservoir of *Yersinia pestis*. In rodents that act as disseminating hosts, for example, rats, and in humans, plague is often fatal. When the domestic rodent reservoir dies off in an epidemic, infected fleas seek alternate hosts in humans.

of *Y. pestis* throughout the body via the bloodstream without the formation of buboes and is so severe that it usually causes death before a diagnosis can be made.

Bubonic plague can be successfully treated with streptomycin or gentamicin, administered by injection. Alternatively, doxycycline, ciprofloxacin, or chloramphenicol may be given intravenously. If treatment is started promptly, mortality from bubonic plague can be reduced to less than 5%. Pneumonic and septicemic plague can also be treated, but these forms progress so rapidly that antibiotic therapy, even if begun when symptoms first appear, is usually too late. *Y. pestis* is an organism that could be used for a bioterrorism attack (🔗 Section 28.8), and oral doxycycline and ciprofloxacin would be used as treatments in such a scenario.

Plague Epidemiology and Control

Sylvatic plague is enzootic in a variety of rodents including ground squirrels, prairie dogs, chipmunks, and mice; rats are the primary hosts in urban communities and were typically the hosts in episodes of sylvatic plague that triggered human pandemics during the Middle Ages. Fleas are intermediate hosts and vectors for plague, spreading the disease between rodent hosts and humans (**Figure 30.19**). Most infected rats or other rodents die soon after symptoms appear, but a small proportion of survivors develop a chronic infection, providing a persistent reservoir of *Y. pestis* cells to fuel new outbreaks.

Plague is endemic in developing countries in Africa, Asia, the Americas, and in south-central Eurasia; most cases occur in sub-Saharan Africa. Pandemic plague was historically associated with unsanitary surroundings, a major factor supporting large rat populations. In sparsely populated rural areas, this is not so great a problem as the disease runs its course when the rodent population dies off, leaving a shortage of hosts. But in urban centers where alternative hosts (humans) are plentiful, an outbreak of sylvatic plague can set the stage for a human plague epidemic. In the United States only a handful of cases of plague are diagnosed annually, mostly in the southwestern states where sylvatic plague is enzootic among wild rodents (Figure 30.19).

Plague control is accomplished through good sanitation practices, surveillance and control of rodent reservoirs and vectors (fleas), isolation of active cases, and imposing quarantine on those who have had contact with diseased individuals. Improved public health practices and the control of rodent populations are the major reasons that outbreaks of plague are extremely rare in developed countries.

MINIQUIZ

- Distinguish among sylvatic, bubonic, septicemic, and pneumonic plague.
- What are the insect vector, the natural host reservoir, and the treatment for plague?

III • Soilborne Bacterial Diseases

30.8 Anthrax

Some human diseases are caused by microorganisms whose major habitat is soil, and anthrax is an excellent example. We covered some aspects of the disease anthrax in Section 28.8 in the context of its use as a potential bioterrorism or biological warfare agent. Here we focus more on the biology of the organism and the disease process.

Discovery and Properties of Anthrax

The famous pioneering medical microbiologist Robert Koch (🔗 Section 1.8) first isolated the causative agent of the disease anthrax, the endospore-forming bacterium *Bacillus anthracis* (**Figure 30.20a**). Using mice caught in the wild as experimental animals, Koch used the disease anthrax to develop his principles for linking cause and effect in infectious disease—Koch's postulates

UNIT 6

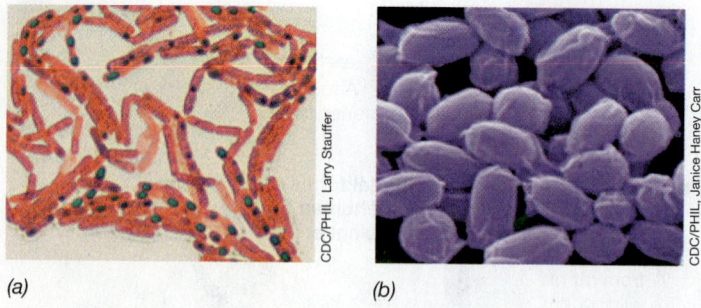

(a)   *(b)*

Figure 30.20 Bacillus anthracis. The anthrax pathogen produces endospores. *(a)* Light micrograph of a malachite green–stained smear of *B. anthracis* cells showing greenish-blue endospores. *(b)* Colorized scanning electron micrograph of *B. anthracis* endospores. Cells of *B. anthracis* are about 1.2 μm in diameter.

(⮌ Figure 1.20). Anthrax is quickly fatal in mice, but in humans, anthrax can take on several different forms, from mild to severe skin infections to respiratory failure and death.

Anthrax is an enzootic disease of worldwide occurrence. *B. anthracis* lives a saprophytic existence in soils, growing as an aerobic chemoorganotroph and forming endospores (Figure 30.20*b*) when conditions warrant. From soil, cells or spores of *B. anthracis* can become embedded in animal hair, hides, or other animal materials, or can be ingested, and from here the disease can develop, allowing *B. anthracis* spores to be transmitted to humans. Anthrax is primarily seen in domesticated farm animals, particularly in cattle, sheep, and goats, and is transmitted from them to humans.

Forms of Human Anthrax

The disease anthrax can manifest itself in one of three forms: cutaneous (on the skin), intestinal, and respiratory (inhalation anthrax). In all forms, disease symptoms are due to a series of toxins whose mode of action was discussed when we considered anthrax as a bioterrorism agent in Section 28.8. The different forms of anthrax show increasing severity, which is primarily a function of where in the body these toxins are excreted. The unusual protein capsule that surrounds cells of *B. anthracis* (**Figure 30.21***a*) is also an important virulence factor as it prevents destruction of the bacterium following ingestion by macrophages. Instead, cells of *B. anthracis* grow within the macrophage, eventually killing it and giving it access to the bloodstream.

Virtually all cases of human anthrax are *cutaneous anthrax*, where spores of *B. anthracis* have entered through a skin lesion, germinate, and form a painless, black and swollen pustule (Figure 30.21*b*); this is highly characteristic of the disease and allows for a firm diagnosis even though human anthrax is rarely seen in clinical medicine. In cutaneous anthrax the bacterium usually remains localized and the disease is easily treatable. Although cutaneous anthrax is fatal for about 20% of those untreated, most cases are treated because of the obvious symptoms, and thus fatalities are rare. *Intestinal anthrax* is very uncommon and is triggered by the ingestion of spores of *B. anthracis* (Figure 30.20*b*) in undercooked meat from diseased animals. Symptoms of intestinal anthrax include abdominal pain, bloody diarrhea, and ulcer-like lesions throughout the intestinal tract. The disease

is still treatable at this stage but because of its rarity, diagnoses are easily missed. As a result, about half of all cases of intestinal anthrax are fatal.

Inhalation anthrax is the most severe form of the disease and is fatal in almost every case (Figure 30.21*c, d*). Inhalation anthrax occurs from the inhalation of endospores of *B. anthracis* and, along with cutaneous anthrax, is an occupational hazard for farm workers that process wool and hides (inhalation anthrax is also known as "woolsorter's disease"). In inhalation anthrax, the organism enters the bloodstream from inhaled dust or animal dander and multiplies to become systemic. The mounting toxemia from this runaway growth of *B. anthracis* triggers septic shock and fluid accumulation in the lungs (Figure 30.21*c*) that can kill a patient in less than a day.

Prevention and Vaccines

Complete prevention of anthrax is impossible because the reservoir of the organism is the soil. However, anthrax is avoidable by limiting close exposure to farm animals and is easily treatable with antibiotics. For the cutaneous form this is a routine treatment, but antibiotic therapy is less effective in intestinal anthrax and especially in inhalation anthrax. By the time the latter is diagnosed, the disease has progressed to the point where it is usually too late to save the patient. An anthrax vaccine is available but because the disease is so rare, it is only recommended for high-risk individuals such as scientists working with the organism, slaughterhouse or livestock workers, and military personnel (for biowarfare reasons). An effective and inexpensive anthrax vaccine is available for vaccinating livestock and is commonly used in cattle, sheep, goats, and horses.

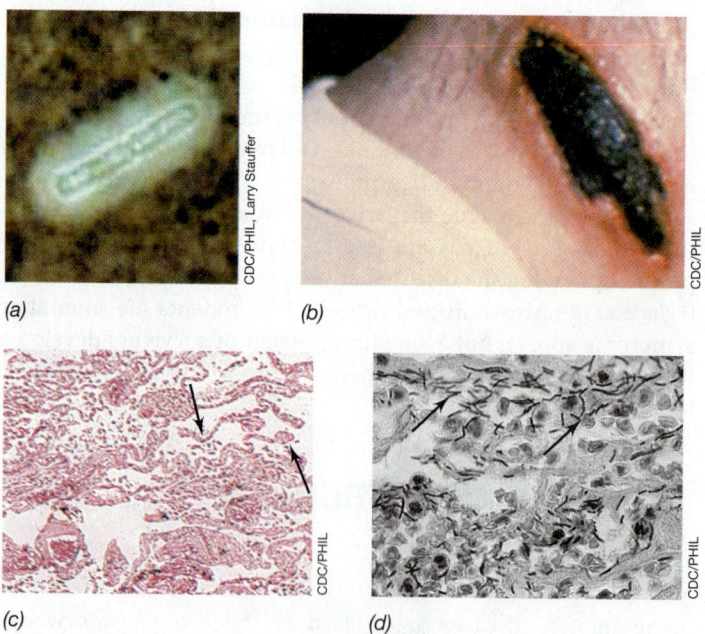

(a)   *(b)*

(c)   *(d)*

Figure 30.21 Anthrax pathology. *(a)* The protein capsule of *Bacillus anthracis* cells is a major virulence factor because it prevents killing by macrophages. *(b)* Cutaneous anthrax, with its characteristic black scabby appearance on the neck of a patient. *(c, d)* Inhalation anthrax. *(c)* The lung fills with bacterial cells (arrows) and fluids (cleared zones). *(d)* From the systemic infection, *B. anthracis* cells can be found almost anywhere, including the lining of the central nervous system (arrows).

MINIQUIZ ---

• What are the major virulence factors of *Bacillus anthracis*?

• What are the three forms of anthrax, and which is most dangerous?

30.9 Tetanus and Gas Gangrene

Tetanus is a serious, life-threatening disease. Although tetanus is completely preventable through immunization, it still causes over 150,000 deaths per year, mostly in countries in Africa and Southeast Asia. *Gas gangrene* is caused by the growth in dead tissues of bacteria related to the tetanus pathogen, leading to a gassy putrefaction and loss of an infected limb or death from systemic shock. Both diseases are caused by clostridia.

Biology and Epidemiology of Tetanus

Tetanus is caused by an exotoxin produced by *Clostridium tetani*, an obligately anaerobic, endospore-forming rod (**Figure 30.22a**; ↩ Section 15.7). The natural reservoir of *C. tetani* is soil, where it is a ubiquitous resident, although it is occasionally found in the gut of healthy humans, as are other *Clostridium* species.

Cells of *C. tetani* normally gain access to the body through a soil-contaminated wound, typically a deep puncture. In the wound, anoxic conditions develop around the dead tissue and allow germination of endospores, growth of the organism, and production of a potent exotoxin, the *tetanus toxin*. *C. tetani* is essentially noninvasive; its sole ability to cause disease is through toxemia, and thus tetanus is observed only as the result of untreated deep tissue injuries. The onset of tetanus symptoms may take from four days to several weeks, depending on the number of endospores inoculated at the time of the injury.

Pathogenesis of Tetanus

We have already examined the activity of tetanus toxin at the cellular and molecular level (↩ Section 23.9). The toxin directly affects the release of inhibitory signaling molecules in the nervous system. These inhibitory signals control the "relaxation" phase of muscle contraction. The absence of inhibitory signaling molecules results in rigid paralysis of the voluntary muscles,

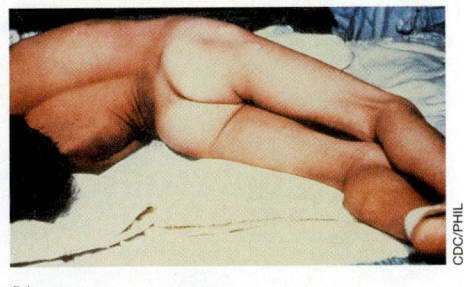

(a)   *(b)*

Figure 30.22 Tetanus. *(a) Clostridium tetani* showing the "drumstick" appearance of sporulating cells with their terminal endospores. Cells of *C. tetani* are about 0.8 μm in diameter. *(b)* A tetanus patient showing the rigid paralysis characteristic of tetanus. Tetanus paralysis typically begins with the facial muscles ("lockjaw") and descends to lower body regions.

often called *lockjaw* because it is observed first in the muscles of the jaw and face. Preceding actual lockjaw, tetanus symptoms typically include mild spasms of facial muscles and muscles of the neck and upper back. Later on, the paralysis extends to the torso and lower body (Figure 30.22b). When tetanus is fatal, death is usually due to respiratory failure. Mortality is relatively high, occurring in about 10% of all reported cases, and up to 50% of cases in which treatment is delayed until generalized full body tetanus has set in.

Diagnosis, Control, Prevention, and Treatment of Tetanus

Diagnosis of tetanus is based on exposure, clinical symptoms (Figure 30.22b), and, rarely, identification of the toxin in the blood or tissues of the patient. The natural reservoir of *C. tetani* is the soil and thus control measures must focus on disease prevention rather than pathogen removal. The tetanus toxoid vaccine is highly effective, and thus virtually all tetanus cases occur in individuals who were inadequately immunized.

A second line of tetanus protection is to administer appropriate medical care to serious cuts, lacerations, and punctures. Even though vaccination against tetanus is widely practiced, any serious wound should be thoroughly cleaned and the damaged tissue removed. If the vaccination status of the individual is unclear or the last tetanus booster was more than 10 years ago, revaccination is recommended. If a deep wound is severe or heavily contaminated by soil, treatment might also include administration of a tetanus antitoxin preparation, especially if the patient's immunization status is unknown or is out of date.

Acute symptomatic tetanus (Figure 30.22b) is treated with antibiotics, usually penicillin, to stop growth and toxin production by *C. tetani*, and antitoxin is injected intramuscularly (or into the sheath surrounding the spinal cord if necessary) to prevent binding of newly released toxin to cells. Supportive therapy such as sedation, administration of muscle relaxants, and mechanical respiration may be necessary to control the effects of paralysis. Treatment cannot provide a quick reversal of symptoms, because toxin that is already bound to tissues cannot be neutralized. Even with antitoxin and antibiotic administration and supportive therapy, tetanus patients show significant morbidity and mortality. A complete recovery from tetanus often takes many months.

Gas Gangrene

Tissue destruction due to the growth of proteolytic and gas-producing clostridia is called **gas gangrene**. In this life-threatening condition, amino acids obtained from the breakdown of muscle proteins are fermented to the gases H_2 and CO_2 plus a variety of foul-smelling organic compounds, including short-chain fatty acids and other putrid molecules; ammonia released during amino acid fermentation (↩ Section 13.13) adds to the stench. In addition, a variety of bacterial toxins are produced that accelerate tissue destruction.

Although *C. tetani* is a proteolytic *Clostridium* species, it does not cause gangrene but can be associated with cases of gangrene triggered by a deep tissue wound. The most common causes of gangrene are *Clostridium perfringens* (**Figure 30.23a**), which is also a common cause of foodborne illness unrelated to gangrene,

UNIT 6

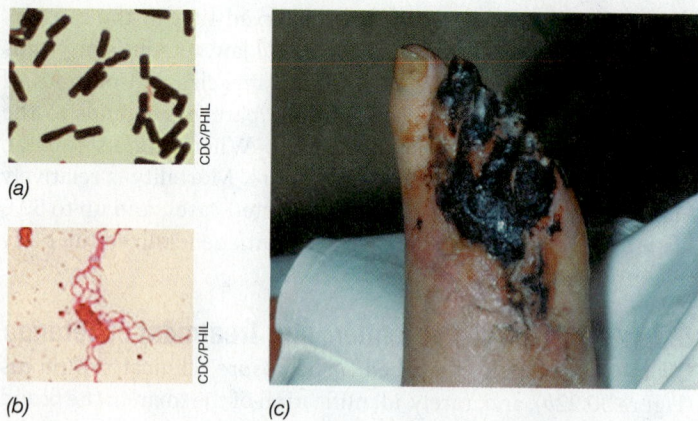

CDC/PHIL

(a)

CDC/PHIL

(b)   (c)

Figure 30.23 Gas gangrene. *(a)* Gram stain of cells of *Clostridium perfringens*, the most common cause of gangrene. *(b)* Flagellum stain showing a cell of *Clostridium novyi*, also an agent of gangrene. *(c)* A case of gas gangrene. Cells of both *C. perfringens* and *C. novyi* are about 1.2 μm in diameter.

C. novyi, and *C. septicum*. These organisms reside in soil and are also part of the normal human intestinal flora. When these species reach deep into tissues, typically from traumatic tissue invasion such as a war wound or other puncture wound, or occasionally from gastrointestinal tract surgery, spores and vegetative cells of proteolytic clostridia are inserted into what are now dead tissues. As the bacteria grow, they release enzymes that destroy collagen and tissue proteins and also excrete a series of toxins. *C. perfringens* (Figure 30.23*a*) *alpha toxin*, which is distinct from the toxins it produces in perfringens food poisoning (↩ Section 31.9), is a major virulence factor in gangrene, as is the general

ability of the pathogens to grow rapidly in the warm, moist environment created by an invasive injury. Alpha toxin is a phospholipase that hydrolyzes the membrane phospholipids of host cells, leading to cell lysis and the typical accumulation of gas and fluids that accompanies gas gangrene (Figure 30.23*c*).

In severe cases of gas gangrene, the toxemia can become systemic and cause death. Antibiotic treatment is taken as a preventive measure in cases of gangrene in addition to the typical though dramatic treatment: amputation of the infected limb. Gangrenous tissues are dead and will not regenerate, and amputation prevents the infection from reaching healthy tissues.

Hyperbaric oxygen treatment of the infected limb is attempted in some cases to try to save it, with the high levels of O_2 inhibiting growth of the obligately anaerobic clostridia. In hyperbaric treatment, the patient sits in an enclosed chamber containing 100% O_2 at about twice atmospheric pressure. This enriches the blood in O_2 and helps still living blood vessels seed the formation of new tissue. Several hyperbaric treatments are administered and may be accompanied by surgical removal of some of the dead tissue. If an adequate blood supply can be established in damaged tissues, a skin graft may also be done to help connect regenerating with damaged tissues.

MINIQUIZ

- Describe infection by *Clostridium tetani* and the elaboration of tetanus toxin.

- Describe the steps necessary to prevent tetanus in an individual who has sustained a puncture wound.

- How does the physiology of *Clostridium perfringens* make it suitable for growing in puncture wounds?

BIG IDEAS

30.1 • Rabies occurs primarily in wild animals but can be transmitted to humans and domestic animals. In the United States rabies is transmitted primarily from the wild animal reservoir to domestic animals or, very rarely, to humans. Vaccination of domestic animals is key to the control of rabies. Most human deaths from rabies occur in developing countries.

30.2 • Hantaviruses are present worldwide in rodent populations and cause zoonotic diseases such as hantavirus pulmonary syndrome and hemorrhagic fever with renal syndrome in humans. Hantavirus is a highly dangerous hemorrhagic fever virus related to Ebola. In the Americas, hantavirus infections have case fatality rates of over 30%.

30.3 • Rickettsias are obligate intracellular parasitic bacteria transmitted to hosts by arthropod vectors. The incidence of spotted fever rickettsiosis and other rickettsial syndromes is increasing due to several factors. Most rickettsial infections can be controlled by antibiotic therapy, but prompt recognition and diagnosis of these diseases remains difficult.

30.4 • Lyme disease is caused by the spirochete *Borrelia burgdorferi* and is the most prevalent arthropod-borne disease in the United States today. Lyme is transmitted from several mammalian host vectors to humans by ticks. Prevention and treatment of Lyme disease are straightforward, but accurate and timely diagnosis of infection is essential.

30.5 • Yellow and dengue fevers are caused by related flaviviruses transmitted to humans by mosquito bites. Both

diseases are widespread in tropical and subtropical countries and both can show mild to very severe symptoms, including hemorrhagic fevers. An effective vaccine for yellow fever is in use, but no vaccine is available to prevent dengue infection.

30.6 • West Nile fever is a mosquito-borne viral disease. In the natural cycle of the pathogen, birds are infected with West Nile virus by the bite of infected mosquitoes. Humans and other vertebrates are occasional terminal hosts. Most human infections are asymptomatic and undiagnosed, but complications of some infections cause about 5% mortality.

30.7 • Plague can be transmitted to individuals who have had contact with rodent populations and their parasitic fleas, the enzootic reservoirs for the plague bacterium, *Yersinia pestis*. A systemic infection or a pneumonic infection leads to rapid death, but the bubonic form is treatable with antibiotics.

30.8 • Anthrax is caused by the endospore-forming bacterium *Bacillus anthracis* and can take on three different forms: cutaneous, intestinal, or inhalation. Cutaneous anthrax is most common and along with inhalation anthrax is an occupational hazard for livestock workers, where *B. anthracis* endospores can be transmitted from animal hides to humans.

30.9 • *Clostridium tetani* is a soil bacterium that causes tetanus, a potentially fatal disease characterized by a toxemia and rigid paralysis. Treatment for acute tetanus includes antibiotics and active and passive immunization, and the disease is preventable by toxoid immunization. Gas gangrene occurs from the growth of various proteolytic clostridia in traumatic wounds, leading to gas and toxin formation.

MasteringMicrobiology® **Review what you know and challenge what you have learned with MasteringMicrobiology!** Access study materials, chapter quizzes, animations, and microbiology lab tutorials in the Study Area to ensure that you have mastered this chapter's content.

REVIEW OF KEY TERMS

Enzootic an endemic disease present in an animal population

Epizootic an epidemic disease present in an animal population

Gas gangrene tissue destruction due to the growth of proteolytic and gas-producing clostridia

Hantavirus pulmonary syndrome (HPS) an emerging, acute disease characterized by pneumonia, caused by rodent hantavirus

Hemorrhagic fever with renal syndrome (HFRS) an emerging acute disease

characterized by shock and kidney failure, caused by rodent hantavirus

Lyme disease a tick-transmitted disease caused by the spirochete *Borrelia burgdorferi*

Plague an enzootic disease in rodents caused by *Yersinia pestis* that can be transferred to humans through the bite of a flea

Rabies a usually fatal neurological disease caused by the rabies virus, which is usually transmitted by the bite or saliva of an infected animal

Rickettsias obligate intracellular bacteria of the genus *Rickettsia* responsible for diseases including typhus, spotted fever rickettsiosis, and ehrlichiosis

Spotted fever rickettsiosis a tick-transmitted disease caused by *Rickettsia rickettsii*, characterized by fever, headache, rash, and gastrointestinal symptoms; also called Rocky Mountain spotted fever

Tetanus a disease characterized by rigid paralysis of the voluntary muscles, caused by an exotoxin produced by *Clostridium tetani*

Typhus a louse-transmitted disease caused by *Rickettsia prowazekii*, characterized by fever, headache, weakness, rash, and damage to the central nervous system and internal organs

West Nile fever a neurological disease caused by West Nile virus, a virus transmitted by mosquitoes from birds to humans

Zoonosis an animal disease transmissible to humans

REVIEW QUESTIONS

1. Identify the animals most likely to carry rabies in the United States. Why is rabies so rare in humans and domesticated animals in developed countries? (Section 30.1)

2. Describe the conditions that may cause emergence of hantavirus pulmonary syndrome (HPS). How can HPS be prevented? (Section 30.2)

3. Identify the three major categories of organisms that cause rickettsial diseases. For typhus, spotted fever rickettsiosis, and ehrlichiosis, identify the most common reservoir and vector. (Section 30.3)

4. Identify the most common reservoir and vector for Lyme disease in the United States. How can the spread of Lyme disease be controlled? How can Lyme disease be treated? (Section 30.4)

5. In what ways are yellow fever and dengue fever similar? In what ways do they differ? (Section 30.5)

6. Describe the spread of West Nile virus infections in the United States. Which animals are the primary hosts? Are humans productive alternate hosts? (Section 30.6)

7. For a potentially serious disease like bubonic plague, vaccines are not routinely recommended for the general population; why not? Identify the public health measures used to control plague. (Section 30.7)

8. Which key feature of the bacterium *Bacillus anthracis* allows this organism to persist for extended periods on animal hides or other environments where growth may not occur? Which form of anthrax is the most serious? (Section 30.8)

9. Discuss the major mechanism of pathogenesis for tetanus and define measures for prevention and treatment. Why is it possible that a traumatic puncture wound could end up causing both tetanus and gas gangrene? (Section 30.9)

APPLICATION QUESTIONS

1. Describe the sequence of steps you would take if your child received a bite (provoked or unprovoked) from a stray dog with no collar and record of rabies immunization. Present one scenario in which you were able to capture and detain the dog and another for a dog that escaped. How would these procedures differ from a situation in which the child was bitten by a dog that had a collar and rabies tag with documented, up-to-date rabies immunizations?

2. Discuss at least three common properties of the disease agents and review the disease process for spotted fever rickettsiosis, typhus, and ehrlichiosis. Why is ehrlichiosis emerging as an important rickettsial disease? Compare its emergence to that of Lyme disease.

3. Devise a plan to prevent the spread of West Nile virus to humans in your community. Identify the relative costs involved in such a plan, both at the individual level and at the community level. Find out if a mosquito abatement program is active in your community. What methods, if any, are used in your area for the reduction of mosquito populations? What is a simple way to limit mosquito numbers around your residence?

31 · Water and Food as Vehicles of Bacterial Diseases

microbiology**now**

Prison Pruno Packs a Life-Threatening Punch

Conditions inside prisons often drive prisoners to extremes. These include making homemade prison wine—a powerful drink called "pruno." However, even if prisoners are relegated to a life behind bars, they are not immune to foodborne illnesses. And in the case of pruno, this can be a potentially fatal foodborne illness.

Pruno is made from a mixture of oranges, grapes, apples, and sugar, with potatoes, corn, or even bread sometimes thrown in. Each batch is unique and is a reflection of the ingredients available at the time. The ingredients (photo) are crushed in warm water, stored in a plastic bag, and allowed to ferment for up to 2 weeks. The fermented product is then strained through a sock or pillowcase to yield a yellowish-orange to red (depending on the ingredients) juice containing anywhere from 2 to 14% alcohol. The latter is similar to the alcohol content of a full-bodied red wine.

Since 2011, prison pruno has been linked to at least 20 cases of botulism, a severe form of food poisoning. In 2012, eight prisoners in the Arizona Correctional System nearly died from drinking pruno containing botulinum toxin.[1] All affected inmates were given botulinum antitoxin (antibodies that neutralize the botulinum toxin), and several required gastrointestinal intubation, a medical procedure used to remove intestinal contents. All inmates were lucky to survive what would have been certain death without medical intervention.

Botulism is caused by the gram-positive endospore-forming anaerobic bacterium *Clostridium botulinum,* commonly found in soil. In the Arizona and other outbreaks, the common ingredient in the tainted batches of pruno was raw potatoes. It is likely that spores of *C. botulinum* entered the fermenting pruno from soil on the potatoes, and the anoxic state of the brew was the perfect environment for growth and toxin production.

[1]Briggs, G., et al. 2013. Botulism from drinking prison-made illicit alcohol—Arizona, 2012. *Morbidity and Mortality Weekly Report. 62:* 88.

In this chapter we consider microbial pathogens whose mode of transmission is either water or food. The diseases these pathogens cause are called "common-source" diseases because they occur only in those who have consumed the same contaminated water or eaten the same contaminated food. Waterborne and foodborne illnesses are common infectious diseases worldwide.

I • Water as a Disease Vehicle

Water is used in enormous quantities, and its microbiological safety rests in the hands of wastewater and drinking water engineers and microbiologists. Indeed, water quality is the single most important factor for ensuring public health. In Chapter 21 we examined the microbiology of wastewater and examined how highly polluted water can be cleaned by microbial activities and reused for many purposes, including for drinking. Here we see what can happen when water intended for human use becomes a vehicle for disease.

Waterborne diseases begin as infections (or occasionally as toxemias), and contaminated water may cause an infection even if only small numbers of the particular pathogen are present. Whether or not exposure causes disease is a function of the virulence of the pathogen and the ability of the host to resist infection.

31.1 Agents and Sources of Waterborne Diseases

Many different microorganisms can cause waterborne infectious diseases, and some of the major ones are summarized in Table 31.1. Waterborne pathogens include bacteria, viruses, and parasitic protists. Here we will consider bacterial pathogens with a major focus on cholera, a waterborne disease of pandemic proportions. We consider parasitic diseases in Chapter 32 and a few waterborne viral pathogens that are also transmitted in food later in this chapter.

We begin by considering the disease vehicle itself—water. Waterborne illnesses can be transmitted through untreated or improperly treated water used for drinking or food preparation or from water used for swimming and bathing (recreational water sources). The major waterborne illnesses traced to drinking water and recreational waters are typically quite different, and these different disease patterns are shown in Table 31.2.

Potable Water

Water supplies in developed countries typically meet rigid quality standards, greatly reducing the spread of waterborne diseases. Drinking water in particular undergoes extensive treatment that includes both filtration and chlorination. Although filtration removes turbidity and many microorganisms, it is *chlorination* that makes drinking water safe. Chlorine gas (Cl_2) is a strong oxidant and oxidizes both organic matter dissolved in the water and microbial cells themselves. Drinking water chlorination facilities add sufficient chlorine to allow a residual level to remain in the water all the way to the consumer. Water suitable for human consumption is called **potable** water (⊘ Sections 21.6–21.9).

Despite filtration and chlorination, waterborne disease outbreaks from potable water occasionally occur. In the United States an average of 25 outbreaks of disease associated with drinking water are recorded in a year (a waterborne outbreak is defined as two or more human illnesses specifically linked to the consumption of the same water at the same time). Nearly 60% of drinking water disease outbreaks are due to *bacterial* pathogens, most notably *Legionella*, the causative agent of legionellosis (Table 31.2 and Section 31.4).

Recreational Waters

Recreational waters include freshwater aquatic systems such as ponds, streams, and lakes, as well as public swimming and wading

Table 31.1 Major waterborne pathogens

Pathogen	Disease
Bacteria[a]	
Vibrio cholerae	Cholera
Legionella pneumophila	Legionellosis
Salmonella enterica serovar Typhi	Typhoid fever
Escherichia coli	Gastrointestinal illness
Pseudomonas aeruginosa	Nosocomial pneumonia, septicemia, and skin infections
Viruses	
Norovirus	Gastrointestinal illness
Hepatitis A virus	Viral hepatitis
Parasites[b]	
Cryptosporidium parvum	Cryptosporidiosis
Giardia intestinalis	Giardiasis
Schistosoma	Schistosomiasis

[a]Except for *S. enterica* serovar Typhi, these bacteria have been associated with major outbreaks of waterborne illness in the United States in recent years, as have the bacteria *Shigella sonnei*, *Campylobacter jejuni*, and *Leptospira* sp.
[b]See Chapter 32.

Table 31.2 Sources of outbreaks of acute gastrointestinal illness in drinking and recreational waters[a]

	Drinking water (%)	Recreational water (%)
Bacteria	58	16
Legionella pneumophilia	33	8
Other	25	9
Parasites	8	51
Viruses	14	4
Chemical/toxin	3	7
Unidentified	11	22
Multiple causes	5	1

[a]Data are for the United States, 2007–2008. All numbers are rounded to the nearest percent and were obtained from the Centers for Disease Control Waterborne Disease and Outbreak Surveillance System.

pools. Recreational waters can be sources of waterborne disease, and on average they cause disease outbreaks at levels slightly lower than those due to drinking water, about 20 outbreaks per year. In contrast to drinking water, where *bacterial* pathogens are most common, disease outbreaks from recreational waters are more frequently linked to *parasitic* pathogens. In addition, recreational waters often transmit gastrointestinal illnesses that are of either unknown microbial origin or due to chemicals or other toxic materials (Table 31.2).

In the United States, the operation of public swimming pools is regulated by state and local health departments. The United States Environmental Protection Agency (EPA) establishes limits for bacterial numbers in both potable and public recreational water sources, but local and state governments can set standards above or below these guidelines for nonpotable sources. By contrast, the water quality of *private* recreational waters, such as swimming pools, spas, and hot tubs is totally unregulated, and these are therefore prime vehicles for waterborne disease outbreaks.

MINIQUIZ

- What is potable water?
- Contrast the major pathogens responsible for disease outbreaks in drinking water versus recreational waters.

31.2 Public Health and Water Quality

Water that looks perfectly transparent may still be contaminated with high numbers of microorganisms and thus pose a risk of disease. It is impractical to screen water for every pathogenic organism that may be present (Tables 31.1 and 31.2), and so both potable and recreational waters are routinely tested for specific *indicator organisms*, the presence of which signals the potential for waterborne disease.

Coliforms and Water Quality

A widely used indicator for microbial water contamination is the **coliform** group of bacteria. Coliforms are useful because many of them inhabit the intestinal tract of humans and other animals. Thus, the presence of coliforms in water indicates likely fecal contamination. Coliforms are defined as facultatively aerobic, gram-negative, rod-shaped, nonsporulating bacteria that ferment lactose with the production of gas within 48 h at 35°C. However, this definition includes several bacteria that are not necessarily restricted to the intestine; for this reason, it is *fecal coliforms* that are important in water safety assessments. *Escherichia coli*, a coliform whose only habitat is the intestine and that survives only a relatively short time outside the intestine, is the key fecal coliform of interest. The presence of cells of *E. coli* in a water sample is taken as evidence of fecal contamination and means that the water is unsafe for human consumption. Conversely, however, the absence of *E. coli* does not ensure that a water source is potable, since other pathogenic bacteria or pathogenic viruses or protists may still be present.

Testing for Fecal Coliforms and the Importance of *Escherichia coli*

Well-developed and standardized methods are in routine use for detecting coliforms and fecal coliforms in water samples. A common method is the *membrane filter* (*MF*) procedure where at least 100 ml of freshly collected water is passed through a sterile membrane filter, trapping any bacteria on the filter surface. The filter is placed on a plate of eosin–methylene blue (EMB) medium, which is selective for gram-negative, lactose-utilizing bacteria. EMB medium is also differential, allowing strongly fermentative species such as *E. coli* to be distinguished from weakly fermentative species such as *Proteus* (**Figure 31.1a**).

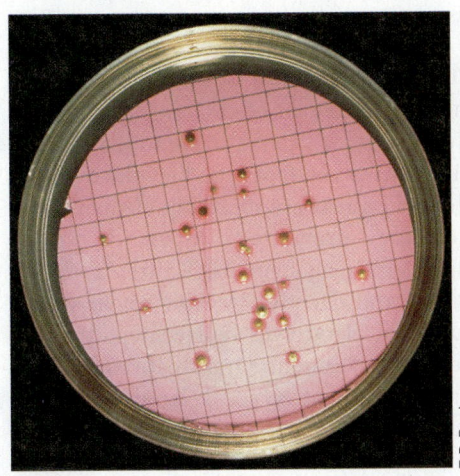

Figure 31.1 Fecal coliforms and their detection in water samples. *(a)* Colonies growing on a membrane filter. A drinking water sample was passed through the filter and the filter placed on an eosin–methylene blue (EMB) agar plate (EMB is both selective and differential for coliforms; more strongly fermentative species form colonies with darker centers). *(b)* Total coliforms and *Escherichia coli*. A filter exposed to a drinking water sample was incubated at 35°C for 24 hours on a medium containing special compounds that fluoresce when metabolized. The filter was then examined under UV light. The single *E. coli* colony in the sample fluoresces dark blue (arrow). Coliforms that do not metabolize the compound form colonies that fluoresce white to light blue. *(c)* The IDEXX Colilert water quality test system. When specific reagents are added to water samples and incubated for 24 h, they develop a yellow color if they contain coliforms (right). Samples containing *Escherichia coli* develop a yellow color but also fluoresce blue (left). Samples negative for coliforms remain clear (center).

Selective media are also available that not only detect total coliforms but also specifically identify *E. coli* simultaneously. These *defined substrate tests* are typically faster and more accurate than EMB-based assays. One popular plate-based test is based on the ability of *E. coli* but not other enteric bacteria to metabolize a combination of two specific chemicals to form a fluorescent blue compound (Figure 31.1*b*). A commonly used liquid method reveals whether coliforms are present and also specifically detects *E. coli* in the water sample (Figure 31.1*c*).

In properly regulated drinking water supply systems, total coliform and *E. coli* fecal coliform tests should be negative. A positive test indicates that a breakdown has occurred in either the purification or distribution system (or both). In the United States, microbiological standards for drinking water are specified in the *Safe Drinking Water Act* and are administered by the Environmental Protection Agency (EPA). Water utilities must report coliform test results to the EPA monthly, and if they do not meet the prescribed standards, the utilities must notify the public and take steps to correct the problem.

Major improvements in public health in the United States beginning in the early twentieth century were largely due to the adoption of water filtration and chlorination procedures in large-scale wastewater and drinking water treatment plants (↺ Figure 28.7). Where drinking water standards have not reached this level, especially in undeveloped countries, a variety of waterborne diseases are common. We turn our attention to these diseases now, beginning with cholera, the most widespread and devastating of all waterborne diseases.

MINIQUIZ --

- Why is *Escherichia coli* used as an indicator organism in microbial analyses of water?

- What procedures are used to ensure the safety of potable water supplies?

II • Waterborne Diseases

31.3 *Vibrio cholerae* and Cholera

Cholera is a severe gastrointestinal diarrheal disease that is now largely restricted to countries in the developing world. Cholera is caused by ingestion of contaminated water containing cells of *Vibrio cholerae*, a gram-negative and motile curved species of *Proteobacteria* (**Figure 31.2**). Cholera can also be contracted from contaminated food, especially improperly cooked shellfish.

The ingestion of a large number ($>10^8$) of *V. cholerae* cells is required to cause disease. The ingested cells attach to epithelial cells in the small intestine where they grow and release *cholera toxin*, a potent enterotoxin (↺ Figure 23.23). Studies with human volunteers have shown that normal stomach acidity (about pH 2) is why the large inoculum of *V. cholerae* cells is needed to initiate disease. In studies with human volunteers, those given bicarbonate to neutralize gastric acidity develop cholera when given as few as 10^4 cells. Even lower cell numbers can initiate infection if *V. cholerae* is ingested with food, presumably because the food protects the vibrios from destruction by stomach acidity.

Cholera enterotoxin causes severe diarrhea that can result in dehydration and death unless the patient is given fluid and electrolyte therapy. The enterotoxin causes fluid losses of up to 20 liters (20 kg or 44 lb) per person per day, causing severe dehydration. The mortality rate from *untreated* cholera is 25–50% and can be even higher under conditions of severe crowding and malnutrition as often occurs in refugee camps or in areas that have experienced natural disasters such as floods, earthquakes, and the like. In these situations there is often a near-complete breakdown in sanitation leading to the contamination of drinking water with feces and the rapid transmission of cholera.

Diagnosis, Treatment, and Prevention of Cholera

At treatment facilities in large outbreaks of cholera, each cholera patient is placed on a "cholera cot," which is a conventional folding cot containing an opening into which feces can be voided (**Figure 31.3***a*). The feces of a cholera patient are more liquid than solid, and confirmation of the disease is straightforward since the pathogen is easily cultured on selective agar media (Figure 31.3*b–d*). Cholera treatment is simple and effective. An oral (or in severe cases, intravenous) liquid and electrolyte replacement therapy is the most effective means of treatment. Oral treatment is preferred because no special equipment or sterile precautions are necessary. The rehydration solution is a mixture of glucose, salt (NaCl), sodium bicarbonate ($NaHCO_3$), and potassium chloride (KCl). If the solution is administered quickly during an outbreak, cholera mortality can be greatly reduced, as rehydration allows patients the time necessary to mount an immune response.

Antibiotics may shorten the course of cholera infection and the shedding of viable cells, but antibiotics are of little health benefit without simultaneous fluid and electrolyte replacement. Public health measures such as adequate sewage treatment and

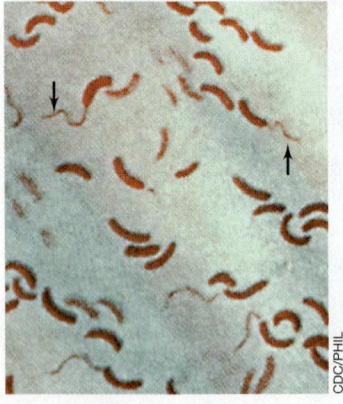
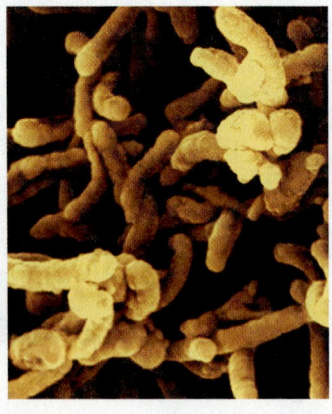

CDC/PHIL

(a)   (b)

Figure 31.2 *Vibrio cholerae*, **the causative agent of cholera.** *(a)* Gram-stained preparation shows the typically curved (vibrio-shaped) cells of this bacterium and polar flagella (arrows). *(b)* Colorized scanning electron micrograph of cells of *V. cholerae*. A single cell is about 0.5 × 2 µm.

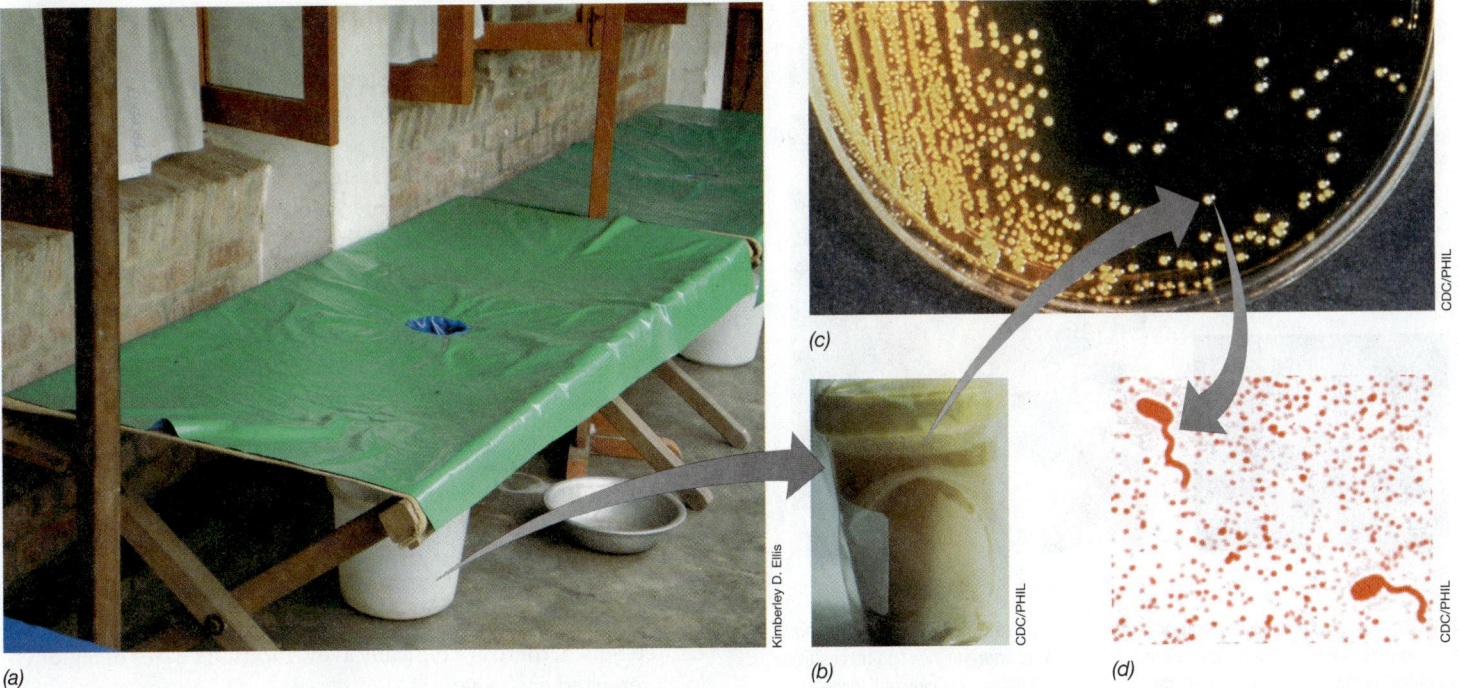

Figure 31.3 Cholera and its diagnosis. *(a)* A cholera cot. The cot allows a person to lie prostrate and void feces directly into a bucket. Cholera cots are used during cholera outbreaks for treating active disease cases with rehydration therapy. *(b)* Feces from a cholera patient. The "rice-water" stool is mostly liquid (the solid material in the bottom is mucus). *(c) Vibrio cholerae* is easily cultured on medium TCBS, which is both selective and differential. TCBS contains high levels of bile salts and citrate, which inhibit enteric bacteria and gram-positive bacteria, and thiosulfate and sucrose, which cells of *V. cholerae (d)* use as a sulfur and carbon/energy source, respectively.

a reliable source of safe drinking water are the keys to preventing cholera. *V. cholerae* is eliminated from wastewater during proper sewage treatment and drinking water purification procedures (Chapter 21). For individuals traveling in cholera-endemic areas, attention to personal hygiene and avoidance of untreated water or ice, raw food, and raw or undercooked fish or shellfish that can feed on phytoplankton contaminated with *V. cholerae* (**Figure 31.4**) can prevent cholera.

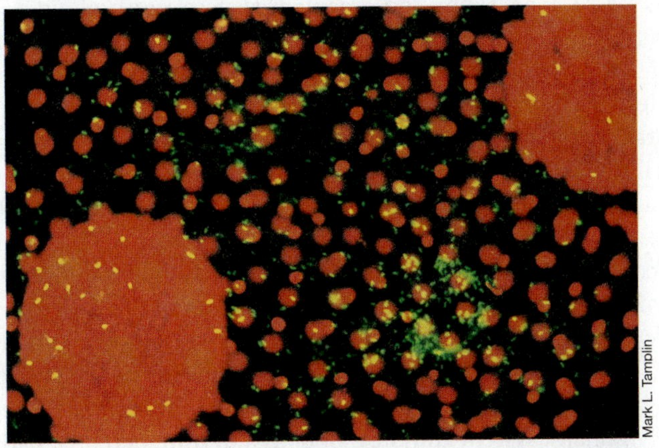

Since 1817, cholera has swept the world in seven major pandemics with an eighth pandemic likely already started (↺ Section 28.10 and Figure 28.17). The World Health Organization estimates that only 5–10% of cholera cases are reported, so the total worldwide incidence of cholera probably exceeds 1 million cases per year. Only a handful of cases of cholera are reported each year in the United States, typically from imported shellfish that are eaten raw or after only minimal cooking.

MINIQUIZ

- What organism causes cholera and what are the symptoms of the disease?
- Describe how cholera can be both prevented and treated.

31.4 Legionellosis

Legionella pneumophila, the bacterium that causes *legionellosis*, is an important waterborne pathogen whose transmission was originally linked to aerosols from evaporative cooling devices. However, *L. pneumophila* (**Figure 31.5**) is now known to be a major pathogen in residential water systems as well, where the organism persists in biofilms that form on water distribution pipe surfaces and also within the cells of certain microbial parasites. In these sites, *L. pneumophila* is protected from the chlorine present in potable waters, and thus biofilms and infected parasites are reservoirs for transmitting legionellosis by a waterborne route (↺ Section 21.9 and Figure 21.21).

Figure 31.4 Cells of *Vibrio cholerae* attached to the surface of *Volvox*, a freshwater alga. The sample was from a cholera-endemic area in Bangladesh. The *V. cholerae* cells are stained green by a fluorescently labeled antibody against a *V. cholerae* cell surface protein. The red color is from chlorophyll *a* fluorescence of the algae.

UNIT 6

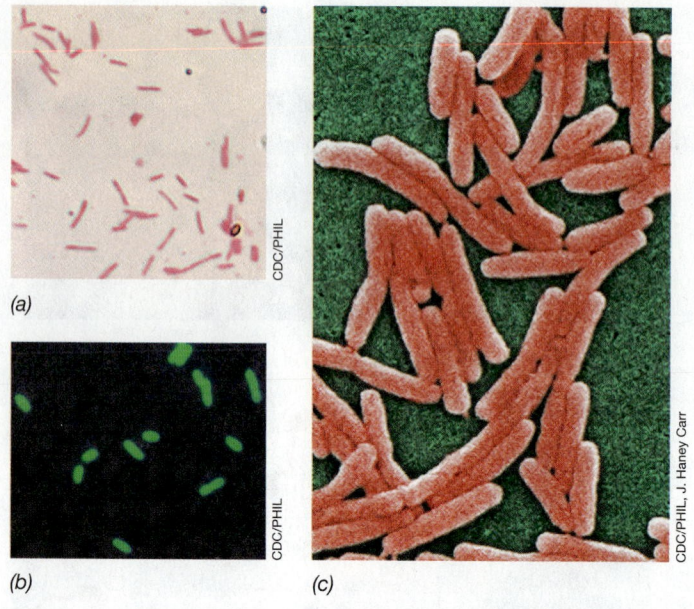

Figure 31.5 *Legionella pneumophila.* *(a)* Gram-stained cells of *L. pneumophila* from lung tissue of a legionellosis victim. *(b)* Cells of *L. pneumophila* can be positively identified using fluorescent anti–*L. pneumophila* antibodies. *(c)* Colorized scanning electron micrograph of *L. pneumophila* cells. Cells are about 0.5 × 2 μm.

Pathogenesis, Diagnosis, and Treatment

Cells of *L. pneumophila* invade the lungs and grow within macrophages and monocytes. Infections are often asymptomatic or produce only a mild cough, sore throat, mild headache, and fever; these self-limiting cases typically resolve themselves in 2–5 days. However, the elderly, whose resistance may be naturally reduced, and those with compromised immune systems often acquire more serious *Legionella* infections resulting in pneumonia. Prior to the onset of pneumonia, intestinal disorders, followed by high fever, chills, and muscle aches, are common. These symptoms precede the dry cough and chest and abdominal pains typical of legionellosis. Up to 10% of cases that reach this stage are fatal, usually due to respiratory failure.

Clinical detection of *L. pneumophila* infection is usually done by culture of the organism from bronchial washings, pleural fluid, or other body fluids or tissues (Figure 31.5*a*). Various serological tests can detect anti-*Legionella* antibodies or *Legionella* cells in these samples and also in patient urine, and this is used to confirm a diagnosis (Figure 31.5*b*). Legionellosis can be treated with the antibiotics rifampin and erythromycin, and intravenous administration of erythromycin is the treatment of choice for life-threatening cases.

Epidemiology

L. pneumophila is a gram-negative, obligately aerobic rod-shaped gammaproteobacterium (Figure 31.5), and shows complex nutritional requirements including an unusually high requirement for iron. The organism can be isolated from terrestrial and aquatic habitats as well as from legionellosis patients. *Legionella pneumophila* was first recognized as the pathogen that caused an outbreak of fatal pneumonia in Philadelphia (USA) in 1976. Besides legionellosis, the same bacterium can also cause a milder syndrome called *Pontiac fever.*

L. pneumophila is present in freshwaters and in soil. It is relatively resistant to heating and chlorination, so it can spread through drinking water distribution systems (↺ Section 21.9). The pathogen is often found in large numbers in improperly sanitized cooling towers and evaporative condensers of large air conditioning systems. The pathogen grows in the water and is disseminated in humidified aerosols. Human infection is by way of airborne droplets, but the infection does not spread from person to person.

Besides evaporative coolers and domestic water systems, *L. pneumophila* has also been detected in hot water tanks and spas; in the latter, it can reach high cell numbers in warm (35–45°C), stagnant water, especially if chlorine (or other sanitizer) levels are not maintained. Many outbreaks of legionellosis have been linked to swimming pools. *L. pneumophila* can be eliminated from water supplies by hyperchlorination or by heating water to greater than 63°C. Although incidence peaks in the summer months, epidemiological studies indicate that *L. pneumophila* infections can occur at any time of year, primarily as a result of aerosols generated from heating and cooling systems and contaminated premise water (↺ Section 21.9) used for showering or bathing. In the United States, there are typically a few thousand cases of legionellosis reported each year.

MINIQUIZ --

- How is legionellosis transmitted?
- Identify specific measures for control of *Legionella pneumophila*.

31.5 Typhoid Fever and Norovirus Illness

Although cholera remains the most widespread and potentially dangerous of waterborne diseases, other waterborne pathogens also cause serious disease. We focus on two major ones here, the causative agents of typhoid (a bacterium) and norovirus gastrointestinal illness (an RNA virus).

Typhoid Fever

On a global scale, probably the most important waterborne bacterial pathogens are *Vibrio cholerae* (Section 31.3) and *Salmonella enterica* serovar Typhi, the organism that causes typhoid fever. *S. enterica* serovar Typhi is a gram-negative peritrichously flagellated bacterium related to *Escherichia coli* and other enteric bacteria (**Figure 31.6***a*). The organism is transmitted in feces-contaminated water, and thus typhoid fever, like cholera, is primarily restricted to areas where sewage treatment and general sanitation are either absent or ineffective. Typhoid today is a well-entrenched endemic disease in sub-Saharan Africa, the Indian subcontinent, and Indonesia, but appears only sporadically in North America, Europe, northern Asia, and Australia.

Typhoid fever progresses in several stages. Cells of the pathogen (Figure 31.6*a*) ingested in contaminated water (or occasionally food) reach the small intestine where they grow and enter the lymphatic system and the bloodstream; from here, the pathogen can travel to many different organs. One to two weeks later, the first symptoms of typhoid appear; these include a mild fever, headache, and general malaise. During this period, the liver and spleen of the typhoid patient become heavily infected. About a week later, the fever becomes more intense (up to 40°C) and the

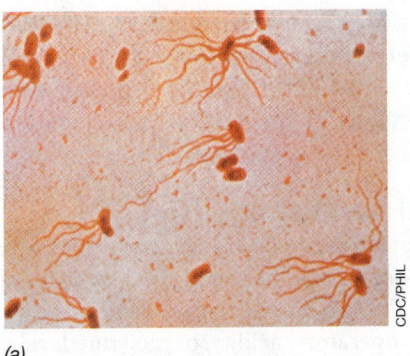
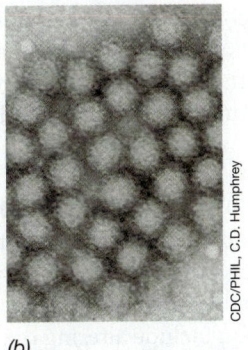

(a)

(b)

Figure 31.6 **Bacterial and viral agents of severe gastrointestinal waterborne diseases.** (a) Flagella-stained cells of *Salmonella enterica* serovar Typhi showing peritrichous flagellation. A single cell measures about 1 × 2 μm. (b) Transmission electron micrograph of virions of norovirus. A single virion is about 30 nm in diameter.

patient typically becomes delirious; diarrhea can occur in this stage and abdominal pain can be severe. Complications can follow, including intestinal bleeding and perforation of the small intestine. The latter releases large numbers of bacterial cells into the abdomen, leading to a condition called *sepsis* (systemic infection and inflammation) and possibly to septic shock; both are potentially fatal conditions (up to 40% of sepsis cases are fatal). After about a week in this crisis stage, the symptoms of typhoid begin to wane and recovery occurs.

In the United States, fewer than 400 cases of typhoid occur per year, but typhoid fever used to be a major public health threat before drinking water was routinely filtered and chlorinated (ᴄ⊋ Figure 28.7). However, breakdown of water treatment methods, contamination of water during floods, earthquakes, and other disasters, or contamination of water supply pipes with leaking sewer lines can propagate epidemics of typhoid fever, even in developed countries.

In some typhoid patients, the gallbladder becomes infected with the pathogen. If these individuals also have gallstones, these can become colonized with *S. enterica* serovar Typhi cells and serve as a long-term reservoir of the pathogen from which it is continuously shed into feces and urine. Such individuals are otherwise healthy "carriers" of typhoid and can transmit the disease over long periods. The notorious cook, "Typhoid Mary," was the classic example of a typhoid carrier (ᴄ⊋ Section 28.4).

Norovirus Illness

Viruses can be transmitted in water and cause human disease. Norovirus (Figure 31.6b) is one of these and is a common cause of gastrointestinal illness due to contaminated water (or food, Section 31.14). Norovirus is a single-stranded plus-sense RNA virus (ᴄ⊋ Section 9.8) and is the leading cause of gastrointestinal illnesses worldwide (see Table 31.5).

Norovirus infection causes symptoms of vomiting, diarrhea, and malaise of relatively short duration. The disease is rarely fatal, although in compromised individuals (very young, elderly, or immune deficient), the significant dehydration that accompanies repeated bouts of norovirus-triggered vomiting and diarrhea can be life-threatening. A clinical diagnosis of norovirus gastrointestinal illness is made by a combination of observing symptoms and the direct detection of either viral RNA by RT-PCR (ᴄ⊋ Sections 11.3 and 27.10) or viral antigens by enzyme immunoassay in samples of feces or vomit.

Norovirus disease is easily transmitted person to person or to food by the fecal–oral route. The infectious dose is very small, as exposure to as few as 10–20 norovirus virions (Figure 31.6b) is sufficient to initiate disease. The sources of waterborne norovirus outbreaks are most often well water or recreational waters that have been contaminated with sewage. Norovirus is also often the culprit when mass common-source gastrointestinal illnesses strike people on cruise ships or in long-term care facilities or other group settings. In these situations, the virus can be transmitted person to person, by contaminated food or water (usually food), or by any combination of these.

MINIQUIZ

- Contrast the causative agents of typhoid and noro gastrointestinal disease.
- What public health conditions allow for outbreaks of typhoid fever?

III • Food as a Disease Vehicle

The foods we eat, whether they are fresh, prepared, or preserved, are rarely sterile. Instead, they are almost always contaminated with spoilage microorganisms of various kinds and occasionally with pathogens. Microbial activities are key to the production of some foods, such as fermented foods, but most of the microorganisms in or on food are unwelcome and diminish either food quality or safety (or both). In this unit, we explore the contrasting processes of food spoilage and food preservation, how food safety is assessed, and the transmission of pathogens in food. In the next two units we focus on major foodborne diseases.

31.6 Food Spoilage and Food Preservation

Many foods provide an excellent medium for the growth of bacteria and fungi. Properly stored food can still undergo food spoilage but is usually not a vehicle for disease assuming that it was free of pathogens to begin with. This is because with rare exception, organisms responsible for food spoilage are not the same as those that cause foodborne illnesses.

Food Spoilage

Food spoilage is any change in the appearance, smell, or taste of a food product that makes it unacceptable to the consumer, whether or not it is due to microbial growth. Foods are rich in organic matter and the physical and chemical characteristics of a food determine its susceptibility to microbial activity. With respect to spoilage, a food or food product falls into one of three categories: (1) **Perishable foods** include many fresh food items such as meats and many vegetables; (2) **semiperishable foods**

UNIT 6

include foods such as potatoes, some apples, and nuts; and (3) **nonperishable foods**, such as sugar and flour. The foods in these categories differ primarily with regard to their *moisture content*, as measured by their water activity (a_w, ↺ Section 5.15). Nonperishable foods have low moisture levels and can generally be stored for long periods without spoilage. Perishable and semi-perishable foods, by contrast, typically have higher moisture levels and hence these foods must be stored under conditions that inhibit microbial growth.

Fresh foods can be spoiled by many different bacteria and fungi (Table 31.3). The chemical properties of foods vary widely, and each food is characterized by its moisture level and the nutrients it contains as well as other factors, such as its acidity or alkalinity. As a result, each susceptible food is typically spoiled by a specific group of microorganisms. The time required for a microbial population to reach a significant level in a given food product depends on both the size of the initial inoculum and the rate of growth during the exponential phase. Microbial numbers in a food product may initially be so low that no measurable effect can be observed, with only the last few cell doublings leading to observable spoilage. Hence, an unconsumed portion of a food product that is palatable and eaten one day can be badly spoiled the next.

The type of food spoilage and the microbial composition of the spoilage community (Table 31.3) are functions of both the food product and the storage temperature. Food spoilage microorganisms are often *psychrotolerant*, meaning that although they grow best at temperatures above 20°C, they can also grow at

refrigeration temperatures (3–5°C) (↺ Section 5.12). However, at any given storage temperature, some species grow faster than others, and thus the composition of the microbial spoilage community of the same food product stored at different temperatures can vary significantly.

Food Preservation and Fermentation

Food storage and preservation methods are designed to slow or stop the growth of microorganisms that spoil food or that can cause foodborne disease. The major methods of food preservation include altering the temperature, acidity, or moisture level of the food, or treating it with radiation or chemicals that prevent microbial growth.

Refrigeration slows microbial growth but a remarkable number of microorganisms, particularly bacteria, can grow at refrigeration temperatures. Storage in the household freezer reduces growth considerably, but slow growth still occurs in pockets of liquid water trapped within the frozen food. In general, a lower storage temperature results in less microbial growth and slower spoilage, but storage at temperatures below −20°C is too expensive for routine use and also can negatively affect food appearance, consistency, and taste.

Heat reduces the bacterial load and can even sterilize a food product, and is especially useful for the preservation of liquids or high-moisture foods. The limited heat treatment of **pasteurization** (↺ Section 5.17) does not sterilize liquids but reduces microbial numbers and eliminates pathogens. *Canning*, by contrast, typically sterilizes the food but requires careful processing in a sealed container at the correct temperature for the correct length of time. If viable microorganisms remain in a can or glass jar, their growth can produce gas, resulting in bulges or even explosions (Figure 31.7). The environment inside a can or sealed jar is anoxic, and an important genus of anaerobic bacteria that can grow in canned foods is the endospore-forming *Clostridium*, one species of which causes botulism (Section 31.9).

Foods can be made drier by either physically removing the water or by adding solutes, such as salt or sugar. Extremely dry or solute-loaded foods help prevent bacterial growth, but spoilage

Table **31.3** Microbial spoilage of fresh food[a]		
Food product	**Type of microorganism**	**Common spoilage organisms, by genus**
Fruits and vegetables	Bacteria	*Erwinia, Pseudomonas, Corynebacterium* (mainly vegetable pathogens; rarely spoil fruit)
	Fungi	*Aspergillus, Botrytis, Geotrichum, Rhizopus, Penicillium, Cladosporium, Alternaria, Phytophthora,* various yeasts
Fresh meat, poultry, eggs, and seafood	Bacteria	*Acinetobacter, Aeromonas, Pseudomonas, Micrococcus, Achromobacter, Flavobacterium, Proteus, Salmonella, Escherichia, Campylobacter, Listeria*
	Fungi	*Cladosporium, Mucor, Rhizopus, Penicillium, Geotrichum, Sporotrichum, Candida, Torula, Rhodotorula*
Milk	Bacteria	*Streptococcus, Leuconostoc, Lactococcus, Lactobacillus, Pseudomonas, Proteus*
High-sugar foods	Bacteria	*Clostridium, Bacillus, Flavobacterium*
	Fungi	*Saccharomyces, Torula, Penicillium*

[a]The organisms listed are the most commonly observed spoilage agents of fresh, perishable foods. Many of these genera include species that are human pathogens (Chapters 29, 30, and 32).

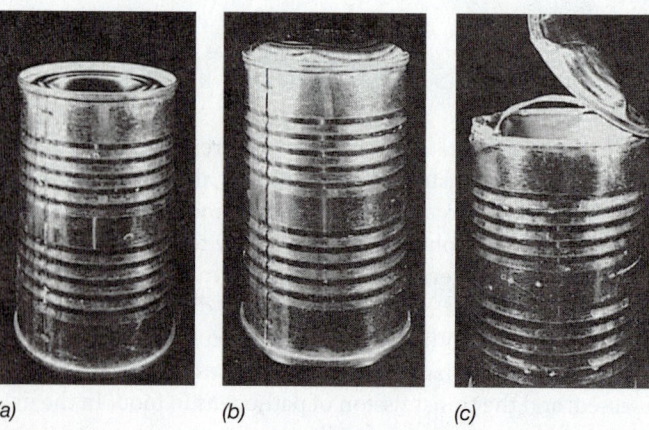

(a) *(b)* *(c)*

Figure 31.7 Changes in sealed tin cans as a result of microbial spoilage. *(a)* A normal can. The top of the can is pulled in a bit due to the normal slight vacuum inside. *(b)* Swelling due to gas production. *(c)* The can shown in b was dropped, and the gas pressure resulted in a violent explosion, tearing the lid apart.

John M. Martinko and Cheryl Broadie

Figure 31.8 Fermented foods. Bread, sausage meats, cheeses, many dairy products, and fermented and pickled vegetables are food products that are produced or enhanced by fermentation reactions catalyzed by microorganisms (see also Table 31.4).

Table 31.4 Fermented foods and fermentation microorganisms

Food category/ Preservative	Primary fermenting microorganisms[a]
Dairy foods/Lactic acid, propionic acid	
Cheeses	*Lactococcus, Lactobacillus, Streptococcus thermophilus, Propionibacterium* (Swiss cheese)
Fermented milk products	
Buttermilk and sour cream	*Lactococcus*
Yogurt	*Lactobacillus, Streptococcus thermophilus*
Alcoholic beverages/Ethanol	*Zymomonas, Saccharomyces*[b]
Yeast breads/Baking	*Saccharomyces cerevisiae*[b]
Meat products/Lactic and other acids	
Dry sausages (pepperoni, salami) and semidry sausages (summer sausage, bologna)	*Pediococcus, Lactobacillus, Micrococcus, Staphylococcus*
Vegetables/Lactic acid	
Cabbage (sauerkraut)	*Leuconostoc, Lactobacillus*
Cucumbers (pickles)[c]	Lactic acid bacteria
Vinegar/Acetic acid	*Acetobacter*
Soy sauce/Lactic acid and many other substances	*Aspergillus*[d], *Tetragenococcus halophilus*, yeasts

[a]Unless otherwise noted, these are all species of *Firmicutes* except for *Micrococcus*, which is in the *Actinobacteria*, and *Zymomonas* and *Acetobacter*, which are in the *Alphaproteobacteria*.
[b]Yeast. Various *Saccharomyces* species are used in alcohol fermentations. *S. cerevisiae* is the common baker's yeast.
[c]Nonfermented pickles are cucumbers marinated in vinegar (5–8% acetic acid).
[d]A mold.

can still occur and when it does, it is typically from fungi. Many foods are preserved by the addition of small amounts of antimicrobial chemicals. These chemicals, which include nitrites, sulfites, propionate, and benzoate along with a few others, find wide application in the food industry for enhancing or preserving food texture, color, freshness, or flavor. Although not widely practiced in many countries, the *irradiation* of food with ionizing radiation is also an effective means for reducing microbial contamination.

Many common foods and beverages are preserved through the metabolic activities of microorganisms; these are *fermented foods* (Figure 31.8 and Table 31.4). The fermentation process (Chapters 3 and 13) yields large amounts of preservative chemicals. The major bacteria important in the fermented foods industry are organic acid–producing bacteria such as the lactic acid bacteria (in fermented milks), the acetic acid bacteria (in pickling), and the propionic acid bacteria (in certain cheeses) (Table 31.4). The yeast *Saccharomyces cerevisiae* produces alcohol as the preservative in the production of alcoholic beverages. The high level of organic acids or alcohol generated from these fermentations prevents the growth of both spoilage organisms and pathogens in the fermented food product.

MINIQUIZ

- List the major food groups as categorized by their susceptibility to spoilage.
- Identify physical and chemical methods used for food preservation. How does each method limit growth of microorganisms?
- List some dairy, meat, beverage, and vegetable foods produced by microbial fermentation. What is the preservative in each case?

31.7 Foodborne Diseases and Food Epidemiology

Foodborne illnesses resemble waterborne illnesses in being *common-source* diseases. Most foodborne disease outbreaks are due to improper food handling and preparation by domestic consumers;

these typically affect only a few people and are rarely reported. However, occasional disease outbreaks due to breakdowns in safe food handling and preparation at restaurants or food-processing and distribution plants can affect large numbers of people in geographically widespread regions.

Foodborne Diseases and Microbial Sampling

The most prevalent foodborne diseases in the United States are *food infections* and *food poisonings*; some foodborne diseases fall into both categories. Food infections are the most common foodborne illnesses in the United States and account for four of the top five leading foodborne illnesses. Table 31.5 lists the major microorganisms that cause food infections and food poisonings in the United States.

Eight microorganisms account for the great majority of foodborne illness, hospitalizations, and deaths in the United States: *Salmonella* species, *Clostridium perfringens*, *Campylobacter jejuni*, *Staphylococcus aureus*, *Listeria monocytogenes*, and *Escherichia coli* (all bacteria); norovirus; and *Toxoplasma* (a protist) (Table 31.5). Four of these—norovirus, *Salmonella*, *C. perfringens*, and *Campylobacter*—account for nearly 90% of all foodborne illness, with norovirus (Sections 31.5 and 31.14) being the most common culprit (60%).

UNIT 6

Table 31.5 Major foodborne pathogens[a]

Organism	Disease[b]	Foods
Bacteria		
Bacillus cereus	FP and FI	Rice and starchy foods, high-sugar foods, meats, gravies, pudding, dry milk
Campylobacter jejuni	FI (4)[c]	Poultry, dairy
Clostridium perfringens	FP and FI (3)[c]	Meat and vegetables held at improper storage temperature
Escherichia coli O157:H7	FI	Meat, especially ground beef, raw vegetables
Other enteropathogenic *Escherichia coli*	FI	Meat, especially ground meat, raw vegetables
Listeria monocytogenes	FI	Refrigerated "ready to eat" foods
Salmonella spp.	FI (2)[c]	Poultry, meat, dairy, eggs
Staphylococcus aureus	FP (5)[c]	Meat, desserts
Streptococcus spp.	FI	Dairy, meat
Yersinia enterocolitica	FI	Pork, milk
All other bacteria	FP and FI	
Protists[d]		
Cryptosporidium parvum	FI	Raw and undercooked meat
Cyclospora cayetanensis	FI	Fresh produce
Giardia intestinalis	FI	Contaminated or infected meat
Toxoplasma gondii	FI	Raw and undercooked meat
Viruses		
Norovirus	FI (1)[c]	Shellfish, many other foods
Hepatitis A	FI	Shellfish and some other foods eaten raw

[a]Data from the Centers for Disease Control and Prevention, Atlanta, Georgia, USA.
[b]FP, food poisoning; FI, food infection.
[c]The number in parentheses is the rank of the top five foodborne pathogens in the United States.
[d]All of these protists are discussed in Chapter 32.

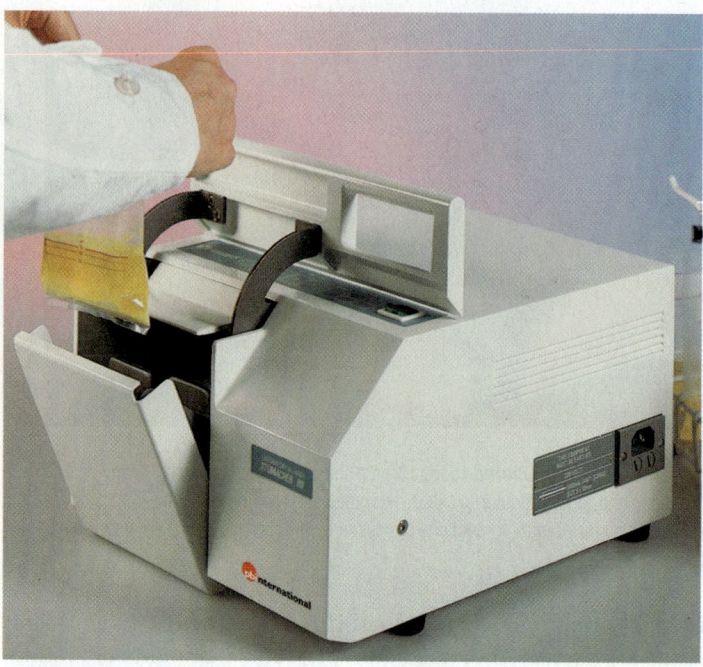

Figure 31.9 **A stomacher.** Paddles in this specialized blender homogenize the food sample in a sealed and sterile bag. The sample is first suspended in a sterile solution to form a uniform mixture.

Food poisoning, also called **food intoxication**, results from ingestion of foods containing preformed microbial toxins. The microorganisms that produced the toxins do not have to grow in the host and may not even be alive at the time the contaminated food is consumed; ingestion and activity of the toxin is what causes the illness. We previously discussed some of these toxins, notably the exotoxin of *Clostridium botulinum* and the superantigen toxins of *Staphylococcus* and *Streptococcus* (⮎ Sections 23.9 and 24.9). In contrast to food poisoning, **food infection** occurs from the ingestion of food containing sufficient numbers of viable pathogens to cause colonization and growth of the pathogen in the host, ultimately resulting in disease.

Rapid diagnostic methods that do not require culturing an organism have been developed to detect important food pathogens and many of these were described in Chapter 27. Isolation of pathogens from foods usually requires preliminary treatment of the food to suspend microorganisms embedded or entrapped within. A standard method for this purpose employs a blender called a *stomacher* (**Figure 31.9**), a device to process food samples

sealed in sterile bags. Paddles in the stomacher crush, blend, and homogenize the samples in a fashion resembling the peristaltic action of the stomach but under conditions that prevent contamination. The homogenized samples are then analyzed for specific pathogens or their products.

In addition to identifying pathogens in the food itself, disease investigators must also recover the foodborne pathogen from the diseased patient in order to establish a cause-and-effect relationship between the pathogen and the illness. In fact, identification of the *same strain* of a particular pathogen in patients and the suspected contaminated food is the "gold standard" for linking cause and effect in a foodborne disease outbreak, and a variety of microbiological, immunological, and molecular techniques are available for these purposes (Chapter 27).

Foodborne Disease Epidemiology

An outbreak of foodborne disease can occur in a home, a school cafeteria, a college dining hall, a restaurant, a military mess hall, or anywhere a contaminated food is consumed by many individuals. In addition, central food-processing plants and distribution centers provide opportunities for contaminated foods to cause disease outbreaks far from where the food was originally processed. It is the job of the food epidemiologist to track disease outbreaks and determine their source, often down to the precise location in which the food was contaminated.

A good example of effective foodborne disease tracking is the outbreak caused by *Escherichia coli* O157:H7 (see Section 31.11 and Figure 31.14) in the United States in 2006. Through culturing and molecular studies, this outbreak was linked to the consumption of contaminated packaged spinach and was quickly traced to a food-processing facility in California. The contaminated spinach was distributed nationwide from the California plant, but most disease cases were in the Midwest. In the summer of 2013, another "packaged" outbreak occurred in the Midwest but in this case was linked to lettuce instead of spinach and to the parasite *Cyclospora cayetanensis* (⮌ Section 32.4) instead of to the bacterium *E. coli.*

To be effective, foodborne disease trackers must work quickly. For example, when the first case in the *E. coli* spinach outbreak appeared in late August, a link to the specific spinach product was made less than a month later. Because *E. coli* O157:H7 has been well studied, public health officials were able to quickly identify the strain contaminating the bagged spinach. Authorities then traced this strain back to the processing plant and eventually identified a specific agricultural field near the processing plant as the source of the pathogen. Although it remains unclear how the spinach was contaminated, domestic animal manure was the likely source. During the outbreak, two foodborne disease surveillance networks, *FoodNet* (Centers for Disease Control and Prevention) and *PulseNet* (an international molecular typing network for foodborne diseases) played important roles in exposing and ending the outbreak.

The spinach *E. coli* epidemic, although serious and even deadly for some, was discovered, contained, and stopped very quickly. However, this incident shows how centralized food-processing facilities can quickly spread disease to distant populations. Because of this, food hygiene standards and surveillance must be maintained at the highest possible level at all times in restaurants and central food-processing and distribution facilities.

MINIQUIZ

- Distinguish between food infection and food poisoning.
- Describe microbial sampling procedures for solid foods such as meat.
- Describe how a foodborne disease outbreak is tracked.

IV • Food Poisoning

Food poisoning can be caused by various bacteria and a few fungi. Here we consider *Staphylococcus aureus*, *Clostridium botulinum*, and *Clostridium perfringens*, the most common causes of bacterial food poisoning. Two of these bacteria—*S. aureus* and *C. perfringens*—are part of the "top five" causes of foodborne illness (Table 31.5).

31.8 Staphylococcal Food Poisoning

A powerful form of food poisoning is caused by enterotoxins produced by the gram-positive bacterium *Staphylococcus aureus* (**Figure 31.10**; ⮌ Section 15.7). This organism is commonly associated with the skin and upper respiratory tract and is a frequent cause of pus-forming wounds (⮌ Section 29.9 and Figure 29.30). *S. aureus* can grow aerobically or anaerobically in many common foods and produces heat-stable enterotoxins. When consumed, the toxins cause gastrointestinal symptoms characterized by one or more of nausea, vomiting, diarrhea, and dehydration. The onset of symptoms is rapid, within 1–6 h of ingestion depending on the amount of enterotoxin consumed, but the symptoms usually pass within 48 h.

Staphylococcal Enterotoxins

S. aureus can produce several related enterotoxins. Many of these are fairly heat-stable and all are stable to stomach acidity. Most strains of *S. aureus* produce only one or two of these toxins, and some strains are nonproducers. However, any one of the staph enterotoxins can cause food poisoning. The toxins pass through the stomach to the small intestine and trigger disease symptoms from there. Besides their normal gastrointestinal activities, staph enterotoxins are also *superantigens* and can lead to potentially lethal toxic shock syndrome (⮌ Sections 23.9 and 24.9).

S. aureus enterotoxins are given acronyms beginning with "SE" (for "staphylococcus enterotoxin"): SEA, SEB, SEC, and SED, which are encoded by the genes *sea*, *seb*, *sec*, and *sed*. Not all of these genes are on the *S. aureus* chromosome but their sequences show them to be highly related. The genes *seb* and *sec* are encoded on the bacterial chromosome, *sea* on a lysogenic bacteriophage (⮌ Section 8.8), and *sed* on a plasmid. The phage- and plasmid-encoded genes can transfer the ability to make toxin to nontoxigenic strains of *Staphylococcus* by horizontal gene transfer (Chapter 10). SEA is the most common cause of staph food poisoning worldwide.

Disease Properties and Prevention

Foods may contain cells of *S. aureus* for several reasons. The organism may have been present on the food source itself; for example, on a meat product. But more commonly, cells of *S. aureus* are introduced to the food by contamination from the food preparer or by contamination of the food product with raw meat or a contaminated sauce or dressing. A common scenario for a staph food poisoning incident is when a food preparer introduces *S. aureus* from nasal secretions or from an uncovered skin

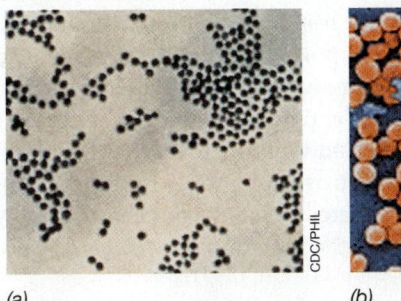
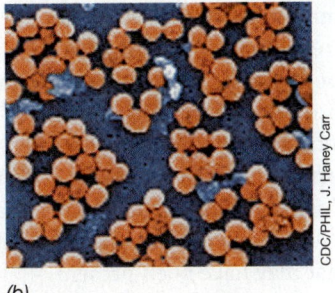

(a)   (b)

Figure 31.10 *Staphylococcus aureus.* *(a)* Gram-stained light micrograph showing the typical "cluster of grapes" morphology of staphylococci. *(b)* Colorized scanning electron micrograph of cells of *S. aureus.* A single cell is about 0.8 μm in diameter.

UNIT 6

wound or leaking bandage into the food during its preparation. If the contaminated food is then stored at room temperature or above, the stage is set for the rapid growth of *S. aureus* and the production of staph enterotoxins.

Each year there are an estimated nearly quarter million cases of staphylococcal food poisoning in the United States. The foods most commonly implicated are custard- and cream-filled baked goods, poultry, eggs, raw and processed meat, puddings, and creamy salad dressings. Salads prepared with mayonnaise-based dressings or those that contain shellfish, chicken, pasta, tuna, potato, egg, or meat, are also common vehicles. Salted foods such as ham can be vehicles because of the ability of *S. aureus* to grow quickly in salty environments (↩ Section 29.9). If any of these foods are contaminated with *S. aureus* but are refrigerated immediately after preparation, they usually remain safe because the organism grows poorly at low temperatures. But if enterotoxin has already been produced, mild heating may not make the food safe, as staph enterotoxins are stable to 60°C.

Treatment of staph food poisoning with antibiotics is not useful because any ingested cells of *S. aureus* have already been killed by the acidity in the stomach and antibiotics have no effect on the enterotoxins. As for any foodborne illness, staphylococcal food poisoning can be prevented by proper sanitation and hygiene in food production, preparation, and storage. In this regard, food preparers should practice thorough and frequent handwashing, prevent foods from coming into contact with nasal tissues and secretions, and routinely wear and frequently change disposable gloves when handling food products, especially if they have a bandaged hand wound.

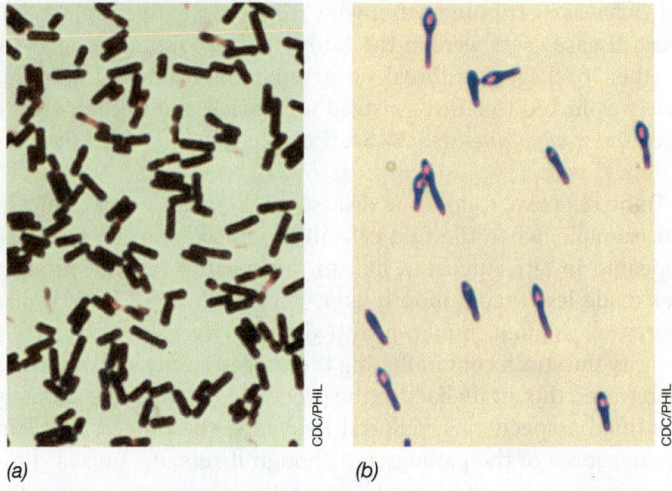

(a)   (b)

Figure 31.11 **Food poisoning clostridia.** *(a)* Gram stain of a growing culture of *Clostridium perfringens*, the bacterium that causes perfringens food poisoning. A cell measures about 1×3 μm. *(b)* Gram stain of a sporulating culture of *Clostridium botulinum*, the agent of botulism. A cell measures about 1×5 μm.

> **MINIQUIZ**
> - Identify the symptoms and mechanism of staphylococcal food poisoning.
> - Why does antibiotic treatment not affect the outcome or the severity of disease with staph food poisoning?

31.9 Clostridial Food Poisoning

The endospore-forming anaerobic bacteria *Clostridium perfringens* and *Clostridium botulinum* (↩ Section 15.8) cause serious food poisoning. Canning and cooking procedures kill vegetative cells of these species but may not kill all endospores. If this occurs, viable endospores in the food can germinate and the resulting cells produce toxins.

There is a clear distinction in the disease process between perfringens food poisoning and botulism. In the case of botulism, the toxin is a neurotoxin and only the toxin is required for disease; growth of *C. botulinum* in the human body is not required but may occur, particularly in cases of infant botulism. By contrast, with perfringens food poisoning, a large number of cells must be ingested in order for the toxin—in this case, an enterotoxin—to be produced.

Clostridium perfringens Food Poisoning

Clostridium perfringens (**Figure 31.11***a*) is commonly found in soil but can also be found in sewage, primarily because it lives in small numbers in the intestinal tract of humans and other animals. *C. perfringens* is the third most often reported cause of food poisoning in the United States behind norovirus illnesses (Sections 31.5 and 31.14) and *Salmonella* infections (Section 31.10 and Table 31.5). In 2011, nearly 1 million perfringens cases were estimated to have occurred in the United States.

C. perfringens is a proteolytic bacterium; proteins are catabolized by fermentation (↩ Section 13.13). Perfringens food poisoning requires the ingestion of a large dose ($>10^8$) of *C. perfringens* cells in contaminated cooked or uncooked foods, usually high-protein foods such as meat, poultry, and fish. *C. perfringens* can grow in meat dishes cooked in bulk where heat penetration is often insufficient. *C. perfringens* grows quickly in the food, especially if left to cool at room temperature. It is when sporulation begins that the perfringens enterotoxin is produced. The toxin alters the permeability of the intestinal epithelium, leading to nausea, diarrhea, and intestinal cramps. The onset of perfringens food poisoning typically begins 7–15 h after consumption of the contaminated food and usually resolves within 24 h; fatalities from perfringens food poisoning are rare.

A diagnosis of perfringens food poisoning is made from isolation of *C. perfringens* from the feces or, more reliably, by an immunoassay that can detect *C. perfringens* enterotoxin in feces. Prevention of perfringens food poisoning requires that cooked foods not be contaminated with raw foods and that all foods be properly heated during cooking and home canning. The perfringens enterotoxin is heat-labile and thus any toxin that may have formed in a food product is destroyed by proper heating (75°C). Cooked foods should be refrigerated as soon as possible to rapidly lower temperatures and inhibit the growth of any *C. perfringens* that may have been present.

Botulism

Botulism is a severe and potentially fatal food poisoning caused by the consumption of food containing the exotoxin produced by *C. botulinum* (Figure 31.11*b* and see page 903). This bacterium normally inhabits soil or water, but its cells or endospores may contaminate raw and processed foods. If viable endospores of *C. botulinum* remain in the food, they may germinate and produce botulinum toxin; ingesting even a small amount of this highly poisonous substance can cause severe illness or death.

Botulinum toxin is a neurotoxin that affects autonomic nerves that control key body functions such as respiration and heartbeat; the typical result is a flaccid paralysis (⮌ Section 23.9). At least seven distinct botulinum toxins are known. Because the toxins are destroyed by heat (80°C for 10 minutes), thoroughly cooked food, even if contaminated with toxin, is harmless. Much foodborne botulism is from improperly processed home-canned foods, especially nonacidic foods such as corn and beans. Any viable *C. botulinum* endospores that remain in the sealed (and now anoxic) jar may germinate during storage and produce toxin. Many of these foods are used without cooking when making cold salads, and hence any botulinum toxin present is not destroyed. Prevention of foodborne botulism thus requires careful attention to canning and related food preservation practices.

Although infants can be poisoned by toxin-contaminated food, the majority of infant botulism cases occur from toxin produced following actual *infection* of the infant with *C. botulinum*. This occurs most commonly in newborns up to about 2 months of age because they lack a well-developed intestinal microflora that can outcompete *C. botulinum*. Ingested *C. botulinum* endospores germinate in the infant's intestine, triggering growth and toxin

production. Wound botulism can also occur from infection, presumably from endospores in contaminating material introduced via a parenteral route. Wound botulism is most commonly associated with illicit injectable drug use.

All forms of botulism are rare. In the United States about 150 cases are observed each year with about 70% being infant, 15% wound, and 15% foodborne. Botulism, however, is very serious because of the high mortality associated with untreated disease. Because most cases are diagnosed and treated, less than 5% of all botulism cases result in death. Botulism is diagnosed when either botulinum toxin or *C. botulinum* cells are detected in the patient (or in the contaminated food) coupled with clinical observations of localized paralysis (impaired vision and speech) beginning 18–24 h after ingestion of the contaminated food. Treatment for botulism is by administration of botulinum antitoxin if the diagnosis is early, and mechanical ventilation if signs of respiratory paralysis have already appeared. If the dose of toxin is not too high, infant botulism is usually self-limiting, and most infants recover with only supportive therapy, such as assisted ventilation.

MINIQUIZ -

- Compare and contrast toxin production and toxemia in botulism and perfringens food poisoning.

- Describe differences in the transmission of botulism in adults versus infants.

V • Food Infection

Food infection results from ingestion of food containing sufficient numbers of viable pathogens to cause growth of the pathogen and disease in the host. Food infections are very common, and in the United States, the sum total of food infections greatly outnumbers cases of food poisoning by nearly 10-fold. Sections 23.1 and 23.6–23.8 review the infection process, summarizing the steps by which microorganisms—both friend and foe alike—attach and become established in host tissues.

31.10 Salmonellosis

Salmonellosis is a gastrointestinal disease typically caused by ingesting food contaminated with *Salmonella* or by handling *Salmonella*-contaminated animals or animal products (**Figure 31.12**). Salmonellosis is the most common bacterial food infection in the United States and second only to norovirus in total number of cases. Symptoms of salmonellosis begin after the pathogen— a gram-negative, facultatively aerobic rod related to *Escherichia coli* (⮌ Section 15.3 and see Figure 31.13)—colonizes the intestinal epithelium. *Salmonella* species normally inhabit the intestine of warm-blooded and many cold-blooded animals (Figure 31.12) and are common in sewage. Thus, some cases of salmonellosis are waterborne rather than foodborne infections and this is especially the case for typhoid fever (Section 31.5).

The accepted species epithet for pathogenic *Salmonella* is *enterica*, and there are seven subspecies of *S. enterica*. Most human salmonellae fall into the *S. enterica* subspecies *enterica*

(a)   (b)

(c)   (d)

Figure 31.12 **Some sources of *Salmonella*.** *(a)* Poultry contain *Salmonella* in their intestines and droppings. *(b) Salmonella* can be transferred to humans from both *(b)* reptiles and *(c)* amphibians. *(d)* Fresh poultry and eggs.

group. Each subspecies is also divided into *serovars* (serological variants). Thus, there is *Salmonella enterica* serovar Typhi, *Salmonella enterica* serovar Typhimurium, and so on. *S. enterica* serovars Typhimurium and Enteritidis are most frequently associated with foodborne salmonellosis.

Pathogenesis and Epidemiology

The most common form of salmonellosis is *enterocolitis*. Ingestion of food containing viable cells of *Salmonella* results in colonization of both the small and large intestines. From here, cells of *Salmonella* invade phagocytic cells and grow intracellularly, spreading to adjacent cells as host cells die. After invasion, pathogenic *Salmonella* deploy several virulence factors including endotoxins, enterotoxins, and cytotoxins that damage and kill host cells (⮔ Sections 23.9 and 23.10). Symptoms of enterocolitis typically appear 8–48 h after ingestion and include a headache, chills, vomiting, and diarrhea, followed by a fever that can last for several days. The disease normally resolves without intervention in 2–5 days. After recovery, however, patients may shed *Salmonella* in their feces for several weeks and some become healthy carriers. A few serovars of *S. enterica* may also cause septicemia (a blood infection) and enteric or typhoid fever, a potentially fatal disease characterized by systemic infection and high fever lasting several weeks (Section 31.5).

The incidence of salmonellosis in the United States has been steady over the last decade, with about a million estimated cases each year. There are several routes by which *Salmonella* may enter the food supply. The bacteria may reach food through fecal contamination from food handlers. Food production animals such as chickens, pigs, and cattle harbor *Salmonella* serovars that are pathogenic to humans, and these may be carried through to fresh foods such as eggs, meat, and dairy products (Figure 31.12). *Salmonella* food infections are often traced to products such as custards, cream cakes, meringues, pies, and eggnog made with uncooked eggs. Other foods commonly implicated in salmonellosis outbreaks are meats and meat products, especially poultry, cured but uncooked sausages and other meats, milk, and milk products. The simple handling of *Salmonella*-contaminated animals (Figure 31.12) can also lead to salmonellosis.

Diagnosis, Treatment, and Prevention

Foodborne salmonellosis is diagnosed from a combination of clinical symptoms, a history of recent consumption of high-risk foods, and culturing of the organism from feces. Selective, differential media are used to isolate *Salmonella* and discriminate it from other gram-negative enteric bacteria (Figure 31.13). Tests for the presence of *Salmonella* are commonly carried out on foods of animal origin such as raw meat, poultry, eggs, and powdered milk. Tests include several rapid tests (Chapter 27), but even rapid tests usually rely on enrichment procedures to increase cell numbers of *Salmonella* to testable levels.

Treatment of enterocolitis is usually unnecessary, and antibiotic treatment does not shorten the course of the disease or eliminate the carrier state. Foods containing *Salmonella* but heated to at least 70°C are generally safe if consumed immediately, held at 50°C or above, or quickly refrigerated. Any foods that become contaminated by an infected food handler can support the growth

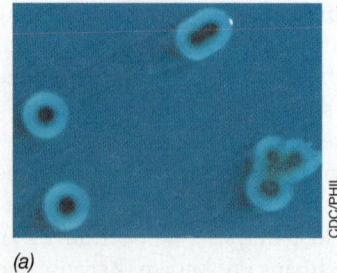 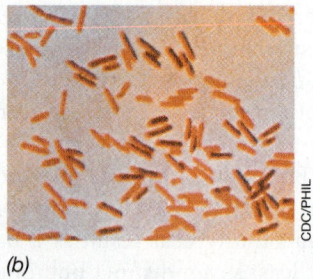

(a) *(b)*

Figure 31.13 **Isolation of *Salmonella*.** *(a)* Colonies of *S. enterica* serovar Typhimurium on Hektoen agar, which contains inhibitors of gram-positive bacteria and both lactose and peptone as carbon sources. Thiosulfate in the medium is reduced to H_2S by *Salmonella* and complexes with iron to form black FeS. *Salmonella* thus forms white colonies (because it does not ferment lactose) with black FeS centers, a pattern unique among enteric bacteria. *(b)* Gram stain of cells of *Salmonella*.

of *Salmonella* if the food is held for a long enough period, especially if it is not kept very warm or refrigerated.

MINIQUIZ

• Describe salmonellosis food infection. How does a food infection differ from food poisoning?

• How might *Salmonella* contamination of food production animals be contained?

31.11 Pathogenic *Escherichia coli*

Most strains of *Escherichia coli* are common microflora in the human colon and are not pathogenic. However, a few strains are potential foodborne (and occasionally waterborne) pathogens (Figure 31.14) and produce potent enterotoxins. These pathogenic strains are grouped on the basis of the type of toxin they produce and their specific disease syndromes. We focus here on Shiga toxin–producing *E. coli* and briefly consider some other toxigenic *E. coli* strains.

Although not in the "top five" in terms of foodborne infection pathogens (Table 31.5), pathogenic *E. coli* strains cause disease symptoms so severe that they often require hospitalization. Indeed, infections with pathogenic *E. coli* may cause life-threatening diarrheal disease and urinary tract distress.

Shiga Toxin–Producing *Escherichia coli* (STEC)

Shiga toxin–producing *Escherichia coli* (STEC) strains produce *verotoxin*, an enterotoxin similar to the Shiga toxin produced by

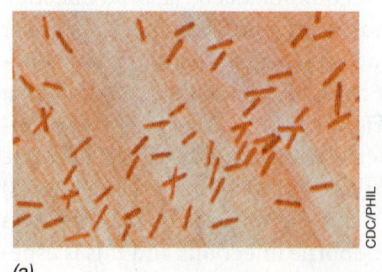 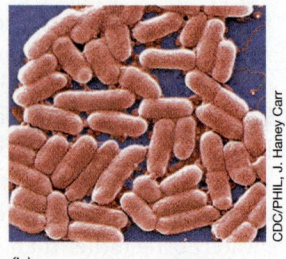

(a) *(b)*

Figure 31.14 **Pathogenic *Escherichia coli*.** *(a)* Gram-stained cells showing the typical gram-negative rod-shaped morphology of *E. coli*. *(b)* Colorized scanning electron micrograph of cells of *E. coli* O157:H7. Cells measure about 1 × 3 μm.

Shigella dysenteriae (⇄ Table 23.5), a close relative of *E. coli*. STEC strains of *E. coli* are also called *enterohemorrhagic E. coli* (EHEC). The most widely distributed STEC is *E. coli* O157:H7 (Figure 31.14*b*). Following ingestion of food or water containing STEC, the bacteria infect the small intestine where they grow and produce verotoxin, which both causes a bloody diarrhea and initiates signs of kidney failure.

Nearly half of STEC infections are caused by the consumption of contaminated uncooked or undercooked meat, particularly mass-processed ground beef. *E. coli* O157:H7 is normally present in the intestines of healthy cattle and enters the human food chain if meat is contaminated with the animal's intestinal contents during slaughter and processing. STEC strains have also been implicated in food infection outbreaks caused by dairy products (especially raw milk products), fresh fruit, and raw vegetables. Contamination of the fresh foods by fecal material, typically from cattle carrying STEC strains, has been implicated in several of these cases (Section 31.7).

Other Pathogenic *Escherichia coli*

Children in developing countries often contract diarrheal disease caused by *E. coli*, and *E. coli* can also be the cause of "traveler's diarrhea," a common infection causing watery diarrhea (as opposed to the bloody diarrhea of STEC strains) in travelers to developing countries. The primary causal agents here are *enterotoxigenic E. coli* (ETEC). These strains infect the small intestine and produce one of two heat-labile, diarrhea-producing enterotoxins.

In studies of United States citizens traveling in Mexico, the infection rate with ETEC is often greater than 50%. The prime vehicles are perishable foods such as fresh vegetables (for example, lettuce in salads) and public water supplies. The local population is typically resistant to the ETEC strains because of long-term contact with the organism. Other pathogenic *E. coli* strains include *enteropathogenic E. coli* (EPEC) strains that cause diarrheal diseases in infants and small children but do not cause invasive disease or produce toxins, and *enteroinvasive E. coli* (EIEC) strains, which invade the colon and cause watery and sometimes bloody diarrhea.

Diagnosis, Treatment, and Prevention

The general pattern established for the diagnosis, treatment, and prevention of STEC infection reflects current procedures used for all pathogenic *E. coli* strains. Laboratory diagnosis requires culture from the feces and identification of the O (lipopolysaccharide) and H (flagellar) antigens and toxins by immunological methods. Identification and typing can also be done using various molecular analyses.

Treatment of STEC infections includes supportive care for dehydration and monitoring of renal function, blood hemoglobin, and platelets. Antibiotics may actually be harmful because they may trigger the release of large amounts of verotoxin from dying *E. coli* cells that would otherwise be voided intact in feces. For other pathogenic *E. coli* infections, treatment includes supportive therapy and, for severe cases and invasive disease, antimicrobial drugs to shorten and eliminate infection.

The most effective way to prevent infection with pathogenic *E. coil* of any types is to wash raw foods vigorously and make sure

that meat, especially ground beef, is cooked thoroughly, which means that it should appear gray or brown with clear juices and have attained a temperature of greater than 70°C. In general, proper food handling, water purification, and appropriate hygiene also prevent the spread of pathogenic *E. coli*. While traveling, diarrhea from pathogenic *E. coli* can be prevented by drinking water only from properly sealed bottled water and avoiding any uncooked foods.

MINIQUIZ -

- How do STEC strains of *Escherichia coli* differ from other pathogenic *E. coli*?

- Why are meats prime vehicles for pathogenic *E. coli*? How can contaminated meat be rendered safe to eat?

31.12 *Campylobacter*

Along with salmonellosis (Section 31.10) and perfringens food poisoning (Section 31.9), *Campylobacter* infections are the most common bacterial foodborne infections in the United States (Table 31.5). Cells of *Campylobacter* are gram-negative and motile spiral-shaped *Epsilonproteobacteria* (⇄ Section 15.5) that grow best at reduced oxygen tension (microaerophilic). Several species of *Campylobacter* are recognized, but *C. jejuni* and *C. fetus* (**Figure 31.15**) are mostly commonly linked to human foodborne illnesses.

Epidemiology and Pathology

Campylobacter is transmitted to humans via contaminated food, most commonly in undercooked poultry or pork, raw shellfish, or

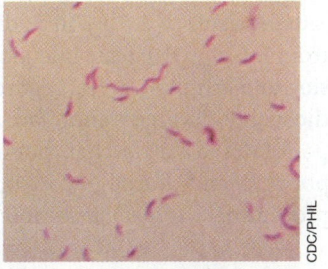

(a)   *(b)*

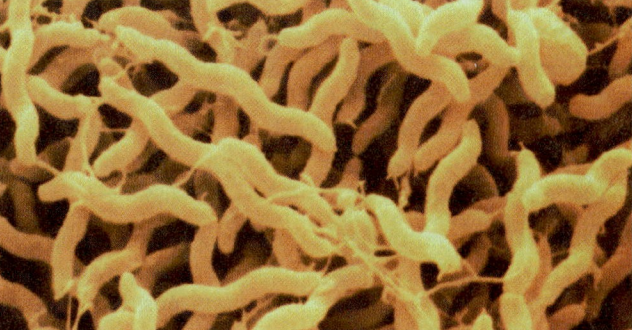

(c)

Figure 31.15 *Campylobacter.* *(a)* Colonies of *C. jejuni* grown on *Campylobacter* agar, a selective medium. The medium contains several antibiotics to which *Campylobacter* species are naturally resistant. *(b)* Gram stain and *(c)* scanning electron micrograph of cells of a *Campylobacter* species. Single cells average 0.4 × 2 μm in size.

occasionally in fecally contaminated water from surface sources. *C. jejuni* is a normal resident of the intestinal tract of poultry, and according to the United States Department of Agriculture, up to 90% of turkey and chicken carcasses are contaminated with *Campylobacter*. Pork can also carry *Campylobacter*, while beef is rarely a vehicle. *Campylobacter* species also infect domestic animals such as dogs, causing a milder form of diarrhea in the animal than that observed in humans. *Campylobacter* infections in infants in particular are often traced to infected domestic animals, especially dogs.

After cells of *Campylobacter* are ingested, the organism multiplies in the small intestine, invades the epithelium, and causes inflammation. Because *C. jejuni* is sensitive to gastric acid, cell numbers as high as 10^4 may be required to initiate infection. However, this number may be reduced to fewer than 500 cells if the pathogen is ingested in food or if the person is taking medication to reduce stomach acid production. *Campylobacter* infection causes a high fever (usually greater than 40°C), headache, malaise, nausea, abdominal cramps, and diarrhea with watery, frequently bloody emissions; symptoms subside in about a week.

Diagnosis, Treatment, and Prevention

Diagnosis of *Campylobacter* food infection requires isolation of the organism from feces and identification by growth-dependent tests, immunological assays, or genomic analyses. Culture media containing multiple antibiotics to which campylobacters are naturally resistant have been developed for selective isolation of this organism (Figure 31.15a). Various immunological methods are also available for diagnosing a campylobacter infection.

Antibiotic treatment with the drug azithromycin is widely practiced if a confirmed diagnosis is made from culture or culture-independent evidence. In addition, severe cases of dehydration from a *Campylobacter* infection may require intravenous perfusion and hospitalization. Rigorous personal hygiene, especially by those in food preparation facilities, proper washing of uncooked poultry (and any kitchenware coming in contact with uncooked poultry), and thorough cooking of meat are the major means of preventing *Campylobacter* infections.

MINIQUIZ

- Describe the pathology of *Campylobacter* food infection. What are the major vehicles for this pathogen?
- How might *Campylobacter* contamination of food production animals be controlled?

31.13 Listeriosis

Listeria monocytogenes causes **listeriosis**, a gastrointestinal food infection that may lead to bacteremia (bacteria in the blood) and meningitis. *L. monocytogenes* is a gram-positive, nonsporulating coccobacillus (*Firmicutes*) that is acid-, salt-, and cold-tolerant and facultatively aerobic (**Figure 31.16**) (⮌ Section 15.7). Although a minor foodborne pathogen in terms of the number of cases observed per year, *Listeria* infections can be very severe and cause an estimated 20% of all deaths from foodborne illness in the United States.

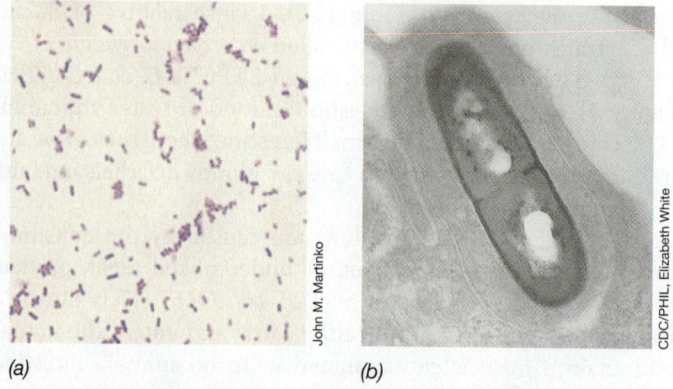

(a)   (b)

Figure 31.16 *Listeria monocytogenes.* *(a)* Gram stain and *(b)* transmission electron micrograph of cells of *L. monocytogenes*, the cause of listeriosis. The *Listeria* cell in part *b* is within host tissues (see Figure 31.17).

Epidemiology

L. monocytogenes is present in soil and water and although it is not common in foods, virtually no food source is safe from possible *L. monocytogenes* contamination. Food can become contaminated at any stage during production or processing. Ready-to-eat meats, fresh soft cheeses, unpasteurized dairy products, and inadequately pasteurized milk are the major food vehicles for *Listeria*, even when these foods are properly stored at refrigerator temperature (4°C). Food preservation by refrigeration, which ordinarily prevents the growth of other foodborne pathogens, is ineffective in the case of *Listeria* because the organism is psychrotolerant. Cells of *L. monocytogenes* produce a series of branched-chained fatty acids that keep the cytoplasmic membrane functional at cold temperatures (⮌ Section 5.12).

Evidence from animal studies and from observations of human cases of listeriosis, along with the high frequency of contamination of both raw and processed foods with *L. monocytogenes*, suggests that the organism is not highly invasive and that a large inoculum is likely required to initiate disease symptoms. Listeriosis is primarily seen in the elderly, pregnant women, newborns, and adults with weakened immune systems. Typically fewer than a thousand cases of listeriosis are reported in the United States each year, but mortality can be as high as 25% of those showing symptoms.

Pathology

Immunity to *L. monocytogenes* is normally conferred by cell-mediated Th1 inflammatory cells (⮌ Section 25.6). However, if cells of *Listeria* evade these immune cells, as they can in those with a compromised immune system, the organism is taken up by intestinal phagocytic cells. Although one might think that this is good from the standpoint of host defense, it is actually not because phagocytic uptake begins the *Listeria* infection cycle.

Listeria cells are taken up by host phagocytic cells into a vacuole called the *phagosome*. This triggers production of a major *Listeria* virulence factor, the toxin *listeriolysin O*, and this protein lyses the phagosome and releases *L. monocytogenes* into the cytoplasm (**Figure 31.17**). Here the bacterium multiplies and produces a second major virulence factor, *ActA*, a protein that induces host cell actin polymerization; the actin coats the bacterial cell and assists

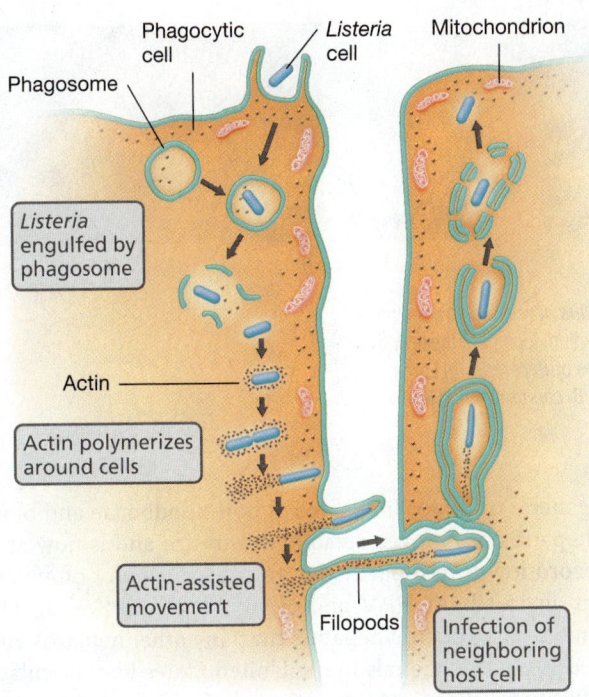

Figure 31.17 Transmission of *Listeria* during listeriosis. Cells of *Listeria* are taken up by phagocytic cells in phagosomes. These are eventually lysed by the virulence factor listeriolysin O to release *Listeria* cells. The bacterial cells then become covered with actin that assists in their movement to the cell periphery. Filopods facilitate transfer of *Listeria* cells to neighboring host cells, where the cycle repeats.

in moving the pathogen to the host cell cytoplasmic membrane. Once there, the complex pushes out, forming protrusions called *filopods* and these are then taken up by surrounding phagocytic cells (Figure 31.17). Filopod formation allows cells of *L. monocytogenes* to move about host tissues without exposure to the major weapons of the immune system: antibodies, complement, and neutrophils (Chapters 24 and 25).

Cells of *Listeria* in the intestine cross the intestinal barrier and are carried by the lymph and blood to other organs, in particular the liver, and multiply there as they do in intestinal phagocytes (Figure 31.17). From here cells of *L. monocytogenes* can infect the central nervous system, where they grow in neurons and lead to inflammation of the meninges (the tissues covering the brain and spinal cord), causing meningitis. In addition to listeriolysin O, which also allows *Listeria* to establish chronic infections in many host tissues, other major virulence factors include phospholipases that can destroy host cell membranes, antioxidants that counter phagocytic cell oxidants, and an array of "stress proteins" common in many bacteria (↩ Section 7.10).

Diagnosis, Treatment, and Prevention

Listeriosis is diagnosed by culturing *L. monocytogenes* (Figure 31.16) from the blood or cerebrospinal fluid. *L. monocytogenes* can be identified in foods by direct culture or by several molecular methods. The latter are also used to subtype clinical isolates in order to track the source(s) of infection. Intravenous antibiotic treatment with penicillin, ampicillin, or trimethoprim plus sulfamethoxazole is used in invasive cases of listeriosis.

Prevention measures include recalling contaminated food and taking steps to limit *L. monocytogenes* contamination at the food-processing site. Because *L. monocytogenes* is susceptible to heat and radiation, raw food and food-handling equipment can be readily decontaminated. However, without pasteurizing or cooking the finished food product, the risk of contamination cannot be eliminated because of the widespread distribution of the pathogen.

MINIQUIZ ---

- What is the likely outcome of *Listeria monocytogenes* exposure in normal healthy individuals?
- Which populations are most susceptible to serious disease from *L. monocytogenes* infection?

31.14 Other Foodborne Infectious Diseases

Over 200 microorganisms, viruses, and other infectious agents can cause foodborne diseases, and we have thus far summarized the major ones. Here we consider a few other bacterial pathogens that are rather uncommon compared with the "top five" (Table 31.5), and we take a second look at norovirus (previously considered as a waterborne pathogen, Section 31.5) in its more frequent context as a foodborne pathogen and overall number one cause of gastrointestinal illness in the United States.

Bacteria

Besides the major *bacterial* foodborne pathogens we have already considered, several other bacteria cause human gastrointestinal illnesses. *Yersinia enterocolitica* is an enteric bacterium commonly found in the intestines of domestic animals and causes foodborne infections from contaminated meat and dairy products. The most serious consequence of *Y. enterocolitica* infection is *enteric fever*, a severe, life-threatening infection. *Y. enterocolitica* can be isolated on the same selective/differential medium used to isolate *Salmonella* (**Figure 31.18a, b**) but is easily distinguished from this organism on plates (compare Figures 31.13a and 31.18b).

Bacillus cereus is responsible for a relatively small number of food poisoning cases. This endospore-producing bacterium (↩ Sections 2.16 and 15.8) produces two enterotoxins that cause different symptoms. In the *emetic form*, symptoms are primarily nausea and vomiting. In the *diarrheal form*, diarrhea and gastrointestinal pain are observed. *B. cereus* grows in foods such as rice, pasta, meats, or sauces that are cooked and left at room temperature to cool slowly. When endospores of this bacterium germinate, toxin is produced. Reheating may kill the *B. cereus* cells, but the toxin is heat-stable and may remain active. *B. cereus* is readily culturable and can be tentatively identified by a combination of microscopy and its typically large, grainy, and spreading colonies (Figure 31.18c, d).

The enteric bacterium *Shigella* causes the food infection *shigellosis*, and species of *Vibrio* can also cause food poisoning, primarily from consumption of contaminated shellfish. Most *Shigella* infections are the result of fecal to oral contamination, but food and water

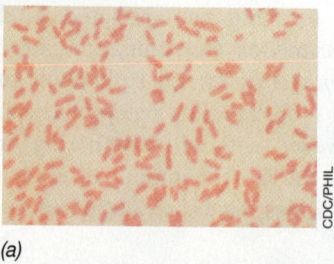
(a)

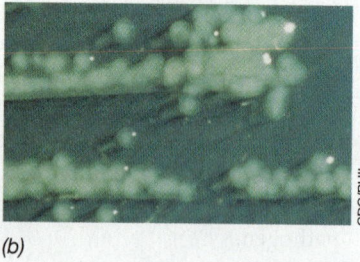
(b)

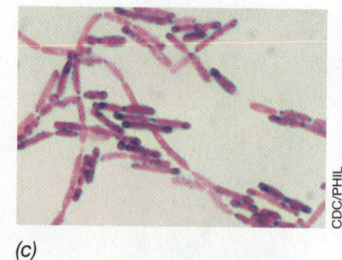
(c)

(d)

Figure 31.18 **Less common foodborne bacterial pathogens:** *Yersinia enterocolitica* **and** *Bacillus cereus.* *(a)* Gram-stained cells of *Y. enterocolitica.* *(b)* Colonies of *Y. enterocolitica* on Hektoen agar, a selective and differential medium (compare these with colonies of *Salmonella* on Hektoen agar in Figure 31.13a). *(c)* Gram-stained cells of a sporulating culture of *B. cereus.* *(d)* Large crystalline-like colonies of *B. cereus* formed on blood agar. Foodborne illness due to *Y. enterocolitica* or *B. cereus* is much less common than illness due to *Salmonella, Campylobacter,* or *Clostridium perfringens.*

are occasional vehicles. We discussed the Shiga-like toxin produced by some pathogenic strains of *Escherichia coli* in Section 31.11.

Viruses

About 70% of annual foodborne infections in the United States are caused by norovirus (**Figure 31.19a**) (Section 31.5). The virus is also known as *Norwalk virus* and is a single-stranded, plus-sense RNA virus related to poliovirus (⮌ Section 9.8). In general, noroviral foodborne illnesses are characterized by diarrhea, often accompanied by nausea and vomiting. Recovery from norovirus infections is typically spontaneous and rapid, usually within 24–48 h (thus the disease is often nicknamed "the 24-hour bug").

Rotavirus, astrovirus, and hepatitis A make up the bulk of the remaining foodborne *viral* infections. These viruses inhabit the gut and are often transmitted in food or water contaminated with feces. Hepatitis A virus (HAV, Figure 31.19b) is an RNA virus that, like norovirus, is related to poliovirus, but it replicates in liver cells. We considered hepatitis viruses transmitted primarily by blood in Section 29.11, but HAV is mainly a foodborne virus. HAV usually triggers mild, and in many cases subclinical, symptoms, but rare cases of severe liver disease from HAV can occur. The most significant food vehicles for HAV are shellfish, usually oysters or clams harvested from water polluted by human feces and then eaten raw. In recent years, HAV has also been seen in fresh produce served without cooking.

The general trend for incidence of both foodborne and blood-borne hepatitis has moved steadily downward and is now at all-time record low levels, partly due to the availability of effective vaccines against both HAV and HAB (⮌ Figure 29.33). HAV causes more cases of viral hepatitis than any other hepatitis virus, and over 30% of individuals in the United States have circulating antibodies to HAV, indicating past subclinical infections.

Protists and Other Agents

Important foodborne protist diseases are listed in Table 31.5. The major pathogens here include *Giardia intestinalis, Cryptosporidium parvum,* and *Toxoplasma gondii. G. intestinalis* and *C. parvum* are spread in foods when contaminated water is used to wash, irrigate, or spray crops. Fresh foods such as fruits are often implicated as vehicles for these protists. *Toxoplasma gondii* is a protist spread primarily through cat feces, but it can also be found in raw or undercooked meat, especially pork. We discuss the diseases giardiasis, cryptosporidiosis, and toxoplasmosis in Chapter 32.

At least one foodborne disease agent is neither cellular nor viral; these are the prions. *Prions* are proteins that adopt novel conformations, inhibiting normal protein function and causing degeneration of host neural tissues (⮌ Section 9.13). Human prion diseases are characterized by neurological symptoms including depression, loss of motor coordination, and eventual dementia. A foodborne human prion disease called *variant Creutzfeldt–Jakob disease* (vCJD) has been linked to consumption of meat products from cattle suffering from *bovine spongiform encephalopathy* (BSE), a disease caused by a prion. Although several thousand cases of vCJD were diagnosed in Great Britain in the mid-1990s, bans on cattle feeds containing rendered cattle parts and bone meal have greatly diminished the incidence of BSE in Europe and have kept the incidence of this disease very low in the United States.

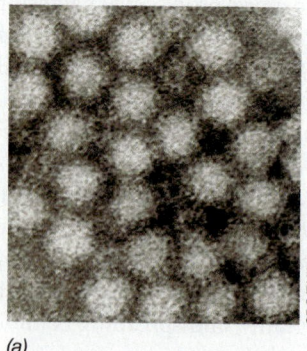
(a)

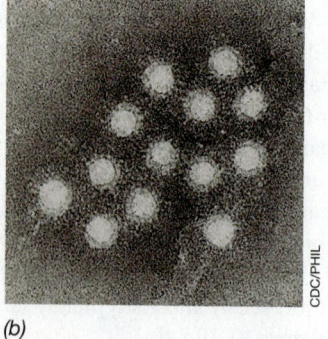
(b)

Figure 31.19 **Viruses transmitted in contaminated foods.** *(a)* Transmission electron micrograph of norovirus; an individual virion is about 30 nm in diameter. *(b)* Transmission electron micrograph of hepatitis A virus; a virion is 27 nm in diameter.

MINIQUIZ

- In what two forms can *Bacillus cereus* food poisoning manifest itself?

- Compared with all other foodborne or waterborne pathogens, what is unique about prions?

BIG IDEAS

31.1 • Contaminated drinking and recreational waters are sources of waterborne pathogens. In the United States, the number of disease outbreaks due to these sources is relatively small in relation to the large exposure the population has to water. Worldwide, lack of adequate water treatment facilities and access to clean water contribute significantly to the spread of infectious diseases.

31.2 • Drinking water quality is determined by counting coliform and fecal coliform bacteria using standardized techniques. Filtration and chlorination of water significantly decreases microbial numbers. Water purification methods in developed countries have been a major factor in improving public health, although in developing countries, waterborne illness is still a significant source of infectious disease.

31.3 • The bacterium *Vibrio cholerae* causes cholera, an acute diarrheal disease associated with severe dehydration. Cholera occurs in pandemics, primarily in developing countries where sewage treatment and sanitation is lacking. Oral rehydration and electrolyte replacement can effectively treat cholera and greatly reduce disease mortality.

31.4 • *Legionella pneumophila* is a respiratory pathogen that causes Pontiac fever and legionellosis, a more serious infection that may result in pneumonia. *L. pneumophila* grows to high numbers in warm waters and is spread via cooling tower aerosols and in domestic water distribution systems where the bacterium develops in biofilms.

31.5 • Typhoid fever, caused by a *Salmonella* species, and norovirus illness are important waterborne diseases. Typhoid is common in developing countries while norovirus illness is seen worldwide. Both of these diseases can be controlled by good sanitation practices and effective water treatment.

31.6 • The potential for microbial food spoilage depends on the nutrients and moisture levels of the food. Growth of microorganisms in perishable foods can be controlled by refrigeration, freezing, canning, pickling, dehydration, chemicals, and irradiation. Microbial fermentations can be used to naturally preserve many foods, including dairy products, meats, fruits and vegetables, and alcoholic beverages.

31.7 • Food poisoning results from the activities of microbial toxins while food infections are due to the growth of the pathogen within the body. Identification of common characteristics of foodborne pathogens from seemingly isolated foodborne outbreaks can pinpoint the origin of foodborne contamination and track the spread of the disease.

The top five foodborne pathogens in the United States in decreasing order of their appearance are: norovirus, *Salmonella* spp., *Clostridium perfringens*, *Campylobacter jejuni*, and *Staphylococcus aureus*.

31.8 • Staphylococcal food poisoning results from the ingestion of a preformed staphylococcal enterotoxin, a superantigen produced by cells of *Staphylococcus aureus* as they grow in food. Proper food preparation, handling, and storage can prevent staphylococcal food poisoning.

31.9 • *Clostridium* food poisoning results from ingestion of toxins produced by microbial growth in foods or from microbial growth followed by toxin production in the body. Perfringens food poisoning is quite common and is usually a self-limiting gastrointestinal disease. Botulism is a rare but serious disease, with significant mortality.

31.10 • More than a million cases of salmonellosis occur every year in the United States. Infection results from ingestion of cells of *Salmonella* introduced into food primarily from animal-derived food products or food handlers.

31.11 • Toxigenic *Escherichia coli* cause many food infections, and of these, STEC strains are the most severe. Contamination of foods from animal feces spreads these pathogenic strains of *E. coli*, but good hygiene practices and specific antibacterial measures such as irradiation or thorough cooking of ground beef, a major vehicle, can control disease outbreaks.

31.12 • *Campylobacter* infection is the third most prevalent foodborne bacterial infection in the United States. Poultry is a major vehicle for *Campylobacter* illness, whereas beef and pork are not. Proper poultry preparation and cooking can prevent *Campylobacter* illness.

31.13 • *Listeria monocytogenes* is a ubiquitous bacterium, and in healthy individuals, it seldom causes infection. However, in immunocompromised individuals, *Listeria* can cause serious disease as it grows as an intracellular pathogen and invades the central nervous system. Listeriosis is uncommon but shows high mortality.

31.14 • Viruses, especially norovirus, cause the most foodborne illness while the bacteria *Bacillus cereus* and *Yersinia enterocolitica* are only occasionally linked to foodborne disease outbreaks. Hepatitis A virus is also a serious foodborne pathogen. Some protists and prions also cause foodborne illness but are far less common foodborne pathogens than are bacteria and viruses.

REVIEW OF KEY TERMS

Botulism food poisoning due to ingestion of food containing botulinum toxin produced by *Clostridium botulinum*

Coliforms gram-negative, nonsporulating, facultatively aerobic rods that ferments lactose with gas formation within 48 hours at 35°C

Food infection a microbial infection resulting from the ingestion of pathogen-contaminated food followed by growth of the pathogen in the host

Food poisoning (food intoxication) a disease caused by the ingestion of food that contains preformed microbial toxins

Food spoilage a change in the appearance, smell, or taste of a food that makes it unacceptable to the consumer

Listeriosis a gastrointestinal food infection caused by *Listeria monocytogenes* that may lead to bacteremia and meningitis

Nonperishable foods foods of low water activity that have an extended shelf life and are resistant to spoilage by microorganisms

Pasteurization the use of controlled heat to reduce the microbial load, including both pathogens and spoilage organisms, in heat-sensitive liquids

Perishable foods fresh foods generally of high water activity that have a very short shelf life due to spoilage by microbial growth

Potable in water purification, drinkable; safe for human consumption

Salmonellosis enterocolitis or other gastrointestinal disease caused by any of several species of the bacterium *Salmonella*

Semiperishable foods foods of intermediate water activity that have a limited shelf life due to potential for spoilage by growth of microorganisms

REVIEW QUESTIONS

1. What are the two main classes of water and how is water from a surface source, for example, from a lake, made safe to drink? (Section 31.1)

2. Define the term fecal coliform and explain the coliform test. Why is the coliform test used to assess the purity of drinking water? (Section 31.2)

3. Why are antibiotics ineffective for the treatment of cholera? What methods are useful for treating cholera victims? (Section 31.3)

4. What are the major reservoirs for the pathogen that causes legionellosis? What aspects of pathogenesis distinguish this disease from other waterborne diseases? (Section 31.4)

5. Contrast the diseases typhoid and salmonellosis. How are they similar and how do they differ? Which is the more serious disease? (Sections 31.5 and 31.10)

6. Identify and define the three major categories of food perishability. Why is milk more perishable than sugar even though both are rich in organic matter? (Section 31.6)

7. Identify the major methods used to preserve food and the major categories of fermented foods. (Section 31.6)

8. Distinguish between food infection and food poisoning and give an example of each. (Section 31.7)

9. What causes the symptoms of staphylococcal food poisoning? Why are cases of staph food poisoning often linked to a food preparer with an open hand wound? (Section 31.8)

10. Identify the two major types of clostridial food poisoning. Which is most prevalent? Which is most dangerous and why? (Section 31.9)

11. What are the possible sources of *Salmonella* spp. that cause food infections? (Section 31.10)

12. How does *Escherichia coli* O157:H7 end up in ground beef? To what class of pathogenic *E. coli* does this strain belong? (Section 31.11)

13. Name a food product that could transmit both *Salmonella* and *Campylobacter* simultaneously. How could this food product be rendered safe to eat? (Section 31.12)

14. Identify the food sources of *Listeria monocytogenes* infections. How does *Listeria* evade the immune system? (Section 31.13)

15. Name two bacteria that are only rare causes of foodborne illness. What agent is the number one cause of gastrointestinal illness? (Section 31.14)

16. What is the causative agent of vCJD? How does the structure of this agent differ from that of the agent of noro foodborne illness? (Section 31.14)

APPLICATION QUESTIONS

1. As a visitor to a country in which cholera is an endemic disease, what specific steps would you take to reduce your risk of cholera exposure? Will these precautions also prevent you from contracting other waterborne diseases? If so, which ones? Identify waterborne diseases for which your precautions may not prevent infection.

2. Argue a case for why perfringens foodborne illness can be considered both a food poisoning and a food infection.

3. Improperly handled potato salads are often the source of both staphylococcal food poisoning and salmonellosis. List some reasons why this might be the case.

CHAPTER

32 · Eukaryotic Pathogens: Fungal and Parasitic Diseases

microbiology**now**

Deadly Fungi

People usually associate fungi with decaying organic matter, or in a medical context, with superficial fungal infections such as athlete's foot. But fungi can cause serious, even deadly infections, as we were reminded when pathogenic fungi triggered a widespread outbreak of fungal meningitis in the United States in 2012.[1]

Glucocorticoids such as methylprednisolone are often prescribed for pain relief, especially in adults suffering chronic lower back pain. The drug is typically injected directly into the outermost part of the spinal canal (epidural injection). Several lots of methylprednisolone formulated by a small pharmaceutical company in Massachusetts were found to be contaminated with fungi, including the mold *Exserohilum rostratum* (photo). *E. rostratum* is common in soil and can naturally infect several human tissues, in particular the cornea, lungs, and lining of the heart (pericardium). By early December of 2012 there were 590 fungal infections—mostly cases of meningitis—linked to injections of the contaminated drug, with 37 of these being fatal.

Using standard epidemiological techniques, the Centers for Disease Control and Prevention (Atlanta, Georgia, USA) and a specially formed multistate response team quickly linked the contaminated medication to the Massachusetts company and more specifically, to particular batches of the drug produced by this company. *E. rostratum* was found in sealed vials of methylprednisolone, and the nonpathogenic yeast *Rhodotorula* and mold *Rhizopus* were found in some others.

This outbreak of fungal disease underscores the critical importance of ensuring that drugs are both sterile and uncontaminated with foreign substances, especially drugs used for injections. In addition, the quick action by public health authorities to identify the disease source and prevent further infections is a testimony to the effectiveness of this branch of the healthcare system in the United States.

[1]Smith R.M., et al. 2012. Fungal infections associated with contaminated methylprednisolone injections—Preliminary report. *N. Engl. J. Med.* DOI: 10.1056/NEJMoa1213978.

In this chapter we focus on pathogenic microorganisms that are eukaryotic. These include several fungi—both molds and yeasts—and various parasitic protists. Some small worms also cause infectious diseases and we consider the most significant of these in the final section.

A common problem in treating diseases caused by eukaryotic pathogens is the fact that their hosts are also eukaryotic. This thwarts many therapeutic strategies and often makes these diseases highly refractory and long-term chronic infections. This is especially true of systemic fungal pathogens.

I · Fungal Infections

Fungi cause a variety of human diseases. Some are mild and self-limiting, whereas others can be firmly entrenched systemic diseases. We begin by considering some of the major fungal pathogens followed by a description of some major fungal diseases, the mycoses.

32.1 Medically Important Fungi and Disease Mechanisms

The fungi include the *yeasts*, which normally grow as single cells, and *molds*, which form branching filaments called *hyphae* with or without septa (cross walls); hyphae eventually intertwine to form visible masses called *mycelia*. The diversity of the molds and yeasts was discussed in Chapter 17.

Common Fungal Pathogens

Fortunately, most fungi are harmless to humans. Most fungi grow in nature as saprophytes on dead organic material; in so doing, fungi are important catalysts in the carbon cycle, especially in oxic environments in soil. Fungi are also important in medicine as both agents of disease and in chemotherapy (antibiotic production). Only about 50 species of fungi cause human diseases, and in healthy individuals, the incidence of serious fungal infections is low although certain superficial fungal infections (for example, athlete's foot) are fairly common. In those with compromised immune systems, however, fungal infections can be systemic, reaching even the deepest of internal tissues. Such infections can cause serious health problems and be life-threatening.

Common fungal pathogens include both yeasts and molds (**Figure 32.1**). However, many pathogenic fungi are *dimorphic*, meaning that they can exist as *either* yeasts *or* in filamentous form. In *Histoplasma*, for example, cells in laboratory culture form hyphae and mycelia and thus exist in the mold form (Figure 32.1*a*). By contrast, when *Histoplasma* causes histoplasmosis, cells grow in the host in the yeast form (see Figure 32.5*a*).

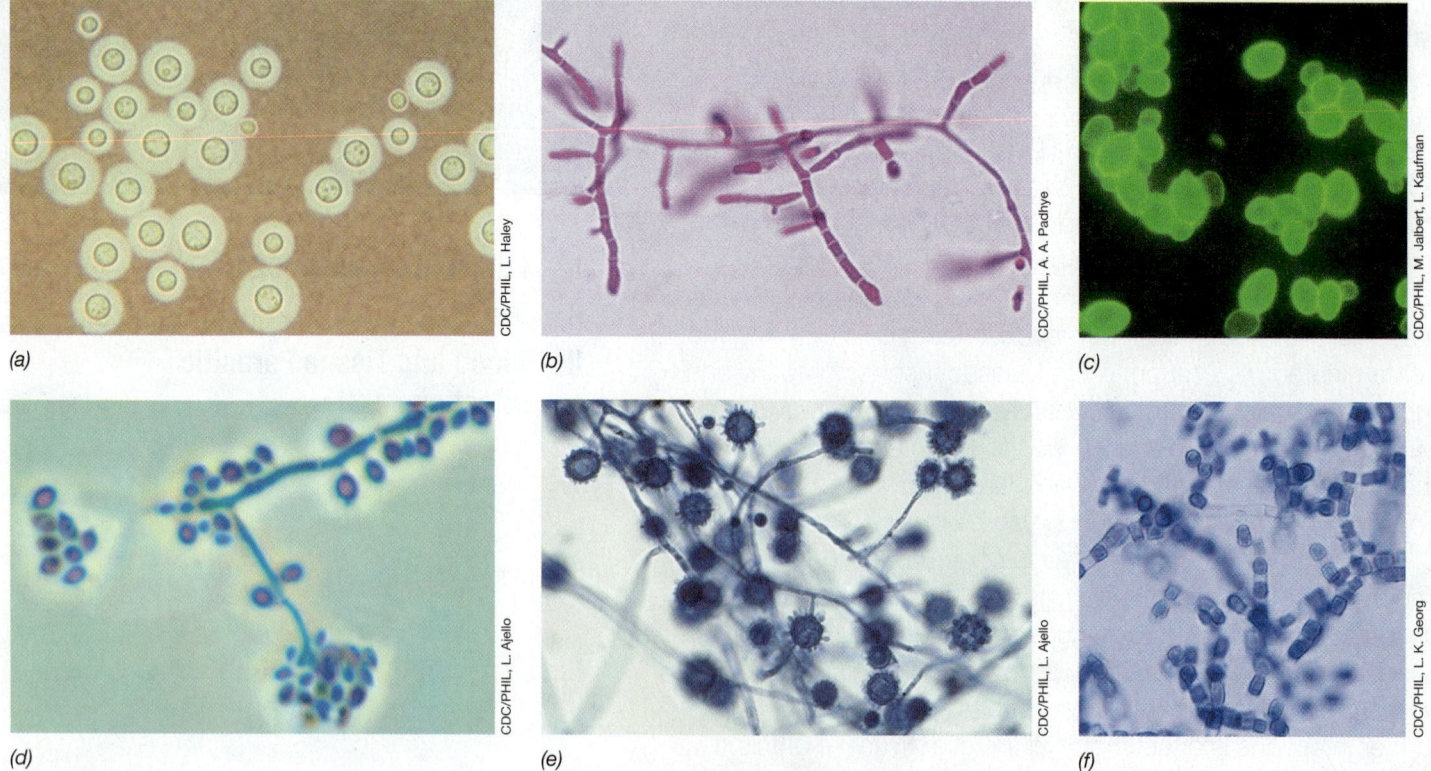

Figure 32.1 Pathogenic fungi. These organisms range from about 4 to 20 μm in diameter. *(a) Cryptococcus neoformans* yeast cells stained to reveal the capsule. *(b) Trichophyton* spp. mycelia and conidia. *(c) Candida albicans* yeast form stained with a fluorescent antibody. *(d) Sporothrix schenckii* mycelia and conidia. *(e) Histoplasma capsulatum* mycelia and large conidia. *(f) Coccidioides immitis* conidia. See fungal disease symptoms in Figure 32.5.

Table 32.1 Major pathogenic fungal diseases[a]

Class and disease	Causal organism	Site
Superficial mycoses		
Athlete's foot	*Epidermophyton, Trichophyton*	Between toes, skin
Jock itch	*Trichophyton, Epidermophyton*	Genital region
Ringworm	*Microsporum, Trichophyton*	Scalp, face
Subcutaneous mycoses		
Sporotrichosis	*Sporothrix schenckii*	Arms, hands
Chromoblastomycosis	*Phialophora verrucosa,* other fungi	Legs, feet, hands
Systemic mycoses		
Aspergillosis	*Aspergillus* spp.[b]	Lungs
Blastomycosis	*Blastomyces dermatitidis*	Lungs, skin
Candidiasis	*Candida albicans*[c]	Oral cavity, intestinal tract, vagina
Coccidioidomycosis	*Coccidioides immitis*[c]	Lungs
Paracoccidioidomycosis	*Paracoccidioides brasiliensis*	Skin
Cryptococcosis	*Cryptococcus neoformans*[c]	Lungs, meninges
Histoplasmosis	*Histoplasma capsulatum*[c]	Lungs
Pneumocystis pneumonia	*Pneumocystis jiroveci*[c]	Lungs

[a]Symptoms of many of these diseases are shown in Figures 32.3–32.5.
[b]*Aspergillus* can also cause allergies, toxemia, and limited infections.
[c]An opportunistic pathogen frequently implicated in the pathogenesis of HIV/AIDS.

In the mold form spores are produced, either asexual spores—*conidia*—or sexual spores (↩ Sections 17.9 and 17.10). When filamentous fungi are cultured from an infection, the morphology of these spore-bearing structures is observed and is often a major clue in reaching a diagnosis. In addition to microscopy, a variety of clinically useful molecular and immunological tools (Figure 32.1c) are also available to diagnose fungal infections. Table 32.1 lists some major fungal pathogens and the types of infections they cause.

Fungal Disease Classes and Treatment

Fungi cause disease through three major mechanisms: inappropriate immune responses; toxin production; and mycoses. Some fungi trigger immune responses that result in allergic (hypersensitivity) reactions following exposure to specific fungal antigens. Reexposure to the same fungi, whether growing on the host or in the environment, may cause allergic symptoms. For example, *Aspergillus* spp. (**Figure 32.2a**), a common saprophyte often found in nature as a leaf mold, produces potent allergens, triggering asthma attacks or other hypersensitivity reactions in susceptible individuals.

Fungal disease may occur from the production of *mycotoxins*, a large and diverse group of fungal exotoxins. The best-known examples of mycotoxins are the *aflatoxins* (Figure 32.2b) pro-

duced by *Aspergillus flavus*, a species that commonly grows on improperly stored dry foods, such as grain. Aflatoxins are highly toxic and are also carcinogenic, inducing tumors in some animals, especially in birds that feed on contaminated grain. Although aflatoxins are known to cause human liver damage including cirrhosis and even liver cancer, adults are not seriously affected by low-level aflatoxin exposure. However, chronic exposure in children can cause serious liver disease and other health effects.

The final fungal disease-producing mechanism is through actual host infection. The growth of a fungus on or in the body is called a **mycosis** (plural, **mycoses**). Mycoses are fungal infections that range in severity from superficial to life-threatening. Mycoses fall into three classes (Table 32.1). **Superficial mycoses** are those in which the fungus infects only the surface layers of skin, hair, or nails (see Figure 32.3). **Subcutaneous mycoses** are infections of deeper layers of skin (see Figure 32.4) and are typically caused by different fungi than superficial infections (Table 32.1). The **systemic mycoses** are the most serious category of fungal infections. These are characterized by fungal growth in internal organs of the body (see Figure 32.5) and can be either primary or secondary infections. A *primary* infection occurs when an otherwise normal, healthy individual is infected with the fungal pathogen; these are rather uncommon. By contrast, a *secondary* infection occurs in a host that harbors a predisposing condition, such as antibiotic therapy or immunosuppression, which makes the individual more susceptible to infection.

Superficial and subcutaneous mycoses are for the most part easily treatable with topical drugs, including tolnaftate (applied topically), various azole drugs (applied either topically or orally), and griseofulvin, a relatively nontoxic drug that can be taken orally but passes through the bloodstream to the skin where it inhibits fungal growth. Chemotherapy against systemic fungal infections is more difficult because of issues with host toxicity (↩ Section 27.16). For example, one of the most effective antifungal agents, amphotericin B, is widely used to treat systemic fungal infections but can also affect kidney function and have other unwanted side effects. Hence, effective treatment of the most serious of the mycoses is sometimes very difficult.

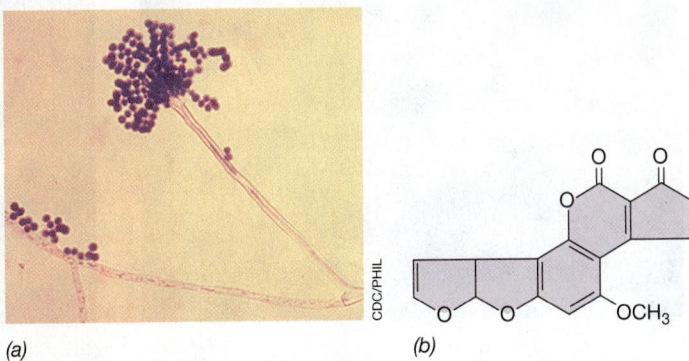

Figure 32.2 Aspergillus and aflatoxin. (a) Mycelia and conidia of an *Aspergillus* species. (b) Structure of aflatoxin B1. This toxin is one of a group of related compounds produced by *Aspergillus flavus*.

UNIT 6

MINIQUIZ

• Differentiate between superficial, subcutaneous, and systemic mycoses.

• What is a dimorphic fungus?

• Distinguish between a primary and a secondary fungal disease.

32.2 Mycoses

The two extremes of fungal infection are the superficial mycoses and the systemic mycoses. *Superficial mycoses* are quite common, and most individuals experience at least one in their lifetime. By contrast, *systemic mycoses* are far less common, and primarily affect the elderly or otherwise immune compromised. As people age, cell-mediated immunity slowly declines due to surgeries, transplantations, immunosuppressive drug treatments for rheumatism and autoimmune diseases, and the onset of other conditions, such as pulmonary decline, diabetes, and cancer. Any of these can predispose the elderly to disease. Systemic mycoses also target those of any age whose immune systems have been impaired or destroyed, for example, by HIV/AIDS (↩ Figure 29.45). Systemic mycoses are thus diseases of **opportunistic pathogens**, microorganisms that cause disease only in those whose immune defenses can no longer fight them off.

Superficial Mycoses

Table 32.1 listed some of the fungi that cause superficial mycoses; collectively, these pathogens are called *dermatophytes*. In general, superficial mycoses can be bothersome and often recurrent infections, but are not serious health concerns. Fungi such as *Trichophyton* (Figure 32.1b) cause infections of the feet (athlete's foot) and other moist skin surfaces, and are quite common (**Figure 32.3a**). These infections cause flaking and itchy skin and are easily transmitted by cells or spores of the pathogen present in contaminated shower stalls, gymnasium and locker room floors, contaminated shared articles such as towels or bed linens, or from close person-to-person contact. Superficial mycoses

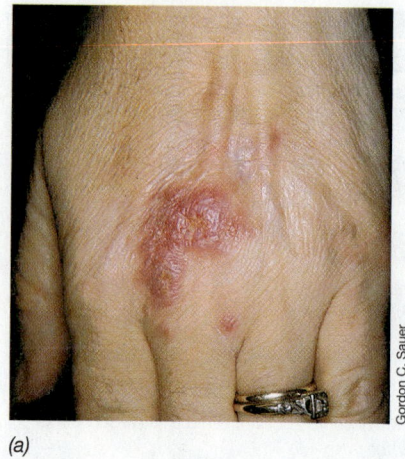
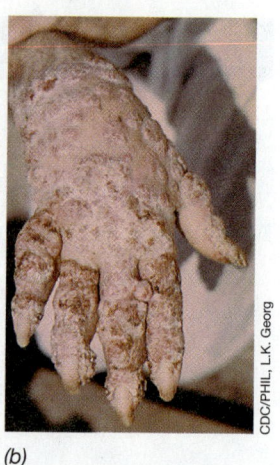

Figure 32.4 **Subcutaneous mycoses.** *(a)* Sporotrichosis, a subcutaneous infection due to *Sporothrix schenckii*. *(b)* Chromoblastomycosis on the hand caused by the fungus *Phialophora verrucosa*. Chromoblastomycosis can also be caused by species of the fungal genera *Fonsecaea* and *Cladosporium*.

can be treated with topical antifungal creams or liquid aerosols, although prophylactic application on a long-term basis may be necessary if constant exposure to the pathogen (for example, to *Trichophyton* on a locker room floor) is unavoidable.

Related surface mycoses include "jock itch," an itchy infection of the groin, skin folds, or anus, and *ringworm* (Table 32.1). Despite the name, ringworm is a fungal infection, typically localized to the scalp or the extremities; the infection causes hair loss and inflammation-like reactions (Figure 32.3b, c). These more severe superficial mycoses are usually treated topically with either miconazole nitrate or griseofulvin.

Subcutaneous Mycoses

Subcutaneous mycoses are fungal infections of deeper layers of skin than those of the superficial mycoses (Table 32.1). One disease in this class is *sporotrichosis* (**Figure 32.4a**), an occupational hazard of agricultural workers, miners, gardeners, and others who

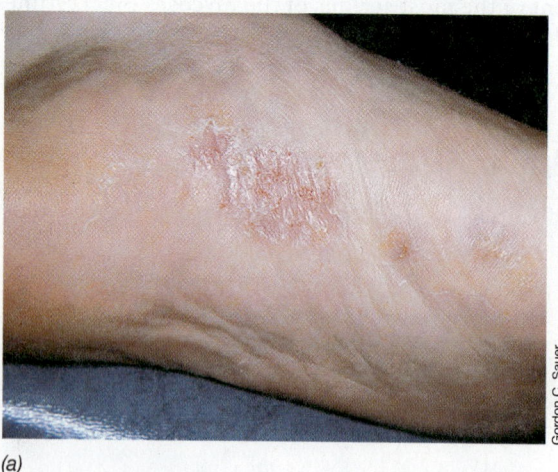
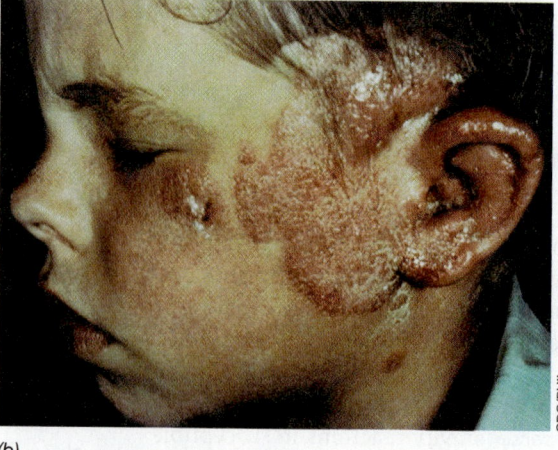
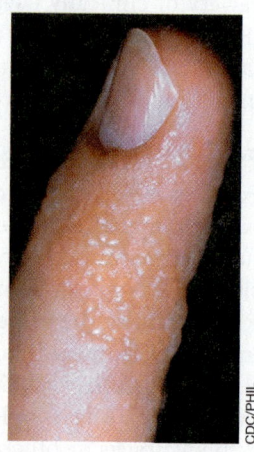

Figure 32.3 **Superficial mycoses caused by *Trichophyton* spp.** *(a)* Athlete's foot. *(b)* Ringworm on a child's face and *(c)* on an adult index finger. "Jock itch" (ringworm of the groin) is another common *Trichophyton* infection and can occur in females as well as males.

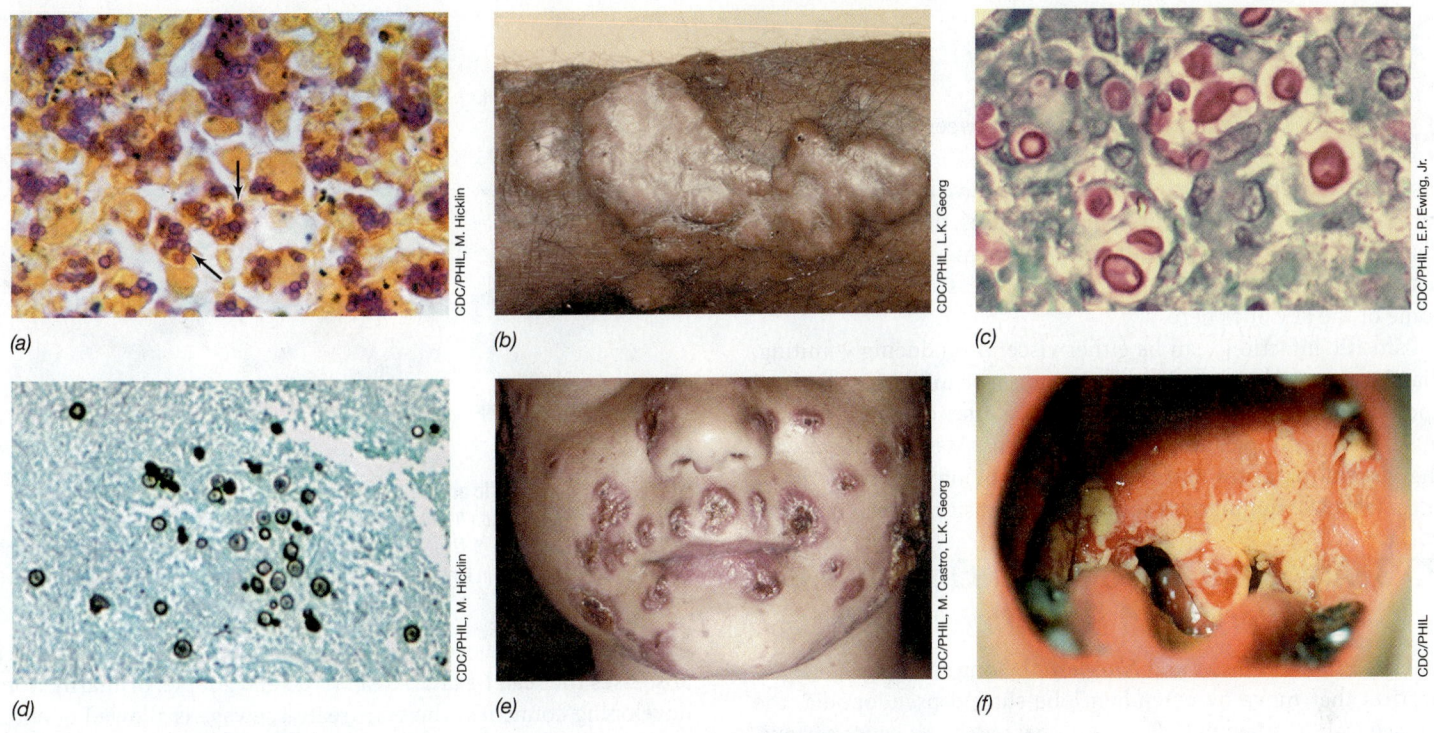

Figure 32.5 Systemic mycoses. *(a)* Histoplasmosis; yeast-form cells of *Histoplasma* (arrows) in spleen tissue. *(b)* Cutaneous blastomycosis on the arm. *(c)* Cryptococcosis; yeast-form cells (stained red) in lung tissue. *(d)* Coccidioidomycosis; yeast-form cells (stained blue-black) in lung tissue. *(e)* Paracoccidioidomycosis lesions on the face. *(f)* Oral thrush. Masses of *Candida albicans* cells (yellow) line the back of the throat. See photomicrographs of cultures of the pathogens causing most of these mycoses in Figure 32.1.

come into close and continual contact with the soil. The causal organism, *Sporothrix schenckii* (Figure 32.1*d*), is a ubiquitous soil saprophyte whose spores can enter through a cut or abrasion and infect subcutaneous tissues (Figure 32.4*a*). *Chromoblastomycosis* is due to pathogenic fungal growth in both surface (cutaneous) *and* subcutaneous skin layers, forming crusty, wartlike lesions on the hand (Figure 32.4*b*) or leg. The disease is primarily one of tropical and subtropical countries and occurs when the fungus becomes implanted under the skin from a puncture wound. Both sporotrichosis and chromoblastomycosis can be treated with oral administration of azoles.

Systemic Mycoses

Systemic fungal pathogens normally live in soil, and humans become infected by inhaling airborne spores that later germinate and grow in the lungs. From there the organism migrates throughout the body, causing deep-seated infections in the lungs and other organs and in the skin. In the United States, the three major systemic mycoses are, in order of decreasing incidence: histoplasmosis, coccidioidomycosis, and blastomycosis. Mortality from these is high, about 10%.

Histoplasmosis (**Figure 32.5*a***) is caused by *Histoplasma capsulatum* (Figure 32.1*e*), and *coccidioidomycosis* (San Joaquin Valley fever, Figure 32.5*d*) is caused by *Coccidioides immitis* (Figure 32.1*f*). Histoplasmosis is primarily a disease of rural areas in midwestern states of the United States, especially in the Ohio and Mississippi River valleys, whereas coccidioidomycosis is generally restricted to the desert regions of the southwestern United States. In more tropical climates *blastomycosis*, caused by *Blastomyces dermatitidis*, is prevalent (Figure 32.5*b*). *Paracoccidioidomycosis*, caused by the fungus *Paracoccidioides brasiliensis*, is primarily a subtropical disease with lesions forming on the face (Figure 32.5*e*) or other extremities.

Cryptococcosis (Figure 32.5*c*), caused by the dimorphic yeast *Cryptococcus neoformans* (Figure 32.1*a*), can occur in virtually any organ of the body and is the major mycosis seen in HIV/AIDS patients. The dimorphic yeast *Candida albicans* (Figure 32.1*c*) is often present as a minor component of the human normal flora. However, this fungus can cause a variety of diseases including mild vaginal infections, more serious oral infections such as thrush (Figure 32.5*f*), and systemic infection of virtually any organ in those with HIV/AIDS. Like *Histoplasma* and *Coccidioides*, *Candida* and *Cryptococcus* are primarily opportunistic pathogens and rarely cause life-threatening infections outside of the immune compromised.

Our discussion transitions now from fungi to pathogenic parasites. Like fungi, parasites are eukaryotic microorganisms, but the pathogenic parasites typically attack quite different body tissues and organs than do the pathogenic fungi.

MINIQUIZ -

- Give an example of a superficial, a subcutaneous, and a systemic mycosis.

- Why are systemic fungal pathogens called "opportunistic"?

UNIT 6

II • Visceral Parasitic Infections

Parasitism is a symbiotic relationship between two organisms, the parasite and the host (Chapter 22). The parasite derives essential nutrients from the host and may have little or no harmful effect on the host. However, in many cases, the parasite causes disease in the host. Many different phylogenetic groups of protists (Chapter 17) cause parasitic human diseases and we examine some of the key ones here.

Parasitic infections can be either visceral—inducing vomiting, diarrhea, and other intestinal symptoms—or infections of blood and internal tissues. Some of the major diseases of human history, malaria for example, are parasitic diseases. We begin here with the visceral parasites and then consider blood and tissue parasites. Table 32.2 summarizes some major parasitic human diseases.

32.3 Amoebae and Ciliates: *Entamoeba*, *Naegleria*, and *Balantidium*

The genera *Entamoeba* and *Naegleria* belong to a large group of protists that move by extending lobe-shaped pseudopodia, the Amoebozoa (↩ Section 17.8). Both parasites can cause serious, even fatal infections, although *Naegleria* infections are very rare. *Balantidium* is a ciliated species of the alveolate group (↩ Section 17.5) and is mainly a disease of tropical countries.

Entamoeba histolytica (**Figure 32.6a**) is transmitted by contaminated water or occasionally through contaminated food. *E. histolytica* is an anaerobe, and the organism's *trophozoites* (the active, motile, feeding stage of the parasite) lack mitochondria. Like another common waterborne pathogen, *Giardia* (Section 32.4), the trophozoites of *E. histolytica* produce cysts, which are the means of transmission. Ingested cysts germinate to form amoebae that grow both on and in intestinal mucosa. This leads to tissue invasion and ulceration that triggers diarrhea and severe intestinal cramps.

With further growth, the amoebae can invade the intestinal wall—a condition called *dysentery*, characterized by intestinal inflammation, fever, and the passage of intestinal blood and mucus. If the infection is not treated, *E. histolytica* can invade the

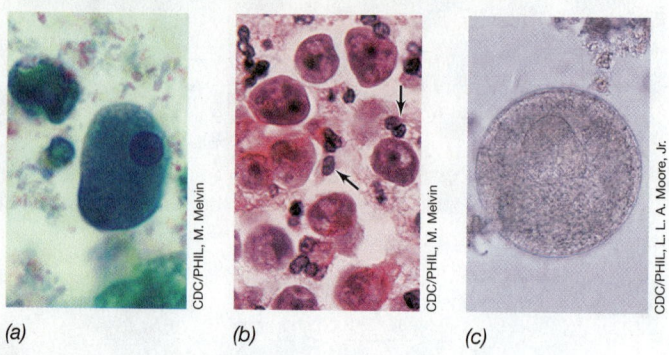

(a)   (b)   (c)

CDC/PHIL, M. Melvin   CDC/PHIL, M. Melvin   CDC/PHIL, L. L. A. Moore, Jr.

Figure 32.6 Parasitic amoebae and ciliates. *(a)* The growing stage (trophozoite) of *Entamoeba histolytica*; these can be up to 60 µm in length. *(b)* Trophozoites (arrows) of *Naegleria fowleri* in sectioned and stained brain tissue; the parasites are 10–25 µm in length. *(c) Balantidium coli* cyst present in a fecal sample.

liver, the lungs, and even the brain. Growth in these tissues causes abscesses that can be fatal. Nearly 100,000 people, primarily from developing countries where untreated sewage is allowed to enter surface waters, die each year from invasive amebic dysentery. *E. histolytica* amebiasis can be treated with a variety of drugs, but the host immune system plays a significant role in recovery as well. However, protective immunity is not conferred from a primary infection, and reinfection is common.

Naegleria fowleri can cause amebiasis, but in a very different form from that of *E. histolytica*. *N. fowleri* is a free-living amoeba present in soil and in runoff waters. *N. fowleri* infections result from swimming or bathing in warm, soil-contaminated waters such as warm springs or lakes and streams in summertime. *N. fowleri* enters the body through the nose and burrows directly into the brain. Here, the organism propagates, causing extensive hemorrhage and brain damage (Figure 32.6b), a condition called **meningoencephalitis**. Diagnosis of an *N. fowleri* infection requires observation of the amoebae in cerebrospinal fluid. If a definitive diagnosis is made quickly, the drug amphotericin B can save the patient; untreated infections are almost always fatal.

Balantidium coli is a ciliated intestinal human and swine parasite that alternates between the trophozoite and cyst (Figure 32.6c) stage; only the cysts are infective. *B. coli* is the only known ciliated parasite of humans. Cysts, typically transmitted in fecally contaminated water, germinate in the colon and infect mucosal tissues, leading to symptoms that resemble those of amebiasis, for which the disease is sometimes mistaken. An infected patient usually experiences a spontaneous recovery or may become an asymptomatic carrier, continuously shedding *B. coli* cysts in the feces. Compared with amebiasis, *B. coli* infections are uncommon, and cases are rarely fatal.

Table 32.2 Major parasitic human diseases

Parasitic diseases by site	Causal organism[a]
Gastrointestinal	
Amebiasis	*Entamoeba histolytica*
Giardiasis	*Giardia intestinalis*
Cryptosporidiosis	*Cryptosporidium parvum*
Toxoplasmosis	*Toxoplasma gondii*
Blood and tissue	
Malaria	*Plasmodium* spp.
Leishmaniasis	*Leishmania* spp.
Trypanosomiasis (African sleeping sickness)	*Trypanosoma brucei*
Chagas' disease	*Trypanosoma cruzi*
Schistosomiasis	*Schistosoma mansoni*

[a]All are protists (Chapter 17) except for *Schistosoma*, a helminth.

MINIQUIZ

- Contrast an *Entamoeba* and *Naegleria* infection in terms of tissues infected and symptoms.
- Describe a scenario for contracting a *Naegleria* infection.

32.4 Other Visceral Parasites: *Giardia, Trichomonas, Cryptosporidium, Toxoplasma*, and *Cyclospora*

The protists *Giardia intestinalis* and *Trichomonas vaginalis* are flagellated anaerobic parasites that contain either mitosomes or hydrogenosomes in place of mitochondria (⟳ Sections 2.21 and 17.3); the parasites cause intestinal and sexually transmitted infections, respectively. The protist *Cryptosporidium* is related to *Toxoplasma*, but unlike *Toxoplasma*, which is primarily transmitted by infected food as is the pathogenic protist *Cyclospora*, *Cryptosporidium* is transmitted primarily by contaminated water. We consider all five of these major human parasites here.

Giardiasis

Giardia intestinalis (also called *Giardia lamblia*) is typically transmitted to humans in fecally contaminated water and causes an acute gastroenteritis, *giardiasis*. The trophozoites of *Giardia* (**Figure 32.7a, c**) produce highly resistant cysts (Figure 32.7b) that function in transmission. Ingested cysts germinate in the small intestine to form trophozoites, and these travel to the large intestine where they attach to the intestinal wall and cause the symptoms of giardiasis: an explosive, foul-smelling, watery diarrhea, intestinal cramps, flatulence, nausea, weight loss, and malaise. The foul-smelling diarrhea and the absence of fecal blood distinguish giardiasis from diarrheas due to bacterial or viral intestinal pathogens.

G. intestinalis causes a significant number of drinking water infectious disease outbreaks in the United States. The thick-walled cysts are resistant to chlorine, and most outbreaks have been associated with water systems that used only chlorination as a means of water purification. Water subjected to proper clarification and filtration followed by chlorination or other disinfection (⟳ Section 21.8)

should be free of *Giardia* cysts. Most surface water sources (lakes, ponds, and streams) contain *Giardia* cysts, as beavers and muskrats are carriers of this pathogen. This is why surface waters should never be drunk untreated but instead should be filtered and disinfected with iodine or chlorine or alternatively, filtered and boiled. The drugs quinacrine, furazolidone, and metronidazole are useful for treating acute giardiasis.

Trichomoniasis

Trichomonas vaginalis (**Figure 32.8**) causes a sexually transmitted infection, *trichomoniasis*. *T. vaginalis* does not produce resting cells or cysts and as a result, *Trichomonas* transmission is typically from person to person, generally by sexual intercourse. However, unlike most sexually transmitted bacterial pathogens, cells of *T. vaginalis* can survive for several hours on moist surfaces and up to a day in urine or semen. Hence, in addition to disease transmission by intimate contact, trichomoniasis can be transmitted by contaminated toilet seats, sauna benches, and towels.

T. vaginalis infects the vagina in women, the prostate and seminal vesicles of men, and the urethra of both males and females. Trichomoniasis is often asymptomatic in males. By contrast, trichomoniasis in females is characterized by a yellowish vaginal discharge (Figure 32.8b) that causes a persistent vaginal itching and burning. The infection is more common in females; surveys have shown that up to 25% of sexually active women are infected with *T. vaginalis* while only about 5% of males are infected. Trichomoniasis is diagnosed by observation of the motile protists in a wet mount of fluid discharged from the patient (Figure 32.8b). The antiprotozoal drug metronidazole is effective for treating trichomoniasis.

Cryptosporidiosis, Toxoplasmosis, and Cyclosporiasis

Cryptosporidium, Toxoplasma, and *Cyclospora* are genera of parasitic Coccidia (which group among the alveolates, ⟳ Section 17.5). These parasites are transmitted to humans in fecally

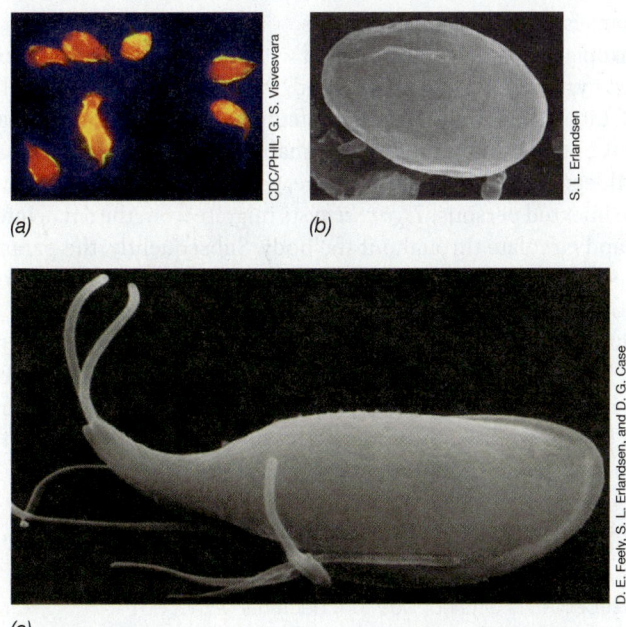

(a) CDC/PHIL, G. S. Visvesvara
(b) S. L. Erlandsen
(c) D. E. Feely, S. L. Erlandsen, and D. G. Case

Figure 32.7 *Giardia. (a)* Fluorescently stained cells of *Giardia intestinalis. (b, c)* Scanning electron micrographs of *(b)* a giardial cyst and *(c)* a motile *G. intestinalis* trophozoite. The trophozoite is 15 μm long and the cyst about 11 μm wide.

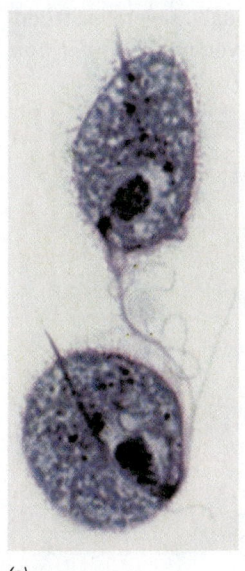
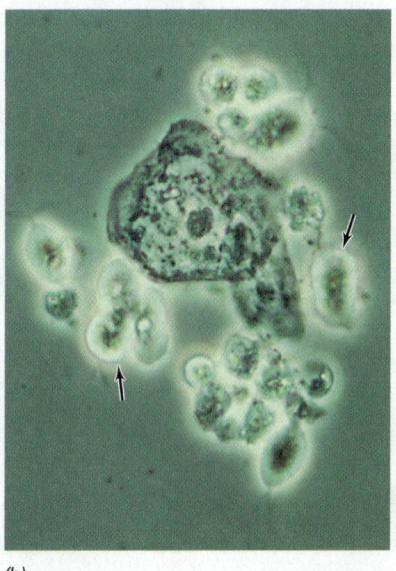

(a)   (b)   CDC/PHIL

Figure 32.8 *Trichomonas vaginalis. (a)* Light micrograph of stained cells; cells vary from 10 to 20 μm in diameter. *(b)* Vaginal discharge from a female with trichomoniasis. *T. vaginalis* cells (arrows) are present along with vaginal secretions and epithelial cells.

contaminated food or water and can trigger serious bouts of diarrhea, or in the case of *Toxoplasma*, serious internal organ damage.

Cryptosporidium parvum infects many warm-blooded animals, in particular cattle. The organism forms small, coccoid cells that invade and grow intracellularly in mucosal epithelial cells of the stomach and intestine (**Figure 32.9a**), resulting in the gastrointestinal illness *cryptosporidiosis*. *C. parvum* produces thick-walled, highly resistant cysts called *oocysts* (Figure 32.9b), which enter water from the feces of infected animals. The infection is then transmitted to other animals and humans when they consume the fecally contaminated water.

Cryptosporidium oocysts are highly resistant to chlorine, and because of this, sedimentation and filtration are the only reliable way to remove them from water supplies. In an average year, *Cryptosporidium* is responsible for the majority of recreational waterborne disease outbreaks in the United States (Chapter 31) but is only occasionally associated with drinking water outbreaks. Nevertheless, *C. parvum* was responsible for the largest single outbreak of disease associated with drinking water ever recorded in the United States. In the spring of 1993, one-quarter of the population of Milwaukee, Wisconsin (USA), developed cryptosporidiosis from consuming water from the municipal water supply. Spring rains and runoff from cattle manure on farmlands had drained into Lake Michigan (the water supply for the city) and overburdened the water purification system, leading to contamination by *C. parvum*.

Cryptosporidiosis typically causes only a mild self-limiting diarrhea, making treatment unnecessary. However, individuals with impaired immunity, such as that caused by HIV/AIDS, or the very young or old can develop serious complications from a *C. parvum* infection. The primary laboratory diagnostic method for cryptosporidiosis is the demonstration of oocysts in the stool (Figure 32.9b). Immunological and molecular tools are also available for more precise identification of strains of the pathogen when such tracking is necessary.

As for *C. parvum*, the parasite *Cyclospora cayetanensis* also forms oocysts and causes a mild to occasionally severe gastroenteritis, which is called *cyclosporiasis*. However, unlike *C. parvum*,

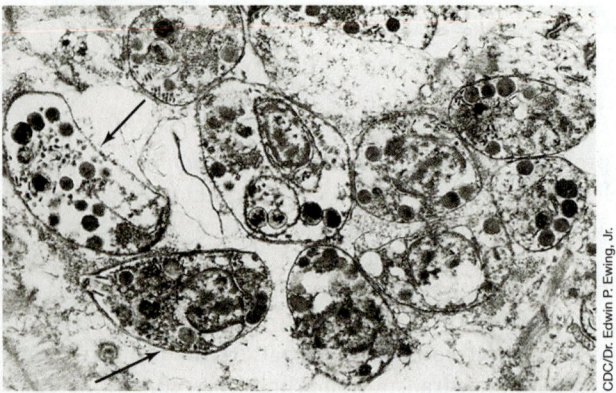

CDC/Dr. Edwin P. Ewing, Jr.

Figure 32.10 *Toxoplasma*. Tachyzoites (rapidly growing cells) of *Toxoplasma gondii*, an intracellular parasite. In this transmission electron micrograph, the tachyzoites (arrows) form a cystlike structure in a host cardiac cell. The *Toxoplasma* tachyzoites are 4–7 μm long.

C. cayetanensis is primarily transmitted by fecally contaminated food products—typically fresh foods—rather than by contaminated water. Most cases of cyclosporiasis have been linked to contaminated fruits or vegetables, with a major outbreak occurring in the United States linked to packaged lettuce in the summer of 2013 (↩ Section 31.7).

Toxoplasmosis is caused by *Toxoplasma gondii* (**Figure 32.10**). This parasite infects many warm-blooded animals, and roughly half of all adults in the United States are infected but asymptomatic because their immune system keeps the organism in check. *T. gondii* is typically transmitted to humans in the form of cysts present in undercooked beef, pork, or lamb, by direct infection from cats, which are major carriers of *T. gondii*, and occasionally from contaminated water. A key step in the *T. gondii* life cycle is completed in felines and thus they are obligate hosts; humans and other animals are only incidental hosts. Most transmission to humans is thus probably from cats.

Toxoplasmosis can be associated with mild to severe symptoms. When cysts of *T. gondii* are ingested, they penetrate the wall of the small intestine. From this initial infection, symptoms can be inapparent or apparent but indistinguishable from those of a mild case of influenza (headache, muscle ache, general malaise). However, in some infected persons, *T. gondii* cysts migrate from the small intestine and circulate throughout the body. Subsequently, the parasite can penetrate nerve cells and infect tissues of the brain and eyes. Although disease symptoms in healthy adults are uncommon, in immune-compromised individuals, toxoplasmosis can damage the eyes, brain, and other internal organ systems. In addition, a first-time infection with *T. gondii* in expectant mothers can lead to birth defects in newborns; thus pregnant women who have not been in contact with cats should avoid cats until after giving birth.

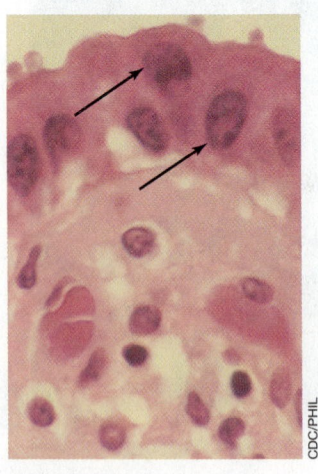
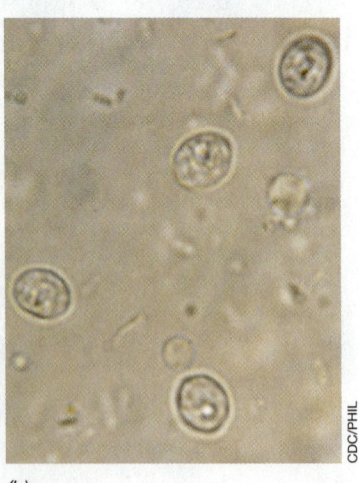

CDC/PHIL

(a) (b)

Figure 32.9 *Cryptosporidium parvum*. (a) Arrows point to intracellular trophozoites of *C. parvum* embedded in human gastrointestinal epithelium. The trophozoites are about 5 μm in diameter. (b) Thick-walled *C. parvum* oocysts are about 3 μm in diameter in this fecal sample.

MINIQUIZ
- What symptoms of giardiasis would suggest that your gastroenteritis was not due to a bacterial pathogen?
- How does one contract a case of trichomoniasis? A case of toxoplasmosis?
- What is unusual about the oocysts of *Cryptosporidium* that facilitates its transmission by a water route?

III • Blood and Tissue Parasitic Infections

Several human parasites infect organs and tissues other than the gastrointestinal tract and are typically transmitted by insect vectors. We begin our consideration of these with malaria, the most devastating and widespread of parasitic diseases and one that remains a major global health problem today.

32.5 *Plasmodium* and Malaria

Malaria is caused by protists of the alveolate group (⮌ Section 17.5). Several species of the protozoal genus *Plasmodium* cause malaria-like diseases in warm-blooded hosts; up to 500 million people worldwide contract malaria annually and about 1 million die from the disease. Malaria is thus one of the most common causes of death worldwide from infectious disease and certainly the most prevalent of parasitic diseases.

In malaria, the complex parasite life cycle requires a mosquito vector. Four species of *Plasmodium—P. vivax, P. falciparum, P. ovale,* and *P. malariae*—cause most human malaria. The most widespread disease is caused by *P. vivax,* whereas the most serious disease is caused by *P. falciparum.* Humans are the only reservoirs for these four species. The protists carry out part of their life cycle in the human and part in the female *Anopheles* mosquito, the only vector that transmits *Plasmodium* spp. The vector spreads the protist from person to person.

Malarial Life Cycle

The life cycle of *Plasmodium* is complex and involves a number of stages (**Figure 32.11**). First, the human host is infected by plasmodial *sporozoites,* small, elongated cells produced in the mosquito that localize in the salivary gland of the insect. The mosquito (Figure 32.11 inset) injects saliva containing the sporozoites into the human when obtaining a blood meal. The sporozoites travel to the liver where they infect liver cells. Here they can remain quiescent for indefinite periods but eventually replicate and become enlarged in a stage called the *schizont* (see Figure 32.12*b*). The schizonts then segment into a number of small cells called *merozoites,* which exit the liver into the bloodstream. Some of the merozoites then infect red blood cells (erythrocytes).

The plasmodial life cycle in erythrocytes proceeds with repeated division, growth, and release of merozoites (**Figure 32.12**); this results in destruction of the host red blood cells. Plasmodial growth in red cells typically repeats at synchronized intervals of 48 h. During this 48-h period, the host experiences the defining clinical symptoms of malaria: chills followed by fever of up to 40°C (104°F). The chill–fever pattern coincides with the release of merozoites from the erythrocytes during the synchronized reproduction cycle. Vomiting and severe headache may accompany the chill–fever cycles, and over the longer term, characteristic symptomatic malaria can alternate with asymptomatic periods. Because of the destruction of red blood cells, malaria typically causes anemia and some enlargement of the spleen (splenomegaly).

Plasmodial merozoites eventually develop into *gametocytes,* cells that infect only mosquitoes. The gametocytes are ingested when an *Anopheles* mosquito takes a blood meal from an infected person, and they mature within the mosquito into *gametes.* Two gametes fuse to form a zygote, and the zygote

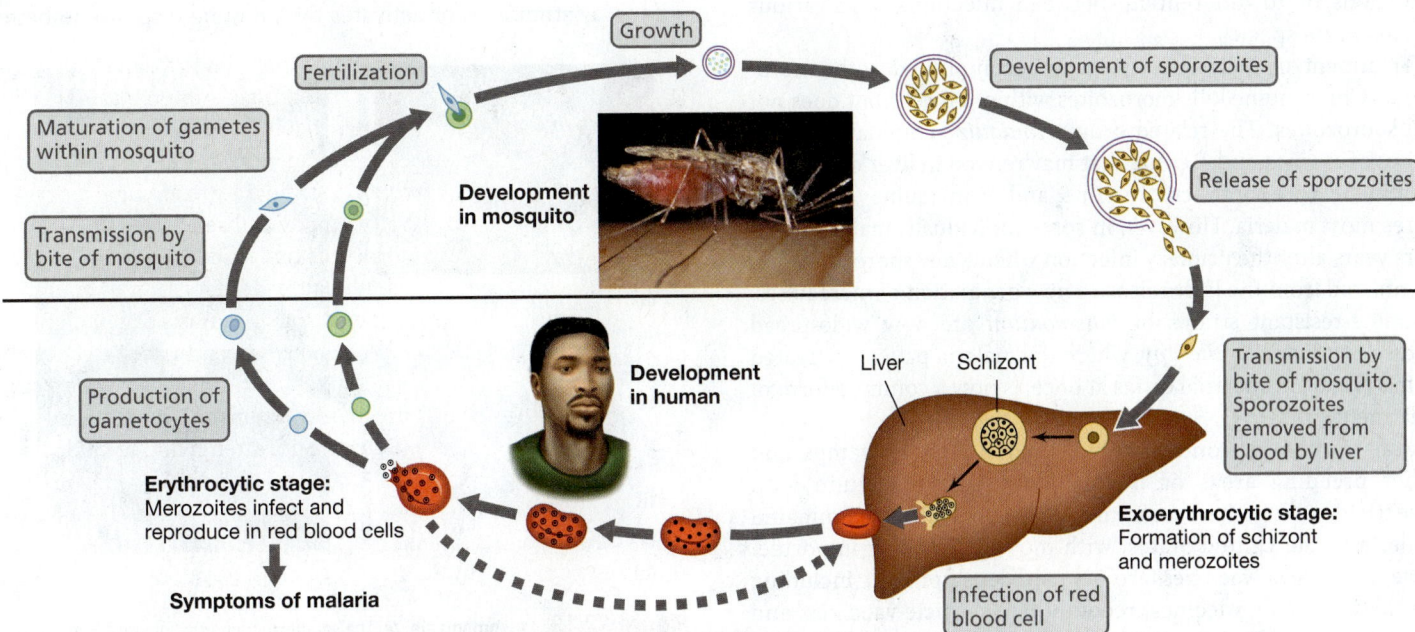

Figure 32.11 **The life cycle of *Plasmodium*.** The life cycle of *Plasmodium* requires both a warm-blooded host and the mosquito vector. Transmission of the protist to and from the warm-blooded host is done by the bite of an *Anopheles gambiae* mosquito (inset). Mosquito photo courtesy of CDC/PHIL, J. Gathany.

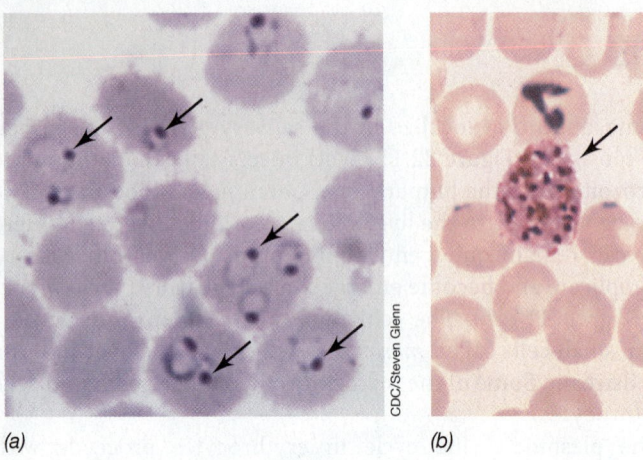

Figure 32.12 ***Plasmodium* and malaria.** *(a)* Merozoites of *Plasmodium falciparum* (arrows) growing within human red blood cells. *(b)* A schizont of *P. vivax* (arrow) along with red blood cells. When released from the schizont, the merozoites infect erythrocytes (Figure 32.11). Red blood cells are about 6 μm in diameter.

migrates by amoeboid motility to the outer wall of the insect's intestine where it enlarges and forms several sporozoites. These are released and reach the salivary gland of the mosquito from where they can be inoculated into another human, and the cycle begins anew (Figure 32.11).

Epidemiology, Diagnosis, Treatment, and Control

Anopheles mosquitoes (Figure 32.11 inset) live predominantly in the tropics and subtropics and are the vector for malaria. Diagnosis of malaria requires the identification of *Plasmodium*-infected erythrocytes in blood smears (Figure 32.12). Fluorescent nucleic acid stains, nucleic acid probes, PCR assays, and various antigen-detection methods are also used to verify *Plasmodium* infections or to differentiate between infections with various *Plasmodium* species.

Treatment of malaria is typically accomplished with *chloroquine*. Chloroquine kills merozoites within red cells but does not kill sporozoites. The related drug *primaquine* eliminates sporozoites of *P. vivax* and *P. ovale* that may remain in liver cells. Thus treatment with both chloroquine and primaquine effectively cures most malaria. However, in some individuals, malaria reoccurs years after the primary infection when a few sporozoites not eliminated from the liver release a new generation of merozoites. Quinine-resistant strains of *Plasmodium* are now widespread and so *combination therapy*, where the malaria patient is treated with several antimalarial drugs at once, is now a common form of treatment.

Malaria can be controlled by either draining swamps and other breeding areas or by eliminating the mosquito with insecticides. Together, these measures have all but eliminated malaria in the United States, with most cases being imported. Several malaria vaccines are also in development, including synthetic peptide vaccines, recombinant particle vaccines, and DNA vaccines (↩ Section 24.7), but thus far no highly effective and reliable malaria vaccine has emerged for use in mass vaccination programs.

32.6 Leishmaniasis, Trypanosomiasis, and Chagas' Disease

Parasites of the genera *Leishmania* and *Trypanosoma* are transmitted by bloodsucking insect vectors. These parasites are *hemoflagellates*, organisms that reside in blood or related tissues such as the liver and spleen, and cause major human diseases, primarily in tropical and subtropical countries.

Leishmaniasis

Leishmaniasis is a parasitic disease of various forms caused by species of the genus *Leishmania*, a flagellated protozoan related to *Trypanosoma*. The disease is transmitted to humans by a bite from the sandfly. *Cutaneous leishmaniasis*, caused by either *L. tropica* or *L. mexicana*, is the most common form of leishmaniasis. Following transmission of the parasite in a blood meal (**Figure 32.13a, b**), the parasite infects and grows within human macrophage cells, leading eventually (weeks or months later) to the formation of a small nodule on the skin. The nodule then becomes ulcerated and can enlarge to form a major skin lesion (Figure 32.13c) that contains active parasites. In the absence of secondary bacterial infections, which are common if the ulcerated tissue is left open, the lesions heal spontaneously over a period of several months but can leave a permanent scar.

Leishmaniasis has historically been treated with injections of pentavalent antimony (Sb^{5+}) compounds. Although the mode of action of these compounds is unknown, it is thought that Sb^{5+} in some way stimulates or activates the immune response to better

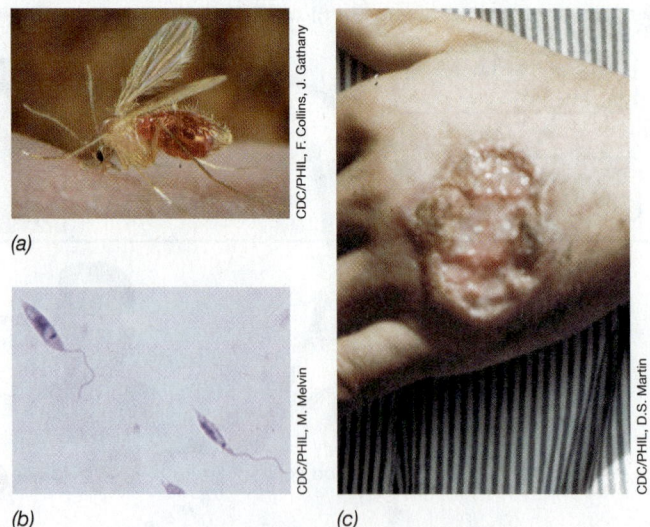

Figure 32.13 **Leishmaniasis.** *(a)* The sandfly (genus *Phlebotomus*) transmits leishmaniasis in a blood meal. *(b)* *Leishmania* spp. are flagellated protozoans and the cause of leishmaniasis. *(c)* Cutaneous leishmaniasis showing an open ulcer on the hand. Secondary bacterial infections of these ulcers are common.

attack the *Leishmania* parasites. At present, however, many *Leishmania* species are resistant to antimony compounds, but a variety of other drugs are available for treating resistant cutaneous forms of the disease. Estimates of cutaneous leishmaniasis prevalence worldwide are about 1 million.

Visceral leishmaniasis is caused by *Leishmania donovani* and is the most severe form of the disease. In visceral leishmaniasis, the parasite travels from the site of infection to internal organs, in particular the liver, spleen, and bone marrow; if left untreated, the visceral disease is almost always fatal. Common symptoms of visceral leishmaniasis include a cycling of fever and chills, a slow reduction in both red and white blood cell numbers, and significant enlargement of the spleen and liver that can lead to major distention of the abdomen. Treatment includes injections of antimony (as for the cutaneous disease), long periods of bed rest, and blood transfusions in acute cases if blood cell counts become dangerously low. Estimates of visceral leishmaniasis prevalence worldwide are about 300,000.

Trypanosomiasis and Chagas' Disease

Flagellated protozoans of the genus *Trypanosoma* (♺ Section 17.4) cause two related forms of **trypanosomiasis**. Two subspecies of *Trypanosoma brucei* native to Africa, *T. brucei gambiense* (**Figure 32.14a**) and *T. brucei rhodesiense*, cause *African trypanosomiasis*, better known as *African sleeping sickness*. The species *T. cruzi* causes *Chagas' disease*, also known as *American trypanosomiasis*. These diseases are transmitted by insect bites from either a fly or a bug.

Sleeping sickness is transmitted by the tsetse fly (genus *Glossina*), an insect similar in dimensions to a housefly and native only to tropical regions of Africa; sleeping sickness is therefore endemic only in countries of sub-Saharan Africa. The disease begins with intermittent fever, headache, and malaise. The parasite multiplies in the blood and later infects the central nervous system and grows in spinal fluid. Neurological symptoms soon begin, including sleep patterns that are no longer diel. The parasite produces the aromatic alcohol *tryptophol*, a derivative of the amino acid tryptophan, which triggers a sleep response. Without treatment, the infection

gradually progresses to a coma, multiple organ failure, and eventually death after months or years depending on the case. A variety of anti-trypanosomal drugs are available for treating sleeping sickness; some are used primarily for treating the blood infection while others are used if the disease has progressed to the neurological stage. About 10,000 new cases of sleeping sickness are reported annually but most cases are thought to go unreported.

Chagas' disease, named for its discoverer, is caused by *T. cruzi*, a close relative of *T. brucei*, and is transmitted by the bite of the "kissing bug" (Figure 32.14*b, c*). Chagas' disease mainly occurs in Latin American countries. The parasite affects several organs including the heart, gastrointestinal tract, and central nervous system, causing inflammatory reactions and tissue destruction. The acute illness is usually self-limiting, but if chronic illness develops, heart disease is significant and is the eventual cause of premature death. About 20,000 deaths due to Chagas' disease occur annually in endemic Latin American countries.

Currently no vaccines are available for prevention of African or American trypanosomiases.

32.7 Parasitic Helminths: Schistosomiasis and Filariases

Some parasitic diseases are caused by helminths, tiny worms that burrow into the human host and cause debilitating diseases and death. We consider the most widespread of these here, schistosomiasis, along with brief coverage of two other less common helminth infections.

Schistosomiasis

Schistosomiasis, also called *snail fever*, is a chronic parasitic disease caused by species of trematodes (flatworms) of the genus *Schistosoma*; the major species is *S. mansoni* and adult worms can be up to a centimeter in length (**Figure 32.15a**). The life cycle of the parasite requires both snails and humans (or other mammals) as hosts. Schistosome eggs (Figure 32.15*b*) released into a freshwater aquatic environment hatch to generate *miracidia*, the form of the worm that infects snails. In the snail, miracidia are transformed into *cercaria* (Figure 32.15*c*), the motile stage of the parasite that is released and infects humans.

Cercaria burrow into the skin, leaving a small surface lesion (Figure 32.15*d*), and then migrate to the lungs and liver; in the process, the worm establishes a long-term infection in the blood vessels. From the liver, the parasite infects the bladder, kidneys, and urethra and the female worm produces large numbers of eggs. The eggs are shed in the urine and also pass through the intestinal wall and are shed in the feces. Large egg masses also become trapped along with fluids in the bladder, liver, and other organs, triggering

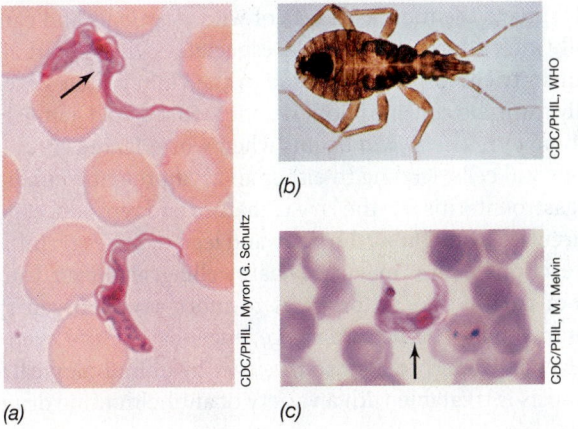

Figure 32.14 African trypanosomiasis and Chagas' disease. (a) Two cells of *Trypanosoma brucei* (arrow), the causative agent of African sleeping sickness (African trypanosomiasis), in a blood smear. (b) The "kissing bug" (*Triatoma infestans*), the vector for Chagas' disease (American trypanosomiasis). (c) A cell of *Trypanosoma cruzi* (arrow), the causative agent of Chagas' disease, in a blood smear.

CDC/PHIL, WHO

CDC/PHIL, Myron G. Schultz

CDC/PHIL, M. Melvin

UNIT 6

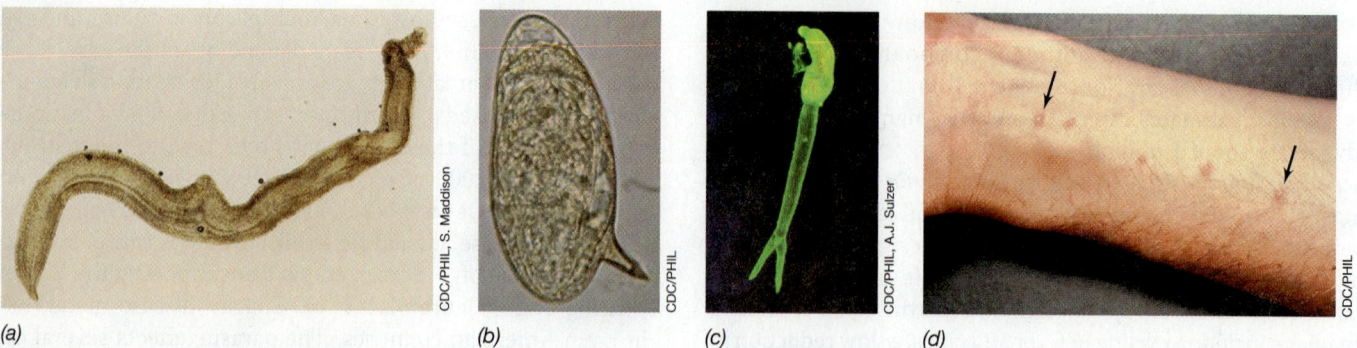

(a)   (b)   (c)   (d)

Figure 32.15 Schistosomiasis. *(a)* Adult worm of *Schistosoma mansoni*; the worm is about 1 cm in length. *(b)* An *S. mansoni* egg, about 0.15 mm long. The lateral spine is characteristic of the eggs of this species. *(c)* Fluorescently stained cercaria, the infective form of *S. mansoni*. From the head (top) to the bifurcated tail is about 1 mm. *(d)* Cercarial infection of the forearm. Five infection sites (arrows) are apparent.

an inflammatory response and a major distention of the abdomen, a condition commonly seen in infected children (**Figure 32.16a**). Other symptoms include bloody urine, diarrhea, and abdominal pain. Eggs as well as adult worms can live in the body for years, causing chronic symptoms that can last from youth into adulthood.

Schistosomiasis is a disease of tropical countries, primarily those in Africa, but some cases also occur in subtropical countries such as Latin America and the Caribbean region. Schistosomiasis can be effectively treated with the drug praziquantel, and diagnosis is relatively easy by assessing symptoms and observing parasite eggs in the urine and feces. Mortality from schistosomiasis is low, about 0.1%, but schistosomiasis is second only to malaria in terms of total parasitic infections worldwide. In 2011, nearly 250 million cases of the disease were treated and many others probably went untreated.

Filariases

Several other parasitic helminth infections are known, and chief among these are the *filariases*, infections by parasitic nematodes (roundworms). Unlike the schistosomiasis parasite, these worms are clearly macroscopic in the adult stage (several centimeters in length, depending on the filariasis).

 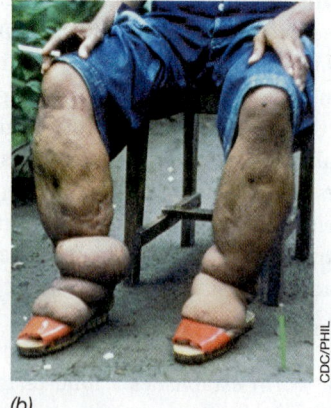

(a)   (b)

Figure 32.16 Symptoms of parasitic helminth infections. *(a)* Schistosomiasis in a small child. The swollen abdomen from the accumulation of fluids and worm eggs is characteristic of the infection. *(b)* Bancroft's filariasis (elephantiasis). The swollen legs are the result of edema from infection of lymph tissues by the roundworm *Wuchereria bancrofti*.

Bancroft's filariasis (also called "elephantiasis") is a chronic infection of the lymphatic system by *Wuchereria bancrofti*. The worm is transmitted to humans in tiny *microfilariae* in a mosquito bite. Once in the host, microfilariae develop into adult worms and these interrupt lymph flow, leading to major accumulation of fluids (edema). Fluid accumulation in lower regions of the body can cause massive enlargement of the legs (Figure 32.16b). Over 120 million people in the tropics suffer from *W. bancrofti* infection, but the microfilarial stage of the disease is readily treatable.

Onchocerciasis (also called *river blindness*) is due to a chronic infection by the large parasitic roundworm *Onchocerca volvulus*. Humans are the only known host for this parasite, but flies are vectors when they become infected with microfilariae in a blood meal and transmit them to uninfected humans in a bite. The microfilariae invade the cornea and from there the iris and retina, triggering an inflammatory response that causes scarring and partial to total loss of vision. *O. volvulus* infection is second only to trachoma (⟲ Section 29.13) as a cause of infectious blindness. It is estimated that about 20 million people are infected with this parasite, primarily in equatorial Africa.

The disease *trichinosis* (also called *trichinellosis*) is caused by species of the parasitic roundworm *Trichinella*. This worm commonly infects the muscle tissues of wild mammals and can occasionally infect domestic animals, especially swine; about 20 cases of human trichinosis are reported in the United States each year, usually from the consumption of undercooked wild game. Human infection with *Trichinella* begins when worm larvae enter intestinal mucosal cells, leading to either an asymptomatic condition or mild gastroenteritis. As the larvae mature and reproduce, new larvae circulate throughout the body and lead to systemic inflammatory reactions such as malaise, facial swelling, and fever. Untreated cases of trichinosis can progress to more severe organ-specific symptoms including heart damage, encephalitis, and even death. However, if properly diagnosed, usually by immunological assays, trichinosis is treatable with a variety of antihelminthic drugs.

MINIQUIZ

- How does the pathogen causing schistosomiasis differ from all other pathogens considered in this chapter?
- From what source are most cases of human trichinosis contracted?

BIG IDEAS

32.1 • Fungi include the molds and yeasts, and some fungi are dimorphic, meaning that both mycelial and yeast phases can occur. Superficial, subcutaneous, and systemic mycoses refer to fungal infections of the skin surface, skin subsurface, and internal organs, respectively. Fungal infections can be mild or serious, depending on the health and immune status of those infected.

32.2 • Superficial mycoses such as athlete's foot or jock itch are mild and easily treatable, whereas subcutaneous mycoses, such as sporotrichosis, or especially systemic mycoses, such as histoplasmosis, are more difficult to treat effectively. The ability of fungi that cause systemic mycoses to infect internal organs makes these pathogens particularly dangerous to the elderly or those otherwise immune compromised.

32.3 • The genera *Entamoeba* and *Naegleria* are amoebic human parasites that cause gastrointestinal and brain infections, respectively. *Entamoeba* is transmitted in fecally contaminated waters, whereas *Naegleria* inhabits warm, soil-contaminated waters. *Balantidium* is a ciliated intestinal parasite transmitted by fecally contaminated water.

32.4 • The protists *Giardia intestinalis* and *Cryptosporidium parvum* are major waterborne pathogenic parasites, whereas *Toxoplasma gondii* is primarily a foodborne or cat-transmitted parasite and *Trichomonas vaginalis* a sexually transmitted parasite. The pathogenic parasite *Cyclospora* is primarily transmitted by fresh vegetables such as lettuce and spinach contaminated with animal feces. None of these parasites cause life-threatening diseases in otherwise healthy individuals, although *T. gondii* can trigger severe and even fatal infections in immune-compromised hosts.

32.5 • Infections with *Plasmodium* spp. cause malaria, a widespread, mosquito-transmitted disease of the blood that causes significant morbidity and mortality in tropical and subtropical regions of the world. Malaria is treatable with quinines and other drugs but is not yet preventable by vaccination.

32.6 • Leishmaniasis is a parasitic disease caused by *Leishmania* species; the cutaneous form of the disease is most common. *Trypanosoma brucei* causes African trypanosomiasis (African sleeping sickness), while the related species *Trypanosoma cruzi* causes Chagas' disease. All of these diseases are transmitted by insect bites from an insect vector, either a fly or a bug.

32.7 • Schistosomiasis is a major parasitic disease caused by a microscopic worm, *Schistosoma mansoni*. The life cycle of the parasite requires both snails and mammals. The worm infects the liver and kidneys and produces large egg masses that accumulate in the body, leading to systemic inflammation and abdominal distention. Other parasitic worm diseases, such as elephantiasis and river blindness, also leave readily visible signs of infection.

UNIT 6

MasteringMicrobiology®   **Review what you know and challenge what you have learned with MasteringMicrobiology!** Access study materials, chapter quizzes, animations, and microbiology lab tutorials in the Study Area to ensure that you have mastered this chapter's content.

REVIEW OF KEY TERMS

Leishmaniasis a disease of the skin or viscera caused by infection with species of a parasitic flagellated protozoan, *Leishmania*

Malaria a disease characterized by recurrent episodes of fever and anemia, caused by the protist *Plasmodium* spp., usually transmitted between mammals through the bite of the *Anopheles* mosquito

Meningoencephalitis the invasion, inflammation, and destruction of brain tissue by the amoeba *Naegleria fowleri* or a variety of other pathogens

Mycosis (plural, mycoses) any infection caused by a fungus

Opportunistic pathogens organisms that cause disease in the absence of normal host resistance

Schistosomiasis a chronic disease caused by a parasitic worm that leads to internal organ damage and accumulation of fluids and worm egg masses

Subcutaneous mycoses fungal infections of deeper layers of skin

Superficial mycoses fungal infections of the surface layers of skin, hair, or nails

Systemic mycoses fungal growth in internal organs of the body

Trypanosomiasis any parasitic disease of the blood and internal tissues caused by species of the flagellated protozoan *Trypanosoma*; African sleeping sickness and Chagas' disease are two major trypanosomiases

REVIEW QUESTIONS

1. Which are more common, superficial or systemic mycoses? Have you had a case of either one? (Section 32.1)

2. What are the most common systemic mycoses in the United States and which populations are most susceptible to such infections? (Section 32.2)

3. If you were to have one or the other, which would you rather have, an *Entamoeba* infection or a *Naegleria* infection? (Section 32.3)

4. In contrast to disease caused by *Trichomonas*, what do giardiasis and cryptosporidiosis have in common? (Section 32.4)

5. Malaria symptoms include fever followed by chills. These symptoms are related to activities of the pathogen. Describe the growth stages of *Plasmodium* spp. in the human host and relate them to the fever–chill pattern. (Section 32.5)

6. Contrast leishmaniasis with the two types of trypanosomiasis in terms of causative agents, symptoms, and transmission vectors. (Section 32.6)

7. Contrast schistosomiasis with all other parasitic infections covered in this chapter. In what major way does it differ? (Section 32.7)

APPLICATION QUESTIONS

1. Malaria eradication has been a goal of public health programs for at least 100 years. What factors preclude our ability to eradicate malaria? If an effective vaccine was developed, could malaria be eradicated?

2. In terms of public health, what is a common problem that unites many of the visceral parasitic infections covered in this chapter? How could this problem be attacked? Why are these diseases rare in developed countries?

3. Explain why the diseases malaria, leishmaniasis, and trypanosomiasis are primarily diseases of tropical regions. How could humans be affecting the future geographical ranges of these diseases?

4. Explain why systemic fungal infections are typically seen only in certain individuals even though many people have contact with the pathogen, whereas an outbreak of giardiasis affects virtually everyone that has come in contact with the pathogen.

Appendix 1
Energy Calculations in Microbial Bioenergetics

The information here is intended to help the reader calculate changes in free energy accompanying chemical reactions carried out by microorganisms. It begins with definitions of the terms required to make such calculations and proceeds to show how knowledge of redox state, atomic and charge balance, and other factors are necessary to calculate free-energy problems successfully.

I. Definitions

1. ΔG^0 = standard free-energy change of the reaction under "standard conditions" (1 atm pressure and 1 M concentrations); ΔG = free-energy change under the conditions specified; $\Delta G^{0\prime}$ = free-energy change under standard conditions at pH 7. The prime mark ($\prime$) throughout this appendix indicates pH 7 (approximate cellular conditions).

2. Calculation of ΔG^0 for a chemical reaction from the free energy of formation, G_f^0, of products and reactants:

$$G^0 = \Sigma \; G_f^0(\text{products}) - \Sigma \; G_f^0(\text{reactants})$$

That is, sum the ΔG_f^0 of products, sum the ΔG_f^0 of reactants, and subtract the latter from the former.

3. For energy-yielding reactions involving H^+, converting from standard conditions (pH 0) to cellular conditions (pH 7):

$$\Delta G^{0\prime} = \; G^0 + m \; G_f^0(H^+)$$

where m is the net number of protons in the reaction (m is negative when more protons are consumed than formed) and $\Delta G_f^0(H^+)$ is the free energy of formation of a proton at pH 7 (-39.83 kJ) at 25°C.

4. Effect of concentrations on ΔG: With soluble substrates, the concentration ratios of products formed to exogenous substrates used are generally equal to or less than 10^{-2} at the beginning of growth and equal to or greater than 10^{-2} at the end of growth. From the relation between ΔG and the equilibrium constant (see item 8), it can be calculated that ΔG for the free-energy yield in these situations differs from the free-energy yield under standard conditions by no more than 11.7 kJ, and so to a first approximation, free-energy yields under standard conditions can be used in most situations. However, if H_2 is a product, H_2-consuming bacteria may consume the H_2 and keep its concentration so low that the free-energy yield of the reaction is significantly affected. Thus, in the fermentation of ethanol to acetate and H_2 by syntrophic bacteria ($C_2H_5OH + H_2O \rightarrow C_2H_3O_2^- + 2H_2 + H^+$), the $\Delta G^{0\prime}$ (at 1 atm H_2) is $+9.68$ kJ, but at 10^{-4} atm H_2 it is -36.03 kJ. With H_2-consuming bacteria present, therefore, ethanol fermentation by syntrophic bacteria converts from an endergonic to an exergonic reaction. (See also item 9.)

5. Reduction potentials: By convention, electrode equations are written as *reductions*; that is, the direction is oxidant + $ne^- \rightarrow$ reductant, where n is the number of electrons transferred. The standard reduction potential (E_0) of the hydrogen electrode, $2H^+ + 2e^- \rightarrow H_2$, is set by definition at 0.0 V at 1 atm pressure of H_2 gas and 1.0 M H^+ at 25°C. $E_0\prime$ is the standard reduction potential at pH 7. See also Table A1.2.

6. Relation of free energy to reduction potential:

$$\Delta G^{0\prime} = -nF\Delta E_0\prime \quad \text{or} \quad \Delta G^{0\prime} = -RT \ln K\prime_{eq}$$

where n is the number of electrons transferred, F is the Faraday constant (96.48 kJ/V), and $\Delta E_0\prime$ is the $E_0\prime$ of the electron-*accepting* couple minus the $E_0\prime$ of the electron-*donating* couple. R and T are constants (see 8) and for K_{eq}, see 7.

7. Equilibrium constant, K_{eq}. For the generalized reaction $aA + bB \leftrightarrow cC + dD$,

$$K_{eq} = \frac{[C]^c[D]^d}{[A]^a[B]^b}$$

where A, B, C, and D represent reactants and products; a, b, c, and d represent number of molecules of each; and brackets indicate concentrations. This is true only when the chemical system is in equilibrium. $K\prime_{eq}$ is K_{eq} at pH 7.

8. Relation of equilibrium constant, K_{eq}, to free-energy change. At constant temperature, pressure, and pH,

$$\Delta G\prime = \Delta G^{0\prime} + RT \ln K\prime_{eq}$$

where R is the universal gas constant (8.315 J/mol/Kelvin) and T is the absolute temperature (in Kelvin).

9. Two substances can react in a redox reaction even if the standard potentials are unfavorable, provided that the concentrations are appropriate.

Assume that normally the reduced form of A would donate electrons to the oxidized form of B. However, if the concentration of the reduced form of A was low and the concentration of the reduced form of B was high, it would be possible for the reduced form of B to donate electrons to the oxidized form of A. Thus, the reaction would proceed in the direction opposite that predicted from standard potentials. A practical example of this is the utilization of H^+ as an electron acceptor to produce H_2. Normally, H_2 production in fermentative bacteria is not extensive because H^+ is a poor electron acceptor; the $E_0\prime$ of the $2H^+/H_2$ pair is -0.41 V. However, if the concentration of H_2 is kept low by its continual removal (for example, by methanogenic *Archaea*, which use $H_2 + CO_2$ to produce methane, CH_4, or by many other anaerobes capable of consuming H_2 anaerobically), the potential will be more positive and then H^+ will be a suitable electron acceptor.

II. Oxidation State or Number

1. The oxidation state of an element in an elementary substance (for example, H_2, O_2) is zero.

2. The oxidation state of the ion of an element is equal to its charge (for example, $Na^+ = +1$, $Fe^{3+} = +3$, $O^{2-} = -2$).

3. The sum of oxidation numbers of all atoms in a neutral molecule is zero. Thus, H_2O is neutral because it has two H at $+1$ each and one O at -2.

4. In an ion, the sum of oxidation numbers of all atoms is equal to the charge on that ion. Thus, in the OH^- ion, $O(-2) + H(+1) = -1$.

5. In compounds, the oxidation state of O is almost always -2 and that of H is $+1$.

6. In simple carbon compounds, the oxidation state of C can be calculated by adding up the H and O atoms present and using the oxidation states of these elements as given in item 5, because in a neutral compound the sum of all oxidation numbers must be zero. Thus, the oxidation state of carbon in methane, CH_4, is -4 (4 H at $+1$ each $= +4$); in carbon dioxide, CO_2, the oxidation state of carbon is $+4$ (2 O at -2 each $= -4$).

7. In organic compounds with more than one C atom, it may not be possible to assign a specific oxidation number to each C atom, but it is still useful to calculate the oxidation state of the compound as a whole. The same conventions are used. Thus, the oxidation state of carbon in glucose, $C_6H_{12}O_6$, is zero (12 H at $+1 = 12$; 6 O at $-2 = -12$) and the oxidation state of carbon in ethanol, C_2H_6O, is -2 each (6 H at $+1 = +6$; one O at -2).

8. In all oxidation–reduction reactions there is a balance between the oxidized and reduced products. To calculate an oxidation–reduction balance, the number of molecules of each product is multiplied by its oxidation state. For instance, in calculating the oxidation–reduction balance for the alcoholic fermentation (glucose $\rightarrow 2\,C_2H_6O + 2\,CO_2$), there are two molecules of ethanol at -4 (for a total of -8) and two molecules of CO_2 at $+4$ (for a total of $+8$), so the net balance is zero. When constructing model reactions, it is useful to first calculate redox balances to be certain that the reaction is possible.

III. Calculating Free-Energy Yields for Hypothetical Reactions

Energy yields can be calculated either from free energies of formation of the reactants and products or from differences in reduction potentials of electron-donating and electron-accepting partial reactions.

Calculations from Free-Energy Values

Free energies of formation are given in **Table A1.1**. The procedure to use for calculating energy yields of reactions follows.

1. **Balancing reactions.** In all cases, it is essential to ascertain that the coupled oxidation–reduction reaction is *balanced*.

Balancing involves three things: (a) the *total number of each kind of atom* must be identical on both sides of the equation; (b) there must be an *ionic balance* so that when positive and negative ions are added up on the right side of the equation, the total ionic charge (whether positive, negative, or neutral) exactly balances the ionic charge on the left side of the equation; and (c) there must be an *oxidation–reduction balance* so that all the electrons removed from one substance are transferred to another substance. In general, when constructing balanced reactions, one proceeds in the reverse of the three steps just listed. Usually, if steps (c) and (b) have been properly handled, step (a) becomes correct automatically.

2. **Examples:** (a) What is the balanced reaction for the oxidation of H_2S to SO_4^{2-} with O_2? First, decide how many electrons are involved in the oxidation of H_2S to SO_4^{2-}. This can be most easily calculated from the oxidation states of the compounds, using the rules given previously. Because H has an oxidation state of $+1$, the oxidation state of S in H_2S is -2. Because O has an oxidation state of -2, the oxidation state of S in SO_4^{2-} is $+6$ (because it is an ion, and using the rules given in items 4 and 5 of the previous section). Thus, the oxidation of H_2S to SO_4^{2-} involves an *eight-electron transfer* (from -2 to $+6$). Because each O atom can accept two electrons (the oxidation state of O in O_2 is zero, but in H_2O is -2), this means that two molecules of molecular oxygen, O_2, are required to provide sufficient electron-accepting capacity. Thus, at this point, we know that the reaction requires 1 H_2S and 2 O_2 on the left side of the equation, and 1 SO_4^{2-} on the right side. To achieve an ionic balance, we must have two positive charges on the right side of the equation to balance the two negative charges of SO_4^{2-}. Thus, 2 H^+ must be added to the right side of the equation, making the overall reaction

$$H_2S + 2O_2 \longrightarrow SO_4^{2-} + 2H^+$$

By inspection, it can be seen that this equation is also balanced in terms of the total number of atoms of each kind on each side of the equation.

(b) What is the balanced reaction for the oxidation of H_2S to SO_4^{2-} with Fe^{3+} as electron acceptor? We have just ascertained that the oxidation of H_2S to SO_4^{2-} is an eight-electron transfer. Because the reduction of Fe^{3+} to Fe^{2+} is only a one-electron transfer, 8 Fe^{3+} will be required. At this point, the reaction looks like this:

$$H_2S + 8Fe^{3+} \longrightarrow 8Fe^{2+} + SO_4^{2-} \text{(not balanced)}$$

We note that the ionic balance is incorrect. We have 24 positive charges on the left and 14 positive charges on the right (16+ from Fe, 2− from sulfate). To equalize the charges, we add $10H^+$ on the right. Now our equation looks like this:

$$H_2S + 8Fe^{3+} \longrightarrow 8Fe^{2+} + 10H^+ + SO_4^{2-} \quad \text{(not balanced)}$$

To provide the necessary hydrogen for the H^+ and oxygen for the sulfate, we add 4 H_2O to the left and find that the equation is now balanced:

$$H_2S + 4H_2O + 8Fe^{3+} \longrightarrow 8Fe^{2+} + 10H^+ + SO_4^{2-} \text{(balanced)}$$

Table A1.1 Free energies of formation (G_f^0) for some substances (kJ/mol)[a]

Carbon compound	Carbon compound	Metal	Nonmetal	Nitrogen compound
CO, −137.34	Glutamine, −529.7	Cu$^+$, +50.28	H$_2$, 0	N$_2$, 0
CO$_2$, −394.4	Glyceraldehyde, −437.65	Cu^{2+}, +64.94	H$^+$, 0 at pH 0; −39.83 at pH 7 (−5.69 per pH unit)	NO, +86.57
CH$_4$, −50.75	Glycerate, −658.1	CuS, −49.02	O$_2$, 0	NO$_2$, +51.95
H$_2$CO$_3$, −623.16	Glycerol, −488.52	Fe^0, 0 Fe^{2+}, −78.87	OH$^-$, −157.3 at pH 14; −198.76 at pH 7; −237.57 at pH 0	NO$_2^-$, −37.2
HCO$_3^-$, −586.85	Glycine, −370.8	Fe^{3+}, −4.6	H$_2$O, −237.17	NO$_3^-$, −111.34
CO$_3^{2-}$, −527.90	Glycolate, −530.95	FeCO$_3$, −673.23	H$_2$O$_2$, −134.1	NH$_3$, −26.57
Acetaldehyde, −139.9	Glyoxylate, −468.6	FeS$_2$, −150.84 FeS, −100.4	PO$_4^{3-}$, −1026.55	NH$_4^+$, −79.37
Acetate, −369.41	Guanine, +46.99	FeSO$_4$, −829.62	Se^0, 0	N$_2$O, +104.18
Acetone, −161.17	α-Ketoglutarate, −797.55	PbS, −92.59	H$_2$Se, −77.09	N$_2$H$_4$, +128
Alanine, −371.54	Lactate, −517.81	Mn^{2+}, −227.93	SeO$_4^{2-}$, −439.95	
Arginine, −240.2	Lactose, −1515.24	Mn^{3+}, −82.12	S^0, 0	
Aspartate, −700.4	Malate, −845.08	MnO$_4^-$, −506.57	SO$_3^{2-}$, −486.6	
Benzene, +124.5	Mannitol, −942.61	MnO$_2$, −456.71	SO$_4^{2-}$, −744.6	
Benzoic acid, −245.6	Methanol, −175.39	MnSO$_4$, −955.32	S$_2$O$_3^{2-}$, −513.4	
n-Butanol, −171.84	Methionine, −502.92	HgS, −49.02	H$_2$S, −27.87	
Butyrate, −352.63	Methylamine, −40.0	MoS$_2$, −225.42	HS$^-$, +12.05	
Caproate, −335.96	Oxalate, −674.04	ZnS, −198.60	S^{2-}, +85.8	
Citrate, −1168.34	Palmitic acid, −305			
o-Cresol, −37.1	Phenol, −47.6			
Crotonate, −277.4	n-Propanol, −175.81			
Cysteine, −339.8	Propionate, −361.08			
Dimethylamine, −3.3	Pyruvate, −474.63			
Ethanol, −181.75	Ribose, −757.3			
Formaldehyde, −130.54	Succinate, −690.23			
Formate, −351.04	Sucrose, −370.90			
Fructose, −951.38	Toluene, +114.22			
Fumarate, −604.21	Trimethylamine, −37.2			
Gluconate, −1128.3	Tryptophan, −112.6			
Glucose, −917.22	Urea, −203.76			
Glutamate, −699.6	Valerate, −344.34			

[a]Values for free energy of formation of various compounds can be found in Dean, J.A. 1973. *Lange's Handbook of Chemistry*, 11th edition. McGraw-Hill, New York; Garrels, R.M., and C.L. Christ. 1965. *Solutions, Minerals, and Equilibria*. Harper & Row, New York; Burton, K. 1957. In Krebs, H.A., and H.L. Komberg. Energy transformation in living matter, *Ergebnisse der Physiologie* (appendix). Springer-Verlag, Berlin; and Thauer, R.K., K. Jungermann, and H. Decker. 1977. Energy conservation in anaerobic chemotrophic bacteria. *Bacteriol. Rev. 41*: 100–180.

In general, ionic balance can be achieved by adding H$^+$ or OH$^-$ to the left or right side of the equation, and because all reactions take place in an aqueous medium, H$_2$O molecules can be added where needed. Whether H$^+$ or OH$^-$ is added generally depends on whether the reaction is taking place under acidic or alkaline conditions.

3. **Calculation of energy yield for balanced equations from free energies of formation.** Once an equation has been balanced, the free-energy yield can be calculated by inserting the values for the free energy of formation of each reactant and product from Table A1.1 and using the formula in item 2 of the first section of this appendix. For instance, for the equation

$$H_2S + 2O_2 \longrightarrow SO_4^{2-} + 2H^+$$

G_f^0 values: $(-27.87) + (0) \longrightarrow$
$(-744.6) + 2(-39.83)$ (assuming pH 7)

$$\Delta G^{0\prime} = [(-744.6) + 2(-39.83)] - [(-27.87) + (0)]$$

$$= -796.39 \text{ kJ}$$

The G_f^0 values for the products (right side of reaction) are summed and subtracted from the G_f^0 values for the reactants (left side of reaction), taking care to ensure that the arithmetic signs are correct. From the data in Table A1.1, a wide variety of free-energy yields for reactions of microbiological interest can be calculated.

Calculation of Free-Energy Yield from Reduction Potential

Reduction potentials of some important redox pairs are given in Table A1.2. The amount of energy that can be released from two half reactions can be calculated from the *differences* in reduction potentials of the two reactions and from the number of electrons transferred. The farther apart the two half reactions are, and the greater the number of electrons transferred, the more energy released.

The conversion of potential difference to free energy is given by the formula $\Delta G^{0\prime} = -nF\Delta E_0{}'$, where n is the number of electrons, F is the Faraday constant (96.48 kJ/V), and $\Delta E_0{}'$ is the difference in reduction potentials. Thus, the $2\,H^+/H_2$ couple has a potential of -0.41 V and the $\frac{1}{2}\,O_2/H_2O$ pair has a potential of $+0.82$ V, and so the potential difference is 1.23 V, which (because two electrons are involved) is equivalent to a free-energy yield (ΔG^0) of -237.34 kJ. On the other hand, the potential difference between the $2\,H^+/H_2$ and the $NO_3{}^-/NO_2{}^-$ reactions is less, 0.84 V, which is equivalent to a free-energy yield of -162.08 kJ (⇄ Figure 3.9).

Because many biochemical reactions are two-electron transfers, it is often useful to give energy yields for two-electron reactions, even if more electrons are involved. Thus, the $SO_4{}^{2-}/H_2$ redox pair involves eight electrons, and complete reduction of $SO_4{}^{2-}$ with H_2 requires $4\,H_2$ (equivalent to eight electrons). From the reduction potential difference between $2\,H^+/H_2$ and $SO_4{}^{2-}/H_2S$ (0.19 V), a free-energy yield of -146.64 kJ is calculated, or -36.66 kJ per two electrons. By convention, reduction potentials are given for conditions in which equal concentrations of oxidized and reduced forms are present. In actual practice, the concentrations of these two forms may be quite different. As discussed earlier in this appendix (Section I, item 9), it is possible to couple half reactions even if the potential difference is unfavorable, provided the concentrations of the reacting species are appropriate.

Table A1.2 Microbiologically important reduction potentials[a]

Redox pair	$E_0{}'$ (V)	Redox pair	$E_0{}'$ (V)
$SO_4{}^{2-}/HSO_3{}^-$	-0.52	Acrylyl-CoA/propionyl-CoA	-0.015
CO_2/formate	-0.43	Glycine/acetate$^-$ + $NH_4{}^+$	-0.010
$2\,H^+/H_2$	-0.41	$S_4O_6{}^{2-}/S_2O_3{}^{2-}$	$+0.024$
$S_2O_3{}^{2-}/HS^- + HSO_3{}^-$	-0.40	Fumarate^{2-}/succinate^{2-}	$+0.033$
Ferredoxin ox/red	-0.39	Cytochrome b ox/red	$+0.035$
Flavodoxin ox/red[b]	-0.37	Ubiquinone ox/red	$+0.113$
$NAD^+/NADH$	-0.32	$AsO_4{}^{3-}/AsO_3{}^{3-}$	$+0.139$
Cytochrome c_3 ox/red	-0.29	Dimethyl sulfoxide (DMSO)/dimethyl sulfide (DMS)	$+0.16$
CO_2/acetate$^-$	-0.29	$Fe(OH)_3 + HCO_3{}^-/FeCO_3$ (Fe^{3+}/Fe^{2+}, pH 7)	$+0.20$
S^0/HS^-	-0.27	$S_3O_6{}^{2-}/S_2O_3{}^{2-} + HSO_3{}^-$	$+0.225$
CO_2/CH_4	-0.24	Cytochrome c_1 ox/red	$+0.23$
FAD/FADH	-0.22	$NO_2{}^-/NO$	$+0.36$
$SO_4{}^{2-}/HS^-$	-0.217	Cytochrome a_3 ox/red	$+0.385$
Acetaldehyde/ethanol	-0.197	Chlorobenzoate$^-$/benzoate$^-$ + HCl	$+0.297$
Pyruvate$^-$/lactate$^-$	-0.19	$NO_3{}^-/NO_2{}^-$	$+0.43$
FMN/FMNH	-0.19	$SeO_4{}^{2-}/SeO_3{}^{2-}$	$+0.475$
Dihydroxyacetone phosphate/glycerolphosphate	-0.19	Fe^{3+}/Fe^{2+} (pH 2)	$+0.77$
$HSO_3{}^-/S_3O_6{}^{2-}$	-0.17	Mn^{4+}/Mn^{2+}	$+0.798$
Flavodoxin ox/red[b]	-0.12	O_2/H_2O	$+0.82$
$HSO_3{}^-/HS^-$	-0.116	$ClO_3{}^-/Cl^-$	$+1.03$
Menaquinone ox/red	-0.075	NO/N_2O	$+1.18$
Adenosine phosphosulfate/AMP + $HSO_3{}^-$	-0.060	N_2O/N_2	$+1.36$
Rubredoxin ox/red	-0.057		

[a]Data from Thauer, R.K., K. Jungermann, and K. Decker, 1977. Energy conservation in anaerobic chemotrophic bacteria. *Bacteriol. Rev.* 41: 100–180.
[b]Separate potentials are given for each electron transfer in this potentially two-electron transfer.

Appendix 2

Bergey's Manual of Systematic Bacteriology, Second Edition

List of Higher-Order Taxa[1]

Domain *Archaea*[2]

Phylum I. *Crenarchaeota*
  Class I. *Thermoprotei*
    Order I. *Thermoproteales*
    Order II. *Caldisphaerales*
    Order III. *Desulfurococcales*
    Order IV. *Sulfolobales*
Phylum II. *Euryarchaeota*
  Class I. *Methanobacteria*
    Order I. *Methanobacteriales*
  Class II. *Methanococci*
    Order I. *Methanococcales*
  Class III. *Methanomicrobia*
    Order I. *Methanomicrobiales*
    Order II. *Methanosarcinales*
  Class IV. *Halobacteria*
    Order I. *Halobacteriales*
  Class V. *Thermoplasmata*
    Order I. *Thermoplasmatales*
  Class VI. *Thermococci*
    Order I. *Thermococcales*
  Class VII. *Archaeoglobi*
    Order I. *Archaeoglobales*
  Class VIII. *Methanopyri*
    Order I. *Methanopyrales*

Domain *Bacteria*

Phylum I. *Aquificae*
  Class I. *Aquificae*
    Order I. *Aquificales*
Phylum II. *Thermotogae*
  Class I. *Thermotogae*
    Order I. *Thermotogales*
Phylum III. *Thermodesulfobacteria*
  Class I. *Thermodesulfobacteria*
    Order I. *Thermodesulfobacteriales*
Phylum IV. *Deinococcus–Thermus*
  Class I. *Deinococci*
    Order I. *Deinococcales*
    Order II. *Thermales*
Phylum V. *Chrysiogenetes*
  Class I. *Chrysiogenetes*
    Order I. *Chrysiogenales*
Phylum VI. *Chloroflexi*
  Class I. *Chloroflexi*
    Order I. *Chloroflexales*
    Order II. *Herpetosiphonales*
  Class II. *Anaerolineae*
    Order I. *Anaerolineales*

Phylum VII. *Thermomicrobia*
  Class I. *Thermomicrobia*
    Order I. *Thermomicrobiales*
Phylum VIII. *Nitrospirae*
  Class I. *Nitrospira*
    Order I. *Nitrospirales*
Phylum IX. *Deferribacteres*
  Class I. *Deferribacteres*
    Order I. *Deferribacterales*
Phylum X. *Cyanobacteria*
  Class I. *Cyanobacteria*
Phylum XI. *Chlorobi*
  Class I. *Chlorobia*
    Order I. *Chlorobiales*
Phylum XII. *Proteobacteria*
  Class I. *Alphaproteobacteria*
    Order I. *Rhodospirillales*
    Order II. *Rickettsiales*
    Order III. *Rhodobacterales*
    Order IV. *Sphingomonadales*
    Order V. *Caulobacterales*
    Order VI. *Rhizobiales*
    Order VII. *Parvularculales*
  Class II. *Betaproteobacteria*
    Order I. *Burkholderiales*
    Order II. *Hydrogenophilales*
    Order III. *Methylophilales*
    Order IV. *Neisseriales*
    Order V. *Nitrosomonadales*
    Order VI. *Rhodocyclales*
    Order VII. *Procabacteriales*
  Class III. *Gammaproteobacteria*
    Order I. *Chromatiales*
    Order II. *Acidithiobacillales*
    Order III. *Xanthomonadales*
    Order IV. *Cardiobacteriales*
    Order V. *Thiotrichales*
    Order VI. *Legionellales*
    Order VII. *Methylococcales*
    Order VIII. *Oceanospirillales*
    Order IX. *Pseudomonadales*
    Order X. *Alteromonadales*
    Order XI. *Vibrionales*
    Order XII. *Aeromonadales*
    Order XIII. *Enterobacteriales*
    Order XIV. *Pasteurellales*
  Class IV. *Deltaproteobacteria*
    Order I. *Desulfurellales*
    Order II. *Desulfovibrionales*
    Order III. *Desulfobacterales*
    Order IV. *Desulfarcales*

    Order V. *Desulfuromonales*
    Order VI. *Syntrophobacterales*
    Order VII. *Bdellovibrionales*
    Order VIII. *Myxococcales*
  Class V. *Epsilonproteobacteria*
    Order I. *Campylobacterales*
Phylum XIII. *Firmicutes*
  Class I. *Bacilli*
    Order I. *Bacillales*
    Order II. *Lactobacillales*
  Class II. *Clostridia*
    Order I. *Clostridiales*
    Order II. *Halanaerobiales*
    Order III. *Thermoanaerobacterales*
  Class III. *Erysipelotrichi*
    Order I. *Erysipelotrichales*
Phylum XIV. *Actinobacteria*
  Class I. *Actinobacteria*
    Order I. *Actinomycetales*
    Order II. *Bifidobacteriales*
    Order III. *Catenulisporales*
    Order IV. *Corynebacteriales*
    Order V. *Frankiales*
    Order VI. *Glycomycetales*
    Order VII. *Jiangellales*
    Order VIII. *Kineosporiales*
    Order IX. *Micrococcales*
    Order X. *Micromonosporales*
    Order XI. *Propionibacteriales*
    Order XII. *Pseudonocardiales*
    Order XIII. *Streptomycetales*
    Order XIV. *Streptosporangiales*
  Class II. *Acidimicrobia*
    Order I. *Acidimicrobiales*
  Class III. *Coriobacteria*
    Order I. *Coriobacteriales*
  Class IV. *Rubrobacteria*
    Order I. *Rubrobacterales*
  Class V. *Thermoleophilia*
    Order I. *Thermoleophilales*
    Order II. *Solirubrobacterales*
Phylum XV. *Planctomycetes*
  Class I. *Planctomycetacia*
    Order I. *Planctomycetales*
    Order II. *Brocadiales*
Phylum XVI. *Chlamydiae*
  Class I. *Chlamydiae*
    Order I. *Chlamydiales*
Phylum XVII. *Spirochaetes*
  Class I. *Spirochaetes*
    Order I. *Spirochaetales*

[1]*Bergey's Manual of Systematic Bacteriology*, second edition, consists of 5 volumes; shown here are phyla and classes of prokaryotes found in this reference. Each class contains one or more orders, each order one or more families, each family one or more genera, and each genus one or more species. The genus name of many important prokaryotes forms part of the names of their higher-order taxa. Because bacterial taxonomy is a work in progress, updates to the list shown here occur regularly. For the most up-to-date prokaryotic taxonomy, refer to http://www.bacterio.net. An overview of prokaryotic diversity is presented in Chapters 14–16 of this text. Select orders of *Bacteria* are described in Chapters 14 and 15, including some not yet in *Bergey's Manual*.

[2]The phyla *Thaumarchaeota*, *Nanoarchaeota*, and *Korarchaeota* (see Chapter 16 and Figure 16.1 of this text) are recently established higher taxa of *Archaea* and should appear along with the *Crenarchaeota* and *Euryarchaeota* in a future edition of *Bergey's Manual*.

Phylum XVIII. *Tenericutes*
  Class I. *Mollicutes*
    Order I. *Mycoplasmatales*
    Order II. *Entomoplasmatales*
    Order III. *Acholeplasmatales*
    Order IV. *Anaeroplasmatales*
Phylum XIX. *Fibrobacteres*
  Class I. *Fibrobacteres*
    Order I. *Fibrobacterales*
Phylum XX. *Acidobacteria*
  Class I. *Acidobacteria*
    Order I. *Acidobacteriales*

  Class II. *Holophagae*
    Order I. *Holophagales*
    Order II. *Acanthopleuribacterales*
Phylum XXI. *Bacteroidetes*
  Class I. *Bacteroidia*
    Order I. *Bacteroidales*
  Class II. *Flavobacteria*
    Order I. *Flavobacteriales*
  Class III. *Sphingobacteria*
    Order I. *Sphingobacteriales*
  Class IV. *Cytophagia*
    Order I. *Cytophagales*

Phylum XXII. *Verrucomicrobia*
  Class I. *Verrucomicrobiae*
    Order I. *Verrucomicrobiales*
  Class II. *Opitutae*
    Order I. *Opitutales*
    Order II. *Puniceicoccales*
  Class III. *Spartobacteria*
    Order I. *Chthoniobacterales*

Glossary

Only the major terms and concepts are included. If a term is not here, consult the index.

ABC (ATP-binding cassette) transporter A membrane transport system consisting of three proteins, one of which hydrolyzes ATP, one of which binds the substrate, and one of which functions as the transport channel through the membrane.

Abscess A localized infection characterized by production of pus.

Acetic acid bacteria Obligately aerobic organisms that produce acetate from alcohols; used for making vinegar.

Acetogen A bacterium that carries out acetogenesis.

Acetogenesis Energy metabolism in which acetate is produced from either H_2 plus CO_2 or from organic compounds.

Acetotrophic Acetate consuming.

Acetyl-CoA pathway A pathway of autotrophic CO_2 fixation and acetate oxidation widespread in various obligate anaerobes including methanogens, acetogens, several clostridia, and sulfate-reducing bacteria.

Acetylene reduction assay A method of measuring activity of nitrogenase by substituting acetylene for the natural substrate of the enzyme, N_2. Acetylene is reduced to ethylene or ethane, depending on the nitrogenase system involved.

Acid-fastness A property of *Mycobacterium* species; cells stained with basic fuchsin dye resist decolorization with acidic alcohol.

Acid mine drainage Acidic water containing H_2SO_4 (sulfuric acid) derived from the microbial oxidation of iron sulfide minerals.

Acidophile An organism that grows best at acidic pH values.

Acridine orange A nonspecific fluorescent dye used to stain microbial cells in a natural sample.

Actinomycete A term used to refer to aerobic filamentous bacteria in the phylum *Actinobacteria*.

Activation energy The energy needed to make substrate molecules more reactive; enzymes function by lowering activation energy.

Activator protein A regulatory protein that binds to specific sites on DNA and stimulates transcription; involved in positive control.

Active immunity An immune state achieved by self-production of antibodies. Compare with *passive immunity*.

Active site The portion of an enzyme that is directly involved in binding substrate(s).

Active transport The energy-dependent process of transporting substances into or out of the cell without chemically changing the transported substances.

Acute infection A short-term infection, usually characterized by dramatic onset.

Adaptive immunity (antigen-specific immunity) The acquired ability to recognize and destroy a particular pathogen or its products; dependent on previous exposure to the pathogen or its products.

Adenosine triphosphate (ATP) A nucleotide that is the primary form in which chemical energy is conserved and utilized in cells.

Adherence The enhanced ability of a microorganism to attach to a cell or surface.

Aerobe An organism that grows in the presence of O_2; may be facultative, obligate, or microaerophilic.

Aerobic anoxygenic phototroph An organism that is an aerobic heterotroph and uses anoxygenic photosynthesis as a supplemental source of energy.

Aerobic secondary wastewater treatment Digestive reactions carried out by microorganisms under aerobic conditions to treat wastewater containing low levels of organic materials.

Aerosol Suspension of particles in airborne water droplets.

Aerotolerant anaerobe An anaerobic microorganism whose growth is not inhibited by O_2.

Agglutination A reaction between antibody and particle-bound antigen resulting in visible clumping of the particles.

Algae Phototrophic eukaryotic micro- and macroorganisms.

Alkaliphile An organism that grows best at high pH.

Allele A sequence variant of a given gene.

Allergy A harmful immune reaction, usually caused by a foreign antigen in food, pollen, or chemicals, which results in immediate-type or delayed-type hypersensitivity.

Allosteric protein An enzyme that contains two combining sites, an active site for binding substrate and an allosteric site for binding an effector molecule such as the end product of a biochemical pathway.

Amoeboid movement A type of motility in which cytoplasmic streaming moves the organism forward.

Amino acid One of the 22 monomers that make up proteins; chemically, a two-carbon carboxylic acid containing an amino group and a characteristic substituent on the alpha carbon.

Aminoacyl-tRNA synthetase An enzyme that catalyzes the attachment of the correct amino acid to the correct tRNA.

Aminoglycoside An antibiotic such as streptomycin, containing amino sugars linked by glycosidic bonds.

Anabolic reactions (anabolism) The biochemical processes involved in the synthesis of cell constituents from simpler molecules, usually requiring energy.

Anaerobe An organism that grows in the absence of O_2; some may even be killed by O_2 (obligate or strict anaerobes).

Anaerobic respiration Use of an electron acceptor other than O_2 in an electron transport–based oxidation leading to a proton motive force.

Anammox Anoxic ammonia oxidation.

Anaphylaxis (anaphylactic shock) A violent allergic reaction caused by an antigen–antibody reaction.

Anergy The inability to produce an immune response to specific antigens due to neutralization of effector cells.

Anoxic Oxygen-free. Usually used in reference to a microbial habitat.

Anoxic secondary wastewater treatment Digestive and fermentative reactions carried out by microorganisms under anoxic conditions to treat wastewater containing high levels of insoluble organic materials.

Anoxygenic photosynthesis The use of light energy to synthesize ATP by cyclic photophosphorylation without O_2 production.

Antenna pigments Light-harvesting chlorophylls or bacteriochlorophylls in photocomplexes that funnel energy to the reaction center.

Antibiogram A report indicating the sensitivity of clinically isolated microorganisms to the antibiotics in current use.

Antibiotic A chemical substance produced by a microorganism that kills or inhibits the growth of another microorganism.

Antibody A soluble protein produced by B lymphocytes and plasma cells that interacts specifically with antigen; also called *immunoglobulin*.

Anticodon A sequence of three bases in transfer RNA that base-pairs with a codon during protein synthesis.

Antigen A molecule capable of interacting with specific components of the immune system.

Antigen-presenting cell (APC) A macrophage, dendritic cell, or B cell that takes up and processes antigen and presents it to T-helper cells.

Antigenic drift In influenza virus, minor changes in viral proteins (antigens) due to gene mutation.

Antigenic shift In influenza virus, major changes in viral proteins (antigens) due to gene reassortment.

Antimicrobial Harmful to microorganisms by either killing or inhibiting growth.

Antimicrobial agent A chemical that kills or inhibits the growth of microorganisms.

Antimicrobial drug resistance The acquired ability of a microorganism to resist the effects of an antimicrobial drug to which it is normally susceptible.

Antiparallel In reference to double-stranded nucleic acids, the two strands run in opposite directions; one strand runs $5' \rightarrow 3'$, the complementary strand $3' \rightarrow 5'$.

Antiseptic (germicide) A chemical agent that kills or inhibits growth of microorganisms and is sufficiently nontoxic to be applied to living tissues.

Antiserum A serum containing antibodies.

Antitoxin An antibody that specifically interacts with and neutralizes a toxin.

Apoptosis Programmed cell death.

Arbuscule Branched or coiled hyphal structure within cells of the inner cortex of plants with a mycorrhizal infection.

Archaea Phylogenetically related prokaryotes distinct from *Bacteria*.

Artificial chromosome A single copy vector that can carry extremely long inserts of DNA and is widely used for cloning segments of large genomes.

Aseptic technique The manipulation of sterile instruments or culture media in such a way as to maintain sterility.

Aspartame A nonnutritive sweetener composed of the amino acids aspartate and phenylalanine, the latter as a methyl ester.

ATP Adenosine triphosphate, the principal energy carrier of the cell.

ATPase (ATP synthase) A multiprotein enzyme complex embedded in the cytoplasmic membrane that catalyzes the synthesis of ATP coupled to dissipation of the proton motive force.

Attenuation In a pathogen, a decrease or loss of virulence. Also, a mechanism for controlling gene expression; typically, transcription is terminated after initiation but before a full-length mRNA is produced.

Autoantibody An antibody that reacts to self antigens.

Autoclave A sealed sterilizing device that destroys microorganisms with temperature and steam under pressure.

Autoimmunity The immune reactions of a host against its own self antigens.

Autoinducer A small signal molecule that takes part in quorum sensing.

Autoinduction A gene regulatory mechanism involving small, diffusible signal molecules that are produced in larger amounts as population size increases.

Autolysis The lysis of a cell brought about by the activity of the cell itself.

Autoradiography Detection of radioactivity in a sample, for example, a cell or gel, by placing it in contact with a photographic film.

Autotroph An organism able to grow with CO_2 as its sole source of carbon.

Auxotroph An organism that has developed a nutritional requirement through mutation. Contrast with *prototroph*.

B cell A lymphocyte that has immunoglobulin surface receptors, may present antigens to T cells, and may form plasma cells, which produce immunoglobulin.

B cell receptor (BCR) A cell-surface antibody that acts as an antigen receptor on a B cell.

Bacteremia The presence of bacteria in the blood.

Bacteria Phylogenetically related prokaryotes distinct from *Archaea*.

Bacterial artificial chromosome (BAC) Circular artificial chromosome with bacterial origin of replication.

Bacteriochlorophyll A pigment of anoxygenic phototrophic organisms consisting of light-sensitive magnesium tetrapyrroles.

Bacteriocidal agent An agent that kills bacteria.

Bacteriocins Agents produced by certain bacteria that inhibit or kill closely related species.

Bacteriocyte A specialized insect cell in which bacterial symbionts reside.

Bacteriome A specialized region in several insect groups that contains insect bacteriocyte cells packed with bacterial symbionts.

Bacteriophage A virus that infects prokaryotic cells.

Bacteriorhodopsin A protein containing retinal that is found in the membranes of certain extremely halophilic *Archaea* and that is involved in light-mediated ATP synthesis.

Bacteriostatic agent An agent that inhibits bacterial growth.

Bacteroid A swollen, deformed *Rhizobium* cell found in the root nodule; capable of nitrogen fixation.

Banded iron formation Iron oxide–rich ancient sedimentary rocks containing zones of oxidized iron (Fe^{3+}) formed by oxidation of Fe^{2+} by O_2 produced by cyanobacteria.

Basal body The "motor" portion of the bacterial flagellum, embedded in the cytoplasmic membrane and wall.

Base composition In reference to nucleic acids, the proportion of the total bases consisting of guanine plus cytosine or thymine plus adenine base pairs. Usually expressed as a guanine plus cytosine (GC) value, for example, 60% GC.

Basic reproduction number (R_0) The number of expected secondary transmissions from each single case of a disease in an entirely susceptible population.

Batch culture A closed-system microbial culture of fixed volume.

β-Lactam antibiotic An antibiotic such as penicillin that contains the four-membered heterocyclic β-lactam ring.

Binary fission Cell division whereby a cell grows by intercalary growth to twice its minimum size and then divides to form two cells.

Binomial system The system devised by Linnaeus for naming organisms by giving them a genus name and a species epithet.

Biochemical oxygen demand (BOD) The amount of dissolved oxygen consumed by microorganisms for complete oxidation of organic and inorganic material in a water sample.

Biofilm Microbial colonies encased in an adhesive, usually polysaccharide material and attached to a surface.

Biofuel A fuel made by microorganisms from the fermentation of carbon-rich feedstocks.

Biogeochemistry The study of microbially mediated chemical transformations of geochemical interest, for example, nitrogen or sulfur cycling.

Bioinformatics The use of computer programs to analyze, store, and access DNA and protein sequences.

Biological warfare The use of biological agents to kill or incapacitate a population.

Bioluminescence The enzymatic production of visible light by living organisms.

Bioremediation The use of microorganisms to remove or detoxify toxic or unwanted chemicals in an environment.

Biosynthesis The production of needed cellular constituents from other (usually simpler) molecules.

Biotechnology The use of organisms, typically genetically altered, in industrial, medical, or agricultural applications.

Black smoker A deep-sea hydrothermal vent emitting superheated 250–400°C water and minerals.

Bone marrow A primary lymphoid organ containing the pluripotent precursor cells for all blood and immune cells, including B cells.

Botulism Food poisoning due to ingestion of food containing botulinum toxin produced by *Clostridium botulinum*.

Brewing The manufacture of alcoholic beverages such as beer and ales from the fermentation of malted grains.

Broad-spectrum antibiotic An antibiotic that acts on both gram-positive and gram-negative *Bacteria*.

Calvin cycle The series of biosynthetic reactions by which many autotrophic organisms convert CO_2 to organic compounds.

Canning The process of sealing food in a closed container and heating to destroy living organisms and endospores.

Capsid The protein shell that surrounds the genome of a virus.

Capsomere The subunit of the virus capsid.

Capsule A dense, well-defined polysaccharide or protein layer closely surrounding a cell.

Carboxysome Polyhedral cellular inclusion of crystalline ribulose bisphosphate carboxylase (RubisCO), the key enzyme of the Calvin cycle.

Carcinogen A substance that causes the initiation of tumor formation. Frequently a mutagen.

Cardinal temperatures The minimum, maximum, and optimum growth temperatures for a given organism.

Carotenoid A hydrophobic accessory pigment present along with chlorophyll in photosynthetic membranes.

Carrier A subclinically infected individual who may spread a disease.

Cassette mutagenesis Creating mutations by the insertion of a DNA cassette.

Catabolic reactions (catabolism) The biochemical processes involved in the breakdown of organic or inorganic compounds, usually leading to the production of energy.

Catabolite repression The suppression of alternative catabolic pathways by a preferred source of carbon and energy.

Catalysis An increase in the rate of a chemical reaction.

Catalyst A substance that promotes a chemical reaction without itself being changed in the end.

CD4 cells T-helper cells. They are targets for HIV infection.

CD4 coreceptor A protein found exclusively on Th cells that interacts with MHC II on an antigen-presenting cell.

CD8 coreceptor A protein found exclusively on Tc cells that interacts with MHC I on a target cell.

Cell The fundamental unit of living matter.

Cell-mediated immunity An immune response generated by interactions with antigen-specific T cells. Compare with *antibody-mediated immunity*.

Cell wall A rigid layer present outside the cytoplasmic membrane that confers structural strength on the cell and protection from osmotic lysis.

Centers for Disease Control and Prevention (CDC) The agency of the United States Public Health Service that tracks disease trends, provides disease information to the public and to healthcare professionals, and forms public policy regarding disease prevention and intervention.

Chaperonin (molecular chaperone) A protein that helps other proteins fold or refold from a partly denatured state.

Chemiosmosis The use of ion gradients, especially proton gradients, across membranes to generate ATP.

Chemokine A small, soluble protein produced by a variety of cells that modulates inflammatory reactions and immunity.

Chemolithotroph An organism that obtains its energy from the oxidation of inorganic compounds.

Chemoorganotroph An organism that obtains its energy from the oxidation of organic compounds.

Chemostat A continuous culture device controlled by the concentration of limiting nutrient and dilution rate.

Chemotaxis Movement toward or away from a chemical.

Chemotherapeutic agent An antimicrobial agent that can be used internally.

Chemotherapy The treatment of infectious disease with chemicals or antibiotics.

Chitin A polymer of *N*-acetylglucosamine commonly found in the cell walls of fungi.

Chloramine A water purification chemical made by combining chlorine and ammonia at precise ratios.

Chlorination A highly effective disinfectant procedure for drinking water using chlorine gas or other chlorine-containing compounds as disinfectant.

Chlorine A chemical used in its gaseous state to disinfect water. A residual level is maintained throughout the distribution system.

Chlorophyll A pigment of phototrophic organisms consisting of light-sensitive magnesium tetrapyrroles.

Chloroplast The chlorophyll-containing organelle of phototrophic eukaryotes.

Chlorosome A cigar-shaped structure enclosed by a nonunit membrane and containing the light-harvesting bacteriochlorophyll (*c*, *d*, or *e*) in green sulfur bacteria and in *Chloroflexus*.

Chromogenic Producing color; for example, a chromogenic colony is a pigmented colony.

Chromosomal island A bacterial chromosome region of foreign origin that contains clustered genes for some extra property such as virulence or symbiosis.

Chromosome A genetic element carrying genes essential to cellular function. Prokaryotes typically have a single chromosome consisting of a circular DNA molecule. Eukaryotes typically have several chromosomes, each containing a linear DNA molecule.

Chronic infection A long-term infection.

-cidal Suffix indicating killing; for example, a bacteriocidal agent kills bacteria. Compare with *-static*.

Ciliate A protist characterized in part by rapid motility driven by numerous short appendages called cilia.

Cilium (plural, cilia) Short, filamentous structure that beats with many others to make a cell move.

Cirrhosis Breakdown of the normal liver architecture, resulting in fibrosis.

Citric acid cycle A cyclical series of reactions resulting in the conversion of acetate to CO_2 and NADH. Also called the tricarboxylic acid cycle or the Krebs cycle.

Clarifier (coagulation basin) A reservoir in which the suspended solids of raw water are coagulated and removed.

Clonal anergy The inability to produce an immune response to specific antigens due to the neutralization of effector cells.

Clonal deletion For T cell selection in the thymus, the killing of useless or self-reactive clone precursors.

Clonal selection A theory that each B or T lymphocyte, when stimulated by antigen, divides to form a clone of itself.

Clone In immunology, a copy of an antigen-reactive lymphocyte, usually in large numbers. Also, a number of copies of a DNA fragment obtained by allowing an inserted DNA fragment to be replicated by a phage or plasmid.

Cloning vectors Genetic elements into which genes can be recombined and replicated.

Coagulation The formation of large insoluble particles from much smaller, colloidal particles by the addition of aluminum sulfate and anionic polymers.

Coccoid Sphere-shaped.

Coccus A spherical bacterium.

Codon A sequence of three bases in messenger RNA that encodes a specific amino acid.

Codon bias The nonrandom usage of multiple codons encoding the same amino acid. Also called codon usage.

Coenocytic The presence of multiple nuclei in fungal hyphae without septa.

Coenzyme A low-molecular-weight molecule that participates in an enzymatic reaction by accepting and donating electrons or functional groups. Examples: NAD^+, FAD.

Coevolution Evolution that proceeds jointly in a pair of intimately associated species owing to the effects each has on the other.

Coliform Gram-negative, nonsporulating, facultatively aerobic rod that ferments lactose with gas formation within 48 hours at 35°C.

Colonial The growth form of certain protists and green algae in which several cells live together and cooperate for feeding, motility, or reproduction; an early form of multicellularity.

Colonization The multiplication of a microorganism after it has attached to host tissues or other surfaces.

Colony A macroscopically visible population of cells growing on solid medium, arising from a single cell.

Cometabolism The metabolic transformation of a substance while a second substance serves as primary energy or carbon source.

Commensalism A type of symbiosis in which only one of two organisms in a relationship benefits.

Commodity chemicals Chemicals such as ethanol that have low monetary value and thus are sold primarily in bulk.

Common-source epidemic An epidemic resulting from infection of a large number of people from a single contaminated source.

Communication Interactions between cells using chemical signals.

Community Two or more cell populations coexisting in a certain area at a given time.

Compatible solutes Organic compounds (or potassium ions) that serve as cytoplasmic solutes to balance water relations for cells growing in environments of high salt or sugar.

Competence The ability to take up DNA and become genetically transformed.

Complement A series of proteins that react sequentially with antibody–antigen complexes, mannose-binding lectin, or alternate activation pathway proteins to amplify or potentiate target cell destruction.

Complement fixation The consumption of complement by an antibody–antigen reaction.

Complementarity-determining region (CDR) A varying amino acid sequence within the variable domains of immunoglobulins or T cell receptors where most molecular contacts with antigen are made. Also called a hypervariable region.

Complementary Nucleic acid sequences that can base-pair with each other.

Complex medium Any culture medium whose precise chemical composition is unknown. Also called undefined media.

Concatemer A DNA molecule consisting of two or more separate molecules linked end to end to form a long, linear structure.

Congenital syphilis Syphilis contracted by an infant from its mother during pregnancy.

Conidia Asexual spores of fungi.

Conjugation The transfer of genes from one prokaryotic cell to another by a mechanism involving cell-to-cell contact.

Consensus sequence A nucleic acid sequence in which the base present in a given position is that base most commonly found when many experimentally determined sequences are compared.

Consortium A two-membered (or more) bacterial culture (or natural assemblage) in which each organism benefits from the others.

Contagious Transmissible.

Convergent evolution A circumstance where a trait or set of traits that are similar in form and/or function between two organisms are not inherited from a shared ancestor (i.e., traits that are similar but not homologous).

Core genome The part of a genome shared by all strains of a species.

Cortex The region inside the spore coat of an endospore, around the core.

Coryneform bacteria Gram-positive, aerobic, nonmotile, rod-shaped organisms with the characteristic of forming irregular-shaped, club-shaped, or V-shaped cell arrangements, typical of several genera of unicellular *Actinobacteria*.

Covalent bond A nonionic chemical bond formed by a sharing of electrons between two atoms.

Crenarchaeota A phylum of *Archaea* that contains both hyperthermophilic and cold-dwelling organisms.

Crista (plural, cristae) An inner membrane in a mitochondrion; a site of respiration.

Crown corrosion The destruction of the upper half, or crown, of concrete wastewater pipes by sulfuric acid produced through the concerted activities of sulfate-reducing and sulfur-oxidizing bacteria.

Culture A particular strain or kind of organism growing in a laboratory medium.

Culture medium An aqueous solution of various nutrients suitable for the growth of microorganisms.

Cutaneous Relating to the skin.

Cyanobacteria Prokaryotic oxygenic phototrophs containing chlorophyll *a* and phycobilins.

Cyclic AMP A regulatory nucleotide that participates in catabolite repression.

Cyst A resting stage formed by some bacteria and protists in which the whole cell is surrounded by a thick-walled, chemically and physically resistant coating; not the same as a spore or endospore.

Cytochrome An iron-containing porphyrin complexed with proteins, which functions as an electron carrier in the electron transport system.

Cytokine A small, soluble protein produced by a leukocyte that modulates inflammatory reactions and immunity.

Cytoplasm The fluid portion of a cell, bounded by the cell membrane.

Cytoplasmic membrane A semipermeable barrier that separates the cell interior (cytoplasm) from the environment.

Cytoskeleton Cellular scaffolding typical of eukaryotic cells in which microtubules, microfilaments, and intermediate filaments define the cell's shape.

DAPI A nonspecific fluorescent dye used to stain microbial cells in a natural sample to obtain total cell numbers.

Decontamination Treatment that renders an object or inanimate surface safe to handle.

Deep sea Marine waters below a depth of 1000 m.

Defective virus A virus that relies on another virus, the helper virus, to provide some of its components.

Defined medium Any culture medium whose exact chemical composition is known. Compare with *complex medium*.

Degeneracy In relation to the genetic code, the fact that more than one codon can code for the same amino acid.

Delayed-type hypersensitivity (DTH) An inflammatory allergic response mediated by Th1 lymphocytes.

Deletion The removal of a portion of a gene.

Denaturation The irreversible destruction of a macromolecule, for example, the destruction of a protein by heat.

Denaturing gradient gel electrophoresis (DGGE) An electrophoretic technique capable of separating nucleic acid fragments of the same size that differ in sequence.

Dendritic cell A type of leukocyte having phagocytic and antigen-presenting properties, found in various body tissues; transports antigen to lymph nodes and spleen.

Denitrification Anaerobic respiration in which nitrate is reduced to nitrogen gases (primarily N_2) under anoxic conditions.

Denitrifier An organism that carries out denitrification.

Dental caries Tooth decay resulting from bacterial infection.

Dental plaque Bacterial cells encased in a matrix of extracellular polymers and salivary products, found on the teeth.

Deoxyribonucleic acid (DNA) A polymer of nucleotides connected via a phosphate–deoxyribose sugar backbone; the genetic material of cells and some viruses.

Desiccation Drying.

Diazotroph An organism that can assimilate N_2 into biomass by activity of the enzyme nitrogenase.

Dideoxynucleotide A nucleotide lacking the 3′-hydroxyl group on the deoxyribose sugar. Used in the Sanger method of DNA sequencing.

Differential media Growth media that allow identification of microorganisms based on phenotypic properties.

Differentiation The modification of a cell in terms of structure and/or function occurring during the course of development.

Dipicolinic acid A substance unique to endospores that confers heat resistance on these structures.

Diploid In eukaryotes, an organism or cell with two chromosome complements, one derived from each haploid gamete.

Disease An injury to a host organism, caused by a pathogen or other factor, that affects host function.

Disinfectant An antimicrobial agent used only on inanimate objects.

Disinfection The elimination of pathogens from inanimate objects or surfaces.

Disproportionation The splitting of a chemical compound into two new compounds, one more oxidized and one more reduced than the original compound.

Dissimilative sulfate-reducer An anaerobic microorganism that conserves energy through the reduction of SO_4^{2-}.

Dissimilative sulfur-oxidizer A microorganism that conserves energy for growth through oxidation of reduced sulfur compounds.

Dissimilative sulfur-reducer An anaerobic microorganism that conserves energy through the reduction of S^0 but cannot reduce SO_4^{2-}.

Distilled alcoholic beverage A beverage containing alcohol concentrated by distillation.

Distribution system Water pipes, storage reservoirs, tanks, and other means used to deliver drinking water to consumers or store it before delivery.

Divisome A complex of proteins that directs cell division processes in bacteria.

DNA Deoxyribonucleic acid, the genetic material of cells and some viruses.

DNA cassette An artificially designed segment of DNA that usually carries a gene for resistance to an antibiotic or some other convenient marker and is flanked by convenient restriction sites.

DNA–DNA hybridization The experimental determination of genomic similarity by measuring the extent of hybridization of DNA from the genome of one organism with that of another.

DNA gyrase An enzyme found in most prokaryotes that introduces negative supercoils in DNA.

DNA helicase An enzyme that uses ATP to unwind the double helix of DNA.

DNA ligase An enzyme that seals nicks in the backbone of DNA.

DNA polymerase An enzyme that synthesizes a new strand of DNA in the $5' \rightarrow 3'$ direction using an antiparallel DNA strand as a template.

DNA vaccine A vaccine that uses the DNA of a pathogen to elicit an immune response.

Domain (1) The highest level of biological classification. The three domains of organisms are the *Bacteria*, the *Archaea*, and the *Eukarya*. (2) A region of a protein having a defined structure and function.

Downstream position Refers to nucleic acid sequences on the 3' side of a given site on the DNA or RNA molecule. Compare with *upstream position*.

Dysbiosis An alteration or imbalance of an individual's microbiome relative to the normal, healthy state, primarily observed in the microbiota of the digestive tract or skin.

Early protein A protein synthesized soon after virus infection and before replication of the virus genome.

Ecology Study of the interrelationships between organisms and their environments.

Ecosystem A dynamic complex of organisms and their physical environment interacting as a functional unit.

Ecotype A population of genetically identical cells sharing a particular resource within an ecological niche.

Effluent water Treated wastewater discharged from a wastewater treatment facility.

Ehrlichiosis One of a group of tick-transmitted diseases caused by rickettsias of the *Ehrlichia* genus.

Electron acceptor A substance that accepts electrons during an oxidation–reduction reaction.

Electron donor A compound that donates electrons in an oxidation–reduction reaction.

Electron transport phosphorylation Synthesis of ATP involving a membrane-associated electron transport chain and the creation of a proton motive force. Also called *oxidative phosphorylation*.

Electrophoresis Separation of charged molecules in an electric field.

Electroporation The use of an electric pulse to enable cells to take up DNA.

ELISA See *enzyme immunoassay*.

Emerging disease Infectious disease whose incidence recently increased or whose incidence threatens to increase in the near future.

Enantiomer One form of a molecule that is the mirror image of another form of the same molecule.

Endemic disease A disease that is constantly present in a population, usually in low numbers. Compare with *epidemic*.

Endergonic reaction A chemical reaction requiring an input of energy to proceed.

Endocytosis A process in which a particle such as a virus is taken intact into an animal cell. Phagocytosis and pinocytosis are two kinds of endocytosis.

Endoplasmic reticulum An extensive array of internal membranes in eukaryotes.

Endospore A differentiated cell formed within the cells of certain gram-positive bacteria that is extremely resistant to heat as well as to other harmful agents.

Endosymbiosis The engulfment of one cell type by another cell type and the subsequent and stable association of the two cells.

Endosymbiotic hypothesis The idea that a chemoorganotrophic bacterium and a cyanobacterium were stably incorporated into another cell type to give rise, respectively, to the mitochondria and chloroplasts of modern-day eukaryotes.

Endotoxin The lipopolysaccharide portion of the cell envelope of certain gram-negative *Bacteria*, which is a toxin when solubilized. Compare with *exotoxin*.

Enriched media Media that allow metabolically fastidious organisms to grow because of the addition of specific growth factors.

Enrichment bias A problem with enrichment cultures in which "weed" species tend to dominate in the enrichment, often to the exclusion of the most abundant or ecologically significant organisms in the inoculum.

Enrichment culture technique The use of selective culture media and incubation conditions to isolate specific microorganisms from natural samples.

Enteric bacteria A large group of gram-negative, rod-shaped *Bacteria* characterized by a facultatively aerobic metabolism and commonly found in the intestines of animals.

Enterotoxin A protein that is released extracellularly by a microorganism as it grows and that produces immediate damage to the small intestine of the host.

Entropy A measure of the degree of disorder in a system; entropy always increases in a closed system.

Enveloped In reference to a virus, having a lipoprotein membrane surrounding the virion.

Enzootic An endemic disease present in an animal population.

Enzyme A catalyst, usually composed of protein, that promotes specific reactions or groups of reactions.

Enzyme immunoassay (EIA) A test that uses antibodies to detect antigens or antibodies in body fluids. Also called enzyme-linked immunosorbent assay (ELISA).

Epidemic A disease occurring in an unusually high number of individuals in a population at the same time. Compare with *endemic*.

Epidemiology The study of the occurrence, distribution, and determinants of health and disease in a population.

Epilimnion The warmer and less dense surface waters of a stratified lake.

Epitope The portion of an antigen that is recognized by an immunoglobulin or a T cell receptor.

Epizootic An epidemic disease present in an animal population.

***Escherichia coli* O157:H7** An enterotoxigenic strain of *E. coli* spread by fecal contamination of animal or human origin to food and water.

Eukarya The phylogenetic domain containing all eukaryotic organisms.

Eukaryote A cell or organism having a unit membrane–enclosed nucleus and usually other organelles; a member of the *Eukarya*.

Euryarchaeota A phylum of *Archaea* that contains primarily methanogens, extreme halophiles, *Thermoplasma*, and some marine hyperthermophiles.

Evolution Descent with modification; DNA sequence variation and the inheritance of that variation.

Evolutionary distance In phylogenetic trees, the sum of the physical distance on a tree separating organisms; this distance is inversely proportional to evolutionary relatedness.

Exergonic reaction A chemical reaction that proceeds with the liberation of energy.

Exoenzyme An enzyme produced by a microorganism and then excreted into the environment.

Exons The coding sequences in a split gene. Contrast with *introns*, the intervening noncoding regions.

Exotoxin A protein that is released extracellularly by a microorganism as it grows and that produces immediate host cell damage. Compare with *endotoxin*.

Exponential growth Growth of a microbial population in which the cell number doubles within a fixed time period.

Exponential phase A period during the growth cycle of a population in which growth increases at an exponential rate.

Expression The ability of a gene to function within a cell in such a way that the gene product is formed.

Expression vector A cloning vector that contains the necessary regulatory sequences allowing transcription and translation of a cloned gene or genes.

Extein The portion of a protein that remains and has biological activity after the splicing out of any inteins.

Extracellular matrix (ECM) Proteins and polysaccharides that surround an animal cell and in which the cell is embedded.

Extreme halophile An organism whose growth is dependent on large concentrations (generally >9%) of NaCl.

Extreme piezophile An organism requiring several hundred atmospheres of pressure for growth.

Extremophile An organism that grows optimally under one or more chemical or physical extremes, such as high or low temperature or pH.

Facultative Indicates that an organism is able to grow in either the presence or absence of an environmental factor (for example, "facultative aerobe").

FAME Fatty acid methyl ester; a technique for identifying microorganisms by their fatty acids.

Fatty acid An organic acid containing a carboxylic acid group and a hydrocarbon chain of various lengths; major components of lipids.

Feedback inhibition A decrease in the activity of the first enzyme of a biochemical pathway caused by buildup of the final product of the pathway.

Fermentation The anaerobic catabolism of an organic compound in which the compound serves as both an electron donor and an electron acceptor and in which ATP is usually produced by substrate-level phosphorylation.

Fermentation (industrial) A large-scale microbial process.

Fermenter An organism that carries out the process of fermentation.

Fermentor A growth vessel, usually quite large, used to culture microorganisms for the production of some commercially valuable product.

Ferredoxin An electron carrier of very negative reduction potential; small protein containing iron–sulfur clusters.

Fever An abnormal increase in body temperature.

Filamentous In the form of very long rods, many times longer than wide.

Filtration The removal of suspended particles from water by passing it through one or more permeable membranes or media (e.g., sand, anthracite, or diatomaceous earth).

Fimbria (plural, fimbriae) Short, filamentous structure on a bacterial cell; although flagella-like in structure, it is generally present in many copies and not involved in motility. Plays a role in adherence to surfaces and in the formation of pellicles. See also *pilus*.

Finished water Water delivered to the distribution system after treatment.

FISH See *Fluorescence in situ hybridization*.

Fitness The capacity of an organism to survive and reproduce as compared to competing organisms.

Flagellum (plural, flagella) A long, thin cellular appendage that rotates (in prokaryotic cells) and is responsible for swimming motility.

Flavoprotein A protein containing a derivative of riboflavin, which functions as an electron carrier in the electron transport system.

Flocculation The water treatment process after coagulation that uses gentle stirring to cause suspended particles to form larger, aggregated masses (flocs).

Flow cytometry A technique for counting and examining microscopic particles by suspending them in a stream of fluid and passing them by an electronic detection device.

Fluorescence in situ hybridization (FISH) A method employing a fluorescent dye covalently bonded to a specific nucleic acid probe for identifying or tracking organisms in the environment.

Fluorescent Having the ability to emit light of a certain wavelength when activated by light of another wavelength.

Fluorescent antibody An antibody molecule covalently modified with a fluorescent dye that makes the antibody visible under fluorescent light.

Fluorescent protein Any of a large group of proteins that fluoresce different colors, including the *green fluorescent protein* (GFP) for tracking genetically modified organisms and determining conditions that induce the expression of specific genes.

Fomite An inanimate object that when contaminated with a viable pathogen can transfer the pathogen to a host.

Food infection A microbial infection resulting from the ingestion of pathogen-contaminated food followed by growth of the pathogen in the host.

Food poisoning (food intoxication) A disease caused by the ingestion of food that contains preformed microbial toxins.

Food spoilage Any change in a food product that makes it unacceptable to the consumer.

Frameshift mutation A mutation in which insertion or deletion of nucelotides changes the groups of three bases in which the genetic code is read within an mRNA, usually resulting in a faulty product.

Free energy (G) Energy available to do work; $G^{0\prime}$ is free energy under standard conditions.

Fruiting body A macroscopic reproductive structure produced by some fungi (for example, mushrooms) and some *Bacteria* (for example, myxobacteria), each distinct in size, shape, and coloration.

FtsZ A protein that forms a ring along the mid-cell division plane to initiate cell division.

Functional diversity The component of biological diversity that deals with the forms and functions of organisms as they relate to differences in physiology and ecology. Compare with *phylogenetic diversity*.

Fundamental niche The range of environments in which a species will be sustained when it is not resource-limited, such as may result from competition with other species.

Fungi (singular, fungus) Nonphototrophic eukaryotic microorganisms with rigid cell walls.

Fungicidal agent An agent that kills fungi.

Fungistatic agent An agent that inhibits fungal growth.

Fusion inhibitor A synthetic polypeptide that binds to viral glycoproteins, inhibiting fusion of viral and host cell membranes.

Fusion protein A protein that is the result of fusing two different proteins together by merging their coding sequences into a single gene.

Gametes In eukaryotes, the haploid germ cells that result from meiosis.

Gas gangrene Human tissue destruction due to the growth of proteolytic and gas-producing clostridia.

Gas vesicle A gas-filled structure made of protein; confers buoyancy on a cell when present in the cytoplasm in large numbers.

GC ratio In DNA (or RNA) from any organism, the percentage of the total nucleic acid that consists of guanine plus cytosine bases (expressed as mol% GC).

Gel An inert polymer, usually made of agarose or polyacrylamide, used for separating macromolecules such as nucleic acids and proteins by electrophoresis.

Gel electrophoresis A technique for separation of nucleic acid molecules by passing an electric current through a gel made of agarose or polyacrylamide.

Gene A unit of heredity; a segment of DNA (or RNA in some viruses) specifying a particular protein or polypeptide chain, or a tRNA or an rRNA.

Gene chip Small, solid supports to which genes or portions of genes are affixed and arrayed spatially in a known pattern. Also called *microarrays*.

Gene cloning See *molecular cloning*.

Gene disruption Also called gene knockout. The inactivation of a gene by insertion of a DNA fragment that interrupts the coding sequence.

Gene expression Transcription of a gene followed by translation of the resulting mRNA into protein(s).

Gene family Genes that are related in sequence to each other as the result of a common evolutionary origin.

Gene fusion A structure created by joining together segments of two separate genes, in particular when the regulatory region of one gene is joined to the coding region of a reporter gene.

General-purpose medium A growth medium that supports the growth of most aerobic and facultatively aerobic organisms.

Generation time The time required for a cell population to double. Also called doubling time.

Gene therapy Treatment of a disease caused by a dysfunctional gene by introduction of a normally functioning copy of the gene.

Genetically modified organism (GMO) An organism whose genome has been altered using genetic engineering. The abbreviation is also used in terms such as GM crops and GM foods.

Genetic code The correspondence between nucleic acid sequence and amino acid sequence of proteins.

Genetic drift A process that results in a change in allele frequencies in a population as a result of random changes in the number of offspring from each individual over time.

Genetic element A structure that carries genetic information, such as a chromosome, a plasmid, or a virus genome.

Genetic engineering The use of in vitro techniques in the isolation, manipulation, alteration, and expression of DNA or RNA and in the development of genetically modified organisms.

Genetic exchange The transfer of genes or the acceptance of genes between prokaryotic cells.

Genetic map The arrangement of genes on a chromosome.

Genetics Heredity and variation of organisms.

Genome The total complement of genetic information of a cell or a virus.

Genomic library (DNA library) A collection of cloned DNA fragments that contains all the genetic information for a particular organism.

Genomics The discipline that maps, sequences, analyzes, and compares genomes.

Genotype The complete genetic makeup of an organism; the complete description of a cell's genetic information. Compare with *phenotype*.

Genus A taxonomic group of related species.

Germicide (antiseptic) A chemical agent that kills or inhibits growth of microorganisms and is sufficiently nontoxic to be applied to living tissues.

Global warming The predicted and ongoing warming of the atmosphere and oceans attributed to anthropogenic release of greenhouse gases, primarily CO_2, that trap infrared radiation emitted by Earth.

Glycocalyx Polysaccharide components outside of the bacterial cell wall; usually a loose network of polymer fibers extending outward from the cell.

Glycolysis Reactions of the Embden–Meyerhof–Parnas pathway in which glucose is converted to pyruvate.

Glycosidic bond A type of covalent bond that links sugar units together in a polysaccharide.

Glyoxylate cycle A series of reactions including some citric acid cycle reactions that are used for aerobic growth on C_2 or C_3 organic acids.

Gonococcus *Neisseria gonorrhoeae*, the gram-negative diplococcus that causes the disease gonorrhea.

Gram-negative cells A major phylogenetic lineage of prokaryotic cells with a cell wall containing relatively little peptidoglycan, and an outer membrane composed of lipopolysaccharide, lipoprotein, and other complex macromolecules; stain pink in the Gram stain.

Gram-positive cells A major phylogenetic lineage of prokaryotic cells containing mainly peptidoglycan in their cell wall; stain purple in the Gram stain.

Gram stain A differential staining technique in which cells stain either pink (gram-negative) or purple (gram-positive), depending upon their structural makeup.

Green fluorescent protein (GFP) A protein that fluoresces green and is widely used in genetic analysis. See also *fluorescent protein*.

Green nonsulfur bacteria Anoxygenic phototrophs containing chlorosomes, a type II photosystem, bacteriochlorophylls *a* and *c* as light-harvesting chlorophyll, and typically growing best as photoheterotrophs.

Green sulfur bacteria Anoxygenic phototrophs containing chlorosomes, a type I photosystem, bacteriochlorophyll *c*, *d*, or *e* as light-harvesting chlorophyll, and typically growing with H_2S as an electron donor.

Group translocation An energy-dependent transport system in which the substance transported is chemically modified during the process of being transported by a series of proteins.

Growth In microbiology, an increase in cell number.

Growth factor analog A chemical agent that is related to and blocks the uptake or utilization of a growth factor.

Growth rate The rate at which growth occurs, usually expressed as the generation time.

Guild Metabolically similar microbial populations that exploit the same resources in a similar way.

HAART (highly active antiretroviral therapy) The treatment of HIV infection with two or more antiretroviral drugs at once to inhibit the development of drug resistance.

Habitat An environment within an ecosystem where a microbial community could reside.

Halophile An organism requiring salt (NaCl) for growth.

Halorhodopsin A light-driven chloride pump that accumulates Cl^- within the cytoplasm.

Halotolerant Capable of growing in the presence of NaCl, but not requiring it.

Hantavirus pulmonary syndrome (HPS) An emerging acute viral disease characterized by pneumonia, caused by rodent hantavirus.

Haploid An organism or cell containing only one set of chromosomes.

Hapten A low-molecular-weight substance not inducing antibody formation itself but still able to combine with a specific antibody.

Healthcare-associated infection (nosocomial infection) An infection contracted in a healthcare-associated setting.

Heat shock proteins Proteins induced by high temperature (or certain other stresses) that protect against high temperature, especially by refolding partially denatured proteins or by degrading them.

Heat shock response Response to high temperature that includes the synthesis of heat shock proteins together with other changes in gene expression.

Heliobacteria Anoxygenic phototrophs that contain bacteriochlorophyll *g*.

Helix A spiral structure in a macromolecule that contains a repeating pattern.

Helper virus A virus that provides some necessary components for a defective virus.

Hemagglutination Agglutination of red blood cells.

Hemolysins Bacterial toxins capable of lysing red blood cells.

Hemolysis Lysis of red blood cells.

Hemorrhagic fever with renal syndrome (HFRS) An emerging acute disease characterized by shock and kidney failure, caused by rodent hantavirus.

Hepadnavirus A virus whose DNA genome replicates by way of an RNA intermediate.

HEPA filter A high-efficiency particulate air filter used in laboratories and industry to remove particles, including microorganisms, from intake or exhaust air flow.

Hepatitis Liver inflammation, commonly caused by an infectious agent.

Herd immunity The resistance of a group to a pathogen as a result of the immunity of a large proportion of the group to that pathogen.

Herpes simplex The virus that causes both genital herpes and cold sores.

Heterocyst A differentiated cyanobacterial cell that carries out nitrogen fixation.

Heteroduplex A DNA double helix composed of single strands from two different DNA molecules.

Heterofermentative Describes lactic acid bacteria capable of making more than one fermentation product.

Heterotroph An organism that requires organic carbon as its carbon source; also a chemoorganotroph.

Hfr cell A cell with the F plasmid integrated into the chromosome.

High GC gram-positive bacteria Describes bacteria in the *Actinobacteria*.

Histones Basic proteins that protect and compact the DNA in eukaryotes and some *Archaea*.

Homoacetogens *Bacteria* that produce acetate as the sole product of sugar fermentation or from $H_2 + CO_2$. Also called *acetogens*.

Homofermentative In reference to lactic acid bacteria, producing only lactic acid as a fermentation product.

Homologs Genes related in sequence to an extent that implies common genetic ancestry; includes both orthologs and paralogs.

Homologous antigen An antigen that reacts with the antibody it has induced.

Homology Having shared ancestry.

Horizontal gene transfer The unidirectional transfer of genetic information between unrelated organisms as opposed to its vertical inheritance from parental organism(s). Also called *lateral gene transfer*.

Host (or host cell) An organism or cell type capable of supporting the growth of a virus or other parasite.

Host-to-host epidemic An epidemic resulting from host-to-host contact, characterized by a gradual rise and fall in disease incidence.

Human artificial chromosome (HAC) An artificial chromosome with a human centromere sequence array.

Human granulocytic anaplasmosis (HGA) Rickettsiosis caused by *Ehrlichia ewingii* or *Anaplasma phagocytophilum*.

Human leukocyte antigen (HLA) An antigen-presenting protein encoded by a major histocompatibility complex gene in humans.

Human monocytic ehrlichiosis (HME) A rickettsiosis caused by *Ehrlichia chaffeensis* or *Rickettsia sennetsu*.

Human papillomavirus (HPV) A sexually transmitted virus that causes genital warts, cervical neoplasia, and cancer.

Humoral immunity An immune response involving antibodies.

Humus Dead organic matter.

Hybridization Base pairing of single strands of DNA or RNA from two different (but related) sources to give a hybrid double helix.

Hydrogenase An enzyme, widely distributed in anaerobic microorganisms, capable of taking up or evolving H_2.

Hydrogen bond A weak chemical bond between a hydrogen atom and a second, more electronegative element, usually an oxygen or nitrogen atom.

Hydrogenosome An organelle of endosymbiotic origin present in certain microbial eukaryotes that oxidizes pyruvate to H_2, CO_2, and acetate, and couples this to ATP synthesis.

Hydrolysis Breakdown of a polymer into smaller units, usually monomers, by addition of water; digestion.

Hydrophobic interactions Attractive forces between molecules due to the close positioning of nonhydrophilic portions of the two molecules.

Hydrothermal vents Warm or hot water–emitting springs associated with crustal spreading centers on the seafloor.

Hydroxypropionate pathway An autotrophic pathway found in *Chloroflexus* and a few *Archaea*.

Hypersensitivity An immune reaction causing damage to the host, caused either by antigen–antibody reactions or cellular immune processes. See *allergy*.

Hyperthermophile A prokaryote having a growth temperature optimum of 80°C or higher.

Hypolimnion The colder, more dense, and often anoxic bottom waters of a stratified lake.

Icosahedron A geometrical shape occurring in many virus particles, with 20 triangular faces and 12 corners.

Immediate hypersensitivity An allergic response mediated by vasoactive products released from tissue mast cells.

Immobilized enzyme An enzyme attached to a solid support over which substrate is passed and converted to product.

Immune Able to resist infectious disease.

Immune memory The capacity to respond more quickly and vigorously to second and subsequent exposures to an eliciting antigen.

Immunity The ability of an organism to resist infection.

Immunization (vaccination) Inoculation of a host with inactive or weakened pathogens or pathogen products to stimulate protective immunity.

Immunoblot (Western blot) The detection of specific proteins by separating them via electrophoresis, transferring them to a membrane, and adding specific antibodies.

Immunodeficiency Having a dysfunctional or completely nonfunctional immune system.

Immunogen A molecule capable of eliciting an immune response.

Immunoglobulin (Ig) A soluble protein produced by B cells and plasma cells that interacts with antigen; also called *antibody*.

Immunoglobulin gene superfamily A family of genes that are evolutionarily, structurally, and functionally related to immunoglobulins.

Incidence The number of new disease cases reported in a population in a given time period.

Induced enzyme An enzyme subject to induction.

Induced mutation A mutation caused by external agents such as mutagenic chemicals or radiation.

Induction Production of an enzyme in response to a signal (often the presence of the substrate for the enzyme).

Industrial microbiology The large-scale use of microorganisms to make products of commercial value.

Infection Growth of an organism within a host.

Infection thread In the formation of root nodules, a cellulosic tube through which *Rhizobium* cells travel to reach and infect root cells.

Inflammation A nonspecific reaction to noxious stimuli such as toxins and pathogens, characterized by redness (erythema), swelling (edema), pain, and heat, usually localized at the site of infection.

Informational macromolecule Any large polymeric molecule that carries genetic information, including DNA, RNA, and protein.

Inhibition In reference to growth, the reduction of microbial growth because of a decrease in the number of organisms present or alterations in the microbial environment.

Innate immunity (nonspecific immunity) The noninducible ability to recognize and destroy an individual pathogen or its products that does not rely on previous exposure to a pathogen or its products.

Inoculum Cell material used to initiate a microbial culture.

Insertion A genetic phenomenon in which a piece of DNA is inserted into the middle of a gene.

Insertion sequence (IS) The simplest type of transposable element, which carries only genes involved in transposition.

Integrase The enzyme that inserts cassettes into an integron.

Integrase inhibitor A drug that interrupts the HIV replication cycle by interfering with integrase, the HIV protein that catalyzes the integration of viral dsDNA into host cell DNA.

Integrating vector A cloning vector that can be inserted into a host chromosome.

Integration The process by which a DNA molecule becomes incorporated into another genome.

Integron A genetic element that collects and expresses genes carried on mobile cassettes.

Intein An intervening sequence in a protein; a segment of a protein that can splice itself out.

Interactome The total set of interactions between proteins (or other macromolecules) in an organism.

Intercalary growth In cell division, enlargement of a cell at several growing points.

Interferons Cytokine proteins produced by virus-infected cells that induce signal transduction in nearby cells, resulting in transcription of antiviral genes and expression of antiviral proteins.

Interleukin (IL) Soluble cytokine or chemokine mediator secreted by leukocytes.

Intermediate filament A filamentous polymer of fibrous keratin proteins, supercoiled into thicker fibers, that functions in maintaining cell shape and the positioning of certain organelles in the eukaryotic cell.

Interspecies hydrogen transfer The process by which organic matter is degraded by the interaction of several groups of microorganisms in which H_2 production and H_2 consumption are closely coupled.

Introns The intervening noncoding sequences in a split gene. Contrast with *exons*, the coding sequences.

Invasion The ability of a pathogen to enter into host cells or tissues, spread, and cause disease.

In vitro In glass, away from the living organism.

In vivo In the body, in a living organism.

Ionophore A compound that can cause the leakage of ions across membranes.

Irradiation In food microbiology, the exposure of food to ionizing radiation to inhibit microorganisms and insect pests or to retard growth or ripening.

Isolation In the context of infectious disease, the separation of persons who have an infectious disease from those who are healthy. Compare with *quarantine*.

Isomers Two molecules that have the same molecular formula but that differ structurally.

Isotopes Different forms of the same element containing the same number of protons and electrons but differing in the number of neutrons.

Isotopic fractionation Discrimination by enzymes against the heavier isotope of the various isotopes of carbon or sulfur, leading to enrichment of the lighter isotopes.

Jaundice The production and release of excess bilirubin in the liver due to destruction of liver cells, resulting in yellowing of the skin and whites of the eye.

Joule (J) A unit of energy equal to 10^7 ergs; 1000 joules equal 1 kilojoule (kJ).

Kilobase (kb) A 1000-base fragment of nucleic acid. A *kilobase pair* (kbp) is a fragment containing 1000 base pairs.

Kinase An enzyme that adds a phosphoryl group, usually from ATP, to a compound.

Koch's postulates A set of criteria for proving that a given microorganism causes a given disease.

Korarchaeota A phylum of *Archaea* that contains the hyperthermophile *Korarchaeum cryptophilum*.

Lactic acid bacteria Fermentative bacteria that produce lactic acid, are found in the *Firmicutes*, and are important in food production and preservation.

Lagging strand The new strand of DNA that is synthesized in short pieces during DNA replication and then joined together later.

Lag phase The period after inoculation of a culture before growth begins.

Laser tweezers A device used to obtain pure cultures by optically trapping a single cell with a laser beam and moving it away from surrounding cells into sterile growth medium.

Latent virus A virus present in a cell, yet not causing any detectable effect.

Late protein A protein synthesized later in virus infection after replication of the virus genome.

Lateral gene transfer The transfer of genes from a cell to another cell that is not its offspring. Also called *horizontal gene transfer*.

Leaching Removal of valuable metals from ores by microbial action.

Leading strand The new strand of DNA that is synthesized continuously during DNA replication.

Leghemoglobin An O_2-binding protein found in root nodules.

Leishmaniasis A disease of the skin or viscera caused by infection with species of a parasitic flagellated protozoan, *Leishmania*.

Leukocidin A substance able to destroy phagocytes.

Leukocyte A nucleated cell found in the blood; a white blood cell.

Lichen A fungus and an alga (or a cyanobacterium) living in symbiotic association.

Lipids Water-insoluble organic molecules important in structure of the cytoplasmic membrane and (in some organisms) the cell wall. See also *phospholipid*.

Lipopolysaccharide (LPS) Complex lipid structure containing unusual sugars and fatty acids found in most gram-negative *Bacteria* and constituting the chemical structure of the outer membrane.

Listeriosis A gastrointestinal food infection caused by *Listeria monocytogenes* that may lead to bacteremia and meningitis.

Lophotrichous Having a tuft of polar flagella.

Lower respiratory tract The trachea, bronchi, and lungs.

Low GC gram-positive bacteria Describes bacteria in the phylum *Firmicutes*.

Luminescence The production of light.

Lyme disease A tick-transmitted disease caused by the spirochete *Borrelia burgdorferi*.

Lymph A fluid similar to blood that lacks red blood cells and travels through a separate circulatory system (the lymphatic system) containing lymph nodes.

Lymph nodes Organs that contain lymphocytes and phagocytes arranged to encounter microorganisms and antigens as they travel through the lymphatic circulation.

Lymphocytes A subset of leukocytes in the blood that are involved in the adaptive immune response.

Lyophilization (freeze-drying) The process of removing all water from frozen food under vacuum.

Lysin An antibody that induces lysis.

Lysis Loss of cellular integrity with release of cytoplasmic contents.

Lysogen A prokaryote containing a prophage. See also *temperate virus*.

Lysogenic pathway After virus infection, a series of steps that leads to a state (lysogeny) in which the viral genome is replicated as a provirus along with that of the host.

Lysogeny A state following virus infection in which the viral genome is replicated as a provirus along with the genome of the host.

Lysosome An organelle containing digestive enzymes for hydrolyses of proteins, fats, and polysaccharides.

Lytic pathway A series of steps after virus infection that lead to virus replication and the destruction (lysis) of the host cell.

Macromolecule A large molecule (polymer) formed by the connection of a number of small molecules (monomers); proteins, nucleic acids, lipids, and polysaccharides in a cell.

Macrophage A large leukocyte found in tissues that has phagocytic and antigen-presenting capabilities.

Magnetosome A small particle of Fe_3O_4 present in cells that exhibit magnetotaxis (magnetic bacteria).

Magnetotaxis The directed movement of bacterial cells by a magnetic field.

Major histocompatibility complex (MHC) A genetic region that encodes several proteins important for antigen presentation and other host defense functions. MHC I proteins are expressed on all cells. MHC II proteins are expressed only on antigen-presenting cells.

Malaria An insect-transmitted disease characterized by recurrent episodes of fever and anemia; caused by the protist *Plasmodium* spp., usually transmitted between mammals through the bite of the *Anopheles* mosquito.

Malignant In reference to a tumor, an infiltrating metastasizing growth no longer under normal growth control.

MAR-FISH A technique that combines identification of microorganisms with measurement of metabolic activities (see *microautoradiography* and *FISH*).

Mast cells Tissue cells adjoining blood vessels throughout the body that contain granules with inflammatory mediators.

Medium (plural, media) In microbiology, the nutrient solution(s) used to grow microorganisms.

Megabase (Mb) One million nucleotide bases (or base pairs, abbreviated Mbp).

Meiosis A specialized form of nuclear division that halves the diploid number of chromosomes to the haploid number, for gametes of eukaryotic cells.

Membrane Any thin sheet or layer. See especially *cytoplasmic membrane*.

Memory (immune memory) The ability to rapidly produce large quantities of specific immune cells or antibodies after subsequent exposure to a previously encountered antigen.

Memory B cell A long-lived B cell responsive to an individual antigen.

Meningitis Inflammation of the meninges (brain tissue), sometimes caused by *Neisseria meningitidis* and characterized by sudden onset of headache, vomiting, and stiff neck, often progressing to coma within hours.

Meningococcemia A fulminant disease caused by *Neisseria meningitidis* and characterized by septicemia, intravascular coagulation, and shock.

Meningoencephalitis The invasion, inflammation, and destruction of brain tissue by the amoeba *Naegleria fowleri* or a variety of other pathogens.

Mesophile An organism living in the temperature range near that of warm-blooded animals and usually showing a growth temperature optimum between 20 and 40°C.

Messenger RNA (mRNA) An RNA molecule that contains the genetic information to encode one or more polypeptides.

Metabolism All biochemical reactions in a cell, both anabolic and catabolic.

Metabolome The total complement of small molecules and metabolic intermediates of a cell or organism.

Metagenome The total genetic complement of all the cells present in a particular environment.

Metagenomics Genomic analysis of pooled DNA or RNA from an environmental sample without first isolating or identifying the individual organisms. Also called environmental genomics.

Metaproteomics The measurement of whole-community protein expression using mass spectrometry to assign peptides to the amino acid sequences coded by unique genes.

Metatranscriptomics The measurement of whole-community gene expression using RNA sequencing.

Metazoa Multicellular animals.

Methanogen A methane-producing species of *Archaea*.

Methanogenesis The biological production of methane (CH_4).

Methanotroph An organism capable of oxidizing methane.

Methylotroph An organism capable of oxidizing organic compounds that do not contain carbon–carbon bonds; if able to oxidize CH_4, also a methanotroph.

MHC class I protein An antigen-presenting molecule found on all nucleated vertebrate cells.

MHC class II protein An antigen-presenting molecule found on macrophages, B lymphocytes, and dendritic cells in vertebrates.

Microaerophile An organism requiring O_2 but at a level lower than that in air.

Microarray Small, solid supports to which genes or portions of genes are affixed and arrayed spatially in a known pattern (also called a *gene chip*).

Microautoradiography (MAR) Measurement of the uptake of radioactive substrates by visually observing the cells in an exposed photograph emulsion.

Microbial community Two or more populations of cells that coexist and interact in a habitat.

Microbial ecology The study of microorganisms in their natural environments.

Microbial leaching The extraction of valuable metals such as copper from sulfide ores by microbial activities.

Microbially influenced corrosion (MIC) The contribution of microbial metabolic activities to accelerating the corrosion of metal and concrete structures.

Microbial mat A thick, layered, diverse community nourished either by light in a hypersaline or an extremely hot aquatic environment, and in which cyanobacteria are essential; or by chemolithotrophs growing on the surface of sulfide-rich marine sediments.

Microbial plastics Polymers consisting of microbially produced (and thus biodegradable) substances, such as polyhydroxyalkanoates.

Microenvironment The immediate, micrometer-scale physical and chemical surroundings of a microorganism.

Microfilament A filamentous polymer of the protein actin that helps maintain the shape of a eukaryotic cell.

Micrometer One-millionth of a meter, or 10^{-6} m (abbreviated μm), the unit used for measuring microorganisms.

Microorganism A microscopic organism consisting of a single cell or cell cluster, also including the viruses, which are not cellular.

Microsensor A small glass sensor or electrode for measuring pH or specific compounds such as O_2, H_2S, or NO_3^- that can be immersed into a microbial habitat at microscale intervals.

Microtubule A filamentous polymer of the proteins α-tubulin and β-tubulin that functions in eukaryotic cell shape and motility.

Middle protein A protein with either a structural or a catalytic function synthesized after the early proteins in a virus infection.

Minimum inhibitory concentration (MIC) The minimum concentration of a substance necessary to prevent microbial growth.

Minus (negative)-strand nucleic acid An RNA or DNA strand that has the opposite sense of (would be complementary to) the mRNA of a virus.

Missense mutation A mutation in which a single codon is altered so that one amino acid in a protein is replaced with a different amino acid.

Mitochondrion A eukaryotic organelle responsible for the processes of respiration and electron transport phosphorylation.

Mitosis The normal form of nuclear division in eukaryotic cells in which chromosomes are replicated and partitioned into two daughter nuclei.

Mixotroph An organism that uses organic compounds as carbon sources but uses inorganic compounds as electron donors for energy metabolism.

Modification enzyme An enzyme that chemically alters bases within a restriction enzyme recognition site and thus prevents the site from being cut.

Molds Filamentous fungi.

Molecular chaperone A protein that helps other proteins fold or refold properly.

Molecular clock A DNA sequence, such as a gene for ribosomal RNA, that can be used as a comparative temporal measure of evolutionary divergence.

Molecular cloning The isolation and incorporation of a fragment of DNA into a vector where it can be replicated.

Molecule Two or more atoms chemically bonded to one another.

Monoclonal antibody A single type of antibody produced from a single clone of B cells. This antibody has uniform structure and specificity.

Monocytes Circulating white blood cells that contain many lysosomes and can differentiate into macrophages.

Monomer A building block of a polymer.

Monophyletic In phylogeny, a group descended from one ancestor.

Monotrichous Having a single polar flagellum.

Morbidity The incidence of illness in a population.

Morphology The shape of an organism.

Mortality The incidence of death in a population.

Most-probable-number (MPN) technique The serial dilution of a natural sample to determine the highest dilution yielding growth.

Motif In antigen presentation, a conserved amino acid sequence found in all peptide antigens that bind to a given MHC protein.

Motility The property of movement of a cell under its own power.

Mucosa-associated lymphoid tissue (MALT) A part of the lymphatic system that interacts with antigens and microorganisms that enter the body through mucous membranes, including those of the gut, the genitourinary tract, and bronchial tissues.

Mucous membrane Layers of epithelial cells that interact with the external environment.

Mucus Soluble glycoproteins secreted by epithelial cells that coat the mucous membrane.

Multilocus sequence typing (MLST) A taxonomic tool for classifying organisms on the basis of gene sequence variations in several housekeeping genes.

Multiple displacement amplification (MDA) A method to generate multiple copies of chromosomal DNA from a single organism.

Mushroom The aboveground fruiting body, or basidiocarp, of basidiomycete fungi.

Mutagen An agent that induces mutation, such as radiation or certain chemicals.

Mutant An organism whose genome carries a mutation.

Mutation An inheritable change in the base sequence of the genome of an organism.

Mutator strain A mutant strain in which the rate of mutation is increased.

Mutualism A type of symbiosis in which both organisms in the relationship benefit.

Myc factors Lipochitin oligosaccharides produced by mycorrhizal fungi to initiate symbiosis with a plant.

Mycorrhizae A symbiotic association between a fungus and the roots of a plant.

Mycosis Any infection caused by a fungus.

Myeloma A malignant tumor of a plasma cell (antibody-producing cell).

Nanoarchaeota A phylum of *Archaea* that contains the hyperthermophilic parasite *Nanoarchaeum equitans*.

Natural killer (NK) cell A specialized lymphocyte that recognizes and destroys foreign cells or infected host cells in a nonspecific manner.

Natural penicillin The parent penicillin structure, produced by cultures of *Penicillium* not supplemented with side-chain precursors.

Negative control A mechanism for regulating gene expression in which a repressor protein prevents transcription of genes.

Negative selection In T cell selection, the deletion of T cells that interact with self antigens in the thymus. See *clonal deletion*.

Negative strand A nucleic acid strand that has the opposite sense to (is complementary to) the mRNA.

Negative-strand virus A virus with a single-stranded genome that has the opposite sense to (is complementary to) the viral mRNA.

Neutralization An interaction of antibody with antigen that reduces or blocks the biological activity of the antigen.

Neutrophil A type of leukocyte exhibiting phagocytic properties, a granular cytoplasm (granulocyte), and a multilobed nucleus. Also called a *polymorphonuclear leukocyte* (*PMN*).

Neutrophile An organism that grows best around pH 7.

Niche In ecological theory, an organism's residence in a community, including both biotic and abiotic factors.

Nitrification The microbial oxidation of ammonia to nitrate (NH_3 to NO_3^-).

Nitrifiers Chemolithotrophic *Bacteria* and *Archaea* that catalyze nitrification.

Nitrogenase The enzyme complex required to reduce N_2 to NH_3 in biological nitrogen fixation.

Nitrogen fixation Reduction of nitrogen gas to ammonia ($N_2 + 8\,H \rightarrow 2\,NH_3 + H_2$) by the enzyme nitrogenase.

Nod factors Oligosaccharides produced by root nodule bacteria that help initiate the plant–bacterial symbiosis.

Noncoding RNA (ncRNA) An RNA molecule that is not translated into protein.

Nonnucleoside reverse transcriptase inhibitor (NNRTI) A nonnucleoside compound that inhibits the action of retroviral reverse transcriptase by binding directly to the catalytic site.

Nonperishable (stable) foods Foods of low water activity that have an extended shelf life and are resistant to spoilage by microorganisms.

Nonpolar Possessing hydrophobic (water-repelling) characteristics and not easily dissolved in water.

Nonsense codon Another name for a stop codon.

Nonsense mutation A mutation in which the codon for an amino acid is changed to a stop codon.

Normal microflora Microorganisms that are usually found associated with healthy body tissue.

Northern blot A hybridization procedure where RNA is the target in the gel and DNA or RNA is the probe. Compare with *Southern blot* and *immunoblot*.

Nosocomial infection (healthcare-associated infection) An infection contracted in a healthcare setting.

Nucleic acid A polymer of nucleotides. See *deoxyribonucleic acid* and *ribonucleic acid*.

Nucleic acid probe A strand of nucleic acid that can be labeled and used to hybridize to a complementary molecule from a mixture of other nucleic acids. In clinical microbiology or microbial ecology, a short oligonucleotide of unique sequence used as a hybridization probe for identifying specific genes.

Nucleocapsid The complete complex of nucleic acid and protein packaged in a virus particle.

Nucleoid The aggregated mass of DNA that makes up the chromosome of prokaryotic cells.

Nucleoside A nucleotide minus phosphate.

Nucleoside reverse transcriptase inhibitor (NRTI) A nucleoside analog compound that inhibits the action of viral reverse transcriptase by competing with nucleosides.

Nucleosome A spherical complex of eukaryotic DNA plus histones.

Nucleotide A monomeric unit of nucleic acid, consisting of a sugar, a phosphate, and a nitrogenous base.

Nucleus A membrane-enclosed structure in eukaryotes containing the genetic material (DNA) organized in chromosomes.

Nutrient A substance taken by a cell from its environment and used in catabolic or anabolic reactions.

Obligate Indicates an environmental condition always required for growth (for example, "obligate anaerobe").

Oligonucleotide A short nucleic acid molecule, either obtained from an organism or synthesized chemically.

Oligotroph An organism that grows only or grows best at very low levels of nutrients.

Oligotrophic (1) Describes a habitat in which nutrients are in low supply. (2) Describes organisms that grow best under low-nutrient conditions.

Oncogene A gene whose expression causes formation of a tumor.

Open reading frame (ORF) A sequence of DNA or RNA that could be translated to give a polypeptide.

Operator A specific region of the DNA at the initial end of a gene, where the repressor protein binds and blocks mRNA synthesis.

Operon One or more genes transcribed into a single RNA and under the control of a single regulatory site.

Operon fusion A gene fusion in which a coding sequence that retains its own translational signals is fused to the transcriptional signals of another gene.

Opportunistic infection An infection usually observed only in an individual with a dysfunctional immune system.

Opportunistic pathogen An organism that causes disease in the absence of normal host resistance.

Opsonization The enhancement of phagocytosis due to the deposition of antibody or complement on the surface of a pathogen or other antigen.

GLOSSARY

Organelle A bilayer membrane–enclosed structure such as the mitochondrion found in eukaryotic cells.

Ortholog A gene found in one organism that is similar to that in another organism but differs because of speciation. See also *paralog*.

Osmophile An organism that grows best in the presence of high levels of solute, typically a sugar.

Osmosis The diffusion of water through a membrane from a region of low solute concentration to one of higher concentration.

Outbreak The occurrence of a large number of cases of a disease in a short period of time.

Outer membrane A phospholipid- and polysaccharide-containing unit membrane that lies external to the peptidoglycan layer in cells of gram-negative *Bacteria*.

Overlapping genes Two or more genes in which part or all of one gene is embedded in the other.

Oxic Containing oxygen; aerobic. Usually used in reference to a microbial habitat.

Oxidation A process by which a compound gives up electrons (or H atoms) and becomes oxidized.

Oxidation–reduction (redox) reaction A pair of reactions in which one compound becomes oxidized while another becomes reduced and takes up the electrons released in the oxidation reaction.

Oxidative (electron transport) phosphorylation The nonphototrophic production of ATP at the expense of a proton motive force formed by electron transport.

Oxygenase An enzyme that catalyzes the incorporation of oxygen from O_2 into organic or inorganic compounds.

Oxygenic photosynthesis The use of light energy to synthesize ATP and NADPH by noncyclic photophosphorylation with the production of O_2 from water.

Oxygen minimum zone (OMZ) An oxygen-depleted region of intermediate depth in the marine water column.

Palindrome A nucleotide sequence on a DNA molecule in which the same sequence is found on each strand but in the opposite direction.

Pandemic A worldwide epidemic.

Pan genome The totality of the genes present in the different strains of a species.

Paralog A gene within an organism whose similarity to one or more other genes in the same organism is the result of gene duplication (compare with *ortholog*).

Parasite An organism able to live in or on a host and cause disease.

Parasitism A symbiotic relationship between two organisms in which the host organism is harmed in the process.

Passive immunity Immunity resulting from transfer of antibodies or immune cells from an immune to a nonimmune individual.

Pasteurization The use of controlled heat to reduce the microbial load, including disease-producing microorganisms and spoilage microorganisms, in heat-sensitive liquids.

Pathogen A disease-causing microorganism.

Pathogen-associated molecular pattern (PAMP) A repeating structural component of a microbial cell or virus recognized by a pattern recognition receptor.

Pathogenicity The ability of a pathogen to cause disease.

Pathogenicity island A bacterial chromosome region of foreign origin that contains clustered genes for virulence.

Pathway engineering The assembly of a new or improved biochemical pathway, using genes from one or more organisms.

Pattern recognition receptor (PRR) A protein in a phagocyte membrane that recognizes a pathogen-associated molecular pattern (PAMP), such as a component of a microbial cell surface structure.

Penicillin A class of antibiotics that inhibit bacterial cell wall synthesis; characterized by a β-lactam ring.

Pentose phosphate pathway A major metabolic pathway for the production and catabolism of pentoses (C_5 sugars).

Peptide bond A type of covalent bond joining amino acids in a polypeptide.

Peptidoglycan The rigid layer of the cell walls of *Bacteria*, a thin sheet composed of *N*-acetylglucosamine, *N*-acetylmuramic acid, and a few amino acids.

Periplasm The area between the cytoplasmic membrane and the outer membrane in gram-negative *Bacteria*.

Perishable food Fresh food generally of high water activity that has a very short shelf life due to potential for spoilage by growth of microorganisms.

Peritrichous flagellation In flagellar arrangements, having flagella attached to many places on the cell surface.

Peroxisome An organelle that functions to rid the cell of toxic substances such as peroxides, alcohols, and fatty acids.

Pertussis (whooping cough) A disease caused by an upper respiratory tract infection with *Bordetella pertussis*, characterized by a deep, persistent cough.

pH The negative logarithm of the hydrogen ion (H^+) concentration of a solution.

Phage See *bacteriophage*.

Phagocyte One of a group of cells that recognizes, ingests, and degrades pathogens and pathogen products.

Phagocytosis A mechanism for ingesting particulate food in which a portion of the cytoplasmic membrane surrounds the particle and brings it into the cell.

Phenotype The observable characteristics of an organism, such as color, motility, or morphology. Compare with *genotype*.

Phosphodiester bond A type of covalent bond linking nucleotides together in a polynucleotide.

Phospholipid A lipid containing a substituted phosphate group and two fatty acid chains on a glycerol backbone.

Photoautotroph An organism able to use light as its sole source of energy and CO_2 as its sole carbon source.

Photoheterotroph An organism using light as a source of energy and organic compounds as a carbon source.

Photophosphorylation The synthesis of energy-rich phosphate bonds in ATP using light energy.

Photosynthesis The series of reactions in which ATP is synthesized by light-driven reactions and CO_2 is fixed into cell material. See also *anoxygenic photosynthesis* and *oxygenic photosynthesis*.

Phototaxis Movement of a cell toward light.

Phototroph An organism that obtains energy from light.

Phycobilin The light-capturing open chain tetrapyrrole component of phycobiliproteins.

Phycobiliprotein The accessory pigment complex in cyanobacteria that contains phycocyanin or phycoerythrin coupled to proteins.

Phycobilisome An aggregate of phycobiliproteins.

Phylogenetic probe An oligonucleotide, sometimes made fluorescent by attachment of a dye, complementary in sequence to some ribosomal RNA signature sequence.

Phylogenetic tree A diagram that depicts the evolutionary history of an organism; consists of nodes and branches.

Phylogeny The evolutionary (natural) history of organisms.

Phylotype One or more organisms with the same or related sequences of a phylogenetic marker gene.

Phylum A major lineage of cells in one of the three domains of life.

Phytanyl A branched-chain hydrocarbon containing 20 carbon atoms, commonly found in the lipids of *Archaea*.

Phytopathogen A microorganism that causes plant disease.

Pickling The process of acidifying food, typically with acetic acid, to prevent microbial growth and spoilage.

Piezophile An organism that lives optimally at high hydrostatic pressure.

Piezotolerant An organism able to tolerate high hydrostatic pressure but growing best at 1 atm.

Pilus (plural, pili) A filamentous structure that extends from the surface of a cell and, depending on type, facilitates cell attachment, genetic exchange, or twitching motility.

Pinocytosis In eukaryotes, phagocytosis of soluble molecules.

Plague An endemic disease in rodents caused by *Yersinia pestis* that is occasionally transferred to humans through the bite of a flea.

Plaque A zone of lysis or growth inhibition caused by virus infection on a lawn of host cells.

Plasma The liquid portion of the blood containing proteins and other solutes.

Plasma cell A large, differentiated, short-lived B lymphocyte specializing in abundant (but short-term) antibody production.

Plasmid An extrachromosomal genetic element that is not essential for growth and has no extracellular form.

Plate count A viable counting method in which the number of colonies on a plate is used as a measure of cell number.

Platelet A noncellular disc-shaped structure containing protoplasm, found in large numbers in blood and functioning in the blood-clotting process.

Plus-strand nucleic acid An RNA or DNA strand that has the same sense as the mRNA of a virus.

Point mutation A mutation that involves a single base pair.

Polar Possessing hydrophilic characteristics and generally water-soluble.

Polar flagellation In flagellar arrangements, having flagella attached at one end or both ends of the cell.

Poly-β-hydroxybutyric acid (PHB) A common storage material of prokaryotic cells consisting of a polymer of β-hydroxybutyrate (PHB) or other β-alkanoic acids (PHA).

Polyclonal antibodies A mixture of antibodies made by many different B cell clones.

Polyclonal antiserum A mixture of antibodies to a variety of antigens or to a variety of determinants on a single antigen.

Polygeny The occurrence of multiple copies of evolutionarily, genetically, structurally, and functionally related genes.

Polymer A large molecule formed by polymerization of monomeric units. In water purification, a chemical in liquid form used as a coagulant to produce flocculation in the clarification process.

Polymerase chain reaction (PCR) Artificial amplification of a DNA sequence by repeated cycles of strand separation and replication.

Polymorphism In a population, the occurrence of multiple alleles for a gene locus at a higher frequency than can be explained by recent random mutations.

Polymorphonuclear leukocyte (PMN) Motile white blood cells containing many lysosomes and specializing in phagocytosis. Characterized by a distinct segmented nucleus. Also called a *neutrophil*.

Polynucleotide A polymer of nucleotides bonded to one another by phosphodiester bonds.

Polypeptide Several amino acids linked together by peptide bonds.

Polyprotein A large protein expressed from a single gene and subsequently cleaved to form several individual proteins.

Polysaccharide A long chain of monosaccharides (sugars) linked by glycosidic bonds.

Polyvalent vaccine A vaccine that immunizes against more than one disease.

Population A group of organisms of the same species in the same place at the same time.

Porins Protein channels in the outer membrane of gram-negative *Bacteria* through which small to medium-sized molecules can flow.

Positive control A mechanism for regulating gene expression in which an activator protein functions to promote transcription of genes.

Positive selection In T cell selection, the stimulation of growth and development of T cells that interact with MHC–self peptide in the thymus.

Positive strand A nucleic acid strand that has the same sense as the mRNA.

Positive-strand virus A virus with a single-stranded genome that has the same complementarity as the viral mRNA.

Potable In water purification, drinkable; safe for human consumption.

Precipitation A reaction between antibody and soluble antigen resulting in visible antibody–antigen complexes.

Prevalence The total number of new and existing disease cases reported in a population in a given time period.

Pribnow box The consensus sequence TATAAT located approximately 10 base pairs upstream from the transcriptional start site. A binding site for RNA polymerase.

Primary adaptive immune response The production of antibodies or immune T cells on first exposure to antigen; the antibodies are mostly of the IgM class.

Primary antibody response Antibodies made on first exposure to antigen; mostly of the class IgM.

Primary disinfection The introduction of sufficient chlorine or other disinfectant into clarified, filtered water to kill existing microorganisms and inhibit further microbial growth.

Primary endosymbiosis Acquisition of the alphaproteobacterial ancestor of the mitochondrion or of the cyanobacterial ancestor of the chloroplast by another kind of cell.

Primary lymphoid organ An organ in which precursor lymphoid cells develop into mature lymphocytes.

Primary metabolite A metabolite excreted during the exponential growth phase.

Primary producer An organism that synthesizes new organic material from CO_2 and obtains energy from light or from oxidation of inorganic compounds. Also an *autotroph*.

Primary structure In an informational macromolecule, such as a polypeptide or a nucleic acid, the precise sequence of monomeric units.

Primary transcript An unprocessed RNA molecule that is the direct product of transcription.

Primary wastewater treatment The physical separation of wastewater contaminants, usually by separation and settling.

Primase The enzyme that synthesizes the RNA primer used in DNA replication.

Primer A short length of DNA or RNA used to initiate synthesis of a new DNA strand.

Prion An infectious protein whose extracellular form contains no nucleic acid.

Probe See *nucleic acid probe*.

Probiotic A live microorganism that, when administered to a host, may confer a health benefit.

Prochlorophyte A bacterial oxygenic phototroph that contains chlorophylls *a* and *b* but lacks phycobilins.

Prokaryote A cell or organism lacking a nucleus and other membrane-enclosed organelles and usually having its DNA in a single circular molecule. Members of the *Bacteria* and the *Archaea*.

Promoter The site on DNA where the RNA polymerase binds and begins transcription.

Prophage The state of the genome of a temperate virus when it is replicating in synchrony with that of the prokaryotic host, typically integrated into the host genome. See *provirus*.

Prophylactic Treatment, usually immunological or chemotherapeutic, designed to protect an individual from a future attack by a pathogen.

Propionic acid bacteria Gram-positive fermentative bacteria that generate propionate as a fermentation end product and are important in the production of cheese.

Prostheca (plural, prosthecae) A cytoplasmic extrusion bounded by the cell wall, such as a bud, hypha, or stalk.

Prosthetic group The tightly bound, nonprotein portion of an enzyme; not the same as a *coenzyme*.

Protease inhibitor A compound that inhibits the action of viral protease by binding directly to the catalytic site, preventing viral protein processing.

Protein A polymeric molecule consisting of one or more polypeptides.

Protein fusion A gene fusion in which two coding sequences are fused so that they share the same transcriptional and translational start sites.

Protein splicing Removal of intervening sequences from a protein.

Proteobacteria A large phylum of *Bacteria* that includes many of the common gram-negative bacteria, including *Escherichia coli*.

Proteome The total set of proteins encoded by a genome or the total protein complement of an organism.

Proteomics The large-scale or genome-wide study of the structure, function, and regulation of the proteins of an organism.

Proteorhodopsin A light-sensitive retinal-containing protein found in some marine *Bacteria* that fuels a proton pump that yields ATP.

Protist A unicellular eukaryotic microorganism; may be flagellate or aflagellate, phototrophic or nonphototrophic, and most lack cell walls; includes algae and protozoa.

Proton motive force (pmf) A source of energy resulting from the separation of protons from hydroxyl ions across the cytoplasmic membrane, generating a membrane potential.

Protoplasm The complete cellular contents, cytoplasmic membrane, cytoplasm, and nucleus/nucleoid of a cell.

Protoplast A cell from which the wall has been removed.

Prototroph The parent from which an auxotrophic mutant has been derived. Contrast with *auxotroph*.

Protozoa Unicellular eukaryotic microorganisms that lack cell walls.

Provirus The genome of a temperate or latent animal virus when it is replicating in step with the host chromosome.

Pseudomonad A term used to refer to any gram-negative, polarly flagellated, aerobic rod able to use a diverse suite of carbon sources.

Psychrophile An organism able to grow at low temperatures and showing a growth temperature optimum of $<15°C$.

Psychrotolerant Able to grow at low temperature but having a growth temperature optimum of $>20°C$.

Public health The health of the population as a whole.

Pure culture A culture containing a single kind of microorganism.

Purine One of the nitrogen bases of nucleic acids that contain two fused rings; adenine and guanine.

Purple nonsulfur bacteria A group of anoxygenic phototrophic bacteria that contain bacteriochlorophyll *a* or *b*, grow best as photoheterotrophs, and have a relatively low tolerance for H_2S.

Purple sulfur bacteria A group of anoxygenic phototrophic bacteria containing bacteriochlorophylls *a* or *b* and characterized by the ability to oxidize H_2S and store elemental sulfur inside the cells (or, in the genera *Ectothiorhodospira* and *Halorhodospira*, outside the cell).

Pyogenic Pus-forming; causing abscesses.

Pyrimidine One of the nitrogenous bases of nucleic acids that contain a single ring; cytosine, thymine, and uracil.

Pyrite A common iron ore, FeS_2.

Pyrogenic Fever-inducing.

Quarantine The separation and restriction of well persons who may have been exposed to an infectious disease to see if they develop the disease. Compare with *isolation*.

Quaternary structure In proteins, the number and arrangement of individual polypeptides in the final protein molecule.

Quinolones Synthetic antibacterial compounds that interact with DNA gyrase and prevent supercoiling of bacterial DNA.

Quorum sensing A regulatory system that monitors the population size and controls gene expression based on cell density.

Rabies A usually fatal neurological disease caused by the rabies virus, which is usually transmitted by the bite or saliva of an infected carnivore.

Radioisotope An isotope of an element that undergoes spontaneous decay with the release of radioactive particles.

Raw water Surface water or groundwater that has not been treated in any way (also called *untreated water*).

Reaction center A photosynthetic complex containing chlorophyll (or bacteriochlorophyll) and other components, within which occur the initial electron transfer reactions of photophosphorylation.

Reading-frame shift See *frameshift mutation*.

Realized niche The range of natural environments supporting a species when that organism is confronted with factors such as resource limitation, predation, and competition from other species.

Recalcitrant Resistant to microbial attack.

Recombinant DNA A DNA molecule containing DNA originating from two or more sources.

Recombination Resorting or rearrangement of DNA fragments resulting in a new sequence combination.

Redox See *oxidation–reduction reaction*.

Reduction A process by which a compound accepts electrons to become reduced.

Reduction potential (E_0') The inherent tendency, measured in volts, of the oxidized compound of a redox pair to become reduced.

Reductive dechlorination An anaerobic respiration in which a chlorinated organic compound is used as an electron acceptor, usually with the release of Cl^-.

Reemerging disease An infectious disease, previously under control, that produces a new epidemic.

Regulation Processes that control the rates of synthesis of proteins, such as induction and repression.

Regulatory nucleotide A nucleotide that functions as a signal rather than being incorporated into RNA or DNA.

Regulon A set of operons that are all controlled by the same regulatory protein (repressor or activator).

Replacement vector A cloning vector, such as a bacteriophage, in which some of the DNA of the vector can be replaced with foreign DNA.

Replication The synthesis of DNA using DNA as a template.

Replication fork The site on the chromosome where DNA replication occurs and where the enzymes replicating the DNA are bound to untwisted, single-stranded DNA.

Replicative form A double-stranded DNA molecule that is an intermediate in the replication of single-stranded DNA viruses.

Reporter gene A gene incorporated into a vector because the product it encodes is easy to detect.

Repression Prevention of the synthesis of an enzyme in response to a signal.

Repressor protein A regulatory protein that binds to specific sites on DNA and blocks transcription; involved in negative control.

Reservoir A source of viable infectious agents from which individuals may be infected.

Resolution In microbiology, the ability to distinguish two objects as distinct and separate under the microscope.

Respiration Catabolic reactions producing ATP in which either organic or inorganic compounds are primary electron donors and organic or inorganic compounds are ultimate electron acceptors.

Response regulator protein One of the members of a two-component system; a regulatory protein that is phosphorylated by a sensor protein (see *sensor kinase protein*).

Restriction enzymes (restriction endonucleases) Enzymes that recognize and cleave specific DNA sequences, generating either blunt or single-stranded (sticky) ends.

Restriction map A map showing the location of restriction enzyme cut sites on a segment of DNA.

Retrovirus A virus whose RNA genome has a DNA intermediate as part of its replication cycle.

Reverse citric acid cycle A mechanism for autotrophy in green sulfur bacteria and a few other autotrophic *Bacteria*, and also in some *Archaea*.

Reverse DNA gyrase A topoisomerase present in all hyperthermophilic prokaryotes that introduces positive supercoils in DNA.

Reverse electron transport The energy-dependent movement of electrons against the thermodynamic gradient to form a strong electron donor from a weaker electron donor.

Reverse transcriptase The retroviral enzyme that makes a DNA copy using RNA as template.

Reverse transcription The process of copying information found in RNA into DNA.

Reversion An alteration in DNA that reverses the effects of a prior mutation.

Rheumatic fever An inflammatory autoimmune disease triggered by an immune response to infection by *Streptococcus pyogenes*.

Rhizosphere The region immediately adjacent to plant roots.

Ribonucleic acid (RNA) A polymer of nucleotides connected via a phosphate–ribose backbone; involved in protein synthesis or as genetic material of some viruses.

Ribosomal RNA (rRNA) The types of RNA found in the ribosome; some rRNAs participate actively in the process of protein synthesis.

Ribosome A structure composed of RNAs and proteins upon which new proteins are made.

Riboswitch An RNA domain, usually in an mRNA molecule, that can bind a specific small molecule and alter its secondary structure; this in turn controls translation of the mRNA.

Ribotyping A means of identifying microorganisms from analysis of DNA fragments generated from restriction enzyme digestion of genes encoding their ribosomal RNA.

Ribozyme An RNA molecule that can catalyze a chemical reaction.

Ribulose monophosphate pathway A reaction series in certain methylotrophs in which formaldehyde is assimilated into cell material using ribulose monophosphate as the C_1 acceptor molecule.

Rickettsias Obligate intracellular bacteria that cause disease, including typhus, spotted fever rickettsiosis, and ehrlichiosis.

RNA Ribonucleic acid; functions in protein synthesis as messenger RNA, transfer RNA, and ribosomal RNA.

RNA editing Changing the coding sequence of an RNA molecule by altering, adding, or removing bases.

RNA interference (RNAi) A response that is triggered by the presence of double-stranded RNA and results in the degradation of ssRNA homologous to the inducing dsRNA.

RNA life A hypothetical ancient life form lacking DNA and protein, in which RNA had both a genetic coding and a catalytic function.

RNA polymerase An enzyme that synthesizes RNA in the $5' \rightarrow 3'$ direction using an antiparallel $3' \rightarrow 5'$ DNA strand as a template.

RNA processing The conversion of a precursor RNA to its mature form.

RNA replicase An enzyme that can produce RNA from an RNA template.

Rocky Mountain spotted fever See *spotted fever rickettsiosis*.

Rolling circle replication A mechanism, used by some plasmids and viruses, of replicating circular DNA, starting by nicking and unrolling one strand. For a double-stranded genome, the unrolled strand is used as a template for DNA synthesis; for a single-stranded genome, the other, still circular strand is used as the template for DNA synthesis.

Root nodule A tumorlike growth on certain plant roots that contains symbiotic nitrogen-fixing bacteria.

RubisCO The acronym for ribulose bisphosphate carboxylase, a key enzyme of the Calvin cycle.

Rumen The forestomach of ruminant animals in which cellulose digestion occurs.

Salmonellosis Enterocolitis or other gastrointestinal disease caused by any of several species of the bacterium *Salmonella*.

Sanitizers Agents that reduce, but may not eliminate, microbial numbers to a safe level.

Scale-up The conversion of an industrial process from a small laboratory setup to a large commercial fermentation.

Scarlet fever A disease characterized by high fever and a reddish skin rash resulting from an exotoxin produced by cells of *Streptococcus pyogenes*.

Schistosomiasis A chronic disease caused by a parasitic worm that leads to internal organ damage and accumulation of fluids and worm egg masses.

Screening A procedure that permits the identification of organisms by phenotype or genotype, but does not inhibit or enhance the growth of particular phenotypes or genotypes.

Secondary adaptive immune response The enhanced production of antibodies or immune T cells on second and subsequent exposures to antigen; the antibodies are mostly of the IgG class.

Secondary aerobic wastewater treatment Oxidative reactions carried out by microorganisms under aerobic conditions to treat wastewater containing low levels of organic materials.

Secondary anaerobic wastewater treatment Degradative and fermentative reactions carried out by microorganisms under anoxic conditions to treat wastewater containing high levels of insoluble organic materials.

Secondary antibody response Antibodies made on second (subsequent) exposure to antigen; mostly of the class IgG.

Secondary disinfection The maintenance of sufficient chlorine or other disinfectant residual in the water distribution system to inhibit microbial growth.

Secondary endosymbiosis The acquisition by a mitochondrion-containing eukaryotic cell of the chloroplasts of a red or green algal cell.

Secondary fermentation A fermentation in which the substrates are the fermentation products of some other organism.

Secondary lymphoid organ An organ at which antigens interact with antigen-presenting phagocytes and lymphocytes to generate an adaptive immune response; these include lymph nodes, spleen, and mucosa-associated lymphoid tissue.

Secondary metabolite A product excreted by a microorganism in the late exponential growth phase and the stationary phase.

Secondary structure The initial pattern of folding of a polypeptide or a polynucleotide, usually the result of hydrogen bonding.

Sediment (1) In water purification, the soil, sand, minerals, and other large particles found in raw water. (2) In large bodies of water (lakes, the oceans), the materials (mud, rock, and the like) that form the bottom surface of the water body.

Selection Placing organisms under conditions that favor or inhibit the growth of those with a particular phenotype or genotype.

Selective medium A growth medium that enhances the growth of certain organisms while inhibiting the growth of others due to an added media component.

Selective toxicity The ability of a compound to inhibit or kill pathogenic microorganisms without adversely affecting the host.

Self-splicing intron An intron that possesses ribozyme activity and splices itself out.

Semiconservative replication DNA synthesis yielding new double helices, each consisting of one parental and one progeny strand.

Semiperishable food Food of intermediate water activity that has a limited shelf life due to potential for spoilage by growth of microorganisms.

Semisynthetic penicillin A natural penicillin that has been chemically altered.

Sensitivity The lowest amount of antigen that can be detected by a diagnostic test.

Sensor kinase protein One of the members of a two-component system; a kinase found in the cell membrane that phosphorylates itself in response to an external signal and then passes the phosphoryl group to a response regulator protein (see *response regulator protein*).

Septicemia (sepsis) A bloodborne systemic infection.

Sequence alignment The insertion of gaps into a set of molecular sequences organized in rows so that homologous positions are organized in vertical columns. Alignment is necessary prior to phylogenetic analysis because deletion and insertion mutations cause variations in the length of molecular sequences.

Sequencing In reference to nucleic acids, deducing the order of nucleotides in a DNA or RNA molecule.

Serine pathway A reaction series in certain methylotrophs in which formaldehyde is assimilated into cell material by way of the amino acid serine.

Serology The study of antigen–antibody reactions in vitro.

Serum The fluid portion of blood remaining after the blood cells and materials responsible for clotting are removed.

Sewage Liquid effluents contaminated with human or animal fecal material.

Sexually transmitted infection (STI) An infection that is usually transmitted by sexual contact.

Shine–Dalgarno sequence A short stretch of nucleotides on a prokaryotic mRNA molecule upstream of the translational start site

GLOSSARY

that binds to ribosomal RNA and thereby brings the ribosome to the initiation codon on the mRNA. Also called ribosome-binding site.

Short interfering RNA (siRNA) Short double-stranded RNA molecules that trigger RNA interference.

Shotgun cloning Making a genomic library by random cloning of DNA fragments.

Shotgun sequencing Sequencing of DNA from previously cloned small fragments of a genome in a random fashion, followed by computational methods to reconstruct the entire genome sequence.

Shuttle vector A cloning vector that can replicate in two or more different organisms; used for moving DNA between unrelated organisms.

Sickle cell trait A genetic trait that confers resistance to malaria, but causes a reduction in the oxygen-carrying capacity of the blood by reducing the life expectancy of the affected red blood cells.

Siderophore An iron chelator that can bind iron present at very low concentrations.

Signal sequence A special N-terminal sequence of approximately 20 amino acids that signals that a protein should be exported across the cytoplasmic membrane.

Signal transduction Indirect transmission of an external signal to a target in the cell (see *two-component regulatory system*).

Silent mutation A change in DNA sequence that has no effect on the phenotype.

Simple transport system A transporter that consists of only a membrane-spanning protein and is typically driven by energy from the proton motive force.

Site-directed mutagenesis Construction in vitro of a gene with a specific mutation.

16S rRNA A large polynucleotide (~1500 bases) that functions as a part of the small subunit of the ribosome of *Bacteria* and *Archaea* and from whose sequence evolutionary relationships can be obtained; eukaryotic counterpart is 18S rRNA.

S-layer A paracrystalline outer wall layer composed of protein or glycoprotein and found in many prokaryotes.

Slime layer A diffuse layer of polymer fibers, typically polysaccharides, that forms an outer surface layer on the cell.

Slime molds Nonphototrophic eukaryotic microorganisms lacking cell walls, which aggregate to form fruiting structures (cellular slime molds) or simply masses of protoplasm (acellular slime molds).

Small subunit (SSU) RNA Ribosomal RNA from the 30S ribosomal subunit of *Bacteria* and *Archaea* or the 40S ribosomal subunit of eukaryotes, that is, 16S or 18S ribosomal RNA, respectively.

Solfatara A hot, sulfur-rich, generally acidic environment commonly inhabited by hyperthermophilic *Archaea*.

Somatic hypermutation The mutation of immunoglobulin genes at rates higher than those observed in other genes.

SOS repair A DNA repair system activated by DNA damage.

Southern blot A hybridization procedure where where DNA is the target in the gel and RNA or DNA is the probe. Compare with *Northern blot* and *immunoblot*.

Species Defined in microbiology as a collection of strains that all share the same major properties and differ in one or more significant properties from other collections of strains; defined phylogenetically as a monophyletic, exclusive group based on DNA sequence.

Species abundance The proportion of each species in a community.

Species richness The total number of different species present in a community.

Specificity (1) The ability of the immune response to interact with individual antigens. (2) The ability of a diagnostic or research test to identify a specific pathogen.

Spheroplast A spherical, osmotically sensitive cell derived from a bacterium by loss of some but not all of the rigid wall layer. If all of the rigid wall layer has been completely lost, the structure is called a *protoplast*.

Spirilla (singular, spirillum) Spiral-shaped cells.

Spirochete A slender, tightly coiled gram-negative bacterium characterized by possession of endoflagella used for motility.

Spliceosome A complex of ribonucleoproteins that catalyze the removal of introns from RNA primary transcripts.

Splicing The RNA-processing step by which introns are removed and exons joined.

Spontaneous generation The hypothesis that living organisms can originate from nonliving matter.

Spontaneous mutation A mutation that occurs "naturally" without the help of mutagenic chemicals or radiation.

Spore A general term for resistant resting structures formed by many prokaryotes and fungi.

Sporozoa Nonmotile parasitic protozoa.

Spotted fever rickettsiosis A tick-transmitted disease caused by *Rickettsia rickettsii*, characterized by fever, headache, rash, and gastrointestinal symptoms; formerly called Rocky Mountain spotted fever.

Stable isotope probing (SIP) A method for characterizing an organism that incorporates a particular substrate by supplying the substrate in ^{13}C or ^{15}N form and then isolating heavy isotope–enriched DNA and analyzing the genes.

Stalk An elongate structure, either cellular or excreted, that anchors a cell to a surface.

Start codon A special codon, usually AUG, that signals the start of a protein.

-static Suffix indicating inhibition of growth. For example, a bacteriostatic agent inhibits bacterial growth. Compare with *-cidal*.

Stationary phase The period during the growth cycle of a microbial population in which growth ceases.

Stem cell A cell that can develop into a number of final cell types.

Stereoisomers Mirror-image forms of two molecules having the same molecular and structural formulas.

Sterilant (sterilizer, sporicide) A chemical agent that destroys all forms of microbial life.

Sterile Free of all living organisms and viruses.

Sterilization The killing or removal of all living organisms and viruses from a growth medium.

Sterols Hydrophobic multiringed structures that strengthen the cytoplasmic membrane of eukaryotic cells and a few prokaryotes.

Stickland reaction The fermentation of an amino acid pair in which one amino acid serves as an electron donor and a second serves as an electron acceptor.

Stop codon A codon that signals the end of a protein.

Strain A population of cells of a single species all descended from a single cell; a clone.

Stratified water column A body of water separated into layers having distinct physical and chemical characteristics.

Stringent response A global regulatory control that is activated by amino acid starvation or energy deficiency.

Stroma The lumen of the chloroplast, surrounded by the inner membrane.

Stromatolite A laminated microbial mat, typically built from layers of filamentous and other microorganisms; may fossilize.

Subcutaneous mycoses Fungal infections of deeper layers of skin.

Substrate The molecule that undergoes a specific reaction with an enzyme.

Substrate-level phosphorylation The synthesis of energy-rich phosphate bonds through the enzymatic reaction of inorganic phosphate with an activated organic substrate.

Superantigen A pathogen product capable of eliciting an inappropriately strong immune response by stimulating greater than normal numbers of T cells.

Supercoil Highly twisted form of circular DNA.

Superficial mycoses Fungal infections of the surface layers of skin, hair, or nails.

Superoxide anion (O_2^-) A derivative of O_2 capable of oxidative destruction of cell components.

Suppressor A mutation that restores a wild-type phenotype without altering the original mutation, usually arising by mutation in another gene.

Surveillance Observation, recognition, and reporting of diseases as they occur.

Suspended solid A small particle of solid pollutant that resists separation by ordinary physical means.

Symbiosis An intimate relationship between two organisms, often developed through prolonged association and coevolution.

Synthetic DNA A DNA molecule that has been made by a chemical process in a laboratory.

Syntrophy The cooperation of two or more organisms to anaerobically degrade a substance neither can degrade alone.

Systematics The study of the diversity of organisms and their relationships; includes taxonomy and phylogeny.

Systemic Not localized in the body; an infection disseminated widely through the body.

Systemic mycoses Fungal growth in internal organs of the body.

Systems biology The integration of data from genomics and other "omics" areas to build an overall picture of a biological system.

Taxis A movement toward or away from a stimulus.

Taxonomy The science of identification, classification, and nomenclature.

T cell A lymphocyte that interacts with antigens through a T cell receptor for antigen; T cells are divided into functional subsets including Tc (T-cytotoxic) cells and Th (T-helper) cells. Th cells are further subdivided into Th1 (inflammatory) cells and Th2 cells, which aid B cells in antibody formation.

T cell receptor The antigen-specific receptor protein on the surface of T lymphocytes.

T-cytotoxic cell (Tc) A lymphocyte that interacts with MHC I–peptide complexes through its T cell receptor and produces cytotoxins that kill the interacting target cell.

T-DNA The segment of the *Agrobacterium* Ti plasmid that is transferred into plant cells.

Teichoic acid A phosphorylated polyalcohol found in the cell wall of some gram-positive *Bacteria*.

Telomerase An enzyme complex that replicates DNA at the end of eukaryotic chromosomes.

Temperate virus A virus whose genome is able to replicate along with that of its host without causing cell death in a state called lysogeny.

Termination Stopping the elongation of an RNA molecule at a specific site.

Tertiary structure The final folded structure of a polypeptide that has previously attained secondary structure.

Tertiary wastewater treatment The physicochemical or biological processing of wastewater to reduce levels of inorganic nutrients.

Tetanus A disease involving rigid paralysis of the voluntary muscles caused by an exotoxin produced by *Clostridium tetani*.

Tetracycline A member of a class of antibiotics characterized by a four-membered naphthacene ring.

Th1, Th2, Th17 See *T-helper (Th) cells*.

Thalassemia A genetic trait that confers resistance to malaria, but causes a reduction in the efficiency of red blood cells by altering a red blood cell enzyme.

Thaumarchaeota A phylum of *Archaea* that contains widespread species capable of aerobic ammonia oxidation.

T-helper (Th) cells Lymphocytes that interact with MHC II–peptide complexes through a T cell receptor and produce cytokines that act on other cells. Th subsets include **Th1** cells that activate macrophages; **Th2** cells that activate B cells; **Th17** cells that activate neutrophils; and **Treg** cells that suppress adaptive immunity.

Thermocline The zone of water in a stratified lake in which temperature and oxygen concentration drop precipitously with depth.

Thermophile An organism with a growth temperature optimum between 45 and 80°C.

Thermosome A heat shock (chaperonin) protein complex that functions to refold partially heat-denatured proteins in hyperthermophiles.

Thylakoid A membrane layer containing the photosynthetic pigments in chloroplasts and in cyanobacteria.

Thymus The primary lymphoid organ responsible for development of T cells.

Ti plasmid A conjugative plasmid in the bacterium *Agrobacterium tumefaciens* that can transfer genes into plants.

Titer (1) In immunology, the quantity of antibody present in a solution. (2) In virology, the number of infectious virions in a viral suspension.

Tolerance The acquired inability to produce an immune response to a specific antigen.

Toll-like receptor (TLR) One of a family of pattern recognition receptors (PRRs) found on phagocytes, structurally and functionally related to Toll receptors in *Drosophila*, that recognize a pathogen-associated molecular pattern (PAMP).

Toxicity The ability of an organism to cause disease by means of a preformed toxin that inhibits host cell function or kills host cells.

Toxic shock syndrome (TSS) The acute systemic shock resulting from host response to an exotoxin produced by *Staphylococcus aureus*.

Toxigenicity The degree to which an organism is able to elicit toxic symptoms.

Toxin A microbial substance able to induce host damage.

Toxoid A toxin modified so that it is no longer toxic but is still able to induce antibody formation.

Transcription The synthesis of an RNA molecule complementary to one of the two strands of a double-stranded DNA molecule.

Transcriptome The complement of all RNAs produced in an organism under a specific set of conditions.

Transduction The transfer of host genes from one cell to another by a virus.

Transfection The transformation of a prokaryotic cell by DNA or RNA from a virus. Used also to describe the process of genetic transformation in eukaryotic cells.

Transfer RNA (tRNA) A small RNA molecule used in translation that possesses an anticodon at one end and has the corresponding amino acid attached to its other end.

Transformation (1) The transfer of genetic information via free DNA. (2) A process, sometimes initiated by infection with certain viruses, whereby a normal animal cell becomes a cancer cell.

Transgenic organism A plant or animal with foreign DNA inserted into its genome.

Transition A mutation in which a pyrimidine base is replaced by another pyrimidine or a purine is replaced by another purine.

Translation The synthesis of protein using the genetic information in a messenger RNA as a template.

Transmissible spongiform encephalopathy (TSE) A degenerative disease of the brain caused by prion infection.

Transpeptidation The formation of peptide bonds between the short peptides present in peptidoglycan, the cell wall polymer of *Bacteria*.

Transporters Membrane proteins that function to transport substances into and out of the cell.

Transposable element A genetic element with the ability to move (transpose) from one site to another on host DNA molecules.

Transposase An enzyme that catalyzes the insertion of DNA segments into other DNA molecules.

Transposon A type of transposable element that carries other genes in addition to those involved in transposition; often these genes confer selectable phenotypes such as antibiotic resistance.

Transposon mutagenesis The insertion of a transposon into a gene; this inactivates the host gene, leading to a mutant phenotype, and also confers the phenotype associated with the transposon gene.

Transversion A mutation in which a pyrimidine base is replaced by a purine or vice versa.

Treg See *T-helper (Th) cells*.

Tuberculin test A skin test for previous infection with *Mycobacterium tuberculosis*.

Turbidity A measurement of suspended solids in water.

Two-component regulatory system A regulatory system consisting of a sensor protein and a response regulator protein (see *sensor kinase protein* and *response regulator protein*).

GLOSSARY

Typhus A louse-transmitted disease caused by *Rickettsia prowazekii*, causing fever, headache, weakness, rash, and damage to the central nervous system and internal organs.

Universal tree of life A phylogenetic tree that shows the evolutionary positions of representatives of all domains of cellular life. Also called universal phylogenetic tree.

Untreated water Surface water or groundwater that has not been treated in any way (also called *raw water*).

Upper respiratory tract The nasopharynx, oral cavity, and throat.

Upstream position Refers to nucleic acid sequences on the 5′ side of a given site on a DNA or RNA molecule. Compare with *downstream position*.

Vaccination (immunization) The inoculation of a host with inactive or weakened pathogens or pathogen products to stimulate protective active immunity.

Vaccine An inactivated or attenuated pathogen, or a harmless pathogen product used to induce artificial active immunity.

Vacuole A small space in a cell that contains fluid and is surrounded by a membrane. In contrast to a vesicle, a vacuole is not rigid.

Vector (1) A self-replicating DNA molecule that carries DNA segments between organisms and can be used as a cloning vector to carry cloned genes or other DNA segments for genetic engineering. (2) A living agent, usually an insect or other animal, able to carry pathogens from one host to another.

Vector vaccine A vaccine made by inserting genes from a pathogenic virus into a relatively harmless carrier virus.

Vehicle A nonliving source of pathogens that transmits the pathogens to large numbers of individuals; common vehicles are food and water.

Viable Alive; able to reproduce.

Viable count A measurement of the concentration of live cells in a microbial population.

Viral load The number of viral genome copies in the tissue of an infected host, providing a quantitative assessment of the amount of virus in the host.

Viricidal agent An agent that stops viral replication and activity.

Virion A virus particle; the virus nucleic acid surrounded by a protein coat and in some cases other material.

Viroid Small, circular, single-stranded RNA that causes certain plant diseases.

Viristatic agent An agent that inhibits viral replication.

Virulence The relative ability of a pathogen to cause disease.

Virulent virus A virus that lyses or in some other way kills the host cell after infection; a nontemperate virus.

Virus A genetic element that contains either RNA or DNA and that replicates only in host cells; has an extracellular form.

Volatile fatty acids (VFAs) The major fatty acids (acetate, propionate, and butyrate) produced during fermentation in the rumen.

Wastewater The liquid derived from domestic sewage or industrial sources, which cannot be discarded in untreated form into lakes or streams.

Water activity (a_w) An expression of the relative availability of water in a substance. Pure water has an a_w of 1.000.

Western blot See *immunoblot*.

West Nile fever A neurological disease caused by West Nile virus, a virus transmitted by mosquitoes from birds to humans.

Wild type A strain of microorganism isolated from nature or one used as a parent in a genetics investigation. The usual or native form of a gene or organism.

Winogradsky column A glass column packed with mud and overlaid with water to mimic an aquatic environment in which various bacteria develop over a period of months.

Wobble In reference to protein synthesis, a less rigid form of base pairing allowed only in codon–anticodon pairing.

Xenobiotic A synthetic compound not produced by organisms in nature.

Xerophile An organism adapted to growth at very low water potentials.

Yeast The single-celled growth form of various fungi.

Yeast artificial chromosome (YAC) A genetically engineered chromosome with yeast origin of replication and centromere sequence.

Zoonosis Any disease primarily of animals that is occasionally transmitted to humans.

Zygote In eukaryotes, the diploid cell resulting from the union of two haploid gametes.

Photo Credits

Front Matter AU.1: Nancy L. Spear; AU.2: Mary Heer; FM.3: Dusan Kostic/Fotolia.

Chapter 1 Opener: Alison E. Murray and top, Peter Glenday; lower left, Emanuele Kuhn; lower right, Christian H. Fritsen and Clinton Davis; 1.1: Paul V. Dunlap; 1.2a: John Bozzola and Michael T. Madigan; 1.2b: Reinhard Rachel and Karl O. Stetter, Archives of Microbiology 128:288-293 (1981). ©1981 by Springer-Verlag GmbH & Co. KG; 1.2c: Samuel F. Conti and Thomas D. Brock; 1.4a: Image produced by M. Jentoft-Nilsen, F. Hasler, D. Chesters (NASA/Goddard) and T. Nielsen (Univ. of Hawaii)/NASA Headquarters; 1.5a: Norbert Pfennig and Michael T. Madigan; 1.5b: Norbert Pfennig, University of Konstanz, Germany; 1.5c: Thomas D. Brock; 1.7a: Douglas E. Caldwell, University of Saskatchewan; 1.7b: From R. Amann, J. Snaidr, M. Wagner, W. Ludwig, and K.-H. Schleifer, 1996. In situ visualization of high genetic diversity in a natural bacterial community. Journal of Bacteriology 178:3496-3500, Fig. 2b. © 1996 American Society for Microbiology. Photo: Jiri Snaidr; 1.7c: Steve Gschmeissner/Photo Researchers; 1.9a: Joe Burton; 1.10.1: Scimat/Photo Researchers, Inc; 1.11.1: mylisa/Fotolia; 1.11.2: M.T. Madigan; 1.11.3: Vankad/Shutterstock; 1.11.4: Pearson Education; 1.12a(t): Stephen Ausmus/USDA Agricultural Research Service; 1.12a(b): lola1960/Shutterstock; 1.12b: U.S. Department of Energy; 1.13: Library of Congress; 1.14a: Thomas D. Brock; 1.14b: Library of Congress; 1.14c: Brian J. Ford; 1.15: Drawing by Ferdinand Cohn, originally published in Hedwigia 5:161-166 (1866); 1.16a: CDC/PHIL; 1.18a: Pearson Education; 1.18b: M.T. Madigan; 1.19: Images from the History of Medicine, The National Library of Medicine; 1.21: Walter Hesse, 1884. "Uber quantitative Bestimmung der in der luft enthaltenen Mikroorganismen," in H. Struck (ed.), Mittheilungen aus dem Kaiserlichen Gesundheitsamte. Verlag August Hirschwald; 1.22: Robert Koch, 1884. "Die Aetiologie der Tuberkulose." Mittheilungen aus dem Kaiserlichen Gesundheitsamte 2:1-88; 1.23a: Photograph by Lesley A. Robertson for the Kluyver Laboratory Museum, Delft University of Technology, Delft, The Netherlands; 1.23b: Paintings by Henriette Wilhelmina Beijerinck, photographed by Lesley A. Robertson for the Kluyver Laboratory Museum, Delft University of Technology, Delft, The Netherlands; 1.24a: From Sergei Winogradsky, Microbiologie du Sol, portion of Plate IV. Paris, France: Masson et Cie Editeurs, 1949. Reproduced by permission of Dunod Editeur, Paris, France; 1.24b: Sergei Winogradsky, Microbiologie du Sol. Paris, France: Masson, 1949; 1.EMW.1: CDC; 1.EMW.2: CDC/William Archibald.

Chapter 2 Opener: Electron Microscope Lab, Berkeley; 2.1a(t): Marie Asao and M.T. Madigan; 2.1a(b): LEO Electron Microscopy; 2.2a: Thomas D. Brock; 2.2b: Norbert Pfennig, University of Konstanz, Germany; 2.3.1: LEO Electron Microscopy; 2.3.2: Marie Asao and M.T. Madigan; 2.4b: Leon J. Le Beau, University of Illinois at Chicago; 2.4c: Molecular Probes; 2.5: M.T. Madigan; 2.6a,b: Richard W. Castenholz, University of Oregon; 2.6c: Nancy J. Trun, National Cancer Institute; 2.7: Linda Barnett and James Barnett, University of East Anglia, U.K; 2.8a: Subramanian Karthikeyan, University of Saskatchewan; 2.8b: Gernot Arp, University of Göttingen, Göttingen, Germany, and Christian Boker, Carl Zeiss Jena, Germany; 2.9: ZELMI, TU-Berlin, Germany; 2.10a: Stanley C. Holt, University of Texas Health Science Center; 2.10b: Robin Harris; 2.10c: F. Rudolf Turner, Indiana University; 2.11.1-2.11.3: Norbert Pfennig, University of Konstanz, Germany; 2.11.4: Ercole Canale-Parola, University of Massachusetts; 2.11.5: Norbert Pfennig, University of Konstanz, Germany; 2.11.6: Thomas D. Brock; 2.12a: Esther R. Angert, Cornell University; 2.12b: Heide Schulz-Vogt/University of CA Davis; 2.14c: Gerhard Wanner, University of Munich, Germany; 2.24b: Leon J. Le Beau, University of Illinois at Chicago; 2.24c: J.L. Pate; 2.24d: Thomas D. Brock and Samuel F. Conti; 2.24e,f: Akiko Umeda and K. Amako; 2.26a: Leon J. Le Beau, University of Illinois at Chicago; 2.29b: Terry J. Beveridge, University of Guelph, Guelph, Ontario; 2.29c: Georg E. Schulz; 2.31: Susan F. Koval, University of Western Ontario; 2.32a: Elliot Juni, University of Michigan; 2.32b: M.T. Madigan; 2.32c: Frank B. Dazzo and Richard Heinzen; 2.33: J.P. Duguid and J.F. Wilkinson; 2.34: Charles C. Brinton, Jr., University of Pittsburgh; 2.35b1: Michael T. Madigan; 2.35b2: Mercedes Berlanga and International Microbiology; 2.36a: M.T. Madigan; 2.36b: Norbert Pfennig, University of Konstanz, Germany; 2.37: CNRS, Karim Benzerara & Stephan Borensztajn; 2.38a: Stefan Spring, Technical University of Munich, Germany; 2.38b: Richard Blakemore and W. O'Brien; 2.38c: Dennis A. Bazylinski, Iowa State University; 2.39: Thomas D. Brock; 2.40a: A.E. Walsby, University of Bristol, Bristol, England; 2.40b: S. Pellegrini and Maria Grilli Caiola; 2.41a: Reproduced from A.E. Konopka et al., Isolation and characterization of gas vesicles from Microcyclus aquaticus. Archives of Microbiology 112: 133-140 (March 1, 1977). © 1977 by Springer-Verlag GmbH & Co. KG; 2.42: Hans Hippe, Deutsche Sammlung von Mikroorganismen und Zellkulturen GmbH, Braunschweig, Germany 2.43: Hans Hippe, Deutsche Sammlung von Mikroorganismen und Zellkulturen GmbH, Braunschweig, Germany; 2.44: Judith F.M. Hoeniger and C.L. Headley; 2.45a: H.S. Pankratz, T.C. Beaman, and Philipp Gerhardt; 2.45b: Kirsten Price, Harvard University; 2.48: Elnar Leifson; 2.49: Carl E. Bauer, Indiana University; 2.50a: R. Jarosch; 2.50b: Norbert Pfennig, University of Konstanz, Germany; 2.51a: David De Rosier; 2.52: Ken F. Jarrell; 2.55a,b: Richard W. Castenholz, University of Oregon; 2.55c,d: Mark J. McBride, University of Wisconsin, Milwaukee; 2.58: Nicholas Blackburn, Marine Biological Laboratory, University of Copenhagen, Denmark; 2.59a: Norbert Pfennig, University of Konstanz, Germany; 2.59b: Carl E. Bauer, Indiana University; 2.61: E. Guth, T. Hashimoto, and S.F. Conti; 2.62: Elisabeth Pierson, FNWI-Radboud University Nijmegen, Pearson Education; 2.63: Don W. Fawcett, M.D., Harvard Medical School; 2.64: Helen Shio and Miklos Muller, The Rockefeller University; 2.65a: Thomas D. Brock; 2.65b: A. Wellma/NaturimBild/Blickwinkel/age footstock; 2.65c: T. Slankis and S. Gibbs, McGill University; 2.66: SPL/Photo Researchers; 2.67a: Rupal Thazhath and Jacek Gaertig, University of Georgia; 2.67b: Michael W. Davidson/The Florida State University Research Foundation; 2.67c: Ohad Medalia & Wolfgang Baumeister; 2.68: Melvin S. Fuller.

Chapter 3 Opener: Daniel H. Buckley; 3.2, 3.4: James A. Shapiro, University of Chicago; 3.7: Richard J. Feldmann, National Institutes of Health; 3.15: Pearson Education; 3.17b: Richard J. Feldmann, National Institutes of Health; 3.21b: Siegfried Engelbrecht-Vandré; 3.32a,b: Wael Sabra, German Centre for Biotechnology, Braunschweig, Germany; 3.32c: Alicia M. Muro-Pastor; 3.T02: Cheryl L. Broadie and John Vercillo, Southern Illinois University at Carbondale.

Chapter 4 Opener: Somenath Bakshi and James Weisshaar; 4.5: Stephen P. Edmondson and Elizabeth Parker; 4.6e: S. B. Zimmerman, J. Struct. Biol. 156:255 (2006); 4.9: Huntington Potter and David Dressler; 4.20: Sarah French; 4.21: Dr. Katsu Murakami, The Pennsylvania State University; 4.35b: Reprinted with permission from M. Ruff et al., Class II aminoacyl transfer RNA synthetases: crystal structure of yeast aspartyl-tRNA synthetase complexed with tRNA(Asp). Science 252:1682-1689 (1991). © 1991, American Association for the Advancement of Science. Photo by Dino Moras; 4.43: Thomas C. Marlovits and Lisa Konigsmaier.

Chapter 5 Opener: Patricia Dominguez-Cuevas; 5.2b: T. den Blaauwen and Nanne Nanninga, University of Amsterdam, The Netherlands; 5.5b: Alex Formstone; 5.5c: Christine-Jacobs Wagner; 5.6b: Akiko Umeda and K. Amako; 5.16: Deborah O. Jung and M.T. Madigan; 5.21a-c: John Gosink and James T. Staley, University of Washington; 5.21d: Michael T. Madigan; 5.22a: Katherine M. Brock; 5.22b, 5.23: Thomas D. Brock; 5.24: Nancy L. Spear; 5.28a: Deborah O. Jung and Michael T. Madigan; 5.28b: Coy Laboratory Products; 5.31: Thomas D. Brock; 5.33, 5.34: John M. Martinko; 5.36: Thomas D. Brock; 5.37: John M. Martinko; 5.38a: Carlos Pedros-Alio and Thomas D. Brock; 5.38b: Janice Carr and Rob Weyant, HIP, NCID, CDC; 5.40: Thomas D. Brock; 5.MS.1: Deborah Jung; 5.MS.2: Soren Molin.

Chapter 6 Opener: José de la Torre and David Stahl; 6.2b: Michael T. Madigan; 6.15: Jonathan Eisen (PLoS Biol. 2006 Jun;4(6):e188. Metabolic complementarity and genomics of the dual bacterial symbiosis of sharpshooters. The Institute for Genomic Research, Rockville, Maryland, USA); 6.18a: GeneChip® Human Genome U133 Plus 2.0 Array, Affymetrix; 6.18b: Affymetrix; 6.20: Jack Parker.

Chapter 7 Opener: Todd Ciche; 7.2: T. Doan, R. Losick, and D. Rudner; 7.4b(l): Stephen P. Edmondson, Southern Illinois University at Carbondale; 7.4b(r): Fenfei Leng; 7.10: Reprinted with permission from S. Schultz et al., Crystal structure of a CAP-DNA complex: The DNA is bent by 90 degrees. Science 253:1001-1007 (1991). © 1991 by the American Association for the Advancement of Science. Photo by Thomas A. Steitz and Steve C. Schultz; 7.21: Timothy C. Johnston, Murray State University; 7.23: Olga E. Petrova and Karin Sauer 2009. PLoS Pathogens 5(11): e1000668; 7.26: C. Fernandez-Fernandez and J. Collier; 7.28: Alicia M. Muro-Pastor.

Chapter 8 Opener: CDC/PHIL, Dr. Fred Murphy, Sylvia Whitfield; 8.3: John T. Finch, Medical Research

Index

Phylogeny of *Bacteria*

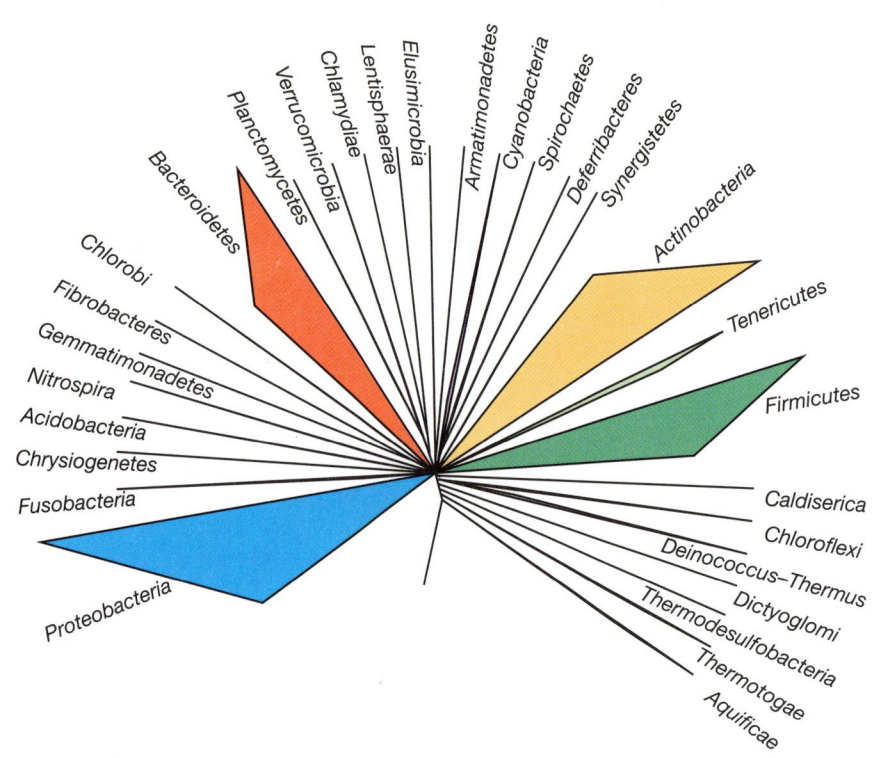